KT-519-785

THE OXFORD COMPANION TO THE THEATRE

The Oxford Companion to the Theatre

Edited by

PHYLLIS HARTNOLL

FOURTH EDITION

Oxford New York Toronto Melbourne

OXFORD UNIVERSITY PRESS

Oxford University Press, Walton Street, Oxford OX2 6DP
Oxford New York Toronto
Delhi Bombay Calcutta Madras Karachi
Petaling Jaya Singapore Hong Kong Tokyo
Nairobi Dar es Salaam Cape Town
Melbourne Auckland

and associated companies in
Berlin Ibadan

Oxford is a trade mark of Oxford University Press

Published in the United States
by Oxford University Press, New York

First edition 1951
Second edition 1957
Third edition 1967
Fourth edition 1983
Reprinted 1985 (with corrections), 1988

The initial scheme of revision for this edition and the commissioning of contributions was undertaken by Simon Trussler.

British Library Cataloguing in Publication Data
Hartnoll, Phyllis
The Oxford companion to the theatre.—4th ed.
1. Theatre—Dictionaries
I. Title
792'.03'21 PN2035
ISBN 0-19-211546-4

Printed in Great Britain by
Butler & Tanner Ltd, Frome and London

PREFACE TO THE FOURTH EDITION

DURING the years that have elapsed since the publication of the Third Edition of this Companion, the theatre throughout the world has undergone some radical changes, particularly in the sphere of technology. Scenery, Costume, and Lighting have always demanded the services of experts; but the application of modern inventions in, for instance, Architecture or Acoustics, have made some aspects of theatre-building or play-production into specialized subjects which lie outside the competence of a general encyclopaedia such as this. We have therefore made no attempt to give a comprehensive view of recent developments in such branches of theatre science, contenting ourselves with a short general history of each topic. In the same way, the growth of what one might almost call the 'theatre industry', with the proliferation of theories about everything from the art of acting to the inner meaning of the play-text, has gone beyond what can be contained in one volume. This has led, among other things, to the almost total omission of literary experts on drama, and of theatre critics, except for those who were also dramatists, or whose books, apart from those of collected criticism, serve some outstandingly useful purpose.

The rapid spread of the 'alternative' or 'experimental' or 'other' theatre, which goes under many names, has produced a world-wide flood of new companies, some—and those not necessarily the least important—short-lived, others with some degree of stability. It would be impossible to list them all. A tentative selection has therefore been made among those that are associated with improvization, collective creation, happenings, audio-visual experimentation, the woman's movement, etc., as examples of an important modern development which needs a book to itself. It is as yet too soon to give a balanced account of any such companies. Time must try them, and it may be many years before an assessment can be made of their true value in the development of modern world theatre. Theatre history provides many examples of modest undertakings which have had a global influence, as well as of those launched with great éclat which have sunk without trace.

There has in recent years been a considerable increase in the publication of works of reference covering almost every imaginable topic. As a result, we have abandoned any attempt to deal with ballet and other forms of dance and with opera, which are much better served by books dealing with each of these theatrical forms. Less happily, current economic stringency has imposed a strict limit on the overall size of the work, and we have elected to concentrate on what is known as the 'legitimate' theatre throughout its history, confining popular genres such as music-hall, vaudeville, and musical comedy to single main articles with biographical entries only for a few major figures. Limitations of space have also led to a determined effort, in dealing with the history of separate countries, to follow the main line of development over the years, however tempting the branch-lines, dead-ends, and side-lines may have seemed. Such space as was then available, once the broad pattern had been set, has been devoted to the inclusion of more, and younger, actors, dramatists, and directors, and in particular to a wider and more representative selection of British provincial and North American regional theatres, in an effort to keep pace with the territorial expansion of the theatre which has been one of the outstanding features of the last few decades. Some omissions there are bound to be; given the benefit of hindsight, in ten or fifteen years one will be able to see what should have been included, what omitted. But it is not always possible to anticipate correctly, and it was thought better to risk omitting a manifestation which may in the future prove to have been important rather than lose sight of the basic essentials in a search after ephemeral novelties.

In the former prefaces to this Companion, we have, we hope, thanked—though one could never do so adequately—the many friends and contributors who helped to disentangle the multifarious strands which go to make up world theatre, past and present. No one could have embarked on such a book, as this one was embarked on in 1939, without complete confidence in the competence and kindness of all those who helped to establish it as the first English-language encyclopaedic survey of world theatre; but with that confidence, which has been fully justified, went also the acceptance of responsibility for all errors and omissions. As the subject under discussion has become more complicated, so the possibilities of error have increased, but the responsibility still remains with the Editor, who is no less ready to accept it than she was at the beginning. The only stipulation she would like to make is that those who, publicly or privately, point out flaws in her work will also help to rectify them. Over the years many people have given freely of their time and knowledge to help solve knotty problems, or establish correctly a name or a date, and often the sternest critics have proved the most helpful. Long may this continue to be the case!

The foundations laid by the contributors to the First Edition, many of whom are now dead, have been on the whole maintained, though much new material has been added over the years. Even those whose articles have now been rewritten or omitted played their part in preparing the ground for the new structures. Among those whose work for the Third Edition has continued into the Fourth, special mention must be made of Miss Dorothy Swerdlove, who, in taking over from our good friend Mr Paul Myers, has proved a tower of strength on the details of the American theatre, which the Editor has not recently been able to explore for herself. Others who have survived the years between and again lent their invaluable support in specialized subjects have been the late Dr James Arnott, Dr Suresh Awasthi, former Director of the National Academy of Music, Dance, and Drama, New Delhi, and Visiting Professor, Graduate Department of Drama, New York University, Alan Deyermond, Dr Sybil Rosenfeld, and Dr J. E. Varey.

New contributors to the Companion include many younger scholars whose reputation has been made during the past fifteen to twenty years. Chief among these was Professor Tom Lawrenson, of Leicester University, whose recent untimely death was a sad blow. His unfailing interest and ready assistance extended far beyond the articles on the French theatre, for which, with Professor W. D. Howarth, of Bristol University, he was mainly responsible. Professor James McFarlane, of the University of East Anglia, made smooth the path of the Scandinavian theatre, while work on the German theatre came under the general guidance of Professor Hugh Rorrison, of Leeds University, and on the Italian theatre under that of Professor Kenneth Richards, of the University of Manchester and Dr Laura Richards of Salford University. Information on recent developments in Italy, Germany, and Czechoslovakia was generously supplied by Mrs M. Barzetti, Dr Ingeborg Krengel-Strudthoff, and Dr J. Milenová respectively. The intricacies of the theatre in Central and Eastern Europe were dealt with by Ossia Trilling, who, though not formerly named as a contributor to the Companion, gave his help and support to the venture from the beginning.

A very present help in time of trouble was Professor Glynne Wickham, of Bristol University, who worked devotedly on the English theatre, as did David Hutchison on the Scottish, Dr David Gardner on the Canadian, Raymond Stanley on the Australian, and David Carnegie on the New Zealand theatre. We are much indebted to Geoffrey Axworthy and Martin Banham for work on Africa, and to Anne Lonsdale for work on the Far Eastern theatres, particularly China. Others who deserve recognition for their most valuable help include Graham Walne, Managing Director of Leisureplan Theatre Consultants, with Richard Cullyer, on Sound; Dr Stanley Wells, on Shakespeare's life and works; Miss Barbara Hancock, Shakespeare Librarian of the City of Birmingham Public Libraries Department, for research on Shakespeare in translation and Shakespeare festivals; and Mrs E. Foster, Librarian of the British Theatre Association, for advice and information on a wide variety of subjects. Michael Thornton supplied not only sympathy and encouragement when needed, but also much useful information on the English theatre; while Theodore and Adele Edling Shank did invaluable work on Collective Creation and related subjects.

All those—and they were numerous—who helped to collect and collate the details of individual places and people deserve our thanks, while chief among the many friends who supplied general information and corrections of small errors Mr Ernest Trehern must take pride of place. The staff of the Oxford University Press, particularly Betty Palmer, Peter Found, and Susan le Roux, have laboured hard, and not, we hope, in vain, to support and encourage the Editor; and on a more personal note, all thanks are due to Miss Jean Heppenstall and Mrs Jenny Lobb for their unfailing help in such practical matters as transport and typing, and Miss Winifred Kimberley for continuing to maintain that domestic peace and plenty which are so necessary to a hardworking editor.

Lyme Regis, 1982.

CONTRIBUTORS

The initial scheme of revision for this edition and the commissioning of contributions was undertaken by Simon Trussler.

Alan Andrews
Peter D. Arnott
Suresh Awasthi
Richard Axton
Geoffrey Axworthy
Arthur H. Ballet
Martin Banham
Jean-Norman Benedetti
Frederick Bentham
C. W. E. Bigsby
Michael Billington
Alan Brody
Malcolm Burgess
Christophe Campos
David Carnegie
Helen E. Chambers
David F. Cheshire
Fay King Chung
F. T. Cloak
David Cook
E. Helen Cooper
Frank Coppetiers
W. A. Coupe
Alan Deyermond
John East
John Elsom
Richard Findlater
T. Gerald Fitzgibbon
Derek Fogg
David Gardner
H. F. Garten
Freda Gaye
Arnold M. Goldman
Adrian Gratwick
Ronald Hawkes
Norman Henley
Nicholas Hern
Ronald Hingley
Diana Howard
W. D. Howarth
Jorge A. Huerta
David Hutchison
N. C. Jain
Mendel Kohansky
T. E. Lawrenson
G. L. Lewis
Anne Lonsdale
James McFarlane
Peter Mackridge
Albert F. Mclean
Philippa MacLiesh
Jennifer Merin
Christopher Murray
Paul Myers
Brian Powell
Kenneth Richards
Laura Richards
Anthony Richardson
J. M. Ritchie
Sue Rolfe
Hugh Rorrison
Sybil Rosenfeld
Ann Saddlemyer
Lawrence Senelick
Robert A. Schanke
Theodore Shank
Adele Edling Shank
Wilfrid Sharp
George Speaight
Roy Stacey
Raymond Stanley
Dorothy L. Swerdlove
John E. Tailby
Terry Theodore
Michael Thornton
Thomas J. Torda
Ossia Trilling
J. E. Varey
Clive Wake
Graham Walne
Irving Wardle
Stanley Wells
George E. Wewiora
Glynne Wickham
Nicholas Worrall
William C. Young

Picture Research
Pat Hodgson

ILLUSTRATIONS

NOTE TO THE READER

Entries are in simple letter-by-letter alphabetical order, with spaces and hyphens ignored; all names beginning with Mc are arranged as though they were prefixed with Mac, and St is ordered as though it were spelt Saint. Cross-references are indicated by the use of small capital letters: if a name or term appears in this form on its first appearance in an article, it will be found to have its own entry. The most frequently occurring name in this volume, that of Shakespeare, has been excepted from the cross-reference treatment. The author's name has been given for every play mentioned, except in a handful of cases where it has proved impossible to trace and except for plays by Shakespeare. When a date appears in brackets after a play's title, this is the year of first production, as far as can be ascertained, unless the performance mentioned is so obviously a revival that brackets have been considered preferable, for the sake of brevity, to a longer form of words. Illustrations have been grouped in subject areas, listed on p. ix, with each section arranged chronologically. They function independently of the text, but further information may be obtained where there is an entry in the main sequence for names mentioned in the captions. In general, the cut-off point for information is the end of 1980, although certain important events of 1981 and 1982 have been accommodated.

ABBA, MARTA, see PIRANDELLO.

ABBEY, HENRY EUGENE (1846–96), early American theatre manager who was one of the first to present good plays and operas outside New York and to import Continental stars into the United States. He first worked as a ticket-seller in the opera house at Akron, and within two years was lessee of the theatre and arranging tours of good companies. In 1876 he was in management in Buffalo, and in the following year went to New York, where the elder SOTHERN played under him, and he brought together for the first time William H. CRANE and Stuart ROBSON. In 1880 he made his first visit to Europe, and on his return presented Sarah BERNHARDT for the first time in New York. He was subsequently responsible for the visits of the company from the GAIETY THEATRE, London, of COQUELIN, and of Henry IRVING, who with Ellen TERRY appeared at Abbey's Theatre (see KNICKERBOCKER THEATRE) when it opened in 1893.

ABBEY THEATRE, Dublin. This opened on 27 Dec. 1904 as the permanent home of the IRISH NATIONAL DRAMATIC SOCIETY. Funds to open the theatre, in the Mechanics' Institute on the site of the New Princess Theatre of Varieties, Abbey Street, were supplied by Miss HORNIMAN, who until 1910 also gave the theatre an annual subsidy. The first directors were Lady GREGORY, SYNGE, and YEATS, and the opening productions were *On Baile's Strand* by Yeats, *Cathleen ni Houlihan* by Yeats and Lady Gregory, and *Spreading the News* by Lady Gregory.

From the first there was pressure from nationalists, within the company and outside, to make the Abbey conform to political ideals. After a walk-out by some actors and writers in 1905 a splinter group was formed, the Theatre of Ireland, with Edward MARTYN as president, and in another dispute Yeats sided with Miss Hornimann against Frank and W. G. FAY, who left the Abbey in 1908. Yeats defended Synge's *The Playboy of the Western World* (1907) with a passion equal to that of audiences who condemned it as a betrayal of national ideals; in 1910, however, he refused to close the theatre during the funeral of Edward VII, according to Miss Horniman's wishes, and her subsidy was withdrawn.

By now the Abbey had achieved an international reputation, chiefly for its naturalistic acting style, which was largely the work of the Fay brothers. Although Yeats had hoped to encourage poetic drama, plays analyzing provincial life in the manner of IBSEN became the staple repertoire, as in the work of Lennox ROBINSON and T. C. MURRAY, though a vein of light satire first struck by William BOYLE reappeared in Robinson's *The White-Headed Boy* (1916).

Foreign tours, organized by the indefatigable Lady Gregory from 1911 to 1914, brought fame if not fortune to the Abbey Theatre, though Irish-American audiences took violent exception to several of the plays, and in Philadelphia the entire cast of Synge's *The Playboy of the Western World* was imprisoned on a charge of obscenity. The actors made a considerable impression on discerning playgoers, however, including the young Eugene O'NEILL. The First World War put a stop to the tours, and for many years the Abbey was on the brink of bankruptcy, but in 1925 an annual subsidy was provided by the newly formed Free State Government, and the worst was over. The plays of O'CASEY brought back the audiences, in spite of rioting sparked off by his treatment of the 1916 rebellion in *The Plough and the Stars* (1926), and new playwrights, such as George SHIELS and Brinsley MACNAMARA, came forward with lively comedies. In 1925 the Peacock Theatre was opened for poetic and experimental productions, and was also made available to outsiders, the GATE THEATRE having its beginnings here in 1928. The late 1920s saw a resurgence at the Abbey, with an excellent company which included F. J. MCCORMICK, Barry FITZGERALD, and Sara ALLGOOD in typically Abbey plays characterized by colourful language, exuberant characters, a mixture of comedy and tragedy, and a realistic urban or rural kitchen setting. The early 1930s were not so happy, with the main company often on tour, and a controversy over the staging in 1935 of O'Casey's *The Silver Tassie*. With the arrival of Hugh HUNT as director in 1936, however, the adventurous spirit of the 1920s revived.

After the death of Yeats in 1939 a new phase began. The Abbey was from 1941 to 1967 managed by Ernest Blythe (1889–1975), who saw its function as being 'to preserve and strengthen Ireland's national individuality.' The cultivation of GAELIC DRAMA became a priority, actors were required to be bilingual, and new plays were allowed to run on. In 1947 there was a public protest in the theatre over its current standards. Ria MOONEY became director, but had hardly begun work when in 1951 the Abbey building was destroyed by fire. The company moved to the QUEEN'S

THEATRE, opening on 24 Sept. 1951 with a revival of *The Silver Tassie*. The Queen's was a large theatre, and an expensive one, which imposed a commercial policy on the Abbey; but a few notable plays emerged, among them Joseph TOMELTY's *Is the Priest at Home?* (1954).

The new Abbey clearly marked a new beginning. It opened 15 years to the day after the fire which had destroyed the old theatre, on 18 July 1966: the year was also the Jubilee of the 1916 rebellion. The opening production, *Recall and Years*, was a review of the Abbey's history. Many new Irish plays have been presented, with significant work from Brian FRIEL—from *The Loves of Cass McGuire* (1967) to *Faith Healer* (1980)—Hugh LEONARD—*Time Was* (1976) and *A Life* (1979)—Thomas Kilroy, and Tom Murphy. A new Peacock Theatre opened in 1967, to be used for plays in Irish and for experimental drama.

ABBOTT, GEORGE (1887–   ), American playwright and director, who became interested in the theatre while still a student, and in 1912 went to Harvard to work under George BAKER. A year later he went on the stage, and continued to act until the success of *The Fall Guy* (1925), which he wrote in collaboration with James Gleason, enabled him to devote all his time to writing and direction. He was responsible for numerous productions, particularly of his own plays, which were mostly written in collaboration; they included *Love 'em and Leave 'em* and *Broadway* (both 1926); *Coquette* (1927), which established Helen HAYES as a star; *Three Men on a Horse* (1935), which was also a success in London a year later; *The Boys from Syracuse* (1938), a musical based on *The Comedy of Errors* which was seen in London in 1963; *Where's Charley?* (1948), a musical version of Brandon THOMAS's *Charley's Aunt*; and three later musicals, *The Pajama Game* (1954), *Damn Yankees* (1955), and *Fiorello* (1959), the first two being also seen in London shortly after their New York productions. Abbott was a resourceful play doctor as well as a competent director, and among his outstanding productions of other dramatists' musicals were *Boy Meets Girl* (1935; London, 1936); *Pal Joey* (1940; London, 1954); *High Button Shoes* (1947; London, 1948); and *A Funny Thing Happened on the Way to the Forum* (1962; London, 1963). In 1966, to mark his long and successful career, the old Adelphi Theatre in New York was renamed the GEORGE ABBOTT THEATRE. He was active into his 90s, directing his 119th show in 1978.

ABBOTT, WILLIAM (1789–1843), English actor, made his first appearance at Bristol in 1805, and by 1813 was at COVENT GARDEN, playing Pylades to the Orestes of MACREADY in a revival of Ambrose PHILIPS's *The Distrest Mother* when Macready made his first appearance at that theatre in 1816. Abbott also created the part of Appius Claudius in Sheridan KNOWLES's *Virginius* (1820), and was a member of the company which visited Paris under Charles KEMBLE in 1827, playing Charles Surface in SHERIDAN's *The School for Scandal*. He later went to America, and died in New York.

ABBOT OF MISRULE, OF UNREASON, see MISRULE, ABBOT OF.

ABBOTS BROMLEY HORN DANCE, see FOLK FESTIVALS.

ABELE SPELEN (sing. *abel spel*), see NETHERLANDS.

ABELL, KJELD (1901–61), Danish dramatist and artist, who worked as a stage designer in Paris and with Balanchine at the ALHAMBRA, London, in 1931. His first play, *Melodien, der blev væk*, was produced in Copenhagen in 1935, and at the ARTS THEATRE, London, a year later as *The Melody That Got Lost*. None of his other plays has been produced in English, though three of them have been published in translation—*Anna Sophie Hedvig* (1939, trans. 1945), widely regarded as his masterpiece; *Dronning gaar igen* (*The Queen on Tour*, 1943, trans. 1955) which, produced during the German occupation of Denmark, was a protest against loss of freedom (after MUNK's murder Abell interrupted a performance at the KONGELIGE TEATER, Copenhagen, to protest and then went underground); and *Dage paa en sky* (*Days on a Cloud*, 1947, trans. 1964). During the 1950s he wrote *Den blå pekineser* (*The Blue Pekinese*, 1954), *Kameliadamen* (*The Lady of the Camelias*, 1959), and *Skriget* (*The Scream*, performed posthumously in 1961). Abell's work represents a sustained attempt to bring an experimental theatre of dream and vision to Denmark.

ABINGTON [*née* Barton], FRANCES (1737–1815), English actress, who was first a flower-girl and street singer but, after being temporarily employed by a French milliner who taught her the elements of refinement and gentility, went on the stage, making her first appearance at the HATMARKET THEATRE in 1755 as Miranda in Mrs CENTLIVRE's *The Busybody*. On the recommendation of Samuel FOOTE she was taken on at DRURY LANE, where she found herself overshadowed by Kitty CLIVE and soon left, going to Dublin for five years. At some point she made a short-lived and unhappy marriage with a music-master, retaining her married name after their separation. On the invitation of GARRICK, who disliked her but thought her an excellent actress, she returned to Drury Lane, and remained there for 18 years. During this time she played many important roles, and was the first Lady Teazle in SHERIDAN's *The School for Scandal* (1777). She was much admired as Beatrice in *Much Ado About Nothing* and as Miss Prue in CONGREVE's *Love for Love*, in which character she was painted by Reynolds. She also revived Anne OLDFIELD's part of Lady Betty Modish in CIBBER's *The Careless Husband*. In 1782 she went to COVENT GARDEN, where she remained until 1790, finally retiring in 1799. She was an ambitious, clever, and witty woman, and in spite of her humble origins she achieved an enviable position in society, where the women paid her the supreme compliment of copying her clothes.

ABOVE, see STAGE DIRECTIONS.

ABSURD, Theatre of the, the name given by Martin Esslin, in a book of that title published in 1962, to the plays of a group of dramatists, among them BECKETT and IONESCO and, in England, PINTER, whose work has in common the basic belief that man's life is essentially without meaning or purpose and that human beings cannot communicate. This led to the abandonment of dramatic form and coherent dialogue, the futility of existence being conveyed by illogical and meaningless speeches and ultimately by complete silence. The first, and perhaps most characteristic, play in this style was Beckett's *Waiting for Godot* (1952), the most extreme—since it has no dialogue at all—his *Breath* (1970). The movement, which liberated playwrights from many outmoded conventions, left a profound and lasting impression on the theatre everywhere.

ACCESI. A company of actors of the COMMEDIA DELL'ARTE, first mentioned in 1590. Ten years later they were under the leadership of Pier Maria CECCHINI and the famous ARLECCHINO Tristano MARTINELLI, with whom they visited France. Among the actors were Martinelli's brother Drusiano, Flaminio SCALA, and possibly Diana da PONTI, formerly of the DESIOSI. On their next visit to France in 1608 they were without their Arlecchino, but were nevertheless much admired by the Court and by Marie de Médicis. Shortly afterwards Cecchini joined forces with the younger ANDREINI, but the constant quarrelling of Cecchini's and Andreini's wives caused the two parties to separate. Cecchini retained the old name of Accesi, but little is known of his subsequent activities. Silvio FIORILLO, the first CAPITANO Matamoros, was with the Accesi in 1621 and 1632.

ACCIUS, LUCIUS (*c*. 170—85 BC), Roman dramatist, and the last important writer of tragedy for the Roman stage. The titles of over 40 of his plays have survived, and show that he dealt with every field of tragedy open to a Latin writer, from the translation of Greek works of the 5th century and later to the composition of FABULA *praetexta*. Characteristic of him are plots of a violent, melodramatic nature, flamboyant personages, majestic utterance, and powerful repartee; when we can compare him with his Greek models we find that he works up the rhetorical possibilities of each situation to the highest degree. Thus, while Eteocles' command to Polynices is simply expressed in the Greek: 'Then get thee from these walls, or thou shalt die', Accius gives us four imperatives in six words: 'egredere, exi, ecfer te, elimina urbe!' The continual search for rhetorical effect, the eagerness to exploit each situation to the full, characteristic of Roman tragedy as a whole, seems to reach its culminating point in Accius. Inevitably it tends to eliminate the half-tones of nature and reduce all portraiture to glaring white and black. Accius's heroes are grand and striking, and do sometimes surpass their Greek prototypes, not merely in Stoic fortitude but in a certain grave humanity and sympathy with misfortune. On the other hand, we have no evidence that he ventured to alter the structure given him by his originals, though he seems to have remodelled certain passages and occasionally to have inserted lines from other sources. Comparing the extant lines from some 40 known titles with the plays of SENECA we see at once that Accius is still under the salutary discipline of having to write for a real stage, and in his *Pragmatica* he attempted a formulation of the dramatist's technique.

ACHARD, MARCEL (1899—1974), French dramatist, whose poetic plays are mainly concerned with insubstantial and ironic love, and show a mingling of burlesque with unexpected pathos, often using clowns and other PANTOMIME characters. The best known of his early plays are *Voulez-vous jouer avec moâ?* (1923), directed by DULLIN (for whom in the same year Achard adapted Ben JONSON's *Epicœne* as *La Femme silencieuse*), *Marlborough s'en va-t-en guerre* (1924), *Jean de la lune* (1929), *Domino* (1931), and *Le Corsaire* (1938), all staged by JOUVET. In later years he wrote for a more indulgent audience; the plays of this period, among which are *Auprès de ma blonde* (1946), *Nous irons à Valparaiso* (1948), *Patate* (1957), and *L'Idiote* (1960), are less interesting and more ephemeral than his excellent early work. S.N. BEHRMAN adapted *Auprès de ma blonde* as *I Know My Love* (1949); Irwin Shaw adapted *Patate* in 1958; and *L'Idiote* was adapted by Harry Kurnitz as *A Shot in the Dark* (1961).

ACHURCH, JANET (1864–1916), English actress, who made her first appearance at the OLYMPIC in London in 1883, and later toured with BENSON, playing leading parts in Shakespeare. It is, however, as one of the first actresses in England to play IBSEN that she is remembered. She was Nora in *A Doll's House* at the Novelty (later the KINGSWAY THEATRE) in 1889, and in 1896 produced *Little Eyolf* at the Avenue (later the PLAYHOUSE THEATRE) with herself as Rita, Mrs Patrick CAMPBELL as the Ratwife, and Elizabeth ROBINS as Asta. She was also seen in the title role of SHAW's *Candida* and as Lady Cecily Waynflete in his *Captain Brassbound's Conversion* (both STRAND, 1900). With her husband Charles Charrington she toured extensively, and was the first English actress to appear in the Khedivial Theatre, Cairo. A beautiful woman, with a superb carriage and lovely voice, she was called by Shaw 'the only tragic actress of genius we now possess'. Some excellent descriptions of her acting can be found in his *Our Theatres in the Nineties*. She retired from the stage in 1913. In 1978 a play based on her correspondence with Shaw was produced at the GREENWICH THEATRE.

ACKERMANN, KONRAD ERNST (1712–71), German actor, who in about 1742 joined SCHÖNEMANN's company, playing mainly in comedy. A handsome man, with a restless, vagabond temperament, Ackermann was well suited to the life of a strolling player, and soon left Schönemann to form his own company, taking with him as his leading lady Sophie SCHRÖDER, whom he married after the death of her husband. Together they toured Europe, being joined eventually by EKHOF and by Sophie's son F. L. SCHRÖDER, until in 1767 Ackermann, feeling the need for a permanent residence and influenced by the ideas of SCHLEGEL and Löwen, established in HAMBURG the first German National Theatre. Here he intended to stage outstanding productions of old and new plays, many of them by German writers, in place of the popular and lucrative ballets and farces of the day; but although the enterprise had the backing of LESSING, whose *Hamburgische Dramaturgie* dates from this period, it failed, mainly owing to dissension between Ackermann and his young stepson, who left the company for a time, returning shortly before Ackermann's death to take over on behalf of his mother. In the company were his two half-sisters, Dorothea (1752–1821) and Charlotte (1757–75) Ackermann, who both played leading roles in his productions, the former being greatly admired as the heroine in Lessing's *Minna von Barnhelm* (1767) and as the Countess Orsina in his *Emilia Galotti* (1772), in which Charlotte, then only 14, played the title role. Dorothea also played Maria in GOETHE's *Götz von Berlichingen* (1773) and the title role in *Stella* (1776), and, also in 1776, was seen as Ophelia to BROCKMANN's Hamlet and as Desdemona in Schröder's production of *Othello*. She retired shortly afterwards on her marriage, but Charlotte, who at 17 was one of the greatest and most admired actresses of Germany, died before her 18th birthday, probably from the strain of too many long and taxing roles and constant public appearances, for which Schröder was blamed.

ACOUSTICS AND SOUND. Communication between actor and audience is the basis of all drama, and although visual elements may enhance this communication, they cannot (except in the case of wordless MIME plays and ballet, which are usually accompanied by music) replace in the theatre the main medium of speech, nor compensate for poor sound reception by the audience. The human voice is produced by the passage of air from the chest cavity through the throat and mouth. The resulting sounds transmit energy to the surrounding atmosphere, shifting air particles successively in a sinusoidal pattern until the energy decays and the sound dies away. This process is three-dimensional and cyclic: the elastic movement of air particles outward from the sound source has a concomitant 'bounce' in the reverse direction, and the number of cycles—the movement out and back again to the centre—per second is expressed in Hertz (Hz). Sound does not spread evenly: lower frequencies tend to be omni-directional and higher frequencies uni-directional. The shape and furnishing of an auditorium with elements that may reflect, resonate, or absorb sound of various frequencies, will affect the audience's reception of an actor's voice, and this effect may be different according to the position of particular audience members in relation to the structural or decorative surfaces via which sounds reach them. Classical Greek theatres, noted for their fine acoustic properties, were in effect modified open-air arenas. A wall behind the actors and a hard floor in front of and slightly below them provided simple sound reflectors, while the lack of opposed plane surfaces eliminated both echo and reverberation and consequent indistinctness. The directional qualities of the human voice were well served by the curved amphitheatre and steeply raked seating, for the audience itself absorbs much of the sound emitted by actors. As early as the 1st century BC, VITRUVIUS recognised that the rake should be a straight incline to avoid any obstacles to sounds reaching each audience member. He also postulated the use of *echeia*, vases, tuned in response to the notes of the tetrachord and placed under seats in the amphitheatre to enhance the resonance of vowel sounds. The sloping roof over the stage of Roman theatres would have provided a further sound-reflecting surface; and Vitruvius also mentions PERIAKTOI, tall triangular prisms set in the wings of the Roman stage, which could rotate about their axes to provide three sets of scenes. These, too, if set at an angle of about 45°, would reflect sound out into the auditorium.

The rhythm of verse contributes to the carrying power of the human voice and it is notable that the two great schools of verse drama, Greek and Elizabethan, developed in unroofed auditoriums. Elizabethan playhouses, based on inn yards, had a hard stage floor that reflected sound up towards an audience seated in raised tiers. There was also a stage roof (the 'shadow' or 'testa') which reinforced the downward movement of sound to the groundlings, although it muffled speech delivered from the stage balcony, so that for listeners in the top gallery Juliet would be heard less well than Romeo. The rhymed alexandrines of French dramatic verse, less flexible in delivery than the verse of Shakespeare and his contemporaries, demanded an enclosed space of comparatively small dimensions for good audibility. The delicate modified vowels of French, too, need the resonance of an interior setting. Although the plays of MOLIÈRE were sometimes performed out of doors, French drama developed mainly in the halls of palaces and converted TENNIS-COURTS. Great halls were also used for plays and MASQUES in 17th-century England. A raised stage and banked seats gave conditions similar to those of the playhouse, an open floor between the front of the audience and the stage (primarily intended for processions or dancing) gave a sound-reflecting surface, and the ceiling trapped the actors' voices to reflect back into the audience.

Of the sound leaving the actor's lips and travelling outward in all directions, only a part travels directly to a given audience member. This direct sound is supplemented by reflected sound arriving shortly after. If the interval between the arrival of direct and reflected sounds is 1/30 second or less, they will be combined into a single resonant tone: a greater interval will give the effect of blurring the actor's delivery or, ultimately, of an echo. The ceiling of an auditorium is a danger point for good acoustic quality, its height, shape, and materials having the power to absorb sound or to bounce it into the audience from many angles and after unacceptable intervals. A theatre designed by VANBRUGH in the Haymarket, opened in 1705, had a high concave vault. It was immediately criticized for sacrificing 'quality and convenience' to visual grandeur, uncontrolled reverberation making speech from the stage virtually inaudible.

The 18th century did however produce all over Europe theatres that were acoustically excellent, owing much to the development of the Italian opera house. Sound reflection was provided by the sides and top of the proscenium arch, the forestage, and the ceiling of the auditorium, usually a flat vault or sloping upwards from the proscenium. Reverberation was minimized by the tiers of boxes with their drapery and by the elaborate ornamentation of surface areas; modern acoustic studies have demonstrated that a mixture of resonant, reflecting, and absorbing materials distributed equally over the surfaces of an auditorium help to achieve the brightness of tone characteristic of these theatres, equally suitable for drama and for the opera of the day.

Although 19th-century theatres maintained the horse-shoe plan and elaborate ornament of earlier years, increasing size brought the need for domed or high-vaulted ceilings, with consequent acoustic faults. Domes produced some notable echoes, such as that of the old ALHAMBRA in Leicester Square, and curved ceilings gave an unequal distribution of reflected sound, so that the myth of the 'blind spot' became current. Later in the century the rise of domestic comedy and drama led to the building of smaller 'comedy' theatres, suited to less oratorical styles of acting and offering intimate acoustic qualities.

The 20th-century revolution in architectural style was naturally carried into theatre design. Ornament and drapery were replaced by large flat surfaces in hard plaster, and the fan-shaped auditorium (which seats more people at a medium distance from the stage with unobstructed sightlines) was widely adopted. Loudness was increased by the even reflecting surfaces, especially in the rear seats; but this increased sound was reflected from surfaces at the back of the auditorium, together with coughs and other noises originating in the cheaper seats, towards the spectators in the front rows. Sound-absorbing materials applied to the walls gave only partial improvement, and it was realised that structural remedies were needed. The large rear wall should be straight or polygonal, not curved; irregular side walls and draped proscenium areas reduce unwanted reverberation; ceilings that are stepped, rather than splayed, will reflet speech into the rear seats while avoiding the return of sound to the stalls. The ceiling of the OLIVIER THEATRE in London's NATIONAL THEATRE, for example, is a concave inversion of the seating area, with acoustic panels (which double as screens for light sources) suspended in an asymmetric pattern; the walls represent five sides of an irregular hexagon.

In rejecting the conventions of the proscenium stage, THEATRE-IN-THE-ROUND has inevitably lost the acoustic benefits of traditional theatre buildings. Complaints from the audience members behind an actor that he is inaudible, and from those facing him that he is shouting, are matched by the players' objections to the lack of target—to speaking into a void. Attempts have been made to meet both creative and economic needs by FLEXIBLE STAGING, which by the use of lifts and flexible seating can offer a variety of stages—open, proscenium, thrust, arena—for all kinds of performance. If drapes, carpets, and well-upholstered seating are dispensed with in the interests of convertibility and economy, sound absorbent material must be applied to walls and ceilings to reduce reverberation.

The development of film and television, together with that of intimate drama, has rendered declamatory styles of acting obsolete and few actors of today are able, or need, to produce the volume of sound needed in the theatre buildings of the past. While there is still a certain resistance to electronic amplification in the straight theatre, it has become commonplace in musicals, especially with the vogue for featuring straight actors or pop singers who lack the vocal technique and training to fill a theatre with their singing. Rock musicals use electronic amplification and effects as an intrinsic element of the genre, and the increasing sophistication of electronic equipment is overcoming its earlier problems such as 'howlaround' or feedback—a high-pitched screech issuing from loudspeakers, picked up by microphones, and setting up a circuit of oscillation. This phenomenon is most likely to start at frequencies emphasized by the theatre's acoustic characteristic, especially when the sound level is high.

Modern PA (public address) systems can compensate for excessive reverberation times and have been used with particular success in large spaces such as cathedrals. Without increasing the volume of the speaker's voice, these systems increase clarity by amplifying its early reflections and thereby achieving an acceptable interval between direct and reflected sound. The directional quality of sound is maintained by siting the loudspeakers at a distance from the audience, or by incorporating into the PA circuit a delay to ensure that direct sound reaches all members of the audience before the amplified reflections. This method of improving a theatre's acoustics, while seemingly no more objectionable than adjustments to the physical elements of the building, has rarely proved acceptable in the straight theatre.

At a higher level of sophistication are such techniques as ambiophony and assisted resonance. Again, increase in volume is not their purpose: they increase the reverberation time by picking up reverberant sound in the auditorium and in part replacing the energy lost by absorption; and because each unit carries a range of settings the ambient resonance can be selected according to the needs of a given production. The rapid development of microprocessors has also been influential, leading to greater efficiency in calculating acoustic equations, setting up sound systems, and controlling levels of volume.

ACT, a division of a play, which may contain one or more scenes. Greek plays were continuous, the only pauses in the action being marked by the chorus. Horace, in his *Ars Poetica* (l. 189), was the first to insist on the importance of five acts in tragedy, a formula adopted by the Italian playwrights of the Renaissance as standard practice. It was accepted by the French writers of tragedy, CORNEILLE and RACINE, and passed into English drama by way of Ben JONSON. There is no evidence that Shakespeare divided his plays in this way, the five-act divisions in the First Folio having probably been introduced by the editors CONDELL and HEMINGE in imitation of Jonson. In comedy more licence was allowed to the individual, two or three acts being quite usual, even in MOLIÈRE. In the 20th century three acts were found to be convenient for actors and audience alike, but two acts are not uncommon, and many revivals of Shakespeare's plays have only one interval. A division into four acts, found mainly in the 19th century, is now seldom used.

A self-contained performance, or turn, on the VARIETY or MUSIC-HALL stage is also called an 'act'.

ACT-DROP, see CURTAIN.

ACTING COMPANY, New York, see CITY CENTRE OF MUSIC AND DRAMA.

ACTOR, ACTRESS, ACTING. As it is probable that the earliest manifestations of drama had no spoken dialogue, the first performers were presumably singers and dancers. Of the very early actors, nothing is known, but in Ancient GREECE, where they were participants in a religious ceremony, they were evidently men of some repute. Tragic actors, who enjoyed personal immunity, were often used as diplomatic envoys, and in the 4th century BC they organized themselves into a guild, the ARTISTS OF DIONYSUS. Of actors in Greek comedy nothing is known beyond a few names. In ROME the social status of actors was low, and they were often recruited from the ranks of slaves, a circumstance which may be linked with the decadence of the theatre there and the decline from classical tragedy and comedy. With the coming of Christianity actors were finally proscribed and sank into obscurity, handing on some mutilated traditions through the little bands of itinerant JONGLEURS and GOLIARDS who catered for the crowds at FAIRS and other populous places in the larger cities, and even through the more highly regarded MINSTRELS. They were not actors, but when the LITURGICAL DRAMA brought the theatre back into Europe they may have taken part in the performances of MYSTERY PLAYS.

The emergence of a vernacular drama in each country was naturally accompanied by the rise of the professional actor, who finally became established in the 16th century—in Italy with the formation of the COMMEDIA DELL'ARTE troupes, in Spain with the work of the first professional actor-manager, Lope de RUEDA, in England with the building in 1576 of the first permanent playhouse in London, the THEATRE, and in France, at the end of the century, with the establishment of the first professional troupe at the Hôtel de BOURGOGNE. Germany, owing to internal dissension and division, had to wait longer for a settled theatre, and not until the 18th century, with Carolina NEUBER, were any German players famous enough to be known by name unless they were also playwrights. Russia can hardly be said to have had a national and professional theatre much before the middle of the 19th century.

Women did not act in Greece at all, and in Rome only if very depraved. The medieval stage may have employed a few women, to play Eve in the Garden of Eden, for instance, but they were still amateurs. The professional actress emerges first in Italy, the best known being Isabella ANDREINI, and in France appears at the same time as the professional actor. Elizabethan drama made no use of women at all: all the heroines of Shakespeare and other playwrights of the time were played by boys and young men, and it was not until the Restoration in 1660 that women were first seen on the London stage. The name of the first English actress is not known. There are several claimants but, whoever she was, she played Desdemona (in *Othello*) on 8 Dec. 1660, and paved the way for the actresses of the PATENT THEATRES. It is probable that the first actress to be seen in Germany before Carolina Neuber was English, since George JOLLY had at least one in his company of ENGLISH COMEDIANS in 1654.

The social status of the modern actor was for a long time precarious. In Catholic countries he was refused the sacraments, and an anomalous situation arose in which an honoured public figure like MOLIÈRE had to be buried in unconsecrated ground. Legally even Shakespeare and his contemporaries were still liable to be classed as 'rogues and vagabonds', and it was not until the 19th century that the English actor achieved a definite place in society, culminating in the knighthood bestowed on Henry IRVING in 1895.

Fashions in acting change as in everything else, and only in the Far East—particularly in JAPAN and to a certain extent in CHINA—have traditions in acting remained unbroken down the centuries; even in those countries they are now in a state of flux. In Greece the tragic actor was static—a voice and a presence; in Rome he was lively, and, in the last resort, an acrobat. The *commedia dell'arte* demanded a quick wit and a nimble body; French tragedy postulated a noble presence and a sonorous voice; Restoration comedy called for polished brilliance in both men and women. The MELODRAMA of the 19th century was based on a ranting delivery and rip-roaring violence of action; the more intimate style of modern drama and comedy helped to develop a NATURALISM strongly influenced by the ideas of STANISLAVSKY. Yet the great actor still needs to be a little of everything—singer, dancer, mimic, acrobat, tragedian, comedian—and to have at his command a good physique, a retentive memory, an alert brain, a clear, resonant voice with good articulation and controlled breathing. Some of the greatest among them, notably Edmund KEAN and David GARRICK, have been able to compensate for their lack of one or other of these essentials by hard work, and the impress of an unusual personality. Much of the actor's art must be born in him; something can be taught. The rest comes by experience.

ACTORS' COMPANY, see MCKELLEN, IAN.

ACTORS' EQUITY ASSOCIATION, see AMERICAN, BRITISH, AND CANADIAN ACTORS' EQUITY ASSOCIATION.

ACTORS' STUDIO, see KAZAN, ELIA; METHOD; and STRASBERG, LEE.

ACTORS' THEATRE, American membership group formed in 1922 for the presentation of good classic and new plays. Dudley DIGGES was one of the directors, and among the first productions were SHAW's *Candida* and IBSEN's *The Wild Duck*. In 1927 the group joined the company at the Greenwich Village Theatre under Kenneth MACGOWAN. A second American group, also known as the Actors' Theatre, was founded in 1939, and produced a number of plays at the PROVINCETOWN PLAYHOUSE. The new group was intended as a training ground for young actors and for the tryout of new plays; most of its activities were suspended on the outbreak of war in 1941, though it continued to function intermittently until 1947.

ACTORS THEATRE OF LOUISVILLE, the official State Theatre of Kentucky, founded in 1964. First housed in a loft over a store, it moved to a converted railway station and in 1972 to its present home which consists of two theatres, the Pamela Brown Auditorium, seating 637 round a thrust stage, and the Victor Jory Theatre, seating 161. The former presents seven productions of classical and modern plays in a season which runs from Sept. to May, while the latter houses the 'Off Broadway' Series, a programme of provocative plays that has included many American and world premières. Both theatres participate in the annual Festival of New American Plays. Among the plays which have had their first production at the Actors Theatre of Louisville are *Tricks* (1971: N.Y., 1973), based on MOLIÈRE's *Les Fourberies de Scapin*; D. L. Coburn's PULITZER PRIZE-winner *The Gin Game* (1977; N.Y., 1977; London, 1979), seen on Broadway and in London with Hume CRONYN and Jessica TANDY; Marsha Norman's *Getting Out* (1977); and James McLure's *Lone Star* (1979), both of which were also seen in New York. The company makes an annual regional tour, and also runs a free children's theatre.

ACT-TUNES, musical interludes between the acts of plays. That these were customary in the Elizabethan theatre is shown by their mention in a number of stage directions. In the Restoration theatre the act-tunes became very important, and composers like Purcell were commissioned to write them. The introductory music was sometimes known as the Curtain-Music or Curtain-Tune.

ADAM DE LA HALLE (*c.* 1245–*c.*1288), French *trouvère* from Arras (like his famous predecessor Jean BODEL), known also as Adam le Bossu (Hunchback) d'Arras, and the virtual founder of the French secular theatre. His first play, *Le Jeu de la feuillée* (*The Play of the Leafy Bower*—but the title may conceal an untranslateable pun), was written in about 1276 for an outdoor performance in Arras. Bawdy and satirical, it seems to anticipate the later anti-clerical and anti-authoritarian foolery of the Parisian law students (*les clercs de la basoche*). Adam, who was in the service of Count Robert II of Artois, accompanied him to Italy, and in about 1283 produced for the amusement of the Court a pastoral, *Le Jeu de Robin et Marion*, which may have been seen first in Naples. It tells of the encounter of a knight and a shepherdess, and by virtue of its music—for Adam was a fine composer as well as a poet—is sometimes considered to be the first French light opera. First printed in 1822, it was revived in a modern version in Arras in 1896.

ADAMOV, ARTHUR (1908–70), Russian-born French dramatist, who in 1924 became a member of the Surrealist group in Paris. His first play, *La Parodie*, written in 1947, was not produced until 1952, two later plays, *La Grande et la Petite Manœuvre* and *L'Invasion*, having been performed two years earlier. Adamov's early works, including *Le Professeur Taranne* and *Tous contre Tous* (both 1953), have much in common with the Theatre of the ABSURD, as does *Ping-Pong* (1955), a satire on the world of commerce and politics; but with *Paolo Paoli*, an exposure of the corruptions of the French social scene first produced by PLANCHON at Lyons in 1957, Adamov moved towards the epic theatre of BRECHT, whose influence was even more evident in *Le printemps '71* (1961), which dealt with the Paris Commune of 1871, and in *La Politique des Reste* (1963). Two of Adamov's later works were based on GORKY—*Les Petits bourgeois*, produced in 1959, and *Les Âmes mortes*, produced in 1960.

ADAMS, EDWIN (1834–77), American actor, who made his first appearance at Boston in 1853 in Sheridan KNOWLES's *The Hunchback*. At the opening performance of BOOTH'S THEATRE, New York, on 3 Feb. 1869, he played Mercutio to the Romeo of Edwin BOOTH. He was an excellent light comedian; his best known role, however, was Enoch Arden in a dramatization of TENNYSON's poem. He toured in it all over the United States, but made his last appearance in San Francisco, as Iago to the Othello of John MCCULLOUGH. His early death was deplored by actors and audience alike.

ADAMS [Kiskadden], MAUDE (1872–1953), American actress, daughter of the leading lady of the Salt Lake City stock company. At the age of 5 she scored a triumphant success as Little Schmeider in Halliday's *Fritz, Our German Cousin* at the San Francisco Theatre, and also played such parts as Little Eva in one of the many dramatizations of Harriet Beecher Stowe's *Uncle Tom's Cabin*. In 1888 she made her first appearance in New York, and three years later was engaged to play opposite John DREW in H. C. de Mille's *The Lost Paradise*. She first emerged as a star with her performance as Lady Babbie in *The Little Minister* (1897), a part which BARRIE rewrote and enlarged specially for her. Her quaint, elfin personality suited his work to perfection, and she appeared successfully in the American productions of his *Quality Street* (1901), *Peter Pan* (1905), *What Every Woman Knows* (1908), *Rosalind* (1914), and *A Kiss for Cinderella* (1916). She was also much admired as the young hero of ROSTAND's *L'Aiglon* (1900), and in such Shakespearian parts as Viola, Juliet, and Rosalind. In 1918 she retired, not acting again until 1931, when she appeared on tour, in *The Merchant of Venice*, as Portia to the Shylock of Otis SKINNER. In 1934 she went on tour again as Maria in *Twelfth Night* and in 1937 she appeared in New York in Rostand's *Chantecler*.

A.D.C., see CAMBRIDGE.

ADDISON, JOSEPH (1672–1719), English politician and man of letters, author of *Cato* (1713), a tragedy on the French classical model seen at DRURY LANE, where it was well received. Although it contains some fine poetry written in unrhymed heroic couplets, it did not prove theatrically effective, and was seldom, if ever, revived. The part of Cato was originally offered to Colley CIBBER, who declined it, and it was finally played by Barton BOOTH, with Anne OLDFIELD as Lucia. Addison's only other play was a comedy, *The Drummer; or, the Haunted House* (1716), also performed at Drury Lane; but his dramatic theories and criticisms, which form an important part of his work, can be found in several papers of *The Spectator*, which he edited jointly with Sir Richard STEELE *The Tatler*, which he also edited, contains in No. 42 (1709) an amusing mock inventory of the properties and furnishings of Drury Lane.

ADE, GEORGE (1866–1944), American journalist, humorist, and playwright famous for his wisecracks, whose 'fables in slang' brought a new and refreshing idiom to American literature. His plays of contemporary life, of which the most successful were *The County Chairman* (1903), *College Widow* (1904), which added a new phrase to contemporary language, *Just Out of College* (1905), and *Father and the Boys* (1908), were full of homely humour and wit. He was also responsible for the books of several MUSICAL COMEDIES, among them *The Fair Co-Ed* (1909), which made Elsie Janis a star.

ADELAIDE FESTIVAL OF ARTS, see AUSTRALIA.

ADELAIDE GALLERY, see GATTI's.

ADELPHI THEATRE, London, in the Strand, was originally built by John Scott, a wealthy business man, for his daughter, and opened as the Sans Pareil on 27 Nov. 1806 with *Miss Scott's Entertainment*, a mixture of songs and recitations, lantern shows, and firework displays. The venture was sufficiently successful for Scott to make a number of alterations to the building, adding a gallery and later building a new façade on the Strand. It changed hands in 1818, and reopened on 18 Oct. as the Adelphi Theatre, playing mainly MELODRAMA and BURLETTA. The first production at this theatre to exceed 100 performances was MONCRIEFF's *Tom and Jerry; or, Life in London* (1821). In 1837 an adaptation of DICKENS's *Pickwick Papers* heralded a series of plays from his novels, which included *Nicholas Nickleby* (1838) and *Oliver Twist* (1839), the last being *The Cricket on the Hearth* (1845). Under the joint management of Mme CÉLESTE and Ben WEBSTER the theatre housed a series of 'Adelphi dramas', mostly written by BUCKSTONE, the best known being

The Green Bushes (1845) and *The Flowers of the Forest* (1847). In 1858, Webster, who had for some years been sole manager, reconstructed the theatre with a four-tier auditorium holding 1,000 people, reopening it officially on 27 Dec. as the Theatre Royal, Adelphi, by which name it had been known since 1829. BOUCICAULT's *The Colleen Bawn* (1860) and *The Octoroon* (1861) drew large audiences; other successes included DALY's *Leah* (1863), with Kate BATEMAN, and *Rip Van Winkle* (1865), with Joseph JEFFERSON. In 1865 the name of the theatre was again changed, to the Royal Adelphi, and in 1879 it was leased to the Gatti brothers (see GATTI's), who presented a series of 'Adelphi melodramas', among them *Harbour Lights* (1885), by G. R. Sims and H. Pettitt, and *The Union Jack* (1888), by Pettitt and GRUNDY; the leading man in all these was William TERRISS, who was assassinated at the stage door of the theatre by a madman in 1897. In 1901 the theatre was completely rebuilt on the same site and reopened as the Century on 11 Sept. with an American musical, *The Whirl of the Town*; a year later the old name was restored, by popular demand.

From 1908 George EDWARDES made the Adelphi a home of MUSICAL COMEDY, beginning with *The Quaker Girl*, and ending when in 1922 a dramatization of Ethel M. Dell's popular novel *The Way of an Eagle* started a four-month run. A mixture of straight plays, musical comedies, and revues followed, finishing with *The House that Jack Built* (1929), after which the theatre was again closed for rebuilding. The new Adelphi—the fourth on the site—opened on 3 Dec. 1930 with the musical *Ever Green*, the first of a series of successful productions by C. B. COCHRAN. It was followed by KNOBLOCK's adaptation of Vicki Baum's novel *Grand Hotel* (1931) and Noël COWARD's *Words and Music* (1932). In 1936 Eric Maschwitz's *Balalaika* began a long run, which was only exceeded by the revival of Ivor NOVELLO's *The Dancing Years*, which opened on 14 Mar. 1942 and ran for 969 performances. Cochran returned after the Second World War with three productions, *Big Ben* (1946), *Bless the Bride* (1947), and *Tough at the Top* (1949), all with libretti by A. P. Herbert and music by Vivian Ellis. The first production to reach 1,000 performances at the Adelphi was the revue *London Laughs* (1952), which was followed by *Talk of the Town* (1954). Later successes were Jerome Lawrence and Robert E. Lee's *Auntie Mame* (1958), based on Patrick Dennis's novel and starring Beatrice LILLIE, and two musicals by Lionel BART, *Blitz!* (1962) and *Maggie May* (1964). On 15 Dec. 1965 *Charlie Girl*, a musical starring Anna Neagle, opened to bad notices but became the Adelphi's longest-running production, completing 2,200 performances by the time it closed on 27 Mar. 1971. The successful Broadway musical by SONDHEIM, *A Little Night Music*, opened in 1975 with Jean Simmons and Hermione Gingold, and there were successful revivals of the musicals *Irene* in 1976 and *My Fair Lady* in 1979, the latter running for two years.

ADELPHI THEATRE, New York, see GEORGE ABBOTT THEATRE.

ADMIRAL'S MEN, company of actors which made its first appearance at Court in 1585 as 'the admiral's players', their patron, Lord Howard, having been made an admiral earlier in the year. By 1590 they were installed, together with some of STRANGE'S MEN, in the THEATRE; but after a dispute involving young Richard BURBAGE, son of the builder of the playhouse, most of them moved to the ROSE THEATRE, under the management of HENSLOWE. The company lost a few members to the CHAMBERLAIN'S MEN, their only serious rival, when they were formed in 1594, but most of them remained at the Rose, with Edward ALLEYN as their leading actor. They had a large and varied repertory of plays, most of which, except for those by MARLOWE, are lost or forgotten. The temporary retirement of Alleyn in 1597 was a great blow, but he returned in 1600, when the company moved to the new playhouse, the FORTUNE, built for them by Henslowe. On the accession of James I in 1603 the company was re-organized and renamed Prince Henry's Men, and on the prince's death in 1612 became the Palsgrave's Men, under the patronage of the Elector Palatine, husband of James's daughter Elizabeth. In 1621 their theatre burned down, and although it was rebuilt and reopened in 1623 with practically the same company as before, it lost some of its popularity. The combination of plague and the death of James I in 1625 also affected it, and in 1631 the company, after a long and honourable career, was disbanded, some of its members joining the second, newly-formed company of PRINCE CHARLES'S MEN.

ADOLPH, JOHANN BAPTIST (1657–1708), Austrian dramatist, successor to AVANCINI as Court dramatist in Vienna. He took his subjects from many sources—*Philemon et Apollonius* from Avancini, *Carnevale seu Voluntas de Carne Triumphans* from CALDERÓN, and *Das Prallhansenspeil* from the literary dramas of GRYPHIUS. In Adolph the sober splendour of Avancini was transmuted into Viennese cheerfulness, and there are leavening scenes of contemporary peasant life.

ADRIAN [Bor], MAX (1903–73), Irish-born actor who first came into prominence during a season at the WESTMINSTER THEATRE in 1938, when he played Pandarus in a modern-dress *Troilus and Cressida* and Sir Ralph Bloomfield Bonnington in SHAW's *The Doctor's Dilemma.* He was then with the OLD VIC company, and in 1944 joined GIELGUD's repertory company at the HAYMARKET THEATRE, playing a variety of parts, including Tattle in CONGREVE's *Love for Love*. He was however an instinctive REVUE artist, and in 1942 appeared successfully in *Light and Shade*, following it with *Tuppence*

Coloured (1947), *Oranges and Lemons* (1948), *Penny Plain* (1951), *Airs on a Shoestring* (1953), which ran for nearly two years, and *Fresh Airs* (1956). He then returned to New York, where he had first appeared in 1934, to play in Bernstein's musical *Candide*, and back in London was seen in FEYDEAU's *Look After Lulu* (1959). In 1960, as a founder member of Peter HALL's ROYAL SHAKESPEARE COMPANY, he gave outstanding performances as a malevolent Machiavellian Cardinal in WEBSTER's *The Duchess of Malfi*, and a hauntingly melancholic Feste in *Twelfth Night*. One of his finest characterizations was the tetchy, wily, egotistical Serebryakov in CHEKHOV's *Uncle Vanya* at the 1963 CHICHESTER FESTIVAL. Later the same year he joined the NATIONAL THEATRE company, playing the Inquisitor in Shaw's *Saint Joan*, Brovik in IBSEN's *The Master Builder*, and Serebryakov again. His one-man shows, on Shaw in 1966 and on GILBERT and Sullivan in 1969, brought him further renown. Adrian was that rare phenomenon in the theatre, a brilliant and fantastic individualist who could nevertheless fit easily into a company.

ADVENT PLAY, see LITURGICAL DRAMA.

ADVERTISEMENT CURTAIN, see CURTAIN.

AE [George William Russell] (1867–1935), Irish poet, who was connected with the early years of the Irish dramatic movement through his one play, *Deirdre*. Acts I and II were first performed, with the author in the cast, in a friend's house in Dublin at Christmas 1901. AE was then persuaded to finish the play for production by the FAY brothers' IRISH NATIONAL DRAMATIC SOCIETY, and it was performed in a double bill with YEATS's *Cathleen ni Houlihan* in 1902. AE was active in the founding of the future ABBEY THEATRE company, but resigned in the spring of 1904 after a disagreement with Yeats concerning artistic control.

AERIAL GARDENS THEATRE, New York, see NEW AMSTERDAM THEATRE.

AESCHYLUS (525/4–456 BC), Greek tragic dramatist, who was born at Eleusis, near Athens, and die at Gela in Sicily, having won distinction both as a soldier in the Persian War and as a playwright at home and abroad. Later generations regarded him as the first great figure of the Athenian theatre and attributed to him, perhaps with greater reverence than accuracy, the introduction of several features characteristic of Greek play production. He was the first playwright whose works were officially preserved, and after his death they were accorded the unique honour of revival in the festivals, to which normally only new plays were admitted.

Aeschylus is said to have written 90 plays; the titles of 79 are known, but only 7 are extant. These are the *Persians* (472), the *Seven Against Thebes* (469), the trilogy known as the *Oresteia* (the *Agamemnon*, the *Choephori* or *Libation Bearers*, and the *Eumenides*) (458), the *Suppliant Women*, and the *Prometheus Bound*. The dates of the last two are still unknown. The *Suppliant Women* was for a long time thought to be Aeschylus' earliest surviving play, on the basis of its predominantly choric content, but recent papyrus discoveries, together with new textual studies, now suggest a date closer to the *Oresteia*. The *Prometheus Bound* also, because of its comparative simplicity of structure, was once attributed to Aeschylus' early career, but most scholars now place it near the time of his death.

Aeschylus presided over a formative stage in the development of Greek tragedy, and even with the re-dating of his plays it is possible to see in his work a transition from what was primarily a choral dance-drama to a form in which the actor plays a much larger part. Greek tradition, accepted by many modern scholars, considered that tragedy developed out of choral song with the introduction of the first actor, or Protagonist, who was still restricted to solo arias and occasional dialogue with the CHORUS. Aeschylus is said to have introduced a second actor, the Deuteragonist, thereby making possible more complex scenes; and in his later plays he followed the precedent set by SOPHOCLES of employing also a third actor, the Tritagonist, though reserving him for special effects such as the bursting into speech of Cassandra in the *Agamemnon*, or the moment in the *Libation Bearers* when Pylades, having been silent for most of the play, suddenly speaks to remind Orestes that he is under a vow to kill his mother. For the most part, Aeschylus' plays are written primarily for two actors who divide a number of roles between them and a chorus which continues to retain considerable importance.

The size of the chorus used by Aeschylus has been the subject of controversy. Traditionally, the number was originally 50, and Aeschylus may well have retained this for his earlier plays; but at some point he reduced the number to twelve or fifteen. One Greek source associates this reduction with the horrifying effect of the chorus of Furies in the *Eumenides*, which was aid to have thrown the audience into such a panic that the authorities decreed a reduction in the size of the chorus thereafter. This simple story undoubtedly conceals more complex reasons, not the least of which would have been the economics of play production.

Although Aeschylus was regarded in his own day as a master of SATYR-DRAMA, only a few fragments of his work in this boisterous genre have survived. In his tragic plays he appears to have been unique in making the three plays which had to be presented at the festival of DIONYSUS a true trilogy, in the modern sense of the word; not always, for the *Persians*, which in any case is exceptional, being the only surviving Greek tragedy to deal with a contemporary event, seems to have stood alone. From the form of the *Oresteia*, the only complete trilogy to survive, a consistent pattern may be assumed, with the first play offering a statement, the second a counter-statement, and the third a resolution. In the *Agamemnon*, Clytemnestra kills her husband for what she holds to be good and valid reasons; in the *Libation Bearers*, Orestes murders her to avenge his father's death; and in the *Eumenides* the arguments on each side are assessed, and judgement given, by a council of the gods. It may be assumed that the Prometheus trilogy, of which *Prometheus Bound* was the first play, ended similarly with a reconciliation between the offending Titan and the gods whose rights he had infringed. The *Suppliant Women* was also the first play of three; the *Seven Against Thebes* the third; and, judging by what has survived, the scale of these trilogies was hardly less majestic than that of the *Oresteia*. This device of connecting the three plays gave the drama of Aeschylus an amplitude which has never been approached since. The gradeur of his conception was matched by bold dramatic technique, immense concentration, a wonderful sense of structure, and magnificent poetry. He made the utmost use of spectacle and colour; and by virtue of the beauty and strength of his choral odes he might well be regarded as one of the greatest of lyric poets, as well as, possibly, the greatest of dramatists. While appealing irresistibly to Athenian patriotism, he also used the scope provided by the trilogy to discuss the place of man in the universe, the conflicting forces to which the individual is subject, and the nature of divine and human justice. His characters are normally removed from the realm of the everyday, and are depicted as gods, demi-gods, or heroic supermen. For Aeschylus, the argument is more important than the individual, and he tends to draw his personages larger than life; but when he wishes he can also create a character of earthy realism to serve as a touchstone of humanity in the weighty issues involved, as with Oceanus in the *Prometheus Bound*, or Orestes' old nurse in the *Libation Bearers*. By virtue of his many talents, Aeschylus imposed a unity on the theatre which it was soon to lose; as well as writing the plays, he was his own director, chief actor, designer, composer, and choreographer. In his plays, perhaps more than in any others, it is possible to hear the poet's voice speaking clearly and with a single purpose.

AESOPUS, CLAUDIUS, Roman tragic actor of the 1st century BC, much admired by Horace. He was a friend of Cicero, who speaks of him as having great powers of facial expression and fluent gesture. During Cicero's exile Aesopus would often allude to him on the stage, in the hope of swaying public opinion in his favour.

AFINOGENOV, ALEXANDER NIKOLAEVICH (1904–41), Soviet dramatist, who began writing in 1926. His first important play (translated into English by Charles Malamuth as *Fear*) was performed at the Leningrad Theatre of Drama in 1931. Dealing with the conversion to socialism of a psychologist who has claimed that fear governs the U.S.S.R., it was one of the first Soviet plays to combine sound technique and dramatic tension with party propaganda. This fusion was even more apparent in a later play, seen at the VAKHTANGOV in 1934, which, as *Distant Point*, was produced at the GATE THEATRE, London, in 1937. In 1940 Afinogenov, who had been in trouble with the authorities, was restored to favour and had a light comedy, *Mashenka*, produced by ZAVADSKY at the MOSSOVIET THEATRE. A year later he was killed in an air raid, leaving his last play *On the Eve*, dealing with the German invasion of Russia, to be performed posthumously. His untimely death deprived the Soviet Union of a dramatist who might have had world-wide appeal.

AFRANIUS, LUCIUS (*fl.* later 2nd cent. BC), Roman dramatist, probably the best of the writers of the FABULA *togata*, or play based on the daily life of Roman citizens: 44 titles and 400 fragments have survived. His plots appear to have been somewhat more complicated than usual, particularly in such a play as *Vopiscus*, which has more than ten characters and at least a couple of intrigues. He is reputed to have introduced the theme of homosexuality into his plays.

AFRICA. Theatre, in the European sense of scripted plays performed by trained actors in special buildings, has a very short history in Black Africa. All kinds of professional entertainers—storytellers, musicians, dancers, jugglers, acrobats—flourished, as in Europe, five centuries ago in the great Mali Empire which stretched from Senegal to Nigeria; and traditional drama, with its associated arts of storytelling, song, and dance, still plays an important part in the life of most Black African communities. The strength of family and tribal loyalties has preserved the traditions in which history, religion, and social duty are all embodied, despite the disapproval of the missionary and the colonial governor. The importance of such traditions was highlighted by the struggle for political independence, particularly in the 1950s when new university colleges, museums, and other cultural agencies began to document material that had hitherto remained exclusively oral. Although indigenous culture is increasingly threatened by western education, urbanization, and the mass media, it provides a rich source of dramatic material and idiom for contemporary West African dramatists, who draw also on political themes inspired by the experience of liberation from the colonial powers.

A performance of *Hamlet* on a British ship off the West Coast of Africa in 1607 illustrates the long-standing British custom of amateur theatricals, which was eagerly maintained by the élite in all British-ruled territories. Lagos was annexed in 1861; within 20 years repatriated Brazilian slaves had formed a theatre company. In 1897 there was a performance of *The Merchant of Venice* in Yoruba. Lagos newspapers had their dramatic critics, and there was talk of an opera house. The Glover Memorial Hall opened in Lagos in 1899 and remained the only large theatrical facility in the capital for 60 years.

French colonial society lacked an amateur tradition, and it was left to the educators to lay the foundations of the Afro-French theatre. Initially the curriculum of the École Normale William Ponty, opened in Dakar in 1913 to train teachers and government officials, was identical to those of similar institutions in metropolitan France. In the early 1930s, however, especially after Charles Béart became director, the curriculum was adapted to African needs; the study of folk-tales and traditions was introduced, and regional identities were respected. The annual Festival of Indigenous Arts led to vacation tours of original plays, including, in 1937, a visit to the COMÉDIE DES CHAMPS-ÉLYSÉES in Paris, where critics noted similarities between African drama, EURIPIDES, and MOLIÈRE. From these roots grew both the drama of French West Africa and the spectacular dance companies found in most former French colonial territories. Bernard DADIÉ wrote one of his early plays while he was a student at the École William Ponty. Another former pupil, the Guinean Fodéba Keïta, created the first professional Afro-French theatre company, the Théâtre Africain, in 1949; after Keïta abandoned drama for dance, the company was renamed in 1958 the Ballets Africains, which became internationally famous. Guillaume OYONO-MBIA has maintained the comic traditions of Molière in an African context, but Jacques RABÉMANANJARA from the Malagasy Republic is unusual in being a literary playwright whose work is more successful on the page than in production.

In the Gold Coast and Nigeria, too, public schools like Achimota, Accra, and Government College, Ibadan, gave future writers and directors their first taste of theatre, and the new university colleges opened at Accra and Ibadan in 1948 built on these foundations. At the Arts Theatre, Ibadan, a broadly-based drama programme ranged from the Greeks to new African plays: over 100 productions were mounted between 1956 and 1967. The theatre became a regular venue for travelling vernacular companies, and when the School of Drama opened in 1962, under the directorship of Geoffrey Axworthy, one of the most popular Yoruba groups, the Ogunmola Theatre Company, was attached to it for a time; their one production there, a folk opera based on the novel *The Palm Wine Drinkard* by Amos Tutuola, became a classic of the Nigerian theatre. The school developed its own Theatre-on-Wheels to tour the country during vacations, and its graduates have played a leading part in the formation of the many theatres and theatre departments that have opened in Nigeria. Major Nigerian playwrights of recent years include Wole SOYINKA, Ola ROTIMI, John Pepper CLARK, and James Ene HENSHAW.

The School of Music and Drama at the University of Ghana at Legon brought together a number of playwrights and directors already active in the development of drama, notably Efua SUTHERLAND, Felix Morriseau-Leroy, and Joe de Graft. Like the Ibadan school, Legon launched a new generation into the African theatre, notably Ama Atta Aido (*Dilemma of a Ghost*, 1965; *Anowa*, 1970), Martin Owusu (*The Story Ananse Told*, 1971), Kofi Awoonor (*Ancestral Power*, 1972), and Patience Addo (*Company Pot*, 1973). Joe de Graft's own plays, dealing with a sophisticated metropolitan society and the pressures that disrupt it, include *Sons and Daughters* (1964) and *Through a Film Darkly* (1970).

Sierra Leone, though early in the field of higher education (Fourah Bay College opened in 1876), was late in the development of drama. A Krio version of *Julius Caesar*, by T.A.L. Decker, was performed in Freetown in 1964, and Raymond Sarif EASMON wrote two plays about contemporary life, but their successors have not found things easy. Pat Maddy, a talented writer with several published plays, now lives abroad.

East Africa, unlike West Africa, had a large settled white population and thus a stronger tradition of amateur and school dramatics in the western style. The first European professional theatre company in East Africa was formed in 1947 by Donovan Maule (1899–   ), OBE 1963, and the Donovan Maule Theatre in Nairobi opened in 1948. The struggle for independence was more traumatic. Kenya, for instance, was the scene of the bitter Mau-Mau campaign: the hero of James Ngugi's *Trials of Dedan Kimathi* is a Mau-Mau martyr. The Ugandan playwrights Robert Serumaga and John Ruganda, both university-based, wrote political satire at the risk of their freedom; Michael Etherton, director of the University Travelling Theatre in Zambia, was deported. In both Tanzania and Malawi artistic initiatives were soon overtaken by political realities. East Africa had two important advantages: unlike West Africa, with its bewildering variety of languages, it has an African lingua franca in Swahili; and, being a decade or more later in the movement towards independence, it could learn from the West African experience. Determined efforts have been made to take the theatre to the people, mostly in the form of spontaneous traditional performances closer to opera than to the conventional drama.

In Black Africa generally, theatre no longer enjoys the freedom it had in the early period of independence. Since almost all theatrical activity of a formal kind is state-funded it is inevitably conservative in utterance, preferring historical pageantry, song and dance, and domestic drama to political statements. The work of writers like the Nigerians Bode Sowande and Femi Osofisan signals a more critical temper, however, and the building of university theatres and the setting up of performing arts departments offers a context for new movements in play-writing and theatrical presentation.

AFRICAN ROSCIUS, see ALDRIDGE, IRA.

AFTER-PIECE, a short comedy or farce, performed after a five-act tragedy in London theatres of the 18th century, partly to afford light relief to the spectators already present and partly to attract those who found the opening hour of 6 p.m. too early. Half-price was charged for admission. The after-piece was often a full-length comedy cut to one act, but many short plays were specially written for the purpose by such writers as GARRICK, MURPHY, and FOOTE.

AGATE, JAMES EVERSHED (1877–1947), English dramatic critic who numbered among his ancestors the actor Ned SHUTER. From 1923 until his death he was dramatic critic of the *Sunday Times*, succeeding S. W. Carroll and being succeeded by Harold HOBSON. His weekly articles, many of which were collected and published in book form, were vigorous and outspoken, and always entertaining, in spite of his refusal to admit greatness in any actor later than IRVING.

AGATE, MAY (1892–1960), English actress, sister of the above. She studied for the stage with Sarah BERNHARDT, on whom she wrote an interesting book, *Madame Sarah* (1945), and with whom she made her first appearances on the stage in Paris and London. In 1916 she joined Miss HORNIMAN's company at the Gaiety, Manchester, and in 1921 appeared in London, where she had a distinguished, though not spectacular, career, making her last appearance as the Duchess Ludoviska in Elizabeth Sprigge's *Elisabeth of Austria* in 1938. With her husband Wilfred Grantham she made an English adaptation of MUSSET's *Lorenzaccio* as *Night's Candles* (1933), in which Ernest MILTON appeared with great success.

AGATHON, of Athens, Greek tragic poet and a younger contemporary of EURIPIDES, who won the first of his two victories in 416 BC. Few fragments of his work survive, but according to ARISTOTLE he was the first to write choral odes, unconnected with the plot, as entr'actes. ARISTOPHANES laughed at him for effeminacy, but on his death lamented his loss. There is a characer sketch of him in PLATO's *Symposium*.

AGGAS [also ANGUS], ROBERT (*c.*1619–79), English scene painter, much esteemed in the Restoration theatre. He worked for KILLIGREW, and with Samuel Towers painted the elaborate scenery used at DRURY LAND for CROWNE's *The Destruction of Jerusalem* in 1677, later suing the theatre for payment.

AGON. A Greek word meaning 'contest', used in dramatic theory to signify the conflict which lies at the heart of Greek TRAGEDY. It is surmised that the *agon* may have been, in pre-dramatic ritual, an actual physical conflict, involving the death of one of the participants. In drama this becomes transmuted into a clash between two principal characters, though often some notion of physical violence is retained, as for instance in the opposition betwen Pentheus and Dionysus in the *Bacchae* of EURIPIDES. Some scholars postulate an early choral *agon*, surviving in such scenes as the attack by the CHORUS on the principal character in the *Acharnians* of ARISTOPHANES, or the rivalry between the semi-choruses of Old Men and Old Women in his *Lysistrata.* Normally, however, the *agon* takes the form of a debate, where the only weapons are words.

AIKEN, GEORGE L. (1830–76), American actor and playwright, who made his first appearance on the stage in 1848. He is chiefly remembered for his adaptation—the best of many—of Harriet Beecher Stowe's *Uncle Tom's Cabin*, prepared for George C. Howard, who wished to star in it his wife, Aiken's cousin, as Topsy, and his daughter as Little Eva. It was first given at Troy in 1852 and in New York a year later, and its popularity assisted the abolitionist cause; it was constantly revived up to 1924. Aiken, who wrote or adapted a number of other plays, continued to act until 1867.

AINLEY, HENRY HINCHLIFFE (1879–1945), English actor, possessed of a remarkably fine voice and great personal beauty and charm. He made his first success as Paolo in Stephen PHILLIPS's *Paolo and Francesca* (1902), at the ST JAMES's THEATRE under George ALEXANDER, and soon became known as a romantic actor in such plays as Justin McCarthy's *If I Were King* (1902) and Bleichmann's *Old Heidelberg* (1903). He was also good in Shakespearian parts, and in 1912 made a great impression as Leontes in GRANVILLE-BARKER's production of *The Winter's Tale*. A year later he showed his versatility by playing Ilam Carve in Arnold BENNETT's *The Great Adventure*. He was associated in the management of several theatres, including His (now HER) MAJESTY'S, where he was seen in FLECKER's *Hassan* (1923), which provided him with one of his finest parts. Illness then kept him from the stage for some years, but in 1929 he returned to score an instantaneous success as James Fraser in St John ERVINE's *The First Mrs Fraser*, which, with Marie TEMPEST in the title role, ran for 18 months. A year later he was seen as a magnificent Hamlet at a Command Performance, and he finally retired in 1932.

AKIMOV, NIKOLAI PAVLOVICH (1901–68), Soviet scene designer and director, who first attracted attention by his designs for the productions of IVANOV's *Armoured Train 14–69* (1929) and AFINOGENOV's *Fear* (1931) at the Leningrad Theatre of Comedy. Moving to Moscow, he worked at the VAKHTANGOV THEATRE and was responsible for the famous 'formalist' production of *Hamlet* in 1932, in which Hamlet faked the Ghost, and Ophelia, a 'bright young thing', was not mad but drunk: the play was taken off in deference to public opinion. In 1936 Akimov became Art Director of the Leningrad Theatre of Comedy, being responsible for a beautifully staged *Twelfth Night* there. In the 1950s he strove to create an original repertory, and fought for the acceptance of new Soviet comedies. His name is particularly associated with the plays of Evgenyi SHWARTZ, three of which he directed in Leningrad. From 1955 until his death he was on the staff of the Leningrad Theatrical Institute.

AKINS, ZOË (1886–1958), American poet and dramatist, whose first play *Déclassée* (1919) provided an excellent part for Ethel BARRYMORE. This was followed in the same year by an amusing comedy, *Papa*, and by *Daddy's Gone A-Hunting* (1921) and *The Texas Nightingale* (1922). In *The Varying Shore* (1921) she showed a desire to experiment, but her most interesting play was perhaps *First Love* (1926). All her work was overshadowed by the success of *The Greeks Had a Word For It* (1930), which was equally successful in London in 1934. In 1935 she was awarded a PULITZER PRIZE for her dramatization of Edith Wharton's novel *The Old Maid*. She adapted several plays from the French, among them Pujet's *Les Jours heureux* as *Happy Days* (1941) and VERNEUIL's *Pile ou face* as *Heads or Tails* (1947), and from the Hungarian, and also dramatized several more American novels. Her last play was *The Swallow's Nest* (1951).

ALARCÓN Y MENDOZA, see RUIZ DE ALARCÓN Y MENDOZA, JUAN.

ALBANESI, MEGGIE [Margharita] (1899–1923), English actress, who as an 18-year-old student at the Royal Academy of Dramatic Art was awarded the Bancroft Gold Medal for her playing of Lady Teazle in SHERIDAN's *The School for Scandal*. She made her first professional appearance later the same year in a revival of GRUNDY's *A Pair of Spectacles*, and then toured with Fred TERRY. She first came into prominence as Sonia Strong in HEIJERMANS' *The Rising Sun* (1919), and consolidated her reputation in the role of Jill Hillcrest in GALSWORTHY's *The Skin Game* (1920). She made a great success as Sydney Fairfield in Clemence DANE's *A Bill of Divorcement* (1921) and as Daisy in Somerset MAUGHAM's *East of Suez* (1922), and made her last appearance on the stage as Elizabeth in J. Hastings Turner's *The Lilies of the Field* (1923), dying the following December. The intensity and truth of her acting had made a deep impression on her audiences and fellow-players alike, and her early death was much regretted. A plaque in her memory, designed by Eric Gill, was placed in the foyer of the ST MARTIN'S THEATRE, and a scholarship in her name was founded at RADA.

ALBANIA. Unbroken cultural links with ancient Greek Illyrian settlements point to the antiquity of the Albanian theatre, but only in 1912, at the end of five centuries of Turkish rule, could it acquire a modern identity. The first recorded plays, acted in Italian or Latin by amateurs, were tragedies by such writers as Gerolamo De Rada (1819–1903), Antonio (Ndon) Sartori (1819–94), author of *Emira*, the first Albanian drama about Arberesh peasants, and Leonard Martini (1830–1923). In 1879 some Catholic schoolboys staged a Nativity play, and from 1890 onwards greater activity in the theatre, still by amateur writers and actors, was apparent; but they were subject to severe censorship, and until independence patriotic dramatists, working in exile, were forced to write plays mainly to be read. Among them were Sami Bey Frashëri [Sami Shemsettin] (1850–1904), whose drama *Besa* (*The Pledge of Honour*), written in 1880, translated from Turkish and published in Bulgaria, enjoyed great popularity in Albania despite official efforts to suppress it. But it was never acted, except by Albanian émigrés in the U.S.A. Other writers active at this time included Andon Cajupi-Zako (1866–1930), considered one of the founders of the Albanian theatre, whose one-act satire on family life, *Katermbëdhjet Vjeg Dhëndërr* (*The Fourteen-Year-Old Son-in-Law*, 1902) and verse-play on Skanderbeg, *Gjergi Kastriota*, written while in exile in Egypt, are now regularly revived. Contemporary with him were Fath Gjergj Fishta (1871–1940), Catholic poet and the translator of EURIPIDES and MOLIÈRE, and Mihail Grameno (1872–1931), who spent most of his life in Romania; his plays include the satirical *Mallkimi i Gjuhës Shqipe* (*The Curse of the Albanian Language*) and *Vdekja e Piros* (*The Death of Pyrrhus*, 1906). Better known to English scholars because of his translations of Shakespeare (and some of the plays of IBSEN) was Fan Stylian Noli (1882–1965), poet, writer, politician, and bishop, who in 1920 returned from exile in Egypt and the U.S.A. to stand for Parliament, becoming Prime Minister in 1924. Among his many dramas *Izraelitë dhe Filistinë* (*Israelites and Philistines*, 1907) enjoyed considerable success.

Between the two World Wars there was a noticeable increase in theatrical activity, though it remained purely amateur, the various companies being catered for by such writers as Mehdi Frashëri (1872–1963) (nephew of Sami Bey), best remembered for a powerful drama on Skanderbeg, *Trathtija* (1926), and a wartime anti-Fascist play, *Prefekti* (*The Prefect*). Outstanding among his contemporaries was Kristo Floqi (1873–1949), a prolific writer of satirical comedies, of which *Fe* e Kombesi (*Faith and Patriotism*, 1914) is typical. After the Italian occupation of 1940 most of the amateur groups went underground; the foundations of a post-war Communist theatre were laid at that time by the dramatists Besim Levonja and S. Vacari, the actors N. Frashëri and L. Filippi, and the director A. Pano. In 1945 Kol Jakova (1917–), playwright, actor, and director, founded the People's Theatre in Tirana, which he directed himself. Among this own plays given there were the satirical *Dom Pjon* and two 'revenge' tragedies—one historical, *Halili e Hajrija*, in which a widow avenges her husband's murder by the Turks, and one contemporary, *Toka Jonë* (*Our Soil*), first performed in 1955. Jakova also wrote a number of plays dealing with the struggle for the liberation of Albania and socialist reconstruction afterwards. Following the example of the People's Theatre, playhouses opened in a number of other towns, among the outstanding actors being Zef Yubani (1910–58), leading member of the Üsküdar Migjeni Theatre, who gave excellent performances as Harpagon in Molière's *L'Avare*, Stockman in Ibsen's *An Enemy of the People*, and Tuç Maku, the villain of Jakova's *Our Soil*, and Pjeter Gjoka (1913–), who made his début at the theatre in Tirana in 1947. Among his outstanding roles have been Bubnov in GORKY's *The Lower Depths*, Willer in SCHILLER's *Kabale und Liebe*, Halili in Jakova's *Halili e Hajrija*, and the title role in *King Lear*.

Before Albania broke with the U.S.S.R., Soviet and Czech theatre specialists assisted the young Albanian theatre, and Soviet plays bulked large in the repertoire, one notable production being that of POGODIN's *Kremlin Chimes* in 1957 in which Filippi gave a fine performance as Lenin. After the break, a new generation of writers who brought to the theatre the aesthetics of Maoism included S. Pitarka, in whose *Trimi i Mire Me Shokë Shumë* (*A Brave Man Among Many*, 1973) Gjoka played both Palush Dibra and Artan.

A flourishing Geg theatre, as against the Tosk dialect used in Albania proper, exists in Priština in the Kosmet Republic of YUGOSLAVIA. It is directed by Azam Shkreli (1938–), who in 1972 staged there *Erveheja*, by A. Qerozi.

ALBEE, EDWARD FRANKLIN (1857–1930), American showman and theatre manager, who in 1885 joined forces with B. F. Keith to promote the new-style VAUDEVILLE introduced by Tony PASTOR and provide entertainment which would appeal to professional and middle-class audiences. He also vastly improved conditions backstage, and built a number of new variety theatres, including the Keith in Boston in 1894, in which he employed the new technologies of electric lighting, steel-frame construction, and fire-proofing. A shrewd and implacable business man, he was very much concerned with improving the conditions of employment of vaudeville performers, but was staunchly opposed to the rise of trade unionism among those he engaged. By 1920 he owned a vaudeville circuit of some 70 theatres, with an interest in about 300 others, and unlike many of his fellow managers, he stoutly refused to convert them into cinemas.

ALBEE, EDWARD FRANKLIN (1928–), American playwright, grandson by adoption of the above, and one of the few major dramatists to emerge in the United States in the 1960s. He had his first play, a one-acter entitled *The Zoo Story*, performed in 1959 in Berlin; a year later it was seen at the PROVINCETOWN PLAYHOUSE and at the ARTS THEATRE in London. It was followed by several other short plays, including *The Death of Bessie Smith* (1960), which also had its première in Berlin, and *The American Dream*. Performed as a double bill in New York and London in 1961, they embodied Albee's recurrent themes—the need to distinguish reality from illusion and to sustain a humane vision in the deteriorating human and physical environment. Albee had his first success on Broadway with his first full-length play *Who's Afraid of Virginia Woolf?* (1962; London, 1964), about a night of conflict between an ineffectual professor and his sharp-tongued wife, involving the 'death' of their imaginary child; Uta HAGEN created the leading role on both sides of the Atlantic. *The Ballad of the Sad Café* (1963), based on a novel by Carson McCullers, was followed by *Tiny Alice* (1964; ROYAL SHAKESPEARE COMPANY, 1970), which caused a great deal of controversy and was declared by six New York critics to be incomprehensible; Irene WORTH played Miss Alice in both New York and London. It was followed in 1966 by what many regarded as Albee's best play to date, the PULITZER PRIZE-winner *A Delicate Balance* (RSC, 1969). Two adaptations, *Malcolm* (1966), from the novel by James Purdy, and *Everything in the Garden* (1967), from Giles Cooper's play, proved disappointing, as did the double bill of *Box* and *Quotations from Chairman Mao Tse-Tung* (1968). *All Over* (1971; RSC, 1972), concerning quarrels at a dying man's bedside, was another caustic look at family relationships. *Seascape* (1975), which the author directed, depicts a seashore encounter between two couples, each undergoing a process of self-assessment, the situation deriving piquancy from the fact that one of the couples is human—the other belongs to the lizard family. *Seascape* won another Pulitzer Prize, but Albee's later work has found little favour. His short play *Counting the Ways* was premièred at the OLIVIER THEATRE in 1976, with Beryl Reid; its American première, in a double bill with *Listening*, was given by the Hartford Stage Company in 1977, and the plays were not performed in New York until 1979, in an off-off-Broadway production directed by the author. *The Lady From Dubuque* (1980), again with Irene Worth, and *Lolita* (1981), adapted from Nabokov's novel, had only brief runs in New York.

ALBERTI, RAFAEL (1902–), Spanish poet and dramatist, who returned to traditional themes in an endeavour to reform the contemporary Spanish theatre. *El hombre deshabitado* (*The Deserted Man*, 1931) is a modern AUTO SACRAMENTAL, dealing with the theme of original sin. *El adefesio* (*The Odd One*, 1944) is a tragic farce, absurd and grim. *El trébol florido* (*Clover in Flower*, 1940) and *La Gallarda* (1944) are rural dramas in verse, powerful and compelling in their presentation of love, death, and fate.

ALBERY, BRONSON JAMES (1881–1971), English theatre manager, son of James (below), knighted in 1949. He made his first ventures into management at the CRITERION THEATRE, where his mother, Mary Moore, was for many years leading lady and, after the death of her second husband Charles WYNDHAM, manager. When she died in 1931, Albery took control, together with his half-brother Howard WYNDHAM, of the Criterion, the NEW THEATRE, and WYNDHAM'S THEATRE. This joint management proved extremely successful and, particularly in the 1930s, many interesting new plays were seen at all three theatres. After Albery's death his son, Donald Arthur Rolleston (1914–), who had been associated with him for many years, successfully continued his policy, and in 1973 renamed the New Theatre the Albery, in memory of his father.

ALBERY, JAMES (1838–89), English dramatist, whose only memorable play, the comedy *Two Roses*, was first seen at the VAUDEVILLE THEATRE in 1870 and provided Henry IRVING, as the 'decayed gentleman' Digby Grant, with one of his earliest successes in London. A year later Irving appeared as Jingle in Albery's adaptation of DICKENS's *The Pickwick Papers*, given as an afterpiece to Leopold Lewis's *The Bells*. Of Albery's other plays the most successful were taken from the French and, like *The Pink Dominoes* (1877), were written mainly for the CRITERION THEATRE where his wife Mary Moore, later Lady WYNDHAM, was playing leading parts. The best of them was probably *The Crisis* (1878), based on AUGIER's *Les Fourchambaults*.

ALBERY THEATRE, London, see NEW THEATRE.

ALCANO, CIELO D', see ITALY.

ALDEN, JOHN, see AUSTRALIA.

ALDRICH [Lyon], LOUIS (1843–1901), American actor, who as an infant prodigy billed as the Ohio Roscius toured America in such parts as Richard III, Macbeth, Shylock (in *The Merchant of Venice*), and Young Norval (in HOME's *Douglas*). After a break for schooling, he returned to the stage as an adult actor under the name of Aldrich, which he retained for the rest of his life. He spent five years in St Louis, and then went to New York, where he appeared in 1866 in Charles KEAN's farewell performance. He was for some years a member of the STOCK COMPANY in Boston, and from 1873 to 1874 was Mrs John DREW's leading man at the Arch Street Theatre in PHILADELPHIA. He then toured for many years in his most successful part, Joe Saunders in Bartley CAMPBELL's *My Partner*, first produced in 1879, which brought him in a fortune. He was also much admired as Shoulders, a drunken creature in Edward J. Swartz's *The Kaffir Diamond* (1888). He made his last appearance in New York in 1899 and died there during rehearsals for a further appearance under BELASCO.

ALDRIDGE, IRA FREDERICK (1804–67), the first great American Negro actor, who in 1863 became a naturalized Englishman. Little is known of his early years, but he is believed to have been born on the west coast of Africa, a member of the Fulah tribe. In New York he was in the service of WALLACK, and acted Rolla in KOTZEBUE's *Pizarro* and Romeo with a Negro company. Billed as the African Roscius, he made his London début as Othello at the ROYALTY THEATRE in 1826. He then went on tour, and in Belfast had as his Iago Charles KEAN. He was also good as Macbeth, Zanga in Edward Young's *The Revenge*, and Mungo in BICKERSTAFFE's *The Padlock*, and was generally regarded as one of the outstanding actors of the day. He was the recipient of many honours, amassed a large fortune, and married a white woman. He was last seen in England in 1865, and then returned to the Continent, where he had first toured in 1853. He was immensely popular in Germany, where he played in English with a supporting cast playing in German. His Lear was much admired in Russia, the only country in which he appeared in the part.

ALDWYCH THEATRE, London. This playhouse, with its three-tier auditorium and a seating capacity of 1,100, was built for Seymour HICKS, who opened it on 23 Dec. 1905 with himself and his wife Ellaline TERRISS in a revival of their 'dream fantasy', *Bluebell in Fairyland* (1901), under the management of Charles FROHMAN. The venture was not a success, but it is worth noting that in 1911 the STAGE SOCIETY, persuaded by Bernard SHAW, put on at the Aldwych the first production in England of CHEKHOV's *The Cherry Orchard*. The building was damaged during the First World War, but after restoration it reopened and in 1923 had its first outstanding success with *Tons of Money*, a farce by Will EVANS and Valentine transferred from the SHAFTESBURY THEATRE. In the cast were Ralph LYNN and Tom WALLS, and both actors stayed on to appear with Robertson HARE and Mary BROUGH in a succession of so-called 'Aldwych farces' written by Ben TRAVERS.

The series ended in 1933 with *A Bit of a Test*, and the next outstanding productions were from America—Lillian HELLMAN's *Watch on the Rhine* (1943), Robert SHERWOOD's *There Shall Be No Night* (1945), with the LUNTS, and Tennessee WILLIAMS's *A Streetcar Named Desire* (1949), in which Vivien LEIGH gave a fine performance as Blanche du Bois. In 1954 came Christopher FRY's *The Dark Is Light Enough*, with Edith EVANS, and a year later Maxwell ANDERSON's *The Bad Seed*. More successful than either of these was the farce *Watch It, Sailor* (1960), by Philip King and Falkland Cary, a sequel to *Sailor, Beware!* (1955), with Kathleen Harrison as Emma Hornett, which moved to the APOLLO to finish its long run. A month later, in Nov. 1960, the Aldwych became the London home of the ROYAL SHAKESPEARE COMPANY from STRATFORD-UPON-AVON. Extensive alterations were made to the interior, including the installation of a completely new lighting system and an apron stage, with a proscenium opening 31 ft wide. At the same time the seating capacity was slightly reduced, to 1,030. The first season opened on 15 Dec. 1960 with Peggy ASHCROFT in WEBSTER's *The Duchess of Malfi*, and the theatre subsequently housed new plays and revivals of English and foreign classics, as well as productions transferred from Stratford. From 1964 to 1975 (except for 1974) the Aldwych also housed, during the absence of the regular company, the annual WORLD THEATRE SEASON organized by Peter DAUBENY. The first new work presented by the RSC was John WHITING's *The Devils* (1961), and in the same year GIRAUDOUX's *Ondine* and ANOUILH's *Becket* represented recent work from abroad, with BRECHT's *The Caucasian Chalk Circle* following in 1962. *The Hollow Crown* (1961), an anthology on the lives and deaths of English sovereigns devised by John BARTON, proved so successful that it won a place in the repertory. Other notable new productions included DÜRRENMATT's *The Physicists* and Rolf HOCHHUTH's *The Representative* in 1963; the epoch-making and controversial *Marat/Sade*, by Peter WEISS, in Peter BROOK's production of 1964; Harold PINTER's *The Homecoming* (1965), *Landscape* and *Silence* (1969), and *Old Times* (1971); Marguerite DURAS's *Days in the Trees* (with Peggy Ashcroft) and Charles Dyer's *Staircase* (which provided a rare modern comedy role for Paul SCOFIELD) in 1966. From America came Jules Feiffer's *Little Murders* in 1967 and *God Bless* in 1968; and Edward ALBEE's *A Delicate Balance* in 1969, followed by his *Tiny Alice* in 1970 and *All Over* in 1972. Tom STOPPARD's *Travesties* (1974), received with critical acclaim, was seen in both New York and the West End, and ARBUZOV's *Old World* (1976) was produced soon after its world première. Two works by Peter NICHOLS, his musical *Privates On Parade* (1977) and *Passion Play* (1981), were also staged, the former having a subsequent West End run.

A major feature of the company's work has been the revival of English and foreign classics from all periods, beginning with GOGOL's *The Government Inspector* in 1965, followed by VANBRUGH's *The Relapse* in 1967 and O'CASEY's *The Silver Tassie* in 1969. In 1970 there was a remarkably popular revival of BOUCICAULT's *London Assurance*, which continued its run a year later at the NEW THEATRE and was also produced in New York. The virtual 'discovery' of GORKY in the English theatre began at the Aldwych with productions of *Enemies* (1971), *The Lower Depths* (1972)—previously produced unsuccessfully by the company at the ARTS THEATRE in 1962—*Summerfolk* (1974), *The Zykovs* (1976), and *Children Of the Sun* (1979). Other revivals have included GILLETTE's *Sherlock Holmes* (1974), which had a long run in New York, O'NEILL's *The Iceman Cometh* (1976), O'KEEFFE's *Wild Oats* (also 1976), and KAUFMAN and HART's *Once In a Lifetime* (1979), the last two both transferring to the West End. The year 1980 was particularly successful, with a fine revival of O'Casey's *Juno and the Paycock* and two mammoth productions: *The Greeks*, a three-part adaptation by John Barton of 10 Greek plays, mainly by EURIPIDES; and David Edgar's eight-hour adaptation of DICKENS's *Nicholas Nickleby*, which was revived in 1981 and produced in New York.

ALECSANDRI, VASILE, see ROMANIA.

ALEICHEM [Rabinovich], SHOLOM (1859–1916), Jewish writer, who in 1888 was owner and editor of a Kiev newspaper. In 1905 he emigrated to the United States, where a number of plays, based on his novels and short stories of life in the Jewish communities of the Ukraine, were performed in the Yiddish Art Theatres, mainly through the efforts of Maurice SCHWARTZ. Aleichem's characters, simple, kindly, but shrewd, offer considerable scope to the actor, and gave both MIKHOELS in the U.S.S.R. and Muni Wiesenfreund (known in the cinema as Paul Muni) in Germany their first successes. Some idea of Aleichem's work can be obtained from Maurice Samuel's *The World of Sholom Aleichem* (1943). In 1959 his centenary was celebrated by a production at the Grand Palais Theatre—the last surviving Yiddish theatre in London—of his three-act comedy, *Hard to be a Jew*. An adaptation of another comedy, *Tevye the Milkman*, as a musical, *Fiddler on the Roof*, was successfully produced on Broadway in 1964, with Zero MOSTEL as Tevye, and in London in 1967, with the Israeli actor TOPOL, who was succeeded in the part by Alfie Bass.

ALEOTTI, GIOVANNI BATTISTA (1546–1636), Italian theatre architect, designer of the Teatro Farnese in Parma which, begun in 1619, opened in 1628. With the rising passion for operatic spectacles in mind, it consisted of a single-storey auditorium with parallel sides and a rounded end enclosing an open space, suitable for processions, tournaments, or sea-fights, behind which rose a high stage with a proscenium opening. On stage was a complete system of WINGS, probably invented and built by Aleotti himself. The building still stands, but is not normally used for dramatic performances.

ALEXANDER, GEORGE [George Alexander Gibb Samson] (1858–1918), English actor-manager, knighted in 1911. He made his first appearance at Nottingham in 1879, and in 1881 appeared in London, joining Henry IRVING's company at the LYCEUM at the end of the year to play Caleb Deecie in a revival of James ALBERY's *Two Roses*. He remained there, except for a short period with KENDAL and HARE at the ST JAMES'S THEATRE in 1883, until 1889, when he entered into management on his own, first at the Royal Avenue (later PLAYHOUSE) Theatre until 1890 and from 1891 until his death at the St James's. Both managements were artistically and financially successful. At a time when the London theatre was cluttered up with adaptations of French farces, Alexander encouraged new writers in English: of over 80 plays which he produced at the St James's only eight were by foreigners. Among his most important early productions were WILDE's *Lady Windermere's Fan* (1892), PINERO's *The Second Mrs Tanqueray* (1893), and Wilde's *The Importance of Being Earnest* (1895), in which he played John Worthing. A man of distinguished appearance and great charm, he succeeded in investing any part he appeared in with romance and dignity. His greatest success was in the dual role of Rudolf Rassendyll and the King in Antony Hope's *The Prisoner of Zenda* (1896), but he was also much admired as the equally picturesque Villon in Justin McCarthy's *If I Were King* (1902). Also in 1902 he introduced Henry AINLEY to the London stage in Stephen PHILLIPS's verse-drama *Paolo and Francesca*, and a year later scored a further success himself as Karl Heinrich in Bleichmann's *Old Heidelberg* (1903). Perhaps the most typical 'St James's play' was Pinero's *His House in Order* (1906), in which Alexander played Hilary Jesson. The last outstanding production under Alexander's management was Jerome K. JEROME's *The Passing of the Third-Floor Back* (1908), starring Johnston FORBES-ROBERTSON, in which he did not appear, mostly confining himself in his last years to revivals of his former successes.

ALEXANDRA THEATRE, Birmingham, see BIRMINGHAM.

ALEXANDRA THEATRE, London. There have been three theatres of this name, all of which have now disappeared. The first was a hall in the grounds of a pleasure resort in Highbury Barn, in which farces and burlesques were given from 1861 to 1871. The second, in Park Street, Camden Town, opened on 31 May 1873 with an operetta by the manager, Thorpe Pede, and in 1877 was leased by a Madame St Claire, who made an unsuccessful appearance as Romeo. In 1879 it became the Park Theatre. It was burnt down on 11 Sept. 1881 and never rebuilt. The third Alexandra, in Stoke Newington, opened on 27 Dec. 1897 with a production of *Dick Whittington* which established a tradition for lavish Christmas pantomimes, and during the rest of the year was used for transfers from the West End. It closed in Oct. 1940.

ALEXANDRINSKY THEATRE, St Petersburg, see PUSHKIN THEATRE, LENINGRAD.

ALEXIS, later 4th-century BC writer of comedies, whose works fall between those of ARISTOPHANES and MENANDER, whose uncle he was. The fragments of his writings and 130 of his reputed 245 titles that survive, suggest that he was instrumental in developing the stock character of the Parasite, a recurrent figure in Greek MIDDLE COMEDY.

ALFIERI, VITTORIO AMEDEO (1749–1803), Italian dramatist. Born in Asti of a noble and wealthy family, he had an unhappy childhood and left home early to travel throughout Europe, returning to Italy in his twenties. His first tragedy, *Cleopatra*, was given at Turin in 1775 with great success, and was followed by some 20 more of which it has been said that their action 'flies like an arrow to its mark'. The best in performance were probably *Saul* (1782) and *Mirra* (1784). His verse is austere, as befits his conception of tragedy as a dramatic presentation of a great theme from which all lesser matters must be banished; but a production by the Teatro Stabile dell'Aquila in 1967 of *Il Divorzio* (1802), one of the half-dozen comedies he wrote towards the end of his life, shows a remarkable gift for satiric humour. His bicentenary was celebrated by productions of his *Oreste* in Rome, directed by VISCONTI, and of his *Filippo* at Asti, by the PICCOLO TEATRO DELLA CITTÀ DI MILANO.

ALHAMBRA, The, a famous London MUSIC-HALL, whose ornate Moorish-style architecture dominated the east side of Leicester Square for over 80 years. Opened unsuccessfully on 18 Mar. 1854 as an exhibition centre, it was bought by E. T. SMITH and converted into a music-hall seating 3,500 on four tiers. As the Alhambra, a name it retained through successive changes of title (Palace, Music-Hall, Theatre, etc.), it opened on 10 Dec. 1860, one of its first spectacular successes being the trapeze artist Léotard, making his first appearance in London in 1861. In 1870 the theatre lost its licence because Frederick Strange, its manager since 1864, presented there 'an indecent dance'—the can-can. However, a year later Strange was able to get a dramatic licence, for the first time, and under him and later managers the theatre proved successful until on 7 Dec. 1882 it was burned down. It was quickly rebuilt, and reopened a year later, its great days as a music-hall being from 1890 to 1910, when the lavish spectacular ballets starring Catherine Geltzer rivalled those of the EMPIRE. In 1912 Charlot inaugurated a series of REVUES, and in 1916, under Stoll, came George GROSSMITH's picture of London life, *The Bing Boys Are Here*, with George ROBEY and Violet Loraine, which, like its successor *The Bing Boys on Broadway* (1918), had a long run. Diaghilev's Ballets Russes company was at the Alhambra in 1919, returning in 1922 with a splendid but unsuccessful production of *The Sleeping Princess*. In 1923 came the revue *Mr Tower of London*, which made Gracie Fields a star, and in 1931 *Waltzes from Vienna*, featuring the music of the Strauss family, gave the theatre, whose fortunes were beginning to decline, a long run. There were seasons of Russian ballet in 1933 and 1935, and on 12 Aug. 1936 the last production at the Alhambra, *Sim-Sala-Bim*, opened, to close on 1 Sept. The building was then demolished, and the Odeon cinema was erected on the site.

ALIENATION, see VERFREMDUNGSEFFEKT.

ALIZON (*fl.*1610–48), French actor who specialized in playing comic elderly maidservants, particularly in farce, or the heroine's nurse in more serious pieces. He appears to have been with the troupe of the Faubourg Saint-Germain in 1634, when he was transferred to the Hôtel de BOURGOGNE. As with JODELET, his own name was often given to the parts intended for him.

ALKAZI, EBRAHIM, see INDIA.

ALLEN, CHESNEY, see CRAZY GANG.

ALLEN, VIOLA (1867–1948), American actress, who made her first appearance on the stage in 1882, succeeding Annie Russell in the title role of William GILLETTE's *Esmeralda* at the MADISON SQUARE THEATRE. A year or two later she was leading lady to John MCCULLOUGH, and in 1886 played with SALVINI. From 1891 to 1898 she was a member of FROHMAN's stock company at the EMPIRE, where she gained a great reputation as one of the leading actresses of the day. On leaving Frohman she toured for some time as a star, her first venture being Gloria Quayle in *The Christian* by Hall Caine. She was also excellent as Viola in *Twelfth Night*, and appeared in *The Winter's Tale, Cymbeline,* and *As You Like It*. In 1915 she toured with J. K. HACKETT as Lady Macbeth, and made her last appearance in 1916 as Mistress Ford in *The Merry Wives of Windsor*.

ALLEN, WILLIAM (? –1647), English actor, a member of LADY ELIZABETH'S MEN in 1624, when he is known to have played a leading role in MASSINGER's *The Renegado*. He may as a child actor have been apprenticed to Christopher BEESTON, as he became a member of QUEEN HENRIETTA'S MEN on its formation in 1625 under the leadership of Beeston, and played in Thomas Nabbes's *Hannibal and Scipio* at the COCKPIT. On the dissolution of the company a year later, he joined the KING'S MEN. He has sometimes been confused with a William Allen who was an officer in the army during the Civil War.

ALLEYN, EDWARD (1566–1626), English actor, one of the first of whom we have any detailed knowledge, and founder of Dulwich College. In 1583 he was one of the Earl of Worcester's players but probably left them in about 1587 to join the ADMIRAL'S MEN, since he is known to have played with them at the ROSE THEATRE which opened in that year. He soon became their leading actor and achieved a great reputation, being highly praised by NASHE and JONSON and considered the only rival of Richard BURBAGE of the CHAMBERLAIN'S MEN. In 1592 he married Joan Woodward, stepdaughter of HENSLOWE, and became part-owner in his father-in-law's theatrical enterprises, eventually owning both the Rose and the FORTUNE THEATRE, as well as the Paris Garden and several other places of entertainment. He retired in 1603, on the accession of James I, having appeared in many outstanding Elizabethan plays including MARLOWE's *Tamburlaine the Great* (*c.* 1589), *The Tragical History of Dr Faustus* (*c.* 1589), and *The Jew of Malta* (*c.* 1590), as well as GREENE's *Orlando Furioso* (*c.* 1591). Two legends later current about him—that he founded Dulwich College in a fit of remorse after the Devil appeared to him during a performance of *Faustus*, and that he died a pauper in his own charitable foundation—are now discredited. He remained a man of substance to the end, and was able to leave a sizeable fortune to his second wife, daughter of the poet John Donne.

ALLEY THEATRE, Houston, see NINA VANCE ALLEY THEATRE.

ALLGOOD, MOLLIE, see O'NEILL, MAIRE.

ALLGOOD, SARA (1883–1950), Irish actress, sister of Maire O'NEILL, who studied for the stage with Frank FAY and his brother. She joined their IRISH NATIONAL DRAMATIC SOCIETY in 1903, playing Princess Buan in YEATS's *The King's Threshold*, and a year later appeared as Cathleen in SYNGE's *Riders to the Sea*. In Dec. 1904 she appeared in the opening productions of the ABBEY THEATRE and remained there until 1914, some of her later parts being Deirdre in Yeats's play, Mrs Delane in Lady GREGORY's *Hyacinth Halvey*, and Widow Quin in Synge's *The Playboy of the Western World*. In Apr. 1908 she played Isabella in William POEL's opening production of *Measure for Measure* for Miss HORNIMAN's company at Manchester. In 1915 she joined a touring company of John Hartley MANNERS's *Peg o' My Heart*, travelling with them to Australia, and during the tour married her leading man, Gerald Henson. In 1920 she returned to the Abbey Theatre, where she was seen as Mrs Geoghegan in a revival of Lennox ROBINSON's *The White-Headed Boy*; but her finest performances were undoubtedly given as Juno Boyle in Sean O'CASEY's *Juno and the Paycock* (1924), in which she was also seen in London and America, and as Bessie Burgess in his *The Plough and the Stars* (1926). In London in 1936 she made a great success as Honoria Flanagan in BRIDIE's *Storm in a Teacup*. She made her last appearance on the stage in New York in 1940, and thereafter appeared only in films.

ALLIO, RENÉ (1921–), influential French scenographer, who worked for a time with PLANCHON at Lyon, and since the early 1960s has helped to redesign several theatres. He considers that in an age of technology the design and functioning of a theatre must be visible to the spectators; that it must allow for creative experimentation and COLLECTIVE CREATION, and that it must avoid restricting itself to any one shape. This has led him to advocate an adjustable theatre, based on two seating areas, one circular and the other, crescent-shaped, embracing part of the first, various sections of which can be changed into acting areas when required. Such a design would allow ten different combinations of acting and seating areas; further combinations would be made possible by dividing the building into two theatres, back to back. Allio's work in England has included the design for the NATIONAL THEATRE's production of FARQUHAR's *The Recruiting Officer* in 1963.

ALL-RUSSIAN THEATRICAL SOCIETY (V.T.O.), organization founded as a charitable body in 1883 to help actors in need and to fight for the recognition of their professional, civic, and social rights. In 1895 the progressive members, who had found themselves in conflict with Imperial officialdom, formed a separate branch, the Union of Stage Workers, which looked after the interests of dramatists and composers as well as of actors. After the Revolution of 1917 the legal and professional side of the work was taken over by the Union of Art Workers, and the Society concentrated on raising the level of acting and production throughout the country by arranging lectures, providing a forum for the discussion of new plays, and giving artistic aid to the provincial theatres. From 1916 until her death in 1964 the great actress YABLOCHKINA was President of the Society. Other republics of the Soviet Union, such as GEORGIA and the Ukraine, have their own Theatrical Societies, run on similar lines.

ALMA-TADEMA, Sir LAWRENCE, see SCENERY.

ALONI, NISSIM (1926–), Israeli playwright and director, born in Tel-Aviv, who worked in Paris under Jean-Marie SERREAU. His first play, *Abzar Mikol Ha'melech* (*The King the Cruelest of All*, 1953), based on a Biblical theme, was produced by HABIMAH, and attracted attention by its pacifist ideas and the beauty of its language. But it was not until 1961, when Aloni himself directed *Bigdey Ha'melech* (*The Emperor's New Clothes*) for Habimah, that he was recognized as Israel's leading playwright. Two years later he was instrumental in founding the 'Theatre of the Seasons', where his *Ha'anesiha Ha'amerikait* (*The American Princess*) was performed; but in 1966 he returned to writing and directing for other companies where he soon became known for his unorthodox methods, writing or completely rewriting the scripts of his plays during rehearsals. While laying great stress on visual beauty, he also employed a variety of clever devices such as slapstick comedy and quick, bewildering cuts from one scene to another. His works, which show the influence of the Theatre of the ABSURD, are complex and enigmatic, full of poetic imagery side by side with an imaginative use of slang and verbal humour, and are based on myths and *commedia dell'arte* masks in contemporary guise. In some of his later plays, notably *Doda* (*Aunt*) *Liza* (1969) which deals with three generations of an Israeli family, he seems to be moving from a world of fantasy into a locally-based reality.

ALTERNATIVE THEATRE, see COLLECTIVE CREATION, COMMUNITY THEATRE, and FRINGE THEATRE.

ALTWIENER VOLKSTHEATER ('Folk-theatre of Old Vienna'), a tradition of popular entertainment inaugurated by STRANITZKY, whose German company was seen in VIENNA in the early years of the 18th century. Their variant of the HAUPT- UND STAATSAKTION integrated HANSWURST into the courtly classical plot, but left him free to improvise dialogue and comic business. In the hands of Stranitzky's successor PREHAUSER, partnered from 1737 by the BERNADON of KURZ, a world of exotic kings and princes gave way to a Viennese setting populated by local doctors, lawyers, and officers, where comedy took precedence over the original historical or mythological plots. In Germany at this time popular comedy was already succumbing to GOTTSCHED's reforms; in Vienna the Empress Maria Theresia decreed in 1752 that comedy, and particularly extempore comedy which the censors found too elusive, should be banned for the moral and artistic improvement of the drama. The decree was short-lived and ineffective, but it led to the first fully scripted Viennese comedies, written by Philipp HAFNER. The tradition of extempore comedy died with Prehauser in 1769, and folk theatre moved into the suburbs with the opening of the Theater in der Leopoldstadt in 1781, the

Theater an der Wien in 1786, and the Theater in der Josefstadt in 1788. In the first two decades of the 19th century BÄUERLE, GLEICH, and MEISL, with a joint total of some 500 plays, provided the scripts for the fairy-tale plays (*Zauberstücke*) and local farces (*Posse*) which made up the repertories of these theatres, reaching a new peak in 1823 when RAIMUND completed Gleich's unfinished script for *Der Barometermacher auf der Zauberinsel*, and followed it with other similar plays before discovering his true vocation in the writing of local farces, in which he created a gallery of period characters whose witty combination of dialect and inventive and untranslateable word play still comes across today. When Raimund's successor and rival NESTROY retired in 1860, operettas took over the function and theatres of the old folk tradition.

ÁLVARES, AFONSO, see PORTUGAL.

ÁLVAREZ QUINTERO, SERAFÍN (1871–1938) and JOAQUÍN (1873–1944), Spanish dramatists, brothers and collaborators in about 200 light comedies, of which the first, *Esgrima y amor* (1888), was seen in Seville while they were still in their teens. These charming trifles, based on the characteristic life and customs of Andalusia and suffused with kindly tolerance and gentle good humour, act well and compensate for their somewhat slight plots by their abundant wit and often racy dialogue. Several of them were translated into English by Helen and Harley GRANVILLE-BARKER, and have been seen in London and New York, among them *Fortunato* and *The Lady from Alfaqueque* (1928; N.Y., 1929); *A Hundred Years Old* (1928; N.Y., 1939); *The Women Have Their Way* (1933; N.Y., 1939); *Doña Clarines* (1934); and *Don Abel Wrote a Tragedy* (1944).

ALVIN THEATRE, New York, at 250 West 52nd Street between Broadway and 8th Avenue. A handsome building with an Adam-style interior seating 1,344, this opened on 22 Nov. 1927 with *Funny Face*, a MUSICAL COMEDY with music by Gershwin. It took its name from the first names of Alex A. Aarons and Vinton Freedley, who built it and retained control until 1932, using it mainly for musical shows with such stars as the Astaires, Ginger Rogers, and Ethel MERMAN, but also for occasional straight plays, among them revivals of O'NEILL's *Mourning Becomes Electra* in 1932 and of the famous stage version of Harriet Beecher Stowe's *Uncle Tom's Cabin*, with Otis SKINNER, in 1933, and the first production of *Mary of Scotland* (also 1933) by Maxwell ANDERSON. In 1935 Gershwin's opera *Porgy and Bess* was first seen, and in 1937 KAUFMAN and HART's *I'd Rather Be Right* inaugurated a series of successful productions, including *The Boys from Syracuse* (1938), a musical based on *The Comedy of Errors*, and the LUNTS in SHERWOOD's *There Shall Be No Night* (1940). Margaret WEBSTER directed *The Tempest* in 1945, with Canada LEE as Caliban, and later successes included Anderson's *Joan of Lorraine* (1946), starring Ingrid Bergman; *Mister Roberts* (1948) by Heggan and Logan; Sidney KINGSLEY's dramatization of Koestler's novel *Darkness at Noon* (1951); *A Funny Thing Happened on the Way to the Forum* (1962), a musical drawn from PLAUTUS and starring Zero MOSTEL; STOPPARD's *Rosencrantz and Guildenstern are Dead* (1967); and *The Great White Hope* (1968) by Howard Sackler, which came to Broadway from the ARENA STAGE in Washington. In the 1970s the theatre staged three successful musicals: SONDHEIM's *Company* (1970); *Shenandoah* (1975), based on the film of that name; and *Annie* (1977), based on the comic strip 'Little Orphan Annie', which ran into the 1980s.

AMATEUR DRAMATIC CLUB, see CAMBRIDGE.

AMATEUR THEATRE, in modern terms an institution for those who regard acting and the staging of plays, however seriously, as a leisure-time occupation, and not as a means of earning a living. It is probably true to say that in most countries the theatre was founded by amateurs: the priests, choirboys, temple dancers, ordinary citizens, and merry-making peasants who originally provided religious spectacles and secular farces for the amusement of their fellows were certainly not professionals, though it was from among them that professional actors first emerged. With the establishment of permanent professional companies, amateur acting became the prerogative of the leisured classes, from the Court down to the moneyed middle classes, and of students, for whom it was often a means of education. In England it reached its height in the 19th century, when many of the best-known amateur societies were founded, among them the Old Stagers, who have given performances closely associated with the Canterbury Cricket Festival since 1842; the Manchester Athenaeum Dramatic Society, which has presented plays regularly since 1854; and the student societies of CAMBRIDGE (A.D.C.) founded in 1855 and OXFORD (OUDS) founded in 1885. The dramatic societies of other universities, though formed much later, have also done excellent work, as can be seen by the productions at the annual student drama festival organized by the National Union of Students.

Amateur organizations can be of four kinds: the occasional group which draws its members from a particular body of people—church, school, or factory—for a particular purpose, often to raise funds for charity; the small but more permanent group connected with a stable organization such as a Women's Institute, a Youth Club, or a Community Centre; the dramatic or operatic societies which flourish mainly in large towns and sometimes reach a high standard, though they are more inclined to play safe with a noted West End success than to embark on experimental work; and the Little Theatres which, though fewer in number, are important for artistic and aesthetic reasons. These last usually own or lease their own theatres, and present every season a number of productions varying from three or four to a dozen or more. Most of them are members of the Little Theatre Guild of Great Britain, founded in 1946, which publishes an annual report of their activities, and arranges conferences and discussions. One of the most interesting is the MADDERMARKET THEATRE at Norwich, which houses the Norwich Players, founded in 1911. The oldest is the Stockport Garrick Society, which dates from 1901.

Although the most significant amateur activity takes place outside London, there are—since the demise of the excellent St Pancras People's Theatre, which lasted from 1926 to 1940—three important centres in the capital—the QUESTORS THEATRE at Ealing, the TOWER THEATRE in Islington, and the MOUNTVIEW THEATRE SCHOOL, Crouch Hill. The left-wing theatre UNITY was something of a hybrid, having during its career been sometimes amateur and sometimes professional.

There are several organizations which form a unifying element in British amateur drama, among them the British Drama League, founded in 1919 by Geoffrey Whitworth (1883–1951) and now known as the British Theatre Association, and the National Operatic and Dramatic Association founded in 1899. Scotland has a Scottish Community Drama Association founded in 1926 and Wales an Amateur Drama Association formed in 1965. Ireland also has an Amateur Drama Council, founded in 1952, and an important amateur drama festival held annually since 1953 in Athlone, with the winning play being seen later in Dublin. In 1952 an International Amateur Theatre Association was founded with E. Martin BROWNE (then director of the B.D.L.) as its first president. It holds conferences every two years in different countries and a four-yearly international amateur theatre festival in Monaco. In 1968 it established its permanent secretariat in The Hague.

The amateur movement has a natural affinity with the use of drama in education, one outcome of which was the foundation in 1956 by Michael Croft, formerly a teacher, of the NATIONAL YOUTH THEATRE. There are also a number of UNIVERSITY DEPARTMENTS OF DRAMA, not only in England but in many European countries. But the country in which the academic amateur drama flourished most strongly was undoubtedly the United States of America, where as early as 1925 YALE had a post-graduate Department of Drama under George Pierce BAKER, and many college playhouses opened in the wake of the amateur academic movement. In a country dominated by the standards of Broadway, and to a certain extent off-Broadway, productions, which attract the major talents in playwriting and acting, the local theatres across the U.S. which adhere to their amateur status may still continue to contribute serious and stimulating work in outlying regions, and, like the amateur theatres outside London in the 1930s, keep alive the spirit of theatre in remote places seldom visited by professionals. That this can be done was proved by the Little Theatre Movement in Canada, a country which has seen its widespread and flourishing amateur movement give rise in a comparatively short time to a soundly-based and rapidly-growing professionals theatre.

AMBASSADORS THEATRE, London, small playhouse in West Street, near St Martin's Lane, which opened on 5 June 1913. It had little success until in 1914 COCHRAN began to stage a series of intimate REVUES there starring Alice DELYSIA. From 1919 to 1930 it was leased by H. M. HARWOOD, whose productions included his own play *The Grain of Mustard Seed* and Lennox ROBINSON's *The White-Headed Boy* (both 1920). In 1925 O'NEILL's *The Emperor Jones* occasioned Paul ROBESON's first appearance in London. Sydney CARROLL took over the theatre in 1932 and under him Vivien LEIGH made her successful West End début in Carl STERNHEIM's *The Mask of Virtue* (1935), translated by Ashley DUKES. A later success was John Perry and M. J. Farrell's *Spring Meeting* (1938), with the inimitable Margaret RUTHERFORD as Bijou. *The Gate Revue*, transferred from the GATE early in 1939, had a long run, followed by a sequel, *Swinging the Gate* (1940), and the intimate revues *Sweet and Low* (1943), *Sweeter and Lower* (1944), and *Sweetest and Lowest* (1946), the work mainly of Alan Melville. On 25 Nov. 1952 the record-breaking run began of Agatha CHRISTIE's *The Mousetrap*, which occupied the theatre until 1973 when it moved to the ST MARTIN's. A year later Athol FUGARD's *Sizwe Bansi is Dead* and *The Island* came from the ROYAL COURT. The Ambassadors has since housed successful productions of Denis Cannan's *Dear Daddy* (1976) and James SAUNDERS's *Bodies* (1979), the latter seen earlier at the HAMPSTEAD THEATRE CLUB.

AMBASSADOR THEATRE, New York, at 215 West 49th Street, between Broadway and 8th Avenue. Under the management of the Shubert brothers, this theatre, with a capacity of 1,121, opened on 11 Feb. 1921 with a musical, *The Rose Girl*, and later the same year scored a success with a romantic musical drawn from the life and music of Schubert, *Blossom Time* (known in England as *Lilac Time*). The theatre several times housed visiting companies, including the ABBEY THEATRE from Dublin in 1937 in a repertory of modern Irish plays. It was given over to radio and television for some years, but reverted to straight plays in 1956, since when it has housed Meyer Levin's *Compulsion* (1957), Rama Rau's *A Passage to India* (1962), based on E. M. Forster's novel, Peter WEISS's *The Investigation* (1966), and four short plays by Robert ANDERSON under the title *You*

Know I Can't Hear You When the Water's Running (1967). Frank DUNLOP's production of *Scapino*, based on MOLIÈRE's *Les Fourberies de Scapin*, was seen here in 1974, and in 1978 the theatre housed *Eubie!*, a REVUE with over 20 songs by Eubie Blake. A Gospel musical, first seen at the LYCEUM in 1976, *Your Arms Too Short to Box With God*, was revived here in 1980.

AMBIGU, Théâtre de l', Paris, opened in 1769 by Audinot as L'Ambigu-Comique, and one of the first theatres to be built on the BOULEVARD DU TEMPLE. To circumvent the monopoly of the COMÉDIE-FRANÇAISE, which lasted until the Revolution of 1789, the theatre was at first used for marionette and children's shows, but later for the melodramas of such writers as PIXÉRÉCOURT and Bouchardy. Burnt down in 1822, it was rebuilt on the nearby Boulevard Saint-Martin, its original site being used for the FOLIES-DRAMATIQUES, and throughout the 19th century it was a popular home of melodrama, with dramatizations of such novels as the elder DUMAS's *Trois Mousquetaires*, HUGO's *Notre-Dame de Paris*, and Sue's *Le Juif errant*, which played to audiences of some 2,000 people. Its fortunes declined as its audiences were drawn away by the cinema and in 1971, after an abortive attempt to revive melodrama as a curiosity, it was demolished.

AMEIPSIAS, ancient Greek comic poet, contemporary and rival of ARISTOPHANES in OLD COMEDY. In one of his plays he is known to have ridiculed Socrates, and his *Revellers* defeated Aristophanes' *Birds* in 414 BC. Only fragments of his works survive.

AMERICAN ACADEMY OF DRAMATIC ARTS, New York, see SCHOOLS OF DRAMA.

AMERICAN ACTORS' EQUITY ASSOCIATION, trade union for professional American actors, founded on 16 May 1913 and affiliated to the American Federation of Labor. Unlike the BRITISH ACTORS' EQUITY ASSOCIATION, it deals only with performers in the legitimate theatre. It called its first strike in Aug. 1919, and was successful in gaining official recognition for itself and better conditions for its members. A second strike in 1960 resulted in further improvements in members' contracts, and in 1961 a policy of racial non-discrimination was set on foot. In recent years the union has suffered some reversals. A new pay code for actors working in off-off-Broadway theatres, enacted in 1979, was disputed by members who felt that the increased wages would force many small theatres to close, leading to more unemployment in the profession. The Equity Library Theatre, which opened in the Master Institute on Riverside Drive in 1941, in association with the New York Public Library, functions as a showcase for actors, directors, and technicians.

AMERICAN COMPANY, small troupe of professional actors, made up of the remnants of the elder HALLAM's company, and a company under David DOUGLASS, with Hallam's widow (whom Douglass married) as leading lady and his son Lewis as leading man. The name was first used in a notice of their presence at Charleston in 1763–4, and was retained during successive changes of management, John HENRY and Lewis HALLAM the younger succeeding Douglass, and being in turn succeeded by HODGKINSON and, in 1796, by DUNLAP. It was the first professional company to produce a play by an American—*The Prince of Parthia*, by Thomas GODFREY—and also staged several plays by Dunlap, who was still with the company when it went from the old JOHN STREET THEATRE to the new PARK THEATRE, opening on 28 Feb. 1798. In 1795 the elder Joseph JEFFERSON became a member of the company, and remained with it until 1803. The company's identity was lost when, in 1805, Dunlap went bankrupt and retired, the Park Theatre being taken over by Thomas Abthorpe COOPER, who had been with the company for some years. Its virtual monopoly of acting in the United States was challenged only by WIGNELL's company at the CHESTNUT STREET THEATRE in Philadelphia.

AMERICAN CONSERVATORY THEATRE, San Francisco, California, originally the Columbia, was opened in 1910, one of the eight theatres built to replace those destroyed in the earthquake and fire of 1906, and the only one still in professional full-time operation. Seating 1,456, it was known briefly in the 1920s as the Wilkes and then the Lurie, and in 1928 became the Geary. Since 1967 it has housed a company, founded in Pittsburgh in 1965, which was invited by the Chamber of Commerce to play in San Francisco, opening its programme with MOLIÈRE's *Tartuffe*. It is now the largest and most active regional theatre in the U.S.A., playing an annual season of 33 weeks in true repertory, and presenting each season 10 plays drawn from classical and modern sources, including the latest offerings of New York and London. In 1972 a programme of Plays in Progress was initiated, offering five or six new plays each season, most of them previously unproduced, in a 'workshop' atmosphere in the small Playroom, seating 49. Since its inception the company has attached great importance to theatre training, and since 1978 it has been empowered to offer a master's degree in acting. It also presents guest production, including musicals such as *Hair* and *Godspell*, and has been host to important companies from overseas, among them the ROYAL SHAKESPEARE and the NATIONAL THEATRE companies from London. Tours of the U.S.A. are undertaken regularly, and the company has visited the U.S.S.R. and Japan.

AMERICAN LABORATORY THEATRE, a training school and production company founded by Richard Boleslavsky (1889–1937) and Maria Ouspenskaya (1876–1949) in New York in the 1920s. Formerly members of the MOSCOW ART THEATRE, they trained actors and directors in the techniques of STANISLAVSKY, of which they became the outstanding American exponents. Their students included Lee STRASBERG and Harold CLURMAN, who together in 1931 founded the GROUP THEATRE. Among the plays which the American Laboratory Theatre produced were WILDER's *The Trumpet Shall Sound* in 1926, Clemence DANE's *Granite* in 1927, and Jean-Jacques BERNARD's *Martine* and SCHNITZLER's *The Bridal Veil* in 1928.

AMERICAN MUSEUM, New York, on the south-east corner of Broadway and Ann Street, was opened as a showplace in 1841 by Phineas Taylor Barnum (1810–91), an astute trickster, self-styled 'the Prince of Humbugs', the discoverer of General Tom Thumb [Charles Stratton] and later part-owner of Barnum and Bailey's Circus, which he publicized as 'the Greatest Show on Earth'. In his new museum, which contained some 600,000 objects, the prize exhibit was a fake mermaid. By 1849 the Museum had lost its appeal and become a theatre with a good stock company and some visiting stars; but it still retained its old name. It was enlarged in 1850, reopening on 17 June with SEDLEY-SMITH's melodrama *The Drunkard*. Barnum sold it in 1855, being short of money, but five years later was able to buy it back. Under his management plays were again gradually ousted by freaks, baby-shows, and boxing contests until on 13 July 1865 the building burned down. Barnum moved temporarily to the Winter Garden, later the METROPOLITAN-THEATRE, and on 6 Sept. 1865 opened his New American Museum at 539–41 Broadway, which since 1859 had been the Chinese Rooms. It was taken over by Van Amburgh for his menagerie in 1867, but plays were evidently still being given there, as it was during the run of a dramatization of Harriet Beecher Stowe's *Uncle Tom's Cabin* that on 3 Mar. 1868 the second Museum was burnt to the ground and never rebuilt.

AMERICAN MUSIC HALL, New York, see AMERICAN THEATRE.

AMERICAN NATIONAL THEATRE AND ACADEMY (Anta), organization founded on 5 July 1935, when President Roosevelt signed a charter permitting the establishment of a nation-wide, tax-exempt institution which was to be 'a people's project, organized and conducted in their interest, free from commercialism, but with the firm intent of being as far as possible self-supporting'. The charter carried with it no government grant, and the existence of the FEDERAL THEATRE PROJECT, together with the subsequent outbreak of the Second World War, made it difficult to raise money privately; it was not until 1945 that any headway could be made. The Board of Directors, formed originally of people interested in, but not connected with, the theatre, was then reorganized to include a number of outstanding theatre personalities and the heads of such organizations as AMERICAN ACTORS' EQUITY, and in 1948 Anta became the U.S. Centre of the INTERNATIONAL THEATRE INSTITUTE. Two years later it acquired the former Guild Theatre as its headquarters, renaming it the Anta Playhouse and running it for some years as an experimental theatre, after which, as the ANTA THEATRE, it was leased to commercial managements as a source of income, Anta retaining part of the building as offices. In 1963, pending the completion of the Lincoln Center, Anta was responsible for the erection of a temporary structure, the WASHINGTON SQUARE THEATRE, to house the VIVIAN BEAUMONT THEATRE repertory company. In 1968 the Anta Theatre was presented to the nation to serve as a performing arts centre for non-profit-making groups under the government-subsidized National Council on the Arts.

AMERICAN NEGRO THEATRE, organization founded in 1940, under the direction of Abram Hill. Its first production, in the New York Public Library Theatre on 135th Street, was Theodore Browne's *Natural Man* (1941), and among its later productions were two plays by Hill, *Walk Hard* (1944) and *On Strivers' Row* (1946), Walter Carroll's *Tin Top Valley* (1947), and, in a programme at the Harlem Children's Center in 1949, SYNGE's *Riders to the Sea*. The great success of this group was Philip Yordan's *Anna Lucasta*, starring the Negro actor Frederick O'Neal, which, after an initial production at the Library Theatre in 1944, moved to Broadway, where it ran for nearly three years. In 1947 it was seen with equal success in London. In 1953 the company toured in England and Europe, but after its return to the United States apparently ceased to function.

AMERICAN OPERA HOUSE, New York, see CHATHAM THEATRE (1).

AMERICAN PLACE THEATRE, New York, at 111 West 46th Street, was founded in 1964 by Wynn Handman and Sidney Lanier, pastor of St Clement's, in which the theatre found its first home. Productions of new and controversial plays, exclusively by American writers, were mounted in the church, services being held in the set of the current production. In 1971 the company moved to a theatre in a newly-built Manhattan skyscraper (as permitted under revised building regulations) seating about 300, and incorporating an experimental studio, the Sub-Plot, in the basement. The opening production was a double bill consisting of Ronald Ribman's *Fingernails Blue As Flowers* and Steven Tesich's *Lake Of the*

Woods; later productions included Sam SHEPARD's two one-act plays *Killer's Head* and *Action* (1975).

AMERICAN REPERTORY THEATRE, Harvard University, see HARVARD UNIVERSITY.

AMERICAN REPERTORY THEATRE, New York, see CRAWFORD, CHERYL; LE GALLIENNE, EVA; and WEBSTER, MARGARET.

AMERICAN SHAKESPEARE THEATRE, Stratford, Connecticut, originally the American Shakespeare Festival Theatre, which houses an annual summer festival, was founded through the initiative of Lawrence LANGNER. Designed by Edwin Howard and seating 1,534, it stands on the bank of the Housatonic river, its octagonal shape being based on that of Shakespeare's GLOBE THEATRE in London, and seeks to combine modern and traditional forms in a flexible, functional whole. It opened on 12 July 1955 with Raymond MASSEY and Christopher PLUMMER as Brutus and Mark Antony in *Julius Caesar*, which shared the repertory with *The Tempest*, in which Massey played Prospero and Plummer Ferdinand. The following year the repertory was extended to include three productions over a longer season, and the nucleus of a permanent company was established. The festival enhanced its reputation under John Houseman, its Artistic Director from 1956 to 1959, and in 1958 Katharine HEPBURN, who had already been seen as Beatrice in *Much Ado About Nothing* and Portia in *The Merchant of Venice* during the 1957 season, played the former role on the company's first national tour. Since 1959 there have been special pre-season programmes for students, and the repertory now includes classic plays other than those of Shakespeare as well as revivals of modern American plays, while the work of new American playwrights is seen in its workshop. A number of famous actors and actresses have been seen in the theatre, including Morris CARNOVSKY since 1956, notably as King Lear, while Katharine Hepburn returned in 1960 as Viola in *Twelfth Night* and Cleopatra in *Antony and Cleopatra*; in 1961 Jessica TANDY appeared as Lady Macbeth and in 1969 Kate REID was seen in CHEKHOV's *Three Sisters* and as Gertrude in *Hamlet*; she returned in 1974 to play the Nurse in *Romeo and Juliet* and Big Mama in Tennessee WILLIAMS's *Cat on a Hot Tin Roof*. In 1981 James Earl JONES and Christopher Plummer played Othello and Iago, the latter also playing the title role and Chorus in *Henry V*. In 1972 the word 'Festival' was dropped from the theatre's name, and in 1977 the Connecticut Center for the Performing Arts was established, the theatre's programme being extended to include appearances by guest artists and national touring companies outside the festival season.

AMERICAN THEATRE, New York, at 260 West 42nd Street, on the east side of 8th Avenue, opened on 22 May 1893, with *The Prodigal Daughter* by Pettit and Augustus HARRIS. In 1908 it was renamed the American Music Hall, and after a somewhat chequered career it became a burlesque house in 1929. On 18 Dec. 1930 it was badly damaged by fire, and was demolished in 1932. Another American Music Hall opened on 14 March 1934, in a converted church at 139–41 East 55th Street, with a production of SEDLEY-SMITH's *The Drunkard*, which ran for 277 performances.

When E. L. DAVENPORT took over the CHAMBERS STREET THEATRE for a season in 1857, it was called the American Theatre; and during HAMBLIN's long period of management the BOWERY THEATRE was for a time re-christened the American Theatre, Bowery.

AMERICAN THEATRE ASSOCIATION (A.T.A.), see UNIVERSITY DEPARTMENTS OF DRAMA.

AMERICAN THEATRICAL COMMONWEALTH COMPANY, see LUDLOW, NOAH, and SMITH, SOL.

AMES, WINTHROP (1871–1937), American director, who used the money inherited from his father, a railroad capitalist, to back non-commercial ventures in the theatre. In 1904 he took over the Castle Square Theatre in Boston, and later, in an effort to establish a true repertory theatre in New York, built three playhouses there: the New Theatre, later known as the CENTURY, in 1909, the LITTLE THEATRE in 1912, and the BOOTH THEATRE in 1913. None of his attempts was successful, but in spite of many setbacks he did good work for the American theatre up to his retirement in 1932. Among his productions were a number of plays by Shakespeare and other classics, some revivals of GILBERT and Sullivan, and such modern plays as HOUSMAN and GRANVILLE-BARKER's *Prunella* in 1913, Alice Brown's *Children of Earth* (1915), winner of a prize of 10,000 dollars offered by Ames for a new American play, MAETERLINCK's *The Betrothal* in 1918, Clemence DANE's *Will Shakespeare* in 1923, and GALSWORTHY's *Old English* in 1924.

AMORIM, GOMES DE, see PORTUGAL.

AMPHITHEATRE (*amphitheatrum*), a Roman building of elliptical shape, with tiers of seats enclosing a central arena. It was not intended for dramatic performances, which in ancient theatres were always given in front of a permanent backscene, but for gladiators and wild beast shows, and mimic sea battles. In plan it might be described as a double theatre, which may account for Pliny's description of an amphitheatre, said to have been built by Curio in 50 BC, which consisted of two separate theatres, back to back, capable of being pivoted to meet along their straight sides and so form a single amphitheatre. The first amphitheatre was probably that built by Julius Caesar in 46 BC. The most famous was the Colosseum in Rome, completed in AD 80, and said to be capable of seating 87,000 spectators, which is still extant.

AMPLEFORTH SWORD PLAY, see MUMMERS' PLAY.

ANCEY [de Curnieu], GEORGES (1860–1926), French dramatist, one of the best of the naturalistic writers who followed in the steps of Henry BECQUE. His plays, which were produced by ANTOINE at the THÉÂTRE LIBRE, were even franker than Becque's, and perhaps more than any other deserve to be described as *comédies rosses*. The best were probably *Les Inséparables* (1889), in which two friends share the same mistress, and *L'École des veufs* (also 1889), in which a father and son find themselves in the same position. Ancey's plays went out of fashion with the decline of NATURALISM and have not been revived.

ANDERSON, JOHN MURRAY, see REVUE.

ANDERSON, JUDITH [Frances Margaret Anderson-Anderson] (1898–), actress from Australia, appointed D.B.E. in 1960. She made her first appearance on the stage in Sydney in 1915, and afterwards toured Australia for 2 years. She then went to the United States, where she first adopted her stage name, and after some experience in stock companies and on tour, made a great success on Broadway as Elise in Martin Brown's *Cobra* (1924). She was later seen in PIRANDELLO's *As You Desire Me* (1931) and O'NEILL's *Mourning Becomes Electra* (1932). In 1936 she played Gertrude to GIELGUD's Hamlet in New York, and a year later she made her first appearance in London, giving an outstanding performance as Lady Macbeth to the Macbeth of Laurence OLIVIER at the OLD VIC. In 1947 she gave a superb rendering of the name-part in EURIPIDES' *Medea*, in a new adaptation by Robinson Jeffers in which she also appeared in Berlin, Paris, and Australia (with the ELIZABETHAN THEATRE TRUST). She was seen again at the Old Vic in 1960, as Madame Arkadina in CHEKHOV's *The Seagull*, and then toured the United States in a recital of scenes from her most famous parts. In 1970, at the age of 71, she toured as Hamlet, in emulation of Sarah BERNHARDT.

ANDERSON, LINDSAY (Gordon) (1923–), English director and critic, whose first productions, at the ROYAL COURT THEATRE, were Kathleen Sully's *The Waiting of Lester Abbs* (1957), Willis HALL's *The Long and the Short and the Tall* and John ARDEN's *Serjeant Musgrave's Dance* (both 1959), and a musical, *The Lily-White Boys* (1960). For the CAMBRIDGE THEATRE he directed *Billy Liar* (also 1960), by Keith Waterhouse and Willis Hall, and in the same year was responsible for the production of FRISCH's *Andorra* by the NATIONAL THEATRE at the OLD VIC. He returned to the Royal Court to direct Frisch's *The Fire Raisers* (1961) and GOGOL's *The Diary of a Madman* (1963), which he adapted himself in collaboration with Richard Harris, who played the sole character. In 1966 Anderson was at the CHICHESTER FESTIVAL THEATRE, where he directed CHEKHOV's *The Cherry Orchard*, and three years later joined William GASKILL and Anthony Page in running the Royal Court, where he later directed David STOREY's *The Contractor, In Celebration* (both 1969), and *Home* (1970). He continued his association with Storey in *The Changing Room* (1973), *The Farm, Life Class* (both 1974), and *Early Days* (1980), the last at the National Theatre, starring Ralph RICHARDSON, whom he had previously directed in William Douglas HOME's *The Kingfisher* (1977); he also directed *The Kingfisher* in New York in 1978, with Rex HARRISON.

ANDERSON, MARY (1859–1940), American actress, who made her first appearance at the age of 16 in Louisville, playing Shakespeare's Juliet. She toured the United States for some years in a wide variety of parts, being much admired as Julia in Sheridan KNOWLES's *The Hunchback*, Pauline in BULWER-LYTTON's *The Lady of Lyons*, and Parthenia in Mrs Lovell's *Ingomar*, and she made her first appearance in London as Parthenia in 1883. Two years later she was seen at Stratford-upon-Avon as Rosalind in *As You Like It*, always one of her best parts, and in 1887 she appeared at the LYCEUM in *The Winter's Tale*, in which she was the first actress to double the parts of Perdita and Hermione. She appeared in London also in several plays by W. S. GILBERT, his *Comedy and Tragedy* (1884) being specially written for her. She retired in 1889, reappearing occasionally to play for charity, and in 1890 married Antonio de Navarro, with whom she settled at Broadway, in Worcestershire. She was part-author with Robert S. Hichens of *The Garden of Allah* (N.Y., 1911; London, 1920), a play based on his best-selling novel.

ANDERSON, MAXWELL (1888–1959), American dramatist, whose first play was *White Desert* (1923), a tragic study of a lonely woman on a farm, written partly in verse. *What Price Glory?* (1924), written in collaboration with Laurence Stallings, portrayed sympathetically but realistically the American soldier in action during the First World War and achieved a great popular success. After two historical plays, also written with Stallings, came *Saturday's Children* (1927), a 'serious comedy' by Anderson alone dealing with the marriage problems of a young couple, which was also successful. He collaborated with Harold

Hickerson in a passionately-written thesis-play, *Gods of the Lightning* (1928), which was not a success, and a realistic play of modern city life, *Gypsy* (1929). Anderson believed that without great poetry there is no great drama: he embarked on a series of historical and pseudo-historical dramas, partly in verse, which included *Elizabeth the Queen* (1930), *Night over Taos* (1932), and *Mary Queen of Scots* (1933); but, in his growing concern over modern social problems, produced also in 1933 *Both Your Houses*, a savage attack on political corruption which was awarded a PULITZER PRIZE. He returned to historical themes with *Valley Forge* (1934) and, after *Winterset* (1935), a modern 'study of conscience' based on the Sacco-Vanzetti case, with *The Wingless Victory* (1936), seen in London in 1943. He combined poetic drama with formal verse, philosophy, and political commentary in a further series, notably *High Tor* (also 1936), *The Masque of Kings* (1937; London, 1938), on the tragedy at Mayerling, *Knickerbocker Holiday* (1938), and *Key Largo* (1939). Among his last plays were *The Eve of St Mark* (1942; London, 1943), *Storm Operation* (1944)—both set in the Second World War—*Truckline Café* and *Joan of Lorraine* (both 1946), *Anne of the Thousand Days* (1948), on Anne Boleyn, and *The Bad Seed* (1954; London, 1955), a study of inherited homicidal tendencies.

ANDERSON, ROBERT WOODRUFF (1917–), American dramatist, whose first play, *Come Marching Home* (1945), was awarded first prize in a National Theatre Conference contest. He is best known for *Tea and Sympathy*, a considerable success on Broadway in 1953 and in London in 1957; his treatment of the victimization of a schoolboy accused of homosexuality was seen as an oblique comment on McCarthyism. His later plays are *All Summer Long* (1957); a double bill, *Silent Night/Lonely Night* (1959); *The Days Between* (1965); *You Know I Can't Hear You When the Water's Running* (1967; London, 1968), consisting of four short plays; *I Never Sang For My Father* (1968; London, 1970); and *Solitaire/Double Solitaire* (1971), another double bill. Anderson's plays deal compassionately with the problems inherent in human relationships, especially within the family, although they sometimes verge on the sentimental.

ANDREINI, FRANCESCO (1548–1624), author and actor of the COMMEDIA DELL' ARTE, who was associated for most of his career with the GELOSI troupe playing lovers' parts; he later abandoned these and achieved success with the role of the CAPITANO, which he made a subtle variation on the braggart-soldier type. He took his troupe to France in 1603, playing at the Hôtel de BOURGOGNE in Paris and at Fontainebleau, but on the death of his wife Isabella ANDREINI a year later he disbanded the company and went into retirement, devoting the rest of his life to writing, his most important work being *Le Bravure del Capitan Spavento da Vall'Inferna* (Part I, 1606; Part II, 1618), an invaluable collection of dialogues between the Captain and his servant Trappola.

ANDREINI, GIOVANN BATTISTA (*c.*1579–1654), actor of the COMMEDIA DELL' ARTE, known as Lelio. He was the eldest son of Francesco (above) and Isabella (below), and first acted in his parents' company, playing young lovers. Some time before his mother's death in 1604 he joined the FEDELI, probably helping in its formation, and provided plays for the troupe. Of these the best known are the *Adamo* (1613) and *La Centaura* (1622), a curious spectacle-play made up of three acts, of which the first is comic, the second pastoral, and the third tragic. It was successfully revived in 1972 in Rome with a student cast under the direction of Luca Ronconi. Giovann was twice married, in 1601 to Virginia Ramponi, who acted with him as Florinda, and after her death to Virginia Rotari, also an actress, known as Lidia.

ANDREINI, ISABELLA (1562–1604), *née* Canali, one of the finest actresses of the COMMEDIA DELL' ARTE, who with her husband Francesco ANDREINI led the GELOSI company on its travels throughout Europe. She seems to have been as highly regarded for her literary works (later edited by her husband) as for her acting. She died in childbirth at Lyon as the troupe was returning from a successful visit to France, leaving seven children of whom the best known was Giovann Battista (above).

ANDREWS, HARRY FLEETWOOD (1911–), English actor gifted with impressive height, strong attack, and a fine, resonant voice. He began his career in 1933 at the LIVERPOOL PLAYHOUSE, and two years later was in London, playing Tybalt in GIELGUD's production of *Romeo and Juliet*. During the next few years he was seen almost entirely in classical parts, notably as Diomedes in the modern-dress *Troilus and Cressida* at the WESTMINSTER THEATRE in 1938 and as Laertes in *Hamlet* at the LYCEUM in 1939, a part he also played in Elsinore later the same year. After war service he joined the OLD VIC company in 1945, playing a variety of parts including Creon in SOPHOCLES' *Oedipus Rex*, Sneer in SHERIDAN's *The Critic*, Tullus Aufidius in *Coriolanus*, and Lucifer in MARLOWE's *Dr Faustus*. He was at Stratford-upon-Avon for several years from 1949, appearing with distinction as Brutus in *Julius Caesar*, Benedick in *Much Ado About Nothing*, the Duke in *Measure for Measure*, and Enobarbus in *Antony and Cleopatra*. He left to make one of his rare appearances in a modern play, as Casanova in Tennessee WILLIAMS's *Camino Real* (1957), but returned to the Old Vic to give an impressive performance in the title role of *Henry VIII*. Later roles included General

Allenby in RATTIGAN's *Ross* (1960), Ekart in BRECHT's *Baal* (1963), and the title role in BOND's *Lear* (1971).

ANDREWS, JULIE, see MUSICAL COMEDY.

ANDREYEV, LEONID NIKOLAIVICH (1871–1919), Russian dramatist, began to write for the theatre with the encouragement of GORKY. He espoused the revolutionary cause but emigrated after the rising of Oct. 1917, and his plays, permeated with pessimism, express the deterioration and bitterness of the period 1905–17. They fall into two groups, realistic and symbolic, the latter group including *The Life of Man* (1906), *The Seven Who Were Hanged* (1908), and *The Sabine Women* (1911), a satire on political compromise. The only play by Andreyev to survive in the repertory is *He Who Gets Slapped* (1914), which was first seen in translation in New York in 1922 and in London in 1927.

ANDRONICUS, LUCIUS LIVIUS (*c*.284–*c*.204 BC), Roman dramatist, probably of Greek origin, and a manumitted slave. He was the first to write and produce plays based on Greek originals for the Roman stage, which up to then had known only a formless medley of dance, song, and buffoonery. The introduction of plays with a regular plot, taken from EURIPIDES, MENANDER, and other Greek dramatists, was successful, and Andronicus continued to produce tragedies, eight of which are known by name, and comedies until his death. From the fragments of his work that remain it is evident also that he introduced into his verse metres which he was, as far as we know, the first to employ, and which later Roman dramatists used for over 200 years. Although his style was uncouth and Cicero pronounced his plays unworthy of a second reading, he is of importance as a pioneer both in style and subject, since he faced for the first time the enormous problems which beset the adaptor of a work in which the aesthetic element predominates.

ANGEL, EDWARD (*fl.* 1660–73), English Restoration actor, referred to in WYCHERLEY's *The Gentleman Dancing-Master* (1671) as 'a good fool'. He was first with RHODES at the COCKPIT, and later joined DAVENANT at DORSET GARDEN, where he paired excellently with the great comedian, NOKES. Angel specialized in parts of low comedy, particularly French valets. He is not heard of after 1673.

ANGELICA, see MARTINELLI, DRUSIANO.

ANGELO, FRANCESCO D', see PARADISO.

ANGLIN, MARGARET (1876–1958), North American actress, accounted one of the finest of her day. Born in Ottawa, the daughter of the then Speaker of the House of Commons there, she was educated in Canada, and was a pupil at the dramatic school attached to the EMPIRE THEATRE, New York, when Charles FROHMAN engaged her to play Madeleine West in Bronson HOWARD's *Shenandoah*, in which she made her first professional appearance in 1894. After touring with James O'NEILL and E. H. SOTHERN, she made an outstanding success as Roxane in Richard MANSFIELD's production of ROSTAND's *Cyrano de Bergerac* in 1898; a year later she became Frohman's leading lady at the Empire where she played a wide variety of parts, being particularly admired in Henry Arthur JONES's *Mrs Dane's Defence*. Among her later successes were SHAW's *The Devil's Disciple* and *Camille* by the younger DUMAS, in both of which she appeared opposite Henry MILLER; William Vaughn MOODY's *The Great Divide*, both under that title and in its original form as *The Sabine Women*; several new translations of classical Greek plays, notably as Antigone, Electra, Iphigenia, and Medea; and such Shakespearian parts as Cleopatra, Rosalind, and Viola. In later years she proved an excellent Mrs Malaprop in SHERIDAN's *The Rivals*, in which she was first seen in 1936.

ANGUS, ROBERT, see AGGAS.

ANIMAL IMPERSONATION. The representation of birds and beasts by human beings played an important part in early FOLK-FESTIVALS, and may be associated with primitive fertility rites in which live animals were sacrificed, or with the wild-beast skins worn by priests and even participants in ritual religious dances, particularly in hunting communities. In general the short playlets in which these 'animals' appeared followed the main plot of the MUMMERS' PLAY, with the death and resurrection of the chief character. At Gèdre, in the Pyrenees, and in Bukovia, the 'animal' was a bear; the Mascarade of La Soule in south-west France included a *gherrero*, or 'pig-man', and a *gatuzain*, or 'cat-man'. The goat was a common folk-disguise in Scandinavia, the stag in Switzerland and Hungary. In England the 'animal' was a horse, giving rise to the HOBBY HORSE of the mummers, though in Derbyshire the Derby Tup, or Derby Ram, was more usual. The 'dragon' killed by ST GEORGE was a composite creature, with several men inside one skin.

Although ARISTOPHANES used animal disguises effectively for satirical purposes in his comedies, in the modern theatre they are mainly confined to 'comic turns' in music-halls and cabarets. They appear occasionally in straight comedies—the animals in OBEY's *Noah*, for instance, or the lion in SHAW's *Androcles and the Lion*—but otherwise find their chief employment in English PANTOMIME, Dick Whittington's cat and the Wolf in *Red Riding Hood* being the ones most often seen. The cow in *Jack and the Beanstalk* is unusual in being played by two actors, controlling respectively the front and back legs of the animal; this applies also to pantomime horses. In some cases, the 'animal' is suggested by one or two costume details—antennae and wings, or heads and tails (as for Bottom the Weaver in *A Midsummer Night's Dream*)—reinforced by appropriate gestures. There is also a form of impersonation limited to sound, as in the music-hall act 'A Day in the Country', where the bird-calls and farmyard noises are usually produced by one man. (See also FEMALE IMPERSONATION and MALE IMPERSONATION.)

ANIMATION CULTURELLE, Centres d', see DÉCENTRALISATION DRAMATIQUE.

ANNUNZIO, GABRIELE D', see D'ANNUNZIO.

ANOUILH, JEAN-MARIE-LUCIEN-PIERRE (1910–), French dramatist, whose first play *L'Hermine* was produced by LUGNÉ-POË in 1932. He was later associated with PITOËFF and with BARSACQ, who produced many of his plays, up to and including *Médée* (1953). The recurring theme of all Anouilh's plays, most of which have been successfully produced in translation both in London and in New York, is the loss of innocence implicit in the struggle for survival in a decadent society. This was treated romantically in what he described as as his *pièces roses—Le Bal des voleurs* (*Thieves' Carnival*, 1938), *Léocadia* (*Time Remembered*, 1940), *Le Rendezvous de Senlis* (*Dinner with the Family*, 1941), and *Colombe* (1951). His *pièces noirs—Le Voyageur sans bagages* (*Traveller Without Luggage*, 1937) and *La Sauvage* (*The Restless Heart*, 1938)—were tinged with a melancholy that was later transmuted into the glittering wit of the *pièces brilliantes—L'Invitation au Château* (*Ring Round the Moon*, 1947) and *La Répétition, ou l'Amour puni* (*The Rehearsal*, 1950)—or the bitter disillusionment of the *pièces grinçantes—Ardèle, ou la Marguerite* (1948), *La Valse des toréadors* (*The Waltz of the Toreadors*, 1952), and *Pauvre Bitos, ou le Diner des têtes* (*Poor Bitos*, 1956). He based other plays on history—*L'Alouette* (*The Lark*, 1953) on Joan of Arc and *Becket, ou l'Honneur de Dieu* (1959) on Thomas à Becket—and on classical themes, as in *Eurydice* (*Point of Departure*, in America *Legend of Lovers*, 1942), *Antigone* (1944), first seen in German-occupied Paris and perhaps the most successful of all Anouilh's plays world-wide, and *Médée*. In the 1950s Anouilh, in association with Roland Pietri, took over the direction of the COMÉDIE DES CHAMPS-ÉLYSÉES, where he was active in the production of his own and other writers' plays, including a revival in 1962 of *Victor, ou les Enfants au pouvoir* by VITRAC whom, with MOLIÈRE and GIRAUDOUX, he considers the main influence on his own work. Later plays by Anouilh himself include *Hurluberlu, ou le Réactionnaire amoureux* (1959), a sequel to *Ardèle* which, as *The Fighting Cock*, was seen at the CHICHESTER FESTIVAL and in London in 1966; *La Grotte* (*The Cavern*, 1961); and *La Foire d'Empoigne* (*The Rat Race*) and *L'Orchestre* (both 1962). Although Anouilh's earlier work was of undeniable originality and importance, he tended in such plays as *Cher Antoine* (1970), *Ne réveillez pas Madame* (1971), and *Le Directeur de l' Opéra* (1973), to be content with regular variations on the theme of characters thirsting for purity but unable to escape the blurring of make-believe and reality, somewhat in the style of PIRANDELLO. He returned to form with *Le Nombril* (*The Navel*, 1981), about an elderly writer regarded more highly by the public than by the critics.

ANSKY [Solomon Rappoport] (1863–1920), Jewish ethnologist and man of letters, who wrote one enduring play, *The Dybbuk, or Between Two Worlds*, a study of demoniac possession drawn from the Hasidic doctrine of pre-ordained relationship. More folk-lore than drama, it owes much of its renown to the manner of its production, first in Yiddish by David Hermann for the Vilna Troupe in 1920, and two years later in Hebrew by VAKHTANGOV for HABIMAH in Moscow. It was first seen in New York in 1925 and in London in 1930. Ansky also wrote an unfinished play, again on a Hasidic theme, and a satirical poem on the popular conception of Heaven and Hell. He was keenly interested in politics, but after the purge consequent on the attempted assassination of Lenin in 1918 he retired to Vilna, where he died.

ANSPACHER THEATRE, New York, see PUBLIC THEATRE.

ANTA, see AMERICAN NATIONAL THEATRE AND ACADEMY.

ANTA THEATRE, New York, at 245 West 52nd Street, between Broadway and 8th Avenue. Built by the THEATRE GUILD to house its own productions, this opened on 13 Apr. 1925 as the Guild Theatre, with Helen HAYES in SHAW'S *Caesar and Cleopatra*. During subsequent seasons further plays by Shaw were given, as well as plays by European and new American playwrights. In 1950 the theatre was taken over by the AMERICAN NATIONAL THEATRE AND ACADEMY, and as the Anta Playhouse was used for experimental productions. Given its present name in 1954, it reverted to commercial use in 1957, some of the productions there since being Archibald MacLeish's *J. B.* (1959), James Baldwin's *Blues for Mr Charlie* (1964), seen in London during the WORLD THEATRE season of 1965, Peter SHAFFER'S *The Royal Hunt of the Sun* (1965), and a documentary on the assassination of John Kennedy, *The Trial of Lee Harvey Oswald* (1967). In 1968 the Anta Theatre was presented to the nation; the building continues to house both the American National Theatre and Academy and the American branch of the INTERNATIONAL THEATRE INSTITUTE. Several famous visiting companies have since appeared there, and other productions have included Charles Gordone's *No Place To Be Somebody* (1969) and William INGE's *Summer Brave* (1975), a revised version of his *Picnic*. The very successful all-Black musical *Bubbling Brown Sugar* was staged here in 1976, and STOPPARD's *Night And Day*, with Maggie SMITH, in 1979.

ANTHONY STREET THEATRE, New York, at 79–85 Worth Street, opened on 20 May 1812 as the Olympic Theatre with a company from Philadelphia, headed by Mrs MELMOTH and Thomas TWAITS. Among the plays, interspersed with circus turns, given during the season was the famous equestrian melodrama *Timour the Tartar* by M. G. LEWIS, then seen for the first time in the United States. In 1813 Twaits again opened the theatre, redecorated and enlarged and renamed the Anthony Street Theatre, with a passable company which included the young PLACIDE and a Mrs Beaumont who was excellent in tragedy. This group survived until 1814, when the theatre passed through various hands and was named successively the Commonwealth and the Pavilion, opening only for the summer. After the destruction by fire of the first PARK THEATRE on 29 May 1820 its company moved to the Anthony Street Theatre, which soon proved too small to hold the crowds that flocked there to see Edmund KEAN's first appearance in New York on 29 Nov. 1820, as Richard III. When the Park Theatre was rebuilt the following year the company moved back, and the Anthony Street Theatre was demolished.

ANTI-(ANTE-) MASQUE, see MASQUE.

ANTOINE, ANDRÉ (1858–1943), French actor, director, and theatre manager, one of the outstanding figures in the theatrical reforms of the late 19th century. In 1887 he founded the THÉÂTRE LIBRE for productions of the new naturalistic drama then coming to the fore in Europe, and in 1890 produced there IBSEN's *Ghosts*, himself playing Oswald, following it with plays by HAUPTMANN, STRINDBERG, BJØRNSON, BECQUE, and BRIEUX, among others. Here, too, he revolutionized French acting and inaugurated a new era of scenic design. Inspired by him, Otto BRAHM found the FREIE BÜHNE in Berlin and GREIN the Independent Theatre in London. In 1896, after working at the ODÉON for a time, he returned to the Théâtre des Menus-Plaisirs, which had been the home of the Théâtre Libre since 1890, renamed it the THÉÂTRE ANTOINE, and made it a rallying-point for many young and experimental dramatists. From 1906 to 1916, when he retired, he was director of the Odéon. His influence all over Europe and in North America was incalculable, and among the outstanding figures of the next generation who owed much to him was Jacques COPEAU, founder of the VIEUX-COLOMBIER.

ANTONELLI, LUIGI (1882–1942), Italian dramatist, whose works developed further CHIARELLI's *teatro del* GROTTESCO. His *L'Uomo che incontrò se stesso* (*The Man Who Met Himself*, 1918) shows that a man is a fool if he hopes to correct the errors of his youth by the wisdom (or experience) of his maturity. Of Antonelli's other plays *L'Isola delle scimmie* (*The Island of Monkeys*, 1922) ironically lays bare the squalor of human civilization; *La bottega dei sogni* (*The Dream Shop*, 1927) analyzes man's use of dreams; and *Il maestro* (1933) is a neatly-turned refutation of PIRANDELLO's theories, built up on the 'master's' own foundations. The world of Antonelli's plays is sombre yet visionary, and essentially truthful.

ANVIL PRODUCTIONS, see OXFORD PLAYHOUSE.

ANZENGRUBER, LUDWIG (1839–89), Austrian dramatist, the first to present realistic peasant life on the modern Austrian stage. Equally successful in tragedy and comedy, he employed a dialect-flavoured dialogue, philologically inaccurate, which lent an air of authenticity without being as difficult to understand as the real dialect literature which came later. His freethinking views emerge in such attacks upon religious intolerance as *Der Pfarrer von Kirchfeld* (1870) and *Der Meineidbauer* (1871), in which he shows that peasant life can furnish matter for true tragedy. His mastery of comic techniques, derived from the Viennese popular theatre, is evident in *Die Kreuzlschreiber* (1872), a comic battle for matrimonial supremacy, and *Der Doppelselbstmord* (1876), a farcical village version of *Romeo and Juliet* which ends happily. Anzengruber's last important play was *Das vierte Gebot* (1877), in which he deserts the country for the town but remains uncompromisingly a realist.

APOLLODORUS of Carystos (*fl.* later 4th cent. BC), Greek comic poet of NEW COMEDY of whose works only a few fragments remain. TERENCE modelled his *Phormio* and *Hecyra* on plays by him.

APOLLO THEATRE, London, in Shaftesbury Avenue, opened on 21 Feb. 1901 with a musical play, *The Belle of Bohemia*, which was not a success; but later productions had good runs and the Apollo, though never the permanent home of a great management, became a consistently successful theatre. It is well situated and, seating 796 in three tiers, the right size for either musicals or straight plays. Among its earlier successes were Charles Hannen's *A Cigarette-Maker's Romance*, transferred from the ROYAL COURT THEATRE in 1901 and starring MARTIN-HARVEY; Owen Hall's *The Girl from Kay's* (1902); and seasons of the *Pélissier Follies* from 1908 to 1912. The most notable production during the First World War was Harold BRIGHOUSE's comedy *Hobson's Choice* (1916), and immediately after the war Ian HAY's *Tilly of Bloomsbury* (1919) had a long run. Later came NOVELLO's *Symphony in Two Flats* (1929) and in 1933 Diana WYNYARD appeared as Charlotte Brontë in Clemence DANE's *Wild Decembers*, Beatrix LEHMANN playing Emily. In 1938, during the Munich crisis, SHERWOOD's *Idiot's Delight* scored a notable success, and good wartime productions were Emlyn WILLIAMS's *The Light of Heart* (1940), John VAN DRUTEN's *Old Acquaintance* (1941), and RATTIGAN's *Flare Path* (1942). *Seagulls Over Sorrento* (1950) by Hugh Hastings scored a long run, and two plays by GIRAUDOUX were of particular interest—*Tiger at the Gates* (1955), with Michael REDGRAVE, and *Duel of Angels* (1958) with Claire BLOOM and Vivien LEIGH. In 1960 an adaptation by Beverley Cross of Marc Camoletti's *Boeing-Boeing* started a four-year run, and in 1968 GIELGUD appeared in Alan BENNETT's *Forty Years On*, returning in 1970 with Ralph RICHARDSON in David STOREY's *Home*, transferred from the Royal Court. Peter NICHOLS's *Forget-Me-Not Lane* transferred from the GREENWICH THEATRE in 1971, and in 1975 Margaret LEIGHTON and Alec GUINNESS starred in *A Family and a Fortune*, based on one of Ivy Compton-Burnett's novels. Later successes included AYCKBOURN's group of short plays *Confusions* (1976), a revival of Rattigan's *Separate Tables* (1977), and John Chapman and Anthony Marriott's *Shut Your Eyes and Think of England* (also 1977), with Donald SINDEN.

APOLLO THEATRE, New York, on 42nd Street between 7th and 8th Avenues. Originally the Bryant, used since 1910 for films and vaudeville, this building opened as the Apollo Theatre 17 Nov. 1920. A year later it housed an interesting but unsuccessful production by Arthur Hopkins of *Macbeth*, with Lionel BARRYMORE in the title role and triangular settings by Robert Edmond JONES. A series of musical comedies followed, and from 1924 to 1931 the house was occupied by *George White's Scandals*. It became a cinema again in 1933.

Three other theatres briefly known as the Apollo were the Third Avenue Variety Theatre in 1885, a playhouse on Chuter Street which opened in Oct. 1926, and a burlesque house on 125th Street.

APPIA, ADOLPHE (1862–1928), French-speaking Swiss artist whose theories on stage design, and particularly on stage LIGHTING, had a great influence on 20th-century methods of play production. Rejecting flat painted scenery as unsuitable for a three-dimensional actor, he took advantage of the introduction of electricity to use light as the visual counterpart of music, enhancing the mood of the play and linking the actor to the setting. He embodied his ideas in extremely simple but effective designs for plays by SHAW and IBSEN, and formulated them so clearly in his book *Die Musik und die Inscenierung* (1899) that other directors were able to put them into practice without the need for special apparatus. Mobile lighting, which breaks up and diversifies the direction, intensity, and colour of light, is used today by many directors who do not realise the extent of their debt to Appia's experiments.

APRON STAGE, see FORESTAGE.

AQUARIUM THEATRE, London, see IMPERIAL THEATRE.

AQUATIC DRAMA, spectacular representations of nautical battles, shipwrecks, and storms at sea, which came to London from the circuses of Paris and became exceedingly fashionable in the early 18th century. In imitation of the flooded circus-rings, many London theatres replaced their stages by large water tanks filled from a neighbouring river and took advantage of the fervour engendered by Nelson's naval victories to put on elaborate reconstructions, mainly anonymous, of the battle of the Nile, the bombardment of Copenhagen, and—a favourite subject, since it terminated with the romantic death of the hero in the moment of victory—the battle of Trafalgar. In 1822 the ROYALTY THEATRE even staged a reconstruction of the shipwreck of the *Grosvenor*, an East Indiaman, and DRURY LANE and COVENT GARDEN themselves fell victim to the popular craze; but the theatre most addicted to Aquatic Drama was SADLER'S WELLS, which even changed its name temporarily to the Aquatic Theatre.

AQUATIC THEATRE, London, see SADLER'S WELLS THEATRE.

ARBUZOV, ALEXEI NIKOLAYEVICH (1908–), Soviet dramatist, one of the few to have found an audience in the West, and compared by some English critics to CHEKHOV. He began writing in the 1920s, his early work including *Class* (1930), but his first major success came with *Tanya* (1939), in which BABANOVA gave an outstanding performance in the title role. Later plays included *European Chronicle* (1953), *The Years Of Wandering* (1954), and *The Twelfth Hour* (1959), produced at the OXFORD PLAYHOUSE in 1964. *City At Dawn* (1957) and *It Happened In Irkutsk* (1959) were performed at the VAKHTANGOV THEATRE. The second, reminiscent in form of WILDER's *Our Town*, became one of the most popular plays in the Soviet Union and made a star of Yulia BORISOVA; it was performed in 1961 in Paris by the Vakhtangov Theatre and in 1967 in Sheffield. By 1963 Arbuzov's plays were running simultaneously at over 70 Russian theatres. *The Promise* (1965) established his reputation abroad, in spite of a sentimental plot covering the lives of three Russians, a girl and two men, from adolescence in 1942 during the siege of Leningrad until 1960. It was produced in London in 1967 with Judi DENCH and Ian MCKELLEN, and in New York in the same year with Eileen ATKINS and McKellen. *Old World* was produced by the ROYAL SHAKESPEARE COMPANY in 1976, the year after its première in Poland. A two-character play about an autumnal romance between a doctor and his patient, an ex-circus performer, it starred Peggy ASHCROFT and Anthony QUAYLE. It was staged in Paris in 1977 as *Le Bateau pour Lipaïa* with Edwige FEUILLÈRE, and a New York production in 1978, under the title *Do You Turn Somersaults?*, starred Mary MARTIN and Quayle.

ARCHER, WILLIAM (1856–1924), critic and playwright, born in Scotland. He had some experience of journalism before migrating to London, where from 1894 to 1898 he was drama critic of the *World*. He took the theatre seriously as an art, and was the first to introduce the plays of IBSEN to the London public. His translation of *Samfundets Støtter* (1877) as *Quicksands; or, the Pillars of Society* was seen in 1880; *Et Dukkehjem* (1879) as *A Doll's House* followed in 1889; *Gengangere* (1881) as *Ghosts* in 1891; *Bygmester Solness* (1892) as *The Master Builder* in 1893; *Lille Eyolf* (1894) as *Little Eyolf* in 1896; and *John Gabriel Borkman* (1896) in 1897. In 1906–8 Archer published the complete works of Ibsen in English in 11 volumes. He had an astringent sense of humour which probably accounted for his long friendship with George Bernard SHAW, who introduced him into *Fanny's First Play* (1911) as Mr Gunn, and like Shaw he was antagonistic to IRVING who, he maintained, had done nothing for the modern British dramatist. He always protested at the critical over-valuation of older plays and under-valuation of modern ones and, being more interested in drama as an intellectual product than as a vehicle for virtuoso acting, he upheld the supremacy of the author's script. On 6 Sept. 1923 his own play *The Green Goddess*, an improbable melodrama which had been seen in New York in 1921, was produced in London at the ST JAMES'S THEATRE and ran for 416 performances.

ARCHETRINGLE, Laupen, see FOLK FESTIVALS.

ARCHITECTURE, see THEATRE BUILDINGS.

ARCH STREET THEATRE, Philadelphia, see CHESTNUT STREET THEATRE and PHILADELPHIA.

ARC LIGHT, see LIGHTING.

ARDEN, JOHN (1930–   ), English dramatist, who has been compared to BRECHT in his preoccupation with moral and social problems and his use of historical themes to illuminate contemporary life. Large sections of his plays are in rhyming verse. Among his early plays, produced at the ROYAL COURT THEATRE, were *Live Like Pigs* (1958), on the unsuccessful rehousing of a gipsy family; *Serjeant Musgrave's Dance* (1959; N.Y., 1966), an anti-military work which was a failure when first produced, but has since been widely revived and is now recognized as an important contribution to modern drama; and *The Happy Haven* (1960; N.Y., 1967), set in an old people's home. In 1963 *The Workhouse Donkey*, a play on English provincial politics, originally seen at the Royal Court for one night in 1957 as *The Waters of Babylon*, was produced at the CHICHESTER FESTIVAL THEATRE, where *Armstrong's Last Goodnight* (based on a Scottish ballad, and first seen at the CITIZENS' THEATRE in Glasgow in 1964) was produced in 1965 with Albert FINNEY, subsequently going on to London; in the same year *Left-Handed Liberty*, written to celebrate the 750th anniversary of the sealing of the Magna Carta, was produced at the MERMAID THEATRE. At Christmas 1967 a play for children, *The Royal Pardon; or, the Soldier Who Became an Actor*, was given in London. This was a collaboration with his wife Margaretta D'Arcy, who was also part-author of a musical featuring Nelson, *The Hero Rises Up* (1968); the outstanding result of their collaboration was *The Island of the Mighty* (1972), a play based on the Arthurian legends. Its production by the ROYAL SHAKESPEARE COMPANY at the ALDWYCH THEATRE led to a violent disagreement with the authors, who have since worked only with small, non-professional groups. Their latest works are overtly Marxist and the poetic qualities of the earlier plays are less evident.

ARDREY, ROBERT (1908–80), American dramatist who had had several plays produced in New York, among them *Star-Spangled* (1936), *Casey Jones*, and *How To Get Tough About It* (both 1938), before the GROUP THEATRE produced his best known work, *Thunder Rock* (1939; London, 1940). This allegory about a young man, disillusioned by the prospect of the coming world war, retiring from the world only to learn from the history of man's earlier struggles that he must return and take part in it as an affirmation of his belief in humanity's ultimate triumph, had great success, particularly in London, where the young man Charleston was played by Michael REDGRAVE. Among Ardrey's later plays were *Sing Me No Lullaby* (1954), which attacked post-war conformism, and *Shadow of Heroes*, a study of the Hungarian uprising of 1956 which was first seen in London in 1958 and, as *Stone and Star*, in New York in 1961. Ardrey then gave up the theatre in favour of anthropology and ethology.

ARENA STAGE, see OPEN STAGE and THEATRE-IN-THE-ROUND.

ARENA STAGE, Washington, D.C., theatre built in 1961 to house a group founded in Aug. 1950 by Zelda Fichandler. The group's first production,

GOLDSMITH's *She Stoops To Conquer*, was given in an adapted cinema in downtown Washington, where the company remained until 1955, when it moved to a disused brewery, the Old Vat. The first production in the purpose-built arena-stage theatre, which seats 827 on four tiers, was BRECHT's *The Caucasian Chalk Circle*. A second theatre, the Kreeger, opened in 1971 and has a seating capacity of 514 arranged in a fan-shaped auditorium round an end-thrust stage. A rehearsal room in the Kreeger's basement was converted in 1976 into the Old Vat Room, seating 180, with a cabaret-style stage. The Arena Stage company, which has become an important force in the modern American theatre, presents a wide range of classic and modern plays, both American and European, during a season which runs from Oct. to June, and has given the first productions of such new American plays as Howard Sackler's *The Great White Hope* (1967) and Weller's *Moonchildren* (1971) as well as of the musical *Raisin* (1973), based on Lorraine Hansberry's play *A Raisin in the Sun*. The Arena Stage also supports Living Stage, a community group which specializes in improvization.

ARENA THEATRE, Cardiff, see WALES.

ARENT, ARTHUR, see LIVING NEWSPAPER.

ARETINO, PIETRO (1492–1556), Italian author and playwright, chiefly remembered for his comedies, which, though written in haste and lacking refinement, are original, amusing, and thoroughly Italian in their realistic and satiric thought and presentation, throwing light on less creditable aspects of the social life of the day. It has been suggested that Ben JONSON's *Epicœne* (1609) owes something to Aretino's comedy *Il Marescalco* (*The Sea Captain*, 1533), which was based on PLAUTUS' *Casina*. His other comedies are *La Cortigiana* (1526), *La Talanta* (1541), *Lo Ipocrito* (1542), a precursor of MOLIÈRE's *Tartuffe*, and *Il Filosofo* (1546). Aretino's one tragedy, *Orazio* (1546), is perhaps the best of those written at the time.

ARGENT, Théâtre de l'Hôtel d', after the Hôtel de BOURGOGNE the second licensed theatre building in Paris. By the end of the 16th century travelling companies were permitted to play at the Paris fairs, and in 1598 Pierre Venier, an actor-manager from the provinces who had been at the Foire St-Germain, took his company, with his daughter Marie VENIER as leading lady, to an improvised theatre in the Hôtel d'Argent in the rue de la Verrerie, where he was allowed to remain on payment of a tax for each performance to the CONFRÉRIE DE LA PASSION, owners of the Hôtel de Bourgogne. This theatre must have remained in use intermittently for many years, as in 1609 and 1610 Marie Venier and her husband Laporte are found there, having temporarily left the Hôtel de Bourgogne where they had been acting with VALLERAN-LECOMTE. The assassination of Henri IV in 1610 caused all the actors to leave Paris, and on their return the two companies went back together to the Hôtel de Bourgogne, while a new company led by LENOIR and MONTDORY leased the Hôtel d'Argent. They later dispersed to other parts of Paris, until in 1634 Montdory opened the famous Théâtre du MARAIS.

ARGYLL ROOMS, London, see TROCADERO PALACE OF VARIETIES.

ARION of Lesbos (*c.*625–585 BC), Greek poet, an important figure in the development of Greek dramatic poetry, since he was the first to give literary shape both to the DITHYRAMB and to the SATYR-DRAMA. He is credited with a miraculous escape from death by drowning at the hands of pirates, being conveyed to land on the back of a dolphin charmed by his singing.

ARIOSTO, LODOVICO (1474–1533), Italian poet and playwright, best known for his epic poem, *Orlando Furioso*, published in 1532, which was adapted for the stage in 1969 by Edoardo Sanguinetti. Ariosto was also one of the first and best writers of the early Italian COMMEDIA ERUDITA, using material taken from Renaissance city life though modelled upon Roman comedy. His first plays, *La cassaria* (1508) and *I suppositi* (1509), were given at the Court of the d'Este family, his patrons, in Ferrara in a theatre built in a classical style under the influence of VITRUVIUS, which survived until 1533. The scenery was by Raphael. Both plays were in prose, but were later rewritten in verse, as was *Il negromante* (written 1520, prod. 1530). An English translation of *I suppositi* by George GASCOIGNE was performed at Gray's Inn in 1566.

ARISTODEMOS, see GREECE, ANCIENT.

ARISTOPHANES (*c.*448–*c.*380 BC), Greek dramatist, author of some 40 comedies of which 11 are extant; the *Acharnians* (425), *Knights* (424), *Clouds* (423), *Wasps* (422), *Peace* (421), *Birds* (414), *Lysistrata* (411), *Women at the Festival* (*Thesmaphoriazousae*) (410), *Frogs* (405), *Women in Parliament* (*Ecclesiazousae*) (392), and *Plutus* (388).

The ancient critics regarded Aristophanes as the greatest of the Attic comic poets, but his direct influence on drama has been slight; the form and spirit of his comedies were too intensely local to offer models or material to comic dramatists of other times and places. On the other hand, his purely literary influence has been great, particularly on Rabelais and FIELDING. Many of his plays take their titles from the disguises assumed in them by the CHORUS—wasps, frogs, clouds, etc. It is difficult to assess his individual contribution to OLD COMEDY as no other complete examples of it survive for comparison. In the earlier plays, there is little development of plot. The action customarily begins with a short farcical episode to catch the audience's attention, after which the main theme of the play is stated and then developed in a series of short scenes interspersed with choral songs. The theme normally had some direct reference to the political situation of the moment or to some urgent social question: the *Acharnians* dramatized the desire for peace in the person of an old countryman who makes a private treaty with the enemy; the *Knights* satirized contemporary politicians; the *Frogs* was inspired by the recent death of EURIPIDES, whom Aristophanes parodies unmercifully. In the same way, characters are often burlesques of contemporary Athenians—people like Cleon or Socrates who were well known to the audience—and even the gods were not immune. Round some small incident Aristophanes weaves a web of fantasy, unsparing and often unfair satire, brilliant verbal wit, literary and musical parody, exquisite lyrics, hard-hitting political propaganda, and uproarious farce. For all his sophistication he was essentially a popular dramatist, fond of slapstick and comic business, with a familiar range of obviously well-loved stock characters; and he panders to a number of Athenian prejudices, particularly the notion that all foreigners are invariably funny.

In his later years, under pressure of war and the growing insecurity of Athens, his plays became milder in tone, with less fantasy, invective, and irreverence, and more emphasis on plot. *Lysistrata* is notably lacking in personal references; *Women at the Festival* and the *Frogs* abandon political for literary targets. Certainly *Women in Parliament* was seen by ancient critics as belonging more to the quieter MIDDLE COMEDY, while *Plutus* is little more than innocuous allegory. (For the 'English Aristophanes', see FOOTE, SAMUEL.)

ARISTOTLE of Stagira (384–322 BC), Greek critic, philosopher, and scientist. His *Poetics* analyses the function and structural principles of tragedy and (to some extent) of epic; a second book, on comedy, has been lost. The *Poetics* is a reply to the criticisms of PLATO and of Socrates, who says (in Plato's *Apology*) that the poets are unable to give a coherent account of what they do. To the latter criticism Aristotle replies by working out a logical theory of poetic composition; to Plato, who condemns poetry and drama for what it does not do (that is, inculcate virtue), he replies by enquiring quite objectively what it does do. It aims at pleasure, but at the rational pleasure which is a part of the good life; by its representation of serious action it does indeed excite emotions, but only to purge them and so leave the spectator strengthened; since art represents universals and not particulars, it is nearer to the truth than actual events and objects are, not further from it, as Plato maintained, 'Poetry is more philosophical than history.'

Aristotle's account of the historical development of tragedy and comedy is extremely brief. His analysis of the form of tragedy is penetrating, but it is something less than an account of the existing types of tragedy; for Aristotle regarded tragedy as an organism which, like biological organisms, developed until it achieved its 'natural' form, and then decayed; its basic laws are therefore to be disclosed by an analysis of the mature, not of the immature or the enfeebled, stages of the art. The 'complete' form of tragedy, in which all its structural features are fully developed and most vigorously used, is that of SOPHOCLES; it is, therefore, the Sophoclean tragedy, in all essentials, that Aristotle analyses, and his dicta are not necessarily true of AESCHYLUS or even EURIPIDES.

The neo-classical critics of the 17th and 18th centuries, especially in France, were anxious to claim Aristotle's authority for their own doctrines: but of the famous three UNITIES Aristotle actually mentions only one and a half. He insists on unity of action, though this is capable of wide interpretation; but his passing reference to 24 hours as the normal time span of a tragedy is descriptive rather than prescriptive, and he says nothing regarding unity of place. The *Poetics* remains uniquely valuable for its summary of artistic practice appropriate to one time and place, rather than for any attempt to lay down universal laws.

ARLECCHINO, one of the comic ZANNI or servant roles of the COMMEDIA DELL'ARTE, who came originally from Bergamo. Wearing a suit patched with scraps of different colours, with a soft cap on his shaven head, he carried a slapstick, or bat, and was above all a dancer and acrobat. His unique blend of stupidity and shrewdness soon singled him out, and he became an important character in many plays. The first actor to be identified with the role, if not the first to play it, was Tristano MARTINELLI; as Arlequin it became popular in Paris when played by Giuseppe BIANCOLELLI, and his successor GHERARDI. It also figured in the plays of MARIVAUX, as in *Arlequin poli par l'amour* (1720) where, though the character remained a servant, the author first showed him as capable of true love. Imported into England, he underwent a metamorphosis, shedding his grosser traits but retaining his excellence as a dancer to emerge as the resourceful and elegant HARLEQUIN.

ARLISS [Andrews], (Augustus) GEORGE (1868–1946), English actor, now chiefly remembered for his films, who also had a successful career on the stage in London and New York. He made his first appearance in 1886, and after some years in the provinces returned to London, where he first came into prominence in 1900–1 when he played the Duke of St Olpherts and Cayley Drummle in revivals of PINERO's *The Notorious Mrs Ebbsmith* and *The Second Mrs Tanqueray* opposite Mrs Patrick CAMPBELL, with whose company he subsequently went to the United States. In 1902 he was seen on Broadway in BELASCO's *The Darling of the Gods* with Blanche BATES, and in IBSEN's *Hedda Gabler* (1903) and *Rosmersholm* (1907) with Mrs FISKE. Among his finest parts at this time were the title roles in MOLNÁR's *The Devil* (1908) and L. N. PARKER's *Disraeli* (1911), and the Rajah in William ARCHER's *The Green Goddess* (1921). In this last part he reappeared in London in 1923, after an absence of over 20 years. He then returned to New York to show his versatility by playing with equal success the elderly gentleman in GALSWORTHY's *Old English* (1924) and, his last stage role, Shylock in *The Merchant of Venice* in 1928.

ARMIN, ROBERT (*c.*1568–*c.*1611), Elizabethan clown, pupil and successor of TARLETON. He appears in the list of actors in Shakespeare's plays, and probably played Dogberry in *Much Ado About Nothing* in succession to William KEMPE. He was also known as a writer. His *Foole upon Foole, or Six Sortes of Sottes* appeared in 1600; on the title-page the author is described as 'Clonnico de Curtanio Snuffe', changed in a later edition, by which time Armin had probably moved from the CURTAIN to the GLOBE, to 'Clonnico del Mondo Snuffe'. His own name is found on an enlarged edition published in 1608 as *A Nest of Ninnies*. He was probably the author of *Quips upon Questions* (also 1600), a collection of quatrains on stage 'themes', or improvisations on subjects suggested by the audience, and is credited with the authorship of one play, *Two Maids of Moreclacke*, produced in 1609.

ARMSTRONG, PAUL (1869–1915), American dramatist who was for a time a purser in the Merchant Navy, and later a journalist. His early plays were not successful, but in 1905 he scored a triumph with *The Heir to the Hoorah*, and became one of the most popular and prolific playwrights of the day. Among his later successes, some written in collaboration, were *Blue Grass* (1908), *Alias Jimmy Valentine* (1909), the first of a long line of crook plays, and *The Deep Purple* (1910). His dialogue was crisp, with strong situations and grim humour, in the melodramatic style later exploited by the cinema.

ARMSTRONG, WILLIAM (1882–1952), English actor and director, who was for many years connected with the LIVERPOOL PLAYHOUSE where his work was of great value not only to the repertory movement but to the English theatre as a whole, many of his company going on to become leading actors in London. He made his first appearance on the stage at Stratford-upon-Avon in 1908, in Frank BENSON's Shakespeare company, and after touring extensively was with both the GLASGOW and the BIRMINGHAM repertory companies. After a tour as leading man with Mrs Patrick CAMPBELL in IBSEN's *Ghosts* in 1922 he went to Liverpool as manager and director, remaining there until the theatre closed in 1941 because of heavy air raids. He then went to London, where his productions included VAN DRUTEN's *Old Acquaintance* (1941) and Thomas Job's *Uncle Harry* (1944), and in 1945 accepted an invitation from Barry JACKSON to go back to Birmingham, where he remained until his death.

ARNAUD, YVONNE GERMAINE (1892–1958), actress born and educated in France, who spent her entire professional life in London. Trained as a concert pianist, she toured Europe as a child prodigy but in 1911, with no previous experience, took over the part of Princess Mathilde in the musical comedy *The Quaker Girl* (1910) at the ADELPHI THEATRE. She was immediately successful, as she was in *The Girl in the Taxi* (1912) at the LYRIC THEATRE. She continued to appear in musical comedy and farce, notably in *Tons of Money* (1922) by Will EVANS and in Ben TRAVERS's *A Cuckoo in the Nest* (1925), her charm and high spirits, added to a musical broken English-French accent, making her a general favourite with the public. Among her outstanding performances were Mrs Pepys in FAGAN's *And So To Bed* (1926), the title role in his *The Improper Duchess* (1931) and Mrs Frail in CONGREVE's *Love for Love* (1943). One of her few failures was Madame Alexandra in the English version of *Colombe* (1951), where her essential kindliness and good nature was ill-suited to the cruelty and egotism of ANOUILH's ageing actress; but in the following year she had another success as Denise in Alan Melville's *Dear Charles*. She lived for a long time near Guildford, Surrey, where the Yvonne Arnaud Theatre, opened in 1965, was named after her.

ARNE, SUSANNA MARIA, see CIBBER, THEOPHILUS.

ARNICHES, CARLOS, see SPAIN.

ARNOLD, MATTHEW (1822–88), English poet and educationist, who was for a short time drama critic of the *Pall Mall Gazette*. Although his own verse plays *Empedocles on Etna* (1852) and *Merope* (1858) were intended for reading rather than acting, he was intensely interested in the theatre as a cultural influence and tried to win official support for it, saying in his essay 'The French Play in London' (1882): 'The theatre is irresistible: organize the theatre!'

ARONSON, BORIS (1900–80), American scene and costume designer, Born in Kiev, he went to New York in 1923 and began to design scenery for the Unser Theatre and for Maurice SCHWARTZ's Yiddish Art Theatre. His early work was influenced by the sets in Cubist-fantastic style designed by Chagall and Nathan Altman for the Jewish Theatre in Moscow, but later they were symbolic, restating by use of form and colour the mood of the play, as in those he designed for ODETS's *Awake and Sing* (1935). In addition to a long series of stage settings for New York, among which that for Archibald MacLeish's *J. B.* (1959) was outstanding, he developed 'projected scenery', consisting of a basic permanent set of interrelated abstract shapes made of neutral grey gauze which can be 'painted' any desired colour by directing light on to it through coloured slides. The method, which is especially applicable to scenery for ballet and opera, was first used in the production of SAROYAN's ballet *The Great American Dream* (1940). Among Aronson's later sets were those for *Coriolanus* at Stratford-upon-Avon in 1959; *Incident at Vichy* by Arthur MILLER and the musical *Fiddler on the Roof* (both 1964); the musical *Cabaret* (1966); and SONDHEIM's *Company* (1970) and *A Little Night Music* (1972).

ARRABAL, FERNANDO (1932–), French dramatist of Spanish extraction, who has lived in Paris and written in French since 1955. His oppressive upbringing during and after the Spanish Civil War is reflected in the plays that have made his reputation, which portray a perverse childish revolt against established beliefs. *La Cimetière des voitures* (written in 1957, but not produced until 1966, when it was directed by GARCIA) begins as a satire on suburban community living and ends in a blasphemous crucifixion scene on a bicycle; as *The Car Cemetery* it was seen in London in 1969. In *La Communion solennelle* (1958), the Mass is parodied in a necrophiliac ceremony; and *Guernica* (1959) derides the 'myth' of the martyrs of Guernica by portraying the events from the point of view of an unsavoury old couple trapped in the débris. In a prolific and varied output (he has written some 30 plays) Arrabal has ventured into formal experiment, as in *Orchestration théâtrale* (1960), a play without actors; social criticism, as in *Pique-Nique en campagne* (written in 1952, produced 1959, seen in London as *Picnic on the Battlefield* in 1964), which juxtaposes banal family life with a military operation; and direct political comment, as in *Et ils passèrent des menottes aux fleurs* (produced and then banned in 1969), which, translated by Charles MAROWITZ as *And They Put Handcuffs on the Flowers* and directed by the author, was seen in New York in 1972 and London in 1973. This was an attack on Spanish political prisons, from which Arrabal had been extracted by international intellectual pressure after a brief return to his own country. His most characteristic vein derives from his very early plays, written before he left Spain, which are peopled by innocent, childish characters producing ferocious acts in the course of their naïve games. In *Le Tricycle* (written in 1953 and seen for one performance in Madrid in 1958) a primitive manhunt is enacted in a park; *Fando et Lis* (written in 1955, produced in Mexico in 1961) tells of the epic struggle of an apparently loving couple to reach a promised city, at the end of which Fando sadistically kills his companion. These simple pieces contain ritual elements that Arrabal later exploited in a more mature way. *Le Grand Cérémonial* (written in 1963, produced in 1966) formulates a ritual battle between men seen as sadistic torturer-victims and women seen as mother-doll-whores. Arrabal's most successful play, seen in translation in London at the NATIONAL THEATRE in 1971, is probably *L'Architecte et l'Empereur d'Assyrie* (1967), a parable of civilization in which two men enter into a role-playing love-hate relationship on a desert island which ends with one eating the other in an attempt to achieve unity. His later plays include *Le jardin des délices*, in which Delphine SEYRIG appeared in 1969, *Jeunes Barbares d'aujourd'hui* (1975), a skit on cycle-racing, *La Tour de Babel* (1976), and *Théâtre Bouffe* (1978).

Arrabal's fantasies, gradually gathering in complexity, accumulating sinister allegoric overtones, and emphasized by an often circular dramatic structure, reach a level of meaning which is ceremonial and quasi-religious. He himself has called his work 'théâtre Pan-ique'.

ART, Théâtre d', see FORT, PAUL, and LUGNÉ-POË.

ARTAUD, ANTONIN (1896–1948), French actor, director, poet, and theorist, whose work has been one of the seminal influences on experimental theatre, in France since the end of the Second World War, and in the rest of the world since his 'discovery' some ten years later. Entering the theatre in 1921, Artaud acted successively with LUGNÉ-POË, DULLIN, and PITOËFF. From 1923 to 1927 he was an active advocate of Surrealism, and with Roger VITRAC and Robert Aron founded the Théâtre Alfred JARRY, where between 1926 and 1929 he produced STRINDBERG's *Dream Play*, his own short incantatory piece *Le Jet de sang*, and Vitrac's *Victor, ou les Enfants au pouvoir* (1928). As early as 1922 he had been impressed by the power of the oriental theatre, and in 1931 a performance at the Colonial Exhibition in Paris by a group of Balinese dancers stimulated him to attempt to redefine the meaning of theatre in a series of essays, collected in 1938 as *Le Théâtre et son double*. Here he formulated his theory of a Theatre of CRUELTY which, by jettisoning language in favour of symbolic gesture, movement, sound, and rhythm, would have the power to disturb the spectator and impel him to action by the inner force of the material presented on stage. Having observed that during the great plagues of history men were liberated from the restraints of morality and reason and returned to a state of primitive ferocity and power, he envisaged the theatre as a similar catalyst, freeing the unconscious and forcing men to view themselves as they really are. Earlier, in 1935, he had managed to give practical expression to his theories in directing his own play *Les Cenci*, based on SHELLEY and Stendhal, with which Roger BLIN was also associated. The production ran for only two weeks, and was Artaud's last practical contribution to the theatre. After travelling in Mexico and Ireland he succumbed to a mental crisis and spent nine years in various asylums before being released in 1946. In France his influence was seen in the work of BARRAULT, whose adaptation of Kafka's *The Trial* (1947) perhaps came nearest to applying Artaud's ideas, VILAR, and Blin; in the early plays of ADAMOV; and in the works of GENET, ARRABAL, and VAUTHIER. More generally, it was a major factor in the shift to COLLECTIVE CREATION in the 1960s, and provided a theoretical explanation for such experiments in audience participation as HAPPENINGS. Peter BROOK's production of WEISS's *Marat/Sade* (1962) has so far been the most notable Artaudian production to achieve conventional theatrical success. Artaud's mystical views of theatre, though not his methods, are akin to those of GROTOWSKI.

ARTISTS OF DIONYSUS, organizations of actors, musicians, mask-makers, writers, stagehands, etc, in the ancient Greek theatre. The Athenian Guild was the oldest, dating from some time in the late 4th century BC, but after about 290 several others were established, among them the Egyptian and the Ionian-Hellespontic. In Roman times there was a world-wide guild. Because of its original association with religion, the profession of acting, confined to men of good repute, gave actors a large degree of immunity; they were often used, as Aristodemos was, as diplomatic envoys in disputes between Greek states. Each guild had its constitution and its main function was to negotiate rates of pay and maintain the right of safe passage from one city to another for theatrical troupes. As Dionysus lacked both official recognition and respectability in Rome, the nearest equivalent to the Artists of Dionysus seems to have been the *collegium scribarum*, established *c.* 200 BC, while the rank and file of theatre people called themselves *artifices scaenici*.

ARTS COUNCIL OF GREAT BRITAIN, the main channel for the distribution of public money in subsidies to the British theatre and other arts. It is in essence a continuation of the Council for the Encouragement of Music and the Arts (CEMA) founded in 1940, early in the Second World War, to take concerts and plays to crowded evacuation areas, factory canteens, and other commercially unprofitable centres. The Arts Council, which took its present name in 1946, receives all its funds from the Treasury, being answerable to Parliament for its expenditure. Funds for dramatic activities are made available mainly through non-profit-distributing companies and charitable trusts, and are intended to help new playwrights and to promote the training of theatre personnel in all branches. This has given a great impetus to the REPERTORY THEATRE movement throughout the country and to improvements in artistic standards as well as backstage working conditions. Of the sum allotted for drama, a large part goes to the NATIONAL THEATRE and the ROYAL SHAKESPEARE COMPANY, and a sum is also earmarked for the assistance of touring drama companies. The Arts Council helps to subsidize 12 Regional Arts Associations in England, together with the Scottish and Welsh Arts Councils, that for Northern Ireland being a separate body. These organizations, which receive additional subsidies from sources such as local authorities, give grants to local dramatic organizations, often CIVIC THEATRES, and coordinate theatre activities within their own areas.

ARTS LABORATORY. This opened in Jan. 1968 in a disused warehouse in Drury Lane, London, under the direction of Charles MAROWITZ and Jim Haynes, founder of the TRAVERSE THEATRE CLUB in Edinburgh. Films, music, and exhibitions of graphic art were presented, and many theatre groups established themselves there, among them the PEOPLE SHOW, the PIP SIMMONS THEATRE GROUP, and Wherehouse LA MAMA. Although the Arts Laboratory closed in Oct. 1969, it exercised a major influence on the London FRINGE THEATRE movement.

ARTS THEATRE, see CAMBRIDGE.

ARTS THEATRE, London, in Great Newport Street, off St Martin's Lane. This opened on 20 Apr. 1927, for the production of unlicensed and experimental plays for members only. Seating 347 in an intimate two-tier auditorium, it has a proscenium width of 20ft and stage depth of 18ft. Its first important production, *Young Woodley* (1928) by John VAN DRUTEN, transferred to a commercial theatre, as did Reginald Berkeley's *The Lady with a Lamp* (1929), Ronald Mackenzie's *Musical Chairs* (1931), Gordon Daviot's *Richard of Bordeaux* (1932), and Norman Ginsbury's *Viceroy Sarah* (1934). In 1942 Alec CLUNES took over, and for ten years made the theatre a vital centre, producing a wide range of plays and winning for it the status of a 'pocket national theatre'. Christopher FRY's *The Lady's Not For Burning* had its first performance here in 1948, with Clunes as Thomas Mendip. The theatre changed hands in 1953, but continued to stage new plays, including BECKETT's *Waiting for Godot* (1955) and ANOUILH's *The Waltz of the Toreadors* (1956). Harold PINTER's *The Caretaker* received its first performance here on 27 Apr. 1960, and ALBEE's *The Zoo Story* was seen later in the same year. In 1962 the theatre was leased for six months by the ROYAL SHAKESPEARE COMPANY for a major experimental season, which opened on 14 Mar. with Giles Cooper's *Everything in the Garden*, followed by five new plays and by revivals of GORKY's *The Lower Depths* and MIDDLETON's *Women*

Beware Women. The theatre was then used for mixed programmes, including films, ballet, and plays for children presented by the Unicorn Theatre under the direction of Caryl Jenner. In 1975 Robert Patrick's *Kennedy's Children* came to this theatre from the King's Head, Islington, the first FRINGE THEATRE production originating in a public house to transfer to the West End. A double bill by STOPPARD, *Dirty Linen* and *New Found Land*, opened in 1976 and ran for four years.

ASCH, SHOLOM (1880–1957), Jewish dramatist and novelist, born at Kutno, Poland, author of several plays in Yiddish, of which the best known is *God of Vengeance*. As it could not be performed in Russian territory because of the ban at the time on Yiddish plays, this was given its first production by REINHARDT in Berlin in 1907 in German, as *Gott der Rache*. It immediately attracted the attention of the literary and theatrical world to the possibilities of Yiddish drama, and has since been translated and produced in many countries. Some of Asch's Yiddish novels have also been dramatized or adapted for the stage. More than anyone else Asch helped to raise the standard of Yiddish writing and place it on a literary basis.

ASCHE, OSCAR [John Stanger Heiss] (1871–1936), English actor of Scandinavian descent, born in Australia. He made his first appearance on the stage in 1893 and was for some time a member of Frank BENSON's Shakespeare company. He joined TREE's company at HER (then His) MAJESTY'S THEATRE in 1902, taking over the management of the theatre in 1907 and producing there *As You Like It* with himself as Jaques; *Othello*, in which he played the title role; and *The Taming of the Shrew*, in which he appeared as Petruchio, always one of his best parts, together with Shylock in *The Merchant of Venice* and Antony in *Antony and Cleopatra*. In 1911 he added Falstaff in *The Merry Wives of Windsor* to his repertoire and in the same year made a great success in the part of Hajj in Edward KNOBLOCK's *Kismet*. He is, however, chiefly remembered for his own oriental fantasy with music *Chu-Chin-Chow* (1916), in which he played Abu Hassan; it ran for five years without a break. A successor to it, *Cairo* (1921), was not successful. Asche's wife, the actress Lily Brayton (1876–1953), who made her first appearance in 1896 under Benson, played opposite her husband in most of his productions, being seen as Marsinah in *Kismet* and Sahrat-al-Kulub in *Chu-Chin-Chow*.

ASHCROFT, PEGGY [Edith Margaret Emily] (1907–   ), DBE 1956, English actress, made her début in 1926 at the BIRMINGHAM REPERTORY THEATRE as Margaret in BARRIE's *Dear Brutus*. A year later she was in London, where she first attracted attention in 1929 playing Naemi in FEUCHTWANGER's *Jew Süss*, in which she showed the simplicity and sense of poetic tragedy that have since made so many of her performances remarkable. She enhanced her reputation with an excellent Desdemona to Paul ROBESON's Othello in 1930, and in the autumn of 1932 joined the OLD VIC, where she played, among other parts, Rosalind in *As You Like It*, Perdita in *The Winter's Tale*, Imogen in *Cymbeline*, and Cleopatra in SHAW's *Caesar and Cleopatra*. She established herself as one of the outstanding young actresses of the time by her Juliet in GIELGUD's production of *Romeo and Juliet* in 1935 and her Nina in CHEKHOV's *The Seagull* a year later. She then made her New York début in Maxwell ANDERSON's *High Tor* (1937), returning to join Gielgud again at the QUEEN's, where she was much admired as Lady Teazle in SHERIDAN's *The School for Scandal* and Irina in Chekhov's *Three Sisters*. She also played one of her best comedy parts, Cecily Cardew in WILDE's *The Importance of Being Earnest*, under Gielgud's direction at the GLOBE in 1939, and in a famous season at the HAYMARKET THEATRE in 1944–5 she was Ophelia to his Hamlet, Titania to his Oberon in *A Midsummer Night's Dream*, and the Duchess to his Ferdinand in WEBSTER's *The Duchess of Malfi*.

After the Second World War she gave two outstanding performances in modern plays, as the alcoholic wife Evelyn Holt in Robert MORLEY's *Edward, My Son* (1947), in which she was also seen in New York, and as Catherine Sloper in the Goetz's *The Heiress* (1949), based on Henry JAMES's *Washington Square*. She returned to Shakespeare in 1950, playing Beatrice to Gielgud's Benedict in *Much Ado About Nothing* and Gordelia to his Lear at Stratford-upon-Avon. After a further excursion into modern drama as Hester Collyer in RATTIGAN's *The Deep Blue Sea* (1952), she gave what was undoubtedly one of the finest performances of her career in IBSEN's *Hedda Gabler* in 1954, for which she was awarded the King's Medal by King Haakon of Norway, and then showed her amazing versatility by playing Miss Madrigal in Enid BAGNOLD's *The Chalk Garden* and the dual role of the prostitute and the prostitute's male cousin in BRECHT's *The Good Woman of Setzuan* (both 1956). In 1961 she became a member of the ROYAL SHAKESPEARE COMPANY (of which she was later made a director), and appeared for them not only in such classic roles as Margaret of Anjou, from youth to old age, in *The Wars of the Roses* (comprising *Henry VI, Edward IV*, and *Richard III*), but also in modern plays—DURAS's *Days in the Trees* (1966), ALBEE's *A Delicate Balance*, and PINTER's *Landscape* (both 1969). In 1972 she was seen in the West End in William Douglas HOME's *Lloyd George Knew My Father*, which had a long run, and in 1975 was at the NATIONAL THEATRE in Ibsen's *John Gabriel Borkman* and BECKETT's *Happy Days*, but returned to the ALDWYCH in 1976 to appear in ARBUZOV's *Old World*. Again at the National Theatre she was in the 1980 revival of Lillian HELLMAN's *Watch on the*

Rhine. In 1962 a theatre named after her opened in Croydon, Surrey, where she was born.

ASHWELL [Pocock], LENĀ (1872–1957), English actress, who made her first appearance in 1891, and in 1900 scored a great success in Henry Arthur JONES's *Mrs Dane's Defence*, in which she was also seen in the U.S. On her return to England she took over the KINGSWAY THEATRE, remaining there until in 1915 she left to organize entertainments for the troops in France and later in Germany, for which she was awarded the OBE. She was active in the foundation of the British Drama League (see AMATEUR THEATRE) in 1919, and in 1925 returned to management at the BIJOU THEATRE in Bayswater, renaming it the Century and producing there a number of new plays, including her own adaptations of DOSTOEVSKY's *Crime and Punishment* and Stevenson's *Dr Jekyll and Mr Hyde* (both 1927). Her autobiography, *Myself a Player*, was published in 1936.

ASOLO STATE THEATRE, Sarasota, since 1965 the State Theatre of Florida (a title also given to the Players Theatre of Miami in 1977), which became fully professional in 1966. It evolved from the summer festival of 18th- and 19th-century comedies first staged in 1960 by the Division of Theatre of Florida State University, and is now housed in an exquisite Court theatre, seating 320, which was erected in 1798 in Asolo, Italy, and reassembled in Sarasota by the Ringling Museums in 1951. The Asolo Theatre company is one of the few American companies to play in true repertory, its season running from mid Feb. to Labor Day Weekend early in Sept. As many as five matinées a week are often given, with different plays at matinée and evening performances, productions ranging from classics to modern plays and musicals. First performances of new plays are also presented, and over a dozen PULITZER PRIZE-winners have been staged ranging from SHERWOOD's *Idiot's Delight* (1980) to Michael Cristofer's *The Shadow Box* (1979), winners in 1936 and 1977 respectively. Since 1966 the theatre has been deeply committed to educational work, the Asolo Touring Theatre, founded in 1975, visiting schools throughout the state with a repertory designed for different age groups. In conjunction with Florida State University, of which it forms a part, the theatre in 1973 established a Conservatory of Acting offering a course leading to a Master's degree and presenting plays at the Asolo Stage Two, which opened in 1977 in downtown Sarasota.

ASPHALEIAN SYSTEM, one of the earliest methods of controlling the stage floor, first used in the Budapest Opera House in 1884, named after the Austrian company which evolved it. It lessened the risk of fire, since steel was mainly used in its construction, and marked the beginning of a new era in stage technique. The whole stage was divided into individual platforms resting on hydraulic pistons, each of which could be separately raised, lowered, or tipped.

ASSEMBLY THEATRE, New York, see PRINCESS THEATRE.

ASSOCIATION OF COMMUNITY THEATRES (A.C.T.), see NEW ZEALAND.

ASTLEY'S AMPHITHEATRE, London, on the south bank of the Thames in Westminster Bridge Road. This theatre, immortalized by Charles DICKENS in *Sketches by Boz*, was originally an open circus ring, used from 1769 onwards by Philip Astley (1742–1814), a retired cavalryman who invented the modern circus and established the popular but short-lived EQUESTRIAN DRAMA. On a site near the ring he built in 1784 a covered amphitheatre with a stage, pit, gallery, and boxes, which was used purely as a circus until 1787, when BURLETTA and PANTOMIME were added to the programmes. Rebuilt after a fire in 1794, the building reopened a year later as the Royal Grove, but was again destroyed by fire in 1803. While it was being rebuilt once more, Astley, who had been responsible for the opening of circuses all over Europe, moved to the OLYMPIC THEATRE, but returned in 1804 to the new building, which he named the Royal Amphitheatre. Among the equestrian spectacles for which it later became famous, the first was Amherst's *The Blood-Red Knight; or, the Fatal Bridge* (1810). After Astley's death the theatre was taken over by Davis, who gave it his own name and continued to present spectacular melodramas starring Edward Alexander Gomersal (1788–1862), an actor of whom Thackeray left a description in *The Newcomes*. After a fire in 1830 Davis retired and his place was taken by the renowned horseman Andrew Ducrow (1793–1848), who was so illiterate that he seldom played a speaking part but excelled in riding, stage management, and production. He remained in charge until 1841, when the theatre was again burnt down. Rebuilt, it was run for a time by William Batty (1801–68), who gave it his own name, followed by William Cooke, under whom *Macbeth* and *Richard III* became equestrian dramas, Richard's horse, White Surrey, playing a leading part. In 1862 Dion BOUCICAULT took over the theatre and renamed it the Theatre Royal, Westminster. It failed dismally, in spite of a spirited revival of Boucicault's own play *The Relief of Lucknow* in 1862 and the first English production of his adaptation of SCOTT's *The Heart of Midlothian* as *The Trial of Effie Deans*, first seen in New York in 1860. Boucicault left, heavily in debt, and was succeeded by E. T. SMITH, who attracted huge audiences with his presentation of Adah Isaacs MENKEN in Lord BYRON's *Mazeppa*. Following its reconstruction in 1872/3 with a four-tier auditorium seating 2,407, the theatre came under the control of the circus proprietor 'Lord' George

Sanger (1827–1911), and was known as Sanger's Grand National Amphitheatre. Closed as unsafe in 1893, it was demolished two years later. No trace of it remains, but in 1951 a memorial plaque was unveiled on the site at 225 Westminster Bridge Road.

ASTON, TONY [Anthony] (*fl.* 1700–50), Irish actor, also known as 'Matt Medley', a wild, irresponsible creature of a mercurial disposition, who tried his hand at many things but soon wearied of them. He was at one time at DRURY LANE, and is believed to have been the first professional actor to appear in the New World, since he is known to have been in Charleston in 1703 and in New York later the same year. There is no information about what he acted and whether he was alone or in company with other actors. He may have appeared in a variety entertainment similar to the 'medley' in which he was seen in the English provinces in 1717. Although he says in the preface to the 1731 printed edition of his play *The Fool's Opera; or, the Taste of the Time*, acted in Oxford by his own company, that he had appeared in 'New York, East and West Jersey, Maryland, Virginia (on both sides Cheesapeek), North and South Carolina, South Florida, the Bahamas, Jamaica, and Hispaniola', he is not thought to have spent much time in America. He was the author of several other plays, of which the first, *Love in a Hurry* (1709), was seen in Dublin, and the second, *The Coy Shepherdess* (1712), at Tunbridge Wells. In 1735 he protested against the proposed bill for regulating the stage, and he was still alive in 1749, when he was spoken of as 'travelling still, and as well-known as the posthorse that carries the mail'.

ASTOR PLACE OPERA HOUSE, New York, one block east of Broadway, was built for Italian opera, and opened on 22 Nov. 1847. On 4 Sept. 1848 the English actor MACREADY appeared there, but owing to the jealousy of Edwin FORREST his visit terminated on 10 May 1849 with the anti-British Astor Place riot, in which 22 people were killed and 36 wounded by shots fired by the militia. The theatre then closed for repairs, and reopened on 24 Sept. 1849 with *Romeo and Juliet*. Actors who played in the theatre between seasons of opera included Charlotte CUSHMAN, Julia DEAN, and George VANDENHOFF, but it never recovered from its reputation as the 'Massacre Opera House', and finally closed on 12 June 1850, being converted into a library and lecture room.

ASTOR THEATRE, New York, at 1537 Broadway, on the north-west corner of 45th Street. It opened on 21 Sept. 1906 with *A Midsummer Night's Dream*, the first notable productions being Eugene Walter's *Paid in Full* and Booth TARKINGTON and H. L. Wilson's *The Man from Home* (both 1908). In 1913 George M. COHAN's dramatization of Earl Biggers's novel *Seven Keys to Baldpate* was also a great success. The theatre's career was thereafter uneventful and in 1925 it became a cinema. It was later demolished and the site now serves as a 'flea market'.

A.T.A. (AMERICAN THEATRE ASSOCIATION), see UNIVERSITY DEPARTMENTS OF DRAMA.

ATELIER, Théâtre de l', Paris, suburban playhouse built in 1822 off a small square (now called Place Charles-Dullin) between Montmartre and Pigalle. It was first named the Théâtre de Montmartre, but became the Théâtre des Élèves when it was used for melodramas and vaudevilles played by young actors, some still pupils at the Conservatoire. It became a cinema in 1914, but in 1922 it was taken over by DULLIN and became one of the recognized centres of AVANT-GARDE drama, with excellent productions of PIRANDELLO, ACHARD, and ROMAINS; a school of acting attached to it trained many well-known French actors, including BARRAULT. When in 1940 Dullin left to go to the Théâtre de PARIS, André BARSACQ moved in with his Compagnie des Quatre Saisons and, in spite of growing financial difficulties, continued to uphold the theatre's reputation as a home of intellectually ambitious drama. After Barsacq's death it was managed by his son André-Alexis.

ATELLAN FARCE, see FABULA (1).

ATHÉNÉE, Théâtre de l', Paris. Built on its present site under a bank and business complex close to the Madeleine in 1893, this playhouse was first called the Comédie-Parisienne, changing its name three years later to Athénée-Comique, from the name of a notoriously frivolous, perhaps immoral, establishment nearby that had had to close ten years earlier. For some forty years it offered a mixed fare of vaudevilles, comedies, and melodramas, but in 1934 it was taken over by Louis JOUVET, who put on there all his famous productions, including plays by ACHARD, MOLIÈRE, and especially GIRAUDOUX. After his death in 1951 several managers tried to continue his work, but without success.

ATKINS, EILEEN (1934–   ), English actress, first appeared in London at the OPEN AIR THEATRE in Regent's Park in 1953, and from 1957 to 1959 was at STRATFORD-UPON-AVON, leaving there to join the BRISTOL OLD VIC company. In 1962 she played leading roles with the OLD VIC, but her performance as Childie, the child-like 40-year-old lesbian in Frank Marcus's *The Killing of Sister George* (1965; N.Y., 1966), first brought her into prominence, particularly when she followed it with a portrayal of a neglected, stonily observant wife in David STOREY'S *The Restoration of Arnold Middleton* (1967). In New York, also in 1967, she appeared in ARBUZOV'S *The Promise*, and back in London was seen as Celia Coplestone in T. S. ELIOT'S *The Cocktail Party* (1968). Her next opportunity to prove her versatility and emotional directness was as Elizabeth I in Robert BOLT's *Vivat! Vivat Regina!* (1970; N.Y., 1972); in 1973, after waging a two-year campaign on its behalf, she brought Marguerite DURAS's *Suzanna Andler* to London, playing the leading role herself. She appeared as Rosalind in *As You Like It* for the ROYAL SHAKESPEARE COMPANY (1973), as Hesione Hushabye in SHAW's *Heartbreak House* for the NATIONAL THEATRE (1975), and for PROSPECT in his *St Joan* (1977), FRY's *The Lady's Not For Burning*, and as Viola in *Twelfth Night* (both 1978). In 1981 she returned to the R.S.C. as Billie WHITELAW's *alter ego* in Peter NICHOLS's *Passion Play*.

ATKINS, ROBERT (1886–1972), English actor and director, whose career was devoted mainly to Shakespeare. He made his first appearance at HER (then His) MAJESTY'S THEATRE in 1906 in *Henry IV, Part I*, and in 1915 joined the OLD VIC company, returning there in 1920 after war service. He remained until 1925, during which time he directed and acted in a number of plays, including IBSEN's *Peer Gynt* (1922), its first production in London, and the rarely-seen *Titus Andronicus* and *Troilus and Cressida* (both 1923). On leaving the Old Vic he took his own company on tour, and in 1936 founded the Bankside Players, directing them in *Henry V, Much Ado About Nothing*, and *The Merry Wives of Windsor* at the RING, Blackfriars, in something approximating to Elizabethan conditions. He also directed at STRATFORD-UPON-AVON, and at the OPEN AIR THEATRE in London's Regent's Park, which he ran for many years. During Oct. 1940, at the height of the bombing of London, he continued to present plays by Shakespeare at the VAUDEVILLE THEATRE. As an actor he was at his best in such parts as Sir Toby Belch in *Twelfth Night*, Touchstone in *As You Like It*, Caliban in *The Tempest*, Bottom in *A Midsummer Night's Dream*, Sir Giles Overreach in MASSINGER's *A New Way to Pay Old Debts*, and James Telfer in PINERO's *Trelawny of the 'Wells'*.

ATKINSON, (Justin) BROOKS (1894–1984), American dramatic critic, educated at Harvard during the influential period of George Pierce BAKER. He became assistant drama critic of the *Boston Daily Evening Transcript* in 1918 and literary editor of the *New York Times* in 1922. From 1926 to 1960 (except during the war years) he was that paper's dramatic critic—an unrivalled tenure. He understood both the literary value of drama and the need for it to be a medium of active communication; his literate, scrupulously honest, and acute writing won a tremendous following. He had great influence: his approval guaranteed a respectable run for a play irrespective of the opinion of other critics. On his retirement the Mansfield Theatre in New York was renamed the BROOKS ATKINSON in his honour.

ATTA, TITUS QUINTIUS (?–77 BC), Roman dramatist, one of the chief exponents of the *fabula togata*, or comedy based on daily life in Rome (see FABULA, 7). Eleven titles and 20 lines are all that survive of his work, but it is evident that in his *Aquae Caldäe* he set the scene in a Roman spa—a fruitful source of farcical comedy. He was said to have excelled in his delineation of women.

AUBIGNAC, FRANÇOIS HÉDELIN, Abbé d' (1604–76), one of the first important French writers on the drama. His *Pratique du théâtre* upheld the UNITIES, and criticized with much acerbity those who departed from them. Its publication in 1657 marked a decisive step in the formation of the French neo-classical style; it was translated into English in 1684. Aubignac's own plays, of which the first was the tragedy *Zénobie* (1640), were intended as models for aspiring dramatists, but were not particularly successful, partly owing to the mediocrity of their style, and partly to the attacks of the many enemies which d'Aubignac's criticisms had made for him; CORNEILLE is known to have been particularly sensitive to his strictures.

AUCKLAND THEATRE TRUST, see MERCURY THEATRE, AUCKLAND, and NEW ZEALAND.

AUDIBERTI, JACQUES (1899–1965), French poet, novelist, and dramatist, who was originally a journalist and did not take up imaginative writing until about the age of 35. His Surrealist world is close to that of ARTAUD, as is evinced in his first plays, *Quoat-Quoat* (1946), in which a French ship bound for Mexico becomes a symbol for organized society at the mercy of atavistic forces, and *Le Mal court* (1947), a tale of innocence corrupted by experience set in the 18th century. Audiberti's next play, *Les Femmes du boeuf* (1948), was produced at the COMÉDIE-FRANÇAISE, as was *Fourmi dans le corps* (1962). Among his other plays were *La Fête noire* (1948), *Pucelle* (1950), a somewhat irreverent treatment of the life of Joan of Arc, *La Hobereaute* (1956), set in medieval Burgundy, *L'Effet Glapion* (1959), and his last work, *L'Opéra du monde* (1965). Audiberti belonged to no particular school of dramatists, but created his own world of fantasy and farce in which his eccentricities found full scope, though his work was often rendered unintelligible by an excess of verbal dexterity.

AUDITION, a trial run by an actor seeking employment, either to display his talents in general by singing, dancing, and reciting, or to demonstrate his fitness for a particular role by reading some part of it to the DIRECTOR and his associates. In earlier times actors were often engaged for a company merely on hearsay, by a letter of recommendation, or after a private audition by the manager. In modern times leading actors may be specifically engaged for a part on their reputation alone, while the supporting cast is chosen only after a number of auditions have been held.

AUDITORIUM, that part of a THEATRE BUILDING designed or intended for the accommodation of the people witnessing the play—the audience. The word in its present usage dates from about 1727, when it gradually replaced the much earlier Auditory. This however continued in use intermittently until the mid-19th century, together with a new coinage, Spectatory, which had a short life. The auditorium can vary considerably in size and shape, placing the audience in front of, part way round, or entirely round the acting area. It can also be in one entity or divided, with galleries above the main area and boxes at the sides or all round.

AUGIER, (Guillaume Victor) ÉMILE (1820–89), French dramatist and one of the first to revolt against the excesses of the Romantics. After a few plays in verse, he found his true vocation in writing domestic prose-dramas dealing with social questions of the moment, of which *Le Gendre de Monsieur Poirier* (*M. Poirier's Son-in-Law*, 1854) is the best known. Among others which were successful in their day were *Le Mariage d'Olympe* (1855), which paints the courtesan as she is and not as idealized by the younger DUMAS; *Les Lionnes pauvres* (1858), which shows the disruption of home life consequent on the wife's adultery; and two political comedies, *Les Effrontés* (*Bold as Brass*, 1861) and *Le Fils de Giboyer* (1862). Under the influence of the new currents of thought running across Europe, he ended his career with two problem plays, *Mme Caverlet* (1876) and *Les Fourchambault* (1878), adapted by James ALBERY as *The Crisis*. Augier's plays were well written in the theatrical conventions of his time, and he was more of a realist than either the younger Dumas or SARDOU.

AUGUSTUS DRURIOLANUS, see HARRIS, Sir AUGUSTUS.

AULAEUM, see CURTAIN.

AUSOULT, JEANNE, see BARON, ANDRÉ.

AUSTRALIA. The first recorded theatrical performance in Australia was a production of FARQUHAR's *The Recruiting Officer*, given in Sydney on 4 June 1789 by a cast of convicts. In 1796 a Mr Sidaway got permission to build a theatre in Sydney which opened with a melodrama, *The Revenge*, but rowdyism among the audience caused it to be closed on the governor's orders, and it was demolished two years later. Although in 1800 another theatre, under the patronage of the governor, again presented *The Recruiting Officer*, together with FIELDING's *The Virgin Unmask'd* (the subtitle of his *An Old Man Taught Wisdom*, 1735), there was very little theatre in Australia apart from convict performances until the 1830s, and it was not until 1832 that the first professional theatre in the country opened in Sydney, under the management of Barnett LEVEY, with JERROLD's *Black-Ey'd Susan*. At this theatre, too, the first native Australian play, the convict Edward Geoghegan's *The Hibernian Father* (1844), was produced. It was set in medieval Ireland and was probably plagiarized: his operetta *The Currency Lass; or, My Native Girl* (also 1844) can be said to be the first Australian play with an Australian setting to be produced in the country. It was followed by productions of David Burns's *The Bushrangers*, first produced in Edinburgh in 1829, but the staple fare was farce and melodrama, mainly imported. Meanwhile, in 1837, the Royal Victoria, now the Theatre Royal, opened in HOBART. Adelaide also had a Theatre Royal, at the Adelaide Tavern; another Royal Victoria opened there in 1839, and in 1842 a Mr Lazar, formerly a theatre manager in Sydney, opened the Queen's. Melbourne had its first theatre in 1841, a wooden building known as the Pavilion, later renamed the Theatre Royal.

In 1842 the English actor George Selth COPPIN arrived in Australia, and soon afterwards settled in Melbourne, where he was the first to import visiting stars from overseas. He engaged James Cassius WILLIAMSON and his wife Maggie MOORE in 1874, and shortly after Williamson founded the firm of J. C. Williamson Theatres Ltd, which became the largest theatrical management in the southern hemisphere. In the 1880s the firm suffered a certain amount of competition from a comedy company run by 'Dot' BOUCICAULT in partnership with the dramatist Robert Brough, but this was disbanded in the 1890s, though Boucicault and his wife Irene VANBRUGH returned to tour Australia again on several occasions.

By this time variety shows were being staged with great success, and many well-known British music-hall stars, among them Marie LLOYD and LITTLE TICH, were being brought to Australia by Harry Rickards [Harry Benjamin Leete] (1843–1911). He was also responsible for the visit in 1891 of the company from the GAIETY THEATRE in London, led by Fred Leslie. Australia now also had its own stars, of whom the most famous, perhaps, was the singer and actress Nellie STEWART, while Gladys MONCRIEFF achieved great popularity in musical comedy and operetta.

Alongside the commercial theatre there sprang up at the turn of the century a number of 'little' theatres and repertory societies, mostly amateur, of which the first was the Adelaide Repertory Society, founded in 1908. In Sydney in the 1930s Doris Fitton founded the Independent Theatre (which closed for lack of substantial subsidies in July 1977), and in the early years of the Second World War May Hollinworth's Metropolitan Theatre and Kathleen Robinson's Minerva Theatre were also active. In Melbourne the Little Theatre, later the ST MARTIN'S THEATRE, was founded in 1931, followed eventually by Frank Thring's Arrow Theatre and Wal Cherry's Emerald Hill Theatre, both now defunct. In Brisbane there was the Twelfth Night Theatre, while Hobart and Perth both had similar companies. During the inter-war period Allan Wilkie (1878–1970), with his wife Frediswyde Hunter-Watts as his leading lady, toured in a number of plays by Shakespeare (whose *Henry IV*, incidentally, had been seen in Sydney as early as the 1790s), and after the Second World War John Alden also toured with his own Shakespeare company.

For many years the Australian theatre relied heavily on plays imported from England and the U.S.A., usually with English and American stars, often past their best or little known in their own countries. Their Australian counterparts, however experienced and talented, were forced to look for success abroad, rarely to return. Plays written by native Australians during the interwar period were, in any case, mainly inferior to the imported products. One exception was *A Touch of Silk* (1930) by Betty Roland, the precursor of such works as K. Brownville's *Sleep to Wake* (1939), C. Duncan's *Sons of the Morning*, Sumner Locke-Elliott's *The Invisible Circus* (both 1946), and Russel Oakes's *Enduring as the Camphor Tree*. Louis Esson has also been generally considered as a neglected Australian playwright, though the revival of two of his plays in 1973 perhaps did little to confirm this judgement. Occasionally an Australian play would cause a stir, as did Locke-Elliott's *Rusty Bugles* in 1948, but no decisive impact was made on the international theatre scene until the production in 1955 by the newly founded ELIZABETHAN THEATRE TRUST of Ray LAWLER's *The Summer of the Seventeenth Doll*, which was later seen in London and in New York. Lawler's success was quickly followed up by Richard Benyon with *The Shifting Heart* (1957), seen in London in 1959, *The Slaughter of St Teresa's Day* (1959) by Peter KENNA, and Alan Seymour's *The One Day of the Year* (i.e., Anzac Day) (1960), also seen in London in 1961. Apart from Patrick WHITE's very specialized talent and Hal PORTER's near-misses, only Alan HOPGOOD upheld the cause of Australian drama in the early 1960s. Later in that decade, and in the early 1970s, a number of new playwrights, aided by government subsidies, came to the fore, among them Alexander BUZO and David WILLIAMSON; the most sophisticated work of its period was probably *As It's Played Today* (1974) by the actor John MCCALLUM, who with his wife Googie WITHERS had returned to Australia (he was born in Brisbane) in 1958 to direct plays and also appear in them, mainly in Melbourne. In 1976 a marked success was scored by the actor Gordon Chater in a one-man play by Steve J. Spears entitled *The Elocution of Benjamin Franklin* (described by Chater as being about 'bigotry, loneliness, and sex'), which opened at the Nimrod Theatre in Sydney, was then seen in Melbourne, and after a long tour of Australia ran for several months at the MAYFAIR THEATRE, London, in 1978.

By the 1970s the Elizabethan Theatre Trust had been largely superseded by the Australian Council for the Arts, founded in 1968 and later renamed the Australia Council. The organization of the theatre in Australia, long divided into 'commercial' and 'non-commercial', now became 'subsidized' or 'unsubsidized', the unsubsidized sector staging mainly productions of overseas successes while the subsidized sector concentrated on more serious, experimental, and naturalistic types of play and new works by Australian writers. Each capital city had its own major subsidized company —the OLD TOTE THEATRE in Sydney, replaced in 1979 by the Sydney Theatre Company, which plays in the SYDNEY OPERA HOUSE; the MELBOURNE THEATRE COMPANY; and the South Australian Theatre Company in Adelaide, where since 1960 a biennial Festival of the Arts has been held, to which in 1980 Peter BROOK took his Paris company in a collective work, *The Conference of the Birds*. Brisbane has the Queensland Theatre Company, Perth the National Theatre Company, and Hobart the Tasmanian Theatre Company. Other minor but fully professional companies, also receiving ever-increasing subsidies, include, in Sydney alone, the Nimrod Theatre, one of Australia's best; the Ensemble, founded by the American actor and singer Hayes Gordon, which plays in-the-round; and the Marian Street, once the Community Theatre. Melbourne has La Mama, the Australian Performing Group, and the Hoopla Theatre Foundation (this last renamed in 1980 the Playbox Theatre Company). For a time Adelaide had John Edmund's now extinct Theatre '62.

With the tightening up of the immigration laws in Britain and the difficulty of obtaining visas to work in the U.S.A., more Australian actors were forced to remain in their own country, and by the mid-1970s Australian Equity had also begun to make it difficult for actors from overseas to be engaged for Australia in the hope that native players would find employment more easily. But at the same time more actors were being drawn away to local television and film-making, so that managements, faced with escalating costs, found it almost impossible to stage good quality work. Nevertheless, it is to be hoped that the Australian theatre 'renaissance' of the early 1970s will continue and in time be sustained and developed to the point of supporting a thoroughly national theatre. By 1980 commercial managements were tending increasingly to co-present productions with the Elizabethan Theatre Trust and organizations such as the Adelaide Festival Centre Trust.

AUSTRIA. The first stirrings of a truly indigenous theatrical art in the Austrian lands can be traced to the small town of Sterzing in the Tyrol, where from 1455 onwards theatrical troupes were formed under the leadership of Stoffl Schopfer and Vigil Raber for the performance of a PASSION PLAY with an original text and for tales of chivalry drawn from the German heroic epics. More important to the development of the Austrian theatre, however, was the work of the humanists of the early 16th century. In the plays of Konrad CELTIS the later Viennese HAUPT-UND STAATSAKTION can be seen in embryo, while his pupil Joachim von WATT first created the Viennese *Posse*, or farce —these two dramatists representing the university, with its atmosphere of secular humanism, and reflecting in the popular elements in their plays something of that earthy realism which had already produced the plays of NEIDHART VON REUENTHAL. Benedictus CHELIDONIUS meanwhile pioneered a more dialectic form of religious drama, while Wolfgang SCHMELTZL wrote plays in the vernacular on Biblical subjects.

The soil of VIENNA proved unfruitful for this kind of evangelical SCHOOL DRAMA; but 1551, the year in which Schmeltzl left Vienna, was also the year of the Jesuits' arrival. Two years later the first Jesuit school opened; within a year it had 300 pupils, by the end of the century 1,000, and nowhere was the Jesuit policy of harnessing all the resources of human art to the task of education more successful than in Vienna, not least in the domain of JESUIT DRAMA. As early as 1562 a report from MUNICH (where a school had been founded in 1559) emphasized that farce was 'the best medium of winning over the Germans, of making friends of heretics, and of filling the schools'. The same insight into the theatrical propensities of the Austrians induced the Jesuits, who also had an important school in Graz, to depart from their own regulations by permitting vernacular, instead of Latin, interludes in plays from 1588 onwards, while from 1601 female roles were allowed in all areas north of the Alps. The first Jesuit play to be performed in Vienna was Lewin Brecht's *Euripus sive de inanitate rerum omnium* (1555), which shows many of the characteristics typical of the best Jesuit drama in the second half of the 17th century: the humanistic preference for classical themes, the moral purpose derived from the medieval MORALITY PLAY, and the setting forth of the Christian message. To these may be added three others which went to make up the unique flavour of the Austrian theatre: a hierarchic view of society which permitted the triumphs and glories of a divinely appointed (Habsburg) ruler over a divinely constituted (and thus symbolic) social order to be celebrated in a dramatic spectacle designed to justify the ways of God to man; a tendency, derived from the gloomy splendours of the Counter-Reformation in Spain, and specifically from CALDERÓN, to emphasize the dreamlike quality of the everyday when set against the reality of the beyond; and the contact with Italian opera, which gave the Jesuit drama in Vienna, and through it the Viennese theatre as a whole, its reliance on music and spectacle: much of the later Austrian theatre developed in imitation of and in competition with Italian opera. The splendours of the Jesuit production *Pietas victrix* (1659), by the greatest writer of Jesuit drama in Austria Nikolaus AVANCINI, were rivalled by those of Cesti's opera *Il pomo d'oro* (1667), to mention only two productions among many. It was pressure from operatic productions which caused the Jesuits to abandon the old MULTIPLE SETTING in favour of a single set with *telari* (see PERIAKTOI) arranged in perspective. The stage thus set out to be a second reality rivalling and transcending the real world; and this second reality could in its turn be transcended and thus made to underline the baroque moral—*vanitas, vanitatis vanitas*—by means of Italian machinery, to which TRANSFORMATION SCENES, trials by fire and water, and the descents of the DEUS EX MACHINA were child's play. These outward splendours made a significant contrast with, say, the purely inward spiritual realities of such plays as *Cenodoxus* (1602), by the Munich Jesuit dramatist Jakob BIDERMANN, in which the outer world is shown to be all sham and hypocrisy.

The Viennese baroque theatre carried within it the seeds of its own decay; when the spiritual impulse flagged, the spectacle became empty rhetoric. And from another quarter too its ultimate downfall was being prepared. The gradual inclusion of the vernacular interlude, aimed originally at the uneducated part of the audience, became a bridgehead for Viennese humour, with its love of cynical comment. This process can be seen at its best in the work of Johann Baptist ADOLPH, whose plays mark a second high-water mark in the history of the Jesuit theatre in Vienna. After him, although it continued high in royal and popular favour, its creative impulse weakened under the rise of the rationalistic spirit which led in 1773 to the suppression of the Order.

During these centuries the popular theatre in Austria was represented by the ENGLISH COMEDIANS, who arrived late in the 16th century. Green's company is recorded as having visited Graz in 1608, where it alternated with seminarists from the Jesuit College in providing a typically mixed Whit-week programme of plays given before the Archduke Ferdinand. The Jesuits performed *Cipriano et Justina*, the English Comedians *The Parable of the Prodigal Son, Faustus* (after MARLOWE), and *King Ludwig and King Frederick of Hungary*, among other pieces. Some of the texts of these plays, in which the vernacular was already being used, are extant, and show the customary Haupt- und Staatsaktion, with comic interludes from the clown Pickelhering created by Robert REYNOLDS. As in the rest of the German-speaking lands, these troupes of wandering players, in which the English members were soon replaced by native-born actors, had no permanent theatre. They set up their stages in courtyards, marketplaces, halls, and palaces—wherever the authorities would tolerate them. At this period in their development they were not especially characteristic of Austria; it was only when they achieved stability in Vienna early in the 18th century that their typical offerings became integrated with the Viennese popular theatre to which, during its first flowering period, they gave a characteristic flavour.

The credit for this achievement goes to Joseph STRANITZKY. The decision of the Vienna city fathers to build a permanent theatre at the Kärntnertor, which was completed in 1710, gave rise to a dispute as to whether it should be left to native or Italian companies; Court influence gained the Italians the victory, but the disgruntled citizenry boycotted them and Stranitzky's company entered the theatre in triumph a year later. They continued to produce the same plays as during their years on tour; what was new about the ALTWIENER VOLKSTHEATER which Stranitzky inaugurated was the replacement of the COMMEDIA DELL'ARTE and English FOOL types by HANSWURST, a new comic figure derived from Austro-Bavarian folk art, who gave rise throughout the German-speaking lands to numerous imitations which departed more and more from the original. Stranitzky, however, being in control of his own theatre, was able to nominate his successor in the part—Gottfried PREHAUSER, in whose hands Hanswurst underwent a mellowing process which toned down much of his original peasant coarseness, leaving him more genial, even a little sophisticated. With the death of Prehauser the age of impromptu burlesque in Vienna drew to its close. The creation of new comic types and forms of popular burlesque, like the BERNADON of Josef von KURZ and the Viennese comedy of manners of Philipp HAFNER, showed that the baroque conception of a social hierarchy with its privileged aristocracy was yielding to a middle-class ethos with a pronounced emphasis upon honesty and civic pride. In 1776 Josef II, in his endeavours to reform Austria in accordance with his rationalist principles, decided to enlist the help of the theatre, which he considered a powerful educational force in the state; he decreed that the Vienna Burgtheater, built in 1741, should become a national theatre under the direction of SONNENFELS, a disciple of GOTTSCHED, and should devote itself entirely to the performance of serious and improving literary drama of the type produced by Cornelius von AYRENHOFF. The popular theatre was banished to the suburbs, where it flourished under Marinelli and Schickaneder, better known as Mozart's librettist, in the well-loved Leopoldstädter and Josefstädter theatres and the Theater an der Wien. Before the extempore farce disappeared for ever it added two more comic types to the gallery of Hanswurst's successors: KASPERL, which Johann LAROCHE made an important figure, and THADDÄDL, the creation of Anton HASENHUT, who already belongs to a different age and, since not improvised, theatre.

As one form of popular theatre faded another flourished. Its foundations were laid by three immensely productive writers, none of whom was primarily a playwright by profession—Joseph GLEICH, Karl MEISL, and Adolf BÄUERLE. They wrote for audiences which had suffered severely under the Napoleonic wars and were now ruled by a government pledged to restore order by reactionary legislation, strict censorship being one of its features. Thus, topical themes had to be avoided, and a powerful impetus was given to the production of *Zauberstücke*—fairy-tale plays with magical figures and decidedly supernatural elements—which appeared innocuous while still giving the actors, skilled in improvization, the chance to slip in topical jibes and parody contemporary institutions. All three authors provided innumerable examples of the genre. Bäuerle also made famous the new stock comic figure STABERL the umbrella-maker, originally played by Ignaz SCHUSTER; the most famous actor-dramatist and impresario of the day, Karl CARL, wrote many *Staberliaden* which popularized the character throughout the German-speaking lands. After Gleich, Meisl, and Bäuerle the Austrian theatre was ready for a shift from a purely external popular entertainment, with no claims to profundity, towards an interior experience which, in the works of Ferdinand RAIMUND, was to achieve the level of great dramatic art. In turn, Raimund's last years were to be overshadowed by the growing popularity of Johann NESTROY, the last great figure in the Viennese popular theatre. If Raimund, with his sensitive, melancholy, elegiac temperament and delicately sensuous humour, was clearly a product on the one hand of a declining Romanticism and on the other of the stable, ordered world of the *ancien régime* which endured longer in Austria than elsewhere, Nestroy, with his irreverence, his acerbity, his love of parody and his fondness for social comment, reflected the rising liberalism and dissatisfaction with the existing order which found an outlet in the revolution of 1848. After Nestroy the forces of disintegration which he both depicted and promoted led to a decline of the popular theatre from which it never recovered. It had derived its strength from a homogeneity of spirit among the citizens of the Austro-Hungarian empire which transcended class barriers. The forces which disrupted that society robbed it of its audience, the truth about the Austrian national character, watered down, sentimentalized, cosmopolitanized, reduced to a few basic clichés, was set to waltz-time, and in the form of Viennese operetta went out to conquer the world.

Literary drama, meanwhile, after the impetus given to it by the reforms of Sonnenfels, received a further influx of vigour with the appointment in 1814 of Josef SCHREYVOGEL to the directorship of the three major Viennese theatres, the Burgtheater, the Kärntnertor Theater, and the Theater an der Wien. To him goes the credit for making the Burgtheater one of the foremost German-speaking stages and to him also is due the discovery, soon after he took office, of Austria's greatest dramatist, Franz GRILLPARZER. Many of his reforms continued to exercise a beneficial influence even after their originator's ungrateful and summary dismissal in 1832. The interregnum was an undistinguished one, but in 1849, with the appointment of Heinrich Laube (1806–84), a new and splendid era began. But it was a long time before Austria could produce another dramatist as distinguished as Grillparzer. His contemporaries Friedrich Halm (1806–71) and Eduard BAUERNFELD provided the stage with thin, 'well-made' tragedies and satirical farces respectively. The trend towards realism in literature which spread throughout Europe during the second half of the 19th century left its mark, though it never reached the heights attained by NATURALISM in BERLIN, largely because it ran counter to the Viennese tradition of the *Gesamtkunstwerk*, which embraced spectators as well as actors in the act of creation. Among the earliest writers to attempt a realistic presentation of peasant life on the stage was Ludwig ANZENGRUBER, whose hedonistic outlook accorded well with the limited image of itself that the Viennese popular operetta was developing at the time. His successor was Karl SCHÖNHERR, something of whose blend of inherited theatrical flair and modernity of spirit is also found in the vastly different social dramas of Arthur SCHNITZLER. While naturalism found its chief Viennese advocate in Hermann BAHR, Schnitzler's Impressionism was better suited to the twilight of the Austrian Empire, and in this he was followed both by Richard Beer-Hofmann (1866–1945) and by Schnitzler's friend and admirer Hugo von HOFMANNSTHAL, who, starting in the impressionistic and somewhat decadent atmosphere of the 1890s, attempted to lead a return to moral values and the roots of the Austrian theatrical tradition. The First World War brought to an end that tradition, as well as the world Hofmannsthal had hoped to reform. The new Austrian Republic of 1919 lacked the diversity of the old monarchy, with its richly flavoured cultural life. Tendencies in literature which derived from outside Austria—from Berlin, for example, with its vigorous postwar theatrical experiments—produced a gradual loss of identity. Worthwhile things were, however, still produced. Oskar Kokoschka had arguably anticipated German EXPRESSIONISM as early as 1910, though Franz WERFEL was to be its major Austrian exponent. Max Mell (1882–1971) followed Hofmannsthal in reviving the morality play, while naturalism blended with Expressionism in the plays of Anton WILDGANS, who also directed the Burgtheater during the 1920s. This theatre, still at the height of its fame during Hofmannsthal's early years, when actors of the calibre of Josef KAINZ and Hermann Müller trod its boards, had declined since the war, but Wildgans's reforms contributed to its rising reputation, which was well established by the 1930s. Actors of this period, who made their reputations in Berlin, included Werner KRAUSS and Fritz KORTNER. Schnitzler and Hofmannsthal continued to write in the inter-war period, other dramatists of note including Alexander LERNET-HOLENIA, Franz CSOKOR, Richard BILLINGER, Ferdinand BRUCKNER, and Stefan ZWEIG. The most notable writer to emerge after the Second World War was Fritz HOCHWÄLDER who, although he lived in Switzerland after 1938, nowhere more clearly revealed his Austrian antecedents than in his assertion that the theatre is the best instrument for repairing the image of the human race, shattered by the Nazis from whom he had been forced to flee.

The Burgtheater was badly damaged during the war, but reopened under the direction of Berthold VIERTEL. Its repertoire revealed the post-war concern with the restoration of traditional Austrian values, the emphasis being on the classics, with new plays conforming to traditional models and expressing a worthy, mercantile philosophy. During the later 1950s the so-called Wiener Gruppe began to react against the confines of traditionalism, Konrad Bayer being its chief theatrical figure with his revival of the folk-hero Kasperl in short black farces. This group created the experimental climate within which Peter HANDKE and Wolfgang BAUER were able to develop in the following decade. Austria's second important city in theatrical terms is Graz, whose theatre dates from 1825 and now seats 580. There is also a 100-seat workshop theatre where experimental work is encouraged. Since 1965 there has been a constant flow of young talent from Graz into the Austrian and German theatres.

AUTHOR'S NIGHTS, see ROYALTY.

AUTO SACRAMENTAL (pl. *autos sacramentales*), the Spanish term for a religious play (*auto*) which derived from the Latin LITURGICAL DRAMA, the earliest known text in the vernacular being the *Auto de los reyes magos* (*c.* 1200). Its development followed in general that of the MYSTERY PLAY in England, France, and Germany, but by the end of the 16th century it had come to be recognized as a dramatic restatement of the tenets of the Catholic faith which embodied the preoccupations and ideals of the Counter-Reformation as the later Mystery plays did in Germany. In this form it provided one of the main elements in the celebrations on Corpus Christi (the Thursday after Trinity Sunday), an important feast day on the Continent. The most distinctive feature of the *auto sacramental* is found in its use of elaborate allegory; its subject matter could be purely sacramental, or Biblical, or hagiological, the first masters of the form being Lope de VEGA and José de Valdivielso (1560–1638). Originally performed on a scaffold before the church, these plays came to be presented in cities such as Madrid on flat moveable waggons (*carros*) on which scenery could be erected, and machinery for elaborate effects installed. By the time of CALDERÓN, the pre-eminent writer of *autos sacramentales*, two pairs of waggon-stages were used, a symmetrical arrangement which appealed to his logical mind since it enabled him to make considerable use of antithetical balance. From 1651 until his death in 1681 Calderón wrote two *autos* a year, and even after his death they were frequently revived for performance in Madrid. These religious plays continued to be performed in Spain long after they had disappeared in France and England; they were finally prohibited by a royal edict of 1765, but isolated examples continued to be given in rural areas as late as 1840.

AVANCINI, NIKOLAUS (1611–86), Austrian nobleman who joined the Society of Jesus in 1627 and attained high office in Vienna, combining his religious functions with the post of Court Poet to Leopold I. He was an outstanding exponent of JESUIT DRAMA, his plays, staged with great splendour, being on subjects taken impartially from the classics, history, or the Bible, but usually ending with a reference to the glories of the House of Habsburg. Thus, in the *Curae Caesarum* (1654), the story of Theodosius ends with a prophecy of Austria's future glory, presaged by the horses, gryphons, elephants, and camels which assemble from the four corners of the earth to do her honour. Avancini's best known play, mainly because it was published with nine engravings showing the stage settings, is *Pietas victrix sive Flavius Constantinus Magnus de Maxentio tyranno victor* (1659) which ends with the conversion to Christianity of Constantine, and a eulogy of Leopold I as his successor.

AVANT-GARDE, term used loosely nowadays to describe any experimental or progressive art. In terms of theatre it originated in France in the 1920s as a description of a specific approach envisaged by the Cartel, an association of four actor-managers—BATY, DULLIN, JOUVET, and PITOËFF—who, having many basic principles in common, determined to combine, each in his own theatre, in an effort to revitalize the French dramatic scene. Reacting against the realism of ANTOINE, they sought to stimulate the imagination of the spectator, placing great emphasis on the role of the author as an artist with a vision, and on the director as the interpreter of that vision. They also emphasized the importance of bodily movement as well as vocal expression in the work of the actor, and set out to educate a new type of spectator, who would become attached to one specific playhouse and demand of it more than mere entertainment. At a time when the cinema was killing the live theatre, they gave it a new impetus and a greater sense of its own dignity and, having entered the COMÉDIE-FRANÇAISE together as joint directors in 1936, they embarked on many reforms, but were stopped in mid-career by the Second World War, which only Jouvet survived. Although many of their ideals, and particularly the consistently high standard of their repertoires, have been overtaken by events, they inspired the work of BARRAULT and VILAR, and of many experimental theatres in the post-war years, which gave unusual emphasis to the work of the director and the actor in the production of a play-text.

AVENUE THEATRE, London, see PLAYHOUSE.

AVIGNON FESTIVAL, founded in July 1947 by Jean VILAR with the first production in France of Shakespeare's *Richard II* in which he himself played the title role. In an effort to re-create the popular festive tradition of outdoor plays, he arranged for productions to be presented in the inner court of the papal palace, whose floodlit façade provided an imposing setting. The festival began modestly, the material for the first stage being borrowed from a local army corps, but when in 1951 Vilar was appointed head of the THÉÂTRE NATIONAL POPULAIRE, he was able to use its resources, including its admirable company, and in the 1950s Avignon sprang into prominence with memorable performances by Gérard PHILIPE in CORNEILLE's *Le Cid* and KLEIST's *Prinz Friedrich von Homburg*. By 1967 the number of spectators had risen to 150,000 and the 17 official performances were regularly sold out. After his resignation from the T.N.P. in 1962 Vilar continued his association with the festival until his death in 1971, widening its cultural basis, increasing the official theatres to 4, extending the season into Aug., and organizing a number of 'fringe' activities, including the engagement of experimental companies. This culminated in 1968 with an invitation to the LIVING THEATRE to provide one of the official offerings, which, following the events in Paris in May of that year, led to trouble, and ended in the expulsion of the Living Theatre by the authorities, who were worried by the subversive tendencies of fringe productions and the growing 'hippy' population attracted by them. By 1970, however, matters had settled down, and in 1973 there were dozens of dramatic performances each day, usually playing to capacity audiences. The very success of the festival seems to have been its undoing, as young directors present highly controversial versions of such classical plays as *Hamlet, King Lear*, and RACINE's *Andromaque*, and more and more small experimental groups fill every available corner, playing to ever smaller audiences. Present developments seem to be tending further away from Vilar's intentions when he founded the festival and it remains to be seen if sufficient audiences will be found to support the type of audio-visual theatre which was so much in evidence in the 1981 festival.

AVON THEATRE, Canada, see STRATFORD (ONTARIO) FESTIVAL.

AVON THEATRE, New York, see KLAW THEATRE.

AXENFELD, ISRAEL, see JEWISH DRAMA.

AXER, ERWIN (1917–), Polish director, pupil and successor of Leon SCHILLER. In 1945 he established a company in Łódź, which was later taken over by DEJMEK, and four years later founded the Contemporary Theatre in Warsaw. From 1955 to 1957 he also directed at the National Theatre, and he has worked in many theatres outside Poland. He has been responsible for the first production of many new Polish plays, including KRUCZKOWSKI's *The Germans* (1949), WITKIEWICZ's *The Mother* (1970), and MROŻEK's *Tango* (1965), *The Happy Arrival* (1973), and *The Tailor* (1979), but has also added a number of foreign classics and new plays to his repertory, including Shakespeare's *Twelfth Night* in 1950; BRECHT's *Der aufhaltsame Aufstieg des Arturo Ui*, which he also staged in Leningrad, in 1962; CHEKHOV's *Three Sisters* in 1963; PINTER's *Old Times* and IONESCO's *Macbett*, both in 1972; BOND's *Lear* in 1974; and the world première of FRISCH's *Triptychon* (1980). In 1964 the company from the Contemporary Theatre was seen in the WORLD THEATRE SEASON in London in FREDRO's *The Life Annuity*, directed by Jerzy Kreczmar, and in two short plays by Mrożek, *What a Lovely Dream* and *Let's Have Fun*, both directed by ŚWINARSKI.

AYALA, A. L., see LOPEZ DE AYALA.

AYCKBOURN, ALAN (1939–), English playwright and director, who worked as stage manager and actor with various repertory companies before directing plays for the first time at the theatre-in-the-round at Scarborough (see STEPHEN JOSEPH THEATRE IN THE ROUND) where he was appointed director of productions in 1970. Most of his plays have been written for this company and then transferred to London. The first to reach the West End was *Mr Whatnot* (1964), originally produced at the VICTORIA THEATRE, Stoke-on-Trent, but the first to be outstandingly successful there was *Relatively Speaking* (1967), which was followed by *How the Other Half Loves* (London, 1970; N.Y., 1971) and *Time and Time Again* (1972). *Absurd Person Singular* (1973; N.Y., 1974) depicts three couples on three successive Christmas Eves, each act being set in the kitchen of one of them. Ayckbourn's next work, *The Norman Conquests* (1974; N.Y., 1975), was a trilogy of full-length plays, *Table Manners, Living Together*, and *Round and Round the Garden*, showing simultaneous events in the dining-room, living-room, and garden of the same house during the same week-end. A musical, *Jeeves* (1975), based on P. G. Wodehouse, for which Andrew Lloyd Webber wrote the music, was not a success, but in the same year *Absent Friends* had a long run. *Bedroom Farce* (1977; N.Y., 1979), one of Ayckbourn's best plays, was first presented by the NATIONAL THEATRE and then moved to the West End; the action passes between the bedrooms of three married couples, which jointly occupy the stage. *Just Between Ourselves* (also 1977) and *Joking Apart* (1979) were more serious and had comparatively short runs. *Sisterly Feelings* (National Theatre, 1980) can be performed in four different variations, the first and last of its four scenes remaining the same but the middle two scenes having alternative versions. *Taking Steps* and *Season's Greetings*, about a family reunion at Christmas, were also seen in London in 1980.

It has been said of Ayckbourn that the more farcical his work is, the more clearly one hears the authentic sound of human desperation. He depicts with humour, accuracy, and an occasional note of cruelty the sexual and other stresses of English middle-class life, and his plays are ingeniously constructed. They have been performed all over the world, and at one time five of them were running in London simultaneously.

AYLMER [Aylmer-Jones], FELIX (1889–), English actor, knighted in 1965. He began his career in London in 1911 under Seymour HICKS, and in 1913 joined the newly-formed BIRMINGHAM REPERTORY THEATRE company, playing Orsino in the opening play, *Twelfth Night*. After serving in the R.N.V.R. in the First World War, he returned to the stage, appearing in several plays by SHAW at the EVERYMAN THEATRE, including the first public performance in London of *The Shewing-Up of Blanco Posnet* (1921) in which he played Sheriff Kemp. From then on he worked almost continuously in London and New York, being seen in many distinguished plays, including DRINKWATER's *Abraham Lincoln* (also 1921) and *Robert E. Lee* (1923), and was first seen in New York in GALSWORTHY's *Loyalties* (1922). He returned to Shaw in 1924 to play Lord Summerhayes in *Misalliance* and Dr Paramore in *The Philanderer*, playing Sir Colenso Ridgeon in *The Doctor's Dilemma* in 1926. Some of his later appearances were in Drinkwater's *Bird in Hand* (1928), FEUCHTWANGER's *Jew Süss* (1929), and, in the 1930s, SHERRIFF's *Badger's Green* (1930), Galsworthy's *Strife* (1933), and GRANVILLE-BARKER's *The Voysey Inheritance* (1934) and *Waste* (1936). He had by now made an enviable reputation as a dependable and immaculate actor with beautiful diction, and was much in demand for parts needing patrician authority and sardonic humour, playing the Earl of Warwick in Shaw's *Saint Joan* (1934). One of his finest performances was as the First Lord of the Admiralty, Walter Harrowby, in Charles MORGAN's *The Flashing Stream* (1938; N.Y., 1939), and the full maturity of his gifts was seen in his portrayals of Sir Joseph Pitts in BRIDIE's *Daphne Laureola* (1949) and the Judge in Enid BAGNOLD's *The Chalk Garden* (1956).

AYNESWORTH, ALLAN [E. Abbot-Anderson] (1865–1959), English actor, who made his first appearance under Beerbohm TREE at the HAYMARKET THEATRE in 1887 and for the next 50 years was associated with all the outstanding players, managements, directors, and writers of the day. He directed many of the plays in which he appeared, played leading parts with Marie TEMPEST from 1903 to 1907, and shortly before the First World War was in management at the NEW, GARRICK, and CRITERION theatres successively. Among the roles he created were Algernon Moncrieff in WILDE's *The Importance of Being Earnest* (1895); Bertram Bertrand in Hope's *The Prisoner of Zenda* (1896); Major Bagnal in Cyril Harcourt's *A Pair of Silk Stockings* (1914); Lord Porteous in MAUGHAM'S *The Circle* (1921), a part he played again in 1931; Dominic in A. A. MILNE's *The Dover Road* (1922); Wilfrid Everitt in Openshaw's *All the King's Horses* (1926), with Irene VANBRUGH; and Lord Crayle in LONSDALE's *The High Road* (1927). He was an impressive actor, with a distinguished appearance, one of his last parts being Lord Conyngham in Laurence HOUSMAN's *Victoria Regina* when it was first presented publicly in 1937, with Pamela Stanley as the Queen and Mabel TERRY-LEWIS as the Duchess of Sutherland.

AYRENHOFF, CORNELIUS VON (1733–1819), Austrian dramatist whose serious dramas, in contrast to the light fare popular in VIENNA at the time, were intended to raise the standard of the theatre. They were given at the Burgtheater, under the able management of Joseph von SONNENFELS, the most successful being a comedy, *Der Postzug* (1769); but they proved unequal to the task of weaning the public from the light diet provided by harlequinade, farce, and operetta.

AYRER, JAKOB (1543–1605), German dramatist, successor to Hans SACHS and like him a prolific writer of FASTNACHTSSPIELE, of which about 60 were published in 1618. Ayrer, who spent most of his life in Nuremberg and was probably one of the MEISTERSÄNGER, was much influenced in his work by the ENGLISH COMEDIANS from whom he took the character of the FOOL, recognizable under his many metamorphoses by his first name of Jan. Although Ayrer's actors were amateurs he aimed at big, spectacular effects and was one of the first German dramatists to write extensive stage directions. Three of his plays were probably derived from the same source as Shakespeare's—the *Comedia von zweyen Brudern aus Syracusn*, like *The Comedy of Errors*, from PLAUTUS' *Menaechmi; Comedia von der schönen Phaenica und Graf Tymbri von Golison aus Arragonien*, like *Much Ado About Nothing*, from a story by Bandello; and *Comedia von der schönen Sidea*, which has scenes resembling those between Miranda and Ferdinand in *The Tempest*, from a source yet to be discovered.

B

BABANOVA, MARIA IVANOVNA (1900–), Soviet actress, who began her career under the direction of KOMISARJEVSKY and from 1920 to 1928 was with MEYERHOLD in his Theatre Workshop, where in 1926 she appeared as the Chinese galley-boy in TRETYAKOV's *Roar, China!* Two years later she was at the Theatre of the Revolution, under DIKIE in FAIKO's *The Man with the Portfolio* and becoming the leading actress of the company; she played Juliet in POPOV's production of *Romeo and Juliet* in 1935 and the title role in ARBUZOV's *Tanya* in 1939. When the theatre was renamed the MAYAKOVSKY she remained with the company, first under OKHLOPKOV and subsequently under Goncharov.

BABO, JOSEPH MARIUS, see RITTERDRAMA.

BACCHUS, see DIONYSUS.

BACK ALLEY THEATRE, see WASHINGTON, D.C.

BACKCLOTH, a flat painted canvas, or a plain surface on which light can be projected. Hanging at the back of the stage, suspended from the GRID, it was used in combination with WINGS to form a wing-and-backcloth scene, later superseded in most straight plays by the BOX-SET and now used only in ballet, opera, and PANTOMIME. Known in America as a backdrop, it is often placed behind a theatrical gauze-cloth (scrim drop in America) to act as a substitute for a CYCLORAMA in small theatres.

BACKING FLAT, see FLAT.

BACK STAGE, a term originally applied to a recess in the back wall of the stage, used for the last pieces of scenery in a deep spectacular vista, and at other times for storage. By extension the word was applied to all parts of the theatre behind the stage, including the actors' dressing-rooms and the GREEN ROOM. The expression 'to go (or work) backstage' is of recent origin, the first recorded use of it to indicate a visit behind the scenes dating only from 1923.

BACON, FRANK (1864–1922), American actor and playwright, who made his first appearance on the stage at the age of 25 in William Pratt's MELODRAMA *Ten Nights in a Bar-Room*, first seen in London at the OLD VIC in 1867. He was for some years in San Francisco, where he played a variety of parts, and was also seen in VAUDEVILLE with his wife and family. After the earthquake there in 1906 he went to New York, where he scored a triumph with his play *Lightnin'* (1918), written in collaboration with Winchell Smith, in which he played the leading character, Lightnin' Bill Jones, a lovable rascal with a taste for strong drink and a wonderful gift for exaggeration, a part well suited to his homely simplicity and engaging personality. The play ran for three years and was seen in London in 1925, with Horace Hodges in the title role. Bacon died during the Chicago run which followed the play's New York success.

BADDELEY, (Madeleine) ANGELA CLINTON- (1904–76), English actress. She was on the stage as a child, making her adult début as Jenny Diver in Nigel PLAYFAIR's production of GAY's *The Beggar's Opera* at the LYRIC THEATRE, Hammersmith, in 1920. She subsequently toured Australia, mainly in plays by BARRIE, and on her return made a great success as the heroine of Allen Harker's *Marigold* (1927). After a tour of South Africa she was seen with her sister Hermione (below) in Zoë AKINS's *The Greeks Had a Word For It* (1934), followed by Emlyn WILLIAMS's *Night Must Fall* (1935) in which she made her New York début the following year. Among other new plays in which she appeared were Dodie SMITH's *Dear Octopus* (1938). Emlyn Williams's *The Light of Heart* (1940) and *The Morning Star* (1941), RATTIGAN's *The Winslow Boy* (1946), and GIRAUDOUX's *The Madwoman of Chaillot* (1951). She was an excellent Miss Prue in GIELGUD's revival of CONGREVE's *Love for Love* in 1943, and appeared with the OLD VIC and STRATFORD-UPON-AVON companies in a variety of parts by Shakespeare under the direction of her second husband Glen BYAM SHAW. In 1961 she was seen in an American musical, *Bye Bye Birdie*, and at the time of her death was appearing in another musical, Stephen SONDHEIM's *A Little Night Music* (1975).

BADDELEY, HERMIONE CLINTON- (1906–), English actress, who was on the stage as a child, made her adult début in 1923, and soon achieved an enviable reputation in REVUE, appearing with *The Co-Optimists* at the PALACE in 1924 and then in COCHRAN's *On With the Dance* (1925) and its sequel *Still Dancing* (1926). In 1927 she made a great success as Ninetta, The Infant Phenomenon, in Nigel PLAYFAIR's *When Crummles Played*, an extravaganza based on the theatrical chapters in DICKENS's *Nicholas Nickleby*, and was also much admired as Polaire in Zoë AKINS's *The Greeks Had a Word For It* (1934), in which her sister Angela (above) also appeared. She returned to revue with *Nine Sharp* (1938) and *The Little Revue* (1939) and made her one excursion into classic comedy with Margery Pinchwife in WYCHERLEY's *The Country Wife* in 1940. After two revues in which she was partnered by Hermione Gingold, *Rise Above It* (1941) and *Sky High* (1942), she returned to straight plays as Ida Arnold in Frank Harvey's stage version of Graham GREENE's *Brighton Rock* (1943), and rejoined Gingold in COWARD's *Fumed Oak* and *Fallen Angels* (1949). She was also excellent as Mrs Pooter in GROSSMITH's *Diary of a Nobody* (1955). She made her New York début in 1961, when she took over the part of Helen in Shelagh Delaney's *A Taste of Honey*, and was later seen there in Tennessee WILLIAMS's *The Milk Train Doesn't Stop Here Anymore* (1964). Her last appearance in the West End was as Mrs Peachum in BRECHT and Weill's *The Threepenny Opera* in 1972.

BADDELEY, ROBERT (1732–94), English actor, who was first a pastrycook and then a valet, in which capacity he toured the Continent for three years acquiring a knowledge of foreign manners and accents that he later turned to good account. He was first seen on the stage in 1760 at DRURY LANE, where he created the parts of Canton in the elder COLMAN and GARRICK's *The Clandestine Marriage* (1766) and Moses in SHERIDAN's *The School for Scandal* (1777), and was dressing for the latter, his most famous part, when he was taken ill and died. He was excellent in such broken-English parts as Fluellen in *Henry V*, and in comic roles like Brainworm in JONSON's *Every Man in His Humour*. On his death he left a sum of money for a cake and wine to be partaken of by the company at Drury Lane annually on Twelfth Night, a custom which is still observed. He was married briefly and unhappily to the actress Sophia SNOW.

BADEL, ALAN FIRMAN (1923–82), English actor, who had made only a few appearances on the stage before joining the army in 1942. He returned to the theatre in 1947, and was seen in Birmingham, Stratford-upon-Avon, and London (at the OLD VIC) in a wide range of classical roles, including Romeo in 1952, to Claire BLOOM's Juliet, and Hamlet in 1956. Among other plays he directed HOCHHUTH's *The Public Prosecutor* (1957), in which he played Fouquier-Tinville and, as part of Furndel Productions (with Lord Furness), was responsible in 1959 for the London production of Marjorie Barkentin's *Ulysses in Nighttown* (based on James JOYCE's *Ulysses*), in which he played Stephen Dedalus. In 1965 he appeared as John Tanner in a revival of SHAW's *Man and Superman* and, after a long absence from the stage spent in films and television, he reappeared at the OXFORD PLAYHOUSE in 1970 as Othello, and in SARTRE's adaptation of the elder DUMAS's *Kean*, which was later seen in London with great success. In 1976 he played Richard III at the ST GEORGE'S THEATRE, Islington.

BAGNOLD, ENID [Lady Roderick Jones] (1889–1981), English author, who was already well known as a novelist when her *Serena Blandish* (1924) was dramatized by S. N. BEHRMAN and produced in New York in 1929. It was seen in London in 1938 with Vivien LEIGH in the title role. Enid Bagnold then dramatized two of her other novels, *Lottie Dundass* (1943) and *National Velvet* (1946); but her most successful play was *The Chalk Garden* (N.Y., 1955), seen in London a year later with Peggy ASHCROFT and Edith EVANS, the latter playing the part created by Gladys COOPER. Three later plays, *The Last Joke* (1960), *The Chinese Prime Minister* (N.Y., 1964; London, 1965), and *Call Me Jacky* (1967), were unsuccessful in spite of excellent casts, the last being seen only on tour in the English provinces. It was however produced in New York in 1976 as *A Matter of Gravity*, with Katharine HEPBURN.

BAHR, HERMANN (1863–1934), Austrian dramatist, critic, and novelist, who after nearly 20 years of journalism and playwriting in Vienna became a director at the Berlin DEUTSCHES THEATER in 1906. He was one of the first to rally to the NATURALISM of IBSEN, and has been described as a 'lesser SCHNITZLER, softening the sharp edges of realism by his scrupulous art'; yet he also wrote essays championing neo-Romanticism (in 1891) and EXPRESSIONISM (in 1916). Nevertheless, he is unmistakably Austrian in his mixture of flippant gaiety, warm-heartedness, and witty acumen: his best play is a sophisticated farce entitled *Das Konzert* (1910), which deals with the matrimonial difficulties of a musician and his wife. Among almost 80 other plays, *Das Tschaperl* (1898) is outstanding.

BAIRD, DOROTHEA, see IRVING, HENRY BRODRIBB.

BAKER, BENJAMIN A. (1818–90), American actor-manager and playwright. Apprenticed to a saddler, he escaped to join a travelling company, and was soon playing Brabantio to the elder BOOTH's Othello. In 1839 he appeared in New York, and was engaged for the company which opened the OLYMPIC. He had already written a number of plays when on 15 Feb. 1848 he produced, for his own benefit night, *A Glance at New York in 1848*, in which Frank CHANFRAU made a great success as the hero Mose, a New York volunteer fireman. The 'Mose' plays were burlettas in the style of MONCRIEFF's *Tom and Jerry*

(1821) and they soon had many imitators in New York and elsewhere, giving rise to a type of play verging on MELODRAMA, with strong action played against a background of local conditions. Further plays by Baker in the same vein were *New York As It Is* (1848), again with Chanfrau, *Three Years After* (1849), and *Mose in China* (1850). Baker was manager successively of several theatres, including the Metropolitan in San Francisco, returning to New York in 1856 as manager of Edwin BOOTH's company. He continued to work as manager and theatrical agent until his death. Known as 'Uncle Ben' Baker, he was a well-loved figure in the American theatre.

BAKER, GEORGE PIERCE (1866–1935), one of the most vital influences in the formation of modern American dramatic literature and theatre. He was educated at HARVARD, where he later became the first Professor of Dramatic Literature, inaugurating in 1905 a course in practical playwriting which led to the foundation of his '47 Workshop' for the staging of plays written under his tuition. One of the immediately successful results of Baker's enterprise was the professional production by Mrs FISKE in 1908 of *Salvation Nell* by Baker's pupil Edward SHELDON. Other American playwrights who attended his special courses were Eugene O'NEILL, Sidney HOWARD, and George ABBOTT. In 1925, by which time he had seen his pioneer work bear fruit in many other centres of learning, often under his own old pupils, Baker left Harvard, where his methods were considered perhaps a little unorthodox, and went to YALE, where he remained as director of the post-graduate department of drama until his retirement in 1933.

BAKER, HENRIETTA, see CHANFRAU, FRANK S.

BAKER, SARAH (1736/7–1816), English theatre manageress, whose activities in Kent continued for more than 50 years. The daughter of an acrobatic dancer, Ann Wakelin, she married an actor in her mother's company in about 1761, and was widowed in 1769. With three small children to support, she went into management on her own account, probably with a puppet-theatre, and then from 1772 to 1777 managed her mother's company. On the latter's retirement, Mrs Baker formed a new company, with an ambitious repertory which included Shakespeare and SHERIDAN, and with it established a regular circuit, visiting between six and ten towns throughout the country. At first she used a portable theatre or any suitable building, but from about 1789 she built her own theatres—ten in all. Among the actors who appeared with her early in their careers were Edmund KEAN, Thomas DIBDEN, and William DOWTON; the last married her daughter Sally, and tried unsuccessfully to run the theatres after Mrs Baker's death.

BAKST, LÉON, see COSTUME and SCENERY.

BALE, JOHN (1495–1563), Bishop of Ossory in Ireland, author of a number of anti-Catholic plays. His surviving works, which include MORALITY PLAYS and a translation of KIRCHMAYER's *Pammachius*, are filled with coarse and incessant abuse of popery and priests. The most important is *Kynge Johan* (1538), which may claim to rank as the first historical drama in English literature. In its mingling of such abstract figures as Sedition, Clergy, and England, and the historical King John and Cardinal Pandolphus, it forms a link between medieval and Elizabethan drama. It may have been first acted at St Stephen's, Canterbury, and was revived at Ipswich in 1561.

BALIEFF, NIKITA (1877–1936), Russian theatre director, who left Moscow after the Revolution and settled in Paris, where he produced an excellent revue known as Le Chauve-Souris (The Bat), from the name of the cabaret with which he had been associated from 1908. This consisted of a number of turns, introduced and often commented upon by Balieff himself, set and costumed with a richness reminiscent of the Russian ballet. Among them were short burlesques, often on topical themes, and a series of mimed sketches based on old ballads, folksongs, prints, engravings, the woodenness of a toy soldier, or the delicacy of a Dresden shepherdess, all rendered amusing by very slight and subtle guying of the material. The show was brought to London in 1921 by C. B. COCHRAN, and eventually became an integral part of the London theatrical scene. In New York, where it was first seen in 1922, its success was instantaneous and unflagging, and both there and in London Balieff, a big burly man with a vast genial moon-face, who eked out his slender store of English with expressive shrugs and grimaces, achieved immense personal popularity. After his death efforts were made to continue the show, but without success; it was last seen in London in 1938.

BALLAD OPERA, a play of a popular and often topical nature, with spoken dialogue interspersed with songs originally set to existing popular tunes, as in the most famous example of the genre, John GAY's *The Beggar's Opera* (1728), the performers being primarily actors rather than singers. Such was the vogue for the ballad opera after 1728 that in the next ten years 95 were presented in London, including *The Gentle Shepherd* by the Scottish poet Allan RAMSAY, at least six by Henry FIELDING, and a couple by Colley CIBBER. In 1731 Charles Coffey (?–1745) adapted a play by Thomas JEVON as a ballad opera, *The Devil To Pay*, which was seen in Germany in 1743 as *Der Teufel ist los*. Its success started a vogue for ballad operas in Germany, to which the name *Singspiele*, originally used for the JIG or afterpiece introduced on the Continent by the ENGLISH COMEDIANS, was applied. The *Singspiele* undoubtedly influenced the development of the great German operas with spoken dialogue exemplified by Mozart's *Die Entführung aus dem Serail* (1782) and Beethoven's *Fidelio* (1805).

A second wave of popular plays with musical numbers, on the lines of the original ballad operas, but with the music specially written for the purpose, was initiated in London by Isaac BICKER-STAFFE's *Love in a Village* (1762), based on Charles Johnson's *The Village Opera*, first produced in the same year as Gay's work. It had music by Thomas Arne, and was musically described as a *pasticcio*, being less a true ballad opera than a forerunner of English operetta and MUSICAL COMEDY.

BALLARD, SARAH, see TERRY, BENJAMIN.

BAŁUCKI, MICHAŁ, see POLAND.

BALZAC, HONORÉ DE (1799–1850), French novelist, who in 1839, in the hope of meeting his many debts, persuaded the manager of the PORTE-SAINT-MARTIN to stage *Vautrin*, based on his own novel *Le Père Goriot*, with FRÉDÉRICK in the chief part. On the first night, 14 Mar. 1840, Frédérick's appearance, made up to resemble Louis-Philippe, caused a riot, and the play was banned. Balzac, deeper in debt than ever, wrote two more plays in collaboration with Frédérick. The first, which was never produced, is lost, but the second, *Mercadet*, was finally put on in 1851, after Balzac's death, in a new version by Dennery. In 1842 Balzac produced an extravaganza, *Les Ressources de Quinola*, for the ODÉON, which ran for 19 performances; *Paméla Giraud* (1843) at the GAÎTÉ, rewritten by two hacks while Balzac was in Russia, managed 21. Both helped to pay a few bills. Balzac might have had a success with his last play *La Marâtre* (*The Stepmother*, 1848), but the troubles of the time closed the theatres. Those who heard Balzac read his own plays said he had great powers of mimicry and a most expressive voice; but it is by his great novels that he is now remembered.

BANBURY, FRITH (1912–), English actor, manager, and director, who made his first appearance on the stage in 1933 and played a wide variety of parts before directing *Dark Summer* (1947) by Wynyard Browne, whose comedy *The Holly and the Ivy* (1950) he also directed. He was then responsible for a number of interesting productions, including N. C. HUNTER's *Waters of the Moon* (1951); RATTIGAN's *The Deep Blue Sea* (1952); John WHITING's *Marching Song* (1954); BOLT's *Flowering Cherry* (1957) and *The Tiger and the Horse* (1960); and *The Wings of the Dove* (1963), based by Christopher Taylor on Henry JAMES's novel. Banbury also directed plays in New York, and in 1968 directed Peter NICHOLS's *A Day in the Death of Joe Egg* for the CAMERI THEATRE in Tel-Aviv, and Robin Maugham's *The Servant* in Paris. Since 1970 he has been mainly involved in revivals —SHAW's *Captain Brassbound's Conversion* in 1971, SHERWOOD's *Reunion in Vienna* for the CHICHESTER FESTIVAL THEATRE in 1971, LONSDALE's *On Approval* in Canada and South Africa in 1976; the last two were seen in London in 1972 and 1977 respectively.

BANCES CANDAMO, FRANCISCO ANTONIO DE (?1662–1704), Spanish dramatist, who, in the *Teatro de los teatros de los pasados y presentes siglos*, wrote the first history of the Spanish theatre. His plays, in the manner of CALDERÓN, were composed mainly for the Court, and were often intended to further the interest of a particular party. His most successful play was *Por su Rey y por su dama* (*For His King and for His Lady*, 1685). Bances also wrote AUTOS SACRAMENTALES.

BANCROFT, SQUIRE (1841–1926), English actor-manager, knighted 1897, who with his wife, Marie Effie (*neé* Wilton) (1839–1921), introduced a number of reforms into the London theatre and started the vogue for drawing-room comedy and drama in place of MELODRAMA. Marie Wilton, the daughter of provincial actors, was on the stage from early childhood, first appearing in London in the title role of William Brough's BURLESQUE on *The Winter's Tale, Perdita; or, the Royal Milkmaid* (1856). She continued to play in burlesque, notably at the STRAND THEATRE in H. J. BYRON's plays, until she decided to go into management on her own account. On a borrowed capital of £1,000, of which little remained when the curtain went up, she leased an old and dilapidated theatre nicknamed the 'Dust Hole'—later the SCALA THEATRE. Renamed the Prince of Wales's, charmingly decorated, and excellently run, it opened on 15 Apr. 1865. In the company was Squire Bancroft, who had made his first appearance in Birmingham in 1861, and had previously played with Marie Wilton in Liverpool. The new venture was a success, and the despised 'Dust Hole' became one of the most popular theatres in London, where the Bancrofts—they married in 1867—presented and appeared in the plays of T. W. ROBERTSON, Bancroft giving one of his finest performances as Captain Hawtree in *Caste* (1867). The Bancroft management did much to raise the economic status of actors, paying higher salaries than elsewhere and providing the actresses' wardrobes. Among other innovations they adopted Madame VESTRIS's idea of practicable scenery, including the BOX-SET. In 1880 they moved to the HAYMARKET THEATRE and continued their successful joint career, retiring in 1885. They made their last appearance on 20 July in that year in a mixed bill consisting of the first act of BULWER-LYTTON's *Money*, a scene from BOUCICAULT's *London Assurance*, and the second and third acts of Tom TAYLOR's *Masks and Faces*.

BANDBOX THEATRE, New York, at 205 East 57th Street, opened on 23 Nov. 1912 as the Adolph Phillip Theatre, taking its new name on 22 Dec. 1914. Here the WASHINGTON SQUARE PLAYERS first appeared in New York in a series of one-act plays from 1915 to 1917. The theatre closed on 28 Apr. 1917, and a cinema was built on the site.

BANKHEAD, TALLULAH BROCKMAN (1902–68), American actress, a strikingly attractive woman with a deep, husky voice, whose off-stage notoriety made her for some years a popular figure in London, but prevented her from being taken seriously as an actress. She made her first appearance in New York, in *The Squab Farm* by Frederick Hatton, in 1918, and in 1923 was seen in London as Maxine in Hubert Parsons' *The Dancer*, playing opposite Gerald DU MAURIER. She then played the title role in KNOBLOCK's *Conchita* (1924) and Julia Sterroll in COWARD's *Fallen Angels* (1925), and was seen in the younger DUMAS's *The Lady of the Camellias* in 1930, afterwards returning to New York. Although she appeared in several outstanding parts, including Sadie Thompson in MAUGHAM's *Rain* in 1935 and Cleopatra in *Antony and Cleopatra*, her first classical part, in 1937, it was her Regina Giddens in HELLMAN's *The Little Foxes* (1939) which first brought her seriously to the attention of the New York critics. A year later she toured in PINERO's *The Second Mrs Tanqueray* and then created two important new roles, Sabina in WILDER's *The Skin of Our Teeth* (1942) and the Queen in COCTEAU's *The Eagle Has Two Heads* (1947). She also appeared in two revivals of plays by Tennessee WILLIAMS—as Blanche du Bois in *A Streetcar Named Desire* in 1956 and as Mrs Goforth in *The Milk Train Doesn't Stop Here Anymore* in 1964. Her autobiography, *Tallulah*, was published in 1952.

BANKS, LESLIE JAMES (1890–1952), English actor, who made his first appearance in the provinces in 1911, and in London, at the VAUDEVILLE THEATRE, in 1914. After serving in the army during the First World War he joined the BIRMINGHAM REPERTORY company, and returned to the West End in 1921, establishing himself as a player of power and restraint in a long series of successful productions. He was also seen several times in New York, notably as Captain Hook in BARRIE's *Peter Pan* in 1924, and Henry in Benn Levy's farce *Springtime for Henry* in 1931. Among his best parts were Petruchio in *The Taming of the Shrew*, in which he played opposite Edith EVANS in 1937; the schoolmaster hero in James Hilton's *Goodbye, Mr Chips* (1938); and the Duke in Patrick Hamilton's *The Duke in Darkness* (1942). His versatility was shown in a repertory season at the HAYMARKET THEATRE in 1944–5, when he played Lord Porteous in MAUGHAM's *The Circle*, Tattle in CONGREVE's *Love for Love*, Bottom in *A Midsummer Night's Dream*, and Antonio Bologna in WEBSTER's *The Duchess of Malfi*. He was also an excellent director, usually of plays in which he was appearing.

BANKSIDE GLOBE PLAYHOUSE, see WANAMAKER, SAM.

BANKSIDE PLAYERS, see ATKINS, ROBERT.

BANNISTER, JOHN (1760–1836), English actor, son of Charles Bannister (1741–1804), who established a modest reputation as a comedian under FOOTE and GARRICK. The son, who trained as an artist at the Royal Academy, was encouraged by Garrick to go on the stage, and in 1778 made his first appearance at the HAYMARKET THEATRE in one of the parts previously played by WOODWARD, who had just died. He proved successful, and was engaged for DRURY LANE where he began in tragedy but, overshadowed by the rising popularity of HENDERSON and John Philip KEMBLE, found his true bent in comedy. He was the first to play Don Ferolo Whiskerandos in SHERIDAN's *The Critic* (1779), in which he later appeared as Sir Fretful Plagiary, and was also admired in such parts as Sir Anthony Absolute in Sheridan's *The Rivals*, Tony Lumpkin in GOLDSMITH's *She Stoops to Conquer*, Scrub in FARQUHAR's *The Beaux' Stratagem*, and Dr Pangloss in the younger COLMAN's *The Heir-at-Law*. He became one of the managers of Drury Lane, where he spent most of his career with occasional visits to the provinces, and in 1783 married Elizabeth Harper (1757–1849), principal singer at the Haymarket, who retired in 1793 to take care of her increasing family. Bannister remained on the stage until 1815, and then retired to spend the rest of his life in the company of his family and friends, among whom were his former teacher Rowlandson, Morland, and Gainsborough.

BANVARD'S MUSEUM, New York, see DALY'S THEATRE.

BAPTISTE, see DEBURAU.

BARANGA, AUREL, see ROMANIA.

BARBICAN NURSERY, see NURSERIES.

BARBIERI, NICCOLÒ (1576–1641), actor and dramatist of the COMMEDIA DELL'ARTE, who played under the name of Beltrame and is credited with the invention of the mask of SCAPINO. He is first found with the GELOSI at Paris in 1600–4 and then joined the FEDELI, becoming joint director with G. B. ANDREINI; finally he was with the CONFIDENTI. One of his plays, *L'Inavertito* (1629), originally only a SCENARIO for improvisation, was later published with dialogue in full, and was made use of by MOLIÈRE for *L'Étourdi*. Barbieri also wrote his memoirs and an account of the stage in his time,

La Supplica (1634), which was edited by Taviani (1971).

BARD, WILKIE, see MUSIC-HALL.

BARKER, a member of the fairground or itinerant theatre company, who stood at the door of the booth and with vociferous and spellbinding patter induced the audience to enter. He is probably as old as the theatre itself and was known in the ancient world. He reappeared, in France particularly, in the 16th century, and from there spread to all the fairs and showgrounds of Europe and America, becoming well known in England in the 19th century.

BARKER, HARLEY GRANVILLE-, see GRANVILLE-BARKER.

BARKER, HOWARD (1946–), English playwright, who had his first plays, *Cheek* and *No One Was Saved*, produced at the ROYAL COURT Theatre Upstairs in 1970. *Alpha Alpha* (1972), about twin gangsters, and *Claw* (1975; N.Y., 1976), about a pimp, were performed at Charles MAROWITZ's Open Space, and *Stripwell* (also 1975) was given a major production at the Royal Court, with Michael HORDERN as a judge whose house is invaded by a criminal he has sentenced. *That Good Between Us* (1977), produced at the WAREHOUSE, is set in a totalitarian Britain of the future; *Fair Slaughter* dates from the same year. *The Hang Of the Gaol* (1978), which concerns a prison governor who becomes an arsonist, was followed by *The Love of a Good Man*, set on a First World War battlefield in 1920, and by *The Loud Boy's Life* (both 1980). *No End of Blame* (1981) deals with the pressures on a Hungarian cartoonist before and after his emigration to England. Like many other FRINGE THEATRE playwrights, Barker is a bitter critic of capitalist society, his moral outrage giving rise both to his outlandish plots and to his frequently excellent dialogue.

BARKER, JAMES NELSON (1784–1858), American dramatist, of whose plays only three have survived: *Tears and Smiles* (1807), a comedy of manners somewhat in the style of TYLER's *The Contrast* (1787), first seen in Philadelphia at the CHESTNUT STREET THEATRE with a brilliant cast which included the WARRENS, the WOODS, and the first Joseph JEFFERSON with his wife; *The Indian Princess; or, La Belle Sauvage* (1808), the first play about American-Indian life to be produced in America, seen at DRURY LANE as *Pocahontas; or, the Indian Princess* in 1820; and *Superstition* (1824), the story of a Puritan refugee from England who leads his village against the Indians, mingled with a tale of religious intolerance and witch-hunting, which was extremely successful in its day. Barker also made a dramatization of SCOTT's *Marmion*, first acted with considerable success in 1812, which held the stage for the next forty years. The script of another play, *The Embargo; or, What News?* (1808), said to have been well received, was taken by an actor in the cast to Baltimore, to be produced there, and lost.

BARLACH, ERNST (1870–1938), German dramatist, novelist, and sculptor, an exponent of EXPRESSIONISM, whose plays, written in North German dialect, use religious and mystical subjects. The first, *Der tote Tag* (*The Dead Day*), written in 1912, was not produced until 1919; it was followed by *Der arme Vetter* (*The Poor Cousin*) in the same year, both plays being concerned with the relationship of father to son. Later plays, of which *Der blaue Boll* (*The Blue Bulb*, 1929), first produced in Berlin, is usually accounted the best, also deal with an individual's struggle to cast off the bonds of the material world in his search for God. This theme was developed with even more complex symbolism in Barlach's unfinished play *Der Graf von Ratzeburg*, published in 1951. Between 1956 and 1960 a cycle of plays by Barlach was produced by Lietzau for the Schillertheater in Berlin.

BARN DOOR SHUTTERS, see LIGHTING.

BARNES, CHARLOTTE MARY SANFORD (1818–63), American actress and dramatist, the daughter of John Barnes (1761–1841), who with his wife Mary went from DRURY LANE to the Park Theatre in New York (see PARK THEATRE (1)), where they were both popular for many years, the husband as a low comedian, the wife in such parts as ROWE's Jane Shore and Isabella in SOUTHERNE's *The Fatal Marriage*. Charlotte first appeared on the stage at the age of four, as the child in 'Monk' LEWIS's *The Castle Spectre*, in which she also made her adult début in 1834, playing the heroine in Boston and New York. She was in London in 1842, where she played Hamlet among other parts, and was well received, though she was never accounted as good an actress in male parts as her mother, who was an excellent Romeo to her daughter's Juliet. In 1846 Charlotte married an actor-manager, Edmond S. Connor, and became his leading lady, being associated with him also in the management of the Arch Street Theatre in PHILADELPHIA. Her first play, *Octavia Bragaldi* (1836), was based on a contemporary murder case but set in Renaissance Italy. It was produced at the National Theatre, New York, with the author as the heroine, and frequently revived, the last time at the BOWERY under her husband's management. Of her other plays, mainly adaptations of French melodramas and popular novels, only *The Forest Princess* (1844), based on the story of Pocahontas, has survived.

BARNES, CLIVE ALEXANDER (1927–), English dance and drama critic, who after leaving Oxford became a freelance journalist, writing mainly on the performing arts until in 1961 he became dance critic of the London *Times* and editor of *Dance and Dancers*, of which he remains New York editor. In 1965 he became dance critic of the *New York Times*, being appointed drama critic also two years later. In 1978 he moved to the *New York Post*, where he holds the same positions. One of the most important and powerful critics in America, he has been attacked both by directors and by other critics for a certain lack of appreciation of American theatrical sensibility, perhaps because he has on occasion supported off-Broadway and international theatre at the expense of commercial Broadway productions; his judgements have on the whole been justified by time.

BARNES, Sir KENNETH, see SCHOOLS OF DRAMA and VANBRUGH, IRENE.

BARNES, THOMAS (1785–1841), English editor and critic, whose love of the theatre is manifest in his early work in the *Examiner* under the name of 'Criticus'. In this journal, while its usual dramatic critic Leigh HUNT was in prison, Barnes described the first appearance of Edmund KEAN as Shylock in *The Merchant of Venice* on 26 Jan. 1814, his account of this momentous event in the history of DRURY LANE, though less well known, bearing comparison with that of HAZLITT. Barnes was editor of *The Times* from 1817 until his death.

BARNES THEATRE, London, in Church Road, Barnes, a two-tier hall built about 1905 which Philip Ridgeway opened as a small theatre on 2 May 1925. After several short runs of new plays came the much-publicized stage version of *Tess of the D'Urbervilles*, prepared by Thomas Hardy himself, with Gwen FFRANGCON-DAVIES as Tess. This ran for several months, and was succeeded by CHEKHOV's *Uncle Vanya*, directed by KOMISARJEVSKY, which opened on 16 Jan. 1926 with Jean FORBES-ROBERTSON as Sonia. It heralded a remarkable run of Russian plays—Chekhov's *Three Sisters* in Feb., with GIELGUD as Tusenbach; ANDREYEV's *Katerina*, again with Gielgud, in Mar.; GOGOL's *The Government Inspector*, with Charles LAUGHTON, in April—all directed by Komisarjevsky. The series continued in Aug. with a dramatization of DOSTOEVSKY's *The Idiot*, but was broken in Sept. by John DRINKWATER's dramatization of Hardy's *The Mayor of Casterbridge*, which he also directed. Chekhov returned with *The Cherry Orchard* at the end of Sept. and a revival of *Three Sisters* in Oct., and this interesting and extremely fruitful venture ended in Nov. 1926 after a revival of Drinkwater's *Abraham Lincoln*.

BARNSTORMERS, a name given in the late 19th century to the itinerant companies whose stages were often set up in large barns and whose work was characterized by ranting, shouting, and general violence in speech and gesture.

BARNUM, P. T., see AMERICAN MUSEUM.

BARON [Boyron], ANDRÉ (*c.*1601–55), French actor, who was taken into MONTDORY's company at the MARAIS in 1634–5, to replace one of the actors sent by Louis XIII to the Hôtel de BOURGOGNE, where Baron himself went in 1641. In the same year he married Jeanne Ausoult (1625–62), the child of strolling players and herself an excellent actress. Her early death, during the run of Mlle DESJARDINS's *Manlius* at the Hôtel de Bourgogne, was much regretted by CORNEILLE, who said he had written a part for her in his next play, probably that of Sophonisbe. Baron played kings and noblemen in tragedy and rustic characters in comedy; he is reported to have died of a self-inflicted wound in the foot from a property sword, received while he was playing Don Diègue (in a revival of Corneille's *Le Cid*) too energetically. They were the parents of Michel BARON, below.

BARON, MICHEL (1653–1729), French actor, son of the above. Orphaned before he was ten, he joined the juvenile Troupe du Dauphin formed by the elder RAISIN and run after his death, not very successfully, by his widow. Touched by her plight, MOLIÈRE allowed her to use his PALAIS-ROYAL for three days and, chancing to visit the theatre during one of the children's performances, he was so struck by the acting of young Baron that he took him into his own company and eventually gave him small parts to play, among them Myrtil in *Melicerte* (1666). Unfortunately Molière's wife took a dislike to the boy, and on one occasion slapped his face, whereupon he ran away and rejoined his former companions. He remained with them until 1670, when Molière asked him to return to play the part of Domitian in CORNEILLE's *Tite et Bérénice*. After Molière's death Baron went to the Hôtel de BOURGOGNE to play the young tragic heroes of RACINE's plays and married a member of the company, Charlotte (1661–1730), daughter of LA THORILLIÈRE. With her he joined the COMÉDIE-FRANÇAISE on its foundation in 1680 and became its leading actor, being endowed with a fine presence, a deep voice, amplitude of gesture, and a quick intelligence. To these gifts, ably fostered by Molière, he joined application, attention to detail, and a firm belief in the dignity of his profession. He did much to raise the status of actors in his day and helped many a struggling dramatist by his fine rendering of a poor part. He was himself a playwright, the best of his comedies being *L'Homme à bonne fortune* (1686). In another of his plays, *Le Rendez-vous des Tuilleries* (1685), he introduced some of the actors of the Comédie-Française under their own names, as well as a number of well-known dandies of Paris. In 1691, at the height of his powers, Baron suddenly retired, but he continued to write plays, and occasionally acted at Court and in private theatricals until 1720, when he returned to the Comédie-Française and remained there until his death. Except for his voice, which was a little quavery, he was as good as ever, and was able to play many of his old parts again. He was taken ill on stage during a revival of ROTROU's *Venceslas*, and died three months later. His son Étienne (1676–1711), a coldly correct actor, was also a member of the Comédie-Française, as were Étienne's son and one of his nieces.

BARRACA, La, Spanish theatre company, more formally known as the Teatro Universitario, formed after the creation of the Spanish Republic in 1931 to take plays from the classical repertoire to rural areas. Headed by GARCÍA LORCA and Eduardo Ungarte, the fit-up company, consisting largely of students, met with an enthusiastic response and influenced the form of the three great tragedies of Lorca's maturity. It disappeared after the Spanish civil war.

BARRAULT, JEAN-LOUIS (1910–), French actor, director, and manager, who began his career in 1931 as a pupil of DULLIN at the ATELIER. His first stage appearance, on his 21st birthday, was as one of the servants in Jules ROMAINS's adaptation of JONSON's *Volpone*. He then began the study of MIME, which was to play such an important role in the development of his technique; his first independent venture was a mime-drama adapted with CAMUS from Faulkner's novel *As I Lay Dying*. In 1937 he directed a revival of CERVANTES's only tragedy *Numancia*, and two years later adapted and directed *Faim* (*Hunger*), based on a novel by Knut HAMSUN. After serving in the army, he was engaged by COPEAU for the COMÉDIE-FRANÇAISE, where Madeleine RENAUD, whom he later married, was already an established star. He made his début in CORNEILLE's *Le Cid*, but his most important work was done in his productions of RACINE's *Phèdre* with Marie BELL, *Antony and Cleopatra*, and CLAUDEL's *Le Soulier de satin*.

When the company was reorganized in 1946, he and his wife left to found their own troupe. For ten years they were at the Théâtre MARIGNY and accomplished the feat, unique for a private company in modern times, of establishing a repertory of modern and classical plays, which formed a stable background to their experiments with new plays, and in which they also toured constantly. They opened their theatre on 17 Oct. 1946 with GIDE's translation of *Hamlet* with Barrault in the title role, in which he was seen two years later at the EDINBURGH FESTIVAL. *Les Fausses confidences* by MARIVAUX was followed by *Baptiste*, a ballet pantomime, and by SALACROU's great play of the Resistance, *Les Nuits de la colère*. Among the important productions of the early years were *Le Procès*, based on Kafka, and a revival of GIRAUDOUX's *Amphitryon 38* (both 1947); a superb production with Edwige FEUILLÈRE of Claudel's *Partage de midi* in 1948 (later seen in London); ANOUILH's *La Répétition, ou, l'Amour puni* (1950), a year which saw also revivals of SARTRE's *Les Mains sales* and MONTHERLANT's *Malatesta*; and in 1951 Gide's version of SOPHOCLES' *Oedipus the King* with Jean VILAR. Much of Barrault's finest work was done with Claudel. In addition to the plays mentioned above, he directed *L'Échange* (1951), *Christophe Colomb* (1951), and *Tête d'Or* (1959), all of which entered the company's repertory together with *Le Soulier de satin*, revived in 1958. The importance of Marivaux in the company's schedule can be attributed partly to the outstanding acting of Madeleine Renaud in this author's works, while Barrault has been largely responsible for the revival and immense popularity of FEYDEAU, whose *Occupe-toi d'Amélie* figured in the repertory as early as 1948. Barrault also directed such important modern plays as FRY's *A Sleep of Prisoners* (1955), VAUTHIER's *Personnage combattant* (1956), and IONESCO's *Rhinoceros* (1960). In 1956 he moved from the Marigny to the Théâtre Sarah-Bernhardt and then to the PALAIS-ROYAL; three years later he was appointed director of the ODÉON, where his work with Madeleine Renaud included BECKETT's *Oh! Les Beaux Jours!* (1960), Ionesco's *Le Piéton de l'air* (1963), and BILLETDOUX's *Il faut passer par les nuages* (1964). He also sponsored such important productions as BLIN's staging of GENET's *Les Paravents* (1966) and organised seasons for the THÉATRE DES NATIONS. In 1968 he was in trouble with the authorities after rioting students took over his theatre and was relieved of his duties. A year later he was at the Élysées-Montmartre, formerly an indoor wrestling stadium near Clichy, and there, with seats tiered around a rectangular stage, he directed his own adaptations of Rabelais's novels and of *Ubu sur la butte*, based on JARRY's plays. *Rabelais* was seen in London, in both French and English, and toured Europe and America. In 1972 Barrault set up a circus tent inside the disused hall of the Gare d'Orsay, where he directed Claudel's *Sous le vent des îles Baléares*, a sequel to *Le Soulier de satin*, and *Zadig*, bassed on a story by VOLTAIRE. This building was taken from him in 1980 to become a museum, and he was given the Palais des Glaces, an old skating rink near the Rond-Point des Champs-Élysées, to which he transported the staging from the Gare d'Orsay, making a larger auditorium seating 920 spectators, and a smaller for 200, the planned repertoire to include a number of new plays including *L'Amour de l'amour*, which he adapted from three classic tales of Psyche, and revivals of some of the company's earlier successes.

As an actor Barrault, elegant and outwardly nonchalant, gives the impression of strong passions firmly controlled by disciplined intelligence and physically sustained by the strenuous exercise of mime. As a mime-actor he is best known through his appearance as DEBURAU in the film *Les Enfants du Paradis* (1945); but the rigour of his training can most effectively be judged on stage, where his slightest gesture speaks volumes.

BARREL, a counterweighted tube of metal (or, rarely, wood) 1½–2 in. in diameter, known in

America as a pipe batten. It is hung on wire-rope or hemp lines from the GRID, and is used to attach scenery or lighting equipment, being referred to in the latter case as a spot bar, or, in America, a light pipe.

BARRETT, GEORGE HORTON (1794–1860), American actor, son of an English actor who went to New York in 1797 and of his wife, Mrs Rivers. They both joined the company at the PARK THEATRE (1) under DUNLAP, and there in 1798 young Barrett made his first appearance as the child in *The Stranger* by KOTZEBUE, later being seen at the age of eleven as Young Norval in HOME's *Douglas*. As an adult actor he became one of the best light comedians of his day, making his reputation in SHERIDAN's plays—as young Absolute in *The Rivals*, Charles Surface in *The School for Scandal*, and Puff in *The Critic*—and as Aguecheek in *Twelfth Night*. He was for a long time stage manager of the BOWERY THEATRE under Gilfert, and later of the Tremont Theatre in Boston and of the BROADWAY THEATRE (1), New York, from its opening in 1847. Familiarly known, from his elegant appearance and gracious manners, as 'Gentleman George', he married in 1825 an American actress named Mrs Henry; he retired in 1855.

BARRETT, LAWRENCE (1838–91), American actor and producer, who was on the stage as a boy of 14, and travelled the United States with many outstanding companies, including that of Julia DEAN. In 1857 he was seen in New York in Sheridan KNOWLES's *The Hunchback* and other plays, and was leading man under BURTON at the METROPOLITAN. In 1858 he was a member of the Boston Museum company, and was in the army during the Civil War. Afterwards he travelled widely, and became extremely popular. He managed BOOTH's THEATRE in 1871, and played Cassius in *Julius Caesar*, the part in which he is best remembered, to the Brutus of Edwin BOOTH. While IRVING was in America in 1884, Barrett took over the LYCEUM in London, and though his visit was not a success financially, he was made welcome and fêted on all sides. He was a scrupulous and competent man of the theatre. All the standard classics were in his repertory, and he tried to encourage American playwrights, though he remained faithful to romantic and poetic drama. Tall, with classic features, and dark, deeply sunken eyes, he was probably at his best in Shakespeare, to whose interpretation he brought dignity, a dominant personality, and intellectual powers somewhat exceptional in an actor at that date.

BARRETT, WILSON (1846–1904), English actor-manager, who in his day had few equals in MELODRAMA, for which he was well fitted by nature; strikingly handsome, with a resonant voice and powerfully developed chest and arms, he lacked only height to make him a fine figure of a man. After a varied career he took over the management of the PRINCESS'S THEATRE, remaining there for five years and making a great success early in his tenancy with G. R. Sims's *The Lights o' London* (1881) which had a long run and then toured the world successfully. Barrett's next success was Henry Arthur JONES and Henry Herman's *The Silver King* (1882), in which he gave a fine performance as Wilfred Denver. His other famous rolė was Marcus Superbus in his own play *The Sign of the Cross*. This melodramatic story of a Roman patrician in love with a beautiful young Christian convert, with whom he goes into the arena to face the lions, was first produced in the U.S.A. in 1895, at a time when Barrett was on tour and badly in need of money. It brought him a fortune. It was first seen in England at the Grand Theatre in Leeds later in 1895 and in London a year later, when it created a sensation. Clergymen preached sermons about it, and people who had never before entered a theatre crowded to see it. It was several times revived and perennially successful on tour. Barrett was less successful in Shakespearian roles, which he sometimes essayed, only his Mercutio in *Romeo and Juliet* being thought passable. He had a brother and nephew in his company, and his grandson was also on the stage.

BARRIE, JAMES MATTHEW (1860–1937), Scottish novelist and dramatist, knighted 1913. Although already well known as a novelist in the 1880s, Barrie's first two plays were not successful, and it was not until TOOLE appeared in his farce *Walker, London* (1892) that he achieved some recognition in the theatre. The success of *The Professor's Love Story* (1894) and *The Little Minister* (1897), the latter based on an earlier novel, consolidated his position as a promising playwright, and he enhanced his reputation further with the romantic costume play *Quality Street* (1902) and the social comedies *The Admirable Crichton* (also 1902) and *What Every Woman Knows* (1908). *Peter Pan*, a whimsical children's play adapted from his own novel *The Little White Bird* (1902), was first produced in 1904 with Nina BOUCICAULT as Peter. It is an admirable example of the blend of fantasy and sentimentality which appears in much of Barrie's work and was again apparent in *Dear Brutus* (1917) and *Mary Rose* (1920), both of which evoke supernatural elements in a realistic setting. Among the shorter plays, the one-act *The Old Lady Shows Her Medals* (1917) and *Shall We Join the Ladies?* (1922), the latter intended as the first act of a full-length murder mystery, have been popular with amateur companies. Barrie's last play *The Boy David* (1936) was specially written for Elisabeth BERGNER, who played David. It was not a success and he is likely to be remembered for his earlier works, especially *The Admirable Crichton*, whose title has passed into everyday speech to describe a man of all-round excellence, and *Peter Pan*, which became a regular Christmas feature of the London theatrical scene with a succession of distinguished young actresses playing the title role.

BARRY, ELIZABETH (1658–1713), the first outstanding English actress. Her first appearances on the stage were lamentable failures, but she persevered and was finally successful in a revival of ORRERY's *Mustapha* in 1675, though she did not attain the full height of her powers until she was past her first youth. She played opposite BETTERTON for many years, and created a number of famous roles—one estimate puts it at 119—including the heroines of *The Orphan* (1680) and *Venice Preserv'd* (1682) by OTWAY, who all his life nourished a hopeless passion for her. She was also the first to play Almeria in CONGREVE's *The Mourning Bride* (1697). Though good in comedy, she was at her best in tragedy, in which she displayed great power and dignity. In private life she was not so estimable, according to contemporary accounts, and many scandals are attached to her name, some perhaps undeservedly. She retired from the theatre in 1710 and, as far as is known, never married.

BARRY, PHILIP (1896–1949), American dramatist, who graduated from Yale in 1919, studied playwriting at HARVARD under G. P. BAKER, and served in the diplomatic service. His first professional production, *You and I* (1923), was a study of a father's effort to realize his artistic ambitions in his son. Later plays included in 1928 *Cock Robin*, in which he collaborated with Elmer RICE, and *Holiday*, a bright comedy concerning the revolt of youth against parental snobbery. A probing psychological drama, *Hotel Universe* (1930), proved less acceptable to audiences than the domestic comedies *Tomorrow and Tomorrow* (1931) and *The Animal Kingdom* (1932); and although Barry won prestige with a mystifying but provocative allegory of good and evil, *Here Come the Clowns* (1938), *The Philadelphia Story* (1939), a deft comedy of manners and character, was more to the public taste. The Second World War inspired *Liberty Jones* (1941), an allegory of the dangers threatening American democracy; *Without Love* (1942), a romantic comedy that strained a parallel between politics and passion; and *The Foolish Notion* (1945), an ingeniously constructed theatrical fantasy concerning a husband's return from the war. A reflective comedy, *Second Threshold* (1951), left unfinished, was completed by Barry's friend Robert SHERWOOD and presented posthumously in New York with some success.

BARRY, SPRANGER (1719–77), Irish actor, first seen on the stage in Dublin in 1744. Two years later he was at DRURY LANE, where he appeared as Othello to the Iago of MACKLIN but soon became known as an outstanding player of young lovers. He remained with the company when GARRICK took over in 1747, and together they alternated such parts as Jaffeir and Pierre in *Venice Preserv'd*, Chamont and Castalio in *The Orphan* (both by OTWAY), Hastings and Dumont in *Jane Shore*, and Lothario and Horatio in *The Fair Penitent* (both by ROWE). In 1750 Barry went to COVENT GARDEN, where he indulged in open rivalry with Garrick, playing Romeo, Lear, and Richard III. It was generally conceded that in *Romeo and Juliet* Barry was better in the balcony scene and in Juliet's tomb, but Garrick showed to greater advantage in the scenes with the friar and the apothecary; as Lear, Barry was impressive, dignified, and pathetic, but inferior to Garrick in the mad scene. He also failed as Richard III, but made an excellent Young Norval, superb in white satin, in HOME's *Douglas*, playing opposite Peg WOFFINGTON as Lady Randolph. In 1767, having ruined himself by the speculative building of a theatre in Dublin, he returned to Drury Lane where, his first wife dying shortly after, he married in 1768 Ann Dancer (*née* Street) (1734–1801), a member of the company. She excelled in high comedy, being particularly admired as Millamant and Angelica in CONGREVE's *The Way of the World* and *Love for Love*, and as Mrs Sullen in FARQUHAR's *The Beaux' Stratagem*. She retired in 1798.

BARRYMORE, ETHEL (1879–1959), American actress, made her first appearance in New York in 1894 under the aegis of her grandmother Mrs John DREW, and in 1898 was in London, where she appeared with Henry IRVING in Laurence IRVING's *Peter the Great*. Back in New York she scored her first major success in Clyde FITCH's *Captain Jinks of the Horse Marines* (1901) and was equally successful in the title role of the comedy *Cousin Kate* (1903) by H. H. DAVIES. She was also much admired in 1905 as Nora in IBSEN's *A Doll's House*, in which her brother John (below) played Dr Rank. After another excursion into light comedy as the heroine of PINERO's *Trelawny of the 'Wells'* in 1911, she was seen as Marguerite Gautier in the younger DUMAS's *The Lady of the Camellias* in 1917; as Juliet in *Romeo and Juliet* in 1922; as Paula in Pinero's *The Second Mrs Tanqueray* in 1924; and as Ophelia in *Hamlet* and Portia in *The Merchant of Venice* in 1925, both with Walter HAMPDEN. Three years later she opened the ETHEL BARRYMORE THEATRE with MARTÍNEZ SIERRA's *The Kingdom of God*, in which she played Sister Gracia, and she appeared there again in 1931 as Lady Teazle in SHERIDAN's *The School for Scandal*. Having been seen in VAUDEVILLE as early as 1915, she returned to it in 1935 with BARRIE's one-act *The Twelve-Pound Look*, in which she was seen at the LONDON PALLADIUM. A beautiful woman with a warm and dignified presence, she gave some of her finest performances in later life, notably as Gran in *Whiteoaks* (1938) by Mazo de la Roche and as Miss Moffat in Emlyn WILLIAMS's *The Corn is Green* (1940), her last outstanding part.

BARRYMORE, JOHN (1882–1942), American actor, made his first appearance in 1903 in Chicago as Max in *Magda*, an adaptation of SUDERMANN's *Heimat.* His success in J. M. BARRIE's *Pantaloon* in 1905 and W. S. GILBERT's *The Fortune Hunter* in 1909 showed him to be an excellent light comedian, and he became something of a matinée idol; but his Falder in GALSWORTHY's *Justice* in 1916 confirmed him as a serious actor and he gave excellent performances in the title role of Geoge Du Maurier's *Peter Ibbetson* in 1917, as Fedor in *The Living Corpse* (a version of TOLSTOY's *Redemption*) in 1918, and as Giannetto Malespini in *The Jest*, based on BENELLI's *La cena delle beffe* (1908), in 1919. In 1922 he electrified New York with his Hamlet, which he played for a record-breaking 101 performances with Blanche YURKA as Gertrude and settings designed by Robert Edmond JONES. He repeated this success in London in 1925, when Gertrude was played by Constance COLLIER and Ophelia by Fay COMPTON. Barrymore's meticulous and scholarly reading of the part was enhanced by his romantic beauty and perfect diction; his subsequent career was a process of decline hastened by alcoholism and he never again approached such heights. After an absence from the stage of 14 years, spent mainly in films, he returned unsuccessfully in *My Dear Children* (1940) by Catherine Turney and Jerry Horwin, which appeared to be partly autobiographical, and after a short tour was not seen again. The eccentricities of the three Barrymores, and the majestic personality of their grandmother Mrs John DREW, were the subject of a play, *The Royal Family* (1927), by Edna Ferber and George S. KAUFMAN, seen in London as *Theatre Royal* in 1934.

BARRYMORE, LIONEL (1878–1954), American actor, brother of Ethel and John BARRYMORE and grandson of Mrs John DREW, under whom he made his first appearance on the stage at the age of 15. After achieving some success, notably in Isaac Henderson's *The Mummy and the Humming Bird* (1902) and the title role of BARRIE's *Pantaloon* (1905), he went to Paris to study art. Returning to the United States in 1909, he reappeared on the stage in Conan Doyle's *The Fires of Fate*, and soon became one of New York's leading actors, being particularly admired in George Du Maurier's *Peter Ibbetson* (1917), Augustus THOMAS's *The Copperhead* (1918), and *The Jest* (1919), based on BENELLI's *La cena delle beffe* (1908), in all of which his brother John also appeared. After John's success as Hamlet, Lionel appeared as Macbeth, but the venture was something of a failure, and in 1925 he withdrew from the theatre and spent the rest of his life in films, in which he had a long and distinguished career.

BARRYMORE, MAURICE [Herbert Blythe] (1847–1905), English-born actor, who took his stage name from an old playbill hanging in the HAYMARKET THEATRE. He studied law and was well known as an amateur boxer before his first appearance on the London stage in 1875; in the same year he went to New York, where he was seen in DALY's *Under the Gaslight*. A handsome, well-built man, he played opposite many leading actresses including MODJESKA, for whom he wrote *Nadjezda* (1886): he later accused SARDOU of having plagiarized it in *Tosca*. He appeared in London in 1886 in Sardou's *Diplomacy* and Charles READE's *Masks and Faces*, and in 1895 starred in BELASCO's *The Heart of Maryland*; he also appeared under the management of William A. BRADY in his own play *Roaring Dick*. Towards the end of his career he became a VAUDEVILLE star. Barrymore married the actress Georgina DREW in 1876; their three children, Ethel, John, and Lionel BARRYMORE, all had distinguished acting careers.

BARRYMORE, RICHARD BARRY, Earl of, see PRIVATE THEATRES IN ENGLAND.

BARRYMORE, WILLIAM. Two actors of this name, no relation to each other or to any of the above, appeared on the London and New York stages. The elder (1758–1830) opened the OLD VIC as the Royal Coburg in 1818 with his own play *Trial By Battle; or, Heaven Defend the Right*, and wrote several more in the same style for the same theatre. He was also the author of one of the versions of the melodrama *The Dog of Montargis; or, the Forest of Bondy*, seen at COVENT GARDEN in 1814. The younger (d. 1845), a sound, useful actor, also wrote a number of plays. His wife, who acted with him in the United States for many years, died in England in 1862.

BARSACQ, ANDRÉ (1909–73), French director and manager of Russian descent, born in the Crimea and taken to Paris at the age of eight. He was studying design when a visit to the ATELIER, then under the direction of DULLIN, turned his thoughts to the theatre. He worked with Dullin on several plays and films, including the famous 1928 adaptation of JONSON's *Volpone* by Jules ROMAINS, and in 1937 founded with Jean DASTÉ the Compagnie des Quatre Saisons, a travelling repertory company which was seen in the same year at the International Exhibition and then spent two years in the French Theatre in New York. During this period Barsacq produced ANOUILH's *Le Bal des voleurs* (1938) and then almost all his later plays until 1953, including the subversive *Antigone* which escaped German censorship in 1944. After the war Barsacq succeeded Dullin at the Atelier and remained there until his death, in spite of illness and financial difficulties, making the theatre a shop-window for new dramatists, and introducing to Parisian audiences BETTI, Félicien MARCEAU,

OBALDIA, and DÜRRENMATT. He also directed several successful adaptations of CHEKHOV, DOSTOEVSKY, and TURGENEV.

BART, LIONEL (1930–), English lyricist and composer who helped to revive English MUSICAL COMEDY after a long period of American domination. His first work, *Fings Ain't Wot They Used T'Be* (1959), was produced by THEATRE WORKSHOP and then ran in the West End for two years. He also wrote the lyrics for *Lock Up Your Daughters* (1959), a play with music based by Bernard MILES on FIELDING's *Rape Upon Rape*. His greatest success was *Oliver!* (1960; N.Y., 1963), based on DICKENS's *Oliver Twist*. It had a long run when first produced and was revived in 1967 and again in 1977, when it ran for almost three years. Two other productions, *Blitz!* (1962) and *Maggie May* (1964), about a Liverpool prostitute between the wars, were also successful, but a Robin Hood musical *Twang!* (1965) was a disastrous failure, as was *La Strada* (1969), a musical based on Fellini's film, which had only one performance in New York.

BARTER THEATRE, Abingdon, Virginia, a proscenium-arch theatre seating 380. Founded during the Depression, it opened on 10 June 1933, admission being by barter, each member of the audience paying in kind for the privilege of seeing a play. In this way the actors, though always short of money, at least had enough to eat. The theatre prospered, and in 1946 was declared the official State Theatre of Virginia (a title also given later to the VIRGINIA MUSEUM THEATRE, Richmond). In 1953 it was refurbished with some of the furnishings from the old EMPIRE THEATRE, New York, and in 1971 a second theatre, the Barter Playhouse, seating 100 round a three-quarters arena stage, was added. Tours are undertaken by the company outside the resident season of classic and contemporary plays, which runs from Apr. to Oct., and special plays for children are presented in the Playhouse.

BARTHOLOMEW FAIR, see FAIRS and JONSON, BEN.

BARTON, JOHN BERNARD ADIE (1928–), English director, who, while a Fellow of King's College CAMBRIDGE, directed plays for the Marlowe Society and the A.D.C. He joined the ROYAL SHAKESPEARE COMPANY on its foundation in 1960, becoming associate director in 1964, his first production there being *The Taming of the Shrew*. He devised for the company the recitals *The Hollow Crown* (1961), *The Art of Seduction*, and *The Vagaries of Love* (both 1962), in all of which he took part, and was also responsible for *The Wars of the Roses* (1963), a condensation of all three parts of *Henry VI* and *Richard III*. In 1969 for the R.S.C.'s schools' programme Theatregoround he prepared a shortened version of *Henry IV* as *When Thou Art King*, concentrating the action on the relationship of Prince Hal and Falstaff. He also made a shortened version of *Henry V* as *The Battle of Agincourt*, and in 1970 added to it two more scenes from the history cycle as *The Battle of Shrewsbury* and *The Rejection of Falstaff*. His non-Shakespearian productions included his own version of MARLOWE's *Dr Faustus* (1974) and *The Greeks* (1980) based on ten plays, mainly by EURIPIDES.

BASSERMANN, ALBERT (1867–1952), German actor, who made his first appearances in Mannheim, where he was born, under his uncle August BASSERMANN. He was a member of the MEININGER ensemble from 1890 to 1895, and in 1899 joined BRAHM's company in Berlin, where his naturalistic, subtly analytic style of acting made him an outstanding interpreter of IBSEN; he was also much admired in plays by HAUPTMANN and SCHNITZLER. From 1909 to 1914 he was with REINHARDT in Berlin, and after the First World War returned there many times in between extensive tours of Germany, during which he extended his range to include such great classical roles as Shakespeare's Shylock and Lear, SCHILLER's Wallenstein, Philip II and the title role in *Don Carlos*, GOETHE's Egmont and Mephistopheles (in *Faust*). In 1934 he left Germany for the United States, appearing on Broadway in 1944 in his first English-speaking part, the Pope in *Embezzled Heaven* (based on a novel by WERFEL). He also went to Hollywood, where he played minor roles in films, and in 1945 returned to Europe, settling in Zurich where he died.

BASSERMANN, AUGUST (1848–1931), German actor, uncle of the above, who made his début at Dresden in 1873 and later appeared in Vienna in such parts as Rolla in SHERIDAN's *Pizarro* and Karl Moor in SCHILLER's *Die Räuber*. In 1886 he became manager of the Mannheim Theatre and from 1904 until his death he was director of a theatre at Karlsruhe. He went to New York several times, and played in the German Theatre there, being much admired in classic and heroic parts.

BASSINGHAM PLAY, see MUMMERS' PLAY.

BATALOV, NIKOLAI PETROVICH (1899–1937), Soviet actor, who joined the MOSCOW ART THEATRE company in 1916 and soon proved an outstanding player of comic parts. In 1927 he gave a lively performance as Figaro in BEAUMARCHAIS's *Le Barbier de Séville*, and later the same year proved his versatility by giving a portrayal of Vasska Okorka in IVANOV's *Armoured Train 14–69* which brought praise from LUNACHARSKY for its warmth and tempestuous gaiety. An optimist by nature, he was also able to give a fine satirical edge to his playing of the anarchist-marauder in Ivanov's *The Blockade* (1929).

BATEMAN, HEZEKIAH LINTHICUM (1812–75), American impresario, who married Sidney Frances Cowell (1823–81), daughter of the English actor Joseph COWELL, and by her had four daughters who were all on the stage from early childhood. Bateman and his family made several visits to England, and in 1871 leased the LYCEUM THEATRE in London to launch his third daughter, Isabel (below), on her adult career. He engaged Henry IRVING as his leading man and after some initial disappointments allowed him to put on Lewis's *The Bells*, which was an immediate success. Bateman continued as lessee of the theatre, though Irving effectively ran and publicized it, until his death, when Mrs Bateman took over. She finally ceded the lease to Irving in 1878 and went with Isabel to SADLER'S WELLS THEATRE, which she continued to manage until her death. She was herself an excellent actress and the author of a number of plays staged by her husband.

BATEMAN, ISABEL EMILIE (1854–1934), American-born actress, third daughter of H. L. BATEMAN. She appeared on the stage as a child, and in 1871 made her adult début at the LYCEUM THEATRE in London, which her father had leased with the intention of launching her on a career as successful as that of her elder sister Kate (below). Her first appearance there was in the title role of *Fanchette, or Will o' the Wisp*, an adaptation by her mother of George Sand's novel *La Petite Fadette*. The production was not a success, as she was unsuited to the somewhat frivolous part, being essentially of a serious turn of mind. This, coupled with her beauty and rigorous training at the hands of her parents, enabled her, however, to give a good account of herself in later parts, including Henrietta Maria in Wills's *King Charles I* and Ophelia, particularly in the mad scene, to IRVING's Hamlet. She also played Desdemona to his Othello in 1874, and Elizabeth I in TENNYSON's *Queen Mary*, with Kate as Mary Tudor. Although she disliked acting, and everything connected with the theatre, she dutifully accompanied her widowed mother to SADLER'S WELLS THEATRE in 1878 and continued to act and assist in running the business until her mother died in 1881. Even then she was forced to go on working until 1898 in order to pay off the heavy debts incurred by her mother's unsuccessful management; but she eventually achieved her life-long desire to become a nun and entered the Community of St Mary the Virgin in Wantage in 1899, becoming Reverend Mother General of the Order in 1920.

BATEMAN, KATE JOSEPHINE (1842–1917), American-born actress, eldest daughter of H. L. BATEMAN. She was on the stage from early childhood, being only 5 when she made her first appearance in Louisville, Kentucky, in *The Babes in the Wood*. For several years she and her younger sister Ellen Douglas (1844–1936) toured the United States in a programme of short sketches which included scenes from several of Shakespeare's plays, Kate playing Richmond, Portia, and Lady Macbeth to Ellen's Richard III, Shylock, and Macbeth. They were seen in these parts in London in 1851. Ellen retired after her marriage to Claude Greppo in 1860, while Kate made her adult début in New York as Geraldine in a dramatization of Longfellow's *Evangeline* (1860), in which her mother also appeared. She then made a sensational success in the title role of DALY's *Leah the Forsaken* (1863), with which she was thereafter always associated. She was seen in it in London in 1864, and played it in Manchester with Henry IRVING as Joseph. When her father took over the London LYCEUM she appeared there several times while Irving and his company were on tour, and also played Emilia in his production of *Othello*, Mary Tudor in TENNYSON's *Queen Mary* in 1876 opposite her sister Isabel (above), and Queen Margaret in *Richard III* in 1877. She also appeared at SADLER'S WELLS during her mother's tenancy of the theatre, but her husband's ill-health caused her to quit the theatre for long periods. Under her married name, Mrs George Crowe, she opened a school of acting in 1892 and ran it successfully for many years. She made one of her rare reappearances as Mrs Dudgeon in the first London production of SHAW's *The Devil's Disciple* (1907).

BATEMAN, VIRGINIA FRANCES, see COMPTON, EDWARD.

BATES, ALAN ARTHUR (1934–), English actor, who made his first appearance in London at the ROYAL COURT in 1956, and later the same year attracted favourable notice as Cliff Lewis in OSBORNE's *Look Back in Anger*, in which he made his New York début the following year. He then played Edmund Tyrone in O'NEILL's *Long Day's Journey into Night* (1958) at Edinburgh and in London; but it was as Mick in PINTER's *The Caretaker* (1960; N.Y., 1961) that he made his first outstanding success. He then went into films and his next notable stage role was as Richard III at the STRATFORD (ONTARIO) FESTIVAL in 1967. Two years later he was back in London, where he appeared in David STOREY's *In Celebration*, and in 1970 was seen as Hamlet at the NOTTINGHAM PLAYHOUSE. This was brought to London in 1971; later the same year (N.Y., 1972) Bates played the title role in Simon GRAY's *Butley*. After an 'effervescently likeable' Petruchio in *The Taming of the Shrew* at the ROYAL SHAKESPEARE THEATRE in 1973 he renewed his association with Storey in *Life Class* (1974) and with Gray in *Otherwise Engaged* (1975). In 1976 he played Trigorin in CHEKHOV's *The Seagull*, and then continued his outstanding film career, returning to the theatre in another play by Simon Gray, the comedy-thriller *Stage Struck* (1979).

BATES, BLANCHE (1873–1941), American actress, whose parents were both in the theatre. She first appeared on tour in California, from 1894 to 1898, and then made a brief appearance in New York under the management of Augustin DALY. It was, however, as the leading lady in plays by BELASCO—*Madam Butterfly* (1900). *The Darling of the Gods* (1902), and *The Girl of the Golden West* (1905)—that she made her reputation. She was also much admired as Cigarette in a dramatization of Ouida's *Under Two Flags* (1901). After leaving Belasco she continued to act, under her own and other managements, but less successfully, until her retirement in 1927.

BATH. The first theatre in Bath was built in 1705 and demolished in 1737, when the company moved to a room under Lady Hawley's Assembly Rooms. In 1750 a second theatre was erected in Orchard Street; the two theatres were rivals until 1756 when they amalgamated under John PALMER and settled in Orchard Street. The theatre there was reconstructed in 1767 and a year later became a Theatre Royal under a patent from George III, the first provincial theatre in England to be so honoured. In 1774 the building was again reconstructed; from 1778 to 1782 Mrs SIDDONS was a member of its STOCK COMPANY: the shell of the Orchard Street Theatre still stands. By 1805 it had become too small and a new and larger theatre was erected in Beaufort Square; it flourished for a time but by the 1820s it was in decline and its interior was destroyed by fire in 1862. Less than 12 months later the present Theatre Royal, seating 615, with a further 250 in the gallery, opened on the same site with a production by the BRISTOL stock company of *A Midsummer Night's Dream* in which Ellen TERRY, aged 15, played Titania. Since 1867 it has been used by touring companies, Henry IRVING's being the first to visit it. It was redecorated in 1892 and 1974, but has otherwise undergone no major renovation. In 1979 the theatre was taken over by a Trust and an appeal was launched for funds to preserve and improve it with the purpose of establishing a base for the NATIONAL THEATRE's middle-scale touring.

BATHYLLUS, Roman freedman of the first century BC, a protégé of Maecenas, referred to by Juvenal (vl. 63). With PYLADES he popularized PANTOMIME as a dramatic form on the Roman stage, devising and performing in many pantomimes himself.

BATTEN, a length of timber used to stiffen a surface of canvas or boards, as by 'sandwich-battening' a cloth, that is, fixing the upper and lower edges between pairs of 3- or 4-inch by 1-inch battens screwed together, or by 'battening-out' a section of boards or a run of FLATS with crossbars. A light batten, known in America as a strip light or border light, is a horizontal trough containing a row of lights, often divided into compartments (and then known as a compartment batten), each containing a lamp, reflector, and coloured filter.

BATTY, WILLIAM, see ASTLEY'S AMPHITHEATRE.

BATY, GASTON (1885–1952), French director, who travelled extensively, and worked in a number of Parisian theatres before opening the Théâtre Montparnasse under his own name in 1930. Here he put on an imposing series of old and new plays, many of them foreign classics, and several dramatizations of novels, of which the best were probably his own version of DOSTOEVSKY's *Crime and Punishment* and *Dulcinée* (1938), based on an episode in CERVANTES' *Don Quixote*. He was sometimes accused of subordinating the text of the play to the décor, which for him was all-important and led him to substitute pictorial groupings for action; but it resulted in some fine work, most of all in his productions of Gantillon's *Maya*, LENORMAND's *Simoun*, and J.-J. BERNARD's *Martine*. In 1936 Baty was appointed a director of the COMÉDIE-FRANÇAISE, where his erudition and feeling for the best in theatre helped to reanimate the classical repertory.

BAUER, WOLFGANG (1941–), Austrian dramatist, one of a group of dialect playwrights who emerged from Graz at the end of the 1960s. His plays are in an Austrian tradition which embraces Horváth's realistic dialogue, SCHNITZLER's *ennui*, and NESTROY's clockwork satire. Like Schnitzler in particular, Bauer presents the comedy inherent in a decadent boredom so severe that in attempting to alleviate it a character may meet his death, as in *Sylvester oder das Massaker im Hotel Sacher* (1971). As in other plays, including the shorter ones like *Film und Frau* (also 1971), briefly seen in London in 1972 as *Shakespeare the Sadist*, the setting is a jaded milieu of pop, pot, and casual sex, among talentless would-be avant-gardists. His control of pace and mood, his convincing theatrical climaxes, even his liking for fast changes of role and clothes, give his plays the impact of farce; but they are true comedies of modern manners and have enjoyed great success in Germany as well as in Austria.

BÄUERLE, ADOLF (1786–1859), Austrian dramatist, creator in *Die Bürger in Wien* (1813) of the comic character STABERL, played by Ignaz SCHUSTER. At first specifically Viennese in his humour, Staberl became the leading character of four more plays by Bäuerle, and was then transmuted by Karl CARL into a figure acceptable all over Austria and Germany. Bäuerle, who wrote nearly 80 plays, was also a successful provider of Viennese *Zauberstücke*, or fairy stories, of which his *Aline*, *oder Wien in einem anderen Weltteile* (1822) is typical. Most of his plays were produced at the Theater in der Leopoldstadt, where he was on the staff from 1822 to 1827.

BAUERNFELD, EDUARD VON (1802–90), Austrian dramatist, friend of Schubert and of GRILLPARZER. His early plays showed the influence of KOTZEBUE, but more successful were the light comedies staged by SCHREYVOGEL at the Burgtheater in Vienna from 1831 to 1835. After the 1848 revolution he had noteworthy successes with *Der kategorische Imperative* (1850) and *Aus der Gesellschaft* (1867), the latter a deft handling of a social problem of the time, *mésalliance*.

BAVARIAN WILD MEN, see FOLK FESTIVALS.

BAYES, NORA, see NORWORTH, JACK, and VAUDEVILLE, AMERICAN.

BAYLIS, LILIAN MARY (1874–1937), theatre manageress, one of the outstanding women of the English theatre. She was only the second woman outside the University to be made an Hon. M. A. at Oxford (in 1924), and in 1929 was appointed C. H. for her services to the arts in Great Britain. Born in London, the eldest child of two musicians, she studied the violin, and had made several appearances as a child prodigy before going with her family to South Africa in 1890. There she toured in a combined musical and dramatic entertainment and later settled in Johannesburg, where she was one of the first music teachers. Returning to London in 1895, she stayed to assist her aunt Emma Cons in running the old Victoria Theatre (later the OLD VIC) as a temperance hall, and when her aunt died in 1912 she took over the management herself. An intensely religious and single-minded woman, she devoted the rest of her life to the founding and running of popular homes for drama and opera in London, working herself and her staff and actors, who were very poorly paid, extremely hard, and making many friends and not a few enemies in the process. When drama, mainly Shakespeare, threatened to oust opera from the Old Vic, she took over and reopened in 1931 the old SADLER'S WELLS THEATRE, where in time ballet became as popular as opera, if not more so; the companies were to evolve into the Royal Ballet and the English National Opera. Under her management all Shakespeare's plays were produced at the Old Vic, from *The Taming of the Shrew* in 1914, under the direction of Matheson LANG, to *Troilus and Cressida* in 1923, directed by Robert ATKINS. *Hamlet*, the second play to appear in the repertory, was several times done in its entirety, always at matinées on account of its length. Plays by other dramatists were sometimes included in the programme, among them IBSEN's *Peer Gynt* in 1922, but it was as the home of Shakespeare, presented very simply with good and in later years excellent actors, that the Old Vic became famous. It says much for the soundness of the foundation on which Lilian Baylis had established it that it survived the shock of her death until 1963, when it became the temporary home of the NATIONAL THEATRE company; it should, in the opinion of many people, have become the National Theatre itself.

BAYREUTH, see THEATRE BUILDINGS.

BEAR GARDENS MUSEUM, see WANAMAKER, SAM.

BEATON, CECIL WALTER HARDY (1904–80), English designer and photographer, knighted 1972. His first work for the stage was the scenery and costumes for the REVUE *Follow the Sun* (1936). After designing several ballets, he made his first venture into the non-musical theatre with the costumes and scenery for GIELGUD's production of WILDE's *Lady Windermere's Fan* at the HAYMARKET THEATRE in 1945. A year later he made his first appearance as an actor as Cecil Graham in an American production of this play, for which he also designed the costumes and scenery. Among the London productions with which he was later associated were revivals of PINERO's *The Second Mrs Tanqueray* in 1950, LONSDALE's *Aren't We All?* in 1953, and *Love's Labour's Lost* at the OLD VIC in 1954. He was also responsible for the décor and costumes of COWARD's *Quadrille* (1952; N.Y., 1954). Thereafter his work was seen mostly in New York, with Enid BAGNOLD's *The Chalk Garden* in 1955, Coward's *Look After Lulu* (based on FEYDEAU's *Occupe-toi d'Amélie*) in 1959, and the musical *Coco* (1969), in which Katharine HEPBURN played the title role. His best, and certainly his best known, work was the lavish and witty costume design for *My Fair Lady* (1956; London, 1958) (based on SHAW's *Pygmalion*), for which he designed both the stage and the 1964 screen versions. He was the author of a play, *The Gainsborough Girls*, which was produced at BRIGHTON in 1951 but never reached London.

BEAUBOUR, PIERRE TRONCHON DE (1662–1725), French actor, third husband of Louise Pitel (*c.* 1665–1740), actress daughter of BEAUVAL, who as a child played the part of Louison in MOLIÈRE's *Le Malade imaginaire*, both her parents being members of his company. Beaubour joined the COMÉDIE-FRANÇAISE in 1691 to replace BARON. Though his good looks and excellent presence made him acceptable in tragedy, he never attained the heights of his predecessor, being somewhat noisy and declamatory in the old-fashioned style: PALAPRAT lamented that Beaubour had realized his potential as a comic actor too late to profit by it. He retired from the theatre in 1718 together with his wife, who had played only secondary roles.

BEAUCHÂTEAU [François Chastelet] (*fl.* 1625–65), French actor who joined the company at the

Hôtel de BOURGOGNE at the same time as BELLEROSE and remained there until 1634, when he went with his wife Madeleine de Pouget (1615–83) to the Théâtre du MARAIS, returning to the Hôtel de Bourgogne in 1642. Husband and wife were parodied as the young lovers in MOLIÈRE's *L'Impromptu de Versailles* (1664); they had their admirers though, like Bellerose, they were usually thought somewhat insipid and sentimental, BOILEAU calling the husband 'an execrable actor'.

BEAUMARCHAIS, PIERRE-AUGUSTIN CARON DE (1732–99), French writer and public figure, whose multifarious activities included writing the most successful comedy of the 18th century. The son of a watchmaker, Beaumarchais began by following his father's trade and first made his mark by litigation over his copyright in an invention. He married a widow, taking over her husband's position at Court and adopting the name Beaumarchais (from a property he acquired on marrying) as an addition to his family surname of Caron. He was soon a widower, established at Court, where he gave music lessons to Louis XV's daughters and directed the music of the Court entertainments. He speculated successfully, indulged in constant litigation, and kept himself continually in the public eye, mainly through his notorious feud with the magistrate Goezmann whom he lampooned in four published pamphlets. He also travelled in Spain, partly to rescue his sister from an unfortunate love-affair (an incident used later by GOETHE in his play *Clavigo*) and partly on secret diplomatic business; undertook to supply arms to the insurgent American colonists during the War of Independence; set up a printing-press at Kehl for the publication of a complete edition of the works of VOLTAIRE; and instituted through the Société des Auteurs, of which he was the founder, the system of payment for plays by means of a fixed percentage on takings (see ROYALTY). During the Revolution he lived an extremely hazardous life. Although he was a member of the Committee of Public Safety, he was suspected of profiteering in arms, had to go into exile, saw his fortune confiscated and his family imprisoned. He returned to France in 1793 and the remaining years of his life saw a partial return of reputation, fortune, and popularity.

Beaumarchais's career as a dramatist began with a series of PARADES written for various private theatres, the resultant skill in manipulating language in order to produce fast-moving comedy standing him in good stead when he came to create the immortal character of Figaro. He wrote two plays in the style of DIDEROT's domestic drama, *Eugénie* (1767) and *Les deux amis* (1770), before achieving an outstanding success with *Le Barbier de Séville*. Originally intended as a play with music (and later used as the basis of an opera by Rossini), this was refused by the COMÉDIE-ITALIENNE, who thought its hero a caricature of their leading actor, himself formerly a barber. It was accepted in a modified form by the COMÉDIE-FRANÇAISE and, after delays imposed by the censor, was seen there in 1775. The sequel, *Le Mariage de Figaro, ou la Folle Journée*, encountered even more difficulties with the censor; it was seen as a provocative political satire, and Beaumarchais had to be content for a number of years with readings of it in society drawing-rooms. The eventual first night, in April 1784, was one of the most eagerly awaited theatrical events of the century. It had 80 performances, a record for the 18th century, and its popularity can be judged by the numerous contemporary parodies. In later years it suffered from competition with Mozart's opera, but it remained in the repertory and there have been notable modern productions. It was first seen in London in 1784, soon after its production in Paris, in an adaptation by HOLCROFT entitled *The Follies of a Day*. In 1926 a new adaptation by Barry JACKSON was produced and in 1974 it was revived by the NATIONAL THEATRE. In its time it represented, as few plays have ever done, the mood of an age—an age critical of governments and of established institutions. Historically it stands as perhaps the most important comedy in France after MOLIÈRE, certainly as one of the most important plays of its century in any language; and in his dramatic craftsmanship Beaumarchais offered an invaluable lesson to such 19th-century writers of comedy and vaudeville as SCRIBE, LABICHE, and FEYDEAU. After the two great Figaro plays Beaumarchais returned to the theatre in 1787 with an opera, *Tarare*, and he completed the Figaro trilogy with *La Mère coupable* (1795), in which he reverted to the moralizing manner of his earlier domestic dramas.

BEAUMÉNARD, ROSE-PETRONELLE LE ROY (1730–99), French actress, who appeared with some success at the Opéra-Comique in 1743 and then toured the provinces, being a member of the company taken on his campaigns by Marshal Saxe. In 1749 she joined the COMÉDIE-FRANÇAISE, where she excelled in such parts as Nicole in MOLIÈRE's *Le Bourgeois Gentilhomme*, though she was less at home with the subtleties of MARIVAUX. She was a great favourite with the public and was variously nicknamed La Rieuse or Gogo, the latter after her successful role in Favart's *Le Coq du village*. In 1761, after a brief retirement, she married the actor BELLECOUR and continued to act until 1791, being the first to play the redoubtable Marceline in BEAUMARCHAIS's *Le Mariage de Figaro* (1784). The Revolution rendered her penniless, and in 1799 she was forced to take up her career again. She attempted to play Nicole once more, but was only a shadow of her former self.

BEAUMONT, Sir FRANCIS (1584–1616), English dramatist, whose name is so closely associated with that of John FLETCHER that scholars are still disentangling their separate contributions to the plays that pass under their joint names. Beaumont had already written *The Woman Hater* (1606) and may have been sole author of *The Knight of the Burning Pestle* (1607), a satire on the audiences that flocked to see the romantic historical plays so popular at the time, before his first known collaboration with Fletcher, *Philaster, or Love Lies Bleeding* (1609). It is now thought that they worked together on no more than six or seven plays, though 53 are assigned to them in collections published in 1647 and 1679. Most of these have since been attributed to JONSON, MASSINGER, MIDDLETON, ROWLEY, SHIRLEY, and TOURNEUR. Beaumont ceased to write for the stage on his marriage in 1613, though he once or twice contributed material to Court entertainments after that date.

BEAUMONT, MURIEL, see DU MAURIER, GERALD.

BEAUVAL [Jean Pitel] (*c.* 1635–1709), French actor, first heard of in 1663 as candle-snuffer to the Prince de Condé's troupe. He married the illegitimate daughter of the actor-manager FILANDRE, who made her first appearances in her father's company, and in 1670 they joined the troupe of MOLIÈRE. Beauval was a poor actor; more by nature than art he made a success as Thomas Diafoirus in *Le Malade imaginaire*, in which his wife played Toinette. Though coarse-featured and bad-tempered, she was an excellent comic actress, given to fits of irresistible laughter which Molière incorporated into her part of Nicole in *Le Bourgeois Gentilhomme*. She also played Zerbinette in *Les Fourberies de Scapin*. After Molière's death the Beauvals went to the Hôtel de BOURGOGNE and eventually became members of the newly established COMÉDIE-FRANÇAISE, retiring in 1704. One of their numerous children married as her third husband the actor BEAUBOUR.

BEAZLEY, SAMUEL (1786–1851), English theatre architect, designer of the LYCEUM, the ST JAMES's, the CITY OF LONDON, that part of the ADELPHI fronting on the Strand, and the colonnade of DRURY LANE. His buildings, though plain and somewhat uninteresting, were good and well adapted for their purpose. A prolific dramatist, mainly of ephemeral farces and short comedies, Beazley was also responsible for poor translations of several operatic libretti.

BECCARI, AGOSTINO (?–1598), Italian dramatist, author of an early PASTORAL, *Il sacrifizio* (*The Sacrifice*), which was performed at Ferrara in 1554–5, before the Duke and his Court. Although it contains some happy imitations of classical pastoral poetry, it is mainly important as having preceded and perhaps inspired the *Aminta* of TASSO.

BECK, HEINRICH (1760–1803), German actor, who began his career in 1777 in his birthplace Gotha under EKHOF, at the same time as IFFLAND who became his close friend and later his associate in theatre management. After Ekhof's death in 1778 he went to Mannheim where he remained many years, playing Ferdinand in the first production there in 1784 of SCHILLER's *Kabale und Liebe* with his wife, Karoline Ziegler, playing Louise. He was also the author of a number of now forgotten plays.

BECK, JULIAN, see LIVING THEATRE.

BECKETT, SAMUEL (1906–), Irish-born dramatist, resident since 1938 in Paris, who has written mainly in French and in 1969 was awarded the NOBEL PRIZE for Literature. He was already well known as a novelist with *Molloy, Malone meurt*, and *L'Innommable* when in 1953 his first play *En attendant Godot* was produced by Roger BLIN. Two years later it was seen in London in the author's own translation as *Waiting for Godot*, and in 1956 it was produced in New York with Bert LAHR. In it Beckett abandoned conventional structure and development in both plot and dialogue in order to present a dramatic vision of the human predicament in a world in which mankind seems to have no place. His two tramps, indecisive and incapable of action, wait hopefully for help which never comes. This epoch-making play in now regarded as one of the masterpieces of the Theatre of the ABSURD. It was followed in 1957 by *Fin de partie*, in which a blind man sits in an empty room flanked by two old people in dustbins. It had its first performance in London, in the original, together with a mime-play *Actes sans paroles* (*Actions Without Words*). As *Endgame* it was produced in English in 1958, in company with a monologue originally written in English, *Krapp's Last Tape*, in which an old man listens uncomprehendingly to recordings he made as a young man; this was seen in Paris in 1960 as *La Dernière bande*. *Endgame* was seen in New York in 1958, *Actions Without Words* in 1959, and *Krapp's Last Tape* in 1960. *Oh! les beaux jours*, which is virtually a monologue for an actress progressively buried in earth until only her head is visible, had its first production in New York in 1961, as *Happy Days*. It was seen in English in London in 1962 and in French, played by Madeleine RENAUD, in 1965. *Comédie*, first produced in Germany in 1963 and in Paris and London (as *Play*) in 1964, was a short dialogue spoken by three heads protruding from urns which was repeated *da capo*. In this and in his later plays, which include *Va et Vient* (*Come and Go*, 1966; London, 1970; N.Y., 1975), *Breath* (1969), which lasts about 30 seconds, beginning with the cry of a newborn child and ending with the last gasp of a dying man, and *Not I* (1971; N.Y., 1972; London, 1973), Beckett moved further and further away from conventional theatre in an effort to convey with the minimum of speech and action man's inability to communicate and his unawareness of his failure to control his destiny. Beckett's plays for radio—*All That Fall* (1957), *Embers* (1959), *Words and Music* (1962), and *Cascando* (1963)—were important contributions to the medium and, together with his novels, must be taken into account in any critical assessment of his work.

BECQUE, HENRY [also Henri] FRANÇOIS (1837–99), French dramatist, an outstanding exponent of NATURALISM in the style of ZOLA. His first important play was *Michel Pauper* (1870), which was underrated on its first production and only appreciated in a revival of 1886. Meanwhile Becque had written the two plays with which his name is usually associated, *Les Corbeaux* (1882) and *La Parisienne* (1885), both naturalistic dramas of great force and uncompromising honesty, which present rapacious or amoral characters who seem unaware of their own degradation. These '*comédies rosses*' (bitter comedies) were not wholly successful until ANTOINE'S THÉÂTRE LIBRE provided an ambience suitable for them.

BEDFORD MUSIC-HALL, London, in Camden High Street. Built on part of the tea garden belonging to the Bedford Arms, it opened in 1861 and had a most successful career, its interior, which then had a seating capacity of 1,168 on three tiers, being the subject in 1890 of a painting by Walter Sickert. Among the variety stars who appeared there were George LEYBOURNE, LITTLE TICH, and Marie LLOYD. In 1949 it became a theatre for six months, housing revivals of such old favourites as Miss Braddon's *Lady Audley's Secret*, JERROLD's *Black-Ey'd Susan*, Leopold Lewis's *The Bells*, and Mrs Henry Wood's *East Lynne*. It then closed and was demolished in 1969.

BEEKMAN STREET THEATRE, New York, see CHAPEL STREET THEATRE.

BEERBOHM, (Henry) MAX(imilian) (1872–1956), English writer, dramatic critic, and dramatist, half-brother of Herbert TREE, knighted in 1939. From 1898 to 1910 he was dramatic critic of the *Saturday Review* in succession to George Bernard SHAW, who introduced him as 'the incomparable Max'. His appreciation of the art of the theatre, which he excitingly communicated to his reader, embraced everything from dignified classical acting to the homely humour of the music-halls. He was the author of the one-act play *The Happy Hypocrite* (1900), based on one of his own short stories and produced at the ROYALTY THEATRE with Mrs Patrick CAMPBELL; a three-act version of the same story, by Clemence DANE, successfully starred Ivor NOVELLO at HER (then His) MAJESTY'S THEATRE in 1936. His other plays were *The Fly on the Wheel* (1902), in which he collaborated with Murray Carson, and a short sketch, *A Social Success* (1913), in which George ALEXANDER made his first music-hall appearance. Beerbohm married in 1908 Florence KAHN, whom he had first seen in that year as Rebecca West in IBSEN's *Rosmersholm*.

BEESTON, CHRISTOPHER (?1570–1628), an important figure in the Jacobean and Caroline theatre. He began his career with STRANGE'S MEN; when they split up in 1594 he may have joined the CHAMBERLAIN'S MEN, as in 1605 he received a legacy of 30*s*. from Augustine Philipps, a member of that company who left a similar sum to Shakespeare. By that time Beeston was with Worcester's Men at the ROSE THEATRE and remained with them when on the accession of James I in 1603 they became QUEEN ANNE'S MEN. In 1612 he succeeded the actor Thomas Greene as business manager to the company but in 1616–17 took over the COCKPIT and ceased to act. On the accession of Charles I in 1625 it became the home of the newly formed company QUEEN HENRIETTA'S MEN with Beeston as manager, and when they disbanded in 1636 he used the theatre for a young company known as Beeston's Boys, several of whom later made their adult reputations in the Restoration theatre. On his death the control of the company and the theatre passed to his son William (below). His eldest daughter Anne married the actor Theophilus BIRD.

BEESTON, WILLIAM (?1606–82). son of Christopher (above) who probably trained him as an actor. On his father's death he took over the control of Beeston's Boys, but found himself in trouble with the authorities over an unlicensed play and was imprisoned, DAVENANT taking his place as manager of the COCKPIT. He returned in 1641 when Davenant went abroad; after the closing of the theatres the following year he is not heard of until 1647 when he managed to acquire SALISBURY COURT, which he used as a training ground for boy actors until in March 1649 it was wrecked by Commonwealth soldiers. He was suspected of being the 'ill Beest' who in 1652 betrayed the actors who were appearing surreptitiously at the VERE STREET THEATRE, but it may have been his brother-in-law Theophilus BIRD. He retained possession of Salisbury Court, which, after he had restored it, became the first theatre to open in 1660. He leased it to several companies and in 1663–4 ran a company there himself, but the granting of Royal PATENTS to Davenant and KILLIGREW put him out of business and nothing more is known of his activities. Beeston was an important link between the Elizabethan and the Restoration stages: his ability in training young actors, which is several times referred to in contemporary records, must have meant that some of the traditional business of the old actors was imparted to those who were to be the leading players of the new age.

BEGOVIČ, MILAN, see YUGOSLAVIA.

BEHAN, BRENDAN FRANCIS (1923–64), Irish dramatist, whose early experiences in prison as a member of the I.R.A. gave him the material for his first play *The Quare Fellow*. First seen in Gaelic, this was produced in Dublin in English in 1954; in London, by Joan LITTLEWOOD, in 1956; and in New York in 1958. Also in 1958 a second play, *The Hostage*, was seen in Dublin in English after an initial production in Gaelic as *An Giall* (1957). It reached London in 1959 and New York in 1961. A tragi-comic account of I.R.A. activities in a seedy brothel which ends with the shooting of an English soldier (the hostage of the title), the play almost by accident reflects in its construction, with interspersed songs and dances and direct addresses to the audience, the powerful influence of BRECHT. This was even more apparent in *Richard's Cork Leg*, left unfinished at Behan's death, adapted and enlarged by Alan Simpson for production at the ABBEY THEATRE's experimental Peacock Theatre in 1972, and seen in London at the THEATRE ROYAL, STRATFORD, a few months later. A short play, *The Big House*, originally written for television, was also adapted by Alan Simpson for the PIKE THEATRE, Dublin, in 1957, and in 1963 was given at the Stratford Theatre Royal as part of a festival of Irish comedy. A dramatization of Behan's first volume of autobiography, *Borstal Boy*, was staged in New York in 1970 and at the Abbey in 1972. This small body of work created a considerable stir. Behan took a Rabelaisian delight in the idiosyncracies of his eloquent characters, none of whom has the staying-power to be a fully-fledged villain and whose gently ironical and frankly irreverent view of life is an unashamedly sentimental plea for life at any price.

BEHN, Mrs APHRA (1640–89), playwright and novelist, the first Englishwoman to earn a living by her pen. Brought up in the West Indies, the scene of her novel *Oroonoko* (dramatized in 1695 by SOUTHERNE), she returned to England in 1658 and married a merchant of Dutch extraction. Soon widowed, she went to the Netherlands during the Dutch war as a spy and did good work for which she was apparently not paid, as she was shortly after imprisoned for debt. She was probably released through the intervention of KILLIGREW. Her first play, *The Forc'd Marriage; or, the Jealous Bridegroom* (1670), was a tragi-comedy produced at LINCOLN'S INN FIELDS THEATRE, with BETTERTON and his wife in the leading roles. It was, however, in comedies of intrigue that she did her best work, and her first outstanding success came with *The Rover; or, the Banish't Cavaliers* (1678). The play was often revived in a modified version, the part of Willmore the Rover being a favourite one with many leading actors. It was followed by several other comedies, of which *The Feign'd Curtizans; or, a Night's Intrigue* (1679) was dedicated to Nell GWYNN, while *The Roundheads; or, the Good Old Cause* (1681) and *The City-Heiress; or, Sir Timothy Treat-all* (1682) owed most of their success to their topicality. Her later plays were less successful, but *The Emperor of the Moon* (1687), a pantomime-farce based on a COMMEDIA DELL'ARTE scenario recently played in Paris, is historically interesting as the forerunner of the HARLEQUINADES which led to the development of the English PANTOMIME. Witty and high-spirited, Aphra Behn was the most prolific dramatist of her day except for DRYDEN, who was one of her many literary friends. Her plays, though coarse, are somewhat better than those of the average playwright of the day. Much of their indecency is due to the fashion of the time and to the necessity a woman was under of writing with masculine pungency if her work was to be given a hearing.

BEHRMAN, S(amuel) N(athaniel) (1893–1973), American dramatist, best remembered for the deft characterization and sparkling dialogue of his comedies, of which the first was *The Second Man* (1927). He had previously collaborated with Kenyon Nicholson in the trifles *Bedside Manner* (1923) and *Love Is Like That* (also 1927), and in 1929 he made an excellent dramatization of Enid BAGNOLD's *Serena Blandish*, which was seen in London in 1938. It was followed by *Meteor* (also 1929), a study of an egotist in the world of business; *Brief Moment* (1931), which dramatized the misalliance of a socially prominent young man; *Biography* (1932), which dealt with the contrasting temperaments of a carefree woman portrait painter and a crusading journalist; *Rain From Heaven* (1934), which made use of an English houseparty to establish the untenability of civilized detachment in a strife-torn world; and *End of Summer* (1936), showing the bankruptcy of the idle rich. He then made for Alfred LUNT and Lynn Fontanne an excellent adaptation of GIRAUDOUX's *Amphitryon 38* (1937) and in his next play, *No Time for Comedy* (1939), dealt with what was in a way his own dilemma, that of a writer who is anxious to come to grips with serious contemporary problems but has a talent only for light comedy. He then embarked on a series of adaptations, two of which, *The Pirate* (1942) from a play by Ludwig Fulda and *I Know My Love* (1949) based on Marcel ACHARD's *Auprès de ma blonde*, again provided excellent vehicles for the Lunts, while Franz WERFEL's *Jacobowsky and the Colonel* (1944) had successful runs both in London and New York. *Jane* (1952) was a straight comedy based on a story, *Miss Thompson*, by Somerset MAUGHAM, and *Fanny* (1945) a musical based on Pagnol's 'Marius' trilogy of films. Behrman's last plays were a comedy of character, *The Cold Wind and the Warm* (1958), based on his own autobiography, and *Blues for Mr Charlie* (1964), seen in the first repertory season of the VIVIAN BEAUMONT THEATRE at the Lincoln Center.

BEIL, JOHANN DAVID (1754–94), German actor who, after several years in small touring companies, went to Gotha to act under EKHOF in 1777, the same year as IFFLAND and BECK. Round-faced and jovial, he was excellent in comic parts, a good mimic and lively observer, but unfortunately he drank too much. He was at MANNHEIM under DALBERG after Ekhof's death, and appeared in SCHILLER's early plays and in the plays of Iffland, to whom he proved a loyal friend and colleague during the great days of the Mannheim stage. His early death was a great loss to Iffland, who had just been appointed director of plays at the theatre and in addition had to take on a number of Beil's roles.

BÉJART, ARMANDE-GRÉSINDE-CLAIRE-ÉLISABETH (1641–1700), French actress, youngest child of a family intimately associated with MOLIÈRE, whom she married in 1662. She was brought up by her eldest sister Madeleine (below), who had herself lost a girl-child at the time of Armande's birth, and contemporary gossip believed her to be Molière's child. His enemies therefore accused him of having married the daughter of his own old mistress and even, according to the actor MONTFLEURY, his own daughter. The calumny has never been expressly disproved, though it is no longer believed, and Louis XIV in effect refuted it by standing godfather to the first child of the union. The marriage was an unhappy one, and after Molière's death his young widow married another actor, Guérin d'Étriché, and was apparently happy with him. An excellent actress who owed all her training to her husband, she first appeared on the stage as Élise in his *Critique de l'École des femmes* and as herself in *L'Impromptu de Versailles* (both 1663). In the following year she played a gipsy woman in *Le Mariage forcé*, when she danced with Louis XIV, and had her first important part as the heroine in *La Princesse d'Élide*. After 1664 she played most of Molière's heroines, which were written expressly for her. She was an able and energetic woman, and contemporaries admired the firm way in which she kept the company together after her husband's death.

BÉJART, JOSEPH (1616–59), French actor, and like his younger brother Louis (below) a member of MOLIÈRE's company from its early days. In spite of a stammer he was a useful actor and a popular member of the company, and his sudden death only a year after Molière had established himself in Paris was a great blow. He was taken ill during a performance of Molière's comedy *L'Étourdi*.

BÉJART, LOUIS (1630–78), French actor, known as L'Éguisé, because of his sharp tongue. He was the younger brother of Madeleine (below) and like her spent the whole of his working life with MOLIÈRE. He was lame in the right leg, a trait Molière incorporated into the part of La Flêche in *L'Avare*, where Harpagon calls him 'lame dog', and the limp has remained traditional for the part. He also played Madame Pernelle in *Tartuffe*. He retired in 1670, being the first of Molière's actors to draw a pension.

BÉJART, MADELEINE (1618–72), French actress, eldest sister of a family intimately associated with MOLIÈRE. She was already an actress of some repute when he met her, and it is generally assumed that he became an actor under her influence. Together they formed the ILLUSTRE-THÉÂTRE, and after it failed joined a provincial troupe. Madeleine returned to Paris with Molière in 1658, shared in his success, and died a year to the day before him—17 Feb. 1672. In her early years she played the heroines of classical tragedy, and in later life created a number of Molière's witty maids who ridicule the follies of their mistresses. At one time she managed the company's finances. It is impossible to assess how much Molière owed to her constant affection and support over thirty years. Her sister, Geneviève (1624–75), also a member of Molière's company, called herself Mlle Hervé after her mother's maiden name; she was a better actress in tragedy than in comedy. For her brothers and youngest sister see above.

BELASCO [Valasco], DAVID (1859–1931), American actor, manager, and playwright, born in California of a Portuguese-Jewish family. His father had appeared as HARLEQUIN at a number of London theatres before emigrating to the U.S. and young Belasco was on the stage as a child, playing, among other parts, the Duke of York in *Richard III* during Charles KEAN's farewell tour. He later toured the Pacific coast and began dramatizing novels, poems, and stories, and adapting old plays, materially assisted by the lack of any copyright laws. As stage manager at various theatres he devised some spectacular effects in MELODRAMA—mainly battles and conflagrations—and at one time directed a Passion Play with real sheep. He made his first appearance in New York with James A. HERNE in *Hearts of Oak* (1879), which they had jointly adapted from an earlier melodrama, *The Mariner's Compass* (1865) by H. Leslie. It was not a success, and Belasco returned to California until in 1882 he was recalled to New York by Daniel FROHMAN to become stage manager of the MADISON SQUARE THEATRE where his first venture was Bronson HOWARD's *Young Mrs Winthrop*. Two years later he directed his own play *May Blossom* and also made his first visit to London. On his return he became stage manager for Steele MACKAYE at the LYCEUM THEATRE (1) where he remained until 1890, continuing to stage a number of his own plays, mostly written in collaboration. The most successful of these were *The Girl I Left Behind Me* (1893), written with Franklyn Fyles, and *The Heart of Maryland* (1895), in which

Maurice BARRYMORE gave a fine performance as the hero. *Zaza* (1899), based on a French play, starred Mrs Leslie CARTER whom Belasco had launched on a spectacular career some years before; it was a success both in New York and in London, as was *Madam Butterfly* (1900), based on a story by John Luther LONG, in which Blanche BATES, another of Belasco's discoveries, gave an excellent performance in the title role, played in London in the same year by Evelyn Millard. In 1901 Belasco began his association with the actor David WARFIELD, whom he took from the VARIETY stage to star in Charles KLEIN's *The Auctioneer*, and in the following year achieved a lifelong ambition by leasing the REPUBLIC THEATRE from Hammerstein and naming it after himself. Completely rebuilt, it opened with a revival of Belasco's *Du Barry* starring Mrs Leslie Carter, which had previously been seen at the CRITERION. It was followed by two more Belasco works, *The Darling of the Gods* (also 1902), another Japanese story written in collaboration with Long and starring Blanche Bates and George ARLISS, and *Sweet Kitty Bellairs* (1903). Warfield scored another triumph with Klein's *The Music Master* (1904) in which he toured for many years; other successes at this first Belasco theatre were *Adrea* (also 1904) with Mrs Leslie Carter; *The Girl of the Golden West* (1905) which, like *Madam Butterfly*, was used as the basis of an opera by Puccini; and Tully's *The Rose of the Rancho* (1906). Also in 1906 Belasco built a new theatre, which opened as the Stuyvesant with Warfield in *A Grand Army Man* (1907), becoming the BELASCO in 1910. Here Belasco remained until his death and his private rooms in the theatre, where *objets d'art* were jumbled up with theatrical properties, the whole presided over by the owner with his deceptively benign and clerical appearance, have passed into legend. At this theatre Belasco produced *The Return of Peter Grimm* (1911), with Warfield as the old man who returns after death to rectify the errors of his life; *The Governor's Lady* (1912), for which an exact replica of a Child's restaurant was shown on stage; *The Case of Becky* (also 1912), starring Francis Starr; *Kiki* (1921), with Lenore Ulric; *Laugh, Clown, Laugh* (1923), based on F. M. MARTINI's *Ridi, pagliaccio* (1919); and, his last production as well as his last play, *Mima* (1928), based on MOLNÁR's *The Red Mill* (1923). Thus he ended as he began, adapting the work of others. His long career spanned a transition period in American drama, during which he perpetuated the older tradition of importing material from Europe rather than encouraging the new American dramatists then beginning to emerge. His skill lay in constructing excellent vehicles for the stars of the day, and in his elaborate, scrupulously exact, scenic displays, in which he employed the latest mechanical inventions and made interesting experiments in the use of lighting. He was a complete man of the theatre of his time, even in his vanity and posturing, and a meticulous stage director; and his long and ultimately successful fight against the stranglehold of the THEATRICAL SYNDICATE involved the whole question of the independence of the American theatre. His large collection of theatrical material is now housed in the New York Public Library at Lincoln Center.

BELASCO THEATRE, New York. The first playhouse of this name, on West 42nd Street, was originally known as the REPUBLIC THEATRE. The second, at 111 West 44th Street, between Broadway and 6th Avenue, with seating for 1,008, opened as the Stuyvesant on 16 Oct. 1907 with BELASCO's *A Grand Army Man*. Renamed the Belasco, it reopened on 3 Sept. 1910 with *The Lily*, Belasco's own adaptation of Leroux's *Le Lys* (1908). Belasco ran the theatre, and lived on the premises, until his death in 1931, after which it was for a time leased by Katharine CORNELL, who produced there a translation of OBEY's *Le Viol de Lucrèce* (1931), as *Lucrece* (1932), and Sidney HOWARD's *Alien Corn* (1933). A year later two plays by Elmer RICE, *Judgement Day* and *Between Two Worlds*, were seen at this theatre, and from 1935 to 1941 it was the headquarters of the GROUP THEATRE, who produced there several plays by Clifford ODETS. *Trio* (1944) by Dorothy and Howard Baker, based on the former's novel, created a scandal because it dealt with lesbianism, and the theatre closed. It was used for broadcasting from 1949 to 1953, but in the latter year housed a successful run of Teichmann and KAUFMAN's *The Solid Gold Cadillac*. In 1960 there was another long run with Tad Mosel's *All the Way Home*, adapted from James Agee's novel *A Death in the Family*. The National Repertory Theatre, with Eva LE GALLIENNE, played a season at the Belasco in 1964 and two years later came Frank Marcus's *The Killing of Sister George*. The nude revue *Oh, Calcutta!* moved here in 1971, and in 1975 the interior was converted to cabaret-style seating for *The Rocky Horror Show*, later reverting to conventional seating. *Ain't Misbehavin'* transferred from the PLYMOUTH THEATRE in 1981. The Belasco is now owned by the Shubert organization.

BELCARI, FEO, see ITALY.

BELFAST CIVIC ARTS THEATRE, founded as the Mask Theatre in 1944 with aims very similar to those of the GATE THEATRE in Dublin. It opened in a converted loft with Charles MORGAN's *The Flashing Stream* and three years later, with many successful productions of European classics behind it, was renamed the Belfast Arts Theatre. Two important American plays had their European premières here, KINGSLEY's *Darkness at Noon* and VAN DRUTEN's *I am a Camera*, and ANOUILH's *Waltz of the Toreadors* was first produced in English by the Belfast Arts. In 1961 the company moved to a new playhouse in Botanic Avenue, which opened on 19 Apr. with Tennessee WILLIAMS's *Orpheus Descending*. Owing to civil

disturbances the theatre was forced to close in 1971, the company's last production being *The Mating Season* by Sam Cree. It reopened in 1976 with the same play presented by the Ulster Actors Company, formed a year earlier by Roy Heayberd. The enterprise has achieved striking success in spite of continuing tensions within the community, gaining a large and enthusiastic following for children's shows, rock musicals, revivals of plays from London and New York, and productions of work by Irish authors including Brendan BEHAN, Brian FRIEL, Hugh LEONARD, Joseph TOMELTY, and Paul Vincent CARROLL. The company is aided by grants from the Belfast City Council and the Arts Council of Northern Ireland. The Arts Theatre's comfortable auditorium seats 500, and there are good front-of-house and backstage facilities.
(See also LYRIC PLAYERS, Belfast, and ULSTER GROUP THEATRE.)

BEL GEDDES, NORMAN (1893–1958), American scene designer, who as early as 1915 had the idea of a theatre without a proscenium, and in 1923 won instant recognition with his magnificent designs for REINHARDT's American production of Volmöller's *The Miracle*. In 1931 he designed a complex of steps and rostrums for his own production of *Hamlet* which was far in advance of anything so far seen; another of his successful experiments was a COMPOSITE SETTING for his own production of KINGSLEY's *Dead End* (1935). His plan for a monumental production of Dante's *Divine Comedy* at Madison Square Garden, for which he designed an immense circular stage, was unfortunately never carried out, nor was his scheme for a THEATRE-IN-THE-ROUND in 1930. Nevertheless his ideas had a great influence on the development of modern American scene design. His activity in the theatre was at its height during the 1930s, after which he concentrated mainly on industrial design. His daughter Barbara (1922–   ) made her stage début in 1940 and five years later was a great success in Arnaud d'Usseau and James Gow's *Deep Are the Roots*; later roles included Margaret in Tennessee WILLIAMS's *Cat On a Hot Tin Roof* (1955) and the title role in Jean Kerr's *Mary, Mary* (1961). She is probably best known in the role of Ellie Ewing in the long-running television series 'Dallas'.

BELGIUM, which occupies roughly the southern half of the territory formerly known as Flanders or the Low Countries, shares its early theatre history with the NETHERLANDS and still has a bi-lingual drama, French and Flemish. That French was a language of importance even in early times is attested by the Mons PASSION PLAY of 1501; the director's notes and list of expenses for this production, discovered and published by Gustave COHEN in 1925, are a valuable source of information on the staging of LITURGICAL DRAMA. It is evident that at least five of the actors were rhetoricians, or members of the local CHAMBER OF RHETORIC, which was more closely associated with similar groups in Burgundy than were those further north. There is ample evidence also of an independent development of plays in French, important in the light of future trends. The small number of such plays which have survived compared to those in Flemish may be due to chance, but more probably to the fact that from an economic and cultural point of view the southern half of the area was less influential at this time than the northern, where the main theatrical development was in Flemish. During the Renaissance, particularly after the reconquest of the area by Spain, the theatre in the south became even more dependent on France and produced little original work. This continued throughout the 16th and 17th centuries, and it was not until towards the end of the 18th century that French-language companies were established in Brussels and Antwerp. Meanwhile the Flemish theatre underwent a period of trial which seriously hindered its development; and it was as a result of the so-called 'Flemish movement' of the mid-19th century that professional Flemish theatres were first founded in Belgium—in Antwerp in 1853, in Ghent in 1871, and in Brussels in 1887.

The French-speaking theatre remained very much under the influence of French dramatists, and by the end of the 19th century NATURALISM had made a noticeable impact on Belgian audiences through the regular performances of ANTOINE'S THÉÂTRE LIBRE in Brussels, and performances by the MEININGER COMPANY in 1888 also had some influence. Though most Belgian dramatists writing in French still had their plays first produced in Paris, the Belgian theatre achieved international recognition with the works of Maurice MAETERLINCK whose first plays, *La Princesse Maleine* (1890), *L'Intruse*, and *Les Aveugles* (both 1891), marked the triumph of SYMBOLISM over naturalism. Maeterlinck's contemporary, the symbolist poet Emile Verhaeren (1855–1916) whose plays are less well known abroad, is nevertheless still remembered for *Le Cloître* (1900), a strong drama of religious passions set in a monastery, and for the fact that *Les Aubes* (1898) was the play with which MEYERHOLD opened his first post-revolutionary theatre in Moscow in 1920, after adapting it and placing it in a contemporary setting. At the same time Belgian drama was once again in the limelight with *Le Cocu magnifique* (1920), an intense study of marital jealousy by Fernand CROMMELYNCK. Like Maeterlinck he was Flemish, and though he wrote in French his work shows much affinity with the basic spirit of Flanders, as does that of the earlier and less well known Charles Van Lerberghe and Georges Rodenbach.

A Belgian author writing in French whose works only became well known after the Second World War was Michel de GHELDERODE, whose work as a whole again shows the influence of his Flemish background. Flemish once again became a force to reckon with in the Belgian theatre under the stimulating influence of Dr Jan Oscar de Gruyter, who was finally responsible for the founding in Antwerp after the First World War of the Vlaamse Volkstoneel, or Flemish Folk Theatre, which won international acclaim during the 1920s, particularly for its work in the field of EXPRESSIONISM under the Dutch director Johan de Meester. The most important of the dramatists connected with it was probably Herman Teirlinck (1879–1967) who in *The Slow-Motion Film* (1922) wrote what is usually considered to be the first Flemish expressionist play. His version of a medieval MIRACLE PLAY, *I Serve* (1923), and his *Man Without a Body* (1925) were also well received, and he continued his search for new forms of dramatic technique after the Second World War with *De Oresteia* (1946) and *Taco* (1958). But his main importance, apart from his novels, was as a brilliant theatre theoretician and teacher of drama. He introduced into the Flemish theatre the ideas of APPIA and Gordon CRAIG, and was active in the foundation of a Flemish National Theatre and of a school of acting in Antwerp which now bears his name. Among other Flemish dramatists, whose work belongs more to the popular tradition of local comedy, the most important was Gaston Martens (1883–1967), many of whose plays of village life portray the Flemish peasant as Breughel saw him.

Among Belgian dramatists writing in French the only one to achieve recognition outside Belgium since the Second World War has been Félicien MARCEAU, and that only for a short time. Hugo CLAUS, writing in Dutch, has also seen his plays incorporated into the repertory of the theatre in the Netherlands.

Brussels has always been the chief centre of theatrical activity in Belgium, and some of its theatres date back to earlier times, the Théâtre de la Monnaie, for instance, having been founded at the beginning of the 18th century. The Théâtre du Parc, which played an important part in the development of Belgian symbolist drama, first opened in 1782, the Théâtre des Galeries in 1846, the Théâtre Molière in 1867. In the 20th century came the foundation in 1945 of the Belgian National Theatre by the Huysman brothers, who before and after the Second World War did excellent work up and down the country with a travelling company known as the Comédiens Routiers, modelled on that of COPEAU. Since its inception the National has shown Belgian-French theatre at its best, in choice of play, production, and acting, and frequently invites guest directors from abroad, among those from England being Michael Langham, Denis Carey, and André van Gyseghem. Important also for the development of the Belgian theatre was the creation in 1950 of the Théâtre de Poche, where many foreign dramatists, including ARRABAL, have been introduced to Belgian audiences. In the 1960s the LIVING THEATRE visited Brussels, and in the 1970s the city saw the foundation of many small experimental theatres of which only the Laboratoire Vicinal, with its policy of COLLECTIVE CREATION, appears to have survived.

BELGRADE THEATRE, Coventry, erected in 1958, received its name in recognition of a gift of timber used in its construction from the City of Belgrade in Yugoslavia. It was the first new theatre to be built in England since the OXFORD PLAYHOUSE in 1938 and also the first CIVIC THEATRE, reflecting in its conception and design the changed role of the theatre within the community as envisaged in the aftermath of the Second World War. Built by the local authority and run by a Trust appointed by them, it receives subsidies from local funds and from the ARTS COUNCIL, and is slowly repaying the original cost of its construction. The members of the company live in flats which form part of the theatre complex; the excellent foyer, with its bookstalls, is large enough for displays of works by local artists and other exhibitions; and its coffee bar is used for lunch-time concerts. There is also a restaurant. The theatre, whose first manager was Brian Bailey, opened on 27 Mar. 1958 with *Half in Earnest*, a musical version of WILDE's *The Importance of Being Earnest*. Its seating capacity is 900 in stalls and circle, and the stage is well equipped for all purposes. Among the new plays presented on it have been several by Arnold WESKER, *Semi-Detached* (1963), by David Turner, all seen later in London, and *Diary of a Desperate Woman* (1979), by Andrew Davies, which, as *Rose*, provided an excellent part for Glenda JACKSON during its London run. As well as encouraging new playwrights, the Belgrade has made serious efforts to attract a young audience through its Youth Theatre, which presents about six plays a year, and also runs a theatre-in-education scheme. Under Warren Jenkins, who took over after the death of Bailey in 1960, the repertory was extended so as to include plays to suit all tastes, works as different as the musical comedy *No, No, Nanette!* and BOND's *Narrow Road to the Deep North* being presented in the same season. This policy drew good audiences, but perhaps prevented the company from developing a distinctive style. Under the present directors, appointed in 1974, there has been a change from two- to three-weekly rep. A studio theatre, the Belgrade Venue, attached to the main theatre, stages experimental and rarely-performed plays, and the Belgrade company is also responsible for annual performances of the Coventry MYSTERY PLAYS in the ruins of the bombed cathedral.

BELGRAVIA THEATRE, London, see ROYAL COURT THEATRE (1).

BELIGAN, RADU, see ROMANIA.

BELL, JOHN (1745–1831), pioneer English publisher and bookseller, who was responsible for *Bell's British Theatre*, a comprehensive selection of plays, each prefaced by an interesting character portrait. His 1773 acting edition of Shakespeare, edited by Francis GENTLEMAN, was based on the prompt books of the Theatres Royal, and is interesting as showing what was actually performed on the stage at the time.

BELL, JOHN JOY (1871–1934), Scottish dramatist, a pioneer writer of plays in the Scottish vernacular. Most of his works were one-act comedies, the most popular of which were *Courtin' Christina*, *The Pie in the Oven*, and *Wee MacGregor's Party*. This last, based on his own sketches of Glasgow working-class life, was produced by the GLASGOW REPERTORY THEATRE. His best known play, *Thread o'Scarlet* (1923), is a tense horror-drama akin to those that Edgar WALLACE and others were providing for the GRAND GUIGNOL programmes of the immediate post-war era. A great favourite with amateur societies, it has been many times revived.

BELL, MARIE [Marie-Jeanne Belon] (1900–), French actress and theatre manager, who appeared from 1921 at the COMÉDIE-FRANÇAISE, notably in a number of tragic roles including RACINE's Phèdre, Bérénice, and Agrippine (in *Britannicus*). She was also seen at other theatres in Paris in modern plays including BARRAULT's production of CLAUDEL's *Le Soulier de satin* (1943). From 1959 she managed the Théâtre du GYMNASE, both acting and directing there, being seen in the first productions of Félicien MARCEAU's *La Bonne Soupe* (1959) and GENET's *Le Balcon* (1960), and in 1973 in a version of William Douglas HOME's *Lloyd George Knew My Father* as *Ne couper pas mes arbres*.

BELLAMY, GEORGE ANNE (*c.* 1727–88), English actress, who received her first names as a result of a mishearing of Georgiana at her christening. She was the child of Lord Tyrawley by a Miss Seal, who eloped with him from boarding-school. He acknowledged her and had her educated, but cast her off when she defied him by going to live with her mother, who had married a sea-captain named Bellamy. Introduced to Christopher RICH, and accepted by him for COVENT GARDEN, she is believed to have made her début there in 1742 as Miss Prue in CONGREVE's *Love for Love* and certainly played opposite QUIN in OTWAY's *The Orphan* two years later. She was at her best in romantic and tragic parts, being an admirable Juliet to the Romeo of GARRICK who made her one of his leading ladies at DRURY LANE, but much of her early success was due to her youth and beauty. Because of the many scandals attached to her name—she was twice married, once bigamously—managers grew chary of engaging her, and an appearance in Dublin in 1780 was a complete failure. Arrogant and extravagant, she alienated many of her friends but continued to act intermittently until her retirement in 1785, when a benefit was given for her at Covent Garden. A 6-volume *Apology* for her life (1785) is clearly by another hand, and is readable rather than reliable.

BELLECOUR [Jean-Claude-Gilles Colson] (1725–78), French actor, husband of Mlle BEAUMÉNARD. After studying for some years as a painter he decided to go on the stage and served his apprenticeship in the provinces. At Bordeaux he attracted the attention of RICHELIEU, who in 1750 encouraged him to try for a place at the COMÉDIE-FRANÇAISE in opposition to VOLTAIRE's protégé LEKAIN. He was accepted, his success being chiefly due to his fine stage presence, for Lekain was undoubtedly the better actor. When Lekain was finally admitted to the company, Bellecour was glad to hand over to him all the tragic roles, reserving for himself the fine gentlemen of comedy, in which he excelled.

BELLEROCHE [Raymond Poisson] (*c.* 1630–90), French actor, who in about 1650 joined a provincial company, and a few years later was at the Hôtel de BOURGOGNE with his wife, who took over the parts formerly played by the wife of BELLEROSE. A big man with an expressive face and a large humorous mouth, Belleroche was an excellent comic actor in spite of a slight stutter, which was inherited by his son and grandson, Paul and François POISSON. Raymond took from SCARRON's *L'Écolier de Salamanque* (1654) the character of the valet CRISPIN and made it peculiarly his own, introducing it into several of his own comedies and playing the part himself. His numerous plays are now forgotten, except for *Le Baron de la Crasse* (1663), which remains of interest because it shows a strolling company performing before some local gentry; the leader may be intended as a satirical portrait of MOLIÈRE.

BELLEROSE [Pierre le Messier] (*c.* 1592–1670), French actor, who was in VALLERAN-LECOMTE's company in 1609, played with him at the Hôtel de BOURGOGNE in 1610, and probably went with him to Holland in 1612. After the death of Valleran-Lecomte he became the leader of the company, and in 1622 joined forces with the troupe then occupying the Hôtel de Bourgogne, becoming head of the united company in 1634 to be succeeded by FLORIDOR in *c.*1646, though he continued to act. He was much admired in both comedy and tragedy, though some critics found him insipid and sentimental. Unlike his rival MONTDORY, at the MARAIS, he was not a ranter, his style being quiet and rhetorical rather than declamatory. The name part in CORNEILLE's *Le Menteur*, with which he became closely associated though he was not the first to play it, was probably well suited to him. He figures as himself in Gougenot's *La Comédie des comédiens* (1652), and created a number of leading roles in pastoral plays and tragi-comedies. In 1638 he married an actress, Nicole Gassot, who died in 1680.

BELLEVILLE, see TURLUPIN.

BELLIDO CORMENZANIA, JOSÉ MARÍA, see SPAIN.

BELL INN, see INNS USED AS THEATRES.

BELLOY, PIERRE LAURENT BUIRETTE DE (1727–75), French dramatist, known by one play only, *Le Siège de Calais* (1765). It owed its great and unexpected success partly to the acting of LEKAIN, MOLÉ, and Mlle CLAIRON, but above all to its topicality. France had just signed a humiliating peace treaty, and the audience were consoled by seeing on stage defeated Frenchmen forcing the admiration of their conquerors by their moral virtues. The play, in which De Belloy followed VOLTAIRE's example by choosing a subject from French national history, was also very much of its time in that it put forward as its heroes bourgeois characters who contrasted favourably with those of higher social rank. After a remarkable reception at the COMÉDIE-FRANÇAISE and at Court it was often put on in provincial theatres, especially in garrison towns, as a means of stimulating the audience's patriotism. In itself it is of little value, and shows only too clearly the continued decline of French neo-classical tragedy. De Belloy's later plays had little success and are now forgotten.

BELLWOOD, BESSIE, see MUSIC-HALL.

BELMONT THEATRE, New York, at 125 West 48th Street. Originally named after its first owner and star, the comedian Jack NORWORTH, it opened on 18 Jan. 1918 and changed its name three months later. It did not achieve a success until it housed Zona Gale's PULITZER PRIZE-winner *Miss Lulu Bett* (1920), which was followed by Gilbert Emery's *The Hero* (1921) and by Philip BARRY's first play *You and I* (1923), which ran for a year. The Belmont's last successful production was VAN DRUTEN's *Young Woodley* in 1925. It became a cinema in 1936 and after a period of disuse was finally demolished in 1951.

BELOW, see STAGE DIRECTIONS.

BEL SAVAGE INN, see INNS USED AS THEATRES.

BELTRAME, see BARBIERI, NICCOLÒ.

BENAVENTE Y MARTÍNEZ, JACINTO (1866–1954), Spanish dramatist, author of over 150 plays, who was awarded the NOBEL PRIZE for Literature in 1922. His plays mark the transition from the comedies of intrigue made popular by ECHEGARAY to the drama of social criticism implicit in such works as *Gente conocida* (*Important People*, 1896), *La noche del sábato* (*Witches' Sabbath*, 1903), *Los malhechores del bien* (*The Evil Doers of Good*, 1905), a plea for tolerance which attacks misguided charity, and *Más fuerte que el amor* (*Stronger than Love*, 1906). The best known of his works outside Spain were *Los intereses creados* (1907) which, as *The Bonds of Interest*, was the first play presented in New York by the THEATRE GUILD in 1919, being seen a year later in London; and *La malquerida* (*The Passion Flower*, 1913), seen in New York in 1920 and in London in 1926. The first is unusual among Benavente's plays in making use of the traditional masks of the COMMEDIA DELL'ARTE to depict modern society as a puppet show in which men are moved by the strings of passion, selfishness, and ambition, worked by the Machiavellian hero Crispin, nominally a lackey but in reality a puppetmaster. The second is a rustic tragedy in which a peasant girl succumbs to evil. Benavente, who had a wide knowledge of European drama, translated several plays by MOLIÈRE and Shakespeare, and in 1909 opened a theatre for children with his own play, *El príncipe que todo lo aprendió en los libros* (*The Prince who learned Everything from Books*). Although his reputation has declined since his death, he was in the 1920s highly thought of in England, where four volumes of his plays, in a translation by J. G. Underhill, appeared between 1917 and 1924.

BENEFIT, a special performance common in the English theatre in the 18th and early 19th centuries, of which the financial proceeds, after deduction of expenses, were given to a member of the company, who was allowed to choose the play for the evening, sometimes one written by himself. In the days of the SHARING SYSTEM and the STOCK COMPANY an actor might rely almost entirely on his benefit night to provide him with ready money, his weekly 'share' barely paying current expenses. In larger companies towards the end of the 18th century a good benefit might bring in as much as £200–£300 for a popular leading actor. Benefits could also be given for an actor who was ill and in need of money, or for his widow and other needy dependants. Before the introduction of the ROYALTY system, a benefit performance was sometimes given for an author.

The first benefit performance in England was probably that given in about 1686 for Mrs BARRY, but it did not become common practice until much later. It was an unsatisfactory arrangement, which exposed the actor to petty humiliations and kept him in a constant state of financial uncertainty. It gradually died out between 1840 and 1870, lingering on in the provinces long after it had been abandoned in London, though MACREADY had a London benefit as late as 1848. There is an interesting account of the way it worked in DICKENS's

Nicholas Nickleby (1838), and the abuses of the system in the early American theatre can be studied in Odell's *Annals of the New York Stage*. A slightly more dignified method of making money was the Bespeak Performance, whereby a wealthy patron or group of friends would buy most of the tickets for one evening and sell or give them away, choosing the play to be performed from the company's current repertory. In that case the proceeds were divided among all the members of the cast.

BENELLI, SEM (1875–1949), Italian dramatist, considered in some ways a successor to D'ANNUNZIO, though with less poetic talent. His first important play, *Tignola* (*The Bookworm*, 1908), had much in common with the works of the CREPUSCOLARI, but the most successful of his later plays, all in blank verse, were historical—*La maschera di Bruto* (*The Mask of Brutus*, also 1908), dealing with the Medici; *La cena delle beffe* (1909), a Renaissance melodrama which, as *The Jest*, was seen in New York in 1919 with John and Lionel BARRYMORE; and *L'amore dei tre rei* (*The Love of the Three Kings*, 1910), a medieval tragedy. Although *Orfeo e Proserpina* (1928) is usually regarded as his last major work, he continued to write until shortly before his death.

BENFIELD, ROBERT (?–1649), actor who appears in the list of those who acted in the plays of Shakespeare. He must already have been experienced when in 1615 he joined the KING'S MEN, probably as a replacement for William OSTLER who had died the previous year, as he played Ostler's part of Antonio in a revival of WEBSTER's *The Duchess of Malfi* in 1619. He is known to have played in BEAUMONT and FLETCHER's plays, and was also seen as Junius Rusticus in *The Roman Actor* (1626) and as Ladislaus King of Hungary in *The Picture* (1629), both by MASSINGER. He seems from the first to have specialized in dignified elderly parts such as kings, counsellors, and noble old men. Although not an original SHARER in the GLOBE or BLACKFRIARS, he later acquired an interest in both.

BENGER, Sir THOMAS, see MASTER OF THE REVELS.

BENNETT, ALAN (1934–), English actor and dramatist, who appeared in 1960 at the EDINBURGH FESTIVAL in the epoch-making REVUE *Beyond the Fringe*, of which he was part-author. It subsequently had a long run in London and in New York. Bennett's first play, *Forty Years On* (1968), also had a successful run, with John GIELGUD (followed by Emlyn WILLIAMS) as the retiring headmaster of a minor public school in whose honour the boys enact a savagely satirical pageant of recent British history. It was followed by *Getting On* (1971), which starred Kenneth More as a disillusioned middle-aged Labour M.P. longing for earlier certainties; a farce, *Habeas Corpus* (1973; N.Y., 1975), with Alec GUINNESS in London and Donald SINDEN in New York—the author himself took over one of the female roles during the play's London run; and *The Old Country* (1977), again with Guinness as a British traitor in exile in Soviet Russia. *Enjoy* (1980) had a disappointingly short run, despite fine performances by Joan PLOWRIGHT and Colin BLAKELY as a working-class married couple in old age.

BENNETT, (Enoch) ARNOLD (1867–1931), English novelist and dramatist, whose plays enjoyed some success in their own day, their plots being ingeniously constructed and the character-drawing having great vitality mellowed by a homely humour. They are now forgotten, except for *Milestones* (1912), written in collaboration with Edward KNOBLOCK, and *The Great Adventure* (1913), based on his own successful novel *Buried Alive* (1908), which owed much of its appeal in the theatre to the acting of Henry AINLEY. Both plays have been several times revived. Bennett's later plays were less successful, to his great disappointment, the best of his post-war works being probably *London Life* (1924), again with Ainley, and *Mr Prohack* (1927), in which Charles LAUGHTON played the title role. In both Bennett again collaborated with Knoblock.

BENNETT, JILL (1931–), English actress, who made her first appearances at Stratford-upon-Avon in 1949, and her London début in Denis Cannan's *Captain Carvallo* (1950). Her first outstanding success was as Isabelle in ANOUILH's *Dinner With the Family* (1957), in which she was perfectly cast as the young girl who comes to dine with her lover's family and meets only actors hired for the occasion. She next appeared in a double bill, *Last Day in Dreamland* and *A Glimpse of the Sea* (1959), by her first husband Willis HALL. Although at her best in roles requiring elegance and languorous wit, she proved her versatility by giving a good account of the disciplined spinster daughter in Donald Howarth's *A Lily in Little India* (1965). She married John OSBORNE in 1968 and was seen in several of his plays including *Time Present* (1968), in which she gave an excellent performance as the bitchy, eloquent, and disdainful actress, Pamela, and *West of Suez* (1968). In 1972 she gave perhaps her best performance, as Hedda Gabler in Osborne's version of IBSEN's play, and in 1976 she played Sally Prosser in Osborne's *Watch It Come Down* for the NATIONAL THEATRE. In 1977 she appeared in a revival of RATTIGAN's double bill *Separate Tables*, and in 1980 played Gertrude in the ROYAL COURT production of *Hamlet*.

BENNETT, RICHARD (1872–1944), American actor, whose handsome appearance and magnificent speaking voice made him a matinée idol, though he also had great acting talent.

Unpredictable both on and off the stage, he would sometimes interrupt a play to criticize the audience. His début in Chicago in 1891 as Tombstone Jake in Elmer E. Vance's *The Limited Mail* was followed by his first New York appearance in the same role later in the year. Other roles were Father Anselm in Robert Marshall's *A Royal Family* in 1900, Hector Malone in SHAW's *Man and Superman* in 1905, Jefferson Ryder in Charles KLEIN's *The Lion and the Mouse* (also 1905) in which he made his London début in 1906, and John Shand in BARRIE's *What Every Woman Knows* (1908). He played George Dupont in *Damaged Goods* (1913), an adaptation of BRIEUX's notorious study of venereal disease, and important later parts included Robert Mayo in O'NEILL's *Beyond the Horizon* (1920), He in ANDREYEV's *He Who Gets Slapped* (1922), and Tony in Sidney HOWARD's *They Knew What They Wanted* (1924). One of his most famous roles came at the end of his career—that of Judge Gaunt in Maxwell ANDERSON's *Winterset* (1935). He was the father of the film stars Joan, Constance, and Barbara Bennett.

BENOIS, ALEXANDRE, see COSTUME and SCENERY.

BENSERADE, ISAAC DE (1613–91), French poet and dramatist, related to Cardinal RICHELIEU, who gave him a pension. He came of a good Norman family, and was intended for the Church, but a passion for the wife of the actor BELLEROSE, contracted while he was studying theology in Paris, turned his thoughts to the theatre. In 1635 he produced his first play, a comedy of little account, followed in the next year by a tragedy, *Cléopâtre*, the best of his not very considerable output. On the death of Richelieu he gave up writing plays, and devoted himself to the composition of libretti for COMÉDIES-BALLETS, then much in favour at Court. He excelled at this, collaborating with MOLIÈRE in *Les Fétes de l'Amour et de Bacchus* (1672).

BENSON, FRANK ROBERT (1858–1939), English actor-manager, knighted by George V in 1916 during the Shakespeare Tercentenary celebrations at DRURY LANE—the only actor ever to receive the accolade in a theatre—with a sword borrowed from a military outfitters' nearby. While at OXFORD Benson had been a prominent member of the OUDS, playing Clytemnestra in his own production of AESCHYLUS' *Agamemnon* in Greek. He made his first appearance on the professional stage in 1882, playing Paris in IRVING's production of *Romeo and Juliet* at the LYCEUM. A year later he formed his own company and took it on tour, directing it himself in due course in all Shakespeare's plays except *Titus Andronicus* and *Troilus and Cressida*. He visited STRATFORD-UPON-AVON annually from 1886 to 1916, leading his company in a repertory of seven or eight plays during the summer months there, and making theatrical history in 1899 by giving festival audiences the uncut version of *Hamlet*. At other times he travelled all over the provinces, thus keeping Shakespeare's plays always before the public and providing a good training school for a number of young players, who were always proud of being 'Old Bensonians'. His success as a manager somewhat obscured his personal performances, but he was a capable actor, and in some parts excellent, having a noble appearance, with handsome aquiline features and much of the air of an 'antique Roman'. He married in 1886 a member of his company, Constance Featherstonhaugh (1860–1946), who continued to play leading roles with him for many years.

BENTLEY, ERIC RUSSELL (1916–   ), British-born critic, director, and playwright, best known for his promotion of the work of BRECHT in the English-speaking world; having begun to translate Brecht's plays in the 1940s he collaborated with the author on a production of *Mutter Courage und ihre Kinder* in Munich in 1950. His work as a director also includes *Him* by e.e. cummings, with Kenneth TYNAN in the lead, in Salzburg and GARCÍA LORCA's *The House of Bernarda Alba* at the ABBEY THEATRE, Dublin, (both 1950). In the same year he co-directed O'NEILL's *The Iceman Cometh*, in a German translation, in Zurich. He was drama critic of the *New Republic* from 1952 to 1956 and was appointed Brander Matthews Professor of Dramatic Literature at Columbia University in 1953. Since 1977 he has also been Katharine Cornell Professor of Theatre at the State University of New York (Buffalo). His own plays include *Are You Now Or Have You Ever Been . . .?* (1972), *The Recantation of Galileo Galilei* (1973), and *Expletive Deleted* (1974); he has adapted and translated plays by PIRANDELLO and SCHNITZLER as well as Brecht and written a number of books on the theatre.

BENZON, OTTO, see DENMARK.

BEOLCO, ANGELO (*c.* 1502–42), actor and playwright connected with the origins of the COMMEDIA DELL'ARTE, though not its originator, as was said by RICCOBONI who probably knew little of his work. A gifted amateur, he appears to have played in Venice, Ferrara, and Padua during Carnival from 1520 onwards under the name of Il Ruzzante (the gossip), a shrewd, talkative, and rebellious young peasant. His plays, which are fully written out in the dialect of his birthplace Padua, show a close and sympathetic observation of country life and provide plenty of opportunities for pantomime and comic turns—BURLE and LAZZI. They are highly thought of today and a number of them have been performed on the Italian stage, notably *La Moschetta* (*The Coquette*, 1528) at Turin in 1960. They were translated into French in 1925–6 by Beolco's biographer Mortier, and an English version of *Il Reduce* (*c.* 1528), as *Ruzzante Returns from the Wars*, is included in Vol. I of Eric

BENTLEY's *The Classic Theatre* (1958). An Italian edition of the plays, edited by Zorzi, was published in 1967.

BÉRAIN, JEAN (1637–1711), French theatrical designer, who replaced VIGARANI at the SALLE DES MACHINES and also as scenic designer to the Paris Opéra. In 1674 he succeeded Gissey as designer to the King, and his costumes and decorations for Court spectacles had a great influence on all forms of contemporary art all over Europe. Their most striking characteristic is the complete synthesis of fantasy and contemporary taste. There was no attempt at realism or archaeological reconstruction, and even in costumes for Romans, Turks, or mythological characters, the exotic elements were absorbed with an intensely personal style. Many of Bérain's designs were preserved, sometimes in copies or tracings, by the pious hand of his son Jean (1678–1726), who succeeded him in his official functions, but, though industrious and inventive, lacked his originality and skill.

BÉRARD, CHRISTIAN (1902–49), French artist and theatre designer, who had a great influence on European stage design in his time, his work being characterized by a wonderful feeling for the visual aspects of the theatre and great skill in the use of colour. After designing ballet sets and costumes for most of the great choreographers of the day, he began in 1934 a fruitful collaboration with JOUVET with the décor for COCTEAU's *La Machine infernale*, continuing with such important and diverse new productions as GIRAUDOUX's *La Folle de Chaillot* (1945) and GENET's *Les Bonnes* (1947), and classics such as BEAUMARCHAIS's *Le Mariage de Figaro* in 1939 and MOLIÈRE's *Don Juan* in 1947. Bérard also designed a number of productions for other theatres, including the COMÉDIE-FRANÇAISE, and for BARRAULT's company at the MARIGNY he designed two revivals of Molière—*Amphitryon* in 1946 and *Les Fourberies de Scapin* in 1949. He died while supervising the lighting of this last play on the night before its production.

BERGBOM, KAARLO, see FINLAND.

BERGELSON, DAVID, see MOSCOW STATE JEWISH THEATRE.

BERGERAC, CYRANO DE, see CYRANO DE BERGERAC.

BERGMAN, HJALMAR FREDERIK (1883–1931), Swedish dramatist and novelist, one of the most influential in the Swedish theatre after the death of STRINDBERG, who, with MAETERLINCK and IBSEN, had a great influence on his early work. Though he began his career as a playwright in 1905, he first came into prominence with two one-act 'Marionette Plays' produced in 1917—*Dödens Arlekin* and the exquisite psychological tragedy *Herr Sleeman kommen*. These were followed by *Ett Experiment* (1918), and by three plays published in 1923—*Vavaren i Bagdad* (*The Weaver of Bagdad*), which he wrote after completing a translation of Sir Richard Burton's *One Thousand and One Nights; Spelhuset* (*The Playhouse*); and *Porten* (*The Door*). He also adapted two of his own regional novels as *Hans Nåds Testamente* and *Markurells i Wadköping* (both 1930). Of his later plays, in which comedy replaced the tragic mood of his earlier work, the best known are *Swedenhjelms* (1925), a realistic comedy about the eccentric family of a Nobel Prize-winner, which was seen at the BIRMINGHAM REPERTORY THEATRE in 1960 as *The Family First*, and *Patrasket* (1928), a Jewish folk comedy.

BERGMAN, INGMAR (1918–), Swedish director, who at the age of 20 was directing with an amateur company. In 1942 he wrote a play, *Kaspers död* (*Kasper's Death*), which he also directed. He was head of the municipal theatre in Hälsingborg from 1944 to 1946 and of Gothenburg's city theatre from 1946 to 1949, where his opening production was CAMUS's *Caligula*, with Anders EK. From this time he worked increasingly in the cinema, developing a distinguished international reputation as a film director; but he concurrently held appointments at the Malmö city theatre from 1952 to 1960 and at the KUNGLIGA DRAMATISKA TEATERN, Stockholm, from 1963 to 1966, where he introduced a number of reforms and attracted to the company some of Sweden's most distinguished players. His intense, often morbid, concentration on psychological interpretation in the tradition of STRINDBERG has been brought to bear on authors ranging from CHEKHOV via PIRANDELLO and BRECHT to ALBEE. A number of his productions, including IBSEN's *The Wild Duck* and Strindberg's *The Dream Play*, were seen in London as part of the WORLD THEATRE SEASON during the 1970s. In 1976, after being wrongly charged with tax irregularities, he left Sweden for West Germany, and from 1977 worked at the Residenztheater in MUNICH.

BERGNER, ELISABETH (1900–), Austrian-born actress, who appeared in a number of German cities before going to BERLIN in the 1920s. There, under Max REINHARDT, she had her first resounding success as SHAW's St Joan in 1924. Her boyish figure and vivacious temperament were admirably suited to the part, and also made her an excellent interpreter of Shakespearian heroines such as Rosalind, Viola, Portia, and Juliet. She was also good in modern roles, among them STRINDBERG's Miss Julie and Queen Christina, Nina in O'NEILL's *Strange Interlude*, and Alcmene in GIRAUDOUX's *Amphitryon 38*. In 1933 she moved to London, where she made her début in an English-speaking role as Gemma Jones in Margaret Kennedy's *Escape Me Never* (1933). She also played the title role in BARRIE's *The Boy David* (1936) and was seen again as St Joan at the

MALVERN FESTIVAL in 1938. She played in New York during the 1940s, her roles including the Duchess in WEBSTER's *The Duchess of Malfi* in 1946. She returned to London in 1951 in *The Gay Invalid*, adapted from MOLIÈRE, and toured Germany and Austria in RATTIGAN's *The Deep Blue Sea* (1954), O'Neill's *Long Day's Journey Into Night* (1957), and Jerome Kilty's *Dear Liar* (1959). Later she played the Countess Aurelia in Giraudoux's *The Madwoman of Chaillot* in both Germany (1964) and England (1967).

BERGOPZOOMER, JOHANN BAPTIST (1742–1804), Austrian actor, who made his début in Vienna in 1764, where he played in comedy and in the old Viennese improvised farces. Later he went with KURZ's company to Germany, where the young SCHRÖDER played the valet to his Don Juan. The two men, who were much the same in age and temperament, became friends, and when Schröder went to Vienna in 1782 he was delighted to find Bergopzoomer there as one of the chief tragedians of the Burgtheater, where in spite of his somewhat old-fashioned and ranting style he was much admired.

BERGSTRØM, HJALMAR (1868–1914), Danish dramatist, one of the best known of his period outside his own country. His plays deal, under the influence of IBSEN, with such social problems as the struggle between the classes, as in *Lynggaard & Co.* (1905), and the emancipation of women, as in *Karen Bornemann* (1907) and *Dame-Te* (*Ladies' Tea*, 1910), an anecdotal analysis of spinsterhood.

BERLIN, formerly the capital of Germany, now divided into the capital of the German Democratic Republic and the walled enclave of West Berlin. In spite of sporadic visits from travelling companies from 1620, when the ENGLISH COMEDIANS first appeared there, the town had no resident German company until 1764, when Franz SCHUCH's son built a theatre in Behrenstrasse which DÖBBELIN took over in 1767. In 1787 Döbbelin established the Nationaltheater in the premises which Frederick the Great had built ten years previously for his favourite French actors. He ran it successfully, mainly with modern German plays, until 1789; but IFFLAND's management, from 1796 to 1814, finally made Berlin the theatrical equal of such towns as HAMBURG and VIENNA. He trained his company in the naturalistic style of acting he had learned in MANNHEIM under DALBERG, and built up a repertory dominated by domestic drama and comedy but with a leavening of Shakespeare, GOETHE, and especially SCHILLER, whose *Die Jungfrau von Orleans* was a spectacular success. When a new theatre was opened on 1 Jan. 1802, Iffland continued his successful career there until his death in 1814, being succeeded by Brühl, who with Ludwig DEVRIENT as his leading man continued to put on notable productions of classic and modern plays. When the theatre burnt down in 1817, with the loss of all its costumes and scenery, it was rebuilt in a neo-classical style by Schinkel, its resident designer. The prestige of the company endured until Devrient died in 1832, when the theatre in Berlin entered on a period of decadence and triviality that was to last until almost the end of the century.

Meanwhile the commercial theatre had begun to make some headway, playing mainly comedy, farce, and opera; but a trend towards good literary drama at modest prices became apparent with the founding in 1883 of the DEUTSCHES THEATER, where L'ARRONGE, with such excellent actors as KAINZ and Agnes SORMA, embarked on a series of classical productions in the style of the MEININGER ensemble. In 1894 Otto BRAHM who at his FREIE BÜHNE since 1884 had been introducing IBSEN, HAUPTMANN, and other naturalistic playwrights to Berlin audiences, took over with a repertory of serious modern drama, remaining there until 1904. In 1905 it was taken over by REINHARDT, who had played there as a young man. Under him the Deutsches Theater (together with the Kammerspiele) dominated the decade before the First World War, and Berlin was able to attract talent from the whole of the German-speaking world, even, when Max PALLENBERG joined the company, from Vienna itself. Brahm's theatre had been literary; that of Reinhardt was spectacular. In 1914 the VOLKSBÜHNE opened its own theatre, where PISCATOR made his first major experiments in political theatre; but with his departure from Berlin, together with that of Reinhardt, JESSNER, who directed the Staatliche Schauspiele from 1919 to 1930, and many more whose admirable acting, designing, and directing talents were lost to Germany by emigration, the town sank to a low level theatrically, though GRÜNDGENS at the Schauspielhaus and Hilpert at the Deutsches Theater managed to maintain a classical repertory of some artistic integrity until the theatres that had survived the air-raids were closed by official decree on 1 Sept. 1944.

One of the first to reopen, almost a year later, was the Hebbeltheater, with a revival of BRECHT's *Die Dreigroschenoper*; and Brecht's return, first to the Deutsches Theater in 1949 and then to the Theater am Schiffbauerdamm in 1954, once again brought Berlin back into the theatrical limelight. In West Berlin the tide turned with the reopening in 1951 of the rebuilt Schillertheater, where good new plays from the international repertory—by BECKETT (who directed some of his own plays), ANOUILH, GIRAUDOUX, Arthur MILLER, Tennessee WILLIAMS, John OSBORNE, ZUCKMAYER, FRISCH, and DÜRRENMATT—together with revivals of standard classics, were seen with distinguished casts, directors, and designers. Under BESSON the Volksbühne became the most artistically enterprising theatre in East Berlin; in West Berlin the Schaubühne, founded in 1962, has since 1970 been under the direction of Peter Stein (1937–   ), whose productions have been widely acclaimed. Both parts of Berlin run theatre festivals, West Berlin having a German-language *Theatertreffen* in May and an international festival in September, and East Berlin an eastern-bloc festival, *Theatertage*, in October.

BERLINER ENSEMBLE, theatre company founded in 1949 in East BERLIN by BRECHT and his wife Helene WEIGEL. It was first housed at the DEUTSCHES THEATER and in 1954 moved to the Theater am Schiffbauerdamm. It achieved pre-eminence in Germany and world-wide acclaim with its productions of Brecht's plays, directed by himself and after his death, when his widow took over, by PALITZSCH, BESSON, and others. By 1974, its 25th anniversary, the company had staged a total of 54 productions and developed a very personal style of acting and presentation designed to prevent the suspension of disbelief and encourage in the audience, as was Brecht's desire, a critical appraisal of the play. To avoid the danger, apparent since the mid-1960s, of turning itself into a Brecht museum, the company has extended its repertoire to include BÜCHNER, SHAW, WEDEKIND, and Heiner MÜLLER, as well as continuing to present selected plays from the entire Brecht canon. The company was received with great enthusiasm in Paris in 1954 and 1955 and visited London in 1956 (just after Brecht died) and 1966. Manfred WEKWERTH became its manager in 1977.

BERMAN, ED, see COMMUNITY THEATRE.

BERNADON, Viennese comic character created by KURZ, a restless and impetuous youth whose adventures lead him into the world of the supernatural, where his comic genius ranges itself on the side of white, as opposed to black, magic. The best of these adventures are found in Kurz's *Die 33 Schelmereien des Bernadon*.

BERNARD, JEAN-JACQUES (1888–1972), French dramatist, son of Tristan BERNARD but a writer in quite a different style, since his work deals with the tragedy of unrequited or unacknowledged love, and derives from the Theatre of SILENCE invented by MAETERLINCK. Several of his plays were seen in London in translation, among them *The Sulky Fire* (*Le Feu qui reprend mal*, 1921) in 1926, *The Unquiet Spirit* (*L'Âme en peine*, 1926) in 1928, *Martine* (1922) in 1929 (New York, 1928), *The Springtime of Others* (*Le Printemps des autres*, 1924) in 1934, and *Invitation to a Voyage* (*L'Invitation au voyage*, 1924) in 1937.

BERNARD, JOHN (1756–1828), American actor and manager, who had already achieved a fine reputation in England as a light comedian before he accepted an invitation from Thomas WIGNELL to join the company at the CHESTNUT STREET THEATRE in Philadelphia, where he first appeared in 1797. He remained there until 1803, playing a wide variety of comic roles, and then went to BOSTON, where in 1806 he became joint manager of the Federal Street Theatre. After five years of unprofitable management he went on tour throughout the United States and Canada, settling in Albany and opening the first regular theatre there in 1813, where he invited Samuel DRAKE and his family to join him. He went back to Boston in 1816 but returned to settle permanently in England in 1819, spending his last years in almost total poverty. His memoirs, published in 1887, are an important source of information on the early American theatre.

BERNARD, TRISTAN (1866–1947), French dramatist, father of J.-J. BERNARD, author of a number of deftly-constructed light comedies in which the chief character becomes entangled in petty intrigues from which the author extracts him with great ingenuity. The most successful were *L'Anglais tel qu'on le parle* (1899) and *Triplepatte* (1905), which, translated by Clyde FITCH as *Toddles*, was played in London by Cyril MAUDE in 1906 and several times revived. It was also seen in New York in 1908.

BERNHARDT, SARAH (-Marie-Henriette) (1844–1923), French actress, who became world famous as much for her eccentricities as for her superb acting. Her voice, likened to a 'golden bell' or the 'silver sound of running water', was one of her chief charms, added to a slim, romantic figure, dark eyes, and a consummate mastery of her art. After some initial training, she made her début in 1862 at the COMÉDIE-FRANÇAISE, where she was destined to make several brief and stormy visits, not accommodating herself easily to the traditions of that venerable institution. After the first break she appeared at the ODÉON, where in 1869 she attracted favourable attention by her performance in Coppée's *Le Passant*. Her career was interrupted by the Franco-Prussian war, and in 1872 she returned to the Comédie-Française, achieving a double triumph as Cordelia in *King Lear* and the Queen in HUGO's *Ruy Blas*. She consolidated her position at the head of her profession by outstanding performances as RACINE's Phèdre and as Doña Sol in Hugo's *Hernani*, one of her finest parts. She made her first appearance in London in 1879 in *Phèdre*, and in New York in 1880 in SCRIBE and Legouvé's *Adrienne Lecouvreur*, scoring an immediate triumph in both capitals. She returned many times, her last appearance in London being in *Daniel* not long before her death. In Paris, after her final departure from the Comédie-Française in 1880, she managed a number of theatres, including the AMBIGU and the PORTE-SAINT-MARTIN, before opening the present Théâtre de la Ville as the Théâtre Sarah-Bernhardt in 1899. There she revived a number of her former successes and appeared in some outstanding new plays, among them the younger DUMAS's *La Dame aux camélias*, SARDOU's *Fédora, Théodora*, and *La Tosca*, and

ROSTAND's *L'Aiglon*, with which her name is always associated. She was also seen on several occasions as Shakespeare's Hamlet. She was an accomplished painter and sculptress and wrote poetry and plays, appearing in some of the latter herself. Among them were *L'Aveu* (1898), *Un Coeur d'homme* (1909), and the unpublished *Dans les nuages*, the manuscript of which is now in the Harvard College library. A volume of reminiscences published in 1907 was translated into English the same year as *My Double Life*. (The American edition, reprinted in 1968, was entitled *Memories of My Life*.)

BERNINI, GIOVANNI LORENZO (1598–1680), Italian architect, artist, and sculptor, who was responsible for the décor and some of the scenery of the theatre built in Rome by the Barberini family, which opened in 1634. He was a great inventor of theatre machines and spectacular effects, some of which were seen in a Roman theatre by a friend of Inigo JONES, Richard Lascelles, who described them in his *Italian Voyage*, published in 1670. They included water pieces with boats floating on them, floods, storms, men and chariots flying through the air, houses falling in ruins, the sudden appearance of temples, forests, and populous city streets, serpents, dragons, and articulated monsters, and TRANSFORMATION SCENES of all kinds.

BERNSTEIN [*née* Frankau], ALINE (1881–1955), American scene designer, and founder in 1937, with Irene Lewisohn, of the Museum of Costume Art in Rockefeller Center, New York. Her first work for the stage was done in 1924, with designs for an Indian play, *The Little Clay Cart*, in a production by the NEIGHBORHOOD PLAYHOUSE. She was responsible for the décor of a number of important productions, including Philip Moeller's *Caprice* (1928) and SHERWOOD's *Reunion in Vienna* (1931) for the THEATRE GUILD; CHEKHOV's *The Cherry Orchard* (1928) and *The Seagull* (1929), *Romeo and Juliet* (1930), and MOLNÁR's *Liliom* (1932) for the Civic Repertory Theatre under Eva LE GALLIENNE; and, for other managements, BARRY's *The Animal Kingdom* and Sidney HOWARD's *The Late Christopher Bean* (both 1932); the dramatization of Christopher Morley's *Thunder on the Left* (1933); Elmer RICE's *Judgement Day* (1934); and two plays by Lillian HELLMAN, *The Children's Hour* (1934) and *The Little Foxes* (1939). Her designs for the American production of COCTEAU's *The Eagle Has Two Heads* (1949) were also much admired.

BERNSTEIN, HENRY (1876–1953), French dramatist, whose early plays were written under the influence of the THÉÂTRE LIBRE. His own, more distinctive manner emerged in plays of action with strong situations, ably interpreted by Lucien GUITRY, among them *La Rafale* (1906) and *Le Voleur* (1907) which, in translation as *The Whirlwind* and *The Thief*, were equally successful on Broadway and in London. In such later works as *Samson* (1910), *L'Assaut* (1912), *Le Secret* (1913), *Judith* (1922), *La Galérie des glaces* (1925), and *Le Venin* (1927), Bernstein attempted to emulate the younger playwrights who were preoccupied with 'le théâtre de l'inquiétude'; but his attempts at profundity were heavy-handed and it is in his earlier plays that his true, if limited, talent is best revealed.

BERSENEV [Pavlishchev], IVAN NIKOLAIEVICH (1889–1951), Soviet actor and director, who in 1907 appeared at the Solovtsev Theatre in Kiev under MARDZHANOV, and later was seen in many other Ukrainian towns. From 1928 to 1938 he acted and directed at the MOSCOW ART THEATRE, leaving there to become director and leading man of the company at the LENKOM THEATRE, where he trained a talented group of young actors, and staged, among other new Soviet plays, K. M. SIMONOV's *And So It Will Be* (1944). His productions were notable for their precision and clarity of ideas, their profound intellectuality, and a lively feeling for the detail of contemporary life.

BERTINAZZI, CARLO (1713–83), known as Carlin, the last of the great Arlequins of the COMÉDIE-ITALIENNE and also one of the last Italians to join the troupe, which was eventually to be composed entirely of French actors. He appears to have acted widely in Italy but particularly in Venice, and according to Casanova was in St Petersburg from 1735 to 1737. He joined the Paris company in 1741, and had the unenviable task of replacing the great Thomassin (Tommaso VICENTINI), in which he succeeded all the more creditably in that the period was undoubtedly that of the sunset of the COMMEDIA DELL'ARTE at the Hôtel de BOURGOGNE. He was much admired by GARRICK, who said his back wore the expression his face would have shown had he not been masked.

BERTOLAZZI, CARLO (1870–1916), Italian dramatist, with PRAGA one of a group of realists in Milan who in the 1890s kept alive the theatre in the local dialect. Among his plays in Milanese dialect, which began with *Mamma Teresa* in 1887, the best was *El nost' Milan* (*Our Milan*, 1893), which was in two parts—*Le povere gente* (*The Poor Ones*) and *I sciori* (*The Rich Ones*). It was successfully revived by Strehler for the PICCOLO TEATRO DELLA CITTA DI MILANO in 1955. Other plays by Bertolazzi include *La gibigianna* (*The Reflection*, 1898), *L'egoista* (1900), and *Lulù* (1903), the last a sensitive portrayal of nymphomania.

BESPEAK PERFORMANCE, see BENEFIT.

BESSENYI, GYÖRGY, see HUNGARY.

BESSON, BENNO (1922–), Swiss director, who after working with SERREAU in Paris became an actor and director with the BERLINER ENSEMBLE in 1949. He proved himself the best and most imaginative of BRECHT's company, helping him to adapt MOLIÈRE's *Don Juan* in 1954 and FARQUHAR's *The Recruiting Officer* (as *Pauken und Trompeten*) in 1955. He left the company in 1958 and in 1961 joined the DEUTSCHES THEATER, being responsible for ingenious and elegant productions of Peter Hacks's adaptations of ARISTOPHANES' *Peace* in 1962 and Offenbach's *La Belle Hélène* in 1964, and of SHWARTZ's modern fairy-tale *The Dragon* in 1965. In 1969 he became director of the VOLKSBÜHNE, which he revitalized, making it an entertaining and cosmopolitan 'people's theatre'. In 1973 and 1974 he staged *Spectacles I and II*, mini-festivals involving as many as twelve different performances under one roof—backstage, on stage, in the foyer, and so on—and in 1975 confirmed the pre-eminence of his theatre in East Berlin with an outstanding production of *Die Schlacht* by Heiner MÜLLER. His productions outside Germany include *As You Like It, Hamlet*, and Brecht's *The Caucasian Chalk Circle*, seen at the AVIGNON FESTIVAL in 1976, 1977, and 1978 respectively. He left the Volksbühne in 1979 and later became director of the Théâtre de la Comédie in Geneva.

BETTERTON, THOMAS (*c.* 1635–1710), English actor, and the outstanding figure of the Restoration stage. Apprenticed to John RHODES, he went with him to the COCKPIT when it reopened in 1660, but was soon taken away with several of his fellow-actors by Sir William DAVENANT, to join the company at LINCOLN'S INN FIELDS. In 1671, after Davenant's death, the company, of which Betterton was by then joint manager with Henry HARRIS, moved to a new theatre in DORSET GARDEN. They remained there until in 1682 they were amalgamated with the actors at DRURY LANE. Although Betterton was the acknowledged leader of the united company, he found it impossible to work with the manager then in charge of the theatre, Christopher RICH; in 1695 he broke away, taking a group of actors to the old Lincoln's Inn Fields theatre and reopening it most successfully with the first performance of CONGREVE's *Love for Love*. In 1705 the company moved again, to VANBRUGH's new theatre in the Haymarket (on the site of the present HER MAJESTY'S THEATRE). Betterton was a good manager and an excellent actor, his Hamlet, Sir Toby Belch (in *Twelfth Night*), and Falstaff (in *Henry IV, Part I*) being equally admired. Though not perhaps so well suited by nature to Restoration comedy, he excelled in the high-flown rhetoric of the HEROIC DRAMA, and created many leading parts by dramatists of the day who were always ready to profit by his advice. He also adapted a number of plays, including some of Shakespeare's, to suit the taste of the time, and in 1690 turned FLETCHER's *The Prophetess* into an opera whose elaborate stage directions show to what a pitch the mechanical effects of the time had been brought, aided by Betterton's own study in Paris of contemporary French scenic machinery.

Betterton married one of the first English actresses, Mary Saunderson (?–1712), whom Pepys always referred to as Ianthe, from her excellent playing of that part in Davenant's *The Siege of Rhodes*. She and her husband were much esteemed by their contemporaries, particularly for their help and kindness to young actors, several of whom lodged in their house and were trained by them.

BETTI, UGO (1892–1953), Italian playwright, much of whose work was written under the influence of PIRANDELLO. A lawyer by profession, he gave many of his plays a legal setting, as in his most important work, *Corruzione al Palazzo di Giustizia* (1949), which, as *Corruption in the Palace of Justice*, was seen in New York in 1963. His other plays—he wrote nearly 30—include light comedies like *Una bella domenica di settembre* (*A Fine Sunday in September*, 1937) and more serious works such as *Frano allo scala Nord* (*Landslide on the North Quay*, 1936); *Ispezione* (1947) which, as *Island Investigations*, was produced at the University of Bristol in 1956; *Delitto all'isola delle capre* (1948), seen in Oxford in 1957 as *Crime on Goat Island; Lotta fino all'alba* (*The Struggle Ends at Dawn,* 1949); *Il giocatore* (1951), seen in New York as *The Gambler* in 1952 in an adaptation by Alfred Drake, who also played the leading role; and the posthumous *La fuggitiva* (*The Fugitive*, 1943). In 1955 three of Betti's most important plays were seen in London in translations by Henry Reed—*Il paese delle vacanze* (1942) as *Summertime; La regina e gli insorti* (1951) as *The Queen and the Rebels*; and *L'aiuola bruciata* (1942) as *The Burnt Flower Bed*. Betti has been called 'the Kafka of drama'. For him the world is on trial and his characters, astray in private nightmares, haunted by visions of a lost Earthly Paradise, find themselves continually up against intangible frontiers. Running away from what they fear, as in 'The Hound of Heaven' they arrive at what they are flying from; for in spite of his realistic portraits of degradation, Betti is an optimist, and for him man's journey leads ultimately to the discovery of Christ.

BETTY, WILLIAM HENRY WEST (1791–1874), a child prodigy, who as the Young Roscius took London by storm in the season of 1804–5. Born in Ulster, he had already played with great success in Ireland and Scotland when on 1 Dec. 1804 he appeared at COVENT GARDEN as the young Achmet in John Brown's tragedy *Barbarossa*. There, and subsequently at DRURY LANE, he was seen in such parts as Hamlet, Romeo, Osman in VOLTAIRE's *Zaïre*, Rolla in SHERIDAN's *Pizarro*, Frederick in Mrs INCHBALD's *Lovers' Vows*, and

Young Norval in HOME's *Douglas*, in which character he was painted by Northcote and Opie. For a short time he ousted even Mrs SIDDONS and John Philip KEMBLE from public favour, but after a couple of seasons of hectic success, interspersed with strenuous summer tours, opinion turned against him and he was hissed off the stage when he attempted Richard III. In 1808 he went as an undergraduate to Cambridge, and three years later again appeared on the stage without success. He was ignored, his father squandered his money, and the rest of his life was passed in obscurity.

BEVERLEY, WILLIAM ROXBY (*c.* 1814–89). English scene-painter, who was for some years an actor, appearing at the Theatre Royal in Manchester under the management of his father, also William Roxby Beverley (1765–1842). His main interest, however, was in scene design, and in 1843 he became chief designer at the theatre, for which he painted a beautiful act-drop which remained in use for 25 years. Some time later he was in London, where he did good work for Madame VESTRIS at the LYCEUM THEATRE, achieving his greatest success with his designs for PLANCHÉ's extravaganza *The Island of Jewels* (1849). His long and fruitful association with DRURY LANE, where his best work was done for the annual Christmas PANTOMIME, began in 1854 and lasted through successive managements until his retirement in 1885. He also worked intermittently elsewhere, and designed the sets for several of Charles KEAN's productions at the PRINCESS'S THEATRE, including *Henry IV, Part 1, King John*, and *Macbeth*, and for an elaborate production of Milton's *Comus*. A one-surface painter, firmly opposed to the innovation of BUILT STUFF, he was next to STANFIELD the most distinguished and influential scene-painter of the 19th-century English theatre.

BIANCOLELLI, GIUSEPPE DOMENICO (*c.*1637–88), actor and playwright of the COMMEDIA DELL'ARTE, who became famous as Dominique, playing the part of ARLECCHINO, gallicized as Arlequin. Son and grandson of actors, he was invited in 1654 by Mazarin to join the Italian players in Paris. At that time they played only in Italian, and in spite of the determined opposition of the established French company, Dominique was able to persuade Louis XIV, a great admirer of his acting, to allow them to play also in French. Dominique later became a naturalized Frenchman and married an actress, one of his daughters, also an actress, marrying Pierre LA THORILLIÈRE, with whom she appeared at the COMÉDIE-FRANÇAISE.

BIANCOLELLI, PIETRO FRANCESCO (1680–1734), Italian actor, youngest son of the above. He acted in Paris both as Arlequin (see ARLECCHINO) and as PIERROT until the Italian players were banished in 1697. As Dominique le Jeune, he was a member of the company which RICCOBONI took back to Paris in 1716, and remained with it until his death, playing in the first productions of some of the plays of MARIVAUX.

BIBBIENA, Cardinal, see ITALY.

BIBIENA [BIBBIENA, DA BIBBIENA], a family of scenic artists and architects, originally from Florence, whose work, in pure baroque style, is found all over Europe, though Parma and VIENNA probably saw their greatest achievements. The family name was Galli, and Bibiena (or Bibbiena) was added later, from the birthplace of Giovanni Maria Galli (1625–65), father of Ferdinando (1657–1743) and Francesco (1659–1739) who together founded the family fortune and renown. While still a young man, Ferdinando worked in the beautiful Teatro Farnese, built by ALEOTTI, which he left to go to Vienna. There, with the help of his brother and his sons, he was responsible for the decorations of many Court fêtes and theatrical performances. His eldest son Alessandro (1686–1748) became an architect, but the three younger ones, Giuseppe (1695–1756), Antonio (1697–*c.* 1774), and Giovanni Maria (1700–74), worked in the theatre, Giuseppe being probably the first designer to use transparent scenery lighted from behind, in 1723. His son Carlo (1721–87), one of whose stage settings is preserved at DROTTNINGHOLM, was associated with his father in the building and decoration of the Bayreuth opera-house. The family made its home in Bologna, but its members can be traced all over Europe. They worked so much in collaboration that it is sometimes impossible to apportion their work individually. They introduced many innovations into scenic design, particularly with their development of the *scena d'angolo*, or perspective scene, which inaugurated a new era in stage design, and made possible the elaborate architectural settings characteristic of the family style.

BIBLE-HISTORIES, see MYSTERY PLAY.

BICKERSTAFFE, ISAAC (1735–1812), English dramatist, considered in his own day the equal of John GAY as a writer of BALLAD OPERAS. The first of these, *Thomas and Sally; or, the Sailor's Return* (1760), was seen at COVENT GARDEN and was followed by *Love in a Village* (1762), based on *The Village Opera* (1728) by Charles Johnson. *The Maid of the Mill* (1765), based on Richardson's novel *Pamela*, held the stage for many years, and is found in the repertory of the TOY THEATRE. Among Bickerstaffe's later productions, many of them written in collaboration with Samuel FOOTE and Charles DIBDIN, the most successful was *Lionel and Clarissa* (1768), which was frequently revived and was last seen at the LYRIC THEATRE, Hammersmith, in 1925. In 1772 Bickerstaffe, who enjoyed the friendship of Dr JOHNSON, GARRICK, and GOLDSMITH, was suspected of capital crime, and fled to the Continent, where he died in poverty. His name was used as a pseudonym by both Jonathan Swift and STEELE.

BIDDLES (BEDELLS), ADELAIDE HELEN, see CALVERT, LOUIS.

BIDERMANN, JAKOB (1578–1639), a Jesuit priest and an outstanding writer of JESUIT DRAMA. The best of his plays to have survived is *Cenodoxus*, the story of a pious hypocrite in Paris whose soul, after death, is tried and cast into Hell. The play ends with the founding by St Bruno, who has witnessed the condemnation, of the religious order of Carthusians. *Cenodoxus* was first performed on 3 July 1602 in Augsburg, and was seen in Munich in 1609. In 1958, in an abridged German text based on a translation from the original Latin published in 1635, it was presented at the Residenztheater in Munich as part of the city's 800th anniversary celebrations.

BIEDERMANN, JOSEPH, see JEWISH DRAMA.

BIJOU THEATRE, London. There were two small theatres of this name in London, the first, in the Haymarket, being a concert hall attached to the Royal Opera House (later HER MAJESTY'S THEATRE). It was used occasionally for plays, Clement SCOTT appearing there as a youth as Fleance in an amateur production of *Macbeth*. In 1862 Charles James MATHEWS and his second wife Lizzie Davenport also appeared there. The theatre was burned down with the opera house on 6 Dec. 1876.

A hall in Archer Street, Bayswater, also used by amateurs, was renamed the Bijou in 1886. It was used for COPYRIGHT performances of several of SHAW's plays—*The Devil's Disciple* in 1897, *The Philanderer* and *You Never Can Tell* in 1898, and *Man and Superman* in 1903. Oscar WILDE's *Salome* was given its first production there on 10 May 1905. In 1925 Lena ASHWELL took over the theatre, renamed it the Century, and appeared there in plays which she directed herself. Among them were St John ERVINE's *The Ship* and MASEFIELD's *Good Friday*, and in 1927 her own dramatizations of DOSTOEVSKY's *Crime and Punishment* and R. L. Stevenson's *Dr Jekyll and Mr Hyde*. From 1933 to 1937 the theatre stood empty, and was then taken over by the Rudolf Steiner Association. It returned briefly to live theatre from 1945 to 1946, as the Twentieth Century, but without success, and eventually became a warehouse.

BIJOU THEATRE, New York. (1) At 1239 Broadway. A small playhouse devoted to light entertainment, it opened as the Brighton in 1878, and was known as Wood's Broadway Theatre and the Broadway Opera House before becoming the Bijou on 31 Mar. 1880. Lillian Russell appeared there in several musical plays in 1882. On 7 July 1883 the building was demolished and a larger theatre of the same name was built on the site, opening on 1 Dec. with Offenbach. It was at this theatre that Julia MARLOWE made her first appearance in New York as an adult actress, playing Parthenia in Mrs Lovell's *Ingomar the Barbarian* in 1887. The building continued in use, mainly as a home of musical and light entertainment, until 1911; it was demolished in 1915.

(2) A second Bijou Theatre, at 209 West 45th Street, was opened by the Shuberts on 12 Apr. 1917. Among the plays produced there were revivals of BARRIE's *What Every Women Knows*, with Helen HAYES, in 1926, and of IBSEN's *The Lady From the Sea*, with Blanche YURKA, in 1929. The last successful production, in 1931, was Benn Levy's *Springtime for Henry*. The theatre then became a cinema, the Toho, showing Japanese films, but it reopened as a theatre in 1945, when Howard LINDSAY and Russel Crouse's *Life With Father* (1939) moved there from the EMPIRE, remaining until 1947. Ten years later Eugene O'NEILL's *A Moon for the Misbegotten* had its first New York production there, and Graham GREENE's *The Potting Shed* had a long run. As the D. W. Griffith, and again as the Toho, the Bijou was a cinema from 1962 until 1972, reverting to live drama with Don Petersen's *The Enemy is Dead* in 1973. *Mummenschanz*, an evening of pantomime, began a long run in 1977.

(3) A Bijou Theatre, seating 200, situated in the off-Broadway Playhouse Theatre at 359 West 48th Street, opened in 1970.

BILL-BELOTSERKOVSKY, VLADIMIR NAUMOVICH (1884–1970), Soviet dramatist who added Bill, his nickname in the U.S.A., where he worked from 1911 to 1917, to his family name. His first play, *Echo* (1924), depicts American dockers refusing to load arms for anti-Soviet use. *Hurricane* (or *Storm*), produced in 1925 at the MOSSOVIET THEATRE, a stirring piece of propaganda about the struggles of a Revolutionary leader in a small village during the Civil War, was a landmark in Soviet theatre history, being the first realistic play about the new Soviet state. It has been produced all over Eastern Europe, either in its original form or in a new version prepared by the author in 1951 which was staged by Yuri ZAVADSKY, also at the Mossoviet. Of Bill-Belotserkovsky's later plays, one, produced in 1934, was published in 1938 in an English translation by Anthony Wixley as *Life is Calling* (it is also known as *Life Goes Forward*). The plot is conventional, but is heightened by the conflict between social duty and personal happiness.

BILLETDOUX, FRANÇOIS (1927–), French journalist and dramatist, who studied under DULLIN and achieved success in 1959 with *Tchin-Tchin*, a play considered at the time a sign of renewed vigour in the Parisian AVANT-GARDE theatre. Billetdoux takes real-life situations in which an unusual event throws things off balance, and ends by throwing doubt on the whole social fabric and suggesting a more essential truth beyond. Thus *Tchin-Tchin* shows a drunken complicity developing between a bricklayer and the wife of a doctor who is having an affair with the bricklayer's wife; *A la nuit la nuit* (1960) revolves round the mysterious identity of a man who visits a prostitute; *Le Comportement des époux Bredburry* (also 1960) is about a woman who tries to sell her husband through the small ads; and *Va donc chez Törpe* (1961) is set in an inn where all the guests have come to commit suicide. Like GIRAUDOUX and AUDIBERTI, Billetdoux writes highly-coloured prose which sometimes captures a throbbing moment, though constantly running the risk of sounding mannered. Among his later plays are *Comment va le monde, môssieu? Il tourne, môssieu!* (1964), which contrasts two characters who, in the manner of VOLTAIRE's Candide, traverse the battlefields of the Second World War; *Il faut passer par les nuages* (also 1964), a play directed by BARRAULT in which a successful middle-class woman is led by a personal crisis to allow her family to disintegrate around her; *Silence, l'arbre remue encore* (AVIGNON, 1967); and *Rintru pa trou tar hin* (1971), where the police try to reconstruct with the help of a family of survivors the ambiguous events leading to an explosion in a block of flats which eventually assumes the aspect of modern society.

BILLINGER, RICHARD (1890–1965), Austrian dramatist and poet, whose first play *Das Perchtenspiel* (1928) was staged by Max REINHARDT. Its theme of conflict between Christian and pagan values was repeated in *Rauhnacht* (*Twelfth Night*) and *Rosse* (*Horses*), both produced in 1931, and in *Die Hexe von Passau* (*The Witch of Passau*, 1935), on which Ottmar Gerster based an opera in 1941. *Der Gigant* (*The Giant*, 1937) exemplified Billinger's other characteristic concern, the struggle between urban and rural lifestyles. *Die Fuchsfalle* (*The Fox Trap*, 1941), *Der Zentaur* (*The Centaur*, 1946), and *Der Galgenvogel* (*The Gallows-Bird*, 1948) were less successful than his earlier work.

BILLY ROSE THEATRE, New York, at 208 West 41st Street, between 7th and 8th Avenues. As the National Theatre, with a capacity of 1,162, this opened on 1 Sept. 1921 with Sidney HOWARD's first play *Swords*. This was a failure, but success came with John Willard's *The Cat and the Canary* (1922), followed by Clemence DANE's *Will Shakespeare* with Katharine CORNELL and Cornelia Otis SKINNER and Walter HAMPDEN in a revival of ROSTAND's *Cyrano de Bergerac* (both 1923). Later successes were Veiller's *The Trial of Mary Dugan* (1927), KNOBLOCK's *Grand Hotel* (1930), O'CASEY's *Within the Gates* (1934), a dramatization of Edith Wharton's novel *Ethan Frome*, Noël COWARD's *Tonight at 8.30* (both 1936), Lillian HELLMAN's *The Little Foxes* (1939), and Emlyn WILLIAMS's *The Corn is Green* (1940), in which Ethel BARRYMORE gave a fine performance as the elderly schoolteacher Miss Moffat. After closing for structural alterations the theatre reopened on 11 Nov. 1941 with Maurice EVANS in *Macbeth*, and in 1959 was bought and redecorated by Billy Rose [William Samuel Rosenberg] (1899–1966), its interior being designed by Oliver MESSEL. It reopened as the Billy Rose Theatre on 18 Oct. with a revival of SHAW's *Heartbreak House*; in 1962 came ALBEE's *Who's Afraid of Virginia Woolf?* starring Uta HAGEN and in 1964 his *Tiny Alice*, with John GIELGUD and Irene WORTH. Later productions included Bradley-Dyne's *The Right Honourable Gentleman* (1965) based on the career of Sir Charles Dilke, a revival of Coward's *Private Lives* in 1969, and two new plays, PINTER's *Old Times* (1971) and STOPPARD's *Jumpers* (1974). In 1978 a London company, in conjunction with the New York producer James Nederlander, bought the theatre as a showcase for successful plays from London. As the Trafalgar Theatre, it opened in 1979 with Brian Clark's *Whose Life Is It Anyway?*, followed by Pinter's *Betrayal* (1980). The London company then withdrew and the theatre was renamed the Nederlander.

BILTMORE THEATRE, New York, at 261–5 West 47th Street, between Broadway and 8th Avenue. Built by the Chanins, with a seating capacity of 994, it opened on 7 Dec. 1925 and had an undistinguished career until 1928, when *Pleasure Man* by Mae WEST was closed by the police after three performances. In 1936 the FEDERAL THEATRE PROJECT presented its experimental LIVING NEWSPAPER *Triple-A Plowed Under* by Arthur Arent and the theatre was then taken over by Warner Brothers, for whom George ABBOTT staged a series of successful shows. Later productions included *My Sister Eileen* (1940), based on stories by Ruth McKenny, and Hugh Herbert's *Kiss and Tell* (1943). From 1952 to 1962 the theatre was used for broadcasting but then reverted to straight plays with Neil SIMON's *Barefoot in the Park* (1963), which had a long run, as did the erotic revue *Hair* (1968). This was revived in 1977 after the run of Jules Feiffer's *Knock Knock* (1976) (originally produced by the CIRCLE REPERTORY COMPANY) and in 1978 there was a production of William Douglas HOME's *The Kingfisher*, starring Rex HARRISON.

BIO-MECHANICS, the name given to MEYERHOLD's method of training actors and directing plays. Its formative influences were the stylized acting of the COMMEDIA DELL'ARTE and, to a certain extent, the *Kabuki* plays of JAPAN; the reduction by Gordon CRAIG of the actor to the status of 'super-marionette'; and, on Meyerhold's own admission, Pavlov's theory of conditioning. Rejecting STANISLAVSKY's method of building a character from the inside outwards, Meyerhold demanded the complete elimination of the actor's personality, and the subjection of his mind and body to the will of the director. He also stripped the stage to its bare essentials, eliminating even 'detail' scenery, and leaving the imagination of the audience to build up the necessary stage picture by association. Through his admiration for Charlie Chaplin, Meyerhold brought athletic and pantomimic skills back into the legitimate theatre and called for a much more plastic use of the actor's body than was common at the time. An intellectual rather than an emotional approach to theatre, Bio-Mechanics served its purpose in the early days of the Revolution by clearing away the inessentials of production and the inertia among actors which had cluttered up the old theatres; but although some of the technical exercises invented by Meyerhold have found their way into the Soviet system of actor training, the method itself was outstripped by the forces it had liberated.

BIRD, ROBERT MONTGOMERY (1806–54), American playwright, leader of the Philadelphia group of dramatists, by profession a doctor. In 1831 Edwin FORREST produced his romantic tragedy *The Gladiator*, playing Spartacus, the hero. It was an immediate success, and Forrest selected it for his first appearance at DRURY LANE in 1836, continuing to act in it until his retirement in 1872. John MCCULLOUGH also played the part, making his last appearance in it in 1884, and it was seen as late as 1892. Bird wrote another romantic tragedy, *Oralloossa* (1832), and a domestic drama, *The Broker of Bogota* (1834), for Forrest and revised for him STONE's *Metamora*. They were all popular and frequently revived; but owing to the chaotic state of the COPYRIGHT laws Bird made no money from their production, and Forrest would not allow them to be printed. Bird therefore withdrew from the theatre, writing several successful novels, later dramatized by other hands, and engaging in journalism and politics.

BIRD, THEOPHILUS (1608–64), English actor, who bridged the gap between the Caroline and Restoration theatres. He was the son of a minor actor, William (?–1624), who appears in HENSLOWE's diary as a member of the ADMIRAL'S MEN. The son is first mentioned as playing female roles with QUEEN HENRIETTA'S MEN and by 1635 he had graduated to such male parts as Massinissa in Thomas Nabbes's *Hannibal and Scipio*. Shortly afterwards he married Christopher BEESTON's eldest daughter Anne and joined the KING'S MEN, remaining with them until the closing of the theatres in 1642. Bird negotiated the transfer of the lease of SALISBURY COURT to his brother-in-law William BEESTON in 1647 and he, not Beeston, may have been the 'ill Beest' who betrayed the actors appearing surreptitiously at the VERE STREET THEATRE in 1652. John DOWNES names him first among the actors who reappeared immediately on the reopening of the theatres in 1660, and PEPYS notes in his Diary a rumour that Bird had broken his leg while fencing in a revival of Suckling's *Aglaura* in 1662.

BIRMAN, SERAFIMA GERMANOVNA (1890–1976), Soviet actress and director, who in 1911 joined the MOSCOW ART THEATRE company. In 1938 she became leading lady at the LENKOM THEATRE, where in 1942 she directed a new production of TOLSTOY's *The Living Corpse*, and in 1943 a new translation of ROSTAND's *Cyrano de Bergerac*, with BERSENEV in the title role. Among a number of new plays she directed was a dramatization of K. M. SIMONOV's novel *The Russian Question* in 1947, and in 1955 her playing of Kashperskaya in Shteyn's *Wheel of Fortune* earned her high praise.

BIRMINGHAM, one of England's most important manufacturing towns, has had a theatrical history typical, especially in its early days, of many provincial cities. Before 1730 two theatre BOOTHS are known to have existed, but the first permanent theatre building was erected in Moor Street in about 1740. Richard Yates brought a company from London there every summer for several years. A theatre opened in King Street in 1751 and another in New Street in 1774. This was burned down and rebuilt in 1792, the father of William MACREADY becoming its manager three years later. It was again destroyed by fire and rebuilt in 1820. The first TOURING COMPANIES visited it in 1849, though the old STOCK COMPANY lingered on until 1878. The building closed in 1901, was rebuilt in 1904, and finally demolished in 1957. Another theatre, the Prince of Wales's, was destroyed by enemy action during the Second World War. Barry JACKSON's BIRMINGHAM REPERTORY THEATRE earned the city a lively theatrical reputation during the 1920s and 30s.

The Alexandra, built in 1901 and for its first year known as the Lyceum, has since 1976 been used by touring companies only. In its heyday it staged mainly MELODRAMA, PANTOMIME, MUSIC-HALL, and REVUE. In 1927 it became a REPERTORY THEATRE, and though rebuilt in 1935 its policy remained unchanged until 1946, when regular spring and autumn visits from touring companies were introduced. In 1964 this theatre housed the first performance by the NATIONAL THEATRE company of *Othello*, with Laurence OLIVIER in the title role. The building was bought by the Birmingham Corporation in 1968, and it became a 'touring only' theatre in 1976. Extensively renovated in 1979, it now has a seating capacity of 1,562.

The Birmingham Hippodrome, which opened in 1900 as the Tivoli, taking its present name in 1903, was for many years a VARIETY theatre. It had a somewhat chequered career and in 1968 was in danger of demolition. It is now a major lyric theatre housing opera, ballet, and other large scale productions, being used extensively by the Welsh National Opera company. It was bought by the local authority in 1979 and is now run by a

Trust, having been extensively altered before its reopening in 1981.

The Midlands Arts Centre, which opened in 1962, includes a studio theatre shared by the resident Cannon Hill Puppet Theatre, visiting companies, and films; the Birmingham Youth Theatre visits frequently. The theatre seats 200, and has a flexible stage. The puppet theatre caters mainly for young children and, like the rest of the centre, has been criticized for failing to attract audiences from the poorer quarters of the city. For this its situation in Edgbaston, on the edge of the town, seems partly responsible.

BIRMINGHAM REPERTORY THEATRE, one of the most significant enterprises in the English theatre during the first half of the 20th century, developed from the work of the Pilgrim Players, an amateur company founded and run by Barry JACKSON. Inspired by the opening of a repertory theatre in MANCHESTER by Miss HORNIMAN in 1907, and of the LIVERPOOL PLAYHOUSE (originally the Liverpool Repertory Theatre) in 1911, Jackson built and equipped a theatre in Station Street, the first in England to be designed specifically as a repertory theatre. It opened on 15 Feb. 1913 with *Twelfth Night* and during the next 10 years a wide variety of uncommercial plays was produced, including DRINKWATER's *Abraham Lincoln* (1918) and the first English production of SHAW's *Back to Methuselah* (1923), one of the many plays by Shaw to be produced by Jackson. In 1924, disheartened by lack of local support, Jackson closed the theatre and went to London. However, the Birmingham Civic Society guaranteed a sufficient number of season-ticket holders to induce him to return and reopen the theatre. The situation was further ameliorated by the subsequent commercial success of such plays as Eden PHILLPOTTS's *The Farmer's Wife* (1924), Drinkwater's *Bird in Hand* (1927), and Besier's *The Barretts of Wimpole Street* (1930), all of which were transferred to London, and by Jackson's controversial productions of *Hamlet* (1925) and *Macbeth* (1928) in modern dress. From 1929 to 1938 the Birmingham Repertory Theatre also provided the nucleus of the company which appeared at the MALVERN FESTIVAL. In 1935 Jackson transferred his theatre to a Board of Trustees, but remained its director until his death in 1961. It became known as a fine training ground for young actors and actresses, including Laurence OLIVIER, Paul SCOFIELD, Margaret LEIGHTON, and, more recently, Albert FINNEY. In 1966 Peter Dews became artistic director, his production of Peter Luke's *Hadrian the Seventh* (1967), with Alec MCCOWEN, being seen later in London and New York. In the late 1960s the Birmingham City Council donated the site for a new and larger theatre, seating 900, which opened in 1971 in the heart of the City Centre, financed by the Council, the ARTS COUNCIL, and public subscription. The size of the auditorium, twice that of the old theatre, which even so was seldom filled to capacity, has under successive directors intensified the normal conflict in a subsidized theatre between artistic and commercial considerations. The programmes today are necessarily less innovatory than those of Jackson's era, and consist mainly of established classics, outstanding modern plays, among them BRECHT's *The Caucasian Chalk Circle*, successfully presented in 1974, and occasional new works, such as Arnold WESKER's *The Merchant* (1978). These, however, are usually presented in the studio theatre, one of the most flexible in the country, which seats 160.

BÍRÓ, LAJOS (1880–1948), Hungarian dramatist, most of whose work was produced abroad. He was working in Berlin when his first play *The Family Circle* (1909), an ironic study of bourgeois domestic hypocrisy, was produced at the FREIE BÜHNE. It was well received, but *The Yellow Lily* (1910), a satire on the Hungarian aristocracy, caused trouble with the authorities in Hungary and was banned. Bíró then worked for a film company and eventually went to the U.S.A., but during the Second World War he was in England, where he wrote and had produced three plays in English—*School for Slavery* (1941), set in Nazi-occupied Poland, *Patricia's Seven Houses*, an amusing variant of the theme of SHAW's *Widowers' Houses* in which an elderly English spinster inherits seven brothels, and *Our Katie* (both 1946).

BITS, see BURLESQUE, AMERICAN.

BJERREGAARD, HENRIK, see NORWAY.

BJØRNSON, BJØRN (1859–1942), Norwegian actor and theatre manager, son of Bjørnstjerne BJØRNSON. After training in Vienna, he joined the MEININGER COMPANY in 1880, and five years later became a director of the KRISTIANIA THEATER, where he remained until 1893. After some years abroad, he returned to Norway in 1899 to take over the newly established NATIONALTHEATRET and for the next 40 years was active in the theatre in Norway in many different capacities.

BJØRNSON, BJØRNSTJERNE (1832–1910), Norwegian poet, dramatist, and novelist, awarded the NOBEL PRIZE for Literature in 1903. In spite of the severity of his childhood as the son of a provincial pastor, he had a sanguine and energetic disposition, applying his talents as creative writer, journalist, and orator to a variety of liberal causes but chiefly to that of Norwegian nationalism.

After spending the years 1857–9 as director of the Norske Teatret in Bergen (see Den NATIONALE SCENE), Bjørnson embarked on a series of dramas on historical and patriotic themes: *Mellem Slagene* (*Between the Battles*, 1858), *Halte-Hulde* (*Lame Hulda*, 1858), *Kong Sverre* (*King Sverre*, 1861), the trilogy *Sigurd Slembe* (1862), and *Maria Stuart i Skotland* (1864); he was to return to such historical subjects with *Sigurd Jorsalfar* and *Kong Eystejn* in 1872. Following his appointment as artistic director of the KRISTIANIA THEATER (1865–7), his plays show an increasing interest in social issues: *De Nygifte* (*The Newly-Marrieds*, 1865) had already dealt with a matrimonial problem in a contemporary setting. *Redaktøren* (*The Editor*, 1874), *En Fallit* (*The Bankrupt*, 1875), *Kongen* (*The King*, 1877), and *Det ny System* (*The New System*, 1879) all debate contemporary themes, especially the urban and commercial concerns of the time.

The plays of Bjørnson's maturity fall into two groups: private and largely individual problems are dealt with in *En Handske* (*A Gauntlet*, 1883), *Geografi og Kjærlighed* (*Love and Geography*, 1885), *Laboremus* (1901), *Daglannet* (1904), and the sparkling comedy *Når den ny vin blomstrer* (*When the New Wine Blooms*, 1909); the individual's relations with established authority are considered in *Over Ævne I* (1883), best known in English as *Pastor Sang* and incomparably Bjørnson's most powerful work, *Paul Lange og Tora Parsberg* (1893), and *Over Ævne II* (1895, seen in London in 1901 as *Beyond Human Power*). Bjørnson's international standing has been overshadowed by that of IBSEN; his treatment of social themes has tended to become dated and the optimism of his tone has perhaps lessened the impact of his observations. In Norway, however, he is honoured as a public figure and patriot, as well as a writer of the first rank.

BLACK, GEORGE, see REVUE.

BLACKFACE, see MINSTREL SHOW.

BLACKFRIARS THEATRE, London. Two theatres were built within the boundaries of the old monastery, part of which was used by the MASTER OF THE REVELS for the storage of costumes and properties for Court entertainments. In 1576 Richard Farrant, master of the choristers at Windsor, adapted part of the building as a theatre where plays were given occasionally by the choirboys up to the time of his death in 1581. Two years later the theatre, whose dimensions and furnishings are unknown, was again being used by a BOY COMPANY made up of choristers from Windsor and the Chapel Royal in London, together with the boys attached to OXFORD'S MEN. It does not seem to have been used after 1584, in which year the young actors were seen in two of LYLY's plays.

In 1596 James BURBAGE bought another part of the old monastery, hoping to use it as a roofed theatre for the CHAMBERLAIN'S MEN in the winter months, but was prevented by opposition from the authorities who did not wish to see yet another playhouse established in London. On James's death in 1597 the property passed to his son Cuthbert BURBAGE, who in 1600 leased it to Henry Evans for the use of the Children of the Chapel Royal. When the lease ran out in 1608 the adult company, renamed the KING'S MEN in 1603, was able to take possession, remaining in occupation until the closing of the theatres in 1642. This second Blackfriars, which was certainly bigger than the first, was 46 ft by 66 ft with galleries on three sides and the stage on the fourth. Its prices were higher than those of the open-air or 'public' theatres, JONSON, some of whose plays were presented there, referring to the stools on its stage as the 'twelvepenny seats'. There was no standing room, the pit being filled with benches. Music was a great feature of the house: musicians paid to play there in the hope of attracting patronage from the nobility in the audience. It may have been the first English playhouse to use scenery, which had for some time previously been used for Court MASQUES, as it is referred to in Suckling's *Aglaura* (1637) and in Habington's *Queen of Aragon* (1640), both presented by the King's Men. An attempt to close the theatre in 1631 was unsuccessful, and it was one of the most popular of the pre-Restoration playhouses. It fell into disrepair after 1642 and was demolished in 1655.

BLAGROVE, THOMAS, see MASTER OF THE REVELS.

BLAKE, WILLIAM RUFUS, see CANADA.

BLAKELY, COLIN GEORGE EDWARD (1930–), Irish-born actor, first appeared in Belfast in 1958, and was first seen in London at the ROYAL COURT THEATRE a year later as the Second Rough Fellow in O'CASEY's *Cock-a-Doodle-Dandy*. In 1963 he joined the newly founded NATIONAL THEATRE company and made an outstanding success as Pizarro in SHAFFER's *The Royal Hunt of the Sun* (1964). After returning to the Royal Court to play Astrov in Christopher HAMPTON's new version of CHEKHOV's *Uncle Vanya* in 1970, he joined the ROYAL SHAKESPEARE COMPANY to appear in PINTER's *Old Times* (1971) and in *Titus Andronicus* (1972). He was in SHAW's *Heartbreak House* (1975) with the National Theatre, Shaffer's *Equus* (1976), AYCKBOURN's *Just Between Ourselves*, and DE FILIPPO's *Filumena* (both 1977). In 1980 he gave an excellent study of old age in Alan BENNETT's *Enjoy* and in 1981 was in a striking revival of Arthur MILLER's *All My Sons*.

BLANCHAR, PIERRE (1896–1963), French actor, of striking presence and impressive delivery, who from 1919 to 1923 played at the ODÉON, and after a short interval in films returned to the stage in such plays as Pagnol's *Jazz* (1925), ACHARD's *Domino* (1931), and SALACROU's *L'Inconnue d'Arras* (1935). From 1939 to 1946 he was with the COMÉDIE-FRANÇAISE, and thereafter continued to appear both on stage and on the screen, some of his notable stage performances being given in MONTHERLANT's *Malatesta* (1950), and CAMUS's dramatization of DOSTOEVSKY's *Les Possédés* (*The Possessed*, 1958).

BLANCHARD, KITTY, see RANKIN, ARTHUR MCKEE.

BLANCHARD, WILLIAM (1769–1835), English actor, who in 1785 joined a travelling company under the name of Bentley. He reverted to his own name in 1789 and was for many years an actor and manager in the provinces. He made his first appearance in London on 1 Oct. 1800, playing Bob Acres in SHERIDAN's *The Rivals* at COVENT GARDEN. He remained there for the rest of his career, except for a short visit to the United States in 1831 when he played at the BOWERY under HAMBLIN, who had married his daughter Elizabeth, an actress as was her mother. Blanchard was at his best in heavy comedy and character parts, particularly drunkards and old men. His son Edward Leman Blanchard (1820–89) became well known for his pantomimes and other dramatic works, of which nothing has survived. He also wrote dramatic criticism for the *Daily Telegraph* and the *Era*.

BLANCHARD'S AMPHITHEATRE, New York, see CHATHAM THEATRE (1).

BLIN, ROGER (1907–84), French actor and director, who made his first appearance in 1935 in an adaptation by ARTAUD of SHELLEY's *The Cenci*. He studied mime with BARRAULT, with whom he appeared many times during the late 1930s, and in 1949 began a successful career as a director with a production of STRINDBERG's *The Ghost Sonata*. His production of BECKETT's *En attendant Godot* in 1953 brought him into prominence; he later directed all Beckett's plays up to and including *Oh! les beaux jours* (1965). Blin was also intrumental in the discovery of ADAMOV and Dubillard and in spite of political opposition to GENET helped to popularize his plays, giving them successfully flamboyant productions. In general he was a discreet director, giving due prominence to text and actors without obtruding himself on either.

BLIZIŃSKI, JÓZEF, see POLAND.

BLOK, ALEXANDR ALEXANDROVICH (1880–1921), Russian symbolist poet, much influenced by the philosophy of Vladimir Solovyov. He had a wide knowledge of Russian and European drama and was anxious to express his ethical and aesthetic ideas in theatrical form. The first of a series of lyric plays, *The Puppet Show*, was staged by MEYERHOLD in 1906; but *The Rose and the Cross*, written in 1912, reveals a totally new conception of the struggle between the forces of good and evil. Blok hoped that STANISLAVSKY would stage it and after two years of hesitation the MOSCOW ART THEATRE did prepare to present it, but the Revolution halted the production. It was staged in 1941, in Russian, by the undergraduate members of the Cambridge Slavonic Society.

BLONDEL, see MINSTREL.

BLONDIN, CHARLES, see MUSIC-HALL.

BLOOD-TUB, see GAFF.

BLOOM, CLAIRE (1931–   ), English actress, who made her first appearance in Oxford in 1946 and was seen briefly in London a year later in WEBSTER's *The White Devil*. After a season at Stratford-upon-Avon she returned to London, and first attracted attention in FRY's *The Lady's Not for Burning* (1949) and ANOUILH's *Ring Round the Moon* (1950). She then joined the OLD VIC company, and for the next six years played leading roles in Shakespeare. After appearing as Cordelia to GIELGUD's King Lear in the West End in 1955 she returned to the Old Vic, touring America with the company and making her début in New York as Juliet, a role in which she was especially admired. In 1958 she starred in London with Vivien LEIGH in GIRAUDOUX's *Duel of Angels* and in 1959 she went to New York again, where she appeared in Fay and Michael Kanin's *Rashomon*, adapted from Akutagawa. She reappeared in London in 1961 in SARTRE's *Altona*, and in 1965 played in CHEKHOV's *Ivanov* with Gielgud. She was seen in New York in 1971 in two plays by IBSEN, *A Doll's House* and *Hedda Gabler* (repeating the former later in London), and also gave a fine performance there as Mary, Queen of Scots, in Robert BOLT's *Vivat! Vivat Regina!* in 1972. In London she was seen as Blanche du Bois in Tennessee WILLIAMS's *A Streetcar Named Desire* in 1974, and as Rebecca West in Ibsen's *Rosmersholm* in 1977. Although a very beautiful woman, she relies not on her looks but on the depth and intensity of her interpretation of arduous and challenging roles.

BLUES, see BORDER.

BLYTHE, ERNEST, see ABBEY THEATRE, DUBLIN.

BOARDS, the component parts of the stage floor, which run up- and down-stage supported on joists running crossways; used also as a phrase indicating the acting profession, to be 'on the boards' or to 'tread the boards' signifying 'to be an actor'. In earlier times the boards were removable, to facilitate the working of TRAPS and other machinery. This meant that each joist had to be supported separately by its own system of vertical posts rising from the floor of the cellar under the stage. With a raked stage a certain stress was exerted which tended to make the stage slide in the direction of the slope, and since the working of the traps and the BRIDGES made it impossible to counteract this by permanent cross-bracing, there can be found, beneath old raked stages, a system of metal tie-bars, linking one row of uprights with the next behind, and so knitting the whole together. These could be unhooked when necessary to allow the passage of a piece of scenery.

BOAR'S HEAD INN, see INNS USED AS THEATRES.

BOAT TRUCK, a platform running on castors, on which scenes or sections of scenes can be moved on and off stage. Two such trucks, pivoted each at the down-stage and off-stage corners so as to swing in and out over the acting area, are known as a Scissor-Stage or Jack-knife Stage. The Segment Stage is a large wedge-shaped wagon pivoted at the up-stage apex, designed to allow two or more sets to be mounted side by side and brought into position with a small rotation of the waggon. A further variant is the Waggon Stage (called a Slip Stage in America), run on a system of rails and lifts. As many as five waggon stages can be found in some large theatres, each capable of moving aside from the acting area, or of rising and sinking in the cellar, with a full load of scenery.

BOBÈCHE [Antoine Mandelot] (1791–*c*. 1840), French farce-player, well-known on the BOULEVARD DU TEMPLE where in his red jacket and grey tricorne with butterfly antennae, he amused the holiday crowds with his PARADES in company with Galimafré [Auguste Guérin, 1790–1870]. Both became extremely popular and were invited to private houses to entertain the guests, but the topical jokes of Bobèche offended Napoleon and he was banished. He returned under the Restoration and was once more successful, but later went into management in the provinces, failed, and was not heard of again. Galimafré, who refused to act after the fall of Napoleon in 1814, joined the stage staff at the GAÎTÉ and elsewhere and died in retirement. A play based by the brothers Cogniard on the lives of the two comedians was produced at the PALAIS-ROYAL in 1837.

BOBO, the rustic clown of early Spanish plays, who in such productions as the PASOS of Lope de RUEDA amused the audience with his naïve witticisms and malapropisms. He may have contributed to the development of the slightly more sophisticated GRACIOSO.

BOCAGE [Pierre-François Tousez] (1797–1863), French actor, who led a hard life in the provinces until 1821 and tried unsuccessfully to join the COMÉDIE-FRANÇAISE before going eventually to the ODÉON. A forceful personality, with a fine physique and a sonorous voice, he found his true vocation in the new dramas and melodramas, which suited him better than the old set forms of comedy and tragedy. He was outstanding in the plays of DUMAS *père*, particularly *Antony* (1831) and *La Tour de Nesle* (1832). He was finally accepted by the Comédie-Française, but its somewhat arid atmosphere proved uncongenial and he returned to the boulevards, where his popularity was greater than ever. In 1845 he became director of the Odéon. He died at the height of his success, much of which he owed to the dramatists of the Romantic movement.

BODEL, JEAN (*c*. 1165–1210), French poet and playwright, whose *Jeu de Saint-Nicolas* is the earliest surviving MIRACLE PLAY in any vernacular language. Its theme, the conversion to Christianity of a Saracen king whose stolen treasure is miraculously restored to him through the intervention of the saint, reflects contemporary enthusiasm for the Fourth Crusade, and it was probably first performed in 1200 in the Guildhall at Arras on 5 Dec., the eve of St Nicholas' Day. Its double plot, in which scenes of dramatic intensity are interspersed with comic episodes showing the thieves drinking and dicing in a tavern, and its racy dialogue, are unlike any of the extant LITURGICAL DRAMAS which preceded it; the realistic details of contemporary life are found again in later dramatists of 13th-century Arras, notably ADAM DE LA HALLE. The play was still being performed by students in Northern France two centuries after the death of Bodel, who in about 1202 contracted leprosy and retired to a lazar-house in Arras where he died.

BOECK, JOHANN MICHAEL (1743–93), German actor, who was the first to play Karl Moor in SCHILLER's *Die Räuber* when it was put on under DALBERG at MANNHEIM in 1782, with IFFLAND, whom Boeck greatly helped and supported in his work, as Franz Moor. He was not a particularly intelligent actor, but he had a fine presence and a passionate style which made him acceptable to the audience. His wife, Sophia Schulze, who had received her early training under SCHÖNEMANN, was also a very popular member of the Mannheim company, and in her youth was excellent in breeches parts (see MALE IMPERSONATION).

BOGUSŁAWSKI, WOJCIECH (1757–1829), Polish actor, director, and playwright, and the virtual founder of the theatre in Poland. He joined the company of the National Theatre in 1778, 13 years after its foundation, and became its director in 1783. During the next 30 years he directed its fortunes, training its actors and encouraging such new Polish playwrights as Franciszek Zabłocki and Julian Niemcewicz. Of his own plays the most successful, which are still in the repertory, were *Henry IV Goes Hunting* (1792) and *Cracovians and Highlanders* (1794): the revival of the latter play—substantially a ballad opera—by Leon SCHILLER at the National Theatre in 1950 proved an important landmark in modern Polish theatre history. Bogusławski was himself a fine actor, and was the first Pole to play Hamlet, appearing in 1797 in his own version of Shakespeare's play based on the German text by F. L. SCHRÖDER. He was responsible for the establishment of theatres in towns throughout Poland—Vilna in 1785, for instance, and Lwów in 1794—and in 1901 a theatre named after him opened in Warsaw.

BOILEAU-DESPRÉAUX, NICOLAS (1636–1711), French critic who exercised a great influence on French literature and drama. He was intended for the Church, but studied law and in 1660 took to literature. His criticism brought a new and invigorating atmosphere into the literary debates of Paris. He was not a poet but a good writer of verse. He had no imagination, no warmth, but plenty of common sense and a flair for the best in art. He appreciated and was the friend of the great men of his time, of RACINE—whom he taught to write verse—of La Fontaine, and particularly of MOLIÈRE, whose satirical genius complemented his own, and whom he called *le contemplateur*.

BOISROBERT, FRANÇOIS LE METEL DE, Abbé (1592–1662), French dramatist, and something of a buffoon, who was a friend of RICHELIEU and used his position to benefit men of letters. He was 40 before he wrote his first play, and its success encouraged him to continue. Of nearly 20 the only one to be remembered is the comedy *La Belle Plaideuse* (1654), set in the fairground of St Germain and other Parisian localities, which gives an excellent picture of contemporary manners in the capital. In it MOLIÈRE found a hint of *L'École des maris* (1661) and a scene for *L'Avare* (1668). Boisrobert was a foundation member of the French Academy and younger brother of the profligate playwright d'OUVILLE.

BOKER, GEORGE HENRY (1823–90), American dramatist, one of the few successful writers in modern times of poetic tragedy, the American counterpart of the later English writer Stephen PHILLIPS. His romantic dramas based on historical incidents include *Calaynos* (1849) and *Leonor di Guzman* (1853), both with Spanish settings; the most popular was *Francesca da Rimini*, first produced in New York in 1855. In 1883 it was revived by Lawrence BARRETT with himself as Lanciotto, Francesca's husband, who in Boker's version is the chief character. It remained in Barrett's repertory for many years and was again revived in 1901 by Otis SKINNER. The play was never acted from the printed version, as Boker himself altered it considerably for production and it was heavily revised by William WINTER for Barrett's revival. Boker was a poet of some standing; he was American envoy to Turkey, 1871–5, and to Russia, 1875–8.

BOLESLAVSKY, RICHARD, see AMERICAN LABORATORY THEATRE.

BOLT, ROBERT OXTON (1924–   ), English dramatist, whose first play *The Critic and the Heart* (1957) was staged at the OXFORD PLAYHOUSE. He first attracted attention with *Flowering Cherry* (also 1957; N.Y., 1959), produced at the HAYMARKET THEATRE with Ralph RICHARDSON as a man with a pipe-dream of owning an apple orchard. It was followed by *The Tiger and the Horse* (1960), in which Michael REDGRAVE appeared with his daughter Vanessa, and *A Man for All Seasons* (also 1960; N.Y., 1961), an outstanding portrait of Sir Thomas More, played by Paul SCOFIELD, with Leo McKern as the commentator who doubles a variety of parts—a Brechtian touch which delighted critics and audiences alike. *Gentle Jack* (1963), in which Edith EVANS appeared, had only a short run, but after a play for children, *The Thwarting of Baron Bolligrew* (1965), and *Brother and Sister* (1967), which was seen only in the provinces, Bolt had another success with *Vivat! Vivat Regina!* (1970), a study of the relationship between Elizabeth I and Mary, Queen of Scots, which was first seen at the CHICHESTER FESTIVAL THEATRE and transferred to London and to New York in 1972. Another historical play, *State of Revolution* (1977), featuring Trotsky and Lenin, was produced at the NATIONAL THEATRE.

BOLTON, GUY, see MUSICAL COMEDY.

BOLTON, Lancs, see OCTAGON THEATRE.

BONARELLI DELLA ROVERE, GUIDOBALDO (1563–1608), Italian dramatist, author of the best known PASTORAL of the 17th century, the *Filli di Sciro*, produced at Ferrara in 1607. As *Scyros* it was given at Cambridge in a Latin translation by Samuel Brooke before Charles, Prince of Wales, on 2 or 3 Mar. 1613. An English translation (printed in 1655) was played before the Court in London in 1630–31 and a second English version, by Gilbert Talbot, in 1657.

BOND, EDWARD (1935–   ), English dramatist, whose plays have aroused fierce controversy. The first, *The Pope's Wedding* (1962), was given a Sunday night 'production without décor' at the ROYAL COURT. *Saved* (1965), at the same theatre (N.Y., 1970), achieved some notoriety, mainly because of a scene in which a baby is stoned to death in its pram. Both these plays were set in contemporary England and their characters were almost inarticulate; *Early Morning* (1968), which was originally banned, was a surrealist work in which one of the characters was Queen Victoria, and *Narrow Road to the Deep North* (also 1968; N.Y., 1972) used Japanese history to attack imperialism. The last three plays were revived in 1969 at the Royal Court, an indication of the importance attributed to Bond by those who first promoted his work. Of his later plays, *Black Mass* (1970) and *Passion* (1971) were anti-apartheid and *Lear* (also 1971) was a rewriting of Shakespeare's play in an attempt to give it contemporary relevance. It was followed by *The Sea* (1973; N.Y., 1975) and *Bingo*

(1974; N.Y., 1976), the latter a portrait of Shakespeare (played in London by John GIELGUD) which condemned him for lack of political commitment. All three plays were presented at the Royal Court. Also in 1974, Bond's adaptation of WEDEKIND's *Frühlings Erwachen* as *Spring Awakening* was seen at the NATIONAL THEATRE; it was produced in New York in 1978. More recent works include *The Fool* and *A-A-America!* (both 1976), the first based on the life of the poet John Clare, the second consisting of two parts, *Grandma Faust* and *The Swing*. *The Bundle* (1978), performed by the ROYAL SHAKESPEARE COMPANY at the WAREHOUSE, was again set in historical Japan, while the characters in *The Woman* (also 1978), seen at the National Theatre, were taken from Greek tragedy. *The Worlds* (1979) was presented by the Royal Court at the Theatre Upstairs.

BONEHILL, BESSIE, see MUSIC-HALL.

BONFILS THEATRE, Denver, Colorado, founded in 1929 as the University Civic Theatre, using stage facilities provided by the University of Denver, a town whose theatre history dates back to 1858, when Charles R. THORNE and his sons appeared at Apollo Hall. By the 1890s it had three theatres, of which the Elitch Gardens Theatre, the oldest summer theatre in the U.S.A., still functions, having celebrated its 90th anniversary in 1981. In 1953 the University Civic Theatre moved to a new 550-seat building erected by Helen Bonfils, an actress and producer who was also the owner and publisher of the *Denver Post*. A non-profit-making organization, it provides amateur actors with a fully staffed professional theatre in which to work. During a season which runs from Sept. to June, the Bonfils Mainstage presents five 'Broadway-style' productions, including many new works; Bo-Ban's Cabaret, situated downstairs, offers five innovative, smaller-scale shows to an audience of 90 seated at tables. The Bonfils Theatre for Children presents four shows every season for young people, while the Festival Caravan has presented free shows in city parks each summer since 1973. A Bonfils Theatre School holds classes every Saturday, and during the summer a Creative Drama Program teaches self-expression to underprivileged children.

In 1972 the Denver Center for the Performing Arts was created out of the Bonfils Theatre organization, including the original Bonfils Theatre and, on a separate downtown location, the Helen Bonfils Theatre Complex, opened in 1980, which houses the Denver Center Theatre Company. The Complex comprises The Stage, seating 650, and The Space, seating 450, which operate a true repertory system whereby it is possible to see five plays, with essentially the same actors, in five nights. The Complex also houses The Lab, an experimental performing space seating 150.

Denver contains a number of other theatrical enterprises, of which the most interesting is possibly Central City Opera House, a beautiful granite building situated in the mountains west of Denver which opened in 1878 in what was then a prosperous mining area. With the end of the town's prosperity the theatre fell into disuse, but in 1932 a group of enthusiasts, including Robert Edmond JONES, helped to restore it, and it is now open in summer as a tourist attraction.

BONSTELLE, JESSIE [Laura Justine Bonesteele] (1872–1932), American actress and theatre manager, nicknamed 'the Maker of Stars'. She began her career in 1890 with a touring company under Augustin DALY, and a year later was managing the Shuberts' theatre in Syracuse. After similar ventures in other towns including Toronto, she leased the Garrick Theatre in Detroit in 1910, remaining there until 1922. She then bought the Playhouse in the same city, opening it in 1925; in 1928, having aroused the interest of the townsfolk, she was able to make it one of America's first civic playhouses, run on the lines of the THEATRE GUILD. Under her control it flourished until her death. She encouraged much native American talent, among her discoveries being Katharine CORNELL. In 1892 she married an actor, Alexander Hamilton Stuart, who died in 1911.

BONTEMPELLI, MASSIMO (1878–1960), Italian dramatist, akin to ROSSI DI SAN SECONDO in his musicality and his recognition of man's creative need for fantasy, and to PIRANDELLO in his vision of a world turned upside down. In *La guardia della luna* (*The Guardian of the Moon*, 1920) he traces with genuine pathos the madness and eventual death of a woman after the loss of her infant daughter, and in *Minnie la candida* (1929) he portrays a woman driven to suicide by her belief that human beings are literally puppets, and that she too may be one. The intuitive sympathy shown in these plays is contradicted by *Nostra Dea* (1925), which is based on the satirical concept that woman can acquire a personality only from her clothes.

BON TON THEATRE, New York, see KOSTER AND BIAL'S MUSIC HALL.

BOOK, the spoken dialogue of a MUSICAL COMEDY or musical play. (The word libretto is sometimes used, but more commonly denotes the text of an opera.) The book and the lyrics of the songs are often written by the same author, but composers such as Irving Berlin and Cole Porter wrote their own lyrics, as does Stephen SONDHEIM, while authors such as Guy Bolton were book rather than lyric writers. The rare feat of writing book, lyrics, and music was achieved by Ivor NOVELLO and Noël COWARD.

BOOK (or **BOOKED**) **CEILING,** see CEILING-CLOTH.

BOOK (or **BOOKED**) **FLAT,** see FLAT.

BOOK-HOLDER, the Elizabethan name for the prompter, not to be confused with the BOOK-KEEPER. In addition to prompting the actors, the book-holder was responsible for seeing that they were ready to enter on cue (a function performed later by the CALL-BOY) and for handing them their PROPS, of which he was in sole charge.

BOOK-KEEPER, an important member of an Elizabethan company, responsible for the custody of the play scripts belonging to it and for the copying and distributing of individual parts, which were supposed to be returned to him when not in use. They were probably taken home by the book-keeper when the theatre closed or the company migrated to another building or went on tour, and the apparently low survival rate of Elizabethan play texts may in part be accounted for by their concentration in so few hands, with the risk of loss by fire, theft, or carelessness.

BOOK WINGS, or Revolving Wings, a method of changing the side pieces in a scenic set used in some early Victorian theatres, among them the Theatre Royal at Ipswich where they are known to have been installed before 1857. There were usually four book wings at each wing position, each quartet being fastened like pages in a book to its own central upright spindle which passed down through a hole in the stage; at its lower end was a grooved wheel. By means of a connecting rope passing over these wheels, all the spindles could be rotated and all the WINGS of a scene changed simultaneously by means of one master handle.

BOOM, BOOMERANG, see LIGHTING.

BOOTH, BARTON (1681–1733), English actor, who while still at Westminster showed great aptitude as an actor in a schoolboy production of TERENCE's *Andria*. In 1700, after gaining experience in the provinces and in Dublin, he was engaged by BETTERTON for LINCOLN'S INN FIELDS THEATRE. He was slow to establish himself, due mainly to the jealousy of WILKS, but played a wide range of tragic parts. The most striking feature of his acting was his adoption of 'attitudes'—the pose, for instance, in which as Othello he listened with appropriate gestures to Emilia's address to the dying Desdemona: even those who were not moved by his acting conceded his effectiveness at such moments. Though not tall he had a dignified appearance and a richly resonant voice, and made a great success as Pyrrhus in PHILIPS's *The Distrest Mother* (1712) and in the title role of ADDISON's *Cato* (1713). He became manager of DRURY LANE with CIBBER and Wilks after DOGGETT retired.

BOOTH, EDWIN THOMAS (1833–93), an outstanding tragedian, and the first American actor to achieve a European reputation. Son of Junius Brutus BOOTH senior, he made his first appearance at 16 in his father's company, playing Richard III at 18 and three years later touring Australia with Laura KEENE. Although at his best in tragedy he was, like his father, much admired as Sir Giles Overreach in MASSINGER's *A New Way to Pay Old Debts*, which he played on his first visit to London in 1861 together with Shylock in *The Merchant of Venice* and the title role in BULWER-LYTTON's *Richelieu*. From 1863 to 1867 he was manager of the Winter Garden (see METROPOLITAN THEATRE), where in 1864 he played 100 consecutive performances of *Hamlet*, a record unbroken until John BARRYMORE's 101 in 1922. The same year saw a performance of *Julius Caesar*, in which Edwin and his two brothers, John Wilkes and Junius Brutus junior (below), appeared together for the only time. After the destruction by fire of the Winter Garden in 1867 he built his own theatre (see BOOTH'S), which opened in 1869 with *Romeo and Juliet* in which he played opposite Mary McVicker, later his second wife. (His first wife was the actress Mary Devlin, 1840–63.) The venture was not a success and Booth went bankrupt in 1873, returning successfully to starring tours in the United States, England, and Germany. In 1881, the year of his second wife's death, he appeared at the LYCEUM in London by invitation of Henry IRVING, alternating with him the roles of Othello and Iago. In 1888 he presented his house in Gramercy Park to the newly founded PLAYERS' CLUB, of which he became the first president, living there until his death. He was an unhappy man, his habitual melancholia aggravated by the madness of his father, second wife, and younger brother, the assassin of Abraham Lincoln.

BOOTH, JOHN WILKES (1839–65), American actor, younger brother of Edwin BOOTH, who at 10.22 p.m. on 14 Apr. 1865 assassinated Abraham Lincoln in FORD'S THEATRE, Washington, during a performance by Laura KEENE's company of Tom TAYLOR's *Our American Cousin*. There are two theories as to the reason for this act, one asserting that Booth, an indifferent actor, was jealous of his elder brother's success and sought notoriety through crime, the other ascribing his action to mistaken patriotism. The whole question is systematically surveyed in *The Mad Booths of Maryland* (1940) by Stanley Kimmel.

BOOTH, JUNIUS BRUTUS, senior (1796–1852), English-born American actor, eldest of a notable theatre family. Trained for the law, he went on the stage at 17 and after touring for some years appeared at COVENT GARDEN as Richard III, almost immediately entering into rivalry with Edmund KEAN. He was seen at Covent Garden as Sir Giles Overreach in MASSINGER's *A New Way to Pay Old Debts*, Posthumus Leonatus in *Cymbeline*, and in 1818 as Shylock. Two years later he went to DRURY LANE, where he played Iago to Kean's Othello, Edgar to his Lear, and Pierre to his Jaffier in OTWAY's *Venice Preserv'd*. In 1821 he deserted his legal wife, by whom he had one son, and with Mary Ann Holmes, a flower-seller in Bow Street, went to America, making his first appearance there as Richard III at Richmond, Va. From then until his death, except for two short visits to Drury Lane, he was constantly seen in America, making his first appearance in New York late in 1821, again as Richard III, and becoming manager of several theatres. Among his other achievements was the playing in French of Oreste in RACINE's *Andromaque*, and he is said to have played Shylock in Hebrew. He made his last appearance at the St Charles Theatre, New Orleans. He was a rough and unpolished actor but full of grandeur and eloquence, with a resonant voice and ample gestures. He and Mary Ann had ten children before in 1851 his wife divorced him and they were able to marry. Several of them died young, a fact which contributed not a little to Booth's habitual melancholy and intemperance; but three of his sons, Edwin and John Wilkes (above) and Junius Brutus junior (below), achieved success on the stage, and his daughter Asia married the actor-manager John Sleeper CLARKE.

BOOTH, JUNIUS BRUTUS, junior (1821–83), American actor, brother of Edwin and John Wilkes Booth (above). He made his first appearance on the stage on tour with his father Junius Brutus Booth senior (above) in 1835, playing Iago to his father's Othello in 1852. Though not as gifted as his father, he was a good manager and director, and was for many years a useful member of the stock company at the BOWERY THEATRE. One of his best parts was Dan Lowrie in Frances Hodgson Burnett's *That Lass o' Lowrie's* (1878). He married three times and had two sons on the stage. The elder shot himself and his wife in a London hotel; the younger, Sydney Barton Booth (1873–1937), who made his début at WALLACK'S THEATRE in 1892, became a successful actor and also played in VAUDEVILLE in 1902 and 1903.

BOOTH, SHIRLEY [Thelma Booth Ford] (1907–), American actress, who first appeared on the stage in 1919 and made her New York début six years later. Her successful career later included such parts as Elizabeth Imbrie in Philip BARRY's *The Philadelphia Story* (1939), Ruth Sherwood in *My Sister Eileen* (1940) by Joseph FIELDS and Jerome Chodorov, and Grace Woods in Fay Kanin's *Goodbye, My Fancy* (1948). She made her greatest impact in 1950, in the role with which she is always associated, the bored, sluttish, unloved Lola in William INGE's *Come Back, Little Sheba*. In 1951 she appeared in her first musical, *A Tree Grows in Brooklyn* based on a book by Betty Smith, and after playing Leona in Arthur Laurents's *The Time of the Cuckoo* (1952) was seen in another musical *By the Beautiful Sea* (1954). This was followed by Bunny Watson in William Marchand's *The Desk Set* (1955) and by the title role in *Juno* (1959), a musical based on O'CASEY's *Juno and the Paycock*. She was away from the stage during most of the 1960s in a television series, but returned in the musical *Look to the Lilies* and a revival of Noël COWARD's *Hay Fever* (both 1970).

BOOTHS, portable theatres which provided travelling companies with an adequate stage, supplanting the adapted inn-yards, barns, or makeshift rooms which were at first the only places available for theatrical productions. Booths were probably first erected in the grounds which accommodated FAIRS in England and on the Continent, and consisted originally of tents housing a small stage on which were presented short sketches and DROLLS interspersed with juggling and rope dancing. Portable buildings, easily dismantled and re-erected, later made their appearance, becoming more elaborate as time went on. In the 19th century they seemed to have conformed to a general pattern, the stage itself being solidly constructed, sometimes on the carts that carried the show around, and having a few simple backcloths and properties. The auditorium consisted of a canvas tent that could easily be rolled up and transported on the wagons, the seats being almost always plain wooden planks. Some of the more successful showmen even contrived to build up a front of gaudily painted canvas FLATS, sometimes with a platform outside on which the performers could appear to tempt the audience in. This was a particular feature of the French fairground theatres which evolved from the early booths, and the first permanent fairground theatres in Paris—which had no counterpart in England—had large balconies for the performance of introductory PARADES.

Within the booths performances were gone through as quickly as possible so as to clear the seats for another audience, often as many as ten or twelve a day. The plays were necessarily simple, farces or strong drama based on popular legends, stories from the classics, the Bible, or history. There were seldom any written scripts, though sometimes a popular play from London or Paris would be adapted for a local audience, particularly if it dealt with a topical and sensational crime. The classic example in English is *Maria Marten, or the Murder in the Red Barn*, an anonymous dramatization of a murder which took place in 1828.

Although travelling companies still transport their own buildings, as the circus does, these bear little resemblance to the humble booths of the fairgrounds, which lasted longest for the accommodation of PUPPET shows, and of which the portable stage of the PUNCH AND JUDY show is a vestigial remnant.

BOOTH'S THEATRE, New York, on the southwest corner of the Avenue of the Americas and 23rd Street. Built for Edwin BOOTH, who hoped to establish it as a national playhouse for poetic drama, it had several unusual features. The stage had no RAKE, the WINGS being supported by braces; there was a tall 'stage house', believed to be the first in New York, and large, hydraulically-powered elevator trays for lowering three-dimensional scenery to the cellar, though during Booth's own management BOX-SETS were often used. The venture began auspiciously on 3 Feb. 1869 with *Romeo and Juliet*, starring Booth and Mary McVicker, later his second wife, followed by *Othello*. During subsequent seasons Booth was seen in many of his finest parts, supported by such outstanding players as Lawrence BARRETT and Charlotte CUSHMAN. It was at this theatre that Adelaide NEILSON made her first appearance in New York, and the younger James WALLACK and Kate BATEMAN their last appearances; the elder James HACKETT was seen there in his last regular New York season; Joseph JEFFERSON III appeared in BOUCICAULT's adaptation of *Rip Van Winkle*; JANAUSCHEK in her own adaptation of DICKENS's *Bleak House*; and the young Minnie Maddern, later famous as Mrs FISKE, made an early appearance as Prince Arthur in *King John*. But Booth was not able to make a success of his venture and in 1873 he went bankrupt leaving the management of the theatre to his elder brother Junius Brutus BOOTH. A year later it passed out of the control of the Booth family and was used for plays by Boucicault and for the farewell of Charlotte Cushman, who ended her last season there with Lady Macbeth on 7 Nov. 1874. The great event of 1875 was the importation from the Theatre Royal, Manchester, of *Henry V* directed by Louis CALVERT, whose production of BYRON's *Sardanapulus*, with scenery by TELBIN, was seen in the same season. In 1879 the lease of the theatre was taken over by Boucicault, under whom many stars including Rose COGHLAN, John BROUGHAM, and Robert B. MANTELL made successful appearances, Charles THORNE being seen there in his last performance before his early death in 1883. The last production, *Romeo and Juliet* with MODJESKA and Maurice BARRYMORE, closed on 30 Apr. 1883, after which the building was pulled down and a large department store was erected on the site.

BOOTH THEATRE, New York, at 222 West 45th Street. Built by Winthrop AMES in association with Lee Shubert, this small playhouse seating 766 opened on 16 Oct. 1913 with *The Great Adventure*, based by Arnold BENNETT on his novel *Buried Alive*. In 1925 there was a production of *Hamlet* in modern dress, and in 1936 KAUFMAN and HART's *You Can't Take It With You* began a run of two years. Another outstanding production was the PULITZER PRIZE-winner *The Time of Your Life* (1939) by William SAROYAN, and in 1946 there was a fantastic production of MOLIÈRE's *Le Bourgeois gentilhomme* as *The Would-Be Gentleman*, with the comedian Bobby Clark as Monsieur Jourdain. Later productions included INGE's *Come Back, Little Sheba* (1950) with Shirley BOOTH, the two-character play *Two for the Seesaw* (1958) by William Gibson, the revue *At the Drop of Another Hat* (1966) with Michael Flanders and Donald Swann, and PINTER's *The Birthday Party* (1967). Leonard Gershe's *Butterflies Are Free* (1969) had a long run and successful productions of the 1970s included two from the PUBLIC THEATRE, Jason Miller's *That Championship Season* (1972) and *For Colored Girls Who Have Considered Suicide: When the Rainbow is Enuf* (1976), based on the poems and prose of Ntozake Shange. Bernard Pomerance's *The Elephant Man* began a long run in 1979.

BOQUET, LOUIS-RENÉ, see COSTUME.

BORBERG, SVEND, see DENMARK.

BORCHERT, WOLFGANG (1921–47), German dramatist, whose only play, *Draussen vor der Tür*, was performed on 21 Nov. 1947, the day after the author's death. A passionate protest against contemporary corruption and decadence, it made a deep impact on war-torn Germany, and has been frequently revived, being considered a link between the plays of EXPRESSIONISM and the later Theatre of the ABSURD. Its central character is a soldier who returns from a prisoner-of-war camp (of which Borchert had personal experience after being captured on the Russian front) and is unable to adapt himself to civilian life. Through him Borchert voiced the nihilism and despair which prevailed in Germany immediately after the Second World War. As *The Man Outside*, it was published in 1952 in a translation by David Porter.

BORDER, a narrow strip of painted cloth, known also as a Top Drop, battened at the top edge only and used to mask the top of the stage from the audience. If the lower edge is cut to represent foliage it is known as a Tree Border. The use of cloud borders, known as CLOUDINGS, to mask the top of almost any scene was formerly common. In Victorian times plain Sky Borders were sometimes known as Blues. Tails, or Legs, on a border, are long vertically-hanging extensions at each end which with the border form an arch over the scene. The furthest downstage border, in front of the front CURTAIN, is known as the Proscenium Pelmet (Grand Drape or Valance in America), while the one directly upstage of it is called the Proscenium Border (Teaser or Valance in America).

BORDER LIGHT, see BATTEN and LIGHTING.

BORISOVA, YULIA KONSTANTINOVNA (1925–), Soviet actress, who in 1949 became a member of the company at the VAKHTANGOV THEATRE. There she played a number of parts portraying the vivid personality of the contemporary Russian girl, among them Natasha in *City at Dawn* (1957) and Valya in *It Happened in Irkutsk* (1959), both by ARBUZOV. One of her finest performances was as Nastasya Filipovna in the dramatized version of DOSTOEVSKY's *The Idiot* in 1958.

BOSTON, Massachusetts. Early efforts to introduce plays into Boston met with fierce puritanical opposition, and *Othello*, on its first showing, had to be presented as a 'Moral Lecture against the Sin of Jealousy'. Plays continued to be disguised in this way even after the building of the first permanent playhouse, the New Exhibition Room, later the Board Alley Theatre, in 1792. Nevertheless, at least two more theatres were built before the end of the century, the Federal Street in 1794 and the Haymarket in 1796. The former, a substantial brick building, was destroyed by fire in 1798, rebuilt immediately, and survived until 1852. The latter, built entirely of wood, was unsuccessful, and was pulled down in 1803. A successful rival to the Federal Street Theatre for many years was the Tremont Street Theatre, which opened in 1827 and was destroyed by fire in 1852, having ceased to be used as a theatre in 1843. In 1832 the American Amphitheatre, later better known as the National Theatre, was opened, and, after changing its name to the Warren, was also burned down in 1852. The best known and most successful theatre in the town was perhaps the Boston Museum. The first playhouse of this name, which stood on the corner of Tremont and Bromfield Streets, opened in 1841, and proved so popular that in 1846 a much larger theatre of the same name was erected on the east side of Tremont Street, between Court and School Streets. It underwent major renovation in 1872, and survived until 1893, reaching the peak of its prosperity between 1873 and 1883, when it housed an excellent resident stock company with good visiting stars. In 1846 the actor J. H. HACKETT opened the Howard Athenaeum, and in 1854 the Boston Theatre was built to replace the Federal Street Theatre. With a seating capacity of 3,140 and an almost circular auditorium, it was one of the most advanced theatre buildings of its time. The centenary of the opening of the Federal Street Theatre saw the opening of the Castle Square Theatre. Situated at the junction of Tremont, Chandler, and Ferdinand Streets, it held 1,800, and was considered one of 'the finest, safest, and best-equipped' playhouses in the country. Boston's pre-eminence in theatre, however, has now waned, and desite its long theatrical history, its excellence in other cultural fields, and its proximity to HARVARD UNIVERSITY, it no longer ranks as one of the outstanding theatrical cities of the United States.

BOTTLE, to, see PUNCH AND JUDY.

BOTTOMLEY, GORDON (1874–1948), English poet and dramatist, whose POETIC DRAMAS *King Lear's Wife* (1915), *Britain's Daughters* (1922), *Gruach* (1923), and *Laodice and Danae* (1930), were all produced in London. Although they show the influence of Shakespeare and of the *nō* play of JAPAN, their themes are taken from legends of the north and he introduced to England, as YEATS did to Ireland, the Celtic twilight, the old world of fear and evil. He wrote also a number of one-act plays, performed mainly by amateurs, and *The Acts of St Peter*, the Exeter Cathedral Festival play for 1933.

BOUCHER, FRANÇOIS, see COSTUME.

BOUCICAULT (BOURCICAULT, BOURSIQUOT), DION(-ysius) LARDNER (1820–90), Irish playwright, actor, director and theatre manager, born in Dublin of Huguenot extraction. He began his career as an actor, first appearing under the name of Lee Moreton in 1838, but reverted to his own name after the success of his play *London Assurance* (1841), which was produced in London by Mme VESTRIS with herself and her husband Charles James MATHEWS in the cast. It was constantly revived during the 19th century, and a highly successful production in 1970 by the ROYAL SHAKESPEARE COMPANY proved the continuing vitality of a playwright who bridges the gap between earlier Irish writers, CONGREVE, FARQUHAR, GOLDSMITH, and SHERIDAN, to whom Boucicault owed so much, and the later Irishmen SHAW, WILDE, SYNGE, and O'CASEY, who all acknowledged their debt to him. During the 1840s and again in the 1860s, Boucicault was closely associated with Ben WEBSTER, both at the ADELPHI, for which he adapted *Don Cesar de Bazan* in 1844, and at the HAYMARKET, where his *Old Heads and Young Hearts* was first seen in the same year. In 1845 he was in France with his first wife, who died there, and on his return was engaged by Charles KEAN as resident dramatist at the PRINCESS'S THEATRE, where his adaptations from the French of *The Corsican Brothers* in 1852 and of *Louis XI* in 1855 were given spectacular production. His successful career in the U.S.A. dates from about the same time with *The Poor of New York* (1857), later known, according to the locality in which it was being produced, as *The Poor* (or *The Streets*) *of Liverpool, London*, etc. It was followed by *Jessie Brown; or, the Relief of Lucknow* (1858), in which the part of Jessie was played by Boucicault's common-law wife Agnes ROBERTSON, who thereafter appeared in many of his plays including the spectacular *The Octoroon; or, Life in Louisiana* (1859), the first play to treat the American negro seriously. Also in 1859 Boucicault, who had already shown great skill in adapting plays as well as writing them, made a successful dramatization of DICKENS's *The Cricket on the Hearth* as *Dot*, in which Joseph JEFFERSON played his first serious part, Caleb Plummer.

In 1860 Boucicault returned to England, but not before producing for Laura KEENE in New York the first of his highly successful Irish plays, *The Colleen Bawn; or, the Brides of Garryowen*. Other plays of 'authentic' Irish life and character, but still cast in the melodramatic mould of the day, were *Arrah-na-Pogue; or, The Wicklow Wedding* (1864) and *The Shaughraun* (1874).

Among Boucicault's later plays were a new version of Washington IRVING's *Rip Van Winkle* for Jefferson, first produced in 1865, *Hunted Down* (1866), originally known as *The Two Lives of Mary Leigh*, in which Henry IRVING appeared early in his career: *The Flying Scud; or, Four-Legged Fortune* (also 1866), one of the first of the popular horse-racing melodramas; and a musical extravaganza, *Babil and Bijou; or, the Lost Regalia* (1872), in which he collaborated with J. R. PLANCHÉ. He then returned to America, where he remained, with occasional visits to England, until his death, and re-entered the New York theatre with *Mimi* (1872) and *Belle Lamar* (1874), one of the first plays to depict the American Civil War.

During his lifetime Boucicault was probably more esteemed as an actor than as a dramatist, in spite of his authorship of about 250 plays. He was highly praised, both by the critics and by his fellow-actors, for his skill in characterization, his timing, and his technical perfection, and towards the end of his life he lectured and wrote on the art of acting as well as running a school for young actors in association with A. M. PALMER and connected with the MADISON SQUARE THEATRE. His facility and inventiveness as a director and his innovations in theatre management made him one of the outstanding personalities of the 19th-century theatre. He was also actively engaged with the American dramatists BOKER and BIRD in their efforts to ensure the passing of the first American dramatic COPYRIGHT law in 1856, more than twenty years after that in England, and by his demands for a more equitable deal for playwrights from managements materially assisted the establishment of the ROYALTY system; he was the first dramatist in England to receive a royalty on his plays, beginning with *The Octoroon* on its London production in 1861 at the Adelphi. In 1888 he married a young American actress Louise Thorndyke, after repudiating Agnes Robertson by whom he had had four children; all were on the stage, the best-known being 'Dot' and Nina (below).

BOUCICAULT, 'DOT' [Darley George] (1859–1929), English actor and dramatist, son of Dion BOUCICAULT and Agnes ROBERTSON. He made his first appearance in his father's company in New York in 1879, and then appeared in London and toured in Australia with Robert Brough. From 1901 to 1915 he directed the plays given at the DUKE OF YORK'S THEATRE in London under the management of Charles FROHMAN. He was an excellent actor, particularly in later life, two of his outstanding roles being Sir William Gower in PINERO's *Trelawny of the 'Wells'* (1898) and Carraway Pim in A. A. Milne's *Mr Pim Passes By* (1901). He married in 1901 the actress Irene VANBRUGH, who appeared in many of his productions.

BOUCICAULT, NINA (1867–1950), English actress, daughter of Dion BOUCICAULT and Agnes ROBERTSON. Born in London, she first appeared in her father's company in America, playing Eily O'Connor in a revival of his *The Colleen Bawn* in 1885. She then toured Australia with him, and was in London in 1892 when she appeared as Flossie Trivett in H. A. Kennedy's *The New Wing* and as Kitty Verdun in the first London production of Brandon THOMAS's *Charley's Aunt*. In 1903 she was much admired as Bessie Broke in a dramatization of Kipling's *The Light that Failed*, and thereafter had a long and successful career, making her last appearance in 1936 as the Countess Mortimer in the first public performance of GRANVILLE-BARKER's *Waste*. She is, however, chiefly remembered as the first actress to play Peter in BARRIE's *Peter Pan* (1904).

BOUFFES DU NORD, Théâtre des, Paris, see INTERNATIONAL CENTRE OF THEATRE RESEARCH.

BOULEVARD DU TEMPLE, fairground in Paris which became a centre of entertainment, with circuses, booths (in which BOBÈCHE and Galimafré revived memories of the earlier PARADES), children's theatres, and puppet-shows. During the Revolution, a number of small permanent theatres were built there, among them the FUNAMBULES, the AMBIGU, and the GAÎTÉ, in which actors like DEBURAU and FRÉDÉRICK appeared in plays, often by the elder DUMAS and later by PIXÉRÉCOURT. From the latter's works it got its nickname of 'the Boulevard of Crime'. The whole picturesque scene, with its sideshows, waxworks, fireworks, museums, cafés, concerts, and perambulating ballad-singers, was swept away in Haussman's rebuilding of Paris in 1862, and the Boulevard Voltaire now occupies most of the site.

BOULEVARD PLAYS (*Théâtre de Boulevard*), term used to describe the type of play usually found in the commercial theatres in Paris, the equivalent of London's West End or New York's Broadway. In the 18th century only licensed theatres were allowed to operate within the city, and unlicensed booths offered holiday entertainment to the crowds outside the walls. Many of these developed into regular theatres, and when in the 1850s the old moats were filled in to provide the foundations for Haussmann's network of strategic and ornamental boulevards, they found themselves advantageously placed, especially along the section which runs between the Boulevard de Strasbourg and the Madeleine on the right bank of the Seine and in the area around Montparnasse on the left. The audiences that filled these luxurious theatres came to be amused and entertained, and were not prepared to accept experimental or too-serious productions: their staple fare was farce, comedy, melodrama, and romance, and they offered substantial financial rewards to those dramatists who were prepared to fulfil their demands. The peak of the Boulevard play was during the 'belle époque' when Paris was the playground of Europe, roughly from 1880 to 1914. After a final successful period in the 1920s the boulevard theatres began to decline, eroded on the one hand by the cinema and television and on the other by the growing influence of the AVANT-GARDE and experimental theatres. By the 1970s there remained only about 20 theatres offering the old-style boulevard plays, among them the Théâtre du PALAIS-ROYAL and the COMÉDIE DES CHAMPS-ÉLYSÉES. In these the salient features of the boulevard plays are preserved, either through revivals of the works of such playwrights as AUGIER, FEYDEAU, and LABICHE, or of contemporary playwrights in the same style—ACHARD, Roussin, and MARCEAU. Some former experimental dramatists have also had their later plays produced in boulevard theatres, among them ANOUILH, OBALDIA, and IONESCO, while the companies attract actors and directors both from the experimental theatres and from such state enterprises as the COMÉDIE-FRANÇAISE. Much of their financial support comes from provincial tours and from foreign and provincial visitors to Paris, and their repertory includes many adaptations of West End and Broadway successes, though their favourite themes—light-hearted attacks on sexual morality, witty topical satire, and whimsical admiration of the power of money—are considered typically Parisian. Those who use the term 'boulevard play' in derision are in danger of forgetting that one of the chief functions of the theatre is to entertain, and that these theatres cater for large audiences which are not attracted by experimental play-writing and production. They may also foster the talents of playwrights whose work may later be judged of permanent value—as witness the post-war reappraisal of Feydeau, once considered a typical purveyor of 'boulevard' material.

BOURCHIER, ARTHUR (1863–1927), English actor-manager, husband of Violet VANBRUGH whom he married in 1894. He was one of the founders of the OXFORD University Dramatic Society (OUDS), and appeared in several of their productions of Shakespeare. On leaving the university he joined Lillie LANGTRY's company, making his first professional appearance as Jacques in *As You Like It* at Wolverhampton. After his marriage he toured with his wife, who played the leading parts in his productions, many of them comedies and farces adapted from the French. From 1900 to 1906 he was in management at the GARRICK THEATRE, where he appeared in Shakespeare, and also in a number of modern plays, among them PINERO's *Iris* (1901), Henry Arthur JONES's *Whitewashing Julia* (1903), and Sutro's *The Walls of Jericho* (1904). In 1910 he joined TREE's company at HER (then His) MAJESTY'S THEATRE and was seen mainly in Shakespeare, being particularly admired as Henry VIII. He later appeared at the Oxford Music-Hall in a sketch based on Bruce Bairnsfather's *The Better 'Ole* (1917). He was at his best in truculent, fiery, or hearty parts, but had little subtlety and resented criticism, spoiling much of his best work by impatience. In 1918, after a divorce, he married the actress Violet Marion Kyrle Bellew.

BOURDET, ÉDOUARD (1887–1944), French dramatist, whose plays on sexual themes, written under the influence of BECQUE, are subtle and penetrating portrayals of social and moral disequilibrium. The first, *Le Rubicon* (1910), was a comedy based on the overcoming of nuptial inhibitions, but with *La Cage ouverte* (1920) and *L'Heure du berger* (1922) he displayed analytical powers shown to their fullest effect in *La Prisonnière* (1926), a frank but compassionate study of lesbianism. Some years later he followed this with a more light-hearted study of homosexuality in *La Fleur des pois* (1932). Of his other plays the most important are *Vient de paraître* (*Just Published*, 1927), a satire on the commercialism of literature, and *Le Sexe faible* (1930), on the corruption of fashionable society. Several of his plays were successfully produced in English translations in London and New York. He was administrator of the COMÉDIE-FRANÇAISE from 1936 to 1940 and many famous actors, including FRESNAY, appeared in his plays.

BOURGOGNE, Théâtre de l'Hôtel de, the first theatre in Paris, built in 1548 in the ruins of the palace of the Dukes of Burgundy in the rue Mauconseil for occupation by the CONFRÉRIE DE LA PASSION, who held the monopoly of acting in Paris. Recent research has established that the theatre was long and narrow, some 102 ft long and about 42 ft wide. The total stage depth from front edge to back wall would have been approximately 43 ft and the acting area of necessity yet more restricted. The greater part of the auditorium was occupied by a pit in which the spectators stood; at the back, on a base of only 10 ft, were sharply rising tiers of benches. There were two rows of boxes, seven down each side and five along the back in each row—38 in all. Both stage and auditorium were lit by candles which had to be snuffed frequently during the performance. The medieval open-air MULTIPLE SETTING was still in vogue, in a cramped and curved indoor version which forced the actors to declaim downstage.

As soon as the new theatre was ready for occupation the Confrérie were forbidden to appear in the religious plays which formed the greater part of their repertory, on the grounds that the mixture of sacred and profane elements which had by now become general in the MYSTERY PLAYS was bringing religion into disrepute. The company struggled along for some time with farces and secular plays but gradually lost their audiences and towards the end of the 16th century were forced to hire out their hall to travelling companies from the provinces. As early as 1578 Agnan SARAT appeared there, and an English company under Jean Thays is traditionally believed to have been there 20 years later. The first permanent company to occupy the theatre was that of VALLERAN-LECOMTE, known as the King's Players; it reigned supreme in Paris, with BELLEVILLE as its leading actor, until in 1634 the Théâtre du MARAIS opened under MONTDORY. His early retirement again left the Hôtel de Bourgogne, under Belleville's successors FLORIDOR and MONTFLEURY, in an unchallenged position until the arrival of MOLIÈRE in Paris in 1658. Many of the outstanding plays of the 17th century were first seen at the Hôtel de Bourgogne, until in 1680 the company was finally merged with the other actors in Paris to form the COMÉDIE-FRANÇAISE. The new combined company moved to the theatre in the rue Guénégaud, and the stage of the Hôtel de Bourgogne was occupied intermittently by the COMÉDIE-ITALIENNE until 1783.

BOURSAULT, EDMÉ (1638–1701), French dramatist, who was almost entirely self-educated, and by study and a natural bent for literature made a name for himself. He had had several short plays produced at the Hôtel de BOURGOGNE when, thinking himself attacked by MOLIÈRE in the character of Lisidas in *La Critique de l'École des femmes* (1663), he wrote for the company there *Le portrait du peintre* (also 1663), in which he accused Molière of putting recognizable contemporary people on stage. This drew from Molière *L'Impromptu de Versailles*, performed before Louis XIV later the same year, which lampooned not only Boursault but all the actors of the Hôtel de Bourgogne. Boursault, fundamentally a modest and good-natured man, was soon reconciled with Molière, whom he much admired, and with BOILEAU and RACINE, with whom he had also quarrelled. For many years he wrote no more plays, but returned to the theatre in 1683 with a comedy, *Le Mercure galant*, named after a famous periodical of the day, and two satires on contemporary manners based on the fables of La Fontaine—*Ésope à la ville* (1690) and *Ésope à la cour* (1701), produced posthumously.

BOUSCHET, JAN [Jean Posset], see ENGLISH COMEDIANS.

BOUWMEESTER, LOUIS, see NETHERLANDS.

BOWERY THEATRE, New York. (1) At 46–8 Bowery, opened on 23 Oct. 1826 as the New York Theatre, Bowery, but was always known simply as the Bowery. It was lit by shaded gas lights, not by the naked jets used elsewhere. The managers were G. H. BARRETT and Gilfert, whose first production was HOLCROFT's *The Road to Ruin*, with a fine cast: they also mounted SHERIDAN's *The School for Scandal*, starring Barrett and Mrs Gilfert. Edwin FORREST made many of his early successes at the Bowery, first appearing there as Othello. On 26 May 1828 the theatre was burnt down, and the actors had to migrate to the Sans Souci at NIBLO'S GARDEN. A second Bowery opened on 20 Aug. of the same year, again with Forrest as its star, and when Gilfert died a year later HAMBLIN entered on 20 years of management. Under him the Bowery was the first theatre in New York to have continuous runs.

The Bowery saw, in the season of 1835–6, the last appearance in New York of Thomas Abthorpe COOPER and the first of Charlotte CUSHMAN. In Sept. 1836 the theatre was once again burnt down, and although a third Bowery was ready by 2 Jan. 1837 it too was destroyed by fire on 18 Feb. 1838 and not reopened until 6 May 1839, once more under the management of Hamblin. A series of unsuccessful plays had brought the fortunes of the theatre to a very low ebb when on 25 Apr. 1845 it was again burnt down and rebuilt, reopening on 4 Aug. 1845.

In 1851 an actor who was to be the idol of the Bowery audiences, Edward EDDY, made his appearance in a series of strong parts. He took over in 1857 but, unable to restore prosperity to the old theatre, closed it after one season. On 7 Aug. 1858 it reopened under George L. FOX and James W. Lingard with plays and pantomimes. When they left for the New Bowery (below), the old Bowery, by now the oldest playhouse in New York, was once again subjected to a series of incompetent managers. During the Civil War it was occupied by the military, and then became a circus. After thorough renovation, Fox reopened it as Fox's Old Bowery and put on a succession of novelties, including a pantomime based on the tale of Old Dame Trot and her Wonderful Cat. Long after old-fashioned farce had vanished from New York's playhouses it continued to flourish at the old Bowery, but MELODRAMA was always its staple fare. The theatre then fell a victim to the prevalent craze for BURLESQUE, and closed in 1878, reopening a year later as the Thalia for plays in German. It was destroyed by fire in 1923 and again, for the sixth and last time, in 1929.

(2) The New Bowery opened on 5 Sept. 1859, under Fox and Lingard, with a good company, some filched from the old Bowery, where the managers had been the previous year. It had a short and undistinguished career, enlivened only by visits from guest stars and the inevitable *Uncle Tom's Cabin* in AIKEN's adaptation. The season of 1866–7 was well under way, mainly with melodrama and pantomime, when on 18 Dec. 1866 the theatre was destroyed by fire. It was never rebuilt.

A 'Zoological Institute' at 37–9 Broadway was in 1835 adapted for circus acts and named the

Bowery Amphitheatre. In 1844 it was run as a theatre, the Knickerbocker, and by 1854 it was called the Stadt Theatre, and German opera was presented there as well as plays in English.

BOXES, small compartments in the AUDITORIUM which formerly ran right round the PIT, each holding up to 20 people. This arrangement can still be found in some opera houses, but elsewhere they are situated only on each side of the stage and hold 4 to 6 people. Though often the most expensive seats in the theatre they were also the most frustrating, since they gave a distorted sideways view of the stage and were usually occupied by those who wished to be seen even more than to see. Even in theatres that retain them they are now seldom used, except to house lighting or television equipment. In many Court and Continental theatres a large Royal or King's Box was placed in the centre of the first tier over the entrance to the pit, directly facing the stage, a position it still retains even in the absence of Royalty, except in England where in Georgian times it was shifted to one side.

BOX-OFFICE, that part of a theatre devoted to the selling of seats. The name dates from the time when the majority of seats in the house, except in the GALLERY, were in BOXES. The box-office staff may consist of from one to six persons, according to the size of the house, some of the larger theatres having separate divisions for differently priced seats, and for advance booking. The term pay-box, used in the 18th century for the recesses at which entrance money was paid, is now generally reserved for those windows which collect money at the door, and do no advance or telephone bookings; but customs may vary from one theatre to another. The box-office is under the control of a Box-Office Manager, an important member of the theatre staff, who may remain attached to one theatre for many years and become, as it were, a repository of its history and traditions. The term 'box-office' is used metaphorically to indicate the appeal of a play, or player, to the public.

BOX-SET, a scene representing the three walls and ceiling of a room, not by means of perspective painting on WINGS, BACKCLOTH, and BORDERS as in earlier times, but by an arrangement of FLATS which form continuous walls, with practicable doors and windows, the whole completely covered in by a CEILING CLOTH. The flats are lined-and-cleated together, edge to edge, on any desired ground plan (or, particularly if they are to be used on a REVOLVING STAGE, joined by hinges), with reveals, or false thickness-pieces, to give solidity to the openings and returns, or setbacks in the walls. The bottoms of openings in a door or arch flat are strengthened by flat metal strips called sill-irons (saddle-irons in America).

In England the box-set was first used in Mme VESTRIS's production of W. B. Bernard's *The Conquering Game* at the OLYMPIC THEATRE on 28 Nov. 1832, and brought to perfection in her production of BOUCICAULT's *London Assurance* at COVENT GARDEN on 4 Mar. 1841. It was in general use for more than a century, especially for contemporary drama; but since it can only function behind a PROSCENIUM arch, the widespread acceptance of various kinds of OPEN STAGE, as well as the cost of changing the scene for each act, has almost led to its disappearance even from traditional theatres, though it is still used occasionally for one-set plays.

BOY BISHOP, the elected ruler of the festivities indulged in by the boys of choir schools and monastic establishments in medieval times, either on St Nicholas's Day (6 Dec.) or on Holy Innocents' Day (28 Dec.). Earlier than the FEAST OF FOOLS with which it was sometimes amalgamated, this festival probably began as a serious church service conducted by the boys themselves, but soon developed a secular, merry-making character, with plays—usually in Latin—acted by the children and considered as part of their educational curriculum. It can be traced all over Europe from the 13th century, and was extremely popular in England where the first mention of it (at York) dates from before 1221. The custom of appointing a Boy Bishop survived until the Reformation; it was briefly revived at Bermondsey in 1963, and later at Boston, Lincs., but has now died out.

BOY COMPANIES, the acting troupes of choirboys attached to the Chapels Royal in London and Windsor (known also as the Children of the Chapel and the Children of Windsor) and to St Paul's Cathedral (the Children of Paul's), who in the 16th and early 17th centuries formed an important part of the theatrical world of London. As in other educational establishments where SCHOOL DRAMA flourished, the performance of plays by children had always been considered part of the curriculum and an isolated note records that the Children of Paul's appeared in a Biblical play as early as Christmas 1378. But it was not until 1517 that the Children of the Chapel, under their then Master William Cornish, emerged as a regularly constituted dramatic group, playing regularly at Court, often for the entertainment of distinguished visitors, and taking part in the many elaborate DISGUISINGS that Cornish organized for the young Henry VIII. Their appearances before the sovereign continued under successive Masters, the most successful being Richard Edwardes who may have written the tragedy *Damon and Pithias* which the Chapel Children performed at Court at Christmas 1565. Meanwhile the Children of Paul's, under Sebastian Westcott, were also performing regularly, not only at Court, where they were seen not less than 17 times between 1557 and 1582 when Westcott died, but before invited audiences in their own cathedral courtyard or in the school hall. The first of the boy companies to act in a theatre was that from Windsor, which under Richard Farrant had been appearing regularly at Court since 1567. In 1576 he adapted part of an old monastic building as the first BLACKFRIARS THEATRE and used it for performances until his death in 1581, after which the Windsor Children do not seem to have acted in London again. Some of them, however, may have been members of the joint company of the Chapel and Paul's Children who, together with the boys attached to OXFORD'S MEN, gave the first productions of LYLY'S *Campaspe* and *Sapho and Phao* at Blackfriars in 1584. The Chapel Children alone, during the next five years, appeared before Elizabeth I in eight of Lyly's plays, while Paul's Children embarked on a final period of activity both at Court and in their own hall with a repertory which included plays by most of the dramatists of the day. Their last recorded performance was in the anonymous and now lost play *The Abuses*, with which James I entertained King Christian of Denmark in July 1606. The Chapel Children, under a new Master Nathaniel Giles with Henry Evans as their business manager, were installed at the second Blackfriars Theatre, which Evans leased from the sons of James BURBAGE in 1600, and there, to all intents and purposes a fully professional company, they enjoyed a popularity which drew audiences away from the adult companies, and caused Shakespeare (who did not write for them) to refer to them disparagingly as 'little eyases' who 'are now in fashion'. Among them were Ezekiel FENN, Nathan FIELD, William OSTLER, John RICE, and Richard ROBINSON who later became well known as adult actors; also Salathiel Pavy, praised by JONSON for his playing of elderly men, who died at the age of 13, and William Penn, a useful member of the KING'S MEN. This company gave the first performances of many outstanding plays, including Jonson's *Cynthia's Revels* (1600), his *The Poetaster* (1601), and CHAPMAN'S *May-Day* (1602). On the accession of James I they were reorganized and renamed the Children of the Queen's Revels. In spite of being implicated in the trouble over *Eastward Ho!* (1605), which lost them the royal favour and caused Jonson and Chapman to be put in prison, they continued to be successful until in 1606 there was further trouble over DAY'S *The Isle of Gulls*. The association between the acting company and the choristers of the Chapel Royal was then broken and the players were renamed the Children of Blackfriars. They managed to survive the king's displeasure over Chapman's *Charles, Duke of Byron* (1608), which offended the French Ambassador, but lost their theatre when Evans's lease ran out in the same year. The King's Men took over Blackfriars, and the children moved to WHITEFRIARS, conveniently empty owing to the demise of the short-lived Children of the King's Revels, founded in 1607. The vogue for boy actors was coming to an end. Under Philip Rosseter, a musician who had himself been a Child of the Chapel, the survivors of the Blackfriars company, as the Children of Whitefriars, continued to appear at Court in plays by BEAUMONT and FLETCHER, and at Whitefriars in such plays as Jonson's *Epicœne, or the Silent Woman* (1609) and Chapman's *The Revenge of Bussy d'Ambois* (1610). Rosseter then planned to build a new theatre for them at PORTER'S HALL, but although they were able to appear there in Beaumont and Fletcher's *The Scornful Lady* (1613) the venture was not a success; by 1615 the company had been disbanded, and the actors had taken up other work or disappeared into the provinces.

BOYLE, ROGER, see ORRERY, LORD.

BOYLE, WILLIAM (1853–1923), Irish dramatist, whose first play, *The Building Fund* (1905), was produced at the newly established ABBEY THEATRE. With this, and with his later plays, *The Eloquent Dempsey* and *The Mineral Workers* (both 1906), Boyle helped to establish a new type of 'realistic' Irish play which materially altered the character of the Abbey Theatre as YEATS and Lady GREGORY had originally envisaged it.

BOZDÊCH, EMANUÊL, see CZECHOSLOVAKIA.

BRACCO, ROBERTO (1862–1943), Italian playwright, who achieved a solid success in the theatre during the 1890s and up to 1914. After the First World War his opposition to Fascism caused his plays to be almost completely neglected, and they are only now becoming known again. Like GIACOSA, he came under the influence of both the younger DUMAS, as in *Una donna* (*A Woman*, 1892) and *Maschere* (*Masks*, 1893), and of IBSEN'S 'theatre of ideas', as can be seen in such plays as *L'infedele* (*The Faithless One*, 1894), *Trionfo* (1895), and the later *Fantasmi* (*Ghosts*, 1906). Though a brilliant writer of comedy, as he showed in *Uno degli onesti* (*One of the Honest Men*, 1900) and *Il perfetto amore* (1910)—the latter a sophisticated vehicle for a virtuoso actor and actress—his acutely developed social conscience is apparent in *Tragedia dell'anima* (*Tragedy of a Soul*, 1899), while his sympathy with feminism is shown in *Maternità* (1903). Some of his best work was done in naturalistic studies of Neapolitan lower life, as in *Don Pietro Caruso* (1895) and *Sperduti nel buio* (*Lost in the Shadows*, 1901), but his most accomplished play, and the most significant historically, is *Il piccolo santo* (*The Little Saint*, 1909), remarkable for having anticipated the Theatre of SILENCE of J.-J. BERNARD and Charles VILDRAC.

BRACE, BRACE JACK, see FLAT.

BRACEGIRDLE, ANNE (*c.* 1673–1748), one of the first English actresses, pupil and protégée of BETTERTON. As a child of six or seven she was already a member of his company, and in DOWNES'S *Roscius Anglicanus* (1708) she is mentioned as one of the leading young actresses at DRURY LANE in 1688. She went with Betterton to LINCOLN'S INN FIELDS THEATRE in 1695, where she was much admired as Angelica in the opening production, *Love for Love* by CONGREVE, whose mistress she was for some years. She was also considered outstanding as Millamant in his *The Way of the World* (1700). She left the stage at the height of her powers in 1707, wisely preferring not to be overshadowed by her younger rival Anne OLDFIELD; she was buried in Westminster Abbey, where in the register of burials her age was mistakenly entered as 85.

BRADY, WILLIAM ALOYSIUS (1863–1950), American actor and theatre manager, who made his first appearance on the stage in San Francisco in 1882. He later toured with his own company, and achieved an outstanding success as Svengali in Du Maurier's *Trilby*. Among the New York theatres with which he was associated were the FORTY-EIGHTH STREET, which he built and opened in 1912, the STANDARD, which he renamed the Manhattan and managed from 1896 until its demolition in 1909, and the PLAYHOUSE, which he built in 1911. He was responsible for a number of outstanding productions, many starring his second wife Grace GEORGE. His daughter by his first wife, Alice Brady (1892–1939), was also on the stage for a time from 1909, but spent her later years in films.

BRAHM [Abrahamsohn], OTTO (1856–1912), German director and literary critic, whose interest in contemporary drama, and particularly in the work of ANTOINE at the THÉÂTRE LIBRE in Paris, led him to found in Berlin in 1889 the FREIE BÜHNE, where he successfully produced a number of new naturalistic dramas, opening with IBSEN's *Ghosts*. In 1894 he took over the DEUTSCHES THEATER where he continued to train his actors, led by Emanuel Reicher, in a naturalistic style based on the teachings of STANISLAVSKY which was well suited to the plays of Ibsen, HAUPTMANN, and SCHNITZLER, less so to the classics. Under him the Deutsches Theater enjoyed ten years of success and popular esteem, but in 1904 he resigned his post to REINHARDT and took over the Lessing Theater in Berlin, where he remained until his death. Although the naturalistic movement had by then lost its impact, Brahm had performed a great service to the German stage by clearing it of outmoded traditions and bringing it into the main current of European drama.

BRAITHWAITE, (Florence) LILIAN (1873–1948), DBE 1943, English actress. She had already had considerable experience as an amateur when in 1897 she made her first professional appearances in South Africa with the Shakespeare company of her husband Gerald LAWRENCE, playing several major roles. She made her first appearance in London as Celia in *As You Like It* to the Rosalind of Julia NEILSON, and then joined Frank BENSON. She later appeared with George ALEXANDER at the ST JAMES'S THEATRE, where she was seen as Francesca in Stephen PHILLIPS's *Paolo and Francesca* (1902) and in Bleichmann's *Old Heidelberg* (1903). Among other parts she created during her long and distinguished career were Mrs Gregory in Vernon and Owen's *Mr Wu* (1913) and Florence Lancaster in Noël COWARD's *The Vortex* (1924). She gave some of her finest performances in her later years, notably as Elizabeth I in the English version of André Josset's *Elisabeth la femme sans homme* (1938) and Abby Brewster in the London production of Joseph Kesselring's *Arsenic and Old Lace* (1942) which ran for over three years. Her daughter Joyce Carey (b.1892), also an actress, is best remembered in a number of plays by Coward. Under the pseudonym Jay Mallory she wrote the popular play *Sweet Aloes* (1934) in which she also appeared.

BRANDANE [MacIntyre], JOHN (1869–1947), Scottish dramatist, born in Bute of working class parents. As a child he worked long hours in a cotton-mill and later became a clerk, but at the age of 32 he qualified as a doctor. While working in England as a general practitioner he became interested in the theatre through seeing the ABBEY THEATRE players on their visits to London, and when he eventually settled in Glasgow he began writing plays for the SCOTTISH NATIONAL PLAYERS, of which he became a director. It was largely due to his influence that James BRIDIE also became a playwright. Brandane's plays, which are all set in the Highlands, are strongly constructed and alive with acute humour and observation. The best known is *The Glen is Mine*, a comedy drawn from the conflict between industrialization and the Highland environment. It was seen throughout Scotland, and in 1930 was produced at the EVERYMAN THEATRE in London.

BRANDÃO, RAUL (1867–1930), Portuguese novelist and dramatist, whose plays, published in 1923, are concerned with the lives of the abjectly poor and the social causes of their poverty. The best of them, which show considerable psychological insight, were *O gébo e a sombre* (*The Hunchback and the Shadow*), *O Diabo e a morte* (*The Devil and Death*), and *Jesus Cristo em Lisboa*. The last, written in collaboration with the poet Teixeira de Pascoais (1877–1952), portrays Christ living in the slums of Lisbon.

BRANDES, GEORG MORRIS COHEN (1842–1927), Danish critic and scholar, who exercised a unique influence upon the literature and culture of Scandinavia. Coming to IBSEN with a mind nourished on Kierkegaard, Taine, Sainte-Beuve, and John Stuart Mill (whose main works he translated), he urged the pursuit of literature which would 'submit problems to debate'; he was the leading spirit of 'det moderne Gennembrud' ('the modern breakthrough') in the field of NATURALISM. Beginning in 1871 with an important series of lectures on European literature at the University of Copenhagen, he eventually completed his *Hovedstrømninger* (*Main Currents in Nineteenth-Century Literature*) in six volumes (1872–90). Among his many monographs were some of the earliest critical studies of Ibsen and BJØRNSON. In the late 1880s his was the first influential voice in Europe to draw attention to the virtually neglected Nietzsche, and he stimulated an exchange of correspondence and views between the latter and STRINDBERG.

His brother Edvard (1847–1931), also a critic, wrote some of the earliest Scandinavian social dramas; he is remembered chiefly for *Et Besøg* (1882) which was translated by William ARCHER as *A Visit* (1892).

BRANDES, JOHANN CHRISTIAN (1735–99), German actor and playwright, who in 1757 joined SCHÖNEMANN's company, of which EKHOF was then a member. He was not a success and did better in the less exalted company of the old HARLEQUIN player SCHUCH, who proved a good friend to him. Although he realised that he was no actor, Brandes prided himself on his plays, now forgotten, and on his MONODRAMAS, which enjoyed an immense vogue for about five years (from 1775 to 1780), the most popular being *Ariadne auf Naxos*. He married an actress, Esther Charlotte Henrietta Koch (1746–84), who was much admired in tragedy, and his eldest daughter Minna (1765–88), the godchild of LESSING, was also on the stage. After she died Brandes retired to Berlin, where he appeared occasionally at the National Theatre under IFFLAND.

BRANNER, HANS CHRISTIAN (1903–66), Danish novelist and dramatist, whose work reveals a mind deeply concerned with the problems of the individual in the modern world. His reputation rests mainly on his fiction, his first successful drama being a dramatization of one of his own novels, *Rytteren* (*The Riding Master*, 1949). Later plays include *Søskende* (*The Siblings*, 1951, trans. as *The Judge*, 1955) and *Thermopylae* (1957), dealing with the period of the German Occupation. He also wrote for radio as early as the 1930s.

BRASSEUR [Espinasse], PIERRE (1905–72), French actor, who was on the stage very young (using his mother's maiden name), and by 1929 was playing leading roles at a number of theatres in Paris. He then went into films, but in 1949 returned to the theatre and thereafter divided his time between stage and screen. Among the important plays in which he appeared were the first productions of CLAUDEL's *Partage de midi* and SARTRE's *Le Diable et le Bon Dieu* (both 1951). A romantic and passionate actor, he also made a great impression as Edmund KEAN in Sartre's recreation of the elder DUMAS's play on the subject. Among his later successes were ANOUILH's *Ornifle* and MONTHERLANT's *Don Juan*, the French version of PINTER's *The Homecoming*, SHAW's *Don Juan in Hell* (from *Man and Superman*), and Shaw himself in Jerome Kilty's *Dear Liar*.

BRAYTON, LILY, see ASCHE, OSCAR.

BRECHT, BERTOLT [Eugen Berthold Friedrich] (1898–1956), German poet and dramatist, who in 1918 was studying medicine in Munich when he wrote his first play *Baal*, which was not performed until 1923. Although he joined in the abortive attempt to give the emergent Weimar Republic a Soviet form, his early writings are indebted rather to EXPRESSIONISM than to the Marxism which was to provide the political mainspring of his work. The first of his plays to reach the stage was *Trommeln in der Nacht* (*Drums in the Night*, 1922). This sober and somewhat cynical study of a soldier returning from the war proved a great success. It was awarded the Kleist prize, and taken to Berlin, where in 1924 Brecht settled as assistant to REINHARDT at the DEUTSCHES THEATER. His next plays, *Im Dickicht der Städte* (1923), *Edward II* (based on MARLOWE, 1924), and *Mann ist Mann* (1926), were less successful, but marked the beginning of his attempt to develop his own form of EPIC THEATRE, with its hotly debated 'distancing techniques' (alienation) or VERFREMDUNGSEFFEKTE. Throughout his career he undertook much theoretical writing, and his aesthetic position was given its most nearly definitive form in *Kleines Organon für das Theater* (1949). In 1928 Brecht married his second wife Helene WEIGEL, who appeared in many of his plays, and had his first great success in the theatre with *Die Dreigroschenoper* at the Berlin Schiffbauerdamm. This very free adaptation of GAY's *The Beggar's Opera* had music by Kurt Weill, who also collaborated with Brecht on the operas *Aufstieg und Fall der Stadt Mahagonny* (1927), *Happy End* (1929), *Der Jasager und der Neinsager* (1929–30), and the opera-ballet *Die sieben Todsünden* (*The Seven Deadly Sins*, 1933). Weill's wife Lotte Lenya appeared in all of these works except *Happy End* and *Der Jasager und der Neinsager*. In the early 1930s Brecht wrote a number of short didactic plays or *Lehrstücke*, the best of which were probably *Die Massnahme* (1930), *Die Ausnahme und die Regel* (*The Exception and the Rule*, 1938), and *Die heilige Johanna der Schlachthöfe* (*St Joan of the Stockyards*), not staged until 1959 but considered by many the first work to show his full stature. When Hitler came to power in 1933 Brecht, with his wife and two children, went into exile—'changing countries more often than his shoes', as he says in one of his poems—first to Switzerland, then to Denmark and Finland, and finally in 1941 to the U.S.A. During these years he wrote what are generally considered to be his best plays—

Mutter Courage und ihre Kinder (1941), *Leben des Galilei, Der gute Mensch von Sezuan* (both 1943), and *Der kaukasische Kreidekreis* (1954)—which combine maturity of vision and depth of expression with a wider sympathy for the human predicament than is found in any of his earlier works, with the exception perhaps of *Die heilige Johanna der Schlachthöfe*. During the same period he also wrote *Herr Puntila und sein Knecht Matti* (1948), *Die Gesichte der Simone Machard* (1957), and *Der aufhaltsame Aufstieg des Arturo Ui* (1958) which, though more overtly political, are no less theatrically potent. In 1949 Brecht returned to East Germany and founded with Helene Weigel, who took charge of it after his death, the BERLINER ENSEMBLE, based from 1954 at the Theater am Schiffbauerdamm, where its work proved a triumphant affirmation of the practical viability of many of his theories. His last years were spent mainly on revivals of his own plays, and on adaptations of foreign plays, among them Shakespeare's *Coriolanus* (as *Koriolan*) and FARQUHAR's *The Recruiting Officer* (as *Pauken und Trompeten*). Like all his adaptations, these were so different from the original that they constituted virtually new works. This was particularly true of his Marxist version of GOZZI's *Turandot*, begun in 1930, taken up again in 1954, and left unfinished at his death.

While few would now dispute Brecht's greatness as a playwright, there remains strong disagreement between those who regard him as a great Marxist writer and those who see him as a great writer in spite of his Marxism; equally, his aesthetic theories are seen both as essential and as obstructive to his creative work. His persuasive presentation of 'epic' as a necessary alternative to 'dramatic' theatre remains, however, dialectically satisfying, and a valuable counterbalance, albeit to STANISLAVSKY rather than to ARISTOTLE (whom Brecht more truly complements than contradicts). Writers described by others as his disciples tend to disclaim his direct influence; but that he has helped to liberate the English-speaking theatre from the constraints of the 'well-made' play in three acts seems undeniable. This influence was belated, in spite of Brecht's residence in the U.S.A. The Berliner Ensemble has yet to visit New York; in London shortly after Brecht's death in 1956 the company was seen in *Mutter Courage und ihre Kinder, Pauken und Trompeten*, and *Der kaukasische Kreidekreis*. On a second visit in 1965 it presented *Die Dreigroschenoper, Koriolan, Die Tage der Kommune* (*The Days of the Commune*) and *Der aufhaltsame Aufstieg des Arturo Ui*. Most of Brecht's plays have now been seen in English, the first being *Senora Carrar's Rifles* (from *Die Gewehre der Frau Carrar*) in 1938 at the UNITY THEATRE which also staged *Mother Courage, Simone Machard*, and *The Exception and the Rule*. These demonstrated, as did productions by Keith Hack with students at Cambridge University in the early 1970s, that the uncomplicated approach of amateurs under a good director can often do greater justice to Brecht than that of professionals with strong prejudices for or against. Interesting productions were seen later in London of *Mother Courage* with Joan LITTLEWOOD in 1955, *The Good Woman of Setzuan* with Peggy ASHCROFT in 1956, *Galileo* in 1963 with Bernard MILES and in 1980 at the NATIONAL THEATRE, and *Baal* with Peter O'TOOLE in 1963. The ROYAL SHAKESPEARE COMPANY at the Aldwych staged *The Caucasian Chalk Circle* in 1962 and *Puntila* in 1965. *St Joan of the Stockyards* was seen in London in 1964, and two adaptations, *Edward II* and *Trumpets and Drums*, in 1968 and 1969 at the National Theatre, London, and the LYCEUM, Edinburgh, respectively. Perhaps the most satisfying British production of Brecht originated in Glasgow, where Michael Blakemore directed *Arturo Ui* at the CITIZENS' THEATRE in 1967, Leonard Rossiter creating a vividly alienated caricature in the title role. This was seen in London at the SAVILLE THEATRE in 1969. In New York, an adaptation of *The Threepenny Opera* by Marc Blitzstein was seen in 1954 at the THEATRE DE LYS and ran for seven years, being followed by *Brecht on Brecht*, compiled by George Tabori from Brecht's writings, which also had a long run; Lotte Lenya appeared in both productions, as well as in the London production of the latter in 1962. Other plays seen in translation in New York are *Galileo* in 1947, *The Good Woman of Setzuan* in 1956, *In the Jungle of Cities* in 1960, *Mother Courage* in 1963, *Arturo Ui* in 1964, and *The Caucasian Chalk Circle* in 1965.

BRÉCOURT [Guillaume Marcoureau] (1638–85), French actor and dramatist, who while playing in the provinces married an actress, Étiennette des Urlis (c. 1630–1713), and with her went to the Théâtre du MARAIS. A year later, probably through the efforts of his sister-in-law Catherine who had been with MOLIÈRE in the ILLUSTRE-THÉÂTRE venture, he joined the company at the PALAIS-ROYAL but left after two years, unregretted owning to his quarrelsome temper. He then joined FLORIDOR at the Hôtel de BOURGOGNE and after Molière's death wrote a one-act play about him, *L'Ombre de Molière* (1674), in which he spoke feelingly of the great actor-manager's good qualities. Brécourt, who was the first to play Britannicus in RACINE's play of the name, had the misfortune in 1680 to kill a coachman during a quarrel, and was forced to take refuge in Holland. Pardoned by Louis XIV, whose life he is said to have saved during a boar-hunt, he became a member of the COMÉDIE-FRANÇAISE, and remained there until his death. Among his plays, some of which are lost, the best was *La Feinte Mort de Jodelet* (1659), in which JODELET himself played the name part.

BREDERO, GERBRAND ADRIAANSZ (1585–1618), Flemish dramatist, whose romantic plays, drawn mainly from tales of chivalry, are remarkable for the realism of the comic interludes, where characters from everyday life mingle with the more stereotyped knights and their ladies. Among them are *Griane* (1612) and *The Dumb Knight* (1618). His best work was done in the farces and comedies which laid the foundations of the Dutch comic tradition, among them the *Farce of the Cow* (1612), based on a German tale which tells how a rogue tricks an old peasant into selling him a cow and then returning him the money paid for it. In spite of its German origin, it is purely Dutch in character and Bredero, himself a man of the people, caught admirably the spirit and dialogue of his times. His last and perhaps his best play was *The Spanish Brabanter* (1617), a series of comic scenes loosely held together by the two central characters, a master and his servant, which vividly creates on stage the widely different worlds of the pretentious but penniless nobleman and the ordinary working people of Amsterdam. It became one of the most successful comedies of its day and was frequently revived.

BREECHES PARTS, see MALE IMPERSONATION.

BRENTON, HOWARD (1942–), English dramatist, who wrote his first play in 1965 while still at Cambridge and had a one-act play, *It's My Criminal*, performed at the ROYAL COURT the following year. His first full-length play, *Revenge* (1969), was also seen at the Royal Court (in the Theatre Upstairs), as was *Magnificence* (1973), which dealt with urban terrorism. *Brassneck* (also 1973), written in collaboration with David HARE, who in 1969 had staged Brenton's *Christie in Love* at his Portable Theatre, was produced at the NOTTINGHAM PLAYHOUSE, where *The Churchill Play* (1974), set in Britain in 1984 under a coalition government, with militant trade unionists being brutally treated by the army, was also seen. *Weapons of Happiness* (1976), the first new play to be produced in the NATIONAL THEATRE's new building, was followed by *Epsom Downs* (1977) at the ROUND HOUSE. In 1980 *The Romans in Britain*, which drew a parallel between the Roman invasion of Britain and the British presence in Northern Ireland, was also produced at the National Theatre and aroused a good deal of controversy because of its violent scenes of bloodshed and homosexual rape (see CENSORSHIP). The same year saw the production of Brenton's translation of BRECHT's *Galileo* at the National Theatre and of *A Short, Sharp Shock*, which satirized Mrs Thatcher's Government.

BRESSELAU, MENAHEM MENDEL, see JEWISH DRAMA.

BRETÓN DE LOS HERREDOS, MANUEL (1796–1873), Spanish playwright, author of about 200 plays of which some 50 were original, the others being translations from French or adaptations of 17th-century Spanish comedies. Most are written in easy, flowing verse, and the early ones include an element of social satire in the style of Leandro Fernández de MORATÍN. In his refusal to be swept away by contemporary literary currents, Bretón perhaps reflects more correctly the taste of his time than either Moratín or ROJAS ZORRILLA. Among his best known plays are *Marcela, o ¿cual de los tres?* (*Marcela, or Which One of the Three?*, 1831), in which a vivacious young woman is confronted by three equally unacceptable suitors, and ends by choosing none of them; *¡Muerete y veras!* (*Die and You Will See*, 1837), an attack on Romanticism; and *El pelo de la dehesa* (*As Green as Grass*, 1840).

BRICE, FANNY, see REVUE.

BRIDGE, a mechanical device (called an elevator in America) for raising heavy pieces of scenery or a tableau of actors from below stage to stage level or about it. The joists and floorboards of the stage are cut and framed, usually with sliders, to allow a large ROSTRUM, framed and tied, to rise in corner-grooves with the aid of counterweights and a winch. Great variety is found in the design of bridges, and today the electrically-controlled bridges of a large theatre may reach a high degree of engineering complexity. (See also TRAP.)

The term is sometimes used for the LIGHTING bridge which is erected over the stage.

BRIDGES, Dr JOHN, see STEVENSON, WILLIAM.

BRIDGES-ADAMS, WILLIAM (1889–1965), English director, who gained his experience in provincial repertory companies, including the BIRMINGHAM REPERTORY THEATRE, before in 1919 he was appointed director of the theatre in STRATFORD-UPON-AVON in succession to Frank BENSON. There he directed 29 plays by Shakespeare and was instrumental in keeping the company together after the theatre was destroyed by fire in 1926, remaining with it until it moved into the new theatre in 1932. He took the company several times on tour to Canada and the U.S.A. After leaving Stratford in 1934 he worked with the BRITISH COUNCIL, initiating the policy of sending theatrical companies abroad, mostly to Europe.

BRIDIE, JAMES [Osborne Henry Mavor] (1888–1951), Scottish doctor and playwright, whose interest in the theatre was aroused by the visits of such companies as the ABBEY THEATRE players to Glasgow, his native city. He was encouraged in his efforts as a dramatist by John BRANDANE, who in the period after the First World War was working to establish an indigenous Scottish drama, and it was through Brandane that Bridie's first plays were produced by the SCOTTISH NATIONAL PLAYERS. His initial effort, *The Sunlight Sonata* (1928), was staged under the pseudonym 'Mary Henderson' and made little impact; but its director was Tyrone GUTHRIE, who not only directed the first play credited to 'James Bridie'—*The Anatomist* (1930)—but also took it in 1931 to London, where most of Bridie's later plays were seen, giving him a wider audience than most contemporary Scottish playwrights have achieved. Among them were *Tobias and the Angel* (also 1930), *Jonah and the Whale* (1932), *A Sleeping Clergyman* (1933), *Colonel Wotherspoon* (1934), *The Black Eye* (1935), *Storm in a Teacup* (1936), an adaptation of Bruno FRANK's *Sturm im Wasserglass* with a Scots setting, *Susannah and the Elders* (1937), *The King of Nowhere* (1938) seen at the OLD VIC with Laurence OLIVIER in the chief part, Vivaldi, and *What Say They?* (1949), first seen at the Golders Green Hippodrome with Yvonne ARNAUD in the lead. In spite of war service Bridie continued to write, with *Mr Bolfry* (1943) in which Alastair SIM scored a great success in London. Sim also played in and directed Bridie's next two plays, *It Depends What You Mean* and *The Forrigan Reel* (both 1944). Bridie had in the meantime been active in the foundation of the CITIZENS' THEATRE in Glasgow, where many of his plays were first seen. In 1949 he scored his greatest success with *Daphne Laureola*, with Edith EVANS in the title role. For the EDINBURGH FESTIVAL he then wrote what many consider his best play, *The Queen's Comedy* (1950); his last two plays were produced posthumously by the Citizens' Theatre, *The Baikie Charivari* in 1952 and *Meeting at Night* in 1954. Bridie remained in medical practice until 1938 and served as an army doctor in both world wars; he probably did more than anyone to establish and encourage Scottish drama and was instrumental shortly before his death in forwarding the foundation of a drama department within the Royal Scottish Academy of Music in Glasgow, now the Royal Scottish Academy of Music and Drama.

BRIEUX, EUGÈNE (1858–1932), French dramatist, whose plays are naturalistic dramas in the tradition of ZOLA and BECQUE. Each of his plays is a plea for the amelioration of a particular evil—a plea which sometimes degenerates into sermonizing; but at his best he combined didacticism with a fierce pity for individuals that produced work of significance, at least in the opinion of his contemporaries, especially SHAW, whose high praise contributed to his somewhat inflated reputation. His most interesting plays are *Les Trois Filles de M. Dupont* (1898), which portrays the dangers of a marriage of convenience; *La Robe rouge* (1900), which exposes the abuses of the judiciary; *Les Avariés* (1902), a study of venereal disease which as *Damaged Goods* created a sensation in London and New York; *Maternité* (1903), which advocates legalized birth-control; and *La Femme seule* (1912), which deals with the financial dependence of married women.

BRIGHELLA, one of the ZANNI or servant roles of the COMMEDIA DELL'ARTE. Originally a thief and a bully, with much in him of the Neapolitan street-corner boy, he later became a lackey, though retaining his love of intrigue and lying. Through the establishment of the COMÉDIE-ITALIENNE in Paris he influenced the development of the French valets of MARIVAUX and so helped to shape the Figaro of BEAUMARCHAIS.

BRIGHOUSE, HAROLD (1882–1958), English dramatist, usually referred to as one of the so-called 'Manchester School', as his early realistic comedies of North-country life were produced at the Gaiety Theatre in Manchester under Miss HORNIMAN. *The Northerners* (1914), on the Luddites, has been favourably compared with TOLLER's *Die Maschinenstürmer* (1922), but has apparently not been seen in London. Of Brighouse's many plays the only one to survive on the stage is *Hobson's Choice* (1916), with which he scored an instant success when it was presented at the APOLLO THEATRE in London with Norman McKinnel as the old cobbler Hobson whose headstrong daughter Maggie insists on marrying her father's timid little employee Willie Mossop. It has been several times revived, notably in 1965 by the NATIONAL THEATRE company at the OLD VIC, with Colin BLAKELY as Hobson and Frank FINLAY as Willie Mossop.

BRIGHTON, Sussex. This famous seaside resort had its first permanent theatre, in North Street, in 1774. In 1790 another theatre was erected in Duke Street at a cost of £500 and later remodelled in horseshoe form, a royal box being added. It closed in 1806, when a theatre was built in New Road. Under the actor-manager John Brunton this opened a year later with Charles KEMBLE as Hamlet. In spite of the patronage of the Prince Regent and the visits of such stars as Mrs SIDDONS, it ruined a number of managers and closed in 1820. Three years later, after being redecorated, it reopened with a PANTOMIME and continued to function until in 1866 Nye Chart, its lessee since 1854, bought it outright, demolished it, and built the present Theatre Royal on the same site. Touring companies first visited the theatre in 1868, but it managed to retain a STOCK COMPANY until 1873. After Chart's death in 1876 it was managed for a time by his widow. It was while playing at this theatre on tour in Oct. 1891 that Fred TERRY and Julia NEILSON were married. The present theatre, seating 1,000, was once the most popular stopping-place for productions on their way to London's West End, but its importance diminished with the reduction in touring and the increased use of the RICHMOND Theatre, Surrey, for pre-London runs. It still presents a varied programme, though it cannot accommodate large-scale opera and ballet or big musicals. It receives no subsidies.

BRIGHTON THEATRE, New York, see BIJOU THEATRE (1).

BRISTLE TRAP, see TRAPS.

BRISTOL, whose university was the first in Britain to have a UNIVERSITY DEPARTMENT OF DRAMA, has a long, though somewhat fragmentary, theatre history. The first entertainments were probably given by itinerant players and puppeteers at the annual fairs—there are apparently no records of LITURGICAL DRAMA in Bristol—and Elizabethan companies on tour are known to have played in the city. Edward ALLEYN and Richard BURBAGE appeared there, and there is some evidence for the existence of a private theatre in the early years of the 17th century. In the early 18th century a company from BATH, with which Bristol was later to be closely linked for a long period, appeared in the town, probably in a room in Tucker (now Bath) Street, and in 1828 a troupe of acrobats including Madame Violante occupied a temporary theatre in St Augustine's Back, followed by a company of actors who engaged in intense rivalry with a company from Bath which appeared at the same time in the Long Room in Hotwells in GAY's *The Beggar's Opera*. In this production the part of Peachum was played by John Hippisley, who remained in the town and built its first permanent theatre at Jacob's Wells outside the city boundary, hoping to escape the hostility of the authorities and some of the leading citizens: it opened on 23 June 1729 with CONGREVE's *Love for Love*. Little is known of its early history, but Mrs Hannah PRITCHARD appeared there in the 1740s, as did Charles MACKLIN. After Hippisley's death in 1748 the fortunes of the theatre declined and it closed during 1757, opening a year later under John PALMER, who was also managing the Orchard Street Theatre in Bath. It soon proved too small for Bristol, which was a thriving and rapidly expanding city, and the quarter in which it was situated became less fashionable. It was used for the last time in 1765 and then abandoned, being demolished some time between 1803 and 1826. Meanwhile a larger theatre had been built in King Street, with a new-style horseshoe-shaped auditorium. It opened on 31 May 1766, and for many years its summer seasons were run by the London actor William Powell and later by James DODD, both of whom imported good players and new plays from London. In 1778 efforts to obtain a royal patent were finally successful and it was henceforth known as the Theatre Royal, being run in conjunction with the Theatre Royal at Bath until 1817, a connection which brought prosperity to both houses. The best known and most respected manager of this period was William Wyatt Dimond and after his death in 1812 the fortunes of the theatre fluctuated considerably, particularly after the break with Bath. In 1819 Charles MACREADY's father took over the management, his son appearing under him, as did Junius Brutus BOOTH. Later, under the elder Macready's son-in-law John Henry Chute, the theatre, which had been on the verge of bankruptcy under his widow, became once more prosperous; her death in 1853 left Chute free to spend a large sum on repairs and redecoration, and to mount a series of excellent Shakespeare productions based on those of Charles KEAN at the PRINCESS'S THEATRE in London. The STOCK COMPANY in Bristol became once more a training-ground for young actors, among them Kate and Ellen TERRY, Marie Wilton, later the wife of Squire BANCROFT, and Madge Robinson, who married William KENDAL. Up to the mid-1860s most of the well known London players appeared at the Theatre Royal, but as the centre of the city ceased to be a residential area, so the status of the theatre declined. On 14 Oct. 1867 Chute opened a new and much larger theatre in Park Row, which after his death in 1878 was used by touring companies only. Originally known as the New Theatre Royal, it was renamed the Prince's Theatre in 1884, and flourished for a time. It was finally vanquished by the popularity of the cinema and the MUSIC-HALLS, of which Bristol had three—the People's Palace (1892), the Empire (1893), and Stoll's BRISTOL HIPPODROME (1911). It existed for many years on a hand-to-mouth basis, until on the night of 24 Nov. 1940 it was completely demolished in an air-raid. The old Theatre Royal too had been going through hard times, playing mainly FARCE and PANTOMIME for a largely working-class audience, while the building itself became little more than a slum. The work it should have been doing was partially undertaken by the LITTLE THEATRE, and later the Rapier Players, and when it closed after the raid of 16–17 Mar. 1941, it seemed as if it was doomed to destruction; but the energy, enthusiasm and unselfish co-operation of many individuals and corporate and civic bodies eventually ensured its preservation, and in 1943 it started a new lease of life (see BRISTOL OLD VIC).

BRISTOL HIPPODROME, The, built by Stoll in 1911–12 and intended to be the largest and best home of VARIETY outside London. It seated 2,000 people (now 1,975), and opened with a spectacular show, *Sands o' Dee*, which utilized the theatre's 40-ft steel water tank. Later productions included opera, concerts, plays, circus, and, at the end of the First World War, a period of twice-nightly family REVUE. It became a cinema in the 1930s but reverted to variety in 1938. In 1948 the stage area was destroyed by fire and the theatre was out of action for nearly a year. When it reopened it became a popular venue for musicals, *Guys and Dolls* and *The Music Man* having their British premières there. It is now a major touring house, with one of the largest stages in the country, 84 ft wide and 60 ft deep. It presents ballet, opera, musicals, and plays by the NATIONAL THEATRE and other touring companies.

BRISTOL OLD VIC. In 1941 the old Theatre Royal in BRISTOL was threatened with final closure and perhaps demolition, but the Council for the Encouragement of Music and the Arts (CEMA, later the ARTS COUNCIL) helped to save it and it became not only the oldest working theatre in England but also the first to be state subsidized. After reconstruction and redecoration the theatre reopened in 1943 with GOLDSMITH's *She Stoops to Conquer*—first performed there in 1773—and three years later, through the joint efforts of CEMA and the London OLD VIC, a new resident company, the Bristol Old Vic, was launched under Hugh HUNT, soon achieving a national and international reputation. One of the theatre's most popular productions was the musical play *Salad Days* (1954) by Dorothy Reynolds and Julian Slade, which moved to London for a long run. A production of *Love's Labour's Lost*, staged in 1964 in honour of Shakespeare's Quatercentenary by Val [Valentine] May (1927–), the theatre's director from 1961 to 1975, toured all over the world gaining golden opinions. Other productions which were later successful in London were *A Severed Head* (1963), based by J. B. PRIESTLEY and Iris Murdoch on the latter's novel, Frank Marcus's *The Killing of Sister George* (1965), which provided a fine part for Beryl Reid, Barry England's *Conduct Unbecoming* (1969), all directed by May; and Peter NICHOLS's *Born in the Gardens* (1979), again with Beryl Reid. When the Arts Council relinquished the lease of the theatre in 1963 it was taken over by a Trust and in the early 1970s extensive redevelopments were carried out, including the building of a new stage, improved backstage facilities, and the acquisition of the adjacent Coopers' Hall, an 18th-century guildhall, to form a new entrance and foyer. Some historical and traditional features were inevitably lost, but both comfort and practicability were improved for audience and actors alike. The main theatre now seats 647. A studio theatre, the New Vic, which opened in 1972, was built in the space occupied by the old entrance; it has a flexible stage suitable for the presentation of modern and experimental drama, and seats about 150. From 1963 to 1980 the Bristol Old Vic also presented plays at the city-owned LITTLE THEATRE in the Colston Hall, and the main company took refuge there during the rebuilding in 1970–72. An excellent Theatre School was founded at the same time as the company, and since its inception the Bristol Old Vic has worked in close conjunction with the Department of Drama at Bristol University.

BRITANNIA MUSIC-HALL, London, see ROTUNDA, THE.

BRITANNIA THEATRE, London, in High Street, Hoxton, originally the Britannia Saloon at the rear of the Britannia Tavern, was opened by Sam LANE on Easter Monday 1841. It was used at first for music-hall entertainments which were free, the profits being made on the sale of refreshments; but the abolition of the monopoly of the PATENT THEATRES in 1843 enabled Lane to turn it into a theatre and stage complete plays, mainly farces and strong drama. 'The Brit.', as it was called, prospered; enlarged in 1850, it was rebuilt in 1858 to seat 3,923 people (reduced in 1866 to 2,972) in three tiers. Lane continued in management until his death in 1871 and was succeeded by his wife Sara, an excellent actress, who managed it until her death 28 years later, no other theatre at that time having been so long under one management. It was a unique institution; authors wrote exclusively for it, actors joined the company as young men and remained until old age. It was supported mainly by the people of the surrounding district, who loved the theatre and revered its manageress, who played in the annual PANTOMIME (which always ran until Easter) until she was in her 70s. After her death the control of the theatre passed to her nephews who were not successful. It was shut from April to Dec. 1903, when it was reconstructed and modernized. It reopened with a Christmas pantomime, and then became a music-hall. John M. EAST ran a stock company there in 1904, unsuccessfully, as did other managers who followed him, and in 1912 it became a cinema. It was destroyed by bombing in 1940.

BRITISH ACTORS' EQUITY ASSOCIATION, the trade union for professional actors founded on 1 Dec. 1929 on a proposal by Ben WEBSTER during a meeting held at the DUKE OF YORK'S THEATRE. It regulates all questions relating to actors' salaries, working conditions, and terms of employment, in London and in the provinces, and advises the Ministry of Labour on the issuing of work permits for foreign actors. Before its foundation there had been various attempts to organize theatre workers in the hope of improving salaries and back-stage conditions, which by the end of the 19th century were often very bad, particularly on tour. Henry IRVING was president of the first Actors' Association, formed in 1891, but like the Actors' Union, founded in 1905 by H. B. IRVING and Seymour HICKS, it had a very short life. The suggestion that the Actors' Association should become a trade union and work to improve minimum rates of pay and raise the status of its members was first made in 1906 and the first actors' strike took place in 1920.

Equity gradually came to represent all performers in the theatre, in films, and on radio and television, though the Variety Artists' Federation, formed in 1906, continued to act independently until it merged with Equity in 1968. Standard 'Equity-approved' contracts and a *de facto* 'closed shop' were early achievements.

During the 1970s Equity achieved recognition of the need for a greatly increased minimum wage, but was riven by internal disputes. In 1972 it was expelled from the Trades Union Congress, to which it had finally agreed to affiliate in 1940, for remaining registered under the controversial Industrial Relations Act in order to retain the privilege of operating a closed shop for the profession. It was readmitted to the T.U.C. in 1973, but remained split between those favouring some degree of political commitment to the left and those advocating a purely professional association. By maintaining a fund to assist professional companies in financial difficulties, Equity has helped to cushion the effects of economic recession on the theatre and, to a degree, to protect its members' jobs.

BRITISH COUNCIL, organization founded in 1934 to promote a wider knowledge of the United Kingdom and the English language abroad. A drama department was founded in 1937 and, after amalgamating with the music department in 1961, again became a separate entity in 1977; it was renamed the drama and dance department in 1980, though dance had in fact always been within its jurisdiction. The department sponsors tours overseas by leading British companies; distributes copies of British plays with a view to their performance abroad; gives advice and information on all aspects of British theatre; sponsors visits to Britain by theatre personalities from other countries for professional study and contacts with representatives of the British theatre; and publishes *Play Bulletin*, a quarterly review of new British plays.

BRITISH DRAMA LEAGUE (B.D.L.), BRITISH THEATRE ASSOCIATION (B.T.A.), see AMATEUR THEATRE.

BRITISH THEATRE MUSEUM ASSOCIATION, see THEATRE MUSEUM.

BRITTON, HUTIN, see LANG, MATHESON.

BRIZARD [Britard], JEAN-BAPTISTE (1721–91). French actor, who had been for ten years in the provinces when in 1756 he attracted the attention of Mlle CLAIRON and Mlle DUMESNIL in Lyon and the following year joined the COMÉDIE-FRANÇAISE. Owing to being prematurely white, he was able to play old men without a wig. Of a dignified presence, with a good voice and a natural style of acting, he was much admired as Henry of Navarre in COLLE's *La Partie de chasse d'Henri IV* when in 1774 it was finally passed by the censor. Brizard was also the first French actor to play King Lear, in the adaptation by DUCIS in 1782. He retired in 1786.

BROADHURST, GEORGE HOWELLS (1866–1952), English-born American dramatist and theatre manager. His early plays were farcical comedies like *What Happened to Jones* (1897) and *Why Smith Left Home* (1899), but *The Man of the Hour* (1906), probably his best work, is in a more serious vein; in *Bought and Paid For* (1911) he reverted to an earlier melodramatic style. On 27 Sept. 1917 he opened a theatre under his own name (below) with SHAW's *Misalliance*, the first of his own plays to be seen there being *The Crimson Alibi* (1919). His later productions included a revival of *Bought and Paid For* and an adaptation of Edgar Rice Burroughs's *Tarzan of the Apes* (both 1921).

BROADHURST THEATRE, New York, at 235 West 44th Street. George BROADHURST opened this theatre, which has a seating capacity of 1,185, on 27 Sept. 1917 with the first production in New York of SHAW's *Misalliance*. In 1926 *Broadway* by George ABBOTT and Philip Dunning started a vogue for gangster plays, and in 1931 *Hamlet* was seen in a vast setting designed by Norman BEL GEDDES. Several successful MUSICAL COMEDIES followed, and in 1933 the PULITZER PRIZE-winner *Men in White* by Sidney KINGSLEY had a long run, as did HOUSMAN's *Victoria Regina* (1935), in which Helen HAYES appeared as the Queen, Thomas Job's *Uncle Harry* (1942), and Agatha CHRISTIE's *Ten Little Indians* (1944). The revival of the musical *Pal Joey* was a big success in 1952, as was Jerome Lawrence and Robert E. Lee's *Auntie Mame*, based on Patrick Dennis's novel, in 1956. Later productions were *Fiorello!* (1959), a musical based on the life of Mayor La Guardia of New York; *Oh, What a Lovely War!* (1964), played by the THEATRE WORKSHOP company in a production by Joan LITTLEWOOD; *Half-a-Sixpence* (1965), a British musical based on H. G. Wells's novel *Kipps*; and *Cabaret* (1966), for the first few months of its long run. Eugene O'NEILL's *More Stately Mansions* (1967), starring Ingrid Bergman and Colleen DEWHURST, had its first New York production at the Broadhurst, as did Robert BOLT's *Vivat! Vivat Regina!* (1972), with Claire BLOOM and Eileen ATKINS. In 1974 there was a revival of GILLETTE's *Sherlock Holmes* with John WOOD, and in 1976 *Sly Fox*, adapted from JONSON's *Volpone* by Larry Gelbart, occupied the theatre. In 1978 the Bob Fosse musical *Dancin'* was a big success, as was Peter SHAFFER's *Amadeus* (1980) with Ian MCKELLEN.

BROADWAY MUSIC-HALL, New York, see WALLACK'S LYCEUM.

BROADWAY OPERA HOUSE, New York, see BIJOU THEATRE (1).

BROADWAY THEATRE, New York. (1) At 326 Broadway. Modelled on the HAYMARKET THEATRE in London, and resplendent in gold and white, this theatre opened on 27 Sept. 1847 with Henry WALLACK in SHERIDAN's *The School for Scandal*, his nephew Lester WALLACK making his first appearance in New York in the afterpiece. When the old PARK THEATRE was destroyed by fire in 1848 the Broadway, which housed a permanent STOCK COMPANY, took its place as a home for visiting stars. Edwin FORREST was appearing there during the riot at the ASTOR PLACE OPERA HOUSE, caused, it was said, by his jealousy of MACREADY, who was playing there. In 1849 Charlotte CUSHMAN, who had been in England for four years, made her reappearance in New York at the Broadway, and in the same year the BATEMAN children, Kate aged 6 and Ellen aged 4, were seen in a wide range of adult parts in Shakespeare's plays. This theatre also saw the first production of BOKER's epoch-making *Francesca da Rimini* (1855). Shortly afterwards excavations next door caused the theatre to collapse, and although it was quickly rebuilt it never regained its former eminence. After a poor season in 1857–8, redeemed only by a successful visit from the younger Charles MATHEWS, it turned to EQUESTRIAN DRAMA and finally closed on 2 Apr. 1859.

(2) At Broadway and West 41st Street. This opened as the Metropolitan Casino on 27 May 1880 and was rebuilt and renamed the Broadway, opening with Fanny DAVENPORT in SARDOU's *Tosca* on 3 Mar. 1888. Several productions, including *Ben Hur* (1899), adapted from Lew Wallace's novel, and an English PANTOMIME imported from London in 1901, had long runs; but in 1907 all the theatre's previous records were broken by the immense popularity of Frances Hodgson Burnett's *Little Lord Fauntleroy*. Among the European visitors who played at this theatre were MODJESKA and SALVINI; Helen HAYES also appeared there as a child. It became a cinema early in 1913 and was demolished in 1929.

(3) A cinema at 1681 Broadway and 53rd Street which opened in 1924 is now used as a theatre, and since 1930 (with the exception of the years 1937–40) has been known as the Broadway. From 1930 to 1935 it housed a musical, *The New Yorkers, Earl Carroll's Vanities*, and VAUDEVILLE. In 1943, after another period as a cinema, it was again used for musicals, with *Carmen Jones* and a revival of Moss HART's *Lady in the Dark* with its original star Gertrude LAWRENCE; in 1959 *Gypsy*, starring Ethel MERMAN, began a long run; *Cabaret* moved in from the IMPERIAL THEATRE in 1968. Later successes included *Candide* (1974), based on VOLTAIRE, from the CHELSEA THEATRE CENTER, *The Wiz* (1977), transferred from the MAJESTIC, and *Evita* (1979) by Tim Rice and Andrew Lloyd Webber.

A building at 410 Broadway, originally called the Euterpean Hall, was named the Broadway Theatre for a brief period in 1837, as were WALLACK'S LYCEUM in its last years, and DALY'S THEATRE at 1221 Broadway in 1877–8.

BROCHET, HENRI, see GHÉON, HENRI.

BROCKMANN, JOHANN FRANZ HIERONYMUS (1745–1812), German actor, friend and pupil of SCHRÖDER to whose company he belonged in Hamburg. He played Hamlet in the first production there of a German version of Shakespeare's play in 1776, with Schröder as the Ghost. He later went to the Burgtheater in VIENNA, where he was highly thought of and remained until his death.

BRÓDY, SÁNDOR, see HUNGARY.

BRODY SINGERS, see JEWISH DRAMA.

BROME, RICHARD (*c.* 1590–1653), English dramatist, at one time in the service of Ben JONSON, whose influence is clearly visible in the comedies on which Brome's reputation chiefly rests, though he also wrote some romantic dramas in imitation of FLETCHER and MIDDLETON. In his later plays he seemed to be working towards a more individual style, but his career was ended by the closing of the theatres in 1642, his last play being *A Jovial Crew, or, the Merry Beggars* (1641), first played by Christopher BEESTON's Boys. It proved extremely popular and was several times revived, being used in 1731 as the basis of a BALLAD OPERA. Among his earlier plays the best are probably *The City Wit; or, the Woman Wears the Breeches* (1628), *The Northern Lass* (1629), and *The Sparagus Garden* (1635). Mrs Aphra BEHN's *The Debauchee; or, the Credulous Cuckold* (1677) was based on Brome's *The Mad Couple Well Matched* (1636), and DRYDEN is also believed to have based two of his early plays, *The Mistaken Husband* and *The Wild Gallant* (both 1663), on plays by Brome.

BRONNEN, ARNOLT (1895–1959), German dramatist, one of the strongest exponents of experimental EXPRESSIONISM, who wrote on overtly sexual themes in the manner of WEDEKIND. He first came into prominence in 1922, when his sensational *Vatermord* (*Parricide*), written in 1915, was seen in Berlin. In this, which portrays incest, as well as the murder of a father by his son, and in *Die Geburt der Jugend*, seen in the same year, he dealt with the contemporary theme of the revolt of youth against age. His later works became increasingly brutal, and although he anticipated his later support of Nazism in *Reparationen* (1926), this was in some measure a rationalization of the movement's crudely emotive appeal. After the war he became a theatre critic in East Berlin, but wrote a few plays, his last being *Die jüngste Nacht* (1958) on the subject of Americans in Germany.

BROOK, PETER STEPHEN PAUL (1925–), English director, who was in his late teens when he directed MARLOWE's *Dr Faustus* and COCTEAU's *The Infernal Machine* on the minute stages of the Torch and Chanticleer theatres in London. In 1945 his production of *King John* at the BIRMINGHAM REPERTORY THEATRE attracted a good deal of attention, and a year later he was at the SHAKESPEARE MEMORIAL THEATRE where he directed an enchanting *Love's Labour's Lost* costumed *à la Watteau*. In the same year he directed SARTRE's *Vicious Circle* with Alec GUINNESS as Garcin, and in 1947 two further plays by the same author—

Men Without Shadows and *The Respectable Prostitute*—in a double bill at the LYRIC THEATRE, Hammersmith. Later that year, at Stratford, he was responsible for a *Romeo and Juliet* which infuriated the critics, mainly on account of the clumsy cutting and speaking of the verse. He then went to direct opera at Covent Garden, but returned to the theatre with excellent productions of ANOUILH's *Ring Round the Moon* (1950), with Paul SCOFIELD, and OTWAY's *Venice Preserv'd* in 1953, with Scofield and GIELGUD. Later productions included FRY's *The Dark is Light Enough* (1954) with Edith EVANS, Anouilh's *The Lark* (1955) with Dorothy TUTIN as Saint Joan, and Shakespeare's *Titus Andronicus* (also 1955) at Stratford. Also in 1955 he directed Scofield in *Hamlet*, the company later going on to Moscow—the first English company to appear there since 1917. In 1962 Brook was appointed co-director, with Peter HALL and Michel SAINT-DENIS, of the ROYAL SHAKESPEARE COMPANY. His first production at Stratford in his new capacity was *King Lear* starring Scofield, and at the ALDWYCH THEATRE in London in 1964 he directed WEISS's *Marat/Sade*, written and directed under the influence of ARTAUD's Theatre of CRUELTY, followed by an experimental documentary, *US* (1966), on the American intervention in Vietnam. After directing SENECA's *Oedipus* for the NATIONAL THEATRE in 1968, and at Stratford in 1970 an internationally-acclaimed *A Midsummer Night's Dream* using stilts, trapezes, and other acrobatic tricks, Brook left England for Paris where, under the influence of GROTOWSKI and with the encouragement of Jean-Louis BARRAULT he opened his INTERNATIONAL CENTRE OF THEATRE RESEARCH. In 1978 he returned to Stratford to direct *Antony and Cleopatra*.

BROOKE, GUSTAVUS VAUGHAN (1818–66), actor of Irish extraction, who appeared in Dublin at the age of 14 and then toured England and Scotland as the Dublin, or Hibernian, Roscius, playing adult roles including Young Norval in HOME's *Douglas*, Rolla in SHERIDAN's *Pizarro*, and Virginius in Sheridan KNOWLES's play of that name. He had a good presence and a fine voice, but his intemperance led to a chequered career during which he alternately triumphed and failed in such Shakespearian roles as Othello, Richard III, Hamlet, and Shylock, in London, the United States, and Australia, and he was finally imprisoned for debt in Warwick jail. On his release he set sail again for Australia and was drowned when his ship sank in the Bay of Biscay. He was twice married, both his wives being actresses.

BROOKLYN THEATRE, New York. After the second PARK, this was the second playhouse to be built in this district. It opened on 2 Oct. 1871 under F.B. CONWAY, whose widow took over when he died and remained there until her death in 1875. Her daughter Minnie then took over, but was unsuccessful and left after a few months. On 20 Sept. 1875 the theatre reopened under PALMER for a brief but exciting season, and on 5 Dec. 1876 it was burnt down during a performance of Oxenford's *The Two Orphans*.

BROOKS ATKINSON THEATRE, New York, at 256 West 47th Street, between Broadway and 8th Avenue. Seating 1,088, it opened as the Mansfield Theatre on 15 Feb. 1926, the first important event in its history being a visit from the Moscow HABIMAH Players in ANSKY's *The Dybbuk* the following Dec. Some successful musical comedies followed, and Marc CONNELLY's *The Green Pastures* (1930) had a great and somewhat unexpected success. Among later productions were Gordon Sherry's *Black Limelight* (1936) and Robert ARDREY's *Thunder Rock* (1939), both of which were less successful than in London, and Ruth Gordon's nostalgic *Years Ago* (1946). The theatre was then used for radio and television shows, but reopened on 12 Sept. 1960 under its present name, given in honour of the critic Brooks ATKINSON. Neil SIMON's *Come Blow Your Horn* (1961) had a long run, and later interesting productions were Rolf HOCHHUTH's *The Deputy* and Tennessee WILLIAMS's *The Milk Train Doesn't Stop Here Anymore* (both 1964); Peter USTINOV's *Halfway Up the Tree* (1967), with Eileen HERLIE and Anthony QUAYLE; Peter NICHOLS's *A Day in the Death of Joe Egg* (1968), with Albert FINNEY; and Ronald MILLAR's *Abelard and Heloise* (1971), with Keith MICHELL and Diana RIGG. In 1973 the theatre housed the Negro Ensemble Company in Joseph A. Walker's *The River Niger* from the off-Broadway St Mark's Playhouse. Later productions included Charles Laurence's *My Fat Friend* (1974), two plays by Bernard Slade, *Same Time, Next Year* (1975) and *Tribute* (1978), Alan AYCKBOURN's *Bedroom Farce* (1979), and Lanford Wilson's *Talley's Folly* (1980).

BROOKS THEATRE, Cleveland, see CLEVELAND PLAY HOUSE.

BROUGH, FANNY WHITESIDE (1854–1914), English actress, daughter of the dramatist Robert Brough (1828–60) and niece of Lionel (below). She made her first appearance in Manchester in 1869 in a pantomime written by Lionel's elder brother William (1826–70), and a year later was seen in London under Mrs John WOOD. Although she sometimes appeared in Shakespeare in her early years, she was at her best in modern comedy and during her long career on the London stage appeared in a number of new plays, creating the parts of Mary Melrose in H. J. BYRON's *Our Boys* (1875), Mary O'Brien in Frances Hodgson Burnett's *The Real Little Lord Fauntleroy* (1888) and Lady Markby in Oscar WILDE's *An Ideal Husband* (1895). In 1901 she appeared with Charles HAW-

TREY in Anstey's *The Man from Blankley's*, and in 1909 made a great success as the Hon Mrs Beamish in the sporting melodrama *The Whip*, by Cecil Raleigh and Henry Hamilton, at DRURY LANE. She made her last appearance, also at Drury Lane, at Christmas 1911, playing the Baroness Chicot in the pantomine *Hop o' My Thumb*.

BROUGH, LIONEL (1836–1900), English actor, who made his first appearance in a play by his elder brother William (1826–70). He was for a time in journalism, but returned to the stage in 1864, and in 1873 was principal low comedian at the GAIETY under HOLLINGSHEAD. He was not so much a character actor as a clown in the best sense, his gift of improvisation and rich sense of humour making him an excellent player in BURLESQUE. Two of his finest parts were Tony Lumpkin in GOLDSMITH's *She Stoops to Conquer* and Bob Acres in SHERIDAN's *The Rivals*, but he was also good in Shakespeare's comic parts, among them Sir Toby Belch in *Twelfth Night* which he played with TREE at HER MAJESTY'S THEATRE. He toured extensively, and was extremely popular in the United States and South Africa. His daughter Mary (below) was an excellent actress.

BROUGH, MARY (1863–1934), English actress, daughter of Lionel (above). She made her first stage appearance in Brighton in 1881 and was seen in London later the same year playing small parts with Lillie LANGTRY. Like her cousin Fanny BROUGH she was at her best in comedy, and had already had a long and successful career when in 1925 she joined the ALDWYCH THEATRE company to play Mrs Spoker in *A Cuckoo in the Nest* by Ben TRAVERS. She appeared in all his subsequent productions up to *Dirty Work* (1932) in which she played Mrs Bugle.

BROUGHAM, JOHN (1810–80), American actor and dramatist, born in Ireland, who made his first appearance in London in July 1830 in Egan's *Tom and Jerry*. After a long engagement with Mme VESTRIS he became manager of the LYCEUM, and in 1842 went to America, making his New York début at the PARK THEATRE. On 23 Dec. 1850 he opened his own theatre on Broadway as Brougham's Lyceum, hoping to rival the success of Mitchell's OLYMPIC. Brougham was a fine actor, an experienced manager, and a jovial popular personality, and his enterprise ought to have been successful; but the short burlesques and farces which had served Mitchell well for so long were now outmoded, and in spite of the hurried importation of such stars as FLORENCE and Charlotte CUSHMAN control of the theatre passed into the hands of the elder WALLACK, who opened it as WALLACK'S LYCEUM on 25 Jan. 1852. Brougham continued to act, spending several years in England before opening his second playhouse, the FIFTH AVENUE THEATRE (1), on 25 Jan. 1869. This again was not a success. Brougham retired from management a few months later and up to the time of his death appeared with various stock companies in New York, making his last appearance on 25 Oct. 1879 at BOOTH'S THEATRE. He was essentially a comedian and his best parts were stage Irishmen such as Sir Lucius O'Trigger in SHERIDAN's *The Rivals* and Dennis Brulgruddery in the younger COLMAN's *John Bull*, as well as Mr Micawber in DICKENS's *David Copperfield*, Captain Cuttle in his *Dombey and Son*, and Dazzle in BOUCICAULT's *London Assurance*. He wrote a number of plays, none of which has survived.

BROWN, PAMELA (1917–75), English actress, who in a distinguished, though comparatively short, career won many admirers. She made her stage début in 1936 at the SHAKESPEARE MEMORIAL THEATRE as Juliet in *Romeo and Juliet*, her other roles during the same season including Cressida in *Troilus and Cressida*. In the same year she made her London début in Clifford Bax's *The King and Mistress Shore*, and later played in repertory at the OXFORD PLAYHOUSE. She was briefly at the OLD VIC in 1939, before returning to Oxford for a further season when her roles included IBSEN's Hedda Gabler and Nina in CHEKHOV's *The Seagull*. She made her first outstanding success at the ST MARTIN'S THEATRE in the title role of Rose Franken's *Claudia* (1942). She played Ophelia in *Hamlet* in 1944 and Goneril in *King Lear* in 1946 for the Old Vic, and in 1947 made her New York début as Gwendolen in WILDE's *The Importance Of Being Earnest*. On returning to London she played leading roles in Aldous Huxley's *The Gioconda Smile* (1948) and FRY's *The Lady's Not For Burning* (1949; New York 1950), her haunting looks, far from conventionally beautiful, and husky voice giving a disturbing ambiguity to both characterizations. After appearing in Charles MORGAN's *The River Line* (1952) she joined GIELGUD's company at the LYRIC, Hammersmith in 1953, being seen as Millamant in CONGREVE's *The Way of the World* and Aquilina in OTWAY's *Venice Preserv'd*. In the same year she appeared opposite Paul SCOFIELD in Wynyard Browne's *A Question of Fact*, returning to New York to play Lady Fidget in WYCHERLEY's *The Country Wife* in 1957 and Lady Utterword in SHAW's *Heartbreak House* in 1959. She made her last stage appearance in Jack Ronder's *This Year, Next Year* (1960).

BROWNE, E(lliott) MARTIN (1900–80), English actor and director, closely connected with the revival of poetic, and particularly religious, drama in England. In 1935 he directed and played in T S. ELIOT's *Murder in the Cathedral*, and thereafter directed all Eliot's plays including *The Family Reunion* (1939), *The Cocktail Party* (1949), *The Confidential Clerk* (1953), and *The Elder Statesman* (1958). He was from 1939 to 1948 director of the Pilgrim Players, who in association with the ARTS COUNCIL toured England in a repertory of religious plays, and in 1945 took over the MERCURY THEATRE in London for the production of new plays by poets, among them FRY's *A Phoenix Too Frequent* (1946). In 1951 he was responsible for the production in its native city, for the first time since 1572, of the York Cycle of MYSTERY PLAYS which he revived several times. In 1945 he had succeeded Geoffrey Whitworth as director of the British Drama League (now the British Theatre Association), and he held this post until 1957, becoming in 1952 the first president of the International Amateur Theatre Association. He then went to the U.S.A. to inaugurate a programme of Religious Drama at the Union Theological Seminary in New York, where he remained as visiting professor until 1962. On his return to England he was appointed Honorary Drama Adviser to the rebuilt Coventry Cathedral; he was also President of Radius (see RELIGIOUS DRAMA).

BROWNE, MAURICE (1881–1955), actor, dramatist, and theatre manager. Born and educated in England, he made his name in the U.S.A., where he is credited with having founded the Little Theatre movement by the establishment in 1912 of the Chicago Little Theatre, which he directed for several years. In 1920 he was directing on Broadway, and in 1927 he made his first appearance in London, at the ARTS THEATRE, as Adolf in STRINDBERG's *The Creditors*. At the beginning of 1929 he took over the management of the SAVOY THEATRE and presented there with remarkable success R. C. SHERRIFF's war play *Journey's End*, which had been tried out by the STAGE SOCIETY at the end of 1928. In the following year he produced *Othello* with Paul ROBESON in the title role, playing Iago himself with Peggy ASHCROFT as Desdemona. He was later responsible for the management of the GLOBE and QUEEN's theatres, presenting among other plays *Hamlet*, transferred from the OLD VIC with GIELGUD in the title role in 1930, and seasons of MOISSI and the PITOËFFS. Among his later successful productions were FAGAN's *The Improper Duchess* (1931) with Yvonne ARNAUD, *Wings Over Europe* (1932), of which he was part-author with Robert Nichols, Norman Ginsbury's *Viceroy Sarah* (1935) with Edith EVANS, and Esther McCracken's *Quiet Wedding* (1939). He retired at the end of 1939.

BROWNE, ROBERT, English actor, first mentioned as one of Worcester's Men in 1583. He became one of the best known of the ENGLISH COMEDIANS travelling on the Continent at the end of the 16th century. In 1592 he toured Holland and Germany at the head of an English company which included in its repertory JIGS, Biblical plays, early English comedies, and cut versions of plays by MARLOWE. In Aug. of that year he made the first of many visits to the great fair at FRANKFURT-AM-MAIN and in 1595 he and his companions took service with Maurice, Landgrave of Hesse-Kassel. At the beginning of the 17th century Browne was intermittently in London, but he continued to visit Frankfurt and other German towns and spent the winter of 1619 at the Court of Bohemia in Prague. The last mention of him is in connection with the Easter fair at Frankfurt in 1620. He may then have returned to England for good, and may be a Robert Browne who toured the English provinces with a puppet-show during the 1630s. His work on the Continent was continued by his friend and pupil John Green.

BROWNE, W. GRAHAM, see TEMPEST, MARIE.

BROWNING, ROBERT (1812–89), English poet, three of whose plays in verse were produced in the theatre: *Strafford* (1837), with MACREADY for whom it was written in the leading role, at COVENT GARDEN; *A Blot on the 'Scutcheon* (1843) at DRURY LANE; and *Colombe's Birthday* (1853) at the HAYMARKET. Helen FAUCIT appeared in all of them. None of them was particularly successful, and they serve only to mark the great gap between poetry and the stage in the 19th century. Browning's other plays were written to be read and belong to literature rather than to the theatre, though the dramatic poem *Pippa Passes* was performed at the NEIGHBORHOOD PLAYHOUSE in New York in 1917, and at the MADDERMARKET THEATRE in Norwich in 1925. It was revived at the OXFORD PLAYHOUSE in 1968.

BROWN-POTTER, Mrs JAMES [*née* Cora Urquhart] (1859–1936), one of the first American society women to go on the stage. She was trained by David BELASCO, though she never appeared under his management, being first seen as a professional, after amateur performances for charity, in London in 1886. With Kyrle Bellew as her leading man she toured Australia and the Far East, preceded by immense publicity, and returned to London in 1892. During the Boer War she raised large sums of money for war charities by her recitation of such poems as Kipling's 'The Absent-Minded Beggar', and in 1904 became manager of the SAVOY. Her venture was unsuccessful, and ended in bankruptcy, after which she went on tour, retiring from the stage in 1912.

BRUCKNER, FERDINAND [Theodor Tagger] (1891–1958), Austrian dramatist, whose first play *Krankheit der Jugend* (*Malady of Youth*, 1926) caused something of a sensation; it dealt with a group of medical students morbidly preoccupied with suicide, death being the one certain alternative to disillusion. Several of his later plays dealt with modern problems under the guise of history, as in *Die Verbrecher* (*The Criminals*, 1928), set in Tudor England but dealing fundamentally with the causes of criminality and the workings of justice. He had his greatest success with a play on the same period, *Elisabeth von England* (1930), which was seen in translation in London in 1931. Like *Die Verbrecher* it made brilliant use of simultaneous action by juxtaposing on stage the Courts of Elizabeth I and Philip II of Spain. His *Timon; Tragödie vom überflüssigen Menschen* (*Timon, the Tragedy of the Superfluous Man*, 1933), ostensibly a reworking of Shakespeare's *Timon of Athens*, lays bare the weakness of the Weimar Republic in allowing the rise of Hitler, as does *Napoleon I* (1937). The earlier *Rassen* (*Races*, 1933) offers a vivid picture of the tragedies caused by racial persecution among German students. Bruckner left Europe in 1936 and lived in America until 1951, his most important plays during this period being *Heroische Komödie* (1938), which deals with Mme de Staël and Benjamin Constant, and *Simon Bolivar* (in two parts, 1943/5). After his return to Germany he attempted to revive classical tragedy in a modern form, using verse and a chorus in *Der Tod einer Puppe* (*The Death of a Doll*) and *Der Kampf mit dem Engel* (*The Fight with the Angel*) (both 1956). His last stage work was an adaptation of the old Indian play *Mṛcchakaṭikā* (known in English as *The Little Clay Cart*) under the title *Das irdene Wägelchen* (1957).

BRÜCKNER, JOHANNES (1730–86), German actor, a member of the company of KOCH, who married his sister. He was trained for the stage by LESSING and EKHOF, and was one of the best tragedians of his time. His Mellefont in Lessing's *Miss Sara Sampson* was considered superior even to Ekhof's. One of Brückner's greatest triumphs was in the title role of GOETHE's *Götz von Berlichingen*.

BRUEYS, DAVID-AUGUSTIN DE (1640–1723), French dramatist, who collaborated with PALAPRAT in several plays for the COMÉDIE-FRANÇAISE, among them *Le Grondeur* and *Le Muet* (both 1691). Brueys also wrote a number of plays on his own, including a version of the old FARCE of *Maistre Pierre Pathelin*, which, as *L'Avocat Pathelin* (1706), remained in the repertory of the Comédie-Française for many years.

BRUN, NORDAHL, see NORWAY.

BRUNELLESCHI, FILIPPO, see PARADISO.

BRUNO, GIORDANO (1548–1600), Italian philosopher, author of one play, *Il Candelaio* (*The Candle-Maker, c.* 1582), a brilliant satire which castigated the corrupt customs of the time with unusual freedom, and was banned. Bruno went to England and visited Oxford at the invitation of Sir Philip Sidney. Among the contemporary English authors who met him may have been Thomas Carew whose MASQUE *Coelum Britannicum*, performed at Whitehall in 1634, shows traces of his influence. Bruno was burnt at the stake by order of the Inquisition. In 1965 *Il Candelaio*, in a cut and modernized version, had what was probably its first professional performance, and in 1970 a play by Morris West, *The Heretic*, based on Bruno's life, had a brief run in London.

BRUNSWICK THEATRE, London, see ROYALTY THEATRE (1).

BRUNTON, ANNE, JOHN and LOUISA, see MERRY, MRS.

BRUNTON, ELIZABETH, see YATES, FREDERICK HENRY.

BRUSCAMBILLE [Jean Deslauriers] (*fl.* 1610–34), French actor at the Paris fairs in the early 17th century who went with Jean FARINE to the Hôtel de BOURGOGNE to play in farce. He won fame as a speaker of witty prologues and harangues to the rowdy audiences, which he composed himself. These are extant, and give an interesting picture of the tribulations of the actor before his profession became respectable, summed up in Bruscambille's oft-quoted epigram, 'une vie sans soucis et quelquefois sans six sous'.

BRUSTEIN, ROBERT, see HARVARD and YALE.

BRYAN, DORA, see MUSICAL COMEDY and REVUE.

BRYANT, MICHAEL (1928–), English actor, who made his first stage appearance in Brighton in 1951, and was first seen in London in IONESCO's *The New Tenant* (1956). He was in the London production of O'NEILL's *The Iceman Cometh* in 1958, in which year he first came into prominence as the young tutor in SHAFFER's *Five Finger Exercise*, the part in which he made his New York début and toured the U.S. a year later. Back in London in 1961, he successfully took over from Alec GUINNESS the part of T.E. Lawrence in RATTIGAN's *Ross*, displaying the solitary and introspective qualities he had shown in previous appearances. He then played Jacko in Robert BOLT's short-lived *Gentle Jack* (1963), and in 1964 joined the ROYAL SHAKESPEARE COMPANY, his roles including Selim Calymath in MARLOWE's *The Jew of Malta* and Teddy in PINTER's *The Homecoming* (1965). He joined the NATIONAL THEATRE company in 1977 and was in John MORTIMER's translation of FEYDEAU's *The Lady from Maxim's*. He has since played a variety of classic roles, including IBSEN's Brand and Sir Paul Plyant in CONGREVE's *The Double Dealer* in 1978; Jaques in *As You Like It* and Gregers Werle in Ibsen's *The Wild Duck* in 1979; and Iago to Paul SCOFIELD's Othello in 1980. His modern roles include Lenin in Robert Bolt's *State of Revolution* (1977) and Julius Caesar in Howard BRENTON's *The Romans in Britain* (1980).

BRYANT'S OPERA HOUSE, New York, see KOSTER AND BIAL'S MUSIC HALL.

BRYANT THEATRE, New York, see APOLLO THEATRE.

BRYDEN, BILL [William] CAMPBELL ROUGH (1942–), Scottish director, who in 1965 was at the BELGRADE THEATRE, Coventry, his first production there being SHAW's *Misalliance*. He moved to the ROYAL COURT THEATRE in 1967, where his first production was Ronald Ribman's *Journey of the Fifth Horse*, and from 1971 to 1974 was at the LYCEUM THEATRE in Edinburgh where he directed his own plays *Willie Rough* (1972) and *Benny Lynch* (1974). In 1975 he became an associate director of the NATIONAL THEATRE, staging such plays as SYNGE's *The Playboy of the Western World* in 1975, OSBORNE's *Watch It Come Down* (1976), and O'CASEY's *The Plough and the Stars* in 1977. In 1978 he was appointed director of the COTTESLOE THEATRE, its flexibility enabling him to stage excellent 'promenade' productions, in which all the seats were removed and the audience mingled with the actors, including *The Passion*, drawn from the York MYSTERY PLAYS, and *Lark Rise* and *Candleford*, based on a book by Flora Thompson. He also directed in 1980 a season of plays by Eugene O'NEILL—*Hughie, The Iceman Cometh*, and *The Long Voyage Home*—as well as Arthur MILLER's *The Crucible*, which transferred to the West End.

BRYLL, ERNEST, see POLAND.

B.T.A. (BRITISH THEATRE ASSOCIATION), see AMATEUR THEATRE.

BUCCO, a stock character in the early Roman FABULA *atellana*, whose name suggests a clown with fat cheeks, perhaps a big eater and talker. He is the chief character in two plays by Pomponius, *Bucco Out on Bail* and *Bucco Adopted*, and appears in his *Soothsayer* of which a fragment has survived.

BUCHANAN, JACK [Walter John] (1890–1957), Scottish actor-manager, singer, dancer, and director, famous for his debonair manner, elegant dancing, and slightly croaky singing voice. He made his professional début at the Edinburgh Empire in 1911, and his first London appearance in a comic opera, *The Grass Widows* (1912). He toured in the musical *To-Night's the Night* and was in several revues before, in 1921, he starred in the Charlot revue *A to Z*, for which he also directed the sketches and staged the musical numbers. The first of his many ventures into management was *Battling Butler* (1922), a musical farce in which he played the title role. In 1924 he made his début in New York, where he was to become well known, co-starring in the *Charlot Revue* with Gertrude LAWRENCE and Beatrice LILLIE, all three being seen there again in the 1926 edition of the revue. He was in the London production of Kern's musical *Sunny* in 1926 and the British musical *That's a Good Girl* in 1928, returning to New York in 1929 to star in another revue, *Wake Up and Dream*, this time with Jessie MATTHEWS. During the 1930s he appeared in London in such native musicals as *Stand Up and Sing* (1931), of which he was part-author, and *This'll Make You Whistle* (1936). In 1944, however, he played a straight role in LONSDALE's *The Last Of Mrs Cheyney*, and thereafter he was seen mainly in non-musical productions, including Lonsdale's *Canaries Sometimes Sing* (1947), *Don't Listen Ladies!* (N.Y., 1948; London, 1949), adapted from a play by Sacha GUITRY, Alan Melville's *Castle In the Air* (1949), and Vernon Sylvaine's *As Long As They're Happy* (1953). He directed many of the productions he acted in or presented, and at the time of his death controlled the GARRICK THEATRE and the KING'S THEATRE, Hammersmith.

BÜCHNER, GEORG (1813–37), German dramatist, by profession a doctor, who died young of typhoid fever. Unknown during his lifetime, his reputation has grown steadily since he was discovered by HAUPTMANN in the 1890s and taken as a model by adherents of EXPRESSIONISM. His first play, *Dantons Tod* (*Danton's Death*), was written and published in 1835 but not acted until 1903 in Berlin; its portrayal of the struggle for power between Robespierre and Danton during the early years of the French Revolution reflects Büchner's own disillusionment with politics after he had narrowly escaped arrest for revolutionary activities. Its enduring quality was proved by a revival in Berlin in 1927 and by its success in translation in a New York production by Orson WELLES in 1938, and in London, directed by Jonathan MILLER for the NATIONAL THEATRE, in 1971. Büchner's only other complete play was a comedy entitled *Leonce und Lena*, written shortly before his death but not published until 1850, first performed in 1895 at Munich. *Woyzeck*, of which only an unfinished version remains, was discovered and published as *Wozzeck* in 1879. Alban Berg made it the subject of his opera under that title in 1925, having seen the first production in Vienna in 1913. Its radically innovatory hero is an uneducated, down-trodden army private, willing to do any work, however humble, to support his mistress and their child: there is a strong element of social criticism in the plot, which shows Woyzeck being driven to madness and murder, and the brief, unconnected scenes make a powerful short drama, usually ending with Woyzeck's suicide. It has several times been revived in Germany, and was seen, as *Woyzeck*, in New York in 1971 and in an adaptation by Charles MAROWITZ at the Open Space in London in 1973.

BUCK, Sir GEORGE, see MASTER OF THE REVELS.

BUCKINGHAM, GEORGE VILLIERS, second Duke of (1628–87), English nobleman and a prominent literary figure of the Restoration. In 1666 PEPYS noted in his diary that he had seen and enjoyed at DRURY LANE a new version by Buckingham of *The Chances* (1623) by FLETCHER, based on a story by CERVANTES. (It was successfully revived at the CHICHESTER FESTIVAL of 1962.) Buckingham was also the author of *The Rehearsal* (1671), which satirized the HEROIC DRAMA of DRYDEN and others and provided a model for many later burlesques of which the best was SHERIDAN's *The Critic; or, a Tragedy Rehearsed* (1779). Dryden responded by depicting Buckingham as Zimri in his satirical poem *Absalom and Achitophel* (1681).

BUCKLEY'S OLYMPIC, New York, see OLYMPIC THEATRE (3).

BUCKSTONE, JOHN BALDWIN (1802–79), English actor and dramatist, who made his first appearance in the famous melodrama *The Dog of Montargis* in a barn in Peckham. Edmund KEAN, who saw him act, encouraged him, and between 1823 and 1827 he appeared at the Coburg (later the OLD VIC) and SURREY theatres. In 1827 he was seen at the ADELPHI in his own play *Luke the Labourer*, which had been given anonymously the previous year because the manager Daniel TERRY had lost the author's name and address. Buckstone was for many years manager of the HAYMARKET THEATRE where many of his plays were given; he is said to haunt the building. Among his many melodramas and farces the best known are *Married Life* (1834), *Single Life* (1839), *The Green Bushes; or, A Hundred Years Ago* (1845), and *The Flowers of the Forest* (1847). He was a popular comedian of great breadth and humour; the mere sound of his voice, a mixture of chuckle and drawl, heard off-stage was enough to set an audience laughing.

BUERO VALLEJO, ANTONIO (1916–), Spanish dramatist, whose *Historia de una escalera* (*Story of a Staircase*, 1949) won the Lope de Vega Prize and marked the ascendancy of social realism in the Spanish post-Civil War theatre. His most important plays are *Hoy es fiesta* (*Today's a Holiday*, 1956), another social drama, *Las meniñas* (*The Ladies-in-Waiting*, 1960), on a historical subject, the symbolic *El concierto de San Ovidio* (1962), and *El sueño de la razón* (*The Dream of Reason*, 1970), a hallucinatory interpretation of Goya. Later works include *La fundación* (1974) and *La detonación* (1977). All his work is marked by a resigned yet hopeful humanism.

BUFFALO BILL, see CODY, W.F.

BUILT STUFF, all specially carpentered three-dimensional objects, from banks and ROSTRUMS to columns, rocks, and complete scenes on BOAT TRUCKS. The most common is perhaps the rostrum, which may vary in size from a small throne-dais to a platform 8 ft high, approached by steps or a ramp. Beyond this, by way of porches and mantelpieces, one reaches all the ingenious applications of light carpentry, chicken-wire reinforcement, glued and shaped canvas, papier-mâché, and all the new plastics, such as high-impact polystyrene and low-density polyethylene sheet for vacuum-forming, rigid urethane and polystyrene foams, Celastic dipped in acetone, and synthetic rubber; all these enable a room complete in all details, or a wood scene with every tree, leaf, and rock in the round, to take shape upon the stage.

BULANDRA, LUCIA and TONY, see ROMANIA.

BULGAKOV, MIKHAIL AFANASEYEV (1891–1940), Soviet dramatist, whose first play was a dramatization of his own novel *The White Guard*, dealing with the Civil War in the Ukraine in 1918. As *The Days of the Turbins* it was produced by STANISLAVSKY at the MOSCOW ART THEATRE in 1926. The play was at first harshly criticized as being too favourable to the 'Whites' (as opposed to the victorious 'Reds') and was withdrawn after a few performances. A second play on the same subject, *The Flight*, was not staged, while two later plays, a comedy entitled *The Red Island* (1928) and *The Cabal of Saintly Hypocrites* (1936), which dealt with MOLIÈRE's difficulties with the censorship over *Tartuffe*, were seen only briefly at the KAMERNY and the Moscow Art theatres respectively. Bulgakov joined the staff of the Moscow Art Theatre; he also prepared for the company there an excellent dramatization of GOGOL's *Dead Souls* (1930). His last play, based on CERVANTES' *Don Quixote*, was produced posthumously at the VAKHTANGOV in 1941. *The White Guard*, adapted by Rodney Ackland and directed by Michel SAINT-DENIS, was seen in London in 1938; it was revived in 1979 at the ALDWYCH by the ROYAL SHAKESPEARE COMPANY, and as *The Days of the Turbins* it was produced in New York in 1977. An adaptation of *The Cabal of Saintly Hypocrites* as *Molière in Spite of Himself* was staged in New York in 1978, as was *The Master and Margarita*, a play by Andrei Serban based on a novel by Bulgakov, which Serban directed.

BULGARIA. The theatre in Bulgaria developed late in the country's history; there is no trace of religious or secular drama in the literature of the Middle Ages, and the long period of Turkish rule, from 1396 to 1878, prevented not only contact with the theatre in other countries, but also all indigenous cultural activity, so much so that the written Bulgarian language almost vanished. The first modern literary work, a patriotic history of the Bulgars by an Orthodox monk Paisii Khilendarski, appeared only in 1762, and the first grammar of the language not until 1835. Drama may be said to have begun tentatively in the 1840s, with a series of moral and patriotic 'dramatic dialogues' by Iordan Djinot, a schoolmaster, which were soon staged and imitated throughout the country. During the 1850s several amateur companies were founded, but they appeared only in translations of such authors as FONVIZIN, GOLDONI, MOLIÈRE, and SCHILLER. Even the first play written directly in the Bulgar language was based, somewhat freely, on a Serbian comedy. This was *Mikhail*, the work of a schoolmaster, S. I. Dobroplodnij (1820–94), some of whose pupils acted it in a café in Shumen on 15 Aug. 1856. Two of Dobroplodnij's pupils themselves became dramatists —Dobri Vojnikov (1833–78) and Vasil Drumen (1841–1901). The latter was the author of *Ivanko the Assassin of Asen I* (1872), a vigorous costume drama of intrigue, which had a immense success when first produced and is still occasionally revived. It was first seen in Brăila, Romania, where both men lived in exile, and where Vojnikov directed from 1866 to 1871 a theatre intended for Bulgar émigrés, writing for it melodramas based on incidents from Bulgaria's medieval past—*Princess Raina* (1866); *The Christianization of the Court at Preslav* (1868); *The Accession of Tsar Krum the Terrible* (1871)—and a few comedies, of which the best was *Civilization Misunderstood* (also 1871), an attack on the current vogue for everything French (as in SHAW's *Arms and the Man* some 20 years later) which ends with the conversion of all the characters to the language, manners, and customs of their own country. An earlier play on much the same theme was *The Crinoline* (1864) by P. R. Slavejkov (1827–95), in which the pretensions of Europeanized Bulgarians are satirically contrasted with local peasant innocence. But on the whole the plays of the pre-Liberation period were either historical, designed to stimulate Bulgarian national consciousness, or melodramatic, as in the dramatization of his own story *Izgubena Stanka* produced in 1870 by the writer Iljia Bluskov (1839–1913), where the sufferings of the heroine at the hands of the Tartars were taken by the audience as symbolic of their own sufferings under Turkish rule.

The years preceding the Liberation in 1878 have become known as the period of *hayduk* drama, plays based on poems and ballads celebrating the exploits of the Bulgarian outlaws (*hayduks*) who emerged briefly from the forests and hills to harry the Turks and their collaborators. The best of these plays, which owe a good deal to BYRON and Schiller, are S. Iordanov's *Golden Stoyan the Voivode* (1865), L. Karavelov's *Hadji Dimitar Asenov* (1871), K. Velichkov's *Svetoslav and Nevenka* (1874), and such post-Liberation works as P. Ivanov's *Apostles of Freedom* (1884), T. Frangov's *An Incident at Drianovskaya Monastery in 1876* (1885), and R. Popova's *On the Day of Revolt* (1900).

As the number of Bulgarian plays increased organized theatrical activity became necessary; by 1880 there were seven companies in the country, the most important being those at Sofia, under the Croatian G. Jovanovič, and at Plovdiv, at that time the cultural capital of Bulgaria, where in 1881 S. Popov (1846–1920) and K. Sapunov (1844–1906) established the Rumelian Company. In 1886 N. Kravarev founded another amateur company there which became professional on moving to Sofia two years later. Also in Sofia was the Laughter and Tears company, founded by I. Popov, V. Nalburov, and R. Kaneli, which functioned until 1906 and was the first group to play Shakespeare (*Othello* in 1897) and CHEKHOV (*Uncle Vanya* in 1904) in Bulgaria. It was from this group that the National Theatre in Sofia, which opened in a fine new building on 3 Jan. 1907, was derived. When the original building burnt down in 1923 it was replaced by an excellent modern theatre designed by the German architect Duelfer.

The dominant figure in Bulgarian drama at the end of the 19th century was Ivan Vazov (1850–1921) whose *Khushove* (1894) remains in the repertory to this day. Like many of his other works it deals with the patriotic fervour preceding the Liberation; the plays he wrote for the newly opened National Theatre—including *Under the Yoke* (1910) and *The Czarina of Kazalar* (1911), both based on his own novels—also dealt with the recent past, while in *Towards the Precipice* (1907), *Borislav* (1909), and *Ivailo* (1914) he returned to Bulgaria's medieval period for his subjects. Vazov was also the author of a number of satirical comedies, among them *The Job Seekers* (1903), which attacked the frenetic careerism that had become a marked feature of Bulgarian public life.

As the years of enforced cultural isolation ended the Bulgarian theatre became more open to European developments and strictly moral or patriotic dramas went out of fashion. The example of IBSEN is seen in Anna Kirima's *The Awakening* (1902) and I. Kirilov's *The Lark* (1906), that of SYMBOLISM in Hristov's tragedy *Boyan the Sorcerer* (1911), the first Bulgarian play in verse. These and other foreign influences are seen in the work of one of the outstanding writers of the period, Anton Strashimirov (1872–1937) whose *Vampire* (1901), a realistic study of peasant passion, and *The Mother-in-law* (1907), which describes a possessive woman's efforts to destroy her son's marriage, are considered classics of the Bulgarian stage. Strashimirov's most able contemporary Petko Todorov (1879–1916) wrote a number of symbolist dramas in which he attempted to combine the mood and techniques of Ibsen's later plays with subjects taken from Bulgarian folk tradition, as in *The Stonemasons* (1901) and *The Serpent's Wedding* (1910), the latter play achieving a curious combination of realistic human psychology with the semi-fantastic setting in which the characters move. Todorov also wrote several dramas of social conflict, among them *The First* (1907), in which a group of villagers struggle against the tyranny of the *chorbadzhiia* (wealthy Turkish collaborators) who feel their power slipping away.

A dramatist who had perhaps a better understanding of theatre technique than any of his contemporaries was Peio Yavorov (1877–1914), who was for several years artistic director of the National Theatre where his realistic tragedies of abnormal sexual passion, *In the Foothills of Mount Vitosha* (1911), *When the Thunder Rolls*, and its sequel *As the Echo Replies* (both 1912), were first produced. Like the hero of his first play, he committed suicide after the death of his wife.

The years 1900–13 saw great advances in Bulgarian theatre and although her alliance with Germany in the First World War cut her off for a time from France and Russia, her most important sources of culture, as soon as the war ended actors and producers were again ready to respond to new technical influences. The National Theatre, directed from 1915 to 1920 by the brilliant actor K. Sarafov and then by the symbolist poet G. Milev, was profoundly affected by the Soviet stage. Many of its actors had studied at the MOSCOW ART THEATRE, and its directors with MEYERHOLD whose methods of staging they followed. In 1925 a Russian actor, N. O. Massalitinov who had directed the Moscow Art Theatre Second Studio, arrived in Sofia to take charge of the National Theatre and its drama school, and thereafter Russian influence was paramount.

The 1920s saw also the arrival of the first Bulgarian who was primarily a playwright and not a poet or novelist first, Stefan L. Kostov (1879–1939). His comedies, of which the best are *The Gold Mine* (1926), a study in credulous cupidity, and *Golemanov* (1927, written in 1920), which satirizes a 'job seeker' in the style of Vazov's earlier play of that name, mark the high point of his career; his later plays are negligible. Between the two wars no single style of writing prevailed in the Bulgarian theatre. *Hayduk* plays and historical dramas continued to be popular, especially those glorifying national heroes, and in *The Millionaire* (1930) Yordan Yovkov (1880–1937) continued the vein of bourgeois satirical comedy mined by Vazov and Kostov with a portrait of a shabby-genteel vet whose non-existent fortune enables him to marry into a greedy and hypocritical rich family. Two other plays by Yovkov, *Albena* (1929) and *Borjuna* (1932), are, like I. Enchev's *Delba* (1931), acute studies of contemporary peasant life. Other plays worthy of note are the gently humorous domestic comedies of S. Mikhailovski (1856–1927)—*When the Gods Laugh* (1922) and *The Tragedy of Conjugal Love* (1927); the symbolist drama *The Golden Chalice* (1922) by Ivan Grozev, a study of the heretic Bogomils of medieval Bulgaria; and *Master Craftsmen* (1927), a well-constructed study of personal and professional rivalry by Racho Stoianov (1883–1951), who apparently wrote nothing else of any importance.

After 1945 the Bulgarian theatre flourished materially, and many new companies were established in their own playhouses, but there was something of a dearth of new playwrights. Some dramatists were able to adapt the old *hayduk* drama to fit tales of communist partisans operating against the Nazis, and other propaganda plays like Gulyashki's *The Promise* (1949) dealt with the struggles between loyal workmen and foreign saboteurs. After the death of Stalin in 1953 there was some relaxation, and such plays as *Fear* (1956) by T. Genov, which portrays a communist opportunist working himself into an important position with the Party, were able to reach the stage. But such playwrights as Kamen Zidarov (1902–) still found it best to confine themselves to historical subjects, as in his *Ivan Shishman* (1960), a verse play set in the years immediately before the Turkish annexation of Bulgaria. Controls soon tightened up again and Bulgarian drama became standardized, with excellent productions of new foreign and Eastern European plays but with no outstanding dramatists of its own.

BULL, OLE (1810–80), Norwegian violinist and patriot whose initiative led to the founding of the first genuinely Norwegian theatre. Det Norske Teatret (later Den NATIONALE SCENE) opened in Bergen, Bull's birthplace, on 2 Jan. 1850, with a policy of appointing Norwegians (and not, as was then customary, Danes) as actors and directors. IBSEN, BJØRNSON, and Gunnar HEIBERG were on the theatre's staff between 1851 and 1888.

BULL INN, see INNS USED AS THEATRES.

BULWER-LYTTON, EDWARD GEORGE EARLE LYTTON, Baron Lytton (1803–73), English novelist and dramatist, whose most successful play *The Lady of Lyons; or, Love and Pride* (1838) was first produced at COVENT GARDEN by MACREADY, who played the hero Claude Melnotte with Helen FAUCIT as Pauline. Its continued popularity led to a number of revivals, notably by Henry IRVING at the LYCEUM in 1879 with himself and Ellen TERRY in the leading roles. It was also constantly burlesqued, most successfully by H. J. BYRON in *The Lady of Lyons; or, Twopenny Pride and Penny-Tence* (1858). Though the plot is romantic and sentimental, it has a touching sincerity and wore well for many years. Bulwer-Lytton's next play *Richelieu; or the Conspiracy* (1839) was also first produced by Macready, and frequently revived, Irving presenting it at the Lyceum no less than four times. *Money* (1840) struck a more serious and contemporary note, seeming to foreshadow the reforms of T. W. ROBERTSON. It too held the stage for many years, being last seen in London in 1911 at a Command Performance with an all-star cast. Bulwer-Lytton's half-dozen other plays are now forgotten.

BUN HILL NURSERY, see NURSERIES.

BUNRAKU, see JAPAN.

BUONTALENTI, BERNARDO (1536–1608), Italian machinist and scene designer, who entered the service of the MEDICI in 1547 and worked for them for the rest of his life. Among his many duties were the organization of firework displays (hence his nickname 'delle Girandole') and the construction of theatrical machinery and SCENERY for Grand Ducal festivities. Drawings and engravings of his work for the Florentine *Intermezzi* of 1589, in honour of the marriage of Ferdinand I of Tuscany, have been preserved and are among the most important early documents of scenic history.

BURBAGE, CUTHBERT (*c.* 1566–1636), son of James (below) and brother of Richard BURBAGE. On his father's death Cuthbert became joint inheritor of his interest in the THEATRE and the BLACKFRIARS THEATRE and having obtained permission to use the latter, unfinished when his father died, he leased it to a children's company, regaining possession in 1608. Meanwhile his connection with the Theatre had embroiled him in financial and other troubles, and after a particularly acrimonious dispute with the ground landlord the Burbages dismantled the building and conveyed the material across the Thames, using it to build the first GLOBE THEATRE.

BURBAGE [BURBADGE, BURBEGE], JAMES (*c.* 1530–97), the builder of the first permanent playhouse in London, the THEATRE. This opened in 1576 and was used by LEICESTER'S MEN with whom Burbage, a carpenter by trade, had begun to act in about 1572. A man of violent temper, stubborn and unscrupulous and beset all his life by financial and other difficulties, he never wavered in his allegiance to the theatre, and though apparently not a good actor himself he was excellent at selecting and training his fellow players, among them his son Richard (below). In 1596 James took over part of an old monastery and adapted it as the second BLACKFRIARS THEATRE, to use in winter in place of the unroofed Theatre, but he died before getting permission to open it and the project was completed by his eldest son Cuthbert (above).

BURBAGE, RICHARD (*c.* 1567–1619), younger son of James (above) and the first outstanding English actor, the creator of most of Shakespeare's great tragic roles—Hamlet, Othello, Lear, and Richard III, the last being accounted by his contemporaries his finest part. It is believed that Shakespeare also wrote for him the part of the Bastard in *King John*. He appeared in a number of other contemporary plays, among his known roles being Hieronimo in the revised version of KYD's *The Spanish Tragedy* (1597) and Ferdinand in WEBSTER's *The Duchess of Malfi* (1614). After his father's death he was associated with his brother Cuthbert in the management of the THEATRE and the building of the GLOBE THEATRE, scene of his greatest triumphs. He began his acting career early, probably under the tuition of his father, the first mention of him being in 1590 when he was with the ADMIRAL'S MEN. He became the leading actor of the CHAMBERLAIN'S MEN on their formation in 1594 and retained that position when in 1603 they were reorganized as the KING'S MEN. Burbage's reputation stood high during his lifetime, and his name long remained synonymous with all that was best in English acting.

BÜRGERLICHES TRAUERSPIEL, the domestic middle-class or bourgeois tragedy in prose inaugurated in Germany by LESSING's *Miss Sara Sampson* (1755). Written under the influence of DIDEROT and LILLO, it refuted ARISTOTLE's assertion that tragedy belonged only to kings and princes and proved that it could equally well arise from a conflict between aristocratic depravity and middle-class morality. This Lessing showed again in *Emilia Galotti* (1772), and SCHILLER even more forcibly in *Kabale und Liebe* (1784) which, while continuing the former criticism of absolutism, also suggests for the first time that the middle classes too may have their faults. One of the best plays of this type, HEBBEL's *Maria Magdalene* (1848), went so far as to dispense with aristocratic villainy; its middle-class heroine is seduced by an ambitious young clerk for the sake of her dowry, but her father's self-righteous obsession with respectability drives her to suicide. Though the term is no longer used, the theme became a staple of social drama.

BURGTHEATER, see AUSTRIA and VIENNA.

BURIAN, EMIL, see CZECHOSLOVAKIA.

BURK, JOHN DALY (*c.*1775–1808), Irish-born dramatist who went to America in 1796 and settled in Virginia around 1800. He is credited with 7 plays: the first and best known, *Bunker-Hill; or, the Death of General Warren*, was first performed in Boston on 17 Feb. 1797, opening in New York at the JOHN STREET THEATRE seven months later. The play, depicting a famous American victory and using bombastic verse rhetoric, was popular with patriotic audiences for 50 years and was regularly performed on holidays such as the Fourth of July. Burk died in a duel with a Frenchman after speaking insultingly of Napoleon.

BURKE, BILLIE, see ZIEGFELD, FLORENZ.

BURKE, CHARLES, see JEFFERSON, JOSEPH.

BURLA (pl. *burle*), a comic interlude or practical joke introduced, usually extempore, into a performance by the servant masks of the COMMEDIA DELL'ARTE. Unlike the LAZZO, which was often no more than a witty aside or an apt gesture, the *burla* involved some horseplay and could be developed at will into a small independent 'turn', the charac-

Theatre at Epidauros, dating from the mid-4th century BC but showing the circular orchestra characteristic of earlier theatres.

Theatre of Dionysus, Athens. The stage was raised above the horse-shoe shaped orchestra, with elaborate sculpted figures forming a frieze, in the rebuilding of AD 61–66. Both theatres have been extensively restored and are regularly used for productions of the Greek classics.

Cart of Thespis, from an Attic vase painting, showing Dionysus attended by satyrs.

Marble copy of a tragic mask; the original masks were probably of wood and painted cloth.

Actor holding a tragic mask and wearing high boots (*cothur*

Actors as birds. Dating from *c*.500 BC, this vase painting probably depicts a religious ceremony, but it suggests a style of costume for the chorus in Aristophanes' *Birds*.

Actors as horses. Vase painting, *c*.550 BC, showing a processional chorus. The riders, as well as the 'horses', are uniformly masked.

Relief panel, 1st century BC, thought to depict Menander in his studio with the masks of three of his chief characters — the young man, the maiden, and the angry father.

Scenes from the *Eumenides*. This krater painting does not show a theatrical performance, but illustrates the legend used by Euripides: Clytemnestra (*left*) wakens the Furies while (*centre*) Orestes is purified of his blood-guilt.

Modern performance of Sophocles' *Oedipus Rex* at Delphi in 1951. Alexis Minotis as Oedipus, Katina Paxinou as Jocasta.

Modern performance at Epidauros, showing the use of the circular orchestra by the chorus in front of the reconstructed stage house.

Roman theatre at Orange, *c.* AD 2, with semi-circular orchestra and elaborate *scaenae frons*.

Roman theatre at Verulamium, probably AD 140–50, but modified several times up to the 4th century. It was originally used for sports as well as dancing and other entertainments.

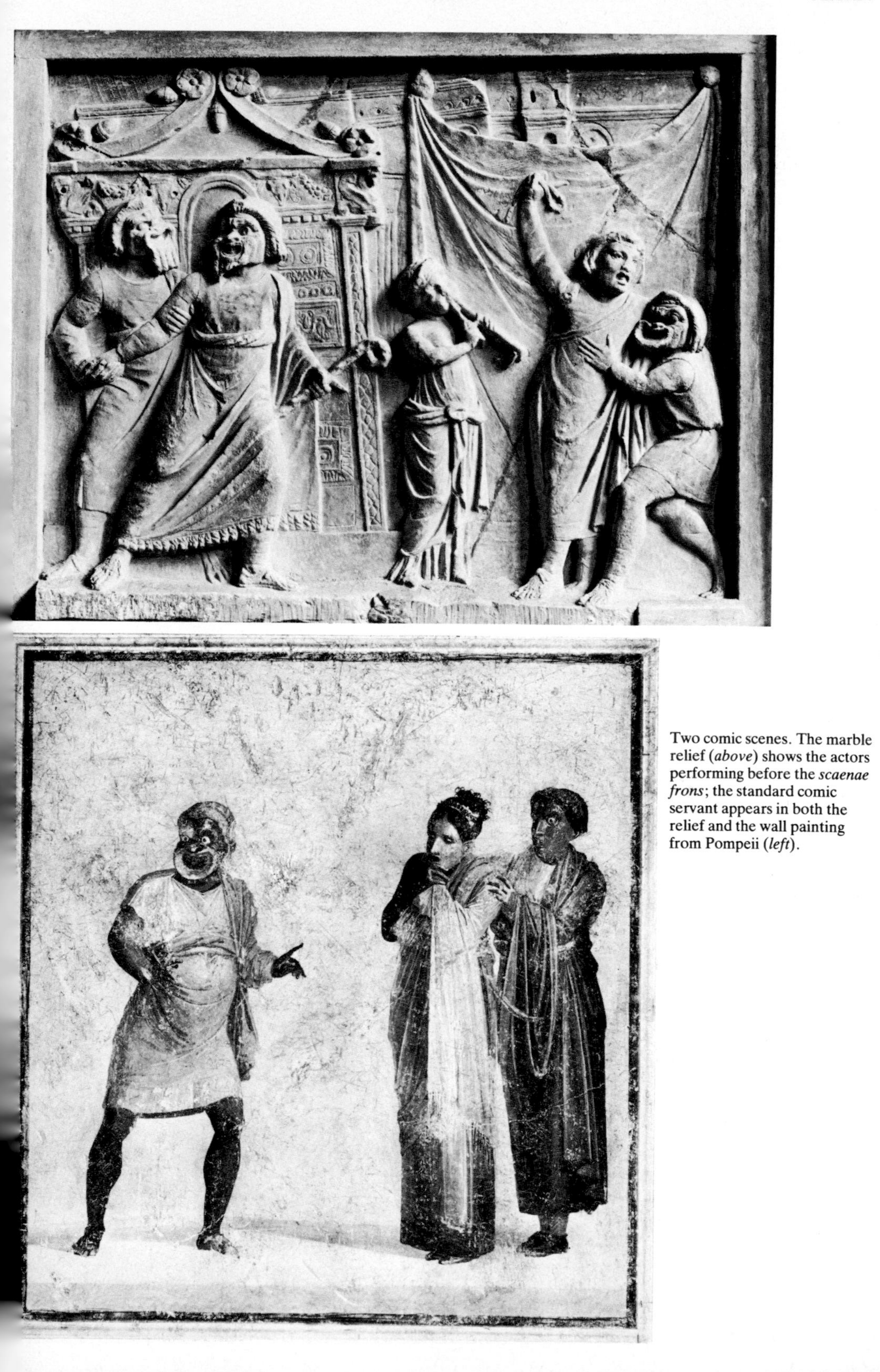

Two comic scenes. The marble relief (*above*) shows the actors performing before the *scaenae frons*; the standard comic servant appears in both the relief and the wall painting from Pompeii (*left*).

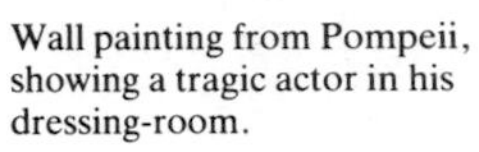

Wall painting from Pompeii, showing a tragic actor in his dressing-room.

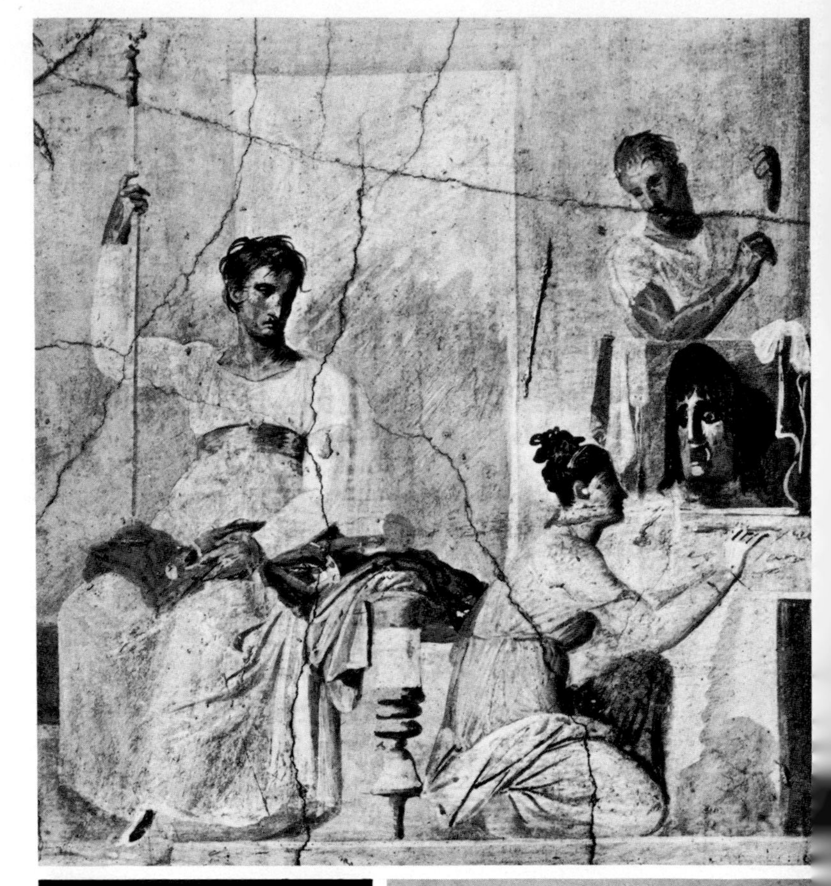

Terracotta figure (*right*) of a comic character from the *fabula atellana*.

Marble figure (*far right*) of an actor in a comic mask, in the role of a slave.

Liturgical drama (*left*): the *Quem quaeritis?* c.1100. The three Marys encounter the angel in the tomb of Christ.
Everyman (*above*), title page of the English version, c.1500, of the old Dutch morality play *Elkerlyc*.

Courtly maskers costumed as satyrs, 1363. According to Froissart, the pitch used to stick animal hair to the performers' costumes accidentally caught fire and four died as a result.

French open-air religious play, *c.*1460: the martyrdom of St Apollonia. Hell mouth is on the right, the ascent to heaven on the left. The booths ranged behind the raised stage are analogous to the mansions of the multiple setting (*see below*).

Valenciennes mystery play, 1547. Part of the multiple setting is shown, with the purification of the Virgin, the flight into Egypt, the massacre of the innocents, and the devil bearing away the soul of Herod, accommodated in various stage areas and mansions.

Terence-stage, 1493. An elaborately decorated booth stage on two levels which is an obvious precursor of the stage-house of Shakespeare's theatres.

Double pageant stage erected in front of the Church of the Holy Innocents in Paris for the wedding of Mary Tudor to Louis XII in 1514. Mechanical flowers opened to reveal performers representing the bride and groom flanked by allegorical figures.

Temporary stage set up by the Rederykers for a procession in Antwerp, 1582. The lavish decoration demonstrates the extensive funds devoted to such open-air entertainments.

Sebastiano Serlio's three basic sets — for tragedy, comedy, and satyr-plays — published in 1545. His theories of perspective stage design had a long-lasting influence all over Europe.

Setting for a school play, 1581, reminiscent of the multiple setting of the medieval mystery play. The playing area is supported on casks.

The Teatro Olimpico at Vicenza, designed by Palladio and completed by Scamozzi in 1585.

Village mummers of the 16th century, performing the story of Valentine and the wild man of the woods.

Booth stage, 1608, typical of the temporary staging used by travelling players. It is set up against the side of an inn; rowdy members of the audience are being driven off the stage while an actor waits to make his entrance.

ters returning at its conclusion to the main theme of the plot. Although there is no adequate English translation of the word, the terms BURLETTA and BURLESQUE derive from it.

BURLESQUE, American, a sex-and-comedy entertainment originally intended for men only. Known popularly as 'burleycue' or 'leg show', it was devised by Michael Bennett Leavitt (1843–1935) in about 1868. Each performance was divided into three parts, the first combining chorus numbers and monologues with comedy sketches known as 'bits'; the second, known as the 'olio', was made up of VARIETY turns—acrobats, instrumentalists, magicians, conjurors, and singers of sentimental songs. The third part, which also had chorus numbers and 'bits', contained the show's only claim to the original name of burlesque in that it might feature a travesty on politics, or a parody of a current theatrical success. The final number, known as 'the Extra Added Attraction', was usually a belly dance, or 'hootchy-kootchy'. One of the early stars of burlesque was Lillian Russell [Helen Louise Leonard] (1861–1922) who first appeared in 1881. In about 1920 strip-tease, an innovation of uncertain origin, was introduced to counteract the competition of the films. This involved a complicated ritual of a woman undressing to music and exposing herself for a second completely nude except for a G-string. The most famous strip-tease artiste was Gypsy Rose Lee [Rose Louise Hovich] (1914–70), who succeeded in establishing her act as a conventional Broadway revue speciality. The musical comedy *Gypsy* (N.Y., 1959; London, 1973) was based on her autobiography. Heavy drinking contributed to the uninhibited atmosphere of burlesque houses, and with the enforcement of prohibition in America burlesque lost its hold over the public and was finally banned from New York in Apr. 1942.

BURLESQUE, English, a satirical play, usually based on some well known contemporary drama or dramatic fashion that offered scope for parody. The prototype was BUCKINGHAM's *The Rehearsal* (1671), which made fun of DRYDEN and his HEROIC DRAMA, and the genre culminated in *The Critic* (1779), in which SHERIDAN amused himself at the expense of the sentimental foibles of his day. In the meantime the traditions of burlesque had been upheld by GAY with *The Beggar's Opera* (1728); by Henry Carey, who burlesqued both opera and drama; and by FIELDING in *The Tragedy of Tragedies; or, the Life and Death of Tom Thumb the Great* (1739).

In the 19th century a new type of burlesque flourished. It retained enough of its origins to choose as its target a popular play, but the element of criticism was lacking. This may have been due to the increase in the size of the audience and the lowering of its general educational level, since the success of the earlier type of burlesque had depended on familiarity with the subject of ridicule. One of the best writers of the new type of burlesque was H. J. BYRON, with *Aladdin; or, the Wonderful Scamp* (1861), *The Corsican 'Bothers'; or, the Troublesome Twins* (1869), and *Robert MacMaire; or, the Roadside Inn Turned Inside Out* (1870). Possibly Byron's execrable puns, as well as the reform of the stage initiated by T. W. ROBERTSON, finally killed the burlesque, though not before it had provided London theatregoers with a good deal of amusement at the old GAIETY, with the famous 'quartet' headed by Nellie FARREN. It survived for a time as sketches in such REVUES as the *Gate* (1938), the *Little* (1939), and the *Sweet and Low* series (1943–6), parodying a current Shakespeare production or some long-running London success.

BURLETTA, originally a farce with interpolated music or a short comic opera. The name was extended in the mid-19th century to include plays put on at the minor theatres which, in order to circumvent the monopoly of the PATENT THEATRES, contained at least five songs in each act. This made any play legally a burletta, and so not subject to the licensing laws. The burletta had much in common with the EXTRAVAGANZA, which led PLANCHÉ, one of the leading exponents of the genre, to subtitle one of his plays 'A Most Extravagant Extravaganza, or Rum-Antic Burletta.' The addition of songs, and sometimes of instrumental interludes or solo dances, allowed the unlicensed theatres to present any 'legitimate' play, including those of Shakespeare, and this way of evading the law helps to explain many of the odd interpolations found in 19th-century productions of 'straight' plays.

BURNABY, DAVY, see PIERROT TROUPES.

BURNACINI, GIOVANNI and LODOVICO OTTAVIO, see COSTUME and SCENERY.

BURNAND, F. C., see CAMBRIDGE.

BURNT-CORK MINSTRELS, see MINSTREL SHOW.

BURTON [Jenkins], RICHARD WALTER (1925–84), Welsh actor, son of a miner, who took his stage name from that of his benefactor Philip Burton. He made his first appearance in Emlyn WILLIAMS's *The Druid's Rest* (1944) and later appeared in three plays by Christopher FRY—*The Lady's Not For Burning* (1949), in which he also made his New York début in 1950, *The Boy with a Cart*, and *A Phoenix Too Frequent* (both 1950). In 1953–4 he was at the OLD VIC, where he played, among other parts, Hamlet, the

Bastard in *King John*, Sir Toby Belch in *Twelfth Night*, Caliban in *The Tempest*, and Othello and Iago alternately with John NEVILLE. In June 1957 he was in New York in ANOUILH's *Time Remembered*, and in 1960 scored a big success there in the musical *Camelot*. From that time he worked mainly in films, but in 1966 he appeared with the OXFORD University Dramatic Society in a production of MARLOWE's *Dr Faustus* directed by his former tutor Nevill Coghill, his second wife, the film star Elizabeth Taylor, playing Helen. Later stage appearances were in New York, where in 1976 he starred in Peter SHAFFER's *Equus*, and in 1980 was seen again in *Camelot*.

BURTON, WILLIAM EVANS (1804–60), English-born actor-manager and dramatist, son of a printer, who inherited his father's business but gave it up to go on the stage. He first appeared in London in 1831 and in the following year was at the HAYMARKET THEATRE where he played with Edmund KEAN. In 1834 he made his American début at the Arch Street Theatre, Philadelphia; he first appeared in New York in 1837. After converting a circus in Philadelphia into a theatre and running it successfully as the National, he returned to New York and took over Palmo's Opera House, later the CHAMBERS STREET THEATRE. Renovated and redecorated, it opened on 10 July 1848 as Burton's and became one of the most important theatres of the day. Burton himself played there a number of richly comic parts, notably Captain Cuttle in BROUGHAM's adaptation of DICKENS's *Dombey and Son*, which was the theatre's first outstanding success. The arrival in 1851 of Mrs Warner from SADLER'S WELLS led to the revival of several plays by Shakespeare, beginning with *The Winter's Tale*. In the following year Burton began to suffer from the success of WALLACK's, which took some of his best actors from him, but before he finally deserted his theatre early in 1856 it had seen fine productions of *A Midsummer Night's Dream* and *The Tempest*, with Burton as Bottom and Caliban. On 8 Sept. Burton took over the METROPOLITAN, opening with SHERIDAN's *The Rivals*, but the competition of Wallack's was too keen and the theatre was too big to stand up to the financial crisis of 1857. In spite of some successful new plays and visits from outstanding guest artists, among them Edwin BOOTH and Charlotte CUSHMAN, it closed on 9 Sept. 1858. Burton made his last appearance in New York at NIBLO'S GARDEN in 1859. He was the author of several plays, of which none has survived.

BURY, JOHN (1925–), stage designer, born in Wales, who began his career with THEATRE WORKSHOP as an actor, his first work as a designer being done for a production of IBSEN's *An Enemy of the People* in 1954. He later did the designs for several of the company's most successful productions, including Brendan BEHAN's *The Quare Fellow* (1956), Shelagh Delaney's *A Taste of Honey* (1958), and the musicals *Fings Ain't Wot They Used T'Be* (1959) and *Oh, What a Lovely War!* (1963; N.Y., 1964). In 1963 he was appointed associate designer with the ROYAL SHAKESPEARE COMPANY, being responsible for the décor of such productions as DÜRRENMATT's *The Physicists* (1963; N.Y., 1964) and John BARTON's adaptation of *Henry VI* and *Richard III* as *The Wars of the Roses* later in 1963; PINTER's *The Homecoming* (1965; N.Y., 1967), *Landscape* and *Silence* (1969), a double bill, and *Old Times* (1971); and ALBEE's *A Delicate Balance* (1969) and *All Over* (1972). In 1973 he moved to the NATIONAL THEATRE as Head of Design, and has since been responsible for the décor of *Hamlet* in 1975, MARLOWE's *Tamburlaine the Great* in 1976, GALSWORTHY's *Strife* in 1978, and Peter SHAFFER's *Amadeus* (1979). His freelance work has included the York Mystery Cycle (1963), Shaffer's *The Battle of Shrivings* (1970) in the West End, and Ibsen's *Hedda Gabler* and *A Doll's House* in New York in 1971.

BURY ST EDMUNDS, Suffolk, a small market town possessing the third oldest working theatre in England, the two older ones being the Theatre Royal, BRISTOL, and the GEORGIAN THEATRE, Richmond, Yorks. The first theatre in the town opened in the 1730s and was remodelled in the 1770s. By the end of the century it formed part of the Norwich CIRCUIT, though it was licensed for only nine weeks in the year and was small and inconvenient. In 1819 it was replaced by a new theatre in Westgate Street, built by William Wilkins and seating 780, which ran into difficulties and closed in 1843. Two years later, after considerable refurbishing, it opened again, taking its present name of Theatre Royal though it does not appear to have had a royal patent. By the 1870s it was in constant use throughout the year but in 1925 it closed and was used as a barrel store. In 1961 a Trust was formed to restore the theatre as close as possible to its original form, and it reopened in 1965. It now seats 352, and no seat is more than 39 feet from the stage in the horseshoe-shaped auditorium. It is used for concerts and amateur and touring productions, including small-scale opera and ballet, and is the only working theatre in the country owned by the National Trust, which purchased it in 1975.

BUSCH, ERNST (1900–80), German actor, who played in TOLLER's *Hoppla, wir leben!* (1927) under PISCATOR and sang in political cabaret. He left Germany in 1933, returning in 1945 to appear at the DEUTSCHES THEATER in Berlin. In 1950 he joined BRECHT's company, playing Semyon Lapkin in *Die Mutter*, Brecht's adaptation of GORKY's *Mother*, in 1951. He also played the Cook in *Mutter Courage und ihre Kinder* and Asdak in *Der* *kaukasische Kreidekreis* in 1954, and the title role in *Leben des Galilei* in 1957.

BUSKER, performer in city streets and on the beaches at seaside resorts; the term included conjurors, jugglers, acrobats, Punch-and-Judy operators, peep-show proprietors, and a wide variety of street musicians, from German bands to organ-grinders and peripatetic one-man bands. The word is derived from 'buskin', the long boot or COTHURNUS worn by Greek actors in tragedy, the equivalent soft heelless shoe of the comic actor, the *soccus*, being known as the sock. In Elizabethan times the two words stood for tragedy and comedy respectively, but sock fell out of use and 'busking' became a general term for any kind of itinerant entertaining. Some buskers worked from a set pitch in a street or marketplace, presenting a show lasting up to half-an-hour, while others specialized in a much briefer turn performed for queues outside theatres and later cinemas, becoming an integral part of urban life throughout the Victorian and Edwardian periods. Most buskers led a hand-to-mouth existence, particularly during the winter months, and the accounts of street entertainers given in Henry Mayhew's *London Labour and the London Poor* suggest that the distinction between busker and beggar was often not very clear. Many of them were Italian and the traditional buskers' language, still spoken by a few old performers, contains a curious mixture of Italian words and phrases with 19th-century Cockney slang. After the 1930s the busker tended to disappear, driven out by traffic congestion and by increasing competition from portable radios, but recently a number of young amateur buskers have appeared in the streets, parks, and shopping precincts of big cities, and on London's Underground stations, and a new wave of street entertaining has started which includes also such experimental groups as the Bread and Puppet Theatre.

BUTLER, SAMUEL, father and son, see GEORGIAN THEATRE, RICHMOND, YORKS.

BUTT, ALFRED, see REVUE.

BUTTI, ENRICO ANNIBALE (1868–1912), Italian dramatist, a constructive pessimist in the Ibsenite sense, who, while recognizing the integrity of the progressives and sharing their views, none the less depicts the final triumph of conservative attitudes. Among his plays the most important are *Il vortice* (*The Vortex*, 1892), produced in the year which saw the first Italian version of IBSEN's *Ghosts; L'Utopia* (1894), a consideration of mercy-killing; *La fine d'un ideale* (1898), in which he juxtaposes free love and the security of marriage; *Lucifero*, on agnosticism, and *La corsa al piacere* (*The Pursuit of Pleasure*), both 1900; and *La tempesta* (1901), a study of social reform through revolution. Butti shows his characters reaching the conclusion—through fearless debate, and not cravenly through inertia and timidity—that man does better to acquiesce in society's rules rather than struggle unavailingly against them.

BUZO, ALEXANDER (1944–), Australian dramatist, whose plays include *Norm and Ahmed* (1968), *The Front Room Boys* (1970), seen in London in 1971, *The Roy Murphy Show* (1971), *Rooted, Macquarie,* and *Tom* (all 1972), a modern version of IBSEN's *An Enemy of the People* set in Australia and entitled *Batman's Beach-Head* (1973), *Coralie Lansdowne Says No* (1974), *Martello Towers* (1976), *Makassar Reef* (1978), and *Big River* (1980). He has also written for radio and films, and was resident playwright for the MELBOURNE THEATRE COMPANY in 1972 and 1973.

BYAM SHAW, GLEN(cairn) ALEXANDER (1904–), English actor and director. He made his first appearance on the stage in 1923, and two years later was seen in London as Yasha in CHEKHOV's *The Cherry Orchard*. He gained further experience in FAGAN's company at the OXFORD PLAYHOUSE and made his New York début in 1927 as Pelham Humfrey in Fagan's *And So To Bed*. In 1937–8 he was a member of GIELGUD's repertory company at the QUEEN'S THEATRE, and in 1939 played Horatio to Gielgud's Hamlet in Elsinore and at the London LYCEUM. After serving in the army during the Second World War he returned to the theatre as a director, having already had some experience in play production before 1939. His first independent production was RATTIGAN's *The Winslow Boy* (1946; N.Y., 1947). He was then associated with the short-lived drama school attached to the OLD VIC, and also directed *As You Like It* (in 1949) and *Henry V* (in 1951) for the first YOUNG VIC. In 1953 he was appointed co-director with Anthony QUAYLE of the SHAKESPEARE MEMORIAL THEATRE company of Stratford-upon-Avon, becoming sole director in 1956. While there he staged a number of plays (including *Macbeth, Hamlet, Othello*, and *King Lear*) in several of which his wife Angela BADDELEY appeared. He resigned in 1959 and returned to London to direct Rattigan's *Ross* in the West End and also on Broadway a year later, and in 1962 became director of productions for the Sadler's Wells Opera company.

BYRON, GEORGE GORDON, Lord (1788–1824), English poet, author of several plays in verse which were staged in the hope of reviving poetry in the English theatre, but with little success. Only one was produced during his lifetime, *Marino Faliero* (1821), seen at DRURY LANE, of which Byron joined the Committee in 1815. His letters are full of references to theatrical matters, but in spite of his undoubted dramatic talents his plays read better than they act. *Werner* (1830) was first produced by MACREADY, also at Drury Lane; it was revived by PHELPS at SADLER'S WELLS in 1844 and by IRVING at the LYCEUM in 1887. *Sardanapalus* (1834) was also given its first production by Macready. *Manfred*, a dramatic poem, and *The Two Foscari* were produced at COVENT GARDEN in 1834 and 1838 respectively. *Cain*, written in 1821, does not appear to have been produced in England, but it has been several times translated into German, in which language it was produced at Frankfurt-on-Main in 1958 and at Lucerne in 1960.

BYRON, HENRY JAMES (1834–84), English actor and playwright, best known in his own day for a long series of BURLESQUES and EXTRAVAGANZAS, usually staged at the smaller London theatres, which began in 1857 with *Richard of the Lion Heart* and ended in 1881 with *Pluto; or, Little Orpheus and his Lute*. In 1865 he contributed a burlesque of Bellini's opera *La Sonnambula* to the opening programme of the BANCROFTS at the Prince of Wales (later the SCALA) Theatre, where two of his more serious comedies, *War to the Knife* (also 1865) and *A Hundred Thousand Pounds* (1866), were also given. His most successful play and the only one now remembered was *Our Boys*. This was produced at the VAUDEVILLE THEATRE on 16 Jan. 1875 and ran till 18 Aug. 1879, setting up a long-lasting record of 1,362 performances. He wrote in all about 150 works, based on themes from mythology, nursery tales, opera, legend, and topical events; his style, though ingenious, was overloaded with puns and in his more serious plays he tended to rely on stock types. He reflected the prevailing taste of the day, against which T. W. ROBERTSON was to rebel, and that, and his own charming personality, doubtless accounted for much of his ephemeral success. As an actor he appeared mainly in his own plays, making his first appearance in London in 1869 in one of his early comedies and his last (exceptionally) in W. S. GILBERT's *Engaged* in 1881.

BYZANTIUM. With the fall of the Roman Empire the greater part of its dramatic achievement passed into oblivion, but some activity continued in Constantinople, which became the new capital of the Empire. Byzantium did not create a genuine theatre or any truly dramatic poetry, but there were undoubtedly dramatic elements in its festivals, games, mummings, and masqueradings. There are records of four early theatres in Constantinople, not counting hippodromes and circuses, and many other cities within the domain of Byzantine culture—Antioch, Alexandria, Brytos, Gaza, Caesarea—had at least one theatre each. According to Ioannes Lydus, a 6th-century Byzantine historian, the main types of Roman comedy—or FABULA—were still played in spite of the Edict of the Trullan Council and the active opposition of the Christian Church. Up to the capture of Constantinople by the Turks in 1453 there is evidence that the mime-player retained his popularity, his material varying from simple buffoonery and scenes of adultery to social satire, one of his main targets being the corruption of State officials: there may also have been some ridiculing of Christian rites and beliefs. The pantomime-player also survived, mainly as a solo dancer, sometimes impersonating several characters; but the playing of full-length tragedies and comedies was discontinued, or survived only in simple declamation of the important speeches. The religious theatre in Byzantium appears to have been closely linked with the sermon or homily, many of which contained a strong dramatic element, so much so that the scholar La Piana called them 'dramatic' homilies. They consisted of fully dramatized episodes from the life of Christ which were accompanied or followed by oratorical passages. The *Encomium* attributed to Proclos also mentions a 'homily-drama', which it defines as one in which the dramatic elements make up a complete play to which the fundamental idea gives unity of purpose and meaning, though it remains linked to the sermon. It appears as part of the *Panegyris* reserved for the greater liturgical feasts. There seem to have been two cycles of such 'homily-dramas', each embracing a trilogy—the Annunciation, Nativity, and Flight into Egypt; and the Baptism of Jesus, his Passion, and the Harrowing of Hell. The Byzantine 'homily-drama' was obviously derived from three sources—the Apocrypha, the Syriac *sogitha* or homily-canticle, in which the hymn writers tended to dramatize sacred history, and the material of the mime-player.

What remains controversial is the relationship between these 'homily-dramas' and the scenario of a Greek Passion Play of unknown date preserved in a Vatican manuscript of the 13th century. According to the detailed scenario, it began with a prologue addressed to the actors followed by the Raising of Lazarus, the Entry of Jesus into Jerusalem, the Last Supper, the Washing of the Disciples' Feet, the Betrayal by Judas, the Denial by Peter, Jesus before Herod, the Crucifixion, the Resurrection, and Doubting Thomas. The play was probably acted outside the church, and whether it was derived from a Western original or the Greek Passion Cycle influenced Western LITURGICAL DRAMA has not yet been decided. There is also a Syrian manuscript in Berlin which may possibly be a translation from a Greek original, and could be the scenario of a religious play called *The Tale of the Actors*. In the first part a company of pagan actors are ridiculing Christianity; in the second they have been converted and are martyred for the faith.

Apart from these scenarios, which may have been written for performance, a number of CLOSET DRAMAS were written for reading during this period, as were poems in dramatic form which may have been intended for public performance. There is also a reference to a play by Stephanus Sabbaita, *The Death of Christ*, and a Greek drama, *Christos Paschon*, variously dated from the 4th to the 11th century, which is a mosaic composed of verses from the Greek tragic poets, particularly EURIPIDES, applied to the Passion, Death, and Resurrection of Jesus, which was probably read in public. It is possible that Western religious drama in the Middle Ages may have been derived from Byzantine material, particularly the homilies, but this is purely conjectural. It is also doubtful if the traditions of the Byzantine stage continued in Istanbul after the Turks took over. There are certain similarities, which may derive from reciprocal exchanges or influences, between the Western religious drama and the guild processions in TURKEY, as well as in certain aspects of Karagöz (the Turkish SHADOW SHOW), and in the rural Dionysiac and other mysteries (notably the Eleusinian) which still survive in Anatolia and are practised by the peasants of modern Turkey.

C

CAECILIUS STATIUS (*c.*219–*c.*166 BC), Roman dramatist, a manumitted slave probably born in Milan, who between the death of PLAUTUS and the advent of TERENCE translated Greek comedies, mainly by MENANDER, for the Roman stage. Some 40 titles and about 300 lines have survived. Caecilius allowed himself considerable freedom in expression and style, as Plautus had done, and many of his 'translations' differ widely from the original, no doubt in an effort to pander to popular Roman taste. His plays, which were not at first well received, finally succeeded because of the steady support given them by Ambivius TURPIO.

CAEN, Compagnie de, see DÉCENTRALISATION DRAMATIQUE.

CAFÉ LA MAMA, see LA MAMA EXPERIMENTAL THEATRE CLUB.

CAFFÉ CINO, New York, at 31 Cornelia Street, a theatre and coffee-house whose opening by Joe Cino in 1958 is often regarded as marking the beginning of the off-off-Broadway movement. Although the production of plays became a regular part of the café's arts programme only in 1961 and was curtailed after Cino's death in 1967, the venture had in that time become a vital part of the city's theatrical scene and had already presented, on a stage only 8 ft square, a considerable body of new works by American dramatists including John GUARE'S *The Loveliest Afternoon of the Year* and *A Day for Surprises* in 1966.

CAHILL, MARIE, see VAUDEVILLE, AMERICAN.

CAJUPI-ZAKO, ANDON, see ALBANIA.

CALDERÓN, MARÍA, Spanish actress, known as *La Calderona*, the mistress of Philip IV and mother of Don John of Austria. After appearing in AUTOS SACRAMENTALES and COMEDIAS, she retired from the theatre and became abbess of a convent in Guadalajara.

CALDERÓN DE LA BARCA, PEDRO (1600–81), Spanish dramatist, successor to Lope de VEGA. He wrote about 200 plays, the best of them dating from between 1625 and 1640. He wrote little while the public theatres of Madrid were closed, from about 1644 to 1650, and after being ordained priest in 1651 he wrote mainly AUTOS SACRAMENTALES, which presented the abstract ideas of Catholic theology on stage with intense lyricism and excellent stagecraft, and spectacle-plays for the Court theatre at Buen Retiro, with which he had been associated since its inception under Philip IV. The best known of his early religious plays are *La cena del rey Baltasar* (*Belshazzar's Feast*, *c.*1634), translated into English in 1969 by R. G. Barnes, and *El gran teatro del mundo* (*The Great World Theatre*, *c.*1641), on which Hugo von HOFMANNSTHAL based his *Grosse Welttheater*, first seen at the SALZBURG festival in 1922 in a production by REINHARDT and frequently revived since. Of the secular plays the finest and best known is undoubtedly *La vida es sueño* (*c.*1638) which, as *Life's a Dream*, was produced in London in 1922 in an adaptation by Edward FitzGerald. It is a story of human regeneration; the Prince who at the start of the play is little better than a beast, having passed all his life shut up in a dungeon, learns when he is set free that he must keep in check his pride and brute passions and comes to realize the value of reason and prudence. FitzGerald also adapted *El alcalde de Zalamea* (*The Mayor of Zalamea*, *c.*1640), a study of prudent virtue and of the vices which spring from uncontrolled passions, set against an idealized vision of the class structure of contemporary society. As in all Calderón's plays, the ideas are transmitted not only through the relationship between the characters, the actors' speeches, significant actions, dress, and gestures, the highly developed use of the stage and of impressive machinery, but also through the images which parallel or subtly contradict the surface movement. The play was produced, in a translation by Adrian Mitchell, at the NATIONAL THEATRE in London in 1981.

Calderón's plays are highly organized; he is always firmly in control of his medium in that the sub-plot becomes an aspect of the main plot and the GRACIOSO serves as much as the hero and heroine to illustrate the moral lesson implicit in the action. No character is superfluous, no action irrelevant, no detail extraneous. Among his other outstanding works are the FAUST-like *El magico prodigioso* (*The Wonder-Working Magician*, 1637), based on the life of St Cyprian, part of which was translated by SHELLEY; and the series based on the 'point of honour' so important in Spanish life, *El medico de su honra* (1635), admirably translated by Roy Campbell as *The Surgeon of His Honour* in 1960; *El pintor de su deshonra* (*The Painter of His Dishonour*, 1637); and *A secreto agravio, secreta venganza* (1635), translated in 1961 by Edward Honig as *Secret Vengeance for Secret Insult*. Much controversy has been aroused by these and by a companion piece, *El mayor monstruo, los celos* (*No Monster Like Jealousy*, *c.*1634) on the subject of Herod and his wife Mariana, some critics maintaining that they are intended to show that Calderón did not by any means accept all the conventional implications of the contemporary code of honour.

Calderón had a considerable influence on European drama, many of his plays becoming known to English Restoration dramatists through French translations, and resulting in TUKE's *Adventures of Five Hours* (1663), based on *Los empeños de seis horas* (at the suggestion of Charles II), and Digby's *Elvira; or, the Worst not Always True* (*c.*1663), based on *No siempre la peor es cierto*. It is also probable that KILLIGREW found some of *The Parson's Wedding* (1664) in *La dama duende*, a cloak-and-sword play written for production at Court in *c.*1629, and that WYCHERLEY took some of *The Gentleman Dancing-Master* (1672) from *El maestro de danzar*. Out of favour during the 18th century, Calderón was rediscovered by the German critics of the Romantic period, notably by SCHLEGEL and TIECK. He has made little impact on the English theatre, probably because of the difficulties of translating his highly personal lyrical style; even in Spain his Court plays, of which one of the best is *Casa con dos puertos mal es de guarda* (*It is Difficult to Guard a House with Two Doors*, *c.*1629), have been little studied.

CALDWELL, JAMES H. (1793–1863), British-born American theatre manager, who developed and dominated the pioneering theatre of the Mississippi Valley. In 1816 he was engaged by Joseph G. Holman to play light comedy roles at the theatre in Charleston, and a year later was managing theatres in North Carolina and Virginia. In 1820 he was in New Orleans, where he opened the American Theatre on Camp Street on 1 Jan. 1824. He established a number of touring companies and built theatres in such cities as Mobile, Nashville, and Cincinnati, finally extending his operations throughout the entire lower Mississippi Valley into parts of Ohio. The high point of his career was the construction of the St Charles Theatre in New Orleans, the first large and important theatre in the South. Caldwell's permanent company was a fine one, and every year attracted a number of visiting stars, but the financial panic of 1837 and his rivalry with Noah M. LUDLOW and Sol SMITH ruined him, and the destruction by fire of the St Charles Theatre in 1842 virtually put an end to his career.

CALDWELL, ZOË (1934–), Australian born actress, who showed unusual maturity early in her career, dominating scenes in which she appeared. An original member of the Union Theatre Repertory Company, Melbourne, she made her professional début with them in 1953. She first appeared in England in 1958 at the SHAKESPEARE MEMORIAL THEATRE, returning in the following year to give outstanding performances as Helena in *All's Well That Ends Well*, and as Cordelia in *King Lear* with Charles LAUGHTON. In 1960 she moved to the ROYAL COURT THEATRE, where in 1961 she played Isabella in MIDDLETON and ROWLEY's *The Changeling*, being seen in the same year as Rosaline in *Love's Labour's Lost* at the STRATFORD (ONTARIO) FESTIVAL.' After a visit to Australia in 1962, where she played SHAW's Saint Joan, she was at the opening season at the GUTHRIE THEATRE, Minneapolis, in 1963 (Frosine in MOLIÈRE's *The Miser*, Natasha in CHEKHOV's *Three Sisters*), and returned in 1965 when her roles included Millamant in CONGREVE's *The Way of the World*. In the same year she made her New York début when she took over the part of the Prioress in John WHITING's *The Devils*. Later roles in New York included Polly in Tennessee WILLIAMS's *The Gnädiges Fräulein* (1966), a highly praised Jean Brodie in Jay Presson Allen's *The Prime of Miss Jean Brodie* (1968), based on Muriel Spark's novel, Alice in STRINDBERG's *The Dance of Death* (1974), and Mary Tyrone in O'NEILL's *Long Day's Journey Into Night* (1976). She returned to Stratford, Ontario, in 1967 to play Cleopatra (in *Antony and Cleopatra*) and other roles, and to London in 1970 to play Emma Hamilton in RATTIGAN's *A Bequest to the Nation*. One of the most eminent actresses to emerge from Australia, she brings a contemporary quality and original characterization to classic plays without any sacrifice in style.

CALL BOARD, see STAGE DOOR.

CALL-BOY, a functionary in the English theatre, usually, though not necessarily, a young boy, whose duty it was to summon the players from their dressing-rooms in time for them to appear on stage on cue. He took the place of the Elizabethan BOOK-HOLDER, who was also the PROMPTER. The word was already current at the end of the 18th century, since MALONE erroneously said that Shakespeare's first job in the theatre was that of 'call-boy'. It became obsolete in the mid-20th century, when the call-boy was replaced by a tannoy or loudspeaker, placed in each dressing-room, thus putting on the actor the onus of being in the right place at the right time.

CALL DOORS, see PROSCENIUM DOORS.

CALLIPIDES, see GREECE, ANCIENT.

CALMO, ANDREA (1509–71), Italian playwright and amateur actor, a gondolier by profession, who seems to have had a considerable influence on the development of the COMMEDIA DELL' ARTE. A contemporary and rival of Angelo BEOLCO, he specialized in the playing of old men somewhat in the style of PANTALONE, though the name was not yet in use. His plays were edited by Rossi in 1888.

CALVERT, LOUIS (1859–1923), English actor, son of a provincial actor-manager Charles Calvert (1828–79) and his wife Adelaide Helen Biddles or Bedells (1837–1921), a well known actress. Louis made his first appearance in Durban, South Africa, in 1879 and then toured Australia. On his return to England in 1880 he joined Sarah THORNE's stock company at Margate and in 1886 joined the company at DRURY LANE. In the following year he was with Henry IRVING at the LYCEUM, and subsequently was associated with most of the leading managements of London, including that of Fred TERRY with whom he gave an outstanding performance as an old actor in KESTER's *Sweet Nell of Old Drury* (1900). He was a leading member of the company at the ROYAL COURT THEATRE during the VEDRENNE and GRANVILLE-BARKER management, playing Broadbent in the first production of SHAW's *John Bull's Other Island* (1904) and William the Waiter in a revival of his *You Never Can Tell* in 1905 with equal effectiveness. His Mercutio in *Romeo and Juliet* and Casca in *Julius Caesar* were much admired, as was his Creon to the Oedipus of MARTIN-HARVEY in SOPHOCLES' *Oedipus Rex* at COVENT GARDEN in 1912. At his best in parts requiring dramatic strength, which gave his fine voice full scope, he could also handle comedy roles with ease and polish. He was an excellent director and a successful manager, several times forming his own company to tour England and the United States.

CALVO-SOTELO, JOAQUÍN (1905–), Spanish playwright, author of about 50 plays. He achieved a major success with *Cuando llego la noche* (*When Night Falls*, 1943), and outstanding among his later plays is *Milagro en la Plaza del Progreso* (*Miracle in Progress Square*, 1953) which brought a deliberate ambiguity to the statement of a complex problem. *La muralla* (*The Impasse*, 1954), a vivid account of one man's attempt to reconcile his Catholic faith with his role in society, was Spain's most successful post-war play, receiving over 5,000 performances.

CÂMARA, JOÃO DA, see PORTUGAL.

CAMBRIDGE, the site of the second oldest university in England, has a long theatrical history. Although LITURGICAL DRAMA was probably seen in churches there as early as elsewhere, the first mention of it dates from 1350 in connection with the staging of a play entitled *The Children of Israel*. There are frequent references during the 15th century to the town MINSTRELS, and by the 16th century the acting of plays in Latin was firmly established as part of the students' curriculum, as is evident from the production of KIRCHMAYER's *Pammachius* at Christ's College in 1546. A few years later the early English comedy *Gammer Gurton's Needle* was seen at the same college, and there are records of plays given for the townsfolk, probably by a visiting company, in the Falcon yard and at the Saracen's Head in 1556. Elizabeth I, on her only visit to Cambridge in 1564, was entertained in the chapel of King's College by a performance of PLAUTUS' *Aulularia* and an original Latin tragedy on the subject of Dido and Aeneas by a Fellow of the College, Edward Halliwell. An English play, UDALL's *Ezechias*, was also revived, but in general the authorities were not in favour of works in the vernacular as 'nothing beseeminge our students', and although by 1575 there was a settled tradition of winter performances, plays in Latin predominated, several of them translated from contemporary Italian works directly or through the French. St John's College seems to have been particularly given to play-acting, and a number of productions are recorded there, among them a Latin tragedy, *Richardus Tertius* (*Richard III*) by Thomas Legge, in 1579. There too the anonymous *Pilgrimage to Parnassus* was first performed in 1598, as were the two parts of the *Return from Parnassus* in 1602 and 1603. In 1605 King's College gave the first performance in England of a Latin translation of GUARINI's famous PASTORAL *Il Pastor Fido*, and in 1615 Trinity men entertained Prince Charles (later Charles I) with *Scyros*, a Latin translation of BONARELLI's *Filli di Sciro*. The visit of James I in 1615 was made memorable by the performance, among other plays, of RUGGLE's *Ignoramus* (based on DELLA PORTA's *La Trappolaria*) which so delighted the king that he returned shortly afterwards to see it again.

All this activity was of course amateur, but there appear to have been occasional visits to Cambridge by professional companies, particularly in summer, when plague closed the London theatres. The CHAMBERLAIN'S (or KING'S) MEN are believed to have played *Hamlet* in Cambridge in 1603, and there were also a number of dramatic performances at nearby Stourbridge Fair, which caused trouble in 1592. On the whole non-academic acting was not encouraged, and as the Commonwealth approached even college productions grew fewer in number. The last recorded play before the closing of the theatres in 1642 was *The Guardian*, an early version of COWLEY's *Cutter of Coleman Street*, given before the eleven-year-old Prince Charles (later Charles II) in 1641.

The Restoration brought no revival of acting in Cambridge, but in the 18th century regular visits from the professional company on the Norwich circuit were well received, though no acting was allowed within the university precincts. There was, however, a commodious wooden theatre at Stourbridge, and another at Barnwell which was successively rebuilt and in the 19th century housed a number of London companies on tour. It then became derelict until in 1926 it was taken over by Terence GRAY to become the Festival Theatre.

Although several efforts were made in the first

half of the 19th century to revive the old tradition of college acting, including the founding of an Amateur Theatrical Society which gave productions of Shakespeare from 1855 to 1868, it was the founding, also in 1855, of the Amateur Dramatic Club (A.D.C.) by F. C. Burnand, later editor of *Punch* and a prolific writer of light plays, which made the theatre again an integral part of undergraduate life. The A.D.C., which has produced a number of outstanding professional actors and directors, celebrated its centenary on 12 June 1955 and is still active, as is the Footlights Club, founded in 1883, whose annual revue has become an important feature of Cambridge theatrical life and is often seen in London for a short season.

In 1907 a group of undergraduates who were interested in reviving Elizabethan and other early plays put on a production of MARLOWE's *Dr Faustus* with sufficient success to warrant the founding of the Marlowe Society a year later, with Rupert Brooke, who had played Mephistophcles, as its first president. An outstanding feature of its work has been its fine speaking of verse, and under the guidance of George Rylands, who directed many of its productions as well as those of the A.D.C., it has recorded for the BRITISH COUNCIL all the plays of Shakespeare. Women's parts were played by men until the production of *Antony and Cleopatra* in 1934, and a tradition of anonymity was maintained until the mid-1960s. Like the A.D.C., the Marlowe Society has given a number of fine actors to the professional stage, and a group of outstanding directors including Peter HALL and John BARTON.

In 1928, before women were admitted to full membership of the A.D.C., the Cambridge Mummers were founded with both men and women members. This group usually gives two performances annually in the A.D.C. theatre, and sometimes one outside Cambridge during the vacation.

The first Cambridge Theatre Royal was adapted in 1882 from the old St Andrew's Hall by the artist William Redfern, who demolished it in 1896 to make way for the New Theatre. This became a cinema in the early 1930s, and in 1936 the economist John Maynard Keynes built at his own expense the Arts Theatre. Under his direction and that of his wife, the Russian ballerina Lydia Lopokova, it did much to stimulate theatrical life in Cambridge. After Keynes's death in 1946 it went through a difficult period, but the skill and devotion of George Rylands finally saved it. Run by a Trust and not, like the OXFORD PLAYHOUSE and other university theatres, under direct university control, it is supported by grants from the local authorities and other sources which have put it on a secure financial basis. It has had a notable record, housing not only the productions of the Footlights Club and the Marlowe Society but also visiting professional companies on tour, and sponsoring organizations such as the Cambridge Theatre Company, one of the country's leading touring companies. The first performances of such modern plays as SHAFFER's *Five Finger Exercise* and PINTER's *The Birthday Party* (both 1958), KOPIT's *Oh Dad, Poor Dad, Mamma's Hung You In the Closet and I'm Feeling So Sad* (1961), and ORTON's *What the Butler Saw* (1969) were given at the theatre.

An important theatrical event in Cambridge is the triennial production of a Greek play in the original which began in 1882 with SOPHOCLES' *Ajax*. The *Birds* of ARISTOPHANES a year later had music by Parry, and from 1921 to 1950 ten plays were directed by J. T. Sheppard, later Provost of King's, beginning with AESCHYLUS' *Oresteia* and ending with Sophocles' *Oedipus Coloneus*. For each of these a modern English composer was commissioned to write or arrange the music. Later productions have been directed by George Rylands.

Since the Second World War many colleges have built their own small theatres, and the number of university productions has increased enormously. A Chair of Drama was established at the university in 1974, though there is as yet no department of drama.

CAMBRIDGE THEATRE, London, in Earlham Street, at Seven Dials. This theatre, which seats 1,275 people on three tiers, opened on 4 Sept. 1930 with *Charlot's Masquerade*, starring Beatrice LILLIE. In 1931 there was a season of BALIEFF's *Chauve Souris* and BRUCKNER's *Elizabeth of England* translated by Ashley DUKES had a good run. In the following year a French company appeared in two plays by Sacha GUITRY, and in 1934 the COMÉDIE-FRANÇAISE gave a season which included CORNEILLE's *Le Cid* and MOLIÈRE's *Le Misanthrope*. Nothing of any importance then happened until 1943, when there was a successful revival of SHAW's *Heartbreak House* with Robert Donat, Edith EVANS, and Isabel JEANS, and in 1944 Strauss's operetta *A Night in Venice* began a long run. In 1946 Jay Pomeroy took over the theatre for a series of artistically successful but financially unrewarding seasons of opera, while the successes of 1949 and 1950 were the revues *Sauce Tartare* and *Sauce Piquante*. After the first performance of Menotti's opera *The Consul* in 1951 and several good seasons by foreign dance companies, a series of successful comedies, inaugurated on 28 Aug. 1952 with Louis VERNEUIL's *Affairs of State*, included William Douglas HOME's *The Reluctant Debutante* (1955). Later productions were T. S. ELIOT's *The Elder Statesman* (1958), *Billy Liar* (1960) by Keith Waterhouse and Willis HALL, with Albert FINNEY, and a dramatization of H. G. Wells's *Kipps* as a musical, *Half-a-Sixpence* (1963), starring Tommy Steele. In 1965 Michael REDGRAVE and Ingrid Bergman had a considerable success in TURGENEV's *A Month in the Country*. In 1968 there were successful revivals of the musical comedies *The Desert Song* and *The Student Prince*, with John Hanson, and in 1970 Maggie SMITH appeared in Ingmar BERGMAN's production of IBSEN's *Hedda Gabler*. Ingrid

Bergman returned in 1971 in a revival of Shaw's *Captain Brassbound's Conversion*, and later in the year John OSBORNE's *West of Suez* opened with Ralph RICHARDSON. Brian RIX took over the theatre for Michael Pertwee's *A Bit Between the Teeth* (1974), the flamboyant musical *The Black Mikado* began a long run in 1975, and Jonathan MILLER's production of CHEKHOV's *Three Sisters* was seen in 1976. The American musical *Chicago* transferred from the CRUCIBLE THEATRE, Sheffield, in 1979 for a highly successful run.

CAMERI (Chamber) **THEATRE,** a company founded in Palestine in 1944 by a group of young actors under Joseph Millo who were in revolt against the conservatism of the two existing theatre groups, HABIMAH and OHEL, which still adhered to the earlier larger-than-life Russian style of acting, with exclusively realistic settings. The new company's first production, on 10 Oct. 1945, was GOLDONI's *The Servant of Two Masters*, followed shortly afterwards by the ČAPEKS' *The Insect Play* and by GARCÍA LORCA's *Blood Wedding*. Audiences were delighted by the light colourful style of acting and the impressionistic settings. The venture proved a success, and on 31 May 1948 the Cameri, which had done good work in presenting a number of contemporary European plays in Hebrew translations, broke new ground with the presentation of Moshe Shamir's *He Walked Through the Fields*, not only the first play to be produced in the independent State of Israel but also the first play in Hebrew to present native-born Israelis in a realistic manner. Cameri continued to present new Jewish plays, but also remained faithful to its plan of bringing in good plays from outside Israel, and scored two of its greatest successes with BRECHT's *The Good Woman of Setzuan*, presented at the THÉÂTRE DES NATIONS in Paris in 1956, and Nathan Alterman's dramatization of Sammy Gronemann's *The King and the Cobbler*, seen in London during the WORLD THEATRE SEASON of 1965. Since 1961, when Millo became artistic director of the newly-opened Haifa Municipal Theatre, the company has been run by Isaiah Weinberg. Its repertory has since included new English plays such as Peter NICHOLS's *A Day in the Death of Joe Egg* and PINTER's *The Homecoming*, and from America Arthur MILLER's *The Price*. In 1961 the company moved into its own premises in the commercial and entertainment centre of Tel-Aviv, and since 1970 it has been the city's official municipal theatre.

CAMERON, BEATRICE, see MANSFIELD, RICHARD.

CAMINELLI, ANTONIO, see SACRA RAPPRESENTAZIONE.

CAMÕES, LUÍS VAZ DE (*c.* 1525–80), Portuguese dramatist, best known as the poet of *Os Lucíadas*. Two of his three plays have classical subjects, but all are treated in a non-classical manner deriving ultimately from VICENTE. *Os Anfitriões* (publ. 1587), based on the *Amphitruo* of PLAUTUS, and *El-Rei Seleuco* (publ. 1645), a story of royal love and self-sacrifice, were performed when Camões was in Lisbon as a young man. *Filodemo*, his best play, has a plot recalling those of chivalresque fiction and, with its neo-Platonic hero and baroque literary style, blends the Italian PASTORAL with contemporary realism. Believed to have been performed in Goa before the Governor-General in about 1555, it may have been written much earlier.

CAMP, see FEMALE IMPERSONATION.

CAMPBELL, BARTLEY (1843–88), American dramatist, and one of the first to make playwriting his only profession. He even gave up journalism when his first play, a melodrama entitled *Through Fire*, was produced in 1871, on the grounds that a playwright could not also be a dramatic critic. He directed for some years a theatre in Chicago where several of his own plays were produced, among them *Fate* (1872) and *Risks* (1873). His most successful work, *My Partner* (1879), a drama of the American frontier, was based on the stories of Bret Harte, and was first produced with Louis ALDRICH in the lead. Among his other works were *The Galley Slave* (also 1879) and *The White Slave* (1882), emotional melodramas which he directed and financed himself, thus losing the fortune which *My Partner* had brought him. He died insane, of overwork and worry.

CAMPBELL, HERBERT, see LENO, DAN, and MUSIC-HALL.

CAMPBELL, KEN, see EVERYMAN THEATRE, LIVERPOOL.

CAMPBELL, Mrs PATRICK [*née* Beatrice Stella Tanner] (1865–1940), English actress, who through her devastating wit and marked eccentricities became a legend in her lifetime. She made her first appearance in London in 1890 in Sheridan KNOWLES's *The Hunchback*, and was already well known when she created the parts of Paula in *The Second Mrs Tanqueray* (1893) and Agnes in *The Notorious Mrs Ebbsmith* (1895), both by PINERO. Also in 1895 she was seen in the title role of SARDOU's *Fédora*, and after playing Juliet to FORBES-ROBERTSON's Romeo in the same year created the title role in *Magda* (1896), the English version of SUDERMANN's *Heimat* (1893). She was to make her New York début in this part in 1902. In 1898 she appeared as Mélisande in the English version of MAETERLINCK's *Pelléas et Mélisande*, repeating the role in French in 1904 to the Pelléas of Sarah BERNHARDT, and in 1907 she was seen as IBSEN's Hedda Gabler. George Bernard SHAW, who called her 'perilously bewitching', wrote for her the part of Eliza Doolittle in *Pygmalion* (1914) and exchanged letters with her over a long period. Their correspondence was published after his death in 1950, and on it the American actor Jerome Kilty based a dramatic dialogue, *Dear Liar* (1959). Much of Mrs Campbell's later career was spent in revivals of her earlier successes, but in 1929 she gave an excellent performance as Anastasia Rakonitz in *The Matriarch*, based by G. B. Stern on one of her own novels. In spite of long and frequent absences from the stage, and a temperament which rendered her somewhat difficult to deal with, she was one of the outstanding stage personalities of her generation.

CAMPEN, JACOB VAN (*c.* 1595–1657), Dutch architect, the designer of the first theatre in Amsterdam, the Schouwburg, which opened on 3 Jan. 1638 with a historical tragedy, *Gijsbrecht van Amstel* by Joost van den VONDEL. Modelled on PALLADIO's Teatro OLIMPICO in Vicenza, it had an elaborate *scena stabile*, or permanent setting, with a balcony on each side and no proscenium arch, one ceiling covering both auditorium and stage, and a central arch reminiscent of the earlier open-air stages designed by the members of the CHAMBERS OF RHETORIC. In 1665 van Campen's theatre was rebuilt by Jan Vos (1615–67) with a contemporary-style Italian proscenium arch flanked by pillars, an orchestra pit in front of the stage, and painted canvas scenery. It was burnt down in 1772.

CAMPION [or CAMPIAN], THOMAS (1567–1629), English poet and composer, who at one time practised as a doctor. He was a friend of Philip Rosseter, through whom he may have had some contact with the public stage, particularly at WHITEFRIARS where the BOY COMPANY run by Rosseter mainly played. His own works, however, of which (unlike Ben JONSON) he wrote both the words and the music, were seen only at Court. The texts of these, together with notes on costume and scenery which add considerably to our knowledge of the staging of MASQUES at the Court of James I, were printed in the 1909 edition of Campion's works edited by Percival Vivian.

CAMPISTRON, JEAN-GALBERT DE (1656–1723), French dramatist, who was befriended by the actor J.-B. RAISIN and had two plays produced at the COMÉDIE-FRANÇAISE. A commission to write the libretto for Lully's opera *Acis et Galatée* (1686) brought him success and a lucrative position in a noble household. He had already had some success with his neo-classical tragedies *Andronic* and *Alcibiade* in 1685, the latter owing much to the excellent acting of BARON, and among his later works *Tiridate* (1691) was also well received. All his subjects were taken from history, but his plays were weak in execution and over-loaded with intrigue. The poverty of French drama at this time is evident by the fact that from 1683 to 1693 Campistron was considered its leading tragic writer. Like RACINE, whose successor he was accounted to be though his work showed more clearly the influence of CORNEILLE, he gave up the theatre to enter the King's service.

CAMUS, ALBERT (1913–60), French dramatist and novelist, who from 1936 to 1939 gained experience of the theatre with a left-wing group in Algiers, the Théâtre du Travail (later de l'Équipe), for which he directed and acted in a wide variety of plays. While editing a clandestine newspaper during the German occupation of France he wrote two important plays, *Le Malentendu* (1944) and *Caligula* (1945), in the second of which the young Gérard PHILIPE appeared with great success. Both plays are concerned with Camus's own concept of absurdity as defined in his essay *Le Mythe de Sisphe* (1942), a definition later applied to the Theatre of the ABSURD. The theme is developed further in *L'État de siège* (1947), based on his novel *La Peste* in collaboration with Jean-Louis BARRAULT, in which the plague is symbolic of the evil that thwarts mankind in its quest for freedom. Among Camus's later plays were *Les Justes* (1949) and adaptations of CALDERÓN, DOSTOEVSKY, and William Faulkner which show him working towards a wider and freer theatrical technique. Three years before his tragically 'absurd' death in a car crash he was awarded the NOBEL PRIZE for Literature.

CANADA. The recorded history of theatre in Canada, excluding the highly developed ritualistic ceremonies of the Amerindians and Eskimos, spans approximately 400 years. Although Sir Humphrey Gilbert is known to have had musicians and a group of MORRIS DANCERS with their HOBBY HORSE on board his ship when he claimed Newfoundland for Elizabeth I in Aug. 1583, the first dramatic performance in the modern sense appears to have been an original aquatic MASQUE in rhymed alexandrines by Marc LESCARBOT entitled *Le Théâtre de Neptune en la Nouvelle France*, which was performed at Port Royal (now Annapolis Royal) in Nova Scotia on 14 Nov. 1606. Three years earlier the first European play about Canada had been published in France. This was *Acoubar, ou la Loyauté trahie* (*Acoubar, or Loyalty Betrayed*), by Jacques du Hamel, based on a romance by Antoine du Périer, who is known to have visited Canada with a French expedition. In 1640 an anonymous tragi-comedy was put on in Quebec City as part of the prolonged celebrations of the first birthday of the future Louis XIV and served to inaugurate a series of plays, staged by local amateurs under the auspices of the Governor, which reflected the Court life of France. Pierre CORNEILLE's tragedy *Le Cid*, for example, was seen in 1646, only nine years after its first performance in Paris, and a year later there was a performance of ballet in a fur-trader's warehouse. In addition there were the more academic productions of the students in the Ursuline and Jesuit colleges, with whom Corneille also proved popular, his *Héraclius* (1647) being presented in 1651 and *Le Cid* again in 1652. Following the precedent established by the widespread JESUIT DRAMA in Europe, there were also a number of original allegorical PAGEANTS devised by the Jesuit fathers to honour various officials, the most notable being *La Réception de Monseigneur le Vicomte d'Argenson*, given on 28 July 1658 to honour the arrival of a Governor-elect.

When Quebec achieved Crown Colony status in 1663 troops were stationed there and a large number of marriageable girls were sent over from France, bringing with them the dances and songs of their own regions. Their presence seems to have led to an increase of rowdyism and bad behaviour among the audiences which gathered to watch the comedies performed at Carnival time, notably on Shrove Tuesday; and one week-long *charivari* (a wedding entertainment resembling the modern REVUE), which involved some pointed dramatic satire directed against the authorities, resulted in a threat of excommunication for the whole of Quebec. During the Carnival of 1668 the Jesuits endeavoured to repair the damage by staging a MORALITY PLAY, *Le Sage visionnaire*, and, at Easter of the same year, a PASSION PLAY in Latin. Hostilities between the temporal and spiritual powers came to a head with the famous scandal known as 'l'affaire du "Tartuffe" '. In the early 1690s Governor Frontenac had inaugurated a series of garrison *soirées* at his château. Corneille's *Nicomède* (1651) and RACINE's *Mithridate* (1673) had been played without incident, but the news that the officers of the garrison were preparing to act in MOLIÈRE's attack on religious hypocrisy provoked a furious response, and drew a series of *mandements* from the Bishop, Saint-Vallier, directed against the play and particularly against the *aide-de-camp* Lieutenant Jacques de Mareuil, a young man of somewhat unsavoury reputation who was cast as the sanctimonious imposter Tartuffe. Bribery was suspected when the Governor accepted money from the Bishop to cancel the production—or, as the more charitably-minded said, to reimburse the actors for expenses already incurred. Mareuil's protracted trial on charges of blasphemy and corrupt morals, with no mention of Molière—or Tartuffe—proved inconclusive; none the less, the Bishop's edicts were interpreted as forbidding all theatrical performances (*Tartuffe*, in fact, was not seen in Canada until the elder COQUELIN played it on tour in 1893); and up to the end of the French colonial era in 1759 there were only a few isolated theatrical events: an opera, *Les Quatre Saisons*, in 1706—probably the first to be performed in Canada; a religious PASTORAL given in 1727; an anonymous comedy acted in Montreal during the Carnival of 1749; and in July 1757, at Fort Niagara, an original one-act farce, *Le Vieillard dupé* (*The Old Man Deceived*).

The first production in Canada of a play in English, also given at Annapolis Royal, dates from the winter of 1743–4 and was probably a translation of Molière's *Le Misanthrope*: the text was found among the papers of Paul Mascarene, Lieutenant-Governor of Nova Scotia, and may have been made by him, though there is no proof of this. A second performance of this play is now known to have been staged on 6 Jan. 1744 in celebration of the birthday of Frederick, Prince of Wales. But it was not until the era of British colonial rule from 1760 to 1867 that plays in English were given regularly. Throughout the whole of that period the officers of the army and navy co-operated with the townsfolk to provide garrison entertainments for the gentry. The comic repertory mirrored that of London's PATENT THEATRES, and all profits were donated to charity. This period in Canadian theatrical history is characterized by the transition from such homespun indigenous amateurism to the commercial domination exercised by the professional touring companies, mainly from the U.S.A. A study of theatre in eastern Canada reveals a gradual movement west from the Atlantic, with the various urban centres coming alive at approximately 10-year intervals. In 1824, with the completion of the Erie Canal connecting Lake Ontario with Albany and the city of New York, a Great Lakes circuit became possible, and by 1856 the water routes were further supplemented by the railways. At Victoria, in British Columbia, there may have been theatrical shows under canvas as early as 1848. Certainly plays were being performed on board British naval vessels in the waters off Victoria in 1853 and 1855, as they had been by the ROYAL ARCTIC THEATRE as early as 1819. Victoria also became the terminus of a north/south touring circuit based on San Francisco, and although Canada's acquisition of the Northwest Territories from the Hudson's Bay Company did not take place until 1869, Winnipeg had enjoyed its first pioneer military productions in 1866. When in 1884 and 1886 the American and Canadian transcontinental railroads linked the coasts the mid-west was opened up, and by 1890 the entire continent was covered by one vast commercial theatre network.

The first professional players to appear in Canada, at Halifax from 26 Aug. to 28 Oct. 1768, were based in North Carolina and led by Messrs Mills and Giffard, of whom nothing further is at present known. Giffard may have been the London actor Henry GIFFARD under whom GARRICK made his first appearances in London, but this seems unlikely, as he would then have been in his 70s; it may have been his son or nephew. The first resident stock companies began to arrive soon after the American War of Independence, to cater for the large influx of English-speaking loyalists. One such group, led by Edward Allen, William Moore, and John Bentley, established itself in Montreal and Quebec in 1786, and on 19 Oct. presented in Quebec City a 5-act tragedy in verse, *The Con-*

quest of Canada; or, The Siege of Quebec (publ. 1766), by an English-born Bostonian, George Cockings (?–1802). This was one of the earliest plays in English on a Canadian theme, the first apparently being *Liberty Asserted* (1704) by the English playwright John DENNIS. Other 18th-century plays on North American themes were Robert Rogers's unacted *Ponteach; or, the Savages of America* (1766); an anonymous Halifax comedy *Acadius; or, Love in a Calm* (1774), also thought to be the first original play in English to be written and performed on Canadian soil; and an elegant comedy with music *Colas et Colinette, ou le Bailli dupé* (1790) by Joseph Quesnel, resident dramatist of Montreal's revived French amateur group, the Théâtre de Société.

In the beginning the assembly-rooms of taverns and inns served as playing-places. Halifax, Nova Scotia, lays claim to Canada's first playhouse, the New Grand Theatre built in 1789, which survived until 1844. Complete with boxes, it seated 500 people. Nova Scotia was also the birthplace of the actress Winetta Montague (1851–77) and the actor and playwright Henry James FINN who with his compatriot William Rufus Blake (1805–63) wrote plays in the 1820s. These were probably the first native-born Canadian playwrights; the first indigenous French-Canadian dramatist, Pierre Petitclair (1813–60), began writing a decade or so later. Born in Quebec, he was the author of several comedies in the style of Molière which made use of the local *patois*.

In Montreal a new theatrical era was heralded by the opening in 1825 of Molson's Theatre Royal. Like the Halifax theatre it survived until 1844. With seating for 1,000 spectators and a strong resident company, it was soon able to attract visiting stars. Edmund KEAN was there in 1826, Charles MACREADY in 1844; and in May 1842 Charles DICKENS, on tour with readings from his own works, deviated from his prescribed programme to spend ten days with the Garrison Amateurs of Montreal as actor and director. Only four years later Canada suffered her worst theatre fire when on 12 June 1846 45 people lost their lives in the burning of Quebec's Théâtre Saint-Louis.

Meanwhile Toronto had come into its own with the opening in Dec. 1848 of the Royal Lyceum Theatre (destroyed by fire on 30 Jan. 1874) which from 1853 to 1858 was successfully managed by John NICKINSON, whose daughter Charlotte went on to manage Toronto's 1,800-seat Grand Opera House from 1874 to 1879. Even more memorable was the 23-year-long management, from its opening in 1852 to 1875, of the second Theatre Royal in Montreal (which closed in 1913) by J. W. Buckland and his wife. E. A. SOTHERN played three summer seasons in Halifax, from 1857 to 1859, when he leased the Spring Garden Theatre there (which opened in 1846 and was demolished in 1885), renaming it temporarily Sothern's Lyceum.

All these theatres, as was usual at the time, had resident companies which could, when necessary, give adequate support to visiting stars; but by 1880 they had mostly vanished, having fallen victim to the completely packaged touring companies which took advantage of the fast and inexpensive travel offered by the railways. Most of these came from outside Canada, to such an extent that in 1911 the Toronto theatre critic B. K. Sandwell was moved to write: 'Canada is the only nation in the world whose stage is entirely controlled by aliens.' This was not entirely correct. Professional Canadian writers, actors, and companies did exist by the time he was writing; the only trouble was that they so rarely worked together.

From the beginning there were in Canada three major streams of playwriting. First, an exotic series of poetic dramas, heavily indebted to Shakespeare, BYRON, and Gothic MELODRAMA and written by authors remembered today as poets rather than as dramatists. Aptly described as 'playwrights in a vacuum', they had little or no connection with the practical stage. However, a few of their plays, set in Canada, have a genuine historical interest, among them *The Female Consistory of Brockville* (1856) by 'Caroli Candidus', a remarkable melodrama based on an actual incident in which a Presbyterian minister had been forced to leave his rectory by a group of militant females; *Laura Secord* (1876) by Sarah Anne Curzon, which tells of the heroine who in the War of 1812 passed through the enemy lines leading a cow to warn the British of an impending American attack; *Tecumseh* (1886) by Charles Mair, on the Indian Chief in the same war who sided with the British; and two plays on French heroes, *Papineau* (1880) by Louis-Honoré Fréchette, about the Frenchman who in 1839 led an early abortive attack on British rule, and *Le Jeune Latour* (1884) by Antoine Gérin-Lajoie, on the heroic defender of a 17th-century fort.

A second group of plays consisted mainly of popular musical shows and political satires. Among them were *Dolorsolatio* (1865) by 'Sam Scribble', predicting Confederation, which was not to come for another two years; *H.M.S. Parliament* (1880) by W. H. Fuller, set to the music of GILBERT and Sullivan's *H.M.S. Pinafore* (1878), which ridiculed the government and its involvement in a political scandal; and *Ptarmigan* (1895) by Jean Newton McIlwraith, which celebrated the return home of a Canadian who had emigrated to the U.S. There was also at this time a third group of minor professional playwrights writing commercially for the Broadway theatres who, although they hailed from Canada, were seldom recognized as Canadian, though their plays were sometimes seen in Canada on tour. The best known was perhaps William A. Tremayne (1864–1939), who wrote several melodramas for Robert MANTELL; another was George V. Hobart (1867–1926), who wrote the book and lyrics for five editions of the ZIEGFELD *Follies*.

It is perhaps not surprising that the U.S. theatre journal *Variety* once called Canada 'the greatest theatrical training-school America has'. Certainly a number of players in the U.S.A. and England were expatriate Canadians, among them the actresses Margaret ANGLIN and, later, Beatrice LILLIE. In the field of light entertainment there were Douglas Brian (1877–1948), Marie Dressler, May Irwin, Eve Tanguay, and Helen Morgan. Julia Arthur, who played with IRVING at the London LYCEUM and on tour and in 1924 was seen on Broadway as Joan of Arc, was also a Canadian. American actors born in Canada include Clara MORRIS, McKee RANKIN, and the younger James HACKETT. Matheson LANG was born in Canada, as were others now remembered as film stars, who began their careers in the theatre—Mary Pickford, Walter Huston, and Walter Pidgeon. Viola ALLEN, Lena ASHWELL, David BELASCO, Henry MILLER, and Annie Russell were also reared and educated in Canada. Two famous MINSTREL stars from Canada were Colin ('Cool') Burgess (1840–1905) and the step-dancer George Primrose (1852–1919), who with Lew DOCKSTADER formed Primrose's Minstrels. George White (1890–1968), whose *Scandals* became an institution in the New York theatre between 1919 and 1939, was a Canadian, as was John Murray Anderson (1886–1954), hailed in New York as 'the King of revues'.

Luckily not everyone went to Broadway and by the opening of the 20th century Canada was slowly, in an atmosphere of awakening national consciousness, beginning to develop an indigenous theatrical life. There were a dozen or so independent Canada-based companies under competent actor-managers travelling the country, and even a solitary classical and Shakespearian troupe under Harold Nelson (Shaw) pioneering in the West. All these companies appeared in the larger towns as well as developing smaller regional circuits of their own. There were also a number of Canadian entrepreneurs, both in the East—C. J. Whitney, C. W. Bennett, Ambrose J. Small, Jacob and Sparrow—and in the West—C. P. Walker, W. B. Sherman, Alexander Pantages—and in 1913 a small group consisting of Leyel, Holles, Lydiatt, and Shaughnessy made a determined bid to break the stranglehold of the American THEATRICAL SYNDICATES in Canada. The idea behind their British-Canadian Theatre Organization (or Trans-Canada Theatres) was to sponsor companies and star actors from England on round-the-world tours. In 1914 and 1921 MARTIN-HARVEY's first two 'grand tours' across Canada were backed by this organization, as was the tour by Laurence IRVING in 1914, which ended in tragedy when the *Empress of Ireland* sank after a collision in the St Lawrence. One of Canada's great unsolved mysteries occurred on 2 Dec. 1919, when Ambrose J. Small sold his holdings to Trans-Canada Theatres for $1,700,000 and disappeared the same day. Trans-Canada was later bought out by movie interests.

Vaudeville continued to flourish in Canada until the advent of the 'talkies', and there was another flurry of British, French, and American stock companies in residence between 1907 and 1930. But essentially professional theatre in Canada collapsed during the depression of the early 1930s, and it was left to the Little Theatre Movement to keep theatre alive. One of the outstanding amateur theatres in late 19th-century Canada was inaugurated on 13 Mar. 1873, when the Governor-General Lord Dufferin opened a new ballroom equipped with a small stage in Government House (Rideau Hall in Ottawa). During his tenure of office this served as an active little playhouse, the Dufferin children's tutor Frederick Dixon collaborating with the Cathedral organist Frederick Mills in the composition of several original operettas which were staged there most successfully. Among them were *The Maire of St Brieux* (1875) and *Maiden Mona, the Mermaid* (1877). Dixon was also the author of a masque, *Canada's Welcome*, performed at the Grand Opera House in Ottawa on 24 Feb. 1879 to celebrate the arrival of Lord Dufferin's successor the Marquis of Lorne and his wife H.R.H. Princess Louise. The patronage of the arts which had always been associated with Government House continued under Earl Grey. He founded the first annual National Drama and Music Festival, which took place from 1907 to 1911 and provided a precedent for Lord Bessborough's Dominion Drama Festival, held for the first time in 1933.

In 1898 the St Jean-Baptiste Society, which had erected the *Monument National* in Montreal as a symbol of French-language culture, began a series of 'family evenings' designed to present French classical plays to middle-class audiences, and in 1908 Paul Cazeneuve, who had for some time been running a company based in Montreal, arranged a *concours* on Montreal Island which attracted 18 entries from the *cercles*, or amateur groups, which were functioning in and around the city. By about 1900 Canadian universities were welcoming drama as an extra-curricular activity, though it did not enter the curriculum as a subject for serious study until after the Second World War. There were, however, schools of 'Elocution and Dramatic Expression' in Toronto in the 1890s, and two Conservatories were founded in Montreal in 1900 and 1907 respectively.

With the example of the ABBEY THEATRE in Dublin and the many European Art Theatres of the time to encourage them, the enlightened amateurs finally determined to challenge the monopoly of the commercial theatre, and after sporadic pre-war attempts the Little Theatre Movement blossomed from coast to coast immediately after the First World War. Some of the best efforts were Carroll Aikins's experimental Home Theatre at Naramata, B.C., which functioned from 1920 to *c.* 1924, and Toronto's HART HOUSE THEATRE where Raymond MASSEY began his career. But by 1928 the first glow of idealism was fading, and with some exceptions the majority of groups reverted to the more mundane aim of 'social diversion' only. However, the 1930s did see the founding of the co-lingual Montreal Repertory Theatre and its School and, in several cities, the establishment of theatres specifically for children. It was a time when one-act plays were much in vogue, perhaps because, like revue sketches, they were easier to handle, and in 1936 the Ottawa Little Theatre sponsored its first one-act play-writing competition. Several of the outstanding Canadian writers of the time showed themselves equally at home in the short or full-length play, among them Merrill Denison (1893–1975), Mazo de la Roche (1885–1961), and Gwen Pharis Ringwood (1910–). Two important innovations of the 1930s were the symphonic dramas of Herman Voaden and Toronto's Theatre of Action, a spin-off from the Workers' Theatre, which flourished from 1935 to 1940.

The amateur movement came into focus with the founding in 1933 of the Dominion Drama Festival, a moveable national feast in which regional festival winners in one-act plays competed before an invited British or French bi-lingual adjudicator. After the Second World War the festival changed to presenting three-act plays; it also appointed Canadian adjudicators, presented an all-Canadian contest for Centennial year in 1967, and began to lay more stress on its role as a training ground for future Canadian actors and playwrights. But this was a bid for survival rather than for leadership, which had begun to pass to the professional theatre that was coming into being all over the country. Spokesmen from the amateur movement had for a long time been urging the creation of a national chain of theatres across Canada, but what they envisaged was not an indigenous professional network, rather an alliance of the Little Theatres, like Hart House and the French Compagnons de Saint-Laurent, with colleges and universities. Of the literally hundreds of community theatres which had sprung up in the wake of the Little Theatre Movement, it is significant that only a few, in cities like Winnipeg, Calgary, Montreal, and London (Ont.), made the transition to professional status. The Dominion Drama Festival, whose biggest mistake had been to close down during the Second World War, officially came to an end in 1978. The Little Theatre Movement did without doubt help to bridge the gap between the end of the commercial age of American touring and the beginning of Canada's own non-profit-making professional theatre, and, being voluntary, it was perhaps the only form of theatre that could have survived during those unsubsidized and economically depressed years.

The rise of the French professional theatre in Canada, which has now come into its own, was a slow business. In 1900 there were in Montreal two important companies, the Théâtre des Nouveautés, which lasted until 1907, and the Théâtre National, which survived until about 1918. In the mid-1920s there were hopes of an all-French Canadian company, which came about in 1927 when Fred Barry and Albert Duquesne teamed up as actor-managers. In 1930 they bought the small Chanteclerc Theatre, renaming it the Stella, and played there professionally until 1935. Similarly, in Quebec City in 1936 a group of professionals founded Les Artistes Associés de Québec, which lasted until 1942. In Montreal the revues of Gratien GÉLINAS, based upon his 'Fridolin' character, spanned the years from 1937 to 1946, and eventually gave rise to his co-lingual Théâtre de la Comédie-Canadienne, which functioned from 1958 to 1969. Father Émile Legault's amateur group Les Compagnons de Saint-Laurent, which did exciting work between 1937 and 1952, was responsible for the emergence of such future key men as Jean GASCON, Pierre Dagenais, founder of the short-lived but highly important L'Équipe (1943–7), which drew its actors from Radio-Canada, and Jean-Louis Roux, who with Gascon founded in 1951 the internationally acclaimed Théâtre du NOUVEAU MONDE. Montreal's oldest established company is the Théâtre du Rideau Vert, founded in 1948 and led by Yvette Brind'Amour and Mercedes Palomino in the historic Stella Theatre. Slightly later were the Théâtre-Club, which lasted from 1954 to 1964; the Théâtre Populaire du Québec, founded in 1963 and still active, having as an offshoot the experimental Grand Cirque Ordinaire; La Nouvelle Compagnie Théâtrale, founded in 1964 and now occupying the Théâtre Denise Pelletier, named in honour of a well-loved Quebec actress; and the Théâtre de Quat'Sous, founded in 1965, which grew out of a children's theatre called La Roulotte, founded in 1952. La Poudrière, since it was established in 1957, has tried to be both commercial and multi-lingual, doing plays in German and Italian as well as in French and English. A newcomer to the established ranks is the Compagnie Jean Duceppe, which started in 1975. There are also over 100 young companies, which, like their English-Canadian counterparts, lean towards COLLECTIVE CREATION. From the beginning, the leading exponent of this 'alternative' theatre has been Jean-Claude Germain's Théâtre d'Aujourd'hui, founded in 1968. Quebec City was slower than Montreal to go professional, the first such company being L'Estoc, founded in 1957. This closed in 1966, but in 1970 the opening of the Grand Theatre complex, with a resident company, Le Trident, once again provided a professional focus.

Playwriting in French has flourished since Gélinas produced his *'Tit-Coq* in 1948. Until the mid-1960s his plays and those of Marcel DUBÉ predominated, documenting sympathetically the disintegration of the older Quebec values rooted in church and family. In 1968, with *Les Belles-Soeurs*, Michel TREMBLAY became the voice of the Québecois. The introduction into his plays of lower-class slang (*joual*) imported on to the stage a new francophone means of expression. The theatre, adding its voice to the emotional rallying-cry of the poets and song-writers, supported the populist movement for French cultural sovereignty and political separation from Canada. Of the many plays which reflected this socio-political ferment, Gélinas's *Hier, les enfants dansaient* (*Yesterday, the children were dancing*, 1966) and Robert Gurik's *Hamlet, Prince de Québec* (1968) were significant. With the election in 1976 of a separatist government, Québecois dramatists relaxed once more into the familiar modes of poetic fantasy and social criticism. Contemporary dramatists of note include Jean Barbeau, Roch Carrier, Jacques Ferron, Michel Garneau, Roland Lepage, Françoise Loranger, and the Acadian Antonine Maillet.

Although the gap left in the English-speaking part of Canada by the demise of the American touring system in the early part of the 20th century was partly bridged by the Little Theatre Movement, the indigenous professional theatre there too began slowly to emerge, and it was the lighter forms of entertainment which flourished first. The First World War saw the rise of the Dumbells, an endearing concert party which eventually provided a series of annual revues that toured throughout the 1920s. Its hit show *Biff, Bing, Bang* was seen at the AMBASSADOR THEATRE in New York, opening on 9 May 1921 and running throughout the summer, thus marking the first successful foray by an all-Canadian production on to the Broadway scene. The Second World War gave further impetus to the expression of national feeling, and the Canadian Navy, Army, and Air Force shows were highly professional and again acclaimed, this time in England. It is interesting to compare the state of the English and French-speaking revues between the wars. Where Montreal had Fridolin, English Canada had the *Winnipeg Kiddies*, a children's touring troupe which lasted from 1915 until about 1935, and Jane Mallett's *Town Tonics*, which started in 1932 and continued until 1942. There was also a single professional stock company, John Holden's Actors' Colony Theatre, which from 1934 to 1941 gave summer seasons in Bala, Muskoka, alternating from 1936 to 1940 with winter stock seasons in Winnipeg's Dominion Theatre.

After the Second World War a new generation of Canadian actors and directors came to the fore, inspired by such devoted teacher-directors as Sydney Risk, Betty Mitchell, Dora Mavor Moore, Robert Gill, and Eleanor Stuart, among many others, and resulting in the foundation of such professional groups as Toronto's New Play Society and Jupiter Theatre; Montreal's Mountain Playhouse; Ottawa's Canadian Repertory Theatre; St John's London Theatre Company; Halifax's Theatre-in-the-Round; Saskatoon's Western Stage Society; and Vancouver's Everyman, Totem, York, and Lancaster theatres. But, lacking subsidies, most of these ventures were doomed to failure. A few survived, undoubtedly because of their commercial appeal, among them Vancouver's Theatre under the Stars, which lasted from 1940 to 1962, and Toronto's annual revue *Spring Thaw*, which ran from 1948 to 1973. Of the many flourishing theatres founded at this time the outstanding example is the STRATFORD (ONTARIO) FESTIVAL theatre, which opened in 1953. There is no doubt that its establishment, together with that of the touring Canadian Players (founded by Tom Patterson and Douglas Campbell a year later), gave a terrific impetus to the development of the performing arts throughout the country. When Parliament, acting on the Massey Report of 1951, created the Canada Council in 1957, the theatre received for the first time government and financial recognition which resulted in an explosion of growth. The Regional Theatre system then established meant the opening of a civic playhouse in every major urban centre in Canada, and the effective buying back of the country's theatre industry from outside control. The pattern was set in 1958 by John HIRSCH's Manitoba Theatre Centre in Winnipeg; the Vancouver Playhouse and the Neptune in Halifax followed in 1963, the Citadel in Edmonton in 1965. Other enterprises, like Toronto's Crest Theatre, which lasted from 1954 to 1966, being privately run and not so well housed, were less easily accommodated into the Regional system. An early alternative theatre was Toronto Workshop Productions, which dates from 1959. To meet the immediate demand thus created for highly trained professional players, the co-lingual National Theatre School in Montreal was founded in 1960, and is still doing fine work.

During this later period new major summer festivals appeared: the Vancouver International Festival, which had an uneven career from 1958 to 1968; the (G.B.) SHAW FESTIVAL at Niagara-on-the-Lake, Ont., in 1962; and in 1965 the Charlottetown Festival, devoted to Canadian MUSICAL COMEDY.

With the arrival of Centennial year in 1967, and the virtual absence of American influence during the war in Vietnam, the Canadian theatre entered a new phase. Many new auditoriums were built to house the Centennial celebrations, notably Charlottetown's Confederation Centre in 1964; the Arts and Culture Centre in St John's in 1967; the National Arts Centre in Ottawa in 1969; Toronto's St Lawrence Centre and the Regina Arts Centre, both in 1970; and the Hamilton Place Theatre, in Hamilton, Ont., in 1973. These theatres reflected the earlier taste for huge and somewhat unwieldy opera houses like the twin Jubilee auditoriums in Calgary and Edmonton, which date from 1957, the Queen Elizabeth in Vancouver, built in 1959, the O'Keefe Centre in Toronto, 1960, and the Place des Arts, Montreal, 1963. Two of Canada's newest theatres are the second Citadel in Edmonton, built in 1976 to rehouse the company formed in 1965, and the renovated Grand in London, Ont., which opened in 1979. In 1980 plans were afoot for the new civic centres in Calgary and Saint John, N. B., as well as a new theatre building for the Neptune company in Halifax.

As might be expected, the Centennial celebrations brought to a head the discontent felt for many years over the dearth of good English-Canadian plays, and stimulated the demand for more of them. The professional machinery for play production was now in good working order, but where was the Canadian product? The post-war years had seen the emergence of several new English-Canadian playwrights: Robertson DAVIES, with *Overlaid* (1949), John Coulter with *Riel* (1950), Patricia Joudry with *Teach Me How To Cry* (1955), and Len Peterson with *The Great Hunger* (1960). Centennial year itself brought a number of important new works, among them *Fortune and Men's Eyes* by John Herbert, *The Ecstasy of Rita Joe* by George RYGA, and *Colours in the Dark* by James REANEY. It also saw the beginning of the 'alternate theatre' movement, a groundswell of small *théâtres de poche*, or 'pocket theatres', for experimental work, designed to encourage the writing of new Canadian plays and to act as 'alternatives' to the 'dinosaurs' or large subsidized playhouses of the establishment. Symbolically, the movement began at Stratford with the Canadian Place Theatre, whose founders would go on to open the Toronto Free Theatre in 1972. In Montreal it was the Centre d'essai, or 'try-out centre', that took up the cause of new writers as early as 1963; in Vancouver it was the New Play Centre, founded in 1970. In the same year the Factory Theatre Lab led the way in Toronto, handing on the torch to the Tarragon Theatre in 1971, while the earlier (English-speaking) Théâtre Passe-Muraille, founded in 1968, opted for collective creation, concentrating especially on themes drawn from Canadian folk-culture. Collective creation has, in fact, become a distinctly Canadian form of theatrical expression and experimentation, adopted in the 1970s by such theatres as Tamahnous in Vancouver, Theatre Network in Edmonton, Twenty-Fifth Street House and Persephone in Saskatoon, and the Mummers Troupe of Newfoundland.

In 1971 the Canada Council convened a conference on playwriting, held at the Gaspé in Quebec, out of which emerged a recommendation that at least half the plays presented by a Canadian company should be by Canadian playwrights. In 1972 an All-Canadian summer festival started at Lennoxville, P.Q., and in the same year the Playwrights' Co-op (now known as Playwrights Canada) was formed, and the business of publishing Canadian plays began in earnest. The 1970s also witnessed the emergence of several theatre and drama magazines, an accelerated study of Canadian theatre history, a Canada Council Touring Bureau, international stardom for many Canadian actors, among them Hume CRONYN, Christopher PLUMMER, Kate REID, and Kate Nelligan, numerous play productions abroad, and awards for achievements at home. But primarily those years saw the recognition of the fact that a repertory of plays by Canadian authors was the only fit and proper backbone for the theatre in Canada. Some of the English-language writers of the 1970s include Carol Bolt (*One-Night Stand*), Peter Colley (*I'll Be Back Before Midnight*), Michael Cook (*Jacob's Wake*), David Fennario (*Balconville*), David Freeman (*Creeps*), David French (*Jitters*), Cam Hubert (*The Twin Sinks of Allan Sammy*), John Murrell (*Waiting for the Parade*), Sharon Pollock (*Walsh*), Rick Salutin (*Les Canadiens*), and George Walker (*Zastrozzi*).

As the 1970s ended, the activity of the 'alternate theatres' appeared to be slowing down, and the pendulum seemed to be swinging back towards the older and larger theatres. Financial cutbacks had already resulted in more small-cast plays, one-person shows, shorter seasons, and the departure or absorption into the larger establishments of a number of players and directors. Tension over the importation of British and American actors led to talk of protectionist policies. On the positive side, the financial pressures of the times were instrumental in promoting increased co-operation between theatres by means of co-productions and exchanges, and a more business-like approach to the writing and marketing of plays. The growth of cabaret and of supper- and lunch-time FRINGE THEATRES, which blossomed in all the major cities, also underlined this new striving after a more commercial base to theatrical enterprise, with less dependence on subsidies. The rapid expansion of the performing arts in Canada is shown by the fact that in 1980 the country had 220 professional theatre groups and 8 opera companies, as well as 3 television networks and a thriving film industry.

CANADIAN ACTORS' EQUITY ASSOCIATION. In 1955 a branch office of the AMERICAN ACTORS' EQUITY ASSOCIATION was opened in Toronto to deal with conditions for Canadian actors, and on 1 Aug. 1976 the Canadians amicably decided to become completely independent of the U.S.A. In addition to the Toronto headquarters, a West Coast office was opened in Vancouver in Feb. 1979. To commemorate 25 years of friendly liaison between the American and Canadian Equity members (a reciprocal 'open border' agreement), the Canadians commissioned a stained glass window for the lobby of the New York office. Designed by Ira Ginsburg and installed in Mar. 1979, it depicts Marc LESCARBOT'S *Le Théâtre de Neptune en la Nouvelle France*, the first play to be performed in CANADA.

CANDLER THEATRE, New York, see SAM H. HARRIS THEATRE.

CANE [CAIN, KEIN, KEYNE], ANDREW (?–1644), a goldsmith by trade, who in 1622 joined the newly formed LADY ELIZABETH'S MEN. He was evidently a comedian, as he is usually referred to as 'Cane the Clown'. He later became one of the former ADMIRAL'S MEN, when under the patronage of the Elector Palatine they were known as the Palsgrave's Men, and remained with them when they were reorganized as PRINCE CHARLES'S MEN, playing Trimalchio, a humorous gallant, in their production in 1631 at SALISBURY COURT of Shackerley Marmion's *Holland's Leaguer*. He was also with them at the FORTUNE THEATRE, and thirty years later was still remembered for his JIGS there, and at the RED BULL THEATRE, where he appeared surreptitiously after the closing of the theatres in 1642. During the Civil War he returned to his old trade and engraved dies at Oxford for the debased coinage of the Royalists.

CAÑIZARES, JOSÉ DE, see SPAIN.

CANKAR, IVAN, see YUGOSLAVIA.

CANTERBURY MUSIC-HALL, London, in the Westminster Bridge Road, Lambeth. The first of the great MUSIC-HALLS, it was erected by Charles MORTON in 1852 on the site of an old skittle-alley adjacent to his Canterbury Tavern, and paid for out of the profits on drink made during free entertainment formerly offered there. The Canterbury Hall, as it was originally called, proved so popular that in 1854 Morton was able to replace it by the larger New Canterbury Music-Hall, which had a large platform stage and accommodation for 1,500 people. By 1867, when Morton left to work elsewhere, the earlier programmes of light music and ballad singing had been dropped and comedy predominated. In 1876 the building was reconstructed as a three-tier theatre, its bar being for many years the favourite rendezvous of music-hall performers. The Canterbury was well patronized by Royalty, being visited regularly by the Prince of Wales (later Edward VII), the Duke of Cambridge, and the Duke and Duchess of Teck. When the popularity of the 'halls' began to decline, drastic reductions in the price of seats brought audiences back for a while; but the heyday of the music-hall had long been over before the Canterbury was destroyed by bombing in 1942.

CANTOR, EDDIE, see REVUE and VAUDEVILLE, AMERICAN.

CAPA Y ESPADA, Comedia de, see COMEDIA.

ČAPEK, KAREL (1890–1938), Czech playwright, whose first play *The Brigand* (1920) was produced at the National Theatre in Prague, where it was revived in 1954. With his next play, *R.U.R.* (1921)—the initials stand for Rossum's Universal Robots—Čapek became widely known outside Czechoslovakia. It was seen both in London and in New York, and introduced the word 'robot'—a mechanical drudge—into the English language. *The Insect Play* (1922), like *R.U.R.*, was much influenced by German EXPRESSIONISM and employed the techniques of revue; both plays depict the horrors of a regimented technological world and the terrible end of the populace if they fail to rise against their oppressors. *The Insect Play*, written by Karel in collaboration with his brother Josef (1887–1945), an artist who died of typhus in Belsen, was immediately seen in London with the sub-title *And So Ad Infinitum* and in New York as *The World We Live In*. Among Čapek's later plays, *The Makropulos Affair* (1923), on which Janáček based his 1926 opera of the same title, argues the case for longevity and unlike SHAW's *Back to Methuselah* (1922) decides against it. A sequel to *R.U.R.*, *Adam the Creator* (1927), again written jointly with Josef, shows man endeavouring to rebuild the world destroyed by the robots; and two anti-Fascist plays, *The White Scourge* (1937), seen in London as *Power and Glory* in 1938, and *The Mother* (1938), seen in London in 1939, deal with the rise of dictatorship and the devastating effects of war. All Čapek's plays reflect the world he lived in and comment on its grosser follies. He also wrote an amusing short monograph, *How a Play is Produced*.

CAPITANO, Il. The braggart soldier of the COMMEDIA DELL'ARTE, vainglorious and cowardly, usually a Spaniard. One of the first to play him was ANDREINI, as Capitano Spavento, and during the first three years of his retirement he collected and published his famous *Bravure del Capitano Spavento del Vall'Inferna*. Fornaris, also a *capitano*, was Andreini's contemporary, while a certain 'Cardone' seems to have been his successor. In the 17th century Silvio FIORILLO played the part as Capitano Mattamoros, 'death to the Moors', anglicized as Captain Matamore. Il Capitano may derive from the MILES GLORIOSUS of PLAUTUS, and perhaps contributed something to Shakespeare's Armando in *Love's Labour's Lost* and Pistol in *Henry IV, Part 2* and *The Merry Wives of Windsor*.

CAPON, WILLIAM (1757–1827), English architect and painter, who in 1791 was appointed scenic director of the new DRURY LANE THEATRE by John Philip KEMBLE. Of a plodding, pedestrian temperament, he was a painstaking antiquarian, which accorded well with Kemble's plans for scenic reform. He did away with the old system of FLATS and WINGS in favour of BUILT STUFF, which, though effective, was cumbersome and difficult to shift, and designed, among other things, streets of ancient houses copied from actual remains of the period, and a 14th-century cathedral with nave, choir, and side aisles, all superbly decorated. He continued to work at Drury Lane until the theatre was burnt down in 1809, involving him in a severe monetary loss since payment was still due for much of the scenery. He then worked for Kemble again at the new COVENT GARDEN Theatre, where many of his sets remained in use for years, serving as set pieces until the arrival of MACREADY. Capon, who had something of Charles KEAN's pedantic inaccuracy in the application of archaeological studies to the stage, delighted the public with his Anglo-Norman hall of *Hamlet*, which was made up of fragments from the periods of Edward the Confessor, William Rufus, and Henry I.

CAPUANA, LUIGI (1839–1915), Italian dramatist, novelist, and critic, who with MARTOGLIO and the Sicilian dialect actors Grasso and Musco helped to keep alive the vernacular theatre in Sicily. Although he argued, in his critical study *Il teatro italiano contemporaneo* (1872), that prose fiction was better suited to the new naturalism than drama, his own plays, under the influence of ZOLA, are vigorous examples of Italian VERISMO. Notable among them are *Malia* (*Enchantment*, 1895) and *Lu Cavalieri Pidagna* (1909).

CARAGIALE, ION LUCA, see ROMANIA.

CARBON ARC, see LIGHTING.

CARDIFF, see WALES.

CAREW, JAMES, see TERRY, ELLEN.

CAREY, JOYCE, see BRAITHWAITE, LILIAN.

CARL, KARL [Karl Andreas von Bernbrunn] (1789–1854), Austrian actor, dramatist, and impresario, who after a chequered army career turned actor and was soon playing romantic leads at the MUNICH Court theatre, where he eventually became director. He was responsible for introducing the figure of STABERL to Munich, the character becoming in Carl's cosmopolitanized version a favourite throughout Germany. He became director of the Theater in der Josefstadt in VIENNA in 1826, the year that marked the beginning of his successful collaboration with the actor-playwright NESTROY, and from 1827 to 1845 managed the Theater an der Wien, where he increasingly depended on the local Viennese *Posse* (farce). In 1847 he built a new theatre on the site of the old Leopoldstädter Theater. A ruthless business man, Carl paid low wages—his contracts were known as 'Korsarenbriefe'—and he died a millionaire.

CARLIN, see BERTINAZZI.

CARMONTELLE [Louis Carrogis] (1717–1806), French painter and dramatist, who wrote amusing sketches for performance in the private theatres of his day. These catered for the 18th-century vogue for dramatic 'proverbs', that is, plays illustrating a proverb which had to be guessed by the audience, and they were to exert a considerable influence on the *Comédies et Proverbes* of Alfred de MUSSET. Carmontelle's little sketches still possess a genuine interest as realistic conversation pieces; he was brought into the repertory of the COMÉDIE-FRANÇAISE in 1938, and one of his playlets, *Le Veuf*, soon established itself as a popular curtain-raiser.

CARNEY, KATE, see MUSIC-HALL.

CARNIVAL PLAY, see FASTNACHTSSPIEL.

CARNOVSKY, MORRIS (1897–   ), American actor, who made his first appearance in New York with the PROVINCETOWN PLAYERS in Sholom ASCH's *God of Vengeance* in 1922. From 1924 to 1930 he was with the THEATRE GUILD, his roles including Alyosha in an adaptation of DOSTOEVSKY's *The Brothers Karamazov* in 1927, in which year he was also seen in SHAW's *The Doctor's Dilemma* and PIRANDELLO's *Right You Are If You Think You Are*. In 1929 he played the title role in CHEKHOV's *Uncle Vanya* and a year later was in Shaw's *The Apple Cart*. During the 1930s he was with the GROUP THEATRE and appeared in a number of modern plays, including Paul GREEN's *The House of Connelly* (1931), and Clifford ODETS's *Awake and Sing* (1935) and *Golden Boy* (1937); he was in the London production of this last in 1938. From 1940 to 1950 he was a member of the Actors' Laboratory Theatre in Hollywood, directing a number of productions there, but he returned to New York in 1950 to appear in IBSEN's *An Enemy of the People*, followed by Maurice Samuels' *The World of Sholom Aleichem* (1953), ANSKY's *The Dybbuk* (1954), GIRAUDOUX's *Tiger at the Gates* (1955), and many other productions. After 1956 he appeared regularly at Stratford, Connecticut (see AMERICAN SHAKESPEARE THEATRE), crowning a distinguished career by playing King Lear there in 1963 and 1965, and appearing in that taxing role again in 1975 at the age of 77.

CARPENTER'S SCENE, an insertion into a PANTOMIME or spectacular MUSICAL COMEDY, played in front of a BACKCLOTH while elaborate scenery is set up behind out of sight of the audience. In Victorian times it was used also in serious plays, but was no longer needed when the practice of dropping the front curtain between scene changes became general.

CARPET CUT, a long narrow stage-opening behind the front curtain, stretching nearly the whole width of the proscenium opening. It is closed by hinged flaps which trap the edge of a carpet or stagecloth, so that the actor cannot be tripped up by it.

CARRETTO, GALLEOTTO DEL, see SACRA RAPPRESENTAZIONE.

CARRIAGE-AND-FRAME (or Chariot-and-Pole), a device for changing the WINGS, devised by

TORELLI in 1641 for the Teatro Novissimo in Venice; it was used extensively on the Continent, and can still be seen in operation at DROTTNINGHOLM. Each wing-piece was suspended just clear of the stage on a rectangular frame (or on a pole) which projected downward through a long slit in the floor, and was borne on a wheeled carriage running on rails in the CELLAR. At each wing position this arrangement existed in duplicate, the two carriages being connected by ropes working in opposite directions to a common drum which served the whole series of wing sets in such a way that as one carriage of each pair moved off stage, its neighbour moved on. The withdrawn wing was then replaced by the wing needed for the next scene, and the process was repeated as often as necessary. In England, where the wings usually ran in GROOVES, a somewhat similar system was used to change the BOOK WINGS.

CARROLL, EARL, see EARL CARROLL THEATRE and REVUE.

CARROLL, PAUL VINCENT (1900–68), Irish dramatist, who attacked the Catholic Church through his plays, and became as well known in England and the United States as he was in Ireland. His first play *The Watched Pot* (1931) roused little interest, and it was with his second, *Things That Are Caesar's* (1932; London and N.Y., 1933), that he first came into prominence. It was produced at the ABBEY THEATRE, Dublin, as were *The Wise Have Not Spoken* (1933; London, 1946; N.Y., 1954) and *Shadow and Substance* (1934; London, 1946). The latter, about the conflict between a canon and a liberal-minded schoolmaster, is perhaps his best play, and had a long run on Broadway (1937–8), as did *The White Steed*, given its first production in New York in 1939 (London, 1947). In 1942 a war play set in Glasgow, *The Strings, My Lord, Are False*, was seen in Dublin at the OLYMPIA THEATRE, where it was well received; it was less successful in New York in the same year. Among Carroll's later plays were *The Old Foolishness* (London, 1943), *The Devil Came from Dublin* (1952; London, 1953), and *The Wayward Saint* (N.Y., 1955). *We Have Ceased to Live* was published posthumously in 1972. Carroll was a schoolmaster in Glasgow from 1921 to 1937; he helped to found the Curtain Theatre there and later, with James BRIDIE, the CITIZENS' THEATRE, for which he wrote *Green Cars Go East* (1940), set in the Glasgow slums. All Carroll's plays are marked by sympathy, humour, and subtle characterization; his studies of Irish priests are especially memorable.

CARTEL, see AVANT-GARDE.

CARTER, Mrs LESLIE [*née* Caroline Louise Dudley] (1862–1937), American actress, who in 1889, when her nine-year-old marriage broke up asked BELASCO to launch her on an acting career. In spite of some initial opposition to her as a divorcée, she became one of his star players, her first outstanding part being Maryland Calvert in Belasco's *The Heart of Maryland* (1895), in which she was seen in London, as she was in his *Zaza* (1899; London, 1900). She then played the title roles in *Du Barry* (1901) by Belasco and *Adrea* (1905) by Belasco and John Luther LONG. On her second marriage in 1906 she severed her connection with Belasco, touring under her own and other managements with some success until she retired in 1917. She returned to the stage in 1921 in MAUGHAM's *The Circle* and continued acting until shortly before her death.

CARTON [Critchett], RICHARD CLAUDE (1856–1928), English dramatist, who had a short career as an actor from 1875 to 1885 and then turned to playwriting. His first plays were much influenced by DICKENS, the best being *Liberty Hall* (1892), but in 1898 he scored a success with *Lord and Lady Algy*, and continued to write comedies, bordering on farce, which poked discreet fun at the aristocracy and were much enjoyed by the occupants of the London stalls. Elegantly staged, they served as starring vehicles for his wife Katherine Mackenzie Compton (1853–1928), who played almost exclusively in his plays from 1885 onwards, and was a great factor in their success. She was adept at portraying the society woman of shrewd wit and few scruples. The daughter of Henry COMPTON, she made her first appearance in London in 1877 as Julia in SHERIDAN's *The Rivals*, and was for some years at the ST JAMES'S THEATRE.

CARTOUCHERIE DE VINCENNES, Paris, see SOLEIL, THÉÂTRE DU.

CARTWRIGHT, WILLIAM. There were two English actors of this name, probably father and son. The elder appears in HENSLOWE's diary, being first mentioned as one of the ADMIRAL'S MEN in 1598 and remaining with the company until 1622. He is believed to have died *c.* 1650. The younger was born *c.* 1606 and may have acted with his father as a boy, but his first recorded appearance was in 1634. He was with PRINCE CHARLES'S MEN at SALISBURY COURT when the theatres closed in 1642, and was one of the actors who appeared surreptitiously thereafter at the COCKPIT. When the theatres reopened in 1660 he joined KILLIGREW's company, and was still appearing in actor-lists in 1684–5, after the two London companies had combined, being then around his 80th year. The date of his death is unknown.

A William Cartwright (1611–43) who was a Proctor at Oxford University was the author of four plays written as academic exercises for the students there and probably performed at Christ Church between 1635 and 1638.

CASARÈS, MARIA (1922–), Spanish-born French actress, permanently resident in Paris since being exiled from her own country during the Spanish Civil War. After success in a French version of SYNGE's *Deirdre of the Sorrows* in 1942, she was associated with several existentialist plays, notably CAMUS's *Le Malentendu* (1944), *L'État de siège* (1948), and *Les Justes* (1949), and SARTRE's *Le Diable et le bon Dieu* (1951), as well as playing Jeannette in ANOUILH's *Roméo et Jeannette* (1946). She was outstanding as Phèdre in RACINE's play while at the COMÉDIE-FRANÇAISE, 1952–4, and also played many leading roles in the early days of VILAR's productions at the AVIGNON FESTIVAL, being particularly admired as Lady Macbeth in 1954, and at the THEATRE NATIONAL POPULAIRE, which she joined in 1955. Later she was notable as the mother in GENET's *Les Paravents* (1966) at BARRAULT's Théâtre de France, and her later roles have included BRECHT's Mother Courage (1968), Queen Victoria in BOND's *Early Morning* (1970), and Shakespeare's Cleopatra (1975).

CASE, CHARLIE, see VAUDEVILLE, AMERICAN.

CASINO THEATRE, New York, on the south-east corner of Broadway and 39th Street, for almost 50 years the leading musical-comedy house of New York. Built in a massively Moorish style, it held 1,300 people, and had a roof garden for summer evening concerts. It opened on 21 Oct. 1883, and among the musical plays produced there were *Florodora* (1900), *A Chinese Honeymoon* (1902), *Wildflower* (1923), and *The Vagabond King* (1925). Its last hit was *The Desert Song* (1926). It was pulled down in 1930.

The EARL CARROLL THEATRE was renamed the Casino from 1932 to 1934.

CASONA, ALEJANDRO, see SPAIN.

CASSETTE, see SLOTE.

CASSON, LEWIS THOMAS (1875–1969), English actor and director, knighted in 1945. After some experience as an amateur, he made his first professional appearance at the ROYAL COURT THEATRE under the VEDRENNE and GRANVILLE-BARKER management, playing mainly in Shakespeare and SHAW. In 1908 he became a member of the repertory company at the Gaiety Theatre in Manchester under Miss HORNIMAN, which he directed from 1911 to 1914; there he met and married Sybil THORNDIKE, with whom his career was later almost entirely associated. After serving in the First World War he returned to the London stage in 1919, and with his wife appeared in seasons of GRAND GUIGNOL at the LITTLE THEATRE from 1920 to 1922, in which year they were also seen in a season of Greek plays in translation at the HOLBORN EMPIRE. Casson then devoted much of his time to direction, though continuing to act occasionally, notably as Stogumber in Shaw's *Saint Joan* (1924), in which his wife played the title role, and as Buckingham in *Henry VIII* at the OLD VIC in 1926. With his wife he made many overseas tours before the Second World War, and during it accompanied her on a tour of the Welsh coalfields in *Macbeth*. After the war they both toured extensively again in dramatic recitals, and Casson gave some excellent performances in such parts as Professor Linden in PRIESTLEY's *The Linden Tree* (1947), Sir Horace Darke in Clemence DANE's *Eighty in the Shade* (1959), and Telyegin in CHEKHOV's *Uncle Vanya* at the CHICHESTER FESTIVAL THEATRE in 1962. He made his last appearance at the age of 90 in a revival of Kesselring's *Arsenic and Old Lace* in 1966. He was a careful and painstaking director, and one of his greatest assets as an actor was his clear and beautifully modulated voice. His son John Casson wrote *Lewis and Sybil* (1972).

C.A.S. THEATRE, Auckland, see COMMUNITY ARTS SERVICE.

CASTRO Y BELLVÍS, GUÍLLEN DE (1569–1631), Spanish dramatist, chiefly remembered for his dramatization of the ballads celebrating the exploits of Spain's national hero the Cid. *La mocedades del Cid*, in two parts, was first published in 1618, and it was on this that CORNEILLE based his own play *Le Cid* (1637). Castro also wrote a number of other plays, including many comedies in the style of his friend and contemporary Lope de VEGA, and made three dramatizations of parts of CERVANTES' *Don Quixote* of which the best concentrates on the episode of Cardenio and Lucinda.

CATWALK, see FLIES.

CAULDRON TRAP, see TRAPS.

CAVE, JOE, see MUSIC-HALL.

CAWARDEN, Sir THOMAS, see MASTER OF THE REVELS.

CECCHI, GIOVANNI MARIA (1518–87), prolific Italian dramatist, whose work ranged from religious plays to comedies of which the best is possibly the *Assiuolo* (1550). The secularization of Italian drama was carried a step further by his interpolation into Bible stories of extraneous characters, usually farcical, borrowed from classical comedy, and incidents taken from contemporary life.

CECCHINI, PIER MARIA (1575–1645), actor and author of the COMMEDIA DELL'ARTE, whose stage name was Fritellino. He joined a professional company in about 1591, touring Italy and visiting Paris, where he returned in 1600 as the leader of the ACCESI. In the company was his wife Orsola, whom he married in 1594. She may have been the daughter of Flaminio SCALA, her stage name being Flaminia, and together with her husband she later joined the FEDELI.

CEILING-CLOTH, a canvas or muslin stretch, battened-out and suspended flat over the top of a BOX-SET. Variations of the ceiling-cloth are the Roll Ceiling, which has two permanent battens along the length of the rectangular stretch and can be rolled up, and the Book (or Booked) Ceiling, which consists of two flats hinged together along their length. When the unit is flown, the two halves fall face to face upon one another, thus taking up less space in the FLIES.

CÉLESTE, Mme CÉLINE (1814–82), French dancer and PANTOMIME player, who appeared with much success on the Parisian stage as a child. In 1827 she went with a troupe of dancers to New York, where she was much admired for her exquisite dancing and expressive gesture, and in 1830 she was seen in London. She scored a triumph in Haines's *The French Spy* (1837). She did not attempt a speaking part in English until 1838, then created a number of parts in new plays, including Madame Defarge in a dramatization of DICKENS's *A Tale of Two Cities* (1860). She was manager of the ADELPHI THEATRE in London for a short time in association with Ben WEBSTER. She made her last appearance in New York in 1865, and in London in 1874, reappearing for one performance in a revival of BUCKSTONE's *The Green Bushes* as Miami, one of her best parts, at DRURY LANE in 1878. From her photographs she appears to have been an extremely plain woman, but contemporary accounts leave no doubt of the beauty and expressiveness of her dancing and miming.

CELESTINA, La, the most important literary work of 15th-century Spain, published anonymously but now known to be the work of Fernando de ROJAS. Although written in dialogue form, it was probably intended to be read aloud rather than acted but a number of plays have been based on it, the earliest English version being John RASTELL's *Calisto and Meliboea* (*c.* 1530). It was the source of Ashley DUKES's *The Matchmaker's Arms* (1931), and of *Celestina* produced by THEATRE WORKSHOP under Joan LITTLEWOOD in 1958. Dramatized versions in French and Italian have also proved successful.

CELLAR, the space under the stage which housed the machinery necessary for TRAPS, scene-changing, and special effects. In English theatres the GROOVES, which allowed tall framed pieces of scenery to be slid to the sides of the stage for a scene-change, meant that the cellar could be shallow. It consisted usually of a room below the stage floor, where most of the machines were housed, and beneath this a well in the central part of the area, into which the traps and BRIDGES descended, and at the bottom of which were the DRUM-AND-SHAFT systems which worked them. In Continental theatres, which were in any case of far greater dimensions than the more intimate English playhouses, any framed backgrounds (as opposed to hanging DROPS) had to be lowered beneath the stage; to accommodate them the cellar (or *dessous*) often descended four or five storeys below the stage.

CELLE, small German town, between Hanover and Hamburg, whose castle contains the oldest existing playhouse in Germany. Designed by Arighini in baroque style, it has a horseshoe-shaped auditorium seating 330. It was first used in 1674, mainly for intimate opera. When the castle became the summer residence of the Hanoverian Court, plays and balls were given there until in 1705 the Elector became George I of England. The theatre was then abandoned until 1772. F. L. SCHRÖDER appeared there early in his career, but during the 19th century it was seldom used. In 1935 it was restored and redecorated, the stage enlarged, and a fly-door added to permit the use of elaborate scenery. It reopened on 13 May. During the Second World War it escaped damage, and in 1950 a permanent repertory company was installed there.

CELTIS (or **CELTES**) [Bickel or Pickel], KONRAD (1459–1508), German humanist, one of the most brilliant of his time. As Professor of Poetry and Rhetoric he made the university of VIENNA, where he settled in 1497, a centre of classical studies, and instituted afternoon performances of plays by TERENCE, PLAUTUS, and SENECA. He was himself the author of two plays in Latin, *Ludus Dianae* and *Rhapsodia, laudes et victoria Maximiliani de Boemannis*, both of which were performed before the Emperor Maximilian, the former at Linz in 1501, the latter at Vienna in 1504. The germ of the later HAUPT- UND STAATSAKTIONEN can be seen in them, with their elevated subjects glorifying the ruling house, their use of music, ballet, and choruses, their exchanges between players and audience, and their traces of the COMMEDIA DELL'ARTE with its stereotyped figures. In 1501 Celtis discovered and published the plays of HROSWITHA.

CEMA (Council for the Encouragement of Music and the Arts), see ARTS COUNCIL.

CENSORSHIP. Specific statutory provision for theatrical censorship is now rare in democratic countries, one of the last to abolish the practice being Great Britain in 1968 (see LORD CHAMBERLAIN). In general, self-censorship is the rule, current levels of public tolerance being shrewdly, if unconsciously, assessed by writers, producers, and directors. Some governments retain the right to intervene in the interest of good order or public morality, as did the French government in the case of GENET's *Les Paravents* (*The Screens*) in 1966, and most federal countries vest powers in regional or state authorities to act against theatrical productions considered offensive to local susceptibilities: the musical *Oh! Calcutta!* was banned in South Australia and Victoria, and plays by COCTEAU, Graham GREENE, and others have been banned in the Swiss canton of Valais. Throughout the world prosecutions may be laid *post hoc* under the laws relating to obscenity, blasphemy, libel, and so on: examples include Tennessee WILLIAMS's *The Rose Tattoo* in Eire in 1957; Mart Crowley's *The Boys in the Band* in Victoria, Australia, in 1969; the musical *Hair* in New Zealand in 1972. It has been rare for such prosecutions to succeed and, as the gulf widens between artistic attitudes to matters of traditional morality and those of the general public, more arcane applications of common law to theatrical performances have been attempted. In 1981 a private prosecution brought against Michael Bogdanov, who had directed Howard BRENTON's *The Romans in Britain* for the NATIONAL THEATRE, was based on Section 13 of the Sexual Offences Act of 1956, the grounds for complaint being the stage enactment of an attempted homosexual rape; the case was withdrawn before any point of law could be established.

In Communist countries and military autocracies pre-production censorship is usual, the severity of scrutiny varying with current orthodoxies. Political and social heresy is the censors' main concern, although attitudes to sexual matters are often more prudish than in libertarian societies. Here, too, an awareness of prevailing levels of acceptability will tend to ensure self-censorship.

CENTLIVRE, Mrs SUSANNAH (1667–1723), English dramatist and actress, a masculine-looking woman who delighted in playing men's parts. She had already written several plays when she achieved her first success with *The Gamester* (1705), a sentimental drama based to some extent on REGNARD's *Le Joueur* (1696) but with the moral tone of Colley CIBBER and Richard STEELE. It was followed in the same year by a somewhat similar play *The Bassett-Table*, which, like all her early works, was published under her second married name, Mrs Carroll. In 1706 she went to Windsor to play Alexander the Great and there met and married as her third husband Joseph Centlivre, cook to Queen Anne, whose name she thereafter retained. Of her many comedies of intrigue, which in verve and ingenuity rivalled those of Aphra BEHN, the best were *The Busie Body* (1709), *The Wonder, a Woman Keeps a Secret* (1714), and *A Bold Stroke for a Wife* (1718). All three were frequently revived, the second, which owed much of its initial success to the acting of Anne OLDFIELD as Violante, later providing an excellent vehicle for GARRICK in the part of Don Felix. The last, perhaps better classified as a comedy of manners, shows the influence of CONGREVE, and may have been partly written by John Mottley.

CENTRAL CITY OPERA HOUSE, Denver, see BONFILS THEATRE.

CENTRAL SCHOOL OF SPEECH AND DRAMA, see SCHOOLS OF DRAMA.

CENTRAL THEATRE, Auckland, see NEW ZEALAND.

CENTRAL THEATRE OF THE SOVIET ARMY, Moscow, see RED ARMY THEATRE.

CENTRE 42, see ROUND HOUSE and WESKER, ARNOLD.

CENTREPOINT THEATRE, Palmerston North, see NEW ZEALAND.

CENTRES DRAMATIQUES, see DÉCENTRALISATION DRAMATIQUE.

CENTURY GROVE, New York, see CENTURY THEATRE (1).

CENTURY THEATRE, London, see ADELPHI THEATRE and BIJOU THEATRE.

CENTURY THEATRE, New York. (1) On Central Park West at 62nd Street, was erected by Winthrop AMES, in association with Lee Shubert, to further his plan to establish a true REPERTORY THEATRE in New York. It opened on 6 Nov. 1909 as the New Theatre, with Julia MARLOWE and E. H. SOTHERN in *Antony and Cleopatra*. Ames's venture was not a success and the theatre closed, to reopen as the Century on 15 Sept. 1911. It housed mainly musical shows, but in 1916 Shakespeare's Tercentenary was celebrated by a fine production of *The Tempest*, and in 1921 MARTIN-HARVEY appeared in *Hamlet*. In 1924 REINHARDT's production of Karl Vollmöller's spectacle-play *The Miracle*, with striking scenery by Norman BEL GEDDES, was a great success. Reinhardt returned in 1927 with his productions of *A Midsummer Night's Dream*, HOFMANNSTHAL's version of *Everyman*, and BÜCHNER's *Dantons Tod*. The theatre closed in 1929 and was demolished a year later. On the roof of the Century there was a small theatre known as the Cocoanut Grove or the Century Grove which staged intimate REVUE and plays for children.

(2) At 932 7th Avenue, between 58th and 59th Streets, a large playhouse opened by the Shuberts as the Jolson on 6 Oct. 1921. In 1923 it housed on epoch-making visit by the company of the MOSCOW ART THEATRE under STANISLAVSKY in plays by CHEKHOV, GORKY, TOLSTOY, and TURGENEV, which had a lasting influence on the American stage. The following year saw a visit

from GÉMIER of the Paris ODÉON and the opening of Romberg's famous musical *The Student Prince*, which had a long run. In 1932 a company called the Shakespeare Theatre gave 15 of Shakespeare's plays at low prices to a mainly student audience. The theatre, which from 1934 to 1937 was known as the Venice, also housed a Negro operetta, an Italian company, the FEDERAL THEATRE PROJECT, and Maurice SCHWARTZ and his Yiddish Players. It was renamed the Century in 1944 and in 1946 saw a visit from the OLD VIC company from London, headed by Laurence OLIVIER and Ralph RICHARDSON. *Kiss Me, Kate*, Cole Porter's musical version of *The Taming of the Shrew*, began a long run in 1948, and in 1952 Margaret WEBSTER's production of SHAW's *Saint Joan* moved there from the CORT THEATRE. The Century was demolished in 1961.

CERNESCU, DINU, see ROMANIA.

CERVANTES SAAVEDRA, MIGUEL DE (1547–1616), Spanish novelist and dramatist, who as a young soldier was present at the battle of Lepanto in 1571, where he was wounded and lost the use of his left hand. Later he was captured by Barbary pirates and spent five years as a slave in Algiers, being ransomed and returning to Spain in 1580. He is best remembered as the author of *Don Quixote de la Mancha*, a satirical romance published in two parts, in 1605 and 1615, which has been widely translated and dramatized in many languages. The first English version for the stage was made by D'URFEY, who probably based it on Sheldon's racy but somewhat inaccurate translation (the best modern English translation is by J. M. Cohen); it was performed at DORSET GARDEN in 1694 with music by Purcell. In 1895 IRVING played Don Quixote at the LYCEUM in a version by W. G. WILLS, and in 1969 *The Travails of Sancho Panza* by James SAUNDERS was the Christmas entertainment by the NATIONAL THEATRE company at the OLD VIC. An American musical *Man of La Mancha* began a three-year run in New York in 1965 and was seen in London in 1968. In France one of the most successful productions by BATY was his own dramatization of an episode from the book, *Dulcinée* (1938).

Cervantes himself said he wrote thirty plays, but apart from eight COMEDIAS, of which the best, *Pedro de Urdemalas*, was translated by Walter Starkie in 1964 as *Pedro the Artful Dodger*, and eight ENTREMESES, all of which were translated in 1948 by S. Griswold Morley and in 1964 by Edwin Honig, the only full-length plays to have survived are *El cerco de Numancia* (*The Siege of Numancia*), a heroic tragedy which was revived by BARRAULT in Paris in 1937, and *El trato de Argel* (*Business Affairs in Algiers*), based on the author's experiences there; these were probably produced between 1580 and 1590.

CÉSAIRE, AIMÉ (1913–   ), West Indian poet and dramatist, writing in French, whose efforts towards the restoration of racial self-confidence have had an impact on many Third World intellectuals. His verse and prose-poems react against rational modes of thought imposed on African peoples by Europe, and hold up the counterbalancing concept of 'negritude'. His use of torrential adjectives and close, assonant rhythms has made his plays difficult to accept on European stages. *Et les chiens se taisaient*, a 'drame-poème' published in 1946, is concerned with the inability of former slaves to express their ancestral history. *La Tragédie du roi Christophe* (1963) was produced by SERREAU. It dealt with the emotional politics and the search for freedom conducted by the first native ruler of a newly emancipated state, and *Une saison au Congo* (1967) took up a similar theme, applying it specifically to the assassinated leader Patrice Lumumba whose fate had been almost exactly predicted in the earlier play. Serreau also produced Césaire's adaptation of Shakespeare's *The Tempest* (1969), in which Caliban and Ariel stand for two kinds of negro revolt against the rational, conquering white man Prospero.

ČESKÝ KRUMLOV, Czechoslovakia, a castle about 100 miles from Prague which contains a small theatre built in 1766—the same year as that at DROTTNINGHOLM. It has a small, horseshoe-shaped auditorium and a magnificent baroque proscenium. The theatre's most exciting asset is ten sets of contemporary scenery, probably the oldest surviving in Europe. Painted by two artists from Vienna, Jan Wetschela and Leo Merkla, these are invaluable in showing how the sumptuous perspective scenes of such artists as the BIBIENAS were translated into wood and canvas. The sets, one of which is normally displayed on the stage, while others are set up in the adjacent riding-school, are for the usual scenes—palace, wood, harbour, and so on—and one, for a street, has for backdrop one of the angled views leading down two asymmetrical perspectives for which the Bibienas were famous. The stage machinery consists of axles, ropes, and winches for instantaneous changes of scenery, as at Drottningholm, and there are four TRAPS. Performances are occasionally given, but the precious and indeed unique scenery is too fragile to be used often.

CHAIKIN, JOSEPH, see OPEN THEATRE.

CHAILLOT, Palais de, Paris, Moresque building on the exhibition site opposite the Eiffel Tower, used by the first THÉÂTRE NATIONAL POPULAIRE under GÉMIER from 1920 to 1934. It was demolished in 1934 to make way for the present building, which was part of the complex built for the International Exhibition in Paris in 1937. It houses two museums as well as two theatres built into the hillside. The larger auditorium, originally seating 3,000 people, was used by the United Nations General Assembly from 1946 to 1952 and then for the next 20 years by the second Théâtre National Populaire under VILAR and then WILSON, whose simple but grandiose style of acting was strongly influenced by the proportions of the stage and auditorium. Closed in 1972, it became under the policy of DÉCENTRALISATION DRAMATIQUE one of the Théâtres Nationaux, the wasteful spaces of auditorium and access halls being restructured to provide a variable-shaped theatre space with dependent exhibition halls, acting areas, and studios.

CHAIRMAN, the presiding genius of the old MUSIC-HALL, and one of the last links with the tavern 'free-and-easy' from which it sprang. In full evening dress, he sat at a table at auditorium level with his back to the stage. A mirror enabled him to see what was going on there; a clock helped him to time the 'turns' and, if necessary, cut them short; and a gavel helped him to keep order. Usually an old 'pro.', of ruddy countenance and commanding presence, with a stentorian voice, he needed a ready wit and a gift for impromptu repartee. There were many famous chairmen, among them Jonghmann of the OXFORD, himself a singer and composer; W. H. Fair, known as 'Punch', famous for his singing of 'Tommy, Make Room for your Uncle'; Charles SLOMAN, who outlived his fame to end as chairman at the MIDDLESEX; and Walter Leaver of Canning Town, who in 1905, at the age of 75, was as magnificent and imposing as ever and probably the last of his kind, for the chairman, though to many synonymous with pre-1923 music-hall, had virtually disappeared by 1890 and had no place in the vast new Theatres of Variety. When in 1937 the PLAYERS' THEATRE tried to revive the atmosphere of the late Victorian music-hall, the chairman was Leonard Sachs, who performed the same function from the 1970s for television's 'The Good Old Days'.

CHALUPKA, SAMO, see CZECHOSLOVAKIA.

CHAMBERLAIN'S MEN, company of actors with which Shakespeare was most closely associated, and for which he wrote most of his plays. Founded in 1594, it took possession of the THEATRE, recently vacated by the ADMIRAL'S MEN, and remained there until the lease ran out in 1599. It then migrated to the new GLOBE THEATRE on Bankside, and with Richard BURBAGE as chief actor had a long and prosperous career apart from a little trouble over Shakespeare's *Richard II* which Elizabeth I, already angered by the Earl of Essex's unsuccessful rebellion, chose to regard as treasonable. However, the company is known to have played at Court not long before the Queen's death in 1603. With the accession of James I, the somewhat chaotic conditions of theatrical life in London were reorganized and the Chamberlain's Men, who were allowed to remain in sole occupation of their theatre, were appointed Grooms of the Chamber and renamed the KING'S MEN, coming under the direct patronage of the sovereign.

CHAMBERS OF RHETORIC, official organizations of rhetoricians which originated in Northern France as early as the 12th century and spread in the 13th century into the Low Countries (see BELGIUM and the NETHERLANDS), where their members were known as Rederykers. They became a forceful element in civic cultural life, some of the large towns having as many as five different chambers. From the beginning of the 15th century the Rederykers were responsible for dramatic contests at which both serious and farcical plays, some of them specially written and afterwards published, were performed. These festivals, in which the various Chambers competed among themselves, sometimes lasted for as much as a week. Most of the plays performed corresponded to the MYSTERY, MIRACLE, and MORALITY PLAYS of England and other European countries, but the farces were mainly indigenous. The Morality Plays gave rise to a genre peculiar to the Low Countries, the *spel van sinne*, which featured, together with the allegorical figures which motivated the main plot, a group of little imps, or *sinnekens*, who incited the characters to evil doing and so provided a satiric and comic element, somewhat akin to the Old VICE of the English theatre. It was the development of the *sinnekens* that eventually brought the rhetoricians in the southern half of the Low Countries into disrepute, but in the Netherlands they prospered until the beginning of the 17th century after which they declined in popularity and disappeared, the last outstanding playwright with whom they were associated being VONDEL.

CHAMBERS STREET THEATRE, New York, at 39–41 Chambers Street. Opened in 1844 as Palmo's Opera House, on the site of the popular Steppani's Arcade Baths, this was taken over by William E. BURTON, who opened it on 10 July 1848 as Burton's Chambers Street Theatre. He had particular success with plays based on novels by DICKENS and later with works by Shakespeare; but by 1852 he was suffering from the competition of WALLACK, who took away some of his best actors. Four years later Burton moved to the METROPOLITAN, and on 15 Sept. 1856 Edward EDDY mounted a season of farce and melodrama at Chambers Street. The house was renamed the American Theatre during the brief tenancy of E. L. DAVENPORT from 13 Feb. 1857: it finally closed on 30 March of that year.

'CHAMPAGNE CHARLIE', see LEYBOURNE, GEORGE.

CHAMPION, HARRY, see MUSIC-HALL.

CHAMPMESLÉ, CHARLES CHEVILLET (1642–1701), French actor and dramatist, who began his career in the provinces, later being seen both at the MARAIS and at the Hôtel de BOURGOGNE where he played mostly in tragedy, creating the role of Antiochus in RACINE's *Bérénice* (1670)—unless, as has been said, the author created it for him. He was a friend of La Fontaine, to whom several of his plays were once attributed. Among his many comedies the best is probably *Crispin Chevalier* (1673), a one-act farce based on the three-act *Les Grisettes* of 1671, but the most popular in its day was *Le Florentin* (1685), which was revived in Paris as late as the 1920s, together with *La Coupe enchantée* (1688). Champmeslé was the husband of one of the finest actresses of the day (below), and with his wife was a founder member of the COMÉDIE-FRANÇAISE.

CHAMPMESLÉ [Marie Desmares], Mlle (1643–98), French actress, wife of the above, whom she married in 1665 as her second husband. They were together in the company at the MARAIS in 1669, where the husband was at first accounted the better player; but his wife profited so much from the teaching of LAROQUE that in six months she was playing leading roles. She went with her husband to the Hôtel de BOURGOGNE in 1670, replacing Mlle DESOEILLETS, and became so popular that the rival troupes competed for her services. In 1679 she joined the amalgamated company formed after the death of MOLIÈRE and so became the leading lady in tragedy of the COMÉDIE-FRANÇAISE on its foundation in 1680, playing opposite BARON and remaining with the company until her death. She created many famous tragic roles, including the title roles in *Bérénice* (1670) and *Phèdre* (1677) by RACINE, whose mistress she was, and in the younger CORNEILLE's *Ariane* (1672). She favoured a chanting, sing-song declamation which she taught to her niece Charlotte DESMARES and to Mlle DUCLOS, the leading actresses of the next generation.

CHAMPS-ÉLYSÉES, Comédie des, see COMÉDIE DES CHAMPS-ÉLYSÉES.

CHANCEREL, LÉON (1886–1965), French dramatist and director, a pupil of Jacques COPEAU. He was also with JOUVET at the COMÉDIE DES CHAMPS-ÉLYSÉES in 1926, and from 1929 to 1939 directed the Comédiens Routiers, a touring company which played to young audiences. In 1935 he founded in Paris a company of young actors to play to children. Known as Le Théâtre de L'Oncle Sebastien, its plays were improvised, in deliberate imitation of the COMMEDIA DELL'ARTE. Chancerel, who succeeded Jouvet as President of the Société d'Histoire du Théâtre, was head of the Centre Dramatique in Paris (see DÉCENTRALISATION DRAMATIQUE). After his death a prize was established in his name to be awarded every two years to the writer of the best new play for children.

CHANFRAU, FRANK S. [Francis] (1824–84), American actor, best remembered for his playing of Mose the New York fireman, first seen at the OLYMPIC in Benjamin A. BAKER's *A Glance at New York in 1848*. Later in the year Chanfrau continued Mose's adventures in *The Mysteries and Miseries of New York*, seen at the CHATHAM THEATRE (2) of which he was lessee for a time, renaming it the National. He continued to appear as Mose for many years, in a series of plays written by Baker or himself. Among his other parts the best was the pioneer Kit Redding in Spencer and Tayleure's *Kit the Arkansas Traveller* (1870), which he constantly revived up to 1882. He married in 1858 the actress Henrietta Baker [Jeannette Davis] (1837–1909), who played Portia in *Julius Caesar* at the METROPOLITAN THEATRE in 1864, the only occasion on which Edwin BOOTH and his two brothers appeared on stage together. Her son Henry revived a number of his father's parts, playing Kit until 1890.

CHANFRAU'S NEW NATIONAL THEATRE, New York, see CHATHAM THEATRE (2).

CHANIN'S 46TH STREET THEATRE, New York, see FORTY-SIXTH STREET THEATRE.

CHANNING, CAROL, see MUSICAL COMEDY.

CHANTILLY [Marie-Justine-Benoiste Duronceray], Mlle (1727–72), French actress, wife of the dramatist and librettist Charles Simon Favart (1710–92), with whom she was at one time in the private company maintained by the Maréchal de Saxe. She played for many years at the COMÉDIE-ITALIENNE and is credited with the first introduction there of historical and local details into her costume, a practice adopted soon after at the COMÉDIE-FRANÇAISE by LEKAIN and Mlle CLAIRON.

CHAPEL STREET THEATRE, New York, on the south side and western end of Chapel Street, later Beekman Street (by which name the theatre was also known). This small theatre, measuring only 90 ft by 40 ft overall, was opened by David DOUGLASS on 19 Nov. 1761 after the failure of his CRUGER'S WHARF THEATRE, and it was here that the first known performance of *Hamlet* in New York was given on 26 Nov. with the younger HALLAM in the title role. After the departure of Douglass's company in May 1762, the theatre was used occasionally by amateurs, and by officers of the British garrison. In May 1766 a professional company was appearing in DODSLEY's *The King and the Miller of Mansfield* when the Sons of Liberty, an unruly band of anti-Britishers to whom all players were suspect, broke up the performance and greatly damaged the building, which was not used again.

CHAPMAN, GEORGE (*c.* 1560–1634), English poet and dramatist, author of the translation of Homer which inspired Keats's sonnet. He may be the original of Shakespeare's Holofernes in *Love's Labour's Lost* and Thersites in *Troilus and Cressi-*

da. Little is known of his early life, but in 1596 he was on HENSLOWE's payroll as accredited dramatist to the ADMIRAL'S MEN. Most of these early plays are lost, but enough remains of his later works, which were given mainly by the BOY COMPANIES so popular at the beginning of the 17th century, to ensure his acceptance as a major playwright. Among his comedies the best are probably *May-Day* (1602) and *All Fools* (*c*. 1604). Of the tragedies that have survived the most important is *Bussy d'Ambois* (also 1604), with its sequel *The Revenge of Bussy d'Ambois* (1610). Chapman was one of the authors, with JONSON and MARSTON, of the comedy *Eastward Ho!* (1605) which gave offence to James I and resulted in a term of imprisonment for himself and Jonson; and he narrowly escaped a second term of imprisonment for his *Charles, Duke of Byron* (1608) which upset the French Ambassador.

CHAPMAN, WILLIAM (1764–1839), one of the earliest and possibly the first of the American SHOWBOAT managers. Born in England, he was as a young man a member of a travelling company under Richardson which appeared mainly at fairs and country merrymakings. In 1803 he made his first appearance on the London stage and in 1827, with his wife and numerous children, went to New York, where he was first seen at the BOWERY THEATRE. He then started on tour for the south-west, and in Pittsburgh persuaded a Captain Brown to build for him a 'floating theatre' on which he travelled up and down the Ohio and Mississippi rivers from 1831 until his death. His widow and sons continued to run the business until in about 1847 they sold out to Sol SMITH. Two years before they had opened the first Showboat in New York, known as Chapman's Temple of the Muses. It was built on a flat-bottomed steam packet, the *Virginia*, 90 ft long, 52 ft wide, and 50 ft high, seating 1,200 people. The proscenium was 27 ft wide, with a stage area 42 ft wide by 45 ft deep. When it opened on 2 Apr. 1845 it was anchored between Canal and Charlton Streets, but it moved from place to place, the audience locating it by the Drummond Light which was lit only on performance nights. It does not appear to have been used for more than about six months. One of the Chapman sons, George, who married a member of his father's company, a widow named Mary Park, headed west in 1851 and with San Francisco as his base established a coastal circuit. He also, in 1860, opened the first Canadian theatre west of Ontario, the Colonial in Victoria.

CHAPPUZEAU, SAMUEL (1625–1701), a French man of letters, author of *Le Théâtre françois* (1674), which consists of three parts: *De l'usage de la comédie*, a defence of the theatre; *Des auteurs qui soutiennent le théâtre*, a dictionary of dramatists; and *De la conduite des comédiens*, an apology for players, with lives of many of the best-known up to the date of publication. This third part contains a long chapter on MOLIÈRE, who had just died. Though somewhat inaccurate, the work as a whole is an important source-book of the 17th-century French theatre. An early farce by Chappuzeau, published in Lyons in 1656, seems to have had some influence on Molière's *Les Précieuses ridicules* (1658).

CHAPTER ARTS CENTRE, Cardiff, see WALES.

CHARING CROSS MUSIC-HALL, see GATTI'S.

CHARING CROSS THEATRE, in King William Street, London, originally the Polygraphic Hall. Nothing of importance happened to it under its new name until in 1872 John Sleeper CLARKE took it over and staged there a revival of SHERIDAN's *The Rivals* in which he played Bob Acres to the Mrs Malaprop of Mrs STIRLING, her first appearance in what was to prove her finest role. Alexander Henderson took the theatre over in 1876, renamed it the Folly, and produced there a series of BURLESQUES starring his wife Lydia Thompson. In 1879 J. L. TOOLE took on the management, giving the house his own name in 1882. He too staged a number of burlesques, mainly by H. J. BYRON, but also some straight plays, among them two of PINERO's early comedies, which were not successful, and BARRIE's first play, *Walker, London* (1895). The last production by Toole was a comedy by Ralph Lumley, *Thoroughbred* (also 1895). After it closed later the same year the building, which Toole had improved and enlarged to hold about 900 people, was demolished to make way for an extension of the Charing Cross Hospital.

CHARIOT-AND-POLE, see CARRIAGE-AND-FRAME.

CHARKE, CHARLOTTE (1713–*c*. 1760), English actress, the daughter of Colley CIBBER. Intensely masculine and scorning all pursuits except hunting and shooting, she made herself conspicuous in London, being even more eccentric than her brother Theophilus. Married at 16 to a violinist at DRURY LANE and widowed eight years later, she went on the stage, quarrelled with the London managers, and ran away, disguising herself as a man and playing both men's and women's parts in strolling companies. She was also a conjuror's assistant in Petticoat Lane, a puppet-master, a performer at fairs, and finally a tavern-keeper in Drury Lane, near the theatre. In 1755 she published a remarkable *Narrative of the Life of Mrs Charlotte Charke*.

CHARLES HOPKINS THEATRE, New York, see PUNCH AND JUDY THEATRE.

CHARLESTON, South Carolina, see DOCK STREET THEATRE.

CHARLOT, ANDRÉ, see REVUE.

CHARON, JACQUES (1920–75), French actor and director, who began his career as assistant to Gaston BATY and later became a member of the COMÉDIE-FRANÇAISE and one of its outstanding directors, specializing in farce and VAUDEVILLE. He was first seen in London in 1964 when he appeared in his own production of FEYDEAU's *Un Fil à la patte* during the WORLD THEATRE SEASON at the ALDWYCH. He directed the same author's *La Puce à l'oreille* in Paris in 1966, and in the same year at the OLD VIC in London in an English translation entitled *A Flea in Her Ear* by John MORTIMER. During the 1967 World Theatre Season the Comédie-Française again appeared in his production of *Feu la Mère de Madame*.

CHÂSSIS A DÉVELOPPEMENT, an extension of the FALLING FLAPS method, used in the 19th century to transform a set piece instantaneously before the eyes of the audience. In its simplest form, the set piece had half its subject painted on the normal surface and the other half on one face of a vertically hinged flap. By means of a trick-line, this flap was swung over, like the page of a book, to lie over the first half of the piece, revealing a new subject painted on the remaining portion and the reverse face of the hinged flap. In trick set-pieces a series of separate flaps were all spring-hinged together to lie compactly flat when folded; upon release all flew out to present not only a totally different painted subject, but one which considerably exceeded the original piece in size.

CHATHAM THEATRE, New York. (1) Between Duane and Pearl Streets. Originally known as the Pavilion and opened in 1823, this was built by Barrière, a pastrycook who had some years previously obtained a licence to provide music and other entertainments in a pleasure resort named Chatham Gardens. It seated 1,300 persons, was the first theatre in New York to be lit by gas, and soon proved a serious rival to the old-established PARK THEATRE (1). The original company included the first Joseph JEFFERSON, who made his last appearance in New York at this theatre on 4 Oct. 1825, and his son, the second Joseph, and also Henry WALLACK who, with James HACKETT, took over the management after Barrière died on 21 Feb. 1826. He was not successful, and in 1830 the theatre, which had in 1829 been renamed the American Opera House, was acquired by BLANCHARD, who ran it with a mixture of straight and EQUESTRIAN DRAMA. Under his successor, HAMBLIN, it finally closed in 1832 and became a Presbyterian chapel.

(2) A New Chatham Theatre, on the south-east side of Chatham Street between Roosevelt and James Streets, opened on 11 Sept. 1839 under the management of James Anderson, an actor from the BOWERY. A large house, seating 2,200, it soon found itself in financial difficulties; it was rescued by the elder Charles THORNE, and prospered for a time with somewhat mixed bills, Shakespeare one week being followed by Negro farce the next. Thorne retired in 1843 and in 1848 the comedian Frank CHANFRAU took the theatre over, renaming it the New National Theatre. It was henceforth known as the National, a name it retained under its next manager, Purdy. It was at this theatre that Edwin BOOTH made his first appearance in New York on 27 Sept. 1850. One of the great successes here was AIKEN's dramatization in 1853 of Harriet Beecher Stowe's anti-slavery novel *Uncle Tom's Cabin* (publ. 1852), and during Purdy's last season in 1859 Adah Isaacs MENKEN, as a young and inexperienced actress, made her first appearance in New York. The building was severely damaged by fire on 9 July 1860, but continued in use as the Union Theatre, the National Concert Hall, and once again the Chatham. It finally became the National Music Hall and in Oct. 1862 it was pulled down.

CHAUVE-SOURIS, see BALIEFF.

CHAYEVSKY, PADDY, see UNITED STATES OF AMERICA.

CHEESEMAN, PETER, see VICTORIA THEATRE, STOKE-ON-TRENT.

CHEKHOV, ANTON PAVLOVICH (1860–1904), Russian dramatist, possibly the one best known outside Russia, whose plays are in the repertory of every country. He came of humble parentage, but graduated as a doctor from Moscow University in 1884, and always thought of himself as more of a doctor than a writer. Well known for his short stories even during his student days, he was early attracted by the theatre, and his first plays were one-act comedies—*The Bear* (1888), *The Proposal* (1889), *The Wedding* (1890)—for which he always retained an amused affection. His first full-length plays, *Ivanov* (1887) and *The Wood-Demon* (1889), were unsuccessful, as was *The Seagull* (1896) when performed at the old-fashioned Alexandrinsky Theatre in St Petersburg (now the PUSHKIN THEATRE, Leningrad); it had nothing in common with the plays currently popular and was incomprehensible to actors trained in the old traditions. Chekhov might have given up the theatre entirely had not NEMIROVICH-DANCHENKO persuaded him to let the newly founded MOSCOW ART THEATRE revive *The Seagull*. The production was a success, and was followed by *Uncle Vanya* (1899) (a revision of *The Wood-Demon*), *Three Sisters* (1901); and *The Cherry Orchard* (1904), in all of which Chekhov's wife Olga KNIPPER played leading parts. Chekhov died shortly afterwards, at the height of his powers.

His plays portray the constant attrition of daily life and the waste, under the social conditions of Old Russia, of youthful energy and talent. At the same time they contain a note of hope for the future which is heavily stressed in modern Russian productions. This hopefulness seems to accord with Chekhov's own view of his plays, but not with STANISLAVSKY's, who wrote that he wept when he first read *The Cherry Orchard* and who conveyed to its first audience his own impressions of regret and impermanence. But however one interprets the plays, they demand a subtlety of ensemble playing which was not available in Russia until the founding of the Moscow Art Theatre. Chekhov may still, particularly in translation, be falsified, but gradually the truth of his work, both for his own time and for ever, is imposing itself on his interpreters. The forces at work in Chekhov's dramas are those which produced the Russian Revolution, and the great speech by the student Trofimov at the end of Act II of *The Cherry Orchard*—the celebrated 'All Russia is our garden'—is not only a concise review of then recent Russian history, but an astonishingly accurate prediction of the Revolution itself.

As far as can be ascertained, the first of Chekhov's plays to be acted in English was *The Seagull* (GLASGOW REPERTORY THEATRE, 1909, and London LITTLE THEATRE, 1912). The STAGE SOCIETY was persuaded by SHAW to produce *The Cherry Orchard* in 1911 and *Uncle Vanya* in 1914. These early productions failed, but helped to make Chekhov's work known to actors and critics alike. *Three Sisters* was produced in London for the first time in 1920, but the first production to be commercially successful was *The Cherry Orchard*, brought to London in 1925 by FAGAN from the OXFORD PLAYHOUSE. Even more successful was a series of productions by KOMISARJEVSKY at the BARNES THEATRE (*Ivanov*, 1925; *Uncle Vanya, Three Sisters, The Cherry Orchard*, 1926), which finally naturalized Chekhov in England. The early unfinished play *Platonov*, first produced in Russia in the 1920s, was seen in London in 1960, after a preliminary production at the NOTTINGHAM PLAYHOUSE. The Moscow Art Theatre, on visits to London, played *The Cherry Orchard* in Russian in 1928 at the GARRICK THEATRE and in 1958 at SADLER'S WELLS, together with *Uncle Vanya* and *Three Sisters. The Cherry Orchard* was seen again in 1964. America's introduction to Chekhov was in Russian, when Paul ORLENEV and Alla NAZIMOVA appeared in *The Seagull* in 1905. In an English translation, this was produced by the WASHINGTON SQUARE PLAYERS in 1916, but American interest in Chekhov dates from the visits of the Moscow Art Theatre in 1923 and 1924, and subsequent American productions have been strongly influenced by Stanislavsky.

There have been a number of translations of Chekhov's plays into English. The standard acting versions for many years were those of Constance Garnett (1923), which have frequently been used as the basis of new adaptations. The Penguin Classics published versions of *The Cherry Orchard, Three Sisters*, and *Ivanov* by Elisaveta Fen. All Chekhov's plays are contained in the first three volumes of the Oxford Chekhov, in translations by Ronald Hingley based on the definitive Moscow edition of 1944–51.

CHEKHOV, MICHAEL ALEXANDROVICH (1891–1955), Russian actor and director, nephew of the above. He joined the company of the MOSCOW ART THEATRE in 1910, having previously had some experience in St Petersburg, and in 1917 came under the influence of VAKHTANGOV, director of the theatre's Studio One. Chekhov himself took over the directorship of this group in 1924, by which time he had moved away from STANISLAVSKY's naturalism, believing that the artist's task was limited to knowing how to copy nature more or less accurately. His ideas were attacked by Communist Party critics, and in 1927 he left Russia for England, where in 1936 he founded a school of dramatic art at Dartington Hall in Devon. He moved to the United States in 1939, where he again founded a school of acting and also appeared in a number of Hollywood films. He published three volumes detailing his ideas on theatre aesthetics—*The Path of the Actor* (1928), *To the Actor* (1935), and *To the Director and Dramatist*, which appeared posthumously in 1963.

CHELIDONIUS, BENEDICTUS, Austrian humanist of the early 16th century and Abbot of the Schottenkloster in VIENNA. His *Voluptatis cum virtute disceptatio* (1515) affords an early example of the drama of the religious orders, with its tendency to present a moral debate in dramatic form: Greek gods and goddesses act out an allegorical tale in which the conflict of virtue and vice is resolved in Christian terms.

CHELSEA THEATRE, London, see ROYAL COURT THEATRE (1).

CHELSEA THEATRE CENTER, New York, an off-off-Broadway company founded in 1965 to stage new American and foreign plays and the lesser known works of classic authors. It began operations in the Chelsea area of Manhattan, but in 1968 moved to the Brooklyn Academy of Music, becoming that institution's resident acting ensemble. From 1973 to 1978 the company was responsible for productions in both Brooklyn and Manhattan, but from 1978 until 1981, when financial difficulties caused it to close down, it functioned only in Manhattan. Notable productions by the group included Imamu Barak's *Slaveship* in 1969, GENET's *The Screens* in 1971, GAY's *The Beggar's Opera* and Allen Ginsberg's *Kaddish* in 1972, David STOREY's *The Contractor* and the revised musical version of VOLTAIRE's *Candide* (which later had a long run on Broadway) in 1973, and KLEIST's *The Prince of Homburg* in 1976.

CHÉNIER, MARIE-JOSEPH (1764–1811), French dramatist (younger brother of the poet André Chénier, who was guillotined in 1794), and one of the few important literary figures of the French Revolution. He had already produced some unsuccessful plays when, after a battle with the censor lasting two years, his *Charles IX* was performed in 1789. It was enthusiastically received, mainly owing to the magnificent acting of TALMA, who later opened the Théâtre de la République (see COMÉDIE-FRANÇAISE) with Chénier's *Henri VIII* (1791), following it in the same year with his *Jean Calas*, and in 1792 with *Caius Gracchus*, another revolutionary play well suited to the temper of the time. Chénier was active in politics and, as a member of the Convention, he voted for the death of Louis XVI. His political career came to an end in 1802, when he opposed the rise of Napoleon. He continued to write, but without success, and his *Timoléon*, first performed in 1792, was proscribed and burnt by order of the censor. Only one copy, saved by the actress Françoise VESTRIS, who much admired him, escaped destruction.

CHERIF, CIPRIANO RIVAS, see SPAIN.

CHERKASSOV, NIKOLAI KONSTANTINOVICH (1903–66), Soviet actor, who began his career at the Leningrad Theatre of Young Spectators in 1926, where he played CERVANTES' Don Quixote. Between 1927 and 1932 he made his name in the music-halls of Moscow and Leningrad as an eccentric comedian, but in 1933 he joined the company at Leningrad's PUSHKIN THEATRE where he remained until his death. During the 1950s and early 1960s he appeared in a number of new Soviet plays, but was best known in classical roles. Internationally he became famous for his appearances as Alexander Nevsky and Ivan the Terrible in Eisenstein's films.

CHERRY LANE THEATRE, New York, at 38 Commerce Street. This early 19th-century building was converted into a theatre in 1924, the first production on 10 Feb. being Robert H. Presnell's *Saturday Night*. In 1927 the New Playwrights Theatre took over the house, gave it their own name, and opened with *The Belt* by Paul Sifton, with settings designed by John Dos Passos. The theatre reverted to its original name in 1928 and managed to survive the years of the Depression by a widespread use of reduced price ticket vouchers. In 1945 the theatre was leased by The Spur, an experimental company which opened with O'CASEY's *Juno and the Paycock*. Since then a number of companies have made the Cherry Lane their headquarters—On-Stage Productions in the late 1940s. Proscenium Productions in the 1950s, and Theatre 1960 and Theatre 1967 in the 1960s, when the first productions in New York of Edward ALBEE's *The American Dream* and *The Death of Bessie Smith*, Samuel BECKETT's *Play*, and Harold PINTER's *The Lover* were presented. Other notable productions have included *To Be Young, Gifted and Black* (1969), adapted by Robert Nemiroff from the speeches, journals, and letters of Lorraine Hansberry; Edward BOND's *Saved* (1970); a double bill in 1973 of *Picnic On the Battlefield* by ARRABAL and *Ionescopade*, adapted from the works of IONESCO; and the long-running *The Passion of Dracula* (1977), adapted by Bob Hall and David Richmond from Bram Stoker's novel.

CHESTNUT STREET THEATRE, Philadelphia, Pennsylvania, America's first theatre in the English PROSCENIUM style, was built in 1793 for a company brought from England by Thomas WIGNELL, but was not opened until 17 Feb. 1794 because of an epidemic of yellow fever. A copy of the Theatre Royal at BATH, it was an elegant and impressive building holding about 2,000, with numerous dressing-rooms and two green-rooms. Wignell remained there until his death in 1803, and was succeeded by William WOOD and the elder William WARREN as joint managers. Under them the theatre continued to prosper, and enjoyed almost a monopoly of acting in Philadelphia until the opening of the WALNUT STREET THEATRE in 1811. Sometimes known, from its pre-eminent position in American theatrical life, as 'Old Drury', it was one of the first theatres in the United States to be lit by gas (on 25 Nov. 1816). Wood's retirement and the success of the Walnut Street and Arch Street theatres, finally forced the theatre into bankruptcy, and the company was disbanded. The building was then occupied by a stock company which supported visiting stars until on 2 Apr. 1820 it was damaged by fire. It reopened on 2 Dec. 1822, but on 1 May 1855 was again burned down and it was not rebuilt until 1863. The last performance was given there in 1910, and the building was finally demolished in 1917.

CHETTLE, HENRY (*c*.1560–1607), prolific Elizabethan playwright, credited by HENSLOWE in his diary with a long list of plays, mostly written in collaboration and now lost. Of those that survive, the only one he appears to have produced unaided was *Hoffman*, known also as *Revenge for a Father* (*c*. 1602), which follows the pattern of the REVENGE TRAGEDY set by KYD. Chettle was also a printer, in which capacity he published the pamphlets of NASHE as well as *A Groatsworth of Wit* (1592), which contains the famous attack on Shakespeare by its author Robert GREENE.

CHEVALIER, ALBERT (1861–1923), one of the outstanding stars of the British MUSIC-HALL, who was originally an actor, appearing on stage as a child, and making his professional début at the age of 13. After a successful and varied career in comedy, BURLÉSQUE, and MELODRAMA, he was persuaded by his friend Charlie Coburn, already a well known name in the entertainment world, to go on the 'halls'. He made his first appearance at the LONDON PAVILION in 1891 singing his own composition 'The Coster's Serenade' and was immediately successful, being seen at the TIVOLI when it opened a year later and finally establishing his reputation as a fine singer of Cockney songs. Although he later presented other characters, including a Chelsea Pensioner, a yokel, and a curate, it was as a Cockney that he gained his immense popularity, with 'Knocked 'Em in the Old Kent Road', 'The Narsty Way 'E Sez It', 'It Gits Me Talked Abaht', 'Appy 'Ampstead', and above all 'My Old Dutch', for which he wrote the words himself, his brother Charles Ingle supplying the music. Chevalier was also the author of a number of sentimental ballads and sketches and of one unsuccessful play, *The Land of Nod* (1898). He was one of the few music-hall stars who never appeared in PANTOMIME.

CHEVALIER, MAURICE, see MUSIC-HALL.

CHIADO, ANTONIO RIBERO, see PORTUGAL.

CHIARELLI, LUIGI (1884–1947), Italian dramatist, the chief exponent of the Teatro GROTTESCO, which took its name from Chiarelli's own description of his best known play *La maschera e il volto* (1916, written in 1913) as '*grottesco in tre atti*'. This study of man in modern society, where hypocrisy is rampant and all true feeling must be hidden under an expressionless mask, was seen in London and then in New York in 1924 as *The Mask and the Face*. In 1931 a play on a similar theme, *Fuochi d'artificio* (*Fireworks*, 1923), was produced in London as *Money, Money!* Chiarelli's other plays include *La morte degli amanti* (*The Death of the Lovers*, 1921), *K.41* (1929), a frankly commercial melodrama, and *La reginetta* (*The Little Queen*, 1931). *Una più due* (*One Plus Two*, 1935), a sophisticated light comedy, has great pace and unexpected depth, while *Il cerchio magico* (*The Magic Circle*, 1937) is a whimsical trifle in the manner of BARRIE or VAN DRUTEN. *Essere* (*To Be*) was produced posthumously in 1953.

CHICAGO, Illinois, was little more than a village until about 1830. Its first theatre, the Rialto, opened in 1838 under the management of John B. Rice and closed in 1840, the second Joseph JEFFERSON being a member of the company. Rice's own theatre was built in 1847 and burned down three years later. In 1857 John H. McVicker built what was described as 'the most substantial, convenient, safe, and costly theatre building standing in the west', which was destroyed in the disastrous fire of 1871 and immediately rebuilt on the same site. Ten years later J. H. Haverly opened a theatre which he hoped would outdo McVicker's, though it seated only 2,000 against its rival's 2,600. In 1885 IRVING's company on its first American tour appeared at Haverly's, which was then rechristened the Columbia Theatre at a ceremony presided over by Ellen TERRY. The Auditorium, opened in 1889, was reputed to be the largest theatre in the United States at the time, seating 4,500 and having outstandingly good acoustics. The vast Spectatorium planned by Steele MACKAYE for the Chicago Exposition of 1893 was abandoned owing to the financial panic of that year, and the unfinished building was sold as junk for a nominal sum. From 1900 Chicago became almost totally dependent on New York for theatrical entertainment, through road shows and through local companies which offered re-runs of Broadway plays. The opening of the GOODMAN MEMORIAL THEATRE in 1925 improved the city's theatrical reputation, but it tended to be regarded mainly as a staging post on the touring circuit until the 1970s, when a big growth in theatre audiences was stimulated by the formation of groups such as the St Nicholas Theatre Company. The Organic Theatre, founded in 1970, introduced its own brand of adaptive, innovative theatre, and the Academy Festival Theatre at Lake Forest, founded in 1973, runs a June-Sept. season. The Auditorium, closed since the 1930s, was restored and reopened in the 1960s and is now used mainly for concerts and conferences. The League of Chicago Theatres, founded in 1979, includes over 30 theatrical organizations, and a number of productions originating in the city have achieved successful runs in New York.

The Chicago-born dramatist David Mamet (1947–) was a co-founder of the St Nicholas Theatre Company, for which he was also playwright in residence and, from 1973 to 1976, Artistic Director; several of his plays were premièred by the company, including the one-act *Duck Variations* and *Sexual Perversity in Chicago* (1974), both seen in New York in 1975. In 1978 he became Associate Artistic Director and playwright in residence at the Goodman. His comedy *American Buffalo*, about the abortive plans of two minor crooks to steal a coin collection, received its world première at the Goodman in 1975; he made his Broadway début with the play in 1977 and it was seen at the COTTESLOE THEATRE in London in 1978. In *A Life In the Theatre* (1977), which also had its world première at the Goodman and went on to New York, two actors of different generations have dressing-room encounters at various stages in their lives, parts of plays in which they are currently appearing being sometimes seen.

CHICANO THEATRE, U.S.A., the name given to the manifestations of the Chicano (Mexican-American) theatre groups which cater for the Chicano community in a bilingual mixture of Spanish and English, with local colloquialisms added for regional flavour. Most of their productions are improvised, in the style of the COMMEDIA DELL'ARTE, and deal with such immediate problems as unemployment, housing, education, drug abuse, low wages, and racial discrimination. The movement began in 1965 with the founding in California of the Teatro Campesino (Fieldworkers' Theatre). Under the direction of Luis Valdéz a group of workers from the vineyards, who were on strike, toured in a programme of short sketches (*actos*) in an effort to recruit other workers to join the strike. Two years later they extended their activities to the whole of the United States, raising funds for the fieldworkers' union by playing on campuses, in theatres, and even on the steps of the Senate House, and eventually breaking off their connection with the union in order to widen their scope and deal with issues that affected the whole Chicano community. In 1969 they took part in the Seventh World Theatre Festival in Nancy, France, and in 1970 organized the first annual Chicano Theatre Festival. A year later the group moved to a permanent home in San Juan Bautista, California, where they devote their whole time to the movement, and in 1972 presented an improvised play based on COLLECTIVE CREATION entitled *La Gran Carpa de los Rasquachis* (*The Great Tent of the Underdogs*).

Although the first Chicano Theatre actors were field workers, later groups have been formed mainly of Chicano college or university students. One of the most important is the Teatro de los Barrios (*barrios* being Chicano communities or neighbourhoods), formed in 1969 by a number of high school students in San Antonio, Texas. Linked at first with a local Chicano political party, they became involved with the wider movement in 1972, when they attended the third annual Chicano Theatre Festival in Orange County, and also participated in the fourth (in San José) and the fifth (in Mexico City). By 1974 most of its members were college students, and the group was selected to act as host to the sixth annual festival in 1975.

Other groups are the Teatro del Piojo (Theatre of the Louse), formed in 1970 by students from the University of Washington in Seattle, which remained active in the American Northwest for ten years; the Teatro de la Gente (Theatre of the People), founded in 1970 in San José, which under the direction of Manuel Martinez and Adrian Vargas later became an important part of the national organization; and the Teatro de la Esperanza (Theatre of Hope), founded in 1971 by students of the University of California at Santa Barbara, under the direction of Jorge Huerta. Members of this group first appeared in songs and *actos* written for the Teatro Campesino, but later developed their own collective creations. They are now professional actors, performing their bilingual programmes on a full time basis throughout the United States, in Mexico, and in various European countries.

In 1971 the various *teatros* of the Chicano Theatre movement formed an organization, the Teatro National de Aztlan, uniting all the groups from coast to coast. Known as Tenaz, this organizes the yearly festivals and regional conferences and provides workshops in an effort to strengthen the lines of communication between its various members. There are now approximately 100 *teatros* in the United States, the majority of them in the South-West.

CHICHESTER FESTIVAL THEATRE. The enterprise shown in the founding of the STRATFORD (ONTARIO) FESTIVAL Theatre in Canada inspired a citizen of Chichester, Leslie Evershed-Martin, to plan a somewhat similar theatre for his own city. The money for its construction was raised mainly from private and commercial sources and from the ARTS COUNCIL. The hexagonal building, which is situated in Oaklands Park, has an auditorium seating 1,394 round a thrust stage, no one being more than 65ft from it. The stage width is 31ft 8ins, its depth 39ft. It has a semi-permanent balcony and a catwalk all round the interior wall; there is good backstage accommodation, together with various stores and small workshops. Four productions are presented annually. The theatre opened on 3 July 1962 under Laurence OLIVIER, his company later forming the nucleus of the first NATIONAL THEATRE. The most successful play of that season was CHEKHOV's *Uncle Vanya*, the others being FORD's *The Broken Heart* and FLETCHER's *The Chances*, in the Duke of BUCKINGHAM's version. The first new play to be given was *The Workhouse Donkey* (1963) by John ARDEN, whose *Armstrong's Last Goodnight* (1965) was also first seen at Chichester as were SHAFFER's *The Royal Hunt of the Sun* (1964) and *Black Comedy* (1965). Olivier resigned at the end of 1965 and was succeeded by John CLEMENTS, who appeared in a number of productions himself and also directed BRECHT's *The Caucasian Chalk Circle*, with TOPOL, PINERO's *The Magistrate*, with Alastair SIM, WYCHERLEY's *The Country Wife*, with Maggie SMITH (all 1969), and BOLT's *Vivat! Vivat Regina!* (1970). In 1974 Keith MICHELL took over as director, plays given under him including TURGENEV's *A Month in the Country* (1974), with Dorothy TUTIN, IBSEN's *An Enemy of the People* (1975) with Donald SINDEN, MAUGHAM's *The Circle* (1976), and N. C. HUNTER's *Waters of the Moon* (1977), with Ingrid Bergman. Peter Dews's seasons as director, 1978–80, brought revivals of Michael REDGRAVE's adaptation of Henry JAMES's *The Aspern Papers* and LONSDALE's *The Last of Mrs Cheyney*. He was succeeded at the end of 1980 by Patrick Garland. Many of the productions at Chichester have subsequently been seen in London. The theatre is used for Sunday concerts and poetry readings, and during the winter by local amateur and visiting companies.

CHIKAMATSU MONZAEMON, see JAPAN.

CHILDREN OF BLACKFRIARS, OF THE CHAPEL, OF PAUL'S, OF THE REVELS, OF WHITEFRIARS, OF WINDSOR, see BOY COMPANIES.

CHILDREN'S THEATRE, term used for performances given by adult professional or amateur actors and puppeteers for young people, either in theatres or in adapted halls. It does not cover the activities of such groups as the BOY COMPANIES, or amateur performances by schoolchildren in public, for which see JESUIT DRAMA, SCHOOL DRAMA, and the WESTMINSTER PLAY; nor has it anything to do with the modern use of drama in schools as part of the educational curriculum. The activities of young people in youth clubs come under the heading of AMATEUR DRAMA, and of those training for the adult theatre under SCHOOLS OF DRAMA.

It is self-evident that to introduce young people to good theatre as early as possible is the surest way of building up an adult audience for the future, but only in countries where such enterprises are fully state-subsidized can stable conditions and steady progress be found; such companies cannot be entirely and continuously self-supporting. In all Communist countries children's companies—such as, in the Soviet Union, the Central Children's Theatre (founded in 1921) and the Moscow Theatre of the Young Spectator (1929)—are heavily subsidized, often having their own theatres and touring organizations, the actors being specially trained for their work; the companies are used not only for entertainment but also for socialist propaganda and mass education. Apart from the Soviet Union, Communist countries active in children's theatre include Czechoslovakia and the German Democratic Republic, where the Theater der Freundschaft in East Berlin is especially notable.

The international organization for children's theatre is ASSITEJ (Association Internationale du Théâtre pour l'Enfance et la Jeunesse), based in Paris, with members in almost 40 countries. Scandinavian countries were early in the field, and Sweden is especially active. France made a good start with Léon CHANCEREL's Théâtre de l'Oncle Sebastien, which lasted from 1935 to 1939, and now has a number of children's theatres such as the Théâtre des Jeunes Années in Lyons and the Théâtre du Gros Caillou in Caen. Children's theatre also flourishes in the Federal Republic of Germany, where the Gripstheater in West Berlin is outstanding, and in Japan. In Germany, as in most non-Communist countries where the theatre is heavily subsidized, young people are able to buy tickets for normal theatre-going very cheaply.

In Britain there have been many attempts to found a permanent children's theatre, but lacking sufficient funds they have usually ended with the retirement or death of the founder. Among them were Bertha Waddell's Scottish Children's Theatre, founded in 1927, which survived until 1968; Jean Sterling Mackinlay's afternoon seasons for six weeks over Christmas from 1914 to 1939; and the English School Theatre (partly educational, since it was recognized by the then Board of Education and supported by some local authorities), which was founded in 1936 but had to be disbanded during the Second World War. Still in existence are the Unicorn Theatre for Children, founded in 1948, run from 1962 until her death in 1973 by Caryl Jenner; the Theatre Centre (formerly the London Children's Theatre), founded in 1954, which tours schools; and the Little Angel Marionette Theatre, founded in 1961. Both the NATIONAL YOUTH THEATRE and the YOUNG VIC do excellent work, and many regional theatres run Theatre in Education schemes. The national association for children's theatre is NADECT (National Association for Drama in Education and Children's Theatre, formerly the British Children's Theatre Association), founded in 1959; the Young Theatre Association is concerned with amateur children's theatre.

Children's theatre in the United States—which, as in Britain, receives very little subsidy—has depended heavily on individuals such as the late Charlotte Chorpenning, who headed the Goodman Children's Theatre in Chicago for 21 years. There are now many children's theatres, among them the Children's Theatre Company and School in Minneapolis, founded in 1961 as the Moppet Players. Many of the regional theatres, like their British counterparts, work in collaboration with the schools. Children's theatre thrives also in Canada.

CHINA. In China, as elsewhere, dramatic performances were of ancient ritual origin. As early as the Han dynasty, 206 BC to AD 220, singing and dancing were apparently combined with the dramatization of a story, and the famous Tang emperor Ming Huang is credited with the foundation of the first official school of drama, known as the 'Pear Garden', in AD 20. Traditional Chinese drama included many types of local theatre (*difnagxi*), classical drama (*kunqu*), and above all popular drama, which in the early 19th century replaced the older classical drama as the dominant form of traditional theatre. Now known as the Peking Opera, this amalgam of spoken dialogue, operatic singing, dancing, and acrobatics, continues to flourish. The plays, roughly classified as civil (*wen*) or military (*wu*) and introduced by the banging of a hand-gong, heralded by a clash of cymbals, are written in a theatrical dialect of combined colloquial and literary Chinese: sung passages are in verse; spoken dialogue breaks the tension, elucidates the plot, and gives the singer a rest. When the protagonist appears he chants a prologue and prefatory poem introducing himself, his names, family, circumstances, motives, and intentions. A character speaks until his gathering emotions plunge him into melody; the last spoken word is a prolonged *crescendo* as a signal to the orchestra. Many plays are performed repeatedly to tireless, knowledgeable audiences. They are divided not into acts but into numerous scenes of varying length, and the UNITIES of time and place are disregarded. This unlocalized drama allows the Chinese playwright, as it allowed the Elizabethan, to indulge in loose, flowing construction, episodic plots, and complex action. The subject-matter of the plays is mainly derived from legends, historical anecdotes, and famous novels, and always makes some pretence of pointing a moral. Goodness is rewarded, wrongs are redressed, and evil punished—eventually. There is no clear distinction between comedy and tragedy, and the majority of traditional plays may be described as melodramas with happy endings.

Over the centuries Chinese actors have developed a technique that renders scenery superfluous. The projecting stage, which is almost square, is flanked by two curtained doors for entrances and exits and has an embroidered or painted backcloth. Stage properties are minimal and, like almost everything connected with the Chinese theatre, have been reduced to an elaborate set of conventional symbols. An ordinary table may stand for an altar, a judge's bench, a bridge, or a banqueting board. It may also represent a mountain to be climbed or battlements to be scaled; by jumping over it the actor makes it a wall. To indicate a change of locality he treads in a circle. Among his important stage properties are a tasselled horsewhip, which enables him to ride an invisible horse; a horsehair duster, which symbolizes spirituality and is carried by deities and religious characters; and an oar, which represents a boat. Large banners in groups of four, inscribed with characters denoting military rank, represent an army; coloured blue, they indicate water; black, a violent wind. A fan is usually a sign of frivolity. Other props, such as military weapons, state umbrellas, shop-signs, lanterns, or candles, are more realistic. Scenic problems are solved by vivid pantomime. Every movement of the actor, standing on a bare, projecting stage and viewed spatially, is sculptured to combine grace and elegance of line with maximum significance. His strenuous training begins at the age of ten, or even earlier, and from his physique and talents he is eventually selected for the type of role to which he is most suited, and then meticulously rehearsed in it for many years before appearing on stage.

The characters which he is trained to interpret are classified under four main types; *sheng*, males in general; *dan*, females in general; *jing* or *hualian*, robust males with faces painted like masks; and *chou*, broad comedians, with a number of precise subdivisions. From 1850 to about 1910 the righteous, elderly, bearded male role (*laosheng*) was the most important, and there were no less than four distinct methods of interpreting the part; but with the meteoric rise to fame of MEI LANFANG in the 1920s, the female impersonator dominated the cast. From the earliest times, as in Elizabethan England, young men played the women's parts. Neo-Confucian prejudice against the mingling of the sexes on stage had encouraged this practice, facilitated by the slender build characteristic of Chinese men and their exiguous growth of whiskers. Even after the advent of the actress characters in a play were at first all represented by one sex, predominantly male; only a few plays were acted by women only. Today, however, it is normal for casts to be mixed, with women playing the female parts.

Except for *chou* and *hualian* roles, characters in Peking Opera speak and sing in falsetto, accompanied by an orchestra seated on stage. Actors walk on and off stage to a tempo set by the orchestra, every movement following a strict convention. Music, which developed in the open-air conditions of primitive theatres resembling the temporary mat-shed structures erected in villages and at temple fairs, consists of a wooden block on which a performer beats time, the resultant sound somewhat resembling that of castanets; a *huqin*, or fiddle, the leading instrument of the vocal accompaniment, with a hollow cylindrical body and two strings bowed horizontally; a *sona*, or clarinet; and several brass percussion instruments. An astonishing volume of sound can be produced by a small Chinese orchestra when all these are played *fortissimo*.

The actors' costumes, adapted from the styles of the Tang, Song, Yuan, and Ming dynasties, indicate social rank, character, and occasion: emperors wear yellow, high officials red, civilian worthies blue, elderly people brown, and rough characters black. On a warrior's costume, embroidered tiger heads are attached to heavily padded shoulders and waist, and a 'heart protecting' mirror glitters on the chest. Four triangular pennons, fixed between the shoulders, are worn by generals, whose head-dress is often surmounted by two pheasant plumes up to ten feet long, which, skilfully manipulated, can express arrogance, triumph, or despair. White silk cuffs, about two feet in length, sewn on to the actor's long flowing sleeves, whose adroit manipulation is a salient part of his technique, accentuate physical grace and the delicacy of hand movements, and enhance dramatic expression. For example, in 'asides' the right hand is raised level with the cheek, and the sleeve hangs down like a curtain; when a character weeps, a corner of the left sleeve is held up to the eyes.

Players of painted-face roles (*hualian*) wear the most complicated make-up, and for them actors shave their foreheads to lengthen or broaden their features. Temperament is expressed by colours and the lines about the eyes, nose, and mouth. White indicates treachery; black, straightforwardness and integrity; red, loyalty and courage; blue, stubbornness and ferocity; dull pink and grey, advanced age. Green is used for outlaws, brigands, and demons; gold for gods and immortals. The great historical roles have been perfected over generations, from facial make-up to the most trivial movements. Cao Cao, for instance, has become a classical monument of villainy. Dou Erdum, a brigand chief with scarlet eyebrows and enormous scarlet beard, wears facial make-up with indigo blue as its basic colour. Curving lines of black and white round the eyes represent the tiger-head hooks which were his most formidable weapon, and another pair of eyes, painted beneath his own, denote his remarkable powers of vision. Beards can indicate age and character. The longest and fullest, covering the mouth, denote heroism and prosperity; tripartite beards, culture and refinement; red and blue beards, supernatural beings (announced also by flashes of fire); while moustaches are generally a sign of coarseness or cunning. In anger the foot is stamped and the beard swept upward. Make-up for the broad comedy characters, with white patches on the nose, is more realistic, particularly in the domestic farces which sometimes act as curtain-raisers.

In modern China the traditional popular drama co-exists with the modern 'spoken drama' (*hua ju*) which developed under foreign influence during the early years of the 20th century. Peking Opera, having survived the vicissitudes of war and revolution, remains firmly established as the most popular form of theatrical entertainment, assisted by certain reforms which have enabled it to ride out the storms of social change. The most notable of these was, of course, the introduction in about 1911 of actresses into casts formerly composed entirely of men. Since 1949 greatly increased state support has also facilitated the survival, along with the Peking Opera, of many types of local drama, which, while sharing the composite song-dance-mime form of all traditional theatre, have each their own repertoire and characteristic styles of music and acting. Local companies frequently visit Peking and other centres, rivalling the Peking Opera in popularity, and in recent years traditional Chinese theatre companies have appeared on tour in most parts of the world.

The 'spoken drama' (*hua ju*), now widespread, originated with a group of Chinese students in Japan who in 1907 founded the Spring Willow Society in Tokyo. Their first production was a Chinese adaptation of the younger DUMAS's *La Dame aux camélias*, but it was not until after the beginning of the 'New Culture' movement in 1917 that the 'spoken drama' gained a serious following. It was then that the works of IBSEN and other modern Western dramatists were introduced into China in translation, and the first attempts were made to write original Chinese plays along Western lines. Prominent in this pioneering activity in the 1920s were the dramatists Tian Han and Hong Shen, who in addition to writing plays organized theatre companies and founded schools of drama. In the late 1930s the plays of Cao Yu, notably *Thunderstorm, Sunrise*, and *Peking Man*, reached a higher technical level than had been attained hitherto, and they continue to be enjoyed by theatre audiences today. After the Japanese invasion of 1937 the theatre also became increasingly a vehicle of patriotic propaganda.

With the establishment in 1949 of the People's Republic, many of China's leading actors and dramatists, including Mei Lanfang, Cao Yu, Tian Han, and Hong Shen, chose to stay on. Traditional plays were increasingly adapted to serve ideological needs, while new material, based on officially-approved models, dramatized the problems of the new society. Amateur performances of standard works were encouraged in communes and factories but strictly regional forms of drama did not receive the same support, though large-scale works, part-opera, part-ballet, with large professional casts and orchestras, were produced on proscenium stages, culminating in five revolutionary modern plays—*Taking Tiger Mountain by Strategy, Sea Harbour, Raid on White Tiger Regiment, Shajiabang*, and *Red Lantern*—and two revolutionary modern ballets—*Red Detachment of Women* and *White-Haired Girl*. These were personally selected by Jiang Qing, the wife of Mao Tse-tung and a former minor film actress, and were the only works officially allowed to be performed in China during the Cultural Revolution from 1967 to 1972. Their ubiquitous presence, not only as plays but as films and concert items, or as the subject of essays, critical studies, or items of folk art, probably served to nullify their didactic and inspirational purpose. All other works were condemned, usually on moral or counter-revolutionary grounds, and many playwrights and actors underwent self-criticism, often disappearing completely. Since the death of Mao and the fall in 1976 of the Gang of Four led by Jiang Qing some of them have reappeared; drama has revived, and traditional Peking and *kunqu* operas are performed again, as well as plays of the *hua ju* type—new, western-style 'spoken dramas' of social comment, often with a marked satirical or political content.

CHIONIDES, see GREECE, ANCIENT.

CHIRGWIN, GEORGE H., see MUSIC-HALL.

CHISWICK EMPIRE, London, see MUSIC-HALL.

'CHOCOLATE-COLOURED COON' [G. H. Elliott], see MUSIC-HALL.

CHODOROV, JEROME, see FIELDS, JOSEPH.

CHOERILUS, see GREECE, ANCIENT.

CHOREGUS. In ancient Greece, the person made responsible by the rotation of the LITURGY for equipping and paying a chorus for a tragic, comic, or dithyrambic contest. Hence the name *choregus*, or chorus-leader. The dramatic contests in the festivals were thus between *choregi* as well as between poets and, later, actors. It was important to the poet that his *choregus* should not be stingy, since the proper presentation of the play depended on him; *choregi* were therefore assigned to the poets by lot. In the Roman theatre the Latin form of the title (*choragus*) came to mean little more than property-man or stage manager.

CHORUS, in Greek drama a group of actors who stand aside from the main action of the play and comment on it, often in general terms, as in the tragedies of AESCHYLUS and SOPHOCLES; EURIPIDES, whose tragedies are more personal than cosmic, made less use of the traditional chorus, concentrating some of its functions in one person who was much more closely concerned with the fate of the protagonists. In the comedies of ARISTOPHANES the Chorus performs the same function as in the tragedies, but points the satire of each play by being dressed in some relevant disguise—birds, frogs, wasps—and by being more closely concerned in its comments with the intimate and bawdy aspects of the plot.

On the Elizabethan stage the Chorus was the speaker of an introductory prologue, a legacy from Euripides handed down via the Roman CLOSET DRAMAS of SENECA. He spoke either at the beginning of the play, as in *Henry VIII*, or before two or more acts, as in *Romeo and Juliet, Henry V*, and, in the person of Gower, in *Pericles Prince of Tyre*. The long gap of 16 years between the third and fourth acts of *The Winter's Tale* is bridged by 'Time, the Chorus'.

In the late 19th-century English theatre the chorus was the troupe supporting singers and dancers in MUSICAL COMEDY, VARIETY, and REVUE. There had always been a number of them to accompany the principal actors in BURLESQUE, EXTRAVAGANZA, and PANTOMIME, but it was not until about 1870 that the 'chorus girl' became a player in her own right. At first she was asked to do no more than wear lovely clothes and move gracefully in unison with her companions, often accompanied by handsome but fairly static 'chorus boys'. By the 1920s chorus girls were wearing fewer clothes, and, like the Tiller Girls in England and the Rockettes at RADIO CITY MUSIC HALL in New York, reached a high standard of precision dancing. C. B. COCHRAN, many of whose chorus girls went on to become distinguished actresses, always referred to them as his 'Young Ladies'. The men in the chorus were often dispensed with at this time, but after the Second World War, under the influence of the American musical, they have returned in full force, both men and women now taking a much larger share in the plot, and being expected not only to sing well, but to excel in all forms of dance, often under the direction of outstanding choreographers.

CHRISTIANIA THEATRE, see KRISTIANIA THEATER.

CHRISTIE [Mrs Max Mallowan, *née* Miller], AGATHA MARY CLARISSA (1890–1976), English novelist and playwright, appointed DBE in 1970, on her 80th birthday, publication day of her 80th book. Although chiefly remembered for her detective fiction, she has a place in theatrical history as the author of *The Mousetrap* (originally written for radio), which opened in London at the AMBASSADORS THEATRE on 25 Nov. 1952 and was still running in 1982. It was seen in New York in 1960. Mrs Christie dramatized several of her novels, including *Ten Little Niggers* (1943), produced in New York in 1944 as *Ten Little Indians; Appointment with Death* (1945); *Murder on the Nile* (1946), seen at Wimbledon the previous year as *Hidden Horizon*; and *The Hollow* (1951). *Witness for the Prosecution* (1953) was still running when *Spider's Web* (1954) was presented, thus giving her, with *The Mousetrap*, three plays running concurrently in the West End. She later wrote *Verdict* and *The Unexpected Guest* (both 1958), and adapted *Go Back for Murder* (1960) from her novel *Five Little Pigs*. Among adaptations of her novels by other hands, the most successful were *Love from a Stranger* (1936), based by Frank Vosper on *Philomel Cottage*, and *Murder at the Vicarage* (1950), adapted by Moie Charles and Barbara Toy. *A Murder Is Announced*, adapted by Leslie Darben, was produced posthumously in 1977.

CHRONEGK, LUDWIG (1837–91), German actor, director, and manager, who made his début at the Kroll Theatre, Berlin, in 1855 as a juvenile comedian and clown and went on to act in various German-speaking cities, notably Zurich, Hamburg, and Leipzig. In 1866 he joined the MEININGER COMPANY, specializing in comic Shakespearian roles. The Duke of Saxe-Meiningen appointed him director in 1871, after which he became successively manager of the company (1877), a councillor (1882), and superintendent of the Court theatres (1884). While the Duke reserved to himself the general supervision of all production, Chronegk was responsible for the execution of his decisions, conducting rehearsals, supervising the company, and arranging their tours. His last illness, brought on by the problems of the troupe's second Russian tour in 1890, ended the Meininger's activities, the ageing Duke deciding that his work would be impossible without Chronegk's help.

CHURCHILL, FRANK, see LEWES, GEORGE HENRY.

CIBBER, CHARLOTTE, see CHARKE, MRS.

CIBBER, COLLEY (1671–1757), English actor, theatre manager, and playwright. Well educated and destined for a learned profession, he went on the stage against the wishes of his family, and with few advantages of voice or person studied his art so diligently that he soon became an excellent comedian, being particularly admired in the fops of Restoration comedy and of his own plays. The pressing needs of a young family turned his thoughts to playwriting and in 1696 his first venture, *Love's Last Shift; or, the Fool in Fashion*, was successfully produced at DRURY LANE. Striking a happy balance between the Restoration comedy of manners and the new vogue for sentiment and morality, this play is now considered the first sentimental comedy. It undoubtedly influenced STEELE and FARQUHAR, as well as setting the pattern for Cibber's own later plays, and it inspired the young VANBRUGH to write *The Relapse; or, Virtue in Danger* (1696), in which Cibber gave a brilliant pertormance as the cynical rake Lord Foppington. In tragedy Cibber was less successful, though he achieved notoriety with his famous adaptation of Shakespeare's *Richard III* (1700), which remained the standard acting text until well into the 19th century. Among his later plays *She Would and She Would Not; or, the Kind Imposter* (1702)—revived as late as 1886 by DALY in New York—and *The Careless Husband* (1704) consolidated his reputation as a writer of comedies. A snob and a social climber. Cibber moved only on the fringe of the company he depicted, and his work is obviously mediocre compared with that of CONGREVE; yet for over a century his comedies were taken as representative of English high society.

The greater part of Cibber's working life was spent at Drury Lane, first as actor, and from 1711 as joint manager with DOGGETT and WILKS and later Barton BOOTH. Although he was in many ways a good manager, and was particularly successful in judging the merits of new plays—choosing them always for their theatrical effectiveness rather than their literary worth—he was decidedly unpopular. Tactless, rude to minor playwrights and small-part actors, conceited, supremely self-confident, and given to posing as an expert on subjects of which he knew nothing, he was savagely ridiculed, notably by Pope in *The Dunciad*, by Dr JOHNSON, and by FIELDING in several of his plays as well as in the opening chapter of his novel *Joseph Andrews*. As a playwright Cibber was competent rather than inspired, but he was good at doctoring other authors' work, as can be judged by his adaptation of MOLIÈRE'S *Tartuffe* as *The Non-Juror* (1717) and his completion of Fielding's *A Journey to London* which as *The Provoked Husband* was produced at Drury Lane in 1728 with Anne OLDFIELD as Lady Townly. In 1730 Cibber was appointed Poet Laureate, which caused no little dismay; he is best remembered for his *Apology for the Life of Mr Colley Cibber, Comedian* (1740), which contains some admirable descriptions of Restoration acting.

CIBBER, THEOPHILUS (1703–58), English actor, son of Colley CIBBER. A wild and undisciplined character, who made his first appearance on the stage at the age of 16, he seemed at first to have the makings of a good actor, particularly in such parts as Ancient Pistol in *Henry IV, Part 2*, but his imprudence and extravagance were his undoing. He was for a time manager of DRURY LANE in succession to his father, but soon forfeited all claim to respect by his insolence and complacency. This, coupled with his shameful treatment of his second wife Susanna Maria Arne (1714–66), sister of Dr Arne and an excellent actress, particularly in such parts as the title role in Aaron HILL'S *Zara* (1736) and Lady Brute in VANBRUGH'S *The Provoked Wife*, in which she made her last appearance on the stage, drove him out of London. He was drowned in the Irish Sea on his way to act at the Smock Alley Theatre in Dublin.

CICOGNINI, GIACINTO ANDREA (1606–60), prolific and popular Italian dramatist, much influenced by the Spanish theatre, which had come into Italy by way of Naples, where, under a Spanish viceroy, the plays of Lope de VEGA and CALDERÓN were frequently performed by visiting Spanish companies. He is credited with over 40 plays, of which perhaps half were by him, and with being the first Italian dramatist to handle the legend of DON JUAN.

CIECO D'ADRIA, Il, see GROTO, LUIGI.

CINCINNATI PLAYHOUSE IN THE PARK, Cincinnati, Ohio, is situated on a hilltop in a park, and is the professional theatre for a three-state region of the Ohio River Valley. Of its two auditoriums, the Thompson Shelterhouse, seating 219, was converted from a gazebo in 1959, the Robert S. Marx Theatre, seating 627, being constructed in 1968. During the Oct.–July subscription seasons, six productions are presented in the larger theatre and four in the smaller. They have included 18 world premières and three American, as well as revivals of works by Shakespeare, MOLIÈRE, Samuel BECKETT, Tennessee WILLIAMS, and other noted playwrights, classical and contemporary. Playhouse productions tour to other American cities, and the subscription seasons are augmented by special productions from all over the United States, and by community productions. The theatre also seeks new audiences through student seasons and special presentations, and is an important training resource.

CINQUEVALLI, PAUL, see MUSIC-HALL.

CINTHIO, Il, see GIRALDI, G. B.

CIRCA THEATRE, Wellington, see NEW ZEALAND.

CIRCLE-IN-THE-SQUARE, New York. The first theatre of this name, at 5 Sheridan Square, was designed specifically for THEATRE-IN-THE-ROUND productions by the Loft Players under José QUINTERO. It opened on 2 Feb. 1951 with a revival of Richardson and Berney's *Dark of the Moon*, and among later productions were revivals of Tennessee WILLIAMS'S *Summer and Smoke* in 1952, Truman Capote's *The Grass Harp* in 1953, and Arthur Hayes's *The Girl on the Via Flaminia* in 1954, after which the building was closed as a fire hazard. When adequate safety measures had been taken it reopened to become one of off-off-Broadway's most popular playhouses, staging in 1956 a production of O'NEILL'S *The Iceman Cometh* directed by Quintero which was considered by some critics superior to the original THEATRE GUILD production of 1946. In 1960 the demolition of the building forced the company to move to the former New Stages Theatre, at 159 Bleecker Street, where they opened on 8 Jan. with WILDER's *Our Town*, and achieved success with productions of Dylan Thomas's *Under Milk Wood* (1961), Wilder's *Three Plays for Bleecker Street* (1962), and, in 1970, Athol FUGARD's *Boesman and Lena*. In 1972 the Circle-in-the-Square moved uptown to 1633 Broadway. The new theatre, in the basement of the URIS THEATRE and called after Joseph E. Levine, was part of the first new theatre building to be erected on Broadway in 44 years. Seating 650, it has an adaptable stage, and the seats are sharply tiered. It opened on 15 Nov. with O'Neill's *Mourning Becomes Electra* and presents four plays annually for a three-month run each. Interesting productions have included Tennessee Williams's *The Glass Menagerie* in 1975; IBSEN's *The Lady From the Sea* in 1976; MOLIÈRE's *Tartuffe* in 1977; and Ibsen's *John Gabriel Borkman* in 1980. The old theatre continues as the Circle-in-the-Square Downtown; the feminist musical *I'm Getting My Act Together and Taking It on the Road* moved there in 1978 from the PUBLIC THEATRE to begin a long run.

CIRCLE REPERTORY COMPANY, an off-off-Broadway company founded in a loft in 1969 by a group who had worked together at Café LA MAMA, CAFFÉ CINO, and elsewhere, and which, under the direction of Marshall W. Mason (1940–), is making an important contribution to the American repertory. Its initial season was remarkable for a double production of CHEKHOV's *Three Sisters* in which the original and an experimental version alternated. The company is best known, however, for its productions of new American plays, including Mark Medoff's *When You Comin' Back, Red Ryder?* (1973), Edward Moore's *The Sea Horse* (1974), Feiffer's *Knock Knock* (1976), and Albert Innaurato's *Gemini* (1977), all of which transferred to the commercial theatre. Originally established at Broadway and 83rd Street, the company moved in 1974 to the former Sheridan Square Playhouse at 99 7th Avenue South, which became the Circle Repertory Company Theatre, a flexible performing space seating 100–150.

The playwright Lanford Wilson (1937–) was one of the founders of the company, and several of his plays were first performed by it, notably *The Hot l Baltimore* (1972), about a group of characters in the lobby of a hotel whose shabbiness is indicated by the absence of the 'e' from its electric sign; *The Fifth of July* (1978), in which a group who were radical students in the 1960s come together in the 1970s at the home of a legless Vietnam veteran; and *Talley's Folly* (1979), which takes a closer look at one of the characters in *The Fifth of July*. All these plays moved on to other New York theatres, *Talley's Folly* being awarded the PULITZER PRIZE in 1980. *A Tale Told*, a third play about the Talley family, was produced at the Circle in 1981.

CIRCUIT, the name given in the United Kingdom, to a group of towns within reasonable distance of each other which were served by the same travelling company. The system began early in the 18th century, when the proliferation of companies led to some unfortunate clashes, as when two groups of actors arrived together in Norwich and two companies gave performances of GAY's *The Beggar's Opera* on the same night in Newcastle. Among the most important circuits, some of which took the name of their chief town instead of that of their patron as in earlier times, were the Aberdeen, which included Perth, Montrose, and Dundee; the Exeter, which covered Plymouth and Weymouth; the Kent, originally including Margate, Canterbury, Dover, Deal, Maidstone, Faversham, and Rochester, which became so unwieldy that it was divided into East and West Kent; the Manchester, which took in Shrewsbury, Chester, Lichfield, and usually Buxton; the Newcastle, one of the largest, since it comprised Lancaster, Chester, Whitehaven, and Preston to begin with, and then took in Scarborough, Durham, Sunderland, South and North Shields, Stockton, Darlington, and Coventry; the Norfolk and Suffolk, which visited Halesworth, Wells, Woodbridge, Sudbury, Eye, Dereham, Swaffham, Lowestoft, Bungay, North Walsham, Thetford, and Newmarket; the Norwich, covering Yarmouth, Ipswich, Bury, Colchester, Cambridge, and Lynn; the Winchester, which took in Portsmouth, Southampton, Chichester, and Newport; and one of the best known, the York, which included Hull and Wakefield, Leeds, Pontefract, and Doncaster. In addition, Bath and Bristol, though not a circuit, were linked together until 1817. The circuit was usually so arranged that the chief towns were visited several times a year, if possible during Race or Assize Weeks, when large audiences could be expected. These towns usually had permanent theatre buildings, called into being by the circuit system; that in BATH dated from 1705, in BRISTOL from 1729, in YORK from 1734, in Ipswich from 1736. Smaller towns might be visited only once or twice a year, the actors playing in temporary accommodation. The companies had large repertories, the York circuit company under Samuel Butler, and later under Tate WILKINSON, who ran it for 30 years, performing as many as 100 plays and afterpieces in a year, while Mrs Sarah BAKER, lifelong manager of the Kent circuit, could reckon on a good 80. All the circuit companies, particularly the better known ones, became recognized training-grounds for young actors, and in their heyday it was rare for a player to begin his career anywhere else.

CIRCUS, in Roman times a place of exhibition for chariot racing and athletic contests. In its modern sense a circus is a kind of entertainment of a particular kind which lies outside the scope of this book. Cosmopolitan and mainly itinerant, it is performed in the central area of a tent (the Big

Top) or in a specially adapted building. The programme is built up of separate turns featuring variations on animal acts and acrobatics, controlled by a ringmaster, and loosely connected by the antics of the CLOWN or Auguste. Owing to the specialized nature of their work, circus performers tend to remain a class apart, with much intermarrying. The great names of the circus in the U.S.A. are Barnum and Sand and in Britain Sanger, Mills, and Chipperfield; but the circus is universal, and since it relies very little on the spoken word can be at home anywhere. It flourishes in Moscow and throughout the Eastern European countries. It was briefly associated with the theatre proper in EQUESTRIAN DRAMA and, to a certain extent, in DOG DRAMA. Its only connection with AQUATIC DRAMA was that the mimic sea-fights often took place in a circular flooded arena reminiscent of the circus ring.

CITIZENS' THEATRE, Glasgow, a repertory company founded in 1943 by James BRIDIE, Paul Vincent CARROLL, and others, who leased from the Corporation the Royal Princess's Theatre built in 1878. The theatre, in the slum area of the Gorbals, suffers from insecurity due to redevelopment plans; seating 1,004 people, it has a large and somewhat clumsy stage, and the whole building is in need of renovation. Nevertheless, the company soon gained a considerable reputation for good productions and continues to play to a predominantly young audience at very low prices. The intention of the founders had been to put on as many Scottish plays as possible, a policy followed by most of their successors. Among the early productions were Bridie's *The Forrigan Reel* (1946) and *The Tintock Cup* (1949) (the latter written in collaboration with George Munro) and after his death in 1951 there were productions of his last two plays, *The Baikie Charivari* (1952) and *Meeting at Night* (1954). Other important productions were William Douglas HOME's *Now Barabbas* in 1947, John ARDEN's *Armstrong's Last Goodnight* in 1964, and Peter NICHOLS's *A Day in the Death of Joe Egg* in 1967. Tyrone GUTHRIE's notable revival of LYNDSAY's *Ane Pleasant Satyre of the Thrie Estaitis* (1552), presented at the EDINBURGH FESTIVAL of 1948, showed not only the theatrical validity of Lyndsay's old MORALITY PLAY but also the high standard reached by the company in its first five years. Since 1969 the theatre has been managed by Giles Havergal, who in 1970 formulated a new policy for it and formed the present company, run by himself, the designer Philip Prowse, and the playwright Robert David MacDonald, all of whom are responsible for directing productions. The theatre embarked on a variety of classical, new, and experimental plays presented in repertory, Havergal himself directing the British première of Tennessee WILLIAMS's *The Milk Train Doesn't Stop Here Anymore* (1969). In 1972 visits were made to four European Arts Festivals with his production of *Antony and Cleopatra*. There have also been notable productions of plays by BRECHT, beginning with *Arturo Ui* directed by Michael Blakemore in 1967, the production being recreated in 1969 for the SAVILLE THEATRE, London. Little overtly Scottish drama found a place in the programmes, and the European commitment was underlined in MacDonald's own plays including *Chinchilla* (1977), drawn from incidents in the life of Diaghilev, and *A Waste of Time* (1980), an audacious adaptation of Proust's work. Productions became vehicles for spectacular theatrical techniques, with frequent emphasis on the decadent, the violent, and the sexually perverse. Experimental works staged at the small Close Studio Theatre from 1965 until its destruction by fire in 1973 included GENET's *The Maids*, Sam SHEPARD's *Icarus's Mother*, and Simon GRAY's *Spoiled*.

The Citizens' Theatre receives subsidies from the Scottish Arts Council and from the Corporation of Glasgow, whose residual control over artistic as well as financial policy has on occasion led to considerable friction. Productions are cast from a pool of actors, all of whom receive the same salary; the Citizens' Theatre for Youth company runs a theatre-in-education scheme.

CITY CENTER OF MUSIC AND DRAMA, New York, at 131 West 55th Street. Built as a Masonic temple in 1924, this building came into the ownership of the city due to a tax default and opened as a theatre under its present name on 11 Dec. 1943 with a concert by the New York Philharmonic Orchestra. Two days later the drama programme was inaugurated with a revival of Rachel CROTHERS's *Susan and God*, starring Gertrude LAWRENCE. The theatre, which has a seating capacity of 2,935, with prices kept as low as possible, served also as the home of the New York City Ballet and the New York City Opera until their activities were transferred in the 1960s to the new Lincoln Center, but the drama section continued for some time to offer a wide variety of plays and musicals under various directors, including José FERRER, Maurice EVANS, and Jean Dalrymple. In July 1972 the City Center Acting Company was formed to foster ensemble acting by a group of Juilliard School graduates; it severed its connection with the City Center in 1975 and since 1980 has been affiliated to the John F. Kennedy Center in Washington, D.C. It is now known as The Acting Company. The City Center still plays host to various dance companies and foreign troupes.

CITY DIONYSIA, see DIONYSIA.

CITY OF LONDON THEATRE, London, in Norton Folgate, adjoining Bishopsgate. This was designed by Samuel BEAZLEY with a seating capacity of 2,500 and opened under the management of Christopher Cockerton on 27 Mar. 1837 with a dramatization of DICKENS's *Pickwick Papers* by Edward Stirling—before the book had completed its first serial publication. In 1838 Osbaldiston became manager in succession to Mrs Honey, a popular actress who had specialized in MELODRAMA, and ran the theatre for three years; but its most successful period was between 1848 and 1868 under the control of Nelson Lee and Johnson. Having learned their trade by running the booth which they took over from John Richardson, they gave their audiences good value for money, kept their prices low, and engaged many leading actors and actresses. After Lee retired the theatre began to go downhill, though Sarah Thorne was there in 1867, as were the Christy Minstrels later in the same year. By 1868 the house had reverted to crude melodrama and romantic spectacle, and in 1871 it was destroyed by fire.

CITY PANTHEON, London, see CITY THEATRE.

CITY THEATRE, London, in Grub Street (later Milton Street). This was a converted chapel which opened as a theatre under the management of John Bedford, a popular comedian, in 1829–30. It functioned for six years, during which time its name was changed to the City Pantheon, Bedford being succeeded by John Kemble Chapman in 1830. Edmund KEAN played Shylock in *The Merchant of Venice* there in 1831, and in the same year Ellen TREE appeared in T. Egerton Wilkes's *Eily O'Connor*. Isaac BICKERSTAFFE's *Love in a Village* and JERROLD's *Black-Ey'd Susan* were also revived there with excellent casts. In 1831 Chapman retired in favour of Davidge, who made an arrangement with the Royal Coburg, later the OLD VIC, by which the same companies appeared at both theatres, being taken to and fro in hackney coaches. Ben WEBSTER ran the theatre for a while, as did MONCRIEFF from Nov. 1833, selling tickets off the premises to evade the monopoly of the PATENT THÉATRES. The City was last used as a theatre in 1836 and then became a warehouse.

CITY THEATRE, New York. (1) A small playhouse at 15 Warren Street which opened on 2 July 1822 with a company of amateurs under the direction of the actress Mrs Baldwin, formerly of the PARK THEATRE. An outbreak of yellow fever brought their first season to a premature end, and the theatre closed in 1823.

(2) In the upper part of the City Saloon on Broadway, between Fulton and Ann Streets. This opened on 13 July 1837 under the management of J. J. Adams, with Joe COWELL as Crack in Thomas Knight's *The Turnpike Gate*, but in spite of Cowell's popularity the venture failed and the theatre closed later the same year.

(3) At 116 East 14th Street, between 3rd Avenue and Park Avenue South. Opening on 18 Apr. 1910 with *Miss Innocence*, this was converted into a VAUDEVILLE house in the following Nov., and after briefly housing Maurice SCHWARTZ's Yiddish Art Theatre in 1928, it became a cinema in 1929. It was demolished in 1952.

CIULEI, LIVIU, see GUTHRIE THEATRE and ROMANIA.

CIVIC REPERTORY COMPANY, New York, see LE GALLIENNE, EVA.

CIVIC THEATRE, in Great Britain, a theatre owned and subsidized by the local authority as a cultural amenity. Clause 132 of the Local Government Act of 1948 empowered local authorities to spend the proceeds of not more than a sixpenny rate on the provision of entertainment 'of any nature'; but it was ten years before the opening of the BELGRADE THEATRE in Coventry paved the way, and true civic theatres are still a rarity. Two main types of theatre enjoy civic support. Touring theatres, often providing facilities for large-scale opera and ballet, may accommodate both subsidized and commercial productions; they include the Alexandra and the Hippodrome in BIRMINGHAM, the Empire in LIVERPOOL, the Grand Theatre and Opera House in LEEDS, the PALACE THEATRE, Manchester, and the Theatres Royal at BATH, LINCOLN, NEWCASTLE, NORWICH, and NOTTINGHAM. Repertory theatres receiving local authority support to a greater or lesser degree, but often managed by a Trust to ensure artistic independence, include the BIRMINGHAM REPERTORY THEATRE; BRISTOL OLD VIC; the CRUCIBLE, Sheffield; the HAYMARKET, Leicester; the LIBRARY THEATRE, Manchester; LEEDS Playhouse; LIVERPOOL PLAYHOUSE; the Mercury, Colchester; NEWCASTLE PLAYHOUSE; the NORTHCOTT, Exeter; NOTTINGHAM PLAYHOUSE; the OCTAGON, Bolton; the PALACE, Watford; the PHOENIX ARTS CENTRE, Leicester; the ROYAL EXCHANGE, Manchester; the Theatre Royal, YORK; and the VICTORIA, Stoke-on-Trent. (See also REPERTORY THEATRES.)

CLAIRON [Claire-Josèphe-Hippolyte Léris de La Tude], Mlle (1723–1803), French actress, who at an early age showed signs of precocious ability and joined the troupe of the COMÉDIE-ITALIENNE, where she played minor parts for a year or so before joining the provincial troupe headed by LA NOUE. It was apparently while she was playing at Lille that GARRICK saw and admired her. Later she developed a fine singing voice, and in 1743 went to the Opéra, but her real talent was for acting and she was transferred to the COMÉDIE-FRANÇAISE to understudy Mlle DANGERVILLE. Her début in the demanding role of RACINE's Phèdre on 19 Sept. 1743 was a triumph, and led to her being entrusted with other important tragic roles. With the great actor LEKAIN, encouraged by MARMONTEL and DIDEROT, she was responsible for introducing some much-needed modifications into the contemporary costumes worn on the stage for every type of play, mainly by adding some historical detail in keeping with the character she was portraying. Also, again following the good advice of Marmontel, she abandoned in about 1753 her somewhat stiff and declamatory method of acting for a freer and more natural style. The naturally tragic depth of her voice, though esteemed too low for high passion and not particularly felicitous in tender scenes, prevented her from lapsing into triviality, and the new departure was a success. In 1765 Clairon, with other members of the Comédie-Française, was imprisoned for refusing to play with an actor who had brought disgrace upon the company; on her release she took refuge with VOLTAIRE, in many of whose plays she had appeared, thereafter acting only in his private theatre at Ferney and on her return to Paris at Court and in private theatricals. In 1773 she was invited to the Court of the Margrave of Anspach, and it was there that she wrote her *Mémoires et réflexions sur l'art dramatique* (1779). She died in poverty in Paris, her actor's pension having ceased with the outbreak of the Revolution and her benefactress being dead.

CLARK, JOHN PEPPER (1935–), Nigerian playwright and poet, born at Kiagbodo in the mid-west of Nigeria and educated at the University of Ibadan. His first three plays make up a trilogy having much in common with Greek tragedy; *Song of a Goat* (1962) is about a childless wife who takes her husband's brother as a lover; their child is the central figure of the second play, *The Masquerade* (1964), and in *The Raft* (also 1964) the threads of the earlier plays are woven together in a discussion between four men adrift on the Niger. Clark's dramatic work arises from the myths and folk-lore of the Ijaw people; his *Ozidi* (1966) deals with the world of the imagination and the magical, being based on an Ijaw saga of the same name of which Clark has prepared an English translation, providing unique evidence of the dramatic nature and potential of such material.

CLARKE, AUSTIN (1896–1974), Irish poet, dramatist, and novelist, who kept interest in verse drama alive in Ireland during the 20 years following the death of YEATS. His first play, *The Son of Learning*, was performed at the Cambridge Festival Theatre in 1927 and three years later, under the title *The Hunger Demon*, was seen at the GATE THEATRE, Dublin. This was followed by two short plays set in a convent, *The Flame* (1932) and *Sister Eucharia* (1939), also staged at the Gate Theatre. *Black Fast*, a one-act farce based on a medieval debate, was produced by the ABBEY THEATRE company in 1941.

In 1944 Clarke established the Lyric Players for the purpose of reviving poetic drama. Among his own plays produced by this group at the Abbey Theatre were *The Viscount of Blarney* (1944), *The Second Kiss* (1946), *As the Crow Flies* (1948), and *The Plot Succeeds* (1950), a 'poetic pantomime'. Although many of Clarke's plays are set in medieval or legendary Ireland and reflect the satirical tone of early Irish writers, his works comment thoughtfully and provocatively on contemporary issues, probing the conflict between faith and reason and between individual conscience and established teaching.

CLARKE, JOHN SLEEPER (1833–99), American actor and theatre manager, who made his début in 1851 in Boston, and later became joint lessee of the Arch Street Theatre, Philadelphia. He married the sister of Edwin BOOTH and with him managed several American theatres. He was first seen in London in Oct. 1867, and in 1872 became manager of the CHARING CROSS THEATRE. He later took over the HAYMARKET and the STRAND, appearing at all three theatres under his own management.

CLAUDEL, PAUL-LOUIS-CHARLES-MARIE (1868–1955), French poet, dramatist, and diplomat, whose early works were published anonymously for fear their ardent Catholicism should harm his diplomatic career. The first, *Tête d'or* and *La Ville* (pub. in 1890 and 1893 respectively), were poetic dramas intended for a form of total theatre far in advance of their time, and the first to be seen on stage were *L'Annonce faite à Marie* (1912) and *L'Échange* (1914), the former directed by LUGNÉ-POË at the Théâtre de L'Oeuvre, the latter by COPEAU at the VIEUX-COLOMBIER. Lugné-Poë also directed *L'Ôtage* (1914). All three were produced in London, as *The Tidings Brought to Mary* and *The Exchange*, directed by Edith CRAIG for the Pioneer Players in 1912 and 1916, and *The Hostage*, in which Sybil THORNDIKE gave an outstanding performance as Synge de Coûfontaine, in 1919. Many of Claudel's later plays were first seen outside France, especially in Germany where their complex structure offered scope for experimental staging. Among them were *Le Pain dur* (1926) in Oldenburg, *Le Père humilié* (1928) in Dresden, and *Christophe Colomb* (1930) in Berlin. *Le Repos du septième jour* (1928) had its première in Warsaw, *Protée* (1929) in Groningen, *La Ville* (1931) in Brussels, and the dramatic oratorio *Jeanne d'Arc au bûcher* (1938), with music by Honegger, in Basle. It was not until the Second World War that Jean-Louis BARRAULT, working in close collaboration with Claudel, first brought his work to the notice of the general public in France with a production at the COMÉDIE-FRANÇAISE on 27 Nov. 1943 of what is usually regarded as Claudel's masterpiece, *Le Soulier de satin, ou le Pire n'est pas toujours sûr*. The incidental music was again by Honegger. The impression it made was reinforced by further productions by Barrault at the MARIGNY—*Partage de midi* (first seen in 1916) in 1948, with Edwige FEUILLÈRE as Ysé, *Christophe Colomb* in 1953, and *Le Soulier de satin* again in 1965. All three productions were seen in London during visits from the Renaud-Barrault company, in 1951, 1956, and 1965 respectively. The company was also seen in New York in 1957 in *Le Soulier de satin*, which, as *The Satin Slipper, or The Worst is not the Surest*, was published in 1931 in an English translation by Father John

O'Connor, *Le Pain dur* and *Le Père humiliè*, sequels to *L'Ôtage*, were translated in 1946 by John Heard. *La Ville*, as *The City*, and *Tête d'Or*, under the same title, have also been translated, and in 1972 *Partage de midi*, in a translation by Jonathan Griffin as *Break of Noon*, was produced at the Arts Theatre in Ipswich. *Jeanne d'Arc au bûcher* was first performed in London in 1954. Claudel's last play, *L'Histoire de Tobie et de Sara*, had its first production at the AVIGNON FESTIVAL in 1947 under Jean VILAR. Most of Claudel's plays were revised and rewritten several times and two or more versions exist of all his major works. All must be regarded as statements of his Christian faith, depicting the unending struggle between good and evil and the redemption of mankind through sacrifice.

CLAUS [Klaus], HUGO (1929–), Belgian poet novelist, and dramatist, writing in Dutch, and probably the leading and most versatile figure of the Belgian post-war literary scene. He came early into contact with the international avant-garde through a chance meeting with ARTAUD in Paris, where he lived for some time, and although many of his plays have a Flemish background his analysis of man's contemporary situation has a universal application. His most successful plays have been *A Bride in the Morning* (1953), a psychological study of a brother-and-sister idyll; *Sugar* (1958), which deals with Flemish seasonal workers in France; and *Friday* (1969), which depicts the confrontation between a man released from prison and his wife, who during his absence has had a child by a neighbour. This was seen in London at the ROYAL COURT Theatre Upstairs in 1971 and in the same year it was first seen in a French version at the Belgian National Theatre, having had its world première in Amsterdam. It was also taken on a Canadian tour by the Belgian National Theatre company in 1973. *Leopold II* (1970) is a satire on Belgian colonialism. In 1976 Claus's version of *Orestes*, based on EURIPIDES, was produced in Antwerp, and later seen in a number of other towns. Several of his plays have been translated and played world-wide, including productions in Japan.

CLAUSEN, SVEN, see DENMARK.

CLEMENTS, JOHN SELBY (1910–), English actor-manager, knighted in 1968. He made his first appearance on the stage in 1930, and five years later founded the Intimate Theatre in Palmers Green, a northern suburb of London, where he directed a weekly repertory of old and new plays, in many of which he appeared himself. It was the first theatre in London to reopen after the outbreak of war in Sept. 1939, with a revival of RATTIGAN's *French Without Tears*. Clements later starred in West End plays, including PRIESTLEY's *They Came to a City* (1943), and for a time managed the ST JAMES'S THEATRE, where he appeared with his second wife Kay Hammond [Dorothy Katharine Standing] (1909–80)—already well known for such roles as Diana Lake in *French Without Tears* (1936) and Elvira in COWARD's *Blithe Spirit* (1941)—in DRYDEN's *Marriage à la Mode* in 1946. He also managed the PHOENIX THEATRE, where FARQUHAR's *The Beaux' Stratagem*, with Clements as Archer and his wife as Mrs Sullen, began a record run in 1949. In 1951 they appeared as John Tanner and Ann Whitefield in a revival of SHAW's *Man and Superman* which Clements presented and directed, eventually including in it once a week the 'Don Juan in Hell' scene which is usually omitted. A year later they were seen in his own play *The Happy Marriage*, based on Jean-Bernard Luc's *Le Complexe de Philémon*, and from 1955 to 1957 he was in management at the SAVILLE THEATRE, where he again presented some excellent revivals of classical comedies, including SHERIDAN's *The Rivals* and CONGREVE's *The Way of the World*, in both of which he starred with his wife. He later appeared in such productions as Benn W. Levy's *The Rape of the Belt* (1957) which he directed, playing Heracles to his wife's Hippolyte, and Ronald MILLAR's *The Affair* (1961) and *The Masters* (1963), based on novels by C. P. Snow, also directing the latter. From 1965 to 1973 he was director of the CHICHESTER FESTIVAL THEATRE, where his productions included PHILLPOTTS's *The Farmer's Wife*, PINERO's *The Magistrate*, and Shaw's *Heartbreak House*, in which he played Shotover; he also played Prospero in *The Tempest* and Antony in *Antony and Cleopatra*. He returned to Chichester in 1979 to appear in Shaw's *The Devil's Disciple* and WILDE's *The Importance of Being Earnest*.

CLEVELAND PLAY HOUSE, Cleveland, Ohio, the first resident professional theatre in the United States. Founded by an amateur group in 1915, it took over a disused church two years later and operated there on a small scale until the appointment in 1921 of Frederic McConnell as its first director. He began to transform it into a professional company, and in 1927 it moved into a new theatre incorporating two auditoriums, the Drury and the Brooks, seating 515 and 160 respectively. A third theatre, the Euclid-77th Street, seating 567, was added in 1949, when a nearby church was converted into a thrust-stage theatre which opened with *Romeo and Juliet*. At the Play House about a dozen productions are presented during a season which runs from Oct. to May, with guest actors making occasional appearances. The repertory covers a wide range and over 70 world and American premières have been given, including the world premières of Paul Zindel's *The Effect of Gamma Rays On Man-in-the-Moon Marigolds* (1969) and Robert E. Lee and Jerome Lawrence's *First Monday in October* (1975). American premières have included Christopher FRY's *A Yard of Sun* (1972), David WILLIAMSON's *The Removalists* (1973), ARBUZOV's *Confession at Night* (1975), and

GORKY's *The Romantics* (1978). The Play House also operates a Youth Theatre and a Play House Comes to School programme, and conducts an annual student festival.

CLIFTON, HARRY, see MUSIC-HALL.

CLINE, MAGGIE, see VAUDEVILLE, AMERICAN.

CLIVE, KITTY [*née* Catherine Raftor] (1711–85), English actress, who at the age of 17 was playing minor parts at DRURY LANE under Colley CIBBER. She made her first success as Phillida in his *Love in a Riddle* (1729) and three years later married a barrister named George Clive. The marriage soon broke up, but merely through incompatibility and no scandal was ever attached to her name. Most of her career was spent at Drury Lane where she and GARRICK were constantly at loggerheads, mainly because he endeavoured to prevent her appearing in parts to which she was unsuited. She was excellent in low comedy, farce, and burlesque, but quite unfitted for tragedy or even genteel comedy; though she made a success of Portia in *The Merchant of Venice*, particularly in the trial scene, where she mimicked famous lawyers of the day. Passionate and vulgar, she was always generous and quite without pride or ostentation. One of her greatest admirers was Horace Walpole, who on her retirement in 1769 presented her with a small house—Clive's-Den—on Strawberry Hill, where her company and conversation were much relished by his friends, particularly Dr JOHNSON. She was frequently painted, notably by Hogarth, and was the author of several short farces.

CLOAK-AND-SWORD PLAYS, see COMEDIA.

CLOSE STUDIO THEATRE, Glasgow, see CITIZENS' THEATRE.

CLOSET DRAMA, plays written to be read as opposed to those intended for production on stage. Among the most important and influential were those of SENECA. Many verse plays by established poets, particularly in the 19th century, though meant to be read, were, like those of Alfred de MUSSET and Lord TENNYSON, successful in production; though Tennyson's have not been revived, de Musset's have joined the permanent repertory. The term 'closet drama' also includes translations of plays, usually classics in their own countries, intended for reading and not for acting.

CLOTH (Drop in America), any large unframed expanse of scenic material, made of widths of canvas or muslin seamed horizontally together and attached top and bottom to a sandwich BATTEN. Among the specialized cloths are the BACKCLOTH; the Cut-Cloth (Cut-Out Drop in America), with cut openings which, if elaborately fretted, need the reinforcement of a piece of netting, glued on behind; the Gauze-Cloth (Scrim Drop in America), used for CYCLORAMA facing and for special effects such as a TRANSPARENCY; the Sky-Cloth (Sky Drop), sometimes faced with gauze, used as a backdrop instead of a cyclorama; and the Stage-Cloth (Ground Cloth in America), with its variants such as the Sand-Cloth for desert and other outdoor scenes, an expanse of painted canvas laid on the stage as a floor covering, a function also performed by the TRAGIC CARPET. The CEILING-CLOTH differs from the others in being battened-out before being suspended over the top of a BOX-SET.

CLOUDINGS, permanent cloud BORDERS used to mask the top of a scene. They could be drawn off sideways by hooked poles, and are mentioned as late as 1743. The detail of their arrangement is not clear, but they recall the form of border used by Inigo JONES in his last masque, *Salmacida Spolia* (1640), which had also 'side-clouds'. These could presumably be drawn off to reveal another set behind, thus changing a stormy sky into a calm one, or vice versa.

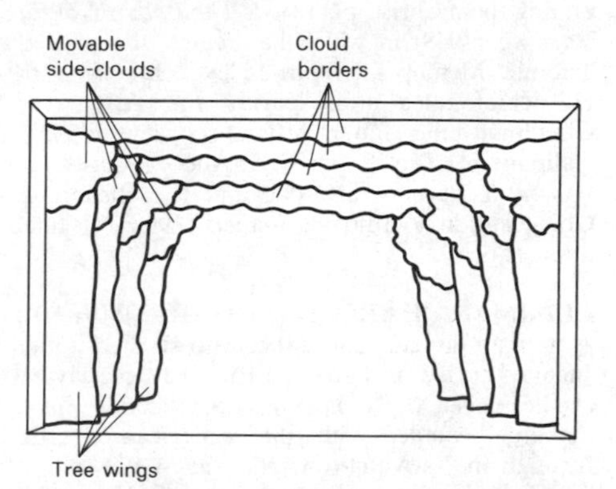

Clouding, as used in *Salmacida Spolia*
(After R. Southern, *Changeable Scenery*, p. 79)

CLOWN, a composite comic character, who may be a simpleton, a knave, or a Court Jester. Shakespeare provides examples of all three with Costard in *Love's Labour's Lost*, Autolycus in *The Winter's Tale*, and Touchstone in *As You Like It*. One of the formative elements of the Elizabethan clown was the Old VICE of the LITURGICAL DRAMA, who tripped up the Devil and played tricks on the serious characters. The best known players of clowns in Shakespeare's day were TARLETON and KEMPE. The COMMEDIA DELL'ARTE Arlecchino first came to England as a clown, but ended up as the young lover HARLEQUIN in the English PANTOMIME, where the part of the Clown, a purely local importation, was immortalized by GRIMALDI. He became known as Joey, a name applied since to all clowns who wear Grimaldi's costume: baggy trousers, white face with bright red nose, and a bald wig with tufts of hair. Today clowns are found only in circuses where Joey is joined by the

Auguste, who wears an ill-fitting dress suit and a bald wig with a ragged fringe of hair, and where their slapstick is elaborated with skilful acrobatics and juggling. In the heyday of the MUSIC-HALL the image of the clown as a dolt was perpetuated by the 'red-nosed' comic, his more polished counterpart by such accomplished players as GROCK.

CLUNES, ALEC [Alexander] S. DE MORO (1912–70), English actor, director, and theatre manager. Both his parents were on the stage, and after considerable experience as an amateur he joined Ben GREET's company in 1934, and was then at the OLD VIC, where he played a wide variety of parts, and at the MALVERN FESTIVAL, later appearing in London in SHAW's *In Good King Charles's Golden Days*. In 1942 he took over the ARTS THEATRE in London and remained there until 1950, producing and directing over 100 plays in many of which he appeared himself. These ranged from his initial production, a revival of ODETS's *Awake and Sing*, through revivals of English classics like FARQUHAR's *The Constant Couple* to new plays, among them Christopher FRY's *The Lady's Not for Burning* (1948) in which he created the part of Thomas Mendip later played by John GIELGUD. His chief roles after leaving the Arts were Claudius to the Hamlet of Paul SCOFIELD in 1955, Caliban in *The Tempest* at the SHAKESPEARE MEMORIAL THEATRE in 1957, and the Bishop of Chichester in HOCHHUTH's *Soldiers* (1968), his final appearance.

CLURMAN, HAROLD EDGAR (1901–80), American director and critic, who studied under Jacques COPEAU in Paris and Richard Boleslavsky at the AMERICAN LABORATORY THEATRE. He made his first contact with the American theatre through the PROVINCETOWN PLAYERS in 1924 and a year later joined the THEATRE GUILD. In 1931, with Cheryl CRAWFORD and Lee STRASBERG, he founded the GROUP THEATRE, directing his first play on Broadway for them—ODETS's *Awake and Sing* (1935). He later directed Odets's other full-length plays for the Group, of which in 1936 he assumed sole management. After it disbanded in 1940 he directed a number of important new plays for other managements, including Carson McCullers's *The Member of the Wedding* (1950), William INGE's *Bus Stop* (1955), ANOUILH's *Waltz of the Toreadors* (1957), O'NEILL's *A Touch of the Poet* (1958), and Arthur MILLER's *Incident at Vichy* (1964). Outside New York he directed Lillian HELLMAN's *Montserrat* in Tel-Aviv in 1949; GIRAUDOUX's *Tiger at the Gates* in London in 1955; O'Neill's *The Iceman Cometh* in Tokyo in 1968; and CHEKHOV's *Uncle Vanya* in Los Angeles in 1969. Clurman, who wrote several books on the theatre, was drama critic of the *New Republic* from 1949 to 1953, when he became drama critic of the *Nation*. From 1959 to 1963 he was guest critic of the London *Observer*, and in 1963–4 served as executive consultant to the VIVIAN BEAUMONT THEATRE before it opened at the Lincoln Center.

COAL HOLE, see EVANS's and TERRY, EDWARD.

COATES, ROBERT (1772–1848), a wealthy eccentric from the West Indies who believed himself to be a superb actor, and in 1810 rented the Theatre Royal, BATH, to display himself as Romeo; hence his nickname 'Romeo' Coates. Taking the hilarity of the audience as a tribute to his genius, he set out to tour the provinces and finally appeared at the HAYMARKET THEATRE in London, again as Romeo, wearing a sky-blue spangled cloak, tight red pantaloons, a muslin vest, a full-bottomed wig, and a tall hat. Later he appeared as Lothario in ROWE's *The Fair Penitent*. He enjoyed a brief blaze of notoriety, but the public soon grew tired of his absurdities and he relapsed into penniless obscurity.

COBORN, CHARLES, see MUSIC-HALL.

COBURG THEATRE, London, see OLD VIC THEATRE.

COBURN, CHARLES DOUVILLE (1877–1961), American actor and manager, who at 17 was in charge of the Savannah Theatre, where he had started work as a programme boy. He acted for a time with a stock company in Chicago and in 1901 made his first appearance in New York, subsequently touring the United States in Hall Caine's *The Christian*. In 1906, with his wife Ivah Wills (1882–1937), he organized the Coburn Shakespearian Players, both husband and wife playing leading parts with it for many years. When it disbanded he returned to New York, where he made a great success as Old Bill in *The Better 'Ole* (1918) by Bruce Bairnsfather and Arthur Eliot, which ran for two years, and as James Telfer in an all-star revival of PINERO's *Trelawny of the 'Wells'* in 1925. Another of his successful parts was Wu Hoo Git in *The Yellow Jacket*, a play 'in the Chinese manner' by Hazelton and Benrimo, which he frequently revived on tour. In 1934 the Coburns inaugurated the Mohawk Dramatic Summer Festival at Union College, Schenectady, which became an annual event, and in the same year Coburn appeared in Emmet Lavery's *The First Legion*. He retired from the stage on the death of his wife but in 1946 returned to play Falstaff in *The Merry Wives of Windsor* for the THEATRE GUILD.

COBURN THEATRE, New York, see DALY'S SIXTY-THIRD STREET THEATRE.

COCHRAN, CHARLES BLAKE (1872–1951), English impresario, known in the theatre as 'Cocky', knighted in 1948. He began his career as an actor in America, where he was also for some years the personal representative of Richard MANS-

FIELD. After gaining experience in practically every branch of popular entertainment, he went into management on his own account, his first venture being a production of IBSEN's *John Gabriel Borkman* in New York in 1897. In London in 1911 he was responsible for REINHARDT's production at Olympia of Vollmöller's *The Miracle*, which he revived at the LYCEUM in 1932, and in 1917 he presented in London BRIEUX's controversial play *Damaged Goods*. He had already presented intimate REVUE with *Odds and Ends* (1914) and similar shows, and in 1918 began a series of remarkable productions at the LONDON PAVILION, beginning with Wimperis's *As You Were* and including Noël COWARD's *On With the Dance* (1925) and *This Year of Grace* (1928), the former being the first show to feature Cochran's famous 'Young Ladies'. He brought many plays and players from abroad to London, arranging Sacha GUITRY's first visit in 1920, the first production in London of an O'NEILL play, *Anna Christie* in 1923, and of Robert SHERWOOD's *The Road to Rome* in 1928 and the Heywards' Negro tragedy *Porgy* in 1929; in that year he also presented O'CASEY's *The Silver Tassie*. A year later he presented MOISSI in *Hamlet* and the PITOËFFS in SHAW's *Saint Joan*. He was closely associated with Coward, being responsible for the production of his *Bitter Sweet* (1929), *Cavalcade* (1931), *Words and Music* (1932), and *Conversation Piece* (1934), and during the 1930s he was associated with a long list of successful new productions. In addition he was manager of the Royal Albert Hall from 1926 to 1938 and promoted rodeo shows and boxing matches. His longest-running production was A. P. Herbert and Vivian Ellis's musical *Bless the Bride* (1947) and his last William Templeton's *The Ivory Tower* (1950).

COCKPIT, London. (1) In Drury Lane. Built for cockfights in 1609, this was converted into a roofed or 'private' theatre in 1616 by Christopher BEESTON for QUEEN ANNE'S MEN. Burnt down by rioting apprentices on Shrove Tuesday 1617, it was rebuilt and renamed the Phoenix, though the old name remained in use for some time. Several different companies occupied it, including a French troupe under FLORIDOR, who for a short season in 1635 appeared in *Mélise*, conjectured to be either CORNEILLE's *Mélite* or *La Mélize* by du Rocher. In 1637 Beeston installed a young company known as Beeston's Boys. On his death in 1638 his son William took over, but was arrested a year later for presenting an unlicensed play which offended Charles I. He was fined and imprisoned, his place being taken by DAVENANT. The Cockpit closed in 1642, in common with all the London theatres, but must have been used for illicit performances as it was raided by Commonwealth soldiers in 1649, the audience found there being fined a total of £3.11*s*.6*d*. on the spot. Two of Davenant's early 'plays with music' (considered the first English operas) were seen at the Cockpit by special permission of Cromwell—*The Cruelty of the Spaniards in Peru* (1658) and *Sir Francis Drake* (1659). When the theatres reopened in 1660, John RHODES occupied the Cockpit with a group of young actors which included BETTERTON. Davenant and KILLIGREW also ran it jointly for a few weeks before separating to take over the PATENT THEATRES. In 1661 George JOLLY was there with a company which remained until 1665, after which the theatre, faced with competition from DRURY LANE, fell out of use.

(2) In Gateforth Street. This was built for the Inner London Education Authority to provide facilities for London schoolchildren to mount plays themselves during their last year at school, or to attend performances by such visiting companies as the NATIONAL YOUTH THEATRE. Its complex of theatre, workshop studios, and rehearsal rooms opened on 19 Jan. 1970. The auditorium holds 120–240, depending on the shape of the stage which can be used either for THEATRE-IN-THE-ROUND or as a thrust or OPEN STAGE.

COCKPIT-IN-COURT, London. The cockpit built for Henry VIII at Whitehall, on the edge of St James's Park, was converted into a playhouse in about 1604. It had a steeply pitched conical roof topped by a lantern and was about 50 ft in diameter. Although primarily intended for private performances before the monarch and his courtiers, particularly by the KING'S MEN, it was later used for some public entertainments. It was reconstructed in 1632 and again when the theatres reopened in 1660, but was not used as a theatre after 1664.

COCOANUT GROVE, New York, see CENTURY THEATRE (1).

COCTEAU, (Clement Eugène) JEAN MAURICE (1889–1963), French dramatist, poet, novelist, critic, artist, and film director, whose many-sided activities involved most of the outstanding artists and musicians of his day. As a dramatist he first came into prominence with a new version of SOPHOCLES' *Antigone* (1922), directed by DULLIN in a setting by Picasso. This was followed by *Orphée* (1924), directed by PITOËFF. In 1934 JOUVET was responsible for the production of another excursion into Greek mythology with *La Machine infernale*, based on the Oedipus legend. Cocteau's other plays include *Les Mariés de la Tour Eiffel* (1921), an exercise in Surrealism first performed at the COMÉDIE-FRANÇAISE; *Les Parents terribles* (1938); *Les Monstres sacrés* (1940), on the nature of the theatrical experience; *La Machine à écrire* (1941), a fairly straightforward thriller; two one-act monologues for an actress—*La Voix humaine* (1930) and *Le Bel Indifférent* (1941), written for PIAF; an excursion into Arthurian legend, *Les Chevaliers de la Table Ronde* (1937); a tragic love-story in verse, *Renaud et Armide* (1943), produced at the Comédie-Française with Marie BELL and Maurice Escande in the title roles; and a romantic costume drama, *L'Aigle à deux têtes* (1946). Although several of Cocteau's plays (*Orphée*, 1927; *The Human Voice*, 1938; *The Infernal Machine*, 1940; *Intimate Relations*, 1951; *The Holy Terrors*, 1952) have been translated into English and performed in London, mainly at club theatres like the ARTS and the GATE, the only one to achieve commercial success was *The Eagle Has Two Heads*, which had a long run in 1946 with Eileen HERLIE as the Queen, the part created by Edwige FEUILLÈRE. It was also seen briefly a year later in New York with Tallulah BANKHEAD; *Intimate Relations* had been produced there in 1932 and *The Infernal Machine* was to come in 1958; but on the whole Cocteau's plays have had less influence outside France than his films and ballets.

CODY, WILLIAM FREDERICK (1846–1917), American showman, better known as Buffalo Bill. Born on a farm in Iowa, he was at the Colorado gold-mines as a boy, and then became a Pony Express rider and a Civil War scout. He appeared between 1872 and 1883 in more than a dozen plays written for him, the theatrical equivalent of Western dime novels. In 1883 he embarked on the Wild West show which made him famous all over the world. For 18 years the show starred the sharpshooter Annie Oakley (1860–1926), and the pair provided the basis for Irving Berlin's musical *Annie Get Your Gun* (1946).

COFFEY, CHARLES, see BALLAD OPERA.

COGHILL, NEVILL, see OXFORD and OXFORD PLAYHOUSE.

COGHLAN, ROSE (1850–1932), American actress, born in England, who made her first appearance on the stage as a child, playing one of the witches in *Macbeth*. In 1869 she made her adult début under HOLLINGSHEAD at the GAIETY THEATRE in London, and two years later was seen in New York with Lydia Thompson. Her success was such that after a few years in London and the English provinces she returned to the United States in 1877 and settled there, becoming an American citizen in 1902. She was for many years leading lady at WALLACK'S THEATRE, where she appeared with her brother Charles Coghlan (1842–99), playing Lady Teazle, one of her best parts, to his Charles Surface in SHERIDAN'S *The School for Scandal*. She also appeared in such modern productions as H. A JONES and Herman's *The Silver King*, Tom TAYLOR'S *Masks and Faces*, and SARDOU'S *Diplomacy*, and was much admired as Rosalind in *As You Like It*. In 1893 she starred in the first production of WILDE'S *A Woman of No Importance*; her career then began to decline, but though her style was outmoded her fine voice, stately presence, and technical ability caused her to be still in demand, and in 1916 she celebrated her stage jubilee. She was frequently seen in VAUDEVILLE, and made her last appearance in 1921 in BELASCO'S production of Sacha GUITRY'S *Deburau*.

COHAN, GEORGE M(ichael) (1878–1942), American actor, dramatist, and manager, who appeared as a child with his parents and sister in an act billed as 'The Four Cohans', and at 15 was writing songs and skits for performance in VAUDEVILLE. In 1901 he made his first appearance in New York in his own play *The Governor's Son* and soon built up an enviable reputation with such shows as *Little Johnny Jones* (1904) and *George Washington, Jr.* (1906). In 1911 he opened his own playhouse, the GEORGE M. COHAN THEATRE, with *Get-Rich-Quick Wallingford*, appearing there later the same year in *The Little Millionaire* and a revised version of *Forty-Five Minutes from Broadway*, first seen in 1905. Among his later successes were *Seven Keys to Baldpate* (1913), based on a story by Earl Biggers, and *The Song and Dance Man* (1923), in which he played a second-rate VARIETY performer who thinks he is good. Apart from appearing in his own plays, he was a success in O'NEILL'S *Ah, Wilderness!* (1933) and KAUFMAN and Moss HART'S *I'd Rather Be Right* (1937). A statue of Cohan was unveiled in 1959 in Duffy Square, New York, and a musical based on his career, *George M.*, was produced on Broadway in 1968.

COHAN AND HARRIS THEATRE, New York, see SAM H. HARRIS THEATRE.

COHEN, GUSTAVE (1879–1958), French theatre scholar, author of a standard work on LITURGICAL DRAMA published in 1906, *Histoire de la mise en scène dans le théâtre religieux français du moyen âge*. This was revised and republished in 1926 in the light of new material found in the notebooks relating to a PASSION PLAY given at Mons in 1501, which Cohen also published separately in 1925 as the *Livre de conduite et compte des dépenses pour le Mystère de la Passion joué à Mons en 1501*. After his appointment to the staff of the Sorbonne he formed a group of students with whom he produced a number of medieval plays, adapted and slightly modernized by himself. The first, in 1933, was RUTEBEUF'S *Le Mystère de Théophile*, from which the little company took its name—Les Théophiliens. It was followed by ADAM DE LA HALLE'S *Jeu de Robin et de Marion* and by a reconstruction of the earliest known French play, the anonymous *Mystère d'Adam*. On the outbreak of war Cohen went to the United States, returning to France in 1945 to continue his researches into medieval texts. He also published a number of books on the theatre of the Renaissance and the 17th century.

COHEN, SHALOM, see JEWISH DRAMA.

COLERIDGE, SAMUEL TAYLOR (1772–1834), English poet, critic, and philosopher, author of several plays in verse, one of which, *Remorse*, written in 1797 as *Osorio*, was produced at DRURY LANE in 1813 with moderate success. The others, which include several translations from the German, remained unacted, except for a Christmas entertainment which, with alterations by DIBDIN, was produced at the SURREY THEATRE in 1818. Coleridge's chief importance in theatre history lies in his critical and editorial work on Shakespeare, though he was handicapped by ignorance of Elizabethan stage conditions, which MALONE was only just bringing to light.

COLISEUM, The, London, in St Martin's Lane, was originally a MUSIC-HALL, which opened under the management of Sir Oswald Stoll on 24 Dec. 1904. It was the first London theatre to have a revolving stage, made up of three concentric circles. Seating 2,358 people in four tiers, it had a proscenium width of 49 ft and a stage depth of 80 ft. Not at first successful, it closed for 18 months from June 1906, but from 1908 onwards its fortunes improved and many famous 'turns' were seen there. Among the actresses of the time who appeared at this theatre were Ellen TERRY and Edith EVANS in 1917 in scenes from Shakespeare, Lillie LANGTRY, and Sarah BERNHARDT. Diaghilev's Russian Ballet company gave three seasons at the Coliseum between 1918 and 1925. In 1931 variety gave way to MUSICAL COMEDY, inaugurated by the long-running *White Horse Inn*. Subsequently the theatre was used for every kind of entertainment, including ice-shows, and a Christmas pantomime became an annual event for some years from 1936. In 1945 a spectacular REVUE, *The Night and the Music* by Robert Nesbitt, began a long run, after which the theatre housed a series of American musicals—*Annie Get Your Gun* (1947), *Kiss Me, Kate!* (1951), *Guys and Dolls* (1953), *Can-Can* (1954), *The Pajama Game* (1955), *Damn Yankees* and *Bells are Ringing* (both 1957), and *The Most Happy Fella* (1960). The theatre then closed, and reopened in June 1961 as a cinema. In 1968 the building was modernized and taken over by the SADLER'S WELLS Opera Company to become the permanent London home of the English National Opera.

COLLÉ, CHARLES (1709–83), French dramatist and librettist, who provided excellent libretti for such musicians as Monsigny and Rameau and also wrote, over a period of some 20 years, a number of plays which were later published in two volumes, *Théâtre de Société* and *Théâtre des boulevards*. Written for his patron the Duc de Chartres (later the Duc d'Orléans, father of Philippe-Égalité) who was an excellent comic actor, they were imitations of the medieval PARADE as it had survived at the Paris fairground theatres—short, witty, and extremely licentious. The best of them is probably *La vérité dans le vin*. Collé also wrote two comedies, *Dupuis et Desronais*, played with some success at the COMÉDIE-FRANÇAISE in 1763, and *La Partie de chasse d'Henri IV*, based on DODSLEY's *The King and the Miller*. This was banned by the censor on the orders of Louis XV and not seen in public until 1774, though it had been printed and played in the provinces much earlier.

COLLECTIVE CREATION, a theatrical process whereby a group of persons working together develop a production from initial concept to finished performance. In contrast with the prevalent 20th-century process of a playwright writing the script of a play in isolation and then depending on a company under a director to stage it, those companies using a collective approach create their productions autonomously through research, discussion, improvisation, writing, and rehearsal involving the entire group.

This collective process, which became the typical method of the alternative theatre movement of the 1960s and early 70s, may in some ways be regarded as a return to earlier theatrical methods, since it appears to have much in common with the way the COMMEDIA DELL'ARTE companies worked; and actor-playwrights such as Shakespeare and MOLIÈRE collaborated with other members of their companies in the productions of their plays. But the new form of collective creation came about in the 1960s as a revolt against the increasing dependence of a technological society on specialization and a hierarchical work structure. The typical professional theatres of the 20th century had come to use a method based on individual specialists: by contrast, the collective theatres of the 1960s and early 70s focused upon the group rather than on individuals or scripts. They were not concerned with the promotion of individual talents nor did they primarily consider their productions commercial products. The script-oriented Theatre of the ABSURD was concerned with the philosophical alienation of the individual, the new collective theatres expressed the social commitment of the group. A collective method seemed most suitable because it would permit the entire company to participate in the creative process.

At first some of the theatre groups of the 1960s were so idealistic in their commitment to the collective process that instead of assigning responsibilities, everyone did everything regardless of skill or interest. The process by which works were created was often considered more important than the product which resulted because it was in keeping with the anti-authoritarian and anti-specialization mood current among young people. However, as idealism gave way to greater pragmatism in the mid-1970s, the strict democratic principles of the method gave way to more practical ways of distributing responsibility.

Theatre collectives have varied greatly in their structures. From the beginning the LIVING THEATRE was directed by Julian Beck and Judith Malina, who made most of the important decisions, but other members were encouraged to participate in the creative process. Joseph Chaikin was the director of the OPEN THEATER throughout its ten years, but the productions were created by the entire group under his guidance. The SAN FRANCISCO MIME TROUPE, after its director was replaced by a collective organization, no longer publicly identified actors, directors, or writers, but one or more members were assigned to write scripts and others to direct within guidelines determined by the group as a whole. The early productions of the PERFORMANCE GROUP were developed under its founder and director Richard SCHECHNER, who guided improvisations which were the chief means used by the group to create material which was then shaped by Schechner. In England members of the PEOPLE SHOW often worked independently on a forthcoming production and then integrated the separate contributions. In WELFARE STATE INTERNATIONAL members of the company are responsible for sculptural elements within the general scheme devised by John Fox or Boris Howarth, the two directors. Simmons, of the PIP SIMMONS THEATRE GROUP, served both as writer and director, but was open to the ideas of others in the company.

COLLIER, CONSTANCE (1878–1955), English actress, whose long and distinguished career began at the age of 3 when she played Peaseblossom in *A Midsummer Night's Dream*. In 1884 she played the child Cissie in H. A. JONES and Herman's *The Silver King* with Wilson BARRETT and in 1893 she made her first appearance in London, where she became one of the famous Gaiety Girls. She made a great success as Chiara the Gypsy in ESMOND's *One Summer's Day* (1898) at the COMEDY THEATRE, and later joined Fred TERRY to play Lady Castlemaine in KESTER's *Sweet Nell of Old Drury* (1900). Engaged by Beerbohm TREE for HER (then His) MAJESTY'S THEATRE, she remained with him from 1901 to 1908, appearing in all his major productions, and then made her first appearance in New York, dividing her time thereafter between London and the United States. One of her outstanding parts was Nancy in Comyns Carr's dramatization of DICKENS's *Oliver Twist*, which she first played under Tree in 1905, and several times revived. She was also much admired as the Duchess of Towers in the dramatization of George Du Maurier's novel *Peter Ibbetson* (1915) and as the Duchesse de Surennes in MAUGHAM's *Our Betters* (1923); in 1925 she played Gertrude to John BARRYMORE's Hamlet in London. The best of her later parts was Anastasia in the New York production of G. B. Stern's *The Matriarch* in 1930. She was part-author with Ivor NOVELLO, who appeared in both plays, of *The Rat* (1924) and *Down Hill* (1926), which she directed herself. She published her reminiscences as *Harlequinade* (1929).

COLLIER, JEREMY (1656–1726), English non-juror and pamphleteer, best known for his *Short View of the Immorality and Profaneness of the English Stage*, published in 1698, in which he attacked Restoration comedy in general and the plays of CONGREVE and VANBRUGH in particular. Congreve replied later the same year in his *Amendments of Mr Collier's False and Imperfect Citations*..., one among many *Replies, Vindications*, and other pamphlets which Collier's work provoked, his own *Defence* following in 1699 and his *Further Vindication* in 1708. Collier's original work made several valid points, but was marred by pedantry and by ignorance of the history and technique of the theatre and he laid himself open to ridicule by his lack of literary ability and his apparently avid search for bawdry everywhere. Nevertheless his work had a salutary effect, and reflected the contemporary middle-class desire for a drama based on its own standards of morality rather than on those of the Court.

COLLIER, JOHN PAYNE (1789–1883), English dramatic and literary historian, who ruined what would have been valuable research work on Shakespeare and his contemporaries by his habit of forging entries in Elizabethan documents, many of which still persist in spite of rectification by later scholars. It is impossible to say how far his forgeries extend. They have been traced in State Papers and in private collections to which he was given access, and it was his manuscript additions to a second folio of Shakespeare which first aroused suspicion. Proof of his guilt, which he never admitted, was found in his papers after his death.

COLLINS, LOTTIE, see MUSIC-HALL.

COLLINS [Vagg], SAM (1826–65), a chimney-sweep who went on the 'halls', and was probably the first, and certainly the best remembered, of the Irish Comedians. He first appeared at EVANS's 'song-and-supper rooms', making his reputation with such ballads as 'The Rocky Road to Dublin', 'Paddy's Wedding', and 'The Limerick Races'. He later starred at the CANTERBURY and most of London's other major halls, and in 1858 opened the Marylebone Music-Hall, later taking over the Lansdowne Tavern in Islington Green, which he opened as COLLINS'S MUSIC-HALL on 4 Nov. 1863. On Collins's grave in Kensal Green cemetery are carved pictures of his billycock hat, his shillelagh, and the shamrock he always wore on stage.

COLLINS, (William) WILKIE (1824–89), novelist and dramatist, whose crime novels *The Woman in White* (1860) and *The Moonstone* (1868) appear to have inaugurated the genre. The first was originally published in *Household Words*, a periodical run by Charles DICKENS who was the author's intimate friend. They collaborated in a drama, *No Thoroughfare* (1867), which was seen at the ADEL-

PHI THEATRE, and Collins's first plays (he wrote about 15, all forgotten), *The Lighthouse* (1857) and *The Frozen Deep* (1866), were first acted by amateurs at Dickens's private theatre in Tavistock House. Collins himself dramatized his best known novels, *The Woman in White* being seen in 1871 and *The Moonstone* in 1877, both at the OLYMPIC THEATRE.

COLLINS'S MUSIC HALL, London, on Islington Green. This was originally a room in the Lansdowne Arms public house, which had been used as a place of entertainment since 1851. In 1863 it was replaced by a new building seating 600, which Sam COLLINS named after himself, running it with great success until his death two years later. It was taken over by his widow and continued for some time to be one of the most popular halls in North London. In 1897 it was rebuilt and enlarged and became a repertory theatre. Later it was known for its rough comedy and striptease acts. It was severely damaged by fire in 1958 and subsequently demolished.

COLMAN, GEORGE the elder (1723–94), English dramatist and theatre manager, who was attracted to the theatre by his friendship with GARRICK, to whom his first play *Polly Honeycombe* (1760) was originally attributed. It was not acknowledged by its author until after the success of *The Clandestine Marriage* (1766), which followed *The Deuce is in Him*, in which Tattle, played by Tom KING, was probably the first of those 'patter parts' in which the elder MATHEWS was later to excel; adaptations of BEAUMONT and FLETCHER's *Philaster* and Shakespeare's *A Midsummer Night's Dream* (all 1763); and some excellent translations of TERENCE published in 1765. Colman had had some help from Garrick in writing *The Clandestine Marriage*, one of the most popular comedies of its day, but its production led to a breach between the two friends as Garrick refused to play the chief part, Lord Ogleby, leaving it to King. Colman, to show his displeasure, immediately took a lease of COVENT GARDEN, where he remained until 1774 in spite of difficulties and lawsuits with his associates, producing a number of good plays, including those of Oliver GOLDSMITH and his own *The Oxonian in Town* (1767), as well as *King Lear* in 1768 with his own alterations instead of those of Nahum TATE. In 1776 he took over the HAYMARKET THEATRE from Samuel FOOTE, engaged a good company headed by John HENDERSON, and did well for several seasons, being an energetic manager of known probity. One of the chief events of his management was the production of John GAY's *Polly* (1777), the sequel to *The Beggar's Opera*, and among the other plays he produced at this time were the first works of his son George (below).

COLMAN, GEORGE the younger (1762–1836), English dramatist and theatre manager, son of the above. His early plays were produced at the HAYMARKET THEATRE by his father, the most successful being *Inkle and Yarico* (1787), and in 1794 he succeeded the elder Colman as manager. He was not a good manager, being reckless and extravagant and constantly involved in lawsuits, and he was also hampered in his career by a secret marriage he had contracted in 1784 with a young actress Clara Morris. But he was a good dramatist, excelling in the creation of comic characters, among them Dr Pangloss in *The Heir at Law* (1797) and Dennis Brulgruddery in *John Bull; or, the Englishman's Fire-side* (1803). Of his other plays, the best were probably *The Iron Chest* (1796), which later supplied KEAN and other tragedians with a fine part in Sir Edward Mortimer, and *Blue-beard; or, Female Curiosity* (1798), based on a French original. These two were first produced at DRURY LANE. In 1805 Colman left the Haymarket, being succeeded eventually by his brother-in-law David Morris, and in the same year wrote for the elder MATHEWS *The Actor of All Work; or, First and Second Floor*, in which two rooms were shown on the stage simultaneously, an innovation. From 1824 until his death he was Examiner of Plays, in which capacity he showed a prudery and strictness which were unexpected in view of the general tenor of his own works, particularly his comic poems. He is believed to have married as his second wife the actress Maria LOGAN.

COLON, JENNY (1808–42), French actress, who made her first appearance at the Opéra-Comique (see COMÉDIE-ITALIENNE), where her elder sister was for many years a useful member of the company, and two years later toured Britain, making a runaway marriage at Gretna Green. Back in Paris she was seen at a number of theatres, and the critic Janin, reviewing her in a new part at the Théâtre des VARIÉTÉS, called her 'queen and fairy, compounded at once of songs and smiles'. She was capricious and self-willed but full of life and vivacity, and she kept her hold on the public until her early and much regretted death.

COLONIAL THEATRE, New York, see HAMPDEN, WALTER.

COLOSSEUM, London, in Albany Street, Regent's Park. This opened in the early 1830s, and was occasionally used for plays. From 1835 to 1838 it was run by Braham, first in partnership with Frederick YATES, later alone; but he abandoned it when he left the ST JAMES'S THEATRE. The last mention of it dates from 1840. It should not be confused with the Colisseum in Regent's Park, which was not used as a theatre.

COLOSSEUM THEATRE, New York, see HERALD SQUARE THEATRE.

COLUM, PADRAIC (1881–1972), Irish dramatist and man of letters, three of whose early plays, together with those of William BOYLE and Lennox ROBINSON, had a strong influence on the development of realistic Irish drama. *Broken Soil*, a moving and imaginative study of the vagrant artist in conflict with the practical demands of the life of a small farmer, was produced by the IRISH NATIONAL DRAMATIC THEATRE in 1903 and at the ABBEY THEATRE in 1905. Colum then revised it, and as *The Fiddler's House* it was seen again in 1907 and at the Abbey in 1919. *The Land* (1905), Colum's second play, is a firmly-drawn study of farming life. Its characters are memorable, and though the speech is plain and literal, there is poetry in the conflict between the two generations, and between the love of Irish soil and the longing for the adventure of emigration in the minds of the young people. The setting of *Thomas Muskerry* (1910) is a small town; the drab life of its inhabitants is as meticulously delineated as was the passionate life of the farming folk in *The Land*.

Colum's relationship with the Abbey directors, particularly YEATS, was always an uneasy one and in 1905 he left the theatre, emigrating to the United States in 1914. He achieved an enviable reputation as a poet and writer, but few of his later plays were successful. *The Desert*, published in 1907 and revised many times, was finally produced at Dublin's GATE THEATRE in 1932 as *Mogu of the Desert*, and in his last years he returned to Irish themes and to the *nō* theatre of JAPAN earlier explored by Yeats. Three of his five 'Plays for Dancers'—*Glendalough*, *Monasterboice*, and *Cloughoughter*—were produced at the LANTERN THEATRE, Dublin, in 1966 as a trilogy under the title *The Challengers*.

COLUMBINE, the young girl of the English HARLEQUINADE, daughter or ward of the old man Pantaloon and in love with HARLEQUIN, with whom she eventually elopes. She evolved from one of the maidservants in the COMMEDIA DELL'ARTE whose name, Colombina, was used by several actresses of the COMÉDIE-ITALIENNE in Paris in the late 17th century and so passed to England through WEAVER's 'Italian Night Scenes'.

COLUMBUS CIRCLE THEATRE, New York, see MAJESTIC THEATRE (I).

COMEDIA, in the Golden Age of Spanish drama a secular play, whether serious or comic, as distinct from the religious and devotional AUTO SACRAMENTAL. There were two main categories—the MACHINE PLAY (*comedia de tramoyas, de apariencia*, or, more expressively, *de ruido*), and the COMEDY of wit (*comedia de ingenio*), which depended on intrigue rather than on spectacle. The cloak-and-sword play (*comedia de capa y espada*) was a sub-division of this latter category, the characters introducing the necessary complications into the plot by disguising themselves in the cloak and fighting the resultant duels with the sword. Spanish 19th-century dramatists were very much attracted to this type of play, and the name then came to mean any romantic costume play with a strong love interest and some sword-play.

In the foreword to his *Propalladia*, published in 1517, TORRES NAHARRO distinguishes further between realistic comedy (*a noticia*) and the purely imaginative (*a fantasía*). The comedy of character (*a figurón*), in which MORETO Y CABANA excelled, is concerned with universal rather than individual characteristics.

COMEDIA A FANTASÍA, A NOTICIA, see TORRES NAHARRO.

COMEDIAS DE CAPA Y ESPADA, see COMEDIA.

COMÉDIE-BALLET, a form of entertainment in vogue at the Court of Louis XIV, mingling speech, usually in rhymed couplets and originally of little importance, songs, and dancing, in which the King and his courtiers usually joined. Although the genre was not invented by MOLIÈRE, he gave it literary form, making the dialogue equal to or even more important than the music. His first *comédie-ballet* was *Les Fâcheux* (1661) but the finest example of the genre is *Le Bourgeois gentilhomme* (1671). Among Molière's collaborators in the preparation of these Court spectacles were Philippe QUINAULT and de BENSERADE, and on one occasion, in *Psyché* (1671), the great CORNEILLE himself. The music was by Lully, who eventually obtained a monopoly of music in Paris which effectively put an end to the development of such 'plays with music', and so hastened the overwhelming popularity of opera and operetta.

COMÉDIE-CANADIENNE, Théâtre de la, see GÉLINAS, GRATIEN.

COMÉDIE DES CHAMPS-ÉLYSÉES, Paris. Built in 1912 in the Avenue Montaigne as an annexe to the concert hall of the same name, the reputation of this theatre was made in the 1920s with GÉMIER's AVANT-GARDE productions of plays by STRINDBERG and SHAW. He was followed briefly by the PITOËFFS, and then by JOUVET, who began his collaboration with GIRAUDOUX at this theatre in 1928 with a production of *Siegfried*. During the 1950s the theatre was owned by ANOUILH, who staged BOULEVARD PLAYS and revivals of his own works.

COMÉDIE-FRANÇAISE, La, the foremost theatre of France, founded in 1680 by Louis XIV, who ordered the fusion of the company at the Hôtel de BOURGOGNE with the already amalgamated troupes of MOLIÈRE (who had died in 1673) and the Théâtre du MARAIS. Also known as Le Théâtre-Français and La Maison de Molière, it was apparently called the Comédie-Française to distinguish it from the Italian company who took over the Hôtel de Bourgogne as the COMÉDIE-

ITALIENNE. At first the new company continued to play in the theatre in the rue Guénégaud, with Mlle CHAMPMESLÉ, Armande BÉJART (Molière's widow) and her husband GUÉRIN D'ÉTRICHE, BARON, HAUTEROCHE, and the elder POISSON as its leading members. In 1689 they moved to a new theatre specially built for them by François d'Orbay in the tennis-court of the Étoile, rue Neuve des Fossés, St Germain-des-Prés, where they remained until 1770. After some years in the SALLE DES MACHINES, at the Tuileries, the company moved in 1781 to a new theatre on the present site of the ODÉON. The Revolution caused a split, the more revolutionary actors headed by TALMA going to the PALAIS-ROYAL as the Théâtre de la République, while the others under MOLÉ remained *in situ* as the Théâtre de la Nation. The second group soon lost the favour of the public, who considered them 'aristos', and first Laya's *Ami des lois* and then Neufchâteau's *Paméla*, based on the novel by Richardson, caused riots which culminated in the arrest and imprisonment of the actors concerned. In 1799 the company of the Comédie-Française was reconstituted in the theatre occupied by Talma, though in the process it lost the monopoly it had enjoyed for so long.

The organization by Louis XIV of the Comédie-Française, formalized by Napoleon in 1812, resembled that of the medieval CONFRÉRIE DE LA PASSION. The company is a co-operative society in which each actor holds a share or, in the case of younger or less important actors, a half or quarter share. Admission depends on merit, and the aspiring player is allowed to choose his or her own part in tragedy and comedy for his début. If successful he is then on probation as a *pensionnaire*, drawing a fixed salary. After a time, which may vary from weeks to years, he may be admitted to the company as a full member or *sociétaire*, taking the place of a former member who has died or resigned. On retirement, which is not usually permitted under 20 years' service, the *sociétaire* is entitled to a pension for the rest of his life. The oldest actor, in years of service not in age, is the nominal head of the company and is known as the *doyen* (though a director under contract now handles its day-to-day running). Some reorganization took place in 1945, after the Second World War, and again under André Malraux in 1959, when the Odéon, up till then the second theatre of France and attached to the Comédie-Française, was separated from it and put under the charge of Jean-Louis BARRAULT until 1968, only to be returned to it in 1971.

COMÉDIE-ITALIENNE, La, Paris, the name given in 1680 on the formation of the COMÉDIE-FRANÇAISE to the Italian COMMEDIA DELL'ARTE actors who then took over the Hôtel de BOURGOGNE. They had used the theatre ever since GANESSA's first visit in 1570–1 and now settled in to play every day except Friday. In the company were Tiberio FIORILLO, known as Scaramouche, an old man but still incredibly active, Domenico BIANCOLELLI, known as Dominique and like Fiorillo a naturalized Frenchman, and his two daughters, together with Angelo COSTANTINI, known as Mezzetin. For some years before 1680 the company had been interlarding its Italian with French songs and phrases, and even whole scenes. This innovation had been opposed by the French actors, but Biancolelli, who was a favourite of Louis XIV, persuaded him to allow the Italians the free use of French in their productions. This soon led to the acting of some plays entirely in French, and contemporary dramatists like DUFRESNY, REGNARD, and PALAPRAT were not slow to take advantage of this new outlet for their works. The acting, however, continued to be purely that of the *commedia dell'arte*, and even in French plays the actors figured under their own names and were allowed ample scope for improvisation. They fell victims ultimately to their high spirits, and after having been warned several times they were expelled from France in 1697 because, it was said, they had offended Mme de MAINTENON by playing Lenoble's *La Fausse Prude*, which the audience took delight in applying to her. The Italians stayed away from Paris during the lifetime of Louis XIV, but after his death in 1716 they returned under the leadership of the younger RICCOBONI, and again settled in the refurbished Hôtel de Bourgogne. They found that harlequinades were out of fashion, the Italian language out of favour, and the French plays in their repertory old-fashioned, but the younger dramatists were ready to write for them, and they appeared in, among others, many of the finest plays of MARIVAUX, no longer playing in the *commedia dell'arte* style but in a new and specialized manner in which foreign and native material were blended. In 1723, on the death of the Regent, they were given the title of *comédiens ordinaires du roi*, with a yearly grant from public funds which is still paid to their successors the Opéra-Comique. From this time they were Italian in name only and shared the theatre-going public with the Comédie-Française, their productions ranging from true comedy to ballet-pantomime and the newly fashionable VAUDEVILLES. French actors finally ousted the Italians, of whom the last —and incidentally the last Arlequin—was Carlin BERTINAZZI whom GARRICK considered one of the best actors in Paris. The success in 1752 of a visiting Italian *opera buffa* company led the Comédie-Italienne to venture into this new territory, with such success that they were able to absorb their rivals but at the expense of their former repertory and individuality. In Apr. 1783 they left the Hôtel de Bourgogne and opened a fine new theatre on the Boulevard des Italiens, called after them. The success in 1789 of a rival theatre (known under the Revolution as the Théâtre Feydeau, while the Italians called themselves the Théâtre Favart) nearly ruined them, but in 1801 the two groups amalgamated as the Opéra-Comique, the name given to their present theatre built in 1835 after the destruction of the old one by fire; and the

Comédie-Italienne, which had held such an important place in French theatre history, ceased to exist in name as well as in fact.

COMÉDIE LARMOYANTE ('tearful comedy'), a type of play popular in 18th-century France, with which the name of LA CHAUSSÉE is especially associated. While retaining the formal characteristics of stylized verse comedy, these plays retained virtually no traditional comic characters, the writing was sententious and sentimental, and the happy ending was reached only after many tears had been shed during five acts over the misfortunes of virtue unjustly persecuted. This type of drama made a strong appeal to contemporary sensibility.

COMÉDIE ROSSE, see ANCEY, GEORGES.

COMEDY, in its modern sense, covers a wide variety of plays. These differ from tragedy in that they have a happy ending, and from farce in that they contain some subtlety and character-drawing. The word, meaning 'revel-song' from the Greek *comos* and *ode*, was applied to the satiric plays of ARISTOPHANES and also to the FABULAE of TERENCE and PLAUTUS. By medieval times it merely indicated any tale with a happy ending, particularly one written in a colloquial style and dealing with the love affairs of lesser folk. The Renaissance brought back the term to the theatre, but without its former satiric connotation; it also lost in course of time its connection with 'comic' and 'comedian', terms now reserved for low humour, though on the Continent the latter term in the generic sense of actor was used later than in England, where from the 18th century it was applied to players of farcical parts, and from the late 19th century to music-hall performers.

Comedy by its very nature resists translation, for it depends far more than tragedy on local and topical interest and innumerable comedies enormously successful in their own day were soon forgotten. This handicap is, of course, subject to the overriding force of genius, and the comedies of Aristophanes, Shakespeare, and MOLIÈRE can still be enjoyed, even in translation, though they demand from the audience a degree of co-operation in recapturing the spirit of their times.

Comedies may be classified under various headings. The Comedy of Humours, as practised by JONSON and FLETCHER, was influenced by classical models. The Comedy of Intrigue, which subordinates character to plot, originated in Spain and was practised in England by Mrs Aphra BEHN. With it may be classed Romantic Comedy, which also came from Spain and reached its highest point in France during the Romantic Revival led by Victor HUGO and the elder DUMAS. It is marked by exaggeration and violence and by an overpowering use of local colour, costume, and scenery. In the hands of a great poet it may give an illusion of greatness but it easily degenerates into melodrama. The Comedy of Manners originated in France with Molière's *Les Précieuses ridicules* (1658), and Molière himself defined it when he said 'the correction of social absurdities must at all times be the matter of true comedy.' Pushing it to its logical extreme, he adventured into the Comedy of Morals—the correction of abuse by the lash of ridicule—of which the greatest exemplar is *Tartuffe*.

In England, the Comedy of Manners is represented by the plays of CONGREVE, FARQUHAR, VANBRUGH, and WYCHERLEY, later classed as Old Comedy but now generally known as Restoration Comedy. These writers created a completely new dramatic genre, stylish, witty, and artificial, and salted with a degree of indecency which by the early 18th century was becoming unacceptable. The Comedy of Manners revived under SHERIDAN with much wit and less indelicacy, but even Sheridan tended to be influenced by the prevalence of Sentimental Comedy, a type of pathetic play best represented by the work of STEELE, which reflected the sensibility of the rising 18th-century middle class. In France this led to the COMÉDIE LARMOYANTE, which reached its height in the plays of LA CHAUSSÉE and, after blurring the distinction between tragedy and comedy and ousting them from the stage, was in its turn eclipsed by the *drame bourgeois*.

In the English-speaking theatre comedy was during the 19th century virtually replaced by FARCE; modern playwrights, among them Alan AYCKBOURN, Michael FRAYN, Neil SIMON, and Tom STOPPARD, use verbal wit and robust humour to examine contemporary topics which are basically serious. The lack of true comedy is supplied by frequent revivals of plays by WILDE and COWARD (perhaps the last writers of English comedy) and, in a more liberal moral climate, of many Restoration comedies.

The playing of comedy makes heavy demands on the actor, who must be able without affectation or pedantry to suggest elegance, leisure, and a nimble wit. Its broader forms may demand great mobility of countenance and a command of dialect. All comedy acting needs an impeccable sense of timing. In the 19th century comedy was regarded as a specialized art and comedian and tragedian rarely trespassed on each other's territory; but the ability to succeed in both fields is now highly regarded.

COMEDY THEATRE, Leningrad, see AKIMOV, N. P.; Moscow, see RUSSIA AND SOVIET UNION.

COMEDY THEATRE, London, in Panton Street, off the Haymarket. This theatre, which holds 820 people on four tiers, opened on 15 Oct. 1881 and housed a series of light operas until in 1887 Beerbohm TREE took it over for his first venture into management with a production of W. O. Tristram's *The Red Lamp*. He was succeeded by Charles HAWTREY and then by Comyns Carr, under whom Winifred EMERY and her husband

Cyril MAUDE appeared successfully in Sydney GRUNDY's *Sowing the Wind* (1893) and *The New Woman* (1894) and PINERO's *The Benefit of the Doubt* (1895). A lean period followed until in 1902 Lewis WALLER produced TARKINGTON and Sutherland's *Monsieur Beaucaire* which ran for over 400 performances. Gerald DU MAURIER had one of his greatest successes at this theatre with Hornung's *Raffles* (1906), and Marie TEMPEST was also successful in Clyde FITCH's *The Truth* (1907) and in Somerset MAUGHAM's *Mrs Dot* (1908) and *Penelope* (1909). One of the longest runs—710 performances—was scored by John Hartley MANNERS's *Peg o' My Heart* (1914), which starred his wife Laurette TAYLOR. During the First World War the Comedy was occupied by a series of REVUES, including COCHRAN's *Half-Past Eight* (1915) and Charlot's *Seesaw* (1916). Interesting productions between the wars included Norman Macowan's *The Infinite Shoeblack* (1929) and Dorothy L. Sayers's *Busman's Honeymoon* (1938). Though slightly damaged by bombing in the Second World War the theatre continued in use, housing the revue *New Faces* (1940) and Philip King's farce *See How They Run!* (1945). In 1956, after THEATRE WORKSHOP's production of Brendan BEHAN's *The Quare Fellow*, the New Watergate Theatre Club made the Comedy its headquarters and staged there for its members a series of plays which had failed to obtain the LORD CHAMBERLAIN's licence, among them Arthur MILLER's *A View From the Bridge* (1956), Robert ANDERSON's *Tea and Sympathy* (1957), and Tennessee WILLIAMS's *Cat on a Hot Tin Roof* (1958). In 1959 the club was disbanded and Peter SHAFFER's first play, *Five Finger Exercise*, was seen, followed in 1960 by Santha Rama Rau's dramatization of E. M. Forster's novel *A Passage to India*. Later *An Evening of British Rubbish* (1963) and *Son of Oblomov* (1964) explored a distinctively British vein of humour, and were followed by Peter NICHOLS's *A Day in the Death of Joe Egg* (1967). Farce kept the theatre open from 1968 to 1972, when Alan AYCKBOURN's *Time and Time Again* (1972), with Tom COURTENAY, had a successful run. Christopher HAMPTON's *Savages* (1973), from the ROYAL COURT, starred Paul SCOFIELD, who returned to the theatre in Athol FUGARD's *Dimetos* (1976) from NOTTINGHAM PLAYHOUSE. The musical *The Rocky Horror Show* (1979) was originally staged at the Royal Court Theatre Upstairs, and two successive productions, David STOREY's *Early Days* (1980), with Ralph RICHARDSON, and Arthur MILLER's *The Crucible* in 1981, were transfers from the COTTESLOE THEATRE.

COMEDY THEATRE, New York, at 108 West 41st Street, between Broadway and Avenue of the Americas. Built by the Shuberts as a small, intimate playhouse, it opened on 6 Sept. 1909 with ZANGWILL's *The Melting Pot* and in 1912 saw the first production in New York of SHAW's *Fanny's First Play*. The WASHINGTON SQUARE PLAYERS made some of their early appearances there, as did Ruth DRAPER, who in 1928–9 created a record with five months' solo playing. In 1937–8, renamed the Mercury, the theatre housed a stimulating season by Orson WELLES which included *Julius Caesar* in modern dress, in which he played Brutus, DEKKER's *The Shoemaker's Holiday*, SHAW's *Heartbreak House*, in which he played Captain Shotover, and BUCHNER's *Danton's Death*, in which he played St Just. In 1939 the Artef Players appeared in GUTZKOW's *Uriel Acosta* and other Yiddish plays. The building was demolished in 1942.

COMELLA Y VILLAMITJANA, LUCIANO FRANCISCO (1751–1812), Spanish dramatist, very successful in his time but now almost forgotten. He wrote over 100 plays, ranging from falsifications of contemporary events to one-act satires: the latter can still hold attention as sketches of contemporary customs. Comella was heavily satirized by FERNÁNDEZ DE MORATÍN in *El café*, which attacks the absurdity of his over-written historical pieces.

COMMEDIA BILE, see GOLIARD.

COMMEDIA DELL'ARTE, the name usually given to the popular Italian improvised comedy first recorded in 1545, which flourished from the 16th to the early 18th centuries. Other names for it are *commedia a soggetto*, because it was acted in accordance with a scenario or pre-arranged synopsis; *all'improvviso*, because the actors made up their speeches as they went along; *dei* ZANNI, from the comic servants who provided most of the humour; *dei maschere*, because most of the actors wore masks; and *all'italiana*, because its home was in Italy. *Dell'arte*, the only phrase to survive in general use, is hard to translate exactly, but means roughly 'of the profession', the actors being trained professionals. To distinguish it from the popular theatre, the written Italian drama of this time was known as the COMMEDIA ERUDITA.

The history of the *commedia dell'arte* is obscure and has to be pieced together from fragments. It has something in common with the Roman FABULA *atellana* or even with the earlier Greek comedies of MENANDER or, further back still, ARISTOPHANES; but there is as yet no proof of a direct link with any of them. Other influences may have been the SACRA RAPPRESENTAZIONE, which found a place for the improvised antics of mimes and acrobats; costumed clowns in masks showing fixed types, who had long participated in Carnival ceremonies and processions; and extemporizing buffoons found by the early years of the 16th century at Courts, in private houses, and in public meeting-places. Possibly such solo professional performers were gradually drawn in to reinforce the amateur productions of the *commedia erudita*, and so realized the possibilities inherent in disciplined group performances as against solo turns: certainly the connections of the later *commedia dell'arte* with the learned drama were substantial, above all in the borrowing of plots for scenarios. But exactly how the *commedia dell'arte* companies developed, and why their professional productions took the particular form of improvized playing, remains obscure. The genesis of such companies appears to have been the result of subtle interactions between many kinds of entertainment and methods of performance, including the 16th-century rustic farces of southern Italy played at Carnival time by amateur groups headed by such men as Angelo BEOLCO or Andrea CALMO; but their plays were written down and have survived, whereas the true *commedia dell'arte* style was marked by the harmonious collaboration of a group of players improvising dialogue and situations around a previously agreed synopsis.

The chief companies of the 16th century were the GELOSI, with the ANDREINI as their mainstay; the DESIOSI, led by Diana da PONTI and sometimes including Tristano MARTINELLI; the CONFIDENTI, led by the PELLESINI; and the UNITI, under Drusiano MARTINELLI. The next generation carried on the tradition in the competing groups of the second Confidenti, directed by Flaminio SCALA; the ACCESI, under Pier Maria CECCHINI; and the FEDELI, under the younger ANDREINI, known as Lelio.

Of the early patrons of the *commedia dell'arte* the Court of the Gonzaga at Mantua was the most important, followed in the latter part of the 16th century by the Courts of Modena and Parma. The companies soon took to the road, and in the 1570s GANASSA was already leading a company in Spain, followed there a decade later by the Martinelli brothers. One of these, Drusiano, is the first Italian comedian known to have appeared in England (in 1577–8), and a troupe which included one Flaminio Curtesse performed at the English Court in 1602. The extent to which the Italian players' style and methods influenced the English theatre of the period is difficult to determine, but it would be reasonable to assume that something of the *brio* and inventiveness of Italian performances were absorbed into English comic stage business. The proximity of London to Paris, where a permanent Italian company was already settled when MOLIÈRE arrived in 1658 to share a theatre with them, meant that visits could easily be made, and Tiberio FIORILLO was there twice in the 1670s, in 1673 and 1675. His success with the public helped to establish a vogue in the London theatres for farces containing heavily anglicized *commedia dell'arte* elements, notably RAVENSCROFT's *Scaramouche* (1677) and Aphra BEHN's *The Emperor of the Moon* (1687). When the COMÉDIE-ITALIENNE, as the troupe was called after 1680, was banished from Paris in 1697, some of the players may have come to England, but any influence they may have had on the development of the English HARLEQUINADE was inextricably mingled with that of native popular entertainers and visitors from the French fairs. After their return to Paris in 1716 they played more often in French, which had already been creeping in before their banishment, than in Italian, and with their productions of MARIVAUX's early plays became more a part of the history of the French theatre than of the *commedia dell'arte*, which finally disappeared towards the end of the 18th century.

In a *commedia dell'arte* company each member had his or her own character or 'mask' and played nothing else, though the player of a youthful part might later graduate to an elderly one. It is still a matter of dispute as to who were the original inventors of the chief masks, which were adapted or altered to suit successive generations of players; but the basic characteristics of each remained unaltered. The young lovers, whose desire for marriage and its constant thwarting by their elders supplied in general the plot of the play, did not wear masks but depended on their youth and personal charm and eloquence to captivate the audience. For this purpose they were recommended to form their style by reading good authors and committing to memory verse or prose passages which could form the basis of the improvised laments, dialogues, conceits, scenes of jealousy, love scenes, and soliloquies, which they had to provide as needed. In league against them were the old men, fathers or guardians, of whom the most important was the Venetian PANTALONE (later the English Pantaloon), while the favourite mask for the second old man was the Bolognese lawyer, the DOTTORE, usually known as Graziano. An independent role, though he could be a rival for the hand of the young girl, was the braggart CAPITANO, a satire on the alien soldier occupying the country, in Italy at this time the Spaniard though in Venice he could be a Greek *stradiotto* or mercenary. Round these revolved the numerous servants who helped or hindered the lovers—the *zanni* (the Bergamask diminutive of Giovanni, John). It was they who gave the *commedia dell'arte* its characteristic flavour and under various names have infiltrated the literature and theatre of the whole of Western Europe. They took over the functions of the slaves of classical comedy, discharging them with the physical skill of acrobats and the impudence of their immediate prototypes, the Bergamask *facchini* or odd-job men who lounged about the piazza. Greed, shrewdness, and a love of mischief for its own sake were among their outstanding characteristics, added to fertility of invention, an eye to the main chance, and a deep-rooted instinct for survival. Some of the lesser masks displayed a bovine stupidity which provided an amusing contrast to the quick wit of their companions, and all found infinite possibilities for surprise and twists of fortune in the BURLA, or practical joke, and the smaller piece of business known as a LAZZO which formed the best part of their stock-in-trade. Not all the *zanni* survived, but among those that handed on their masks and are still remembered—often by other names—are ARLECCHI-

NO (Harlequin), PEDROLINO (Pierrot), and PULCINELLA (Punch); the only female servant to have survived is Colombina (COLUMBINE), originally an attendant on the leading lady. The French theatre adopted and made its own the masks of MEZZETINO (Mezzetin), PASQUINO (Pasquin), SCAPINO (Scapin), and SCARAMUCCIA (Scaramouche), the last having in him originally something of the braggart soldier; and from a compound of other masks now forgotten—Buratino, Francatrippe, Fritellino, Trappolino—it created its own inimitable CRISPIN.

The average company consisted of about 12 to 15 members under an acknowledged leader, though other outstanding performers carried considerable weight in matters concerning the troupe, particularly in choice of scenario and methods of staging, the latter varying according to the status of the company and the nature of the place where they were to play. The smaller travelling troupes carried portable equipment which included basic props, costumes, and canvas scenery, and a platform stage for erection in such playing areas—open spaces and public squares—as opportunity offered. When set up the stage would be at about head-height to standing spectators, the players performing before a painted canvas backcloth on which was usually depicted the traditional comic scene of a piazza or street with houses. For more prosperous companies, who could expect to rent a hall or theatre for a time—as the Andreini did in Paris—a similar background could be provided by portable wings showing rows of houses with practicable doors and windows and a perspective backcloth. Some companies at the highest level might be accommodated in their patron's private theatre when not on the road, and have at their disposal the most sophisticated scenic equipment; but basically, until the end of the 17th century, a variation on the street scene provided all that was necessary for a *commedia dell'arte* production, in which performance rather than *mise en scène* was the determining factor. And the performance was ruled by the choice of scenario. Although the essence of the Italian style was improvization a skeleton plot, with indications of possible tricks and cues for music and dance, helped to ensure that the improvization developed in a way which would lead to the desired denouement. Successful extemporizing depended on the players' innate theatrical sense, their ability to supply and pick up cues readily, and a constant awareness of audience response. Although actors always played the same part, the details of it could vary enormously; preparation for it, even after years of experience, entailed study, including the accumulation of relevant material from any accessible sources, particularly the *zibaldoni* or commonplace books which contained speeches suitable for all occasions, and the acrobatic and other spectacular comic scenes demanded constant testing and rehearsing. A number of collections of such skeleton plots survive, of which Flaminio Scala's *Teatro* (1611) is perhaps the most unusual: it is printed; it represented his personal taste, if not always his own composition; and it is closely associated with the Gelosi. It contains 39 comedies, one tragedy, a tragi-comedy, a mixed entertainment, a pastoral, and a few fairytale plays. Other scenarios exist in manuscript in many Italian libraries, in Paris, and in Leningrad. In the larger collections farce, with some comedy, forms the largest part; much is drawn at second or third hand from classical and neo-classical plays; and in the 17th century the taste for Spanish drama (see COMEDIA) opened the way for a larger proportion of melodramatic and sentimental plots. Pastorals with a strong infusion of buffoonery were also evidently popular. The way in which these 'dried plays' could be put to soak and swell is explained by Perucci in his *Arte rappresentativa* (1699), an invaluable handbook much of which has been reprinted by Petraccone in his *La commedia dell' arte, storia, tecnica, scenari* (1927). Other hints on production can be found in the semi-critical writings of such players as Cecchini, BARBIERI, and RICCOBONI, and it is sometimes possible to deduce details of technique from the texts of the plays themselves.

By the 18th century the vitality of the *commedia dell'arte* was beginning to flag. France had absorbed much of it into her own drama, the German-speaking countries had drawn on it for their own clowns, among them HANSWURST, and general decadence had set in. In an attempt to revive the splendour of the old days GOLDONI substituted a written text for improvization. His contemporary and rival GOZZI preferred to use the old masks and methods for his own purposes, and in 1761 provided SACCHI's company with the first of a series of *fiabe*, in which he mingled fairytales with farce: with him the *commedia dell'arte* made a good end. Improvisation soon became a lost art and actors were no longer expected to be acrobats, dancers, and singers. Something of the tradition lingered on in puppet-shows, in mask names, in the English harlequinade, and in the pantomime company which still plays in the Tivoli Gardens in Copenhagen. The idea of *commedia dell'arte* skills, even though no longer practised, remains a vital force in the modern theatre, which with COLLECTIVE CREATION is turning to a new form of improvisation. Many directors, notably COPEAU and REINHARDT, drew inspiration from the Italian comedy, and in 1964 and 1973 Peppino DE FILIPPO, in the *Metamorphoses of a Wandering Minstrel* seen during the WORLD THEATRE season, gave London audiences some idea of what a *commedia dell'arte* performance may have been like. Similar intimations have been given by Vito Pandolfi in Jean Renoir's film *The Golden Coach*, by the work of Giacomo Oreglia with the actors of the KUNGLIGA DRAMATISKA TEATERN in Stockholm, and by certain of Giorgio Strehler's productions at the PICCOLO TEATRO DELLA CITTÀ DI MILANO.

COMMEDIA ERUDITA, the early 16th-century Italian 'learned' counterpart to the 'popular' COMMEDIA DELL'ARTE. Based on the Roman comedies of PLAUTUS and TERENCE, it tended towards the creation of types rather than individuals, as is clearly shown by the list of comic types appended to the prologue of ARETINO's comedy *Il Marescalco* (*The Sea Captain*, 1533). Typical of the genre are the comedies of DELLA PORTA, which are all reducible to a single plot—two young people crossed in love, whose difficulties are resolved by a servant; the plays usually end with a recognition scene, after it has been revealed that one of the characters has previously been carried of by pirates, sold into slavery, or captured by the Turks. The writers of learned comedies were convinced that they had breathed new life into classical comedy—as indeed they had, but mainly thanks to their observation of contemporary life and manners. Among the masterpieces of the *commedia erudita*, apart from the works of Aretino, are the comedies of ARIOSTO, *La Calandria* (1506) by Cardinal Bibbiena, and *La Mandragola* (1520) by MACHIAVELLI.

COMMONWEALTH THEATRE, New York, see ANTHONY STREET THEATRE.

COMMUNITY ARTS SERVICE (C.A.S.), Auckland, a theatre company sponsored by the Adult Education Centre of Auckland University, which toured the North Island of New Zealand playing in theatres and halls of every description, often with the slenderest of stage resources, and bringing live theatre to many small country communities. It also provided an excellent training ground for many young actors. During its lifetime, from 1947 to 1962, its most notable productions included *Twelfth Night*, PRIESTLEY's *An Inspector Calls*, Somerset MAUGHAM's *The Circle*, Michael REDGRAVE's adaptation of Henry JAMES's *The Aspern Papers*, and in 1958 the first production in New Zealand of BECKETT's *Waiting for Godot*, a landmark in the country's theatre history.

COMMUNITY THEATRE, which began to flourish in England in the early 1970s, is closely allied to political theatre and tends to have a socialist commitment which may also embrace feminism and gay liberation, though there are also specialist groups devoted to these causes. Community theatre groups, which usually perform in such non-theatrical places as working-men's clubs, public houses, village halls, streets, and open spaces, seek to relate their activities to a particular locality by presenting material of local interest and by creating a closer rapport with their audiences than is possible in a conventional theatre. The leading community theatre group in England is Inter-Action, founded in 1968 by an American, Ed [Edward David] Berman, who was also responsible for the foundation of the London FRINGE THEATRES Ambiance and Almost Free. The group runs Dogg's Troupe, which presents plays for children with audience participation on housing estates, and also a Fun Art Bus for other theatrical activities.

Other community theatre groups include the Covent Garden Community Theatre in London, the Interplay Community Theatre in Leeds and West Yorkshire, the Second City Theatre Company in Birmingham and the West Midlands, and the Solent People's Theatre in Hampshire. A form of community theatre which has no political aims, involving professionals, amateurs, and local schoolchildren in plays based on subjects of local interest, is emerging, particularly in the Dorset-Devon area under Ann Jellicoe.

In the United States community theatre is largely in the hands of the regional theatres (see UNITED STATES OF AMERICA) and off-Broadway and off-off-Broadway groups. (See also COLLECTIVE CREATION and HAPPENING.)

COMPAGNIA DEI GIOVANI, see ITALY and VALLI, ROMOLO.

COMPAGNIA REALE SARDA, Italian theatre company founded in 1821 in Turin. Its repertory was initially dominated by 18th-century drama and spectacle, particularly the works of GOLDONI and Metastasio; this gradually gave way to contemporary 'WELL-MADE' PLAYS in the French style. The company attracted the talents, at some stage in their careers, of many of the leading actors and actresses of the day, partly by virtue of the standing and stability due to its receiving a regular official grant. Perhaps the most distinguished of these players were Tommaso SALVINI and Adelaide RISTORI. The latter acted with the company at the age of 15 and made her Paris début with it in 1855, the year the company broke up because its grant was withdrawn. Its history was written by G. Costetti in 1893.

COMPAGNIE DES QUINZE, French theatre company, formed by several of the actors from COPEAU's itinerant troupe, Les Copiaus, after its dissolution in 1929. With Copeau's nephew Michel SAINT-DENIS at its head, the company achieved a distinguished reputation, particularly in the plays of André OBEY—*Noé, Le Viol de Lucrèce, La Bataille de la Marne*—written specifically to display the actors' expertise in diction, gesture, mime, and movement; but it did not long survive the departure of some of its members in 1934 and of Saint-Denis himself, who left for England in the following year.

COMPARTMENT BATTEN, see BATTEN.

COMPOSITE SETTING, the modern equivalent of the medieval MULTIPLE SETTING, where the action of a play takes place in several distinct areas, such as rooms, gardens, and streets, shown on stage at the same time. The lighting in each area intensifies as the action moves into it, dimming as it leaves.

Usually only one area is lighted at any one time, but two or three areas may be lit and in use at the same time. Excellent examples of composite settings were seen in London in 1935 in Rodney Ackland's *The Old Ladies*, designed by Motley, and in 1949 in Tennessee WILLIAMS's *The Glass Menagerie*, designed by Jo MIELZINER for the American production of 1947. In America the composite setting is known as the Simultaneous-Scene Setting; it may be regarded as one form of the permanent SET (unit setting or formal setting in America).

COMPTON, EDWARD (1854–1918), English actor-manager, son of Henry COMPTON, and the founder of the Compton Comedy Company, which from 1881 to 1918 toured the English provinces in a repertory of plays by Shakespeare, SHERIDAN, GOLDSMITH, and other classic dramatists. An excellent actor, perhaps insufficiently appreciated in his day, he appeared several times in London, playing opposite his wife, Virginia Frances Bateman (1855–1940), the youngest daughter of H. L. BATEMAN. With her sisters she appeared on stage as a child in the United States, and in 1871 played under her parents' management at the LYCEUM THEATRE in London. After her marriage in 1882 she appeared exclusively under her husband's management. She was the mother of the novelist Compton Mackenzie (1883–1972), who reverted to the original family name, and of Francis, known as Frank (1885–1964), Viola (1886–1971), Ellen, and Fay (below), who were all on the stage.

COMPTON, FAY [Virginia Lillian Emmeline] (1894–1978), English actress, daughter of Edward COMPTON. She had a long and distinguished career, making her début in the *Follies* of her first husband H. G. Pélissier and being the first to play the title role in BARRIE's *Mary Rose* (1920) and Phoebe Throssel in his *Quality Street* (1921), as well as Elizabeth in MAUGHAM's *The Circle* (1921) and Constance in his *The Constant Wife* (1927). She also appeared with her third husband Leon QUARTERMAINE in Ashley DUKES's *The Man With a Load of Mischief* (1925). Among her later successes were Fanny Grey in *Autumn Crocus* (1931) and Dorothy Hilton in *Call It a Day* (1935), both by Dodie SMITH, Ruth in COWARD's *Blithe Spirit* (1941), and Gina in IBSEN's *The Wild Duck* (1948). At the first CHICHESTER FESTIVAL in 1962 she played Marya in CHEKHOV's *Uncle Vanya* and on the opening of the Yvonne Arnaud Theatre in Guildford in 1965 was seen as Anna in TURGENEV's *A Month in the Country*. In the course of her career she played Ophelia to the Hamlets of John BARRYMORE (1925) and John GIELGUD (1939) and was seen in a number of Shakespearian parts at the OPEN AIR THEATRE, Regent's Park, and with the OLD VIC company, for which she also played Lady Bracknell in WILDE's *The Importance of Being Earnest* in 1959. She appeared, too, in pantomime and variety. In 1926 she published a volume of reminiscences, *Rosemary*.

COMPTON, HENRY [Charles Mackenzie] (1805–77), actor and theatre manager, a collateral descendant of the Scottish actor-manager David Ross (1728–90). The son of a Scottish minister, he took his grandmother's maiden name when he decided to go on the stage. He had a long and successful career, married an actress, Emmeline Montague (d. 1910), and had seven children connected with the stage, the most important being Edward (above). His only daughter Katherine married the dramatist R. C. CARTON.

CONCERT PARTY, see PIERROT TROUPES and REVUE.

CONCERT THEATRE, New York, see JOHN GOLDEN THEATRE (1).

CONDELL, HENRY (?–1627), Elizabethan actor, first mentioned in 1598 as playing in JONSON's *Every Man in His Humour*. His only other known role is the Cardinal in WEBSTER's *The Duchess of Malfi* (1614), but he is believed to have appeared in some of Shakespeare's plays in such supporting roles as Horatio in *Hamlet*. Evidently a well-known and respected figure in Jacobean theatre society, and one of the SHARERS in both the GLOBE and BLACKFRIARS, he left the stage to devote himself to business matters in about 1616. Shakespeare made no provision for the printing of his plays, and the quarto copies of single plays were more often than not incomplete and badly mutilated. Condell and his fellow-actor HEMINGE were jointly responsible for the collecting and printing in one volume—the so-called First Folio—of 36 plays, published in 1623, price 20 shillings. Although there is no record of how many copies were printed, a second edition was not called for until 1636.

CONFEDERATE THEATRE, Richmond, Va., see OGDEN, R.D'O.

CONFIDENTI, a company of the COMMEDIA DELL'ARTE, first mentioned in 1574, which toured Italy, France, and Spain, and by 1580 was under the control of Giovanni PELLESINI. Some of the GELOSI may have joined this company after the death of Isabella ANDREINI in 1604, but its history remains obscure until it re-emerges in about 1610, under the patronage of Giovanni de' Medici, with Flaminio SCALA as its leader. It toured mainly in Italy and was disbanded in 1621, some of the actors joining the FEDELI in Paris.

CONFRÉRIE DE LA PASSION (Confraternity of the Passion), an association of the burghers of Paris, formed in 1402 for the performance of religious plays. Their first permanent theatre was in the disused hall of the Guest-House of the Trinity outside the walls of Paris, in the direction of the Porte Saint-Denis. In 1518 they were given a monopoly of acting in Paris which later proved a serious hindrance to the establishment of a permanent professional theatre. When the Confraternity left their first home they built themselves a theatre in the ruins of the palace of the Dukes of Burgundy, which became known as the Théâtre de l'Hôtel de BOURGOGNE. No sooner was it ready for their occupation in 1548 than they were forbidden to act religious plays, though still retaining their monopoly. Deprived of the major part of their repertory, they gave up acting, and from about 1570 onwards leased their theatre to travelling and foreign companies, always retaining, however, a couple of boxes and the right of free entry. The Confraternity kept a jealous eye on any company that tried to establish itself in Paris, and usually succeeded in having it sent away. When a company finally became their permanent tenants, they still insisted on the payment of a levy for every performance, and were constantly engaged in recriminations and lawsuits with other companies as well as with their own. The first breach in their privileges was made when in 1595 the FAIRS of St-Germain and St-Laurent were thrown open to provincial actors, but the monopoly lingered on until 1675, after the death of MOLIÈRE and only a few years before the foundation of the COMÉDIE-FRANÇAISE.

CONGREVE, WILLIAM (1670–1729), the greatest English writer of the Restoration comedy of manners. Born in England, he was educated in Ireland, where Jonathan Swift was one of his schoolfellows, and in 1689 returned to England. Four years later he had his first play, *The Old Bachelor*, produced at DRURY LANE with a fine cast headed by BETTERTON and Mrs BRACEGIRDLE. It was well received, its dialogue giving a witty edge to the theatrical conventions of the time, and was quickly followed by *The Double Dealer* (1694), played by the same company. This was less successful, perhaps because of the complexities of the plot, but the commendation of Queen Mary improved its fortunes. When in 1695 Betterton seceded from Drury Lane and reopened the old LINCOLN'S INN FIELDS THEATRE his first production was Congreve's *Love for Love*, his most successful play and one which calls for a high degree of skill in the acting. Betterton himself played Valentine, with Mrs Bracegirdle as Angelica. Congreve's one tragedy *The Mourning Bride* (1797) failed to please the critics, though it was a success with the public and the players, the part of Almeria, first played by Bracegirdle, being long a favourite with tragedy queens. Congreve's last, and in some critics' opinion his best, play was *The Way of the World* (1700); it was poorly received, and a combination of pique, laziness, and ill health kept him out of the theatre. He died of injuries received when his coach overturned on the way to Bath, and was buried in Westminster Abbey.

Although *The Mourning Bride* has not been seen in London since 1804 except for a brief appearance at the SCALA THEATRE in 1925, the three best known comedies continue to hold the stage. *Love for Love*, in which GIELGUD played Valentine in 1943–4, was revived in 1965 by the NATIONAL THEATRE, which also revived *The Double Dealer* in 1978. *The Way of the World*, seen in 1924 with Edith EVANS as Millamant, in 1953 with Gielgud as Mirabell and Margaret RUTHERFORD as a superb Lady Wishfort, and in 1956 with John CLEMENTS and Kay Hammond as Mirabell and Millamant, was also revived by the ROYAL SHAKESPEARE COMPANY in 1978 with Judi DENCH as Millamant.

CONNELLY, MARC(us) COOK (1890–1980), American dramatist, who was working in theatrical journalism in New York when he met George S. KAUFMAN, with whom he collaborated in a number of successful plays including *Dulcy* (1921), *To the Ladies* and *Merton of the Movies* (both 1922), and *Beggar on Horseback* (1924). He is chiefly remembered for his own play *The Green Pastures* (1930), which was awarded a PULITZER PRIZE. Based on Roark Bradford's Southern sketches *Ol' Man Adam an' His Chillun*, with a cast of Negro actors led by Richard B. HARRISON as De Lawd, this endeavoured to describe the image that black plantation workers in the Southern states had of God and Heaven. In spite of some original misgivings it proved immensely popular and ran for over five years, being several times revived. Connelly wrote several other plays, including *The Farmer Takes a Wife* (1934) and *Everywhere I Roam* (1938), but never again achieved such a success. His memoirs, *Voices Offstage*, were published in 1969.

CONQUEST, GEORGE AUGUSTUS (1837–1901), English dramatist, acrobat, and pantomimist, the only son of the theatre manager, Benjamin Oliver (1805–72), who had adopted the name of Conquest when he first went on the stage. George Augustus acted under his father's management at the GARRICK THEATRE in Leman Street as a child and was then sent to Paris to study, his parents intending him for a musical career. He, however, preferred to return to the theatre, and alone or in collaboration wrote some 40 PANTOMIMES, celebrated for their brilliant flying ballets and acrobatic effects, and about 100 MELODRAMAS, many of them adaptations from the French. These were mainly produced at the GRECIAN THEATRE, which he took over when his father died while in management there. A competent artist, Conquest designed, painted, and made scenery, properties, and masks for his productions, in many of which he appeared himself. He was a powerful character actor, and an excellent animal impersonator. Offstage he stammered badly but this was never apparent in his acting. In 1879 he sold the Grecian and embarked on what was intended to be a lengthy tour of the United States, but a serious accident on stage soon after his first appearance forced him to return to England. In 1881 he took over the SURREY THEATRE and again made it famous for pantomimes and melodramas. He had three sons, all pantomimists. George (1858–1926) inherited the Surrey, sold it in 1904, and was for a time in 1910 manager of the BRITANNIA, Hoxton; Fred (1871–1941) was well known for his pantomime Goose; Arthur (1875–1945) played his MUSIC-HALL act 'Daphne the Chimpanzee' for nearly 20 years in pantomime.

CONSERVATOIRE, Paris, see SCHOOLS OF DRAMA.

CONTACT THEATRE COMPANY, see MANCHESTER.

CONTAT, LOUISE (1760–1813), French actress, who made her first appearance at the COMÉDIE-FRANÇAISE in 1776. In tragedy she was, like her tutor and fellow actor PRÉVILLE, distinguished but cold, and it was soon evident that her bent was for high comedy. Her success as Suzanne in the first production of BEAUMARCHAIS's *Le Mariage de Figaro* (1784), in which her younger sister, aged 13, appeared as Franchette, set the seal on her growing reputation. She was also particularly admired in the plays of MARIVAUX when they were eventually admitted to the repertory. She retired at 50, having given up youthful parts in favour of elderly matrons owing to her increasing size.

CONTEMPORARY THEATRE, A, see SEATTLE REPERTORY THEATRE.

CONTI, ITALIA (1874–1946), English actress, who made her first appearance on the stage in 1891 at the LYCEUM and later toured extensively in England and Australia. In 1911 she was asked by Charles HAWTREY to train the children, among them Noël COWARD, who were to appear with him in the fairy play *Where the Rainbow Ends* by Clifford Mills and John Ramsey. This led her to found the Italia Conti School for the stage training of children, which still survives, and to which she devoted the rest of her life, one of her outstanding pupils being Gertrude LAWRENCE. She continued to act intermittently, being seen annually from 1929 to 1938 at the Holborn Empire as Mrs Carey in the Christmas production of *Where the Rainbow Ends.* (See also SCHOOLS OF DRAMA.)

CONTOUR CURTAIN, see CURTAIN.

CONTRASTO, see ITALY.

CONWAY, FREDERICK BARTLETT (1819–74), American actor, son of William (below). He was an important figure in the development of the theatre in Brooklyn, New York. With his wife, the actress Sarah Crocker (1834–75), he opened in 1863 the first theatre to be erected there, and remained in control very successfully until 1871, when he built and opened the larger BROOKLYN THEATRE. There, with a good stock company and frequent visits from American and overseas stars, he continued to provide excellent entertainment until he died. His daughter Minnie [Marianne] (1854–96) was also on the stage, making her first appearance under his management in 1869. She married the actor Osmond TEARLE.

CONWAY [Rugg], WILLIAM AUGUSTUS (1789–1828), Irish-born actor, known from his exceptional good looks and fine carriage as 'Handsome' Conway. He first appeared on the stage in Dublin in 1812 and then went to England. He was a good actor but morbidly sensitive, and is said to have thrown up an excellent part in London because of adverse criticism. In 1824 he went to New York and appeared at the PARK THEATRE in several Shakespearian parts with great success. He also played Edgar to the Lear of Thomas COOPER, then at the height of his fame, Faulconbridge to his King John, and Joseph Surface, in SHERIDAN's *The School for Scandal*, to his Charles Surface. An actor of the KEMBLE school, Conway seemed on the threshold of a brilliant career; but he threw himself overboard on a voyage to Charleston and was drowned.

COOK, PETER, see REVUE.

COOKE, GEORGE FREDERICK (1756–1812), eccentric and unstable English actor, who made his first appearance at Brentford in 1776 and then, except for a fleeting engagement at the HAYMARKET THEATRE, spent more than 20 years as a strolling player in the provinces. On 31 Oct. 1800 he appeared as Richard III at COVENT GARDEN. He was immediately successful and remained there for 10 years, playing a wide range of parts but constantly in trouble with the management, undisciplined, dissipated, usually in debt, and often in prison. He seemed to act better when drunk and was probably somewhat insane from constant inebriation. In 1810 he went to New York and appeared before an enthusiastic audience at the PARK THEATRE, but soon proved as undependable as in London and lost his popularity. With DUNLAP as his manager he toured the United States, already a dying man. He was buried in New York where Edmund KEAN, who had a high opinion of his capabilities, erected a monument to his memory. Careless in studying his parts, he picked them up quickly and played them intuitively. A powerful actor with fiery eyes and a lofty forehead, he was unequalled at expressing the darker passions.

COOKE, THOMAS POTTER (1786–1864), English actor, known as 'Tippy' Cooke, who in 1804 forsook the Royal Navy in order to appear on stage at several minor theatres in London. He made his first outstanding success in 1820 as Ruthven in PLANCHÉ's *The Vampire; or, the Bride of* *the Isles* and was then seen as the Monster in *Presumption; or, the Fate of Frankenstein* (1823), based on Mary Shelley's novel; but it was as William in the 400 consecutive performances of JERROLD's nautical melodrama *Black-Ey'd Susan* (1829) and as Harry Halyard in Edward Blanchard's *Poll and My Partner Joe* (1857) that he was best remembered. He retired in 1860 after a farewell performance at COVENT GARDEN.

COOPER, GLADYS (1888–1971), DBE 1967, English actress, who excelled in drawing-room comedy but was by no means limited to this style. She made her first appearance in London in 1906 in *The Belle of Mayfair*, and was subsequently seen in both MUSICAL COMEDY and straight plays with equal success. In 1916, with Frank Curzon who remained with her until 1928, she took over the management of the PLAYHOUSE, where she presented and often played in a varied programme of old and new plays. She was particularly admired in revivals of PINERO's *The Second Mrs Tanqueray* in 1922 and SUDERMANN's *Magda* in 1923 and was in the first production of MAUGHAM's *The Sacred Flame* (1929). She occasionally appeared for other managements, playing opposite Gerald DU MAURIER at the ST JAMES'S THEATRE in the first production of LONSDALE's *The Last of Mrs Cheyney* (1925). She remained at the Playhouse until 1933 and a year later made her first appearance in New York, in Keith Winter's *The Shining Hour*, repeating the role in London in the same year. Her later stage career was divided between London and New York in roles ranging from Desdemona in *Othello* and Lady Macbeth in *Macbeth* (N.Y., 1935) to Fran Dodsworth in Sidney HOWARD's *Dodsworth* (London, 1938), adapted from Sinclair Lewis's novel. She was seen in London in Noël COWARD's *Relative Values* (1951), returning to New York to appear as Mrs St Maugham in Enid BAGNOLD's *The Chalk Garden* (1955), in which she also made her last stage appearance in London shortly before her death. By the marriage of her daughter Joan by her first husband H. J. Buckmaster (she was three times married) she became the mother-in-law of Robert MORLEY.

COOPER, THOMAS ABTHORPE (1776–1849), British-born American actor, and one of the first to become an American citizen. He made his début at Edinburgh in 1792, and three years later appeared in London, at COVENT GARDEN as Hamlet and Macbeth. In 1796 he went with WIGNELL to the CHESTNUT STREET THEATRE in Philadelphia and a year later was seen in New York as Jaffier in OTWAY's *Venice Preserv'd*, always one of his best parts. After a quarrel with Wignell he joined the AMERICAN COMPANY under DUNLAP and in 1806 took a lease of the PARK THEATRE (1) in New York, but without much success. With Stephen PRICE as his partner he then toured the eastern circuit—New York, Philadelphia, and Charleston, South Carolina—in the big tragic roles of Shakespeare, being much admired as Macbeth. A handsome man, with a fine voice, much eloquence, and a dignified presence, he went on acting too long, and towards the end of his life his popularity declined.

CO-OPTIMISTS, see PIERROT TROUPES.

COPEAU, JACQUES (1879–1949), French actor and director, whose work had an immense influence on the theatre in Europe and America. Although interested in the innovations of ANTOINE, he disliked the realistic theatre and in 1913 took over the Théâtre du VIEUX-COLOMBIER in an effort to bring back, as he said, 'true beauty and poetry' to the French stage. His repertory was based mainly on MOLIÈRE and Shakespeare, and he had one of his earliest successes in 1914 with *Twelfth Night*, translated by Théodore Lascaris as *Nuit des rois*. In 1921 he produced *The Winter's Tale* as *Conte d'hiver*, and later published French versions of Shakespeare's major tragedies. At the ATELIER in 1937 he staged, as *Rosalinde*, his own free adaptation of *As You Like It.* He trained his actors himself; in his original company were Charles DULLIN and Louis JOUVET who later founded their own theatres. During the First World War Copeau spent two years (1917–19) at the GARRICK THEATRE in New York, and returned to the Vieux-Colombier convinced that the training of young actors was an essential part of the reforms he envisaged in the theatre. In 1924 he withdrew to his native Burgundy with a group of youngsters, later known as 'les Copiaus', with whom he worked on various facets of technical training, giving performances at irregular intervals in the surrounding villages. In 1930 several of his actors joined the COMPAGNIE DES QUINZE, directed by his nephew and collaborator Michel SAINT-DENIS. By 1936 the importance of Copeau's work had been recognised and he became one of the directors of the COMÉDIE-FRANÇAISE, retiring in 1941. Copeau's ideas on the theatre are summed up in the two *Cahiers du Vieux-Colombier* (1920, 1921) which he edited. He also published an annotated edition of Molière.

COPPIN, GEORGE SELTH (1819–1906), English-born actor-manager, usually referred to as 'the father of the Australian theatre'. The son of strolling players, he arrived in Australia in 1842 and at various times was in actor-management in Sydney, Tasmania, and Adelaide, finally settling in Melbourne. A low comic and clown in Shakespeare, he brought over 200 entertainers to Australia and built six theatres, the most famous, the Olympic in Melbourne (known as 'the Iron Pot'), having been pre-fabricated in Manchester. Actors brought to Australia by Coppin included Gustavus Vaughan BROOKE, Charles KEAN, Barry SULLIVAN, Edwin BOOTH, Madame CÉLESTE, and the one who was to have the greatest effect upon the Australian theatre, J. C. WILLIAMSON.

COPYRIGHT IN A DRAMATIC WORK. 1. *Great Britain*. By the beginning of the 19th century, the right of the dramatist (as of other writers) to prevent the making of copies of his work by unauthorized persons for a limited period of time had been established by various Acts of Parliament. There was as yet, however, no statutory protection for stage productions. It was the Dramatic Copyright Act of 1833, commonly known as BULWER-LYTTON's Act, which finally gave to the author of 'any tragedy, comedy, play, opera, farce, or other dramatic entertainment', again for only a limited period, the sole right to perform it, or authorize its performance. In 1842 the Literary Copyright Act consolidated the law relating to the protection of dramatic as well as literary and musical property and brought within the terms of a single statute the two rights so far recognized, 'copyright' or the right of 'multiplying' copies, and 'performing right' or the right of representation. Performances of dramatizations of non-dramatic works were not covered by the Act, and the only way in which a novelist, for instance, could protect himself from unauthorized stage productions of his works was by dramatizing those works himself and so acquiring the right of representation in them as 'dramatic pieces'. Another institution brought into being by the ambiguous phrasing of the Act was the 'copyright performance'. It was generally believed, though with no clear legal support, that if a play was published before being performed the 'performing right' was lost. This led to the practice, much indulged in by SHAW for instance, of hiring actors to give a public reading of a manuscript play in a hall, or other 'place of public entertainment', usually without costumes or scenery. These 'copyright performances' have often led to considerable confusion in the dating of first performances of well known plays. Meanwhile the Berne Convention of 1886, which had established the International Copyright Union, having been amended in 1896 and in 1908, it was essential that Great Britain should keep abreast of international developments, particularly those covered by the 1908 revisions. A committee was therefore appointed, many of whose recommendations were incorporated in the Copyright Act of 1911, which for the first time merged the right of multiplying copies and the right of representation in the general term 'copyright'; and for the first time no formalities were necessary to obtain copyright. This Act remained in force until the Copyright Act of 1956, which was made necessary by the introduction of films, records, radio, and television, and by the provisions of the 1948 Brussels Convention and the Universal Copyright Convention signed at Geneva in 1952.

Under the 1956 Act copyright subsists in *inter alia* every original dramatic work written by a British subject or other 'qualified person'. Section 2 (5) lists the 'acts restricted by copyright' as reproducing the work in any material form; publishing the work; performing the work in public; broadcasting the work; causing the work to be transmitted to subscribers to a rediffusion service; and making an adaptation of the work. 'Reproduction' is defined as including, in the case of a dramatic work, reproduction in the form of a record or cinematograph film and 'performance' as covering 'any mode of visual or acoustic presentation' (Section 48). 'Adaptation' in the case of a non-dramatic work means 'a version of the work (whether in its original language or a different language) in which it is converted into a dramatic work' and, in the case of a dramatic work, means 'a version of the work (whether in its original language or a different language) in which it is converted into a non-dramatic work' (Section 2 (6)).

'Dramatic work' includes 'a choreographic work or entertainment in dumb show if reduced to writing in the form in which the work or entertainment is to be presented but does not include a cinematograph film, as distinct from a scenario or script for a cinematograph film' (Section 48).

The phrases 'original dramatic work' and 'in public' are not defined in the Act and their interpretation has therefore been a matter for the Courts. However, for the avoidance of doubt about the nature of school performances, Section 41 (3) of the Act states that where a dramatic work

(a) is performed in class, or otherwise in the presence of an audience, and
(b) is so performed in the course of the activities of a school, by a person who is a teacher in, or a pupil in attendance at, the school

'the performance shall not be taken for the purposes of this Act to be a performance in public if the audience is limited to persons who are teachers in, or pupils in attendance at, the school, or are otherwise directly connected with the activities of the school.' A parent or guardian of a pupil is 'not connected with the activities of the school' for this purpose (Section 41 (4)).

It should also be noted that 'The reading or recitation in public by one person of any reasonable extract from a published literary or dramatic work, if accompanied by a sufficient acknowledgement' (Section 6 (5)), does not constitute an infringement of copyright. This subsection does not apply to broadcasting.

In addition to the general liability for infringement by public performance, a special liability falls upon the person who lets for hire a place of entertainment. Section 5 (5) provides that 'The copyright in a literary, dramatic or musical work is also infringed by any person who permits a place of public entertainment to be used for a performance in public of the work, where the performance constitutes an infringement of the copyright in the work' but the subsection does not apply if the 'person' 'was not aware, and had no reasonable grounds for suspecting, that the performance would be an infringement of the copyright' or 'gave the permission gratuitously, or for a consideration which was only nominal or (if more than nominal) did not exceed a reasonable estimate of the expenses to be incurred by him in consequence of the use of the place for the performance'. There need be no express permission or authorization, but a general authority to use a theatre for public performance is not sufficient (*The Performing Right Society Ltd.* v. *Ciryl Syndicate* (1924) *I.K.B.I.*).

The acts 'restricted by copyright' under Section 2 (5) apply as much to 'a substantial part' of a work (Section 49 (1)) as to a complete work. The phrase 'substantial part' is a third phrase which is not defined in the Act. In the various cases that came before the courts when the identical phrase in the 1911 Act was under consideration, stress was laid on the fact that quantity is not the sole criterion by which to judge whether a substantial part of a work has been copied. The quality and importance of the passage taken must also be given consideration. For this reason, four lines from Kipling's thirty-two-line poem *If* were held to amount to a substantial part of the poem in *Kipling* v. *Genatosan Ltd.*, Macg. C.C. (1920), 203, and in *Hawkes & Son Ltd.* v. *Paramount Film Services Ltd.*, Macg. C.C. (1933), 473, twenty-eight bars, the playing time of which was twenty seconds, were held to amount to a substantial part of a musical work, the playing time of which in its entirely was four minutes.

Where there is no verbatim copying 'both the plot (including in that word the idea and the arrangement of the incidents) and the dialogue and working out of the play must be regarded in order to see whether one play is a reproduction of the other or of a substantial part of it, and regard must also be had to the extent to which both plays include stock incidents' (*Rees* v. *Melville*, Macg. C.C. (1911–16), 168). In this case, it was decided that the basic idea of a young man marrying a beggar girl to comply with the terms of a will and acquire a fortune did not constitute a substantial part of a play.

It has been emphasized in many cases that, particularly in the field of melodrama, there are certain stock characters and stock situations the inclusion of which in two plays does not of itself give the author of the earlier play a good cause of action against the author of the later play, 'though the combination of these ordinary materials may nevertheless be original, and when such combination has arrived at a certain degree of complexity it becomes practically impossible that they should have been arrived at independently by a second individual' with the result that there is a presumption that there has been infringement.

As copyright is not a monopoly, proof that a later author had no knowledge of an earlier author's work and arrived at his results independently of it will be a good defence, however similar the two works may be.

The normal period of copyright protection is 50 years from the end of the calendar year in which the author dies. In the case of works of joint authorship, the period of protection dates from the end of the calendar year of the death of the author who dies last (Schedule 3, para. 2). If the work has not been published, performed in public, offered for sale on gramophone records, or broadcast during the author's lifetime, the period of protection is fifty years from the end of the calendar year which includes the earliest occasion on which one of these acts was done (Section 2).

The first owner of the copyright in a dramatic work is normally the dramatist. However, in those rare cases where a dramatic work is made in the course of the dramatist's employment by another person (who is not the proprietor of a newspaper or magazine) under a contract of service or apprenticeship, that other person is entitled to the copyright in the work in the absence of agreement to the contrary.

Infringement of copyright is primarily a civil offence for which the remedies are an injunction and damages. Proceedings must be instituted within six years of the date of the infringement. Criminal proceedings lie only against a person who makes, sells, lets for hire, exhibits by way of trade, imports, or distributes articles which he knows to be infringing copies of the work (Section 21 (1) and (2)) or 'causes a literary, dramatic or musical work to be performed in public' knowing that copyright is infringed thereby.

In 1973 a Government Departmental Committee was set up under the Chairmanship of Mr Justice Whitford to consider amendment of the copyright law required by developments since the 1956 Act was passed, in particular in the fields of reprography, audio and video recording, and computer technology. Amendment of the law was also needed to enable the UK to ratify the 1971 Paris text of the Berne Convention (see below).

2. *The United States*. American copyright derived initially from a British Act of 1710. The first comprehensive American Act was passed in 1790. During the 19th century various amendments were made to the Act, but by the early years of the 20th century the mass of piecemeal copyright legislation and case law had reached very much the same confused state as that obtaining in Great Britain at the end of the 19th, and a report on the position, made by the first Recorder of Copyrights in 1903, prepared the way for the passing of the Copyright Act of 1909.

The principal improvements made by this Act were the extension of the term of copyright protection, a slight relaxation in the copyright formalities and the expansion of copyright protection to 'all the writings of an author' (Copyright Act 1909, Section 4).

The Act also provided, as previous United States copyright legislation had done, that all books and periodicals written in the English language, with the exception of dramatic or dramatico-musical compositions, seeking copyright protection in the U.S.A., should be manufactured in the United States.

During and after the Second World War there were modifications of the regulations governing the 'manufacturing clause' requirements but since 1957 the 'manufacturing clause' has been removed altogether in the case of new works by British authors, who can obtain copyright protection by complying with the provisions of the Universal Copyright Convention which both the United States and the United Kingdom have ratified (see below). The 1909 Act provided for two separate terms of copyright: a period of 28 years from publication, followed by a renewal period of a further 28 years.

However, the comprehensive new Copyright Act 1976, which came into force on 1 Jan. 1978, makes fundamental changes in the duration of protection for new works, and also contains some complicated transitional provisions as follows:

(*a*) *Works in their second 28-year period on 1 Jan. 1978*: Copyrights, orginally registered with the U.S. Register of Copyrights before 1950 and renewed before 1978, are automatically extended by the new Act until 31 Dec. of the 75th calendar year from the original date they were secured. In effect, this means that all existing copyrights in their second term were extended for 19 years. This extension applied not only to copyrights less than 56 years old, but also to older copyrights that had previously been extended in duration under a series of Congressional enactments beginning in 1962.

(*b*) *Works in their first 28-year period on 1 Jan. 1978*: Works originally copyrighted between 1 Jan. 1950 and 31 Dec. 1977 must still be renewed. Application for renewal must be made to the U.S. Register of Copyrights within one year prior to the expiry of the original 28-year term, but under the new Act the renewal period is extended to 47 years, making up the total period of 75 years.

Works by British authors published on or after 27 Sept. 1957, which acquired their first 28-year term of protection under the U.C.C. simply by bearing the prescribed copyright notice, cannot be renewed unless there has already been an original registration for the first term. However, as long as the necessary applications, copies and fees are all received by the end of the first term, original and renewal registrations may be made simultaneously.

(*c*) *Works created on or after 1 Jan. 1978*: Works created after the new law came into force are automatically protected for the author's lifetime, and for an additional 50 years after the author's death. For works made for hire, and for anonymous and pseudonymous works, the new term is 75 years from publication or 100 years from creation, whichever is shorter.

(*d*) *Works in existence but not copyrighted on 1 Jan. 1978*: Such works automatically receive protection under the new law. The copyright will generally last for the same life-plus-50 or 75/100-year terms provided for new works. However, all works in this category are guaranteed at least 25 years of statutory protection; the law specifies that in no case will copyright in a work of this sort expire before 31 Dec. 2002, and if the work is published before that date the term is extended by another 25 years, to the end of 2027.

Proceedings for infringement of the copyright in a play cannot be instituted in the United States if the play has not been registered.

3. *International Copyright*. The Berne Convention of 1886, amended in 1896, 1908, 1914, 1928, 1948, 1967, and in Paris in 1971, established an International Copyright Union and in 1980 had more than 70 member states. As well as the U.K. and Commonwealth and all European countries, it includes among its membership Japan, Mexico, and a number of states in Africa and South America. Although individual countries in joining the Union have made certain special reservations and stipulations, all adhere to the principle that a national of any member state of the Union enjoys in every other country of the Union the rights and privileges of a national of that country. No registration or other formality is required.

The most prominent absentees from the Union are the United States of America, the U.S.S.R., and China. The first two of these are, however, members of the Universal Copyright Convention signed at Geneva in 1952 (the effective date for the U.K. is 27 Sept. 1957) which has almost the same number of members as the Berne Union. It requires a minimum period of copyright protection of 25 years from the death of the author and no formalities are necessary beyond the publication on all copies of works claiming protection of the symbol © accompanied by the name of the copyright proprietor and the year of first publication 'placed in such manner and location as to give reasonable notice of claim of copyright'. This Convention is supplementary to and not in substitution of the Berne Union. Countries, therefore, which are members of both Conventions are bound in their relations with each other by the Berne Union, which provides for more comprehensive protection and for a longer period.

COQUELIN, CONSTANT-BENOÎT (1841–1909), French actor, known as Coquelin *aîné* to distinguish him from his brother Ernest-Alexandre-Honoré (1848–1909), also an actor, known as Coquelin *cadet*. They were both at the COMÉDIE-FRANÇAISE, Coquelin *aîné* finally leaving the company in 1892 after a long absence on tour in Europe and America. Six years later he appeared at the PORTE-SAINT-MARTIN as CYRANO DE BERGERAC in the play of that name by ROSTAND, his son Jean (1865–1944) playing Ragueneau the pastrycook. In 1900 he toured with Sarah BERNHARDT, with whom he had appeared at her own theatre in Rostand's *L'Aiglon* (1898), and towards the end of his life he was seen frequently in London where he was very popular. He died suddenly while rehearsing Rostand's *Chantecler*, in which he was to have played the lead, Coquelin *cadet* dying insane a fortnight later. A big man with a fine voice and presence, Coquelin *aîné* was outstanding in the great comic roles of MOLIÈRE and in romantic and flamboyant modern parts. He was the author of two books, *L'Art et le comédien* (1880) and *Les Comédiens par un comédien* (1882).

CORCORAN, KATHARINE, see HERNE, JAMES A.

CORK, see IRELAND.

CORNEILLE, PIERRE (1606–84), French tragic dramatist. He wrote his first play *Mélite*, a comedy, for a strolling company under LENOIR, with MONTDORY as its star, which was appearing in Rouen, Corneille's birthplace, in some of Alexandre HARDY's plays. In 1629 the company was in Paris, and *Mélite* was given in a converted tennis-court near the PORTE-SAINT-MARTIN. After a slow start it was a success, in spite of the fact that it contained none of the stock personages of farce, and that it violated the UNITIES, then coming into fashion under the influence of Jean MAIRET.

Piqued by some of the criticisms levelled at *Mélite*, Corneille next wrote a tragicomedy, *Clitandre* (1631), and it was said 'the critics were then ready to implore him to return to his earlier style.' This he did, and his next four plays were comedies—*La Veuve* (1631–2), *La Galerie du palais* (1632), *La Suivante* (1633), and *La Place royale* (1633). In 1635 came his first tragedy, *Médée*, written probably as a result of the success of Mairet's *Sophonisbe* at the MARAIS the previous year, followed by another comedy *L'Illusion comique* (1636). Meanwhile Corneille, who had settled in Paris, had been chosen by RICHELIEU as one of the five authors commissioned to write the Cardinal's plays for him. Corneille was not temperamentally fitted for such servitude and after incurring the wrath of Richelieu by altering some part of the plot allotted to him he returned to Rouen. There he dipped into Spanish literature and on *Las mocedades del Cid* by CASTRO Y BELLVÍS he based *Le Cid*. Its production in 1636, or more probably the early days of 1637, is regarded as marking the beginning of the great age of French drama. First seen at the Marais with Montdory in the title role, it was an immediate success; having been translated into English by Joseph Rutter, it was played in London by Beeston's Boys (see BEESTON, William) in 1637. A Spanish version was played in Madrid and *Le Cid* was soon translated into several other languages.

The success of *Le Cid* created a number of enemies for Corneille, chief among them Mairet and SCUDÉRY, and a fierce pamphlet war, the *querelle du Cid*, was waged. The public, however, remained faithful to Corneille and was ready to applaud his *Horace* (1640), *Cinna* (1641), and *Polyeucte* (1642). These were all given at the Marais with FLORIDOR in the title roles. The year 1643 saw the production of another tragedy *Le Mort de Pompée* and Corneille's finest comedy *Le Menteur*, based on RUIZ DE ALARCÓN's *La verdad sospechosa*, in which Floridor played Dorante with JODELET as Cliton. Its great success was not repeated with its sequel *La Suite du menteur* the following year. Corneille's next group of plays, *Rodogune, Théodore* (both 1645), and *Héraclius* (1646), were also staged at the Marais before Floridor's move to the Hôtel de BOURGOGNE in 1647.

Corneille was now established as France's major dramatist, and in 1647 he was elected a member of the French Academy. *Nicomède* (1651), one of his best and most popular plays which MOLIÈRE chose for his reappearance in Paris in 1658, was preceded by the rather weak *Don Sanche d'Aragon* (1649) and by *Andromède* (1650), a MACHINE PLAY written at the request of Mazarin to show off TORELLI's new scenic devices at the PETIT-BOURBON. By now Corneille was showing signs of the fatigue which led to the poor work of his later years and which, with the upheavals of the Fronde, may help to account for the disastrous failure of *Pertharite* (1652), after which he abandoned the theatre for some years.

It was not until 1659 that Paris again saw a play by its veteran dramatist. This was *Oedipe*, at the Hôtel de Bourgogne where Floridor was now established in succession to BELLEROSE. It was followed by another machine play, *La Toison d'or*, written for the marriage of Louis XIV, and played by the actors of the Marais at the castle of the Marquis de Sourdéac in Nov. 1660. Later, having been given the scenery and machines by their patron, the actors were able to put the play on in their own theatre. *Sertorius* (1661) was also given there, though *Sophonisbe* (1663) and *Othon* (1664) were done at the Hôtel de Bourgogne, without much success.

In 1663 Corneille finally deserted Rouen for Paris, but his plays became less and less successful, partly because of the increasing popularity of RACINE. *Agésilas* (1666) was hampered by the author's employment of a new verse form, while *Attila* (1667) was overshadowed by the success of Racine's *Andromaque* in the same year. It was done by Molière, who followed it by a production of *Tite et Bérénice* (1670), given at the same time as Racine's play on the same subject at the rival Hôtel de Bourgogne. In Corneille's play the young BARON had his first adult part as Domitien, LA THORILLIÈRE played Tite, and Molière's wife Armande BÉJART was Bérénice. The play was moderately successful, and was followed by one of Corneille's most charming works, *Psyché* (1671) in collaboration with Molière and Quinault. *Pulchérie* (1672), at the Marais, and *Suréna* (1674), at the Hôtel de Bourgogne, are

Corneille's last plays, and both, though successful for a short time, soon fell out of the repertory.

Corneille was brusque and shy with strangers, had none of Racine's easy graces—but was probably a far finer character—and was sometimes too pleased with himself and his work. Yet even Racine at his best could not easily prevail against the popularity of Corneille in his decline, which did not prevent his making a memorable eulogy of the older dramatist at the Academy on the reception of Thomas CORNEILLE in his brother's stead. Corneille's fame was a little eclipsed in the 18th century, but the 19th restored him to his true place, even though the Corneille canon was reduced, for practical purposes, to the four great plays written around 1640—*Le Cid, Horace, Cinna*, and *Polyeucte*—with occasional revivals of some others including *Rodogune, Nicomède*, and *Le Menteur*. In the 20th century both academic critics and directors in the theatre have rescued from neglect some of the earlier and later plays: JOUVET, for instance, mounted a memorable production of *L'Illusion comique* at the COMÉDIE-FRANÇAISE in 1937 which helped to establish the play as a minor masterpiece of the pre-classical baroque tradition.

The comparison between Corneille and Racine has been a commonplace of dramatic criticism since the 17th century. While the former's heroic characters compel admiration, the latter creates psychologically convincing individuals with human weaknesses. In Racine's plays the women's roles stand out; Corneille's most rewarding parts are for men—Horace and his father, Augustus in *Cinna*, Polyeucte. His most memorable female roles are those of 'masculine' women such as Médée, Émilie in *Cinna*, Cléopâtre in *Rodogune*, and Sophonisbe; and while his more 'feminine' heroines like Camille in *Horace* and Pauline in *Polyeucte* have attracted some of the greatest French actresses, among them Mlle CLAIRON and RACHEL, they do not possess the stature of either the great Racinian heroines or their counterparts in Corneille.

CORNEILLE, THOMAS (1625–1709), French dramatist, usually known as Corneille de l'Isle to distinguish him from his elder brother (above). He wrote over 40 plays, mostly successful though he had none of his brother's genius. *Timocrate*, his first tragedy, performed at the MARAIS in 1656, was immensely popular. His best plays are considered to be *Ariane* (1672), in which Mlle CHAMPMESLÉ and later RACHEL were outstanding, and *Le Comte d'Essex* (1678), given at the Hôtel de BOURGOGNE again with Champmeslé. Like his brother, the younger Corneille also wrote a number of comedies, some adapted from the Spanish and some on topical subjects. None of his plays are remembered today, but with QUINAULT he filled the gap between the zenith of Pierre Corneille's career and the rise of RACINE. Thomas Corneille was commissioned by MOLIÈRE's widow to produce a versified and less provocative adaptation of his *Don Juan*; it remained in the repertory of the COMÉDIE-FRANÇAISE until 1847.

CORNELL, KATHARINE (1893–1974), American actress, daughter of a theatre manager in Buffalo, who made her first appearance with the WASHINGTON SQUARE PLAYERS in 1916. In 1919 she was in London, where she made a great success as Jo in a dramatization of Louisa M. Alcott's *Little Women*, and back in New York first came into prominence with her performances in Clemence DANE's *A Bill of Divorcement* (1921) and *Will Shakespeare* (1923). She was also seen in the title role of SHAW's *Candida* in 1924. In 1931 she appeared under her own management as Elizabeth Moulton-Barrett in Besier's *The Barretts of Wimpole Street*, a part she played many times and with which her name is always associated. It was directed by her husband Guthrie MCCLINTIC, who was responsible for many other productions in which she appeared, notably Sidney HOWARD's *Alien Corn* (1933), *Romeo and Juliet* (1934), Shaw's *The Doctor's Dilemma* (1941), *Antony and Cleopatra* (1947), FRY's *The Dark is Light Enough* (1955) and *The Firstborn* (1959), and Jerome Kilty's *Dear Liar* (1959), in which she played Mrs Patrick CAMPBELL. She retired on the death of her husband in 1961, the recipient of many honorary degrees and other awards. She wrote two books of reminiscences, *I Wanted to be an Actress* (1939) and *Curtain Going Up* (1943).

CORNER TRAPS, see TRAPS.

CORNISH, WILLIAM, see BOY COMPANIES and DISGUISING.

CORNISH ROUNDS. There are in Cornwall, particularly in the western part of the county, remains of circular earthworks which are now known to have been used as open-air theatres for annual performances of medieval MYSTERY PLAYS. These could accommodate on banks (usually seven) cut in the rising ground, and sometimes faced with granite, a large number of spectators grouped round a central area or 'playing-place' (*plên an gwary*). Some were still in use in the 17th century. The surviving plans for a trilogy consisting of a play on the Creation, one on the Passion, and one on the Resurrection, and for a two-day play on the life of St Meryasek, show that Heaven was at the eastern end, with Hell on the north, probably a survival from the days when performances were given in church. There must also, from the stage directions, have been rostrums or stages of varying levels on the central area. One feature of the Cornish 'playing-place' which has given rise to some controversy is the 'conveyour', originally thought to be a person, i.e., a stage-manager, charged with the duty of shepherding or 'conveying' a character to his appointed place. It is now thought to mean a central covered pit with a tunnel running to it from the side of the 'playing-place', by which a character such as the Devil could appear unexpectedly, as from a 19th-century TRAP. This seems to be confirmed by the existence at Perran Round of the so-called Devil's Spoon, a long shallow trench running from a depression in the first bank of seats to a central pit. The study of Cornish 'playing-places' has taken on an added urgency since Professor Wickham's suggestion (*Early English Stages*, II, p. 165 *seq*) that it is to them that we must look for the origin of the first London theatres.

CORONET THEATRE, New York, see EUGENE O'NEILL THEATRE.

CORPUS CHRISTI PLAYS, see MYSTERY PLAYS.

CORREIA, ROMEU, see PORTUGAL.

CORRIE, JOE (1894–1968), Scottish dramatist, who supplied the modern Scottish dramatic movement, and especially the amateur section, with some of its best one-act plays, several of which have been translated into French and Russian. He began writing while working in the coal-mines in Fife, and had his first plays, *The Shillin'-a-Week Man* and *The Poacher*, performed by the Bowhill Village Players during the General Strike of 1926. This group of actor-miners, of which Corrie himself made one, toured the music-halls of Scotland and Northern England with his plays during 1929 and 1930, and for the first time in the history of the music-hall put on a three-act play, *In Time o' Strife*, in one of them. Corrie was also the author of a full-length play on Robert Burns and of *Master of Men* (1944), produced at the Glasgow CITIZENS' THEATRE.

CORSICAN TRAP, see TRAPS.

CORTÊS, ALFREDO, see PORTUGAL.

CORT THEATRE, New York, at 138 West 48th Street, between the Avenue of the Americas and 7th Avenue. This opened on 20 Dec. 1912 with Laurette TAYLOR in the immensely successful comedy *Peg o' My Heart* by her husband John Hartley MANNERS. In 1916 Charles COBURN and his wife appeared in a revival of Hazelton and Benrimo's *The Yellow Jacket*, and in 1919 came the successful production of DRINKWATER's *Abraham Lincoln*. In 1930 there was a notable revival of CHEKHOV's *Uncle Vanya*, and in 1933 *The Green Bay Tree* by Mordaunt Shairp, with Laurence OLIVIER and his first wife Jill Esmond in an elegant setting by Robert Edmond JONES, had a good run. Also successful were the fine war play *A Bell for Adano* (1944), based on a novel by John Hersey; ANOUILH's *Antigone* (1946), with Katharine CORNELL; Joseph Kramm's *The Shrike* (1952); *The Diary of Anne Frank* (1955) adapted by Frances Goodrich and Albert Hackett; Doré Schary's *Sunrise at Campobello* (1958), with Ralph Bellamy as Franklin D. Roosevelt; and *Purlie Victorious* (1961) by Ossie Davis. In 1969 it was announced that the theatre had been leased for television productions but it reopened as a legitimate theatre in April 1972 with *All The Girls Came Out To Play*, by Richard T. Johnson and Daniel Hollywood. William Douglas HOME's *The Jockey Club Stakes* was seen here in 1973, and the successful musical *The Magic Show* in 1974. An important production was *Home* (1980), by Samm-Art Williams, about the Black experience in America.

COSMOPOLITAN THEATRE, New York, see MAJESTIC THEATRE (1).

COSSA, PIETRO (1830–81), Italian playwright, who endeavoured to evolve a new form of tragedy in verse by applying the formulae of bourgeois REALISM to plots taken from Roman and Italian history. In spite of his poor poetry his plays reveal a lively sense of the theatre and there is considerable forcefulness in his writing; but his work as a whole shows that it was impossible for the old classical tragic form to accommodate the new realistic spirit. His most successful play was *Nerone* (1872), which reveals beneath its fustian a shrewd psychological insight, showing Nero as a frustrated mediocre actor rather than as evil incarnate. It was translated into English shortly after its publication by Frances E. Trollope, mother of the novelist, but her version does not appear to have been acted. Cossa, who was the product of a liberal, middle-class environment, vented his anticlericalism in *Cola di Rienzo* (1874), and to a certain extent also in *Giuliano l'apostata* (1877). Among his other numerous plays *Messalina* (1876) stands out for its remarkable evocation of the sensuality and lasciviousness of the period under review. Also successful in their day were *Pushkin* (1870) and *Plauto e il suo secolo* (1873).

COSTA, FRANCISCO DA (1533–91), Portuguese dramatist, who died in captivity in Morocco, and, during his long imprisonment there, wrote many poems and seven plays on biblical or hagiographic subjects. Some at least were performed by the Christian captives, and have survived in manuscript. The only examples of their kind, they have now been published.

COSTANTINI, ANGELO (*c.* 1655–1729), actor of the COMMEDIA DELL'ARTE, who first appeared with the troupe of the Duke of Modena. In 1683 he joined the COMÉDIE-ITALIENNE in Paris, ostensibly to share the role of Arlequin with BIANCOLELLI, but played it so seldom that he finally adopted and enlarged that of MEZZETINO. In the company with him was Tiberio FIORILLO whose life he wrote, rather inaccurately, as *La Vie de Scaramouche* (1695). When the Italians were banished from Paris in

1697 Costantini went to Brunswick where he had the misfortune to be the successful rival in love of the Elector of Saxony, which cost him 20 years in prison. His brother Giovan Battista (?–1720) was also an actor, who as Cintio played young lovers in the Paris troupe.

COSTELLO, TOM, see MUSIC-HALL.

COSTER, SAMUEL, see NETHERLANDS.

COSTUME. From earliest times costuming has been an essential part of the theatre. In Greek tragedy the actors wore MASKS and long robes with sleeves, quite unlike the dress of the day. In the Old Comedy of ARISTOPHANES the chorus wore symbolic details like horses' heads and tails, or feathered wings, while the chief characters wore loose tunics, grotesquely padded, and a large red leather phallus. This was later abandoned, and in the New Comedy of MENANDER actors wore the ordinary clothes of the time. When the Romans took over Greek tragedy and comedy they adopted the existing costumes, with the addition of the *toga* and *stola* and also, for tragedy, the high boot (COTHURNUS) and the exaggeratedly high peak (*onkos*) over the forehead. By the time of the Caesars, stage costumes, except for the popular pantomimes, which were played almost in the nude, and the patchwork rags of the Atellan farce (see FABULA 1: ATELLANA), had become very elaborate and colourful. This trend continued as the Empire moved towards its dissolution, and in the 6th century actresses incurred the disapproval of the Church for wearing sumptuous dresses of cloth of gold enriched with pearls.

When the theatre, which had disappeared with the collapse of the civilized world, was reborn in LITURGICAL DRAMA, the priests and choirboys who acted the first short Easter and Nativity plays wore their usual robes, with some simple additions, such as veils for the women characters, crowns for the Three Kings, or a rough cloak for a shepherd. But once the plays were moved from the church to the market-place and acted by laymen, costuming again became important, and sometimes the MYSTERY PLAY of the later Middle Ages was expensively dressed, with golden robes for God and his angels, and fantastic leather garments for the Devil and his attendant imps. Although no attempt at historical accuracy was made, a slight oriental touch seems to have crept into the costumes of the Magi. In the MORALITY PLAY the allegorical figures representing the virtues and vices wore richly fantasticated versions of contemporary dress, except for the Devil, who kept the costume he had worn in the Miracle play. His attendant, the Old VICE, a new character, was usually dressed as a fool or jester, with a long-eared cap decorated with bells, a cockscomb, and a parti-coloured close-fitting tunic and leggings, reminiscent of the costumes of the Roman farce-players. Much the same type of costume was worn by the masked actors of the COMMEDIA DELL' ARTE, though except for the lovers, who wore contemporary dress, each character was recognizably from a different part of Italy: PULCINELLA from Naples, ARLECCHINO from Bergamo, PANTALONE from Venice, while the CAPITANO was a Spaniard, Spain at this time still holding considerable tracts of territory in Italy, or sometimes a Greek. The gaudy rags of Arlecchino soon became stylized into the diamonds of brightly contrasted silk worn by HARLEQUIN today, the white belted and bloused suit of Pulcinella was stereotyped by Watteau into the familiar garb of PIERROT, but otherwise the characters retained their individuality, cropping up in unexpected places but having little influence on the evolution of theatrical costume in general.

The main line of development was through the academic theatre of Renaissance Italy and the Court play. The Italian *intermezzi*, triumphs, and pageants, the French *ballets de cour*, the English MASQUES, gave immense scope for fantastic mythological costumes, often, for the men, based on the 'Roman' pattern, with a plumed headdress, a breastplate moulded to the body, and some variation of the Roman kilt; this fashion was adopted on the public stage and continued to be used for tragic heroes for over 200 years. Women's costumes, as always, followed contemporary fashions. In the Court entertainments and in the public theatre, traces of medieval costuming can be found in the grotesque elements—the Wild Men in the masque, the Old Vice, referred to by Shakespeare —together with some borrowings from the *commedia dell'arte*. Many of the designs for costumes worn during the extravagant entertainments devised on the Continent, particularly in France to enhance the prestige of Louis XIV, still survive. Some of the earliest were by the great Italian stage designer TORELLI, but French designers soon took over, the most influential being Jean BÉRAIN, whose work, not only for the theatre but over the whole field of decorative art, synthesizes all the tendencies of the time, as did that of his contemporary Burnacini in Vienna. After the death of Louis XIV the elegant balance of *le style Bérain* declined into the fantasies of rococo, best studied in the work of the artists Gillot and Watteau, in the costume designs of Boucher, and in the shepherds and shepherdesses of J. B. Martin. It achieved its finest development on stage in the designs of Louis-René Boquet, who from 1760 to 1782 worked both for the Paris Opéra and for the Court. Like Bérain, he was content to work mainly in a contemporary style, suggesting character or period by some small decorative detail. His costumes, both male and female, are characterized by his use of wide paniers, forming a kind of ballet skirt covered with rococo detail. This panier-skirt reached as far as England, where QUIN wore it in Thomson's *Coriolanus* (1749). Among the reforms of GARRICK was the abolition of such garments in favour of contemporary styles, Macbeth, for instance, being acted in the scarlet of the

King's livery; and in 1758, when playing the part of an ancient Greek, Garrick insisted on wearing the costume of a Venetian gondolier on the grounds that the majority of Venetian gondoliers were of Greek origin. His leading actresses in tragedy also wore contemporary costume, including the high head-dresses of the 1770s with a crown or flowing veil, or, for Eastern potentates, a turban with a waving plume. Twenty years earlier VOLTAIRE, helped by the actor LEKAIN, had begun a campaign for correct costuming, at least in classical plays. In his *Orphelin de la Chine* (1755) Lekain as the hero wore an embroidered robe in place of the usual panier-skirt, and the heroine, played by Mlle CLAIRON, a simple sleeveless dress which, though not noticeably Chinese, was less incongruous than the exaggerated skirt of the time, held out by hip-pads; and when playing Roxane in a revival of RACINE's *Bajazet*, she appeared in a fairly close approximation to Turkish costume. A great advance towards accuracy was made by the actor TALMA when in 1789 he appeared as Brutus, in Voltaire's play of that name, with bare legs and arms. The reaction of the public was unfavourable, but by the end of the century the Revolutionary passion for anything remotely connected with antiquity had made classical costumes acceptable on the stage.

During the early 19th century the popularity of SCOTT's novels, many of which were dramatized, led to a growth of interest in 'historical' costumes, which meant in practice the addition of Elizabethan, Stuart, and other details to contemporary dress. The resulting mixture is not without charm, but bears very little relation to historical accuracy. The first step towards 'antiquarian' detail in dress was made by PLANCHÉ with his costume designs for Charles KEMBLE's production of *King John* at COVENT GARDEN in 1823. Charles KEAN strove for accuracy in his productions of Shakespeare at the PRINCESS's in the 1850s, though as Hermione in *The Winter's Tale* Mrs Kean wore a perfectly correct Grecian costume over a crinoline. Mrs BANCROFT, playing Peg Woffington in what she fondly imagined was 18th-century dress, now looks, in the photographs that survive, purely contemporary. Towards the end of the 19th century the triumph of REALISM in the theatre meant on the one hand an almost pedantic accuracy in dress which was sometimes very untheatrical, and on the other an increase in the number of modern plays in which the actors wore the dress of the day, often their own clothes. Consequently the designer, excluded from the legitimate stage, let his fancy run riot in lighter musical productions and in opera.

From about this time costume, scenery, and lighting began to be seen as interlocking components of the overall design plan. Diaghilev's Russian Ballet, which startled Europe and America in the years 1911–16, introduced a brilliant non-naturalism in both settings and costumes (notably in the designs of Léon Bakst) which influenced all departments of theatre design. In contrast, APPIA and CRAIG had been arguing for simplicity, grouping figures statuesquely against austere three-dimensional settings or neutral screens. Abstraction and symbolism swept through the AVANT-GARDE theatres of France, Germany, and, especially, Soviet Russia, where the Constructivists of the post-Revolutionary years favoured an impersonal approach with the actor treated almost as an icon. The more extreme examples of modernist costume design were self-defeating: the shape and the movements of the human body conflict with the elements of 'pure' design. Max REINHARDT instituted a form of stylized realism that was strikingly successful in his large-scale productions, but the main line of development ran from PISCATOR through the anti-illusionist work of BRECHT and his designer Caspar NEHER.

As in scenic design (see SCENERY) the boom in new materials since the 1960s, especially plastics and adhesives, has greatly increased the costume designer's range, although the demands of actors must restrain costume design from the more extreme experimental forms. For a contemporary piece the designer will often choose ready-made clothes; a play set in the past is rarely dressed exactly in the fashions of the time, but in versions modified to highlight character and mood. Plays by Shakespeare have been subjected to costuming in the style of almost every period and culture, from ancient Greek to futuristic and from Japan to the Wild West.

The production designer will often be responsible for both scenery and costumes, or each may have a separate designer. In either case, close collaboration with the lighting designer (see LIGHTING) as well as with the director is a vital part of the design plan.

COTHURNUS, from the Greek *kothornos*, a loose-fitting, thick-soled boot, coming high up the calf. It was worn by Greek women in everyday life, and in the Greek and Roman theatre came to be a distinctive feature of the tragic actor's costume. Evidence from the early theatre is lacking, but in the Hellenistic period and later the soles became very much exaggerated, adding inches to the actor's stature and contributing to his padded, unnatural appearance.

COTTESLOE THEATRE, the smallest of the three playhouses which make up the NATIONAL THEATRE in London, the others being the LYTTELTON and the OLIVIER. Named after the first Chairman of the South Bank Theatre Board, which is responsible for the structure of the building, it is a flexible theatre in which the position and shape of the stage can be changed, the seating capacity—maximum 400—varying according to the arrangement. The Cottesloe, used mainly for testing new techniques and presenting new or little known plays, opened on 4 Mar. 1977 with Ken Campbell and Chris Langham's *Illuminatus*, previously seen in Liverpool (see EVERYMAN THEATRE). Plays which have had their first productions in the Cottesloe include Julian Mitchell's *Half-Life* (1977), with John GIELGUD, Charles WOOD's *Has 'Washington' Legs?* (1978), with Albert FINNEY, and David STOREY's *Early Days* (1980), with Ralph RICHARDSON. David Mamet's *American Buffalo* also had its British première there in 1978. In 1978 *Lark Rise* and *Candleford*, based by Keith Dewhurst on Flora Thompson's book, were presented as 'promenade' productions, for which the seats were removed and the audience mingled with the actors. The same technique was adopted for Bill BRYDEN's productions of *The Passion* (1978, 1980), based on parts of the York cycle of MYSTERY PLAYS. In 1980 an O'NEILL season consisting of *Hughie, The Iceman Cometh*, and *The Long Voyage Home* was seen at the Cottesloe, as well as a revival of *The Crucible* by Arthur MILLER, which, like *Half-Life* and *Early Days*, was transferred to the West End.

COULDOCK, CHARLES WALTER (1815–98), American actor, born in London, who decided to go on the stage after seeing MACREADY act. In 1836 he joined a provincial company and appeared with most of the famous players of the day. Among them was Charlotte CUSHMAN, who in 1849 engaged him to go with her to New York where he made an immediate success in KOTZEBUE's *The Stranger*. He spent several seasons at the WALNUT STREET THEATRE in Philadelphia and was also a great favourite in Canada, where he first appeared at John NICKINSON's Royal Lyceum Theatre in Toronto in 1851, returning regularly until 1858 and again from 1874 to 1879 to star with his daughter Charlotte's stock company at the Grand Opera House. He is said to have coached young Henry MILLER for his stage début there in about 1875. Among his finest parts were Abel Murcott in Tom TAYLOR's *Our American Cousin* (1858) and Dunstan Kirke in Steele MACKAYE's *Hazel Kirke* (1880). A man of great vitality and energy and a witty, genial companion (he was a close friend of Joseph JEFFERSON and Edwin BOOTH), he was on the stage for 60 years.

COUNCIL FOR THE ENCOURAGEMENT OF MUSIC AND THE ARTS, see ARTS COUNCIL.

COUNSELL, JOHN, see WINDSOR.

COUNTERWEIGHT SYSTEM, the modern method of flying scenery by means of endless lines, pulleys, and counterweights, as opposed to the traditional system of lines from a fly-floor used in HAND WORKING. In a counterweight house the lines, usually made of rope, are attached to BARRELS from which the scenery is hung. From there they pass over blocks to tie-off loops on the top of a counterweight cradle (known in America as an arbor) which holds the counterweights for the scenery load. This moves up and down on wire guides, and is raised and lowered by means of a purchase line (also known as an endless, overhaul, or working line) of ¾-in manila hemp, which eventually passes back to the bottom of the cradle. The counterweights are loaded high off the stage on a special loading floor or loading platform.

The above is the normal one-power system; but several two-power ('two-in-one') systems have been devised and are in widespread use. Instead of being tied off to the counterweight cradle, the lines pass down and under a block at the top of it and back up to the bottom of a stationary pulley attached to a head block, around which the purchase line passes. In this system the fly-floor, with its locking rail, tension block, etc., is located halfway up to the GRID. The main advantage of the 'two-in-one' is that the barrel moves twice as fast and twice as far as the purchase line; its principal disadvantage is that the counterweight cradle needs twice as much weight to balance the load as in the one-power system.

COURTELINE [Moineau], GEORGES (-Victor-Marcel) (1861–1929), French dramatist, author of a number of amusing farces, some of which were produced by ANTOINE before finding their way into the repertory of the COMÉDIE-FRANÇAISE. They deal mainly with the humour of military life, as in *Les Gaîtés de l'escadron* (*Regimental Fun and Games*, 1886), and of the law, as in *L'Article 330* (1901); or, as in what is perhaps his best play, *Boubouroche* (1893), with episodes in the life of ordinary people, salted with much wit and a certain gross brutality which recalls the farces of the early French theatre.

COURTENAY, TOM [Thomas] DANIEL (1937–), English actor whose haggard good looks were admirably suited to Konstantin in CHEKHOV's *The Seagull*, in which part he made his professional début with the OLD VIC company in Edinburgh in 1960. After taking over from Albert FINNEY as the fantasizing hero of Keith Waterhouse and Willis HALL's *Billy Liar* in 1961, he appeared with the NATIONAL THEATRE company in Max FRISCH's *Andorra* (1964), and two years later began a long association with the 69 Theatre Company in Manchester (see ROYAL EXCHANGE THEATRE), his roles including Lord Fancourt Babberley in Brandon THOMAS's *Charley's Aunt* (1966), Hamlet in 1968, Young Marlow in GOLDSMITH's *She Stoops to Conquer* in 1969, and the title role in IBSEN's *Peer Gynt* in 1970. He was seen in London in two new works by Alan AYCKBOURN—*Time and Time Again* (1972) and the trilogy *The Norman Conquests* (1974)—returning to Manchester in 1976. He made his New York début in Simon GRAY's *Otherwise Engaged* (1977), playing Simon, a publisher.

In 1980 he was seen in Manchester and in London (N.Y., 1981) in the title role of Ronald Harwood's *The Dresser*. Courtenay's voice lacks the richness needed for the great poetic roles, but he is an actor of great integrity and versatility.

COURT FOOL. A member of the Royal Household, also known as the King's Jester, not to be confused with the humbler FOOL of the FOLK FESTIVALS. His origin has been variously traced to the Court of Haroun-al-Raschid, to the classical dwarf-buffoon, and to the inspired madman of Celtic and Teutonic legend. At some point in his career he adopted the parti-coloured costume of his rival, which led to confusion between the two types, but the Court Fool has nothing else in common with the folk tradition, whose Fool is nearer to the CLOWN. Shakespeare's Fools derive from the Court Fool, already a tradition in his time. In dramatic use they serve as vehicles for social satire.

COURTNEIDGE, (Esmeralda) CICELY (1893–1980), DBE 1972, English actress, who made her first appearance on the stage in Manchester at the age of 8 playing Peaseblossom in *A Midsummer Night's Dream*. After touring Australia with her father, the actor Robert Courtneidge (1859–1939), she made her first appearance in London in 1907 in the MUSICAL COMEDY *Tom Jones* and was seen in a variety of musical plays, including *The Arcadians* (1909) and *The Mousmé* (1911). She then went on the variety stage and into PANTOMIME and REVUE, and in 1925 scored a success in the revue *By-the-Way*, in which she made her New York début in the same year; it was directed by her husband Jack Hulbert (1892–1978) who also appeared in it. They were together in many subsequent productions including *Clowns in Clover* (1927), *The House That Jack Built* (1929), *Under Your Hat* (1938), a musical comedy that ran over the outbreak of the Second World War until 1940, and *Something in the Air* (1943). In 1956 she appeared in Ronald MILLAR's comedy *The Bride and the Bachelor* and in 1960 was seen in its sequel *The Bride Comes Back*; in 1971 she played another straight comic part in *Move Over, Mrs Markham*, by Ray Cooney and John Chapman—her last appearance in the West End, though she later toured in several productions. Her autobiography *Cicely* was published in 1953. Jack Hulbert's brother Claude (1900–64) was also on the stage; they toured together in *The Hulbert Follies* (1941) and were in other productions together.

COURT STREET THEATRE, see MILWAUKEE REPERTORY THEATRE.

COURT THEATRE, Christchurch, see NEW ZEALAND.

COURT THEATRE, London, see ROYAL COURT THEATRE.

COVENT GARDEN, Theatre Royal, London, in Bow Street. There has been a theatre on the site of the present Royal Opera House, Covent Garden, since 1732, when John RICH, holder of the patent granted to Sir William DAVENANT (see PATENT THEATRES), built one on land owned by the Duke of Bedford that had once been part of a convent garden. Designed by Edward Shepherd, it seated 1,897 in pit, amphitheatre, two galleries, and three tiers of side-boxes. The proscenium opening was 26 ft and the stage depth 42 ft, with an apron stage giving an extra 13 ft. The new theatre opened on 7 Dec. 1732 with a revival of CONGREVE's *The Way of the World*, acted by the company headed by James QUIN which Rich had brought with him from LINCOLN'S INN FIELDS. For several years plays alternated with the operas of Handel, until in 1740 Peg WOFFINGTON made a brilliant success as Silvia in FARQUHAR's *The Recruiting Officer*, following it almost immediately with her famous BREECHES PART Sir Harry Wildair in the same author's *The Constant Couple*. She made her last appearance at this theatre, in 1757 as Rosalind in *As You Like It*. Other famous players who appeared under Rich, who was himself seen only in his own PANTOMIMES, were George Anne BELLAMY, Spranger BARRY, and for a short time GARRICK. Rich's son-in-law John Beard, who inherited the Patent in 1761, was interested only in opera and in 1767 sold it to George COLMAN the elder and three partners, of whom one, Thomas Harris, became sole manager in 1774 after a good deal of wrangling and some physical violence. In the previous year GOLDSMITH's *She Stoops to Conquer* had had its first performance, and MACKLIN had appeared on 23 Oct. as Macbeth in something approaching realistic Scottish garb, though not in the kilt. He was to be seen at Covent Garden again in his own play *The Man of the World* (1781) and made his last appearance there as Shylock in *The Merchant of Venice* in 1789. On 17 Jan. 1775 SHERIDAN's *The Rivals* had its first performance. In 1782 the auditorium was gutted and reconstructed, the whole building being again reconstructed within the original walls by Henry Holland in 1792. George Frederick COOKE, during his meteoric career, was mainly associated with this theatre, making his London début there as Richard III on 31 Oct. 1800. Three years later John Philip KEMBLE bought a sixth share of the Patent, and appeared in company with his sister Sarah SIDDONS. One of his first importations was the child prodigy Master BETTY. On 20 Sept. 1808 the theatre was burnt down. Rebuilt by Smirke with a pit and five galleries seating 2,800, the new theatre opened with *Macbeth*. Owing to the high cost of the rebuilding Kemble abolished the shilling gallery but this, together with the appearance of Madame Catalani who was unpopular with a section of the audience, caused trouble which culminated in the famous O.P. (Old Prices) Riots. After disturbances every night for about two months, Kemble was forced to submit and restore the shilling gallery.

Between 1809 and 1821 most of the famous actors of the day, and many singers, appeared at Covent Garden, as did pantomimists like Farley and GRIMALDI. On 29 June 1812 Mrs Siddons made her farewell appearance, and on 16 Sept. 1816 MACREADY his first. Kemble retired on 23 June 1817 (in which year the theatre was first lit by gas) and his younger brother Charles KEMBLE took over, staging in 1823 a 'historically accurate' production of *King John* with costumes and scenery designed by PLANCHÉ—probably the most important innovation in costuming since Macklin's *Macbeth*. In 1829 Charles's daughter Fanny KEMBLE appeared reluctantly as Juliet, to save her father from bankruptcy, and playing a wide range of parts filled the theatre for three years. Edmund KEAN made his last appearance on the stage on 25 Mar. 1833 and later in the same year Alfred Bunn took over the management of the theatre, to be succeeded in 1835 by Osbaldiston, who engaged a strong company including Charles KEAN, Macready, and Helen FAUCIT, making her London début. Macready himself took over in 1837, introducing limelight on to the stage long before it was in general use as a LIGHTING effect. His reign was marked by much internal dissension and he left in 1839, being succeeded by Madame VESTRIS who with her husband Charles James MATHEWS put on a number of beautifully staged productions of Shakespeare; but financially her greatest success was BOUCICAULT's first play *London Assurance* in 1841. After Madame Vestris left in 1842 the theatre failed and was finally closed, to reopen on 6 Apr. 1847 as the Royal Italian Opera House, its seating having been increased following reconstruction to over 4,000. Smirke's building was burnt down on 5 Mar. 1856 and the new theatre, designed by E. W. Barry, with pit, stalls, three tiers of boxes, amphitheatre, and gallery, opened on 15 May 1858 with a performance of opera. Theatrical entertainments were in future limited to a handful of pantomimes, some revues, and in 1912 REINHARDT's production of SOPHOCLES' *Oedipus the King* starring MARTIN-HARVEY. Otherwise the house was entirely given over to opera, except for a brief existence during the Second World War as a Palais de Danse. In 1946 it became the joint home of London's chief opera and ballet companies, the latter coming from SADLER'S WELLS. Major alterations took place in 1964, involving the abolition of the gallery, and in 1970 the seating capacity was 2,158. The removal of the neighbouring fruit and vegetable market later provided the opportunity for long-needed improvements to back-stage facilities.

COVENTRY, see BELGRADE THEATRE and HOCKTIDE PLAY.

COWARD, NOËL PIERCE (1899–1973), English actor, director, composer, librettist, and playwright, knighted in 1970. He was on the stage from childhood, making his first appearance on 27 Jan. 1911 in a fairy play called *The Goldfish* and later playing small parts in Charles HAWTREY's company, as well as Slightly in BARRIE's *Peter Pan* several years running, and eventually Charley in a touring production of Brandon THOMAS's *Charley's Aunt*. His early plays, some of which aroused controversy because of their characters' sexual sophistication, included *The Young Idea* (1923; N.Y., 1932), *The Vortex* (1924; N.Y., 1925), *Fallen Angels* (1925; N.Y., 1927), and *Hay Fever* (London and N.Y., 1925). He then wrote several REVUES for C. B. COCHRAN, including *On With the Dance* (also 1925, in which year Coward had five productions running in London) and *This Year of Grace* (London and N.Y., 1928), the success of the latter making up for the failure of *Sirocco* (1927) which caused a riot on its first night. Further successes were scored by the musical romance *Bitter Sweet* (London and N.Y., 1929), *Private Lives* (1930; N.Y., 1931), in which Coward starred memorably opposite Gertrude LAWRENCE, the patriotic *Cavalcade* (1931), and a new revue *Words and Music* (1932). *Design for Living* (N.Y., 1933; London, 1939) aroused further controversy because of its light-hearted approach to a *ménage à trois* (played in New York by Coward and the LUNTS); it was followed by *To-Night at 8.30* (London and N.Y., 1936), consisting of six (later ten) one-act plays performed in groups of three. After the musical play *Operette* (1938) came the record-breaking *Blithe Spirit* (London and N.Y., 1941), which ran for 1,997 performances in London. In 1943 *Present Laughter* (N.Y., 1946) and *This Happy Breed* were presented in London in repertory, and Coward then returned to revue with *Sigh No More* (1945). *Peace in Our Time* (1947) was a serious play about England after a Nazi victory, but *Relative Values* (1951) marked a welcome return to sophisticated comedy and *Quadrille* (1952; N.Y., 1954) provided an excellent vehicle for the Lunts. John GIELGUD starred in the London production of *Nude With Violin* (1956; N.Y., 1957) and Vivien LEIGH in *South Sea Bubble* (also 1956) and the London production of *Look After Lulu* (London and N.Y., 1959), an adaptation of FEYDEAU's *Occupe-toi d'Amélie*. After *Waiting in the Wings* (1960), set in a home for retired actresses, Coward produced his last original musical *Sail Away* (N.Y., 1961; London, 1962) and then wrote the lyrics and music for *The Girl Who Came to Supper* (N.Y., 1963), a musical based on RATTIGAN's *The Sleeping Prince*. His last work (in which he also made his last stage appearance) was *Suite in Three Keys* (1966), consisting of *A Song at Twilight, Shadows of the Evening*, and *Come Into the Garden, Maud*, the last two forming a double bill; *A Song at Twilight* and *Come Into the Garden, Maud* were seen in New York in 1974 as *Noël Coward in Two Keys*. Coward was an excellent actor who as well as taking the lead in many of his own plays appeared occasionally in plays by other authors, notably as Lewis Dodd in

Margaret Kennedy's *The Constant Nymph* (1926) and King Magnus in SHAW's *The Apple Cart* in 1953. He wrote two volumes of autobiography, *Present Indicative* (1937) and *Future Indefinite* (1954), and his plays are frequently revived.

COWELL, JOE LEATHLEY [Joseph Hawkins-Witchett] (1792–1863), English actor, who served in the navy for some years before making his first appearance on the stage in Plymouth, playing Belcour in CUMBERLAND's *The West Indian*. He became a fine low comedian, particularly noted for his playing of Crack in Knight's *The Turnpike Gate*, which he constantly revived both in England and in the United States where he became one of the most popular comedians of the day. He made his last appearance in New York in 1850 as Crack and then returned to England, where he remained until he died. Through his first wife he was connected with the SIDDONS family, one of his sisters-in-law marrying Sarah Siddons's son Henry; his daughter by his second wife married the American impresario H. L. BATEMAN. His second son Sam (below) became a MUSIC-HALL star.

COWELL, SAM(uel) HOUGHTON (1820–64), second son of the comedian Joe COWELL and one of the earliest stars of the British MUSIC-HALL. Born in the United States, he appeared as a child with his father, billed as the Young American Roscius. As an adult he went first to Edinburgh and then on to London, where he appeared in E. L. Blanchard's BURLESQUE *Nobody in London* (1851). His success as a singer of comic songs between the acts led to his appearance at EVANS's 'song-and-supper rooms', and in 1852 Charles MORTON engaged him for the newly opened CANTERBURY MUSIC-HALL, after which he deserted the 'legit.' for the 'halls'. An excellent mimic with an engaging personality, he toured the music-halls with a repertory which included Blanchard's 'Villikins and his Dinah', first popularized by Frederick ROBSON, E. Bradley's 'The Ratcatcher's Darter', and the anonymous 'Billy Barlow', for which he appeared in shabby workman's clothes, with a brimless hat, one badly blacked eye, and a clay pipe in his hand. He returned to America in 1860 but after a strenuous tour his health gave way and he returned to England to die, leaving nine children. His daughter Florence, better known as Mrs A. B. TAPPING, was an excellent actress, as were his daughter Sydney Cowell (1846–1925) and his granddaughter Sydney FAIRBROTHER.

COWL, JANE (1884–1950), American actress and playwright, who made her first appearance on the stage under BELASCO in his *Sweet Kitty Bellairs* (1903). She subsequently played in a number of his productions, including Tully's *The Rose of the Rancho* (1906) and Belasco's own *A Grand Army Man* (1907), and soon established a reputation as one of the leading actresses of the day. Her finest part was probably Juliet, which she first played in 1923, but she was also seen in other Shakespearian plays, including *Antony and Cleopatra* in 1924 and *Twelfth Night*, in which she played Viola, in 1930; she was also much admired in revivals of the younger DUMAS's *The Lady of the Camellias*, as Camille, in 1931, and of SHAW's *Captain Brassbound's Conversion* in 1940 and *Candida* in 1942. Among the new plays in which she appeared were COWARD's *Easy Virtue* (1925), in which she made her London début a year later, Robert SHERWOOD's *The Road to Rome* (1927), Edward SHELDON's *Jenny* (1929), Thornton WILDER's *The Merchant of Yonkers* (1938), and VAN DRUTEN's *Old Acquaintance* (1940). She appeared also in her own plays *Lilac Time* (1917), written in collaboration with Jane Murfin, *Information, Please* (1918), and *Smilin' Through* (1919). *Hervey House*, written in collaboration with Reginald Lawrence, was produced in London in 1935, directed by Tyrone GUTHRIE.

COWLEY, ABRAHAM (1618–67), English poet and dramatist, whose first volume of verse was published when he was only 15. He was also the author of a pastoral comedy *Love's Riddle*, which though published in 1638 was not acted until 1723, when it was 'arranged' for a performance at a young ladies' boarding school, and of a Latin play, *Naufragium Joculare*, based on PLAUTUS. This was performed at Trinity College, Cambridge, in 1638, and in 1704 was translated by Charles Johnson as *Fortune in Her Wits*. Cowley is, however, best remembered as the author of *The Guardian* acted before Prince Charles (later Charles II) in Cambridge in 1642. As *Cutter of Coleman Street* it was one of the first plays performed publicly at the Restoration, being given at LINCOLN'S INN FIELDS THEATRE by DAVENANT's company in Dec. 1661. Samuel PEPYS was present at the performance and very much enjoyed it. A comedy of contemporary manners which satirized both the Royalists and the Puritans, it was also seen at Court and in spite of arousing some controversy was revived in 1668 with James NOKES as Puny, one of his best parts.

COWLEY [*née* Parkhouse], HANNAH (1743–1809), English playwright, whose works mark the transition from Restoration comedy to the 18th-century comedy of manners. Her first play *The Runaway* (1776) was said to have been 'improved' by GARRICK, who produced it at DRURY LANE; the most successful was *The Belle's Stratagem* (1780), based on DESTOUCHES's *La Fausse Agnès* (1759). First produced at COVENT GARDEN with Mrs JORDAN as Letitia, it was many times revived, notably by Henry IRVING in 1881 with himself as Doricourt and Ellen TERRY as Letitia. It was last seen in London in 1913. It was one of the first English comedies to be seen in the New World, being in the repertory of the HALLAMS and HODGKINSON in New York in 1794. It was often revived there, too, Ada REHAN playing Letitia in 1893 to the Doricourt of Arthur BOURCHIER. Mrs Cowley's later plays, which included *Which Is the Man?* (1782), *A Bold Stroke for a Husband* (1783), and *The Town Before You* (1794), were all given their first productions at Covent Garden.

COWLEY, RICHARD (?–1619), Elizabethan actor, who is first mentioned as acting with the ADMIRAL'S MEN in the 1590 revival of TARLETON'S *The Seven Deadly Sins*. In 1593 he was with STRANGE'S MEN in the provinces, and a year later joined the CHAMBERLAIN'S MEN on their formation. He figures in the list of actors in Shakespeare's plays, and was probably the first to play Verges in *Much Ado About Nothing* (1598) to the Dogberry of KEMPE. Tall and thin, with a long, pale face, he may also have played Aguecheek in *Twelfth Night* and Slender in *The Merry Wives of Windsor*.

COX, ROBERT (?–1655), English actor, who was on the stage before 1639, being one of BEESTON'S Boys at the COCKPIT. After the closing of the playhouses in 1642 he evaded the ban on acting by appearing in DROLLS, or short farcical pieces taken from longer plays (for example 'Bottom the Weaver' from *A Midsummer Night's Dream*), interspersed with rope-dancing and conjuring, at country fairs and in London. He was arrested while acting surreptitiously at the RED BULL playhouse in 1653, and imprisoned. The chief items in his repertory were published in 1662 by Francis Kirkman as *The Wits; or, Sport upon Sport*. The frontispiece of the second edition (1672) has the earliest known illustration of footlights on the English stage.

CRABTREE, CHARLOTTE (1847–1924), American actress, known as Lotta. Born in New York, she was taken at the age of six to California, where she was taught to dance by Lola Montez. An attractive child, with black eyes and a mop of red hair, she toured the mining camps from the age of eight, singing, dancing, and reciting, and becoming a well known and much loved figure. In 1865 she went to New York and made a great success as Little Nell and the Marchioness in BROUGHAM'S dramatization of DICKENS'S *The Old Curiosity Shop* (1867). Throughout her career she preserved a look of youthful innocence, even in her most daring dances and by-play. She was outstanding in BURLESQUE and EXTRAVAGANZA and in slight plays specially written for her, like Fred Marsden's *Musette* in which she toured indefatigably. She retired unmarried in 1891, having amassed a large fortune which she left to charity.

CRAIG, EDITH GERALDINE AILSA (1869–1947), English actress and stage director, daughter of Ellen TERRY and E. W. GODWIN, and sister of Gordon CRAIG. She first appeared on the stage as a child, and as a young woman acted with her mother and Henry IRVING at the LYCEUM. She also appeared with Ellen Terry in HEIJERMANS' *The Good Hope* (1903) and BARRIE'S *Alice Sit-By-The-Fire* (1905) and acted as her stage-manager during her tour of the U.S.A. in 1907. Later she withdrew from the theatre and studied music in London and Berlin, but on her return to England she turned her attention to production, and from 1911 directed the Pioneer Players, for whom she also designed costumes and scenery, in some 150 plays up to 1921; one of the most interesting was CLAUDEL'S *L'Ôtage* as *The Hostage* (1919), in which Sybil THORNDIKE played the chief character Synge de Coûfontaine. After 1921 Edith Craig directed several plays in London and elsewhere and in 1929, on the first anniversary of her mother's death, inaugurated an annual Shakespeare matinée in memory of her, the performances being given in a converted barn adjacent to the house at Small Hythe where Ellen Terry had spent her last years.

CRAIG, (Edward Henry) GORDON (1872–1966), English scene designer and theorist. The son of Ellen TERRY and E. W. GODWIN, he was for a time an actor, but later turned to the theory and practice of scene design and direction. In 1903 he prepared some interesting sets for Fred TERRY'S production of Calvert's *For Sword or Song* and for his mother's productions of *Much Ado About Nothing* and IBSEN'S *The Vikings*. He was also responsible for the designs for OTWAY'S *Venice Preserv'd* (*Das gerettete Venedig*) in Berlin in 1905, and Ibsen's *Rosmersholm* for DUSE in 1906. In 1908 he settled in Florence, where he founded and edited *The Mask*, a journal devoted to the art of the theatre, and also ran a school of acting in the Arena Goldoni. In 1911 he designed the costumes for YEATS'S *The Hour Glass* at the ABBEY THEATRE, where his invention of screens as a background for lights to play on, intended for the MOSCOW ART THEATRE *Hamlet* of 1912, were seen in public for the first time. In 1926 he designed and produced Ibsen's *The Pretenders* in Copenhagen. His last designs for the theatre were for a New York production of *Macbeth* in 1928. His theories on acting, which have been criticized as reducing the actor to nothing more than a puppet working under the direction of a 'master-mind', and his theories on stage settings are impossible to summarise: they can best be studied in his numerous publications. They had an immense influence on the theatre in Europe and the U.S.A. (though not so much in England), more because of their originality and proliferation of ideas than through Craig's actual achievements. Shortly before his death his vast theatrical library was bought by the French government for the Rondel Collection in Paris.

CRAIG THEATRE, New York, see GEORGE ABBOTT THEATRE.

CRANE, RALPH (*c.* 1550/60–*c.* 1632), an underwriter in the Privy Seal Office who added to his income by copying plays for authors and actors. He was responsible for the copy of MIDDLETON's *The Witch* (1615) now in the Bodleian, and two copies made by him of Middleton's *A Game at Chess* (1624) are in the Bodleian and the British Library respectively. His copy of *Sir John van Olden Barnavelt* (1619), ascribed to Middleton and FLETCHER, made for the KING'S MEN as were those mentioned above, was evidently a prompt copy. An immense amount of such copying must have been done for the playhouses over the years and it is curious that so little is known about it. It was probably poorly paid and mostly done by hacks and hangers-on of the literary and theatrical professions.

CRANE, WILLIAM HENRY (1845–1928), American actor, who was for some years an opera singer, but proved so good in light comedy that he finally deserted singing in favour of acting. In 1877, after 14 years on the stage, he joined forces with the comedian Stuart ROBSON and together they appeared in *Our Boarding House*, specially written for them. One of their most successful joint appearances was as the two Dromios in *The Comedy of Errors* in 1878. Crane was also much admired as Sir Toby Belch in *Twelfth Night* in 1881 and as Falstaff in *The Merry Wives of Windsor* in 1885. His last appearance with Robson was in *The Henrietta* (1887), which Bronson HOWARD wrote with them in mind. They parted amicably in 1889 and Crane went on to appear in a long list of new plays, among them *Business is Business* (1904), an adaptation of MIRBEAU's *Les Affaires sont les affaires*, until he retired in 1916.

CRANKO, JOHN, see REVUE.

CRATES, Greek writer of OLD COMEDY, who won his first prize in 450 BC. He was a contemporary of ARISTOPHANES, who praised him in the *Knights* for his wit and graceful style, and he was said by ARISTOTLE to have been the first to abandon the 'lampoon-form' and to write 'generalized plots'. Only fragments of his work survive.

CRATINUS (*c*. 520–*c*. 423 BC), Greek dramatist, one of the masters of OLD COMEDY, and an elder contemporary of ARISTOPHANES, who in the *Knights* makes fun of Cratinus as a worn-out drunkard; in the following year (424 BC) Cratinus had his revenge by defeating Aristophanes' *Clouds* with his own *Wine-Flask*, probably his last play. Aristophanes also speaks of Cratinus' torrential style, and of the popularity of his lyrics. Fairly extensive fragments of his work survive, and some 20 titles of his plays are known.

CRAVEN, Lady ELIZABETH, Margravine of Anspach, see PRIVATE THEATRES IN ENGLAND.

CRAVEN, FRANK (1880–1945), American actor, dramatist, and producer, who made his first appearance on stage as a child in Boston, where his parents were members of a stock company. His first New York success was in BROADHURST's *Bought and Paid For* (1911), in which he also made his London début in 1913. During the long run of this play he wrote his first comedy *Too Many Cooks* (1914). The most successful of his later works was *The First Year* (1920), a modern domestic comedy. Apart from appearing in his own plays, he was seen as the Stage Manager in Thornton WILDER's *Our Town* (1938). Much of his later work was in films.

CRAVEN, HAWES [Henry Hawes Craven Green] (1837–1910), English scene-painter, son of a PANTOMIME actor and an actress, who served his apprenticeship under the scene-painter at the BRITANNIA THEATRE, London. His first outstanding work was done for Wilkie COLLINS's *The Lighthouse* (1857) at the OLYMPIC THEATRE, and he also worked at COVENT GARDEN, DRURY LANE, and at the Theatre Royal, Dublin, from 1862 to 1864. His finest work was done for IRVING at the LYCEUM, where he was considered the equal of STANFIELD and BEVERLEY in craftsmanship, and their superior in his grasp of theatrical essentials. His scenery for *Hamlet, Romeo and Juliet*, W. G. Wills's *Faust*, and TENNYSON's *Becket* was particularly admired. As an innovator he ranks with de LOUTHERBOURG, and he was a pioneer in the skilful use of the newly-introduced electric LIGHTING.

CRAWFORD, CHERYL (1902–), American actress and theatre director, who made her first stage appearance in 1923 for the THEATRE GUILD, appearing in many of its later productions and being its casting manager from 1928 to 1930. She was also active in the foundation and running, with Harold CLURMAN and Lee STRASBERG, of the GROUP THEATRE in 1930; of the American Repertory Theatre, in association with Eva LE GALLIENNE and Margaret WEBSTER, in 1946; and of the Actors' Studio (see METHOD) in 1947. In 1950 she became one of the directors of the Anta play series produced annually by the AMERICAN NATIONAL THEATRE AND ACADEMY, and among her other activities presented, often in association with other producers, a number of plays. These included Tennessee WILLIAMS's *The Rose Tattoo* (1951), *Camino Real* (1953), and *Sweet Bird of Youth* (1959); FRISCH's *Andorra* and BRECHT's *Mother Courage and her Children* (both 1963); *Doubletalk* (1964), two one-act plays by Lewis John Carlino; the musical *Celebration* (1969); *Colette* (1970) by Elinor Jones; and ARBUZOV's *Do You Turn Somersaults?* (1978).

CRAZY GANG, The, an association of seven British comedians, consisting of three double-act teams—Bud Flanagan [Robert Winthrop, originally Chaim Reuben Weintrop] (1896–1968) and Chesney [William] Allen (1894–1982); Jimmie Nervo [James Holloway] (1897–1975) and Teddy

Knox [Albert Edward Cromwell-Knox] (1896–1974); Charlie Naughton (1887–1976) and Jimmy Gold [James McConigal] (1886–1967)—and 'Monsewer' Eddie Gray (1897–1969), who appeared together in a number of 'crazy' shows at the LONDON PALLADIUM in the 1930s and at the VICTORIA PALACE from 1947 to 1960. All the Gang had wide experience of variety and circus comedy techniques, and were able to make specific individual contributions to their shows, which included knockabout, satirical, sentimental, and farcical incidents, with a good deal of improvization.

CRÉBILLON, PROSPER JOLYOT DE (1674–1762), French dramatist, author of a number of neoclassical tragedies on subjects taken from Greek mythology and ancient history, of which the first was *Idoménée* (1705) and the best *Rhadamiste et Zénobie* (1711). In his own day he was considered the successor of RACINE in tragedy, to the detriment of CAMPISTRON, his immediate predecessor, but he had his detractors: BOILEAU called him 'Racine drunk' and 'a Visigoth in an age of good taste'. His verse was crude, but his audiences enjoyed the romantic element in his plays and the atmosphere of terror he succeeded in imparting. He said himself: 'CORNEILLE had taken the superhuman, Racine the human; all I was left with was the infernal regions.' His plays are not tragedies so much as melodramas, and they remained popular until ousted by the even more melodramatic works of the elder DUMAS. Crébillon became a member of the French Academy in 1731 and was later appointed one of the Court censors, using his position to oppose VOLTAIRE, one of his more determined critics. To prove his superiority, Voltaire took over the subjects of five of Crebillon's plays and successfully reworked them. The only advantage the older man, who was always short of money, derived from this feud was a pension, granted him at the age of 72 through the intrigues of the Court party hostile to Voltaire.

CREED PLAY, see LITURGICAL DRAMA.

CREPIDATA, see FABULA 2: PALLIATA.

CREPUSCOLARI, I, Italian dramatic movement of the early 20th century, the 'twilight' school, with which is linked the work of the *intimisti*. Characterized by a quiet pessimism, in conscious contrast to the excesses of D'ANNUNZIO, it found its chief exponents in Fausto Maria MARTINI and Cesare Vico LODOVICI, though both BENELLI's *Tignola (The Bookworm*, 1908) and the anti-myths of MORSELLI have by some critics been assigned to the *crepuscolari* canon.

CRISPIN, a character in French comedy who derives from the COMMEDIA DELL'ARTE mask SCARAMUCCIA, gallicized as Scaramouche. As originally played in Paris by Raymond Poisson (see BELLEROCHE) he had in him something also of the braggart soldier, the CAPITANO, but successive generations of the POISSON family playing the part until 1735 made it more that of a quick-witted, unscrupulous valet. The name was introduced into theatrical literature by SCARRON, and was used by LESAGE for the hero of his first successful one-act comedy *Crispin rival de son maître* (1707). It was also the name of a character in REGNARD's *Le Légataire universel* (1708).

CRITERION THEATRE, London, in Piccadilly Circus. Originally an adjunct to Spiers and Pond's Criterion restaurant, this theatre was built entirely underground, and at the time was considered a great novelty, as air had to be pumped into it to ensure adequate ventilation. Designed by Thomas Verity with a three-tier auditorium seating 675 people, it opened on 21 Mar. 1874 with H. J. BYRON's *An American Lady* starring Mrs John WOOD and had its first outstanding success with *Pink Dominoes* (1877), adapted by James ALBERY from a French farce. In the cast was Charles WYNDHAM who took over the theatre in 1879, inaugurating his management with Bronson HOWARD's *Truth*. In 1883 the theatre was found to be unsafe and was closed and reconstructed, electricity being installed at the same time. It reopened on 16 Apr. 1884 with a revival of Howard's *Brighton*. Many of Henry Arthur JONES's plays were first produced at the Criterion, including *The Case of Rebellious Susan* (1894), *The Physician*, and *The Liar* (both 1897). Another success in 1896 was *Rosemary* by L. N. PARKER and M. Carson. Wyndham left in 1899 to open WYNDHAM'S THEATRE, but remained lessee of the Criterion until his death, when it passed to his widow Mary Moore who made many outstanding appearances there. In 1902 the theatre was remodelled and Frank Curzon took over the management, opening on 10 Feb. 1903 with R. C. CARTON's *A Clean Slate*. In 1907 Wyndham returned temporarily to score a great success in Hubert H, DAVIES's *The Mollusc*. One of the greatest successes of the First World War, Walter Ellis's *A Little Bit of Fluff*, began its run here in 1915, while post-war successes included Sydney Blow's *Lord Richard in the Pantry* (1919), HACKETT's *Ambrose Applejohn's Adventure* (1921), and *Advertising April* (1923), a farce by Herbert Farjeon (see REVUE) and Horace Horsnell mainly remarkable for the presence of Sybil THORNDIKE in the title role. Between the wars the chief successes were Ronald Mackenzie's *Musical Chairs* (1932), with John GIELGUD, RATTIGAN'S *French Without Tears* (1936), and Lesley Storm's *Tony Draws a Horse* (1939). The theatre then became a BBC studio, reopening in Sept. 1945 with a revival of SHERIDAN'S *The Rivals* with Edith EVANS as Mrs Malaprop. The post-war years brought a run of successes, including Warren Chetham-Strode's *The Guinea Pig* (1946), Arthur Macrae's *Traveller's Joy* (1948), with Yvonne ARNAUD, and *Intimacy at 8.30* (1954), a revue which ran for 550 performances. Samuel

BECKETT's *Waiting for Godot* (1955) and ANOUILH's *Waltz of the Toreadors* (1956) were transferred here from the ARTS THEATRE, and N. F. Simpson's *One-Way Pendulum* (1960) from the ROYAL COURT. In 1963 *A Severed Head*, adapted by J. B. PRIESTLEY and Iris Murdoch from the latter's novel, began a long run, and it was followed by *The Owl and the Pussycat* by Bill Manhoff and Joe ORTON's *Loot* (both 1966). In 1969 Roy DOTRICE scored a tremendous success with his solo performance in Aubrey's *Brief Lives*, and in 1971 Simon GRAY's *Butley* began a long run, with Alan BATES in the title role. *Absurd Person Singular* gave Alan AYCKBOURN another West End success in 1973, and Ray Cooney and John Chapman's *There Goes the Bride* was popular in the following year. Other notable productions were a revival in 1975 of STOPPARD's *Rosencrantz and Guildenstern Are Dead* from the YOUNG VIC, Michael Pertwee's *Sextet* (1977), Martin Sherman's *Bent* (1979), from the Royal Court, and *Tomfoolery* (1980), a collection of the songs of the American satirist Tom Lehrer.

CRITERION THEATRE, New York, on the east side of Broadway, between 44th and 45th Streets. Built as part of Oscar Hammerstein's OLYMPIA complex to seat 2,800, with a roof garden, this opened as the Lyric Theatre in 1895. In 1899, after being sold at auction, it reopened under Charles FROHMAN as the Criterion. Among its early productions were Clyde FITCH's *Barbara Frietchie* (1899) with Julia MARLOWE in the title role and PINERO's *The Gay Lord Quex* (1900) starring John HARE and Irene VANBRUGH. The last production before it became a cinema was a version of BRIEUX's *La Robe rouge* as *The Letter of the Law*, with Lionel BARRYMORE. The building was demolished in 1935, together with the old Olympia Music-Hall (see NEW YORK THEATRE (2)).

The HERALD SQUARE THEATRE, originally the Colosseum, was named the Criterion from 1882 to 1885.

CROFT, MICHAEL, see NATIONAL YOUTH THEATRE.

CROMMELYNCK, FERNAND (1888–1970), French-language Belgian dramatist, who was for a time an actor. His early plays, which include *Nous n'irons plus au bois* (1906) and *Le Sculpteur de masques* (1908), were written under the influence of his compatriot Maurice MAETERLINCK, and though praised by the critics had little success. His international reputation, however, was assured by *Le Cocu magnifique* (1920), a study in obsessive marital jealousy which had its first performance in Paris under LUGNÉ-POË, who played the part of the husband Bruno. It was widely acted in translation, MEYERHOLD staging it in a memorable production in Moscow in 1922. It was seen in London in 1932, in English but under its original title, with Peggy ASHCROFT as Stella, the wife pushed into infidelity and finally flight with a cowherd by her husband's unreasoning jealousy, and was given a successful revival in Paris in 1945 under Jacques HÉBERTOT. Crommelynck's later plays, which included *Tripes d'or* (1925), a satire on snobbery, *Carine ou la jeune fille folle de son âme* (1929), a complicated mixture of romantic and symbolic characters, and *Une Femme qui a le coeur trop petit* (1934), which included ballet, music, and pantomime, were not successful, and he is now remembered only for *Le Cocu magnifique*.

CRONYN, HUME (1911–), Canadian actor, who trained for the stage in New York and made his début in Washington, D.C., in 1931. A notable character actor, he made his name on Broadway in the 1930s with a series of outstanding performances including Elkus in Maxwell ANDERSON's *High Tor* (1937) and Andrei in CHEKHOV's *Three Sisters* in 1939. In 1942 he married Jessica TANDY and after the Second World War, during which he organized shows for the Canadian and American forces, appeared with her in a number of plays including Jan de Hartog's *The Fourposter* (1951) and N.C. HUNTER's *A Day by the Sea* (1955). They also appeared together in the London production of Hugh Wheeler's *Big Fish, Little Fish* (1962); she did not appear in the 1961 New York production. In 1963 Cronyn starred with his wife in the first season of the GUTHRIE THEATRE in Minneapolis, playing Harpagon in MOLIÈRE's *The Miser*, Tchebutykin in *Three Sisters*, and Willy Loman in Arthur MILLER's *Death of a Salesman*. They were together again in the 1965 season, in which he played Richard III, and in the New York productions of DÜRRENMATT's *The Physicists* (1964) and ALBEE's *A Delicate Balance* (1966). He returned to Canada in 1969 to play the title role in Peter Luke's *Hadrian VII* at the STRATFORD (ONTARIO) FESTIVAL, where in 1976 he also played Shylock in *The Merchant of Venice* and Bottom in *A Midsummer Night's Dream*, his wife acting in the same season. In New York in 1972, and later on tour, he was the solo performer in BECKETT's *Krapp's Last Tape*, being also seen in New York with Jessica Tandy in the same author's *Happy Days*. They also appeared together in the double bill *Noël Coward in Two Keys* (1974) and in D. L. Coburn's *The Gin Game* (1977; London, 1979).

CROOKED MIRROR, Theatre of the [Krivoe Zerkalo], Leningrad, the best satirical theatre of pre-revolutionary Russia. It was founded in St Petersburg in 1908, its first outstanding success being *Vampuka, or the Bride of Africa*, a hilarious skit on grand opera. It attracted a number of brilliant young talents, including EVREINOV, whose *Fourth Wall* which ridiculed VERISMO opera was produced there, as was ANDREYEV's *The Pretty Sabines*, a biting social satire. One of the theatre's most successful productions was GOGOL's *The Government Inspector* shown as it might have been staged by a number of contemporary directors, including MEYERHOLD and STANISLAVSKY. The theatre closed in 1918, but managed to reopen in 1922 and in the following year the company made a brief visit to Moscow. After the death in 1928 of the critic A. P. Kugel, one of its original founders, it continued to function, but lost its special flavour and finally closed in 1931.

CROSS, BEVERLEY, see SMITH, MAGGIE.

CROSS KEYS INN, London, see INNS USED AS THEATRES.

CROTHERS, RACHEL (1878–1958), American dramatist, whose first short plays were produced while she was a student at a drama school. Her full-length plays, which she directed herself, include *The Three of Us* (1906); *A Man's World* (1909), an attack on the moral double standards which was regarded as highly significant in its time; *He and She* (also known as *The Herfords*; 1911), in which she herself played the lead in a revival in 1920; *A Little Journey* (1918); *Expressing Willie* (1924); *Nice People* (1920), a study of post-war youth; *As Husbands Go* (1931), which contrasts English and American marriages; and *When Ladies Meet* (1932), a deft study of feminine psychology. Her most important work, *Susan and God* (1937), starred Gertrude LAWRENCE as a woman devoted to a new religious cult. Always in the vanguard of public opinion, she never allowed her feminist viewpoint to weaken the theatrical effectiveness of her writing.

CROUSE, RUSSEL, see LINDSAY, HOWARD.

CROWDER'S MUSIC-HALL, see GREENWICH THEATRE (3).

CROWE, EILEEN, see MCCORMICK, F. J.

CROWE, Mrs GEORGE, see BATEMAN, KATE.

CROWNE, JOHN (?1640–?1713), English Restoration dramatist, a favourite of Charles II. He is believed to have been born in Shropshire, the eldest son of the William Crowne to whom Cromwell later gave title to half Nova Scotia. He accompanied his father to North America in 1657, and studied for three years at Harvard. The Restoration in 1660 put the Crownes' title in jeopardy, and they hastened back to England, only to lose it. Crowne is said to have turned to playwriting in the hope of mending his fortunes and regaining his lost lands; but to no avail, and he died in obscurity. Among his forgotten plays, which included several tragedies, a MASQUE entitled *Calisto* (1675), seen at Court, and two plays based on parts of Shakespeare's trilogy *Henry VI*, the only one to hold the stage for a time was *Sir Courtly Nice; or, It Cannot Be* (1685), based on a comedy by MORETO Y CABAÑA; the title role provided a favourite part for many actors. A two-part heroic drama, *The Destruciton of Jerusalem by Titus Vespasian* (1676/7), seems to have owed much of its success to its scenery, painted by AGGAS and Towers, who later sued DRURY LANE for payment for their work. Six of Crowne's plays, published between 1675 and 1698, are in the Lawrence Lande Collection of Canadiana at McGill University, Montreal.

CROW STREET THEATRE, Dublin, see IRELAND.

CRUCIBLE THEATRE, Sheffield, opened in 1971, replacing the old Sheffield Playhouse—a converted British Legion hall—where a repertory company was started in 1923. By the 1960s the company had gained an outstanding reputation, being recognized by the ARTS COUNCIL as a major repertory group, and in 1967 a Trust was formed to organize the building of a new theatre, the Crucible. It was financed mainly by the Arts Council, the City Council, and public donations. The only full-time professional theatre in South Yorkshire, it seats just over 1,000 in one semi-circular tier round three sides of the thrust stage, no seat being more than 59 ft from its centre. There is also a studio theatre seating 250, which is used for work of less popular appeal—many new plays have been seen there—and for productions by Theatre Vanguard, the Crucible's touring unit, which takes plays and projects to the surrounding community, mainly to schools. Under the theatre's first Artistic Director Colin George the excellent permanent company presented such plays as CHEKHOV's *The Cherry Orchard* and BRECHT's *The Caucasian Chalk Circle* and also local documentaries such as *The Stirrings in Sheffield on Saturday Night* by Alan Cullen. Peter James succeeded as Artistic Director in 1974, since when notable productions have included Tennessee WILLIAMS's *The Glass Menagerie* and the British premières of two American musicals, *Chicago*, which began a long run in London in 1979, and *The Wiz*, an all-black version of *The Wizard of Oz*. James was succeeded by Clare Venables in 1981.

CRUELTY, Theatre of, a form of drama which seeks to shock the spectator into an awareness of the underlying primitive ruthlessness of man's precarious existence, stripped of the artificial restrictions of conventional behaviour. Stemming from the works of ARTAUD, particularly *Le Théâtre et son double* (1938), the movement was eagerly adopted by members of the post-war AVANT-GARDE, including the dramatists GENET, ORTON, and WEISS and the director Peter BROOK.

CRUGER'S WHARF THEATRE, New York, below Water Street, between Cuyler's Alley and Old Slip; sometimes referred to as the Wharf Theatre. This was built in 1758 by David DOUGLASS for a company which included the widow of Lewis HALLAM and her three children. They opened on

28 Dec. with ROWE's *Jane Shore*, and after a season of two months ended with *Richard III*. They left New York in February 1759, after which 'the Theatre on Mr Cruger's Wharff', which is known to have had a 'box, pit, and gallery', was used no more.

CRUZ CANO Y OLMEDILLA, RAMÓN FRANCISCO DE LA (1731–94), Spanish writer of lively one-act SAINETES, vivid farces or satirical sketches which owed nothing to the prevailing neo-classical fashion of his time but were the descendants of the PASOS and ENTREMESES of the 16th and 17th centuries and forerunners of the 19th-century sketches of daily life and realistic drama. Ramón de la Cruz also wrote libretti for ZARZUELAS depicting working-class life in Madrid, an innovation in a genre which had hitherto concerned itself largely with the love intrigues of classical gods and goddesses.

CSIKY, GERGELY, see HUNGARY.

CSOKOR, FRANZ THEODOR (1885–1969), Austrian dramatist, whose first plays, of which the best was *Die rote Strasse* (1918), were written under the influence of EXPRESSIONISM, and contain certain technical characteristics still discernible in such later works as *Gesellschaft der Menschenrechte* (1929) on the life of BÜCHNER, whom Csokor much admired, and *Dritter November 1918* (1937), which portrays the collapse of the Austro-Hungarian army at the end of the First World War. He left Austria when the Nazis took over in 1938 but was later interned until 1945, after which he returned to Vienna where he remained until his death. Two other plays, which he later grouped into a *Europäische Trilogie* with *Dritter November 1918*, were *Besetztes Gebiet* (1930), on the French occupation of the Ruhr, and *Der verlorene Sohn* (1947). A similar religious trilogy contains the notable play *Cäsars Witwe* (1952). His last plays were *Die Erweckung des Zosimir* (1960) and *Das Zeichen an der Wand* (1962).

CUEVA, JUAN DE LA (1543–1610), Spanish dramatist whose most important plays were produced in Seville between 1578 and 1581. Writing in the period when classical influence was still paramount, Cueva owes much to SENECA, but he was also the first writer of plays based on Spanish history. His dramas on Bernardo el Carpio and King Sancho and his tragedy *Los Siete infantes de Lara* (*The Seven Princes of Lara*) paved the way for the great dramas of Lope de VEGA and his contemporaries. Cueva was also an innovator in the Spanish comedy of manners with his *Comedia del viejo enamorado* (*Comedy of the Infatuated Old Man*) and *El infamador* (*The Backbiter*).

CUMBERLAND, RICHARD (1732–1811), English dramatist, grandson of the great Richard Bentley, Master of Trinity College, Cambridge. He spent some years in politics before embarking on a literary career, mainly because he was in need of money. He became a prolific playwright, his tragedies including an adaptation of Shakespeare's *Timon of Athens* with a new fifth act, seen in 1771, and new versions of MASSINGER's *The Bondman* and *The Duke of Milan* (both 1779) which have not survived; but his best plays were sentimental domestic comedies like *The Brothers* (1769) and *The West Indian* (1771), the latter proving his most successful work when produced by GARRICK at DRURY LANE. *The Fashionable Lover* (1772), in the same style, was less well received and of Cumberland's many later plays the only one of interest is *The Jew* (1794), one of the first to plead the cause of Jewry on stage. It was frequently revived and translated into several languages, the title role providing a fine part for some of the outstanding actors of the day. Cumberland, who was extremely sensitive to criticism, figures in SHERIDAN's *The Critic* (1779) as Sir Fretful Plagiary.

CUMMINGS, CONSTANCE (1910–), American-born actress, who made her début in 1926 in San Diego and her first appearance in New York in 1928 in the chorus of the musical show *Treasure Girl*. After her marriage to the English playwright Benn Wolfe Levy (1900–73) she lived and worked mainly in England. She had roles in several of her husband's works, including the title roles in *Young Madame Conti* (1936; N.Y., 1937) (written in collaboration with Hubert Griffith), based on a play by Bruno FRANK, and, in New York only, *Madame Bovary* (1937), adapted from Flaubert's novel. She returned to London in 1938 to appear in James Hilton's *Goodbye, Mr Chips*, following it with Levy's production of his own play *The Jealous God* (1939). Later in the same year, for the OLD VIC, she showed great poise and emotional power as Juliet in *Romeo and Juliet* and as SHAW's St Joan. She was seen in the London production of Robert SHERWOOD's *The Petrified Forest* in 1942, and after the Second World War appeared in more plays by Levy as well as the London production of Archibald MacLeish's *J.B.* (1961). In 1964 she successfully took over from Uta HAGEN the role of Martha in ALBEE's *Who's Afraid of Virginia Woolf?* in London, and in 1969 she played Gertrude in *Hamlet*, directed by Tony RICHARDSON, both in London and New York. In 1971 she joined the company of the NATIONAL THEATRE in London, where she gave a superb performance as the drug-addicted Mary Tyrone in O'NEILL's *Long Day's Journey into Night*, and was also much admired as Ranevskaya in CHEKHOV's *The Cherry Orchard*. She went to the ROYAL COURT THEATRE to appear with Michael HORDERN in Howard BARKER's *Stripwell* (1975), and back in New York in 1979 was highly praised for her performance (mainly a monologue) as a woman recovering from a stroke in Arthur KOPIT's *Wings*, seen at the COTTESLOE THEATRE in London later in the same year.

CUP-AND-SAUCER DRAMA, see ROBERTSON, T.W.

CURTAIN. The open-air Greek theatre had no need of curtains, though a small one may occasionally have been used instead of doors to hide the entrance through the back wall. In the Roman theatre two curtains were apparently used, the *siparium*, which served as a screen or backcloth in an otherwise permanent set, and the *aulaeum*, which hid the stage before the play began. Although early European theatres may have used curtains to cover the entrances to houses and temples, the curtain to hide the stage returned to the theatre only with the introduction of enclosed theatre buildings in the 16th and 17th centuries. In the English theatre after 1660 it rose at the conclusion of the PROLOGUE, which was spoken on the FORESTAGE, and remained out of sight until the play was over and the epilogue spoken, when it fell to indicate the end of the performance. It was at first, and remained for many decades, green in colour. It was occasionally, and for special effects only, dropped during a performance, and it was not until about the mid-18th century that it began to fall regularly to mark the end of an act and hide the stage during the interval. This function was shortly after transferred to the Act-Drop, a painted cloth which descended in the PROSCENIUM opening and was painted with a permanent and

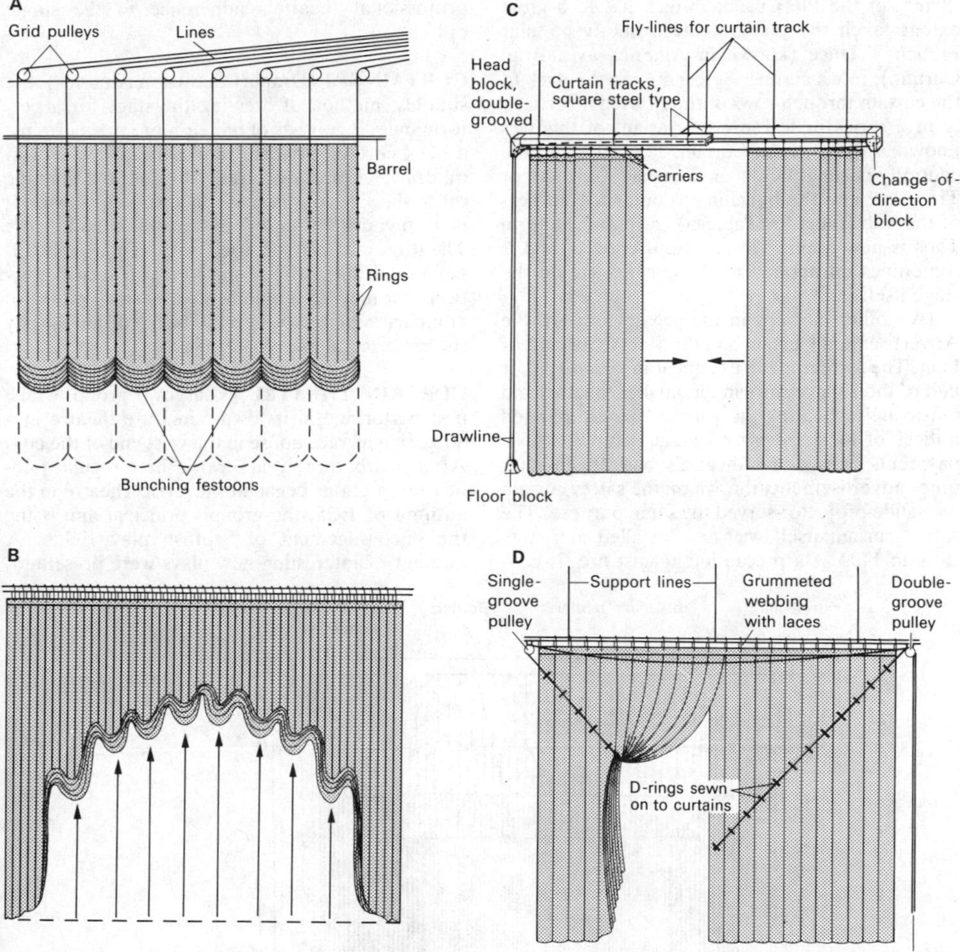

A French Valance or Festoon, rear view. **B** Contour Curtain, front view. Operation similar to A. **C** Draw or Traverse Curtain, or Traveller. A Draw Curtain may be flown by detaching the drawline from the floor block. The carriers have a double wheel, not shown, which rides in the track. **D** Tab (Tableau) Curtain, rear view

decorative picture which had no connection with the play being performed. Some of these act-drops became closely associated with the theatre in which they were placed, and by 1895 had become important enough for articles to be written in the popular press about their correct design. The act-drop, however, marked only the intervals between the acts; it was not until IRVING revived BOUCICAULT's *The Corsican Brothers* in 1880 that a crimson velvet curtain was introduced to hide changes of scene during an act.

The Front, or House, Curtain, so long a feature of all post-Restoration playhouses, is now being discarded in a number of theatres as part of the reaction against the BOX-SET and the proscenium picture-frame, and no provision is made for it in many modern theatre buildings. It can be worked in a variety of ways, including the straightforward 'flying' of the Fly Curtain, which has to a great extent taken the place of the formerly popular French Valance (known in America as a Brail Curtain), in which lines descending at the back of the curtain through rows of rings raised it vertically in a series of festoons. A variant of this was known as the Contour Curtain. There are also the centrally-parting Draw or Traverse Curtain, or Traveller, and the bunching to outer top corners of the Tab (short for Tableau) Curtain. The term Tabs is now applied to any front curtain, and is sometimes misapplied to the CURTAIN SET on the stage itself.

Two other curtains in the proscenium are the Advertisement Curtain and the Safety Curtain, or Iron. The former made its appearance in the latter half of the 19th century in the smaller theatres and music-halls, bearing in various panels painted notices of local shops and manufacturers whose payments helped the theatre's budget. In later times advertisements thrown on the safety curtain by a slide projector served the same purpose. The safety curtain itself was first installed at DRURY LANE in 1794 as a precaution against fire. It consists of an iron or fireproof sheet dividing the stage area from the auditorium, which must by law be lowered once during every performance. This is usually done during an interval. The device is now being made obsolete by modern water curtains and sprinklers.

CURTAIN-MUSIC, CURTAIN-TUNE, see ACT-TUNES.

CURTAIN-RAISER, a one-act play, usually farcical, which in the 19th century served to whet the appetite of the audience before the main five-act drama of the evening, the last relic of the days when a full evening's entertainment included several plays to which latecomers were admitted on payment of a reduced fee. Like the AFTERPIECE, it disappeared in the 20th century with the professional theatre's adherence to the single bill.

CURTAIN SET (Drapery Setting in America), the simplest method of dressing the stage for a performance. It consists of one or more side curtains, a back curtain, BORDERS, and perhaps a traverse, or draw, CURTAIN, centrally divided and running off to the sides of the stage on a wire or railway. It is a favourite standby of amateurs and Little Theatres and can be used with remarkable ingenuity. It has, strictly speaking, no scenic function, but may be supplemented by small pieces of BUILT STUFF placed before the back curtain and by the resourceful use of LIGHTING.

CURTAIN THEATRE, Glasgow, group which first performed in its own miniature theatre in a Victorian terrace house in the west end of the city. After nearly three years' work there regular public presentations began at the Lyric Theatre in the autumn of 1935, the group's principal aim being the encouragement of Scottish playwrights. A number of interesting new plays were presented,

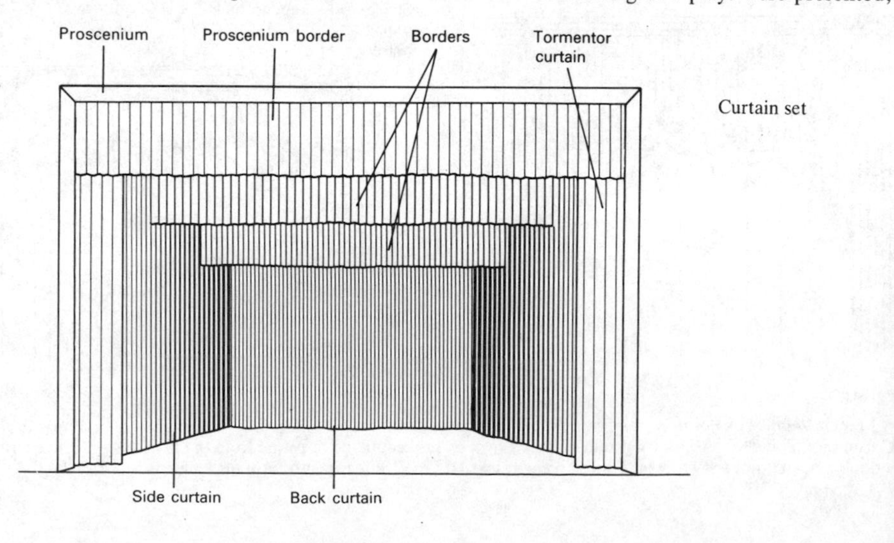

Curtain set

including works by Paul Vincent CARROLL (one of the founders) and Robert MCLELLAN. Although the Curtain disbanded in the spring of 1940, in the following Jan. John Steward, who had been its music adviser, opened in the house next door the Park, a club theatre which offered a new production once a month until it closed in 1949 when Stewart founded the PITLOCHRY FESTIVAL THEATRE.

CURTAIN THEATRE, London, in Curtain Close, Finsbury Fields, Shoreditch. This playhouse opened in 1577, a year after its close neighbour the THEATRE, which it resembled; but its dimensions are not known, nor are there any details of its actual construction. It housed a number of distinguished companies including the CHAMBERLAIN'S MEN, among whose members were TARLETON, ARMIN, and Thomas POPE, who had a share in the theatre. Shakespeare's *Romeo and Juliet* (*c.* 1595) and *Henry V* (*c.* 1599) may have been given their first productions here, in which case the 'wooden O' referred to by Chorus in the latter play would be the Curtain and not the GLOBE as formerly believed. It may also indicate that the auditorium of the Curtain was circular, a point still in dispute. In 1598 JONSON'S *Every Man in His Humour* was first seen at this theatre. Several attempts were made to suppress the Curtain, particularly after the building of the FORTUNE, with the aim of limiting the number of playhouses in London. From 1603 to about 1609 QUEEN ANNE'S MEN occupied it and its last recorded use was in 1622 by PRINCE CHARLES'S MEN, though the building is known to have been standing in 1642 and perhaps as late as 1660.

CURTO, RAMADA, see PORTUGAL.

CUSACK, CYRIL (1910–), Irish actor, director, and playwright, who was on the stage as a child and later toured for several years throughout Ireland. In 1932 he joined the company at the ABBEY THEATRE, Dublin, and remained there for 13 years, appearing in more than 65 plays including SYNGE'S *The Playboy of the Western World*, SHAW'S *Candida* (in which he played Marchbanks), and many of the major works of Sean O'CASEY, George SHIELS, Lennox ROBINSON, and Paul Vincent CARROLL. During this time he made his first appearance in London in O'NEILL'S *Ah, Wilderness!* (1936) and was subsequently seen there in a number of parts including Michel in COCTEAU'S *Les Parents terribles* (1940) and Dubedat (playing opposite Vivien LEIGH) in a revival of Shaw's *The Doctor's Dilemma* in 1942. In the same year he directed the Gaelic Players in Dublin in his own play, *Tareis an Aifreann* (*After the Mass*). In 1945, dissatisfied with the artistic policy of the Abbey, he left to form his own company with which he toured in Ireland and abroad in a repertory of Irish and European plays, giving the first performance of O'Casey's *The Bishop's Bonfire* in Dublin in 1955. He made several visits to the United States, and in 1960 was in Paris at the THÉÂTRE DES NATIONS, where he received the International Critics' Award for his solo performance in *Krapp's Last Tape* by Samuel BECKETT. In 1963, after 20 years' absence from the London stage, he returned to play in the ROYAL SHAKESPEARE COMPANY'S production of DÜRRENMATT'S *The Physicists*, and the following spring appeared with the NATIONAL THEATRE company at the OLD VIC in FRISCH'S *Andorra*. He returned to the ALDWYCH for the 1968 WORLD THEATRE SEASON in the Abbey Theatre's production of *The Shaughraun*, fulfilling a longstanding ambition to bring BOUCICAULT back on to the stage; in that year he played Gaev in CHEKHOV'S *The Cherry Orchard* and in 1969 the title role in Peter Luke's *Hadrian VII*. He took the part of Fluther in the Abbey's 1974 production of O'Casey's *The Plough and the Stars*, which toured the United States in 1976. He was seen at the OLIVIER THEATRE in London in the same role in 1977, and two years later played the title role in Chekhov's *Uncle Vanya* at the Abbey. He returned with the company to the Old Vic in 1980 in Hugh LEONARD'S *A Life*, in which he scored a personal triumph as the ageing, terminally ill civil servant Drumm.

CUSHMAN, CHARLOTTE SAUNDERS (1816–76), American actress, good in such parts as Lady Macbeth, which she played in 1836 for her début in New York, and later in London with MACREADY, Mrs Haller in KOTZEBUE'S *The Stranger*, and above all Meg Merrilies in a dramatization of SCOTT'S *Guy Mannering*. She also played Lady Gay Spanker in the first American production of BOUCICAULT'S *London Assurance* in 1841. In 1845 she was seen in London at the PRINCESS'S in the above parts, as well as Rosalind in *As You Like It*, Beatrice in *Much Ado About Nothing*, and Portia in *The Merchant of Venice*. Being of a somewhat masculine cast of countenance, she also played a number of male roles, including Hamlet, Romeo to the Juliet of her younger sister Susan (1822–59), Oberon in *A Midsummer Night's Dream*, and Claude Melnotte in BULWER-LYTTON'S *The Lady of Lyons*. During the last years of her life she gave a great many Shakespeare readings, which proved very successful. In 1907 a Charlotte Cushman Club was established in Philadelphia. Its clubroom contains many interesting theatrical relics, paintings, and material contemporary with her career.

CUT-CLOTH, CUT-OUT DROP, see CLOTH.

CWMNI THEATR CYMRU, see WALES.

CYCLORAMA, a curved wall, or section of a dome, built at the back of the stage, with an unbroken surface upon which light can be thrown. The effects thus achieved can be striking, but since to present a perfect surface a cyclorama must be rigidly and heavily built, generally of hard cement, it has several drawbacks, interfering with access to the stage from the sides and with the suspension of scenery from above. Being immobile, it also becomes a potential obstacle when not in use, and limits the scenery that can be used on that particular stage to one style only. A moveable cyclorama has been invented, but never widely adopted. A partial, or shallow, cyclorama, or even a plain, uncurved, distempered back wall, has often been used instead of a full cyclorama, and proved just as effective and much less restricting. In Germany, where the cyclorama was first used, it is called a *Rundhorizont*.

CYDER CELLARS, see EVANS'S.

CYRANO DE BERGERAC, SAVINIEN (1619–55), French dramatist, soldier, and celebrated swordsman, always ready to take offence if anyone dared to remark on his enormous nose. On leaving the army he settled in Paris and cultivated his undoubted gifts for literature and science. His friendship with MOLIÈRE and SCARRON turned his thoughts to the stage, but his tragedy *La Mort d'Agrippine* (1653) was not a success, and a comedy, *Le Pédant joué*, written between 1645 and 1649, does not appear to have been performed; it must have circulated in manuscript since both Scarron and Molière—the latter in a scene from *Les Fourberies de Scapin* and one in *Don Juan*—appear to have been influenced by it. The hero of ROSTAND's great Romantic drama *Cyrano de Bergerac* (1898) is a mixture of fact and fiction, a swashbuckling Gascon with the soul of a poet, whereas the real Cyrano was a Parisian and probably much less given to altruism and introspection than Rostand's hero.

CZECHOSLOVAKIA. The bilingual character of Czechoslovakia makes it necessary to deal separately with the two main currents of the theatre there—the Czech and the Slovak. The Czech language is found in the texts of early LITURGICAL DRAMA in the 13th century, and it was not long before the use of the vernacular in some productions led to the introduction into the action of secular comic and satirical scenes such as that of the quack doctor. The tradition of the native PASSION PLAY, however, lapsed during the Hussite wars of religion in the early part of the 15th century, and was not revived until 1970, in Brno. When in the 16th century the theatre made a fresh start it relied mainly on the SCHOOL DRAMA in Latin of such educational writers as Comenius (Jan Amos Komenský, 1592–1670), though plays by a few dramatists, among them Pavel Kyrmezer (?–1589), author of *The Czech Comedy of Dives and Lazarus* (1566), portrayed scenes from contemporary life in the vernacular. The Thirty Years War, and the subsequent tyranny of the Hapsburgs, put an end to the promising beginnings of both humanist and local drama, and after the defeat of the Czechs at the battle of Bílá Hora (The White Mountain) in 1620, the nobility and intelligentsia were either executed or forced to emigrate. For more than 150 years the Czech theatre seemed dead, while Italian opera, German plays, and the school productions of JESUIT DRAMA reigned supreme. Itinerant foreign companies did occasionally play in Czech, while the tradition of the vernacular Nativity play, usually written by a priest, or a Saint's play, like the late 17th-century anonymous *Play About the Holy Maid Dorothy* in which the conflict between the pagan king and the Christian saint typified that between the Czech people and their oppressors, continued, giving rise to an occasional performance inside or outside the church; while in the villages folk comedies still survived. Among these were the comic interludes of Václav František Kozmánek (1607–79) and later the plays of Karel Kolčava (1656–1717), which survived into the 18th century and were to be revived as typical folk dramas by E. F. Burian in the 1930s. From the 18th century also came such comedies in the vernacular as *The Comedy of Frances, the King of England's Daughter and Jock the Merchant's Son*, in which the young artisan Jock (Honzíček) rescues the princess Frances (Františka) from the Turks and eventually marries her.

The earliest theatres—the first in Prague was built in 1737—were intended for the use of foreign companies; although a few plays in Czech were produced in Prague and Brno between 1767 and 1783, it was not until 1785 that František Bulla, manager of the theatre built in 1783 by Count Nostitz-Rienick (now the Tyl Theatre), was able to put them on more frequently, with the intention of contributing to the revival of Czech national culture by means of the spoken word. The outstanding personalities of the Czech theatre in this period of the 'enlightenment' were the Thám brothers—Karel Ignác (1763–1816) and Václav (1765–1816)—whose work was connected more especially with the Bouda (or Hut) Theatre from 1786 to 1789, and Prokop Šedivý (1764–1810), who inaugurated a type of local Prague farce which later became popular throughout Europe. At the turn of the century the chief dramatists were Jan Nepomuk Štěpánek (1783–1859) and Václav Kliment Klicpera (1792–1859), whose lively comedies burlesqued country bumpkins and small-town philistines impartially, while their poetic dramas inaugurated the era of Czech Romanticism; but the most influential figure of the time was undoubtedly Josef Kajetán TYL, himself an actor and the reputed founder of the modern Czech theatre. His work was complemented by that of Josef Jiří Kolár (1812–96), who was also an actor and director; Ferdinand Břetislav Mikovec (1826–62); Vítězslav Hálek (1835–74), who, like Kolár, was much influenced by Shakespeare in his writing of plays based on Czech history; and Emanuěl Bozděch (1841–89), whose WELL-MADE PLAYS, in the manner of SARDOU, though now forgotten, were popular in their time. The drama of contemporary Czech life, inaugurated by Tyl, was continnued by František

Věnceslav Jeřábek (1836–93), with such plays as *His Master's Servant* (1870), and František Ferdinand Šamberk (1839–1904), author of *The Eleventh Commandment* (1883). Two poetic dramatists, who continued the work of the Romantics, were Julius Zeyer (1841–1901), with such plays as *Radúz and Mahulena* (1898), based on Slovak folk-lore, and Jaroslav Vrchlický [Emil Frída] (1855–1912), whose *Night at the Karlstein* (1884), set in the time of the Emperor Charles IV, and *The Trial of Love* (1886) were well constructed comedies of great charm.

While these and other dramatists were producing a fine body of work in the Czech language, the material side of the theatre was not neglected. By the mid-19th century there were resident companies, almost fully professional, in Prague and Brno, and the first Czech touring company, Prokopš, had already set out on its travels. Another potent influence was puppetry, particularly that of the famous Czech puppeteer Kopecký, who travelled with his family all over the country. Nor should the influence of the many amateur groups of the time be underrated. Inevitably in the early days actors, among them Jan Kaška (1810–69), Karel Šimanovský (1825–1904), and the young Francesca JANAUSCHEK, who after two years in Prague left to continue her career elsewhere, modelled their style on that of the world-famous VIENNA Burgtheater company; but they were soon joined, and superseded, by a younger generation who played in a freer, more realistic style; among them were Josef Šmaha (1843–1922), Vendelín Budil (1847–1928), and Marie Bittnerová (1854–98), working under the playwright and director František Adolf Šubrt (1849–1915), who from 1883 to 1900 ran the National Theatre. Among his own plays, which contributed not a little to the movement towards REALISM, were *Jan Výrava* (1886), *The Great Freeholder* (1891), and the starkly realistic *Drama of Four Bare Walls* (1893), based on an actual strike by Bohemian miners. Other realist dramatists of the time were Josef Štolba (1846–1930) and Václav Štech (1859–1947), who began by writing for amateur and provincial companies, but after the success of the folk-comedy *Our Upstart Villagers* (1887) by Ladislav Stroupežnický (1850–92) achieved recognition with such comedies as *The Wife* (1888) by Štech and *The Will* (1889) by Štolba. Other playwrights who depicted life in Czech and Moravian villages as it was, comic or tragic, and not as the Romantics would have liked it to be, were Alois Jirásek (1851–1930) with *Father* (1894); Gabriela Preissová (1862–1946), whose *The Step-daughter* (1891) provided the libretto of Janáček's opera *Jenufa* (1904); and the Mrštik brothers, Alois (1861–1925) and Vilém (1863–1912), joint authors of *Maryša* (1894), set in a Moravian village. Jirásek also brought realism to the historical play with his trilogy set in Hussite times, *Jan Žižka* (1903), *Jan Hus* (1911), and *Jan Roháč* (1918).

The reaction against realism, apparent at the turn of the century and after, owed much to the current trend in Europe towards SYMBOLISM, as can be seen in Jirásek's *The Lantern* (1905), and in several plays by Jaroslav Hilbert (1871–1936), among them *The Outcasts* (1900), *Job* (1928), and *Twin Spirits* (1930). Fráňa Šrámek (1877–1952) introduced Impressionism into Czech drama with such excellent plays as *Summer* (1915) and *Moon Over the River* (1922), which remain in the repertory; while Arnošt Dvořák (1880–1931), in such works as *Wenceslas IV* (1910) and *The Hussites* (1919), exemplifies the growing influence of EXPRESSIONISM. But the outstanding man of the theatre of the first half of the 20th century was undoubtedly Jaroslav KVAPIL, actor, playwright in many moods, and from 1900 to 1918 manager of the National Theatre. Under him a new generation of actors arose, headed by the overwhelming personality of Eduard Vojan (1853–1920); among them were Kvapil's wife, Hana Kvapilová (1866–1907), who appeared in many of his plays; Marie Hübnerová (1862–1931), unsurpassed in the interpretation of women of the people in popular drama; and Anna Sedláčková (1887–1967), an elegant and entrancing comedienne.

Kvapil was succeeded as manager of the National Theatre by Karel Hugo HILAR, under whom the plays of Karel ČAPEK, the first Czech dramatist to achieve international fame, were produced, as were those of František LANGER. Meanwhile socialist and experimental theatres were being run by such writers and actors as Jiří Voskovec [Wachsmann] (1904–81) and Jan Werich (1905–80), who appeared in their own political skits and sketches at the Free (or Unfettered) Theatre from 1927 until the Nazi invasion in 1938 drove them to New York, whence only Werich returned. One of their joint works, *The Rag Ballad* (1935), enjoyed a considerable success when revived at the East Berlin Festival in 1958. Another dramatist connected with the Free Theatre was the fine novelist Vladislav Vančura (1891–1942), younger brother of Antonín Vančura (1882–1939), who as Jiří Mahen wrote a number of plays reflecting all the misery and turmoil of Central Europe after the First World War, as in *Heaven, Hell, Paradise* (1919), *Progeny* (1921), and *The Deserter* (1924). He committed suicide after Munich; Vladislav, best remembered for his allegorical play *Teacher and Pupil* (1927) and his study of contemporary trends in medicine in *The Sick Girl* (1928), was taken hostage after the assassination of Heydrich and executed in a concentration camp. Vitězslav Nezval (1900–58), though arrested in 1941, survived to work in the post-war theatre and in 1954 produced his prophetic and utopian allegory *Tonight the Sun Still Sets on Atlantis*, and Emil František Burian (1904–59), founder in 1933 of his own experimental theatre the D34, did excellent work until he was deported in 1941. Returning after the war, Burian again raised his theatre, now called after him, to a high pitch of excellence, one of his most interesting and controversial productions being an idiosyncratic version of BRECHT's *Die Dreigroschenoper* in 1957, with František Vnouček as Mackie Messer. One of his young associates at the D34, Alfréd Radok (1914–76), also deported in 1941, returned in 1945 to become a director at the National Theatre and internationally known through his connection with Josef SVOBODA and his 'Laterna Magika' multi-media show, first seen at the Brussels Exhibition in 1958. Himself the author and adapter of several plays, Radok left Czechoslovakia in 1968 and worked abroad until his death. Since 1938 the Czech theatre had been in complete disarray, as war, occupation, death, and emigration took their toll, reinforced by the nationalization of the theatres in 1945, and the imposition of SOCIALIST REALISM in 1948. For a time didactic drama prevailed, as in Vašek Káňa's *Grinder Karhan's Gang* (1949), widely seen throughout Eastern Europe; in the 1950s and 1960s some loosening of control permitted the emergence of new talents. Jiří Suchý and Jiří Šlitr, working in the tradition of Voskovec and Werich, produced satirical musical plays with much use of improvisation and their own idiosyncratic version of the chanson. Like the playwright Pavel Kohout, who also directed at the Realistic Theatre, and Ladislav Smoček, whose Drama Club Theatre opened in 1965 with his own *Picnic* and other plays written under the influence of the Theatre of the ABSURD, they encountered official opposition after the Soviet invasion of 1968, as did Josef Topol, whose *Hour of Love* (1967), directed by Otomar KREJČA, was seen in London during the WORLD THEATRE SEASON of 1969, and Václav Havel, who was for a time resident dramatist at the Theatre on the Balustrade, and from 1955 at the Theatre Behind the Gate, with its founder Krejča, until it was closed in 1972. Since then all Havel's plays have been produced abroad. Apart from Krejča, other directors who remained active during the period were the veteran actor Zdeněk Štěpánek (1896–1968), who became manager of the National Theatre in 1954, where he also appeared in leading roles in several of Čapek's plays, and Jan GROSSMAN, who returned to Czechoslovakia in 1974 after some years abroad. Many of the country's best actors have turned to films, but one who still makes the theatre his chief concern is Ladislav Pešek, who in the 1966 World Theatre Season in London was seen as The Tramp in Čapek's *The Insect Play*.

In Slovakia the theatre was slow to establish itself, and although there may have been something similar to the development elsewhere of liturgical drama, there is little written evidence of it. For a long time plays were produced only in Latin or German, except for the vernacular folk-dramas of the villages, and all were acted by amateurs, a state of affairs which continued until the establishment of the Czechoslovak Republic in 1918. Drama, in fact, advanced more quickly than theatre, with the pioneer Slovak dramatists Samo Chalupka (1791–1871), whose comedies of small-town life proved that the Slovak language could be used on stage, and his successor Jan Palárik (1822–70), who set out to use drama as a means of furthering the cause of Slovak nationalism. One of his satirical comedies, *The Tinker* (1860), was chosen by the Croatians to open the 1863 season of their new National Theatre in Zagreb. Among the Slovaks it was less popular, but with it and his other plays Palárik had at least assured a firm foundation for the development of Slovak drama. His vein of comedy was developed by other writers, and some progress towards realism was made in the plays of Jozef Gregor-Tajovský (1874–1940); but the most popular playwright of the early 20th century among the amateur groups with their often limited capabilities was Ferko Urbánek (1859–1934), whose many plays appear to have been straightforward melodramas, with virtue triumphant and vice properly punished. First performed by village groups, they were later seen at the Municipal Theatre in Bratislava when in 1920 it became the National Theatre, with professional actors playing both in Czech and Slovak. During the period between the two world wars the leading playwrights were Ivan Stodola (1888–1942), author of satirical comedies of provincial life and also of historical dramas, and Július Barč-Ivan (1909–55), who drew his characters mainly from the modern business world and looked for the solution of social problems in the light of his own religious faith. During the Second World War both he and Stodola wrote allegorical satires directed against fascism, as did Peter Zvon (1913–42), while Mária Mazusová-Martáková in *Jánošík* (1941) appealed yet again to Slovak patriotism with the story of Slovakia's Robin Hood, who plundered rich Hungarian merchants for the benefit of oppressed Slovak peasants, a subject treated many times, not only by Slovak playwrights but also by Jiří Mahen in Czech.

The most important playwright of the post-war years is Peter Karvaš, author of *Midnight Mass* (1959), set in the time of the Slovak uprising of August 1944, and *Antigone and Others* (1961), set in a concentration camp. Among his comedies the most successful was *The Bigwig* (1964), which satirized the continuing 'cult of personality'. Two other post-war dramatists are Štefan Králik, with *Saint Barbara* (1953), and Igor Rusnák, who in *Foxes, Good Night* (1964) dramatized the problems of the younger generation. Since 1968, however, little good new work seems to have been produced, Karvaš in particular having had nothing staged since *Experiment Damocles* (1968). An outstanding actor of the Slovak stage was Andrej Bagár, who died in 1966. He was already well known before the war and during the German occupation his young company performed for and supported the partisans, the group forming the nucleus of the Theatre of the Slovak Uprising still active in Martin. He was deported by the Nazis in 1939. In 1949 he became head of the National

Theatre in Bratislava, where his finest roles were considered to be Shakespeare's Hamlet and Macbeth, and Herod in *Herod and Herodias* by the poet Pavol Országh Hviezdoslav (1849–1921). Josef Budský, some years younger, was originally with the Czech company at the National, where he played Dr Galen in Čapek's *The White Scourge*, but moved to the Slovak section in 1936, where he gave impressive performances as SOPHOCLES' Oedipus and MOLIÈRE's Tartuffe. After the war he did more directing than acting, his production of *As You Like It* in 1946 being considered remarkable.

Czechoslovakia as a whole has a remarkable record in theatre design, the two leading designers being Josef Svoboda in Prague and Ladislav Vychodil in Bratislava. It also has excellent and numerous puppet-theatres, its chief puppet-master for many years being Josef Skupa (1892–1957); and its strong tradition of MIME (the great DEBURAU was born in Bohemia) has been upheld by Ladislav Fialka, with MARCEAU one of the great mime artists of modern times. A lively FRINGE THEATRE includes groups such as Divadlo na Provázku (Theatre on a String) in Brno, Ypsilonka in Prague, and the popular duo Lasica and Satinský in Bratislava, who all contrive to present a remarkable amount of social criticism. Czechoslovakia is fortunate in possessing two 18th-century theatres, with scenery and working parts still intact, at ČESKÝ KRUMLOV and Litomyšl.

D

DADA, art movement of the 1920s which was later to have a vital influence on European drama. Irrational, nihilistic, and anarchic, it attacked the complacency of so-called 'society', and set out to arouse in its audiences the outrage later evoked by ARTAUD'S Theatre of CRUELTY, exploiting also the sense of futility which gave rise to the Theatre of the ABSURD.

DADIÉ, BERNARD BINLIN (1916–), the outstanding dramatist of French-speaking Africa, who became Ivory Coast Minister of Culture. His first play *Les Villes* (1933) was written while he was still at school in Bingerville, Ivory Coast, his second, the historical drama *Assémien Déhylé, roi du Sanwi* (1936), during his time at the École Normale William Ponty in Dakar. Dadié then established his reputation as a novelist, poet, and satirist, returning to the theatre in the late 1960s, *Monsieur Thôgô-gnini* being seen at the pan-African cultural festival in Algiers in 1969; it has since been performed several times in various parts of French-speaking Africa. Dadié's satire is based on a perceptive appreciation of human behaviour; Thôgô-gnini and Nahoubou, the chief character in *Les Voix dans le vent,* have a flavour of JARRY'S *Père Ubu*, linking Dadié's work with the Theatre of the ABSURD.

DAGERMAN, STIG (1923–54), Swedish novelist and dramatist whose work, though somewhat static, reflects the disillusion of the early post-war years, using the techniques of EXPRESSIONISM to convey a compelling sense of evil. Thus in *Den dödsdömde* (*The Condemned*, 1947) the reprieved hero gains existential strength from his confrontation with death. *Skuggan av Mart* (*The Shadow of Mart*, 1948) and *Ingen går fri* (*Nobody is free*, 1949) combine realistic and symbolic elements, and *Streber* (*The Upstart*, 1949) shows the betrayal of the Syndicalist movement's ideals. *Den yttersta dagen* (*Judgement Day*), written in the year of his suicide, concerns the expiation of guilt.

DAGMARTEATRET, Copenhagen, Danish theatre established in 1883 and for some decades the main rival in Copenhagen of the KONGELIGE TEATER. Bjørn BJØRNSON held an appointment there for two years in the 1890s and did much to improve its status. After the First World War it ran into financial difficulties and finally closed in 1937, the building being demolished.

DALBERG, WOLFGANG HERIBERT, Baron von (1750–1806), German aristocrat whose interest in the theatre led to his appointment in 1778 as honorary director of the newly opened National Theatre at MANNHEIM. After the death of EKHOF later the same year Dalberg arranged for the actors in his company to be transferred to Mannheim, and with them and IFFLAND, whose first plays were produced there, the theatre flourished. Dalberg prepared for it the first German version of Shakespeare's *Julius Caesar*, but his most memorable achievement was the production in 1782 of *Die Räuber* by SCHILLER. The National Theatre continued to prosper until the rigours of war reached Mannheim in 1796, when the company was disbanded and Iffland left to go to BERLIN.

DALIN, OLOF VON (1708–63), Swedish poet, dramatist, and essayist, best known for his editorship and authorship of the *Swänska Argus* (*Swedish Argus*, 1732–4), a 'moral weekly' modelled on the *Tatler* of ADDISON and STEELE. Dalin was the first dramatist to introduce the French classical style into Sweden, and his comedies show the influence of both MOLIÈRE and HOLBERG. Among the plays he wrote for the Svenska Dramatiska Teatern were the comic *Den afvuns juke* (*Envy*, 1738) and the tragic *Brynhilda* (1739).

DALLAS THEATER CENTER, Dallas, Texas, which opened in 1959, is the only theatre building in the United States designed by the architect Frank Lloyd Wright. It houses two separate theatres. The larger, the Kalita Humphreys, seats 416 in 11 rows, bringing all spectators into close proximity to the modified thrust stage. It presents a balanced programme of classical works, important contemporary plays, and premières of new plays. The smaller theatre, the Down Center Stage, which opened in 1964, seats 56 and has a proscenium-arch stage. It is devoted mainly to the production of promising new plays. The Center, which has a season running from Oct. to July, combines regional professional theatre with an academic programme leading to a graduate degree in theatre, and also runs the Magic Turtle Children's Theater.

DALSUM, ALBERT VAN, see NETHERLANDS.

DALY, (Peter Christopher) ARNOLD (1875–1927), American actor, who made his first stage appearance in 1892. In 1903 he came into prominence with the first public performance in New York of *Candida*, which he directed as well as playing Marchbanks. This was the first production in the United States of a play by SHAW since MANSFIELD's tour of *The Devil's Disciple* in 1897, and Daly followed its success with a production of *Mrs Warren's Profession* which led to an uproar and prosecution by the police. In spite of this he produced *You Never Can Tell, Arms and the Man*, and a double bill consisting of *The Man of Destiny* and a trifle written specially for him by Shaw, *How He Lied to Her Husband*. He also played this last sketch in vaudeville. Opposition to the new 'theatre of ideas' finally defeated him and he was forced back into the run-of-the-mill life of the theatre, dying in a fire in his early fifties.

DALY, (John) AUGUSTIN (1839–99), American manager and dramatist, who while working as a dramatic critic on various papers wrote or adapted a number of plays, among them *Leah the Forsaken* (1862), based on Mosenthal's *Deborah*, and a melodrama, *Under the Gaslight; or, Life and Love in These Times* (1867). In 1869 he took over the management of the FIFTH AVENUE THEATRE, which prospered until it was burnt down in 1873. He then opened another theatre, also on Fifth Avenue, which he left in 1879 to open the New Broadway as DALY'S, with a fine company headed by John DREW and Ada REHAN, which in 1884 was seen in London, at the CHARING CROSS THEATRE (then Toole's), with such success that after further visits between 1886 and 1890 Daly decided to open his own theatre there. Both in New York and in London Daly's first nights were important events, and he set a high standard in spite of his tendency to tamper with the text of established classics.

DALY'S SIXTY-THIRD STREET THEATRE, New York, at 22–6 West 63rd Street, between Central Park West and Broadway. This was built as the Davenport in 1909, but was not completed until 1919 when it became a cinema for children. It was renamed Daly's on 3 Oct. 1922, and staged a notable revival of CONGREVE's *Love for Love* in 1925. Known successively as the Coburn (1928), the Recital (1932), the Park Lane (1932), and Gilmore's (1934), it was leased as the Experimental Theatre in 1936 by the FEDERAL THEATRE PROJECT, who gave there the first New York production of SHAW's *On the Rocks* in 1938. It stood empty for some years, being used by the Shubert management for storage before being demolished in 1957.

DALY'S THEATRE, London, in Cranbourn Street, off Leicester Square. This theatre, which had a seating capacity of about 600 in three tiers, was leased to Augustin DALY by George EDWARDES, and opened on 27 Jan. 1893 with *The Taming of the Shrew*, Ada REHAN playing Katharina. This was followed by Sheridan KNOWLES's *The Hunchback*, with Violet VANBRUGH, and in 1894 by *Twelfth Night* and *As You Like It*. In the same year Daly presented Eleonora DUSE in the younger DUMAS's *La Dame aux camélias*, and later Edwardes transferred the musical comedy *A Gaiety Girl* to Daly's from the PRINCE OF WALES THEATRE. During the next 15 years Daly's became one of the most fashionable and successful theatres in London, among the eleven productions which occupied it being *San Toy* (1899), *A Country Girl* (1902), *The Merry Widow* (1907), and *The Dollar Princess* (1909). Edwardes's last production there was *The Marriage Market* (1913). The theatre had a further musical smash hit when in 1917 Oscar ASCHE produced *The Maid of the Mountains* which, with José Collins as Teresa, ran for 1,352 performances; it was followed by *A Southern Maid* (1920). In 1922 Daly's was bought by the financier James White, whose productions before he committed suicide in 1927 included *The Lady of the Rose* (1922), *Madame Pompadour* (1923), and *Cleopatra* (1925). At the end of 1927 Noël COWARD's *Sirocco* proved a failure, and a number of old musicals were revived with little success. The last production at this famous theatre was Emmet Lavery's *The First Legion*, which closed on 25 Sept. 1937. The site is now occupied by the Warner Cinema.

DALY'S THEATRE, New York, at 1221 Broadway, below 30th Street. Originally known as Banvard's, and later Wood's, Museum, this opened in 1867 and changed its name a year later. Although the repertory was mainly burlesque, variety, and melodrama, straight plays were occasionally performed after 1872, the year in which Laura KEENE made her last appearance there. After several seasons as the Broadway and the New Broadway, the theatre was remodelled and redecorated and opened on 17 Sept. 1879 as Daly's, to become one of the leading playhouses of New York, particularly after the loss by fire of the NEW PARK THEATRE (1) in 1882 and the UNION SQUARE THEATRE in 1888. One of its most successful productions was *The Taming of the Shrew*, which opened on 17 Jan. 1887 with Ada REHAN, who had first been seen at this theatre in 1875 in Lumley's *The Thoroughbred*. Augustin DALY continued to manage the theatre until his death in 1899, and it retained his name until it was demolished in 1920, having been a cinema since 1915.

DAME, a female character in the English PANTOMIME, traditionally played by an actor, sometimes one recruited from the MUSIC-HALLS like Dan LENO and George ROBEY. Among the familiar Dame parts are Aladdin's mother Widow Twankey (a name taken by H. J. BYRON from a Chinese tea-exporting port), Idle Jack's mother in *Jack and the Beanstalk*, usually known as Dame Durden or Dame Trot, and Cinderella's ugly sisters, who go under many pairs of names. If the ugly sisters are played by women, then their mother the Baroness is played by a man. Other Dame parts are the Cook in *Dick Whittington*, the Queen of Hearts, Mother Goose, and Mrs Robinson Crusoe.

D'AMICO, SILVIO (1887–1955), Italian theatre scholar, appointed in 1923 Professor of the History of the Theatre at the St Cecilia Academy in Rome, who inaugurated and until his death edited the monumental *Enciclopedia dello Spettacolo*. He was dramatic critic of several Italian papers, founding and editing the theatre review *Scenario* in 1932 and the *Rivista italiana del dramma* (later *del teatro*) in 1937. He also founded in 1935 the academy of dramatic art now named after him. He wrote many books on the theatre, of which the most important was his *Storia del teatro drammatico* (1939–40; 3rd edition in 4 vols., 1953). A great lover of the living theatre, he did not confine his attention to its literary and academic aspects but encouraged his students to act and direct, and a number of them entered the professional theatre.

DANCER, ANN, see BARRY, SPRANGER.

DANCHENKO, VLADIMIR NEMIROVICH-, see NEMIROVICH-DANCHENKO.

DANCOURT, FLORENT CARTON (1661–1725), French dramatist and actor, who in 1680 married Marie Thérèse Lenoir (1663–1725), daughter of LA THORILLIÈRE, giving up the study of law to join her on the stage. He had a good face and figure, and a natural liveliness of disposition which was an asset in comedy, though in tragedy he was judged cold and monotonous. After some years in the provinces he and his wife joined the COMÉDIE-FRANÇAISE and remained there until 1718. During this time he wrote more than 50 comedies based on small scandals or other topical events, his better plays showing some skill in etching the contemporary scene. He was particularly good in his delineation of the rising middle class, with its love of money, its desire for political power, its shrewdness in taking advantage of a corrupt government, and its avidity for easy pleasure. There is no bitterness in Dancourt's writing, but under the superficial wit and good humour much brutality and selfishness are apparent. His best play is *Le Chevalier à la mode* (1687), written in collaboration with Saint-Yon, a portrait of contemporary life with much satire at the expense of parvenu financiers. Others worthy of note are *La Maison de campagne* (1688) and the one-act *Les Vendanges de Suresnes* (1695), the most frequently revived of all his works. Dancourt was not as versatile as DUFRESNY nor as poetic as REGNARD, but more than anyone after MOLIÈRE he was a dramatist who was also a man of the theatre.

DANCOURT, MARIE-ANNE-ARMANDE (1684–1745) and MARIE-ANNE-MICHELLE (1685–1780), daughters of the above, known respectively as Manon and Mimi. They were both on the stage as children, making their first appearance together in their father's play *La Foire de Bezons* in 1695. They were precocious and talented and their father continued to write plays for them in which they were much admired. In 1701 they made their adult débuts as members of the COMÉDIE-FRANÇAISE. Manon retired a year later on her marriage, but Mimi remained and was still drawing her pension at the age of 95.

DANE, CLEMENCE [Winifred Ashton] (1888–1965), English dramatist and novelist, who took her pseudonym from the church of St Clement Danes in the Strand. She acted for a few years as Diana Cortis, making her first appearance in London in H. V. ESMOND's *Eliza Comes to Stay* (1913), but the success of her first play *A Bill of Divorcement* (1921) led her to give up acting and concentrate entirely on writing. It was in this play, which deals sympathetically with the problem of divorce on the grounds of insanity, that Katharine CORNELL made one of her first successful appearances in New York. Among Clemence Dane's later plays were *Will Shakespeare* (also 1921), *Naboth's Vineyard, Granite* (both 1926), and *Wild Decembers* (1932), a play on the Brontës in which Diana WYNYARD played Charlotte, Beatrix LEHMANN Emily, and Emlyn WILLIAMS Branwell. Her last play, *Eighty in the Shade* (1958), was specially written for Sybil THORNDIKE and Lewis CASSON, who played the leading parts. Several of her novels deal with the theatre; she was also an excellent sculptor, and her bust of Ivor NOVELLO stands in the foyer of the Theatre Royal, DRURY LANE.

DANGEVILLE [Marie-Anne Botot], Mlle (1714–96), French actress, member of a family that served the COMÉDIE-FRANÇAISE for three generations. After playing elsewhere as a child, she was admitted there in 1730 to play comedy roles in which she excelled. Some of her greatest triumphs were in MARIVAUX's plays, in which she was ably supported by PRÉVILLE, and GARRICK considered her superior even to Mlle CLAIRON. A warm-hearted woman whom no breath of scandal ever touched, she adopted in old age a granddaughter of the great actor BARON, whom she found living in poverty. Her retirement in 1763 was much regretted, particularly by her fellow actors who paid frequent surprise visits to her house in Vangirard. In the garden there on her birthday some of them first performed COLLÉ's *La Partie de chasse d'Henri IV*, which Louis XV had banned from the public stage.

DANIEL, ISAAC, see HEBREW THEATRE, NEW.

DANIEL, SAMUEL (*c.* 1563–1619), English poet and dramatist, author of an unacted tragedy in the classical manner on the subject of Cleopatra, and of *Philotas* (1604), which caused trouble on account of a fancied resemblance in its plot to the unhappy fate of the Earl of Essex. Daniel was at this time in charge of one of the BOY COMPANIES and was implicated in the trouble over their production of *Eastward Ho!* (1605) by JONSON, CHAPMAN, and MARSTON, which gave offence to James I. He was also the author of two pastorals, and of a number of Court MASQUES which contain some excellent poetry, little relished by Jonson, Daniel's rival in this field, but highly praised by many of his contemporaries.

D'ANNUNZIO [Rapagnetta], GABRIELE (1863–1938), Italian poet, novelist, and dramatist, who in 1919 captured the port of Fiume (now Rijeka in Yugoslavia) for Italy, and thereafter became a political rather than a literary figure. His plays, which are simple in structure but rich in poetry and sensuality, aroused much controversy when first produced, but none is the product of an emotionally mature mind and none, except perhaps *La figlia di Jorio* (1904), is truly dramatic. His stage directions reveal the extent and accuracy of his archaeological knowledge, but his people are puppets, driven by elemental passions, and his plays live mainly by their poetry. Among the best known are *La città morta* (1898), in which worship of the Nietzschean superman is already apparent; *La Gioconda* (1899), in which DUSE, who spent much time and money in furthering the production of his plays, starred opposite SALVINI; and *Francesca da Rimini* (1902). None of these has remained in the repertory, but a religious play written in French, *Le Martyre de San-Sébastien* (1911), is sometimes revived in a cut version for the sake of the incidental music by Debussy. His medieval verse tragedies, such as *La fiaccola sotto il moggio* (*The Light under the Bushel*, 1905) best demonstrate his tempestuous yet curiously static style. Both in life and in art he possessed a certain grandiose vitality; but his search for heroic transcendence, and frequent underlying sadism, reflect all too tragically the mood which gave rise to Fascism.

DANSKE SKUEPLADS, Den (The Danish Theatre) Copenhagen. This opened in Lille Grønnegade on 23 Sept. 1722 and Ludvig HOLBERG wrote 28 comedies for the theatre before it closed in 1727, thus establishing a truly Danish drama for the first time. The theatre was re-established under royal patronage in 1747, with Holberg acting informally as consultant. The following year the company moved to a new building in Kongens Nytorv. In 1770 the monarchy accepted financial responsibility for it, and two years later the building was officially designated as the KONGELIGE TEATER.

DANVERS, JOHNNIE, see LENO, DAN.

D'ARCY, MARGARETTA, see ARDEN, JOHN.

DARLINGTON, W(illiam) A(ubrey) (1890–1979), English dramatic critic and dramatist, who was drama critic of the *Daily Telegraph* from 1920 to 1968, continuing to write theatrical articles for the paper after his retirement until his death, and also drama correspondent of the *New York Times* from 1936 to 1960. His first play, based on one of his own novels, was *Alf's Button* (1924), an extravaganza which had a great success and was several times revived and filmed. It was followed by *Carpet Slippers* (1930), *A Knight Passed By* (1931), a burlesque version of BOUCICAULT's *The Streets of London* (1932), and *Marcia Gets Her Own Back* (1938). He was also the author of lives of SHERIDAN (1932) and J. M. BARRIE (1938) and of several critical works on the theatre. His autobiography *I Do What I Like* was published in 1947.

DASTÉ, JEAN (1904–); French theatre director and actor, who began his acting career in 1919. He was with COPEAU in Burgundy, married his daughter, and was later with the COMPAGNIE DES QUINZE. Before the Second World War he helped BARSACQ to found the Compagnie des Quatre Saisons, which pioneered provincial tours with specially designed productions, and endeavoured to continue this work under the German occupation. In 1945, after an unsuccessful attempt to set up a company in Grenoble, he moved to St Étienne, where in 1947 he became director of the newly formed dramatic centre set up under the new policy of DÉCENTRALISATION DRAMATIQUE. Over the years his programme of two or three classics and one new play each season was modified until it carried only one classic a year and several plays by new authors, notably GATTI, BRECHT, and FRISCH. He resigned in 1970 after a disagreement with the municipal authorities, but continued to take an interest in the company. As an actor he was at his best in MOLIÈRE and as the comic characters in ANOUILH's plays. He wrote *Voyage d'un comédien* (1977).

DATT, UTPAL, see INDIA.

DAUBENY, PETER LAUDERDALE (1921–75), English impresario and theatre manager, knighted in 1973. He started his theatrical career as an actor under William ARMSTRONG at the LIVERPOOL PLAYHOUSE, but soon moved to London, where he was responsible for a long series of productions, early successes including WERFEL's *Jacobowsky and the Colonel* (1945) and Hugh Mills's *The House by the Lake* (1956). He also arranged, often under great difficulties, visits by many outstanding foreign theatrical groups, notably the BERLINER ENSEMBLE in 1956, a visit which had immense repercussions in the English theatre, and the MOSCOW ART THEATRE in 1958. Later productions, among them Michael REDGRAVE's version of Henry JAMES's *The Aspern Papers* (1959) and BILLETDOUX's *Chin-Chin* (1960), met with general approval, though the LIVING THEATRE's production of Jack GELBER's *The Connection* (1961) caused an uproar. Daubeny was also responsible for an annual WORLD THEATRE SEASON from 1964 to 1973, which without official support played in London the role undertaken in Paris by the Government-subsidized THÉÂTRE DES NATIONS, and was the author of *Stage by Stage* (1952) and *The World My Theatre* (1971).

DAUVILLIERS [Nicolas Dorné] (*c.* 1646–90), French actor, who was at the MARAIS in 1670, marrying two years later an actress in the company, Victoire-Françoise (*c.* 1657–1710), daughter of Raymond POISSON. After the death of MOLIÈRE in 1673 husband and wife joined the depleted company run by Molière's widow, but were excluded from it a year later for three months because of their opposition to the costly revival of Thomas CORNEILLE'S MACHINE PLAY *Circé*. On the formation of the COMÉDIE-FRANÇAISE in 1680 Dauvilliers became a member of the company, remaining with it until his death; his wife retired from acting because of ill health, but remained as prompter. It was said of Dauvilliers that he had no advantages of person and was poor in comedy, but that in tragedy he sometimes showed a genius akin to madness. Indeed, tradition has it that he went mad on the stage and tried to kill himself with a property sword while playing Eros in a revival of LA CHAPELLE's *Cléopâtre*, subsequently dying in an asylum.

DAVENANT, Sir WILLIAM (1606–68), English dramatist and theatre manager, reputed to be the natural son of Shakespeare by the hostess of the Crown Inn, Cornmarket, Oxford. There is no proof of this, though Shakespeare may have been his godfather. Certainly Davenant had a great love of the theatre and before the closing of the playhouses in 1642 had written a number of plays and Court MASQUES in the style of Ben JONSON whom he succeeded as Poet Laureate in 1638. During the Civil War he fought on the King's side, and was knighted by Charles I in 1643. Towards the end of the Commonwealth he managed to evade the ban on stage plays and by presenting them as 'music and instruction' got permission to put on *The Siege of Rhodes* (1656), considered by some the first English opera, *The Spaniards in Peru* (1658), and *Sir Francis Drake* (1659). On the Restoration in 1660 he obtained from Charles II a patent under which he opened LINCOLN'S INN FIELDS THEATRE as the Duke's House, with a company led by Thomas BETTERTON. Later he built a new theatre, DORSET GARDEN, but died before it opened. Davenant, who was responsible for several adaptations of Shakespeare to suit the tastes of the time, was also one of the first to encourage the vogue for machinery, dancing, and music which came in with the proscenium theatre and elaborate scenery, and his work, though not always to the taste of later critics, gave a great impetus to the development of the English theatre. He was originally the person aimed at in the character of Bayes in BUCKINGHAM's *The Rehearsal* (1671), which was altered after his death to satirize DRYDEN instead.

DAVENPORT, EDGAR and HARRY, see DAVENPORT, EDWARD; JEAN and THOMAS, see LANDER, JEAN; LIZZIE, see MATHEWS, C. J.; MAY, see SEYMOUR, WILLIAM.

DAVENPORT, EDWARD LOOMIS (1815–77), American actor, son of an innkeeper, who first appeared on the stage, billed as Mr E. Dee, at Providence, R. I., playing a small part to the elder Junius Brutus BOOTH's Sir Giles Overreach in MASSINGER's *A New Way to Pay Old Debts* in 1837. After some years on tour he was seen in New York in 1843 in Mrs John DREW's company, and in 1848, with Mrs MOWATT, visited England, where he remained for several years, being much admired in such roles as Othello, Richard III, Sir Giles Overreach, Claude Melnotte in BULWER-LYTTON's *The Lady of Lyons*, and the dual title roles in BOUCICAULT's *The Corsican Brothers*. On his return to New York he played Hamlet, and then formed his own company. Most of his subsequent career was spent outside New York, and he made his last appearance in 1877 in W. S. GILBERT's *Dan'l Druce, Blacksmith*, at the National Theatre, Washington. He married an English actress, Fanny Elizabeth Vining (1829–91), who regularly played opposite him; of their nine children five were on the stage, the best known being Fanny (below). Another daughter, Blanche, became an opera singer; May (1856–1927), a successful actress, retired on her marriage to William SEYMOUR. Edgar Longfellow (1862–1918) and Harry George Bryant (1866–1949) both played in their father's company as children, Edgar as an adult appearing with Julia MARLOWE and Viola ALLEN, among other stars; Harry married McKee RANKIN's daughter Phyllis, and for some years appeared with her in VAUDEVILLE before going into films.

DAVENPORT, FANNY LILY GIPSY (1850–98), American actress, daughter of Edward DAVENPORT, who was on the stage as a child in her father's company. She made her adult début in 1865 as Mrs Mildmay in Tom TAYLOR's *Still Waters Run Deep*, and from 1869 to 1877 was Augustin DALY's leading lady in New York. She then formed her own company, starring with it in the principal theatres of the United States. Her range of parts was wide, including Shakespeare's heroines and such contemporary women as Polly Eccles in ROBERTSON's *Caste* and Lady Gay Spanker in BOUCICAULT's *London Assurance*. Between 1883 and 1895 she also appeared in four plays by SARDOU: *Fedora, Tosca, Cleopatra*, and *Gismonda*.

DAVENPORT THEATRE, New York, see DALY'S SIXTY-THIRD STREET THEATRE.

DAVIDSON, GORDON, see MARK TAPER FORUM.

DAVIES, HUBERT HENRY (1869–1917), English dramatist, who was for a time a journalist in New York, where his *Cousin Kate* (1903) was produced with Ethel BARRYMORE in the title role. It was seen at the HAYMARKET in London the same year under the management of Cyril MAUDE, who appeared in it himself with Ellis Jeffreys as his leading lady. Among Davies's other plays the best were *Mrs Gorringe's Necklace* (also 1903; N.Y., 1904) and *The Mollusc* (1907; N.Y., 1908), both produced by Charles WYNDHAM, the first at Wyndham's own theatre and the other at the CRITERION, which he also owned, with himself and Mary Moore in the cast. Davies's last play, *Outcast* (London and N.Y., 1914), which dealt with a high-grade courtesan and the prevailing double standard of morality, was commercially his most successful, though *Cousin Kate* and *The Mollusc* were better constructed and dramatically more satisfying.

DAVIES, ROBERTSON (1913–), Canadian author and dramatist, who was a member of the OLD VIC company from 1938 to 1940. He is the author of two books on Shakespeare—*Shakespeare's Boy Actors* (1939) and *Shakespeare for Young Players* (1942)—and, in collaboration with Tyrone GUTHRIE, of *Renown at Stratford* (1953), *Twice Have the Trumpets Sounded* (1954), and *Thrice the Brinded Cat Hath Mew'd* (1955) on the first three seasons of the STRATFORD (ONTARIO) FESTIVAL. His first plays, written in about 1948, included the one-act *Overlaid* and *Eros at Breakfast*, and were followed by the full-length *Fortune my Foe* (1949), *King Phoenix*, and *At My Heart's Core* (both 1950). Among his later works were two plays for boy-actors, *A Masque of Aesop* (1952) and *A Masque for Mr Punch* (1962), *A Jig for the Gypsy* (1954), and *Hunting Stuart* (1955). *Love and Libel* (1960), based on his own novel *Leaven of Malice*, was first seen in New York, and under its original title was revived at the SHAW FESTIVAL at Niagara-on-the-Lake in 1975. In 1963 Davies was appointed Master of Massey College; while teaching drama and theatre history there he wrote several social comedies with a campus setting, though *Question Time* (1975), first seen at the St Lawrence Centre in Toronto, is concerned with political responsibility and the use and abuse of power.

DAVILÁ, ALECSÁNDRU, see ROMANIA.

DAVIS, FAY, see LAWRENCE, GERALD.

DAVIS, HALLIE FLANAGAN, see FLANAGAN, HALLIE.

DAVIS, JOHN (1822–75), British-born actor, manager, and dramatist, who had already established himself in England as a good actor before in 1855 he went to America, where he joined the company of his brother-in-law W. H. Crisp. Shortly after the outbreak of Civil War in Mar. 1861, Davis took over the Concert Hall in New Orleans, renamed it the Confederate Theatre, and almost single-handed kept it open until the city was captured by Union troops at the end of April 1862, after which he left New Orleans and became stage manager for the Queen Sisters in Augusta. He was the author of a number of plays of which the most popular was *The Roll of the Drum* (1861), which became such a favourite with Confederate audiences that seldom a month passed during the war without its being performed somewhere in the Southern States.

DAVIS, OWEN (1874–1956), American dramatist who, finding no market for his tragedies in verse, wrote over 100 ephemeral but remunerative melodramas. In 1921, however, he returned to his earlier ideals with a sincere and moving play, *The Detour*, which was recognized as one of the best productions of the year, in spite of its lack of financial success when compared with his earlier works. It was followed by *Icebound*, a study of New England farming folk which was awarded the PULITZER PRIZE for 1923. Unfortunately his later work failed to maintain the promise of these two plays, though in 1936 he made with his son a good dramatization of Edith Wharton's novel *Ethan Frome*.

DAVIS'S ROYAL AMPHITHEATRE, London, see ASTLEY'S AMPHITHEATRE.

DAY, JOHN (*c.* 1574–*c.* 1640), English dramatist, noted in HENSLOWE'S diary as writing plays for the ADMIRAL'S MEN, mainly in collaboration with CHETTLE, with whom he wrote the first part of *The Blind Beggar of Bethnal Green*, also known as *Thomas Strowd* (1600). A year later, possibly with a different collaborator, he added two further parts to this. His later works include the ill-fated *Isle of Gulls* (1606), played by the BOY COMPANY at BLACKFRIARS; because of its satire on the uneasy relations between England and Scotland under James I, it lost the children the patronage of the royal family and led to the imprisonment of some of those connected with its production. Day, however, continued to write, alone or in collaboration, being author with William ROWLEY of *The Travails of the Three English Brothers* (1607), on his own of *The Parliament of Bees* (1608), and with DEKKER of *The Bellman of Paris* (1623), apparently his last play.

DEAN, BASIL HERBERT (1888–1978), English actor, dramatist, and theatre director, who made his first appearance on the stage in 1906, and was then for four years a member of Miss HORNIMAN's company at the Gaiety Theatre, Manchester. In 1911 he became the first director of the repertory company at the LIVERPOOL PLAYHOUSE. From 1919 onwards he was active in the London theatre, both as manager (with Alec Rea as ReandeaN until 1926) and as director. Among the many important plays which he directed for his own and other managements were Clemence DANE's *A Bill of Divorcement* (1921), James Elroy FLECKER's *Hassan* (1923), Margaret Kennedy's *The Constant Nymph* (1926), John VAN DRUTEN's *Young Woodley* (1928), Dodie SMITH's *Autumn Crocus* (1931), and J. B. PRIESTLEY's *Johnson Over Jordan* (1939). He was a pioneer of stage LIGHTING, importing much new equipment from Germany and the U.S.A., and devising some of his own. He had already organized entertainment for the troops during the First World War, and during the Second he became director of the Entertainments National Service Association (ENSA) with overall responsibility for sending every form of entertainment to all the theatres of war. He gave a graphic description of his experiences in both wars in *The Theatre at War*, published in 1955. He returned to the London theatre in 1946 with a production of Priestley's *An Inspector Calls*, and among his later assignments were *The Diary of a Nobody* (1954), based on the GROSSMITHS' book, and Henry JAMES's *The Aspern Papers* (1959), with Michael REDGRAVE, who made the adaptation for the stage. Dean also organized the first British Repertory Theatre Festival in 1948.

DEAN, JULIA (1830–69), American actress, granddaughter of Samuel DRAKE. With her father and step-mother (her mother Julia Drake, also an actress, having died when she was 2) she appeared as a child under the management of LUDLOW and SMITH, and in 1846 made her adult début in New York as Julia in Sheridan KNOWLES's *The Hunchback*. A beautiful woman, with a gentle personality and great charm of manner, she was at her best in roles of tenderness and pathos, such as SCRIBE's Adrienne Lecouvreur and Mrs Haller in KOTZEBUE's *The Stranger*. She made an unhappy marriage in 1855, and her acting declined. A tour of California in 1856 was a success, but she never regained her position in New York, where she returned after her divorce. She died in childbirth following her second marriage in 1867.

DE ANGELIS, (Thomas) JEFFERSON (1859–1933), American actor, son of one of the original members of the San Francisco MINSTREL SHOW. With a younger sister he appeared on the stage as a child, touring around small cities and mining camps from St Louis to the coast, a life he portrayed vividly in his autobiography *A Vagabond Trouper* (1931). He was never a great actor but always a reliable and likeable one, who appeared in light opera, MUSICAL COMEDY, straight plays, and VAUDEVILLE with equal facility. His greatest personal success was the song 'Tammany', which he sang in his own production of *Fantana* (1905), by Robert B. Smith and Sam S. Shubert. In 1927 he appeared in Edna Ferber and George S. KAUFMAN's *The Royal Family* (known in England as *Theatre Royal*), and he made his last appearance later the same year in Dorrance Davis's *Apron Strings*.

DE BAR, BENEDICT (1812–77), British-born equestrian actor, manager, and theatre owner who, through his association with Noah M. LUDLOW and Solomon F. SMITH, did much to help establish New Orleans and St Louis as leading theatre cities in America. He made his American début in New Orleans in 1835 as Sir Benjamin Backbite in SHERIDAN's *The School for Scandal*, and acted there for the next 20 years except for a few years spent in New York and London. He eventually became stage manager of the St Charles Theatre where he had first appeared. In 1856 he moved to St Louis where he remained for the rest of his life, becoming owner of the influential Bates Theatre, renaming it first the St Louis Theatre and then simply De Bar's. Six years later he also acquired the St Louis Varieties Theatre and rechristened it De Bar's Opera House. He had a fine comic gift and was highly praised as Shakespeare's Falstaff and CHANFRAU's Mose the Fireman.

DE BRIE [Catherine Leclerc du Rozet], Mlle (*c.* 1630–1706), French actress, who as Mlle de Rose joined MOLIÈRE's company in 1650, and was destined to create many of his finest women's parts, including Cathos in *Les Précieuses ridicules* (1658) and Agnès in *L'École des femmes* (1662). She continued to play Agnès for over 20 years, and when in later life she was replaced by a younger actress the audience clamoured for her return. She retired on pension in 1685, having been one of the original members of the COMÉDIE-FRANÇAISE. She married in 1651 Edmé Villequin de Brie (1607–76), a mediocre actor who had been in Molière's provincial company and later played such small parts as the fencing master in *Le Bourgeois gentilhomme* (1671) and the river-god in *Psyché* (1671).

DEBURAU, JEAN-GASPARD [Jan Kašpar Dvořák] (1796–1846), famous French player of PANTOMIME, creator of a new conception of PIERROT based on the COMMEDIA DELL'ARTE mask PEDROLINO, which has since remained a popular and constant figure in the public imagination. Born in Bohemia, member of an acrobatic family with whom he toured the Continent, he arrived in Paris in 1814 and in 1820 was engaged with his companions for the Théâtre des FUNAMBULES on the BOULEVARD DU TEMPLE where he remained until his death. He was at first an inconspicuous member of the company, whose Pierrot was Felix Chiarini, from whom Deburau, eventually known as Baptiste, learned his art. He took over the role of Pierrot in 1826, and as he developed, with great subtlety and many delicate touches, his concept of the character as the white-clad, ever-hopeful, always disappointed lover, as child, prince, poet, and eternal seeker, all Paris flocked to see him and his praises were sung by all the critics, particularly Jules Janin, who in 1832 published a volume devoted entirely to him and his work. A later biography by Tristan Rémy, published in 1954, did much to enhance his fame in modern times. Deburau's appeal to the working-class public of Paris was based on a thorough knowledge of their lives and working conditions, which he studied minutely as he wandered alone among the craftsmen of the Faubourg du Temple; with his lightning wit and occasional flashes of cruelty and malice he expressed their spirit of revolt. He was the inspiration behind all subsequent attempts to re-establish the art of MIME in the modern theatre. After his death his son Charles (1829–73) carried on the Pierrot tradition, without his father's genius but with a vast store of goodwill and popularity. He remained at the Funambules until 1858, when he opened the Théâtre MARIGNY under his own name but without success.

DE CAMP, MARIA THERESA, see KEMBLE, CHARLES.

DÉCENTRALISATION DRAMATIQUE, a term used to describe the renewal of theatrical life in the French provinces, begun in 1945 as a result of private and government ventures and increasingly supported by local authorities. Since the spread of the cinema, the only attempts to establish a decentralized theatre company not based on the capital, as all touring companies were at that time, had been those of GÉMIER before the First World War and of COPEAU, whose Copiaus lived and worked in Burgundy between 1924 and 1929. Plans for decentralization, interrupted by the Second World War, were taken up by Jeanne Laurent, an official in the Ministry of Beaux Arts, and implemented with the help of a group of directors, actors, producers, and authors who between them made great changes in the theatrical map of France. The whole pattern of dramatic innovation has now altered radically, with many new authors and theatrical groups appearing in the provinces and on the edges of the Paris urban area, all publicly sponsored and forming a national network which is almost entirely independent of the commercial theatre in central Paris, and within which productions and personnel circulate freely. As these ventures receive their subsidies directly from the State, they are in the main independent of small-town pressures, and this independence has often allowed them to establish themselves even in the face of local opposition.

The organizations grouped under the overall cover of decentralization are of several different kinds. The first, and perhaps most important, are the Centres Dramatiques, of which the first, covering Eastern France, was founded under André Calvé in 1946, giving its first production in Jan. 1947. Based originally on Colmar, it later moved to Strasbourg, and includes Haguenau, Metz, and Mulhouse, becoming known as the Théâtre National de Strasbourg. The Comédie de Saint-Étienne, which serves the Lyonnais, the Valley of the Rhône, Burgundy, and the Alps, was founded in 1947 under Jean DASTÉ, who remained with it until 1970. It was followed in 1949 by the Grenier de Toulouse, which was for a long time the only active company in the south-western quarter of France, touring from the Loire to the Rhône and in North Africa. Based on a company founded in 1945 by SARRAZIN, it was staffed by amateurs until 1950, but in 1965 moved into its own theatre, built by the municipality and named in honour of Daniel SORANO. Also in 1949 the Comédie de l'Ouest was founded, serving Brittany, Normandy, and the Valley of the Loire, with its centre in Rennes under the direction of Hubert Gignoux, who remained with it until 1957. In 1952 Gaston BATY, convalescing in Aix-en-Provence, was instrumental in founding the Centre Dramatique du Sud-Est, serving the eastern Midi and the coast from Perpignan to Nice. After his death later the same year it came under the direction of Georges Douking and eventually of Antoine Bourseiller who, after a disagreement with the town council, moved the company to Marseille and renamed it Action Culturelle du Sud-Est. Two further centres date from the early 1960s—the Théâtre de la Commune d'Aubervilliers, directed by Gabriel Garran, and housed by a generous town council in a theatre rebuilt under the supervision of René ALLIO; and the Théâtre des Amandiers at Nanterre, directed by the Belgian actor Pierre Debacuhe, who has been particularly successful in creating a focus of culture in an extremely refractory immigrant, impoverished, and rehoused area. In 1968, under a fresh impetus given to the policy of decentralization by André Malraux as Minister of Culture, the Compagnie de Caen was founded, based on a group run since 1950 by Jo Tréhard who remained with it until his death in 1972.

In addition to the Centres Dramatiques, which are charged with the responsibility of producing three plays and giving at least 100 performances annually and of sending productions out on tour, or alternatively organizing coach sevices to bring in spectators from outlying areas, there are a number of other enterprises, such as the Maisons de la Culture established in a number of provincial towns and financed equally by the local council and the State, where theatre is only part of their activities; on a smaller scale are the Centres d'Animation Culturelle, of which by 1975 there were 19. Also under the aegis of the Ministry of Beaux Arts are the Troupes Permanentes, a name given by Malraux to provincial companies which have a permanent playhouse but are committed to constant touring in their area; among them are the Théâtre Populaire de Flandres, founded in 1953; the Théâtre de Bourgogne, founded in 1955; and the Comédie des Alpes, founded in 1960. The Théâtres Nationaux, which are owned and wholly funded by the State, include, in addition to the COMÉDIE-FRANÇAISE, the ODÉON, the Théâtre National de CHAILLOT, and the Théâtre de l'EST PARISIEN in Paris, the Théâtre National de Strasbourg, and the present THÉÂTRE NATIONAL POPULAIRE at Villeurbanne.

DE COURVILLE, ALBERT, see REVUE.

DEEVY, TERESA (1894–1963), Irish dramatist, whose sensitive portrayals of Irish provincial society and psychological studies of ambition placed her among the outstanding Irish dramatists of the 1930s. Her first play, *The Reapers* (1930), was produced at the ABBEY THEATRE, and was followed in 1931 by *A Disciple* and *Temporal Powers*. With *The King of Spain's Daughter* (1935) and *Katie Roche* (1936)—usually considered her best work—her plays became part of the Abbey's standard repertory; but *The Wild Goose* (also 1936) was the last to be produced there as she then turned to the writing of equally successful radio plays.

DE FILIPPO, EDUARDO (1900–84), Italian actor and dramatist, illegitimate son of the Neapolitan actor Eduardo Scarpetta, in whose company he and his brother Peppino (1903–80) and sister Titina [Anastasia] (1898–1963) acted as children. In 1932 they opened their own playhouse in Naples—the *Teatro Umoristico*—and soon achieved a great reputation for their productions in the style of the COMMEDIA DELL'ARTE. Many of their plays, particularly those in the Neapolitan dialect, were the work of Eduardo, including his first play *Natale in casa Cupiello* (*Christmas at the Cupiello's*, 1931). In 1945 Peppino left to found his own company, with which he was seen in London during the WORLD THEATRE SEASON of 1964, in his own play *Metamorphoses of a Wandering Minstrel*. Titina remained with Eduardo until her death, appearing in many of his comedies, of which the finest is probably *Filumena Marturano* (1946). Other outstanding works are *Napoli milionaria!* (1945), with which Eduardo visited the 1972 World Theatre Season in London; *Questi fantasmi!* (*These Ghosts!*, 1946); *Le voci di dentro* (*Voices from Within*, 1948); *La paura numero uno* (*Terror Number One*, 1950); *Bene mio e core mio* (*My Goods and My Heart*, 1955); and two plays directly connected with one of the old masks of the *commedia dell'arte*—*Pulcinella in cerca della sua fortuna per Napoli* (*Pulcinella in Search of His Fortune in Naples*, 1958) and *Il figlio di Pulcinella* (*The Son of Pulcinella*, 1959). Always topical, his plays developed an increasingly social slant, their language reflecting the evolving Neopolitan dialect and the vocabulary of modern life. In 1973 Laurence OLIVIER and Joan PLOWRIGHT appeared with the NATIONAL THEATRE company in London in Eduardo's *Saturday, Sunday, Monday* (written in 1959), directed by ZEFFIRELLI. This had a successful West End run, as did Zeffirelli's production of *Filumena* in 1977, again starring Joan Plowright. The last two plays were also seen in New York—the first, again directed by Zeffirelli but with a different cast, in 1974; the second in 1980, with Joan Plowright and Frank FINLAY. *Grande magia* (1951), was seen in New York as *Grand Magic* in 1979.

DEFRESNE [Abraham-Alexis Quinault], see DUFRESNE.

DÉJAZET, PAULINE-VIRGINIE (1798–1875), French actress, who was already well known for her MALE IMPERSONATIONS in VAUDEVILLE when in 1821 she appeared with some success in a vaudeville by SCRIBE at the GYMNASE, where she remained for some time, playing male roles. Her reputation really dates from the opening of the PALAIS-ROYAL in 1831, where she stayed for 13 years and became one of the most popular actresses in Paris. After an argument over her salary she continued her triumphant career at the VARIÉTÉS and the GAÎTÉ, still playing masculine roles—soldiers, collegians, students—as well as great ladies and pretty peasant girls. Some idea of her versatility is given by the fact that her parts included Voltaire, Rousseau, Napoleon, Henri IV, and Ninon de l'Enclos. In 1859 she took over the FOLIES-NOUVELLES, which she renamed Théâtre Déjazet, making a great hit at the age of 62 in a male part in SARDOU's *Monsieur Garat*. She made her last appearance in Paris in 1870, and in the same year was seen in London at the OPÉRA COMIQUE in a season of French plays.

DEJMEK, KAZIMIERZ (1924–), Polish director, who worked under Leon SCHILLER in Łódź, and in 1949 took over the theatre there. His success, particularly in the staging of new plays, led to his being appointed in 1960 director of the National Theatre in Warsaw, where he remained until his revival of MICKIEWICZ's *Forefathers' Eve* in 1968 offended the authorities and he was dismissed. After working abroad for a time he returned to Warsaw in 1973, and then went back to Łódź, where he directed a number of new Polish plays, including the first Polish production of GOMBROWICZ's *Operette* (first seen in Paris in 1967) in 1975, and MROŻEK's *The Hunchback* (1976). He also revived in 1978 FREDRO's comedy *Vengeance*. He is probably best known outside Poland for his reconstruction and production of two 16th-century religious plays, *The Glorious Resurrection of Our Lord* by Mikotaj of Wilkowiecka and Rej's *The*

Life of Joseph, both being seen in London and Paris.

DEKKER, THOMAS (*c.* 1572–*c.* 1632), English dramatist, who worked mainly in collaboration and is believed to have had a hand in more than 40 plays of which about 15 survive, several having been destroyed by WARBURTON's cook. The most important of his own works is *The Shoemaker's Holiday* (1599), first played at the ROSE THEATRE by the ADMIRAL'S MEN, which materially assisted the evolution of English comedy. Robust and full-blooded, it tells how Simon Eyre, a master shoemaker, became Lord Mayor of London, and shows a promise which was not fulfilled in Dekker's later works, particularly in two of which he was again sole author—*If It be not Good, the Devil is in It* (1610) and *Match Me in London* (1611). In collaboration with MARSTON he wrote *Satiromastix* (1601), in which Ben JONSON, who had ridiculed the authors in his *The Poetaster* earlier the same year, is satirized under the name of Horace; with MIDDLETON he wrote *The Honest Whore* (1604) and *The Roaring Girl* (1610); and with MASSINGER a tragedy, *The Virgin Martyr* (1620), which may however have been only a revision of one of his own early plays *Diocletian* (1594). He is also believed to have been part-author of William ROWLEY's *The Witch of Edmonton* (1621). In 1609 he published *The Gull's Handbook*, a satirical account of the fops and gallants of the day which gives some interesting information about the contemporary theatre.

DELACORTE THEATRE, New York, see PAPP, JOSEPH.

DE LA MOTTE, H., see HOUDAR DE LA MOTTE.

DELAUNAY, LOUIS-ARSÈNE (1826–1903), French actor, who after attracting attention by his excellent acting in a classical comedy at the ODÉON made his first appearance at the COMÉDIE-FRANÇAISE on 25 Apr. 1848 and had a long and brilliant career, playing young lovers until he was 60 and never losing his hold on the affections of the public. He made his first outstanding success when he appeared with RACHEL in a one-act play by Bartet, *Le Moineau de Lesbie* (*Lesbia's Sparrow*). He was the original Fortunio in Alfred de MUSSET's *Le Chandelier* (1848), and was so much admired as Coelio in the same author's *Les Caprices de Marianne* (1851) that after his retirement in 1886 it went out of the repertory for many years. He also appeared in HUGO's *Hernani* (1831) when it finally passed into the repertory of the Comédie-Française, and in the course of his long career appeared in many new plays as well as in revivals of MARIVAUX and REGNARD.

DELAVIGNE, CASIMIR (1793–1843), French dramatist, in whose *L'École des vieillards* (1823) TALMA made one of his rare appearances in comedy, playing opposite Mlle MARS. Delavigne's tragedies, of which the best were *Les Vêpres siciliennes* (1821), successful because of its political implications, and *Marino Faliero* (1829), in which FRÉDÉRICK played, show a somewhat cautious adoption of Romanticism while retaining much of the form of neoclassicism. One of his plays, *Les Enfants d'Édouard* (1833), was based on English history, dealing with the death of the Princes in the Tower. He was elected a member of the French Academy in 1825.

DELLA PORTA, GIAMBATTISTA (1538–1615), Italian scientist, philosopher, and dramatist, exponent of the COMMEDIA ERUDITA, 14 of whose plays have survived out of a possible 33. They are in prose, on subjects taken mainly from Boccaccio and PLAUTUS, though *Il due fratelli rivali* is based on the story by Bandello which also supplied Shakespeare with the plot of *Much Ado About Nothing*. Two of Della Porta's plays, *La fantesca* and *La Cintia*, were translated by Walter Hawkesworth and performed at Trinity College, Cambridge, as *Leander* (in 1598) and *Labyrinthus* (in 1603) respectively, and *La trappolaria* was used by George RUGGLE as the basis of his *Ignoramus*, which James I much enjoyed when he visited Cambridge in 1615.

DE LOUTHERBOURG, PHILIP JAMES, see LOUTHERBOURG.

DELYSIA [Lapize], ALICE (1889–1979), French actress and singer, who made her first appearance at the Moulin Rouge in Paris in 1903 in the chorus of the musical *The Belle of New York*. She was later seen at the VARIÉTÉS and at the FOLIES-BERGÈRE with Yvonne PRINTEMPS. In 1905 she made her first appearance in the United States, being seen with Edna May at DALY's in *The Catch of the Season*, and in 1914 C. B. COCHRAN engaged her for London, where she made a great success in the revue *Odds and Ends*. She appeared for many years under Cochran's management, notably in *On With the Dance* and *Still Dancing* (both 1925), and in his memoirs he constantly pays tribute to her loyalty and good nature. In 1930 she appeared in London in Pagnol's *Topaze*, adapted by Benn Levy, and in 1933 was in *Mother of Pearl*, a musical in which she also toured Australia. She made her last appearance in London at the CRITERION THEATRE in 1939, as Hortense in *The French for Love*, a light comedy by Marguerite Steen and Derek Patmore in which she also toured in 1944, having worked for ENSA from 1941 to 1943 entertaining the troops in the Near East. She then retired to Brighton, where she died.

DEMETAR, DIMITRIJE, see YUGOSLAVIA.

DENCH, JUDI [Judith] OLIVIA (1934–   ), English actress, who made her début in 1957 with the OLD VIC company as Ophelia to the Hamlet of

John NEVILLE. She remained with the company for four years, then joined the ROYAL SHAKESPEARE COMPANY to play Anya in CHEKHOV's *The Cherry Orchard* in 1961 followed by Isabella in *Measure for Measure* in 1962, and was then seen in her first modern play as Dorcas Bellboys in John WHITING's *A Penny for a Song*. She made another excursion into modern drama in ARBUZOV's *The Promise* in 1967, developing convincingly from a pert and tubby adolescent to the resigned wife of 13 years later. She also scored a success as Sally Bowles in her first musical *Cabaret* (1968), based on VAN DRUTEN's *I Am a Camera*. She returned to the classics in 1969 to play Viola in *Twelfth Night* for the R.S.C. followed by the title role in SHAW's *Major Barbara* in 1970; in that year she was also seen as the bluestocking Grace Harkaway in an acclaimed revival of BOUCICAULT's *London Assurance* and in 1971 played the title role in WEBSTER's *The Duchess of Malfi*. Again with the R.S.C., her roles included Beatrice in *Much Ado About Nothing* and Lady Macbeth (1976), Millamant in CONGREVE's *The Way of the World* (1978), and Juno in a highly praised production of O'CASEY's *Juno and the Paycock* in 1980. Like many English actresses who have proved their worth in the great classical parts, her problem is to find challenging contemporary roles.

DENISON, (John) MICHAEL TERENCE WELLESLEY (1915–), English actor, who made his first appearance on the stage in Brandon THOMAS's *Charley's Aunt* in 1938 and his London début in *Troilus and Cressida* later the same year. In 1939 he married Dulcie Gray (1919–), who made her London début at the OPEN AIR THEATRE, Regents Park, in 1942, playing Maria in *Twelfth Night*, and then appeared in several West End plays, including an adaptation of Graham GREENE's *Brighton Rock* (1943). Husband and wife appeared together for the first time in Peter Watling's *Rain on the Just* (1948), and then in several other plays including Jan de Hartog's *The Fourposter* (1950) and PRIESTLEY's *Dragon's Mouth* (1952). They toured South Africa together in 1954–5, and on returning to England Denison appeared at the SHAKESPEARE MEMORIAL THEATRE; they were together again in Dulcie Gray's own play *Love Affair* (1955), which Denison directed. Among later plays in which they co-starred were revivals of LONSDALE's *Let Them Eat Cake* in 1959, SHAW's *Candida* in 1960 and *Heartbreak House* in 1961, and a new play *Where Angels Fear to Tread* (1963), based on a novel by E. M. Forster. After touring in a joint Shakespeare recital they were separated for a time, to be reunited in a succession of productions, including revivals of WILDE's *An Ideal Husband* in 1965, Lonsdale's *On Approval* in 1966, and IBSEN's *The Wild Duck* in 1970. In 1974 Denison was in BARRIE's *Peter Pan* and in 1975 he was Pooh-Bah in *The Black Mikado*, which ran for a year, while Dulcie Gray was in Agatha CHRISTIE's *A Murder Is Announced* in 1977.

DENMARK. The documented history of Danish drama begins in the 16th century, where the presence of relatively late medieval forms suggests that there had been, to some extent, an earlier drama also; but only the mid-16th century SCHOOL DRAMA of Christiern Hansen—*Den utro hustru* (*The Unfaithful Wife*) and *Dorothiae Komedie*—and Hieronymus Justesen Ranch—*Karrig Niding* (*The Miser*)—and the anonymous *Ludus de Sancto Kanuto Duce* (1530) survive to indicate varying degrees of independence in the use of native sources. The absence of vernacular plays throughout the 17th century, coupled with the habit of producing French and German plays in the Copenhagen theatres in the early 18th century, suggests that, with the disappearance of the ecclesiastical and scholastic Latin or vernacular school drama, dramatic writing ceased for a time in Denmark.

It was recreated at a stroke by Ludvig HOLBERG who, though a Norwegian by birth, worked in Copenhagen and was the first dramatist to use the Danish language. His fecund comic genius provided a rich repertoire of Danish plays for Den DANSKE SKUEPLADS, which opened on 23 Sept. 1722. Holberg had a strong and lasting influence on Danish drama, providing a model for realistic comedy combining the technique of the classics and of MOLIÈRE—the more important in that Danish literature was soon afterwards exposed to strong German influences.

After Holberg's death in 1754 the Danish theatre went into something of a decline and the repertoire was filled with translations of second-rate French and Italian pieces—works which WESSEL's *Love Without Stockings* (1772) devastatingly parodied. In 1772 Holberg's old theatre became the KONGELIGE TEATER (the Royal Theatre) and the following year the buildings were renovated, inaugurating a very much livelier period of theatre in Copenhagen. Johannes EWALD began to display his power as a dramatic poet, and with *The Fishermen* in 1778 produced what is probably the finest of his works. There were frequent performances of Holberg and Ewald, and of important foreign dramatists like Molière, BEAUMARCHAIS, and SHERIDAN.

Denmark was also one of the first countries to translate and act the plays of Shakespeare. The famous Danish actor Peter Foersom (d.1817) made a number of versions, beginning with *Hamlet* in 1813 in which he also played the title role. In 1873 Edvard Lembcke translated all the plays except *Titus Andronicus* and *Pericles*; his versions are still used in the theatre. Valdemar Österberg published a number of more literary translations with excellent introductions and notes, intended mainly for reading. The best of the 19th-century actors in Shakespeare was probably the great comedian Olaf Poulsen, whose Falstaff and Bottom were famous.

At the death of Ewald, Denmark was on the verge of the Golden Age—*Guldalderen*—in which her poets were from time to time dramatists and her dramatists poets. After the turn of the century, the emphasis in the theatre was firmly on Adam OEHLENSCHLÄGER, Shakespeare, and the German *bürgerliches Drama*. Very soon this was augmented by the Danish vaudevilles of Johan Ludwig HEIBERG and Jens Christian Hostrup (1818–92), the comedies of Henrik HERTZ, and the dramas of Johannes Carsten HAUCH. Thomas Overskou (1798–1873), a dramatic critic and historian, also left a number of plays of which *Pak* (1845) is probably the best known. In the later 19th century a movement originating with the eminent critic Georg BRANDES stimulated Danish interest in the work of IBSEN. In 1874 the Kongelige Teater acquired new premises and soon made a speciality of presenting plays by Ibsen and BJØRNSON. With Edvard Brandes (the brother of Georg) and Otto Benzon (1856–1927), characteristically modern drama, analysing and criticising contemporary society, became predominant, but the most popular plays being written in Denmark were the poetic and lyrical dramas of Holger DRACHMANN. In the post-Ibsen era Danish theatre developed a healthy balance between translated and indigenous works.

The end of the 19th and the beginning of the 20th centuries witnessed a blossoming of Danish theatrical talent. Among the distinguished players were Betty Hennings, Emil and Olaf Poulsen followed by Emil's son Johannes POULSEN, Bodil IPSEN, Poul REUMERT, Bodil KJER, and Mogens WIETH. The best known playwrights of the early 20th century were Hjalmar BERGSTRØM, Gustav WIED, and Helge Rode. From the 1920s to the end of the Second World War the dominant writers were Kaj MUNK, Kjeld ABELL, and SOYA, with lesser contributions from Svend Borberg (1888–1947)—*Ingen* (*Nobody*, 1923)—Sven Clausen (1893–1961)—*Naevningen* (*The Juryman*, 1929)—and Leck Fischer (1904–56), with a series of somewhat drab Communist dramas from 1936 to 1950. The immediate post-war period was linked with the names of H. C. BRANNER and Knut SØNDERBY; the 1960s witnessed a flowering of the musical theatre, particularly in the satires of Erik KNUDSEN and Klaus RIFBJERG. Ernst Bruun Olsen was only one of a number of writers who turned to live theatre after a fruitful apprenticeship in radio drama. The experimental Odin Teatret moved in 1966 from Norway to Hostelbro, Denmark.

Today, apart from the Kongelige Teater, Copenhagen has several other first-class theatres, including the DAGMARTEATRET, Casinoteater, Folketeater, and Det Nye Teater; it has also acquired its own city theatre, Det Ny Scala, which offers a balanced repertoire of modern and classic pieces. Provincial theatres of high standing include those of Aarhus, Odense, and Aalborg; and the Kongelige Teater also regularly sends a travelling company round the provinces.

DENNIS, JOHN (1657–1734), English critic and dramatist, whose first play, a comedy entitled *A Plot and No Plot* (1697), was followed by a bombastic tragedy *Rinaldo and Armida* (1698), satirized by Pope in his *Essay on Criticism*. Dennis also figures as Sir Tremendous in *Three Hours After Marriage* (1717), a satirical farce by John GAY and others. He was responsible for an adaptation of Shakespeare's *The Merry Wives of Windsor* as *The Comical Gallant; or, the Amours of Sir John Falstaff* (1702) and for a comedy, *Gibraltar; or, the Spanish Adventure* (1705) seen at DRURY LANE, as were his last two pseudo-classical tragedies, *Appius and Virginia* (1709) and *The Invader of his Country; or, the Fatal Resentment* (1719) on Coriolanus. None of his work has survived on the stage, but he deserves mention as his *Liberty Asserted* (1704), first seen at LINCOLN'S INN FIELDS THEATRE, is believed to be the first play on a Canadian theme. He was the author of a defence of the stage against those who attacked it on moral grounds, and engaged in several pamphlet wars with theatrical persons whom he disliked.

DENNY, FRANCES ANN (1797–1875), American actress who at the age of 13 so impressed Samuel DRAKE that, with her mother's consent, he made her his protégée and took her with him on a tour of the West and South, where she quickly blossomed into an exceptionally fine actress. In 1816 she married Drake's son Alexander. She made her first starring tour in 1820 and performed in all the principal cities in the United States and Canada before making her début in New York that year, going on to frequent starring engagements at the PARK, CHATHAM, and BOWERY Theatres. Billed as America's great 'tragedy queen', she was a powerful actress acclaimed for her intensity in such roles as Shakespeare's Lady Macbeth and Evadne in FLETCHER's *The Maid's Tragedy*.

DERWENT, CLARENCE (1884–1959), actor and producer. Born and educated in London, he made his first appearance on the stage in 1902, and was for five years with Frank BENSON's company. He subsequently spent two years at the Gaiety, Manchester, under Miss HORNIMAN's management, and then appeared in London in 1910 under TREE. In 1915 he went to America, where he had a distinguished career on Broadway, making only one further appearance in London, in *The Late Christopher Bean* (1933) by Emlyn WILLIAMS. His last appearance was in GIRAUDOUX's *The Madwoman of Chaillot* (1948). He was responsible for many productions, and in 1945 instituted the Clarence Derwent awards given in London and New York annually for the best performances by players in supporting roles. He was the author of several plays, and in 1946 was elected President of the AMERICAN ACTORS' EQUITY ASSOCIATION.

DÉRY, TIBOR, see HUNGARY.

DESCLÉE, AIMÉE-OLYMPE (1836–74), French actress, who studied at the Conservatoire, but finding the discipline irksome left to make her début at the GYMNASE. There and at other theatres she met with little success, so she joined a company which was touring Europe with a repertory of French plays. The younger DUMAS, seeing her in Brussels in 1867 in his own play *Diane de Lys*, persuaded the Gymnase to re-engage her. She returned to Paris, and the hard years of touring bore fruit when she made an immediate success in MEILHAC and Halévy's *Frou-Frou*. Her gratitude to Dumas knew no bounds, but he being by this time a fervent moral reformer, she could only show it by her excellent performances in his plays, being particularly admired as the heroines of his *Visite de noces* (1871) and *La Femme de Claude* 1873). She died suddenly at the height of her success.

DESEINE, Mlle, see DUFRESNE.

DESIOSI, a company of COMMEDIA DELL'ARTE actors who may have been led by Diana da PONTI. They are first noted in 1581, and in 1595 Tristano MARTINELLI, who had been with PELLESINI's troupe, joined them to play ARLECCHINO. The group broke up some years later, the chief actors then joining the ACCESI.

DESJARDINS, MARIE-CATHERINE-HORTENSE (1632–83), one of the first women playwrights of France. She left home when young, and may have been an actress in MOLIÈRE's provincial company. Settling in Paris she wrote poetry, novels, and plays, including a comedy, *Le Favory* (1665), produced by Molière at the PALAIS-ROYAL. She is best remembered, however, for her tragedy *Manlius* which was produced at the Hôtel de BOURGOGNE in 1662. It was while appearing in this play that the mother of the famous actor BARON died and was replaced by Mlle DESŒILLETS. It was successful on its first appearance and went into the repertory of the COMÉDIE-FRANÇAISE where, as *Manlius Torquatus*, it remained until the outbreak of the French Revolution.

DESMARES, CHARLOTTE (1682–1753), French actress, daughter of Nicolas DESMARES, who profited so well by the tuition of her aunt Mlle CHAMPMESLÉ that she eventually succeeded to many of her parts, sharing leading feminine roles at the COMÉDIE-FRANÇAISE with Mlle DUCLOS. At her best in pathetic and tender roles, she was particularly admired in VOLTAIRE's *Oedipe*. She retired in 1721, eclipsed by the young and lovely Adrienne LECOUVREUR.

DESMARES, NICOLAS (*c*.1645–1714), French actor and dramatist, brother of Mlle CHAMPMESLÉ and father of Charlotte (above). He spent some years in a French company at the Court of Copenhagen, and was also a member in 1680 of a provincial company under the patronage of Le Grand Condé. Five years later he joined the company of the COMÉDIE-FRANÇAISE, where he remained until his retirement in 1712. His range was limited, but he was unsurpassed in the playing of peasant characters.

DESMARETS DE SAINT-SORLIN, JEAN (1595–1676), French novelist, poet, and dramatist, a foundation member of the French Academy and a frequenter of the Hôtel de Rambouillet. Urged by his patron RICHELIEU to attempt the theatre, he wrote *Aspasie* (1636), a mediocre tragedy given shortly before CORNEILLE's *Le Cid*; but he is best remembered for his comedy *Les Visionnaires* (1637), a witty comment on the foibles of fashionable society which was produced at the MARAIS with MONTDORY as the hallucinated old-fashioned poet who believes himself to be a great modernist. It had some influence on MOLIÈRE, who revived it twice at the PALAIS-ROYAL. Desmarets wrote several other plays, including the greater part of *Mirame* (1641), the first play to be given in Richelieu's private theatre, in which the Cardinal himself had a hand.

DESŒILLETS [Alix Faviot], Mlle (1621–70), French actress, who served a long apprenticeship in the provinces, where she married an actor. She was in her late 30s before she was seen in Paris, and in 1661 was at the Théâtre du MARAIS, where she played Viriate in the first production of CORNEILLE's *Sertorius*. From there she went to the Hôtel de BOURGOGNE a year later, to take the place of Mlle Baron, mother of the famous actor Michel BARON. Neither young nor pretty, and short of stature, she was nevertheless an excellent actress, most moving in tragedy and in high favour with the audience. She created the title role in Corneille's *Sophonisbe* (1663) and that of Hermione in RACINE's *Andromaque* (1667), being replaced in the latter part during an illness by Mlle CHAMPMESLÉ. The younger actress was so good that on seeing her Mlle Desœillets retired from the theatre in tears and never acted again.

DESSOUS, see CELLAR.

DESTOUCHES [Philippe Néricault] (1680–1754), French dramatist, important in the development of 18th-century *drame* from 17th-century comedy. He was a successor to MOLIÈRE, and it has been said that his titles—*L'Ingrat* (1712), *Le Médisant* (1715)—read like subtitles to the latter's plays. Destouches had already written several moderately successful comedies before in 1716 he visited London, where he contracted a secret marriage with an Englishwoman which later supplied him with the material for one of his best works, *Le Philosophe marié* (1727), adapted by Mrs INCHBALD as *The Married Man* (1789). On his return to France he retired to the country and gave himself up wholly to writing. His stay in England may account for the mingling of sentiment and tragedy with comedy in his later plays; many, particularly towards the end of his career, were spoilt by sententiousness. His best play is *Le Glorieux* (1732), which was translated into English by HOLCROFT as *The School for Arrogance* (1791). It pictures the struggle between the old nobility and the newly rich who are rising to power, and some traits of the central character are said to have been taken from the actor DUFRESNE who played it. Destouches became a member of the French Academy in 1723, in succession to CAMPISTRON. He was a man of a serious, even a religious, turn of mind, and at 60 turned his attention entirely to theology, though he left several plays in manuscript, one of which, *La Fausse Agnès*, was produced posthumously in 1759 with some success.

DETAIL SCENERY, small, changeable pieces of scenery used for a particular scene in or before a formalized Permanent Setting (known in America as a Unit, or Formal, Setting).

DEUS EX MACHINA, literally 'the god from the machine', in classical drama the character, usually a god, who enters at the end of a play to resolve the complexities of a plot which would otherwise be insoluble. In Greek tragedy he was lowered from above by a crane (MECHANE). The use of this device was later criticized as showing the dramatist's lack of skill in resolving his plot naturally. By extension, the term has come to mean any arbitrary form of plot resolution.

DEUTERAGONIST, see PROTAGONIST.

DEUTSCH, ERNST (1890–1969), German actor, born in Prague, who in 1916 proved himself an ideal interpreter of EXPRESSIONISM when he played the title role in HASENCLEVER's *Der Sohn* (1916). For REINHARDT in Dresden he played in SORGE's *Der Bettler* (1917) and was seen in many outstanding plays and films up to 1933, when he emigrated to the U.S. He returned to Germany in 1947, where among his more important parts were the title roles in LESSING's *Nathan der Weise* in 1954 and SCHNITZLER's *Professor Bernhardi* in 1955, and Clausen in HAUPTMANN's *Vor Sonnenaufgang* in 1961.

DEUTSCHES THEATER, a private play-producing society founded in Berlin in 1883 for the purpose of staging good plays in repertory, as a protest against the deadening effect of long runs and outmoded theatrical tradition. Under Adolf L'ARRONGE a group of actors led by Josef KAINZ and Agnes SORMA presented classical historical plays in the style of the MEININGER COMPANY. In 1894 the enterprise was given a new direction by its affiliation with the FREIE BÜHNE and it enjoyed another period of fame under Max REINHARDT, who went there in 1905 from the Neues Theater with a band of young actors trained in his own methods. After the First World War the theatre was in the doldrums but revived somewhat under Heinz HILPERT. It was closed in 1944, reopened the following year, and in 1946 became the National Theatre of East Berlin. It housed the BERLINER ENSEMBLE from its foundation until 1954. From 1961 to 1969 the theatre knew a further period of success under BESSON.

DE VILLIERS, see VILLIERS.

DEVINE, GEORGE ALEXANDER CASSADY (1910–65), English actor and theatre director, who while at OXFORD was instrumental, as President of the OUDS in 1932, in inviting John GIELGUD to direct *Romeo and Juliet* for the society, he himself playing Mercutio. After making his professional début later the same year he joined Gielgud's company at the QUEEN'S THEATRE, and from 1936 to 1939 was on the staff of the London Theatre Studio founded by Michel SAINT-DENIS. After six years in the army he joined Saint-Denis at the OLD VIC school, being also a director of the YOUNG VIC company. In 1954 he gave an outstanding performance as Tesman in IBSEN's *Hedda Gabler* to the Hedda of Peggy ASHCROFT, and in 1955 at the SHAKESPEARE MEMORIAL THEATRE directed an interesting *King Lear* with décor by the Japanese artist Isamu Noguchi, Gielgud playing the title role. A year later he was appointed artistic director of the newly-formed ENGLISH STAGE COMPANY at the ROYAL COURT, of which he was one of the founders, and under his expert and enthusiastic guidance it proved a galvanizing force in the London theatre. In addition to directing a number of plays he gave some fine performances, particularly as Mr Shu Fu in BRECHT's *The Good Woman of Setzuan* (1956), Mr Pinchwife in WYCHERLEY's *The Country Wife*, and the Old Man in IONESCO's *The Chairs* (both 1957). In the early 1930s he was associated with the firm of Motley, the stage designers, and in 1940 married Sophia Harris, one of the partners. In 1966 a George Devine Award was instituted in his memory to provide financial encouragement to young workers in the theatre.

DEVRIENT, (Philipp) EDUARD (1801–77), German actor, nephew of Ludwig DEVRIENT. He was an opera singer in Berlin until 1834, when his voice failed and he turned to acting. In 1844 he went to the Court Theatre in Dresden, where the pre-eminence of his younger brother Emil (below) led him to concentrate more on directing. He also began the publication of his history of the German theatre, the first of its kind, which appeared in 5 vols. between 1848 and 1874 (a new edition in 2 vols was published in 1967). From 1852 until he retired in 1870 he was director of the Court

Theatre at Karlsruhe; after his retirement he devoted much of his time to translations of Shakespeare's plays, which, with the help of his son Otto, he published between 1873 and 1876 as the *Deutsche Bühnen- und Familien-Shakespeare*. Though somewhat bowdlerized, his versions were better suited to the stage than other more literary translations such as those of SCHLEGEL.

DEVRIENT, (Gustav) EMIL (1803–72), German actor, the youngest of the three nephews of Ludwig DEVRIENT, who made his début in 1821, and played in a number of provincial towns before going to the Dresden Court Theatre in 1831, where he was joined in 1844 by his brother Eduard (above). Emil remained there until 1868, being outstanding in such parts as GOETHE's Tasso, Orestes (in *Iphigenia*), and Egmont, and in the plays of SCHILLER, to which his fine speaking voice and heroic gestures were well suited. He was frequently given leave of absence from Dresden to star elsewhere, including London where his Hamlet was well received.

DEVRIENT, KARL AUGUST (1797–1872), German actor, eldest nephew of Ludwig DEVRIENT, who made his stage début as Rudenz in SCHILLER's *Wilhelm Tell* and later became a leading actor at the Court Theatre in Dresden, where he was much admired as Posa in Schiller's *Don Carlos*, as Hamlet, GOETHE's Egmont, and Schiller's Wilhelm Tell and Wallenstein. In later years, playing in Karlsruhe and Hanover, he excelled in character parts such as Goethe's Faust, Shylock in *The Merchant of Venice*, and King Lear.

DEVRIENT, LUDWIG (1784–1832), German actor, the first of a famous 19th-century theatrical dynasty. He had already had a good deal of experience before in 1814 he was engaged by IFFLAND for BERLIN, where he became the leading German actor of the Romantic period, evoking comparisons with Edmund KEAN by his wild-eyed, unbridled, passionate acting. His rendering of King Lear's madness was celebrated, as was his incandescently evil Franz Moor in SCHILLER's *Die Räuber*, in which he made his first appearance in Berlin, eclipsing Iffland's quieter though equally evil interpretation. He was also an excellent comedian, Falstaff in *The Merry Wives of Windsor* being considered his finest part; he was much admired as Shylock in *The Merchant of Venice* and as Harpagon in MOLIÈRE's *L'Avare*.

DEVRIENT, MAX (1857–1929), German actor, younger son of Karl DEVRIENT by his second wife, Johanna Block. He made his début in DRESDEN in 1878 and after appearing all over Germany was for many years at the VIENNA Burgtheater, making his first appearance there in SCHILLER's *Die Räuber* in 1882. A handsome man of commanding presence, he excelled in big tragic roles, particularly in the plays of GOETHE, SCHILLER, and Shakespeare; he could also play comedy to good effect, and was much admired as Petruchio in *The Taming of the Shrew*.

DEVRIENT, OTTO (1838–94), German actor, director, and dramatist, son of Eduard DEVRIENT, under whom he played at the Karlsruhe Court Theatre until in 1873 he went as a character actor to WEIMAR, where in 1876 his production of both parts of GOETHE's *Faust* in the style of the old MYSTERY PLAY, on a stage with three levels, aroused much interest. He was the author of three tragedies which have not been revived.

DEWHURST, COLLEEN (1926–   ), Canadian-born actress, who made her New York début in 1946 while still a drama student, and in 1952 had a small part in the Anta production of O'NEILL's *Desire Under the Elms*: in 1963 she was to play the leading role of Abbie Putnam. A passionate and powerful actress, she starred in other plays by O'Neill—*A Moon for the Misbegotten* (in 1958 at the Spoleto Festival in Italy and in 1973 with Jason ROBARDS); in the first New York production of *More Stately Mansions* in 1967; and in a 1972 revival of *Mourning Becomes Electra*. She was seen to advantage in several Shakespearian roles for Joseph PAPP: Tamora in *Titus Andronicus*, Katharina in *The Taming of the Shrew* (both 1956), Lady Macbeth (1957), Cleopatra (1959 and 1963), and Gertrude in *Hamlet* (1972). She also appeared in plays by ALBEE—*The Ballad of the Sad Café* (1963), *All Over* (1971), and as Martha in *Who's Afraid of Virginia Woolf?* in 1976. Other outstanding appearances were in Athol FUGARD's *Hello and Goodbye* in 1969, BRECHT's *The Good Woman of Setzuan* in 1970, and as Lillian HELLMAN in a revival of *Are You Now or Have You Ever Been . . . ?* (1978) by Eric BENTLEY.

DEXTER, JOHN (1925–   ), English director, originally an actor. After joining the ENGLISH STAGE COMPANY at the ROYAL COURT THEATRE in 1957 he came into prominence with his productions of Arnold WESKER's trilogy *Chicken Soup with Barley* (1958), *Roots* (1959), and *I'm Talking About Jerusalem* (1960). Also in 1960 he directed the West End production of Lillian HELLMAN's *Toys in the Attic*, and after directing another Wesker play *Chips With Everything* (1962; N.Y., 1963) was responsible for the production of a musical, *Half-a-Sixpence* (1963) with Tommy Steele. From 1963 to 1966 he was an associate director of the NATIONAL THEATRE, where he directed revivals of SHAW's *Saint Joan*, with Joan PLOWRIGHT, Harold BRIGHOUSE's *Hobson's Choice*, and *Othello*, with Laurence OLIVIER. He also directed Peter SHAFFER's *The Royal Hunt of the Sun* (1964), in London and a year later in New York, where he remained to direct the American productions of Shaffer's double bill *Black Comedy* and *White Lies* and USTINOV's *The Unknown Soldier and His Wife* (both 1967). Back in London he directed Alec GUINNESS in Simon GRAY's *Wise Child* (also 1967). Among his later productions were RATTIGAN's double bill *In Praise of Love* (1973); Trevor Griffiths's *The Party* and Shaffer's *Equus* (both 1973) for the National Theatre (the latter also in 1974 in New York); and a revival of Shaw's *Pygmalion* in 1974. From 1974 to 1981 he was Director of Production at the Metropolitan Opera House in New York, but he also directed plays, including *Phaedra Britannica* (1975) based on RACINE and BRECHT's *Life of Galileo* (1980), both for the National Theatre.

DIAS, BALTASAR, see PORTUGAL.

DIBDIN, CHARLES (1745–1814), English actor, dramatist, and composer of many ballads, of which the popular 'Tom Bowling' was written in memory of his brother Thomas, a naval captain. He also wrote a number of BALLAD OPERAS, one of which, *The Waterman* (1774), was successful enough to pass into the repertory of the TOY THEATRE. He was a good actor, making his first success as Mungo in BICKERSTAFFE's *The Padlock* (1768), for which he wrote the music, and from 1788 to 1793 gave a series of one-man entertainments, playing, singing, and reciting monologues. Handsome, but quarrelsome and bad-tempered, he was constantly at odds with the London theatre managers, particularly GARRICK. He left his wife after he had spent all her money, and by his association with the actress Harriet Pitt had three children of whom two were on the stage (see below).

DIBDIN [Pitt], THOMAS JOHN (1771–1833), English actor and playwright, illegitimate son of Charles DIBDIN. Under his mother's name of Pitt he made his first appearance on the stage at the age of four, playing Cupid to the Venus of Mrs SIDDONS. He was later apprenticed to an upholsterer but ran away to become an actor, scene-painter, and dramatist, taking the name of Dibdin to annoy his father, whom he admired but resented, accusing him of neglect of him and his two brothers. He composed about 2,000 songs in the style of the elder Dibdin, to whom they are often attributed, and his most successful work was the PANTOMIME *Harlequin and Mother Goose* (1806), in which GRIMALDI first played CLOWN. Dibdin's NAUTICAL DRAMAS, such as *The Mouth of the Nile* (1798) and *Nelson's Glory* (1805), were also extremely popular. He was the author of one of the many translations of Caigniez's *La Pie voleuse*, as *The Magpie; or, the Maid of Palaiseau* (1815). He married an actress named Nancy Hilliar and had four children who, under the name of Pitt, were all connected with the stage, their children going to the United States, where they were to be found in management for many years. Thomas's elder brother Charles Isaac Mungo Pitt (1768–1833), who retained his mother's name, was also a popular and successful writer of plays and pantomimes, and was for a time the proprietor and manager of SADLER'S WELLS THEATRE.

DICKENS, CHARLES JOHN HUFFAM (1812–70), the great English novelist, was all his life intimately connected with the stage and had an immense influence on it through the numerous dramatizations of his books. Although there is no proof that he was ever an actor, there is a stubborn tradition that he was once in the Portsmouth company of T. D. Davenport, who may have been the prototype of the actor-manager Vincent Crummles, his little daughter Jean (later Mrs LANDER) being the Infant Phenomenon. Certainly both *Nicholas Nickleby* and *Great Expectations* show an intimate knowledge of the details of an actor's life between 1837 and 1844. Those who saw Dickens in his many amateur appearances, which included the leading roles in three farces performed in 1842 in Montreal with the Garrison Amateurs of the Coldstream Guards, particularly admired him as Shakespeare's Justice Shallow (in *Henry IV, Part 2*) and JONSON's Captain Bobadil (in *Every Man in his Humour*), and thought he would have made a fine eccentric comedian. His famous readings from his own works were in a way solo dramatic performances, as has been clearly shown by Emlyn WILLIAMS's reconstruction of them. Early in his career he wrote several operatic burlettas which he did not wish to see revived, and in later years he had in his London home, Tavistock House, a small private theatre, perfectly fitted up, where with his friends and family he gave private performances before a distinguished audience. Two of Wilkie COLLINS's plays were produced there before they were seen at the OLYMPIC, and Dickens also collaborated with Collins in the writing of *No Thoroughfare* (1867), in which FECHTER and Ben WEBSTER appeared.

It would be impossible to catalogue here all the plays based on Dickens's novels, many of which were seen before the books had finished appearing in fortnightly parts. The most persistent adapters were W. T. MONCRIEFF and Edward Stirling, but many versions were performed anonymously. Dickens entrusted Albert SMITH with the dramatization of *The Cricket on the Hearth* at Christmas 1845, when twelve different versions were being given at London theatres, all to be superseded later by Dion BOUCICAULT's excellent *Dot* (1862), and Dickens himself wrote the script for *The Old Curiosity Shop*, not seen until 1884 at the OPERA COMIQUE. His own version of *Great Expectations*, in which he hoped TOOLE would appear, was never used, and the first to be staged was that prepared by W. S. GILBERT, seen at the ROYAL COURT THEATRE in May 1871. Many dramatizations were also prepared for the American stage, but owing to the absence of copyright laws, Dickens received nothing for them.

Dickens's characters are so vivid, his plots so dramatic, that it is not surprising they did well on the stage. Many famous actors had their favourite

Dickens characters, even IRVING appearing as Jingle (from *The Pickwick Papers*), TREE as Fagin (from *Oliver Twist*), and, most successful of all, MARTIN-HARVEY as Sidney Carton in *The Only Way* (based on *A Tale of Two Cities*). Bransby WILLIAMS had a whole gallery of Dickens characters, while Betsey Prig and Sairey Gamp (from *Martin Chuzzlewit*) were for a long time acted by men. In the 1960s a new turn was given to adaptations of Dickens's novels (many of which have been serialized as plays on radio and television) by the conversion of *Oliver Twist, The Pickwick Papers*, and *A Tale of Two Cities* into musicals as *Oliver!* (1960) by Lionel BART, *Pickwick* (1963), with Harry Secombe, and *Two Cities* (1969). It is symptomatic of Dickens's appeal even in translation that a version of *The Pickwick Papers* was an outstanding success on the Soviet stage, being seen at the MOSCOW ART THEATRE in 1934.

The ROYAL SHAKESPEARE COMPANY's production of a two-part, 8-hour adaptation of *Nicholas Nickleby* was a phenomenal success at the ALDWYCH in London in 1980 and later in New York.

DIDASCALIA, a Greek word meaning teacher, and, by extension, dramatic poet, since in the earliest times of Greek drama the poet taught the CHORUS its parts and produced his own play. Hence *didascalia*, meaning production, and *didascaliae*, or catalogues, made by ARISTOTLE and others, of productions, giving the title, author, date, actor-manager's name, and, in the case of Roman plays, details of the music and musicians. Only fragments of such works remain, but where they do survive, or where they are noted as having been consulted by ancient scholars, they are invaluable to the theatre historian as contemporary and authentic evidence of productions.

DIDEROT, DENIS (1713–84), French man of letters. Compared with his Encyclopaedia (1751–77), and his numerous works on any and every subject, Diderot's dramatic writings are only a minor feature of his literary life; but they were an important expression of the Enlightenment, and had a considerable influence on LESSING and through him on the European drama of the 19th century. Diderot was an exponent of bourgeois drama, that offshoot of COMÉDIE-LARMOYANTE whose mixed sentiment, virtue, and sheer priggishness appealed so strongly to the middle-class audiences of the 18th century. The titles of his plays—*Le Fils naturel, Le Père de famille*—are sufficiently revealing, and, in spite of some feeling for dialogue, they degenerate into didactic expositions of philosophical theories. They were in some cases published years before they were acted, while some, like *Est-il bon, est-il méchant?*, were seen only on private stages. Yet the extent of his influence may be gauged by the fact that during his lifetime they were translated into German, English, Dutch, and Italian. The best of Diderot's work for the theatre must be looked for in his *Observations sur Garrick, Essai sur la poésie dramatique, Entretien avec Dorval*, and especially *Paradoxe sur le comédien*, published posthumously in 1830, a dialogue on the art of acting which deals notably with the relative merits of spontaneous sensibility and intellectual control in the actor's performance, the paradox consisting for Diderot, himself the most spontaneous and emotional of men, in the view that intellectual control and the ability to simulate emotion are the gifts of a great actor, while only the mediocre performer allows genuine emotion to govern his work. One of Diderot's stories in dialogue, *Le Neveu de Rameau* written between 1761 and 1774, was staged in Paris in the 1960s with Pierre FRESNAY in the title role; it seems to show that Diderot had the makings of a very different sort of playwright from the one revealed in his extant plays.

DIGGES, (John) DUDLEY WEST (1720–86), English actor, served his apprenticeship in Dublin and Edinburgh, where he was the first to play Young Norval in HOME's *Douglas*. He appeared at the HAYMARKET in London between 1777 and 1781 as Macbeth, Lear, Shylock in *The Merchant of Venice*, Wolsey in *Henry VIII*, ADDISON's Cato, and HOME's Old Norval. He then returned to Dublin and acted there until he was incapacitated by paralysis in 1784.

DIGGES, DUDLEY (1879–1947), Irish-American actor, who made his first appearances with the IRISH NATIONAL DRAMATIC SOCIETY in 1901–3. He was trained for the stage by Frank J. FAY, and was in the original production of AE's *Deirdre* (1902) and some early plays by YEATS, being particularly good as the Wise Man in *The Hour-Glass* (1903). In 1904 he went with the ABBEY THEATRE company to New York, and remained there until his death, making his first appearance on Broadway on 16 Oct. 1904 in SHAW's *John Bull's Other Island*. For 7 years he acted as stage manager for George ARLISS and then joined the THEATRE GUILD company, appearing in their first production, BENAVENTE's *The Bonds of Interest* (1919). He remained with the Guild until 1930, being seen in most of their productions, notably MOLNÁR's *Liliom* in 1921, RICE's *The Adding Machine* (1923), and O'NEILL's *Marco Millions* (1928), and also directing, among other plays, Shaw's *Pygmalion* in 1926 and *The Doctor's Dilemma* in 1927. In 1929 he directed an all-star revival of *Becky Sharp* (a dramatization of Thackeray's *Vanity Fair*), and in 1937 rejoined the Theatre Guild company to play the Emperor Franz Joseph in Maxwell ANDERSON's *The Masque of Kings*. His last and one of his finest parts was Harry Hope in O'Neill's *The Iceman Cometh* (1946).

DIKIE, ALEXEI DENISOVICH (1889–1955), Soviet actor and director, who began his theatrical career at the MOSCOW ART THEATRE in 1910, where he came under the influence of STANISLAVSKY. After the Revolution he concentrated mainly on directing, being responsible at various theatres for productions of such plays as FAIKO's *The Man with the Portfolio* (1928) and Wolf's *Sailors of Cattaro* (1930). He continued to act, and in 1942 gave an unforgettable performance as Ivan Gorlov in KORNEICHUK's *Front* at the VAKHTANGOV THEATRE. Two years later he went to the MALY THEATRE, where he produced SOFRONOV's *Moscow Character* (1948) and a revival of OSTROVSKY's *The Warm Heart*. In 1951 he went to the PUSHKIN THEATRE, Moscow, where he produced in 1953 a revival of *Shadows* by SALTIKOV-SHCHEDRIN.

DILLINGHAM, CHARLES BANCROFT (1868–1934), American theatre manager who, through his friendship with Charles FROHMAN, became the manager of Julia MARLOWE and of many other famous actors, some of them making their first visit to New York under his aegis. In 1905 he introduced SHAW's *Man and Superman* to America, and later the comedies of Frederick LONSDALE. In 1910 he opened the Globe Theatre (later the LUNT-FONTANNE) and for the next 20 years was responsible for all the productions there, mainly musical shows and revues. He also managed other New York theatres, and at the height of his prosperity had as many as six plays running at once. In 1914 he took over the HIPPODROME THEATRE for lavish vaudeville shows, which included trained elephants, Anna Pavlova on her first visit to New York, Gaby Deslys, and performing seals. He produced over 200 plays, but became bankrupt in 1933.

DIMMER, see LIGHTING.

DIONYSIA, ancient Greek festival in honour of the god DIONYSUS, closely associated with the development of comedy and tragedy, which, under the influence of THESPIS, evolved from the DITHYRAMB or ceremonial hymn. The most important of the Greek festivals was the City or Great Dionysia, held in Athens in March-April, when the city was normally full of visitors. It was reorganized on a grand scale by the tyrant PISISTRATUS in the 6th century BC, and lasted for five or six days, opening with a splendid Dionysiac procession to the Parthenon. On the next three days a tragic TRILOGY, with its attendant SATYR-DRAMA, was presented in the morning, while from the early 5th century onwards, comedies, first seen only at the smaller and more intimate winter festival, the LENAEA, were given in the afternoon. The festivities concluded with a competitive contest for ten Dithyrambs. Plays given at the City Dionysia were usually revived later at the RURAL DIONYSIA.

DIONYSUS, a Greek nature-god (Bacchus being the roughly equivalent Latin name), associated mainly but not exclusively with wine. His worship took many forms, the most remarkable being the orgiastic revels in which his votaries (the women in particular) withdrew for a time into the wild and experienced a mystical communion with nature. As Dionysus was a vegetation-spirit, who died and was reborn each year, he was associated not only with all kinds of rites designed to promote fertility but also with mystery-religions, of which an important part was their teaching about death, purgation, and rebirth. The death, sufferings, and rebirth of Dionysus were commonly presented in a quasi-dramatic ritual (the same sequence of events is also found in the English MUMMERS' PLAY, which may have evolved from a similar ritual, though no evidence of this has ever been found), and in fact drama in Athens, whether tragic, satyric, or comic, was always strictly associated with the DIONYSIA and similar festivals. In OLD COMEDY the phallus (as a symbol of fertility) and the Dionysiac *komos* or revel (whence, perhaps, 'comedy') are constant features; another element, the AGON, is by some scholars derived from the 'agony' of Dionysus. In SATYR-DRAMA, too, direct Dionysiac influence is obvious, the satyrs, or 'horse-men', imaginary creatures of the wild, coming to be regarded as attendants on Dionysus, though apparently they were not that originally.

With TRAGEDY the connection is less clear and may be more indirect. Attempts have been made to trace a Dionysiac ritual-sequence in the basic form of tragedy, but they are not very convincing. It is possible, though it cannot be proved, that the earliest (lost) forms of tragedy in Athens dealt exclusively or mainly with Dionysiac subjects; but although, among later tragedies, plays with Dionysiac subjects are common (of which only one, the *Bacchae* of EURIPIDES, happens to have survived), they are not more common than the dramatic nature of the Dionysus-legends would lead us to expect. When ARISTOTLE says that tragedy originated in the DITHYRAMB, he meant only that its form developed from a choral performance, not that the new art was Dionysiac in spirit or content, though it may have been. He states also that tragedy developed 'out of the satyr-drama' (which was undoubtedly very closely connected with the more boisterous side of Dionysus worship) by discarding its rollicking metres, its ridiculous diction, and its insignificant plots. It might therefore be argued that tragedy actually became itself by getting rid of the Dionysiac elements.

DIPHILUS OF SINOPE (d. 290 BC), Greek poet of NEW COMEDY, contemporary with MENANDER. He was frequently imitated by TERENCE and PLAUTUS. Only fragments of his plays and 60 out of about 100 titles survive.

DIRECTOR, the person responsible for the general interpretation of the play, and for the conduct of rehearsals; known on the Continent as the *régisseur*. In England until 1956 he was called the PRODUCER, but it was then officially decided to adopt the usage of the American stage and the cinema.

It was not until the beginning of the 20th century that play-direction became a recognized profession; before that rehearsals were conducted by the author; by the chief actor who, like MOLIÈRE, might also be the author and manager; by the STAGE-MANAGER; or by the PROMPTER; and in any case they were brief and perfunctory. It is evident from memoirs and letters that many actors appeared on stage with no preliminary rehearsal, trusting to the older members of the company to carry them until they found their feet. The first to insist on long detailed rehearsals was the actress-manageress Madame VESTRIS, when she took over the OLYMPIC THEATRE in 1830. Dion BOUCICAULT, who worked with Vestris, took her methods to the United States, where he directed his own plays and influenced not only the work of Augustin DALY but more especially that of David BELASCO, who may be considered America's first outstanding director in the modern style. Meticulous rehearsals, especially of crowd scenes, had already found favour in Germany with the MEININGER COMPANY, working under Ludwig CHRONEGK, whose influence was apparent on Otto BRAHM in his own country, on ANTOINE in France, and on STANISLAVSKY in Russia. Among their disciples were REINHARDT in Germany, and in Russia MEYERHOLD. Meanwhile in England GRANVILLE-BARKER had taken over the management of the ROYAL COURT THEATRE and trained actors in the new meticulous way, arousing the admiration of SHAW, ahead of his time as always, who called him 'the best possible director of my plays'. He was also much admired by the French actor-manager COPEAU, the first of a new generation which also produced JOUVET, BATY, DULLIN and, from Russia via Switzerland, PITOËFF. In Germany the work of the director was extended by PISCATOR and JESSNER; in Russia Meyerhold gave way to TAÏROV, OKHLOPKOV, and VAKHTANGOV; and in England French influence was brought to bear when Copeau's nephew Michel SAINT-DENIS settled in London. America, which with the emergence of O'NEILL as a major playwright no longer had to depend on English classics for great productions, developed a body of truly American directors with Eva LE GALLIENNE, followed by STRASBERG, KAZAN, and CLURMAN. England was represented by Basil DEAN and by the brilliant Shakespearean director Tyrone GUTHRIE, who also worked in the United States and Canada; later directors are too numerous to mention individually, but outstanding among them are Peter BROOK, Peter HALL, and Joan LITTLEWOOD. On the Continent the main currents of play production were influenced by BRECHT, VILAR, and later PLANCHON. There is no lack of young and original directors in the theatres of all countries today.

The style of each director is nowadays so personal that it may be immediately recognized, and in some cases plays are referred to as being not by the author but by the director—Reinhardt's *Hamlet*, Stanislavsky's *Three Sisters*—a form of journalistic shorthand which should be corrected wherever possible. There are nowadays courses for directors, as distinct from those for actors, but most outstanding directors still appear to be born, not made, and it has been said that the ideal director should be an actor, an artist, an architect, an electrician, an expert in geography, history, costume, accessories, and scenery, and have a thorough understanding of human nature—the last being perhaps still the most essential.

DISCOVERY SPACE, see INNER STAGE.

DISGUISING, a term used in the 15th and 16th centuries to describe a form of Court entertainment in England which had something in common with the MUMMERS' PLAY and, through the use of masks, with the later Court MASQUE. The word was already falling out of use by 1544 and being replaced by 'mask'. The Disguising was essentially an amateur production, involving the royal household and attendant nobility, and was often given in honour of a distinguished guest, ending with an exchange of gifts and a dance in which everyone joined. Music and dancing were originally more important than the text, and the task of arranging a Disguising usually fell to a professional musician in the royal household; many of the Disguisings enjoyed by the young Henry VIII were the work of William Cornish, a Gentleman of the Chapel Royal and the Master of the choirboys there, who, as an organized BOY COMPANY, played an important part in the festivities.

DITHYRAMB, a hymn originally in honour of the Greek god DIONYSUS, and then of other gods. It was performed by a CHORUS of 50 and would normally relate some incident in the life of the deity to whom it was addressed. The leader of the chorus later became the solo PROTAGONIST, and in the ensuing dialogue between him and the rest of the chorus is believed to lie the origin, or one of the origins, of drama.

DMITREVSKY, IVAN AFANASYEVICH (1733–1821), Russian actor, who appeared with the amateur company founded by Fedor VOLKOV, and went with him to play before the Court in Jan. 1752. He was later a member of the company organized by SUMAROKOV. After the death of Volkov in 1763 Dmitrevsky was appointed Inspector of Theatres, and took a leading part in the running of the State playhouses. Between 1765 and 1768 he twice went abroad to complete his theatrical education, spending most of the time in Paris with the leading French actors of the day. On his return he occupied the highest position in the St Petersburg theatre, both as actor and administrator, appearing with equal success in tragic and comic parts. His best performances were considered to be the title-roles of MOLIÈRE's *Le Misanthrope* and

Sumarokov's *Dmitri the Impostor*, and Starodum in *The Minor* by FONVIZIN.

DMITRIEV, VLADIMIR VLADIMIROVICH (1900–48), Soviet stage designer whose early death was much regretted. Only the previous year he had designed a most evocative setting for Virta's *Our Daily Bread* at the MOSCOW ART THEATRE, in which a quiet autumnal landscape formed a poignant contrast to the feverish activity within the house, seen through its lighted windows. Other important plays for which he designed the décor were OSTROVSKY's *The Last Sacrifice* at the Moscow Art Theatre in 1944, and GORKY's last plays, *Yegor Bulychov and Others* and *Dostigayev and Others*, at the VAKHTANGOV in 1932 and 1933.

DMITRI OF ROSTOV, Saint [Daniel Tuptalo] (1651–1709), Russian bishop, canonized in 1757. His six sacred dramas are very similar to medieval European MYSTERY and MORALITY PLAYS, and are based on the sacred SCHOOL DRAMAS or dialogues acted at the Kiev Academy in his time, which in turn were based on JESUIT DRAMA. In rhyming syllabic metre, they include *The Nativity Play, The Penitent Sinner*, and *The Resurrection of Christ*. Biblical characters appear together with such allegorical figures as Jealousy, Hope, Despair and Death, and humorous peasants who comment irreverently on the sacred action. It was in St Dimitri's *Esther and Ahasuerus* that Ivan DMITREVSKY made his first appearance before the Russian Court in 1752.

DÖBBELIN, KARL THEOPHIL (1727–93), German actor-manager, who began his career in 1750 with Carolina NEUBER and was later with ACKERMANN, but much preferred the strolling harlequin-players with whom he toured from time to time. In 1756, encouraged by GOTTSCHED who had much admired his acting in a revival of his *Der sterbende Cato*, he formed his own company, and later joined SCHUCH's, eventually taking it over and settling in BERLIN, where he had a great success with his production of LESSING's *Minna von Barnhelm* in 1768 and played the title role in the first unsuccessful production of his *Nathan der Weise* in 1783. In 1787 his company was installed in Berlin's Nationaltheater. He remained manager until 1789, and continued to act and direct until his death. He was a ranting actor, remaining always, whether in tragedy or comedy, the undisciplined though resourceful strolling player. In spite of his faults he was a cheerful, hard-working man who kept his company well under control, provided them with a good repertory, and several times forced the critics to take his productions seriously.

DOCKSTADER, LEW [George Alfred Clapp] (1856–1924), American VAUDEVILLE artist and one of the foremost 'blackface' performers in the MINSTREL SHOW, known for his engaging manner, quick wit, and absurd impersonations. Born in Hartford, Conn., he first appeared there as an amateur, but made his professional début in 1873 with the Emmet and Wilde minstrels, still under his real name. His stage name came through his partnership in a minstrel troupe with Charles Dockstader, billed as the Dockstader Brothers. When Charles died in 1883 Clapp kept the name, even after the troupe had disbanded. He then opened his own theatre in New York, but without success as he was not a good business man, and from 1898 to 1913 ran a minstrel troupe in partnership with George Primrose (1852–1919). When this folded up he went on the Keith circuit until 1923, when he retired. He gave a new dimension to the declining plantation show by turning it into an instrument of political and social satire, parodying politicians, polar explorers, aviators, and other men in the news, his basic innovation being to portray not comic Negro characters but comic whites in 'blackface'. His performances invariably contained at least one of his most popular comic songs, such as 'Everyone Works But Father', or 'He Used To Breakfast With Us Every Morning', and one of the highlights of his act was an impression of Theodore Roosevelt in Rough Rider uniform, complete with persistent smile, walrus moustache, and a black face.

DOCK STREET THEATRE, Charleston, South Carolina. The first theatre of this name opened with FARQUHAR's *The Recruiting Officer* in 1736, a year after the city had its first known theatrical performance, of OTWAY's *The Orphan*. The theatre was not a success, and no performances are recorded there after 1737, though the building survived until at least 1749. The site was later occupied by a hotel, but in 1937 a second Dock Street Theatre, seating 463, a re-creation of the earlier one, was built there, and again opened with *The Recruiting Officer*, which was produced by the Footlight Players, Charleston's oldest community theatre group founded in 1931. The Players remained in Dock Street until 1941, running the theatre during the last few years with its then lessee, the Carolina Art Association. When the latter formed its own producing group the Footlight Players withdrew to a converted cotton warehouse seating 295, but they returned to Dock Street in 1952 and since 1958 have been sole lessee of the theatre, which is owned by the City of Charleston.

During the interval between the two Dock Street theatres Charleston had several other theatres. The Church Street opened in 1773 and housed David DOUGLASS and his company on their last engagement before the American Revolution; the first Charleston opened in 1793 and was visited by Junius Brutus BOOTH and Edmund KEAN; and the second Charleston, built in 1837, had a good stock company run by William ABBOTT from the London HAYMARKET. It was burned down in 1861.

and its successor was the Academy of Music, which opened in 1869. It closed in 1936 after a production by the Footlight Players of COWARD's *Hay Fever*.

DOCUMENTARY DRAMA, see LIVING NEWSPAPER and FACT, THEATRE OF.

DÓCZY, LAJOS, see HUNGARY.

DODD, JAMES WILLIAM (1734–96), English comedian, the last of the fops who began with Colley CIBBER. After playing in the provinces, he went to DRURY LANE in 1765, and in 1777 created the parts of Lord Foppington and Sir Benjamin Backbite in SHERIDAN's *A Trip to Scarborough* and *The School for Scandal* respectively. He was also the first to play Dangle in Sheridan's *The Critic* (1779). One of his finest parts was Aguecheek in *Twelfth Night*, in which Charles LAMB saw and admired him, but he was also good as Bob Acres in Sheridan's *The Rivals* and as Tattle in CONGREVE's *Love for Love*. Off-stage he was serious and a great student, leaving at his death a fine library.

DODD, LEE WILSON (1879–1933), American poet, novelist, and playwright. The most important of his stage works were *Speed* (1911), a satire on the early motor-car craze, and *The Changeling* (1923). Another satirical comedy, *A Stranger in the House*, was abandoned when its leading man, Henry MILLER, died during rehearsals in 1926. After some financial losses in the slump of 1929, Dodd became a teacher of English, and at the time of his death had just been invited to succeed George Pierce BAKER, whose assistant he had been for some time, in his playwriting course at YALE.

DODSLEY, ROBERT (1703–64), English playwright and publisher. Having run away from home, he was working as a footman in London when his literary gifts attracted the attention of Pope and Defoe. Helped by them, and by the success of his first play, *The Toy Shop* (1735), he established himself as a bookseller and publisher at the sign of Tully's Head in Pall Mall, where he issued works by such authors as Pope and Dr JOHNSON, and also published a *Select Collection of Old Plays*, later revised and edited by HAZLITT. His best known play was *The King and the Miller of Mansfield* (1737), which, with its sequel, *Sir John Cockle at Court* (1738), was first given at DRURY LANE and frequently revived. It provided the basis for COLLÉ's *Partie de chasse d'Henri IV*, and became part of the repertory of the 19th-century TOY THEATRE. Dodsley also wrote the libretto of a BALLAD OPERA, *The Blind Beggar of Bethnal Green* (1741). His last play, *Cleone* (1758), was a tragedy which owed much of its success to the acting of George Anne BELLAMY and was revived by Mrs SIDDONS in 1786.

DOG DRAMA, a type of spectacular entertainment which probably had its origin in the performing dog troupes of the circuses. It became extremely popular in Europe in the 19th century, the immediate cause of the craze for dogs on stage in London being a little after-piece at DRURY LANE in which a real dog, Carlos, rescued a child from a tank of real water—an echo of the equally fashionable AQUATIC DRAMA. The most famous Dog Drama was PIXÉRÉCOURT's *Le Chien de Montargis, ou la Fôret de Bondy*, first seen at the Théâtre de la GAÎTÉ on 18 June 1814, and in an English adaptation as *The Dog of Montargis; or, the Forest of Bondy*, by William BARRYMORE, at COVENT GARDEN on 30 Sept. As *The Forest of Bondy*, it was revived at the LYCEUM in 1824 and at Drury Lane in 1835. It is said that the Duke of Weimar's insistence on seeing a German version of this play, with a performing poodle, caused GOETHE to retire from the management of the Weimar Court Theatre.

DOGGETT, THOMAS (*c.* 1670–1721), English actor, a fine low comedian for whom CONGREVE wrote the part of Fondlewife in *The Old Bachelor* (1693) and Ben in *Love for Love* (1695). He was himself the author of a comedy, *The Country-Wake* (1690), which Colley CIBBER later used as the basis of his *Hob, or the Country Wake* (1711). In 1711 Doggett became joint manager of DRURY LANE with WILKS and Cibber, who in his *Apology* has left an excellent pen-portrait of him, but retired in disgust when Barton BOOTH, whose politics he disapproved of, was given a share in the patent. In 1714, in honour of the accession of George I, Doggett instituted the Doggett Coat and Badge for Thames watermen, a trophy which is still rowed for annually and which plays an important part in the plot of *The Waterman* (1774), a BALLAD OPERA by Charles DIBDIN.

DOGG'S TROUPE, see COMMUNITY THEATRE.

DOHERTY, BRIAN, see SHAW FESTIVAL.

DOLCE, LODOVICO (1508–68), Italian playwright of whom Walker in his *Historical Memoir on Italian Tragedy* (1799) says scathingly 'little is known that can be related with pleasure.' Most of his plays, on subjects drawn from classical antiquity, were considered bloodthirsty even in an age which applauded the notorious *Canace* (1543) of SPERONI; but his retelling of the story of Herod and Mariamne, *Marianna* (1565), was, for its period, unusually subtle, and was received with great applause on its first performance. His *Giocasta*, based on EURIPIDES' *Phoenician Women*, was translated by George GASCOIGNE, and as *Jocasta* was performed at Gray's Inn in 1655.

DOMINION DRAMA FESTIVAL, see CANADA.

DOMINIQUE, see BIANCOLELLI, GIUSEPPE.

DON JUAN, character derived from an old Spanish legend, who first found vital expression in TIRSO DE MOLINA's *El burlador de Sevilla y convidado de piedra* (*The Trickster of Seville and the Stone Guest*, c.1630) and has since become a constantly recurring figure in European literature. There is no evidence for the historical existence of Don Juan, though tradition usually connects him with Seville. Tirso's play combines two plots derived from separate sources, the first being concerned with the character and activities of the hero, the second with his mocking invitation to dinner given to the marble statue, who accepts it and brings retribution by supernatural means upon Don Juan in punishment of his many crimes. Don Juan is not portrayed merely as a sensual man; he is the embodiment of self-will, unable to curb his desires although he knows they are evil. There is no lack of Catholic belief in him, as there is in some later versions; he does not doubt that retribution will come, but he continually puts off repentance, hoping through God's mercy and long-suffering to remain immune as long as possible.

Among the many works on the same theme are Mozart's opera *Don Giovanni* (1787), BYRON's poem *Don Juan* (1819–24), MOLIÈRE's *Le Festin de pierre* (1665), an Italian version by GOLDONI, and a Russian one by PUSHKIN, several works in Spanish, the best being Zorrilla y Moral's *Don Juan Tenorio* (1844), ROSTAND's *La Dernière Nuit de Don Juan* (c.1910), FRISCH's *Don Juan, oder Die Liebe zur Geometrie* (1953), and Ronald Duncan's *Don Juan* and *The Death of Satan* (both 1956). Don Juan also appears in the third act of SHAW's *Man and Superman* (1905), and in Tennessee WILLIAMS's *Camino Real* (1953).

DOONE, RUPERT, see GROUP THEATRE, LONDON.

DOORS OF ENTRANCE, see PROSCENIUM DOORS.

DORSET GARDEN THEATRE, London. This theatre, known also as the Duke's, or Duke of York's, House, fronted the Thames, to the south of Salisbury Court. It was planned in c.1669 by DAVENANT for his company the Duke's Men, but he died before it opened and his widow, with Henry HARRIS and BETTERTON, took control. The building, by tradition attributed to Wren, was 57ft wide by 140ft long; it was uncomfortable and inconvenient, its long narrow auditorium giving a tunnel-like view of the stage, which was completely separated from the audience, and unlike most contemporary theatres it had no side boxes. It had, however, a highly ornate proscenium and a stage capable of displaying elaborate scenery. The theatre opened on 9 Nov. 1671 with DRYDEN's *Sir Martin Mar-All*, a tried favourite. The first new play was CROWNE's *King Charles VIII of France*; opera, for which Dorset Garden later became famous, began with Davenant's adaptation of *Macbeth* (1673). An operatic version of *The Tempest* by SHADWELL was given a spectacular production in 1674, as was his own play *Psyche* in the same year. This was specially designed to show off splendid scenery and complicated machines newly imported from France. Betterton's performances in *The Libertine* (1675) and *The Virtuoso* (1676), both by Shadwell, enhanced his already considerable reputation, while Mrs BARRY made her reputation in *The Orphan* (1680) and *Venice Preserv'd* (1681), specially written for her by OTWAY.

Under Betterton's capable management the theatre flourished and from 1672 until 1674, while DRURY LANE was out of action, it was London's only first-class playhouse. Among the dramatists who wrote for it were D'URFEY, SETTLE, Aphra BEHN, ETHEREDGE, and RAVENSCROFT. By 1682, however, both Dorset Garden and Drury Lane were in financial difficulties, and the two companies combined, choosing Drury Lane as their headquarters. Dorset Garden was deserted, except for occasional performances of opera, though in 1689 it was renamed the Queen's Theatre as a compliment to Queen Mary who had just ascended the throne with William of Orange. In 1706 the Drury Lane company returned to Dorset Garden for an autumn season, and thereafter the theatre was used for acrobatic and wild animal shows. It was demolished in 1709.

DORST, TANKRED (1925–   ), German dramatist, who has moved from an early association with the Theatre of the ABSURD, with such farces as *Die Kurve* and *Freiheit für Clemens* (both 1960), to a politically committed deployment of the techniques of documentary drama. He began work in a puppet-theatre, and his first play, *Gesellschaft im Herbst* (*Company in Late Autumn*, also 1960), showed the influence of puppet-drama, while *Grosse Schmährede an der Stadtmauer* (*Great Vituperation at the City Wall*, 1961) has been compared to the work of BRECHT in its portrayal of a woman's impassioned but ultimately futile protest against war. In *Toller* (1968) he uses the dramatist's own technique of EXPRESSIONISM to examine the failure of the Munich revolt in 1918, in which TOLLER was implicated, incorporating scenes from Toller's *Masse-Mensch* (1920). *Eiszeit* (*Ice Age*, 1973) is about Knut HAMSUN.

DORVAL [*née* Delaunay], MARIE-THOMASE-AMÉLIE (1798–1849), French actress, child of strolling players, who was on the stage from her earliest years, and as an adult actress first came into prominence as Amélie in the melodrama *Trente Ans, ou la Vie d'un joueur* (1827), playing opposite FRÉDÉRICK. On 21 April 1835 she made her first appearance at the COMÉDIE-FRANÇAISE, giving a remarkable performance as Kitty Bell in *Chatterton* by Alfred de VIGNY, whose mistress she was at the time. Later in the same year she appeared with success in HUGO's *Angelo* and might have remained with the company, but she found the restrictions, after her free life, too irksome, and the jealousy of Mlle MARS insupportable. She therefore returned to the popular theatres, where she again played with Frédérick. In 1842 she was seen at the ODÉON in RACINE's *Phèdre*, with marked success, but her health failed after a gruelling provincial tour and she returned to Paris to die in poverty.

DOSSENNUS, one of the stock characters of the FABULA *atellana*, whose name, from a connection with *dorsum*, 'back', may indicate that he was humpbacked. He shared some of the characteristics of MANDUCUS, and the names may have been interchangeable. Dossennus gave his name to one of the plays of Novius, *The Two Dossennuses*, and also figures in a fragment from Pomponius' *Campani* which shows him to have been something of a scrounger.

DOSTOEVSKY, FYODOR MIKHAILOVICH (1821–81), Russian novelist, several of whose novels have been dramatized, notably *The Idiot* and *The Brothers Karamazov*. These were produced in Russia before 1917 and have been revived since. *The Insulted and Injured* (also known in English as *The Despised and Rejected*) was adapted for the MOSCOW ART THEATRE by V. A. Solovyov, and in 1946 an adaptation of *Crime and Punishment*, by Rodney Ackland, was given with some success in London, John GIELGUD playing Raskolnikov. *The Idiot*, in an adaptation by Simon GRAY, was seen at the OLD VIC in 1970 in a production by the NATIONAL THEATRE company directed by Anthony QUAYLE.

DOTRICE, ROY (1923–), British actor, who became interested in the theatre while a prisoner of war in Germany in the Second World War, making his first appearance in 1945 in a revue performed by ex-prisoners-of-war, after which he played in various repertory companies. In 1958 he joined the company at the SHAKESPEARE MEMORIAL THEATRE, and there and at the ALDWYCH THEATRE in London played a wide variety of parts, including Father Ambrose in WHITING's *The Devils* (1961), Simon Chachava in BRECHT's *The Caucasian Chalk Circle* (1962), Caliban in *The Tempest* (1963), Shallow in *Henry IV, Part II* (1964), and the title role in Brecht's *Puntila* (1965). He is best known for his solo performance as John Aubrey in *Brief Lives*, seen at the HAMPSTEAD THEATRE CLUB in 1967 and on Broadway later the same year. In 1969 it was presented in the West End, where it had a long and well-deserved run, being an admirable re-creation of the malicious 17th-century diarist and his age. Dotrice has since given this performance all over the world, while intermittently appearing in other roles, notably at the CHICHESTER FESTIVAL THEATRE in 1970 as Peer Gynt in a new version of IBSEN's play by Christopher FRY. In New York in 1980 (London, 1981), he gave another one-man performance, as Abraham Lincoln, and he was also seen in New York in Hugh LEONARD's *A Life* in 1980.

DOTTORE, Il, the second elderly character of the COMMEDIA DELL'ARTE, forming a pair with PANTALONE. He was usually depicted as a Bolognese lawyer named Graziano, given to long expositions and pedantic utterances which in the coarser playing of the character lapsed into *spropositi* (exaggerations) and nonsensical tongue-twisters, the original use of local dialect giving way to the habit of 'saying everything the wrong way round'. He was distinct from the stage pedant, who belonged to the COMMEDIA ERUDITA, but they had many characteristics in common and were later often confounded by foreigners. Unlike Pantalone, whom he resembled in his greed, gullibility, and amorous pretensions, he did not pass into the English HARLEQUINADE, but he was adopted on to the French stage through the plays of MOLIÈRE.

DOUBLE MASQUE, see MASQUE.

DOUGLASS, DAVID (?–1786), American actor-manager who in 1758 met and married the widow of the elder HALLAM in Jamaica. Amalgamating his actors with hers, he took them back to New York, named them the AMERICAN COMPANY, and built first a temporary theatre on CRUGER'S WHARF, then another in Beekman Street—also known as the CHAPEL STREET THEATRE—and a third in JOHN STREET. He was also responsible for the erection of the first permanent theatre in the United States, the SOUTHWARK in PHILADELPHIA, which opened in 1766 in spite of some puritanical opposition. It was under Douglass's management that the American Company staged Thomas GODFREY's *The Prince of Parthia* (1767), the first American tragedy to have a professional production, and that John HENRY, later to succeed Douglass as manager, first joined the company in New York.

DOWN CENTER STAGE, see DALLAS THEATRE CENTER.

DOWNES, JOHN (*fl.* 1661–1710), author of *Roscius Anglicanus*, a volume of scattered theatrical notes which is one of the rare sources of information on the early Restoration theatre. Downes wanted to be an actor, but his first appearance in DAVENANT's *The Siege of Rhodes* in 1661 was such a fiasco that he gave up, and worked back-stage as PROMPTER and BOOK-KEEPER. *Roscius Anglicanus*, first published in 1708, was reprinted in 1886 in an edition by Joseph Knight, and in 1930 appeared in a new edition by Montague Summers.

DOWN STAGE, see STAGE DIRECTIONS.

DOWNSTAGE THEATRE, Wellington, New Zealand, company founded in 1964 by actors and writers seeking a professional outlet for their work. Starting as a small café-theatre, it developed over the years into a highly expert ensemble which staged many plays by local authors, including Peter Bland, Robert Lord, Roger Hall, Joseph Musaphia, and Bruce Mason. In 1970 Sunny Amey returned from working with the NATIONAL THEATRE in London to become the theatre's artistic director, supervising the move into its new flexible Hannah Playhouse. He was later succeeded by Anthony Taylor. Raymond Boyce, brought over from England by the NEW ZEALAND PLAYERS, was first a consultant and later resident designer to the Downstage Theatre company, his contributions to stage design in New Zealand reaching a high standard and exerting a great influence. In 1974 Downstage put on the first full season of plays entirely by New Zealand playwrights. The company has also appeared in several plays written by its members and in a series of late-night shows reminiscent of its café-theatre antecedents.

DOWTON, WILLIAM (1764–1851), English actor, who in 1791 joined Sarah BAKER's travelling company in Kent, and soon became her leading man, marrying her daughter Sally (1768–1817) in 1794. Two years later he was seen by SHERIDAN while playing Sheva in CUMBERLAND's *The Jew*, and was engaged for the winter seasons at DRURY LANE. He continued to tour with Mrs Baker in the summer and in 1815 took over the management of her company but without success, possibly because all his interests lay in London. His elder son William Paton (1797–1883) took over from him, but he too went to Drury Lane in 1832 and the Kent circuit was finally abandoned in 1838. Meanwhile the elder William had been recognized as an outstanding character actor in such parts as Sir Anthony Absolute in Sheridan's *The Rivals*, Old Hardcastle in GOLDSMITH's *She Stoops to Conquer*, and Old Dornton in HOLCROFT's *The Road to Ruin*. In 1815 he essayed Shylock in *The Merchant of Venice*, but was not well received by a public accustomed to seeing him in comic roles such as Falstaff, in which he made his New York début at the PARK THEATRE in 1836. His later years were unhappy, and by 1840 he was destitute; a benefit was organized for him at HER MAJESTY'S THEATRE, on the proceeds of which he went into retirement.

DRACHMANN, HOLGER HENRIK HERHOLDT (1846–1908), Danish poet, novelist, and dramatist, who for many years devoted himself equally to painting and writing and travelled widely on the Continent and in England (which he visited in 1871). He came for a time under the influence of Georg BRANDES, but a passionate and volatile temperament took him through successive stages of radical socialism, romantic traditionalism, and ultimately to the 1890s style of Bohemianism. He wrote much prose and verse before attempting dramatic composition in the early 1880s, and to the end his work was lyrical (somewhat in the manner of Swinburne) rather than dramatic. *Der var en Gang* (*Once Upon a Time*, 1885) established him as a dramatist; and his melodramas, the best known of which are *Vølund Smed* (*Wayland the Smith*, 1894) and *Renæssance* (1894), were collected in *Melodramaer* (1895). *Brav Karl* (1898) added to his popularity as a playwright; and the neo-romantic *Gurre* (1898), the lyric drama *Hallfred Vandraadeskjald* (1900), and *Der grønne Haab* (*Green Hope*, 1903) finally consolidated it.

DRAG, see FEMALE IMPERSONATION.

DRAKE, ALFRED, see MUSICAL COMEDY.

DRAKE, SAMUEL [Samuel Drake Bryant] (1768–1854), American actor-manager, born in England, who left for America in 1810 with his large and talented theatrical family. They performed in Boston for several years, moving to Albany in 1813 to join the company of John BERNARD. They left for Kentucky in 1815, and after a long and dangerous journey performed in Louisville, Lexington, and Frankfort. Drake visited Missouri in 1820, and then went to Cincinnati where in 1826 his son Alex and daughter-in-law Frances Ann DENNY took over until his death. He may not have been the first actor to perform in the West, but he brought the most talented and professional company to appear there and established playhouses in Kentucky and elsewhere on a firm basis. His granddaughter Julia DEAN became an outstanding American actress.

DRAMA. 1. A term applied loosely to the whole body of work written for the theatre, as English drama, French drama, or to a group of plays related by their style, content, or period, as Restoration drama, realistic drama.

2. A term applicable to any situation in which there is conflict and, for theatrical purposes, resolution of that conflict with the assumption of character. This implies the cooperation of at least two actors, or, as in early Greek drama, a PROTAGONIST and CHORUS, and rules out narrative and monologue. The dramatic instinct is inherent in man, and the most rudimentary dialogue with song and dance may be classed as drama. In a narrower sense the word is applied to plays of high emotional content, which at their best may give us literary masterpieces, and at their worst degenerate into MELODRAMA. The term dramatist is not necessarily restricted to a writer of such dramas, but serves, like playwright, to designate anyone writing for the theatre.

DRAMA AND COMEDY, Theatre of, Moscow, see TAGANKA THEATRE.

DRAMA SCHOOLS, see SCHOOLS OF DRAMA.

DRAMATIS PERSONAE. The characters in a play, usually listed in the introductory pages of a printed text, or on programmes. In Latin, *persona* (from the Etruscan) replaced the Greek *prosopon*, or MASK, which concealed the identity of the actor and allowed him to 'become' the personage he was portraying.

DRAME, the name given by DIDEROT to a type of play of which his own works are significant examples. He envisaged it as a blending of the outmoded forms of comedy and tragedy which would deal seriously with the domestic problems of the middle-class audiences who now frequented the theatre. It differed from the earlier *tragédie bourgeoise* as exemplified by some of the plays of VOLTAIRE in that it usually ended happily, or at least peacefully, with a reconciliation after repentance for past errors. Rooted in the COMÉDIE LARMOYANTE of LA CHAUSSÉE, it was strongly influenced from England by the novels of Richardson and the domestic dramas of LILLO and MOORE. Among the exponents of the *drame*, apart from Diderot, were SEDAINE and MERCIER.

DRAPER, RUTH (1884–1956), American actress who achieved world-wide fame with dramatic monologues which she wrote herself. She first employed her gift for mimicry in short sketches performed at private parties and charity performances, and it was not until 29 Jan. 1920 that she made her first professional appearance at the Aeolian Hall in London. She quickly established herself as an international figure and for the rest of her life toured continuously, elaborating and adding to her repertory but never changing the basic formula—a bare stage, a minimum of props, and herself as one person responding to invisible companions (as in 'Opening a Bazaar' or 'Showing the Garden') or as several people in succession (as in 'Three Generations', 'Mr Clifford and Three Women', or 'An English House Party'). Her career was a long series of triumphs on the Continent, in New York, where she often remained in the same theatre for four or five months, and in England. She was seen for the last time in 1956 in London at the ST JAMES'S THEATRE and in New York at the PLAYHOUSE. The basic texts of some of her best known monologues were included in *The Art of Ruth Draper* (1960), by M. D. Zabel, but they are *aides-mémoire* only, since she varied her dialogue at every performance.

DRAPERY SETTING, see CURTAIN SET.

DRESDEN, the capital of the Electorate (later the Kingdom) of Saxony until 1918, and now in the German Democratic Republic. It was visited by travelling companies during the 17th century, but German drama played a secondary role to Italian opera, for which several opera-houses were built, the first in 1667, and it was not until the 1770s that subsidized German companies, under such managers as DÖBBELIN and Abel SEYLER, settled in the town, playing in the theatre built in 1761. Ludwig TIECK, who was attached to the theatre from 1835 to 1841, established a literary repertoire with plays by KLEIST, CALDERÓN, and above all Shakespeare. Carl Zeiss, coming in 1901 to the Hoftheater built by Semper in 1841, renovated not only the repertory but also the building, installing Linnebach's elaborate lifts and revolving stage. In 1916, when Ernst DEUTSCH appeared in the first expressionist play seen in Germany, HASENCLEVER'S *Der Sohn* (see EXPRESSIONISM), Dresden became one of the major theatrical centres of the Weimar Republic, Oskar Kokoschka directing there the first German production of his *Mörder, Hoffnung der Frauen* (*Murderer, Hope of Women*) in 1917. The Dresden Staatstheater is now one of the major theatres of the German Democratic Republic.

DRESSLER, MARIE, see VAUDEVILLE, AMERICAN.

DREW, GEORGIANA (1856–93), American actress, daughter of Mrs John DREW. At the age of 16 she appeared with her mother's company in Philadelphia, and was later with Augustin DALY in New York. An actress of great ability whose career was hampered by ill health, she married in 1876 Maurice BARRYMORE, with whom she appeared in *Diplomacy*, a version of SARDOU'S *Dora* by B. C. Stephenson and Clement SCOTT, and also in MODJESKA'S company. She also played with Lawrence BARRETT and Edwin BOOTH, and under the management of Charles FROHMAN. Her three children, Ethel, John, and Lionel BARRYMORE, were all on the stage.

DREW, JOHN (1853–1927), American actor, son of Mrs John DREW, under whom he first appeared on the stage in Philadelphia. In 1875 he was engaged by Augustin DALY to play opposite Fanny DAVENPORT and, later, Ada REHAN. During the 1880s he was several times seen in London, being much admired as Petruchio in *The Taming of the Shrew* and appearing in other classical comedies, including *As You Like It* and SHERIDAN'S *The School for Scandal*. From 1892 to 1915 he was in a series of modern comedies under the management of Charles FROHMAN, often playing opposite Maude ADAMS and making frequent tours of the United States. A handsome, distinguished-looking man, he gave a fine performance in 1916 as Major Arthur Pendennis in a dramatization of Thackeray's novel. He was last seen on tour in PINERO'S *Trelawny of the 'Wells'*. He was the third president of the PLAYERS' CLUB, and in 1903 presented the library of the theatre historian and bibliographer Robert W. Lowe, which he had acquired, to HARVARD, thus laying the foundation of the fine theatre collection there. His daughter Louise (?–1955), who married the actor Jack Devereaux (?–1958), and his grandson John were both on the stage.

DREW, MRS JOHN [*née* Louisa Lane] (1820–97), American actress and theatre manager, daughter of English actors who could trace their theatrical ancestry back to Elizabethan days. She went on the stage in London as a small child and played with such actors as MACREADY and COOKE. In 1827 she was taken by her widowed mother to New York, where she appeared as many characters in one play, in the style of Clara FISHER, with much success. She was also seen with the elder BOOTH, with the first Joseph JEFFERSON, and with Edwin FORREST, who much admired her precocious talent. By the age of 16 her roles included Lady Macbeth and the Widow Melnotte in BULWER-LYTTON's *The Lady of Lyons*, and she was appearing all over the United States. In 1850 she married as her third husband John Drew (1827–62), an Irish actor who was admired in eccentric parts such as Sir Lucius O'Trigger in SHERIDAN's *The Rivals* and Handy Andy in the stage version of Samuel Lover's novel; she was thereafter known as Mrs John Drew. From 1861 to 1892 she managed the stock company at the Arch Street Theatre, PHILADELPHIA, and from 1880 to 1892 was constantly seen on tour as Mrs Malaprop in *The Rivals*, one of her best parts, with Joseph JEFFERSON as Bob Acres. A woman of strong, almost masculine, personality, she ruled her theatre and family with firmness and energy, and contributed not a little to the establishment of the American theatre during the 19th century. Two of her children, John and Georgiana (above), were on the stage. *The Royal Family* (1927) by George S. KAUFMAN and Edna Ferber was drawn from the personalities of the Drew family and the succeeding BARRYMORE generation; it was produced in London as *Theatre Royal* (1934) with Marie TEMPEST playing the formidable matriarch, a role taken in the New York production by Haidee Wright.

DRINKWATER, JOHN (1882–1937), English poet and dramatist, who was for some years actor and general manager at the BIRMINGHAM REPERTORY THEATRE, where most of his plays were first produced. The most successful of the early ones was *X=O; a Night of the Trojan War* (1917), but it was in *Abraham Lincoln* (1919) that he did his best work. Transferred to the LYRIC, Hammersmith, later the same year, it ran for over a year and in 1920 was seen in America, where Drinkwater, who had played Burnet Hook and later the title role in London, played the Chronicler. Of his later plays the most successful was a comedy, *Bird in Hand* (1927), seen at the ROYALTY THEATRE in London in 1929.

DROLL, a short comic sketch, usually a scene taken from a longer play. It originated in London during the Puritan interregnum (1642–60), when the actors, deprived of the right to act, of scenery, of costumes, and often of their playhouses, nevertheless managed to give a certain amount of entertainment. For their illicit purposes they invented the 'droll'—short for Droll Humours or Drolleries—rounding it off with dancing in the manner of the JIG. Some of the most famous drolls are taken from Shakespeare—'Bottom the Weaver' from *A Midsummer Night's Dream* and 'The Gravemakers' from *Hamlet*. Others were from biblical sources. The best-known player of drolls was Robert COX. Droll was also the name applied to early puppet shows, and was given to collections of humorous or satiric verse, as in *Westminster Drolleries* (1672). It was sometimes used to designate actors, particularly players of humorous parts, and men of quick wit and good company: Pepys uses it in this sense of KILLIGREW. In the late 19th century it was applied to such comediennes as Louie Freear (1871–1939), who was equally successful in minstrel shows, music-halls, musical comedy, and Shakespeare.

DROP, the early English name for CLOTH, still retained in America. In England the word is first found in about 1690 to indicate an unframed canvas BACKCLOTH which offered a plain surface for painting, free from the central join which marked the earlier 'pair of flats' (see FLAT). It was rolled up on a bottom roller, which ascended by means of lines, furling up the drop as it rose. Records show that swords, cloaks, and trains frequently got caught up in the ascending roller, to the detriment of dramatic dignity, and an alternative method known as 'tumbling' was then used. A BATTEN was fixed across the back of the drop a third of the way up, and it was taken away in bights, with a loose roller, or 'tumbler', inside to weight it and keep the bend straight. As it was not possible to have a practicable door or window in an unframed drop, it was later provided with battens at the back to which doors and windows could be fixed, thus making it, in effect, a single flat which could be flown. The lack of height above the stage made this impracticable in the early theatres.

DROTTNINGHOLM THEATRE AND MUSEUM, Sweden, part of a royal palace on an island near Stockholm, built in 1766. Until 1771 it was used during the summer by a French company which during the winter played in the capital, and it was also occasionally used by courtiers for amateur productions. With the palace it became State property in 1777 and had its most brilliant period during the reign of Gustaf III (1772–92), when a Frenchman, Louis-Jean Desprez, designed scenery and costumes for the productions given there. During the 19th century it fell into disuse, but its employment as a lumber-room saved it from demolition or modernization, and in 1921 it was restored under Dr Agne Beijer, the only alteration being the substitution of electric light for the former wax candles. The stage is about 57 ft deep and 27 ft wide at the footlights. The 18th-century stage machinery on the CARRIAGE-AND-FRAME system is still in working order, and there are more than 30 sets of usable scenery of the same period. The theatre is now used for short summer seasons of early opera. The museum exhibits, which include a rich deposit of 17th-century French stage designs, are housed in the former royal apartments.

DRUM-AND-SHAFT SYSTEM, an early method of moving theatre scenery by means of a rope fixed to a cylindrical shaft. A lever was inserted through the shaft and the whole twisted like a corkscrew, drawing the piece of scenery to the required position. An improvement on this method was achieved by the use of a circular drum, or BARREL, built round the shaft and considerably larger in diameter. The shaft was then rotated by pulling on a line wound round the drum, and in this way the piece was moved more easily and steadily. The system was used to move BORDERS, BRIDGES, CLOUDINGS, DROPS, and TRAPS. By using drums of different diameters on the same shaft several pieces of scenery could be moved at the same time, and by attaching them singly to different drums they could be made to move at different speeds. The Glory, a scenic effect frequently employed in Renaissance and Baroque spectacles in which a number of clustered cloudings gradually expanded into a great aureole, was controlled by the drum-and-shaft method. It is now obsolete, all pieces of flown scenery today being worked independently.

DRURIOLANUS, AUGUSTUS, see HARRIS, AUGUSTUS HENRY GLOSSOP.

DRURY LANE, Theatre Royal, London's most famous theatre, the present building being the fourth on the site. The first was erected by KILLIGREW under a charter granted by Charles II in 1662, making it one of London's two PATENT THEATRES. Seating about 700, it occupied a site 112 ft by 58 ft, and being hemmed in by other buildings could be reached only by a narrow passage from Brydges Street. It opened on 7 May 1663 (as the Theatre Royal, Brydges Street) with BEAUMONT and FLETCHER's *The Humorous Lieutenant*, the company including Charles HART and Michael MOHUN. In 1665 Nell GWYNN made her first appearance in DRYDEN's *The Indian Queen*, and from June 1665 to Nov. 1666 the theatre remained closed because of plague and the Great Fire. It was itself burnt down on 25 Jan. 1672. The second theatre on the site, known as the Theatre Royal in Drury Lane, is believed to have been designed by Sir Christopher Wren, though there is no documentary proof. It had a pit, amphitheatre, and two galleries, holding 2,000 people; the stage

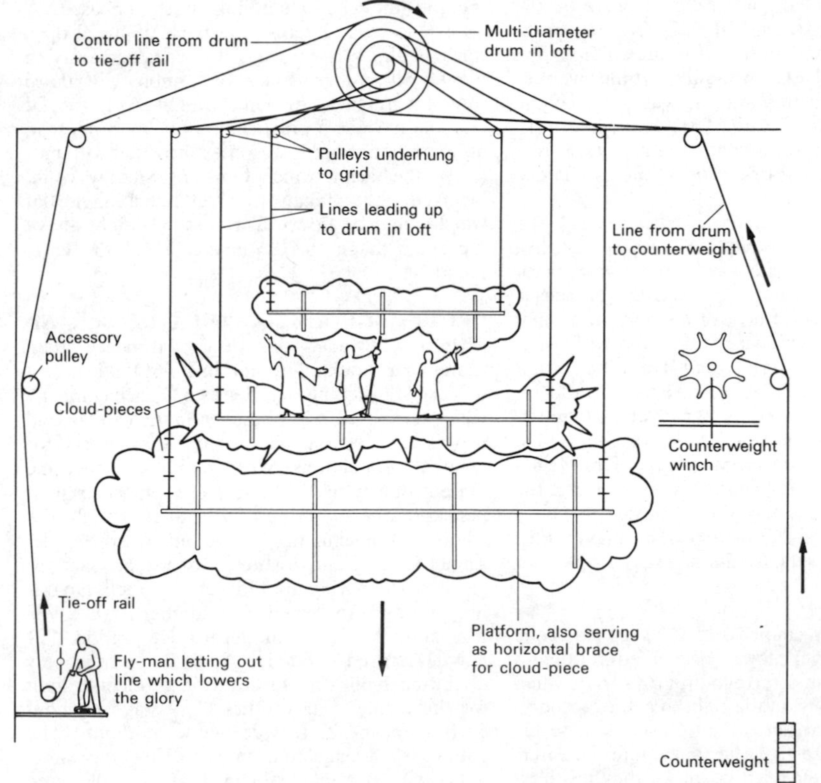

Schematic rear elevation of a Glory and its rigging (After Sonrel, *Traité de Scénographie*, p. 188)

depth from the front of the apron was no less than 130 ft. It opened on 26 Mar. 1674 with Fletcher's *The Beggar's Bush* and prospered for a while, but then lost its younger actors to the company at the more popular DORSET GARDEN THEATRE and was forced to close in 1676. By 1682 it was evident that London could support only one theatre, so a combined company under BETTERTON settled at Drury Lane. In 1690 the patent under which the theatre operated was bought by Christopher RICH; his clashes with some of the actors led to their departure to LINCOLN'S INN FIELDS, again headed by Betterton. Rich lost his charter in 1709, and the theatre closed. A triumvirate consisting of Colley CIBBER, Robert WILKS, and Thomas DOGGETT then took over, and with Anne OLDFIELD as their leading lady inaugurated a period of prosperity which lasted until 1733. Serious difficulties then arose between the theatre's management and the leaseholders, which were only resolved when Charles FLEETWOOD gained complete control. He ran into difficulties himself when he abolished the Footmen's GALLERY, an action which led to rioting on 5 May 1737, and being an inveterate gambler he soon plunged the theatre into debt. The one noteworthy event of this unfortunate period was MACKLIN's performance of Shylock in *The Merchant of Venice* on 11 Jan. 1741, in effect inaugurating the new school of interpretive acting which GARRICK was later to popularize. Garrick himself made his first appearance at Drury Lane on 11 May 1742, as Chaumont in OTWAY's *The Orphan*, and was then seen as Lear and Richard III. Five years later, with John LACY, he took over the theatre and rescued it from bankruptcy, instituting a number of reforms—regular rehearsals, careful casting, improved lighting, and more accurate texts, particularly of Shakespeare. He even managed, with some difficulty, to remove members of the audience from the stage. His first production, on 15 Sept. 1747, was again *The Merchant of Venice* with Macklin as Shylock; and with a company which included Kitty CLIVE, Peg WOFFINGTON, Mrs Cibber, Mrs PRITCHARD, Spranger BARRY, YATES, and SHUTER, he brought prosperity to the theatre for the next 30 years. He was responsible for major alterations to the building by Robert Adam in 1775, the year in which the 20-year-old Mrs SIDDONS made an unsuccessful début as Lady Anne to his Richard III. Garrick himself retired a year later, and was succeeded by R. B. SHERIDAN, whose first production was his own play *The School for Scandal* on 8 May 1777. In 1780 the theatre was damaged during the Gordon Riots, and a company of Guards was then posted there nightly until 1896. On 10 Oct. 1782 Mrs Siddons returned to play the title role in SOUTHERNE's *Isabella; or, the Fatal Marriage* with outstanding success, being joined the following year by her brother John Philip KEMBLE, to whom Sheridan, preoccupied with politics, handed over the effective management of the theatre in 1788 while still retaining his share of the patent. In 1791 the theatre was rebuilt by Holland, on a site 155 ft by 300 ft, its capacity being increased to 3,611. This third Drury Lane opened on 12 Mar. 1794 with a concert. Plays began on 21 Apr. with Kemble and Mrs Siddons in *Macbeth*, followed by an epilogue during which an iron safety curtain was lowered to prove that the theatre was now protected against fire. This device, however, did not prevent it burning down again on 24 Feb. 1809, after a stormy period during which Sheridan had so mismanaged affairs that Kemble had left for COVENT GARDEN, taking his sister with him, and their place had been taken by MELODRAMAS and spectacles which brought real elephants and performing dogs on to the Drury Lane stage. After the fire there were no funds available for rebuilding, but the situation was saved by the brewer Samuel Whitbread, a shareholder with Sheridan in the patent, who raised £400,000 to pay for the new building. This fourth Drury Lane, which still stands, is a four-tier house with a capacity of 2,283, a proscenium opening of 42 ft and a stage depth of 80 ft. It opened on 10 Oct. 1812 with a performance of *Hamlet*. On 26 Jan. 1814 Edmund KEAN made his first appearance at Drury Lane as Shylock, and continued to play there until 1820; but even his success could not keep pace with rising costs. Whitbread committed suicide in 1815, and ELLISTON, who took over in 1819 and under whose management the portico was added in 1820 and the interior reconstructed in 1822, finally went bankrupt. He was succeeded in 1826 by the American impresario Stephen PRICE, under whom Charles KEAN made his first, and GRIMALDI his last, appearance. From then onwards the history of Drury Lane was one of unmitigated artistic and financial disaster, apart from the brief reign of MACREADY, which lasted from 1841 to 1843. This unhappy period ended with the withdrawal in 1878 of the then manager, F. B. Chatterton, who from sad experience laid it down that 'Shakespeare spells ruin, and Byron bankruptcy.' The theatre remained closed until on 1 Nov. 1879 Sir Augustus HARRIS reopened it with a revival of *Henry V*, and embarked on a series of spectacular shows, realistic melodramas, and an annual PANTOMIME of great splendour in which Dan LENO—who first appeared at 'the Lane' in 1889—and Herbert Campbell played together for many years. Harris was succeeded on his death in 1896 by Arthur Collins, who was to remain in charge until 1923. He continued Harris's policy, staging explosions, earthquakes, avalanches, chariot races, shipwrecks, and, in *The Whip* (1909), a horse race with real horses; but he also had to his credit IRVING's last London season, Ellen TERRY's Jubilee (both 1905), and FORBES-ROBERTSON's farewell appearance (1913). In 1916 Shakespeare's Tercentenary was celebrated by a performance of *Julius Caesar* after which Frank BENSON, who played Caesar, was knighted by George V in the royal box with a property sword.

In 1921–2 the interior of the theatre was reconstructed. From 1924 to 1931 Alfred Butt presented a run of great MUSICALS including *Rose Marie* (1925), *The Desert Song* (1927), *Show Boat* (1928), and *New Moon* (1929). *The Land of Smiles* introduced Richard Tauber to London in 1931, and Ivor NOVELLO then occupied the theatre for several years with a series of highly successful musicals—*Glamorous Night* (1935), *Careless Rapture* (1936), *Crest of the Wave* (1937), and *The Dancing Years* (1939), which was still running when the theatres closed on the outbreak of the Second World War. Drury Lane then became the headquarters of the Entertainments National Service Association (ENSA), under Basil DEAN, reopening with Noël COWARD's *Pacific 1860* (1946), followed by a series of popular musicals by Rodgers and Hammerstein—*Oklahoma!* (1947), *Carousel* (1950), *South Pacific* (1951), and *The King and I* (1953). *My Fair Lady* (1958) ran for nearly five years, and the musicals *Camelot* (1964) and *Hello, Dolly!* (1965) were also successful. *The Great Waltz* (1970) ran for two years, being followed in 1972 by a musical version of *Gone With the Wind*, in 1973 by Anna Neagle in a revival of *No, No, Nanette!*, in 1974 by *Billy* with Michael Crawford. The musical *A Chorus Line* (1976) ran for over two and a half years, but the run of SONDHEIM's *Sweeney Todd* (1980) was disappointingly short.

DRURY THEATRE, Cleveland, see CLEVELAND PLAY HOUSE.

DRUTEN, JOHN VAN, see VAN DRUTEN.

DRYDEN, JOHN (1611–1700), English critic, poet, satirist, and one of the outstanding dramatists of the Restoration period, though his best work was done in other fields. He was responsible, alone or in collaboration, for some 30 plays of all kinds. The first, *The Wild Gallant* (1663), was seen at DRURY LANE, as was *The Rival Ladies* (1664), based on a Spanish original. His first HEROIC DRAMA, of which he was the chief exponent, was *The Indian Queen* (also 1664), written in collaboration with his brother-in-law Sir Robert Howard. Its sequel, *The Indian Emperor* (1665), was written by Dryden alone, and was followed by one of his most successful plays, *Secret Love; or, the Maiden Queen* (1667), based partly on Mlle de Scudéry's famous novel *Le Grand Cyrus*; the parts of Florimel and Celadon, played by Nell GWYNN and Charles HART, have sometimes been considered the prototypes of CONGREVE's Millamant and Mirabell in *The Way of the World* (1700). Later in 1667 LINCOLN'S INN FIELDS THEATRE saw the first production of an excellent comedy, *Sir Martin Mar-All; or, the Feign'd Innocence*, while at Drury Lane a year later came a comedy, *An Evening's Love; or, the Mock Astrologer*, which combined elements from CORNEILLE, MOLIÈRE, and QUINAULT. Another heroic drama, *Tyrannic Love; or, the Royal Martyr* (1669), was followed by Dryden's finest work in this style, *Almanzor and Almahide*, usually known by its subtitle *The Conquest of Granada*. This vast and complicated play, of which the first part was performed in 1670 and the second in 1671, contains all the elements, good and bad, of the heroic drama—rant, bombast, poetry, vigour, battle, murder, and sudden death. It was satirized unmercifully in BUCKINGHAM's *The Rehearsal* (1671) and the genre soon died a natural death. Dryden returned to comedy with *Marriage à la Mode* (also 1671) and *The Assignation; or, Love in a Nunnery* (1672), but produced two more heroic tragedies in *Amboyna; or, the Cruelties of the Dutch to the English Merchants* (1673) and *Aureng-Zebe* (1675). From the restraints of rhymed couplets he then returned to blank verse for the play which, in the opinion of posterity no less than in that of his contemporaries, was his masterpiece—*All For Love; or, the World Well Lost* (1677), a retelling of the story of Antony and Cleopatra which takes only its plot from Shakespeare: all the rest is Dryden's own. It is well constructed, contains some fine poetry, and observes more strictly than any other English tragedy the UNITIES of time, place, and action. It was frequently revived in the 18th century, and in 1922 was seen in London in a production by the PHOENIX SOCIETY which starred Edith EVANS and Ion Swinley. Dryden's later plays were less important; they included two comedies, *The Kind Keeper; or, Mr Limberham* (1678) and *The Spanish Friar: or, the Double Discovery* (1680), re-writings of *Troilus and Cressida* (1679) and PLAUTUS' *Amphitryon* (1690), and, his last play, *Love Triumphant; or, Nature Will Prevail* (1693), a tragicomedy produced at Drury Lane. Dryden also wrote a large number of prologues and epilogues, then very much in fashion, both for his own and for other people's plays, which are not only mines of information about theatrical matters but also notable contributions to English poetry. In his prefaces and critical writings he contributed largely to literary and theatrical controversies; but his most important contribution to the English theatre was undoubtedly the heroic drama, a genre which rose, flourished, and fell with him.

DRŽIĆ, MARIN, see YUGOSLAVIA.

DUBÉ, MARCEL (1930–), French-Canadian playwright, born in Montreal and educated there and in Paris. His early plays, which established him as a popular writer strongly influenced by contemporary American social realism, included *Le Bal triste* (1950), *De l'autre côté du mur* (1952), and *Zone* (1953), *Chambres à louer* and *Le Barrage* (both 1955), and *Un Simple Soldat* and *Le Temps des lilas* (both 1958). He was then responsible for the French adaptations of several modern American plays, among them Arthur MILLER's *Death of a Salesman* and *Two for the Seesaw* by William Gibson. His later works, in a more elegant and poetic style, include *Florence* (1961), *Les Beaux Dimanches* (1965), based on one of his own television plays, and a musical comedy, *Il est une saison* (also 1965). One of the few of his plays to be translated into English (as *The White Geese*) was *Au Retour des oies blanches* (1966), a near-classical tragedy which many consider his best work. It was followed by *Virginie, Pauvre Amour, Un Matin comme les autres* and *Manuel* (all 1968). Dubé was for several seasons resident playwright with the Théâtre du NOUVEAU MONDE.

DUBLIN ROSCIUS, see BROOKE, G.V.

DUBLIN THEATRE FESTIVAL, an annual event which takes place usually in October, when for about two weeks all the theatres of Dublin combine to participate in a festival of new plays, revivals, musicals, mime shows, and fringe events. The first festival, in 1957, was an ambitious affair, involving operas, concerts, ballet performances, and international folk-dancing, as well as theatrical events which included the THÉÂTRE NATIONAL POPULAIRE from Paris, and a visit from an English company in WILDE's *The Importance of Being Earnest* with Margaret RUTHERFORD as Lady Bracknell. At local level there were revivals of JOHNSTON's *The Old Lady Says 'No'!* at the GATE THEATRE; O'CASEY's *Juno and the Paycock* and SYNGE's *The Playboy of the Western World* at the ABBEY THEATRE; seven of YEATS's plays by the Globe Theatre company; and, at the PIKE THEATRE, a notorious production of Tennessee WILLIAMS's *The Rose Tattoo*. No festival was held in 1958, following a row over O'Casey's *The Drums of Father Ned* which resulted in O'Casey banning professional productions of his plays in Ireland, and from 1959 the festival narrowed in scope, concentrating mainly on Irish material and performers, though visits from such foreign companies as the LIVING THEATRE and the Black Theatre of Prague have continued to form an important part of the programme. Since 1960 the policy of encouraging new Irish writing has led to the discovery of many new playwrights, among them Brian FRIEL, John B. KEANE, and Hugh LEONARD.

DUCHESS THEATRE, London, in Catherine Street, off the Aldwych. This small theatre, seating 491 in two tiers, with a proscenium opening of 25 ft, opened on 25 Nov. 1929, and housed some early productions by the People's National Theatre company under Nancy PRICE. In 1934 J. B. PRIESTLEY, whose *Laburnum Grove* had had a successful run at the theatre the previous year, took it over and produced his own plays *Eden End* (1934) and *Cornelius* (1935). Emlyn WILLIAMS then appeared in his own thriller *Night Must Fall* (also 1935), which ran for a year and was followed by the first West End staging of T. S. ELIOT's *Murder in the Cathedral*. Another long-running Priestley play was *Time and the Conways* (1937). In 1938 Emlyn Williams returned in his own play *The Corn is Green*, which was still running when the theatre closed on the outbreak of war in Sept. 1939. It reopened shortly afterwards, but closed again until 1942, when COWARD's *Blithe Spirit*, first produced at the PICCADILLY THEATRE, began a long run. Later productions included Priestley's *The Linden Tree* (1947), RATTIGAN's *The Deep Blue Sea* (1952), William Douglas HOME's *The Manor of Northstead* (1954), and Ronald MILLAR's *The Bride and the Bachelor* (1956). Two plays by Agatha CHRISTIE, *The Unexpected Guest* (1958) and *Go Back for Murder* (1960), were seen there and Bill Naughton's *Alfie* transferred from the MERMAID in 1963. *The Reluctant Peer* by William Douglas Home began a long run in 1964 and Donald Howarth's *Three Months Gone* moved to the Duchess from the ROYAL COURT in 1970. The nude revue *Oh! Calcutta!* ran from 1974 to 1980 and was followed by a successful revival of Coward's *Private Lives* transferred from the GREENWICH THEATRE.

DUCIS, JEAN-FRANÇOIS (1733–1816), French dramatist, the first adapter of Shakespeare for the French stage. As a young man he served in the Seven Years War and returned home to live quietly and amuse himself with plays and poetry. His own plays were not successful and he consoled himself with those of Shakespeare, which he probably read in the deplorable translations of LAPLACE and LETOURNEUR since he almost certainly knew no English. Encouraged by the current French Anglomania, he adapted several plays for the COMÉDIE-FRANÇAISE, often so drastically that nothing remained but the title; but he knew the taste of his audience and realised that they would only accept Shakespeare with modifications. Beginning with *Hamlet* in 1769, in which MOLÉ appeared in the title role, he dealt in his own way with *Romeo and Juliet, King Lear, Macbeth*, and *Othello*, which, given in 1792, owed much of its success to the fine acting of TALMA, as did a revised version of *Hamlet* in 1804. In connection with Ducis' adaptation of *King John* a contemporary critic deplored his wasting his undoubted talents on such rubbish; none of his own plays has however survived.

DUCLOS [Marie-Anne de Châteauneuf], Mlle (1668–1748), French actress, who in 1693 was taken into the company at the COMÉDIE-FRANÇAISE to play tragic roles, in which she later replaced Mlle CHAMPMESLÉ, sharing feminine leads with Mlle DESMARES. A woman of ungovernable temper and strong passions (at 50 she married a young boy, who soon left her), she was unpopular among her fellow actors, venting her fury particularly on the young and lovely Adrienne LECOUVREUR when the latter was given some of her roles. Her strength lay in declamation and her acting was stiff and artificial; by the time she retired at the age of 72 it seemed sadly old-fashioned, compared to the freer methods of the younger generation.

DU CROISY [Philibert Gassot] (*c.* 1626–95), French actor, who with his wife, a mediocre actress who retired in 1665, joined MOLIÈRE's troupe at the PETIT-BOURBON in 1659, after having played in the provinces. Heavily built, with a noble paunch, he was essentially a character actor, and among the parts he played in Molière's plays were Covielle in *Le Bourgeois gentilhomme*, Harpin in *La Comtesse d'Escarbagnas*, Oronte in *Le Misanthrope*, and probably Vadius in *Les Femmes savantes*. He was the first actor to play Tartuffe, both in 1667, when the character was named Panulph, and as Tartuffe in 1669. His step-daughter Marie-Angélique, who was also in Molière's troupe, married the actor Paul POISSON.

DUCROW, ANDREW, see ASTLEY'S AMPHITHEATRE and EQUESTRIAN DRAMA.

DUFF, Mrs [*née* Mary Ann Dyke] (1794–1857), American actress, born in London. With her sister Elizabeth, later the wife of the Irish poet Tom Moore, she appeared in Dublin, and then married William Murray of the Theatre Royal, Edinburgh, brother-in-law of Mrs SIDDONS's son Henry. He died almost immediately, and she married an Irish actor, John Duff (1787–1831), with whom she went to America. She made a great reputation in Boston and Philadelphia, but was never wholly accepted by audiences in New York, where she made her first appearance in 1823 as Hermione in PHILIPS's *The Distrest Mother* to the Orestes of the elder BOOTH. A tall, dark, graceful woman, she was at her best in tragic or pathetic parts. The early death of her second husband left her with seven small children, and in a moment of financial stress she married an American actor, Charles Young (?–1874). The marriage was never consummated and was annulled. In 1835 she took as her fourth husband a lawyer named Seaver from New Orleans, where she appeared a year later as Jane Shore in ROWE's tragedy and as Portia in *The Merchant of Venice*. Although she then announced her retirement she is noted as having appeared at the Royal Lyceum in Toronto as late as the summer of 1850, billed as 'the American Siddons'.

DUFRESNE (or DEFRESNE) [Abraham-Alexis Quinault] (1693–1767), French actor, sometimes known as Quinault-Dufresne. With his wife, Mlle Deseine [Catherine-Marie-Jeanne Dupré] (1705–67), he was a member of the COMÉDIE-FRANÇAISE, as were his father, his elder brother, and his three sisters, the eldest, who died young, acting as Mlle de Nesle, and the youngest, known as Quinault la Cadette, sharing soubrette roles with Mlle DANGEVILLE. Dufresne made his formal début in 1712, playing Oreste in CRÉBILLON's *Électre*, and was engaged to play young heroes and lovers. He was extremely handsome and, as a contemporary said of him, had the good fortune to please the ladies in an age when they took little trouble to fight against their inclinations. At first the simplicity of his acting was against him, but he came into his own after the retirement of BEAUBOUR in 1718. He had a good voice and a fine presence, reminding many in the audience of BARON, whose traditions he had inherited through the teaching of Ponteuil. Dufresne was the first to play the name-part in VOLTAIRE's *Œdipe* (1718) and DESTOUCHES wrote for him the comedy *Le Glorieux* (1732) in which he hardly had to act, so completely was he the person Destouches was satirizing. His wife, who had great charm and a natural style of acting, suffered much from ill health and finally retired in 1736, four years before her husband.

DUFRESNE, CHARLES (*c.* 1611–*c.* 1684), French actor-manager who had been leader of a provincial touring company for some years when in about 1644–5 he was joined by the remnants of MOLIÈRE's ill-starred ILLUSTRE-THÉÂTRE company, which included Madeleine BÉJART with her sister and two brothers. Dufresne soon ceded his leadership to Molière, but remained with the company and returned with it to Paris in 1658, retiring a year later. Even before being joined by Molière his company had been considered one of the best in France.

DUFRESNY, CHARLES-RIVIÈRE (1654–1724), French dramatist, reputed to be a great-grandson of Henri IV. He began by writing for the Italian actors established at the Hôtel de BOURGOGNE, and after their departure from Paris in 1697 turned his talents to the service of the COMÉDIE-FRANÇAISE. Of his numerous plays the best was the one-act *Esprit de contradiction* (1700). Others, successful when first produced, were *Le Double veuvage* (1702), *La Coquette du village* (1715), and *Le Mariage fait et rompu* (1721), all of which kept their place in the repertory for some time. Dufresny was very conscious of the weight of tradition in comedy and made some effort to shake it off, as may be seen from his prologue to *Le Négligent* (1692), in which he complains that a good comic writer is blamed for copying MOLIÈRE and a bad one for not doing so. He was, however, too indolent to produce work of lasting value.

DUGAZON [Jean-Baptiste-Henri Gourgaud] (1746–1809), French actor, who became a member of the COMÉDIE-FRANÇAISE in 1771 and proved himself an excellent comedian, at his best in farcical roles, particularly in revivals of plays by SCARRON. In 1786 he joined the staff of the newly founded School of Declamation, which in 1793 became the Conservatoire. One of his pupils there was TALMA, whom he later supported in the upheavals of the Revolution, joining him at the Comédie-Française again when it was reconstituted under Napoleon. He was the brother of the actresses Françoise VESTRIS and Marie-Marguerite Gourgaud (1742–99), who played soubrette roles at the Comédie-Française from 1767 onwards.

DUKE OF YORK'S THEATRE, London, in St Martin's Lane, built for Frank Wyatt and his wife the actress Violet Melnotte. As the Trafalgar Square Theatre, seating 900 in three tiers, it opened on 10 Sept. 1892, and in 1893 saw the first performances in England of IBSEN's *The Master Builder*, with Elizabeth ROBINS. In 1895 the theatre took its present name, and in 1896 it had its first success with the musical comedy *The Gay Parisienne*, in which the quaint DROLL Louie Freear (1871–1939), who had been on the stage since early childhood, first made a name for herself. In 1897 Charles FROHMAN began a successful tenancy, introducing many well known American actors to London, among them Maxine ELLIOTT in 1899. 'Dot' BOUCICAULT was appointed resident manager in 1901, and under him several plays by BARRIE had their first production, including *The Admirable Crichton* (1902), *Peter Pan* (1904), *Alice Sit-By-The-Fire* (1905), and *What Every Woman Knows* (1908). In 1910 Frohman tried to introduce a REPERTORY system to London with an outstanding programme of modern plays which included GALSWORTHY's *Justice*, SHAW's *Misalliance*, and GRANVILLE-BARKER's *The Madras House*, but without success, and the theatre reverted to straight runs, Somerset MAUGHAM's *The Land of Promise* (1914) being the last new play put on by Frohman before he died. Two later successes at this theatre were Jean Webster's *Daddy Long-Legs* (1916) and Lady Arthur Lever's *Brown Sugar* (1920). In 1923 Violet Melnotte took control, her first success being Charlot's REVUE *London Calling*, written mostly by Noël COWARD whose *Easy Virtue* was produced there three years later. In 1929 Peggy ASHCROFT made a considerable impact in a small part, Naemi, in *Jew Süss*, adapted by Ashley DUKES from FEUCHTWANGER's novel for Matheson LANG, who had made a success at this theatre a year previously with *Such Men Are Dangerous* (1928) adapted, again by Dukes, from a German play by Alfred Neumann. In 1931 VAN DRUTEN's *London Wall* had a successful run, followed by a season of non-stop Grand Guignol, and from 1933 to 1936 Nancy PRICE's People's National Theatre used the theatre occasionally, but up to the outbreak of the Second World War little of any importance was presented. Violet Melnotte died in 1935, and various managements then came and went. As a result of enemy action the theatre closed in 1940, reopening in May 1943 with Paul Vincent CARROLL's *Shadow and Substance*. Since then it has had a fairly stable career, numbering among its productions Roland Pertwee's *Is Your Honeymoon Really Necessary?* (1944), with Ralph LYNN, *The Happy Marriage* (1952), with John CLEMENTS (who adapted it from a French play) and Kay Hammond, Hugh Mills's *The House by the Lake* (1956), with Flora ROBSON, and *A Scent of Flowers* (1964) by James SAUNDERS. Further successes were scored by Frank Marcus's *The Killing of Sister George* (1965) from BRISTOL OLD VIC, Alan AYCKBOURN's *Relatively Speaking* (1967), John OSBORNE's *Time Present* (1968), from the ROYAL COURT, and Arthur MILLER's *The Price* (1969). In 1978 there were two successful transfers, of Julian Mitchell's *Half-Life*, with John GIELGUD, from the COTTESLOE THEATRE, and Michael FRAYN's *Clouds*, seen at the HAMPSTEAD THEATRE CLUB in 1976. In Nov. 1978 the theatre was bought by Capital Radio, and major structural changes were made, including the replacement of the gallery by a recording and broadcasting studio. The theatre reopened in 1980 with two highly popular productions, Andrew Davies's *Rose*, with Glenda JACKSON, originally produced at the BELGRADE THEATRE, Coventry, and Tom Kempinski's *Duet For One*, transferred from a FRINGE THEATRE, the Bush.

DUKES, ASHLEY (1885–1959), English dramatist, theatre manager, and dramatic critic, in which last capacity he worked for *Vanity Fair*, the *Star*, and the *Illustrated Sporting and Dramatic News*. He was also for many years English editor of the American *Theatre Arts Monthly*. He had a wide knowledge of modern Continental drama and adapted a number of French and German plays for the London stage, among them Neumann's *Der Patriot* as *Such Men Are Dangerous* (1928) and FEUCHTWANGER's *Jew Süss* (1929), both for Matheson LANG. His very free version of an episode from Fernando de ROJAS's *La Celestina* as *The Match-maker's Arms* (1930) provided an excellent part for Sybil THORNDIKE, and Vivian LEIGH made her first outstanding appearance in his *The Mask of Virtue* (1935), based on STERNHEIM's *Die Marquise von Arcis*. In 1933 he opened the MERCURY THEATRE where he did excellent work with productions of new and foreign plays, particularly verse plays. His translation of MACHIAVELLI's *Mandragola* was seen there in 1939. Among his own plays the most successful was *The Man With a Load of Mischief* (1924), first produced by the STAGE SOCIETY, in which Fay COMPTON made a great success as the Lady when it was revived at the HAYMARKET THEATRE a year later.

DUKE'S HOUSE, DUKE OF YORK'S HOUSE, DUKE'S MEN, London, see DORSET GARDEN THEATRE and LINCOLN'S INN FIELDS THEATRE.

DUKE'S THEATRE, London, see HOLBORN THEATRE.

DULLIN, CHARLES (1885–1949), French actor and producer, a pupil of GÉMIER, who, after some appearances in melodrama, joined COPEAU when he first opened the VIEUX-COLOMBIER. In 1919 Dullin formed his own company and took it on a long provincial tour. Back in Paris, confronted by many difficulties and always short of money, he finally succeeded in establishing the company in the Théâtre de l'ATELIER, which soon gained a great reputation as one of the outstanding experimental theatres of Paris. The list of plays produced there covers the classics of France, the comedies of ARISTOPHANES, translations of famous foreign plays, among them CALDERÓN's *La Vida es sueño* (1922), Shakespeare, Ben JONSON, PIRANDELLO for the first time in France, and such new French plays as COCTEAU's *Antigone* (also 1922). Himself an excellent actor, Dullin ran a school of acting connected with his theatre, and in 1936 was invited to become one of the directors at the COMÉDIE-FRANÇAISE. During the occupation of France he toured the unoccupied zone with MOLIÈRE's *L'Avare*, and in 1943 he was responsible for the first production of *Les Mouches by* SARTRE.

DUMAS [Davy de la Pailleterie], ALEXANDRE, *père* (1802–70), French novelist and playwright, of Creole parentage, Dumas being the name of his West Indian grandmother. He is now chiefly remembered for his novels, but his dramas played an important part in the French Romantic movement, his historical drama *Henri III et sa cour* (1829) being the first triumph of the Romantic theatre. In the prevailing torpor of the French stage the colour and movement of Dumas's play delighted the audience and brought him the friendship and admiration of Alfred de VIGNY and Victor HUGO. A play on Napoleon was followed by *Antony* (1831), seen at the Théâtre de la PORTE-SAINT-MARTIN, where several of Dumas's more melodramatic pieces were first produced, including the famous *La Tour de Nesle* (1832), which for terror and rapidity of action, not to mention the number of corpses, surpassed anything seen on the French stage since the days of Alexandre HARDY. In *Antony* Dumas had presented a plea for one type of misfit in contemporary society, the illegitimate child; in *Kean, ou Désordre et génie* (1836), based on the career of the great English actor who had died three years earlier, he portrayed another, the man of genius, alternately idolized and ostracized by his contemporaries. With FRÉDÉRICK in the title role, this play had a tremendous success (as did an adaptation of it by SARTRE in 1953, intended for another flamboyant, extrovert actor, Pierre BRASSEUR). Dumas then proceeded, alone or in collaboration, to dramatize his own historical novels, with varying success. Some of them were produced at the Théâtre Historique, which he built and financed himself and which nearly ruined him when it failed in 1850; for in spite of his enormous earnings, Dumas was extravagant and an easy prey to harpies and hangers-on. He was eventually rescued from his creditors by his daughter, who, with the help of her brother Alexandre (below), took over the management of her father's affairs in 1868 and remained with him until his death.

DUMAS, ALEXANDRE, *fils* (1824–95), French dramatist, natural son of the above, who approached the theatre by way of a dramatization of his own novel *La Dame aux camélias* (1848). First acted in 1852, this became one of the outstanding theatrical successes of the second half of the 19th century, and is still occasionally revived. It was equally popular in England, America, and Italy, where it was originally known as *Camille*. In spite of its success it was destined to remain the younger Dumas's only Romantic play; he turned to social drama and, though himself an agnostic, sought to enforce Christian virtues and conventional morality by using the stage as a pulpit as in *Les Idées de Madame Aubray* (1867). Dumas had little liking for the bohemian society in which his childhood had been passed and to which he gave a permanent label in the title of his play *Le Demi-Monde* (1855). The bitterness of his own illegitimacy found expression in *Le Fils naturel* (1858) and *Un Père prodigue* (1859), while the question of sexual morality was ventilated in such plays as *La Femme de Claude* (1873), *L'Étrangère* (1873), and his last play *Francillon* (1887). Only occasionally, as in *La Question d'argent* (1857), did he deal with social issues of a more general scope, and much of his theatre has disappeared with the conditions which gave rise to it. In his own day a popular and powerful social dramatist, he is now remembered only by his least typical work, mainly because the consumptive and pathetic figure of his heroine Marguerite Gautier offers as fine a part for an ambitious and passionate actress as does her counterpart Violetta for a singer in *La Traviata* (1853), the opera which Verdi based on the play.

DU MAURIER, GERALD HUBERT EDWARD (1871–1934), English actor-manager, knighted in 1923. He was first seen on the stage in London in 1894 and a year later was with TREE at the HAYMARKET THEATRE playing a small part in *Trilby*, a dramatization of the novel by his father George Du Maurier (1834–96). Gerald remained with Tree for some years, first coming into prominence when he went to the DUKE OF YORK'S THEATRE, under the management of Charles FROHMAN, to appear in BARRIE's *The Admirable Crichton* (1902). In the cast was Muriel Beaumont (1881–1957), whom he married shortly afterwards. He was the first to play Mr Darling and Captain Hook in Barrie's *Peter Pan* (1904), and made a great success at the COMEDY THEATRE in E. W. Hornung's *Raffles* (1906), following it with *Arsène Lupin* in 1909, in which year he directed a production of *An Englishman's Home* by his elder brother Guy (1865–1916) who was killed in the First World War. Produced anonymously, this was a stirring patriotic play which stimulated Territorial Army recruitment. In 1910 Du Maurier again scored an immense success with Paul ARMSTRONG's *Alias Jimmy Valentine*, and then took over the management of WYNDHAM'S THEATRE in partnership with Frank Curzon, making the theatre a home of light comedy for many years. His range of parts was limited; but within those limits he was seldom excelled, and he could on occasion step beyond them, as was proved by his portrayal of Mr Dearth in Barrie's *Dear Brutus* (1917). A more typical part, and one which brought him solid success, was the title role in 'Sapper's *Bull-Dog Drummond* (1921). Shortly afterwards he left Wyndham's and went to the ST JAMES'S THEATRE, where he made his last outstanding appearances in LONSDALE's *The Last of Mrs Cheyney* (1925) and Roland Pertwee's *Interference* (1927). His daughter Daphne (1907–), best known as a novelist, is also the author of three plays, *Rebecca* (1940), based on one of her own novels, *The Years Between* (1945), and *September Tide* (1949).

DUMB BALLET, see TRICKWORK.

DUMESNIL, MARIE-FRANÇOISE (1713–1803), French actress, who joined the COMÉDIE-FRANÇAISE in 1737 and soon showed herself an excellent actress, particularly in passionate roles. She was considered by VOLTAIRE superior even to Adrienne LECOUVREUR in high pathos, and he attributed much of the success of his *Mérope* (1743) to her acting. Unlike her contemporary and rival Mlle CLAIRON, she took no interest in the reform of theatrical costume and aimed at magnificence rather than correctness, being always robed in rich stuffs made in contemporary styles and loaded with jewels. She retired in 1776 but remained in full possession of her faculties until the end of her long life, and was able to pass on to younger players traditions temporarily lost during the upheavals of the French Revolution.

DUNLAP, WILLIAM (1766–1839), American dramatist and theatre manager, the dominating force of the American stage from 1790 to 1810. He went to England in 1784 to study art under Benjamin West, but neglected his work for the theatre, going to see most of Shakespeare's plays, a good many contemporary comedies, and such actors as Charles KEMBLE and Mrs SIDDONS. Late in 1787 he returned to the United States and, inspired by the success of TYLER's *The Contrast*, wrote a comedy for the AMERICAN COMPANY which was accepted but never acted, mainly because of the lack of suitable parts for the manager John HENRY and his wife. A second comedy, *The Father; or, American Shandyism*, was produced at the JOHN STREET THEATRE, New York, on 7 Sept. 1789. Dunlap continued to write for the American Company, and in 1796 became one of its managers in partnership with the younger HALLAM and HODGKINSON, strengthening the company by the inclusion of the first Joseph JEFFERSON. In 1797 Hallam withdrew from active management, and Hodgkinson and Dunlap opened the first PARK THEATRE on 29 Jan. 1798 with *As You Like It.* One of the first plays staged there was *André*, a tragedy which Dunlap based on an incident in the War of Independence—the first native tragedy on American material. Hodgkinson played André, and the part of his friend Bland was taken by Thomas Abthorpe COOPER, who was to succeed Dunlap as lessee and manager of the theatre.

In 1798 Hodgkinson left the Park and Dunlap continued on his own, producing adaptations of French and German plays, mainly those of KOTZEBUE of which the most successful were *The Stranger, Lovers' Vows, The Wild Goose Chase, The Virgin of the Sun*, and its sequel *Pizarro*, as well as some of his own plays, among them *Leicester* and *The Italian Father*. Many of them were performed also in Boston, where Dunlap had leased the Haymarket Theatre, and in Philadelphia under Warren and Wood at the CHESTNUT STREET THEATRE. Meanwhile Dunlap struggled on at the Park, hampered by temperamental actors and recurrent epidemics of yellow fever, until in Feb. 1805 he went bankrupt. A year later he became assistant stage manager at the Park under Cooper, in which capacity he engaged in 1809 the parents of Edgar Allan Poe for parts in M. G. LEWIS's *The Castle Spectre*. In 1812 he accompanied George Frederick COOKE on his American tour, and then retired from the theatre to devote himself to literature and painting. He planned the publication of his plays in a uniform edition but produced only the first volume; he also published an invaluable *History of the American Theatre* (1832).

DUNLOP, FRANK (1927–), English director, who had already had a good deal of experience in England and abroad when in 1961 he became director of the NOTTINGHAM PLAYHOUSE where he remained until 1964, seeing the new theatre building through its first season. In 1966 he directed the Pop Theatre in *The Winter's Tale* and EURIPIDES' *Trojan Women* for the EDINBURGH FESTIVAL, and in the following year he joined the NATIONAL THEATRE company in London, where his early productions included BRECHT's *Edward II,* MAUGHAM's *Home and Beauty* (both 1968), and WEBSTER's *The White Devil* (1969). In 1970 he founded the YOUNG VIC, initially as part of the National Theatre, and directed its first production *Scapino* (N.Y., 1974), which he adapted with Jim Dale, who played the title role, from MOLIÈRE's *Les Fourberies de Scapin*. Among his other productions for the company were *The Comedy of Errors* in 1971, GENET's *The Maids* in 1972, and the musical *Joseph and the Amazing Technicolor Dreamcoat* (also 1972; N.Y., 1976). When the Young Vic became independent in 1974 he continued as Director. He has also worked extensively freelance, staging the ROYAL SHAKESPEARE COMPANY's revival of William GILLETTE's *Sherlock Holmes* in 1974, which he also directed in New York, other productions there including Alan BENNETT's *Habeas Corpus* in 1975. In 1980 he directed a revival of the musical *Camelot* and returned to the Young Vic to direct *King Lear*.

DUNNOCK, MILDRED (1900–), American actress, formerly a schoolteacher, who after studying with Lee STRASBERG and Elia KAZAN made her New York début in 1932. From 1937 she was seen in a variety of roles including Queen Margaret in *Richard III* in 1943 and Lavinia Hubbard in Lillian HELLMAN's *Another Part of the Forest* (1946) before scoring a big success as Linda Loman in Arthur MILLER's *Death of a Salesman* (1949). This was followed by major roles in other important plays such as Gina in IBSEN's *The Wild Duck* in 1951 and Big Mama in Tennessee WILLIAMS's *Cat on a Hot Tin Roof* (1955). In 1962 she appeared at the Festival of Two Worlds in Spoleto, Italy, in another Tennessee Williams play, *The Milk Train Doesn't Stop Here Anymore*, and played the same part in New York in 1963, a process that was repeated with the role of Hecuba in EURIPIDES' *Trojan Women* also in 1963. Madame Renaud in ANOUILH's *Traveller Without Luggage* followed in 1964. She continued to make regular stage appearances in her seventies, including leading roles in Marguerite DURAS's *A Place Without Doors* (1970) and *Days in the Trees* (1976).

DUNSANY, EDWARD JOHN MORETON DRAX PLUNKETT, Lord (1878–1957), Irish man of letters, who was connected with the early years of the ABBEY THEATRE, where his first plays were produced—*The Glittering Gate* (1909) and *King Argimenes* (1911). These were also seen in London, as were *The Gods of the Mountain* (1911) and *The Golden Doom* (1912), both at the HAYMARKET. Most of his later plays were seen only in Ireland or in amateur productions, but *If* had a long run at the AMBASSADORS in 1921. A versatile writer, his plays range from one-act farces (*Cheezo*), fantasy (*The Old King's Tale*), and satire (*The Lost Silk Hat*) to full-length tragedy (*Alexander*) and comedy (*Mr Faithful*).

DUNVILLE [Wallen], THOMAS EDWARD (1868–1924), an eccentric comedian who as a young man came to London with a small provincial troupe of acrobatic entertainers. He made his first solo appearance in 1889, and was seen at GATTI's-Under-the-Arches and at the MIDDLESEX MUSIC-HALL with immediate success, soon becoming well known for his short comic songs, consisting of telegraphic phrases delivered in an explosive manner, in the style of 'And the Verdict Was'—'Little Boy, Pair of Skates, Broken Ice, Heaven's Gates.' For many years he was top of the bill wherever he appeared, wearing a long black coat which accentuated his height, with a white Puritan collar, a small bowler hat, baggy trousers, and a Dutch-doll wig over a red-nosed face. When he tried to vary his appearance audiences forced him to return to his usual style. The posters announcing his arrival in a town read 'I'm Sticking Here for a Week', and showed him spreadeagled, suspended from the wall. When in the early 1920s music-hall began to lose its appeal he became very depressed and, after overhearing himself referred to as a fallen star, he drowned himself in the Thames.

DUODRAMA, see MONODRAMA.

DU PARC [Marquise-Thérèse de Gorla], Mlle (1633–68), French actress, who had already had some experience in a travelling company before she married in 1653 a comic actor, René Berthelot (*c.* 1630–64), known as Du Parc and, from his size, as Gros-René. He had been in MOLIÈRE's company since 1648 and his wife now joined it, both going to Paris with Molière in 1658 and playing small parts in his productions. After her husband's death Mlle Du Parc, a woman of great beauty and majestic presence far better in tragedy than in comedy, in which Molière's troupe specialized, left for the Hôtel de BOURGOGNE, the home of tragedy, where she played the title role in RACINE's *Andromaque* (1667). She died suddenly, probably in childbirth; and Racine, whose mistress she was at the time, was later accused by the infamous Cathérine Voisin of having poisoned her to make way for Mlle CHAMPMESLÉ.

DURANTE, JIMMY, see VAUDEVILLE, AMERICAN.

DURAS, MARGUERITE (1914–), French dramatist, film-maker, and novelist, whose plays belong basically to the Theatre of the ABSURD but temper the vision of life as fundamentally ridiculous by means of a surface realism. Her first play *Le Square* (1956) dealt with the impossibility of direct communication between human beings. In *Les Viaducs de la Seine-et-Oise* (1963) and *L'Amante anglaise* (1968) she dramatized the same true murder story (in which two old people throw portions of a dismembered body into the waggons of passing goods trains) apparently fascinated by its combination of fastidiousness and inconsequence. As *The Viaduct*, the first of these was seen at the Yvonne Arnaud Theatre in 1968, with Sybil THORNDIKE and Max ADRIAN as the old couple. The second version, as *A Place Without Doors*, was seen in New York in 1970 and, as *The Lovers of Viorne*, in London in 1971, starring Mildred DUNNOCK and Peggy ASHCROFT respectively. The original version was followed in 1965 by *Les Eaux et les Fôrets*, a bitter comedy on the theme of ingratitude, and by a study of divorce, *La Musica*, which with *Le Square* was seen briefly in London in 1966. In that year Duras's first full-length play, *Des Journées entières dans les arbres* (also 1965), an account of a curious, abortive reconciliation between an estranged mother and son, originally produced by BARRAULT at the ODÉON, was staged by the ROYAL SHAKESPEARE COMPANY at the ALDWYCH as *Days in the Trees* with Peggy Ashcroft as the Mother, played in New York in 1976 by Mildred Dunnock. In 1973 the R.S.C. also put on *Suzanna Andler*, which portrays the anguish of a woman, married for 17 years, as she debates whether to continue an affair. The play's presentation was due largely to the advocacy of Eileen ATKINS, who gave a superb performance in the title role. Duras's later plays, not produced outside France, include *L'Eden cinéma* (1977) with Madeleine RENAUD, who had also starred in *Des Journées entières dans les arbres*.

D'URFEY, THOMAS (1653–1723), Restoration dramatist and songwriter, author of a number of plays, mainly based on the works of earlier English or foreign dramatists, none of which has survived on the stage. The earlier ones are purely farcical, but later ones are tinged with the sentimentality which was soon to bulk so large in English drama. D'Urfey was one of the writers most savagely attacked for indecency by Jeremy COLLIER in his *Short View of the Immorality and Profaneness of the English Stage*, and in 1698 he was prosecuted for profanity.

DÜRRENMATT, FRIEDRICH (1921–), Swiss dramatist, whose work, influenced by BRECHT, WEDEKIND, and EXPRESSIONISM, has nevertheless an unmistakable flavour of its own, with its mixture of the grotesque and the macabre and its clever use of modern dramatic techniques. He first came into prominence with *Es steht geschrieben* (*All As It Is Written*, 1947) which was revised in 1968 under the title of *Die Wiedertäufer* (*The Anabaptists*); but his first outstanding success was a mock-heroic comedy *Romulus der Grosse* (1949), set in a single day, the Ides of March AD 476: the Emperor calmly receives the news of the barbarians' final victory and, deserted by his men, goes out alone to give himself up to the invader. Paradoxically, those who deserted him are killed while he is condemned to live. *Die Ehe des Herrn Mississippi* (1952), seen briefly in London in 1959 as *The Marriage of Mr Mississippi*, and *Ein Engel kommt nach Babylon* (1953) made little impact outside the German-speaking countries, but he won international fame with *Der Besuch der alten Dame* (1956), which as *The Visit* was seen widely in America in 1958 and in England in 1960, in a production by Peter BROOK with Lynn Fontanne and Alfred LUNT. The story concerns a fabulously wealthy woman who returns to her impoverished native town, from which she had been driven out in disgrace as a girl, and offers the citizens a vast sum of money if they agree to kill her seducer. After some argument, they yield to temptation and the murder is committed. After *Frank V*, a musical play which was not a success, Dürrenmatt achieved another world-wide success with *Die Physiker* (1962), seen in London in 1963 (N.Y., 1964) as *The Physicists* again directed by Peter Brook. The scene is a lunatic asylum in which a nuclear physicist has taken refuge after burning his papers to prevent his researches being used for destructive purposes. Two other inmates, also posing as lunatics, are secret agents trying to kidnap him on behalf of their respective governments. Eventually all three men decide to stay in the asylum, but the woman doctor in charge has obtained copies of the papers and intends to exploit them: the fatal progress of science cannot be reversed.

Dürrenmatt's plays since *Die Physiker* have been less successful. *Der Meteor* (1966) shows the impact on the living of a man who has risen from the dead, and *Porträt eines Planeten* (1971) presents a synopsis of man's life on earth from its beginnings to its destruction by a cosmic catastrophe. Dürrenmatt has also adapted several plays, among them STRINDBERG'S *The Dance of Death* under the title *Play Strindberg* (1969), seen in New York in 1971 and London in 1973, and Shakespeare's *King John* and *Titus Andronicus*. Although he has written in his essay 'Theaterprobleme' (1955) that our disintegrating world is a subject for comedy rather than tragedy, all his plays are fundamentally pessimistic, reflecting the anxieties of their time.

DU RYER, PIERRE (*c.* 1600–58), French dramatist, whose earliest plays were TRAGI-COMEDIES in the style of HARDY, three being presented in the year 1628–9. They were spectacular, calling for elaborate staging in the old-fashioned MULTIPLE SETTING, and ignored the UNITIES of time and place. Several comedies followed, containing good parts for the comedian GROS-GUILLAUME, of which the best was *Les Vendanges de Suresne* (1633). The most important of several tragedies, which included some on Biblical subjects, was *Scévole* (1644). This was produced at the Hôtel de BOURGOGNE, as were most of Du Ryer's plays; it was in the repertory of MOLIÈRE'S short-lived ILLUSTRE-THÉÂTRE, was produced by him several times in the provinces before he went to Paris, and remained in the repertory of the COMÉDIE-FRANÇAISE for over 100 years. Du Ryer did more than anyone except MAIRET and CORNEILLE to establish French classical tragedy.

DUSE, ELEONORA (1858–1924), Italian actress, born into a theatrical family. She was on the stage at the age of 4, and at 14 appeared as Juliet in Shakespeare's *Romeo and Juliet*, a performance which D'ANNUNZIO, working on her recollections of it, incorporated into his novel *Il fuoco*. As an adult actress her first notable success came in 1879 when she was seen in ZOLA'S *Thérèse Raquin*. Equally successful was her performance as Santuzza in VERGA'S *Cavalleria rusticana* in 1884, a challenging and rewarding role which reinforced her growing interest in modern drama. In 1885 she made her first foreign tour, in Latin America, and on her return to Italy founded her own company, the Città di Roma. During a tour of Russia in 1891 CHEKHOV saw her as Shakespeare's Cleopatra and was captivated by her acting, which was a revelation to him. It has been suggested that he had her in mind while writing the part of Madame Arkadina in *The Seagull*, and some of the enthusiasm which led to the founding of the MOSCOW ART THEATRE may also have been generated by her.

She made the first of four visits to the United States in 1893 and first visited England in 1894, where on 18 May she played Mirandolina in GOLDONI's *La locandiera* in a command performance at Windsor before Queen Victoria. A year later she and Sarah BERNHARDT were both in London playing in SUDERMANN's *Heimat* (known in English as *Magda*), which gave the London critics the opportunity of comparing their styles, SHAW vigorously championing Duse while Clement SCOTT much preferred Bernhardt. Their rivalry was renewed in Paris in 1897. From her late thirties Duse devoted much of her time and money to promoting the plays of D'Annunzio, who first came into prominence as a dramatist when in 1898 she appeared in his *La Gioconda* and *La città morte*. She retired in 1909 but financial difficulties forced her back to the stage, and she reappeared in Turin in 1921 as Ellida in IBSEN's *The Lady from the Sea*. Two years later she started on her last international tour and died in Pittsburgh on 21 Apr. 1924; her body was returned to Italy and buried at Asolo. A slender, graceful woman, with dark eyes and expressive, mobile features, melancholy in repose, she was noted for the beauty of her gestures. She had a statuesque way of playing, a slowness and subtlety which was not always to the taste of an audience that preferred a more flamboyant manner, and her excessively nervous and overwrought temperament led at times to too much restlessness on stage; her rare moments of immobility revealed her greatness. There is abundant evidence of her dedication and of the sustained application that went into the preparation of her roles; above all, her technical expertise and electrifying presence enabled her to transcend the handicap of a somewhat frail physique. She was noted for her refusal to wear make-up on stage, having apparently the ability to blush or turn pale at will. She was much admired in such big emotional parts as SARDOU's Fédora, Tosca, or Théodora, and Camille in the younger DUMAS's *La Dame aux camélias*, and was outstanding in Ibsen —particularly as Hedda Gabler, and as Rebecca West in *Rosmersholm* in the production which Gordon CRAIG designed for her in Florence in 1906.

DUST HOLE, London, see SCALA THEATRE.

DVOŘÁK, ARNOŠT, see CZECHOSLOVAKIA.

DYBWAD, JOHANNE (1867–1950), Norwegian actress who made her first appearance in 1887 in Bergen. After the opening of Norway's NATIONALTHEATRET in 1899 she became its leading actress, a position she retained for the next 40 years. She appeared frequently in other Scandinavian countries, and also in Hamburg, Berlin, and Paris, and was particularly esteemed for her roles in IBSEN, which included Solveig and Aase (both in *Peer Gynt*), Nora in *A Doll's House*, Mrs Alving in *Ghosts*, Rebecca West in *Rosmersholm*, and Hilda Wangel in *The Master Builder*.

E

EAGLE SALOON, London, see GRECIAN THEATRE.

EAGLE THEATRE, New York, see STANDARD THEATRE.

EARL CARROLL THEATRE, New York, on the south-east corner of 7th Avenue and 50th Street. Built and managed by the impresario after whom it was named, who staged there from 1923 several editions of his REVUE *Earl Carroll's Vanities*, on the lines of ZIEGFELD'S *Follies*, this opened on 27 Feb. 1922. A number of straight plays were also seen there, including Leon Gordon's *White Cargo* in 1923. A second Earl Carroll Theatre, designed by George Keister, opened on the same site on 27 Aug. 1931. It seated 3,000 and boasted one of the most technically innovatory stages in the country. The economic depression, however, forced Carroll to close it, and it reopened on 19 May 1932 as the Casino with the musical *Show Boat*. In 1934 it became a theatre restaurant, known as the French Casino, and again changed its name in 1938, when it became the Casa Mañana.

EASMON, RAYMOND SARIF (*c.* 1930–   ), a doctor in Sierra Leone, a former British colony which achieved independence in 1960. He is also a playwright, his works satirising the social pretensions of the Europeanized African ruling class and corruption in politics. His *Dear Parent and Ogre* (1964) and *The New Patriots* (1968) were popular with audiences throughout Black Africa, mainly because of their simple language and characterization and their easily identifiable targets.

EAST, JOHN MARLBOROUGH (1860–1924), English actor, manager, and director, who began his career as an actor in 1874, gaining wide experience on tour and appearing in London in the first productions of Sims and Pettitt's *In the Ranks* (1883) and Wills and Collingham's *A Royal Divorce* (1891). In Feb. 1892 he became manager of the LYRIC THEATRE in Hammersmith, where he directed a resident stock company in MELODRAMA and annual PANTOMIMES which attracted a loyal audience from the locality. He appeared in many of his productions, some of which he also wrote; one of his best parts was the Giant in *Jack and the Beanstalk*. After leaving Hammersmith in 1904 he was connected with several theatres, including the BRITANNIA at Hoxton, and in 1908 he was responsible for a dramatization of *Sexton Blake*, seen in London. A year later he devised and directed at the Crystal Palace a mammoth pageant, *Invasion*, which with a cast of 600 and an airship depicted the destruction of an English village by aerial bombardment. His first wife was the actress Leah Marlborough (1868–1953), for many years the leading actress of Mrs KIMBERLEY'S touring company. His brother Charles (1863–1914), with his wife Miss East Robertson (1866–1916), played mainly in melodrama and directed seasons at London suburban theatres.

EAST LONDON THEATRE, see ROYALTY THEATRE (1).

EAST SEVENTY-FOURTH STREET THEATRE, EASTSIDE PLAYHOUSE, New York, see PHOENIX THEATRE.

EBERLE, OSKAR (1902–56), Swiss director, who in the 1930s achieved an international reputation with his revival of the 16th-century Lucerne PASSION PLAY, and of CALDERÓN'S *El gran teatro del mundo*, which he staged in the square in front of the church at Einsiedeln, one of the earliest centres of medieval LITURGICAL DRAMA. He also directed the official Festival play at the Swiss national exhibition in Zürich in 1939. In 1955 he directed the Festival play at Vevey, where a wine festival with drama and music has been held intermittently since 1797, and in the year of his death revived the *Tellspiel* (1512), a play based on the exploits of William Tell, in a version by Jacob Ruf dating from 1545.

ECCLESIASTICAL DRAMA, see LITURGICAL DRAMA.

ECHEGARAY, JOSÉ (1832–1916), Spanish dramatist, awarded the NOBEL PRIZE for Literature in 1905. His plays retain the verse-form and much of the imagery of the Romantics, but deal mainly with social problems, and though enthusiastically received, caused fierce controversy. They had a great influence, not only in Spain but on the European theatre generally. The best-known are *O locura o santidad* (*Madman or Saint*, 1877), *El loco Dios* (*The Divine Madman*, 1900), *El hijo de Don Juan* (*The Son of Don Juan*, 1892), a study of inherited disease which owes something to IBSEN'S *Ghosts*, and, most important of all, *El gran*

Galeoto (1881), produced in England as *Calumny* and in the United States as *The World and His Wife*, in which a woman wrongfully accused of being a poet's mistress finally becomes so.

ECKENBERG, JOHANN CARL (1685–1748), German actor, an acrobat and juggler of great dexterity, who with his wife, a rope-dancer, led a company of acrobats and actors up and down Europe. He was once the successful rival of Carolina NEUBER at HAMBURG, his varied entertainment proving more acceptable to the public than her classic plays.

EDDY, EDWARD (1822–75), American actor, long popular at the BOWERY THEATRE, where he first appeared on 13 Mar. 1851 in BULWER-LYTTON's *Richelieu*. He was also seen as Othello, as Claude Melnotte in Bulwer-Lytton's *The Lady of Lyons*, and in the name-part of one of the many versions of *Belphegor*. He was at his best in youthful, melodramatic parts, earning for himself the sobriquet of 'robustious Eddy', and his most popular role was Edmond Dantès in DUMAS *père*'s *The Count of Monte Cristo*. He was at Burton's (later called the METROPOLITAN) in 1856, but his style was not suited to the fashionable theatres, and he went back to triumph at the Bowery as Richard III and in similar parts until his popularity waned.

EDEN PALACE OF VARIETIES, London, see KINGSWAY THEATRE.

EDEN THEATRE, New York, see PHOENIX THEATRE.

EDESON, ROBERT (1868–1931), American actor, son of an actor-manager and his wife Marion Taliaferro. He started work at 16 in the office of the PARK THEATRE, appearing there a year later as an understudy. By 1892 he was leading man at the Boston Museum, where he was seen by Charles FROHMAN in H. J. BYRON's *Our Boys*. Frohman took him to the EMPIRE in New York where he made a success as the Revd Gavin Dishart in BARRIE's *The Little Minister* (1897) with Maude ADAMS. Two years later he made his first appearance in London, as David Brandon in ZANGWILL's *Children of the Ghetto*, and appeared there again in 1907 in his most successful part, Soangataha in *Strongheart* by William C. de Mille. After further successes, including two plays by himself, he made his last stage appearance as the Vagrant in the ČAPEKS' *The Insect Play* (as *The World We Live In*) in 1922, spending the rest of his career in films. He had four wives, the third being the actress Mary Newcomb.

EDINBURGH CIVIC THEATRE COMPANY, see LYCEUM THEATRE, EDINBURGH.

EDINBURGH FESTIVAL OF MUSIC AND DRAMA, an international event presented annually since 1947 for three weeks, opening usually in mid-August. Although the emphasis is on music, some distinguished productions have been seen there over the years, including those of a number of foreign companies, since it has always been the policy of the Festival's directors to invite visitors from overseas. The first, in 1947, was JOUVET's company from the ATHÉNÉE in MOLIÈRE's *L'École des femmes* and GIRAUDOUX's *Ondine*. In 1948 the RENAUD-BARRAULT company appeared in GIDE's translation of *Hamlet* and MARIVAUX's *Les Fausses Confidences*; in 1955 Edwige FEUILLÈRE was seen in the younger DUMAS's *La Dame aux camélias*; and in 1957 the Renaud-Barrault company returned in ANOUILH's *La Répétition*. Other companies which have visited Edinburgh include the COMÉDIE-FRANÇAISE, the THÉÂTRE NATIONAL POPULAIRE, and the PICCOLO TEATRO DELLA CITTÀ DI MILANO. Marcel MARCEAU has also appeared there several times. New plays in English that have had their first productions at the Festival include T. S. ELIOT's *The Cocktail Party* (1949), *The Confidential Clerk* (1953), and *The Elder Statesman* (1958); Charles MORGAN's *The River Line* (1952); Thornton WILDER's *The Matchmaker* (1954) and *A Life in the Sun* (1955) and Athol FUGARD's *Dimetos* (1975). The last was produced on a specially constructed stage in the Assembly Hall, where Tyrone GUTHRIE directed in 1948, 1949, and 1959, in an adaptation by Robert KEMP, the old Scottish morality play by Sir David LYNDSAY, *Ane Pleasant Satyre of the Thrie Estaitis*, last seen in 1552. It was revived again for the 1973 Festival in a new version by Tom Wright directed by Bill BRYDEN. Other performances at the Assembly Hall have included the OLD VIC company in *Romeo and Juliet*, JONSON's *Bartholomew Fair*, and SCHILLER's *Mary Stuart* (translated by Stephen Spender), and the STRATFORD (ONTARIO) FESTIVAL Theatre company in SOPHOCLES' *Oedipus the King*. The latter company also appeared at the Festival in 1956 with a production of *Henry V* in which French-Canadian actors played the King of France and his courtiers. In 1949 the candle-lit Regency Hall of the Royal High School was the setting for an enchanting revival by Guthrie of RAMSAY's *The Gentle Shepherd*, first seen in 1729; and a year later the Glasgow CITIZENS' THEATRE was seen in BRIDIE's *The Queen's Comedy* and in a revival of the Scottish tragedy *Douglas* by John HOME, first seen in 1756. The same company returned in 1968 with Michael Blakemore's outstanding production of BRECHT's *Arturo Ui*. Other Scottish companies to have appeared regularly at the Festival include the GATEWAY, who were responsible for the first production of Robert MCLELLAN's *Young Auchinleck* (1962), and the Edinburgh Civic Theatre Company from the LYCEUM THEATRE, with John McGrath's *Random Happenings in the Hebrides* (1970). Other repertory companies which have visited the Festival include those of BIRMINGHAM, Perth, and Dundee; the BRISTOL OLD VIC; the ABBEY THEATRE from Dublin; and the NOTTINGHAM PLAY-

HOUSE company. The ENGLISH STAGE COMPANY from the ROYAL COURT THEATRE in London has made several visits, presenting the British premières of O'CASEY's *Cock-a-Doodle-Dandy* (1959) and IONESCO's *Exit the King* (1963), and in 1962 the ROYAL SHAKESPEARE COMPANY made its first visit to the Festival with the British première of FRY's *Curtmantle*; they returned in 1974 with MARLOWE's *Dr Faustus*. The Actors' Company, led by Ian MCKELLEN and others, also chose the Festival for their début in 1972, appearing in FEYDEAU's *Ruling the Roost* and FORD's *'Tis Pity She's a Whore*: they were to return several times. A memorable event of later years was the adaptation of *Richard III* staged in 1979 by the Rustaveli company from GEORGIA, and the following year the NATIONAL THEATRE company made its first visit to the Festival.

The Festival has from the beginning attracted to Edinburgh a 'fringe', not paralleled on the musical side, of plays and revues outside the official programme, of which the best known and most successful was the revue *Beyond the Fringe* in 1960. While the standard of drama varies enormously, several good plays have been given their first performances by 'fringe' companies, notably Tom STOPPARD's *Rosencrantz and Guildenstern Are Dead* (1966). Much experimental work has been produced, particularly by new companies working on COLLECTIVE CREATIONS, and the TRAVERSE THEATRE, which expands its year-round programme of events during the Festival, has acted always as a cultural catalyst to the 'official' proceedings.

EDISON THEATRE, New York, at 240 West 47th Street, which seats 499, opened in 1970, its first production being *Show Me Where the Good Times Are*, a musical based on MOLIÈRE's *Le Malade imaginaire*. Later productions have included Kurt Vonnegut's *Happy Birthday, Wanda June* (also 1970) and Athol FUGARD's *Sizwe Bansi is Dead* and *The Island* (1974). A revival of the nude revue *Oh! Calcutta!* opened there in 1976 and was still running in the early 1980s.

EDOUIN, WILLIE [William Frederick Bryer] (1846–1908), English comedian, who appeared in London as a child, mainly in PANTOMIME. As an adult actor he played in Australia and then went to New York, where he was associated with Lydia Thompson and her English troupe, one of whom he married. At WALLACK'S THEATRE in 1871 he was seen in his most popular character, Wishee-Washee in a burlesque of *Bluebeard*. He made his first adult appearance in London in 1874, and in 1884 was associated with the management of TOOLE'S THEATRE; he also managed the STRAND THEATRE for a time from 1888, but being a poor business man was not successful financially. He was excellent in certain grotesque and whimsical parts, but his main success was in BURLESQUE and EXTRAVAGANZA.

EDWARDES, GEORGE (1852–1915), Irish-born theatre manager, who introduced MUSICAL COMEDY to London. He first worked as a theatre manager in Ireland in 1875, and in 1881 became manager of the newly opened SAVOY THEATRE in London. In 1885 he went into partnership with John HOLLINGSHEAD at the GAIETY THEATRE, a year later becoming sole manager and producing a fine series of musical comedies with the famous chorus of 'Gaiety Girls', whom he chose and trained himself. In 1893 he built a theatre in London for Augustin DALY of which he later took control, making it as successful as the Gaiety, and managing also at various times the PRINCE OF WALES' and the APOLLO. Always known as 'The Guv'nor', he was tall, good-looking and burly, with a very soft voice. He had an extraordinary flair for knowing what the public wanted, and spared no cost in providing it, with a meticulous care for detail. He was rewarded by crowded houses and the trust and affection of the public and the profession alike.

EDWARDES, RICHARD, see BOY COMPANIES.

EDWARDS, HILTON (1903–82), actor and director born in London, who began his career at the OLD VIC in 1922, where he appeared in all but two of Shakespeare's plays. In 1927 he went with Anew MCMASTER to Ireland, where he remained and with Michéal MACLIAMMÓIR founded in 1928 the Dublin GATE THEATRE. There and at the GAIETY THEATRE he directed over 300 plays, in many of which he appeared himself, including new works by Irish authors, among them Brian FRIEL, Denis JOHNSTON, and W. B. YEATS. In 1937 he directed *Hamlet* at Elsinore, playing Claudius. He took the Gate Theatre company on many tours in Europe, North Africa, and the United States and in 1947 was seen again in London with his company, notably in MacLiammóir's *Ill Met By Moonlight*. In 1970, while the Gate was closed for repairs, he directed a production of CHEKHOV's *The Seagull* at the ABBEY THEATRE. His association with the Gate continued until his death.

EDWARDS, RICHARD, see ENGLAND.

EDWIN, JOHN, the elder (1749–90), English comedian, whose friendship with Ned SHUTER led to his playing the latter's parts in Manchester and Dublin. During summer seasons he also appeared at the HAYMARKET THEATRE in plays by O'KEEFFE, who wrote for him the comic songs in which he excelled. On the death of Shuter in 1776 Edwin replaced him at COVENT GARDEN where he remained until his death, being himself succeeded by MUNDEN. A good reliable actor who managed to be both humorous and handsome, he played with restraint and subtlety, among his best parts being Dogberry in *Much Ado About Nothing*, the First Grave-Digger in *Hamlet*, Launcelot Gobbo in *The Merchant of Venice*, and Sir Hugh Evans in *The*

Merry Wives of Windsor. He was also much admired as Sir Anthony Absolute in SHERIDAN'S *The Rivals*, which Shuter had been the first to play.

EDWIN, JOHN, the younger (1768–1803), English actor, son of the above. He was first on the stage as a boy, playing with his father in Bath and at the HAYMARKET THEATRE. As an adult, he made his début at COVENT GARDEN in 1788, and then came under the patronage of the Earl of Barrymore, for whom he devised private theatricals at Wargrave; he died of dissipation without having realised his full powers as an actor. His wife, Elizabeth Rebecca Richards (*c.* 1771–1854), who had been in the provincial company of Tate WILKINSON, joined the company at DRURY LANE after her husband's death.

EFTÍMIU, VICTOR, see ROMANIA.

EISENHOWER THEATER, Washington, D.C. This theatre, which seats 1,142, forms part of the John F. Kennedy Center for the Performing Arts. It opened on 16 Oct. 1971 with a revival of IBSEN'S *A Doll's House* starring Claire BLOOM. It houses plays on tour, often with major stars, many of which are also seen in New York. Highlights include the world premières of the popular musical *Annie* (1977) and Arthur KOPIT'S *Wings* (1978), and a visit in 1981 by Elizabeth Taylor on her way to her Broadway début in Lillian HELLMAN'S *The Little Foxes*.

EK, ANDERS (1916–79), Swedish actor, who from 1946 to 1950 worked in the city theatre of Gothenburg, from 1952 to 1960 with the KUNGLIGA DRAMATISKA TEATERN, and subsequently with the city theatre of Stockholm. A powerful actor, he was seen mainly in such parts as Othello, Macbeth, Jean in STRINDBERG'S *Miss Julie*, and the title roles in Eugene O'NEILL'S *The Emperor Jones* and CAMUS'S *Caligula*. In 1965 he returned to the K.D.T., and one of his last roles was Tessier in BECQUE'S *Les Corbeaux*, directed by Alf SJÖBERG. He worked often with Ingmar BERGMAN in both theatre and films.

EKHOF, KONRAD (1720–78), German actor, who in 1740 joined the newly-formed company of SCHÖNEMANN, where he remained for 27 years. Being short and ungraceful with no pretensions to good looks, he was not at first thought much of, but he had a fine speaking voice and quickly perfected his art, discarding the stiff declamatory style of Carolina NEUBER for a more natural style of acting hitherto unknown in Germany. During his years with Schönemann he met and married a young actress, and trained her in his own methods. He was the first professional theorist on German dramatic art—since GOTTSCHED was no actor—and in 1753 opened a short-lived Academy of Acting for the discussion and production of plays. As time went by Schönemann, who was more interested in horse-racing, left the running of the company to Ekhof, who however had no taste or talent for business management and called in KOCH to take control. This led to constant friction, and in 1767 Ekhof left to join ACKERMANN'S company. He was at the height of his powers and left the company after five years only because he could no longer endure the rudeness and arrogance of the youthful SCHRÖDER. After several miserable years spent touring with Abel SEYLER he joined the Court Theatre company in WEIMAR, where in later years he became a friend of GOETHE, imparting to him some of the theatrical reminiscences that figure in *Wilhelm Meister*. A disastrous fire at Weimar in 1775 led Ekhof to become chief actor and director at Gotha, where one of his last official acts was to engage IFFLAND for the company. He made his last appearance on stage as the Ghost in Schröder's adaptation of *Hamlet*. Ekhof excelled in the mingled tragi-comic and pathetic roles of the new drama. Some of his most admired parts were Old Barnwell in LILLO'S *The London Merchant*, the father in DIDEROT'S *Le Père de famille*, and Odoardo in LESSING'S *Emilia Galotti*. In spite of his physical shortcomings, he was outstanding in tragedy and in comedy was subtle and discreet, notably in his portrayal of north-German peasants. He lived to see his fellow actors, in great part through his own exertions, raised from the misery of strolling players to the dignity of an assured profession under noble patronage.

EKKYKLEMA. A piece of ancient Greek stage machinery which has occasioned a great deal of controversy among modern scholars. Meaning literally something 'rolled out', it was obviously used to bring forward some object, character, or grouping important in the play's context. This was once thought to have been a movable platform which was either pushed on stage or revolved on a turntable to show an interior scene. It is now thought to have been nothing more than a couch on wheels, or a grouping arranged within a pair of double doors which opened to reveal it as the doors open in AESCHYLUS' *Libation Bearers* to reveal the bodies of Clytemnestra and Aegisthus.

EKMAN, GÖSTA (1890–1938), Swedish actor, director, and manager who, after holding appointments in various Stockholm and Gothenburg theatres between 1908 and 1913, was closely associated with the Svenska Teater in Stockholm between 1913 and 1925. After the destruction of the theatre by fire he spent periods at the Oscarsteater (1926–30) and the Vasateater (1931–5), and made several tours. He had something of the clown in him, and brought an irrational brilliance to the interpretation of such roles as Hamlet, IBSEN'S Peer Gynt and Hjalmar Ekdal (in *The Wild*

Duck), STRINDBERG's Gustav Vasa, and Marchbanks in SHAW's *Candida*.

ELEN, GUS, see MUSIC-HALL.

ELEPHANT AND CASTLE THEATRE, London, a playhouse devoted to 'transpontine' MELODRAMA and PANTOMIME, built in 1872 on the site of NEWINGTON BUTTS. After a steady but unspectacular career catering mainly for a local audience, it closed in 1928 to reopen four years later, after partial rebuilding, as a cinema under its old name.

ELEVATOR, see BRIDGE.

ELIOT, T(homas) S(tearns) (1888–1965), poet and dramatist, American by birth but English by adoption, who initiated a revival of POETIC DRAMA in England with a play on the murder of Thomas à Becket, *Murder in the Cathedral* (1935). First acted in the Chapter House of Canterbury Cathedral, it was subsequently revived several times with great success in commercial theatres in Britain and the United States. A later play, *The Family Reunion* (1939), based on the *Oresteia* of AESCHYLUS, was considered less successful, mainly because Eliot failed to integrate the ritualism of the CHORUS with the realism of the setting, but it has had occasional revivals. Of his last three plays, commissioned for performance at the EDINBURGH FESTIVAL, *The Cocktail Party* (1949) and *The Confidential Clerk* (1953) were based on EURIPIDES' *Alcestis* and *Ion* respectively; *The Elder Statesman* (1958) on SOPHOCLES' *Oedipus at Colonus*. In them Eliot moved closer to a mannered realism, disguising his serious purpose under the form of modern drawing-room comedy, and discarding the closely-wrought poetic style of the earlier plays for a plain undecorated verse.

ELITCH GARDENS THEATRE, Denver, see BONFILS THEATRE.

ELIZABETHAN STAGE SOCIETY, see POEL, WILLIAM.

ELIZABETHAN THEATRE TRUST, Australian, body founded in 1955 to mark the visit of Queen Elizabeth II to Australia the previous year, and to establish the principle of government subsidies throughout the country for the performing arts. It was hoped that among its activities would be the training of young actors, the encouragement of promising young playwrights, and eventually the founding of an Australian National Theatre. Its initial preoccupation was however to provide touring companies covering the whole country, and with this in mind a Board of Trustees under Hugh HUNT (succeeded in 1960 by Neil Hutchinson) took over the former Majestic Theatre in Melbourne (built in 1917 but later used as a cinema) as a base for the Trust's work. Renovated and redecorated, it opened as the Elizabethan Theatre on 27 July 1955 with a cast of guest artists from England in RATTIGAN's *The Sleeping Prince*. The Trust's own company, headed by Judith ANDERSON, made its début in Canberra the following Sept. in Robinson Jeffers's version of EURIPIDES' *Medea*, and then embarked on a long tour. Among the new Australian plays presented by the Trust were Ray LAWLER's *The Summer of the Seventeenth Doll*, Alan Seymour's play about Anzac Day *The One Day of the Year*, and Douglas STEWART's verse drama *Ned Kelly*. The Trust also had its own ballet and opera companies which later became autonomous; the theatre company was then disbanded and the subsidy work of the Trust largely taken over by what is now the Australia Council.

ELLIOTT, DENHOLM MITCHELL (1922–), English actor, who trained at the Royal Academy of Dramatic Art, then joined the R.A.F. and spent three years as a prisoner of war. He made his first appearance on the stage in 1945 and a year later was seen in London in Warren Chetham-Strode's *The Guinea Pig*. He first attracted attention as Edgar in Christopher FRY's *Venus Observed* (1950), and later that year appeared in New York in the dual role of Hugo and Frederic in ANOUILH's *Ring Round the Moon*. On returning to London he appeared in another play by Fry, *A Sleep of Prisoners* (1951), and was then seen in T. S. ELIOT's *The Confidential Clerk* (1953). He continued to appear in plays of unusual quality, displaying a haunting vulnerability in Julien GREEN's *South* (1955) and Anouilh's *Traveller Without Luggage* (1959). At Stratford-upon-Avon in 1960 his parts included Troilus in *Troilus and Cressida* and Valentine in *The Two Gentlemen of Verona*, and in 1963 he toured America in CHEKHOV's *The Seagull* and Arthur MILLER's *The Crucible*, both plays being seen in New York, where he also appeared as Cornelius Melody in O'NEILL's *A Touch of the Poet* (1967). Back in London, he played four roles in John MORTIMER's *Come as You Are* (1970) and in 1972 was seen as Judge Brack in IBSEN's *Hedda Gabler*. Among his later roles were Dick in Peter NICHOLS's *Chez Nous* (1974) and the title role in Graham GREENE's *The Return of A. J. Raffles* (1975) with the ROYAL SHAKESPEARE COMPANY.

ELLIOTT, (May) GERTRUDE (1874–1950), American actress, sister of Maxine ELLIOTT. She made her first appearance in 1894 in Rose COGHLAN's company at Saratoga in WILDE's *A Woman of No Importance*, appearing in New York later the same year, and in London in 1899, when she was seen with her sister in Clyde FITCH's *The Cowboy and the Lady*. In 1900 she was engaged by FORBES-ROBERTSON, with whom she went on tour, marrying him at the end of the year. She was his leading lady until he retired in 1913, and then appeared in London under her own management, later undertaking extensive tours of South Africa, Australia, and New Zealand. She returned to New York in 1936 to play the Queen to Leslie Howard's Hamlet. She was the mother of Jean FORBES-ROBERTSON.

ELLIOTT, G. H., see MUSIC-HALL.

ELLIOTT, MAXINE [Jessie Dermot] (1868–1940), American actress, who made her stage début with E. S. WILLARD in 1890 in H. A. JONES's *The Middleman*, adopting her stage name at the suggestion of Dion BOUCICAULT. From Willard's company she went to that of Rose COGHLAN, playing among other parts Dora in *Diplomacy*, a version of SARDOU's *Dora* by B. C. Stephenson and Clement SCOTT. In 1895 she joined Augustin DALY's company, appearing in his productions both in New York and in London, where she was first seen as Silvia in *The Two Gentlemen of Verona*. After touring Australia with her husband Nat GOODWIN, she appeared with him in a number of plays by Clyde FITCH. It was in the latter's *Her Own Way* (1903) that she made her first independent starring venture, following it by *Her Great Match* (1905). She opened her own New York playhouse, MAXINE ELLIOTT'S THEATRE, in 1908. A woman of great beauty and charm, and a personal friend of Edward VII, she was extremely popular in England and retired there shortly before the First World War, during which she was active in war work. In 1920, impoverished by her relief work for Belgium in particular, she returned to the stage and again became one of the best-loved stars of the American theatre until her retirement, when she returned to live in England.

ELLIOTT, MICHAEL, see ROYAL EXCHANGE THEATRE, MANCHESTER.

ELLISTON, ROBERT WILLIAM (1774–1831), English actor, who after touring the provinces made his London début in 1796 at the HAYMARKET THEATRE. Charles DIBDIN wrote a number of entertainments for him, and he was frequently seen at DRURY LANE, being one of the most popular actors of the day, second only to GARRICK in tragedy. He was also much esteemed in comedy, playing Doricourt in Mrs COWLEY's *The Belle's Stratagem*, Charles Surface in SHERIDAN's *The School for Scandal*, Rover in O'KEEFFE's *Wild Oats*, and Ranger in HOADLY's *The Suspicious Husband* with equal success. In later life one of his finest parts was Falstaff in *Henry IV, Part I* in which he had previously played Hotspur. He managed a number of provincial theatres and in 1819, after ten years as lessee of the SURREY THEATRE, he achieved his ambition of managing Drury Lane where, surrounding himself with a fine company, he opened with Edmund KEAN in *King Lear* and in 1821 put on BYRON's *Marino Faliero* in the face of much opposition. But his resources could not stand up to his expenditure and his outside speculations, and in 1826 he went bankrupt. He then returned to the Surrey Theatre, where he made a substantial profit from JERROLD's *Black-Ey'd Susan* (1829) starring T. P. COOKE. He made his last appearance about a fortnight before his death. Contemporary critics were loud in his praise: though eccentric, extravagant, and a heavy drinker he was a fascinating personality, unrivalled in the playing of gentlemanly rakes and agreeable rattles.

ELTINGE, JULIAN [William Dalton] (1883–1941), American actor, who developed the modern style of FEMALE IMPERSONATION. Abandoning the grotesque representations of comic servants and elderly women, he presented himself as a convincingly glamorous young lady, paying particular attention to his walk, hand movements, make-up, and clothes. He made his first appearance in a female part in the musical comedy *Mr Wix of Wickham* in 1905, and from 1911 to 1914 toured in the dual roles of Mrs Monte and Hal Black in *The Fascinating Widow*, a musical comedy specially written for him. He continued to appear on stage and in films until his retirement in 1930, returning briefly in 1940 to work in night-clubs.

ELTINGE THEATRE, New York, at 236 West 42nd Street, between 7th and 8th Avenues. Named after Julian ELTINGE, this opened on 11 Sept. 1912 with a melodrama, *Within the Law* by Bayard Veiller, which established Jane COWL as a star. The theatre housed a number of successful productions before, in 1930, it became a home of American BURLESQUE, notorious for the daring of its strip-tease acts and its dubious jokes. It was closed in 1942, and a year later became a cinema.

EMANTS, MARCELLUS, see NETHERLANDS.

EMBASSY THEATRE, Hampstead, London. This opened as a try-out theatre for new plays on 11 Sept. 1928 with *The Yellow Streak* by M. E. Hope and sent many successful plays to the West End, among them Anthony Armstrong's *Ten-Minute Alibi* (1933) which went to the HAYMARKET THEATRE and Michael Egan's *The Dominant Sex* (1934) which opened at the Prince's (now the SHAFTESBURY THEATRE) at the beginning of 1935. The Embassy was damaged by bombing during the Second World War but continued in use, Anthony Hawtrey running it successfully from 1945 until his death nine years later. It then closed and in 1957 was taken over by the Central School of Speech and Drama to house student productions.

EMERY, JOHN (1777–1822), English actor, son of the actor Mackle Emery (1740–1825). Even as a boy he had a remarkable talent for playing old men and at 15 was already considered an outstanding actor. In 1798 he was engaged by COVENT GARDEN to take the place of the comedian John

A scene from *Titus Andronicus* sketched by Henry Peachum in 1595. Aaron is depicted as a black man.

The second Blackfriars Theatre, 1597. This reconstruction shows Richard Burbage's adaptation of a hall in the old Blackfriars Monastery as an indoor theatre.

The Swan Theatre, 1596, the only known contemporary representation of an Elizabethan theatre.

The title-page of Francis Kirkman's *The Wits: or Sport upon Sport*, 1672, probably a composite impression of the type of stage clandestinely used when the theatres were closed between 1642 and 1660, and the earliest English illustration to indicate the use of footlights. The characters depicted include the Hostess and Falstaff in a droll taken from *Henry IV*.

The second Globe Theatre (*below*), built in 1614 and pulled down in 1644, from Hollar's engraving published in 1647 (where it is wrongly labelled the Bear Garden). The building was almost certainly polygonal, not circular as it appears here.

Two clowns of Shakespeare's time: Richard Tarleton (*above left, c.* 1590) and Richard Kempe (*above right*, 1600) were especially known for the jigs they interpolated into their performances.

Henry Harris as Wolsey (*left*) in *Henry VIII, c.* 1661. The decline in popularity of Shakespeare's plays after the Restoration accounts for the paucity of visual records from this period.

Garrick and Mrs Pritchard as Macbeth and Lady Macbeth, Drury Lane, 1768. Although historical costume was not attempted, the setting has a medieval flavour.

Hamlet, presented by Karl Döbbelin's company, Berlin, 1778, with Johann Brockmann as Hamlet. The costuming is still contemporary in essence.

Henry VIII, 1817. This composite group shows Mrs Siddons as Queen Katharine, with her brothers John Philip, Charles, and Stephen Kemble as Wolsey, Cromwell, and the King; there is no evidence that they ever appeared together in the play.

Macbeth, 1853. Setting by I. Dayes for Charles Kean at the Princess's Theatre, showing the solid 'archeological' detail demanded by Kean. Banquo's ghost appears at the banquet by means of a transparency.

Design for *Hamlet* by William Telbin, for Charles Fechter's production at the Lyceum Theatre, 1864.

The Taming of the Shrew, 1887, Augustin Daly's production starring Ada Rehan at Daly's Theatre, New York.

Julius Caesar, 1898, the Forum scene, from one of Beerbohm Tree's spectacular productions at Her Majesty's Theatre.

As You Like It, 1897, produced by Barry Sullivan at the Shakespeare Memorial Theatre. The stuffed stag (formerly an inhabitant of Charlecote Park) became an obligatory feature of the play at Stratford until 1919.

Henry Irving in the play scene from *Hamlet*, Adelphi Theatre, 1905.

Ellen Terry in a scene from *Much Ado About Nothing* at her jubilee performance, Drury Lane, 1906. Setting and costumes were designed by her son Gordon Craig for her production of 1903.

King Lear, Haymarket Theatre, 1909, design by Charles Ricketts, reflecting the influence of Gordon Craig in its combination of simplification and massiveness.

Hamlet, 1911, Moscow Art Theatre, the final scene. The controversial sets were by Gordon Craig.

Antony and Cleopatra, Leningrad, 1924. A Constructivist design by V. Shchuko.

Design by Komisarjevsky for his own production of *The Merchant of Venice* at the Shakespeare Memorial Theatre, 1932.

Romeo and Juliet, 1930. Eva le Gallienne's production for the Civic Repertory Theatre, with settings by Aline Bernstein.

A Midsummer Night's Dream, 1933, at the Open Air Theatre, Regent's Park. Phyllis Neilson-Terry as Oberon, Jessica Tandy as Titania.

Twelfth Night, 1933, at the Moscow Art Theatre. Sergei Obraztsov as Feste, S. Giasintova as Maria.

The Tempest, 1934, the Old Vic Company at Sadler's Wells, directed by Tyrone Guthrie, designs by John Armstrong. Charles Laughton as Prospero, Elsa Lanchester as Ariel.

Macbeth, 1934, at the Old Vic, directed by Guthrie, costumes by Armstrong, settings by Wells Coates.

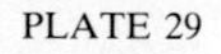

Henry V, 1937, Ben Iden Payne's production at the Shakespeare Memorial Theatre.

Othello, 1937, design by Robert Edmond Jones for a production at the New Amsterdam Theatre, New York.

The Merry Wives of Windsor, 1955, at the Shakespeare Memorial Theatre, directed by Glen Byam Shaw, designs by Motley.

Henry IV, 1969, staged in fairground style at the Leningrad Bolshoi Gorky Theatre by Georgyi Tovstonogov.

As You Like It (*left*), 1967, an all-male modern-dress production by the National Theatre, London, designed by Ralph Koltai.

A Midsummer Night's Dream (*below*), 1970, Peter Brook's historic 'circus' production for the Royal Shakespeare Company.

The open-air Delacorte Theatre in New York's Central Park, where productions of Shakespeare's plays are mounted with free admission.

Richard III, 1977, at the Stratford (Ontario) Festival Theatre, directed by Robin Phillips, designed by Daphne Dare. Brian Bedford as Richard, with Peter Brikmanis and Stephen Hunter.

QUICK and apart from a short engagement at the HAYMARKET remained there until his sudden death. His Caliban in *The Tempest* was highly praised, and he was good as Sir Toby Belch in *Twelfth Night*, the First Grave-digger in *Hamlet*, and Dogberry in *Much Ado About Nothing*. He was also an artist, and between 1801 and 1817 exhibited frequently at the Royal Academy. His son Sam and grand-daughter Winifred (below) were also on the stage.

EMERY, SAM(uel) ANDERSON (1817–81), English actor, son of John EMERY, much of whose talent he inherited. He was with the KEELEYS at the LYCEUM THEATRE from 1844 to 1847, where he created the parts of Jonas and Will Fern in Edward Stirling's dramatizations of DICKENS's *Martin Chuzzlewit* (1844) and *The Chimes* (1845) respectively, and also played John Peerybingle in Albert SMITH's version of *The Cricket on the Hearth* (also 1845). In 1853 he went to the OLYMPIC THEATRE and in the following year became lessee of the Marylebone (later WEST LONDON) Theatre making his last appearance in London in 1878. One of his best parts was Sir Peter Teazle in SHERIDAN's *The School for Scandal*, and he appeared in many new plays of the period including Tom TAYLOR's *Plot and Passion* (1853) and *Still Waters Run Deep* (1855). He was the father of Winifred EMERY.

EMERY, (Isabel) WINIFRED MAUD (1862–1924), English actress, daughter of Sam EMERY. She began her long and distinguished career in 1870 at Liverpool where she appeared as the child Geraldine in BUCKSTONE's *The Green Bushes*. Four years later she was in PANTOMIME in London, and in 1879 made her adult début at the IMPERIAL THEATRE. She was later with Wilson BARRETT, John HARE, and the KENDALS, and then joined the LYCEUM THEATRE company under Henry IRVING whom she accompanied on his American tours in 1884 and 1887, playing leading roles in Shakespeare and other productions. She returned to London in 1888 and married Cyril MAUDE, becoming his leading lady when he went into management at the HAYMARKET THEATRE in 1896. A beautiful woman of great charm, she was one of the most versatile and popular actresses of her day; among her finest parts were the title role in WILDE's *Lady Windermere's Fan* (1892) and Lady Babbie in BARRIE's *The Little Minister* (1897), both of which she was the first to play.

EMMET, ALFRED, see QUESTORS THEATRE.

EMNEY, FRED (1900–80), a leading comedian in MUSICAL COMEDY and REVUE, son of the MUSIC-HALL star Fred Emney (1865–1917), whose best known sketch was 'A Sister to Assist 'Er' and who died of injuries sustained during the PANTOMIME *Cinderella* at DRURY LANE after slipping on the soapsuds during a knockabout comedy scene. The younger Fred made his first appearance on the stage in London as the page-boy in Edward SHELDON's *Romance* (1915), and from 1920 to 1931 played in VAUDEVILLE in the United States. On his return to England he played at a few 'halls', but then turned to light opera and musical plays, being part author of *Big Boy* (1945) and *Happy as a King* (1953), in both of which he appeared. In later life he gave several excellent performances in straight comedy, playing Admiral Ranklin in PINERO's *The Schoolmistress* in 1950 and Ormonroyd, the photographer, in PRIESTLEY's *When We Are Married* in 1970. He was an excellent raconteur and in his solo performance of 'songs at the piano' his heavyweight personality, imperturbable face and manners, his eyeglass, and his eternal cigar made up an unforgettable image of good-tempered lethargy.

EMPIRE, The, London, in Leicester Square. This famous MUSIC-HALL, originally a theatre which from 1884 to 1886 housed BURLESQUES and EXTRAVAGANZAS, stood on the site of Saville House, the home of George II when Prince of Wales. In 1887 it opened as the Empire Theatre of Varieties under the joint management of George EDWARDES and Augustus HARRIS, whose spectacular ballets, with such stars as Adeline Genée and Phyllis Bedells, rivalled those of the ALHAMBRA. One of the Empire's most popular features was its promenade, which was attacked as a 'haunt of vice' by Mrs Ormiston Chant in her 1894 'Purity Campaign'. Screens were erected to separate the promenade and its prostitutes from the auditorium but the audience rioted and tore them down, egged on by Winston Churchill, then a cadet at Sandhurst. Shortly before the First World War REVUE made its appearance, and on 21 Feb. 1918 the first of a series of successful MUSICAL COMEDIES, *The Lilac Domino*, began a long run, followed by *Irene* (1920), *The Rebel Maid* (1921), and *Lady, Be Good!* (1926), with music by George Gershwin. On 22 Jan. 1927 the Empire closed and was demolished. It was replaced by a cinema which from late 1949 to 1952 featured spectacular revues with a mixed programme of films and live entertainment. This was itself closed and gutted in 1961, and the interior was reconstructed as a cinema and dance hall.

EMPIRE, Islington, see GRAND THEATRE.

EMPIRE THEATRE, Liverpool, see LIVERPOOL.

EMPIRE THEATRE, New York, on the south-east corner of Broadway and 40th Street. Built for Charles FROHMAN, this handsome edifice with its red and gold rococo interior opened on 25 Jan. 1893 with BELASCO's *The Girl I Left Behind Me*, which had a long run. Let in the summer months to visiting touring companies, it housed during the winter for many years not only Frohman's STOCK COMPANY but also a company led by John DREW, who in 1899 scored a great personal success in the title role of *Richard Carvel*, based on a novel by the American writer Winston Churchill, and another led by Maude ADAMS who, also in 1899, was seen as Juliet to the Romeo of William FAVERSHAM and in 1905 as BARRIE's Peter Pan. One of the most popular plays seen at the Empire, which relied mainly on light modern comedies, was KESTER's *When Knighthood was in Flower* (1901), with Julia MARLOWE. Other famous actresses who visited this theatre were Ellen TERRY, seen in 1907 in SHAW's *Captain Brassbound's Conversion* and HEIJERMANS's *The Good Hope*; Ethel BARRYMORE, who appeared in PINERO's *Mid-Channel* in 1910 and Zoë AKINS's *Déclassée* in 1919; and Jane COWL, seen in 1925 in COWARD's *Easy Virtue*. In 1926 the Empire was temporarily closed by the authorities after the production of *The Captive*, a translation of Edouard BOURDET's controversial play *La Prisonnière*. It soon reopened and continued its successful career, Katharine CORNELL appearing in 1928 in *The Age of Innocence*, adapted from Edith Wharton's novel; she returned in 1931 in Besier's *The Barretts of Wimpole Street* and in 1937 in Shaw's *Candida*. Two important productions seen during the 1930s were Elmer RICE's *We, The People* (1933), with settings by Aline BERNSTEIN, and, in 1936, IBSEN's *Ghosts*, with NAZIMOVA. Also in 1936 *Hamlet*, with John GIELGUD, ran for 132 performances, breaking the records set up by Edwin BOOTH and John BARRYMORE. On 8 Nov. 1939 came the first night of Crouse and LINDSAY's *Life With Father*, based on the book by Clarence Day, which occupied the theatre for 6 years. After its transfer to the BIJOU THEATRE in 1946, the Empire saw the LUNTS in RATTIGAN's *O Mistress Mine* (produced in London as *Love in Idleness*). The last full-scale production was *The Time of the Cuckoo* (1952) by Arthur Laurents, which closed on 30 May 1953. The previous Sunday evening a memorial programme, *High Lights of the Empire*, had been presented by the AMERICAN NATIONAL THEATRE AND ACADEMY, with Cornelia Otis Skinner as compère. The theatre was pulled down in the same year.

ENCINA, JUAN DEL (1469–*c*. 1539), Spanish playwright, who with his later contemporaries TORRES NAHARRO and Gil VICENTE founded and secularized Spanish Renaissance drama. He was attached to the household of the Duke of Alba for several years from about 1492 as dramatist, musician, actor, and director of entertainments, and for his patron wrote eight eclogues, or pastoral dialogues, in the Italian style, though the content was purely Spanish. Three of these are on religious themes—Christmas, the Passion, and the Resurrection—the others are secular, with simple but well constructed plots and amusing dialogue, often in peasant dialect. Into all of them Encina introduced *villancicos*, or rustic songs, and his work in general owed much to his musical talent. In 1513 one of Encina's plays, the *Égloga de Plácida y Vitoriano* was given in Italy before a brilliant audience which included the Spanish ambassador. It is probable that they were all first acted by talented amateurs of high rank, or their servants, either at Court or in the private houses of the nobility; but with the rise of the professional theatre they eventually found their way into the repertories of the strolling players.

ENGLAND. Modern research is revealing that, contrary to earlier belief, it is safer to regard drama in England as having developed in the early Middle Ages from two separate sources—one ecclesiastical, the other secular—and in two separate environments—the Church and the Hall—rather than to suppose that having originated within the liturgies of the Christian Church it subsequently became secularized.

What makes this probable is the invariable association of all dramatic plays and games between the 10th and 15th centuries of which we have any knowledge with the main festivals, or Holy Days, of the Christian calendar. Given an occasion of such special significance to the community as to warrant a public holiday, it would be normal to celebrate it both by a simple re-enactment in the church of the religious event being commemorated, and by secular sports and games elsewhere. The first gave rise to the LITURGICAL DRAMA enacted mainly in churches, the second to those festivities which are now grouped under the general heading of FOLK FESTIVALS and FOLK-PLAYS which usually took place in summer in the open air; but in winter a hall, whether of castle or palace, offered shelter from the weather for the scôps who sang or recited sagas like *Beowulf*, and for the assorted jugglers, bearwards, acrobats, singers, and clowns who under the generic name of 'gleemen' supplied the feudal nobility with entertainments ranging from the serious to the frivolous.

In England, as on the Continent, liturgical drama developed from a simple scene played in the church at Easter to a wide range of plays connected with all the major Church festivals; some of these plays, together with others on subjects taken from the Bible or from the lives of saints, were written by monastic clerks or students at the newly-established universities, not necessarily for performance inside the church. Some, like the Resurrection Play acted at Beverley in Yorkshire in about 1220, were meant to be played in the churchyard, or alternatively in monastic refectories and in the banqueting halls of the nobility on such festive occasions as Christmas and Carnival time (Shrove Tuesday).

From what is known of the development of the folk and ROBIN HOOD plays it seems unlikely that any important change took place much before the latter part of the 12th century, since audiences for social recreation of any kind remained virtually the same until the growth of towns brought into existence a middle class whose wealth lay in merchandise and money rather than in land. It was for this new audience that during the 13th century such plays as the anonymous English *Interludium de clerico et puella* (*The Interlude of the Student and the Maiden*) were written. Elaborations of such traditional folk features as the MORRIS DANCE and the MUMMERS' PLAY also offered dramatic possibilities, as did the dramatization of popular romances devoted to love and chivalry—the *chansons de geste*—and the tournaments, street PAGEANTS, and DISGUISINGS which were the staple entertainment of the Royal household. By the 14th century the idea of a play as existing in its own right, for amusement as well as instruction, was widely accepted. The way was thus open for the development of the Miracle or MYSTERY PLAY, a dramatization in rhyming English verse of episodes from the Old and New Testaments chosen to reinforce the central doctrine of Christianity—the redemption of fallen mankind, in the persons of Adam and Eve and their descendants, through Christ's Crucifixion and Resurrection. This achieved an even greater intensity with the establishment in 1311 of the feast of Corpus Christi—the Body of Christ—symbolically present in the Host which was the focal point of the new festival. Devised, financed, and cast locally, the preparation and production of a Mystery play involved everyone in the town where it was being presented, and it therefore became not only a corporate act of worship but also a community project run by local people—clergy and laity alike—for their own amusement. Practically complete texts of such cycles of plays survive from a number of English towns, among them Chester, Coventry, Lincoln, Wakefield, and York, and fragments from others. Evidence establishing the existence of many more, whose texts probably perished during the Reformation, exists in Account Rolls and Minute Books from all over Britain, including London. Just as the surviving texts differ in structure, characterization, and dialect, so conditions of performance must have varied to suit local conditions. In Lincoln the entire cycle of plays was presented on a single fixed stage; Chester, Wakefield, and York appear to have preferred to stage each scene on a separate pageant-wagon, which perambulated the streets, stopping in turn at fixed points in the town with sufficient open space for the accommodation of an audience. The time-span also varied. One day, from dawn to dusk, seems to have sufficed for a performance at York, while in Chester three days were needed; in London the time was extended to seven days. Dates of performance also differed, Chester preferring Whitsun and Lincoln St Anne's Day (26 July). An even more important variation was the occasional substitution of another long but wholly different play. Beverley sanctioned a Paternoster play, as did Lincoln. York preferred a Creed play (based on the Apostles' Creed). No texts of these have survived, but from other evidence it may be presumed that they resembled the didactic MORALITY PLAY which, with its shorter version the MORAL INTERLUDE, had developed alongside the Mystery play, as had the specialized SAINT PLAY, often written to celebrate a patronal festival. The only surviving text of a play which may have been used as an alternative to the more general Mystery play is *The Play of the Sacrament* from Croxton, Norfolk, which dates from about 1475.

It was from among the players in the Moral Interludes, which were not tied to any particular festival but could be performed at any time with a small cast, that the first professional actors arose in the course of the 15th century. Direct descendants of the multifarious itinerant entertainers of the early Middle Ages, they were probably employed in menial capacities as household servants in the palaces and castles of the nobility and the abbeys and priories of the superior clergy. They therefore had a trade to fall back on in hard times; but once they had decided to turn their mimetic skill to personal advantage, they had a readymade audience and could probably rely on the household chaplain to provide them with scripts. By the end of the century resident companies, usually consisting of four men and a boy, were being maintained not only by the King, whose *Lusores Regis*, or 'Players of the King's Interludes', flourished from about 1490 to 1580, but by the nobility at large. The rush-strewn floor of a hall or refectory, with its dais and high table at one end and its minstrels' gallery and serving screen at the other, provided their playing-place. When they travelled, as they were allowed to do in summer, wearing their master's livery and carrying his letters of permission to avoid arrest for vagrancy, wooden planks mounted on barrels in front of a canvas changing-room served the same purpose, whether at fairgrounds, in market squares, or on village greens. In halls they could claim a fixed reward for each performance, but elsewhere they were at the mercy of the collecting-box. In such a small company, doubling and trebling of parts was a prime requirement, as was the versatility needed to switch swiftly from one role to another.

These private companies maintained by wealthy individuals were quickly imitated in the larger towns by groups of artisans, who also toured their plays, at least within their own county boundaries, as did many villagers with their humbler folk-games and dances. Churchwardens' accounts of the time abound in entries recording payments made to, or sums received from, 'players' and 'gamesters' who journeyed to and from their local parish on public holidays to entertain their neighbours. The word 'game' in this context is as important as 'play', since it helps towards an understanding of the normal holiday spectacle of the late Middle Ages, which included, besides a fully dramatic play, game, or interlude, whether religious or secular, such activities as tilting and jousting, sword-play, bull- and bear-baiting, cock-fighting, wrestling, tumbling, and puppetry, a situation which obtained from the 12th century until well into the Jacobean period.

The cross-fertilization of clerical and secular forms and methods of entertainment was a constant source of strength to both genres and helped to preserve dramatic art as the most popular and widespread of community recreations. It injected a sense of moral or social purpose into FARCE, as can be seen in Lydgate's *Mumming at Hertford* (*c.* 1425), and saved dramatized sermons from becoming boring, as is evident in the Morality play *Everyman* (*c.* 1495). Sometimes, more especially during the FEAST OF FOOLS and at Carnival time, buffoonery, parody, and satire passed beyond tolerable limits. The players were then open to charges of obscenity, blasphemy, and sedition, which inevitably provoked retribution from either the ecclesiastical or the civic authorities, or both, in the form of injunctions and prohibitions supported by threats of excommunication, fines, and imprisonment. Fully documented examples of this survive from places as widely separated as London, Lincoln, Bristol, and Exeter; and on occasion such objections could be extended to blanket condemnation of all acting, as is exemplified by the *Tretis against Miraclis* (*c.* 1380).

The accession of Henry VII in 1485 coincided with a revival of classical learning in Italy which led to practical attempts to present Latin plays in the light of contemporary studies. English scholars brought back to their own country news of this renaissance of classical drama, and both Henry VII and Henry VIII actively encouraged similar revivals at Court and in the universities and schools throughout the country. The influence of Latin comedy showed itself more quickly in farces presented in secular environments than in plays on religious subjects, and is first apparent in such works as Henry Medwall's *Fulgens and Lucrece* (1497), John RASTELL's *Four Elements* (*c.* 1517), John Redford's *Wit and Science* (*c.* 1539), and particularly in the farces of John HEYWOOD. Added impetus stemmed from the work of dons and schoolmasters who had an interest in encouraging the study of Latin. Implicit in this situation, however, was the acceleration of a split in the audience, first occasioned by social and educational advantages, which would lead eventually to the building of 'private' as distinct from 'public' playhouses in Jacobean London. The process was gradual and did not, as happened elsewhere in Europe, result in the abrupt relegation of popular theatrical traditions to the untutored realms of folk-culture. The reason for this is to be found in the effect of the Reformation on English playwrights and actors, which after 1530 served as a counterweight to any Latin or Italian innovations. What had until then been regarded as the voice of Apollo speaking from Parnassus became that of Anti-Christ speaking from Hell. The traditional Morality play, already shifting under humanist influence from an exclusive concern with the salvation of the individual soul to a broader interest in the corporate welfare of the State, as exemplified in John SKELTON's *Magnyfycence* (*c.* 1515), became, under the influence of Swiss and German playwrights, an instrument of propaganda. Also admirably fitted for this purpose, as is shown by John BALE's *Kynge Johan* (1536), was the Saint Play, concerned as it was with persecution and martyrdom. Not surprisingly, this sudden translation of traditional religious drama into the arena of politics caused a sharp reaction on Catholic spectators, who responded to satirical, in their view blasphemous, attacks on their faith by recourse to their fists and swords.

The inevitable result of such breaches of the peace was government intervention, leading to legislation, censorship, and ultimately, under James I, to the absolute control of actors, plays, and playhouses by the monarch (see LORD CHAMBERLAIN); but, as successive governments found, it was one thing to legislate, another to enforce, and the further a community lay from London the easier it was to ignore or defy the new laws with impunity. Changes therefore occurred piecemeal, particularly as the country as a whole was constantly fluctuating between Protestant and Roman Catholic policies under Edward VI, Mary I, and Elizabeth I successively. The gainers in all these upheavals were the adult professional players and the established companies of choir-boys, schoolboys, and students attached to corporate bodies. The losers were necessarily the amateurs. Henry VIII's sharp curtailment of the number of feast days, and the suppression in 1548 of the festival of Corpus Christi, cut the occasions on which they could act; and the official animus against religious plays (which, incidentally, had been banned in France in 1548) deprived them of the whole of their repertory. Even those Mystery cycles which survived longest in the Midlands and the North, notably at Chester, York, Wakefield, and Coventry, were finally forbidden between 1574 and 1581.

The suppression of religious drama had far-reaching results, and by 1590 it was no longer possible for playwrights to discuss religious issues in their plays, the last to do so being Christopher MARLOWE in *Dr Faustus* (*c.* 1592). In future politics and religion would have to be approached by way of history or mythology, by allegorized devices, and by indirect allusion. By and large the new rules were respected, since economic ruin faced the players if they were not; but on occasion the forbidden subjects were treated too openly, bringing their authors before the courts and often landing them, together with the actors, in prison. With the loss of religious subjects for plays went the loss of churchmen as the normal suppliers of new works and a corresponding growth in the demand for scripts from other sources. This was met in part by schoolmen like Thomas Ingelend with *Nice Wanton* (1550) and *The Disobedient Child* (1560), or Richard Edwards with *Damon and Pythias* (1565) and *Palamon and Arcite* (1566). The first English comedy written under classical influence, *Ralph Roister Doister* (*c.* 1552), was by a school-master, Nicholas UDALL, and was performed by schoolboys, as was probably *Gammer Gurton's Needle* (*c.* 1560) ascribed to William STEVENSON; from the Inns of Court in London came Thomas NORTON and Thomas Sackville, joint authors of the first regular English tragedy *Gorboduc* (1562). These plays were not intended for 'popular' audiences; those were catered for by men like Thomas PRESTON, author of the hybrid tragi-comedy *Cambyses* (*c.* 1570). It was from a fusion of these two types of play that English drama was to develop rapidly in the last 20 years of the 16th century, through the work of the so-called 'University Wits' like GREENE, LYLY, NASHE, and PEELE and the humbler journeymen-playwrights like CHETTLE, DEKKER, JONSON, MUNDAY, and Shakespeare, the last being considered in his own day only one among many.

The great advance in modern English drama that began with the plays of Marlowe, particularly *Tamburlane* (1587), has sometimes been accounted for by saying that Marlowe was a great poet and dramatist while his immediate predecessors were only well-meaning purveyors of entertainment. Even if this were true, the advance in the physical conditions under which plays were performed in public must be taken into account, and in this the legitimizing of regular weekday performances from 1574 onwards, together with the subsequent erection of permanent playhouses, played an important part. Two factors governed the establishment and later development of Elizabethan theatres—the professional actor-manager's desire to possess a home of his own in which to perform regularly and the availability of adequate capital with which to lease or buy a suitable building. As long as companies lacked these resources they had to rely on finding somewhere to rent, sometimes a disused chapel or a hall but usually an inn-yard or, as is now thought more probable, a large room in an inn. Among the London INNS USED AS THEATRES the best known were probably the Bell, the Bel Savage, the Bull, the Cross Keys, and the Red Lion. Wherever they went the actors ran the risk not only of failing to find a suitable place in which to act, whether in London or in the country, but, when they did, of being forbidden to use it by the City Council. When in 1576 James BURBAGE leased land on which to build England's first playhouse, the THEATRE, for LEICESTER'S MEN, and when Richard Farrant adapted rooms in an abandoned monastery on Crown land near St Paul's Cathedral and made it into the first BLACKFRIARS THEATRE for the use of the Children of the Chapel Royal, both were careful to choose sites which did not come within the jurisdiction of the City of London, and both had to borrow money for building or conversion work. The pattern thus set for alternative types of playhouse continued. One type, the so-called 'public' playhouse, like the Theatre and its successors the CURTAIN, the FORTUNE, the GLOBE, the ROSE, and the SWAN, all erected before the end of 1600, and the later RED BULL (1606) and HOPE (1614), was open to the sky (and the weather) and, with standing room round the stage and galleries for seated patrons, held a large number of people; its admission prices were correspondingly low, ranging between one penny and sixpence. The other, of which the Blackfriars was the prototype, included the COCKPIT, the SALISBURY COURT, and the WHITEFRIARS; these 'private' theatres, having sacrificed size for the protection afforded by a roof, had to compensate for smaller audiences by charging higher prices. Thus financial considerations were added to the academic and social factors which had already split the originally homogeneous audience of medieval times into two sections—what Ben Jonson called 'the sharp and the learned' and those with 'grounded judgement'. The first group supported the plays presented by the BOY COMPANIES, in which refined verbal wit was allied to music and spectacle in the smaller and more select 'private' theatres, while for the larger, less discriminating audiences of the 'public' playhouses, bustling stage-action, matched by rousing blank verse and presented by professional adult actors, became the primary attraction. The boys, as long as the fashion for them continued, were safely ensconced under the patronage of the Court, being mainly connected with the Chapels Royal; the adult companies had from their inception been attached to the households of members of the nobility. Among them, roughly in order of their foundation, were the 16th Earl of OXFORD'S MEN, who in 1580 amalgamated with the Earl of Warwick's Men; the Earl of LINCOLN'S MEN, known also as Clinton's Men, from the Earl's family name; STRANGE'S MEN, maintained by the heir to the 4th Earl of Derby, who also had his own company, as did his grandson the 6th Earl who succeeded to the title within eighteen months of his grandfather's death; and, a latecomer, the Earl of PEMBROKE'S MEN, who, together with Strange's Men, are believed to have employed Shakespeare when he first arrived in London. The two most important companies, however, were undoubtedly the CHAMBERLAIN'S MEN, active from about 1594, with whom Shakespeare was to be intimately connected for much of his working life, and their only serious rival the ADMIRAL'S MEN, active from about 1585. By the end of Elizabeth I's reign the rivalry between the two types of play and presentation had become a subject for bitter and angry comment within the text of the plays themselves, and a veritable 'war of the theatres' began, as can be seen from Shakespeare's reference in *Hamlet* II.2 to the 'little eyases' who 'are now in the fashion'.

The chief provider of plays for the Boy Companies was John Lyly, who did for comedy what Marlowe had already done for tragedy. The latter had given the English theatre the resonance and amplitude of blank verse, allied to a sense of the grandeur and inevitability of tragic events; Lyly's delicate and sensitive prose gave comedy a finer edge and showed how humour could be combined with sentiment, wit with romance. Few of his contemporaries achieved his unique blend of fantasy and reality, but something of it shows in Peele's *Old Wives' Tale* (*c.* 1590) and again in Dekker's *Old Fortunatus* (1599).

Thus, when in about 1590 Shakespeare began writing for the stage, he found several types of play already firmly established. The oldest, and in many ways the most popular, was the long rambling 'chronicle history', with its roots far back in the Saint play. Shakespeare used it for his first effort, the three-part *Henry VI*. Rivalling the chronicle history in popularity was the 'revenge' tragedy, of which the worthy prototype was KYD's *The Spanish Tragedy* (*c.* 1585–9). This introduced to the English stage the horrors of SENECA's full-blooded drama; in his youth Shakespeare wrote *Titus Andronicus* in this style, and in his maturity *Hamlet*. The influence of Marlowe's more subtle tragedy can be seen in *Richard III*; and that of Lyly in such early comedies as *Love's Labour's Lost*, and in Shakespeare's love of wordplay.

If Shakespeare may be said to have succeeded by going with the tide, his contemporary Ben Jonson spent most of his energies swimming against it, aiming to write realistic comedies that would expose and correct the follies of his age. He was not always successful; but in some of their highly effective scenes, his comedies based on the 'humours' of mankind—envy, greed, malice, melancholy—show an unusual power of concentrated satire; and his conscientious craftsmanship is in striking contrast to the careless and often incoherent plotting of many of his less painstaking contemporaries.

By the end of the 16th century the theatre in London was in a somewhat uncertain situation. To the rivalry between the boy and adult companies was added the constant bickering of the adults among themselves, or with the owners of the playhouses, or with the civic authorities over the perennially vexed issues of the control of theatres and the censorship of plays, all aggravated by the rising tide of hostility to any form of amusement consequent upon the increasing strength of Puritanism. With the accession of James I in 1603 stern measures were taken to reduce and contain the opposing forces which threatened to tear the theatre apart. The companies were put under the direct control of the royal household and allotted to specific playhouses, the Boy Companies, as the King's Revels and the Queen's Revels, staying in Whitefriars and Blackfriars respectively. The adult companies were reorganized, the Chamberlain's Men becoming the KING'S MEN at the Globe, and the Admiral's Men, at the Fortune, Prince Henry's Men. After the death of their patron in 1612 they were allotted as Palsgrave's Men to the Elector Palatine, husband of James's daughter Elizabeth. A third company, made up of Oxford and Worcester's Men, became QUEEN ANNE'S MEN, and took over the Curtain. Up to the closing of the theatres in 1642 these arrangements remained in force, though they necessarily became somewhat relaxed, new playhouses being eventually licensed and new companies formed—the LADY ELIZABETH'S MEN in 1611 and PRINCE CHARLES'S MEN (usually known as the Prince's Men) in 1612. After the accession of Charles I this company was disbanded, the former Palsgrave's Men taking over the name under the control of the future Charles II while a new company, QUEEN HENRIETTA'S MEN, was formed to play at the Cockpit. Queen Anne's Men had already been renamed the Red Bull company and occupied that theatre on the death of their patron in 1619. One great change which had already taken place by then was the demise of the Boy Companies, the adult companies having put them out of business by taking over both their repertoires and their theatres. The King's Revels lasted only a year at Whitefriars, which was then taken over by the Queen's Revels who survived until 1615, their former theatre at Blackfriars having been occupied by the King's Men in 1609.

Whatever reservations London's city fathers may have had about James I's autocratic assumption of control over the theatres, they continued to enjoy good enough relations with the leading playwrights and actors to employ them to write and act in the pageants prepared for his coronation and annually thereafter to mark the installation of a new Lord Mayor. Dekker, Jonson, MIDDLETON, WEBSTER, and Thomas HEYWOOD all wrote texts for such occasions, and adult professionals as well as boy actors were paid to perform in them. Towering above all others in this period of transition were the actors Richard BURBAGE and Edward ALLEYN, the former closely associated with the plays of Shakespeare, the latter with those of Jonson. Shakespeare, as chief playwright to the King's Men and one of the part-owners of their theatre, needed no other patron. Jonson, having failed to win that prize, looked for support to Queen Anne and Prince Henry and to the universities and Inns of Court. While Shakespeare continued to try and meet the tastes of all playgoers, Jonson, determined to be recognized as a 'man of letters'—and supported in this ambition by CHAPMAN, BEAUMONT, FLETCHER, and others—sought to wean English audiences away from the traditional chronicle plays and 'mongrel tragi-comedies' towards Aristotelian principles of play construction that deliberately separated 'comedy' from 'tragedy' and both from 'pastoral'. Court audiences (more especially under Charles I and his French consort Henrietta Maria) welcomed the PASTORAL for the excuse it offered for elaborate scenic and musical decoration of the kind that Jonson and others, in collaboration with the designer Inigo JONES and the composers Thomas CAMPION, Matthew Locke, and Henry Lawes, were pioneering in Court MASQUES; and it was along this route that lyric drama advanced, from the homespun fairyland of Shakespeare's *A Midsummer Night's Dream* via Fletcher's *The Faithful Shepherdess* (1609) and Milton's *Comus* (1634) towards opera.

The substitution of satire for romance as the basis of the Jonsonian 'comedy of humours' served similarly to shift attention away from lyrical subject matter, settings, and poetry towards a harsher and more prosaic depiction of mercantile and domestic concerns in city life conspicuous in the plays of Middleton and MASSINGER, or the intrigues of the leisured few in those of James SHIRLEY and Richard BROME.

Tragedy of the kind envisaged but not achieved by Jonson failed to win a popular following; it was not until Milton wrote *Samson Agonistes* (1671) that its potential was fully realised in England. Only John FORD succeeded in writing tragedies that merit continued revival.

The steady shift in the tone of the drama away from imaginative, poetic visions of life towards rational explanations of it is marked by a gradual downgrading of the importance attached by writers of the period to the 'public' playhouses and a corresponding increase of interest, by writers and actors alike, in the manners and opinions of the select audiences of the intimate and exclusive 'private' playhouses; also by a steady decline of support for plays of any sort in provincial cities, where after 1623 Account Books reveal that even the royal acting companies were, with increasing frequency, 'paid not to play'. The desire to cultivate French and Italian models of neoclassical dramatic theory and practice is reflected in the Palladian architectural features of the last theatres to be opened in London before the Civil War—the former Cockpit in Whitehall Palace transformed into a Court Theatre by Inigo Jones in 1629/30, and the Salisbury Court which opened in 1631.

When the Civil War finally broke out in 1642 the theatre was one of the first casualties. The London playhouses, long frowned upon by the Puritans, were closed by Act of Parliament—the logical end to the processes of control and censorship begun following the Reformation in 1531—and dramatic performances virtually ceased. Only a few desperate or rebellious actors tried to evade the ban by putting on surreptitious productions in some of the abandoned theatres, mostly the smaller ones such as the Cockpit; they were speedily discovered and suppressed, being either fined or imprisoned for their flouting of the law. Not until a few years before the ending of the PURITAN INTERREGNUM was there a partial relaxation of the ban on acting, when Sir William DAVENANT was permitted in 1656 to stage *The Siege of Rhodes*, a musical play with changeable scenery, at Rutland House.

With the return of Charles II in 1660 the theatre quickly revived; but the habit of theatre-going had been lost under the Commonwealth, and for many years to come theatre audiences would consist mainly of courtiers, upper-class fribbles and their hangers-on, rakes, bullies, and ladies of pleasure, young Inns-of-Court men, and a few *nouveaux riches*. The two PATENT THEATRES which were all London could support—Davenant's King's Men at LINCOLN'S INN FIELDS and later at DORSET GARDEN, KILLIGREW's Duke's Men at what was to become DRURY LANE—had to draw on pre-Commonwealth drama for their plays, and it is significant that the playwrights most acceptable to the new audience were not Shakespeare and Jonson but men like Beaumont and Fletcher and Shirley, amusing, easy to listen to, and making no intellectual demands.

One of the most striking innovations of the newly founded theatre was the introduction of actresses in place of the former boy players. The early ones are somewhat shadowy figures except for Nell GWYNN, who was with Killigrew for a time; but in the persons of Mrs BARRY and later Anne BRACEGIRDLE, who both played opposite Davenant's leading man Thomas BETTERTON, they became a force to be reckoned with. There can be no doubt that they contributed greatly to the success of the Restoration 'comedy of manners', the one important contribution of the age to English drama. (In fact only a small part of this comedy falls within the Restoration period proper; it reached the height of its success under William and Mary and came to an end under Queen Anne.) The first of its impertinent young men and witty assured young women, who were to reappear with little variation in the comedies of the next forty years or so, are met with in George ETHEREGE's *Love In A Tub* (1664) and *She Would if she Could* (1668), where they have an air of innocence that disarms criticism and leaves the spectator hardly aware of the implications of their intrigues. It is far otherwise in the plays of William WYCHERLEY where, in the world of inverted moral values depicted in *The Country Wife* (1673) and *The Plain Dealer* (1676), cuckolding has become the main business of life, everything else being subordinated to the battle of the sexes, the successful management of an intrigue, the ridicule of marriage, and a nice discrimination between fools and men of sense. All this was summed up by CONGREVE in his last play, *The Way of the World* (1700), where his aristocratic detachment, flawless style, and a certain remoteness from reality enable him to skate over the thinnest of ice; this is intellectual comedy at its best. John VANBRUGH, whose best plays *The Relapse* (1696) and *The Provoked Wife* (1697) have stood the test of time almost as well as *The Way of the World*, was more of a realist. His characters are no less indecent than Congreve's but we feel the indecency more because they are flesh and blood and are not the exquisitely artificial phantoms that Charles LAMB later declared all these Restoration characters to be. George FARQUHAR, with whose death in 1707 Restoration comedy may be said to have come to an end, has little of Congreve's deliberate perfection. In *The*

Recruiting Officer (1706) and *The Beaux' Stratagem* (1707), intellectual comedy is beginning to dissolve; his young men are still rakes and scamps, but they are not heartless. The change is no doubt due in part to Farquhar's warm-hearted Irish temperament; but it is probable too that the celebrated attack by Jeremy COLLIER on the profaneness of the English stage, published in 1698, was already having an effect, and audiences were beginning to be dissatisfied with, perhaps even a little shocked by, the dry wit of inconstant lovers. With the turn of the century comedy became progressively more decent, if also a good deal less amusing.

Of the other comedies written during the period the most interesting are probably SHADWELL's attempts to revive the 'comedy of humours'. He was a shrewd observer of the contemporary scene, and such plays as *The Squire of Alsatia* (1688) and *Bury Fair* (1689) have point and good sense. The 'comedy of intrigue' (usually borrowed from Spanish sources) was the speciality of Mrs Aphra BEHN. In bulk it becomes excessively tedious; it is rarely leavened with wit, though it has plenty of impudence, and relies for most of its interest on the comic vicissitudes of the plot. Otherwise only an occasional comedy such as TUKE's *Adventures of Five Hours* (1663), OTWAY's *The Soldier's Fortune* (1680), or CROWNE's *Sir Courtly Nice* (1685), stand out from the rest, and there is little interest today in the works of such writers as D'URFEY or TATE.

The late 17th century was not a heroic age and it is not surprising that on the whole its tragic drama is inferior to its comedy. The period produced a form peculiar to itself, the rhymed HEROIC TRAGEDY wittily satirized in BUCKINGHAM's *The Rehearsal* (1671). John DRYDEN's *The Conquest of Granada* (1670/1) is typical of the genre, full of rant and hyperbole, characterized by the constant juxtaposition of Love and Honour, by violent action, and by startling reversals of fortune. Luckily Dryden also wrote a number of comedies. *Marriage à la Mode* (1671) is among the more memorable Restoration plays and the rest are deservedly forgotten. In blank verse tragedy he was more successful, particularly with *All For Love* (1677) (based on Shakespeare's *Antony and Cleopatra*) and *Don Sebastian* (1689). In the general dearth of tragedy two by Thomas Otway stand out—*The Orphan* (1680) and *Venice Preserv'd* (1682). They were graced by the presence of Betterton and Mrs Barry when first produced and continued to move audiences until well into the 19th century. Both have been revived in the present century, *Venice Preserv'd* notably in 1953 with Paul SCOFIELD and John GIELGUD as Pierre and Jaffier and Pamela BROWN as Belvidera. Their pathos must have come as a welcome change from the rant and bluster of such writers as Nathaniel LEE or the vapidity of John Banks.

At the beginning of the 18th century London still had only two theatres, and from 1683 to 1695 only one, when quarrels between actors and management led Betterton to take the best actors from the two Patent Theatres into a United Company at Drury Lane, from which he was ousted by Christopher RICH. He then settled in the old theatre in Lincoln's Inn Fields until in 1705 a new theatre, built by Vanbrugh, opened as the QUEEN's (on the present site of HER MAJESTY'S) with Vanbrugh's own play *The Confederacy*. It soon proved acoustically unsuitable for drama, and became an opera house. Meanwhile John RICH, who had inherited Davenant's patent from his father, remained at Lincoln's Inn Fields where in 1728 he made a popular and financial success with GAY's *The Beggar's Opera*, until in 1732 he opened a new theatre in Bow Street, COVENT GARDEN, which, after extensive rebuilding, still stands today as the London home of opera and ballet. Rich, who was an excellent dancer and mime and a renowned HARLEQUIN, had transformed WEAVER's 'Italian Night Scenes' into the forerunner of the modern English PANTOMIME. This was so popular that it became a threat to the serious drama as great as that of Italian opera, which in London under Queen Anne drew away from the playhouses to the opera houses large audiences composed of most of the nobility and people of fashion. The actors struggled on, as did the playwrights, though compared with the previous century, English drama showed a sharp decline. Tragedy in the first decade was represented mainly by the works of Nicholas ROWE, who followed at a respectful distance behind Otway with *The Fair Penitent* (1703) and *Jane Shore* (1714). Joseph ADDISON showed the influence of French neoclassical tragedy in his *Cato* (1713) which, like the plays of several tragic poets of the time, is now forgotten. *Irene* (1749) is remembered only because it was written by Dr JOHNSON; and HOME's *Douglas* (1756), though it proved its viability in a revival at the CITIZENS' THEATRE, Glasgow, in 1950 with Sybil THORNDIKE as Lady Randolph, is best known as the source of a once-popular passage for recitation, 'My name is Norval.' A more interesting development was the 'domestic tragedy' best represented by the plays of George LILLO and Edward MOORE, particularly *The London Merchant* (1731) and *The Gamester* (1753) respectively. In them there is a good deal of sentiment and a strong insistence on the middle-class virtues; but no great dramatist arrived to lift the genre above interesting mediocrity.

In comedy the record is not quite so depressing. At the turn of the century Colley CIBBER, in *Love's Last Shift* (1696), was allowing the rake as much freedom as before for four acts, but pulling him up short in the fifth and dismissing him as a reformed character; while Mrs CENTLIVRE, in *The Busybody* (1709) and *A Bold Stroke for a Wife* (1718), considerably toned down the bawdiness of Mrs Aphra Behn's 'comedy of intrigue'. The moral tone of English drama under Queen Anne was distinctly less libertine than before, giving evidence of a new, if modified, Puritanism, and STEELE, whose essays in the *Tatler* and *Spectator* had done much to amend contemporary morals and manners, used comedy for the same end. His four sentimental comedies, culminating in *The Conscious Lovers* (1722), were all written to render vice contemptible and virtue attractive. They suffer from the faults inherent in the genre: pious sentiments fall flat, laughter is drowned in tears of repentance or forgiveness, and while the characters are enunciating good resolutions reality unobtrusively escapes. Yet they are robust when compared with the mawkish plays of Hugh KELLY and Richard CUMBERLAND, where 18th-century sentimentalism sobs itself to a standstill; the success of *She Stoops to Conquer* (1773) by Oliver GOLDSMITH indicated that public taste was turning against too much sentimental refinement. In the plays of SHERIDAN the theatre returned to the comedy of manners, with all its traditional wit and perfection of phrase but none of the indecency. He had many emulators; good examples of the more ordinary run of 18th-century comedy are *The Suspicious Husband* (1747) by Benjamin HOADLY and *The Jealous Wife* (1761) by George COLMAN the elder.

The early part of the 18th century saw the introduction and subsequent popularity of the AFTER-PIECE, often a full-length comedy cut short to provide late-comers to the theatre (usually middle-class business men and their families) with solid fare in place of the previous light diversions of singing and dancing given after a five-act tragedy. As the popularity of the after-piece grew, it came to be specially written, often by actors like FOOTE or MURPHY. Other genres in which the 18th century excelled were the BURLESQUE—Carey's *Chrononhotonthologos* (1734), FIELDING's *Tom Thumb* (1730), and above all Sheridan's *The Critic* (1779)—and the BALLAD OPERA. A word too should be said of the continuing popularity of the 17th-century prologue and epilogue throughout the century; they point to an age in which the theatre was still very much a social rendezvous and the audience could be addressed directly and made pleasantly conscious of itself. The century brought some interesting changes in theatrical representation and acting, some of which must be credited to David GARRICK who, with his leading ladies Peg WOFFINGTON, Kitty CLIVE, Mrs BELLAMY, and Mrs ABINGTON and such actors as Tom KING, Harry WOODWARD, and Richard YATES, reigned supreme at Drury Lane for many years. He rescued English acting from the stylized declamation into which it had fallen with QUIN and banished the audience from the stage, where they had sat intermittently since Elizabethan times. Under him new lighting was installed, more attention was paid to the supporting actors, and towards the end of the century some attempt was made to replace the contemporary costumes worn for every type of play by something more in keeping with the period in which the play was set. The main difficulty the actors had to contend with was the enormous size of the two rebuilt Patent Theatres, which at the beginning of the 19th century were still the only ones licensed to play LEGITIMATE DRAMA. Intimate or naturalistic acting was impossible, and both playhouses concentrated on providing their large and often undiscriminating audiences, swollen by the influx of a moneyed middle class, with plays that combined the maximum amount of spectacle, music, colourful costumes, and swift, violent action with as little dialogue as possible—in short with MELODRAMA, the most popular product of the time. Introduced by Thomas HOLCROFT at the turn of the century, it took the place of both tragedy and comedy and reigned supreme for more than 50 years. Serious writers withdrew from the theatre, and the smaller, unlicensed theatres which sprang up—among them the ADELPHI, the HAYMARKET, the LYCEUM, the OLYMPIC, and the SURREY—housed the lighter fare provided by such journeyman entertainers as Charles DIBDEN, J. R. PLANCHÉ, and Frederick REYNOLDS. Other theatres throughout London and the suburbs relied on productions which evaded the LICENSING LAWS by adding songs even to straight plays; and together with farce and burlesque such new genres as EXTRAVAGANZA and operetta flourished, until in 1843 a new act was passed regulating the theatres and the old monopoly of the Patent Theatres came to an end. The loosening of the stranglehold exerted by Covent Garden and Drury Lane, jealous guardians of their privileges under the patents of 1660, undoubtedly contributed not a little to the dramatic revival in the later decades of the 19th century; but meanwhile the serious dramatists of the first half of the century, among them BYRON and TALFOURD who both wrote plays for MACREADY, were working in an outworn tradition, and their plays are forgotten. More successful at the time but equally forgotten were James Sheridan KNOWLES, with *The Hunchback* (1832), and BULWER-LYTTON, with *The Lady of Lyons* (1838) and *Money* (1840), the latter being one of the first plays to attempt to deal with a contemporary theme. Most of the poets of the 19th century tried their hands at playwriting, but all of them, including COLERIDGE, SHELLEY, BROWNING, and TENNYSON, dwindling at last to Stephen PHILLIPS, continued to write for the Victorians as if they were Elizabethans or ancient Greeks. When a poetic play had any success on the stage it was due to the actor rather than the dramatist, as when IRVING put on Tennyson's *Becket* (1893) or ALEXANDER staged Phillips's *Paolo and Francesca* (1902). In the second half of the century realism began to creep in, with Tom TAYLOR's *The Ticket-of-Leave Man* (1863) and the cup-and-saucer dramas of T. W. ROBERTSON, who provided a number of plays for the BANCROFTS at the PRINCE OF WALES', in which recognizable contemporary characters found themselves involved in contemporary dilemmas. Neither achieved the immediate success of the prolific Dion BOUCICAULT, whose plays, melodramatic in form and content as they often were, nevertheless showed some excellence in character study and a vast amount of technical skill. His first success, *London Assurance* (1841), brought financial stability to the management of Mme VESTRIS and Charles James MATHEWS at Covent Garden, where the KEMBLE family, led by John Philip and Mrs SIDDONS, had reigned supreme for so long while MACKLIN and Edmund KEAN bolstered up the fortunes of Drury Lane. In a revival by the ROYAL SHAKESPEARE COMPANY in 1970 *London Assurance* again proved a sparkling comedy of contemporary manners.

Meanwhile the influence of the French 'WELL-MADE' PLAYS of SARDOU and SCRIBE had been felt in London, and for a time dominated the work of English dramatists. Henry Arthur JONES and Arthur Wing PINERO were, however, opening windows on to modern life and letting some much-needed fresh air into the run-of-the-mill playhouses. Jones was a serious dramatist and a conscious pioneer whose *Saints and Sinners* (1884), in spite of some melodramatic features, introduced the Victorians to a naturalism they had not yet met in the theatre, though they were already familiar with it in prose fiction. Pinero's *The Second Mrs Tanqueray* (1893) gave them another shock. As George Bernard SHAW later pointed out, both Jones and Pinero tended to run away at the last moment from the issues they had raised, and the behaviour of their characters is still governed more by the conventions of the theatre than by those of life itself. They had the success that comes to men of talent who walk ahead, but not too far ahead, of their contemporaries. In contrast, Oscar WILDE was writing brilliant artificial comedies of which *The Importance of Being Earnest* (1895) is the most likely to survive, and Victorian wit was finding its most typical expression in the comic operas provided by W. S. GILBERT in collaboration with the musician Arthur Sullivan, from *Thespis* (1871) to *The Grand Duke* (1896). He also wrote a number of now forgotten plays for outstanding actresses of the day, among them Madge KENDAL and Mary ANDERSON.

New theatres sprang up all over London, and many new ones were also built in the provinces, where the old STOCK COMPANY and CIRCUIT system were breaking down under the impact of TOURING COMPANIES. These took advantage of the network of railways which soon spread across the countryside to present replicas of London successes to provincial audiences. A new phenomenon which arose roughly in the 1850s was the MUSIC-HALL, which soon created its own stars, most of whom never appeared in a play but like Marie LLOYD, Vesta TILLEY, George ROBEY, or Albert CHEVALIER relied entirely on solo turns. Soon every town had its own music-hall, the larger ones very often having local stars, some of whom went to London and returned on tour as nationally-known entertainers while others were content to remain at home and build up a personal relationship with their faithful audiences. From SCOTLAND, which had struggled for many years to build up its own indigenous theatre, came a special brand of Scottish comedians, of whom the best known outside Scotland was probably Harry LAUDER. Irish comedians, of whom the first was Sam COLLINS, were also popular, and from America came the MINSTREL SHOWS, at first played by white men with blackened faces but later by true Negroes. Their performances, unlike those at the music-halls, were intended for family entertainment, and when the fashion for them died out some of the participants, like George Chirgwin, Eugene Stratton, and G. H. Elliott, migrated to the 'halls' to start afresh on their own.

Many of the new theatres built in London at the turn of the century were controlled by a newcomer to the theatrical scene, the actor-manager who starred in his own company in plays chosen and directed by himself, usually spending the winter in London and the summer on tour. The greatest was Irving, who for many years made the Lyceum London's leading theatre; but equally popular and hardworking were Alexander at the ST JAMES'S and TREE, first at the Haymarket, later at Her Majesty's. In most of the theatres of London could be seen members of the TERRY family, from Kate and Ellen, the eldest daughters, who first appeared on stage in the late 1850s, down to Fred, the youngest son who with his wife Julia NEILSON was still active in the 1920s. Their children and grandchildren continued the family tradition, one of the outstanding actors of the mid-20th century, John Gielgud, being Kate's grandson.

By the end of the 19th century the theatre in all its ramifications was the main public amusement; but the glamour and scenic splendour of most of the productions did not blind some people to the fact that there was little serious drama, and what there was was often lost in the flood of meretricious entertainment. In the 1890s some enthusiasts for the 'theatre of ideas' which was already making headway on the Continent, particularly with the advent of such dramatists as IBSEN and STRINDBERG, started a reaction against what they regarded as the despotism of actor-managers, the policy of the long run, the overpowering use of scenic spectacle, the escapism of romantic melodrama, and the stereotyped form and conventional morality of the 'well-made' play. The modern movement may be said to have begun with Shaw, whose first play *Widowers' Houses* was presented privately by J. T. GREIN's Independent Theatre. Neither this venture nor the New Century Theatre founded by William ARCHER survived the 1890s, but as a result of their pioneer work Grein and others were able to establish the STAGE SOCIETY, a private play-producing group which for the next 40 years arranged Sunday performances of experimental and banned plays in West End theatres. Harley GRANVILLE-BARKER staged a series of brilliant productions at the ROYAL COURT THEATRE between 1904 and 1907 which set new standards in ensemble acting and in the use of simple but significant décor. His innovation in banning long runs and adopting the repertory system already in use in Germany and elsewhere (which was in essence only a return to the old theatre habit of constant changes of programme) had important repercussions in the provinces, where in 1907 Miss HORNIMAN established in MANCHESTER the first REPERTORY THEATRE, to be followed in 1911 by that in LIVERPOOL and in 1913 by that in BIRMINGHAM.

Among the writers of new drama Shaw stands somewhat apart. His range of interest was wider than theirs, his plots enlivened by touches of the fantastic and the absurd, and the characteristic incisiveness of his dialogue is blended with the soaring but finely organized rhetoric of his long speeches. Granville-Barker, GALSWORTHY, St John HANKIN, and other dramatists of the new age were more naturalistic. They appealed above all to the intellect, and were specially concerned with the emancipation of women and the influence of money on moral principles. Among the other dramatists of the time, Alfred Sutro was an agreeable purveyor of comedies of manners well suited to the Edwardian age, as was Somerset MAUGHAM in his early plays though he later became a penetrating satirist. J. M. BARRIE, immensely popular for a time, was *sui generis*. His fantasy and sentimentality, curiously blended with pessimism, as exemplified in the early years of the century in *The Admirable Crichton* (1902), *Quality Street* (1903), and *Peter Pan* (1904), founded no school, though the last has achieved a kind of immortality. During the early part of the 20th century tentative efforts were made to revive POETIC DRAMA. James Elroy FLECKER's *Hassan* (1923) was successful in its time, but did not stand up in revival; BOTTOMLEY's dramatizations of Celtic and Northern legends fall short of the symbolic wisdom of the *nō* play of JAPAN where YEATS and MASEFIELD also sought inspiration, the latter combining it with Greek classicism. None of their plays remains in the repertory.

Major changes in the theatrical scene came with the outbreak of war in 1914. Although the actor-managers had been accused of giving little or no encouragement to the new drama, they had maintained a certain continuity of policy in their theatres and helped to buttress the drama proper against the competition of the music-hall, MUSICAL COMEDY, and REVUE. During the First World War, however, theatre rents and the costs of production quadrupled, and the actor-manager was ousted by financial combines whose one idea was to make money by providing light entertainment for a war-weary audience swollen by the influx of overseas soldiers. Musical comedy, melodrama, and farce became the staple fare of most theatres, in London and in the provinces; the Manchester venture succumbed in 1917, Liverpool and Birmingham surviving only with difficulty, the latter under the enlightened patronage of Barry JACKSON, who also founded the MALVERN FESTIVAL. In London, under the able and thrifty management of Lilian BAYLIS, the OLD VIC became a home for Shakespeare's plays, played in true repertory style.

Between 1918 and 1939 the practice of sub-letting the principal London theatres made productions more costly than ever and, with a few honourable exceptions, managements preferred long runs of obvious appeal to serious but experimental plays. During the 1920s Fred Terry could still tour the provinces successfully, as could such actor-managers as F. R. BENSON and Matheson LANG; but the arrival of the films, and particularly talking films, caused many theatres to be converted into cinemas. The staging of new plays became more and more the prerogative of a few small theatres, often clubs with a limited private membership. There were memorable productions of Strindberg, Shaw, and O'NEILL at the EVERYMAN THEATRE, Hampstead, under Norman MACDERMOTT; of German expressionist drama at the GATE THEATRE under Peter Godfrey and Norman MARSHALL; of CHEKHOV at the BARNES THEATRE under KOMISARJEVSKY; and of poetic drama by ELIOT and Auden at the MERCURY THEATRE under Ashley DUKES. In OXFORD, FAGAN revived a number of European classics, and in CAMBRIDGE, at the Festival Theatre, Terence GRAY was responsible for some highly experimental productions on a stage of revolutionary design.

With the disappearance of the actor-managers went also the flamboyant style of acting and the melodramatic type of play they had favoured. The disillusioned and iconoclastic audience of the 1920s and 1930s preferred the adroit understatement of such actors as Gerald DU MAURIER and the critical realism of the problem play and the drama of social purpose. Schooled by the cinema, it found no fault with the episodic structure of Galsworthy's last important play *Escape* (1927) and applauded the characters who defied the law by helping a convict on the run. The indictment of the inhuman laws governing divorce in Clemence DANE's *A Bill of Divorcement* (1921), the disquisition on eugenics in James BRIDIE's *A Sleeping Clergyman* (1933), and the discussion of Christianity and birth control in St John ERVINE's *Robert's Wife* (1937) were all to its taste. The conflict between the younger and the older generations was a favourite theme for problem plays, notably in the works of Noël COWARD, a deft craftsman, as was Frederick LONSDALE, who in *On Approval* (1927) wrote the nearest approach of the day to the comedy of manners, which Coward also attempted in *Private Lives* (1930).

Although NATURALISM was the prevailing dramatic technique of the inter-war period, some playwrights searched for a method which would permit a more imaginative presentation of life without a return to sentimental romanticism. In this spirit DRINKWATER wrote *Abraham Lincoln* (1918), in which history was related to contemporary problems, with verse commentaries by two choric characters between the prose acts; Sutton Vane tackled the problem of life after death in the allegorical *Outward Bound* (1923); Sean O'CASEY, in *The Silver Tassie* (1929), blended poetry, song, and expressionist devices more successfully than in any of his later plays; and J. B. PRIESTLEY moved from the efficient naturalism of *Eden End* (1934) to the dramatization of new theories of the circularity of time in *I Have Been Here Before* and *Time and the Conways* (both 1937), and to a mixture of Morality Play and expressionist techniques in *Johnson Over Jordan* (1939). In many respects the anti-naturalistic movement of the time reached its climax in T. S. Eliot's *Murder in the Cathedral* (1935). In this and in his later plays he endeavoured to show that modern verse, in various syllabic lengths, can be made to serve contemporary situations as effectively as Shakespeare's iambic pentameter.

The upheaval of the Second World War, which closed many theatres in London and the larger provincial cities for lengthy periods between 1939 and 1942, nevertheless saw an important revival of dramatic art in Britain, chiefly because for the first time the state began to subsidize it through the creation of the Council for the Encouragement of Music and the Arts (CEMA), later the ARTS COUNCIL. This resulted in the establishment of the BRISTOL OLD VIC, in the formation of several excellent touring companies, and in sponsored repertory seasons in London comparable to that given by the Old Vic company at the NEW THEATRE in London. Memorable short runs of English and foreign plays of high quality were also staged by Alec CLUNES during his management of the ARTS THEATRE from 1942 to 1953. Peter USTINOV and Terence RATTIGAN emerged as young dramatists of promise, while such pre-war writers as Emlyn WILLIAMS and particularly Priestley continued their successful careers. Most of the outstanding players of the 1930s were still active, and were able to make the transition from the classical and romantic repertory in which they had made their names to the more realistic atmosphere of the post-war theatre. This was made easier by the sudden and somewhat unexpected interest aroused by the poetic plays of Christopher FRY: Gielgud starred in *The Lady's Not for Burning* (1949), Laurence OLIVIER in *Venus Observed* (1950), Edith EVANS in *The Dark is Light Enough* (1954). Peggy ASHCROFT and Ralph RICHARDSON were seen together in *The Heiress* (1949), based on Henry JAMES, and Alec GUINNESS appeared in T. S. Eliot's *The Cocktail Party* (1949). All of them were also to be seen later in the plays of the younger dramatists who came to the fore after interest suddenly shifted from the modern verse plays of Fry and Eliot to plays of protest and satire which exalted the nonconformist, the misfit, and the martyr, and showed sympathy with and understanding of the frustrations and fears of the common man. The new movement was sparked off by the violent reaction of young people against the stereotyping processes of mass civilization, the regimentation of the Welfare State, and the anxieties of the atomic age, as well as by the feeling that contemporary democracy was only a façade concealing an oligarchical 'Establishment' associated with middle-class morality, imperialism, the use of nuclear weapons, and capital punishment. It began quietly with the production of John WHITING's *Saint's Day* (1951), a vision of nuclear doom, and exploded in John OSBORNE's *Look Back in Anger* (1956), at the Royal Court Theatre, the first of the naturalistic 'kitchen-sink' dramas, in which the hero is a provincial graduate turned street vendor who hurls invective at class distinctions, Kiplingesque patriotism, suburban ennui, Sunday newspapers, and his mother-in-law, indiscriminately. Correspondingly, the older generations of provincial families were criticized in Shelagh Delaney's *A Taste of Honey* (1958) and in WESKER's *Roots* (1959), while the patriotic shibboleths of the Irish establishment were mocked in Brendan BEHAN's *The Hostage* (also 1958).

The influence of BRECHT, whose plays were beginning to be seen in translation on the English stage, showed itself in the episodic narrative methods of Robert BOLT in *A Man for All Seasons* (1960), which starred Paul Scofield, one of the best of the post-war actors; while that of the Theatre of the ABSURD as developed in France by BECKETT and IONESCO, which had already reached London with the former's *Waiting for Godot* (1955) and the latter's *The Bald Prima Donna* (1956), *The Chairs* (1957), and other plays, showed clearly in PINTER's *The Caretaker* (1960). These plays, and many others, made clear the cosmic absurdity of man, his imperfect powers of communication with his fellows, and his inevitable fears and loneliness. The new movement brought a refreshing variety of contemporary idioms and dialects to the stage and made effective use of functional and symbolic settings, breaking down the 'fourth wall' convention. This was matched architecturally by new THEATRE BUILDINGS derived from medieval, Elizabethan, and Greek models, which jettisoned the proscenium arch in favour of a thrust stage that aimed to provide a more intimate association between actor and audience than existing theatres allowed. One of the younger playwrights, John ARDEN, was the most eager to exploit this opportunity, reverting deliberately in *The Happy Haven* (1960), *The Workhouse Donkey* (1963), and *Armstrong's Last Goodnight* (1964) to earlier conventions of play construction and theatrical representation. The logical extension of this movement was naturally THEATRE-IN-THE-ROUND, the keenest advocate of which was Stephen JOSEPH. Almost equally popular with the 'arena' theatre, as it was also known, was the 'end stage', as at the MERMAID in London, and the three-sided stage, as at the CHICHESTER FESTIVAL THEATRE.

The British theatre was fortunate in that both the innovative and the naturalistic spirit among dramatists, architects, and designers was found also in the new generation of young players and directors, all eager to experiment with new ideas in varied combinations, among them Albert FINNEY, Peter O'TOOLE, and Tom COURTENAY and, as directors, Joan LITTLEWOOD with her THEATRE WORKSHOP, Peter BROOK with the ROYAL SHAKESPEARE COMPANY, founded in 1961, and Tony RICHARDSON with the ENGLISH STAGE COMPANY, founded in 1956 at the Royal Court. As the new movement, typified by its youth, grew in strength, it fuelled the rebellion against all forms of authority by directing attention to the taboos of sex and religion, and by channelling its energies into a co-ordinated battle for the freedom of the theatre, led by the artistic directors of the Royal Court headed by George DEVINE, who had from the beginning been in the forefront of the battle. In 1968 Parliament finally repealed the old Licensing Act and abolished the Lord Chamberlain's powers of censorship. The heady sense of freedom that followed generated a brief wave of plays in the early 1970s that sought to test public reaction to verbal obscenity and physical nudity and eventually to prove that playwrights, actors, and managers between them could be trusted to know where artistic necessity ended and pornography began. Of the playwrights first confronted with this new liberty those most successful in finding compromises acceptable to audiences were Peter SHAFFER and John MORTIMER. With censorship no longer an issue drama in the 1970s showed two main trends, towards sardonic comedy and political polemic, the former best exemplified in the plays of Alan AYCKBOURN, Alan BENNETT, Michael FRAYN, and Tom STOPPARD, the latter by those of Howard BARKER, Howard BRENTON, and David HARE.

The opening of the NATIONAL THEATRE complex in 1976, together with the continuing strength of the Royal Shakespeare Company and the excellence of the new regional theatres, helped to emphasize the precarious state of the commercial theatre. The West End theatre relied increasingly on transfers from the subsidized sector, the FRINGE, and theatres such as those at GREENWICH and HAMPSTEAD; and subsidized productions such as the R.S.C.'s adaptation of DICKENS's *Nicholas Nickleby* and the National Theatre's production of Shaffer's *Amadeus* have triumphed on Broadway. It became accepted that the survival of the theatre—whether in London or the provinces—at any but the most basic level entailed public subsidy, especially in the severe economic climate of the 1980s. (See also CIVIC THEATRE.)

ENGLISH ARISTOPHANES, see FOOTE, SAMUEL.

ENGLISH COMEDIANS, troupes of English actors who toured the Continent during the late 16th and early 17th centuries. Although a group of English musicians is known to have visited the Danish Court in 1579–80, the first actors to appear abroad seem to have been those who accompanied the Earl of Leicester to Holland and Denmark in 1584–6. Among them was the famous clown Will KEMPE who with his fellow actors was invited to DRESDEN. Thereafter references to visits from English actors can be found in the archives of many German towns. The first to become widely known was Robert BROWNE, who toured from 1592 to 1619. He was joined in 1606 by John Green, who later succeeded him as head of the company; together they went to the private theatre built by Count Moritz of Hesse in that year, visiting Poland a year later. In 1610 a popular newcomer to the company was Richard Jones (?–1615), formerly with the ADMIRAL'S MEN in London, who remained with it until shortly before his death. One of Browne's original actors, Thomas Sackville, broke away to form a company of his own which was for a time at WOLFENBÜTTEL in the household of Heinrich Julius, Duke of Brunswick. Sackville was probably the first to create an indigenous comic character, Jan Bouschet (Jean Posset), designed specifically for an audience whose knowledge of English was very slight, who combined the often bawdy antics of the CLOWN with the wit of the FOOL and relied mainly for his effects on pidgin German laced with English and Dutch phrases. He was the forerunner of such German-speaking clowns as Hans Stockfisch, created by John Spencer who with headquarters in Berlin travelled as far afield as The Hague and Dresden, and Robert REYNOLDS's Pickelhering who was the prototype of such purely indigenous characters as HANSWURST and THADDÄDL.

It seems likely that the English Comedians first appeared in JIGS and short comic sketches whose humour was broad enough to appeal even to a foreign audience, but a collection of texts from their repertory, printed in 1620, shows that by that time they were relying more on pirated editions of full-length plays, the dialogue pruned to the minimum and the plots reduced to a series of dramatic incidents. The titles in this collection are all of English plays, but a further collection printed in 1630 shows a preponderance of German titles, and it is probable that by then the plays were given mainly in German. Even in serious plays a good deal of fooling by the clowns was added for the benefit of the groundlings, with additional music and dancing for the more sophisticated. From the literary point of view the influence of the English Comedians on German drama was negligible, but they helped to give German audiences the habit of theatre-going, and the brevity of their dialogue may have done something to counteract the native tendency to excessive discussion. Once the companies began to engage German actors the use of English was discontinued and indigenous drama, in the shape of the HAUPT- UND STAATSAKTION, took over. In spite of the troubles of the Thirty Years War, some English actors continued to tour the Continent, particularly during the PURITAN INTERREGNUM, the last authentic record of them dating from 1659 and the last notable player in the tradition being George JOLLY; but such was the prestige of the 'Englische Komödianten' that the name was used in Germany for publicity purposes as late as the middle of the 18th century.

ENGLISH STAGE COMPANY, organization formed in 1956, mainly to present plays by young and experimental dramatists but also to put before the public the best contemporary plays from abroad. After an unsuccessful attempt to reopen the bomb damaged KINGSWAY THEATRE, the company, under the artistic direction of George DEVINE, took over the ROYAL COURT THEATRE in Sloane Square, where it still functions. For a time it played an indispensable role, promoting the work of such playwrights as OSBORNE, WESKER, and ARDEN and making the Royal Court the most exciting theatre in London. Its function is now shared by the NATIONAL THEATRE, the ROYAL SHAKESPEARE COMPANY, FRINGE THEATRES, and some of the regional theatres, but it still plays a useful part in the London theatrical scene. For individual productions see under Royal Court Theatre.

ENNIUS, QUINTUS (239–169 BC), one of the greatest of Latin poets, and an important figure in the history of Latin drama. Born at Rudiae in the south of Italy, he went to Rome in 204 BC. In the absence of a reading public at Rome it was inevitable that Ennius should turn to drama; the death of ANDRONICUS and the disgrace of NAEVIUS left the way clear for him in tragedy. He seems to have taken many of his plays from EURIPIDES, in whom he found a questioning spirit and humanitarian outlook like his own. Like other Roman dramatists Ennius kept, or tried to keep, close to the sense of the Greek he was translating, though allowing himself considerable freedom in expression and metre; of original construction we have no evidence, except that in his *Iphigenia* he introduced a chorus of soldiers. Surviving fragments (400 lines from some 20 tragedies) illustrate his poetic power; compared with the Greek they may seem rhetorical, but they are free from the excesses of later Roman tragedy in this respect, and they contain much that is beautiful and moving. His contemporary PLAUTUS burlesques his style—a proof that his plays were well known—and in the next generation TERENCE refers to him as one of the 'careless' but nevertheless admirable dramatists of the past. His plays continued to be read down to the end of the Republic, though Roman opinion seems to have ranked PACUVIUS and ACCIUS higher as writers of tragedy. He also attempted comedy, apparently without much success, and two FABULAE *praetextae* are doubtfully assigned to him—the *Sabinae*, dealing presumably with the Rape of the Sabines, and the *Ambracia*, which perhaps told of the conquest of that region by Ennius' patron, M. Fulvius Nobilior.

ENTERTAINMENTS NATIONAL SERVICE ASSOCIATION (ENSA), organization formed in 1938–9 to provide entertainment for the British and allied armed forces and war workers during the Second World War. It was directed by Basil DEAN, and had its headquarters at DRURY LANE THEATRE, London. Working closely in collaboration with the Navy, Army, and Air Force Institute (NAAFI), which was responsible for the financial side, it provided all types of entertainment, from full-length plays and symphony orchestras to concert parties and solo instrumentalists, not only in the camps, factories, and hostels of Great Britain, but on all war fronts, from the Mediterranean to India and from Africa to the Faroes.

ENTHOVEN [*neé* Romaine], GABRIELLE (1868–1950), English theatre historian, who in 1924 presented to the Victoria and Albert Museum in London the vast collection of theatre material which bears her name. It includes manuscripts, playbills, engravings, prints, books, models, and newspaper cuttings, covering the history of theatrical production in London from the end of the 17th century onwards, and is constantly being added to by gifts, bequests, and a systematic acquisition policy. Mrs Enthoven, who for many years administered the collection herself, was a good amateur actress and the author of several plays, including an English adaptation of D'ANNUNZIO's *Le Chèvrefeuille* (1913) which as *The Honeysuckle* was produced in New York in 1921. In 1948 she became the first President of the newly founded SOCIETY FOR THEATRE RESEARCH.

ENTREMÉS, a Spanish term, deriving originally from the French *entremets*, applied to a diversion, dramatic or otherwise, which took place between the courses of a banquet. The name was first applied in Spain to the short dramatic interludes which enlivened the Corpus Christi processions in Catalonia, and was later applied in Castilian to a short comic interlude, often ending in a dance, which was performed, sometimes incongruously, between the acts of a play given in the public theatres. The *entremés* reached the height of its popularity with QUIÑONES DE BENAVENTE, and most of the well known dramatists of the Golden Age wrote them, including Lope de VEGA, CERVANTES, and CALDERÓN. In modern times the genre was revived by the brothers ÁLVAREZ QUINTERO.

EPHRATI, JOSEPH, see JEWISH DRAMA.

EPICHARMOS (*c.* 550–460 BC), Greek comic poet of Syracuse, whose work survives only in a few fragments. With SOPHRON he represents the literary offshoot of the early popular and crudely realistic MIME, in which mythological burlesque was an important element. His plays had some influence on the later development of Athenian OLD COMEDY and, according to Horace, PLAUTUS imitated his fast-moving 'patter' style.

EPIC THEATRE, a phrase taken from ARISTOTLE, where it implies a series of incidents presented without regard to theatrical conventions, and used in the 1920s by such pioneers as BRECHT and PISCATOR of episodic productions designed to appeal more to the audience's reason than to its emotions, thus excluding sympathy and identification with the drama being portrayed on stage (see also VERFREMDUNGSEFFEKT). It employs a multi-level narrative technique and places the main emphasis on the social and political background of the play. Typical of Epic Theatre was a production by Piscator of a dramatized version of Leo TOLSTOY's vast novel *War and Peace*. This was first seen in Berlin, and in 1962 made a brief appearance at the BRISTOL OLD VIC and in London.

EPIDAUROS, one of the finest surviving examples of a Greek theatre building, dating from the 4th century BC. In its original form it had a full circular ORCHESTRA, an elaborate stone stage building or skene, and a raised stage with ramps leading to ground level. The orchestra and the seating, though not the buildings, have been restored, and the theatre is in regular use for annual summer festivals of ancient drama performed by the Greek National Theatre. The auditorium holds about 14,000 spectators and is notable for its superb acoustics.

EPILOGUE, see PROLOGUE.

EPIPHANY PLAY, see LITURGICAL DRAMA.

EQUESTRIAN DRAMA, a form of entertainment, popular in London in the first half of the 19th century, which evolved out of the feats of horsemanship shown at ASTLEY'S AMPHITHEATRE. The first was *The Blood-Red Knight; or, the Fatal Bridge* (1810) by J. H. Amherst, who was also responsible for such later spectacular shows as *The Battle of Waterloo* (1824) and *Buonaparte's Invasion of Russia; or, the Conflagration of Moscow* (1825), in both of which horses and their riders figured largely. Among the outstanding actors in equestrian drama were Edward Alexander Gomersal (1788–1862) and Andrew Ducrow (1793–1848), the latter appearing not only at Astley's but also at COVENT GARDEN, where in 1823 he and his company were seen in PLANCHÉ's *Cortez; or, the Conquest of Mexico* and a revival of 'Monk' LEWIS's *Timour the Tartar*, first seen in 1811. Among the many plays that were adapted for Astley's one of the most successful was Shakespeare's *Richard III*, with the leading role allotted to Richard's horse White Surrey; but the most famous of all equestrian dramas was *Mazeppa*, based by H. M. Milner on BYRON's poem and first seen in London in 1823. The part of Mazeppa was afterwards associated entirely with Adah Isaacs MENKEN, who first appeared in it in the United States in 1863. Although the fashion for equestrian drama soon died out, real horses were used on stage at DRURY LANE as late as 1909 in *The Whip*, a sporting drama by Cecil Raleigh and Henry Hamilton.

EQUITY, see AMERICAN, BRITISH, and CANADIAN ACTORS' EQUITY ASSOCIATION.

ERCKMANN-CHATRIAN, pseudonym of two French authors who wrote novels and plays in collaboration, Emile Erckmann (1822–99) and Louis-Gratien-Charles-Alexandre Chatrian (1826–90). They are best remembered for their melodrama, *Le Juif polonais* (*The Polish Jew*, 1869) which, in an English adaptation by Leopold Lewis as *The Bells*, provided IRVING with the fine part of Mathias, in which he made his first great success at the LYCEUM in 1871. He played the part many times, and after his death it was played also by his elder son, H. B. IRVING, and by MARTIN-HARVEY.

ERLANGER'S THEATRE, New York, see ST JAMES THEATRE.

ERNST, PAUL (1866–1933), German dramatist, associated in the 1890s with HOLZ and Schlaf. An austere writer, he forsook NATURALISM, which failed to satisfy his innate sense of form, for Neoclassicism. In his tragedies, written for the theatre at Düsseldorf, he isolated the inner action and reaction in a timeless sphere and so reduced drama to dialogue, as in *Demetrios* (1905) and *Canossa* (1908). They were not successful and there is more movement in his comedies, *Eine Nacht in Florenz* (1904), *Der heilige Crispin* (1910), and *Pantaloon und seine Söhne* (1916), where the characters, though traditional, are seen from a new angle, while the comic situation, often arising out of a confusion of identity, provokes reflection rather than laughter.

ERVINE, (John) ST JOHN GREER (1883–1971), Irish dramatist and critic, who later settled in England, where he was dramatic critic for a number of papers including the *Morning Post* and the *Observer*. In his early years he was for a short time manager of the ABBEY THEATRE in Dublin, where most of his early plays were produced, among them *Mixed Marriage* (1911) and *John Ferguson* (1915). *Jane Clegg* (1913), in which Sybil THORNDIKE gave a moving performance in the title-role, was first seen at the Gaiety Theatre in Manchester, under Miss HORNIMAN. These were serious plays dealing with social problems, but Ervine later wrote for London a number of light comedies, of which the most successful were *Anthony and Anna* (1926) and *The First Mrs Fraser* (1929). He returned to serious matters with *Robert's Wife* (1937), in which Edith EVANS gave an outstanding performance, and *Private Enterprise* (1947), the last of his plays to be seen in London.

ESCAMILLA [Vazquez], ANTONIO DE (?–1695), Spanish actor and manager of a touring company from 1661 to 1690. He appeared in many of the plays of CALDERÓN and sang in ZARZUELAS. He was particularly admired for his playing of

GRACIOSO parts. His daughter Manuela (1648–1721) was also an excellent actress.

ESMOND [Jack], HENRY VERNON (1869–1922), English actor-manager and dramatist, who made his first appearance in London in 1889 and was for a time associated with E. S. WILLARD and Edward TERRY. He then joined George ALEXANDER's company at the ST JAMES'S THEATRE, where he scored a success as Cayley Drummle in PINERO's *The Second Mrs Tanqueray* (1893) and played Little Billee in Du Maurier's *Trilby* (1895). While at the St James's he began writing plays, mainly sentimental comedies which had a great vogue, and in which he later toured with his wife Eva Moore (1870–1955). The only one to be remembered is *Eliza Comes to Stay* (1913), though his last play *The Law Divine* (1918) had a considerable success at the time. His daughter Jill was also on the stage, becoming the first wife of Laurence OLIVIER.

ESPERT I ROMERO, NÚRIA (1936–), Spanish actress, who founded her own theatrical company in 1959 with the aim of performing works by classical Spanish dramatists and contemporary Europeans. She played Claire in GENET's *Les Bonnes* in 1969 and the title role in a striking production of GARCÍA LORCA's *Yerma* in 1971, both directed by Victor GARCIA, who also directed her in VALLE INCLAN's *Divinas palabras* (*Divine Words*). *Yerma* was seen in the London WORLD THEATRE SEASON of 1972 and universally acclaimed.

ESPY, L', see L'ESPY.

ESRIG, DAVID, see ROMANIA.

ESSLAIR, FERDINAND (1772–1840), German actor, who made his début at Innsbruck and for many years toured the provinces, until in 1820 he was appointed leading actor and manager of the Court Theatre in MUNICH. His passionate, inspirational acting, allied to exceptional graces of person and voice, made him one of the most popular actors in Germany. He was seen at his best in SCHILLER's plays, particularly as Wilhelm Tell and Wallenstein.

EST, Centre Dramatique de l', see DÉCENTRALISATION DRAMATIQUE.

ESTCOURT, DICK (1668–1712), English actor immortalized by Sir Richard STEELE in No. 468 of *The Spectator*, which also recorded his death. He does not appear to have been an outstanding actor—he played Kite in the first production of FARQUHAR's *The Recruiting Officer* (1706)—but he was an amazing mimic, and his natural good humour and vivacity made him a favourite in any company. He was the author of a comedy, *The Fair Example; or, the Modish Citizens* (1703), seen at DRURY LANE, where he was for many years a member of the company and for a short time one of the managers.

ESTÉBANEZ, JOAQUÍN, see TAMAYO Y BAUS, MANUEL.

ESTELLE R. NEWMAN THEATRE, New York, see PUBLIC THEATRE.

EST PARISIEN, Théâtre de l' (T.E.P.), Paris, playhouse in the eastern, working-class area of the city. Its director, Guy Rétoré, founded in 1950 a company called La Guilde which operated in a church hall. In 1957 they won a prize in a competition for young companies and used it to set themselves up in a larger hall which they called the Théâtre de Ménilmontant. Here they acted classics and contemporary works in repertory, until in 1959 they became an officially subsidized company under the policy of DÉCENTRALISATION DRAMATIQUE and finally one of the Théâtres Nationaux. In 1963 they moved into a converted cinema, under their present name, and made a successful start, attracting 18,000 subscribers and 140,000 spectators in their first year. Popular support in the area was sufficient, during an internal financial crisis in 1964, to persuade the government to increase the theatre's subsidy.

ETHEL BARRYMORE THEATRE, New York, at 243 West 47th Street, between Broadway and 8th Avenue. Built by Lee Shubert, with a seating capacity of 1,099, it opened on 20 Dec. 1928 with Ethel BARRYMORE, in whose honour it was named, in MARTÍNEZ SIERRA's *The Kingdom of God*. Its subsequent history, though not sensational, has been one of almost unvaried success. Among its outstanding productions have been COWARD's *Design for Living* (1933) and *Point Valaine* (1935); Elsie Schauffler's *Parnell* (also 1935); *Night Must Fall* (1936) by Emlyn WILLIAMS, who also appeared in it; and Clare Boothe's scathing comedy *The Women* (also 1936). In 1947 Tennessee WILLIAMS's *A Streetcar Named Desire*, with Marlon Brando and Jessica TANDY, won a PULITZER PRIZE, as did Ketti Frings's adaptation of Wolfe's novel *Look Homeward, Angel* ten years later. Other productions have included Frederick Knott's *Wait Until Dark* (1966); Peter SHAFFER's double bill *Black Comedy* and *White Lies* (1967); and Christopher HAMPTON's *The Philanthropist* (1971) with Alec MCCOWEN. A musical about Black life, *Ain't Supposed to Die a Natural Death* (1971), scored a great success, and later productions were the double bill *Noël Coward In Two Keys* (1974), with Jessica Tandy and Hume CRONYN; STOPPARD's *Travesties* (1975), with John WOOD; the musical *I Love My Wife* and David Mamet's *American Buffalo* (both 1977); and Bernard Slade's *Romantic Comedy* (1979).

ETHEREGE, SIR GEORGE (1634–91), English dramatist, the first to attempt the social comedy of manners developed by CONGREVE and later perfected by SHERIDAN. Etherege spent part of his early years in France, and was doubtless influenced by memories of MOLIÈRE when writing his first play, *The Comical Revenge; or, Love in a Tub* (1664), a serious verse drama with a comic prose sub-plot. It was this latter style that Etherege explored further in his two later comedies, *She Would if She Could* (1668) and *The Man of Mode* (1676). The latter, a typical Restoration comedy, is Etherege's best play, a picture of a society living exclusively for amusement, with a tenuous plot of entangled love-affairs offering an opportunity for brilliant dialogue and character-drawing. It contains the 'prince of fops', Sir Fopling Flutter, and the heartless, witty Dorimant, often considered a portrait of Lord Rochester, just as the poet Bellair is supposed to be Etherege himself.

ETHIOPIAN OPERA, a turn in the MINSTREL SHOW, consisting of burlesques on Shakespeare and opera, with Negro melodies inserted. It was modelled on the early burlesques of T. D. RICE (Jim Crow), of which the best were *Bone Squash Diavolo* (on 'Fra Diavolo'), *Jumbo Jim, Jim Crow in London*, and a burlesque of *Othello*.

ETTINGER, SOLOMON, see JEWISH DRAMA.

EUCHEL, ISAAC, see JEWISH DRAMA.

EUCLID-77TH STREET THEATRE, Cleveland, see CLEVELAND PLAY HOUSE.

EUGENE O'NEILL THEATRE, New York, at 230 West 49th Street, between Broadway and 8th Avenue. Seating 1,075, this opened as the Forrest on 24 Nov. 1925. Its history was uneventful until on 17 Sept. 1934 *Tobacco Road*, based on a novel by Erskine Caldwell, was transferred there from the JOHN GOLDEN THEATRE where it had opened on 4 Dec. 1933; it ran until 31 May 1941. In 1945 the theatre closed, and after extensive alterations reopened on 14 Dec. as the Coronet, its first success being a revival of Elmer RICE's *Dream Girl*, followed by Arthur MILLER's *All My Sons* (1947). A revival of Eugene O'NEILL's *The Great God Brown* was presented in 1959 and the theatre then took its present name, reopening with INGE's *A Loss of Roses*. In 1969 it housed *The Last of the Red Hot Lovers* by Neil SIMON who became owner of the theatre, several more of his plays being later seen there: *The Prisoner of Second Avenue* (1971), *California Suite* (1976), *Chapter Two*, transferred from the IMPERIAL in 1979, and *I Ought To Be In Pictures* (1980).

EUGENE O'NEILL THEATRE CENTER, see O'NEILL, EUGENE.

EUPHORION, ancient Greek dramatist, the son of AESCHYLUS and apparently a tragic poet of no mean order, since in 431 BC he defeated both EURIPIDES and SOPHOCLES. Nothing of his work survives.

EUPOLIS (*c*. 446–411 BC), ancient Greek dramatist, who with ARISTOPHANES and CRATINUS was considered one of the best writers of comedy in Athens. He is known to have written 18 plays, reputedly winning seven victories, and in brilliant satire and lewdness was said to outdo even Aristophanes. Many fragments of his work survive, including 200 lines of the *Demes* in which great statesmen of the past return to rule once more over Athens.

EURIPIDES (484–406 BC), ancient Greek tragic dramatist, born probably on the Athenian island of Salamis. His parents were well-to-do, his mother apparently of good family in spite of ARISTOPHANES' recurring joke about her having a vegetable stall in the market. Little is known of Euripides' life, but tradition represents him as a recluse living in a cave on Salamis; this may have arisen from his reluctance to take an active part in civic life. Unlike his contemporaries AESCHYLUS and SOPHOCLES he seems to have given all his time to writing plays, and he stands out as an individualist in an age which still venerated the ideal of duty towards one's fellow citizens and the holding of public offices. He was also labelled a misogynist, in spite of the obvious sympathy with women's rights and problems displayed, for example, in his *Medea*; and the dislike his audiences sometimes felt for the heretical statements which he put into the mouths of his characters no doubt rebounded on to him.

Euripides is said to have written 92 plays, of which there survive 16 tragedies, one SATYR-DRAMA (the *Cyclops*, a burlesque version of Odysseus's adventures in Sicily), and a large number of fragments, testimony to his later popularity. The extant tragedies, with dates where known or conjectured, are: *Medea* (431), *Hippolytus* (428), the *Children of Heracles* (?also 428), *Hecuba* (?426), the *Trojan Women* (415), *Iphigenia in Tauris* (?414), *Helen* (412), the *Phoenician Women* (411), *Orestes* (408), and, of unknown date, the *Madness of Heracles*, the *Suppliant Women* (with a plot different from that of Aeschylus' tragedy of the same name), *Ion*, and *Electra*. The *Bacchae* and *Iphigenia in Aulis* were both produced posthumously. One other play, *Alcestis* (438), though usually classed among the tragedies, was entered for competition at the festival in place of the customary satyr-drama and contains pronounced satyric elements. The *Rhesus*, a curious play with several puzzling features, is thought by some to be by Euripides and by others to be a 4th-century imitation of his style.

Euripides' unpopularity with his contemporaries appears to have affected his career. He won only five victories compared to Sophocles' 18 and in 408 left Athens, perhaps not entirely of his own free will, to visit the Court of Archelaus, King of

Macedonia, where he died. Sophocles, who outlived him by only a few months, was said to have dressed his next chorus in mourning as a sign of respect, and in the *Frogs* Aristophanes, though he lampoons Euripides unmercifully, makes it clear that he regards his death as marking the end of an era, the only hope for the theatre's future lying in revivals.

In his lifetime Euripides aroused great interest and great opposition with his realism, his interest in abnormal psychology, his portraits of women in love, his new and emotional music, his unorthodoxy, and his argumentativeness. Unlike Sophocles, who carried into a democratic age the outlook and traditions of an aristocratic society, Euripides was critical, sceptical, interested less in the community than in the individual, dealing less with broad questions of morality and religion than with personal emotions and passions—love, hate, revenge—and with specific social questions like the suffering of the individual in war. He tried to bring tragedy down to earth by using more colloquial language, popular forms of music, and characters who, though still drawn from myth and legend, had recognizable counterparts in 5th-century Athens. His fondness for violence and stage spectacle and his reworking of familiar material into unexpected shapes shocked his audiences into a new awareness and made them think for themselves. After his death his reputation increased immeasurably, for his plays could speak to the new HELLENISTIC AGE established after Alexander's conquests in a way in which those of Aeschylus and Sophocles could not; and they continued to be performed and to influence new generations of writers for many centuries.

The plays of Euripides fall into two clearly marked categories—tragedies in the modern sense, and plays which may be variously called tragi-comedies, black comedies, romantic drama, melodrama, or even high comedy. The tragedies have often been much misunderstood and criticized because of their apparently episodic plot structure. Sometimes (as in *Hecuba* and the *Madness of Heracles*) the play is based on two stories which seem to have little in common except the leading characters, but in this way Euripides was able to achieve in one play much of what Aeschylus had earlier accomplished in a TRILOGY. The first half of the play illuminates a set of incidents from one point of view; the second treats them from a different point of view; and the audience is asked to judge between the two. The cumulative effect of the play often negates the 'new morality' which individual characters appear to be advocating in their speeches, and self-interest is revealed as ultimately destructive.

Euripides was described by ARISTOTLE as 'the most tragic of the poets', and it is in such tragedies as *Medea, Hippolytus*, the *Bacchae*, the *Trojan Women*, and *Hecuba* that his best work is probably to be found. *Electra* and *Orestes*, in which some critics have found a shrewish heroine and a pusillanimous hero, are powerful studies in morbidity and insanity; *Alcestis, Ion*, and *Iphigenia in Tauris* are excellent tragi-comedy or romantic drama; *Helen* is delightful high comedy; and the *Phoenician Women* is perhaps best described as a pageant-play. All these except the last show Euripides as a master of non-tragic dramatic writing, and through them he, and not the contemporary writers of OLD COMEDY, opened the way for the writers of NEW COMEDY, particularly MENANDER who acknowledged Euripides as his master. Other aspects of his work—emotionalism, sententiousness, pathos, and rhetoric—were easily imitated and through SENECA those characteristics, absent in Aeschylus and Sophocles, were passed on to the dramatists of the Renaissance.

It is interesting to note that Euripides continued to use the three actors and CHORUS as finally established by Sophocles, with one important innovation. The opening song of the chorus as used by Aeschylus, or the scene in dialogue with which all the extant plays of Sophocles begin, were replaced by a formal *prologos* spoken sometimes by a character in the play, sometimes by an external god, which summarizes the story up to the point at which the action begins. Together with the progressive detachment of the chorus from the main action, inevitable once the play was concerned with private rather than public issues, this eventually, via Seneca, gave rise to the Elizabethan idea of an extraneous person called the Chorus speaking a PROLOGUE.

EUSTON MUSIC-HALL, London, see REGENT THEATRE.

EUTERPEAN HALL, New York, see BROADWAY THEATRE.

EVANS, EDITH MARY (1888–1976), D.B.E. 1946, distinguished English actress. She made her first appearance on the stage in 1912 as an amateur, playing in Cambridge under POEL's direction, and making her London début as Cressida in his production of *Troilus and Cressida* later the same year. She then turned professional, toured with Ellen TERRY, and after varied experience, mainly in modern plays—though in 1922 she played Cleopatra in DRYDEN's *All for Love* for the PHOENIX SOCIETY—she created the parts of the Serpent and the She-Ancient in SHAW's *Back to Methuselah* (1923) at Birmingham. A fine performance as Millamant in CONGREVE's *The Way of the World* at the LYRIC, Hammersmith, in 1924 established her as one of the outstanding actresses of her generation; and in 1925–6 she was at the OLD VIC, playing Rosalind in *As You Like It*, Katharina in *The Taming of the Shrew*, and the Nurse in *Romeo and Juliet* (all of which she was to repeat successfully in later years) and several other roles including Beatrice in *Much Ado About Nothing* and Kate Hardcastle in GOLDSMITH's *She Stoops To Conquer*. In 1926 she was seen in her first and only IBSEN role, Rebecca West in *Rosmersholm*, and a year later returned to Restoration comedy as Mrs Sullen in FARQUHAR's *The Beaux' Stratagem*. In 1929 she made a great success in Reginald Berkeley's *The Lady With the Lamp*, playing Florence Nightingale, the part in which she made her New York début two years later; in 1929 she also created at the MALVERN FESTIVAL the role of Orinthia in Shaw's *The Apple Cart* and played Lady Utterword in a revival of his *Heartbreak House*. She returned to the Old Vic in 1932, playing Emilia in *Othello* and Viola in *Twelfth Night*, and was then seen in a series of leading roles in new plays, including Irela in *Evensong* (1932; N.Y., 1933) by Edward KNOBLOCK and Beverley Nichols; Gwenny in Emlyn WILLIAMS's *The Late Christopher Bean* (1933); the title role in Norman Ginsbury's *Viceroy Sarah* (1934); and Agatha Payne in Rodney Ackland's *The Old Ladies* (1935), based on a novel by Hugh Walpole. In 1936 she was seen as Madame Arkadina in KOMISARJEVSKY's production of CHEKHOV's *The Seagull*, and later in the year, with the Old Vic company, played Lady Fidget in WYCHERLEY's *The Country Wife*, which was followed in 1937 by Katharina in *The Taming of the Shrew* to the Petruchio of Leslie BANKS and a long run in St John ERVINE's *Robert's Wife*. In 1939 she first played her most famous role, Lady Bracknell in WILDE's *The Importance of Being Earnest*, in which her beautifully modulated voice was used to excellent comic effect. During the Second World War she was seen in VAN DRUTEN's *Old Acquaintance* (1941) and again in Shaw's *Heartbreak House* in 1943, this time as Hesione Hushabye. She proved an excellent Mrs Malaprop in SHERIDAN's *The Rivals* in 1945, and a superb Katerina Ivanovna in DOSTOEVSKY's *Crime and Punishment* (1946), adapted by Ackland. Her 'old peeled wall' of a Lady Wishfort in *The Way of the World* in 1948, again with the Old Vic, was as outstanding as her earlier Millamant, and she was much admired, in the same year, as Madame Ranevskaya in Chekhov's *The Cherry Orchard*. In her sixties she was seen in several new plays, which provided her with some of her finest roles—Lady Pitts in BRIDIE's *Daphne Laureola* (1949; N.Y., 1950); Helen Lancaster in N. C. HUNTER's *Waters of the Moon* (1951); Countess Rosmarin in FRY's *The Dark is Light Enough* (1954); and Mrs St Maugham in Enid BAGNOLD's *The Chalk Garden* (1956). In 1958 she was at the Old Vic to play Queen Katherine in *Henry VIII*, and in 1959 at the SHAKESPEARE MEMORIAL THEATRE as the Countess of Rousillon in *All's Well That Ends Well*. She returned to modern drama with BOLT's *Gentle Jack* (1963), Enid Bagnold's *The Chinese Prime Minister* (1965), and ANOUILH's *Dear Antoine* (CHICHESTER, 1971), and made her last appearance on the stage at the HAYMARKET THEATRE in 1974 in a programme entitled *Edith Evans... and Friends*.

EVANS, MAURICE HERBERT (1901–), English-born actor, who became an American citizen in 1941. After some amateur experience he made his professional début in 1926 in Cambridge under Terence GRAY and then went to London, where he played a variety of small parts before coming into prominence as Lieutenant Raleigh in SHERRIFF's *Journey's End* (1928). He joined the OLD VIC company in 1934, playing a wide variety of parts including Benedick in *Much Ado About Nothing*, Petruchio in *The Taming of the Shrew*, Richard II, and Hamlet (in its entirety). He then went to the United States, where he remained, playing Romeo in 1935 to the Juliet of Katharine CORNELL and then appearing in a succession of Shakespearian plays directed by Margaret WEBSTER. During the Second World War he entertained the troops with his so-called G.I. version of *Hamlet*, which was published in 1947, and in the post-war theatre played a number of parts by SHAW including John Tanner in *Man and Superman* in 1947, Dick Dudgeon in *The Devil's Disciple* in 1951, King Magnus in *The Apple Cart* in 1956, and Captain Shotover in *Heartbreak House* in 1959. Among his other parts were Crocker-Harris in RATTIGAN's *The Browning Version* (1949), Hjalmar Ekdal in IBSEN's *The Wild Duck* (1951), and, an uncharacteristic role, Tony Wendice in Frederick Knott's *Dial 'M' for Murder* (1952), which had a long run. In 1962 he played the lead in *The Aspern Papers*, based by Michael REDGRAVE on Henry JAMES's novel.

EVANS, WILL (1875–1931), star of British MUSIC-HALL and PANTOMIME, son of an acrobat and clown, Fred(erick) William Evans (1840–1909), with whom he made his first appearance in 1881 at the age of 6, in *Robinson Crusoe* at DRURY LANE. He then toured with his father's troupe, and returned to London in 1890 to appear on the 'halls' with his wife Ada Luxmore and, after her death, as a solo eccentric comedian. He specialized in 'scenes of domestic chaos', of which the most popular were 'Building a Chicken House', 'Whitewashing the Ceiling', and 'Papering a House'. For many years he imported the same style of knockabout humour into kitchen scenes in Drury Lane pantomimes. He was part-author (with 'Valentine', Archibald Thomas Pechey) of *Tons of Money*, written in 1914, but not staged until 1922. It was first seen at the SHAFTESBURY THEATRE and transferred to the ALDWYCH a year later to become the first of a long series of Aldwych farces. Evans also collaborated with R. Guy Reeve in *The Other Mr Gibbs* (1924) and was the author of many of his own music-hall songs and sketches.

EVANS'S, in King Street, Covent Garden, probably the best known of the convivial 'song-and-supper rooms' which preceded the MUSIC-HALL, and were themselves the successors of the old singing-clubs at such resorts as the 'Cyder Cellars' and the 'Coal Hole', where entertainment of a mixed kind but mostly comic songs and monologues was provided free of charge for those who came to eat and drink, in the style of the modern cabaret. Evans's was first situated in the basement of a large tavern, whose lessee before W. C. Evans, a comedian from COVENT GARDEN, took over in about 1820 was a Mr Joy; hence its first name 'Evans's, late Joy's'. It had at first a modestly successful career, Evans being succeeded in 1844 by Paddy Green [Greenmore], who kept the old name even when in about 1856 new and larger premises were built over the garden; by this time, under the influence of the new music-halls, the original simple 'sing-song' had given way to highly organized entertainment, though retaining the old separate tables and the presiding CHAIRMAN. In the 1840s, Evans's heyday, the programme included such entertainers as Charles SLOMAN, then at his best, and Sam COWELL, who was to be one of the early stars of the music-halls, and varied from the somewhat *risqué* songs of such singers as J. W. Sharpe to the boys of the Savoy Chapel in glees and extracts from opera. It also included Harry Clifton in his 'motto' songs, such as 'Paddle Your Own Canoe', and one of the earliest Negro MINSTREL SHOWS, the 'Æthiopian Serenaders', in 'Swanee River' and other 'coon songs'. But by the 1860s Evans's was feeling the competition of the new music-halls, and in 1880 it lost its licence and closed. Its premises were taken over in 1934 by the PLAYERS' CLUB, who later punned on the original name by labelling their 'Victorian-style' show *Ridgeway's Late Joys*.

EVELING, STANLEY, see SCOTLAND.

EVERAC, PAUL, see ROMANIA.

EVERYMAN THEATRE, Liverpool, originally a chapel, opened in 1964. It was intended by its founders, who included Terry HANDS, for the production of serious modern plays which it was felt were being ignored by the LIVERPOOL PLAYHOUSE, and among the dramatists presented during the first six years were BECKETT, BOND, Peter NICHOLS, and Harold PINTER. Later the theatre became known for its local musical documentaries; one on the Beatles, *John, Paul, George, Ringo . . . and Bert* (1974) by a local author Willy Russell, was transferred to the West End for a long run. The theatre has supported other local authors such as Bill Morrison, Alan Bleasdale, John McGrath, and Mike Stott. Plans were made for rehousing the theatre, but in the mid-1970s, when they failed to materialize, it was rebuilt on its existing site. Owned by an independent Trust, it now seats 430 and has an open-end stage with a small thrust capacity.

In 1980 Ken Campbell (1941–) became artistic director. He was already well known as director of the Science Fiction Theatre of Liverpool, which, in makeshift premises, presented works which combined imagination, fantasy, and outlandish comedy. Its first production, *Illuminatus* (1976), a group of five plays written and directed jointly by Campbell and Chris Langham, was seen in London in 1977 as the opening production at the COTTESLOE THEATRE. Though the authorities eventually forbade the use of its premises for theatrical purposes, the Science Fiction Theatre of Liverpool retained its title, presenting Neil Oram's *The Warp* (1979) at the ICA Theatre in London. It later became Campbell's first offering at the Everyman, being shown in 10 weekly instalments; the seats were removed and the audience found itself in the midst of the action. Other productions, for which the seats were replaced, and which, like *The Warp*, were experimental, included Hrant Alianak's *Lucky Strike* and an adaptation of Karel ČAPEK's novel *The War With the Newts*, later seen at the RIVERSIDE STUDIOS. Though not specifically a youth theatre, the Everyman attracts a predominantly young audience.

EVERYMAN THEATRE, Hampstead, London. This small theatre, now well known as a cinema, had a brief blaze of glory from 1920 to 1926 under Norman MACDERMOTT. Originally a Volunteer Drill Hall, it was converted and opened on 15 Sept. 1920 with BENAVENTE's *The Bonds of Interest*. The success of SHAW's *You Never Can Tell* a fortnight later led to the appearance of many more Shaw plays, including the first public production in London of *The Shewing-Up of Blanco Posnet* (1921). Other foreign plays first seen at this theatre include BJØRNSON's *Beyond Human Power* (1923), CHIARELLI's *The Mask and the Face* (1924), and PIRANDELLO's *Henry IV* (1925). Chiarelli's play transferred to the West End, as did Noël COWARD's *The Vortex* (1924). Eugene O'NEILL was seen for the first time in London with *In the Zone* and *Different* (1921), *Ile* (1922), and *The Long Voyage Home* (1925). Among other productions were IBSEN's *A Doll's House, John Gabriel Borkman* (both 1921), *Hedda Gabler* (1922), *The Wild Duck, Ghosts* (both 1925). The venture ended in January 1926 with a dramatization of Chesterton's *The Man Who Was Thursday*. The scenery and costumes for most of the plays were designed by MacDermott himself, with the help of John Garside, and Rudolph Schwabe. After MacDermott's departure the Everyman was taken over by Raymond MASSEY, followed by Malcolm Morley, under whom the first English production of OSTROVSKY's *The Storm* was given on 3 Dec. 1929. It was then used intermittently by students until 1947 when the building was converted into a cinema intended for non-commercial foreign and experimental films.

EVREINOV, NIKOLAI NIKOLAIVICH (1879–1953), Russian dramatist, an exponent of SYMBOL-

ISM, whose anti-naturalistic theatre is exemplified in *The Theatre of the Soul*, a MONODRAMA in which various aspects of the same person appear as separate entities. His *Revisor* (1912) illustrated amusingly what such directors as STANISLAVSKY and REINHARDT might do to GOGOL's masterpiece *The Government Inspector*, and in *The Fourth Wall* (1915) he portrayed an ultra-realistic director let loose on GOETHE's *Faust. The Chief Thing* (1921), which deals with the power of illusion, has some affinity with the work of PIRANDELLO. Evreinov's theories were swept aside by the SOCIALIST REALISM of the Soviet era. Two of his short plays were translated by C. E. Bechofer and published in *Five Russian Plays* (1916) as *A Merry Death* and *The Beautiful Despot*.

EWALD, JOHANNES (1743–81), Danish lyric poet, and author of the first tragedies to be written in Danish, who strangely fused in his work the traditional qualities of 18th-century rationalism with an anticipation of Romanticism. He came to share the enthusiasms of the German STURM UND DRANG school for Shakespeare and Ossian, and his eager reading also included Milton, ROUSSEAU, and Percy's *Reliques. Adam og Eva* (*Adam and Eve*, 1769) is essentially a dramatic poem rather than a drama; his tragedy *Rolf Krage* (1770) draws inspiration from Norse legend; *Balders Død* (1773, trans. by G. Borrow as *The Death of Balder*, 1779), based on Scandinavian mythology, is in blank verse; and *Fiskerne* (*The Fishermen*, 1780), his most famous play, strikes a new note of patriotism.

EXETER, Devon. Although there is evidence of acting by waits, mummers, and minstrels in Exeter as early as the mid-14th century, and excerpts from MIRACLE PLAYS were given annually on Corpus Christi (the Thursday after Trinity Sunday) in the early 16th century, there is little theatrical activity to record until the early 18th century. A room in the Seven Stars Inn, just outside the city boundary, is known to have been used as a theatre in about 1735, at the time when the first small theatre was erected by a company from BATH. Closed by the Theatres Act of 1737, this became a Methodist chapel but reverted later to its original use as part of the Bath CIRCUIT. It was managed for a time in the 1760s by Thomas Jefferson, great-grandfather of the famous American actor Joseph JEFFERSON. In 1787 a new theatre was built, based on the plans of SADLER'S WELLS, the London stars who appeared there including Stephen KEMBLE, Mrs SIDDONS, Master BETTY, and Edmund KEAN. It was destroyed by fire on 7 Mar. 1820, the only part to survive being the colonnade which was incorporated into a new building. This opened on 10 Jan. 1821 with *The Merchant of Venice*, and was later visited by Mme VESTRIS, Charles and Fanny KEMBLE, MACREADY, and PHELPS. The first performance in 1880 of *Hearts of Oak* by Henry Arthur JONES (the first play by a local man) had in its cast the future playwright and humorist Jerome K. JEROME. This theatre, which although it had no royal patent was known locally as the Theatre Royal from about 1828, also burned down on 7 Feb. 1885. A new Theatre Royal, which opened on 13 Oct. 1886, was also destroyed in one of the worst conflagrations in the history of the English theatre when on 5 Sept. 1887, during a performance of G. R. Sims's *Romany Rye*, the building was entirely gutted with the loss of 186 lives. It was not until 1889 that a new theatre opened. It survived the financial vicissitudes of the 1920s and 1930s and escaped damage by bombing during the Second World War, but although after the war it became well known for its long-running Christmas pantomimes, it closed and was demolished in 1962. Exeter was then left without a theatre until the opening in 1967 of the NORTHCOTT THEATRE, closely connected with the university.

EXIT, see STAGE DIRECTIONS.

EXPERIMENTAL THEATRE, New York, see DALY'S SIXTY-THIRD STREET THEATRE.

EXPRESSIONISM, a movement that began in Germany in about 1910, and is best typified by the plays of Georg KAISER and Ernst TOLLER. The term was first used in 1901 by Auguste Hervé to describe some of his paintings conceived in reaction against Impressionism in art. It was later used of art, music, and literature, as well as plays that displayed reality as seen by the artist looking out from within, instead of, as with Impressionism, reality as it affects the artist inwardly. The Expressionist theatre was a theatre of protest, mainly against the contemporary social order and the domination of the family. Most of its dramatists were poets who used the theatre to further their ideas, and it was partly their use of poetic language that led to the collapse of the movement. It was too personal, as was the concentration of attention on the central figure, the author-hero whose reactions are 'expressed' in the play. Among the forerunners of Expressionism were STRINDBERG and WEDEKIND; the first drama of the Expressionist movement is usually considered to be *Der Bettler*, by Reinhard SORGE. Other dramatists important in the movement were Walter HASENCLEVER, Fritz von UNRUH, and Ernst BARLACH. One of the few dramatists outside Germany to be influenced by Expressionist drama was Eugene O'NEILL, particularly in *The Emperor Jones* (1920) and *The Hairy Ape* (1922).

EXTRAVAGANZA, a spectacular and brilliantly costumed dramatic entertainment which flourished in England in the mid-19th century. Distinguished from BURLESQUE by its lack of a satiric target, it was usually based on a well known story from mythology or folk-lore and characterized by its witty, punning use of rhymed couplets and clever songs. It was first offered as an alternative to PANTOMIME at Easter and Whitsun in 1824, and with a HARLEQUINADE attached and sometimes billed as a pantomime, was gradually introduced into the theatre on Boxing Night (Dec. 26). Its leading exponents were J. R. PLANCHÉ, H. J. BYRON, and E. L. Blanchard.

EYSOLDT, GERTRUD (1870–1955), German actress, who made her first appearance with the MEININGER COMPANY in 1890 in *Henry IV*. After touring in Germany and Russia she appeared in BERLIN in 1899, and later played under REINHARDT. An extremely clever and subtle player, she was at her best in modern realistic parts, particularly in the works of WEDEKIND, in which she played opposite the author, and was also good in IBSEN, in such parts as WILDE's Salome, and in the plays of MAETERLINCK. She was in the first German-language production of SCHNITZLER's *Reigen* at the Kleines Schauspielhaus in Berlin in 1921, and at the subsequent court hearing defended the play against the charge of obscenity.

EYTINGE, ROSE (1835–1911), American actress, who made her début in 1853 as Melanie in BOUCICAULT's *London Assurance*. She was a beautiful and temperamental player, greatly admired in both high comedy (as Lady Gay Spanker in *London Assurance* and as Beatrice in *As You Like It*) and tragedy (as Ophelia in *Hamlet*, Desdemona in *Othello*, and especially Cleopatra in *Antony and Cleopatra*). Perhaps her most popular role was that of Nancy in DICKENS's *Oliver Twist* at the STAR THEATRE with E. A. DAVENPORT as Bill Sikes and the younger James WALLACK as Fagin. Her autobiography, *The Memories of Rose Eytinge* (1905), is a racy account of her colourful life.

F

FABBRI, DIEGO (1911–80), Italian dramatist influenced both by PIRANDELLO and by his older contemporary Ugo BETTI. His plays unlike those of the dialect dramatists of his day were written in Italian; they are markedly religious and discuss contemporary social and moral problems in the light of the Catholic faith. His best known work, *Processo a Gesù* (*The Trial of Jesus*, 1955), which as *Between Two Thieves* was seen in New York in 1960, was first produced for the PICCOLO TEATRO DELLA CITTÀ DI MILANO by Orazio Costa. Among his other plays, which include *Inquisizione* (*The Inquisition*, 1950), *Il sedutore* (*The Seducer*, 1951), *Processo di famiglia* (*The Family on Trial*, 1953), and *Veglia d'armi* (*The Armed Vigil*, 1956), *La Bugiarda* (*The Liar*, 1953) was seen in London during the WORLD THEATRE SEASON of 1965, in a production by VALLI's Compagnia dei Giovani. Fabbri also made excellent dramatizations of DOSTOEVSKY's *The Devils* and *The Brothers Karamazov*, published in 1961 in a volume entitled *I demoni*.

FABRE D'ÉGLANTINE [Philippe-François Nazaire] (1755–94), French dramatist and revolutionary, who took the name Églantine after winning a prize for his verses at the Floral Games in Toulouse. He was for some time an actor but in spite of a good voice and presence he was unsuccessful, and soon forsook the stage in order to devote himself to the Revolution in which he took an active part, being a member of the Convention with Danton, Marat, and Robespierre. He was guillotined in April 1794. He wrote several poor plays and a number of librettos for the little theatres; the only play by which he is remembered is *Le Philinte de Molière* (1790), a sequel to *Le Misanthrope* which illustrates the views of J.-J. ROUSSEAU on MOLIÈRE's play.

FABRICIUS, JAN, see NETHERLANDS.

FABULA, the generic Latin name for a play, under which many different types of drama were grouped.

1. **Atellana**, a form of popular farce, traditionally connected with the small town of Atella nine miles from Capua on the road to Naples. It seems to have been designed to cater for the taste of the crowds which gathered on market days in country towns, and was always sharply distinguished by the Romans from Greek performances of a farcical nature such as the MIME. The essential feature of the *atellana* was the appearance in each one of certain traditional characters —BUCCO, DOSSENNUS, MACCUS, MANDUCUS, and PAPPUS—all coarse, greedy clowns, wearing stock costumes and masks. A recurring feature of the *atellana* was disguise and masquerade, and hence arose those complications, called *tricae atellanae* (the origin of our word 'intrigue'), indicated by such titles as *Marcus the Maiden*. Atellan pieces were short, with few characters; their use as INTERLUDES would seem to suggest that they corresponded to the later CURTAIN-RAISER. Originally performed in the Oscan tongue, and always impromptu, they were given literary form by Novius and Lucius in the opening decades of the first century BC, and were then written down in Latin. In the process they must have lost something of their traditional character and coarse jesting but, to judge by the titles and fragments which have survived, they retained their primitive, rustic atmosphere—*The Pig*, *The Sow*, *The She-Goat*, *The She-Ass*—but could be set either in town—*The Candidate*, *The Inspector of Morals*—or in the country—*The Farmer*, *The Vine-Gatherers*, *The Wood-Pile*, *A-Hoeing*. The language remained homely: a peasant defines wealth as 'a short-lived blessing, like Sardinian cheese', and the wit clownish; when told to 'make a clean job of it', Bucco replies 'I've just washed my hands.' But the jokes were no less popular for their familiarity, and when discussing the nature of jest, Cicero took his examples from the *atellanae* of Novius. The few fragments of these rustic farces that survive from the period after Pomponius and Novius suggest that there may have been a return to improvisation. On the stages of the early Roman Empire the atellan performers, though social outcasts, enjoyed considerable licence. They retained their popularity for some time but eventually had to give way to the equally vulgar but more sophisticated mime players.

2. **Palliata**, a play directly translated from Greek NEW COMEDY, from *pallium*, the Greek cloak of everyday wear. Authors of *palliatae* included ANDRONICUS, CAECILIUS STATIUS, NAEVIUS, PLAUTUS, and TERENCE. The last writer of *palliatae* for the stage was Turpilius, who died in 103 BC. The *fabula crepidata*, a term which is sometimes found, may be synonymous with the *fabula palliata*, since the *crepida* was a Greek shoe always worn with the *pallium* and quite unlike the COTHURNUS worn by tragic actors.

3. **Praetexta**, an original Latin play of a serious

character on a theme taken from Roman legend or contemporary history: from *toga praetextata*, the purple-bordered toga worn by Roman magistrates. The form was created by Naevius: only some half-dozen examples of it survive from the time of the Republic, though ACCIUS, ENNIUS, and PACUVIUS are thought to have written *praetextae*. In 43 BC L. Cornelius Balbus illustrated his lack of good taste by producing at Gades a *praetexta* on his own achievements in the civil war. Other *praetextae* intended for reading only were written under the Empire: the only one extant is the anonymous *Octavia*, once attributed to SENECA.

4. **Rhinthonica**, a burlesque of Greek tragedy named after the Greek writer Rhinthon of Tarentum (*fl.c.*300 BC), the chief exponent of this type of play. Another name for it was *hilarotragoedia*, 'merry tragedy'. The popularity of such plays among the Greeks of southern Italy is illustrated by the so-called PHLYAX vases, which show, for example, Zeus on a love-adventure about to climb a ladder to the window from which a lady is looking out, while Hermes holds a lantern to assist him. The nearest example of such a play in Latin is the *Amphitruo* of Plautus.

5. **Riciniata**, a synonym for the ancient Roman mime play, from the *ricinium*, or characteristic hood worn by actors in such plays.

6. **Saltica**, in the Roman theatre, a 'play for dancing', that is, the libretto or text sung by the CHORUS while the solo actor—the PANTOMIMUS—mimed the action.

7. **Togata**, a form of Roman comedy which arose from the irrepressible Italian instinct for social satire and topical allusion, which audiences failed to find in the imported *palliata*, especially in its later and more strictly Hellenic form. As the ruling classes in Rome would have reacted strongly to any political or personal attacks the *togata* took its themes from the life of ordinary folk in Italian towns or the lower classes in Rome, as can be seen from its alternative name, *tabernaria*, from *taberna*, a poor man's house. Three writers of *togatae* are known by name: AFRANIUS, ATTA, and TITINIUS, of the 2nd century BC. The cast of a *togata*, like that of an *atellana*, was apparently smaller than that of a *palliata*, and the plot simpler, less restrained by social conventions and more overtly sexual, even extending to adultery and to the portrayal of homosexuality, a theme said to have been introduced into the theatre by Afranius. Usually the world of the *togata* is that of business, amusement, matchmaking, and family relationships, with very little romance, but as slaves in the *togata* were not allowed to be cleverer than their masters, there are none of those tricky, swindling servants who play so large a part in the plots of the *palliatae*. No complete play of this type survives, and it is impossible to tell from the fragments that remain how the plot was built up. Once acted, it appears that a *togata* was seldom revived, though literary *togatae* continued to be written down to Juvenal's day.

8. **Trabeata**, a form of *fabula togata* dealing with Roman knights, from *trabea*, the official robe of the equestrian order. Invented by Melissus, a freedman of Maecenas, it was probably short-lived, and nothing is known of it except its name.

FACT, Theatre of, term first applied in the 1950s to documentary plays which derive in part from the technique of the American pre-war LIVING NEWSPAPER. Based entirely on the 'facts' of history, used either unadorned, as in KIPPHARDT's *In the Matter of J. Robert Oppenheimer* (1964) and *Die Ermittlung* (1965) by Peter WEISS, or with some attempt at theatrical coherence and conscious arrangement, as in *Oh, What a Lovely War!* (1963), evolved by Joan LITTLEWOOD, the genre was typified in England by *US* (1966), directed by Peter BROOK for the ROYAL SHAKESPEARE COMPANY at the ALDWYCH in London, which, on its first production, was hailed as 'the ultimate non-play so far'.

FAGAN, J(ames) B(ernard) (1873–1933), playwright and director, born in Ulster, who began his stage career in Frank BENSON's company in 1895. He then played two seasons under Beerbohm TREE at HER MAJESTY'S THEATRE, and later became manager of the ROYAL COURT THEATRE, where he gave some notable productions of Shakespeare, including *The Merchant of Venice* in 1919 with the great Jewish actor Maurice Moskovitch as Shylock. In 1923 he took over the old Big Game Museum in Oxford and made it into a small theatre, the OXFORD PLAYHOUSE, installing there a repertory company, mainly of young players, which included at various times Flora ROBSON, John GIELGUD, Tyrone GUTHRIE, and Raymond MASSEY. In 1929 he became director of the Festival Theatre, CAMBRIDGE, and he was also responsible for many productions by the Irish Players. His excellent dramatization of R. L. Stevenson's *Treasure Island* was first seen in London in 1922, and then annually as a Christmas play up to 1931. Of his own plays, the most successful were *And So To Bed* (1926), based on PEPYS's Diary, and *The Improper Duchess* (1931), in both of which Yvonne ARNAUD gave fine performances.

FAIKO, ALEXEI MIKHAILOVICH (1893–), Soviet dramatist, whose early plays *Lake Lyul* (1923) and *Bubus the Teacher* (1925) were produced by MEYERHOLD. *The Man with the Portfolio* (1928) dealt for the first time with the problem of the intellectual at odds with the new Soviet régime. Its chief character, a professor at a Moscow college, tries to make the best of both worlds by conforming in public and abusing the Government in private, but finally, a spiritual bankrupt, he commits suicide. The play was revived several times after the Second World War in Leningrad and Moscow. A comedy produced at the LENKOM THEATRE in Moscow in 1957, *Don't Set Yourself up as God!*, was well received, but on the whole

Faiko confined himself in later years to writing for theatre journals.

FAIR, WILLIAM B., see MUSIC-HALL.

FAIRBROTHER [Parselle], SYDNEY (1872–1941), English actress, grand-daughter of the MUSIC-HALL star Sam COWELL and daughter of Florence TAPPING by her first husband. Sydney took her great-grandmother's name when she went on the stage, making her first appearance in 1889, and subsequently coming into prominence playing Wally in G. R. Sims and Arthur Shirley's *Two Little Vagabonds* (1896). She was later seen in a great variety of roles, including Proserpine in SHAW's *Candida* at the ROYAL COURT THEATRE in 1904, and during 1912 to 1914 toured the music-halls very successfully with the elder Fred Emney in a sketch, 'A Sister to Assist 'er'. In 1916 she played Mahbubah in Oscar ASCHE's *Chu-Chin-Chow*, remaining in the part until 1920 when she left to make one of her most successful appearances as Mrs Badger in Gertrude E. Jennings's *The Young Person in Pink*, followed by a similar character part, Mrs Alfred Butler in a musical farce *Battling Butler*, and, in contrast, Maria in *Twelfth Night*. She continued active until shortly before her death, her last part being Alice Matheson in N. C. HUNTER's *A Party for Christmas* (1938).

FAIRS. In England and Europe the big fairs, usually held in the spring and autumn, were from the beginning associated with theatrical enterprise, just as the small country fairs were never without their groups of acrobats, dancers, and singers. The assembling of a number of people in one place provided a ready-made audience for the travelling companies of the 15th and 16th centuries, and finally offered sites for permanent theatres. This was most noticeable in the case of the Paris fairs, particularly Saint-Germain in early spring and Saint-Laurent in late August and September. Some of the best known farce-players of 17th-century Paris are believed to have graduated from the fairs or to have been the children of fairground actors, but the main development of the *forains*, as they were called, came at the end of the 17th and in the 18th centuries, when they gradually extended their season beyond that of the fairs themselves and replaced their wooden BOOTHS with permanent playhouses. Their popularity, and the fact that during the fairs provincial companies were allowed to play in Paris without further permission, gave constant annoyance to the COMÉDIE-FRANÇAISE and the COMÉDIE-ITALIENNE, until the latter made common cause with the *forains* and combined to form the Opéra-Comique, which during the summer months occupied the Théâtre de la Foire Saint-Laurent, built in 1721 and destroyed in 1761. The old French farce survived at the fairs in the form of the PARADE, which formed a link between the COMMEDIA DELL'ARTE and the various spectacles offered on the BOULEVARD DU TEMPLE up to the mid-19th century.

In England the fairs of Saint Bartholomew (immortalized by Ben JONSON), Smithfield, and Southwark, as well as the smaller Greenwich and May fairs were always connected with theatrical entertainments, particularly PUPPET shows, and there were also booths for living actors. Elkanah SETTLE is believed to have ended his days as a green dragon in Mrs Myn's booth at Southwark, and he also wrote DROLLS and short sketches for Bartholomew Fair. The theatrical development of the English fairs was, however, less than in France, perhaps owing to the inclemency of the weather, and there is no record of any permanent playhouses being built on fair-sites.

Of the German fairs, those of LEIPZIG and FRANKFURT-AM-MAIN are best known in connection with theatrical matters, as it was at Frankfurt that the ENGLISH COMEDIANS appeared most frequently and had if anywhere a permanent home, while Leipzig saw the reforms of GOTTSCHED first put into action by Carolina NEUBER. Fairs also played a big part in the history of the early Russian theatre, which, however, developed somewhat differently from the rest of Europe. Little information seems to be available on the fairs of Italy and Spain, but there can be little doubt that in Italy the itinerant *commedia dell'arte* troupes, and in Spain the travelling companies of such actor-managers as Lope de RUEDA were in evidence wherever there was a chance of finding a ready-made audience.

FALCKENBERG, OTTO (1873–1947), German director, who first made his name with a production at the MUNICH Kammerspiele in 1915 of STRINDBERG's *Spöksonaten*. From 1915 to 1944 he directed this theatre, and made it an outstanding home of contemporary drama with such plays as Johst's *Der Einsame* (1917), BARLACH's *Der tote Tag* (1924), and BILLINGER's *Rauhnacht* (1931). He also staged BRECHT's first play, *Trommeln in der Nacht* (1922), with settings by Reigbert, and revived a number of classics, among them *As You Like It* in 1917, *Troilus and Cressida* in 1925, and *Cymbeline* in 1926. He was adept at bringing out the particular atmosphere—and notably any visionary qualities—of each play. He was above all an actor's director, and many of those who played under him achieved fame.

FALLING FLAPS, a method of making quick changes of scenery and trick effects which dates from about the beginning of the 19th century. It consists of a series of double flat scenes, framed and moving on hinges. One side of these was painted to represent, for example, an indoor scene and the other an outdoor. The stage was thus set with a complete picture in sections, each piece being kept in place by catches. When these were simultaneously released, the flaps fell by their own weight, presenting an immediate change of scene. The method is documented in the

Victorian TOY THEATRE sheets, some of which are labelled 'Trick Scene'. These have dotted lines across them, and on a separate sheet labelled 'Tricks' are smaller pieces, which when cut out and mounted on cardboard fit into the areas marked by the dotted lines. They can be hinged into position and dropped to effect a sudden change of scene.

FALSE, or Inner, **PROS(CENIUM),** temporary structure set behind the PROSCENIUM arch to diminish the size of the opening. In America it is known as a Portal Opening, and in France as *le manteau d'Harlequin*. It is particularly useful on tour, when scenery has to be accommodated on stages of varying sizes, and consists usually of a pair of WINGS, opened to a right angle or more and set on either side of the stage, with a deep BORDER connecting the two sides across the top. The whole is cut and painted to simulate hanging drapery, with folds, swags, fringes, cords, and tassels. In some modern proscenium theatres the wings are often replaced by a fixed pair of narrow FLATS covered in black velvet (Tormentors) linked by a plain border (a Teaser). This arrangement can be used with a CYCLORAMA to reduce the amount of scenery needed to dress the stage. A false pros, consisting of a pair of flats and a decoratively profiled border (Show Portal in America) is often used in REVUE and MUSICAL COMEDY.

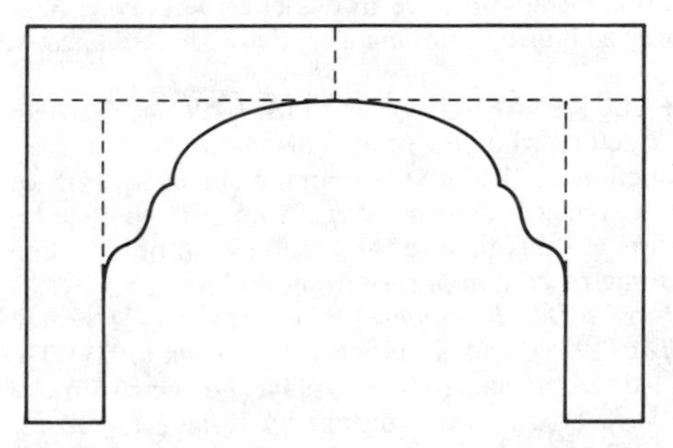

False proscenium: dotted lines indicate assembly from smaller units

FAN EFFECT, see TRANSFORMATION SCENE.

FARCE, a form of popular comedy in which laughter is raised by horseplay and bodily assault in contrived and highly improbable situations. It must, however, retain its hold on humanity, even if only in depicting the grosser faults of mankind, otherwise it degenerates into BURLESQUE. It deals with the inherent stupidity of man at odds with his environment, and belongs in its origins to the great submerged stream of folk-drama, of which few written records remain. It stands at the beginning of classical drama (see FABULA) as well as of modern European drama, and was especially popular in France in the later Middle Ages. Among the many medieval farces that were long current, the greater part no doubt transmitted orally, the best known is *Maître Pierre Pathelin* (*c*.1470), the portrait of a rascally lawyer which has survived revision, adaptation, and even translation without losing its savour and robust humour. Traditional farce survived in France until well into the 17th century, particularly in the provinces, but died out in Paris in the 1640s with the disappearance of the gifted trio of farce-players TURLUPIN, GROS-GUILLAUME, and GAULTIER-GARGUILLE of the Hôtel de BOURGOGNE. In his early career as an actor MOLIÈRE played in farces, particularly in the neighbourhood of Lyon, and not only do his own early plays follow the same pattern but the tradition of the farce exercised a great influence on his whole career as a dramatist. There were elements of farce in early English Biblical plays, and farcical interludes were later written by scholars for production in schools and other places, but as in Italy and Germany the influence of the French farce was paramount, culminating in the works of John HEYWOOD.

In the 18th and 19th centuries short one-act farces were popular on the English and American stages, usually as part of a bill which also included a five-act TRAGEDY. They were ephemeral productions, though some of them achieved a great success, mainly through the acting of some particular comedian.

In modern usage the word farce is applied to a full-length play dealing with some absurd situation, generally based on extra-marital adventures—hence 'bedroom farce'. An early exponent of modern farce in England was PINERO, several of whose early plays in this genre have been successfully revived. A famous example of a full-length farce which still holds the stage is *Charley's Aunt* (1892) by Brandon THOMAS. In the 1920s and 1930s there was a series of successful farces at the ALDWYCH THEATRE, mostly written by Ben TRAVERS, and in the 1950s and 1960s a similar series was produced by Brian RIX at the WHITEHALL. Thanks to its robust character, farce survives translation better than COMEDY, as is shown by the recent success in England of the farces of FEYDEAU, some of the best of their kind.

FARINE, JEAN (*fl*.1600–35), French actor in the style of TABARIN, who is believed to have been originally a travelling quack doctor. After serving his apprenticeship at the Paris fairs he went to the Hôtel de BOURGOGNE with BRUSCAMBILLE, and together they played improvised farces in the manner of TURLUPIN and his associates, with whom they are often mentioned approvingly in contemporary doggerel.

FARJEON, HERBERT, see REVUE.

FARQUHAR, GEORGE (1678–1707), Irish-born dramatist who is usually classed among the writers of English Restoration comedy, though he stands a little apart from them, both chronologically and in showing more variety of plot, depth of feeling, and, in his later and best plays, a more conscious effort to adapt the licentiousness of the early comedy of manners to the changing taste of the time. He was for a short time an actor at the Smock Alley Theatre in Dublin, but believing himself more fitted for writing than acting he went to London, where his first play *Love and a Bottle* (1698) was produced at DRURY LANE. With his second, *The Constant Couple; or, a Trip to the Jubilee* (1699), he established himself as one of the foremost dramatists of the day. The play held the stage throughout the 18th century, the hero Sir Harry Wildair, first played by WILKS, the friend who had encouraged Farquhar to leave Dublin, being a favourite BREECHES PART with Peg WOFFINGTON and Dorothy JORDAN. A sequel, *Sir Harry Wildair* (1701), proved less successful, as did a revision of BEAUMONT and FLETCHER's *The Wild Goose Chase* (1652) as *The Inconstant; or, the Way to Win Him* (1702), and two new farces, *The Twin Rivals* (1703) and *The Stage Coach* (1704). Having contracted a marriage with a penniless lady whom he believed to be an heiress, Farquhar left London for a time, but returned to the theatre in 1706 with a fine comedy *The Recruiting Officer*, based on his own experiences in that capacity in Shropshire. An immediate success, it was followed by his finest play *The Beaux' Stratagem* (1707), in which Wilks played Archer and Anne OLDFIELD Mrs Sullen. A realistic comedy with a wholesome, open-air humour, it was Farquhar's last play, written in six weeks while he was lying ill and penniless in mean lodgings. Both it and *The Recruiting Officer* are still revived. The latter, on which BRECHT based his *Pauken und Trompeten* (1955), was seen at the OLD VIC in 1963 with OLIVIER as Brazen; the former at the Old Vic in 1970 with Maggie SMITH as Mrs Sullen, which was also one of Edith EVANS's best parts.

FARR, FLORENCE [Mrs Edward Emery] (1860–1917), English actress and director, chiefly connected with the introduction of plays by IBSEN, SHAW, and YEATS to the London stage. She played Rebecca West in the first London production of Ibsen's *Rosmersholm* in 1891 and in 1892 appeared as Blanche in Shaw's *Widowers' Houses*, the first of his plays to be staged. Financed by Miss HORNIMAN, she then produced at the Avenue Theatre (later the PLAYHOUSE) his *Arms and the Man*, creating the part of Louka. It was preceded by Yeats's one-act play *The Land of Heart's Desire*, again the first of his plays to be seen in London. She was later associated with Yeats in Dublin where she played Aleel the Minstrel in *The Countess Cathleen* (1899); back in London she arranged the music and trained the chorus for GRANVILLE-BARKER's productions of EURIPIDES' *The Trojan Women* in 1905 and *Hippolytus* in 1906, in which she played the Leader of the Chorus and the Nurse respectively. A keen student of theosophy and Indian philosophy, she learned in 1912 that she had cancer and went to Ceylon, where she spent her last years as principal of a Hindu girls' college.

FARRANT, RICHARD, see BLACKFRIARS THEATRE and BOY COMPANIES.

FARREN, ELIZABETH (1759–1829), English actress, the child of strolling players and on the stage from her earliest years. As an adult she made her first appearance in London on 9 June 1777 at the HAYMARKET THEATRE as Kate Hardcastle in GOLDSMITH's *She Stoops to Conquer*, and then joined the company at DRURY LANE. She was not immediately successful, owing to the outstanding popularity of Mrs ABINGTON, but she became recognized as an excellent player of fine ladies, for which her natural elegance, tall, slim figure, and beautifully modulated voice rendered her particularly suitable. At one time she organized private theatricals for the Duke of Richmond. She made her last appearance on the stage in 1797 as Lady Teazle in SHERIDAN's *The School for Scandal* and then married the recently-widowed Earl of Derby, who had been in love with her for many years.

FARREN, NELLIE [Ellen] (1848–1904), English actress, great-granddaughter of William FARREN the elder. Being small and slight she specialized for many years in boys' parts—Smike in DICKENS's *Nicholas Nickleby*, Sam Willoughby in Tom TAYLOR's *The Ticket-of-Leave Man,* and the cheeky Cockney lads in H. J. BYRON's burlesques and extravaganzas. From 1868 until her retirement in 1891 she was a great favourite with the audiences at the old GAIETY THEATRE, where she was a member of the famous burlesque quartet with Edward TERRY, Kate VAUGHAN, and Edward Royce. Her husband Robert Soutar (1827–1908), also an actor, was for many years stage-manager at the Gaiety, and her two sons were on the stage, Joseph, whose career stretched over 50 years, dying in 1962 at the age of 91.

FARREN, WILLIAM, the elder (1725–95), English actor, founder of a distinguished theatrical family. He went on the stage as a young man, but it was not until 1776 that he joined a major company, going to DRURY LANE where he created the parts of Careless in *The School for Scandal* (1777) and Leicester in *The Critic* (1779), both by SHERIDAN. He was also good in Shakespeare, playing such parts as Hotspur in *Henry IV, Part 1*, the Ghost in *Hamlet*, and Buckingham in *Henry VIII*. In 1784 he went to COVENT GARDEN, where he remained until his death. His elder son Percival (1784–1843), was for a short time an actor, but deserted the boards for stage management. He was the friend and tutor of Helen FAUCIT. For his son William, see below.

FARREN, WILLIAM, the younger (1786–1861), English actor, younger son of William FARREN. He made his first appearance in Plymouth under the management of his elder brother Percival, and was first seen in London at COVENT GARDEN in 1818, playing Sir Peter Teazle in SHERIDAN's *The School for Scandal*, generally considered one of his best parts. He was also much admired as Lord Ogleby in the elder COLMAN's *The Clandestine Marriage* and during his long career played a wide range of Shakespearian parts, being seen in his early years with the Covent Garden company as Sir Andrew Aguecheek in *Twelfth Night* and in later years as Malvolio. Other parts in which he was considered outstanding were Shallow in *Henry IV, Part 2*, Dogberry in *Much Ado About Nothing*, and Polonius in *Hamlet*. He retired in 1853 after a stroke. His sons William (1825–1908), and Henry (1826–60), his grandson William Percival (1853–1937), and his grand-daughter Nellie (above) were all on the stage.

FASTNACHTSSPIEL, the German Carnival or Shrovetide play of the 15th century, in one act, performed mainly in Nuremberg by students and artisans. It shows, in its somewhat crude couplets, a mingling of religious and popular elements interesting in the light of later developments in German drama. In its earliest form it had no plot but developed, independently of religious drama, from a sequence of narrated episodes in which each character endeavoured to outdo the others in absurdity or obscenity. The earliest known writer of *Fastnachtsspiele* was Hans ROSENPLÜT. The subjects of the Carnival plays are those which would appeal to the mainly urban audiences before whom they were usually acted—the weaknesses and venial sins of lawyers and their clients, doctors and their patients, clerics and their female parishioners. In most of these farces the NARR or fool is the central character, sometimes with a dull-witted companion to serve as a butt for his practical jokes. At first the plays were presented in the simplest possible way, rather in the style of the English MUMMERS' PLAY but later, when the town guilds took over, a raised stage and hangings, with a few properties, became general. Many of the Carnival plays were written by the MEISTERSÄNGER, of whom the best known is Hans SACHS.

FATE DRAMA or **TRAGEDY,** see SCHICKSALSTRAGÖDIE.

FAUCIT, HELEN(a) SAVILLE (1817–98), English actress, who was trained by Percival Farren, great-uncle of Nellie FARREN. A woman of great charm and beauty, her best work was done in revivals of Shakespeare and in new verse plays, many of them specially written for her. She first appeared in the provinces in 1833, and in 1836 made her début in London as Julia in Sheridan KNOWLES's *The Hunchback*. A year later she was seen in BROWNING's poetic drama *Strafford* and then had a great success in one of her best parts, Pauline in BULWER-LYTTON's *The Lady of Lyons* (1838), playing opposite MACREADY. She was also in Browning's *A Blot in the 'Scutcheon* (1843) and *Colombe's Birthday* (1853). It was said that her refusal to appear in Matthew ARNOLD's *Merope* caused him to abandon its production. Her finest Shakespearian roles were considered to be Juliet in *Romeo and Juliet*, Portia in *The Merchant of Venice*, and Desdemona in *Othello*. She married in 1851, becoming Lady Martin when her husband Theodore Martin was knighted in 1880. In later life she acted almost entirely for charity and became the friend and guest of Queen Victoria.

FAUST, JOHANN (*c*. 1480–1540), a wandering conjurer and entertainer whose name became linked with an earlier medieval legend of a man in league with the devil. The story of his adventures was published in a Frankfurt chapbook in 1587, which in an English translation provided the material for a play, *The Tragical History of Doctor Faustus* (*c*. 1589–92), by Christopher MARLOWE. The legend returned to Germany via the ENGLISH COMEDIANS, who stressed the spectacular and farcical elements, and lived on in a puppet-show until once more taken up seriously by LESSING in 1759. Only fragments of his work remain, but it is evident that he envisaged Faust as a scholar whose inquiring mind finds itself in conflict with the limits imposed by God on human knowledge. During the period of STURM UND DRANG the Faust legend made a strong appeal, being used by Friedrich Müller, KLINGER (in a novel), and particularly by GOETHE, whose play on the subject, in two parts, engaged his attention from 1774 to his death in 1832. In it Faust, though tempted by Mephistopheles and guilty of the death of Gretchen, defies the devil and escapes him, his soul, as in Lessing's version, being eventually borne up to heaven. Since Goethe, the subject has been treated in German by Lenau, Heine, GRABBE, and others; in French by Paul Valéry; and in English by Wills (1885), and Rawson (1924) (from Goethe), and by Stephen PHILLIPS and Comyns Carr in collaboration. It has also supplied the librettos for operas by Boito, Busoni, Gounod, and others.

FAUTEUIL, see STALL.

FAVART, CHARLES SIMON, see CHANTILLY, MLLE.

FAVART, Théâtre, see COMÉDIE-ITALIENNE.

FAVERSHAM, WILLIAM (1868–1940), American actor-manager, born in London, where he made a brief appearance on the stage before going to New York in 1887. There he played small parts at the LYCEUM with Daniel FROHMAN's company, and was for two years with Mrs FISKE. In 1893 he was engaged by Charles FROHMAN for the EMPIRE and remained there for eight years, playing a wide variety of parts including Romeo to the Juliet of

Maude ADAMS. He appeared in a number of modern plays, and in 1909 made his first independent venture with the production at the LYRIC THEATRE, New York, of Stephen PHILLIPS's *Herod*, in which he played the title role. It was, however, his production of *Julius Caesar* in 1912, in which he played Mark Antony, that set the seal on his reputation both as actor and director. With a fine cast, the play ran for some time in New York and then went on tour, being followed by productions of *Othello* and *Romeo and Juliet*. After a further series of new plays, Faversham toured Australia, and on his return was seen in several more plays by Shakespeare, and as Jeeter Lester in Kirkland's long-running *Tobacco Road* (1933).

FAY, FRANK J. (1870–1931) and WILLIAM GEORGE (1872–1947), Irish actors, who in 1892 formed the Ormonde Dramatic Society, playing in Dublin and the surrounding country in a repertory of sketches, short plays, and farces. Among their fellow actors were Dudley DIGGES and Sara ALLGOOD, who were associated with them in 1898 in the IRISH NATIONAL DRAMATIC SOCIETY and in 1904 went with them to the ABBEY THEATRE. Frank, who was interested in verse-speaking, was responsible for the company's speech training, while W. G. acted as stage-manager. Both brothers appeared in most of the plays produced at the Abbey at this time, W. G. playing Christy Mahon in SYNGE's epoch-making *The Playboy of the Western World* (1907) and Frank his rival Shawn Keogh. In 1908, after a disagreement with YEATS and Lady GREGORY over artistic control, the Fays left the Abbey and went to America, where they directed a repertory of Irish plays for Charles FROHMAN, W. G. making his first appearance on Broadway in Yeats's *The Pot of Broth*. Back in London in 1914, W. G. was seen in several new plays and was successively a director at the Nottingham and BIRMINGHAM REPERTORY theatres. Among his later parts were the Tramp in Synge's *In the Shadow of the Glen* when it was revived in London in 1928; Mr Cassidy in BRIDIE's *Storm in a Teacup* (1936), which he also directed; and the title role in Paul V. CARROLL's *Father Malachy's Miracle* (1945). Frank returned to Dublin in 1918 and became a teacher of elocution, going back briefly to the Abbey in 1925 to play in a revival of Yeats's *The Hour-Glass*.

FEAST OF FOOLS, the generic name for the New Year revels in European cathedrals and collegiate churches, when the minor clergy usurped the functions of their superiors and burlesqued the services of the Church. The practice may have arisen spontaneously, as an outlet for high spirits, or may be an echo of the Roman Saturnalia. It appears to have originated in France in about the 12th century, and from the beginning evidently included some form of crude drama. The proceedings opened with a procession headed by an elected 'king'—in schools a BOY BISHOP—riding on a donkey, a detail which was taken over by the LITURGICAL DRAMA for the scenes of Balaam's Ass, the Flight into Egypt, and perhaps Christ's Entry into Jerusalem. The Feast of Fools lingered on in France until the 16th century, by which time the festivities had moved out of the church, and was eventually absorbed into the merrymaking of the *sociétés joyeuses*; but in England, where it is known to have taken place at St Paul's in London as well as at Lincoln, Beverley, and Salisbury, it died out some time in the 14th century. Part of its functions, though without the burlesque church services, survived during the Christmas season, either at Court or in the colleges of the universities (particularly at St John's and Merton in Oxford), as late as 1577, the 'king' being replaced by an Abbot or Lord of MISRULE; at the Inns of Court in London the custom of appointing a Christmas Lord of Misrule lingered on intermittently until the 1660s.

FEATHERSTONHAUGH, CONSTANCE, see BENSON, FRANK.

FECHTER, CHARLES ALBERT (1824–79), actor of French parentage, who played in French and English, both in America and in Europe, with equal success. He made his début at the COMÉDIE-FRANÇAISE in 1840, but soon left to make his reputation elsewhere, and became an outstanding *jeune premier*, being the first actor to play Armand Duval in the younger DUMAS's *La Dame aux camélias* (1852). At that time he was co-director of the ODÉON but chafing at the restrictions imposed on him in favour of the Comédie-Française he returned to London, where he had appeared some years earlier, in an English translation of HUGO's *Ruy Blas*. Though his accent was never very good, he played with so much fire and fluency that the audience was carried away, and he became very popular. He followed this success with a revolutionary production of *Hamlet* in 1861, in which the subtlety and depth of his performance impressed even those who clung to the traditional style. His Othello was not so successful, and in a revival he played Iago. In 1863 he took over the management of the London LYCEUM, where he appeared in a series of melodramas, and in 1870 went to New York. After touring for some time, he opened the old Globe Theatre under his own name, making many improvements and introducing many innovations, and appeared there in new plays as well as in some of his former successes. His quarrelsome and imperious nature made him many enemies and after a last visit to England he retired in 1876 to a farm near Philadelphia where he died.

FEDELI, a company of actors of the Italian COMMEDIA DELL'ARTE formed in about 1598 by Giovann Battista ANDREINI. It first came into prominence in 1603, and later took in actors from the GELOSI and ACCESI. For some time the company's ARLECCHINO was the famous Tristano MARTINELLI who paid prolonged visits to Paris, his acting, and that of Andreini, being much admired by Louis XIII. The Fedeli also toured extensively in Italy, and seems to have continued in existence until shortly after 1644. The company was officially under the protection of the Duke of Mantua in succession to the Gelosi.

FEDERAL THEATRE PROJECT, the first theatre scheme in the United States to be officially sponsored and financed by the Government, administered through the Works Progress Administration, and intended to give socially useful employment to members of the theatrical profession who were out of work. It was inaugurated on 1 Oct. 1935, being directed from Washington by Hallie FLANAGAN and operating through regional assistants. At its peak the Project employed some 10,000 people, ran theatres in 40 different states, published a nationally-distributed magazine, conducted a play and research bureau which served its own theatres and also about 20,000 schools, churches, and community groups throughout America, and played to audiences totalling many millions. From the first it offered an ambitious programme which included classical and modern plays, from EURIPIDES to IBSEN, dance dramas, musical comedies, children's plays, religious plays, puppet shows, and cycles of plays by both new and established modern playwrights. Between 1936 and 1939 it had a number of successful productions running on Broadway, where it occupied the Adelphi (later the GEORGE ABBOTT) Theatre, as well as in its various regions. Among its artistic achievements were the LIVING NEWSPAPER, a cinematic-type production which dealt with social and economic problems; a Negro theatre responsible for such productions as the Haitian *Macbeth* (1936), *Haiti* (1938), and *The Swing Mikado* (1939); the simultaneous production in 21 cities of Sinclair Lewis's *It Can't Happen Here* (1936); regional productions, such as Paul GREEN's *The Lost Colony* (1937), which is still played annually to thousands of people on Manteo Island in North Carolina; and the American première of SHAW's *On the Rocks* (1938). It had a great following of young people never before (or since) able to afford theatre-going; but its outspoken comments on economic issues, especially in some editions of the Living Newspaper, led to criticism by witnesses before the House Committee on unAmerican Activities and the sub-committee of the House Committee on Appropriations, and on 1 July 1939 in spite of a public outcry the Federal Theatre was disbanded by the United States Government.

FEHLING, JÜRGEN (1885–1968), German director, whose career was spent in BERLIN where, rejecting the influence of REINHARDT, he infused passion, violence, and pathos into the REALISM inherited from Otto BRAHM and created his own visionary realism. Among the plays he introduced to the public, as well as many interesting revivals, were BRECHT's *Eduard II* (1924), JAHNN's *Medea*, and BARLACH's *Der blaue Boll* (both 1929), all at the Staatstheater where he worked from 1922 to 1932, having been at the VOLKSBÜHNE from 1918. He then deserted EXPRESSIONISM, and with the aid of such designers as Traugott MÜLLER, evolved a highly personal version of the BOX-SET, with just a few props on a bare and steeply raked stage. He made some notable productions of Shakespeare, including *The Merchant of Venice* with KORTNER as Shylock and *Richard III* with KRAUSS as Richard, both in 1927; in 1939 he directed *Richard II*, with GRÜNDGENS in the title role.

FELIŃSKI, ALOJZY, see POLAND.

FELLOWES, AMY, see TERRISS, WILLIAM.

FEMALE IMPERSONATION. In the Greek and Elizabethan theatres female parts were invariably played by boys and men, as they were for centuries in China and Japan, where the actress is a comparatively late importation. Some actors became renowned for their skilled portrayal of women—BATHYLLUS in Rome, ALIZON in Paris, Ned KYNASTON in London, MEI LANFANG in China. In England the playing of the DAME in PANTOMIME by a man is more a form of BURLESQUE or caricature than an expression of the art of impersonation, though some famous Dames, notably Dan LENO, raised their characterization of elderly women to great heights of artistry. Today the term 'female impersonation' is reserved for those actors, often solo performers, whose appearance in woman's dress, or 'in drag' as it is commonly called, is so perfect as almost to convince the spectators that they are in fact women. Among the first modern exponents of 'drag' were the Americans Tony Hart [Anthony Cannon] (1855–91) and Julian ELTINGE. In England Arthur Lucan toured for many years as Old Mother Riley, and an acknowledged master of the art at present is Danny LA RUE. Outstanding also are the singing duo Hinge and Bracket (Dr Evadne Hinge and Dame Hilda Bracket, really George Logan and Patrick Fyffe), and the Australian Barry Humphries, whose *alter ego* is known as Dame Edna Everage.

FENN, EZEKIEL (1620–?), a boy actor with QUEEN HENRIETTA'S MEN under Christopher BEESTON at the COCKPIT, where in 1635 he played the chief female part in Nabbes's *Hannibal and Scipio*. He must already have had a good deal of experience with a BOY COMPANY, as at about the same time he played an even more exacting role, Winifred in DEKKER and ROWLEY's *The Witch of Edmonton*. When the Queen's Men disbanded in 1637 Fenn stayed on with Beeston, probably as one of the older members of his new company Beeston's Boys, and played his first male role in 1639. Fenn's history after the closing of the theatres in 1642 is unknown.

FENNELL, JAMES (1766–1816), English actor who gained his experience in the provinces and in 1792 went to America, where he was very popular. He was a member of the AMERICAN COMPANY and also of WIGNELL's company in Philadelphia. His best part was always considered to be Othello, though he appeared in most of the tragic roles of the current repertory, including Jaffier in OTWAY's *Venice Preserv'd*, which he played at the PARK THEATRE, New York, in 1799, in company with COOPER and Mrs MELMOTH. An excellent actor, over 6 ft tall, with a mobile, expressive face, he retired in 1810. Four years later he published his autobiography, *An Apology for the Life of James Fennell*.

FENTON, LAVINIA (1708–60), English actress, daughter of a naval lieutenant named Beswick, but always known by her stepfather's name. She made her first appearance on the stage at the HAYMARKET THEATRE in 1726, playing Monimia in OTWAY's tragedy *The Orphan* followed immediately by Cherry in FARQUHAR's comedy *The Beaux' Stratagem*, and was equally admired in both, being pretty, witty, and only eighteen. Two years later, on 29 Jan. 1728, she created the part of Polly in GAY's *The Beggar's Opera* at LINCOLN'S INN FIELDS THEATRE, with such success that she became one of the best known actresses in London. She made her last appearance on the stage on 28 June in the same year, and then retired to live with the Duke of Bolton whom she married in 1751.

FERBER, EDNA, see KAUFMAN, GEORGE S.

FERNÁNDEZ, LUCAS (*c*. 1474–1542), Spanish playwright, born like his contemporary ENCINA in Salamanca, where he taught music at the university from 1522 until his death. He was the author of three religious plays of which the *Auto de la Pasión*, written to be performed in Holy Week, tells of the conversion to Christianity of Dionysius the Areopagite by St Peter—the first known Spanish play to portray a Christian convert. Fernández, whose work is known only from one manuscript of 1514, wrote also three secular comedies, of which the *Farsa o cuasi comedia del soldada* derives from the MILES GLORIOSUS of PLAUTUS by way of Centurio (a character in the CELESTINA), while the *Farso o cuasi comedia de une doncella, un pastor y un caballero* introduces in rustic speech a sad shepherd who suffers from unrequited love for a lady betrothed to a knight.

FERNÁNDEZ DE MORATÍN, L. and N., see MORATÍN, FERNÁNDEZ DE.

FERRARI, PAOLO (1822–89), Italian dramatist, who was practising law when the first of his numerous plays was produced. He first came into prominence in 1853 when a play on GOLDONI—*Goldoni e le sue sedici commedie*—won a prize offered by a Florentine dramatic academy. Of his later works, which deal with social problems but are intended to reinforce rather than criticize prevailing attitudes towards them, the most successful were *Il duello* (1868), which argues in favour of duelling; *Il suicido* (1875), which held the stage for many years; and *Due dame* (*The Two Ladies*, 1877), which supports the prevailing prejudice against marriages between gentlemen and women of the lower classes, however noble and virtuous such women may be. Ferrari's best known work is *La satira e Parini* (1856), on the life of the poet Giuseppe Parini and his great satirical 4-part poem *Il Giorno* (1863–81). One character in this play, the Marchese Colombi, is as familiar to Italian audiences as Mrs Malaprop in SHERIDAN's *The Rivals* is to English.

FERREIRA, ANTÓNIO (1528–69), Portuguese dramatist, author of the finest play of the Portuguese Renaissance and also the first to combine classical models with native material, the *Tragédia de Dona Inêz de Castro* (pub. 1587), which tells of the murder in 1355 of the mistress of Pedro, eldest son of Alfonso IV, a subject which has inspired many other dramatists, including CAMÕES and, in recent times, Henri de MONTHERLANT in *La Reine Morte* (1942). Ferreira also wrote two moral but witty comedies, called after their chief characters, *O Cioso* and *Bristo* (pub. 1622).

FERRER, JOSÉ VICENTE (1912–), Puerto Rican-born American actor, producer, and director, who originally studied architecture. He made his stage début in a show-boat melodrama touring Long Island Sound in 1934, and a year later was seen in New York in *A Slight Case of Murder* by Howard LINDSAY and Damon Runyon. It was followed by a series of Broadway appearances culminating in Maxwell ANDERSON's *Key Largo* (1939) and a revival of Brandon THOMAS's *Charley's Aunt* in 1940, in which he played Lord Fancourt Babberley. In 1943 he made an outstanding success as Iago to Paul ROBESON's record-breaking Othello, and in 1946 gave a highly praised performance in the title role of ROSTAND's *Cyrano de Bergerac*. He played a number of roles at the CITY CENTER in 1948, among them the title role in JONSON's *Volpone*, Jeremy and Face in his *The Alchemist*, and Fat Joe in O'NEILL's *The Long Voyage Home*. He directed and starred in a revival of Hecht and MacArthur's *Twentieth Century* (1950) and in Joseph Kramm's *The Shrike* (1952), played the title role in *Richard III* (1953), and produced and directed Milton Geiger's *Edwin Booth* (1958), in which he also played the title role. The last part he created in New York was the Prince in *The Girl Who Came to Supper* (1963), a musical based on RATTIGAN's *The Sleeping Prince*, but he twice—in 1966 and 1967—took over the role of Don Quixote in the musical *Man of La Mancha*. As a producer his most successful work was Donald Bevan and Edmund Trzcinski's *Stalag 17* (1951).

FESTIVAL THEATRE, Cambridge, see GRAY, TERENCE.

FEUCHTWANGER, LION (1884–1958), German dramatist and novelist, best remembered for his historical novel *Jud Süss* (1925), which he dramatized himself. As *Jew Süss*, it was seen in London in a translation by Ashley DUKES in 1929, with Matheson LANG in the title role, and Peggy ASHCROFT as Naemi, her first outstanding success. Feuchtwanger's other plays include three written in collaboration with BRECHT, among them *Die Gesichte der Simone Machard* (1957) based on his own novel *Simone* (1944), and a number of historical and political works on Jewish themes. He left Germany in 1933, was interned by the French during the Second World War, and escaped to America, where he died. Among his later plays were *Wahn oder der Teufel in Boston* (*Madness, or the Devil in Boston*, 1948) and *Die Witwe Capet* (1956), on Marie-Antoinette.

FEUILLÈRE, EDWIGE [Caroline Cunati] (1907–), French actress, who, as Cora Lynn, first appeared in light comedies at the PALAIS-ROYAL. She made her début at the COMÉDIE-FRANÇAISE in 1931 but left to go into films, returning to the Paris stage in 1934. Her quality as an actress became apparent in 1937, when she appeared in BECQUE's *La Parisienne* and in 1940, when she played Marguerite Gautier in the younger DUMAS's *La Dame aux camélias*, a role she returned to in 1952. She created the parts of Lia in GIRAUDOUX's *Sodome et Gomorrhe* (1943) and the Queen in COCTEAU's *L'Aigle à deux têtes* (1946) and a year later joined the BARRAULT-RENAUD company to play Ysé, one of her finest parts, in CLAUDEL's *Partage de midi*. In 1957 she played the Queen in a French version of BETTI's *La regina e gli insorti*, and she was in Paris productions of KOPIT's *Oh Dad, Poor Dad* in 1963, Giraudoux's *La Folle de Chaillot* in 1965, Tennessee WILLIAMS's *Sweet Bird of Youth* in 1971, James Goldman's *The Lion in Winter* in 1973, and DÜRRENMATT's *Der Besuch der alten Dame* in 1976. In 1977 she made an oustanding success in ARBUZOV's *Old World*, known in French as *Le Bateau pour Lipaïa*.

She was seen in London in 1951, and again in 1968, in *Partage de midi*, in 1955 in *La Dame aux camélias*, and in 1957, with her own company at the PALACE THEATRE, in a repertory that included *La Parisienne* and RACINE's *Phèdre*. Her autobiography, *Les Feux de la mémoire*, was published in 1977.

FEYDEAU, GEORGES-LÉON-JULES-MARIE (1862–1921), French dramatist, son of the novelist Ernest-Aimé Feydeau (1821–73), and author of more than 60 farcical comedies which were not only lively but often well constructed. He had his first success with *Tailleur pour dames* (1887) and continued writing until his death, his last play, *Cent million qui tombent*, being produced posthumously. Among his more notable plays were *Le Système Ribadier* (1892), *Le Dindon* (1896), *La Dame de Chez Maxim* (1899), seen in London in 1902 as *The Girl from Maxim's*, and some one-act farces, including *Feu la mère de Madame* (*My Late Mother-in-law*, 1908), seen during the WORLD THEATRE SEASON of 1967; *On purge bébé* (*Time for Baby's Medicine*, 1910); *Mais n'te promène pas toute nue!* (*For God's Sake, Get Dressed!*, 1912); and *Hortense a dit: J'm'en fous!* (*Hortense Couldn't Care Less*, 1916). Regarded in his lifetime as nothing more than an adroit purveyor of light entertainment, Feydeau has come to be regarded as an outstanding writer of classic farce, and his reputation has been consolidated by productions of his best plays at the COMÉDIE-FRANÇAISE and elsewhere. Several of his masterpieces have been seen in London in translation, among them *L'Hôtel du Libre Échange* (1899) as *Hotel Paradiso* in 1956, in an adaptation by Peter Glenville starring Alec GUINNESS; *Occupe-toi d'Amélie* (1908) as *Look After Lulu* (1959), by Noël COWARD, with Anthony QUAYLE and Vivien LEIGH; *Une puce à l'oreille* (1907) as *A Flea in Her Ear* (1965), and *Un Fil à la patte* (1908) as *Cat Among the Pigeons* (1969), both adapted by John MORTIMER, the former for the NATIONAL THEATRE at the OLD VIC, the latter for the PRINCE OF WALES'. *Un Fil à la patte* was also seen in French during the World Theatre season of 1964. In 1967 the Tavistock Repertory Company produced a translation of *Le Ruban* (1894) as *Honours Even*, and in 1969 *Monsieur Chasse* (1892), as *The Birdwatcher*, was seen at the Yvonne Arnaud Theatre in Guildford. The attention paid to Feydeau as a classic *farceur*, in both France and England since 1945, has been one of the most interesting developments in the post-war theatre.

FEYDEAU, Théâtre, see COMÉDIE-ITALIENNE.

FFRANGCON-DAVIES, GWEN (1896–), English actress of Welsh extraction, who was trained as a singer and appeared with great success as Etain in Rutland Boughton's 'music drama' *The Immortal Hour*, first seen in London in 1920. She was, however, attracted by straight acting, and had already had considerable experience before appearing as Phoebe Throssel in a London revival of BARRIE's *Quality Street* in 1921. In the same year she joined the BIRMINGHAM REPERTORY THEATRE company where she played many leading roles including Eve and the Newly Born in the first English production of SHAW's *Back to Methuselah* (1923). Back in London she embarked on a distinguished career, being particularly admired as Shakespeare's Juliet in 1924 and Eliza in Shaw's *Pygmalion* and Ann Whitefield in his *Man and Superman* in 1927. She also played Elizabeth Moulton-Barrett in Rudolf Besier's *The Barretts of Wimpole Street* (1930), Anne of Bohemia in Gordon Daviot's *Richard of Bordeaux* (1932), Mary Stuart in her *Queen of*

Scots (1934), and Mrs Manningham in Patrick Hamilton's *Gas Light* (1939). In John GIELGUD's memorable production of WILDE's *The Importance of Being Earnest*, also 1939, she played Gwendolen and Lady Macbeth to his Macbeth in 1942, then spent several years acting in South Africa, but was at the SHAKESPEARE MEMORIAL THEATRE in 1950 and with the OLD VIC company in 1953. Among the best of her later roles was Mary Tyrone in O'NEILL's *Long Day's Journey into Night* in 1958. Her first New York appearance was in 1963 in SHERIDAN's *The School for Scandal*, she played Amanda in a London production of Tennessee WILLIAMS's *The Glass Menagerie* in 1965, and in 1970 gave a beautifully controlled and appealing performance as Madame Voynitsky in CHEKHOV's *Uncle Vanya* at the ROYAL COURT THEATRE.

FIABE, see GOZZI, CARLO.

FICHANDLER, ZELDA, see ARENA STAGE, WASHINGTON.

FIELD, NATHAN (1587–1620), English actor and playwright, who was taken from St Paul's School to become a boy-actor with the Children of the Chapel at the second BLACKFRIARS THEATRE where he appeared in *Cynthia's Revels* (1600) and *The Poetaster* (1601), both by JONSON in whose *Epicoene; or, the Silent Woman* (1609) he was later to appear. He was an excellent actor, particularly admired in such parts as the title role in CHAPMAN's *Bussy d'Ambois* (*c.* 1604), and in 1615 joined the KING'S MEN, possibly in succession to Shakespeare. He wrote two plays, *A Woman is a Weathercock* (1609) and *Amends for Ladies* (1611), and collaborated in several works with MASSINGER and FLETCHER.

FIELD, SID, see MUSIC-HALL.

FIELDING, HENRY (1707–54), English novelist and dramatist. His first three plays were comedies aimed at contemporary and literary follies, as was *Tom Thumb the Great; or, the Tragedy of Tragedies* (1730), which satirized the conventions of the HEROIC DRAMA. With his BALLAD OPERA *The Welsh Opera; or, the Grey Mare the Better Horse* (1731) he openly attacked both political parties, and also brought the Royal Family, thinly disguised, on to the stage. The repercussions were sufficiently alarming for Fielding to drop politics for a time and write for DRURY LANE five plays dealing light-heartedly with the contemporary social scene. The hated Excise Bill, coupled with a parliamentary election, induced him to write *Don Quixote in England* (1734), in which he continued his attacks on Walpole and on corrupt electioneering practices.Encouraged by the success of this satire, he staged his two most audacious attacks on the government, *Pasquin* (1736), which had almost as great a success as GAY's *The Beggar's Opera*, and *The Historical Register for the Year 1736* (1737). He also put on plays by other authors which attacked Walpole's administration, and it was probably the cumulative effect of all these that finally decided the government to curtail the liberty of the theatres. Adopting as a pretext the scurrility of a play entitled *The Golden Rump*, which had nothing to do with Fielding but had been submitted to GIFFARD at GOODMAN'S FIELDS THEATRE and found its way to Walpole's desk, the authorities rushed through the Licensing Act, which limited to two—COVENT GARDEN and DRURY LANE—the legitimate theatres in London, thus creating a monopoly which lasted until 1843 and creating a censorship over new plays which continued until 1968 (see LORD CHAMBERLAIN). The closing of the unlicensed HAYMARKET THEATRE, where all Fielding's plays had been produced, hit him hard, and with a wife and growing family to support he deserted the stage for the more lucrative career of novel-writing, producing his two best works in the genre, *Joseph Andrews* and *Tom Jones*, in 1742 and 1749 respectively.

It was once believed that Fielding had run a booth at Bartholomew Fair, but this actually belonged to Timothy Fielding (?–1738), an actor who played a small part in his namesake's play *The Miser* (1733).

FIELDS, GRACIE, see MUSIC-HALL.

FIELDS, JOSEPH (1895–1966), son of Lew Fields (see WEBER AND FIELDS), who collaborated with Jerome Chodorov (1911–) in several plays, including *My Sister Eileen* (1940; London, 1943), *Wonderful Town* (1953; London, 1955), and *Anniversary Waltz* (1954; London, 1955). He was also part-author with Anita Loos of the dramatization of her book *Gentlemen Prefer Blondes* (1949), and with Oscar Hammerstein of the book of the musical *Flower-Drum Song* (1958), which he directed in New York and also in London in 1960. His brother Herbert (1897–1958) and sister Dorothy (1905–74) were responsible for the lyrics and books of many shows, among them the musicals *Mexican Hayride* (1944) and *Annie Get Your Gun* (1946; London, 1947).

FIELDS, LEW, see WEBER AND FIELDS.

FIELDS [Dukenfield], W(illiam) C(laude) (1879–1946), American actor and VAUDEVILLE star, born in Philadelphia, where he made his first appearance at an open-air theatre as a 'tramp juggler', in 1897. A year later he was in New York, where he juggled and appeared in a slapstick comedy routine, soon becoming a 'topliner' on the Keith circuit. In 1900 he was in London and thereafter toured Europe. He gradually began to introduce comic monologues into his juggling act and finally established himself as a comedian, appearing in 1905 in a loosely connected series of vaudeville sketches entitled *The Ham Tree*, in which he toured for a couple of years, appearing in 1908 at the FOLIES-BERGÈRE in Paris. Evolving slowly, he abandoned vaudeville permanently in 1915 to appear as an eccentric comedian in the *Ziegfeld Follies* and *George White's Scandals*. Stout, with small, cold eyes and a bulbous nose, a rasping voice, and a grandiose vocabulary and manner, he usually played petty and inept crooks, or anarchic characters devoid of any conventionally decent feelings. In 1923, after appearing as a strolling carnival swindler in the musical *Poppy*, he abandoned the theatre for films, in which he had a brilliant career.

FIFTH AVENUE THEATRE, New York. (1) On 24th Street near Broadway. Originally the Fifth Avenue Opera House, specializing in Negro MINSTREL SHOWS, it opened as a theatre on 2 Sept. 1867, but closed abruptly after a fight in the auditorium in which a man was killed. It reopened on 25 Jan. 1869 with John BROUGHAM in one of his own plays, and on 16 Apr. became DALY's Fifth Avenue Theatre, with a good company that included Mrs GILBERT and Fanny DAVENPORT. Its first outstanding success was Daly's own version of *Frou-Frou* by MEILHAC and Halévy, but being anxious to encourage American dramatists he staged on 21 Dec. 1870 Bronson HOWARD's *Saratoga*, which had a long run. In 1872 the theatre was redecorated and much improved, but on 1 Jan. 1873 it was burnt down and not rebuilt. The site was later used for the Fifth Avenue Hall which became the MADISON SQUARE THEATRE.

(2) Daly opened his second Fifth Avenue Theatre, at Broadway and 28th Street, with an elaborate and costly production on 21 Feb. 1874 of *Love's Labour's Lost*, which had not previously been seen in New York. It failed, and the financial panic of 1873–4 also had a bad effect. Success came with the production on 17 Feb. 1875 of Daly's own play *The Big Bonanza*, in which John DREW made his New York début. The following season saw the New York production of the popular London success H. J. BYRON's *Our Boys*, in which Georgiana Drew, later the wife of Maurice BARRYMORE, made her first appearance in New York. The first night of SHERIDAN's *The School for Scandal* on 5 Dec., with Charles COGHLAN, was marred by the disastrous BROOKLYN THEATRE fire. By 1878 Daly was finding the financial loss on the theatre too great, and left to take over Banvard's Museum, which he opened as DALY'S THEATRE. The Fifth Avenue then came under the management of Stephen Fiske, under whom Mary ANDERSON and Helena MODJESKA made their first appearances in New York. After that the theatre was leased to various travelling companies, and after many changes of name was pulled down in 1908.

FIFTY-EIGHTH STREET THEATRE, New York, see JOHN GOLDEN THEATRE (1).

FIFTY-FIRST STREET THEATRE, New York, see MARK HELLINGER THEATRE.

FIFTY-FOURTH STREET THEATRE, New York, see GEORGE ABBOTT THEATRE.

FIGUEIREDO, MANUEL DE, see PORTUGAL.

FILANDRE [Jean-Baptiste Monchaingre *or* Jean Mathée *or* Paphetin] (1616–91), French actor-manager, whose long and active life was spent mostly in the provinces, where he led a company which toured Northern France, Holland, and Belgium. He was however at the MARAIS from 1647 to 1648, replacing FLORIDOR who had gone to the Hôtel de BOURGOGNE, after which he returned to the provinces. He is believed to have served as the model for Léandre, the leader of a travelling company in SCARRON's novel *Le Roman comique*. Filandre retired officially in 1667, but was glimpsed at Grenoble in 1670 and may have continued to act for some years after that.

FILIPPO, E. and P. DE, see DE FILIPPO.

FINLAND. Although traditions of ritual performance can be traced far back into Finnish prehistory, the first attempt to introduce formal theatre into Finland seems to have been made at Åbo Akademi (Turku) in 1640, when a student company acted translations of Renaissance dramas, moralities, and Latin comedies on a raised stage with classical columns and tapestry hangings. Owing to the prevalence of war and plague, however, this had little lasting influence, and during the 18th century Finland depended mainly on Swedish travelling actors for dramatic entertainment. Much was owed to the Swedish actor Carl Gottfried Seuerling (1727–92) who, with his wife, brought Shakespeare to Finland, beginning with *Romeo and Juliet* in 1780, as well as RACINE, MOLIÈRE, CALDERÓN, and HOLBERG. In the early 19th century, after the annexation of Finland by Russia, German influence became particularly strong.

Finland acquired its first permanent theatre in 1860, when the Swedish theatre opened in Helsinki with a performance of Topelius's *Prinsessen av Cypern* (*Princess of Cyprus*). Shortly afterwards, the inauguration of a Finnish-language theatre was marked by the production, also in Helsinki, of Aleksis KIVI's drama *Lea* on 10 May 1869; theatre in Finland has subsequently developed bilingually, with separate Finnish-speaking and Swedish-speaking institutions though with occasional shifts of actors and technicians between them.

Regular performances in Finnish followed the establishment of a Finnish National Theatre company in 1872, the moving spirit of which was Kaarlo Bergbom (1842–1906), who gathered round him a company which included the great actress Ida Aalberg Kivekäs (1857–1915). Bergbom was a great lover of IBSEN, and the first performance of *John Gabriel Borkman* was given at his theatre in 1897. In 1902 the company moved to a new and splendid building. The development of theatre went hand in hand with the emergence of a vigorous native Finnish drama, in which the conspicuous names are Kivi, Minna Canth (1844–97), whose realistic plays deal with contemporary problems, Juhana Henrik Erkko (1849–1906), who drew on the *Kalevala* for two of his more important plays, and the prolific Eino Leino (1878–1926), who was much influenced by Nietzsche. Later dramatists included Maria Jotuni (1880–1943), author of satirical comedies of rural life, the expressionist-influenced Lauri Haarla (1890–1944), and Hella VUOLIJOKI.

In the earlier years of the Swedish-speaking theatre, much reliance was placed on plays translated from other European languages; but an indigenous Swedish drama also began to emerge, associated with the names of Runeberg, Topelius, and Wicksell. In 1916, the Swedish-speaking National Theatre was established, which brought with it the acceptance of Fenno-Swedish accents from its actors in place of the earlier insistence on 'High Swedish'. Since 1955, the Lilla Teatern, first under the direction of Viveca Bandler and from 1967 under Lasse Pöysti, has (as Helsinki's second Swedish theatre) done much experimental work. Among Fenno-Swedish dramatists of significance are Walentin Chorell and Benedict Zilliacus.

There are in Finland (with a population of four and a half million, of which one-tenth is Swedish speaking) 30 professional Finnish theatres (eight in Helsinki) and five Swedish (two in Helsinki, one at Åbo, one at Vasa, and a travelling company which tours the provinces). There are also innumerable amateur groups. A new and ultra-modern Helsinki City Theatre was opened in 1968.

FINLAY, FRANK (1926–), English actor, who first appeared on the stage in 1957 and a year later came into prominence as Harry Kahn, the ageing Jewish paterfamilias in Arnold WESKER's *Chicken Soup With Barley*, which he played at the BELGRADE THEATRE in Coventry and later at the ROYAL COURT in London, where he also appeared as Attercliffe in John ARDEN's *Serjeant Musgrave's Dance* (1959), Libby Dobson in Wesker's *I'm Talking About Jerusalem* (1960), and Corporal Hill in his *Chips With Everything* (1962). In 1963 Finlay scored a personal triumph as the thieving Alderman Butterthwaite in another Arden play, *The Workhouse Donkey*, at CHICHESTER, and he then joined the NATIONAL THEATRE company, where his roles included Willie Mossop in BRIGHOUSE's *Hobson's Choice* and Iago to OLIVIER's Othello in 1964, Giles Corey in Arthur MILLER's *The Crucible* in 1965, and Joxer Daly in O'CASEY's *Juno and the Paycock* in 1966. After playing in David MERCER's *After Haggerty* (1970) for the ROYAL SHAKESPEARE COMPANY he returned to the National Theatre to appear as the silently suffering father in DE FILIPPO's *Saturday, Sunday, Monday*, with Joan PLOWRIGHT, and Sloman in Trevor Griffiths's *The Party* (both 1973). Later productions in which he appeared include OSBORNE's *Watch It Come Down* and Howard BRENTON's *Weapons of Happiness* (both 1976 for the National Theatre), the short-lived musical *Kings and Clowns* (1978), and de Filippo's long-running *Filumena* (also 1978), succeeding Colin BLAKELY and again with Joan Plowright, which had a brief run in New York with the same stars in 1980. In 1981 he succeeded Paul SCOFIELD as Salieri in SHAFFER's *Amadeus* at the National Theatre, continuing the role in the successful West End run. Essentially a fine character actor, he is especially strong in moments of watchful stillness.

FINN, HENRY JAMES (1785–1840), actor and playwright, born on Cape Breton Island, Nova Scotia, who studied law at Princeton but deserted it for the stage. After working in Charleston and Boston he joined the company at the PARK THEATRE, New York. When it was burnt down in 1820 he went to the ANTHONY STREET THEATRE, playing Hamlet there, as he did at the CHATHAM THEATRE in 1824. After making a success as Aguecheek in *Twelfth Night* he gave up tragedy for comedy in which he was inimitable, and in 1825 appeared in his own play, *Montgomery; or, the Falls of Montmorency*, as Sergeant Welcome Sobersides, an amusing Yankee character who was later incorporated into *The Indian Wife* (1830) and played by James H. HACKETT. Finn died in the destruction by fire of the SS *Lexington* in Long Island Sound.

FINNEY, ALBERT (1936–), English actor, who gained his experience at the BIRMINGHAM REPERTORY THEATRE, where he was a member of the company from 1956 to 1958, playing among other parts Macbeth, Henry V, and Face in JONSON's *The Alchemist*. After appearing with Charles LAUGHTON in Jane Arden's *The Party* (1958) he went to the SHAKESPEARE MEMORIAL THEATRE to play Edgar to Laughton's Lear, and also took over the part of Coriolanus from Laurence OLIVIER. He then returned to London, where he first came into prominence in a new musical *The Lily-White Boys* and in Willis HALL and Keith Waterhouse's *Billy Liar* (both 1960). His reputation was further enhanced by his appearance in OSBORNE's *Luther* (1961), his thickset physique and a hint of surly stubbornness perfectly fitting the part; he made his New York début in the same role in 1963. In 1965 he joined the NATIONAL THEATRE company, creating at the CHICHESTER FESTIVAL THEATRE the title role in ARDEN's *Armstrong's Last Goodnight*, seen in London later in the year at the OLD VIC. During the same season his roles also included the antique collector in SHAFFER's *Black Comedy* and the mutinous servant Jean in STRINDBERG's *Miss Julie* in a double bill, and an impotent bourgeois and a dim-witted hotel porter in FEYDEAU's *A Flea in Her Ear*. In 1967 (N.Y., 1968), he co-presented Peter NICHOLS's *A Day in the Death of Joe Egg*, previously seen at the CITIZENS' THEATRE in Glasgow, also playing the leading role in New York. In 1972 he gave a powerful performance as the embittered husband in E. A. Whitehead's *Alpha Beta* at the ROYAL COURT, of which he was Associate Artistic Director from 1972 to 1975. He appeared in the West End in another play by Nichols, *Chez Nous* (1974), and in 1975 rejoined the National Theatre company to play Hamlet during the last days at the Old Vic and in the first appearance of the company on the new LYTTELTON stage, also playing the title role in MARLOWE's *Tamburlaine the Great*, the opening production at the OLIVIER. He was later seen at the National as Horner in WYCHERLEY's *The Country Wife* (1977), as Macbeth, and in CHEKHOV's *The Cherry Orchard* and Charles WOOD's *Has 'Washington' Legs?* (all 1978). Finney is often credited with having introduced into English acting a new spirit of working-class earthiness, but he has several times proved himself an actor in the true heroic tradition.

FIORILLO, SILVIO (?–*c*.1633), an actor of the COMMEDIA DELL'ARTE, the original CAPITANO Matamoros and probably the first PULCINELLA. He had a company of his own in Naples, his birthplace, in the last years of the 16th century, and later appeared with other companies, notably the ACCESI on their visits to France in 1621 and 1632. He was the author of several plays and of scenarios in the *commedia dell'arte* tradition. His son Battista (*fl.* 1614–51) was married to the actress Beatrice Vitelli (*fl.* 1638–54), and is known to have played SCARAMUCCIA. Silvio may also have been the father of Tiberio, below.

FIORILLO, TIBERIO (1608–94), a COMMEDIA DELL'ARTE actor who may have been the son of Silvio, above, but whose name is also found as Fiorilli and Fiurelli. Born in Naples, he joined an itinerant company at Fano, with which he toured Italy, eventually graduating to more distinguished troupes. He was already well known before making several visits to Paris in the 1640s; in 1658 he was with the Italian company which shared the PETIT-BOURBON with MOLIÈRE, an arrangement repeated at the PALAIS-ROYAL when he was again in Paris, where he spent the latter part of his life, in 1661 with a new company. Although he was not the first to play SCARAMUCCIA, who as Scaramouche became a stock character in French comedy, he was certainly its most skilled interpreter, playing it without a mask. He visited London at least twice (in 1673 and 1675) and was popular with Court and public alike, displacing French puppets in the affections of fashionable society. His picaresque 'autobiography', *La Vie de Scaramouche* (1695) written by Angelo COSTANTINI, is of dubious accuracy. It was translated into English in 1696 and again by Cyril Beaumont in 1924. A musician, dancer, singer, acrobat, and pantomimist, Fiorillo retained his agility well into his 80s, when he was still so supple that he could tap a fellow-actor's cheek with his foot. He was twice married, first to the actress Isabella del Campo (?–1687), known as Marinetta, who played servant roles, and then to his former mistress Marie-Robert Duval.

FIRES IN THEATRES. Both the London Patent Theatres, DRURY LANE and COVENT GARDEN, have been burnt down twice, the first in 1672 and 1809, the second in 1808 and 1856. The first recorded theatre fire in America was that of the Federal Street Theatre in Boston in 1798, and between then and 1876, when the BROOKLYN THEATRE went up in flames during the last act of Oxenford's *The Two Orphans* with the loss of about 300 lives including two members of the cast, over 75 serious fire disasters were reported. In Richmond, Va., 70 people were killed on 26 Dec. 1811 when the candles of a stage chandelier set fire to the scenery. The introduction of gas lighting, coupled with the vogue for very large theatres, caused the heavy death-rolls in 19th-century disasters: in the Lehman Theatre in St Petersburg in 1836 there were 800 casualties; in Quebec in 1846, 50; in Karlsruhe a year later, 631; in Leghorn in 1857, 100. The greatest disaster of all time, however, was probably the fire in a Chinese theatre in 1845 which killed 1,670 persons.

Even after the introduction of more stringent fire regulations in Britain in 1878, the Theatre Royal, EXETER, was burnt down in 1885, rebuilt, and again destroyed by fire in 1887, with 186 people killed. In America, where strong safety measures were taken after the Brooklyn fire, the futility of strict regulations without equally stringent enforcement was demonstrated on 30 Dec. 1903, when the supposedly fireproof Iroquois Theatre in Chicago was the scene of the worst such catastrophe in the history of the American theatre. A fire in an overcrowded house, though quickly controlled, led to a panic which resulted in the loss of over 600 lives.

With the introduction of electric lighting, fireproofing of stage materials, and a comprehensive code of fire regulations which laid the onus for prevention of fire on theatre managers, conditions improved, and fires in theatres are now rare, though the evasion of fire regulations caused 492 deaths at the Cocoanut Grove night club in Boston on 28 Nov. 1942 and on 6 July 1944 a circus fire at Hartford, Conn., caused 168 deaths, mostly of children.

FISCHER, LECK, see DENMARK.

FISHER, CLARA (1811–93), a child prodigy who made her first appearance on the stage in London at the age of 6, being much admired as Richard III, Shylock in *The Merchant of Venice*, and Young Norval in HOME's *Douglas*, as well as in *The Actress of All Work* (1819), a series of sketches in which she impersonated a number of characters. At the age of 16 she went to New York, making her début at the PARK THEATRE as the four Mowbrays in the farce *Old and Young*, and for the next few years toured in light opera and vaudeville. In 1834 she married a musician, James G. Maeder, but remained on the stage, though with less success. She retired in 1844 but returned later to star in *opéra bouffe*, notably at the Theatre Royal, Montreal, from 1852 to 1855, finally retiring in 1880.

FISKE, MINNIE MADDERN [Marie Augusta Davey] (1865–1932), American actress, who appeared with her parents at the age of three under her mother's maiden name of Maddern, and at five went to New York, where she played juvenile parts which included the Duke of York in *Richard III* and Prince Arthur in *King John*. At 13 she graduated to adult parts, being seen as the Widow Melnotte in BULWER-LYTTON's *The Lady of Lyons*. One of her most successful parts was Mercy Baxter in *Caprice* (1884), which Henry P. Taylor wrote specially for her. She retired from the stage in 1890 on her marriage to Harrison Grey Fiske (1861–1942), but three years later returned to star in her husband's play *Hester Crewe* and in 1894 was seen as Nora in IBSEN's *A Doll's House*. Her playing of Tess in a dramatization of Hardy's novel in 1897 established her reputation, and she was also much admired as Becky Sharp in Langdon MITCHELL's dramatization of Thackeray's *Vanity Fair* (1899), with Maurice BARRYMORE as Rawdon and Tyrone POWER as Steyne. Since her husband's opposition to the THEATRICAL SYNDICATE prevented her from appearing in their theatres, she rented the Manhattan (formerly the STANDARD THEATRE) in 1901, and remained there for five years with an excellent company in a series of fine plays, including Ibsen's *Hedda Gabler* in 1903. She was also responsible for the first professional production of a play from George Pierce BAKER's '47 Workshop', SHELDON's *Salvation Nell* (1908), which she staged at Hackett's Theatre, formerly the second WALLACK'S THEATRE. After an unsuccessful spell in films she returned to the New York stage in a series of light comedies, and then made a long tour in SHERIDAN's *The Rivals*, in which she gave an excellent performance as Mrs Malaprop. In 1927 she again went on tour, as Mrs Alving in Ibsen's *Ghosts*, and she made one of her last appearances as Beatrice in *Much Ado About Nothing*. She had considerable ability as a stage director, and during her years in management gave encouragement to new dramatists, proving a potent force in the battle for theatrical realism. As an actress, although effective in serious roles, she was most admired in comedy.

FITCH, (William) CLYDE (1865–1909), American dramatist, author of about 50 plays. The first, *Beau Brummell* (1890), was commissioned by Richard MANSFIELD who retained it in his repertory until he retired. Fitch poured out plays, the best of his early work being *Nathan Hale* (1898) and *Babara Frietchie* (1899), based on American history and mingling personal and political problems in marked contrast to his subsequent light comedies. He was at the height of his popularity in 1901, when *The Climbers* and *Lovers' Lane*, social comedies of life in New York, and *Captain Jinks of the Horse Marines*, in which Ethel BARRYMORE first appeared as a star under Charles FROHMAN, were running simultaneously in New York, while in London TREE was producing *The Last of the Dandies*. Among Fitch's other plays, many of which were written to order for certain stars, were *The Moth and the Flame* (1898) and *The Cowboy and the Lady* (1899), both melodramas, *The Stubbornness of Geraldine*, dealing with the American abroad, *The Girl with the Green Eyes*, a drama of jealousy (both 1902), *The Truth* (1906), which is sometimes considered to be his best play, and *The Woman in the Case* (1909). One of America's best-loved playwrights, Fitch was excellent at pinpointing certain aspects of contemporary and domestic life; but his social observation was marred by his deference to the prevailing fashion for melodrama.

FITZBALL [Ball], EDWARD (1792–1873), English dramatist, author of a vast number of immensely popular melodramas, now forgotten. Several, notably *The Red Rover; or, the Mutiny of the Dolphin* (1829), based on a novel by Fenimore Cooper, and *Paul Clifford* (1835), from a novel by BULWER-LYTTON, were successful enough in their day to be included in the repertory of the TOY THEATRE. He also helped forward the development of NAUTICAL DRAMA with the dramatization of another of Cooper's novels as *The Pilot; or, a Tale of the Sea* (1825) and such works as *Nelson; or, the Life of a Sailor* (1827), also subtitled *Britannia Rules the Waves*. His *Jonathan Bradford; or, the Murder at the Wayside Inn* (1833), based on a sensational murder case, made a fortune for the manager of the SURREY THEATRE, where it was first produced. Fitzball's autobiography, *Thirty-Five Years of a Dramatic Author's Life* (1859), contains interesting material on the theatrical life of his time.

FITZGERALD, BARRY [William Joseph Shields] (1888–1961), Irish actor, brother of Arthur SHIELDS. He first walked on for the ABBEY THEATRE company in 1914, gradually taking larger roles in plays by Lennox ROBINSON, Bernard SHAW, Lady GREGORY, George SHIELS, Brinsley MACNAMARA, Lord DUNSANY, W. B. YEATS, and others. But it was not until Sean O'CASEY wrote for him the role of Captain Boyle in *Juno and the Paycock* (1924) that Fitzgerald finally committed himself to acting as a profession. From 1929, when he played Sylvester Heegan in the London production of O'Casey's *The Silver Tassie*, he alternated between Dublin and London, joining the Abbey players again for their American tour in 1934. In 1936 he went to Hollywood and from that time made his home in the United States, ceasing to appear on the stage after 1941.

FITZMAURICE, GEORGE (1877–1963), Irish dramatist and short story writer, whose unusual qualities set him somewhat apart from other ABBEY THEATRE playwrights. Nearly all his plays, written in the strong, harsh dialect of his native Kerry, draw their plots from folk-fantasy. The first, *The Country Dressmaker* (1907), a three-act peasant drama swinging from bitter comedy to pitiless tragedy, was produced at the Abbey, as was *The Pie-Dish* (1908), a one-act Faustian tragedy which, like his next one act play *The Magic Glasses* (1913), blends pagan and Christian beliefs, realism and the supernatural, to create a poetic but relentlessly fantastic world unlike anything else produced in Ireland until the work of John B. KEANE.

Only one more play by Fitzmaurice was produced at the Abbey during his lifetime, a slight farce entitled *'Twixt the Giltinans and the Carmodys* (1923); but *The King of the Barna Men* was produced posthumously in 1967, and an eerily poetic folk-fantasy of great power, *The Dandy Dolls* (published in 1914 but rejected by both YEATS and Lady GREGORY) was finally seen at the Abbey in 1969; it had been produced in 1945 by Austin CLARKE at the Lyric Theatre, where *The Linnaun Shee* was also presented in 1949. In 1948 the Earlsfort Players produced Fitzmaurice's nationalist drama *The Moonlighter*, written over 35 years earlier, and in 1952 the Studio Theatre Company, Dublin, presented an early play about Dublin, *One Evening Gleam*, which owed much to the dramatist's lifelong love of the music-hall.

FITZROY THEATRE, London, see SCALA THEATRE.

FLAMINIA, see CECCHINI, P. M.

FLANAGAN, BUD, see CRAZY GANG.

FLANAGAN, HALLIE [*née* Ferguson, later Mrs Philip H. Davis] (1890–1969), American theatre historian and organizer. After a year as assistant to George Pierce BAKER at Harvard, she taught at Grinnell College, Iowa, from 1924 to 1925, then went to Vassar until 1941, founding experimental theatres in both colleges. *Dynamo* (1943) is an account of her work at Vassar, where she co-wrote and directed *Can You Hear Their Voices?* (1931), a forerunner of LIVING NEWSPAPER drawing on her visits to Europe and the Soviet Union; she had already published *Shifting Scenes of the Modern European Theatre* (1928). As director of the FEDERAL THEATRE PROJECT, 1935–9, she was responsible for its nationwide activities comprising more than 1,000 productions: her book *Arena* (1940) is an account of these years. She became Dean of Smith College, Northampton, Mass., in 1942, resigning after four years because of ill-health but remaining as Professor of Drama until her retirement in 1955.

FLANDERS, MICHAEL, see REVUE.

FLAT, a frame made of 3 in. × 1 in. timber, generally covered with canvas, though modern stage designers use a variety of surfaces. It is the most important single element of the designer's equipment, and the solution to the problem that scenes must be made up of separate parts which should be as light as possible. The standard full-sized flat is 18 ft high; in small theatres it may be less and in large ones may reach 24 ft. Its width may vary from 1 ft to 6 ft, exceptionally to 8 ft. For widths above this two or more flats hinged together are used; these are known as Book (or Booked) Flats. In America they are called Two- or Three-folds. Flats may be plain or contain openings; they may be straight-edged or bear Profiling Boards, fixed to the side and sawn to the required shape. They may be used as wings, and then carry on their on-stage edge an extension of profiling known as a Flipper. This is hinged so that it faces the audience even if the flat is set at an angle, and can be folded flat for packing. Flats are used to form the three walls of a BOX-SET. When those at the back are battened together so that the whole wall can be flown it is called a French Flat. Variants of the Flat are the GROUNDROW, which may be likened to a flat on its side lying across the stage, the SET PIECE, a low flat profiled all round and probably cut to the silhouette of some representational or decorative shape, and the Backing Flat, set outside a door or other opening to block the view beyond.

Traditionally, the frame of an English flat consists of two Stiles, or side pieces, and a top and bottom Rail. The corners are mortised-and-tenoned, the stiles bearing the tenons and the rails the mortises. No glue or screws are used, the parts being put together with wooden pins, so that any flat can be altered without the waste of cutting away glued joints. To counteract the draw of size paint drying on the canvas, two or three Toggle Rails of slightly thinner timber are 'jumped in' between the stiles as stays, each rail tenoned and pinned at either end into the mortise of a triangular Toggle, or cross-piece, about 18 in. long. To preserve squareness at the corners a diagonal Brace, about 3 ft long, is shouldered across at one or two of the corners. If the flat is to be profiled, the profile board (usually fireproof plywood) is attached to the stile at this stage. The canvas is then tacked or stapled to the face of the flat, the waste canvas is roughly trimmed off, the edges beyond the tacks are laid back while the frame is glued, and then pressed back in position on the glue. Finally a sharp knife is run round ½ in. from the edge of the flat, cutting the canvas clean and at the same time making a slight incision into which the cut edge of the glued canvas can be pressed and so sealed. If the flat is to be used in a box-set it is provided with a Throwline (Lashline in America), fixed near the top of the left-hand stile. The flat is now ready for priming and painting. There are quicker and cheaper ways of constructing a flat, but this traditional method gives the lightest and strongest result. A free-standing flat is supported by a Stage Brace—a wooden rod, generally extensible, hooked to the back of the flat at one end and screwed into the floor at the other, or held down by a weight made to fit over the foot of the brace. Sometimes the flat is supported by a French Brace (Brace Jack in America), which is a framework of wood shaped like a right-angled triangle and hinged to the back of the piece, ready to be opened out and weighted as needed.

Small pieces of scenery that belong to the 'flat' family, in that they are canvas-covered frames, include rostrum-fronts, balustrade-pieces, staircase sides, and raking pieces (that is, pieces with a sloping top edge, used either independently as small groundrows or as fillings before ramps).

Historically the flat is one of the most interesting pieces of stage equipment, dating back to the 'flat scenes' of the Court MASQUES mounted by Inigo JONES. These consisted of a painted cloth or a pair of painted 'panels' which could be opened to show another scene behind. Some were divided horizontally into two halves, the upper ones opening independently to show a view on to the sky or a mountain top, the bottom ones revealing a cellar or a cave in the mountain. These continued in use for many years, and as late as 1736 Henry FIELDING alludes to an imaginary scene painter as Mynheer Van Bottom-Flat.

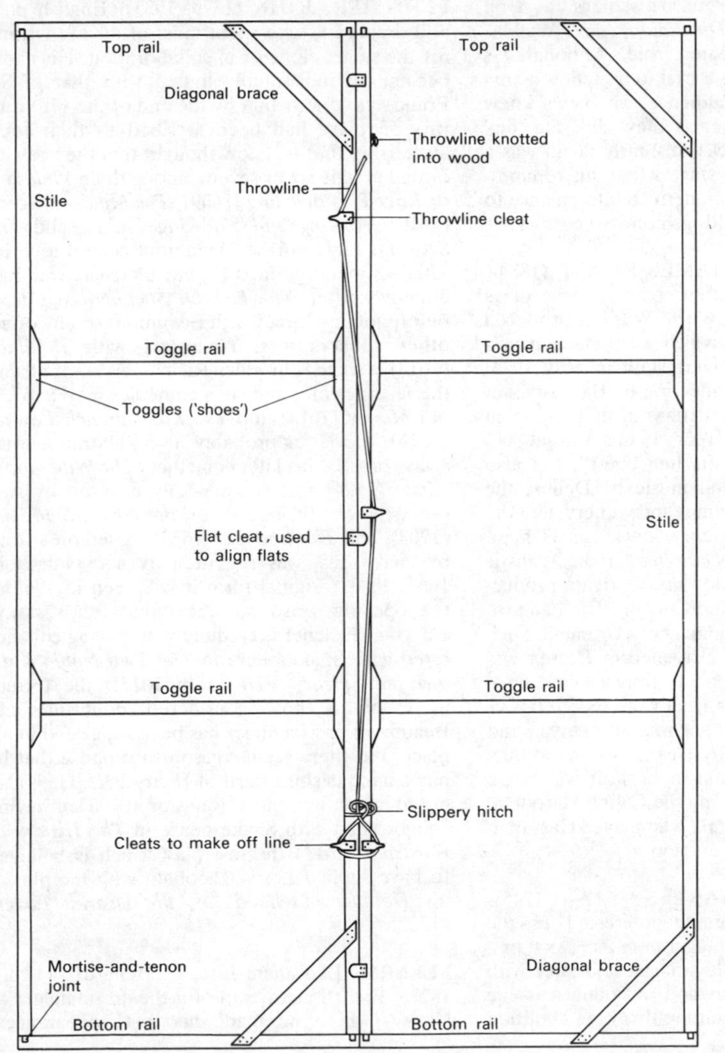

Pairs of flats, rear view

From earliest times the two halves of the 'pair of flats', as they continued to be called until the mid-19th century, might contain openings for doors, windows, arches, or the interstices between the trees in a forest scene. With such cut flats another piece of scenery, later known as an Inset, or Inset Scene, was needed behind them to mask the opening. (See SCENERY.)

FLAVIO, see SCALA, FLAMINIO.

FLECK, JOHANN FRIEDRICH FERDINAND (1757–1801), German actor, who joined the company at the BERLIN National Theatre shortly before IFFLAND arrived in 1796. He first made his mark as Gloucester to SCHRÖDER's Lear, his fine presence, resonant voice, and fiery artistic temperament also making him an ideal interpreter of the stormy heroes of the STURM UND DRANG period. He was especially admired in SCHILLER's plays, as Don Carlos, Karl Moor in *Die Räuber*, and Wallenstein—his last and greatest role. Although his capricious character was a trial to his fellow actors and an irritation to his audience (who never knew whether they would see, as they said, the 'big' Fleck or the 'little' Fleck), his early death was a great loss to the Berlin stage where his romanticism was a promisingly modern counterbalance to Iffland's cautious and old-fashioned repertory.

FLECKER, (Herman) JAMES ELROY (1884–1915), English poet, author of two verse plays written before the First World War but produced posthumously. *Hassan*, which was first seen in a German translation at Darmstadt in May 1923, was given a splendid production by Basil DEAN at HER (then His) MAJESTY'S THEATRE in London in September of the same year. It was a great success, partly because of its fine poetry, but also because of the incidental music by Delius, the sumptuous oriental costumes and scenery, and the excellent cast led by Henry AINLEY as Hassan. Flecker's second play, *Don Juan*, more realistic and modern in style, had only a private production by the Three Hundred Club in 1926. In a production by the Oxford University Dramatic Society in 1931 the part of Pervaneh in *Hassan* was played by Peggy ASHCROFT. Both plays have been revived, *Don Juan* unofficially at the GATEWAY THEATRE during the 1950 EDINBURGH FESTIVAL, and *Hassan*, again directed by Dean, in London during the Festival of Britain in 1951. It was not a success. In the cast, playing the Caliph Haroun al Rashid, was Frederick Valk who played Hassan in the original German production.

FLEETWOOD, CHARLES (?–*c.* 1745), wealthy English gentleman interested in the theatre, who became manager of DRURY LANE, first in partnership with Colley CIBBER and later with Charles MACKLIN. Fleetwood introduced some much-needed reforms, among them the abolition in 1737 of the free entry of lackeys to the Footmen's GALLERY which caused some turmoil at the time but eventually ended a constant source of annoyance and disorder. Unfortunately both Fleetwood and Macklin were gamblers by temperament and not even Macklin's epoch-making Shylock in *The Merchant of Venice* in 1741 and GARRICK's arrival in 1742, fresh from his triumphs at GOODMAN'S FIELDS THEATRE, could save Drury Lane from financial disaster. Fleetwood, having ruined himself and everyone who came into contact with him, finally gave up, and sold his share of the Patent in 1744.

FLEMISH THEATRE, see BELGIUM and NETHERLANDS.

FLESCHELLES, see GAULTIER-GARGUILLE.

FLETCHER, JOHN (1579–1625), English poet and dramatist, who spent most of his life writing for the stage, alone or in collaboration. His name became so indissolubly linked with that of Sir Francis BEAUMONT that by the end of the 17th century 53 plays had been ascribed to their joint authorship, but it is now thought that they collaborated in only six or seven, among them *Philaster, or Love Lies Bleeding* (1609), *The Maid's Tragedy* (1610), *A King and No King* (1611), and *The Scornful Lady* (1613). Beaumont ceased to write after his marriage in 1613, but Fletcher, who had already written *The Faithful Shepherdess* (1608) before joining forces with Beaumont or any of his other collaborators, continued with *Bonduca* (1613), a tragedy in which Richard BURBAGE played the leading role, and such comedies as *Wit Without Money* (1614) and *The Little French Lawyer* (1619), the latter probably in collaboration with MASSINGER. Of his later comedies, *The Wild Goose Chase* (1621) was successfully adapted by FARQUHAR as *The Inconstant; or, the Way to Win Him* (1703), and *The Chances* (1625), based on a story by CERVANTES, was rewritten by BUCKINGHAM in 1666. In its original form it was seen in 1962 at the opening season of the CHICHESTER FESTIVAL THEATRE. Fletcher is credited with having collaborated with Shakespeare in *The Two Noble Kinsmen* and *Henry VIII* (both 1613): the former ascription is now considered doubtful, and Beaumont or Massinger has been suggested in his place; but there seems reason to suppose that he had a hand in some parts of *Henry VIII.* He is also noted in the Stationers' Register in 1653 as having collaborated with Shakespeare in *The History of Cardenio* (1613) the lost play which is believed to have suplied Lewis Theobald with the plot of his *Double Falsehood; or, the Distrest Lovers* (1727).

FLEURY [Abraham-Joseph Bénard] (1750–1822), French actor, son of a theatre manager at Nancy, where he made his first appearances.

Encouraged by VOLTAIRE, who discerned great promise in him, he made his first attempt to join the COMÉDIE-FRANÇAISE in 1774, helped by LEKAIN who had known his father. After further experience in the provinces he was accepted in 1778, remaining with the company until his retirement in 1818 and becoming its 17th doyen. After imprisonment during the Revolution he was one of the members of the company reconstituted in 1799. An excellent actor, he owed his position to hard work and an innate feeling for the theatre, being almost totally uneducated. He had a natural nobility of carriage and character and was a master of polished comedy, being particularly admired as Alceste in MOLIÈRE's *Le Misanthrope*.

FLEXIBLE STAGING. In an effort to break away from the conventions of the PROSCENIUM or picture-frame stage, directors have experimented for many years with various types of staging which go back, in essentials, to the Greek and Elizabethan theatres. The initial aim of these experiments was to re-create a sense of intimacy and immediacy between actors and audience by abolishing the proscenium arch. It was soon realized that new or experimental groups need not rely on acquiring or building a conventional theatre, but could adapt any hall to their purpose. This led to the evolution of the all-purpose theatre which provides a variety of seating plans and adaptable levels and shapes for the stage. Such theatres can be used for THEATRE-IN-THE-ROUND, for productions with the audience on three sides of the acting area (the OPEN, or Thrust, STAGE), for staging at one end of a hall, usually with a large FORESTAGE, or for proscenium productions. Pioneering English examples include the university theatre at BRISTOL, the QUESTORS THEATRE at Ealing, in London, and the experimental theatre belonging to the London Academy of Music and Drama. The installation of flexible facilities in the small COTTESLOE THEATRE which forms part of the NATIONAL THEATRE has endorsed its usefulness. René ALLIO has been a major advocate of flexible staging in France.

FLIES, the space above the stage, hidden from the audience, where scenery can be lifted clear from the stage, or 'flown', by the manipulation of ropes. In the traditional system of HAND WORKING, the stage hands in charge of the flies were known as fly-men and worked on fly-floors, or galleries running along each side of the stage. Between the fly-floors there were formerly catwalks, or narrow communicating bridges about 2 to 3 ft wide over each set of WINGS, slung on iron stirrups from the GRID to which the ropes, or lines, were attached. Usually the rope-ends were tied off on the rail of the fly-floor on the prompt side. Large Continental theatres and opera houses sometimes had as many as three pairs of fly-floors, but English theatres rarely had more than one pair. Scenery is now generally flown by the COUNTERWEIGHT SYSTEM.

FLIPPER, see FLAT.

FLOATS, see LIGHTING.

FLOODLIGHT, see LIGHTING.

FLORENCE, WILLIAM JERMYN (*or* JAMES) [Bernard Conlin] (1831–91), American actor, who made his New York début in 1850 at NIBLO'S GARDEN in dialect impersonations. Two years later, while appearing at the BROADWAY THEATRE, he married the actress Malvina Pray (1831–1906), sister of Mrs Barney WILLIAMS, and with her toured North America in a repertory of Irish plays, being seen in 1852 with John NICKINSON's company at the Royal Lyceum in Toronto. Husband and wife were in London in 1856 when they appeared at DRURY LANE in a farce, *The Yankee Housekeeper*, with great success. Among Florence's best parts were Captain Cuttle in DICKENS's *Dombey and Son*, Bob Brierly in Tom TAYLOR's *The Ticket-of-Leave Man*, which he was the first to play in America, and Bardwell Slote in Woolf's *The Mighty Dollar* (1876), a play which remained in his repertory for many years. He was good in burlesque and as a comedian ranked with Joseph JEFFERSON, playing Sir Lucius O'Trigger to his Bob Acres in SHERIDAN's *The Rivals* and Zekiel Homespun to his Dr Pangloss in the younger COLMAN's *The Heir-at-Law*. Florence was responsible for the first production in New York of ROBERTSON's *Caste*, which he had memorized in London, staging it at the Broadway Theatre on 5 Aug. 1867, only four months after its first appearance, with the realistic scenery and atmosphere of the BANCROFT production. This caused a good deal of controversy, as Lester WALLACK had bought the American rights of the play, but in the absence of COPYRIGHT laws he had no redress against piracy. Probably fearing to challenge comparison with Florence's excellent production, Wallack did not himself stage the play until 1875.

FLORIDOR [Josias de Soulas, Sieur de Primefosse] (*c.* 1608–71), French actor, leader of a troupe of strolling players whom he took to London in 1635, appearing before the Court and at the COCKPIT in Drury Lane. Three years later he toured the French provinces with FILANDRE and in 1638 joined the company of the Théâtre du MARAIS. Some of his best performances were given in the plays of Pierre CORNEILLE, and it may have been Floridor's move to the Hôtel de BOURGOGNE in Apr. 1647 which induced Corneille to give his later plays to that theatre rather than the Marais. Floridor soon became the leader of the company, his quiet, authoritative acting being in marked contrast to the bombastic style of his colleague MONTFLEURY, and he was the only actor spared by

MOLIÈRE in his mockery of the rival troupe in *L'Impromptu de Versailles* (1663).

FLORINDO, see ANDREINI, G. B.

FLY-FLOOR, FLY-MEN, see FLIES.

FLYING EFFECTS have been achieved in the theatre since the earliest times. The Greeks had the DEUS EX MACHINA; in LITURGICAL DRAMA, and later in the Renaissance theatre, flying effects varied from the simple rise and fall of a figure or group of figures on a movable platform concealed behind CLOUDINGS, as in the PARADISO, to the elaborate undulating flight of a character across the stage. The mechanism needed to produce such effects was well known in England by the Restoration period, when references to 'flyings' are frequently found, and by the end of the 18th century there were in existence diagrams of the procedure to be followed for complicated flights. By the mid-19th century the machines in use included a complicated arrangement of pulleys and counterweights for controlling the circulatory gyrations of a pair of flying figures. In the PANTOMIMES of the early 20th century a group known as Kirby's Flying Ballet gave many brilliant and graceful performances; the lines to which players are attached are still known as Kirby Lines. In such effects, where no clouding or chariot is used for the character being flown, a line is attached to a hook at the player's back which forms part of harness worn under the costume. It is easily fixed and discarded, but a safety device prevents it from leaving the hook during flight. One of the few 20th-century plays in which flying is essential is J. M. BARRIE's *Peter Pan*, first produced in 1904 and still frequently revived.

FO, DARIO (1926–), Italian dramatist, actor-manager, and mime, who with his wife Franca Rama founded in 1957 a company which staged mainly his own short plays. He began his career by collaborating with Franco Parenti in satirical revues, among them *Il dito nell'occhio* (*A Finger in the Eye*, 1953) and *I sani da legare* (*Certified in Possession of their Faculties*, 1954). Among his later plays, some of which were first produced in Belgium, were *Gli arcangeli non giocano al flipper* (*Archangels Don't Play the Pin-Tables*, 1959) and *Aveva due pistole con gli occhi bianchi e neri* (*He had Two Pistols with Black-and-White Eyes*, 1960). After the abolition of censorship in 1962 Fo's comedies adopted more explicitly political themes, among them *Settimo; ruba un po'meno* (*Seventh: Thou Shalt Steal a Little Less*, 1964) and *La colpa è sempre del diavolo* (*It's Always the Devil's Fault*, 1965). In 1967 he established his own group, La Nuova Scena: his one-man show *Mistero buffo* (1969) was a reworking of old MYSTERY PLAYS. For the co-operative group La Comune he wrote *Morte accidentale di un anarchico*, which as *Accidental Death of An Anarchist* was seen at the Half Moon Theatre in Whitechapel, London, in 1979 and in 1980 began a long run at WYNDHAM'S. *Non si paga non si paga* (1975), as *We Won't Pay! We Won't Pay!* was seen in New York in 1980 and as *Can't Pay? Won't Pay!* at the CRITERION THEATRE in London in 1981.

FOERSOM, PETER, see DENMARK and SHAKESPEARE IN TRANSLATION.

FOGERTY, ELSIE, see SCHOOLS OF DRAMA.

FOIRE SAINT-GERMAIN, SAINT-LAURENT, see FAIRS.

FOLGER, HENRY CLAY (1857–1930), American businessman, founder of the great Shakespeare Library in Washington which bears his name. Beginning with a modest set of Shakespeare's plays in 13 volumes, he went on to acquire the finest collection of Shakespeariana in the world, which he left to the nation. Under Dr Joseph Quincy Adams, the great Shakespearian scholar who became its first director, the collection, administered by Amherst College, increased rapidly in size and scope and now contains a series of English manuscripts of the 15th and 16th centuries comparable with those in the British Museum and the Bodleian Library.

FOLIES-BERGÈRE, the first major Parisian music-hall, which opened on 1 May 1869 as a *café-spectacle*, with a mixed bill of light opera and pantomime in which Paul Legrand for many years played PIERROT. Its main attraction, however, apart from its delightful winter garden, was the immense promenade (copied at the EMPIRE, London) where young men about town could meet their friends in comfort. Its international reputation was achieved under Paul Derval, who managed it from 1919 to 1966. It catered to a large extent for visitors, whether French or foreign, and one of the main features of its programmes was a bevy of beautiful young women who paraded either stark naked or clad only in inessentials, with enormous feathered head-dresses and long spangled trains. For the rest its turns consisted of acrobats, singers, and sketches, the last either extremely vulgar or incredibly beautiful, the scenic resources of the theatre being vast, though its stage is only 11 by 6½ metres. Its name has become synonymous with French 'naughtiness', now somewhat out of date.

FOLIES-DRAMATIQUES, London, see KINGSWAY THEATRE.

FOLIES-DRAMATIQUES, Théâtre des, Paris, on the BOULEVARD DU TEMPLE, built on the site of the first Théâtre de l'AMBIGU which was destroyed by fire in 1822. It opened with MELODRAMA on 22 Jan. 1831 and had a successful career, catering mainly for a local audience with short runs of patriotic and melodramatic plays by the Cogniard brothers, Comberousse, de Kock, Théaulon, and others. In 1834 FRÉDÉRICK had one of his first successes at this theatre in his own play *Robert Macaire*. When the Boulevard du Temple was demolished in 1862 a new Folies-Dramatiques was built in the rue de Bondy, which became a home of light musical shows.

FOLIES-MARIGNY, see MARIGNY, THÉÂTRE.

FOLIES-NOUVELLES, Théâtre des, Paris, a small theatre which had a short but brilliant career. It opened in 1852 and under several successive names was a home of pantomime and light opera. Two years later it became the Folies-Nouvelles, with programmes like those of the old fairground theatres, simple, crude, and naïvely charming. It had a great vogue until 1859 and then became a straight theatre, being renamed Théâtre DÉJAZET after the actress of that name who appeared there in plays by SARDOU and others under the management of her son. Before it finally disappeared it had reverted to light entertainment and fallen out of favour.

FOLK FESTIVALS, seasonal celebrations connected with the activities of the agricultural year, particularly seed-time and harvest. They appear to have arisen spontaneously among primitive peoples from the time of the first organized communities and vary considerably in form. Some consist merely of a processional dance, as in the MORRIS DANCE and the Horn Dance, unique to Abbot's Bromley in Staffordshire and held annually on the first Monday after 4 Sept., in which groups of all-male dancers carry huge sets of reindeer horns. The New Year's Eve Archetringle at Laupen in Switzerland contains a mock combat between two bands of masked men, one side wearing ribbons and bells reminiscent of Morris dancers; there is more horseplay among the Bavarian Wild Men, who appear on St Nicholas's Day (6 Dec.) on skis and wearing animal skins and horns or antlers. A rudimentary form of drama is found in the Mascarade of the Soule region of south-west France, which contains elements seen in the most elaborate form of folk festival, covering the life-cycle of birth, death, and resurrection, the MUMMERS' PLAY. In England the simpler forms of folk festival were connected with PLOUGH MONDAY; May-Day (1 May), with its sports, Maypole, May Queen, and ROBIN HOOD play; Midsummer Day (24 June), with its fires derived from the Celtic festival of Beltane; and Harvest Home, which celebrated the final gathering into barns of the year's grain harvest. Such festivals as these, found all over Europe, survived the rise and fall of Greece and Rome and the coming of Christianity, although the Church, considering them an undesirable pagan survival, tried either to suppress them or to graft them on to its own festivals of Christmas, Easter, Corpus Christi, and local saints' days. They may have been influenced from time to time by wandering minstrels or other nomadic entertainers, but in the main they relied on local talent.

FOLK PLAY. Under this heading may be grouped the rough-and-ready dramatic entertainments given at village festivals by the villagers themselves. These were derived from the dramatic tendencies inherent in primitive FOLK FESTIVALS, and were given on May-day, at Harvest Home, or at Christmas, when to the central theme of a symbolic death and resurrection, which comes from remotest antiquity, were added the names and feats of local worthies. Later, though not before 1596, these were replaced by the Seven Champions of Christendom or other heroes, probably under the influence of the village schoolmaster (cf. Holofernes in *Love's Labour's Lost*). As patron saint of England, ST GEORGE may have figured among them from the earliest times. With some dramatic action went a good deal of song and dance, of which the MUMMERS' PLAY, and the MORRIS DANCE are the main survivals. Practically no written records of the folk play survive, though a number of texts have been collected orally by such scholars as E. K. Chambers and R. J. E. Tiddy, and it contributed very little to the main current of modern drama; but its influence should not on that account be entirely disregarded.

FOLLIES, Ziegfeld, see REVUE and ZIEGFELD, FLORENZ.

FOLLY THEATRE, London, see CHARING CROSS THEATRE.

FOLZ, HANS (*c*.1435/40–1513), German poet, dramatist, printer, publisher, and producer. By profession a surgeon, he was in Nuremberg at least from 1459. After ROSENPLÜT, he was the second writer of FASTNACHTSSPIELE from the 15th century known by name, and he also carried through an important reform of the Nuremberg schools of MEISTERSÄNGER.

FONTANNE, LYNN, see LUNT, ALFRED.

FONVIZIN, DENIS IVANOVICH (1744–92), Russian dramatist, whose work bridges the gap between the neo-classical literary plays of SUMAROKOV and the social comedies of OSTROVSKY. His first attempt at playwriting was a comedy, *The Minor*, which pointed the contrast between the uneducated noblemen of the provinces and the cultured nobility of the city. He put it aside unfinished in favour of *The Brigadier-General*, which he read before the Court in 1766 with great success. It satirized the newly-rich illiterate capitalists who were at that time forcing their way into Russian society and also attacked the current craze for praising everything from Western Europe at the expense of everything Russian. It was not until

1781 that Fonvizin again took up *The Minor*, and virtually rewrote it, sharpening his satire on the landowners and their politics, and this version is still in the repertory of the Soviet theatre. As the landowning party was then in the ascendant, Fonvizin's daring was his downfall and he was forced into premature retirement. Although he wrote in the comic tradition of MOLIÈRE and the French 18th century, he imparted to his work a native element of Russian folk comedy which was to be further exploited by his successors.

FOOL, the licensed buffoon of the medieval FEAST OF FOOLS, later an important member of the *sociétés joyeuses* of medieval FRANCE, not to be confused with the COURT FOOL. The traditional costume of the Fool, who is associated with such FOLK FESTIVALS as the MORRIS DANCE and the MUMMERS' PLAY (especially the Wooing Ceremony), is a hood with horns or ass's ears, and sometimes bells, covering the head and shoulders; a parti-coloured jacket and trousers, usually tight fitting; and sometimes a tail. He carries a *marotte* or bauble, either a replica of a fool's head on a stick, or a bladder filled with dried peas, and he sometimes has a wooden sword or 'dagger of lath', like the 'Old VICE' in the MORALITY PLAYS, from whom he may have borrowed it. It is surmised that this costume is a survival of the old custom of animal disguising, when worshippers in primitive religious ceremonies wore the head and skin of a sacrificial animal, while the ass's ears were taken from the donkey who figures in the Feast of Fools.

FOOLS, Feast of, see FEAST OF FOOLS.

FOOTE, SAMUEL (1720–77), English actor and dramatist, well born and well educated, but so extravagant that in 1744 lack of money drove him to adopt the stage as a profession. His first appearances at the HAYMARKET THEATRE were not particularly successful, and he left London to act in Ireland where he was well received. On his return to England he joined the company at DRURY LANE, and in 1747 took over the Haymarket Theatre. There, with great ingenuity, he evaded the Licensing Act of 1737 by inviting his friends to a dish of tea or chocolate, their invitation card giving them admittance to an entertainment in which Foote mimicked his fellow-actors and other public figures. In 1749, having inherited a second fortune, he went to Paris, spent it, and returned to London to take up in earnest a life of hard work as actor-manager and playwright. He had already written a few farces but his first successful play was *The Englishman in Paris* (1753), with its sequel *The Englishman Returned from Paris* (1756). He then took over the Haymarket again and staged there his best play *The Minor* (1760), a satire on Whitefield and the Methodists in which he himself played Shift, a character intended to ridicule Tate WILKINSON. He remained at the still unlicensed Haymarket, appearing in his own comedies in afternoon performances until in 1766 he lost a leg through some ducal horseplay and the Duke of York, who was present at the accident, procured him a Royal Patent in compensation. He was then able to present summer seasons of 'legitimate' plays, the first in 1767 with Spranger BARRY and his wife in leading parts. He evaded trouble for many years but was finally forced to dispose of his Patent to the elder COLMAN, dying shortly afterwards on his way to France.

Foote had a bitter wit, and his plays were mainly devised to caricature people he disliked, particularly GARRICK, but he had such wonderful powers of mimicry and repartee that even his victims found themselves laughing at him. Although he had a keen eye for character, and wrote brilliant sketches of contemporary manners which caused him to be nicknamed 'the English Aristophanes', his plays were successful through their topicality, and have not survived. Short, fat, flabby, with an ugly but intelligent face, he was at once feared and admired by his contemporaries. Portraits of him were painted by both Zoffany and Joshua Reynolds.

FOOTLIGHT PLAYERS, see DOCK STREET THEATRE, CHARLESTON.

FOOTLIGHTS, see LIGHTING.

FOOTLIGHTS CLUB, see CAMBRIDGE.

FOOTLIGHTS TRAP, a long rectangular opening at the front of the stage before the curtain, with a post below at either end. Between the posts a framework slid up and down, and upon this the lamps of the footlights were arranged. They were lowered (according to an account of 1810) not only to enable a stage hand in the cellar to trim them, but so that the stage might, upon occasion, be darkened. For this purpose a pair of lines was taken up from the ends of the framework over pulleys at the heads of the posts and down again to a DRUM AND SHAFT in the cellar. A line led from the drum in the opposite direction to a counterweight, designed to balance the framework, and from the shaft an endless line was brought by pulleys to a point at the side under the stage floor, where it ascended to a winch in the prompter's corner, so that the whole operation of dimming the footlights could be conducted from the stage without going into the cellar.

FOOTMEN'S GALLERY, see GALLERY.

FORBES-ROBERTSON, JEAN (1905–62), English actress, second daughter of Johnston FORBES-ROBERTSON. She made her first appearance in 1921 in South Africa, touring with her mother under the name of Anne McEwen and in 1926, back in London, was seen as Sonia in KOMISARJEVSKY's production of CHEKHOV's *Uncle Vanya* under her own name which she thenceforth retained. In the same year she scored a success as Helen Pettigrew in John L. Balderston and J. C. Squire's *Berkeley Square*, based on Henry JAMES's novel *The Sense of the Past*. From 1927 until 1934 she played the title role in the Christmas production of BARRIE's *Peter Pan*, for which her slight, boyish figure made her eminently suitable, as it did also for Puck in *A Midsummer Night's Dream* in 1937 and Jim Hawkins in Stevenson's *Treasure Island* in 1945. She was for some time with the OLD VIC company, where she played a number of Shakespeare's heroines and was much admired in the plays of IBSEN, particularly as Hedda Gabler and as Rebecca West in *Rosmersholm*, both in 1936. Among her other oustanding parts were Lady Teazle in SHERIDAN's *The School for Scandal* and Marguerite in the younger DUMAS's *The Lady of the Camellias*. She also appeared with great success in a number of modern plays, among them Rodney Ackland's *Strange Orchestra* (1932) and PRIESTLEY's *Time and the Conways* (1937). After the war she was seen less often in London, but returned to make her last appearance as Branwen Elder in Priestley's *The Long Mirror* (1952) at the ROYAL COURT THEATRE.

FORBES-ROBERTSON, JOHNSTON (1853–1937), English actor-manager, knighted in 1913. He first studied art, but gave it up for the stage, learning his perfect elocution from Samuel PHELPS. He made his first appearance in 1874, his first outstanding success being at the HAYMARKET THEATRE in W. S. GILBERT's *Dan'l Druce, Blacksmith* (1876). Two years later he joined the BANCROFTS, appearing successfully in SARDOU's *Diplomacy* (1878), ALBERY's *Duty* (1879), and a revival of T. W. ROBERTSON's *Ours*, also in 1879. He was then engaged by Wilson BARRETT to play opposite MODJESKA, and in 1882 went to the LYCEUM under IRVING, later touring with Mary ANDERSON with whom he made his first appearance in New York as Orlando to her Rosalind in *As You Like It*. Returning to England he went back to the Lyceum to give an outstanding performance as the Duke of Buckingham in Irving's production of *Henry VIII* in 1892, and three years later played Romeo there to the Juliet of Mrs Patrick CAMPBELL. At the same theatre he was seen as Hamlet for the first time in 1897, and proved himself to be one of the finest players of the part in his generation, his voice, fine ascetic features, and graceful figure being ideally suited to the part. Outstanding later roles included Mark Embury in *Mice and Men* (1902) by M. L. Ryley; Dick Heldar in Kipling's *The Light That Failed* (1903), and above all the Stranger in Jerome K. JEROME's *The Passing of the Third-Floor Back* (1908), with which he inaugurated his last season at the ST JAMES'S THEATRE. He made his last appearance at DRURY LANE as Hamlet on 6 June 1913. He married in 1900 Gertrude ELLIOTT, who acted with him until 1913. Their daughter Jean FORBES-ROBERTSON was also on the stage, as were Forbes-Robertson's three brothers Ian, Norman, and Eric.

FORD, JOHN (1586–1639), English dramatist, several of whose known plays are lost, destroyed by WARBURTON's cook. Little is known of his life, except that he was born and died in Devonshire, and in 1602 was admitted to the Inner Temple. He may have been part author with ROWLEY and DEKKER of *The Witch of Edmonton* (1621) and with Dekker alone of *The Sun's Darling* (1624), which may however have been only a revision of one of Dekker's earlier plays, *Phaethon* (1598). Ford's most important plays are those he wrote alone —*Love's Sacrifice*, *'Tis Pity She's a Whore* (both 1627), *The Lover's Melancholy* (1628), and *The Broken Heart* (1629). The best known, *'Tis Pity She's a Whore*, was seen after the reopening of the theatres at Salisbury Court in 1661 and not revived again until 1923; but it has been played several times since, most recently by the ROYAL SHAKESPEARE COMPANY at Stratford-upon-Avon in 1977 and in London in 1978. In 1961 VISCONTI directed in Paris a French version as *Dommage qu'elle soit putain*. *The Broken Heart* was revived in 1962 in the first season at the CHICHESTER FESTIVAL THEATRE. Ford displays a totally different attitude towards morality from that of Shakespeare or JONSON: he has been called 'a romantic rebel, a believer in the divine impulse of human passion which leads in the end to self-destruction'.

FORDE, FLORRIE, see MUSIC-HALL.

FORD'S THEATRE, Washington, D.C., was originally a Baptist church, being taken over by John T. Ford (1829–94) and opened on 19 Mar. 1862 as Ford's Atheneum. Destroyed by fire, it was rebuilt and reopened on 27 Aug. 1863 as Ford's Theatre. It then became an immediate success, and some of the finest artists of the American stage soon appeared there. Its success came to an abrupt end, however, on 14 Apr. 1865, when President Abraham Lincoln was assassinated there during a performance by Laura KEENE's company of Tom TAYLOR's *Our American Cousin*. The assassin was John Wilkes BOOTH, whom Lincoln had once seen in a performance at the theatre. After the assassination Ford and his brother were imprisoned for 39 days, but later acquitted of complicity in the crime. The theatre was closed by official order and prevented from reopening by public outcry; the government bought the property and used it for storage and office space. Part of the theatre collapsed in 1893 and the building remained derelict for some time, but in 1932 it became a Lincoln museum, which it still remains. In 1954 Congress began to vote funds to restore the theatre as closely as possible to its original form. The remodelled building, outwardly resembling the original structure but with its capacity reduced to 741 from almost 2,000, opened on 12 Feb. 1968 with Stephen Vincent Benét's *John Brown's Body*, performed by the CIRCLE-IN-THE-SQUARE company. Ford's Theatre now presents four to six new or touring plays each season, many of which have gone on to Broadway success, including the Black musicals *Don't Bother Me, I Can't Cope* (1971 and 1974) and *Your Arms Too Short to Box With God* (1976).

FORESTAGE, a small area of the stage in front of the PROSCENIUM arch and the front CURTAIN, which served the same purpose as the unlocalized *platea* of the medieval MYSTERY PLAY, actors approaching it either from the main stage or through the PROSCENIUM DOORS. It was the last vestige of what had been the main acting area in Restoration times, the apron stage, itself a modified form of the Elizabethan platform stage, and its loss later proved a great handicap in staging revivals of Elizabethan and Restoration plays. In some new theatre buildings, particularly those intended for experimental or academic work, forestages were built out in front of the proscenium, or, as at the OLD VIC and the SHAKESPEARE MEMORIAL THEATRE, added to existing stages. In the LYTTELTON THEATRE in the NATIONAL THEATRE building, the proscenium arch can be adjusted to give a narrow forestage.

FORK, see GROOVES.

FORMALISM, a theatrical method popular in Russia soon after the October Revolution, and practiced by MEYERHOLD, by AKIMOV in his earlier period, and to a lesser degree by TAÏROV. It entailed the subjection of the actor so that he became nothing more than a puppet in the hands of the director and insistence on exterior symbolism at the expense of inner truth. Though it originally helped to clear the stage of the old falsities and conventions of pre-Revolutionary days, when pushed too far it resulted in a complete lack of harmony between actors and audience which eventually caused it to be abandoned.

FORMAL SETTING, see DETAIL SCENERY.

FORMBY, GEORGE, see MUSIC-HALL.

FORREST, EDWIN (1806–72), American tragedian, who appeared in an amateur performance at the age of 10, and four years later played Young Norval in HOME's *Douglas* at the WALNUT STREET THEATRE in Philadelphia, his birthplace. His early years were overshadowed by poverty and thwarted ambition, but in the end he became the acknowledged head of his profession for nearly 30 years. Even then the defects of his character made him as many enemies as friends, and no one received more abuse mingled with the praise which was his due, yet Mrs John DREW, who at the age of 8 appeared with him in Sheridan KNOWLES's *William Tell*, praised his unselfishness in allowing other actors to gain the attention of the audience during their important moments. He had expressive features and a powerful voice, which he used unsparingly, but his acting lacked delicacy, and in his early years he was much criticized for 'ranting'. This he later cured to some extent, and was then outstanding as Lear, Hamlet, Macbeth. Othello, and Mark Antony in *Julius Caesar*, and in two parts especially written for him: Spartacus in *The Gladiator* (1831) by BIRD and the title role in STONE's *Metamora* (1829). Among his other parts were Jaffier in OTWAY's *Venice Preserv'd*, Rolla in SHERIDAN's *Pizarro*, and the title role in Knowles's *Virginius*. He appeared in London in 1836 with some success, but in 1845 met with a hostile reception that he attributed to the machinations of MACREADY. Their quarrel led eventually to a riot at the ASTOR PLACE OPERA HOUSE in New York in 1849, when Macready barely escaped with his life. This caused Forrest to be ostracized by more sober members of the community, but he was the idol of the masses, who looked on him as their champion against the tyranny of the English.

In 1837 Forrest married Catharine Norton Sinclair (1817–91), who was on the stage and later became manager of a theatre in California; they were divorced in 1850. From 1865 his career declined and he died lonely and unhappy; his last performance was at the Globe Theatre, Boston, on 2 Apr. 1872 in BULWER-LYTTON's *Richelieu*.

FORREST THEATRE, New York, see EUGENE O'NEILL THEATRE.

FORT, PAUL (1872–1960), French poet, director, and theatre manager, who at the age of 18, in reaction against the REALISM of ANTOINE'S THÉÂTRE LIBRE, founded first the short-lived Théâtre Mixte and then the Théâtre d'Art for the production of poetic plays. His theories of symbolic and abstract art led him to look on his actors also as abstractions, and defeated by their solid reality he gave up his theatre after only two years, being succeeded in its management in 1893 by LUGNÉ-POË.

FORTUNE THEATRE, Dunedin, see NEW ZEALAND.

FORTUNE THEATRE, London. (1) In Golden Lane, Cripplegate. This was built in 1600 for Edward ALLEYN and Philip HENSLOWE to house the ADMIRAL'S MEN, of which Alleyn was the leading actor. It was erected by Peter Street, who also built the first GLOBE, and the contract for it is still in existence; it gives a good deal of valuable information, though a number of details are dismissed with the words 'like unto the Globe'. It is clear, however, that it was an 80-ft-square timber building, almost certainly constructed within the walls of a former inn or dwelling house, and it had two galleries reputed to hold 1,000 people. It took its name from a statue of the Goddess of Fortune over the entrance. It was a popular playhouse, and drew a fashionable audience of nobles and distinguished foreign visitors. On 9 Dec. 1621 it was burnt down, and with the building perished also the wardrobe and all the playbooks. Two years later, rebuilt in brick and with a repertory of 14 new plays, it reopened, still tenanted by the same company. Though not as successful as before, the Fortune continued in use until all the theatres were closed in 1642. Even then it was used occasionally for illicit performances, until in 1649 it was raided by Commonwealth soldiers, who dismantled the interior. The building was finally demolished in 1661.

(2) The old name was revived when on 8 Nov. 1924 the present Fortune Theatre, in Russell Street, Covent Garden, opened under Laurence Cowen with his own play *The Sinners*. It seats 424 people in three tiers, and has a proscenium width of 25 ft. It had few successes in its early years, except for LONSDALE's *On Approval* (1927) which ran for 469 performances and a season of plays by O'CASEY presented by J. B. FAGAN. It was the headquarters of Nancy PRICE's People's National Theatre for ten months from Oct. 1930, and was then used by amateur companies. During the Second World War it was occupied by ENSA. It reopened in 1946 with Kenneth Horne's *Fools Rush In*, but during the next 10 years housed more amateur productions than professional, though *Joyce Grenfell Requests the Pleasure* (1954) was popular. In 1957 Michael Flanders and Donald Swann presented their musical entertainment *At the Drop of a Hat*, which had a long run, as did the revue *Beyond the Fringe* (1961), with Peter Cook, Dudley Moore, Alan BENNETT, and Jonathan MILLER; it was succeeded by another revue *Wait a Minim* (1964). A further success was scored by *The Promise* (1967) by Alexei ARBUZOV, with Ian MCKELLEN and Judi DENCH. The theatre then housed a number of unsuccessful *avant-garde* plays interspersed with musicals and revues, including *Close the Coalhouse Door* (1968), Alan Plater's documentary musical on the Durham miners. In 1971 a thriller, *Suddenly at Home* by Francis Durbridge, began a long run which lasted until June 1973; it was succeeded in the same year by Anthony Shaffer's *Sleuth* transferred from the GARRICK, and in 1976 by Agatha CHRISTIE's *Murder at the Vicarage* transferred from the SAVOY, both of which had long runs. In 1980 MARLOWE's *Dr Faustus* and *Jeeves Takes Charge*, based on P. G. Wodehouse, moved there from the LYRIC, Hammersmith. The theatre was taken over in 1981 by Fortune Theatre Management Ltd, but its policy of six- to eight-week runs of plays from provincial and FRINGE theatres was not a success.

FORTUNY, MARIANO, see LIGHTING.

FORTY-EIGHTH STREET THEATRE, New York, at 157 West 48th Street. This opened on 12 Aug. 1912 and housed opera and musical comedy, including GILBERT and Sullivan, as well as a number of successful straight plays. In 1922 came George KELLY's delightful satire on pretentious 'little' theatres *The Torchbearers*, which ran for several months. Later successes where Jean Bart's *The Squall* (1926), with Blanche YURKA, a revival of BOUCICAULT's *The Streets of New York* (1931), *Harvey* (1944) by Mary Chase, and *Stalag 17* (1951) by Donald Bevan and Edmund Trzcinski. Robert ANDERSON's *Tea and Sympathy*, which opened at the ETHEL BARRYMORE THEATRE in 1953, finished its long run here in June 1955. Two months later, when the theatre was unoccupied, a water tank fell through the roof, causing considerable damage to the auditorium, and the building was demolished later the same year. From 1937, when it was leased to Labor Stage who opened on 20 Nov. with *Work Is For Horses*, until 1943 it was known as the Windsor Theatre.

FORTY-FOURTH STREET THEATRE, New York, at 216 West 44th Street. This opened on 21 Nov. 1912 as the New Weber and Fields Music-Hall, seating 1,463 people. It was taken over a year later by the Shuberts, who staged there a successful musical revival, *The Geisha*. In 1915 Robert B. MANTELL appeared in an extensive classical repertory, but after he left the theatre reverted to light opera and musicals. The Marx Brothers had a long run in 1928 in *Animal Crackers*; the outstanding production of 1930 was Gilbert Seldes's adaptation of ARISTOPHANES' *Lysistrata*. Highlights of the 1930s were J. B. PRIESTLEY's *The Good Companions* in 1931, a four-week season by Walter HAMPDEN in 1935, and the shortlived but memorable *Johnny Johnson* (1936) by Paul GREEN with incidental music by Kurt Weill. One of the last outstanding productions at this theatre, which was demolished in 1945, was Moss HART's musical documentary on aviation *Winged Victory* (1943).

On the roof of the Forty-Fourth Street Theatre was a smaller theatre, which underwent several changes of name but in 1918 was called the Nora Bayes after the wife of Jack NORWORTH. It reverted to that name when it was taken over by the FEDERAL THEATRE PROJECT in 1937, and was demolished with the main theatre. The basement housed the café famous in the era of prohibition as the Little Club.

FORTY-NINTH STREET THEATRE, New York, at 235 West 49th Street. This small house opened on 26 Dec. 1921 and early in the following year Morris Gest presented BALIEFF's La Chauve-Souris for a season which continued during the summer on the roof of the old CENTURY THEATRE. A series of successful new plays followed, though WEDEKIND's *Lulu* and COWARD's *Fallen Angels* both failed in 1927. In 1928 came a fine revival of IBSEN's *The Wild Duck* with Blanche YURKA followed by *Hedda Gabler* in 1929, and a year later the public were delighted by the great Chinese actor, MEI LANFANG. Among later productions were DRINKWATER's *Bird in Hand* (1930), a revival of STRINDBERG's *The Father* (1931) with Robert LORAINE, a season of Yiddish plays with Maurice SCHWARTZ, and three productions by the FEDERAL THEATRE PROJECT. The last play seen in the theatre was *The Wild Duck* played in modern dress in Apr. 1938, after which it became a cinema; it was later demolished.

FORTY-SIXTH STREET THEATRE, New York, at 226 West 46th Street. Owned by the Chanin Construction Company and intended for musical shows, with a seating capacity of 1,338, it opened as Chanin's Forty-Sixth Street Theatre on 24 Dec. 1925 and took its present name on 10 Oct. 1932. Straight plays produced there include *The First Legion* (1934), a sensitive portrayal of Jesuit life by Emmet Lavery, and a GROUP THEATRE production of *Weep for the Virgins* (1935) by Nellise Child; but it is best known for such successful musicals as *Hellzapoppin* (1938), *Du Barry Was a Lady* (1939), *Panama Hattie* (1940), *Finian's Rainbow* (1947), *Guys and Dolls* (1950), and *Damn Yankees* (1955). In 1958 GIELGUD occupied the theatre with his one-man recital of passages from Shakespeare *The Ages of Man*, and later successes included *How to Succeed In Business Without Really Trying* (1961), *I Do! I Do!* (1966), *1776* (1969), a revival of *No, No, Nanette* (1971), *Chicago* (1975), and *The Best Little Whorehouse in Texas* (1978), which ran into the 1980s.

FORUM THEATRE, Manchester, see LIBRARY THEATRE.

FORUM THEATRE, New York, see VIVIAN BEAUMONT THEATRE.

FOSS, WENCHE (1917–), Norwegian actress who, after an early career in comedy and farce in the 1930s, developed into a player of great versatility. At the NATIONALTHEATRET (1952–67) she played (among other parts) Célimène in MOLIÈRE's *Les Précieuses Ridicules* in 1957 and IBSEN's Hedda Gabler in 1960. In 1967 she returned to the OSLO NYE TEATER, with which she had spent the years 1949–51, to devote herself to comedy and musicals, returning to the Nationaltheatret in 1978. There she played leading roles in ARBUZOV's *Old World* in 1978, Ibsen's *John Gabriel Borkman* in 1979, and William Douglas HOME's *The Kingfisher* in 1980.

FOUR SEASONS THEATRE, Wanganui, see NEW ZEALAND.

FOURTEENTH STREET THEATRE, New York, at 105–9 West 14th Street. Opening as the Théâtre Français on 26 May 1866, it staged French-language plays for three years, until purchased and refurbished by Charles FECHTER (who never, however, assumed its management) in 1871. As the Lyceum it reopened on 11 Sept. 1873 with an English company in a version of HUGO's *Notre-Dame de Paris*. Its subsequent history was mainly concerned with light opera, and it was known as the Fourteenth Street Theatre from 1886 until 1926, when until 1932 it enjoyed a brief distinction as the home of Eva LE GALLIENNE's Civic Repertory Company. It was demolished in 1938.

FOX, G. L. [George Washington Lafayette] (1825–77), American actor and pantomimist, was on the stage as a child, and from 1850 to 1858 was at Purdy's National Theatre (see CHATHAM THEATRE (2)) where in 1853 he persuaded the management to put on AIKEN's adaptation of Harriet Beecher Stowe's *Uncle Tom's Cabin*, which was a great success, proving the best of the many versions current. In 1858, in partnership with J. W. Lingard, he took over the management of the BOWERY, and from 1862 to 1867 he staged a long series of successful pantomimes there. In 1867 he gave a fine performance as Bottom in James E. Hayes's production of *A Midsummer Night's Dream* at the OLYMPIC (2), and a year later he produced at the same theatre his pantomime *Humpty-Dumpty* and a travesty of *Hamlet* which Edwin BOOTH is said to have enjoyed. Considered the 'peer of pantomimists', he continued to appear in successive editions of *Humpty-Dumpty* until his death. Much of the pantomime 'business' in his shows was devised by his brother Charles Kemble Fox (1835–75), who was also a child actor. He was in *Uncle Tom's Cabin*, as was his mother, and played Pantaloon in his brother's pantomimes.

FOX WEDGES, see RAKE.

FOY, EDDIE [Edwin Fitzgerald] (1856–1928), American actor and VAUDEVILLE player. He was a singer and entertainer from childhood, and in 1878 toured the Western boom towns with a MINSTREL SHOW. He was later seen in comedy and MELODRAMA, and from 1888 to 1894 played the leading part in a long series of EXTRAVAGANZAS in Chicago. He was acting in the Iroquois Theatre there in 1903 when fire broke out, and did his best to calm the audience, but without success, the ensuing panic resulting in the loss of over 600 lives. Foy, who was an eccentric comedian with many mannerisms and a distinctive clown make-up, played in MUSICAL COMEDY until 1913 and then went into vaudeville, accompanied by his seven children, with whom he made his last appearance in 1927. He subsequently wrote his autobiography, *Clowning Through Life* (1928).

FRAGSON, HARRY, see MUSIC-HALL.

FRANCE. As elsewhere in Europe, a national drama evolved in France from the LITURGICAL

DRAMA of the Christian Church, with important and definite accretions from the FOLK PLAY and MINSTREL traditions. The evolution of the play based on the liturgy and performed wholly or mainly by the clergy inside the church was complete by the middle of the 13th century, and subsequent developments intensified the secular elements which had already crept in. This process began when exigencies of space and a growing sense of the limitations imposed by the church walls led to the removal of the play from the nave and choir to the churchyard, and eventually to the marketplace. In France this move was made early, for a 12th-century Anglo-Norman play, *Le Mystère* (or *Jeu*) *d'Adam*, was clearly intended for outdoor performance.

From the early days of secularization plays were closely connected with the Guilds, who first supplied the extra actors needed and later were responsible for financing and producing each performance. They fell into two groups—MYSTERY PLAYS (*Mystères*) based on the Bible, and SAINT PLAYS (*Miracles*) based on the lives of Christian saints and martyrs and, in France particularly, on the life of the Virgin Mary. These remained linked with the main Church festivals—Whitsun and, particularly, the feast of Corpus Christi (the Thursday after Trinity Sunday) being the favourite times for theatrical activity.

One outstanding result of the interaction of religious and secular elements in drama was the extended use of the vernacular, already present in the *Mystère d'Adam*. This inevitably produced a national, as opposed to a cosmopolitan, drama, and encouraged the comic and rowdy elements which had begun to make their appearance inside the church, mainly in connection with the extraneous characters grafted on to the Bible story—wives, merchants, servants—and with the imps and devils attendant on Satan.

By the early 13th century, as can be seen from *Le Jeu de Saint-Nicolas* by Jean BODEL, scenes of everyday life were already being incorporated into the religious plays, thus further weakening their original liturgical character—though the great four-day PASSION PLAYS of MARCADÉ and GREBAN were not written until the early 15th century. Another influence at work in the direction of humour was that of the purely secular FARCE, a twofold inheritance from the almost submerged traditions of the classical theatre preserved in the repertory of the Minstrels and from the licensed buffoonery of the FEAST OF FOOLS. Just as the serious religious play became the concern of the town guilds and literary societies, so the farcical horseplay of the minor clergy became the perquisite of the *sociétés joyeuses*, bands of light-hearted, irreverent youths who, when ecclesiastical merrymaking fell into disrepute, took it upon themselves to provide amusement for their fellow townsmen. They and such similar societies as *les clercs de la basoche*, formed by law students in the universities under Philippe le Bel (1285–1314), and the student society of the Parisian *enfants sans souci*, formed under Charles VI (1380–1422), were primarily associations of amateurs, not to be confused with the professional corporations of minstrels. These latter, however, were not excluded from the revels, and some of them may have been employed by the guilds to assist in the production of religious plays, particularly in the provision of music. They certainly played a part in secular plays, where their experience and ready pens would have been most welcome. To ADAM DE LA HALLE goes the honour of having written in the later 13th century the earliest known French secular plays; and one work by another minstrel, RUTEBEUF, is also extant.

By the 15th century drama had utterly forsaken the Church and been forsaken by it. The CONFRÉRIE DE LA PASSION, a guild of amateurs formed in 1402 to act religious plays, was the model for many such associations up and down the country. The literary societies (the *puys*) and student companies performed allegorical MORALITY PLAYS and evolved the typical French SOTIE, of which the best known writer was Pierre GRINGORE. Farce, a perennial form of popular drama, flourished also, the most famous example being the anonymous 15th-century *Maître Pierre Pathelin*, performed in 1470 by *les enfants sans souci*, the robust humour of which has survived modernization and translation. During the whole of the 15th and 16th centuries in France, miracle, mystery, and morality plays, *soties* and farces, continued to be performed, often by the same actors in the same programme.

This mingling of the genres had its dangers, the most obvious being the final swamping of the religious element by the secular. A decree of 17 Nov. 1548, while confirming the Confrérie de la Passion in its monopoly of acting in Paris, also forbade the performance of anything but secular pieces, on the grounds that the mixture of sacred and profane elements brought religion into disrepute. The Confrérie was thus deprived of the greater part of its repertory, and with the ban the early religious drama in France came to an end.

Owing to the ravages of civil war the secular theatre took a long time to establish itself, and the 16th century, which saw the emergence of a national drama in SPAIN and ENGLAND, was still in France a period of confusion and experiment. The influence of the Renaissance triumphed in the end over the old farces and *soties* as it did over the complicated and outmoded Bible histories. Drawing not only their inspiration but also their form and content from classical antiquity, the early writers of the French Renaissance, such as Lazare de Baïf and Jean de la Péruse, were students first and playwrights almost by accident. It was Étienne JODELLE who opened the way for the great neoclassical drama of the 17th century. Among his contemporaries, who all looked to Greece and Rome (in particular SENECA) for their models, the best was probably Robert GARNIER; Antoine de

MONTCHRÉTIEN later continued the vogue for tragedy.

True to what they believed to be the principle of their classical models, these writers allowed themselves no mingling of the serious and the comic. The medieval mixture of high tragedy and low farce was no longer permissible, and the genres were sharply defined. An author might adventure in both, as Jodelle did, and Jacques GRÉVIN; but their comic writing owed no more to the popular farce of old France than did the comedies of the Italian-born Pierre de LARIVEY which were directly indebted to the COMMEDIA DELL'ARTE, an early association of French with Italian comedy which foreshadowed that between the *commedia dell' arte* scenarios and the early plays of MOLIÈRE. Also from Italy came the PASTORAL, the delight of a courtly society, which shared the popularity of tragedy.

The years which saw the gradual shaping and perfecting of French drama in a form which was to reach its height in the plays of CORNEILLE and RACINE saw also the emergence of a body of professional actors to interpret it. The early writers had relied on amateurs, on themselves and their friends, or on the schools, particularly those which produced plays in the tradition of JESUIT DRAMA. Jodelle himself played the heroine of his *Cléopâtre captive* (1552), and the plays of Grévin were produced at the Collège de Beauvais.

While the new drama was establishing its literary status, the old discredited popular theatre was still being played in the provinces by bands of strolling actors who by the 1570s were beginning to form themselves into permanent professional companies. But the country was in too troubled a state to permit the establishment of permanent theatre buildings, and even after peace had been restored with the accession of Henri IV in 1594, Paris had still to be content for many years with visits from foreign companies, mainly Italian, brought in expressly for the amusement of the Court, and from those provincial companies which included the town in their itineraries. One such company was that of Agnan SARAT, first noted in 1578; with VALLERAN-LECOMTE as its chief actor and the prolific Alexandre HARDY as its dramatist, it became the first professional company to settle permanently at the Hôtel de BOURGOGNE, leased from the Confrérie de la Passion. Since the Confrérie still clung to its monopoly of acting in Paris, no other company could legally act in the city, and for some time Sarat's company had no rivals. A breach was, however, made in the Confrérie's monopoly in 1595, by the publication of an edict permitting other companies to play at the Paris FAIRS of St Germain in the spring and St Laurent in the autumn. The father of Valleran-Lecomte's leading lady Marie VENIER was quick to take advantage of this, and having established a company at the spring fair, he moved it to the Hôtel d'ARGENT, where he was allowed to play for a time on payment of a levy to the Confrérie. On a later visit he was joined for a time by his daughter and her husband Laporte.

The opening years of the 17th century thus found Paris with two professional companies, which were allowed to function as long as the Confrérie de la Passion allowed them to. The fortunes of both groups fluctuated, as did their composition, for actors constantly migrated from one company to another, while a political upheaval such as the assassination of Henri IV in 1610 was enough to uproot them all and send them back into the provinces again. But by 1629 the *Comédiens du roi*, as the company once headed by Valleran-Lecomte was now called, had established itself firmly at the Hôtel de Bourgogne, while a rival company, originally under Charles LENOIR, was competing for public favour elsewhere, settling at the Théâtre du MARAIS under Lenoir's companion MONTDORY in 1634.

Paris had not yet attained complete control of theatrical activity and good companies could still be found in the provinces, particularly in Normandy, while an increasing number of plays were published in Rouen. A study of the repertory, where particulars of it can be found, shows a curious mingling of the old medieval plays with those written under classical or Italian influence. The pastoral continued in high favour with TRAGI-COMEDY not far behind, but comedy, neglected by Hardy who was to have a great influence on many later dramatists, was practically non-existent. Farces were seldom printed, perhaps hardly ever written down, so few have survived, but their popularity may be judged by the fact that the outstanding actors of the period whose names have come down to us were farce-players. Among them were TABARIN, whose clowning on the trestle-stage of the Pont-Neuf was long a feature of Parisian life, and BRUSCAMBILLE and Jean FARINE, who were said to have begun their careers in the fair booths. The famous comic trio TURLUPIN, GAULTIER-GARGUILLE, and GROS-GUILLAUME, who as Belleville, Fleschelles, and Lafleur also played serious parts, were the delight of audiences at the Hôtel de Bourgogne for many years, though the leading actor there, BELLEROSE, did much to raise the status of his profession and oust the farce-players from public favour. Under him such popular clowns as GUILLOT-GORJU had to be content with a secondary place, while a notable feature of the period was the sudden flowering of feminine talent with such actresses as Mlles BEAUCHÂTEAU and BEAUPRÉ, both married to actors in the company. At the Théâtre du Marais Montdory, with Mlle de VILLIERS, led an equally able and experienced company, which gave the first performances of many of the early plays of the French classical period, including Corneille's youthful comedies and his epoch-making 'tragi-comedy' *Le Cid* (1637).

Many influences had been at work, however, before Corneille produced his masterpiece, the most important being the establishment of the neo-classical UNITIES of time, place, and action.

These were first formulated in the preface to the published edition of MAIRET's *Silvanire* (1630), a pastoral written like his *Sylvie* (1624) under the influence of *Les Bergeries* (1620) by RACAN, a Court poet who with TRISTAN L'HERMITE not only raised the prestige of the dramatist but introduced for the first time into French dramatic writing some distinction in style and versification. The success of *Silvanire* induced other writers to follow Mairet's example in adhering to the unities, just as the success of his *Sophonisbe* (1634) brought about a return to popular favour of tragedy (after its partial eclipse by tragi-comedy) strengthened by the later success of Tristan L'Hermite's *La Mariane* (1636). It is therefore understandable that when *Le Cid*, now regarded as one of the outstanding works of French literature, was first put on it was not, on the whole, favourably received. The main criticism of it was that it violated the unities, and a fierce quarrel broke out among the literary figures of Paris—the so-called *Querelle du Cid*—with Mairet and his friends supporting their beloved rules, while those who already had some perception of Corneille's stature argued that genius makes its own rules and cannot be fettered by those of other people. The question was finally brought for arbitration to Cardinal RICHELIEU, who added to his many other pre-occupations an interest in the theatre, an urge to organize it, and a desire to shine as a playwright. He very sensibly handed the problem over to his newly founded French Academy, and meanwhile, undeterred by the storm he had aroused, Corneille continued to write. His next plays, however, were mainly tragedies, as though he wished to prove himself a master of the recognized vehicle for classical French drama. Among other writers who followed his example were Pierre DU RYER, Jean ROTROU, who became official dramatist to the Hôtel de Bourgogne in succession to Hardy, and LA CALPRENÈDE, who was unusual in that he looked to English rather than Roman history for his tragic subjects.

The political troubles of the Fronde, from 1648 to 1653, caused a slowing down of theatrical activity and a slight but noticeable change in the development of French drama. Corneille's temporary retirement in 1652 after the production of *Andromède* (1650), the first MACHINE PLAY embellished by the mechanical inventions of TORELLI, coincided with that of Du Ryer and the death of Rotrou. The public taste was now for lighter fare and new dramatists came forward to satisfy it, among them Corneille's younger brother Thomas CORNEILLE whose first play was produced in 1647; Philippe QUINAULT, the friend and protégé of Tristan L'Hermite; and the elderly but prolific Abbé BOISROBERT, whose main activity was confined to the 1650s.

Although all these dramatists wrote tragedy, their more important work was mainly in comedy, a genre which now came into prominence for the first time and was to culminate in the plays of the actor-dramatist Molière. In the early years of the 17th century playgoers had relied for humour mainly on the horseplay of the farce, or on the incidental comic scenes and characters in the pastorals and tragi-comedies; the 1630s saw a few isolated examples of comedy by such established writers as Rotrou, Du Ryer, and SCUDÉRY; but it was Corneille, with the series of plays which preceded *Le Cid*, who liberated true comedy from the entanglements of farce and melodrama. Although he wrote no more comedies after *La Suite du Menteur* (1644), which failed to achieve the success of its predecessor *Le Menteur* in the previous year, he had succeeded in making comedy a genre which merited the attention of experienced and serious playwrights. Among the few notable examples of the type before *Le Menteur*, decidedly the best of its period, was *Les Visionnaires* (1637) by DESMARETS DE SAINT-SORLIN, one of Richelieu's protégés. It might have led French comedy to concentrate on character rather than on intrigue had it not been for the overwhelming influence of those writers who turned to Spanish comedies of intrigue for their plots, among them not only Thomas Corneille but Antoine d'OUVILLE, elder brother of Boisrobert, and Paul SCARRON, friend of Pierre Corneille and first husband of the future Mme de MAINTENON. The revival of true French comedy, in the line of the elder Corneille but with its roots in the indigenous farce, was the work of Molière, whose influence it is almost impossible to overestimate, not only in his own country but throughout Europe; he alone made comedy a genre worthy to rank with tragedy. After an unsuccessful start in Paris the ILLUSTRE-THÉÂTRE, the first company to which Molière belonged, worked and travelled unceasingly in the provinces, and the experience gained there, mainly in farce, proved useful when in 1658 Molière brought his company back to Paris. In addition to himself and Madeleine, Joseph, and Louis BÉJART from the original company, he had acquired two good actresses in Mlles DU PARC and DE BRIE, a staunch veteran in Charles DUFRESNE, and a good comedian in GROS-RENÉ.

Paris now had three permanent troupes of actors—at the Hôtel de Bourgogne, whose company under Bellerose's successors FLORIDOR and MONTFLEURY had achieved supremacy in tragedy; at the Marais, which under LAROQUE had fallen somewhat from its former eminence, and was now devoting itself mainly to 'machine plays'; and at the PETIT-BOURBON, where Molière's company shared the stage with a *commedia dell'arte* troupe until in 1660 he moved to the PALAIS-ROYAL.

Molière did not neglect the work of his contemporaries, and among the plays he produced at the Palais-Royal was *La Thébaïde* (1664) by Jean RACINE, a young writer whose fame was soon to surpass even that of Corneille. For many years before this important event, tragedy had been in the doldrums, able to offer only the works of the ageing Corneille; of his brother Thomas, whose *Timocrate* (1656) heralded a timid return to normality after the troubles of the Fronde; the mediocre works of Jean MAGNON; and the one tragedy, *Manlius* (1662), of Mlle DESJARDINS: the time was ripe for a new talent. *La Thébaïde* was followed by *Alexandre et Porus* (1665), which Racine, with a callous disregard for Molière's kindness that cost him the older man's friendship, gave to the Hôtel de Bourgogne a fortnight after it had opened at the Palais-Royal. Even worse was the defection of Molière's leading actress Mlle Du Parc, who became Racine's mistress and went to the rival house to play the title role in his *Andromaque* (1667). Its success established him as the leading dramatist of the day, for Corneille's recent plays, *Agésilaus* (1666) and *Attila* (1667), had been coolly received. The rivalry between them came to a head in 1670, when both produced plays on the same subject, Racine's *Bérénice* being given at the Hôtel de Bourgogne a week before Corneille's *Tite et Bérénice* was seen at the Palais-Royal. Racine's play, with the great tragic actress Mlle CHAMPMESLÉ in the title role, was judged the better of the two, in spite of the excellence of Molière's wife Armande BÉJART as Corneille's heroine and the fine Domitien of young Michel BARON, later to be the foremost actor of France.

Jealous of Racine's immense prestige, a powerful cabal now extolled a mediocre play on the same subject as his *Phèdre* (1677) by Nicolas PRADON. The temporary success of this inferior work and Racine's official appointment as historiographer to the King combined to make him virtually forsake the theatre, and his last two plays, *Esther* (1689) and *Athalie* (1690), were written for performance by young girls. Racine wrote only one comedy, *Les Plaideurs* (1668); Molière's chief rivals in the field were his brother actors, among them Claude de VILLIERS, Charles CHAMPMESLÉ, and BELLEROCHE, who was the first to immortalize the valet CRISPIN. Guillaume BRÉCOURT also wrote several comedies, of which *La Feinte Mort de Jodelet* (1659) was one of a series intended to display the comic powers of the actor JODELET. The only one to approach Molière, in either quality or quantity, was an actor at the Hôtel de Bourgogne, Noël de HAUTEROCHE. A non-actor, Antoine de MONTFLEURY, was in constant competition with Molière, mainly owing to the pressing need for the company at the Hôtel de Bourgogne to put on a new comedy every time Molière produced one. Finally there was Edmé BOURSAULT, a redoubtable enemy, whose attack on Molière, *Le Portrait du peintre*, was seen at the rival theatre in 1663; he was however to do his best work after Molière's death, with two comedies based on Aesop's fables.

The great age of French drama, which opened with *Le Cid* in 1637, closed with *Phèdre* in 1677. Between lay the comedies of Molière, who in his lifetime saw great changes in the structure and staple fare of the French theatre, many of which he had helped to bring about. He lived through three eras of stage-craft. In his youth he saw, and may have used, the medieval system of MULTIPLE SETTING. This gradually gave way to an unlocalized set scene suitable for the formal acting of French tragedy, usually summed up as *palais à volonté*. The setting of comedy, influenced by the needs of the *commedia dell'arte* company which controlled Molière's first stage, showed a street or square in perspective, with houses grouped round from which the actors issued to transact their most intimate affairs in the open air—domestic quarrels, signing of wills, consultations with doctors, and so on. This fairly simple setting, which employed stylized scenery of painted and folding screens, was taken by Molière to Court for his productions there, but he soon came under the influence of the spectacular effects achieved in opera by Lully, and one of the last things he did was to rebuild the stage of the Palais-Royal to accommodate elaborate scene-shifting and machines. After his death the new stage proved very useful to the wily Lully, who managed to get it assigned to him, under his monopoly of music, for productions of opera.

The change in stage settings is easily illustrated, but it is more difficult to assess the change which had come over acting during the same period. Bellerose and Montdory, the one gentle and somewhat insipid, the other forceful and declamatory, had set a high standard, which was maintained by Floridor and Montfleury. Molière, though never good in tragedy, excelled in comedy; it was a far cry from the rough buffoonery of the early farce-players to his subtle and more natural style of acting, while his discarding of the mask, which he originally wore in imitation of the *commedia dell'arte* (as did the farce players), allowed full play to facial expression. Other actors in his company worthy of note were LA GRANGE, his close friend, editor of his works, and keeper of a register of the theatre which has proved invaluable to later students of the period; DU CROISY, the first Tartuffe; and LA THORILLIÈRE, founder of a large theatrical family. Lesser parts were played by HUBERT who, in the old tradition, was the first to play Mme Jourdain in *Le Bourgeois gentilhomme*; L'ESPY, brother of Jodelet; and BEAUVAL, whose one great success was the part of Thomas Diafoirus in *Le Malade imaginaire*. His wife, however, was an excellent actress worthy to rank with Mlle de Brie, who played the part of Agnès in *L'École des femmes* for nearly 50 years, and Mlle Du Parc, who had the distinction of being admired by Corneille, trained by Molière, and loved by Racine. At the Hôtel de Bourgogne the outstanding actor in tragedy was LA TUILLERIE, who married a daughter of Belleroche and wrote a number of comedies about the valet Crispin for his father-in-law; and Mlle DESOEILLETS, who created Corneille's Sophonisbe and Racine's Andromaque, the role later taken over by Mlle Champmeslé.

Although Paris had by the end of the century attained complete supremacy in theatrical matters, as in all else, provincial companies continued to circulate, and some of their actor-managers achieved a certain reputation. Among them was FILANDRE, who is credited with the early training of Mlle Beauval and possibly of Floridor; DORIMOND, whose career resembles that of Molière in his early years; and ROSIMOND, who eventually went to the Marais and played several of Molière's roles in the combined company run by Molière's widow. This mixture of the players of the Marais with the former companions of Molière was in 1680, as a result of Louis XIV's passion for centralization, merged with the actors at the Hôtel de Bourgogne to form what eventually became the COMÉDIE-FRANÇAISE, still the national theatre of France. The name seems to have arisen in an effort to distinguish the French actors from those of the COMÉDIE-ITALIENNE, who, after intermittent but increasingly lengthy visits to Paris during the past century, had taken possession of the disused Hôtel de Bourgogne, and began to mingle French with their Italian. The Comédie-Française had the sole right of acting and producing plays in Paris and first call upon the services of provincial actors. Its most important members on its foundation were Baron, Belleroche, Hauteroche, Mlle Beauval, Mlle Champmeslé, with her husband, and Molière's widow and her second husband GUÉRIN D'ÉTRICHÉ. A newcomer who was to prove so good in comedy that he was nicknamed 'le petit Molière' was Jean-Baptiste RAISIN, whose brother, sister, and wife were also members of the company. A few years later a second generation of actors was represented by Paul POISSON, son of Belleroche, his wife, daughter of Molière's Du Croisy, and Pierre de LA THORILLIÈRE.

At the one theatre with which they now had to be content, Parisian audiences could see revivals of Molière, Corneille, and Racine, together with some older plays and a host of inferior imitations. Tragedies on classical and exotic subjects continued to be written, often by those whose aptitudes were for some other form of art, but they had begun to lose their hold upon the public. In the same way comedy, while appearing to flourish in new hands, degenerated into a comedy of manners which no longer interested an audience whose own mode of life had undergone some drastic changes. Among the few writers of tragedy who made some small showing are LA CHAPELLE, Racine's immediate successor, and the prolific CAMPISTRON. The playwright whose best tragedy marked the end of the 17th century was LA FOSSE, whose *Manilius Capitolinus* (1698), based to some extent on OTWAY's *Venice Preserv'd* (1682), remained in the repertory after a resounding initial success for many years. His later and less interesting plays were overshadowed by the works of the infant prodigy LA GRANGE-CHANCEL. In the hands of CRÉBILLON tragedy turned towards melodrama, its force and dignity giving way to the horrors of incest and infanticide, its psychological realism to easy intrigue and romance.

Unlike Crébillon, who confined himself entirely to tragedy, La Grange-Chancel and Campistron both produced comedies, but they were isolated and not very successful productions and again the actors were doing the best work in the genre. Baron, famous for his acting in tragedy, nevertheless attempted to keep alive a comedy of character—an attempt frustrated by the concentration of his fellow-actor and brother-in-law DANCOURT on the comedy of manners. Dancourt dramatized, with some skill and much gaiety, the social struggle implicit in the changing conditions of his day; another actor LEGRAND based his own ephemeral comedies on contemporary events, while other writers were catering for the Italians, who were becoming increasingly important in the theatrical life of Paris as their use of French was extended. Permission to use French at all had only been achieved with a struggle, and was strenuously opposed by the Comédie-Française. With the help of Giuseppe BIANCOLELLI, who himself wrote several plays for his companions, the Italians won their case after appealing to Louis XIV, and then looked increasingly to French dramatists to supply them with material. Among their collaborators the most important were DUFRESNY and REGNARD, who both contributed also to the repertory of the Comédie-Française after the King dismissed the Italian troupe in 1697, mainly at the insistence of Mme de Maintenon (who accused them of holding her up to ridicule in *La Fausse Prude* by Étienne LENOBLE) but also perhaps to lighten his budget by withdrawing their large subsidy.

Two authors who collaborated in comedy for the French actors, though one at least wrote for the Italians also, were PALAPRAT and de BRUEYS. The latter, who also wrote several tragedies on his own, refurbished the old farce of Pathelin in a version which, as *L'Avocat Pathelin* (1706), became one of the most popular plays in the repertory of the French theatre. The finest comedy of the time was to come from a novelist, LESAGE, whose early works had been based on Spanish cloak-and-sword plays. His most important contribution to the theatre was undoubtedly *Turcaret* (1709), a scathing attack on the financiers and *nouveaux riches* who embittered the last years of the reign of Louis XIV. Lesser writers of comedy were Boindin, whose *Bal d'Auteuil* (1702) caused a tightening-up of the censorship laws, and HOUDAR DE LA MOTTE, who also made one notable contribution to post-classical tragedy with his *Inès de Castro* (1723). The last outstanding writer of tragedy was, however, VOLTAIRE. His universal genius applied itself to the theatre as to every other aspect of 18th century thought, but his influence in itself contributed to the disintegration of the tragic form; for while his lifelong enemy Crébillon, now censor of plays, kept in however pedestrian a style to the old ways in form and content, Voltaire followed Racine for the form of his plays, but found no difficulty in reconciling that form with the liberal ideas of his age. At the same time his use of spectacular effects, and the intrusion of sentiment and emotion, contributed to the development of the DRAME (or *tragédie bourgeois*) which, with the COMÉDIE LARMOYANTE, constituted the main offering of the 18th century to the evolution of the French theatre. Both were the logical outcome of the influx into the auditorium of the middle class, with its passion for moral teaching expressed through strong emotional situations placed in the recognizable décor of modern society. The germ of *comédie larmoyante*, and so of *drame*, can be found in the plays of DESTOUCHES, who detracted from the vigour of his comedies by insistence on the moral virtues. His best play, *Le Glorieux* (1732), which dramatized once more the struggle between the old nobility and the wealthy parvenu, provided an excellent part for DUFRESNE, one of the few outstanding players of his time. Though good in some parts, BEAUBOUR could not replace Baron, while Mlle Champmeslé had been succeeded, but not equalled, by her niece Mlle DESMARES and her pupil Mlle DUCLOS. The Comédie-Française had to suffer the competition of the unlicensed actors in the illegal theatres of the fairs, now open nearly all the year round. In spite of constant complaint and litigation, these *forains*, as they were called, flourished, and evaded each new law as it was passed with an ingenuity which merely served to enhance their popularity; and by the 1770s their actors had joined with the rejuvenated Comédie-Italienne to form the Opéra-Comique.

The Italians had been recalled to Paris by the Regent in 1716, a year after the accession of the 5-year-old Louis XV, and under the younger RICCOBONI and Pietro BIANCOLELLI, son of Giuseppe, had achieved new triumphs, becoming completely assimilated into French theatrical life. Indeed, the finest plays of MARIVAUX were written for and performed at the Comédie-Italienne, for he found the actors at the Comédie-Française overbearing and narrow-minded, and much preferred the gaiety and freedom of the Italians.

Marivaux, however, was not much appreciated by his contemporaries, and he remains an isolated phenomenon in French dramatic literature, the one great exponent of that fusion of French and Italian art which arose from the acclimatization of the *commedia dell'arte*. Much more to the taste of 18th-century audiences were the 'tearful comedies' of LA CHAUSSÉE, the writer most closely associated with the *comédie larmoyante*; others who essayed it included PIRON, best known for his comedy *La Métromanie* (1738); GRESSET; and even Voltaire with *L'Enfant prodigue* (1736) and *Nanine* (1749). Voltaire, however, also continued to write tragedies in his own manner, and was fortunate in being able to call on a new generation of actors to appear in them. First among them was the lovely Adrienne LECOUVREUR, whose early death was a great loss, although she was replaced by Mlle DUMESNIL, for whom Voltaire wrote *Mérope* (1743), and by Mlle DANGEVILLE, who, with PRÉVILLE, excelled in the plays of Marivaux when they eventually reached the Comédie-Française. The greatest players of the period were Mlle CLAIRON and LEKAIN, who appeared opposite one another in many of Voltaire's plays, and were instrumental in effecting certain reforms in the theatre, notably in the adoption of a modified form of historical costume. Also, under the influence of MARMONTEL in whose *Denys le tyran* she appeared in 1748, Mlle Clairon dropped her declamatory style for a more natural tone, and induced her companions to do likewise. Lekain and Voltaire together were also responsible for the long-overdue clearance of the stage, which had been cluttered up with members of the audience for nearly a century. This encroachment had been tolerated in French classical tragedy, but had proved impossible in the 'machine plays'; matters finally came to a head with the introduction, under the influence of Shakespeare, of crowd scenes and spectacular effects in some of Voltaire's later tragedies, notably *Sémiramis* (1748).

Another actor associated with Voltaire was LA NOUE, director of a provincial company in which Mlle Clairon played for a short time. He later joined the Comédie-Française to play secondary roles, the leaders of the company at the time being GRANDVAL, preferred by some to Lekain, and BELLECOUR, who was excellent in comedy.

Voltaire, though living in exile, was still writing for the French stage, and had just sent *Tancrède* (1760) to the Comédie-Française when DIDEROT saw his first play *Le Père de famille*, written and published two years earlier, in production. Diderot's work was typical of the style to which he himself gave the name *drame*, its characteristics being an exalted, declamatory prose, weighed down with moral sentiments which now appear somewhat bathetic in view of the usual domestic settings. Yet both Diderot and MERCIER introduced into French drama a concern with man in relation to his social surroundings. The plays of SEDAINE, though equally moral, are written with greater simplicity and some humour, a quality noticeably lacking in the work of SAURIN, whose tragedy *Blanche et Guisard* was coldly received. The actors were consoled for their disappointment by the sudden success, on patriotic grounds, of a poor play entitled *Le Siège de Calais* (1766), by de BELLOY. Another dramatist who had some success at this time was LA HARPE, mainly remembered as a critic; and at the same time DUCIS conceived the idea of refashioning the plays of Shakespeare to the taste of the time.

In the welter of tears, piety, blood, sentimentality, and sheer dullness that characterized most of the theatre of the second half of the 18th century only one writer of comedy stands out, BEAUMARCHAIS, though Collin d'HARLEVILLE deserves mention for his charming *Les Châteaux en Espagne* (1789).

In 1770 the Comédie-Française had moved, first

to the Tuileries and then, in 1781, to a new theatre on the present site of the ODÉON. Mlle Clairon had been replaced by two sisters, Mlles DUGAZON and VESTRIS, the latter related by marriage to the great English actress Mme VESTRIS. At this new theatre the first night of Beaumarchais's *Le Mariage de Figaro* took place in 1784, after many delays. Figaro was played by DAZINCOURT and the charming Mlle CONTAT was Suzanne, the cast also including Préville, who had played Figaro in *Le Barbier de Séville* (1775), and MOLÉ. Beaumarchais's play was a brilliant success; attracted to it by curiosity, dazzled by its wit, and blind to its political significance, the audiences that applauded it represented the very people who were later to suffer under the Revolution it presaged. When it finally came, in 1789, it put an end to the monopoly the Comédie-Française had enjoyed for so long in spite of the activities of the unlicensed booth theatres on the BOULEVARD DU TEMPLE, where the veteran showman NICOLET with his performing monkeys was capable of drawing a larger audience than the King's players. Many of the early booths had already been replaced by more permanent structures, the first, the Ambigu-Comique (now the AMBIGU) which opened in 1769, circumventing the monopoly by the use of life-size wooden puppets. This was only one of the many subterfuges invented to flout authority, others being the insertion of songs and dances into straight plays, and a lavish use of dumb-show and even written placards to keep the audience informed of the plot.

Although the actors in general welcomed the freedom conferred on them by the Revolution, it had its opponents, among them some of the members of the Comédie-Française. A number of those who favoured the new régime broke away from their reactionary colleagues and under TALMA, the finest actor of the time, went to the Palais-Royal to open the Théâtre de la République. Those who remained, as the Théâtre de la Nation, soon lost the favour of the public and were nearly all arrested and imprisoned; and it was not until 1803 that Talma was able to return in triumph to the Comédie-Française at the head of a reconstituted company. Meanwhile, with the execution of Louis XVI in 1792 and the rise to power of Robespierre, drama had been used by the authorities as an effective instrument, not in theatres but in mass spectacles in the open air or in such buildings as Notre-Dame, where the 'Feast of Reason' was celebrated in 1793. After Robespierre's fall the liberated theatre, sharing the general revulsion from bloodshed, withdrew from the political arena and offered only easy and escapist fare to a populace sadly in need of escape.

The Revolution produced only one dramatist of note, Marie-Joseph CHÉNIER, though FABRE D'ÉGLANTINE showed promise with his one comedy *Le Philinte de Molière ou la Suite du Misanthrope* (1790). PICARD, founder of the Odéon, provided light amusing plays for his company under the Consulat and Empire, which successfully kept the audience in a good humour while avoiding giving offence to the authorities. For Napoleon, though interested in the theatre and a good friend to Talma, was quick to suppress any manifestations of originality and re-established the censorship in 1804. His friend and admirer LEMERCIER, after an effort to re-establish historical tragedy with a mediocre *Agamemnon* (1797) and comedy with *Pinto* (1800), lacked the force and vitality necessary to battle with the new restrictions and turned away from Napoleon when he saw where ambition was taking him. The chief and most popular genre of the time, apart from operetta, was the VAUDEVILLE, which took over the theatres of the old Boulevard du Temple, the most important, the FUNAMBULES, opening its permanent building in 1816. These theatres provided light entertainment for the vast new audiences thrown up by the Revolution and at the same time served as the echo and mirror of the times. They flourished under the Empire and the Restoration, their authors adapting their subjects to the prevailing ideology with constant backward glances to the seemingly heroic ages of the past; but the greatest delight of the often illiterate crowds who thronged the popular theatres were the melodramas of PIXÉRÉCOURT which, with their direct appeal to the emotions and their spectacular effects, showed virtue triumphing after all the pleasurable excitements of villainy. Meanwhile, the learned and elderly formed a fast-diminishing audience for the rigid tragedies and dull comedies which were all that the Comédie-Française could offer. The time was ripe for a revival of literary drama, and it came with the onslaught of Romanticism. The leader of the new movement was Victor HUGO, its manifesto was set out in the Preface to his *Cromwell*, published in 1827, and its battle-cry was his *Hernani*, whose first night in 1830 was one of the most tempestuous the national theatre had ever seen.

A better dramatist than Hugo was the elder DUMAS, whose *Henri III et sa cour* was produced in 1829. Alfred de VIGNY, author of the poignant *Chatterton* (1835), marked a step further in the acclimatization of Shakespeare in France with his adaptation of *Othello* (1829); the plays of Alfred de MUSSET were also written at this time, but did not reach the stage till much later, after the disastrous reception accorded to his first play *La Nuit vénitienne* in 1830.

The life of the Romantic theatre was short. The public soon wearied of its excesses, and in 1843 Hugo's *Les Burgraves* failed completely, while the calm neo-classicism of PONSARD's *Lucrèce* was acclaimed. The real taste of the time was soon revealed in the success of the WELL-MADE PLAYS of the prolific Eugène SCRIBE, who during the whole of the Romantic period and after continued to turn out an inexhaustible supply of bourgeois comedies, beautifully constructed, optimistic, platitudinous, and no longer acted. His superb though desiccated technique brought him a vast fortune and undisputed pre-eminence in the theatre of his time, as well as a host of imitators of whom the best was Legouvé.

The Comédie-Française had ceased to be the only theatre whose actors were worthy of notice. Great reputations could be made elsewhere, as witness the careers of such players as FRÉDÉRICK, the first Ruy Blas, DEBURAU, the beloved Pierrot of the Funambules, Mlle DORVAL, star of the PORTE-SAINT-MARTIN, and many other favourites of the Parisian public. Even the members of the Comédie-Française itself were not averse to seeking remunerative employment elsewhere: SAMSON signed a lucrative contract with another theatre which he was not, however, allowed to implement, and RACHEL, who had succeeded Mlle MARS as the leading lady of the Comédie-Française, was more often in the provinces or abroad than in Paris. Yet her influence was considerable, and she revived the fine tradition of French classic acting which had almost been lost sight of in the upheavals of the Revolution and the vagaries of Romanticism. She could not, however, sustain alone the burden of a classic repertory, and received little support from her comrades, many of whom were jealous of her prestige, and the Comédie-Française was, for almost the first time in its history, glad to take over plays which had proved popular elsewhere.

The Revolution of 1848 brought new preoccupations to the theatre, and the comedies of Scribe were replaced by a more serious type of social drama, though it was long before French dramatists broke away from the traditions of the 'well-made play' in matters of form. The first exponent of the new drama was AUGIER, who had been associated with Ponsard in the short-lived classical reaction against Romanticism. The younger DUMAS too chose contemporary themes, and accustomed Parisian audiences to discussions of social and ethical problems in the theatre, thus creating a favourable climate for the later exponents of social drama. These efforts to reform society by showing morality in action in the theatre were seconded rather faintly by Théodore Barrière (1823–77), while farcical comedy was kept alive by LABICHE, who set out frankly to amuse and succeeded so well that he was compared by his grateful contemporaries to Molière. But the spirit of the Second Empire found its truest expression in light opera and in the musical burlesques of MEILHAC and his friend Halévy, whose collaboration gave theatrical expression to the trivial, corrupt, but amusing world which came to an end in 1870.

The work of Scribe had been continued by SARDOU, who was at home in all types of play—comedy of contemporary manners, historical romance, political satire, social drama. He must be credited with having provided some excellent parts for Sarah BERNHARDT who, like Rachel, was a member of the Comédie-Française, but had her greatest triumphs elsewhere. Indeed, much of the history of the French theatre from this time onwards must be looked for outside the national theatre, which tended to lag further and further behind the times, in spite of the excellence of such actors as GÔT and the elder COQUELIN, who appeared together in an amusing comedy by Édouard Pailleron (1834–90), *Le Monde où l'on s'ennuie* (1881), and were responsible for the first visit of the Comédie-Française to London; of MOUNET-SULLY, the greatest tragic actor of his day; and of WORMS and the younger Coquelin, and many excellent actresses. From time to time the venerable institution suffered the shocks of modernization, and its most outspoken critics sometimes became its directors, but in the main it served mostly as a repository of tradition.

Like many others, the fine *comédienne* RÉJANE carried the glories of French art and acting all over Europe and North America without ever appearing at the Comédie-Française. Theatrical life was now flowing into the experimental theatres of the end of the century. The impulse towards a new realism came first from the novelists BALZAC, Flaubert, and the GONCOURTS, and ZOLA took their ideas into the theatre, rejecting with equal fierceness the problem plays of Augier and the younger Dumas and the well-conducted intrigues of Scribe and Sardou, substituting for their artifices the naturalistic ideal of a 'slice of life'. Zola's chief follower was BECQUE, whose plays were uncompromisingly naturalistic; the movement was strengthened by the efforts of ANTOINE, who in 1887 founded the THÉÂTRE LIBRE where he staged plays both by contemporary foreign dramatists and by his own countrymen, notably ANCEY, MIRBEAU, ROMAINS, and BRIEUX. Among other exponents of NATURALISM in France, though in a somewhat diluted form, were François de Curel (1854–1928), Paul Hervieu (1857–1915), a writer in the tradition of the younger Dumas, and LAVEDAN.

In the inevitable reaction against naturalism the leading figures to emerge were the Belgian MAETERLINCK, whose work was staged by Paul FORT and LUGNÉ-POË; and ROSTAND, whose *Cyrano de Bergerac* (1898) is still in the repertory. Defying any rigid classification was CLAUDEL, whose works bore traces of symbolist influence but were deeply impregnated with his fervent Catholicism. In 1894 the production of *Amoureuse* by PORTO-RICHE set a new fashion for 'daring' plays on sexual themes, of which other exponents were Henri Bataille (1872–1922) and BERNSTEIN, while a more light-hearted approach to the subject is to be found in the comedies of Tristan BERNARD. Two outstanding masters of French farce at this time were FEYDEAU, whose work enjoyed a tremendous popular revival in the years following the Second World War, and COURTELINE, whose first plays were staged by Antoine. Lucien GUITRY and his son Sacha continued the tradition of many French actors by appearing in their own light comedies, now forgotten.

In 1913 Jacques COPEAU started an experimental theatre known as the VIEUX-COLOMBIER, where important work was done in introducing new authors, among them VILDRAC, and with new productions of established authors. In the previous year Lugné-Poë had returned to the Théâtre de l'Oeuvre and, after the cataclysmic interruption of the First World War, these two theatres provided a thread of continuity into the interwar period. In 1929 disciples of Copeau led by Michel SAINT-DENIS formed the COMPAGNIE DES QUINZE, which was to be intimately associated with the work of André OBEY.

A very different response to the upheaval of the war years was expressed in the movement towards Surrealism, anticipated in the work of JARRY. Given formal definition by André Breton, it found its chief theatrical exponents in the visionary ARTAUD and VITRAC, while the work of COCTEAU, though not avowedly indebted to the movement, displayed more fully the force that had been generated by Jarry's seminal *Ubu-roi* (1896). The actor Firmin GÉMIER, who played the original Ubu, was destined, like Artaud and Jarry, to have a more enduring influence after the Second World War, when the THÉÂTRE NATIONAL POPULAIRE, which he directed in its first manifestation from 1920 to 1934, was at last to find a permanent home. Gémier also managed the Odéon from 1921 to 1928, introducing there the work of SALACROU. It was, however, the CARTEL formed in 1927 by BATY, DULLIN, JOUVET, and PITOËFF which represented the mainstream of the AVANT-GARDE, in the studio theatres which were increasingly recognized as a major third force to be added to the Comédie-Française and the Odéon, on the one hand, and the popular productions of BOULEVARD PLAYS on the other. Dullin had settled at the ATELIER in 1922, creating there an impressive classical repertory and introducing to the public the work of ACHARD, while Jouvet had gone in the same year to the CHAMPS-ÉLYSÉES under Baty, who himself went in 1930 to the Montparnasse, giving it his own name. Pitoëff did his best work after 1925 at the Arts (now named after the active and experimental impresario HÉBERTOT) and the MATHURINS, including in his programme plays by Claudel, Cocteau, and the newcomers LENORMAND and ANOUILH.

The partnership of Jouvet with the dramatist GIRAUDOUX was of immense importance to the work of both. Giraudoux, with Anouilh, had introduced a type of serious drama that would (at least potentially) appeal also to boulevard audiences, but many conventional plays for such audiences continued to be written by such authors as Jean-Jacques BERNARD, SARMENT, and VERNEUIL. Henri GHÉON, with his religious plays, remained outside the main stream, as did the anachronistic but individual RAYNAL, and Gabriel MARCEL, whose concern with EXISTENTIALISM was ahead of its time and would find more understanding audiences after the Second World War.

The war years, 1939 to 1945, were indeed to see a polarization between the Christian standpoint of Claudel (who at last achieved recognition in productions at the Comédie-Française under BARRAULT) and of Henri de MONTHERLANT, and the existentialist philosophy of Jean-Paul SARTRE, the one outstanding talent to emerge during the war. Meanwhile Anouilh put on under the German occupation in 1944 an *Antigone* which provided a rallying-point for the aspirations of insurgent youth, and followed it with a steady stream of successful plays, all brilliantly stageworthy but increasingly revealing a certain shallowness of inspiration. His only rival in popularity after the Second World War was André Roussin, a prolific author of light comedies; but the most important figure of the immediate post-war period was undoubtedly the actor-director Barrault, who with his wife Madeleine RENAUD from 1946 to 1956 at the Théâtre MARIGNY staged a memorable series of productions, carrying on the traditions of the Cartel not only in his intelligent choice of classical and modern plays, but in his boldly experimental use of staging and lighting. Even Barrault, with his excellent company to which the fine actress Edwige FEUILLÈRE belonged in the late 1940s, was unable to stimulate any real creative surge in playwriting, and most of his best work was done in classical revivals or in works by writers already established, some, like CAMUS and GIDE, being already known as novelists. Other writers of note in the early post-war period were AUDIBERTI, the American-born Julien GREEN, and, in a lighter vein, Marcel Aymé.

The Comédie-Française, which had been administered from 1915 to 1936 by the minor dramatist Émile Fabre, had received an infusion of fresh blood in 1936, when (although another writer, Édouard BOURDET, rather than the widely canvassed Copeau, succeeded Fabre) Dullin, Jouvet, Baty, and Copeau were at least invited to direct. But in spite of the excellence of such players as Marie BELL during the war years, the company went into a decline and apart from the work of the Renaud-Barrault company the most notable acting of the period was seen at the Atelier under BARSACQ's direction; in the company formed by Sacha PITOËFF in 1949; in the work of the mime-actor Marcel MARCEAU; and in the work of two clowns, Les GRENIER-HUSSENOT, whose interest in popular forms of entertainment was to be reflected in a more general concern to widen the appeal of the theatre.

Thus in 1951 the actor-director Jean VILAR was appointed to the directorship of the second Théâtre National Populaire, Daniel SORANO and Gérard PHILIPE being among its members, and soon turned it into one of the best companies in Paris, supplementing Barrault's work at the Marigny. Vilar tapped on entirely new, young, and popular audience, staging his productions with startling simplicity. He resigned in 1963, and after a period under Georges WILSON the T.N.P. was transferred to Villeurbanne under Roger PLANCHON. This move would hardly have been possible before the implementation of the post-war policy of DÉCENTRALISATION DRAMATIQUE, intended to reverse the situation which had followed on the disappearance of the provincial theatres and old touring system and the centralization of all theatrical activity in Paris. Between 1946 and 1961 seven *Centres Dramatiques*, all run with state and local financial aid with the intention of providing genuinely regional theatres, were formed, and permanent companies committed to touring were instituted, as were the more broadly based *Maisons de la Culture*. Vilar's annual festival at AVIGNON was another important regional contribution to the theatre, while the THÉÂTRE DES NATIONS in Paris helped to widen the audiences' international horizons.

In 1959, when André Malraux as Minister of Culture announced the reorganization of the national theatres, the two houses of the Comédie-Française were separated and the Odéon was given to Barrault. The Comédie-Française's subsidy was cut by approximately a quarter, this sum being awarded to Barrault, and Vilar was given a second home, the Théâtre Récamier. As his subsidy was not increased he was forced to abandon this second theatre within 18 months, but not before he had mounted notable productions of plays by Armand GATTI, Boris VIAN, and René de OBALDIA. Both Vian and Obaldia showed some debt to the movement which can be linked to the earlier existentialist drama by Camus's coinage of the concept of the ABSURD, which is most often used to describe it. Samuel BECKETT (whose best known plays were all directed by Roger BLIN) and Eugene IONESCO were the earliest writers to achieve international reputations for work in the absurdist idiom, though the imprecision of the label may be gauged from the number and variety of writers to whom it has been applied, from Vian to ADAMOV, ARRABAL, and Marguerite DURAS. Even the work of Jean GENET has been so described, though it reflects more truly the Artaudian impulse which was to find a belated but widespread expression chiefly outside France.

In the continuing but somewhat atrophied boulevard style, meanwhile, some success had been achieved by Félicien MARCEAU, while BILLETDOUX sustained the tradition of Giraudoux. The plays of Gatti, Planchon, and Kateb YACINE showed a new, stylistically fresh attempt to come to terms with the reality of post-imperial France, as did those of Aimé CÉSAIRE from the point of view of the 'exploited'; but the events of May 1968, the political expression of that reality, contributed rather to the impulse towards COLLECTIVE CREATION, which had already been shown in the activities of such experimental groups as the Grand Magic Circus, the Théâtre du Chêne Noir, and most notably in the Théâtre du SOLEIL.

The enduring strength of the actor-director in the French theatre was demonstrated by the response of Barrault to his expulsion from the Odéon following its take-over by students in 1968—a burst of creative energy in a new direction. Other directors who made and sustained reputations during the 1960s, often in the regional centres, include Marcel MARÉCHAL and Antoine VITEZ, while Victor GARCIA has stretched the director's influence over a given text to the utmost.

FRANCE, Théâtre de, Paris, see ODÉON.

FRANK, BRUNO (1887–1946), German dramatist, who in reaction from the extreme drama of REALISM and EXPRESSIONISM wrote a number of light, sophisticated comedies. *Sturm im Wasserglass* (1930) was successful in England in an adaptation by James BRIDIE as *Storm in a Teacup* (1936) and in America as *Storm over Patsy* (1937). Frank also wrote some historical dramas, of which the best and most characteristic is *Zwölftausend* (1927) which, as *Twelve Thousand*, was successful in New York in 1928 and was seen at the EMBASSY THEATRE, London, in 1931.

FRANKFURT-AM-MAIN, a German city whose Easter and Autumn FAIRS made it attractive to early touring companies, the ENGLISH COMEDIANS playing there many times between 1592 and 1658. It had, however, an undistinguished theatrical history until Carl Zeiss made it a stronghold of EXPRESSIONISM with the production at the Schauspiel in 1917 of Kornfeld's *Die Verführung* (*The Seduction*), followed by UNRUH's *Ein Geschlecht* (*A Generation*, 1918), STERNHEIM's *1913* (1919), and BRONNEN's *Vatermord* (*Parricide*, 1922). In 1919 Weichert directed the first performance of HASENCLEVER's *Antigone* and with first productions at the Neues Theater of plays by KAISER—*Die Bürger von Calais* (1917), *Gas I* and *II* (1918/1920), and *Hölle Weg Erd* (*Hell, Path, Earth*, 1919)—Frankfurt became a serious rival to BERLIN. Between 1951 and 1968 productions of a number of BRECHT's plays when they were mainly being ignored by other West German theatres gave Frankfurt back part of its earlier importance, and in the 1970s there were experiments in COLLECTIVE CREATION under PALITZSCH and Rainer Werner Fassbinder.

FRANKLIN THEATRE, New York, at 175 Chatham Street, a small playhouse seating 550, which opened on 7 Sept. 1835 with Thomas MORTON's *The School of Reform*, followed by some classic comedies. It soon went over to MELODRAMA and FARCE, in spite of a visit by Junius Brutus BOOTH in 1836–7 in several tragedies from his repertoire. The Franklin is memorable as being the first theatre in New York at which Joseph JEFFERSON (the third) appeared in 1837, at the age of eight. Successive managers failed to make it profitable over the next three years, partly because it suffered from the competition offered by the neighbouring CHATHAM THEATRE (2), but in 1841 the company from the PARK THEATRE, which was up for sale following the death of its manager

Stephen PRICE, gave a successful season. The same year saw the first production in New York of the anonymous *Fifteen Years of a New York Fireman's Life*, followed in 1842 by BULWER-LYTTON's *Money*. The theatre then gave up legitimate drama and under various names became the home of PANTOMIME and MINSTREL SHOWS. It was closed by the authorities in 1854 after a scandal involving some *tableaux vivants*.

FRANZ, ELLEN, see MEININGER COMPANY.

FRASER, CLAUDE LOVAT (1890–1921), English artist and stage-designer, whose designs for Nigel PLAYFAIR's productions of *As You Like It* and, particularly, GAY's *The Beggar's Opera* at the LYRIC THEATRE, Hammersmith, in 1920 inaugurated a new era in stage design. He took his inspiration from the 18th and early 19th centuries, and his work embodied a gay, brightly-coloured romanticism. His influence, considering the brevity of his career, was phenomenal.

FRAYN, MICHAEL (1933–), English playwright and novelist, who was already well known for his novels when his collection of four short plays *The Two of Us* (1970) was produced in London. His first full-length play *The Sandboy* was seen at the GREENWICH THEATRE in the following year, and he achieved his first outstanding success with *Alphabetical Order* (1975), set in the cuttings library of a provincial newspaper. *Donkeys' Years* (1976) dealt with the reunion after 20 years of six former university students and *Clouds* (also 1976), starring Tom COURTENAY, was set in Cuba, its two main characters being rival male and female press reporters. After the comparative failure of *Balmoral* (1978), which as *Liberty Hall* was seen at Greenwich in 1979 but did not reach the West End, he was again successful with *Make and Break* (1980), about a work-obsessed salesman at a foreign trade fair and, like his other plays, a comedy with serious undertones. His translations of CHEKHOV's *The Cherry Orchard* and TOLSTOY's *The Fruits of Enlightenment* were produced at the NATIONAL THEATRE in 1978 and 1979 respectively.

FRAZEE THEATRE, New York, see WALLACK'S THEATRE (2).

FRÉDÉRICK [Antoine-Louis-Prosper Lemaître] (1800–76), French actor, who embodied all the glory and excesses of the Romantic drama, many of whose heroes he created. He never appeared at the COMÉDIE-FRANÇAISE, which was still clinging to the last vestiges of the classical tradition, but spent most of his career in the theatres on the BOULEVARD DU TEMPLE, having made his first appearance at the Variétés-Amusantes as the lion in a pantomime, *Pyrame et Thisbe*, at the age of 15. He then went to the FUNAMBULES, where he was a contemporary of the great mime DEBURAU, and also attended classes at the Conservatoire. In 1820 he became a member of the company at the ODÉON but, finding the audiences apathetic and opportunities for sustained acting too few, returned thankfully to the popular stage, replacing Fresnoy at the AMBIGU-COMIQUE where he made his first appearance as Robert Macaire in *L'Auberge des Adrets* (1823), a part ever after associated with him. The play had been written as a serious melodrama, but Frédérick, sensing that the public was beginning to want something different, carried it to success by burlesquing it. He was equally successful in a sequel, *Robert Macaire* (1834), much of which he wrote himself and which he took to London in 1835. Like BEAUMARCHAIS's *Le Mariage de Figaro*, *Robert Macaire* had political repercussions; it was said to have contributed not a little to the downfall of Louis-Philippe. At the PORTE-SAINT-MARTIN in 1827, Frédérick had made a tremendous impression in *Trente Ans, ou la Vie d'un joueur*, a play on the evils of gambling which he took to London on his first visit in 1828. Among his later successes were Othello, in DUCIS's translation, and several of the leading roles in the plays of the elder DUMAS, notably the title role in *Kean, ou Désordre et génie* (1836), which was specially written for him, as was BALZAC's *Vautrin* (1840). When Dumas and Victor HUGO took over the old Théâtre Ventadour and renamed it the Théâtre de la RENAISSANCE, Frédérick opened it in 1838 as the home of Romantic drama with an electrifying performance in Hugo's *Ruy Blas*, a play about a lackey who falls in love with a queen, which gave great offence to Queen Victoria when she saw it in London in 1852, though she had been one of Frédérick's greatest admirers on his visits in 1845 and 1847. His last years were unhappy, for the taste of the public changed and he could find only trivial or unsuitable plays to appear in; but he left his mark on the theatre, and when BATEMAN first saw IRVING in Leopold Lewis's *The Bells* (1871), he could bestow no higher praise on him than to say that his acting was equal to that of Frédérick.

FREDRO, ALEKSANDER (1793–1876), Polish poet and playwright, one of the few who remained in Poland and saw his plays produced there. The son of an aristocratic family, he served in Napoleon's Grand Army from 1809 to 1814, and then settled down on his family estates. His comedies, written in strict classic form and impeccable verse, deal with the life of the country gentleman as he knew it or as he had heard of it from his elders. Among them the best are *Mr Moneybags* (1821), the story of a social climber which somewhat resembles MOLIÈRE's *Le Bourgeois gentilhomme*; *Husband and Wife* (1822), seen in London in 1957 in a production by the Polish National Theatre company; *Ladies and Hussars* (1825), an excellent farce about the invasion of officers' quarters by a group of determined women, seen in translation in New York in 1925; *Maidens' Vows* (1833), splendidly revived in Warsaw in 1963; *Vengeance*

(1834), which a company from Cracow took to the THÉÂTRE DES NATIONS in Paris in 1955; and *The Life Annuity* (1835), again a comedy in the style of Molière about a usurer's troubles with a young rake. This was seen in London during the WORLD THEATRE SEASON of 1964 in a production by the Contemporary Theatre from Warsaw. After a silence of some years, due to unjustified attacks on him by the young Romantics of the time, Fredro resumed writing, but none of his last dozen or so plays approach the excellence of his early comedies and they were not staged in his lifetime, though several, including *The Candle Has Gone Out*, first seen in the 1880s, have been revived recently.

FREEAR, LOUIE, see DROLL and DUKE OF YORK'S THEATRE.

FREEDLEY, GEORGE REYNOLDS (1904–67), American theatre historian, founder and director of the Theatre Collection of the New York Public Library, now housed in the Lincoln Center for the Performing Arts. A graduate of George Pierce BAKER's '47 Workshop' at Yale, Freedley was for a short time an actor and stage manager on Broadway, where he first appeared in 1928, being associated a year later with the THEATRE GUILD. In 1931 he joined the staff of the New York Public Library to administer the theatre section based on the recently presented David BELASCO Collection. He was also one of the founders of the Theatre Library Association in 1939 and was closely associated with the running of the AMERICAN NATIONAL THEATRE AND ACADEMY. From 1938 until his death he was drama critic and drama feature writer of the New York *Morning Telegraph*, and lectured and wrote extensively on theatre history.

FREEHOLD, The, see LA MAMA EXPERIMENTAL THEATRE CLUB.

FREIE BÜHNE [Free Stage], private theatre club founded in BERLIN in 1889 by ten writers and critics under Otto BRAHM, for the production of plays by the new writers of the school of NATURALISM, on the lines of ANTOINE'S THÉÂTRE LIBRE in Paris. It had no home of its own, and played only matinées in different theatres. Its first production was IBSEN'S *Ghosts*, and in 1890 it produced HOLZ'S *Die Familie Selicke*, a sordid picture of lower middle-class life played in a realistic manner against an equally realistic background. The greatest achievement of the Freie Bühne was the discovery of Gerhard HAUPTMANN, whose first play *Vor Sonnenaufgang* was produced on 20 Oct. 1889. Several more of his plays, up to *Die Weber* (1893), also had their first performances here. Among other authors staged were BJØRNSON, TOLSTOY, ZOLA, STRINDBERG, and ANZENGRÜBER. To help publicize its work the Freie Bühne started in 1890 a periodical under its own name (changed later to *Die neue Rundschau*), which remains one of the leading German literary reviews. Along with the promotion of contemporary drama, the Freie Bühne developed a new, realistic style of ensemble playing, with such actors as Emanuel Reicher, Rudolf Rittner, and Agnes SORMA. When in 1894 Brahm was appointed director of the DEUTSCHES THEATER, the Freie Bühne was attached to it as an experimental theatre. It continued to present new plays up to the turn of the century, but had by then fulfilled its mission as a platform for naturalistic drama.

FREIE VOLKSBÜHNE, Berlin, see VOLKSBÜHNE.

FRENCH BRACE, FRENCH FLAT, see FLAT.

FRESNAY [Laudenbach], PIERRE (1897–1975), French actor, trained at the Conservatoire, who made his début at the COMÉDIE-FRANÇAISE in 1915 in MARIVAUX'S *Le Jeu de l'amour et du hasard*. He remained with the company until 1927 but thereafter played at a number of theatres, not only in Paris but in London, where in 1934 he took over Noël COWARD'S role in the first production of *Conversation Piece*, playing for the first time opposite his wife Yvonne PRINTEMPS. With her he took over in 1937 the management of the Théâtre de la Michodière, where they appeared together in a long succession of light comedies.

FRESNEL SPOT, see LIGHTING.

FREYTAG, GUSTAV (1816–95), German writer who, although best known as the author of sociological and historical novels, began his literary career as a dramatist, and became the German exponent of the WELL-MADE PLAY. His best work was a comedy, *Die Journalisten* (1852), a good-humoured portrayal of party politics in a small town during an election; his attempts at serious problem-plays and a historical tragedy in verse are less attractive. In 1863 he published his *Technik des Dramas*, with its famous pyramid, or diagrammatic plot, of a 'well-made' play.

FRIDOLIN, see GÉLINAS, GRATIEN.

FRIEL, BRIAN (1929–), Ulster dramatist, whose favourite theme is the interaction between various kinds of institutional failure—of government, Church, class, and family—and that of the individual personality. His play *This Doubtful Paradise* was produced by the ULSTER GROUP THEATRE, Belfast, in 1959; *The Enemy Within* (1962) and *The Blind Mice* (1963) were both produced in Dublin, but his first outstanding success was *Philadelphia, Here I Come!* (Dublin, 1964; N.Y., 1966; London, 1967), in which a despairing young Irishman contemplating emigration to America seeks some sign of regret from an unresponsive father. It was followed by *The Loves of Cass McGuire* (N.Y., 1966; Dublin, 1967); *Lovers*

(Dublin, 1967; N.Y., 1968; London, 1969), consisting of two plays *Winners* and *Losers*; *Crystal and Fox* (Dublin, 1968; N.Y., 1973); and *The Mundy Scheme* (Dublin and N.Y., 1969). After *The Gentle Island* (Dublin, 1971) came *The Freedom of the City* (London and Dublin, 1973; N.Y., with Kate REID, 1974), in which three innocent people are shot dead during a Londonderry demonstration. *Volunteers* and *Living Quarters* were seen in Dublin in 1975 and 1977 respectively, while *Faith Healer* was produced in New York in 1979 before being seen in Dublin in 1980 and London, with Helen MIRREN, in 1981. *Aristocrats* (Dublin, 1979) was about the decline of a family of Catholic gentry, and in the following year his most highly praised play *Translations* was premièred in Londonderry. Produced in New York in 1981, it was seen in the same year in London, first at the HAMPSTEAD THEATRE and later at the NATIONAL THEATRE. Set in a 'hedge school' in Donegal in 1833, it tells with humour and compassion of the resentment caused by the arrival of a British Army unit to make the first maps, the rationalization of Celtic place names into English symbolizing the cultural rape of Ireland. Friel's translation of CHEKHOV's *Three Sisters* was staged in Londonderry in 1981.

FRINGE THEATRE in Britain, like off-off-Broadway in America, offers a platform for productions that, for example because of their political content or experimental nature, are unlikely to succeed in more conventional surroundings. The term dates from the late 1960s and probably derives from the activities on the 'fringe' of the EDINBURGH FESTIVAL. There are well over 50 fringe theatres in London, mostly away from the centre where rents are lower. They are usually small, their seating ranging from 40 to 200, and few were built as theatres: mostly they are found in converted warehouses or factories, in basements, or in rooms in public houses. Performances are usually given in the evenings and sometimes at lunchtime, though lunchtime theatres were more common in the early 1970s. Fringe theatres are less formal, less expensive, and usually less comfortable than ordinary theatres, and their audiences tend to be younger and more anti-Establishment. Being always short of money and often housed in temporary accommodation (see ARTS LABORATORY), they tend to be transitory, but some, such as the King's Head at Islington, the Soho Poly, and the Orange Tree at Richmond (where many of James SAUNDERS's plays have been presented), have achieved a certain stability. Other well established fringe theatres are the Bush at Shepherds Bush in West London, which first presented Tom Kempinski's *Duet for One* (1980), later to have a West End run, and the Half Moon at Tower Hamlets in East London, whose revival of the Rodgers and Hart musical *Pal Joey* in 1980 was also a West End success. Outside London fringe theatre is housed in arts centres, university and college theatres, and the studio theatres attached to many CIVIC THEATRES, as well as in more makeshift quarters. About 70 fringe groups, including Foco Novo, Shared Experience, Belt and Braces, and Hull Truck, tour these locations. Some, like the PIP SIMMONS THEATRE GROUP, Joint Stock, and the 7:84 Theatre Companies, work mainly by COLLECTIVE CREATION, and some of the others use it at times.

The fringe has produced a number of interesting playwrights, though their passionate left-wing stance often leads to distortion and caricature. Notable among them are Howard BARKER; Howard BRENTON; David HARE; and John McGrath (1935–), author of such plays as *Events While Guarding the Bofors Gun* (1966), *Soft Or A Girl* (1971), *The Cheviot, the Stag, and the Black, Black Oil* (1973), and *Yobbo Nowt* (1975). He is also the founder of the 7:84 Theatre Companies in England (1971) and Scotland (1973) for which several of his plays were written. (See also COMMUNITY THEATRE.)

FRISCH, MAX RUDOLF (1911–), Swiss playwright, a disciple of BRECHT, whose influence is apparent in the early plays, *Nun singen sie wieder* (*Now They Sing Again*, 1945), *Die chinesische Mauer* (*The Great Wall of China*, 1946), *Als der Krieg zu Ende war* (*When the War was Over*, 1949), and *Graf Öderland* (1951). In complete contrast to these moral dramas are *Santa Cruz* (1946), a blend of dream and reality, and *Don Juan, oder Die Liebe zur Geometrie* (1953), in which Frisch gives a new twist to the old legend, since DON JUAN's only interest in life is geometry. He has no time for women, so they are attracted to him and in the end he finds that his only way of escape is through marriage. In 1958 Frisch produced a short play, *Biedermann und die Brandstifter*, which, as *The Fire Raisers*, had a successful run at the ROYAL COURT in London in 1961, and as *The Firebugs* was seen in New York in 1963. Frisch's next play, *Andorra* (1961), the story of a non-Jewish illegitimate child brought up as a Jew, who is murdered by an anti-Semitic invader, had a resounding success. It was seen in New York in 1963, and was twice performed in London in 1964, in German by the Schillertheater company at the ALDWYCH during the WORLD THEATRE SEASON, and in English by the NATIONAL THEATRE company. *Biografie* (1967), seen in New York as *Biography: a Game* (1979), is a play on one of Frisch's recurrent themes, the problem of identity. The world première of his *Triptychon* was directed by AXER in 1980.

FRITELLINO, see CECCHINI, P. M.

FROHMAN, CHARLES (1860–1915), American manager, who like his brothers Daniel and Gustave (below) inherited a passion for the theatre. From selling souvenirs and programmes he graduated to various managerial posts. He first visited

England in 1880 as business manager of Haverley's Minstrels, for whom he achieved Royal patronage; back in America he became a manager, a theatrical agent, and a manager of touring companies. His first great success came in 1888–9 with Bronson HOWARD's *Shenandoah*: this laid the foundations of his fortune, since he had shrewdly bought the American rights. In 1893 he opened the EMPIRE THEATRE with a fine STOCK COMPANY which he ran for many years and he soon had a controlling interest in many other theatres, in New York and elsewhere. In 1896 he had his first London success with KLEIN's *A Night Out* at the VAUDEVILLE and then joined forces with the Gattis, with George EDWARDES, and with BARRIE, taking Edwardes's musical comedies and Barrie's plays to America. He took a long lease of the DUKE OF YORK'S THEATRE and mounted many notable productions there, including Barrie's *Peter Pan* (1904). At one time he controlled five London theatres, although he failed to establish a repertory system at the Duke of York's in 1910. Frohman died when the *Lusitania* was torpedoed in 1915: a memorial to him stands in the churchyard at Great Marlow, Bucks. A small, odd-looking man, charming and kindly, who never used a written contract, he left a name for fair dealing and a record of remarkable theatrical achievements.

FROHMAN, DANIEL (1851–1940), American manager, who started work as a journalist but in 1880 became business manager of the MADISON SQUARE THEATRE. He went into independent management in 1885 with his brother Gustave (1855–1930) at the LYCEUM THEATRE; here he brought together an excellent STOCK COMPANY and was responsible for a number of outstanding productions, including plays by PINERO and H. A. JONES. E. H. SOTHERN scored his earliest successes under Daniel Frohman's management. When the old Lyceum closed in 1902 Frohman became owner of the new theatre bearing the same name and he was also manager of DALY's New York theatre from 1899 to 1903. On the death of his youngest brother Charles (above) he took over the administration of his affairs in America. He published two volumes of reminiscences, *Memoirs of a Manager* (1911) and *Daniel Frohman Presents* (1935).

FRONT OF HOUSE, all those parts of a theatre which are used by the audience, as distinct from the performers, who are BACK STAGE. They include the auditorium, passages, lobbies, foyers, bars, cloakrooms, refreshment-rooms, and the box-office or pay-box for the booking of seats, the whole being under the control of a front-of-house manager who also acts as host to important guests and is at all times concerned with the comfort and well-being of the audience.

Front-of-House (FOH) LIGHTING is any stage-lighting equipment which is placed on the auditorium side of the PROSCENIUM arch. Its purpose was originally to illuminate the FORESTAGE or far down-stage area by means of carbon arcs and, later, spotlights mounted on the front of the balcony or gallery. Modern theatre buildings have in addition provision for spotlights suspended from the ceiling of the auditorium, and many have permanent lighting positions in the walls as well as a lighting bridge in the ceiling outside the proscenium line.

FRY [Harris], CHRISTOPHER (1907–), English dramatist, who took his stage name from his grandmother's maiden name. Originally a schoolmaster, he gained his early experience of the theatre in Oxford, and had already written a couple of pageants, a religious play about St Cuthbert entitled *The Boy with a Cart* (1937), and, for Tewkesbury, *The Tower* (1939), when he first attracted attention with *A Phoenix Too Frequent* (1946), a one-act *jeu d'esprit*, based on Petronius's story of the widow of Ephesus. In 1948 *The Firstborn* (on the early life of Moses) was seen at the EDINBURGH FESTIVAL, *Thor, With Angels* at Canterbury, and *The Lady's Not For Burning* at the ARTS THEATRE in London, with Alec CLUNES as Thomas Mendip. This last (the 'spring' play in what was planned as a tetralogy of the seasons), transferred to the GLOBE, with GIELGUD as Thomas and Pamela BROWN as Jennet Jourdemayne, consolidated Fry's reputation, and seemed to herald a renaissance of poetic drama on the English stage. It was followed in 1950 by *Venus Observed*, the 'autumn' play of the tetralogy, which starred Laurence OLIVIER, and a translation of ANOUILH's *L'Invitation au Château* as *Ring Round the Moon*, in which Paul SCOFIELD played the dual role of the hero and the villain. Fry then reverted to his earlier biblical vein with *A Sleep of Prisoners* (1951), planned for performance in a church and first seen at St Thomas's, Regent Street, and saw the 'winter' play of his tetralogy, *The Dark Is Light Enough* (1954), produced at the ALDWYCH. In spite of the superb acting of Edith EVANS as the Countess Rosmarin, this did not have the success hoped for, and for some time Fry busied himself with translations, among them Anouilh's *L'Alouette* (on Joan of Arc) as *The Lark* and GIRAUDOUX's *La Guerre de Troie n'aura pas lieu* as *Tiger at the Gates* (both 1955), the latter, with Michael REDGRAVE as Hector, having long runs in London and New York. A translation of Giraudoux's *Pour Lucrèce* as *Duel of Angels* (1958) was also seen in New York in 1960.

It was by now obvious that the trend of the English theatre was away from poetry towards what was known as 'kitchen-sink' realism, and Fry's next play *Curtmantle* (on Henry II and Becket) was first produced in the Netherlands, at the opening in 1961 of the Stadsschouwburg in Tilburg. It was seen in London a year later, but had only a qualified success, as did the 'summer' play *A Yard of Sun* (1970), first produced at the

NOTTINGHAM PLAYHOUSE and seen at the OLD VIC for a week during the company's visit there. Two further translations, of IBSEN's *Peer Gynt* and ROSTAND's *Cyrano de Bergerac*, were performed at the CHICHESTER FESTIVAL THEATRE in 1970 and 1975 respectively: both employed Fry's characteristic form of free but regularly stressed verse with richly imaginative word-play.

FUGARD, ATHOL (1932–), South African playwright, director, and actor, born in the remote, arid Karroo district, of Anglo-Irish and Afrikaner parents. He studied philosophy and social anthropology at the University of Cape Town. His first play, *No Good Friday* (1959), was closely followed by *Ngogo* (translatable as 'a 35-cent woman'), the story of a mine-worker's whore. *The Blood Knot* (1961; London, 1963; New York, 1964, with James Earl JONES) tackled the bitter topic of the temptation to pass for white in a racially-segregated society.

In 1963 Fugard and his wife, the actress Sheila Meiring, were invited to help a group of non-white actors in the African township of New Brighton, Port Elizabeth, in forming an amateur drama group, the Serpent Players. His first production there, MACHIAVELLI's *The Mandrake*, was followed by productions of plays by BRECHT, BÜCHNER, SOPHOCLES, and others, and work with the group concurrently enriched his own writing. *Hello and Goodbye* (1965; N.Y., 1969; London, 1973) and *Boesman and Lena* (1969; N.Y., 1970 with Jones; London, 1971) form with *The Blood Knot* a trilogy about the lives of poor people in Port Elizabeth; *People Are Living There* (1969; N.Y., 1971; London, 1972) is set in a cheap boarding house in Johannesburg.

Working with the Serpent Players, Fugard explored IMPROVISATION and COLLECTIVE CREATION to put on plays presenting the Black South African point of view. This resulted in *The Coat* (1971), *Sizwe Bansi Is Dead* (1972), dealing with the effects of the pass laws, and *The Island* (1973), a dialogue between two prisoners on Robben Island. The last two, played by the two fine actors credited as co-authors, John Kani and Winston Ntshona, were seen at the ROYAL COURT THEATRE, London, in 1974, as was *Statements After an Arrest under the Immorality Act*, which was receiving its first performance and was performed in New York in 1978; the same productions of *Sizwe Bansi is Dead* and *The Island* were also seen in New York in 1974.

The EDINBURGH FESTIVAL of 1975 commissioned *Dimetos*, a poetic allegory with an unlocalized setting which was less well received than Fugard's work to date; although a production at the NOTTINGHAM PLAYHOUSE in 1976, with Paul SCOFIELD, transferred to London. *A Lesson from Aloes*, produced in 1980 at the COTTESLOE THEATRE in London and in New York, again with James Earl Jones, was a triumphant return to the South African environment that has inspired his best plays.

FULHAM THEATRE, London, see GRAND THEATRE (2).

FULLER, ISAAC (1606–72), English scene painter, who studied in Paris under François Perrier, probably at the new Academy there. He worked for the Restoration theatre and in 1969 painted a scene of Paradise for DRYDEN's *Tyrannic Love*, later suing DRURY LANE for payment. He was awarded £335 10*s*. 0*d*.—a large sum in those days—but his scene may also have been used for other plays.

FULLER, ROSALINDE (1892–1982), English actress, who made her first success in America, where in 1922 she played Ophelia to the Hamlet of John BARRYMORE. She first appeared in London in 1927, among her outstanding parts being the Betrothed in RAYNAL's *The Unknown Warrior* and Irina in CHEKHOV's *Three Sisters*. She was also seen in a number of SHAW plays, and in 1938–40 she was with Donald WOLFIT's Shakespeare company, her roles including Viola in *Twelfth Night*, Katharina in *The Taming of the Shrew*, Portia in *The Merchant of Venice*, Desdemona in *Othello*, and Beatrice in *Much Ado About Nothing*. In 1950 she first appeared in a solo programme entitled *Masks and Faces*, consisting of her own adaptations of a number of short sketches from such authors as DICKENS, Maupassant, Henry JAMES, and Katherine Mansfield, performed in costume but with a minimum of scenery. She developed and specialised in this form of entertainment, touring all over the world, sometimes under the auspices of the British Council, and adding new programmes to her repertoire, among them *Subject to Love* (1965) and *The Snail Under the Leaf* (1975), the latter based on the life and work of Katherine Mansfield.

FULL SCENERY, a stage set where all the parts of the stage picture belong to the current scene only and must be changed for another scene, as opposed to DETAIL SCENERY, where some elements only are changed against a Permanent Setting. Because of its cost, both in materials and labour, full scenery is now generally confined to single-set plays.

FULTON THEATRE, New York, see HELEN HAYES THEATRE.

FUNAMBULES, Théâtre des, Paris, a playhouse on the BOULEVARD DU TEMPLE which derived its name from the Latin for rope-dancers. It began as a booth for acrobats and pantomime, but in 1816 a permanent theatre was built on the site, which, under the existing laws, had to have a BARKER outside while the actors, even in slightly serious roles, had to indulge in somersaults and handsprings. It was at this theatre that in the year it opened the famous actor FRÉDÉRICK first appeared, in HARLEQUINADES, but its great days began in the 1830s with the appearance of DEBURAU as PIERROT, and lasted until his death. He was always surrounded by a good company, and the pantomimes in which he starred were given with wonderful scenery, transformation scenes, and tricks, the stage being excellently equipped; it even had the apparatus necessary to produce a real waterfall. After Deburau's death in 1846 his son continued to play his parts for a time. The theatre was finally demolished by Haussmann in 1862 in his building of the great boulevards.

FURTENBACH, JOSEF, see LIGHTING.

FUTURISM, an artistic movement which emerged in Italy during the first decade of the 20th century, and of which the principal theorist was MARINETTI. Embracing painting, sculpture, poetry, and the theatre, its main concern was to introduce contemporary ingredients into art, and notably to reflect the dynamic impact of technology. During the 1920s its theories became identified with Fascism, though in many of its experiments with language, and in its breaking-up of action, it prefigured the Theatre of the ABSURD. A number of futurist plays were performed at the Teatro degli INDIPENDENTI. Futurism was essentially a paradox, a right-wing revolt against right-wing decadence. It was childishly romantic, its chief fallacy deriving from its 19th-century glorification of the machine while it believed that it spoke for the 20th century. Its value lay in accidentally bringing about an examination of several neglected aspects of theatre—the role of the fragment; the function of the arena, especially the circus ring; the use of white light (reaching forward to BRECHT); the interpenetration of spheres of consciousness and desire, actuality and vision; the masks provided by parts of the body other than the face; the tonality of colour; the codes of syntax, by no means thoroughly explored by the exponents of VERISMO; and the deployment of disintegrated humanity in search of a unified whole, however transitory and illusory.

FUZELIER, LOUIS (1672–1752), French dramatist, whose plays were written mainly for the small theatres of the Paris FAIRS, with PIRON, LESAGE, and others. The COMÉDIE-FRANÇAISE having tried to stop their performances, regarding them as an infringement of their monopoly, Fuzelier and his companions solved their difficulties by writing the verses on long placards, held up by two children dressed as cupids, so that the audience could sing them to a popular air while the actors mimed the action. Many of Fuzelier's plays written for this convention are very charming and have been reprinted in numerous collections. He also wrote several serious plays, done at the Comédie-Française, of which none has survived. He was editor of the *Mercure* from 1744 to 1752, and wrote widely for it on the theatre and other subjects.

FYFFE, WILL, see MUSIC-HALL.

G

GABRIELLI, FRANCESCO (1588–*c.* 1636), an actor of the COMMEDIA DELL'ARTE, son of Giovanni, a virtuoso solo player. Francesco is sometimes credited with the invention of the character of SCAPINO, which he may have taken up and enlarged from BARBIERI. He is first found with the ACCESI in 1612, for many years he was one of the outstanding members of the CONFIDENTI, and in 1624 he was with G. B. ANDREINI and the FEDELI in Paris, later rejoining the Accesi. He appears to have been a skilful musician, and a number of his letters which have been preserved give interesting information on the theatrical life of the time. He had several children, all of whom went on the stage.

GAELIC DRAMA. No plays were written in IRELAND in the native Gaelic tongue until the 20th century, when there were several attempts to create a distinctively native drama, mainly as part of a policy for restoring Gaelic as the vernacular. The first important production in Gaelic was *Casadh an tSúgáin* (*The Twisting of the Rope*, 1901) by Douglas HYDE, billed with YEATS and MOORE's *Diarmuid and Grania* as the final presentation by the IRISH LITERARY THEATRE at the GAIETY THEATRE in Dublin.

In 1926 the Free State Government provided a subsidy for the staging of plays in Irish, and subsequently an amateur group called An Comhar Dramaíochta presented plays at the ABBEY THEATRE, the Peacock, and the GATE THEATRE until 1942. At the Gate, An Comhar found both an ally and an enthusiast in Micheál MACLIAMMÓIR, who wrote some of their plays and directed or acted in others. In 1942 An Comhar was absorbed into the Abbey Theatre by Ernest Blythe, the manager, who hoped to make Gaelic drama an integral part of the repertoire. One-act plays in Irish were performed as afterpieces to plays in English, and from 1945 an annual pantomime in Irish was staged, cast from the regular Abbey company, which is now required to be bilingual. The Abbey's Peacock Theatre, opened in 1967, was intended as a centre for Gaelic drama, and a few plays in Irish are staged there annually, though the main support for Irish plays comes from local amateur groups. There is, however, a national Gaelic drama festival in Dublin every spring.

The work which has emerged from this movement consists of translations of plays by Irish writers such as Lady GREGORY, O'CASEY, and SYNGE; translations of foreign classics; and some original plays. The Damer Hall, Dublin, a converted Unitarian Church, is generally used and here Brendan BEHAN's *An Giall* (1957) was first staged, but even to Irishmen it is better known as *The Hostage* (1958).

Outside Dublin there are regional Gaelic theatres, notably in Gweedore, Co. Donegal, and in Galway. Compántas Corchaí, founded in Cork in the late 1940s, led to the staging of plays by Sean O'Tuama in the University's Granary Theatre. The Siamsa Tire, or Irish Folk Theatre, was founded in Co. Kerry in 1974, with plans for a central state-supported theatre in Tralee. An Taibhdhearc in Galway is a semi-professional theatre founded in 1928 with the help of Hilton EDWARDS and MacLiammóir, whose *Diarmuid agus Grainne* (1928) was its first production. Managers since MacLiammóir include Frank Dermody, who directed Behan's *An Giall* and went on to be a director at the Abbey, and Walter Macken, actor, novelist, and playwright. Many well-known actors and actresses began at An Taibhdhearc, notably Siobhán MCKENNA, but, as with the Gaelic theatre movement as a whole, An Taibhdhearc relies heavily on translations for its repertoire.

GAFF, 19th-century term for an improvised theatre, in the poorer quarters of London and other large towns, on whose stage an inadequate company dealt robustly with a repertory of MELODRAMA. The entrance fee was a penny or twopence. The lowest type of gaff was known as a blood-tub. In Scotland it was called a 'geggie'.

GAIETY THEATRE, Dublin, one of the two surviving 19th-century theatres in Dublin, the other being the OLYMPIA. The Gaiety opened in 1871 but had no resident company, relying on visits from well known English companies in a variety of plays from Shakespeare to SHAW. Somewhat incongrously, the theatre was chosen by the IRISH LITERARY THEATRE for the presentation of seasons of Irish plays in 1900 and 1901, but there was little other native fare before the arrival of Jimmy O'DEA in variety in 1937. The outbreak of war in 1939 put an end to visits from England, and the gap was bridged by the local operatic society and by the GATE THEATRE, which was first seen in 1940 in a new play by Micheál MACLIAMMÓIR, *Where Stars Walk*, and for several years played more often at the Gaiety than at the Gate. After the war the visiting companies returned but important Irish contributions were also made, among them Anew

MCMASTER's production of GOLDSMITH's *She Stoops to Conquer* in 1946 and the first production, by Cyril CUSACK, of O'CASEY's *The Bishop's Bonfire* (1955). The theatre was structurally altered in 1955 and the old gallery removed. A threat of demolition in 1966 was averted when the television personality Eamonn Andrews took over the lease so that 'there would always be a Gaiety Theatre in Dublin.' The theatre's programme now includes a PANTOMIME and a summer show, each lasting about 12 weeks, international opera seasons in spring and autumn, and ballet.

GAIETY THEATRE, London, at the east end of the Strand. This opened as the Strand Musick Hall on 15 Oct. 1864, but its mixed bill of serious music and music-hall turns proved unsuccessful and it closed on 2 Dec. 1866. A new building was then erected, seating 1,126 people in three tiers and covering a larger area than the former hall. It opened as the Gaiety on 21 Dec. 1868 after a series of mishaps, including the loss of some of the scenery in a fire. The company included Nellie FARREN, Madge Robertson (the future Mrs KENDAL), and Alfred Wigan. The manager, John HOLLINGSHEAD, instituted a number of reforms, establishing regular morning matinées and doing away with the old BENEFIT SYSTEM by paying his actors better salaries. One of the theatre's early succcesses was H. J. BYRON's *Uncle Dick's Darling* (1869) in which Henry IRVING played Reginald Chenevix, and on 26 Dec. 1871 came the first joint work by GILBERT and Sullivan—*Thespis; or, the Gods Grown Old*.

The great feature of the Gaiety was its BURLESQUES, many of them written by Burnand and Byron, featuring the famous 'quartet'—Nellie Farren, E. W. Royce, Edward TERRY, and Kate VAUGHAN—first seen in Byron's *Little Don Cesar de Bazan* (1876). Fred Leslie joined the company in 1885 and his partnership with Nellie Farren lasted until 1891. Just before Hollingshead retired in 1885 he was joined by George EDWARDES who took over the management, his first production being the comic opera *Dorothy*. Burlesques, however, continued to occupy the theatre: *Ruy Blas; or, the Blasé Roué* (1889) featured a *pas de quatre* with Fred Leslie, C. Danby, Ben Nathan, and Fred Story dressed as ballet girls and made up to look like Irving, TOOLE, Edward Terry, and Wilson BARRETT. Irving protested and the make-up was altered, but not before a number of people had enjoyed it. In Dec. 1892 Edwardes transferred from the PRINCE OF WALES' THEATRE a new type of show called *In Town*, now considered the first MUSICAL COMEDY. It was followed by a series of similar shows—*The Shop Girl* (1894), starring Ada Reeve, *The Circus Girl* (1896), *A Runaway Girl* (1898)—all featuring in the chorus the 'Gaiety Girls', chosen for their good looks as well as their singing and dancing ability.

The Gaiety was a popular and successful theatre, but it was destined to fall victim to the Strand widening scheme. The last outstanding production there was *The Toreador* (1901) in which Gertie MILLAR made a hit and after a farewell performance entitled *The Linkman*, in which many Gaiety stars appeared, the theatre closed on 4 July 1903, its place being taken by a second Gaiety nearby. This was built by Edwardes on an island site at the west end of the Aldwych. A four-tier house holding 1,267 spectators, it opened on 26 Oct. 1903 in the presence of Edward VII and Queen Alexandra, with a musical comedy, *The Orchid*, again starring Gertie Millar. Among its later successes were *The Girls of Gottenberg* (1907) and *Our Miss Gibbs* (1909). When Edwardes died the fortunes of the Gaiety declined, in spite of the success of a new comedian, Leslie HENSON, who appeared in *Tonight's the Night* (1915), in which he had previously starred in New York. The long run of musical comedies was broken when in 1921 MAETERLINCK's *The Betrothal* (a sequel to *The Blue Bird*) was staged, but began again later in the year with *The Little Girl in Red*. A later success was *Love Lies* (1928), with Laddie Cliff, and in 1936 and 1937 Fred EMNEY and Leslie Henson had good runs in *Swing Along* and *Going Greek*. The last production was *Running Riot* (1938) which closed on 25 Feb. 1939. The building then stood empty until in 1945 Lupino LANE bought it, hoping to reopen it; but in 1950 he was forced to sell and in 1957 it was demolished.

GAIETY THEATRE, Manchester, a dilapidated music-hall with a somewhat dubious reputation, which in 1907 was bought by Miss HORNIMAN to house her REPERTORY COMPANY, the first to be established in England. To provide a more spacious auditorium with better sight-lines, the interior of the theatre was remodelled, cutting the original seating of 2,500 by half, and many improvements were made backstage. The newly decorated theatre opened on 9 Sept. 1908, and among its early productions was a one-act play, *Reaping the Whirlwind* by Allan MONKHOUSE, which is usually considered the starting-point of the so-called 'Manchester School of Drama', a phrase first used apparently in the Manchester Grammar School magazine of Oct. 1909. Other dramatists connected with it were Harold BRIGHOUSE and Stanley HOUGHTON, who both had their first though not their best known plays produced at the Gaiety, and St John ERVINE, whose *Jane Clegg* was seen there in 1913. The chief characteristic of the Manchester dramatists was their REALISM, and their works formed part of the 'new drama' which came into prominence shortly before the First World War. The Gaiety, which was directed for some time by Lewis CASSON, was artistically a great success but financially it failed, and Miss Horniman disbanded the company in 1917, though she maintained her interest in the theatre until 1921 when it was sold and became a cinema.

GAIETY THEATRE, New York, at 1547 Broadway. This attractive playhouse opened on 31 Aug. 1908; a year later W. S. GILBERT's *The Fortune Hunter*, with John BARRYMORE, had a good run, as did a dramatization of Jean Webster's *Daddy Long-Legs* in 1914. In 1918 *Lightnin'* brought fame to Frank BACON, its star and part-author, running for three years, then a Broadway record. Later successes included GALSWORTHY's *Loyalties* in 1922, LONSDALE's *Aren't We All?* in 1923, Philip BARRY's *The Youngest* in 1924, and a revival of *Rain*, based on Somerset MAUGHAM's short story, in 1926. In 1943 the theatre, which since 1932 had housed mainly BURLESQUE and films, was renamed the Victoria.

GAÎTÉ, Théâtre de la, Paris. This playhouse became famous in the mid-19th century, when FRÉDÉRICK performed there regularly. Until 1862 it was situated on the BOULEVARD DU TEMPLE, having been opened by Nicolet in 1764 for the performance of puppet-shows. Later it specialized in mime-plays and acrobatics. Renamed Théâtre de la Gaîté in 1795 after the upheavals of the French Revolution, and rebuilt in 1805, it was assigned to MELODRAMA by Napoleon's decree on theatres, and staged productions of many works by PIXÉRÉCOURT, its manager from 1823 to 1834. When the Boulevard du Temple disappeared in the building of Haussmann's boulevards in the 1860s, the Gaîté was rebuilt nearby, adjacent to the Boulevard Sebastopol, where it still stands. The Paris Council, which owns it, renamed it the Théâtre de la Musique.

GALANTY SHOW, see SHADOW-SHOW.

GALDOS, BENITO PÉREZ, see PÉREZ GALDOS.

GALIMAFRÉ, see BOBÈCHE.

GALLACHER, TOM, see SCOTLAND.

GALLERY, in the 19th-century theatre building the highest and cheapest seats in the house, usually unbookable. The seating generally consisted of wooden benches, in some cases without backs. The occupants of the gallery were in about 1752 nicknamed 'the Gods'; they often formed the most perceptive and certainly the most vociferous part of the audience. In the Restoration period the usual charge for the upper gallery was one shilling, but at DORSET GARDEN and later at DRURY LANE footmen waiting for their masters were admitted free at the end of the fourth act. From 1697 onwards Christopher RICH, hoping to curry favour with the rougher element in the audience, allowed footmen to occupy the Drury Lane gallery without payment from the opening of the performance, which led to its being known as the Footmen's Gallery. The custom led to so much noise and disorder that it was finally abolished by Fleetwood in 1737.

GALLI, see BIBIENA.

GALLIENNE, EVA LE, see LE GALLIENNE.

GALLINA, GIACINTO (1852–97), Italian dramatist, who wrote mainly in the Venetian dialect, thus keeping alive there the tradition of vernacular drama. His early plays, of which *Le Barufe in famegia* (*Family Quarrels*, 1872) is typical, were written very much in the style of GOLDONI; but he moved towards an increasing awareness of and hatred for contemporary materialistic society, expressing his view of it in terms of the conflict between succeeding generations, as can be seen in the bitter yet sympathetic *Così va il mondo, bimba mia* (*So Wags the World, My Child*, 1880) and the acrid *Serenissima* (1891). His major work is now considered to be *La famegia del santolo* (*The Godfather's Family*, 1892), a grey but tender tragedy, reminiscent in its patient acceptance of cuckoldry of GIACOSA's *Tristi amori* (1888).

GALLO THEATRE, New York, see NEW YORKER THEATRE.

GALSWORTHY, JOHN (1867–1933), O.M. 1929, English novelist and dramatist, awarded the NOBEL PRIZE for Literature in 1932, mainly for his novels. His plays deal with questions of social justice in the fashion of his time. *The Silver Box* (1906; N.Y., 1907) highlights the inequality before the law of a rich thief and a poor one and *Strife* (London and N.Y., 1909) explores the effect of an industrial strike. Both were well received and several times revived up to 1933, as was his best known play, *Justice* (1910; N.Y., 1916), which includes a scene showing the effect of solitary confinement on its chief character, Falder, and was at the time credited with having led to a reform of prison practice in this respect. Of his later plays, in which the reformer seems to have triumphed over the dramatist, the most successful were *The Skin Game* (London and N.Y., 1920), *Loyalties* (London and N.Y., 1922), and *Old English* (London and N.Y., 1924); his last play *The Roof* (1929; N.Y., 1931) failed. *Justice* was seen in London in 1968, and *Strife* at the NATIONAL THEATRE in London in 1978.

GANASSA, ZAN (?–*c.* 1583), actor-manager of the COMMEDIA DELL'ARTE, whose real name was probably Alberto Naseli. He is first heard of in about 1568, and his company was one of the earliest to leave Italy and travel abroad. He was in Paris in 1571, and a year later took part in the festivities held in honour of the marriage of the King of Navarre (later Henri IV) to the sister of Charles IX, Marguerite de Valois. A curious painting now in the Bayeux Museum, probably painted at this time and ascribed to either Paul or Frans Porbus, is believed by Duchartre, who reproduced it in his *The Italian Comedy* (p. 84), to represent a performance by Ganassa's company assisted by the French king and members of the royal household. It was, however, in Spain that Ganassa made his most successful tours, being found there frequently during the 1570s, and there seems to be no doubt that he exerted a considerable influence on the nascent Spanish professional theatre.

GARCÃO, PEDRO ANTÓNIO CORREIA, see PORTUGAL.

GARCIA, VICTOR (1934–), Argentinian-born theatre director who has worked mainly in France since 1961, being with SERREAU from 1962 to 1964. His association with the work of ARRABAL began in 1966, and with that of GENET in 1969, when he directed *Les Bonnes* for Núria ESPERT I ROMERO in Spain, following it in 1970 with *Le Balcon* in Brazil. He directed Arrabal's *The Architect and the Emperor of Abyssinia* for the NATIONAL THEATRE in London in 1971, in which year he also directed GARCÍA LORCA's *Yerma* for Espert. His work generally revolves around an overall environmental concept, like the massive raked trampoline on which his production of *Yerma* took place, and he has been accused of subordinating the text of the play to the stage picture.

GARCÍA DE LA HUERTA, VICENTE (1734–87), Spanish dramatist, who endeavoured to rectify the contemporary neglect of Spanish classics in favour of translations from the French by publishing in 1785–6 a selection of forgotten Spanish plays. Unfortunately he did not always choose the best, omitting all works by Lope de VEGA, TIRSO DE MOLINA, and RUÍZ DE ALARCÓN. In his own work he was much influenced by the theory of the three UNITIES, imposed on Spain by LUZÁN; his most successful play *La Raquel* (1778) was modelled on classical French lines, but its inspiration was entirely Spanish and in the general poverty of Spanish drama of the period it stands out as a not unworthy successor to the plays of the Golden Age.

GARCÍA GUTIÉRREZ, ANTONIO (1813–84), Spanish playwright of the Romantic period, whose play *El Trovador* (1836) achieved worldwide popularity as the basis of Verdi's opera *Il Trovatore*. García Gutiérrez wrote a number of other plays, of which the last two, *Venganza catalana* (1864) and *Juan Lorenzo* (1865), were far superior to his earlier works. He was also the author of several ZARZUELAS, set to music by Arrieta, the best being *El grumete* (1864) and its sequel *La vuelta del corsario* (1865). His work shows the influence of DUMAS *père*, two of whose melodramas he translated into Spanish as well as several plays by SCRIBE.

GARCÍA LORCA, FEDERICO (1898–1936), Spanish poet and playwright, executed by firing squad at the opening of the Spanish Civil War. As a child he made and played with puppets in his own miniature theatre, and later produced puppet-plays in his native Granada. His first full-length play *El maleficio de la mariposa* (*The Butterfly's Curse*) was produced by MARTÍNEZ SIERRA in 1920 at the Teatro Eslava. It was followed by a historical drama, *Mariana Pineda* (1927), and several light comedies, among them *La zapatera prodigiosa* (*The Shoemaker's Amazing Wife*, 1930). His fame rests on his tragic folk trilogy *Bodas de sangre* (1933), *Yerma* (1934), and *La casa de Bernarda Alba* (produced posthumously in 1945 in Buenos Aires), a powerful indictment of a society in which natural impulses are frustrated by the dead hand of convention from which there is no escape. These have been presented all over the world, the first to be seen in English being *Bodas de sangre*, as *Bitter Oleander* in New York in 1935 and as *The Marriage of Blood* in London in 1939; it has since been widely revived as *Blood Wedding*. *Yerma* was produced in London in 1957 and *The House of Bernarda Alba* in New York in 1951. In his lifetime García Lorca's influence on the Spanish theatre was most important, both through his own plays and productions and through his association with the itinerant amateur company La BARRACA, which did much to familiarize the working people of Republican Spain with their great dramatic heritage.

GARDEN THEATRE, New York, at 61 Madison Avenue. This opened on 27 Sept. 1890 and in the following year saw the reappearance in New York of Sarah BERNHARDT in SARDOU's *La Tosca*. Other productions included GILBERT's *The Mountebanks* (1893) (with music by Alfred Cellier), a dramatization of Ouida's *Under Two Flags* (1901), and HAUPTMANN's *The Weavers* (1915). In Jan. 1910 a season of Shakespeare's plays was given by Ben GREET and his company, and in 1919 the theatre became the Jewish (later Yiddish) Art Theatre. It was demolished in 1925.

GARDIN, VLADIMIR (1877–1965), Russian actor, who while playing in the provinces attracted the attention of Vera KOMISARJEVSKAYA. She engaged him for her theatre in St Petersburg, where he made an outstanding success in such parts as Krogstad in IBSEN's *A Doll's House* and Shalimov in GORKY's *The Summer Guests*. He left in 1907 to found a small private theatre in Finland for the production of plays banned by the Russian authorities, among them HAUPTMANN's *Die Weber* and two plays by ANDREYEV. A year later he was with a company which toured Paris, London, and other capital cities with a repertory of Russian plays, and in 1912 was in Moscow where he joined the company at the KORSH THEATRE, his finest and favourite part there being Potrassov in TOLSTOY's *The Living Corpse*. From 1913 his career was entirely devoted to films.

GARNIER, ROBERT (*c.* 1545–90), French Renaissance dramatist, by profession a lawyer, who with his seven tragedies on classical models prepared the way for the great tragic writers of the next generation. His first play *Porcie* (1568) dealt in the style of SENECA with the suicide of Brutus's wife Portia; *Hippolyte* (*c.* 1568) may be considered a forerunner of RACINE's *Phèdre*; but his best play is usually considered to be *Les Juives* (*c.* 1580), based on the Old Testament story of Nebuchadnezzar's cruelty to Zedekiah after the fall of Jerusalem. All his plays give evidence of wide reading, keen perceptions, and great lyric gifts. Garnier's choruses, a traditional feature which he retained from classical tragedy and which helped to give his plays a strong moral flavour, are particularly fine; and at its best his writing is forceful, imaginative, and fluent. He was also the author of the first French tragi-comedy, *Bradamante* (1582) on a theme from ARIOSTO's *Orlando Furioso*, which was still being acted in the 17th century.

GARRETT, JOÃO BATISTA DE ALMEIDA (1799–1854), Portuguese dramatist of the Romantic movement, the first of whose plays to be produced was *Lucrécia* (1819). He became involved in Liberal politics—his early neo-classical play *Catão* (1821) has clear political implications—and for some years took refuge in England. After the Revolution of 1836 he returned to Portugal and was appointed Inspector-General of Theatres, being responsible for the foundation of the Teatro Nacional (later the Teatro de D. Maria II) and the Conservatório. Determined to provide a repertoire of Portuguese plays, he produced his first major dramatic work *Um auto de Gil Vicente* (1838), a love story in a historical setting which provided the starting-point for Romantic drama in Portugal. His masterpiece was *Frei Luís de Sousa* (1843), a personal tragedy set in a time of national tragedy, the late 16th century.

GARRICK, DAVID (1717–79), one of the greatest English actors, who effected a radical change in the style of acting in his day when he replaced the formal declaration favoured by James QUIN with an easy, natural manner of speech. Of Huguenot descent, he showed an early inclination for the stage, and at the age of eleven appeared with some success as Sergeant Kite in a schoolboy production of FARQUHAR's *The Recruiting Officer*. Later he was sent to study under Dr JOHNSON at Lichfield, accompanied him to London, and there indulged in amateur theatricals at the expense of his business career in the wine trade, which he soon abandoned. Some obscurity surrounds his early apppearances on the professional stage, but in 1741 he was playing small parts at GOODMAN'S FIELDS THEATRE and went with the manager GIFFARD to Ipswich where, under the name of Lyddal, he played a variety of roles with a success which seemed to justify his continuing in the theatrical profession. Rejected by the managers of both DRURY LANE and COVENT GARDEN, he returned to Goodman's Fields and on 19 Oct. 1741 made his formal début as Shakespeare's Richard III with such success that he was soon drawing crowds to the theatre. He was then engaged by FLEETWOOD for Drury Lane, where on 11 May 1742 he embarked on a triumphant career which continued until his retirement in 1776. A small man, with a clear though not resonant voice, he appeared unsuited to tragedy, but was nevertheless unsurpassed in the tragic heroes of contemporary works as well as in such great parts as Hamlet, Macbeth, and Lear. (Garrick himself related how he modelled the old king's madness on that of an unfortunate man who had accidentally killed his two-year-old child by dropping it from a window.) He was not at his best as Romeo, a part he soon resigned to Spranger BARRY, nor as Othello, partly because blacking his face deprived him of that marvellously expressive play of mobile features which constituted one of his greatest assets. He was much admired in comedy, one of his earliest and most acclaimed performances being Abel Drugger in JONSON's *The Alchemist*; he was good as Benedick in *Much Ado About Nothing* and as Ranger in HOADLY's *The Suspicious Husband*; and as Bayes in BUCKINGHAM's *The Rehearsal* he scored a triumph with his mimicry of well known actors of the time; however he exempted Quin, being always careful to give the older actor his due and admitting his excellence in such parts as Falstaff and ADDISON's Cato while taking from him all the other great parts in which he had long reigned supreme. Garrick's fiery temper, vanity, and snobbishness, as well as his sudden rise to fame, brought him many enemies, among them Samuel FOOTE who lampooned him mercilessly. He also had to contend with the petulance of unacted authors and disappointed small-part actors. He was not always to blame for the quarrels in which he found himself involved, notably with Dr Johnson over the failure of his *Irene* (1749) and with the elder COLMAN whom he offended by refusing to play the part of Lord Ogleby in *The Clandestine Marriage* (1766), a comedy on which they had collaborated.

In 1747 Garrick first became involved in the management of Drury Lane, where the major part of his working life was spent. On the death of LACY he became sole manager, passing on his share in the Patent on retirement to SHERIDAN, Ford, and Linley. During his management many reforms were introduced, both before and behind the curtain, the most important being the introduction of stage lighting concealed from the audience and the banishment of spectators from the stage, a measure long overdue and achieved at about the same time by VOLTAIRE in Paris. Garrick was always careful to gather a good company round him, his leading ladies being Peg WOFFINGTON (his mistress for many years), Kitty CLIVE, Mrs BELLAMY, who played Juliet to his Romeo, Mrs ABINGTON, and Mrs Cibber, the wife of Theophilus CIBBER, who was said to resemble him like a sister.

Among the men, apart from intermittent appearances by Quin and Barry, his chief supporters were the unfortunate MOSSOP, WOODWARD, who wrote the Drury Lane pantomimes and appeared in them as HARLEQUIN (a part once played by Garrick in his early years), and Tom KING, who was the original Sir Peter Teazle in Sheridan's *The School for Scandal* (1777) and took over the part of Ogleby in *The Clandestine Marriage* when Garrick refused to play it.

Garrick's management was marred by two serious riots, the first in 1755, occasioned by the appearance of French dancers in Noverre's ballet 'The Chinese Festival' just as war between France and England was about to break out, and the second in 1762, when the concession of 'half-price after the third act' was abolished but had to be restored. This led Garrick to retire for a time, and from 1763 to 1765 he travelled on the Continent with his wife Eva Maria Veigel (or Faigel) (1724–1822), a Viennese dancer whom he had married in 1749 after her appearances at the HAYMARKET as Mlle Violette (also found as Violetta and Violetti). They were well received everywhere, particularly in France, and Garrick's reputation, enhanced by a number of recitals which he gave in private houses in Paris, shed much lustre upon the contemporary English theatre. He returned to London greatly refreshed and a public surfeited with musical spectacles was glad to see him in a succession of his greatest parts. One of his most publicized achievements, and the one which occasioned most malice at his expense, was his 'Shakespeare Jubilee' at STRATFORD-UPON-AVON in 1769. It was remarkable for the number of odes, songs, speeches, and other effusions by Garrick, and for the complete absence of anything by Shakespeare. In spite of the cares of management and constant appearances on stage, Garrick was a prolific dramatist, vivacious and competent, at his best about equal to Colley CIBBER. The best of his own works were the farces *Miss in her Teens; or, the Medley of Lovers* (1747), in which he himself played Fribble with great success, and *Bon Ton*, usually known from its subtitle as *High Life Above Stairs* (1775). Much of his energy was expended on re-writing old plays, the most successful being *The Country Girl* (1766), a bowdlerization of WYCHERLEY's *The Country Wife* (1675) which held the stage for many years. He was also a great writer of prologues and epilogues, for both his own and other men's plays; these were published with his other works in a three-volume edition in 1785. It has been asserted that he restored the original texts of Shakespeare, freeing them from the gross corruptions of the 17th century, but this claim must be treated with caution; he himself was responsible for a production of *Hamlet* with the Grave-diggers omitted (a bad precedent followed in our own day by Maurice EVANS in his *G. I. Hamlet*); of *King Lear* without the Fool; and of a *Romeo and Juliet* which allowed the lovers a scene together in the tomb before dying. He also concocted a *Katharine and Petruchio* and a *Florizel and Perdita* (both 1756) from *The Taming of the Shrew* and *The Winter's Tale* respectively, and did not scruple to add to or alter scenes and speeches in any play by Shakespeare he presented. He had, however, to contend with the taste of the time and the ignorance and lack of appreciation of Shakespeare's genius then prevalent, and even in his worst excesses he seems to have retained a deep and sincere appreciation of the dramatist whose works he was tampering with.

Garrick made his last appearance on the stage on 10 June 1776 as Don Felix in Mrs CENTLIVRE's *The Wonder, a Woman Keeps a Secret*, which he had first played 20 years earlier and in which he was unequalled for spirit and vivacity. He then retired to Hampton to enjoy the companionship of his wife and friends. His death, which was felt as a personal loss by many admirers, drew from Dr Johnson the memorable epitaph: 'I am disappointed by that stroke of death which has eclipsed the gaiety of nations and impoverished the public stock of harmless pleasure.' He was buried in Westminster Abbey, where his wife later joined him, and the carriages of the mourners reached as far as the Strand. His younger brother George (1722–79), who had for many years been his right-hand man at Drury Lane, died a few days later because, said the wits of the time with rueful humour, 'Davy wanted him.'

GARRICK CLUB, London, gentleman's club, with membership restricted to 700, which has strong theatrical associations. Its first committee meeting was held on 17 Aug. 1831, the club, with the Duke of Sussex as its patron, opening in Nov. of the same year, though its premises in Probatt's Family Hotel, King Street, Covent Garden, were not ready for the use of members until Feb. 1832. The present club-house, which opened on 4 July 1864, stands on part of old Rose Street which ran through the warren of crowded alleys between King Street and St Martin's Lane. It has a fine collection of theatrical portraits, of which an annotated catalogue was prepared in 1908 by Robert Walters, and other stage memorabilia.

GARRICK THEATRE, London. (1) At 70 Leman Street, Whitechapel. This theatre, which took its name from its proximity to the old GOODMAN'S FIELDS THEATRE where GARRICK made his début, opened in 1831 under Benjamin Conquest. Burnt down in 1846, it was rebuilt and reopened in 1854. It held a very low position even among East End theatres, and was practically a GAFF. In 1873 J. B. Howe, a popular local actor, took it over and attempted to raise its standards, but within two years he was bankrupt. After the theatre had again been reconstructed, to hold 462 people on two tiers, May Bulmer took it over and staged Bazin's *opéra bouffe Le Voyage en Chine* (1879) as *A Cruise to China* in which Beerbohm TREE made his first professional appearance. The building fell into disuse after 1881.

(2) In the Charing Cross Road, near Trafalgar Square. This theatre, which has a three-tier auditorium seating 800, opened on 24 Apr. 1889 under John HARE with PINERO'S *The Profligate*. Grundy's comedy *A Pair of Spectacles* had a long run in 1890, and five years later *The Notorious Mrs Ebbsmith*, also by Pinero, with Mrs Patrick CAMPBELL in the title role, caused a sensation, particularly after a woman named Ebbsmith was found drowned in the Thames with the counterfoil of a ticket for the play in her pocket. Hare left after this production, and standards declined until in 1900 Arthur BOURCHIER took over, and with his wife Violet VANBRUGH inaugurated a long and brilliant period of productions ranging from Shakespeare to farce, among them KNOBLOCK'S *Kismet*, starring Oscar ASCHE, and *The Merry Wives of Windsor*. After Bourchier left in 1915 the theatre had no regular policy or management, and was used mainly for REVUE. Closed in 1939, it opened again in 1941 and in 1944 scored a success with Thomas Job's *Uncle Harry* starring Michael REDGRAVE. Later hits included Vernon Sylvaine's *Madame Louise* (1945), *One Wild Oat* (1948), and *As Long As They're Happy* (1953), Garson Kanin's *Born Yesterday* (1947), and the revues *La Plume de ma Tante* (1955) and *Living for Pleasure* (1958). A transfer from THEATRE WORKSHOP of the musical *Fings Ain't Wot They Used T'Be* in 1960 was followed by long runs of Charles Dyer's *Rattle of a Simple Man* (1962) and George Ross and Campbell Singer's *Difference of Opinion* (1963). In 1967 Brian RIX began a series of farces, which he presented and appeared in, with *Stand By Your Bedouin* by Roy Cooney and Tony Hilton; later ones included *Let Sleeping Wives Lie* (1967) by Harold Brooke and Kay Bannerman and Michael Pertwee's *She's Done It Again* (1969) and *Don't Just Lie There, Say Something* (1971). Ira Levin's thriller *Death Trap* ran from 1978 until 1981.

GARRICK THEATRE, New York, at 65–9 West 35th Street. This opened as Harrigan's Theatre on 29 Dec. 1890. In 1895 Richard MANSFIELD took it over and renamed it, opening on 23 Apr. with SHAW'S *Arms and the Man*, the longest run under his management being that of William GILLETTE in his own play *Secret Service*. Gillette also appeared as Sherlock Holmes, a part always associated with him. Among later productions were *Captain Jinks of the Horse Marines* (1901) with Ethel BARRYMORE, *The Stubbornness of Geraldine* (1902), and *Her Own Way* (1903), all by Clyde FITCH, while 1905 saw a successful run of Shaw's *You Never Can Tell*.

From 1917 to 1919 the theatre was occupied by a French company under Jacques COPEAU, who named it after his theatre in Paris, the VIEUX-COLOMBIER, and inaugurated his tenancy on 20 Nov. 1917 with a performance of MOLIÈRE'S *Les Fourberies de Scapin*. On 19 Apr. 1919 the THEATRE GUILD opened at the Garrick, which had reverted to its old name, with BENAVENTE'S *The Bonds of Interest*; the Guild's productions were presented there until it opened its own playhouse, now the ANTA THEATRE, in 1925. The PROVINCETOWN PLAYERS made their last appearance at the Garrick in 1929; it was demolished in 1932.

GASCOIGNE, BAMBER, see REVUE.

GASCOIGNE, GEORGE (*c.* 1542–77), a scholar of Cambridge who helped to prepare the entertainments given before Elizabeth I at Kenilworth and Woodstock in 1575. He had earlier prepared translations of Lodovico DOLCE'S *Giocasta* (based on EURIPIDES' *Phoenician Women*) as *Jocasta*, and of ARIOSTO'S *I suppositi* (based on PLAUTUS'S *Captivi*) as *The Supposes*, both performed at Gray's Inn in 1566. *The Supposes*, which provided Shakespeare with the subplot of Bianca and her suitors in *The Taming of the Shrew*, was revived at Trinity College, Oxford, in 1582.

GASCON, JEAN (1921–), French-Canadian actor and director, born in Montreal. He began acting with the amateur companies Les Carabins and Les Campagnons de Saint-Laurent, and then went to Paris, where he spent a year at the VIEUX-COLOMBIER. Returning to Montreal, he worked with Ludmilla PITOËFF. He was the co-founder in 1951 of the Théâtre du NOUVEAU MONDE and its artistic director until 1966, during which time he appeared in MOLIÈRE'S *L'Avare* and *Don Juan* and STRINDBERG'S *The Dance of Death*, but became best known for his excellent productions of plays by European playwrights, among them, apart from Molière and Strindberg, BEAUMARCHAIS, BRECHT, FEYDEAU, MUSSET, and John WEBSTER. As artistic director of the STRATFORD (ONTARIO) FESTIVAL from 1967 to 1974 he repeated many of these productions in English versions, as well as directing some of Shakespeare's lesser-known plays such as *Cymbeline* and *Pericles*. In 1975 he directed a revival of John Coulter's play on a Canadian folk-hero *Riel* for the National Arts Centre in Ottawa, and at the Centre's Studio in 1978 he played the title role in Strindberg's *The Father*. He became the first director-general of the National Theatre School of Canada in 1960.

GASKILL, WILLIAM (1930–), English theatre director, who spent several years as an actor and stage manager in repertory before going to the ROYAL COURT THEATRE, where in 1958 he directed N. F. Simpson's double bill *A Resounding Tinkle* and *The Hole*, and *Epitaph for George Dillon* by John OSBORNE and Anthony Creighton. He was also responsible for the production of Simpson's *One-Way Pendulum* (1959), and directed *Richard III* (1961) and *Cymbeline* (1962) at the ROYAL SHAKESPEARE THEATRE. Back in London, he directed the ROYAL SHAKESPEARE COMPANY in BRECHT's *The Caucasian Chalk Circle* (also 1962), Peter O'TOOLE in Brecht's *Baal* (1963) at the PHOENIX, and, as associate director of the NATIONAL THEATRE company, FARQUHAR's *The Recruiting Officer* (also 1963), Brecht's *Mother Courage and her Children*, and ARDEN's *Armstrong's Last Goodnight* (both 1965). Returning to the Royal Court as artistic director, 1965–72, he was responsible for the production of several controversial new plays, including Edward BOND's *Saved* (1965), *Early Morning* (1968), *Lear* (1971), and *The Sea* (1973). In 1974 he helped to found the Joint Stock Company with the aim of involving dramatists in collaborative preparatory work with their actors; he co-directed David HARE's *Fanshen* for the company in 1975, and his later productions for it included Howard BRENTON's *Epsom Downs* (1977) and Stephen Lowe's adaptation of Robert Tressell's novel *The Ragged Trousered Philanthropists* (1978). He was responsible for the National Theatre productions of GRANVILLE-BARKER's *The Madras House* in 1977 and MIDDLETON and ROWLEY's *A Fair Quarrel* in 1979, in which year he also directed Nicholas Wright's *The Gorki Brigade* at the Royal Court.

GASPAR, ENRIQUE (1842–1902), Spanish dramatist, whose later plays made a considerable contribution to the acceptance of REALISM on the Spanish stage. In the best of them—*Las personas decentes* (1890), *La huelga de hijos* (1893), and *La eterna cuestión* (1895)—he explored, in the manner of his immediate predecessor TAMAYO Y BAUS, the moral decadence of the middle classes and, like PÉREZ GALDÓS in the novel, satirized their greed, selfishness, and overriding concern for respectability at all costs.

GASSMAN, VITTORIO (1922–), Italian actor-manager, who made his first appearance on the stage in 1943 and soon proved himself an excellent actor in both comedy and tragedy, having a good presence and a finely modulated voice. In companies run by Squarzina and VISCONTI in the late 1940s he played roles as diverse as Antony in the elder DUMAS's play of that name, Orlando in *As You Like It*, Kowalski in Tennessee WILLIAMS's *A Streetcar Named Desire*, and the title role in ALFIERI's *Oreste*, this last being accounted one of his best parts. Among his own productions, in most of which he starred himself, were IBSEN's *Peer Gynt* at the Teatro Nazionale in 1950–1 and *Hamlet* at the Teatro dell'Arti in 1952–3. In 1954–5, with his own company the Teatro Popolare Italiano, he played two of his most famous roles, SOPHOCLES' Oedipus and KEAN in SARTRE's version of the elder Dumas's play; in 1956 he alternated the parts of Othello and Iago with Randone. The company, modelled somewhat on VILAR's T.N.P., was one of the first in Italy to perform in unconventional locations: in 1960 he appeared with it in a circus tent in Rome in Manzoni's *Adelchi*, and it was seen in London in 1963 in a programme entitled *The Heroes*, incorporating excerpts from his extensive repertory. In later years Gassman excelled in comedy, especially in films.

GASSNER, JOHN [Jenö] WALDHORN (1903–67), American theatre critic, historian, and playwright, born in Hungary. He had already held a number of academic posts when in 1956 he became Professor of Dramatic Literature at YALE. In 1928 he began adapting foreign plays for the THEATRE GUILD among them Emil Ludwig's *Versailles* as *Peace Palace* (1931) and Stefan ZWEIG's *Jeremiah* (1939), and in 1940 he dramatized Robinson Jeffers's poem, *The Tower Beyond Tragedy*, for Judith ANDERSON. He also made modern acting versions of SOPHOCLES' *Antigone* and *Oedipus Rex*, and of *Everyman* and *The Second Shepherd's Play*. He wrote several books on the theatre, and edited a number of play anthologies, including ten volumes of American plays and about ten of English and Continental plays. He was drama critic successively of *New Theatre Magazine, Forum, Time, The Educational Theatre Journal*, and *Theatre Arts*.

GATE THEATRE, Dublin, was founded in 1928 by Michéal MACLIAMMÓIR and Hilton EDWARDS, to fill the gap left by the ABBEY THEATRE's concentration on Irish drama in the naturalistic mode. The company played for two seasons at the Abbey's experimental Peacock Theatre, and then moved to its present home in Rotunda Buildings, which seats 400. During their first season, they presented seven plays, including IBSEN's *Peer Gynt*, O'NEILL's *The Hairy Ape*, and WILDE's *Salome*. Plays by ČAPEK, GALSWORTHY, RICE, and TOLSTOY were staged the following year, together with a new Irish play, Denis JOHNSTON's *The Old Lady Says 'No!'* (1929), which had been refused by the Abbey.

The first production in the Rotunda Buildings, on 17 Feb. 1930, was GOETHE's *Faust*. Over the next few years the productions were equally ambitious, including SHAW's *Back to Methuselah* in its entirety and AESCHYLUS' *Oresteia*. The latter was translated by the Earl of LONGFORD, who with his wife became patron of the Gate in 1931. In 1936 a split developed between the founders of the theatre and the Longfords, as a result of which two separate companies played at the Gate for six months each until the death of Lord Longford in 1961.

In 1940 Edwards and MacLiammóir, unable to make their usual overseas tours to Egypt and elsewhere because of the Second World War, presented their plays at the larger GAIETY THEATRE, Dublin. After the war, the Gaiety continued to be used, though overseas tours were resumed. Meantime, the Gate, under the general management of Lady Longford, was let out to various companies all through the 1960s, and for a time was even closed.

In 1969 a government subsidy was awarded for the renovation of the Gate Theatre under Edwards, MacLiammóir, and Lady Longford, and on 15 March 1971 it officially reopened, with a production of ANOUILH's *Ornifle* as *It's Later than You Think*. With the aid of an annual Government subsidy the former policy was again implemented at the Gate, where the resident company mounts productions for six or eight months and makes the theatre available to other companies for the remainder of the year. Particularly popular at the Gate Theatre in the early 1970s were the vintage productions of GOLDSMITH, Shakespeare, Shaw, and Wilde in which the company's style is probably seen at its best; Irish authors are far from dominant in the Gate's programmes, which are drawn from the full range of classic and contemporary drama. The theatre's golden jubilee, 1978 (also the year of MacLiammóir's death), was marked by the publication of a history of the enterprise, *Enter Certain Players* by Peter Luke.

GATE THEATRE, London. The first home of this small experimental theatre club was in Floral Street, Covent Garden, on the top floor of a warehouse. Accommodating 96 people, it opened on 30 Oct. 1925 under the direction of Peter Godfrey with Susan GLASPELL's *Bernice* and had its first success a year later with KAISER's *From Morn to Midnight*, translated by Ashley DUKES. The company mounted 32 productions in two years before moving to new premises in Villiers Street, off the Strand, where it occupied part of the site formerly known as GATTI'S Under-the-Arches (also the Hungerford or Charing Cross Music-Hall), the rest being taken over later by the PLAYERS' THEATRE. The second Gate Theatre opened on 22 Nov. 1927 with Gantillon's *Maya*. The venture began well, with such productions as TOLLER's *Hoppla* and O'NEILL's banned play *Desire Under the Elms*, with Eric PORTMAN and Flora ROBSON; but by 1933 it was obviously failing and Godfrey withdrew. In 1934 Norman MARSHALL took over, opening on 1 Oct. with Toller's play on Mary Baker Eddy, *Miracle in America*. This was followed by J.-J. BERNARD's *Le Feu qui reprend mal* (as *The Sulky Fire*) and by *This Year, Next Year*, the first of a series of witty, intimate Gate revues starring Hermione Gingold. Among Marshall's outstanding productions were Elsie T. Schauffler's *Parnell* with Margaret Rawlings as Kitty O'Shea, a part she had previously played on Broadway; Leslie and Sewell Stokes's *Oscar Wilde* with Robert MORLEY as Wilde; Lillian HELLMAN's *The Children's Hour* (all 1936); AFINOGENOV's *Distant Point* (1937), one of the first and best Soviet plays to be seen in London; Josset's *Elisabeth la femme sans homme* (1938), with Lilian BRAITHWAITE giving a fine performance as Elizabeth I; STEINBECK's *Of Mice and Men* (1939); and COCTEAU's *Les Parents Terribles* (1940), with Martita Hunt and Cyril CUSACK. Several of these plays, including *Oscar Wilde*, *Of Mice and Men*, and the 1938 Gate Revue, were successfully transferred to commercial theatres. The last play to be produced at the Gate was Beckwith's *Boys in Brown*, a picture of Borstal life, which opened on 13 June 1940; on 16 April 1941 the theatre was extensively damaged by bombing and was not reopened.

GATEWAY THEATRE, Edinburgh, a small theatre, previously the Broadway cinema, which in 1946 was opened by the Church of Scotland for films and plays, the latter being provided by visiting companies or by a company directly engaged by the management. From 1953, however, the theatre was rented for about eight months in the year to an independent company under the chairmanship of Robert KEMP, some of whose plays were produced there. The Church's policy was liberal, though there were a few clashes over productions, and in 1961 the conduct of the Gateway was approved by the General Assembly of the Church of Scotland after a debate which recalled the controversy over HOME's *Douglas* two centuries earlier. In 1965 the company was wound up, passing on to the newly formed Edinburgh Civic Theatre Trust at the LYCEUM THEATRE 'its limited assets and its unlimited good wishes'.

GATEWAY THEATRE, Tauranga, see NEW ZEALAND.

GATHERER, a functionary in the Elizabethan playhouse whose task it was to collect the spectators' entrance fees from that part of the theatre, usually the upper gallery and the boxes, allotted to the HOUSEKEEPERS, or owners of the actual theatre building, as distinct from the SHARERS in the assets of the company.

GATTI, ARMAND (1924–), Monacan-born dramatist and film-maker, whose plays, directly political in content and purpose, created a stir in the years leading up to the upheavals in France in 1968. *Le Crapaud-bufle*, directed by VILAR in 1959, was for example a series of sketches dealing with France's attitude to international affairs. His best play, *Chant publique devant deux chaises électriques* (1966), deals with the execution of the two Italian anarchists Sacco and Vanzetti. *La Passion du général Franco*, which was to have opened at the THÉÂTRE NATIONAL POPULAIRE in 1968, was prohibited by the French government after protests from the Spanish ambassador. Gatti's work demands complicated sets, often involving puppets, mechanical devices, and complicated lighting plots, and has been most successful in these public or semi-public theatres in France which are interested in good political drama and are able to spend lavishly on them.

GATTI'S, London. In 1856 the brothers Carlo and Giovanni Gatti, with Giuseppe Marconi, had a restaurant in Hungerford Market. This was demolished to make way for Charing Cross Station. With the money they received in compensation they opened another restaurant in the Westminster Bridge Road, which in 1865 was licensed as a music-hall. Meanwhile the arches under the new station were being let, and in 1866 the Gattis acquired two of them, which they opened a year later as another music-hall. To avoid confusion it was known as Gatti's-Under- (or In-) the-Arches, the other being nicknamed Gatti's-Over-the-Water or in-the-Road. This was rebuilt in 1883 to hold 1,183 people in a two-tier auditorium, and reopened as Gatti's Palace of Varieties; it was there that Harry LAUDER made his first London appearance in 1910. It closed in 1924, and was demolished in 1950. Gatti's-Under-the-Arches was renamed the Hungerford Music-Hall in 1883, and later became the Charing Cross Music-Hall. It closed in 1903 and in 1910 became a cinema. From 1927 to 1941 part of the site was occupied by the GATE THEATRE and in 1946 another part of the premises was taken over by the PLAYERS' THEATRE.

The Gattis also ran a restaurant, the Adelaide Gallery, a former theatre for which they were unable to get a performing licence, and for some years owned the VAUDEVILLE THEATRE.

GAULTIER-GARGUILLE [Hugues Guéru] (*c.* 1573–1633), French actor, chief farce-player, with GROS-GUILLAUME and TURLUPIN, of the company at the Hôtel de BOURGOGNE to which he probably graduated from the Paris FAIRS. He was a tall, thin man with a dry humour much appreciated by Parisians. Under the name of Fleschelles he also played serious parts, particularly kings in tragedy, but is best remembered as a low comedian. He figures with other members of the company in Gougenot's *La Comédie des comédiens*, produced at the Hôtel de Bourgogne in 1631.

GAUSSIN [Gaussem], JEANNE-CATHÉRINE (1711–67), French actress, daughter of BARON's valet who made her first appearance at the COMÉDIE-FRANÇAISE in 1731 as Junia in RACINE's *Britannicus* and later succeeded Mlle DUCLOS in tragedy. She was at her best in roles demanding tenderness and grief rather than the portrayal of the sterner passions; with her dark, languorous eyes and rich voice (which, LA HARPE said, 'had tears in it'), she never lost her youthful look, and at the age of 50 could still play young girls. She appeared in several of VOLTAIRE's plays, notably *Zaïre* (1732), and was considered unrivalled in the sentimental comedies of LA CHAUSSÉE. In 1759 she made an unhappy marriage with a dancer at the Opéra, and retired from the stage four years later.

GAUTIER, THÉOPHILE (1811–72), French poet and journalist, and an enthusiastic supporter of the Romantic movement, who led the applause at the first night of Victor HUGO's *Hernani* (1830). He himself wrote a number of melodramas and comedies, but is best remembered as the author of librettos for Romantic ballets such as *Giselle*. His lifelong connection with the Paris theatres, and his loyalty to his Romantic contemporaries, make him one of the most interesting dramatic critics of his day.

GAUZE-CLOTH, see BACKCLOTH, CLOTH, and TRANSPARENCY.

GAVELLA, BRANKO, see YUGOSLAVIA.

GAY, JOHN (1685–1732), English poet and satirist, author of *The Beggar's Opera*, first given at LINCOLN'S INN FIELDS in 1728 under John RICH, thus, it was said, 'making Gay rich and Rich gay'. A light-hearted mixture of political satire and burlesque of Italian opera, then a fashionable craze, this BALLAD OPERA, one of the best of its kind, had music selected by Pepusch from well known airs, subsequently arranged by Linley. It was constantly revived and the music rearranged, notably by Frederic Austin for a production at the LYRIC, Hammersmith, in 1920, and by Benjamin Britten for SADLER'S WELLS in 1948. In 1928 a German adaptation was made by BRECHT as *Die Dreigroschenoper*, with music by Kurt Weill. *Polly*, a sequel to *The Beggar's Opera*, was not produced for many years owing to political censorship, but was finally given at the HAYMARKET in 1777 with alterations by the elder George COLMAN. Gay was also the author of several comedies and of the libretto of Handel's *Acis and Galatea* (1731).

GEGGIE, see GAFF.

GELBER, JACK (1932–), American dramatist, whose first play *The Connection* (1959) was produced by the LIVING THEATRE. A brutally realistic study of a group of drug addicts, it aroused vigorous controversy, both in the U.S. and in England, where it was seen in 1961 at the DUKE OF YORK'S THEATRE; but it helped to establish off-Broadway as a source of innovation in both form and subject-matter. *The Apple* (1961), *Square in the Eye* (1965), *The Cuban Thing* (1968), and *Sleep* (1972) attracted less attention, though they display considerable inventiveness. He directed WESKER's *The Kitchen* in New York in 1966 and KOPIT's *Indians* (1968) for the ROYAL SHAKESPEARE COMPANY in London, as well as *Jack Gelber's New Play: Rehearsal* (1976) at the AMERICAN PLACE THEATRE, where he also directed Sam SHEPARD's *Seduced* (1979).

GÉLINAS, GRATIEN (1909–), the first outstanding actor of the French-Canadian theatre, known as Fridolin from the leading character in the 'Fridolinons', revues which he wrote for production at the Monument National, Montreal, from 1938 to 1946. Two years later his play *'Tit-Coq* (*Lil' Rooster*) broke all Canadian records with a run of more than 450 performances in French and English; it was filmed in 1952. In 1958

Gélinas founded the Théâtre de la Comédie Canadienne, housing it in a former burlesque theatre, bought and redecorated with a donation from a brewery. The opening production was ANOUILH's *L'Alouette* (*The Lark*) in both French and English. Another of his own plays, the immensely successful *Bousille et les justes* (1959), was also played in both languages in Montreal and on tour, as was *Hier, les enfants dansaient* (1966), about a family divided by separatism. The company also staged plays by other Canadian authors, among them Marcel DUBÉ and Guy DuFresne, as well as a number imported from abroad. In 1956 Gélinas appeared with other French-Canadian actors at the STRATFORD (ONTARIO) FESTIVAL playing Charles VI in *Henry V* and Dr Caius in *The Merry Wives of Windsor*. He left the live theatre in 1970 to join the Canadian Film Development Corporation.

GELOSI, one of the earliest and best known of the COMMEDIA DELL'ARTE companies, which after an initial visit to France in 1571 was summoned to play before the French king, Henri III, at Blois in 1577, and from there went to Paris, thus inaugurating the visits of the Italian players which later resulted in the establishment of the COMÉDIE-ITALIENNE. In the company, which was attached to the household of the Duke of Mantua, were Francesco ANDREINI and his wife Isabella. After constant travelling, the Gelosi returned to Paris in 1603, but on their way back to Italy in 1604 Isabella died at Lyons and her husband disbanded the troupe.

GÉMIER [Tonnerre], FIRMIN (1869–1933), French actor, director, and manager, who played a major part in the theatrical revival of the 1920s. A pupil of ANTOINE, he made his first big success as the original Ubu in LUGNÉ-POË's production of JARRY's *Ubu Roi* (1896). Between 1892 and 1914 he played some 200 parts, and from 1906 onwards also managed the Théâtre Antoine. He was an inspired, if somewhat slapdash, actor, with great powers of rhetoric and ample gestures, and he had a remarkable ability for directing crowd scenes in which he also appeared. His interest in popular theatre was first shown in his production of Romain ROLLAND's *Le 14 juillet* in 1902, and in 1911 he founded the Théâtre Ambulant—a scheme which entailed moving around France by rail what was in effect the entire equipment of a 1700-seat theatre. During the First World War he helped to cement the Franco-British alliance by producing Shakespeare in Paris, and his sense of public responsibility, heightened by an important meeting with Max REINHARDT, led him to attempt with the help of Gaston BATY a series of gigantic dramatic events at the Cirque d'Hiver. In 1920 he became manager of the first THÉÂTRE NATIONAL POPULAIRE; at the same time he was managing the COMÉDIE DES CHAMPS-ÉLYSÉES, and from 1921 to 1928 the ODÉON. During this last period he successfully directed spectacular performances of foreign classics by such authors as GORKY, KLEIST, and TOLSTOY, as well as Shakespeare, and presented for the first time an eclectic assortment of new playwrights, including Paul FORT, SALACROU, STRINDBERG, O'NEILL, and SHAW. Towards the end of his life he left the Odéon and returned to his original idea of a travelling theatre. Much of his pioneer work later bore fruit in the emergence of a second Théâtre National Populaire under Jean VILAR and in the work of Jean-Louis BARRAULT.

GENERAL UTILITY, see STOCK COMPANY.

GÉNERO CHICO, or *teatro por horas*, generic term applied in Spain to a form of light dramatic entertainment dating from about 1868, which reached the height of its popularity towards the end of the 19th century. Deriving from the SAINETE, it consisted at first of one-act scenes of everyday life, usually set in Madrid and heightened to the point of caricature. As the fashion for musical accompaniment became more widespread, *género chico* became synonymous with the one-act ZARZUELA. Though still to be seen in Madrid and the provinces, it has been largely superseded by the fantastic *astracanadas*.

GENET, JEAN (1910–   ), French dramatist and poet, whose view of theatre as an act of revolt against society has been conditioned by an early life spent largely in correctional institutions and prisons. Although his novels reveal some self-indulgence in their glorification of the criminal and the pederast, his dramatic work is now generally recognized as important, not only for its extreme beauty of language but also for its creation of a disciplined, realistic form akin to that of the Theatre of CRUELTY advocated by ARTAUD. His first play, *Les Bonnes* directed by JOUVET in 1947, introduced his conception of a play as ceremony and masquerade. Under the mask the characters act out their dreams and secret desires, thus demonstrating the nullity of what is usually termed 'reality'. The ceremony, or ritual, imposed on the masquerade is designed, on the analogy of the Catholic mass, to unite the spectators in a metaphysical experience beyond normal conceptions of good and evil. This conception, implemented by the brilliant use of language, is developed and strengthened in *Haute surveillance* (1949); *Le Balcon* (produced as *The Balcony* in London in 1957; not seen in France until 1960, when it was also seen in New York), a sexual-political ceremony in which a brothel becomes the focus for a revolution; and *Les Nègres* (1959), which as *The Blacks* had a long run in 1961 off-Broadway, and was seen in London in the same year. *Les Paravents*, first performed in Berlin in 1961, deals obliquely yet perceptively with the Algerian struggle for independence; it had a stormy reception at the ODÉON in 1966. As *The Screens* it was given an abbreviated workshop production by Peter BROOK in 1964. Of Genet's earlier plays, *Les Bonnes*, as *The Maids*, was seen in New York in 1955 and in London in

1956; *Haute surveillance*, as *Deathwatch*, appeared in New York in 1958 and in London in 1961. In 1972 *The Balcony* was revived by the ROYAL-SHAKESPEARE COMPANY.

GENGENBACH, PAMPHILUS (*c*. 1480–1525), a printer by trade, resident in Basle and a supporter of the Reformation, whose *Totenfresser* (*The Corpse-Eaters*, 1521) is a savage attack on the Roman Catholic Church, which battens on the dead, growing rich on the money paid by credulous relatives in order to save the souls of the departed from Purgatory. It depicts the Pope, a Bishop, a lay brother, a monk, and a nun, sitting round a table carving up a corpse, while the Devil plays the fiddle. In the foreground the laity, led by a parson, pray for deliverance. The play is primitive in form, but anger imparts fire to its uncouth speeches. Gengenbach was also the author of a MORALITY PLAY, *Das Spiel von den zehn Altern* (1515), the first significant 16th century drama to be written in German.

GENTLEMAN, FRANCIS (1728–84), Irish critic, dramatist, and actor, the best of whose criticisms were published anonymously in 1770 as *The Dramatic Censor* in two volumes, the first dedicated to GARRICK, the second to Samuel FOOTE. Gentleman's remarks on contemporary actors at the end of each article are more valuable than his somewhat verbose criticisms of the plays of Shakespeare and of his own day. From 1752 to 1755 he acted as 'Mr Cooke' in Bath, where two of his tragedies, *Osman* and *Zaphira*, were produced. He was seen as Othello in Edinburgh, where his most successful play, *The Tobacconist* (*c*. 1760), an adaptation of JONSON's *The Alchemist*, was first staged, being seen at the HAYMARKET in London in 1771. In 1761 he appeared in his own play *The Modist Wife* at Chester, where he also played opposite MACKLIN; probably his best work, this was also seen at the Haymarket in 1773.

GEORGE, Mlle [Marguerite-Josephine Weymer] (1787–1867), French tragic actress, who made her début at the COMÉDIE-FRANÇAISE in 1802 as RACINE's Iphigénie. Her majestic bearing and fine voice assured her instant success in tragic parts, but in 1808 she suddenly eloped with a dancer named Louis Laporte and went to Russia, where she acted with a French company for five years. Back at the Comédie-Française, she was again successful until in 1817 her fellow actors, tiring of her ungovernable temper, asked her to resign. She went to London and on tour, and in 1822 returned to Paris to star at the ODÉON. She was reputed to have been successively the mistress of Napoleon, Talleyrand, Metternich, and Ouvrard, and for many years lived with the manager Harel, following him into the provinces and to the PORTE-SAINT-MARTIN in Paris, where she appeared with great success in romantic drama. Increasing stoutness led her to retire and she became a teacher of elocution, but her extravagance forced her back to the stage, where she found herself outmoded and forgotten. She finally retired to die in obscurity.

GEORGE, GRACE (1879–1961), American actress, wife of the manager William A. BRADY. She played the lead in many of her husband's productions, and was particularly admired in such parts as Lady Teazle in SHERIDAN's *The School for Scandal*, Barbara Undershaft in the first American production of SHAW's *Major Barbara* (1915), and the title-role in St John ERVINE's *The First Mrs Fraser* (1929). Her only appearance in London was at the DUKE OF YORK's in 1907 as Cyprienne in SARDOU's *Divorçons*, which many considered her finest part.

GEORGE ABBOTT THEATRE, New York, at 152 West 54th Street, between 7th and 8th Avenues. Seating 1,401, it opened on 24 Dec. 1928 as the Craig, but with little success, and after standing empty for some time it reopened on 27 Nov. 1934 as the Adelphi. It was still unsuccessful, and two years later was taken over by the FEDERAL THEATRE PROJECT, which retained control until its own dissolution in 1939. The most successful production of this period was Arthur Arent's LIVING NEWSPAPER on housing, *One-Third of a Nation* (1938). In 1947 the THEATRE GUILD put on a musical version of Elmer RICE's *Street Scene*, and the theatre was then used for radio and television. When it reopened as a playhouse in 1958 it was renamed the Fifty-Fourth Street Theatre, among its productions being CAMUS's *Caligula* (1960) and a musical, *No Strings* (1962). On 1 Jan. 1966 the theatre was renamed again, this time in honour of the playwright George ABBOTT, and a year later housed the American production of Peter USTINOV's *The Unknown Soldier and his Wife*. It was demolished in 1970.

GEORGE M. COHAN THEATRE, New York, at 1482 Broadway. This was opened on 13 Feb. 1911 by COHAN, who on 25 Sept. appeared there with his parents in his own play, *The Little Millionaire*, as he did again in *Broadway Jones* (1912). Among later successes were the Jewish comedy *Potash and Perlmutter* (1913) by Charles KLEIN and Montague Glass, the farces *It Pays to Advertise* (1914) by Walter HACKETT and *Come Out of the Kitchen* (1916) by A. E. Thomas, and several musical comedies, including *Two Little Girls in Blue* (1921). Clemence DANE's *A Bill of Divorcement* (also 1921) established Katharine CORNELL as a star and was followed in the same year by Ed Wynn in his own musical comedy, *The Perfect Fool*, from which he took his nickname. The last years of the theatre were uneventful, the final production being a musical, *The Dubarry* (1932), with Grace Moore. A year later it became a cinema, and in 1938 it was demolished.

GEORGIA, a constituent republic of the Soviet Union, which has had a long history of folk-drama and regional dances, but until the 19th century no professional actors, though plays were given by amateurs as part of the struggle against foreign domination. It was not until 1821, when GRIBOYEDOV went to live in Tbilisi (Tiflis), that with help from Russia a professional theatre emerged in Georgia. From then on it remained in close contact with both the Russian and Soviet theatres, one of its early leaders, Abashidze, having been a student at the MALY THEATRE in Moscow. A National Theatre, which in 1921 was renamed the Rustaveli, was founded in the 1880s, and was modernized and reorganized on Soviet lines by the Georgian-born Soviet actor MARDZHANOV, whose production of Lope de VEGA's *Fuente Ovejuna* soon after his return to Georgia in 1922 inaugurated a period remarkable for its productions of new and classic plays from Russia and other parts of Europe, staged and acted by those whom Mardzhanov had himself trained in modern methods. In 1979 the Rustaveli company was seen at the EDINBURGH FESTIVAL in a new Georgian version by Z. Kiknadze of *Richard III*, the title role being superbly enacted by Georgia's outstanding actor Ramaz Chkhikvadze. The production, directed by Robert Sturua in a set by Mirian Shvelidze, was seen in London at the ROUND HOUSE early in 1980. There are now a number of Georgian theatres throughout the country, a second State Theatre in Tbilisi founded in 1930 by Mardzhanov now being named after its founder. The repertory there, as elsewhere in Georgia, is international, and the acting, much influenced by STANISLAVSKY, shows the national characteristics of vigour and feeling so noticeable in Georgian dancing and also great beauty of movment, the result of basic training in ballet.

GEORGIAN THEATRE, Richmond, Yorkshire, one of the three oldest working theatres in England, the others being in BRISTOL and BURY ST EDMUNDS. Small, with a rectangular auditorium originally seating 400, it is unique in still having its original proscenium. Built by Samuel Butler, a provincial manager whose CIRCUIT included Harrogate, Kendal, Ripon, and Whitby, in all of which he built theatres, it opened on 2 Sept. 1788. In 1811 it assumed the title of Theatre Royal apparently without justification. Butler died in 1812, his wife and his son Samuel Butler (1803–45) continuing his work for a time; but when the lease of the theatre ran out in 1830 Butler dismissed the company and went to America, where he became well known as a tragedian. In 1848 the theatre finally closed, the lower part becoming a wine vault, the upper an auction room, and the sunken pit being boarded over. In 1943 the unrestored building was used for a performance in commemoration of the 850th anniversary of the enfranchisement of the borough, and in 1960 a Trust was formed to restore and redecorate it. It reopened in 1963, and now seats 238, the stage being 24 ft wide and 27 ft 8 ins deep, with a proscenium opening of 17 ft. There are two seasons of about eight weeks each in spring and autumn and a Christmas season, and the building is also used for concerts and by amateurs. A theatre museum was opened in 1979.

GERMANOVA, MARIA NIKOLAEVNA (1884–1940), Russian actress, who made her début at the MOSCOW ART THEATRE in 1904 as Calpurnia in *Julius Caesar*. Her intensity and delicate appearance struck NEMIROVICH-DANCHENKO as ideal for the more expressionist dramas with which he tried to counterbalance STANISLAVSKY's naturalism, and among her many successful roles were Agnes in IBSEN's *Brand* in 1906, the Fairy in MAETERLINCK's *The Blue Bird* in 1908, Grushenka in DOSTOEVSKY's *The Brothers Karamazov* in 1910, and the title role in ANDREYEV's *Katerina Ivanovna* (1912), which was written for her. From 1922 to 1929 she was on tour with the Prague Group of the Moscow Art Theatre, visiting Paris in 1926 and London in 1928. She retired in 1930.

GERMANY. Indigenous pagan rites are now thought to have been of far greater importance in the growth of religious as well as of secular German drama than was formerly conceded, a theory which helps to explain the presence of grotesque and comic elements in LITURGICAL DRAMA. Thus the FASTNACHTSSPIEL or Carnival Play, which emerged in the 15th century, far from being a mere offshoot of the *Mysterienspiel* or MYSTERY PLAY, frequently shows cognate native elements in an unchristianized form. No doubt the wandering MINSTREL was also instrumental in the propagation of medieval farce, at any rate before it became the preserve of the burghers, while the Church, by admitting the practice of the FEAST OF FOOLS, also furthered its development.

The Carnival plays which form the pre-Reformation secular drama of Germany attacked what the townsman disliked or despised, but lacked the satirical pungency and the political element of the French SOTIE as well as the charm of the English INTERLUDE, for they had little contact before the 16th century with either SCHOOL DRAMA or with Courts, petty or imperial. This is an instance of the way in which the German theatre has, like the Italian, been affected from the first by the absence of a central focus of culture such as Paris or London.

Religious subjects still appeared in these secular farces, as for example in a 'Debate between Church and Synagogue', where the punishments meted out to the losing party are so revolting as to preclude ecclesiastical patronage. Contemporary events are reflected, as in a play about the Turkish wars, and there are links with ancient pagan festivals, as in the plays of the Austrian dramatist NEIDHART VON REUENTAL. In these plays the weaknesses of lawyers and their clients, or doctors and their patients, are realistically portrayed, and the illicit sexual relations of the clergy are shown up with gusto. Indeed, all human failings that undermine domestic peace—unfaithfulness, quarrelling, gluttony, avarice, bullying, lack of physical control—are described with disconcerting frankness, but not usually shown in action; rather, the problem is ventilated before a magistrate or some other person in authority, or the help of a doctor or friend is sought (in which case the advice given often redounds on the giver). A delightful example of the former type is found in *Rumpolt und Marecht*, which dramatizes a breach-of-promise case in which Rumpolt desires to rid himself of Marecht, who declares that she has just cause to demand the marriage she eventually achieves.

A variable feature of many 15th and 16th century German farces is the NARR or fool, who is not necessarily funny but in the Carnival plays is usually made to appear so. These early farces, in one act of variable length, were probably performed by youths, usually artisans, from the neighbourhood, and their jokes were doubtless largely personal. Gradually the plays assumed a more general significance, and groups of such actors, having gained a certain competence, occasionally visited other neighbourhoods; but no organized companies on the scale of the Parisian *enfants sans souci* are recorded in Germany. At first the presentation of the plays must have been extremely simple, performers and audience being almost one, and both joining in the final dance. Originally the actors stepped forward to speak from a seated row or semi-circle; but as more action was introduced, properties such as chairs, tables, and so on would have been required, so that it became more convenient to raise the actors above the level of the audience on tables and trestles. But the practice of having all the actors on stage from start to finish continued, and as the action grew more elaborate accommodation had to be found for more than one group at a time. The stage was therefore erected in the open air, or, if permission could be obtained, in the hall of a public building. As the process of elaboration continued there was no doubt some idea of emulating the magnificence of the religious plays or the richer setting of the SCHOOL DRAMA, but not until the middle of the 15th century, when the guilds undertook the performances, was public money forthcoming under the stimulus of local pride. The most active of these guilds were of course the MEISTERSÄNGER, who flourished in the large towns of southern Germany.

Serious secular drama was hardly known in Germany at this period. The *Tellspiel* from Uri, celebrating the winning of Swiss independence, probably belonged to the 16th century, and since the unique work of HROSWITHA in the 10th century Latin drama had lain fallow. In 1450 the University of Heidelberg acquired some manuscripts of TERENCE and SENECA, and intensive study of these authors ensued; but for a time admiration for the early Renaissance drama in Latin as it developed in Italy, and the use of the dialogue for argument, retarded the adoption of classical dramatic form.

Latin comedy and its derivatives were first in the field, notably in the work of Johannes Reuchlin (1455–1522), and tragedy did not on the whole flourish in 16th-century Germany. Seneca was republished, commented on, and translated, but only for educational purposes. Of the Greek tragedians EURIPIDES was the first to evoke interest, Erasmus and Melanchthon making translations of which a few performances are recorded. The German humanist drama was thus important rather as a model of form, and for the importation of classical themes, than for positive achievement, though Nikodemus Frischlin (1547–90), author of several plays in Latin on Biblical subjects, was also known for his spirited but indiscreet comedies, mainly based on ARISTOPHANES. By the end of the 15th century religious drama, though widespread, had largely deteriorated into pageantry, and was beginning to yield to the onslaught of the Reformation, while the traditional farces relied on the rehandling of old material and on a somewhat offensive coarseness. The Reformation brought new life and fresh material, however, certain practices of the Church being attacked, particularly in the free cities of Switzerland and in the work of Pamphilus GENGENBACH, with a violence hitherto unknown in German drama. The work of Niklaus MANUEL of Berne was also strongly Protestant in sentiment, and often more subtle in its methods. Elsewhere, too, the old dramatic forms were used in support of the Reformation—in Riga the former monk Burkhart WALDIS turning even the parable of the Prodigal Son against the Catholics. Thus it was only to be expected that Protestant dramatists would make use of the *mise en scène* of the PASSION PLAY, the dominant German form of the *Mysterienspiel*, and a striking example of this occurs in the *Pammachius* (1538) of the humanist Thomas KIRCHMAYER. Nothing of comparable vigour was produced by the Catholics until the second half of the 16th century, when JESUIT DRAMA employed play-acting as an effective method of training and propaganda. The increasing number of printed copies of both Protestant and Catholic plays indicates that their authors could now count on a reading public.

The early dramatists of the Reformation had been content to use popular or traditional forms, but sporadic division into acts and scenes reveals the influence of classical studies. The humanist movement had been gaining ground in Germany towards the end of the 15th century, and the mere fact that Latin plays were acted with much pomp and ceremony in the school and universities led to a wider acquaintance with the form of classical drama. Religious drama had required a more elaborate type of staging than the Fastnachtsspiel, as the action moved from one part of the stage to another, but for performances of Terence a special stage had been evolved in Renaissance Italy, ostensibly on a classical pattern, and this TERENCE STAGE was a deliberate attempt to counteract both elaboration and naturalism. Later in the 16th century the desire for realism again prevailed, and a series of scenic houses on both sides of a sharply converging prospect was used, terminating in a painted backcloth. This technique was also of Italian origin, and it was rendered more flexible by the use of *telari* or triangular prisms revolving on pivots. Such a stage may have been used by Paul REBHUN, the major dramatist of the period, whose natural sense of act-division was in marked contrast to the purely mechanical divisions of most of the longer plays. Hans SACHS, for example, recognized no special virtue in the five-act play as distinct from the two- , seven-, or even eleven-act play. To him belongs, however, the honour of having instituted the first German theatre building, when in 1550 he took over the Marthakirche, disused since Nuremberg had adopted Protestantism. His stage was certainly a simple one: curtains with slits screened the side and back; visible steps led from the floor-level to the sides, hidden steps to the back; there was a floor TRAP and a pulpit; and existing steps may have been incorporated. Isolated features such as doors were probably of painted wood, and properties were brought on to the stage and removed by the actors in the course of the play. As in the case of School Drama, the audience sat facing the stage, and not on three sides of it. In the Carnival play, which Sachs purged of indecency without abating its fun, the acting must have been fairly realistic but in serious plays he could hardly have followed the example of the School Drama, which laid its principal emphasis on delivery and facial expression, and so probably fell back on the stylized gestures of the old religious play. As his stage was so small, crowd scenes must have been resolved into single incidents. In the case of persons of exalted rank, the costume seems to have been traditional, while the lesser characters wore their ordinary clothes.

Sachs clearly gave some training to his actors, although they could in no sense be regarded as professionals, and by the 1580s German drama was gropingly making headway, hampered by tradition. There is evidence that by slightly raising one portion of the stage and using a partition with an exit to a lower level, interior scenes were practicable alongside the usual scenes in public places. There were other experiments too; in the *Comedi vom Crocodilstechen* (*c.* 1596) a stone crocodile in effigy on a wall scares the beholders and sets the whole of Nuremberg, including contemporary notabilities, by the ears. Apart from its personal element, this racy Carnival play in two parts is well in the Sachs tradition.

The real quickening of dramatic life in Germany came late in the 16th century from the ENGLISH COMEDIANS. Many German princes, among them Heinrich Julius, Duke of Brunswick, himself a playwright, engaged these strolling companies for periods varying in length and lodged them in their castles. Otherwise, for performances in towns, permission had to be obtained from the authorities, usually after much bargaining as to the price to be charged for entrance and seats. The players carried all their apparatus with them, so the scenery must have been modest; but their Shakespearian stage clearly offered them greater possibilities than that used by Hans Sachs. Their repertory, in which most of the Elizabethan dramatists were represented, was characterized by the complete absence of the moralizing tendency prevalent in contemporary German drama, and the vivid acting of these professionals—masters of gesture and facial expression, supported by superb fooling, expert dancing, fencing, and acrobatics—raised the standard of acting in Germany for all time. German authors could not afford to ignore this new importation. In the south Sachs still had a great following, and Jakob AYRER, his successor, retained much of his manner, including the doggerel verse and the old verbosity, but he learned many tricks of the trade from the English Comedians, and his stage must have been of a fair size to accommodate his large casts.

Given the popularity of the English actors and the slow progress of the Renaissance in Germany owing to the Reformation and the wars of religion, it is not surprising that the general trend of drama about 1600 was towards noisy hilarity or gory sensationalism. It was not until the turn of the century that comparison with the achievements of other countries opened the eyes of patriots to the backwardness of the German stage. The opening years of the 17th century were, however, even before the outbreak of war in 1618, a period of disintegration. The well-meant efforts of numerous literary societies based on the Italian *Accademia della Crusca* were unavailing until Martin Opitz (1597–1639) advised young writers to look to the classics and to France for guidance. His almost unchallenged authority resulted in a complete break with national tradition, and in an academic literature, smooth, not ungraceful at times, but devoid of life and action. Towards the middle of the 17th century, under the stress of suffering and religious perplexity, the more highly coloured exuberance of the Italian seicento gained ground in Germany, mainly through the influence of the Jesuit Counter-Reformation. It produced a rhetorical style, expressive of pent-up emotion controlled by rationalistic stoicism; and this baroque mentality produced in its turn German tragedy. At the lower end of the scale the strolling players struggled on, being rapidly eclipsed by their more moneyed rivals, the Court opera and the Jesuit productions; their decline is indicated by the fact that the middle of the 17th century saw managers resorting to marionettes. Protestant School Drama, also active, did not allow itself to be entirely outshone by its Catholic rivals but it evinced greater sobriety, both of necessity and on purpose, as is evinced by the plays of Andreas GRYPHIUS, the major dramatist of his time, who typified the clash of imperious desires and stoic endurance. His stage, like that of the Jesuits, was no longer the simple Terence stage, but a deeper area divided into three sections by curtains, with painted scenery that moved on ropes, the front portion being used for intimate or indoor scenes, the full depth for important or crowded scenes. With the increased interest in the past and in other countries characteristic of this age, costumes were more carefully differentiated, and lighting effects were freely used. The prevalent martyr-plays demanded flaming stakes, descending clouds, and so on. Avoiding the ultra-rhetorical gestures of the Jesuits, Gryphius appears to have aimed at as much expressiveness as was consistent with dignity. Comedy, of course, allowed of more realistic acting; and women, who had for some time appeared in strolling companies, were now allowed on the serious stage.

The baroque drama proper reached its peak in the work of Daniel von LOHENSTEIN and his imitators, and towards the close of the century a sharp reversal of taste began to occur—in the direction of realism on the one hand and simplicity on the other. The former tendency was manifest in the plays of Christian WEISE, in which all the characters, whether historical, Biblical, fanciful, or farcical, are made to speak in the same vein of homespun prose; the second was marked by an increasing number of translations and adaptations from French classical drama, and was first revealed in the work of Johannes VELTEN, who strove to improve the status of actors and to educate them and the public by such translations. Although Velten was forced eventually to revert to the popular HAUPT- UND STAATSAKTION to replenish his coffers, in the last decade of the century French classical tragedies were being performed at the Court of Brunswick in imitation of the Parisian scene.

The 17th century had seen the emergence of the professional actor, whose improvisations had invaded the domain of the dramatist; in contrast the 18th century saw the dramatist reinstated and an advance made towards the establishment of the German theatre on a permanent basis. Johann GOTTSCHED joined forces with the philosopher Wolff and other advocates of enlightenment in their endeavours to educate the German public; the French theatre, with its decorum and its easily formulated rules was his ideal as it was that of all the petty German princelings who set out to emulate VERSAILLES, and of BERLIN, where VOLTAIRE had brought in a French company to play before Frederick the Great. It is undoubtedly true that the German theatre as Gottsched found it was in a poor way. The public was dazzled by the splendour of opera, its taste vitiated by crude dramatic fare, and there were few German plays of any note. Gottsched, ably assisted by his wife, tried to fill the gap, both with translations and with original plays, all written in sober prosaic alexandrines—a far cry from the improvised fooling of HANSWURST and his companions. Good fortune brought him into contact with the best theatrical company of his day, that of Carolina NEUBER, and his alliance with her lasted from 1727 to 1739. Although Gottsched's preoccupation with detail and rigid outlook made their eventual separation inevitable, 'la Neuberin' did succeed in combining French stateliness with German emotion and so inaugurated a new era in German acting.

Gottsched found compensation for her loss in the companies of Heinrich KOCH and Johann SCHÖNEMANN, both of whom had worked under Carolina Neuber; but his day was over. By 1740 everyone was acclaiming the superiority of English imaginative poetry, and even the most gifted of his disciples, J. E. SCHLEGEL, was championing the *Julius Caesar* of Shakespeare in the face of Gottsched's condemnation. Yet Schlegel himself in his plays adhered strictly to Gottsched's ideal of French classical drama, and the force of his authority may be judged by the violence with which LESSING and others attacked it. Schlegel's own tragedies are chiefly remarkable for the fact that several of them deal with episodes from German history; and this vein was further exploited by the poet KLOPSTOCK, whose emotional scenes lacked dramatic life but were symptomatic of a rising national consciousness. The playgoing public was at last being diverted from the stilted heroes and heroines of French origin not only by the new regard for Shakespeare, but by the converging streams of the COMÉDIE LARMOYANTE and the DRAME, in which a strong middle-class element was asserting itself. The moralizing sentimentality of Richardson's novels, then very popular in Germany, fused with the sterner accents of LILLO's *The London Merchant* to produce a type of drama truly German in character, and of vivid actuality, in which verse was to give way to prose—the BÜRGERLICHES TRAUERSPIEL.

The first notable play of this genre was Lessing's *Miss Sara Sampson* (1755), but between this and his only other essay in the form, *Emilia Galotti* (1772), appeared *Minna von Barnhelm* (1767), a comedy in which the middle classes are made to inspire affection and respect. For the first time in German drama the characters were truly three-dimensional and German to the core. The play was also new in that it dealt with a topical subject, the aftermath of the Seven Years War.

From the time of Carolina Neuber until the late 1760s the German theatre presents a confused picture of companies formed, broken up, and reshuffled, and all homeless, all dependent on the whim of the public or a patron, all wandering incessantly from Court to town, and town to Court, mostly in Germany, but with excursions abroad to Zurich, Vienna, even Russia, under the leadership of such players as Christian Speigelberg and Franz SCHUCH. Yet not long after came a number of eminent actors who in the course of the century were to bring the German theatre to the front rank of European art, notably under Konrad EKHOF (who also founded the first academy of acting in Germany in 1753, thus raising his calling to the status of a profession) and Konrad ACKERMANN, who sought to establish in HAMBURG the first German 'national' theatre. Members of his distinguished company included Sophie HENSEL, Karoline KUMMERFELD, Friedrich SCHRÖDER, and Johann BOECK. His venture was a failure, as was a second attempt at Gotha under Ekhof, who went there when fire destroyed the theatre at WEIMAR in 1774; but under him the great actor IFFLAND had done excellent work, and in 1779 he was invited by Count DALBERG to establish yet a third 'national' theatre at MANNHEIM. This met with some success, the outstanding events of its early years being the first performance of SCHILLER's epoch-making play *Die Räuber* (1782) and the efforts made by Schröder to establish Shakespeare in the repertory. Some attempts had already been made to force *Hamlet* and *Romeo and Juliet* into the prevailing mode, notably by Christian Weisse, but it was not until GOETHE's *Götz von Berlichingen* (1773) had broken down the barriers of the old 'regular' play that Shakespeare's plays could be acted and appreciated. Schröder, who had seen *Hamlet* in Vienna (though arranged *à la viennoise*), had already taken the bold step of putting it on for an astonished Hamburg audience in 1776, with BROCKMANN in the title role and himself playing the Ghost. Certainly the influence of Shakespeare—in the boldness with which the characters are sketched, their wide range, the utter disregard of the UNITIES, the historical perspective—is clear in *Götz von Berlichingen* as in all the plays of the STURM UND DRANG movement. In many of its productions the poetic licence of Shakespeare was actually exceeded; and even where the form was not irregular, the quickened pace swept the audience into a different world where the action was spotlit, or a mood reflected, in short, vivid scenes, as in the work of Jakob LENZ. The break with classicism was complete, and the informality, even slovenliness, of the language also called for increased realism in acting. An incidental change brought about by the historical RITTERDRAMA, an offshoot of *Sturm und Drang* which Goethe and KLINGER pioneered, was the acquisition by a number of companies of wardrobes of medieval costumes, which led to a greater use of the picturesque element in production.

By the time of the French Revolution in 1789 the literary upheaval in German had spent itself. In *Egmont* (1788) Goethe had abandoned the use of short scenes and rough-hewn language, returning to a regular form and blank verse of exquisite mellowness. Schiller, still avid for freedom, sought it now in the realm of thought and conscience, and gave it expression in verse even more vibrant that Goethe's. It was Iffland, ably assisted by FLECK, who first presented Schiller's later work to the public beyond Weimar where it had first appeared. In the 1790s the campaign against the French revolutionaries had caused much disorganization in the Mannheim theatre; Iffland, by then in charge of production, went on tour and appeared with great success in 1796 in Berlin, until then much in the rear of theatrical development. He was promptly appointed director, by royal decree, of yet another 'national' theatre. During the French occupation of the city he had to adapt his repertory to the demands of the authorities, but with the return of the royal family he once more received a subsidy, and music and ballet as well as drama were entrusted to him.

When in 1792 Goethe was entrusted by the Grand Duke with the entire responsibility for the Weimar Court theatre, he went to work with his usual thoroughness and circumspection. By this time the naturalism fostered by the domestic conversation-piece and the *Sturm und Drang* plays had become popular among dramatists to the exclusion of all else, the worst offender in Goethe's eyes being August von KOTZEBUE whose works had to be included in the Weimar repertory to swell the box-office takings. As the years went by Goethe's horror of realism increased, the Greece of SOPHOCLES becoming his mental refuge. He sought to inculcate in his actors a technique of dignified gesture and declamation, very different from the rhetorical style prevalent in France at the beginning of the century. They were not allowed to turn away from the audience, or to diverge from carefully prescribed groupings, or to attract undue attention to themselves by departing from a prescribed cadence of phrasing in their lines. For the first time an artistic ensemble was made the prime purpose of dramatic presentation; but unfortunately the artificiality which so much careful preparation engendered increased after the death of Schiller, and the rugged humour of such a play as KLEIST's *Der zerbrochene Krug* (1808) was smoothed out, with disastrous results.

The achievement of the 18th century in the German theatre had been the employment, in the absence of centralization, of drama as one of the main vehicles of culture and as a potentially important factor in social and literary life. On the other hand, the profession of the actor was still hampered by insecurity and lack of funds, and there was a disquieting cleavage between the intellectual leaders and the general public. Nor could the Romantic movement, which began in about 1798, bridge the gap, since the highly intellectual and individualistic outlook of its writers was not favourable to drama. They produced no dramatic works of lasting value, their main characteristics being frequent use of the supernatural, mythical, and mysterious, the accentuation of abnormality, and indefiniteness of form. The ironical twist which Ludwig TIECK gave to his dramatized folk-tales robbed them of their vigour; but his public readings from Shakespeare's plays probably did more to make the latter's genius known and loved in Germany than all the theorizing of the Romantics. Meanwhile the authors of the so-called SCHICKSALSTRAGÖDIE, or Fate Drama —Zacharias WERNER, Adolf Müllner, and the young GRILLPARZER with *Die Ahnfrau* (1817)—turned into sheer malignancy the power Schiller had revered in Sophocles. Only one dramatist of genius appeared—Heinrich von Kleist who, neglected by his contemporaries, strove to unite in his work the grandeur of the fate element in antiquity with Shakespearian character-drawing.

The German theatre in the opening decades of the 19th century seemed to be in a state of stagnation. The Napoleonic wars hampered enterprise, and after the battle of Waterloo the reactionary Metternich régime, which excluded social, political, and religious subjects from public discussion, reduced the theatre to a mere place of entertainment in which WELL-MADE PLAYS by SCRIBE, the works of French writers of VAUDEVILLE, and the dramas of OEHLENSCHLÄGER from Denmark occupied the major part of the repertory. The chief German purveyors of light entertainment were Eduard von BAUERNFELD with his conversation pieces; Roderich Benedix (1811–73) with his bourgeois domestic interiors; Ernst Raupach (1784–1852), would-be successor to Schiller; and Friedrich Halm (1806–71). To these may be added Elias Niebergall (1815–43) in Frankfurt and Charlotte Birch-Pfeiffer (1800–68), wholesale purveyor of other people's goods (her *Die Waise von Lowood* of 1856 is based on Charlotte Brontë's *Jane Eyre*), who outshone even Bauernfeld in popularity. The only writers to rebel against this emasculation of the stage were the politically-minded members of the *Junge Deutschlands* (Young Germany) movement, with the result that official vigilance increased until the troubles of 1848 produced some relaxation. The movement produced at least two dramatists of vigour and vitality, Christian GRABBE and Georg BÜCHNER. Karl GUTZKOW was also connected with it, while the work of his contemporary Gustav FREYTAG marked the transition from unrest to the solid bourgeois endeavour characteristic of the first half of the 19th century—the so-called *Biedermeierzeit*.

If the written drama in Germany during the inter-revolutionary years was on the whole mediocre, some enterprise in the theatrical field was evident with the able, often brilliant acting of Ferdinand ESSLAIR and Ludwig DEVRIENT; but the individual actor still tended to display his powers to the detriment of artistic unity. Moreover, the dispersion of theatrical companies among so many small states meant that only in a few large towns possessing several theatres could they cultivate a style of their own. Thus, in addition to Tieck's work in Dresden, Karl IMMERMANN founded a colony of artistically-minded men at the Städttheater in Düsseldorf, based partly on the Weimar style but avoiding its later exaggeration. Though he was not on the whole successful, certain technical improvements which he helped to sponsor became widespread at this time. Scenery was lighter and more easily moved, gas-lighting was introduced after 1830—though Cologne still considered it too dazzling in 1837—and in Berlin, the most highly subsidized of the Court theatres, Iffland's successor von Brühl insisted on greater accuracy than hitherto in historical costume.

The work of Richard Wagner, although belonging mainly to the domain of opera, exercised a great influence over the theatre in general. He may be regarded as the fulfilment of the Romantic ideal—the total fusion of poetry, music, and the pictorial arts. His most notable contemporaries in the field of pure drama were Friedrich HEBBEL, whose first play *Judith* was produced in 1840, and Otto LUDWIG, whose *Erbförster* (1850) is linked by some of its characters to the traditional domestic tragedy while anticipating the future in its vigorous realism. By the time of Ludwig's death in 1865 a great change was coming over Germany. Successful wars had brought her political supremacy, unification had stimulated industrial enterprise, and the sudden rise in prosperity had created extravagant hopes, naïve national pride, and rampant materialism. All this was expressed in the plays of Ernst von Wildenbruch (1845–1909), but no other dramatist of note marked the opening years of the Empire, founded in 1871. Among the outstanding actors of the time were August BASSERMANN and the younger members of the DEVRIENT dynasty, while the Duke of Saxe-Meiningen combined at his Court theatre the earlier endeavours of von Brühl in Berlin and Dingelstedt in Weimar with a new desire for verisimilitude, and for historical accuracy in scenery and costume (see MEININGER COMPANY). The chief difficulty of the new method was the necessity for avoiding frequent changes of scene, and this was not finally overcome until in 1896 Karl Lautenschläger introduced the REVOLVING STAGE.

Although in 1876 the opening of the new opera-house at Bayreuth had raised high hopes of artistic progress, in Berlin at any rate there was by 1883 a feeling of stagnation, intensified by the increasing practice of long runs. The new movement in drama, NATURALISM, being a movement of revolt was in bad odour socially and politically. Yet the more thoughtful of the younger generation were shocked by the national complacency and by the misery of the working class under the new industrialism, and IBSEN's plays gave further impetus to their discontent. The DEUTSCHES THEATER in Berlin, founded by Adolf L'ARRONGE with the actors Josef KAINZ and Agnes SORMA, was anxious to open its doors to the new dramatists, but there was trouble with the censor and in 1880 a private company, the FREIE BÜHNE modelled on the THÉÂTRE LIBRE in Paris, came into being under the aegis of the critic Otto BRAHM and the actor Rudolf Rittner. There the work of the young Gerhart HAUPTMANN, vibrating with actuality, found a home, his chief rival in popularity being Hermann SUDERMANN, a clever playwright with an unfailing if self-conscious sense of the theatre.

Lesser exponents of naturalism included Arno HOLZ and Max Halbe (1865–1944).

The founding of the Freie Bühne was followed, when the need for providing good plays at low prices for the working class was realized, by the founding of the Neue Freie VOLKSBÜHNE in 1890 and the Schillertheater in 1894. In the new drama that was sweeping across Europe Berlin was well to the fore. The Meiningen methods of staging could be adapted to the requirements of naturalism, provided such conventions as facing the audience, speaking standard German, and so on were scrapped, for the stress laid on the darker side of life and on the frailties of human nature demanded from the actor, as from the dramatist, the most minute observation and the subordination of the actor to his role, making the play as a whole a 'slice of life'.

A reaction against naturalism originated in MUNICH and was a sign of the gradual ascendancy of the southern capital. Frank WEDEKIND acted his own heroes, ably supported by Gertrud EYSOLDT, with flamboyant irony, in garish settings which suggested the cabaret, that product of modernity which, notably in the Berlin of the 1920s, was to become a serious rival to the legitimate theatre. By the turn of the century neo-Romantic drama, written mostly in verse and offering opportunities for spectacular scenery and gorgeous costumes, was beginning to rival the naturalistic drama in popularity, due in part to the European success of the plays of MAETERLINCK and of Oscar WILDE's *Salome* (1894). Herbert Eulenberg's *Ritter Blaubart* (1905), Ernst Hardt's *Tantris der Narr* (1907), and Wilhelm Schmidtbonn's *Graf von Gleichen* (1908) are examples of the German equivalent of this genre. Ironically enough, a neo-classicism which began to manifest itself in these years seems to have been an offshoot of neo-Romanticism rather than a reaction to it, although Paul ERNST, and Wilhelm von Scholz in *Der Jude von Konstanz* (1905), followed their models Schiller and Hebbel in writing plays classical in form and with a clearly expressed ethical content.

With the greater opportunities for scenic variety and splendour offered by these different types of drama, and with the invention or increased efficiency of various technical devices, the role of the director as a co-ordinator in the theatre became of increasing importance. The age found its man in Max REINHARDT, who was director of the Deutsches Theater from 1905 to 1920, with Max PALLENBERG, Alexander MOISSI, and Albert BASSERMANN in his company, and again from 1924 to 1933. His production of *The Merchant of Venice* in 1905 marked the beginning of one of the most brilliant eras in Berlin's theatrical history.

Two authors whose works marked the beginning of a change in the dramatic climate at this time were Carl STERNHEIM, who elaborated the element of savage social satire found in the plays of Wedekind, and Georg KAISER, who may be described as the inheritor of Wedekind's glorification of sex as representative of the life-force in a decadent world—a theme he later combined with a slightly irrevent attitude to the Scriptures and the German classics. Kaiser is generally regarded as the leading dramatist of EXPRESSIONISM, but the pioneer of the movement as far as drama is concerned was Reinhard SORGE, whose first play *Der Bettler* (1912) had all the elements later considered typical of the genre, which now found their way into the plays of a number of Sorge's contemporaries. Meanwhile, Kaiser's *Die Bürger von Calais* (1914) developed the idea of 'der neue Mensch' in highly experimental language. Short explosive phrases separated by dashes replaced grammatically constructed sentences, articles and pronouns were omitted as unessential, and verbs and nouns stood starkly without adjectives—a 'telegram style' typified also by the later plays of August Stramm. In the work of Fritz von UNRUH these linguistic experiments developed out of an attempt to convey the mood peculiar to the clipped speech of army officers; but where Unruh's backgrounds became increasingly metaphysical, those of Ernst BARLACH, for instance, were often localized. Reinhard GOERING's choice of a particular contemporary incident as the subject of his *Seeschlacht* (1917) illustrated the division during the First World War between those dramatists mainly interested in the spiritual rebirth of the individual and those more concerned with the concrete problems of political and social reform thrown up by the Russian revolution of 1917. The first tendency, is represented in an already well developed form, by the *Himmel und Hölle* (1919) of Paul Kornfeld, which shows the spiritual progress of the characters from devils to angels in a series of stages—a form already present in embryo in earlier expressionist plays. This technique was also employed by the main representative of the opposite or 'activist' tendency, Ernst TOLLER, while the contemporary political and social scene was more specifically criticized in the adaptation of Sophocles' *Antigone* by Walter HASENCLEVER. Such plays were indicative of a coming change in the tone and mood of expressionist drama. The idealism expressed, sometimes grimly, in the adverse years of the war, and the high hopes placed in the future and a new generation of men, barely outlasted the peace of 1919, which was followed by social upheaval and political revolution. The second part of Kaiser's *Gas* (1920) is typical of the pessimistic turn taken by expressionism from that time onwards, and prepares the way for the decline of its 'ecstatic' form. The treatment by Hanns Johst of the *poète maudit* in *Der Einsame* (1917) was very popular at the time, but now appears ranting and emotional; the *Spiegelmensch* (1920) of Franz WERFEL complicates the stages of a soul's journey to redemption by giving it a reflection and making it play hide-and-seek with itself; while Toller's genius for conveying the physical horrors of reality was carried to embarrassing extremes in *Hinkemann* (1924).

By this time, however, a generation of writers

too young to have taken part in the inception of the movement was beginning to make its presence felt, its members including Ferdinand BRUCKNER, Arnold BRONNEN, and Hans Henny JAHNN—in whose *Kronung Richards III* (1922) the excesses of expressionism in decline became an end in themselves. Oskar Kokoschka's (1886–1980) theme, the revolt of man's animal nature in a society ordered with unnatural severity, really belongs to the pre-expressionist period of Wedekind; but when his *Mörder, Hoffnung der Frauen* (*Murderer, Hope of Women*, 1909) was revived in 1919 it seemed to reflect the moral and social chaos of post-war Germany in the callous frigidity with which horror is piled on horror.

After these years of crisis the need was expressed for a return to a more realistic, less subjective style of drama, coupled with a more orthodox approach to dramatic form and technique. Most of the expressionist writers of the previous decade were affected by this, but the works of the new generation gave it authority. The *Heimkehrerdrama*—plays on the predicament of the soldier returning from the war—had already been the subject of BRECHT's first play *Trommeln in der Nacht* (1922), for which he was awarded the Kleist Prize. It expressed the bleak reality of the present, in contrast to the attempt of Carl ZUCKMAYER in *Schinderhannes* (1927) to place the spectacle of a world run mad in its proper perspective by aiming at sanity and sobriety of vision. A year later Brecht, more cynically inclined, produced *Die Dreigroschenoper*, a version of GAY's *The Beggar's Opera* in which Macheath, in league with a corrupt chief of police, battens on the poor whom Brecht sees as the real victims. Historical drama was extremely popular at this period, particularly in the late 1920s, when a historical setting was used to present contemporary political problems and crises, as for example in Unruh's *Bonaparte* and Werfel's *Paulus unten den Juden* (both 1926) and the works of Lion FEUCHTWANGER. The irony implicit in these plays became more savage a few years later with Brecht's *Die heilige Johanna der Schlachthöfe* (1929) and Zuckmayer's bitterest (and most successful) play, *Der Hauptmann von Köpenick* (1931). Among the outstanding actors who interpreted the expressionist plays of the interwar period were Ernst DEUTSCH and Alexander Granach, both members of the company with which the most notable director of the period, Erwin PISCATOR, achieved eminence as an exponent of EPIC THEATRE, inaugurated with the production of Brecht's *Im Dickicht der Städte* in 1924. Also in Piscator's company were Fritz KORTNER, Ernst BUSCH, and Brecht's future wife Helene WEIGEL. Other expressionist directors of the interwar period were Otto FALCKENBERG and Leopold JESSNER, inventor of the Spiel- or Jessnertreppe; while in Berlin Jurgen FEHLING and Heinz HILPERT explored the heritage of REALISM.

With the coming to power of the Nazis in 1933 many German writers and theatre artists went into exile. The dramatists who remained, or who made a name for themselves in the period up to the close of the Second World War, wrote ideological dramas, nationalistic in tone, such as Johst's *Schlageter* (1933), the best exponent of such works being Erwin Kolbenheyer (1878–1962). The only form of drama to flourish during this generally barren period was the Thingspiel, a massive pageant played in the open air before large audiences, and dedicated to the glorification of German nationalistic ideals in general and Nazism in particular. Although German theatre in exile did not equal the volume and power of fiction and poetry composed in such circumstances, it was during his years outside Germany that Brecht's art reached its summit, and the prolific Friedrich WOLF continued while in Russia to follow in the footsteps of Toller. Recovery after the war was slow, and it was in Switzerland that new German-speaking dramatists of interest began to appear—notably Max FRISCH and Friedrich DÜRRENMATT—while Fritz HOCHWÄLDER continued to enhance his reputation in his adopted country. The perpetuation of the allied zones of occupation to form the Federal Republic (West Germany) and the Democratic Republic (East Germany), combined with the physical destruction of the war and the philosophical disorientation of the German peoples, inhibited new writing in the late 1940s and 1950s. Although theatres were rebuilt, and foreign plays and revivals of German classics mounted prolifically, only the return of Brecht and the creation in 1949 of the BERLINER ENSEMBLE in East Berlin made any international impact. Zuckmayer returned to Europe to stage *Des Teufels General* in 1946 and Wolfgang BORCHERT became known posthumously for his anti-war play *Draussen vor der Tür* (1947), but it was only in the 1960s that new dramatists of any distinction began to emerge. Günter Grass and Tankred DORST were indebted in their early plays to the Theatre of the ABSURD, but soon responded to the political impulse evident in the work of such dramatists as Peter WEISS, which also lay behind the DOCUMENTARY DRAMA of the controversial Rolf HOCHHUTH and Heinar KIPPHARDT.

If the Democratic Republic has produced few dramatists distinguished by more than political conformism, the work at the Berliner Ensemble of Brecht and his fellow directors Manfred WEKWERTH, Benno BESSON, and Peter PALITZSCH did establish a tradition in which forms of dialectical drama might flourish, as in the work of Heiner MÜLLER and Peter Hacks, who moved to East Berlin after the success of his *Die Eröffnung des indischen Zeitalters* (1955), a Marxist interpretation of the discoveries of Columbus. But the brief cultural thaw after 1956 effectively ended with the departure of Kipphardt for the West in 1959, though the same year saw the beginning of government support for theatre groups in factories and on farms. Interesting work has been done in East Germany at this level, but its participatory excitement has only served to emphasize the paucity of good writing for the professional stage.

The physical separation of Germany divided Berlin, which up to the outbreak of the Second World War had increasingly become the focus of German theatrical activity, and heavy subsidies from local municipalities encouraged a return to importance of the other major metropolitan centres in the Federal Republic. They often became identified with the work of a particular director —such as GRÜNDGENS in Hamburg or the 'pop art' interpretations of classic dramas by Peter Zadek (1926-   ) in Bremen.

The 'events' of May 1968 had a profound effect throughout the Federal Republic, as in Western Europe as a whole. They contributed to a sense of fragmentation which the plays of the Austrian Peter HANDKE and Rainer Werner Fassbinder anticipated and reflected, and which gave a new impetus to the new theatrical form of COLLECTIVE CREATION: both theatrically, as in the work of a number of new and experimental troupes, and in Peter Stein's seminal revival of Goethe's *Torquato Tasso* in Bremen in 1969; and administratively, in the increased participation of theatre workers in the running of the previously hierarchical subsidized companies. At the Schaubühne in Berlin from 1970, Stein attracted increasing attention with productions of works by Kleist, GORKY (*Summerfolk* in 1974, seen at the NATIONAL THEATRE in London in 1977), Shakespeare, and AESCHYLUS, whose *Oresteia* was the opening event of the new Schaubühne in 1981. Meanwhile a revival of the old folk-play tradition, associated with such dramatists as Martin SPERR and Franz Kroetz, and the director Peter Zadek, has attempted from another direction to bring the theatre to a wider and less sophisticated audience.

GHELDERODE, MICHEL DE (1898–1962), Belgian dramatist, who created in his plays a world which recalls the Flemish fairgrounds painted by Breughel and Bosch. Although Ghelderode's early plays, *La Mort regarde à la fenêtre* (1918) and *Le Repas des fauves* (1919), were first acted in French, as were the later *Christophe Colomb* (1927) and *Les Femmes au tombeau* (1928), several of his most important works, beginning with *Barabbas* (also 1928), were originally produced in Flemish by the Théâtre Populaire Flamand. Ghelderode was little known outside Belgium until after the Second World War, when he was 'discovered' by the directors of the AVANT-GARDE, who revived *Hop! Signor* (1935) in Paris in 1947, followed by *Escurial* (1927) in 1948, and *Mademoiselle Jaïre* (1934) and *Fastes d'enfer* (1929) in 1949. The last caused such a scandal that it was eventually withdrawn, though others of Ghelderode's plays were later seen in theatres on the Left Bank. His work, which includes early plays for puppets, and shows the influence of MAETERLINCK, is almost unknown in England and the United States.

GHÉON [Vanglon], HENRI (1875–1943), French dramatist, an important figure in the revival of religious drama in the inter-war period. His early plays, among them *Le Pauvre sous l'escalier* (1913), were produced by COPEAU at the VIEUX-COLOMBIER; but from 1920 onwards he was engaged in writing and directing almost 100 plays on religious and Biblical themes for schools, colleges, and parish churches. The best known of these are *L'Histoire du jeune Bernard de Menthon* (1925), translated by Barry JACKSON as *The Marvellous History of St Bernard* and produced at the BIRMINGHAM REPERTORY THEATRE and the MALVERN FESTIVAL in 1926, and *Le Noël sur la place* (1935), translated by Eric Crozier as *Christmas in the Market Place*, which has had many amateur productions in England, the first being by the Pilgrim Players of Canterbury in 1943. It was originally presented by Les Compagnons de Jeux, a semi-amateur company organized in 1932 by Henri Brochet (1898–1952), an excellent actor and religious dramatist, to take the place of Ghéon's earlier troupe, Les Compagnons de Notre-Dame. Like the best of Gheon's work, it combines poetry and theatricality with a simplicity appropriate to his intended audiences.

GHERARDI, EVARISTO (1663–1700), an actor of the COMMEDIA DELL'ARTE and a famous ARLECCHINO, whose career was spent entirely with the COMÉDIE-ITALIENNE in Paris, where he made his début in 1689 in a revival of REGNARD's *Le Divorce*. He is chiefly remembered for his publication of the repertory of scenarios used by the Italians in Paris.

GHOST EFFECTS, see LIGHTING; PEPPER'S GHOST; SLOTE; and TRAPS.

GHOST GLIDE, see TRAPS.

GIACOMETTI, PAOLO (1816–82), Italian dramatist, a prolific author who wrote mainly on historical and social subjects. Among the first the most successful was *Elisabetta regina d'Inghilterra* (1853); of his social plays the best known is *La morte civile* (*The Outlaw*, 1861), which helped to further a shift in the Italian theatre away from the rhetorical tradition towards a more naturalistic mode. Like many of Giacometti's plays, which were designed to appeal to the new bourgeois audiences, it was essentially an actor's vehicle, being concerned with an escaped convict who returns to his wife and daughter after 15 years in prison; he finds them comfortably settled in a new life and rather than bring shame on them he commits suicide. A fusion of realistic and romantic elements, this remained a stock piece in the Italian theatre for some fifty years, and provided a major part for SALVINI, who toured in it extensively.

GIACOSA, GIUSEPPE (1847–1906), Italian playwright, after VERGA the most important dramatist of the Italian school of VERISMO.

Although an accomplished craftsman and an alert observer of contemporary life, his range was narrow, his psychological insight limited, and his ideas essentially derivative; but, seeking to assert the value of middle-class morality, he unconsciously painted an accurate and pathetic portrait of the decadence of the Italian bourgeoisie. His *Diritti dell'anima* (*The Rights of the Soul*, 1894) is typical of his limitations. It shows clearly the influence of IBSEN's *A Doll's House*, but lacks firmness of purpose. More forceful presentations of social problems are found in *Come le foglie* (*Like the Leaves*, 1900) and *Il più forte* (*The Stronger*, 1904), the latter being reminiscent of SHAW's *Mrs Warren's Profession* in its handling of the estrangement between parent and child. A kindly and humane man, Giacosa was more at home in such charming and successful comedies as *Una partita a scacchi* (*A Game of Chess*, 1871), *La zampa del gatto* (*The Cat's Paw*, 1883), and *L'Onorevole Ercole Malladri* (1884), which was successfully revived in Turin in 1956 to mark the 50th anniversary of his death. With the historical play *Signora di Challant* (1891) he provided an excellent vehicle for both BERNHARDT and DUSE, and in collaboration with Luigi Illica he was responsible for the librettos of Puccini's operas *La Bohème* (1896), *Tosca* (1900), and *Madama Butterfly* (1904).

GIBBONS' TENNIS COURT, London, see VERE STREET THEATRE.

GIBBS, Mrs, see LOGAN, MARIA.

GIBSON, WILLIAM, see UNITED STATES OF AMERICA.

GIDE, ANDRÉ (1869–1951), French novelist and dramatist, whose early plays *Le Roi Candaule* (1899) and *Saül* (1902) made little stir, but whose *Œdipe* (1932) successfully reinterpreted SOPHOCLES' drama as a quest for humane values in a clerically-dominated society. Although he reworked several other classical themes, it was with his translations of Shakespeare that he did his best work for the theatre, with *Antony and Cleopatra* (1920) and particularly *Hamlet* (1946). He had published a brilliant first act for the latter in 1928, and a chance meeting with BARRAULT in 1942 led him to complete the play, with which the Renaud-Barrault company opened their first season at the Théâtre MARIGNY on 17 Oct. 1946 and two years later visited the EDINBURGH FESTIVAL. It was also for Barrault that Gide prepared *Le Procès*, a dramatization of Kafka's novel, which opened on 10 Oct. 1947 and was revived by Barrault at the ODÉON IN 1962.

GIEHSE, THERESE (1898–1975), German actress, who from 1923 to 1933 was at the Kammerspiele in MUNICH, where she played Mutter Wölffen in a revival of HAUPTMANN's *Der Biberpelz* in 1925, directed by FALCKENBERG. From 1933 to 1945 she was in ZURICH, and from 1952 onwards alternated between there and Munich. Technically a very fine actress, she built up her roles with a wealth of well-observed realistic details. She was the first to play the title roles in BRECHT's *Mutter Courage und ihre Kinder* (1941) and in *Die Mutter* (1949), his adaptation of GORKY's novel *The Mother*; among her other outstanding roles were Clara Zachanassian in DÜRRENMATT's *Der Besuch der alten Damen* (1956) and Dr Mathilde von Zahnd in his *Die Physiker* (1962). She was also seen in 1970 as Juno Boyle in a German adaptation of O'CASEY's *Juno and the Paycock*.

GIELGUD, (Arthur) JOHN (1904–), English actor and director, knighted in 1953. The grand-nephew of Ellen TERRY and through his father the great-grandson of the Polish actress Madame Aszperger, he made his first appearance on the stage at the OLD VIC in 1921 as the Herald in *Henry V*. He then toured in J.B. FAGAN's *The Wheel* (1922) and returned to London to play Felix in the ČAPEKS' *The Insect Play* (1923). In 1924 he became a member of Fagan's repertory company in Oxford (see OXFORD PLAYHOUSE), where he acted in a wide variety of plays, mainly English and foreign classics. In 1925 in London he took over the part of Nicky Lancaster in COWARD's *The Vortex* from the author, also succeeding Coward as Lewis Dodd in Margaret Kennedy's *The Constant Nymph* in 1926. He made a brief appearance in New York in 1928 in Neumann's *The Patriot*, and back in London later in that year he played Oswald in IBSEN's *Ghosts*. In 1929 he joined the company at the Old Vic to play, among other parts, Romeo, Richard II, Macbeth, and Hamlet, his greatest role, which he was eventually to play over 500 times. In the summer of 1930 he played another part which was to be important in his career—John Worthing in WILDE's *The Importance of Being Earnest*—and he then returned to the Old Vic to play such roles as Prospero in *The Tempest* and Antony in *Antony and Cleopatra*. On the reopening of SADLER'S WELLS THEATRE in Jan. 1931 he played Malvolio in *Twelfth Night* and was subsequently seen as Sergius in SHAW's *Arms and the Man*, Benedick in *Much Ado About Nothing*, and King Lear. On leaving the Old Vic he appeared later in 1931 in PRIESTLEY's *The Good Companions*, and the following year achieved his first popular success with Gordon Daviot's *Richard of Bordeaux*, which he also directed. He then undertook the production of several new plays, including Rodney Ackland's *Strange Orchestra* (also 1932) and Somerset MAUGHAM's *Sheppey* (1933), and appeared in Ronald Mackenzie's *The Maitlands* (1934) and OBEY's *Noah* (1935). Later in 1935 he alternated the parts of Romeo and Mercutio with Laurence OLIVIER in a production of *Romeo and Juliet* which he also directed, achieving the longest run on record for this play, and he followed it, after a brilliant performance as Trigorin in CHEKHOV's *The Seagull* in 1936 (in which in 1925 he had played Treplev), with *Hamlet* in New York. In 1937 he took over the QUEEN'S THEATRE, where he headed a distinguished repertory company with which he played Richard II, Joseph Surface in SHERIDAN's *The School for Scandal*, Vershinin in Chekhov's *Three Sisters*, and Shylock in *The Merchant of Venice*, also directing the first and last plays. At the same theatre in the autumn of 1938 he was seen in Dodie SMITH's *Dear Octopus*. Before and after playing Hamlet at the LYCEUM THEATRE before it finally closed in 1939, he again appeared in *The Importance of Being Earnest*, which he also directed, and which was still running when the Second World War broke out on 3 Sept. During the war he toured extensively for ENSA, and was also seen in London in a repertory season at the HAYMARKET, during which he played Oberon in *A Midsummer Night's Dream* and Felix in WEBSTER's *The Duchess of Malfi*, as well as Hamlet, Valentine in CONGREVE's *Love for Love*, and Arnold Champion-Cheney in Maugham's *The Circle*. After a tour of Burma he returned to London to give an outstanding performance as Raskolnikoff in DOSTOEVSKY's *Crime and Punishment* (1946; N.Y., 1947), and then scored another personal triumph as Thomas Mendip in FRY's *The Lady's Not for Burning* (1949; N.Y., 1950). His first appearance at the SHAKESPEARE MEMORIAL THEATRE in 1950, during which he played for the first time Angelo in *Measure for Measure*, was followed in 1951 by a memorable Leontes in *The Winter's Tale* in the West End, and an excellent repertory season at the LYRIC, Hammersmith, in 1952–3, his roles there including Jaffier in OTWAY's *Venice Preserv'd* and Mirabell in Congreve's *The Way of the World*; the latter he also directed. Having played most of the great classic roles, he turned his attention to modern plays, and during the 1950s appeared in N.C. HUNTER's *A Day By The Sea* (1953) and Coward's *Nude with Violin* (1956), both of which he also directed, and Graham GREENE's *The Potting Shed* (1958), displaying in all of them new facets of his consummate art. In 1958 he also gave a superb performance at the Old Vic as Wolsey in *Henry VIII*. In the same year he first presented his Shakespeare recital *Ages of Man* in North America, bringing it to London in 1959 to reopen the QUEEN'S THEATRE, and subsequently taking it on extensive world tours. He was responsible for the production of Enid BAGNOLD's *The Chalk Garden* (1956), Terence RATTIGAN's *Variations on a Theme* (1958), Peter SHAFFER's *Five Finger Exercise* (also 1958; N.Y., 1959), and Greene's *The Complaisant Lover* (1959). He also directed two productions of Chekhov—*The Cherry Orchard* (1961), in which he played Gaev, and *Ivanov* (1965; N.Y., 1966), in which he played the title role—and appeared in new plays, among them Kilty's *The Ides of March* (1963), ALBEE's *Tiny Alice* (N.Y., 1964), Alan BENNETT's *Forty Years On* (1968), Peter Shaffer's *The Battle of Shrivings* (1970), and David STOREY's *Home* (London and N.Y., 1970). In 1974 he inaugurated Peter HALL's directorship of the NATIONAL THEATRE by appearing as Prospero in *The Tempest*, but returned to modern drama with Edward BOND's *Bingo* (also 1974), and gave a much-acclaimed performance in PINTER's *No Man's Land* (1975; N.Y., 1976), also for the National Theatre. He was seen at the National in 1977 in the title role of *Julius Caesar*, in JONSON's *Volpone*, and in Julian Mitchell's *Half-Life*, which later had a long run in the West End. Although the foremost Shakespearian actor of his day, he has also proved outstanding in Chekhov; and he has surprised many of his admirers by his versatility in modern roles, in many of which he has displayed an unexpected vein of humour.

He is the author of an autobiography, *Early Stages* (1938), a collection of essays and speeches, *Stage Directions* (1963), and two volumes of theatrical reminiscences, *Distinguished Company* (1972) and *An Actor and his Time* (1979).

GIFFARD, HENRY (*c.* 1695–1772), English actor and manager, who in 1733 opened the GOODMAN'S FIELDS THEATRE with a revival of LILLO's *George Barnwell*, the excellent company being headed by Richard YATES. In 1737 Giffard took to Walpole, who rewarded him with £1,000, the script of a scurrilous play entitled *The Golden Rump*, which the government used as an excuse for the passing of the Licensing Act of 1737. Since this led to the closing of all unlicensed playhouses—that is, all except DRURY LANE and COVENT GARDEN—Giffard found himself without a theatre. However, like a number of other managers, he found ingenious ways of evading the prohibition on play-acting, and successfully reopened the Goodman's Fields Theatre, where in 1741 David GARRICK made his first appearances on the professional stage. When in the following year FLEETWOOD engaged Garrick for Drury Lane, Giffard and his wife, a fine actress who played Lady Macbeth and other parts opposite Garrick, went with him. Giffard for a short time held part of the Patent of Drury Lane, but relinquished it because he could not stomach Fleetwood's extravagances. Nothing more is known of him; it has been conjectured that he might be the Henry Giffard who appeared in North America in 1768, though in view of his advanced age this seems unlikely.

GILBERT, GABRIEL (*c.* 1620–80), French dramatist, author of a number of plays, but chiefly remembered for his connections with CORNEILLE and RACINE. His *Rhodogune* was virtually contemporary with the *Rodogune* (1645) of Corneille, and the similarities between them are too obvious to have been the result of chance. It seems certain that Gilbert's play must have been based on Corneille's. Gilbert's *Hippolyte* (1646), on the other hand, was certainly one of the versions of the story of Phaedra known to Racine when he wrote his *Phèdre* (1677), and he may have made some use of it. Gilbert's other plays are negligible, though *Téléphonte* (1641) may have been performed in the private theatre of Cardinal RICHELIEU.

GILBERT, Mrs GEORGE H. [*née* Ann Hartley] (1821–1904), American actress, who was in her youth a ballet dancer in London, where in 1846 she married George Gilbert, an actor and dancer with whom she toured England and Ireland. They emigrated to Wisconsin, but failed to make a success of farming and in 1850 returned to the stage, touring the larger cities. Gilbert was injured in a fall through a TRAP, and though he continued to work in the theatre, he was unable to appear on the stage, and died in 1866. His wife found her niche playing comic elderly women, in which capacity she was with Mrs John WOOD's company at the OLYMPIC, New York, in 1864. Her first important part was the Marquise de St Maur in FLORENCE's pirated production of ROBERTSON's *Caste* in 1867. The period of her greatest fame, however, was from 1869 to 1899, when she was in DALY's company, with Ada REHAN and John DREW. When Daly died she was engaged by FROHMAN, with whom she remained until her death. Her angular body and homely features were assets in her skilled playing of eccentric spinsters and elderly dowagers.

GILBERT [Gibbs], JOHN (1810–89), American actor, who made his début in Boston in 1828. He then toured the Mississippi river towns until 1834 and returned to the Tremont Theatre, Boston, remaining there until it closed, playing character roles. An excellent comic actor, he was particularly good as Sir Anthony Absolute and Sir Peter Teazle in SHERIDAN's *The Rivals* and *The School for Scandal* respectively. In 1847 he had a successful season in London and then returned to play at the PARK THEATRE, New York, until it was destroyed by fire. He was with Lester WALLACK's company from 1861 until it was disbanded in 1888, dying the following year while touring with JEFFERSON in *The Rivals*. He was a man of some erudition, and after his death his widow presented his fine collection of books to the Boston Public Library.

GILBERT, WILLIAM SCHWENCK (1836–1911), English dramatist, knighted in 1907, whose name is always associated with that of Sir Arthur Sullivan, for whose music he wrote the libretti of the 'Savoy operas', so called because they were mainly produced at the SAVOY THEATRE in London. Their collaboration began with *Thespis; or, the Gods Grown Old* (1871), and lasted for over 20 years, ending with *The Grand Duke* (1896), though *The Gondoliers* (1889) was the last of those which still hold the stage. The others are: *The Sorcerer* (1877), *H.M.S. Pinafore* (1878), *The Pirates of Penzance* (1880), *Patience* (1881), *Iolanthe* (1882), *Princess Ida* (1884), *The Mikado* (1885), *Ruddigore* (1887), and *The Yeoman of the Guard* (1888). *Utopia Limited* (1893) was never revived. Gilbert wrote libretti for other composers but without success, just as Sullivan wrote music for other librettists but failed to recapture the brilliance of the Savoy operas. Yet their partnership was not a happy one, Gilbert being a man of irascible temperament and a martinet at rehearsals. He was already known as a dramatist before joining forces with Sullivan, having been encouraged by T.W. ROBERTSON to write for the stage. His early works were mainly BURLESQUES and EXTRAVAGANZAS, the first, *Dulcamara; or, the Little Duck and the Great Quack* (1866), being produced at the ST JAMES'S THEATRE. Among his more serious plays the most successful were *The Palace of Truth* (1870), produced by BUCKSTONE at the HAYMARKET THEATRE with himself and the KENDALS in the cast; *Pygmalion and Galatea* (1871), in which the part of Galatea was played by several young actresses of the day, including Mary ANDERSON and Julia NEILSON; *Sweethearts* (1874); *Broken Hearts* (1875); *Dan'l Druce, Blacksmith* (1876), whose title role was long a favourite with character actors; and *Engaged* (1877). He continued writing almost up to his death, but of all his works only the librettos for Sullivan have survived.

GILCHRIST, CONNIE (1865–1946), English MUSICAL COMEDY star, who made her first appearances on the stage as a child in PANTOMIME and in later years was a skipping-rope dancer at the GAIETY THEATRE, where in 1880 she made an immense hit as Libby Ray in the American comedy *The Mighty Dollar* (1861) by B.E. Woolf. She preferred musicals to straight comedy, however, and returned to them to have a brief but glorious career before she left the stage in the late 1880s to become the Countess of Orkney. Her name is best remembered in connection with the supreme example of 'judicial ignorance'; at the height of her fame, Mr Justice Coleridge in Scott v. Sampson asked: 'Who is Miss Connie Gilchrist?'

GILDER, (Janet) ROSAMOND DE KAY (1891–), American drama critic, daughter of the distinguished poet and editor Richard Watson Gilder, whose letters she edited in 1916. She was actively concerned with the foundation and work of the FEDERAL THEATRE PROJECT, the AMERICAN NATIONAL THEATRE AND ACADEMY (Anta), and the INSTITUTE FOR ADVANCED STUDIES IN THE THEATRE ARTS, and for many years headed the United States delegations to the Congresses of the INTERNATIONAL THEATRE INSTITUTE, of which she was one of the founders. She was also its President from 1963 to 1967, and in 1968 was nominated first President of the independent International Theatre Institute of the United States, Inc., housed in the ANTA THEATRE. From 1924 to 1945 she was on the staff of the distinguished journal *Theatre Arts Monthly*, of which she was editor from 1945 to 1948, when it changed hands, and wrote copiously for it on all aspects of American and international theatre. Her books include *A Theatre Library* (1932), *Theatre Collections in Libraries and Museums* (in collaboration with George FREEDLEY, 1936), *John Gielgud's Hamlet* (1937), and *Enter the Actress* (1961). In 1950 she edited a *Theatre Arts Anthology*, and in 1969 *Theatre I*. She received awards from France and Czechoslovakia, as well as from the United States, in recognition of her work for the theatre. As a dynamic administrator with a broad experience of theatre practice, she was one of the outstanding American theatre personalities of her generation.

GILES, NATHANIEL, see BLACKFRIARS THEATRE and BOY COMPANIES.

GILL, CLAES (1910–73), Norwegian actor, director, and author. After relatively brief periods with the Studioteatret in Oslo and the Rogaland Teater, he became in 1968 artistic adviser to the Riksteatret, the State Travelling Theatre. As an actor he displayed great versatility in parts that ranged from the title role in HOLBERG's *Jeppe paa Bjerget* in 1954 and Dr Stockmann in IBSEN's *An Enemy of the People* in 1953 to Becket in T.S. ELIOT's *Murder in the Cathedral* in 1958 and the title roles in BRECHT's *Galileo* and PIRANDELLO's *Enrico IV* in 1959.

GILL, PETER (1939–   ), Welsh director, was an actor from 1957 to 1965, the first play he directed being D.H. LAWRENCE's *A Collier's Friday Night* at the ROYAL COURT in 1965, one of a trilogy of Lawrence's plays he directed there. He worked mostly at the Royal Court until in 1977 he was appointed the first director of RIVERSIDE STUDIOS, his productions there including the highly praised opening production of CHEKHOV's *The Cherry Orchard* (1978), MIDDLETON and ROWLEY's *The Changeling* (also 1978), and *Measure For Measure* (1979). In 1980 he moved to the NATIONAL THEATRE, becoming Joint Director with Peter HALL of the OLIVIER and the COTTESLOE, his first production at the National being TURGENEV's *A Month in the Country* (1981). He has also written several plays.

GILLETTE, WILLIAM (1855–1937), American actor and dramatist, author of a number of adaptations and dramatizations of novels in most of which he appeared himself, among them Conan Doyle's *Sherlock Holmes* (1899), with which his name is always associated. He played Holmes with outstanding success both in England and America, and frequently revived it up to his retirement in 1932. Of his original plays the best were melodramatic spy stories of the Civil War, *Held by the Enemy* (1886) and *Secret Service* (1895). He also wrote a comedy, *Too Much Jonson* (1894). He appeared in BARRIE's *The Admirable Crichton* and *Dear Brutus*; but his best work was done in his own plays, the only one of which to have been revived is *Sherlock Holmes*, which had a notably successful production by the ROYAL SHAKESPEARE COMPANY at the ALDWYCH in 1974, seen in New York in the same year.

GILMORE'S THEATRE, New York, see DALY'S SIXTY-THIRD STREET THEATRE.

GILPIN, CHARLES SIDNEY (1878–1930), black American actor, who spent many years in VAUDEVILLE. In 1916 he became manager of the first all-Negro stock company in New York, at the Lafayette Theatre, Harlem. Among his later Broadway parts were the Negro clergyman in DRINKWATER's *Abraham Lincoln* (1919) and Brutus Jones in O'NEILL's *The Emperor Jones* (1921). The latter provided him with a great emotional part in which he was at once powerful, terrifying, and extremely moving. He retired in 1926, but occasionally reappeared in revivals of *The Emperor Jones* until his death.

GINGOLD, HERMIONE, see REVUE.

GIRALDI, GIOVANNI BATTISTA (1504–73), Italian humanist, known as 'il Cinthio'. He was the author of a number of 'horror' tragedies written under the influence of SENECA, of which the first, *Orbecche* (1541), was performed before the Duke of Ferrara at the house of Alfonso Della Viola, composer of the music specially commissioned for the occasion. Two of Giraldi's later plays, *L'Altile* (1543) and *Epizia* (1547), were based on stories which he later included in his *Hecatommithi* (1565), a collection of *novelle* in the style of Boccaccio's *Decameron*. It was from this collection that Shakespeare, who had already used the story of Epizia, translated into English by George Whetstone as *Promos and Cassandra* (1578), in his *Measure for Measure*, took the plot of *Othello*. For him, as for other English dramatists, among them GREENE, SHIRLEY, and BEAUMONT and FLETCHER, who all used material from the *Hecatommithi*, Giraldi opened up chivalric sources and provided romantic situations suitable for tragedy and tragi-comedy; but in Italy his work served to strengthen the stranglehold of the neoclassical form. He was content to further the moral and sensational aspects of tragedy, making only minor concessions to his audience by avoiding strict imitation of the classical chorus and treating his women with unusual sympathy.

GIRAUD, GIOVANNI, see ITALY.

GIRAUDOUX, (Hippolyte) JEAN (1882–1944), French novelist and dramatist, whose fruitful collaboration with the actor-director JOUVET produced some of the finest plays of the period. Giraudoux was already a well-known novelist before his first play, *Siegfried* (1928), was produced at the COMÉDIE DES CHAMPS-ÉLYSÉES, and as a dramatist his reputation grew with each succeeding work. After *Amphitryon 38* (1929), which, in an English adaptation by S.N. BEHRMAN, had a great success in England and America with the LUNTS as Jupiter and Alcmena, and *Judith* (1931), a biblical tragedy, came the enchanting *Intermezzo* (1933), a mixture of fantasy and realism seen in London in 1956 in a production by the RENAUD-BARRAULT company. In *La Guerre de Troie n'aura pas lieu* (1935), Giraudoux traced the causes of the Trojan War to a lie and a misunderstanding. As *Tiger at the Gates* (with Michael REDGRAVE as Hector) this play made a belated but successful appearance in London in 1955 in a translation by Christopher FRY, going on to achieve equal success in New York.

After the short *Supplément au Voyage de Cook* (1935), *Électre* (1937), *L'Impromptu de Paris* (1937), a burlesque on contemporary French attitudes to the theatre in the style of MOLIÈRE, and the one-act *Cantique des Cantiques* (1938), Giraudoux produced what many believe to be his finest work, *Ondine*, a retelling of the legend of the water-nymph on which he brought to bear all the poetry and imagination of which he was capable and (in the second act) all the theatrical tricks which he knew Jouvet could stage for him. Produced on 3 May 1939, its run was cut short in 1940 and it was not revived until 1949. It was seen in French in London in 1953 (at the LYRIC, Hammersmith) in a production by the Théâtre National de Belgique, and in an English translation was one of the successes of the ROYAL SHAKESPEARE COMPANY's first season at the ALDWYCH in 1961, Leslie Caron playing Ondine.

In 1942 *L'Apollon de Bellac* was produced in Rio de Janeiro by Jouvet, who directed its first production in France in 1947. In 1943 came the première in Paris of *Sodome et Gomorrhe*, in which Lia, the leading role, was played by Edwige FEUILLÈRE. But it was with *La Folle de Chaillot*, produced posthumously on 19 Dec. 1945, that Giraudoux's magic re-established itself. The play was an instant success in Paris and, as *The Madwoman of Chaillot*, was seen in New York in 1948 and in London in 1951 and has had several revivals. Giraudoux's last play, *Pour Lucrèce* (1953), directed by Barrault at the MARIGNY and staged in London as *Duel of Angels* in 1958 with Vivien LEIGH as Paola, played originally by Feuillère, was successful. Revivals of some of Giraudoux's earlier plays have shown that their distinctive blend of verbal and theatrical fluency continues to be an attractive combination.

GITANA, GERTIE, see MUSIC-HALL.

G.I.T.I.S., Moscow, see SCHOOLS OF DRAMA.

GJOKA, PJETER, see ALBANIA.

GLASGOW REPERTORY THEATRE. This opened in 1909, under the management of Alfred Wareing, with a production of SHAW's *You Never Can Tell* in the Royalty Theatre, built in 1879. For five seasons the excellent company played in works by many English and Continental dramatists, including *The Seagull* in 1909, claimed as the first performance of CHEKHOV in English. In its initial efforts to foster a purely Scottish drama, it produced a few plays in the vernacular, but its Scottish commitment was never very strong. The venture was well on its way to success when the outbreak of the First World War forced it to close down. The remaining funds were transferred to the St Andrew Society, and later used to help launch the SCOTTISH NATIONAL PLAYERS.

GLASPELL, SUSAN (1882–1948), American novelist and dramatist, active in the formation of the PROVINCETOWN PLAYERS, who produced several of her one-act plays and, in 1915, her first three-act play, *Bernice*, in which she herself played the devoted servant Abbie. Her finest work for the theatre was *Alison's House* (1930), based partly on the life of the American poet Emily Dickinson. Produced by Eva LE GALLIENNE at the Civic Repertory Theatre in New York, this was awarded the PULITZER PRIZE for drama. It was seen in London in 1932.

GLASS CRASH, see SOUND EFFECTS.

GLEICH, JOSEPH ALOIS (1772–1841), Austrian playwright, author of some 220 plays and one of the founders of the Viennese popular theatre which succeeded the old improvised burlesque, a theatre compounded of magic, farce, and parody. Gleich, who was at one time official playwright to both the Leopoldstädter and the Josefstädter theatres in VIENNA, did his best work in the so-called '*Besserungsstück*', the moral tale which forces the sinner to recognize the error of his ways but always by way of slapstick comedy. The best example of this is *Der Eheteufel auf Reisen* (1822), in which the chief character, who blames his wife for their matrimonial difficulties, is 'magicked' into becoming a partner in five different marriages and ends by confessing that man is usually the guilty party.

GLOBE PLAYHOUSE, London, see WANAMAKER, SAM.

GLOBE THEATRE, London. (1) On the south side of Maiden Lane, Bankside, in Southwark, the theatre most intimately associated with Shakespeare, who was one of the HOUSEKEEPERS, or owners of a share in it. It was built in 1599 by Cuthbert BURBAGE of materials taken from the THEATRE, built by his father, and was the largest and best known of the Elizabethan playhouses. It housed the company known as the CHAMBERLAIN'S MEN, led by Richard BURBAGE. Being largely open to the elements, it was used only during the summer months, the company, which in 1603 was renamed the KING'S MEN, transferring after 1613 to the BLACKFRIARS THEATRE for the winter. Its chief rival was HENSLOWE's company at the FORTUNE THEATRE, where ALLEYN was the leading actor. All that is known of its interior is that it had a thatched roof over the upper gallery and that some of its dimensions were the same as those of the Fortune. In 1613, during or shortly after a performance of *Henry VIII*, it caught fire and was burned down. Rebuilt in substantially the same style, but with a tiled roof in place of the thatch in which the fire was believed to have originated, it reopened in 1614 and remained in constant use until all the London theatres were closed in 1642. When the Burbages' lease ran out in 1644 the building was demolished. A conjectural replica, designed by the Shakespearian scholar Dr John Cranford Adams, was erected in 1950 at Hofstra College, Long Island, N.Y. The plan of ST GEORGE'S THEATRE is based on the Globe, and Sam WANAMAKER proposes to build a reconstruction close to the original site.

(2) A four-tiered auditorium holding about 1,800 people which stood in Newcastle Street at the east end of the Strand near the OPERA COMIQUE, the two theatres being known as the Rickety Twins. It opened on 28 Nov. 1868 with H. J. BYRON's *Cyril's Success*, which lived up to its name; subsequent productions were less fortunate, and there were frequent changes of management. In 1876 Jennie Lee scored a personal triumph in the title role of *Jo*, a play based on DICKENS's *Bleak House*, and in 1884 HAWTREY's *The Private Secretary*, with PENLEY in the lead, transferred from the PRINCE OF WALES'S THEATRE, where it had not done well, to achieve an immense success and become one of the classic stage farces of all time. In 1893 Penley returned to the Globe, which had in the meantime housed MANSFIELD in 1889 and Frank BENSON's company in 1890, with another classic farce, Brandon THOMAS's *Charley's Aunt*, transferred from the ROYALTY, which ran for four years. In 1898 John HARE became manager, his most interesting productions being J. S. Ogilvie's *The Master* (1898), which brought Kate TERRY out of retirement, and PINERO's *The Gay Lord Quex* (1899), with Irene VANBRUGH. The last production, in March 1902, was a revival of KESTER's *Sweet Nell of Old Drury*, with Fred TERRY and Julia NEILSON, after which the theatre, for long considered a fire hazard, was demolished as part of the Strand-widening scheme.

(3) In Shaftesbury Avenue and built for Seymour HICKS, whose name it first bore. Having a three-tier auditorium, a proscenium width of 30 ft, and stage depth of 36 ft, it opened under the management of Charles FROHMAN on 27 Dec. 1906 with Hicks and his wife Ellaline TERRISS in a musical play *The Beauty of Bath*, transferred from the ADELPHI. Renamed in 1909, it remained Frohman's London headquarters until his death in 1915: he was succeeded by Alfred Butt, who transferred J. H. MANNERS's successful *Peg o' My Heart* to the Globe from the COMEDY. From 1918 to 1927 the theatre had a number of successes under Anthony Prinsep, in most of which Prinsep's wife Marie LÖHR appeared, including A. A. Milne's *The Truth About Blayds* (1921), LONSDALE's *Aren't We All?*, MAUGHAM's *Our Betters* (both 1923), COWARD's *Fallen Angels* (1925), and Charles Openshaw's *All the King's Horses* (1926), with Irene Vanbrugh. After Prinsep's retirement the theatre changed hands several times, Maurice BROWNE being responsible for the appearance in 1930 of MOISSI in *Hamlet*, followed by the PITOËFFS in SHAW's *Saint Joan*. A year later came FAGAN's *The Improper Duchess*, with Yvonne ARNAUD, and in 1935 Dodie SMITH's *Call it a Day* had a long run. In Feb. 1937 the theatre was taken over by H. M. Tennent, opening with a revival of Shaw's *Candida*. Among the successful productions of this management were St John ERVINE's *Robert's Wife* (1937), with Owen NARES and Edith EVANS, RATTIGAN's *While the Sun Shines* (1943), FRY's *The Lady's Not For Burning* (1949), with John GIELGUD, and his adaptation of ANOUILH's *Ring Round the Moon* (1950), in which SCOFIELD scored his first West End success. Gielgud returned in Coward's *Nude With Violin* (1956), and again in *The Potting Shed* (1958) by Graham GREENE, whose *The Complaisant Lover* was seen in 1959 with Ralph RICHARDSON and Scofield. The lease of the theatre was then taken by the Prince Littler organization, and in 1960 BOLT's *A Man for All Seasons* was seen, with Scofield as Sir Thomas More. Anouilh's *Becket* (1961) and Peter SHAFFER's double bill *The Private Ear* and *The Public Eye* (1962) ended the Globe's long run of successes until the opening on 15 June 1966 of Terence Frisby's *There's a Girl in my Soup* with Donald SINDEN. Other successful productions have been Woody Allen's *Play It Again, Sam* (1969); SARTRE's *Kean* (1971), in which Alan BADEL gave a virtuoso performance; Peter NICHOLS's *Chez Nous* (1974) with Albert FINNEY and *Born In the Gardens* (1980), transferring from the BRISTOL OLD VIC; and Michael FRAYN's *Donkeys' Years* (1976). The theatre has also staged several plays by Alan AYCKBOURN—the trilogy *The Norman Conquests* (1974) starring Tom COURTENAY, from GREENWICH THEATRE, *Ten Times Table* (1978), and *Joking Apart* (1979).

GLOBE THEATRE, New York, see LUNT-FONTANNE THEATRE and NEW YORK THEATRE (1).

GLORY, see DRUM-AND-SHAFT SYSTEM.

GLOVE-PUPPET, see PUNCH AND JUDY and PUPPET.

GNESSIN, MENAHEM, see HEBREW THEATRE, NEW.

GOBO, see LIGHTING.

GODFREY, CHARLES, see MUSIC-HALL.

GODFREY, THOMAS (1736–63), American dramatist, who became the first playwright of the United States when in 1759 he wrote a tragedy entitled *The Prince of Parthia* which he sent to DOUGLASS, manager of the AMERICAN COMPANY. It was received too late for production during the season in Philadelphia, and Godfrey died before it was given by Douglass for one night at the SOUTHWARK THEATRE in 1767. It was then not acted again until its revival at the University of Pennsylvania in 1915. It was published in 1765 with some poems by Godfrey, and shows the influence of the plays which were in the repertory of the elder HALLAM in about 1754.

GODS, see GALLERY.

GODWIN, EDWARD WILLIAM (1833–86), archaeologist, architect, and theatrical designer, who was living in Bristol in 1863 when he first met the 15-year-old Ellen TERRY. After the breakdown of her shortlived marriage to the painter F. G. Watts, she and Godwin met again, and from 1868 to 1875 they lived together, their children Edith and Gordon CRAIG being born during this time. After their separation Godwin continued to advise Ellen Terry on theatrical matters, and he supervised the production of *The Merchant of Venice* in 1875 in which she played Portia. From then onwards he was much absorbed by work for the theatre, and wrote a good deal on archaeology in relation to the contemporary stage. His most notable achievement was probably the production of John Todhunter's *Helena in Troas* in 1886, for which he designed a classical theatre which was erected inside the existing structure of Hengler's Circus.

GOERING, REINHARD (1887–1936), German dramatist, whose first play *Seeschlacht* (1918), based on the naval battle of Jutland, is one of the key works of German EXPRESSIONISM. The cast consists of seven unnamed sailors in the gun-turret of a battleship; there is no attempt at realism, and with stylized language and ballet-like movements the impression, in REINHARDT's first production in BERLIN, was one of extreme depersonalization, with all the rigidity of classical tragedy. It was an immediate success and was frequently revived. Goering's later plays, which included one on the scuttling of the German battleships in Scapa Flow and one on the tragic journey of Captain Scott to the South Pole, were also produced in Berlin, the latter by JESSNER, but were less successful.

GOETHE, JOHANN WOLFGANG (1749–1832), German man of letters, statesman, scientist, and philosopher, who devoted a not inconsiderable portion of his time and genius to the theatre. Born in Frankfurt of a respected and well-to-do family, he had a good education, the French occupation of the town from 1759 to 1762 enabling him to perfect himself in the language and also to see a number of French plays. He had already written a couple of unimportant comedies when he was introduced to the works of Shakespeare by Herder, whom he met while studying law in Strasbourg. Immediately captivated by them, his enthusiasm expressed itself in the first German play to be written in the Shakespearian style, *Götz von Berlichingen mit der eisernen Hand*. Produced in Berlin in 1773, this somewhat idealized portrait of a robber baron was an immediate success. Though diffuse, with the action badly broken in the middle, its wide range of characters, its disregard for the UNITIES, and its historical perspective made it the spearhead of the STURM UND DRANG movement, a model for young dramatists, particularly SCHILLER, and the prototype of a wave of RITTERDRAMA which swept across Germany. In 1774 the success of his short novel *Die Leiden des jungen Werthers* confirmed Goethe's position as the leader of the young Romantics. Plays written at this time include *Stella* (1776); the domestic tragedy *Clavigo* (1779), set in Spain and based on a true incident in the life of BEAUMARCHAIS, which was seen in London in the WORLD THEATRE SEASON of 1964; and *Egmont*, which was not completed until 1787. Dealing with the revolt of the Netherlands against Spain in 1567, Goethe portrays his hero not as a circumspect Fleming of heroic mould—a role reserved for William of Orange—but as a warm-hearted believer in the fundamental goodness of mankind. With sublime indifference he walks into the snare set for him and goes to the scaffold dreaming of his country's liberation. The play was first performed, with extensive alterations by Schiller, in 1791. The original version was not seen until May 1810, without the overture and nine songs commissioned for it from Beethoven which were not ready for performance until the following June.

In 1775 Goethe accepted an invitation from the reigning Duke to settle in WEIMAR, where he remained for the rest of his life. One of his duties was to organize ducal entertainments, and among the plays he directed with a cast of young courtiers was the first (prose) version of his *Iphigenie auf Tauris* (1779). Being himself a good amateur actor, he appeared in this production as Orestes to the Iphigenia of Caroline Schröter. He soon found himself adding to his original duties in the Duchy, becoming responsible for agriculture, mining, and forestry, with a seat in the Cabinet and the chairmanship of the Treasury, and acquitting himself in all these undertakings with tact and efficiency. His situation was not, however, without its tensions and frustrations, and in 1786, in need of rest and refreshment, he went to Italy where he underwent a fundamental change, losing the last vestiges of his Romanticism and becoming an avowed classicist. During his two years' sojourn there he recast *Iphigenie auf Tauris* in verse, superimposing on the situation of EURIPIDES' Iphigenia the moral vision of the 18th century. Believing in the goodness of the gods, she avoids human sacrifice until the time comes for her brother Orestes to be killed. Then, trusting in the fundamental humanity of Thoas and in divine clemency, she reveals their relationship and wins from the king an unequivocal pardon. The play, one of the masterpieces of European drama, is an expression of Goethe's belief that the salvation of mankind can come only through humanity and renunciation. In its new form it was produced in 1802. The visit to Italy was also responsible for *Torquato Tasso* (1807), in which is portrayed a poetic temperament in conflict with the world of action. These two classical plays are cast in iambic verse of exquisite mellowness and have an intimate, personal appeal lacking in Goethe's earlier works. After his visit to Italy the vitality in his nature which had made him feel akin to all things human abated, as the first signs of the Olympian detachment which was such a feature of his old age began to appear. In 1791, the time for amateur dramatics being over, he was appointed director of a professional company established in the Court Theatre, where he found the practical knowledge of the theatre imparted to him earlier by SCHRÖDER extremely useful; with the help of Schiller, his co-director for a time, and IFFLAND, he established a company which became known throughout Europe for the excellence of its acting and the high standard of its repertory. He also took up again a novel set in the world of the strolling player, based on Schröder's life, of which the first draft, as *Wilhelm Meisters theatralische Sendung*, was not published until 1910. Under the influence of Italy, and of the immense widening of his personal outlook, he enlarged its scope, turning the quest for perfect theatre (which was to have ended with an ideal production of *Hamlet*) into the quest for true citizenship. Published in 1795 as *Wilhelm Meisters Lehrjahre*, this was translated into English in 1824 as *Wilhelm Meister's Apprenticeship* by Thomas Carlyle, who also translated its sequel *Wilhelm Meisters Wanderjahre* as *Wilhelm Meister's Travels* in 1827.

The crowning achievement of Goethe's career, as both playwright and poet, and in some ways as philosopher, was his *Faust*. Begun in the early 1770s and inspired by a puppet-play seen in his youth, this was destined to occupy him throughout his whole life. The original draft, which he abandoned, was discovered and published as *Ur-Faust* in 1887. A second version, which had not been acted, was included by Goethe in an edition of his works published in 1790. But it was Schiller who in about 1799 persuaded Goethe to return to the play and finish it. Part I, which contains the seduction and desertion of the innocent Gretchen and her execution for infanticide, appeared in 1808; Part II, in which occurs the scene of the raising of Helen of Troy used by MARLOWE in his *Dr Faustus*, was completed in the last year of Goethe's life and appeared posthumously. Taken together, the two parts are a distillation of an old man's wisdom, accumulated over a long and active career and embracing all aspects of human life, political, social, economic, and even fantastic, and insisting throughout on unremitting activity and endeavour as the true aim of man's life on earth. Owing to its complexity the play has always proved difficult to stage and many directors fall back on the simpler *Ur-Faust*. Part I was not seen in Germany until 1829; Part II in 1854. The first production of both parts was seen in 1876 at Weimar, where Otto DEVRIENT overcame some of the problems by using the old medieval MULTIPLE SETTING, which was subsequently widely adopted. In Berlin in 1909 REINHARDT produced Part I, and in 1911 Part II, with all the resources of the new revolving stage and striking lighting effects. For Salzburg in 1933 he again used a multiple setting of small scenes on different levels built into a 60-ft-high cliff. The best of a number of recent productions was by GRÜNDGENS, who also played Mephistopheles, at Hamburg—Part I in 1957, Part II in 1958.

Although most of Goethe's works, including his plays, have been translated into English, very little of his dramatic work has been seen on stage; but in 1963 the BRISTOL OLD VIC put on a version of *Götz von Berlichingen*, by John ARDEN, as *Ironhand*, and *Iphigenia in Tauris* was seen in Manchester and London in 1975.

GOGOL, NIKOLAI VASILIEVICH (1809–52), Russian writer and dramatist and the first great realist of the Russian theatre. He was already known for his short stories when in 1832, after making the acquaintance of the actor SHCHEPKIN, he began work on a play, but abandoned it when he realized that as a satire on bureaucracy it would not pass the censor; the manuscript was later destroyed. Two other plays, both satires, were begun but left, to be finished in 1842; Gogol's dramatic masterpiece *Revizor* was produced at the Court theatre in the presence of the Tsar in 1836. Its theme was official corruption in a small town, where an impecunious impostor is mistaken for a government official and fêted accordingly. It came at an opportune moment, when the authorities were busily engaged in reorganizing local municipal affairs, but the satire bit too deep, and the play was viciously attacked. Gogol, already broken in health, left Russia, not to return until 1848. The play had an immense influence in Russia and has also became well known in translation, being first seen in London in 1920 as *The Government Inspector* and in New York in 1923 as *The Inspector-General*. In 1966 an outstanding revival of it was directed by Peter HALL at the ALDWYCH THEATRE, with Paul SCOFIELD as Khlestakov. An English translation of an amusing but less important comedy *Marriage* (1842) was seen as *The Marriage Broker* at the MERMAID THEATRE in 1965. Gogol's other dramatic works include the farce *Gamblers* (1842) and a sketch entitled *On Leaving the Theatre After a Performance of a New Comedy*, which analyses audience reaction in St Petersburg in the same year.

In 1928 Gogol's great novel *Dead Souls* was dramatized by BULGAKOV and presented by STANISLAVSKY at the MOSCOW ART THEATRE, and this version was seen in London in 1964 when the company appeared during the WORLD THEATRE SEASON. On the occasion of the 150th anniversary of Gogol's birth the Moscow Transport Theatre was renamed the Gogol in his honour.

GOLD, JIMMY, see CRAZY GANG.

GOLDEN, JOHN, see JOHN GOLDEN THEATRE, NEW YORK.

GOLDEN THEATRE, New York, see JOHN GOLDEN THEATRE and ROYALE THEATRE.

GOLDFADEN [Goldenfodim], ABRAHAM (1840–1908), the first important Yiddish dramatist and the first to put women on the stage in Yiddish plays. In 1862, while a student at Zhitomir, he played the title role in the memorable first performance of Ettinger's *Serkele*, written in about 1825. He soon became known as a writer, and a number of his folk songs and dramatic sketches were performed by the Brody Singers, who in Oct. 1876 performed in Simon Marks's Wine Cellar in Jassy a two-act musical entertainment in Yiddish which Goldfaden had prepared for them. Its success encouraged him to form a company for the production of his own plays, of which he wrote about 400. Among the best known are *The Recruits* (1877), *The Witch* (1879), *The Two Kune Lemels*, a version of *Romeo and Juliet, Shulamit* (both 1880), *Dr Almosado* (1882), and *Bar Kochba* (1883). At the time of his death his last play, *Son of My People* (1908), was running at the Yiddish People's Theatre in New York, where he had settled in 1903, opening a school of drama. The great Jewish actor Maurice Moskovitch [Marskopf] (1871–1940) was one of his pupils. Goldfaden was also the author of the first Hebrew play seen in New York, *David at War* (1904).

GOLDONI, CARLO (1709–93), Italian dramatist, born in Venice and destined for the law. Although he completed his studies, and even practised for a while, he had always since childhood had a passion for the theatre and on 24 Nov. 1734 his first play, *Belisario*, was produced in Verona. Like many plays which he had seen or read, this was a tragi-comedy; but Goldoni's main purpose in writing plays, which he embarked on almost immediately, was to reform the now moribund COMMEDIA DELL'ARTE by substituting fully written comedies of character for the old-style improvisation based on an outline plot. In this he was bitterly opposed, not only by some of the actors but by his contemporary Carlo GOZZI, who was also anxious to reform the Italian stage, but in an entirely different way.

Goldoni began his reforms in 1738 with *Momolò Cortesan*, which had one part fully written out. In what later became one of his most popular works, *Il servitore di due padroni* (*The Servant of Two Masters*, 1743), he left some of the scenes to be improvised by Truffaldino, the 'servant' of the title, but wrote them out in full, no doubt incorporating much of the comic business of the actor Sacchi who first played the part, when the text was published in 1755. He achieved his first real success with the public with *La vedova scaltra* (*The Wily Widow*, 1748), produced by Girolamo Medebac in Venice at the Teatro Sant'Angelo, which also saw the first nights of *La buona moglie* (*The Good Wife*), *Il cavaliere e la dama* (both 1749), and *Il teatro comico* (1750), in which he set out his plans for the reform of the stage. Like PIRANDELLO's *Sei personaggi in cerca d'autore* (*Six Characters in Search of an Author*, 1921) this opens with a bare stage on which a group of actors is assembling for a rehearsal. It was well received, and soon Goldoni was at the height of his fame, with the success of *Il bugiardo* (*The Liar*), *La bottega del caffè* (*The Coffee House*, both 1750), and particularly what is sometimes regarded as his masterpiece, *La locandiera* (1751), a picture of feminine coquetry which delighted audiences all over Europe and later America, more especially when Mirandolina, the hostess of the title, was played by Eleonora DUSE.

In 1751 Goldoni left Medebac and took his plays to the Teatro San Luca, where he soon found himself in difficulties. The theatre was too large for his intimate comedies, and he had also to content with the hostility not only of Gozzi but of a mediocre playwright Pietro Chiari (1711–85) who had taken his place at the Sant'Angelo. Wearied by a rivalry which threatened to split Venetian audiences into two camps, he left Venice in 1761 for Paris, where he wrote both in Italian and in French for the COMÉDIE-ITALIENNE. But here again he found the older exponents of improvised comedy opposed to his reforms, and his plays, among them *Il ventaglio* (*The Fan*, 1763) and *Le Bourru bienfaisant* (*The Kindly Curmudgeon*, 1773), were not as successful as he had hoped. Caught up in the turmoil of the French Revolution, he died in poverty, leaving three volumes of memoirs and more than 150 plays. His work has been both over- and underrated. He had none of the elegance or profoundity of MOLIÈRE, with whom he has often been compared, but his technical competence in stagecraft and dialogue reached high levels, his comic invention seldom flagged, and many of his characters, like the fishermen in *Le baruffe chiozzotte* (*Squabbles in Chioggia*, 1770), are based on close and sympathetic observation. By employing the masks of the old *commedia dell'arte* in a realistic manner he paved the way for the later bourgeois drama of such writers as DIDEROT, who based his *Le Père de famille* (1761) on Goldoni's *Il padre di famiglia* (1750). Much of his work survives on the Italian stage, notably the plays in Venetian dialect like *I rusteghi* (*The

Boors, 1760) and *Sior Tòdero Brontolon* (1762), and the three plays produced in 1954 as *Triologia della villeggiatura* (*Trilogy on Country Holidays*) by Giorgio Strehler for the PICCOLO TEATRO DELLA CITTÀ DI MILANO (who had already made an international success with a revival of *Il servitore di due padroni* in 1947). Others, among them *L'impresario delle Smirne* (*The Impresario of Smyrna*, 1760), have joined *La locandiera* in the international repertory, and several have been used for opera librettos.

GOLDSMITH, OLIVER (1730–74), English poet, novelist, and dramatist, whose two plays, *The Good-Natured Man* (1768) and *She Stoops to Conquer* (1773), stand, like the works of SHERIDAN his contemporary, far in advance of the drama of his time. The first, produced by the elder COLMAN at COVENT GARDEN, had a cool reception, but the second, also at Covent Garden, was an immediate success. It has little in common with the genteel comedy of the day and has been constantly revived. Goldsmith wrote nothing more for the theatre, but in 1878 his novel *The Vicar of Wakefield* was made into a charming play, *Olivia* by W. G. Wills, with Ellen TERRY as the young heroine.

GOLIARD, a name given to the wandering scholars and clerks of the early Middle Ages who, unamenable to discipline, joined themselves to the itinerant entertainers of the time and were often confused with them, as in an order of 1281 that 'no clerks shall be jongleurs, goliards or buffoons'. They imparted a flavour of classical learning to the often crude performances of their less erudite fellows, and even when, as happened in the 14th century, the word was used for 'minstrel' without any clerical association, the goliard is still shown rhyming in Latin, as in Langland's late 14th-century poem *Piers Plowman*. Some idea of the entertainment offered by the goliards during the Middle Ages can be seen in the *Commedia Bile*, a short play in Italian for three characters. The text dates from the 14th century, but the subject is taken from the *Deiphosophisti* of Athenaeus (195 AD) and concerns a host who gives his guest the smaller fish, while on a specious pretext he reserves the larger one for himself.

GOMBROWICZ, WITOLD (1904–69), Polish novelist and playwright, who spent most of his later life in exile, going first to Argentina on the outbreak of the Second World War, to Berlin in 1963, and finally to Paris, where he died. His three 'grotesque allegories', which presage the Theatre of the ABSURD, were not seen on the Polish stage until 1957, when *Iowna, Princess of Burgundy* was produced in Cracow; a German version of this play was presented by the Berlin Schillertheater in London during the WORLD THEATRE SEASON of 1971. A second play, *Marriage*, written in 1946, was first produced in 1963 in Paris, as was *Operetta*, presented by the THÉÂTRE NATIONAL POPULAIRE in 1967. *Marriage*, which has affinities with *Hamlet*, was produced at Zurich in 1972, with a remarkable surrealist décor by Krystyna Zachwatowicz, seen again at the Warsaw production in 1975. *Operetta*, which portrays the dying Western civilization in the form of a musical comedy, was directed by DEJMEK in Łódź, also in 1975.

GOMERSAL, EDWARD ALEXANDER, see ASTLEY'S AMPHITHEATRE.

GONCOURT, EDMOND-LOUIS-ANTOINE HUOT DE (1822–96) and JULES-ALFRED HUOT DE (1830–70), French novelists and men of letters, who lived and wrote in collaboration all their lives. The theatre held an important place in their work, though their own plays are negligible. *Henriette Maréchal* (1865) failed in an atmosphere of political recrimination, and *La Patrie en danger*, written in 1873, had to wait until 1889 before it was put on at ANTOINE'S THÉÂTRE LIBRE. Several plays based on their novels fared no better; however, these, and their authors' theoretical ideas, had a great influence on such writers as ZOLA at the end of the 19th century.

GONZAGA, PIETRO GOTTARDO [Pyotr Fyodorovich] (1751–1831), Italian painter and scenic artist, who from 1772 to 1778 worked in the studio of the famous Galliari brothers. In 1779 he was appointed principal scene designer at the Scala, Milan, and later worked for a number of other theatres in Italy. In 1792 he was invited by Prince Yusupov to work in Russia, where he spent the rest of his life. Appointed décor designer for the Imperial Theatres, he worked for the Bolshoi and Hermitage theatres in St Petersburg, for theatres in Gatchina, Pavlovsk, and Peterhof, and also for the Petrovsky Theatre opened by Mikhail MEDDOKS in Moscow. The bulk of his designs were for opera and ballet, but in 1804 he designed the scenery for Ozerov's drama *Oedipus in Athens*. In 1817 Prince Yusupov commissioned him to design and build a theatre on his country estate at Arkhangelskoye; of the 12 changes of scenery which he designed for this theatre, four have survived. He retired in 1828. Throughout his life he was a partisan of the neo-classical style as opposed to the Baroque, favouring 'the liberation of art from distorting vices and its restitution to the simple principles of imitation', and his ideas were keenly followed by Russian stage designers of the early 19th century.

GOODMAN MEMORIAL THEATRE, Chicago, Illinois. This theatre, which opened on 22 Oct. 1925, is after the CLEVELAND PLAY HOUSE the oldest regional theatre in the United States. Donated to the Art Institute of Chicago by the parents of the playwright Kenneth Sawyer Goodman in memory of their son, it was built mainly underground to avoid spoiling the view across Grand Park and seats 683. It was intended to combine a school of acting with a resident professional acting company, but the Depression forced the company to disband in the 1930s, though the school continued to flourish. In the 1950s it was reorganized, and a subscription series of plays introduced during which professional guest artists performed with students. In the 1960s the Goodman once again housed a fully professional company, the student productions being moved to the adjacent Goodman Theatre Studio, seating 135, until in 1977 the theatre was incorporated as the Chicago Theatre Group Inc. and the school ceased to be connected with it. The Mainstage theatre now offers five productions a year, classic revivals and works by new authors, while the studio theatre houses smaller productions. There is also a Young People's Drama Workshop.

GOODMAN'S FIELDS THEATRE, London. There were two theatres of this name, both in Ayliffe Street, Whitechapel. The first opened on 31 Oct. 1727, in a converted shop, under the management of Thomas Odell, deputy Licenser of Plays, with a performance of FARQUHAR's *The Recruiting Officer*. It was here that FIELDING's second play *The Temple Beau* was produced on 26 Jan. 1730. Not having much knowledge of theatrical affairs Odell soon retired, handing over to Henry GIFFARD, whose last production opened on 23 May 1732. The theatre was then used for exhibitions of rope-walking and acrobatics, though there was a brief return to drama before it finally closed in 1751. It became a warehouse and was burned down in 1802. Meanwhile Giffard had built a new theatre nearby to plans drawn up by Edward Shepherd, the architect of the first COVENT GARDEN theatre. This opened on 2 Oct. 1732 with *Henry IV, Part I*, the company including Thomas Walker (1698–1744), the original Macheath in GAY's *The Beggar's Opera*, Richard YATES, Harry WOODWARD, and a low comedian Christopher Bullock. The passing of the Licensing Act of 1737, for which he was mainly responsible, lost Giffard his licence, and the theatre had to close, Giffard renting LINCOLN'S INN FIELDS for a time; but he was able, by various subterfuges, to reopen Goodman's Fields in 1740, and on 15 Jan. 1741 put on *The Winter's Tale*, 'not seen in London for over 100 years', with himself as Leontes. At this theatre on 9 Oct. of the same year GARRICK made his first professional appearance, as Richard III. On 27 May 1742 the theatre closed, never to reopen.

There may have been another, older theatre in the vicinity, for in 1703 a periodical called *The Observator* stated that 'the great playhouse has calved a young one in Goodman's Fields, in the passage by the Ship Tavern, between Prescot Street and Chambers Street.' No other record of this has yet been found.

GOODWIN, NAT [Nathaniel] CARL (1857–1919), American actor, who made his début in Boston, his birthplace, in 1874 at the Howard Athenaeum in Joseph Bradford's *Law in New York*. His career as a vaudeville comedian began at Tony PASTOR's Opera House, New York, in 1875, and the following year he scored an immense success in a show called *Off the Stage* at the New York LYCEUM with imitations of the popular actors of the day. Although best known as an actor of light comedy, he was also successful in such serious plays as Augustus THOMAS's *In Mizzoura* (1893) and Clyde FITCH's *Nathan Hale* (1899), which was written for him and in which he played opposite his third wife, Maxine ELLIOTT. In 1901 he played Shylock to her Portia and in 1903 appeared as Bottom, but Shakespeare proved beyond his range and he returned to modern comedy. The trade mark of his mature years was a drily humorous manner which made him popular as a comedian and as a vaudeville raconteur. His reminiscences, *Nat Goodwin's Book*, appeared in 1914.

GORCHAKOV, NICOLAI MIKHAILOVICH (1899–1958), Russian producer, who began his career as a student of VAKHTANGOV soon after the Revolution. In 1924 he joined the company of the MOSCOW ART THEATRE as assistant director under STANISLAVSKY, his productions including KATAYEV's *Squaring the Circle*, Kron's *An Officer of the Fleet*, and SHERIDAN's *The School for Scandal*. From 1933 to 1938 he was director of the Moscow Theatre of Drama, and from 1941 to 1943 of the Theatre of SATIRE. In 1939 he was appointed to the Chair of Theatre Production at the Lunacharsky State Institute of Theatre Art in Moscow. He wrote widely on theatrical subjects; one of his books, based on shorthand notes taken during Stanislavsky's rehearsals, was published in an American translation as *Stanislavsky Directs* (1962). One of his last productions was Rakhmanov's *The Troubled Past*, which the Moscow Art Theatre company performed in London during their visit in May 1958.

GORDIN, JACOB (1853–1909), Jewish dramatist, born in the Ukraine, who in 1891 emigrated to New York and was drawn into the orbit of the newly founded Yiddish theatre there. The success of his first play encouraged him to continue, and he wrote about 80 plays, of which *The Jewish King Lear* (1892), *Mirele Efros* (1898), which portrays a feminine counterpart of Shakespeare's Lear, and *God, Man and Devil* (1900), based to some extent on GOETHE's *Faust*, are the best known. Like most of his contemporaries, Gordin took much of his material from non-Jewish plays, adapting and rewriting them with Jewish characters against a Jewish background, setting his face against improvisation and insisting on adherence to the written text. In its simplicity, seriousness, and characterization, his work marks a great advance on what had gone before. He was one of the directors of GOLDFADEN's drama school in New York, and towards the end of his life fought against the 'star' system which he foresaw would be a danger to the theatre.

GORDON-LENNOX, COSMO, see TEMPEST, MARIE.

GORKY, MAXIM [Alexei Maximovich Pyeshkov] (1868–1936), Russian dramatist, and the only one to belong equally to the Tsarist and Soviet epochs, one of his major works being staged before the Revolution and one after. He had a hard and unhappy childhood and a youth overshadowed by brutality: his pseudonym, 'bitter', was well chosen. Painfully, during a series of menial jobs, he educated himself, and in 1892 published his first short story. It was followed by a succession of works in which he became the outspoken champion of the underdog. They brought him fame, and some money, but also imprisonment and eventually exile to Italy; he did not return to Russia for good until 1931. It was CHEKHOV who in 1902 persuaded the MOSCOW ART THEATRE to stage his first play, *Scenes in the House of Bersemenov* (known also as *The Smug Citizens*). This was followed by what is generally considered to be his best play, *The Lower Depths*, which depicts with horrifying realism the lives of some of the inhabitants of Moscow's underworld, huddled together in the damp cellar of a doss-house. In a translation by Laurence IRVING this was seen in London in 1903 and has been revived several times, notably by the ROYAL SHAKESPEARE COMPANY at the ALDWYCH in 1972. In New York it was first produced in 1930 as *At the Bottom* and revived in 1964 as *The Lower Depths*. It was followed by *Summerfolk* (1904, seen at the Aldwych in 1974); *Children of the Sun* (1905; Aldwych, 1979); *Enemies* (written in 1906, but not staged in Russia until 1933, or in London—again at the Aldwych—until 1971; N.Y., 1972); *Mother Zheleznova* (or *The Mother*) (1910); *The Zykovs* (1913; N.Y., 1975; Aldwych, 1976); and the posthumous *Yakov Bogolomov* (1941). In 1931 Gorky began a trilogy on the decay of the Russian bourgeoisie; of this only two parts were staged—*Yegor Bulychov and Others* at the VAKHTANGOV THEATRE in 1932 and *Dostigayev and Others*, also at the Vakhtangov, in 1933. The last part, *Somov and Others*, was left unfinished. Some of Gorky's novels were successfully dramatized by other hands, notably *Foma Gordeyev* (1899) and *Mother* (1907). Though he later suffered some disillusionment, Gorky welcomed the Revolution and by his work helped to bring about the establishment of the new régime. In acknowledgement of this his birth-place Nizhny-Novgorod was renamed Gorky, as was the theatre in Leningrad where his plays were first seen outside Moscow (see below).

GORKY THEATRE, Grand, Leningrad, founded in 1919 by Alexandr BLOK, and named after GORKY, who took a great interest in it. Its first production was SCHILLER's *Don Carlos*, followed by a number of other foreign classics and some new plays. A period of EXPRESSIONISM featured plays by TOLLER, KAISER, SHAW, and O'NEILL produced in Constructivist style. From 1925 onwards Soviet plays were introduced into the repertory and in 1933 Gorky's *Enemies* was first seen at this theatre, being produced in Moscow two years later. Among the theatre's scene designers were Alexandre Benois, who left Russia to join Diaghilev, and AKIMOV. After the Second World War the theatre, which had been severely damaged, was repaired, and the company embarked on an ambitious programme, including the revival of a fine production of Shakespeare's *King Lear* which had been cut short by the outbreak of war. The company appeared in London during the WORLD THEATRE SEASON of 1966 in a musical entitled *Grandma, Uncle Iliko, Hilarion, and I*, and in DOSTOEVSKY's *The Idiot*, dramatized and directed by TOVSTONOGOV, with SMOKTUNOVSKY as Prince Myshkin. During the 1960s and 1970s a number of Russian classics were presented at this theatre including GRIBOYEDOV's *Woe From Wit* (or *Wit Works Woe*) in 1962, CHEKHOV's *Three Sisters* in 1965, and GOGOL's *The Government Inspector* in 1972. Productions of foreign plays included BRECHT's *The Resistible Rise of Arturo Ui* in 1963, O'Neill's *A Moon for the Misbegotten* and MILLER's *The Price* in 1968, and a composite version of the two parts of Shakespeare's *Henry IV* in 1969.

GOSET, see MOSCOW STATE JEWISH THEATRE.

GÔT, EDMOND-FRANÇOIS-JULES (1822–1901). French actor, who passed the whole of his long and honourable career at the COMÉDIE-FRANÇAISE, of which he was the 29th *doyen*. He made his début in comedy roles on 17 July 1844, playing Mascarille in MOLIÈRE's *Les Précieuses ridicules*, and became a member of the company in 1850. He was one of the finest and most dependable actors of his day, playing innumerable new parts, as well as most of the classic repertory.

GOTTLOBER, ABRAHAM BER, see JEWISH DRAMA.

GOTTSCHED, JOHANN CHRISTOPH (1700–66), German literary critic who tried to reform the German stage on the lines of the French classical theatre. He was helped by the actress Carolina NEUBER, for whom he prepared a model repertory, later published in six volumes as the *Deutsche Schaubühne nach den Regeln der alten Griechen und Römer eingerichtet* (1740–45). This consisted of adaptations of French plays by himself, his wife, and his friends, with a few original works of which the best were those of J. E. SCHLEGEL. Some of them were acted by Carolina Neuber's company in place of the traditional farces featuring HANSWURST, whom she banished from the stage. She also produced Gottsched's own plays, of which *Der sterbende Cato* (1732), based on ADDISON's *Cato* (1713), was at first successful. By 1740 Gottsched, who had quarrelled with Carolina Neuber, no longer had any contact with or influence on the theatre. However, his *Nöthiger Vorrath zur Geschichte der deutschen dramatischen Dichtkunst* (1757–65) is a well-documented bibliography of German drama from the 16th century onwards. The oft-quoted remark by LESSING, that it would have been better if Gottsched had never meddled with the German theatre, need not now be taken seriously. The discipline he imposed was salutary and necessary, and first led German audiences to recognize artistic merit in serious drama.

GOUGH [GOFFE], ROBERT (?–1624), English actor, first heard of in *c.* 1590/1, playing a woman's part, possibly Aspasia in the revival of TARLETON'S *The Seven Deadly Sins*. He joined the CHAMBERLAIN'S MEN on their formation in 1594, and remained with them when they became the KING'S MEN in 1603, his known roles including Memphonius in the anonymous *The Second Maid's Tragedy* (1611) and Leidenberch in *Sir John van Olden Barnavelt* (1619), a tragedy attributed to FLETCHER and MASSINGER. He appears in the list of actors in Shakespeare's plays affixed to the First Folio of 1623, and is believed, on slender evidence, to have created the role of Portia in *The Merchant of Venice* and possibly Juliet. In 1605 he inherited shares in the GLOBE and BLACKFRIARS from his brother-in-law Augustine PHILLIPS, who was also a member of the King's Men. Gough's son Alexander was for a time a boy player but later became a publisher.

GOURGAUD, FRANÇOISE, see VESTRIS, FRANÇOISE; MARIE-MARGUERITE, see DUGAZON.

GOW, RONALD, see HILLER, WENDY.

GOWARD, MARY ANN, see KEELEY, MARY ANN.

GOZZI, CARLO (1720–1806), Italian dramatist, who tried to reform the moribund COMMEDIA DELL'ARTE in the middle of the 18th century by using its characters and methods, but not its subject-matter, for a new type of play which he called *fiabe*—a mixture of fantasy and fooling in a set text which nevertheless allowed plenty of room for improvisation. In opposition to the realistic and bourgeois comedies of his contemporary and rival GOLDONI, Gozzi used stories of magicians, fabulous animals, and fairytale characters, and wrote in 'pure Tuscan' as a counterblast to Goldoni's use of local Italian dialects. Most of Gozzi's plays were produced by the actor SACCHI, whose leading lady was Teodora Ricci. They were for a time successful and had some influence in Germany, on such dramatists as TIECK, and in France, particularly on Alfred de MUSSET. The best were *L'amore delle tre melarance* (*The Love of Three Oranges*) and *Il corvo* (*The Raven*), both 1761; *Il re cervo* (*King Stag*), 1762; *Turandot* (based on a Chinese fairytale and perhaps Gozzi's best work) and *L'augellino belverde* (*The Beautiful Green Bird*), both 1765. Perhaps because his vein of fantasy has become a contemporary preoccupation (he anticipated PIRANDELLO in his use of myth and the working of the subconscious), there have in this century been some notable productions, particularly *Turandot* directed by VAKHTANGOV in Moscow in 1922, and *King Stag* directed by George DEVINE for the YOUNG VIC in London in 1946 and later produced by several children's theatres. Gozzi's plays have appealed particularly to musicians, who have written a good deal of incidental music for them and used them as librettos for opera, the best known being Prokofiev's *The Love of Three Oranges* (1921) and Puccini's *Turandot* (1926).

GRABBE, CHRISTIAN DIETRICH (1801–36), German poet and dramatist who with BÜCHNER was the most notable playwright associated with the 'Young Germany' movement in the early 19th century. His grotesque satirical play *Scherz, Satire, Ironie und tiefere Bedeutung*, written in 1822 but not produced until 1892, is now seen as a forerunner of the Theatre of the ABSURD. The first of his plays to be produced was the ambitious *Don Juan und Faust* (1829), in which he strove to emulate both Mozart and GOETHE; it has some striking scenes, as has his *Napoleon* (1835) which, though consisting of little more than a series of sketches loosely strung together in the style of the STURM UND DRANG movement, was the first German drama in which the mob was the main character. Grabbe left several unfinished plays and his *Hermannschlacht*, written in 1836, was not produced for 100 years; he led an unhappy, harassed life and died young. He was later made the hero of a play, *Der Einsame* (1925) by Hanns Johst.

GRACIOSO, the comic servant or peasant in Spanish plays of the Golden Age. Though he may ultimately derive from Latin comedy, his immediate ancestor appears to be the BOBO of Lope de RUEDA's interludes, or PASOS. He first appears in the plays of Gil VICENTE and his contemporaries, particularly TORRES NAHARRO. In works by Lope de VEGA, who did not as he claims introduce the *gracioso* to the Spanish stage, he is nevertheless an important element, parodying the actions of his master in lively and popular language. By CALDERÓN he is used to present yet another facet of the moral or doctrinal lesson implicit in the play, whether COMEDIA or AUTO SACRAMENTAL. The *gracioso* reached his apotheosis in the plays of Agustín de MORETO, where he and his female counterpart set on foot and maintain the complicated intrigues.

GRAHAM, MARY ANN, see YATES, RICHARD.

GRAMATICA, EMMA (1875–1965), Italian actress, the child of strolling players, and on the stage from her earliest years. Though originally considered less promising than her elder sister Irma (below) she later gave notable interpretations of the leading characters in the realistic plays of IBSEN, PIRANDELLO, and particularly SHAW, appearing in the Italian versions of his *Mrs Warren's Profession, Pygmalion, Saint Joan*, and *Caesar and Cleopatra*. She was also seen as Marchbanks in *Candida*. An actress of great versatility, she was equally admired as Sirenetta in D'ANNUNZIO's *La Gioconda*; in the title role of ROSTAND's *La Samaritaine*; and in ROSSO DI SAN SECONDO's *Tre vestiti che ballano* (*Three Dresses which Dance*). In the 1930s she toured Europe, and she went to America in 1945. She was an excellent linguist and played in German and Spanish.

GRAMATICA, IRMA (1873–1962), Italian actress, elder sister of the above, and like her on the stage from childhood. As a young girl she became well known for her skilful playing of the heroines of contemporary French plays in translation, among them MEILHAC and Halévy's *Froufrou* and SARDOU's *Odette*. She was also much admired in such new Italian plays as GIACOSA's *Come le foglio* (*Like the Leaves*, 1900) and VERGA's *Dal tuo al mio* (*From Thine to Mine*, 1903). She was admirable as Katharina in *The Taming of the Shrew*, and made her last appearance in 1938 as Lady Macbeth to the Macbeth of RUGGERI, after which she retired. An actress of great charm and vivacity, she could make even trivial parts appealing.

GRANACH, ALEXANDER, see JEWISH DRAMA.

GRAND GUIGNOL, see GUIGNOL.

GRAND MASTER SYSTEM, see LIGHTING.

GRAND OPERA HOUSE, New York, on the north-west corner of 8th Avenue and 23rd Street. This opened on 9 Jan. 1868 as Pike's Opera House, from the name of its first owner. Its early years were uneventful and it was often closed between visits from touring companies. In Feb. 1869 it changed its name to the Grand Opera House during a season of *opéra bouffe*, and from 1872 to 1874 it was managed by DALY. Under him Mrs John WOOD made her last apperance in New York, playing Peachblossom in a revival of his melodrama, *Under the Gaslight; or Life and Love in These Times*. C. A. FECHTER was seen as the elder DUMAS's Count of Monte Cristo, and Fay Templeton, later a variety star, played Puck in *A Midsummer Night's Dream*. After a visit in 1875 from E. L. DAVENPORT the theatre closed, to reopen after redecoration as a local 'family' house, with inexpensive seats, for Broadway successes on tour with a sprinkling of stars. In 1938 it became a cinema and the building was demolished in 1961.

GRAND THEATRE, Leeds, see LEEDS.

GRAND THEATRE, London. (1) In Upper Street (now Islington High Street). As the Philharmonic Music-Hall this opened in 1870, being started on its way by the 'father of the halls', Charles MORTON. It was used for some years for *opéra bouffe*, and was destroyed by fire in 1883. As the Grand Theatre, it reopened in the same year, being itself burned down in 1887 and again in 1900. Rebuilt both times, it became in 1901 a staging post for provincial and London companies on tour. Its Christmas PANTOMIMES were famous, the music-hall star Harry Randall playing in them for many years. In 1908 it became a music-hall again as the Islington Empire, giving its last live 'variety' show in 1932, after which it became a cinema.

(2) In Putney Bridge Approach. As the Fulham Grand, this opened in 1897 as a home of MUSICAL COMEDY, Gertie MILLAR making her London début there as Dandini in *Cinderella* at Christmas 1899. Renamed the Fulham Theatre in 1906, it failed to find good audiences and its fortunes declined. In 1933, as the Shilling Theatre, it tried to attract local audiences at 5p a head, but the experiment failed, and except for a few amateur productions the building remained closed until it was demolished in 1958.

GRANDVAL, FRANÇOIS-CHARLES RACOT DE (1710–84), French actor, who made his début at the COMÉDIE-FRANÇAISE at the age of 19 as Andronic in CAMPISTRON's tragedy of that name. He played some of BARON's roles and succeeded DUFRESNE in tragic parts with a force and intelligence considered by some to be the equal of LEKAIN himself. He retired at 52, but lack of money brought him back to the stage; again successful, though not to the same extent, this proved his undoing and some jealous fellow actors were suspected of hiring a gang of thugs to howl him down in VOLTAIRE's *Alzire*, whereupon he retired for good. He wrote a certain amount of witty verse, and some scurrilous but amusing comedies are attributed to him.

GRANOVSKY, ALEXANDER [Abraham Ozark] (1890–1937), Jewish Yiddish-speaking actor and theatre director, who in 1919 founded in Leningrad the Jewish Theatre Studio, which later moved to Moscow and became the MOSCOW STATE JEWISH THEATRE. His first production in Leningrad was MAETERLINCK's *Les Aveugles*, with the text freely adapted to his own theories of dramatic effect, but his opening evening in Moscow in 1921 was devoted to short sketches by Sholom ALEICHEM, and he then concentrated on the production of plays by Yiddish dramatists. His methods were perhaps most clearly shown in his own dramatization of Isaac PERETZ's poem *Night in the Old Market* (1925). With no more than a thousand words of text to work on, he made music the basic element, while a subtle use of lighting evoked the presence of the dead who, with the market people and the *badchan*, or professional jester, made up the characters of the play. In 1927, while on a tour of Europe, Granovsky resigned his directorship of the company and was succeeded by his chief actor MIKHOELS. In 1930 he produced GUTZKOW's *Uriel Acosta* for the HABIMAH company, who appeared in it on tour.

GRANVILLE-BARKER, HARLEY (1877–1946), English theatre scholar, actor, and director, one of the outstanding figures of the progressive theatre at the beginning of the 20th century. In 1891 he joined the stock company at Margate, and later toured with Lewis WALLER, Ben GREET, and Mrs Patrick CAMPBELL. In London he appeared in such diverse productions as *Under the Red Robe* (1896), based on Stanley Weyman's romantic novel, and *Richard II* in William POEL's production for the Elizabethan Stage Society in 1899. In 1900 SHAW chose him to play Marchbanks in the first London production, by the STAGE SOCIETY, of *Candida*, and he appeared also in their productions of Shaw's *Captain Brassbound's Conversion* (also 1900), *Mrs Warren's Profession* (1902), and *Man and Superman* (1905), in which his first wife Lillah MCCARTHY played Ann Whitefield to his John Tanner. In 1904, with J. E. VEDRENNE, he assumed the management of the ROYAL COURT THEATRE, where he embarked on an extensive programme of productions of new plays, many of them by contemporary European authors, including HAUPTMANN, SCHNITZLER, and IBSEN. New English dramatists included GALSWORTHY and St John HANKIN, and he also put on his own play *The Voysey Inheritance* (1905), in which the hero finds that the firm he has inherited achieved its wealth dishonestly. An earlier play, *The Marrying of Ann Leete* (1901), had been successfully produced by the Stage Society, but their later production of Barker's *Waste* in 1907 fell foul of the LORD CHAMBERLAIN because it contained an abortion, and the play was not licensed for public production until 1936. His only other play of any importance, *The Madras House* (1910; N.Y., 1921), was first seen at the DUKE OF YORK'S THEATRE, which he was then directing for Charles FROHMAN. His experiences at the Royal Court, which had been artistically rather than financially rewarding, made him a fervent advocate for a subsidized theatre in England, for which he campaigned ceaselessly. His own approach to Shakespeare, whom he naturally considered as the foundation stone of any English NATIONAL THEATRE, was conditioned by his early association with Poel, and his productions at the SAVOY THEATRE in 1912–14 of *The Winter's Tale, Twelfth Night*, and *A Midsummer Night's Dream* were later considered epoch-making in their simplicity and poetic beauty. Barker was at the height of his career in England when a visit to America directed his energies into fresh channels. Divorced from Lillah McCarthy, he married as his second wife the American Helen Huntingdon. At her instigation he gave up all contact with the theatre backstage, hyphenated his name, and settled down to translate, with her help, the plays of MARTÍNEZ SIERRA and the ÁLVAREZ QUINTERO brothers, and to write the *Prefaces to Shakespeare* (1927–47) on which his posthumous fame chiefly rests. *The Marrying of Ann Leete* was revived by the ROYAL SHAKESPEARE COMPANY in 1975, and *The Madras House* by the National Theatre in 1977.

GRASSI, PAOLO, see PICCOLO TEATRO DELLA CITTÀ DI MILANO.

GRASSO, GIOVANNI (1873–1930), Italian actor, son and grandson of Sicilian puppet-masters, who was encouraged to go on the stage by the actor-manager ROSSI, under whom he trained. He was seen all over Europe, and was considered a fine actor of the VERISMO school, being at his best in the plays of PIRANDELLO. He was several times seen in London, where in 1910 his Othello was considered outstanding.

GRAU, JACINTO (1877–1958), Spanish dramatist, whose slow-moving but powerful works, such as *El Conde Alarcos* (1917) and *El hijo pródigo* (*The Prodigal Son*, 1918), were meant to restore the dignity of true poetic drama to the Spanish stage. A later, and better-known, play, *El señor de Pigmalión* (1923), which owes something to PIRANDELLO, is a bitter satire which makes excellent intellectual and theatrical use of the metaphor of the human being seen as a puppet.

GRAVE TRAP, see TRAPS.

GRAY, DULCIE, see DENISON, MICHAEL.

GRAY, 'MONSEWER' EDDIE, see CRAZY GANG.

GRAY, SIMON JAMES HOLLIDAY (1936–), English dramatist, whose first stage play *Wise Child* (1967), starring Alec GUINNESS in a travesty role (played by Donald Pleasence in New York in 1972), was followed by *Dutch Uncle* (1968) and *Spoiled* (1971). He had his first outstanding success with *Butley* (also 1971), about a university lecturer facing the breakdown of both his marriage and his homosexual relationship. It starred Alan BATES, as did *Otherwise Engaged* (1975), depicting a publisher who avoids emotional entanglements. Both plays were seen in New York, the first with Bates in 1972, the second starring Tom COURTENAY in 1977. *Molly* (1977), in which Billie WHITELAW appeared, and *The Rear Column* (London and N.Y., 1978) had only short runs. *Close of Play* (NATIONAL THEATRE, 1979; N.Y., 1981) concerns a family reunion at which a distinguished academic, played by Michael

REDGRAVE, remains silent while relationships disintegrate. Gray resumed his successful partnership with Bates in the comedy thriller *Stage Struck* (1979), while *Quartermaine's Terms* (1981), considered by some to be his best play, is a study of anguish and non-communication in a Cambridge language school. His adaptation of DOSTOEVSKY's *The Idiot* was produced by the National Theatre company in 1970.

GRAY, TERENCE (1895–   ), co-founder in 1926 of the Festival Theatre, Cambridge (formerly the Barnwell), which during his brief management had an influence, particularly on the Continent, out of all proportion to the work actually done there. Like CRAIG, whose theories on lighting and stage-craft were the basis of his experiments, Gray fertilized the theatre more by his ideas than by his achievements. He abolished the proscenium arch and footlights, and built out a forestage connected with the auditorium by a staircase, broken by platforms on different levels which offered exceptional opportunities for significant groupings. With Maurice EVANS as his leading man, and with interesting experiments in lighting by Harold Ridge, who later became an authority on stage lighting, Gray produced the *Oresteia* of AESCHYLUS, following it with a number of English and foreign classical and modern plays, including Shakespeare. The choreography was by Ninette de Valois, Gray's cousin, whose dancers later formed the nucleus of the Vic-Wells ballet. In 1929 Anmer Hall brought a company headed by Flora ROBSON and Robert Donat to the theatre, profiting by the experience gained there to build the WESTMINSTER THEATRE, and in 1932 Norman MARSHALL ran a season there. Gray returned intermittently but finally abandoned his project in 1933, after the first performance in English of Aeschylus' *Suppliants*. The theatre was then bought by a commercial management and restored in a conventional style. It is now owned by the Trustees of the CAMBRIDGE Arts Theatre and used as a workshop and costume store.

GRAZIANO, see DOTTORE, IL.

GREAT DIONYSIA, see DIONYSIA.

GREATHAM SWORD PLAY, see MUMMERS' PLAY.

GREAT LAKES SHAKESPEARE FESTIVAL, Lakewood, Ohio, see SHAKESPEARE FESTIVALS.

GREAT QUEEN STREET THEATRE, London, see KINGSWAY THEATRE.

GREAT VANCE, The, see VANCE, ALFRED.

GREBAN, ARNOUL (*c.* 1420–*c.* 1471), French musician, organist, and dramatist, who was choirmaster at Notre-Dame in Paris from about 1447 to 1452, during which time he wrote one of the earliest and perhaps the best of the French PASSION PLAYS. The music for the numerous songs or organ pieces in the play has unfortunately not survived, but the text, which runs to over 34,500 lines, mainly in octosyllabics but with a wide variety of lyric metres inset, is remarkable for the quality of its poetry and its observation of everyday life. It follows the same pattern as the earlier Passion Play of MARCADÉ, beginning with a prologue depicting the Fall, followed by the events of Christ's life from the Annunciation to the Ascension, the whole performance taking four days. The text, or adaptations of it, notably one by the Angevin Jean Michel performed in sumptuous style at Angers in 1486, was widely used all over France in the late 15th century. Greban probably collaborated also in a MYSTERY PLAY by his brother Simon, *Actes des Apôtres*, which followed at length the wanderings of the twelve Apostles after the death of Christ. This was still being played in Paris by the CONFRÉRIE DE LA PASSION as late as 1540.

GRECIAN THEATRE, London, in Shepherdess Walk, City Road, Shoreditch. This opened under Thomas Rouse in *c.* 1830 as a two-tier concert hall for the production of light opera. It was situated in the pleasure grounds of the Eagle Saloon, immortalized in the nursery rhyme 'Pop goes the weasel', where open-air entertainments had already been given for some time. In spite of excellent actors and singers, of whom Frederick ROBSON was one, from 1844 to 1849, the venture was a failure and Rouse was kept solvent only by the profits from the tavern. In 1851 he was succeeded by Benjamin Conquest, who obtained a full theatre licence and presented a wide range of plays, including Shakespeare. The theatre was reconstructed in 1858 with four tiers to hold 3,400 persons. Drama and pantomime formed an important part of the entertainment presented under the Conquests, drawing audiences from all parts of London, and ballet was under the direction of Mrs Conquest, herself an excellent dancer. Their son George CONQUEST, an outstanding player of PANTOMIME, was responsible for many years for the Grecian Christmas pantomimes. He succeeded to the management of the theatre on the death of his father in 1872, and four years later the theatre was again rebuilt with a three-tier auditorium to hold about 1,850 people. It was sold in 1879 to Thomas Clark, one-time lessee of the ADELPHI THEATRE, who lost a great deal of money on it, and eventually sold out to the Salvation Army in 1881.

GREECE, Ancient. The classical Greek drama which reached its maturity in the 5th and 4th centuries BC was in fact Athenian drama: for although every Greek city and many a large city elsewhere came to have its own theatre, and although some dramatic forms, such as MIME, originated and flourished elsewhere, Athens established and maintained a complete pre-eminence among the Greek states, and all the Greek drama that we possess was written by Athenians for Athens. The question of the origin of TRAGEDY and COMEDY is obscure and difficult, but of little practical importance when compared with the major fact that both forms of drama were, from their inception, part of the religious celebrations in honour of DIONYSUS—that is, since the word 'religious' in this context can be misleading to modern ears, a serious and splendid civic and national occasion. Both forms of drama were, from the beginning, addressed to a whole community, which came to the theatre as a community, not as individuals; a community which was its own political master and its own government.

Tragedy was formed while the Athenian democracy was being formed; it ennobled itself while the Athenian people were ennobling themselves by the part they played in repelling the Persians (490–479); it lost its vitality and vigour at the end of the 5th century, under the double strain of the long Peloponnesian War and an age of criticism and self-consciousness. During this century it moved, on the whole, from communal to individual or private themes. The typical concern of AESCHYLUS, the first great Athenian writer of tragedy, is the moral government of the universe: sin produces suffering and counter-sin; that produces more sin and suffering; until a resolution is reached in Justice. The tragedy of the individual hero was brought to perfection by SOPHOCLES, who won his first victory in 468, this being the form of tragedy that ARISTOTLE analyses in the *Poetics*. In EURIPIDES' plays, and in some of the later works of Sophocles, there is a more analytical and critical view of society, some interest in the abnormal, and a more complex exploitation of dramatic situation, which eventually produces romantic drama, melodrama, even high comedy, as in Euripides' *Helen* (412).

Comedy followed a roughly parallel course, but some 50 years later: OLD COMEDY (*c.* 435–405) was a riotous burlesque and criticism of current personalities and movements in political life (this is represented for us by the first nine plays of ARISTOPHANES); MIDDLE COMEDY was still strongly satirical, but quieter, more coherent, and with a social rather than a political background; NEW COMEDY (*c.* 350–292), which ended with the death of its chief exponent MENANDER, was a delicate, sometimes sentimental, comedy of manners.

In both forms of drama the CHORUS, the communal element, is originally very prominent and the movement of plot is restricted; but in both forms the histrionic element and the plot grow in importance, while the chorus becomes more and more out of place and finally disappears.

Although it is beyond question that tragedy as written and performed in Athens in the 5th century BC developed out of the choral lyric, an art which had reached a high degree of perfection in the preceding century, particularly among the Dorian peoples of the Peloponnese, nothing further can be said without qualification. The account of the origin of tragedy given by Aristotle in his *Poetics* is bald and perfunctory. It was, he says, at first an improvisation; then, passing out of the stage of SATYR-DRAMA and abandoning short plots and ludicrous diction, it gradually attained dignity. These statements accord with the fact that tragedies were performed only during the festivals in honour of Dionysus, and that the statutory tragic trilogy was regularly followed by a satyr-play. It has been suggested that what Aristotle says is only an inference drawn from these facts. In modern times many detailed theories have been advanced; 'ritual-sequences' of the death and rebirth each year of a nature-god, such as Dionysus was, have been found throughout the world, but nowhere did they lead to anything comparable to Greek tragedy. Its Dionysiac origin is generally accepted, but attempts to link tragedy with any ritual worship of the god cannot profitably look back further than the DITHYRAMB. It would have been a natural development, and in accord with Aristotle's statement, if the leader of the dithyrambic chorus had engaged in a simple form of dialogue with his fellows, and so created a histrionic precedent. But if ever tragedy had a special association with Dionysiac subjects, this was lost before the genre reached maturity. Nor is the word 'tragedy' itself any help; it simply means a 'goat-song'. The role of the goat is equally obscure; perhaps it was the prize for the best song.

Other formative influences can, however, be recognized. The legendary poet ARION is credited with the invention of a form of lyric tragedy distinct from the choral ode, and the scarcely less legendary THESPIS was apparently the first person to take a touring company round the villages of Attica, travelling in a cart which provided a raised stage for semi-improvised performances. In these he himself was the chief and only actor, as distinct from the former chorus-leader. It was, appropriately, Thespis who won the prize for tragedy at the first of the tragic contests introduced by the tyrant PISISTRATUS when he reorganized, with great splendour, the City DIONYSIA in 534 BC. His successor Choerilus was said to have written 160 plays, the first being produced about 520 BC.

The first great Athenian writer of tragedy Aeschylus (it is not known how many predecessors he had) introduced a second actor, found in the earliest extant Greek tragedy, the *Persians*. Although there is no known example of the one-actor play, it is interesting to note that in his *Suppliant Women*, attributed to his later career, Aeschylus for some reason returned to the earlier form. In this the actor playing Danaus is silent practically throughout, and the play is virtually carried by a single actor (the King) and the Chorus of Suppliants. With very little plot, it nevertheless contâins a most effective intensification on quasi-musical lines of an already existing situation and, apart from the dancing, must have borne some resemblance to modern oratorio.

Aeschylus, with the introduction of a second

actor, had intensified the histrionic side of his work at the expense of the chorus, which he himself reduced at some time from 50 to 12 or 15. The introduction of a third actor by Sophocles produced a different type of drama, based on the tragic interplay between different characters, between a character and circumstance, or even between two different aspects of a single character. This demanded a more subtle plot and greater complexity of characterization treated in a detailed and naturalistic way. It was therefore inevitable that Sophocles should abandon the vast canvas used by Aeschylus for his trilogies and present three quite distinct plays. He also reduced and altered the scope of the chorus, whose chief function became purely lyrical, as in the ecstatic ode in the *Antigone* which immediately precedes the Messenger's tale of death.

After Sophocles there were no important changes in the externals of tragedy, though in the late *Oedipus at Colonus* he introduced a fourth actor. The innovations of Euripides were concerned more with style and treatment—more realism and pathos, a new and more emotional style of music. The number of actors increased—in the *Phoenician Women* there are 11—while the role of the chorus declined. The last we hear of it is Aristotle's statement that it was AGATHON, a younger contemporary of Euripides, who first introduced choral odes that had no connection with the plot. These acted as interludes, dividing the action, which had previously run without a break through dialogue and choral ode, into separate sections, and so eventually giving rise to Horace's idea of a play being necessarily 'in five acts'.

Although the worship of Dionysus had its jovial, bawdy, indeed indecent, aspects, which seem to have been given free rein in the satyr-drama, tragedy itself remained entirely disassociated from comedy, which had its own, equally obscure, but quite different origins. It appears to have evolved from two separate elements. Old Comedy, represented now only by the works of Aristophanes, exhibits with variations a complex and elaborate structure which is elaborately symmetrical; the chorus enters, there is a dispute between it and an actor or between two actors each supported by a semi-chorus; there is a formal 'contest' or debate, and finally the *parabasis* or 'coming-forward', in which the chorus directly addresses the spectators. All this is essentially choral in form. But there are also scenes for actors alone which precede the entry of the chorus and follow the *parabasis*. It thus seems clear that the formal part of comedy developed out of some ritual, while the histrionic scenes have obvious affinities with the Sicilian mime. The use of the phallus certainly makes clear some connection with early fertility-rites, which were a mixture of singing, dancing, scurrilous jesting involving the bystanders, and ribaldry, while the disguises of the chorus as animals or birds points to an influence from the popular masquerade often depicted on vase-paintings. The first contest for comedy was added to the City Dionysia in 486 BC and was won by a poet called Chionides; it came to maturity during the second half of the 5th century. Old Comedy, in which anything or anyone prominent in city life was unsparingly ridiculed, was a unique mixture of fantasy, criticism, wit, burlesque, obscenity, parody, invective, and exquisite lyricism. It gave way in Aristophanes' last two plays, the *Women in Parliament* and *Plutus*, to Middle Comedy, in which little is left of the former spirit of revelry, and by the middle of the 4th century this in turn had passed into New Comedy, which, in the capable hands of Menander, became the model and quarry for Roman comedy, and was transmitted by way of the Renaissance to the great comic writers of the European theatre.

Drama in ancient Greece was always a civic preoccupation, never a private venture, and preparations for it were in the hands of the State. Plays were given only at the city festivals, the Dionysia and the LENAEA, and early in the official year, which in Athens began at midsummer, officials in charge of the festivities would begin their preparations by choosing from among the many applicants three poets whose works would be performed in the festival. It is surmised that the lesser known, who might already have served their apprenticeship to playwriting in the RURAL DIONYSIA, were asked for a complete script, established writers for a synopsis only. The chosen three were then assigned a CHOREGUS, who became responsible for all the expenses in connection with the production, except for the chorus and the three statutory actors who were paid by the State. Since this left the poet dependent on the generosity or otherwise of his choregus, it was thought fairer to assign them by lot. The poet not only wrote the play, but also composed the music and arranged the dances. In earlier times he trained the chorus and acted the chief part himself, but later these functions were handed over to specialists, and in the 5th century individual actors seem to have been associated closely with particular poets—Cleander and Mynniscus with Aeschylus, for instance, and Cleidimides and Tlepolemus with Sophocles. When later the importance of the actor increased it was again thought fairer, at least in the case of the PROTAGONIST, to assign them also by lot.

Very little is known about individual Greek actors, but with the decline of tragedy in the 4th century they became more prominent. The most famous was POLUS, who is said to have taught Demosthenes elocution. Other actors known by name include, in the 4th century, Theodorus, who had a reputation for adapting plays to suit his own personality, and Aristodemos, who is known to have been sent as an envoy from Athens to the Court of Macedonia. Actors, who were usually men of good repute and members of guilds like the ARTISTS OF DIONYSUS, were often employed on secret missions, since their semi-religious function gave them a degree of diplomatic immunity. From the 5th century have survived the names of Nicostratos, famous for his delivery of messenger speeches, and Callipides, often the butt of comic writers because of his high-flown style.

Playgoing in Greece was a civic duty, and still retained traces of its religious origins, as did the theatre. These were all open-air, cut into the side of a hill, and had to be big enough to contain a vast number of spectators during a day-long session—EPIDAUROS could seat 14,000 people—while seats had to be provided for distinguished visitors and officials, the seat of honour in the centre of front row being reserved for the priest of Dionysus. Originally all other seats were free; later a charge of 2 obols (about 4p or 8 cents) was made. Those citizens who were too poor to pay even that small sum were given a grant from the THEORIC FUND. The acting area, with its *skene* or stage-building against which the actors played, and its vast circular ORCHESTRA for the chorus, probably duplicated the original playing-place in front of a temple wall, and the orchestra still retained its *thymele* or altar to Dionysus. When the importance of the chorus diminished and a raised stage or *logeion* was placed against the back wall for the actors, that too had its altar, which could also be used if necessary as a tomb or other holy shrine.

There was no scenery in the early Greek theatres, colour and splendour being supplied by the rich robes of the actors, who all wore MASKS, and the multifarious costumes of the chorus, particularly in comedy. Later, easily changed backcloths helped to diversify the permanent set, and PERIAKTOI, or movable screens, indicated a change of scene. There were also mechanical devices, such as the MECHANE, or crane, and the EKKYKLEMA, or wheeled platform.

GREECE, Modern. After the virtual disappearance of the theatre under the rule of BYZANTIUM, Greek drama resurfaced in the 16th and 17th centuries in Venetian-occupied Crete, where tragedies, comedies, and other dramas, all in verse, were written according to Italian conventions but in a lively demotic language; the chief dramatist was G. Chortatsis. This tradition was continued in the Ionian Islands, particularly Cephalonia and Zante, during the 17th and 18th centuries. Outside the islands, Greek theatrical activity was resumed on the eve of the War of Independence, the first new Greek play to be performed in modern times being presented at Jassy in 1805, while the first production of ancient drama in modern Greek was mounted at Odessa in 1818. In Greece itself during the Ottoman occupation the only theatrical tradition was that of the Karaghiozis SHADOW-SHOW. This theatre, with its cardboard puppets representing various stereotyped Greek and Turkish characters, was adapted to Greek needs from Turkish models, and survives today as a popular entertainment for children, parents, and intellectuals.

Soon after Athens became the capital of the new kingdom in 1834 temporary outdoor theatres were set up for the summer season. At first these were used by Italian acrobats; the first play performed in Athens since ancient times was Metastasio's *Olimpia*, translated by Rigas Pheraios, in 1836. The first 'winter' theatre in Athens, built in 1840, put on chiefly Italian opera.

During most of the 19th century Greek theatre was stultified by the imposition of an artificial official form of the language (*katharevousa*) on all serious writing. This went with a barren adherence to the forms (but not the spirit) of classical Greek culture. Thus, tragedies were written in ancient or Byzantine settings, mainly to be read rather than performed. It was usually only comedies that met with any success in the theatre; one of these was *Vavylonia* (1836) by D. Vyzandios, a satire on the Greek language question which still preserves its popularity. Meanwhile in Zante a precocious bourgeois drama in demotic prose, *The Basil Plant* (1830) by A. Matesis, was produced; but it was to have no successor.

The only two playwrights of any note during this period were S. Vasiliadis (1845–1874) and D. Vernardakis (1833–1907). The former produced successful tragedies of cold perfection combining ancient myth with motifs from modern Greek folklore, while the latter wrote more romantic historical dramas. During the 1880s the Greek theatre was revitalized by the emergence of a popular new genre, the *komeidyllio* or comic idyll, which combined comedy with sentimentality; these light plays, written in the demotic language and often containing songs, were set either in an idealized folksy countryside or among the poorer sectors of urban society. This movement brought a marked increase in the popularity of theatre in Athens which led to the foundation of the Royal Theatre, built in 1901. The *komeidyllio* opened the way for serious drama in the demotic, set in a contemporary Athens milieu; the chief exponent was Yannis Kambysis (1872–1901), who was influenced by IBSEN, STRINDBERG, and HAUPTMANN.

Since the Royal Theatre was seen very much as an establishment institution, K. Christomanos founded a rival New Stage, also in 1901, aimed at a younger audience. Its leading ladies Kyveli Adrianou and Marika Kotopouli were to become legendary figures in the modern Greek theatre. The New Stage put on productions of young playwrights such as Kambysis, K. Palamas, and G. Xenopoulos (1876–1951), the last of whom became one of the most prolific and popular writers of 'serious comic' plays and sentimental dramas in Greece.

It was about this time that performances of ancient drama in demotic translation began to be mounted. Hitherto these plays had been staged in academic performances in ancient Greek, particularly in the Theatre of Herod Atticus on the Acropolis. Even the staid Royal Theatre staged performances at the Herod Atticus of AESCHYLUS'

Oresteia in demotic translation in 1903, which gave rise to the famous 'language riots', and an audience of 1,500 was reported at a single performance of ARISTOPHANES' *Clouds* in 1900 at the Athens Municipal Theatre, founded in 1888. But the Royal Theatre closed down after a few years, and dramatic activity in Greece generally slackened off until in 1927 Angelos Sikelianos (1884–1951) mounted the first Delphic Festival, which included performances of Aeschylus' *Prometheus Bound* in the ancient theatre of Delphi. The Delphic Festivals received international attention and Sikelianos went on to state his belief in tragedy as the highest form of social and spiritual creation. In 1930 the Royal Theatre was refounded as the National Theatre by Fotos POLITIS, whose influence on the production of ancient tragedy was to last well after the end of his brief career. Politis also directed performances of Cretan tragedy, and of a play based on the Karaghiozis shadow-show. In 1938 the National Theatre first staged a production (starring Katina PAXINOU) in the great ancient theatre of EPIDAUROS, where since 1954 there has been an annual festival in which the National Theatre has performed the ancient repertory before 10,000 people at a time.

In the years since the Second World War the National Theatre has kept up the quality of its productions, particularly of ancient tragedies and comedies directed by Alexis Solomos, Takis Mouzenidis, and others, and with Katina Paxinou and Alexis MINOTIS as their leading players. The increasingly conservative nature of these performances has contrasted with the more innovative productions of the Art Theatre, founded in 1942 by Karolos KOUN, which has usually outshone the National in its productions of ancient plays at the Athens and Epidauros festivals as well as on tours to major centres abroad. Especially successful have been its productions of Aristophanes, with much deliberate anachronism and the use of music by the leading popular composers Hadjidakis and Theodorakis. The Art Theatre has also been far more adventurous in its choice of plays, mounting memorable performances of PINTER, BECKETT, SHAFFER, as well as contemporary Greek playwrights. Theatrical life has also thrived outside Athens since the war. In 1961 the National Theatre of Northern Greece was founded at Salonica, while a National Theatre touring company puts on plays in the remoter corners of the Greek mainland and islands. Ancient theatres such as Dodone in Epirus and Philippi in Macedonia have their own festivals, and an increasing number of less conventional locations are used, such as medieval castles.

Experimental theatre companies were founded in the 1960s and 1970s, and there is now a whole generation of dramatists who write solely for the theatre, unlike earlier playwrights such as Sikelianos, N. Kazantzakis, A. Terzakis, and G. Theotakas, whose best work lies in other genres. The new drama of Greece is political and absurdist, with much influence from BRECHT, as well as Pinter and Beckett. The settings are contemporary, and these plays treat social and existential problems with a bitter humour.

Theatregoing in Greece today is a popular activity not confined to any particular class of people and is central to Greek cultural life. There are about 40 'winter' theatres in Athens putting on a wide range of serious plays, from ancient drama and Shakespeare to the latest avant-garde works by Greek and foreign playwrights. In summer most plays are performed in the open air, and the majority of the repertoire then consists of revues and light comedies. One of the most interesting companies today is Spyros Evangelatos' Amphitheatro, which revolves not round its leading player but, like the Art Theatre, round its director. Evangelatos has been responsible for some original and amusing productions of plays which are outside the standard repertoire, such as Cretan comedy and oddities like *Iphigenia in Lixouri* by the 18th-century writer Petros Katsaitis, a version of the ancient myth which ends in farce in a contemporary Cephalonian town.

GREEN, JOHN, see ENGLISH COMEDIANS.

GREEN, JULIEN (1900–   ), French novelist of American parentage, whose three plays were written in the mid-1950s. Of these *Sud* (*South*, 1953), a 'strong' drama set in the American Civil War of an officer's discovery of his own homosexuality, was seen in London, at the ARTS THEATRE, in 1955. His other plays, *L'Ennemi* (1954) and *L'Ombre* (1956), which were staged only in Paris, deal with the spiritual conflict between good and evil.

GREEN, PAUL ELIOT (1894–1981), American dramatist, born on a farm, who gained a knowledge of the Negro from working in the fields. Early in his career he wrote almost 40 one-act plays, mainly produced by the Carolina Playmakers, on the problems of Negroes and poor whites in the American South. His first full-length play, *In Abraham's Bosom* (1926), which deals with the frustrated attempts of an ambitious but illiterate Negro, son of a white man, to start a school for Negro children, culminating in his murder at the hands of an infuriated mob, was awarded a PULITZER PRIZE for its imagination, sympathy, and power. Other full-length plays include *The Field God* (1927), on religious repression, *Tread the Green Grass* (1929), *The House of Connelly* (1931), the first independent production by the GROUP THEATRE, *Johnny Johnson* (1936), with music by Kurt Weill, and an adaptation of Richard Wright's novel *Native Son*, for a production by Orson WELLES in 1941. In 1937 Green wrote *The Lost Colony* (produced at Roanoke Island, North Carolina), the first of 15 'symphonic dramas' celebrating American history, all designed to be performed out-of-doors in specially built amphitheatres and using professionals with local amateurs. They include *Wilderness Road* (Berea, Kentucky, 1955), *The Founders* (Williamsburg, Virginia, 1957), *Cross and Sword* (St Augustine, Florida, 1965), *Texas* (Palo Duro Canyon, Texas, 1966), *Trumpet in the Land* (New Philadelphia, Ohio, 1970), and *Louisiana Cavalier* (Natchitoches, Louisiana, 1976). Many of these productions have become annual events in their localities.

GREEN-COAT MEN, footmen in green liveries who, in the early Restoration theatre, placed or removed essential pieces of furniture in full view of the audience.

GREENE, (Henry) GRAHAM (1904–   ), distinguished English novelist and dramatist, appointed C.H. in 1966. His first contact with the theatre was through the successful dramatization by Frank Harvey of his novel *Brighton Rock* in 1943. Two of his other novels, *The Heart of the Matter* and *The Power and the Glory*, were also dramatized, the first by himself and Basil DEAN in 1950, the second by Denis Cannan and Pierre Bost in 1956. Greene's first play written directly for the theatre was *The Living Room* (1953; N.Y., 1954), in which Dorothy TUTIN as a young girl in love with an older married man gave a fine performance. It was followed by *The Potting Shed* (N.Y., 1957; London, with John GIELGUD, 1958); *The Complaisant Lover* (London, with Ralph RICHARDSON, 1959; N.Y., with Michael REDGRAVE, 1961); and *Carving a Statue* (1964; N.Y., 1968), in which Richardson again played the lead in London. In 1975 *The Return of A. J. Raffles*, with Denholm ELLIOTT in the title role, was presented by the ROYAL SHAKESPEARE COMPANY in London and *For Whom the Bell Chimes* was presented by the HAYMARKET THEATRE, Leicester, in 1980. Greene's work bears the strong impress of his Roman Catholicism, though his expression of it is not always acceptable to the authorities.

GREENE, ROBERT (*c.* 1560–92), English dramatist, who led a wild and dissipated life, and shortly before his death wrote his famous recantation *A Groatsworth of Wit Bought with a Million of Repentance*, now chiefly remembered for its malicious attack on Shakespeare—the earliest allusion to his standing as a dramatist—as 'an upstart crow beautified with our feathers . . . in his own conceit the only Shake-scene in a country.' Greene, who was one of a group of writers known as the UNIVERSITY WITS, is thought to have had a hand in the *Henry VI* later re-written by Shakespeare, as well as in KYD's *The Spanish Tragedy* (*c.* 1587) and several other plays of the time. Among those that can be ascribed to him with some certainty the most important is the comedy *The Honorable History of Friar Bacon and Friar Bungay* (1589), a study of white magic which may have been intended as a counterblast to MARLOWE's play of black magic *Doctor Faustus* (*c.* 1588); it proved sufficiently popular to be revived many times up to about 1630. Others are *Alphonsus, King of Aragon* (1587), *James IV* (1590), and an *Orlando Furioso* (1591) based on ARIOSTO.

GREEN MAN, a figure common to many spring FOLK FESTIVALS, particularly that of May-Day. Dressed always in green and sometimes hung about with leaves and small blossoming branches, he appears to typify the spirit of the renascent earth. With the HOBBY HORSE, he can sometimes be found in the MORRIS DANCE, and ROBIN HOOD, in his suit of Lincoln green, may be a later development of the character, which was also known as the Woodman or Jack-in-the-Green.

GREEN ROOM, after the Restoration in 1660, a room behind the stage in which actors and actresses gathered before and after the performance to chat or to entertain their friends. It has almost disappeared from the modern English theatre, but still exists in a modified form at DRURY LANE. The first reference to it occurs in SHADWELL's *A True Widow* (1678) and it is mentioned in Colley CIBBER's *Love Makes a Man* (1700). It seems probable that it got its name simply because it was hung or painted in green. It was also known as the Scene Room, a term later applied to a room where scenery was stored, and it has been suggested that 'green' is a corruption of 'scene'. In the larger early English theatres there was sometimes more than one green room; they were then strictly graded in use according to the salary of the player, who could be fined for presuming to use a green room above his rank.

GREEN ROOM CLUB, London, in Adam Street, a social club formed in 1866 when the professional members of the old Junior GARRICK CLUB were outvoted. It remains a notable meeting place for the actors and actresses who comprise 75 per cent of its membership, and maintains reciprocal arrangements with the LAMBS in New York.

GREENWICH THEATRE, London. (1) The first permanent theatre in Greenwich was opened in 1709, probably in Church Street, by William PENKETHMAN. Colley CIBBER's *Love Makes a Man* is known to have been produced there the following year, but there is no further mention of it after 1712.

(2) A second Greenwich Theatre, at 75 London Road, opened in 1864 under a Mr Noble. It held 721 people in a three-tier auditorium, and later came under the control of William Morton, who gave it his own name, as did his successor Arthur Carlton in 1902. It was converted into a cinema in 1910 and demolished in 1937.

(3) The present Greenwich Theatre, in Stockwell Street, Crooms Hill, was originally a concert-hall attached to the Rose and Crown public house, which opened in 1855. It was rebuilt in 1871 and then became Crowder's Music-Hall. After several

changes of name and management, it became the Greenwich Hippodrome in 1911, and finally closed in 1924. Later it reopened as a cinema and then became a warehouse, and by 1962 was derelict. On the initiative of a group of enthusiasts from the BRISTOL OLD VIC, headed by Ewan Hooper, the present theatre was erected within the walls of the old building, with a new façade. It holds 426 in a two-tier auditorium, and has an open stage. Its first production was *Martin Luther King* by Hooper, which opened on 21 Oct. 1969. Many notable new works have since been staged there, including John MORTIMER's *A Voyage Round My Father* (1970), Peter NICHOLS's *Forget-Me-Not Lane* (1971), and Alan AYCKBOURN's trilogy *The Norman Conquests*, which came from Scarborough in 1974; all were successfully transferred to the West End. Jonathan MILLER has directed several plays there, and the theatre gave the British premières of O'NEILL's *More Stately Mansions* (1974) and Hugh LEONARD's *Da* (1979). Later West End transfers have included COWARD's *Private Lives* (1980) and *Present Laughter* (1981). In 1978 Hooper was succeeded as director by Alan Strachan, who controls not only the main theatre but also Tramshed, a community theatre project in Woolwich, and the Greenwich Young People's Theatre in Plumstead.

GREENWICH VILLAGE THEATRE, New York, see MACGOWAN, KENNETH, and PROVINCETOWN PLAYERS.

GREET, BEN [Philip Barling] (1857–1936), English actor-manager, knighted in 1929. He first appeared on the stage in 1879, as Philip Ben, the nickname 'Benjamin' having been given him as the youngest of eight children, and three years later was in London, where he appeared with a number of outstanding players including Lawrence BARRETT and Mary ANDERSON. In 1886 he gave the first of his many open-air productions of Shakespeare, and formed a company with which he toured incessantly in the United Kingdom and the U.S.A., rivalling Frank BENSON as a trainer of young actors. He spent several years in America, but returned in 1914 and was one of those responsible for the foundation of the OLD VIC, where between 1915 and 1918 he produced 24 of Shakespeare's plays, including *Hamlet* in its entirety, as well as a number of other English classics. Many London schoolchildren owed their introduction to the theatre to his visits in the 1920s and 1930s to L.C.C. and other centres with a repertory in which Shakespeare predominated. In his later years he concentrated mainly on productions for schools and open-air performances. He celebrated his stage jubilee in 1929, and continued working until his death.

GREGORY, (Isabella) AUGUSTA, Lady (1852–1932), Irish landowner, who entered the theatre in middle age with an unsuspected gift for comedy-writing which, but for her contact with YEATS and the Irish dramatic movement, might never have been realized. She proved an indefatigable worker in the movement which led to the founding of the ABBEY THEATRE in 1904, and whose early years she described in *Our Irish Theatre* (1914). In 1909 she won a notable victory for the Abbey by frustrating attempts to suppress the production of SHAW's *The Shewing-Up of Blanco Posnet*, and in 1911 took the company on its stormy and triumphant visit to America. The best known of her numerous masterly short plays are comedies of peasant life: *The Pot of Broth* (with Yeats, 1902), *Spreading the News* (1904), *Hyacinth Halvey* (1906), and *The Workhouse Ward* (1908); but she is also known for two fine patriotic plays, *Cathleen ni Houlihan* (with Yeats, 1902) and *The Rising of the Moon* (1907), and for one brief peasant tragedy, *The Gaol Gate* (1906). Later she wrote fantasies of mingled humour, pathos, and poetic imagination—*The Travelling Man* (1910), *The Dragon* (1919), *Aristotle's Bellows* (1921), and others. She also contributed to the Abbey repertory many translations, of which *The Kiltartan Molière*, a version of several of MOLIÈRE's plays transplanted to the west of Ireland, is the best known.

GREGORY, JOHANN GOTTFRIED (1631–75), German pastor, who in 1658 was appointed to the Lutheran Church in Moscow where ten years later he founded a school. On 17 Oct. 1672 he was responsible for the first organized dramatic entertainment given before the Russian Court on the orders of Alexei, father of Peter the Great. This was *The Play of Artaxerxes*, taken from the repertory of the ENGLISH COMEDIANS and performed in German by German students in Moscow. It lasted ten hours, and was enlivened by songs, music, dancing, and comic interludes. Gregory later produced *The Comedy of Young Tobias* (1673) and, with Russian student-actors whom he had trained himself, *The Comedy of Holofernes* (1674). He is portrayed in *A Comedian of the Seventeenth Century* by OSTROVSKY. His theatre, a hastily erected wooden building in the summer palace at Preobrazhen, survived until Alexei's death in Jan. 1676, when it was pulled down.

GREIN, JACK [Jacob] THOMAS (1862–1935), a Dutchman who became a naturalized Englishman in 1895 and as playwright, critic, and manager, did much to further productions of the new 'theatre of ideas' in London at the turn of the century. In 1891, inspired by the example of ANTOINE, he founded the Independent Theatre Club, which he inaugurated at the ROYALTY THEATRE with an 'invitation' performance of IBSEN's *Ghosts* in William ARCHER's translation. The critics, particularly Clement SCOTT, damned the play, but the Club survived and in 1892 put on *Widower's Houses*, the first of SHAW's plays to be performed. Grein's interest extended also to the STAGE SOCIETY, out of which came the epoch-making management of VEDRENNE and GRANVILLE-BARKER at the ROYAL COURT THEATRE.

GRENFELL [née Phipps], JOYCE IRENE (1910–79), English *diseuse*, who made her first appearance on the professional stage in *The Little Revue* (1939) in a selection of the monologues with which she had for some time previously been entertaining her friends privately. She was immediately successful, proving herself an excellent mimic and an accurate though kindly satirist of contemporary manners, particularly of schoolmistresses and middle-class wives and daughters. During the Second World War she toured the service hospitals, and returned to London in 1945 to appear in Noël COWARD'S REVUE *Sigh No More*, which was followed by *Tuppence Coloured* (1947) and *Penny Plain* (1951). She made her first appearance in New York in 1955 in *Joyce Grenfell Requests the Pleasure*, seen the previous year in London, and thereafter toured world-wide in her one-woman entertainments of sketches and songs, many of which she wrote herself. She was the author of *Nanny Says* (with Sir Hugh Casson) (1972); an autobiography, *Joyce Grenfell Requests the Pleasure* (1976); and *George, Don't Do That* (1977).

GRENIER DE TOULOUSE, see DÉCENTRALISATION DRAMATIQUE.

GRENIER-HUSSENOT, Les, French theatre company formed in 1946 by two actors in their early thirties, Jean-Pierre Grenier and Olivier Hussenot, who were members of CHANCEREL's touring company Les Comédiens Routiers until 1939. During the Second World War some of the group continued to perform in Provence, and it was the communal experience of popular entertainment during this period that Grenier and Hussenot brought to the wealthier and more sophisticated audiences in post-war Parisian cabarets. As Bobèche and Lapopie, they headed a company which established a successful compromise between circus acts and light comedy, with a mixture of acrobatics, mimicry, song, dance, and satirical sketches. Talented designers and song-writers were inspired to work for them, and one of the long-surviving offshoots of the venture is the four-man music-hall song-act, Les Frères Jacques. The company disbanded in 1957.

GRESSET, JEAN-BAPTISTE-LOUIS (1709–77), French dramatist, whose poems 'Vert-Vert' (1734), a charming trifle dealing with the misadventures of a parrot in a convent, and 'La Chartreuse' (1735) caused him to be dismissed from the Jesuit college where he was teaching. Thrown on his own resources, he wrote several plays, of which *Edouard III* (1740), though subtitled 'a tragedy', is in essence a sentimental bourgeois drama and shows clearly the artificial cult of the day for the simple life. This is even more noticeable in *Sidneï* (1745), set in an English village and a good example of the Änglomania prevalent in 18th-century France. Gresset's best and most successful play was undoubtedly *Le Méchant* (also 1745), the portrait of a cynical mischief-maker which gives a vivid picture of the corrupt society of the time. Gresset, who was admitted to the French Academy in 1748, later renounced the theatre as being incompatible with religion and morality.

GRÉVIN, JACQUES (1538–70), French poet, a humanist of the school of Ronsard. His tragedy *La Mort de César* (1561)—important as the first original French tragedy to be performed and published—shows the influence of SENECA, and portrays idealized Roman virtues in the style later adopted by the neo-classical CORNEILLE and his contemporaries. It is, however, for his comedies *La Tresorière* (1559) and *Les Esbahis* (1561) that Grévin is chiefly remembered.

GRIBOV, ALEXEI NIKOLAYEVICH (1902–1977), Soviet character actor, who joined the MOSCOW ART THEATRE in 1924, and was the first member of the company to portray Lenin, in 1942. Exceptionally popular with Moscow audiences for his depiction of working-class and peasant roles, he was also an excellent performer in serious classic plays.

GRIBOYEDOV, ALEXANDER SERGEIVICH (1795–1829), Russian diplomat and dramatist, who was assassinated in Teheran while acting as Russian Minister to Persia. His literary works were the fruit of his leisure hours, and with one exception were comedies translated from the French or written in collaboration with his friends. The exception is his classic play *Gore ot Uma*, whose Russian title has been variously rendered into English as *Woe From Wit, Wit Works Woe, Too Clever By Half, The Misfortunes of Reason, The Trouble With Reason*, and *The Disadvantages of Being Clever*. It deals with the struggles of a young man, arriving in Moscow full of liberal and progressive ideas, against the stupidity and trickery of a corrupt society. Classic in form yet realistic and satiric in content, this first dramatic protest against Tsarist society is one of the great plays of the Russian theatre. Griboyedov worked on it for many years, and although it was banned from the stage during his lifetime, it circulated freely in manuscript and was printed in a cut version four years after his death. It was first performed at the Bolshoi Theatre in St Petersburg in 1831, published in full in 1861, and revived by the MOSCOW ART THEATRE in 1906 with KACHALOV and MOSKVIN as Chatsky and Famusov, parts which they played again in a production in 1938. The part of Chatsky is to young Russian actors what Hamlet is to English, and Famusov, the heroine's conservative father, has long been a favourite role with older actors.

GRID, or Gridiron, an open framework from 5 ft to 10 ft below the stage roof from which scenery or lights can be hung. It can be made of wood or metal, and should be built at a height three times that of the proscenium opening. The modern grid is constructed of 3 to 5 pairs of heavy channel-iron beams running from the proscenium wall to the back wall, with a 10-ft space between each pair, which allows for the descending lines from the Loft Blocks, bolted to the top of each pair. The grid floor is composed of strips of channel-iron, 4 to 6 ins apart, which run parallel to the main channels (i.e., up-stage and down-stage) and which are supported by I-beams perpendicular to the main channels. The Synchronous Winch System has now made possible a newer, multi-channelled grid design, with a single main channel every 4 ft and with loft blocks underhung to the stage roof, thus freeing the grid floor from sheaves and lines.

GRIEG, NORDAHL (1902–43), Norwegian dramatist, poet, and novelist, whose untimely death in a bombing raid over Berlin during the Second World War was a great loss to the European theatre. After some years as a student (which included a year at Oxford), he began by publishing volumes of prose and poetry. His first drama, an experimental play called *Barabbas*, was written in 1927. *Vår ære og vår makt* (*Our Honour and Our Power*, 1935) was an anti-war play of overwhelming force and originality; *Nederflaget* (*The Defeat*, 1937), which takes as its subject the Paris Commune of 1871, explores the same theme more deeply; it is believed to have inspired BRECHT's *Die Tage der Kommune* (1956).

GRIEVE, WILLIAM (1800–44), English scene designer, who as a young man worked at COVENT GARDEN, where his father, John Henderson Grieve (1770–1845), was responsible for the scenery of spectacle plays and pantomimes under John Philip KEMBLE. William later went to DRURY LANE, where he did his best work, being considered after the retirement of STANFIELD the finest scenic artist of the day. His moonlit scenes were particularly admired, and he was the first theatre artist to be called before the curtain by the applause of the audience. His elder brother Thomas (1799–1882), who in 1839 was at Covent Garden under Madame VESTRIS, also worked at Drury Lane, being assisted by his son, Thomas Walford Grieve (1841–82), whose painting was remarkable for the brilliance of its style and the artistic beauty of its composition.

GRIFFITH, LYDIA, see SEYMOUR, WILLIAM.

GRILLPARZER, FRANZ (1791–1872), Austrian dramatist, the major figure of the Romantic period, whose first play, *Die Ahnfrau* (*The Ancestress*, 1817), was produced by SCHREYVOGEL at the Theater an der Wien. It was followed by two plays on classical themes, *Sappho* (1818) and *Das goldene Vliess* (1820), and a historical play in the style of SCHILLER, *König Ottokars Glück und Ende*. Written in 1823, this was banned by the censor for two years because of the resemblance between the career of its hero and that of Napoleon. It was then produced at the Burgtheater, but not until it was transferred to the more spacious stage of the Theater an der Wien were its qualities fully realized, and it became one of the acknowledged masterpieces of the German-speaking theatre. Of Grillparzer's other plays, which include *Ein treuer Diener seines Herrn* (*His Master's Faithful Servant*, 1826) and *Des Meeres und der Liebe Wellen* (1829), on the story of Hero and Leander, the most important is his adaptation of *La vida es sueño* by CALDERÓN as *Der Traum ein Leben* (*Life Is a Dream*, 1834). Grillparzer's only comedy, *Weh' dem, der lügt* (*Thou Shalt Not Lie*, 1838), was somewhat too sophisticated for an audience conditioned to the lighter forms of popular theatre, and failed on its first production at the Burgtheater, with the result that Grillparzer turned from the theatre and wrote only for his own amusement. His last three plays, *Ein Bruderzwist in Habsburg, Die Jüdin von Toledo*, and *Libussa*, were published posthumously. The first was produced at the Burgtheater in the autumn of 1872; *Libussa*, which some consider Grillparzer's best work, was also seen at the Burgtheater two years later, with Charlotte Wolter in the title role. *Die Jüdin von Toledo* had to wait till 1888 for its first production.

Grillparzer was not unlike RAIMUND, for whom he had a high regard, in temperament and theatrical instinct; but his sophistication and matter-of-factness were in complete contrast to Raimund's essentially naïve genius.

GRIMALDI, JOSEPH (1778–1837), English actor, the creator of the English CLOWN, the only character in the HARLEQUINADE not derived from the Italian COMMEDIA DELL'ARTE. In his honour all later clowns were nicknamed Joey. The illegitimate son of Giuseppe Grimaldi, ballet-master at DRURY LANE, and a chorus-girl named Rebecca Brooker, he made his first appearance on the stage at SADLER'S WELLS at the age of 2½. As a boy he played with Dubois in PANTOMIME, and by the time he became a regular member of the Sadler's Wells company in 1792 he was already a seasoned actor, well trained in acrobatic and pantomimic skills. In 1796 he was at Drury Lane where he played a number of parts besides Clown, being seen as Aminadab in Mrs CENTLIVRE's *A Bold Stroke for a Wife* and as Robinson Crusoe and Blue Beard in pantomime. But Drury Lane was not the right milieu for his activities and in 1805, on the recommendation of Charles DIBDIN who admired him greatly, he was engaged for COVENT GARDEN, where, with occasional returns to Sadler's Wells, he remained until his retirement in 1823. At these theatres his inventive comic genius —making a man out of vegetables, a coach out of four cheeses, a cradle, and a fender, a hussar's uniform out of a coal-scuttle, a pelisse, and a muff—was given full scope; and he was also the designer of a number of trick scenes, changing a quack's pill into a duck, a drum into a temple, or a post-chaise into a wheelbarrow. Making Clown a rustic booby on the model of Pantaloon's servant, Grimaldi gave him his traditional costume—baggy breeches, a livery coat with scarlet patches, fantastic wigs, usually a turned-up pigtail, and a white face with scarlet triangles and exaggerated eyebrows. His acrobatics were characterized by dynamic energy which finally wore him out, and his place at Covent Garden was taken by his dissolute and sottish son, Joseph S. (1802–32). He made his last appearance at Sadler's Wells on 17 Mar. 1828, and on 27 June was given a benefit at Drury Lane, where, seated before the footlights, he sang 'Hot Codlins' (i.e., roasted apples), which, with the tale of the oyster crossed in love, was one of his most popular turns. He then retired to Pentonville where he died, and was buried in the churchyard of St James's Chapel on Pentonville Hill, now a public garden where his tombstone can still be seen.

GRINGORE, PIERRE (*c.* 1475–1538), French actor, the chief FOOL, or *mère-sotte*, of the Parisian company *les enfants sans souci* for whom he wrote a number of topical and satirical SOTIES, or short farces, between 1502 and 1515, playing in them himself. His best known play is *Le Jeux du prince du Sots*, a political satire on Pope Julius II, acted in the market-place of Les Halles on Shrove Tuesday 1511. Under Louis XII's successor, François I, Gringore was suspected of being a political agitator and left Paris to enter the service of the Duke of Lorraine. He wrote only one more play, a *Mystère de Saint-Louis* (Louis IX), the first to be written on a French national theme. The spelling of his name as Gringoire, sometimes found nowadays, proceeds from an error in Victor HUGO's novel *Notre-Dame de Paris* (1831).

GRIPSHOLM THEATRE, Sweden, a small and extremely beautiful playhouse in one of the round towers of Gustavus Vasa's fortress, built in 1535. The theatre opened in 1782 and its scenery and machinery still remain intact, one set of scenes reproducing exactly the pillared décor of the semicircular auditorium. The interior has been carefully restored, and the theatre is used for occasional performances in the summer months.

GROCK [Charles Adrien Wettach] (1880–1959), for many years a much-loved star of the British MUSIC-HALL, and the supreme CLOWN of his generation. Born in Switzerland, he toured as a boy with a circus and in 1903 joined forces with a clown named Brick, changing his own name to Grock. Later, with a different partner, he appeared in Berlin, and in 1911 was engaged by COCHRAN for the PALACE THEATRE in London. His clowning was so expressive that its meaning, even without words, was immediately apparent to everyone. He was constantly at the mercy of inanimate objects: any chair he sat down on would collapse and entangle him; his enormous double-bass case would be found to contain a tiny violin; when he began to play the piano he discovered he was wearing thick gloves, and his delight at the improvement in his performance when he removed them was a joy to watch. An accomplished performer on at least 20 instruments, his hour-long act consisted of a variety of musical excerpts, which at first he failed to grapple with but finally, to his own simple delight, managed to master. One of the highlights of his performance was the moment when he discovered that the grand piano was too far from the piano-stool. He was already pushing it painfully when his assistant pointed out, to his amazement, that it would be easier to move the stool. In his excitement he would slide across the top of the piano in an attempt to prevent his hat falling off and so revealing his bald head. He always wore the same make-up and costume, a pale, egg-shaped face with a large red mouth, a massive tail-coat with collar and tie, tight trousers, clumsy boots, a bowler hat, and a bald wig. He played in London almost continuously up to 1924, mainly at the COLISEUM, after which he returned to the Continent, giving his farewell performance at Hamburg on 31Oct. 1954.

GROOVES, a characteristic of English, as opposed to Continental, stage machinery, by means of which WINGS and FLATS were slid on and off stage in full view of the audience, the top and bottom of each flat running in a groove between two strips of timber built into the stage structure. The origin of the groove can be found in the Court MASQUE, as seen in the designs of Inigo JONES. For the pastoral *Florimène* in 1635 only the back shutters ran in grooves, but by 1640, for William DAVENANT's *Salmacida Spolia*, there were not only long grooves at the back of the stage for the back scenes but also a series of short grooves on each side for the wings, allowing for up to four changes of scenery at each place. During the Commonwealth a simplified system of grooves, for back shutters only, was used for Davenant's *The Siege of Rhodes* (1656). Although stage directions in plays from 1660 onwards refer indirectly to the use of grooves, the first direct mention of them dates from 1743, at COVENT GARDEN. They were also used in early theatres in America, where in 1897 they were referred to as 'old-fashioned'. During the time they remained in use, several innovations were made to enable the scenery to be changed more quickly, the most efficient being the DRUM-AND-SHAFT SYSTEM. Even so, there were many disadvantages attached to the use of grooves. As they always had to run parallel to the front of the stage, because of the difficulty of placing them obliquely on a raked floor, masking was poor, and spectators in side boxes could see deeply into the wings. Also because of the rake, the wing flats became shorter as one advanced upstage, so each could be used only in one position and interchangeability was impossible. Sometimes the scenes stuck in the grooves, or moved raggedly, and when in about 1820 the bottom grooves were removed and the upper ones cut and hinged, the noise of the falling arms, and of the chains used to check their fall at an exactly correct level, became a nuisance. All these factors combined to bring about the abolition of the grooves system in favour of the Continental CARRIAGE-AND-FRAME, which was first installed at Covent Garden in 1857. The last London theatre to use grooves was the LYCEUM, where they were removed in 1880 by IRVING, who also began at the same time the practice of dropping the front CURTAIN to cover scene changes.

Grooves remained, however, for some years in smaller theatres, and a pivoted variant is found in the 1880s which surmounted the objection to their rigidly enforcing a position parallel to the footlights on all wings, for wing and groove could now be twisted to any angle. Eventually even this modification gave way to forks, in which the tops of the wings were held as by an inverted garden-fork. The modern system of supporting flats by braces has superseded all these earlier methods.

GROPIUS, WALTER, see THEATRE BUILDINGS.

GROS-GUILLAUME [Robert Guérin] (*fl.* 1598–1634), French actor, who under the name of La Fleur played serious parts in tragedy, but is best remembered as a farce-player, with TURLUPIN and GAULTIER-GARGUILLE, in the permanent company at the Hôtel de BOURGOGNE. He probably played at the Paris fairs before going there, as he was already known as an actor by 1630. A fat man, with black eyes and a very mobile face (which he covered with flour for comic parts, thus giving rise to the tradition that he was a baker before going on the stage), he figures as himself with his companions in Gougenot's *La Comédie des comédiens* (1631). His wife and daughter were both actresses, the latter marrying an actor, François Juvenon (*c.* 1623–74), who took his father-in-law's stage name and as La Fleur played kings in tragedy in succession to MONTFLEURY, and Gascons or the ranting CAPITANO in comedy. His son became an actor under the name of LA TUILLERIE.

GROS-RENÉ, see DU PARC.

GROSSES SCHAUSPIELHAUS, Berlin, see POELZIG, REINHARDT, and THEATRE BUILDINGS.

GROSSMAN, JAN (1925–), Czech director, who in 1959 joined the company of the Theatre on the Balustrade in Prague, which had been founded in 1958 by Ladislav Fialka. In 1962 Grossman became responsible for the drama section, Fialka continuing to direct the mime company. Both were with the company in 1967 when they visited London during the WORLD THEATRE SEASON, Fialka with his mime-play *The Fools* and Grossman with his own version of Kafka's *Der Prozess*. In 1968 they returned, with Fialka's *The Fools* and *The Clowns* and with Grossman's production of *King Ubu*, in which he adapted and amalgamated JARRY's *Ubu-Roi* and *Ubu enchaîné*. In the same year, after a long period of official disapproval, Grossman was forced to leave Czechoslovakia and worked abroad until 1974, when he returned as director of the City Theatre in Cheb. It was under his directorship of the Theatre on the Balustrade that several new plays by the young Czech dramatist Václav Havel were first produced.

GROSSMITH, a family of English actors, of whom George (1847–1912) and Walter Weedon (1852–1919) were joint authors of the inimitable *The Diary of a Nobody* (1892) which was dramatized in 1954. George was primarily an entertainer with sketches at the piano but appeared in GILBERT and Sullivan's operettas at the SAVOY THEATRE from 1881 to 1889. Walter wrote several plays; the most successful were *The Night of the Party* (1901), in which he toured England and the United States, and *The Duffer* (1905). George's sons, George (1874–1935) and Lawrence (1877–1944), both had long careers on the stage, the former being well known for his impersonation of the 'dude' in musical comedy and revue. He was part-author with Fred Thompson of the revue *The Bing Boys are Here* (1916) and with P. G Wodehouse of *The Cabaret Girl* (1922).

GROTO, LUIGI (1541–85), Italian dramatist and poet, known as *Il cieco d'Adria*—the blind man of Adria, the town in which he lived. His play *Adriana* (1578) was probably the first dramatization of the story by Bandello on which Shakespeare based his *Romeo and Juliet* (*c.* 1595), though it is unlikely that Shakespeare knew it. Groto also wrote a number of PASTORALS and comedies, and several popular horror-tragedies, of which *La Dalida* (1572) was translated into Latin by a clergyman, William Alabaster, and as *Roxana* was produced at Trinity College, Cambridge, in about 1592. Another of Groto's plays, either *Amoroso* or *Caliste*, was anonymously translated and probably given in Cambridge in about 1610.

GROTOWSKI, JERZY (1933–), Polish director, whose experimental work has exerted a great influence outside his own country. He became director of a theatre in Opole in 1959 and in 1965 moved to Wrocław, where he established a Theatre Laboratory which later became well known through its tours abroad, its three most important productions being WYSPIAŃSKI's *Akropolis*, CALDERÓN's *El príncipe constante*, translated and adapted by SŁOWACKI, and *Apocalypsis cum* *figuris*, based on a collection of Biblical texts and liturgical chants interspersed with quotations from such modern authors as DOSTOEVSKY, T. S. ELIOT, and Simone Weil. Trained in the methods of STANISLAVSKY, Grotowski did not, as some critics appear to think, reject them, but developed them in a psychophysical rather than a purely psychological direction. For him the actor is paramount, and must make use of all the physical and mental powers at his disposal. Apparent spontaneity and even violence must be rigorously controlled, and in reaction against the 'wealth' of the contemporary scene, with its lighting, scenery, costumes, and music, and its all-powerful director, Grotowski has envisaged a 'poor' theatre, stripped of inessentials and relying entirely on the brain and body of the actor. This idea has been set out in a volume entitled *Towards a Poor Theatre*, published in 1968 with a preface by Peter BROOK, whose own work has been much influenced by Grotowski. The Theatre Laboratory gave its first performance outside Poland in 1966, touring Scandinavia with *Akropolis*, and has since travelled throughout Europe and the near East. In 1968 *Akropolis* was seen at the EDINBURGH FESTIVAL, and the group then visited London and several provincial cities, going also to New York, where its influence was apparent in the work of the choreographer Jerome Robbins. In 1976, after ten years' touring, Grotowski disbanded his company and instead launched a novel programme, which involved actors and students working behind closed doors with no audiences.

GROTTESCO, Teatro, Italian dramatic movement which emerged during the First World War, and took its name from CHIARELLI's *La maschera e il volto* (*The Mask and the Face*), described as a *grottesco in tre atti*, which, though written in 1913, was not staged until 1916. As the title implies, the central concern of the movement, which arose in opposition to the heroics of D'ANNUNZIO, was with the contradiction between social behaviour and personal reality, between the hypocrisy of the bourgeois conformist hiding behind his mask and the primitive passions which sometimes tore it from his face, an idea developed by such writers as Enrico Cavacchioli, BONTEMPELLI, and ROSSO DI SAN SECONDO. The movement was short-lived, but proved an important element in the development of PIRANDELLO.

GROUNDROW, originally a strip of gas lights, laid flat along the stage to illuminate the foot of a back scene. As a term in stage LIGHTING it was later applied to the rows of electric bulbs fixed on a BATTEN and sunk into the stage floor, or masked by a scenic groundrow, which is now the name given to all low cut-out strips of scenery, made of canvas stretched on wood, like a FLAT laid on its side, and cut along its upper edge to represent, for instance, a hedge with a stile in it, or a bank topped by low bushes. Series of groundrows, set one

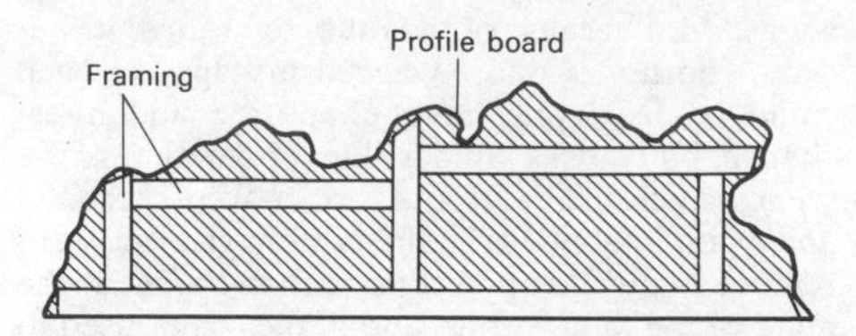

Ground row, rear view
The framing is made of 3″ × 1″ timber, the profile board from material such as 3-ply timber or masonite; a large groundrow may have the outer shape only made of hard profile board, with canvas or muslin filling the inner areas

behind the other, with their top edges cut to represent waves on a lake, river, or sea, so arranged as to allow the passage of a stage boat between them, are known as Set Waters, or Sea or Water Rows.

GROUP THEATRE, London, a private play-producing society founded in 1933 with the object of presenting modern non-commercial plays and revivals of experimental work. It had its headquarters at the WESTMINSTER THEATRE, where most of its productions were staged. These included *Timon of Athens* in modern dress, T. S. ELIOT's *Sweeney Agonistes*, and Jean Giono's *Sowers of the Hills* (all 1935); the poetic plays of W. H. Auden and Christopher Isherwood—*The Dog Beneath the Skin* (1936), *The Ascent of F.6* (1937), and *On the Frontier* (1938); and Stephen Spender's *Trial of a Judge* (1938). Most of the plays were directed by Rupert Doone (1904–66) on a bare stage with little scenery and few props but with the occasional use of masks. Some of the incidental music for the productions was written by Benjamin Britten. During the Second World War the society lapsed, but in 1950 it was re-formed and gave as its first production a translation of SARTRE's *Les Mouches* as *The Flies*. It continued to function spasmodically for a couple of years, but finally disappeared in about 1953.

GROUP THEATRE, New York, production company, formed in 1931 by Harold CLURMAN, Cheryl CRAWFORD, and Lee STRASBERG, which evolved from the THEATRE GUILD. It had high ideals and a democratic mode of operation, and was intended for the presentation of works of serious social content free from the pressures of commercial theatre. Its first production, in Sept. 1931, was Paul GREEN's *The House of Connelly*, and among its other early and successful ventures were John Howard Lawson's *Success Story* (1932), Sidney KINGSLEY's *Men in White* (1933), and Melvin Levy's *Gold Eagle Guy* (1934). The most important dramatist discovered by the Group Theatre was, however, Clifford ODETS, who was a member of the company. In 1935 it produced his *Awake and Sing!*, *Waiting for Lefty*, a one-act play which later formed a double bill with *Till the Day I Die*, and *Paradise Lost*; they were followed by *Golden*

Boy (1937), *Rocket to the Moon* (1938), and *Night Music* (1940). Most of these plays were directed by Clurman, who assumed responsibility when the Group was reorganized and Strasberg and Crawford ceased to direct. Other major productions included Paul Green's *Johnny Johnson* (1936), with music by Kurt Weill, Irwin Shaw's *The Gentle People*, and William SAROYAN's *My Heart's in the Highlands* (both 1939). A permanent repertory company was built up, dedicated to the principles of group acting as formulated by STANISLAVSKY, which produced a number of outstanding actors. In 1941, however, beset by financial difficulties and worn down by conflicts of personality, the Group Theatre ceased production.

GROVE THEATRE, New York, in Bedlow Street, now Madison Street, east of Catherine Street. A small playhouse which opened in 1804 with a company of little known actors including George Bland and Mr and Mrs Frederick Wheatley, who were later at the PARK THEATRE. The Grove was open only on Tuesdays, Thursdays, and Sundays, when the all-powerful Park was closed, and it lasted only a couple of seasons. Among its attractions was a pantomime staged by Signor Bologna from London's COVENT GARDEN (after which the theatre was optimistically renamed) in which he himself played CLOWN.

GRÜNDGENS, GUSTAV (1899–1963), German actor and director, who made his first appearances in HAMBURG and from 1928 until the end of the Second World War was at various theatres in BERLIN, where he directed and acted in a wide range of classical plays. Tall, blond, and extremely good-looking, he made a striking Hamlet in a German version of Shakespeare's play seen at Elsinore in 1938. He was also much admired in plays by SHAW. After a brief period in a Soviet internment camp he was released and went to Düsseldorf, where he remained until 1955, making it one of the outstanding theatre centres of the Federal Republic. He spent his last years in Hamburg, where he added many modern works to the repertory, including BRECHT's *Die heilige Johanna der Schlachthöfe* in 1959, and John OSBORNE's *The Entertainer* in 1957. After a long period of neglect in the theatre, he revived both parts of GOETHE's *Faust* in 1957/8, playing Mephistopheles, the role in which he had first attracted attention in 1922. This was received with acclaim in Edinburgh, New York, Moscow, and Leningrad, and he was negotiating for it to be seen in South America when he died, apparently by his own hand, in Manila.

GRUNDY, SYDNEY (1848–1914), English dramatist, for some years a barrister in his native town of Manchester. His numerous plays, of which the first, *A Little Change* (1872), was seen at the HAYMARKET THEATRE in London, were mainly comedies and farces, many of them adapted from the French; the only one to have survived is *A Pair of Spectacles* (1890), which provided John HARE, as Benjamin Goldfinch, with a part he played to perfection and revived many times. In later years Grundy was influenced by the prevailing taste for sentimental melodrama, and such plays as *Sowing the Wind* (1893) show a curious mingling of old and new fashions in playmaking: he lacked the skill to handle realistic situations with any depth of insight, and was compelled to cramp them into the conventions of an outworn method.

GRYPHIUS, ANDREAS (1616–64), German baroque dramatist, author of a number of tragedies written in lofty poetic prose with scenes of horror and bloodshed which violate all the canons of classical restraint. A firm conviction that earthly existence is vanity pervades all his work, and his heroes range from Papinianus to Charles I of England, who had just been beheaded. This play is particularly interesting for its treatment of a contemporary event (1649) viewed with strong royalist sympathies. Gryphius was sufficiently in advance of his time to write what deserves to rank as the first domestic drama with *Cardenio und Celinde* (1647), which centres on the passions of ordinary men and women and the final miraculous transformation of the repentant sinners. He was also the author of a number of comedies, all in prose except for the doggerel of the Pyramus and Thisbe scenes in *Peter Squentz* (1663). The best of them was *Horribilicribrifax* (also 1663), which gives a lively picture of contemporary follies—bombastic conceit, pedantry, self-seeking—as seen by a naturally austere mind with a sense of humour. Although a staunch Protestant, Gryphius owed much to the JESUIT DRAMA, as well as to the contemporary secular plays in France and elsewhere which he had seen on his travels between 1644 and 1646, and a comedy of intrigue with farcical scenes in Low German shows that he was acquainted with the work of the ENGLISH COMEDIANS. He was the first German dramatist to handle his material with conscious artistry; but there was no permanent theatre in which his plays could be given, and only sporadic performances, mostly by schoolboys, are recorded.

GUAL, ADRIÁ (1872–1943), Catalan dramatist and theatre director, founder of the Teatro Intim in Barcelona, which from 1903 to 1904 presented masterpieces of world drama from AESCHYLUS to IBSEN as well as the works of such Spanish dramatists as BENAVENTE. The technical and artistic innovations which Gual introduced, his careful training of his actors, and his experiments in scenic design had little immediate result outside Catalonia but bore fruit in the dramatic renaissance in Madrid in the 1920s and 1930s inaugurated by MARTÍNEZ SIERRA. Gual, like Martínez Sierra, was much influenced in his early writings by MAETERLINCK, but none of his plays, of which the best is probably *Misteri de Dolor*, has survived on the stage.

GUARE, JOHN (1938–), American dramatist, who first attracted attention when *The Loveliest Afternoon of the Year* and *A Day for Surprises* were produced at the CAFFÉ CINO in 1966. His professional début was in 1968, when *Muzeeka* was performed in a double bill off-Broadway; it was seen at Charles MAROWITZ's Open Space in London in 1969, in which year Guare's double bill of *Cop-Out* and *Home Fires* was seen in New York. He became widely known with *The House of Blue Leaves* (1970), his first full-length play. Centring on the effect of the Pope's visit to New York on a middle-aged zookeeper's family and friends, it showed his gift for savage farce; the play was produced in Paris in 1972 as *Un Pape à New York*. In the following year he helped to adapt and wrote the exuberant lyrics for a successful musical version of *The Two Gentlemen of Verona* produced by Joseph PAPP, which in 1973 was also seen in London. Later works, both produced by Papp, were *Rich and Famous* (1976), in which a playwright fantasizes on his first opening night, and *The Landscape Of the Body* (1977), about a young woman encountering madness and murder in the big city. In *Bosoms and Neglect* (1979) with Kate REID (a comedy in spite of its theme) a couple have a long discussion about themselves after finding that the man's mother is dying of cancer.

GUARINI, GIOVANNI BATTISTA (1537–1612), Italian dramatist, author of *Il pastor fido* (1598), an idyllic tragi-comedy which stands with TASSO's *L'Aminta* (1573) as the outstanding achievement of the Italian PASTORAL. Begun in 1569, published in 1590, it was first produced in Mantua, in great splendour and with much success. Frequently reprinted and translated, it had a great influence on the pastoral and romantic literature of England and France in the 17th century. Some idea of its appeal to a sophisticated English audience can be gained by the fact that it was first seen at Court in English, under its original title, in 1601, and in Latin, as *Pastor Fidus*, at Cambridge in 1604/5. It was revived before the Court in 1630 and 1635 in different translations and in 1647, as *The Faithful Shepherd*, in a translation by Richard Fanshawe. In 1676/7 a new translation, by Elkanah SETTLE, was produced at DORSET GARDEN.

GUÉRIN D'ÉTRICHÉ, ISAAC FRANÇOIS (*c.* 1636–1728), French actor, one of the troupe at the Théâtre du MARAIS, who in 1677 married MOLIÈRE'S widow. He became a member of the COMÉDIE-FRANÇAISE on its foundation and was its third *doyen*, continuing to act until he was over eighty. He was much admired in such elderly parts as Harpagon in Molière's *L'Avare*, and was also the first to play a number of leading roles in early 18th-century plays.

GUERRERO, MARÍA (1868–1928), Spanish actress who studied under Teodora Lamadrid and later in Paris. In her first appearances in 1885 she was associated with Ricardo Calvo. Her father, a wealthy businessman, bought for her the Teatro de la Princesa (now named after her as the Teatro Nacional María Guerrero), and from 1894 she dominated the Spanish theatre, displacing Antonio Vico as the most influential player of the day. Her repertoire was large, including classical plays of the Golden Age and modern works by authors such as BENAVENTE, PÉREZ GALDÓS, GUIMERÀ, and the brothers ÁLVAREZ QUINTERO, but few foreign works.

GUIGNOL, the name of a French puppet which originated in Lyons, probably in the last years of the 18th century, and may have been invented by a puppet-master named Laurent Mourquet, grafting native humour on to POLICHINELLE. In Paris the name attached itself to cabarets which, like the Théâtre du Grand Guignol, specialized in short plays of violence, murder, rape, ghostly apparitions, and suicide. In a modified form these made their appearance in London in 1908 and have been seen sporadically ever since, notably in the seasons of Grand Guignol at the LITTLE THEATRE in 1920, at the Granville, Walham Green, in 1945, and at the Irving in 1951. English Grand Guignol never reached the intensity of the French, however, and its true home is in the small theatres of Montmartre.

In the French theatre a 'guignol' is also a QUICK-CHANGE ROOM.

GUILBERT, YVETTE [Emma Laure Esther] (1865–1944), French singer, known as 'la diseuse fin-de-siècle', and an outstanding figure of the MUSIC-HALL. She made her début in 1887, and soon became an established favourite, bringing to the popular stage a new style of delivery, precise diction, and an intelligent appraisal of the potential of vocal material. Thin to the point of emaciation, her hennaed hair, white mask-like face with a vivid gash of a mouth, and wide range of facial expressions and gestures were immortalized by Toulouse-Lautrec in many paintings and sketches, as were the long black gloves, originating in her early poverty, which became a mark of distinction and were retained by her until the end of her career. Her first songs, on themes from Parisian low life, full of crude language, were delivered with an air of utter innocence—Arthur Symons, one of her fervent admirers, described her as the exponent of 'depraved virginity'—but later she added to her repertoire some thousands of *chansons*, ranging from the 14th to the 19th century, with which she illustrated the many lecture-recitals she gave on tour between 1901 and 1914. From 1915 to 1918 she was in the United States, reappearing in London in a series of recitals in the 1920s, when she settled again in France. She was also an accomplished actress, her outstanding role being Mrs Peachum in the French version of BRECHT's *Threepenny Opera* in 1937.

GUILD THEATRE, New York, see ANTA THEATRE.

GUILLOT-GORJU [Bertrand Hardouin de St Jacques] (1600–48), French actor, who earned a precarious living as a BARKER for quack-doctors at the Paris FAIRS before BELLEROSE took him into the company at the Hôtel de BOURGOGNE, where he played in farce in succession to TURLUPIN and his companions. He was excellent as the pedantic doctor, which he played masked. He married the sister of Bellerose in 1636, and in 1641 retired from the stage to take up his profession of medicine.

GUIMERÀ, ÀNGEL (1849–1924), Catalan poet and dramatist, born in Tenerife but resident for most of his life in Barcelona, where his plays were produced. The first, a poetic tragedy *Gala Placidia* (1879) which showed the influence of Shakespeare and the French Romantics, particularly Victor HUGO, was followed by several more in the same style, of which the best was perhaps *L'ànima morta* (*The Dead Soul*, 1892); but with *La boja* (*The Box-Tree*, 1890) he moved into the realm of contemporary REALISM, depicting the loves and hates of the fishermen and peasants of Catalonia. Of these modern plays the best known are *Maria Rosa* (1894) and *Terra baixa* (1896), seen in New York in 1903 as *Marta of the Lowlands*. His later plays, in which he attempted to dramatize abstract ideas, were less successful.

GUINNESS, ALEC (1914–), English actor, knighted in 1959. He was first seen on the stage in 1934, and from his earliest appearances, notably as Osric to John GIELGUD's Hamlet and as Yakov in CHEKHOV's *The Seagull*, gave signs of future excellence. He first came into prominence, however, as Sir Andrew Aguecheek in *Twelfth Night* at the OLD VIC in 1937, in which year he joined Gielgud's repertory company at the QUEEN'S THEATRE, his parts there including an unusually tender and poetic Lorenzo in *The Merchant of Venice*. Returning to the Old Vic, he was seen in 1938 as Bob Acres in SHERIDAN's *The Rivals*, and then played Hamlet in an uncut modern-dress production by Tyrone GUTHRIE. During the Second World War he served in the Royal Navy, being temporarily released in 1942 to appear in New York in RATTIGAN's *Flare Path*. Returning to London in 1946, he played Mitya in his own adaptation of DOSTOEVSKY's *The Brothers Karamazov* and Garcin in SARTRE's *Vicious Circle* (*Huis-Clos*), and then rejoined the Old Vic company at the NEW THEATRE, where in addition to a number of Shakespearian parts including Richard II and the Fool in *King Lear*, he was seen as Khlestakov in GOGOL's *The Government Inspector* and Abel Drugger in JONSON's *The Alchemist*. At the EDINBURGH FESTIVAL in 1949 (and in New York a year later) he appeared in T. S. ELIOT's *The Cocktail Party*, and in 1951 he played Hamlet in London in his own Elizabethan-style production of the play which, though not commercially successful, provided lovers of Shakespeare with an interesting and valuable experience. After being seen as Richard III in the inaugural production at the STRATFORD (ONTARIO) FESTIVAL in 1953 he appeared in several modern plays—as the imprisoned cardinal in Bridget Boland's *The Prisoner* (1954), in FEYDEAU's farce *Hotel Paradiso* (1956), as T. E. Lawrence in Rattigan's *Ross* (1960), and in IONESCO's *Exit the King* (1963). He went to New York in 1964 to play Dylan Thomas in Sidney Michaels's *Dylan*, which was followed by Arthur MILLER's *Incident at Vichy* (1966) in London. He returned briefly to Shakespeare with *Macbeth* at the ROYAL COURT (also 1966), and then continued to display his protean versatility in a wide range of modern plays: Simon GRAY's *Wise Child* (1967), in which he played a travesty role, John MORTIMER's *A Voyage Round My Father* (1971), as the Father, Alan BENNETT's hilarious farce *Habeas Corpus* (1973), and Julian Mitchell's adaptation of Ivy Compton-Burnett's *A Family and a Fortune* (1975). In 1976 he played Dean Swift in *Yahoo*, of which he was co-deviser with Alan Strachan, and in 1977 he was seen in Alan Bennett's *The Old Country* as a British defector living in Soviet Russia.

GUITRY, LUCIEN-GERMAIN (1860–1925), French actor and dramatist, who made his first appearance as Armand in a revival of *La Dame aux camélias* by DUMAS *fils* in 1878, at the GYMNASE. He was later in Russia, and toured the Continent for many years, returning to the ODÉON in 1891, and appearing with Sarah BERNHARDT at the PORTE-SAINT-MARTIN in 1893. He was for several years manager of the RENAISSANCE, where he also appeared in many of his own productions and plays.

GUITRY, SACHA [Alexandre-Pierre-George] (1885–1957), French actor and dramatist, son of the above. He began writing for the stage at 17 and produced 130 plays and some 30 film scripts. From his first success *Le veilleur de nuit* (1911) until his death, he enjoyed uninterrupted popularity, except for a short period when he was suspected of pro-German sympathies during the Second World War occupation of France. His work is typical of French light BOULEVARD PLAYS during their last great period. In his lightweight social comedies—*La Jalousie* (1915) or *Mémoires d'un tricheur* (1935)—he shows himself heir to SARDOU and FEYDEAU, while his quasi-historical comedies, dealing with anecdotal episodes on the fringe of great events—*Histoires de France* (1929) or *N'écoutez pas, Mesdames* (1942), seen in London in 1948 as *Don't Listen, Ladies!*—and his plays on the private lives of great men—Pasteur, Mozart, DEBURAU—continue a tradition established by AUGIER and DUMAS *fils*. He was five times married to actresses, his third wife Yvonne PRINTEMPS appearing with him in revivals of several of his plays, notably *Nono* (1920) and *Mozart* (1926) in London and New York.

GUIZARDS, GUIZERS, see MASQUE and MUMMERS' PLAY.

GUNDERSEN, LAURA (1832–98), the first really great actress in the Norwegian theatre. She made her début at the KRISTIANIA THEATER in 1850, and remained there with a short break until her death. She is remembered above all for her playing in Shakespeare, BJØRNSON, and IBSEN.

GUNNELL, RICHARD (?–1634), English actor, dramatist, and theatre manager, a friend of ALLEYN in whose diary he often figures. He was with the ADMIRAL'S MEN from 1612, when they became the Palsgrave's Men, and wrote for them two plays, now lost, *The Hungarian Lion* (1623) and *The Way to Content all Women, or How a Man may Please his Wife* (1624). He was a shareholder in the new FORTUNE THEATRE, rebuilt after a disastrous fire in 1618, and seems eventually to have taken over its management, but with little success as he was often in debt. In 1629 he was associated with Blagrove in the building of SALISBURY COURT THEATRE, where he remained in management until his death.

GUSTAV III (1746–92), King of Sweden from 1771 to 1792, a period known as 'the Gustavian Era'. He took a great interest in the theatre, had much influence on its development in Sweden in the latter part of the 18th century, and was himself the author, or part-author, of a number of plays given at Court, chief of which is *Gustav Adolphs ädelmod* (*Gustav Adolph's Magnanimity*, 1783). He founded the KUNGLIGA DRAMATISKA TEATERN in Stockholm in 1788 and encouraged the work of young dramatists.

GUTHRIE, TYRONE (1900–71), English actor and director, knighted in 1961, who through his mother Norah Power was the great-grandson of the Irish actor Tyrone POWER. He made his first appearance on the stage under J. B. FAGAN in Oxford in 1924, and from 1929 to 1930 directed plays at the Festival Theatre, Cambridge, for Anmer Hall, for whom he also directed his first play in London, BRIDIE's *The Anatomist* (1931) with which the WESTMINSTER THEATRE opened. In the same year he produced in Montreal a radio series called *The Romance of Canada*. Much of his finest work was done in Shakespeare. At the OLD VIC in 1933 he directed an interesting *Measure for Measure* with Charles LAUGHTON as Angelo, a production which he repeated in 1937 with Emlyn WILLIAMS in the role. Also in 1937 he directed Laurence OLIVIER in *Hamlet* in a production which was later seen at Elsinore; at Christmas 1937 and 1938 he was responsible for delightful productions of *A Midsummer Night's Dream* with Robert HELPMANN as Oberon, and Mendelssohn's music. Among his other productions at this time were *Hamlet* in 1938 in modern dress, starring Alec GUINNESS, and IBSEN's *Peer Gynt* in 1944, with Ralph RICHARDSON in the title role. For the Edinburgh Festivals of 1948 and 1949 he directed the old Scottish plays LYNDSAY's *Ane Pleasant Satyre of the Thrie Estaitis* and Allan RAMSAY's *The Gentle Shepherd*. From 1953 to 1957 he directed productions at the STRATFORD (ONTARIO) FESTIVAL theatre, which was largely his creation, and in 1963 he directed CHEKHOV's *Three Sisters* and another modern-dress *Hamlet* for the Minneapolis Theatre, later named after him and designed somewhat on the lines of the Stratford, Ontario, theatre. His later productions in Minneapolis included *Henry V* and JONSON's *Volpone* in 1964, and Chekhov's *The Cherry Orchard* and *Richard III*, with Hume CRONYN as Richard, in 1965, together with a *Twelfth Night* which made Feste the chief character in the play. A creative artist who was not afraid to experiment, Guthrie was at his best in the handling of crowd scenes. He worked in many European countries, including Germany and Finland, and in Israel. In 1967 he returned to the Old Vic to direct the NATIONAL THEATRE company in MOLIÈRE's *Tartuffe* with GIELGUD as Orgon and Robert Stephens in the title role.

GUTHRIE THEATRE, Minneapolis, Minnesota, originally the Minneapolis Theatre, was planned in 1958 by Tyrone GUTHRIE as a fully professional classical repertory theatre free from commercial pressure. It opened under his direction on 7 May 1963, seating 1,441 spectators, of whom none was more than 52 ft from the centre of the seven-sided thrust stage designed by Guthrie in conjunction with Tanya MOISEIWITSCH, who was closely associated with the theatre for several years. The theatre originally presented a summer repertory season only, but now offers nine plays during an 11-month season. Many outstanding actors have appeared there, including Hume CRONYN, his wife Jessica TANDY, and Zoë CALDWELL, who were all seen in the opening season, when the plays produced were *Hamlet*, CHEKHOV's *Three Sisters*, Arthur MILLER's *Death of a Salesman*, and MOLIÈRE's *The Miser*. In 1971 the theatre was renamed in honour of Guthrie, who was succeeded by the English director Michael Langham (1919–), who had also succeeded him as director of the STRATFORD (ONTARIO) FESTIVAL. Under his direction the theatre continued to prosper, among his own productions being ROSTAND's *Cyrano de Bergerac* in 1971, SOPHOCLES' *Oedipus Rex* in 1972, *King Lear* in 1974, and WILDER's *The Matchmaker* in 1976. He was succeeded for a year (1978–9) by Alvin Epstein, and then in 1980, after a long search, by the Romanian director Liviu Ciulei (1923–), whose opening season in 1981–2 included *The Tempest*, *As You Like It*, and Thomas Bernhard's *Eve of Retirement*, all directed by himself, and the American première of Susan Cooper and Hume Cronyn's *Foxfire*. For this season

Ciulei covered the original stage by a rectangular end stage which increased the performing area by 30 per cent.

GUTZKOW, KARL FERDINAND (1811–78), German writer, a prominent member of the 'Young Germany' movement, mainly remembered as the author of the great Jewish play *Uriel Acosta*, a moving and terrible picture of the struggle for intellectual freedom, written in 1847, which has become a recognized classic of world drama and is in the repertory of almost all Jewish theatres. It was first seen in England in 1905 and has been played, both in the original and in translation, in most European countries and in America.

GWYNN, NELL [Eleanor] (1650–87), English actress, who was an orange-girl, probably under Mrs MEGGS at DRURY LANE, when she attracted the attention of Charles HART. Profiting by his help, she made her first appearance on the stage in DRYDEN's *The Indian Emperor* (1665). She was not a good actress, and owed her success in comedy to her charm and vivacity. Her best part seems to have been Florimel in Dryden's *Secret Love* (1667), in which she was much admired in male attire. She was first noticed by Charles II when speaking the witty epilogue to Dryden's *Tyrannic Love, or the Royal Martyr* (1669). She then became his mistress and left the stage, her last part being Almahide in Dryden's *The Conquest of Granada* (1670). Tradition has it that the founding of Chelsea Hospital was due to her influence. She became the subject of a number of plays, one of the best known being Paul KESTER's *Sweet Nell of Old Drury* (1900), in which the title role was played many times by Julia NEILSON.

GYLLENBORG, CARL (1679–1746), Swedish dramatist and statesman, whose satirical *Den Swenska sprätthöken* (*The Swedish Dandy*, 1740) was the first genuinely Swedish play to be performed at the newly established Svenska Dramatiska Teatern.

GYMNASE, Théâtre du, Paris. Built in 1820 on what is now the Boulevard Bonne-Nouvelle, this theatre was originally intended as a training-ground for young actors studying at the Conservatoire, who appeared there in one-act adaptations of the classics which it was hoped would attract young audiences from the schools and colleges. The scheme proved impractical, and in 1830 the theatre began to present the plays of Eugène SCRIBE with such success that he was offered a life annuity and bonuses for the sole and permanent rights in his work. The leading lady of the Gymnase from 1842 until her death in 1861 was Rose Chéri, who in 1847 married its manager Monsigny. Scribe was succeeded by SARDOU, AUGIER, and others as resident playwrights, and in 1852 the younger DUMAS's *La Dame aux camélias* had its first night there. From then until the end of the century the theatre ranked third in esteem among the theatres of Paris, surpassed only by the COMÉDIE-FRANÇAISE and the ODÉON. Henri BECQUE's first plays were staged there, BERNHARDT, Lucien GUITRY, and Henry BERNSTEIN all began their careers there, the last managing the theatre between the two world wars, when it continued to offer popular comedy and melodrama. In recent times, under the management of Marie BELL, it has tended more towards social drama and the established classics of comedy.

GYURKÓ, LÁSZLÓ, see HUNGARY.

H

HABIMAH (Stage) **THEATRE,** company founded in Moscow in 1917 by Nahum Zemach to perform plays in Hebrew with the intention of assisting in the revival of the language among the Jewish community there. Its first performance took place on 6 Oct. 1918 and attracted the attention of STANISLAVSKY, who sent his pupil VAKHTANGOV to direct the young company in ANSKY's *The Dybbuk*. This opened on 31 Jan. 1922, after more than three years of intermittent rehearsal, and became world famous, being played by Habimah on tour more than a thousand times. Other early productions were David Pinsky's *The Eternal Jew* (1919) and LEIVICK's *The Golem*. In 1926 Habimah left Moscow for a world tour, and achieved a sensational success in Europe, the critics marvelling as much at the company's artistry as at their command of a language until then considered dead. In the U.S.A. they were less successful and a split developed, a few of the actors remaining in America with Zemach, while the rest returned to Europe and set up a temporary base in Berlin. They were never to return to Russia, but having always intended to settle in Palestine, as it then was, they made their first visit there in 1928, being seen in ALEICHEM's *The Treasure* and CALDERÓN's *The Hair of Absolom*, under the direction of Alexander Diky who had worked with Vakhtangov. After a further tour of Europe, during which Habimah was seen in its first Shakespeare production, *Twelfth Night* staged by Michael CHEKHOV in Berlin, the company finally settled in Tel-Aviv in 1932 and in 1945 moved into its own theatre there, where it remained until 1970, being declared a National Theatre in 1958. In 1969, after some dissension, the government of Israel, which had hitherto subsidized the company to some extent, took over full responsibility for what now became a State Theatre. The 'collective' management hitherto adhered to was disbanded, and a manager and artistic director with full powers were appointed by the authorities. On 2 Apr. 1970 a performance of DEKKER's *The Shoemaker's Holiday* in Hebrew inaugurated a new, comfortable, and well equipped National Theatre building.

HACKETT, JAMES HENRY (1800–71), American actor, who made his first appearance in 1826, and became famous for his portrayal of Yankee characters, many of which he interpolated into new or existing plays. One of his finest parts was Nimrod Wildfire in James Kirke Paulding's *The Lion of the West* (1830), a play which owed much of its success to his acting. The manuscript, which was believed lost—Hackett objected to the printing of plays in which he starred—was found and published in 1954. Hackett also appeared as Rip Van Winkle several years before JEFFERSON made his remarkable success in the part. He was the first American actor to star in London, in 1833, and was manager of the ASTOR PLACE OPERA HOUSE at the time of the rioting by rival admirers of MACREADY and FORREST in 1849. He was a keen student of Shakespeare, Falstaff being one of his best parts, and did much to encourage the development of a native drama in the United States. By his second wife (he was first married to the actress Catherine Lee Sugg, 1797–1848), he was the father of the romantic actor James Keteltas HACKETT.

HACKETT, JAMES KETELTAS (1869–1926), American actor, son of the above. Born in Canada, he was taken to the U.S.A. as a child, and by 1892 was playing leading roles in Shakespeare and SHERIDAN under DALY. In 1895 he joined Daniel FROHMAN's company at the LYCEUM, where one of his best parts was Rassendyll in Hope's *The Prisoner of Zenda*. In youth his tall slim figure, dark hair, and fine features exactly suited the heroes of melodrama and romance popular at the time, but a certain artificiality and lack of basic training prevented him from doing serious work in his later career, though in 1906, with the profits of a production of Sutro's *The Walls of Jericho* the year before, he opened his own theatre in New York, later renamed the WALLACK, and in 1914, with the proceeds of a legacy, put on an *Othello* with sets by Joseph URBAN which marked an important step forward in the history of American stagecraft and scenic design.

HACKETT, WALTER (1876–1944), American dramatist and theatre manager, who in 1915 went to London with his wife Marion LORNE for the production of *He Didn't Want To Do It*, of which he was joint author. For more than 23 years he remained in London, producing a series of light comedies in which his wife starred with great success among them *Ambrose Applejohn's Adventure* (1921), *77 Park Lane* (1928), and *Sorry You've Been Troubled* (1929). In 1930 he opened the WHITEHALL THEATRE with *The Way to Treat a Woman*, transferred from the DUKE OF YORK's, where he had briefly been in management, and remained there until 1934, his productions including *The Gay*

Adventure (1931), *Road House* (1932), and *Afterwards* (1933). Among his later successes were *Hyde Park Corner* (1934), *Espionage* (1935), *The Fugitives* (1936), and *London After Dark* (1937), all at the APOLLO. Hackett's last play for London was *Toss of a Coin* (1938), after which he and his wife returned to the United States.

HAFNER, PHILIPP (1736–64), Austrian playwright and director, usually regarded as the originator of the *Volksstück*, the typical Viennese comedy of manners. As an admirer of LESSING he endeavoured to replace the impromptu burlesque by plays of serious value, and insisted on faithfulness to the author's text; but he was also quick to defend the rights of the living theatre against the over-zealous 'improver'. His own plays, of which the most important are *Megära die förchterliche Hexe* (1755), *Der geplagte Odoardo* (1762), and *Der Furchtsame* (1764), strike a judicious balance between literary and theatrical elements, the latter owing much to his close friendship with the actor PREHAUSER.

HAGEN, UTA THYRA (1919–), German-born American actress, who first appeared on the stage in 1935 in Noël COWARD's *Hay Fever*, and in 1937 played Ophelia to the Hamlet of Eva LE GALLIENNE in Dennis, Mass. She made her New York début with the LUNTS in 1938 as Nina in CHEKHOV's *The Seagull*, and appeared with her first husband José FERRER in Maxwell ANDERSON's *Key Largo* (1939). In 1943 she was seen as Desdemona to Ferrer's Iago and Paul ROBESON's Othello, and she played, in German, in GOETHE's *Faust* in 1947 and IBSEN's *The Master Builder* in 1948. Later in 1948 she took over from Jessica TANDY the part of Blanche Dubois in Tennessee WILLIAMS's *A Streetcar Named Desire*, which she played for two years. In Nov. 1950 she gave a highly praised performance in the title role of ODETS's *The Country Girl*, other notable roles during the next few years including Joan in SHAW's *Saint Joan* in 1951, Tatiana in SHERWOOD's *Tovarich* in 1952, Natalia in TURGENEV's *A Month in the Country* in 1956, and the dual leading role in the first New York production of BRECHT's *The Good Woman of Setzuan* (also 1956). In 1962 she was first seen in her most famous role, Martha in ALBEE's *Who's Afraid of Virginia Woolf?*, making her London début in the same part in 1964. In 1968 she appeared as Ranevskaya in Chekhov's *The Cherry Orchard*. Since 1947 she has taught acting at the studio run by her second husband Herbert Berghof, and she is the author of a handbook, *Respect for Acting* (1973).

HAINES, JOHN THOMAS (1797–1843), English actor and dramatist, author of a number of MELODRAMAS, now forgotten, in many of which he appeared himself. His first work for the stage, seen at the Coburg (later the OLD VIC) Theatre, was *Quentin Durward* (1823), one of the earliest dramatizations of a novel by Walter SCOTT. He was then in Manchester, where he wrote and acted in several local dramas, including *Hulme Hall; or, Manchester in the Olden Times* (1828), *The Spinner's Dream* (1829), perhaps the earliest of the 'factory' dramas to which he later contributed *The North Pole* (also 1829) and *The Factory Boy; or, the Love Sacrifice* (1840). In 1830 he returned to London, where his most successful play, the sentimental NAUTICAL DRAMA *My Poll and My Partner Joe* (1835), was given at the SURREY THEATRE, providing an excellent vehicle for T. P. COOKE, as did *The Ocean of Life; or, Every Inch a Sailor* (1836). Haines's later plays, which were mainly hack-work, include several other nautical dramas, some domestic dramas, one tragedy, *The Life of a Woman; or, a Curate's Daughter* (1840), and a spectacular entertainment in the fashion of the time, *Aslar and Ozines; or, the Lion Brothers* [or *Hunters*] *of the Burning Zaara* (1843).

HAINES, JOSEPH (?–1701), English actor, an excellent comedian, inveterate practical joker, and writer of scurrilous verse and lampoons which on several occasions brought him into conflict with the authorities. After running a booth at Bartholomew Fair he joined KILLIGREW's company at DRURY LANE to play clowns and buffoons, and was one of the first English actors to play HARLEQUIN, in RAVENSCROFT's adaptation of MOLIÈRE's *Les Fourberies de Scapin*. This had already been announced when it was forestalled by OTWAY's *The Cheats of Scapin*, seen at DORSET GARDEN in 1676. It was therefore renamed and remodelled on the lines of a COMMEDIA DELL'ARTE version by Tiberio FIORILLO then running in Paris, and as *Scaramouch a Philosopher, Harlequin a Schoolboy, Bravo, Merchant and Magician. A Comedy after the Italian Manner* was finally produced at Drury Lane in May 1677, after Haines had made a special journey to Paris to study the methods and machinery of the French stage.

HALE, LOUISE [*née* Closser] (1872–1933), American actress and author, who made her stage début at Detroit in Charles T. Dazey's *In Old Kentucky* (1894). Her first great success came as Prossy in SHAW's *Candida*, with Arnold DALY, in 1903. In 1907 she was seen in London as Miss Hazy in Alice Hegan Rice and Anne Crawford Flexner's *Mrs Wiggs of the Cabbage Patch*, and she appeared in many other excellent productions, including MAETERLINCK's *The Blue Bird* (1910), *Ruggles of Red Gap* (1915), adapted from H. L. Wilson's humorous novel, O'NEILL's *Beyond the Horizon* (1920), and IBSEN's *Peer Gynt* (1923). Her later career was entirely in films. She wrote a number of novels, many of them dealing with theatrical life, and some travel books, the latter illustrated by her husband Walter Hale (1869–1917).

HÁLEK, VÍTĚZSLAV, see CZECHOSLOVAKIA.

HALÉVY, LUDOVIC, see MEILHAC, HENRI.

HALEVY, MOSCHE, see HEBREW THEATRE, NEW, and OHEL THEATRE.

HALL, PETER REGINALD FREDERICK (1930–), English theatre manager and director, knighted in 1977. After directing several amateur productions while at CAMBRIDGE, he did his first professional production in 1953 at the Theatre Royal, WINDSOR. A year later he was in London, at the ARTS THEATRE, where among other new plays, he directed BECKETT's *Waiting for Godot*, BETTI's *The Burnt Flower-Bed* (both 1955), and ANOUILH's *The Waltz of the Toreadors* (1956). Also in 1956 he directed Colette's *Gigi* at the NEW THEATRE and was responsible for an excellent *Love's Labour's Lost* at the SHAKESPEARE MEMORIAL THEATRE, where he later produced *Cymbeline* (1957), *Twelfth Night* (1958), *A Midsummer Night's Dream* (1959), and *The Two Gentlemen of Verona* (1960). In 1960 he became managing director of the newly formed ROYAL SHAKESPEARE COMPANY, which took over the ALDWYCH THEATRE as its London home. His productions included Anouilh's *Becket* (1961), *The Wars of the Roses* (based on *Henry VI* and *Richard III*, 1963), Henry LIVINGS's *Eh?* (1964), and PINTER's *The Homecoming* (1965; N.Y., 1967). In 1968 he resigned as Managing Director, though he remained on the Board and continued to direct plays for the company, including Pinter's *Old Times* (London and N.Y., 1971). He became Director of the NATIONAL THEATRE in 1973, having to deal three years later with the many problems entailed by the transfer from the OLD VIC to its present home. His productions at the National have included Pinter's *No Man's Land* (1975; N.Y., 1977); *Hamlet* (1975) and MARLOWE's *Tamburlaine the Great* (1976), the opening productions at the LYTTELTON and OLIVIER theatres respectively; AYCKBOURN's *Bedroom Farce* (1977); and SHAFFER's *Amadeus* (1979; N.Y., 1980). In 1980 he became Joint Director of the Olivier and the COTTESLOE with Peter GILL while retaining overall control of the National Theatre organization.

HALL, ROGER, see NEW ZEALAND.

HALL, WILLIS (1929–), English playwright, most of whose later plays have been written in collaboration with Keith Spencer Waterhouse (1929–). The first of Hall's own plays to be staged, *The Royal Astrologers* (1958), was written for children, his first play for adults being the highly successful *The Long and the Short and the Tall* (1959; N.Y., 1962), a study of the diverse members of a patrol lost in 1942 in the Malayan jungle. After a double bill, *Last Day in Dreamland* and *A Glimpse of the Sea* (also 1959), both plays set in a seaside resort at the end of the season, Hall began his collaboration with Waterhouse, their first joint work—and one of the their most successful—being *Billy Liar* (1960; N.Y., 1963), based on Waterhouse's novel about a fantasizing North Country boy. It was followed by *Chin-Chin* (also 1960), based on a play by BILLETDOUX, *Celebration* (1961), *England, Our England* (1962), a revue, and (also in 1962) another double bill, *The Sponge Room* and *Squat Betty* (seen in New York in 1964). More typical of their collaboration, however, were the comedy *All Things Bright and Beautiful* (1962) and the farce *Say Who You Are* (1965), seen in America in 1966 as *Help Stamp Out Marriage*. They then wrote the book for the musical *The Card* (1973), based on Arnold BENNETT's novel, and had a big success at the NATIONAL THEATRE with *Saturday, Sunday, Monday* (also 1973), based on a play by Eduardo DE FILIPPO, which was transferred to the West End and seen in New York in 1974. In the next few years Hall wrote several plays on his own, including *Kidnapped at Christmas* (1975) and other plays for children, but in 1977 he again collaborated with Waterhouse in *Filumena* (N.Y., 1980), based on another play by de Filippo.

HALLAM, LEWIS, the elder (1714–56), British-born American actor, son of Adam Hallam, an actor at COVENT GARDEN from 1734 to 1741, and later at DRURY LANE. In 1752 Lewis Hallam, already an experienced actor, took his wife and children (with the exception of Isabella, 1746–1826, later famous on the English stage as Mrs Mattocks) and ten other actors to the United States, where they appeared in WILLIAMSBURG in *The Merchant of Venice* and JONSON's *The Alchemist*. A year later he refurbished the NASSAU STREET THEATRE in New York, where his company, in the face of some opposition, appeared in a wide-ranging repertory. Later, in Philadelphia, they played in a building previously occupied by KEAN and Murray. After a visit to Charleston the company went to Jamaica, where Hallam died. His widow (?–1773) then married David DOUGLASS, and the remnants of Hallam's company, including his son Lewis (below), joined Douglass's company.

HALLAM, LEWIS, the younger (*c.* 1740–1808), American actor, son of the above, who went with his father to Williamsburg in 1752, and in 1757 became leading man of the combined companies of his mother and David DOUGLASS, going with them to New York, where, billed in 1758 as the AMERICAN COMPANY, they played in the temporary theatre on CRUGER'S WHARF. Lewis, an excellent actor, appeared in GODFREY's The *Prince of Parthia* (1767), the first American play to be given a professional production. After the death of Douglass in 1786 he took over the management of the American Company, in partnership with John HENRY, and later with HODGKINSON and DUNLAP, reopening the SOUTHWARK THEATRE in Philadelphia and the JOHN STREET in New York, playing also in Baltimore and Annapolis. He married, as his second wife, an actress named Miss Tuke, who became an important but quarrelsome member of his company. Though he retired from management in 1797, he continued to act until his death.

HALLE, ABRON, see JEWISH DRAMA.

HALL KEEPER, see STAGE DOOR.

'HALLS', The, see MUSIC-HALL.

HALLSTRÖM, PER (1866–1960), Swedish dramatist, poet, novelist, essayist, and the author of some significant aesthetic criticism. His plays belong almost entirely to the early 20th century, the first, *Grefven af Antwerpen*, being produced in 1898. It was followed by a tragedy, *Bianco Capello* (1900); two comedies, *En veneziansk komedi* (1901) and *Erotikon* (1908), the first set in Venice; two legend-plays, *Alkestis* and *Ahasverus* (also 1908); two saga-plays, *Önskningarna* (*The Wishes*) and *Tusen och en natt* (*A Thousand and One Nights*, both 1910); two historical plays, *Karl XI* and *Gustaf III* (both 1918); and, his last play, *Nessusdräkten* (*The Garment of Nessus*, 1919). Hallström's chief contribution to drama after 1919 was his translation of the plays of Shakespeare, made between 1922 and 1931.

HAM, a term of derision applied to the old-fashioned rant and fustian which is supposed to have characterized 19th-century acting, particularly in MELODRAMA. The word appears to derive from 'ham-fatter', an American slang term for an incompetent performer, itself perhaps derived from the lard used to remove grease-paint: 'ham' was current in America from the 1880s, and seems to have found its way to England after the First World War. In essence, 'ham' acting is tragic or dramatic acting which is devoid of inner truth or feeling, covering its deficiencies with a veneer of over-worked technical tricks, bombast, and showy but meaningless gestures.

HAMBLIN, THOMAS SOWERBY (1800–53), British-born actor and theatre manager, who after some years in the provinces, appeared at DRURY LANE in leading parts. He went to America in 1825, and made his first appearance at the PARK THEATRE as Hamlet, later touring the United States as a tragedian. He was a fine, though somewhat melodramatic, actor, but in later years his acting was hindered by frequent bouts of asthma. He became lessee of the BOWERY THEATRE in 1830, rebuilt it after the disastrous fire of 1836, and only relinquished it in 1850 after further fires in 1838 and 1845. Ill luck seemed to dog him, for in 1848 he rented and redecorated the old Park Theatre and opened it on 4 Sept., only to see it destroyed by fire three months later, whereupon he retired. He was twice married; his first wife Elizabeth, the daughter of William BLANCHARD, was for many years her husband's leading lady.

HAMBURG, a city important in German theatrical history, since it was there that the first National Theatre was established with LESSING as its accredited dramatic critic and EKHOF as its leading actor under ACKERMANN. It opened in 1767 and closed two years later, its lack of success being due to poor plays, backstage intrigues, and public apathy. The one outstanding event was the first production of Lessing's *Minna von Barnhelm* (1767), the first masterpiece of German comedy. It was SCHRÖDER, leader of Ackermann's company after 1771, who with his powerful acting established Shakespeare on the German stage and introduced the plays of the STURM UND DRANG, making Hamburg a vital theatrical centre where from 1785 onwards IFFLAND's plays were also staged. Schröder's successor F. L. Schmidt successfully staged KLEIST's *Der zerbrochene Krug* in 1820, after its failure in WEIMAR, and in 1827 took his company to a new theatre, now the Staatsoper. A second playhouse, the Thaliatheater, opened in 1843 for the production of popular comedy, and a fresh impulse to theatrical life in Hamburg came in 1905 when JESSNER took over the theatre where he remained until 1915, introducing into the repertory plays by WEDEKIND, BÜCHNER, and IBSEN. Reconstructed in 1950, the Thalia now competes successfully with the Deutsches Schauspielhaus (opened in 1900), which had its greatest period to date from 1955 to 1963, under GRÜNDGENS, who had made his first appearance in the 1920s at the Hamburg Kammerspiele. Originally built in 1907 as the Volksschauspielhaus, this theatre was reconstructed and reopened in 1918 as the Kammerspiele with a series of plays by Wedekind. The following year saw the first productions of KAISER's *Brand im Opernhaus* and BARLACH's *Der arme Vetter*, and subsequently a number of foreign and German plays, by such authors as SHAW, STRINDBERG, HASENCLEVER, STERNHEIM, and GOERING, were seen there.

HAMMERSTEIN II, OSCAR, see MUSICAL COMEDY.

HAMMERSTEIN'S THEATRE, New York, see MANHATTAN THEATRE.

HAMMERTON, STEPHEN (?–*c*. 1648), English actor, who as a boy played women's parts with PRINCE CHARLES'S MEN at SALISBURY COURT. In 1632 he was evidently persuaded to join the KING'S MEN; although his former manager, Christopher BEESTON, petitioned the king for his return, he remained with them, playing adult roles such as King Ferdinand in SHIRLEY's *The Doubtful Heir* (1640). He was still alive in 1647, when he signed the dedication of the BEAUMONT and FLETCHER Folio, but nothing further is known of him.

HAMMOND, KAY, see CLEMENTS, JOHN.

HAMPDEN, WALTER [Walter Hampden Dougherty] (1879–1955), American actor, born in New York, who first appeared on the stage in England, where he was for some years a member of Frank BENSON's company and later played leading parts at the ADELPHI THEATRE in London. In 1907 he returned to the United States, appearing with NAZIMOVA in a series of plays by IBSEN and other modern dramatists. Among his later successful parts were Manson in C. Rann Kennedy's *The Servant in the House* (1908), seen in London a year previously, and the title role in ROSTAND's *Cyrano de Bergerac*, in which he first appeared in 1923, reviving it several times later. He was also seen in a wide range of Shakespearian parts, which included Caliban, Hamlet, Macbeth, Oberon, Othello, Romeo, and Shylock. In 1925 he took over the Colonial Theatre at 1887 Broadway, which he renamed Hampden's, and appeared there in an interesting repertory which included *Henry V*, BENAVENTE's *The Bonds of Interest*, and BULWER-LYTTON's *Richelieu*. He remained in his own theatre until 1930, and then toured, mainly in revivals of his previous successes. In 1939 he played the Stage Manager in Thornton WILDER's *Our Town*, and in 1947 was seen in *Henry VIII*, the first production of the American Repertory Theatre, playing Cardinal Wolsey to the Queen Katharine of Eva LE GALLIENNE. His last Broadway appearance was in Arthur MILLER's *The Crucible* in 1953.

HAMPSTEAD THEATRE CLUB, London, opened at the Moreland Hall, Hampstead, on 24 Sept. 1959, the first London productions of PINTER's *The Room* and *The Dumb Waiter* being seen there in 1960, and of IONESCO's *Jacques* a year later. In 1962 the club moved to its own prefabricated premises at Swiss Cottage, and among many notable productions on its open-end stage were Laurie Lee's *Cider With Rosie* (1963) and several which were later transferred to the West End, notably Donald Howarth's *A Lily in Little India* (1965), John Bowen's *After the Rain* (1966) and *Little Boxes* (1968), and Aubrey's *Brief Lives* (1967), with Roy DOTRICE. In 1970 the club moved to new premises near by, seating 157, where many interesting productions have been staged. Numerous transfers to the West End have included Michael FRAYN's *Alphabetical Order* (1975), Pam Gems's *Dusa, Fish, Stas, and Vi* (1976), James SAUNDERS's *Bodies* (1978), and Pinter's *The Hothouse* (1980); Brian FRIEL's *Translations* (1981) transferred to the LYTTELTON. The theatre also staged the British premières of such foreign plays as Tennessee WILLIAMS's *Small Craft Warnings* (1973), Peter HANDKE's *The Ride Across Lake Constance* (also 1973), both of which transferred to the West End, DÜRRENMATT's *Play Strindberg* (1973), and Sam SHEPARD's *Buried Child* (1980).

HAMPTON, CHRISTOPHER JAMES (1946–), English dramatist, whose first play, *When Did You Last See My Mother?* (1966), was written while he was still an undergraduate. It was so successful when given a Sunday night production at the ROYAL COURT THEATRE that it was immediately transferred to the West End, and was seen in New York a year later. Hampton then became resident dramatist at the Royal Court, where his next play, *Total Eclipse* (1968), dealing with the relationship between Rimbaud and Verlaine, was also produced. It was followed by *The Philanthropist* (1970; N.Y., 1971), in which Alec MCCOWEN gave an excellent performance as Philip, an amiable but dispirited don, whose good intentions are constantly defeated by stronger personalities. *Savages* (1973; N.Y., 1977), which starred Paul SCOFIELD in London, was an ambitious non-naturalistic play combining a plot about the kidnapping of a British diplomat by South American guerrillas with another play about the mass murder of American Indians in Brazil. It was followed by the more conventional *Treats* (1976; N.Y., 1977). Hampton is also a notable translator of plays, having been responsible for English versions of Isaac Babel's *Marya* (1967), CHEKHOV's *Uncle Vanya* and IBSEN's *Hedda Gabler* (both 1970), Ibsen's *A Doll's House* (1971), in which Claire BLOOM appeared in New York (in repertory with *Hedda Gabler*) and London; MOLIÈRE's *Don Juan* (1972), Horváth's *Tales From the Vienna Woods* (1977) and *Don Juan Comes Back from the War* (1978), and Ibsen's *The Wild Duck* (1979), the last three all seen at the NATIONAL THEATRE in London.

HAMPTON COURT PALACE THEATRE, London. The Great Hall of Hampton Court Palace was regularly used for theatrical performances from 1572 onwards. A stage was erected beneath the minstrels' gallery, a chamber off the hall was converted into a dressing room, and the Great Watching Chamber was used for rehearsals. Records show that the cost of costumes and scenery was considerable. Elizabeth I's habit of spending the Christmas season at Hampton Court was continued by both James I and Charles I. The programme for Christmas 1636/7 included Shakespeare's *Othello* (as *The Moor of Venice*) and *Hamlet*, and DAVENANT's *Love and Honour*. There was no permanent theatre in the palace, however, until 1718, when George I invited a company of actors which included Colley CIBBER and Anne OLDFIELD to play there from 23 Sept. to 22 Oct. The performances, which were directly under the patronage of the King, were open to the public, and admission was free. The plays chosen again included *Hamlet*, together with FARQUHAR's *The Constant Couple* and JONSON's *Volpone*; but the most notable choice was Shakespeare's *Henry*

VIII, performed in its actual setting. The last recorded performance in the palace took place on 18 Oct. 1731 in honour of the Duke of Lorraine, but the stage survived until 1798.

HAMSUN, KNUT (1859–1952), Norwegian novelist, poet, and dramatist, awarded the NOBEL PRIZE for Literature in 1920 and widely regarded, by virtue of his novels in the first instance, as the greatest literary figure of his country since IBSEN. His plays are relatively unimportant when set alongside his fiction but they have had some success abroad, *Munken Vendt*, written in 1902, having its first production in Heidelberg in 1926, and a modern comedy, *Livet ivold* (*In the Grip of Life*, 1910), being seen at the MOSCOW ART THEATRE in 1911 and in Germany in 1914, under REINHARDT, as *Vom Teufel geholt.* His other plays are: *Ved rigets port* (*At the Gate of the Kingdom*, 1895), *Livets spil* (*The Game of Life*, 1896), and *Aftenrøde* (*Afterglow*, 1898), a trilogy which portrays three stages in the life of its hero Ivar Karenko; and the historical *Dronningen Tamara* (*Queen Tamara*, 1903). Tankred DORST's *Eiszeit* (1973) is about Hamsun.

HANCOCK, SHEILA (1933–   ), English actress, who made her first appearance in London in Peter Coke's *Breath of Spring* (1958), and was then seen in the revue *One to Another* and the musical *Make Me an Offer* (both 1959). Another revue, *One Over the Eight* (1961), was followed by her first outstanding success, the prostitute in Charles Dyer's *Rattle of a Simple Man* (1962), and she made her New York début in Joe ORTON's *Entertaining Mr Sloane* in 1965. On her return to London she was seen in a series of interesting modern plays, including Bill McIlwraith's *The Anniversary* (1966), Charles WOOD's *Fill the Stage With Happy Hours* (1967), and two plays by ALBEE for the ROYAL SHAKESPEARE COMPANY, *A Delicate Balance* (1969) and *All Over* (1972). After appearing in AYCKBOURN's *Absurd Person Singular* (1973) she returned to revue with *Déjà Revue* (1974), and later starred in the London productions of two major American musicals, *Annie* in 1978 and SONDHEIM's *Sweeney Todd* in 1980. In 1981 she played Tamora in *Titus Andronicus* and Paulina in *The Winter's Tale* at the ROYAL SHAKESPEARE THEATRE, also directing OTWAY's *The Soldier's Fortune* for the Cambridge Theatre Company (see CAMBRIDGE), of which she is an associate director.

HANDKE, PETER (1942–   ), Austrian dramatist and novelist, who first attracted attention with his provocatively anti-theatrical *Publikumsbeschimpfung* (*Offending the Audience*, 1966). Other undramatic, plotless, characterless one-act pieces followed, until *Kaspar* (1968) and *Der Ritt über den Bodensee* (1971) dispelled suspicions of charlatanism, and established Handke's serious concern with the problem of individual expression in a world overstocked with readymade concepts. In *Kaspar*, which was seen in London in 1973 at the Almost Free Theatre and was also used by Peter BROOK as a practice text at his Paris centre in 1971–2, the chief character, an innocent *tabula rasa*, is imprinted with conformist language and behaviour; in *Der Ritt über den Bodensee*, seen in New York in 1972 and at the HAMPSTEAD THEATRE CLUB in 1974 as *The Ride Across Lake Constance*, the characters try to communicate through stereotyped speeches and gestures. In the more conventional *Die Unvernünftigen sterben aus* (*The Foolish Ones Die Out*, 1974) it is the banality of his business life that stifles the sensitivity of the capitalist protagonist. A dramatic monologue reflecting on his mother's suicide was seen in New York in 1977 as *A Sorrow Beyond Dreams*.

HAND-PROPS, see PROPS.

HAND-PUPPET, see PUNCH AND JUDY and PUPPET.

HANDS, TERRY [Terence] DAVID (1941–   ), English director, one of the founders in 1964 of the EVERYMAN THEATRE in Liverpool, where his productions included OSBORNE's *Look Back in Anger*, WESKER's *The Four Seasons*, and ARRABAL's *Fando and Lis*. In 1966 he joined the ROYAL SHAKESPEARE COMPANY, initially as artistic director of Theatregoround, its travelling company, becoming an associate director of the main company in 1967 and its joint artistic director, with Trevor NUNN, in 1978. Outstanding among his productions at the ROYAL SHAKESPEARE THEATRE have been *The Merry Wives of Windsor* in 1968 and 1975, *The Merchant of Venice* in 1971, the Stratford Centenary productions of *Henry IV* and *Henry V* in 1975, and *Henry VI* in 1977, the last three all starring Alan HOWARD. After *Richard II* and *Richard III* in 1980, again with Howard, he had directed all Shakespeare's history plays in five years. His productions at the ALDWYCH THEATRE have included Triana's *The Criminals* (1967), the first Cuban play to be seen in London, GENET's *The Balcony* in 1971, ARBUZOV's *Old World* in 1976, with Peggy ASHCROFT and Anthony QUAYLE, and GORKY's *The Children of the Sun* in 1979. From 1975 to 1977 he was also consultant director of the COMÉDIE-FRANÇAISE and he directed several plays there, his productions of *Richard III* in 1972 and *Twelfth Night* in 1976 being much admired. He has also directed plays by Shakespeare at the Burgtheater in Vienna.

HAND WORKING, the traditional practice of raising scenery in the theatre by hand-lines from a fly-floor, as against the modern COUNTERWEIGHT SYSTEM. The piece of scenery to be flown is hung by a set of lines, generally three in number, which pass over three pulley blocks in the GRID. They are taken to a triple head-block at the side of the stage, and their ends descend together to be made fast to a cleat on the fly rail. A complete BOX-SET can be battened together, with property furniture attached to its walls, and the whole unit flown entire on a number of sets of lines.

A hand-worked theatre was formerly known as a Rope House, and sometimes as a Hemp House.

HANKIN, EDWARD CHARLES ST JOHN (1869–1909), English dramatist who, in revolt against the sentimentalism of the 19th-century theatre, attacked social abuses but without suggesting any remedies. Among his somewhat cynical plays is *The Return of the Prodigal* (1905), in which the central character is supported by his family in case he should harm their reputation: it had a short run in 1948 with John GIELGUD as the young wastrel. None of Hankin's other plays has been revived.

HANLON-LEES, a troupe of acrobatic actors who became internationally famous for their outstanding TRICKWORK in Dumb Ballets of which the best known was probably *Le Voyage en Suisse*, seen in Paris at the Théâtre des VARIÉTÉS in 1879, and in London, at the GAIETY THEATRE, a year later. It included a bus smash, a chaotic scene on board ship during a storm, an exploding Pullman car, a banquet which turned into a juggling party after one of the Hanlons had crashed through the ceiling on to the table, and, in the opinion of several critics, one of the cleverest drunk scenes ever presented on the stage. The Hanlons, six brothers, began their career some time after 1860 in partnership with a famous acrobat, 'Professor' John Lees, and in 1883 (by which time Thomas had died) were in New York in a show called *Fantasma*, presented at the Fifth Avenue Theatre. This was revived in 1889, without Frederick, who died in 1886, and Alfred, who had retired, dying in 1892. Of the three survivors, William died in 1923, George, the eldest, in 1926, and Edward in 1931.

HANNAH PLAYHOUSE, Wellington, see DOWNSTAGE THEATRE and NEW ZEALAND.

HANNEN, NICHOLAS, see SEYLER, ATHENE.

HANSEN, CHRISTIERN, see DENMARK.

HANSWURST, a comic character from German folk-lore, combined with something of the ZANNI in the COMMEDIA DELL'ARTE, who in the 16th century became the equivalent of the Pickelhering of the ENGLISH COMEDIANS, finding his way into all the improvised comedies of the day and often disrupting the plot with his fooling and acrobatic tricks. It was against the worst of his antics, when they began to creep into serious plays, that GOTTSCHED and Carolina NEUBER waged war in the 1730s. In VIENNA the character had been immortalized by STRANITZKY, who turned him into an astute and witty knave with a knack of wriggling his way out of awkward situations. Unlike Pickelhering, who appeared in a variety of grotesque costumes, the Viennese Hanswurst was instantly recognizable by his outfit—a loose red jacket, a blue smock with a huge green heart on the chest, sometimes embroidered with the letters H. W., yellow pantaloons, red braces with green edgings, a conical green hat, and a wooden sword stuck into a wide leather belt. He had a short black beard, heavy eyebrows, and plenty of padding to make him even more of a contrast to the tall, slim leading man. When Stranitsky retired in 1725 the part was taken over by PREHAUSER, in whose interpretation the character became more sophisticated and less coarse in his verbal and visual clowning. On Prehauser's death in 1769, KURZ took over the character, but changed it into BERNADON, and as the popularity of improvised farce declined in the late 18th century, Hanswurst disappeared from the stage.

HANUSZKIEWICZ, ADAM, see POLAND.

HAPPENING, form of theatrical or artistic expression, popular in the 1960s, which consisted of an unannounced demonstration or tableau. In the theatre, it took the form of an interruption to what was happening on the stage, ostensibly taking the actors by surprise. Happenings could also take place in a non-theatrical context, such as meetings, or among crowds in the street. The purpose might be to evoke greater audience participation than was possible in a normal theatrical situation or simply propaganda. Such happenings were planned in greater or lesser detail according to the desired nature of the audience response. Alternatively, the purpose of a happening might be purely aesthetic, and many of the leading exponents were originally painters. Though it influenced such groups as the LIVING THEATRE, the phenomenon was a transient one.

HARDWICKE, CEDRIC WEBSTER (1893–1964), English actor, knighted in 1934. He made his first appearance on the stage in 1912, joined Frank BENSON's company on tour a year later, and in 1914 was at the OLD VIC. His career was then interrupted by war service, and he was not seen again on stage until in 1922 he joined the company at the BIRMINGHAM REPERTORY THEATRE, where one of his most successful parts was Churdles Ash in PHILLPOTTS's *The Farmer's Wife* in 1924, in which he was subsequently seen in London. He had already appeared in a number of SHAW's plays when in 1925 he was seen as Caesar in a revival of *Caesar and Cleopatra*, and at the MALVERN FESTIVAL in 1929 he created the part of King Magnus in *The Apple Cart*. A year later he appeared as Edward Moulton-Barrett in Besier's *The Barretts of Wimpole Street*, which he then played in London during its long run and on tour. Among his later parts were the Burglar in Shaw's *Too True to be Good* (1932), Dr Haggett in Emlyn

WILLIAMS's *The Late Christopher Bean* (1933), and Prince Mikhail in SHERWOOD's *Tovarich* (1935). In 1936 he went to New York to play the title role in Barré Lyndon's *The Amazing Dr Clitterhouse* and Canon Skerritt in Paul Vincent CARROLL's *Shadow and Substance* (1938) and after spending several years in films in Hollywood, returned to England to tour in a revival of Phillpotts's *Yellow Sands* in 1945. In 1948 he again joined the company at the Old Vic, where he was seen as Sir Toby Belch in *Twelfth Night*, Faustus in MARLOWE's *Dr Faustus*, and Gaev in CHEKHOV's *The Cherry Orchard*. He then settled permanently in New York, where he made a great success as Koichi Asano in Spigelgass's *A Majority of One* (1959).

HARDY, ALEXANDRE (*c.* 1575–*c.* 1631), the first professional French playwright. Though there is no evidence that he was ever an actor, he was certainly attached as paid dramatist to a provincial company, that of VALLERAN-LECOMTE, which settled in Paris at the Hôtel de BOURGOGNE in the early years of the 17th century. Hardy, who had begun writing in about 1595, was a most prolific author, and was said to have written or adapted about 600 to 700 plays, of which 34, published between 1623 and 1628, are all that survive. They are of all types, and include some tragedies in a restrained classical style, others, still called tragedies, which present characters from low life in eventful plots, and a number of tragi-comedies and pastorals, with complicated plots verging on melodrama. With a spark of genius he might have changed the course of French dramatic literature, but in spite of the strong dramatic instinct which shows in some of his works his facility and easy success told against him. It is to his credit that by eliminating the influence of SENECA he helped to bridge the gap between the medieval drama and the frigid tragedies of the Renaissance on one side, and the classical drama of the great age of French tragedy on the other. He lived just long enough to see the triumphs of CORNEILLE, who first made contact with the theatre through his plays, and of the UNITIES, which he had cheerfully ignored.

HARE, DAVID (1947–   ), English playwright, co-founder of the Portable Theatre, a travelling FRINGE group, in 1968. He became literary manager (1969–70) and resident dramatist (1970–71) at the ROYAL COURT THEATRE, where his play *Slag* (London and N.Y., 1971), about three mistresses in a girls' school, was well received, having been produced the previous year at the HAMPSTEAD THEATRE CLUB, where *The Great Exhibition* (1972) was also seen. *Brassneck*, written in collaboration with Howard BRENTON, was first produced at the NOTTINGHAM PLAYHOUSE in 1973, when Hare was resident dramatist there. It covers three generations of an unscrupulous Midlands family, highlighting its commercial and political corruption. *Knuckle* (1974; N.Y., 1975) uses a conventional thriller format to make another attack on capitalist corruption. In 1974 Hare became co-founder of another fringe company, Joint Stock, which in the following year presented *Fanshen*, his adaptation of a book by William Hinton about the Chinese revolution. *Teeth 'n' Smiles*, seen at the Royal Court in 1975 and WYNDHAM's in 1976, starred Helen MIRREN as the drunken lead singer of a disintegrating rock group performing at a Cambridge May Ball and antagonistic to the class privilege of its audience; it was produced in New York in 1979. The protagonist in *Plenty* (1978), seen at the NATIONAL THEATRE, is again a woman, an intelligent and honest character who seeks in vain in the post-war world for an outlet for her wartime idealism.

HARE [Fairs], JOHN (1844–1921), English actor-manager, knighted in 1907. He made his first appearance in Liverpool in 1864, and a year later was in London with the BANCROFTS at the Prince of Wales (later the SCALA) Theatre, where he appeared in the plays of T. W. ROBERTSON, playing Sam Gerridge in *Caste* (1867) with conspicuous success. He left the Prince of Wales in 1875 to take over the management of the ROYAL COURT THEATRE, remaining there until 1879. During this time he directed a number of new plays, among them Charles COGHLAN's *A Quiet Robber* and Tom TAYLOR's *New Men and Old Acres* (both 1876), in both of which he also appeared, and *Olivia* (1878), based by W. G. Wills on GOLDSMITH's novel *The Vicar of Wakefield*, in which Ellen TERRY played the title role. He then went into partnership with the KENDALS at the ST JAMES'S THEATRE, where for 8 years he was the chief actor and director, producing among other new plays PINERO's *The Money Spinner* (1881), in which he also appeared, and *The Ironmaster* (1884). On 24 Jan. 1889 he opened the GARRICK THEATRE, which W. S. GILBERT had built for him, his first outstanding success there being GRUNDY's *A Pair of Spectacles*, in which he played Benjamin Goldfinch, a part thereafter associated with him which he many times revived. He remained at the Garrick until 1895, and then made his first appearance in New York in Pinero's *The Notorious Mrs Ebbsmith*, following it with other parts from his repertory. On his return to England he toured as Old Eccles in *Caste*, and then created the title role in Pinero's *The Gay Lord Quex* (1899), playing opposite Irene VANBRUGH. After playing many of his old parts in revivals, he retired in 1911.

HARE, J. ROBERTSON (1891–1979), English comedian and a consummate player of farce, who made his first appearance on the stage in 1911 in Carton's *The Bear Leaders*, and spent many years touring in the provinces before becoming associated with Ralph LYNN at the ALDWYCH THEATRE in a series of farces by Ben TRAVERS, which succeeded each other from 1925 to 1933. Hare always played the 'little' man, swept along by the succession of outrageous mishaps which constituted the typical Aldwych farce. He was seen in Vernon Sylvaine's *Aren't Men Beasts!* (1936), but a year later was back with Travers in *Banana Ridge*, followed by *Spotted Dick* (1940), *She Follows Me About* (1943), *Outrageous Fortune* (1947), and *Wild Horses* (1952), being reunited with Ralph Lynn in the last two. He also appeared in Sylvaine's *One Wild Oat* (1948), and in Ronald MILLAR's *The Bride and the Bachelor* (1956) and *The Bride Comes Back* (1960). In 1963 he played Erronius in the American musical *A Funny Thing Happened on the Way to the Forum*, and in 1968 appeared in John Chapman's *Oh, Clarence!*, based on stories by P. G. Wodehouse.

HARLEQUIN, the young lover of COLUMBINE in the English HARLEQUINADE. His name, though not his part in the play, comes from the ARLECCHINO of the COMMEDIA DELL' ARTE, where he was only one of the ZANNI or quick-witted, unscrupulous servingmen, dressed usually in a patched suit, with a soft cap on his shaven head, and a small black cat-faced mask. With the passing of time he became one of the chief buffoons in comedy, and passed into the French theatre as Arlequin. When he first arrived in England he was a magician who had the power of turning himself into someone else; but in the opening scenes of the PANTOMIME it was the persecuted lover who was transformed into Harlequin; by this time fantastically dressed in a silken suit of brightly diamond coloured patches, and retaining from his origins his black mask and his magic wand, or bat (see SLAPSTICK). With this he could perform such marvels as the TRANSFORMATION SCENE, and so remove himself and his betrothed from the power of the elderly PANTALONE, anglicized as Pantaloon.

HARLEQUINADE, an important element in the development of the English PANTOMIME. It resulted from the fusion of the dumbshow of the actors at the Paris fairgrounds, where dialogue was forbidden, with the current convention that the trickster Arlequin, derived from the Italian COMMEDIA DELL'ARTE mask ARLECCHINO, could by magic turn himself into someone else. When the 'Italian Night Scenes' in which he figured were brought to London by WEAVER, this convention was not understood, and it was the unhappy lover, or later the hero of the fairy-tale opening, who turned into HARLEQUIN and so gave his name to the entire performance. Once the change had been effected the rest of the evening was devoted to the various stratagems by which the young lovers Harlequin and COLUMBINE managed to escape from Pantaloon (from PANTALONE). Not until the early 19th century, when GRIMALDI made the purely English character CLOWN into the chief personage of the Harlequinade, did Harlequin lose his premier position, and the love scenes between him and Columbine dwindled into short displays of dancing and acrobatics between bouts of horseplay and practical jokes. As these in their turn began to pall, the fairy-tale opening increased in length, and the Harlequinade dwindled into a short epilogue. For a time there was a pretence of changing the fairy-tale hero (the PRINCIPAL BOY, played always by a woman) into Harlequin by means of traps, but at last this too was abandoned, and the Harlequinade, placed after the Grande Finale of the pantomime, lost all meaning. It finally disappeared altogether during the Second World War.

HARLEVILLE, (Jean-François) COLLIN D' (1755–1806), French dramatist, author of a number of successful comedies, of which the first, *L'Inconstant* (1786), was performed first at VERSAILLES and later at the COMÉDIE-FRANÇAISE. It was followed by *L'Optimiste* (1788), which criticizes ROUSSEAU's *homme sensible*, showing that such a man can be happy only if he refuses to face facts, and by D'Harleville's best known play *Les Châteaux en Espagne* (1789), an amusing study of a man who, like VOLTAIRE's Pangloss, thinks everything is for the best in the best of all possible worlds. Jealousy over this play on the part of FABRE D'ÉGLANTINE nearly cost the author his life. He was arrested as an enemy of the Republic but, unlike Fabre, escaped the guillotine and lived to write more plays, of which the best is *Le Vieux célibataire* (1792). D'Harleville was not a particularly good dramatist, but his light-hearted comedies were welcomed by an audience caught up in the turmoil of the Revolution.

HARPER, ELIZABETH, see BANNISTER, JOHN.

HARRIGAN, NED [Edward] (1845–1911), American VAUDEVILLE actor, manager, and dramatist, whose partnership with the female impersonator Tony Hart [Anthony Cannon] (1855–91) in 1872 first brought him into prominence in New York. As Harrigan and Hart they produced many successful shows, notably *The Mulligan Guards*, a series in which Harrigan played Dan Mulligan and Hart his wife Cordelia. They took over Wood's Music Hall, a converted synagogue at 514 Broadway, calling it the Théâtre Comique, then in 1881 moved uptown to the NEW YORK THEATRE, which became their New Théâtre Comique. They parted company in 1885, but Harrigan continued to act, appearing in revivals of a number of his own plays, such as *Old Lavender* (1877) and *The Major* (1881). He opened the GARRICK THEATRE under his own name in 1890, remaining there for five years, and retired in 1908. He wrote a number of songs and over 80 vaudeville sketches, on which some of his later full-length plays were based. His characters were recognizable types of old New York life, chiefly Irish- and German-Americans and Negroes.

HARRIGAN'S PARK THEATRE, New York, see HERALD SQUARE THEATRE; **HARRIGAN'S THEATRE,** New York, see GARRICK THEATRE.

HARRIS, AUGUSTUS HENRY GLOSSOP (1852–96), English theatre manager, son of the Augustus Harris who was for many years manager of COVENT GARDEN. The younger Harris took over DRURY LANE in 1879 and remained there until his death, his devotion to his theatre earning him the nickname of 'Augustus Druriolanus'. He specialized in spectacular MELODRAMA and elaborate Christmas shows, and although some critics held him responsible for the vulgarization of the PANTOMIME by his introduction of MUSIC-HALL artistes, particularly knockabout comedians, he had a feeling for the old HARLEQUINADE, providing it with lavish scenery and machinery and engaging for it excellent CLOWNS and acrobats. He was knighted in 1891, not for his undoubted services to the theatre but because he happened to be Sheriff of the City of London when the German Emperor paid a visit there.

HARRIS, HENRY (*c.* 1634–1704), English actor of the Restoration period, accounted by some contemporary critics superior even to BETTERTON. He joined DAVENANT's company at the LINCOLN'S INN FIELDS THEATRE in 1661, where he played the title roles in ORRERY's *Henry V* (1664) and *Mustapha, Son of Solyman the Magnificent* (1665), and also several Shakespearian parts, including Romeo, Sir Andrew Aguecheek in *Twelfth Night*, and Wolsey in *Henry VIII*. A pastel drawing of him in this last part hangs in Magdalen College, Oxford. In 1668, on the death of Davenant, Harris became joint manager of the theatre with Betterton, and he was also a shareholder in the new DORSET GARDEN THEATRE, built in 1670–1. He made his last appearance on the stage in 1681, playing the Cardinal in CROWNE's *Henry VI, The First Part*.

HARRIS, JULIE [Julia] ANN (1925–), American actress, who made her first appearance in New York in 1945, and a year later, as a member of the OLD VIC company in New York, was seen in *Henry IV, Part 2* and SOPHOCLES' *Oedipus*. It was, however, as the adolescent Frankie Addams in Carson McCullers's *The Member of the Wedding* (1950) that she made her first outstanding success, and in the following year she scored another hit as the bohemian Sally Bowles in VAN DRUTEN's *I Am a Camera*. She was later in two plays by ANOUILH, *Mademoiselle Colombe* in 1954 and (as Joan of Arc) *The Lark* in 1955; she was also seen in classical roles, playing Margery Pinchwife in WYCHERLEY's *The Country Wife* in 1957, Juliet in *Romeo and Juliet* at the STRATFORD (ONTARIO) FESTIVAL in 1960, and Ophelia in *Hamlet* for Joseph PAPP in 1964. She has since been seen in Jay Allen's *Forty Carats* (1968), Paul Zindel's *And Miss Reardon Drinks a Little* (1971), James Prideaux's *The Last of Mrs Lincoln* (1972), and RATTIGAN's *In Praise of Love* (1974). She also gave a highly praised solo performance as Emily Dickinson in William Luce's *The Belle of Amherst* (1976), making her London début in this role in 1977.

HARRIS, ROBERT (1900–), English actor, who made his first appearance in 1923 in BARRIE's *The Will*, his fine voice and excellent presence later proving valuable assets in such parts as Oberon in *A Midsummer Night's Dream* in 1924 and the title role in GHÉON's *The Marvellous History of St Bernard* in 1925, in which year he also made his New York début in COWARD's *Easy Virtue*. In 1931 he joined the OLD VIC company, playing Hamlet and Mark Antony in *Julius Caesar*. He appeared as Marchbanks in SHAW's *Candida* in New York in 1937, and in the same year as Orin Mannon in O'NEILL's *Mourning Becomes Electra* in London; and as Robert Caplan in *Dangerous Corner* (1938) and David Sheil in *Music at Night* (1939), both by J. B. PRIESTLEY. In 1946 and 1947 he was at the SHAKESPEARE MEMORIAL THEATRE, where he played Prospero in *The Tempest*, Richard II, and the title role in MARLOWE's *Dr Faustus*. After two important IBSEN roles, Gregers Werle in *The Wild Duck* in 1948 and Rosmer in *Rosmersholm* in 1950, and an appearance in HOCHWÄLDER's *The Strong are Lonely* (1955), he returned to the Shakespeare Memorial Theatre and then to the Old Vic. In 1963–4 he toured the U.S.A. as More in BOLT's *A Man For All Seasons*, and in 1966 he returned to the West End to play the title role in KIPPHARDT's *In the Matter of J. Robert Oppenheimer*. He made his last appearance in William Templeton's *Make No Mistake* (1971) at the Yvonne Arnaud Theatre in Guildford.

HARRIS, ROSEMARY (1930–), British-born actress, who was seen in New York in HART and Mittelholzer's *The Climate of Eden* (1952) before appearing for the first time in London in George Axelrod's *The Seven Year Itch* (1953). After playing major roles with the BRISTOL OLD VIC and the OLD VIC in London, she went with the Old Vic company to New York in 1956 to play Cressida in Tyrone GUTHRIE's production of *Troilus and Cressida* in modern dress, and has since remained domiciled in the United States, where she toured from 1960 to 1962 in such roles as Ann Whitefield in SHAW's *Man and Superman* and Madame Arkadina in CHEKHOV's *The Seagull*. She returned to England in 1962 to appear at the CHICHESTER FESTIVAL THEATRE in its first season, being seen as Constantia in FLETCHER's *The Chances* and Penthea in FORD's *The Broken Heart*. She returned there in 1963, and later in the year played Ophelia in the NATIONAL THEATRE's inaugural production of *Hamlet*. After appearing in New York in a revival of KAUFMAN and Hart's *You Can't Take It With You* in 1965 and as Eleanor of Aquitaine in James Goldman's *The Lion in Winter* in 1966, she scored a big success back in London, playing three different characters in Neil SIMON's *Plaza Suite* (1969). In America she continued to show her versatility by appearing in PINTER's *Old Times* in 1971, as well as in revivals of Shaw's *Major Barbara* in 1972 (at the AMERICAN SHAKESPEARE THEATRE), Tennessee WILLIAMS's *A Streetcar Named Desire* in 1973, and Kaufman and Edna Ferber's *The Royal Family* in 1975. Back in London in 1981, she gave a highly praised performance in a revival of MILLER's *All My Sons*.

HARRISON, REX CAREY (1908–), English actor, who made his first appearance at the LIVERPOOL PLAYHOUSE in 1924 and was first seen in London in 1930. His good looks and elegant presence led almost inevitably to his being cast in a succession of featherweight parts in comedy, though his performances in RATTIGAN's *French Without Tears* (1936) and COWARD's *Design for Living* (1939) hinted at reserves of irony and anger. During the Second World War he was in the R.A.F., and he then went into films. He returned to the stage in 1948 to play Henry VIII in New York in Maxwell ANDERSON's *Anne of the Thousand Days*, and was in London in 1950, playing the Uninvited Guest in T. S. ELIOT's *The Cocktail Party*. He then returned to New York to star in VAN DRUTEN's *Bell, Book and Candle* (in which he was seen in London in 1954), and in the American productions of FRY's *Venus Observed* in 1952 and USTINOV's *The Love of Four Colonels* in 1953. In 1956 he was seen in his most famous role, Henry Higgins in *My Fair Lady*, a musical based on SHAW's *Pygmalion* by Lerner and Loewe, which he played for two years on Broadway before repeating the part in London. Employing a kind of *Sprechgesang* for the musical numbers, since he had very little singing voice, he showed exactly the right tetchy authority and charming intolerance for Shaw's language-loving Professor. Since then his stage appearances have been intermittent; he played the General in ANOUILH's *The Fighting Cock* in New York in 1959 and the title role in CHEKHOV's *Platonov* in London in 1960, and was later seen in Nigel Dennis's *August for the People* (1961), PIRANDELLO's *Henry IV* (N.Y., 1973; London, 1974) (in the title role), Rattigan's *In Praise of Love* (N.Y., 1974), and Shaw's *Caesar and Cleopatra* (N.Y., 1977). His second wife, the German-born actress Lilli Palmer [Peiser] (1914–), starred with him in *Bell, Book and Candle*, *Venus Observed*, and *The Love of Four Colonels*, and was also seen in London, where she first appeared in 1938, in Coward's *Suite in Three Keys* (1966). He later married the actress Kay Kendall (1926–59), and after her early death Rachel Roberts (1927–80), who appeared with him in *Platonov* and *August for the People*, her other roles including the title role in the musical *Maggie May* (1964) and the embittered wife in E. A. Whitehead's *Alpha Beta* (1972). His autobiography, *Rex*, was published in 1974.

HARRISON, RICHARD BERRY (1864–1935), American Negro actor, the son of slaves who escaped to Canada. He returned as an adult to Detroit, was befriended by L. E. Behymer, and after some training in elocution toured the Behymer and Chautauqua circuits with a repertory of Shakespearian and other recitations. He was working as a teacher of drama and elocution when he was persuaded to play 'De Lawd' in Marc CONNELLY's *The Green Pastures* (1930), in which he made an immediate success, appearing in the part nearly 2,000 times. Of medium build, with a soft but resonant voice, he was for the greater part of his life a lecturer, teacher, and arranger of festivals for Negro schools and churches. A modest man, he was surprised by his success as 'De Lawd'; but he enjoyed his contact with the professional theatre, his one great regret being that he never appeared in Shakespeare, whose works he knew so well.

HARRIS THEATRE, New York, see SAM H. HARRIS THEATRE and WALLACK'S THEATRE (2).

HART, CHARLES (?–1683), English actor, who as a boy actor before the closing of the theatres played the important role of the Duchess in SHIRLEY's *The Cardinal* (1641) with great success. He was a soldier during the Commonwealth and returned to the theatre in 1660, joining KILLIGREW's company and playing Celadon in DRYDEN's *Secret Love* (1667) opposite Nell GWYNN as Florimel. He was excellent in heroic parts, particularly as Alexander in LEE's *The Rival Queens* (1677), and it was said that once on stage he became so absorbed in his part that it was impossible to distract his attention. He retired on pension when the DORSET GARDEN and DRURY LANE companies were amalgamated in 1682.

HART, CHRISTINE, see SCHRÖDER, F. L.

HART, LORENZ, see MUSICAL COMEDY.

HART, MOSS (1904–61), American dramatist and director, who began his career as office boy to Augustus Pitou, a theatre impresario to whom he sold his first play *The Beloved Bandit*. His second, *Once in a Lifetime* (1930; London, 1933), was bought by Sam Harris, and successfully produced after extensive rewriting by George S. KAUFMAN. This led to a long collaboration between the two which resulted in such successes as *Merrily We Roll Along* (1934); *You Can't Take It With You* (1936; London, 1937), which gained a PULITZER PRIZE; and *The Man Who Came to Dinner* (1939; London, 1941). Hart also wrote plays on his own, of which the most interesting were *Winged Victory* (1943), about the Air Force,

Christopher Blake (1946), and *Light Up the Sky* (1948). His last play was *The Climate of Eden* (1952) based on Edgar Mittelholzer's novel *Shadows Move Among Them* and set on a missionary colony in British Guiana. His work for the musical stage included the sketches for the Irving Berlin revue *As Thousands Cheer* (1933), the libretto for the operetta *The Great Waltz* (1934), and the book for Rodgers and Hart's *I'd Rather Be Right* (1937), with Kaufman, and Kurt Weill's *Lady In the Dark* (1941). He directed a number of his own works, and, among those by others, Krasna's *Dear Ruth* (1944), Jerome Chodorov and Joseph FIELDS's *Anniversary Waltz* (1954), and the musicals *My Fair Lady* (1956; London, 1958) and *Camelot* (1960). In 1959 he published an outstanding autobiography, *Act One*, which unfortunately ended with the production of *Once in a Lifetime*. This play was revived with great success in London in 1979 by the ROYAL SHAKESPEARE COMPANY.

HART, TONY, see FEMALE IMPERSONATION and HARRIGAN, NED.

HART HOUSE THEATRE, a small playhouse, seating 500, situated within the students' building of the University of Toronto. It was built by the Massey Foundation under the direction of Vincent Massey, who was its chairman from its opening in 1919 until 1935. Its first director was Roy Mitchell, who had gained an outstanding reputation for his work in Toronto, first with the Arts and Letters Club and then with the Players Club, founded in 1913. Subsequently he had been with the PROVINCETOWN PLAYERS in the U.S.A. for two years. He mounted an adventurous series of productions at Hart House during its first two seasons, making use of the company from the Players' Club, and staged a number of new one-act plays by Canadian authors. He was succeeded as director by Bertram Forsyth (1921–5), Walter Sinclair (1925–7), Carroll Aikins (1927–9), Edgar Stone (1929–34), and Nancy Pyper (1935–7). When in 1935 Vincent Massey left to become Canadian High Commissioner in London, Hart House lost something of its importance, and after being used for two years by visiting companies it closed until 1946. It then became a university theatre and, under the direction of Robert Gill who remained there for 20 years, it trained a new generation of players who were to help build Canada's post-war professional theatre. The university's Graduate Centre for the Study of Drama was founded in 1966, and in 1971 Anne Saddlemyer became director of both the Centre and the theatre, with Martin Hunter as supervisor of productions. An interesting appointment to the Centre in 1976 was that of Ronald Bryden, formerly dramatic critic of the London *Observer* and *New Statesman* and literary manager of the ROYAL SHAKESPEARE COMPANY, who took an active part in the work of the theatre. To celebrate the university's 150th anniversary a season of four plays, including two new works—*Pontiac and the Green Man* by Robertson Davis and *The Dismissal* by James REANEY—was staged in 1977, with casts drawn from Hart House graduates who had become professional actors, and current students. In 1978 Michael Sidnell took over the direction of the theatre and the Drama Centre, but with the departure of Hunter in 1979 Hart House was left without a resident professional to direct its productions, so faculty members and graduate students assumed these responsibilities, and for the 1979–80 season, during Professor Sidnell's absence on sabbatical leave, Ronald Bryden was in charge.

HARTLEY [*neé* White], ELIZABETH (1751–1824), a beautiful English actress who became the favourite model of Sir Joshua Reynolds, three of whose portraits of her hang in the GARRICK CLUB. She made a brief appearance at the HAYMARKET THEATRE under Samuel FOOTE in 1769 and then went into the provinces, to reappear in 1772 at COVENT GARDEN, where the rest of her career was spent, in the title role of ROWE's *Jane Shore*, following it later the same year with the title role in William Mason's dramatic poem *Elfrida*. She was considered better in such tender and pathetic roles than in tragedy, her Lady Macbeth and Cleopatra (in DRYDEN's *All For Love*) not being highly thought of. She must, however, have been good in comedy, as she was the first to play Lady Touchwood in Mrs COWLEY's *The Belle's Stratagem* (1780). Her husband may have been a convenient fiction, as she was buried at Woolwich under her maiden name which also appears on her will.

HARTOG, JAN DE, see NETHERLANDS.

HARTZENBUSCH, JUAN EUGENIO (1806–80), Spanish dramatist, of German parentage, who was quite unknown when his second play, *Los amantes de Teruel*, a romantic tragedy based on an old Spanish legend, had an immense success. Produced at the Teatro del Príncipe in 1837, it remained in the repertory for many years. Although he continued to write plays, some adapted from old Castilian dramas, and to translate foreign plays into Spanish, he never again achieved such a success, though one play, on the great Spanish hero El Cid, *La jura en Santa Gadea* (1845), was well received, as was a play dealing with Judas and Dimas entitled *El mal apóstol y el buen ladrón* (*The wicked apostle and the good thief*, 1860). Hartzenbusch's admiration for the dramatists of the the Golden Age led him to edit, for the *Biblioteca de Autores Españoles* (1846–1880), the plays of TIRSO DE MOLINA, CALDERÓN DE LA BARCA, and RUIZ DE ALARCÓN and a selection of the plays of Lope de VEGA. He became director of the National Library in Madrid in 1862, retiring in 1875.

HARVARD UNIVERSITY, Cambridge, Mass., the oldest institution of higher learning in the U.S.A. The first plays known to have been acted there were ADDISON's *Cato*, Whitehead's *The Roman Father*, and OTWAY's *The Orphan*, performed surreptitiously by students. Acting was later encouraged, and the Hasty Pudding Club was formed, its productions now being mainly musicals. The Harvard Dramatic Club was started in 1908 and for many years produced only plays written by students or graduates of Harvard or Radcliffe (the women's college in Cambridge). It later concentrated on foreign plays. It was at Harvard that Professor G. P. BAKER inaugurated in 1905 his influential '47 Workshop' to stimulate the writing and production of new American plays. The Loeb Drama Center was built for Harvard in 1960. Its 556-seat auditorium was the first fully automated flexible theatre in the United States, offering a proscenium, thrust, or arena stage. There is also an experimental theatre seating 120. The Center is used by the Harvard-Radcliffe Dramatic Club and by the American Repertory Theatre, a professional company under the directorship of the distinguished critic Robert Brustein (1927–) which took up residence at Harvard in 1980, its opening seasons including highly praised productions of *A Midsummer Night's Dream*, BRECHT and Weill's musical *Happy End*, and Jules Feiffer's *Grown-ups*. The Harvard Theatre Collection, begun in 1901, is one of the finest performing arts research libraries in the world. It is a part of the Houghton Library. A descriptive catalogue of the engraved portraits in the collection was issued in four volumes (1930–4) under the editorship of Lillian A. Hall.

HARVEY, JOHN MARTIN-, see MARTIN-HARVEY.

HARWOOD, HAROLD MARSH (1874–1959), English dramatist and theatre manager, whose first play *Honour Thy Father* (1912) was successful enough to warrant his deserting medicine for the theatre. With his wife, Fryniwyd Tennyson Jesse (1889–1958), a great-niece of Lord TENNYSON and a well known author in her own right, he wrote *The Mask* (1913) and *Billeted* (1917), and then achieved a great success with his own play *The Grain of Mustard Seed* (1920). Of his later plays the best known is *The Man in Possession* (1930), long a favourite with repertory companies. It was first produced at the AMBASSADORS THEATRE, of which he was lessee from 1919 to 1932, producing there not only his own works but also such new plays as Lennox ROBINSON's *The White-Headed Boy* (1920) and Sacha GUITRY's *Deburau* (1921). In 1951 he again collaborated with his wife in the dramatization of her book *A Pin to See the Peepshow*, based on the Bywaters-Thompson murder case.

HARWOOD, JOHN EDMUND (1771–1809), American actor. He was for some years at the CHESTNUT STREET THEATRE, Philadelphia, under WIGNELL, and was with the company when it appeared in New York in 1797, being much admired in low-comedy parts. He was engaged by DUNLAP for the PARK THEATRE in 1803, and remained there until his death, making his first great success as Dennis Brulgruddery in the younger COLMAN's *John Bull; or, an Englishman's Fireside*. He was celebrated for his portrayal of Falstaff in *Henry IV*, which he first played at the Park Theatre in 1806 with COOPER as Hotspur. He later changed his style of acting, appearing as polished gentlemen, for which his fine presence and handsome countenance made him eminently suitable. He married the grand-daughter of Benjamin Franklin.

HARWOOD, RONALD, see ROYAL EXCHANGE THEATRE, MANCHESTER.

HASENCLEVER, WALTER (1890–1940), German dramatist and novelist, whose first play *Der Sohn* (1916) was one of the outstanding works of German EXPRESSIONISM, its conflict between father and son typifying the revolt of the younger generation against the life-denying authority of their elders. Equally outspoken were his anti-war plays *Antigone* (1917) and *Der Retter* (1919). His last play in this style was *Mord* (1926). He then turned successfully to comedy with *Ein besserer Herr* (*A Man of Distinction*, also 1926) and *Ehen werden im Himmel geschlossen* (*Marriages are Made in Heaven*, 1928). In 1933, because of his pacifist views, he was deprived of his German citizenship and drifted about Europe. A comedy, *Münchhausen*, written in 1934, was not performed until some years after his death, but an unpublished work, *Ehekomödie*, was seen in London in 1935 as *What Should a Husband Do?* On the outbreak of the Second World War he was interned in France, and committed suicide a year later.

HASENHUT, ANTON (1766–1841), Austrian actor, creator of the comic character THADDÄDL which, despite its roots in Viennese comic tradition, was essentially a sophisticated literary conception. Hasenhut, one of the most popular actors of the day, was much admired by GRILLPARZER, who wrote a poem in celebration of his art, but with the lack of suitable new vehicles and the declining demand for comic types he faded into obscurity after 1817.

HASSE, OTTO EDUARD (1903–78), German actor, who studied at the Max REINHARDT school in Berlin and began his career in Munich, going on to become one of the most popular figures of the German-speaking stage. In Berlin after the Second World War he was seen in a number of distinguished performances, among them Mephistopheles in GOETHE's *Faust* in 1945, Mr Antrobus in Thornton WILDER's *The Skin of Our Teeth* in 1946, Harras in ZUCKMAYER's *Des Teufels General* in 1948, and Domenico Soriano in DE FILIPPO's *Filumena Marturano* in 1953. One of his outstanding appearances was as Bernard SHAW in Jerome Kilty's *Dear Liar*, in which he toured throughout Germany in 1959 with Elisabeth BERGNER as Mrs Patrick CAMPBELL. He was also much admired as Winston Churchill in HOCHHUTH's *Soldaten* (1967), which he was the first to play.

HATTON GARDEN NURSERY, see NURSERIES.

HAUCH, JOHANNES CARSTEN (1790–1872), Danish dramatist, poet, and novelist, who as a young man was discouraged from writing by his friend OEHLENSCHLÄGER, whose work he greatly admired. Instead, he became first an academic scientist, then a student (and eventually a professor) of Aesthetics. His creative writing therefore came relatively late in his career, though it was to make him one of the most important figures of Danish Romanticism. His main achievements were in lyric poetry and the novel; among his more significant plays were *Svend Grathe* (1841), *Søstrene paa Kinnekullen* (*The Sisters of Kinnekullen*, 1849), *Marsk Stig* (1850), and *Tycho Brahes Ungdom* (*The Youth of Tycho Brahe*, 1852).

HAUPTMANN, GERHART (1862–1946), German dramatist, the chief playwright of NATURALISM, who in 1912 was awarded the NOBEL PRIZE for Literature. A Silesian by birth, he often used his native dialect to heighten the authenticity of his plays, of which the first, *Vor Sonnenaufgang (Before Dawn*, 1889), was produced in Berlin by the FREIE BÜHNE. Its depiction of a farmer's family which suddenly becomes wealthy, but sinks into degradation through alcoholism and sexual promiscuity, delighted some sections of the audience but outraged others by its uncompromising presentation of human and social misery. It was followed by two middle-class psychological dramas, *Das Friedensfest* (*The Coming of Peace*, 1890), the study of an unhappy and neurotic family, and *Einsame Menschen* (*Lonely People*, 1891), which examines the problem of marital incompatibility. Hauptmann's next play *Die Weber* (1892), based on the revolt of the Silesian weavers in 1844, was unusual in having as its hero a group of men instead of a single individual. It adheres closely to the theoretical tenets of naturalism, and displays dramatic qualities to a degree seldom achieved in this form: it is probably Hauptmann's best, as well as his best known, play. It was followed by a satirical comedy *Der Biberpelz* (*The Beaver Coat*, 1893), in which, and in its sequel *Der rote Hahn* (*The Red Rooster*, 1901), Hauptmann created some fine character parts, notably Mutter Wolffen, a scurrilous but appealing Berlin washerwoman who cunningly outwits Prussian bureaucracy. *Florian Geyer* (1896), a naturalistic excursion into historical drama based on the Peasants' Revolt in the time of Luther, presents a picture of Götz von Berlichingen very different from that of GOETHE. It was a failure when first performed, but was successfully revived in 1962. Of the other naturalistic plays written at this time, *Rose Bernd* (1903) shows a simple girl driven by circumstances and instinct to infanticide; *Führmann Henschel* (1898) and *Die Ratten* (1911) paint even darker pictures of man's enslavement to environment and circumstance, both ending with the suicide of the main character—a common feature of Hauptmann's dramas. But his considerable dramatic output, of about 40 plays, was not confined to naturalism. *Hanneles Himmelfahrt* (*The Assumption of Hannele*, 1893), with its interpolated dream sequence, marks the beginning of a transition to a more poetic and symbolist approach, while *Die versunkene Glocke* (*The Sunken Bell*, 1896) explores the Romantic theme of the creative artist's relationship with reality. Hauptmann's last dramatic work, written in the shadow of the Second World War, was a blank verse tetralogy on the doom of the Atrides. The themes of senseless bloodshed exacted by inscrutable powers and the subsequent madness of Orestes and Electra, induced by their enforced participation in the destruction of their family, reflect Hauptmann's pain at witnessing Europe disembowelling itself for the second time in his lifespan. The blackness of the tragedy is only relieved by Iphigenia's self-sacrifice in the first play *Iphigenie in Delphi* (1941), but the suggestion of reconciliation is tentative and ambiguous. A shortened version of the tetralogy was produced by PISCATOR at the Freie Volksbühne in 1962 in celebration of Hauptmann's centenary.

Several of Hauptmann's plays have been performed in English translation, *The Sunken Bell* in New York in 1900 and in London in 1907; *Hannele* (N.Y., 1910; London, 1924); *The Weavers* (N.Y., 1915; London, 1980). In 1933 Miles MALLESON's adaptation of *Vor Sonnenuntergang* (1932) as *Before Sunset* was seen in London with Werner KRAUSS as Matthew Clausen and Peggy ASHCROFT as Mrs Peters.

HAUPT- UND STAATSAKTIONEN, plays acted in German by the ENGLISH COMEDIANS and later by professional strolling players between 1685 and 1720, so named by GOTTSCHED to distinguish them from the comic after-pieces and to emphasize their historical aspect. They were usually concerned with the strong passions and elaborate intrigues of those in high places, kings, emperors, great conquerors. The plots were taken from all available European sources and rewritten to provide an unsophisticated audience with a feast of rhetoric, often crude and bombastic, and spectacular effects featuring executions, ghostly apparitions, weddings, and coronations, the whole interrupted at frequent intervals by comic interludes featuring the clown Pickelhering, later replaced by HANSWURST, whose colloquial chatter, vulgarity, and bawdy improvisations were intended to provide a comic contrast to the formal speech and stiff manners of the noble and courtly characters. The texts were not printed, the scripts being the property of the leader of the troupe that produced them. Fourteen Viennese examples of the genre survive from 1724, mostly on classical subjects and all featuring Hanswurst, whose inroads into the more serious parts of the play finally led to his condemnation by Gottsched and his banishment by Carolina NEUBER.

HAUSER, FRANK, see OXFORD PLAYHOUSE.

HAUTEROCHE, NOËL-JACQUES LE BRETON DE (*c.* 1616–1707), French actor and dramatist, who was at the MARAIS before going to the Hôtel de BOURGOGNE. Though MOLIÈRE made fun of him in *L'Impromptu de Versailles* (1663), he was a good actor and became one of the original members of the COMÉDIE-FRANÇAISE on its foundation in 1680, retiring in 1684 and being later stricken with blindness. He wrote a number of comedies reminiscent of Molière, some of which remained in the repertory until late in the 19th century. Among them were *Crispin médecin* (1670) and *Crispin musicien* (1674), written for BELLEROCHE, and *La Dame invisible* (1684), based on *La dama duende* by CALDERÓN, which many years later provided PRÉVILLE with one of his greatest successes.

HAVEL, VÁCLAV, see CZECHOSLOVAKIA.

HAVERGAL, GILES, see CITIZENS' THEATRE, GLASGOW.

HAWTREY, CHARLES HENRY (1858–1923), English actor-manager, knighted in 1922. A far better actor than his popularity in what became known as a 'Hawtrey' part ever allowed him to demonstrate, he was a fine light comedian and had no equal in the delineation of the man-about-town of his day. He always wore a moustache (the only time he shaved it off the play failed), and had perfect poise and sang-froid. He made his first appearance on the stage in 1881, under the name of Bankes, but soon went into management on his own account. In 1883 he adapted a German comedy by Von Moser as *The Private Secretary*. When tried out at the Prince of Wales (later the SCALA) Theatre in 1884 it was not a success, but transferred to the GLOBE THEATRE, with W. S. PENLEY in the title role first played by TREE, it ran for two years and was frequently revived. Hawtrey himself played Daniel Cattermole in this production. Among other plays which he directed and appeared in the most successful were R. C. CARTON's *Lord and Lady Algy* (1898), Ganthony's *A Message from Mars* (1899), Anstey's *The Man from Blankley's* (1901), George A. Birmingham's *General John Regan* (1913), and Walter HACKETT's *Ambrose Applejohn's Adventure* (1921). Hawtrey was for many years responsible for the production at Christmas of the children's play *Where The Rainbow Ends* by Clifford Mills and John Ramsey, first seen at the SAVOY THEATRE in 1911.

HÁY, GYULA [Julius] (1900–75), Hungarian playwright, who spent most of his life in exile. His early plays were produced in Berlin, where REINHARDT directed in 1932 *Gott, Kaiser, Bauer*, which was seen at the National Theatre in Budapest in 1945, the year in which Háy returned from Moscow. A village tragedy *Tiszazug*, first seen in Budapest in the same year, was frequently revived in Germany as *Haben*, and in 1955 was produced by Joan LITTLEWOOD at the Theatre Royal, Stratford, London, as *The Midwife*; it was also broadcast by the B.B.C. in 1969 as *To Have and To Hold*. *Az élet hidja*, staged in Budapest in 1951, was seen in London two years later as *The Bridge of Life*. Jailed for his part in the 1956 uprising, together with his fellow-playwright Tibor Déry, Háy was released in 1960 and again left Hungary, settling in Switzerland where he died. His later plays were all produced abroad, in German-speaking theatres; among them were *Das Pferd* (Salzburg, 1964), on the mad emperor Caligula, which was seen in Oxford later as *The Horse*; *Gaspar Varros Recht,* a savage attack on Communist bureaucracy, written in 1956 but not staged until 1965; *Attilas Nächte* (Bregenz, 1966), and *Mohács* (Lucerne, 1970).

HAY, IAN [John Hay Beith] (1876–1952), English novelist and playwright, and a distinguished army officer, whose first plays were dramatizations of his own light novels *Happy-Go-Lucky*, as *Tilly of Bloomsbury* (1919), and *A Safety Match* (1921). Much of his work was done in collaboration—*Good Luck* (1923) with Seymour HICKS, *A Damsel in Distress* (1928) and *Leave It to Psmith* (1930) with P. G. Wodehouse, *The Middle Watch* (1929) and *The Midshipmaid* (1931) with Stephen King-Hall, and *Orders are Orders* (1932) with Anthony Armstrong. In 1936 he dramatized Edgar WALLACE's novel *The Frog* and his own *The Housemaster*, and in 1947 again collaborated with King-Hall in *Off the Record*. He never attempted a serious note in his plays, which often bordered on farce, but his high spirits and good humour carried the plot along without a hint of offensiveness.

HAYDUK DRAMA, see BULGARIA.

HAYE, HELEN (1874–1957), a highly accomplished English actress of the old school, with superb carriage and diction. She was on the stage for 55 years, making her first appearance in 1898 in Hastings in T. W. ROBERTSON's *School*. After touring extensively, part of the time with Frank BENSON's Shakespeare company, she joined TREE's company at HER (then His) MAJESTY'S THEATRE, and made her first London appearance in 1910 as Gertrude in *Hamlet*, followed by

Olivia in *Twelfth Night*. In 1921 she toured Canada, and made her first appearance in New York in 1925 in LONSDALE's *The Last of Mrs Cheyney*. She excelled in costume and aristocratic parts such as Lady Sneerwell in SHERIDAN's *The School for Scandal*, which she played at the OLD VIC in 1935, and the Dowager Lady Monchesney in T. S. ELIOT's *The Family Reunion* (1939). She taught for many years at the Royal Academy of Dramatic Art, where her pupils included Flora ROBSON, Celia JOHNSON, Charles LAUGHTON, and John GIELGUD. In the 1940s she appeared in a number of new plays, including Margery Sharp's *The Nutmeg Tree* (1943) and Agatha CHRISTIE's *Murder on the Nile* (1946), and she made her last appearance as the Dowager Empress of Russia in Guy Bolton's adaptation of Marcelle Maurette's *Anastasia* (1953) at the ST JAMES'S THEATRE.

HAYES [Brown], HELEN (1900–   ), American actress, who made her first appearance on the stage at the age of five as Prince Charles in Robert Marshall's *A Royal Family*, played the lead in Frances Hodgson Burnett's *Little Lord Fauntleroy* at seven, the dual lead in Mark Twain's *The Prince and the Pauper* at eight, and graduated easily to adult roles from about 1920 onwards by way of Eleanor Porter's *Pollyanna* and Margaret in BARRIE's *Dear Brutus*. She made a great success in KAUFMAN and CONNELLY's *To the Ladies* (1922) and appeared for the THEATRE GUILD as Cleopatra in SHAW's *Caesar and Cleopatra* in 1925. In 1933 she again had an outstanding success as Mary Queen of Scots in Maxwell ANDERSON's *Mary of Scotland*, and she was much admired in 1940 as Viola in *Twelfth Night*. Her greatest triumph, however, was the title role in the American production of Laurence HOUSMAN's *Victoria Regina* in 1935, in New York and later throughout the United States. She made her first appearance in London in 1948 as the mother, Amanda, in Tennessee WILLIAMS's *The Glass Menagerie*. Back in New York she gave a moving performance as Nora Melody in O'NEILL's *A Touch of the Poet* (1958) at the Fulton Theatre, renamed the HELEN HAYES in her honour, and in 1961 undertook a world tour with *The Glass Menagerie*, WILDER's *The Skin of Our Teeth,* and William Gibson's *The Miracle Worker*. In the summer of 1962 she and Maurice EVANS appeared at the AMERICAN SHAKESPEARE THEATRE in Stratford, Connecticut, in *Shakespeare Revisited*, a programme of scenes from the plays which they then took on a nationwide tour, and in 1964 she formed the Helen Hayes Repertory Company to sponsor tours of Shakespeare readings in universities. She published her autobiography *On Reflection* in 1968.

HAYMARKET THEATRE, Leicester. This theatre, which was financed by the City Council, opened in 1973, and was intended to take the place of the Phoenix (see PHOENIX ARTS CENTRE), which had opened ten years earlier. Both theatres worked together, however, under the Leicester Theatre Trust, until 1979, when the Phoenix became independent. The Haymarket is larger than the Phoenix, seating 710, and in its first season presented two highly successful musicals, *Cabaret* and *Joseph and the Amazing Technicolor Dreamcoat*, together with a revival of FARQUHAR's *The Recruiting Officer*, John Hopkins's *Economic Necessity*, BRECHT's *The Caucasian Chalk Circle*, and a pantomime, *Aladdin*. Later seasons showed a similarly wide range, including productions of three notable musicals, *Oliver!, My Fair Lady*, and *Oklahoma!*, all of which later had long London runs, as well as two vastly different trilogies—SOPHOCLES' *Theban Plays* and AYCKBOURN's *The Norman Conquests*. The studio theatre has presented mainly modern plays, including PINTER's *Old Times*, BOND's *Saved*, Trevor Griffiths's *Comedians*, and David STOREY's *Home*. In 1980 it gave the first performance of Graham GREENE's *For Whom the Bell Chimes*. The Haymarket company tours for about 5 or 6 weeks annually, during which time the theatre houses visiting companies.

HAYMARKET THEATRE, London. In 1720 John Potter, a carpenter, built a 'Little Theatre in the Hay' on the site of the old King's Head tavern, the first recorded performance being given on 19 Dec. 1720 by a visiting French company. In 1729 the theatre, which was not licensed and could not stage 'legitimate' drama, had an unexpected success when Samuel Johnson of Cheshire put on a wild burlesque entitled *Hurlothrumbo; or, the Supernatural*, which proved so popular that it ran for 30 nights. In the 1730s the satires of Henry FIELDING, which attacked the Government and the Royal Family, brought notoriety to the theatre and led indirectly to the passing of the Licensing Act of 1737, which caused it to be closed down. It stood empty until in 1747 Samuel FOOTE took it over, evading the law by various ingenious methods. A feature of his entertainments was his mimicry of well known persons, which soon became all the rage. He had set his heart on obtaining a patent for his theatre, which by accident he did in 1766, and although it was valid for the summer months only the Haymarket became a Theatre Royal, a title which it still retains. Foote sold out in 1776 to the elder COLMAN, who made many improvements and launched the theatre on a period of prosperity, all the great actors of the day appearing there in the summer when DRURY LANE and COVENT GARDEN were closed. On 3 Feb. 1794 Colman was succeeded by his son, who unfortunately proved an indifferent manager, always in financial difficulties. In 1817 he was imprisoned for debt; his brother-in-law and partner, David Morris, carried on alone, and in 1820 built the present Haymarket a little to the south of the old building, which was replaced by shops and a café. The new building, designed by Nash, whose pedimented portico of six Corinthian columns extending over the pavement has survived reconstructions.

opened on 4 July 1821 with SHERIDAN's *The Rivals* and in 1825 had a great success with LISTON, always a prime favourite in comedy, in Poole's farce *Paul Pry*. In 1833 Julia Glover, who had already played Hamlet at the LYCEUM with some success, appeared as Falstaff in *The Merry Wives of Windsor*, with unfortunate results. In 1837 Benjamin WEBSTER became manager, and under him the theatre was substantially altered, gas lighting being installed (the Haymarket was the last theatre in London to use candles) and the forestage and proscenium doors abolished. The theatre prospered, PHELPS making his début there in 1837 and Barry SULLIVAN in 1853, while between those two dates most of the great players of the day were seen. One of the good new plays, in which Webster himself appeared, was *Masks and Faces* (1852) by Tom TAYLOR and Charles READE. A year later Webster was succeeded by BUCKSTONE, an excellent comedian whose ghost is said still to haunt the theatre. As Drury Lane at this time was little better than a showbooth and Covent Garden was given over to opera, the Haymarket became the leading playhouse of London. It was under Buckstone's management that the American actor Edwin BOOTH made his first appearance in London in 1861, the year in which E. A. SOTHERN also came from the United States to appear as Lord Dundreary in Taylor's *Our American Cousin*. Sothern made a further success in 1864 in *David Garrick*, by the then unknown T. W. ROBERTSON. Buckstone retired in 1879 and the BANCROFTS took possession. They remodelled the interior of the theatre, doing away with the pit and taking the stage back behind the proscenium arch, thus making it the first picture-frame stage in London, and opened on 31 Jan. 1880 with a revival of BULWER-LYTTON's *Money*. They ran the theatre most successfully, adding immensely to its prestige, until they retired in July 1885. In the autumn of 1887 the theatre passed into the hands of TREE, under whom Oscar WILDE's *A Woman of No Importance* (1893) and *An Ideal Husband* (1895) were first produced. His greatest success, however, was George Du Maurier's *Trilby* (also 1895) with himself as Svengali and Dorothea Baird, later the wife of H. B. IRVING, as Trilby; George's son Gerald DU MAURIER played a small part in this production. A year later Tree moved to HER MAJESTY's and Cyril MAUDE took over the Haymarket with Frederick Harrison, opening on 17 Oct. 1896 with *Under the Red Robe* (from Stanley Weyman's novel). After a long period of success marked by fine acting in good plays with splendid settings, Maude withdrew in 1905, having overseen the reconstruction of the interior of the theatre, but Harrison carried on until his death in 1926, being succeeded by Horace Watson, and he in turn by his son and grandson. Among the outstanding successes of this period were *Bunty Pulls The Strings* (1911), by Graham Moffat; BARRIE's *Mary Rose* (1920), with Fay COMPTON, who returned in 1925 in Ashley DUKES's *The Man With a Load of Mischief*; *Yellow Sands* (1926) by Eden and Adelaide PHILLPOTTS; St John ERVINE's *The First Mrs Fraser* (1929), with Marie TEMPEST; and *Ten-Minute Alibi* (1933) by Anthony Armstrong, transferred from the EMBASSY. The theatre escaped damage from enemy action during the Second World War and in 1944–5 housed a fine company in repertory under John GIELGUD. In 1948 the American actress Helen HAYES made her first appearance in London in Tennessee WILLIAMS's *The Glass Menagerie*. Later successful productions included *The Heiress* (1949), a dramatization of Henry JAMES's novel *Washington Square*; N. C. HUNTER's *Waters of the Moon* (1951) and *A Day by the Sea* (1953); Enid BAGNOLD's *The Chalk Garden* (1956); Robert BOLT's *Flowering Cherry* (1957); RATTIGAN's *Ross* (1960), with Alec GUINNESS as T. E. Lawrence; and Peter Luke's *Hadrian VII* (1969), transferred from the MERMAID. All-star revivals then held the stage until in 1971 Guinness returned in John MORTIMER's *A Voyage Round My Father*. A year later Royce Ryton's *Crown Matrimonial*, on the abdication of Edward VIII, opened to mixed reviews but had a long run. The theatre housed transfers of the CHICHESTER FESTIVAL THEATRE revivals of MAUGHAM's *The Circle* (1976), *Waters of the Moon* (1978), and SHAW's *The Millionairess* (also 1978). Michael FRAYN's *Make and Break* (1980) was a highly successful transfer from the LYRIC, Hammersmith, and in 1981 Maggie SMITH was seen in Edna O'Brien's *Virginia*, about Virginia Woolf, a role she had created at the STRATFORD (ONTARIO) FESTIVAL in 1980.

HAZLITT, WILLIAM (1778–1830), English writer, the first of the great dramatic critics, a contemporary of Leigh HUNT, COLERIDGE, and LAMB. From 1813 to 1818 he reviewed plays for the *Examiner*, the *Morning Chronicle*, the *Champion*, and *The Times*, and it was his good fortune to live in an age of great acting, displayed chiefly in the plays of Shakespeare, since the period coincided with a dearth of good playwrights. In his zeal for bygone dramatists Hazlitt was once led to say he loved the written drama more than the acted, but he nevertheless took a vivid delight in acting and much enjoyed the society of actors. He was the author of several books on Elizabethan drama and a selection of his criticisms was published in 1818.

HEAVY FATHER, LEAD, WOMAN, see STOCK COMPANY.

HEBBEL, FRIEDRICH (1813–63), German dramatist, uneasily poised between idealism and realism. Obsessed by the tragedy of life, he probed ceaselessly for the cause of that tragedy, finding it less in the realm of guilt and human frailty than in the very process of life and progress. He was greatly influenced by Hegel, and much of his work, particularly his studies of wronged women, anticipates the plays of IBSEN and STRINDBERG. His first play, *Judith* (1840), shows his heroine torn between her duty to kill Holofernes and the sexual passion which he arouses in her. It was followed by a powerful middle-class tragedy, *Maria Magdalena* (1844), a fierce tragedy of jealousy in *Herodes und Mariamne* (1849), *Agnes Bernauer* (1852), again a study of unhappy love, and by *Gyges und sein Ring* based on the legend of King Cambyses, pub. in 1855 but not performed until 1869. Hebbel's last work was a trilogy, *Die Nibelungen* (1861), which was revived by FEHLING in 1924. *Judith* was revived by REINHARDT in 1910 with Paul Wegener and Tilla Durieux, and in 1972 there was a revival of *Maria Magdalena* in Cologne. Hebbel was helped in his early, difficult years by his mistress Elise Lensing, who supported him and bore him two children. Ironically, in view of his preoccupation with the wrongs of women, he left her to marry in 1846 the actress Christine Enghaus (1817–1910), in whom he found the ideal interpreter of his heroines.

HÉBERTOT, JACQUES (1886–1970), French theatre director, manager, and critic, who introduced many important foreign plays to Paris. In 1919 he took over the COMÉDIE DES CHAMPS-ÉLYSÉES and the adjacent Théâtre des Champs-Élysées, appointing JOUVET director of both in 1922. A year later he founded, under BATY, the Studio des Champs-Élysées. He abandoned management for criticism in 1926, but in 1940 returned, taking over the Théâtre des Arts, which until 1906 had been known as the Théâtre des Batignolles from its opening in 1838. As the Théâtre Hébertot it continued the excellent work done by the Théâtre des Arts, particularly under PITOËFF in the mid-1920s, and remains a well-known Paris theatre, having seen, among other outstanding productions, the first performance of CAMUS's *Caligula* in 1945, of MONTHERLANT's *Le Maître de Santiago* in 1947, and of Georges Bernanos's *Dialogues des Carmélites* in 1952.

HEBREW THEATRE, The New. Although plays in Hebrew had formed part of the repertory of the JEWISH DRAMA since the 16th century, there was no permanent Hebrew theatre until the 20th. The popularity of Yiddish drama made it clear that the revival of spoken Hebrew aimed at by the Zionists could be encouraged by the provision of Hebrew drama on the stage, and in 1907 two groups were founded for the purpose of providing such drama, one in Warsaw, under Isaac Katzenelson (1886–1943), the other in Jaffa. This second group, consisting of amateurs who called themselves 'Lovers of Dramatic Art', hoped to encourage the use of the Hebrew language among the small Jewish community in their city; they flourished until the outbreak of the First World War in 1914, and the consequent expulsion of the Jews from Jaffa by the Turkish authorities. Meanwhile, in Bialystok, another Hebrew company had a short but active life under Nahum Zemach (1887–1939), who in 1917 founded in Moscow, with Jewish actors from the old Warsaw and Bialystok companies, the important and still flourishing HABIMAH THEATRE. The ultimate intention of the group was to settle in Palestine, but before it did so in 1932 several tentative efforts had been made after the war to establish a Hebrew company there. The most successful of these was the Teatron Ivri, founded in 1921 by David Davidov, a Yiddish-speaking actor from England. With actors who had had little or no training, few designers and technicians, and a potentially small audience, Davidov nevertheless produced a new show every ten days, drawing his plays from the international repertory and hastily translating them into Hebrew. His company survived until 1927, by which time Menahem Gnessin (1882–1953), who had been one of the founder-members of Habimah, had brought the Teatron Erez Israeli from Berlin to Tel-Aviv, where in 1923 he opened the first permanent theatre building in the country, and Moshe Halevy (1895–1974) had founded in 1926 the OHEL THEATRE, which was destined to play an important part in future developments. There was also a small experimental company run by Isaac Daniel (1895–1935), which with very limited resources worked along lines suggested by its founder's experiences in the theatres of France and Germany, particularly under Max REINHARDT.

Most of these undertakings had fairly short lives, as did the Kumkum (Teapot) group, a satirical theatre founded in 1927, which took as its targets the British mandatory government and the Zionist leadership. After it closed some of the members of the group founded Mat'ateh (the Broom), which also acted as the spokesman of the Jewish community against the British. After 1948, its chief purpose in life having been taken from it by the establishment of the State of Israel, it tried unsuccessfully to become a straight comedy theatre and disbanded in 1954.

When in 1928 Habimah paid its first visit to Palestine it was already world famous and had become, in the eyes of the Jewish people, the symbol of the revival of Jewish culture and its ancient Hebrew language. Initially the contemporary Hebrew theatre had perforce to be almost solely a theatre of production, playing international and Jewish classics. It was not until the 1930s that a few new plays appeared; and these took their subjects mainly from the Old Testament. Typical of many of them was Aaron Ashman's *Mihal, the Daughter of Saul* (1941), with its melodramatic plot, stereotyped characters, and long monologues in pseudo-classical style. These plays were followed after the Second World War by works written under the influence of SOCIALIST REALISM, glorifying the pioneers who were fighting to establish a new country, but the War of Independence in 1948 brought forth a new type of play, in which heroics were replaced by a matter-of-fact approach to the problems faced by young people on whom a war had been imposed. The first and most significant of these was Moshe Shamir's *He Walked Through the Fields* (1948), the first to put on the stage a *sabra*, or native-born Israeli, who disdains the romanticism of his Zionist parents and performs acts of self-sacrifice without considering himself a hero. Writers who followed Shamir in portraying contemporary problems included Igal Mossinsohn, whose *Casablan* (1954) dealt with the difficulties facing a Moroccan immigrant labourer; while humour in daily life provided the material for Ephraim Kishon's *The Marriage Contract* (1961). Among the dramatists of the late 1950s and 1960s who were influenced by the European Theatre of the ABSURD, the most important was Nissim ALONI. There was also a revival of Biblical plays, but the tone was mainly a mocking one, as if the young playwrights wanted to lighten the burden of being the 'People of the Book'. In Yosef Shabtai's *A Crown in the Head* (1970), for example, King David is presented as a senile old man, impatient with affairs of state and chiefly interested in the young girl who warms his bed. The early 1970s also produced Hanoch LEVIN, who is sharply critical of Israeli society and in general displays a profound pessimism. In spite of the increase in the number of new playwrights, the majority of plays presented by the Israeli theatres are translations from the international repertory, classical and contemporary.

Along with the change in the content of Israeli plays went a profound change in the style of playing and presentation. For many years the theatre in Palestine lay under the shadow of VAKHTANGOV's production of ANSKY's *The Dybbuk* for Habimah. By 1939 the so-called 'STANISLAVSKY' style, which actually appeared more closely related to that of MEYERHOLD before the Revolution, had degenerated into formalism, and by the end of the Second World War the need for a new orientation had become obvious. The need was answered by the foundation in 1944, by Joseph Millo, of the CAMERI THEATRE in Tel-Aviv, whose light, improvisatory style, and fluent, colloquial Hebrew enchanted the younger generation, as well as the immigrants from Western Europe who found the Russian-inspired declamatory acting and old-fashioned classical Hebrew of the established companies hard to take. The Cameri flourished, and is now the municipal theatre of Tel-Aviv.

The early 1960s, which also saw the establishment of a municipal theatre in Haifa, saw the rise of the commercial theatre on a fairly large scale. The Godik Theatre achieved a great success in 1961 with its production of the musical *My Fair Lady*, and followed it with a series of other musicals and light plays. Other commercial theatres on a smaller scale came into being, and the Zavta Club in Tel-Aviv, sponsored by the left-wing Mapam party, presented avant-garde and experimental plays for small audiences, while other small groups gave occasional performances for young people in out-of-the-way places. In 1970 the municipality of Jerusalem erected a magnificent theatre building, which has no company of its own but houses productions from Tel-Aviv and Haifa, and a year later the Jerusalem Khan, built in the ruins of an ancient caravanserai, began to present under the direction of British-born Michael Alfreds an imaginative programme of documentaries, adaptations, and straight plays which stressed the physical rather than the verbal aspects of play production.

The most interesting aspect of the Hebrew theatre is the size of its audience. A survey made in 1972 revealed that as many as two-thirds of the population attend the theatre regularly, 15% of them at least once a month. One explanation of this phenomenon is the system which calls on all theatre companies to travel; Israel being a small country, any town, no matter how far removed from the centre, can be visited by one group or another. A contributory factor to this support for the theatre is the nature of Israel's rural population, which consists for the most part of members of *kibbutzim*, collective settlements which lay great stress on cultural life, and therefore readily organize visits from visiting companies or to theatres in neighbouring towns.

HEDBERG, TOR (1862–1931), Swedish dramatist and critic, son of Franz Hedberg, a popular dramatist of the mid-19th century, whose interest in the lives of the working population of Stockholm the younger man shared. Tor's earliest works were novels and stories, but he began writing plays in 1886, the best of these early efforts being perhaps *En Tvekamp* (*A Duel*, 1892). His later work ranges from comedies like *Nattrocken* (*The Dressing Gown*, 1893) to dramatic poems—*Gerhard Grim* (1897)—and tragedies such as *John Ulfsjerna* (1907) and *Borga gård* (1915). In 1910 he became a director of the KUNGLIGA DRAMATISKA TEATERN in Stockholm, but continued a steady output of plays, of which the most notable was probably *Perseus och vidundret* (*Perseus and the Monster*, 1917).

HEDGEROW THEATRE, Moylan, Pennsylvania, founded in 1923, second oldest repertory company in the United States, presenting a wide range of classical and modern works. With a semi-professional company, it performs all the year round in a converted grist mill dating from 1840 which seats 136. It staged the world premières of such American plays as *Winesburg, Ohio* (1934), a dramatization of Sherwood Anderson's novel of that name; Jane Bowles's *In the Summer House* (1951); Alex Newell's *White Clouds, Black Dreams* (1973); and Bertrand Russell's *Low on High* (1978). It has toured throughout the United States, and runs a theatre school for actors.

HEDQVIST, IVAN (1880–1935), Swedish actor, active in Stockholm's Svenska Teatern (1906–9 and 1922–5), and in the KUNGLIGA DRAMATISKA TEATERN (1909–19 and 1927–32). He was a powerful representative of the newly emergent naturalistic acting of his day, his more important roles including SOPHOCLES' Oedipus and the King in *Hamlet*.

HEIBERG, GUNNAR EDVARD RODE (1857–1929), Norwegian dramatist, whose work shows the influence of IBSEN's later style. He was for some years (1884–8) director of Det Norske Teatret in Bergen, and in addition to his dramas wrote a number of volumes of dramatic criticism. His first play, *Tante Ulrikke* (1884), was a deftly satirical contribution to contemporary 'problematic' literature. This was followed by *Kong Midas* in 1890. The two plays by which he is best known—*Balkonen* (*The Balcony*, 1894) and *Kjærlighedens Tragedie* (*The Tragedy of Love*, 1904)—explore in a rather heavy-handed fashion the nature of sexual passion. His ability to write light witty dialogue is perhaps best in evidence in *Kunstnere* (*The Artists*, 1893) and *Gerts have* (*Gert's Garden*, 1894). His six later plays did not greatly add to his stature.

HEIBERG, JOHAN LUDVIG (1791–1860), Danish poet, dramatist, and critic, son of P. A. Heiberg (1758–1841), a writer preoccupied with revolutionary causes. After graduating, he spent six years abroad before settling in Copenhagen, where he became one of the leaders of the city's intellectual and cultural life and was for a time director of the KONGELIGE TEATER. In criticism he was influential both as a theorist (drawing heavily on Hegel) and as a practical journalist on the *Kjøbenhavns flyvende Post* (*Copenhagen Flying Mail*). He introduced and established the style of the French VAUDEVILLE in Denmark, laying the theoretical foundations with *Om Vaudevillen som dramatisk Digtart* (*On Musical Comedy as a Dramatic Genre*, 1826).

Although Heiberg's own dramatic output began as early as 1814 with *Marionetteater*, and included over the following years popular successes such as *Julespøg og Nytaarsløjer* (*Christmas and New Year Fun and Games*, 1816) and the vaudeville *Kong Salomon og Jørgen Hattemager* (*King Solomon and Jørgen the Hatter*, 1825), his reputation as a dramatist was finally made secure in 1828 with *Elverhøj* (*Elfin Hill*), a romantic blend of realism and fantasy, which retains its popularity today. Of his later works, *Fata Morgana* (1838), *Syvsoverdag* (*Day of the Seven Sleepers*, 1840), and *En Sjæl efter Døden* (*A Soul After Death*, 1841)—the last an elegantly satirical verse drama and the pinnacle of his achievement—are the most noteworthy.

HEIBERG [*née* Pätges], JOHANNE LUISE (1812–90), Danish actress and dramatist, who married J. L. HEIBERG in 1831. After her début at the KONGELIGE TEATER in Copenhagen in 1826, she revealed a very versatile talent, ranging from high tragedy—she was a notable Lady Macbeth—to comedy and VAUDEVILLE. In her own generation, until her retirement from the stage in 1864, she was supreme among Danish actresses. Her own vaudevilles—*En Søndag paa Amager* (*A Sunday in Amager*, 1843), *Abekatten* (*The Monkey*, 1849), and *En Sommeraften* (*A Summer Evening*, 1853)—were slight pieces, but popular.

HEIJERMANS, HERMAN (1864–1924), Dutch dramatist, and the first since VONDEL to become well known outside his own country. Under the influence of the new NATURALISM, he set out to depict the hypocrisy of contemporary bourgeois morality and the miseries of racial minorities and the working class. His first successful play *Ahasverus* (1893), produced by ANTOINE at his THÉÂTRE-LIBRE in Paris soon after the Amsterdam première, and *Ghetto* (1898), which was seen in London and New York in 1899, both dealt with the sufferings of the Jewish people in Russia and Amsterdam—Heijermans was himself a Jew. Although many of his plays aroused considerable controversy, they were eagerly awaited by Dutch audiences. Among the later one-act plays was *In der Jong Jan* (1903), written for the Dutch actor Henri de Vries, who took it to New York in 1905 as *A Case of Arson*, playing all seven witnesses of the crime himself. Heijermans's masterpiece was undoubtedly the three-act drama of the sea, *Op Hoop van Zegen* (1900), which dealt with the exploitation of poor Dutch fishermen by greedy shipowners. As *The Good Hope* it was seen in London in 1903, with Ellen TERRY as the old mother Kniertje, and in 1927 was produced for the Civic Repertory Company by Eva LE GALLIENNE, who played Jo.

HEINRICH JULIUS, Duke of Brunswick, see WOLFENBÜTTEL.

HELBURN, THERESA, see LANGNER, LAWRENCE.

HELD, ANNA, see JEWISH DRAMA and ZIEGFELD, FLORENZ.

HELEN BONFILS THEATRE COMPLEX, Denver, see BONFILS THEATRE.

HELEN HAYES THEATRE, New York, at 210 West 46th Street, between Broadway and 8th Avenue. Originally a theatre restaurant which failed, it was remodelled and opened as the Fulton Theatre, with a seating capacity of 1,160, on 20 Oct. 1911. It had its first success with Hazelton and Benrimo's *The Yellow Jacket*; equally successful, though in a more serious vein, was BRIEUX's *Damaged Goods* (1913). *Abie's Irish Rose* (1922) by Anne Nichols, began its long run at the Fulton before moving to the REPUBLIC, and among later productions were LONSDALE's *The High Road* (1928), the Stokes's *Oscar Wilde* (1938), with Robert MORLEY in the title role, and Kesselring's *Arsenic and Old Lace* (1941). A play on the Negro problem, *Deep Are the Roots* (1945) by Arnaud d'Usseau and James Gow, was also seen at the Fulton, which on 21 Nov. 1955 was given its present name in recognition of the stage jubilee of Helen HAYES, who appeared there in 1958 as Nora in O'NEILL's *A Touch of the Poet*. In 1961 *Mary, Mary* by Jean Kerr started a three-year run, to be followed by the same author's *Poor Richard* (1964). An adaptation of Muriel Spark's novel *The Prime of Miss Jean Brodie* (1968) starred Zoë CALDWELL, and Alec MCCOWEN played the title role in Peter Luke's *Hadrian VII* (1969). *The Me Nobody Knows* (1970) was an unusual musical based on the writings of New York tenement children. Later productions included Royce Ryton's *Crown Matrimonial* (1973), a revival of George S. KAUFMAN and Edna Ferber's *The Royal Family* (1975), and *Strider: the Story of a Horse* (1979), a musical based on a short story by TOLSTOY, which transferred from the off-Broadway Westside. The theatre was demolished in 1982, together with the nearby MOROSCO THEATRE.

HELLENISTIC AGE, the period following the conquests of Alexander. In the Greek or Hellenizing cities which sprang up everywhere in the Near and Middle East in the later 4th and early 3rd centuries BC, the decline in the political influence of Athens was matched by the increase of its cultural prestige. A theatre became an indispensable public building, and it was often of great beauty: it is from this period that most of the extant theatre buildings date. Little original drama of any merit was produced in them except, early in the period, the NEW COMEDY in Athens, but MIME flourished, with HERODAS and THEOCRITUS as its most notable exponents, and in Alexandria there was an artificial revival of tragedy among a group of seven writers known as the Pleiad (from the seven bright stars in the Pleiades) of whom LYCOPHRON is the best known. The great contribution of this age to drama, and indeed to literature as a whole, was the work of Alexandrian scholars in collecting, annotating, and preserving the texts of the great dramatists of the 5th century BC. The tragedies (not the comedies, which were too purely Athenian) became established as classics and were regularly performed. More attention was paid to acting, the use of MASKS was modified for the new audience, and purely spectacular productions became common.

HELLINGER, MARK, see MARK HELLINGER THEATRE.

HELLMAN, LILLIAN (1905–84), American dramatist, who entered the theatre as a press agent and as a play-reader for the Broadway producer Norman Shumlin. Her first play, *The Children's Hour* (1934; London, 1936), aroused extraordinary interest with its story of a neurotic schoolgirl's defamation of her teachers. *The Little Foxes* (1939; London, 1942), the study of a predatory family of industrial entrepreneurs, and *Watch on the Rhine* (1941; London, 1942), which implied that America would soon have to join the fight against Fascism, fulfilled the promise of her début, as did *The Searching Wind* (1944), which exposed the errors of an American career diplomat. *Another Part of the Forest* (1946) returned to the antecedent history of her 'little foxes' with a Jonsonian picaresque comedy of villains outsmarting one another. *The Autumn Garden* (1951) was a powerful group study in frustration and *Toys in the Attic* (New York and London, 1960) a searing study of failure and possessiveness. She also adapted a number of novels for the stage, among them VOLTAIRE's *Candide* (1956; London, 1959) and Blechman's *How Much?* as *My Mother, My Father and Me* (1963). In the 1950s she was summoned before the House Committee on UnAmerican Activities: her book *Scoundrel Time* (1976) describes this era. She wrote two other autobiographical works, *An Unfinished Woman* (1969) and *Pentimento* (1973). Revivals of *Watch on the Rhine* were seen in London (at the NATIONAL THEATRE) and New York in 1980, while Elizabeth Taylor starred in a revival of *The Little Foxes* in New York in 1981 and London in 1982.

HELPMANN, ROBERT MURRAY (1909–), Australian-born ballet dancer, choreographer, actor, and director, knighted in 1968, who successfully made the transition from dancing to acting, bringing to both careers the same meticulous attention to detail and artistic integrity. He made his first appearances, both as dancer and actor, in Sydney, and in 1933 was in London, where he joined the Vic-Wells Ballet, being principal dancer of the Sadler's Wells Ballet from its inception until 1950. He first came into prominence as an actor in London when he played Oberon in *A Midsummer Night's Dream* at the OLD VIC at Christmas 1937, repeating the part the following Christmas. In 1944 he appeared with the company at the NEW THEATRE as Hamlet, and in 1947, in partnership with Michael Benthall, took over the DUCHESS THEATRE, where he was seen as Flamineo in WEBSTER's *The White Devil* and Prince in ANDREYEV's *He Who Gets Slapped*. He then went to the SHAKESPEARE MEMORIAL THEATRE, Stratford-upon-Avon, where in 1948 he again played Hamlet, as well as Shylock in *The Merchant of Venice*, and King John. He returned to the London stage in 1951 to play Apollodorus in SHAW's *Caesar and Cleopatra* and Octavius Caesar in Shakespeare's *Antony and Cleopatra*, with Laurence OLIVIER and Vivien LEIGH, the two plays being given in repertory as part of the Festival of Britain and then transferring to New York. On his return he was seen as the Doctor in Shaw's *The Millionairess*

(1952) on its first West End appearance, playing opposite Katharine HEPBURN and again going with the company to New York. With the Old Vic company, which he directed in a number of plays including ELIOT's *Murder in the Cathedral* in 1953, he appeared in a wide variety of Shakespearian parts, his versatility encompassing in 1955 Petruchio in *The Taming of the Shrew* and Angelo in *Measure for Measure*, and in 1956–7 Launce in *The Two Gentlemen of Verona* and Pinch in *The Comedy of Errors* as well as the title role in *Richard III*. Among his non-Shakespearian parts were Georges de Valera in SARTRE's *Nekrassov* (in Edinburgh and London) in 1957, and Sebastien in COWARD's *Nude With Violin*, which he took over from GIELGUD in the same year and in which he toured Australia. He also directed GIRAUDOUX's *Duel of Angels* (1960) and the musical *Camelot* (1964). From 1965 to 1976 he was director of the Australian Ballet, and in 1970 was appointed artistic director of the Adelaide Festival, returning briefly to London in 1971 to direct a new production, several times revived, of BARRIE's *Peter Pan*.

HELTAI, JENÖ, see HUNGARY.

HEMINGE [HEMINGES, HEMMINGS], JOHN (1556–1630), English actor, a member of the CHAMBERLAIN'S MEN and probably the first to play Falstaff. When the company was reorganized in 1603 as the KING'S MEN he appears to have acted as their business manager, and probably for this reason was instrumental with his fellow-actor Henry CONDELL in arranging for the printing of Shakespeare's plays in the one-volume edition—the First Folio, containing 36 plays—published in 1623. He and his joint editor have been criticized on several counts, including their arbitrary division of each play into five acts, on the classic model favoured by Ben JONSON, and other sins of omission and commission which have been at the root of much literary controversy; but without their efforts the plays might well have been lost, as the author himself made no move to preserve them and no manuscript indisputably by Shakespeare has been found.

HEMP HOUSE, see HANDWORKING.

HENDERSON, JOHN (1747–85), English actor, whose passion for the stage was fostered by GARRICK. He was unsuccessful in his application to join the company at DRURY LANE, probably because of his weak voice and unprepossessing appearance, due mainly to his extreme poverty. He was, however, determined to act, and after he had appeared in several amateur productions in which he imitated Garrick to the life, the latter gave him a letter of recommendation to John PALMER at Bath. He made his first appearance there, as Hamlet, on 6 Oct. 1772, and was successful enough to be retained in the company. Several years of hard work followed, during which Gainsborough painted him as Macbeth, a portrait which shows him to have been a stoutly-built, fair-haired man with a strong, determined face—not handsome, but commanding. His determination and hard work were rewarded when in 1777 the elder COLMAN engaged him to play Shylock in *The Merchant of Venice* at the HAYMARKET THEATRE in London; even MACKLIN, the greatest Shylock of the day, congratulated him. He then returned to Bath, but a year later SHERIDAN decided that he wanted him for Drury Lane, so badly that in order to persuade Palmer to release him he granted the manager the sole right of playing *The School for Scandal* in Bath. Henderson stayed at Drury Lane for two years, going from there to COVENT GARDEN, and never again left London except on tour. One of his greatest parts was Falstaff, both in *Henry IV* and in *The Merry Wives of Windsor*. He died young, of overwork and early privations, and was buried near Garrick in Westminster Abbey.

HENRY, JOHN (1738–94), Irish-born American actor who was for many years one of the leading men of the AMERICAN COMPANY, which he joined in 1767 at the JOHN STREET THEATRE, New York, under DOUGLASS. On the return of the American Company from the West Indies, where they took refuge during the War of Independence, Henry assumed the management, jointly with the younger HALLAM. He accepted and played in the first of DUNLAP's plays to be produced, *The Father; or, American Shandyism* (1789), and was responsible in 1792 for the importation from England of John HODGKINSON, by whom he was soon forced into the background, retiring shortly before his death. He was twice married, his first wife, an actress named Storer, being lost at sea before the company came back from the West Indies. After living for some time with her sister Ann or Nancy (1749–1816) (later Mrs Hogg), by whom he had a child, Henry married a younger member of the family, Maria STORER.

HENRY MILLER'S THEATRE, New York, at 124 West 43rd Street east of Broadway. Built for the actor Henry MILLER, this opened on 1 Apr. 1918 with *The Fountain of Youth*, a translation of *Flor de la vida*, by the brothers ÁLVAREZ QUINTERO. After achieving a modest success, the theatre embarked on the 'sophisticated' drama with which its name was usually connected, beginning with Noël COWARD's *The Vortex* (1925). In 1929 SHERRIFF's *Journey's End* had a long run, and *The Late Christopher Bean* by Sidney HOWARD had its first production in 1932. The THEATRE GUILD produced O'NEILL's *Days Without End* in 1934; a revival of WYCHERLEY's *The Country Wife*, with designs by Oliver MESSEL, was seen in 1936; and in 1937 RATTIGAN's *French Without Tears* failed to repeat its London success. More successful were Thornton WILDER's *Our Town* (1938), Norman Krasna's *Dear Ruth* (1944), T. S. ELIOT's *The Cocktail Party*

(1950), Saul Levitt's *The Andersonville Trial* (1961), and the ROYAL SHAKESPEARE COMPANY's recital *The Hollow Crown* in 1963. In 1969 the building became a cinema.

HENSEL [*née* Sparmann], SOPHIE FRIEDERIKE (1738–89), German actress, in the company of ACKERMANN at the Hamburg National Theatre. Built on generous lines, with a face and figure of majestic beauty, she was admirable as the noble heroines of German classical drama. Even LESSING, who detested her, had to admit that she was a fine actress. In private life she was malicious and intriguing, her character contributing to the downfall of the Hamburg enterprise. She married as her second husband Abel SEYLER, and effectively prevented him from making any kind of success.

HENSHAW, JAMES ENE (1924–), a physician in Eastern Nigeria, who wrote the first Nigerian play to be published abroad, *This Is Our Chance* (1956), like his next play *The Jewels of the Shrine* (also 1956) based on a simple village folktale. His later works, which include *A Man of Character* (1960), about an official who chooses poverty rather than accept bribes, and *Medicine for Love* (1964), in which a politician, in the middle of an election, finds on his doorstep three 'wives' sent him by well-intentioned relatives, utilize conventional European dramatic techniques, as does *Dinner for Promotion* (1967), about two young clerks trying to to outdo each other in order to impress their employer. Henshaw's plays, simple in construction and unpretentious in language, are perfectly adapted to the talents and resources of the school and amateur groups by which they were first performed, while their didacticism and moral tone recommend them to African audiences.

HENSLER, KARL FRIEDRICH, see RITTERDRAMA.

HENSLOWE, PHILIP (?–1616), English impresario, who unlike most theatre people of his time was never an actor but derives his importance in the history of the Elizabethan stage from being the owner of the FORTUNE, HOPE, and ROSE playhouses. His stepdaughter Joan Woodward married Edward ALLEYN, who on his father-in-law's death inherited his property and papers, the latter now being housed in Dulwich College. Among them is Henslowe's 'diary', a basic document for the study of Elizabethan theatre organization, in which he entered accounts for his various theatres, loans made to actors, payments to dramatists, and various private memoranda. Since some of the actors in the companies which used his theatres were contracted to Henslowe personally, and not, as was usually the case in Elizabethan companies, to their fellow actors, and as he paid the dramatists for their work, it follows that he had a large say in the choice of play and method of presentation. That his relations with his actors were not always cordial is proved by a document headed *Articles of Grievance, and Articles of Oppression, against Mr Hinchlowe*, drawn up in 1615, in which he is accused of embezzling their money and unlawfully retaining their property. There is no note of how the controversy ended, but evidently Henslowe kept actors and dramatists in his debt in order to retain his hold over them. This arrangement was not as good, nor did it make for such stability, as that in force among other companies like the CHAMBERLAIN'S MEN, where the actors, led by their chief player Richard BURBAGE, were joint owners of their own theatre, responsible only to each other.

HENSON, LESLIE LINCOLN (1891–1957), English comedian, best remembered for his appearances in MUSICAL COMEDY, who had been on the stage for some time before appearing in the West End in *Nicely, Thanks* (1912) at the STRAND THEATRE. In 1914 he went to New York, playing Albert in *To-Night's The Night*, and returned to London to play the same part a year later, scoring an instantaneous success. After serving in the First World War he returned to the stage in another musical, *Kissing Time* (1919), which opened the newly converted WINTER GARDEN THEATRE, followed by a series of musicals at the same theatre. He made a great success in *Funny Face* (1928), an apt description of his mobile indiarubber features, and was also seen in *It's a Boy* (1930) and *It's a Girl* (1931). From 1935 to 1939 he appeared in and directed a series of musicals at the GAIETY containing parts specially written for him. He spent most of the Second World War entertaining the troops, though he also appeared in two London revues, *Up and Doing* (1940) and *Fine and Dandy* (1942). After the war he was seen in *Bob's Your Uncle* (1948), and as Pepys in a musical version of FAGAN's *And So To Bed* (1951); he made his last appearance in London as Mr Pooter in the GROSSMITHS' *Diary of a Nobody* (1955). after which he played Old Eccles in a musical version of ROBERTSON's *Caste* at Windsor. He wrote two autobiographical works, *My Laugh Story* (1926) and *Yours Faithfully* (1948).

HEPBURN, KATHARINE HOUGHTON (1909–), American star best known for her work in films but also an excellent stage actress. She made her first appearance in 1928 in Baltimore and her New York début the same year. From 1932 she worked mainly in Hollywood until 1939, when she made a triumphant return to Broadway in Philip BARRY's *The Philadelphia Story*. She again worked in films until 1950, when she played Rosalind in *As You Like It* in New York. Two years later she made her London début as the Lady in SHAW's *The Millionairess*, repeating the role in New York in the same year. In 1955 she was seen in a series of Shakespearian parts in Australia with the OLD VIC company; and in 1957 she played Portia in *The Merchant of Venice* and Beatrice in *Much Ado About Nothing* at Stratford, Connecticut (see AMERICAN SHAKESPEARE THEATRE), returning in 1960 to play Viola in *Twelfth Night* and Cleopatra in *Antony and Cleopatra*. In New York in 1969 she played the title role in *Coco*, a musical based on the life of the French fashion designer Chanel, and in 1976 she starred in Enid BAGNOLD's *A Matter of Gravity*.

HERALD SQUARE THEATRE, New York, at the north-west corner of Broadway and 35th Street. This opened as the Colosseum Theatre on 10 Jan. 1874. Renamed the Criterion in 1882, Harrigan's Park in 1885, and the Park in 1889, it had an undistinguished career until, rebuilt as Herald Square Theatre, it opened on 17 Sept. 1894 with Richard MANSFIELD in SHAW's *Arms and the Man*. It was for some years an important link in the opposition to the powerful THEATRICAL SYNDICATE, but was converted to VAUDEVILLE by the Shuberts, who had taken control in 1900, and was demolished in 1915.

HERBERT, A. P., see MUSICAL COMEDY.

HERBERT, Sir HENRY (1596–1673), younger brother of the poet George Herbert, and from 1623 to 1642 MASTER OF THE REVELS in succession to Sir George Buck. The office-book which he kept, recording details of plays licensed by him for public performance (at £2 for the first performance and £1 for subsequent revivals), and also lists of actors in the different companies and their relations with the authorities, has been lost, but extant passages from it were collected and published by Joseph Quincy Adams in 1917 as *The Dramatic Records of Sir Henry Herbert*. They represent an important source of information on the stage history of his time. At the Restoration Sir Henry made strenuous efforts to revive the powers of his office, but was routed by the monopoly granted by Charles II to DAVENANT and KILLIGREW, the latter assuming the title of Master of the Revels on Herbert's death and passing it on to his son. But it was no longer of any importance, and on the passing of the LICENSING ACT in 1737 its main function, the CENSORSHIP of plays, passed to the LORD CHAMBERLAIN, where it remained until it was abolished in 1968.

HERCZEG, FERENC, see HUNGARY.

HERLIE, EILEEN (1920–), Scottish-born actress, who made her first London appearance in 1942 as Mrs de Winter in Daphne Du Maurier's *Rebecca*. After a season in Liverpool with the OLD VIC she played a number of roles at the LYRIC, Hammersmith, in the last of which, the Queen in COCTEAU's *The Eagle Has Two Heads*, she achieved perhaps the biggest success of her career, transferring to the West End in the part in 1947. In 1948 she had another big success as EURIPIDES' Medea, followed by PINERO's *The Second Mrs Tanqueray* in 1950. Her role in WILDER's *The Matchmaker* in 1954 led to her first appearance in New York in the same play in 1955, after which she remained in America. In 1958 she played Paulina in *The Winter's Tale* and Beatrice in *Much Ado About Nothing* at the STRATFORD (ONTARIO) FESTIVAL. Her varied New York career includes roles in OSBORNE's *Epitaph for George Dillon* (1958) and USTINOV's *Photo Finish* (1963). A return to the classics as Gertrude in John GIELGUD's production of *Hamlet* (1964) was followed by another Ustinov play, *Halfway Up the Tree* (1967), and an appearance in Chicago as Martha in ALBEE's *Who's Afraid of Virginia Woolf?* (1971). In 1973 she played Queen Mary in Royce Ryton's *Crown Matrimonial*.

HER MAJESTY'S THEATRE, London, in the Haymarket. The first theatre on this site was called the Queen's, after Queen Anne. Designed by Sir John VANBRUGH, it opened on 9 Apr. 1705 under the management of William CONGREVE, one of the first plays to be given there being Vanbrugh's *The Confederacy*. The house proved unsuitable for drama, and became London's first opera house, many of Handel's operas, from *Rinaldo* in 1711 to *Jupiter in Argos* in 1739, being first produced there. His *Esther*, the first oratorio to be heard in England, was also given there in 1732. On the death of Queen Anne in 1714 the theatre had changed its name to the King's, in honour of George I, and it retained that name when rebuilt after a fire on 17 June 1789. The new theatre opened on 26 Mar. 1791, and was devoted entirely to opera and ballet. On the accession of Queen Victoria in 1837 it was renamed Her Majesty's, a name it retained until the accession of Edward VII in 1901, when it became His Majesty's again, reverting to Her Majesty's on the accession of Elizabeth II in 1952. It had a consistently successful career until it was again burnt down in Dec. 1867. Although it was rebuilt between 1868 and 1869, it did not reopen until 1877, and never regained its former popularity. It finally closed in 1891 and was demolished, only the Royal Opera Arcade being left standing.

Some years later Beerbohm TREE acquired part of the site (the remainder being occupied first by the Carlton Hotel and then by New Zealand House), and with the profits from his production of Du Maurier's *Trilby* (1895) at the HAYMARKET built a new theatre there, retaining the old name. Designed by C. J. Phipps, it held 1,283 people in four tiers. It opened on 28 Apr. 1897 with Gilbert Parker's *The Seats of the Mighty*, followed by a series of excellent productions, including a number of Shakespeare's plays and new works, played in repertory. It was in his rooms in the dome of this theatre that Tree instituted a drama school in 1904 which eventually moved to other premises to become the Royal Academy of Dramatic Art. After Tree's departure in 1915 the theatre achieved a new kind of success with the opening on 31 Aug. 1916 of *Chu-Chin-Chow*, a musical fantasy by Oscar ASCHE which ran for 2,238 performances. Later productions were *Cairo* (1921), also by Asche, FLECKER's *Hassan* (1923), with music by Delius, COWARD's *Bitter Sweet* (1929), *The Good Companions* (1931), a dramatization of J. B. PRIESTLEY's bestselling novel of that name, and a musical, *The Dubarry* (1932). In 1935 George ROBEY played Falstaff in *Henry IV, Part I*, and in 1936 Elisabeth BERGNER was seen briefly in the title role of BARRIE's last play, *The Boy David*. In June 1939 the Greek actress Katina PAXINOU made her first appearance in London in SOPHOCLES' *Electra*, with the Greek National Theatre company. subsequently playing Gertrude to the Hamlet of her husband Alexis MINOTIS. During the Second World War the theatre housed a series of revivals, but in 1947 Robert MORLEY had a considerable success there with his own play *Edward, My Son*, written in collaboration with Noel Langley. The Lerner and Loewe musical *Brigadoon* (1949) inaugurated a successful series of American productions which included John Patrick's *The Teahouse of the August Moon* (1954) and the musicals *Paint Your Wagon* (1953) and *West Side Story* (1958). Plays returned with M. Bradley-Dine's *The Right Honourable Gentlemen* (1964) and Keith Waterhouse and Willis HALL's *Say Who You Are* (1965), and a five-year run was achieved by a new musical, *Fiddler on the Roof*, based on stories by Sholom ALEICHEM, which opened in 1967, and in which the Israeli actor TOPOL scored a great personal success. Later musicals included *Company* by SONDHEIM and *Applause* (both 1972), the British musical version of *The Good Companions* (1974), and the African musical *Ipi Tombi* (1975). Terence RATTIGAN's *Cause Célèbre* (1977) was followed by a series of disappointingly short runs until Peter SHAFFER's *Amadeus*, with Frank FINLAY, which transferred from the NATIONAL THEATRE, opened in 1981.

HERMANN, DAVID, see VILNA TROUPE.

HERNE [Ahearn], JAMES A. (1839–1901), American actor and dramatist, who made his first appearance on the stage in 1850. He was for a time manager of Maguire's New Theatre in San Francisco, where he appeared in a number of adaptations of DICKENS's novels which appear to have had a great influence on his own writing. He was also leading man at the Baldwin Theatre in the same town, where he worked with BELASCO, collaborating with him in a number of plays including *Hearts of Oak* (1879). This was pure melodrama, but Herne's later works showed more realism and sobriety, mainly under the influence of his wife Katharine Corcoran (1857–1943), a fine actress who played most of his heroines. His first important work, and the only one still to be remembered, was *Margaret Fleming* (1890), a sombre drama of infidelity which was too far ahead of its time to appeal to the public of its own day, in spite of fine acting by Mrs Herne in the title role. It was revived in 1894, with Herne's daughter Chrystal as Margaret Fleming, and in 1915, and was then more sympathetically received. In 1892 Herne put on *Shore Acres*, which after a slow start became one of the most popular plays of the day, mainly through the character of Uncle Nathaniel Berry. It was followed by *The Reverend Griffith Davenport*, of which no complete copy has survived, and *Sag Harbour* (both 1899), in which Herne was acting when he died.

HERO-COMBAT PLAY, see MUMMERS' PLAY.

HERODAS (or HERONDAS) (*c*.300–250 BC), Greek dramatist of the HELLENISTIC AGE, a writer of semi-dramatic MIMES, nine of which were discovered, whole or in part, on an Egyptian papyrus in 1891. They represent very vividly, and with some coarseness, scenes from ordinary life, and are thought to have been written for solo performances.

HEROIC DRAMA, a style of playwriting imported into England during the Restoration in imitation of French classical tragedy. Written in rhymed couplets, and observing strictly the UNITIES of time, place, and action, it deals mainly with the Spanish theme of 'love and honour' on the lines of CORNEILLE's *Le Cid*. Its chief exponent was DRYDEN, whose *Conquest of Granada* (1670/1) best displays both its faults and virtues. Its vogue was short-lived and it was finally killed by BUCKINGHAM's satire *The Rehearsal* (1671).

HERON, MATILDA AGNES (1830–77), American actress, who made her début in Philadelphia in 1851 and in the same year appeared in Toronto and on the Great Lakes circuit with her parents. She later achieved recognition as Marguerite Gautier in DUMAS *fils*'s *The Lady of the Camellias*, which she had seen while on a visit to Paris, subsequently making a fairly accurate version of it in which she toured all over the United States. She was not, however, the first to play the part in America, Jean LANDER having forestalled her in 1853 with an innocuous adaptation entitled *Camille; or, the Fate of a Coquette*. Miss Heron's version was particularly successful in New York and she made a fortune out of it, most of which she spent or gave away. She later played Medea in EURIPIDES' tragedy and Nancy in a dramatization of DICKENS's *Oliver Twist*, and was seen in several of her own plays. Among the actresses she trained for the stage was her daughter by her second marriage, Hélène Stoepel, known as Bijou Heron (1862–1937), who married Henry MILLER.

HERTZ, HENRIK (1798–1870), Danish poet and dramatist, author of nearly 50 plays, light comedies and romantic pieces, as well as national and historical dramas written for production at the KONGELIGE TEATER by his friend and close associate J. L. HEIBERG. The only plays by Hertz to have survived are *Sparekassen* (*The Savings Bank*, 1836), *Sven Dyrings Hus* (*The House of Sven Dyring*, 1837), and the one-act *Kong Renes Datter* (1843). The last, as *King René's Daughter*, was seen at the HAYMARKET in London in 1855, with Helen FAUCIT as the blind heroine. Under the name of the heroine, *Iolanthe*, in 1880, it provided an excellent part for Ellen TERRY playing opposite IRVING at the LYCEUM.

HEVESI, SÁNDOR (1873–1939), Hungarian director, who in 1904 founded the Thália Society, on the lines of the THÉÂTRE LIBRE and the FREIE BÜHNE, and produced a number of new Hungarian plays, among them the early works of Melchior Lengyel, as well as new versions of classical European plays. He was much influenced by the theories of Gordon CRAIG and STANISLAVSKY, which he endeavoured to put into effect when in 1916 he began to direct plays at the National Theatre in Budapest. He became manager of the theatre in 1928, and remained there until 1933, when he moved to the Magyar Theatre.

HEYWOOD, JOHN (*c.* 1497–1580), English dramatist, author of a number of interludes which mark the transition from medieval plays to the comedy of Elizabethan times. He married Elizabeth, the daughter of John RASTELL, and probably owed much to the influence of his wife's uncle Sir Thomas More, who took a great interest in plays and playing. Heywood's best known work is *The Playe called the foure P.P.; a newe and a very mery enterlude of a palmer, a pardoner, a potycary, a pedler*, each of whom tries to outdo the others in lying. The palmer wins when he says that in all his travels he never yet knew one woman out of patience. This was acted in about 1520, probably at Court, and was published some 20 years later by Heywood's brother-in-law William Rastell, as were *The Play of the Wether* and *The Play of Love* (both 1533). *The Dialogue of Wit and Folly*, which probably dates from the same year, remained in manuscript until 1846, when it was issued by the Percy Reprint Society. Two further interludes are sometimes attributed to Heywood, *The Pardoner and the Frere* and *Johan Johan*, both published anonymously by Rastell in 1533, while he may also be the author of *Thersites*, sometimes attributed to UDALL.

HEYWOOD, THOMAS (*c.* 1570–1641), English actor and dramatist, who may have been related to John HEYWOOD. He was with the ADMIRAL'S MEN, and later, until they disbanded in 1619, with QUEEN ANNE'S MEN, after which he does not seem to have acted again. As a dramatist he is first mentioned in 1596, when HENSLOWE recorded in his diary an advance payment made to him for an unnamed play which may have been *Captain Thomas Stukeley*, acted that year by the Admiral's Men. In 1599 he was apparently part-author with CHETTLE and others of a chronicle play, *Edward IV*, and the following year he produced a somewhat absurd romantic drama *Four Prentices of London*, which was later satirized by BEAUMONT in *The Knight of the Burning Pestle* (1607). From then onwards he produced a vast quantity of work, mainly for Henslowe, and said himself that he had written or had a hand in over 200 plays, most of which are lost as he did not trouble to publish them until forced to it by piracy. His subject-matter ranged from classical myth, as in the four plays *The Golden*, *The Silver*, *The Brazen*, and *The Iron Age* (*c.* 1611–*c.* 1613), to realistic contemporary drama, in which he did some of his best work. His masterpiece is undoubtedly the domestic tragedy *A Woman Killed With Kindness* (1603) which was been many times revived, most recently by the NATIONAL THEATRE in 1971. His other extant plays include *The Wise Woman of Hogsdon* (1604); a rambling account of the early years of Elizabeth I's reign entitled *If You Know Not Me, You Know Nobody* (1604/5); *The Rape of Lucrece* (1607), obviously designed to profit from the popularity of the poem of the same name by Shakespeare, from whom Heywood made frequent borrowings; and *The Fair Maid of the West* (1610). After his retirement from acting he wrote nothing for some years, but returned in 1625 with *The English Traveller*, which contains echoes of *Macbeth*, and in 1631 added a second part to *The Fair Maid of the West* as *The Girl Worth Gold*. He also wrote the first of the civic pageants which he contributed annually to the Lord Mayor's Show, from *London's Jus Honorarium* (1631) to *Londini Status Pacatus* (1639). The last play to be attributed to him was a comedy, *Love's Masterpiece*, now lost. Among his non-dramatic works the most interesting is an *Apology for Actors*, published in 1612.

HIBERNIAN ROSCIUS, see BROOKE, G.V.

HICKS, (Edward) SEYMOUR (1871–1949), English dramatist and actor-manager, knighted in 1935. He began his stage career by walking on at the Grand, Islington, in 1887, and was for a long time with the KENDALS, both in England and America. With Charles Brookfield he produced the first REVUE seen in London, *Under the Clock* (1893). In the course of his long and varied life he topped the bill in the MUSIC-HALLS and appeared with equal success in MUSICAL COMEDY and straight plays, being particularly admired as Valentine Brown in BARRIE's *Quality Street* (1902). Among his numerous plays the most successful were *Bluebell in Fairyland* (1901), which was frequently revived at Christmas up to 1937; *The Gay Gordons* (1907), in which he played Angus Graeme; *Sleeping Partners* (1917), in which his own performance was a *tour de force* of silent acting; and *The Man in Dress Clothes* (1922), which, like many of his plays, was taken from the French. He was manager of several London theatres, including the ALDWYCH, which he built and opened in 1906; the GLOBE (first named the Hicks), which he built for Charles FROHMAN and opened in 1907 with his own play *The Beauty of Bath*; and DALY's, where he inaugurated his management by playing Charles Popinot in *Vintage Wine* (1934), a play taken from the German in which he collaborated with Ashley DUKES. He was the first actor to take a party of entertainers to France in the First World War, being awarded the Legion of Honour by the French Government in 1931 both for his work then and for his services generally to French drama in London. He was also the first actor to go to France during the Second World War. He married the actress Ellaline TERRISS, who appeared with him in many of his productions.

HICKS THEATRE, London, see GLOBE THEATRE.

HILAR [Bakule], KAREL HUGO (1885–1935), Czech director, who was first at the Municipal Theatre in Prague, and in 1921 took over at the National. He was responsible for a number of excellent productions, particularly of foreign classics, combining the REALISM of STANISLAVSKY with the visionary imagination of Gordon CRAIG. It was under him that the plays of Karel and Josef ČAPEK, including *R.U.R.* (1921) and *The Insect Play* (1922), were first staged, and also some of the plays of František LANGER.

HILARIUS, see LITURGICAL DRAMA.

HILAROTRAGOEDIA, see FABULA 3: RHINTHONICA.

HILL, AARON (1685–1750), English playwright, satirized by Pope in *The Dunciad* (1728). He is now chiefly remembered for his association with Handel, whose first London opera *Rinaldo* (1711)—for which Hill wrote the libretto—was performed under his management, and for his translations of three plays by VOLTAIRE—*Zaïre* (1732), *Alzire, ou les Américains* (1736), and *Mérope* (1743). The first, as *The Tragedy of Zara*, was produced at DRURY LANE in 1736, the second, as *Alzira*, at LINCOLN'S INN FIELDS later the same year, and the third, as *Meropé*, again at Drury Lane in 1749. Hill's own plays include several tragedies in the French style and a farce, *The Walking Statue; or, the Devil in the Wine Cellar* (1710), which was extremely successful and frequently revived up to 1845. His letters and journals contain a good deal of entertaining information about the theatre of his time.

HILL, JENNY, see MUSIC-HALL.

HILLBERG, EMIL (1852–1929), Swedish actor, one of the first realistic players of any stature in his own country during the 19th century. He made his first appearance in 1873 at the KUNGLIGA DRAMATISKA TEATERN, and later became well known for his excellent interpretations of parts by IBSEN (notably the title role in *Brand* and Stockmann in *An Enemy of the People*), and STRINDBERG (particularly the title role in *Gustav Vasa* and the Captain in *The Father*).

HILLER, WENDY (1912–), DBE 1975, English actress. She began her career at the MANCHESTER Repertory Theatre in 1930, and scored an instant success five years later in the London production of *Love on the Dole*, adapted from Walter Greenwood's novel by the author and Ronald Gow (1897–), later her husband and the author of several other plays, including *Ma's Bit o' Brass* (1938). She repeated the role of Sally Hardcastle in New York in 1936, and later the same year was at the MALVERN FESTIVAL, where she played the title role in SHAW's *Saint Joan* and Eliza in his *Pygmalion*. After some years in films she returned to the West End in 1944 in a revival of MARTÍNEZ SIERRA's *The Cradle Song*, and a year later was seen as Princess Charlotte to Robert MORLEY's Prince Regent in Norman Ginsbury's *The First Gentleman*, which had a long run. At the BRISTOL OLD VIC in 1946 she played Tess in Gow's adaptation of Hardy's novel, repeating the part in London and in 1947 appearing on Broadway as Catherine Sloper in *The Heiress*, based by Ruth and Augustus Goetz on Henry JAMES's novel *Washington Square*. Back in London, she was seen in Gow's adaptation of H. G. Wells's *Ann Veronica* (1949) and was in the all-star cast of N. C. HUNTER's *Waters of the Moon* (1951), in which she gave an outstanding performance as Evelyn Daly. This play ran for two years, after which she joined the OLD VIC company to play a wide variety of Shakespearian parts. In 1957 she was in New York, playing Josie Hogan in O'NEILL's *A Moon for the Misbegotten*. Among her later parts were Carrie Berniers in Lillian HELLMAN's *Toys in the Attic* (1960); Miss Tina in Michael REDGRAVE's adaptation of Henry James's *The Aspern Papers* (N.Y., 1962); and Susan Shepherd in James's *The Wings of the Dove* (1963), adapted by Christopher Taylor. In 1967 she was seen in a revival of MAUGHAM's *The Sacred Flame*, and three years later in Peter SHAFFER's *The Battle of Shrivings*. The finest part of her later career was a remarkably faithful portrait of Queen Mary in Royce Ryton's *Crown Matrimonial* (1972), on the abdication of Edward VIII. For the NATIONAL THEATRE, in 1975 she played Gunhild Borkman opposite Ralph RICHARDSON in IBSEN's *John Gabriel Borkman*, and two years later she was at the CHICHESTER FESTIVAL THEATRE in a revival of *Waters of the Moon*, this time as Mrs Whyte, a part she also played in London a year later.

HILPERT, HEINZ (1890–1967), German director in the tradition of Berlin REALISM, which he laced with humour and common sense. He first came into prominence with his production of BRÜCKNER's *Der Verbrecher* (1928), which he followed with the first productions in 1931 of Horváth's *Geschichten aus dem Wiener Wald* (*Tales from the Vienna Woods*) and ZUCKMAYER's *Der Hauptmann von Köpenick*. From 1934 to 1945 he managed the DEUTSCHES THEATER with complete integrity in the face of Nazi intransigence; after the war he directed the first performances of Zuckmayer's *Des Teufels General* (1946) in Zurich and of his *Der Gesang im Feuerofen* (1950) in Göttingen, where from 1950 to 1966 he was director of the Deutsches Theater.

HIPPODROME, London, in Cranbourne Street, Leicester Square. This opened on 15 Jan. 1900 as a circus with a large water-tank which was used for aquatic spectacles. It later became a MUSIC-HALL and in 1909, when it was taken over by Edward Moss as his London headquarters, was reconstructed internally, the circus arena being covered by stalls. Ballet and variety were seen there, and from 1912 to 1925 Albert de Courville staged a number of successful REVUES. These were followed by an equally successful series of musical comedies, among them *Sunny* (1926), *Hit the Deck* (1927), *Mr Cinders* (1929), and *Please, Teacher* (1935). Among later productions were the revue *The Fleet's Lit Up* (1938), Ivor NOVELLO's musical comedy *Perchance to Dream* (1945), and Herman Wouk's play *The Caine Mutiny Court-Martial* (1956). In 1958, after complete reconstruction of the interior, the building opened as a combined restaurant and cabaret, the Talk of the Town, which closed in 1982.

HIPPODROME THEATRE, New York, on the Avenue of the Americas, between 43rd and 44th Streets. This theatre, the largest in America, seating 6,600 people, opened on 12 Apr. 1905 with a lavish spectacle entitled *A Yankee Circus on Mars*, and a year later was taken over by the Shuberts, who were succeeded by Charles DILLINGHAM. Every kind of entertainment was given, including grand opera. In 1923, as B. F. Keith's Hippodrome, it became a VAUDEVILLE house, and in 1928, as the R.K.O. Hippodrome, a cinema. Closed in 1932, it reopened in 1933 as the New York Hippodrome, and in 1935 was taken over by Billy Rose, whose spectacular musical *Jumbo* marked the end of the Hippodrome as a theatre. It was finally demolished in Aug. 1939.

HIRSCH, JOHN STEPHEN (1930–   ), Canadian director, born in Hungary. He emigrated to Canada in 1948 and after graduating from the University of Manitoba became interested in children's theatre, writing several plays for his own Muddiwater Puppets Theatre. He later became artistic director of Rainbow Stage, an outdoor summer MUSICAL COMEDY theatre, and helped to establish a professional group, Theatre 77. In 1958 he was appointed artistic director of the Manitoba Theatre Centre, which with its resident stock company set the pattern for a chain of regional theatres across Canada. His productions there included BRECHT's *Mother Courage* in 1964 and FRISCH's *Andorra* in 1965, and on one of his many return visits he directed, in 1973, ANSKY's *The Dybbuk*. From 1967 to 1969 he was co-director with Jean GASCON of the STRATFORD (ONTARIO) FESTIVAL theatre, where he was responsible for staging *Henry VI, Richard III, A Midsummer Night's Dream*, and *Hamlet*, as well as CHEKHOV's *The Cherry Orchard*, a spoof version of *The Three Musketeers*, and a production based on the *Satyricon* of Petronius. He also directed the first production of *Colours in the Dark* (1967) by the Canadian author James REANEY, and has since worked on Broadway, at the VIVIAN BEAUMONT THEATRE in New York, and at the HABIMAH THEATRE in Israel. At the end of 1980 he was reappointed as director of the Stratford theatre for a three-year term.

HIRSCH, ROBERT PAUL (1925–   ), French comic actor, a member of the COMÉDIE-FRANÇAISE from 1948 to 1974. His inventiveness and agility predisposed him at first towards such roles as Arlequin in MARIVAUX's *La Double Inconstance* and *Le Prince travesti*, in 1950, Sosie in MOLIÈRE's *Amphitryon* in 1951 and Scapin in his *Les Fourberies de Scapin* in 1956, the valet in FEYDEAU's *Le Dindon*, and, probably his best part, Bouzin in Feydeau's *Un Fil à la patte*, in which he was seen in London in the first WORLD THEATRE SEASON in 1964. Later he brought a touch of burlesque to more serious roles such as that of Nero in RACINE's *Britannicus* (in 1961), Raskolnikov in an adaptation of DOSTOEVSKY's novel *Crime and Punishment* (1963), and the title roles in Molière's *Tartuffe* and BRECHT's *Arturo Ui*. In the fantasies which highlight certain aspects of any character he plays, there is always a touch of Surrealism.

HIRSCHBEIN, PERETZ (1880–1949), Jewish actor and dramatist, and founder of the first Yiddish Art Theatre in Odessa. Although his first play, *Miriam* (pub. 1905), was written in Hebrew, he founded his company in 1908 for the production of plays in Yiddish, in which language he wrote his later works, translating them into Hebrew himself. Among the most important are *The Blacksmith's Daughters* (1915) and *Green Fields* (1919), both idylls of Jewish country life, the latter being considered one of the finest works of Yiddish dramatic literature.

HIRSCHFELD, KURT (1902–64), German theatre manager and director, who in 1933 left Darmstadt for exile in Zurich, where he worked from 1938 until his death. Though noted for the first productions of such plays as BRECHT's *Herr Puntila und sein Knecht Matti* (1948) and FRISCH's *Andorra* (1961) as well as importations from abroad, among them T. S. ELIOT's *The Elder Statesman* in 1960, his main achievement lay in assembling a cosmopolitan repertoire during the Second World War which became a model for many post-war German theatres. His constructive criticism proved helpful to many playwrights, among them Frisch and DÜRRENMATT in their early days.

HIS MAJESTY'S THEATRE, London, see HER MAJESTY'S THEATRE.

HOADLY, BENJAMIN (1706–57), well-known English physician, son of a bishop, who in 1747 offered GARRICK a comedy entitled *The Suspicious Husband*. Though it reads poorly, with Garrick in the part of Ranger—in which he was much admired—it made an unexpected success and was often revived. It was Hoadly's only known contribution to the theatre, though he is believed to have written another comedy, now lost, and to have collaborated in a third.

HOBART, the capital of Tasmania, whose Theatre Royal is the oldest playhouse in Australia. It opened officially on 8 Mar. 1837 with MORTON's *Speed the Plough*, which was revived in 1937 for the theatre's centenary celebrations. Although the external walls and stage remain much as they were originally, a gallery and a handsome classical façade were added to the building in 1860 and in 1911 the auditorium was reconstructed and the seating accommodation enlarged. Many international stars touring Australia played at the Royal, but in recent years the cost of transporting scenery and large casts from the mainland has become too high, though visits from solo performers are encouraged. For most of the year the theatre now houses the nationally-subsidized Tasmanian Theatre Company.

HOBART, GEORGE V., see CANADA.

HOBBY HORSE, form of ANIMAL IMPERSONATION common to many FOLK FESTIVALS throughout Europe, and particularly popular in England, the most famous being the Hooden Horse of Kent, the Old Hoss of Padstow, in Cornwall, and the Minehead Soulers' Horse. It is probably a survival of the primitive worshipper dressed in the skin of a sacrificial animal, and usually appears as a wicker framework in the shape of a horse covered with a long green cloth which conceals the legs of the man animating it. It can, however, be merely a carved horse's head attached to a long stick, in which form it has always been a popular toy for children. The Hobby Horse usually accompanies the MORRIS DANCE, and is often found in the MUMMERS' PLAY. In the Mascarade of La Soule a horse is the chief character, and dies and is resurrected in token of the return of spring after the death of winter.

HOBSON, HAROLD (1904–), English dramatic critic, knighted in 1977, who began his career in 1931 as dramatic critic to the *Christian Science Monitor*. He always championed the cause of the AVANT-GARDE and experimental theatre and was also passionately devoted to the modern French theatre; in recognition of his services to French dramatic literature he was appointed to the Legion of Honour in 1960. From 1944 to 1976 he was on the staff of the *Sunday Times*, becoming chief dramatic critic in 1947.

HOCHHUTH, ROLF (1931–), German dramatist, resident in Switzerland. His first play, *Der Stellvertreter* (1963; London, as *The Representative*, also 1963; N.Y., as *The Deputy*, 1964), first seen in Berlin in a production by PISCATOR, indicted Pope Pius XII for criminal nonintervention in the Nazis' extermination of the Jews. The Jesuit priest, whose confrontation with the Pope and subsequent self-immolation in Auschwitz form the core of the play, is an invented character, but most of the other figures are taken from real life. This fact, together with the immense amount of research undertaken by Hochhuth, much of it clumsily incorporated into the stage directions and the dialogue, caused the play to be labelled a 'documentary drama'. Though far too long (it had to be cut by half for performance), over-ambitious, and muddled in its aims, it remains an impressive dramatic treatment of the Nazi era, and can be seen as the beginning of a post-war renaissance in German drama. Hochhuth's second play, *Soldaten: Nekrolog auf Genf* (1967), also too long and overlaid with documentation, accuses Winston Churchill of causing the death of his Polish ally General Sikorski and of criminal inhumanity in the bombing of Dresden. The play became a *cause célèbre* in Britain long before it was produced in London as *Soldiers* in 1968, following the abolition of stage censorship. It had previously been seen in English in Toronto, and briefly in New York, also in 1968. The first production, in Berlin, was to have been directed by Piscator, but he died while working on it. The lack of his expert advice may have been one of the reasons for the comparative failure of Hochhuth's next plays, *Guerillas* (1970), a wholly fictional account of a modern revolution in America, *Die Hebamme* (*The Midwife*, 1972), a comedy based on the real life of a social worker, and *Lysistrate und Nato* (1974), none of which have had any impact abroad.

HOCHWÄLDER, FRITZ (1911–), Austrian dramatist, born in Vienna, but resident since 1938 in Switzerland, where he had his first success in the theatre with *Das heilige Experiment* (1945), which as *The Strong are Lonely* was seen in New York in 1953 and in London, with WOLFIT in the chief part, in 1955. It deals with the destruction of the Jesuit Community in Paraguay in the 18th century, and displays considerable theatrical skill in dramatizing a moral issue. After *Der Flüchtling* (*The Fugitive*, 1945), set in modern times, Höchwalder again used a historical setting for *Der öffentliche Ankläger* (1947), seen in London in 1947 as *The Public Prosecutor*, with Alan BADEL as Fouquier-Tinville who, briefed to conduct a case against an anonymous 'enemy of the people', finds that he himself is the accused. Of Hochwälder's other plays, *Donadieu* (1953) depicts a man, after an agonizing mental struggle, giving shelter to his wife's murderer; *Die Herberge* (*The Inn*, 1956) shows the investigation of a theft leading to the discovery of a far greater crime; *Der Unschuldige* (*The Innocent Man*, 1958), is a comedy with serious undertones, when a man realises that though innocent of the murder of which he is accused, he might very well have committed it; *Der Himbeerpflücker* (*The Raspberry Picker*, 1965) satirizes the latent Nazism of a small Austrian town whose inhabitants welcome a suspected Nazi criminal, only to find that he is a petty thief on the run; and in *Der Befehl* (*The Command*, 1968), Hochwälder returns to the situation of *The Public Prosecutor*: an Austrian police inspector, ordered to investigate a former Nazi crime, discovers that he is himself the culprit. In all his plays Hochwälder is concerned with a conflict of ideas, emphasized not by the use of modern stage techniques but rather by a strict adherence to the classical UNITIES.

HOCKTIDE PLAY, a survival of one of the early English FOLK FESTIVALS, given in Coventry on Hock Tuesday (the third Tuesday after Easter Sunday), and revived as a pleasant antiquity in the Kenilworth Revels prepared for the visit of Queen Elizabeth in July 1575. It began with a Captain Cox leading in a band of English knights, each on a HOBBY HORSE, to fight against the Danes, and ended with the leading away of the Danish prisoners by the English women. It was intended to represent the massacre of the Danes by Ethelred in 1002, but this is probably a late literary assimilation of an earlier folk-festival custom, traceable in other places (Worcester, Shrewsbury, Hungerford), by which the women 'hocked' or caught the men and exacted a forfeit from them on one day, the men's turn coming the following day. The practice was forbidden at Worcester in 1450. This ceremony may have been a symbolic re-enactment of the capture of a victim for human sacrifice.

HODGE, MERTON, see NEW ZEALAND.

HODGKINSON [Meadowcroft], JOHN (1767–1805), English actor, who had had some experience in the provinces when in 1792 he accepted an offer from John HENRY to join the AMERICAN COMPANY at the JOHN STREET THEATRE in New York, and spent the rest of his life in the United States. He soon became extremely popular, ousting Henry from management as well as from public favour, and becoming joint manager with the younger HALLAM and DUNLAP of the PARK THEATRE when it first opened. A handsome man, with a good memory and a fine stage presence, he excelled both in tragedy and comedy, among his best parts being the title role in Dunlap's *André* and Rolla in his adaptation of KOTZEBUE's *Pizarro*.

HODSON, HENRIETTA (1841–1910), English actress, daughter of a London publican, who made her first appearances in Scotland in a company which included the young Henry IRVING. She went with him to Manchester, and then joined the stock company at BRISTOL, where she appeared in burlesque and extravaganza under the management of Madge Robertson (later Mrs KENDAL) together with Kate and Ellen TERRY. She retired on her marriage to a young solicitor, but being soon widowed returned to the theatre and in 1866 was seen in London, where she became known as an actress of highly individual style and technical accomplishment, at her best in humorous or even farcical parts; pathos and deep sentiment lay outside her range. In 1868 she married the journalist Henry Labouchère (1831–1912), who had rented for her the QUEEN'S THEATRE where she again appeared with Irving, playing Nancy in DICKENS's *Oliver Twist* to his Bill Sikes. She continued to act under her maiden name and later became manageress of the ROYALTY THEATRE, where she produced and played in a number of burlesques. She retired in 1878 and three years later was instrumental in introducing Lillie LANGTRY on to the London stage.

HOFMANNSTHAL, HUGO VON (1874–1929), Austrian poet and dramatist, who as a young man was in the forefront of the reaction against NATURALISM. His early verse-plays, of which *Der Tor und der Tod* (*Death and the Fool*) was produced at Munich in 1898, reveal his delight in beauty and in poetic and mystical intuitions, but in his later plays his exquisite poetry was diluted in an effort to enhance the dramatic content of his dialogue and relate his characters to a social framework. He turned to subjects which were obviously theatrical, and his first success in the theatre came with *Elektra* (1903), produced in Berlin by REINHARDT, who directed most of his plays. It was followed by *Ödipus und die Sphinx* (1905), *König Ödipus* (1907), and *Alkestis* (1909). Also to this period belongs the first of his adaptations, which in his hands became almost new plays—*Das gerettete Venedig* (1905), based on OTWAY's *Venice Preserv'd*, seen at the Lessingtheater in Berlin with sets by Gordon CRAIG. In 1911 Hofmannsthal produced what is perhaps his best known play, *Jedermann*, based on the old Dutch MORALITY PLAY *Elckerlyc* (*Everyman*) and incorporating some elements from Hans SACHS. After its first production in Berlin it was transferred to the Salzburg Festival, founded in 1917 by Reinhardt and Hofmannsthal, and it has since been seen at all the festivals there, in an open-air setting in front of the Cathedral. For the festival in 1922 Hofmannsthal adapted CALDERÓN's *El gran teatro del mundo* as *Das Salzburger grosse Welttheater*. In two plays of the same period, set in and near Vienna, *Der Schwierige* (*The Difficult Man*, 1921) and *Der Unbestechliche* (*The Incorruptible Man*, 1923), Hofmannsthal reverted to the vein of comedy which had been apparent in *Der Abenteurer und die Sängerin* (*The Adventurer and the Singer*, 1898) and *Christinas Heimreise* (*Cristina's Journey Home*, 1909). Both are pleas for marital fidelity as the natural and desirable sequel to the equally natural, though occasionally extended, philanderings of immaturity, but *Der Schwierige* in particular stands out as a masterpiece, combining high comedy of subtle human relationships with irony and social satire. Hofmannsthal's last play was again based on Calderón, an adaptation, as *Der Turm*, of *La vida es sueño*, which takes up once more the conflict between material power and spiritual integrity. The first version of 1925 ends with a Utopian picture of a new and purified world; but the revised version of 1927, produced a year later, closes in tragedy, showing the spirit as indestructible but powerless and isolated in a tyrannical world.

Hofmannsthal's plays are little known in the English-speaking world, where he is remembered mainly as the librettist of Richard Strauss's operas *Elektra* (1909, based on his play); *Der Rosenkavalier* (1911); *Ariadne auf Naxos* (1912); *Die Frau ohne Schatten* (1919); *Die ägyptische Helen* (1928); *Arabella* (1933).

HOIST, see SLOTE.

HOLBERG, LUDVIG (1684–1754), historian, philosopher, satirist, and playwright, who, though born in Norway, spent most of his working life in Denmark, being for many years Professor of Metaphysics at the University of Copenhagen. As a young man he travelled widely in Europe, coming into close and sympathetic contact with the literature of France, Italy, and England, where he studied for a time at the University of Oxford. His connection with the theatre was brief, but far-reaching in effect. When in 1721 he was appointed director of the DANSKE SKUEPLADS, he brought to his task a knowledge of and love for the drama unusual in Denmark at that time, his favourite playwrights being PLAUTUS and MOLIÈRE. There were at this time no Danish plays, the company at the theatre when he took over playing only in French and German. To remedy this he wrote a number of plays in Danish during his six years as director, bringing on to the stage Danish and Norwegian characters and situations not seen there before, and creating in the vernacular a tradition of comedy which was upheld by his successors. His first two productions, in 1722, were a translation of Molière's *L'Avare* and his own *Den politiske Kandestøber* (*The Political Tinker*). He wrote in all 32 comedies, six of them, not so good as his earlier work, after 1747, when the Danske Skueplads reopened after nearly 20 years of inactivity. The best of his plays belong to the earlier period—*Jean de France* and *Mester Gert Westphaler* (*The Babbling Barber*), both given in 1722; *Den Vægelsindede* (*The Weathercock*) and *Barselstuen* (*The Lying-in Room*), both 1723; *Jacob von Tyboe* and *Henrik og Pernille*, both 1724; and *Den Stundesløse* (*The Fussy Man*), 1726. *Erasmus Montanus*, though written in 1723, was not performed until 1742. Although many of Holberg's plays became well known in translation in Germany, Holland, and France, they have not made much impact in England or the U.S.A. Apparently the only one to be professionally produced is *Jeppe paa Bjerget, eller den forvandlede Bonde* (1722), which, as *Jeppe of the Mountains*, in a translation by Michael Meyer, was seen at the PITLOCHRY FESTIVAL THEATRE in 1966, although a number of his plays have been published in English translation.

HOLBORN EMPIRE, London, in High Holborn. As Weston's Music-Hall, this opened on 16 Nov. 1857, and proved the first serious rival to the CANTERBURY. It had a consistently successful career, being renamed the Royal Music-Hall in 1868. Rebuilt in 1887, it opened as the Royal Holborn and saw the débuts of many famous music-hall stars, among them Bessie Bellwood and J. H. Stead; its CHAIRMAN for many years was W. B. Fair, singer of 'Tommy, Make Room for Your Uncle.' Round about 1900 its popularity began to decline, and it closed in June 1905. Completely rebuilt, it reopened on 29 June 1906 as the Holborn Empire, again with great success. It was occasionally used for plays, Sybil THORNDIKE appearing there early in 1920 in the title roles of SHAW's *Candida* and EURIPIDES' *Medea*, and as Hecuba in the latter's *Trojan Women*. In 1928 William POEL produced a revival of JONSON's *Sejanus*, and from 1922 to 1939 the children's play *Where the Rainbow Ends* was given an annual Christmas production by Italia CONTI. In 1932 Nellie WALLACE appeared in *The Queen of Hearts*, a REVUE by Tom Arnold, and at the end of 1939 another revue, *Haw-Haw!*, was staged by George Black. On the night of 11–12 May 1941 the building was severely damaged by enemy action; it was finally demolished in 1960.

HOLBORN THEATRE, London, at 42 High Holborn. Built by Sefton Parry, this opened on 6 Oct. 1866 with BOUCICAULT's *The Flying Scud*, which was a success. Subsequent productions and several managements failed until in 1875 Horace Wigan assumed control, reopening the theatre on 24 Apr. as the Mirror. In Oct. of the same year he had his greatest success with *All For Her*, the first dramatization of DICKENS's *A Tale of Two Cities*, starring John Clayton. In 1876 the theatre again changed hands, opening on 8 Jan. under F. C. Burnand as the Duke's, with his own burlesque of JERROLD's *Black-Ey'd Susan*, transferred from the OPERA COMIQUE. After a further undistinguished period a modern drama by Paul Merritt, *The New Babylon* (1879), proved acceptable, and had just returned to the theatre after a provincial tour when the building was destroyed by fire on 4 June 1880.

HOLCROFT, THOMAS (1744–1809), English dramatist, who is usually credited with the introduction of MELODRAMA on the London stage with his *A Tale of Mystery* (1802), an adaptation of *Coelina, ou l'Enfant de mystère* (1800) by PIXÉRÉCOURT, which was several times revived. His best known play is *The Road to Ruin* (1792), with its excellent roles of Goldfinch and Old Dornton, the latter a favourite part with many elderly character actors both in London and in America, where the play was frequently revived. It was last seen in London in 1937. Holcroft, who was entirely self-educated, was a very good French scholar and had a phenomenal memory, a combination which enabled him while in Paris to learn by heart BEAUMARCHAIS's *Le Mariage de Figaro*, and to put it on the London stage in 1784 as *The Follies of a Day*. He also translated DESTOUCHES's *Les Glorieux* (1732) as *The School for Arrogance* (1791). Among his other comedies the most successful was *Love's Frailties* (1794), based on a German original, *Der deutsche Hausvater* by Gemminger. Holcroft, who was a friend of Charles LAMB, was editor of the *Theatrical Recorder*, which appeared monthly from 1805 to 1806 and contained plays translated from the French and Spanish. His *Memoirs* (1816) were edited by HAZLITT.

HOLLAND, GEORGE (1791–1870), English actor, son of a dancing-master, who after seven years on the London stage went to New York and founded a family of American actors, his sons Edmund Milton (1848–1913) and Joseph Jefferson (1860–1926) both being well known as light comedians; his daughter was also on the stage, dying just as she was beginning her career under Augustin DALY. George Holland made his first appearance in New York at the BOWERY THEATRE in 1827, and soon became a popular comedian. He toured extensively and was well known in the South, being for some time in management with LUDLOW and Sol SMITH, and he was for six years at Mitchell's OLYMPIC in BURLESQUE. From 1855 to 1867 he played character parts in Lester WALLACK's company, being outstanding as Tony Lumpkin in GOLDSMITH's *She Stoops to Conquer*, which he was still playing at the age of 75. In 1869 he was engaged by Daly, dying soon after.

HOLLINGSHEAD, JOHN (1827–1904), English theatre manager, whose name is chiefly associated with the GAIETY THEATRE. He opened it on 21 Dec. 1868 and remained there for 18 years, being succeeded by George EDWARDES, who was to make it the home of MUSICAL COMEDY. Under Hollingshead it had been used mainly for BURLESQUE. Hollingshead is credited with the introduction of matinées and with being the first manager to use electric light outside the theatre (in 1886). In 1880 he staged a translation by William ARCHER of *Samfundets Støtter* as *Quicksands; or, the Pillars of Society*, the first play by IBSEN to be seen in London.

HOLLOWAY, JOSEPH (1861–1944), Irish architect, dramatic critic, and theatre historian, who in 1904 was commissioned by Miss HORNIMAN to renovate the old Mechanics Theatre on Abbey Street, Dublin, for the use of the IRISH NATIONAL DRAMATIC THEATRE. An inveterate 'first-nighter', his theatre criticisms were as frequently subtle in their judgement of the art of acting as they were erratic in their assessment of playwrights. On his death he left to the Municipal Gallery of Dublin and the National Library of Ireland his collection of theatre memorabilia, including portraits, playbills, photographs, and newspaper clippings, and his 25,000,000-word diary, covering the 45 years from the founding of the IRISH LITERARY THEATRE in 1899, out of which a four-volume history of the ABBEY THEATRE in his time was compiled by Robert Hogan and Michael K. O'Neill.

HOLM, IAN [Ian Holm Cuthbert] (1931–), English actor who was with the company at Stratford-upon-Avon in 1954 and 1955, playing small parts, and was then seen in London. He returned to Stratford after touring Europe in 1957 with OLIVIER in *Titus Andronicus*, and remained there for many years, playing a wide variety of parts, among them a tetchy, senile Gremio in *The Taming of the Shrew* and a romantic Lorenzo in *The Merchant of Venice* (both 1960). His combination of sardonic humour and technical precision, backed by an impressive vocal technique, was given full play in the trilogy *The Wars of the Roses* (1963–4), adapted by John BARTON and directed by Peter HALL, in which he played a shrewdly calculating Prince Hal, a mud-stained, anti-heroic Henry V, and a Richard III who balanced venom with open self-mockery. His rare appearances in modern drama have included the dapper pimp in PINTER's *The Homecoming* (1965; N.Y., 1967), Nelson in RATTIGAN's *A Bequest to the Nation* (1970), in which he gave the character considerable surface charm and a sense of inner torment, and Hatch in BOND's *The Sea* (1973).

HOLT, BLAND (1853–1942), English-born actor-manager who, taken to Australia at the age of four, began his stage career in Sydney in 1876, eventually becoming known as 'Australia's Monarch of Melodrama'. He was himself a fine comedian, but was equally admired for his spectacular productions, which sometimes included circus acts and early motor-cars on stage. He was also a skilful adapter of old scripts, heightening the drama and bringing the entire play up to date. He retired in 1909.

HOLT, HELEN, see TREE, HERBERT BEERBOHM.

HOLZ, ARNO (1863–1929), German novelist and dramatist, whose *Die Familie Selicke* (1890), written in collaboration with Johannes Schlaf, became the manifesto of the new school of NATURALISM. Produced by the recently founded FREIE BÜHNE, this dreary catalogue of misery, rape, disease, and death, set against a sordidly realistic background and devoid of all theatrical tricks, proved a rallying point for the younger dramatists, among them Gerhart HAUPTMANN. Holz wrote a number of other plays but, lacking the collaboration of Schlaf, without success.

HOME, the Revd JOHN (1722–1808), Scottish minister, author of the tragedy *Douglas* (1756), produced in Edinburgh with Dudley DIGGES as the hero, Young Norval. It caused a great deal of controversy, as many members of the Church of Scotland were horrified that one of their number should write for the theatre; but it was rapturously received by the audience. In 1757 it was accepted by John RICH for COVENT GARDEN, where Spranger BARRY ('six feet high and in a suit of white puckered satin,' says Doran) played Young Norval to the Lady Randolph of Peg WOFFINGTON. The play was constantly revived, Lady Randolph being a favourite part with Sarah SIDDONS, while many young actors in England and America, including Master BETTY and John Howard PAYNE, delighted in Young Norval. The speech beginning 'My name is Norval' became a regular recitation piece, and the play eventually found its way into the repertory of the TOY THEATRE. It was revived at the EDINBURGH FESTIVAL in 1950, with Sybil THORNDIKE as Lady Randolph. Home wrote other tragedies, none of them successful.

HOME, WILLIAM DOUGLAS (1912–), English dramatist, who was for a time on the stage, making his first appearance in London in Dodie SMITH's *Bonnet Over the Windmill* (1937). After service in the Second World War he devoted his energies to playwriting and acted only occasionally in his plays, which are mainly light comedies with an upper-class background. He first achieved success with a political comedy, *The Chiltern Hundreds* (1947), which starred A. E. MATTHEWS, following it with a serious play about prison life, *Now Barabbas . . .* (also 1947). Among his later plays were *Ambassador Extraordinary* (1948), *The Bad Samaritan* (1953), *The Manor of Northstead* (1954), a sequel to *The Chiltern Hundreds* which again starred Matthews, and the highly popular *The Reluctant Debutante* (1955; N.Y., 1956). It was some time before he had another outstanding success, but *The Reluctant Peer* (1964), with Sybil THORNDIKE, did well, as did *The Secretary Bird* (1968). *The Jockey Club Stakes* (1970; N.Y., 1973), a comedy about racing starring Alastair SIM in London, also had a long run. During the next few years Home's plays continued to attract major stars—Ralph RICHARDSON and Peggy ASHCROFT were in *Lloyd George Knew My Father* (1972), Michael DENISON and Dulcie Gray in *At the End of the Day* (1973); Celia JOHNSON in *The Dame of Sark* (1974); and Richardson with Celia Johnson in *The Kingfisher* (1977), which starred Rex HARRISON in New York in 1978.

HOOFT, PIETER CORNELISZ (1581–1647), Flemish dramatist, author of several historical plays of some importance, among them *Geeraerdt van Velsen* (1613), which shows the continuing influence of SENECA, as do several of his other plays, all on classical themes. The allegorical characters common to many of the Rederykers' plays still appear (see CHAMBERS OF RHETORIC), but in Hooft they have more lite and are endowed with the spirit of his time. He is however, chiefly remembered for his comedy *Warenar* (1616), which, though based on PLAUTUS's *Aulularia*, is firmly situated in Amsterdam.

HOPE THEATRE, London, on Bankside. This brick and wood building, of the same shape and size as the SWAN THEATRE, was built by Philip HENSLOWE for the LADY ELIZABETH'S MEN. It replaced the old Bear Garden, and had a removeable stage so that it could still be used for bull- and bear-baiting, and for cockfights. The contract for its construction still survives; it is dated 29 Aug. 1613 and the building, which was roofed over, was in use in the autumn of the following year, when JONSON's *Bartholomew Fair* was shown there before being presented at Court. After Henslowe's death in 1616 a new agreement was made between ALLEYN, who had replaced Nathan FIELD as leading man, and the company, now known as PRINCE CHARLES'S MEN; but, having for a time attracted the audience which had formerly frequented the GLOBE, burnt down in 1613, the Hope lost it again when the rebuilt Globe opened in 1614, and its fortunes continued to decline. It was used by a few minor companies until 1617, after which it reverted to bear-baiting and to its old name of Bear Garden. It was finally demolished in or after 1682.

HOPGOOD, ALAN (1934–), Australian actor and dramatist, whose first play *Marcus* was staged at Melbourne University in 1959. In 1963 his *And The Big Men Fly* was given its first performance by the Union Theatre Repertory Company, now the MELBOURNE THEATRE COMPANY, and has since been revived and televised all over Australia. The U.T.R.C. also staged his *The Golden Legion of Cleaning Women* (1964) and *Private Yuk Object* (1966), about the Vietnam War. *Terr-*

ibly Terribly was first produced in Davis, California, in 1968 and was seen in Melbourne in 1974.

HOPKINS, PRISCILLA, see KEMBLE, JOHN PHILIP.

HORDERN, MICHAEL MURRAY (1911–), English actor, knighted 1983, was an amateur with the St Pancras People's Theatre for several seasons before making his professional début in 1937. After serving in the Navy during the Second World War, he returned to the stage in 1946 as Torvald Helmer in IBSEN's *A Doll's House* and was also seen as Bottom in Purcell's masque *The Fairy Queen* at Covent Garden. He was at Stratford-upon-Avon at Christmas 1948 and 1949 to play Mr Toad in *Toad of Toad Hall*, A. A. Milne's adaptation of Kenneth Grahame's *The Wind in the Willows*, and was then seen in the title role of CHEKHOV's *Ivanov* in 1950, and as Paul Southman in John WHITING's *Saint's Day* (1951). In 1952 he joined the Shakespeare company at Stratford, establishing himself as a first-rate classical actor and going from there to the OLD VIC, where in 1954 he gave an excellent performance as Malvolio in *Twelfth Night*. After appearing in the West End as Sir Ralph Bloomfield Bonington in SHAW's *The Doctor's Dilemma* in 1955, he greatly enhanced his reputation as the senile barrister, cavorting with mincing pomposity, in John MORTIMER's *The Dock Brief* (1958), and then returned to the Old Vic to play Macbeth and Pastor Manders in Ibsen's *Ghosts*. He joined the ROYAL SHAKESPEARE COMPANY in 1962 to appear in London in PINTER's *The Collection* and DÜRRENMATT's *The Physicists*, and was seen at the DUKE OF YORK'S THEATRE in Alan AYCKBOURN's *Relatively Speaking* (1967), followed by Tom STOPPARD's *Enter a Free Man* (1968) at the ST MARTIN's. He then returned to the ALDWYCH to appear in ALBEE's *A Delicate Balance* (1969), and a year later was seen at the CRITERION as the lecherous, agnostic clergyman in David MERCER's *Flint*. He gave for the NATIONAL THEATRE what was perhaps his finest performance as George, the word-spinning, God-obsessed metaphysician in Stoppard's *Jumpers* (1972), and was also excellent in an adaptation of Evelyn Waugh's *The Ordeal of Gilbert Pinfold* (Manchester, 1977; London, 1979) and as Prospero in *The Tempest* at Stratford (1978). He is a remarkable actor who has few peers in the portrayal of quixotic eccentrics.

HORN DANCE, Abbotts Bromley, see FOLK FESTIVALS.

HORNIMAN, ANNIE ELIZABETH FREDERICKA (1860–1937), theatre patron and manager, and one of the seminal influences in the Irish and English theatres at the beginning of the 20th century. The daughter of a wealthy Victorian tea-merchant, she had no connections with the theatre, but from her travels abroad came to realize the important part played in the cultural life of various countries, particularly Germany, by a subsidized repertory theatre. She therefore made funds available in 1894 for a repertory season at the Avenue Theatre (later the PLAYHOUSE) which included *Arms and the Man*, the first play by SHAW to be seen in the commercial theatre, and *The Land of Heart's Desire*, the first play by YEATS to be seen in London. Her introduction to Yeats through this season led her to act for some time as his unpaid secretary, and to take an interest in the new IRISH LITERARY THEATRE, and this in turn led her to build the ABBEY THEATRE in Dublin in 1904. In 1908 she bought and refurbished the GAIETY THEATRE, Manchester, where from 1908 to 1917 she maintained an excellent repertory company, putting on more than 200 plays, at least half of which were new works, among them the plays of the so-called 'Manchester School' of which BRIGHOUSE, MONKHOUSE, and HOUGHTON were the chief exponents, and an early play by St John ERVINE, *Jane Clegg* (1913). Most of the productions were directed by Lewis CASSON, who married a member of the company, Sybil THORNDIKE. The venture was not a success financially and the company was disbanded in 1917, the building becoming a cinema in 1921. Miss Horniman lived long enough, however, to see her pioneer work bear fruit with the spread of the REPERTORY THEATRE movement, and players who had served their apprenticeship under her at the head of their profession in London. On her death she left her extensive library of plays to the British Drama League, now the British Theatre Association.

HÔTEL D'ARGENT, Théâtre de l', see ARGENT.

HÔTEL DE BOURGOGNE, Théâtre de l', see BOURGOGNE.

HOUDAR DE LA MOTTE, ANTOINE (1672–1731), French dramatist, who almost gave up the theatre after the failure of his first play *Les Originaux* (1693), but returned to write opera librettos and lyrics for ballets, some of them the best since QUINAULT. His one important dramatic work is a tragedy, *Ines de Castro* (1723), which was given with great success at the COMÉDIE-FRANÇAISE; his comedies, which include *La Matrone d'Ephèse* (1702), though now forgotten, were successful enough at the time to be parodied at the unlicensed theatres of the Paris fairs. He was an advocate of modernism in the controversy over the UNITIES, and held advanced views for his time about the composition of tragedy, emphasizing the spectators' pleasure rather than obedience to academic rules, and advocating that it should be written in prose. In practice, however, he conformed to the rules like everyone else.

HOUGHTON, (William) STANLEY (1881–1913), English playwright, one of the best of the so-called 'Manchester School' and much influenced by IBSEN, whose plays, dealing with the revolt against parental authority and the struggle between the generations, were first seen at the GAIETY THEATRE in Manchester under Miss HORNIMAN. They include *The Dear Departed* (1908) and *The Younger Generation* (1910), both of which later proved popular with amateur and repertory companies. His best known play *Hindle Wakes* (1912) was, however, first seen in London, played by actors from the Gaiety company and directed by Lewis CASSON, their resident producer. It was an immediate success, though its plot, in which Fanny Hawthorn, a working girl, refuses to marry the cowardly, vacillating rich man's son who has seduced her, portrayed a reversal which took contemporary playgoers by surprise and caused a good deal of controversy. It was several times revived, and was seen in New York soon after its London production. Houghton subsequently had time to write only a couple of one-act plays before his untimely death.

HOUSEKEEPERS, the name given in the Elizabethan theatre to the men, whether actors or not, who owned part (or in some cases the whole) of the actual building in which the company was working. They were responsible for the maintenance of the building and for payment of the ground rent, but, unlike the SHARERS, were not part-owners of the wardrobe and playbooks. Shakespeare was one of the housekeepers of the GLOBE THEATRE, the others being Robert and Cuthbert BURBAGE, who owned a quarter each, the rest being divided between Shakespeare and his fellow-actors John HEMINGE, Augustine PHILLIPS, Thomas POPE, and Will KEMPE.

HOUSEMAN, JOHN, see AMERICAN SHAKESPEARE THEATRE; MARK TAPER FORUM, LOS ANGELES; and WELLES, ORSON.

HOUSE OF OSTROVSKY, see MALY THEATRE, MOSCOW.

HOUSMAN, LAURENCE (1865–1959), English author and playwright, brother of the poet A. E. Housman. His first plays, *Bethlehem* (1902), directed by Gordon CRAIG, and *Prunella* (1904), a COMMEDIA DELL'ARTE fantasy with music, were full-length, but most of his work consisted of one-act plays. The best known of them, dealing with Queen Victoria and her Court, were banned by the censor, a selection of them, as *Victoria Regina* (1935), being seen privately at the GATE THEATRE with Pamela Stanley in the title role, in which Helen HAYES scored a big success in New York in the same year. It was not until Edward VIII intervened that they were licensed for public performance in Britain, and produced at the LYRIC THEATRE in 1937, again with Pamela Stanley as the Queen. A further series of one-act plays entitled *The Little Plays of St Francis*, though often performed by amateurs, does not appear to have been given professionally.

HOWARD, ALAN MACKENZIE (1937–), English actor, nephew of the film star Leslie Howard, whose good looks he inherited. He made his first appearance in the initial production at the BELGRADE THEATRE, Coventry, the musical *Half in Earnest* (1958), going with the company to London to appear in WESKER's *Roots* (1959) at the ROYAL COURT, where in 1960 he was seen in the complete Wesker trilogy. He first attracted attention in Julian Mitchell's adaptation of Ivy Compton-Burnett's novel *A Heritage and Its History* (1965), playing the heir presumptive Simon with the true touch of arrogant, aristocratic pride. A year later he joined the ROYAL SHAKESPEARE COMPANY, remaining with it virtually ever since, and playing a wide variety of parts, among them Lussurioso in TOURNEUR's *The Revenger's Tragedy* in 1966, Benedick in *Much Ado About Nothing* in 1968, Hamlet in 1970, and doubling Theseus and Oberon in Peter BROOK's production of *A Midsummer Night's Dream*, also in 1970, achieving a triumph of physical dexterity and meticulous verse-speaking. A year later he made his début in New York in the same two parts, and on returning to London was seen in GORKY's *Enemies*. Over the next few years a series of Shakespearian parts—Prince Hal, Henry V and VI, Richard II, Richard III, Coriolanus, and Antony (to the Cleopatra of Glenda JACKSON)—brought him great critical praise, and he was also much admired as Jack Rover in the revival in 1976 of O'KEEFFE's *Wild Oats*. One of his most acclaimed performances was in the title role of all three parts of *Henry VI* in 1977. He appeared in Gorky's *The Children of the Sun* in 1979, and in 1981 was seen in the British première of OSTROVSKY's *The Forest* at the OTHER PLACE and the world première of C. P. TAYLOR's *Good* at the WAREHOUSE.

HOWARD, BRONSON (1842–1908), American playwright, one of the first to make use of native material with any skill, and also the first to earn his living solely by playwriting. He worked as a journalist in Detroit until success came to him with *Saratoga* (1870), a farcical comedy produced by Augustin DALY. As *Brighton* (1874), it was adapted for London, Charles WYNDHAM playing the hero Bob Sackett. Howard wrote several other comedies, including *The Banker's Daughter* (1878), which had previously been seen in 1871 as *Lilian's Last Love*, and, as *The Old Love and the New*, was successfully produced in London in 1879. His most important play, however, was *Young Mrs Winthrop* (1882), the first to be produced in London (in 1884) without alteration or adaptation. The most successful of his later plays were *The Henrietta* (1887), a satire on financial life, and *Shenandoah* (1888), a drama of the Civil War.

HOWARD, EUGENE and WILLIE, see VAUDEVILLE, AMERICAN.

HOWARD, SIDNEY COE (1891–1939), American dramatist, who studied playwriting under George Pierce BAKER at Harvard. The first of his plays to be produced, a romantic verse-drama entitled *Swords* (1922), was a failure, but success came with the PULITZER PRIZE-winner *They Knew What They Wanted* (1924), a comedy set in grape-growers' country in Howard's native state of California. It was followed by *Lucky Sam McCarver* (1925), the portrait of a night-club proprietor, *Ned McCobb's Daughter* and *The Silver Cord* (both 1926), the first a sympathetic tale of a New England woman at odds with rum-runners, the second a study of maternal possessiveness. The position of the artist in an unsympathetic community was the theme of *Alien Corn* (1933), a somewhat melodramatic piece which owed much of its success to the fine acting of Katharine CORNELL; *Yellow Jack* (1934), which has often been revived in the amateur theatre, was a factually accurate account of the fight against yellow fever; *The Ghost of Yankee Doodle* (1937), though only moderately successful in production, was perhaps the most satisfactory of Howard's plays, showing how in all classes of society economic considerations overcome the normal aversion to war. Howard had just finished the first draft of *Madam, Will You Walk?* when he was killed in an accident. It was produced in 1953 but, lacking the author's revisions, was not a success.

Howard was a prolific translator and adaptor, being responsible for the American versions of VILDRAC's *Le Paquebot Tenacity* in 1922, Pagnol's *Marseilles* in 1930, René Fauchois's *Prenez-garde à la peinture* in 1932, among others. As *The Late Christopher Bean*, the last had a great success, and was further adapted for British audiences by Emlyn WILLIAMS. Howard also dramatized Sinclair Lewis's novel *Dodsworth* in 1934 and Humphrey Cobb's *Paths of Glory* in 1935.

HOYT'S THEATRE, New York, see MADISON SQUARE THEATRE.

HROSWITHA [HROTSVITHA, ROSWITHA], a Benedictine abbess of Gandersheim in Saxony, who in the 10th century, finding herself drawn by the excellence of his style to read the pagan plays of TERENCE and fearing their influence on a Christian world, set out to provide a suitable alternative. This she did in six original prose plays modelled on those of Terence, but dealing with subjects drawn from Christian history and morality—*Paphnutius, Dulcitius, Gallicanus, Callimachus, Abraham*, and *Sapientia*. They were intended for reading rather than production, but the use of miracles and abstract characters links them with the later MYSTERY and MORALITY PLAYS. The Latin is poor, but the dialogue is vivacious and there are some elements of farce. The plays were published in 1923 in an English translation by H. J. W. Tillyard, and *Paphnutius*, which deals with the conversion of Thaïs, was produced in London in 1914 by Edith CRAIG in a translation by Christopher St John.

HUBERT, ANDRÉ (*c*. 1634–1700), French actor, who in 1664 left the MARAIS to join MOLIÈRE's company, where he first played such parts as Damis in *Tartuffe*, Acaste in *Le Misanthrope*, and Cléante in *L'Avare*. On the retirement of Louis BÉJART in 1670 he took over the playing of elderly women (which actresses of the time did not wish to play), and created the parts of Mme Jourdain in *Le Bourgeois gentilhomme*, Mme de Sotenville in *George Dandin*, and Philaminte in *Les Femmes savantes*. He also played Mme Pernelle in *Tartuffe* when the play was revived, Louis Béjart having played it originally. In Molière's last play *Le Malade imaginaire* he played Monsieur Diafoirus. After Molière's death he became responsible, with LA GRANGE, for the finance and administration of the company, until he retired in 1685. His wife also served the theatre in a minor capacity, probably in the box-office.

HÜBNER, ZYGMUNT, see POLAND.

HUBRIS, literally, 'insolence'; in Greek tragedy the type of pride or presumption in a mortal which offends the gods and causes them to punish the hubristic hero by encompassing his downfall—a situation exemplified in the English proverb 'Pride goes before a fall.'

HUDSON THEATRE, New York, at 139 West 44th Street, between Broadway and the Avenue of the Americas. This handsome and elegant playhouse opened on 19 Oct. 1903 with Ethel BARRYMORE in H. H. DAVIES's *Cousin Kate*. Among its early successes were the first productions in New York of SHAW's *Man and Superman* (1905), Henry Arthur JONES's *The Hypocrites* (1906), and Somerset MAUGHAM's *Lady Frederick* (1908). Later productions included the sharply contrasted *Pollyanna* (1916), from the children's book by Eleanor Porter, and Maugham's *Our Betters* (1917). In 1919 Booth TARKINGTON's comedy *Clarence*, with Helen HAYES and Alfred LUNT, had a long run, and in 1927 Sean O'CASEY's *The Plough and the Stars* had its American première. From 1934 to 1937 the building was used for broadcasting, but returned to live theatre with Cedric HARDWICKE in Barré Lyndon's *The Amazing Dr Clitterhouse*, and a year later Ethel Barrymore returned in Mazo de la Roche's *Whiteoaks*. In 1940 there was a revival of CONGREVE's *Love for Love* and in 1945 came *The State of the Union*, a PULITZER PRIZE-winner by Howard LINDSAY and Russel Crouse. The theatre was again used for broadcasting from 1949 until Lillian HELLMAN's *Toys in the Attic* was produced there in 1960. In 1963 the Actors' Studio presented their initial production, a revival of O'NEILL's *Strange Interlude*, and five years later the

Hudson became a cinema. Renovated and adapted as a 1,000-seat cabaret in 1978, and renamed the Savoy, it reopened in 1981 with a rock concert.

HUGHES, GLENN, see THEATRE-IN-THE-ROUND.

HUGO, VICTOR-MARIE (1802–85), one of France's greatest poets, and the leader of the French Romantic movement. He was also a dramatist, whose plays mark the entry of MELODRAMA into the serious theatre. All alike suffer from overwhelming rhetoric, too much erudition, and not enough emotion, yet by their vigour and their disregard of outworn conventions they operated a revolution in French theatre history. The best is *Ruy Blas* (1838), which has two excellent acts, the second and the fourth, and a superb ending. It still retains the power to move an audience by the passion of its lyric poetry, but remains unconvincing, since Ruy Blas kills himself because he is 'only a lackey', yet has nothing of the lackey in his composition, which is that of the usual well-born romantic hero. Of Hugo's other plays, *Cromwell* (published in 1827) was not intended for the stage and would take six hours to act. It was a battle-cry, and its preface became the manifesto of the new Romantic movement. *Marion Delorme* was banned on political grounds and not acted until 1831, a year after *Hernani*, whose first night led to a riot at the COMÉDIE-FRANÇAISE. *Lucrèce Borgia, Marie Tudor* (both 1833), and *Angelo, Tyran de Padoue* (1835) are prose melodramas, of which the second was revived with Maria CASARÈS as Mary in 1954 at the THÉÂTRE NATIONAL POPULAIRE. *Le Roi s'amuse* (1832), banned after one performance, was used for the libretto of Verdi's opera *Rigoletto* (1851). The vogue for Romantic theatre was bound to be short-lived. The failure of Hugo's last play, *Les Burgraves*, in 1843 showed that the tide had turned in favour of prose and common sense, and he withdrew from the stage, but not before he had brought back to it dramatic verse of a quality unknown since RACINE, and of a totally different inspiration—lyrical, elegiac, colourful, and moving, conveying tragic and comic effects with equal skill.

HULBERT, CLAUDE and JACK, see COURTNEIDGE, CICELY.

HUNEKER, JAMES GIBBONS (1860–1921), American dramatic critic, born in Philadelphia, of Irish-Hungarian extraction. He studied law for a short time, but gave it up to become a concert pianist, which he did not achieve, though he later became a music critic of international renown. In 1890 he became music and drama critic of the *Morning Advertiser* and the *New York Recorder* and in 1902 joined the staff of the *Sun*, leaving in 1912 for the *New York Times*. He was perhaps more of an interpreter than a critic, battling in print with William WINTER over IBSEN and SHAW, and his gusto and worldly knowledge shocked the puritans of the day. He edited a two-volume edition of Shaw's criticisms from the *Saturday Review*, and through his contacts with Europe—as a young man, he had lived in Paris for some years—was able to introduce and explain foreign dramatic literature to his compatriots, particularly through his studies of BECQUE, HAUPTMANN, D'ANNUNZIO, and others. His own dramatic criticisms were collected into a number of volumes, and in 1920 he published an amusing autobiography, *Steeplejack*.

HUNGARY. The oldest written survivals of Hungarian dramatic art are the texts of LITURGICAL DRAMA in Latin, which date from about the beginning of the 11th century. There are also references in later documents to the acting of MYSTERY and MIRACLE PLAYS in some of the larger towns, and to the visits of troupes of dancers and musicians. In the 15th century entertainments by Italian COMMEDIA DELL'ARTE actors were given at the Court of Matthias I, and in the course of the 16th century many plays of all kinds—tragic, comic and satirical—are known to have been performed. Details of presentation, even for those texts which have survived, are sadly lacking, but much can be surmised from a study of JESUIT DRAMA, which flourished in Catholic schools at this time. Plays given in Protestant and Piarist schools first encouraged the emergence of drama in the vernacular, since their pupils were not so deeply committed to the study of Latin. Two of the earliest plays by known Hungarian writers, the *Magyar Elektra* (1558), based on SOPHOCLES' play by Péter Bornemisza (1535–84), and *Theofania* (1575), based on the story of Cain and Abel by Lörenz Szegedi (?–1594), are known to have been first performed in Protestant schools. Some 15 play-texts from the 17th century have survived in manuscript, and in 1695 György Felvinczi (*c.* 1650–1716), author of a *Comico-Trageodia* (1693), attempted to found a professional company. He was not successful, and plays continued to be written mainly by scholars for performance by boys and students; but the use of the vernacular by amateurs led to the introduction of recognizable peasant types and of plots taken from indigenous FOLK FESTIVALS; and the secularization of Hungarian drama proceeded apace.

The first plays in Hungarian not intended for student performances were those of Hungary's most important playwright to date György Bessenyi (1747–1811), author of a comedy, *The Philosopher* (1777), and of several historical tragedies which may be said to mark the beginning of modern Hungarian drama.

Although throughout the 17th and 18th centuries there had been private theatres in many palaces and noble houses in Hungary similar to that maintained by the Esterházys, they were used mainly for music and opera and for the visits of foreign troupes. The first professional Hungarian company was founded in 1790 by Ferenc Kazinczy (1759–1831), the first translator of Shakespeare's plays into Hungarian, and one of the leading figures in the crusade to revitalize the Hungarian language. Led by the actor László Kelemen (1760–1814), this group opened in Buda with a comedy by Kristóf Simai (1742–1833), *Igazházi, a Good Pious Father*. Owing to some local prejudice and the opposition of a well-established German company the venture failed, but a second company which opened in Kolozsvár (Cluj) in Transylvania in 1792 was more successful and by 1801 was already touring the countryside, one of their earliest successes being *Mary Batóri* (1793) by András Dugonics. The great actress Róza Déry-Szeppataki (*née* Rozalia Schenback, 1793–1872), after whom the present Déryné touring theatre is named, acted with this company and may have made her début under Kelemen.

Hungarian playwrights were not slow to grasp the opportunities offered by the demand for new plays; one of the most interesting was the lyric poet Mihály Csokonai Vitéz (1773–1805), author of three *commedia dell'arte*-type comedies of which the best was *The Widow Karnyó and the Two Coxcombs* (1799). A little later Karóly Kisfaludy (1788–1830) was writing, for the new companies which were being established, light comedies and musical pieces which earned for him the title of 'the father of Hungarian comedy.' Among his most popular plays were *The Suitors* (1820) and *Disappointments* (1829), in which he brought on to the professional stage for the first time Hungarian peasant types which had previously figured only in the early vernacular folk-plays. The Kisfaludy Society, named after him, was responsible for a complete edition of Shakespeare's plays in translation, published in 1864, which long remained in use on the Hungarian stage.

Although permanent theatre buildings were being erected in some provincial towns—notably Koloszvár and Miskolc—in the 1820s, it was not until 1837 that the first opened in Pest, supported by the leading dramatists and actors of the day. It became the National Theatre in 1840, its leading actor being Gábor Egressy (1808–66). The first play to be performed there was *Arpád's Awakening* by Mihály Vörösmarty (1800–55), whose later plays *Csongor and Tünde*, a romantic fairy-tale, *Blood Wedding*, a tragedy (both 1831), and *Czillei and the Hunyadis* (1844), a historical drama, were also seen there. The last was revived in 1966 as a grandiose spectacle play by the director of the National Theatre, Endre Marton.

It was at this theatre in Pest, soon after its opening, that one of the masterpieces of Hungarian literature, *Bánk Bán*, usually translated as *The Viceroy*, was first seen. It was written by Jószef Katona (1791–1839) in about 1814; he was unable to get it produced and it was published in 1821. It has a dual theme—a private struggle in the heart of the viceroy Bánk between loyalty to his King and the desire to avenge his wife's honour, which ends with his assassination of the (German) Queen who connived at the wife's seduction; and a public confrontation between the Hungarian nobles, who side with the common people, and their tyrannical German overlords. In the late 1840s this play was used as the libretto of an opera by Ferenc Erkel, musical director of the National Theatre, but again the unsettled political situation prevented its production and it was not staged until 1861. As far as is known Katona, who was by profession a lawyer, wrote no more plays, but his one tragedy, whether as a play or an opera, is constantly revived on the Hungarian stage and a theatre in his birthplace, Kecskemét, has been named after him.

The first professional Hungarian playwright was Ede [Eduárd] Szigligeti (1814–78), whose whole life was spent in the service of the National Theatre. He was its secretary when it was first founded and later its director, and also the author of over 100 plays, many of which he directed himself. He began by writing historical dramas, of which the best is *The Captivity of Ferenc Rákóczi II* (1848), but soon turned to comedy and to the invention of a popular type of folk-play (*népszinmü*) dealing with the adventures of a romantic and idealized type of peasant-hero, enlivened with colourful Hungarian music and folk-dances. Among the best of these are *The Deserter* (1843), *The Cattleman* (1846), *Liliomfi* (1849), and *The Foundling* (1863). In the hands of some of his imitators the folk-hero became somewhat over-romantic and unreal, but two of them maintained the standard—Ferenc Csepreghy (1842–80), author of *The Yellow Colt* (1877) and *The Red Wallet* (1878), two comedies which are often revived; and Ede Tóth (1844–76), for most of his life a provincial actor and the author of some of the most popular Hungarian folk-dramas, among them *The Organ-Grinder's Family* (1875), *The Village Wastrel* (1875), which provided a fine peasant role for the outstanding actress Luiza Blaha (1850–1926), and *The Vagabond* (1878).

Jószef Szigeti (1822–1902), for many years a popular character actor at the National Theatre, was also the author of several comedies which had a great success in their day, among them *The Old Infantryman and His Hussar Son, The Engagement Ring* (both 1856), and *The Fair Shepherdess* (1857). But on the whole the mood of the Hungarian people in the 1840s and 1850s was sombre, particularly after the failure of the 1848 uprising, and although László Teleki (1811–61) had some success with a historical play *The Favourite* (1841) and Sigmund Czakó (1820–47) with comedies, particularly *The Will* (1845), *Leona* (1846), and *The Frivolous One* (1847), both committed suicide. The climate of despair was further exemplified in one of the most powerful works in the Hungarian repertory, *Az ember tragédiája*, twice translated into English as *The Tragedy of*

Man. This great cosmic drama, in which the chief characters are Adam, Eve, and Lucifer and whose theme is the eternal battle of good and evil and Man's part in it, was the work of the poet and philosopher Imre Madách (1823–64). Like several of Madách's other plays, it was considered impossible to stage, until in 1883 the manager of the National, Ede Paulay, gave it its first production and proved its theatrical quality. It has remained in the repertory ever since, and become internationally famous. Its success led to the staging of two other plays by Madách dating from the 1860s —*The Last Days of Csák* in 1886 and *Moses* in 1888.

It was also in the 1880s that REALISM first came to the Hungarian stage with the plays of Gergely Csiky (1843–91), which depict the problems of the middle class, seeking security in government jobs or in winning rich husbands for their daughters from the ranks of the new industrialists: *The Parasites* (1880), *Glittering Misery* (1881), *A Match for Cecile*, *The Stomfay Family* (both 1883), *Soap Bubbles* (1884), and *The Grandmother* (1891). In sharp contrast to Csiky's mood of satirical pessimism was the neo-romantic style of Jenö Rákosi (1842–1929) exemplified in his light comedies *School of Love* (1873) and *The Quarrelsome Queens* (1890). He also wrote a number of historical dramas of which the most successful was *Andrew and Joanna*, based on the unhappy love-affair of a Hungarian king with Joanna of Naples. Another writer of neo-romantic comedy was Lajos Dóczy (1845–1919), whose most successful play and one often revived was *The Kiss* (1871). The novelist Sándor Bródy (1863–1924), already known as the founder of modern bourgeois literature in Hungary, had a great success in 1901 with a dramatization of his own novel *Snow White* and advanced the cause of NATURALISM in Hungary with such plays as *The Nurse* (1902) and *The Schoolmistress* (1908).

At the turn of the century a number of new theatres were established in Budapest, where in the 1860s a second Hungarian-language People's Theatre had been opened under the direction of György Molnár (1830–91). Meanwhile, under Ede Paulay (1836–94), the National Theatre flourished and began to make an international reputation, its leading players being Vidor Kassai (1840–1928), Mari Jaszái (1850–1926), and Emilia Markus (1860–1949). The first of the new theatres was the Vígszínház (Gaiety Theatre), which opened in 1896 and was rebuilt in 1947, after extensive war damage. Both in comedy and tragedy it followed the new Naturalism, whereas the Magyar Theatre, founded in 1897, housed mainly light comedies and the Karăly Theatre, founded in 1903, was devoted to operetta. Fine experimental work was done between 1904 and 1907 by the Thália Society, founded by Sándor HEVESI, the first to recognize the talent of Melchior [Menyhért] Lengyel (1880–1947), whose portrait of the moral weakness of a Japanese scientist, *Typhoon* (1909), was later to achieve an international success.

In the early years of the 20th century Hungarian drama reached a peak of excellence. Ferenc Heřczeg (1863–1954), though German by extraction, interpreted the true Hungarian spirit of his time, both in his comedies, like *The Gyurkovics Family* (1903), and his historical dramas. Other playwrights of note were Dezsö Szomory (1869–1944), author of a number of social comedies; Ernő Vajda (1887–1954), whose *Fata Morgana* (1915) is a touching drama of first love and disillusionment; and the satirical writer Frigyes Karinthy (1888–1938); while Zsigmond Móricz (1879–1942) achieved what many dramatists had been struggling to do since Kisfaludy, with his truthful and realistic depictions of Hungarian farmers, villagers, and herdsmen, in such plays as *Judge Sarí* (1909), *The Village* (1911), *Love* (1913), and *Gentlemen on a Spree* (1928).

The First World War was followed by economic depression and an inevitable decline in the material prosperity of theatrical life; but good actors were not lacking, among them Gyula Csortos and Gábor Rajnay at the Vigszínház, and Árpád Odry and Gizi Bajor at the National, where from 1916 onwards Hevesi directed some outstanding productions. New playwrights contributed good work, among them Jenö Heltai (1871–1957), whose romantic verse-play *The Silent Knight* (1936) was seen in London in 1937, and Lajos Ziláhy (1891–1974), author of many plays dealing with post-war problems, among them *The General* (1928), *The Firebird* (1932), and above all *The Twelfth Hour* (1933). Better known outside their own country were Ferenc MOLNÁR, on whose *Liliom* (1909) the successful musical *Carousel* (1945) was based, and Lajos BÍRÓ, some of whose plays were written in English and first seen in London. Another exile, Gyula HÁY, had most of his plays produced in Germany, while a number of his contemporaries, though they remained in Hungary, had to wait until after the Second World War for their most successful productions, among them László Németh (1901–75), with his *Galilei* (1956) and *The Two Bolyais* (1962), and Gyula Illyés (1902–), better known as a poet, with *Dozsá György* (1956) and *The Brothers* (1972), both based on the Peasants' Revolt of 1514.

The nationalization of the Hungarian theatre in 1949 raised hopes of greater freedom for writers, and for a time drama flourished, with many new theatres being built and audiences increasing everywhere. Stalinism caused a setback for a time, and the unsuccessful uprising of 1956 sent authors like Háy back into exile. It also silenced many others, among them Tibor Déry (1894–1977), who only began to write plays after the Second World War, and whose one-act satire *The Bootlicker* (1953) had fiercely attacked the 'personality cult' of the time. Some dramatists survived, among them István Örkényi (1912–), whose *The Töt Family* (1966) had a great success; Lajos Mesterházi [Hofstetter] (1916–   ), author of *People of Budapest* (1958), produced by the outstanding director Károly Kasimír, and *The Eleventh Commandment* (1960); Miklos Hubay (1918–   ), whose *Nero Playing* (1968) reveals in a series of one-act plays the several aspects of tyranny; Imre Sarkadi (1921–61), whose *Simeon Stylite*, a sympathetic study of a non-conformist, was not seen until six years after his death by suicide; Ferenc Karinthy (1921–   ), son of Frigyes, some of whose one-act satires exemplify the so-called new 'black comedy', particularly *The Grand Piano* (1966); and László Gyurkó (1930–   ), author of a study in tyranny, *Electra, My Love* (1968), and founder in 1969 of the Huszönötodik (25th) Theatre, where he has directed a number of new plays. Among other outstanding directors of the period are Tamás Major (1910–   ), also one of the leading actors at the National Theatre together with Ferenc Bessenyi (1919–   ), and Zoltán Várkonyi (1912–   ), head of the Vígszınház since 1971, where he has directed many of the plays of the younger generation of writers. Two fine actresses of the present day are Éva Ruttkai, who in 1974 appeared in Örkényi's *Kith and Kin*, and Mari Töröcsik. Materially the theatre in Hungary is flourishing, with more than 15 theatres in Budapest and a similar number in the regions, with permanent companies which employ actors on fixed salaries with the possibility of a pension on retirement. There are also a number of puppet-theatres, and in 1976 the State Theatre Institute, with its excellent theatrical archives, was turned into the world's first international multilingual research centre.

HUNGERFORD MUSIC-HALL, see GATTI'S.

HUNT, HUGH SYDNEY (1911–   ), English theatre director and playwright. He had already had some varied experience in play-production when in 1935 he became a director of the ABBEY THEATRE in Dublin, where his first play *The Invincibles*, written in collaboration with Frank O'Connor, was produced in 1938. After serving throughout the Second World War, he returned to the theatre in 1945 to become the first director of the BRISTOL OLD VIC, leaving it in 1948 to go to the OLD VIC in London. In 1955 he was again involved in pioneer work when he was appointed director of the ELIZABETHAN THEATRE TRUST in Australia, where he remained for five years. From 1961 to 1973 he was the first Professor of Drama at Manchester University, and from 1969 to 1971 also Artistic Director of the Abbey Theatre. His *In the Train* (1958), again written in collaboration with O'Connor, was produced in London. He is the author of a number of books on the theatre. (See also MANCHESTER.)

HUNT, (James Henry) LEIGH (1784–1859), English poet and essayist, and one of the pioneers of modern dramatic criticism. He probably had a keener appreciation of acting than any of his contemporaries, and his criticisms—the best of which are contained in *Dramatic Essays* (1894) edited by William ARCHER—recreate the art of the great players who brought distinction to the theatre of his day. He was the first regular critic of any importance to report upon all the principal theatrical events of his time, both in the *News* from 1805 to 1807 and in his own paper the *Examiner*, which he edited from 1808 to 1821, continuing to supervise it even while in prison in 1813 for having published in it some criticisms of the Prince Regent. In 1840 his only play *A Legend of Florence* was produced at COVENT GARDEN, and in the same year he published an edition of the dramatic works of SHERIDAN and other Restoration dramatists with biographical notes, which inspired Macaulay to publish in the *Edinburgh Review* his famous essay on 'The Comic Dramatists of the Restoration.'

HUNTER, N(orman) C(harles) (1908–71), English dramatist, who wrote a number of light comedies, of which *All Rights Reserved* (1935) and *A Party for Christmas* (1938) were seen in the West End, before he scored an outstanding success with two plays which showed the influence of CHEKHOV in their atmosphere and characterization—*Waters of the Moon* (1951), starring Sybil THORNDIKE and Edith EVANS, and *A Day by the Sea* (1953), in which Sybil Thorndike again appeared, partnered by Irene WORTH, John GIELGUD, and Ralph RICHARDSON. In Hunter's next play, *A Touch of the Sun* (1958), Michael REDGRAVE and his daughter Vanessa appeared together for the first time. Of his later plays, the only one to be seen in London was *The Tulip Tree* (1962). *Waters of the Moon* was revived at the CHICHESTER FESTIVAL THEATRE in 1977 and the HAYMARKET THEATRE in 1978.

HURRY, LESLIE (1909–78), English stage designer, most of whose work was done for ballet and opera. He was, however, responsible for the décor of many Shakespeare plays given at the OLD VIC and at the SHAKESPEARE MEMORIAL THEATRE, and also for productions of MARLOWE'S *Tamburlaine the Great* (1951), Graham GREENE'S *The Living Room*, OTWAY'S *Venice Preserv'd* (both 1953), Tennessee WILLIAMS'S *Cat on a Hot Tin Roof* (1958), and WEBSTER'S *The Duchess of Malfi* (1960). His work was characterized by a sombre magnificence which imparted a brooding air of tragedy to his settings, shot through with sudden gleams of red and gold, and he was at his best in designing for plays which like *Venice Preserv'd* called for the conjuring-up of mystery and a sense of space, together with poetic imagery.

HURWITCH, MOSES, see JEWISH DRAMA.

HUSSENOT, OLIVIER, see GRENIER-HUSSENOT, LES.

HUTT, WILLIAM (1920–), Canadian actor-director, who was a student at the HART HOUSE THEATRE in the University of Toronto, his birthplace, before playing in summer stock in Ontario and with the Canadian Repertory Theatre in Ottawa. In 1953 he joined the company at the STRATFORD (ONTARIO) FESTIVAL theatre in its first season, and remained there to play a number of leading roles in Shakespeare, as well as in MOLIÈRE, CHEKHOV, and GOGOL. He has toured extensively, not only in Canada and the U.S.A., where he was first seen in New York in 1956 in MARLOWE's *Tamburlaine*, but in Europe and Australia, being seen in England at the BRISTOL OLD VIC in 1959 in O'NEILL's *Long Day's Journey Into Night*, and in London in *Waiting in the Wings* (1960) by Noël COWARD, in whose *Sail Away* he toured North America a year later. In 1964 he created the part of the lawyer in the Broadway production of ALBEE's *Tiny Alice*, and in 1968 he was seen in New York in *Saint Joan* by SHAW, a play which he directed also at Stratford (Ontario) in 1975. He returned to London in 1969 to play Caesar in Shaw's *Caesar and Cleopatra*. Among other plays he has directed for Stratford (Ontario) are BECKETT's *Waiting for Godot* (1968), *Much Ado About Nothing* (1971), *As You Like It* (1972), and TURGENEV's *A Month in the Country* (1973). In a notable season in 1978 he appeared there in the title roles of *Titus Andronicus* and Chekhov's *Uncle Vanya* and as Falstaff in *The Merry Wives of Windsor*.

HYDE, DOUGLAS (1860–1949), Irish folklorist, historian, poet, translator, and dramatist, and from 1938 to 1945 the first President of Eire. His devotion to the Irish language and GAELIC DRAMA, and his address as president of the National Literary Society in 1892 on 'The Necessity for De-Anglicizing Ireland', did much to create a favourable climate for the Irish dramatic movement, while at the same time indirectly leading to later political disturbances at the ABBEY THEATRE.

In 1901 he contributed to the third season of the IRISH LITERARY THEATRE a play in Gaelic, *Casadh an tSúgáin* (*The Twisting of the Rope*), which was directed by W. G. FAY, with the author playing the leading role. Collaboration with Lady GREGORY (who drafted the scenarios, usually from stories or ideas supplied by YEATS) led to further plays, written in Irish but later translated by Lady Gregory into English—*An Posadh* (*The Marriage*, 1902); *Ar Naom ar Iarraid* (*The Lost Saint*, 1903); *Teach na mBocht* (*The Poorhouse*, 1903), which was later revised by Lady Gregory as *The Workhouse Ward*; and *Drama Breite Criosta* (*The Nativity Play*), produced at the Abbey in 1911. Hyde also collaborated with George MOORE, who worked with him on *An Tincear Agus an tSideog* (*The Tinker and the Fairy*, 1902). Other plays by Hyde include the bilingual *Pleusgadh na Bulgoide* or *The Bursting of the Bubble* (1903); *Rig Seumas* (*King James*, 1903); and *An Magistir Sgoile* (*The Schoolmaster*, 1904).

I

IBSEN, HENRIK (1828–1906), Norwegian dramatist and poet. After a hard childhood he was apprenticed to an apothecary in the provincial town of Grimstad until in 1850 he moved to Christiania to follow his calling as a writer. In his early years, following the publication of his first drama *Catilina* (1850), he experimented in a range of different styles—historical, nationalistically romantic, and comic—both in prose and in verse. In 1851 he was invited by Ole BULL to join as 'dramatic author' the newly established Norske Teatret in Bergen (see NATIONALE SCENE), on the understanding that he would write at least one new play a year for performance there; inevitably, however, he became involved on the production side. In 1852 he went on a study tour of theatres in Denmark and Germany, which contributed greatly to his sense of theatre. Between 1851 and 1857 five of his early plays—*Sankthansnatten* (*St John's Night*, 1853), *Kjæmpehøjen* (*The Burial Mound*, 1854), *Fru Inger til Osteraad* (1855), *Gildet paa Solhoug* (*The Feast at Solhoug*, 1856), and *Olaf Liljekrans* (1857)—were performed in Bergen; he seems also to have had a hand in no fewer than 145 different plays produced during these years.

In 1857 Ibsen returned to the Norwegian capital to become the first artistic director of the Christiania Norske Teatret, and then (after its bankruptcy in 1862) of the other theatre there, the KRISTIANIA THEATER. Between 1857 and 1863 he completed three further plays: *Hæmændene paa Helgeland* (*The Vikings at Helgeland*, 1857), a severe, simple, and moving tragedy set in the world of the sagas; *Kjælighedens Komedie* (*Love's Comedy*. 1862), a satirical verse-play of contemporary life; and *Kongsemnerne* (*The Pretenders*, 1863), an impressive historical play whose interest is psychological and poetic.

In 1864, after the award of a modest travelling scholarship, Ibsen left Norway for Italy. For the next 27 years he lived abroad. His years in Italy brought a breakthrough in his career; his verse dramas *Brand* (1866) and *Peer Gynt* (1867) were immediate cultural and commercial successes, and the works which did most initially to establish his European reputation. *Brand* revealed his characteristic sternness, strength, and searching questioning of motive and deed in their full power; and his method of implying the nature of truth by a series of unanswered questions and seeming negations is displayed here in all its bleakness. *Peer Gynt* is in many ways complementary to *Brand*, a study of the Norwegian character conducted in a mood of mordant irony.

After *Peer Gynt* Ibsen abandoned verse as a dramatic medium. *De Unges Forbund* (*The League of Youth*, 1869) is an uncharacteristically light-hearted satire on dishonesty and insincerity. The enormously long ten-act *Kejser og Galilæer* (*Emperor and Galilean*, 1873) is a complex study, in prose dialogue, of the struggle between paganism and early Christianity under the Emperor Julian the Apostate, in which the concept of 'the third realm' is advanced, a modern synthesis for the new age.

Beginning in 1877, there followed the twelve 'contemporary' dramas which finally consolidated Ibsen's wider international fame. *Samfundets Støtter* (*Pillars of Society*, 1877) is a study of public life rooted in a lie; *Et Dukkehjem* (*A Doll's House*, 1879) explores the insidious destruction worked by deceit and deception in domestic life; *Gengangere* (*Ghosts*, 1882) uses hereditary disease as a symbol for the moral corruption in society; and *En folkefiende* (*An Enemy of the People*, 1882) considers the position of the man of truth in conflict with a polluted society. All four plays are realistic accounts of the small-town life of Ibsen's own day; and in mercilessly revealing the lies upon which certain societies, self-righteous and self-contained, are founded, they present an ageless and universal parochialism. On the threshold of the last group stands *Vildanden* (*The Wild Duck*, 1884). In this group SYMBOLISM plays an increasingly large part, and the interest shifts gradually from the individual in society to the individual exploring strange areas of experience, isolated and alone. *Rosmersholm* (1885–6) traces the growth of a mind in contact with a tradition of nobility; *Fruen fra Havet* (*The Lady From the Sea*, 1888) is concerned with the overcoming of obsession by freedom and responsibility. *Hedda Gabler* (1890) subtly and skilfully contrasts the virtue of nature and normality with the effects of an artificial society.

Ibsen's last four plays, beginning with *Bygmester Solness* (*The Master Builder*, 1892), form a coherent sub-group within his work, in which his obsession with patterns and problems of human relations in conditions of crisis and stress finds expression. The symbolism of *The Master Builder* is stronger than that of any earlier play, much of it concerned with the relation of the artist and the man within an individual. *Lille Eyolf* (1894) is a study of marital relations, of the nature of love and the distinctions between its kinds; *John Gabriel Borkman* (1896), of unfulfilled genius and the relation of the genius to society. *Naar vi Døde Vaagner* (*When We Dead Awaken*, 1900) is Ibsen's last pronouncement on the artist's relation to life and to truth.

Ibsen's achievement was little regarded outside Scandinavia until 1878, when *Pillars of Society* was widely played in Germany, especially in Berlin. In the 1890s, however, he finally burst upon the European scene. The new wave of independent theatres—the THÉÂTRE LIBRE in Paris, the FREIE BÜHNE in Berlin, and GREIN's Independent Theatre in London—all seized upon *Ghosts* as the work which best expressed the spirit of the age. During this period the appearance of a new Ibsen play was a profoundly European occasion, the concerted launching, in book form and upon the stage, of a single dramatic work in a wide range of Europe's cultural capitals. Never before had a single dramatist so monopolized public debate or so dominated the European theatre. Ibsen is now widely recognized as the creator of modern European drama. As well as providing a new and stimulating means of expression for contemporary social and psychological problems, he revolutionized the potential of prose as a dramatic medium. His consummate technical mastery, his pitiless insight into character, and his subtle exploitation of symbolism gave a lead to 20th-century drama which is still operative and which, both by the intrinsic achievement of his work and by the stimulus it provided, places him among the world's greatest dramatists.

The impact of Ibsen on the English and American theatres was so decisive that some account of the first productions of his plays in translation is essential. Their reception was very different in England and in America, as the influence of a large immigrant population from Scandinavia prevented the emergence in the United States of the opposition which Ibsen's plays had to face in England. The first to be seen in London, in 1880, was *Samfundets Støtter*, translated by William ARCHER (who quickly established himself as the main translator of Ibsen) as *Quicksands*; under the better known title of *Pillars of Society* it was staged in 1889. *Et Dukkehjem*, as *Breaking a Butterfly*, with TREE as Krogstad (renamed Dunkley), was seen in 1884; as *Nora*, in 1885; and as *A Doll's House*, with Janet ACHURCH as Nora, in 1889. Florence FARR and Frank BENSON appeared in *Rosmersholm* in 1891, and later in the same year came the private production of *Gengangere*, as *Ghosts*, which aroused a storm of abuse, especially from Clement SCOTT. The play was not seen in public until 1914. The year 1891 saw also the first performances of *The Lady from the Sea* and in 1893 came *The Master Builder* and *An Enemy of the People*, again with Tree. *The Wild Duck* in 1894, *Little Eyolf* in 1896, and *The League of Youth* in 1900 were followed by a STAGE SOCIETY production of *When We Dead Awaken* in 1903. In the same year *The Vikings at Helgeland* was produced by Gordon CRAIG with his mother Ellen TERRY as Hiordis. *Lady Inger* came in 1906, *Peer Gynt* and *Olaf Liljekrans* in 1911, *The Burial Mound* (as *The Hero's Mound*) and *Brand* in 1912, *The Pretenders* in 1913, *St John's Night* in 1921. In 1936 Donald WOLFIT appeared in *Catiline*. A production outside London deserving mention is *Love's Comedy*, produced as early as 1909 at the GAIETY THEATRE in Manchester. Most of Ibsen's plays have been revived, often in outstanding productions with splendid casts and in new and excellent translations. Apart from Archer, the main providers of modern versions have been Una Ellis-Fermor, whose scholarly texts were not very theatrical; Ann Jellicoe, who was responsible for a new version of *The Lady from the Sea* in 1961, in which Margaret LEIGHTON appeared; Max Faber, in whose new *Hedda Gabler* in 1954 Peggy ASHCROFT gave a superb performance; and Michael Meyer, whose version of *Brand* at the LYRIC, Hammersmith, in 1959 had an unexpected success, followed by *An Enemy of the People* at Nottingham in 1962 and *The Pretenders* at Bristol in 1963. New translations of all Ibsen's plays under the editorship of J. W. McFarlane (1962–77) are accompanied by introductions, commentaries, and notes.

The world première of *Gengangere* was given in Chicago in the original Norwegian in 1882, a year before its first production in Europe, and in the same year America saw its first translated Ibsen play, *Et Dukkehjem* as *The Child Wife*; it was presented in 1883 in Louisville, Kentucky, as *A Doll's House* with MODJESKA as Nora. *Ghosts* followed in 1894 and *John Gabriel Borkman* in 1897. The critical response was generally hostile, *Ghosts* arousing in William WINTER much the same reactions as Scott's in London, but leading ladies of the calibre of Mrs FISKE and Beatrice CAMERON persisted in playing *A Doll's House*, *Ghosts*, and *Hedda Gabler* (first produced in America in 1898) even at some risk to their careers. *The Master Builder* was seen in 1900, *Rosmersholm* and *The Pillars of Society* in 1904, and *When We Dead Awake(n)* in 1905. By the year of Ibsen's death, 1906, opinion had changed in his favour; Richard MANSFIELD presented *Peer Gynt* in that year, and when NAZIMOVA mounted a season of Ibsen plays in 1906–7 the important literary critics James HUNEKER and William Dean Howells came to his defence. The first American productions of *Brand*, *Little Eyolf* (with Nazimova), and *The Lady from the Sea* came in 1910, and of *The Wild Duck* in 1918. Eleonora DUSE's national tour with *Ghosts* and *The Lady from the Sea* in 1923 paved the way for the ACTORS' THEATRE and Eva LE GALLIENNE's Civic Repertory Company, both in New York, to champion Ibsen's plays alongside other important European work. Walter HAMPDEN gave *An Enemy of the People* its American première in 1927 and by 1929 the plays had become popular box-office attractions: in that year there were six Ibsen productions on Broadway, with Eva Le

Stage settings by Buontalenti, *c*. 1589 (*top*), and Furtenbach, 1628 (*bottom*). The influence of Serlio's symmetrical perspective settings was to continue throughout the 17th century.

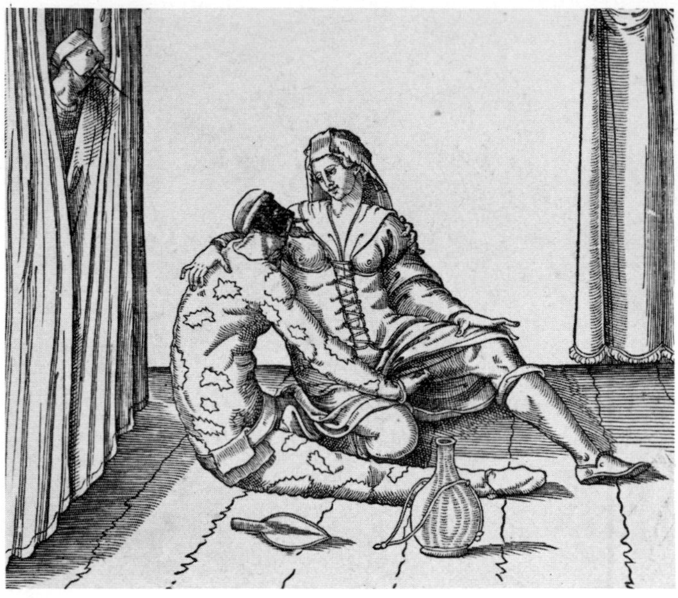

Commedia dell'arte: (*above*) spied on by Pantalone, Arlecchino woos Francisquina, a variable character (on occasion a *travesti* role) who was to disappear from the standard troupe; (*right*) Coviello, wearing a costume and mask typical of the period and dramatic style. This character too was a variable and comparatively short-lived type.

La Farce des gueux, early 17th century, played in one of the booth theatres popular at the Paris fairs.

Mahelot's set for Théophile's *Pyrame et Thisbé* as played at the Hôtel de Bourgogne in the early 1620s, reconstructed in the Nuffield Drama Theatre and Workshop at Lancaster University, 1972.

Farce at the Hôtel de Bourgogne, *c.*1630, with three of the era's most popular comic characters, Gros-Guillaume, Gaultier-Garguille, and Turlupin.

Inigo Jones, design for *The Temple of Love*, 1635. The two-tiered scene of relief is concealed by pairs of shutters drawn back in grooves to reveal the upper or lower part as required.

Setting by Torelli for Corneille's *Andromède*, 1650, a machine play with complex trick effects.

The arrival of Neptune in *Pietas Victrix* by Nikolaus Avancini, a sumptuous Jesuit school production, Vienna, 1659.

Stage of the Schouwburg, Amsterdam, 1637, basically a permanent setting but with pairs of painted wings

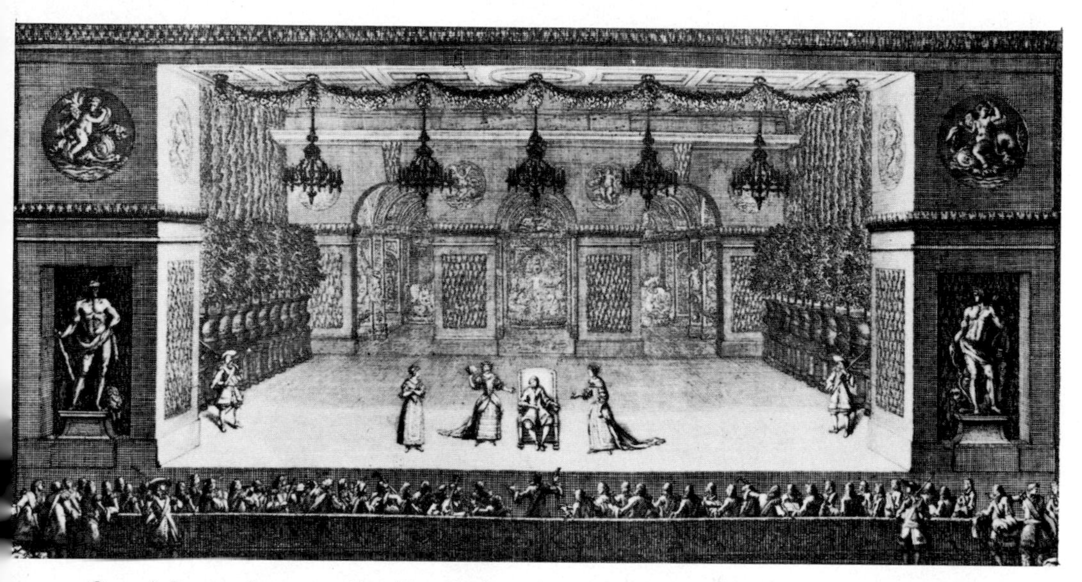

Open-air Court performance at Versailles of Molière's *Le Malade imaginaire*, 1674. A line of large chandeliers in the proscenium opening was the customary stage lighting in the permanent theatres of Paris at this time.

Design by James Thornhill for Motteux's *Arsinoë, Queen of Cyprus*, early 18th century.

Stage design by Ferdinand Galli di Bibiena, early 18th century. The use of angled perspective became a hallmark of the Bibiena family's work for the theatre.

Theatre at the Foire St-Laurent, Paris, built in 1721. The balcony accommodated the *parade*, held to attract audiences into the theatre.

Machinery designed *c.*1730 by Gillot for the Comédie-Italienne. The Italian actors had settled at the permanent theatre at the Foire St-Laurent

The first Burgtheater, Vienna, completed in 1756. The handsomely decorated auditorium is straight-sided, although the horse-shoe plan was by now becoming common.

The John Street Theatre, the first permanent playhouse in New York, opened in 1767. Simpler in style than its European counterparts, it retained the proscenium doors introduced in the English Restoration.

Garrick's production of Sheridan's *The School for Scandal*, Drury Lane. 1777; the screen scene. Older features include the large forestage, stage boxes, and proscenium doors, in striking contrast with the naturalistically painted backcloth and wings and the imaginative use of oblique lighting.

Setting for a farmhouse scene, Drottningholm, late 18th century. Every detail, including the furniture, is painted on the wings and backcloth.

Drury Lane Theatre, 1808. The new theatre, opened in 1794 and burned down in 1809, retained many of the features of the stage used by Garrick.

Sadler's Wells Theatre in 1809, showing one of the wildly popular aquatic dramas performed in a tank installed on the stage.

The stage of the Georgian Theatre (*above*), Richmond, Yorkshire, with a woodland setting of *c.*1820. The theatre was reopened in 1963 with the stage and auditorium restored to their earlier form.

The Park Theatre (*left*), New York, 1822, opened a year after its predecessor had been destroyed by fire.

Bäuerle's *Die Räuber in den Abruzzen, oder der Hund seines Herrn Relber*, produced in Vienna, *c.*1830. Dramas featuring performing animals, especially dogs, had a vogue in Britain, too, at this time.

A production of Holberg's *Jacob von Thyboe* at the Royal Theatre, Copenhagen, 1830. Dating from 1748, the original building was used for a century.

Design for Kleist's *Der Prinz von Homburg, c.* 1875, by the Duke of Saxe-Meiningen. Crowd scenes were a striking feature of the Meininger company's productions.

The Haymarket Theatre, 1880, as remodelled by the Bancrofts. The pit was replaced by stalls, most of the boxes by circles, and the proscenium redesigned to contain the stage picture within a complete frame.

The second Burgtheater, Vienna, opened in 1888. The design by Gottfried Semper and Karl Hasenauer retained the straight-sided auditorium and tiers of boxes, in contrast to the amphitheatral plan becoming common in Europe.

Gallienne and Blanche YURKA appearing simultaneously as Hedda Gabler. *The Vikings* had its first American production in 1930.

With the trauma of the economic depression Americans began to reject foreign classics, including Ibsen's plays which were viewed as museum pieces, and with the notable exception of tours and Broadway productions by Eva Le Gallienne they were rarely seen for the next 20 years. This period of comparative indifference ended with Arthur MILLER's adaptation of *An Enemy of the People* in 1950 and Lee STRASBERG's production of *Peer Gynt* in 1951, and even more attention was gained by television productions featuring popular actresses such as Tallulah BANKHEAD, Julie HARRIS, and Jessica TANDY; the 1960s saw successful Broadway productions of *Hedda Gabler, Rosmersholm*, and *The Wild Duck*, as well as the American première of *The Pretenders* in 1960. Ibsen's plays were once more accepted as popular classics: the EISENHOWER THEATER in Washington's Kennedy Center opened in 1971 with Claire BLOOM in *A Doll's House*, translated by Christopher HAMPTON. As in Britain the outstanding plays have been frequently revived, often in new versions with leading players, for whom Ibsen provided so many varied and rewarding parts.

IBSEN, LILLEBIL (1899–   ), Norwegian ballet dancer and actress, who in 1919 married IBSEN's grandson Tancred (1893–1978), a film director best known for his screen versions of his grandfather's plays. She began her dancing career at the age of 12, and before long was internationally known, appearing often in Berlin and London as well as in the Scandinavian capital cities. Later she devoted herself entirely to acting, and from 1928 until 1956 was attached mainly to the OSLO NYE TEATER. Her great strength lay in comedy, but she successfully appeared in a number of serious roles, including Nora in *A Doll's House* and Mrs Alving in *Ghosts*; and in her later years with the NATIONALTHEATRET, from 1956 to 1969, she was also seen in plays by such modern dramatists as SHAW, ALBEE, and DÜRRENMATT.

ICELAND. The history of the theatre in Iceland can be traced back only to the 18th century, to SCHOOL DRAMA performances, in Latin and in the vernacular, at the Grammar School in Skáholt, later in Reykjavik. By the following century, plays by HOLBERG were frequently performed in translation, and over the years amateur companies began to supplement the School Drama.

In the second half of the 19th century, many French and Danish classics were performed in translation, and in 1892 IBSEN's *The Vikings at Helgeland* was produced at one of the two rather primitive theatres in Reykjavik—the first performance of an Ibsen play in Iceland. Under the inspiration of the dramatist Indriði Einarsson, 1897 saw the founding of the Reykjavik Theatre Association (Leikfélag Reykjavíkur), which over the succeeding decades worked energetically for the establishment of a good theatre; gradually a modern Icelandic drama began to emerge, particularly from the pens of Jóhan SIGURJÓNSSON and Guðmundur KAMBAN, both writing in Danish, and Davið Stefánsson.

A National Theatre building, to plans by the achitect Guðjon Samúelsson, was begun in Reykjavik in 1930, but owing to financial and other difficulties was not completed until 1950. It opened on 20 Apr. with a repertory season which included a number of European plays in translation as well as three plays by Icelandic authors, including *The Bell of Iceland*, adapted from his trilogy of novels by Halldór LAXNESS.

In 1961 a State drama competition brought new vigour to the scene. An experimental theatre group Gríma (Mask) made its appearance alongside the two existing professional companies, the National Theatre and Reykjavik Drama Society, and two young writers, Jökull JAKOBSSON and Oddur Björnsson (1932–   ), firmly established their reputations. Since this breakthrough, a number of younger talented playwrights have had their work performed, notably Erlingur E. Halldórsson (1930–   ), Guðmundur Steinsson (1925–   ), Magnus Jonsson (1938–   ), Svava Jakobsdóttir (1930–   , sister of Jökull Jakobsson), and Nina Björk Árnadóttir (1941 –   ). The permanent company of the National Theatre consists of about 15 actors on long-term contracts, with others engaged for single seasons or performances. A drama school, a ballet school, and a choir are attached to the theatre. Although the population of Reykjavik is only about 70,000, audiences at the theatre have averaged 100,000 a season.

IDEN, ROSALIND, see PAYNE, BEN IDEN, and WOLFIT, DONALD.

IFFLAND, AUGUST WILHELM (1759–1814), German actor and dramatist, who in 1777 was taken on by EKHOF in Gotha, and a year later went with the company, on Ekhof's death, to MANNHEIM, where DALBERG had recently taken charge of the newly opened National Theatre. Here his first plays were produced with much success, and he also played the part of Franz Moor in the first production of SCHILLER's *Die Räuber* (1781). The simple, unsophisticated characters, well-made plots, and superficial nobility and ecstasy of his own ephemeral plays caught the taste of public and critics alike, and the best known, *Die Jäger* (1785), was translated into English as *The Foresters* (1799). He toured in them with much success, and even acted at HAMBURG with the great SCHRÖDER, who, however, thought little of him. Iffland preferred the atmosphere of WEIMAR, where, inspired by the kindness of GOETHE, he played 16 of his best parts to great applause. As an actor he developed a fine technique, but no depth: it was said of him that he was capable of great moments but not of sustained effort. He was at his best in dignified comedy, in the parts of retired officers, elderly councillors, indulgent parents, reverend and witty old men. He left his mark on the German theatre by his work at Mannheim, which he practically controlled throughout its heyday. In 1796 he went to BERLIN, where he remained until his death. His cautious policy and restricted repertory led to trouble with the extremists among the rising generation, and he was often embittered by their attacks, but he trained a number of young actors, not in his own virtuosity, but in the serious, sober style of Schröder, and shortly before his death took on Ludwig DEVRIENT.

ILLINGTON, MARGARET [Maude Ellen Light] (1879–1934), American actress who made her first appearances in Chicago. Engaged by Daniel FROHMAN, who coined her stage name, she first appeared in New York at the CRITERION THEATRE in Sept. 1900. Three years later she married Frohman, from whom she was divorced in 1909. She remained on the stage, and later toured under the management of her second husband, Major Bowes, making her last appearance in 1919. A woman of strong personality, she was at her best in forceful, passionate parts; one of her biggest successes was Marie Louise Voysin in Gordon-Lennox's *The Thief*.

ILLUSTRE-THÉÂTRE, the company with which MOLIÈRE, drawn into it by his friendship with Madeleine BÉJART, made his first appearance on the professional stage. The contract drawn up between the first members, among whom were three of the Béjart family, is dated 30 June 1643, and was modelled upon that of the CONFRÉRIE DE LA PASSION. In essentials, it has remained the basic constitution of the COMÉDIE-FRANÇAISE. The company played for a time in the provinces, possibly at Rouen, opening on 1 Jan. 1644 in Paris, but without much success. By Aug. 1645 the Illustre-Théâtre had come to an ignominious end, and vanished without leaving a trace in contemporary records. Its repertory included plays by CORNEILLE, DU RYER, and TRISTAN L'HERMITE, and some specially written by a member of the company, Nicholas Desfontaines, all of which had the word *illustre* in the title.

ILLYÉS, GYULA, see HUNGARY.

ILYINSKY, IGOR VLADIMIROVICH (1901–), Soviet actor, who made his first appearance in Petrograd in 1918, in which year he also played in the first 'Soviet' play MAYAKOVSKY'S *Mystery-Bouffe*. Two years later he joined MEYERHOLD's company, remaining with him with one or two breaks until 1935 and appearing in all his major productions, including CROMMELYNCK's *The Magnificent Cuckold* in 1922 which made his name. In 1938 he went to the MALY THEATRE, and established himself as a fine interpreter of leading roles in the plays of OSTROVSKY. He was also seen in two new plays, as Chesnok, the co-operative farm president in KORNEICHUK's *In the Steppes of the Ukraine* in 1942, and as the sailor Shibayev in VISHNEVSKY's *The Unforgettable 1919* in 1951. In *John Reed* (1967), an adaptation of Reed's *Ten Days That Shook the World* staged for the 50th anniversary of the Revolution, he played Lenin. He also became a notable director, being responsible in 1958 for a production based on Thackeray's *Vanity Fair*, and in 1962 for SOFRONOV's *Honesty*.

IMMERMANN, KARL (1796–1840), German director, from 1834 to 1837 of the Düsseldorf theatre. Under the influence of GOETHE's work at WEIMAR he tried to raise the standard of theatrical production, insisting on clear diction, and made many technical improvements in scenery, costume, and lighting. The public, however, was apathetic, and his experiments ended in failure.

IMPERIAL THEATRE, London, in Tothill Street, originally the Aquarium Theatre, part of an amusement and exhibition palace known as the Royal Aquarium Winter Garden, which stood on the site of the present Central Hall, Westminster. The theatre opened on 15 Apr. 1876, and PHELPS made his last appearance there in 1878. On 21 Apr. 1879, under the management of Marie Litton (Mrs Wybrow Robertson), the name was changed to the Imperial. It closed in 1899 and Lillie LANGTRY, who had already appeared there in 1882 in Tom TAYLOR's *An Unequal Match*, took over, virtually rebuilding it, and opened it on 22 Apr. 1901 with Berton's *A Royal Necklace*. In spite of good reviews of her acting in the dual roles of Marie Antoinette and Mlle Olivia and of the sumptuous costumes and scenery the play was not a success, and in 1903 she withdrew. Ellen TERRY then presented IBSEN's *The Vikings at Helgeland*, and appeared herself as Beatrice in *Much Ado About Nothing*, both productions being designed and directed by her son Gordon CRAIG. Lewis WALLER was at the Imperial from 1903 to 1906, presenting a series of romantic plays of which the most successful, apart from the perennial *Monsieur Beaucaire* by Booth TARKINGTON, was Conan Doyle's *Brigadier Gerard* (1906). The last play seen at this theatre was Dix and Sutherland's *Boy O'Carrol* (also 1906), with MARTIN-HARVEY. The theatre was then dismantled and taken to Canning Town, where it was re-erected as the Imperial Palace. It later became a cinema, and was destroyed by fire in 1931.

IMPERIAL THEATRE, New York, at 249 West 45th Street, between Broadway and 8th Avenue. This opened on 24 Dec. 1923, with a seating capacity of 1,452, and was intended by the Shuberts for musical shows, among its early successes being *Rose Marie* (1924) and *Oh, Kay!* (1926). In 1936 Leslie Howard appeared there in *Hamlet*, competing unsuccessfully with the record-breaking production starring GIELGUD at the EMPIRE. It then reverted to musicals, Ethel MERMAN starring in *Annie Get Your Gun* (1946) and *Call Me Madam* (1950). Later successful musicals were *The Most Happy Fella* (1956), *Carnival!* (1961), *Oliver!* (1963), *Fiddler on the Roof* (1964), and *Zorba* (1968). John OSBORNE's *A Patriot for Me* (1969) was followed by more musicals, including *Pippin* (1972), and by Neil SIMON's *Chapter Two* (1977) and the musical *They're Playing Our Song* (1979), for which Simon wrote the book.

IMPERSONATION, see ANIMAL, FEMALE, and MALE IMPERSONATION.

IMPROVISATION, an impromptu performance by an actor or group of actors, which may be an element in actor-training, a phase in the creation of a particular role, or part of a staged production. For much of its history the theatre relied heavily on the actor's ability to improvise on a given theme, as in the comic scenes of the MYSTERY PLAY, in early farces and comedies deriving from folk tradition, which gave ample scope for the actor's initiative, and above all in the productions of the COMMEDIA DELL'ARTE. The practice continued in MELODRAMA and PANTOMIME and in the MUSIC-HALL, and even now 'business' and 'ad-libbing' by adroit comedians is permissible in certain types of productions, as long as it is not carried too far. That Shakespeare suffered from the over-enthusiastic improvisation of some of his actors is shown by his admonition in *Hamlet* (III. ii): 'Let those that play your clowns speak no more than is set down for them', and by the end of the 19th century the battle for the supremacy of the author's text had resulted in the actor becoming an interpreter rather than a creator of situations. But even STANISLAVSKY recognized the value of improvisation during lengthy rehearsal periods to help an actor explore a character's background and motivation, resulting in a richer characterization. His ideas were adopted all over the world and exercises in improvisation aimed at the release of personal inhibitions and the development of physical and vocal skills became a part of all actors' training. Meanwhile, in a reaction against the set text, the influence of DADA and Surrealism was encouraging spontaneous activity and the rejection of the achieved 'work of art'. This philosophy resulted in the emergence of COLLECTIVE CREATION, based entirely on group improvisations elaborated from minimal outlines. The movement spread rapidly and from 1960 onwards became worldwide. From it developed HAPPENINGS and similar environmental events, which by incorporating chance occurrences and audience reaction into the show created a unique theatrical experience, considered by many preferable to the controlled repetition of a conventional playtext.

INCHBALD [*née* Simpson], ELIZABETH (1753–1821), English actress and one of the first English women dramatists. She was for a time on the stage, in spite of a slight impediment in her speech which she never wholly overcame. She was first seen in the provinces, where she acted Cordelia to the King Lear of her husband Joseph Inchbald (?–1779), an inoffensive little man who painted and acted indifferently, and survived his marriage to the beautiful and spirited Elizabeth by only seven years. She was acting in Tate WILKINSON's company when her husband died, and later appeared at COVENT GARDEN, retiring in 1785 to devote herself to playwriting after the moderate success of her first plays, which included *I'll Tell You What* (1785), *The Widow's Vow* (1786), and, from the French, *The Midnight Hour; or, War of Wits* (1787). She was a capable writer of sentimental comedy, and though none of her plays has been revived, they were successful in their own day. Among the best were *Wives as They Were and Maids as They Are* (1797), and her last comedy *To Marry, or Not to Marry* (1805). She also made English versions of DESTOUCHES's *Le Philosophe Marié* as *The Married Man* (1789) and of KOTZEBUE's *Das Kind der Liebe* as *Lovers' Vows* (1798).

INCIDENTAL MUSIC, music written expressly for a dramatic performance, which seldom survives the play for which it was intended. Mendelssohn's Overture to *A Midsummer Night's Dream*, written in 1826, was not incidental music but a symphonic poem for concert performance based on the subject of Shakespeare's play. The music Mendelssohn wrote for a production of the play in 1843 was incidental music, properly speaking, and the overture became part of it.

The origin of incidental music must not be looked for in the stage entertainments of Renaissance Italy nor in the MASQUE which derived from them in England, for in both music is an integral part of the work. Traces of it can be seen, however, in Elizabethan drama and also in classical Spanish drama. It is clear that Shakespeare's plays demanded a good deal of music, not only for interpolated songs, for sonnets (a word probably derived from *sonata*), and for tuckets (from *toccata*), but also for interludes and dances. There must have been music, perhaps for a consort of viols or a broken consort of wind and strings, at the opening of *Twelfth Night* and in the last acts of *A Midsummer Night's Dream* and *Much Ado About Nothing*. In Spain, the plays of CALDERÓN, Lope de VEGA, and many other playwrights continually call for music of various sorts; yet they have too much action and spoken dialogue to be classed as operas. Their music is not casual but it is clearly incidental. In English Restoration plays it has been said to be casual, but this is to ignore the important contributions made by Purcell to the production of plays by such dramatists as Shakespeare, BEAUMONT and FLETCHER, CONGREVE, and DRYDEN. His music for plays must be called incidental, and it marks an important historical advance in the use of music in the theatre: nowhere else in Europe in the 17th century did incidental music result in anything as fine as the stage works of Purcell. The reason for this is twofold—the flourishing condition of the spoken drama in England and the failure of opera to thrive there as it did on the Continent, except in Spain where the spoken drama as well as opera was in decline.

There was not much call for incidental music anywhere during the 18th century, mainly owing to the popularity of BALLAD OPERA and other types of light operatic entertainment in which the music, though often trivial, was none the less essential and not incidental. But it did at least prepare the way for developments in the 19th century by enlarging and improving the orchestral resources of the theatre. The opera orchestras in London and Paris were the best of their time, while in Italy, except at one or two ducal courts, they were the only ones. In Germany conditions were much more favourable to the development of incidental music. Although the German Court theatres vied with each other in lavish productions of opera, the majority of them were situated in small provincial towns like MANNHEIM and WEIMAR and depended for their audiences on the Court's household and, in some places, the university. Drama provided a change of bill but, except in a few cities like VIENNA, had to be given in the same theatre. The necessity for keeping an orchestra in readiness for operatic performances led to an increase in the amount of music used in plays, primarily in order to provide employment for the musicians. Overtures and musical interludes between the acts became customary, and it was not long before playwrights made the most of the musical facilities available by inserting into their new plays not only songs and dances and processions to music, but scenes of excitement or pathos in which the spoken words were accompanied by an orchestral undercurrent. The MELODRAMA, popular in England later in the century, in which murder, robberies, suicides, and furtive escapes took place to appropriate music, developed from this earlier practice.

From the beginning of the 19th century, major composers were supplying scores which have survived the dramas for which they were intended, including Beethoven's overture for the 1810 production of GOETHE's *Egmont*, Schubert's music for Helmine von Chézy's ephemeral *Rosamunde* (1823), Schumann's score (1848–9) for BYRON's *Manfred*, and Mendelssohn's full incidental music for *A Midsummer Night's Dream*. By the second half of the century incidental music was established as a separate category of some importance, with such works as Bizet's music for Daudet's *L'Arlésienne* (1872) and Grieg's for IBSEN's *Peer Gynt* (1876). Everywhere poetic drama, in particular, was provided with some form of music, often by famous composers. Both Fauré in 1898 and Sibelius in 1905 wrote incidental music for MAETERLINCK's *Pelléas et Mélisande*, a play which has now been superseded by the music-drama based on it by Debussy. French composers of the inter-war years who wrote incidental music include Darius Milhaud, Arthur Honegger, and Jacques Ibert, whose *Divertissement* (1930), derived from his music for LABICHE's *Un Chapeau de paille d'Italie*, has proved very popular in its own right. In the lavish conditions of pre-revolutionary Russia, Balakirev wrote music for *King Lear* (1861) and Tchaikovsky for *Hamlet* (1891); and the Soviet Union has maintained the traditions of using large-scale orchestral forces in the serious theatre. The cinema has claimed the best of modern American incidental music, although early productions of plays by Eugene O'NEILL had interesting scores.

In England incidental music was rarely taken seriously until the 20th century, though music was commissioned from Sir Arthur Sullivan for productions of Shakespeare between 1862 and 1888, as well as for TENNYSON's *The Foresters* in 1892 and Comyns Carr's *King Arthur* in 1895. A revival of *The Tempest* in London in 1921 in which some of Sullivan's music was used, together with some new and very striking music by Arthur Bliss, showed how good Sullivan could be at theatre music of this kind. Edward German had a talent for simple melodies and agreeably picturesque, if distant, period imitation, but of his scores for plays nothing remains but his dances for Shakespeare's *Henry VIII* (1892) and for Anthony Hope's *English Nell* (1900). Little produced before 1900 was of much more value than the music provided by Jimmie Glover to accompany the villain's misdeeds in Drury Lane melodramas. There were some exceptions, but they were mainly for academic performances—Stanford's music for AESCHYLUS' *Eumenides* (1885) and SOPHOCLES' *Oedipus Tyrannus* (1887) in CAMBRIDGE; Parry's for productions of ARISTOPHANES in OXFORD between 1883 and 1914; and Vaughan Williams's for Aristophanes' *Wasps* (1909), also in Cambridge. Stanford also provided good incidental music for two London plays, Tennyson's *Queen Mary* (1876) and *Becket* (1893), in which Henry IRVING appeared at the LYCEUM.

With the turn of the century things began to improve. The incidental music by Norman O'Neill for plays at the HAYMARKET, where he was musical director for many years, was slight, but combined a special aptitude for the requirements of the stage with graceful, sometimes fanciful invention, particularly in the scores for Maeterlinck's *The Blue Bird* (1909), BARRIE's *Mary Rose* (1920), and Ashley DUKES's *The Man With a Load of Mischief* (1925). Elgar wrote music for the London production of YEATS's *Grania and Diarmid* (1902) and for *The Starlight Express* (1915), a play for children by Violet Pearn based on Algernon Blackwood's novel *A Prisoner in Fairyland*. Other outstanding composers of incidental music at this time were Armstrong Gibbs—for Maeterlinck's *The Betrothal* (1921); Eugene Goossens—for MAUGHAM's *East of Suez* (1922), Margaret Kennedy's *The Constant Nymph* (1926), and Dodie SMITH's *Autumn Crocus* (1931); Frederic Austin—for the ČAPEKS' *The Insect Play* (1923) and a revival of CONGREVE's *The Way of the World* in 1924; and, most important of all, Delius—for FLECKER's *Hassan* (1923). No less a composer than Benjamin Britten wrote the music for the London productions of PRIESTLEY's *Johnson Over Jordan* (1939), Ronald Duncan's *This Way to the Tomb*, and WEBSTER's *The Duchess of Malfi* (both 1945); and before turning his attention to film music during the war years, William Walton provided a score for John GIELGUD's production of *Macbeth* in 1942.

It has for many years been impossible for theatres presenting straight plays to maintain a pit orchestra, and companies who regularly use incidental music have tended to employ a small instrumental ensemble or some form of recorded score. Until it closed in 1963 the OLD VIC had a flexible chamber group, and in its later years commissioned music from a number of distinguished contemporary composers. These include John Gardner (MARLOWE's *Tamburlaine*, 1951; *Hamlet, King John*, 1953); Malcolm Arnold (*The Tempest*, 1954); Peter Maxwell Davies (*Richard II*, 1959); Thea Musgrave (*A Midsummer Night's Dream*, 1960); Peter Racine Fricker (*King John*, 1961); Michael Tippett (*The Tempest*, 1962); Elizabeth Lutyens (*Julius Caesar*, 1962; also Aeschylus' *Oresteia* for the OXFORD PLAYHOUSE company at the Old Vic, 1961). The company's resident composer George Hall wrote incidental music for Ibsen's *Peer Gynt* in 1962: the National Theatre in Oslo had also commissioned a score from Harald Saeverud in 1947, to replace the well known music by Grieg. When Peter HALL became director of the ROYAL SHAKESPEARE COMPANY in 1960, Raymond Leppard was appointed music adviser and there too outstanding composers were employed: for example, Lennox Berkeley (*The Winter's Tale*, 1960); Humphrey Searle (*Troilus and Cressida*, 1960); and Roberto Gerhard (*Macbeth*, 1962). As resident composer Stephen Oliver provided an effective score for the R.S.C.'s adaptation of DICKENS's *Nicholas Nickleby* (1980) as well as for a number of Shakespeare productions. Marc Wilkinson was succeeded as composer at the NATIONAL THEATRE by Harrison Birtwistle.

Electronic music assembled on tape, being economical as well as often highly effective, has proved increasingly useful. Peter BROOK's chilling accompaniment to his production of *Titus Andronicus* (Stratford, 1957) and Raymond Leppard's enchanted music for *The Tempest* (Stratford, 1963) are notable examples. Like the specially-composed scores for small atmospheric instrumental ensembles, electronic incidental music is usually integral to the stage production and is rarely heard in concert versions.

INCORPORATED STAGE SOCIETY, see STAGE SOCIETY.

INDEPENDENT THEATRE, London, see GREIN, J. T.; **INDEPENDENT THEATRE CLUB**, London, see KINGSWAY THEATRE.

INDIA. There is very little reliable information about the origin and early history of the classical Hindu (Sanskrit) drama of India, but it seems to be generally agreed that the earliest surviving play-texts are those discovered in the city of Trivandium in 1910 and ascribed to Bhāsa (*fl.* 3rd century BC) which show that theatrical art had already achieved a high standard. Little is known of the nature of such early dramatic performances, but it would seem that, as in many other countries, they developed out of dramatic dances and mimetic representations whose traditions go back indefinitely. Controversy as to whether the drama was in origin religious or secular has little meaning in a country like India, where the two are so closely intermingled, though it seems likely that it had nothing to do with the official Vedic religion of the ruling orders. The suggestion that it developed under Greek influence has no evidence to support it; nevertheless it is convenient in point of time, and it is possible that the knowledge that plays were performed at the Courts of Greek invaders subsequent to the conquests of Alexander the Great may have stimulated the development of Indian drama, though direct influence of classical models seems highly improbable.

Most of the plays of Bhāsa, of which 13 in all have survived, are on subjects taken from the great Sanskrit epics, the *Mahābhārata* and the *Rāmāyaṇa*, in both of which there are passing references to plays and players from which it appears that music, dance, and mime were at first more important than the spoken word. Sanskrit drama, of which another little known early exponent was Aśvaghosa, who is now believed to have been writing in the 2nd century BC, some three or four centuries earlier than the date formerly assigned to him, rose to its peak in about 700 AD and continued to flourish until about 1000 AD, after which very little original work of any value was produced.

Among the early Sanskrit plays the best known outside India is *Mṛcchakaṭikā* (*The Little Clay* (or *Toy*) *Cart*), formerly ascribed to King Śūdraka, *c.* 150 AD, but now believed to be of unknown authorship and indeterminate date. It has several times been translated into English, and in 1924 was produced in New York as the first offering of the NEIGHBORHOOD PLAYHOUSE. It was seen in London in 1964 at the HAMPSTEAD THEATRE CLUB. It is unusual in having an original plot, not taken from mythology or history, and in being closer in spirit to Western comedy than any other Sanskrit play.

More highly regarded in India, however, are the plays of Kālidāsa (late 4th to early 5th century AD), author of *Shakūntalā*, usually regarded as the masterpiece of Sanskrit drama and known in English as *The Recovered Ring*. It was a translation of this play by Sir William Jones in 1789 which first aroused the interest of European writers, particularly GOETHE, in Sanskrit literature. Among later playwrights the most important seem to have been King Harsha of Kanauj, who wrote two comedies of Court intrigue and one play with a Buddhist theme; Bhattanārāyaṇa, who in such plays as the *Veṇīsamhāra* drew his inspiration from incidents in the *Mahābhārata*; Viśākhadatta, whose *Mudrārākśasa* (*The Signet Ring of Rakshasa*) is a tale of complicated political intrigue; and Bhāvabhūti, whose works, drawn mainly from the *Rāmāyaṇa*, show already the lack of spontaneity so apparent in later Sanskrit drama, which often appears to be intended more for reading than for performance.

The earliest Sanskrit treatire on the theory and practice of drama is the *Nāṭya-śāstra* of Bharata, which may have been written as early as the 1st century AD, though the earliest copy to survive probably dates from the 8th century AD. It deals exhaustively with every branch of the subject, including the erection of theatres, the composition and production of plays, the music, the dancing, and the costumes. Later treatises on the drama contain essentially the same material and add little that is new, though some are valuable as preserving fragments of plays which would otherwise be unknown. Some of the commentaries on these extracts contain useful details about their production.

Regular theatre buildings do not appear to have existed at this time. Bharata's account of play production implies that performances took place in the open air, in specially erected buildings, or in temples and palace rooms adapted for the occasion, usually in connection with a festival or public celebration. Sometimes the king or some rich patron would summon the actors to play for the private amusement of himself and his guests. Whatever space was used for the performance was divided into two sections, stage and auditorium, in accordance with certain specified measurements, the auditorium being again divided by pillars into four sections to be occupied by the four castes. The central position among the seats of brick or wood was reserved for the patron and his entourage. The stage (*raṅga*) was decorated with pictures and reliefs. At the back was a curtain which separated the acting area from the performers' dressing-room and from behind which came 'noises off', sounds of tumult or rejoicing, and the voice of gods, who could not suitably be represented on stage. There was apparently no scenery, and very few props, though it is recorded that in the *Mṛcchakaṭikā* of Śūdraka the toy cart which gives the play its name actually appeared on stage; and in a play based on the life of Udayana, the legendary king of Vatsa and a popular subject for drama, an artificial elephant was used. In general, however, the players depended on conventional gestures to indicate, for instance, mounting a horse, getting into a chariot, arranging flowers, or rowing a boat. Costumes indicated clearly the class, profession, and nationality of the wearer; gods, demi-gods, kings, and princes wore magnificent and brightly coloured robes; ascetics, garments made of bark or rags; girls of the cowherd class wore dark blue; dirty and ragged clothing indicated madness or extreme poverty; sober-coloured garments were worn by priests and those engaged in religious offices. Stylized symbolic make-up also conveyed useful information to the audience; people of the north-west were painted reddish-yellow, as were Brahmins and kings. Those from the Ganges valley, as well as members of the two lower castes, Vaisyas and Sudras, were painted dark brown; men from the south and from primitive tribes were blackened. Decorations, jewellery, garlands, and other accessories, also helped to distinguish one character from another.

The actors (*naṭa*) formed a distinct caste of their own, and to judge from numerous references their social status was not very high, though some enjoyed the friendship of distinguished personages. They were mainly itinerant, wandering in groups from one city to another seeking a patron. They worked under the leadership of a *sūtradhāra*, whose business it was to supervise the construction or adaptation of the theatre and the production of the play, as well as acting one of the chief parts. He was assisted by a deputy or right-hand man, the *pāripārśvika*, and there was sometimes a third official, the *sthāpaka*, whose duties seem to have been connected with the actual construction and management of the stage. Male parts were normally played by men, female by women.

The play proper was preceded by a series of introductory performances (*purvaranga*), consisting of instrumental music, songs, and dances. This part of the programme seems to have been lengthy and complicated, and in addition certain ritual acts of worship and prayers to the gods had to be performed. These were brought to an end by the *nāndi*, or benedictory stanza, after which the *sūtradhāra* made his first appearance and the play began.

From the earliest times characters appear to have been stereotyped. Chief among them was the *nāyaka*, or hero, usually a king or other exalted personage, noble, handsome, brave; the *nāyikā*, or heroine, who could be either a lady of noble birth or in certain types of drama a great courtesan, was always beautiful and accomplished; the *pratināyaka*, or villain, often represented as courageous and resourceful, was none the less violent, stubborn, and wicked. Other types frequently found are the *pīthamarda*, friend or hanger-on of the hero; the *viṭa*, or rake, who assists the hero in his less reputable adventures; and the *ceṭa*, servant to the hero. One of the most interesting and curious characters of Sanskrit drama was the *vidūshaka*, or clown, often a disreputable and illiterate Brahmin with an insatiable appetite for food, bald, deformed, with staring eyes and protruding teeth. Like other low-class characters, he speaks in Prākit, the vernacular current in the early centuries AD, Sanskrit being reserved for kings, nobles, and Brahmins, in accordance with the prevailing social conditions at the time the plays were written. Most of the dialogue is in prose, poetry being reserved for moments of passion and extreme tension.

Among the types of drama found in all regions, the most important, the *samavakara*, dealt exclusively with the activities of the gods and divine persons. Almost equally important was the *nāṭaka*, based on mythological or heroic legends, the chief personage being again a divine person or a king. There were never more than five chief characters in this type of play, which was in not less than five acts and not more than ten. The *nāṭikā*, which resembled the *nāṭaka* in many ways, was slighter in subject and extent, and usually dealt with the hero's amorous adventures. The *prahasaṇa*, or bourgeois comedy, centred on less exalted persons—ministers, Brahmins, merchants—and the plot was usually invented by the author as in the *Mṛcchakaṭikā*, the most famous example of the type. The *prahasaṇa* was a short piece portraying the tricks and intrigues of various low-class characters, while the *bhāṇa* was a monologue in which the *viṭa*, or rake, a thoroughly disreputable character, walks through the less respectable parts of the town holding imaginary conversations with the people he meets.

After the Muslim invasions Sanskrit drama underwent a rapid decline. The new rulers were hostile to it, and for an art which depended so largely on patronage this was a crushing blow. It remained popular for some time in 11th-century Kashmir, but even there it eventually died out. In the end the tradition almost totally vanished, so that knowledge of it can be gained only through the surviving texts of plays and accounts of performances, and to some extent through illustrations. Only in remote Malabar have some traces of the old tradition been preserved among the Chākkiyar, a caste of actors and performers, but they are slight and not very informative.

The vacuum left by the disappearance of the classical drama was eventually filled by the emergence of an equally literary but more diverse form of 'variety' theatre, which retained some of the features of earlier times, but depended far more on spectacle, dance, and music both vocal and instrumental. Episodes from the great epic and narrative poems of India had already provided material for solo dramatic recitations. These, though not intended for performance, could easily be adapted for presentation, and from them developed the big processional pageants, first performed in the temples and then outdoors in the countryside and in the city streets. Most of them were based on the Rama and Krishna legends and on stories from the *Mahābhārata*. Other new types of drama drew on romantic stories, medieval legends, historical episodes, local and topical events, and the stories formerly told by street-singers in the old ballad form, all accompanied by music and mime. In the structure of such dramas the chorus, which in Sanskrit drama had been of little importance, took on an outstanding role, singing the entry-songs and alternating dialogue, either in speech or song, with the chief actor. Its singing also accompanied the actors' dance sequences, and enabled the audience to keep track of the plot.

In this new form of drama, which quickly became traditional, the characters were divided into three main types—*satvika*, or godly; *rajasika*, or worldly; and *tamasika*, or evil-minded. Intermediary types represented different combinations of these qualities, and leading actors tended to specialize during their working life in one type or another. Many of the conventions governing the actors' entrances and exits, the delivery of dialogue, and the treatment of time and place were related to the classical Sanskrit theatre, but with far greater freedom and a wide-ranging culture which embraced every aspect of poetry, music, dance, mime, graphic and plastic arts, religious and civic pageantry, and even the decorative arts and crafts. All this activity benefited greatly from the great cultural and creative renaissance of the 15th and 16th centuries, which also gave a fresh impetus to the already established folk-drama, both forms of entertainment developing quickly across the country with the rapid spread of the vernacular language. Folk-drama, which still flourishes in India, is the last repository of all the medieval arts, with its unlocalized platform or multi-level stage, multiple settings, stylized acting and costumes, the use of a narrator (*sūtrahāra*) with chorus, the antics of the clown (*vidūshaka*), the close relationship between actor and audience, and above all the loosely constructed plots with their mixture of narrative and dramatic elements which leave room for improvisation. Although the major folk dramas are still taken from the Rama and Krishna play-cycles, there are a number of purely localized forms. The *jatra* of Bengal, perhaps the oldest, from being purely operatic developed into a prose drama with songs. The *yakshagana* of South India is also operatic in character, with choreographic acting, the players wearing gorgeous and fantastic costumes. The *bhavai* of Gujarat, the *tamasha* of Maharashtra, and the *nautanki* of Uttar Pradesh are all secular plays, some dealing with contemporary social themes, others based on historical romances or medieval legends. *Nagal, swang*, and *bhandaiti* are short plays, lighter in character, dealing with topical events, and dependent for much of their effectiveness on improvization and ready wit.

It was not until the middle of the 19th century that modern drama made its appearance in India, mainly as a result of the direct impact of the West. Though some earlier playwrights had managed to combine Western-style picture-frame staging with the indigenous conventions of the classical and folk drama, it was the professional Parsee companies which flourished during the latter half of the 19th and early part of the 20th centuries which first juxtaposed local farces and dramas with adaptations of such Western dramatists as Shakespeare and MOLIÈRE. The outstanding Parsee playwrights of the time, Agha Hashar Kashmiri and Radheshyam, both prolific writers, made use of alliterative prose, rhyming couplets, and a Shakespearian plot-structure with Indian classical and folk-stage conventions. In the early years of the 20th century the main influences on modern Indian playwrights were SHAW, IBSEN, and CHEKHOV. Among the many talented writers of the time Rabindranath TAGORE was perhaps the most successful in combining certain aspects of Western drama with indigenous forms of theatre. Some of his plays, at one time considered unsuitable for stage production, have now taken a permanent place in the repertory, mainly owing to the efforts of Sombhu Mitra in Calcutta, one of a group of talented modern directors which includes Utpal Datt (also of Calcutta), Ebrahim Alkazi, head of the National School of Drama, and Habib Tanvir, both of Madras. Since Independence in 1947 theatre activities in India have greatly expanded and intensified, and the mainly amateur theatre is moving towards semi-professionalism, with hard-working and deeply committed groups in most of the larger cities. There is also an active and well-organized commercial theatre in West Bengal and in Maharashtra, two states which have always had a strong theatrical tradition and were the first language areas to support modern theatre during the mid-19th century. Unfortunately in the vast Hindi-speaking region of north India the only professional group, the Prithvi Theatre founded in 1944, closed in 1960, but not before it had had a successful career under the actor-director Prithvi Raj Kapoor, who bridged the gap between Hindi literary drama and the demands of a modern popular audience with such long-running plays as *Deewar* and *Pathan*.

During the mid-1960s a number of playwrights, led by Adya Kangacharya with his *Suno Janamejava* (published in Kannad in 1960 and in Hindi in 1965), set out to reflect in their works the stresses and tensions of a society passing through great social and economic changes. They realized the need to explore and rediscover the heritage of the past, and so evolved a new dramatic form which has become an important factor in modern play production. Among them were Vijay Tendulkar with *Shantata Court Chalu Ahe* (*Silence, the Court is in Session*) and *Ghasiram Kotwal*; Badal Sircar with *Ebam Indrajit* and *Baki Ithas* (*Another Story*); Girish Karnad with *Hayavadan* and *Tughlaq*; and Kohan Rakesh with *Ashadha Ka Ek Din* (*A Day in the Month of Ashadha*) and *Adhe-Adhure* (*Halfway House*). Ten years later a new generation came forward to strengthen and enrich the repertory, each in his own language—Mohit Chatterjee and Manoj Mitra in Bengali, Khanolkar and Mahesh Elhunchwar in Marathi, Chandra Shekbar Kambarr in Kannad, Laxi Marayan Lal, Surendra Verma, and Mani Madhukar in Hindi.

One important aspect of the contemporary theatre has been the revival of Sanskrit drama. Plays in Sanskrit are now regularly performed by specialized groups in such towns as Bombay, Calcutta, and Madras, and a number of Sanskrit plays have been translated and produced in modern Indian languages. This renaissance of India's great theatrical heritage has been matched by the growing popularity of Indian music and dance-drama, led by Uday Shankar, a pioneer of modern Indian dancing and music.

India is traditionally the home of puppetry, and there are references in early literary works which suggest the existence of some form of PUPPET theatre in India in the first centuries of the Christian era. Its religious origins are attested by the rituals which form part of the proceedings. A performance opens with a prayer to the Lord Ganesh, the elephant-headed god who is the presiding deity of the traditional theatre, and closes with a prayer for the welfare of the performers, the audience, and the community. Special rites are also connected with the making, preserving, and handling of the figures representing gods and goddesses. Puppet-plays none the less embody a social art deeply rooted in the everyday life of the people, and new material is continually being added. Puppet companies are often family-based, with six to eight members, and if the group is not complete the chief puppeteer, who is also the head of the family, may decide to take another wife. The art is passed on from one generation to another, and in any given region certain villages are allotted to a specified group, which performs for six or eight months on tour, and then returns to its village, often a tent-village, to work on allied crafts.

Almost all forms of puppet are found in India, but the main types are hand- or glove-puppets, rod puppets, string puppets (marionettes), and leather (or shadow) puppets. Of these, the first is rapidly dying out and now survives only in Uttar Pradesh and Orissa in the North and Kerala in the South. In the former locality it is called *Gulabo-Sitabo*, taking its name from the two co-wives who air their domestic differences in witty dialogue and knockabout comedy. In Orissa it is *Gopi-Leela*—'the sport of the milkmaid'—and deals with the love affairs of Krishna and Radha; one puppeteer manipulates the two figures while another sings the songs, and both improvise prose dialogue, often with local and topical allusions. Sometimes there is only one puppeteer, who manipulates the figures and plays the drum while singing the songs. The glove puppet theatre is most highly developed in Kerala, where the puppets wear the same costumes, facial make-up, headdresses, and ornaments as the Kathakali dancers, and draw on the same material, the epics *Rāmāyaṇa* and *Mahābhārata*. While one puppeteer manipulates the figures a group of instrumentalists and vocalists provide the words and the music.

Rod puppets also are becoming rarer and are found now only in West Bengal where they appear in the *Putuḷ Nauch*, or dance of the dolls. The large figures, 3–4 ft high, wear traditional costume and are manipulated by means of two bamboo rods, one fixed to the back and one to the head of the puppet. The puppeteer stands behind a curtain, keeping up a continual movement which animates the figure, and behind him stand the musicians. The plays are based entirely on episodes from the *Rāmāyaṇa*.

More widely distributed throughout the country are the string puppets, figures made of wood in a variety of sizes and designs. Their performances follow closely the style and content of the local traditional and folk drama, and they may vary in size from a single block of wood about 18 ins high, with one string at the head and another at the waist, to the elaborately carved, rounded figures, 2–3 ft high, of the Tamil *nadu* marionettes used in the so-called *Bomalattam*, or dance of the dolls. They are exceedingly difficult to manipulate, as their hands are moved by two iron rods held in the puppeteer's hand and their other limbs by strings attached to an iron ring on his head. To animate the puppets the puppeteer constantly moves and shakes his head while also moving the iron rods in time to the music.

One of the oldest forms of puppetry is the leather puppet used in the SHADOW SHOW, which is found in Orissa, Andhra Pradesh, Karnataka, and Kerala. They vary in size, those from Orissa being only up to 18 ins high while those from Andhra Pradesh may be as much as 6 ft, and in the material used for their construction, divine characters being made of deerskin, others of goat or buffalo hide. As many as 200 characters, each one drawn and cut out in dramatic postures and brightly coloured, may be used for one performance, when their shadows are projected from behind on to a screen by the light of rush torches or castor-oil lamps. Their manipulation calls for great skill, both in animating the figures and also in adjusting their distance from the screen so as to sharpen or soften their outlines. Movement is not restricted to joints only; the whole puppet can be tilted, made to advance or recede, fall, rise, turn, hover, or descend from above. Some characters like Ravana and Hanuman are characterized by an intense shaking of the whole body and sudden leaps and dips.

Since Independence great efforts have been made to revive and revitalize the puppet theatre as part of the growing interest in and use of traditional arts and crafts. Festivals have been organized and newly formed groups have been given Government support. There is also a growing use of puppets as a medium of education and to publicize development plans for economic growth and social change. Such programmes as family planning, prohibition, health, hygiene, and adult education have been popularized through the medium of puppet plays, and this modern development is in its turn helping to keep alive and revitalise the traditional Indian puppet theatre.

INDIPENDENTI, Teatro sperimentale degli, Italian theatre company, founded in 1922 by Anton Giulio Bragaglia, active until 1930. Working in an underground room in the old Roman baths by Virgilio Marchi, the group laid special emphasis on the visual side of its productions, which included plays by such authors as JARRY, SCHNITZLER, and WEDEKIND, as well as some of the new works of FUTURISM.

INGE, WILLIAM (1913–73), American dramatist, who was drama critic of the *St Louis Star Times* from 1943 to 1946. His first success was *Come Back, Little Sheba* (1950), in which Shirley BOOTH memorably played the garrulous, pathetic, inadequate Lola, lonely wife of an alcoholic. *Picnic* (1953), which won the PULITZER PRIZE, concerns the effect of an unemployed wanderer with powerful sexual magnetism on the women he meets in a Kansas town. It was followed by *Bus Stop* (1955), Inge's most cheerful work, again set in Kansas and bringing together a group of characters in a café used by bus passengers; and *The Dark At the Top of the Stairs* (1957), originally produced in Dallas in 1947 as *Farther Off From Heaven*, and dealing with the sexual and other problems of a travelling salesman and his family in a small town in Oklahoma in the early 1920s. After these four plays—all hits in New York, though little known in Britain—Inge's career went into decline. *A Loss of Roses* (1959) was a failure, as were *Natural Affection* (1962), *Where's Daddy?* (1966), and *Summer Brave*, a revised version of *Picnic*, written in 1962 and produced in New York in 1973. Inge had a vivid sense of the theatre, though his approach to psychological problems was perhaps somewhat elementary.

INGEGNERI, ANGELO, see LIGHTING.

INGELAND, THOMAS, see ENGLAND.

INNER PROS(CENIUM), see FALSE PROS(CENIUM).

INNER STAGE, a presumed feature of the Elizabethan theatre which has aroused a great deal of controversy. It was formerly thought to be either a large curtained recess behind the back wall or a structure projecting on to the stage from the back wall as in the TERENCE STAGE. Some form of concealment must have existed, as stage directions in a number of Elizabethan plays demand 'a discovery' by the drawing of a curtain (for example, Ferdinand and Miranda playing chess in *The Tempest*); but it is now thought that the 'inner stage' was nothing more than part of a narrow corridor behind the stage-wall which could be made visible through an opening usually closed by a door or curtain. As it would not have been possible for many people in the audience to see a scene played inside this recess under the canopy, it seems probable that actors so 'discovered' came forward on to the main stage to join in the action of the play, as Ferdinand and Miranda do when they see Alonso and Prospero. There may also have been small structures—caves, tents, monuments, even simple rooms—either standing free on the stage, as they did in the MASQUE, or abutting on to the back wall. Some such arrangement would have been necessary for the monument scene in *Antony and Cleopatra*. Stage directions prove the existence of a similar 'inner stage' in the Spanish AUTO SACRAMENTAL, where a curtain might be drawn back to reveal a crib, or a statue of the Virgin, or the empty tomb of the Resurrection. Because of the uncertainty over its size and position, the term 'inner stage' is now being abandoned in favour of the expression 'discovery space', which allows for a wide variety of interpretations.

INNS USED AS THEATRES. Before the erection of permanent theatres, English actors performed in inn yards, which provided a convenient playing-place, with floor-space and galleries for the spectators. Even after the building of London's THEATRE in 1576, the practice continued sporadically in the city. Among the best known London inns to give hospitality to the players were
(1) The Bell, in Gracious (now Gracechurch) Street, which continued in use until after 1583.
(2) The Bel Savage, on Ludgate Hill, where the QUEEN'S MEN are known to have played. It continued in use until at least 1588.
(3) The Boar's Head, in Whitechapel beyond Aldgate. The first reference to its use as a playhouse is found in 1557, when a 'lewd' play called *A Sack Full of News* gave offence and caused the actors to be arrested for 24 hours. Later occasional performances were given until in 1595 Oliver Woodlif took over. A surviving document shows that at that time the stage and tiring house still existed and in the next four years a considerable amount of work was carried out, the galleries and stage being improved and covered over for use all the year round. It continued to be used until after 1616. Another Boar's Head, somewhere in Middlesex, was also in use between 1602 and 1608.
(4) The Bull, in Bishopsgate. This was used for plays from about 1576 until after 1594. The Queen's Men played there in 1583 and probably later.
(5) The Cross Keys, also in Gracechurch Street, near the Bell. It was used from before 1579 until about 1596 and STRANGE'S MEN played there in 1589 and 1594.
(6) The Red Bull, in Upper Street, Clerkenwell. This was used for plays before it became the RED BULL THEATRE in about 1605.
(7) The Red Lion, in Stepney, used occasionally for plays, as there is a record of one called *Samson* having been performed there in 1567.
(8) The Saracen's Head, in Islington. This is one of the earliest known inns to be used for plays. According to a reference in Foxe's *Book of Martyrs* (1563), a play was in progress here when the dissenter John Rough was arrested prior to his martyrdom. Topographical references subsequent to the inn's reconstruction early in the 17th century suggest that it may have been situated at the junction of Goswell Street and St John's Street.

INSET, a small set, usually a corner of a room or an attic, lowered from the FLIES, or set inside (i.e., in front of) a full set that does not have to be struck. An Inset, or Inset Scene, is a similar set piece put behind an opening in a FLAT to mask the view beyond.

INSTITUTE FOR ADVANCED STUDIES IN THE THEATRE ARTS (IASTA), New York, an educational venture founded in 1958 by Dr and Mrs John Mitchell to provide opportunities for professional theatre workers to study the style and technique of foreign theatres by working under a director from abroad and attending courses on subjects relating to the plays being produced. The first visitor was Willi Schmidt, from Berlin, who directed a new English version of SCHILLER's *Kabale und Liebe*. Subsequent productions included MOLIÈRE's *Le Misanthrope* under Jacques CHARON, CONGREVE's *The Way of the World* under George DEVINE, and SOPHOCLES' *Electra* under Dimitrios RONDIRIS, as well as *Kabuki* and *Nō* plays directed by leading actors from JAPAN, COMMEDIA DELL'ARTE scenarios, and Indian dance-dramas.

INTERLUDE, in early English drama the name for a short dramatic sketch, from the Latin *interludium*. It appears to have some affinity with the Italian *tramesso*, signifying something extra inserted into a banquet and so an entertainment given during a banquet. By extension it came to indicate short pieces played for light relief between the acts of a long play. For these Renaissance Italy adopted the term *intermedio* or INTERMEZZO, the former term giving rise to the French *entremets* or *intermède*, meaning a short comedy or farce. In Spain the ENTREMÉS, while having a somewhat similar origin, became a distinct dramatic genre. The first English dramatist to make the Interlude an independent dramatic form was

John HEYWOOD. The Players of the King's Interludes (Lusores Regis) were first recorded in 1493 and disappeared under Elizabeth, the last survivor dying in 1580.

INTERMEZZO (*Intermedio*), an interpolation of a light, often comic, character performed between the acts of serious dramas or operas in Italy in the late 15th and early 16th centuries. They usually dealt with mythological or classical subjects, and could be given as independent entertainments for guests at royal or noble festivals, on the lines of the English DISGUISING and dumb-show, the French *momerie* and *entremets*, or the Spanish ENTREMÉS.

INTERNATIONAL AMATEUR THEATRE ASSOCIATION, see AMATEUR THEATRE.

INTERNATIONAL CENTRE OF THEATRE RESEARCH was founded as a result of BARRAULT's invitation to Peter BROOK to conduct in Paris in 1968 a workshop for actors, writers, and directors from diverse backgrounds and cultures. This experience inspired Brook to open two years later, in a former tapestry factory, a theatrical centre for questioning, experimentation, and discovery, his colleagues being chosen from over 150 theatre workers from all over the world. One of the Centre's first projects was a retelling of the myth of Prometheus in a new language called Orghast invented by the poet Ted Hughes. It was staged on a mountain top in front of the tomb of King Artaxerxes, overlooking the ruins of Persepolis. In 1972 Brook took his company to Africa, where they toured through five countries, playing on a carpet in temples, houses, squares, forests, and on dirt roads. Much of their work was improvisational, as in *The Conference of the Birds*, based on a work by the 12th-century Persian poet Farid Uddin Attar, which was revived in the USA in the following year. In 1974 the Centre moved to the derelict Théâtre des Bouffes du Nord in Paris, which was deliberately left in the dilapidated state in which it was acquired. The first production there, a French version of *Timon of Athens*, was followed in 1975 by *The Ik*, the true story of an African tribe, seen in London in 1976. Alfred JARRY's *Ubu-Roi* was presented in 1977, and in 1979 *The Conference of the Birds* was revived in a production which differed greatly from the earlier improvised versions. The company also stages improvised performances in non-theatrical settings in schools, colleges, old people's homes, and small towns and villages.

INTERNATIONAL FEDERATION FOR THEATRE RESEARCH, founded in July 1955 at a meeting in London attended by delegates from 22 countries at the invitation of the English SOCIETY FOR THEATRE RESEARCH. The Federation, which now has 27 member countries, is devoted to the collection, preservation, and dissemination of theatrical material throughout the world. Its constitution was drafted at a meeting in Paris in 1956 and accepted at a world conference in Venice in 1957. It publishes three times a year a journal, *Theatre Research International*, with résumés in French of the articles, which are in English. Full membership is open to institutions engaged wholly in theatre research (those only partly so engaged can become Associate members), and has also a category of individual members. Meetings of its committee are held annually in various European and North American cities (Paris, Vienna, Stockholm, Munich, Prague, New York, Toronto, etc.), and are accompanied by a symposium at which experts are invited to speak on a subject chosen by the host country. Every four years a world conference open to the general public is held, the proceedings of which are published in volume form. The Federation has established an International Institute for Theatre Research in the Casa Goldoni in Venice, where work on the cataloguing of theatre material in libraries, museums, and collections is carried on, and where international summer courses are held.

INTERNATIONAL THEATRE, New York, see MAJESTIC THEATRE (1).

INTERNATIONAL THEATRE INSTITUTE (I.T.I). Founded as a branch of Unesco in Prague in July 1948, after discussions in Paris a year earlier, the I.T.I. exists to promote international co-operation and exchange of ideas among all workers in the theatre. It works through national centres, with headquarters in Paris, and publishes a quarterly illustrated journal in French and English, *International Theatre Information*, formerly *World Theatre*. A world congress is held every two years, together with conferences and colloquia on such questions as the training of the actor or theatre architecture. The British Centre, one of the first to be opened, helps those who come from abroad to study the theatre in Britain and also serves as a central information bureau, particularly about the theatre abroad, for people at home. The American Centre, now an independent body housed in the ANTA THEATRE, covers much the same ground, as do all the other centres; most are extremely active, one of the best being the Polish. A number of international organizations connected with the theatre are affiliated to the I.T.I.

INTIMA TEATERN, Stockholm, a small theatre established in 1907 and directed by August Falck. It was intended exclusively for the presentation of plays by STRINDBERG, who was closely associated with it during the years 1907–10. Despite its cramped stage, which measured 7 metres wide by 4 metres deep (20 ft × 13 ft), the theatre presented no fewer than 24 of his 'Chamber Plays' during the three seasons it was active, including *The Ghost Sonata, The Burnt House*, and *The*

Pelican. Strindberg's *Open Letter to the Intimate Theatre* (1908) gives his mature views on the art of the actor. The theatre's name was later changed to Lilla (Little) Teatern and it closed in 1913. In 1911 a different Stockholm theatre, in Engebreksplan, was called Intima Teatern. This was taken over in 1921 by the KUNGLIGA DRAMATISKA TEATERN and renamed Mindre (Minor) Dramatiska Teatern, changed again in 1923 to Komediteatern. It was converted into a cinema in 1938. The name Intima Teatern was revived once more in 1950 for a small theatre in Odengatan, Stockholm.

INTIMATE THEATRE, London, see CLEMENTS, JOHN.

ION, a Homeric reciter, who gave his name to a dialogue by PLATO which provides valuable evidence about the art of the reciter and hence of the Greek actor.

IONESCO [Ionescu], EUGÈNE (1912–), French dramatist of Romanian origin, exponent and virtual founder of the Theatre of the ABSURD. Rejecting both the realistic and the psychological theatre, Ionesco's plays stress the impotence of language as a means of communication, the oppression of physical objects, and the incapacity of man to control his own destiny. Many are in one act: they include *La Cantatrice chauve* (1950), *La Leçon* (1951), and *Les Chaises* (1952), seen in London as *The Bald Prima Donna* (1956; N.Y., as *The Bald Soprano*, 1958), *The Lesson* (1955; N.Y., 1958), and *The Chairs* (1957); *Victimes du devoir* (*Victims of Duty*, 1953; London and N.Y., 1960); *Jacques, ou la soumission* (written in 1950, but not staged until 1955; N.Y., 1958), *L'Impromptu de l'Alma* (1956; London and N.Y., as *The Shepherd's Chameleon*, 1960), *L'Avenir est dans les oeufs* (1958), and *Le Nouveau Locataire* (written in 1955, seen in London as *The New Tenant* in 1956 and in New York in 1960, but not staged in Paris until 1967). Ionesco's first full-length play, *Amédée, ou Comment s'en débarrasser* (1954), seen in New York in 1955 and in London in French in 1963 together with *L'Avenir est dans les oeufs* in a production by the company of Jean-Marie SERREAU, was only partially successful. Ionesco endeavoured to rectify this by introducing into his next full-length play *Tueur sans gages* (1959), seen in New York in 1960 and in London in 1968 as *The Killer*, a well-intentioned if ultimately impotent 'little man', Élie Berenger, who reappears in *Rhinocéros* (1960), *Le Roi se meurt* (1962), and *Le Piéton de l'air* (1963). The first two were produced at the ROYAL COURT THEATRE in London with Laurence OLIVIER in *Rhinoceros* in 1960 and Alec GUINNESS in *Exit the King* in 1963. They were staged in New York in 1961 and 1968 respectively, the first starring Zero MOSTEL. In 1966 a new play, *La Soif et la faim*, was produced at the COMÉDIE-FRANÇAISE. Ionesco's later work, which has received less attention in the English-speaking world, has included *Jeux de massacre* (1971), *Macbett* (1972; London, 1973), a reworking of Shakespeare's play in his own image, *The Mire* (1972), an experimental montage of speech, image, and sound which broke new ground without departing from Ionesco's characteristic *angst*-ridden themes, and *Ce Formidable Bordel* (*This Dreadful Mess*, 1973).

ION OF CHIOS, Greek tragic dramatist of the 5th century BC. Longinus compares him with SOPHOCLES, saying that he was elegant and faultless but not powerful. He wrote, besides tragedies, other forms of poetry and the first recorded volume of memoirs.

IPSEN, BODIL LOUISE JENSEN (1889–1964), Danish actress, who made her first appearance at the KONGELIGE TEATER, Copenhagen, in 1909, and soon became an outstanding figure on the Danish stage. She is particularly remembered for her performances as Nora in IBSEN's *A Doll's House*, Mrs Alving in his *Ghosts*, as STRINDBERG's Miss Julie, and as Alice in his *The Dance of Death*.

IRELAND. Although certain extant medieval manuscripts of Irish origin suggest that Ireland played her part in the development of LITURGICAL DRAMA, there are few records of dramatic performances there before the 17th century. The first public theatre, in Werburgh Street, Dublin, was built in 1637 by John Ogilby, a Scots dancing master who brought a company from London to stage *St Patrick for Ireland* (1639), the first historical play on an Irish subject, in the presence of the author, James SHIRLEY. This theatre was closed under the Commonwealth but in 1661, soon after the Restoration, Ogilby obtained a royal patent to build the Smock Alley (or Orange Street) Theatre, opening on 18 Oct. 1662 with FLETCHER's *Wit Without Money*. Smock Alley became the third theatre in the British Isles after DRURY LANE and COVENT GARDEN, reaching its peak under the management of Thomas Sheridan, father of the playwright R. B. SHERIDAN, from 1744 to 1758; it was finally demolished in 1815. A rival theatre, the Crow Street, opened in 1758 under Spranger BARRY and Henry WOODWARD who, like the managers of the Smock Alley Theatre and other Dublin rivals, opened a sister house in Cork. It closed in 1820, and the following year Henry HARRIS opened the Theatre Royal in Hawkins Street, which became Dublin's centre of legitimate drama and opera, visited by all the English stars until it was burned down in 1880. The QUEEN'S THEATRE, which opened in 1844, concentrated on melodrama, especially in the later years of the 19th century; like the OLYMPIA, opened in 1855 and rebuilt in 1897, the GAIETY, dating from 1871, and the Theatre Royal, rebuilt in 1897, it survived well into modern times.

The repertoire of Irish theatres until the end of the 19th century was derived from the English stage; Irish writers, from FARQUHAR to SHAW, went to London to establish their reputations. Indigenous theatre began with the movement initiated in 1899 by Lady GREGORY and W. B. YEATS, which offered from the beginning plays by native dramatists upon native subjects (though performed at first by English actors) at the IRISH LITERARY THEATRE. After the union with the FAY brothers' IRISH NATIONAL DRAMATIC SOCIETY in 1901 the players were Irish too, and included such outstanding actors as Sara ALLGOOD and her sister Maire O'NEILL, Maire NIC SHIUBHLAIGH, Dudley DIGGES, and Arthur SINCLAIR. In 1904 Miss HORNIMAN provided the money to found the ABBEY THEATRE, and in 1910 the Irish National Theatre Society became financially independent. After 1904 the ULSTER GROUP THEATRE (as it became later) also formed an important part of the movement, independent but fundamentally akin.

The ideals of the movement were revolutionary; fervently and fundamentally national, the Irish Theatre was to be independent of European fashions, its material native and poetic rather than derived from IBSEN; though only scantily subsidized (if at all), it was to be independent of the box office and popular control. Besides Yeats and Lady Gregory, the poet AE, John Millington SYNGE, George MOORE, Edward MARTYN, and many writers of the Ulster Theatre, drew upon the legendary and historical material which the great Gaelic scholars of their own and the preceding generation had made accessible in translations from Old and Middle Irish; while Lady Gregory's comedies and most of the work of Martyn, and later of Synge, found much of their material in contemporary Irish life. Lady Gregory, followed in this by Synge, used from the first in her dialogue the 'language of the folk', the English idiom of the Irish-speaking peasants of the West, also employed by Douglas HYDE in his translations of Irish poetry.

Direction, setting, and acting at the Abbey Theatre were also revolutionary and in their turn established a continuing tradition. The actors were at first amateurs, as they continued to be for many years (even in recent times Barry FITZGERALD and F. J. MCCORMICK became fully professional only after some years as part-time actors). From the beginning they were encouraged to remain free of the conventions of the European stage, and the resulting naturalness has had an influence reaching far beyond Ireland. Setting and costume were also (necessarily) strictly economical; but this economy was exercised in Dublin and Ulster by highly intelligent artists and craftsmen, resulting in a beauty and simplicity little known elsewhere at that time.

Some of the original leaders—AE, Martyn, Moore, and Padraic COLUM—were either men of letters who contributed an occasional play, or writers who began as playwrights and afterwards developed along other lines. As the movement gathered power new writers were drawn to it, and it gradually developed into something quite other than its original founders had intended; but it never abandoned the early principles expounded by Yeats and Lady Gregory in their 'Advice to Playwrights' and in the early issues of their theatre journal *Samhain*.

It was a sign of the vitality of the movement that the mood of its drama began to change within the first ten years of its life. With the entry of new writers naturalism and objectivity, sometimes gay, sometimes bitter, sometimes satirical, infiltrated the new drama. Between 1903 and 1912 plays by such newcomers as William BOYLE, T. C. MURRAY, and Lennox ROBINSON in Dublin, St John ERVINE and Rutherford MAYNE in Belfast, found their material in the struggles against adversity and the hard daily life of the small farmer or small-town man everywhere in Ireland, including the people of the Dublin or Belfast suburbs. Finally Sean O'CASEY, whose *Shadow of a Gunman* was produced at the Abbey in 1923, extended the setting of new plays to the Dublin slums.

There were other dramatists writing before 1916 who tended in general to belong to the later tradition rather than that of the original founders, most notably George FITZMAURICE and Seumas O'KELLY. During these years Lord DUNSANY began his connection with the Irish theatre with *The Glittering Gate* (1909), and Bernard SHAW his when Lady Gregory produced *The Shewing-Up of Blanco Posnet* in Dublin in 1909, while it was still banned in England. The Drama League, founded in 1919, also brought European plays to the Abbey, thereby paving the way for the founding of the GATE THEATRE by Hilton EDWARDS and Michéal MACLIAMMÓIR in 1928.

New playwrights of the 1920s and 1930s included Brinsley MACNAMARA, George SHIELS, Denis JOHNSTON, Paul Vincent CARROLL, Teresa DEEVY, and Joseph TOMELTY. The work of Lennox Robinson, with its variety of mood, theme, and subject, continued to bridge the gap between periods, no less than did his lifelong association with the Abbey Theatre as playwright, actor, manager, producer, and director.

In 1944 Austin CLARKE founded the Lyric Players for the purpose of reviving Irish poetic drama, which had suffered somewhat from the Abbey's preoccupation with realistic works, and for several years his company staged revivals of poetic plays by Yeats and a number of other dramatists. This work was paralleled in Belfast by Mary O'Malley, who founded the LYRIC PLAYERS there in 1951. In 1961 another venture was established in Ulster when Hubert Wilmot founded the BELFAST ARTS THEATRE, modelled after the Dublin Gate Theatre.

Meanwhile in Dublin the establishment of the DUBLIN THEATRE FESTIVAL in 1957 not only stimulated national interest in the theatre but also encouraged experiments in playwriting and production. The new Abbey Theatre, built with government help on the site of the old one (which, with the Peacock Theatre, had been burnt down in 1951), opened in 1966, and the following year a smaller theatre opened in the same building to replace the Peacock, specifically for the production of plays in Irish, and poetic, experimental, and small-cast plays. Since the retirement of Ria MOONEY as director of the Abbey in 1963 various artistic advisers, among them Hugh HUNT, have encouraged the adoption of a more flexible repertory and also the appearance of more visiting actors, many of whom played with the company in its early days, notably Siobhán MCKENNA, Cyril CUSACK, and Jack MACGOWRAN. In turn, audiences at the Olympia and Gaiety Theatres, traditionally the homes of imported productions from England and the Continent, have seen more plays by Irish writers, including Brian FRIEL, John B. KEANE, and Hugh LEONARD, while the comedy shows established at the Gaiety by Jimmy O'DEA continue with such comic artists as his former partner Maureen Potter.

A number of small theatres, seating from 30 to 50 people, opened in Dublin during the 1950s, on the lines of similar 'théâtres de poche' or pocket theatres in Paris. They were known locally as 'basement theatres', and a number of them were indeed housed in the basements of Georgian houses. Most notable both for repertoire and survival are the PIKE THEATRE, where Brendan BEHAN's work was first seen, and the LANTERN THEATRE, although the work of other groups such as the Players Theatre and Globe Theatre Productions should also be noted.

Elsewhere in Ireland theatre activities have also expanded, helped not a little by the touring over 35 years of the Shakespearian company run by Anew MCMASTER and the upsurge of interest in AMATEUR THEATRE, resulting in the founding of the Athlone all-Ireland amateur festival in 1953. Cork, which has always contributed a steady stream of dramatists and performers to the modern Irish theatre, among them Robinson, Murray, and Keane, has a particularly strong amateur tradition. The Drama League, started by Frank O'Connor in the 1920s, and other active groups including Father James O'Flynn's The Loft, which produced mainly Shakespeare, helped to fill the gap created by the withdrawal of the English touring companies, until the late 1940s saw not only their return but also the founding of the Everyman Theatre, formed by the amalgamation of some of the more adventurous amateur societies. This, and the Southern Theatre Group (later Theatre of the South), which became well known for its production of Keane's popular folk-dramas, quickly developed into semi-professional groups, the Everyman in particular presenting 6 or 8 international plays of high quality each season, making it a focal point for serious drama and enabling it to establish itself in its own theatre building. The founding of Compántas Corchaí, also in the late 1940s, helped greatly to encourage the production of GAELIC DRAMA resulting in the staging of Sean O'Tuama's plays in Gaelic by Cork's most recent venture, the University's Granary Theatre, situated in a converted malthouse.

IRELAND, KENNETH, see PITLOCHRY FESTIVAL THEATRE.

IRELAND, WILLIAM HENRY (1775–1835), a brilliant but eccentric Englishman, who at the age of 19 forged a number of legal and personal papers purporting to relate to Shakespeare which for a time deceived even the experts. He persuaded SHERIDAN to put on at DRURY LANE, on 2 Apr. 1796, a forged play, *Vortigern and Rowena*. Although John Philip KEMBLE and Mrs JORDAN played the title roles it failed, thus preventing Ireland from bringing forward further forgeries, of which *Henry II* was already written and *William the Conqueror* nearly completed. All Ireland's forgeries were published by his father, a dealer in prints and rare books who never ceased to believe in them, saying his son was too stupid to have composed them. They were exposed by MALONE and Ireland was forced to confess. He spent the rest of his life in ignominious retirement, doing nothing but hack work.

IRISH COMEDIANS, see COLLINS, SAM.

IRISH LITERARY THEATRE, enterprise founded by Lady GREGORY and W. B. YEATS in 1899, the first manifestation in drama of the Irish literary revival. The first performances were given, with English actors, on 8–9 May 1899 at the Antient Concert Rooms, Dublin, and consisted of Yeats's *The Countess Cathleen* and MARTYN's *The Heather Field*. In Feb. 1900 the GAIETY THEATRE was hired for the production of George MOORE's *The Bending of the Bough*, Martyn's *Maeve*, and Alice Milligan's *The Last Feast of the Fianna*. On 21 Oct. 1901 a cast headed by Frank BENSON was imported from England to appear in *Diarmuid and Grania*, by Yeats and Moore, and the first Gaelic play, Douglas HYDE's *Casadh an tSugáin*, was performed by Gaelic-speaking amateurs. The Irish Literary Theatre was then taken over by the IRISH NATIONAL DRAMATIC SOCIETY.

IRISH NATIONAL DRAMATIC SOCIETY, company founded by Frank and W. G. FAY, which in 1902 produced AE's *Deirdre*, YEATS and Lady GREGORY's *Cathleen ni Houlihan* and *The Pot of Broth*, and four other plays, including one in Gaelic. These were given in small halls in Dublin, but in 1903 and 1904 the company appeared in a larger hall, where, as the Irish National Theatre Society, they were seen in Yeats's *The Hour-Glass*, *The King's Threshold*, and *The Shadowy Waters*, as well as in Lady Gregory's *Twenty-Five*, SYNGE's *In the Shadow of the Glen* and *Riders to the Sea*, and Padraic COLUM's *Broken Soil*. The company was invited to London in 1903, and on 2 May appeared there in two performances of 5 short plays each.

It is believed that these productions finally decided Miss HORNIMAN to finance the ABBEY THEATRE. In 1904 an Ulster branch of the society was formed with the support of the parent company, whose members took part in an initial production of *Cathleen ni Houlihan*. This group merged with the ULSTER GROUP THEATRE in 1939.

IRISH THEATRE, New York, see PROVINCETOWN PLAYERS.

IRON, see CURTAIN.

IRVING, HENRY [John Henry Brodribb] (1838–1905), English actor-manager, knighted in 1895, the first actor to be so honoured. Of Cornish extraction but born at Keinton Mandeville in Somerset, he went to London at the age of 10 and while at school had elocution lessons to help overcome a stutter. At 14 he became a clerk in a counting-house, but in his spare time frequented the theatres and did some amateur acting. In 1856, on payment of a fee of three guineas, he played Romeo in an amateur production at the Soho Theatre, using the name Irving for the first time, and encouraged by his success he accepted an invitation to join the stock company at the new Royal Lyceum Theatre in Sunderland, where he made his first professional appearance on 29 Sept. 1856. In Jan. 1857 he was in Edinburgh and in the autumn of 1859 returned to London, where he played four small parts including Osric in *Hamlet* at the PRINCESS'S THEATRE. Feeling that he still had a lot to learn, he returned to the provinces and was not seen in London again until in Oct. 1866 he appeared at the ST JAMES'S THEATRE as Doricourt in Mrs COWLEY's *The Belle's Stratagem* and Rawdon Scudamore in BOUCICAULT's *Hunted Down*, a part he had acted earlier that year in Manchester, when the play was called *The Two Lives of Mary Leigh*. His success in both parts was sufficient to keep him in London, and in 1867 he went to the Queen's Theatre in Long Acre to play for the first time with Ellen TERRY in GARRICK's *Katharine and Petruchio*, a one-act version of *The Taming of the Shrew*. He later appeared successfully as Reginald Chevenix in H. J. BYRON's *Uncle Dick's Darling* (1869) and Digby Grant in ALBERY's *Two Roses* (1870), and on 11 Sept. 1871 appeared for the first time at the LYCEUM THEATRE, under the management of the American impresario H. L. BATEMAN. This theatre had long been considered unlucky, and Irving's Jingle, in Albery's adaptation of DICKENS's *The Pickwick Papers*, did nothing to restore its fortunes. Bateman, almost in despair, agreed to let Irving appear in *The Bells*, a study in terror adapted by Leopold Lewis from ERCKMANN-CHATRIAN's *Le Juif polonais*. The theatre was almost empty on the first night; but by next morning Irving was famous and he was to dominate the London stage during the last 30 years of Queen Victoria's reign. *The Bells* was followed by Wills's *Charles I* (1872), which gave Irving as the ill-fated king a chance to explore all the pathos and romantic associations of the Stuart legend, while in the same author's *Eugene Aram* (1873), about another haunted man, he repeated the triumph of his Polish Jew. In the same year he was seen in BULWER-LYTTON's *Richelieu*. In all these productions he deliberately pitted his own conception of acting against that of the current school of MACREADY and, in spite of some dissenting critics, won. In 1874 he appeared as Hamlet, presenting him as a gentle prince who fails to act not from weakness of will but from excess of tenderness. His reading, so different from that generally current at the time, puzzled the audience at first but eventually won them over, and it was with the revival of this play that he inaugurated his own management at the Lyceum on 30 Dec. 1878.

Although he was sometimes blamed for his indifference to modern playwrights, Irving was a good manager as well as a great actor, and employed only the best players, painters, and musicians of his day. Ellen Terry, the first and finest of his leading ladies, was only one of the glories of a theatrical reign that finally overcame the remains of a puritan prejudice against the theatre and made each new production at the Lyceum a universal topic of late-Victorian conversation. Irving was not without his detractors and it was his fate to gain immense prestige and yet never to be free from critical attack. At the height of his renown there were still people who found his mannerisms unsympathetic and even faintly ludicrous; they were drawn to see him because his acting was overwhelming in its intensity and held his audience spellbound, not so much through emotional sympathy as through intellectual curiosity. Once under his spell, they found his peculiar pronunciation, crabbed elocution, halting gait, and the queer intonations of his never very powerful or melodious voice, part of the true expression of a strange, exciting, and dominating personality. Occasionally he chose to depict tenderness, as when he played Dr Primrose in Wills's dramatization of GOLDSMITH's *The Vicar of Wakefield* as *Olivia* in 1885, but in his finest and most powerful parts, Charles I, Richelieu, Wolsey in *Henry VIII*, which he first played in 1892, and TENNYSON's Becket in 1893, he was able to give free rein to his individual genius. Although his tenancy of the Lyceum is mainly remembered for his productions of Shakespeare, from *The Merchant of Venice* in 1879 to *Cymbeline* in 1896, he included in his repertory revivals from his earlier days—Bulwer-Lytton's *The Lady of Lyons* in 1879, Boucicault's *The Corsican Brothers* in 1880, Mrs Cowley's *The Belle's Stratagem* in 1881—and also a number of new plays, usually written for him and now forgotten, among them Wills's adaptation of GOETHE's *Faust* (1885), Merivale's adaptation of SCOTT's *The Bride of Lammermoor* as *Ravenswood* (1890), and Wills's *Don Quixote* (1895). Some of these, together with revivals of *The Bells* and

Charles I, were seen during his tours of America and Canada, which he visited eight times, from 1883 to 1903, returning in 1904 to make his last appearance in New York in Boucicault's *Louis XI*. His last years were unhappy. His health was failing and he was beset by financial difficulties, exacerbated by a disastrous fire in 1898 which destroyed the costumes and settings for 44 plays. He gave up the Lyceum a year later after appearing there in *Peter the Great* by his son Laurence IRVING. Under another management he returned in 1901 as Coriolanus, and made his last appearance there in 1902 as Shylock. His last appearance in London, on 10 June 1905, was as Tennyson's Becket. He then set out on a farewell tour which ended with his death in Bradford on 13 Oct., after appearing in the same part. He was buried in Westminster Abbey. The two sons of his unhappy marriage in 1869 to Florence O'Callaghan, which resulted in their separation after only two years, were both on the stage.

IRVING, HENRY BRODRIBB (1870–1919), English actor-manager, elder son of Henry IRVING, usually known as H. B. to distinguish him from his father. Intended for the law, he left it to go on the stage, making his first appearance in 1891 with John HARE at the GARRICK THEATRE as Lord Beaufoy in ROBERTSON's *School*. He was for a time with Ben GREET, and later with George ALEXANDER, playing Captain Hentzau in Anthony Hope's *The Prisoner of Zenda* (1896), in which he was later seen in the title role, and making a great success as Loftus Roupell in R. C. CARTON's *The Tree of Knowledge* (1897). He scored a further success in the title role of BARRIE's *The Admirable Crichton* (1902) and was also much admired as Nevill Letchmere in PINERO's *Letty* (1903). In 1906 he formed his own company with which he toured England and America, reviving many of his father's famous parts, including Charles I, Louis XI, Mathias in *The Bells*, and the dual role of Lesurques and Dubosc in READE's *The Lyons Mail*. Although he resembled his father in many ways, he was not as great an actor, and his performances proved something of a disappointment. He later managed both the QUEEN'S and the SAVOY theatres, but his last years were on the whole unadventurous. He married in 1896 the actress Dorothea Baird (1875–1933), who a year previously had created the title role in Du Maurier's *Trilby*. She was for some time with Henry Irving at the LYCEUM, and later accompanied her husband on tours of the United States and Australia, retiring in 1913.

IRVING, LAURENCE HENRY FORSTER (1897–   ), English stage designer and theatre historian, son of H. B. IRVING, who in 1951 published the definitive biography of his grandfather, *Henry Irving: The Actor and his World*, continuing the story of the Irving family to the present day in *Successors* (1967). As artist and scenic designer he was responsible for the décor of many London productions, including T. S. ELIOT's *Murder in the Cathedral* (1935) and PRIESTLEY's *I Have Been Here Before* (1937); he was also the first chairman of the British Theatre Museum Association.

IRVING, LAURENCE SIDNEY BRODRIBB (1871–1914), English actor and playwright, younger son of Henry IRVING. He was for a time in the diplomatic service but left it to go on the stage, making his first appearance in 1891 in Frank BENSON's company. In 1898 he was at the LYCEUM with his father, for whom he wrote *Peter the Great* (1898), an epic poem rather than a play, and also adapted SARDOU's *Robespierre* (1899) and *Dante* (1903). Of his later plays the most successful was *The Unwritten Law* (1910). In 1913 he made a great success as Earle Skule in IBSEN's *The Pretenders* at the HAYMARKET THEATRE, and a year later left with his wife, the actress Mabel Hackney, to tour Canada and the U.S.A. They were both drowned when the *Empress of Ireland* sank after a collision in the St Lawrence.

IRVING, WASHINGTON (1783–1859), the first American author to gain recognition abroad. Chiefly remembered as a historian and as the writer of romantic sketches and tales, he also served as a diplomat. In 1802–3 he published, in the New York *Morning Chronicle*, a series entitled 'The Letters of Jonathan Oldstyle', several of which give a vivid picture of the contemporary New York stage. Together with his brother William, and with J. K. Paulding, he wrote *Salmagundi* (1807–8), which includes a number of satiric letters on the state of the theatre. He later collaborated with John Howard PAYNE in several plays. Their *Charles II* (1824) and *Richelieu* (1826), both adapted from French originals, were particularly successful. His short story 'Rip Van Winkle' (first published in 1819) was adapted for the stage by James HACKETT in 1825. Other adaptations followed, the most successful being that of Joseph JEFFERSON III and Dion BOUCICAULT in 1865, in which Jefferson scored his greatest success.

IRVING PLACE THEATRE, New York, see SCHWARTZ, MAURICE.

IRWIN, MAY [Georgia Campbell] (1862–1928), American VAUDEVILLE comedienne and singer, who began her career at 13 in a variety act with her sister Flo [Ada Campbell], their most popular number being 'Sweet Genevieve'. In 1877 they were engaged by Tony PASTOR, and after some years in BURLESQUE they separated, May joining the company of Augustin DALY to play supporting roles with such fine actors as John DREW, Otis SKINNER, Ada REHAN, and Mrs GILBERT. After extensive touring, she returned to vaudeville to play jolly, buxom matrons in a series of short farces specially written for her by John J. McNally, and also the title role in his full-length play *The Widow*

Jones (1895). In 1897 she formed her own troupe, directing and starring in a lively comedy, *The Swell Mrs Fitzwell*, and in *Mrs Peckham's Carouse*, a playlet written for her by George ADE. A minor classic on an anti-prohibitionist theme, this was first performed as a curtain-raiser to the play *Mrs Black is Back*, and after May had played it in a number of vaudeville houses it was passed on to her sister Flo, who also made a success of it. May Irwin wrote a number of her own songs, the best known being 'Mamie, Come Kiss Your Honey Boy'.

ISAACS [*née* Rich], EDITH JULIET (1878–1956), editor of the American international theatre magazine *Theatre Arts* from 1919 to 1945 (and for many years its chief stockholder). For 25 years she was a leading force in the American theatre, exercising not only her remarkable editorial faculties but also her own considerable gifts as a critic of theatre, dance, and music. She was also a clear-headed business woman who understood the economic and aesthetic problems of the theatre; she played a leading part in the founding of the National Theatre Conference and in the campaign for better theatre buildings in New York which led to improvements in the building code. She also assisted in the difficult first days of the FEDERAL THEATRE PROJECT and helped to organize and direct the AMERICAN NATIONAL THEATRE AND ACADEMY, of which she was the first Vice-President. The many writers whose work appeared in *Theatre Arts*, ranging from Bernard SHAW and D. H. LAWRENCE to Robert SHERWOOD and William SAROYAN, to all of whom she gave encouragement and often practical guidance, included many important theatre personalities from the 67 foreign countries to which the magazine found its way.

ISLINGTON EMPIRE, London, see GRAND THEATRE.

ISRAEL, see HEBREW DRAMA, NEW.

ITALIAN NIGHT SCENES, see WEAVER, JOHN.

ITALY. As a political unity Italy is not much more than a century old. For many hundreds of years it was a collection of independent regions and city states, linked by a common tradition but widely differing in many ways, both in life style and in daily spoken language. Thus, Rome has never been a cultural capital in the manner of Paris or London, and much important drama has been written in the local dialects. In the 16th century, for instance, Angelo BEOLCO wrote in the Paduan dialect, in the 18th GOLDONI in the Venetian, and in recent times Eduardo DE FILIPPO has worked principally in the Neapolitan. The most famous opera house, La Scala, is in Milan, as is one of the best modern theatre companies, the PICCOLO TEATRO DELLA CITTÀ DI MILANO. The evolution of the theatre in Italy is therefore the result of the interplay of a number of forces, historical, cultural, and ethnographical—notably a popular tradition of mime, song, and dance going back to the Etruscans; and the heritage of Graeco-Roman culture. Throughout its evolution Italian theatre has thus operated between two polarities—the popular tradition and the classical inheritance—while the two most innovative and indigenous forms, the COMMEDIA DELL'ARTE and the opera, have exerted a continuing influence over the 'literary' drama which is not subject to 'literary' critical analysis.

The disintegration of the Empire of ancient ROME brought with it the disappearance of formal theatre; but the dramatic impulse remained strong at local level. The popular MIME of Etruria continued to exert its influence and, as in other European countries, new forms of theatre developed out of the LITURGICAL DRAMA of the Christian Church; but these were by no means the only sources of drama, for the plays of the late medieval and early Renaissance periods are the result of an interplay of liturgical and popular forces. Local festivals and entertainments also persisted; itinerant performers still wandered from one town to another; the wandering scholars—the GOLIARDS—wrote their own monologues and songs in dog Latin, and sometimes formed themselves into companies: the later *commedia bile* gives some idea of their work. In the universities students presented short plays, usually as part of their graduation ceremonies, and the Church was not without its secular entertainments—those surrounding the election of the BOY BISHOP, for instance, and the FEAST OF FOOLS, when it was not unknown for monks to dance naked upon the altar. Religious drama was constantly invaded by pagan elements, some of which persisted and developed to the exclusion of the religious elements.

The earliest known form of religious drama in Italy was the LAUDA, which developed from a simple choric song of praise into an elaborate stage production. This in turn evolved into the more complex SACRA RAPPRESENTAZIONE, whose subject matter was originally drawn from the Scriptures but was later extended to include the lives of the saints. It achieved its most elaborate form in Florence during the 15th century, where for the first time individual dramatists were to emerge, among them Antonio Meglio and Feo Belcari (1410–84), who collaborated in a play on the Last Judgement, and Lorenzo de' MEDICI. As the simple *lauda* developed into genuine narrative or historical drama, so methods of staging moved from the interior of the church to the streets and public squares; and this brought with it a secularization of the subject matter. The action proper was increasingly broken up by the INTERMEZZO, interludes showing scenes of daily life, which with the passing of time acquired great importance. This invasion of popular, often local, material distinguished medieval drama in Italy from kinds found elsewhere, and the collapse of static forms under the pressure of a popular tradition is a persistent feature of the history of Italian theatre. Thus, parallel with religious drama during the later Middle Ages, there were popular forms which might contain religious elements but which were mainly concerned with contemporary life. One such was the *contrasto*, or debate, which could be a clerical disputation or a domestic row; a 13th-century secular example by the Sicilian Cielo d'Alcano is the earliest extant Italian dialect play, and the form was used by later writers of *laudi*. Another was the *maggio*, an entertainment which, as the name indicates, took place in May and whose origins may well go back to pre-history. The *maggio*, based on legends, with spectacular representations of battles, duels, and processions, would appear to have been as elaborate in its way as some of the later *sacre rappresentazioni*, but unfortunately there are no texts of the period, the surviving examples not having been written down until the 19th century. Both these forms were to be absorbed into the drama of the 16th century.

Various kinds of dramatic presentations, often containing carnival elements, surrounded graduation ceremonies in Italian universities. The *Janus Sacerdos*, first performed at Pavia in 1427, is an example of a complete dramatic text which can stand independently of the occasion for which it was intended. It indicates also the emergence of a subject which was to become popular on the stage—student life. Pavia has also provided two other texts written specifically for the university, the *Philogenia* and the *Repetitio egregii Zanini coqui* (*The Lesson of the Good Cook Zanino*) by Ugolino Pisani. *Philogenia*, with its story of a girl first seduced then abandoned, is the first of a long line of comedies on similar subjects.

Medieval drama had always been free in form—episodic, unconscious of the notion of genre, mixing tragic, comic, religious, and farcical elements in one flowing narrative. The development of the cultural phenomenon which we call the Renaissance, with its notion of the totally created and pre-planned art object, consciousness of form, and formulation of aesthetic principles, brought about a change in the composition and presentation of plays. It was taken for granted that the works of antiquity were necessarily superior to those of the Middle Ages. The rediscovery of Roman drama produced great intellectual excitement; it was felt that a vital contact had been established with the classical world, thus somehow reviving a lost golden age. It was not realized that the plays of PLAUTUS and TERENCE were themselves only imitations of Greek NEW COMEDY, whose original texts were then unknown; so were the works of ARISTOPHANES, which would have provided an apt model for the period, perfectly matching the satirical impulses of the Italian theatre.

The first genuine example of a classically-derived play is the *Paulus* of VERGERIO, whose model was Terence, but the work already demonstrates a tension which was to be a permanent feature of Renaissance comedy, between the desire to imitate a classical model and at the same time to draw an accurate portrait of contemporary society. In 1429 a manuscript containing 12 comedies by Plautus was discovered, providing still further models for new writers; and during the last half of the 15th century a number of performances of plays by both Plautus and Terence were given, so authors were subjected also to the influence of living drama. The first known example was a performance of Terence's *Andria* given in 1476 by students in Florence.

Another major source of material for Italian dramatists was the *Decameron* of Boccaccio, whose *novelle* provided excellent plots for comedy, their close, satirical observation of contemporary life being not unlike the ambience of Roman comedy with its stories of dissolute young men and wily servants. For example, in his comedy the *Calandria*, first seen in Urbino in 1506, Cardinal Bibbiena [Bernardo Dovisi] (1470–1520) successfully combined a contemporary story with the *Menaechmi* of Plautus.

The use of contemporary material naturally entailed the use of the vernacular. Vergerio and the other humanists had written in Latin, but by the beginning of the 16th century contemporary speech, either 'official' Tuscan or a local dialect, had become the language of the theatre.

The Italian theatre of the Renaissance produced no major tragic author, and the period is best considered as a laboratory for the investigation and testing of the recently discovered tragic material. The encounter with Roman and, later, Greek tragedy faced Italian dramatists with problems of adaptation and staging which were to be met with in the rest of Europe, many of which are still with us. The impetus towards the re-creation of classical tragedy may conveniently be dated from the discovery by Lovato de' Lovati (1240–1309) of a manuscript in the library of the Abbey of Pomposa containing the tragedies of SENECA. Medieval scholars had been aware of Seneca's writings, and fragments of his work—usually aphorisms detached from their context—are to be found in much creative writing of the period, but for most contemporary writers Seneca was principally a moralist, a noble pagan, the presumed writer of letters (later known to be forged) to St Paul. They had no idea of him as a dramatic poet. The discovery of his manuscript caused a radical revision of the traditional view of Seneca, and the texts provided a model for classical tragedy as we understand it.

It was not until five years after Lovati's death that the first direct imitation of Seneca appeared; this was the *Ecerinis* of Albertino MUSSATO (1314). In writing it he had no thought of its being performed, though it was given public readings; but it may reasonably be considered the first modern European tragedy. It established the nature of tragedy as being essentially concerned with the reversal of fortune (from good to bad) of great men and princes, and its material, preferably drawn from contemporary history, as being violent and revengeful—thus giving rise to the idea that Senecan tragedy was littered with maimed and slaughtered bodies. Mussato also used a variety of verse forms in his work to convey various emotions, so providing an acceptable diversity in his dialogue. His play was taken as an exemplary pattern for neo-classical tragedy, and many similar plays, also in Latin, were written by later dramatists, among them Giovanni Manzini in his tragedy on the life of Antonio della Scala produced in 1387. Nearly a hundred years later the *De captivitate ducis Jacobi* (1464) by Laudivio de' Nobili, written in Latin and probably never performed, still shows an interesting mingling of Senecan tragedy with the *sacra rappresentazione*.

The discovery of classical Greek drama was slow and difficult, but led ultimately to a fresh examination of the nature of tragedy. The study of the Greek language, which had lapsed during the Middle Ages, was revived in the late 14th century, due mainly to the arrival of many famous Greek teachers, among them Manuel Chrysolaras who drew many scholars to Florence in 1396. The great Italian humanist Giovanni Aurispa brought back from his travels abroad 238 volumes in Greek, including six plays by AESCHYLUS and seven by SOPHOCLES. A fresh impetus to the study of Greek was also provided by the refugees fleeing from the Turks after the fall of Constantinople in 1453. By the beginning of the 16th century editions of the Greek classics had been published in Venice—Aristophanes in 1498, Sophocles in 1502, EURIPIDES in 1503, Aeschylus in 1518. As with the Roman comedies, these Greek tragedies provoked a series of imitations, often written under great difficulties. Dramatists had to master a classical language taught by men who spoke little or no Italian, and many passages in the original plays remained obscure. Both in translation and imitation there were two major problems to be solved: the function of the CHORUS and the use of varying metres in the dialogue. Many writers tackled these problems in both Latin and Italian, TRISSINO, in his *Sofonisba*, producing the most satisfactory compromise by retaining the Chorus throughout the action and by using a stanza form for the choruses and an 11-syllable unrhymed line for the dialogue. Although the play, published in 1515, was not performed until 1524, at Vicenza, its influence is already discernible in works like the *Rosmunda* of Giovanni RUCELLAI (1516). In a broad sense it may be said that Trissino, like Mussato before him, provided a basic model for his contemporaries, particularly as he emphasised that he was not using any one classical play as his source, but rather following the so-called precepts laid down by ARISTOTLE in his *Poetics*. Consequently the serious drama of the 16th century in Italy was characterized by a constant exploration of classical models, either by direct imitation or by adaptation, and this exploration extended to all aspects of the theatre, from the actual writing of plays to the formulation of the aesthetics of dramatic composition and the minutiae of stage presentation. The new definition of theatrical art necessarily entailed restrictions, and the free flow of the liturgical drama gave way to the fixed rules of classical composition, particularly for tragedy, the latter based more often than not on a misunderstanding of Aristotle, as in GIRALDI's treatise *Intorno al comporre delle commedie e delle tragedie* (*Concerning the Writing of Comedies and Tragedies*, 1543). This was the first to insist on the UNITIES of time and action, to which Giulio Cesare Scaligero (1484–1558) in his *Poetices libri septem* (*The Seven Books of the Poetics*, 1561) added the unity of place. Theory went hand in hand with practice, and by the time the first genuinely Italian neo-classical comedies began to appear, not only were the unities being observed, but other rules also were being respected; the setting, as in Roman comedy, was a street or a square—that is, a public place, which meant that no respectable woman could appear, since it was not the custom for them to leave their houses; a pseudo-Aristotelian five-act division was imposed; no more than three speaking characters could appear on stage at the same time; no character could appear more than five times; and a certain number of stock characters became obligatory.

The first of such comedies is credited to ARIOSTO, and the restraints he imposed upon himself illustrate the efforts which many dramatists had to make to conform to the new ideas. The metre which he normally used in his poetry was precisely that used for the *sacre rappresentazioni* given in Florence, but in his comedies he abandoned it in favour of imitations of classical metres. He was also the first Italian dramatist to provide a permanent set for his plays, having one specially built in the Court Theatre in Ferrara, where he was working in 1530/1. Unfortunately it was burned down within a year, but not before it had established the tradition of the comedy in a single set. Both this and his written texts exercised a great influence on contemporary dramatists, particularly ARETINO, and by the end of his life the COMMEDIA ERUDITA, or learned comedy, as it was called, had achieved both a literary form and a stage. It did not, however, have an exclusive hold on the theatre of its time. Popular plays continued to be produced, and some writers continued to ignore the unities. In the anonymous *La Venexiana*, for instance, written some time during the first half of the 16th century, the action moves freely in and out of doors and may well have been presented in the old MULTIPLE SETTING; and in spite of all rules to the contrary, the impulse to present contemporary life and speech was as strong as, or sometimes stronger than, the Renaissance desire to imitate classical antiquity. The most famous comedy of the Italian Renaissance, MACHIAVELLI's *La Mandragola* (1520), illustrates this perfectly.

Although Machiavelli in his other plays worked within the tradition of the *commedia erudita*, translating the *Andria* of Terence and imitating the *Clouds* of Aristophanes, in *La Mandragola* he achieved a perfect fusion of contemporary material (taken from a *novella*) with a structure based on classical models.

Outside the *commedia erudita*, written in Tuscan or in the Florentine dialect, there was also a flourishing, popular dialect theatre of considerable vitality—the *commedia dell'arte* whose continuing existence is one of the most important factors in the history of the theatre in Italy. It is true to say that in the Renaissance period the dialect plays—*La Venexiana* was written in the Venetian dialect—often coarser and broader in tone than the learned comedies, have a freshness and directness which the stricter neo-classical comedies lack. Some writers, like Alione, used the Piedmontese dialect; in the introduction to his first plays he dared to say that he was not going to 'fish around in Plautus and Terence' for his material. Others like Angelo Beolco, known as Il Ruzzante (the gossip), wrote in the Paduan dialect; and in many places local companies confined their activities to the speech commonly used by their audiences. What mainly characterizes Italian comedy in the 16th century, whether learned or popular, is, in the hands of the best writers, its sophistication of tone, whether dealing with urban or rural life. It is the first genuinely modern, as opposed to medieval, theatre: the texts that are still performed are, like *La Mandragola*, a perennial favourite, far from being imitations of the classics. After centuries of neglect the plays of Il Ruzzante have now been revived in Italy and France with considerable success, though they remain unknown in England, but of Ariosto's disciplined efforts at playwriting, for example, nothing remains, and it was his epic poem, *Orlando Furioso*, in which his creative impulse was able to express itself freely outside the straitjacket of the neo-classical rules, that provided one of the most influential spectacles in recent years, in an adaptation by Edoardo Sanguinetti in 1969.

So far as tragedy was concerned, by the mid-16th century the classical texts were available, imitations of them had been written, Italian had been established as the normal language of the theatre, and certain problems of staging had been tackled. But there remained large areas of disagreement on problems of adaptation and style. Giraldi had proclaimed the superiority of Seneca over all other classical authors, and had insisted on the necessity of adapting Aristotle's precepts to contemporary taste. His *Orbecche*, a Senecan catalogue of horrors, was the first neo-classical tragedy to be performed, being seen in Ferrara in 1541. It was constructed according to the rules in five acts—exposition, development, apogee, peripeteia, catastrophe—and there were interludes between the acts, with music by Alfonso della Viola, at whose house the first performance of the play was given in the presence of the Duke of Ferrara, Ercole (II) d'Este. The most significant tragic production of the period, however, was of Sophocles' *Oedipus the King*, which, in a translation by Orsate Giustiniani, inaugurated the Teatro OLIMPICO in Vicenza on 3 Mar. 1583. The permanent set was designed by PALLADIO, the choruses were set to music by Andrea Gabrieli, and the staging was supervised by Ingegneri, whose ability to combine historical concern and scholarship with a sophisticated mode of presentation was unmatched anywhere else in Europe. Significant, too, was his insistence on the importance of music, for with musical settings and interludes, often calling for a large number of instrumentalists, becoming an important part of many stage productions, the theatre of this time was laying the foundations of that fusion of music and drama which culminated in the typically Italian genre of Grand Opera, in which costume and scenic design were also to play a large part.

Meanwhile religious drama was declining into the *mescidato*, in which the form of the *sacra rappresentazione* was used for purely secular subject-matter; in the later 16th century this was to contribute largely to the development of the PASTORAL, an elaborate entertainment concerned with the loves of nymphs and shepherds, situated in an idyllic countryside far removed from reality. Its literary origins lay in the pastoral poems and eclogues of Virgil and Theocritus, and there can be no doubt that it helped the theatre to survive during a difficult period by its capacity to encompass the best in various fields. It was the most directly poetic and lyrical of all 16-century dramatic forms, and so brought the genius of the Italian Renaissance poets into the world of the theatre. It also had a great influence on the development of opera, and it is perhaps significant that the first operatic masterpiece, Monteverdi's *Orfeo* (1607), is based on the same subject as the first play to be written in the vernacular, the *Favola d'Orfeo* (1472) by POLIZIANO.

By contrast with this learned, courtly, and aristocratic theatre which flowered during the 16th and early 17th centuries, the *commedia dell'arte*, in which the actor rather than the author or the theorist was the dominating figure, continued to flourish and, like the *commedia erudita* and the pastoral, left traces in the modern repertory; nothing of Renaissance tragedy now survives on the stage. During the period 1580–1680 in England, Spain, and France, a succession of outstanding dramatists produced fine examples of tragedy in the classic sense; but no such body of work is found in Italy, apart from the Latin productions of JESUIT DRAMA. The few tragic plays by such writers as Michelangelo Buonarotti were mechanical and lifeless. It seems as if the tragic spirit had flowed into opera, and it was in great part the achievements of opera in costume, staging, and machinery that laid the foundations of future developments in the Italian theatre. The 17th century was a time of technical consolidation. There were no major innovations in the 'literary' drama, and the main contribution of the period was the introduction and development of the PROSCENIUM arch theatre as it has survived up to the present day. Stage directors were seized with a passion for perspective scenery and for machines of all kinds. Giacomo TORELLI, for example, managed to get no less than 44 scene changes into one of his productions, and set designs themselves became more and more elaborate as the theatre conformed to the complex and convoluted tastes of the baroque era. Torelli and ALEOTTI were active in Parma, BERNINI in Rome, and from Venice, where he worked later, Torelli went to Paris where his pupil VIGARANI carried on the tradition.

The 18th century saw the emergence of three major dramatists—ALFIERI, Goldoni, and GOZZI. The first grappled once more with the problem of writing tragedies based on classical models, inspired perhaps by the example of Scipione MAFFEI, whose *Merope* (1713) had enjoyed an enormous success and had been translated into several languages but was destined to stand alone until Alfieri inaugurated a series of classical tragedies with *Cleopatra* (1775). This, though well received at the time, demonstrates how much the pure classical tradition was a hindrance to Italian authors; it was a straitjacket which many of them willingly put on, but which ultimately excluded all that was most vital and original in the Italian tradition. There is a certain sense of heroic achievement in the way in which Alfieri forced his passionate nature into a rigid structure, but it is the artist's struggle with his material which rivets the attention, not the dilemma of the protagonists, and all Alfieri's art could not bring the tragic form to vivid life. Writers of comedy experienced the same difficulty, to a lesser degree. The *commedia dell'arte* was still popular, but it had become tired, vulgar, and repetitious; comic invention had given way to mechanical imitation. Literary comedy too had become unimaginative, following established lines, using a small number of stereotypes, and again basing itself mainly on classical models. Rival solutions to this state of affairs were produced by Goldoni and Gozzi, the first seeking to reform the improvised comedy by replacing its *scenarii* or outlined plots by fully written-out dialogue, while Gozzi chose to preserve the old forms and infuse them with new life. Both authors, firmly rooted in the Italian tradition, could on occasion trespass on each other's territory; some of Gozzi's texts were composed with considerable care, leaving only a few episodes for improvization, while some of Goldoni's, notably *Il servitore di due padrone* (*The Servant of Two Masters*, 1746), used several of the old *commedia dell'arte* masks, leaving ample room for the insertion of a quick LAZZO or a longer BURLA. To his contemporaries Goldoni appeared to be a 'realist', and at times they found him somewhat unflattering in his portrayal of the life of the middle (and sometimes lower-middle) class. In the theatre war which developed between the two writers it was Goldoni who was obliged to retreat, taking up his residence in France and leaving Gozzi to triumph. Posterity has reversed this judgement. Goldoni's reputation has risen steadily in his own country and his plays have been seen in translation elsewhere, whereas Gozzi's work is now better known in the adaptations made for operatic librettos. Even he could not save the moribund *commedia dell'arte*; Goldoni on the other hand is now seen to have anticipated the arrival of the bourgeois comedy, and to belong to the mainstream of European drama of the later 18th century.

The battle over the *commedia dell'arte* marked the end of an era, and the political, social, and intellectual upheavals of the French Revolution and the Napoleonic Wars brought a new world into being. For the Italians specifically one dominant idea emerged—that of achieving national unity and, as a necessary concomitant, a national literature. Many of the characteristic features of Italian drama—regional and dialect theatre, realistic observation of urban and rural life—persisted, but were seen as part of the complex pattern of what was coming to be regarded as a distinctively Italian culture. Napoleon's stepson Eugène de Beauharnais made a significant gesture in 1806 when he attempted to found a permanent, though not yet national, theatre in Naples, the Compagnia Reale Italiana. It did not long survive, but neither did it pass unnoticed, and in 1821 came the foundation of the COMPAGNIA REALE SARDA, which continued to function until 1855.

It was inevitable that the nationalistic ideals of the Risorgimento should be reflected in the tragic drama of the period, which developed almost into historical frescoes. Long before, Alfieri had dedicated a play about Brutus to 'the Italian people of the future', and now the public was avid for plays of this kind. Initially subjects were drawn as before from classical antiquity, but in accordance with the prevalent romantic taste medieval settings became gradually more popular, while the influence of Shakespeare and contemporary German writers led to a spate of historical romances, remarkable more for high-flown sentiment and inflated rhetoric than anything else, in which a parallel between a known historical event and a contemporary situation could be presented to the audience emotionally, but without fear of censorship. Most of the writers of such works are now forgotten; typical of them was NICCOLINI, a university professor who used drama as a vehicle for his own libertarian ideals. For the greater part of the 19th century the Italian theatre looked increasingly to France for models—HUGO, SCRIBE, AUGIER, DUMAS *fils*, ZOLA—and, later, to Scandinavia, with IBSEN. The greater part of such derivative work vanished with the fashions which gave rise to it and even great literary gifts were no guarantee of dramatic success. The only two authors of distinction in the period, Alessandro MANZONI and Silvio PELLICO, failed to adapt their talents satisfactorily to the requirements of the theatre, and the problem of producing a major Italian tragic masterpiece remained unsolved.

In the field of comedy no one emerged to take the place of Goldoni, and only Giovanni Giraud (1776–1834), whose *L'aio dell'imbarazzo* (1807) was used in 1824 as the basis of a comic opera by Donizetti, produced anything of interest. In Italy, as elsewhere in Europe, there was a movement towards drama which reflected the background and concerns of the rising middle classes, who wished to see their own life and opinions depicted on the stage; and this coincided with a reaction against the romantic excesses of the revolutionary period. The result was a series of commonsense plays concerned with social ethics and 'correct' behaviour by such writers as Paolo GIACOMETTI, who was much influenced by the French writers of the time, as was Paolo FERRARI, perhaps the best of the bourgeois dramatists, not only in his presentation of social problems but in his defence of the traditional bourgeois responses to them and his affirmation of belief in accepted values. Both these writers were at their best in comedy, but efforts were also made, without much success, to bring tragedy into line with the routine daily life of respectable citizens. All this amounted to nothing more than efficient commercial playmaking, and although it seemed as if the theatre might take a genuine step forward with the well-constructed and highly successful comedy *I mariti* (1867) by Achille Torelli, it proved to be an isolated event; when renewal came it was from another source.

Despite the absence of major authors, the theatre in Italy as an institution was flourishing, due in no small measure to the excellence of contemporary actors. The 19th century was the great age of the Italian virtuoso in acting as in music; for although the names of many great players of the *commedia dell'arte* have come down to us, they are usually connected with one role and not renowned as performers able to cover a wide range of parts. The first and perhaps the finest of the new actors was Gustavo MODENA, who dominated the stage of the mid-19th century. His chief concern was to establish the psychological truth of his characters, and in this he not only expressed the growing contemporary interest in REALISM, but also reiterated the desire, often expressed in the course of Italian theatre history, that acting and production should be 'natural'. When he was a young man his acting profoundly shocked his father Giacomo, also an actor, who described his son's performance in Alfieri's *Saül* as 'sloppy'. But he was soon joined by a number of other players—SALVINI, who inspired STANISLAVSKY, Ernesto ROSSI, Adelaide RISTORI, Ermete NOVELLI, Ermete ZACCONI, and finally the legendary Eleonora DUSE. Between them they brought fine acting to the entire repertory of plays, good, bad, and indifferent, from high romantic tragedy to domestic realism.

The cumulative impact of so many fine performers lent an even greater urgency to the continual demand for a national theatre, or for at least a number of permanent, state-assisted companies in the major cities. Such a demand was a natural corollary to the desire for a politically united nation and a national literature; but the government ignored it. In consequence there were a number of private attempts to form companies, employing as many leading actors as possible. In 1860 Luigi Bellotti-Bon, adopted son of the author Francesco Bon, launched a company which survived for 13 years; when it broke up Bellotti-Bon was unable to meet his obligations and committed suicide. Another company formed in 1860 with Paolo Ferrari as its leader survived for five years. In Rome the Teatro Dramatico Nazionale was established, but for its inaugural production was able to offer no more than an operetta. It was perhaps no accident that a period of sterility in the spoken drama should have coincided with an immense outpouring of creative energy in the lyric theatre, as witness the works of Rossini, Bellini, Donizetti, and above all Verdi. It was in the opera house that the tragic impulse again found its best outlet, though musicians, in common with dramatists, now turned away from purely classical subjects, taking as the basis for their librettos plays by Shakespeare, SCHILLER, Victor Hugo, and, as in Donizetti's *Lucia di Lammermoor*, the novels of Sir Walter SCOTT.

It was the coming of VERISMO which breathed new life into the spoken drama, both in the writing and in the style of production, and the first and most famous example of this movement is the one-act play *Cavalleria rusticana* (1884) by the Sicilian novelist Giovanni VERGA. Influenced by the French exponents of NATURALISM, by the realism of the modern Russian novelists, and most of all perhaps by a captain's log-book which came into his possession and impressed him deeply by the rough truthfulness of the writing, Verga sought to suppress the personality of the writer (so dear to the heart of the Romantics) and to present a sequence of incidents as objectively as possible. His new departure and his choice of subject were significant in a number of ways. First, he was essentially a regional writer and his works were rooted in intimate local knowledge. Second, he turned to peasant life for his material and his method was based upon close and accurate observation of it; like many Italian authors before him, his desire to escape from the accepted conventions portrayed in the 'problem' play caused him to return to the life around him. Third, behind the purely anecdotal surface of a work like *Cavalleria rusticana* (now better known as the subject of an opera by Mascagni first produced in 1890), lie a profound sense of religion and an awareness of tragic undercurrents. The same technique of objective presentation was applied by Giuseppe GIACOSA to the portrayal of middle-class life in the cities, particularly in Milan.

The achievements of these authors coincided with a flowering of dialect theatre, noted writers in this genre being, in Sicily, Luigi CAPUANA and Nino MARTOGLIO; in Milan, Carlo BERTOLAZZI; and in Venice, Giacinto GALLINA. Many of their plays were concerned with working- and middle-class life, notably Bertolazzi's *El nost Milan* (1893) and Vittorio Bersezio's *Le miserie d'Monnsù Travet* (1863), which had such a resounding success that the word *travetto* entered the Italian language as a synonym for 'employee'. The situation in Naples was somewhat different. The *commedia dell'arte* tradition had survived there strongly, in a somewhat debased form, and gradually there appeared a new form of theatre, partly written, partly improvised, based on the old forms and depicting life in the popular quarters of Naples. The dramatists responsible for the best of these entertainments, among them Raffaele Viviani, two of whose plays in Neapolitan dialect under the title of *Naples by Day, Naples by Night* were given in London by the Rome Stabile during the WORLD THEATRE SEASON of 1968, could perhaps better be described as creators of spectacles rather than as authors in the strict sense. Some of their works remained in the repertory, two created by Eduardo Scarpetta having been included in his programmes by his son, Eduardo De Filippo. Meanwhile realistic descriptions of the life of the under-privileged continued to feature in such plays as Roberto BRACCO's *Sperduti nel buio* (*Lost in the Shadows*, 1901), while other writers like Marco PRAGA and, in his earlier works, Gerolamo ROVETTA attempted to place the 'problem' play in a more realistic setting. All these tentatives reflected the beginnings of a reaction against the more narrowly photographic aspects of *verismo*. The intellectual climate was changing. Italy was united at last: since 1871 it had had a king and a central government, and in terms of the theatre it had explored newly developing society at both local and national level; but the prestige of modern Italy, successor to ancient Rome, and its role in the modern world were still in question. In 1901 Rovetta abandoned orthodox *verismo* for *romanticismo*, in which he contrasted contemporary decadence with the moral idealism of the Risorgimento. This captured the public's imagination; afterwards they were almost, but not quite, ready for a man like Gabriele D'ANNUNZIO, who was soon to give theatrical expression to the Nietzschean search for heroic transcendence.

Three important schools of writing emerged in Italy in the first quarter of the 20th century—the CREPUSCOLARI, the Teatro del GROTTESCO, and FUTURISM. All three demonstrated a turning away from the objective rationalism of the previous century; the world of appearances had become suspect, and, at least in part, was believed to obscure rather than reveal the truth. This belief was exemplified in the main movements of European thought—the rise of psychoanalysis, with its insistence on the non-rational basis of much human behaviour; the development of the social sciences; and the growing suspicion that language itself might not be the best, or even the only, means of communication between people. These ideas found their fullest expression in the works of Luigi PIRANDELLO, whose own intensely unhappy personal life provided him with insights that both confirmed and went beyond the purely philosophical. His characters have a life of their own, passionate and at times violent, transcending any extrapolated philosophy. At its best his dramatic language is vigorous and economical, yet with its own kind of poetry.

Though Pirandello overshadows his contemporaries, other writers like Massimo BONTEMPELLI also continued to investigate the tragic contradiction between man's inner nature and social necessity. During the interwar years, 1919–39, a number of small experimental, or art, theatres were started: Pirandello himself founded one. The most famous was the Teatro Sperimentale degli INDIPENDENTI directed by Anton Giulio Bragaglia. Meanwhile, after a century of agitation, state support for the theatre was still not forthcoming. Mussolini, shortly after his rise to power in 1922–5, had expressed to Eleonora Duse his intention of tackling this problem, but it was only in the 1930s that he redeemed his promise. A Corporazione dello Spettacolo was set up in 1930 and a Direzione Generale del Teatro, under the control of the Ministero della Cultura Popolare, in 1935. In the same year an Academy of National Dramatic Art was established in Rome, to train actors and directors. An attempt was also made to take theatre to small towns and villages by means of travelling fit-up companies—the *carri di Tespi*—by an organization concerned with leisure-time activities. Unfortunately, like all totalitarian régimes, Mussolini's government was only concerned with the purveying of 'official' culture, and writers opposed to Fascism or insufficiently enthusiastic in their support of it were excluded. The remaining talent was minimal. With the collapse of the Fascist régime its attendant institutions also disappeared, and in 1945 the Italian theatre had to begin all over again in its search for unity and government assistance. In the event the post-war period was extremely rich in all forms of cultural activity. As the economy expanded, so did the theatre and the cinema; as Italian design became fashionable, so Italian actors and directors assumed an increasingly important role on the international scene. During the immediate post-war period, three writers who had lived through the Fascist experience—Eduardo De Filippo, Ugo BETTI, and Diego FABBRI—stressed the need for a renewal of values, a rediscovery of purity and honesty. Typical of this revival was De Filippo's

Napoli milionari! (1945) (seen in a memorable production in London in the World Theatre Season of 1972), which shows the morally corrosive effects of black marketeering on a Neapolitan family and ends with a 'long night' of self-examination.

Once effective government had been established after the end of the Second World War, the Italian government set up a department responsible for cinematic and theatrical activities of all kinds, which in 1959 became part of the Ministry of Tourism and the Fine Arts. By this time the cinema had in a sense taken over the functions of the popular theatre, bringing drama to a wide public who, in Italy as elsewhere, had come to regard the theatre as an exclusively middle-class domain. The so-called neo-realist school preceded the English school of realism in the theatre and cinema by several years and created a sensation abroad. The cinema also established international reputations for such artists as Luchino VISCONTI and Vittorio GASSMAN, who were also able to use their gifts in the theatre, for an interesting feature of the post-war period in Italy was the two-way traffic from one medium to another. Major directors like ZEFFIRELLI worked in opera, theatre, and film, not only in their own country but all over the world, the expansion of the Italian film industry based on Cinecittà providing a firm economic base from which many artists could work. Security for many came too with the establishment in the major cities of a TEATRO STABILE, or permanent company, to present works in the Italian and international repertory. A dream which had endured a long time was thus realized at last, though the earlier tradition of a company built around a group of celebrated artists still persisted. One of the best of these, the Compagnia dei Giovani founded by Giorgio di Lullo and Romolo VALLI, was seen in London in 1965 and 1966 in Pirandello's *Sei personaggi in cerca d'autore*; this group, and others like the Piccolo Teatro della Città di Milano under Paolo Grassi and Giorgio Strehler, achieved an international reputation. Dialect theatre too continued to flourish, the outstanding examples being the companies of Eduardo De Filippo and his brother Peppino, both seen in London, where the success too of Eduardo's *Saturday, Sunday, Monday* in translation at the NATIONAL THEATRE during the 1973 season, and his *Filumena* in 1977, showed how well such work can succeed outside its own locality.

The later 1960s and early 1970s nevertheless witnessed a certain unease in the Italian theatre. The success of the permanent companies brought its own dissatisfactions—the fear of becoming too staid, too static, and a corresponding desire for artistic renewal. Both Strehler and Grassi left the Milan company and, as with Dario FO, the very acceptability of a dramatist in the classic sense became a matter of dispute. Individual spectacles of great distinction have, however, been presented, such as Luca Ronconi's production of MIDDLETON's *The Changeling* in 1966 and of Ariosto's *Orlando Furioso*, adapted by Edoardo Sanguinetti, which sent shock-waves all through European theatre. The crisis, if such it can be called, is one of belief or 'relevance', and was brought into focus for many by the Paris 'events' of 1968. A number of experimental groups, among them La fabbrica dell'attori, Nuova scena, and the Teatro maschera were all founded in the years immediately following 1968, and together with the Gruppo sperimentazione teatrale, founded in 1965, have explored through COLLECTIVE CREATION new ways of approaching contemporary problems and reaching wider audiences.

I.T.I., see INTERNATIONAL THEATRE INSTITUTE.

IVANOV, VSEVOLOD VYACHESLAVOVICH (1895–1963), one of the first Soviet dramatists. After running away from home at the age of 15 he had a hard and adventurous youth. He began writing while in the army and in 1920, with the help of GORKY, settled in Leningrad. His first play, *Armoured Train 14–69*, was an adaptation of one of his own novels, dealing with the capture of a trainload of ammunition during the Civil War. It was produced by the MOSCOW ART THEATRE in 1927 with KACHALOV in the lead and was the first successful Soviet play to be seen there. It was several times revived and now has a permanent place in the Russian repertory. Ivanov produced nothing comparable to it, though a later play about events in the Far East, *The Doves See the Cruisers Departing* (1938), had some success, as did a play on the life of the 'father' of Russian science, *Lomonosov*, also produced at the Moscow Art Theatre in 1953.

J

JACK-IN-THE-GREEN, see ROBIN HOOD.

JACKKNIFE STAGE, see BOAT TRUCK.

JACKSON, ANNE, see MATHEWS, CHARLES.

JACKSON, BARRY VINCENT (1879–1961), English director and wealthy amateur of the theatre, knighted in 1925. He was trained as an architect, but in 1907 founded an amateur company, The Pilgrim Players, which became professional when in 1913 he built and opened for it the BIRMINGHAM REPERTORY THEATRE in his birthplace. Classics and new plays, tragedy and farce, pantomime and ballet, opera, and even marionettes were seen on its stage, and Jackson maintained it with his own money for 22 years as a creative force in the English theatre, often in the face of local hostility and indifference. Among the many plays he directed were several of his own, among them *The Christmas Party* (1913), a children's play which was many times revived. He helped to establish the reputation of George Bernard SHAW with his production of *Back to Methuselah* (1923), and also presented his own versions of GHÉON's *The Marvellous History of St Bernard* (1925), BEAUMARCHAIS's *The Marriage of Figaro*, and ANDREYEV's *He Who Gets Slapped* (both 1926). Considering the theatre as a workshop for artistic experiment rather than a museum for the preservation of tradition, Jackson produced *Cymbeline* in 1923, *Hamlet* in 1925, and *Macbeth* in 1929 in modern dress and in 1929 founded the MALVERN FESTIVAL, mainly as a shop window for Shaw's plays. In 1935 he transferred the Birmingham Repertory Theatre, whose company had proved an excellent training ground for many young players, to a Board of Trustees, but remained associated with it, among his later plays seen there being versions of Wyss's *The Swiss Family Robinson* (1938), DICKENS's *The Cricket on the Hearth* (1941), and FIELDING's *Jonathan Wild* (1942). In 1945 he was appointed Director of the SHAKESPEARE MEMORIAL THEATRE at Stratford-upon-Avon, where he remained until 1948, helping the theatre over the difficult years which followed the Second World War and instituting several salutary reforms. Among the highlights of his management were *Love's Labour's Lost* directed by Peter BROOK; *Hamlet* in Victorian dress, with Paul SCOFIELD and Robert HELPMANN alternating in the title role; *The Winter's Tale* directed by Anthony QUAYLE, who was to succeed Jackson as director of the theatre; and *Othello* with Godfrey TEARLE, who also directed the play, in the name part.

JACKSON, GLENDA (1936–), English actress of strong personality and high intelligence, who made her first appearance in 1957 and was seen in London later the same year. In 1964 she joined the ROYAL SHAKESPEARE COMPANY, appearing in their Theatre of CRUELTY season in London, and first came into prominence as a strikingly erotic Charlotte Corday in the company's production of WEISS's *Marat/Sade*, directed by Peter BROOK, which sprang directly from that season; in 1965 she made her New York début in the same role. Her Masha in CHEKHOV's *Three Sisters* at the ROYAL COURT THEATRE in 1967 was remarkable for its tough, uninhibited passion. She then appeared only in films for some years, but returned to the stage as the waspish wife in John MORTIMER's *Collaborators* (1973), and was also seen at Greenwich in 1974 in a revival of GENET's *The Maids*. In the following year she toured Britain, the U.S.A., and Australia for the R.S.C. in the title role of IBSEN's *Hedda Gabler*, and in 1976 was seen at the OLD VIC in WEBSTER's *The White Devil*, following it with the title role in Hugh Whitemore's *Stevie* (1977), on the life of the poet Stevie Smith. She was in Stratford in 1978 to play Cleopatra in *Antony and Cleopatra*, returning to London to score a personal success as the discontented, married middle-aged school-teacher in Andrew Davies's *Rose* (1980; N.Y., 1981).

JACKSON, JOE, see VAUDEVILLE, AMERICAN.

JACOBI, DEREK GEORGE (1938–), English actor, who had some experience with the NATIONAL YOUTH THEATRE and at CAMBRIDGE before he made his professional début at the BIRMINGHAM REPERTORY THEATRE in 1960 in N. F. Simpson's *One Way Pendulum*. He remained at Birmingham until 1963, and after a season at the CHICHESTER FESTIVAL THEATRE in that year joined the NATIONAL THEATRE company, making his London début as Laertes in *Hamlet*. He was later seen as Cassio in *Othello* (1964), Brindsley Miller in Peter SHAFFER's *Black Comedy* (1966), Tusenbach in CHEKHOV's *Three Sisters* (1967), and Lodovico in WEBSTER's *The White Devil* (1970). In 1972 he was with the PROSPECT THEATRE COMPANY, playing the title roles in Chekhov's *Ivanov* in 1972 and in *Pericles* in 1973, and repeating for the same company in 1975 his performance of Rakitin in TURGENEV's *A Month in the Country*, previously seen at Chichester. In 1978, again for Prospect, he gave an acclaimed performance as Hamlet, later seen at Elsinore and on a world tour; he was also seen as Thomas Mendip in a revival of FRY's *The Lady's Not for Burning*, and in 1980 made his New York début in Nikolai Erdman's *The Suicide*.

JACOPONE DA TODI, see LAUDA.

JAHNN, HANS HENNY (1894–1959), German dramatist and novelist, by profession an organ-builder. He was a devout exponent of EXPRESSIONISM, and most of his plays deal with violence and dark sexual instincts. The first, *Pastor Ephraim Magnus* written in 1919, caused an uproar when it was first produced in BERLIN in 1923. The year before *Die Krönung Richards III* had been seen at LEIPZIG, and in 1926 *Medea* was produced in Berlin. Among Jahnn's later works are *Armut, Reichtum, Mensch und Tier* (*Poverty, Wealth, Man and Beast*, 1948), *Thomas Chatterton*, a tragedy based on the life of the 18th-century English poet which was staged by GRÜNDGENS at Hamburg in 1956, and *Der staubige Regenbogen* (*The Dusty Rainbow*), produced posthumously in Frankfurt in 1961 by PISCATOR.

JAKOBSSON, JÖKULL (1933–78), Icelandic novelist and dramatist, who examines in his work the dreams and aspirations of people who are otherwise fettered by poverty. *Hart i bak* (*Hard Aport*, 1962) and *Sjóleiðin til Bagdad* (*The Seaway to Bagdad*, 1965) are set against a totally authentic and naturalistically drawn Icelandic background. *Sumarið '37* (*The Summer of '37*, 1968) is possibly less fully resolved in its structure. *Songur skóarans og dóttir bakarans* or *Söngurinn fra My Lai* (*Song of the Shoemaker and the Baker's Daughter* or *The Song from My Lai*, 1978), his last play, is about a group of people in an Iceland fishing village.

JAKOVA, KOL, see ALBANIA.

JAMES, DAVID (1839–93), English actor, uncle of David BELASCO. He made his first appearance in London in 1857 under Charles KEAN, but later made such a success in BURLESQUE and EXTRAVAGANZA—being particularly admired as Mercury in Burnand's *Ixion; or, the Man at the Wheel* (1863)—that with two other actors, Henry MONTAGUE and Thomas THORNE, he built and opened the VAUDEVILLE THEATRE, hoping to make it a home of burlesque. The first productions proving unsuccessful, the management put on ALBERY's *Two Roses* (1870), which helped to launch the career of Henry IRVING who played Digby Grant. Later the theatre had an outstanding success with H. J. BYRON's *Our Boys* (1875), which made a fortune for James and his associates. After leaving the Vaudeville in 1882 James was seen at a number of other theatres, and in 1886 returned to burlesque in a revival of *Little Jack Sheppard*, with Nellie FARREN. In his later years two of his best straight parts were Old Eccles in T. W. ROBERTSON's *Caste* and Stout in BULWER-LYTTON's *Money*.

JAMES, HENRY (1843–1916), American novelist, who spent most of his life in England, and in 1915 became a British subject. He had a great love for the theatre, but after the hostile reception which greeted his *Guy Domville* (1895) when it was produced by George ALEXANDER at the ST JAMES'S THEATRE, he withdrew, and nothing further of his was seen on the stage until FORBES-ROBERTSON played Captain Yule in *The High Bid* (1909) at HER (then His) MAJESTY'S THEATRE; it was revived in 1967 at the MERMAID THEATRE where *The Other House*, written in 1893, had its first performance in 1969. *The Outcry*, written in 1909, was performed posthumously by the STAGE SOCIETY in 1917 and was revived at the ARTS THEATRE in 1968. The theatrical success James had longed for came after his death to dramatizations of his novels by other hands—*Berkeley Square* (1928), based on the unfinished *The Sense of the Past*; *The Heiress* (1947), based on *Washington Square*; *The Innocents* (1950), based on *The Turn of the Screw*; *The Aspern Papers* (1959), adapted by Michael REDGRAVE, who also played 'H.J.' in the London production, Maurice EVANS playing the role in New York in 1962; *The Wings of the Dove* (1963); *A Boston Story*, based on *Watch and Ward*, and *The Spoils*, based on *The Spoils of Poynton* (both 1968). A second version of the latter novel was also seen briefly in London in 1969.

JANAUSCHEK, FRANCESCA ROMANA MADDALENA (1830–1904), Czech actress, born in Prague, where she made her début in 1846 in light comedy, leaving two years later to become leading lady at the theatre in Frankfurt, where she appeared mainly in classic revivals. After ten years, during which time she also toured extensively in Germany and Austria and even as far afield as Russia, she went to the Dresden Court Theatre, and within a short time had built up an enviable reputation as a tragic actress in such parts as EURIPIDES' Medea and Iphigenia, SCHILLER's Joan of Arc and Mary Queen of Scots, and Shakespeare's Lady Macbeth. In 1867 she went for the first time to the United States, where, in the fashion of the time, she played in German opposite Edwin BOOTH and other leading American actors. She studied English intensively, however, and from 1873 played mainly in that language, usually in parts which she had appeared in already. She retired in 1900, after suffering a paralytic stroke, and died in the U.S.A. She was one of the last of the great international actresses in the grand tragic style.

JANIS, ELSIE, see REVUE and VAUDEVILLE, AMERICAN.

JAPAN. The classical theatre of Japan, dating from the late 14th century and still regularly performed today, has never limited itself to spoken dialogue between characters. In performances of *nō, jōruri* puppet theatre (conventionally referred to as *bunraku*, from the name of the theatre which has been its only home since 1909), and *kabuki*, music or dance or both have always been important elements: the major part of the training of a *nō* or *kabuki* actor is in the dance and the puppets are accompanied by musical recitative. As this is what *nō*, puppet *jōruri*, and *kabuki* had become when they reached mature form, it is usual to seek their origins in various earlier dance forms. The latter were introduced to Japan from the Asian mainland during the early decades of the 7th century and by the Nara period (710–794) were an established part of court and religious entertainment. Chief among these was *bugaku*, dance pieces accompanied by music which were sometimes performed as individual items in a form of musical entertainment known as *gagaku* (literally, 'elegant music').

Bugaku dances were subdivided according to their place of origin, and in content ranged from re-enactments of primitive folk practices to illustrations of Buddhist rites. During the Heian period (794–1158) *bugaku* dances can be seen to have had many of the features which characterised *nō* several centuries later. Wooden masks were worn and the costumes were rich in colour and fabric. The structure of the dances followed a set pattern of introduction (*jo*), development (*ha*), and climax (*kyū*) which, in a refined and more complicated form, provides the basic framework for all *nō* plays. *Bugaku* also shared with *nō* a dual audience. It was performed at court and its development owed much to court patronage, but it was also highly popular at shrine and temple festivals and was enjoyed by large audiences from the local populace. The effects of such dual patronage are more marked in the case of the dance forms which superseded *bugaku*, when the Court declined and Japan was governed by military rulers. Two rather rougher forms came into prominence from the early 13th century. Named *sarugaku* and *dengaku*, they were essentially variety shows with singing, dancing, acrobatics, and juggling, *dengaku* being mainly distinguished from *sarugaku* by its closer connection with rural entertainment forms. The players formed themselves into groups and were attached to the larger shrines and temples, which used them to entertain and sometimes to instruct the crowds who gathered at festivals. The military nobility (*shōgun*) also patronized them and would often invite their favourite players to give private performances: *sarugaku* and *dengaku* players, of low social origin themselves and used to unlicensed, even ribald, performances before holiday crowds, thus came into contact with the highly refined culture of the ancient Imperial capital.

Nō as we know it today is the creation of two *sarugaku* players who both enjoyed such patronage, Kan'ami (1333–1384) and his son Zeami (1363–1443). Kan'ami was only exposed to Court culture during the last few years of his life, whereas Zeami spent many of his most creative years under its influence. Kan'ami was an innovator, and Zeami describes himself rather modestly as only developing his father's work. Kan'ami introduced into his *sarugaku* a strongly rhythmic dance known as *kuse*, which gave both dramatic tension and a focal point to the pieces. Enactments of stories from the past had already become an established part of *sarungaku*, and under Kan'ami a form recognizable as danced drama developed. From Zeami's time we may speak of *nō*, although the word, which means 'accomplishment', had been used previously to refer to both *sarugaku* and *dengaku*. Zeami was actor, playwright, and theorist. As head of the house of actors which was clearly the *shōgun*'s favourite and as the author of comprehensive treatises on *nō*, Zeami had unrivalled influence over the development of the form. About a third of the plays currently in the *nō* repertory are attributed to him and every aspect of *nō* performances since his time follows his prescriptions for the art.

Zeami established that the general tone of *nō* should be noble. There should be nothing in it to offend the ears or eyes of court society. Hence the language is honorific and sonorous, whatever the character for whom it is written. The subject matter was mainly to be taken from Japan's classical literature, which included the novels and diaries of court life and war romances, as well as a large corpus of lyric poetry. The most famous passages from this literature would be known to many in the audiences and could be quoted verbatim. The plays were to follow the basic three-part pattern *jo-ha-kyū*, the *ha* section itself subdivided into its own introduction, development, and climax. The rhythmic *kuse* dance should usually be performed as the third part of the *ha*, with a break afterwards before the lively dance which would bring the play to an end. This set structure was also reflected in the sequence of the five plays which made up a complete programme. The first was a stately god play, to give the programme an auspicious start. Next followed a warrior play, in which the ghost of a famous soldier of the past described its sufferings since death. The third play would have a beautiful woman as the main character, and elegant beauty was the effect aimed at by the players. The themes of the fourth group of plays were quite varied, but in general presented more opportunities for heightening the dramatic atmosphere. The programme reached a crescendo in the fifth group of plays, where the main character, usually a god or demon, performed a vigorous dance.

In a typical play this main character would

appear as an ordinary person at a historical site and after revealing himself as the spirit of, for example, a warrior who had fought in a famous battle there centuries before, would reappear in the *kyū* section in his original shape. Players in the *nō* are all male and are supported by a chorus of eight and an orchestra of three or four consisting of drums and the all-important flute. The *nō* chorus plays a passive, narrative role and may even chant the lines of the principal characters while the latter are executing certain dances. This lack of specificity in the attribution of lines of dialogue further distinguishes *nō* from Western theatre forms. To a Westerner the performance of a *nō* play may seem long-drawn-out, even monotonous. If, however, attention is paid to the rhythms provided by the drums and the tension suggested by the flute, the shifting drama underlying the apparently stately movements of the actors is more satisfactorily sensed. *Nō* is now much slower than it was in Kan'ami and Zeami's day, and this was due to the type of patronage under which it flourished: but its essence as theatre has not been lost.

Comic interludes called *kyōgen* ('mad words') were sometimes performed between *nō* plays. These often portrayed a servant outwitting his slow-thinking master, and in many other ways provide interesting comment on the society of the time. *Kyōgen* is now regarded as a theatre form in its own right.

By contrast with *nō*, the other two main forms of Japanese classical theatre are lively, noisy, and emotional. In this they too were reflecting the taste of the patrons on whom they depended. Both the puppet theatre and *kabuki* were products of the rapid urbanization which took place from the beginning of the 17th century, especially in the areas of modern Tokyo and Osaka. A diversified economy needed a cash system and a merchant class to make it work, and both developed rapidly. The merchants used their increasing cash wealth to buy consumer goods and entertainment, and theatre was part of an extensive sub-culture which sought to satisfy the demand for the latter.

Puppetry had been known in Japan since the 7th century: the earliest puppeteers were probably Koreans. Using simple, hand-held puppets, these popular entertainers would present re-enactments of folk tales, while their wives often worked as prostitutes. During the first century of the Edo period (1615–1868) there was a considerable advance in the techniques of puppet manipulation, and the addition of feet to male figures in 1678 led the way to the complicated puppets that are in use today. Apart from the puppets themselves, two other factors were important in the development of this type of theatre. The first was the popularity of the *shamisen*, a new three-stringed instrument whose sound caught the mood of the merchant culture with its combination of traditional pathos and gentle eroticism, and the second was the great vogue for recitation of romantic love stories which began with that of a certain Princess Jōruri celebrated in the *Tales of Heike* (early 13th century). Japan's puppet theatre took its place alongside *kabuki* as a major theatre form when *jōruri* chanters, *shamisen* players, and puppeteers came together in the second half of the 17th century. One master chanter dominated the puppet theatre during the 1680s, Takemoto Gidayū. In 1685 he invited a young *kabuki* playwright, Chikamatsu Monzaemon (1653–1725), to work with him in his theatre in Osaka and together they created what was known as *Gidayū jōruri*, in essence the puppet theatre that can still be seen in Japan today.

Chikamatsu, Japan's most famous playwright, wrote for Takemoto Gidayū between 1685 and 1688, returned to *kabuki* for 15 years, and finally abandoned it for *ningyō jōruri* in 1703. Most of his masterpieces were written after this move. He wrote in a heightened prose that is always near to poetry and at certain points in the plays becomes poetry in a formal sense. He was a master of the *kakekotoba* or pivot-word, which is used to link very different images by common sounds (e.g., in English, 'wain*scot-free*'), and he has left many passages of quite extraordinary poetic beauty and complexity. Chikamatsu wrote two main types of play, the *sewa-mono* or domestic dramas and the *jidai-mono* or historical dramas. The former dramatized merchant life, and many of them ended in a lovers' double suicide. *Sonezaki Shinjū* (*The Love Suicides at Sonezaki*, 1703) is his most famous play in this category. The *jidai-mono* featured warrior heroes of the past, *Kokusen'ya Kassen* (*The Battles of Coxinga*, 1715) representing the genre. The merchant audiences could identify with the characters of the *sewa-mono* and escape into the world of the history plays. A major theme running through both types was the irresolvable clash between the deep desire to act honourably towards the other members of one's social group (*giri*—literally 'duty') and the pull of one's own heart (*ninjō*—'human feelings'). Society gave precedence to the former; the latter had a liberating purity about it which encouraged some to seek it regardless, a course which inevitably led to death.

No comparable *jōruri* playwright emerged after Chikamatsu, but the puppets became bigger and more complicated. In 1730 they acquired movable eyes, in 1733 articulated fingers, and from 1734 their 1.2 metres height and sometimes 20 kilograms weight were supported and worked by three puppeteers, the master, in full view of the audience, working the head and right arm, and his two assistants, sometimes in black and hooded, working the left arm and the feet respectively. Constantly vying with *kabuki* for audiences, and indeed for playwrights, during its early history, the puppet theatre reached its zenith around the middle of the 18th century. There was then a gradual decline, with the number of companies dwindling until only a single one was left. Since 1909 *ningyō jōruri* has been regularly performed only at the Bunraku-za in Osaka, with occasional tours to other large cities.

Ningyō jōruri's great rival was *kabuki*, which by the heyday of Edo art and culture—the Genroku period of 1688–1703—had developed into a part-danced part-acted theatre form able to perform plays as complicated as any being written for the puppets. It had originated in narrative dances performed by women in the early decades of the 17th century. The close connection between these simple dance dramas and prostitution forced the authorities first to ban the women (1629), then to ban the young men who took their place (1652), and finally to require the adult men, whose preserve *kabuki* then became, to include elements of plot development in their performances. *Kabuki* was unashamedly commercial in outlook, adopting from other theatre forms and creating for itself whatever would draw audiences. Thus the *kabuki* of Edo, the military capital, tended to be more robust than that of Kyoto and Osaka. Plays were borrowed from *nō* and *jōruri* and adapted to suit *kabuki*'s own patrons. New stage effects were constantly sought after: TRAPS and REVOLVING STAGES were already in regular use by the second half of the 18th century. The art of *kabuki* was controlled and developed almost entirely by its actors. *Kabuki* actors have traditionally thought of their whole bodies as the medium of expression and they quickly refined techniques of expressing complex emotions through stylized and exaggerated, but to their audiences deeply compelling, full-body techniques. Training began in childhood, mainly through dance, and until the 19th century an actor would usually specialize in either male or female roles. There were subdivisions of the former into virtuous, villainous, or comic. The players of female roles (*onnagata* or *ōyama*) added a subtle eroticism to their stylized portrayal of feminine beauty, and to this day *kabuki* is unthinkable without their presence. *Kabuki* actors were early organised into families, who came to exert such power over the form that birth, not ability, could often ensure stardom. There was thus both a deep conservatism and a certain dynamism about *kabuki* acting. Traditions and conventions, beloved by the audiences, had to be observed, but a real master could break through with his own innovations.

Chikamatsu wrote for *kabuki*, and although there have been celebrated *kabuki* playwrights since his time their number is relatively few considering the long history of *kabuki* and the frequency with which it has always been performed. In general playwrights were the servants of the players, who gave directions on how a play was to be written and altered it at will afterwards. There were, and are, no directors in the Western sense. The plays are divided into groups, as in the case of *nō* and the puppet theatre. There are many sub-groups, but the major division is into historical plays, domestic dramas, and dance pieces. A *kabuki* programme, which regularly consisted of individual scenes or acts from several different plays, would be constructed to preserve a balance between the different types. The most famous *kabuki* play is *Kanadehon Chūshingura* (*The Treasury of Loyal Retainers*, originally written in 1748 for the puppets but adapted immediately for *kabuki*), a historical play based on a vendetta which had caught the public's imagination in 1703.

A visit to a *kabuki* theatre in the Edo period was an exciting event that engaged one's physical and emotional energies for a whole day or more. Theatres were set well apart from the centres of cities and to be on time for the morning start (performances could only be in daylight hours) one might have to set out at dawn or even the night before. The plays would continue until dusk and one could eat, drink, and gossip during the performance. There was always sound coming from the (apron) stage area—music, the actors' voices, or the two types of clappers which were used to orchestrate the performance and highlight the climaxes. Some actors would enter from behind the audience along a raised platform (*hanamichi*) running through the auditorium. The costumes would delight the eye with their brilliance, and however intent on one's food, gossip, or company, one would not miss the high spots of the show, especially the climactic poses (*mie*) executed by the leading actors of male roles.

Kabuki was the dominant theatre form when Japan began modernizing in the late 19th century: puppet theatre had declined and *nō* had been discredited along with its patrons, the military rulers. It was inevitable that forces for change in the theatre would seek either a reform of *kabuki* or a new variation of it, or would try to break away altogether, but *kabuki* survived attempts to make it more respectable and realistic in the 1880s and it flourishes today with many of its artistic traditions intact. *Kabuki* acting is much as it always was, although many actors now appear in films or on television, even in musicals, displaying a versatility that would have been inconceivable a century ago. New playwrights have emerged, and their plays combine elements of psychological and historical realism with dialogue and actions that are only possible for *kabuki* actors. Modern audiences sit in numbered seats before a proscenium and are requested not to eat in the auditorium, but they are just as likely today to emerge from the theatre having been both thrilled and moved by the performance. *Kabuki* is now the monopoly of the Shōchiku entertainment company and it retains its traditional commercialism. During the modern period new theatre forms have developed which have emulated the commercialism and certain elements of the *kabuki* style while acknowledging the advent of a new age. One such was *shimpa* ('new school'), which developed out of political plays performed by amateurs in the 1880s. *Shimpa* in its early political and later (from the early 1900s) non-political forms proved that theatre need not be the exclusive preserve of specially trained actors from established families. While some *kabuki* actors have joined *shimpa*, the general acting style is less exaggerated, and actresses may be used instead of, or even sometimes together with, *onnagata*. The staple of *shimpa* from the time when it achieved its first maturity in about 1905 was dramatizations of popular novels serialized in newspapers. Melodramatic, sentimental, and full of convenient coincidences though many of these were, reviews of *shimpa* were regularly carried in the main theatrical journals in the 1910s, and in spite of several sharp vicissitudes during the past 50 years, *shimpa* still has an important place in contemporary commercial theatre.

By contrast *shingeki* ('new drama') set itself against the commercialism of *kabuki* and *shimpa*. Taking its inspiration from the West (the first two groups opening with IBSEN's *John Gabriel Borkman* and *Hamlet* in 1909 and 1911 respectively) and supported by progressive young intellectuals, *shingeki* has never been able to shake off the charge of being un-Japanese. After financially unsuccessful experiments in the 1910s concerted *shingeki* activity began with the formation of the Tsukiji Shōgekijō (Tsukiji Little Theatre) company in 1924. Its actors were trained in the Western manner, its productions for the first two years were all of Western plays, its directors had seen STANISLAVSKY's and MEYERHOLD's work in the West and its theatre incorporated the latest Western thinking on theatre design. Technically the early efforts of Tsukiji Shōgekijō were poor, but it made a gradual improvement before being engulfed in the passion for Marxism to which its intellectual patrons succumbed in the late 1920s. 'Proletarian drama' required new performance modes, which were enthusiastically learnt, but it also attracted heavy reaction from the police and by 1934 little was left of this briefly flourishing movement. Political drama was impossible in the 1930s and three *shingeki* companies (two with left-wing antecedents and one consciously apolitical) continued Tsukiji Shōgekijō's work (the original company having fragmented in 1928). In the post-war period too there have been three main *shingeki* companies, and they have been joined by a host of smaller groups. There is now a significant corpus of works written by playwrights for *shingeki*, some of whom, such as Kinoshita Junji, have been translated into English and other Western languages. *Shingeki* has sought to create a modern realism that can express the cultural and social complexity of contemporary Japan, and the wide variety of experiments conducted only testifies to the difficulty of the task. It was taking too long for some young *shingeki* actors who, around the time of the world-wide student unrest in 1968, created Japan's underground theatre. Deliberately eschewing the realism of *shingeki* and bringing music (this time modern popular music) and dance back into the theatre, they established a vibrant new form. The underground (or 'little theatre') movement achieved its dramatic effect by kaleidoscopes of images that reflected both Japan's traditional past and her intense present.

The rest of the world has mostly seen Japan's classical theatre and this avant-garde drama, which has elements in common with Western fringe theatre. The foreigner in Tokyo will find a bewildering array of dramatic entertainment available to him. Everything that has been mentioned here, from *nō* onwards, can be seen at almost any time. There are also the Takarazuka Girls Opera (musical plays performed only by girls), commercial realistic theatre, traditional story-telling (*rakugo*), traditional (*yose*) and modern variety shows, and many others. There are more different types of theatre being performed simultaneously in Tokyo than in any other city in the world.

JAROCKI, JERZY, see POLAND.

JARRY, ALFRED (1873–1907), French poet and dramatist, whose *Ubu-roi* is now considered the founder-play of the modern AVANT-GARDE theatre, and a seminal influence on French Surrealism. Written when Jarry was 15 years old and first performed in 1888 as a marionette play, it was given its first live stage production by GÉMIER at the Théâtre de l'Œuvre in 1896. A savagely funny, anarchic revolt against society and the conventions of the naturalistic theatre, it scandalized audiences in 1896 and still has considerable impact, as VILAR's successful revival in 1958 at the THÉÂTRE DES NATIONS amply demonstrated. Père Ubu, vicious, cowardly, coarse, pompously cruel, and unashamedly amoral, is a farcical prototype of the later anti-hero of the nuclear age. Many of the marionette elements in the play, expressly demanded by Jarry in his stage instructions to Gémier, have become common currency in the work of playwrights like GENET and IONESCO and directors like BRECHT and PLANCHON: the use of masks, skeleton sets, crude pantomime and stylized speech to establish character, gross farce and slapstick elements, placards indicating scene changes, cardboard horses slung round actors' necks, and similar unrealistic props. Jarry returned to Mère and Père Ubu in other plays—*Ubu cocu* (completed in 1898, but not published or performed until 1944), *Ubu enchaîné* (1899), and the marionette play, *Ubu sur la butte* (1901)—but without recapturing the same creative spark. *Ubu-roi* was produced at the ROYAL COURT THEATRE in 1966 in an English translation by Iain Cuthbertson. In his novel *Gestes et opinions du Docteur Faustroll* (1911) Jarry elaborated his theory of 'Pataphysics, which he ironically defined as 'the science of imaginary solutions.' IONESCO, VIAN, and others formed a 'College of 'Pataphysics' in 1949, which institutionalized its distinctive brand of erudition, but as a movement it was largely superseded by the Theatre of the ABSURD.

J. C. WILLIAMSON THEATRES, LTD, see WILLIAMSON, JAMES CASSIUS.

JEANS, ISABEL (1891–   ), English actress, who was first seen on the stage in 1909, and had had a good deal of varied experience before she began an important association with the PHOENIX SOCIETY, appearing for them as Celia in JONSON's *Volpone* and Aspatia in BEAUMONT and FLETCHER's *The Maid's Tragedy* in 1921; the First Constantia in Fletcher's *The Chances* and Abigail in MARLOWE's *The Jew of Malta* in 1922; Cloe in Fletcher's *The Faithful Shepherdess* in 1923; and Margery Pinchwife in WYCHERLEY's *The Country Wife* and Laetitia Fondlewife in CONGREVE's *The Old Bachelor* in 1924. She also played in such modern works as Ivor NOVELLO and Constance COLLIER's *The Rat* (1924) and Robert E. SHERWOOD's *The Road to Rome* (1928). In 1931 she starred in *Counsel's Opinion* by her second husband Gilbert Wakefield, and she was seen in revivals of LONSDALE's *On Approval* in 1933, Somerset MAUGHAM's *Home and Beauty* in 1942, and SHAW's *Heartbreak House* in 1943. In 1945 she had a great success as Mrs Erlynne in WILDE's *Lady Windermere's Fan*, which ran for almost 18 months, and in 1949 was much admired as Mme Arkadina in CHEKHOV's *The Seagull*. She appeared in ANOUILH's *Ardèle* (1951), Wilde's *A Woman of No Importance*, T. S. ELIOT's *The Confidential Clerk* (both 1953), and after a ten-year absence from the stage returned in two more plays by Wilde, playing the Duchess of Berwick in *Lady Windermere's Fan* in 1966 and Lady Bracknell in *The Importance of Being Earnest* in 1968. She was last seen in two plays by Anouilh, a revival of *Ring Round the Moon* in 1968 and *Dear Antoine* in 1971. Her exuberant vitality and gift for high comedy kept her in constant demand for both classic and modern plays throughout her career.

JEANS, RONALD, see REVUE.

JEFFERSON, JOSEPH (1774–1832), British-born American actor, whose father Thomas Jefferson (1732–1807) was an actor at DRURY LANE under GARRICK, and for some years manager of a theatre in Plymouth. Four of his five children were on the stage. Joseph, the second son, was trained by his father, and in 1795 went to America. After a short stay in Boston he joined the company at the JOHN STREET THEATRE in New York and later went to the PARK THEATRE, where he was popular with the company and the public, being particularly admired in the roles of humorous elderly gentlemen. He was, however, somewhat held back by the pre-eminence of HODGKINSON, and in 1803 went to the CHESTNUT STREET THEATRE in Philadelphia, where he remained until about 1830. He married an actress, Euphemia Fortune (1774–1830), whose sister Esther was married to William WARREN, thus uniting two families of great importance in American stage history. His seven children all went on the stage, including a second Thomas, John, and four daughters; the best remembered is the second Joseph (1804–42), who inherited his father's happy temperament but not his theatrical talent, being a better scene-painter than actor. By his marriage to the actress Cornelia Thomás (1796–1849), mother by a former marriage of the actor Charles Burke (1822–54), he became the father of the third Joseph (below).

JEFFERSON, JOSEPH (1829–1905), American actor, third of the name, son and grandson of the above. He made his first appearance on stage in Washington, at the age of 4, being tumbled out of a sack by T. D. RICE whose song and dance he then mimicked. With his family he toured extensively, losing his father at 13 and living the hard life of pioneer players. By 1849 he had achieved some eminence, and in 1856 made his first visit to Europe. On his return he joined the company of Laura KEENE, making a great success as Dr Pangloss in the younger COLMAN's *The Heir-at-Law* and as Asa Trenchard in Tom TAYLOR's *Our American Cousin*. He was later at the WINTER GARDEN THEATRE in New York, where in 1859 he played Caleb Plummer in *Dot*, BOUCICAULT's version of DICKENS's *The Cricket on the Hearth*, and Salem Scudder in the same author's *The Octoroon; or, Life in Louisiana*. After the death of his first wife in 1861 he went on a four-year tour of Australia, and on his return to London in 1865 played the part with which he is always associated—Rip Van Winkle. There had been several dramatizations of Washington IRVING's story before this, but it was in Boucicault's version that Jefferson made his greatest success, altering the text so much as he continued to appear in it that in the end it was virtually his own creation, and lived only as long as he did. Not until 1880 did he appear in something else, reviving SHERIDAN's *The Rivals* and making Bob Acres a little more witty and a little less boorish than his predecessors in the part had done. With Mrs John DREW as Mrs Malaprop and a good supporting company this production toured successfully for many years. Jefferson, whose charming, humorous personality made him typical of all that was best in the America of his time, made his last appearance on 7 May 1904, as Caleb Plummer, and then retired, having been 71 years on the stage. In 1893 he had succeeded Edwin BOOTH as President of the PLAYERS' CLUB, and so become the recognized head of his profession. He did a good deal of lecturing, and in 1890 published a delightful autobiography, reissued in 1949. As his second wife he married Sarah Isabel Warren, granddaughter of the first William WARREN, and so strengthened the ties between the two families. His eldest son by his first marriage, Charles Burke Jefferson (1851–1908), was for many years his manager and also an actor, as was Thomas (1857–1931), the third of the name. His only daughter Margaret, by her marriage with the English novelist B. L. Farjeon, was the mother of the composer Harry Farjeon and the novelists and playwrights Eleanor, Joseph, and Herbert Farjeon. By his second marriage Jefferson had four sons, of whom two went on the stage.

JEFFORD, BARBARA MARY (1930–), English actress, who made her first appearance in Brighton in 1949, and in 1950 played Isabella in *Measure for Measure* to GIELGUD's Angelo at Stratford-upon-Avon. After further seasons at Stratford in 1951 and 1954, during which she added roles such as Helena in *A Midsummer Night's Dream* and Katharina in *The Taming of the Shrew* to her repertory, she appeared in the West End as Andromache in GIRAUDOUX's *Tiger at the Gates* (1955), making her New York début in the same play later in the year. From 1956 until 1962 she was with the OLD VIC company, establishing herself as an outstanding classical actress in roles including Beatrice in *Much Ado About Nothing*, Portia in *The Merchant of Venice* (both 1956), Tamora in *Titus Andronicus* (1957), and Viola in *Twelfth Night* (1958). She showed versatility when in 1960 she toured with the company as Lady Macbeth, Joan of Arc in SHAW's *Saint Joan*, and Gwendolen Fairfax in WILDE's *The Importance of Being Earnest*, being particularly memorable as St Joan, communicating both the peasant earthiness of the character and a radiant sense of the ineffable. She appeared at the Old Vic in 1961 as Lavinia Mannon in a revival of O'NEILL's *Mourning Becomes Electra*, and was seen a year later in Oxford (and in 1963 in London) as Lina in Shaw's *Misalliance*; she also played the Stepdaughter in a revival of PIRANDELLO's *Six Characters in Search of an Author* in 1963. In David MERCER's *Ride a Cock Horse* (1965) she made one of her rare excursions into modern drama, and she was much admired as Patsy Newquist in Jules Feiffer's *Little Murders* (1967) at the ALDWYCH THEATRE. In 1976 she joined the NATIONAL THEATRE company, appearing as Gertrude in the initial production of *Hamlet* and in MARLOWE's *Tamburlaine*; in 1977–8 she was with the PROSPECT company at the Old Vic, and in 1979 she took over Joan PLOWRIGHT's part in DE FILIPPO's *Filumena* in the West End before returning to Prospect. Since 1962 she has also been seen regularly at the OXFORD PLAYHOUSE, the NOTTINGHAM PLAYHOUSE, and the BRISTOL OLD VIC.

JEŘÁBEK, FRANTIŠEK VĚNCESLAV, see CZECHOSLOVAKIA.

JEROME, JEROME KLAPKA (1859–1927), English humorist, novelist, playwright, and for a short time an actor. He is now chiefly remembered for his humorous novel *Three Men in a Boat, not to Mention the Dog*, published in 1899, but he was also the author of a number of plays. Several of these were successful in their own day, but the only one remembered now is *The Passing of the Third Floor Back* (1908), in which FORBES-ROBERTSON scored a signal triumph as the mysterious and Christlike stranger whose sojourn in a Bloomsbury lodging-house changes the lives of all its inhabitants. It was revived several times up to 1929.

JERROLD, DOUGLAS WILLIAM (1803–57), English actor, playwright, and journalist, who from its foundation in 1841 until his death was associated with the humorous journal *Punch*. In his early years he appeared at a number of London theatres, but had little taste or talent for acting and soon gave it up in favour of writing. As a playwright he had a good deal of contemporary success, though none of his plays has survived on the stage. Among them were the farce *Paul Pry* (1827), the melodrama *Fifteen Years of a Drunkard's Life* (1828), both produced at the Coburg (later OLD VIC) Theatre, and the NAUTICAL DRAMA *Black-Ey'd Susan; or, All in the Downs* (1829), in which he drew on his early experiences of life at sea, and provided an excellent vehicle for T. P. COOKE. Some of his later plays, notably *The Rent Day* (1832), were first produced at DRURY LANE. In 1836 he took over the STRAND THEATRE, where he put on a number of his own plays, including *A Gallantee Showman; or, Mr Peppercorn at Home* (1837). His last play, a comedy entitled *St Cupid; or, Dorothy's Fortune* (1853), was first seen at Windsor Castle and later the same year at the PRINCESS'S THEATRE. His son William Blanchard Jerrold (1826–84) was the author of a farce, *Cool as a Cucumber* (1851), which provided the younger MATHEWS with one of his best parts.

JESSE, FRYNIWYD TENNYSON, see HARWOOD, HAROLD MARSH.

JESSNER, LEOPOLD (1878–1945), German theatre director, who abandoned the use of representational scenery in favour of a bare stage on different levels connected by stairways (*Jessnertreppe* or *Spieltreppe*). During his years as director of the Staatliche Schauspiele in Berlin (1919–30), where he had Pirchan as his designer and Fritz KORTNER as his leading actor, he was considered one of the most advanced exponents of EXPRESSIONISM and his work greatly influenced the contemporary theatre. Among his outstanding productions were SCHILLER's *Wilhelm Tell* in 1919, Shakespeare's *Richard III* and WEDEKIND's *Der Marquis von Keith* in 1920, and HAUPTMANN's *Die Weber* in 1928. He left Germany in 1933, and after a short stay in London went to the U.S.A., where he died.

JESTER, see COURT FOOL.

JESUIT DRAMA, term used to describe a wide variety of plays, mainly in Latin, written to be acted by pupils in Jesuit colleges. Originally, as in other Renaissance forms of SCHOOL DRAMA, these were simple scholastic exercises, but over the years, particularly in VIENNA under the influence of opera and ballet, they became full-scale productions involving elaborate scenery, machinery, costumes, music, and dancing, as well as an almost professional technique in acting and diction.

The earliest mention of a play produced in a Jesuit college dates from 1551, when an unspecified tragedy was performed at the Collegio Mamertino in Messina, founded as the first Jesuit school three years earlier. In 1555 the first Jesuit play was seen in Vienna, *Euripus sive de inanitate verum* by Levinus Brechtanus (Lewin Brecht), a Franciscan from Antwerp. This was followed by productions at Cordoba in 1556, at Ingoldstadt in 1558, and in MUNICH in 1560. There were already 33 Jesuit colleges in Europe when Ignatius de Loyola, the founder of the Order, died in 1556: by 1587 there were 150 and by the early 17th century about 300. For over two centuries at least one play a year, and often more than one, was performed in each college. The total of plays specially written for these performances was enormous. Only the best were published, but recent researches have brought to light a great many manuscripts. The early plays were based mainly on classical or biblical subjects—Theseus, Hercules, David, Saul, Absalom—but later, stories of saints and martyrs—Theodoric, Hermenegildus—were also used, as well as personifications of abstract characters—Fides (Faith), Pax (Peace), Ecclesia (the Church). The popularity of plays based on the stories of women—Judith, Esther, St Catharine, St Elizabeth of Hungary—led to the early abolition of the rule against the portrayal on stage by the boy pupils of female personages. The use of Latin was less easily disregarded, bound up as it was with its use in class and in daily conversation between masters and pupils. The vernacular seems first to have been used, in conjunction with Latin, in Spain, but the *Christus Judes* (1569) of Stefano TUCCIO was translated into Italian in 1584 and into German in 1603. During the 17th century many plays appeared in French or in Italian, and by the beginning of the 18th century most Jesuit plays were written in the language of the country in which they were to be produced. Parallel with the increased use of the vernacular went the introduction of operatic arias, interludes, and ballets. Of all the splendid productions given in Vienna the most memorable appears to have been the *Pietas Victrix* (1659) of Nikolaus AVANCINI, which had 46 speaking characters as well as crowds of senators, soldiers, sailors, citizens, naiads, Tritons, and angels. The technical development reached by Jesuit stagecraft can be studied in the illustrations to the published text of the play, which was acted on a large stage equipped with seven TRANSFORMATION SCENES. Lighting effects became increasingly elaborate; though the plays often began in daylight, which came through large windows on each side of the stage, they usually ended by torchlight, while in the course of the action sun, moon, and stars, comets, fireworks, and conflagrations were regularly required. All this, added to the splendour of the costumes and the large choruses and orchestras—often employing as many as 40 singers and 32 instrumentalists—made the Jesuit drama a serious rival to the public theatres. In Paris in the 17th century the three theatres in the Lycée Louis-le-Grand, where Louis XIV and his Court often watched the productions, were better equipped than the COMÉDIE-FRANÇAISE and almost on a par with the Paris Opéra. Jesuit drama continued to flourish in such conditions all over Catholic Europe until the Order was suppressed in 1773. It left its mark on the developing theatres wherever it was played, notably through the works of such authors as Avancinus and ADOLPH in Austria, BIDERMANN in Germany, and Tuccio and SCAMMACCA in Italy, and through its influence on pupils who were to become playwrights, among them CALDERÓN, CORNEILLE, GOLDONI, LESAGE, MOLIÈRE, and VOLTAIRE.

JEVON, THOMAS (?–1688), English actor and dancer, credited in a contemporary satire with 'heels of cork and brains of lead'. He was one of the first English actors to play HARLEQUIN, being seen in that part in Aphra BEHN's *The Emperor of the Moon* at DORSET GARDEN in 1687. The previous year his only known play, a farce entitled *The Devil of a Wife; or, a Comical Transformation*, had been given at the same theatre with some success. As a three-act BALLAD OPERA by Charles Coffey, billed as *The Devil to Pay; or, the Wives Metamorphos'd* (1731), later cut to one act by Theophilus CIBBER, it became extremely popular, and in 1743 appeared as *Der Teufel ist los* in Germany, where it had an undoubted influence on the development of the *Singspiel*.

JEWISH ART THEATRE, New York, see GARDEN THEATRE and SCHWARTZ, MAURICE.

JEWISH DRAMA. Any study of Jewish drama, including that of the new HEBREW THEATRE of Israel, must be approached in the light of the Jewish attitude to the theatre in general. Unlike other dramas, that of the Jewish communities had originally no territorial limits. Its boundaries were linguistic, comprising Hebrew, the religious and historical language which never ceased to be written and has now been reborn as a living tongue in Israel; Yiddish, the vernacular of the vast Jewish communities which lay between the Baltic and the Black Seas, spread by emigrants all over the world; and Ladino (Judaeo-Spanish), the speech of the Sephardic Jews of the Middle East, who never achieved a permanent stage, their plays being for reading only. The linguistic frontiers were not clearly defined; Israel ZANGWILL wrote Jewish plays in English; Alexander Granach (1890–1945) began as a Yiddish actor, won fame in the German theatre, and after 1933 returned to the Yiddish stage in Poland.

Drama was not indigenous to the Jew. Deuteronomy xxii.5 expressly forbade the wearing of women's clothes by men, and the connection between early drama and the idol-worship of alien religions was a strong argument against the theatre as late as the 19th century. Yet the classical theatre, low as it had fallen by the time the Jew came into contact with Hellenism, exercised a strong attraction, and Jewish actors were found in Imperial Rome, while Ezekiel of Alexandria, taking EURIPIDES as his model, wrote a Greek tragedy on the Exodus, though there is no reason to suppose that this was ever acted, or indeed intended for the stage. It was from the Jewish itinerant musicians and professional jesters (the *badchans* of modern Yiddish literature), and from the questions and responses in the synagogue services, that the Jewish theatre slowly evolved. An important part was also played by the Hebrew philosophical dialogues dating back to Ibn Ezra in the 12th century and popular down to the end of the 18th—such works as Beer's *Conversations of the Spirit of Poverty with that of Good Reputation* (1674), Fiammetta's *Duet between Grace and Truth* (1697), and Norzi's *Conversation with Death* (1800).

By the 15th century Jewish festivals in Europe, particularly in Germany, had long been enlivened by the songs and dances of itinerant entertainers, and under the influence of German Court drama and the plays of the MEISTERSÄNGER vernacular songs were soon being dramatized to provide such farces as *The Play of the Devil, the Doctor, and the Apothecary* and *A Play of Food- and Drink-Loving Youth*. Soon religious plays in Yiddish appeared also, on such subjects as Adam and Eve, the sacrifice of Isaac, or the death of Moses, modelled on the Carnival plays of Hans SACHS. The earliest extant manuscript of such a play, on Jonah, dates from 1582, and is based on a similar work by the mastersingers Simon Rothen and Balthasar Klein. Many of the elements of these plays were later fused with those developed in Italy to form the PURIM PLAYS, which continued well into the 19th century.

While Yiddish plays were developing in Germany and other parts of Central Europe, plays in Hebrew were being given in Italy and among the Spanish Jews in Holland. The first Hebrew play to survive is *The Comedy of a Marriage*, ascribed to Leone di Somi, which shows clearly the influence of the COMMEDIA DELL'ARTE; the characters include a pining lover, his beloved, her scheming maid, a broad-humoured manservant, and an unscrupulous lawyer. The introduction to the earliest extant manuscript of 1618 says that the play was intended for presentation at Purim (roughly the middle of March).

Throughout the 17th and 18th centuries the Jewish communities in Italy were closely connected with those in the Netherlands. Refugees from Spain settled in Amsterdam; their leaders saw in the drama one way of fostering a community spirit, but knowing no Dutch, and afraid of exciting animosity by the use of Spanish in a country which had only just shaken off the yoke of its Spanish oppressors, they resorted to Hebrew for the language of their plays. The form of the new drama followed that of the Spanish COMEDIA and *auto*, but its content was influenced by the religious plays of the Dutch dramatist Joost van den VONDEL, as can be seen in a play on the delivery of Abraham from the fiery furnace, *Yesod Olam* by Moses Zacuto (1625–97). This was probably first performed in Italy, where a number of Jewish actors could be found; it is known that another of Zacuto's plays, *Tofteh Aruch*, which describes a journey to the next world, was acted by worshippers at the synagogue in Ferrara in about 1700. Its influence can be seen in *Eden Aruch* (*c.* 1720) by Jacob Olmo (1690–1755). Although Italy saw the first performances of many Hebrew plays, it was in Amsterdam that the first was printed. This was *Asiré Hatiqva*, a MORALITY PLAY by Joseph Penço de la Vega (1650–1703), published in Amsterdam in 1673 and reprinted in Leghorn in 1770. In it a king is led astray by Satan, his wife, and his own unbridled passions; his reason and his good angel guide him back to virtue. The play, which contains a number of songs, is believed to have been written for a festivity at a religious school.

The outstanding Jewish dramatist of the early 18th century was undoubtedly Moses Hayim LUZZATO, an Italian Jew resident in Amsterdam, whose two most important plays, *Migdal Oz* (1727) and *Tehilla Layesharim* (pub. 1743), have placed him among the great Hebrew poets. But neither these plays nor those of an increasing number of Hebrew playwrights, among whom was Samuel Romanelli (1757–1814), a Jew resident in Mantua who translated several of Metastasio's librettos, led to the establishment of a permanent Jewish theatre, and apart from plays written for and presented on religious holidays and special festive occasions drama in general remained a literary exercise remote from stage presentation. The Haskala (Enlightenment) Movement which resulted from the period of tolerance following the Thirty Years War meant, however, that German Jews were able to enjoy a relatively peaceful life, and under the influence of the German-Jewish philosopher Moses Mendelssohn, Jews were introduced to the language, literature, and sciences of other nations, which resulted in a revival of their own culture. The principles of the Haskala Movement spread slowly eastward, reaching the more remote regions only in the second half of the 19th century, and for over 100 years provided the dominant feature in the history of Central and Eastern European Jewry. One of the results of the movement was a renewed interest in the writing of plays in Hebrew. Imitations of Luzzato are to be found in such works as *Yaldut Ubahrut* (1786) by Menahem Mendel Bresselau (1760–1827) and *Amal and Tirza* (1812) by Shalom Cohen (1772–1845), the first written for a bar mitzvah, the second for a wedding. Knowledge of other European dramas was extended by the translation of such plays as RACINE's *Esther* and *Athalie*, popular both for their Jewish subjects and because their choral structure appealed to the Jewish love of combined dialogue and song; of GOETHE's *Faust*, and the plays of LESSING and SCHILLER; and of the comedies of MOLIÈRE. Even Shakespeare was drawn on by Joseph Ephrati (1770–1849) of Toplowitz in his *The Reign of Saul* (pub. 1794). All these plays, however, were still intended mainly for reading, and for an occasional performance. The establishment of a truly Hebrew Theatre was not to come until well into the 20th century, with the founding of the State of Israel, and the Yiddish theatre was destined to go ahead first, mainly in an attempt to check the growing vulgarity of the Purim plays. Two of these early Yiddish plays were written by followers of Moses Mendelssohn in Germany: *Reb Henoch* (*c*. 1793) by Isaac Euchel (1756–1804) and *Leichtsin und Frommelei* (1796) by Abron Halle [Wolfsohn] (1754–1835). Others, by Joseph Biedermann (1800–?), were performed by amateurs in Vienna in about 1850. They flourished best in Russia, where a long tradition of wandering musicians, the Brody Singers who took their name from a Polish town, had resulted by about 1850 in the evolution of a type of popular entertainment, featuring comic and sentimental Yiddish songs linked by dialogue and dance, which spread widely throughout Galicia and the Carpathians, and eventually resulted in the creation of a series of one-act sketches, partly written and partly improvised. While such writers as Israel Axenfeld (*c*. 1795–1868), Solomon Ettinger (*c*. 1803–56), author of the very popular *Serkele* (1825), Abraham Ber Gottlober (1811–99), and Ludwig Levinsohn (1842–1904) were laying the foundations of a Yiddish repertory, these sketches kept alive the spirit of the theatre among Russian Jews until in 1876 Abraham GOLDFADEN founded the first permanent Yiddish theatre, probably between 5 and 8 Oct., when, with the help of two Brody Singers, he presented a two-act musical sketch in a tavern in Jassy (Romania). From these humble beginnings the new theatre progressed rapidly. Goldfaden enlarged his company, employing women on the stage for the first time in Jewish history, trained his players, and progressed from the short sketches of his early days to the writing of full-length plays, some of which still remain in the repertory. Although he lacked what the established theatres of Europe could have given him—training, tradition, and experience—and much of his undoubted genius was wasted in the struggle to establish and maintain a theatre under primitive conditions, he gave his audiences what they wanted—a mixture of song and dance, with plots and music borrowed from all over Europe, racy dialogue and broad characterization, much action and little analysis. He succeeded so well that he eventually raised up rivals in his own field. One of his original actors, Israel Gradner, established his own company in the late 1870s, with Joseph Lateiner (1853–1935), who in 1883 was to found a Yiddish Theatre in New York, as its accredited playwright; and when the sophisticated Jews of Odessa, to whom the theatre was no novelty, found the plays of Goldfaden, on a visit there in 1879, too rough, they discovered a playwright more to their taste in Joseph Yehuda Lerner (*c*. 1849–1907), who took most of his material from non-Jewish works dealing with Jewish heroes, and is chiefly remembered for his translation into Yiddish of K. F. GUTZKOW's *Uriel Acosta* (1847).

All this activity was brought to a sudden stop by the anti-semitic measures which followed the assassination of the Tsar Alexander II on 13 Mar. 1881. All plays in Yiddish were forbidden, and the Yiddish theatre existed precariously in Russia until the Revolution of 1917. Most of the actors and dramatists left the country for England and America, and New York became the new centre of Yiddish drama. The first attempt at its establishment there was made in 1882 by Boris Thomashefsky (1868?–1939), and it found its feet with the arrival of Lateiner and Moses Hurwitch (1844–1910). But the stock themes of old Jewish life in Europe, which had been acceptable to the first bewildered and largely illiterate immigrant audiences, together with the broad farce and sentimental melodrama which had provided the only alternatives in the early days, became increasingly outmoded as Americanization proceeded apace; even Goldfaden, when he arrived in New York in 1887, found himself out of touch with the new audiences. It was left to Jacob GORDIN to revitalize the American Yiddish theatre, by free adaptations of plays by the great European dramatists, by discouraging improvisation, and by seeking his material in such modern social phenomena as the breakdown of traditional Jewish family life under the stress of new surroundings. Among those who followed him in the path of social drama were: Solomon Libin [Israel Zalman Hurwitz] (1872–1955), author of some 50 plays on the life of the immigrant Jewish worker in New York, of which *Broken Hearts* (1903) was the best; Moses Nadir (1885–1943), author of *The Last Jew* (1919); Harry Sackler (1883–1973), author of *Yizkor* (*In Memoriam*, 1923); Leo Kobrin (1872–1946), whose *Riverside Drive* was produced in 1927; and Halper LEIVICK, considered by many critics the best Yiddish writer of his time. A number of other companies were founded under the influence of SCHWARTZ, among them that of Rudolf Schildkraut (1862–1930) and Artef, the workmen's studio theatre, which adopted the methods of the MOSCOW STATE JEWISH THEATRE and staged works by

Soviet-Jewish authors. The wide-spread adoption of English in Yiddish-speaking homes, and the slackening in immigration, were potent factors in the continued decline of the Yiddish theatre, which was not arrested by Schwartz's efforts to enlarge the repertory by playing European classics in Yiddish and Yiddish plays in English.

Nevertheless the period between the two world wars was a flourishing time for the Yiddish theatre. The Argentine, home of a large Jewish community, had two permanent Yiddish theatres; London, which had its first Yiddish theatre in Whitechapel in 1888, had two in the 1930s, both in the East End and both playing Yiddish classics. Paris, where Goldfaden founded a company in 1890 with Anna Held (1865–1918) as its leading lady, also had a company which played the more popular operettas from the New York Yiddish stage, giving one performance a week. Vienna, home of the Biedermann company, had in the 1920s and 1930s several more which carried on the earlier tradition, and in the 1920s New York alone had twelve theatres devoted to Yiddish drama, as well as several others scattered throughout the States.

The Nazi holocaust of Jews on the European continent, and the progress of assimilation in Western countries not so affected, led inevitably after 1946 to the decline of the Yiddish theatre, and though isolated pockets may have lingered on, it finally disappeared from the international scene. Yiddish actors migrated to the national stage of the country in which they found themselves or went into films, and most of the theatres were closed, one of the last being the Polish State Jewish Theatre, run by Ida KAMIŃSKA.

JEWISH DRAMA ENSEMBLE, see MOSCOW STATE JEWISH THEATRE.

JIG, an Elizabethan after-piece, given in the smaller public theatres only. (Shakespeare evidently had a low opinion of it, making Hamlet say dismissively of Polonius: 'He's for a jig, or a tale of bawdry.') It consisted of rhymed dialogue, usually on the frailty of women, sung and danced to existing tunes by three or four characters, of whom the CLOWN was always one. The best known exponents of the Jig were KEMPE and TARLETON; it disappeared from the legitimate theatre at the Restoration, but remained in the repertory of strolling players and actors in fair-booths, and from the late 16th century onwards became increasingly popular in Germany, being taken there by the ENGLISH COMEDIANS who toured the Continent. Very few texts survive, and those mostly in German translations, among them *Mr Attowel's Jigge*, first published in 1595 as *Frauncis new Jigge . . . to the tune of Walsingham* and translated as *Die tügende Bawrin*. It was probably played in Germany during a visit there by STRANGE'S MEN, the company to which George Attwell belonged. The Jig is thought to have been a formative element in the development of the German *Singspiel* (see BALLAD OPERA).

JIM CROW, see RICE, T. D.

JIRÁSEK, ALOIS, see CZECHOSLOVAKIA.

JODELET [Julian Bedeau] (?1590–1660), French comedian, brother of L'ESPY. He had already had some experience in the provinces when he joined the company of MONTDORY shortly before the opening of the Théâtre du MARAIS in 1634. With several of his companions he was transferred by order of Louis XIII to the Hôtel de BOURGOGNE but he returned to the Marais in 1641, where he played Cliton in CORNEILLE's *Le Menteur* (1643). He also appeared in a series of farces written for him, mostly with his name in the title, by, among others, SCARRON (*Jodelet, ou le maître-valet*, 1645) and Thomas CORNEILLE (*Jodelet Prince*, 1655). He was extremely popular with Parisian audiences—he had only to show his flour-whitened face to raise a laugh—and he frequently added jokes of his own to the author's lines. When MOLIÈRE established himself in Paris in 1658 he induced Jodelet to join his company, thus assuring the co-operation of the one comedian he had reason to fear, and wrote for him the part of the Vicomte de Jodelet to his own Marquis de Mascarille in *Les Précieuses Ridicules* (1659). He probably intended to give him the title role in *Sganarelle, ou le Cocu imaginaire* (1660), but Jodelet died just before the production opened and Molière played the part himself.

JODELLE, ÉTIENNE (1532–73), French Renaissance poet and dramatist, a member (with Ronsard and others) of the famous 'Pléiade'. His *Cléopâtre captive* (1552) was the first French tragedy to be modelled on SENECA. Together with a comedy, *Eugène*, also an attempt to acclimatize a classical dramatic form, it was given before Henri II and his Court with Jodelle, not yet 21, as Cleopatra. The Pléiade, delighted at the success of one of their members, organized a festival in Jodelle's honour, presenting him with a goat garlanded with ivy; the Church took umbrage at this pagan revival and Jodelle bore the brunt of its displeasure. Of his later plays only *Didon* (1558) survives. Jodelle's tragedies are static, elegiac commemorations of a striking catastrophe. They lack dramatic action, and depend overmuch on soliloquy and narrated event; but he deserves credit as the precursor of a great dramatic tradition.

JODRELL THEATRE, London, see KINGSWAY THEATRE.

JOHN BULL PUNCTURE REPAIR KIT, see PEOPLE SHOW.

JOHN F. KENNEDY CENTER FOR THE PERFORMING ARTS, see EISENHOWER THEATRE, WASHINGTON, D.C.

JOHN GOLDEN THEATRE, New York. (1) At 202 West 58th Street, between Broadway and 7th Avenue. Named after the actor, song-writer, and theatre director John Golden (1874–1955), for whom it was built, it opened on 1 Nov. 1926 and was almost immediately taken over by the THEATRE GUILD, who staged there before the end of the year two plays by Sidney HOWARD—*Ned McCobb's Daughter* and *The Silver Cord*. In 1928 the Theatre Guild again occupied the theatre with a successful run of O'NEILL's *Strange Interlude*, but after a short period as the Fifty-Eighth Street Theatre in 1935–6 it became a cinema, making a brief return to live entertainment as the Concert Theatre in 1942, with intimate REVUE. It is now used for television shows.

(2) At 252 West 45th Street, between Broadway and 8th Avenue. A small theatre, seating 799, this opened as the Masque on 24 Feb. 1927 with a translation of ROSSO DI SAN SECONDO's *Marionette, che passione!* as *Puppets of Passion*. Kirkland's long-running *Tobacco Road* opened on 4 Dec. 1933, and four years later John Golden took over, naming the theatre after himself. His first successful productions were Paul Vincent CARROLL's *Shadow and Substance* (1938) and Patrick Hamilton's *Angel Street* (1941). Later the theatre housed a series of revues which included *At the Drop of a Hat* (1959), *Beyond the Fringe* (1962), and the White African revue *Wait a Minim* (1966), all imported from London. More recent productions have included a double bill by Robert ANDERSON, *Solitaire/Double Solitaire* (1971), STOPPARD's *Dirty Linen*, D. L. Coburn's *The Gin Game* with Jessica TANDY and Hume CRONYN (both 1977), and the musical *A Day in Hollywood—A Night in the Ukraine* (1980), which had more success than in London.

JOHNNY JACKS' PLAY, see MUMMERS' PLAY.

JOHNSON, CELIA (1908–82), English actress, created DBE in 1981. She made her first appearance in SHAW's *Major Barbara* in Huddersfield in 1928 and was first seen in London in 1929. She had a long run as Anne Hargraves in Merton Hodge's *The Wind and the Rain* (1933) and was much admired in 1936 for her playing of Elizabeth Bennet in Helen Jerome's adaptation of Jane Austen's *Pride and Prejudice*. She was playing with equal success in Daphne Du Maurier's *Rebecca* (1940) when the theatres closed because of enemy action. In 1947 she played the title role in Shaw's *Saint Joan* for the OLD VIC company, and in 1951 gave a fine performance as Olga in CHEKHOV's *Three Sisters*. She made a successful appearance in modern comedy in William Douglas HOME's *The Reluctant Débutante* (1955) and gave a sensitive portrayal of Isobel Cherry in Robert BOLT's *Flowering Cherry* (1957). Over the next few years she was seen in a succession of new plays—Hugh and Margaret Williams's *The Grass is Greener* (1958), BILLETDOUX's *Chin-Chin* (1960), N. C. HUNTER's *The Tulip Tree* (1962), and Giles Cooper's *Out of the Crocodile* (1963). She then broadened her range by joining the NATIONAL THEATRE company to play Mrs Solness in IBSEN's *The Master Builder* in 1964 opposite OLIVIER, and took over the part of Judith Bliss in its revival of COWARD's *Hay Fever* the following year, which gave her another opportunity to display her virtuosity in light comedy, as did Alan AYCKBOURN's *Relatively Speaking* (1967). In contrast, she was seen at the CHICHESTER FESTIVAL THEATRE in 1966 as Ranevskaya in Chekhov's *The Cherry Orchard*; and in Nottingham in 1970 and London a year later as Gertrude to Alan BATES's Hamlet. She was later in three plays by William Douglas Home, *Lloyd George Knew My Father* (1972), *The Dame of Sark* (1974), and *The Kingfisher* (1977).

JOHNSON ELIZABETH (*fl.* 1790–1810), American actress, who made her first appearance in Boston in 1795 with the AMERICAN COMPANY, and went with them to the JOHN STREET THEATRE, New York, the following year. A tall, elegant woman, she played Rosalind in *As You Like It* on the opening night at the PARK THEATRE in 1798 and was later seen as Juliet and as Imogen in *Cymbeline* to the Romeo and Iachimo of Thomas A. COOPER. She was much admired in the fashionable ladies of high comedy, and was one of the first actresses in New York to play male parts seriously, appearing in 1804 as Young Norval in HOME's *Douglas*. Her husband John was a good utility actor specializing in old men, and was for a short time manager of the Park, where his daughter Ellen, later Mrs Hilson, made her first appearance as a child of 5.

JOHNSON, Dr SAMUEL (1709–84), the great English lexicographer, was the author of a five-act tragedy, *Irene*, which his friend and fellow townsman David GARRICK produced at DRURY LANE in 1749, with little success. After its failure Johnson never again essayed the stage, though he made more money from the proceeds of the third, sixth, and ninth nights of his play than by anything he had previously done. His edition of the plays of Shakespeare is valuable for the light it throws on the editor rather than on the author, since Johnson had little knowledge of Elizabethan drama or stage conditions, and was not temperamentally a research worker. He should not be confused with Samuel Johnson of Cheshire, author of *Hurlothrumbo* (1729) and other burlesques.

JOHNSTON, (William) DENIS (1901–84), Irish dramatist, who was joint manager of the GATE THEATRE, Dublin, and also appeared in early

Dublin Drama League productions at the ABBEY THEATRE. His first play, *The Old Lady Says 'No!'* (1929), is a satirical review of certain dominant elements in Irish life, exposing the sentimentality inherent in some of them. Refused by Lady GREGORY for the Abbey (hence its title), it was produced by the Gate Theatre during its second season. *The Moon in the Yellow River* (1931) earned wide popularity for the richness of its characterization and the precision with which the author diagnosed the mood of the mid-1920s in Ireland. *A Bride for the Unicorn* (1933) showed the influence of EXPRESSIONISM, but *Storm Song* (1934) was a straightforward play on an original theme. *Blind Man's Buff*, Johnston's adaptation of TOLLER's *Die blinde Göttin*, had a successful run at the Abbey Theatre in 1936. Later works were *The Golden Cuckoo* (1939), a comedy of individual rebellion; *The Dreaming Dust* (1940), on the life of Dean Swift; *A Fourth for Bridge* (1948), a one-act war comedy; *Strange Occurrence on Ireland's Eye* (1956), a successful court-room drama; and *The Scythe and the Sunset* (1958). His autobiography, *Nine Rivers From Jordan*, was published in 1953.

JOHNSTON, HENRY ERSKINE (1777–1845), Scottish actor, who made his first appearance on the stage at the Theatre Royal, Edinburgh, playing Hamlet with no experience or training. He then created a sensation as Young Norval in a revival of HOME's *Douglas* and was nicknamed the Scottish Roscius. Too much undeserved adulation went to his head and prevented him from taking his work seriously, but his youth and handsome presence took him to COVENT GARDEN in 1797, where he created the parts of Henry in MORTON's *Speed the Plough* (1800) and Ronaldi in HOLCROFT's *A Tale of Mystery* (1802), after which he lapsed into obscurity.

JOHN STREET THEATRE, New York, at 15–21 John Street, west of Broadway. The first permanent playhouse in New York and the third to be built by David DOUGLASS, this opened on 7 Dec. 1767 with FARQUHAR's *The Beaux' Stratagem*; it was described by DUNLAP, whose first play *The Father; or, American Shandyism* was produced there in 1789, as 'principally of wood, an unsightly object, painted red.' There is also a reference to it in Royall TYLER's *The Contrast*, a landmark in American theatre history produced at John Street two years earlier, where Jonathan, the country bumpkin, describes his first visit to a theatre. Until the War of Independence the AMERICAN COMPANY gave regular seasons there, and mounted the first productions in New York of such plays as *The Merchant of Venice, Macbeth, King John*, JONSON's *Every Man In His Humour*, and DRYDEN's *All For Love*, as well as a large repertory of contemporary plays and after-pieces. During the war the playhouse was rechristened Theatre Royal and used for productions by the officers of the English garrison, among them Major John André—later the subject of a play by Dunlap—whose scene-painting was much admired. Just before the British evacuated New York a professional company under Dennis Ryan came from Baltimore and stayed for a time at the John Street Theatre, but without much success.

Two years after the British evacuation, in 1785, the American Company, now under the control of the younger HALLAM and John HENRY, took possession of the theatre again, the company being reinforced shortly afterwards by Thomas WIGNELL and Mrs Owen MORRIS. During the next few years regular seasons were given, with productions for the first time in New York of *The School for Scandal* and *The Critic* by SHERIDAN, *Much Ado About Nothing, As You Like It*, and GARRICK's version of *Hamlet*. George Washington visited John Street three times in 1789, the year of his inauguration. On 6 May he saw *The School for Scandal* and a popular farce, *The Poor Soldier* by John O'KEEFFE, in which Wignell was much admired as Darby; on 5 June he saw the elder COLMAN's *The Clandestine Marriage*; and on 24 Nov. he heard himself alluded to on the stage in Dunlap's *Darby's Return*.

Soon after this Wignell and Mrs Morris left John Street to found their own CHESTNUT STREET THEATRE in Philadelphia and John HODGKINSON, newly arrived from England, joined the John Street company, soon becoming so popular, and so grasping, that he ousted both Hallam and John Henry from management and from the affections of the public. Henry and his wife withdrew from the company in 1794, Hallam in 1797, leaving Hodgkinson in command with Dunlap. He had been added to the management the previous year, which had also seen the first appearance in New York with the John Street Company of the first Joseph JEFFERSON, who remained with it until 1803 when he went to Philadelphia.

In the autumn of 1797 Sollee, a theatre manager from Boston, rented the John Street Theatre for his own company and there entered on an intense rivalry with Wignell's company from Philadelphia, established for a season in Rickett's Amphitheatre. In Sollee's company were Miss Arnold, later the mother of Edgar Allan Poe, and Mrs Whitlock, sister of Sarah SIDDONS. The venture was not a success, and the old company returned while waiting to move into their new PARK THEATRE (1), built by Dunlap. The John Street Theatre was used for the last time on 13 Jan. 1798, and was later sold by Hallam for £115.

JOINT STOCK COMPANY, London, see GASKILL, WILLIAM, and HARE, DAVID.

JOLLY, GEORGE (*fl.* 1640–73), English actor, the last of the notable ENGLISH COMEDIANS in the German theatre. He may have been at the FORTUNE THEATRE in London in 1640, and is first found in Germany in 1648. He was particularly active in

FRANKFURT, where Prince Charles (later Charles II) probably saw him act. He appears to have anticipated DAVENANT's use of music and scenery on the public stage, and already had women in his company in 1654. He returned to England at the Restoration, and got permission to open the COCKPIT, where the French theatre historian Chappuzeau saw him in 1665; but KILLIGREW and Davenant managed to deprive him of his patent and he had to content himself with overseeing the NURSERY.

JOLLY BOYS' PLAY, see MUMMERS' PLAY.

JOLSON, AL, see MUSICAL COMEDY and REVUE.

JOLSON THEATRE, New York, see CENTURY THEATRE (2).

JONES, HENRY ARTHUR (1851–1929), English dramatist, who had had several one-act plays performed in the provinces before his farce *A Clerical Error* (1879) was put on in London at the ROYAL COURT THEATRE by Wilson BARRETT who later made an immense success in *The Silver King* (1882), a melodrama written by Jones in collaboration with Henry Herman which was many times revived. Although regarded by his contemporaries as one of the new school of dramatists whose plays belonged to the so-called 'theatre of ideas', it was the melodramatic element rather than the social criticism which drew the public to such plays by Jones as *Saints and Sinners* (1884), *The Dancing Girl* (1891), in which Julia NEILSON made a spectacular success, *The Case of Rebellious Susan* (1894), and particularly *Michael and His Lost Angel* (1896), which had to be withdrawn after 10 performances, mainly on account of the scene in a church in which a priest, standing before the altar, makes a public confession of adultery after having some years before exacted a similar penance from a young woman in his congregation. *The Liars* (1897) and *Mrs Dane's Defence* (1900) also contain a strong melodramatic strain, though the third act of the latter is still considered a masterpiece. With PINERO and SHAW, Jones was one of the most considerable playwrights in the period when IBSEN's influence was beginning to make itself felt in the English theatre, but there is no reason to doubt his own assertion that he was in no way consciously indebted to Ibsen. Shaw praised Jones at the expense of Pinero on the ground that he drew faithful portraits of men and women in society, whereas Pinero merely flattered them with reflections of their own imaginings; posterity has reversed this judgement, and Pinero's plays still hold the stage while those of Jones are forgotten. His skill in naturalistic dialogue and in the creation of dramatic tension remains impressive, but his social and moral criticism lacked a firm philosophical basis and the redeeming gift of humour.

JONES, INIGO (1573–1652), English architect and artist, the first to be associated with scenic decoration in England. Before his day the designing and decorating of Court MASQUES had been arranged by the Office of the Revels, which employed for the purpose any artist who happened to be about the Court. Jones, having studied in Italy and worked in Denmark, was in 1604–5 attached to the household of Prince Henry, and in addition to his work as an architect undertook complete control of the artistic side of the masques. Of the 13 given at Court from 1605 to 1613, 9 were certainly of his devising, the others probably so, the first being JONSON's *Mask of Blackness*. Jones was also in charge of the plays given at Oxford in Christ Church Hall in Aug. 1605, where he first used revolving screens in the Italian manner. He later used as many as five changes of scenery in one play, with backcloths, shutters, or FLATS painted and arranged in perspective, which ran in GROOVES and were supplemented by a turntable (*machina versatilis*) presenting to the audience different facets of a solid structure. He also introduced into England the picture-stage framed in the proscenium arch. His increasing power and responsibility brought him into conflict with the Court poets, particularly Jonson, who satirized him in many of his plays. During the Civil War Jones fell out of favour, was heavily fined, and died in poverty. Many of his designs are preserved in the library of the Duke of Devonshire at Chatsworth.

JONES, JAMES EARL (1931–), American actor, considered the finest Negro player of his generation. He made his stage début at the University of Michigan in 1949 in James Gow and Arnaud d'Usseau's *Deep Are the Roots*, his first engagement in New York being as an understudy in Molly Kazan's *The Egghead* (1957). During his subsequent career he has managed to combine roles in New York and elsewhere with a close association with Joseph PAPP's New York Shakespeare Festival, for which he appeared as Oberon in *A Midsummer Night's Dream* in 1961, Caliban in *The Tempest* in 1962, Othello in 1964, Macbeth in 1966, Claudius in *Hamlet* in 1972, and King Lear in 1973. For other managements he has appeared in GENET's *The Blacks* (1961), James SAUNDERS's *Next Time I'll Sing To You* (1963), Athol FUGARD's *The Blood Knot* (1964), and BÜCHNER's *Danton's Death* (1965). He also scored an outstanding success as the boxer in Howard Sackler's *The Great White Hope* (1968), which ran for over a year. Later appearances have included Boesman in Fugard's *Boesman and Lena* (1970), Lopahin in an all-Black production of CHEKHOV's *The Cherry Orchard* in 1973, Hickey in a revival of O'NEILL's *The Iceman Cometh* in 1973, and Lennie in a revival of STEINBECK's *Of Mice and Men* in 1974. He appeared in the two-character play *Paul Robeson* by Phillip Hayes Dean in New York and London in 1978. In 1980 he was in Fugard's

A Lesson from Aloes and in 1981, at the AMERICAN SHAKESPEARE THEATRE, he again played Othello.

JONES, JOSEPH STEVEN (1809–77), American actor and author, creator of a number of Yankee characters, of whom Solon Shingle in *The People's Lawyer* (1839) was the most popular, John E. OWENS making his final appearance in the part in New York in 1884. The Honorable Jefferson S. Batkins, another Yankee character in *The Silver Spoon* (1852), was first played by the younger William WARREN at the Boston Museum, and the play survived on the stage until well into the 20th century.

JONES, MARGO (1913–55), American director and producer. After studying at the Southwestern School of the Theatre in Dallas she worked with the Ojai Community Players and at the PASADENA PLAYHOUSE. In 1939 she was associated with community and university drama in Houston, Texas, and in 1943 she staged an early play by Tennessee WILLIAMS, *You Touched Me*, at the CLEVELAND PLAY HOUSE. In 1945 she founded an experimental theatre in Dallas, where she encouraged the work of new playwrights and gave experimental productions of older plays, described in her book *Theatre-in-the-Round* (1951). Her work was first seen on Broadway in 1945, when she directed, with Eddie Dowling, Tennessee Williams's *The Glass Menagerie*. Later productions included Williams's *Summer and Smoke*, Maxwell ANDERSON's *Joan of Lorraine*, and Owen Crump's *Southern Exposure*. After her death the theatre she founded continued to function under different directors until the end of 1959, the last production being *Othello*.

JONES, RICHARD, see ENGLISH COMEDIANS.

JONES, ROBERT EDMOND (1887–1954), American writer, lecturer, director, and above all scene designer, whose first designs, for Ashley DUKES's *The Man Who Married a Dumb Wife* in 1915, began a revolution in American scene design. His ability to integrate his designs with all the aspects of the play made his work memorable, particularly when he also directed the actors, as he did in *The Great God Brown* (1926) by O'NEILL, with all of whose early plays he was connected through his association with the PROVINCETOWN PLAYERS. He was also responsible for the décor of a number of Shakespearian productions, his *Othello* in 1937 being much admired, and of such modern plays as Carson and Parker's *The Jest* (1919), Marc CONNELLY's *The Green Pastures* (1930), Maxwell ANDERSON's *Night over Taos* (1932), and Sidney HOWARD's adaptation of a Chinese play, *Lute Song* (1946). He was also part-author with Kenneth MACGOWAN of *Continental Stagecraft* (1922).

JONGLEURS, medieval itinerant entertainers, who flourished throughout Europe, often along the traditional pilgrim routes. The term embraced most kinds of performers—acrobats, jugglers, bear-leaders, ballad-singers—much of their patter and songs being provided by the *trouvères*, who usually performed only their own works. Although some jongleurs worked in family or friendly groups, most were independent; and women, as well as men, followed the profession.

JONSON, BEN(jamin) (1572–1637), English dramatist, friend and contemporary of Shakespeare, and perhaps the only one worthy to rank with him. He was at Westminster School, but was deprived of the university education his attainments there warranted by his stepfather, who apprenticed him to his own trade of bricklaying. Finding this little to his taste, Jonson decamped and went soldiering in the Low Countries, returning to London in about 1595 to seek a living in and around the theatres. As an actor, though by all accounts a bad one, he was implicated in the trouble over NASHE's *The Isle of Dogs* (1597), of which he may also have been part-author, and found himself, not for the last time, in prison. He does not appear to have acted after that, but a year later his comedy *Every Man in His Humour* was given by the CHAMBERLAIN'S MEN, with Shakespeare, who was a member of the company, as Knowell. It was followed by *Every Man out of His Humour* (1599), also played by the Chamberlain's Men, and by *Cynthia's Revels* (1600), acted by the Children of the Chapel at BLACKFRIARS. This BOY COMPANY also played in *The Poetaster* (1601), in which Jonson, a quarrelsome man who despised most of his fellow-playwrights (always excepting Shakespeare) as uneducated hacks, satirized DEKKER and MARSTON. Dekker replied in *Satiromastix* (also 1601); but the quarrel with Marston was evidently patched up in time for him to collaborate with Jonson and CHAPMAN in *Eastward Ho!* (1605). Because of its satirical reflections on the Scottish policy of James I this landed Jonson in prison again: he had already been in trouble with the authorities over his tragedy *Sejanus* (1603), which was considered seditious. This seems to have sobered him somewhat, and a fruitful and less tempestuous phase of his life began. In 1605, in collaboration with Inigo JONES, he prepared for the entertainment of the king and his courtiers the first of his eight MASQUES, in one of which—*Oberon, the Fairy Prince* (1611)—the young Prince Henry, eldest son of James I, appeared shortly before his untimely death. A year later one of Jonson's best plays, *Volpone, or the Fox*, was given by the KING'S MEN, and in 1609 *Epicoene, or the Silent Woman* was seen at WHITEFRIARS in a production by the Children of the Queen's Revels. It was followed by two excellent comedies, *The Alchemist* (1610) and *Bartholomew Fair* (1614), the former a satire on cupidity and credulousness, the latter a portrayal, in short scenes linked by a tenuous plot, of a London crowd in holiday mood; a tragedy, *Catiline* (1611), was no more successful, as the author himself admitted, than *Sejanus* had been. The comparative failure of his next play, *The Devil is an Ass* (1616), caused Jonson to leave the theatre for some years, though he continued to function as Court poet. He finally returned to the stage with *The Staple of News* (1625), *The New Inn* (1629), *The Magnetic Lady* (1632), and a revised version of an earlier work, *A Tale of a Tub* (1596; 1633). These were not to be compared to his previous plays, and are little known today. His reputation rests on a handful of excellent comedies which have never lost their popularity and are still frequently revived, *Volpone* being the one most often seen. It provided Donald WOLFIT with one of his best parts; was seen in New York in 1928; in Paris with DULLIN in the title role in 1931; as a musical, *Foxy*, with Bert LAHR, on Broadway in 1964; and has twice been produced by the NATIONAL THEATRE in London, in 1968 with Colin BLAKELY and in 1977 with Paul SCOFIELD. *The Alchemist*, whose Abel Drugger was one of GARRICK's best comic roles, was revived at the OLD VIC in 1947, with Alec GUINNESS, and again in 1962; it was seen at the CHICHESTER FESTIVAL THEATRE in 1970; and was again revived by the ROYAL SHAKESPEARE COMPANY at the ALDWYCH THEATRE, with Ian MCKELLEN, in 1977. Its first production in New York was in 1948. *Bartholomew Fair* has not been produced professionally in New York, but in 1950 it was seen at the Old Vic and it was produced by the Royal Shakespeare Company in 1969. *Epicoene*, on which Stefan ZWEIG based the libretto of Richard Strauss's opera *Die schweigsame Frau* (1935), has not been seen in London since the PHOENIX SOCIETY revival of 1924. *The Devil is an Ass* was revived in 1977 by the BIRMINGHAM REPERTORY COMPANY.

JORDAN, DOROTHY [Dorothea] (1761–1816), English actress, who excelled as high-spirited hoydens and in BREECHES PARTS. She was the illegitimate daughter of an actress by a gentleman named Francis Bland; her brother, also an actor, called himself George Bland (?–1807), but she went on the stage as Miss Francis. She made her first appearance in Dublin in 1777, at the Crow Street Theatre, playing Phoebe in *As You Like It*, and was later much admired in *The Governess* (a pirated version of SHERIDAN's *The Duenna*) when it was given for the benefit of O'KEEFFE. In 1780 she was engaged by Daly for the Smock Alley Theatre, but two years later, after being seduced by him, she fled secretly to England with her mother and sister. There she was befriended by Tate WILKINSON, who had acted with her while on a visit to Dublin. On his advice, since she was clearly pregnant, she changed her name to Mrs Jordan, though she was never married, and acted with his company up to, and for some time after, the birth of Daly's child. In 1785, probably on the recommendation of William ('Gentleman') SMITH, who saw and admired her at York during race week, she was engaged by Sheridan for DRURY LANE, and in spite of the preferences of audiences of the time for tragedy in the style of Mrs SIDDONS (who was also in the company, and for a time thought poorly of her acting), she chose to make her début on 15 Oct. as Peggy in GARRICK's *The Country Girl*. She was immediately successful and reappeared in the part many times, wisely abandoning tragedy altogether and continuing to delight audiences in such parts as Priscilla Tomboy in *The Romp* (based on BICKERSTAFFE's *Love in the City*), in which character she was painted by Romney, as well as Miss Hoyden in Sheridan's *A Trip to Scarborough*, Sir Harry Wildair in FARQUHAR's *The Constant Couple*, and Miss Prue in CONGREVE's *Love for Love*. Early in her career at Drury Lane Mrs Jordan became entangled with a young man named Richard Ford by whom she had four children. She left him in 1791 to become the mistress of the Duke of Clarence, later William IV, by whom she had 10 children. She continued to act intermittently; she was in the company which performed IRELAND's Shakespeare forgery *Vortigern and Rowena* (1796) at Drury Lane, and in 1800 she appeared as Lady Teazle in Sheridan's *The School for Scandal*. She parted from the Duke, to whom she had been a faithful and affectionate companion, in 1811, and made her last appearances in London in 1814, when she was seen in a revival of *As You Like It* and in a new play by James Kenney, *Debtor and Creditor*. After a final appearance at Margate she retired to Paris, where she died in poverty. Her grave was swept away during rebuilding in the early 1930s.

JORNADA, name given in Spain to each division of a play, corresponding to the English 'act'. It probably comes from the Italian *giornata*, found occasionally in a SACRA RAPPRESENTAZIONE. The word in its present form was first used by TORRES NAHARRO.

JŌRURI, see JAPAN.

JOSEPH, STEPHEN (1921–67), English actor and director, son of the REVUE artist Hermione Gingold and the publisher Michael Joseph. He began his theatrical career as a director at the Lowestoft Repertory Theatre, and then went to America, where he graduated in drama at the University of Iowa. On his return to England in 1955 he founded a company for the express purpose of presenting plays 'in the round', which was to become his main preoccupation. In that year it began to present summer seasons in the public library in Scarborough (see STEPHEN JOSEPH THEATRE IN THE ROUND), and in the autumn it began monthly Sunday night performances in London. From 1956 it toured all over England with a portable theatre-in-the-round, continuing until 1962, when Joseph founded the VICTORIA THEATRE, Stoke-on-Trent, in conjunction with Peter Cheeseman. In the same year he was appointed the first Fellow of the newly founded Department of Drama at Manchester University, which involved both teaching and research, and the Victoria Theatre was run by Cheeseman. Joseph continued in his work and in his books to press the claims not only of his own arena stage but also of the end stage (as at the MERMAID THEATRE) and the three-sided stage (as at the CHICHESTER FESTIVAL THEATRE).

JOUVET, LOUIS (1887–1951), French actor and director, one of the most important figures of the French theatre in the years before the Second World War, who in 1913 joined COPEAU at the VIEUX-COLOMBIER as actor and stage manager. His best work there was done in Shakespeare, as Aguecheek in *Twelfth Night* and Autolycus in *The Winter's Tale*. Jouvet went with Copeau to America in 1917–19, but in 1922 left him to take over the COMÉDIE DES CHAMPS-ÉLYSÉES. Here he was responsible for the success in 1923 of ROMAINS's *Knock, ou le Triomphe de la médecine*, which he directed, playing also the chief part himself and frequently reviving it. In 1928 he began his long association with GIRAUDOUX, in *Siegfried*. In 1934 he moved to the larger Théâtre de l'ATHÉNÉE, to which he added his own name. He joined the staff of the Conservatoire in 1935, and a year later was appointed one of the advisors to the COMÉDIE-FRANÇAISE, becoming a director in 1940, only to be forcibly retired under the German occupation. In 1945 he returned to the Athénée, where he staged GENET's first play *Les Bonnes* in 1947. The great passion of his life was for MOLIÈRE, his finest part being Géronte in *Les Fourberies de Scapin*. He wrote two books, *Réflexions du comédien* (1939) and *Témoignages sur le théâtre*, published posthumously in 1952.

JOYCE, JAMES (1882–1941), Irish novelist, and the author of one play, *Exiles* (1918), written in the manner of IBSEN, whom Joyce much admired. Concerned less with adultery than with the nature of marital fidelity, the play was long considered negligible even after its first production by the STAGE SOCIETY in 1926, but in 1970 it was revived at the MERMAID THEATRE in a notable production by Harold PINTER which he re-created for the ROYAL SHAKESPEARE COMPANY at the ALDWYCH THEATRE a year later. A number of adaptations of Joyce's novels have been made for the stage, including Marjorie Barkentin's *Ulysses in Nighttown* (N.Y., 1958), and *Stephen D*, based on *A Portrait of the Artist as a Young Man* by Hugh LEONARD. This was first seen during the DUBLIN THEATRE FESTIVAL in 1962, and in London the following year. Siobhán MCKENNA had great success with her one-woman recital *Here Are Ladies*, which included characters from Joyce's novels.

JUELL-REIMERS, JOHANNE (1847–82), Norwegian actress, married first to Mathias Juell (their daughter was Johanne DYBWAD), and second to Arnoldus REIMERS. She made her début in 1865, moved in 1866 to the KRISTIANIA THEATER, and was beginning to create a distinctive reputation for herself as an interpreter of IBSEN and BJØRNSON when she died.

JULIAN ELTINGE THEATRE, see ELTINGE THEATRE.

JURČIČ, JOSIP, see YUGOSLAVIA.

JUVARRA, FILIPPO (1676–1736), Italian architect, who designed scenery for several private theatres in Rome, his most important work being done for Cardinal Ottoboni, a great lover of plays and opera. For him Juvarra designed and built a small theatre (probably for rod PUPPETS) in his Palazzo della Cancelleria. Two plans for this are extant; it is not known which was finally used, as the theatre was demolished when Ottoboni died and no trace of it remains. There are, however, still in existence a number of scene-designs which Juvarra made for productions there between 1708 and 1714, including those in an album now at the Victoria and Albert Museum in London which probably dates from 1711. Juvarra left Rome in 1714, and then devoted himself entirely to civil and church architecture.

JUVENILE DRAMA, see TOY THEATRE.

JUVENILE LEAD, TRAGEDIAN, see STOCK COMPANY.

K

KABUKI, see JAPAN.

KACHALOV [Shverubovich], VASILI IVANOVICH (1875–1948), Russian actor, whose career covered the transition from Tsarist to Soviet Russia. After three years' apprenticeship in the provinces he joined the company at the MOSCOW ART THEATRE and made his first appearance in a revival of A. K. TOLSTOY's *Tsar Feodor Ivanovich*. He later played many leading roles, including the name parts in *Julius Caesar* and *Hamlet* and IBSEN's *Brand*. He was also seen as Ivan Karamazov in DOSTOEVSKY's *The Brothers Karamazov* and as the Reader in Leo TOLSTOY's *Resurrection*, considered one of his best parts. He created the role of Vershinin, the hero in IVANOV's *Armoured Train 14–69* (1927), and in 1938 appeared as Chatsky in GRIBOYEDEV's *Woe From Wit*, the part he had played when it was first given at the Moscow Art Theatre in 1906.

KAHN, FLORENCE (1877–1951), American actress, who first came into prominence as leading lady to Richard MANSFIELD. She later became a notable interpreter of IBSEN's women, being seen throughout the United States as Rebecca West in *Rosmersholm*, Mrs Elvsted in *Hedda Gabler*, the Strange Lady in *When We Dead Awaken*, and other parts. In 1908 she was at Edward TERRY'S theatre in London, playing Rebecca West, when she was first seen by Max BEERBOHM, whom she married shortly afterwards. She then left the stage, but made a few guest appearances, the most noteworthy being as Åse in the OLD VIC production of *Peer Gynt* in 1935.

KAINZ, JOSEF (1858–1910), German actor, famed for the richness of his voice and the purity of his diction. He was trained with the MEININGER COMPANY and made his first appearance on the stage in 1874, in Vienna, where in 1899 he returned to end his days as a leading member of the Imperial Theatre. He was for some time in MUNICH, where he was the friend and favourite actor of King Ludwig II of Bavaria, and in 1883 played opposite Agnes SORMA in the newly founded DEUTSCHES THEATER in Berlin. He toured extensively in America in many of his best parts, which included Romeo, Hamlet and the heroes of GRILLPARZER. He was also good as MOLIÈRE's Tartuffe, Oswald in IBSEN's *Ghosts*, and ROSTAND's Cyrano de Bergerac.

KAISER, GEORG (1878–1945), German dramatist, a major exponent of EXPRESSIONISM. His early plays were satirical comedies directed against Romanticism, but the first of his works to attract attention was *Die Bürger von Calais*, written in 1913 but not performed until early in 1917, and generally considered his best play, though not so well known as *Von morgens bis mitternachts*, seen later the same year. This sombre history of a bank clerk whose bid for freedom from the futility of modern civilization leads to suicide was translated by Ashley DUKES as *From Morn to Midnight*, and produced by the STAGE SOCIETY in 1920. It was given its first public production in 1926 under the direction of Peter Godfrey, who again produced it at the GATE THEATRE in 1932, and was seen in New York in 1922 in a production by the THEATRE GUILD. It was followed by the powerful trilogy, *Die Koralle* (1917) and *Gas, I and II* (1918 and 1920), a symbolic picture of modern industrialism crashing to destruction and taking with it the civilization it has ruined. Except for the melodramatic *Der Brand im Opernhaus* (1919), Kaiser's later works—he wrote about 70 plays—made less impact, but he collaborated with the musician Kurt Weill in several operas; in 1938 he left Germany for Switzerland, where he died.

KĀLIDĀSA, see INDIA.

KALITA HUMPHREYS THEATER, see DALLAS THEATER CENTER.

KAMBAN, GUÐMUNDUR JÓNSSON (1888–1945), Icelandic novelist and dramatist, most of whose work was written in Danish. *Hadda-Padda* (1914) and *Vi mordere* (*We Murderers*, 1920), set in New York, were among the earliest plays of the modern Icelandic theatre.

KAMBYSES, YANNIS, see GREECE, MODERN.

KAMERI THEATRE, Israel, see CAMERI THEATRE.

KAMERNY THEATRE, Moscow, Chamber, or Intimate, Theatre, founded in 1914 by Alexander TAÏROV as an experimental theatre for those to whom the naturalistic methods of the MOSCOW ART THEATRE no longer appealed. Here he sought to work out his theory of 'synthetic theatre', which, unlike the 'conditioned theatre' of MEYERHOLD, made the actor the centre of attention, combining in his person acrobat, singer, dancer, pantomimist, comedian, and tragedian. Taïrov's first successful production was VISHNEVSKY's *An Optimistic Tragedy* (1934), in which his wife, Alisa KOONEN, played the heroine. The theatre then became important for its productions of non-Russian plays, providing a link with Western drama at a time when it was badly needed. After Taïrov's death in 1950 the theatre was reorganized and its identity lost, many of the company joining the newly opened PUSHKIN THEATRE.

KAMIŃSKA, IDA (1899–1980), Polish-Jewish Yiddish-speaking actress, who made her début in 1916, and from 1921 to 1939 managed the Polish State Jewish Theatre in Warsaw, founded by her mother Ester Rachel Kamińska (1868–1925). During the Second World War she was in Russia, but returned to Poland in 1949, and rejoined the Jewish Theatre at Łódź, removing with it to Wrocław, and eventually back to Warsaw, where it reopened under her mother's name. For some years it toured Poland under her management, visiting Belgium, the Netherlands, and France in 1956, Germany, France, and England in 1958, and Israel in 1959–60. It closed in 1968, probably owing to the decimation of the Jewish population which formed its main audience. Ida Kamińska then emigrated to the United States. Among her outstanding roles were Nora in IBSEN's *A Doll's House* and the title role in BRECHT's *Mutter Courage und ihre Kinder*.

KANTOR, TADEUSZ (1915–   ), Polish director and scene designer, who from 1945 to 1955 was responsible for the décor of many outstanding productions at the Stary Theatre in Cracow, and in 1955 founded there the experimental theatre Cricot 2. There he gave performances of several of WITKIEWICZ's surrealist plays, including *The Cuttlefish* (1957), *In a Small Country House* in 1961; *The Madman and the Nun* in 1963; *The Water Hen* in 1967; and *Lovelies and Dowdies* in 1973. He employs the HAPPENING in his productions, and has achieved international fame with his *Dead Class* (1975), with which he has toured Europe, America, and Australia. Played mostly in mime, it depicts a group of corpses sitting in a schoolroom, obviously doomed to repeat in death the mistakes they made when alive. Kantor's production of his own *Wielopole, Wielopole* (Florence, 1979) was seen at the EDINBURGH FESTIVAL and RIVERSIDE STUDIOS in 1980; its grotesque-tragic picture of a dying past was made up of childhood memories, family and religious myths, and nightmare images from the First World War.

KARATYGIN, VASILY ANDREYEVICH (1802–53), Russian tragic actor, who made his professional début in 1820. He was noted for the care with which he studied his roles, returning where possible to the original sources, and labouring for historical accuracy in costume and décor, though he was opposed to the realistic style of acting and the innovations of SHCHEPKIN and was criticized for a certain frigidity and mechanical quality. In contrast to MOCHALOV he developed a subtle and calculated technique which enabled him to play the most varied roles, though his preference was always for classical tragedy, and he was especially admired in the patriotic drama of the day.

KARNAD, G. R., see INDIA.

KARVAŠ, PETER, see CZECHOSLOVAKIA.

KASPERL(E), a stock character in the ALTWIENER VOLKSTHEATER, derived largely from HANSWURST. Created about 1764 by LAROCHE, he had no stock costume but appeared in a variety of disguises. He survives as a puppet akin to PUNCH, the Kasperltheater being the Austrian equivalent of the English PUNCH AND JUDY show.

KATAYEV, VALENTIN PETROVICH (1897–   ), Soviet dramatist, whose most successful play, which has been produced in many countries, was *Squaring the Circle* (1928), a comedy about two ill-assorted couples who, owing to the housing shortage, are compelled to live in one room and finally change partners. First produced at the MOSCOW ART THEATRE, it was seen in New York in 1935 and in London three years later. Katayev wrote a number of other comedies, including *The Primrose Path* (1934), which the FEDERAL THEATRE PROJECT produced in New York in 1939, and an amusing trifle called *The Blue Scarf* (1943). Among his more serious plays are *Lone White Sail* (1937), which, in a revised version, had a successful run in 1951; *I, Son of the Working People* (1938); a play for children, *Son of the Regiment* (1946); and *All Power to the Soviets* (1954), the last two being based on his own novels. *Violet* (1974) deals with the moral problems of those who lived through Stalin's rule.

KATONA, JÓSZEF, see HUNGARY.

KATZENELSON, ISAAC, see HEBREW THEATRE, NEW.

KAUFMAN, GEORGE S. (1889–1961), American dramatist and director, whose first plays were written in collaboration with Marc CONNELLY, the most notable being *Merton of the Movies* (1922), about a film-crazy innocent in Hollywood, and *Beggar on Horseback* (1924; London, 1925), a satire on contemporary American philistinism based on a German play by Paul Apel. After writing on his own *The Butter and Egg Man* and *The Cocoanuts* (both 1925), Kaufman joined Edna Ferber to write *The Royal Family* (1927) based on the lives of the DREWS and BARRYMORES, seen in England in 1934 as *Theatre Royal*; they later wrote other plays together, such as *Dinner at Eight* (1932; London, 1933) and *Stage Door* (1936). His collaboration with Moss HART produced a number of light-hearted comedies, among them *Once in a Lifetime* (1930; London, 1933), another satire on Hollywood; *Merrily We Roll Along* (1934); *You Can't Take It With You* (1936; London, 1937), the most successful, which won the PULITZER PRIZE; and *The Man Who Came to Dinner* (1939; London, 1941), in which the leading character is reputedly based on Alexander WOOLLCOTT. He also wrote (usually in collaboration) many works for the musical stage, including sketches for the revue *The Band Wagon* (1931) and the book for the Gershwins' *Of Thee I Sing* (also 1931), Rodgers and Hart's *I'd Rather Be Right*, with Moss Hart (1937), and Cole Porter's *Silk Stockings* (1955). Known as 'the Great Collaborator', Kaufman was an expert technician and an excellent director, with a keen sense of satire and a thorough knowledge of the theatre. He directed not only many of his own works but such varied productions as Hecht and MacArthur's *The Front Page* (1928), STEINBECK's *Of Mice and Men* (1937), Loesser's musical *Guys and Dolls* (1950), and USTINOV's *Romanoff and Juliet* (1957).

KAZAN, ELIA (1909–   ), American actor and director, who made his first appearances on stage with the GROUP THEATRE, playing Agate Keller in *Waiting for Lefty* (1935), Eddie Fuseli in *Golden Boy* (1937)—both by Clifford ODETS—and Ficzur (the Sparrow) in MOLNÁR's *Liliom* in 1940. It was, however, as a director that he became known, among his important productions being Thornton WILDER's *The Skin of Our Teeth* (1942), Tennessee WILLIAMS's *A Streetcar Named Desire* (1947) and *Cat on a Hot Tin Roof* (1955), Arthur MILLER's *Death of a Salesman* (1949), and Archibald MacLeish's *J. B.* (1958). In 1947 he helped to found the Actors' Studio, a workshop where professional actors could experiment and study their art according to the tenets of the METHOD, and remained with it until 1962, when he was appointed co-director of the Lincoln Center repertory company. He resigned in 1965, shortly before the company moved to the VIVIAN BEAUMONT THEATRE, afterwards devoting his energies to novel-writing.

KEAN, CHARLES JOHN (1811–68), English actor-manager, son of Edmund KEAN, who sent him to Eton with the idea of detaching him from the stage. But when, at time of his father's break with DRURY LANE, he was offered an engagement there, he accepted it and made his first appearance in 1827 as Young Norval in HOME's *Douglas*. Realising that he lacked experience he then went into the provinces, and in 1828 was seen on stage with his father in Glasgow. They were not to play together again until Edmund Kean's last appearance, when he collapsed after the third act while playing Othello to his son's Iago. Charles had none of his father's genius, but was a serious, hardworking man, somewhat priggish, but with plenty of application and common sense. With his wife Ellen Tree (1806–80), a good actress who played opposite him in many important productions, he ran an excellent company, which from 1850 to 1859 appeared at the PRINCESS'S THEATRE in a series of carefully-chosen and well-rehearsed plays, set and costumed lavishly, as the fashion of the time demanded, but with some attempt at historical accuracy. Queen Victoria was a frequent visitor to the Princess's, as was the Duke of Saxe-Meiningen, husband of one of the Queen's nieces, and Kean's work probably inspired many of the reforms attributed to the MEININGER COMPANY. It was under Kean that Ellen TERRY, as a child of nine, made her first appearance on the stage.

KEAN, EDMUND (1789–1833), English tragedian, whose undoubted genius was offset by wild and undisciplined behaviour, an ungovernable temper, and habitual drunkenness. Very little is known about his early life. He was apparently the illegitimate son of Ann Carey (?–1833), daughter of the author and entertainer George Savile Carey (1743–1807) and so granddaughter of the Henry Carey remembered as the composer of 'Sally in Our Alley' and author of the burlesque *The Tragedy of Chrononhotonthologos* (1734), himself the natural son of Henry Savile (Lord Eland). Kean's father was reputedly a small-part actor and drunkard, Edmund Kean (1770–92), who committed suicide when his son was about three years old. The child was taken into the care of Charlotte Tidswell (1760–1846), a minor member of the company at DRURY LANE and for a time the mistress of Edmund Kean the elder's brother Moses, a ventriloquist and entertainer, who also died in 1792. Knowing no life but that of the playhouse, she planned to make an actor of him, and had him taught singing, dancing, fencing and elocution. By the age of eight he was already something of a prodigy and had appeared in several small parts at Drury Lane. Anxious to exploit his talents for her own benefit, his mother then reclaimed him, and with her he became a strolling player, acting Hamlet before Nelson and Lady Hamilton in Carmarthen, and giving a command performance in Windsor Castle before George III. In 1801 he played for his mother's benefit in London, and in 1802 was seen in a mixed programme, in which he recited a speech from SHERIDAN's *Pizarro*, at COVENT GARDEN. In 1804, tired of earning money for his mother to spend, he broke away and set out on his own, going from one provincial company to another. During the next nine years he endured not only all the privations of a strolling player's life, but added to his burdens a wife—Mary Chalmers (*c.* 1789–1808), a mediocre actress whom he met in Gloucester—and two sons, of whom the elder died young.

On 26 Jan. 1814 Kean finally achieved his ambition and appeared at Drury Lane in a major role, playing Shylock in *The Merchant of Venice* not in the traditional red wig and beard of the low comedian, which even MACKLIN had not dared to discard, but as a swarthy embittered fiend with a butcher's knife in his grasp and blood-lust in his eyes. The audience acclaimed him, and he continued for a time to delight them in a series of villainous parts. The technical novelty of his acting is revealed by a contemporary comment that 'by-play' was one of his greatest excellencies; he relied less on his naturally harsh voice than on facial expression. In spirit the change was even greater. Deficient in dignity, grace, and tenderness, he needed a touch of the malign, of murderous frenzy, to inspire him. His Lear was a disappointment; as Romeo he stood beneath Juliet's balcony like 'a lump of lead'; his Hamlet showed a severity amounting to virulence. Comedy he seldom essayed; he was seen as Abel Drugger—GARRICK's favourite comic part, in JONSON's *The Alchemist*—only three times. Mild villainy made no appeal to him, and he rejected the part of Joseph Surface, in Sheridan's *The School for Scandal*, with scorn. Even his magnificent Othello was too often overplayed, too constantly on the rack. But as Macbeth he was heartrending, and he gave his finest performances as the arch-villains Richard III and Iago. Two of his other great masterpieces were Sir Giles Overreach, with his ruthless frenzy of miserliness, in MASSINGER's *A New Way to Pay Old Debts*, and the barbarous fiend Barabas, in MARLOWE's *The Jew of Malta*. After his enthusiastic reception at Drury Lane, Kean seemed about to embark on a career of unequalled splendour; but the wildness in his blood and the intemperate habits of the lost years overpowered him. He was seldom sober; his frequent absences from the stage and his poor showing when drunk, added to the many scandals which attached themselves to his name—particularly his affair with the wife of Alderman Cox, one of the members of the Drury Lane general committee—alienated the audiences who had been attracted by his superb acting; he could win them back time and again with a fine performance, only to lose them through his bad behaviour. In the United States, where he made his first appearance in 1820, the same thing happened. Audiences who had awaited him on tiptoe with expectation, and applauded his first appearances with fervour, turned against him when he fell back into his usual state of arrogance and unreliability. He returned to England, but his powers were waning, his wife had left him, and his health was deteriorating. He continued to act intermittently, and in 1831 added to his financial burdens the management of the King's Theatre at Richmond, near London. He even went on tour occasionally, and on his better days returned to the stage of Drury Lane, where on 25 Mar. 1833, while playing Othello to the Iago of his son Charles KEAN, he finally collapsed, dying a few weeks later.

KEAN, THOMAS (*fl.* mid-18th century), American actor, who in 1749 was manager with Walter Murray of a company which acted ADDISON's *Cato* and other plays in a converted warehouse in PHILADELPHIA, later the home of HALLAM's company. In the following year Kean and his players appeared in a theatre in Nassau Street in a repertory which included *Richard III* as altered by Colley CIBBER, CONGREVE's *Love for Love*, OTWAY's *The Orphan*, and LILLO's *George Barnwell*. From there, with a group known as the Virginia Company of Comedians, Murray and Kean set out to tour the major centres of Virginia, starting in Williamsburg, in opposition to Hallam. Nothing is known of them after this, nor is it clear whether they were amateur or professional; if the latter, then theirs was the first such company in the New World.

KEANE, DORIS, see SHELDON, EDWARD.

KEANE, JOHN BRENDAN (1928–   ), Irish dramatist, poet, and essayist, whose plays, popular with amateur companies, have been given a mixed reception by the critics. *Sive* (1959), his first stage play—he had previously written for radio—was followed by *Sharon's Grave, The Highest House on the Mountain* (both 1960), *No More in Dust* (1961), *The Man from Clare* (1962), *The Year of the Hiker* (1963), *The Field* (1965), *Big Maggie* (1969), *The One-Way Ticket, The Change in Mame Fadden* (both 1972), *The Crazy Wall* (1973), *Values* (consisting of *The Springing Of John O'Dorey*, *Backwater*, and *The Pure Of Heart*) (also 1973), *Matchmaker* (1975), *The Good Thing* (1976), *The Buds Of Ballybunion* (1978), and *The Chastitute* (1980). Keane's plays are marked by a primitive strength of situation and character frequently bordering on the grotesque or the melodramatic, and, as with the work of an earlier Kerryman, George FITZMAURICE, the mood and diction are gratingly poetic and powerful.

KEDROV, MIKHAIL NIKOLAYEVICH (1893–1972), Russian actor and director, a pupil of STANISLAVSKY, some of whose works he edited. He joined the company at the MOSCOW ART THEATRE in 1924, and remained there until his death. He directed the production of CHEKHOV's *Uncle Vanya* in which the company was seen at SADLER'S WELLS in London in 1958, and also appeared at the ALDWYCH in the 1970 WORLD THEATRE SEASON. Among his many productions one of the most acclaimed was a version of *The Winter's Tale* which was three years in rehearsal.

KEELEY [*née* Goward], MARY ANN (1806–99), English actress, who was trained as a singer, but turned to acting and had already made a name for herself when in 1829 she married Robert KEELEY with whom she thereafter appeared, being in fact the better player of the two. A small, neatly-made person, she was at her best in pathetic, appealing parts such as Nydia in BUCKSTONE's *The Last Days of Pompeii* (1834), or Smike in Stirling's *The Fortunes of Smike; or, a Sequel to Nicholas Nickleby* (1840), adapted from DICKENS. Her greatest success was scored in the title role of Buckstone's version of *Jack Sheppard* (1839), in which the highwayman was portrayed as a wild youngster, defrauded of his heritage and driven to bad ways by the animosity of Jonathan Wild the thief-taker. She had two daughters, both on the stage, of whom one, Mary, married Albert SMITH.

KEELEY, ROBERT (1793–1869), English actor, who ran away from his apprenticeship to join a strolling company, and in 1818 was in London, at the OLYMPIC THEATRE and later at the ADELPHI. There he made a great success as Jemmy Green in MONCRIEFF's dramatization of Pierce Egan's *Tom and Jerry; or, Life in London* (1821). In later life he was a good low comedian, his stolid look and slow, jerky speech adding much to the humour of his acting. Among his best parts were Dogberry in *Much Ado About Nothing*, Jacob Earwig in Selby's farce *Boots at the Swan* (1842), and Sarah Gamp in a dramatization of DICKENS's *Martin Chuzzlewit*. In his professional life he was somewhat overshadowed by the excellence of his wife Mary Ann (above), with whom he took over the management of the LYCEUM THEATRE from 1844 to 1847, appearing with her later at the HAYMARKET THEATRE under Ben WEBSTER.

KEENE, LAURA (?–1873), actress and theatre manager, born in England. The date of her birth has been variously given as 1820, 1826, 1830, and 1836, and her real name may have been Moss, Foss, or Lee. She is said to have been trained for the stage by an aunt, Mrs Yates, and to have made her first appearance as Juliet at the RICHMOND THEATRE, Surrey, on 26 Aug. 1851: if so, it seems likely that she was born in 1830 or 1836. Shortly after her first appearance she was seen at the OLYMPIC THEATRE in London, and in 1852 went to New York on her way to tour Australia. She returned to New York in 1855 and spent the rest of her life in the United States, where she was the first woman to become a theatre manager. On 18 Nov. 1856 she opened her own theatre in New York, later the OLYMPIC THEATRE (2), with *As You Like It*, in which she played Rosalind. For some time she presented a repertory of good foreign and American plays with an excellent stock company, eschewing the destructive practice of importing visiting stars. Among her actors were the third Joseph JEFFERSON and E. A. SOTHERN, who were jointly responsible for the outstanding success of Tom TAYLOR's *Our American Cousin* (1858), whose long run helped to establish New York as the theatre centre of the United States. Two months after the outbreak of the Civil War in 1861 Laura Keene's was the only theatre open in New York, but she was forced to lower her standards and give poor, showy spectacles. She never recovered her prestige or buoyancy, and relied more and more on MELODRAMA and spectacle until her last season in 1862. She continued to act on tour and in other New York theatres, but never again attained the heights of her previous management. Her company was playing *Our American Cousin* at FORD'S THEATRE in Washington on the night Abraham Lincoln was assassinated there—14 Apr. 1865.

The LYCEUM THEATRE (2), New York, was known as the Laura Keene from 1871 to 1873.

KEITH, BENJAMIN FRANKLIN, see VAUDEVILLE, AMERICAN.

KELLY, CHARLES, see TERRY, ELLEN.

KELLY, FANNY [Frances] MARIA (1790–1882), English actress and singer, the subject of the essay 'Barbara S—', by Charles LAMB, who was in love with her, proposing marriage in a letter dated 20 July 1819. She made her first appearance on the stage at the age of seven, playing with her uncle Michael Kelly, the singer and composer, at DRURY LANE, where as an adult actress she worked for over 30 years, reviving many of the parts associated with Mrs JORDAN. She also acted with Edmund KEAN, playing Ophelia to his Hamlet in 1814. On her retirement in 1840 she built a theatre in Soho, later the ROYALTY, intending to use it as a training-school for young actresses, but it was not a success financially, and after struggling on for several years she was forced to give it up, confining her activities to Shakespeare readings and private tuition.

KELLY, GEORGE EDWARD (1887–1974), American dramatist, whose first full-length play, *The Torchbearers* (1922), was a satire on pretentious amateur theatricals. He expanded a VAUDEVILLE skit, *Poor Aubrey*, into a hilarious satire entitled *The Show-Off* (1924), and was awarded the PULITZER PRIZE for *Craig's Wife* (1925), an exposé of feminine possessiveness and lovelessness. Among his later plays, none of which was quite so successful, were *Daisy Mayme* (1926), *Behold the Bridegroom* (1927), and *Maggie the Magnificent* (1929). After the failure of *Philip Goes Forth* (1931), Kelly withdrew from the theatre for some years, returning unsuccessfully in 1936 with a comedy, *Reflected Glory*. It was not until 1945 that his work was seen on stage again with *The Deep Mrs Sykes*, a satire on feminine intuition which was followed in 1946 by his last play, *The Fatal Weakness*, a study in feminine romanticism.

KELLY, HUGH (1739–77), English playwright, whose sentimental comedy *False Delicacy* (1768) was produced at DRURY LANE by GARRICK in the hope of emulating the success of GOLDSMITH's *The Good-Natur'd Man* at COVENT GARDEN. Though now forgotten, for a time it eclipsed its rival: it was produced in the provinces, revived in London, and translated into French and German. Of Kelly's other plays the best was *The School for Wives* (1773), which most nearly approached the true spirit of the comedy of manners.

KEMBLE, CHARLES (1775–1854), English actor-manager, youngest brother of John Philip KEMBLE. As a child he acted in the provincial company of his father Roger Kemble and at the age of 17 was seen at Sheffield as Orlando in *As You Like It*, later one of his best parts. He made his first appearance in London on 21 Apr. 1794, playing Malcolm to the Macbeth and Lady Macbeth of his brother and sister (Mrs SIDDONS). He was not at first a good actor, being somewhat awkward, with a weak voice, but in time he became an accomplished player, not only as Orlando, but in such parts as Mercutio and Romeo in *Romeo and Juliet*, Benedick in *Much Ado About Nothing*, Young Absolute in *The Rivals* and Charles Surface in *The School for Scandal*, both by SHERIDAN, and Mirabell in CONGREVE's *The Way of the World*. By temperament poetic rather than emotional, he was quite unfitted for tragedy, which he wisely left to John Philip, with whom he went to COVENT GARDEN in 1803, taking over on the latter's retirement in 1817. He was not a good manager, and was only saved from bankruptcy in 1829 by the excellent acting of his daughter Fanny (below), but one important event which took place under his management was a revival of *King John* for which PLANCHÉ designed the costumes. This reinforced the reaction against the conventional costuming of historical plays which John Philip Kemble had already started and which culminated in the 'historical accuracy' insisted on later by Charles KEAN.

Troubled by increasing deafness, Charles retired in 1832 to become Examiner of Plays and to give Shakespeare readings. He also visited America, where his courtesy and affable manner caused him to be considered a typical 'English gentleman'. He married in 1806 the actress Maria Theresa De Camp (1774–1838), who retired in 1819, having been much admired as Edmund in Kenney's *The Blind Boy* (1807) and as Lady Elizabeth Freelove in her own one-act comedy *The Day After the Wedding; or, a Wife's First Lesson* (1808). She was also good as Mrs Sullen in FARQUHAR's *The Beaux' Stratagem* and as Madge Wildfire in Daniel TERRY's musical version of SCOTT's *The Heart of Midlothian* (1818).

KEMBLE, FANNY [Frances] ANNE (1809–93), English actress, daughter of Charles KEMBLE. With no particular desire for a theatrical career, she was persuaded to appear at COVENT GARDEN in 1829 by her father, who hoped she would help to avert the bankruptcy which threatened his management. She was first seen as Juliet in *Romeo and Juliet*, and was immediately successful. For three years she brought prosperity to the theatre and everyone connected with it, being seen as Portia in *The Merchant of Venice*, Beatrice in *Much Ado About Nothing*, Lady Teazle in SHERIDAN's *The School for Scandal*, and a number of tragic parts formerly associated with her aunt Mrs SIDDONS, among them Isabella in SOUTHERNE's *The Fatal Marriage*, Euphrasia in MURPHY's *The Grecian Daughter*, Calista in ROWE's *The Fair Penitent*, and Belvidera in OTWAY's *Venice Preserv'd*. She also created the part of Julia in Sheridan KNOWLES's *The Hunchback*. Unlike the rest of her family she seems to have been good in both tragedy and comedy. In 1832 she went with her father to America, and was received everywhere with acclamation, but in 1834 left the stage on her marriage to an American, whom she divorced in 1848. After living in retirement in Lennox, Mass., for 20 years, she returned to London (where she had last been seen in a series of Shakespeare readings in 1848), where she died.

KEMBLE, JOHN PHILIP (1757–1823), English actor, eldest son of Roger Kemble (1722–1802) and his wife Sarah Ward (?–1807), strolling players. He appeared on the stage as a child with his parents, but was then sent to Douai to study for the priesthood. He returned to the theatre, however, and after acting in the provinces for several years made his London début in 1783 at DRURY LANE as Hamlet, in which character he was painted by Lawrence. He gave an unusual rendering of the part which at first puzzled but finally captivated the audience by its gentleness and grace. During his long career he steadily improved, playing all the great tragic parts of the current repertory—Wolsey in *Henry VIII*, Brutus in *Julius Caesar*, the title role in *The Stranger*, Thompson's adaptation of KOTZEBUE's *Menschenhass und Reue*, Rolla in SHERIDAN's *Pizarro*, ADDISON's Cato, and above all Shakespeare's Coriolanus, the part in which he took leave of the stage on 23 June 1817. He managed both Drury Lane, where he introduced a number of reforms before and behind the curtain, and COVENT GARDEN, where his raising of the prices of admission after the disastrous fire of 1808 caused the famous O.P. (Old Prices) riots. Financially his managements were not a success and as an actor he was limited in his choice of parts, since his somewhat pedantic approach to his work, together with his harsh voice and stiff gestures, rendered him unfit for comedy, which he seldom attempted. Nor was he good in romantic parts, in spite of his handsome presence. Even in tragedy, in which he excelled, he eschewed sudden bursts of pathos or passion, achieving his effects by a studied and sustained intensity of feeling. More respected than loved, he was nevertheless much admired, and exercised a salutary influence on the theatre of his time. He married in 1787 the actress Priscilla Hopkins (1755–1845), widow of the actor William Brereton (1751–87). In her early years she had been a member of GARRICK's company. John Philip's brothers Charles (above) and Stephen (below) were also on the stage, as were his brother Henry and his four sisters: the eldest became famous as Mrs SIDDONS; Frances and Ann had unremarkable careers; Elizabeth emigrated with her husband to America, and became well known there as Mrs Whitlock (1761–1836). Descendants of Roger and Sarah Kemble are still to be found in the theatrical and literary worlds of England and the United States.

KEMBLE, STEPHEN (1758–1822), English actor, son of Roger and Sarah Kemble, strolling players. He was born in a theatre in Herefordshire, his mother having just completed her part in the evening's entertainment, and as a child acted with his parents. He then left the theatre, but returned when his elder sister Mrs SIDDONS became famous, hoping to profit by her popularity. He was a useful actor, but always overshadowed by his elder brother John Philip (above). In later life he became very fat, which enabled him to play Falstaff without padding, so that it was said that DRURY LANE, under the management of John Philip, had the great Kemble, and COVENT GARDEN, where Stephen was appearing, the big Kemble. He led a roving life, being at various times manager of a provincial theatre, of a company in Ireland, and, from 1792 to 1800, of a theatre in Edinburgh. He returned to London in 1818 to manage Drury Lane, with little success, and retired shortly afterwards. He married the actress Elizabeth Satchell (1763–1841), by whom he had a son and a daughter, both on the stage.

KEMP, ROBERT (1908–67), Scottish dramatist, until 1947 a producer with the B.B.C. His first play, *Seven Bottles for the Maestro*, was staged by the Dundee Repertory company in 1945 and his last, *Scotch on the Rocks*, at the Theatre Royal in WINDSOR in 1967. He wrote fluently in both Scots and English, and his work ranged from pageants—*The Saxon Saint* (1949) and *The King of Scots* (1951), both written for performance in Dunfermline Abbey—through history—*John Knox* (1960), written to mark the fourth centenary of the Reformation—to a number of light pieces. The most effective of these was possibly *The Penny Wedding*, seen at the Edinburgh GATEWAY THEATRE in 1957, which satirizes those Scottish nationalists who insist on the revival of the Scots language. Kemp was also responsible for the adaptation of LYNDSAY's *The Thrie Estaitis*, performed under GUTHRIE's direction at the 1948 and subsequent EDINBURGH FESTIVALS. His best play is *The Other Dear Charmer* (1951), in which he makes use of Robert Burns's affair with Mrs MacLehose to explore the problem of identity that still bedevils Scottish life and art.

KEMPE, WILLIAM (?–*c.* 1603), one of the best known of Elizabethan clowns and a great player of JIGS. He was a member of the company which went with the Earl of Leicester to Holland and Denmark in 1584–6, and on his return to London may have joined QUEEN ELIZABETH'S MEN, perhaps in succession to TARLETON who died in 1588. He was with STRANGE'S MEN in 1592 in their production of the anonymous *A Knack to Know a Knave*, of which he may have been part-author. On his return from a provincial tour with them in 1594 he joined the CHAMBERLAIN'S MEN on their formation and was the original Dogberry in Shakespeare's *Much Ado About Nothing* and Peter in *Romeo and Juliet*. He also appeared in Ben JONSON's *Every Man in His Humour* (1598). He was famous for his improvisations, and it may have been Shakespeare's dislike of extempore gagging (as shown in his slighting reference in *Hamlet* to clowns that speak 'more than is set down for them') that caused him to leave the company in 1600. In that year, for a wager, he danced his famous nine-day MORRIS DANCE from London to Norwich, of which he published an account in his *Kemps morris to Norwiche*. He then went back to the Continent, but returned to London in 1602, when he borrowed some money from HENSLOWE (duly recorded in the latter's diary) and is last heard of as one of Worcester's Men at the ROSE THEATRE.

KEMPSON, RACHEL, see REDGRAVE, MICHAEL.

KENDAL [Grimston], WILLIAM HUNTER (1843–1917), English actor-manager, who made his first appearance on the stage in 1861, was a member of the Glasgow stock company, and for some years toured the provinces with Charles KEAN, G. V. BROOKE, and Helen FAUCIT. In 1866 he was engaged by BUCKSTONE for the HAYMARKET THEATRE where he remained for eight years, playing leading parts in plays old and new, among them W. S. GILBERT's *The Palace of Truth* (1870) and *Pygmalion and Galatea* (1871). There he met and married in 1874 the actress Madge [Margaret] Sholto Robertson (1848–1935), appointed DBE in 1926, the twenty-second child of an actor-manager and the younger sister of T. W. ROBERTSON. She had first appeared on the stage at the age of 5, and in 1865 made her adult début as Ophelia in *Hamlet*. After her marriage her career was inseparable from that of her husband. Together they went to the ROYAL COURT THEATRE under HARE, where Mrs Kendal's acting gave a new lease of life to the old play *A Scrap of Paper*, based on SARDOU's *Une Patte de mouche*. After appearing with the BANCROFTS at the Prince of Wales (later the SCALA) Theatre in a revival of BOUCICAULT's *London Assurance* in 1877 and in *Diplomacy* (1878) based by Clement SCOTT and B. C. Stephenson on Sardou's *Dora* the Kendals went to the ST JAMES's in partnership with Hare, remaining with him until 1888 and playing leading roles in many productions. Under other managements they played together until they retired in 1908. Kendal was somewhat overshadowed by his wife's brilliance but was none the less a good reliable actor, better in comedy than in tragedy, his best part being Frank Maitland in Godfrey's *The Queen's Shilling* (1879), and an excellent business manager. Mrs Kendal was a fine *comédienne*, though she could on occasion play in a more gentle mood, as in *The Elder Miss Blossom* (1898) by Ernest Hendrie and Metcalfe Wood, which she frequently revived and took on tour with her husband playing Andrew Quick to her Dorothy Blossom. The Kendals formed an ideal partnership, both on and off stage, and did much to raise the status of the acting profession. The companies with which they were connected were always admirably managed, and provided an invaluable training-ground for many young actors and actresses.

KENNA, PETER (1930–), Australian actor and dramatist, whose best known play *The Slaughter of St Teresa's Day* won an Australian national playwriting competition in 1959. His *Talk to the Moon* had its world première at the HAMPSTEAD THEATRE CLUB, London, in 1963, and its Australian première in Melbourne at the ST MARTIN'S THEATRE in 1972. Kenna directed his own play *Muriel's Virtues* at the Independent Theatre, Sydney, in 1966, and his country comedy *Listen Closely* was given its first production at the same theatre in 1972. In 1973 *A Hard God*, the first part of a proposed trilogy, was staged at the Nimrod Theatre in Sydney, and the one-act *Mates* was also seen there in 1975. *A Hard God* was revived at the Adelaide Festival of Arts in 1978 and the trilogy, under the title *The Cassidy Album*, was completed with *Furtive Love* and *An Eager Hope*, all three plays being directed by John Tasker.

KENNY, SEAN (1932–73), stage designer and director, born in Ireland, where he trained as an architect. He then moved to London, where he designed his first stage set in 1957 for a revival of O'CASEY's *The Shadow of a Gunman* at the LYRIC THEATRE, Hammersmith: but it was his collaboration with Joan LITTLEWOOD on Brendan BEHAN's *The Hostage* (1958) and with Lindsay ANDERSON on Cookson's *The Lily-White Boys* (1960) that launched him on a career which in the next 14 years resulted in over 35 major West End productions. As the first resident Art Director at the MERMAID THEATRE, he helped to arrange the stage and auditorium and designed its first four shows, including the successful musical *Lock Up Your Daughters* (1959). Other musicals on which he worked included Lionel BART's *Oliver!* (1960; N.Y., 1963), for which he designed ingenious moving sets, and Anthony Newley and Leslie Bricusse's *Stop the World—I Want to Get Off* (1961; N.Y., 1962) and *The Roar of the Greasepaint, the Smell of the Crowd* (N.Y., 1965). For the CHICHESTER FESTIVAL THEATRE he designed the sets for CHEKHOV's *Uncle Vanya* in 1962 and IBSEN's *Peer Gynt* in 1970. In 1963 he replanned the interior of the OLD VIC THEATRE to suit the needs of the new NATIONAL THEATRE company and designed the décor for their opening production of *Hamlet*. He was also attracted by the role of director, and in 1968 returned to the Mermaid Theatre to adapt, design, and direct Swift's *Gulliver's Travels*, in a production which made ingenious use of film and sound to achieve shifts in perspective. In 1970 he directed Siobhán MCKENNA's one-woman show *Here Are Ladies*, and when he died he was directing a revival of O'Casey's *Juno and the Paycock* for the Mermaid. Kenny's ingenious set designs and revaluation of the relationship between playwright, designer, and audience did much to revitalize the art of scene design in Britain; many of these ideas were expressed in his design of the NEW LONDON THEATRE, which opened in the year of his death.

KERR [Kempner], ALFRED (1867–1948), German theatre critic, and one of the most influential between 1895 and 1920. A fervent champion of NATURALISM, and so of the plays of IBSEN, HAUPTMANN, and SHAW, he published in 1917 five volumes of his criticisms as *Die Welt in Drama*, notable for their highly subjective approach and idiosyncratic, somewhat mannered style. He expressed his dislike of EXPRESSIONISM and the EPIC THEATRE of BRECHT in two works, *Was wird aus Deutschlands Theater?* in 1932 and *Die Diktatur des Hausknechts* in 1934. He left Germany in 1933, settling in London for the rest of his life.

KESTER, PAUL (1870–1933), American dramatist, whose first play, *The Countess Roudine* (1892), was produced by Mrs FISKE. In 1900 Fred TERRY and his wife Julia NEILSON had a great success with *Sweet Nell of Old Drury*, which remained in their repertory for over 30 years. When it was produced in America later the same year the part of Nell GWYNN was played by Ada REHAN. Kester adapted a number of foreign plays for the American stage, but two of the most successful of his adaptations were of novels by Charles Major, *When Knighthood Was in Flower* (1901) and *Dorothy o' the Hall* (1903), the latter being also a favourite with Fred Terry and his wife, who first appeared in it in London in 1906.

KHMELEV, NIKOLAI PAVLOVICH (1901–45), Soviet actor and director, who joined the MOSCOW ART THEATRE in 1919, where his first role was Fire in MAETERLINCK's *The Blue Bird*. He subsequently played many important parts, including Alexei Turbin in BULGAKOV's *Days of the Turbins*, Karenin in an adaptation of TOLSTOY's *Anna Karenina*, Peklevanov in V. IVANOV's *Armoured Train 14–69*, Zabelin in POGODIN's *Kremlin Chimes*, and Tusenbach in CHEKHOV's *Three Sisters*. In 1937 he became director of the YERMOLOVA THEATRE, where he later produced FLETCHER's *The Woman's Prize; or, the Tamer Tamed*. On the death of NEMIROVICH-DANCHENKO in 1943, Khmelev replaced him at the Moscow Art Theatre.

KIELLAND, ALEXANDER (1849–1907), Norwegian novelist and dramatist. An 'angry young man' of the 1880s, he wrote a succession of realistic novels sustained by a fine sense of indignation at contemporary society and its ways; his dramas, a number of them adaptations of his narrative works, are on a lower level of achievement: *Garman og Worse* (*Garman and Worse*, 1883); *Tre Par* (*Three Couples*, 1886): *Bettys Formynder* (*Betty's Guardian*, 1887); and *Professoren* (*The Professor*, 1888).

KILLIGREW, THOMAS (1612–83), English dramatist and theatre manager. Before the closing of the theatres in 1642 he had already written several tragi-comedies, including *The Prisoners* (1635), *Claracilla, The Princess; or, Love at First Sight* (both 1636), and *The Parson's Wedding* (1641), based on a play by CALDERÓN. This last, when revived in 1664 with a cast consisting of women only, made even PEPYS blush. It is not, however, as a dramatist that Killigrew is important in the history of the English theatre, but as the founder of the present DRURY LANE, which he opened as the Theatre Royal in 1662 under a Charter granted by Charles II. With Sir William DAVENANT, holder of a similar Charter for the Duke's House, later transferred to COVENT GARDEN, Killigrew thus held the monopoly of serious acting in Restoration London, his company including MOHUN, HART, and, for a short while, Nell GWYNN. He also established a training school for young actors at the Barbican, and in 1673, on the death of Sir Henry HERBERT, was appointed MASTER OF THE REVELS. He was, according to Pepys, a 'merry droll' and a great favourite at Court, but he was not as good a business manager as his rival Davenant and was often in financial difficulties. His brother Sir William (1606–95) and his son Thomas (1657–1719) both wrote plays, while another son, Charles (1665–1725), took over the management of the Theatre Royal in 1671, assisted by his half-brother Henry, and became Master of the Revels in succession to his father.

KIMBERLEY, Mrs CHARLOTTE (1877–1939), English theatre manageress and playwright, who toured with her husband F. G. Kimberley, making the Grand Theatre, Wolverhampton, the base from which she directed the other theatres in her CIRCUIT—the Grand, Brighton; the Grand, Llandudno; the County, Shrewsbury; the Palace, Bordesley; and the Playhouse, Swansea. She was one of the most successful supporters of MELODRAMA in its declining years, and wrote over forty popular melodramatic plays, including *Tatters* (1919) and *Kiddie o' Mine* (1921) for her star actress Leah Marlborough (1868–1953).

KINCK, HANS (1865–1926), Norwegian novelist and dramatist, whose reputation as a sensitive observer and analyst of Norwegian peasant life was made chiefly by his novels and short stories. Of his nine published plays, incomparably the finest are the verse-play *Driftekaren* (*The Drover*, 1908) and its sequel *Paa Rindalslægret* (*At Rindal Camp*, 1925), sometimes compared to IBSEN's *Peer Gynt*.

KING [Pratt], DENNIS (1897–1971), British-born actor and singer who spent most of his life in America. He entered the theatre as a call-boy with the BIRMINGHAM REPERTORY THEATRE, making his stage début at that theatre in 1916 as Dennis in *As You Like It*. His first London appearance, in 1919, was in Booth TARKINGTON and Mrs E. G. Sutherland's *Monsieur Beaucaire*, based on the former's novel, and two years later he made his New York début with Ethel and John BARRYMORE in *Clair de Lune*, based on Victor HUGO's novel, *L'Homme qui rit*. He played several further roles in New York, including Mercutio to Jane COWL's Juliet in *Romeo and Juliet* in 1923, before achieving stardom in two famous musicals by Friml: as Jim Kenyon in *Rose-Marie* (1924) and François Villon in *The Vagabond King* (1925), which he played for over two years. In 1928 he was seen as D'Artagnan in another Friml musical *The Three Musketeers*, based on the novel by DUMAS *père*, returning to London to repeat the role at DRURY LANE in 1930. Back in New York he played the title role in a revival of *Peter Ibbetson* (1931) based on the novel by George Du Maurier, and thereafter appeared mainly in straight plays, though he was seen in such musical roles as Gaylord Ravenal in a revival of Kern's *Show Boat* in 1932, Count Willy Palaffi in Rodgers and Hart's *I Married An Angel* (1938), and Bruno Mahler in a revival of Kern's *Music In the Air* in 1951. His non-musical roles included Richard II in the American production of Gordon Daviot's *Richard of Bordeaux* (1934), Torvald Helmer in IBSEN's *A Doll's House* in 1937, Vershinin in CHEKHOV's *Three Sisters* in 1942, and Alexander Hazen in Lillian HELLMAN's *The Searching Wind* (1944). He played General Burgoyne in SHAW's *The Devil's Disciple* in 1950 and was later seen in such plays as Sidney KINGSLEY's *Lunatics and Lovers* (1954), N. C. HUNTER's *A Day By the Sea* (1955), USTINOV's *Photo Finish* (1963), and Brian FRIEL's *The Loves of Cass McGuire* (1966). He made his last appearance in OSBORNE's *A Patriot For Me* (1969).

KING, HETTY, see MUSIC-HALL.

KING, TOM (1730–1804), English actor, who at 17 was a strolling player with Ned SHUTER and made his first appearance at DRURY LANE under GARRICK in Oct. 1748. He was not suited to tragedy, and deciding to confine himself entirely to high comedy he went to Dublin, worked for a time under Thomas Sheridan, and returned to Drury

Lane a finished comedian. His Malvolio in *Twelfth Night* and Touchstone in *As You Like It* were both admirable, but it was as Lord Ogleby, in *The Clandestine Marriage* (1766) by the elder COLMAN and Garrick, that he made his reputation. He was the first to play Sir Peter Teazle in *The School for Scandal* (1777) and Puff in *The Critic* (1779), both by R. B. SHERIDAN, and it was as Sir Peter that he took his leave of the stage in 1802. He was then a wealthy man, but a passion for gambling and an unfortunate venture into management of the BRISTOL and SADLER'S WELLS theatres caused him to die penniless.

KING OF MISRULE, see MISRULE, KING OF.

KING'S CONCERT ROOMS, London, see SCALA THEATRE.

KING'S JESTER, see COURT FOOL and LENO, DAN.

KINGSLEY, SIDNEY (1906–), American dramatist, whose first play *Men in White* (1933; London, 1934), set in a hospital, was awarded a PULITZER PRIZE. The economic depression of the 1930s led him to write *Dead End* (1935), a bleak but provocative study of crime-breeding slum conditions, while his *Ten Million Ghosts* (1936) excoriated the international munitions cartels that had profited from the First World War. *The World We Make* (1939), based on a novel by Millen Brand, gave a moving account of a neurotic rich girl's discovery of comradeship and hope among the poor; *The Patriots* (1949) was a chronicle of the formative years of American democracy; *Detective Story* (1949), a powerful indictment of excessive righteousness covering one day in a New York police station. In 1951 he dramatized Arthur Koestler's anti-Communist novel *Darkness at Noon*, and in 1954 he broke new ground with a farce *Lunatics and Lovers* set in a hotel suite. His last play was *Night Life* (1962). He produced or directed a number of his own plays.

KING'S MEN, company of actors formerly known as the CHAMBERLAIN'S MEN. They received their new name on the accession of James I in 1603, but remained in their own theatre, the GLOBE, which they rebuilt after a fire in 1613. They continued to appear in the plays of Shakespeare, a member of the company, as they were written, their chief actor Richard BURBAGE usually playing the leading parts. On his death in 1619 his place was taken by Joseph TAYLOR, who with John LOWIN also replaced HEMINGE and CONDELL as business managers. Among the plays produced by the King's Men before the death of Shakespeare in 1616 were *Philaster* (*c.* 1610) by BEAUMONT and FLETCHER and WEBSTER'S *The Duchess of Malfi* (1614). Other notable events in the company's history were the taking over of the second BLACKFRIARS in 1608 for use as a winter indoor theatre, and the publication in 1623 of the First Folio of Shakespeare's plays. The following year saw the production of MIDDLETON'S *A Game at Chess*, a popular success which gave offence to the Spanish ambassador. The players were admonished and fined and the play shelved. On the death of James I in 1625 the company came under the patronage of his son Charles I, who with his queen continued the interest shown in them by his father, and frequently commanded them to appear at the COCKPIT-IN-COURT. Among their later dramatists were MASSINGER and SHIRLEY. From the time of their foundation they were the leading London company, their only serious rival being the company under ALLEYN at the FORTUNE. In spite of growing Puritan opposition they continued to act at the Globe until in 1642 all the theatres were closed and the company disbanded.

KING'S THEATRE, London, in the Hammersmith Road. This theatre, which opened on 26 Dec. 1902, was run by the Mulholland family, for whom it was built, for over 50 years, in conjunction with the Wimbledon Theatre. Seating 1,700, it was used mainly by touring companies, Donald WOLFIT'S company making regular appearances there, but in its last few years it became a repertory theatre. It closed in Feb. 1955 and was sold to the B.B.C., who used it as a television rehearsal room until 1958. After standing empty for some time, it was finally demolished in 1963.

KING'S THEATRE, London, in the Haymarket, see HER MAJESTY'S THEATRE.

KINGSTON [Konstam], GERTRUDE (1866–1937), English actress and theatre manager, who after some amateur experience made her first appearance on the stage at Margate in 1887. A year later she was with TREE at the HAYMARKET THEATRE, and in 1910, after a consistently successful career, built and opened the LITTLE THEATRE in John Adam Street as a home of repertory. Her first production was Laurence HOUSMAN'S version of ARISTOPHANES' *Lysistrata*, in which she played the title role, and in 1912 she revived SHAW'S *Captain Brassbound's Conversion*, playing Lady Cicely Waynflete. The venture was not a success, though her efforts, like those of Lena ASHWELL at the KINGSWAY THEATRE, helped to establish the REPERTORY THEATRE system in England, mostly in the provinces. In 1913 Miss Kingston was seen at the VAUDEVILLE THEATRE in Shaw's *Great Catherine*, which he wrote specially for her. She wrote several plays and a number of books, including her reminiscences, *Curtsey While You're Thinking* (1937).

KINGSWAY THEATRE, London, in Great Queen Street, Holborn. This opened on 9 Dec. 1882 as the Novelty Theatre, and closed the same month. Renamed the Folies-Dramatiques, it reopened in March 1883 with a little more success, and in 1888 was called the Jodrell after its manageress.

It had reverted to its original name when on 7 June 1889 the first production in England of IBSEN's *A Doll's House*, translated by William ARCHER, was given with Janet ACHURCH as Nora. It became the New Queen's Theatre in 1890, and the Eden Palace of Varieties in 1894. It then stood empty until it was taken over by W. S. PENLEY, who reopened it on 24 May 1900 as the Great Queen Street Theatre with a revival of Ambient and Heriot's *A Little Ray of Sunshine*. An important event was the first London performance of SYNGE's *The Playboy of the Western World* by the ABBEY THEATRE company on 10 June 1907, and in the following Sept. Lena ASHWELL took over the management, reopening the theatre after further reconstruction as the Kingsway on 9 Oct. with Wharton's *Irene Wycherley*. Outstanding events of the next few years were Arnold BENNETT's *The Great Adventure* (1913) with Henry AINLEY, under the direction of GRANVILLE-BARKER, and a visit in 1925 by the BIRMINGHAM REPERTORY THEATRE company, under Barry JACKSON, during which his modern-dress production of *Hamlet* was seen. The theatre then continued on its erratic course until 1932, when it became the home of the Independent Theatre Club for the production of plays banned by the LORD CHAMBERLAIN. Although these included LUDWIG's *Versailles* and SCHNITZLER's *Fräulein Elsa*, both directed by KOMISARJEVSKY and the latter starring Peggy ASHCROFT, the venture was not a success, and the theatre reverted to normal use. In Feb. 1940 Donald WOLFIT brought his touring company to London for the first time in a season of Shakespeare's plays, and the last play to be produced at this theatre was a revival of Anthony Kimmins's *While Parents Sleep*. This opened on 30 Apr. and closed on 11 May 1940 when the building suffered considerable damage from bombing. It was demolished in 1956.

KIPPHARDT, HEINAR (1922–82), German dramatist, whose first plays were produced in East Berlin; in 1959 he moved to the Federal Republic, where he became the leading exponent of DOCUMENTARY DRAMA. The first of his works to attract attention, *Der Hund des Generals* (1962), was based on one of his own short stories, and is a bitter satire on war; he achieved international fame with *In der Sache J. Robert Oppenheimer* (1964), which deals with the injustice of McCarthy's 'witch-hunts' in the U.S.A. in the 1950s and the consequent trial of the well known American nuclear physicist for suspected treason. As *In the Matter of J. Robert Oppenheimer* it was seen in London in 1966, and in New York in 1969. A further documentary, *Joel Brand* (1965), recalls the offer of the Nazis during the Second World War to spare the lives of a million Jews in exchange for 10,000 lorries. In all these plays, which call in question the integrity of political action and depict the anguish of minority groups caught between contending world powers, Kipphardt showed remarkable skill in turning documented fact into effective theatre. A later comedy, *Die Nacht, in der Chef geschlachtet wurde* (*The Night the Boss was Slaughtered*, 1967), portrayed the technologically advanced society of the West as a nightmare.

KIRBY'S FLYING BALLET, see FLYING EFFECTS.

KIRCHMAYER, THOMAS (1511–63), German Protestant humanist, author, under the pseudonym Naogeorg, of several anti-Catholic plays, of which the most important was *Pammachius* (1538). Written and first performed in Latin, it was translated into German for a production at Zwickau, and some time between 1538 and 1548 an English version in four parts, now lost, was prepared by John BALE. A performance in the original Latin, probably of a shortened version, was seen at Cambridge in 1545.

KIRSHON, VLADIMIR MIKHAILOVICH, see RUSSIA AND SOVIET UNION.

KISFALUDY, KARÓLY, see HUNGARY.

KISIELEWSKI, JAN AUGUST, see POLAND.

KITCHEN-SINK DRAMA, term applied fleetingly in the London theatre after John OSBORNE's *Look Back in Anger* (1956) and particularly plays by Arnold WESKER, to plays using working-class settings rather than the drawing-rooms of polite comedy. It was used somewhat contemptuously, and as modern dramatists enlarged their settings to cover almost as many different places as the WELL-MADE PLAY, it fell out of use as being no longer appropriate.

KIVI [Stenvall], ALEXSIS (1834–72), Finnish playwright and novelist, the first to establish a distinctively national tradition in both forms. His finest play is a comedy of country life, *Nummisuutarit* (*The Village Cobblers*), published in 1864 but not produced during his lifetime. With a one-act comedy *Kihlaus* (*The Betrothal*), published in 1867, it is now frequently revived, and provides an excellent part for a comic actor. The production of Kivi's short Biblical and lyric play *Lea* in 1869 marked the inauguration of a Finnish-language theatre in Helsinki. Among his other plays the tragedy *Kullervo*, written in 1864 but not produced until 1885, is based on the Finnish epic poem *Kalevala*, and a full-length romantic play *Karkurit* (*The Refugees*), published in 1867, shows the influence of *Romeo and Juliet*.

KJÆR, NILS (1870–1924), Norwegian dramatist and essayist, who in his plays combined social comment with a deftly satirical wit. Notable are *Regnskabets Dag* (*Day of Reckoning*, 1902), *Mimosas Hjemkomst* (*Mimosa's Return Home*, 1907), *Det lykkelige Valg* (*The Happy Choice*,

1913), and *For Træet er det Haab* (*There is Hope for the Tree*, 1917).

KJER, BODIL (1917–), Danish actress who began her career in 1937 in Copenhagen and became one of the leading players at the KONGELIGE TEATER. Her roles ranged over Greek tragedy, the comedies of MOLIÈRE, and STRINDBERG, as well as plays by modern Danish writers such as SØNDERBY, ABELL, and RIFBJERG.

KLAW THEATRE, New York, at 251–7 West 45th Street, between Broadway and 8th Avenue. This opened on 2 Mar. 1921 with Tallulah BANKHEAD, Katharine CORNELL, and a fine supporting cast in Rachel CROTHERS's *Nice People*, followed by W. J. Hurlbut's *The Lilies of the Field* (also 1921) and *Meet the Wife* (1923) by Lynn Starling. Henry Hatcher's *Hell-Bent for Heaven* (also 1923), a mountaineering drama which won the PULITZER PRIZE, was first seen at the Klaw for four matinées. In 1925–6 the THEATRE GUILD occupied the theatre with a double bill by SHAW, *Androcles and the Lion* and *The Man of Destiny*. The theatre was renamed the Avon in 1929, and Constance COLLIER and a fine cast appeared in 1931 in a revival of COWARD's *Hay Fever*. The last legitimate production at this theatre was *Tight Britches* (1934) by John Taintor Foote and Hubert Hales, after which it became a broadcasting studio. It was pulled down in Jan. 1954.

KLEIN, CHARLES (1867–1915), British-born American dramatist, who went to the United States at the age of 15, and because of his short stature was able for some years afterwards to continue playing juvenile roles. He first wrote for the theatre when he revised a play he was appearing in, but came into prominence with *Heartsease* (1897), not to be confused with a version of the younger DUMAS's *La Dame aux camélias* made by Mortimer for MODJESKA under the same title in 1880. Klein then had an immense success with two sentimental plays written for and produced by BELASCO, *The Auctioneer* (1901) and *The Music Master* (1904), in which David WARFIELD became famous in the title roles. In 1913, in collaboration with Montague Glass, Klein dramatized some Jewish short stories which as *Potash and Perlmutter* were successful both in New York and in London a year later. Klein was a play reader for Charles FROHMAN and was drowned with him in the sinking of the *Lusitania*.

KLEIST, (Bernd Wilhelm) HEINRICH VON (1777–1811), German dramatist, born into a Prussian military family, who was himself in the army and saw active service. In 1799, unable to accept the discipline and sterility of barrack life, he resigned his commission in order to devote himself to literature and philosophy. Disappointed in his hopes of success, certainly as far as his plays were concerned, he committed suicide at the age of 34. His first play, *Die Familie Schroffenstein* (1804), was a SCHICKSALSTRAGÖDIE involving the destruction of two feuding families; *Die Hermannsschlacht*, a patriotic drama written in 1808 but not performed until 1871, dealt with the defeat of the Romans by Arminius, and was intended to serve as a warning to Napoleon; *Penthesilea*, written at about the same time and produced in a shortened adaptation in 1876, was based on the story of the Amazon Queen; *Käthchen von Heilbronn* (1810), a RITTERDRAMA with a Griselda-like heroine, was first seen at Bamberg and revived by REINHARDT for the opening of the DEUTSCHES THEATER in 1905; and *Prinz Friedrich von Homburg* (1821), in which the hero on impulse disobeys a military command in an hour of national peril and is brought by wise handling to accept the necessity of discipline and his own death sentence (whereupon he is reprieved), was seen in Berlin with Josef KAINZ in the chief part. It was not a success at the time, but is now considered an important part of the German repertory. In 1951, in a French translation, with Gérard PHILIPE as Friedrich and produced by Jean VILAR, it was one of the outstanding successes of the AVIGNON FESTIVAL, its affinity with the then fashionable philosophy of Existentialism making it highly acceptable to French audiences. It was also used as the libretto of an opera by Hans Werner Henze, first performed in 1960. As *The Prince of Homburg* the play was produced in New York in 1976 and at the NATIONAL THEATRE in London in 1982.

Kleist was also the author of two comedies, an adaptation of MOLIÈRE's *Amphitryon*, published in 1807, and *Der zerbrochene Krug*, in which a village magistrate with a Falstaffian virtuosity in lying tries a case in which he is himself the culprit. It was produced at WEIMAR by GOETHE in 1808, but was a complete failure, and not until F. L. SCHMIDT revived it at HAMBURG in 1820 were its merits apparent. It is now regarded as one of the best short comedies in the German language, and has been seen in English as *The Broken Jug* (or *Pitcher*) and in Scots as *The Chippit Chantie*.

KLICPERA, VÁCLAV KLIMENT, see CZECHOSLOVAKIA.

KLINGER, FRIEDRICH MAXIMILIAN (1752–1831), German dramatist, author of *Der Wirrwarr* (1777), a version of the Romeo and Juliet story set in America at the time of the War of Independence, whose subtitle, under which it was subsequently published, gave its name to the STURM UND DRANG movement. Under the combined influence of Shakespeare and GOETHE, Klinger had already written half a dozen plays in about two years—1774 to 1776—of which *Die Zwillinge* (*The Twins*) was awarded first prize in a competition organized by SCHRÖDER and was produced by him in Hamburg in 1776. Rich in action, but weak in plot, it was written in broken sentences and ejaculations intended to express the passionate intensity of the characters' feelings. Later Klinger wrote more conventional plays for the troupe led by Abel SEYLER, and turned from the extravagances of the *Sturm und Drang* to historical tragedy and to comedies resembling those of LESSING, as in *Die falschen Spieler* (*The Cardsharpers*, 1782).

KLOPSTOCK, FRIEDRICH GOTTLOB (1724–1803), German poet, who in the intervals of writing his great religious epic *Der Messias*, inspired by Milton's *Paradise Lost*, produced also several religious plays—*Der Tod Adams* (1757), *Salomo* (1769), and *David* (1772)—as well as a trilogy celebrating German history in an Ossianic or 'bardic' style. This dealt with the victory of Arminius over the Romans in 9 AD—a subject treated later by KLEIST—and consisted of *Hermanns Schlacht* (1769), *Hermann und die Fürsten* (1784), and *Hermanns Tod* (1787).

KNEPP, MARY (?–1677), one of the first English actresses. Trained by KILLIGREW, she was in his company at the first Theatre Royal, a friend and contemporary of Nell GWYNN. She was also a friend of PEPYS, in whose diary she often appears, usually as the source of back-stage gossip. According to him she was a merry, lively creature, at her best in comedy and an excellent dancer. She was also much in demand by authors to speak the witty prologues and epilogues which the fashion of the time demanded.

KNICKERBOCKER THEATRE, New York, at 1396 Broadway, on the north-east corner of 38th Street. As Abbey's Theatre, named after its first manager Henry E. ABBEY, it opened on 8 Nov. 1893 with IRVING and Ellen TERRY in TENNYSON's *Becket*, and later saw the New York débuts of such overseas stars as MOUNET-SULLY, RÉJANE, and John HARE. On 14 Sept. 1896 it was taken over by Al Hayman and renamed the Knickerbocker, but continued its policy of housing famous visitors, among them Wilson BARRETT in his own melodrama *The Sign of the Cross* and Beerbohm TREE in Gilbert Parker's *The Seats of the Mighty*. Among the American stars who appeared there were Maude ADAMS in ROSTAND's *L'Aiglon* in 1900 and BARRIE's *Quality Street* in 1901, and Otis SKINNER, who in 1911 scored an instantaneous success in KNOBLOCK's *Kismet*. The last production at this famous theatre was Philip Dunning's *Sweet Land of Liberty*, which opened on 23 Sept. 1929 and closed after 8 performances. The theatre was demolished in 1930.

The BOWERY THEATRE was called the Knickerbocker for a short time in 1844.

KNIPPER-CHEKHOVA, OLGA LEONARDOVNA (1870–1959), Russian actress, who joined the company of the MOSCOW ART THEATRE on its foundation in 1898 and became famous for her interpretations of the heroines of CHEKHOV, whom she married in 1901. When the Moscow Art Theatre revived *Ivanov* and *The Seagull*, which had previously been unsuccessful elsewhere, she was seen as Elena Petrovna and Madame Arkadina, and she created the roles of Elena Andreyevna in *Uncle Vanya* (1899), Masha in *Three Sisters* (1901), and Madame Ranevskaya in *The Cherry Orchard* (1904). She played this last part again in 1943, on the occasion of the 300th performance of the play. After Chekhov's death in 1904 she remained with the Moscow Art Theatre, and though mainly considered at her best in poetic and delicately psychological roles, she was also much admired in such comedy parts as Shlestova in GRIBOYEDOV's *Woe From Wit*, which she played in a revival in 1925.

KNOBLOCK [Knoblauch], EDWARD (1874–1945), playwright born and bred in the United States, who spent much of his life in England and on the Continent and eventually became a British citizen. He was for a short time an actor, and had a thorough knowledge of the stage, which he applied with much skill to the dramatization of other writers' novels, sometimes in collaboration with them, and was considered an admirable and reliable 'play carpenter' rather than an original dramatist. Of his own fairly numerous plays, some of them based on the French, the most successful were *Kismet* (1911), an oriental fantasy produced in London by Oscar ASCHE and in New York (also 1911) by Otis SKINNER, and *Marie-Odile* (1915), a tale of the Franco-Prussian war which in New York was directed by David BELASCO. In 1910 he translated MAETERLINCK's *Sister Beatrice*, and began his work of collaboration by writing *Milestones* (1912) with Arnold BENNETT, with whom he later wrote *London Life* (1924) and *Mr Prohack* (1927). He also collaborated with Seymour HICKS in *England Expects*. The novels of which he made stage versions include J. E. Goodman's *Simon Called Peter* (1924), PRIESTLEY's *The Good Companions*, Vicki Baum's *Grand Hotel* (both 1931), Beverley Nichols's *Evensong*, and A. J. Cronin's *Hatter's Castle* (both 1932).

KNOWLES, JAMES SHERIDAN (1784–1862), Irish-born playwright, a cousin of R. B. SHERIDAN and a close friend of COLERIDGE, HAZLITT, and Charles LAMB. In 1808 he was a member of the Crow Street Theatre in Dublin, where he proved himself a passable actor and also wrote a melodrama, *Leo; or, the Gypsy* (1810), for Edmund KEAN, who was a member of the same company. He later offered Kean *Virginius; or, the Liberation of Rome* (1820), but it was refused and MACREADY accepted it and played it with great success at COVENT GARDEN. Knowles's most popular play was *The Hunchback* (1832), whose heroine Julia, first played by Fanny KEMBLE, was a favourite part with many young actresses. He himself appeared in it, after many years off the stage, playing Master Walter, the part in which he made his first appearance in New York in 1834. Knowles wrote his tragedies on classical subjects in the light of 19th-century domesticity, and was more concerned with the emotions of his characters than with their actions. None of his works has survived on the stage.

KNOWLES, R. G., see MUSIC-HALL.

KNOX, TEDDY, see CRAZY GANG.

KNUDSEN, ERIK (1922–), Danish dramatist and poet, a trenchant critic of the more modish aspects of contemporary society, whose satirical revues exerted a major influence during the 1960s. *Frihed det bedste guld* (*Freedom the Best Gold*, 1961) criticized the breakdown of social democratic values under 'admass' pressures. *Nik, Nik, Nikolaj* (1966) sardonically portrays a Russian artist's escape to the West, where he finds the values as constricting to his work as those of Communism.

KOBRIN, LEO, see JEWISH DRAMA.

KOCH, ESTHER, see BRANDES, J. C.

KOCH, FREDERICK HENRY (1877–1944), American university professor, who, like George Pierce BAKER, introduced the serious study of playwriting and play production into the curriculum of the university student. He worked with the Carolina Playmakers, students of the University of North Carolina, who toured from Georgia to Washington carrying their scenery and props with them and produced plays, mainly in one act, written by themselves and their fellow drama students on themes of Southern folklore, superstition, and local history. Unlike Baker Koch did not aim to prepare playwrights for Broadway, but the first good plays about the Southern States, which began to appear about 1923, are probably due to him, and he can be credited with the training of at least one outstanding dramatist, Paul GREEN. The Carolina Playmakers certainly influenced the amateur and Little Theatres, and Koch's work gave an added impetus to the teaching of drama in American schools and colleges. He also founded and directed a school of playwriting at Banff in Canada.

KOCH, HEINRICH GOTTFRIED (1703–75), German actor, who in 1728 joined Carolina NEUBER's company. He was a good scene-painter, a good translator and adapter of plays, and a competent actor in the new style sponsored by GOTTSCHED and Neuber, particularly popular in classical comedy and acceptable in tragedy as long as the declamatory style remained in fashion. After the break-up of the Neuber company he started on his own, quarrelling violently in the process with his old companion SCHÖNEMANN, whose company he took over some years later, enjoying a free hand after the departure of EKHOF in 1767. He treated his actors well and was one of the few managers of his time, apart from ACKERMANN, to be generally esteemed. He travelled continuously, but made LEIPZIG his headquarters.

KOLÁR, JOSEF JIŘÍ, see CZECHOSLOVAKIA.

KOLTAI, RALPH (1924–), German-born stage designer, who worked initially in the field of opera, but in 1962 was engaged to design the sets for the ROYAL SHAKESPEARE COMPANY's production of BRECHT's *The Caucasian Chalk Circle*. He was appointed associate designer to the company a year later, working on its production of HOCHHUTH's *The Representative* (1963) and revivals of MARLOWE's *The Jew of Malta* in 1964, SHAW's *Major Barbara* and *Too True To Be Good* in 1970 and 1975 respectively, and O'KEEFFE's *Wild Oats* in 1977. He has also designed sets for several plays given by the NATIONAL THEATRE, among them the all-male *As You Like It* in 1967, BOLT's *State of Revolution* in 1977, IBSEN's *Brand* in 1978 and *The Wild Duck* in 1979, and Brecht's *Baal*, also in 1979. His work elsewhere has included designs for Hochhuth's *Soldiers* (London and N.Y., 1968), and for the musical *Billy* (1974). Although based in England, he has worked in many parts of Western Europe, in the U.S.A., Canada, and elsewhere. From 1965 to 1973 he was head of the School of Theatre Design at the Central School of Art and Design in London.

KOMISARJEVSKAYA, VERA FEDOROVNA (1864–1910), Russian actress and theatre manager, sister of Theodore KOMISARJEVSKY. She made her début in 1891 as Betsy in TOLSTOY's *The Fruits of Enlightenment* and in 1896 went to the Alexandrinsky (see PUSHKIN) Theatre where she played Nina in the ill-fated first production of CHEKHOV's *The Seagull*. She left to found her own theatre, where in the midst of the upheavals of 1905 her productions included plays by GORKY, Chekhov, and IBSEN. In the years of reaction which began in 1906 Komisarjevskaya came under the influence of SYMBOLISM and invited MEYERHOLD to produce in her theatre, but soon broke with him. On a last tour she caught smallpox and died. She never appeared in England, but in 1908 she played a season at DALY's in New York. She was at her best in such parts as Gretchen in GOETHE's *Faust*, Rosy in SUDERMANN's *The Battle of the Butterflies*, and Ibsen's Nora, in *A Doll's House*, and Hedda Gabler.

KOMISARJEVSKAYA THEATRE, Leningrad, founded in 1904 by Vera KOMISARJEVSKAYA with a repertoire of plays by GORKY, IBSEN, CHEKHOV, and others. In 1906 MEYERHOLD was invited to direct the company, and a number of symbolist productions were staged, including MAETERLINCK's *Soeur Beatrice* and BLOK's *The Puppet Show*. When Komisarjevskaya broke with Meyerhold in 1907, he was replaced by her brother Theodore KOMISARJEVSKY and by Nikolai EVREINOV. The company disbanded in about 1908. In 1959 the Leningrad Drama Theatre took over the building used by Komisarjevskaya's company from 1904 to 1906 and the newly opened theatre was officially named after her.

KOMISARJEVSKY, THEODORE [Fedor] (1882–1954), Russian director and designer, brother of Vera (above). Born in Venice, he was brought up in Russia and gained his initial experience in the pre-Revolutionary theatre, where from 1907 he produced a number of plays and operas, first at his sister's theatre and then at his own. In 1919 he emigrated to England, where he worked initially as a theatre designer, and was first noticed during the Russian season at the little BARNES THEATRE in 1926. Although his best work was done in productions of and designs for Russian plays, particularly those of CHEKHOV, he was also responsible for a number of controversial productions at the SHAKESPEARE MEMORIAL THEATRE, notably *Macbeth* (1933) with aluminium scenery, *The Merry Wives of Windsor* (1935) in the style of a Viennese operetta, a widely-acclaimed *King Lear* (1936), *The Comedy of Errors* (1938), and *The Taming of the Shrew* (1939). In London during these years he also produced a wide variety of plays, the last being BARRIE's *The Boy David* (1936). He subsequently went to the United States, where he remained until his death. He was the second husband of Peggy ASHCROFT, who played in many of his productions.

KOMISSAROV, ALEKSANDR MIKHAILOVICH (1904–79), Russian actor, who joined the MOSCOW ART THEATRE in 1924, after studying for the stage with the theatre's Second Studio, and remained there throughout his career. One of his earliest successes was as Chérubin in BEAUMARCHAIS's *Le Mariage de Figaro* in 1927, but later in life he specialized in comic roles, playing them with great charm and liveliness. Among them were Mr Winkle in an adaptation of DICKENS's *Pickwick Papers* in 1934, Careless in SHERIDAN's *The School for Scandal* in 1940, and Petrischev in TOLSTOY's *The Fruits of Enlightenment* in 1951. In 1964 Komissarov visited London with the Moscow Art Theatre company during the WORLD THEATRE SEASON, and was seen as Epihodov in CHEKHOV's *The Cherry Orchard*, the Man with Boots in POGODIN's *Kremlin Chimes*, and Koukou in GOGOL's *Dead Souls*. In this last play he later played Chichikov, and in later years was seen as Alyosha in GORKY's *The Lower Depths*, Golotvin in OSTROVSKY's *The Diary of a Scoundrel*, and Bobchinsk in Gogol's *The Government Inspector*. For many years he was a teacher at the Moscow Art Theatre Drama School.

KONGELIGE TEATER (Royal Theatre), Copenhagen. The original building, designed for the DANSKE SKUEPLADS by Niels Eigtved and completed in 1748, was extensively altered and enlarged in 1772, two years after the monarchy had assumed financial responsibility for the theatre's affairs. The present building, close to the original site in Kongens Nytorv, dates from 1874. In the late 19th century it achieved a European reputation under the direction of E. Fallesen as the theatre most closely associated with IBSEN. An annexe theatre called Stærekassen (Starling Nest Box) was acquired in 1931.

KOONEN, ALISA GEORGIEVNA (1889–1974), Soviet actress, who began her career with the MOSCOW ART THEATRE before joining MARDZHANOV's Free Theatre in 1913, where she met her future husband, Alexander TAÏROV. A year later they opened the KAMERNY THEATRE, and in a remarkable partnership which lasted until Taïrov's death in 1950 produced a wide range of plays, mostly international, in which Koonen, who combined great beauty with exceptional plasticity of movement, played leading parts. Her performance as Adrienne Lecouvreur in a revival of SCRIBE's play in 1919 was said by a critic to have raised the quality of the play from melodrama to Shakespearian tragedy. She was also outstanding as Shakespeare's Juliet, RACINE's Phèdre, SHAW's Saint Joan, Abbie Putnam in O'NEILL's *Desire Under the Elms*, and the Commissar in the first production of VISHNEVSKY's *An Optimistic Tragedy* (1934).

KOPIT, ARTHUR (1938–), American dramatist, whose first plays were produced at HARVARD while he was a student there. They included *On the Runway Of Life You Never Know What's Coming Off Next* (1957), *Across the River and Into the Jungle* (1958), *Sing To Me Through Open Windows* (1959), and *Oh Dad, Poor Dad, Mamma's Hung You In the Closet and I'm Feelin' So Sad* (1960), which achieved an international reputation. A brilliant black farce parodying the Oedipus complex, it was seen briefly in London in 1961 and revived in 1965 with Hermione Gingold as the possessive mother, Madame Rosepettle. It was also seen in New York in 1962, where it had a long run, and in Paris in 1963 with Edwige FEUILLÈRE. Later works included *The Day the Whores Came Out To Play Tennis*, satirizing social-climbing middle-class Americans, in which a group of whores demolish an exclusive country club with their tennis balls. It was presented in New York in 1965 with a revised version of *Sing To Me Through Open Windows*. *Indians* (1968), set in the Wild West with Buffalo Bill (W. F. CODY) as its central character, attacked America's treatment of its Indian population. Its world première was given in London by the ROYAL SHAKESPEARE COMPANY, and a revised version was produced in New York in 1969. In *Wings* (1979) Constance CUMMINGS gave an outstanding performance, first in New York and then in London, as a woman suffering from the effects of a stroke.

KORNEICHUK, ALEXANDER EVDOKIMOVICH (1905–72), Ukrainian dramatist, whose first successful play, *The Wreck of the Squadron* (1934), dealt with the sinking of their fleet by Red sailors to prevent its capture by White Russians. It was followed by *Platon Krechet* (1935), the story of a young Soviet surgeon, and by *Truth* (1937), which shows a Ukrainian peasant led by his search for truth to Petrograd and Lenin at the moment of the October Revolution. *Bogdan Hmelnitsky* (1939) dealt with a Ukrainian hero who in 1648 led an insurrection against the Poles. A war play which proved very popular was *The Front* (1943), while a satirical comedy, *Mr Perkins's Mission to the Land of the Bolsheviks*, in which an American millionaire visits Russia to discover for himself the truth about the Soviet regime, was produced in 1944 by the Moscow Theatre of SATIRE. Later plays included *Come to Zvonkovo* (1946), *Makar Dobrava* (1948), an inimitable portrait of an old Donetz miner, *The Hawthorn Grove* (1950), *Wings* (1954), *Why the Stars Smiled* (1958), and *On the Dnieper* (1961).

KORSH THEATRE, Moscow. Founded in 1882 by the Russian businessman and entrepreneur Fyodor Adamovich Korsh (1852–1923), following the liquidation of the Imperial theatres' monopoly in the capital, the heyday of this theatre was from about 1885 to 1912, when an outstanding company—which at different times included ORLENEV, MOSKVIN, and LEONIDOV—presented a number of fine productions of Russian and foreign classic plays. Later the theatre's policy became indecisive, and it closed in 1932.

KORTNER, FRITZ (1892–1970), Austrian actor, who after some experience in the provinces achieved an immediate success when he appeared in BERLIN in 1919 as the obsessed young hero of TOLLER's *Die Wandlung* (*Transfiguration*). In the same year he consolidated his position with a satanic, sadistic portrayal of Gessler in SCHILLER's *Wilhelm Tell*, and in the 1920s established himself as the ideal interpreter of the new EXPRESSIONISM, beginning at this time a long and fruitful partnership with the director JESSNER. After the energy and menace of his interpretations of the title roles in *Richard III* and WEDEKIND's *Der Marquis von Keith*, in which he used a rapid, high-pressure delivery exploiting to the full the strident power of his voice, he began in 1921, in *Othello*, to make use of quieter and more controlled speech, with greater variety of gesture. This new realism was apparent in his Macbeth, Coriolanus, and Shylock, as well as in IBSEN's *John Gabriel Borkman*, in a production directed by VIERTEL in 1923, and in SOPHOCLES' *Oedipus the King* in 1929. He left Germany in 1933 and played small parts in a number of American films, returning home in 1949. Working mainly in Berlin and MUNICH, he then became one of the most important directors in postwar Germany, directing a series of meticulously detailed productions, informed by a passion for truth and realism, which entailed long and arduous rehearsals. Notable among them were plays by Shakespeare and MOLIÈRE, STRINDBERG's *The Father* in 1950, BECKETT's *En attendant Godot* in 1953, and such German classics as Schiller's *Kabale und Liebe* in 1965, GOETHE's *Clavigo* in 1969 and, in the year of his death, LESSING's *Emilia Galotti*.

KOSTER AND BIAL'S MUSIC HALL, New York, at the corner of West 23rd Street and the Avenue of the Americas. Originally Bryant's Opera House for MINSTREL SHOWS and VARIETY, it was enlarged by Koster and Bial for musical entertainments in general and reopened on 5 May 1879. In 1881 it began the policy of importing outstanding VAUDEVILLE stars from abroad and became a famous house of light entertainment. It closed on 26 Aug. 1893, when Koster and Bial moved to their New Music Hall on the north side of 34th Street, between Broadway and 7th Avenue, which had been opened in 1892 by Oscar Hammerstein as the Manhattan Opera House. They intended to present there a better-class and more expensive brand of variety, but their fortunes declined, and on 21 July 1901 the theatre closed, Macy's department store being built on the site. Meanwhile their old music hall had reopened as the Bon Ton, but it too was unsuccessful, and soon closed, being eventually demolished in 1924.

KOSTIČ, LAZA, see YUGOSLAVIA.

KOSTOV, STEFAN L., see BULGARIA.

KOTT, JAN (1914–), Polish critic and literary historian, who became Professor of Drama at Warsaw University in 1946. Like his compatriot of an earlier generation, WYSPIAŃSKI, he developed a solid knowledge of Shakespeare's plays, and has over the years, particularly after seeing a production by Peter BROOK in 1955 of *Titus Andronicus*, evolved his own somewhat unusual view of them. His conclusions were set out in a volume of essays translated into English as *Shakespeare, Our Contemporary* (1964), in which he argues that we like the Elizabethans live in an age of transition and that in any such era the fool holds the stage, illustrating his thesis by an interesting comparison between *King Lear* and BECKETT's *Endgame* (1957). Kott's ideas on Lear, and even more on the sexual aspect of *A Midsummer Night's Dream*, appear to have had a great influence on productions of these plays by Brook, who wrote the introduction to the English version of Kott's book. Kott's views have also become well known in the U.S.A., where in 1966 he was visiting professor at YALE University.

KOTZEBUE, AUGUST FRIEDRICH FERDINAND VON (1761–1819), German dramatist, who in his day was more popular than SCHILLER with the new audiences of the Revolutionary period, not only in Germany but all over Europe. He wrote over 200 melodramas, the most successful being *Menschenhass und Reue* (1789), in which an erring wife obtains forgiveness from her husband by a life of atonement. As *The Stranger*, in an adaptation by Benjamin Thompson, it was produced at DRURY LANE in 1798, the heroine, Mrs Haller, providing an excellent part for Mrs SIDDONS, playing opposite her brother, John Philip KEMBLE. It was frequently revived up to the end of the 19th century. Equally successful in the following year, with the same leading players, was *Pizarro*, an adaptation by R. B. SHERIDAN of *Die Spanier in Peru*. In America Kotzebue's plays, in adaptations by DUNLAP, led to a vogue for melodrama which tended to eclipse more serious works and pandered to a craving for sensationalism. The best of his comedies is *Die deutschen Kleinstadten* (1803), an entertaining skit on provincialism.

KOUN, KAROLOS (1908–), Greek director, who while teaching English in Athens during the 1930s mounted a remarkable series of productions of the classics performed by young people. In 1942, during the German occupation, he founded the Art Theatre in Athens, where he has been responsible for productions of Greek and foreign plays, including IBSEN's *The Wild Duck* in 1942, GARCÍA LORCA's *Blood Wedding* in 1948, Tennessee WILLIAMS's *The Glass Menagerie* in 1947 and *A Streetcar Named Desire* in 1949. His work for other companies includes Arthur MILLER's *Death of a Salesman* in 1950 and for the Greek National Theatre PIRANDELLO's *Henry IV* and CHEKHOV's *Three Sisters*. In 1962 he took the company from the Art Theatre to the THÉÂTRE DES NATIONS in Paris, where their performance of ARISTOPHANES' *Birds* was much admired. It was seen in London during the WORLD THEATRE SEASON of 1964 and, together with AESCHYLUS' *Persians*, in 1965. In 1967 he directed *Romeo and Juliet* for the ROYAL SHAKESPEARE COMPANY. Koun's school of drama, attached to the Art Theatre, has supplied the Greek stage with a number of leading actors.

KRANNERT CENTER FOR THE PERFORMING ARTS, Urbana, Illinois. One of the largest university cultural centres in the United States, it opened on 24 Apr. 1969 and consists of four indoor theatres and one outdoor amphitheatre. The Great Hall, used for concerts, seats 2,092; the Festival Theatre, used for opera, ballet, modern dance, and Kabuki productions, 979; the Playhouse, used for plays and dance recitals, 678; the Studio Theatre, used for experimental theatre and music, 200; and the outdoor amphitheatre, 560.

KRASIŃSKI, ZYGMUNT (1812–59), Polish romantic poet and playwright, most of whose works were written in exile and published anonymously. His romantic epic drama *The Undivine Comedy*, which owes something to GOETHE's *Faust*, was written when he was only 21, and although it was published in 1835 it was not seen on the stage until 1902 in Cracow. Dealing with an abortive revolution in 1832, it underlines the disastrous consequences of class warfare; an English version by H. E. Kennedy was published in 1924 and it has also been acted in Russian and German. It was revived in Warsaw in 1926 in a production by Leon SCHILLER and, after the era of SOCIALIST REALISM, in Łódź in 1959, where ŚWINARSKI again revived it at the Stary Theatre in 1965. Of Krasiński's other plays the most important is *Irydon*, written in the early 1830s and first staged in 1908 in Łódź. Although less frequently revived than *The Undivine Comedy*, it was given a splendid production in Cracow in 1962 under Jerzy Kreczmar.

KRASOWSKI, JERZY, see SKUSZANKA, KRYSTYNA.

KRASNYA PRESNYA THEATRE, Moscow, see REALISTIC THEATRE.

KRAUSS, WERNER (1884–1959), Austrian actor, who made his first appearance on the stage in 1904. He appeared in BERLIN and VIENNA in many leading classical roles, including Macbeth, Richard III, Julius Caesar, King Lear, and in modern parts, among them the Crippled Piper in REINHARDT's *The Miracle*, both in Germany and in New York, and King Magnus in SHAW's *The Apple Cart*. He was seen in London in 1933 as Matthew Clausen in Miles MALLESON's adaptation of HAUPTMANN's *Vor Sonnenuntergang*.

KREJČA, OTOMAR (1921–), Czech director, who worked for some time with Burian, and from 1946 to 1951 was at the Vinohrady Theatre, where he was responsible for a number of interesting productions. He then went to the National Theatre, becoming its artistic director in 1956. The sensitive style of his first productions there set the tone for all his later work. He staged some remarkable revivals of the classics, including *Hamlet* and CHEKHOV's *The Seagull*, and also set up an experimental workshop to help young writers, one of them being Josef Topol who went with him as resident dramatist when he founded the Theatre Behind the Gate in 1965 and opened it with Topol's *Cat on the Rails*. The company was seen in London in the WORLD THEATRE SEASON of 1969 during a crowded week which included performances of Chekhov's *Three Sisters*; *The Single-Ended Rope*, based on several farces by NESTROY, particularly *Der Zerrissene*; SCHNITZLER's *Der grüne Kakadu*; and a new play by Topol, *An Hour of Love*. The Theatre behind the Gate was officially disbanded in 1972, Krejča having been dismissed the previous year; in 1976 he took up residence in Germany.

KRISTIANIA THEATER, Norway. The theatre in Christiania (now Oslo) traces its origins back to

1827, when the first permanent playhouse was established in the Teatergata by a Swede, J. P. Strömberg; he was quickly succeeded as director by a Dane. In 1835 the theatre burnt down and reopened in 1837 in the Bankplassen. Until well after the mid-19th century, Danish influence and control over the theatre was supreme, despite demonstrations, some of them vigorous, from patriotic Norwegians.

After 1850, a genuinely Norwegian drama of very considerable power began to emerge, linked with the names of IBSEN and Bjørnstjerne BJØRNSON, which encouraged the emergence of Norwegian players and directors. In 1852, Det Norske Teatret began operating in Møllergata, in Christiania, in opposition to the older theatre, but in 1863 the two were combined. Ibsen briefly held an appointment there, succeeded from 1865 to 1867 by Bjørnson. Thereafter the theatre survived (though not without vicissitudes) until 1899, with Hans Schrøder as its director over the last 20 years. This was the period when Ibsen's and Bjørnson's plays dominated the Norwegian theatrical scene, and when between 1884 and 1893 Bjørn BJØRNSON was an energetic director there. The last performance took place at the theatre on the Bankplassen on 15 June 1899; thereafter virtually the whole company moved to the new NATIONALTHEATRET, with Bjørnson as its director.

KRLEŽA, MIROSLAV, see YUGOSLAVIA.

KROG, HELGE (1889–1962), Norwegian dramatist and essayist, an acute and subtle psychologist whose perception of the undertones of human relations remained his chief characteristic whatever material he used and whatever the mood of his play. His dialogue had the skill and fineness of finish associated with certain schools of modern French dramatists. Among his most successful plays, some of which have been translated into English, are: *Det store Vi* (*The Great We*, 1919); *På solsiden* (*On Life's Sunny Side*, 1927); *Blåpapiret* (*The Copy*, 1928); *Konkylien* (*Happily Ever After*, 1929); *Underveis* (*On the Way*, 1931); and, possibly the most notable of all, *Oppbrudd (Break-Up*, 1936).

KRONES, THERESE (1801–30), Austrian actress, who in 1821 secured a contract at the Leopoldstädter in VIENNA, having been discovered by RAIMUND who saw her acting in Hungary. Her voice and person were small, but her dynamism and overflowing temperament more than compensated for these defects, and she had a genius for provocative improvisation delivered in a disarmingly innocent style which, according to contemporary reports, enabled her to sail far closer to the wind than a male actor could have done. She did not hesitate to make indirect allusions to her own notoriously hectic private life, and by this means, and by her inspired creation of a series of comic roles from GLEICH, MEISL, BÄUERLE, and, above all, Raimund (her playing of Youth in *Der Bauer als Millionär*, 1826, to the author's Fortunatus Wurzel was one of the great moments of the Viennese theatre), she became a legend within her own short lifetime.

KRUCZKOWSKI, LEON (1900–62), Polish novelist and playwright, whose dramatization of his own novel *Kordian i cham*, on the Polish uprising of 1931, was staged by Leon SCHILLER in 1935. An anti-Nazi play, *A Hero of Our Time*, was seen in the same year, and after the Second World War three anti-fascist plays, *The Requital* (1948), *The Germans* (1949), and *The First Day of Freedom* (1959), were seen in Warsaw, the last two being directed by AXER at the Contemporary Theatre. *Julius and Ethel* (1954), on the American spy case, argued forcibly for the innocence of the Rosenbergs. Kruczkowski's last play *The Death of the Governor* (1960) was directed by DEJMEK at the Polski Teatr in Warsaw.

KUKOLNIK, NESTO VASILYEVICH, see RUSSIA AND SOVIET UNION.

KUMMERFELD, KAROLINE (1745–1815), German actress, who in 1758 joined the company of ACKERMANN with whom she remained until he went to HAMBURG. There her popularity excited the animosity of Sophie HENSEL, who soon engineered her dismissal. She and her brother, a ballet-master, then joined KOCH in LEIPZIG where she was much admired by the young GOETHE. Later she went to Gotha under EKHOF, after whose death she accompanied her colleagues to MANNHEIM, at the invitation of DALBERG. Her memoirs give an interesting picture of the theatrical life of the period.

KUNGLIGA DRAMATISKA TEATERN, Stockholm, Sweden's Royal Dramatic Theatre, which can trace its origins back to 1737, when the Swedish Dramatic Theatre was established by A. F. Ristell. Over the years it underwent many changes of name and played in a succession of different buildings under various administrations, until in 1788, when it was housed in the Stora Bollhuset where it remained until 1792, it was brought into association with the Swedish Royal Opera, founded in 1773, the two being placed under a single controlling body in 1813. In 1794, after a season in the opera house, the theatre company moved to the Kungliga Mindre (Small Royal) Teatern where it remained until the building burned down in 1825, when it returned to the opera house. In 1863 it took over a theatre built in 1842 which was called successively the Nye (New) and then the Mindre, finally becoming known simply as 'Dramaten'. The present building on Nybroplan was built in 1904–7 and is a fine example of Swedish Jugendstil architecture; it opened on 18 Feb. 1908 with a performance of STRINDBERG's *Mäster Olof*. Extensive alterations were carried out in 1936, when a studio theatre was added to the main building, the latter being again modernized in 1957–60. The theatre is run as a limited company, with financial support from State lotteries.

KUPPELHORIZONT, see LIGHTING.

KURBAS, ALEXANDER STEPANOVICH (1887–1942), Ukrainian actor and director, whose work for the development of a Soviet theatre in the Ukraine paralleled that of MEYERHOLD, TAÏROV, and others in Russia itself. His main efforts were centred on the organization of the Berezil Theatre in Kharkov from 1922 to 1934, when it was renamed the Shevchenko Theatre. His approach to his texts was essentially that of the dedicated director, the actor's role being subordinated to the overall conception realized in terms of rhythm, sound, and form. In spite of this, his unique qualities attracted some of the finest actors in the Ukrainian theatre of the day, who were happy to work with him. Accused of FORMALISM, he was a victim of repression in the late 1930s, but was later rehabilitated.

KURZ, JOSEPH FELIX VON (1715–84), Austrian actor, who developed the typical Viennese peasant-clown HANSWURST into a personal type to which he gave the name BERNADON. He was a staunch champion of the old improvised comedy in its battle against the new neo-classic drama, and when the latter proved victorious in Vienna he and his wife took a company to Germany, where they were joined by the young F. L. SCHRÖDER. Kurz later separated from his wife, who continued to lead the company while he returned to the Burgtheater in VIENNA. The time for his 'Bernadoniades' was, however, over, and he was forced into retirement.

KVAPIL, JAROSLAV (1868–1953), Czech dramatist and director, author of a number of light comedies and fantasies on folk themes in many of which his wife Hana Kvapilová appeared. He became director of the National Theatre in Prague in 1900, remaining there until 1918. During this time he produced no less than 21 plays by Shakespeare, of which 16 were given in 1916 to mark the tercentenary of the dramatist's death. He was thus instrumental in continuing the work begun by Kolár and others in making Shakespeare available to Czech audiences, greatly helped by his leading actor Eduard Vojan.

KYD, THOMAS (1558–94), English dramatist, best remembered for *The Spanish Tragedy* (*c.* 1585–9), one of the most popular plays of its day and the prototype of many later REVENGE TRAGEDIES. It was constantly revived and revised, on one occasion by Ben JONSON, and its strong resemblance to *Hamlet* has led to the theory that Kyd was the author of an earlier play of that name which Shakespeare used as the basis for the one he wrote in about 1600 for the CHAMBERLAIN'S MEN. Kyd has also been suggested as the possible author of a lost play, *The Taming of a Shrew* (1589), believed to have been used by Shakespeare, and as part-author of the latter's *Titus Andronicus* and of the anonymous *Arden of Feversham* (1591). Of the other plays formerly assigned to him, *Soliman and Perseda* (*c.* 1590) is now thought to be by PEELE and the author of *The First Part of Jeronimo* (printed in 1605), whose action precedes that of *The Spanish Tragedy*, remains unknown. Kyd was an intimate friend of MARLOWE, with whom he was implicated in accusations of atheism, extricating himself in a not altogether creditable manner.

KYNASTON, NED [Edward] (*c.*1640–1706), one of the last boy-players of women's roles. After seeing him as the Duke's sister in a revival of FLETCHER's *The Loyal Subject*, PEPYS said: 'He made the loveliest lady that ever I saw in my life', and it was the delight of fashionable ladies to take him, in his petticoats, driving in the Park after the play. Colley CIBBER recounts that on one occasion Charles II had to wait for the curtain to rise at the theatre because Kynaston, who was playing the tragedy queen, was being shaved. In later life he fulfilled the promise of his youth and made many fine appearances in dignified heroic roles. He was particularly admired as the ageing king in *Henry IV*.

L

LABERIUS, DECIMUS (106–42 BC), the first Roman dramatist to give the MIME literary form. His outspoken political criticism incurred the wrath of Julius Caesar, who forced on him the humiliation (for a Roman knight of good standing) of having to act in one of his own mimes in competition with the professional actor PUBLILIUS SYRUS. The prologue which Laberius wrote for this occasion has survived, and contains the line, obviously applicable to Caesar: 'Needs must he fear many whom many fear' and also the subversive cry: 'Roman citizens! We are losing our liberties.' The titles of 43 of his mimes have survived, but only a few fragments from them, not enough to determine the plot; they seem, however, to have had a close affinity with the FABULA *atellana*, which they eventually absorbed or replaced. Among them are *The Soothsayer, The Fuller, The Gossips, The Fisherman*, and *The Salt-Miner*.

LABICHE, EUGÈNE (1815–88), French dramatist, who between 1838 and 1877 wrote more than 150 light comedies, alone or in collaboration, of which the most successful were *Un Chapeau de paille d'Italie* (1851), *Le Voyage de M. Perrichon* (1860), *La Poudre aux yeux* (1861), and *La Cagnotte* (*The Kitty*, 1864). The best known in English is *Un Chapeau de paille d'Italie*, which was twice translated by W. S. GILBERT. As *The Wedding Guest* it was produced at the ROYAL COURT in 1873 and as *Haste to the Wedding*, with music by George Grossmith, at the CRITERION in 1892. A new version, *An Italian Straw Hat* by Thomas Walton, was seen in London in 1945, 1952, and 1955, and also in New York in 1957, where the play had previously been produced in 1936 as *Horse Eats Hat*. Labiche's success with his contemporaries may have been due to a revolt against the serious problem-plays of such authors as the younger DUMAS. With his broad humour and predictable but well-presented situations, he gave new life and gaiety to the VAUDEVILLE inherited from SCRIBE, and raised French FARCE to a height which it has since attained only with FEYDEAU.

LABOR STAGE, New York, see PRINCESS THEATRE.

LA CALPRENÈDE, GAUTIER DE COSTES DE (1614–63), French nobleman, well received at Court, who was also a novelist and dramatist. His tragedies, of which the first, *La Mort de Mithridate* (1635), was produced at the Hôtel de BOURGOGNE, were contemporary with the early plays of CORNEILLE and contributed to the development of the classical tradition in France. Three of his subjects were taken from English history, the most interesting being *Le Comte d'Essex* (1637), in which he introduces the episode of the ring given by Elizabeth to Essex, based on current tradition. This was very successful, and in 1678 was rewritten by Thomas CORNEILLE with equal success.

LACEY, CATHERINE (1904–79), English actress, who made her first appearance in Brighton with Mrs Patrick CAMPBELL in 1925 and her London début later the same year. She first came into prominence playing Leonora Yale in *The Green Bay Tree* (1933) by Mordaunt Shairp, and in 1935 was at the SHAKESPEARE MEMORIAL THEATRE, where she was seen as Cleopatra in *Antony and Cleopatra* and Katharina in *The Taming of the Shrew*. A sensitive actress with great reserves of emotional strength, she made a deep impression as Amy O'Connell in GRANVILLE-BARKER's *Waste* (1936), and was also much admired as the heroine of J.-J. BERNARD's *The Unquiet Spirit* (1937). She played Agatha in the first production of T. S. ELIOT's *The Family Reunion* (1939) and again in 1946. In 1951, with the OLD VIC company, she played Clytemnestra in SOPHOCLES' *Electra*, and after appearing in Lesley Storm's *The Day's Mischief* (also 1951) was seen as Hecuba in GIRAUDOUX's *Tiger at the Gates* (1955), both in London and New York. She returned to the Old Vic to play Elizabeth I in SCHILLER's *Mary Stuart* in 1958, and was then seen in BOLT's *The Tiger and the Horse* (1960) before playing Clytemnestra again in AESCHYLUS' *Oresteia* in 1961. She was back with the Old Vic during their last season, one of her parts being Aase in IBSEN's *Peer Gynt* in 1962, and was last seen in 1970 in the London production of Robert ANDERSON's *I Never Sang For My Father*.

LA CHAPELLE, JEAN DE (1655–1723), French nobleman who as a young man wrote four tragedies much influenced by RACINE. These were performed by the newly founded company of the COMÉDIE-FRANÇAISE, with BARON and Mlle CHAMPMESLÉ in the leading roles. The most successful was *Cléopâtre* (1681), in which Baron was outstanding as Antony. The play, which was sufficiently well known to be parodied soon after its production, remained in the repertory until 1727.

LA CHAUSSÉE, PIERRE CLAUDE NIVELLE DE (1692–1754), French dramatist, and the chief exponent of 18th-century COMÉDIE LARMOYANTE. Though he was already a well known figure in literary society La Chaussée was 40 before he wrote his first play, *La Fausse Antipathie* (1733). This and *Le Préjugé à la mode* (1735) were both well received, and may be said, with their mingling of pathos and comedy, to prefigure the end of the old French comedy and the beginnings of *le* DRAME *bourgeois*, or 'domestic drama'. La Chaussée was a wealthy man with somewhat frank and licentious tastes, and it is clear from his prologue to *La Fausse Antipathie* that he wrote as he did less out of conviction than because he saw that this was what his audience wanted. He produced a number of later plays in the same vein, including *Mélanide* (1741), the most successful example of *comédie larmoyante; Paméla* (1743), adapted from Richardson's novel; and *La Gouvernante* (1747), a foretaste of *East Lynne*. In his own day La Chaussée was immensely popular; he was elected to the French Academy in 1736, and his plays were translated into Dutch, English, and Italian.

LACKAYE, WILTON (1862–1932), American actor, who was intended for the Church, but adopted the stage after a chance visit to the MADISON SQUARE THEATRE on his way to Rome. He was appearing with an amateur company when Lawrence BARRETT gave him a part in his revival of BOKER's *Francesca da Rimini* at the STAR THEATRE, New York, on 27 Aug. 1883. He later appeared many times with Fanny DAVENPORT, and in 1887 made a success in a dramatization of Rider Haggard's novel *She*. He was thereafter constantly in demand, appearing in a number of Shakespeare plays and in many new productions. He also played Jean Valjean in his own dramatization of HUGO's *Les Misérables*. But his greatest part was undoubtedly Svengali in Du Maurier's *Trilby* (1895), which he played for two years and then in many revivals. He founded the Catholic Actors' Guild and helped to organize the AMERICAN ACTORS' EQUITY ASSOCIATION.

LACY, JOHN (?–1681), English actor, originally a dancing-master, who took to the stage when the theatres reopened in 1660 and became a great favourite with Charles II. A painting by Michael Wright now in Hampton Court shows him in three different parts—as Teague in Sir Robert Howard's *The Committee* (1662), as Monsieur Galliard in a 1662 revival of William Cavendish's *The Variety*, and as Mr Scruple in John Wilson's *The Cheats* (1663). He was the first to play Bayes in BUCKINGHAM's *The Rehearsal* (1671), and was judged to have hit DRYDEN off to the life. His Falstaff was also much admired. He was the author of four plays, of which *Sauny the Scot* (1667) was based on *The Taming of the Shrew* and *The Dumb Lady; or the Farrier made Physician* (1669) on MOLIÈRE's *Le Médecin malgré lui*.

LADY ELIZABETH'S MEN, a company under the patronage of James I's daughter. Formed in 1611, they were seen at Court a year later and in 1613 were merged with the Revels company under Philip Rosseter, one of the young men in that company, Nathan FIELD, becoming their leading actor. They were at the HOPE under the management of HENSLOWE, who built it for them, in 1614, and on his death two years later several of the company joined PRINCE CHARLES'S MEN under ALLEYN; but Field was taken in by the KING'S MEN, probably in the place of Shakespeare, who died in the same year. The rest of the Lady Elizabeth's Men went into the provinces, but in 1622 a new company under the same name took possession of the COCKPIT under Christopher BEESTON. They were successful for a time, and appeared in a number of plays by distinguished new dramatists—*The Changeling* by MIDDLETON and ROWLEY, MASSINGER's *The Bondman*, and Thomas HEYWOOD's *The Captives, or the Lost Recovered* among them. The great plague of 1625 caused them to break up and on the accession of Charles I their place was taken by a new company, QUEEN HENRIETTA'S MEN.

LA FLEUR, see GROS-GUILLAUME.

LA FOSSE, ANTOINE D'AUBIGNY DE (1653–1708), French dramatist, who was over forty when his first play *Polixène* (1696) was produced. A second play, *Manlius Capitolinus* (1698), remained in the repertory of the COMÉDIE-FRANÇAISE until 1849: TALMA, in particular, was later very good in the name-part. It was a frank imitation of CORNEILLE and RACINE, and under the guise of Roman names treated of contemporary history, being an account of the conspiracy of the Spaniards against Venice, a subject used some years earlier by OTWAY in *Venice Preserv'd* (1682). La Fosse's later plays were less successful, though many of his contemporaries rated him as highly as Racine.

LAGERKVIST, PÄR (1891–1974), Swedish poet, novelist, and dramatist, who was awarded the NOBEL PRIZE for Literature in 1951. Although best known outside Sweden for his novels and short stories, he was probably the most remarkable playwright of the modern Swedish theatre as well as an influential theorist and critic. His *Ordkonst och bildkonst* (*Verbal Art and Pictorial Art*, 1913) and *Modern teater* (1918) repudiate NATURALISM as a means of expression in modern life and commend a Strindbergian mode of EXPRESSIONISM. He had already written *Sista mänskan* (*The Last Man*) and *Den svåra stunden* (*The Difficult Hour*) before his first play to be performed, *Himlens hemlighet* (*The Secret of Heaven*, 1921), written in 1919, was put on with Harriet Bosse in the leading role. These and the later *Den osynlige* (*The Invisible One*, 1923) demonstrate his devotion to STRINDBERG; but the despair engendered by the First World War and the period immediately following it is none the less conveyed with originality both of thought and technique. The gloom of his mood lessened in the second phase of his career, and such plays as *Han som fick leva om sitt liv* (*He Who Lived His Life Again*, 1928), *Mannen utan själ* (*The Man Without a Soul*, 1936), *Seger i mörker* (*Victory in the Dark*, 1939), and *Midsommardröm i fattighuset* (*Midsummer Dream in the Workhouse*, 1941) show a sensitive, compassionate vision expressed through an increasingly realistic style, though rarely without symbolical or allegorical overtones. His most successful play in the theatre was *De vises sten* (*The Philosopher's Stone*, 1948), set in the Italian Renaissance, and dealing with problems of faith, knowledge, and social responsibility. It was given its first production at the KUNGLIGA DRAMATISKA TEATERN under Alf SJÖBERG and was revived in 1952. The best known plays of his last years are *Låt människan leva* (*Let Man Live*, 1951) and *Barabbas* (1953), the latter based on one of his own novels.

LA GRANGE [Charles Varlet] (1639–92), French actor, who joined the company of MOLIÈRE in 1659, playing young lovers. He also appeared in the title role of RACINE's *Alexandre* (1665). A methodical man, he kept a register of all the plays presented by Molière at the PALAIS-ROYAL and the receipts from each, interspersed with notes on the internal affairs of the company which have proved invaluable to later students of the period. In 1664 he took over Molière's function as Orator to the troupe, and was active in forwarding its affairs after Molière's death. As an act of piety in memory of his friend he edited and wrote a preface to the first collected edition of Molière plays, published in 1682. His wife Marie (1639–1727), known as Marotte from the part she played in *Les Précieuses ridicules*, was the daughter of the pastry-cook Cyprien Ragueneau, immortalized by ROSTAND in his play *Cyrano de Bergerac* (1898). She created the title role in Molière's *La Comtesse d'Escarbagnas* (1671), and was one of the original members of the COMÉDIE-FRANÇAISE, together with her brother-in-law Achille (1636–1709) who acted under the name of Verneuil.

LA GRANGE-CHANCEL, FRANÇOIS-JOSEPH DE (1677–1758), French dramatist, whose first play was written when he was about 13. With the help of RACINE it was put on in 1694, without much success. It was followed by several more tragedies which show clearly the decline of the neo-classical ideal and an increasingly marked tendency to sensationalism: the most successful were *Amasis* (1701) and *Ino et Mélicerte* (1713). La Grange-Chancel, who later became embroiled in controversy with VOLTAIRE, seems to have been an embittered man who suffered from his own precocity and failed to fulfil his youthful promise.

LA HARPE, JEAN-FRANÇOIS DE (1739–1802), French dramatist, whose plays, modelled on those of VOLTAIRE, show the continued decline of classical tragedy during the 18th century. The first, *Le Comte de Warwick* (1763), is usually accounted the best, though *Philoctète* (1783) and *Coriolan* (1784) were well received. It is, however, as a critic that La Harpe is best remembered. His *Cours de littérature ancienne et moderne*, based on lectures given in 1786, was for long a standard work, at its best when dealing with the 17th century, and he wrote excellent commentaries on the plays of RACINE.

LAHR, BERT [Irving Lahrheim] (1895–1967), American VAUDEVILLE and BURLESQUE player, who became a leading actor in musical shows and also in straight plays. Basically a clown, adept at lugubrious leers and with a wide range of expressive grunts and bellows, he first played the vaudeville circuits at the age of 15, and within a few years had developed his own solo act. After serving in the Navy during the First World War he teamed up in a double act with Mercides Delpino, later his wife, and made his Broadway début in the revue *Delmar's Revels* (1927), his first outstanding success being the prizefighter Gink Schiner in the musical *Hold Everything* (1928). During the 1930s he was in a number of musical shows, including the revues *Life Begins at 8.40* (1934), *George White's Scandals of 1936* (1935), and *The Show Is On* (1936), and the Cole Porter musical *Du Barry Was A Lady* (1939), in which year he also became internationally famous for his performance as the Cowardly Lion in the film *The Wizard of Oz*. His subsequent stage career included Skid, a *passé* burlesque comic, in *Burlesque* (1946) by Arthur Hopkins and D. G. M. Walters, Gogo (Estragon) in BECKETT's *Waiting for Godot* (1956), and Boniface in *Hotel Paradiso* (1957), based on a farce by FEYDEAU. He gave fine performances as Bottom in *A Midsummer Night's Dream* in the American Shakespeare Festival production of 1960, and in *Foxy*, a musical version of JONSON's *Volpone*, in 1964. A mainly sympathetic but highly perceptive biography by his son John appeared in 1969 as *Notes on a Cowardly Lion*.

LA MAMA EXPERIMENTAL THEATRE CLUB, New York, AVANT-GARDE theatre group formed in 1961 by Ellen Stewart and originally known as the Café La Mama, from the basement room under an Italian restaurant where new and uncommercial plays by young playwrights were presented for a week's run. After two moves the group settled in its own premises and took its present name, supporting a permanent company formed in 1964 by Tom O'Horgan, who first directed the seminal rock musical *Hair* for La Mama in 1967 and on Broadway a year later. Further experiments, after Horgan's departure in 1969, led to the merging of classical forms of music with rock genres, employing also such skills as tap-dancing and acrobatics. In 1973 financial stringencies forced the company to stop working full-time and its members formed a pool from which performers could be elected for specific projects. La Mama then became an international movement, with off-shoots in England, Europe, Australia, Canada, and South America, all founded by people who at one time worked with the group in New York. La Mama has also produced such experimental groups as Mabou Mines, which (taking its name from a disused coal mine in Nova Scotia) began work in 1970, preparing by COLLECTIVE CREATION productions which the participants describe as Animations or Collaborations the former, like *Red Horse Animation* (1971), being non-literary presentations using images of animals as metaphors for human conditions. La Mama also housed the productions of Chaikin's OPEN THEATER.

LAMB, CHARLES (1775–1834), English critic and essayist, who wrote about the theatre with a warm affection which shows him to have been a scholar who enjoyed the busy traffic of the stage as well as the calm seclusion of the study. An anthology of dramatists contemporary with Shakespeare, published in 1808, revealed to his own contemporaries the little-known beauties of such playwrights as MARLOWE, WEBSTER, and FORD, and although in one of his essays he argued the case for reading Shakespeare in preference to seeing him acted, in the *Essays of Elia* (1823–33) he recalls many happy moments spent in the theatre and in the company of actors. His love for Shakespeare in particular led him to write in collaboration with his sister Mary *Tales from Shakespeare* (1807), which, though now considered out of date, served an a pleasant and easy introduction to many of the plays for several generations of schoolchildren. Lamb was himself the author of four plays, the first of which, a five-act tragedy called *John Woodvil*, was offered to John Philip KEMBLE in 1799 under the title of *Pride's Cure*. It was apparently never acted, but was printed in 1802. A farce, *Mr H—* (1806), was produced unsuccessfully at DRURY LANE with ELLISTON in the title role. *The Wife's Trial; or, the Intruding Widow* (1828) and *The Pawn-broker's Daughter* (1830) were published in *Blackwood's Magazine*; they were never acted.

LAMBS, The, a London supper club founded in the 1860s by a group which included Squire BANCROFT, John HARE, and Henry IRVING. Limited to 24 members under a chairman known as the Shepherd, it had no premises of its own but met regularly for many years at the Gaiety Restaurant and subsequently at the Albemarle Hotel. It survived until the late 1890s. Meanwhile an American actor, Henry MONTAGUE, who had been one of the founder-members, had returned to New York, where in 1875 he started a similar club, with himself as the first Shepherd. On the dissolution of the London club the Shepherd's crook, bell, and badge were presented to the American Shepherd. In 1904, after two moves, the club settled in premises at 128 West 44th Street, which had to be enlarged several times to accommodate a growing membership. These were disposed of in 1945, and the club moved to new quarters in the Women's Republican Club on 51st Street. The Lambs in its present form is roughly the equivalent of the London Savage Club, as the PLAYERS is of the GARRICK CLUB.

LAMDA, see SCHOOLS OF DRAMA.

LA MOTTE, see HOUDAR DE LA MOTTE.

LAMP, see LIGHTING.

LANCHESTER, ELSA, see LAUGHTON, CHARLES.

LANDER [*née* Davenport], JEAN MARGARET (1829–1903), American actress, daughter of the English actor-manager Thomas Donald Davenport (1792–1851), on whom DICKENS is believed to have based Vincent Crummles in *Nicholas Nickleby*—in which case Jean, who at the age of eight was playing such unsuitable parts as Richard III and Shylock, would be the original Infant Phenomenon. She went with her parents to America in 1838 and was exploited as a child prodigy for several years, making her adult début in 1844. She played the title roles in English versions of SCRIBE and Legouvé's *Adrienne Lecouvreur* and, in 1858, the younger DUMAS's *La Dame aux camélias* in America, the latter with the young Edwin BOOTH as Armand, and toured for many years in a repertory which included *The Wife, The Hunchback*, and *Love*, all by Sheridan KNOWLES, Mrs Lovell's *Ingomar*, BULWER-LYTTON's *The Lady of Lyons*, and Charles READE's *Peg Woffington*. She retired from the stage on her marriage in 1860 to General Lander, but on his death in the Civil War two years later returned, being first billed as Mrs Lander at NIBLO'S GARDEN in 1865. She made her last appearance on the stage in Boston on 1 Jan. 1877 as Hester in her own dramatization of Hawthorne's *The Scarlet Letter*. A small, well-formed woman with a sweet face, clear voice, and graceful figure, she was an actress of great talent and intellectual judgement, but lacked fire.

LANDSCAPES AND LIVING SPACES, see PEOPLE SHOW.

LANE, LOUISA, see DREW, MRS JOHN.

LANE, LUPINO [Henry George Lupino] (1892–1959), English actor, nephew of Barry and Stanley LUPINO, who took his stage name from his great-aunt Sara (below). As Nipper Lane he made his first appearance on the stage at the age of four and as an adult actor toured extensively in variety, MUSICAL COMEDY, and PANTOMIME. He made a great success as Bill Snibson in *Me and My Girl*, a musical comedy by Arthur Rose and Douglas Furber, in which he created the dance known as 'The Lambeth Walk.' In the cast with him was his brother Wallace Lupino, also an actor, dancer, and pantomimist, and his son Lauri, who later went into films. Surviving silent films of Lupino Lane's stage act are a good record of the tumbling and acrobatics of the HARLEQUINADE.

LANE, SAM (1804–71), English theatre manager, who for many years ran the Britannia Saloon, later, after the abolition of the Patents in 1843 (see PATENT THEATRES), the BRITANNIA THEATRE. After his death it was taken over by his widow Sara (1823–99) who continued to run it successfully until she died. She was an excellent actress, playing PRINCIPAL BOY in the famous 'Brit.' PANTOMIMES until well into her 70s. By the marriage of her niece Charlotte Robinson to Harry Lupino, uncle of Barry and Stanley LUPINO, she was the great-aunt of Lupino LANE.

LANG, (Alexander) MATHESON (1879–1948), actor-manager and dramatist, born in Canada but brought up in Scotland. Coming from a clerical family—his cousin Cosmo Lang became Archbishop of Canterbury—and himself destined for the Church, his determination to go on the stage was strengthened by visits to the productions of BENSON and IRVING when they toured Scotland. He made his first appearance with CALVERT in Wolverhampton in 1897, and later joined Benson, appearing with him in London in 1900 and going on tour with him to the West Indies. In 1904 he was back in London, where he played under the management of VEDRENNE and GRANVILLE-BARKER at the ROYAL COURT THEATRE in IBSEN and SHAW. He scored his first outstanding success in Hall Caine's *The Christian* (1907) at the LYCEUM, where he also gave good performances as Romeo and Hamlet. He then took his own company to South Africa, Australia, and India, playing Shakespeare and modern romantic drama with much success. On his return to London he appeared in a play with which his name was for a long time associated—*Mr Wu* (1913), an improbable Anglo-Chinese melodrama by Harry Vernon and Harold Owen, which was seen all over the world and frequently revived. In 1914 Lang, with his wife Hutin [Nellie] Britton (1876–1965), who had been with him in Benson's company and subsequently toured as his leading lady, inaugurated the Shakespeare productions at the OLD VIC under Lilian BAYLIS with *The Taming of the Shrew, Hamlet*, and *The Merchant of Venice*. Four years later he was at the LYRIC THEATRE with his own adaptation of a French romantic comedy as *The Purple Mask*, and in 1920 was seen in *Carnival*, written in collaboration with H. C. M. Hardinge, and E. Temple Thurston's *The Wandering Jew*, the latter running for a year and being several times revived. *The Chinese Bungalow* (1925) by Marion Osmond and James Corbet was followed by two adaptations of German plays by Ashley DUKES—*Such Men Are Dangerous* (1928), based on Neumann's *Der Patriot*, and *Jew Süss* (1929), based on a novel by FEUCHTWANGER, in which the young Peggy ASCHROFT played Naemi, the Jew's daughter. In his later years Lang appeared mainly on tour in his most successful parts, and in 1941 he moved to South Africa.

LANGENDIJK, PIETER, see NETHERLANDS.

LANGER, FRANTIŠEK (1888–1965), Czech playwright, author of a number of comedies, of which *The Camel Through the Needle's Eye* (1923) had a considerable success on Broadway when produced there in 1929 by the THEATRE GUILD. Among his more serious plays, written under the influence of EXPRESSIONISM, *Periphery* (1925), in which a man who has escaped punishment for a crime he committed is executed for a crime he did not commit, was seen in London as *The Outskirts* in 1934 and in New York as *The Ragged Edge* in 1935. Langer's other plays include *Cavalry Patrol* (1930), somewhat similar in theme to SHERRIFF'S *Journey's End* (1928), and *The Angel in Our Midst* (1931), which reflects man's search for new spiritual values in modern life.

LANGHAM, MICHAEL, see GUTHRIE THEATRE, MINNEAPOLIS, and STRATFORD (ONTARIO) FESTIVAL.

LANGNER, LAWRENCE (1890–1962), a successful patent agent in New York who was also a potent force in the American theatre. Born in Wales, he was educated in London, and emigrated to the U.S.A. in 1911, becoming an American citizen. In 1914 he helped to organize the WASHINGTON SQUARE PLAYERS, who produced his first play, the one-act *Licensed*; he was also instrumental in helping the group to re-form as the THEATRE GUILD, of which he became a director, and for which, with the playwright Theresa Helburn (1887–1959), he supervised the production of over 200 plays. He was the founder and first president of the AMERICAN SHAKESPEARE THEATRE at Stratford, Conn., and also founded and directed from 1931 to 1933 the New York Repertory Company, building for it the Country Playhouse, Westport, Conn. His autobiography *The Magic Curtain* was published in 1951, and he wrote or translated a number of plays in collaboration with his wife, Armina Marshall.

LANGTRY [*née* Le Breton], LILLIE [Emilie] CHARLOTTE (1853–1929), English actress, daughter of the Dean of Jersey, known from her surpassing beauty as 'the Jersey Lily'. She married at the age of 22 a wealthy Irishman and became prominent in London society, being an intimate friend of the Prince of Wales (later Edward VII). She was one of the first society women to go on the stage, making her début in 1881 under the BANCROFTS at the HAYMARKET THEATRE as Kate Hardcastle in GOLDSMITH'S *She Stoops to Conquer*. She caused a great sensation, but more on account of her looks and social position than by her acting which for many years was not taken seriously by the critics. She eventually organized her own company and proved herself a good manageress, playing at the IMPERIAL and other London theatres and touring the provinces and the United States. Although never a great actress, she became in time a good one, particularly in such parts as Rosalind in *As You Like It* and Lady Teazle in SHERIDAN's *The School for Scandal*.

LA NOUE, JEAN SAUVE DE (1701–*c.* 61), French provincial actor, who in 1739 wrote *Mahomet II*, which was given at the COMÉDIE-FRANÇAISE. VOLTAIRE, who was indebted to it for some of his own *Mahomet, ou le fanatisme* (1742), acknowledged his debt by allowing La Noue to perform the play at Lille before it was given at the Comédie-Française, and then only allowed the latter to have it if La Noue were imported into the company to play the title role, subject to his making a satisfactory début. This he did in 1742, remaining with the company until 1757. In the year before his retirement he wrote a comedy, *La Coquette corrigée*, given first at the COMÉDIE-ITALIENNE, which remained in the repertory for many years.

LANSBURY, ANGELA, see MUSICAL COMEDY and SONDHEIM, STEPHEN.

LANTERN, see LIGHTING.

LANTERN THEATRE, Dublin. The most successful of Dublin's basement theatres, this opened on 2 Nov. 1957 with McDonagh's verse play *Happy as Larry*, and quickly made a name for itself. After overcoming many difficulties it found a permanent home in Dec. 1963 next door to the British Embassy and subsequently acquired a reputation for its productions of poetic drama, though its policy was to stage everything from Shakespeare to revue. Seven of Shakespeare's plays were seen there, as well as five by YEATS, two by Austin CLARKE, and one by George FITZMAURICE. Among the successful productions at the Lantern were *The Road Round Ireland* (1964), a programme compiled from the poetry of Padraic COLUM which so impressed the author that he presented the company with a new play, *The Challengers* (1966); Jack MACGOWRAN in his BECKETT show, *Beginning to End*, in 1965; and O'CASEY's autobiography, dramatized in three parts and staged in a mammoth production in 1972. Following the destruction by fire of the British Embassy, insurance problems forced the Lantern to close, the last production being O'NEILL's *A Moon for the Misbegotten* in Sept. 1972.

LAPLACE, PIERRE ANTOINE DE (1707–93), French man of letters, who in 1745–8 published *Théâtre Anglais*, an 8-volume set of English plays in translation. Four of the volumes are devoted to Shakespeare's plays, and though few of them are complete (a number of scenes are summarized in prose) they mark the first effort to present Shakespeare to French readers.

LAPORTE, see VENIER, MARIE.

LARIVEY, PIERRE DE (*c.* 1540–1612), French dramatist, of Italian extraction—his name is a pun on the family name Giunti, 'the newly arrived'. In 1577 he saw the Italian COMMEDIA DELL'ARTE company the GELOSI play at Blois, and inspired by them he wrote 9 comedies, 6 of which were published in 1579, and 3 in 1611. Though based on Italian models, they are in no sense translations but adaptations which often contain much new material. They were played extensively in the provinces, and also in Paris. The best known, *Les Esprits* (*The Ghosts*), is taken from a comedy by Lorenzino de' Medici, itself based on material from PLAUTUS and TERENCE; in its turn it provided material for both MOLIÈRE—*L'École des maris* (1661) and *L'Avare* (1668)—and REGNARD—*Le Légataire universel* (1708). Larivey was the most substantial writer of comedy in France before Pierre CORNEILLE.

LAROCHE, JOHANN (1745–1806), Austrian actor, the creator of the comic character KASPERLE, which he first played in Graz in 1764. Five years later he went to VIENNA, where he remained until his death, becoming principal comedian of the Leopoldstädter Theater under MARINELLI. In peasant costume, with a short thick beard inherited from HANSWURST, he excelled in improvisation in broad Viennese dialect, the sound of his offstage lament '*Anwedl, anwedl!*' evoking laughter from the audience even before he showed himself on stage.

LAROQUE [Pierre Regnault Petit-Jean] (*c.* 1595–1676), French actor, who was at the MARAIS under MONTDORY, and later became leader of the company. Little is known of his abilities as an actor, but he was much esteemed as a fearless and efficient man of the theatre, and he piloted the Marais through many difficult years, even succeeding for a time in detaching the DU PARCS from MOLIÈRE. He could not, however, stand up to the combined rivalry of Molière and the Hôtel de BOURGOGNE, and in the end his theatre was mainly used for the spectacular new MACHINE PLAYS. When Molière died he and his companions were amalgamated with the remnants of the PALAIS-ROYAL company, and so formed part of the COMÉDIE-FRANÇAISE when it was founded in 1680. He was the first to recognize and encourage the talent of Mlle CHAMPMESLÉ, who joined his company in 1669.

LARRA Y SÁNCHEZ DE CASTRO, MARIANO JOSÉ DE (1809–37), Spanish journalist and satirist, adapter of a number of contemporary French plays for the Spanish stage, and author of an interesting play, *Macías* (1834), which reflects the literary preoccupations of his time. While still neo-classical in form, the emotions exhibited in it are clearly Romantic: passion overrides the bounds of honour and even of religion. Larra's theatrical criticism is of great interest, reflecting not only his own personal preoccupations and the despair which was to end finally in suicide, but also the contemporary association of literature with political and social progress.

L'ARRONGE, ADOLF (1838–1908), German dramatist and theatre director, who from 1873 to 1878 was in charge of the Breslau Lobetheater. During this time he wrote and staged a number of successful plays, including *Mein Leopold* (1873), *Hasemanns Töchter* (1877), and *Doktor Klaus* (1878). In 1883 he founded the DEUTSCHES THEATER in Berlin, destined to become one of the leading theatres in Germany. Although L'Arronge was not wholly in sympathy with the new NATURALISM, he helped to prepare the way for it by gathering together a group of highly trained actors, including Josef KAINZ and Agnes SORMA, who were able to interpret its subtleties under his successor Otto BRAHM.

LA RUE, DANNY [Daniel Patrick Carroll] (1928–), Irish-born actor specializing in FEMALE IMPERSONATION, and largely responsible for the re-establishment of its popularity in England. After appearing in regular series of nightly cabaret performances, many of the later ones being built round his act and called by his name, he made his first appearance in a straight play as Danny Rhodes in Bryan Blackburn's comedy *Come Spy With Me* (1966) which ran at the WHITEHALL THEATRE for over a year. He then appeared in the 1967 pantomime *The Sleeping Beauty* at Golders Green, having previously appeared in pantomime as early as 1954, and in 1968 gave one of his best performances in the long-running *Queen Passionella and the Sleeping Beauty* at the SAVILLE. He has since appeared in a number of excellent variety shows, of which *Danny La Rue at the Palace* (1970) had a long run.

LASHLINE, see BOX-SET and FLAT.

LASHWOOD, GEORGE, see MUSIC-HALL.

LA TAILLE, JEAN DE (1540–1608), early French writer of tragedies in the classical manner. His first play, a Biblical drama entitled *Saül le furieux* (1572), was prefaced by a short treatise on the art of tragedy which contains the first French exposition of ARISTOTLE's theory, and formulates the UNITIES of time and place which were not to be generally accepted in France until the 1660s. His *Les Corrivaux* (*The Rivals*, 1574) was the first French comedy to be written in prose.

LATEINER, JOSEPH, see JEWISH DRAMA.

LATERNA MAGICA, see LIGHTING and SVOBODA.

LA THORILLIÈRE, FRANÇOIS LENOIR DE (1626–80), French actor, who had already played in the provinces before he went to the MARAIS in 1659, having married the niece of LAROQUE, the company's leader. He later joined MOLIÈRE at the PALAIS-ROYAL, where he played leading parts and had a tragedy on the subject of Cleopatra produced with some success. After Molière's death he went to the Hôtel de BOURGOGNE, where he opposed the amalgamation of the company there with two other companies to form the COMÉDIE-FRANÇAISE, which consequently was not done until he died. His daughter Charlotte married Michel BARON, while Marie-Thérèse married DANCOURT. His son Pierre (1659–1731), who became a member of the Comédie-Française, married the daughter of the Italian actor BIANCOLELLI, and their son Anne-Maurice (*c*. 1697–1759) was at the Comédie-Française in 1722, being its *doyen*, or oldest member in years of service, at the time of his death.

LA TUILLERIE [Jean-François Juvenon] (1650–88), French actor and dramatist, the grandson of GROS-GUILLAUME. He was at the Hôtel de BOURGOGNE in 1672, in which year he married Louise Catherine (*c*. 1657–1706), the actress daughter of BELLEROCHE. A tall, stately man, he was at his best in tragedy; in comedy he played minor roles, usually those requiring a fine physique. His plays included tragedies, which were mainly rewritings of older dramas, and farces, of which two, *Crispin précepteur* (1680) and *Crispin bel esprit* (1681), were produced at the Hôtel de Bourgogne with his father-in-law in the title roles. On the foundation of the COMÉDIE-FRANÇAISE his wife retired, but he remained with the company until his death.

LAUDA, literally 'praise'; a medieval dramatic form peculiar to Italy which developed from a fusion between groups who practised two forms of religious observance. The first, founded in Florence in the 12th century and known as the *Laudesi della Beata Vergine*, had as its function the chanting at specific times of the day of hymns in praise of the Virgin. These were originally in Latin, but later in Italian, and in the 13th century they became merged with the songs of praise sung in procession by groups of wandering or mendicant orders, of which the best known was the *Confraternità dei disciplinati*. The movement spread rapidly throughout Italy and into the rest of Europe and led to the creation of groups whose sole function was to form a choir to perform on feast days.

From a simple choric song the *lauda* developed into a full-scale production which demanded soloists, divided choirs, interspersed dialogue, and narration. The surviving examples of such *laudi*, which were often collected to form a *laudario* or *Book of Lauds*, contain many interesting production details and show that the choice of episode, its elaboration, and grouping, was designed to follow the seasons of the Church's year. The purpose of the authors, mainly anonymous, was still religious rather than artistic, and the main emphasis was still on choral music, which serves to differentiate the *lauda* from the later SACRA RAPPRESENTAZIONE. The only known master of the *lauda* was Jacopone da Todi (*c*. 1230–1306), who wrote at least 88, of which many take the form of a *contrasto*, or debate. His *Pianto della Madonna* had a more complex dramatic form; it was set to music in 1921 by the Italian composer Ghedini.

LAUDER, HARRY [Hugh MacLennan], (1870–1950), one of the most famous stars of the British MUSIC-HALL and the first to be knighted, in 1919. Although he was to become the best-known and loved of the Scottish comedians on the 'halls', he made his début as a singer of Irish songs, first at the Argyle in Birkenhead and then in London in 1900, at GATTI's Palace of Varieties (Gatti's-Over-The-Water). When in response to a demand for encores he ran out of material, he turned to Scottish songs and found them even more successful. Thereafter he remained faithful to them, appearing invariably in a kilt and glengarry and carrying a crooked stick. Among his most successful numbers were 'I Love a Lassie', 'Roamin' in the Gloamin'', 'A wee Deoch-an'-Doris', 'It's nice to get up in the morning', and 'Stop yer Ticklin', Jock!' He was well known in all the 'halls' of London and the provinces, and made numerous tours of the United States, South Africa, and Australia. He also appeared in REVUE and in at least one straight play, Graham Moffat's *A Scrape o' the Pen* (1909).

LAUGHTON, CHARLES (1899–1962), stage and film actor, an American citizen since 1940, but born in England, where he made his first appearance on the stage in 1926 in GOGOL's *The Government Inspector*. He then played the title role in *Mr Prohack* (1927) by Arnold BENNETT and Edward KNOBLOCK and in 1931 starred in Jeffrey Dell's *Payment Deferred*, based on a novel by C. S. Forester, making his New York début in the play in the same year. He quickly established himself as an excellent character actor, and his season at the OLD VIC in 1933, when he played Lopakhin in CHEKHOV's *The Cherry Orchard*, Henry VIII, Macbeth, Prospero in *The Tempest*, and Angelo in *Measure for Measure*, greatly enhanced his reputation. After playing Captain Hook in BARRIE's *Peter Pan* at the LONDON PALLADIUM in 1936 he was seen in 1937 at the COMÉDIE-FRANÇAISE in MOLIÈRE's *Le Médecin malgré lui*, being the first English actor to appear there. He was then seen only intermittently on stage, being engrossed in his outstanding film career, but he returned to the theatre in Los Angeles in 1947 in BRECHT's *Galileo*, having adapted the English text in collaboration with the author. In New York in 1951 he directed and played the Devil in the *Don Juan in Hell* section of SHAW's *Man and Superman* and in 1952 he directed his own adaptation of Stephen Vincent Benét's poem *John Brown's Body*. He directed Wouk's *The Caine Mutiny Court Martial* (1954), and both directed and appeared in Shaw's *Major Barbara* in 1956. He was seen once more on the London stage in Jane Arden's *The Party* (1958), which he also directed, and in the following year played King Lear and Bottom in *A Midsummer Night's Dream* at Stratford-upon-Avon.

His wife Elsa Lanchester (1902–) appeared with him in *Payment Deferred*, *The Tempest*, *Peter Pan* (in the title role), and *The Party*, and gave the solo performance *Elsa Lanchester—Herself* (1961) directed by her husband. She wrote *Charles Laughton and I* (1938).

LAURA KEENE'S VARIETIES, New York, see KEENE, LAURA and OLYMPIC THEATRE, NEW YORK (2).

LAUTENSCHLÄGER, KARL, see SCENERY.

LAVEDAN, HENRI (1859–1940), French dramatist, whose plays dealt with social problems and contemporary manners, somewhat in the style of BECQUE and the naturalistic writers, but in a less serious and didactic manner. The most successful was *Le Prince d'Aurec* (1894), in which a decadent young nobleman is saved from the consequences of his folly by the sacrifices of his bourgeois mother. In the sequel, *Les Deux Noblesses* (1897), the hero restores the family fortunes by going into trade. Among Lavedan's other plays the comedies of manners *Le Nouveau Jeu* (1907) and *Le Goût du vice* (1911) had a breezy vitality which made them popular with the audiences of his day.

LAVINIA, see PONTI, DIANA DA.

LAWLER, RAY (1921–), Australian actor and dramatist, who had already written several light comedies when in 1949 his *Cradle of Thunder* was awarded first prize in a national competition. He was directing a repertory company in Melbourne when in 1956 the newly founded ELIZABETHAN THEATRE TRUST put on his *Summer of the Seventeenth Doll* with himself as the leading character, Barney Ibbot. This was also successfully produced in London in 1957 and in New York a year later. Lawler's next play, *The Unshaven Cheek*, was performed at the LYCEUM, Edinburgh, in 1963, and was followed by *Piccadilly Bushman*. In 1970 *A Breach in the Wall*, dealing with the supposed discovery of Thomas à Becket's tomb, was performed at the Marlowe Theatre, Canterbury, and *The Man Who Shot the Albatross* (1971) and *Kid Stakes* (1976) were both staged by the MELBOURNE THEATRE COMPANY.

LAWRENCE, D(avid) H(erbert) (1885–1930), English novelist, poet, and playwright, whose novels, considered shocking and highly controversial when first published, made him famous during his lifetime. Of his eight plays, *The Widowing of Mrs Holroyd*, written in about 1914, and *David*, a biblical drama, were given Sunday-night productions by the STAGE SOCIETY in 1926 and 1927 respectively. The others remained unproduced until in 1965 *A Collier's Friday Night* (written in about 1906) was seen at the ROYAL COURT THEATRE in a 'production without décor', and in 1967 *The Daughter-in-Law* (written during, and based on, the coal strike of 1912) was given a public showing, also at the Royal Court. It was named as one of the best new plays of the year, and led to a D. H. Lawrence season at the Royal Court in 1968, when the three plays, done in repertory to emphasize their common setting of family life in a mining background, revealed Lawrence as a starkly realistic playwright of great subtlety and vigour.

LAWRENCE, GERALD (1873–1957), English actor, who served his apprenticeship with BENSON and later joined Henry IRVING at the LYCEUM. He was playing Henry II the night Irving made his last appearance as Becket (in TENNYSON's play) in 1905. He then went to America, and on his return in 1909 directed a number of Shakespeare's plays at the ROYAL COURT THEATRE; in 1912 he gave an outstanding performance as Brassbound in a revival of SHAW's *Captain Brassbound's Conversion*. After serving in the Royal Navy during the First World War, he returned to the stage in 1919, playing de Guiche to the Cyrano of Robert LORAINE in ROSTAND's *Cyrano de Bergerac*. He also directed and played in Louis N. PARKER's *Mr Garrick* (1922), and in 1923 toured the provinces in a revival of Booth TARKINGTON's *Monsieur Beaucaire*. He made his last appearance on the stage in Oct. 1938. By his first wife Lilian BRAITHWAITE he was the father of the actress Joyce Carey. His second wife, who appeared with him in many of his later productions, was Fay Davis (1872–1945), an American actress who came to London in 1895 and made a great success as Flavia in Anthony Hope's *The Prisoner* of Zenda.

LAWRENCE [Klasen], GERTRUDE (1898–1952), English actress, who trained under Italia CONTI and made her first appearance in PANTOMIME at the age of 12. Continuing her career uninterruptedly, mainly as a dancer and later as a leading lady in REVUE, she appeared in several editions of *Charlot's Revue* in London and New York in 1924–5, and was seen as Kay in the musical *Oh, Kay!* (N.Y., 1926; London, 1927). She also essayed a few straight roles, mainly on tour, and made her first outstanding success as Amanda Prynne in Noël COWARD's *Private Lives* (1930; N.Y., 1931) with Coward himself as her leading man, appearing with him again in his *To-Night at 8.30* (London and N.Y., 1936). After playing in New York in the latter she remained there for some time, starring in Rachel CROTHERS's *Susan and God* (1937), Moss HART and Kurt Weill's musical *Lady in the Dark* (1941 and 1943), and a revival of SHAW's *Pygmalion* (1945). She returned to London in Daphne Du Maurier's *September Tide* (1948) but was back in New York in 1951, and was starring in Rodgers and Hammerstein's enormously popular musical *The King and I* when she became ill, dying shortly afterwards. She wrote a volume of reminiscences, *A Star Danced* (1945), and her second husband, the American theatre manager Richard Aldridge, published a memoir, *Gertrude Lawrence as Mrs A.* (1955).

LAWRENCE, SLINGSBY, see LEWES, GEORGE HENRY.

LAWSON [Worsnop], WILFRID (1900–66), English actor, who after 12 years' experience in the provinces was first seen in London in 1928. He was a member of Charles Macdona's SHAW company at the ROYAL COURT THEATRE from 1929 to 1930, playing a wide variety of parts, and was then seen in ZUCKMAYER's *Caravan* and Edward KNOBLOCK and Beverly Nichols's *Evensong* (both 1932). An unpredictable actor with a gritty voice, he was extremely successful as John Brown in Ronald Gow's *Gallows Glorious* (1933), but a disappointing Mark Antony in the OLD VIC production of *Antony and Cleopatra* in 1934, his only appearance in Shakespeare. He made his first appearance in New York in Ward Dorane's *Libel* (1935), and on returning to London made a great success in J. B. PRIESTLEY's *I Have Been Here Before* (1937; N.Y., 1938). After serving in the R.A.F. from 1940 to 1942 he was in a revival of BOUCICAULT's *The Streets of London* in 1943, but was absent from the theatre from 1943 until 1947, when he returned to play Dan Hillboy in SAROYAN's *The Beautiful People*. He toured in the U.S. in the same year as Edward Moulton-Barrett in Besier's *The Barretts of Wimpole Street*, a part he had played in the London revival of 1935, but was not seen on stage again until 1953, when he reappeared in London as the Captain in STRINDBERG's *The Father* and Korrianke in Zuckmayer's *The Devil's General*, and scored a personal success as the grandfather in Edmund Morris's *The Wooden Dish* in 1954. He appeared at the Royal Court in ARDEN's *Live Like Pigs* (1958) and O'CASEY's *Cock-a-Doodle-Dandy* (1959), and in 1962 at the Old Vic gave an effective and imaginative performance as the Button-Moulder in IBSEN's *Peer Gynt*. His well known intemperance, which hampered his career, in no way affected the power of his acting.

LAXNESS, HALLDÓR KILJAN (1902–), Icelandic novelist who was awarded the NOBEL PRIZE for Literature in 1955. His plays include an adaptation of his trilogy of novels *Islandsklukken* (*The Bell of Iceland*) for the opening of the Icelandic National Theatre in Reykjavik in 1950. He turned from historical to contemporary matters in *Silfurtunglið* (*The Silver Moon*, 1954), protesting against the establishment of American air bases in Iceland; this was followed by *Strompleikurinn* (*The Chimney Play*, 1961) and *Prjónastofan Sólin* (*The Sun Knitting Works*, 1962), experimental attempts to fuse allegory, realism, satire, and even farce. *Dufnaveizlan* (*The Pigeon Banquet*, 1966) is perhaps his best play.

LAYE, EVELYN (1900–), English actress, who made her first appearance on the stage in 1915 and from 1923 to 1927 was at DALY'S THEATRE in a series of musical comedies, creating the title role in *Madame Pompadour* and being seen in revivals of *The Dollar Princess* and *Lilac Time*. She was first seen in New York in 1929 as Sari Linden in COWARD's *Bitter Sweet*, a part she also played in London a year later in succession to Peggy Wood. She was later in VARIETY and PANTOMIME, playing PRINCIPAL BOY in *The Sleeping Beauty* (1938) and *Cinderella* (1943 and 1948), and after the Second World War toured extensively in a number of plays with her second husband Frank Lawton (1904–69) (she had previously been married to the actor Sonnie Hale), with whom she appeared in London in *Silver Wedding* (1957), a comedy by M. C. Hutton. In 1959 she gave an excellent performance as Lady Fitzadam in Anthony Kimmins's *The Amorous Prawn* and in 1969 appeared in a new musical, *Phil the Fluter*. She was in the London run of Marriott and Foot's comedy *No Sex Please—We're British* from 1971 to 1973 and in 1979 toured in a revival of SONDHEIM's musical *A Little Night Music*.

LAZZO (pl. *lazzi*), a slight piece of byplay indulged in by the comic servants of the COMMEDIA DELL'ARTE, consisting of a small, unexpected ornamentation by word or gesture superimposed on the main plot. There is no satisfactory English translation of the word, which can be rendered as 'antic', 'trick', 'gambol', 'pun', 'patter', according to the context. The longer comic episode involving a practical joke or some horseplay was known as the BURLA.

LEAL, JOSÉ DA SILVA MENDES (1818–86), Portuguese dramatist and diplomat, who largely followed French models, his early plays being historical melodramas such as *Os Dois Renegados* (1839) in the style of DUMAS *père*. He later turned to social dramas such as *Homem de Ouro* (*The Golden Man*, 1855) and *Pobreza Envergonhada* (*Embarassed Poverty*, 1858), after DUMAS *fils*. His rhetorical excesses were admired by GARRETT among others.

LEAP, see TRICKWORK.

LEAVITT, MICHAEL BENNETT, see BURLESQUE, AMERICAN.

LEBLANC, GEORGETTE, see MAETERLINCK, MAURICE.

LECOUVREUR, ADRIENNE (1692–1730), French actress, who made her first appearance at the COMÉDIE-FRANÇAISE in 1717 in the title role of CRÉBILLON's *Électre*. Her immediate popularity aroused much jealousy among her fellow actresses, but she continued to triumph with the public. She was better in tragedy than comedy and, disliking the declamatory style which Mlle DUCLOS and Mlle DESMARES had learned from the great Mlle CHAMPMESLÉ, she succeeded, in the teeth of their opposition, in introducing a much simpler and more natural form of delivery. Her reign was a brief one, and she died suddenly after only 13 years. As an actress she was refused Christian burial, and was interred secretly at night in a marshy corner of the Rue de Bourgogne. Anne OLDFIELD, who died in the same year, was buried in Westminster Abbey; VOLTAIRE, in some of whose plays she had appeared, contrasted bitterly the treatment accorded her with the respect shown to the English actress. In 1849 SCRIBE and Legouvé wrote a play on the life of Adrienne Lecouvreur which, though very inaccurate, provided an excellent part for RACHEL and later BERNHARDT.

LEDERER, GEORGE W., see REVUE.

LEDERER THEATRE COMPANY, Providence, see TRINITY SQUARE REPERTORY COMPANY.

LEE, CANADA [Leonard Lionel Cornelius Canegata] (1907–52), American Negro actor, who received his stage training with the Negro Unit of the FEDERAL THEATRE PROJECT, making his first appearance in its all-Negro production of *Macbeth* in 1936. In 1939 he played with Ethel Waters in Dorothy Heyward's *Mamba's Daughters*, and he was also in Paul GREEN's outstanding tragedy of Negro life, *Native Son* (1941), based on the novel by Richard Wright. Among other productions in which he appeared were the musical *South Pacific* (1943), Philip Yordan's *Anna Lucasta* (1944), first produced by the AMERICAN NEGRO THEATRE, Maxine Wood's *On Whitman Avenue* (1946), and Dorothy Heyward's *Set My People Free* (1948). In 1945 he appeared as Caliban in *The Tempest*, directed by Margaret WEBSTER, and in 1946 played Bosola, in a white make-up, in WEBSTER's *The Duchess of Malfi*, directed by George Rylands.

LEE, GYPSY ROSE, see BURLESQUE, AMERICAN.

LEE, NATHANIEL (*c*. 1653–92), English dramatist, author of a number of tragedies on subjects taken from ancient history, of which the best and most successful was *The Rival Queens; or, the Death of Alexander the Great* (1677), dealing with the jealousy of Alexander's wives Roxana and Satira. It owed much of its initial success to the acting of BETTERTON and Mrs BARRY, and held the stage for over 100 years, being in the repertory of John Philip KEMBLE, Edmund KEAN, and Mrs SIDDONS. Lee, who collaborated with DRYDEN in *The Duke of Guise* (1682), was one of the most popular dramatists of his day and his plays were frequently revived and reprinted. Their ranting verse and plots which left the stage encumbered with corpses or lunatics betrayed, however, a streak of insanity which later led to his being confined in Bedlam where he died.

LEEDS, Yorkshire, which in 1876 was left without a theatre after the destruction by fire within a year of both the old Theatre Royal and Joseph Hobson's Amphitheatre, now has the Grand, which opened in 1878 under Wilson BARRETT, who remained its lessee until 1895, and Leeds Playhouse, which opened in 1970 on a site provided by the university. The Grand, which was built as a touring theatre, seats 1,554 and the large orchestra pit accommodates up to 60 musicians. The theatre presents a varied programmes of plays, ballet, opera, musicals, pantomime, and concerts. Leeds Playhouse, the regional repertory theatre for West Yorkshire, opened after six years of vigorous campaigning. It seats 750 in a steeply raked semicircular auditorium, no seat being more than 75 ft from the stage. Its repertory includes classical plays—a production of *The Tempest* with Paul SCOFIELD was later seen in London—musicals, modern plays, including numerous premières such as WESKER's *The Wedding Feast*, and the works of such northern writers as Alan AYCKBOURN, J. B. PRIESTLEY, and Alan Plater. Several of Tom STOPPARD's plays have also been given successful productions, including the regional première of *Every Good Boy Deserves Favour*, as have plays by Howard BRENTON and Howard BARKER. There is a Theatre-in-Education team which works exclusively with and for children, and the building is used for films, concerts, and art exhibitions. Leeds is also fortunate in possessing Britain's only surviving full-time MUSIC-HALL, the City Palace of Varieties, which developed from modest beginnings in 1865 as a room attached to the White Swan public-house. It now seats 713 and is used for the television programme 'The Good Old Days', which re-creates Edwardian music-hall.

LEE SUGG, CATHARINE, see HACKETT, JAMES HENRY.

LE GALLIENNE, EVA (1899–), American actress and director, daughter of the poet Richard Le Gallienne. Born in London, she appeared there in several small parts before going to New York in 1915, where her great success as Julie in the first American production of MOLNÁR's *Liliom* (1921) was followed by five years of starring roles on Broadway. In 1926 she founded the Civic Repertory Company to present important foreign plays at low admission prices, opening on 26 Oct. with *Saturday Night*, a translation of BENAVENTE's *La noche del sábado* (1903). Alla NAZIMOVA appeared with the company, her roles including Ranevskaya in CHEKHOV's *The Cherry Orchard*. Before the enterprise collapsed in the depression of 1935, Le Gallienne had mounted 37 productions, including plays by Shakespeare, ROSTAND, TOLSTOY, MOLIÈRE, GOLDONI, GIRAUDOUX, BERNARD, WIED, and especially IBSEN: she also toured the United States in 1934–5 in her own versions of *A Doll's House, Hedda Gabler*, and *The Master Builder*. With Cheryl CRAWFORD and Margaret WEBSTER she founded the American Repertory Theatre, which was active from 1946 to 1948, playing Queen Katharine in *Henry VIII*, the initial production, and Ella Rentheim in Ibsen's *John Gabriel Borkman*. From 1961 to 1966 she directed and acted with the National Repertory Theatre, touring productions of classic plays coast to coast and receiving great critical acclaim for her performances in Maxwell ANDERSON's *Elizabeth the Queen* and EURIPIDES' *Trojan Women*. In New York in 1968 she appeared in IONESCO's *Exit the King* and directed *The Cherry Orchard*, and two years later she was in the AMERICAN SHAKESPEARE THEATRE's production of *All's Well That Ends Well*. In 1975 she directed *A Doll's House* for the SEATTLE REPERTORY THEATRE, returning in 1977 to appear in a new play, *The Dream Watcher* by Barbara Wersba, after starring on Broadway in a revival of George KAUFMAN and Edna Ferber's *The Royal Family* in 1976.

Her numerous published works include two volumes of memoirs, *At 33* (1934) and *With a Quiet Heart* (1953); translations of Ibsen and Hans Christian Andersen; and *The Mystic in the Theatre* (1973), a study of DUSE with whom Le Gallienne has often been compared.

LEGISLATION, Theatre, see COPYRIGHT IN DRAMATIC WORKS, DRAMATIC CENSORSHIP, FIRES IN THEATRES, and LICENSING ACTS.

LEGITIMATE DRAMA—sometimes abbreviated to 'legit'—term which arose in the 18th century during the struggle of the Patent Theatres COVENT GARDEN and DRURY LANE against the upstart and illegitimate playhouses which were springing up all over London. It covered in general those five-act plays (including Shakespeare's) which depended entirely on acting, with little or no singing, dancing, and spectacle. In the 19th century the term was widely used by actors of the old school as a defence against the encroachments of FARCE, MUSICAL COMEDY, and REVUE.

LEGRAND, MARC-ANTOINE (1673–1728), French actor and dramatist, who in 1702 joined the COMÉDIE-FRANÇAISE to play comic and rustic parts. He occasionally insisted on appearing in tragedy, for which he was quite unsuited, being short and ugly, but his good humour and wit made him popular with everyone. Of his many plays, based mainly on contemporary events, the most successful was *Cartouche*, which dealt with the career of a notorious footpad, Louis-Dominique Bourguignon. Hurriedly written after the man's arrest, it was first performed a week later, on 21 Sept. 1721, and repeated 13 times, the last on the eve of his execution. The play later provided a fine part for FRÉDÉRICK. The success of Legrand's works, in which he sometimes collaborated with BIANCOLELLI, helped to bring back to the Comédie-Française the audiences tempted away by the unlicensed theatres of the FAIRS. Legrand was the teacher of Adrienne LECOUVREUR and his son and daughter were both members of the Comédie-Française.

LEGS, see BORDER.

LEG-SHOW, slang term for a spectacular musical play, largely designed to display chorus-girls in a series of scanty costumes and energetic dances.

LEHMANN, BEATRIX (1903–79), English actress of great power and intensity, who made her first appearance in 1924, and in 1929 attracted attention by her playing of Ella Downey in O'NEILL'S *All God's Chillun Got Wings*. Other notable appearances were as Susie Monican in O'CASEY'S *The Silver Tassie* in the same year, Emily Brontë in Clemence DANE'S *Wild Decembers* (1933), and Stella Kirby in PRIESTLEY'S *Eden End* (1934). In 1936 she consolidated her position as one of London's major actresses when she appeared as Winifred in the OLD VIC production of William ROWLEY'S *The Witch of Edmonton*, and she was seen in a number of important productions, including O'Neill's *Mourning Becomes Electra* (1937) and *Desire Under the Elms* (1940), IBSEN'S *Ghosts* (1943), and Thomas Job's *Uncle Harry* (1944). In 1947 she was at the SHAKESPEARE MEMORIAL THEATRE, where she gave one of her finest performances as Isabella in *Measure for Measure*, also playing Viola in *Twelfth Night*. Returning to contemporary plays, she was seen in Ted Willis's *No Trees in the Street* (1948), Lesley Storm's *The Day's Mischief* (1951), USTINOV'S *No Sign of the Dove* (1953), and ANOUILH'S *The Waltz of the Toreadors* (1956). She was in Harold PINTER'S first play *The Birthday Party*, played Lady Macbeth at the Old Vic (both 1958), and gave a remarkable performance as the 100-year-old Miss Bordereau in Michael REDGRAVE'S *The Aspern Papers* (1959), based on a story by Henry JAMES; but she was thereafter seen less frequently and mostly in supporting roles. She appeared in the revival of SHERWOOD'S *Reunion in Vienna* at the CHICHESTER FESTIVAL THEATRE in 1971 and in London in 1972, and joined the ROYAL SHAKESPEARE COMPANY to play in John ARDEN and Margaretta D'Arcy's *The Island of the Mighty* (also 1972), the Nurse in *Romeo and Juliet*, and the Duchess of York in *Richard II* (both 1973). Her last appearance on the stage was in T. S. ELIOT'S *The Family Reunion* at the ROYAL EXCHANGE THEATRE, Manchester, in 1979.

LEICESTER, see HAYMARKET THEATRE and PHOENIX ARTS CENTRE.

LEICESTER'S MEN, the earliest organized company of Elizabethan actors, founded in 1559 and noted as playing at Court a year later. In the company in 1572 was James BURBAGE, who built for it the first THEATRE in London, and William KEMPE, who later joined the CHAMBERLAIN'S MEN. Leicester's Men remained in high favour with the Queen from about 1570 until the formation in 1583 of QUEEN ELIZABETH'S MEN, to which several of them were drafted. They formed part of the Earl of Leicester's household and continued to act under his patronage until his death in 1588 when they were amalgamated with the provincial company belonging to the household of Lord Strange, later the 6th Earl of Derby.

LEIGH, VIVIEN [Vivian Mary Hartley] (1913–67), English actress, who first came into prominence as Henriette in *The Mask of Virtue* (1935), an adaptation by Ashley DUKES of Carl STERNHEIM'S *Die Marquise von Arcis*. She was then seen as Jenny Meere in Clemence DANE'S adaptation of Max BEERBOHM'S *The Happy Hypocrite* (1936), and in 1937 played Ophelia to the Hamlet of Laurence OLIVIER at Elsinore and Titania in *A Midsummer Night's Dream* to Robert HELPMANN'S Oberon at the OLD VIC. In 1940 she made her New York début as Juliet to the Romeo of Olivier, whom she married the same year as her second husband. Back in London, she gave an excellent performance as Jennifer Dubedat in SHAW'S *The Doctor's Dilemma* in 1942, and as Sabina in WILDER'S *The Skin of Our Teeth* (1945). It was however as Blanche du Bois in Tennessee WILLIAMS'S *A Streetcar Named Desire* (1949) that she gave proof of greater power on the stage than she had hitherto shown. In 1951 she appeared with Olivier in Shaw's *Caesar and Cleopatra* and Shakespeare's *Antony and Cleopatra* in London and New York, the general of opinion being that she was more successful in the former. She again appeared with Olivier in RATTIGAN'S *The Sleeping Prince* (1953), and two years later was at the SHAKESPEARE MEMORIAL THEATRE, where she played Lady Macbeth, Viola in *Twelfth Night*, and Lavinia to the Titus Andronicus of Olivier, both their performances in this little known play being accounted outstanding. In 1957 the production was taken on tour in

Europe and then seen in London. She had further successes in COWARD's *South Sea Bubble* (1956) and *Look After Lulu* (1959) and in GIRAUDOUX's *Duel of Angels* (1958; N. Y., 1960) and in 1961 undertook an extensive tour with the Old Vic which included a new version of the younger DUMAS's *The Lady of the Camellias*. In 1963 she was in New York in a musical version of SHERWOOD's *Tovarich*; her last appearance was in New York in 1966 in CHEKHOV's *Ivanov*, with John GIELGUD in the title role. She also had an outstanding career in films.

LEIGHTON, MARGARET (1922–76), stylish and elegant English actress, who first appeared on the stage at the BIRMINGHAM REPERTORY THEATRE in 1938 and later joined the OLD VIC company, making her début in London in 1944 as the Troll King's Daughter in IBSEN's *Peer Gynt*. In 1950 she emerged as a leading actress with her Celia Coplestone in T. S. ELIOT's *The Cocktail Party*, which was followed by Masha in an all-star production of CHEKHOV's *Three Sisters* (1951) and a season at the SHAKESPEARE MEMORIAL THEATRE (1952), where she played Lady Macbeth, Ariel in *The Tempest*, and Rosalind in *As You Like It*. In a revival of SHAW's *The Apple Cart* in 1953 she played Orinthia to the King Magnus of Noël COWARD, and in the same year was seen as Lucasta Angel in Eliot's *The Confidential Clerk*. She was much admired in RATTIGAN's *Separate Tables* (1954), in which she starred also in New York in 1956, and in the same author's *Variations on a Theme* (1958). After her Beatrice in *Much Ado About Nothing* to the Benedick of John GIELGUD in New York in 1959, she returned to London as Elaine Lee in John MORTIMER's *The Wrong Side of the Park* (1960) and in a revival of Ibsen's *The Lady from the Sea* (1961), and back in New York gave a fine performance as Hannah Jelkes in Tennessee WILLIAMS's *The Night of the Iguana* (also 1961). She remained in the United States for some years, appearing in BILLETDOUX's *Chin-Chin* (1962), Enid BAGNOLD's *The Chinese Prime Minister* (1964), and a revival of Lillian HELLMAN's *The Little Foxes* (1967). Returning to England, she was seen at the CHICHESTER FESTIVAL THEATRE in 1969 as Cleopatra in *Antony and Cleopatra* and again in 1971 as Mrs Malaprop in SHERIDAN's *The Rivals* and Elena in SHERWOOD's *Reunion in Vienna*, repeating the latter part in London a year later. She made her last appearance in Julian Mitchell's dramatization of Ivy Compton-Burnett's novel *A Family and a Fortune* (1975).

LEIPZIG, the largest town in Saxony, now in the German Democratic Republic, which attracted much theatrical activity by its annual FAIRS and in 1731 saw the beginnings of serious German drama with the production by Carolina NEUBER's company of GOTTSCHED's *Der sterbende Cato*, the result of their collaboration aimed at the reform of the lurid, bawdy histrionics of the strolling players. Its success was sufficient to put the HAUPT- UND STAATSAKTIONEN and HANSWURST on the defensive, and to cause the split between literary drama, seen as a form of instruction, and popular theatre, seen purely as entertainment, which was later to bedevil much German writing for the stage. In the 19th century the Stadttheater, established by Theodor Küstner in 1817, was used for productions by him of new literary plays, while elsewhere the productions of the romantic operas of Weber, Marschner, and Lortzing drew large audiences. During the period of the Weimar Republic (1918–33) the rivalry between Kronacher at the Altes Theater and Veihweg at the Schauspielhaus produced some good work; among the plays that were given their first performances in Leipzig were JAHNN's *Der Krönung Richards III* (1922), BRECHT's *Baal* (1923), and TOLLER's *Der deutsche Hinkermann* (1923), all works of black EXPRESSIONISM. Leipzig now has a famous political cabaret Die Pfeffermühle (The Pepper-Mill).

LEISEWITZ, JOHANN ANTON (1752–1806), German dramatist, author of one of the best of the STURM UND DRANG plays, *Julius von Tarent*, produced by SCHRÖDER in Hamburg in 1776 after it had taken second place to KLINGER's *Die Zwillinge* in a drama competition. He was also the author of two short prose plays, *Die Pfandung* (*The Sequestration*), an essay in social realism, and *Der Besuch um Mitternacht* (*The Midnight Visit*), published in the same year. He married the daughter of the travelling manager Abel SEYLER.

LEIVICK, HALPER [Leivick Halpern] (1888–1962), Jewish dramatist, considered by many critics the outstanding Yiddish writer of his time in the United States. His most important play, *The Golem*, based on the ancient Jewish legend of an artificial human being created by a wonder-working medieval rabbi, was produced by HABIMAH in Moscow in Hebrew in 1925, and was later acted in Yiddish, Polish, and English. He also wrote a number of social dramas—*Rags* (1921), *Shop* (1926), and *Chains* (1930)—and in 1945 produced *The Miracle of the Warsaw Ghetto*, which dealt with the struggle of the Polish Jews against the Nazis.

LEKAIN [Caïn], HENRI-LOUIS (1729–78). French actor, who in 1748 organized some amateur productions, playing the leading parts himself with much success. In one of these he was seen by VOLTAIRE, who built a theatre for him, playing there himself with his two nieces. The little company soon achieved an enviable reputation and society clamoured for admission. Lekain stayed six months with Voltaire, and always said he owed him everything. Before Voltaire left Paris for Berlin in 1750, he had the pleasure of seeing Lekain at the COMÉDIE-FRANÇAISE, playing Titus in a revival of his benefactor's *Brutus*. Though he made a good impression, through intrigue and jealousy he was not immediately accepted for the company, preference being given to RICHELIEU's protégé BELLECOUR and to the elegant GRANDVAL. The public, however, wanted Lekain, in spite of his low stature, bow legs, and harsh voice. He knew how to overcome his faults, and on stage they were forgotten. He also worked hard at his roles, as is evident from the *Mémoires et reflexions sur l'art dramatique* by Mlle CLAIRON, his leading lady for many years. He finally wore himself out; when he fell ill the bulletins of his progress were as eagerly awaited as those of a film star today and when he reappeared in 1770 it was remarked by many critics that he acted better than ever. He then went to Berlin on the invitation of Frederick II, and Voltaire, whose memories of French acting went back to BARON, called him the only truly tragic actor. He was also highly praised by Grimm and LA HARPE, and frequently compared to GARRICK (his grandfather was English). He collaborated in many reforms in the theatre, notably the suppression of seats on the stage and the introduction of some trace of historical costume, in which he was nobly supported by Mlle Clairon. Together they did away with hip-pads and panniers, and on 20 Aug. 1755 introduced some touches of *chinoiserie* into Voltaire's *Orphelin de la Chine*. Lekain also insisted on more mobility on stage, doing away with the old tradition of delivering long speeches down stage centre front. After giving a magnificent performance as Vendôme in Voltaire's *Adélaide du Guesclin*, he went out into the chill night air, caught cold, and died just as his benefactor and greatest admirer was returning to Paris after 30 years of exile. The news of his funeral was the first thing Voltaire heard on his arrival.

LEKO, see LIGHTING.

LELIO, see ANDREINI, G. B., and RICCOBONI, LUIGI.

LEMAÎTRE, A.-L.-P., see FRÉDÉRICK.

LEMBCKE, EDVARD, see DENMARK and SHAKESPEARE IN TRANSLATION.

LENAEA, annual festival held in Athens in Jan.–Feb. in honour of the god DIONYSUS. Though briefer and less splendid than the City DIONYSIA, it also included dramatic contests, originally for comedies only, with five comic poets entering one play each. When early in the 5th century comedy was admitted to the City Dionysia, the Lenaea added tragedy to its festivities, two tragic poets being allowed to compete with two plays each.

LENGYEL, MELCHIOR, see HUNGARY.

LENKOM THEATRE, Moscow, founded in 1927. Its original members were amateurs who wrote their own topical plays. They then turned professional, and the importation from the MOSCOW ART THEATRE of experienced directors including SIMONOV gave the theatre new impetus. Under Ivan BERSENEV and Serafima BIRMAN it mounted many notable productions. Evacuated during the Second World War, it eventually returned to Moscow and embarked during the 1960s and early 1970s on a policy of staging plays for young people, the repertory including works by ARBUZOV, ROZOV, and BRECHT.

LENO, DAN [George Galvin] (1860–1904), one of the best loved and most famous stars of the English MUSIC-HALL and PANTOMIME. Born in London of itinerant entertainers, he made his first appearance on the stage at the age of four, and for many years played in the provinces in slapstick sketches with 'the Leno family'—the stage name of his stepfather, whose real name was Wilde. He later did a double dancing act with his elder brother Jack and then with his young uncle Johnny Danvers (1860–1939), later a Mohawk Minstrel and a player in pantomime. Working mainly in the north of England, he specialized in clog-dancing and in 1880, at Leeds, won the World Championship. Returning to London in 1885 with his young wife Lydia Reynolds, a 'comedy vocalist' whom he married in 1883 and by whom he had six children, he found that his dancing was of no account, whereas the comic songs and patter with which he enlivened his act were rapturously received. He therefore abandoned the clogs to concentrate on such comic songs as 'Going to Buy Milk for the Twins' and 'When Rafferty Raffled His Watch' and was soon playing at several 'halls' every night. He first appeared in pantomime at the SURREY in 1886, as Jack's mother in *Jack and the Beanstalk*, and two years later achieved his ambition of playing at DRURY LANE when he was offered the part of the Baroness in *The Babes in the Wood*. He was then seen in every Drury Lane pantomime up to 1903, when he made his last appearance as Queen Spritely in *Humpty-Dumpty*, partnered by Harry Randall (1860–1932), who had been in the pantomimes at the GRAND THEATRE, Islington, for ten years, and was to prove a worthy successor to Dan Leno at Drury Lane. Leno had previously been partnered by another music-hall comedian, Herbert Edward Campbell [Story] (1844–1904), whose vast bulk proved the perfect foil to Leno's diminutive quicksilver figure. Leno's best parts as a pantomime DAME were usually thought to be Widow Twankey in *Aladdin* (1896), Sister Anne in *Bluebeard* (1901), and the title role in *Mother Goose* (1902). The pantomimes usually ran until the end of March or early April, and during the rest of the year Leno, in common with other pantomime stars, returned to the 'halls'. He created a wide range of music-hall characters, two of the most popular being the Shop-Walker and the Beefeater, though some audiences preferred the Railway Guard. He always remained a man of the people, never seeking to emulate the *lions comiques* and their 'man-about-town' successors, but, with his husky voice and worried little face, carried always the stamp of the poverty and privations from which he had emerged. He usually made his entrance in a rush, stopping suddenly and darting suspicious looks at the audience while starting a song, and then leaving it to indulge in long, rambling monologues, with muttered asides and sudden bursts of step-dancing. He worked alone, without much in the way of props, and built up his characters with brilliant use of mime and rapid gestures. In 1901 he was commanded by Edward VII to Sandringham, which earned him the nickname of 'the King's Jester'. Towards the end of his life he broke down from overwork and died insane. Three of his children went on the 'halls'. Sidney Paul Galvin (1891–1962), who looked very like him, using some of his material and billing himself as Dan Leno junior.

LENOBLE, EUSTACHE (1643–1711), French lawyer, who was imprisoned for forgery, and later became a literary hack. Among his miscellaneous works were some unsuccessful plays, including three produced by the COMÉDIE-ITALIENNE. One of them, *La Fausse Prude*, was taken as a reflection on Mme de MAINTENON and was used by the authorities as an excuse to banish the Italian troupe from Paris in 1697.

LENOIR, CHARLES (*fl.* 1610–37), French actor-manager, who in 1620 was leader of a troupe playing in Lille and from 1622 to 1626 was with a company under the patronage of the Prince of Orange, which appeared intermittently in Paris. The actor MONTDORY was also a member of this troupe and at the end of 1629 he and Lenoir brought a new group of actors to Paris, carrying with them the first play of an unknown dramatist, Pierre CORNEILLE. This was the comedy *Mélite*, in which they appeared in 1630. In 1634, just as the company was about to settle permanently at the Théâtre du MARAIS, Lenoir, his wife (who had appeared in MAIRET's plays), and several other actors were sent by Louis XIII to join the company at the Hôtel de BOURGOGNE. Little more is known of Lenoir, and his name is not mentioned after 1637, when his wife, who outlived him, retired from the stage.

LENOIR, MARIE-THÉRÈSE, see DANCOURT, F.C.

LENORMAND, HENRI-RENÉ (1882–1951), French dramatist, who in reaction against the REALISM made popular by ANTOINE exploited in his work Freud's theory of the unconscious. The first of his 20-odd plays was *Le Temps est un songe* (1919), produced by PITOËFF in Geneva and then in Paris. As *Time is a Dream*, it was seen in New York in 1924 and in London in 1950. Pitoëff also produced *Les Ratés* (1920), seen in New York in 1924 as *The Failures*, and *Le Mangeur de rêves* (1922), which, as *The Eater of Dreams*, was produced at the GATE THEATRE in London in 1929. Lenormand's best play *Simoun*, staged by BATY at the Théâtre Montparnasse in 1920, was seen in London in 1927. All his work is marked by a fatalistic vision of man as the victim of a fate determined in his infancy and by the force of his unconscious. He achieved some of his most interesting effects by the use of short scenes played in a spotlight on a blacked-out stage.

LENSKY, ALEXANDER PAVLOVICH (1847–1908), Russian actor, who in 1876 joined the company at the MALY THEATRE in Moscow, and eventually became its director, in which capacity he was responsible for introducing the plays of IBSEN to Russian audiences. A many-sided man, being both actor and artist, he was also an inspired teacher and trained many actors for the Maly. During the difficult days after the abortive rising of 1905 he supported YERMOLOVA in her efforts to revive the classical repertory, but was enlightened enough, under the influence of the MOSCOW ART THEATRE, to further many reforms in methods of production and rehearsal and in the replacement of old-fashioned declamation by a more natural style of acting.

LENS SPOT, see LIGHTING.

LENYA, LOTTE, see BRECHT and MUSICAL COMEDY.

LENZ, JAKOB MICHAEL REINHOLD (1751–92), German dramatist, who became a friend of the young GOETHE, and in his *Anmerkungen übers Theater* (1774) set out the principles of the STURM UND DRANG movement: all rules are rejected; the purpose of drama is to reveal 'naked nature' as Shakespeare had done; and it should deal above all with strong ruthless characters. His first play *Der Hofmeister* (also 1774), a partly autobiographical treatment of the story of Abelard and Héloïse, is marred by lack of unity of action and a grotesque mixture of genres, but testifies to Lenz's power to observe accurately and write convincingly, a form of realism which marks also his best work *Die Soldaten*, which, though written in 1774–5, was not produced until 1863, when it was seen at the VIENNA Burgtheater as *Soldatenliebchen*. Its theme is the seduction and disgrace of a middle-class girl at the hands of an aristocratic officer, and it contains Lenz's suggestion (put forward seriously in a letter to Goethe's patron, Duke Karl August of WEIMAR) that such situations should be avoided in future by the recruiting of a contingent of volunteer harlots attached to the army. The play had a great influence on BÜCHNER's *Woyzeck*, and was used for the libretto of an opera by Zimmermann.

LEONARD, HUGH [John Keyes Byrne] (1926–), Irish dramatist and critic, whose first play, *The Italian Road* (1944), was produced by an amateur company in Dublin. His second, *The Big Birthday* (1956), was seen professionally at the ABBEY THEATRE, as was *A Leap in the Dark* (1957). Two further plays, *Madigan's Lock* (1958) and *Walk on the Water* (1960), were first seen at the Dublin Globe Theatre. For the DUBLIN THEATRE FESTIVAL, of which he was director in 1978, Leonard wrote *The Passion of Peter Ginty* (1961), based on IBSEN's *Peer Gynt*, and two plays based on the works of James JOYCE, *Stephen D.* (1962; London, 1963; N.Y., 1967) and *Dublin One* (1963). Later plays include *The Poker Session* (also 1963; London, 1964; N.Y., 1967), in which a young man is released from a mental hospital and tries to discover why he was sent there; *The Family Way* (1964); *When the Saints Go Cycling In* (1965); and *The Au Pair Man* (1968; London, 1969; N.Y., 1973), a two-character allegory about Britain and Ireland. After *The Patrick Pearse Motel* (Dublin and London, 1971) and *Thieves* (1972) came *Da* (1973; London and N.Y., 1978), an autobiographical play in which a middle-aged man remembers his father on the day of the latter's death; *Summer* (1974: N.Y., 1980); *A Suburb of Babylon*, three one-act plays previously called *Irishmen* (1975); and *Time Was* (1976). *A Life* (1979; London, with Cyril CUSACK, and N.Y., 1980) is a companion piece to *Da* in which one of its characters with six months to live assesses his life.

Leonard is a prolific writer whose technical facility is sometimes given free rein at the expense of depth of characterization; he is by his own admission obsessed with the theme of betrayal, frequently presented through clever satires of the Irish *nouveaux riches* and a re-examination of what he has called the 'four great F's' of Ireland: Faith, Fatherland, Family, and Friendship.

LEONIDOV, LEONID MIRONOVICH (1873–1941), Russian actor and director, who joined the KORSH THEATRE in Moscow in 1901 and in 1903 became a member of the company at the MOSCOW ART THEATRE. His most brilliant performance as a tragic actor was given as Dmitri Karamazov in a dramatization of DOSTOEVSKY's *The Brothers Karamazov*. Among his other outstanding roles were IBSEN's Peer Gynt, Lopakhin in CHEKHOV's *The Cherry Orchard* and Solyony in his *Three Sisters*, Professor Borodin in AFINOGENOV's *Fear*, and Plushkin in the dramatization of GOGOL's *Dead Souls*. At the time of his death he was engaged with NEMIROVICH-DANCHENKO on the production of a new play about Lenin, POGODIN's *Kremlin Chimes*, which was seen at the Moscow Art Theatre in 1942.

LEONOV, LEONID MAXIMOVICH (1899–), Soviet dramatist, whose *Untilovsk* (1928), one of the first Soviet plays to be produced by the MOSCOW ART THEATRE, was not entirely successful; its successor, *Skutarevski* (1934), which dealt with the problems of an elderly scientist torn between his work and his family, and between the old and new régimes, was however warmly received when it was produced at the MALY THEATRE. Among Leonov's later plays was *The Orchards of the Polovtsi* (1938), which as *The Apple Orchards* was produced at the BRISTOL OLD VIC in 1948. *The Wolf* (1939), which dealt with the impact of the Soviet régime on personal problems, was well received, but the play which set the seal on Leonov's growing reputation was *Invasion* (1942), which recounts with great force and pathos the story of a Soviet village under Nazi rule. In 1957 *Gardener in the Shade* was produced at the MAYAKOVSKY THEATRE by OKHLOPKOV, and a revised version of an earlier play, *Golden Chariot*, was seen at the Moscow Art Theatre. *The Snowstorm*, first written in 1939, but officially banned, was revised and finally published in 1963.

LEOPOLDSTÄDTER THEATER, see MARINELLI and VIENNA.

LÉOTARD, JULES, see MUSIC-HALL.

LERMONTOV, MIKHAIL YUREVICH (1814–41), Russian lyric poet and the author of three romantic plays, most of whose work fell foul of the censor. His first play *The Spaniards*, written in 1830, dealt ostensibly with the Spanish Inquisition but was in reality aimed at the despotism of the Tsar; it was banned, not to be performed in Russia until after the Revolution. His next, written two years later, was given the German title *Menschen und Leidenschafter* to indicate its kinship with the STURM UND DRANG movement and particularly with the works of SCHILLER. Based on a family conflict which recalls Lermontov's own unhappy home life, it is again an indictment of contemporary society. In a later rewritten version Lermontov gave it a wider application, with less stress on the family and more on the struggles of the rebellious hero confronting a hostile world. His most important play and the only one by which he is now remembered is *Masquerade*, written in 1835. Showing clearly the influence of Shakespeare—the theme is very similar to that of *Othello*—and BYRON, it depicts the tragedy of a man who murders the wife he loves because he suspects her of infidelity. In deference to the censor the play was later given a happy ending, but it was not produced until 1852 and then only in a mutilated text. The full version was first seen in 1864, and then not again until MEYERHOLD produced it in 1917 at the Alexandrinsky, where it was the last play to be performed before the October Revolution. He later produced it at his own theatre, and again at the Alexandrinsky in 1938. It now forms part of the permanent repertory of the Soviet theatre. In a translation by Robert MacDonald it was seen at the CITIZENS' THEATRE in Glasgow in 1976.

LERNER, JOSEPH YEHUDA, see JEWISH DRAMA.

LERNET-HOLENIA, ALEXANDER (1897–), Austrian poet and dramatist, author of two historical plays in the tradition of the old HAUPT- UND STAATSAKTION, and of a number of comedies of which the first, *Österreichische Komödie* (1926), a cynical assessment of the breakdown of moral standards after the First World War, was followed by the SCHNITZLER-like sexual roundabout of *Ollapotrida* (also 1926) and the equally theatrical *Die Frau des Potiphar* (1934), *Spanische Komödie* (1948), and *Finanzamt* (1955).

LESAGE, ALAIN RENÉ (1668–1747), French novelist and dramatist, orphaned when young and left penniless. Little is known of his early years; by 1694 he was married and established in Paris. His brilliant literary career began with translations of Lope de VEGA and ROJAS, and it was under the influence of Spanish literature that he wrote the two novels which constitute his main claim to fame: *Le Diable boiteux* (1707) and *Gil Blas* (1715). Though pre-eminently a novelist, Lesage is by no means negligible as a dramatist. His first success in the theatre was a one-act comedy, *Crispin rival de son maître* (1707). Two years later a one-act play, *Les Étrennes*, which the actors had refused, was remodelled as *Turcaret* (1710), one of the outstanding comedies in the history of French drama. It is a satire on the vulgar parvenu, the tax-farmer who has enriched himself by exploiting the poor, and it reflects that bitterness against taxation which came to a head under Louis XVI. It met with considerable opposition, and those whom it satirized tried to bribe the author and the actors to suppress it; but it was eventually put on, and was not only successful at the time but has remained in the repertory. It was produced, for instance, in 1960 by Jean VILAR for the THÉÂTRE NATIONAL POPULAIRE, with a musical accompaniment by Duke Ellington. A quarrel between Lesage and the actors of the COMÉDIE-FRANÇAISE, for some unknown reason, led him to break off his association with the official theatre, and for many years he wrote for the theatres of the Paris FAIRS, alone or in collaboration with PIRON, FUZELIER, and others. Lesage's life, though laborious and unremunerative—he was too proud to accept patronage—was mainly a happy one, as he had a devoted wife and four children, and enjoyed the esteem of his fellow-writers. His greatest sorrow was the decision of two of his sons, René-André (1695–1743) and François-Antoine (1700–?), to become actors. The elder, known on the stage as Montménil, became reconciled with his father when he saw him play Turcaret, one of his best roles; the second, known as Pitténec, was not such a good actor, and after his father's death retired to live with his elder brother (Lesage's third son) who was a canon of Boulogne cathedral.

LESCARBOT, MARC (*c.* 1570–1642), French lawyer, author of the first play to be performed in CANADA, which he directed himself. This was a marine masque, *Le Théâtre de Neptune en la Nouvelle France*, seen on 14 Nov. 1606 at Port Royal, Acadia (today Annapolis Royal, Nova Scotia). Lescarbot had been persuaded by his friend Jean de Poutrincourt to accompany him to New France in July 1606. He served there as an instructor in religious matters, and was left in charge of the settlement when De Poutrincourt went with Samuel de Champlain on an exploratory trip to the south. He prepared the masque to welcome them home. Subsequently he quarrelled with Champlain about the future of the colony, advocating agriculture rather than mining and trapping, and returned to France to resume his legal career. In 1609 he published a collection of poems and writings, *Les Muses de la Nouvelle France*, that contained the script of *Le Théâtre de Neptune*, which in 1611 was translated into English by P. Errondelle and published in London.

L'ESPY [François Bedeau] (?–1663), French actor, also known as Gorgibus. The elder brother of JODELET, he is first heard of in a provincial company at Angers in 1603. In 1634 he joined the company of LENOIR and MONTDORY, only to be transferred just before they opened the MARAIS to the Hôtel de BOURGOGNE, on the orders of Louis XIII. He returned to the Marais in 1641, and in 1659 joined the company of MOLIÈRE, with whom he remained until his death. He acted as director of works for the renovation of the PALAIS-ROYAL stage when Molière's company moved there at the end of 1660. Among the parts in Molière's plays which he is known to have played were Gorgibus in *Les Précieuses ridicules* and *Sganarelle*—whence his nickname—and possibly Ariste in *L'École des maris*.

LESSING, GOTTHOLD EPHRAIM (1729–81), German playwright and dramatic critic, the opponent of the reforms of GOTTSCHED, which he attacked in his *Briefe, die neuste Litteratur betreffend* (1759). He later expanded his ideas on aesthetics in *Laokoon* (1766), of which only Part I was completed, and in 1767–8, as official critic to ACKERMANN's short-lived National Theatre at HAMBURG, published a series of papers entitled the *Hamburgische Dramaturgie*. In these he again sought to replace the convention of French classical drama by a freer approach which paid lip-service to Shakespeare, whom he had not, however, fully understood, though he admired his supreme craftsmanship, his truth to nature in character-drawing, and his unerring sense of the theatre. Lessing was more at home with such writers as DRYDEN, whose *Essay of Dramatick Poesie* he translated into German, and Richardson, whose influence is clearly discernible in his first major work as a dramatist, *Miss Sara Sampson* (1755), in which a young girl, betrayed by her lover, is poisoned by his mistress, whereupon he commits suicide. Lessing as a young man studied the theatre at first hand behind the scenes with the company run by Carolina NEUBER, who produced his early plays. These were light comedies written in the prevailing French fashion, but he was also the author of an admirable prose comedy, *Minna von Barnhelm* (1767), and of a second tragedy, *Emilia Galotti* (1772), in which a young girl, abducted by a licentious prince, prefers death at the hand of her own father to dishonour. Towards the end of his life Lessing engaged in a prolonged struggle against narrow-minded orthodoxy, which finds expression in his final work, *Nathan der Weise*, a noble plea for religious tolerance, written in blank verse. First produced two years after his death, its merits were not recognized until GOETHE revived it at WEIMAR in 1801. Translated into many languages, it was given its first professional production in English in 1967 at the MERMAID THEATRE.

LETOURNEUR, PIERRE (1736–88), French man of letters, who made a great contribution to the discovery of Shakespeare by the French with his 20-volume prose translation of all the plays, published between 1776 and 1783. A faithful but staid version, it was nevertheless a good deal more complete than that of LAPLACE, and caused proportionately more resentment on the part of VOLTAIRE, provoking from him a particularly virulent anti-Shakespearian outburst.

LEVEY, BARNETT (1798–1837), London-born merchant, who arrived in Australia in 1821 and in 1829 began to present at his Royal Hotel in Sydney musical evenings consisting of songs and sketches in the style of an 'At Home', many of them performed by himself. The authorities refused to grant him a theatrical licence until the arrival of Governor Bourke in 1832: in December of that year professional theatre began in Australia with a production at Levey's Theatre Royal of JERROLD's *Black-Ey'd Susan*. From then until his death Levey held together, often with some trouble, a professional company which included from time to time most of the leading actors then in Australia.

LEVIN, HANOCH (1943–), Israeli playwright and director, whose first play *The Adventures of Solomon Grip* (1969) passed almost unnoticed, although it contained most of the elements which were to make his later works—*Hefetz* (1972) and *Jacoby and Leidentahl—Temporary Title* (1973)—fiercely controversial. In the first Levin attacks the bourgeoisie for its complacency, smugness, self-satisfaction, and cruelty to those who have not succeeded in making a place for themselves in society; in the second he presents marriage as a trap in which the woman uses her sexuality in order to enslave a man and set herself up as the autocratic ruler of the household. Although he writes under the influence of the Theatre of the ABSURD, Levin is very much an Israeli in his imagery, irony, deceptively simple language, and most of all in his humour. *Jacoby and Leidentahl*, which he directed himself, was a sad and deeply pessimistic play which none the less kept the audience constantly laughing.

LEVINSOHN, LUDWIG, see JEWISH DRAMA.

LEVSTIK, FRAN, see YUGOSLAVIA.

LEVY, BENN W., see CUMMINGS, CONSTANCE.

LEWES, GEORGE HENRY (1817–78), philosopher, dramatist, and dramatic critic. He was the grandson of the actor (Charles) Lee Lewes (1740–1803), who created the role of Young Marlow in GOLDSMITH's *She Stoops to Conquer* (1773), and was himself an amateur actor, appearing with a group formed by Charles DICKENS in such parts as Sir Hugh Evans in *The Merry Wives of Windsor* and Old Knowell in JONSON's *Every Man in His Humour*. In 1849 he appeared in Manchester as Shylock in *The Merchant of Venice* with some success, anticipating IRVING's conception of the part as 'a noble nature driven to outlawry by man'. He also played the chief part in his own play *The Noble Heart* when it was produced in Manchester in 1849 before its appearance in London, where he does not appear to have acted professionally. Under the pseudonyms of Slingsby Lawrence and Frank Churchill he wrote several other plays, of which the first, *The Game of Speculation* (1851), was based on BALZAC's *Mercadet*. *A Chain of Events* (1852) and *A Strange History in Nine Chapters* (1853) were written for and in collaboration with the younger Charles MATHEWS and were produced at the LYCEUM, as were all Lewes's other plays except *Buckstone's Adventure with a Polish Princess* (1855), seen at the HAYMARKET, and *Stay at Home* (1856), produced at the OLYMPIC. Lewes, who in 1854 began a lifelong liaison with the novelist George Eliot, was the founder and editor of *The Leader*, for which as 'Vivian' he wrote dramatic criticism from its foundation in 1850 to 1854, sometimes reviewing his own plays. He particularly disliked Charles KEAN, whom he harried whenever possible, but was one of the first to recognize the genius of Henry Irving.

LEW FIELDS THEATRE, New York, see WALLACK'S THEATRE (2).

LEWIS, MABEL TERRY-, see TERRY-LEWIS.

LEWIS, MATTHEW GREGORY (1775–1818), English novelist and dramatist, known as 'Monk' Lewis from the title of his most famous novel,

Ambrosio; or, the Monk (1795). This provided material for a number of sensational plays, which, together with *The Castle Spectre* (1797) and *Timour the Tartar* (1811), found their way into the repertory of the 19th-century TOY THEATRE. Lewis's work, which was deliberately concocted to appeal to the prevailing taste for MELODRAMA and spectacle, was somewhat crude but offered great scope for effective acting and lavish scenery enhanced by incidental music. He was very much influenced by KOTZEBUE, two of whose plays he translated. Most of his work has vanished with the fashion that gave rise to it.

LEWISOHN, ALICE and IRENE, see NEIGHBORHOOD PLAYHOUSE, NEW YORK.

LEYBOURNE, GEORGE [Joe Saunders] (1842–84), famous MUSIC-HALL star, and the first to be hailed as a *lion comique*. He gained his initial experience in the 'free-and-easys' of provincial and East End taverns, and after appearing in one or two minor London halls was engaged for the CANTERBURY, probably about 1865. Tall and handsome, always immaculately dressed as a 'man-about-town', with monocle, whiskers, and a fur-collared coat, he represented the second generation of music-hall entertainers, whose 'heavy swell' was a complete contrast to the working-class characters of their predecessors like Sam COWELL. Leybourne was the friend and rival of The Great VANCE, also a *lion comique*, and when his singing of 'Champagne Charlie' sent up the bar takings at the Canterbury, Vance at the OXFORD replied with a song in praise of Veuve Cliquot. Leybourne riposted with one in praise of burgundy, and between them they went through the whole wine list. Unfortunately Leybourne's nickname of 'Champagne Charlie' was well deserved, and his last years were a constant struggle with disillusionment and ill-health. He made his last appearance at the Queen's in Poplar shortly before his death.

LIBERTY THEATRE, New York, at 234 West 42nd Street, between 7th and 8th Avenues. This opened on 4 Oct. 1904 with *The Rogers Brothers in Paris*, featuring the popular musical comedy team of the title, for whom the theatre was built. Among its early productions was the great horse-racing drama *Wildfire* (1908), written by George BROADHURST and George Hobart for Lillian Russell. The season of 1909–10 included TARKINGTON'S *Springtime* and the famous musical comedy *The Arcadians*, seen in London the previous year. A further successful import from England was *Milestones* (1912), by Arnold BENNETT and Edward KNOBLOCK. The great Negro musical, *Blackbirds of 1928*, had part of its long run at this theatre, and the last legitimate production was given there on 18 Mar. 1933, after which the building became a cinema.

LIBIN, SOLOMON, see JEWISH DRAMA.

LIBRARY THEATRE, Manchester. In 1934 the basement of the Public Library was adapted by the City Council to serve as a lecture hall seating 308, and from 1946 it was used for a varied programme of entertainments including drama. In 1952 the Libraries Committee formed its own resident repertory company which opened with WILDE'S *The Importance of Being Earnest*. In 1971 a second theatre, the Forum, was opened at Wythenshawe eight miles away as part of a leisure complex. It seats 483 and is run in tandem with the Library Theatre under one director: but after an initial period during which productions at the two theatres were interchangeable they now house separate productions, the Library concentrating on modern plays, such as Hugh Whitemore's *Stevie* and Peter NICHOLS'S *Born in the Gardens*, and the Forum on more commercial shows, including musicals such as *Piaf, Godspell,* and *Sweet Charity*. Costumes, props, and sets for both theatres are made at Wythenshawe. This is the only theatrical enterprise in Britain to be operated directly by a local authority.

LIBRETTO, see BOOK.

LICENSING ACTS, see LORD CHAMBERLAIN.

LICENSING OF THEATRES. Stage plays in Britain may not be publicly performed except on premises which are licensed as suitable for the purpose. Under the Theatres Act of 1843 the monopoly of the PATENT THEATRES was abolished, and the powers of the LORD CHAMBERLAIN to license theatres (as distinct from his responsibility for the CENSORSHIP of plays) were limited to London, Brighton, Windsor, and other places of royal residence. Detailed requirements for licensing were evolved, mainly relating to the safety and comfort of the public: satisfactory means of ingress and egress, reasonable means of escape in case of fire, and the suitable positioning of an adequate number of lavatories. These regulations were confirmed by the Theatres Act of 1968, which transferred the licensing authority for the central London area to the Greater London Council which already exercised, *de facto*, the supervisory function that the London County Council (its predecessor) had taken over some 50 years earlier. The two former patent theatres, COVENT GARDEN and DRURY LANE, although nominally exempt from the need to be licensed by the GLC, were made subject to the same requirements as other London theatres. The authority for licensing in Windsor was transferred to the local authority, in effect confirming appropriate local authorities as the responsible bodies. Licensing authorities are forbidden under the Act to stipulate conditions concerning the nature of the plays performed or the manner of their performance.

LIDIA, see ANDREINI, G. B.

LIGHT BATTEN, see BATTEN.

LIGHT CONSOLE, see LIGHTING.

LIGHTING. The open-air theatres of Greece and Rome, the outdoor stages of the medieval LITURGICAL DRAMA, and the unroofed public theatres of Elizabethan and Jacobean England and of Spain were all illuminated by natural light, torches, candles, and lanterns being used from earliest times as stage properties to indicate darkness during daytime performances. In Elizabethan playhouses some artificial lighting was provided on winter afternoons by cressets—knots of tarred rope in small iron cages. The enclosed private theatre, which spread rapidly from Renaissance Italy to France and so to England, was lit by large windows during the hours of daylight and by torches, candle-sconces, and chandeliers after dark. Little or no distinction was at first made between the lighting of the acting area and that of the auditorium, though a more realistic impression of night or gloom than the public playhouses could manage was sometimes achieved by covering the windows or extinguishing some of the lights.

Stage lighting proper, designed to enhance the stage spectacle, became an integral part of the sumptuous Court entertainments of the Italian Renaissance; its very costliness emphasized their conspicuous extravagance. Writings which have survived from the 16th and early 17th centuries give full accounts of such contemporary lighting methods and resources. Sebastiano SERLIO, in a book on theatre perspective published in 1545 and translated into English as *The Second Book of Architecture* in 1611, describes the stage as 'adorned with innumerable lights, great, middle, and small,' with a 'great part of the lights in the middle, hanging over the scene'; the windows of the lath-and-canvas scenery houses also had lights behind them. To make coloured lights, large lamps or torches with reflectors made from barbers' basins were placed behind glass containers filled with coloured liquids; natural phenomena such as the rising moon were also successfully counterfeited.

The Jewish dramatist Leone Ebreo di Somi (1527–92), in his *Dialogues on Stage Affairs*, written *c.* 1565 and published in an English translation in 1937, gives the fullest account of lighting as practised by an expert in the 1560s. He insists on the careful placing of candles and lamps, on the shading or concealing of most of the lights, and on reducing the amount of light in the auditorium. In common with others, he placed reflectors on the backs of the wings, and also behind lights concealed by columns or set in the openings between the wings, and to enhance the brightness of the stage placed only a few lamps in the auditorium, well towards the back of the hall. He also experimented with the contrast between light and darkness to create atmosphere. When producing a tragedy he lit the stage brightly until the first tragic incident—the death of a queen—and then swiftly and dramatically reduced the illumination.

Angelo Ingegneri (*c.* 1550–*c.* 1613), in *Della poesia rappresentativa e del modo di rappresentare le favole sceniche* (1598), stated that lighting was of 'supreme theatrical importance'. He believed in concealing the stage lights, and, to light the actors' faces and cast a glow over the scene, recommended the equivalent of a concert BATTEN 'fitted with many lighted lamps, having tinsel reflectors to direct the beams upon the actors', and with a valance to prevent light from spilling into the audience. He also urged that the auditorium should be completely darkened, but this had to wait for many years: fashionable audiences wished to be seen as well as to see.

Nicola Sabbattini (1574–1654), designer of a theatre in Pesaro, pointed out in his *Pratica di fabricar scene e machine ne'teatri* (1638) that side-lighting gives more brightness and contrast than lighting from in front or behind. To darken the scene instantaneously he used tin cylinders suspended on wires over every lamp. He also described the equivalent of concealed footlights with reflectors, but added that the smoke and smell of these were drawbacks at stage level. They were screened from the audience by a parapet a foot or so higher than the stage and at a distance from it of anything from 1 to 10 feet. The German architect Josef Furtenbach (1591–1667), in his *Architectura Civilis* (1628), gives a similar description of footlights and of oil lamps in a sunk strip 6 feet wide at the very back of the acting area. These could be used to light the backcloth, and also for special trick effects by rotating the lamps on vertical poles.

As well as candles, two kinds of lamp were used in the Italian theatre, one a simple cruse lamp with a floating wick and the other a *bozze*, or glass globe, which had a short handle opposite the hole left for the wick in its metal holder. The handles could be mounted in rings or in holes drilled in a board, and two *bozzi* could be used with containers of liquid to give coloured light. Burnished reflectors and mirrors were also used, both singly and in lengths.

The lighting of the 17th-century English MASQUE was based on Italian practice. Inigo JONES used an abundance of coloured lights, which he called 'jewel glasses', also concealed lighting and all kinds of reflected light, as well as transparencies and such effects as moons, sunrises, and sunsets. Footlights (recorded in France as early as 1588, on a temporary stage in the Salle de Diana at Montbrison) were adopted more slowly in England, possibly because the Court masque retained the earlier tradition of bringing the performers down to the floor of the hall for the concluding dance. The frontispiece to Francis Kirkman's *The*

Wits; or, Sport upon Sport is the earliest English illustration to suggest the use of footlights: dating from 1672, it may depict a makeshift stage of the Commonwealth period.

The new DORSET GARDEN THEATRE, which opened in 1671, was equipped with up-to-date French machinery, as was the new DRURY LANE, opened in 1674. Although no description of their lighting exists, the fact that both theatres, on the evidence of plays in the repertoire, were able to arrange for 'sudden darkness' during the course of the action indicates the use of concealed lights, probably footlights and lights on wing-ladders which could be controlled. But the main source of illumination was still chandeliers and sconces, a pattern which persisted well into the 18th century. Under John RICH the LINCOLN'S INN FIELDS THEATRE, which he reopened in 1714, had six chandeliers, apparently iron rings hung on chains. Theatrical prints of the 18th century show either a small chandelier or two-branched candelabra over the proscenium doors and double branches at regular intervals round the fronts of the circles. The better known of the two 'Fitzgiggo' illustrations of the COVENT GARDEN stage in 1763 shows double branches between each of the stage boxes, a lofty chandelier over the middle of the stage carrying six candles, and four rings of some 16 candles each, two hung, apparently, in the proscenium arch and two upstage of the central lustre.

Similar conditions prevailed in France at the end of the 17th century. In 1688 the Hôtel de BOURGOGNE had six large chandeliers hung in front of the proscenium arch and behind the arch three on each side, in line with the perspectives, illuminating each wing. At VERSAILLES in the 1660s and 1670s there were five chandeliers hung in a line in the deep arch, lighting the front of the stage very brightly, in contrast to the upstage area which remained in shadow. An unidentified, probably French, painting of 1670 shows a stage illuminated by six chandeliers and a row of 34 footlights arranged in three groups.

Unconcealed chandeliers hung over the stage—particularly in a line in the proscenium arch—impeded the view from the second circle of the upstage area, and the glare from the naked candle-flames was very trying to the eyes (as PEPYS noted in his diary on 12 May 1669). The famous lighting reforms made by GARRICK at Drury Lane in 1765 consisted of removing the six overhead chandeliers from the stage area and massing his lights behind the wings to cast a bright horizontal radiance on to the stage. These innovations were apparently based on contemporary developments in France, and Garrick may have imported some new and brighter lamps from Paris, although it was stated at the time that Drury Lane was far more brilliantly lit than the COMÉDIE-FRANÇAISE. Also in 1765, the lighting at Covent Garden was similarly improved, but instead of candles oil lamps were used, which gave less light as well as smelling unpleasantly. The well known engraving of the 'screen scene' from SHERIDAN'S *The School for Scandal*, as performed at Drury Lane on 8 May 1777, shows the effect of the new lighting. The apron stage is brilliantly lit by footlights well masked from the pit; directed lighting from the prompt side falls on a considerable area of the main stage; and, although a good deal of illumination is provided by the house-lights, all the stage lighting proper except for the footlights comes from behind the proscenium arch. The scene designer Philip de LOUTHERBOURG, who joined Garrick at Drury Lane in 1773, further improved both the naturalistic and spectacular qualities of the stage lighting: his conflagrations and fogs were especially popular. Basic methods changed little, however, until the second decade of the 19th century, except for the 'reverberators' (reflector lamps) which may have been installed in smaller theatres (*see* George Saunders, *A Treatise on Theatres*, 1790). These were fixed to each tier of stage boxes to light the apron area, and may be considered the precursors of the modern floodlight.

Experiments with gas lighting had been going on since the 1790s, and the new fuel excited much interest from the turn of the century. In 1803 and 1804 a German entrepreneur, F. A. Winsor, demonstrated 'coal-gas illumination' at the LYCEUM THEATRE, and early in 1807 he lighted a part of Pall Mall by gas; with the incorporation of the London Gaslight and Coke Company in 1812, gas lighting of thoroughfares and public buildings became commonplace in the metropolis. At first the theatres adopted it to light their exteriors, foyers, and staircases, with Covent Garden leading on 11 Sept. 1815; the OLYMPIC apparently had gas lights in the auditorium from 30 Oct. 1815; Drury Lane was lighted throughout by gas from 6 Sept. 1817, with Covent Garden and the Lyceum following two days later. All the major cities of Britain had gas supplies by 1825 and most theatres quickly adopted the innovation, although a few clung to oil lamps and candles: the Olympic even boasted on 28 Oct. 1822 that it had reverted to 'wax-lights' and gas had been completely removed from its interior.

The increased brilliance offered by gas was not universally welcomed, and throughout the 19th century stage lighting was criticized as being crude and garish. Perhaps in part because gas had increased the amount of light in the auditorium along with that on stage, few attempts were made to explore the dramatic capabilities of the new lighting, although some magical illusions were achieved by, for example, PHELPS in his 1853 production of *A Midsummer Night's Dream*, and in the 1860s FECHTER eliminated much of the objectionable glare by sinking the footlights below stage level. Facilities existed to create a wide range of effects. Apart from footlights, overhead battens, and wing-ladders, all efficiently masked, gas points in various positions around the stage allowed for additional lights as required; coloured light was produced by cloth stretched over wire guards in front of naked gas jets. Limelight, developed by Drummond in 1816, was in general use in the theatre by the 1850s; it consisted of a block or cone of lime which became incandescent when heated by the flame from a gas jet, giving an illumination that was unprecedentedly brilliant and yet mellow. With a reflector the limelight could project a directional beam which, like the modern spotlight, could follow an actor's movements (giving the phrase 'in the limelight' to our language) and could produce realistic effects of moon- and sunbeams. The most important advance offered by gas, however, was its ease of control. Each pipe supplying a group of jets had its own tap; these were sited on a 'gas-table' operated by the prompter, who could dim or extinguish any set of lights at will, using a master tap for complete blackouts; relighting was also under remote control by means of an electric spark circuit. Although control was adequate safety measures were not, and with the introduction of gas FIRES IN THEATRES doubled their number in ten years. Some protection was given by glass chimneys, but the open fish-tail burners used for wing-lights were often left without even a wire guard.

As with Garrick in the 18th century, it was IRVING in the 19th who gave proper attention to the dramatic use of lighting, and the foundation of his achievement was his insistence on darkening the auditorium, a practice advocated and intermittently attempted over the past 300 years. Within the defined stage picture thus presented he used colour with particular subtlety, directing blends of tinted light on to the stage (in contrast to the simple dominant tones then customary) and using coloured paper or transparent lacquer on glass in place of cloth filters. His boldly non-naturalistic lighting was on occasion scoffed at; he seems to have been in advance of his time, for ANTOINE, after seeing Irving's production of *Macbeth* on 9 Feb. 1889, noted in his journal that the lighting was beyond anything then known in France. It is said that Irving resisted the introduction of electricity and, after the majority of theatres had abandoned gas, continued to take his own gas lighting apparatus with him on tour.

Electric arc lights were in use in theatres as early as 1846, at the Paris Opéra, and were first used throughout a theatre at the Paris Hippodrome in 1878. Carbon arcs gave brilliant illumination but, like the limelights which they never entirely ousted, needed constant attention; they were, moreover, noisy and they flickered. Equipped with a hood, lens, and parabolic reflector, they could be used for floodlighting or in the manner of the modern spotlight, and a sliding shutter between the light source and the lens provided a rudimentary form of dimming. Both arcs and limes continued in use long after the introduction of the incandescent bulb had made electricity a practical proposition for all kinds of theatre lighting.

According to David BELASCO, the first American theatre to use electricity was the California, in San Francisco, early in 1879. Over the next decade it was installed in most theatres in Britain, Europe, and the United States, a development that was stimulated by two disastrous fires in 1887, at the Opéra-Comique in Paris and at the Theatre Royal, Exeter (the latter having been burnt down and rebuilt only two years earlier).

The SAVOY THEATRE, built by D'Oyly Carte for the performance of GILBERT and Sullivan operettas, was the first public building in London to be lit throughout by incandescent electric-light bulbs. On 10 Oct. 1881, to resounding cheers, the electric lights in the auditorium were turned on and on 28 Dec. those on the stage. However, unlike the Lyceum, the Savoy auditorium was still fully illuminated throughout the performances, which probably strengthened complaints of glare, echoing those that had greeted the introduction of gas.

The Savoy's stage lighting established a pattern that became standard in succeeding decades. Overhead battens (called 'border lights' in America) and footlights (still commonly called 'floats', harking back to the cruse lamp) were fitted with bulbs lacquered with a variety of colours and backed by simple reflectors. This basic floodlighting was supplemented by carbon arcs and limelights, usually positioned on 'perch platforms' behind the proscenium wall or mounted on the ends of the upper circle or gallery. Lighting GROUNDROWS eliminated shadows from the backcloth and were used to create lighting effects. Dimmers were used from the beginning: the Savoy had six (compared with 88 gas taps controlling the lighting at the Paris Opéra), four shunt regulators operating on the theatre's own steam-driven generators and two series resistances.

The high-intensity gas-filled lamp, introduced in about 1914, provided increased brilliance, but its greater heat melted coloured lacquers applied directly to the glass; gradually the floodlights, both battens and floats, were housed in rows of individual compartments, each with a set of colour-runners for frames holding glass or gelatine colour media. Equipped with lenses to concentrate and focus the beam, the new lamps developed into spotlights, the main lighting resource of the modern theatre. Remotely controlled from the central switchboard, they could be positioned anywhere in the house. At first, hung on the front of the upper balconies, they were used to fill in the dark area between the footlights and the No. 1 batten immediately up-stage of the proscenium arch; the greater throw required of the beam led to rapid developments in design. Mechanisms for varying the distance between lamp and lens; improved reflectors made of various materials, including mirror-glass, chromium, and rhodium; plano-convex and, later, stepped lenses, all increased the efficiency of the light output. Irises and 'barn door' shutters adjusted the size and shape of the beam, and colour wheels, controlled by electro-magnets, were mounted in front of the lens. The positions available to spotlights spread both in front of the proscenium (FRONT OF HOUSE or FOH lighting) and behind it.

Concurrently with developments in lighting units or lanterns (also called 'luminaires' and, in America, 'lighting instruments') dimmers increased in both numbers and refinement. Liquid or wire variable resistances inserted into each circuit were supplemented by auto-transformers which regulate the voltage and are independent of load, and all these dimmers were controlled by sliding keys mounted in vertical channels on the switchboard. To aid the synchronization of large numbers of dimmers, manual operation of the individual keys was replaced by the Grand Master system of interlocked keys connected to a single lever or wheel, so that a complete lighting change could be achieved by a single technician.

In the 1930s these increasingly cumbersome manual methods were replaced in some larger theatres by remote control servo-operated systems, in which resistance dimmers were connected via Mansell electro-magnetic clutches to motor-driven shafts activated by 3-position switches (Up-Off-Down). Covent Garden pioneered this innovation (at first combined with some manual control) and its installation remained in use from 1934 to 1964. In 1935 Frederick Bentham of Strand Electric, the chief influence on the technical development of stage lighting in Britain, used Mansell clutches in his Light Console. This adaptation of a Compton organ console used the organ stops to pre-select the dimmers, the pedals to regulate the motor speed, and the keys to raise or lower the lighting level. A number of Light Consoles were built after the Second World War, the largest being installed at the London COLISEUM and Drury Lane; but the invention was overtaken, first by George Izenour's 44-channel Thyratron system, installed at Yale in 1947, then by applications of silicon chip technology in the 1960s. The Thyratron and the Silicon Controlled Rectifier (SCR), now called the Thyristor, have made possible all-electric dimmers with no moving parts, except for the control levers. The late 1960s saw the introductions of punched card 'memory' systems which could store an unlimited number of preset lighting cues, but as in other industries punched cards were swiftly made obsolete by coded commands stored on miniaturized components such as integrated circuits. A computer programme, usually recorded magnetically on a tape cassette, can activate the successive changes of a complete lighting plot.

Both technical and aesthetic development in stage lighting continued to vary from country to country during the 20th century. In Germany, for example, the opera-house tradition led to massive stages equipped with a series of elaborate lighting bridges below the grid, the open framework above the stage from which lights can be suspended. German theatres pioneered the use of the *Kuppelhorizont* invented by Mariano Fortuny (1871–1949) in 1902. This 'sky-dome' of coloured silk reflected the beams of high-powered arcs down on to the stage, giving a bright, diffused light; it was expensive in terms of current and was replaced by the *Rundhorizont*, a half-dome of silk or plaster which developed into the CYCLORAMA. Adolphe APPIA and Gordon GRAIG did their most innovative work in Germany. They rejected the conventions of painted scenery in favour of a three-dimensional setting: the flat illumination of the late 19th-century stage had consequently to be replaced by directional lighting, casting shadows in the manner of real light to convey the mass and moulding of both actor and set. Appia insisted on the importance of lighting as an integral part of a unified artistic conception, as did Craig, who used changing lights against severely simple settings to build up an effective non-naturalistic stage picture.

In America David Belasco was almost unique in his adventurous use of lighting. As early as 1879 he lit Morse's *The Passion Play* from the front with old locomotive bull's-eye lanterns, anticipating the ideas of REINHARDT and GRANVILLE-BARKER by a quarter of a century. The latter was a pioneer in the use of FOH lighting in Britain, a necessary solution to the lighting problems created by his extension of the forestage acting area over the orchestra pit for productions of Shakespeare.

In general, however, British and American practice lagged behind that of Europe. Commercial productions tended to be lit by footlights, banks of compartment battens, a few FOH spots, and more spots mounted vertically on booms (boomerangs) and spot ladders in the wings and behind the proscenium wall. In Britain floodlighting was radically improved in the 1930s by the Acting Area and Pageant Lanterns, which threw a comparatively narrow, highly-concentrated beam; these could be massed to give overlapping areas of varying brightness, or used singly for sharp accents such as direct 'sunlight'. The technique became popular in revue and musical comedy, but was also used effectively in a non-naturalistic manner by directors such as George DEVINE and Peter BROOK.

Improved spotlights, which are more precise and economical than any flood, began to oust the compartment batten from the late 1930s onwards. The Focus Lamp, or Lens Spot, with a fixed plano-convex lens and moveable bulb and reflector, is now uncommon except in Germany, where it is called a *Linsenscheinwerfer*; elsewhere it has been replaced by the Fresnel Spot. Here the fixed lens is moulded in concentric stepped rings which give a soft-edged beam that can be grouped with others in overlapping clusters to give either an overall even glow or areas of varying brightness. The short focus available with the Fresnel lens produces a particularly high light output. The Profile Spot, which has a moveable objective lens and stationary lamp and reflector, gives a hard-edged beam with negligible scatter; both lamps can be fitted with shutters to confine the beam to a desired shape. The increasing use of THEATRE-IN-THE-ROUND and the various facilities offered by FLEXIBLE STAGING were aided by the development of spots: where the dramatic action is wholly or partly surrounded by the audience, light spill and dazzle present problems that can be solved only by flexible and precise light sources. Pin-spotting, where a tight circle of brightness is achieved by adding a long, narrow hood to a spotlight, probably encouraged the vogue for COMPOSITE SETTINGS, in which a number of locations are successively illuminated as the action moves from one part of the set to another.

Modern theatre lighting equipment generally consists of a range of Fresnel and Profile spots (the latter known in America as Lekos, from the original manufacturers Levy and Kook). The Focus Lamp is still to be found in small studio theatres. Small, highly economical Tungsten-Halogen lamps came into wide use in the 1960s; these have allowed lanterns to be reduced in size while increasing their efficiency, and Baby Spots (in America even Embryo Spots) are available in all three types of lantern. The PAR (Parabolic Aluminized Reflector) Lamp, similar to the spotlights for domestic use, is a lamp and reflector in one available as either spot or flood.

Lanterns can be hung over the acting area on horizontal bars (pipes in America); clamped to vertical booms or ladders in the wings; or mounted on perch (tormentor in America) platforms or booms behind the proscenium walls. Except on the largest stages and in spectacular musicals, floodlights in 6-ft battens tend to be restricted to footlights and to the electric groundrow which lights the backcloth or cyclorama. FOH lighting is still found on the balcony fronts, but booms and pipes suspended from the ceiling and along the side walls of the auditorium are more usual; new theatre buildings have permanent side-slots in the walls, backed by platforms and booms. All lamps are equipped with colour filters made of non-inflammable plastics and available in several dozen tints.

Special effects in the field of lighting can call on a multiplicity of techniques. Apart from imaginative colour changes, one of the simplest uses the Gobo, a cut-out mask slotted into a lantern housing to throw the shadow of leaves, architectural units, or other off-stage features on to the scene. Shadow projection, by the direct-beam projector, is a refinement of the technique introduced in 1916 by Adolph Linnebach (1876–1963) at the Court Theatre, Dresden. The Linnebach and direct-beam projectors create magical effects by means of shadowgraphs. Optical lens projectors, or stereopticons, can create up to a complete setting using large format glass slides specially painted or photographed for the purpose. Moving effects such as clouds, snow, or flames are achieved by effects projectors (sciopticons in America) carrying slides in a motor-driven turntable or drum. Such effects can be particularly successful when projected on to a gauze CLOTH, or unbroken stretch of fine net, in front of the action. Film projection has a place in stage production and has been used to great acclaim in the Laterna Magica of Prague, devised by Josef SVOBODA, and based on the work which he, Alfred Radok, and others created for the Czech Pavilion at the Brussels International Exhibition of 1958. Here slide and film projection are wittily combined with a few live performers to convey the illusion of a stage packed with scenery and action. Ghostly effects (once dependent on devices such as PEPPER'S GHOST and the various kinds of TRAP) can be created with ultra-violet light, under which only objects coloured by fluorescent paint or dye can be seen. The pulsing flashes of strobe (stroboscopic) light can 'freeze' action into a series of jerks; the technique became very popular as a result of the disco vogue and can be powerful when used with discretion. Similar brilliant flashing is produced by the Lobsterscope, a punched metal disc revolving in front of a spotlight.

The complexity of modern resources has made stage lighting a distinct craft, and the role of the lighting designer has since the 1950s become increasingly important. His work with costume and set designers, and with both sound and lighting technicians, in shaping a production to the director's needs is a vital part of theatrical creation today.

LIGHT PIPE, see BARREL.

LILLIE, BEATRICE GLADYS [Lady Robert Peel] (1898–   ), Canadian actress and entertainer, born and educated in Toronto, but equally well known and admired in England and the United States. She made her first appearance in England in 1914, at the Chatham Music Hall, and in the same year was seen at the LONDON PAVILION. She then played in a series of REVUES, and in 1924 made her début in New York in a production by André Charlot, alternating between London and New York in his revues until 1926. In 1928 she appeared in New York in Noël COWARD's revue *This Year of Grace*, which had a long run. In 1932 she played the straight part of the Nurse in the première in New York of SHAW's *Too True to be Good*, but she was later seen in the revues *At Home Abroad* (1935), *The Show Is On* (1936), and Coward's *Set to Music* (1939), all in New York, and *All Clear* (also 1939) in London. During the Second World War she entertained the troops tirelessly with a programme of revue sketches and one-act plays, including Coward's *Tonight at 8.30*, returning to New York to star in the revues *The Seven Lively Arts* (1944) and *Inside U.S.A.* (1948). From 1952 to 1956 she toured the world in a solo entertainment, *An Evening with Beatrice Lillie*. Past mistress of the elegant *double entendre* and tongue-in-cheek riposte, heightened by the manipulation of an enormously long cigarette-holder, she returned to London in 1958 to give a superb performance in Jerome Lawrence and Robert E. Lee's adaptation of Patrick Dennis's novel *Auntie Mame*, a part she had already played in New York in succession to Greer Garson. Her last appearance was in New York in 1964 as Madame Arcati in *High Spirits*, a musical version of Coward's *Blithe Spirit*.

LILLO, GEORGE (1693–1739), English dramatist, best remembered for his play *The London Merchant; or, the History of George Barnwell* (1731). Based on an old ballad, it tells how a good young man's passion for a bad woman leads him to murder his old uncle for money, the murderer and his accomplice being subsequently hanged. It was immensely successful and was frequently revived, notably by Mrs SIDDONS. It was known well enough to be the butt of several burlesques, and was also the play performed by the Crummles family in *When Crummles Played* (1927), based by Nigel PLAYFAIR on DICKENS's novel *Nicholas Nickleby*. It had a great vogue on the Continent where it influenced the development of domestic tragedy, particularly in Germany. Lillo wrote several other plays, of which the most important was *Guilt Its Own Punishment; or, Fatal Curiosity* (1736), also based on an old ballad about a murder done in Cornwall. First produced at the HAYMARKET THEATRE by Henry FIELDING, it was chosen by Mrs Siddons for her BENEFIT in 1797, her brothers Charles and John Philip KEMBLE appearing with her. This play also had a great influence on the German SCHICKSALSTRAGÖDIE, inspiring WERNER's *Der 24 Februar* (1810). Lillo also wrote a version of the anonymous *Arden of Feversham*, produced in 1759.

LIMELIGHT, LIMES, see LIGHTING.

LIMERICK, MONA, see PAYNE, BEN IDEN.

LINCOLN, in the late 17th century, was the headquarters of the Lincoln CIRCUIT, which included Grantham, Boston, Spalding, Peterborough, Huntingdon, Wisbech, and Newark-on-Trent. From 1802 to 1847 the circuit was controlled by the Robertson family, to which T. W. ROBERTSON and his sister Madge KENDAL belonged. The first permanent theatre in the city was built in about 1731 by Erasmus Audley in Drury Lane (so called after a local business man and not after the theatre in London), and in 1764 a second theatre was formed by the adaptation of some buildings in King's Arms Yard, nearer the centre of the town. This was managed up to about 1783 by William Herbert, who had also been manager since 1750 of the Drury Lane theatre. The large collection of playbills in Lincoln Public Library has many announcing performances by 'Mr Herbert's Company of Comedians.' A new theatre was built in King's Arms Yard in 1806. It was burned down in 1892 and another arose on the same site, opening on 18 Dec. 1893 with Brandon THOMAS's *Charley's Aunt*. This, the present Theatre Royal, was partially adapted to serve as a repertory theatre in 1930, and continued to be so used until 1976. It then closed, and reopened later as a 482-seat theatre for touring companies only.

LINCOLN CENTER FOR THE PERFORMING ARTS, New York, see VIVIAN BEAUMONT THEATRE.

LINCOLN'S INN FIELDS THEATRE, London, in Portugal Street. This was originally Lisle's Tennis-Court, built in 1656. It was leased in March 1660 by Sir William DAVENANT, who enlarged it for use as a theatre, making it the first playhouse in England to have a proscenium arch behind the apron stage. It opened in June 1661 (probably on the 28th) with Davenant's own play *The Siege of Rhodes, Part I*, the second part appearing shortly afterwards. Thomas BETTERTON, who was to be closely associated with this theatre, made his first appearance there on 28 Aug. playing Hamlet. There was also a revival of *Romeo and Juliet*, with the original ending one day and the alternative happy ending by James Howard the next. Among the outstanding plays first seen at this theatre were TUKE's *The Adventures of Five Hours* (1663), based on CALDERÓN, and DRYDEN's *Sir Martin Mar-all* (1667), in which the comedian NOKES scored a great success. Davenant died in 1668, and his widow, with the assistance of Betterton, kept the theatre going until DORSET GARDEN, the new playhouse begun by her husband, was completed. The company gave its last performance at Lincoln's Inn Fields Theatre on 9 Nov. 1671. Two months later KILLIGREW took over the empty building, following a fire which had destroyed the first DRURY LANE THEATRE, and remained there until the end of Mar. 1674, after which the building reverted to its former use as a tennis-court, until in 1695 Betterton, who had seceded from the United Company formed by the amalgamation of the actors at Drury Lane and Dorset Garden, took a company which included Elizabeth BARRY and Anne BRACEGIRDLE to the old theatre. He financed the restoration of the building by public subscription, and reopened it with CONGREVF's new play *Love for Love*. Though somewhat handicapped by the smallness of the stage and the limited accommodation in the two-tier auditorium, the company remained in occupation for ten years, moving in 1705 to the new Queen's Theatre in the Haymarket, built by VANBRUGH. The old building then ceased to be used as a theatre until in 1714 Christopher RICH took it over and put in hand extensive renovations, dying before they were completed. It was left to his son John RICH to finish the alterations, which gave the theatre a handsome auditorium seating more than 1,400 spectators and lighted by six overhead chandeliers, and a stage, with mirrors each side, larger than that at Drury Lane. All the scenery was new.

The opening production, on 18 Dec. 1714, was a revival of FARQUHAR's *The Recruiting Officer*, and on 29 Jan. 1728 came the first night of Rich's most important new production, GAY's *The Beggar's Opera*. It was also at this theatre that Rich first appeared as HARLEQUIN in a new-style entertainment which led to the development of the English PANTOMIME. In 1732, for reasons that are not yet fully understood, Rich undertook the building of a new theatre in COVENT GARDEN, and moved there in the autumn of that year. Lincoln's Inn Fields was then used mainly for music and opera, except in the season of 1736–7 and again in 1742–3 when Giffard was there after the closure of GOODMAN'S FIELDS THEATRE. The final performance took place on 11 Dec. 1744, and the old theatre then became, among other things, a barracks, an auction room, and finally the Salopian China Warehouse, used by Spode and later by Copeland. It was pulled down in 1848 to make way for an extension of the museum of the Royal College of Surgeons.

LINCOLN'S MEN, small company of Elizabethan actors, led by Laurence Dutton, who were in the service of the first Earl of Lincoln and of his son Lord Clinton, whose name they sometimes took. They appeared at Court before Queen Elizabeth I several times between 1572 and 1575 and were active in the provinces up to 1577, as was a later company of the same name from 1599 to 1610.

LINDBERG, AUGUST (1846–1916), Swedish actor, whose travelling company did much to bring IBSEN to provincial audiences all over Scandinavia. He was the first actor to play Oswald in *Ghosts* in 1883 and among his other Ibsen roles were Brand, Peer Gynt, John Gabriel Borkman, and Solness in *The Master Builder*. He also appeared as Hamlet and King Lear. From 1906 he worked with the KUNGLIGA DRAMATISKA TEATERN in Stockholm.

LINDSAY, HOWARD (1889–1968), American actor, dramatist, and director, who already had a long list of successes to his credit when he collaborated with Russel Crouse (1893–1966) in a dramatization of Clarence Day's book *Life With Father*. Produced in 1939, with Lindsay as Father, this ran for seven years. They continued their collaboration in the PULITZER PRIZE-winner *State of the Union* (1945) and a dramatization of Day's *Life With Mother* (1948), which also had a long run, Lindsay again playing Father.

LINDSTROM, ERIK (1906–74), Swedish actor, who after training at the Swedish National Theatre's drama school, joined the company there, making his début in 1927 in SHAW's *You Never Can Tell*. For a time he specialized in comedy, one of his earliest successes being Sir Peter Teazle in SHERIDAN's *The School for Scandal*. He was also much admired in a revival of HOLBERG's *Den Stundesløse* (*The Fussy Man*) in 1968. But even in 1929 he had shown that he could play serious drama, appearing as Stanhope in SHERRIFF's *Journey's End* and later in such parts as IBSEN's Peer Gynt, GOETHE's Faust, PIRANDELLO's Enrico IV, and Shakespeare's Macbeth and Iago (with Mogens WIETH as Othello). In 1948, with the company from the Swedish National Theatre, to which he belonged during the whole of his active career, he appeared at Elsinore as Hamlet. One of his outstanding successes in later years was the title role in STRINDBERG's *Gustav III*.

LINLEY, ELIZABETH ANN, see SHERIDAN, RICHARD BRINSLEY.

LINNEBACH, ADOLF, see LIGHTING and SCENERY.

LINSENSCHEINWERFER, see LIGHTING.

LIONS COMIQUES, see LEYBOURNE, GEORGE, and VANCE, ALFRED.

LISLE'S TENNIS-COURT, London, see LINCOLN'S INN FIELDS THEATRE.

LISTON, JOHN (1776–1846), English actor, who in spite of a nervous and melancholic turn of mind in private life had only to appear on stage to set the audience laughing. After extensive experience in the provinces he was seen in London in 1805, under the younger George COLMAN, where he made his début as Sheepface in *The Village Lawyer*, an adaptation by MACREADY of the old French farce *Maître Pierre Pathelin*. He was immediately successful and for more than 30 years was one of the leading players of London, being the first comedian to command a salary greater than that of a tragedian. He occasionally essayed tragedy himself, but without success. His Paul Pry, in John Poole's comedy of that name, was so popular that it was copied, dress and all, by all later performers of the part. After making a fortune for such dramatists as POCOCK and the younger Charles Dibdin by appearing in their comedies, Liston retired in 1837.

LISTON, VICTOR, see MUSIC-HALL.

LITTLE ANGEL THEATRE, Islington, see CHILDREN'S THEATRE and PUPPET.

LITTLE DRURY LANE THEATRE, London, see OLYMPIC THEATRE (1).

LITTLE THEATRE, Bristol, repertory company founded to meet the demand in the city for good modern plays, since both Bristol's existing theatres—the Theatre Royal and the Prince's—were at this time given over to light entertainment only. Under the guidance of an experienced actor, Rupert Harvey (1887–1954) from the OLD VIC in

London, the newcomers leased from the Corporation the Lesser Colston Hall and launched their venture on 17 Dec. 1923. In spite of the small stage and cramped conditions backstage, it was an artistic success, and the company included at various times Laurier Lister, Sebastian Shaw, Marjorie Fielding, and the future playwrights Arthur Macrae and Philip King. Financially matters were less satisfactory, and in spite of the popularity of such authors as BARRIE and SHAW, whose *Back to Methuselah* was staged in its entirety in 1929, and the inclusion each season of at least one new play by a local author, the company was forced to disband in June 1934. The theatre was then taken over by the Rapier Players, a company which functioned from 1935 to 1961, doing excellent work under great difficulties. After they left the theatre was leased to the BRISTOL OLD VIC which used it for special productions, among them plays by Charles WOOD and PINTER's *The Homecoming* (1967), and as its main theatre during the rebuilding of the old theatre in 1970–71, until in 1980 it was again taken over by an independent company.

LITTLE THEATRE, London, in John Adam Street, Adelphi. This small theatre, which held only 250 people until balcony and boxes added in 1912 increased its capacity to 309, opened on 11 Oct. 1910 with Laurence HOUSMAN's adaptation of ARISTOPHANES' *Lysistrata*, under the management of Gertrude KINGSTON, who played the title role. On 27 Jan. 1911 Noël COWARD, aged 11, made his first appearance on the stage as Prince Mussel in a fairy play called *The Goldfish* by a Miss Lila Field and on 19 Apr. SHAW's *Fanny's First Play*, directed by himself, was performed anonymously and then ran for a year, as did G. K. Chesterton's *Magic* (1913). The theatre was severely damaged by bombing on 4 Sept. 1917, but was repaired and reopened on 24 Feb. 1920 with KNOBLOCK's *Mumsee*. Later in the year there was a season of Grand Guignol with Sybil THORNDIKE, followed by *The Nine O'Clock Revue* (1922) and *The Little Revue Starts at Nine* (1923). Two 'horror' plays were produced, both based on novels, Bram Stoker's *Dracula* (1927) and Mary Shelley's *Frankenstein* (1930). In 1932 Nancy PRICE made the theatre the headquarters of her People's National Theatre, the most successful productions being *Lady Precious Stream* (1934) by S. I. Hsiung and Mazo de la Roche's *Whiteoaks* (1936). Herbert Farjeon's *Nine Sharp* (1938) and *Little Revue* (1939) were both successful, the latter being adapted to war conditions by playing from Sept. 1939 onwards in the afternoon. The last production, in Apr. 1940, was a revival of WYCHERLEY's *The Country Wife* with Alec CLUNES. The theatre was severely damaged by enemy action on 16 Apr. 1941 and demolished in 1949.

LITTLE THEATRE, New York, at 240 West 44th Street, between Broadway and Eighth Avenue. Built by Winthrop AMES as a try-out theatre, it opened on 12 Mar. 1912. Productions in the first year included SCHNITZLER's *Anatol* with John BARRYMORE, *Prunella* by Laurence HOUSMAN and Harley GRANVILLE-BARKER, SHAW's *The Philanderer*, and a revival of Clyde FITCH's *Truth*. The Shuberts later took control of the theatre, enlarging its seating capacity from 300 to 600, but it was never very successful and after several changes of name and use the *New York Times* bought it in 1941 and used it for lectures, recitals, and television. On 2 Nov. 1963 it reverted to live theatre, opening with Langston Hughes's adaptation of his own novel *Tambourines to Glory*. In 1964, when Frank D. Gilroy's *The Subject was Roses* was transferred to the Little from the ROYALE, the name was changed to the Winthrop Ames Theatre, in honour of its founder, but a year later it reverted to its original name and was used for television. It became a theatre again in 1974 with the production of Ray Aranha's *My Sister, my Sister*; in 1977 Albert Innaurato's *Gemini*, originally produced by the CIRCLE REPERTORY COMPANY, moved in, and it was still running in the early 1980s.

LITTLE THEATRE GUILD OF GREAT BRITAIN, see AMATEUR THEATRE.

LITTLE THEATRE IN THE HAY, London, see HAYMARKET THEATRE.

LITTLE THEATRE OF THE DEAF, see O'NEILL, EUGENE.

LITTLE THEATRES, see AMATEUR THEATRE and CANADA.

LITTLE TICH [Harry Relph] (1868–1928), English MUSIC-HALL comedian, who got his name from his supposed likeness as a baby to the claimant in the famous Tichborne Case. He made his first appearance as a child of 12 in blackface at one of London's last pleasure grounds, the Rosherville near Gravesend, and was first seen on the 'halls' in 1884 at the MIDDLESEX. After playing in VAUDEVILLE in the United States for some years he returned to England and was seen in a number of PANTOMIMES and BURLESQUES. In 1902 he was at the TIVOLI and there, and in a number of other 'halls' throughout the country, made a great success with his impersonations, which ranged from grocers, blacksmiths, and sailors on leave to fairy queens and Spanish dancers, and invariably ended, at least until his last years, with a fantastic dance in which he balanced on the tips of his preposterous boots, which were as long as he was high. He toured extensively, and was as popular in Paris as in London. His act was essentially pantomimic, and his whole personality was reminiscent of the Court Fool or dwarf of medieval aristocratic households. Though unusually small (just over 4 ft) he was not deformed except for an extra finger on each hand.

LITTLEWOOD, (Maudie) JOAN (1914–), English director, born in London of working-class parents, who studied for the stage at the Royal Academy of Dramatic Art. Although she did well there, the effect of her training, combined with a naturally aggressive and experimental nature, left her impatient with what she regarded as the inanities of the normal West-End theatrical routine and, turning her back on the success which might have attended her in the commercial theatre, she went to Manchester. While working there in radio she founded with her husband the folk-singer Ewan McColl [Jimmy Miller], whom she married in 1935, an amateur group, Theatre Union, which soon made a name for itself with unconventional productions, in halls and in the open air, of experimental plays. The group dispersed on the outbreak of war in 1939, but came together again in 1945 as THEATRE WORKSHOP, with Joan Littlewood as artistic director, and in 1953 took over the lease of the THEATRE ROYAL at Stratford, London. There, working on a system entirely her own (though it seems to have some affinities with those of STANISLAVSKY and BRECHT), she became responsible for a series of productions which were successfully transferred to the West End and were also acclaimed abroad, especially in Paris. The constant drain on the company's resources, and a morbid dread of publicity, led Joan Littlewood to leave Theatre Workshop in 1961 to work elsewhere, though she returned in 1963 to undertake the production of *Oh, What a Lovely War!* which was again a resounding success. From 1965 to 1967 she was at the Centre Culturel, Hammamet, Tunisia, returning again to Theatre Workshop to direct Barbara Garson's political skit based on *Macbeth, Macbird* (first seen off-Broadway), *Mrs Wilson's Diary*, and *The Marie Lloyd Story*, as well as a successful pastiche of VANBRUGH's *The Provoked Wife* as *Intrigues and Amours*. Her last production at Stratford was Peter Rankin's *So You Want to be in Pictures?* (1973) and since 1975 she has worked in France.

LITTMANN, MAX, see THEATRE BUILDINGS.

LITURGICAL DRAMA. The modern theatre in Europe owes much of its early development to the observances of a cult that had been the main factor in abolishing the last vestiges of classical drama, and which in its later history was to battle with varying degrees of success against the manifestations of its secularized offspring. Yet considering the drama inherent in the life of Christ it is perhaps not surprising that the mimetic instinct in man, driven underground by the Early Fathers, should have broken out again in the celebration of the Mass, particularly at Easter and, soon after, at Christmas. The service itself provided the bare bones of a plot, and the introduction of antiphonal singing, which may have owed something to the memory of the Greek CHORUS, paved the way for the use of dialogue. As church services became more elaborate, the earlier antiphons were supplemented by additional melodies, first sung to vowel sounds only (*neumes*) later to specially written texts (*tropes*), many of which took on dialogue form. The best known and most important of these *tropes*, from a dramatic point of view, was the QUEM QUAERITIS? (Whom Seek Ye?) sung on Easter morning. From a short scene in front of the empty tomb (a temporary erection in which a cross was laid on Good Friday and removed on Easter morning) sung by four male voices, this soon developed into a small drama of three or four scenes covering the main events of the Resurrection. Later detached from the Mass and performed separately, it was followed by the *Te Deum*, and so merged into Matins.

An important step forward in dramatic evolution was taken with the introduction of extraneous lyrics to be sung by the three Marys as they approached the tomb, and an even greater one when comic characters, with no Scriptural basis, were introduced in the persons of the merchants who sold spices to the women for the embalming of Christ's body. They probably appeared first in Germany, and were a counter-influence on the late liturgical play from the secular vernacular drama which had developed alongside it and partly under its influence.

A further Easter play, enacted at Vespers and perhaps modelled on the *Quem Quaeritis?*, was the *Peregrinus*, which showed the Risen Christ with the disciples at Emmaus, sometimes accompanied by the three Marys and Doubting Thomas. In a version preserved at Tours, the scene at Emmaus is preceded by the scenes at the tomb; later additions, including the Lamentations of those stationed by the Cross on Good Friday, led by degrees to a drama which extended from the preparations for the Last Supper to the burial of Christ. This was given in a rudimentary form as long as it remained within the church, mostly in dumb show with passages from the Vulgate, but once performed outside, it coalesced and took on more substance to form the all-important and widespread PASSION PLAY.

Meanwhile the services of Christmas had given rise to a play on the Nativity, centring on the crib, with Mary, Joseph, the ox and ass, shepherds, and angels. (The practice of dressing a crib in church, which was the focal point of the Nativity play, never ceased on the Continent and has been widely revived in England during the present century.) The Christmas play never attained in liturgical drama the importance of the Easter play, and, together with a short scene dealing with Rachel and the Massacre of the Innocents, was soon absorbed into an Epiphany play, in which the interest centred on the Wise Men and their gifts. It depicted the arrival of the Magi in Jerusalem, their interview with Herod, their meeting with the shepherds, the presentation of their gifts at the manger, their return home by a dfferent route after a warning from an angel, Herod's rage at being outwitted, the massacre of the infant children, and the Flight into Egypt, Sometimes a further scene was added, showing the death of Herod and the return from exile of the Holy Family. Herod, who was to be such an important figure in later secular plays on religious subjects, was from the first a noisy, blustering fellow, whence Hamlet's phrase 'to out-Herod Herod'. He was probably played by the 'king' chosen from among the minor clergy to preside over the FEAST OF FOOLS, and may be an importation from extra-ecclesiastical gaieties at Christmastide. The textual evolution of the Epiphany play can be studied from collated fragments, and begins with antiphons and prose sentences based on Scripture only. The influence of the wandering scholars, or GOLIARDS, led later to the writing of new metrical texts with occasional tags from Sallust or Virgil, débris thrown up by submerged classical learning and used by the Church for her own purposes.

Another Christmas play, and the most important for the future development of the drama, was based on a narrative sermon attributed to St Augustine and known as tbe Prophet play. It listed all the prophecies from the Old Testament concerning the coming of the Messiah, and included also those of Virgil and the Erythraean Sibyl. At some time in the 11th century it became a metrical dramatic dialogue, introducing Balaam and his Ass and the Three Children in the Fiery Furnace. The ass was probably another importation from the Feast of Fools, and its use may mark a determined effort by the Church to canalize the irrepressible licence of Christmas merry-making by incorporating into its own more orderly proceedings a slight element of buffoonery.

Materials for a study of the evolution of liturgical drama are fragmentary, but are found all over Europe. It seems to have developed most fully in Germany and France, but Spain, Italy, and England were not far behind, and traces of it have been found in Eastern Europe also. Even in the early days some attempt at costume was made, with robes for the angel, the Marys, and the Apostles, and properties such as wings and a palm for the angel and boxes for the spice-sellers, but as long as the play remained within the church it was part of the liturgy, and the actors were priests, choirboys, and perhaps, later on, nuns. The dialogue, entirely in Latin, was chanted, not spoken, and the musical interludes were sung by the choir alone, with no participation by the congregation. An Advent play from Germany, *Antichristus* dating from about 1160, had a large cast and could only have been enacted in the nave of a sizeable church. It is well written, dramatically conceived, and probably had some significance, now forgotten, in church politics of the time. One interesting fact is that it introduces allegorical as well as Biblical figures. It is a far cry from the initial simplicity of the three Marys at the tomb, and among the participants may have been laymen who had experience of acting, even if only in ROBIN HOOD plays or vernacular FARCE. But nothing is known for certain. All that can be said is that the liturgical drama evolved from a simple Easter *trope*; that its evolution was complete by the end of the 13th century, even though in some cities plays in churches were given up to the 15th; and that in its final phase it was no longer necessarily connected with the liturgy, being based on subjects taken from the whole range of the Bible and on the lives and legends of the Saints. From it came the vernacular MYSTERY PLAY and the SAINT PLAY. Eventually the Church, having given back to Europe a regular and coherent form of drama, withdrew and prepared to do battle with the art form which it had engendered, and whose development henceforward was to become part of the theatrical history of each separate European country.

Successful revivals of liturgical plays have encountered few difficulties with the texts which could not be surmounted by careful study and collation, and some inspired guesswork; but they have given rise to much controversy over the music, which usually appears in the extant manuscripts as nothing more than a simple line of melody for the singers. It seems reasonable to suppose that conditions varied from one performance to another, having regard to the size of the church in which it was being given and the expertise of the musicians on whatever instruments were available and likely to be used in a liturgical context.

LITURGY, in ancient Greece, a public service required of wealthy citizens. One such duty was the staging of a play by a CHOREGUS. In Europe the liturgy, or form of worship, of the early Christian Church gave rise to the performance of plays in Latin, or LITURGICAL DRAMA.

LIVANOV, BORIS NIKOLAYEVICH (1904–72), Soviet actor and director, who joined the MOSCOW ART THEATRE company in 1924, playing a wide range of parts, from GOGOL, CHEKHOV, and BEAUMARCHAIS to POGODIN and AFINOGENOV. In 1960 he became the first Soviet director to stage a dramatization of DOSTOEVSKY's *The Brothers Karamazov* (which had not been seen since NEMIROVICH-DANCHENKO's pre-revolutionary production at the Moscow Art Theatre), in which he played the part of Mitya Karamazov. During the visit of the Moscow Art Theatre to London in the WORLD THEATRE SEASON of 1964, Livanov appeared as Nozdryov in BULGAKOV's dramatization of Gogol's *Dead Souls*, one of his finest parts. His own production of Chekhov's *The Seagull* was seen in London during the 1970 World Theatre Season.

LIVERPOOL, which in the 1740s had theatrical entertainments given by visiting Irish players who appeared in a converted room known as the Old Ropery Theatre, had its first permanent theatre building, in Drury Lane, in 1750. Used by actors from London during the summer months, it had originally no boxes. These were added in 1759, and in 1767 a green room and dressing-rooms were also provided. It became a Theatre Royal in 1771, the name being transferred to a new theatre erected a year later, one of its lessees being George Mattocks, whose father-in-law Lewis HALLAM took the first theatrical company to the New World. The interior was rebuilt in 1803 to provide a horseshoe-shaped auditorium which was later adapted as a circus, and eventually the building became a storage depot. In its heyday Liverpool, which was independent of any CIRCUIT, had a theatre season which lasted practically all the year round. The Star Theatre, which opened in 1866, became the home of lurid melodrama until 1911, when it was taken over as a REPERTORY THEATRE, known since 1916 as the LIVERPOOL PLAYHOUSE. Also in 1866, the present Empire Theatre, which in 1979 was taken over by the Merseyside County Council and extensively renovated, opened as the New Prince of Wales, changing its name to the Alexandra a year later. Seating 2,312, it is the largest two-tier theatre in the country, providing entertainment ranging from touring drama, opera, and ballet to variety and pop music. The enterprising EVERYMAN THEATRE has achieved a high reputation, and appeals particularly to young people.

LIVERPOOL PLAYHOUSE, Britain's oldest surviving REPERTORY THEATRE. As the Liverpool Repertory Theatre it opened in 1911, partly in emulation of the venture into repertory being so ably run at the GAIETY THEATRE in Manchester by Miss HORNIMAN. When an experimental season run by Basil DEAN at another theatre showed that Liverpool was willing to accept good plays under repertory conditions, the old Star Theatre, a home of lurid melodrama which had been built in 1866, was taken over and completely reconstructed. It opened on 11 Nov. 1911 and has since been closed only for short summer vacations, being given its present name in 1916. From its inception it set out to cater for every type of playgoer, though under William ARMSTRONG, its artistic director from 1922 to 1944, its programmes were perhaps somewhat less adventurous than those of Barry JACKSON at the BIRMINGHAM REPERTORY THEATRE. In recent years, however, classical and established works have been offset by a number of controversial modern plays by such writers as David HARE, Harold PINTER, Tom STOPPARD, and Alan BENNETT. A new play for children is presented annually at Christmas. The Playhouse, which seats 762, was extensively renovated in 1968, but the auditorium still retains the spaciousness of earlier times, with wide circles and balconies curving round the stage. The company, which is subsidized by the ARTS COUNCIL and the Merseyside County Council, has been a notable school of acting, many well known players starting their careers there, among them Michael REDGRAVE, Rex HARRISON, and Diana WYNYARD before the Second World War, and more recently Ian MCKELLEN. There is also a studio theatre, the Playhouse Upstairs, seating 120.

LIVING NEWSPAPER, a form of stage production which employed documentary sources to present subjects of current social importance, usually in a sequence of short scenes with individualized dialogue alongside more abstract, often didactic, presentation. Many antecedents have been claimed for the Living Newspaper technique, from Russia, Germany, China, and particularly from Vienna, whose Spontaneity Theatre, founded by J. L. Moreno in 1921, brought psychodrama to New York in 1925. But the term is primarily associated with a unit of the FEDERAL THEATRE PROJECT, whose members were theatre and newspaper workers. Together they produced six Living Newspapers, of which the first, *Ethiopia*, a montage of newspaper reports about the current war in Abyssinia, was cancelled before its public showing under strong pressure from the U.S. State Department. Three of the others were by Arthur Arent—*Triple-A Plowed Under* (1936), which dealt with farming conditions during a widespread drought and profiteering from food distribution during the depression; *Power* (1937), which advocated government ownership of the electrical power industry to save it from the menace of financial manipulation; and *One-Third of a Nation* (1938), which advocated federal and state development of low-cost housing. Less successful were *1935*, the most satirical of the Unit's productions, which lampooned the indifference of the general public to social issues, and *Injunction Granted*, which attacked the collusion of the courts with capitalist big-business in applying federal 'anti-trust' laws against labour unions. Following the example of the Unit, other Federal Theatre companies outside New York wrote and presented their own Living Newspapers, often with local or regional relevance. The outspokenness of the Living Newspaper productions was among the complaints made against the Federal Theatre Project as a whole, and one of the main causes of its being disbanded by the Government in June 1939. The work of the Unit ended at the same time, but its theatrical concepts influenced other companies and productions, and during the Second World War the British army used the Living Newspaper technique to keep the troops informed about the wider issues of the conflict and conditions at home. Post-war documentary drama, or Theatre of FACT, also owes a considerable debt to the form.

LIVINGS, HENRY (1929–), English actor and dramatist, who appeared in THEATRE WORKSHOP's London production of Brendan BEHAN's *The Quare Fellow* (1956). His first plays, *Stop It, Whoever You Are*, about a lavatory attendant in a factory, and *Big Soft Nellie*, about a 'mother's boy', were both seen in London in 1961. *Nil Carborundum* (1962), set in the kitchen of an R.A.F. station, was produced by the ROYAL SHAKESPEARE COMPANY, which also, after the performance of *Kelly's Eye* (1963) at the ROYAL COURT THEATRE, presented *Eh?* (1964), which had a New York production in 1966. Livings's later plays include *The Little Mrs Foster Show* (1966), *Honour and Offer* (1969), *The Finest Family in the Land* (1972), and *Jug* (1975). He uses comedy, farce, and fantasy to depict the trials of people who appear to be underdogs but are not necessarily so, and though he was born and lives in Lancashire and his plays have a Lancashire 'tone', they are hardly ever specifically set there.

LIVING STAGE, see ARENA STAGE, WASHINGTON.

LIVING THEATRE, The, experimental off-Broadway group formed in New York in 1951 by Julian Beck and his wife Judith Malina, who had been one of PISCATOR's students. Their company first appeared in public at the Cherry Lane Theatre, one of the first productions of their original and iconoclastic career being Gertrude Stein's *Dr Faustus Lights the Lights*. From 1954 until 1956 they performed in a loft at Broadway and 100th Street, but it was closed by the Building Department because its 65 seats were considered too many for safety. In 1959 they opened a theatre seating 162 at 14th Street and Sixth Avenue, and used it until 1963, when they were evicted for non-payment of taxes. During this period they produced Jack GELBER's *The Connection* (1959), a play about drug addiction involving some improvisation in performance. In Jackson MacLow's *The Marrying Maiden* (1960) an element of chance was introduced during the performance by allowing the throwing of dice and the drawing of cards to determine the order of lines and actions. The final production before the company left for Europe was Kenneth Brown's *The Brig* (1963), based on his experiences in a U.S. Marine Corps prison.

The Living Theatre had already visited Europe in 1961, winning three first prizes at the THÉÂTRE DES NATIONS in Paris, and from 1964 to 1968 it toured Europe with four pieces made by COLLECTIVE CREATION in which they played themselves and interacted with the spectators. *Mysteries and Smaller Pieces* (1964) was a series of exercises and improvisations; *Frankenstein* (1965) dealt with the creation of an artificial man through acculturation; *Antigone* (1967), based on BRECHT, justified civil disobedience as a protest against the war in Vietnam; and *Paradise Now* (1968) was intended to inaugurate a non-violent anarchist revolution by freeing the individual. In 1968 and 1969 the group toured the U.S.A. with these four pieces, and on returning later to Europe it disbanded. In 1970 Beck, Judith Malina, and several other members of the group went to Brazil with the intention of working with oppressed minorities, but were arrested and imprisoned on drug charges and deported to the U.S.A. They have continued developing new performance techniques, mainly in the service of political manifestations. During the 1960s, working in the U.S.A. and Italy and touring throughout Europe, The Living Theatre was the most influential of the experimental groups which emerged from America. Among its members was Joseph Chaikin, founder of the OPEN THEATER.

LLOYD, MARIE [Matilda Alice Victoria Wood] (1870–1922), the idol of British MUSIC-HALL audiences for many years. She first appeared in 1885, billed as Bella Delmere, a name she soon discarded for the one under which she became famous. She made her first great success at the MIDDLESEX, singing Nellie Power's song, 'The Boy I Love is Up in the Gallery', and she was then engaged for a year at the OXFORD. For three years, from 1891 to 1893, she appeared in PANTOMIME at DRURY LANE, but the 'halls' were her true home and she eventually returned to them for good. In her work she was wittily improper, but never coarse or vulgar, and her reputation for *double entendre* lay less in her material than in her delivery of it, with appropriate actions and an enormous wink. Her cheery vitality and hearty frankness won over all but the most captious critics, particularly with such serio-comic numbers as 'As I Take My Morning Promenade', which enabled her to display her alluring walk and elegant figure in well-cut, fashionable gowns designed and made by herself. In her late 40s she lost her looks, and went into semi-retirement, emerging in 1920 to take part in a revival of old-style music-hall. Among her best-loved songs were 'Oh, Mr Porter!', 'My Old Man Said Follow the Van','A Little of What You Fancy Does You Good', and 'I'm One of the Ruins That Cromwell Knocked Abaht a Bit', which she sang on her last appearance at the Edmonton Empire a few days before her death. Two musicals based on her life, *The Marie Lloyd Story* (1967) and *Sing a Rude Song* (1971), were unsuccessful, probably owing to the difficulty of reproducing her unique personality.

LOA, the prologue or compliment to the audience which preceded the early Spanish theatrical performance. It ranged from a short introductory monologue to a miniature drama having some bearing on the play which was to follow. Águstin de ROJAS, in his *El viaje entretenido* (*The Entertaining Journey*, 1603), says that a strolling company generally had a variety of *loas* which could be fitted to any play. In the 17th century the *loa* appears to have been retained in the public theatres for the first performance only of a new play, although allegorical *loas sacramentales* were written for performance on Corpus Christi day as an introduction to an AUTO SACRAMENTAL, and an introductory 'praise of the audience' continued to be an integral feature of Court performances.

LOBANOV, ANDREI MIKHAILOVICH (1900–59), Soviet director, who began his career in the theatre-studio run by Reuben SIMONOV. While working there during the 1930s he was responsible for some interesting productions, among them OSTROVSKY's *Talent and Its Admirers* in 1931, CHEKHOV's *The Cherry Orchard* in 1934, and GORKY's *Children of the Sun* in 1937. His success with these plays showed where his true talent lay, and although he directed a number of contemporary Soviet plays, particularly during his directorship of the YERMOLOVA THEATRE from 1945 to 1956, he did his best work in revivals of pre-Revolutionary Russian classics.

LOBSTERSCOPE, see LIGHTING.

LOCATELLI, BASILEO (?–1650), an Italian who does not appear to have had any professional connection with the theatre, but was so enamoured of the COMMEDIA DELL'ARTE productions that he collected and copied out over 100 scenarios. These are preserved in two manuscript volumes dated 1618 and 1622, now in the Biblioteca Casanatense in Rome.

LOCATELLI, DOMENICO (1613–71), actor of the COMMEDIA DELL'ARTE, who as Trivellino played a ZANNI or comic servant role somewhat akin to that of ARLECCHINO, which he may also have played. He spent many years in Paris, going there first in about 1644, and again from 1653 until his death. His first wife, Luisa Gabrielli (*fl.* 1644–53), was also an actress and accompanied him to France. They were playing in Modena together shortly before her death.

LOCKE-ELLIOT, SUMNER, see AUSTRALIA.

LODOVICI, CESARE VICO (1885–1968), Italian dramatist of the CREPUSCOLARI or 'twilight' school, also highly regarded as a translator of Shakespeare. Greatly influenced by CHEKHOV, he exploited to the full the range of human communication, in which silence plays so large a part. One of his earliest plays, *L'idiota* (1915), was based on DOSTOEVSKY's novel *The Idiot*, but the first to bring him to the attention of the public was the forceful *La donna di nessuno* (*Nobody's Woman*, 1919). His finest work, *L'incrinatura, o Isa, dove vai?* (*The Flaw, or Isa, Where are You Going?*, 1937), is a sensitive portrayal of the tragedy of non-communication resulting from unfounded jealousy. Although Lodovici was not entirely at ease in the writing of historical dramas, *Vespro siciliano* (1940) and *Catherina da Siena* (1949) had some success.

LOEB DRAMA CENTRE, see HARVARD UNIVERSITY.

LÖFFLER, PETER, see SWITZERLAND.

LOFT BLOCKS, see GRID.

LOGAN, MARIA (1770–1844), English actress, probably the second wife of George COLMAN the younger. Of unknown parentage, she was the godchild of the actor John PALMER, with whom she appeared at the HAYMARKET THEATRE at the age of 13. As an adult actress she appeared at GOODMAN'S FIELDS in 1787 billed as Mrs Gibbs, and retained that name for over 50 years. For many years she was one of the leading ladies of the Haymarket company, eventually taking over the elderly ladies of Old Comedy from Mrs Mattocks. On the younger Colman's death in 1836 she retired to Brighton, where she died. Fair, plump, with blue eyes and a sweet singing voice, she was as much admired for the kindness of her heart and the cheerfulness of her disposition as for her undoubted ability as an actress.

LOHENSTEIN, DANIEL CASPAR VON (1635–83), German dramatist, in whose work German baroque drama reached its highest point. Himself a scholar and a man of quiet, abstemious life, his plays are bloodthirsty melodramas, couched in extravagant language and overburdened with farfetched similes. It is not certain that they were produced, but as literary or CLOSET DRAMA they were widely read and acclaimed, satisfying as they did the prevailing taste for Gothic horrors. They include *Ibrahim Bassa* (1650), *Cleopatra* (1661, rev. 1680), *Agrippina* and *Epicharis* (both 1665), based on the life of Nero, and *Sophonisbe* (written in 1666, pub. 1680). In his last play, *Ibrahim Sultan* (1673), Lohenstein returned to the oriental setting of his first, but with less success.

LÖHR, MARIE (1890–1975), Australian-born actress who became a respected and much-loved star of the London stage. She made her first appearance in Sydney in 1894, and was first seen in London in 1901. Between 1902 and 1907 she toured several times with the KENDALS, and she played Mrs Reginald Bridgenorth in the first production of SHAW's *Getting Married* (1908) at the HAYMARKET. Later the same year she was seen as Margaret in *Faust* by J. Comyns Carr and Stephen PHILLIPS, with Beerbohm TREE, subsequently attracting much attention by her performance as Lady Teazle in his revival of SHERIDAN's *The School for Scandal* (1909). She appeared with John HARE in *The Marionettes* (1911), adapted by Gladys Unger from the French, and continued to play leading parts with the foremost managers of the day, including Charles FROHMAN, George ALEXANDER, and Gerald DU MAURIER, until in 1918 she went into management at the GLOBE THEATRE with her husband Anthony Prinsep, remaining there until 1927, her own roles during that period including Lady Caryll in Robert S. Hichens's *The*

Voice from the Minaret (1919), in which she made her début in New York in 1922, and the Hon. Margot Tatham in LONSDALE's *Aren't We All?* (1923). At Christmas 1927 she made the first of several appearances as Mrs Darling in BARRIE's *Peter Pan*, and she was later seen in John L. Balderston and J. C. Squire's *Berkeley Square* (1929), MAUGHAM's *The Breadwinner* (1930), Dodie SMITH's long-running *Call It a Day* (1935), and Esther McCracken's *Quiet Wedding* (1938). After appearing in Lynne Dexter's *Other People's Houses* (1941) she was absent from the stage for almost five years, but returned with her lively wit and warm humanity undimmed to play in such productions as John WHITING's *A Penny for a Song* (1951), William Douglas HOME's *The Manor of Northstead* (1954), and—after a rather surprising appearance in a musical, John OSBORNE's *The World of Paul Slickey* (1959)—COWARD's *Waiting in the Wings* (1960). She made her last appearance as Mrs Whitefield in a revival of Shaw's *Man and Superman* in 1966.

LONDON ACADEMY OF MUSIC AND DRAMATIC ART, see SCHOOLS OF DRAMA.

LONDON CASINO, see PRINCE EDWARD THEATRE.

LONDON COLISEUM, see COLISEUM.

LONDON HIPPODROME, see HIPPODROME.

LONDON MUSIC-HALL, Shoreditch, see MUSIC-HALL.

LONDON PALLADIUM, in Argyll Street. The first building on this site was a circus which ran successfully under Hengler, who enlarged the building considerably, from 1871 to his death in 1887. Its fortunes then declined, and it was reconstructed as a MUSIC-HALL in three tiers, with a proscenium width of 47 ft and a stage depth of 40 ft. Its seating capacity of 2,325 is the largest of any live theatre in the West End. (The COLISEUM, which has more seats, now ranks as an opera house.) The theatre opened as the Palladium on 26 Dec. 1910 with a VARIETY bill which included Nellie WALLACE and also MARTIN-HARVEY in a one-act play. Variety was soon replaced by REVUE, and shows such as *Rockets* (1922) and *The Whirl of the World* (1923) had a considerable success. The theatre was used as a cinema for three months in 1928, but then came under the control of George Black, who brought there the CRAZY GANG in a series of shows which included *Life Begins at Oxford Circus* (1935), *O-Kay for Sound* (1936), *These Foolish Things* (1938), and *The Little Dog Laughed* (1939). From 1930 until 1938 there was also a Christmas revival of BARRIE's *Peter Pan*, usually with Jean FORBES-ROBERTSON as Peter. The theatre was officially renamed the London Palladium in 1934. During the Second World War more revues followed, including Irving Berlin's *This Is the Army*, but under Val Parnell, who succeeded Black in 1946, there was a policy of twice-nightly variety with a number of acts. Later, following a general trend, the Palladium changed to once-nightly performances by a top international star with only one or two supporting acts. Leslie MacDonnell took over in 1960, and in 1968 there was a short break in continuity when Sammy Davis Junior appeared in a musical version of Clifford ODETS's *Golden Boy*. Under Louis Benjamin, who replaced MacDonnell in 1970, musicals and star appearances both featured in the schedules. In 1974 the Christmas pantomime was replaced by a musical based on the life of Hans Andersen, with Tommy Steele in the title role. There was a highly successful revival of *The King and I* in 1979, and in 1981 *Barnum*, with Michael Crawford, began a run which was to become the longest in the Palladium's history. The theatre has been used more than any other for the annual Royal Variety Show since its inception in 1930.

LONDON PAVILION, famous MUSIC-HALL which began as a 'song-and-supper' room attached to the Black Horse Inn in Tichborne Street at the top of the Haymarket, the stable-yard being roofed in for the purpose. It closed in 1860, and after extensive renovation reopened on 23 Feb. 1861 as a music-hall seating 2,000 people. Many famous music-hall stars appeared there, including the Great MacDermott, whose patriotic song 'We don't want to fight, but, by Jingo, if we do . . .', by G. W. Hunt, added the expression 'Jingoism' to the English language. The hall was demolished on 26 Mar. 1885 and a new Pavilion arose on the site, opening on 30 Nov. the same year. It retained the old-style separate tables and chairs, with a presiding CHAIRMAN. They were abolished a year later, when normal theatre seating was installed, but music-hall 'turns', sometimes as many as 20 in one evening, still reigned supreme, until in 1918 COCHRAN took over and the Pavilion became a theatre, famous for its REVUES. Among the best were *Fun of the Fayre* (1921), starring Evelyn LAYE; *Dover Street to Dixie* (1923), with Florence Mills, who reappeared with the all-black *Blackbirds* (1926); COWARD's *On With the Dance* (1925), *This Year of Grace* (1928) (with Jessie MATTHEWS), and *Cochran's 1931 Revue*; *One Dam Thing After Another* (1927) and *Wake Up and Dream* (1929), both also starring Jessie Matthews. After Cochran left in 1931 the theatre turned to non-stop VARIETY, the last performance being on 7 Apr. 1934, after which the building was gutted and became a cinema. It was closed for demolition in 1981.

LONDON THEATRE STUDIO, see SAINT-DENIS, MICHEL.

LONG, JOHN LUTHER (1861–1927), American novelist, playwright, and librettist, whose short story *Madame Butterfly* was dramatized by

BELASCO with great success in 1900. It was used in 1904 as the basis of an opera by Puccini. Long later collaborated with Belasco on *The Darling of the Gods* (1902), a romantic melodrama set in ancient Japan which TREE produced in London in 1903, and on *Adrea* (1905), a tragedy in which Mrs Leslie CARTER successfully played the title role. Long wrote several other plays but, lacking Belasco's technical expertise, these were not successful.

LONGACRE THEATRE, New York, at 220 West 48th Street, between Broadway and 8th Avenue. With a seating capacity of 1,115, this opened on 1 May 1913 and was used mainly for MUSICAL COMEDY, though KAUFMAN's farce *The Butter and Egg Man* (1925) had a long run there. Later the theatre housed short runs of several notable plays, among them G. B. Stern's *The Matriarch* (1930) and OBEY's *Noé* (1935) with Pierre FRESNAY in the title role. Several plays by ODETS were produced under the auspices of the GROUP THEATRE, but in 1944 the building was taken over for broadcasting. It returned to use as a theatre in 1953, Lillian HELLMAN's adaptation of ANOUILH's *The Lark* being staged there in 1955 with Julie HARRIS as Joan of Arc. Emlyn WILLIAMS was seen as Dylan Thomas in his own compilation from the poet's works *A Boy Growing Up* in 1957, and in 1961 Zero MOSTEL starred in IONESCO's *Rhinoceros*. Robert ANDERSON's *I Never Sang For My Father* (1968) was followed in the 1970s by Julie Harris's solo performance as Emily Dickinson in William Luce's *The Belle of Amherst* and GIELGUD and RICHARDSON in PINTER's *No Man's Land* (both 1976); a musical tribute to Fats Waller, *Ain't Misbehavin'* (1978); and John GUARE's *Bosoms and Neglect* (1979). Mark Medoff's *Children of a Lesser God* was a big success in 1980.

LONGFORD, 6th Earl of [Edward Arthur Henry Pakenham] (1902–61), Irish dramatist and director, who became chairman of the GATE THEATRE in Dublin, and founder of Longford Productions, his company alternating at the theatre with Hilton EDWARDS and Michéal MACLIAMMÓIR's Gate Theatre Productions in classic plays by English, Greek, and European writers, as well as translations from modern Continental dramatists. The best of Lord Longford's own plays was probably *Yahoo* (1933), which, with Hilton Edwards as Jonathan Swift, had a great success both in Dublin and in London, where the Longford company played seasons at the WESTMINSTER THEATRE in 1935, 1936, and 1937. With his wife, Christine Patti Trew, he translated plays from classical Greek and French; Lady Longford also wrote a number of plays and stage adaptations of 19th-century novels.

LONGPIERRE, HILAIRE-BERNARD DE ROQUELEYNE (1659–1731), French dramatist, and the immediate successor in neo-classical tragedy of RACINE, whose works inspired him to write for the stage. He was well educated and a student of antiquity, as such study was understood at the time, and he retained enough of the form and psychological conflict of the great days in his plays, of which the best was *Médée* (1694), for his contemporaries to rate him highly. He had everything of Racine but his genius, and his plays are now forgotten.

LONG WHARF THEATRE, New Haven, Connecticut, is situated in a meat and produce terminal. A thrust-stage playhouse seating 484, it opened in 1965 and now has one of the most highly regarded resident companies in the United States. During seasons running from Oct. to June it presents classic and modern plays, including world and American premières. Some of its productions have moved practically unaltered to Broadway, among them David STOREY's *The Changing Room* in 1973; Athol FUGARD's *Sizwe Bansi Is Dead* in 1975; Michael Cristofer's *The Shadow Box* and D. L. Coburn's *The Gin Game*, both in 1977; and David Mamet's *American Buffalo* and Lillian HELLMAN's *Watch on the Rhine*, both in 1980. It has an intimate Stage II.

LONSDALE [Leonard], FREDERICK (1881–1954), English dramatist, who was responsible for the librettos of several MUSICAL COMEDIES, among them *The Maid of the Mountains* (1917), *Monsieur Beaucaire* (1919), and *Madame Pompadour* (1923). He is best remembered, however, as the author of a number of comedies of contemporary manners somewhat in the style of Somerset MAUGHAM, though with less subtlety. The best known are *Aren't We All?* (London and N.Y., 1923), *Spring Cleaning* (N.Y., 1923; London, 1925), *The Last of Mrs Cheyney* (London and N.Y., 1925), *On Approval* (N.Y., 1926; London 1927), *The High Road* (1927; N.Y., 1928), and *Canaries Sometimes Sing* (1929; N.Y., 1930). All gave scope for good, brittle, sophisticated acting, and stars of the calibre of Gladys COOPER, Marie LÖHR, and Yvonne ARNAUD appeared successfully in them. Their amusing situations, easy and effective dialogue and rich, worldly, and well-bred characters made them immensely popular in their day. Lonsdale's later plays, which include *Another Love Story* (N.Y., 1943; London, 1944), *The Way Things Go* (1950), and *Let Them Eat Cake* (produced posthumously in 1959), were not so popular.

LOPE DE VEGA, see VEGA CARPIO, LOPE DE.

LOPE DE RUEDA, see RUEDA, LOPE DE.

LÓPEZ DE AYALA, ADELARDO (1829–79), Spanish dramatist and with TAMAYO the chief representative of the transition from romanticism to realism. His early plays, among them *Un hombre de estado* (*The Statesman*, 1851), are still basically romantic, but his later comedies, particularly *El* *tanto por ciento* (*The Percentage*, 1861) and *Consuelo* (1878), are bitter attacks on the materialistic tendencies of his time.

LORAINE, ROBERT (1876–1935), English actor-manager, who made his first appearance on the stage in 1889 and subsequently made a hit as D'Artagnan in a dramatization of the elder DUMAS's *The Three Musketeers* (1899). In 1911, after a visit to America, he took over the CRITERION THEATRE in London, opening with a revival of *Man and Superman* by SHAW, whose Don Juan he had played in the first production of *Don Juan in Hell* (1907). During the First World War he made a great reputation as an aviator and was awarded the M.C. and the D.S.O. for gallantry in action. Essentially a romantic actor, he returned to the stage in 1919 in the title role of ROSTAND's *Cyrano de Bergerac*, which had a long run. Among his later parts were Deburau in Sacha GUITRY's play of that name in 1921, the dual role of Rassendyl and Prince Rudolf in a revival of Hope's *The Prisoner of Zenda* in 1923, the Nobleman in the New York production of Ashley DUKES's *The Man with a Load of Mischief* in 1925, Adolf in STRINDBERG's *The Father* in 1927 in a double bill with BARRIE's *Barbara's Wedding*, in which he played the Colonel, and a number of Shakespearian parts, including Petruchio in *The Taming of the Shrew* and Mercutio in *Romeo and Juliet*, both in 1926.

LORCA, FEDERICO GARCÍA, see GARCÍA LORCA, FEDERICO.

LORD CHAMBERLAIN, an officer of the British Royal Household under whom the MASTER OF THE REVELS was first appointed in 1494 to supervise Court entertainments. After the Restoration in 1660 the Lord Chamberlain himself began to intervene directly in the regulation of theatres and in CENSORSHIP, mainly in relation to political and religious issues, his powers being legally formulated by the Licensing Act of 1737. These powers, particularly in regard to the LICENSING OF THEATRES, were modified by the Theatres Act of 1843.

New plays, and new matter added to existing plays, had to be submitted to the Lord Chamberlain's office for scrutiny by the Examiner of Plays. At the time of the abolition of his powers the Lord Chamberlain had three English readers and one Welsh, who reported on the work submitted, with particular reference to indecency, impropriety, profanity, seditious matter, and the representation of living persons. Plays written before the passing of the 1843 Act could be suppressed under Section 14 of the Act. Play-producing societies and theatre clubs came within the Lord Chamberlain's powers, although it was customary to treat bona fide societies with lenience. His power to withhold or withdraw a licence was absolute and he was under no legal obligation to disclose the reasons for his readers' decisions; but in practice the Lord Chamberlain's office, when refusing a licence, was normally ready to indicate changes in the text that would enable a licence to be issued. There was no appeal against his decision.

The Theatres Act of 1968 took all theatrical matters out of the hands of the Lord Chamberlain, vesting theatre licensing in local authorities and abolishing theatrical censorship in England, since when stage plays have been subject to common law such as those Acts pertaining to obscenity, blasphemy, libel, breach of the peace, and so on.

LORD OF MISRULE, OF UNREASON, see MISRULE, LORD OF.

LORNE, MARION (1888–1968), American actress who, made her début on Broadway in 1905, and later played a wide variety of parts with the Hartford stock company, including Dora in SARDOU's *Diplomacy*, Lady Babbie in BARRIE's *The Little Minister*, and the title role in ZANGWILL's *Merely Mary Ann*. In 1915 she accompanied her husband Walter HACKETT to London, where she appeared in *He Didn't Want To Do It*, written by Hackett in collaboration with George H. BROADHURST, and during the next 23 years starred in a series of comedies, written by her husband, at various London theatres, including the WHITEHALL, which he opened in 1931 and ran until 1934. Her last appearance in London was in *Toss of a Coin* (1938), in which she played the dual roles of Dollie Goring and Dollie Fairlight, directed by her husband, with whom she then returned to America.

LOS ANGELES, California, see MARK TAPER FORUM.

LOTTA, see CRABTREE, CHARLOTTE.

LOUTHERBOURG, PHILIP JAMES DE (1740–1812), German painter, who had for some time made a special study of stage illusion and mechanics before visiting London in 1771. There he met GARRICK, who two years later appointed him scenic director at DRURY LANE, a position he held also under SHERIDAN. He introduced a number of new devices, including a series of border-battens behind the PROSCENIUM which discouraged the actors from stepping too far outside the picture frame, and enhanced the importance of the scenery by a flood of illumination. He was particularly successful in producing the illusion of fire, volcanoes, sun, moonlight, and cloud-effects, and invented strikingly effective devices for thunder, guns, wind, the lapping of waves, and the patter of hail and rain. He was the first designer to bring a breath of naturalism into the artificial scenic conventions of the day, and paved the way for the realistic detail and local colour favoured by

Charles KEMBLE. A visit to the Peak district in 1779 resulted in an act-drop depicting a romantic landscape—possibly the earliest example of a scenic curtain in Western Europe—which remained in use until Drury Lane was destroyed by fire in 1809. For Sheridan's *The Critic* (1779), in which Mr Puff refers to him by name, he executed a striking design of Tilbury Fort, and he was also responsible for some excellent new transparencies used in a revival of *The Winter's Tale* in the same year. He was probably the first designer in England to break up the scene by the use of perspective, and also the first to use BUILT STUFF, though somewhat sparingly. Much of his best work was lavished on unremarkable plays, to which he gave a momentary popularity. Shortly after preparing the scenery for the first dramatization of Defoe's *Robinson Crusoe* (1781) by Sheridan—the first act alone had 8 changes—Loutherbourg left the theatre, mainly on account of a dispute over his salary, and devoted much of his time to a remarkable scenic exhibition, the 'Ediophusikon', whose influence lingered on until 1820, when ELLISTON attempted to reproduce in *King Lear* the powerful effects of storm and tempest which Loutherbourg had created for his exhibition.

LOVINESCU, HORIA, see ROMANIA.

LOW COMEDIAN, see STOCK COMPANY.

LOWIN, JOHN (1576–1653), English actor, and an important link between the Elizabethan and Restoration stages, since he was said to have passed on to DAVENANT the instructions he had received for the playing of Henry VIII from Shakespeare himself, Davenant passing them on to his chief actor under Charles II, Thomas BETTERTON. It is possible that Lowin, who is always referred to as a big, bluff man, may have been the first to play the part in 1613; he is also believed to have been much admired as Falstaff in revivals of *Henry IV*, and as the eponymous hero of Ben JONSON's *Volpone* (1606), though he may not have been its creator. He is first mentioned as an actor in 1602, and a year later joined the KING'S MEN on their foundation, his name being found in the cast-lists of Jonson's *Sejanus* (1603) and MARSTON's *The Malcontent* (1604), probably as a player of small parts. He created the part of Melantius in BEAUMONT and FLETCHER's *The Maid's Tragedy* (1610), and probably that of Bosola in WEBSTER's *The Duchess of Malfi* (1614); he also appeared in several of MASSINGER's plays. He remained with the King's Men until the closing of the theatres in 1642, and was one of the actors caught and punished under the Commonwealth for playing surreptitiously at the COCKPIT in 1649. In his old age he kept the Three Pigeons at Brentford.

L.T.D., see O'NEILL, EUGENE.

LUCAN, ARTHUR, see FEMALE IMPERSONATION.

LUCILLE LA VERNE THEATRE, New York, see PRINCESS THEATRE.

LUCY RUSHTON'S THEATRE, New York, see NEW YORK THEATRE (1).

LUDLOW, NOAH MILLER (1795–1886), American actor-manager of the pioneering theatre, who in 1815 was engaged by Samuel DRAKE to go to Kentucky and later founded his own company, with which in 1817 he gave English plays in New Orleans. He travelled extensively, with his own or other companies, often being the first actor to penetrate to some of the more remote regions in the South and West. In 1828 he was induced to take over the old CHATHAM THEATRE in New York with COOPER, but failed to make it pay, and moved to St Louis. From 1835 to 1853 he was in partnership with Sol SMITH in the American Theatrical Commonwealth Company, and ran several theatres simultaneously in St Louis, New Orleans, Mobile, and other cities, often engaging outstanding stars. He was himself an excellent actor, particularly in comedy, and the author of an entertaining volume of reminiscences, *Dramatic Life as I Found It*.

LUDWIG, OTTO (1813–65), German novelist and dramatist, a contemporary of HEBBEL, whom he resembles in his progress from prose to verse. Most of his plays were not performed until long after his death, but *Der Erbförster* (*The Hereditary Forester*), a realistic study of bourgeois life, was seen at Dresden in 1850, and an apocryphal drama, *Die Makkabäer*, at the Vienna Burgtheater in 1852. Ludwig, who studied music under Mendelssohn, was an ardent admirer of Shakespeare, on whom he wrote a number of essays published in 1871.

LUGNÉ-POË, AURÉLIEN-FRANÇOIS (1869–1940), French actor, director, and theatre manager, who in 1893 took over the Théâtre d'Art from Paul FORT, renaming it the Théâtre de l'Oeuvre. He remained there until 1899 and returned to it from 1912 to 1929. In his early years he continued Fort's reaction against the realism of the THÉÂTRE LIBRE by staging poetic and symbolist plays, using transparent curtains against a painted backcloth, abolishing the footlights and the box-set, and replacing realistic speech by formal intoning. His first production was *Pelléas et Mélisande* by MAETERLINCK, several of whose later plays were also first seen at this theatre, and among the other foreign authors he presented were IBSEN, BJØRNSON, HAUPTMANN, D'ANNUNZIO, and ECHEGARAY. In 1895 he staged WILDE's *Salome*, and he brought CLAUDEL before the public with *L'Annonce faite à Marie* in 1912 and *L'Ôtage* in 1914. In 1896 he was responsible for the riotous first performances of JARRY's *Ubu-roi*. Between 1919 and 1929 he introduced SHAW and STRINDBERG to Paris, and encouraged such new French playwrights as ACHARD. After he left the Théâtre de l'Oeuvre he continued to direct elsewhere, being associated in 1932 with the production of ANOUILH's first play *L'Hermine*, and in 1935 with *L'Inconnue d'Arras* by SALACROU, whose *Terre à Terre* he had presented as early as 1925. Himself an excellent actor, Lugné-Poë appeared in many of his own early productions, and in 1908 was seen in London in Jules Renard's *Poil de Carotte*.

LUN, see RICH, JOHN; **LUN** junior, see WOODWARD, HARRY.

LUNACHARSKY, ANATOLI VASILEVICH (1875–1933), the first Commissar for Education in Soviet Russia, an able and cultured man to whom the U.S.S.R. owes the preservation and rejuvenation of those Imperial theatres—notably the MOSCOW ART THEATRE—which survived the Revolution. He was also responsible for the organization of the new Soviet theatres which sprang up in vast numbers, many of which are named after him. He realized that the new audiences, many of whom had never been in a theatre before, would inevitably demand new plays and new methods of production, but that the old plays, both Russian and European, were part of the cultural heritage of the new world and must not be discarded. With this end in view he endeavoured, while supporting the move towards SOCIALIST REALISM, to counter some of the post-revolutionary experimental excesses by a return to the past, skilfully adapted to meet contemporary requirements. He was himself the author of a number of plays mainly on revolutionary themes, one of which, *Oliver Cromwell* (1921), pays tribute to a fellow-revolutionary and draws an interesting parallel between the establishment of the Commonwealth in England in the 17th century and the Russian Revolution of 1917.

LUNACHARSKY STATE INSTITUTE OF THEATRE ART, see SCHOOLS OF DRAMA.

LUNCH-TIME THEATRE, see FRINGE THEATRE.

LUNDEQUIST, GERDA (1871–1959), Swedish actress, one of the foremost tragic players of her day, who was a member of the company at the KUNGLIGA DRAMATISKA TEATERN from 1889 to 1891, and then went to the Svenska Teatern, where she remained until 1906, rejoining the Kungliga Dramatiska company in that year and remaining with them until 1911. She was particularly admired as SOPHOCLES' Antigone and as Lady Macbeth.

LUNT, ALFRED (1892–1977), American actor, who made his début in Boston in 1912 and had his first big success in 1919, playing the title role in Booth TARKINGTON's *Clarence*. In 1922 he married LYNN [Lillie Louise] **FONTANNE** (1887–1983), who after studying with Ellen TERRY and making her début in pantomime in 1905, had pursued a successful stage career in London and New York. The Lunts first appeared together in KESTER's *Sweet Nell of Old Drury* (1923). They then joined the THEATRE GUILD company, and over the next 5 years appeared in a succession of distinguished plays, including SHAW's *Arms and the Man* and *Pygmalion*, MOLNÁR's *The Guardsman*, and WERFEL's *Goat Song*. They were first seen together in London in 1929 in G. Sil-Vara's *Caprice*, adapted by Philip Moeller. Later successes were Robert SHERWOOD's *Reunion in Vienna* (1931; London, 1934); COWARD's *Design for Living* (1933), GIRAUDOUX's *Amphitryon 38*, adapted by S. N. BEHRMAN (1937; London, 1938); and RATTIGAN's *Love in Idleness* (1944), seen in New York in 1946 as *O Mistress Mine*. They were at their best in intimate modern comedy, playing together with a subtlety and sophistication which gave life and strength even to the flimsiest dramatic material. In 1958 they appeared at the Globe Theatre, New York, renamed the LUNT-FONTANNE in their honour, in DÜRRENMATT's *The Visit* (with which they had previously toured in England under the title of *Time and Again*). This was the first production seen at the opening of the new ROYALTY THEATRE (3) in London in 1960, their last appearance. In 1964 the Lunts were awarded the US Medal of Freedom by President Johnson.

LUNT-FONTANNE THEATRE, New York, at 205 West 46th Street. This attractive playhouse was opened by Charles DILLINGHAM as the Globe on 10 Jan. 1910, mainly for musical shows, though at the end of the year it housed the company of Sarah BERNHARDT in a repertory of French plays. The *Ziegfeld Follies* were there in 1921 and *George White's Scandals* in 1922 and 1923. After a successful musical, *The Cat and the Fiddle* (1931), it became a cinema; but in 1958, completely remodelled and redecorated, with a seating capacity of 1,714, it reopened as the Lunt-Fontanne with Alfred LUNT and Lynn Fontanne in DÜRRENMATT's *The Visit*. In 1963 it saw a brilliant but short-lived run of BRECHT's *Arturo Ui* and from 9 Apr. to 8 Aug. 1964 Richard BURTON appeared in *Hamlet*, setting up a new record for the run of the play in New York. In the 1970s the theatre had several musical successes, including *The Rothschilds* (1970) and revivals of *My Fair Lady* (1976), from the ST JAMES, *Hello, Dolly!* (1978), and *Peter Pan* (1979), based on BARRIE's play. In 1981 it had another success with *Sophisticated Ladies*, a compilation of Duke Ellington numbers.

LUPINO, BARRY (1882–1962), English comedian, one of a numerous family (including Lupino LANE) descended from an Italian puppet-master who emigrated to England early in the 17th century. He made his first appearance as a baby in *Cinderella*, and as an adult actor was for some years the leading comedian and pantomimist at the BRITANNIA THEATRE, being connected by marriage with the lessees Sam and Sara LANE. He toured extensively, was seen in MUSICAL COMEDY, and played the DAME in innumerable PANTOMIMES, of which he himself wrote more than fifty. Two of his finest parts were Dame Durden in *Jack and the Beanstalk* and the Widow Twankey in *Aladdin*, which he was still playing in his seventies. He made his last appearance as Dame Sarah in *Dick Whittington* in 1954.

LUPINO, STANLEY (1893–1942), English comedian, younger brother of Barry (above). He was on the stage as a child, as a member of an acrobatic troupe, and appeared for many years in PANTOMIME at DRURY LANE. He was also seen in MUSICAL COMEDY and REVUE and was part author, with Arthur Rigby, of *So This Is Love* (1928), in which he played Potty Griggs, *Love Lies* (1929), in which he played Jerry Walker, and *Room for Two* (1932), which he directed. He wrote a number of plays on his own, notably *The Love Race* (1930), *Hold My Hand* (1931), *Over She Goes* (1936), and *Crazy Days* (1937), in all of which he also appeared. In 1934 he published a volume of reminiscences, *From the Stocks to the Stars*.

LUSCIUS LANUVINUS, Roman comic dramatist of the mid-2nd century BC, and a contemporary of CAECILIUS STATIUS. He is now mainly remembered for his attack on TERENCE, whom he accused of plagiarism, poor writing, and falsification (*contaminare*) of the Greek originals of his comedies. Terence called him a 'malicious mischiefmaker' and defended himself hotly against these attacks in the prologues to several of his plays. Luscius is known to have translated MENANDER's *Thesauros* (*The Treasure*) and *Phasma* (*The Ghost*), and although nothing of his work has survived there is reason to believe that he was a much more literal translator than any other known dramatist of the time.

LUTÈCE, Théâtre de, Paris. Theatre in the rue de Jussieu, which under the direction of Lucie Germain became a home for modern experimental work, notably the Theatre of the ABSURD. Roger BLIN's production of GENET's *Les Nègres* was seen there in 1959, and other authors whose work has been presented include Dubillard and OBALDIA.

LUZÁN Y CLARAMONT, IGNACIO (1702–54), Spanish man of letters, the most important theorist of the Spanish neo-classical period. His *Poética*, published in 1737, showed the influence of the classical doctrines of the Italian Renaissance, under whose yoke of quasi-Aristotelian UNITIES Luzán endeavoured to subjugate Spanish poetry and drama. Luzán mentions with admiration the 16th-century poetry of Garcilaso, Camoëns, the Argensola brothers, and Herrera. He regrets that the dramatists of the 17th century, Lope de VEGA, CALDERÓN, and others, did not write according to the rules, for otherwise their work would have been the envy and admiration of other nations. The posthumous second edition (1789) shows a certain mellowing of his views: the dramatist is free to follow his instincts, his whims, or to write for the delight of the common people, but in all periods enlightened writers will turn to the rules which are based on reason and authority.

LUZZATO, MOSES HAYIM (1707–47), an Italian Jew, resident in Amsterdam, author of a number of plays in Hebrew, of which the earliest, written when he was only 17, dealt with Samson among the Philistines. For his uncle's wedding in 1727 he wrote *Migdal Oz*, a pastoral modelled on the *Pastor Fido* of GUARINI, which contains many passages of great lyric beauty. Among his later plays the most important, written for a pupil's wedding and published in 1743, was *Tehilla Layesharim*. Luzzato was one of the first Hebrew dramatists to attempt dramatic criticism, one section of a grammar for which he was responsible containing an attempted definition of drama.

LYCEUM THEATRE, Edinburgh. Built in 1883 to accommodate touring companies, mainly from England, this large theatre in Grindley Street was named after the London home of Henry IRVING, whose company was the first to appear in it in a repertory which included *Hamlet, Much Ado About Nothing*, and Lewis's *The Bells*. In 1964 it was bought by the Edinburgh Corporation to house the newly-founded Edinburgh Civic Theatre company, which had just been formed with the help of grants from the ARTS COUNCIL and the local authority; the opening production on 1 Oct. 1965 was a Scots version of GOLDONI's *Servant of Two Masters*. Two years later Clive Perry became artistic director of the company and built up a large and enthusiastic audience by a skilful mingling of traditional and modern productions, and when in 1971 Bill BRYDEN became associate director a greater commitment to contemporary Scottish drama was foreseen. An offshoot of the Lyceum, the Young Lyceum company, formed in 1975, occupies a studio theatre in Market Street. The artistic directorship of Stephen Macdonald from 1976 to 1979, which promised much, was marred by public disagreements with the Edinburgh District Council over funding and the new company's role in the Scottish theatrical scene.

LYCEUM THEATRE, London, in Wellington Street, just off the Strand. There was a place of entertainment on the site as early as 1765 but it was not until 1809, when the DRURY LANE company moved there after the destruction by fire of their own theatre, that it was licensed for plays. When the new Drury Lane opened in 1812 the Lyceum retained its licence for the summer months only, being used for mixed entertainments of opera and plays. In 1816 it was rebuilt and renamed the English Opera House, opening on 15 June 1817 with two plays starring Fanny KELLY. The elder Charles

MATHEWS introduced the first of his entertainments, 'Mathews at Home', on 2 Apr. 1818; PLANCHÉ's melodrama *The Vampire; or, the Bride of the Isles* opened on 9 Aug. 1820; and in 1821 Julia Glover appeared as Hamlet. On 16 Feb. 1830 the theatre was burnt down and on 14 July 1834 a new building, whose frontage still stands on Wellington Street, opened as the Royal Lyceum and English Opera House. This was the building in which Henry IRVING made his name. It started well, but had a chequered career until the passing of the Licensing Act of 1843 enabled it to go over to the legitimate drama. Robert KEELEY ran the theatre with his wife from 1844 to 1847; among their successes was *Mrs Caudle's Curtain Lectures* (1845) by Stirling. After Keeley left Mme VESTRIS and her husband, the younger Charles MATHEWS, took over, and a series of brilliant productions followed, mostly of EXTRAVAGANZAS by Planché. In one of them, *The Island of Jewels* (1849), the scene-painter BEVERLEY introduced the first TRANSFORMATION SCENE to be seen in London. Vestris's management was not a success, owing to her extravagance, and eventually she went bankrupt. The theatre was then occupied for two years by the company from COVENT GARDEN, which had been burnt down in 1856. Little of note happened, except for seasons by Mme CÉLESTE from 1858 to 1861 and by FECHTER from 1863 to 1867, until in 1871 the American impresario H. L. BATEMAN took the theatre in order to present his three daughters, Kate, Virginia, and Isabel, in a London season, engaging as his leading man the young Henry Irving. Seven years later Irving took control, and with Ellen TERRY as his leading lady inaugurated a series of fine productions which made the Lyceum the most notable theatre in London. They made their last appearance there together (though both continued to appear elsewhere separately) on 19 July 1902 in *The Merchant of Venice*, and after their departure the fortunes of the theatre declined. It was partly demolished and rebuilt as a MUSIC-HALL which opened on 31 Dec. 1904, with no success. Under the MELVILLE brothers from 1910 to 1938 it became famous as the home of MELODRAMA and PANTOMIME. In 1939 the building was scheduled for demolition and six farewell performances of *Hamlet*, with John GIELGUD, took place, ending on 1 July. The outbreak of the Second World War two months later led to the abandonment of the demolition scheme and the theatre stood empty until 1945 when it became a dance hall.

LYCEUM THEATRE, New York. (1) At 312–16 4th Avenue, between 23rd and 24th Streets, a small but lavishly appointed theatre built by Steele MACKAYE after his failure at the MADISON SQUARE THEATRE, which opened on 6 Apr. 1885 with his *Dakolar*, starring Robert B. MANTELL and Viola ALLEN; it was not a success, and Mackaye withdrew from management. The theatre was then taken over by Daniel FROHMAN, who had many successes there until he left in 1902, when it was pulled down.

(2) At 149 West 45th Street, between Broadway and the Avenue of the Americas. This theatre opened on 2 Nov. 1903, having been built by Daniel FROHMAN, who for many years lived in apartments on the top floor. Luxuriously appointed, seating 995, it is one of the few New York theatres to have a GREEN ROOM. The first new play produced there was BARRIE's *The Admirable Crichton*, and later Charles WYNDHAM brought his London company for an eight-week season. From 1916 a number of productions by BELASCO, Frohman's first stage manager at the Lyceum, preceded a fine performance by David WARFIELD as Shylock in *The Merchant of Venice* (1922); the same season brought Ethel BARRYMORE in SHERIDAN's *The School for Scandal. Antony and Cleopatra* (1924), starring Jane COWL, was not a success. In 1929 *Berkeley Square*, by John Balderston and J. C. Squire, ran for 227 performances, and further successes were achieved in 1940 by KAUFMAN and HART's *George Washington Slept Here*; in 1941 by Joseph FIELDS and Jerome Chodorov's *Junior Miss* (in which year SAROYAN's *The Beautiful People* had an artistic though not commercial success); and in 1946 by Garson Kanin's *Born Yesterday*. Later productions included ODETS's *The Country Girl* (1950) and three plays imported from London—OSBORNE's *Look Back in Anger* (1957), Shelagh Delaney's *A Taste of Honey* (1960), and PINTER's *The Caretaker* (1961). From 1964 to 1969 the theatre housed the company from the PHOENIX THEATRE. In 1971 it staged Daniel Berrigan's *The Trial of the Catonsville Nine* and in 1976 a musical, *Your Arms Too Short to Box with God*, based on the Gospel according to St Matthew. Arthur KOPIT's *Wings* was seen in 1979, with Constance CUMMINGS, and in 1980 Paul Osborn's *Morning's at Seven* had a successful revival.

An earlier Lyceum, at the corner of Broadway and Warren Street, survived for a single season in 1808. When it first opened in 1850, WALLACK'S LYCEUM was known as the Lyceum, as was the FOURTEENTH STREET THEATRE from 1873 to 1886.

LYCOPHRON (b. *c.* 324 BC), learned Alexandrian poet of the HELLENISTIC AGE, composer of tragedies, none of which survives. He was one of the original Pleiad.

LYCOPODIUM FLASK, a blow-pipe of vegetable brimstone which added a white flame to the terrors of red fire in the conflagrations of MELODRAMA.

LYLY, JOHN (*c.* 1554–1606), English novelist and dramatist, important as the first writer of sophisticated comedy and for his use of prose in drama. He is perhaps best known for his novels *Euphues; the Anatomy of Wit* (1579) and *Euphues and his England* (1580) whose peculiarly involved and allusive style gave rise to the term 'euphuism'; but by his contemporaries he was regarded as an outstanding dramatist, and the elegance of his style had a salutary effect on some of the more full-blooded dramatists of the day. He wrote almost exclusively for a cultured and courtly audience which delighted in the grace and artificiality of his dialogue and in his many sly allusions to contemporary scandals. All his plays were acted by BOY COMPANIES, the first two, *Campaspe* and *Sapho and Phao* (1584), at BLACKFRIARS and at Court by a combined group of choristers from St Paul's and the Chapel Royal. The later *Galatea* (1585), *Endymion the Man in the Moon* (1588), *Midas and Mother Bombie* (both 1589) (this last based on TERENCE), all like their predecessors comedies on classical themes, were performed, probably at Court and in the hall of their song school, by Paul's Boys, of which Lyly became vice-master in about 1590. They also appeared in the pastoral *Love's Metamorphosis* (1590). *The Woman in the Moon* (1593), again a comedy, may not have been acted. Lyly is believed to have had a hand in a number of other plays and in some of the Court entertainments, but nothing can be ascribed to him with any certainty. He outlived the popularity of his work, but may have had the satisfaction of knowing that he had materially helped to lay the foundations of the great age of Elizabethan drama.

LYNDSAY, Sir DAVID (1490–*c.* 1554), Scottish poet, and the author of the only surviving example of a Scottish medieval MORALITY PLAY *Ane Pleasant Satyre of the Thrie Estaitis*. First performed at Cupar in 1552 in the presence of James V, this was a serious attack on the established Church and the authority of the Pope, interspersed with comic episodes highlighting clerical follies and abuses of the time. The complete play, which was added to and revised in 1554, is very long, but an abbreviated and modernized version was successfully produced at the EDINBURGH FESTIVAL in 1948 under the direction of Tyrone GUTHRIE and has since been revived several times.

LYNN, RALPH (1882–1962), English actor, who specialized in 'silly ass' parts, with monocle, protruding teeth, a winning smile, and sweet though asinine reasonableness in the most trying circumstances. He made his first appearance on the stage at Wigan in 1900, and after many years in the provinces and a visit to New York in 1913 was seen in London in 1914, making his first outstanding success as Aubrey Henry Maitland Allington in Will EVANS's farce *Tons of Money* (1922). First seen at the SHAFTESBURY THEATRE, it was transferred to the ALDWYCH. There, after Walter HACKETT's farce *It Pays to Advertise* (1924), in which he played Rodney Martin, Lynn was to remain for nearly ten years, starring with Robertson HARE and Tom WALLS in the series of so-called 'Aldwych farces' which began with *A Cuckoo in the Nest* (1925), included *Rookery Nook* (1926) and *Thark* (1928), and ended with *A Bit of a Test* (1933), all by Ben TRAVERS. After leaving the Aldwych, Lynn appeared in the long-running farcical comedy *Is Your Honeymoon Really Necessary?* (1944) by E. Vivian Tidmarsh, which he also directed. It was followed by two more farces by Travers, in which Lynn again teamed up with Robertson Hare—*Outrageous Fortune* (1947) and *Wild Horses* (1952). He made his last appearance in London in 1958, but continued to tour in some of his old successes until shortly before his death.

LYRIC HALL, LYRIC OPERA HOUSE, London, see LYRIC THEATRE, HAMMERSMITH.

LYRIC PLAYERS, Belfast, amateur group founded in 1951 by Dr and Mrs Pearse O'Malley for the performance of poetic drama, somewhat on the lines of Austin CLARKE's group of the same name in Dublin. Its first productions, mainly of short plays by YEATS and other Irish writers, were given in a private house, but in 1952 a small theatre was built in some disused stables and there nearly 200 plays, covering the whole range of poetic drama from the Greeks to modern times, were performed on a stage 8 by 10 ft. In 1956 the theatre was enlarged to accommodate a drama school, and reopened with CHEKHOV's *The Seagull*, and in 1968 a new theatre was built for the company, which is now run as a non-profit-making association by the Lyric Players Theatre Trust. The theatre is subsidized by the Arts Council of Northern Ireland and produces approximately ten plays each year, mainly European and Irish classics with some contemporary works. Visits have been paid to Eire, England, France, Germany, and the U.S.A. There is a policy of combining actors, directors, and designers from Northern Ireland, Eire, and England. (See also BELFAST CIVIC ARTS THEATRE AND ULSTER GROUP THEATRE.)

LYRIC PLAYERS, Dublin, see CLARKE, AUSTIN.

LYRIC THEATRE, Hammersmith, West London, in King Street, opened as the Lyric Hall on 17 Nov. 1888. It was reconstructed and reopened as the Lyric Opera House in 1890, and from 1892 to 1904 was managed by John M. EAST, under whom it was rebuilt in 1895, its seating capacity being increased from 550 to 800, and further improved in 1899. The home of a resident stock company, it drew a large local audience to its MELODRAMAS and annual PANTOMIME, but its fortunes declined when East left and it housed mainly touring companies until in 1918 it was taken over by Nigel PLAYFAIR. He redecorated it, renamed it the Lyric Theatre, and made it prosperous and fashionable, drawing large audiences from the West End. The opening production, on 24 Dec. 1918, was A. A. Milne's *Make-Believe*; in 1919 DRINKWATER's chronicle play *Abraham Lincoln*, brought from the BIRMINGHAM REPERTORY THEATRE for two weeks, began a run of 466 performances. A notable revival of GAY's *The Beggar's Opera*, with décor by Lovat FRASER, opened on 5 June 1920 and ran for 1,463 performances. Edith EVANS was a fine Millamant in CONGREVE's *The Way of the World* in 1924 and returned as Mrs Sullen in FARQUHAR's *The Beaux' Stratagem* in 1927. In between Ellen TERRY had made her last appearance on the stage in de la Mare's *Crossings*, and there had been a notable revival of CHEKHOV's *The Cherry Orchard*, with GIELGUD as Trofimov, both in 1925. A number of new light operas were staged, including A. P. Herbert's *Riverside Nights*, and in 1930 WILDE's *The Importance of Being Earnest* was seen in a black-and-white décor by Michael Weight. Playfair left in 1933, after which the theatre stood empty for a considerable time, or was used only intermittently, but in 1946 it achieved a success with a production by Peter BROOK of DOSTOEVSKY's *The Brothers Karamazov* starring Alec GUINNESS. A number of new plays and revivals were then staged, several of which were transferred to the West End, among them COCTEAU's *The Eagle Has Two Heads* (also 1946), VANBRUGH's *The Relapse* (1947), Arthur MILLER's *All My Sons*, and SARTRE's *Crime Passionel* (both 1947). From 1952 to 1953 Gielgud ran a successful repertory season, directing Paul SCOFIELD in *Richard II* and appearing himself as Mirabel in CONGREVE's *The Way of the World* and Jaffier in OTWAY's *Venice Preserv'd*. Among later interesting productions were ANOUILH's *The Lark* (1955) with Dorothy TUTIN as Joan of Arc; MONTHERLANT's *The Master of Santiago* and *Malatesta* (both 1957) with Donald WOLFIT; Bamber Gascoigne's revue *Share My Lettuce* (also 1957), in which Maggie SMITH made her London début; PINTER's first full-length play *The Birthday Party* (1958); and in 1959 BÜCHNER's *Danton's Death*, IBSEN's *Brand*, and Alun Owen's *The Rough and Ready Lot*. A period of stagnation then set in and the theatre finally closed in 1966. Efforts were made to save it, but it was demolished in 1972, some of the fine Victorian plasterwork from the auditorium being preserved and used in the construction of a new theatre 20 yards away, seating 540, which retained the old name and opened on 18 Oct. 1979 with SHAW's *You Never Can Tell*. The theatre had a big success with Michael FRAYN's *Make and Break* (1980) starring Leonard Rossiter and Prunella Scales, which was transferred to the HAYMARKET; other productions in 1980 included Noël COWARD's *Hay Fever*, with Constance CUMMINGS, and IBSEN's *The Wild Duck*. The theatre incorporates a small studio theatre seating 130 from which in 1980 three productions—MARLOWE's *Dr Faustus*, *Jeeves Takes Charge*, based on P. G. Wodehouse, and Joe ORTON's *Loot*—transferred to the West End.

LYRIC THEATRE, London, in Shaftesbury Avenue, a four-tier house holding 1,306 persons (later reduced to 948). The theatre opened on 17 Dec. 1888, and was devoted mainly to musicals, except when in 1893 DUSE made her London début in the younger DUMAS's *La Dame aux camélias*. It had its first outstanding success in 1896, when Wilson BARRETT appeared as Marcus Superbus in his own play *The Sign of the Cross*. Seasons of French plays were given in 1897 by RÉJANE and in 1898 by Sarah BERNHARDT, and in 1902 FORBES-ROBERTSON appeared in *Hamlet* and *Othello*. Successful musical comedies seen at this theatre included *Florodora* (1899), *The Duchess of Dantzig* (1903), and *The Chocolate Soldier* (1910). During the First World War SHELDON's *Romance*, with Owen NARES and Doris Keane, was transferred here from the DUKE OF YORK'S THEATRE and enjoyed a long run, and in 1919 *Romeo and Juliet* was staged with Ellen TERRY as Juliet's nurse. Later productions included *Lilac Time* (1922), with Schubert's music, and during the 1930s a succession of good modern plays occupied the theatre, among them Dodie SMITH's *Autumn Crocus* (1931), PRIESTLEY's *Dangerous Corner* (1932), Robert SHERWOOD's *Reunion in Vienna* (1934), HOUSMAN's *Victoria Regina* (1936) on its first public showing, with Pamela Stanley as the Queen, GIRAUDOUX's *Amphitryon 38* with the LUNTS, and Charles MORGAN's *The Flashing Stream* (both 1938), the latter with Godfrey TEARLE and Margaret Rawlings. During the Second World War the Lunts were seen again in *Love in Idleness* (1944) by Terence RATTIGAN, whose *The Winslow Boy* (1946) provided the theatre's first post-war success. Equally successful was Roussin's *The Little Hut* (1950) with Robert MORLEY, which ran until T. S. ELIOT's *The Confidential Clerk* took over on 16 Sept. 1953. Two musicals, *Grab Me a Gondola* (1956), from the LYRIC Hammersmith, and *Irma la Douce* (1958), did well and in 1964 a succession of short runs was broken by *Robert and Elizabeth*, a musical version of Besier's *The Barretts of Wimpole Street*, with John CLEMENTS as Edward Moulton-Barrett. Clements returned in 1967 in the CHICHESTER FESTIVAL revival of SHAW's *Heartbreak House*, and in 1970 Alan AYCKBOURN's *How the Other Half Loves*, with Robert Morley, started a long run. Alan BENNETT's farce *Habeas Corpus* (1973), with Alec GUINNESS, was highly successful, as was *John, Paul, George, Ringo... and Bert* (1974), Willy Russell's musical reconstruction of the Beatles' career. In 1975 an attempt was made to found a repertory company at the theatre; it was abandoned after only two productions, CHEKHOV's *The Seagull* and Ben TRAVERS's *The Bed Before Yesterday* (both directed by Lindsay ANDERSON), though the latter had a long run on its own. Later successes were William Douglas HOME's *The Kingfisher* with Ralph RICHARDSON and Celia JOHNSON, DE FILIPPO's *Filumena* (both 1977), and Roger Hall's *Middle-Age Spread* (1979).

LYRIC THEATRE, New York, at 213 West 42nd Street. This opened on 12 Oct. 1903 with *Old Hei-*

delberg, starring Richard MANSFIELD, who for some years returned regularly with his company, the theatre being used also for opera and musical plays. One of its earliest successes in the field of legitimate drama was *The Taming of the Shrew*, with Ada REHAN and Otis SKINNER, and in its second season RÉJANE and NOVELLI appeared with their companies. Plays by SUDERMANN and HAUPTMANN were seen and Oscar Straus's musical comedy *The Chocolate Soldier* (1909) had a long run. In 1911 came the first production in the United States of IBSEN's *The Lady From the Sea*. Florenz ZIEGFELD filled the theatre for many years with a series of musical comedies, and the last production in 1933, before the building became a cinema, was a Negro drama with music, Hall Jonson's *Run, Little Chillun*.

New York's CRITERION THEATRE was known as the Lyric for its first four years.

LYTTELTON THEATRE, the first of the three theatres inside the NATIONAL THEATRE building to be opened, the others being the COTTESLOE and the OLIVIER. Named after Lord Chandos (Oliver Lyttelton), the first Chairman of the National Theatre Board, it is a traditional picture-frame theatre seating 890 in two tiers, and has an adjustable proscenium with an opening which can range from 34 to 45 ft, a stage height of from 16 up to 29 ft, and a depth of 51 ft. Although its opening production, on 16 Mar. 1976, was *Hamlet* with Albert FINNEY, it is used mainly for new plays and modern classics, among new plays seen there being AYCKBOURN's *Bedroom Farce* (1977), later seen in the West End, David HARE's *Plenty*, PINTER's *Betrayal* (both 1978), and GRAY's *Close of Play* (1979). Outstanding revivals have included Ben TRAVERS's *Plunder* (1978), MAUGHAM's *For Services Rendered*, Arthur MILLER's *Death of a Salesman* (both 1979), and Terence RATTIGAN's *The Browning Version* (1980) with Alec MCCOWEN and Geraldine MCEWAN. There have also been notable productions of older classics such as SHAW's *The Philanderer* (1978) and VANBRUGH's *The Provok'd Wife* (1980) with Dorothy TUTIN and Geraldine McEwan.

LYTTON, EDWARD GEORGE EARLE LYTTON BULWER-LYTTON, Baron, see BULWER-LYTTON.

LYUBIMOV, YURI PETROVICH (1917–), Soviet director, who joined the company at the VAKHTANGOV THEATRE after the Second World War. A production of BRECHT's *The Good Woman of Setzuan*, which he staged there with a group of third-year students, was so successful that he was made director of the TAGANKA THEATRE, which he made a thriving centre for AVANT-GARDE productions. Although he was nominally the company's artistic director, it functioned as a collective in the true sense, co-operating on rehearsal procedures, pooling ideas, and emerging with a finished production staged with a rare combination of intelligence and vitality in a tradition of COLLECTIVE CREATION that owed much to MEYERHOLD, Vakhtangov, and Brecht, and revived for its audiences a sense of the panache and daring of the experimental productions of the 1920s.

LYUBIMOV-LANSKOY, YEVSEI OSIPOVICH (1883–1943), Soviet actor and director, notable for his work in establishing a specifically Soviet repertory in the early 1920s. He joined the Trades Union (now the MOSSOVIET) Theatre in 1923, and there directed first performances of work by young Soviet writers: his production of BILL-BELOTSERKOVSKY's *Storm* in 1926 is considered an important landmark in the history of the Soviet theatre. His early training in the tradition of the 19th century gave him a thorough grounding in the realist tradition and, fascinated by the way in which the Russian Revolution had brought people in the mass into literature and history, he achieved an enviable reputation as a brilliant arranger of crowd scenes. In 1941 he became a member of the company at the MALY THEATRE, where he remained until his death.

M

MABOU MINES, see LA MAMA EXPERIMENTAL THEATRE CLUB.

McCALLUM, JOHN (1914–), Australian-born actor, who studied for the stage at the Royal Academy of Dramatic Art in London, but made his first appearance in Brisbane in 1934. He then returned to England, where from 1937 to 1940 he appeared with the OLD VIC company. After serving with the Australian forces during the Second World War he returned to Australia, being seen at the Theatre Royal, Sydney, before returning to London for five years, during which time he played a wide variety of parts. In 1948 he married Googie WITHERS, who subsequently appeared with him on tour in Australia and New Zealand before they returned to England, where he played Mr Darling and Captain Hook in BARRIE's *Peter Pan* in 1956. This was followed by a year's run as Lord Dungavel in Lesley Storm's *Roar Like a Dove* (1957), in which he and his wife later toured Australia, together with MAUGHAM's *The Constant Wife*. McCallum then took up an appointment with J. C. Williamson Theatres Ltd and remained with the firm for some years, exercising an important influence on the development of the modern Australian theatre, and directing a number of new plays, including LAWLER's *Piccadilly Bushman*. His own play, *As It's Played Today*, was staged in Melbourne in Oct. 1974. The McCallums then returned to England for a time, but in 1978 again toured Australia and New Zealand in William Douglas HOME's *The Kingfisher*, in which they scored a great success. He is the author of *Life With Googie* (1979).

McCARTER THEATRE, Princeton, see UNIVERSITY DEPARTMENTS OF DRAMA.

McCARTHY, LILLAH (1875–1960), English actress, who made her first appearance on the stage in 1895 and later in the year joined Ben GREET's company to play leading roles in Shakespeare. In 1896 she was with Wilson BARRETT, playing Mercia in his play *The Sign of the Cross*, in which part she made her first appearance in the United States. She remained with him until 1904, and then joined TREE at HER (then His) MAJESTY'S THEATRE. It was then that she first met SHAW, who became her ardent admirer and a great friend; and for whom she played Nora in *John Bull's Other Island*, Ann Whitefield in the first public performance of *Man and Superman* (both 1905), Gloria in a revival of *You Never Can Tell*, and Jennifer Dubedat in *The Doctor's Dilemma* (both 1906). Also in 1906 she married GRANVILLE-BARKER, who had directed all these plays at the ROYAL COURT THEATRE, and among other parts played for him Nan in MASEFIELD's *The Tragedy of Nan* and Dionysus in EURIPIDES' *Bacchae*. In 1911 she took over the LITTLE THEATRE, where she presented IBSEN's *The Master Builder*, appearing in it as Hilda Wangel, and played Margaret Knox in the first production of Shaw's *Fanny's First Play* (1912). In the same year she was seen as Jocasta in SOPHOCLES' *Oedipus Rex*, with MARTIN-HARVEY, and as Iphigenia in Euripides' *Iphigenia in Tauris*, and in 1913 created the part of Lavinia in Shaw's *Androcles and the Lion*. After divorcing Granville-Barker in 1918 to enable him to marry Helen Huntington, she took over the management of the KINGSWAY THEATRE and was later seen with Matheson LANG in E. Temple Thurston's *The Wandering Jew* (1920) and Tom Cushing's *Blood and Sand* (1921). Her second marriage, to Sir Frederick Keeble, took her to Oxford and she left the theatre, returning for one performance as Euripides' Iphigenia in London in 1932, and for a short run in the title role of Binyon's *Boadicea* at Oxford in 1935.

McCLINTIC, GUTHRIE (1893–1961), American actor, producer, and director, who was for a time a member of Jessie BONSTELLE's stock company in Buffalo and later became assistant stage director to Winthrop AMES at the LITTLE THEATRE, New York. He then went into management on his own account, and from 1921, in which year he married the actress Katharine CORNELL, until his death he was active in the New York theatre. He directed many of the plays in which his wife starred, notably Michael Arlen's *The Green Hat* (1925), which first brought him into prominence, Rudolf Besier's *The Barretts of Wimpole Street* (1931), *Romeo and Juliet* (1934), and SHAW's *Candida* (1937). In 1936 he directed *Hamlet* on Broadway with John GIELGUD. He was also instrumental in presenting to New York in 1952 the company of the Greek National Theatre in a season of classical plays.

McCOLL, EWAN, see LITTLEWOOD, JOAN.

McCORMICK, F. J. [Peter Judge] (1891–1947), Irish actor, considered the most versatile and intelligent player of character parts ever to appear at the ABBEY THEATRE in Dublin, where he was first seen in 1918, and remained (except for a brief period in 1921–2) until his death. Among the many roles he created were Seumas Shields in *The Shadow of a Gunman* (1923), Joxer Daly in *Juno and the Paycock* (1924), and Jack Clitheroe in *The Plough and the Stars* (1926), all by Sean O'CASEY; Oedipus in YEATS's versions of SOPHOCLES' *Oedipus the King* (1928) and *Oedipus at Colonus* (1934), and the leading role in *The King of the Great Clock Tower* (1935), also by Yeats; and the title role in George SHIELS's *Professor Tim* (1925). In 1925 he married the Abbey actress Eileen Crowe, who joined the company in 1921 and created, among other roles, Mary Boyle in *Juno and the Paycock*.

McCOWEN, ALEC [Alexander] DUNCAN (1925–), English actor, who gained his early experience in repertory and was first seen in London in 1950, coming into prominence with his playing of Daventry in Roger Macdougall's *Escapade* (1953). He was particularly admired also as Claverton-Ferry in T. S. ELIOT's *The Elder Statesman* (1958) and from 1959 to 1961 was with the OLD VIC company, where he played such varied parts as Richard II, Touchstone in *As You Like It*, and Malvolio in *Twelfth Night*. He was also seen as the Fool in *King Lear* at the SHAKESPEARE MEMORIAL THEATRE in 1962. He returned to modern drama with his portrayal of Father Riccardo Fontana in HOCHHUTH's *The Representative* (1963), the Author in ANOUILH's *The Cavern* (1965), and Arthur Henderson in John Bowen's *After the Rain* (London and N.Y., 1967). In 1968 he made an international reputation in the title role of Peter Luke's *Hadrian VII*, which he played in New York a year later, and followed it with an outstanding performance of Hamlet at the BIRMINGHAM REPERTORY THEATRE in 1969. A year later he was back in London in Christopher HAMPTON's *The Philanthropist* (N.Y., 1971), and in 1973 he played Alceste in MOLIÈRE's *The Misanthrope* and Martin Dysart in Peter SHAFFER's *Equus* for the NATIONAL THEATRE, repeating the former role in New York in 1977. In 1974 at the Albery (see NEW THEATRE) he was an excellent Higgins to Diana RIGG's Eliza in SHAW's *Pygmalion*. In 1977 he played Antony in *Antony and Cleopatra* for PROSPECT and in 1978, in London and New York, gave a remarkable solo performance in which he recited the whole of St Mark's Gospel from memory. He returned to the National Theatre in 1980 to play Andrew Crocker-Harris in a revival of RATTIGAN's *The Browning Version*, in which he was much admired. His autobiography *Young Gemini* was published in 1979.

McCULLOUGH, JOHN (1832–85), American actor, born in Ireland, who emigrated to the United States at the age of 15. He acted with amateur clubs and later in stock and touring companies until in 1861 he was taken by FORREST to play second lead, and later went with him to San Francisco, where he ran the California Theatre in partnership with Lawrence BARRETT, and by himself, until 1875. He then toured, making his last appearance as Spartacus in Robert BIRD's *The Gladiator* in 1884. A big man with a forceful personality, McCullough had a high reputation in his own line of old-fashioned tragedy, in melodrama, and in some of Shakespeare's leading roles, though his Hamlet was a failure.

MACCUS, one of the oldest stock characters in the early Roman FABULA *atellana*, taking his name from a Greek word meaning 'stupid'. Unlike DOSSENNUS, who appears to have had a sharp wit, he was a booby, probably a fat, guzzling rustic. Judging by the numbers of titles which survive containing his name, he must have been extremely popular. Both Pomponius and Novius used him as their chief character, as in *Maccus the Soldier* and *Maccus the Maiden* or *Maccus the Innkeeper* and *Maccus in Exile*.

MACDERMOTT, The Great, see MUSIC-HALL.

MACDERMOTT, NORMAN ALEXANDER (1890–1977), English theatre director who after some experience in the provinces opened the EVERYMAN THEATRE in Hampstead, which he ran from 1920 to 1926 as an experimental and non-commercial playhouse, producing many new English and unknown foreign plays. His first production was *The Bonds of Interest*, an adaptation of BENAVENTE's *Los intereses creados* (1907). Several of his later productions were transferred to West End theatres, including Munro's *At Mrs Beam's*, Sutton Vane's *Outward Bound* (both 1923), *The Mask and the Face*, a translation of CHIARELLI's *La maschera e il volto*, and Noël COWARD's *The Vortex* (both 1924). He also revived eight of SHAW's plays. After leaving the Everyman, Macdermott continued to direct plays in London at other theatres, his last important venture being O'CASEY's *Within the Gates* (1934) at the ROYALTY.

McEWAN [McKeown], GERALDINE (1932–), English actress, who first appeared with the repertory company at Windsor, where she was born, and from 1951 to 1956 was seen in and around London in a series of light comedies. In 1957 she gave an interesting performance as the unhappy and tiresome child Frankie Addams in Carson McCullers's *The Member of the Wedding*, and in 1958 she joined the company at the SHAKESPEARE MEMORIAL THEATRE, going with the company to Russia and returning to Stratford in 1961 to play Beatrice in *Much Ado About Nothing* and Ophelia to the Hamlet of Ian Bannen. In complete contrast was her role as the suburban prostitute-wife Jenny Acton in Giles Cooper's *Everything in the Garden* (1962). In the same year she proved her versatility by taking over the part of Lady Teazle in SHERIDAN's *The School for*

Scandal, in which she made her New York début a year later, starring there also in Peter SHAFFER's double bill *The Private Ear* and *The Public Eye*. Back in London, she joined the NATIONAL THEATRE company at the OLD VIC, where she appeared in ARDEN's *Armstrong's Last Goodnight* and gave an excellent and spirited performance as Angelica in CONGREVE's *Love for Love* (both 1965). She remained with the company for some years, playing in FEYDEAU's *A Flea in Her Ear* in 1966 with the precision and neatness essential to French farce, in STRINDBERG's *Dance of Death* in 1967, in MAUGHAM's *Home and Beauty* in 1968, and in Congreve's *The Way of the World* in 1969, giving an excellent account of herself as Millamant. In WEBSTER's *The White Devil* (also 1969), she radiated flamboyant sexuality. After appearing with the National Theatre company again in 1971 in GIRAUDOUX's *Amphitryon 38*, she was seen in a series of plays in the West End, including Jerome Kilty's *Dear Love* (1973), in which she played Elizabeth Barrett Browning, Peter NICHOLS's *Chez Nous* (1974), and a revival of Feydeau's *Look After Lulu* in 1978 in which she appeared both at the CHICHESTER FESTIVAL THEATRE and in London. She returned to the National Theatre in 1980 to play in RATTIGAN's double bill *The Browning Version* and *Harlequinade*, and in VANBRUGH's *The Provok'd Wife*.

MACGOWAN, KENNETH (1888–1963), American theatre director, who was for some years a dramatic critic, and also worked for the international theatre magazine *Theatre Arts* from 1919 to 1925. He was the author of a number of books on the theatre, some written in collaboration with other scholars. From 1923 to 1926 he was associated with the PROVINCETOWN PLAYERS and the Greenwich Village Theatre, producing for them a number of new and unusual plays, among them some of O'NEILL's. In 1927 he briefly directed the ACTORS' THEATRE, and later he produced several plays on Broadway.

MacGOWRAN, JACK (1918–73), Irish actor, who made his first appearance at the Dublin GATE THEATRE in 1944 in Hilton EDWARDS's revival of John DRINKWATER's *Abraham Lincoln*. During the next six years he appeared in numerous productions both at the Gate and at the ABBEY THEATRE, and also directed plays at the Abbey's experimental Peacock Theatre. In 1954 he was responsible, with Cyril CUSACK, for a production of SYNGE's *The Playboy of the Western World* seen in Paris at the THÉÂTRE DES NATIONS, and in the same year made his London début, playing Young Covey in a revival of O'CASEY's *The Plough and the Stars*. He was later seen in a number of plays by Samuel BECKETT, including *Endgame* (1958), in which he played Clov, and a revival of *Waiting for Godot* in 1964, in which he appeared as Lucky (playing Vladimir in a later revival), both at the ROYAL COURT THEATRE. His association with Beckett, on stage, radio, and television, resulted in three solo programmes derived from the playwright's works—*End of Day* (DUBLIN THEATRE FESTIVAL, 1962), *Beginning to End* (LANTERN THEATRE, 1965), and a final and much revised version of *Jack MacGowran in the Works of Samuel Beckett*, produced by Joseph PAPP in New York in 1970. Although mainly remembered for his work in Beckett and contemporary Irish plays, he also gave a number of intelligent and sensitive performances in such parts as the title role in IONESCO's *Amédée* (1957) and Harry Hope in O'NEILL's *The Iceman Cometh* (1958), and in 1960 spent a season with the ROYAL SHAKESPEARE COMPANY at Stratford-upon-Avon.

McGRATH, JOHN, see FRINGE THEATRE and SCOTLAND.

MACHADO, SIMÃO, see PORTUGAL.

MACHIAVELLI, NICCOLÒ DI BERNARDO DEI (1469–1527), Florentine statesman and political philosopher, whose most famous work is *Il Principe* (*The Prince*, 1513). Exiled from the service of the Medicis on suspicion of conspiracy, he gave some of his time and genius to writing comedies of which the best is *La Mandragola* (*The Mandrake*), written between 1513 and 1520 and considered one of the masterpieces of the COMMEDIA ERUDITA. Its sharp, precise prose contains pungent criticism of Florentine society: it portrays the gradual betrayal of its lovely heroine by her credulous husband, her ardent but unscrupulous lover, and her scheming mother, aided by the evil machinations of the corrupt priest, Fra Timoteo. In a translation by Ashley DUKES it was successfully given at the MERCURY THEATRE in London in 1940, and in 1965, in a modernized version by Carlo Terron, it was revived by Peppino DE FILIPPO, who played Fra Timoteo, at the Sant' Erasmo in Milan. Machiavelli's *Clizia* (1525), derived from the *Casina* of PLAUTUS, was more strictly neo-classical in form. Its theme of amorous dotage may have been an ironic self-criticism of the author's own feelings for the actress Barbara Salutati.

MACHINE PLAY, name given to a type of 17th-century French spectacle which made excessive use of the mechanical contrivances and scene-changes developed in connection with the evolution of opera, particularly by TORELLI. The subjects were usually taken from classical mythology, the first French example being CORNEILLE's *Andromède* (1650); the genre reached its peak with MOLIÈRE's *Amphitryon* (1668) and *Psyché* (1671). Most of the machine plays, though not necessarily the best, were produced at the Théâtre du MARAIS, which had a large and excellently equipped stage suitable for their production.

MACKAY, CHARLES (*c.* 1785–1857), Scottish comic actor, who began his career in theatres in the West of Scotland, and in 1818 joined the company of the Edinburgh Theatre Royal. In Feb. 1819 he appeared as Bailie Nicol Jarvis in Isaac POCOCK's version of Sir Walter SCOTT's *Rob Roy*: his performance created a sensation, and he was hailed as a star. Thenceforth he specialized, and had his greatest successes, in Scots character parts, many of them in further adaptations of Scott's novels. In 1970 *The Bailie*, a one-act play based on Mackay's life, written by Donald McKenzie and performed by Callum Mill, had its first performance at the EDINBURGH FESTIVAL as a fringe production and subsequently toured throughout Scotland.

MACKAYE, (James Morrison) STEELE (1842–94), American actor, playwright, theatre designer, pioneer, and inventor. After studying in London and Paris he played Hamlet in London in 1873, repeating the role later in the year in Manchester with Marion TERRY, who was making her first stage appearance, as Ophelia. Later he appeared in New York with a group of students whom he had trained in the methods of Delsarte. In order to carry out his ideas he remodelled the old Fifth Avenue Theatre, opening it in 1879 as the MADISON SQUARE THEATRE. Although the venture failed, it was at this theatre that Mackaye's own play *Hazel Kirke* (1880) was produced for a long run. It was frequently revived and was seen in London in 1886. After leaving the Madison Square Theatre Mackaye built his own theatre, the LYCEUM, and established there the first school of acting in New York, later known as the American Academy of Dramatic Art. Mackaye, who was thin, dark, nervous, and dynamic, was everything by turns—actor, dramatist, teacher, lecturer. Shortly before his death he planned a vast 'Spectatorium' for the Chicago World Fair, which was never built; but later theatre architects were indebted to its plans for the introduction of many new ideas and methods. He influenced the theatre more by what he thought, dreamed of, and fought for than by what he actually achieved, and none of his plays has survived. His life-story is told in *Epoch*, by his son Percy Wallace Mackaye (1875–1956) who, like his father, was something of a distinguished failure, much of his best work being done in poetic drama, modern masques, and spectacles, nothing of which has survived.

McKELLEN, IAN MURRAY (1939–), English actor, who made his first appearance at the BELGRADE THEATRE in Coventry as Roper in Robert BOLT's *A Man for all Seasons* (1961). In the provinces he played such major roles as Henry V, Luther in John OSBORNE's play, and Arthur Seaton in a dramatization of Alan Sillitoe's novel *Saturday Night and Sunday Morning*. His first appearance in London, in James SAUNDERS's *A Scent of Flowers* (1964), was followed by a season with the NATIONAL THEATRE, after which he gave a remarkable performance as the young hero of Donald Howarth's *A Lily in Little India* (1966), wrestling with his own nascent sexuality. He was then seen in ARBUZOV's *The Promise* (1967), in which he also made his first appearance in New York later the same year, and gave excellent performances in Peter SHAFFER's double bill *White Liars* and *Black Comedy* (1968). He set the seal on his growing reputation by playing Richard II in 1968 and MARLOWE's Edward II in 1969 with the PROSPECT THEATRE COMPANY, showing Richard gradually coming to terms with his own vulnerability and Edward as a wayward and impetuous youth hungering for sensual contacts. He played both parts on tour and in London, as he did Hamlet with the same company in 1971, and in 1972 he helped to found the Actors' Company, in which the actors chose their own plays and shared equal pay, billing, and leading roles. With this group he gave memorable performances in Iris Murdoch's *The Three Arrows* (1972) and CHEKHOV's *The Wood Demon* (1973), and he then joined the ROYAL SHAKESPEARE COMPANY to play leading roles in Marlowe's *Dr Faustus* in 1974, SHAW's *Too True To Be Good* in 1975, *Romeo and Juliet* and *Macbeth* in 1976, and IBSEN's *Pillars of the Community* in 1977. He was seen at the ROYAL COURT and in the West End in 1979 in Martin Sherman's play *Bent*, about two homosexuals in a Nazi prison camp, and scored a big success in New York in 1980 as Salieri in Shaffer's *Amadeus*.

McKENNA, SIOBHÁN (1923–), Irish actress, who made her first appearance in 1940 with the An Taibhdearc Theatre in Galway, where for two years she played leading roles in Gaelic, sometimes in English plays translated by herself (see GAELIC DRAMA). From 1944 to 1946 she played in both Gaelic and English at the ABBEY THEATRE, Dublin, and in 1947 appeared for the first time in London, as Nora Fintry in CARROLL's *The White Steed*. By 1951, when she appeared at the EDINBURGH FESTIVAL in Lennox ROBINSON's *The White-headed Boy* and as Pegeen Mike in SYNGE's *The Playboy of the Western World*, she had established herself as one of the leading interpreters of Irish parts. Avril in O'CASEY's *Purple Dust* followed in 1953, but her outstanding role at this time was SHAW's Saint Joan in 1954 in London. She made her New York début in 1955 as Miss Madrigal in Enid BAGNOLD's *The Chalk Garden*, and at the STRATFORD (ONTARIO) FESTIVAL in 1957 played Viola in *Twelfth Night*.

At the DUBLIN THEATRE FESTIVAL in 1960, and later on a European tour and in London, she again played Pegeen Mike, and a year later she was seen as Joan Dark in BRECHT's *Saint Joan of the Stockyards*, repeating the role at the QUEEN'S THEATRE in London in 1964. In the 1967 Festival she played Cass in Brian FRIEL's *The Loves of Cass*

McGuire, and the following year at the Abbey Theatre she was an excellent Madame Ranevskaya in CHEKHOV's *The Cherry Orchard*. In 1970 in London, and again at the GATE THEATRE, Dublin, in 1975, she presented her one-woman show *Here Are Ladies*, in which she added to her repertoire parts by YEATS and BECKETT and characters from the novels of James JOYCE. In 1973 at the MERMAID THEATRE in London she gave a moving portrayal of Juno in O'Casey's *Juno and the Paycock*, which she had previously played at the GAIETY THEATRE, Dublin, in 1966. Later roles have included Josie in O'NEILL's *A Moon For the Misbegotten* at the Gate in 1976, in which year she also played Bessie Burgess in O'Casey's *The Plough and the Stars* at the Abbey and in New York. In 1980 at the Abbey she again played Juno. Perhaps the outstanding characteristic of this strongly individualistic actress is her ability to capture the common humanity beneath the poetic fabric of her Irish heroines, reminiscent of Sara ALLGOOD at her best.

MACKINLAY, JEAN STERLING, see WILLIAMS, HARCOURT.

MACKLIN [M'Laughlin], CHARLES (*c*. 1700–97), Irish actor, who in 1716, after a wild and restless boyhood, joined a company of strolling players. Four years later he was in Bath, and in 1725 was engaged by Christopher RICH for the LINCOLN'S INN FIELDS THEATRE. There his natural manner of speaking, which preceded GARRICK's reforms in stage delivery, was unacceptable in the high-toned tragedies of the day, and he returned to the provinces and minor theatres, playing HARLEQUIN and CLOWN at fairs and at SADLER'S WELLS. At this period of his life he was known as the Wild Irishman, with the reputation of a jovial boon companion and a good boxer. In 1730 he again appeared at Lincoln's Inn Fields, but this time in comedy, being seen in FIELDING's *The Coffee-House Politician; or, the Justice Caught in his own Trap*, a revised version with an entirely new act of *Rape Upon Rape*, seen at the HAYMARKET THEATRE earlier the same year. In 1732 he was engaged by FLEETWOOD for DRURY LANE, where he played secondary comic parts—Touchstone in *As You Like It*, Scrub in FARQUHAR's *The Beaux' Stratagem*, and Peachum in GAY's *The Beggar's Opera*—until he persuaded the management to revive *The Merchant of Venice*, in which he appeared for the first time on 14 Feb. 1741 as Shylock and became famous overnight. He rescued the character from the clutches of the low comedian, to whom it had been assigned since Restoration times, and made him a dignified and tragic figure, drawing from Alexander Pope the memorable couplet: 'This is the Jew/That Shakespeare drew.' There is an excellent description of him in the part by the German critic Lichtenberg.

With advancing years Macklin became extremely quarrelsome and jealous (he had already killed another actor in a fight over a wig), and moved erratically from one theatre to another, causing trouble backstage and engaging in constant litigation. Apart from Shylock and his Iago to the Othello of Garrick and Spranger BARRY, his most memorable part was Macbeth, which he first played at COVENT GARDEN on 23 Oct. 1773 in something approximating to the dress of a Highland chieftain in place of the red military coat favoured by Garrick. He was the author of several plays, of which two survived well into the 19th century—*Love à la Mode* (1759), in which he himself played the leading role, Sir Archy McSarcasm, and *The Man of the World* (1781), in which, in spite of his great age, he again played the lead, Sir Pertinax McSycophant. He made his last appearance on the stage on 7 May 1789, when he essayed Shylock but was unable to finish it. He was twice married, his second wife surviving him, though all his children, of whom two daughters (one illegitimate) were also on the stage, died before him.

McLELLAN, ROBERT (1907–), Scottish dramatist, whose first play, *Jeddart Justice*, a one-act comedy, was produced at the CURTAIN THEATRE in Glasgow in 1934. Several of his full-length plays were also presented there, including *Tom Byres* (1936) and *King Jamie the Saxt* (1937), in which Duncan MACRAE played the leading role. *Torwatletie* (1946) and *The Flouers o' Edinburgh* (1948) were first produced by Glasgow's UNITY THEATRE, and the CITIZENS' THEATRE later revived all these plays as well as giving *The Road to the Isles* (1954) its first production. *Young Auchinleck* (1962), first seen at the EDINBURGH FESTIVAL in 1962, was produced by the GATEWAY THEATRE, and the Edinburgh Civic Theatre presented *The Hypocrite* (1967) at the LYCEUM THEATRE as well as reviving *The Flouers o' Edinburgh* in 1975 in a stylish production by Bill BRYDEN. McLellan has been closely associated with the Lallans movement, which seeks to revive Lowland Scots as a literary medium, and it can be argued that without his vigorous use of dialect as a theatrical medium John ARDEN might not have written *Armstrong's Last Goodnight* (1975).

MacLIAMMÓIR, MICHÉAL (1899–1978), Irish actor, designer, and dramatist, who as Alfred Willmore appeared on the London stage as a child. He then studied art, travelled widely in Europe, and in 1927 joined Anew MCMASTER's Shakespeare company. In 1928 he founded with Hilton EDWARDS the Galway Theatre (which he directed until 1931) for the production of GAELIC DRAMA, and later the same year the Dublin GATE THEATRE, which opened at the ABBEY's experimental Peacock Theatre with IBSEN's *Peer Gynt*. In addition to acting in and designing over 300 productions for the Gate, including Denis JOHNSTON's first play, *The Old Lady Says 'No!'* (1929), in which he played Emmett, he acted in London, New York (where at the invitation of Orson WELLES he played Hamlet in 1934), and on tour with the Gate company in Egypt and the Balkans. In 1960 came his first one-man show, on Oscar WILDE, *The Importance of Being Oscar*, in which he toured widely, and in 1963 he presented a second one-man entertainment *I Must Be Talking to My Friends*; a third, *Talking about Yeats*, was first seen in 1965. He also wrote a number of plays, including *Diarmuid agus Grainne* (1929), seen in Irish and English; *Ill Met by Moonlight* (1946), seen in London in 1947; and *Prelude in Kazbek Street*, which received its première at the DUBLIN THEATRE FESTIVAL in 1973. MacLiammóir was also the author of several volumes of memoirs: *All for Hecuba* (1946); *Put Money in My Purse* (1954), an account of the filming of *Othello*; *Two Lights on Actors* (1960); and *Each Actor on his Ass* (1961).

McMAHON, GREGAN (1874–1941), Australian theatre director, who made his first appearance on the stage in H. A. JONES's *The Liars* in 1900, and in 1906 joined the firm of J. C. Williamson Theatres Ltd as a play-producer. In 1911 he persuaded them to form the Melbourne Repertory Company, the first of such companies in Australia, which gave the first Australian productions of several plays by SHAW. The company was disbanded during the First World War, and in 1929 McMahon formed his own semi-professional company in Melbourne. He was awarded the C.B.E. for his services to the Australian stage in 1938.

McMASTER, ANEW (1894–1962), Irish actor, who first appeared on the stage in 1911 in Fred TERRY's company, with which he remained for three years. He was leading man to Peggy O'Neill in *Paddy the Next Best Thing* (1920), and in 1921 went to Australia, where among other parts he played Iago to the Othello of Oscar ASCHE. In 1925 he founded a company to present Shakespeare on tour, acting, managing, and directing plays himself. A tall, handsome man, with an imperious face crowned, in his later years, by a fine head of white hair, he 'took the stage' with great dignity, his Shylock, Richard III, Coriolanus, Lear, and Othello being outstanding. He appeared at the SHAKESPEARE MEMORIAL THEATRE in 1933, playing Hamlet and Coriolanus, and then took his company on a tour of the Near East. After the Second World War he toured indefatigably throughout Ireland, and at the time of his death was preparing to play Othello at the DUBLIN THEATRE FESTIVAL to the Iago of Micheál MACLIAMMÓIR. In 1968 Harold PINTER, who toured with McMaster in Ireland in the early 1950s, published an appreciation of him entitled *Mac*.

MACNAMARA, BRINSLEY [John Weldon] (1890–1963), Irish dramatist and novelist, who in 1910 made his first appearance on the stage with the ABBEY THEATRE company. He left in 1912 to devote all his time to writing, and his first play, *The Rebellion at Ballycullen*, was staged at the Abbey in 1919. This was followed by a series of successful plays, including *The Land for the People* (1920), *The Glorious Uncertainty* (1923), *Look at the Heffernans!* (1926), *The Master* (1928), *Margaret Gillan* (1933), *The Grand House in the City* (1936), *The Three Thimbles* (1941), and *Marks and Mabel* (1945). Macnamara became a director of the Abbey Theatre in 1925, but resigned after a disagreement over O'CASEY's *The Silver Tassie* when it was finally produced, against his wishes, in 1935. He was successor to Andrew Malone as drama critic of the *Irish Times*, and also compiled a useful guide to plays produced by the Abbey Theatre company from 1899 to 1948.

MACOWAN, MICHAEL (1906–80), English actor and director, son of the actor-dramatist Norman Macowan (1877–1961), whose plays included *The Blue Lagoon* (1920), *The Infinite Shoeblack* (1929), and *Glorious Morning* (1938). He made his first appearance in 1925 in Charles Macdona's repertory season of SHAW, but in 1931 abandoned acting for direction, producing plays at the OLD VIC, the SHAKESPEARE MEMORIAL THEATRE, and, from 1936 to 1939, at the WESTMINSTER THEATRE. He was associated with the short-lived Old Vic Theatre School, and directed some of FRY's early plays, as well as PRIESTLEY's *The Linden Tree* (1947) and Charles MORGAN's *The River Line* (1952). From 1954 to 1966 he was head of the London Academy of Music and Dramatic Art (LAMDA), but continued to work in the commercial theatre, directing, among other plays, Graham GREENE's *The Potting Shed* (1958).

MACRAE, (John) DUNCAN (1905–67), Scottish actor, who scored his first major success in the Glasgow CURTAIN THEATRE's production of Robert MCLELLAN's *King Jamie the Saxt* (1937), in which he played the title role. In 1943 he joined the newly formed Glasgow CITIZENS' THEATRE, and soon established himself as one of its leading actors, his performance in James BRIDIE's plays *The Forrigan Reel* (1944) and *Gog and Magog* (1949) being particularly memorable, as was his DAME in Bridie and Munro's pantomime *The Tintock Cup* (1949). Macrae was also in Tyrone GUTHRIE's production of LYNDSAY's *Ane Pleasant Satyre of the Thrie Estaitis* at the EDINBURGH FESTIVAL of 1948. Although better known in Scotland than in England, he appeared several times in London, notably as John in IONESCO's *Rhinoceros* (1960), and also worked extensively in film and television. He took an active interest in the affairs of the Scottish branch of BRITISH ACTORS' EQUITY, of which he was chairman until shortly before his death. As an actor, his talent was at its finest in the portrayal of comic characters with more than a touch of the grotesque, helped by the subtle use of his gangling physique.

MACREADY, WILLIAM CHARLES (1793–1873), English actor, one of the finest tragedians of his day. The son of a provincial actor-manager, he was forced by his father's financial difficulties to leave Rugby and go on the stage. He made his first appearance in 1810 in Birmingham, as Romeo, and then toured the provinces, playing Hamlet, later one of his finest parts, for the first time in 1811 at Newcastle. In 1816 he joined the company at COVENT GARDEN to play Orestes in PHILIPS's *The Distrest Mother*, and by 1819 was firmly established both there and at DRURY LANE as the only rival of the great Edmund KEAN. He was much admired in such new plays as Thomas MORTON'S *The Slave* (1816), POCOCK'S *Rob Roy Macgregor* (1818), based on SCOTT's novel, and Sheridan KNOWLES's *Virginius* (1820), but it was his Hamlet, Lear, Macbeth, and, later, Othello, that were universally acclaimed. In 1826 he visited America, where he made his first appearance at the PARK THEATRE in New York as Virginius, and in 1828 he played Macbeth in Paris, returning later the same year to play Hamlet and Othello. In 1837 he appeared at Covent Garden with Helen FAUCIT in BROWNING's poetic play *Strafford*, and a year later created the part of Claude Melnotte to her Pauline in BULWER-LYTTON's *The Lady of Lyons*. Their fine acting did much to ensure the success of the play, which was frequently revived. Another new play which owed its appearance to Macready's encouragement and initiative was Lord BYRON's *The Two Foscari* (1838), and he also scored a personal triumph in Bulwer-Lytton's *Richelieu* (1839).

In the late 1830s Macready, who had an ungovernable temper, became the implacable rival of the American actor Edwin FORREST, their mutual animosity leading eventually to the ASTOR PLACE riot in New York in 1849, when several people were killed. Macready never acted in America again. He made his last appearance on the stage at Drury Lane on 26 Feb. 1851 as Macbeth. Although he never wavered in his dislike of the profession into which he had been forced, he was scrupulous in performing his theatrical duties both on stage and as manager at various times of both PATENT THEATRES, where he instituted a number of reforms and for the first time insisted on full rehearsals, particularly for supers and crowd-scenes. His managements were artistically if not financially successful, and he sought always to improve his productions by subordinating scenery and costume to the integrity of the play as a whole. He also rescued Shakespeare's texts from many of the emendations of the Restoration period. A cultured man, he enjoyed the friendship of some of the finest writers of the day, but made many enemies in the theatre by his constant disparagement of his profession.

MacSWINEY, OWEN, see SWINEY, OWEN.

MADÁCH, IMRE, see HUNGARY.

MADDERMARKET THEATRE, Norwich, a replica of an Elizabethan theatre interior, built inside a dilapidated hall in 1921 by (Walter) Nugent Bligh Monck (1877–1958), to house the Norwich Players, an amateur company which he founded in 1911. It was altered and enlarged in 1953 and again in 1966 and now holds about 300 people. Monck, whose career as a professional actor had been cut short by the First World War, directed and appeared in most of the Maddermarket productions himself up to 1952, and since its inception the company, still amateur, has appeared in all 37 of the plays associated with Shakespeare, as well as other early English plays, foreign plays, Greek tragedies, and the works of SHAW and his contemporaries. A new production is put on every month for nine performances, and the actors are, and always have been, anonymous.

MADDERN, MINNIE, see FISKE, MINNIE MADDERN.

MADDOX, MICHAEL, see MEDDOKS, M. E.

MADISON SQUARE THEATRE, New York, on the south side of 24th Street, west of 5th Avenue. This was built on the site of DALY's first FIFTH AVENUE THEATRE and was adapted by Steele MACKAYE for use as a repertory theatre on Continental lines, with a company made up of students whom he had trained himself. It opened on 23 Apr. 1879, but soon failed and was taken over by Daniel FROHMAN who reopened it on 4 Feb. 1880 with Mackaye's *Hazel Kirke*, a domestic drama which ran for nearly two years. Viola ALLEN made her first New York appearance here in 1882, in HUGO's *The Hunchback of Notre Dame*, and in the same year Bronson HOWARD's *Young Mrs Winthrop*, one of the first good American plays, was seen. Albert M. PALMER took over the management in 1885, and in 1889–90 Richard MANSFIELD appeared in a series of revivals and new plays, including the first straightforward translation of *A Doll's House* which, with Beatrice Cameron as Nora, first brought IBSEN to the attention of the American public. From 1891 until his death in 1900 the theatre was taken over by Hoyt: it was renamed Hoyt's Theatre on 1 Feb 1905. *A Case of Arson* by HEIJERMANS was produced there in 1906. The building was demolished in Mar. 1908.

MAETERLINCK, MAURICE (1862–1949), Belgian poet and dramatist, awarded the NOBEL PRIZE for Literature in 1911. An exponent of SYMBOLISM and a forerunner of the Theatre of SILENCE, he depicts his characters as the instruments of some hidden force emanating from the unseen reality which lies all around us. Most of his plays were first produced in Paris by LUGNÉ-POË, a close personal friend, though *L'Oiseau Bleu* (1909) was first seen in Moscow in a production by STANISLAVSKY. It was produced in London, as *The Blue Bird*, in the same year; its sequel *Les Fiançailles*, as *The Betrothal*, was seen there in 1921. *Le Bourgmestre de Stilmonde* (1919), which some critics consider Maeterlinck's finest play, was seen in London the same year with MARTIN-HARVEY in the title role. It had a great success, and was frequently revived up to 1933, but was less well received in New York, where the WASHINGTON SQUARE PLAYERS had already presented several of Maeterlinck's works. In 1922 Eva LE GALLIENNE appeared in *Aglavaine and Selysette* playing Aglavaine, the part originally written for Georgette Leblanc (1867–1941), the actress and singer who created most of Maeterlinck's heroines from 1896 to 1910. His best known play is *Pelléas et Mélisande* (1893), seen in London in 1898 and in New York in 1902, with incidental music by Fauré. In both these productions Mélisande was played by Mrs Patrick CAMPBELL, who also played the part in London in French in 1904 to the Pelléas of Sarah BERNHARDT. In 1902 Debussy used the play as the basis of an opera. Other dramatic works by Maeterlinck include *La Princess Maleine* (1899), *L'Intruse* (1890), *Les Aveugles* (1891), *Intérieur* (1895), *Monna Vanna* (1902), *Le Miracle de saint Antoine* (1919), and *Ariane et Barbe-bleu* (1901) used as the libretto of an opera by Dukas.

MAFFEI, FRANCESCO SCIPIONE (1675–1755), Italian dramatist, educated by the Jesuits from whom he gained his knowledge of classical drama. Wealthy and cultivated, he was the author of a number of tragedies of which *Merope* (1713) was the most successful. Tightly constructed, it dispensed with narrative passages, messenger speeches, and exposition through minor characters, and drove straight to the heart of its plot in expressive and finely-turned verse. Written as a challenge to the alleged superiority of French classical tragedy, it was first seen in Italy and almost immediately afterwards at the COMÉDIE-ITALIENNE in Paris. It was much admired by VOLTAIRE, who dedicated his own *Merope* (1743) to Maffei and was also discussed by LESSING in his *Hamburgische Dramaturgie* (1767–9). An English translation of Maffei's play by William Ayre was published in 1740. though apparently never acted.

MAGGIO, see ITALY.

MAGNES (*fl.* second half of 5th cent. BC), Greek dramatist, one of the earliest writers of OLD COMEDY. He is mentioned by ARISTOPHANES in the *Knights*, and won a victory in 472 BC. The titles of several of his plays are known, including the *Birds*, the *Insects*, and the *Frogs*, but very few fragments have survived.

MAGNON, JEAN (1620–62), French dramatist, of whom it was said that his works were more easily written than read. He deserves to be remembered, if only because his first play *Artaxerce* (1644) was acted by a small company which included Jean-Baptiste Poquelin, later MOLIÈRE, whose friend Magnon remained all his life. He wrote seven more plays, and a vast encyclopaedia called *La Science universelle*. He was murdered on the Pont-Neuf by the hired bravos of his wife's marquis-lover. His son opened a theatre in Copenhagen where Molière's plays were acted, and exercised a great influence on the development of the dramatist HOLBERG.

MAHELOT, scenic designer, scene painter, and director at the Hôtel de BOURGOGNE. whose *Mémoire*, now in the Bibliothêque Nationale in Paris, contains notes on and sketches for the settings of plays produced at the theatre in 1633 and 1634, at a time when the old MULTIPLE SETTING derived from the LITURGICAL DRAMA was giving way to the single set of French classical comedy and tragedy. It gives valuable information on about 71 published plays and also the titles and dates of many unpublished ones which might otherwise have remained unknown. The *Mémoire*, which was continued by Laurent, was published in 1920 with introduction and notes by H. C. Lancaster, and from its detailed drawings it has been possible to reconstruct some of the settings used for specific plays, as was done for a production of THÉOPHILE's *Pyrame et Thisbé* at Lancaster University's Nuffield Theatre in 1972.

MAHEN, JIŘÍ, see CZECHOSLOVAKIA.

MAINTENON, Madame de [*neé* Françoise d'Aubigné] (1635–1719), the second wife of Louis XIV, whom she married secretly in about 1684. Her first husband was the dramatist SCARRON, on whose household and dramatic works she had a salutary effect. Her ascendancy over the Court was held to be largely responsible for the austerity which reigned there during the 1680s and 1690s, when, in contrast to the encouragement earlier given to MOLIÈRE and others, the public theatre came in for severe disapproval, the King even banishing the actors from the COMÉDIE-ITALIENNE in 1697 for having put on *La Fausse Prude* by Étienne LENOBLE, which offended her. Private theatricals, however, were another matter, and Madame de Maintenon not only persuaded RACINE, after twelve years' abstention from the theatre, to write the poetic and religious dramas *Esther* (1689) and *Athalie* (1691) for performance at her school in Saint-Cyr for impoverished young ladies, but herself composed for them a number of one-act sketches illustrating well-known proverbs, a type of entertainment very popular in society at the time and later made famous by Alfred de MUSSET. These, preserved in manuscript, were not published until 1820.

MAÍQUEZ, ISIDORO (1768–1820), Spanish actor, a pupil of TALMA, who in 1802 introduced true tragic acting into Spain through his interpretation of Shakespeare's Othello. Liberal in his views, he found himself involved in the political troubles of the day, fighting against the French troops in 1805, yet performing in the presence of Napoleon's brother Joseph, King of Naples. He was finally exiled from Madrid shortly before his death.

MAIRET, JEAN (1604–86), French dramatist, whose reputation suffered in later years from his opposition to CORNEILLE in the literary war over *Le Cid* (1637). He was, however, the foremost dramatist of his time and the first to formulate and use the theory of the UNITIES, newly developed in Italy. His first play, *Chryséide et Arimand* (1625), a tragi-comedy, figured in the repertory of both the Hôtel de BOURGOGNE and the MARAIS. Under the influence of THÉOPHILE DE VIAU and RACAN, it was followed by two pastorals, *Sylvie* (1626) and *Silvanire* (1630). It was in the preface to the published edition of the second of these, in 1631, that Mairet first put forward his theory of the unity of time and place, and in his most important play, *Sophonisbe* (1634), a tragedy on a Roman theme, he laid the foundations of French classical tragedy. The play already possessed the simplicity, refinement, and concentration needed for such a work, and it had an immediate influence. Mairet was attached as dramatist to MONTDORY's troupe until 1640, then retired from the theatre and entered the diplomatic service.

MAISONS DE LA CULTURE, see DÉCENTRALISATION DRAMATIQUE.

MAISON DE MOLIÈRE, La, see COMÉDIE-FRANÇAISE.

MAJESTIC THEATRE, New York. (1) At 5 Columbus Circle. As the Cosmopolitan, this opened on 29 Jan. 1903 with a musical version of the famous children's book *The Wizard of Oz*, followed by the almost equally successful *Babes in Toyland*. In 1911 it was renamed the Park, reopening on 23 Oct. with *The Quaker Girl*. Three years later Mrs Patrick CAMPBELL was seen there in SHAW's *Pygmalion*, and in 1917 Constance COLLIER and Herbert TREE appeared in a notable revival of *The Merry Wives of Windsor*. A season of light opera, which included a long run of GILBERT and Sullivan's *Ruddigore*, followed, but from 1923 to 1944 the building was used as a cinema. Renamed the International in 1944, it opened again as a theatre and in 1945, as the Columbus Circle Theatre, housed Maurice EVANS in his *G. I. Hamlet*. It was again known as the International from 1946 to 1949, after which it became a television studio. It was pulled down in June 1954, an exhibition hall, the New York Coliseum, being built on the site.

(2) At 245 West 44th Street, between Broadway and 8th Avenue. With a seating capacity of 1,655, this opened on 28 Mar. 1927 with an ephemeral production which soon gave way to MUSICAL COMEDY, and on 19 Jan. 1928 John GIELGUD made his first appearance in New York as the youthful Grand Duke Alexander in Neumann's *The Patriot* (seen in London as *Such Men Are Dangerous*), which had only 8 performances. After several more failures the theatre reverted to musical comedy, for which it was eminently suitable, though in 1935 it housed Michael CHEKHOV and his Moscow Art Players in a series of Russian plays, and it was also used for the production of several thrillers. In 1945 the THEATRE GUILD presented a musical version of MOLNÁR's *Liliom*, as *Carousel*, which had a long run, and other successful musicals presented here included *South Pacific* (1949) and *The Music Man* (1957). In 1963 Gielgud returned successfully as Joseph Surface in his own production of SHERIDAN's *The School for Scandal*, and in 1967 WEISS's *Marat/Sade* had its New York première, the musical *Fiddler on the Roof* moving in from the IMPERIAL later the same year. Further musicals included *Sugar* (1972); the all-Black *The Wiz* (1975), based on *The Wizard of Oz*; *The Act* (1977); and the highly successful *42nd Street*, which transferred from the WINTER GARDEN in 1981.

MAJOR, TAMÁS, see HUNGARY.

MAKE-UP. The use of make-up in the theatre dates from the earliest times, and may originally have been a survival of the custom of smearing the faces of the participants in a religious rite with the blood of the sacrificial victim or with ash from the sacred fire. This latter usage may posibly account for the blackened faces of the performers in the MORRIS DANCE. Little is known about make-up in classical times, since most of the actors wore MASKS, but the followers of Dionysus are known to have smeared their faces with the lees of wine, to indicate their bibulous habits. The main purpose of make-up in later times was to disguise the actor's face, to alter it in some way so as to make him appear older, more ferocious, less human, or more god-like. The two extremes of make-up used in this way were the gilding of God's face in LITURGICAL DRAMA, and the elaborate painting of the actors' faces in the *Kabuki* plays of JAPAN. The use of everyday cosmetics to enhance personal beauty was probably brought into the European theatre by actresses; the use of make-up by actors in 16th-century Italy is attested by a passage in Leone di Somi's *Dialogues on Stage Affairs* (1565). He was writing of a torch- or candle-lit theatre, but even in the open-air theatre of the Elizabethans there is evidence of some form of character make-up—black for Negroes, umber for sunburnt peasants, red noses for drunkards, chalk-white for ghosts. In the Restoration theatre the grotesque make-up of such a character as Lady Wishfort in CONGREVE's *The Way of the World* (1700) was probably nothing more than a caricature of that of the contemporary lady of fashion. According to RICCOBONI in his *Historical Account of the Theatres in Europe* (1738), the English actor James Spiller added 40 years to his appearance by drawing lines on his face and painting his eyebrows and eyelids. But the general standard of make-up in London theatres appears to have been somewhat low, and one of the excellences attributed to GARRICK was his skill in making up his face to suit the age and character of his part, particularly when he played old men. Before the introduction of modern grease-paints, all make-up was basically a powder, compounded with a greasy substance or with some liquid medium, which was often harmful to the skin and sometimes extremely dangerous, particularly if white lead was used in its composition. Powder make-up also had the disadvantages of drying up and so hindering the mobility of the actor's features, or of melting and streaking in the heat. From the first comprehensive account of make-up in the English theatre, in Leman Thomas Rede's *The Road to the Stage* (1827), it is evident that the introduction of gas-lighting had recently led to changes in the art of making-up, and although the paints used were still powder-based, some form of grease—usually pomatum, though butter and lard are also mentioned—was used as a foundation. Also, grease or oil was used to remove the paint after the performance. Not everyone, however, approved of the use of greasy substances. T. H. Lacy, in *The Art of Acting* (1863), says make-up should be put only on a dry non-greasy surface, a practice also recommended in the anonymous *How to 'Make-Up'*... by 'Haresfoot and Rouge' (1877) though the use of cold cream after the performance seems to have been generally adopted by about 1866. A revolution in make-up was achieved in the second half of the 19th century by the introduction of grease-paint, invented in about 1865 by Ludwig Leichner (1836–?), a Wagnerian opera-singer. He opened his first factory in 1873, and his round sticks, numbered and labelled from 1, light flesh colour, to 8, a reddish brown for Indians (later increased to 20, and by 1938 to 54), were soon to be found in practically every actor's dressing-room. The first sticks were imported into London between 1877 and 1881, and a London branch of L. Leichner Ltd opened in 1928. As well as the thick sticks of grease-paint which the actor was instructed to use over a slight coating of cocoa butter, which was also used to clean the face afterwards, there were thin sticks, or liners, in black, brown, blue, and white, used for painting in fine lines. The use of grease-paint seems to have been pretty general by 1890. and even those actors who still used powder mixed it with a harmless cold cream in place of the pearl powder (subchloride of bismuth) or hydrated oxide of bismuth which were apt to turn grey or black when exposed to fog or the fumes given off by coal fires. The introduction of electric light again caused fundamental changes in theatrical make-up, which has also more recently been influenced by the techniques of film and television make-up. Grease-paint is available now in tubes and tins as well as sticks, and has in some cases been superseded by liquid make-up applied with a sponge, or by a 'water-moist' make-up, greaseless and packed in tubes or, in the case of Max Factor's 'Pancake' make-up, in cake form, packed in plastic containers. In the modern theatre, where the make-up that embellishes and the make-up that disguises are of equal importance, the art of making-up has reached a high standard, and covers more than the simple painting and lining of the face. To age an actor or actress from 20 to 60 in one evening is now a commonplace, though in many cases such a transformation is achieved not by the player but by a make-up expert. This is perhaps truer of films and television than of the theatre, where the individual actor usually still attends to his own make-up, inventing and discarding his own methods within the limits of the materials available.

MALE IMPERSONATION. Since in the early theatre parts were mainly played by men and boys, the question of male impersonation did not arise until actresses had become firmly established on the stage. In England this dated from 1660, and it was not long before young actresses, inspired perhaps by their success in the temporary assumption of male attire when playing such parts as Rosalind in *As You Like It*, Viola in *Twelfth Night*, or Silvia in FARQUHAR's *The Recruiting Officer* (1706), took over some of the male leads in contemporary comedy, which became known as 'breeches parts'. The most famous of these was Sir Harry Wildair in Farquhar's *The Constant Couple*, which Peg WOFFINGTON played with immense success in 1740. Others who were successful in breeches parts were Nell GWYNN, Mrs BRACEGIRDLE, and particularly Mrs MOUNTFORT, the last being considered outstanding as Lothario in ROWE's *The Fair Penitent* and Macheath in GAY's *The Beggar's Opera*. This fashion for playing *en travesti*, as it was later called, formed an essential part of Regency spectacle and Victorian EXTRAVAGANZA, the great exponent of the latter being Madame VESTRIS, and it was one of the formative elements in the development of the PRINCIPAL BOY in pantomime.

Apart from their appearances in such light male roles there were also a number of intrepid actresses who essayed such tragic roles as Hamlet, Romeo, and Richard III, among them Charlotte CUSHMAN and Sarah BERNHARDT, who in 1900 also appeared in a 'serious' breeches part, the young Duc de Reichstadt in ROSTAND's *L'Aiglon*. In a more realistic age such courageous feats are seldom attempted, although Frances de la Tour played Hamlet at the Half Moon Theatre in 1980. In the 19th century and afterwards the term 'male impersonator' was used mainly of women on the MUSIC-HALLS who sang comic songs in a variety of male costumes, from the man-about-town to the Cockney urchin, with particular emphasis on the more glamorous uniforms of the armed forces. Outstanding among them were Vesta TILLEY, Ella Shields, and Hetty King. The woman dressed as a man is still an effective feature of satirical European cabaret, though the other aspects of male impersonation do not seem to have flourished outside England, the U.S.A., and, occasionally, France.

MALINA, JUDITH, see LIVING THEATRE.

MALLESON, (William) MILES (1888–1969), English actor, who made his first appearance at the Liverpool Repertory Theatre (see LIVERPOOL PLAYHOUSE) in 1911, and two years later was in London. In 1918 he was seen as Sir Andrew Aguecheek in *Twelfth Night*, and soon established a reputation as an eccentric comedian with his playing of Sir Benjamin Backbite in SHERIDAN's *The School for Scandal* and Launcelot Gobbo in *The Merchant of Venice* in 1919; Peter Quince in *A Midsummer Night's Dream* in 1920; and Trinculo in *The Tempest* in 1921. In 1925 he was seen as Filch in a long run of GAY's *The Beggar's Opera*, and played Scrub in FARQUHAR's *The Beaux' Stratagem* in 1927 and Wittol in CONGREVE's *The Old Bachelor* in 1931. One of his best parts was Foresight in Congreve's *Love for Love*; he was also much admired as Sir Fretful Plagiary in Sheridan's *The Critic* at the OLD VIC in 1945, and, in more serious vein, as Old Ekdal in IBSEN's *The Wild Duck* in 1948. His finest work was done in his own very free adaptations of MOLIÈRE, which began in 1950 with *The Miser* (*L'Avare*), and continued with *Tartuffe, Sganarelle, The School for Wives* (*L'École de femmes*), *The Slave of Truth* (*Le Misanthrope*), *The Imaginary Invalid* (*Le Malade imaginaire*), and *The Prodigious Snob* (*Le Bourgeois gentilhomme*). Although they aroused some controversy, they had the merit of bringing Molière to the English stage in versions which appealed to the average playgoer, and proved invaluable to repertory and provincial theatres. He made one of his rare appearances in a modern play when he was seen as Mr Butterfly in IONESCO's *Rhinoceros* (1960), and he also appeared as Merlyn in the musical *Camelot* on its London production in 1964. Apart from his adaptations of Molière, he was the author of a number of plays, of which the most successful was *The Fanatics* (1927).

MALONE, EDMUND (1741–1812), one of the first scholars to study and annotate the works of Shakespeare. Born in Ireland, he came as a young man to London, intending even then to dedicate himself to literary criticism. He became the friend of Dr JOHNSON, and was the first to perceive and denounce the Shakespeare forgeries of the young William IRELAND. In spite of the many new facts which have been brought to light by later research, and an entirely new orientation in the study of Shakespeare as a dramatist, Malone's works, which include a biography, a chronology of the plays, and a history of the Elizabethan stage, are still of value. The Malone Society, formed in 1907 to further the study of early English drama by reprinting texts and documents, was named after him in recognition of his eminence in the world of theatre scholarship.

MALVERN FESTIVAL, founded in 1929 by Barry JACKSON, who provided much of the money to initiate and support it, as well as actors from his BIRMINGHAM REPERTORY THEATRE company, distinguished players being imported for special parts. Jackson's long association with Bernard SHAW led him to devote the first year's programme entirely to his plays, with the first English production of *The Apple Cart* (with Cedric HARDWICKE and Edith EVANS) and revivals of *Back to Methuselah, Caesar and Cleopatra*, and *Heartbreak House*. Shaw became the patron-in-chief of the festival, and more than 20 of his plays were presented there, *Geneva* and *In Good King Charles's Golden Days* having their first productions in 1938 and 1939 respectively, with Ernest THESIGER playing the chief part in both. *Too True to Be Good* (1932) and *The Simpleton of the Unexpected Isles* (1935) were also given their first English productions at Malvern, having been seen previously in the United States. Apart from Shaw, the productions of the festival ranged over 400 years of English drama, from *Hickscorner* (*c.* 1513) to contemporary plays by BRIDIE, DRINKWATER, PRIESTLEY, and others, the first non-Shavian play to be seen being Besier's *The Barretts of Wimpole Street* (1930). At the end of the 1937 season Jackson withdrew and the festival was run by Roy Limbert, manager of the Malvern Theatre. During the Second World War Limbert managed to maintain a skeleton organization at Malvern, but after the war the festival was revived only once, in 1949, when Shaw's *Buoyant Billions*, first seen in Zurich in 1948, had its first English production. In 1964 the theatre, seating 799, was taken over by a Trust, which leases it from the local authority, and since 1977 the festival has been revived as a Shaw and Elgar Festival each spring, the rest of the year being taken up by visits from touring companies and productions by a resident amateur company.

MALY THEATRE, Moscow. This theatre (*maly*, small, as opposed to *bolshoi*, big) opened on 14 Oct. 1824 with a company which had been in existence since 1806. It is the oldest theatre in the city, and the only one to keep the traditional drop-curtain. With its unbroken history it has played an important part in the development of Russian drama, particularly in the 1840s, when SHCHEPKIN was appearing in such plays as GOGOL's *The Government Inspector* and GRIBOYEDOV's *Woe from Wit*, and MOCHALOV in translations of Shakespeare's tragedies made directly from the original and not, as hitherto, from the French. In 1854 the Maly first produced a play by OSTROVSKY, and so began a brilliant partnership which lasted until 1885. The actor who first played many of the leading roles was Prov SADOVSKY, whose descendants continued the connection with the theatre. Other great names connected with the Maly are Alexander LENSKY and Maria YERMOLOVA. Having survived the Russian Revolution, the theatre took its place in the theatrical life of Soviet Russia with the production of TRENEV's *Lyubov Yarovaya* in 1926. It has since included many new Soviet plays in its repertoire, but has not neglected the classics, one of the outstanding productions having been *Othello* with the veteran actor Alexander Ostushev (1874–1953) in the title role. For many years the theatre was known as the House of Ostrovsky, and the 150th anniversary of its opening was celebrated in 1974 with a revival of Ostrovsky's *The Storm*.

MAMET, DAVID, see CHICAGO.

MANAGER, see PRODUCER.

MANCHESTER. The first permanent theatre in the city was erected in 1758 in Marsden, and from 1760 to 1775 was used regularly by a company under the provincial manager James Augustus Whitley (*c.* 1724–81) on its way from Leeds to Worcester. It closed when a new theatre was built at the junction of York Street and Spring Gardens, and was put to other uses, being finally demolished in 1869. Having obtained a royal patent, Joseph Younger and George Mattocks, son-in-law of Lewis HALLAM, opened the new building as the Theatre Royal in 1775, and ran it successfully for several years. Among the well known players who appeared there were John Philip KEMBLE, who at 19 years of age was seen as Othello to the Desdemona of his elder sister Mrs SIDDONS, and Mrs Elizabeth INCHBALD. The building was burned down in 1789, but rebuilt and reopened in 1790. It proved to be too small for the new mass audiences, and in 1807 was replaced by a much larger theatre, which soon ruined its first manager, the father of William MACREADY. It too was destroyed by fire, in 1844, and the last Theatre Royal to be built in Manchester, in Peter Street, opened in 1845, became a cinema in 1929, and is now used for bingo. In 1891 the PALACE THEATRE was erected for the use of touring companies, and in 1908 the GAIETY THEATRE, a former music-hall, was opened as the first repertory theatre in Britain by Miss HORNIMAN. The dramatists connected with this theatre, including Harold BRIGHOUSE, Stanley HOUGHTON, and Allan MONKHOUSE, were known as the Manchester School of Drama. The theatre closed in 1917. The LIBRARY THEATRE has had its own company since 1952.

A university theatre, directly connected with the Manchester UNIVERSITY DEPARTMENT OF DRAMA, opened under Hugh HUNT in 1966, and for some years housed the 69 Theatre Company, now the ROYAL EXCHANGE THEATRE Company. Since 1973 the university theatre has had its own resident company, the Contact Theatre Company, founded by Hugh Hunt, which also visits schools and youth clubs and runs educational drama workshops. The theatre is used for part of the year by amateur groups from the university. It has an adaptable auditorium seating between 250 and 350.

MANDUCUS, a stock character in the Roman FABULA *atellana*, whose name indicates that he had a big mouth with champing teeth, perhaps because he was a gluttonous eater, or merely to appear frightening. Like most of the clowns in the Atellan farce he wore a belted tunic to the knee and a short full cloak. He is also known to have had a large hooked nose with a wart on it, in common with DOSSENNUS, and the two names may have been used for the same character indifferently.

MANET, see STAGE DIRECTIONS.

MANHATTAN OPERA HOUSE, New York, see KOSTER AND BIAL'S MUSIC HALL.

MANHATTAN THEATRE, New York, at 1697 Broadway, between 53rd and 54th Streets. This opened on 30 Nov. 1927 as Hammerstein's Theatre, and was used almost entirely for musical shows. In 1931 it was renamed the Manhattan, but adhered to its musical policy until in 1934, after a long period of idleness, it became a theatre-restaurant, the Billy Rose Music-Hall, and soon afterwards the Manhattan Music-Hall. This was a failure, and on 21 Feb. 1936 the FEDERAL THEATRE PROJECT reopened the building, which had reverted to its former name, with *American Holiday*, by E. L. and A. Barker. On 30 Mar. there began a limited run of T. S. ELIOT's *Murder in the Cathedral*, but in Sept. of the same year the theatre was taken over for broadcasting. It was renamed the Ed Sullivan in 1967.

From 1897 to 1901 the STANDARD THEATRE was known as the Manhattan.

MANNERS, JOHN HARTLEY (1870–1928), Irish-born American dramatist, who made his first appearances on the stage in Australia, and until 1902 was in London, where he appeared with ALEXANDER and played Laertes to the Hamlet of FORBES-ROBERTSON. He went to the United States in the company of Lillie LANGTRY, for whom he wrote his first play. In 1908 he settled permanently in America, and from then until his death contributed more than 30 plays to the New York stage. The best known is *Peg o' My Heart* (1912; London, 1914), which was translated into several European languages, and was at one time being played by five touring companies at once through several seasons. Its success overshadowed all his other works, and prevented his being taken seriously as a modern dramatist. The part of the heroine in this and many of his other plays was taken by his wife Laurette TAYLOR.

MANNHEIM, a city important in the history of the German stage. It already had a long tradition of opera behind it when the first Mannheim National Theatre opened in 1778 under DALBERG, who made it one of the foremost playhouses in the country, particularly when after EKHOF's death in Gotha later the same year he took over his troupe with IFFLAND at its head. Dalberg's greatest service to the German theatre was undoubtedly his support of the young SCHILLER, whose *Die Räuber* had its first production at Mannheim in 1782, followed by *Fiesko* and *Kabale und Liebe* in 1784, the year which saw also the production of one of the best plays of the STURM UND DRANG movement, LEISEWITZ's *Julius von Tarent*. In 1796, partly owing to the rigours of war, the fortunes of the Mannheim theatre declined and the company was disbanded. Little was done until in 1884 the National Theatre reopened and there was an upsurge of theatrical activity, which culminated in 1916 with the first production of HASENCLEVER's *Der Sohn*, inaugurating a period of EXPRESSIONISM. The National Theatre was completely destroyed in 1943, but a new building opened in 1957 with an arena production of *Die Räuber*, directed by PISCATOR, since when the repertory has included a wide selection of new European plays as well as German classics.

MANOEL THEATRE, Malta, 18th-century playhouse built by the Grand Master Antonio de Vilhena in 1731. It opened a year later with a performance of MAFFEI's *Merope*, and flourished for a century or more with a nine-month season of mixed opera and plays. Renamed the Royal Theatre by the British occupying powers, it continued pre-eminent until the building of the much larger Opera House, after which it sank into disrepair. In the 1950s it was bought by the Maltese Government, and in 1957 a committee was set up to rebuild and modernize it backstage while retaining the historic auditorium and façade. On 27 Dec. 1960 it reopened as the National Theatre of Malta with a season by the Ballet Rambert. It is still used for occasional summer seasons and gala performances.

MANRIQUE, GÓMEZ (*c*. 1412–*c*. 1490), early Spanish poet and playwright, author of a short Nativity play, *La representación del Nacimiento de Nuestro Señor*, written between 1467 and 1481 for performance by nuns. It combines interestingly the essentials of the secular courtly play with those of the LITURGICAL DRAMA, showing the Infant Christ being presented not with gold, frankincense, and myrrh by the Magi, but with the instruments of the Passion by the archangels Gabriel, Michael, and Raphael. Manrique wrote also some purely secular plays, in one of which, intended to celebrate the birth of his nephew, he introduced the seven Virtues, while in another, produced at Arévalo in 1467 to celebrate the 14th birthday of Alfonso, brother of the Infanta Isabella, the nine Muses descend from Helicon to bestow their several gifts upon the young prince.

MANSFIELD, RICHARD (1854–1907), American actor, born in Berlin, the son of an opera singer and a London wine merchant. Educated in England and on the Continent, he made his first appearance on the stage in London, and then toured in light opera. In 1882 he went to New York, making his first appearance there on 27 Sept., again in light opera; but it was as Baron Chevrial in Feuillet's *A Parisian Romance* (1883) at the UNION SQUARE THEATRE under A. M. PALMER that he first attracted attention and embarked on a successful career. Some of his outstanding parts were the dual title roles in a dramatization of Stevenson's *Dr Jekyll and Mr Hyde* (1887), and the title roles in Clyde FITCH's *Beau Brummell* (1890), specially written for him, ROSTAND's *Cyrano de Bergerac* (1898), and Booth TARKINGTON's *Monsieur Beaucaire* (1901). He also played the leading parts in his own plays *Monsieur* (1887), *Don Juan* (1891), and *The First Violin* (1898), based on a novel by Jessie Fothergill. During a visit to London in 1889 he first played Richard III, later accounted one of his finest parts, and among his other Shakespearian roles were Shylock in *The Merchant of Venice*, Brutus in *Julius Caesar*, and Henry V, making a spectacular appearance after Agincourt on a white horse. He was instrumental in introducing SHAW to America, playing Bluntschli in *Arms and the Man* (1894) and Dick Dudgeon in *The Devil's Disciple* (1897), Raina in the first and Judith Anderson in the second being played by his wife Beatrice Cameron (1868–1940), whom he married in 1892. She had already played Nora in IBSEN's *A Doll's House* in 1889, and it may have been under her influence that Mansfield, who in general had little sympathy with the 'new drama', though he much admired Ibsen's poetic prose, put on in his last season, 1906–7, the first production in English of *Peer Gynt*, with himself in the title role.

MANSFIELD THEATRE, New York, see BROOKS ATKINSON THEATRE.

MANTEAU D'HARLEQUIN, see FALSE PROS(CENIUM).

MANTELL, ROBERT BRUCE (1854–1928), American actor, born in Scotland, who was first on the stage in Belfast and London using the stage name Robert Hudson. He resumed his real name when he visited the United States with MODJESKA in 1878. He then returned to England, but after several years of hard work with little recognition he went back to the States for good, and in 1886 took his own company on tour. As a young man he was at his best in such romantic melodramas as BOUCICAULT's *The Corsican Brothers*, Charles

Selby's *The Marble Heart*, and BULWER-LYTTON's *The Lady of Lyons*. Later he became somewhat heavy and uninspired, but remained popular outside New York, where his careful studies of the leading Shakespearian roles won him respectful admiration.

MANTLE, ROBERT BURNS (1873–1948), American dramatic critic, who in 1898 became dramatic editor of the *Denver Times*. Later he worked as Sunday editor of the *Chicago Tribune*, and from 1922 until his retirement in 1943 he was dramatic critic of the *New York Daily News*. He edited till his death a series of play anthologies which began in 1920 with *The Best Plays of 1919–1920*; each volume contains extracts from ten of the season's productions, together with an annotated index of every play produced in New York and at leading regional theatres during the year. This useful chronicle of the modern American theatre was supplemented by three more volumes covering the years 1894 to 1919 and continues with the sub-title *The Burns Mantle Theater Yearbook*. Mantle was also the author of *American Playwrights of To-Day* (1935) and edited *A Treasury of the Theatre* (1938) with John GASSNER.

MANUEL, NIKLAUS (1484–1530), Swiss poet, playwright, and painter, of Berne, where his plays were performed in the market place. They were written in support of the Reformation and attack the abuses of the Catholic Church. One, seen in 1522, shows two processions, one of papal pomp and splendour, the other of Christ riding humbly on an ass, and draws the obvious moral. Manuel's best play, which gives evidence of a violence hitherto unknown in German drama, is *Der Ablasskrämer* (1528), in which a seller of indulgences, Ricardus Hinderlist (Dick Trickster), returns to a village where the peasants, recognizing him, attack him with rusty billhooks and torrents of abuse, avenging themselves for the tricks he played on them during his previous visit. In the end they string him up and force him to confess his sins and disgorge his ill-gotten gains.

MANZINI, GIOVANNI, see ITALY.

MANZONI, ALESSANDRO FRANCESCO TOMASO ANTONIO (1785–1873), Italian writer, best known for his outstanding novel *I promessi sposi* (*The Betrothed*). He was also the author of two tragedies, *Il conte di carmognola* (pub. 1820) and *Adelchi* (pub. 1822). Both are historically more accurate than most romantic tragedies, and both entirely disregard the UNITIES, in accordance with the principles set forth by the author in his treatise *Lettre sur les unités de temps et de lieu dans les tragédies* (1823). Despite their lyrical impulse—the choruses in *Adelchi* are some of the finest lyrics in 19th-century Italian writing—they are somewhat frigid and have not survived on the stage, though they served their purpose of proving that dialogue need not be unnatural because it happens to be in verse.

MARAIS, Théâtre du, Paris, early French theatre, which is thought to have opened on 31 Dec. 1634 in a converted tennis-court in the rue Vieille-du-Temple, with a company under MONTDORY which had been appearing in CORNEILLE's early comedies, and was responsible early in 1637 for the production of his great tragedy *Le Cid*. Among other notable productions at this time was TRISTAN L'HERMITE's *La Mariane*, in which Montdory gave a powerful rendering of Herod which may have contributed to his breakdown the following year. After he left the theatre went through some bad times. Its best actors joined the rival company at the Hôtel de BOURGOGNE, to which Corneille also gave his new plays, and those that were left were forced to revert to the playing of popular farces. They were however lucky in the return of JODELET, for whom some excellent new farces were written, and later they specialized in spectacular performances with a good deal of imported Italian scenery and machinery. The theatre never regained the place it had held previously in public esteem, particularly after a bad fire in 1644 burnt down the old building. The contract for its replacement on the same site indicates that its overall dimensions were about 114 ft by 36 ft. The newly-housed company survived, mainly owing to the efforts of FLORIDOR and LAROQUE, until 1673, when it was amalgamated with the company of MOLIÈRE, who had just died, and by the combined companies' fusion in 1680 with that of the Hôtel de Bourgogne became part of the COMÉDIE-FRANÇAISE, the Marais stage being finally abandoned.

MARANGA MAI, see NEW ZEALAND.

MARBLE, DANFORTH (1810–49), American actor, originally a silversmith. He made his first appearance on the professional stage in 1831, after some experience as an amateur, and proved himself an excellent mimic of the Yankee dialect. He made a success in the title role of *Sam Patch* (1837), an anonymous play which he may have written or arranged himself, and it proved to be so popular, especially along the Mississippi and at the BOWERY THEATRE in New York, that it was followed by two other 'Sam Patch' plays. In 1844 Marble visited London and the English provinces, being received with enthusiastic applause and going also to Glasgow and Dublin, in such typical pieces from his repertory as the younger COLMAN's *Jonathan in England* and Joseph Jones's *The People's Lawyer*, which gave full scope to his inimitable assumption of Yankee characteristics. He married an actress, one of the four daughters of William WARREN the elder, and died young at the height of his popularity.

MARCADÉ [Mercadé], EUSTACHE (?–1440), French theologian and rhetorician, who became Dean of the Faculty of Ecclesiastical Law at the Sorbonne in Paris shortly before his death. He is thought to be the author of a MYSTERY PLAY on the destruction of the Jews, *Le Mystère de la vengeance de notre seigneur Jhesuscrist sur les Juifs par Vespasien et Titus*, and also of the PASSION PLAY that precedes it in the manuscript, sometimes called the *Passion d'Arras*. This contains some 25,000 lines, and took four days to act. It opens with scenes in Paradise, then dramatizes the life of Christ from the Annunciation to the Ascension. A preacher speaks a prologue and epilogue to each day's play.

MARCEAU, FÉLICIEN [Louis Carette] (1913–), Belgian novelist, dramatist, and critic, whose first play *L'Oeuf* (1956) was a notable box-office success, and was seen in London in 1957 as *The Egg*. His second play *La Bonne soupe* (1959), which was also well received, had a good run in London in 1961 as *Bonne Soupe*, translated by Kitty Black. In spite of considerable inventiveness in devising witty representations of contemporary issues and a successful formula, similar to that devised by Peter NICHOLS, which involves a 'little man' as narrator, discovering how to trick the world, and taking the audience into his confidence, Marceau's popularity has since waned.

MARCEAU, MARCEL (1923–), French actor, the finest modern exponent of mime or, as he prefers to call it, mimodrama. He studied under DULLIN and Decroux and in 1945 joined the company of Jean-Louis BARRAULT, but a year later abandoned conventional acting to study mime, basing his work on the character of the 19th-century French Pierrot and evolving his own Bip, a white-faced clown with sailor trousers and striped jacket. In this part, which he first played at the tiny Théâtre de Poche in Paris in 1946, he has toured all over the world, accompanied by supporting players whom he has trained himself. He has also evolved short pieces of concerted mime, including one based on GOGOL's *The Overcoat*, and longer symbolic dramas like his own *The Mask-Maker*, and with the aid of a screen has contrived to appear almost simultaneously as two sharply contrasted characters—David and Goliath, or the Hunter and the Hunted. His work has given immense impetus to the study of mime by young actors, who have also benefited by the many demonstrations of his technique which he has given to students. His École de mimodrame de Paris opened at the Théâtre de la PORTE-SAINT-MARTIN in 1978.

MARCEL, GABRIEL (1889–1964), French philosopher, dramatist, and critic, whose writings introduced the term 'existentialism' into the language. His plays, which are largely dramas of conscience, include *Le Quatuor en fa dièze* (*The Quartet in F sharp*, 1920), *Le Coeur des autres* (1921), *L'Iconoclaste* (1923), *Un Homme de Dieu* (1925), and his most important work for the theatre *Le Chemin de crète* (1936). His last play, *La Dard*, was produced in 1938.

MARDZHANOV [Mardzhanishvili], KONSTANTIN ALEXANDROVICH (1872–1933), Soviet director, and the virtual founder of the modern theatre in his native GEORGIA. After some experience in the provinces, he joined the MOSCOW ART THEATRE in 1910, remaining there for three years and directing, among other plays, IBSEN's *Peer Gynt* in 1912. In 1913 he organized the so-called Free Theatre, having TAÏROV and his future wife Alisa KOONEN among his actors. His methods of staging had a noticeable influence on Taïrov's subsequent work at the KAMERNY THEATRE. After the Revolution Mardzhanov was in Kiev, but in 1922 he returned to Tbilisi to reorganize the Georgian theatre along Soviet lines. An intensely national artist, though in no sense nationalistic, Mardzhanov encouraged the writing and production of new Georgian plays, many of which he directed himself, but did not neglect the older classics, one of his most successful productions being Shakespeare's *The Merry Wives of Windsor*. In 1928 he went to Kutaisi, where he founded the Second State Georgian Theatre, which in 1930 was transferred to Tbilisi and renamed the Mardzhanov after his death. Among the plays which he directed there were works by POGODIN, AFINOGENOV. TOLLER, and, as before, Shakespeare.

MARÉCHAL, MARCEL LOUIS-NOËL (1937–), French actor and director, an important figure among those who have led the revival of the theatre in the French provinces. His early successes were made with plays by VAUTHIER and AUDIBERTI, but his later work has been on a larger scale, and although he came late to staging Shakespeare, the word 'Elizabethan' has often been used to describe his work. A production of *La Moschetta* in 1970 led to a revival of interest in BEOLCO's plays; revivals of ROLLAND's *Danton* in 1969 and BRECHT's *Herr Puntila und sein Knecht Matti* in 1971 were much acclaimed; and he produced in 1972 an astonishing *Capitaine Fracasse* (adapted from Gautier's novel), using professional masters-at-arms, acrobats, and a comic-strip style aimed at bringing back certain elements of popular culture into the modern theatre. In 1975 he became the first director of the Théâtre National in Marseilles. He is the author of *La mise en théâtre* (1974).

MARIGNY, Théâtre, Paris, a small theatre built in 1850 in the gardens of the Champs-Élysées. The son of DEBURAU took it over in 1858 and opened it under his own name, appearing there in many of his father's old parts, but it was not a success and soon became the Théâtre des Champs-Élysées (not to be confused with the later COMÉDIE

DES CHAMPS-ÉLYSÉES). Success continued to elude it until as the Folies-Marigny it became a home of VAUDEVILLE after the demolition in 1862 of the BOULEVARD DU TEMPLE. It was pulled down in 1881, and a circular building intended for a panorama was erected on the site. This became a music-hall in 1896 and took its present name in 1901. In 1925 it was completely redecorated under a new owner and became the most elegant playhouse in Paris, with spacious approaches, a luxurious auditorium, comfortable seating, and an impressive cupola. The COMÉDIE-FRANÇAISE used it for matinées and special performances, and moved there temporarily during the rebuilding of its own theatre in 1937 and again in 1974. From 1946 to 1956 it housed the company of Jean-Louis BARRAULT and his wife Madeleine RENAUD in a repertory of distinguished plays.

MARINELLI, KARL (1744–1803), Austrian actor, dramatist, and impresario, who in his play *Der Ungar in Wien* (*The Hungarian in Vienna*, 1773) first introduced the figure of the light-hearted romantic Magyar, later a stereotype of Viennese folk-comedy and operetta which survived long enough to be ridiculed by SHAW in *Arms and the Man* (1894). When Joseph II, as part of his rationalist reforms, banished the old impromptu burlesque from the Burgtheater, Marinelli provided a new home for it at the Leopoldstädter Theater, which opened in 1781, and there the great comedian Johann LAROCHE established the comic character of KASPERLE. Marinelli was also the first to recognize the talent of the young actor-dramatist Anton HASENHUT, later famous as THADDÄDL; and with his own plays and adaptations, of which *Die Liebesgeschichte von Hirschau oder Kasperl in sechselei Gestalten* (*The Love Story of Hirschau or Kasperl in Six Shapes*, 1782) and *Dom Juan, oder der steinerne Gast* (1783) are typical, he helped to establish the reputation of the traditional Viennese folk theatre.

MARINETTI, FILIPPO TOMMASO (1876–1944), Italian poet and dramatist, and the principal exponent of FUTURISM. His first play, *Le Roi Bombance* (in Italian *Il re baldoria*), was written in 1905, but not produced until 1909, when it was seen in Paris in French, and created a scandal by its use of the digestive system as a symbol of human corruption. But this, and his other plays, some of which were published in 1920 under the title of one of them *Elettricità sessuale* (written in 1909), were of less importance than his theoretical writings in support of the short-lived futurist movement. The first of these appeared in *Le Figaro* in 1909, and others followed up to the final manifesto of 1921. Marinetti spent his later years in France preaching the cause of Fascism, to which the anarchic individualism of futurism was an apt artistic pendant.

MARIONETTE, see PUPPET.

MARIVAUX, PIERRE CARLET DE CHAMBLAIN DE (1688–1763), French dramatist, friend of Fontenelle and HOUDAR DE LA MOTTE, with whom he frequented the salons of Mme de Tencin and Mme de Lambert. This no doubt helped to develop in him the peculiarly paradoxical and sensitive style which characterizes the dialogue of his plays, a style later known as *marivaudage*, first contemptuously, then in admiration of its superb subtlety. His first plays, both produced in 1720, were an unsuccessful tragedy *Hannibal*, seen at the COMÉDIE-FRANÇAISE, and a successful one-act comedy *Arlequin poli par l'amour*, seen at the COMÉDIE-ITALIENNE. Marivaux's talents were much better suited to the Italian comedy, whose traditional characters, simple plots, and stock situations left him free to concentrate on the developing emotional awareness of his young lovers. He therefore continued to write mainly for the Comédie-Italienne, whose leading lady SILVIA contributed not a little to the success of his plays. The most important of these were *La Surprise de l'amour* (1722); *La Double Inconstance* (1723); *Le Jeu de l'amour et du hasard* (1730); *Les Fausses Confidences* (1737); *and L'Épreuve* (1740). He also wrote occasionally for the Comédie-Française, where *La Seconde Surprise de l'amour* (1727) and *Le Legs* (1736) were well received.

Marivaux's delicate, psychological theatre, in which the major emphasis is on the female roles, was not on the whole popular with his contemporaries, who preferred the cruder emotions of LA CHAUSSÉE's *comédie larmoyante*—though he had a certain following among the representatives of that cultured, refined society in which his own gifts had developed. Lost sight of during the period which led up to the French Revolution, Marivaux came back into favour again with the Restoration, when his plays had a considerable influence on the work of Alfred de MUSSET. It was not, however, until the 20th century that he was fully appreciated. His plays were then frequently revived, not only at the Comédie-Française, where he holds fourth place, in number of performances, after MOLIÈRE, CORNEILLE, and RACINE, but at other theatres, particularly at the MARIGNY under BARRAULT, where Madeleine RENAUD proved herself the perfect interpreter of Marivaux's heroines. Although French companies have performed Marivaux's plays both in London and in New York with some success, the subtlety of his dialogue makes it almost impossible to translate him adequately into English.

MARK HELLINGER THEATRE, New York, at 237 West 51st Street, between Broadway and 8th Avenue; capacity 1,581. Originally a cinema which opened in 1930, it became the Fifty-First Street Theatre on two occasions, first in 1936 with *Sweet River*, a new version of *Uncle Tom's Cabin* by George ABBOTT which was taken off after five performances, and again in 1940 for a short run of *Romeo and Juliet* with Laurence OLIVIER and

Vivien LEIGH. On 22 Jan. 1949 it became a theatre again and was named after Mark Hellinger (1903–47), a newspaper columnist, dramatist for stage and screen, and film producer, reputedly the first newspaperman to write a column solely concerned with Broadway. From 1956 to 1962 the theatre was occupied by the musical *My Fair Lady*. Later musicals seen there included *On a Clear Day You Can See Forever* (1965); *Coco* (1969), starring Katharine HEPBURN as Chanel; and Tim Rice and Andrew Lloyd Webber's *Jesus Christ Superstar* (1971). In 1979 *Sugar Babies*, a tribute to the great days of BURLESQUE, began a long run.

MARKISH, PERETZ, see MOSCOW STATE JEWISH THEATRE.

MARK TAPER FORUM, Los Angeles, California, one of the most important and enterprising theatre organizations in Los Angeles, a city which with several notable commercial theatres, including the Huntington Hartford and the Shubert, and over 60 other theatrical enterprises, ranks as one of the most exciting theatrical areas in the United States outside New York. Since 1967 the Forum, which forms part of the Music Center of Los Angeles County, where the 2,071-seat Ahmanson Theatre is also situated, has housed the Center Theatre Group, a non-profit-making organization created jointly in 1959 (as the Theatre Group) by the University of California Extension and a number of interested theatre, film, and television personalities. Its first full productions, in 1960, were T. S. ELIOT's *Murder in the Cathedral* and CHEKHOV's *Three Sisters*, both directed by John Houseman. The Forum, which has a pentagonal thrust stage and a semicircular auditorium seating 750, has no permanent company, but assembles excellent casts and technicians from the nearby film and television colonies. Under Gordon Davidson (1933–), its Director since 1965, the Group has been responsible for the world premières of several important plays, including Daniel Berrigan's *The Trial of the Catonsville Nine* (1970; N.Y., 1971), Michael Cristofer's *The Shadow Box* (1975; N.Y., 1977), and Mark Medoff's *Children of a Lesser God* (1980; N.Y., 1980; London, 1981), all of which were directed by Davidson himself. The 99-seat Forum Lab presents experimental work to invited audiences.

MARLOWE, CHRISTOPHER (1564–93), playwright of the English Renaissance and an important figure in the development of the Elizabethan stage, which he helped to liberate from the influence of medieval drama and the Tudor INTERLUDE. The son of a Canterbury shoemaker, Marlowe was educated at Cambridge. Though highly thought of by his contemporaries as a poet and a scholar, he was often in danger of arrest owing to his atheistical and outspoken opinions, and he died young in a tavern brawl, perhaps assassinated because of his secret service activities. His quatercentenary was overshadowed by that of Shakespeare, ironically, since there is a theory that he was not killed, but only concealed for a time, returning later unrecognized to write plays under Shakespeare's name.

Marlowe's first appearance in the theatre was probably as part-author with NASHE of a classical tragedy, *Dido, Queen of Carthage*, performed by one of the BOY COMPANIES in 1587–8. At about the same time the ADMIRAL'S MEN produced his *Tamburlane the Great, Part I*, with Edward ALLEYN in the title role. The second part was given about a year later. There can be no doubt that these two plays, which were highly successful, had a great influence on Shakespeare, who probably saw them soon after his arrival in London a few years later, since they continued to hold the stage up to the closing of the theatres in 1642. They were followed, though it is not certain in what order, by *The Jew of Malta, The Tragical History of Dr Faustus, Edward II*, and *The Massacre at Paris*, the last being an inferior piece which has not survived on the stage. *The Jew of Malta*, probably written as early as 1589–90, survives only in a printed version of 1633 which was heavily revised, probably by Thomas HEYWOOD. It seems to have been the most popular of Marlowe's plays, and Alleyn was again much admired in the title role. It may have contributed something to Shakespeare's Shylock in *The Merchant of Venice*. *Dr Faustus*, written round about 1590, was based on the old German legend of FAUST; this again survives only in a fragmentary and much mutilated condition. It was not printed until 1604, and then with the addition of a number of comic scenes featuring the Devil which were probably added for a revival, possibly by BIRD and ROWLEY at the instigation of HENSLOWE. As originally planned it seems to have had much in common with the earlier MORALITY PLAY, and while retaining all the fine poetry of *Tamburlane* it is more consistently dramatic and shows a great advance in stagecraft over the earlier work. *Edward II*, a chronicle play which has affinities with Shakespeare's *Richard II*, though lacking the fine lyricism of some of Marlowe's earlier works, marks the high point of his development as a dramatist. It was printed during his lifetime, and the text may therefore be considered reasonably correct. It maintained its popularity for a few years, and then fell out of the repertory. It was revived by the PHOENIX SOCIETY in 1923, and in 1969 was performed in repertory with Shakespeare's *Richard II* by PROSPECT at the MERMAID, Ian MCKELLEN playing the two title roles on alternate nights; he was seen as Dr Faustus in a revival by the ROYAL SHAKESPEARE COMPANY in 1974, and the play was seen again in the West End in 1980. *The Jew of Malta*, in repertory with *The Merchant of Venice*, was revived at the ROYAL SHAKESPEARE THEATRE in 1965. Donald WOLFIT gave a fine performance as Tamburlane at the OLD VIC in 1951; the most recent revival of this play was at the NATIONAL THEATRE in 1976 with Albert FINNEY.

MARLOWE, JULIA [Sarah Frances Frost] (1866–1950), British-born American actress, who was taken to the United States at the age of 4, and made her first appearance on the stage at the age of 12. She made her adult début in New York in 1887 as Parthenia in Mrs Lovell's *Ingomar*, and was immediately successful. She was at her best in the heroines of Shakespeare's plays, but was also much admired in standard comedy, playing Lydia Languish in SHERIDAN's *The Rivals* with Joseph JEFFERSON and Mrs John DREW, as well as Julia in Sheridan KNOWLES's *The Hunchback* and Pauline in BULWER-LYTTON's *The Lady of Lyons*. She married as her second husband E. H. SOTHERN, and played Juliet to his Romeo in 1904. Three years later she made her first appearance in London, and toured for some time with her husband in a Shakespearian repertory, playing Lady Macbeth for the first time in 1913. She retired for a time in 1915, but returned to play mainly in Shakespeare until her final retirement in 1924.

MARLOWE SOCIETY, see CAMBRIDGE.

MARMONTEL, JEAN-FRANÇOIS (1723–99), French man of letters, whose first play, *Denys le Tyran*, was produced in 1748. Owing to the lack of good contemporary plays this pale reflection of neo-classical French tragedy was well received mainly because of the acting of Mlle CLAIRON, who later became Marmontel's mistress and was persuaded by him in about 1753 to discard her declamatory manner for a more natural style of acting. Marmontel continued to write plays until 1753, but none of them has been revived, and it is mainly as a critic that he is now remembered. Indeed, he may be regarded as the founder of French journalistic dramatic criticism. He also wrote a number of librettos for light operas.

MAROWITZ, CHARLES (1934–), American-born theatre director, who from 1956 worked in England, and was connected with Peter BROOK in a production of *King Lear* in 1962 and in a programme entitled 'Theatre of Cruelty' which in 1964 set out to demonstrate the theories propounded by ARTAUD. He also worked with the TRAVERSE THEATRE CLUB in Edinburgh and London, and in 1968 opened his own theatre, the Open Space, a small adaptable playhouse in Tottenham Court Road seating 130 round a completely flexible stage. (It later moved to Euston Road.) There he introduced several new American writers to English audiences; he attracted notice with 'collage' versions of Shakespeare, *Hamlet* in 1966, *Macbeth* in 1969, *Othello* in 1972, *The Taming of the Shrew* in 1974, and *Measure for Measure* in 1975. The last production at the theatre in the Euston Road was an adaptation of STRINDBERG's *The Father* in 1979. For a time the company continued to present plays elsewhere—*Hedda*, based on IBSEN's *Hedda Gabler*, was seen at the ROUND HOUSE in 1980—but in 1981 it went into liquidation.

MARS, Mlle [Anne-Françoise-Hippolyte Boutet] (1779–1847), French actress, younger daughter of the actor-dramatist Monvel (1745–1812) by a provincial actress. She appeared on the stage as a child, playing in Paris and at Versailles under Mlle MONTANSIER, and in 1795, encouraged by Mlle CONTAT, she made her first appearance at the COMÉDIE-FRANÇAISE. When the company was reconstituted after the upheaval of the Revolution, she was again a member of it, and embarked on a long and successful career. She was at her best in the comedies of MOLIÈRE, but was also much liked in such dramas as *Henri III et sa cour* (1829) by DUMAS *père* and HUGO's *Hernani* (1830). She retired in 1841, making her last appearances on 31 Mar. as Elmire in Molière's *Tartuffe* and Silvia in MARIVAUX's *Le Jeu de l'amour et du hasard*. A beautiful woman with a lovely voice, she continued to play young parts until she was over 60.

MARSHALL, NORMAN (1901–80), English director and theatre manager. After gaining some experience of the stage on tour, he was appointed in 1926 one of the directors under Terence GRAY of the CAMBRIDGE Festival Theatre, where he was responsible for some interesting productions; in 1932, under his own management, he directed there the first production in England of O'NEILL's *Marco Millions*. Two years later he took over the direction of the GATE THEATRE in London, where he presented an annual Gate REVUE, as well as a varied programme of new and uncommercial plays, several of which—among them Elsie Schauffler's *Parnell* (1936), HOUSMAN's *Victoria Regina* (1937), and STEINBECK's *Of Mice and Men* (1939)—were later seen in the West End. After serving in the army during the Second World War he returned to London to direct Robert SHERWOOD's *The Petrified Forest* (1942), and then formed his own company, with which he toured extensively, at the same time continuing to direct in London such new plays as Beckwith's *A Soldier for Christmas* (1944), Norman Ginsbury's *The First Gentleman* (1945), with Robert MORLEY as the Prince Regent, and Peter USTINOV's *The Indifferent Shepherd* (1948). Under the auspices of the British Council he then toured Europe and India (where he was born) in abridged versions of Shakespeare's plays, and in 1952 was invited to direct Ben JONSON's *Volpone* at the CAMERI THEATRE in Tel Aviv. He continued for some years to direct plays in London, and was active in the planning of the NATIONAL THEATRE, as well as lecturing and writing on the modern theatre.

MARSHALL THEATRE, Richmond, Virginia, see RICHMOND THEATRE.

MARSHFIELD PLAY, see MUMMERS' PLAY.

MARSTON, JOHN (*c.* 1575–1634), English dramatist, satirized by JONSON as Crispinus in *The Poetaster* (1601). His mother was Italian, which may help to account for the Italian influence discernible in his first two tragedies, *Antonio and Mellida* and *Antonio's Revenge*, which were performed by the Children of Paul's in 1599. A comedy, *What You Will* (1601), may also have been given by them, and was followed by the best of Marston's plays, *The Malcontent* and *The Dutch Courtesan* (both 1604). A DROLL extracted from the latter, *The Cheater Cheated*, was published in Kirkman's *The Wits* (1662), and was later adapted by Aphra BEHN as *The Revenge; or, a Match in Newgate* (1680). In 1605 Marston was implicated in the trouble over *Eastward Ho!*, in which he seems to have been the chief culprit, and only escaped imprisonment with his co-authors Jonson and CHAPMAN by an ignominious flight. He returned to the theatre with another Italianate tragedy, *The Insatiate Countess* (1610), and then renounced the stage and took holy orders, the immediate cause of his retirement being apparently a play now lost in which he satirized James I and so found himself in prison. He is believed to have had a hand in the writing of Shakespeare's *Troilus and Cressida*.

MARTIN, MARY (1913–), American singer and actress, who first attracted attention in the Broadway musical *Leave It To Me* (1938), in which she sang 'My Heart Belongs to Daddy', a song with which she was ever after associated. After several years in Hollywood she returned to Broadway in another musical, *One Touch of Venus* (1943), and made her first appearance in London at DRURY LANE as Elena Salvador in COWARD's *Pacific 1860* (1946). After touring the United States in *Annie Get Your Gun* (1947–8), she created the role of Ensign Nellie Forbush in *South Pacific* (N.Y., 1949; London, 1951), and was then seen in her first non-singing role, in Krasna's *Kind Sir* (1953). She subsequently starred in a musical version of BARRIE's *Peter Pan* (1954), and in a revival of Thornton WILDER's *The Skin of Our Teeth* (1955). In 1959 she created another famous role, Maria Rainer in the musical *The Sound of Music*, and in 1965 she created in London, at Drury Lane, the role of Dolly Levi in *Hello, Dolly!*, the musical version of Wilder's *The Matchmaker*. In the following year she starred on Broadway in *I Do! I Do!*, the musical version of Jan de Hartog's two-character play *The Fourposter*, which ran for over a year. After a year's tour, 1968–9, she virtually retired from the stage, though she was seen briefly in New York in 1978 in ARBUZOV's *Do You Turn Somersaults?* Her autobiography, *My Heart Belongs*, was published in 1976.

MARTIN BECK THEATRE, New York, at 302 West 45th Street, between 8th and 9th Avenues. With a seating capacity of 1,280, it opened on 11 Nov. 1924 with *Madame Pompadour*, and has since housed many other successful musicals. It was also used by the THEATRE GUILD for Robert Nichols and Maurice BROWNE's *Wings Over Europe* (1928); for the first American production of SHAW's *The Apple Cart* (1930); and by the GROUP THEATRE for its initial productions, GREEN's *The House of Connelly* and SHERWOOD's *Reunion in Vienna* (both 1931). Katharine CORNELL was seen in 1934–5 in a repertory which included *Romeo and Juliet*, Besier's *The Barretts of Wimpole Street*, and BUCKSTONE's *The Flowers of the Forest*. Later productions included Maxwell ANDERSON's *Winterset* (1935) and *High Tor* (1937), Lillian HELLMAN's *Watch on the Rhine* (1941), and O'NEILL's *The Iceman Cometh* (1946). After the success of *The Teahouse of the August Moon* (1953) by John Patrick, Leonard Bernstein's musical version of VOLTAIRE's *Candide* (1956), though highly praised, had only a short run, as did BRECHT's *Mother Courage and her Children* (1963), which was followed by *The Ballad of the Sad Café*, adapted from a novella by Carson McCullers by Edward ALBEE, whose own plays *A Delicate Balance* (1966) and *All Over* (1971) were produced here. Other notable productions of the 1970s were Alan BENNETT's *Habeas Corpus* (1975), with Donald SINDEN, and a new version of Bram Stoker's *Dracula* (1977). In 1981 the theatre had an outstanding success with a revival of Lillian Hellman's *The Little Foxes* starring Elizabeth Taylor and Maureen STAPLETON.

MARTINELLI, DRUSIANO (?–1606/8), actor of the COMMEDIA DELL'ARTE, brother of Tristano (below). He was probably in London in 1577–8 with the first regular Italian company to cross the Channel. His name is found in several actor-lists after this, but his reputation was overshadowed by that of his wife Angelica Alberigi (or Alberghini), who with her husband directed for some time a company known as the UNITI.

MARTINELLI, TRISTANO (*c.* 1557–1630), brother of the above, actor of the *commedia dell'arte*, probably the first to play ARLECCHINO. He was originally a member of the CONFIDENTI under PELLESINI, but appears to have been of a roving and somewhat quarrelsome disposition and is found with many different companies, including the DESIOSI and, in 1600, the ACCESI. He was very popular in Paris, where he appeared on several occasions, and specimens of his extempore wit were preserved in a publication entitled *Compositions de rhétorique de M. Don Arlequin* (1600).

MARTÍNEZ DE LA ROSA, FRANCISCO (1787–1862), Spanish dramatist, virtually the founder of the Romantic movement in the Spanish theatre. His early plays, in the prevailing neo-classical style, were of little interest, and in his comedies, of which the best is *La niña en la casa y la madre en la máscara* (*The Daughter at Home and the Mother at the Masquerade*, 1821), he followed the traditions of FERNÁNDEZ DE MORATÍN. Exiled to Paris in 1823 for his political activities, he came under the influence of Victor HUGO and the young Romantics. His next two plays. *Aben-Humeya* and *La conjuración de Venecia, año de 1310* (*The Venetian Conspiracy of 1310*), both produced in Paris in 1830, show clearly the result of his contact with the Romantic movement. *La conjuración de Venecia* was the first Romantic drama to be seen in Madrid, being produced there in 1834, several months before LARRA's *Macías*, and a year before the *Don Álvaro* of SAAVEDRA, both masterpieces in the genre. *Aben-Humeya*, which deals with the revolt of the Moors under Philip II, followed in 1836. Both plays are markedly Romantic in style, with lavish use of local colour, crowd scenes, violence on-stage, and a characteristic mingling of comedy and tragedy. After the death of Ferdinand VII, Martínez de la Rosa played an important part in the political life of Spain, becoming Prime Minister in 1834.

MARTÍNEZ SIERRA, GREGORIO (1881–1948), Spanish dramatist, whose works are more notable for delicacy and quiet humour than for action or excitement, and have lately been neglected. He was much influenced by BENAVENTE, and by MAETERLINCK, whose plays he translated into Spanish. He was himself fortunate to find sympathetic translators in Harley and Helen GRANVILLE-BARKER, whose version of Sierra's best known play *Canción da cuna* (1910), as *The Cradle Song*, was seen in New York in 1921, and in London in 1926 in a double bill with *The Lover* (*El enamorado*, 1913). *El Reino de Dios* (1916), which as *The Kingdom of God* was seen in London in 1927, is best remembered as the play with which Ethel BARRYMORE, in a fine and moving performance, opened in 1928 the theatre in New York named after her. Other productions in English include *Madame Petipa* (1912; N.Y., 1927; London, 1932); *The Two Shepherds* (*Los pastores*, 1913; London, 1935); *Wife to a Famous Man* (*La mujer del héroe*, 1914; London, 1924); *The Romantic Young Lady* (*Sueño de una noche de agosto*, 1918; N.Y., 1926; London, 1931): and *Take Two From One* (*Triángulo*, 1930; London, 1931). More important than his plays was Sierra's work as a director in charge of the Teatro Eslava from 1917 to 1925, when he introduced to Madrid audiences the new techniques imported into the peninsula by the Catalan Adriá GUAL. He also staged for the first time many contemporary plays in translation, and new or little known Spanish works, including the first of GARCÍA LORCA's plays *La maleficio de la mariposa* (*The Butterfly's Curse*, 1920).

MARTIN-HARVEY, JOHN (1863–1944), English actor-manager, knighted in 1921. He made his first appearance on the stage in 1881 and a year later joined IRVING's company at the LYCEUM, where he remained for 14 years, playing a wide range of parts including some of Irving's leading roles on tour during the summer months. He left the Lyceum in 1896 to appear with other managements, notably in MAETERLINCK's *Pelleas and Melisande* (1898) with Mrs Patrick CAMPBELL, but returned there as manager in 1899. His first production was the overwhelmingly successful *The Only Way*, a dramatization by Wills and Langbridge of DICKENS's *A Tale of Two Cities*, in which he played the hero, Sydney Carton. He became closely identified with the part, and found himself constantly forced to revive it to satisfy the demands of the public, thus limiting the time and energy available for other and perhaps more worthwhile roles. He was none the less associated with a number of new plays, among them Charles Hannan's *A Cigarette Maker's Romance* (1901) and Rutherford's *The Breed of the Treshams* (1903). In 1904 he played Hamlet for the first time, and he was later seen as Richard III, Henry V, and Petruchio in *The Taming of the Shrew*. In 1912 he gave a magnificent performance as Oedipus in REINHARDT's production of SOPHOCLES' *Oedipus Rex at* COVENT GARDEN. In later years he revived many of his old parts, but was also seen in Maeterlinck's *The Burgomaster of Stilemonde* (1918), and in the title role of the old morality play *Everyman* (renamed *Via Crucis*) in 1923. During the 1920s he toured North America, particularly Canada, with great success, and on his return to London added to his repertory two plays by SHAW, being seen as Blanco Posnet in *The Shewing-Up of Blanco Posnet* in 1926 and as Richard Dudgeon in *The Devil's Disciple* in 1930. His last appearances, up to 1939 when he retired, were mainly in revivals of his best known parts. A handsome man, with clear-cut features and a distinguished presence, he was regarded by many as the lineal descendant of Irving, and his death broke the last link with the Victorian stage. He married in 1889 the actress Angelita Helena Margarita de Silva Ferro (1869–1949), who as Nina de Silva was his leading lady for many years.

MARTINI, FAUSTO MARIA (1886–1931), Italian dramatist, an exponent of the CREPUSCOLARI or 'twilight' school of drama, whose plays reflect an increasingly hopeless acceptance of bourgeois values. In *Il giglio nero* (*The Black Lily*, 1913) he portrays the disturbing effect of city attitudes on a quiet provincial couple, and in *Il fiore sotto gli occhi* (*The Bloom Behind the Eyes*, 1921) he preaches the value of monotony and drabness. Between these two plays came a very successful venture into the commercial theatre with *Ridi, pagliaccio!* (1919), a commonplace tale of frustration leading to suicide which as *Laugh, Clown, Laugh!* was staged in New York in 1923; but with *L'altra Nanetta* (*The Other Nanette*, 1923) Martini reveals the influence of PIRANDELLO, and in his next two plays *La facciata* (*The Face*, 1924) and *La sera del 30* (*The Night of the 30th*, 1926), he approaches the 'theatre of silence' of J.-J. BERNARD and Charles VILDRAC.

MARTINSON, HARRY (1904–), Swedish poet, novelist, and dramatist, jointly awarded the NOBEL PRIZE for Literature in 1974. He came late to the theatre, achieving international success with a science-fiction opera adapted from his narrative poem *Aniara* (1959), and consolidating his dramatic reputation with *Tre Knivar från Wei* (*Three Knives from Wei*, 1964).

MARTINSON HALL, New York, see PUBLIC THEATRE.

MARTOGLIO, NINO (1870–1921), Italian dramatist, who, with Luigi CAPUANA, kept alive the Sicilian dialect theatre. His masterpiece, first seen in 1915, is *L'aria del continente*, a hilarious comedy on the old theme of an islander converted to sophisticated 'mainland' ways and his attempts to convert his neighbours back home. Martoglio collaborated in dialect plays with PIRANDELLO, and in the latter's work one can see this idea being taken up and used in different ways.

MARTYN, EDWARD (1859–1923), Irish dramatist, one of the founders of the IRISH LITERARY THEATRE, which as its second production, on 9 May 1899, performed his play *The Heather Field*. *Maeve*, a psychological drama on the clash between England and Ireland, was staged in 1900, as was *The Bending of the Bough*, an adaptation by George MOORE and W. B. YEATS of his play *The Tale of a Town*. None of his other plays, though all were published, was produced. He was president of the Theatre of Ireland, formed by a splinter group from the ABBEY THEATRE company in 1905, and his last active participation in theatre was in 1914, when he founded the Irish Theatre in Hardwicke Street, Dublin.

MARTYNOV, ALEKSANDR EVSTAFEVICH (1816–60), Russian actor, with SADOVSKY one of the early exponents of a natural style of acting, under the influence of the reforms introduced by SHCHEPKIN. Originally trained for ballet, he studied for the stage under KARATYGIN, whose formal technique he soon outgrew. His early successes were in VAUDEVILLE, to which he brought great originality and a comic flair which helped to redeem the poorness of his material; but as he acquired more authority and a surer technique he began to play leading roles in GOGOL, TURGENEV, and OSTROVSKY. Although he had hitherto excelled in comedy, critics of his later work observed in his characterizations a note of suffering which gave him great pathetic power; this may have been in part due to the tuberculosis from which he died while undergoing treatment in Kharkov.

MARYLEBONE MUSIC-HALL, London, see MUSIC-HALL.

MARYLEBONE THEATRE, London, see WEST LONDON THEATRE.

MASCARADE, La Soule, see FOLK FESTIVALS.

MASEFIELD, JOHN (1878–1967), English poet, novelist, and dramatist, who became Poet Laureate in succession to Robert Bridges and was appointed O.M. in 1935. His plays, which contain some of his best poetry, combine the traditions of Greek classical tragedy with those of the *nō* play of JAPAN; though some of them have been successfully staged, they are more poetic than theatrical. Among them are *The Campden Wonder* (1907), on an unsolved murder case; *The Tragedy of Nan* (1908), in which Lillah MCCARTHY played the title role; *The Witch* (1911), based on a Norwegian tragedy, which was first seen at the ROYAL COURT THEATRE and several times revived up to 1944; *The Faithful* (1915), which shows more than any of the others the Japanese influence; *Good Friday* (1917); *Melloney Holtspur* (1923); *The Trial of Jesus* (1926); *A King's Daughter* (1928), on the story of Jezebel; and *The Empress of Rome* (1937), based on an old French Miracle Play.

MASK, a covering for the face with openings for the eyes and mouth. It was originally made of carved wood or painted linen, later of painted cork, leather, or canvas, and later still of papier-mâché or light-weight plastics. The wearing of masks in the theatre derives from the use of animal skins and heads in primitive religious rituals. In the Greek theatre, masks (*prosopon*) served, in an all-male company, to distinguish between the male and female characters and to show the age and chief characteristic of each—hate, anger, fear, cunning, stupidity. (The suggestion that the opening for the mouth served as an amplifier for the voice is no longer tenable.) In tragedy the mask gave dignity and a certain remoteness to demi-gods and heroes, and also enabled one actor to play several parts by changing his mask. In comedy the mask helped to unify the members of the chorus (who, as can be seen in the plays of ARISTOPHANES, wore identical masks of such creatures as frogs, birds, horses, etc.) and served as an additional source of humour, particularly with the comic masks of slaves.

The Roman theatre took over the tragic masks from the Greeks (though not in the MIME), adopting the later exaggerated form with a high peak (*onkos*) over the forehead. Many fine copies of classical masks in marble still survive and they are also shown in several wall-paintings and bas-reliefs. The golden masks worn by God and the archangels in some versions of the medieval MYSTERY PLAY may have been a survival of the Greek tragic mask or an independent discovery of the new European theatre; but the devils' masks, though often comic in intention, seem by their horrific animal forms to be linked to early primitive religious usage. The comic actors of the COMMEDIA DELL'ARTE always wore masks, usually the small black 'cat-mask' which left the lower part of the face bare. Otherwise masks, which continued to be an essential factor in the *nō* play of JAPAN and in other Far Eastern theatres, were discarded in Europe, and they are seldom seen on stage, though they are sometimes used for special effects by such writers as YEATS (*At the Hawk's Well*, 1917; *The Only Jealousy of Emer* and *The Dreaming of the Bones*, both 1919); O'NEILL (*The Great God Brown*, 1926; *Lazarus Laughed*, 1928); and more recently by John ARDEN in *The Happy Haven* (1960), by Peter SHAFFER in *The Royal Hunt of the Sun* (1964) and *Equus* (1973), and in the NATIONAL THEATRE production of AESCHYLUS' *Oresteia* (1981). Apart from such isolated examples, the main use of masks at present is in the training of drama students, on whom they seem to have a liberating effect, particularly in IMPROVISATION.

The old name for the black mask used by the early Tudor actor or 'guizard' in the Court MASQUE was 'visor'. The Latin word for 'mask', *persona*, was used by TERENCE in the sense of character, whence *dramatis personae*, 'the characters in the play'. In the *commedia dell'arte* the word *maschera* was used both of the mask and of the person wearing it.

MASK, see MASQUE.

MASKING UNITS, single elements of SCENERY specially designed to conceal the backstage area and fly-gallery in a theatre from the audience. The same effect can be achieved by the use of black or dark-coloured draperies. An actor is said to be 'masking' another if, inadvertently or otherwise (and if done on purpose, except on the instructions of the director, it is a serious fault), he gets between him and the audience, so that he cannot be seen properly.

MASK THEATRE, Belfast, see BELFAST ARTS THEATRE.

MASON, BRUCE, see NEW ZEALAND.

MASON, MARSHALL, see CIRCLE REPERTORY COMPANY.

MASQUE (originally Mask, the French spelling being first used by Ben JONSON), spectacular entertainment, which combined music and poetry with scenery and elaborate costumes. It derived originally from a primitive folk ritual featuring the arrival of guests, usually in disguise, bearing gifts to a king or nobleman, who with his household then joined the visitors in a ceremonial dance. The presentation of the gifts soon became an excuse for flowery, flattering speeches, while the wearing of outlandish or beautiful costumes and MASKS, or visors, led to miming and dancing as a prelude to the final dance. The early, relatively simple, form of the masque was known as a DISGUISING, and is part of the folk tradition that includes the MUMMERS' PLAY. In Renaissance Italy, mainly under the influence of Lorenzo de' MEDICI, it became a vehicle for song, dance, scenery, and machinery, one of its non-dramatic offshoots being the elaborate Trionfo, or Triumph. At the French Court it gave rise to the simple *ballet de cour* and the more spectacular *mascarade* (from which is derived 'masquerade'), and eventually the COMÉDIE-BALLET. In the 16th century it came back under its new name to Tudor England, where maskers played before the king in elaborate dresses, with all the appurtenances of scenery, machinery, and rich allegorical speech. In Elizabethan times the formula proved useful for the entertainment of the Queen, either in her own palace or during her 'progresses' throughout the land. Shakespeare makes fun of a simple country masque in *Love's Labour's Lost*, and uses the form seriously for typical early 'disguisings' in *Timon of Athens* and *The Tempest*. This latter already shows some of the elaboration reached by the Court masques prepared for James I and Charles I by Ben Jonson (appointed Court Poet in 1603) and the scenic designer Inigo JONES. Their first joint work was the Twelfth Night masque of 1605, their best probably *Oberon the Fairy Prince* in 1611. One of Jonson's innovations was the anti-masque, known also as the ante-masque, because it preceded the main entertainment, or the antic masque, because it employed earlier elements of antic or grotesque dancing. First introduced in 1609, the anti-masque provided a violent contrast to the main theme, as Hell before Heaven, War before Peace, Storm before Calm. The simplicity of the early masque, in which the performers appeared in one guise only—as blackamoors, wild men, or shepherds—later gave way to the double masque, in which they were seen in two different groups of characters—fishermen and market-women, for instance, or sailors and milkmaids. In time the literary content of the masque diminished, and the spectacular aspect, particularly the dancing, in which Charles I and Henrietta Maria became performers after the fashion of Louis XIV, became more important. This led Jonson, after constant altercations with Inigo Jones, to withdraw, his last masque being performed in 1634. He was succeeded by James SHIRLEY, who found himself called on to provide nothing more than a scenario suitable for elaborate effects, and a few dull speeches. The Civil War put an end to the masque, which was never revived, but it had provided the means of introducing into England the new Italian scenery, and the Restoration theatre was to take over many of its spectacular effects. The decorative frame set up for the masque in a ballroom became the PROSCENIUM arch, behind which Inigo Jones's movable shutters or WINGS, trebled or quadrupled, ran in GROOVES to open or close in front of a painted backcloth, or, less often, what Jones called a 'sceane of releave', consisting of cut-out pieces on various planes. As this had to be prepared in advance and shown to the audience by drawing back the shutters, it was termed a Set Scene, whence the modern use of the word SET for the scenic components of a play.

Milton's *Comus* (1634), though entitled 'a masque', is in reality a PASTORAL, and was probably called a masque to distinguish it from the plays given in the public theatre.

MASSEY, CHARLES (?–1625), English actor, who was a friend of ALLEYN and of the actor-dramatist Samuel ROWLEY. He first appeared with the ADMIRAL'S MEN in 1597 and remained with them during their successive renamings until his death. He was one of the actors who leased the FORTUNE THEATRE from Alleyn in 1618, and he was a shareholder in the new Fortune, built in 1622, where he ranked as one of the chief members of the company. He was also a dramatist, two of his plays, *Malcolm King of Scots* (1602) and *The Siege of Dunkirk* (1603), now lost, being given by the Admiral's Men.

MASSEY, RAYMOND HART (1896–1984), American actor and director, born in Toronto. Through his family he became associated with the HART HOUSE THEATRE after the First World War, in which he served with the Canadian forces, and in 1922 he made his début there as Rosmer in IBSEN'S *Rosmersholm*. He then went to London and made his professional début as Jack in O'NEILL'S *In the Zone* at the EVERYMAN THEATRE and in 1924 played two small parts in the first London production of SHAW'S *Saint Joan*. In 1926 he was back at the Everyman as joint manager. He made his New York début as Hamlet in 1931 and for several years was seen in both New York and London, his roles including David Linden in Winter's *The Shining Hour* (N.Y. and London, 1934), Ethan Frome in the dramatization of Edith Wharton's novel of that name (N.Y., 1936), and Harry Van in SHERWOOD'S *Idiot's Delight* (London, 1938). His finest performance was as Abraham Lincoln in Sherwood's *Abe Lincoln in Illinois* (N.Y., 1938), a part which made good use of his dark brooding looks, rangy physique, and distinctive voice, and which he played until 1940. His later stage work, confined almost entirely to the United States, included notable appearances in New York in revivals of Shaw's *The Doctor's Dilemma* (1941), *Candida* (1942), and *Pygmalion* (1945), and STRINDBERG'S *The Father* (1949). In the opening season at Stratford, Conn., in 1955 (see AMERICAN SHAKESPEARE THEATRE) he played Brutus in *Julius Caesar* and Prospero in *The Tempest*, and among his later roles were Mr Zuss in Archibald Macleish's *J.B.* (N.Y., 1958) and Tom Garrison in Robert ANDERSON'S *I Never Sang For My Father* on a return to London in 1970. He also directed many plays, and was the author of the play *The Hanging Judge* (1952), based on Bruce Hamilton's novel, and an autobiography entitled *When I Was Young* (1976).

His children Daniel Raymond (1933–) and Anna (1937–) were born in England and are both well known on the stage. Daniel's roles have included Charles Surface in SHERIDAN'S *The School For Scandal* (1962), Captain Absolute in his *The Rivals* (1966), and John Worthing in WILDE'S *The Importance of Being Earnest* (1968), all at the HAYMARKET THEATRE, London. He was later seen in several roles at the NATIONAL THEATRE, among them the title role in Horváth's *Don Juan Comes Back From the War* (1976), Robert in PINTER'S *Betrayal* (1978), John Tanner in the full-length version of Shaw's *Man and Superman*, and the title role in MOLIÈRE'S *The Hypochondriac* (both 1981). Anna Massey achieved a big success in her first stage appearance, in William Douglas HOME'S *The Reluctant Débutante* (London, 1955; N.Y., 1956). Notable later roles were Annie Sullivan in William Gibson's *The Miracle Worker* (1961), Lady Teazle in Sheridan's *The School For Scandal* (1962), Jennifer Dubedat in Shaw's *The Doctor's Dilemma* (1963), and Laura Wingfield in Tennessee WILLIAMS'S *The Glass Menagerie* (1965), the last three all at the Haymarket Theatre, London. She appeared for the National Theatre in Shaw's *Heartbreak House* (1975) and Simon GRAY'S *Close of Play* (1979).

MASSINGER, PHILIP (1583–1640), English dramatist, author of some 40 plays, about half of which are lost, eight having been destroyed by WARBURTON'S cook. Of those that survive the earliest is *The Duke of Milan* (1620), a tragedy presented by the KING'S MEN. It was followed by *The Maid of Honor* (1621), *The Bondman* (1623), *The Renegado* (1624), and *A New Way to Pay Old Debts* (1625), this last being Massinger's best and best known work. Allowed to lapse during the Restoration, it returned to the stage in the 18th century and has since been constantly revived, the part of Sir Giles Overreach being a favourite with many actors including Edmund KEAN. It was played in the most recent revivals in 1950 and 1953 by Donald WOLFIT. Later plays by Massinger include *The Roman Actor* (1626), also revived by Kean; *The Great Duke of Florence* (1627), with its charming idyll between Giovanni and Lidia; and two comedies, *The City Madam* (1632) and *The Guardian* (1633). Much of Massinger's work was done in collaboration—*The Fatal Dowry* (1619) with FIELD, *The Virgin Martyr* (1620) with DEKKER—and he also had a hand in several of the plays ascribed to BEAUMONT and FLETCHER; he may also have worked with the latter on Shakespeare's *Henry VIII* and possibly on *The Two Noble Kinsmen* (1613).

MASTER OF THE REVELS, an official of the Royal household, first appointed in 1494 to serve under the LORD CHAMBERLAIN in connection with entertainments at the Court of Henry VII. It was at first a temporary appointment, held by several people in succession, until in 1545 Sir Thomas Cawarden, who was to supervise the revels for the coronation of Elizabeth I in 1558, became the first Master for Life, and the Revels Office was officially established as the major instrument of censorship of dramatic and other entertainments. On Cawarden's death in 1559 he was succeeded by Sir Thomas Benger, during whose tenure the powers of the Master were somewhat restricted, matters of finance and production being controlled by other departments and the Master being concerned only with censorship. No new appointment was made after Benger's death in 1572, but work continued smoothly under the permanent under-official Thomas Blagrove, who served the Revels Office faithfully for 57 years. In 1579 his hopes of the Mastership were dashed by the appointment, probably through influence in high places, of Sir Edmund Tilney, who remained Master until his death in 1610. He seems to have done very little work, though he continued to act as censor of plays, and retained all the fees paid for plays publicly performed. By the establishment of the Licensing Commission in 1589, the Revels Office was given virtually exclusive powers to licence plays, companies, playhouses built on Crown lands, and provincial touring companies. Under Elizabeth, the Master exercised these powers on behalf of the Privy Council; but under James I and Charles I he behaved in all such matters as if he were the sovereign's deputy, issuing licences, conducting interrogations, and levying fines as directed by his master. Tilney was succeeded by his nephew, Sir George Buck, on whose death in 1622 the office passed to its most famous holder, Sir Henry HERBERT. By the Licensing Act of 1737 the censorship of plays (abolished in 1968) became the direct responsibility of the Lord Chamberlain, and the old office of Master of the Revels became extinct.

MASTERSINGERS, see MEISTERSÄNGER.

MATHEWS, CHARLES (1776–1835), English actor and entertainer, who from childhood showed amazing powers of mimicry, joined to a most retentive memory and an intense desire to go on the stage. He made his first appearance in Dublin in 1794 and after some years in the English provinces, mostly under Tate WILKINSON at York, he appeared in London, at the HAYMARKET THEATRE, in 1803, soon gaining an enviable reputation as an eccentric comedian. Among his most successful parts at this time were Sir Fretful Plagiary in SHERIDAN's *The Critic* and Risk in the younger George COLMAN's *Love Laughs at Locksmiths* (1803), which was specially written for him. He later appeared at both COVENT GARDEN and DRURY LANE, and in addition to the many new and ephemeral parts which he created, he was much admired as Falstaff in *Henry IV*, Sir Archy Mac-Sarcasm in MACKLIN's *Love à la Mode*, and Sir Peter Teazle in Sheridan's *The School for Scandal*. In 1808 he conceived the idea of the one-man entertainment with which he is chiefly associated and in which he appeared in London and all over the provinces, as well as in America, for more than 20 years. The first of these was *The Mail Coach Adventure; or, Rambles in Yorkshire*, in which Mathews's second wife, the actress Anne Jackson (1782–1869), played a small part; after 1811 she was not seen again, Mathews taking as his partner in his second entertainment, *The Travellers; or, Hit and Miss* (1811), the singer Charles Incledon. Their relationship was not a happy one, and from 1812 Mathews appeared alone in a series of entertainments which from 1817, when Mathews was in Brighton, became known as *Mr Mathews at Home*. These 'At Homes', which were given in London in the winter and throughout the provinces during the summer, became immensely successful. From their early beginnings, with a number of comic songs linked together with depictions of eccentric characters based partly on observation, partly on intuition, they developed into short plays to which many writers of comedy contributed; though the overall product was always attributable to Mathews himself, except in such cases as the younger Colman's *The Actor of All Work; or, First and Second Floor* (1817), in which a manager is shown interviewing a number of applicants for a place in his company. This gave Mathews the opportunity of displaying his powers of mimicry in a bewildering series of totally dissimilar characters. Among the early scripts for which Mathews was himself responsible were *Mr Mathews' Trip to Paris* (1819), *Mr Mathews' Trip to America* (1821), and *The Youthful Days of Mr Mathews* (1822), in which he gave an imitation of MACKLIN. He continued to appear on stage from time to time, in spite of being lame from a carriage accident which occurred in 1814, among his later parts being Goldfinch in HOLCROFT's *The Road to Ruin* and Dr Pangloss in the younger Colman's *The Heir at Law*. He returned to America for the last time in 1834, but was already in poor health and died at Liverpool on the return journey. By his second wife he was the father of the actor Charles James MATHEWS.

MATHEWS, CHARLES JAMES (1803–78), English actor-manager, the only son of Charles MATHEWS. Trained as an architect, he had little connection with the theatre, beyond acting with an amateur company, until on the death of his father in 1835 he briefly succeeded him as joint manager with Frederick YATES of the ADELPHI THEATRE. Later the same year he made his first professional appearance at the OLYMPIC, playing in his own comedy *The Humpbacked Lover*, and partnering John LISTON in William Leman Rede's farce *The Old and Young Stager*. Like his father he was an excellent mimic, and one of his most popular sketches was *Patter v. Clatter* (1838), in which he played five parts. Also in 1838 he married Madame VESTRIS, and on their return from a visit to New York they took over COVENT GARDEN, where they staged some fine productions including BOUCICAULT's *London Assurance* (1841), in which Mathews played Dazzle, always one of his best parts. The venture was not a success financially, and hoping to recover their losses they moved to the LYCEUM THEATRE, which proved even more disastrous. Mme Vestris died in the midst of their bankruptcy, and Mathews continued to act, but for other managements. He also visited America again, returning with his second wife, the actress Lizzie Davenport (?–1899). With her help he eventually extricated himself from his financial embarrassments, and embarked on a more successful, though less adventurous, career which lasted until his death. Tragedy and pathos were outside his range, but he was inimitable not only as Dazzle but in such parts as Sir Fopling Flutter in ETHEREGE's *The Man of Mode*, Puff in SHERIDAN's *The Critic*, Young Wilding in FOOTE's *The Liar*, as well as in such contemporary parts as Affable Hawk in LEWES's *The Game of Speculation* and Plumper in W. B. JERROLD's *Cool as a Cucumber* (both 1851). He also appeared with his second wife in an entertainment, reminiscent of his father's, called *Mr and Mrs Mathews at Home*. He had not the solid gifts of the older Mathews, but much charm and delicacy tempered his high spirits and made him, within certain limits, one of the best light comedians of his day.

MATHURINS, Théâtre des, Paris, in the rue des Mathurins. This was built in 1906 for Sacha GUITRY, and became a leading house for BOULEVARD PLAYS until 1934, when it was taken over by Georges and Ludmilla PITOËFF, who had already had a season there in 1927–8. They remained there until Georges's death in 1939, one of the first plays they introduced to Paris being PIRANDELLO's *Questa sera si recita a soggetto* as *Ce Soir, on improvise*; Pitoëff himself played the part of the director Hinkfuss. He and his wife, who was associated with him in all his work, appeared frequently in the productions at the Mathurins, which they made one of the outstanding theatres of Paris, Ludmilla being particularly admired in a revival of SHAW's *Saint Joan*. The last part Georges played at the Mathurins was Doctor Stockman in IBSEN's *An Enemy of the People*. After his death the theatre was taken over by Jean Marchat, whose notable revival of MOLIÈRE's *Tartuffe* was among the successful classic productions seen there, and from 1951 to 1981 the theatre was directed by the widow of the actor Harry Baur.

MATTHEWS, A(lbert) E(dward) (1869–1960), English actor, son of a Christy Minstrel, and grand-nephew of GRIMALDI's pupil, the Clown Tom Matthews. He began his career as a call-boy at the PRINCESS'S THEATRE in 1886, and continued to act almost up to the time of his death. He never appeared in Shakespeare or in classical plays, which he rightly considered outside his range, but in his own line he was inimitable, with a sure technique which enabled him to seem at his most careless when he was most in control. He excelled in farce, and in his early years toured England, South Africa, and Australia in such plays as PINERO's *Dandy Dick* and *The Magistrate,* HAWTREY's *The Private Secretary*, Hamilton Aidé's *Dr Bill*, and Brandon THOMAS's *Charley's Aunt*. He returned to England in 1896, and was in the first production of R. C. CARTON's *Lord and Lady Algy* (1898), then created a wide range of parts, including Cosmo Grey in BARRIE's *Alice Sit-By-the-Fire* and Eustace Jackson in St John HANKIN's *The Return of the Prodigal* (both 1905). He made his first appearance in New York in 1910, where he returned many times, and was as popular there as in London. Among his last and most successful parts were the Earl of Lister in *The Chiltern Hundreds* (1947) and *The Manor of Northstead* (1949), both by William Douglas HOME. In private life Matthews, who was known affectionately as 'Matty', was an eccentric of dry humour, who refused to take his success seriously, and posed as the bluff country gentleman. In youth he added ten years to his age for fear of seeming too young for the parts he wanted, and in old age cut off ten years in case he was thought too old to go on acting, but it now seems certain that he was 90 at the time of his death.

MATTHEWS, (James) BRANDER (1852–1929), American playwright and theatre historian, the first Professor of Dramatic Literature in the United States—at Columbia, from 1900 to 1924. He was the author of a number of books on the theatre, and through his writings and lectures exercised a great influence on the professional theatre, on contemporary dramatic critics, and on the attitude of the general theatrical public in America. He had a wide knowledge of European drama, and a keen feeling for all that was best in the dramatic literature of his own country.

MATTHEWS, JESSIE [Margaret] (1907–81), English singer and dancer. She made her first appearance at the age of 12 in the children's play *Bluebell in Fairyland* by Seymour HICKS, and in her teens was seen in revue in London and New York. She then starred for COCHRAN in the revues *One Dam Thing After Another* (1927), *This Year of Grace* (1928), and *Wake Up and Dream* (1929), in the last of which she was also seen in New York, returning to London to appear in her first musical comedies, *Ever Green* (1930), by Rodgers and Hart, and *Hold My Hand* (1931). During this period she introduced such famous songs as COWARD's 'A Room With a View', Cole Porter's

'Let's Do It', and Rodgers and Hart's 'My Heart Stood Still' and 'Dancing on the Ceiling'. By the early 1930s she had become Britain's highest paid theatrical star, and she then became an international film star. She returned to the West End in 1940 in the revue *Come Out to Play*, and in 1942 played Sally in Jerome Kern's musical *Wild Rose*. After the Second World War she continued to appear in revue—*Maid to Measure* (1948), *Sauce Tartare* (1949)—but began to be seen more often in straight parts, mostly outside London, playing in RATTIGAN's *The Browning Version* and *Harlequinade* in 1949, revivals of SHAW's *Pygmalion* in 1950 and COWARD's *Private Lives* in 1954, and in Peter SHAFFER's *Five Finger Exercise* in 1960. Later roles included June Buckridge in Frank Marcus's *The Killing of Sister George* (1971), Mrs Doasyouwouldbedoneby in the musical version of Kingsley's *The Water Babies* (1973), and the Duchess of Berwick in WILDE's *Lady Windermere's Fan* (1978). Her autobiography *Over My Shoulder* was published in 1974.

MAUDE, CYRIL FRANCIS (1862–1951), English actor-manager, who came of a military family and was intended for the army or the Church. He, however, preferred the theatre, and while travelling in America in 1884 made his first appearance on the stage at Denver, Colorado, in a stage version of Mrs Henry Wood's *East Lynne*. After appearing in New York he returned to England, where he scored his first outstanding success as Cayley Drummle in PINERO's *The Second Mrs Tanqueray* (1893) at the ST JAMES'S THEATRE with George ALEXANDER. In 1896, in partnership with Frederick Harrison, he took over the HAYMARKET THEATRE, where he put on a number of excellent productions with a distinguished company led by his wife (Isabel) Winifred Maud Emery (1862–1924), who made a great success as Lady Babbie in BARRIE's *The Little Minister* (1897) to his Gavin Dishart. On leaving the Haymarket in 1905 Maude took over the Avenue Theatre (later the PLAYHOUSE), where he was seen in Clyde FITCH's *Toddles* (1907) and Austin Strong's version of *Rip Van Winkle* (1911). While on tour in America in 1913 he made a great success as Andrew Bullivant in Hodges's *Grumpy*, which he repeated in London a year later. He gave up the Playhouse in 1915, and again went on tour, reappearing in London in Sydney Blow and Douglas Hoare's *Lord Richard in the Pantry*. He retired in 1927, returning briefly from time to time and making his final appearance in 1933. On his 80th birthday he emerged from retirement in Devon to play Sir Peter Teazle in SHERIDAN's *The School for Scandal*, always one of his best parts, in aid of the R.A.F. Benevolent Fund. He was for many years President of the Royal Academy of Dramatic Art.

MAUGHAM, W(illiam) SOMERSET (1874–1965), English novelist and dramatist, appointed C.H. in 1954. He trained as a doctor and was well known as a novelist before his first play, *A Man of Honour*, was produced in 1904. The height of his achievement as a playwright was reached in 1908, when he had four plays running in London—*Lady Frederick* (N.Y., 1908), *Jack Straw* (N.Y., 1908), *Mrs Dot* (N.Y., 1910), and *The Explorer* (N.Y., 1912). For the next 25 years he was prolific and, at least with the comedies of manners which formed the bulk of his output, fashionable. This period included *The Land of Promise* (N.Y., 1913; London, 1914), *Caroline* (London and N.Y., 1916), *Our Betters* (N.Y., 1917; London, 1923), *Home and Beauty* (1919), *The Circle* (London and N.Y., 1921), usually considered his best play, *The Constant Wife* (N.Y., 1926; London, 1927), *The Letter* (London and N.Y., 1927), *The Sacred Flame* (N.Y., 1928; London, 1929), *The Breadwinner* (1930; N.Y., 1931), and *For Services Rendered* (1932; N.Y., 1933), a serious play about post-war disillusionment. After the comparative failure of *Sheppey* (1933; N.Y., 1944) Maugham stopped writing for the theatre, but several of his short stories were adapted for the stage by other writers, notably *Rain* (1922) by John Colton and Clemence Randolph and *Jane* (1947) by S. N. BEHRMAN. The most frequently revived of his plays have been *The Circle, The Constant Wife*, and *The Sacred Flame*. Maugham achieved popularity without being good-natured or soothing: his humour was sardonic and his attitude to virtue mistrustful; but he told good stories with neatness and polished malice.

MAULE, DONOVAN, see AFRICA.

MAURSTAD, ALFRED (1896–1967), Norwegian actor, who made his début in 1920, and from 1931 worked with the NATIONALTHEATRET in Oslo as one of its leading players. He is remembered mainly for his interpretations of roles in HOLBERG, BJØRNSON, and IBSEN.

MAURSTAD, TORALV (1926–), Norwegian actor, son of the above, who from his début in 1949 until 1967 worked mainly at the OSLO NYE TEATER and the NATIONALTHEATRET, in plays which ranged from those of Shakespeare, HOLBERG, and IBSEN to those of more recent dramatists. In 1967 he became manager of the Oslo Nye Teater, where he directed a number of productions and where his performances included the Master of Ceremonies in the musical *Cabaret* in 1968 and the young James Tyrone in O'NEILL's *A Moon for the Misbegotten*. He became manager of the Nationaltheatret in 1978, his productions there including Peter SHAFFER's *Amadeus* in 1980.

MAURSTAD, TORDIS (1901–), Norwegian actress, wife of Alfred (above), who worked for the greater part of her career with the NORSKE TEATRET. She was always a player of passion, and found a natural place for herself in such strongly defined classical roles as the Antigone of

SOPHOCLES and the Medea of EURIPIDES, as well as many parts in the dramas of Shakespeare, IBSEN, STRINDBERG, CHEKHOV, and O'NEILL.

MAX, (Alexandre) ÉDOUARD DE (1869–1925), French actor, a pupil of WORMS at the Paris Conservatoire, where he took first prizes for comedy and tragedy in 1891. He made his début at the ODÉON, and was already considered one of the foremost actors of his day when in 1915 he first appeared at the COMÉDIE-FRANÇAISE, playing Néron in RACINE's *Britannicus*. He had a short but glorious career there, dying of heart failure after playing Oreste in Racine's *Andromaque*.

MAXINE ELLIOTT'S THEATRE, New York, at 109 West 39th Street, between Broadway and 6th Avenue. This was built for Maxine ELLIOTT, and opened on 30 Dec. 1908. The first outstanding success, in the following April, was Jerome K. JEROME's *The Passing of the Third Floor Back*. SYNGE's *The Playboy of the Western World*, played by the ABBEY THEATRE company on their first American tour, caused a riot in 1911. In 1913 Doris Keane was seen in SHELDON's *Romance*, which later had a long run in London. In 1922 a dramatization of MAUGHAM's short story *Rain*, by John Colton and Clemence Randolph, ran for 648 performances, and another Maugham success at this theatre was *The Constant Wife* (1926), with Ethel BARRYMORE.

Later productions were *Twelfth Night* (1930), with Jane COWL as Viola, and PIRANDELLO's *As You Desire Me* (1931), with Judith ANDERSON. Lillian HELLMAN's *The Children's Hour* was staged in 1934, and two years later the theatre was taken over for a season by the FEDERAL THEATRE PROJECT with *Horse Eats Hat* (a translation of LABICHE's *Un Chapeau de paille d'Italie*) and MARLOWE's *Doctor Faustus*, both directed by Orson WELLES. *Separate Rooms* (1940) by Carole and Dinehart began its long run at this theatre, but from 1941 it was used only for broadcasting, and it was demolished late in 1959.

MAY, EDNA, see MUSICAL COMEDY.

MAY, VAL, see BRISTOL OLD VIC and NOTTINGHAM PLAYHOUSE.

MAYAKOVSKY, VLADIMIR VLADIMIROVICH (1894–1930), Soviet poet, dramatist, painter, actor, director, and film scenarist, after whom many public institutions, including the former Theatre of the Revolution (below), have been named. As a youth he was several times arrested for revolutionary activity, and in 1917 placed himself and his undoubted talents at the disposal of the Bolsheviks, writing poetry and painting posters in support of the new régime. His first full-length play in verse, *Mystery-Bouffe*, was staged by MEYERHOLD in Petrograd on the first anniversary of the October Revolution, and is usually regarded as the first Soviet play. It presents the end of the old world and the discovery of the new in the form of a futurist mystery-play based on the Biblical story of the Flood. His next important play, *The Bed-Bug*, was commissioned by Meyerhold in 1929. A satire portraying the Soviet world of the future in which a pre-Revolutionary bourgeois and a bed-bug alone survive from the old world, and depicting their struggles to acclimatize themselves, it was staged by Meyerhold at his own theatre with music by Shostakovich, but was not an unqualified success, though ILYINSKY's performance as Prisipkin was greatly admired. Mayakovsky's next play, *The Bath House*—'Drama in Six Acts with a Circus and Fireworks'—again produced by Meyerhold, in 1930, was even less well received. Worn out by the constant strain of meeting deadlines—his literary output in his last years was prodigious—and by an unhappy love affair, he committed suicide. Much of his poetry is semi-dramatic and intended for public declamation; some of it has been translated into English, and his *Collected Plays*, translated by G. Daniels, appeared in 1969. In the Soviet Union thousands of copies of his work are sold every year, and his three important plays have often been revived, notably by TAÏROV. *The Bed-Bug* was first produced in English by the London University Drama Society, and had its first professional production at the MERMAID THEATRE in 1962.

MAYAKOVSKY THEATRE, Moscow, founded in 1922 as the Theatre of the Revolution for the production of propaganda plays. Its first outstanding director was A.D. POPOV, who raised its standards considerably, introducing new and worthwhile Soviet plays, among them POGODIN's *Poem About an Axe* (1931), as well as excellent productions of the classics. His production of *Romeo and Juliet* (1936) remained in the repertory for many years. His policy was continued by OKHLOPKOV, under whom in 1954 the theatre was renamed in honour of MAYAKOVSKY. During his directorship, from 1943 until his death in 1967, Okhlopkov was responsible for such successful new plays as Virta's epic of Stalingrad, *Great Days* (1947), and also for controversial productions of *Hamlet* (1954) and BRECHT's *Mother Courage* (1960). He was succeeded as artistic director by Andrei Goncharov, and later productions include Tennessee WILLIAMS's *A Streetcar Named Desire* (1970) and OSTROVSKY's *It's All in the Family* (1975).

MAY-DAY, MAYING, see FOLK FESTIVALS and ROBIN HOOD.

MAY FAIR THEATRE, London, in Stratton Street, the smallest commercial theatre in London, forming part of the May Fair Hotel. Seating 310 in a single raked tier, it opened on 17 June 1963 with a revival of PIRANDELLO's *Six Characters in Search of an Author* with Ralph RICHARDSON and

Barbara JEFFORD. *All In Love* (1964), a musical based on SHERIDAN's *The Rivals*, followed and was replaced by the REVUE *Beyond the Fringe* transferred from the FORTUNE THEATRE for the rest of its long run. *The Philanthropist* by Christopher HAMPTON, with Alec MCCOWEN, opened in Sept. 1970 after a season at the ROYAL COURT and ran until Oct. 1973, after which Roy DOTRICE was seen in 1974 in a revival of Aubrey's *Brief Lives*. There were two successful transfers from the HAMPSTEAD THEATRE CLUB—Michael FRAYN's *Alphabetical Order* (1975) with Billie WHITELAW and Pam Gems's *Dusa, Fish, Stas, and Vi* (1977). In 1978 Gordon Chater gave a remarkable performance in Steve J. Spears's one-man play from Australia *The Elocution Of Benjamin Franklin*. Two short musicals presented as *A Day In Hollywood—A Night In the Ukraine* (1979), which transferred from the New End Theatre, Hampstead, later achieved a much bigger success in New York.

MAYNE, RUTHERFORD [Samuel Waddell] (1878–1967), Irish dramatist, chiefly associated with the ULSTER GROUP THEATRE for which he began to write soon after its foundation in 1904. Most of his plays have been seen only in Ulster, although *The Drone* (1908) was produced in the U.S.A. in 1913; *The Troth* (1908) had its first production in London; and *Red Turf* (1911), *Peter* (1930), and *Bridgehead* (1934) were first seen at the ABBEY THEATRE in Dublin. Like most Ulster dramatists Mayne concentrated on local issues, but some of his plays achieve a broad view of the problems of rural life and have been translated into Dutch, Norwegian, and Swedish.

MAZARINE FLOOR, see MEZZANINE FLOOR.

MEADOW PLAYERS, see OXFORD PLAYHOUSE.

MECHANE, in the ancient Greek theatre, a piece of stage machinery in the form of a large crane which enabled characters, chariots, etc., to appear to fly through the air. It was also used to lower a character, usually a god, from the top of the stage building so that he could resolve the complexities of the plot—the DEUS EX MACHINA.

MEDDOKS, MIKHAIL EGROVICH [Michael Maddox] (1747–1825), British-born Russian impresario, who in 1767 arrived in St Petersburg with an exhibition of mechanical dolls. Ten years later he was in Moscow, where he was engaged by Prince Ourusov to manage the theatre, built under a licence from the Tsar which gave him the sole right to provide dramatic entertainments in the capital for a period of ten years. After the destruction of his theatre by fire in 1780, Ourusov ceded his monopoly to Maddox, who built a theatre on Petrovsky Square (where the Bolshoi Theatre now stands), and opened with a company drawn mainly from the private theatres of the nobility. It flourished until its destruction by fire in 1795; it was not reopened. One unexpected result of Maddox's work was the adoption of the name Vauxhall (*voksal*), which he gave to the amusement park he opened in Moscow, as the Russian word for a railway station, the first one in the city being built on part of Maddox's park. His own name, first transmuted to Meddoks, later became Medok; one of his direct male descendants bearing that name was killed while serving in the Soviet army during the Second World War.

MEDICI, LORENZO DE' (1449–92), a member of the famous Florentine family, all of whom were patrons of the arts. Himself a fine poet, Lorenzo, known as 'The Magnificent', was the founder of Laurentian Library, and was also the author of a SACRA RAPPRESENTAZIONE produced in Florence in 1499. A younger member of the family, Lorenzino di Pier Francesco, was one of the first Florentine writers of comedy, his *Aridosia*, based on PLAUTUS and TERENCE, being performed in 1536 with fine settings by San Gallo. It was on this play that LARIVEY based his comedy *Les Esprits* (1579).

MEDLEY, MATT, see ASTON, TONY.

MEDWALL, HENRY (*fl.* 1490–1514), early English dramatist whose work was practically unknown until in 1919 the manuscript of his *Fulgens and Lucrece* came to light in a London saleroom. An INTERLUDE, performed in two parts as an entertainment at a banquet, it was probably acted in 1497, and as an example of secular drama is much earlier than anything hitherto known. With its story of the wooing of Lucretia and comic sub-plot of the wooing of her maid, it foreshadows the mingling of romantic and comic elements which was to be a feature of later Elizabethan drama.

MEGGS, Mrs MARY (?–1691), known as Orange Moll, a well known figure in the early days of the Restoration theatre. A widow, living in the parish of St Paul's, Covent Garden, she was granted a licence on 10 Feb. 1663 to sell oranges and other eatables in the Theatre Royal, DRURY LANE. For this she paid £100 down and 6*s*. 8*d*. for every day the theatre was open, which seems to show that the business was a lucrative one. The fire which destroyed the theatre in 1672 was believed to have started under the stairs where she kept her wares. In his *Diary* PEPYS refers to Orange Moll and her orange-girls, of whom Nell GWYNN was one, several times, usually in connection with items of theatrical scandal. Towards the end of her life she was frequently involved in trouble with the management, and when in 1682 the companies of Drury Lane and COVENT GARDEN were amalgamated a new orange-woman was appointed. This led to constant disputes, and the matter was still unsettled when Mrs Meggs died.

MEGLIO, ANTONIO, see ITALY.

MEI LANFANG (1894–1961), Chinese actor, the only one to become known outside his own country. The son and grandson of actors, he began his stage training at the age of 9 and made his first public appearance at 14. Between 1919 and 1935 he visited Japan, the United States, Europe, and Russia, where he enjoyed the friendship of STANISLAVSKY and NEMIROVICH-DANCHENKO. Renowned for his exquisitely delicate playing of *dan*, or female characters, he was the first to combine the dramatic techniques of the five roles in Peking Opera into which they are divided, and his meteoric rise to fame after the First World War gave the female role the predominant place formerly held by the *laosheng*, or elderly male role. During the Chinese civil war Mei Lanfang lived in retirement, signifying his resolution not to appear on stage by allowing his beard and moustaches to grow; but in 1949 he returned to Peking and was appointed President of the Research Institute of Chinese Drama. He then returned to the theatre and in 1958 celebrated his stage jubilee, retiring a year later.

MEILHAC, HENRI (1831–97), French dramatist, whose first play was produced at the PALAIS-ROYAL, for which theatre and for the GYMNASE he wrote many comedies and vaudevilles. He also collaborated with the novelist and dramatist Ludovic Halévy (1834–1908) in writing amusing librettos for the operettas of Offenbach—among them *La Belle Hélène* (1865) and *La Vie parisienne* (1867)—and light comedies such as *Froufrou* (1869). Meilhac, who was for many years one of the outstanding figures of the French theatre, had all the ephemeral gifts and none of the durable ones; his work epitomizes the witty and cynical spirit of the Second Empire.

MEININGER COMPANY, troupe of actors resident at the Court of George II, Duke of Saxe-Meiningen, led by his morganatic wife the actress Ellen Franz (1839–1923). The Duke, who directed the plays himself and also designed the costumes and scenery, was ably assisted by the actor Ludwig CHRONEGK, who joined him in 1866 and became responsible for the general direction and discipline of the company. The innovations for which the Meininger Company later became famous had a great influence throughout Europe, firmly establishing the creative and interpretive role of the director and pioneering ensemble acting. Historically accurate scenery and costumes were used, and the setting of the chief actors within the scene took the place of formal groupings, while the handling of crowd scenes was revitalized by making the crowd a personage of the drama—every member an actor in his own right, yet the whole responding to the needs of the moment in a unified way. The Duke followed the example of Charles KEAN in dividing the supers into small groups, each with a competent actor at its head, and in working out the relationship of the various groups to one another and to the set, ensuring that all gestures should be within the period of the play and related to the style of the time. By the use of steps and rostrums the action was kept moving on different levels, and the inadequacy of the conventional painted set was overcome by moving from two-dimensional to three-dimensional scenery, making use of the BOX-SET. Though the Duke never sought to abolish stage waits and so give Shakespeare's plays in one continuous, rapid flow, his other reforms were exemplary, and even included the requirement that star actors should from time to time play minor roles. In 1874 the Meininger Company appeared for the first time outside its own town when it visited BERLIN and in 1881 it was seen in London, appearing at DRURY LANE in *Julius Caesar, Twelfth Night*, and *The Winter's Tale*, all in German, as well as in a number of German and other classics. In the following years up to 1890, when Chronegk's health broke down and the tours were discontinued, the company visited 38 cities in Europe; STANISLAVSKY saw the Meiningers in Moscow on their second visit there in 1890 (their first was in 1885), and ANTOINE saw them in Brussels; thus the two men who were to become the greatest exponents of stage REALISM, at the MOSCOW ART THEATRE and at the THÉÂTRE LIBRE in Paris respectively, both came under the Meininger influence, which through them spread far into the 20th century. Among the actors trained by Chronegk with the Meiningers were Albert BASSERMANN, Gertrude EYSOLDT, and Josef KAINZ.

MEISL, KARL (1775–1853), Austrian playwright, who with GLEICH and BÄUERLE represents the popular escapist theatre of VIENNA during the troublesome post-Napoleonic era. His numerous plays depended very much on the skill with which they were produced and the accessories of costume, music, and magic. Meisl was at his best in parody, of which *Der lustige Fritz* (1818) was a good example.

MEISTERSÄNGER, name applied originally in the 14th and 15th centuries to German itinerant poets who were also musicians. It was used later to describe groups in certain South German towns who formed Guilds of Mastersingers which competed against each other in musical and poetic contests, mainly on religious subjects. The strict rules governing the compositions submitted for these contests, and their enclosed nature, prevented them from having much influence on the development of music and poetry in Germany, but it was Hans SACHS, the famous Mastersinger celebrated in Wagner's opera *Die Meistersinger von Nürnberg* (1868), who founded the first permanent German theatre in the secularized church of St Martha, where at least from 1551 public performances of Biblical and other, usually historical, plays were given before a paying audience up to the end of the century. Other

Mastersingers who are known to have written plays—mainly short farces—are Hans ROSENPLÜT and Hans FOLZ. Dramatic societies organized on lines similar to that run by Sachs are recorded elsewhere, notably at Freiburg in the late 17th century, and at Augsburg and Memmingen during the 18th century, where productions included a version of MOLIÈRE's *Le Médecin malgré lui* and a parody of *Hamlet*, known in Germany through the activities of the ENGLISH COMEDIANS, entitled *Hammelprinz von Dannemarkt* (*The Muttonprince of Denmarket*).

MELBOURNE THEATRE COMPANY. Founded in 1953 by John Sumner as the Union Theatre Repertory Company for the University of Melbourne, at whose Union Theatre it first appeared, this group was responsible in 1955 for the first production of Ray LAWLER's *Summer of the Seventeenth Doll*. It has since staged a number of interesting plays by such Australian dramatists as Alexander BUZO, Alan HOPGOOD, Patrick WHITE, and David WILLIAMSON. It was first associated with the ELIZABETHAN THEATRE TRUST in 1959, and a year later the Council of Adult Education enabled it to appear at the Russell Street Theatre, which it took over full-time when in 1966 the Union Theatre was no longer available. Two years later it took its present name, and in 1970 Tyrone GUTHRIE visited the theatre to direct a notable production of *All's Well That Ends Well*. In 1973 the company took over the ST MARTIN'S THEATRE, which it relinquished (while still retaining the Russell Street Theatre) when in 1977 it went to the Athenaeum Theatre, where in 1979 it opened the Athenaeum 2, in the former art gallery, for the presentation of 'alternative' experimental plays. Lawler, who had rejoined the company in 1975 as artistic adviser and director, returned to the stage in 1979 to play the lead in Ron Elisha's *In Duty Bound*, which was revised and revived in the following year, when Lawler's title was changed to 'literary adviser'. The English directors Peter James and Frank Hauser have twice been guest directors for the company, and Michael Blakemore was engaged to direct O'NEILL's *Mourning Becomes Electra* in 1981.

MELLON, Mrs ALFRED [*née* Sarah Jane Woolgar] (1824–1909), English actress, who was on the stage as a child and made her adult début in London on 9 Oct. 1843 at the ADELPHI THEATRE under Ben WEBSTER, with whom she remained for many years. She appeared in a number of plays based on DICKENS's novels, being seen at different times as Dot, Tilly, and Bertha in *The Cricket on the Hearth*, Mercy in *Martin Chuzzlewit*, and Mrs Cratchit, one of her best parts, in *A Christmas Carol*. She was also much admired in BUCKSTONE's *The Flowers of the Forest* (1847) and as Mrs Vane in *Masks and Faces* (1852) by Tom TAYLOR. Although she lacked the elegance needed for Old Comedy, she had plenty of high spirits and piquancy. She played opposite T. P. COOKE in his last appearance in JERROLD's *Black-Ey'd Susan*, and was Anne Chute in the first productions of BOUCICAULT's *The Colleen Bawn* (1860). She retired in 1883.

MELLON, HARRIOT (1777–1837), English actress, who was with a strolling company at Stafford in 1795 when SHERIDAN saw her and engaged her for DRURY LANE, where she remained until her retirement in 1815 on her marriage to the banker Thomas Coutts. She was at her best as the light impertinent chambermaids of comedy, in which parts Leigh HUNT much admired her. After her first husband's death she married the Duke of St Albans, leaving the vast Coutts fortune to the daughter of Sir Francis Burdett, later the Baroness Burdett-Coutts, friend and patroness of Sir Henry IRVING.

MELMOTH, Mrs CHARLOTTE (1749–1823), American actress. Born in England (her maiden name is unknown), she ran away from school with an actor named Courtney Melmoth [Samuel Jackson Pratt] (1749–1814), and appeared with him in the provinces. They soon separated, but she continued to act under her married name, and was seen both at COVENT GARDEN and DRURY LANE before leaving for New York in 1793, where she made her début with the AMERICAN COMPANY as Euphrasia in Arthur MURPHY's *The Grecian Daughter*. The excellence of her acting, particularly as Lady Macbeth, made her a universal favourite, and caused many more tragedies to be added to the current repertoire. She was one of the leading actresses at the PARK THEATRE, New York, when it first opened in 1798, and after DUNLAP's bankruptcy was at the CHESTNUT STREET THEATRE in Philadelphia. She retired in 1812, and became a teacher of elocution in New York until her death.

MELO, FRANCISCO MANUEL DE (1608–66), Portuguese statesman and man of letters, whose *Auto do fidalgo aprendiz* (*The Apprentice Nobleman, c.* 1646) has been called the only dramatic work of value on the Portuguese stage of its time. Combining classical, Italian, and Spanish elements, it continued the tradition of comedy established by VICENTE, and has been cited as one of the sources of MOLIÈRE's *Le Bourgeois gentilhomme* (1670); but in spite of its popularity and its astringent wit, it failed to revive the dormant vernacular drama of 17th-century Portugal.

MELODRAMA, type of play popular all over Europe in the 19th century. The term derives from the use of INCIDENTAL MUSIC in spoken dramas, which became customary in German theatres during the 18th century, and from the French *mélodrame*, a dumb show accompanied by music; its application to Gothic tales of horror and mystery, vice, and virtue triumphant, stems from the early works of GOETHE (*Götz von Berlichingen*,

1773) and SCHILLER (*Die Räuber*, 1782), and its most important authors on the Continent were KOTZEBUE and PIXÉRÉCOURT. It was first introduced into England through translations of their plays, particularly those made by Thomas HOLCROFT, whose *A Tale of Mystery* (1802), based on Pixérécourt's *Coelina, ou l'Enfant de mystère* (1800), was the first work in England to be labelled a melodrama. Gradually the music became less important, and the setting of the plays less Gothic. *The Brigand* (1829) by PLANCHÉ was one of the last of the old-fashioned musical melodramas; the setting of JERROLD's *Fifteen Years of a Drunkard's Life* (1828) heralded an era of domestic melodrama, which ran concurrently with a vogue for plays based on real-life or legendary crimes—the anonymous *Maria Marten; or, the Murder in the Red Barn*, which became a classic of melodrama in the 1830s; FITZBALL's *Jonathan Bradford; or, the Murder at the Roadside Inn* (1823); and Dibdin Pitt's *Sweeney Todd; or, the Fiend of Fleet Street* (1847).

The growth of a middle-class audience produced a new type of melodrama, notably at the ADELPHI THEATRE under BUCKSTONE. While the rougher elements on the Surrey side enjoyed the horrors of real life borrowed from *Les Bohémiens de Paris* (1843), with its glimpses of the Paris or London underworld in slums and sewers, prosperous merchant families enjoyed the equally spectacular but less violent domestic tragedies of the elder DUMAS, among them *Pauline* (1840), seen by Queen Victoria at the PRINCESS'S THEATRE in 1851, and *The Corsican Brothers* (1852), the latter adapted by BOUCICAULT. Among his other adaptations was one of *Les Pauvres de Paris*, which was first seen in Liverpool in 1864 as *The Poor* (or *The Streets*) *of Liverpool*, the name later being altered to London or New York according to the town in which the play was being presented. A new phenomenon at this time was the sudden success of the numerous dramatizations of popular novels by women writers—Harriet Beecher Stowe's *Uncle Tom's Cabin* (1852), Mrs Henry Wood's *East Lynne* (1861), and Miss Braddon's *Lady Audley's Secret* (1862). These rivalled in popularity such plays as Tom TAYLOR's *The Ticket-of-Leave Man* (1863), based like so many melodramas on a French original, *Le Retour de Melun* by Brisbarre and Nus. Few of the prolific dramatists of the time bothered to concoct their own plots, though the exercise of the COPYRIGHT laws in the 1860s began to inhibit their wholesale piracy. None the less, all the melodramas staged by IRVING at the LYCEUM from Leopold Lewis's *The Bells* in 1871 to Boucicault's *The Corsican Brothers* in 1880, originated on the Continent. Other actor-managers had their greatest successes with dramatizations of novels—TREE with Du Maurier's *Trilby* (1895), ALEXANDER with Anthony Hope's *The Prisoner of Zenda* (1896), MARTIN-HARVEY with DICKENS's *A Tale of Two Cities*, retitled *The Only Way* (1899), and Fred TERRY with Baroness Orczy's *The Scarlet Pimpernel* (1903). Some exceptions over the years were *The Silver King* (1882) by H. A. JONES and H. Herman; *The Sign of the Cross* (1895) by Wilson BARRETT; and the nautical melodramas popularized by William TERRISS at the Adelphi. The turn of the century saw the spectacular melodramas staged at DRURY LANE, with shipwrecks, railway accidents, earthquakes, and horse-racing, and the joint productions of the MELVILLE brothers with *The Worst Woman in London* (1899) and *The Bad Girl of the Family* (1909). Melodrama had come a long way from its original simplicity, which equated poverty with virtue and wealth with villainy. The day of true melodrama was over, and occasional revivals of such classic examples as *Maria Marten, East Lynne*, and *The Streets of London* have been played as comic caricatures, but melodramatic elements continue to flourish in the theatre as they have done since the time of EURIPIDES.

MELODRAMMA, Italian play with music, each element being equally important. It evolved during the 18th century from the earlier PASTORAL, the chief writers connected with it being Apostolo Zeno (1668–1750) and Metastasio (1698–1782), whose librettos have since been used by innumerable composers of opera. The term is sometimes used for the MONODRAMA popular in Germany during the late 18th century.

MELPOMENE, see MUSES.

MELUCHA THEATRE, see MOSCOW STATE JEWISH THEATRE.

MELVILLE, ALAN, see REVUE.

MELVILLE, WALTER (1875–1937) and FREDERICK (1879–1938), two brothers who for 25 years were joint proprietors of the LYCEUM THEATRE in London, where they produced annually a spectacular PANTOMIME, usually written by Fred. Elaborately produced, these filled the theatre to capacity for several months. The Melvilles, who jointly built the Prince's (later the SHAFTESBURY) Theatre, were also successful writers of highly coloured MELODRAMAS, simple, direct stories with virtue triumphant, much to the taste of their time. Walter was responsible for such masterpieces as *The Worst Woman in London* (1899) and *The Girl Who Took the Wrong Turning* (1906), and Fred for *Her Forbidden Marriage, The Ugliest Woman on Earth* (both 1904), and *The Bad Girl of the Family* (1909), all equally lurid, and equally successful in their day.

MÉMOIRE DE MAHELOT, see MAHELOT.

MENANDER (*c.* 342–293 BC), Ancient Greek poet, son of Diopeithes, and an Athenian dramatist whose plays are classified as NEW COMEDY. For a time he studied philosophy with ARISTOTLE's pupil and successor Theophrastus, whose division of human nature into fixed character-types has obvious affinities with Menander's own works. He was invited to the Court of Ptolemy I of Egypt, but preferred to remain in Athens, where he was recognized as the leading comic writer of his time.

In spite of his great reputation in antiquity, very little of his work has survived, and until this century he was known only through fragments and passages quoted by later writers, including a line quoted by St Paul (Corinthians 15:33): 'Evil communications corrupt good manners.' But in 1905 a papyrus (now in Cairo) was discovered which contained considerable parts of four plays by Menander, with smaller fragments of a fifth. These were the *Heros*, the *Samia* (the *Samian Women*), the *Perikeiromene* (completed, translated, and published by Gilbert Murray in 1941 as *The Rape of the Locks*), and the *Epitrepontes* (also published by Murray in English in 1945 as *The Arbitration*). In 1955 a further find in Egypt brought to light a complete play, the *Dyskolos* (published in an English version by Vellacott as *The Bad-Tempered Man*, or *The Misanthrope*). With this were further portions of the *Samian Women* and about half of the *Aspis* (*The Shield*). In 1965 substantial parts of *Misoumenes* (*The Man She Hated*) were found.

To some extent, these discoveries and other known fragments merely reinforced what could already be surmised about Menander's work from extant Latin versions of his plays, for he was able to give the comic dramatists of Rome what they could never have found in OLD COMEDY—excellent models and useful material. It is in Menander that we first meet the originals of some of the stock characters of later comedy—the irascible old man, the young rake with a good heart, the officious slave—all portrayed with great delicacy and liveliness. The plays are comedies of manners and intrigue set against a background of urban life, designed for the cosmopolitan audiences of the HELLENISTIC AGE, and completely lacking the topicality, fantasy, bawdiness, and invective of ARISTOPHANES. They concern such things as mistaken identities, long-lost children, and hasty marriages. The characters are immediately identifiable by their MASKS; and the motivating force behind the plot is no more than chance or coincidence which brings the characters together. Settings are more realistic than in Old or MIDDLE COMEDY, in that they utilize the standard three doors of the Hellenistic scene-building to represent houses fronting on to a street. As there is no place in a drama which dealt exclusively with the fortunes of private individuals for a CHORUS, which must in some sense represent the community, it is replaced by groups of singers and dancers who provide entertainment between the acts; sometimes they are explained away as 'a band of tipsy revellers', whose songs are quite irrelevant to the action, and their function is indicated in the papyri only by 'something for Chorus'.

Menander was praised by ancient critics for his fidelity to nature. If nowadays this judgement is questioned on the grounds that his plots are contrived and his characters stereotypes, it should be remembered that he stands at the head of a long succession of comic writers, beginning with the Roman playwrights to whom he bequeathed models which they could usefully adapt for their own audiences, and which in turn have inspired a whole succession of brilliant writers of polite comedy continuing through MOLIÈRE and GOLDONI to PINERO, COWARD, and Neil SIMON. Menander is, in a very real sense, the father of modern comedy.

MENKEN, ADAH ISAACS [Dolores Adios Fuertes] (1835–68), American actress, whose theatrical reputation rests entirely on her playing of Mazeppa, in a dramatization of BYRON's poem of that name, 'in a state of virtual nudity while bound to the back of a wild horse'. This equestrian drama was first seen at the Coburg (later the OLD VIC) in London in 1823, and continued to be popular for the next 50 years, though it was not until 1859 that a woman, Charlotte Crampton, played the part of the hero. Menken, who in 1856 had married John Isaacs Menken, and kept his name through all her subsequent matrimonial and other adventures, first played Mazeppa in 1861 in Albany, and after appearing in the part in New York was seen at ASTLEY'S AMPHITHEATRE in London in 1864. She appears to have exercised a fatal fascination over 'literary gentlemen', including DICKENS, the elder DUMAS, and Swinburne. She died in Paris, where she had made her first appearance in 1866.

MEN'S DRAMATIC CEREMONY, see MUMMERS' PLAY.

MERCER, DAVID (1928–80), English playwright, whose plays reflect an interest in both politics—he was a disillusioned Marxist—and schizophrenia and madness. He was already well known for his television plays when in 1965 the ROYAL SHAKESPEARE COMPANY at the ALDWYCH THEATRE staged his one-act political play *The Governor's Lady*. In the same year Peter O'TOOLE appeared in *Ride a Cock Horse* as the working-class anti-hero alienated from his background by his success as a writer; it was produced in New York in 1979. Two further plays, *Belcher's Luck* (1966) and *After Haggerty* (1970), the latter about a dramatic critic with a sense of professional impotence, were produced by the R.S.C., and also in 1970 Michael HORDERN was seen at the CRITERION THEATRE in the title role of *Flint*, playing a lecherous and agnostic clergyman who elopes with a pregnant Irish girl. Following *Duck Song* (1974) came *Cousin Vladimir* (1978), both presented by the R.S.C., the eponymous hero of the second, a fugitive from Soviet Russia confronted by the moral degeneracy of modern England, perhaps mirroring Mercer's own perplexities. His last two plays were *Then and Now* (1979) and *No Limits to Love* (1980), the latter again presented by the R.S.C.

MERCHANT, VIVIEN, see PINTER, HAROLD.

MERCIER, LOUIS-SÉBASTIEN (1740–1814), French dramatist, an exponent of the *drame bourgeois* popularized by DIDEROT. His plays, in which he gives elaborate directions for scenery, were more popular outside France, particularly in Germany, on account of their unimpeachable morality and declamatory style. A characteristic example of his work is *La Brouette du vinaigrier* (1784), the story of a marriage between a wealthy young girl and the son of a working-class man. He had earlier adapted LILLO's *The London Merchant* (1731) as *Jenneval* (1768), allowing the hero to escape punishment by a last-minute conversion. In the same spirit of optimism he gave his translation of Shakespeare's *Romeo and Juliet* a happy ending and reduced *King Lear* to the tale of a bourgeois household quarrelling over the misdeeds of the servants. Having taken refuge in Switzerland because of his political views, Mercier returned to Paris on the outbreak of the Revolution; as a deputy he voted against the execution of Louis XVI and in favour of a life sentence. He was later imprisoned, but was saved from the guillotine by the fall of Robespierre in 1794.

MERCURY THEATRE, Auckland, New Zealand, erected as a cinema in 1910 and formerly known as the King's, the Prince Edward, and the Playhouse. Bought in 1968 to serve as a home for the company run by the Auckland Theatre Trust, it was converted to provide a main auditorium seating 660 people, with a smaller studio theatre. From its foundation up to 1975 its artistic director was Anthony Richardson. The appointment in 1980 of Jonathan Hardy as the first New Zealand artistic director gave hopes of an increase in the number of new local plays in the theatre's repertory. New Zealand plays which have had their first production at the Mercury include James McNeish's *The Rocking Cave* and the musical *Mister King Hong* (both 1973), Robert Lord's *Heroes and Butterflies* (1974), and *The Naval Officer* (1979) by Bruce McNeill.

MERCURY THEATRE, London, at Notting Hill Gate, a small but well-equipped playhouse holding about 150 people, opened in 1933 by Ashley DUKES for the production of new and uncommercial plays and to serve as a centre for his wife's Ballet Rambert. The first production, on 19 Oct., was an adaptation of MOLIÈRE's *Amphitryon*. The most important event of the theatre's early years was the first London production, on 1 Nov. 1935, of T. S. ELIOT's *Murder in the Cathedral*, transferred from the Chapter House at Canterbury Cathedral. Two years later Auden and Isherwood's poetic play *The Ascent of F.6* was first seen in London, and in 1943 O'NEILL's *Days Without End*. In 1945 and 1946 the Mercury became the home of POETIC DRAMA with Norman Nicholson's *The Old Man of the Mountains*, Ronald Duncan's *This Way to the Tomb!*, with music by Benjamin Britten, Anne Ridler's *The Shadow Factory*, and Christopher FRY's *A Phoenix Too Frequent*, all directed by E. Martin BROWNE. In 1947 SAROYAN's *The Beautiful People* and O'Neill's *SS Glencairn* both had their London premières at the Mercury, but after the production of two children's plays by Nicholas Stuart Gray there were no further productions for some time. The theatre was then used only by the Ballet Rambert, until in 1966 the International Theatre Club presented a season of new and experimental plays, beginning with OBALDIA's *Jenusia* on 7 July. The theatre has since been used intermittently by visiting companies, among them the LA MAMA troupe from New York in 1967.

MERCURY THEATRE, New York, see COMEDY THEATRE.

MERMAID SOCIETY, play-producing society founded by Philip Carr, which in the 1900s staged a number of Elizabethan and Stuart plays, by MARLOWE, JONSON, WEBSTER, FORD, and others. It was also responsible for a production of Milton's *Comus* in the Botanical Gardens at Oxford in 1903, and of Jonson's MASQUE *A Hue and Cry After Cupid*, written for the marriage of Lord Haddington in 1608, at Stratford-upon-Avon in 1908. Information about the society is difficult to obtain, and some research into its work and influence is still needed.

MERMAID THEATRE, London, originally a small private theatre in the garden of Bernard MILES's house in St John's Wood. Designed on Elizabethan lines, it opened on 9 Sept. 1951 with a performance of Purcell's opera *Dido and Aeneas*, followed a week later by *The Tempest* and in 1952 by MIDDLETON's *A Trick to Catch the Old One*. In 1953, to celebrate the coronation of Elizabeth II, the Mermaid was re-erected in the City of London for performances of *As You Like It*, *Macbeth*, and JONSON, MARSTON, and CHAPMAN's *Eastward Ho!* The success of his venture encouraged Miles to build a permanent professional theatre, which opened on 28 May 1959 with his own musical adaptation of FIELDING's *Rape upon Rape* as *Lock Up Your Daughters*. The new Mermaid, financed by public subscription and erected within the walls of an old bombed warehouse at Puddle Dock, near Blackfriars Bridge, seated nearly 500 people in one steeply raked tier, and had an open Elizabethan-style stage. Among the productions seen there were BRECHT's *Galileo* (1960), several revivals of SHAW in 1961–2, and a series of plays by O'CASEY, also in 1962. On 31 Jan. 1963 Antrobus's *The Bed Sitting Room*, with Spike Milligan, began a successful run which was completed at the DUKE OF YORK'S, and on 6 Mar. the first of Bill Naughton's plays, *All in Good*

Time, opened, transferring later to the PHOENIX, to be followed by his *Alfie* and *Spring and Port Wine* (both 1963). Two other plays which later went to the West End were a revival of SARDOU's *Let's Get a Divorce* in 1966 and Peter Luke's *Hadrian VII* in 1968. Ian MCKELLEN repeated his much-acclaimed performances in the title roles of *Richard II* and MARLOWE's *Edward II* in 1969, and two interesting revivals were JOYCE's *Exiles* (1970) and SHERRIFF'S *Journey's End* (1972). A tribute to Noël COWARD, a musical biography entitled *Cowardy Custard*, opened on 10 July 1972 and ran for a year, being followed by a similar treatment of Cole Porter (*Cole*, 1974) and Stephen SONDHEIM (*Side by Side by Sondheim*, 1976). In Sept. 1978, at the end of the run of STOPPARD's *Every Good Boy Deserves Favour*, the theatre closed for reconstruction. It reopened on 7 July 1981 with extra seating, a much larger stage, better front-of-house facilities, and a studio theatre for children, the Molecule Club, seating 250. The opening production, a musical version of *Eastward Ho!*, was a failure, as was *Shakespeare's Rome*, an attempt to combine shortened versions of *Julius Caesar* and *Antony and Cleopatra* into a single evening. The consequent financial problems caused the curtailment of future plans, in spite of the successful transfer of Mark Medoff's *Children Of a Lesser God* to the Albery (see NEW THEATRE).

MERMAN [Zimmermann], ETHEL (1909–), American actress and singer, who was seen in cabaret in 1928 and in VAUDEVILLE a year later with Clayton, Jackson, and Durante. Her first role on Broadway was Kate Fothergill in George Gershwin's musical *Girl Crazy* (1930). Her performance as Reno Sweeney in the Cole Porter musical *Anything Goes* (1934) established her as a star, and she went on to play the leading roles in four more Porter musicals—*Red, Hot and Blue!* (1936), *Du Barry Was a Lady* (1939), *Panama Hattie* (1940), and *Something For the Boys* (1943). Her biggest success was as Annie Oakley in Irving Berlin's *Annie Get Your Gun* (1946), which ran for three years, and she had another long run as Sally Adams, American ambassador to a mythical European country, in Berlin's *Call Me Madam* (1950). *Happy Hunting* (1956) was less successful, but she scored a big hit as Gypsy Rose Lee's mother in *Gypsy* (1959). She has not been seen since in a new musical, though she starred in Broadway revivals of *Annie Get Your Gun* (1966) and *Hello, Dolly!* (1970). She was seen in London in 1964 in cabaret. An extrovert personality, with a loud, brassy voice, she brings tremendous attack to every role she plays. In 1955 she published her autobiography *Who Could Ask For Anything More?*

MERRY (*née* Brunton), ANNE (1769–1808), English actress, eldest child of a provincial theatre manager named John Brunton (1741–1822). She first appeared in Bath in her father's company in 1784 with such success that a year later she was engaged for COVENT GARDEN, where she remained until she left the stage in 1792 to marry Robert Merry. Soon after he lost all his money, and she accepted an offer from Thomas WIGNELL to go to the CHESTNUT STREET THEATRE in Philadelphia. She made her first appearance there on 5 Dec. 1796 as Juliet in *Romeo and Juliet*, and was soon acclaimed as the finest actress of the time, both in Philadelphia and in New York, where she first appeared in 1797. Many contemporary accounts testify to her beauty and fine acting, and she was particularly admired in Mrs SIDDONS's great part of Belvidera in OTWAY's *Venice Preserv'd*. Widowed in 1798, in 1803 she married Wignell, who died shortly afterwards. Three years later she married William WARREN, made several successful appearances in New York under William DUNLAP, who had been in the audience when she made her début at Covent Garden, and died in childbirth at the age of 40. Her brother John Brunton (1775–1849) was also an actor, as was her sister Louisa (1779–1860), who retired in 1807 to marry the Earl of Craven.

MERSON, BILLY, see MUSIC-HALL.

MESCIDATO, see ITALY.

MESQUITA, MARCELINO, see PORTUGAL.

MESSEL, OLIVER HILARY SAMBOURNE (1905–78), English artist and stage designer, who first attracted attention with his masks and costumes for COCHRAN's revues in the late 1920s. In 1932 he was given the task of converting the stage and auditorium of the LYCEUM THEATRE into a cathedral for a revival of Vollmöller's *The Miracle* directed by REINHARDT, and three years later he designed the settings for Ivor NOVELLO's musical *Glamorous Night*. He was also responsible for the settings of productions of WYCHERLEY's *The Country Wife* in 1936, *A Midsummer Night's Dream* in 1937, and SHERIDAN's *The Rivals* in 1945. One of his most memorable settings was that for ANOUILH's *Ring Round the Moon* (1950), translated by Christopher FRY, and he provided the sets for two of Fry's own plays, *The Lady's Not For Burning* (1949) and *The Dark Is Light Enough* (1954). He also designed the setting for the production of the latter play in New York, where his work was later seen in Fay and Michael Kanin's *Rashomon* (1959), based on Ryunosuke Akutagawa's Japanese stories, and Anouilh's *Traveller Without Luggage* (1964). He was active also in opera, ballet, and film, and was responsible for the interior decoration of Sekers's private theatre in Whitehaven, and of the BILLY ROSE THEATRE in New York.

MESSENIUS, JOHANNES (1579–1636), Swedish historian and dramatist, who drew his material from Swedish history and saga, using dialogue as the vehicle for secular instead of sacred history. His intention was to cover Swedish history in 50 plays, but only six are extant, of which *Disa* (1611) is the first.

METHOD, name given to an introspective approach to acting based on the system evolved by STANISLAVSKY for the actors at the MOSCOW ART THEATRE and set out by him in such books as *An Actor Prepares* (1926). The Method first came into prominence in the United States during the 1930s, when it was adopted by the GROUP THEATRE in its reaction against what were considered the externalizing, stereotyped techniques current on the contemporary New York and London stages. Its notoriety rested mainly on its adoption by the Actors' Studio, founded in 1947 by Elia KAZAN and others, later including Lee STRASBERG. Strong feelings were aroused by the Method: its upholders maintained that it enabled the actor to give a true interpretation of any part based on personal exploration of the character; its opponents pointed out that it intensified the actor's self-absorption, to the exclusion of the audience and even of other actors, and that it was suitable only for a certain type of play, and certainly not for Elizabethan or Restoration plays. It achieved its greatest success in the plays of some modern American playwrights, particularly those of Tennessee WILLIAMS. It was introduced into England in 1956, but with little result, perhaps because good English actors, like good actors everywhere, were already Method actors without realizing it. In the hands of a teacher such as Stanislavsky it could give excellent results, but once codified it was open to partial or even complete misunderstanding, and often did more harm than good. The controversy originally aroused by the Method has long since died away, and it is now seen as only one of many valid methods of approaching a part, all of which are simply convenient ways of making actors work hard and concentrate on their job. From the audience's point of view the niceties of Method or non-Method acting have always been unimportant. To them it was the end result that mattered.

METROPOLITAN CASINO, New York, see BROADWAY THEATRE (3).

METROPOLITAN MUSIC-HALL, London, in the Edgware Road. In 1862 a small concert hall attached to the White Lion Inn was replaced by a new building, which as Turnham's, named after its proprietor, opened as a MUSIC-HALL on 8 Dec. and was renamed the Metropolitan two years later. It gradually achieved popularity, and under a number of different managers had a prosperous career until 1897 when it was rebuilt. The new three-tier house, holding 1,855 people, opened on 22 Dec. with a star-studded bill and prospered for a time. When the popularity of VARIETY began to decline, seasons of opera and visits by touring companies were tried, but without success, and the hall closed except for occasional shows and wrestling matches on Saturdays. It eventually became a television studio, which shut down on 6 Dec. 1962, the building being demolished a year later.

METROPOLITAN THEATRE, New York, at 667–77 Broadway. Built on the site of Tripler Hall, which was burnt down on 7 Jan. 1854, this theatre opened unsuccessfully on 20 Sept., and soon became a circus. It was occasionally used for plays, and in Sept. 1855 RACHEL, with a French company, appeared there in RACINE's *Phèdre* and SCRIBE and Legouvé's *Adrienne Lecouvreur*. In the following December Laura KEENE took over and in spite of prejudice against a woman manager was doing well when she lost her lease on a technicality. She was succeeded by BURTON, and in 1859 by BOUCICAULT, who reopened the theatre as the Winter Garden on 14 Sept. with *Dot*, his own dramatization of DICKENS's *The Cricket on the Hearth*. He also presented a stage version of *Nicholas Nickleby* and several of his own plays during his first season. An early version of Joseph JEFFERSON's *Rip Van Winkle* was seen at the Winter Garden, and important American actors who appeared there included John Sleeper CLARKE, Charlotte CUSHMAN, SOTHERN, and the FLORENCES. Edwin BOOTH took over the theatre in 1864, appearing there on 25 Nov. in a performance of *Julius Caesar* to raise funds for the Shakespeare statue in Central Park. He played Brutus, with his brothers Junius Brutus and John Wilkes BOOTH as Cassius and Antony respectively; this was the second and final appearance on the New York stage of Lincoln's assassin. Edwin Booth's record run of 100 performances of *Hamlet* took place here in 1864–5, and here, too, he was greeted by an enthusiastic audience when he appeared in public for the first time after Lincoln's murder. On 23 Mar. 1867, when Booth was about to appear as Romeo, the theatre was burnt down: it was not rebuilt.

MEYERHOLD, VSEVOLOD EMILIEVICH (1874–?1940/3), Russian actor and director, who after training under NEMIROVICH-DANCHENKO joined the MOSCOW ART THEATRE on its foundation in 1898, making his first appearance as Treplev in CHEKHOV's *The Seagull*. In 1902 he left to found his own company, with which he toured the provinces untill 1905, in which year he was invited by STANISLAVSKY to take charge of a newly opened studio theatre. It was soon apparent that his conception of the actor as a puppet to be controlled from outside by the director (a concept he formulated under the influence of Gordon CRAIG) was totally opposed to the NATURALISM of the Moscow Art Theatre style, and the studio closed. From 1906 to 1907 Meyerhold worked with Vera KOMISARJEVSKAYA, but again came into conflict with the actors when he tried to put his theories into practice, and was forced to leave. For some years he directed productions at the Imperial theatres in St

Petersburg, and at the same time continued his experiments, based on the stylized methods of the COMMEDIA DELL'ARTE, with a new group of young actors. After the Revolution Meyerhold was the first theatre director to offer his services to the new government; in 1920 he was appointed by LUNACHARSKY as head of the Theatre Section of the People's Commissariat for Education and provided with the old Sohn Theatre for his workshop. It was officially known as the Meyerhold Theatre from 1926. His early post-Revolutionary work, based on a synthesis of his experimental work known as BIO-MECHANICS, proved comparatively short-lived but produced some important results. He was the first director to stage a Soviet play, *Mystery-Bouffe* (1918) by MAYAKOVSKY, whose *The Bed-Bug* (1929) and *The Bath House* (1930) he also directed, as well as new plays by FAIKO, TRETYAKOV, VISHNEVSKY, and others. In 1936, with the establishment of SOCIALIST REALISM, Meyerhold, who had already suffered from official criticism of his FORMALISM, incurred the displeasure of the authorities and was charged with having a pernicious 'foreign' influence on the Soviet theatre. His theatre was closed, and he was arrested in 1938. The circumstances of his death, and that of his wife Zinaida Raikh, are not known, though the official date is 1940. He was rehabilitated in 1955, and has since been recognized, both in his own country and abroad, as one of the most important directors of his time. Two volumes of his writings on the theatre, which reveal him as an original and creative thinker, were published in Moscow in 1968, and a selection from these, with a critical commentary by Edward Braun, appeared in English in 1969.

MEZZANINE (or MAZARINE) **FLOOR,** in large theatres the first level of the CELLAR, which housed most of the TRAPS and other machinery, and also served as a retiring-room for the members of the orchestra.

MEZZETINO, MEZZETIN, ZANNI or servant of the COMMEDIA DELL'ARTE, with many of the characteristics of BRIGHELLA or SCAPINO, though more polished. Towards the end of the 17th century (*c*. 1682) the role was altered and elaborated by Angelo COSTANTINI, who adopted red and white as the distinguishing colours of his costume as opposed to the green and white stripes of Scapino. The character passed into the French theatre by way of the COMÉDIE-ITALIENNE, and one of the finest French Mezzetins was the actor PRÉVILLE of the COMÉDIE-FRANÇAISE, who was painted by Van Loo in the part.

MICHELL, KEITH (1928–), Australian-born actor and singer, formerly an art teacher, who made his stage début in Adelaide in 1947, and then trained at the OLD VIC Theatre School in London. After playing Bassanio in *The Merchant of Venice* (1950) for the YOUNG VIC he appeared as Charles II in the musical version of FAGAN'S *And So To Bed* (1951), and then toured Australia with the company from the ROYAL SHAKESPEARE THEATRE, returning to Stratford for the seasons of 1954 and 1955. At the Old Vic in 1956 he was seen in such roles as Benedick in *Much Ado About Nothing* and Antony in *Antony and Cleopatra*. He then went to the West End to star in another musical, *Irma La Douce* (1958), in which he made his début in New York in 1960, returning there after playing Don John in FLETCHER'S *The Chances* at the first CHICHESTER FESTIVAL in 1962 to appear in ANOUILH'S *The Rehearsal* in 1963. Two other musical roles were Robert Browning in the long-running *Robert and Elizabeth* (1964), based on Besier's *The Barretts of Wimpole Street*, and Don Quixote in *Man of La Mancha* (1968), based on CERVANTES. He scored a great success in Ronald MILLAR'S *Abelard and Heloise* (1970; N.Y., 1971), and again played Browning in Kilty's *Dear Love* (1973). He was Artistic Director of the Chichester Festival from 1974 to 1977, his roles there including the title role of ROSTAND'S *Cyrano de Bergerac*, Iago in *Othello*, and King Magnus in SHAW'S *The Apple Cart*, in which he was also seen in London. After leaving Chichester he appeared in another musical, *On the Twentieth Century* (1980).

MICKERY THEATRE, see NETHERLANDS.

MICKIEWICZ, ADAM (1798–1855), the outstanding poet of Poland, and the founder of the Romantic movement in Polish literature. Born and educated in Vilna, he was exiled in 1829, and lived thereafter mainly in Paris, where in 1840 he became Professor of Slavonic Literature at the newly-founded Collège de France. His finest dramatic work, *Forefathers' Eve*, was composed in a fragmentary manner, not being intended for the stage. Parts II and IV were published as part of his poetic works in 1823; Part III in 1832. Part I, which he left unfinished, was not published until after his death. The four parts were first co-ordinated and staged by WYSPIAŃSKI in Cracow in 1901, and the play was many times revived before being officially banned after the Second World War, during the ascendancy of SOCIALIST REALISM. It was not seen again until 1955, when it was directed by Aleksander Bardini at the Polski Teatr in Warsaw, and continued to be revived. An outstanding arena production by ŚWINARSKI at the Stary Theatre in Cracow was brought to London during the WORLD THEATRE SEASON of 1975 and staged in Southwark Cathedral. In 1979 two productions of it were current in Warsaw at the same time, one directed by Jerzy Kreczmar and the other by Adam Hanuszkiewicz.

MIDDLE COMEDY, term applied to the last two plays of ARISTOPHANES, the *Ecclesiazusae* (*Women in Parliament*) and *Plutus*, and those of his immediate successors in the early and middle 4th century BC, from which much of the spirit of revelry present in OLD COMEDY has disappeared. The chorus shrinks, the importance of the plot grows, the parabasis disappears, and dramatic illusion is taken more seriously. The earlier plays are still concerned, though to a lesser degree, with politics—Aristophanes ridicules both feminism and communism—but they are less personal and fantastic, and even the obscenity becomes less obvious. Judging from the remaining fragments, later Middle Comedy in general was social rather than overtly political, with a background of private life. By the middle of the 4th century it had passed into NEW COMEDY.

MIDDLESEX MUSIC-HALL, London, in Drury Lane, originally the Mogul Saloon, known as the 'Old Mo', a nickname it subsequently retained. It had been a place of entertainment since the early years of the 19th century, but it was not until 27 Dec. 1847 that it began to feature the new-style MUSIC-HALL turns, becoming the Middlesex Music-Hall in 1851. Many famous music-hall stars made their débuts there, including, it is said, Dan LENO. Reconstructed in 1872, it was rebuilt in 1891 and again in 1911, when, as the New Middlesex Theatre of Varieties, it reopened on 30 Oct. with a capacity of 3,000, a far cry from its original 500. Music-hall turns continued to predominate in the programme, though interspersed with a series of sensational French REVUES, until the theatre closed on 1 Feb. 1919, to reopen as the WINTER GARDEN THEATRE. The site is now occupied by the NEW LONDON THEATRE.

MIDDLETON, THOMAS (*c.* 1570–1627), English dramatist, first noted as working for HENSLOWE in collaboration with others, among them DEKKER, with whom he wrote *The Honest Whore* (Part I, 1604; Part II, 1605) and *The Roaring Girl* (1610). His first independent play, *The Phoenix* (1604), was played by the Children of Paul's, who also appeared in his best known comedy, *A Trick to Catch the Old One* (1605)—to which MASSINGER may have been indebted for the idea of his *A New Way to Pay Old Debts* (1625)—and *A Mad World, My Masters* (1606). This comedy of London life was followed by another, *Your Five Gallants* (1607), and by *A Chaste Maid in Cheapside* (1611), *No Wit No Help Like a Woman's* (1613)—which Aphra BEHN rewrote in 1677 as *The Counterfeit Bridegroom, or the Defeated Widow*—and by a tragedy, *Women Beware Women* (1621), a penetrating study of feminine psychology which was successfully revived in 1962 and 1969 by the ROYAL SHAKESPEARE COMPANY. In 1624 a political satire, *A Game at Chess*, on the fruitless attempts being made to unite the royal houses of England and Spain, proved popular with its audiences but led to Middleton being severely admonished by the authorities and perhaps imprisoned. Two of Middleton's best plays were written in collaboration with ROWLEY: *A Fair Quarrel* (1617), revived by the NATIONAL THEATRE in 1979, and *The Changeling* (1622), of which two revivals were current in London in 1978. He was also responsible for a number of MASQUES and Civic Pageants, now lost, and has been credited with *The Revenger's Tragedy* (1607), now attributed to TOURNEUR.

MIELZINER, JO (1901–76), American scene designer, who was for a short time an actor. His first stage designs were done for the LUNTS in MOLNÁR's *The Guardsman* (1924), after which he was responsible for the décor of a wide variety of plays, among them *Romeo and Juliet* (1934) for Katharine CORNELL and *Hamlet* (1936) for John GIELGUD; *Winterset* (1935), *The Wingless Victory* (1936), and *High Tor* (1937) by Maxwell ANDERSON; *The Glass Menagerie* (1945), *A Streetcar Named Desire* (1947), *Summer and Smoke* (1948), and *Cat on a Hot Tin Roof* (1955), by Tennessee WILLIAMS; *Death of a Salesman* (1949), by Arthur MILLER; and a number of musicals, including *The Boys from Syracuse* (1938), *Finian's Rainbow* (1947), *Guys and Dolls* (1950), and *The King and I* (1951). He was also responsible for the design of the WASHINGTON SQUARE THEATRE and for the décor of its productions of Miller's *After the Fall* and BEHRMAN's *But for Whom Charlie* (both 1964); with the Finnish architect Eero Saarinen he designed the VIVIAN BEAUMONT THEATRE at the Lincoln Center.

MIKHOELS [Vovsky], SALOMON MIKHAILOVICH (1890–1948), Jewish actor, who in 1919 joined the Jewish Theatre Studio founded in Leningrad by GRANOVSKY, which later moved to Moscow to become the MOSCOW STATE JEWISH THEATRE. Mikhoels, who first came into prominence with a fine performance in 1921 in Sholom ALEICHEM's *Agents*, became the company's leading actor and in 1927 took over the running of it during a European tour. He remained in charge until his death; one of his finest performances was given as King Lear in a production by RADLOV in 1935.

MILES, BERNARD (1907–   ), English actor and director, knighted in 1969. He made his first appearance on the stage in 1930, and spent several years in repertory before joining the company at the PLAYERS' THEATRE in *Late Joys*. He appeared several times with the OLD VIC company, both in London and on tour, and in 1951 founded the first MERMAID THEATRE, where he was seen as Caliban in *The Tempest*. A year later he played Macbeth to the Lady Macbeth of his wife Josephine Wilson, who has ably assisted him in all his enterprises, and in 1953 was seen at the second Mermaid in *As You Like It*, which he also directed, and as Slitgut in JONSON's *Eastward Ho!* Since opening the third Mermaid with his own adaptation of FIELDING's *Rape Upon Rape* as a musical, *Lock Up Your Daughters*, he has devoted all his time to it, directing many of the plays seen there, and playing, among other parts, Long John Silver in his own dramatization of Stevenson's *Treasure Island*, first seen in 1959 and revived many times. He also played the title roles in BRECHT's *Galileo* (1960), IBSEN's *John Gabriel Borkman* (1961), Brecht's *Schweyk in the Second World War* (1963), and SOPHOCLES' *Oedipus the King* and *Oedipus at Colonus* (1965). He then appeared as the Archbishop in ARDEN's *Left-Handed Liberty* (also 1965), and in 1970 was seen as Falstaff in both parts of *Henry IV*. The replanning and rebuilding of his theatre occupied much of the next decade. He is the author of *The British Theatre* (1947).

MILES GLORIOSUS. The braggart soldier, a stock character in ancient Greek NEW COMEDY, who found his way into Roman comedy and gave his name to one of the plays of PLAUTUS. He reappears among the COMMEDIA DELL'ARTE figures as the Capitano, and in English drama as part of the make-up of such characters as Shakespeare's Pistol and Ben JONSON's Captain Bobadil (in *Every Man in His Humour*).

MILES'S MUSICK HOUSE, London, see MUSIC-HALL and SADLER'S WELLS THEATRE.

MILLAR, GERTIE (1879–1952), one of the most famous of George EDWARDES's 'Gaiety Girls', an excellent dancer, with a beautiful face and figure and a small but sweet voice. Born in Bradford, the daughter of a mill-worker, she first appeared on the stage at the age of 13, playing the Girl Babe in *The Babes in the Wood* in Manchester. In 1899 she appeared as Dandini in *Cinderella* at the Grand, Fulham, and two years later made her first appearance at the GAIETY, singing 'Keep off the Grass' in *The Toreador*. She remained there till 1908, when she made her first appearance in New York in *The Girls of Gottenburg* marrying the composer Lionel Monckton in the same year, and then returned to the Gaiety until 1910, when she was seen at the ADELPHI as Prudence in *The Quaker Girl*; two years later she was at DALY's in *Gipsy Love*, and in 1914 she appeared at the COLISEUM in VARIETY. She made her last appearance on the stage at the PRINCE OF WALES's in *Flora* (1918), and then retired, marrying the 2nd Earl of Dudley after Monckton's death in 1924.

MILLAR, RONALD (1919–   ), English actor and playwright, who made his first appearance on the stage in 1940, and was seen in a number of productions before devoting himself entirely to playwriting. His first play was a thriller, *Murder From Memory* (1942), and *The Other Side* (1946) was based on Storm Jameson's novel of the same title. His own play *Frieda* was also produced in 1946, and was followed by *Champagne for Delilah* (1949) and *Waiting for Gillian* (1954), based on Nigel Balchin's novel *A Way Through the Wood*. He first achieved an outstanding success with *The Bride and the Bachelor* (1956) and its sequel *The Bride Comes Back* (1960), farces in which Cicely COURTNEIDGE and Robertson HARE appeared. He then dramatized three novels by C. P. Snow, *The Affair* (1961; N.Y., 1962), *The New Men* (1962), and *The Masters* (1963), featuring academic life in Cambridge, and wrote a musical on the Brownings, *Robert and Elizabeth* (1964), following it with *Number Ten* (1967), set in Downing Street with Alastair SIM as the Prime Minister, and *They Don't Grow on Trees* (1968), a comedy in which Dora Bryan appeared in nine different roles. His *Abelard and Heloise* (1970; N.Y., 1971), with Keith MICHELL and Diana RIGG in the title roles, had a long run, and he then returned to Snow's novels, adapting *In Their Wisdom* as *The Case in Question* (1975).

MILLER, ARTHUR (1915–   ), American dramatist, who while a student at Michigan University won three drama prizes, one of which, the THEATRE GUILD National Award of 1937, he shared with Tennessee WILLIAMS. His first play, *The Man Who Had All the Luck* (1944), received only five performances, and his first success was with *All My Sons* (1947; London, 1948), an exposure of wartime profiteering. Miller's next play, *Death of a Salesman* (N.Y. and London, 1949), received the PULITZER PRIZE for Drama. Containing some of his best work, it depicts the destructive power of illusion, as evinced by the refusal of the leading character, the salesman Willy Loman, to face the failure of his career and family relationships. In 1950 Miller prepared a new translation of *An Enemy of the People* by IBSEN, a dramatist by whom he is believed to have been much influenced. It was followed by *The Crucible* (1953), a powerful play dealing with the witch-trials of the 17th century, when innocent people were condemned on the evidence of unbalanced, hysterical accusers; it was widely interpreted as an attack on the political witch-hunting then prevalent in the U.S.A. Two one-act plays, *A View from the Bridge* and *A Memory of Two Mondays*, were seen in 1955. Rewritten in three acts, *A View from the Bridge* was produced in London in 1956, as was *The Crucible*, directed by George DEVINE at the ROYAL COURT THEATRE. It was widely believed, though Miller denied it, that his marriage to the film star Marilyn Monroe, which ended in divorce, provided the background to his next play *After the Fall*, the opening production of the Vivian Beaumont Repertory Company at the WASHINGTON SQUARE THEATRE in 1964. Later in the same year the company appeared in *Incident at Vichy*, about a group of men undergoing examination by the Nazis to see if they are Jews, which was seen in London in 1966. Neither of these plays pleased the critics, and in 1968 Miller returned to the scene of his earlier successes—American family life—with *The Price*, seen in London in 1969. His later works—*The Creation of the World and Other Business* (1972), about Adam and his family, *The Archbishop's Ceiling* (1977), not given a New

York production, and *The American Clock* (1980), set in the Depression—have found little favour.

MILLER, HENRY (1860–1926), British-born American actor and manager, who made his début in Toronto in 1878, and then went to the United States, where after touring with many leading actresses of the time and with BOUCICAULT he became leading man of the EMPIRE THEATRE stock company in New York. In 1899 he appeared as Sydney Carton in *The Only Way*, an adaptation by Wills and Langbridge of DICKENS's *A Tale of Two Cities* which had a long run, and in 1906 he went into management, producing and playing in William Vaughn MOODY's *The Great Divide*, in which he was also seen in London the same year. Among his later successes were Charles Rann Kennedy's *The Servant in the House* (1908) and Moody's *The Faith Healer* (1910). In 1916 he opened HENRY MILLER'S THEATRE in New York; he died during rehearsals for Dodd's *A Stranger in the House*. He married Bijou, the daughter of the actress Matilda HERON, and their son Gilbert Heron Miller (1884–1969) was also a theatre manager in London and New York, fostering the exchange of new plays across the Atlantic. From 1918 until its demolition in 1958 he was lessee of the ST JAMES'S THEATRE in London.

MILLER, JONATHAN WOLFE (1934–), English director, who qualified as a doctor at CAMBRIDGE, where he also acted in several Footlights revues. After the unexpected success at the EDINBURGH FESTIVAL of the 1960 revue *Beyond the Fringe*, of which he was co-author, he appeared in *Beyond the Fringe* in London and New York. The first play he directed was John OSBORNE's *Under Plain Cover* (1962) at the ROYAL COURT THEATRE. He then returned to New York, where he directed Robert Lowell's triple bill *The Old Glory* (1964) and Minoff and Price's *Come Live With Me* (1967), returning to London later in 1967 to direct *Benito Cereno*, based on a story by Herman Melville, for the MERMAID THEATRE. Among his later productions were *King Lear* in 1969 at Nottingham, with Michael HORDERN, and *The Merchant of Venice* (1970), with OLIVIER as Shylock, at the NATIONAL THEATRE, where he directed BÜCHNER's *Danton's Death* in 1971 and SHERIDAN's *The School for Scandal* in 1972. In the latter year he also directed *The Taming of the Shrew* for the CHICHESTER FESTIVAL THEATRE, following it with CHEKHOV's *The Seagull* in 1973. Appointed associate director of the National Theatre, 1973–5, he directed BEAUMARCHAIS's *The Marriage of Figaro, Measure for Measure*, and Peter NICHOLS's *The Freeway* (all 1974). He has directed several productions at GREENWICH, including *Hamlet* and IBSEN's *Ghosts* in 1974, WILDE's *The Importance of Being Earnest* in 1975, and ETHEREGE's *She Would If She Could* in 1979. In 1976 he was responsible for a memorable production of Chekhov's *Three Sisters* at the CAMBRIDGE THEATRE. He has also directed several opera productions.

MILLER, MARILYN, see MUSICAL COMEDY and REVUE.

MILLO, MATTEI, see ROMANIA.

MILLS, FLORENCE, see LONDON PAVILION and REVUE.

MILTON, ERNEST (1890–1974), American-born actor who spent most of his life in England. He made his first appearance in the U.S.A. in 1912, and was first seen in London in Montague Glass and Charles KLEIN's *Potash and Perlmutter* (1914) and *Potash and Perlmutter in Society* (1916). After some varied experience, during which he played Oswald in IBSEN's *Ghosts* and Marchbanks in SHAW's *Candida*, he joined the company at the OLD VIC in 1918, and then and on subsequent visits gave interesting and highly idiosyncratic performances of many of Shakespeare's leading characters, especially Hamlet, Shylock, Macbeth, and Richard II. He was at his best in roles which had something fantastic or sinister about them, and was admirable as PIRANDELLO's Henry IV in 1925, a part he played again in 1929 when the play was renamed *The Mock Emperor*. He was also excellent as Channon in ANSKY's *The Dybbuk* (1927) and as Rupert Cadell in Patrick Hamilton's *Rope* (1929), a part he played in New York later the same year when it was retitled *Rope's End*; he was a gaunt and somewhat frightening Pierrot in Laurence HOUSMAN's *Prunella* (1930), and gave a memorable interpretation of Lorenzino de' Medici in *Night's Candles* (1933), an adaptation by May Agate of Alfred de MUSSET's *Lorenzaccio*. Also in 1933 he was seen in New York in *The Dark Tower* by George S. KAUFMAN and Alexander WOOLLCOTT, and after returning to London appeared in the title roles of his own play *Paganini* and *Timon of Athens* (both 1935). After playing King John for the Old Vic at the NEW THEATRE in 1941 he was less often seen in the West End, though he gave some interesting performances elsewhere, including Sir Giles Overreach in MASSINGER's *A New Way To Pay Old Debts* at the CITIZENS' THEATRE, Glasgow, in 1944. At the LYRIC THEATRE, Hammersmith, he was seen as Lorenzo Querini in HOCHWÄLDER's *The Strong Are Lonely* (1955) and as Pope Paul in MONTHERLANT's *Malatesta* (1957). He joined the ROYAL SHAKESPEARE COMPANY in 1962 to play the Cardinal in MIDDLETON's *Women Beware Women*, and in 1963 was seen at the MERMAID THEATRE as Bishop Tibon in CAMUS's *The Possessed*, based on DOSTOEVSKY.

MILWAUKEE REPERTORY THEATRE, Milwaukee, Wisconsin, founded in 1954 as the Fred Miller Theatre, was originally a community theatre using local actors supplemented by stars.

Gradually a fine company evolved which in 1969 moved into a new civic centre home, the Wehr Theatre, which seats 504 and has a thrust stage. The company's programme covers a wide range, including Shakespeare and other classics (*Othello*, 1967, *King Lear*, 1975, JONSON's *Volpone*, 1977), European plays such as IBSEN's *A Doll's House* (1971) and MOLNÁR's *The Play's the Thing* (1973), and established American plays (Tennessee WILLIAMS's *A Streetcar Named Desire*, 1967, KAUFMAN and HART's *You Can't Take It With You*, 1970). The plays of contemporary authors are also presented, among them David RABE's *Sticks and Bones* (1973), Sam SHEPARD's *La Turista* (1974), and Peter NICHOLS's *A Day in the Death of Joe Egg* (1975) and *The Freeway* (1978), the last receiving its American première at the theatre. In 1973 a second, experimental theatre, the Court Street, was established, and since that year new plays have been fostered and their authors given financial support.

MIME, in Ancient Rome a spoken form of popular, farcical drama which, unlike the FABULA *atellana* and the silent acting of the PANTOMIMUS, was played without masks. Among the Dorians of Sicily and south Italy the mime was popular from an early period. Linguistic barriers were no hindrance to the art of gesture and facial expression, and the simple requirements of a troupe of strolling mime-players—a platform and a curtain—could be found anywhere. Soon after the introduction of Greek drama to the Roman stage in the mid-3rd century BC the Floralia, or festival of Flora, was founded, and this became a favourite occasion for the performance of mimes, where licence even went so far as to sanction the appearance of mime-actresses naked on the stage. The influence of mime on the development of Latin comedy must have been considerable; much of the jesting and buffoonery which PLAUTUS introduced into his adaptations of Greek comedy would have been quite appropriate in the mime.

Even at its highest level the mime differed from more conventional forms of drama by its preoccupation with character-drawing rather than plot, a necessary consequence of its more or less improvised nature. The distinctive costume of the mime-player was a hood or *ricinium*—whence the name *fabula riciniata* for a mime—which could be drawn over the head or thrown back, a patchwork jacket, tights, and the phallus; the head was shaven and the feet bare. The companies were small; Ovid speaks of a cast of three to take the roles of the foolish old husband, the erring wife, and the dandified lover, who, unlike the comic *stupidus* or clown in his mime costume, was probably dressed as a man of fashion. A terracotta lamp from Athens also shows three *mimi* acting *The Mother-in-Law*. The plots were simple, the endings often abrupt. When the climax was reached, the curtain (*siparium*) would descend, and the play was over.

In the 1st century BC the mime achieved a precarious status as a literary form when Decimus LABERIUS and later PUBLILIUS SYRUS adopted it and wrote fixed scripts for mime-players. The fragments of their works that survive show that they had much in common with the *fabulae atellanae* which the new mime-plays seem to have absorbed or replaced. But the popular mimes which all but drove other forms of spoken drama from the stage under the Roman Empire were sub-literary, unmetrical, and largely impromptu, with dialogue in prose which the chief actor—the *archimime*—was free to cut or expand at will. The sordid themes and startling indecency of the language, judged by some later fragments which have survived, seem to be characteristic of the mime in general. Not only was adultery a stock theme, but the Emperor Heliogabalus appears to have ordered its realistic performance on stage, and if the plot included an execution it was possible, by substituting a condemned criminal for the actor, to give the spectators the thrill of seeing the execution actually take place. It is hardly surprising that however popular the actors were, socially they were beneath contempt. They countered the opposition of the Christian Church by mocking Christian sacraments; but gradually the Church got the upper hand and in the 5th century succeeded in excommunicating all performers in mime. In the 6th century the emperor Justinian closed all the theatres. Yet the mime lived on. Its simple requirements could be supplied in any public place or private house, and in such settings it continued to entertain audiences who were now officially Christian. Though forced to drop its habit of burlesquing the sacraments, it still scandalized the Fathers of the Church by its indecency and the immorality of its performers.

How far the mime survived the fall of Rome and the onslaught of the barbarians is doubtful. So simple a type of performance might arise independently at different ages and in different countries; yet precisely because of its primitive character it is hard to be sure that the classical mime ever became wholly extinct in Europe. The Middle Ages had their mime-players, who may have taken over from the *mimi*, those last representatives of classical drama, something of their traditions, and handed them on to their descendants in the modern world.

MIME, Modern, of which the outstanding exponent is Marcel MARCEAU, has nothing in common with the Roman mime (above), and approximates far more closely to the art of the Roman PANTOMIMUS, being entirely dependent on gesture and movement, usually accompanied by music, but wordless. The word PANTOMIME has now taken on such an entirely different meaning from its original one that modern exponents of dumb-show call their work 'mime' to differentiate it from the popular English Christmas show. There is nothing intrinsically new in the use of mime in entertainment; the actors of the COMMEDIA DELL'ARTE relied heavily upon it and it is an essential part of ballet. It should be part of any actor's equipment, and is now taught in most drama schools. The work of DEBURAU, which culminated in the famous mime-play *L'Enfant prodigue*, popularized mime in France in the 19th century and the vogue for it spread to England, Again a more general interest in the subject was aroused by the performance of Jean-Louis BARRAULT as Deburau in the famous film *Les Enfants du Paradis* (1945); but in spite of the fact that Marceau can command enthusiastic audiences for his mime-plays wherever he goes, and that there are now a number of pure mimes, pupils of both Barrault and Marceau whose performances are generally acceptable, mime remains, at least in England, a minority interest.

MINACK THEATRE, Porthcurno, Cornwall. This open-air theatre, hewn out of the granite cliffs near Penzance, was privately owned by Miss Rowena Cade. Work on it began in 1932, and from 1933 to 1939 a play was produced there every two years, always by local amateur companies. During the Second World War the theatre was covered with barbed wire, but productions began again in 1949 with a performance of EURIPIDES' *Trojan Women* by pupils from two Penzance schools. From 1951 to 1954 the theatre was run by the Cornwall branch of the National Council of Social Service, but heavy losses were made and it was returned to its owner. In 1959 a Minack Theatre Society was formed to assist and publicize the venture which has since flourished with weekly productions from June to September by companies from all over Britain, plays ranging from Shakespeare and Restoration comedy to the works of such writers as ANOUILH, Christopher FRY, and Alan AYCKBOURN. Many improvements have been made, and the theatre, which seats 600, is probably one of the finest and most skilfully adapted open-air playhouses in existence.

MINOTIS, ALEXIS (1900–), Greek actor and director, who made his first appearance on the stage in 1925, and in 1930 made his New York début as Orestes in a Greek production of EURIPIDES' *Electra*. Returning to Greece, he played a wide range of ancient and modern parts for the National Theatre, with which he was associated from 1931 to 1967, succeeding Fotos POLITIS as director and directing many of the plays as well as acting in them. He was first seen in London in 1939, when he played Hamlet. Among his other Shakespearian roles have been Shylock in *The Merchant of Venice*, King Lear, and Richard III. Since 1955 he has been responsible for the production of many classical tragedies in the open-air theatre at EPIDAUROS. Among the modern plays with which he has been connected are FRY's *The Dark is Light Enough* in 1957 and DÜRRENMATT's *The Visit* and *The Physicists* in 1961 and 1963 respectively. He appeared again in London during the 1966 WORLD THEATRE SEASON, playing Oedipus to the Jocasta of his wife Katina PAXINOU in SOPHOCLES' *Oedipus Rex*, Talthybius to her Hecuba in Euripides' play of that name, and Oedipus again in Sophocles' *Oedipus at Colonus*. From 1967 to 1974 he ran his own company, presenting mainly modern European plays, returning to the National Theatre, as General Director, from 1974 to 1980. In 1981 he gave a memorable performance as the Cardinal of Spain in MONTHERLANT's play.

MINSKOFF THEATRE, New York, on the third floor of 1 Astor Plaza, on Broadway between 44th and 45th Streets, the site of the former Astor Hotel. This theatre was designed under new building code laws which allowed theatres to be built off the street level and inside another type of structure. Named after its managers, and seating 1,621, it opened on 13 Mar. 1973 with a revival of the musical comedy *Irene*, first produced in 1919, which was a big success. Later productions have included *Clarence Darrow*, by David W. Rintels, an all-Black version of *Hello, Dolly!* (both 1975), and in 1980 a revival of *West Side Story*.

MINSTREL, MÉNESTREL, a term dating from the 12th century, derived from the Latin *minister* (servant, official), and originally used to distinguish those performers who were in the regular employment of a particular lord from their itinerant fellows the JONGLEURS, but later descriptive of both. The origin of these professional entertainers, who flourished from the 11th to the 15th centuries, must be sought in the fusion of the Teutonic *scôp*, or bard, with the floating débris of the Roman theatre, particularly the *mimus* (see MIME). The process went on obscurely from the 6th to the 11th century, helped by the GOLIARDS, wandering scholars who brought to the mixture a measure of classical erudition. Dressed in bright clothes, with flat-heeled shoes, clean-shaven face, and short hair—legacies of Rome—and with their instruments on their backs, they tramped, alone or in company, all over Europe, often harassed by the hostility of the Church and the restrictions of petty officialdom. In spite of this, they enlivened the festivities of religious fraternities, and performed wherever they could gather an audience—in noblemen's halls, in market places, along pilgrim routes—and kept alive many traditions handed down from Greece and Rome. They may even have had a share in the development of LITURGICAL DRAMA. Even before guilds were organized to regulate the profession, in France in 1321 and in England in 1469, different grades of minstrels had evolved. At the top were the individual, learned poet-musicians attached permanently to royal and noble households, like Richard I's Blondel; then came those who (often in large groups) presented more or less literate fare—plays, songs, debates on learned topics, and short farces; finally came those who presented the full range of popular entertainments to mass audiences, the ancestors of the circus and fairground showmen of later years.

MINSTREL SHOW, an entertainment which originated in the Negro patter songs of T. D. RICE (known as Jim Crow), and from his burlesques of Shakespeare and opera, to which Negro songs were added. From 1840 to 1880 the Minstrel Show was the most popular form of amusement in the United States, whence it spread to England. Unlike the MUSIC-HALL, which was intended for adults only, it was essentially a family entertainment, given in a hall and not in a theatre. The performers were at first white men with their faces artificially blacked, whence the name Burnt-cork Minstrels, but later they were true Negroes. Sitting in a semi-circle with their primitive instruments, banjos, tambourines, one-stringed fiddles, bones, etc., they sang plaintive coon songs and sentimental ballads interspersed with soft-shoe dances and outbursts of back-chat between the two 'end-men', Interlocutor and Bones. Their humour was simple and repetitive, and after a great burst of popularity the Minstrels gradually faded away, some, like Chirgwin and Stratton, to the music-halls, some to stroll along the beach at seaside resorts in the summer in traditional minstrel costume—tight striped trousers and waistcoat and tall white hat or straw boater—singing and playing their banjos. Among the most famous troupes were the Christy Minstrels, the Burgess and Moore, and the Mohawks. The original formula of the Burnt-cork Minstrels was successfully revived by B.B.C. Television in the late 1940s with the Kentucky Minstrels and in the 1960s with the Black and White Minstrels, who as a live show had a long run at the VICTORIA PALACE in London from 1962 onwards.

MIRACLE PLAY, in English medieval times a synonym for MYSTERY PLAY; in France the term was used for plays in which the central incident was a miraculous intervention in human affairs by the Virgin Mary or one of the saints, St Nicholas being a favourite subject. The chief surviving examples of these anonymous SAINT PLAYS, as they were called in England, are found in a 14th-century collection of forty *Miracles de Notre-Dame par personnages*.

MIRA DE AMESCUA, ANTONIO (*c.* 1574–1644), Spanish dramatist, whose versatility enabled him to write religious and historical plays, as well as comedies of local life. Stylistically, particularly in his comedies, he stands midway between the simple clarity of Lope de VEGA and the complexity of CALDERÓN. He is best remembered for his theological play *El esclavo del demonio* (*The Devil's Slave*, 1612), on a story which parallels that of FAUST, but where damnation is tempered by the doctrine of ultimate redemption which characterizes much of his religious work.

MIRANDA, FRANCISCO DE SÁ DE (1485–1558), Portuguese dramatist, approximately contemporaneous with Gil VICENTE. During his travels in Italy, from which he returned in 1526, he learned to admire and imitate the classic forms of poetry and drama, and his two neo-classical comedies, *Estrangeiros* and *Vilhalpandos* (both written between 1528 and 1538), laid the foundations of the classic theatre in Portugal. Unlike the plays of Vicente they were logically shaped and divided neatly into acts and scenes, with some attention to the UNITIES, and they helped to give the drama in Portugal a sense of dramatic form. Sá de Miranda wrote at least one tragedy, *Cleopatra*, but only a few lines of it have survived.

MIRBEAU, OCTAVE (1848–1917), French novelist, critic, and dramatist, whose plays deal with contemporary problems in the style of Henri BECQUE, as in *Les Mauvais Bergers* (1897), a study of the struggle between labour and capital, and *Le Foyer* (1908), which examines the ambiguous motivations of official charity. His best play, and the only one now remembered, was *Les Affaires sont les affaires* (1903), a mordant satire on the big business man who is a slave to his wealth. As *Business is Business*, this was seen in New York in 1904 and in London a year later.

MIRODAN, AL(exandru), see ROMANIA.

MIRREN, HELEN (1946–   ), English actress, whose early appearances with the NATIONAL YOUTH THEATRE led to her playing an extremely youthful Cleopatra, in *Antony and Cleopatra*, at the OLD VIC in 1965. Two years later she joined the ROYAL SHAKESPEARE COMPANY and remained with it for five seasons, her roles during that time including Cressida in *Troilus and Cressida*, Ophelia in *Hamlet*, Tatyana in GORKY's *Enemies*, and the title role in STRINDBERG's *Miss Julie*. After working with the INTERNATIONAL CENTRE OF THEATRE RESEARCH she played Lady Macbeth for the R.S.C. in 1974, and a year later was seen at the ROYAL COURT in David HARE's *Teeth 'n' Smiles*, in which she gave an outstanding performance as an alcoholic singer with a rock group. In complete contrast were her excellent Nina in CHEKHOV's *The Seagull* and Ella in Ben TRAVERS's farce *The Bed Before Yesterday*, both seen at the LYRIC THEATRE in 1975. She returned to the R.S.C. in 1977 to play in the three parts of *Henry VI*, and was seen as Isabella in *Measure for Measure* at the RIVERSIDE STUDIOS in 1979 and in the title role of WEBSTER's *The Duchess of Malfi* with the ROYAL EXCHANGE THEATRE company in 1980 (ROUND HOUSE, 1981). In 1981 she also appeared in Brian FRIEL's *Faith Healer* at the Royal Court Theatre.

MIRROR THEATRE, London, see HOLBORN THEATRE.

MISS KELLY'S THEATRE, London, see ROYALTY THEATRE (2).

MISRULE, Abbot, King, or **Lord of**, a special officer (*dominus festi*) appointed to oversee the Christmas entertainments at Court and elsewhere in England in the late 15th and early 16th centuries. In Scotland he was also known as the Abbot of Unreason. His appointment was temporary until he was replaced by the MASTER OF THE REVELS, who became a permanent Court official. Lords of Misrule were common in the colleges of the universities, particularly at Merton and St John's, Oxford, where appointments were made as late as 1577. They are also found in the Inns of Court, intermittently until the time of Charles II. It was at the Gray's Inn Christmas Revels of 1594 that Shakespeare's *Comedy of Errors* was first given. The Lord of Misrule is a direct descendant of the 'king' in the FEAST OF FOOLS, and the equivalent in adult circles of the BOY BISHOP of the choir and other schools.

MISTINGUETT [Jeanne-Marie Bourgeois] (1875–1956), Flemish-born French music-hall artist, with beautiful (and highly insured) legs, mischievous good looks, and a smart line in repartee. 'Miss', as she was nicknamed, was a symbol of Paris to millions of tourists, and even to Parisians themselves, for over half a century. In her early years she was an eccentric comedienne, specializing in character sketches of low-life Parisian women, but, reversing the usual pattern, she later became almost exclusively a dancer and singer, appearing first at the MOULIN ROUGE, of which she was for some time part-proprietor, and, with Maurice Chevalier as her partner, at the FOLIES-BERGÈRE, where she was seen in some sensational dances, descending with superb panache vast staircases, wearing enormous hats and trailing yards of feathered train. A hard worker, with a compelling personality, she became world famous in spite of hardly ever appearing outside Paris, her one appearance in London being at the London Casino (later the PRINCE EDWARD THEATRE) in 1947 at the age of 72.

MITCHELL, LANGDON ELWYN (1862–1933), American poet and playwright, who in 1899 dramatized Thackeray's *Vanity Fair* for Mrs FISKE under the title *Becky Sharp*. It was a great success, and was frequently revived. His finest play, however, was *The New York Idea* (1906), a satire on divorce, also produced by Mrs Fiske with a remarkable cast. It was translated into German (directed by Max REINHARDT), and into Dutch, Swedish, and Hungarian. In the same year Mitchell translated *The Kreutzer Sonata* from the Yiddish of Jacob GORDIN, and in 1916 dramatized Thackeray's *Pendennis* for John DREW. As a successful playwright, he was invited to lecture at the University of Pennsylvania, and in 1928 became the first Professor of Playwriting there. He always retained his interest in the professional theatre, and continued to write for it, but his high standards made him very self-critical, and he left a number of unfinished plays in manuscript.

MITCHELL, MAGGIE [Margaret] JULIA (1832–1918), American actress, who first appeared on the stage at 13, playing children's parts at the BOWERY THEATRE and being particularly admired as DICKENS's Oliver Twist. After some years on tour she became leading lady of BURTON's company in 1857 and in 1861 played the part with which she is mainly associated—Fanchon in an adaptation of George Sand's novel *La Petite Fadette*. Although she was good in other parts, notably as Charlotte Brontë's Jane Eyre and as Pauline in BULWER-LYTTON's *The Lady of Lyons*, she was constantly forced by an admiring public to return to Fanchon, which she played for over 20 years, remaining to the end a small, winsome, sprite-like figure, overflowing with vitality.

MITCHELL, ROY, see HART HOUSE THEATRE.

MITCHELL, WILLIAM (1798–1856), British-born actor and theatre manager, who had had a good deal of experience before in 1836 he emigrated to New York, where he took over the OLYMPIC THEATRE (1), opening it with Townley's *High Life Below Stairs*, and embarking on a long series of excellent productions which brought the theatre, known as Mitchell's Olympic, fame and prosperity. He was himself an excellent actor, being particularly admired as Vincent Crummles in a dramatization of DICKENS's *Nicholas Nickleby* and as Hamlet in John Poole's burlesque *Hamlet Travestie*. He retired in 1850.

MITCHELL'S OLYMPIC, New York, see OLYMPIC THEATRE (1).

MITRA, SOMBHU, see INDIA.

MITZI E. NEWHOUSE THEATRE, New York, see VIVIAN BEAUMONT THEATRE.

MNOUCHKINE, ARIANE, see SOLEIL, THÉÂTRE DU.

MOBERG, VILHELM (1898–1973), Swedish novelist and dramatist. His best work for the stage has often been adapted from his own novels, such as *Rid i natt* (*Ride Tonight*, 1941), a historical drama, and *Mans Kvinna* (*Fulfilment*, 1943), about the conflict between sexual and social morality. His original work for the stage ranges from farce to biblical drama, and includes the rustic tragedy *Hustrun* (*The Wife*, 1929) and *Vär ofödde son* (*Our Unborn Son*, 1945), about the the effect of an illegal abortion on the guilt-ridden mother.

Moberg's work shows a deep concern for traditional social values, and is immensely popular in Sweden.

MOCHALOV, PAVEL STEPANOVICH (1800–45), Russian tragic actor, the leading exponent in Moscow of the 'intuitive' school of acting. He made his début in 1817, and was soon playing leading roles, his finest parts being Hamlet, King Lear, and the heroes of SCHILLER's *Die Räuber* and *Kabale und Liebe*. With great gifts of temperament and passion, Mochalov was antipathetic to any rational method, relying entirely on inspiration and so being extremely uneven in his acting and in his effect on his audiences. With the advance of realism his romantic appeal became out of date, and when heavy drinking impaired his powers of concentration and weakened the flow of inspiration, he had no technique to fall back on. His influence persisted for a time, but was eventually to succumb to the reforms of SHCHEPKIN, which he had resisted to the end.

MODENA, GUSTAVO (1803–61), Italian actor, accounted one of the finest of his day. The son of Giacomo Modena (1766–1841), who in his youth was well known for his excellent acting in GOLDONI's comedies, he first came into prominence in 1824, when he played David to his father's Saul in ALFIERI's tragedy of that name with the travelling company of the actor-manager Salvatore Fabbrichesi (1760–1827). He soon established himself as a leading actor in tragedy, comedy, and romantic drama, and in 1829 formed his own company with which he toured Italy. During the 1830s he became involved in revolutionary activities and left Italy, ending up in London, where he supported himself by teaching Italian and giving Dante recitals. He returned to Italy in 1839 and acted where he could, until in 1843 he was able to form a company of young actors in Milan which included the 14-year-old SALVINI. He was now at the summit of his profession, and in 1847 he produced *Othello*, the first play by Shakespeare to be seen in Italy. It was not a great success, but undoubtedly paved the way for the later triumphs in the part of Salvini and Ernesto ROSSI, also his pupil. He made his last appearance in 1860, having done much to forward the reform of the Italian stage and give it a social purpose. Though not handsome, he had a fine figure and a nobly rugged and expressive face.

MODJESKA [Modrzejewska, *née* Opid], HELENA (1840–1909), Polish actress, who in 1860 eloped with an actor named Gustave Zimajer, and with him joined a company in Cracow, adopting the name by which she is best known and retaining it after her second marriage in 1868 to Charles Chłapowski. In the same year she moved to Warsaw, where for eight years she played a wide range of parts, from classical tragedy to the contemporary productions in the style of SARDOU. In 1876 she emigrated to the United States, making her first appearance there a year later in San Francisco, and scoring an immense success, in spite of her poor command of the English language. She had a great love for Shakespeare, having 14 parts in his plays in her repertory of 260 roles, some of which she had already played in Warsaw. She first appeared in London from 1880 to 1882, and again in 1890, and in 1881 achieved her ambition of playing Shakespeare in English in London, when after intensive study of the language she appeared at the ROYAL COURT THEATRE as Juliet to the Romeo of FORBES-ROBERTSON. One of her finest parts was Lady Macbeth, particularly in the sleep-walking scene. She toured extensively in America and Europe, and before she gave her farewell performance at the Metropolitan Opera House in New York in 1905 she had achieved an outstanding reputation as one of the great international actresses of her time, at her best in tragic or strongly emotional parts.

MOGUL SALOON, London, see MIDDLESEX MUSIC-HALL.

MOHOLY-NAGY, LASZLO, see SCENERY.

MOHUN, MICHAEL (*c.* 1620–84), English actor, who as a boy acted under William BEESTON, and was playing adult parts when the theatres closed in 1642. He then joined the Royalist army, and returned to the stage in 1660. With Charles HART, he was leading man of KILLIGREW's company in 1662 when it took possession of the first DRURY LANE theatre. His Iago in *Othello* was much admired, and he created many roles in Restoration tragedy, including Abdelmelech in DRYDEN's *The Conquest of Granada* (1670) and the title role in LEE's *Mithridates, King of Pontus* (1678).

MOISEIWITSCH, TANYA (1914–   ), British stage designer, daughter of the pianist Benno Moiseiwitsch and his first wife, the violinist Daisy Kennedy. She laid the foundations of her career at the ABBEY THEATRE in Dublin, where between 1935 and 1939 she was responsible for the settings of more than 50 productions. After working as resident designer for the OXFORD PLAYHOUSE from 1941 to 1944 she joined the OLD VIC, and in 1949 moved to Stratford-upon-Avon, where her first work for the SHAKESPEARE MEMORIAL THEATRE was a permanent set with no front curtain for *Henry VIII*. She also provided the settings for a cycle of history plays performed there in 1951 for the Festival of Britain, and two years later was in Canada, where she designed the first, tentlike theatre for the STRATFORD (ONTARIO) FESTIVAL and also the settings for its first productions, *Richard III* and *All's Well That Ends Well*. Her association with the theatre continued, and she designed over 20 of its productions. She was closely associated also with the GUTHRIE THEATRE in Minneapolis from its opening season in 1963, to which she contributed the settings for *Hamlet*, MOLIÈRE's *The Miser*, and CHEKHOV's *Three Sisters*, until 1967. Her work for the NATIONAL THEATRE in London has included settings for revivals of JONSON's *Volpone* in 1968, Molière's *The Misanthrope* in 1973 (seen in New York in 1975), and CONGREVE's *The Double Dealer* in 1978. She has also been responsible for the décor of a number of West End productions, among them the musical *Bless the Bride* (1947), R. C. SHERRIFF's *Home at Seven* (1950), RATTIGAN's *The Deep Blue Sea* (1952), and John MORTIMER's *The Wrong Side of the Park* (1960).

MOISSI, ALEXANDER (1880–1935), German actor of Italian extraction, born in Albania where a drama school has been named after him. He played his first speaking part in German at Prague in 1902, and remained there until 1905, when he went to BERLIN. There his fine presence, and above all his rich, musical, speaking voice, soon brought him into prominence. At the DEUTSCHES THEATER under REINHARDT he played a number of Shakespearian parts, including Romeo, the Fool in *King Lear*, Oberon in *A Midsummer Night's Dream*, and Touchstone in *As You Like It*. He was also seen as both Faust and Mephistopheles in GOETHE's *Faust*; as Posa in SCHILLER's *Don Carlos*; as Oswald in IBSEN's *Ghosts*; as Louis Dubedat in SHAW's *The Doctor's Dilemma*, and as Marchbanks in his *Candida*. His playing of SOPHOCLES' Oedipus and AESCHYLUS' Orestes in productions by Reinhardt in Vienna was accounted outstanding. In 1930 he visited London, appearing in a German version of *Hamlet* at the GLOBE THEATRE while GIELGUD was playing it in English at the QUEEN'S THEATRE next door.

MOLANDER, OLOF (1892–1966), Swedish director and manager and member of a notable theatrical family in Sweden. He joined the KUNGLIGA DRAMATISKA TEATERN as an actor in 1914, subsequently becoming a director and head of the theatre. In the 1920s he mounted a series of productions of the classics in the REINHARDT tradition which, together with his later productions of STRINDBERG, made him to a great extent the architect of Sweden's post-naturalistic style of theatre.

MOLÉ [Molet], FRANÇOIS-RENÉ (1734–1802), French actor, who first appeared at the COMÉDIE-FRANÇAISE in 1754, and after being sent into the provinces to gain experience, was received into the company in 1760, to play young heroes and lovers. After a few successes, often in parts specially written for him, he became the idol of the public, and when GRANDVAL retired shared leading roles with BELLECOUR. He excelled in comedy, and was usually content to leave the tragic roles to others, but ironically he was destined to be the first French Hamlet, in 1769, in an adaptation made by DUCIS. This was nothing like Shakespeare's play, but the French liked it, and Molé was considered good in the part. In later life he became fat, but lost nothing of his agility of mind and body. A contemporary said of him that his wit was so quick, and his fooling so excellent, that he was capable of playing a part he did not know, relying entirely on the prompter without the audience realizing it. An adherent of the Revolution, he was not imprisoned with the other actors in 1793, going instead to play revolutionary drama at the theatre of Mlle MONTANSIER. When the Comédie-Française reopened in 1799, he returned with his wife, who was also a member of the company, and remained there until his death.

MOLIÈRE [Jean-Baptiste Poquelin] (1622–73), French dramatist and actor, author of some of the finest comedies in the history of the theatre. He was the eldest son of a prosperous upholsterer of Paris attached to the service of the king. Little is known of his early years, though he is reported to have been an excellent mimic and a great frequenter of theatres. He probably saw BELLEVILLE at the Hôtel de BOURGOGNE, while MONTDORY at the MARAIS introduced him to the tragedies of CORNEILLE. In about 1631 he became a pupil at the Jesuit College of Clermont (later Louis-le-Grand), and left in 1639 a good scholar, particularly in Latin, and an omnivorous reader of poetry and plays. While at school he probably took part in the productions which made up the repertory of the scholastic JESUIT DRAMA. Among his fellow students was the young Prince de Conti, who much later became his patron, and it may have been at Clermont that he met CYRANO DE BERGERAC, a life-long friend and champion.

On leaving school young Poquelin, who was supposed to inherit his father's position at Court, made a half-hearted attempt to study law, and in 1642 went to Narbonne in the suite of Louis XIII, a move which may have been designed to break up a friendship he had formed with a family of actors, the BÉJARTS. If so, it was unsuccessful, for a year later the young man renounced his succession, and with the Béjarts and some of their actor-friends formed a small theatrical company, known as the ILLUSTRE-THÉÂTRE, which took over and adapted a disused tennis-court. The driving force behind this enterprise was undoubtedly Madeleine Béjart, the eldest of the family and already an experienced actress. Molière, as he had now become—no reason for the adoption of this name has yet been found—was merely a fledgling actor, whose aspirations tended towards the playing of tragedy for which he was physically and temperamentally unsuited. It has been surmised that at this time he was the lover of Madeleine, a relationship which was to have tragic repercussions in his later life.

The new company failed completely and Molière was imprisoned for debt. On his release he had some of his former companions gathered together again and set off to act in the provinces, remaining there from 1645 to 1658. These were the formative years of Molière's career, during which he developed his powers as an actor and also supplied the company, of which he soon became the virtual leader, with partly improvised farces in the style of the COMMEDIA DELL'ARTE. These appear to have been successful, though nothing of them survives except a few titles, and stray fragments used in later plays. Little is known, too, of the company's wanderings beyond a list of the towns they visited, under the patronage successively of the Duc d'Épernon and the Prince de Conti, but it is evident that they were reputable, usually well lodged and well received; and the troupe was further strengthened by the later accession of the veteran actor DUFRESNE, of Gros-René and his wife Mlle DU PARC, and of Mlle DE BRIE, both excellent actresses.

Encouraged by his success in the provinces, Molière decided to return to Paris under the patronage of Monsieur, the king's brother, and on 24 Oct. 1658—a date memorable in French theatrical history—he appeared at the Louvre before the 20-year-old King Louis XIV and his Court in Corneille's tragedy *Nicomède*. Most of the audience had probably seen the bombastic and ranting MONTFLEURY (who was present at the production) at the Hôtel de Bourgogne in this part, and Molière's quieter and more natural style of acting made little impression on them. For a moment the fate of the company, and of French drama, hung in the balance. Then Molière came forward and in a modest speech asked to be allowed to perform, as Grimarest says: 'One of the trifling entertainments with which they had gained a certain renown in the provinces. He was confident of success in this, for his actors were trained to extemporize short comic pieces in the Italian manner.' *Le Docteur amoureux*, which followed, was an immediate success, and the company was given permission to share the PETIT-BOURBON with the Italian company already established there under FIORILLO. The relationship between the two companies was most cordial, and Molière was always ready to acknowledge how much he had learned from the Italians.

Though they still continued to appear in tragedy, the growing reputation of the 'Troupe de Monsieur', as it was now called, was almost entirely due to their playing of Molière's own comedies *L'Étourdi* and *Le Dépit amoureux*, both of which had already been seen in the provinces. But it was the production of the one-act satire *Les Précieuses ridicules* (1658) which set the seal on their success. As one play followed another—*Sganarelle, ou le cocu imaginaire* (1660), *L'École des maris* (1661), *L'École des femmes* (1662)—tragedy almost fell out of the repertory, and it is significant that Molière's only real failure among his early plays was his one attempt at high comedy in the heroic style, *Dom Garcie de Navarre* (1661).

By the time Molière had been back in Paris for four or five years, his pre-eminence both as a comic actor and as a comic dramatist was virtually unchallenged, and in the ten years that remained before his death from overwork in 1673, discerning critics hailed him as the greatest comic playwright of all time. For most of his contemporaries, however, he remained primarily a hard-working entertainer and manager, one among many, as witness the painting, executed in 1670, which hangs today in the COMÉDIE-FRANÇAISE, and shows Molière among his fellow-actors TURLUPIN, Fiorillo, and others. Nevertheless the king continued to patronise the company, which in 1661 was allowed to move to the PALAIS-ROYAL, and in 1665 became 'the King's Players'. The royal favour had, however, to be paid for in bouts of feverish activity connected with the COMÉDIE-BALLETS for which Molière's services were in constant demand. His main function was to provide and produce a series of these plays with singing and dancing, for which the Court musician Lully composed the music. The chief parts were played by Molière and his actors, while the notables of the Court, including the king himself, danced in the ballets.

The first of these Court entertainments was *Les Fâcheux* (*The Bores*), produced at Vaux in 1661. Two years later Molière launched a biting satire on the actors of the Hôtel de Bourgogne who had sought to undermine his influence at Court by attacking him not only in his professional but also in his private life. It was a vulnerable spot, for in Jan. 1662 Molière had made an injudicious marriage with the youngest sister of Madeleine, Armande Béjart, a spoilt child, only 20 years old, capricious, flirtatious, and apparently quite lacking in affection. The affair was complicated by the belief of most people, even Molière's friends, that Armande was the daughter, and not the sister, of Molière's former mistress. Montfleury, always jealous of Molière, went so far as to accuse him before the king of having married his own daughter. The king's reply to this was not only to commission *L'Impromptu de Versailles*, which ridiculed Montfleury and his company, but also to stand godfather to Molière's first child, born in 1664, who died the same year. A second child, Esprit Magdeleine (1665–1723), was born before the parents separated; a third child, a son born in 1672 after a temporary reconciliation, lived only a few weeks.

Meanwhile, whatever their private matrimonial troubles, the professional life of this ill-assorted pair continued with unabated success, and Armande, who had made her first appearance as Élise in *La Critique de l'École des Femmes*, provided her husband with an excellent leading lady in all his later plays. In 1664, after the production at Court of *Le Mariage forcé*, in which the king danced the part of a gipsy, Molière was put in charge of the six-day entertainment given at VERSAILLES under the collective title of *Les Plaisirs de l'île enchanté*, for which he hurriedly wrote *La Princesse d'Élide*. It was during these celebrations that Molière also produced the first three acts of a play which was to cause him much distress—*Tartuffe*. This attack on hypocrisy roused against him the religious bigots of Paris, who succeeded in getting the play suppressed for many years. Though frequently performed in private houses, it was not seen in public until it was put on at the Palais-Royal on 5 Aug. 1667 as *L'Imposteur*.

Another play which roused almost as much opposition was *Don Juan, ou le festin de pierre*, given at the Palais-Royal in the same year as *L'Amour médecin* (1665). It was followed by *Le Misanthrope* and *Le Médecin malgré lui* (both 1666) and by *Amphitryon* and *L'Avare* (both 1668), while for the Court Molière provided *Le Sicilien, ou l'amour peintre* (1667) and *George Dandin, ou le mari confondu* (1668). These were first seen at Versailles. *Monsieur de Pourceaugnac* (1669) had its first production at Chambord, and *Les Amants magnifiques* (1670) at Saint-Germain. It was in this last that Louis XIV, as Neptune and Apollo, finally ended his theatrical career.

The best known of the Court plays is undoubtedly *Le Bourgeois gentilhomme*, produced at Chambord in 1670. It was written to provide the king with an oriental entertainment, a genre much in vogue at the time. It was followed by *Psyché* (1671), in which Corneille and QUINAULT collaborated, written, it is said, to make use of some elaborate scenery representing Hell, which Louis hated to see lying idle. The last play written for a Court performance was *La Comtesse d'Escarbagnas* (also 1671).

All the Court entertainments, shorn of much of their splendour, had also been given publicly at the Palais-Royal, which in 1671 was rebuilt and enlarged to take the scenery and machines needed for *Psyché*. In the same year Molière reverted to his earlier farcical manner with the light-hearted *Les Fourberies de Scapin*. In 1672 came another of the great satires, *Les Femmes savantes*, followed by *Le Malade imaginaire*, Molière's last play, in which he was acting on the night of his death, 17 Feb. 1673.

Molière's great achievement was that by his own efforts he raised French comedy to the height attained by French tragedy. He rescued it from the domain of farce, and in his hands it became a vehicle not only for social satire, but also for a universally applicable exposure of human folly and pretentiousness. Though he took his material from many sources—Greek, Latin, Italian, Spanish—by far the most important influence on him was that of the popular comic tradition in which he was trained. As a comic actor he soon abandoned the extrovert, active *commedia dell'arte* character of Mascarille (returning to it only once with his successful creation of Scapin) for the more passive and introspective Sganarelle who, after appearing in half a dozen plays under his own name, was capable of elaboration and diversification into such characters as Arnolphe in *L'École des femmes*, Orgon in *Tartuffe*, Alceste in *Le Misanthrope*, or Monsieur Jourdain in *Le Bourgeois gentilhomme*.

At the Comédie-Française, known as 'La Maison de Molière', his plays account for one-seventh of the total performances given between 1680 and 1920; of the ten plays most frequently performed by 1920, eight were by Molière; and *Tartuffe* is the one play that has been given during every year of the theatre's history. Roles like Alceste, Célimène, Tartuffe, Orgon, and Elmire have served to demonstrate the excellence of countless actors and actresses; and in more recent times Louis JOUVET, the outstanding actor and director of Molière in the 20th century, gave memorable performances of the roles of Don Juan, Arnolphe, and Géronte in *Les Fourberies de Scapin*.

Although not easy for English audiences to understand, either in French or in translation, Molière has always been popular in England. His influence was very marked during the Restoration, and most dramatists borrowed freely from him, usually without acknowledgement. It would be impossible to enumerate all the plays based on his works: DRYDEN made use of *L'Étourdi* and *Amphitryon*; SHADWELL of *L'Avare, Psyché*, and *Don Juan*; WYCHERLEY of *L'École des femmes* and *Le Misanthrope*; BETTERTON of *George Dandin*. In 1671 RAVENSCROFT adapted *Le Bourgeois gentilhomme* as *Mamamouchi; or, the Citizen turned Gentleman*, and also made use of *Monsieur de Pourceaugnac* for his own play *The Careless Lovers* (1672). In 1676 OTWAY made a translation of *Les Fourberies de Scapin* as *The Cheats of Scapin*, which was revived in London as late as 1959. In 1717 CIBBER based his successful *The Non-Juror* on *Tartuffe*, and FIELDING in 1732 translated *L'Avare* and *Le Médecin malgré lui*. After a lull during the 19th and early 20th centuries, a fresh impetus to the acting of Molière in English was given by the very free versions of Miles MALLESON, in which he himself appeared. Other good and more scholarly, though perhaps less actable, versions have been made by George Graveley, and by John Wood (in the Penguin Classics series). In 1973 Tony Harrison's version of *Le Misanthrope* was performed at the NATIONAL THEATRE in London with Diana RIGG as a memorable Célimène, and in 1981 a translation by Alan Drury of *Le Malade imaginaire* as *The Hypochondriac* was produced there.

MOLINA, TIRSO DE, see TIRSO DE MOLINA.

MOLNÁR [Neumann], FERENČ (1878–1952), Hungarian playwright, and one of the best known outside his own country. He first attracted attention with some light-hearted farces in which he exploited the humours of Hungarian city life. These were followed by a variation on the theme of FAUST, *The Devil* (1907), which a year later was seen in translation at two theatres in New York simultaneously. His best known play is *Liliom*

(1909), which was a failure when first produced in Budapest but a success on its revival some ten years later. It was seen in New York in 1921 and in London in 1926, and was used as the basis of the successful musical *Carousel* (1945). Among Molnár's later plays, several of which starred his third wife the Hungarian actress Lili Darvas (1902–74), were *The Guardsman* (1910), which later provided an excellent vehicle for the LUNTS in a production by the THEATRE GUILD; *The Swan* (1914), a satirical comedy of Court life; *The White Cloud* (1916), a fantasy; *The Red Mill* (1923), seen in New York in 1928 as *Mima*; and the Cinderella-story of a Budapest servant-girl's romance, *The Glass Slipper* (1924). *The Play in the Castle* (also 1924), seen in New York in 1926 as *The Play's The Thing*, in an adaptation by P. G. Wodehouse, again showed his mastery of sophisticated comedy, as did *Olimpia* (1927), while in *The Good Fairy* (1931) he reverted to his former whimsical vein with the adventures of a romantic usherette in a Budapest theatre. In 1940 Molnár, who had emigrated some years before to the U.S.A., where he died, became an American citizen. His success abroad for a time militated against his acceptance in Hungary, where his work was largely ignored, but after his death a reaction set in; a revival of *Olimpia* at the Madách Theatre in 1965, with Lili Darvas, set the seal on his growing reputation, and many of his plays have since been revived in Budapest.

MOMUS, the Greek god of ridicule, and so by extension of CLOWNS, who appears in the 'secular masque' provided by DRYDEN for a revival of FLETCHER's *The Pilgrim* in 1700. The name was later used frequently to denote a clown, as in GRIMALDI's reference to himself as 'the once Merry Momus', and became attached to one of the characters in the HARLEQUINADE.

MONAKHOV, NIKOLAI FEDOROVICH (1875–1936), Russian actor, who appeared mainly in the theatres of St Petersburg (Leningrad). He began his career in a café-chantant and had been on the stage for 17 years, mostly in operetta, before he joined the company of the Free Theatre founded in 1913 by Konstantin MARDZHANOV, where his versatility was much appreciated. When the company went bankrupt, Monakhov was invited to the MOSCOW ART THEATRE, but preferred to return to operetta in St Petersburg for another five years. After the Revolution he joined the newly founded Leningrad Theatre, making his first appearance there as Philip II in SCHILLER's *Don Carlos* with great success, and remaining to play leading roles with the company until his death.

MONCK, NUGENT, see MADDERMARKET THEATRE, NORWICH.

MONCRIEFF, GLADYS (1893–1976), Australian actress in operetta and musical comedy, affectionately known as 'Our Glad'. She made her début in 1915 in Sydney as Josephine in GILBERT and Sullivan's *H.M.S. Pinafore*, and played the lead in the original Australian productions, as well as in many revivals, of *The Maid of the Mountains*, *The Merry Widow, The Southern Maid, The Chocolate Soldier, The Belle of New York*, and *Rio Rita*.

MONCRIEFF, WILLIAM GEORGE THOMAS (1794–1857), English dramatist and theatre manager, author of about 200 plays, mainly burlesques and melodramas written for the minor theatres. The most successful was *Tom and Jerry; or, Life in London* (1821), the best of many adaptations of the book by Pierce Egan. In 1823 Moncrieff's *The Cataract of the Ganges* was produced at DRURY LANE with the added attraction of real water. He adapted DICKENS's *Pickwick Papers* for the stage as *Sam Weller; or, the Pickwickians* (1837) almost before the last instalment had appeared, which aroused the anger of the author, who satirized him as 'the literary gentleman' in *Nicholas Nickleby* (1838).

MONKHOUSE, ALLAN NOBLE (1858–1936), dramatic critic and dramatist, who was on the staff of the *Manchester Guardian* when Miss HORNIMAN opened her repertory theatre at the GAIETY THEATRE, Manchester. He supported her enterprise in his column and wrote for her his first play, *Reaping the Whirlwind* (1908). This was followed by *Mary Broome* (1911) and *Nothing Like Leather* (1913), a one-act satire on the Gaiety company in which Miss Horniman appeared fleetingly as herself. Monkhouse's best play, produced in London after the Gaiety company had been disbanded, was *The Conquering Hero* (1924). The story of a soldier who goes unwillingly to war, finds humiliation in battle, and returns to the irony of a triumphal welcome, it anticipated SHERRIFF's *Journey's End* (1928).

MONODRAMA, short solo piece for one actor or actress supported by silent figures or by a chorus. It is sometimes referred to as a MELODRAMA on account of its musical accompaniment. It was popularized in Germany between 1775 and 1780 by the actor Johann Christian BRANDES, who developed the lyrical element introduced into German drama by the poet KLOPSTOCK and intensified by the influence of ROUSSEAU's *Pygmalion*. The Duologue, a similar compilation, had two speaking characters. Both types of entertainment, which were useful in filling out the triple bill then in vogue, frequently consisted of scenes extracted and adapted from longer dramas, somewhat in the style of the earlier English DROLL.

MONTAGUE [Mann], HENRY JAMES (1844–78), American actor who appeared in London in the 1860s, and in 1870 opened the VAUDEVILLE THEATRE in partnership with David JAMES and

Thomas THORNE. He returned to New York in 1874, and spent the rest of his short career with Lester WALLACK, his success in such contemporary comedies as *Diplomacy* (Clement SCOTT's version of SARDOU's *Dora*), T. W. ROBERTSON's *Caste*, and Tom TAYLOR's *The Overland Route* being due, perhaps, more to his good looks, youthful assurance, and personal magnetism than to the quality of his acting. In 1875 he founded the LAMBS, in imitation of the London club of which he had been a founder-member.

MONTALAND, CÉLINE (1843–91), French actress, who at the age of seven appeared at the COMÉDIE-FRANÇAISE in children's parts and was much admired. Going from there to the PALAIS-ROYAL she continued to attract the public in childish parts specially written for her and in 1860 made her adult début in the famous fairytale *Pied de mouton* at the PORTE-SAINT-MARTIN. She was also seen at the GYMNASE, where one of her finest parts was the mother in a dramatization of Daudet's *Jack*. In 1888 she returned to the Comédie-Française, remaining with the company until her death.

MONTANSIER [Marguerite Brunet], Mlle (1730–1820), French actress-manageress who, having fallen in love with a young actor, Honoré Bourdon de Neuville (1736–1812), decided to go on the stage. With the help of powerful friends she took over the management of the theatre at Rouen, with such success that she was soon managing several others, Neuville acting as her business manager. While in charge of the theatre at VERSAILLES, she was presented to Marie-Antoinette, who was so charmed with her gaiety and wit that she invited her to act at Court. In 1777 she built a new theatre at Versailles (demolished in 1886) which was used to try out aspirants to the COMÉDIE-FRANÇAISE. On the outbreak of the Revolution she went to Paris and presided over a salon, where Napoleon, then an officer in the artillery, first met TALMA and formed a lasting friendship with him. Accused by FABRE D'ÉGLANTINE of Royalist sympathies because of her appearances at Court, she was arrested, but was saved from the guillotine by the fall of Robespierre. She immediately married Neuville and returned to the theatre she had recently opened at the PALAIS-ROYAL, giving it her own name and remaining there until 1806.

MONTCHRÉTIEN, ANTOINE DE (*c.* 1575–1621), French dramatist, who by 1600 had written several plays on classical themes which were produced at the Hôtel de BOURGOGNE. In 1605, at the height of his success, he had the misfortune to kill a man in a duel, and fled to England. James I, to whom he had dedicated his play on Mary Queen of Scots, *L'Écossaise* (1603), one of the first French plays to deal with modern history, secured a pardon for him and he returned to France in 1611. In 1621 he joined the Huguenots and was killed by royalist soldiers while taking refuge in an inn. His plays are technically weak and lack action but contain passages of great lyric beauty, and his choruses, like those of GARNIER, are particularly fine, while the note of heroism which he often sounded is similar to that of CORNEILLE.

MONTDORY [Guillaume Desgilberts] (1594–1651), French actor-manager, friend and interpreter of CORNEILLE. He is first heard of in 1612, touring Holland in the company of VALLERAN-LECOMTE. In 1622 he was with LENOIR in a company under the patronage of the Prince of Orange, which several times played in Paris, and two years later he toured in the plays of HARDY through northern France and Holland with his own troupe. He returned to Paris towards the end of 1629 with Lenoir and a new company of actors, bringing with him Corneille's first play *Mélite*, which was probably produced in January 1630. It was an outstanding success, and eventually the company decided to settle permanently in a new theatre, later known as the MARAIS. Just before they opened there Lenoir and his wife, an excellent actress who had appeared in the plays of MAIRET, were sent on the orders of Louis XIII to the Hôtel de BOURGOGNE. This was probably intended to annoy RICHELIEU, who had too openly expressed his preference for the newcomers over the King's Players under BELLEROSE. Under Montdory alone the new theatre continued to flourish, for he was an excellent business man as well as a fine actor in the old declamatory style, and to him goes the honour of having been the first to play the hero Rodrigue in Corneille's great play *Le Cid* (1637). Among his other outstanding roles was Herod in TRISTAN L'HERMITE's *La Mariane*, in which he was appearing before Richelieu when he was stricken with paralysis of the tongue and forced to retire from the stage.

MONTEIRO, LUÍS DE STTAU, see PORTUGAL.

MONTFLEURY [Zacharie Jacob] (*c.* 1600–67), French actor, who was at the Hôtel de BOURGOGNE, where he was second only to BELLEVILLE and later FLORIDOR. He was also the author of a poor tragedy, *La Mort d'Asdrubal* (1647), in which he himself played the leading part. An enormously fat man, with a loud voice and pompous delivery, he was considered a fine tragic actor by his contemporaries and was much sought after by aspiring authors. With his fellow-actors he was satirized by MOLIÈRE in *L'Impromptu de Versailles* (1663). There was no love lost between them, for it was Montfleury who accused Molière before Louis XIV of having married his own daughter. He was also much disliked by CYRANO, who is said to have ordered him off the stage on one occasion, an incident made use of by ROSTAND in his play *Cyrano de Bergerac* (1898). His son (below) was a dramatist, and his daughters

Françoise (*c.* 1640–1708), known as Mlle d'Ennebault, and Louise (1649–1709) were actresses.

MONTFLEURY, ANTOINE (1639–85), French dramatist, son of the above. A lawyer by profession, he wrote a number of plays, first coming into prominence with *L'Impromptu de l'Hôtel de Condé* (1663), intended as a reply to *L'Impromptu de Versailles*, in which MOLIÈRE had ridiculed the company to which his father belonged. Most of his later comedies were probably written to provide the actors at the Hôtel de BOURGOGNE with comedies to rival Molière's, many of them having somewhat similar titles, but though he wrote with much vivacity and showed shrewd traits of wit and observation, his work was soon forgotten. His one serious play, *Trasibule* (1663), bears some resemblance to *Hamlet*; Montfleury was unlikely to have known Shakespeare's play and may merely have used the same source in translation. He married the daughter of FLORIDOR.

MONTHERLANT, HENRI DE (1896–1972), French writer, known chiefly for his novels until in 1942 the success of his first play, *Le Reine morte* (based on the story of Inés de Castro), at the COMÉDIE-FRANÇAISE turned his thoughts seriously to the theatre. Three more plays were produced at the Comédie-Française—*Port Royal* (1954), *Brocéliande* (1956), and *Le Cardinal d'Espagne* (1960). Of his other plays, which include *Fils de personne* (1943), *Le Maître de Santiago* (1948), *Demain il fera jour* (1949), *Celles qu'on prend dans ses bras* and *Malatesta* (both 1950), *Don Juan* (1958), and *La Guerre civile* (1965), the most interesting, *La Ville dont le prince est un enfant*, was written in 1951 but not performed until 1967. It deals with a platonic friendship between two boys which is destroyed by a priest not, as he thinks, out of kindness but out of jealousy. This portrayal of spiritual agony, of the conflict between love and religion, is typical of Montherlant's plays, which contain little external action and are written in a sonorous prose that makes few concessions to realism. Although they have a religious context, Montherlant, unlike CLAUDEL, did not consider himself a 'Catholic' writer, preferring to describe himself as a 'psychological' dramatist. The first of his plays to be seen in English translation was *La Reine morte* as *Queen After Death* (Dundee Repertory Theatre, 1952). In 1957 Donald WOLFIT appeared at the LYRIC, Hammersmith, in *The Master of Santiago* and *Malatesta*, and in 1969 Max ADRIAN was seen in the title role of *The Cardinal of Spain* at the Yvonne Arnaud Theatre, Guildford.

MONTMÉNIL, see LESAGE.

MONTPARNASSE, Théâtre, see BATY, GASTON.

MOODY, WILLIAM VAUGHN (1869–1910), American dramatist, whose work is important in the development of indigenous drama in the United States. Poet, scholar, and educationist, Moody had been for many years a teacher when he decided that his real vocation lay in the theatre. The first of his plays to be staged, by Henry MILLER, was *The Great Divide* (1906), originally known as *The Sabine Woman*. It was successfully produced in London in 1909, the year in which another of Moody's plays, *The Faith Healer*, was seen in New York. Both these plays are written in a dignified, poetic style, and mark the arrival on the American scene of the serious social dramatist, still somewhat melodramatic but moving away from French farce and adaptations of sentimental novelettes. None of his long poetic plays was produced in his lifetime.

MOONEY, RIA (1903–73), Irish actress, teacher, and director for many years with the ABBEY THEATRE, Dublin, where she first appeared in SHIELS's *The Retrievers* (1924), achieving prominence as Rosie Redmond in O'CASEY's *The Plough and the Stars* (1926). In the following year she played Mary Boyle with Arthur SINCLAIR's Irish Players in their New York production of O'Casey's *Juno and the Paycock* and by 1932 was once again with the Abbey company, appearing with them on their American tour in the winter of 1932–3. From 1933 to 1935 she played at the GATE THEATRE, Dublin, in such roles as Catherine Earnshaw in her own adaptation of Emily Brontë's *Wuthering Heights*, and Bride in Denis JOHNSTON's *Storm Song* (1934). She returned to the Abbey in 1935, and two years later was in charge of the Abbey's experimental Peacock Theatre where, among other plays, she directed YEATS's *The Dreaming of the Bones*. In 1948 she was appointed the first woman director of the Abbey Theatre, a position she held until her retirement in 1963.

MOORE, DUDLEY, see REVUE.

MOORE, EDWARD (1712–57), English dramatist, author of the fashionably sentimental plays *The Foundling* (1747) and *The Gamester* (1753). The first bridges the gap between the works of Colley CIBBER and Richard STEELE and those of CUMBERLAND and Mrs INCHBALD. The second, partly written by GARRICK who staged it at DRURY LANE, is a domestic tragedy in the style of LILLO. Translated into French, it had a marked influence on the development of the *tragédie bourgeoise*.

MOORE, EVA, see ESMOND, HENRY VERNON.

MOORE, GEORGE (1852–1933), Irish novelist and dramatist, who spent many years in London and Paris, and in Paris particularly learned something of the theatre. His play *The Strike at Arlingford* was produced by J. T. GREIN's Independent Theatre in London in 1893, and when, early in 1899, plans were started for the production of the first two plays to be given by the IRISH LITERARY

THEATRE—YEATS and Lady GREGORY's *The Countess Cathleen* and Edward MARTYN's *The Heather Field*—Moore helped to recruit the players and directed the rehearsals. He subsequently returned to Ireland for some years, his and Yeats's play *The Bending of the Bough*, based on a play by Edward Martyn, being staged in 1900. Moore collaborated with Yeats again on *Diarmuid and Grania*, which was produced in 1901—the first play of the movement to be taken directly from Irish legend. But in spite of an excellent company, and incidental music by Elgar, the reception was lukewarm, and with the winding up of the Irish Literary Theatre and the arrival of the FAY brothers with a permanent company of Irish actors, Moore withdrew, though he continued to collaborate in the writing of plays.

MOORE, MAGGIE [Margaret Sullivan] (1851–1926), American-born actress, who in 1874 migrated with her husband J. C. WILLIAMSON to Australia and appeared there with him in *Struck Oil; or, the Pennsylvanian Dutchman*, an American comedy which was seen two years later at the ADELPHI THEATRE in London. The play was an immediate success, and the Williamsons remained in Australia, the wife, who was a comedienne, a step-dancer, and a mimic without malice, appearing in a number of light operettas to which her sweet voice was well suited. Divorced from Williamson in 1899, she returned to the U.S.A. and toured there and in England from 1903 to 1908, returning eventually to Australia, where she was occasionally seen in her original role in revivals of *Struck Oil*, with her second husband H. R. Roberts as her leading man.

MOORE, MARY, see ALBERY, JAMES, and WYNDHAM, CHARLES.

MORAL INTERLUDE, a short play, intended mainly for educational purposes, which in the 16th century developed from the longer MORALITY PLAY. It marked the transition from purely abstract characters to individuals, sometimes bringing both types on stage together. It also had a good deal of robust humour, some of the earlier Vices, notably Pride and Gluttony, and such abstract personifications as Free-Will and Self-Love, being transformed into purely comic characters. More space was also given to the antics of the Devil and the old VICE, or buffoon. Notable English examples of the Moral Interlude are the anonymous *Hickscorner* (*c.* 1513) and R. Wever's *Lusty Juventus* (*c.* 1550).

MORALITY PLAY, medieval form of drama which, together with the Creed and Paternoster plays (whose provenance is self-evident), aimed to teach through entertainment. There is reason to believe that the Orders of preaching friars which established themselves early in the 13th century, particularly the Dominicans and Franciscans, were among the first to realize the value of a *ludus*, or play, similar to those from which LITURGICAL DRAMA had evolved and which were even then being used for non-liturgical subjects as a means of instruction. Their sermons already contained within themselves the elements of drama, and it was not difficult for actors to inject into them the necessary three-dimensional animation. In this way complicated and abstract points of doctrine—on original sin, atonement, the need for repentance—could be translated into vivid and concrete verbal images readily understood by the largely illiterate audiences to which they were addressed. Even supernatural and miraculous events could be made comprehensible, and moral teaching reinforced, by the personification of Vices and Virtues, led respectively by a recognizable Devil and a Good Angel, battling for man's soul. Unlike the Christmas and Easter (and many other religious) plays, the Morality play, which in its shorter form became known as an INTERLUDE, was not tied to any specific Church festival; a debate in dramatic form on the salvation of mankind could be presented at any time in any place where sufficient spectators could be found. It therefore provided suitable material for the professional players who by the middle of the 15th century were beginning to emerge from among the amateurs. It marked a big step forward in the evolution of regular drama all over Europe, and many more such plays must have been written than have survived, even in manuscript. The only one still to be seen on the stage is the Dutch *Elckerlyc* (*c.* 1495), known in English as *Everyman*. In an adaptation by William POEL, this was seen in London in 1903 and has been frequently revived. As *Jedermann*, in a German version by Hugo von HOFMANNSTHAL, it has played an important part in the SALZBURG festival ever since its inception.

MORATÍN, LEANDRO FERNÁNDEZ DE (1760–1828), Spanish poet and playwright, son of Nicolás (below). In his plays he combined the Venetian comedy of GOLDONI and the contemporary neo-classicism of France, where he spent some years in exile after supporting Joseph Bonaparte, with traditional Spanish elements, in the presentation both of character and of social background. One of his early successes was a satire on the type of extravagant drama popular at the time, the victim of *La comedia nueva o el café* (1792) being recognizably COMELLA. Moratín's most famous play was *El sí de las niñas* (*When a girl says Yes*, 1806), in which he defends the principle of a woman's freedom to marry the man she loves. Moratín was a great admirer of MOLIÈRE and made excellent though somewhat free translations of *L'École des maris* and *Le Médecin malgré lui*. He was also the first to translate Shakespeare's *Hamlet* directly from the English text, and wrote a history of the Spanish theatre which was published two years after his death.

MORATÍN, NICOLÁS FERNÁNDEZ DE (1737–80), Spanish poet and playwright, father of the above. Like his son, he admired French neoclassical tragedy, and disapproved of the 'excesses' of Lope de VEGA and his followers, and particularly of the religious play, or AUTO SACRAMENTAL. His one comedy was negligible, but in his three tragedies—*Lucrecia* (1763), *Hormesinda* (1770), and *Guzmán el Bueno* (1777)—he upheld the neo-classical theories of Ignacio de LUZÁN and spoke out strongly against despotism and in favour of moral standards. Inferior to his son as a playwright, he was a far better poet.

MORDVINOV, NIKOLAI DMITRIEVICH (1901–66), Soviet actor, who from 1936 to 1940 was at the Gorky State Theatre in Rostov, where his outstanding role was Petruchio in *The Taming of the Shrew*. In 1940 he went with a group of leading actors from Rostov to the MOSSOVIET THEATRE in Moscow, where he greatly influenced the style of acting, particularly in heroic roles. Among the parts in which he was much admired were Arbenin in LERMONTOV's *Masquerade*, and the title roles in Shakespeare's *Othello* and *King Lear*, the latter in 1958. He was seen also in a number of new plays, playing Ognev in KORNEICHUK's *Front* (1943), and Petrov in SOFRONOV's *In One Town* (1946). He also gave an outstanding comic performance as the Cavaliere di Ripafratta in GOLDONI's *La locandiera*.

MORETO Y CABAÑA, AGUSTÍN (1618–69), Spanish playwright, author of a number of good religious and historical plays, but mainly remembered for his COMEDIAS, of which the best are considered to be *El desdén con el desdén* (*Scorn for Scorn*, pub. 1654) and *El lindo Don Diego* (pub. 1662). The first, based on several comedies by Lope de VEGA, who with TIRSO DE MOLINA was the main influence on Moreto's work, shows a young noblewoman who has disdained her suitors being in turn disdained by the man she loves—a situation later used by MOLIÈRE in *La Princesse d'Élide* (1664). The second is an amusing trifle about a foppish young man who spends most of the day getting dressed, but whose arrival in Madrid nevertheless threatens the happiness of two young lovers. It is an excellent example of the *comedia de figurón*, in which Moreto excelled. An inventive and engaging GRACIOSO figures in many of his plays.

MORETTI, MARCELLO, see PICCOLO TEATRO DELLA CITTÀ DI MILANO.

MORGAN, CHARLES LANGBRIDGE (1894–1958), Engish novelist and essayist, who succeeded A. B. WALKLEY as dramatic critic of *The Times* in 1926, a position he held until 1939. He was also the author of three plays—*The Flashing Stream* (1938), *The River Line* (1952), and *The Burning Glass* (1954)—written with the fastidious care and nervous vitality that distinguished all his work.

MÓRICZ, ZSIGMOND, see HUNGARY.

MORLEY, ROBERT (1908–   ), English actor and playwright, son-in-law of Gladys COOPER. He made his first appearance on the stage in 1928, and was in FAGAN's repertory company at Oxford. Later he joined the Cambridge Festival Theatre company under Norman MARSHALL, under whose direction he made his first outstanding success in the title role of Leslie and Sewell Stokes's *Oscar Wilde* (1936) at the London GATE THEATRE, making his New York début in the same part in 1938. In 1937 he played Professor Higgins in SHAW's *Pygmalion* at the OLD VIC, and among his later successes were Sheridan Whiteside in KAUFMAN and HART's *The Man Who Came to Dinner* (1941), the Prince Regent in Norman Ginsbury's *The First Gentleman* (1945), and Arnold Holt in his own play *Edward My Son* (1947; N.Y., 1948), written in collaboration with Noel Langley. He was then seen in two plays by André Roussin, *The Little Hut* (1950), adapted by Nancy Mitford, which ran for three years, and *Hippo Dancing* (1954), which he adapted himself. He made his début in a musical play as Panisse in *Fanny* (1956), and played the Japanese Mr Asano in Spigelgass's *A Majority of One* (1960). Later he appeared in USTINOV's *Half-way Up the Tree* (1967), AYCKBOURN's *How the Other Half Loves* (1970), in which he was also seen in North America and Australia, *A Ghost on Tiptoe* (1974), which he wrote in collaboration with Rosemary Anne Sisson, and a revival of Ben TRAVERS's *Banana Ridge* (1976). Of his own plays, the best known is *Goodness, How Sad!* (1938), which had a good run at the VAUDEVILLE THEATRE. He is limited by his burly form and rich, booming voice to mainly arrogant, domineering roles, but he is an excellent actor and widely and deservedly popular. His autobiography, *Responsible Gentleman*, written in collaboration with Sewell Stokes, was published in 1966.

MOROSCO THEATRE, New York, at 217 West 45th Street, between Broadway and 8th Avenue. Built by the Shuberts, with a seating capacity of 1,009, it opened on 5 Feb. 1917 with *Canary Cottage* by Elmer Harris and Oliver Morosco, the latter a well known West Coast play producer after whom the new theatre was named. Eugene O'NEILL's *Beyond the Horizon* was seen there in 1920, and in the same year the success of *The Bat* by Mary Roberts Rinehart and Avery Hopwood started a fashion for mystery thrillers. George KELLY's *Craig's Wife* (1925) won a PULITZER PRIZE, and in 1927 Katharine CORNELL had a success in MAUGHAM's *The Letter*. The THEATRE GUILD's production of *Call It a Day* by Dodie SMITH ran for six months in 1935. An interesting experiment was WILDER's revision of IBSEN's *A Doll's House* (1937). Later successes include two plays by John VAN

DRUTEN, *Old Acquaintance* (1940) and *The Voice Of the Turtle* (1943), and COWARD's *Blithe Spirit* (1941). Arthur MILLER's *Death of a Salesman* (1949) and Tennessee WILLIAMS's *Cat on a Hot Tin Roof* (1955) were also Pulitzer Prize-winners, and Gore Vidal's comedy of American political life *The Best Man* (1960) did extremely well. In 1963 Arthur KOPIT's *Oh Dad, Poor Dad, Mamma's Hung You in the Closet and I'm Feeling So Sad* was transferred here from the PHOENIX, and later the same year Peter SHAFFER's double bill *The Private Ear* and *The Public Eye* was seen. Arthur Miller's *The Price* (1968) had a long run, and David STOREY's *Home* (1970), with GIELGUD and RICHARDSON, repeated its London success. Other plays from London were Simon GRAY's *Butley* (1972) with Alan BATES and David Storey's *The Changing Room* (1973). After a successful revival of O'Neill's *A Moon For the Misbegotten*, also in 1973, came more imports from London—RATTIGAN's *In Praise of Love* (1974) and AYCKBOURN's *The Norman Conquests* (1975). In 1977 Cristofer's *The Shadow Box* was yet another Pulitzer Prize-winner, and in 1980 ALBEE's *The Lady from Dubuque*, with Irene WORTH, had a brief run. The theatre was demolished in 1982, together with the nearby HELEN HAYES THEATRE.

MOROZOV, MIKHAIL MIKHAILOVICH (1897–1952), Russian Shakespearian scholar, who encouraged the production of Shakespeare's plays all over the U.S.S.R., and early in his career translated *All's Well That Ends Well* and *The Merry Wives of Windsor* (with Samuel Marshak). He also wrote (in English) a short book entitled *Shakespeare on the Soviet Stage*, with an introduction by Dover Wilson, and was a contributor to several volumes of *Shakespeare Survey*.

MORRIS [Morrison], CLARA (1846–1925), American actress who is believed to have been on the stage as a child, appearing in pantomime at the Royal Lyceum in Toronto, her birthplace. At 16 she joined the stock company in Cleveland, Ohio, and a few years later toured North America, being seen in Halifax in the summer of 1870 before going to join DALY's company in New York, where she was much admired in a dramatization of Wilkie COLLINS's *Man and Wife* and as Cora the Creole in his adaptation of *Article 42* (1872). She later played in a wide range of parts, and in spite of a strong accent and an extravagant, unrestrained manner was popular both in New York and on tour. Though not accounted a good actress, she had an extraordinary power of moving an audience, and could always be relied on to fill the theatre, particularly in such parts as Marguerite Gautier in the younger DUMAS's *The Lady of the Camellias*. This led to her being nicknamed 'the Queen of Melodrama', but she also essayed more serious roles, among them Lady Macbeth, Julia in KNOWLES's *The Hunchback*, and a number of modern heroines. She appeared for some years under the management of A. M. PALMER, but after 1885 ill-health forced her into semi-retirement, though she was seen in Canada again in 1889, and in 1892 appeared for the first time in Montreal in SARDOU's *Odette*. She was in California in 1890 and 1892 and made her last appearance on the stage in New York in the title role of *Claire* (1894), which she herself adapted from a novel by Richard Voss, then retired to devote herself to the writing of fiction and an autobiography, *Life on the Stage* (1902). She made a return to the theatre in 1904, playing Sister Geneviève in an all-star revival of Oxenford's *The Two Orphans*, and subsequently appeared in vaudeville until her eyesight failed in 1909.

MORRIS, Mrs OWEN (1753–1826), American actress, whose maiden name is apparently unknown. She was the second wife of the low comedian Owen Morris (?–*c.* 1810), who was with the AMERICAN COMPANY from 1759 to 1790 playing Sir Oliver Surface in SHERIDAN's *The School for Scandal*, Dogberry in *Much Ado About Nothing*, and Polonius in *Hamlet* when they were all seen for the first time in New York. His first wife was also with the company, and is known to have played Ophelia. She was drowned in a ferry accident in New York in 1767 and by 1773 her husband had married again. The second Mrs Owen Morris was a remarkable actress, one of the outstanding members of the American Company, playing Charlotte in Royall TYLER's *The Contrast* (1787), the first indigenous American comedy, and leading roles in Shakespeare. With her husband she left New York after the season of 1789 to join WIGNELL's new company at the CHESTNUT STREET THEATRE, Philadelphia, where she found herself somewhat eclipsed by Mrs MERRY. She remained with the company after her husband's death, and is believed to have been the Mrs Morris who appeared at the Commonwealth Theatre (formerly the ANTHONY STREET THEATRE) in New York in 1815.

MORRIS (MORRICE) DANCE, English dance popular at village FOLK FESTIVALS. Different authorities derive it either from the Germanic sword-dance or from the Morisco, or Moorish dance, the latter on account of the blackened faces of the dancers or their companions. Chambers, however, sees in this custom a relic of the old pagan rite of smearing the face with ash from the sacrificial fire. Others see it simply as a form of disguise. The FOOL who accompanies the dancers, with his bladder and cow's tail, is probably a survival of the primitive worshipper in sacrificial costume. The dancers, usually six in number, wear white suits, have bells on their legs, and carry handkerchiefs or short staves in their hands. The music is traditionally supplied by a pipe and tabor or by bagpipes, though in recent times a fiddle has been used. With the dancers and the fool go the HOBBY HORSE and, at Maytime, a Jack-in-the-

Green; sometimes also a dragon and a Maid Marian, the latter showing an early association, through the May-Day revels, with the ROBIN HOOD legend. The Morris Dance can be traced all over England and Scotland, and reached the height of its popularity outside the folk tradition under Henry VIII, when it appeared at Court. It still survives in the countryside, notably in Hampton in Oxfordshire, and in the early 20th century it had a fostered revival under the auspices of the English Folk Dance Society. The Morris dance can be either stationary or processional, and both forms can be seen in the streets of Oxford on May Morning.

MORRISON, Mrs CHARLOTTE, see NICKINSON, JOHN.

MORSELLI, ERCOLE LUIGI (1882–1921), Italian dramatist of the CREPUSCOLARI or 'twilight' school of drama. In his reworking of three legendary tragic stories—*Orione* (1910), *Glauco* (1919), and the posthumous *Belfagor* (1930)—he explored the nature of sensuality in a more critical and restrained manner than D'ANNUNZIO, his anti-mythic approach anticipating many later 'psychoanalytical' reworkings of classical themes.

MORTIMER, JOHN CLIFFORD (1923–), English dramatist, by profession a barrister, whose aim, in his own words, is to chart 'the tottering course of British middle-class attitudes in decline'. His first plays, the one-act *Dock Brief* and *What Shall We Tell Caroline?* (1958), were produced in that year in a double bill at the LYRIC, Hammersmith and later at the GARRICK THEATRE, and also in New York in 1961; they were followed by another one-acter, *Lunch Hour* (1961). His first full-length plays were *The Wrong Side of the Park* (1960), in which Margaret LEIGHTON played a woman whose illusions about her deceased first husband prevent her from making a success of her second marriage, and *Two Stars for Comfort* (1962). Mortimer then prepared an English version of FEYDEAU's farce *Une Puce à l'oreille* as *A Flea in Her Ear* (1965) for the NATIONAL THEATRE, and after a somewhat non-naturalistic play, *The Judge* (1967), and a second translation from Feydeau—*Un Fil à la patte* as *Cat Among the Pigeons* (1969)—returned to the one-act form with *Come As You Are* (1970), four playlets set in different parts of London showing groups of middle-aged people in the grip of sexual fantasies. They were followed by an autobiographical play, *A Voyage Round My Father*, produced at the GREENWICH THEATRE in 1970 and at the HAYMARKET THEATRE in 1971, where the role of the Father was played first by Alec GUINNESS and then by Michael REDGRAVE. In the same year an adaptation of ZUCKMAYER's *The Captain of Köpenick* was seen at the National Theatre. *I, Claudius* (1972) was based on two novels by Robert Graves, and later plays are *Collaborators* (1973), in which a husband and wife collaborate in writing a film about their marriage, *Heaven and Hell* (1976), another double bill, and *The Bells of Hell* (1977), which depicts the Devil, disguised as an ex-R.A.F. padre, arriving at the home of a trendy South London vicar. A third translation from Feydeau, *The Lady from Maxim's*, was also seen in 1977 at the National Theatre.

MORTON, CHARLES (1819–1904), early and extremely able MUSIC-HALL manager, who from his association with the early pioneering days of music-hall was known as 'the father of the halls'. He opened the first of the major London music-halls, the CANTERBURY, in 1852, and in 1861 inaugurated a new era with the opening of the OXFORD. He was later called in to revive the flagging fortunes of several other well known halls, among them the Philharmonic (later the GRAND THEATRE) in Islington in 1870, the PALACE in Cambridge Circus in 1892, and the TIVOLI in the Strand in 1893. He was successful mainly because he was a hard worker, set a high standard, was good at choosing his performers, and though appreciating honest vulgarity was opposed to innuendo and salacity.

MORTON, JAMES J., see VAUDEVILLE, AMERICAN.

MORTON, JOHN MADDISON (1811–91), English dramatist, son of Thomas (below). He began his dramatic career with a PANTOMIME, *Harlequin and Margery Daw; or, the Saucy Slut and the See-Saw* (1833), which was seen at the ADELPHI THEATRE, but became a prolific writer of short farces, mainly taken from the French, which helped to build up the reputations of such actors as BUCKSTONE, the KEELEYS, and others. The most popular were *The Double-Bedded Room* (1843), which, rewritten as *Box and Cox* (1847), provided the libretto for Sullivan's first operetta *Cox and Box* (1867), and *Lend Me Five Shillings* (1846), first seen at the HAYMARKET THEATRE during the management of Ben WEBSTER.

MORTON, THOMAS (*c.* 1764–1838), English dramatist, whose first play, a MELODRAMA entitled *Columbus; or, a World Discovered* (1792), was produced at COVENT GARDEN, and was followed by *The Children in the Wood* (1793), which was popular enough to be included in the repertory of the TOY THEATRE. He then wrote several sentimental comedies—*The Way to Get Married* (1796), *A Cure for the Heart-Ache* (1797), and *Secrets Worth Knowing* (1798)—before writing the play by which he is best remembered, *Speed the Plough* (1800). It became famous for the character of Mrs Grundy, who never appears but is frequently referred to as the embodiment of British respectability. Among Morton's later plays the most important were *The School of Reform; or, How to Rule a Husband* (1805), an amusing comedy which was frequently revived up to 1883, *The*

Slave (1816), in which MACREADY made an early appearance at Covent Garden, and *Henri Quatre; or, Paris in the Olden Time* (1820), in which Macready again played the title role with great success. One of Morton's most spectacular melodramas was *Peter the Great; or, the Battle of Pultawa*, seen at DRURY LANE in 1829.

MOSCOW ART THEATRE, famous theatre, now dedicated to Maxim GORKY, founded by STANISLAVSKY and NEMIROVICH-DANCHENKO in 1898. It opened with A. K. TOLSTOY'S *Tsar Feodor Ivanovich*, and soon after gave a successful production of CHEKHOV'S *The Seagull*, which had recently failed at the Alexandrinsky Theatre, St Petersburg (now the PUSHKIN THEATRE, Leningrad). This was followed by *Uncle Vanya* (1899), *Three Sisters* (1901), and *The Cherry Orchard* (1904). The ferment which produced the abortive revolution of 1905 was reflected in the production of *The Lower Depths* (1902), by Gorky, whose *Children of the Sun*, based on an incident in that uprising, was staged late in 1905. The repertory of pre-Revolutionary plays included a number of European classics, but the only play by Shakespeare to be performed at this time was *Julius Caesar*. During the upheavals of the Revolution the theatre was saved from extinction by the efforts of LUNACHARSKY, and after a long tour of Europe and America in 1922 and 1923, the company returned to Moscow. Some tentative productions of new plays followed, including BULGAKOV's *The Days of the Turbins* (1926), and the theatre finally found its feet in the new world in 1927 with the production of IVANOV'S *Armoured Train 14–69*. The theatre was now firmly established, many actors, among them KACHALOV, MOSKVIN, and Chekhov's widow Olga KNIPPER, spending the rest of their lives with the company, and being succeeded by actors trained in the theatre's own dramatic school in Stanislavsky's METHOD. From this school developed also several important individual groups, among them those at the REALISTIC and VAKHTANGOV THEATRES. The Moscow Art Theatre visited London in 1958, 1964, and 1970, its productions of plays by Chekhov being of particular interest to English theatre-goers. In 1971 the artistic directorship of the theatre was assumed by Oleg YEFREMOV, and two years later the company moved into new premises on the Tverskoy Boulevard, with a large auditorium seating nearly 1,400. The company also retained its old premises, which, with the dramatic school, gives it three bases from which to work.

MOSCOW STATE JEWISH THEATRE, known in Yiddish as Melucha, and in Russian as Goset. This theatre was founded in Leningrad in 1919 by Alexander GRANOVSKY for the production of plays in Yiddish; it later moved to Moscow, opening at the small Chagall Hall in 1921 in a programme of short comedy sketches by Sholom ALEICHEM, and finally achieved its own theatre, seating 766. Granovsky was succeeded in 1927, during a European tour, by his leading actor Salomon MIKHOELS, who continued much of Granovsky's work with the addition of new plays by such authors as David Bergelson (1884–1952) and Peretz Markish. The leading stage designer at this theatre was for some time Isaac Rabinovich, who later worked for the VAKHTANGOV and MALY theatres, and one of its artistic directors was RADLOV, who in 1935 was responsible for an interesting production of Shakespeare's *King Lear* in which Mikhoels gave a fine performance. After the latter's death in 1948 the company disbanded, but was re-formed in 1962 under Vladimir Shvartser as the Jewish Drama Ensemble, and toured extensively in a production of Aleichem's *Tevye the Milkman* (better known in English as the musical *Fiddler on the Roof*). Among its later productions were revivals of Aleichem's *Two Hundred Thousand*, GORDIN's *Over the Ocean*, and GOLDFADEN's *The Witch*.

MOSCOW TRANSPORT THEATRE, see GOGOL, N. V., and SUDAKOV, I. Y.

MOSKOVITCH, MAURICE, see GOLDFADEN, ABRAHAM.

MOSKVIN, IVAN MIKHAILOVICH (1874–1946), Russian actor, who joined the MOSCOW ART THEATRE on its foundation in 1898, playing the lead in its initial production of A. K. TOLSTOY's *Tsar Feodor Ivanovich* and Epihodov and Lvov in CHEKHOV's *The Cherry Orchard* and *Ivanov*. Among his later roles were Luka in GORKY's *The Lower Depths*, Nozdrev in GOGOL's *Dead Souls*, and the merchant Pribitkov in OSTROVSKY'S *The Last Sacrifice*. In 1942 he played a leading part in POGODIN's play about Lenin, *Kremlin Chimes*.

MOSSOP, HENRY (1729–74), Irish actor, who after some experience in Dublin went to DRURY LANE, where he was successful in tragedy, proving at first an excellent ally to GARRICK in his rivalry with Spranger BARRY. He took over some of the parts of QUIN, and was also seen as Richard III, Horatio to Garrick's Lothario in ROWE's *The Fair Penitent*, Coriolanus, and the title role in John Brown's *Barbarossa* (1754), the last two being considered his best parts. A foolish insistence on playing young lovers, to which he was quite unsuited, led to some failure and ridicule, and the hot-tempered Mossop vented his resentment on Garrick, of whom he had become insanely jealous. He went back to Ireland and opened the Smock Alley Theatre in Dublin in 1761, in opposition to Barry and WOODWARD at the Crow Street Theatre, but was soon in financial difficulties, which he made worse by gambling and general dissipation. Ruined in pocket and health, he crept back to London. Too proud to appeal to Garrick, and impatient of all advice and correction, he was gradually deserted by all his friends and died of starvation.

MOSSOVIET THEATRE, Moscow. Founded in 1923 with the object of encouraging young playwrights to tackle Soviet and European political themes, this theatre's earliest and best dramatist was BILL-BELOTSERKOVSKY, whose *Storm* (also known as *Hurricane*) was produced there in 1926. The company was led by LYUBIMOV-LANSKOY until 1940, when it came under the direction of ZAVADSKY, who in 1951 directed a revised version of *Storm* which proved a landmark in the theatre's history, as were his productions of *The Merry Wives of Windsor* in 1957 and of LERMONTOV's *Masquerade* in 1965. An excellent production of *King Lear* in 1958, with MORDVINOV in the title role, further enhanced the reputation of the theatre, which continued also to produce new Soviet plays, among them Virta's *In Summer the Sky is High* (1961), as well as plays by such European dramatists as SHAW, GARCÍA LORCA, and Heinrich Böll. In 1965 the company was seen at a theatre festival in Paris in plays by Lermontov and ROZOV, and in an adaptation of DOSTOEVSKY's *Uncle's Dream*.

MOSTEL, ZERO [Samuel Joel] (1915–77), American actor, a huge, moon-faced man of enormous talent, who gained his theatrical experience in Greenwich Village night clubs before appearing in VAUDEVILLE on Broadway in 1942. In 1946 he was seen as Hamilton Peachum in *Beggar's Holiday*, an adaptation of GAY's *The Beggar's Opera* with music by Duke Ellington, and in 1952 played Argan in *The Imaginary Invalid*, his own adaptation of MOLIÈRE's *Le Malade imaginaire*. Among his later parts were Shu Fu in BRECHT's *The Good Woman of Setzuan* in 1956, Leopold Bloom in Marjorie Barkentin's dramatization of part of JOYCE's *Ulysses* as *Ulysses in Nighttown* in 1958, John in IONESCO's *Rhinoceros* in 1961, and Prologus in the highly successful musical *A Funny Thing Happened on the Way to the Forum* in 1962. In 1964 he scored a great success in the musical *Fiddler on the Roof*, based on a short story by Sholom ALEICHEM. He was later seen in New York in revivals of both *Ulysses in Nighttown* (1974) and *Fiddler on the Roof* (1976), and died during the pre-Broadway tour of WESKER'S *The Merchant*. A volume of memoirs, *Zero by Mostel*, was published in 1965.

MOTA, ANRIQUE DA (*fl.* early 16th century), Portuguese poet, four of whose dialogue works are extant, together with one in which he apparently collaborated with Gil VICENTE and others. The keynote of Mota's work is his satirical treatment of easily recognizable types, and the language which he puts into the mouth of a young Negress foreshadows the 'special vocabularies' evolved by Vicente. There is no record of the staging of any of Mota's works, but some at least would have been readily adaptable, and one of them, the 260-line *Farsa do Alfaiate* (composed between 1496 and 1506) may be earlier than any of Vicente's work.

MOTION, name given in the 16th and 17th centuries to the puppet-plays of the itinerant showmen. The earliest dealt with biblical subjects, and Shakespeare refers in *The Winter's Tale* to 'a motion of the Prodigal Son'. Later the range of subjects was extended, and episodes were used from medieval romance, mythology, and contemporary history.

MOULIN ROUGE, Paris, well known dance-hall which opened on 5 Oct. 1889, including in its amenities a large garden used in the summer for dancing and entertainments. A constant feature of its programmes has been a cabaret show, in which the cancan made its first appearance, the dancers in 1893 being Grille d'Égout, la Goulue, la Môme Fromage, and Nini-patte-en-l'air. MISTINGUETT was for some years part-proprietor of the Moulin Rouge and frequently appeared there, as did most of the stars of variety and music-hall.

For the Moulin Rouge, New York, see NEW YORK THEATRE (2).

MOUNET-SULLY [Jean Sully Mounet] (1841–1916), famous French actor who made his début at the COMÉDIE-FRANÇAISE in 1872 as Oreste in RACINE's *Andromaque*. His fine physique, beautifully modulated voice, and sombre, penetrating gaze, added to fiery, impetuous acting and great originality, soon brought him into prominence. His career was one of unclouded success and he appeared in all the great tragic roles of the French classical repertory. He was also outstanding in the plays of Victor HUGO when they were finally given at the Comédie-Française. His younger brother Paul (1847–1922) was also a member of the company, which he joined in 1889 from the ODÉON.

MOUNTFORT, WILLIAM (1664–92), English actor, who specialized in 'fine gentlemen' parts, being an excellent Sparkish in WYCHERLEY'S *The Country Wife* and creating the title role in John CROWNE's *Sir Courtly Nice; or, It Cannot Be* (1685). He was also the author of several plays, including a HARLEQUINADE 'in the Italian Manner', *The Life and Death of Doctor Faustus . . . With the Humours of Harlequin and Scaramouche* (1685), and a comedy, *Greenwich Park* (1691), which was extremely successful and frequently revived. He was brutally murdered at the instigation of a certain Captain Hill, who suspected him of being a successful rival in the affections of Mrs BRACEGIRDLE, with whom Hill was passionately but unsuccessfully in love. Mountfort's wife Susanna Percival (1667–1703) was also on the stage, and in her youth was much admired in BREECHES PARTS, playing Bayes in BUCKINGHAM'S *The Rehearsal* with great success. Colley CIBBER has left a description of her playing of Melantha in DRYDEN'S *Marriage à-la-Mode*, in which he considered her unequalled. She married as her second husband the actor J. B. Verbruggen.

MOUNTVIEW THEATRE SCHOOL, at 104 Crouch Hill, London, N.8., an amateur group founded in 1947 by Peter Coxhead and Ralph Nossek, who while serving in the Royal Navy had run an amateur group in Sri Lanka. The group's first production, WILDE'S *The Importance of Being Earnest*, was staged in Nov. 1947 after five months' work had been done to convert a derelict hall into a small theatre. This was held on lease until 1949, and then purchased outright, undergoing major alterations after a bad fire in 1963. Originally the group presented one play a month and from 1950 a production was mounted every three, and later every two, weeks. A high standard was achieved, and in 1970 the company toured the U.S. for six weeks with a repertory which included *Hamlet*, BRECHT'S *Mother Courage*, GENET'S *The Maids*, and Edward BOND'S *Saved*. The drama school, which began operations in 1969 and has now superseded the club, offers full and part time training in acting and technical skills to adults, children, and handicapped people. Student performances are presented in the Mountview Theatre and in the studio theatre, opened in 1971 and named in honour of Judi DENCH: admission is free. These productions are regarded as an integral part of London's FRINGE THEATRE and are regularly taken on tour in Britain and overseas.

MOURQUET, LAURENT, see GUIGNOL.

MOWATT [*née* Ogden], ANNA CORA (1819–70), American author and actress, mainly remembered for her social comedy *Fashion* (1845), considered to be one of the best of the early satires on American life. It was first produced at the PARK THEATRE, so successfully that Mrs Mowatt was encouraged to go on the stage herself, making her début as Pauline in BULWER-LYTTON'S *The Lady of Lyons* in 1845. She then formed her own company, with E. L. DAVENPORT as her leading man, with which she visited London in 1848–51. In 1854 she published her *Autobiography*, and then retired to live quietly in London, where she died.

MROŻEK, SŁAWOMIR (1930–   ), Polish playwright, who with WITKIEWICZ and GOMBROWICZ represents in Polish drama the Theatre of the ABSURD, though with more political content than is usual with exponents of this style elsewhere. Some of his early plays were in one act, and in 1964 two of them *What a Lovely Dream!* and *Let's Have Fun!* (both 1962), were seen in London during the first WORLD THEATRE SEASON, directed by Konrad ŚWINARSKI. Mrożek's longer plays, which include *The Turkey Cock* (1961) and *The Enchanted Night* (1963), culminated in the world-famous satire on totalitarianism, *Tango*, first produced at Belgrade in 1965, and later the same year in Warsaw by AXER. This was the first modern Polish play to make a decided impact on the theatres of Western Europe and America; it was seen in London in 1966 at the ALDWYCH directed by Trevor NUNN, and in New York in 1969. In 1968 Mrożek went into voluntary exile, but most of his recent plays have been seen in Poland, *The Hunchback* (1976) having been given its first production simultaneously in Cracow and Łódź.

MÜLLER, HEINER (1929–   ), playwright of the German Democratic Republic, whose first plays, *Der Lohndrücker* (*The Scab*, 1950) and *Die Korrektur* (*The Correction*, 1958), were propaganda works in the style of BRECHT designed to instil socialist attitudes in his worker-audiences. He then turned to classical subjects, taken mainly from SOPHOCLES, which depicted with less overt socialism the conflicting claims which the community makes on the individual; the best was probably *Philoktet* (1968). Müller also translated and adapted Shakespeare's *Macbeth* and SHWARTZ'S *The Dragon* and made a deft adaptation, in *Cement*, of Gladkov's novel on the Russian Revolution. All his works, with their terse, dense, finely chiselled language, are highly thought of throughout Germany.

MÜLLER, TRAUGOTT (1895–1944), German stage designer, who worked initially with PISCATOR, for whom he evolved multi-purpose constructions with steps, ramps, and levels on a revolving stage. For SCHILLER'S *Die Räuber* in 1926 he designed a fort in battleship-grey, and for TOLLER'S *Hoppla, wir leben!* in 1927 a two-storey house in cross section. For FEHLING'S productions of Shakespeare's *Richard III* in 1937 and *Richard II* in 1939 he made effective use of a bare stage with a steep rake, and for LESSING'S *Emilia Galotti*, directed by GRÜNDGENS in 1937, he paved the stage with stone. His Chinese set for a production of GOZZI'S *Turandot* in 1941 radiated menace.

MULTIPLE SETTING, a term applied to the stage décor of the medieval play (known in France as *décor simultané* and in Germany as *Standort-* or *Simultanbühne*), inherited from the LITURGICAL DRAMA with its 'mansions' or 'houses' representing different sites of the action, disposed about the church. When Biblical dramas were first performed out of doors, the 'mansions' were positioned on three sides of an unlocalized *platea* or acting space, but by the 16th century, at any rate in France, they were set in a straight line or on a very slight curve at the back of the stage. In England the different scenes of a MYSTERY PLAY were on perambulating pageant-waggons, and the multiple setting was not needed, but it continued on the Continent for a long time, and was still in use

Symbolism and naturalism were the predominant and contrasting styles of theatrical production in the years 1880–1920: (*above*) Ibsen's *Peer Gynt* at the Dagmartheatret, Copenhagen, 1886; (*below*) Pinero's *Lady Bountiful* at the Garrick Theatre, 1891.

Two productions of Chekhov at the Moscow Art Theatre. *The Seagull*, Act I (*above*), in 1898 and *Uncle Vanya*, Act III (*below*), in 1899. Both productions were by Stanislavsky and Nemirovich-Danchenko with décor by V. Simov.

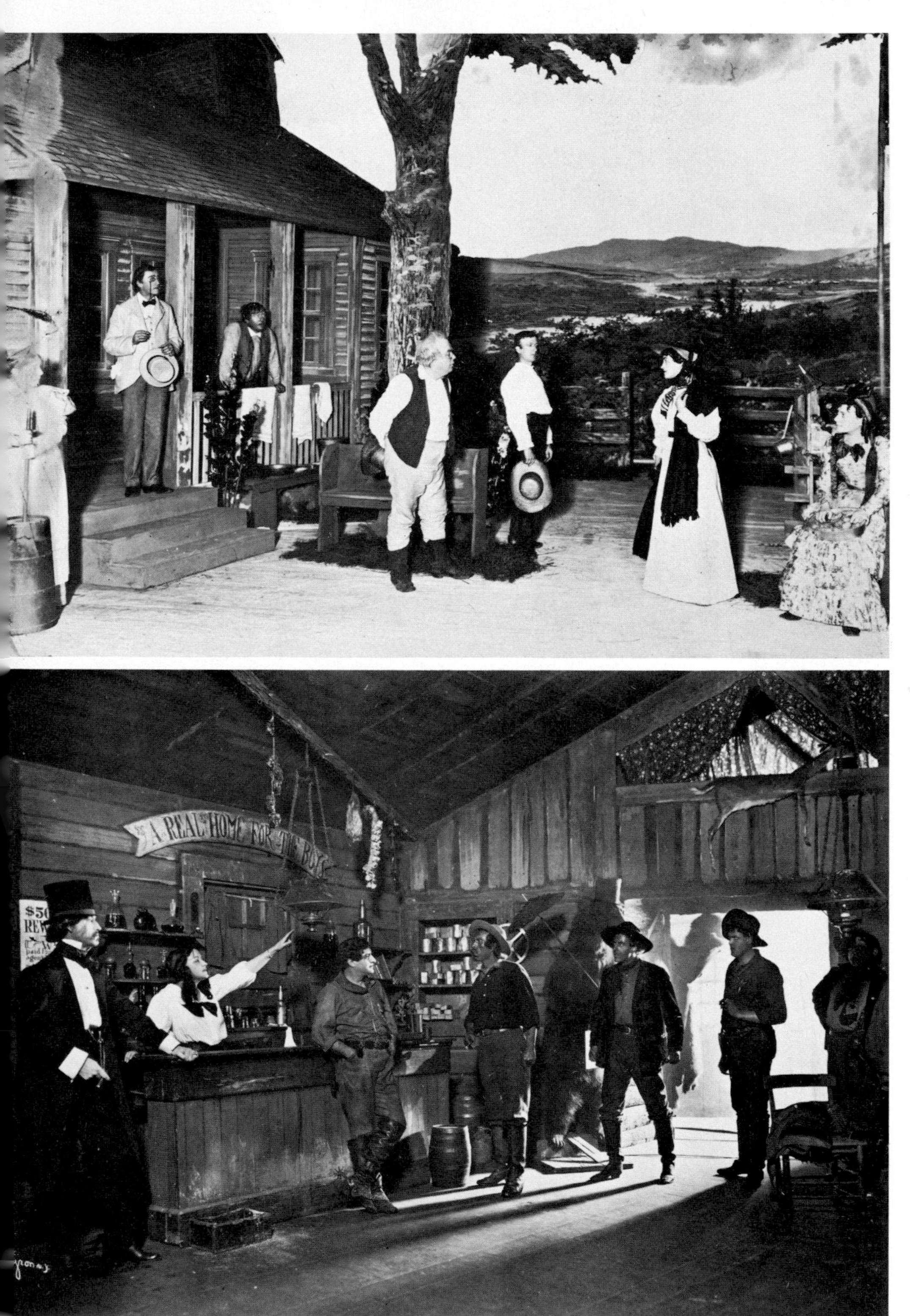

Popular drama in New York. *Way Down East* by Lottie Blair Parker (*above*) at the Manhattan Theatre in 1898 and Belasco's production of *The Girl of the Golden West* (*below*) at his own theatre in 1905.

Ibsen's *Hedda Gabler*, an American production of 1904 starring Minnie Maddern Fiske and George Arliss.

Eleonora Duse in *The Lower Depths* by Maxim Gorky at the Théâtre de l'Oeuvre, Paris, 1905.

Reinhardt in Berlin (*above*): *A Midsummer Night's Dream*, 1912, with a characteristically large-scale architectural setting.

The Grosses Schauspielhaus (*left*), Berlin, designed for Reinhardt by Hans Poelzig, 1919.

Constructivist setting for Calderón's *La dama duende*, designed by Alexandra Exter for the Moscow Art Theatre, 1924.

Design by Reigbert for Brecht's *Trommeln in der Nacht*, Munich, 1922.

Ibsen's *Little Eyolf*, 1924. A design for Act II by Adolph Appia which achieves a brooding atmosphere by very simple means.

Ivanov, *Armoured Train 14–69* at the Moscow Art Theatre, 1927. The studied naturalism of socialist realism contrasts with the expressionism of Reigbert's sketch (*opposite page*).

Two all-black successes on Broadway: Paul Green's *In Abraham's Bosom* (*above*) by the Provincetown Players in 1926 and Marc Connelly's *The Green Pastures* (*below*) at the Mansfield Theatre in 1930.

Dynamo by Eugene O'Neill, Martin Beck Theatre, New York, 1929.
The Constructivist setting was by Lee Simonson.

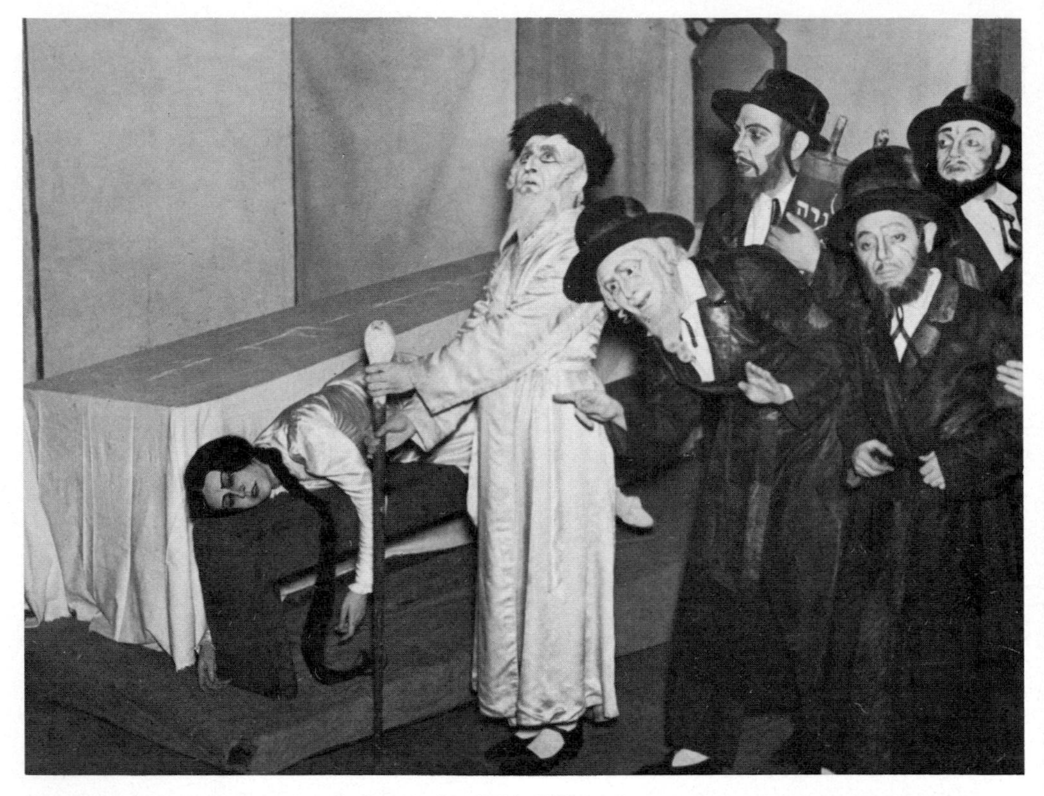

Ansky's *The Dybbuk* was a regular feature of the Habimah Theatre's repertory throughout the 1920s and 1930s.

Strindberg's *The Dream Play*, directed by Olaf Molander at the Royal Dramatic Theatre, Stockholm, 1935.

Our Town by Thornton Wilder, produced in 1938 at Henry Miller's Theatre, New York, combined naturalism and non-naturalism in a new way.

The Abbey Theatre, Dublin; a revival of O'Casey's *The Plough and the Stars* in 1942.

Lyndsay's *The Thrie Estaitis*, revived by Tyrone Guthrie for the Edinburgh Festival in 1948.

Louis Jouvet and Dominique Blanchar in *Ondine* by Giraudoux at the Athénée, Paris, in 1959.

Setting designed by Josef Svoboda for Gogol's *The Government Inspector* at the National Theatre, Prague, 1948.

… à Marie at the Comédie Française, Paris, 1955. Setting for Act II by Georges Wakhevitch

Laterna Magika, presented by Radok and Svoboda of the National Theatre, Prague, at the Brussels Exposition of 1958, combined two or more films, projected simultaneously, with live actors and dancers; witty and innovative lighting effects enhanced the stage illusion.

The first London production of Samuel Beckett's *Waiting for Godot*, Arts Theatre, 1955; Timothy Bateson as Lucky, Paul Daneman as Vladimir, Peter Bull as Pozzo.

John Osborne's *Look Back in Anger* at the Royal Court Theatre, 1956: (*l. to r.*) Mary Ure, Alan Bates, Helena Hughes, Kenneth Haigh. This was another trail-blazing première, labelled the first 'kitchen-sink drama'.

at the Hôtel de BOURGOGNE in Paris in the early 17th century, as is shown by the so-called *Mémoire* of MAHELOT. It is even possible that CORNEILLE's early plays, produced at the MARAIS, were staged in a multiple setting, which was finally ousted by the development of the single set used for classical tragedy, as in the plays of RACINE, and the successive scenes of the spectacular MACHINE PLAY. The set scenes of the Renaissance stage in Italy, notably the TERENCE-STAGE, were single composite backgrounds and not multiple settings in the original sense, though the unlocalized acting space in front of the tragic or comic houses corresponded to the medieval *platea*. The Elizabethan public stages, like that at the first GLOBE THEATRE, did not employ multiple settings, though something of the kind may have been used in the early days of the private roofed playhouses like BLACKFRIARS, and was certainly a feature of the elaborate Stuart Court MASQUES. The modern equivalent of the multiple setting is the COMPOSITE SETTING.

MUMMERS' PLAY, or MEN'S DRAMATIC CEREMONY, the best known type of English folk-play, which appears to derive from the FOLK FESTIVALS of primitive agricultural communities. There is a strong likeness between the mummers' play and such survivals of a spring festival play in remote corners of Europe as the ST GEORGE or Haghios Gheorgios Play of Thrace and Thessaly, the Kalusari or Little Horse Dancers' Play from Romania, and the Little Russia Easter Celebration described by Sir James Frazer in *The Golden Bough*. The common central theme is the death and resurrection of one of the characters, an obvious re-enactment in human terms of the earth awakening from the death of winter.

Texts of the mummers' play have been collected from all over England, from Scotland, and from Ireland; they show a remarkable similarity, though no common prototype has been traced. The play is first mentioned towards the end of the 18th century and flourished until the mid-19th. It seems to have died out in Scotland, but can still be seen in a handful of English villages, in Northern Ireland, and in the outposts of Newfoundland, where it is first recorded from the late 18th century. A play performed at Revesby Abbey in Lincolnshire on 20 Oct. 1779 was thought to be the oldest authentic version of an English folk-play, but is now known to have been assembled by Joseph Banks from various elements of folk drama: the earliest example to have been written down is the Bassingham Play, also from Lincolnshire, collected in 1823. This is unusual in having two versions, one for men and one for children, of which the latter is the more ribald. One of the best survivors is the Symondsbury Play from Dorset, a particularly lively, bawdy, and inventive version.

The extant texts have been divided into three groups: the Hero-Combat Play, the Sword Play, and the Wooing Ceremony. The largest group is the Hero-Combat Play, and many elements of this type are present in the other two groups. The all-male cast, dressed in rags or in costumes of shredded paper, as in the Paper Boys' Play from Marshfield in Gloucestershire, their faces often blackened with soot—showing some affinity with the MORRIS DANCE—go from one house or farm to another. Once admitted, their spokesman, usually Father Christmas, clears a roughly circular playing-place among the spectators ('Room, room, brave gallants all' or, by corrupt oral tradition, 'Room, room, gallons of room!') and promises them a 'dreadful battle'. The hero, St George or in later versions King George, delivers a boastful speech celebrating his slaying of the dragon and other exploits and challenges all comers to single combat. The challenge is taken up by the Turkish Knight, corrupted in some versions into the Turkey Snipe and in others replaced by the Bold Slasher or Captain Slasher, who in turn boasts of his valiant deeds. The two men fight a comic duel with wooden swords and one of them is killed. In the longer versions of the play there follows a lament, usually from Father Christmas who claims the dead man as his son and calls for a doctor. This character, doubtless modelled on the travelling mountebank, has affinities with the DOTTORE of the COMMEDIA DELL'ARTE which are probably coincidental. In this section of the play improvised comedy reaches its height. The Doctor, often accompanied by his servant Jack Finney or Johnny Jack, with whom he indulges in knockabout slapstick, details the remarkable cures he has effected, recites a garbled list of diseases he can treat, and with his famous 'remedy' (usually elecampane, a root widely used in folk medicine) restores the dead man to life. Sometimes he fights, dies, and is again resurrected. The play ends with the *Quête*, a procession and the taking up of a collection, into which a number of extraneous characters are introduced, each one identifying himself in a few lines of doggerel verse. They may include such legendary figures as Beelzebub and Little Devil Dout, who recall the Old VICE of the MYSTERY PLAY. Little Devil Dout leads the *Quête* and carries a broom to brush away the old year's dust, his name being a corruption of 'Do out'. According to the district and the date of the surviving text, historical figures may appear, often Charlemagne and Oliver Cromwell, and even caricatures of contemporary local worthies. Skirmishing on the fringes of the main action there may also be a FOOL and a HOBBY HORSE and there is usually a Female, sometimes known as the Bessie, played by a man in woman's clothes.

The Sword Play, found mainly in north-east England is essentially defined by the slaying of the victim not in single combat but by a group of dancers. They first perform an intricate dance with swords, which ends with the weapons linked so that they can be raised by one man and then placed around the neck of the victim, usually the Fool. A particularly fine example has been recorded from Greatham in Durham. Two survivals from Piedmont in Italy, the Fenestrelle Sword Dance and the Vicoforte Sword Dance, have affinities with the English Sword Play, although both incorporate characters from the *commedia dell'arte*.

The Wooing Ceremony from the East Midlands is characterized by having two women (still played by men) as the chief characters. The Lady is chosen by the Fool in preference to the Old Woman, or Dame Jane; he is killed by rival suitors and then resurrected. The Wooing Ceremony, which seems to have been an occasion rich in ribald vitality and spontaneous folk humour, is no longer performed.

Whatever its origins, the mummers' play, in the more than 3,000 texts that have been recovered, certainly came under the influence of puppet plays, chapbooks, and popular ballads. Literary accretions, too, have made difficulties for scholars: the Wooing Ceremony became conflated with an anonymous playlet entitled *The Recruiting Sergeant* during the 18th century, and the Ampleforth Sword Play contains garbled snatches from CONGREVE's *Love for Love*. Although the mummers' play seems to be mainly associated with Christmas—due perhaps to the influence of Christianity—there are traces of amalgamation with older folk festivals in such rare survivals as the Pace- (or Pasch-) Egg Play, in which the players collect Easter eggs, and the Soul-Cakers' (or Soulers') Play from Cheshire, which takes place on All Souls' Eve (1 Nov.) and for which special cakes are baked and collected, usually by a troupe of children.

The mummers' play has a number of alternative names, such as the Johnny Jacks' Play, the Jolly Boys' Play, and the Tipteerers' Play, while the players were sometimes known as Guizards or Guizers, from their disguise.

MUNDAY, ANTHONY (*c*. 1553–1633), English pamphleteer, ballad-maker, translator, and dramatist, who may have been a boy-actor with OXFORD'S MEN. He is known to have collaborated with CHETTLE and others on several plays, now lost, on ROBIN HOOD and Sir John Oldcastle, and worked mainly for HENSLOWE. The first play attributed to him, *Two Italian Gentlemen, or Fedele and Fortunio* (1584), translated from the Italian, may have been used by Shakespeare for *Much Ado About Nothing* (1598), and parts of his *John a Kent and John a Cumber* (*c*. 1594), first given by the ADMIRAL'S MEN, may have suggested the Bottom scenes in *A Midsummer Night's Dream*, produced by their rivals the CHAMBERLAIN'S MEN a year later. Munday certainly had a hand in the controversial *Sir Thomas More* (1593/5), partly attributed to Shakespeare, since most of the manuscript is in his handwriting. *A Knack to Know a Knave* (1594) has also been attributed to him, and in 1605 he was responsible for *The Triumphs of Reunited Britannia*, the first of several civic pageants which he wrote and devised.

MUNDEN, JOSEPH SHEPHERD (1758–1832), English comedian, who after several years' hard work in the provinces joined the company at COVENT GARDEN in 1790. His first outstanding success was as Old Dornton in the initial production of HOLCROFT'S *The Road to Ruin* (1792). He was the favourite actor of Charles LAMB, who left an admirable description of him, and seems to have been particularly admired in drunken scenes, for which he invented new 'business' at every performance. The broadness of his acting would scarcely be conceivable today and even in his own time he was censured for caricature, but he made the fortune of many a poor play. He remained at Covent Garden for over 20 years, and then went to DRURY LANE, retiring in 1824.

MUNICH, the capital of Bavaria, had a rich cultural life from the 16th century onwards, but although JESUIT DRAMA, first seen in 1560, flourished from 1606 to 1614 with plays written and directed by Jakob BIDERMANN, who in 1609 staged a splendid revival of his *Cenodoxus*, the first half of the 17th century was mainly given over to opera. It was not until 1769 that a German company first became resident in the city, and began gradually to oust the French players who had been so influential there. A National Theatre opened in 1811, and during the 19th century Dingelstedt there and POSSART at the Court Theatre tried to build up a repertoire of German classics in the teeth of considerable opposition, the first importing stars from all over the German-speaking lands to show his audiences contemporary theatre at its best, while the latter became well known for his productions of Shakespeare and SCHILLER with enormous casts in opulent sets. During his years as artistic director of the Court Theatre Josef KAINZ appeared there—from 1881 to 1883—often in private performances commanded by the eccentric King Ludwig II, who insisted on preparing him for an appearance as Schiller's Wilhelm Tell by dragging him round Switzerland to declaim the appropriate lines on the historic sites. Possart also engaged Lautenschläger, the inventor of the revolving stage, and Jocza Savits, a pioneer director who staged Shakespeare without scenery on a stage divided by a curtain into up- and down-stage areas. The trend away from elaborate illusionistic sets was further advocated in Fuchs's theories, which in turn produced Littman's Künstlertheater, founded in 1908. This theatre was leased by REINHARDT from 1911 to 1913. Under the direction of FALCKENBERG the Kammerspiele, founded in 1911 as the Lustspielhaus and already known as the home of EXPRESSIONISM, became the first theatre to stage a play by BRECHT, *Trommeln in der Nacht* (1922), and a year later his *Im Dickicht der Städte* (*In the Jungle of the Cities*) was also first produced in Munich. In 1923 Brecht made his début as a director by producing at the Kammerspiele his free adaptation of MARLOWE'S *Edward II.* Among the other theatres in Munich one of the most interesting, and certainly the most beautiful, is Cuvillies' Residenztheater, built between 1751 and 1753. Bombed in 1943–4, its superb rococo interior was later salvaged and re-erected in the palace, where it reopened in 1951 and is now used for festival productions of intimate drama and opera. Ingmar BERGMAN joined the company in 1977.

MUNICIPAL THEATRE, see CIVIC THEATRE.

MUNK [Petersen], KAJ HARALD LEININGER (1898–1944), Danish priest and playwright, whose first substantial play, *En Idealist* (*Herod the King*, 1928), was disastrously received on its first production and only fully established itself as a major work at its revival in 1938. One of his earliest successes was *Cant* (1931), a verse play about Henry VIII and Anne Boleyn. It was followed by *Ordet* (*The Word*, 1932), which takes as its theme the nature of miracles. After a brief flirtation with Fascism in the early 1930s, Munk became increasingly and openly antagonistic to the movement, as was shown in his next plays, *Sejren* (*The Victory*, 1936), which condemned Italian aggression in Abyssinia, and *Han sidder ved Smeltediglen* (*He Sits by the Melting Pot*, 1938), which was an attack on Nazi anti-Semitism. During the Second World War Munk continued fearlessly to write with *Niels Ebbensen*, *Før Cannae*, and *Ewalds Død*. He was shot by the Nazis to silence his outspoken hostility to their occupation of his country.

MURCELL, GEORGE, see ST GEORGE'S THEATRE, LONDON.

MURDOCH, JAMES EDWARD (1811–93), American actor, considered by many of his contemporaries the finest light comedian of his day. He was especially noted for his clear diction. He began his career at the Arch Street Theatre in Philadelphia in 1829 and appeared at the CHESTNUT STREET THEATRE with Fanny KEMBLE in 1833. Joseph JEFFERSON, who played Moses to his Charles Surface in SHERIDAN's *The School for Scandal* in 1853, admired his acting immensely. Among his other parts were Benedick in *Much Ado About Nothing*, Orlando in *As You Like It*, Mercutio in *Romeo and Juliet*, Mirabell in CONGREVE's *The Way of the World*, and the Rover in Aphra BEHN's play of that name. He remained on the stage until 1858, appearing in England in 1856 in several of his best known parts in which he was well received. On the outbreak in 1861 of the Civil War, in which his only son was killed, Murdoch came out of retirement to give readings and lectures to and in aid of the wounded.

MURPHY, ARTHUR (1727–1805), English actor and dramatist, who gave up acting, for which he had no particular talent, to devote all his energies to playwriting. He adapted VOLTAIRE's *L'Orphelin de la Chine* (1755) for the English stage as *The Orphan of China* (1759) and based some of his best comedies on MOLIÈRE and other French writers, among them *The Way to Keep Him* (1760), *All in the Wrong* (1761), *The School for Guardians* (1767), and *Know Your Own Mind* (1777). His most successful tragedy, which later gave Mrs SIDDONS one of her favourite parts as Euphrasia, was *The Grecian Daughter* (1772), first seen at DRURY LANE. Murphy had little originality but was adept at choosing and combining the best elements from the work of others, and had his place in the revival of the comedy of manners which culminated in SHERIDAN's *The School for Scandal* (1777).

MURRAY, ALMA (1854–1945), English actress, who made her first appearance in London in W. S. GILBERT's *The Princess* (1870) and was later at the LYCEUM with IRVING. She was much admired by BROWNING when she appeared in revivals of his plays—*In a Balcony* in 1884, *Colombe's Birthday* in 1885, and *A Blot in the 'Scutcheon* in 1888. It was, however, as Beatrice in the single private performance of SHELLEY's *The Cenci* on 7 May 1888 that she scored her greatest triumph, and it was much regretted that the LORD CHAMBERLAIN would not allow the play to be given a normal run. The other oustanding part she created was Raina in SHAW's *Arms and the Man* (1894). She continued to play leading parts in many West End productions, and retired in 1915 after playing Mrs Maylie in a dramatization of DICKENS'S *Oliver Twist*.

MURRAY, (George) GILBERT AIMÉ (1866–1957), English classical scholar, poet, humanist, and philosopher, whose verse translations of the plays of EURIPIDES were staged in London at the beginning of the 20th century. Though they have now been superseded, they were in their day superior theatrically to anything heard previously, and revealed the beauties of classical tragedy to those who had no Greek. The first to be staged, under the direction of Harley GRANVILLE-BARKER at the ROYAL COURT THEATRE, was *Hippolytus* in 1904. It was followed by the *Trojan Women* in 1905 and *Electra* in 1906. *Medea*, with the same director, was seen at the SAVOY THEATRE in 1907, the *Bacchae*, in which Lillah MCCARTHY played Dionysus, again at the Court, and *Iphigenia in Tauris*, with Lillah McCarthy in the title role, at the KINGSWAY. The *Alcestis* and the *Rhesus* have had only amateur productions. Murray also translated SOPHOCLES' *Oedipus Rex, Antigone*, and *Oedipus at Colonus*, and AESCHYLUS' the *Oresteia*, the *Suppliant Women, Prometheus Bound*, the *Persians*, and the *Seven Against Thebes*, somewhat less successfully; but his versions of ARISTOPHANES' the *Birds,* the *Frogs*, and the *Knights* were eminently actable, as were his reconstructions of MENANDER's *Perikeiromenê* as *The Rape of the Locks* in 1941 and of the *Epitrepontes* as *The Arbitration* in 1945.

MURRAY, THOMAS CORNELIUS (1873–1959), Irish dramatist, whose subjects were drawn from the peasant and farming life of his native county of Cork, and were distinguished for their sympathetic perception of the deeply religious and sometimes mystical quality of the peasant mind. He made his name with *Birthright*, a tragedy of rivalry and family jealousy produced at the ABBEY THEATRE in 1910. This was followed by *Maurice Harte* (1912) which portrays the conflict between spiritual honesty and family affection in the mind of a young peasant. *The Briery Gap* (published in 1917, but not produced until 1948 at the Abbey's experimental Peacock Theatre) is an exquisite brief tragedy; *Spring* (1918) is a one-act study of poverty and the greed engendered by it; *Aftermath* (1922) is a full-length tragedy on the theme of an arranged marriage; and *Autumn Fire* (1924), another tragedy of mismating, is probably the author's best three-act play. *The Pipe in the Fields* (1927) is a fine, brief study of the sudden flowering of the mind of an artist. *The Blind Wolf* (1928), the first of Murray's plays to be set outside Ireland, is again a peasant tragedy. His last plays were *A Flutter of Wings* (1929), *Michaelmas Eve* (1932), *A Stag at Bay* (1934), *A Spot in the Sun* (1938), and *Illumination* (1939).

MURRAY, WALTER, see KEAN, THOMAS.

MUSAPHIA, JOSEPH, see NEW ZEALAND.

MUSES, The. Three of the nine Muses in Greek mythology, daughters of Mnemosyne the goddess of memory, were particularly connected with the theatre—Melpomene, the Muse of tragedy; Terpsichore, the Muse of dancing; and Thalia, the Muse of comedy.

MUSICAL COMEDY, or MUSICAL, entertainment in which a story is told by a combination of spoken dialogue and musical numbers. Originally the plot of a musical comedy was very slight, but with the importation of more serious themes, particularly in the U.S.A., the word 'comedy' was dropped, and the genre is now known simply as the 'musical'. In the earlier examples the music might or might not be relevant, at some times carrying the story forward or portraying character, at others serving merely as a vehicle for the singers or the chorus. Later it became more integrated with the plot. From the beginning most musical comedies had some dancing, but later it played a more important role. The borderline between the original musical comedy and the operetta or light opera is difficult to define, but the latter tended to be Central European in origin or setting, the music was largely traditional in its rhythms, and there was far less emphasis on dancing.

The musical was ultimately to reach its peak in the U.S.A., where it is generally considered to have originated with *The Black Crook* (1866), an enormously successful spectacle put on at NIBLO'S GARDEN in New York. In it a conventional melodrama by Charles M. Barras was interspersed with songs and ballet. Initially, however, the better examples of the genre originated in England, with productions by George EDWARDES at the GAIETY THEATRE and elsewhere. Unlike plays with INCIDENTAL MUSIC, which continue to be known by the names of their authors, musical comedies and musicals are generally referred to under the names of the composers of the music. Osmond Carr's *In Town* (1892; N.Y., 1897) is often considered the first English musical comedy. It was followed by Sidney Jones's *A Gaiety Girl* (1893; N.Y., 1894), Ivan Caryll's *The Shop Girl* (1894; N.Y., 1895), and Lionel Monckton's *A Country Girl* (London and N.Y., 1902) and *The Quaker Girl* (1910; N.Y., 1911). As musical comedy became established, productions were increasingly exchanged across the Atlantic; Gustave Kerker's *The Belle of New York* (1897), though American in origin, achieved its major success a year later in London, where it became the first American musical to run for over a year, Edna May [Pettie] (1878–1948), who had a short but brilliant career, retiring on marriage in 1907, starring in both productions. On the other hand, Leslie Stuart's *Florodora* (1899), which began its successful career in London, ran even longer in New York after opening there in 1900. A landmark in the early development of the musical comedy was Cook's *In Dahomey* (1903), the first full-length musical written and performed by Negroes to be seen in a major Broadway theatre. It was transferred to London later the same year.

Although the prestige of the American musical stage was materially enhanced by the success of such entertainments as Victor Herbert's *Naughty Marietta* (1910) and Rudolf Friml's *The Firefly* (1912), which, like the 'Savoy operas' of GILBERT and Sullivan in London, deserve perhaps to be classed as light operas rather than musical comedies, musical shows from Europe were still dominant in the early years of the 20th century. Monckton and Talbot's *The Arcadians* and Monckton and Caryll's *Our Miss Gibbs*, both seen in London in 1909, demonstrated the continuing strength of the English musical comedy. Viennese musicals, however, reigned supreme, Franz Lehár's *The Merry Widow* (Vienna, 1905) being seen in London and New York in 1907, and Oscar Straus's *The Chocolate Soldier* (Vienna, 1908)—based without permission on SHAW's *Arms and the Man*—appearing on the New York stage in 1909 and in London in 1910.

The outbreak of the First World War in 1914, with its subsequent wave of anti-German feeling, combined with a somewhat unproductive period on the English musical scéne to bring to an end the domination of the American musical stage by European, or European-style, composers. A number of talented American musicians, writing in a distinctively native idiom, made their appearance, among them Irving Berlin, Jerome Kern, and Cole Porter. Berlin composed the score for the musical-comedy-cum-revue *Watch Your Step* (1914; London, 1915), and in the same season Kern's melodic gifts were shown in the additional songs which he wrote for the American production of Paul Rubens and Sidney Jones's *The Girl From Utah* (London, 1913), including the classic 'They Didn't Believe Me', and for *Nobody Home*, the New York version of Rubens's *Mr Popple of Ippleton*, first seen in London in 1905. Kern's first outstanding success was *Very Good, Eddie!* (1915), which marked a trend towards everyday characters and realistic situations in musical comedy that was to set a pattern for many years to come. One of the collaborators on the book of *Very Good, Eddie!* was Guy Reginald Bolton (1884–1979), born in England of American parents, who was later to collaborate in many famous musicals, including *Kissing Time* (1919), *She's My Baby* (1927), *The Fleet's Lit Up* (1938), and *Follow the Girls* (1945). Cole Porter, whose *See America First* was produced in 1916, was, like Irving Berlin, his own lyricist, but Emmerich Kalman's *Miss Springtime* (also 1916) had lyrics by one of Bolton's most eminent collaborators, P. G. [Sir Pelham Grenville] Wodehouse (1881–1975), a famous English writer of humorous novels, who also worked with Kern and Bolton on *Oh, Boy!* (1917), known in England as *Oh, Joy!* (1919).

The pre-war type of musical comedy still found an audience with the success of *Maytime* (1917) by the Hungarian-born Sigmund Romberg who had settled in America. He also wrote most of the music for *Sinbad* (1918), in which the great star of musical comedy and REVUE, Al Jolson, well known for his singing, in blackface, of coon songs, interpolated two of his best-known melodies, 'My Mammy' and 'Swanee'. The latter was by George Gershwin, later to become one of America's outstanding composers. In 1919 Harry Tierney's *Irene*, with its hit song 'Alice Blue Gown', was the musical success of the season, being seen in London in 1920, in which year Broadway saw *Always You*, the first musical to have book and lyrics by Oscar Hammerstein II (1895–1960). Grandson of the impresario Oscar Hammerstein I (1847–1919), who built theatres in London and New York, he was to become one of the most successful lyric-writers of his generation. The same year also saw the production of *Sally* (London, 1921) by Kern and Bolton, which during its three-year run on Broadway starred Marilyn Miller [Mary Ellen Reynolds] (1898–1936), who became one of Broadway's leading actresses in musical comedy.

A comparative dearth of musicals during the early 1920s was followed by a period of enormous activity during which the American musical achieved a lasting supremacy, even though the first outstanding success of the time, Friml's *Rose-Marie* (1924; London, 1925), seemed to mark a return to the Viennese tradition, as did Romberg's *The Student Prince* (1924; London, 1926), *The Desert Song* (1926; London, 1927), and *The New Moon* (1927; London, 1929). (Hammerstein was co-author of all of these shows except *The Student Prince*.) There were, however, distinct traces of American influence in the music of *Rose-Marie*, and Gershwin's *Lady, Be Good!* (1924; London, 1926) was already employing the jazz rhythms which were to figure so prominently in later musicals of the 1920s. Starring the dancers Fred Astaire and his sister Adèle, with a book by Bolton and Fred Thompson, this work marked the beginning of Gershwin's long association with his brother Ira [Israel] (1896–   ) as lyricist, a collaboration which continued with *Oh, Kay!* (1926; London, 1927), starring Gertrude LAWRENCE, with a book by Bolton and Wodehouse.

In September 1925 four outstanding musical shows opened on Broadway within a week. The first was Vincent Youmans's *No! No! Nanette!*, already seen in London in Mar. 1925, with its hit songs 'I Want To Be Happy' and 'Tea for Two'; then came *Dearest Enemy*, which marked the first collaboration on a full musical score of the composer Richard Rodgers with Lorenz Hart (1895–1943); Friml's *The Vagabond King* followed, with Dennis KING as François Villon, the part played in London two years later by Derek Oldham (1892–1968); and finally Jerome Kern's *Sunny* (London, 1926), with book and lyrics by Harbach and Hammerstein, which again starred Marilyn Miller in the part played in London by Binnie Hale.

In the theatrical season of 1927–8 the number of musicals opening in New York reached a record total of over 50; among them were Henderson's *Good News* (London, 1928); Rodgers and Hart's *A Connecticut Yankee* (seen in London in 1929 as *A Yankee at the Court of King Arthur*), based on a story by Mark Twain; the Gershwins' *Funny Face* (London, 1928) with the Astaires; and, most successful of all, *Show Boat*, with a book and lyrics by Hammerstein based on a novel by Edna Ferber and some of Kern's finest music, including 'Ol' Man River', sung in the 1928 London production by Paul ROBESON.

In spite of the Depression and the advent of the film musical, the heyday of the American musical continued in the 1930s. From the Gershwins came *Strike Up the Band* and *Girl Crazy* (both 1930), the latter marking the first appearance in musical comedy of Ethel MERMAN. Another musical by the Gershwins, *Of Thee I Sing* (1931), written in collaboration with George S. KAUFMAN and Morris Ryskind, received the first PULITZER PRIZE to be awarded to a musical; but the finest work the Gershwins wrote for the stage was undoubtedly the folk opera *Porgy and Bess* (1935; London, 1952), based on a play by Dorothy and Du Bose Heyward. Among the other outstanding musicals of the time were Kern's *The Cat and the Fiddle* (1931; London, 1932), *Music in the Air* (1932; London, 1933), in which he again collaborated with Hammerstein, and *Roberta* (1933); Irving Berlin's *Face the Music* (1932); Rodgers and Hart's *On Your Toes* (1936; London, 1937), with a book by George ABBOTT, *Babes In Arms* (1937), and *I'd Rather Be Right* (also 1937), an amusing political satire by Kaufman and Moss HART in which the part of Franklin D. Roosevelt was played by the vaudeville actor George M COHAN; and Cole Porter's *The Gay Divorce* (1932; London, 1933), with Fred Astaire, and *Anything Goes* (1934; London, 1935), with a book by Bolton and Wodehouse, and starring Ethel Merman.

By 1937, the year in which Gershwin died, the first great phase of the American musical was over. The film musical was already proving an alluring alternative for some of the most talented exponents of the genre, and the political situation had a depressing effect. There was still much to enjoy, however. In 1938 Cole Porter's *Leave It To Me* introduced Mary MARTIN to Broadway, and Rodgers and Hart scored a great success with *The Boys From Syracuse*, adapted by George Abbott from *The Comedy of Errors*; curiously enough, it was not seen in London until 1963. Cole Porter scored again with *Du Barry Was a Lady* (1939; London, 1942) and *Panama Hattie* (1940; London, 1943), both of which provided Ethel Merman with fine starring parts on Broadway, played in the London productions by Frances Day (as the Du Barry) and Bebe Daniels (as Hattie). In other hands the American musical was showing signs of greater sophistication, with Rodgers and Hart's *Pal Joey* (1940; London, 1954), a cynical study of a gigolo, and particularly with *Lady In the Dark* (1941), which starred Gertrude Lawrence in a plot featuring psychoanalysis and dreams. The libretto was by Maxwell ANDERSON, the lyrics by Ira Gershwin, and the music by Kurt Weill, who with his Austrian-born wife, the singer and actress Lotte Lenya [Karoline Blamauer] (1900–81), had emigrated to America from Germany in the 1930s; both are known best for their association with Bertolt BRECHT. A year later Rodgers and Hart were to collaborate for the last time in *By Jupiter*, Hart dying shortly after and Rodgers beginning a new and fruitful career in association with Hammerstein.

Their first work together was *Oklahoma!* (1943; London, 1947), based on Lynn Riggs's play *Green Grow the Lilacs*. Its success made a star of Alfred Drake [Capurro] (1914–   ), who played the part of Curly in New York, and it marked for the American musical as a whole a transition to stronger and more serious plots, greater integration of words and music, more elaborate choreography with the collaboration of such choreographers as Agnes de Mille and Jerome Robbins, and the use of non-contemporary settings. These trends were reinforced by the success of the next Rodgers and Hammerstein musical, *Carousel* (1945; London, 1950), based on MOLNÁR's *Liliom*, though the light-hearted atmosphere of earlier musicals was maintained by the first venture into the theatre of the composer Leonard Bernstein, *On the Town* (1944; London, 1963), which followed the adventures of three American sailors on 24-hour leave in contemporary New York, and by Irving Berlin's *Annie Get Your Gun* (1946), which, with its superb score, provided another spectacular role for Ethel Merman as the sharpshooting Annie Oakley, a role played on tour by Mary Martin and in London in 1947 by Dolores Gray.

Meanwhile the pre-war years had produced little of note in London apart from the musical plays of Noël COWARD and Ivor NOVELLO, who successfully continued the old tradition of European operetta. British musical stars tended to be seen in revue, though Jessie MATTHEWS appeared in Rodgers and Hart's *Ever Green* (1930), which never reached Broadway, and Jack BUCHANAN, Leslie HENSON, Evelyn LAYE, and the husband-and-wife team of Cicely COURTNEIDGE and Jack Hulbert all starred in musical comedy. *Bless the Bride* (1946), which had music by Vivian Ellis and a libretto by A. P. [Sir Alan Patrick] Herbert (1890–1971), had a long run in London, starring the French singer Georges Guétary and the English actress Lizbeth Webb. Ellis and Herbert were also responsible for *Big Ben* (1946), *Tough at the Top* (1949), and *The Water Gipsies* (1955), based on one of Herbert's novels.

After the Second World War, however, the pre-eminence of the American musical went unchallenged, and it was a rare theatrical season that did not produce at least one popular and critical success. *Brigadoon* (1947; London, 1949), about a Scottish village that comes to life for one day every century, was the work of a new team, Alan Jay Lerner (1918–   ) as author of the book and lyrics and Frederick Loewe as composer, who were to have several successes together. Cole Porter's *Kiss Me, Kate* (1948; London, 1951) starred Alfred Drake in New York as an actor feuding with his co-star (who is also his ex-wife) while appearing in a musical version of *The Taming of the Shrew*. (Drake was later to star in 'straight' Shakespeare, his roles including Benedick in *Much Ado About Nothing* at Stratford, Conn., and Claudius to Richard BURTON's Hamlet.) Rodgers and Hammerstein's *South Pacific* (1949; London, 1951) starred Mary Martin (as Ensign Nellie Forbush) in New York and London, and became the second musical to be awarded a Pulitzer Prize. Also in 1949 Jule Styne, who had scored a success in 1947 with *High Button Shoes* (London, 1948), produced *Gentlemen Prefer Blondes* (based on Anita Loos's novel of the same name), which made a star of Carol Channing, her part of Lorelei being played in London in 1962 by Dora Bryan. Irving Berlin's *Call Me Madam* (1950; London, 1952) provided yet another triumph for Ethel Merman as Sally Adams, the part played in London by Billie Worth, and was soon followed by one of the best of modern musicals, *Guys and Dolls* (London, 1953), which had superb music and lyrics by Frank Loesser and an amusing and literate book based on the stories and characters of Damon Runyon which retained all the idiosyncrasies of Runyonesque speech.

During the 1950s the continued success of American composers and librettists successfully concealed the fact that few new talents seemed to be emerging. One of the longest runs was achieved by Brecht and Weill's *The Threepenny Opera* (first produced in Berlin in 1928) with Lotte Lenya in her original role as Jenny. The long-established partnership of Rodgers and Hammerstein had two outstanding successes with *The King and I* (1951; London, 1953), based on Margaret Langdon's biography *Anna and the King of Siam*, which starred Gertrude Lawrence and Yul Brynner in New York and Valerie Hobson and Herbert Lom in London, and *The Sound of Music* (1959; London, 1961), starring Mary Martin in New York, based on Maria Augusta von Trapp's *The Trapp Family Singers*. Lerner and Loewe did well with *Paint Your Wagon* (1951; London, 1953), and then produced one of the greatest musicals of all time, *My Fair Lady* (1956; London, 1958), based on Shaw's *Pygmalion*, in which Rex HARRISON and Julie Andrews starred both in New York and in London. In *Kismet* (1953; London, 1955), a musical version of KNOBLOCK'S play starring Alfred Drake in both New York and London, music by Borodin was used for the songs. The same year saw the first of three outstanding musicals by Bernstein in the 1950s—*Wonderful Town* (London, 1955), in which he again collaborated with the librettists of *On the Town*, Betty Comden (1915–) and Adolph Green (1915–). It was followed by *Candide* (1956; London, 1959), based by Lillian HELLMAN on VOLTAIRE's satire—a failure when first produced, but revised and revived with acclamation in 1973—and by the notable *West Side Story* (1957; London, 1958), a retelling of the story of *Romeo and Juliet* in terms of gang warfare in contemporary New York, with superb choreography by Jerome Robbins, whose idea it was. The lyrics for the last marked the début on the musical comedy scene of Stephen SONDHEIM, who was also responsible for the lyrics of another outstanding musical, Styne's *Gypsy* (1959; London, 1973), in which the leading role of Gypsy Rose Lee's mother was played by Ethel Merman. Cole Porter's last two stage musicals also date from the 1950s, *Can-Can* (1953; London, 1954) and *Silk Stockings* (1955), the latter based on Greta Garbo's film *Ninotchka*. *The Pajama Game* (1954; London, 1955), by two comparative newcomers, Richard Adler and Jerry Ross, which had a libretto by George Abbott, marked the début as a choreographer of Bob Fosse, who was to play an important part in many Broadway musicals. Another stalwart, Loesser, contributed the book and music of *The Most Happy Fella* (1956; London, 1960), based on Sidney HOWARD's play *They Knew What They Wanted*. *The Music Man* (1957; London, 1961), with its popular song 'Seventy-Six Trombones', reached heights never again achieved by its composer Meredith Willson, and the third Pulitzer Prize to be given to a musical was awarded to Jerry Bock's *Fiorello!* (1959; London, 1962), based on the career of New York's famous mayor La Guardia.

Meanwhile the English musical had been showing signs of revival with a gentle parody of the 1920s musical in *The Boy Friend* (1953) by Sandy Wilson, seen in New York in 1954 with Julie Andrews as Polly Browne, and *Salad Days* (1954) by Julian Slade. Neither composer, however, was able to achieve such a success again, and it was not until 1960 that Lionel BART was able to raise English musicals from the doldrums with *Oliver!*, an adaptation of Charles DICKENS's *Oliver Twist*. It was seen in 1963 in New York, where the dearth of new talent was at last making itself felt. After Hammerstein died in 1960 Rodgers wrote five more musicals, for the first of which, *No Strings* (1962), he wrote the lyrics himself, while for *Do I Hear a Waltz?* (1965) the lyric writer was Sondheim. His last score was *I Remember Mama* (1979). Loewe collaborated again with Lerner in the Arthurian *Camelot* (1960; London, 1964), based on T. H. White's *The Once and Future King*, starring Richard BURTON and Julie Andrews in New York and Laurence Harvey in London. Loewe's only further stage work was a collaboration with Lerner on a stage version of their film musical *Gigi* in 1973. Neither Rodgers nor Lerner did so well with new partners.

Loesser's last work to be seen in New York, *How To Succeed in Business Without Really Trying* (1961; London, 1963), won the fourth Pulitzer Prize to be awarded to a musical, and Sondheim scored an outstanding success with *A Funny Thing Happened on the Way to the Forum* (N.Y. and London, 1962), based on PLAUTUS and starring Zero MOSTEL in New York and Frankie Howerd in London. His next work, *Anyone Can Whistle* (1964), though not a great success, marked the début in musicals of the London-born actress Angela Lansbury, who was later to appear in several more, including the long-delayed London production of *Gypsy* in 1973 and a 1974 New York revival. Styne's *Funny Girl* (1964; London, 1966) was based on the career of the revue star Fanny Brice, and made a star of Barbra Streisand. Jerry Herman's *Hello, Dolly!* (1964; London, 1965), based on WILDER's play *The Matchmaker*, owed much of its success to its title song, elaborate staging, and the appeal of Carol Channing, the first actress to play the part of Dolly, later played in America by Ginger Rogers, Ethel Merman, Betty Grable, Mary Martin, and, in an all-black production, Pearl Bailey. In London the part was first played by Mary Martin and then by Dora Bryan. The greatest success of the 1960s, however was Jerry Bock's *Fiddler on the Roof* (1964; London, 1967), which starred Zero Mostel on Broadway and TOPOL in London. Based on Sholom ALEICHEM's *Tevye the Milkman*, it ran in New York for almost eight years, the longest run ever achieved by a Broadway musical (though

Harvey Schmidt's *The Fantasticks*, based on ROSTAND's *Les Romanesques*, which opened in 1960 in a small Greenwich Village theatre, the Sullivan Street Playhouse, was still running in the 1980s).

Other interesting musicals of the 1960s included Mitch Leigh's *Man of La Mancha* (1965; London, 1968), based on CERVANTES'S *Don Quixote*, and Jerry Herman's *Mame* (1966; London, 1969), which starred Angela Lansbury in New York and Ginger Rogers in London. The popular playwright Neil SIMON wrote the books for two adaptations of films, Cy Coleman's *Sweet Charity* (1966; London, 1967), based on Fellini's film *Notti di Cabiria*, and Burt Bacharach's *Promises, Promises* (1968; London, 1969), based on Billy Wilder's film *The Apartment*. Perhaps the most satisfying musical of the decade artistically, however, was John Kander's *Cabaret* (1966; London, 1968), based on VAN DRUTEN's play *I Am a Camera*, itself a dramatization of Isherwood's novel *Goodbye to Berlin*. The show, in which Sally Bowles was played in New York by Jill Haworth and in London by Judi DENCH, evoked most successfully the atmosphere of Berlin in the 1930s. Some of the songs showed the influence of Kurt Weill, whose widow, Lotte Lenya, was in the American production. *Cabaret* was directed by Hal [Harold] Prince (1928–), a producer who became a well-known director, especially of musicals, including many of Sondheim's. The most unorthodox success of the decade was probably Galt MacDermot's *Hair* (N.Y. and London, 1968), which brought rock music and total nudity to the stage.

During the 1960s English writers too made some interesting contributions to the musical scene. *Stop the World—I Want To Get Off* (1961; N.Y., 1962), by Anthony Newley and Leslie Bricusse, was a decided success, and in 1963 (N.Y., 1965) *Pickwick*, with music by Cyril Ornadel and lyrics by Bricusse, proved to be another acceptable Dickens adaptation. David Heneker's *Half-a-Sixpence* (1963; N.Y., 1965), based on H. G. Wells's novel *Kipps*, provided a lively vehicle, both in London and in New York, for Tommy Steele [Hicks] (1936–), a former pop star who was to become an all-round entertainer, even appearing at the OLD VIC in GOLDSMITH's *She Stoops to Conquer* in 1960; the same composer's *Charlie Girl* (1965), which opened to very poor reviews, nevertheless ran for over 2,000 performances with Anna Neagle as Lady Hadwell.

By the late 1960s the American musical had lost much of its former appeal. Though the cinema had long been a serious competitor for talent in the musical field, creators of successful stage musicals could normally assume that a film version, providing extra earnings and a much larger audience, would follow. Hollywood's financial troubles, however, brought about a decrease in the number of film musicals, which were expensive to produce. In any case the music of stage musicals, which had once provided many of the popular songs of the day, no longer held any appeal for film audiences addicted to the fashionable rock and beat music, while the latter was not much liked by theatre audiences and not thought suitable for theatrical purposes. Theatre music almost disappeared from the hit parade, and musical shows therefore received less publicity. These factors, combined with a shortage of talent and the mounting cost of production, made successful musical shows increasingly rare. The desperate need to fill empty theatres on Broadway led to an increasingly bizarre choice of subject. Musicals were made from films—*Ilya Darling* (1967) from Jules Dassin's *Never On Sunday, Applause* (1970; London, 1972) from Joseph Mankiewicz's *All About Eve*; plays—*Sherry* (1967) from Kaufman and Hart's *The Man Who Came To Dinner, Cyrano* (1973) from Rostand's *Cyrano de Bergerac*; novels—*Billy* (1969) from Melville's *Billy Budd, Gantry* (1970) from Sinclair Lewis's *Elmer Gantry*; even comic strips—*You're A Good Man, Charlie Brown* (1967; London, 1968) based on 'Peanuts', *Annie* (1977; London, 1978) on 'Little Orphan Annie'. When such adaptations failed to work, as most of them did, producers turned in desperation to revivals.

Fortunately there was still Stephen Sondheim, the only major composer still writing regularly for the musical stage. One of the stars of his *Company* (1970; London, 1972), who remained in England after appearing in the London production, was Elaine Stritch (1926–), who had already been seen in a number of musicals, such as Coward's *Sail Away* (1961; London, 1962) and revivals of *Pal Joey* (1952) and *On Your Toes* (1954), as well as in straight plays including INGE's *Bus Stop* (1955) and ALBEE's *Who's Afraid of Virginia Woolf?* (1963). Sondheim was to write a number of other musicals during the decade, but otherwise there were few productions of any importance. Stephen Schwartz's rock musical *Godspell* (N.Y. and London, 1971) was based on the Gospel according to St Matthew, and Jim Jacobs and Warren Casey's *Grease* (1972; London, 1973) looked back nostalgically to the rock-and-roll heyday of the 1950s. John Kander's *Chicago* (1975; London, 1979), choreographed by Fosse, who also helped to adapt it from a 1926 play of the same name by Maurine Dallas Watkins, amusingly satirized American justice in the 1920s, using dance-rhythms of the period. Probably the best musical of the 1970s not by Sondheim was Marvin Hamlisch's *A Chorus Line* (1975; London, 1976), in which each person arriving for an audition had to give a potted autobiography as well as a demonstration of dancing; it won the Pulitzer Prize.

For once it was the English musical which was to produce the only major new talent of the decade. Tim Rice (1944–) as lyricist and Andrew Lloyd Webber as composer collaborated on a musical, first seen in New York, based on the life of Christ, *Jesus Christ Superstar* (1971; London, 1972), and on *Evita* (1978; N.Y., 1979), based on the career of Eva Perón. Other English musicals of the decade were John Barry's *Billy* (1974), based on the play *Billy Liar* by Willis HALL and Keith Waterhouse (in turn based on Waterhouse's novel); *The Good Companions* (also 1974), based on J. B. PRIESTLEY's best-selling novel of the 1930s, with music by the conductor André Previn and starring John Mills and Judi Dench; and Monty Norman's *Songbook* (1979), which cleverly parodied various styles of popular music, and had a brief run in New York in 1981 as *The Moony Shapiro Songbook. Cats* (1981), based on T. S. ELIOT's *Old Possum's Book of Practical Cats*, with music by Andrew Lloyd Webber, was a big success.

By the 1980s, however, the musical was at a low ebb on both sides of the Atlantic, and the economic recession made it even harder to stage spectacular shows and attract audiences. It is significant perhaps that the big hit of the 1980 Broadway musical stage was an adaptation—admittedly with dazzling new choreography—of the 1933 film musical *42nd Street*.

MUSIC BOX, New York, at 239 West 45th Street between Broadway and 8th Avenue. Seating 1,010, this theatre opened on 22 Sept. 1921 with a REVUE which ran into several editions. After a further series of revues came KAUFMAN and HART's *Once in a Lifetime* (1930) and *Of Thee I Sing* (1931), a collaboration between Kaufman and Morris Ryskind which was the first musical to win a PULITZER PRIZE. Later successes were Kaufman and Ferber's *Dinner at Eight* (1932); STEINBECK's *Of Mice and Men* (1937); *The Man Who Came to Dinner* (1939), again by Kaufman and Hart; *I Remember Mama* (1944) by John VAN DRUTEN; and Maxwell ANDERSON's *Lost in the Stars* (1949) with music by Kurt Weill. Since then the theatre has mainly been used for straight plays, including INGE's *Picnic* (1953), RATTIGAN's *Separate Tables* (1956), *Rashomon* (1959), based on a play by Akutagawa, *A Far Country* (1961) by Henry Denker, and *Any Wednesday* (1964) by Muriel Resnik. Anthony Shaffer's *Sleuth* (1970), imported from London, was followed by two other successes from the same source, AYCKBOURN's *Absurd Person Singular* (1974) and Trevor Griffiths's *Comedians* (1976). The musical *Side by Side by Sondheim* (1977) was succeeded in 1978 by Ira Levin's *Deathtrap* with John WOOD, which continued into the 1980s.

MUSIC-HALL, a type of entertainment which flourished in Great Britain during the second half of the 19th and early 20th centuries. It arose from the desire of tavern landlords to emulate the success of the music programmes put on in such pleasure resorts as 'Miles's Musick House' at SADLER'S WELLS and in the 'song-and-supper rooms' of which EVANS'S was typical. There had always been 'free-and-easy' sing-songs during convivial evenings in London taverns, but in the 1840s some landlords, hoping to attract more and better-class customers, adapted a room on their premises, or even built one on, specially for organized musical evenings. Among the earliest of these 'saloons' were those attached to the Britannia in Hoxton and the Grapes in Southwark Bridge Road. The latter, renamed in 1856 the WINCHESTER MUSIC-HALL, was already by the mid-1840s displaying many of the features of the later music-halls, as were the Mogul Saloon, or 'Old Mo', in Drury Lane, which in 1851 became the MIDDLESEX MUSIC-HALL, and the Sun Music-Hall in Knightsbridge, also licensed in 1851, which flourished until 1890. But the history of the modern music-hall is generally held to have begun in 1852, when the astute and far-sighted Charles MORTON, who three years earlier had opened a hall for musical programmes attached to his tavern, the Canterbury Arms, built and opened adjacent to it the CANTERBURY MUSIC-HALL. There he exploited and improved on all the tentative efforts of the last ten years, to provide a pattern of entertainment which was seized on and copied by tavern keepers not only in London but all over the country, until the old 'free-and-easy' became a new genre in its own right which eventually demanded its own purpose-built Theatres of Variety. The word 'variety' was contemporary with 'music-hall', the two existing side-by-side for many years until with increasing sophistication 'variety' almost ousted the former 'music-hall'; but both words described the same type of programme, and the same line of development, from the simple stage at one end of a hall furnished with tables and chairs and a bar, the whole presided over by a CHAIRMAN, to the elaborate two- or three-tier auditorium with a fully-equipped stage behind a proscenium arch, which began to take shape in the 1890s.

Once Morton had made a success of his venture others were not slow to follow, and during the next ten years innumerable music-halls opened up everywhere, particularly in the East End of London. Some lasted only a few years, but others survived to enjoy considerable popularity. Typical were the hall, later known as Crowder's, on the site of the present GREENWICH THEATRE (3), and that attached to the Griffin in Shoreditch High Street, which opened in 1856, was renamed the London Music-Hall in 1894, and from 1916 to 1935 was even better known as the Shoreditch Empire, now demolished. In 1857 Weston's Music-Hall, later the HOLBORN EMPIRE, was the first to challenge the supremacy of the Canterbury; a year later a room attached to the Rose of Normandy Tavern in Marylebone High Street was converted by the Irish comedian Sam COLLINS (who in 1863 was to open a music-hall under his own name in Islington Green) into the Marylebone Music-Hall, which flourished until 1894, when it was closed as being unsafe. In 1858 the concert room at Wilton's, in Grace's Alley, Wellclose Square, opened as a music-hall, to become Fredericks' Royal Palace of Varieties before it closed in 1877 after a disastrous fire. The building survived to become one of the oldest relics of the music-hall era in Great Britain, used in 1966 for filming and television while plans were made to restore it to its original use. A close contemporary of Wilton's, the South London, built and decorated in imitation of a Roman villa, opened on 30 Dec. 1860 with a strong bill. It was burned down on 28 Mar. 1869, but soon rebuilt and reopened, and in 1873 an enterprising manager, J. J. Poole, produced excellent shows there, among his 'discoveries' being Connie GILCHRIST. The building was badly damaged by enemy action in 1941 and demolished in 1955.

The music-halls of London set the pattern for those in the provinces, which were visited regularly by all the 'stars', who considered Birmingham, Manchester, and other cities as important as London and used the same material in them all. Their history can best be followed by considering some of the best known, like the ALHAMBRA in Leicester Square, which opened shortly before the South London on 20 Dec. 1860 and survived until 1 Sept. 1936; or the LONDON PAVILION, which opened in 1861 and retained until 1886 the old-style separate tables and chairs which were elsewhere being abolished in favour of conventional theatre seating; or the BEDFORD in Camden High Street, which also opened in 1861 and in 1890 was painted by Sickert.

A landmark in the history of the music-hall was the opening on 26 Mar. 1861, under Charles Morton, of the OXFORD MUSIC-HALL; it marked the beginning of a new era, during which the last links with the old 'free-and-easys' were broken, and the Palaces of Variety began to compete on equal terms with the established theatres, even importing into their programmes stars of the straight stage in dramatic sketches and excerpts from current productions. Among the new-style music-halls were the METROPOLITAN, which as Turnham's opened in 1862; the Royal Standard, later the VICTORIA PALACE, which opened in 1863; and the short-lived Strand Musick-Hall, later the GAIETY THEATRE, which opened in 1864. The last important 'halls' to open in the 1860s were those run by the GATTIS—Gatti's-Over-the-Water in 1865 and Gatti's-Under-the-Arches in 1867. During the next 20 years the music-hall consolidated its position as one of the most important elements in contemporary entertainment, and its stars became highly paid and world famous. Many 'halls' came and went and are now forgotten, but one whose name is still remembered—the TROCADERO, later famous for its cabaret—opened in 1882, and in 1887 another very popular hall, the EMPIRE, close neighbour and rival of the Alhambra, opened in Leicester Square. It was almost the last of the great music-halls, except for the PALACE THEATRE of Varieties in Cambridge Circus, which opened in 1892, and the TIVOLI, in the Strand, which opened in 1893, both being creations of the ever-active Charles Morton, now close to the end of his career as 'the father of the halls'. The music-hall was already feeling the competition of the cinema at the turn of the century, many 'halls' indeed helping the newcomer by showing films in their programmes, and the only additions to the 'halls' of London, all fairly unimportant, were destined to be the Euston, later the REGENT THEATRE, in 1900; Stoll's COLISEUM, in 1904; the Olympia, Shoreditch, which had for many years previously functioned as the STANDARD THEATRE, in 1907; and the Islington Empire, which as the GRAND THEATRE (1) had been famous for its PANTOMIMES, in 1908. In 1909 the HIPPODROME went over to variety, losing its circus ring in the process, and a year later the present LONDON PALLADIUM also became a home of variety mixed with dramatic sketches and revues; and in 1912 there opened a suburban 'hall', the Chiswick Empire, typical of many all round London. Its three-tier auditorium, holding 2,000 people, was often filled to capacity, especially on Saturday nights. Finally, in the same year, the opera house which Oscar Hammerstein had built in 1911 hoping to rival COVENT GARDEN, was given over to variety and revue. It later became the STOLL THEATRE, which Oswald Stoll (1866–1942) made the headquarters of his 'variety' empire, while his former partner Edward Moss (1852–1912) based himself on the Hippodrome. After that no new music-halls were built; the genre had passed its peak and was to decline, in spite of many efforts to revive it.

The stars who made the 'halls' and were made by them were as many and diverse at the buildings themselves. At first singers like Sam COLLINS and Sam COWELL came in from the old 'song-and-supper rooms', which also gave the music-hall its first comedian in 'blackface', Joe [Joseph] Arnold Cave (1823–1912), later manager of the Marylebone Music-Hall, as well as Harry Clifton (1832–72), who wrote the words and music of 'Pretty Polly Perkins of Paddington Green' and sang 'motto' songs like 'Paddle Your Own Canoe', and Victor Liston (1838–1913), who popularized Fred Albert's 'Shabby Genteel' ('Too proud to beg, too honest to steal'), which was much admired by Edward VII when Prince of Wales. It was not long before the music-hall was making its own stars and its own stereotypes. One of the first, which replaced for a time the earlier stereotype of the tough, good-hearted working-man, was the 'man-about-town', the 'West End toff'. This was the invention of George LEYBOURNE, the first of the *lions comiques*, and The Great VANCE, whose versatility covered also Cockney and 'motto' songs, which did not hinder him from being hailed as a *lion comique* on a par with Leybourne.

One of the interesting features of the early music-hall was its employment of women singers, and later *comédiennes*, in its programmes. Women had been seen rarely, if ever, at the Coal Hole and Cyder Cellars and suchlike male resorts, probably not even at Evans's, which reluctantly, in the late 1860s, had allowed women with a male escort to sit in the auditorium behind a grille, so that they could not be seen. But from the early days of the Canterbury, and later at the Oxford, Morton engaged Emily Soldene (1840–1912) to sing excerpts from light opera. She was the forerunner of such almost forgotten stars as Bessie Bellwood [Elizabeth Ann Katherine Mahony] (1847–96), who was turned down at the Holborn for being 'too quiet', but later learned to belt out such Cockney songs as 'Wot Cheer, 'Ria!', and of Jenny Hill [Elizabeth Pasta] (1851–96), a diminutive, sharp-featured little creature, billed as 'the Vital Spark', whose early privations, combined with the frenzy with which she sang and danced once she became famous, contributed to her early death. The work of such performers as these came to perfection in Marie LLOYD, to many the finest of all music-hall's comic women singers.

Although, understandably, it is the great solo performers of the music-hall who are mainly remembered, the many 'turns' on a normal evening's bill—sometimes as many as 20 to 25 at a time—were all different, comprising alone or in groups every type of acrobat, juggler, conjurer, ventriloquist, speciality dancer, slapstick or knockabout comedian, and singers who ran the gamut from vulgar, often suggestive, comic songs to serious ballads. The old 'song-and-supper rooms' had featured little but singers, with an occasional acrobat, conjurer, or ventriloquist, and it is to the music-hall of the second half of the 19th century onwards that the greatest of these belong, among them Jules Léotard (1830–70) and Charles Blondin [Jean François Gravelet] (1824–97), Paul Cinquevalli [Kestner] (1859–1918), and Fred Russell [Thomas Frederick Parnell] (1862–1957), who changed his name because of the current scandal involving the Irish politician Charles Parnell and Kitty O'Shea. Léotard, who was first seen in London in 1861 at the Alhambra, where his grace and agility captured all hearts, has been immortalized in the song 'The Daring Young Man on the Flying Trapeze', and his name is now given to the sober one-piece practice costume, adopted by acrobats and ballet-dancers, which he always wore in contrast to the gaudily spangled accoutrements of many of his fellow trapeze-artists. The only person to approach him in daring and popularity was Blondin, who also appeared at the Alhambra, in 1862. Of a later generation Cinquevalli, a German by birth, also began as an acrobat, but after an accident became probably the greatest juggler of all time, his skill with everything from a billiard ball to a cannon ball being universally recognized. His contemporary Fred Russell, who made his first appearance with his doll 'Coster Joe' in 1896, soon became 'top of the bill' wherever he appeared, touring extensively and retaining his faculties to the end of his long life. He too was known as 'the father of the halls' because of the part he played in founding the Variety Artists' Federation, now part of the BRITISH ACTORS' EQUITY ASSOCIATION, of which he became president when he was 90. No fewer than 14 members of his family were connected with variety, among them his sons Russ [Frederick Russell] Carr (1889–1973), also a ventriloquist, and Val(entine) Charles Parnell (1894–1972), for many years associated with music-hall management.

The 1870s and 1880s saw the rise to popularity of many stars now forgotten, or remembered perhaps for one song. Such a one was James Henry Stead (?–1886), who unaccountably achieved fame for several years with 'The Perfect Cure', which he sang in a costume with broad stripes, his pale face painted with red cheeks like a Dutch doll and adorned with a small moustache and imperial, the whole surmounted by a clown's hat. As he was primarily an eccentric dancer, the song was merely an excuse for complicated contortions, and for a series of leaps—some 1,600 in all—during which he held his arms tightly by his sides. When the act fell out of favour, he had nothing to replace it and died in poverty in an attic in Seven Dials. More versatile in their time, but still remembered for one or two songs only, were William B. Fair (1851–1909), with 'Tommy, Make Room for Your Uncle', which he sang at as many as six 'halls' in one night, and Charles Coborn [Colin Whitton McCallum] (1852–1945), with 'Two Lovely Black Eyes' and 'The Man Who Broke the Bank at Monte Carlo'.

Among those who came to the music-halls from the MINSTREL SHOW, retaining their 'blackface' make-up, were George H. Chirgwin (1854–1922), originally known as 'the White-Eyed Musical Moke', but from 1877 billed as 'the White-Eyed Kaffir' because of the white lozenge-shaped patch round his right eye, who accompanied himself on the one-string fiddle while he sang sentimental 'coon' songs, of which the favourites were 'The Blind Boy' and 'My Fiddle is My Sweetheart', sung in a high-pitched, piping voice; and slightly later Eugene Stratton [Eugene Augustus Ruhlman] (1861–1918), who also sang 'coon' songs, the best remembered being 'Lily of Laguna', to which he whistled a refrain while dancing a 'soft-shoe shuffle' in a spotlight on a darkened stage, a noiseless, moving shadow. Harry Champion (1866–1942), on the other hand, soon discarded his 'blackface', and after making a hit with such songs as 'Ginger, Ye're Barmy' and 'I'm 'Enery the Eighth I am, I am', is mainly remembered for his songs in praise of food—'Boiled Beef and Carrots', 'Hot Tripe and Onions'—which he sang at terrific speed and with great zest and vitality up to the day of his death.

At a time when Britain was at the height of her prosperity, patriotic songs, particularly when sung by a handsome man in army or naval uniform, formed part of the staple fare of the halls. Among the best known singers of such songs was The Great Macdermott [Gilbert Hastings Farrell] (1845–1901), whose 'We Don't Want to Fight, but By Jingo! if we do' added the word 'Jingoism' to the English language. He was something of a *lion comique*, as was Charles Godfrey [Paul Lacey] (1851–1900), who came from the theatre to sing 'Masher King' in 1880, dressed in silk knee-breeches and buckled shoes, but also put on uniform for the descriptive *scena* songs in which he excelled, among them 'Poor Old Benjamin', about a veteran of the Crimean War who when turned away from a casual ward with the words 'Be off. You are not wanted here!' replied 'No, I am not wanted *here*; but at Balaklava—I was wanted *there*!' He reversed the process adopted by Arthur Roberts (1852–1933), who left the music-halls in 1880 to go on the stage, but returned to them in 1903 to become a 'Veteran of Variety'. Another handsome singer of naval songs, Tom Costello (1863–1945), was famous in his own day for 'Comrades', but is now better remembered as the henpecked husband in 'At Trinity Church I Met Me Doom'. Undoubtedly one of the outstanding figures of the 'halls' in the last 15 years of the 19th century was Dan LENO, who made his first adult appearance in a London music-hall in 1885, the same year as Marie Lloyd. For many people these two epitomize the spirit of the 'halls' at their best, and also highlight that connection between music-hall and pantomime which was to be a marked feature of the next 30 years or more. Marie Lloyd was seen in only three pantomimes, but Leno made an annual appearance at DRURY LANE for 15 years, his tiny, mercurial figure set off by the vast bulk and jolly red face of Herbert Campbell [Story] (1844–1904), his partner for many years, and later by the clowning of Harry Randall (1860–1932), who made his first appearance at the Oxford Music-Hall in 1883, and for ten years was the mainstay of the pantos at the Grand Theatre, Islington. He moved to Drury Lane in 1902, and also appeared in Leno's last pantomime in 1903, proving a worthy successor to him in the years that followed. Like Leno and Campbell, he spent the time between pantos (which usually lasted till late March or early April) on the 'halls' and often the three men would find themselves billed as separate turns on the same programme.

The 1890s saw the arrival of many stars who were to survive into the 20th century, among them Mark Sheridan (?–1917), who in top hat, frock coat, and bell-bottomed trousers strapped below the knee, sang Cockney songs with rousing choruses, of which the best-remembered are 'Oh, I Do Like to be Beside the Seaside' and 'Here We Are, Here We Are, Here We Are Again', to which the troops marched during the First World War. He was first seen in London in 1891, as were Richard George Knowles (1858–1919), born in Canada, who made his first appearance the Trocadero, billing himself as 'the very peculiar American comedian', and wore a shabby black frock-coat, opera hat, and white duck trousers as he strode about the stage singing at the top of his voice songs like 'Girly-Girly' and 'Brighton', and the great comedian Albert CHEVALIER, who brought back the homespun Cockney comedian whom the *lions comiques* had almost ousted from the 'halls'. In the same year Gus [Ernest Augustus] Elen (1862–1940), known till then as a 'blackface' comedian, left off his Negro make-up and reinforced Chevalier's picture of the rollicking costermonger with 'Never Introduce Yer Donah to a Pal'. Some of his characterizations were truly Dickensian, among them ''E Dunno Where 'E Are', 'If It Wasn't For the 'Ouses In Between' and 'Wait Till the Work Comes Round'. Among the feminine equivalents of Chevalier was the much-loved Kate Carney (1868–1950), who, dressed in a coster suit of 'pearlies' and a vast hat with towering feathers, sang 'Liza Johnson' and 'Three Pots a Penny'.

As Cockney songs returned to the 'halls', the elegant silhouette of the 'man-about-town', which had been in partial eclipse since the days of Leybourne and Vance, also returned again, but this time in the form of MALE IMPERSONATION, of which Nellie Power (1855–87) and Bessie Bonehill (?–1902) were early examples, and Vesta TILLEY the supreme exponent. She was the first to wear male attire realistically and like a true aristocrat, and outshone even her excellent younger contemporaries who worked in the same style, Ella Shields (1879–1952), who took Vesta Tilley's West-End Burlington Bertie, and made him a pathetic East-Ender striving after gentility, and Hetty King (1883–1972), an exuberant artiste with impeccable timing, who continued to perform up to the time of her death. In her early years she suffered from the hostility of Vesta Tilley's husband, who was reluctant to let her perform in the music-halls which he controlled, but she nevertheless achieved considerable eminence, one of her best-loved songs being 'All the Nice Girls Love a Sailor', and, in the popular image of the young man of the day, 'Follow the Tramlines' and 'I'm Afraid to Come Home in the Dark.'

The last decade of the 19th century, which saw the initial impact of Vesta Tilley, though she was probably at her best from 1900 to 1905, was not without its share of eccentric humorists, among them Thomas Edward DUNVILLE; the ineffable LITTLE TICH, who may have resembled the Tichborne Claimant in features, but in action was like no one but himself; and, proving once again that a career could be built up on one act, Lottie Collins (1866–1910), whose 'Ta-Ra-Ra-Boom-De-Ay', which had failed when first produced in New York, became the rage of London. She first performed it in *Dick Whittington*, the Islington pantomime for 1891, and it was then introduced into the burlesque *Cinder-Ellen Up Too Late* at the Gaiety, while the pantomime was still running. Rushing from one theatre to another, Lottie would begin her song on a low note, gently waving a handkerchief in time to the music. Suddenly, she would place her hands on her hips and whirl into a swift, high-kicking dance, a vision of tossing curls, and swirling skirts below a tiny waist, gyrating faster and faster as the orchestra quickened its pace. She performed her act for years afterwards in English music-halls, and was eventually paid £200 a week when she revived it in New York.

Out-topping them all were a pair as famous as Dan Leno and Marie Lloyd—George ROBEY and Nellie WALLACE. He first appeared in London in 1891, she a year or two later after touring the provinces, and between them they covered a vast range of humorous, and sometimes pathetic, characters in songs and sketches; both appeared in pantomime, Nellie Wallace being probably the only woman ever to play the DAME satisfactorily; and in revue, which was already taking over many of the music-hall's best features. But they continued to carry the spirit of the true music-hall into the 20th century, as did such stars as Bransby WILLIAMS, with his musical monologues; Harry Tate [Ronald Macdonald Hutchison] (1872–1940), who took his stage name from the firm of Henry Tate & Sons, Sugar Refiners, who once employed him, and from his first appearance in 1895 built up a series of sketches on golfing, motoring, fishing, and so on; Wilkie Bard [Billie Smith] (1870–1944), with his high, domed forehead, reminiscent of Shakespeare (hence the 'Bard'), fringed with sparse hairs, and with two black spots over the eyebrows, who in 1908 popularized the vogue for tongue-twisters with 'She Sells Sea-Shells on the Sea-Shore', and played Pantaloon in the HARLEQUINADE in pantomime, reviving much of its old spirit in company with Will EVANS; Florrie Forde [Florence Flanagan] (1876–1940), who came from Australia and made her first appearance in London on August Bank Holiday 1897, being best remembered as a massive PRINCIPAL BOY and as the singer of 'Down at the Old Bull and Bush', 'Has Anybody Here Seen Kelly?', and 'Hold Your Hand Out, Naughty Boy'; and finally George Lashwood (1864–1943), who came from the provinces in 1899, and Charles R. Whittle (*c.* 1870–1947), who, like Lashwood, favoured songs with 'Girl' in the title, but is best remembered for 'Let's All Go Down the Strand'.

The new century opened auspiciously with the first appearance in London of Harry LAUDER, finest of the Scottish Comedians, master of the daft eccentricity, the pawky humour, and the occasional streak of pathos which later marked the work of Will Fyffe (1885–1947), creator of a whole picture-gallery of Scottish worthies, and singer of 'I Belong ta Glasgae', who made his first appearance in London at the Palladium on 18 July 1921. It was in a series of sketches written for but refused by Lauder that Fyffe first went on the 'halls' after many years in melodrama and revue, and established himself as a prime favourite. Among Lauder's more immediate contemporaries were Albert Whelan [Waxman] (1875–1961), who came from Australia to appear at the Empire in 1901, and was the first to use a signature tune ('Lustige Brüder'), which he whistled as he strode on stage in immaculate evening dress, nonchalantly placing on the top of the piano his stick, top-hat, overcoat, white gloves, scarf, and wristwatch, the whole process being reversed at the end of his act; G(eorge) H(enry) Elliott (1884–1962), another recruit from the Minstrel Show who was first seen in London in 1902 and, billed as 'the Chocolate-Coloured Coon', became the successor of Eugene Stratton, equally admired for his soft-shoe dancing and with a similar repertory of 'coon' songs to which he added 'I Used to Sigh for the Silvery Moon'; Billy Merson [William Henry Thompson] (1881–1947), a circus clown and acrobat, who put on an auburn wig with a bald patch at the back and two coy curls across the top of his high forehead to sing about Alphonso Spagoni, 'The Spaniard That Blighted My Life,' and 'The Good Ship Yacki-Hicki-Doola'; Gertie Gitana [Gertrude Mary Ross, *neé* Astbury] (1889–1957), who first appeared on stage at the age of four with a troupe dressed as gipsies from whom she took her stage name, and by the time she was 16 was topping the bill all over the country with songs like 'Nellie Dean'; old Fred Emney, stalwart of slapstick comedy, whose son, also Fred EMNEY, delighted the 'halls' with his songs at the piano and his talent as a raconteur; and Harry Fragson [Potts] (1869–1913), son of an English mother and a Belgian father (who shot him in a fit of insanity); he played in Paris from 1887 onwards with a Cockney accent and in 1905 came to London to play with a French accent; when he appeared as Dandini in *Cinderella* the character was rechristened Dandigny as a tribute to his Anglo-French reputation.

They were all singers and comedians; but when it came to music and comedy combined, none of them was as great as GROCK, the supreme clown of his generation, skilled in the playing of 20 instruments, and constantly surprised by his own accomplishments. From 1903 to 1924 he delighted audiences in London, and then returned to the Continent, where he died in 1959. The music-hall, attacked on all sides by new forms of entertainment, was already dying when he left London, and only a few names are attached precariously to its last flickers, since what had been a full-time profession now had to be split up, not only between REVUE and MUSICAL COMEDY, but between the cinema, the radio, and above all television. Among them are Maurice Chevalier (1888–1972), MISTINGUETT's partner in Paris before the First World War, whose debonair straw hat, raffish appearance, quizzical smile, and seductive voice with a charming broken-English accent made him irresistible; Gracie Fields [Grace Stansfield] (1898–1979), who sang the praises of 'The Biggest Aspidistra in the World' and became a legend in her lifetime, her warm personality, ready wit, and technical ability to hold an immense audience making her the natural successor to Marie Lloyd; her fellow-Lancastrian George Formby (1905–61), who sang and pattered to the ukulele; and Sid Field (1904–50), a DROLL who was essentially a comic actor, in sketches—the best involving the 'spiv', Slasher Green—which relied more on characterization and situation than on slapstick and verbal gags. He had already built up an immense reputation outside London before in 1943 he first appeared in London in the revue *Strike a New Note*, and he made his last appearance on stage as Elwood P. Dowd in Mary Chase's *Harvey* (1949). He was, like his contemporaries, a man of many parts; the dwindling music-halls could no longer contain them, and they sought other worlds to conquer.

In its heyday Music-Hall presented the type of entertainment most loved by the ordinary people. It was gay, raffish, and carefree, vulgar but not suggestive, dealing amusingly with the raw material of their own lives, their emotions, their troubles, their rough humour. Sophistication and subtlety was its undoing. Yet, paradoxically, the main force ranged against it, television, has revived something approaching the earlier form of music-hall, which seems to be springing up again as informal entertainment in the place where it all began, the local public-house. The only surviving full-time music-hall, the City Palace of Varieties in LEEDS, was restored to its earlier glory in order to house B.B.C. Television's music-hall series 'The Good Old Days'.

MUSIC IN THE THEATRE, see INCIDENTAL MUSIC.

MUSSATO, ALBERTINO (1261–1329), Italian diplomat and historian, author of a Latin tragedy, *Ecerinis* (1314), in the style of SENECA but on a contemporary subject—the tyrannous rule of Ezzelino da Romano in Padua—and intended as a warning to his fellow citizens. For this, and for other political activities, he was exiled to Chioggia, where he died.

MUSSET, (Louis-Charles) ALFRED DE (1810–57), French poet and playwright of the Romantic era, and the one man who might have been able to fuse the new Romantic drama with the best of the classical tradition in the French theatre. Unfortunately the failure of his first play *La Nuit vénitienne* (1830) turned him against the stage, and for nearly 20 years he wrote only plays to be read. It was not until 1847 that the COMÉDIE-FRANÇAISE, at the insistence of Mme Allan who had appeared in a Russian translation of the play in St Petersburg, put on *Un Caprice*, which had been published ten years earlier. It was an immediate success and was followed in 1848 by *Il faut qu'une porte soit ouverte ou fermée* (*A Door Should Be Either Open or Shut*) and *Il ne faut jurer de rien* (*One Can Never Be Sure of Anything*). Although this encouraged Musset to write further plays for performance, and to revise some of his earlier ones for the same purpose, his real importance as a dramatist was realized only after his death. His plays, which brought back to a theatre under the dual influence of SCRIBE and BALZAC the poetry it was in danger of losing, show a delicacy and restraint quite unlike the work of his contemporaries Victor HUGO and DUMAS *père*, suggesting rather the influence of MARIVAUX. Published eventually in three volumes as *Comédies et proverbes*, they consist of on the one hand bitter-sweet comedies with a large element of fantasy, as in *Le Chandelier* (*The Decoy*, perf. 1848), *Les Caprices de Marianne* (perf. 1851), *On ne badine pas avec l'amour* (*There's No Trifling With Love*, perf. 1861), and *Fantasio* (perf. 1866), and on the other short scenes of social and salon life usually illustrating some well-known proverb, such as had been popularized by CARMONTELLE in the private theatres of the 18th century and which Musset took up again and made very much his own, as in *On ne saurait penser à tout* (*It's Impossible to Think of Everything*, 1849). Musset's most important dramatic work, however, is the historical drama *Lorenzaccio*, written in 1834 after his tragic liaison with George Sand, but banned because of its provocative theme—the assassination of the unpopular Alexander de' Medici by his young cousin Lorenzo. In a drastically cut and rearranged version it was first staged in 1896 with BERNHARDT in the leading role, a part often compared to Hamlet. Although the size and shapelessness (by French standards) of this play has deterred many directors, there have been some important 20th-century productions, notably in 1952 at the AVIGNON FESTIVAL, when Gérard PHILIPE both directed the play and appeared as Lorenzo, thus breaking the tradition which assigned the part to a woman—Marie-Thérèse Pierat in 1927 at the Comédie-Française and Marguerite Jamois in 1945 in a production by Gaston BATY. In 1933 an English version was produced in London as *Night's Candles*, with Ernest MILTON in the title role. Generally speaking, however, Musset's plays remain little known in England probably because, like Marivaux, he does not translate well.

MYSTERY PLAY, medieval religious play which derives from LITURGICAL DRAMA, but differs in being wholly or partly in the vernacular and not chanted but spoken. Also, it was performed out of doors, in front of the church, in the market square, or on perambulating PAGEANTS. The earlier English name for it was MIRACLE PLAY, now seldom used, and a better name would be Bible-histories, since each play was really a cycle of plays based on the Bible, from the Creation to the Second Coming. Substantial texts of English 'cycles' of such plays have survived from Chester, Coventry, Lincoln, Wakefield, and York.

The genesis of the Mystery play may be looked for in the short dramatic works of well-educated, theatrically-minded monastic clerks and students at the newly-founded universities, who applied the liturgical approach to other Biblical material and adapted poetic and musical traditions of secular origin to dramatic compositions. Some of their plays, like the extant early 13th-century *Daniel* by

Abelard's pupil Hilarius, had no direct connection with the liturgy and were intended for production either in the churchyard, as with the 13th-century Anglo-Norman *Mystère* (or *Jeu*) *d'Adam*, or in monastic refectories and the banqueting halls of the nobility on such festive occasions as Christmas and Carnival time (Shrove Tuesday). There they found themselves in company with, and influenced by, the secular entertainments of the time derived from the old FOLK FESTIVALS and FOLK PLAYS. From the few texts of these early monastic plays to survive it is evident that by the beginning of the 14th century the idea that a play, or *ludus*, could exist in its own right as entertainment as well as in the context of worship was firmly established throughout Western Europe. Once the Church had recognized the value of the *ludus* as a reinforcement of its teaching within the church, it was seen as an admirable means of setting before an unlettered audience outside the fundamental doctrines of Christianity. The institution in 1311 of the Feast of Corpus Christi (the Body of Christ, symbolized in the Host), on the Thursday following Trinity Sunday, gave laymen the opportunity to participate more closely than before in the Church's festivities. The new festival began with a formal procession escorting the Host to form the focal point of celebrations of Mass in the principal thoroughfares of every sizeable town in Christendom. From the beginning the citizens, represented by their guilds, followed the procession with torches, banners, and images, and it was not long before they supplemented these with dramatic performances depicting fallen man's redemption and the salvation of his soul through the miraculous power of the Host (whence 'Miracle play'). Scripted in verbal and visual language that was intelligible to everyone who acted in them or saw them performed, these 'Christian epics', with costumes and properties that became progressively more elaborate, proliferated throughout Europe, and having lost their unifying text in Latin developed entirely in the vernacular, basically identical but with a wide variety of incidental differences to meet the needs and the taste of local audiences. Thus arose simultaneously the English Mystery play, the French *Mystère*, the German *Mysterienspiel*, the Italian SACRA RAPPRESENTAZIONE, and the Spanish AUTO SACRAMENTALE, to name only the most important. Traces of similar plays are found in Russia, in the states of Central Europe, and also in Denmark.

N

NADIR, MOSES, see JEWISH DRAMA.

NAEVIUS, GNAEUS (*c.* 270–*c.* 201 BC), Roman dramatist, a contemporary of ANDRONICUS, and like him a writer of both tragedy and comedy. The titles of 6 of his tragedies and about 28 of his comedies are known, but few fragments have survived. The best known is his description of a flirt from *The Girl from Tarentum*. He was the inventor of the FABULA *praetexta*, a serious play on a legendary or contemporary Roman theme; he is known to have written one on Romulus and one on the battle of Clastidium, won by his famous contemporary, M. Claudius Marcellus, in 222 BC. A man of daring and energy, his gift for social and political satire brought him into conflict with the powerful family of the Metelli, and after a period of imprisonment he is said to have died in exile in Africa.

NAHARRO, BARTOLOMÉ DE TORRES, see TORRES NAHARRO.

NAOGEORG, see KIRCHMAYER, THOMAS.

NARES, OWEN (1888–1943), English actor, for many years one of the most popular 'matinée idols' of the London stage, his charm and good looks somewhat obscuring his gifts as an actor. He made his first appearance in 1908, and came into prominence with an excellent performance as Lord Monkhurst in Arnold BENNETT and Edward KNOBLOCK's *Milestones* (1912). He was also much admired in the title role of Du Maurier's *Peter Ibbetson* and as Thomas Armstrong in the London production of Edward SHELDON's *Romance* (both 1915). In 1923 he was at the ST JAMES'S THEATRE (where he had first appeared with ALEXANDER in 1909), giving an interesting performance as Mark Sabre in a dramatization of A. S. M. Hutchinson's novel *If Winter Comes* . . ., and then toured South Africa with *Romance* and SARDOU's *Diplomacy*. In later years he gave signs of greater depth of feeling and reserves of strength than he had previously been credited with, particularly in Dodie SMITH's *Call It A Day* (1935), St John ERVINE's *Robert's Wife* (1937), in which he played opposite Edith EVANS, and Daphne Du Maurier's *Rebecca* (1940), but he died suddenly before he had had time to develop his newfound powers.

NARR, the German equivalent of the early English FOOL, who appears in the 16th-century FASTNACHTSSPIEL or Carnival Play. Though not originally a comic character (the fools in Sebastian Brant's poem *Das Narrenschiff* (*The Ship of Fools*, 1494) are those who live foolishly), he was made to appear so on the stage. He could also be a comic peasant, like the CLOWN in Shakespeare's plays, as opposed to the more sophisticated Court Jester, and his assumed stupidity could often cover slyness. He bequeathed most of his characteristics to Jan Bouschet (originally John Posset), the creation of one of the ENGLISH COMEDIANS, Sackville, who allowed him, though a secondary character in the play, to hold the stage and improvise at will. Similar types were created by other leading actors—Hans Stockfisch, Pickelhering, and eventually the native HANSWURST and KASPERLE.

NARYMSKI, JÓZEF, see POLAND.

NASHE, THOMAS (1567–1601), English pamphleteer and playwright, friend of GREENE and LYLY, who collaborated with MARLOWE in *Dido, Queen of Carthage* (1587/8) and was also author, or part-author with JONSON who appeared in the play, of the ill-fated 'seditious' comedy *The Isle of Dogs* (1597), now lost, which caused Jonson, and perhaps Nashe also, to be put in prison. Nashe's only extant dramatic work is *Summer's Last Will and Testament* (1592/3), designed for performance in the house of a nobleman, probably Archbishop Whitgift at Croydon. It was used by Constant Lambert in 1936 as the basis of a work for orchestra, chorus, and baritone solo.

NASSAU STREET THEATRE, New York, at 64-6 Nassau Street. Opened as the New Theatre on 11 December 1732, this was one of the earliest New York theatres. Nothing is known of activities there until 5 Mar. 1750, when Walter Murray and Thomas KEAN brought a company from Philadelphia to appear in a repertory of tragedies and farces which opened with CIBBER's version of *Richard III*. In 1751 an actor from London named Robert Upton appeared unsuccessfully in *Othello* and other plays. The elder Lewis HALLAM and his company dismantled, enlarged, and refurbished the old theatre, opening their season on 17 Sept. 1753 with STEELE's *The Conscious Lovers*, and ending it in March 1754 with MOORE's *The Gamester*, after which the theatre seems to have been abandoned.

NATHAN, GEORGE JEAN (1882–1958), American dramatic critic, who in 1905 joined the staff of the *New York Herald* and began his fight for the 'drama of ideas', denouncing the works of such contemporary American playwrights as David BELASCO and Augustus THOMAS, whose popular plays occupied the playhouses to the exclusion of almost everything else. Nathan introduced to America the modern dramatists of Europe—HAUPTMANN, IBSEN, SHAW, STRINDBERG, and a host of others whose early work he published in *The Smart Set*, a magazine which he edited with H. L. Mencken. His most important discovery was Eugene O'NEILL, some of whose earliest plays also appeared in *The Smart Set*. Later Nathan championed Sean O'CASEY, being largely responsible for the New York production of *Within the Gates* (1934), and William SAROYAN, whose first work for the theatre, *My Heart's in the Highlands* (1939), he praised enthusiastically when there was still much doubt as to its value. He wrote over 30 books on the theatre, and for many years produced an annual volume on each New York theatre season.

NATION, Théâtre de la, see COMÉDIE-FRANÇAISE.

NATIONAL CONCERT HALL, NATIONAL MUSIC HALL, New York, see CHATHAM THEATRE (2).

NATIONAL CRITICS INSTITUTE, see O'NEILL, EUGENE.

NATIONALE SCENE, Den (The National Theatre), Bergen, was established in 1850 under its original name, Det Norske Teatret (The Norwegian Theatre), on the initiative of Ole BULL. His aim was to present a truly Norwegian theatre in opposition to the Danish influence which had hitherto dominated the theatre in Norway. IBSEN was the Norske Teatret's first resident dramatic author from 1851 to 1857, and BJØRNSON was its director from 1857 to 1859; both gained valuable experience of dramatic practice there. The theatre was forced to close for financial reasons in 1863, reopening under its present name in 1876. Gunnar HEIBERG was director from 1884 to 1888. The theatre enjoyed a particularly lively period from 1934 to 1939, when Hans Jacob NILSEN was its director. Its Lille Scene (Studio Stage) provided the opportunity for experimental work and contributes to its continuing vigour.

NATIONAL OPERATIC AND DRAMATIC ASSOCIATION (NODA), see AMATEUR THEATRE.

NATIONAL PLAYWRIGHTS CONFERENCE, see O'NEILL, EUGENE.

NATIONAL THEATRE, London. The establishment of a permanent state-subsidized theatre in London, on the lines of the COMÉDIE-FRANÇAISE, was first suggested by David GARRICK in the 18th century, and in the 19th century both IRVING and BULWER-LYTTON were enthusiastic supporters of the idea. It was not until 1908, however, that a committee was set up to investigate the possibility of opening such a theatre in 1916 to celebrate the tercentenary of Shakespeare's death. A large sum of money had been subscribed and a foundation stone laid on a site in Gower Street when the outbreak of the First World War in 1914 brought the project to a standstill. In 1938 another site in South Kensington was acquired and a second foundation stone laid by G. B. SHAW. The outbreak of the Second World War a year later caused further delay, and it was not until 1951 that the idea was taken up again, with Geoffrey Whitworth, founder of the British Drama League (see AMATEUR THEATRE), as an active supporter. A more ambitious plan was launched, and the site moved to the South Bank of the Thames, where Queen Elizabeth, deputizing for her husband George VI, laid a third foundation stone on 13 June of that year. In 1961 a decision was taken to found a National Theatre company under Laurence OLIVIER, to be housed in the OLD VIC THEATRE pending the erection of a new building on a site downstream from the third foundation stone, below Waterloo Bridge.

This building, designed by Denys Lasdun, is a vast complex housing three theatres with extensive backstage accommodation, rehearsal rooms, and workshops, dressing rooms for 150 actors, offices, bars, buffets, a restaurant, and a large foyer for exhibitions and informal concerts. Work began on the site in Nov. 1969 and on 16 Mar. 1976 the first theatre, the LYTTELTON, gave its opening performance, followed on 25 Oct. 1976 by the OLIVIER and on 4 Nov. 1977 by the COTTESLOE. In Nov. 1973 Olivier, who had resigned owing to ill-health, was succeeded by Peter HALL, who still exercises overall control with the assistance of Michael Rudman, who became Director of the Lyttelton in 1980, and Peter GILL, who since 1980 has been Joint Director, with Hall, of the Olivier and the Cottesloe.

NATIONAL THEATRE, New York, see BILLY ROSE THEATRE.

NATIONAL THEATRE CONFERENCE, U.S.A., an honorary professional association limited to 120 members drawn from academic, community, and non-profit-making professional theatres throughout the United States. Its purpose is the exchange of experience and ideas in the field of non-commercial theatre, especially at the Conference's annual meeting. The organization edited the book *Theatre in America* (1968), a survey of professional, community, educational, and repertory theatre activities in the U.S.A.

NATIONAL THEATRE OF THE DEAF, see O'NEILL, EUGENE.

NATIONALTHEATRET, Oslo, playhouse which on 1 Sept. 1899 took over from the old KRISTIANIA THEATER in newly-built premises, and has since held pride of place as the premier theatre of Norway. Its first director was Bjørn BJØRNSON, with a company headed by Johanne DYBWAD, and during the first quarter of the 20th century it contributed a brilliant page to the history of the Norwegian theatre with such players as Ragna Wettergren, Hauk Aabel, and Sofie REIMERS. A second generation of exceptional talent came to the fore in the 1930s, with Tore SEGELCKE, Alfred MAURSTAD, and Olaf Havrevold, with Agnes Mowinckel and Gerda Ring on the production side. During the Second World War the theatre suffered many setbacks. Its leading members were sent to concentration camps, and its official productions, approved by the Occupation authorities, were boycotted by the general public. After the war its problems were mainly economic; eventually the State and the municipality of Oslo had to come to the rescue and subsidies from public funds now play a major role in the economic life of the theatre. In 1959 it acquired an experimental studio, the Lille Scene, which survived only two years. It was succeeded in 1963 by the Amfiscenen, housed in the Nationaltheatret's own premises, which has since developed a healthy life of its own. With the appointment of Arild Brinchmann as director in 1967, the Nationaltheatret took on a new lease of life, technically, administratively, and architecturally, and it continues to flourish under Toralv MAURSTAD, who became manager in 1978.

NATIONAL YOUTH THEATRE, organization founded in 1956 by Michael Croft, a master at Alleyn's School, Dulwich, who had formerly been an actor, for the production of Shakespeare's plays. Drawn originally from Alleyn's and Dulwich College, the players eventually came from schools all over Britain. The first production, at Toynbee Hall, was *Henry V*. It was followed by *Troilus and Cressida* (at the EDINBURGH FESTIVAL), *Hamlet* (in London and on tour), and *Antony and Cleopatra* (at the OLD VIC). The first contemporary play to be presented was David Halliwell's *Little Malcolm and His Struggle Against the Eunuchs* at the ROYAL COURT in 1965. The company formed a particularly productive relationship with Peter TERSON, who has written several plays for it, beginning with *Zigger-Zagger* (1967), about a football fan. In 1971 the N.Y.T. took over the Shaw Theatre, originally erected as a conference centre and adapted by the Camden local authority as a theatre to hold 510 spectators; the actors share the premises with a public library. The N.Y.T. established at the Shaw Theatre a professional company which aims to provide high quality theatre related to the needs of schools and the interests of young people. The company opened with SHAW's *The Devil's Disciple*, and has since presented classic revivals and plays by contemporary dramatists, including, in 1978, the first full revival of WESKER's trilogy. The Shaw Theatre Company encourages new writers, having given the first performance of plays by such authors as David Cregan, Ken Campbell, and Barrie Keeffe. The N.Y.T. company itself, which uses the Shaw Theatre for its summer productions and makes regular tours abroad, remains amateur, continuing to present revivals and the work of new writers. Many well known actors, among them Derek JACOBI and Helen MIRREN, gained their first experience with the N.Y.T. In 1981 its grant from the ARTS COUNCIL was withdrawn, but it was eventually replaced by commercial sponsorship.

NATIVITY PLAY, see LITURGICAL DRAMA.

NATURALISM, a movement in the theatre of the late 19th century which carried a step further the revolt against the artificiality of contemporary forms of playwriting and acting initiated by the selective REALISM of IBSEN. *Thérèse Raquin* (1873), dramatized by ZOLA from his own novel, was the first consciously conceived naturalistic drama, STRINDBERG's *Miss Julie* (1888) its first masterpiece. But it was ANTOINE who established it in the theatre. The influence of his THÉÂTRE-LIBRE led to the foundation of the FREIE BÜHNE in Germany and GREIN's Independent Theatre in London, and the movement finally attained world recognition in the work of STANISLAVSKY, particularly with his production of GORKY's *The Lower Depths* (1902). In Spain naturalism is represented by BENAVENTE's *La Malquerida* (1913) and in the United States by the early works of O'NEILL and the dramatized novels of STEINBECK.

NAUGHTON, CHARLIE, see CRAZY GANG.

NAUTICAL DRAMA, a type of romantic MELODRAMA popular in the late 18th and early 19th centuries, which had as its hero a 'Jolly Jack Tar', a lineal descendant of the sailor characters in the novels of Smollett, who was himself the author of one of the earliest plays of this kind, *The Reprisal; or, the Tars of Old England* (1757). The Jack Tar was further popularized by the elder Charles DIBDIN and by the naval victories of Nelson, before being given its final form in the noble-hearted William, hero of JERROLD's *Black-Ey'd Susan; or, All in the Downs* (1829), in which T. P. COOKE made his name. The character continued to flourish in the minor 'transpontine' theatres, particularly at the SURREY, until well into the 1880s, in spite of being burlesqued in DICKENS's *Nicholas Nickleby* (1838) and GILBERT and Sullivan's *H.M.S. Pinafore* (1878). Unlike AQUATIC DRAMA, in which the water itself often seems to have played the chief part, nautical drama usually took place on dry land, and had nothing in common with the *naumachiae*, or mimic sea-fights, of Roman times, given in an amphitheatre specially built for the purpose by Augustus on the right bank of the Tiber, or with the splendid water-pageants given in Renaissance Italy, often with the help of complicated machinery.

NAZIMOVA, ALLA (1879–1945), Russian actress, who studied with NEMIROVICH-DANCHENKO, acted for a season with the MOSCOW ART THEATRE, and in 1904 was the leading lady of a theatre company in St Petersburg. She toured Europe and America with a Russian company and, having learnt English in less than six months, made her first appearance in an English-speaking part—IBSEN's Hedda Gabler—in 1906 at the PRINCESS THEATRE in New York under the management of the Shuberts, who built and named for her the Nazimova Theatre, later the THIRTY-NINTH STREET THEATRE, which she opened on 18 Apr. 1910 with Ibsen's *Little Eyolf*. A superb actress, vibrant, passionate, yet subtle, she astonished Broadway audiences with the variety of her characterizations; but by 1918 her popularity began to wane and she spent the next ten years in films. She then returned to the stage, appearing with Eva LE GALLIENNE's Civic Repertory Company and for the THEATRE GUILD in Ibsen, CHEKHOV, TURGENEV, and O'NEILL. In 1935 she directed and starred in her own version of Ibsen's *Ghosts*, on Broadway and then on a national tour, after which she went back to Hollywood.

NAZIMOVA THEATRE, New York, see THIRTY-NINTH STREET THEATRE.

NEDERLANDER THEATRE, New York, see BILLY ROSE THEATRE.

NEHER, CASPAR (1897–1962), German stage designer, closely associated with BRECHT, for whom he designed sets for the first productions of *Im Dickicht der Städte* (1923), *Eduard II* (1924), *Baal* (1926), and *Die Dreigroschenoper* (1928). For all these plays he developed a sleazy, speakeasy style, depicting a world of leering primitiveness and low-life characters. He was for ten years, 1934 to 1944, at the DEUTSCHES THEATER in Berlin, where under HILPERT he did some fine work. His settings, in which he used gradations of subdued colours with brilliant stage lighting, were always in tune with the spirit of the play. He returned to Brecht in 1948, designing the set for *Antigone* (1948) in Switzerland, and for the BERLINER ENSEMBLE he designed *Herr Puntila und sein Knecht Matti* in 1949 and *Leben des Galilei* in 1957. Some of his finest work was done for operatic productions, particularly of Mozart.

NEIDHART VON REUENTAL (*c.* 1185–*c.* 1240), Middle High German poet, whose work is remarkable for its realism and the vigour of its style. Many of his poems are in reality dance-songs; typical of them is the so-called *Neidhartspiel* from the Tyrol, in which the author, having seen the first violet of the year, hastens to call his lady and her friends so that they may admire it. The local peasants, who hate him for mocking their ways and decoying their women, pick the flower and, after putting something offensive in its place, tie it to a pole and caper round it. Neidhart, discomfited in the presence of his mistress, suspects the trick and pounces on his foes. A scrimmage ensues in which the peasants are routed, whereupon Neidhart leads his lady to the dance. The *Neidhartspiel* had two actions going on simultaneously, one with the courtiers, one with the peasants, the speeches sometimes coming from one group of actors, sometimes from the other.

NEIGHBORHOOD PLAYHOUSE, New York, at 466 Grand Street, on the Lower East Side. This was built and endowed by Alice and Irene Lewisohn, who designed, choreographed, and directed most of the productions seen there. It opened on 12 Feb. 1915 with a dance-drama entitled *Jephthah's Daughter*, and before it closed in 1927 with the fifth annual edition of a review entitled *Grand Street Follies* it had been responsible for a long list of productions, including short plays by DUNSANY, CHEKHOV, Sholem ASCH, YEATS, and SHAW, and the first dramatic rendering of BROWNING's *Pippa Passes* (1917). Later productions included a number of new dance-dramas and ballets and such varied plays as *Fortunato* by the ÁLVAREZ QUINTERO brothers, GALSWORTHY's *The Mob*, GRANVILLE-BARKER's *The Madras House*, O'NEILL's *The First Man*, SHERIDAN's *The Critic*, the Hindu drama *The Little Clay Cart*, and ANSKY's *The Dybbuk*. After the closing of the theatre, a school of acting under the same name opened at 340 East 54th Street.

NEILSON, (Lilian) ADELAIDE [Elizabeth Ann Brown] (1846–80), English actress, daughter of a strolling player, who had an unhappy childhood. In 1865 she went on the stage, making her first appearance as Julia in Sheridan KNOWLES's *The Hunchback*, always one of her favourite parts. After several years in London and the provinces, where she was much admired in Shakespearian parts and in dramatizations of SCOTT's novels, she made her first visit to the U.S.A. in 1872, touring the country with a fine repertory. She became exceedingly popular with American audiences, and had just returned from a second extended and highly successful tour of the States when she died suddenly. A beautiful woman, with dark eyes and a most expressive countenance allied to a fine speaking voice, she was considered to be at her best as Juliet in *Romeo and Juliet* though her Viola in *Twelfth Night* was also much admired. She made an unhappy marriage, divorcing her husband in 1877.

NEILSON, JULIA EMILIE (1868–1957), English actress, wife of Fred TERRY, whom she married while they were both appearing in Henry

Arthur JONES's *The Dancing Girl* (1891), in which Julia scored a sensational success as Drusilla Ives. From 1900 until Fred's retirement in 1929 she played opposite him, in London and on innumerable tours, in such romantic costume comedies as the Baroness Orczy's *The Scarlet Pimpernel, Sweet Nell of Old Drury* and *Dorothy o' the Hall*, both by Paul KESTER, *Matt o' Merrymount* by Beulah Dix and Mrs Sutherland, and William Devereux's *Henry of Navarre*. Originally intended for a musical career, she was studying at the Royal Academy of Music when she was persuaded by W. S. GILBERT to go on the stage, her first appearances being in revivals of several of his plays. One of her outstanding performances before becoming her husband's leading lady was as Hester Worsley in Oscar WILDE's *A Woman of No Importance* (1893), and she was also much admired as Princess Flavia in George ALEXANDER's production of Anthony Hope's *The Prisoner of Zenda* (1896). After Fred's death in 1932 she was seen with Seymour HICKS in his *Vintage Wine* (1934), and in 1938 she celebrated her stage jubilee. After some years in retirement she made a final appearance in 1944, playing Lady Rutven in *The Widow of Forty*, by Heron Carvic, second husband of Julia's daughter Phyllis Neilson-Terry (1892–1977), who, like her brother Dennis (1895–1932), was also on the stage.

NÉMETH, LÁSZLÓ, see HUNGARY.

NEMIROVICH-DANCHENKO, VLADIMIR IVANOVICH (1859–1943), Russian dramatist and director, who worked under both the Imperialist and Soviet régimes. In his early years he wrote about a dozen light comedies, which were successfully produced at the MALY THEATRE, and he was in charge of the Drama Course of the Moscow Philharmonic Society, where Olga KNIPPER, MEYERHOLD, and MOSKVIN were among his pupils, when in 1897 a meeting with STANISLAVSKY resulted in the founding of the MOSCOW ART THEATRE a year later. He was responsible for the literary quality of the theatre's repertory, as Stanislavsky was for the acting, and it was he who persuaded CHEKHOV to allow *The Seagull* to be revived after its disastrous first performance at the Alexandrinsky (later PUSHKIN) Theatre. He himself directed a number of the theatre's outstanding successes, both classic and modern, the last being POGODIN's play about Lenin *Kremlin Chimes* (1942). He wrote an account of the founding of the Moscow Art Theatre, and also expounded his own philosophy of the drama, in *My Life in the Russian Theatre* (1937).

NERO, Roman Emperor from AD 54 to 68. He was passionately fond of the theatre and appeared frequently on stage, not only as a dancer in PANTOMIME but as a solo tragic actor in such parts as the Mad Hercules, the Blind Oedipus, the Matricide Orestes, even Canace in Travail. On such occasions he wore a mask, but the features were always modelled on his own or on those of his current mistress. From his famous theatrical tour of Greece in AD 66–7 he returned with 1,808 triumphal crowns. Even his worst crimes do not seem to have shocked conservative opinion in Rome as much as these antics—a fact which illustrates the low status of professional entertainers under the EEmpire.

NERONI, BARTOLOMEO, see THEATRE BUILDINGS.

NERVO, JIMMY, see CRAZY GANG.

NESBITT, CATHLEEN (1888–1982), English actress, who made her first appearance, in London, in 1910 in a revival of PINERO's farce *The Cabinet Minister*, and a year later joined the Irish Players, with whom she made her New York début in SYNGE's *The Well of the Saints*. In London she played Deirdre in the same author's *Deirdre of the Sorrows* and Phoebe Throssel in BARRIE's *Quality Street* (both 1913), but she returned to America in 1915 to appear in Vachell's *Quinney's*, in which she had already been seen in Liverpool. She remained in the U.S.A. until 1919, playing a number of parts, among them Ruth Honeywill in GALSWORTHY's *Justice* in 1916 and Varinka in SHAW's *Great Catherine* in 1917. Back in London, she was seen mainly in such classic roles as WEBSTER's Duchess of Malfi in 1919, Doralice in DRYDEN's *Marriage à la Mode* and Belvidera in OTWAY's *Venice Preserv'd* in 1920, and Amarillis in FLETCHER's *The Faithful Shepherdess* in 1923; but she also appeared in such modern plays as HARWOOD's *The Grain of Mustard Seed* (1920), Galsworthy's *Loyalties* (1922), and James Elroy FLECKER's *Hassan* (1923), and being extremely versatile as well as intelligent moved easily from comedy to drama, from verse to prose. She made outstanding appearances in such plays as LONSDALE's *Spring Cleaning* (1925), Margaret Kennedy's *The Constant Nymph* (1926), and Clemence DANE's *A Bill of Divorcement* (1929). She was equally good as Katharina in *The Taming of the Shrew* in 1935 and as Thérèse Raquin in ZOLA's *Thou Shalt Not*... in 1938, and in 1940 she played Goneril to GIELGUD's King Lear for the OLD VIC. Two long-running productions were Edward Percy's *The Shop at Sly Corner* (1945) and the New York production of T. S. ELIOT's *The Cocktail Party* (1950). After the latter she was seen more in the United States than in England, appearing on Broadway in 1956 as Mrs Higgins in the musical *My Fair Lady* and as the Grand Duchess in RATTIGAN's *The Sleeping Prince*, and in 1957 touring the United States as Mrs St-Maugham in Enid BAGNOLD's *The Chalk Garden*. She later toured in Tennessee WILLIAMS's *Suddenly Last Summer* and *The Glass Menagerie* and ANOUILH's *Time Remembered*. Her last London role was the Dowager Lady Headleigh in Robin Maugham's *The Claimant* (1964), but in New York she played Madame Voynitsky in

CHEKHOV's *Uncle Vanya* (1973). At nearly 90 she appeared at the CHICHESTER FESTIVAL THEATRE in Michael REDGRAVE's dramatization of Henry JAMES's *The Aspern Papers* and in her nineties again played Mrs Higgins in an American revival of *My Fair Lady*. Her autobiography, *A Little Love and Good Company*, was published in 1974.

NESBITT, ROBERT, see REVUE.

NESTROY, JOHANN NEPOMUK (1801–62), Austrian actor and playwright, who in his satirical comedies reflected the rising tide of liberalism and social discontent which was to result in the Revolution of 1848. The foundations of his success were laid in VIENNA, where he was born, and to which he returned, after some years on tour, in 1829, playing first at the Theater in der Josefstadt, and subsequently appearing at the Theater an der Wien under Karl CARL, whom he succeeded in 1854 as director of the Leopoldstädter. His gift for improvization, which several times got him into trouble with the authorities, and his immense facility (he wrote at least 83 plays) soon made him a popular figure. He excelled in parody, his main target being Wagner, and he was the last outstanding exponent of Viennese popular theatre, which after him declined into operetta. His first successful play was *Der böse Geist Lumpazivagabundus* (1833), in which two young women in male attire disport themselves in local taverns. Less farcical and of more lasting value are such plays of social comment and political satire as *Zu ebener Erde und im ersten Stock* (*On the Ground Floor and on the First Floor*, 1835) and *Freiheit in Krähwinkel* (1848), of which a translation, as *Freedom Comes to Krahwinkel*, was published in the *Tulane Drama Review* in 1961. But most of his work is untranslatable, as much of his wit depends upon the skilful use of Viennese dialect. In 1842 he adapted John Oxenford's *A Day Well Spent; or, Three Adventures* (1836) as *Einen Jux will er sich machen*, which Thornton WILDER later used as the basis for his play *The Merchant of Yonkers* (1938). He later rewrote it as *The Matchmaker* (1954), on which again was based the musical, *Hello Dolly!* (1965). Nestroy's play was also used in an adaptation by Tom STOPPARD, *On the Razzle*, presented at the NATIONAL THEATRE in London in 1981. Nestroy, who was an omnivorous reader, and could write in the morning what he had to act in the evening, adapted two other English plays, Poole's *Patrician and Parvenu; or, Confusion Worse Confounded* (1835) as *Liebesgeschichten und Heiratssachen* (1843) and BOUCICAULT's *London Assurance* (1841) as *Nur keck!* (1855). Unlike his immediate predecessor RAIMUND, whose themes were mostly original, he took his plots wherever he could find them, but altered them so much that they were hardly recognizable, and he was always ready to sacrifice verisimilitude to theatrical effectiveness. Nothing was sacred to him, and with more wit than Raimund, but less imagination and less heart, he showed considerable courage in his attacks on social and political targets. Among his best plays are *Der Zettelträger* (*The Billposter*, 1827), the first in which he himself appeared, *Das Mädel aus der Vorstadt* (*The Girl from the Suburbs*, 1841), and *Judith und Holofernes* (1849).

NETHERLANDS, the northern part of the territory known in earlier times as Flanders or the Low Countries, the southern part now comprising the independent kingdom of BELGIUM. As in the rest of Europe, the earliest manifestation of theatre in this area was LITURGICAL DRAMA, based on the main services of the Christian Church. The first mention of such plays, written and acted in Latin, dates from about 1275; but parallel with them there developed also plays in the vernacular, known as the *abele spelen*, based mainly on the romances of chivalry. Three of them from the mid-14th century have survived: *Esmoreit*, on the theme of a royal foundling; *Gloriant*, which portrays the love of a Christian prince for a Saracen princess; and, the most highly developed of them all, *Lanseloet van Denemarken*, with its subtle characterization of a princely hero who falls in love with one of his mother's serving-women and through his folly causes the death of them both. There is also an interesting allegorical play from the same period, *The Play of Winter and Summer*, which may provide a clue to the origin of the *abele spelen* in the antithetical debates on abstract personifications conducted in the vernacular early in its development.

The *abele spelen*, and also the liturgical plays even before they ceased to be played in Latin, were usually followed by rough farces in Flemish—*kluchten* or *sotternien* (jokes)—based on scenes of daily life, usually in the form of quarrels between husband and wife, some of which are shown in the later paintings of Pieter Brueghel. Their popularity may have helped the development of religious plays in the vernacular, of which the earliest known are the mid-15th century MYSTERY PLAYS on the Seven Joys of the Virgin Mary: the first and last, dealing with the Annunciation and the Assumption, have survived. The best of the surviving MIRACLE PLAYS is *Mary of Nijmegen*, with its vividly realistic tavern scene. It tells of a young girl who sells herself to the devil for beauty and power and leads a profligate life, repenting after watching scenes from a religious play; this is possibly the first example in European drama of a play within a play. Of the allegorical and didactic MORALITY PLAYS of the time, the only one known nowadays is *Elckerlyc* by the Chartreuse monk Petrus Dorlandus. Printed in Delft in 1405 under its full title, *Den Spieghel der Salicheit van Elckerlyc* (*The Play of the Summoning of Everyman*), it is probably the original of the English play *Everyman*, first seen in London in about 1500. Rediscovered in 1901 by William POEL, who produced it in a modernized version in the grounds of the Charterhouse, it has been revived many times since and as *Jedermann* was one of the great productions of REINHARDT in Salzburg.

From the early Middle Ages to the beginning of the 17th century Flemish—and by extension Dutch—theatre grew in importance. The SCHOOL DRAMA, imported into Flanders from Germany, was already feeling the influence of the Humanist discovery of the Latin dramatists PLAUTUS, SENECA, and TERENCE, and flourished at the hands of such writers as Willem de Volder (1493–1568), who wrote under the name of Gulielmus Gnapheus and with his *Acolastus* (1529), on the theme of the prodigal son, inaugurated the *comicotragoedia* of the Biblical school drama, as distinct from that on classical themes. This was translated into English, and gave rise to a number of plays on the same subject, one of which was *The Glass of Government* (1575) by George GASCOIGNE. More important for the development of the Flemish theatre was the influence of the Rederykers, or members of the CHAMBERS OF RHETORIC, founded in imitation of those in northern France. These became the dominating cultural force in many of the larger cities, and from the beginning of the 15th century were active in the organization of drama festivals, during which the various groups competed among themselves, often with plays specially written for the occasion. Among these were the works of Cornelius Everaert (1485–1556), author of a farce entitled *About the Fisherman*, in which the fisherman's wife confesses her unfaithfulness to her husband during a storm at sea but manages to regain her ascendancy over him once the storm has passed; the anonymous *Mirror of Love*, the first domestic tragedy in Dutch drama, since it is concerned with the love of a rich merchant's son for a poor girl which ends in tragedy for them both and points the moral that one should not seek to marry outside one's own class; and the plays of Jan van Hout (1542–1609), which show traces of early Renaissance influence. The Rederykers were also responsible for the civic pageants and entertainments devised for the reception of royal and noble personages, and the scenic backgrounds built for some of their open-air productions, which recall the triumphal arches of the Romans, are believed to have had some influence on the development of the Elizabethan playhouse.

The war with Spain in the second half of the 16th century seriously disrupted the cultural development of the southern half of the Low Countries, and the theatre there took a long time to recover; when it did, the main influence was no longer Flemish, but French. Meanwhile the province of Holland in the North became an important cultural centre, mainly due to the number of Protestant refugees who flocked to it from the south. The famous scholar and poet Daniel Heinsius (1580–1655), whose dramatic theories, based on ARISTOTLE, were known to JONSON and DRYDEN, still wrote in Latin; but the major playwrights of the time, Pieter Cornelisz HOOFT, Gerbrand Adriaansz BREDERO, and Joost van den VONDEL, all wrote in the vernacular. In 1617 their friend Samuel Coster (1579–1665), a doctor who was also a playwright, opened the first Dutch Academy in Amsterdam as a centre for theatre, literature, and higher learning. This was something new for the Netherlands, which had never had a Court Theatre, and one of its main objectives was attained 20 years later when Jacob van CAMPEN built the first theatre in Amsterdam, the Schouwburg. This opened in 1638 with Vondel's *Gijsbrecht van Amstel*, and in 1665 was remodelled by Jan Vos (*c.* 1620–67), a popular playwright whose neo-classical tragedies owed much to Seneca. He did away with van Campen's permanent set (*scena stabile*), replacing it with changeable scenery and elaborate Italian machinery. This, however, did nothing to halt the decline of Dutch drama, which was rapidly succumbing to the influence of French neo-classicism, resulting in a lessening of creative force and originality. No great playwrights arose to take the place of Vondel, who was the last to be associated with the Rederykers, and the plays of the end of the 17th century, of which those of the prolific Pieter Bernagie (1656–99) were typical, were stiff and colourless. The 18th century, which saw only such imitative tragedies as *Eneas en Turnus* (1705) and *Scilla* (1709) by Luca Rotgans (1654–1710), *Achilles* (1719) by Balthazar Huydecoper (1695–1778), and *Jacoba van Beieren* (1736), on a subject taken from Dutch history by Jan Harmenszoon de Marre (1696–1763), saw good work being done in comedy. The comic tradition created by Bredero and Vondel, and upheld by Thomas Asselyn (*c.* 1620–1701) in such plays as *Jan Klaaz* (1682), culminated in the work of Pieter Langendijk (1683–1756). Some of his comedies, written under the invigorating influence of MOLIÈRE, are excellent. *The Mutual Marriage Hoax* (1714), in which two people pretend to be rich in order to entice each other into matrimony, was the best Dutch comedy of manners of the 18th century, and long remained in the repertory.

During the 19th century the influence of French *tragédie bourgeoise* on the Dutch theatre was reinforced by translations of plays from Germany, particularly those of KOTZEBUE and IFFLAND which were widely imitated. Little of importance emerged, however, nor did the exponents of Romanticism, of whom Henrik Jan Schimmel (1823–1906), with a number of historical plays in the style of Victor HUGO, is typical, add anything of value to the repertory. On the technical side, however, great strides were being made. The destruction by fire of the second Schouwburg in 1772 had led to the building of a new theatre two years later, only a year after the opening of the first theatre in Rotterdam. By the beginning of the 19th century the art of acting had immensely improved, and the new century made a good start with the opening of a Theatre Royal in The Hague in 1804, and the introduction of a new and simpler style of acting by the actor Marten Corver, the first Dutch director. In 1870 the Dutch Theatre Association was founded by prosperous citizens who wanted to improve the quality of acting and production by the creation of a theatre school, and as a result a national theatre was brought into being, its company consisting of the best actors available, among them Louis Bouwmeester (1841–1925), best known for his portrayal of Shylock in *The Merchant of Venice*. With his sister Theo (1850–1939) he also appeared in the plays of the younger DUMAS and of SARDOU, and of those who still modelled their style on imports from France. The guest performances of the MEININGER COMPANY in Amsterdam in 1880 and in Rotterdam in 1888 aroused much controversy but proved on the whole stimulating, and had an important influence on two complementary theatre personalities of the rising generation, Willem Royaards (1867–1929), who brought poetry back to the Dutch stage with productions of Vondel and Shakespeare, and Eduard Verkade (1878–1961), who was educated partly in England, where he came under the influence of Gordon CRAIG, and was the first director to stage SHAW's *Saint Joan* in the Netherlands. A pupil of Verkade, the actor, director, and manager Albert van Dalsum (1889–1971), was responsible for bringing EXPRESSIONISM on to the Dutch stage between the two World Wars.

It was the impact of REALISM which in the 20th century brought the Netherlands back into the main stream of European drama, first with the works of Marcellus Emants (1848–1923), whose most successful play was *The Power of Stupidity* (1904), in which a rising politician is destroyed by the jealousy of his wife and the folly of his worthless son, and secondly with those of Herman HEIJERMANS, who made himself the mouthpiece of the oppressed and explored in his plays the miseries and inequalities of the contemporary scene. His contemporary Josine Adriana Simons-Mees (1863–1948), less well known outside her own country, revealed in her work a sympathetic understanding of the stresses of family life, as did Willem Frederik Schürmann (1879–1915). Better known was Jan Fabricius (1871–1966), many of whose plays were set in the Dutch East Indies, the best being the pathetic *Lonely* (1907), which depicts the despair and desolation of a man stationed on a remote island in the Moluccas. A later dramatist who attracted attention outside the Netherlands is Jan de Hartog (1914–), seaman and actor, who settled in the United States. His *Skipper Next To God* (1942) was seen in London in 1945, with himself in the chief part, and in New York in 1948. *The Fourposter*, a two-character play, had its first performance in English in London in 1950, opening in New York the following year, where it had a long run. The most interesting Dutch-language playwright to emerge since the Second World War is Hugo CLAUS.

In recent years a number of new theatres and theatre schools have opened, the one at Tilburg on 3 Mar. 1961 with the world première of Christopher FRY's *Curtmantle*. There is also in Amsterdam, whose theatre school dates from 1874, a most interesting Theatre Museum, founded in 1959. A notable event in the Dutch theatre, which, like the Belgian, relies heavily on imports from other European countries and from America, has been the founding in 1965 of the Mickery Theatre, which specializes in avant-garde plays. Most of the important experimental groups of Europe and America have appeared at this theatre, which in 1972 moved to Amsterdam, where it continues to exercise a profound influence on all aspects of Dutch theatre.

NETHERSOLE, OLGA ISABEL (1870–1951), English actress and theatre manager, who made her first appearance in London in 1887 and two years later was at the GARRICK THEATRE, where her portrayal of the betrayed country girl Janet Preece in PINERO's *The Profligate* (1889) quickly brought her recognition as an actress of unusual emotional power. In 1893 she scored a further triumph with her Countess Zicka in a revival of SARDOU's *Diplomacy*, and then took over the management of the ROYAL COURT THEATRE, where she directed and played in A. W. Gattie's *The Transgressor* (1894), in which she made her New York début later the same year. For the next 20 years she divided her time between London and the U.S.A., being equally popular in both. Her intense and realistic characterizations of fallen women in such plays as the younger DUMAS's *The Lady of the Camellias* (*Camille*), SUDERMANN's *Magda*, Pinero's *The Second Mrs Tanqueray* and *The Notorious Mrs Ebbsmith* shocked some older play-goers, but to the younger generation she became a symbol of the revolt against prudery, particularly when in 1900 she was arrested by the New York police for alleged indecency while appearing as Fanny Legrand in Clyde FITCH's *Sapho*. Defended by William WINTER, she was eventually acquitted. She managed several London theatres, including HER MAJESTY's in 1898, the ADELPHI in 1902 and the SHAFTESBURY in 1904. One of her last outstanding parts was the title role in MAETERLINCK's *Mary Magdalene* (1910). She retired in 1914.

NEUBER [*née* Weissenborn], (Frederika) CAROLINA (1697–1760), German actress and manager. After an unhappy childhood she eloped with a young clerk, Johann Neuber (1697–1759), and with him joined a theatrical company. Ten years later they formed a company of their own. 'Die Neuberin' was at this time at the height of her powers, being much admired in BREECHES PARTS, and had already attracted the notice of GOTTSCHED, who enlisted her help in his projected reform of the German stage, persuading her in 1727 to stage French classical plays in the place of the old improvised comedies, farces, and harlequinades. She soon found herself in conflict with his rigid principles and they parted in 1739, after which she offended him further by ridiculing his *Der* *sterbende Cato*, which was performed in Roman dress, and by depicting him in a curtain-raiser as a bat-winged censor. The fortunes of her company declined: even the collaboration of the young LESSING was of no avail, and after struggling along until the outbreak of the Seven Years War, which reduced them to poverty, husband and wife died within a year. In spite of its unfortunate conclusion, Carolina Neuber's association with Gottsched is generally regarded as the starting-point of the modern German theatre. She did a great deal for her profession, ruling her company with a firm hand, and insisting on regularity and order. Her style of acting, in its day, was a vast improvement on the old clowning and farcical horse-play and prepared the way for the subtle and more natural methods of EKHOF and SCHRÖDER.

NEUMES, see LITURGICAL DRAMA.

NEUVILLE, HONORÉ DE, see MONTANSIER, MLLE.

NEVILLE, JOHN (1925–), English actor, director, and manager, who gained his early experience in repertory and from 1950 to 1953 was with the BRISTOL OLD VIC. He then spent some years with the OLD VIC in London, being seen in a wide variety of parts—in 1956 he alternated Othello and Iago with Richard BURTON—and going with the company to the United States in 1956 and 1958. After leaving the company in 1959 he made several appearances in London before going in 1961 to the NOTTINGHAM PLAYHOUSE where he remained until 1968, though in the first season at the CHICHESTER FESTIVAL THEATRE he was seen in FLETCHER'S *The Chances* and FORD'S *The Broken Heart*, and in 1963 he was at the MERMAID in the title role of Naughton's *Alfie*. He became joint artistic director at Nottingham in 1963, taking sole charge in 1965, and under him the theatre gained an excellent reputation. His roles there included Sir Thomas More in BOLT'S *A Man for All Seasons* in 1961; Coriolanus in the opening production at the new Playhouse in 1963; Corvino in JONSON'S *Volpone* in 1965; and Willy Loman in Arthur MILLER'S *Death of a Salesman* in 1967. Arout's *Beware of the Dog* (1967), based on short stories by CHEKHOV, in which he played several parts, was transferred to London. He also directed a number of plays, including an adaptation of CALDERÓN'S *The Mayor of Zalamea* in 1964. In 1968 he returned to London, directing LIVINGS'S *Honour and Offer* (1969) and appearing as King Magnus in SHAW'S *The Apple Cart* in 1970. Since 1972 he has worked mainly in Canada, running a theatre in Edmonton from 1973 to 1978 and then becoming artistic director of the Neptune Theatre in Halifax. He was, however, in a revival of BECKETT'S *Happy Days* at the NATIONAL THEATRE in London in 1977.

NEW AMERICAN MUSEUM, New York, see AMERICAN MUSEUM.

NEW AMSTERDAM THEATRE, New York, at 214 West 42nd Street, between 7th and 8th Avenues. This opened on 26 Oct. 1903 with Nat GOODWIN in the Klaw and Erlanger production of *A Midsummer Night's Dream*, while later in the same year came the DRURY LANE pantomime *Mother Goose* which ran for three months. Among visiting stars who appeared at this theatre in its early years were Mrs Patrick CAMPBELL in SARDOU'S *The Sorceress* (1904) and the FORBES-ROBERTSONS in SHAW'S *Caesar and Cleopatra* (1906). *Brewster's Millions* (also 1906) by McCutcheon, Smith, and Ongley and *The Merry Widow* (1907) were both successful productions here. Beerbohm TREE came in *Henry VIII* (1916), but the theatre was mainly occupied by musical comedy and revue, being best known for the *Ziegfeld Follies*, seen annually from 1913 to 1924. In 1933 Eva LE GALLIENNE brought an adaptation of Carroll's *Alice in Wonderland* and CHEKHOV'S *The Cherry Orchard* from the Civic Repertory Theatre for a successful run. The last production was *Othello* in 1937, with Walter Huston in the title role and sets by Robert Edmond JONES, after which the building became a cinema.

On the roof of the building was the Aerial Gardens Theatre, a small playhouse which had frequent changes of names. It was a radio or television studio for much of its existence, though it mounted plays for a brief period from 1943, taking the New Amsterdam's name.

NEW BOWERY THEATRE, New York, see BOWERY THEATRE (2).

NEWCASTLE PLAYHOUSE, Newcastle-upon-Tyne. The first theatre of this name was originally called after Flora ROBSON, who appeared there in Emlyn WILLIAMS'S *The Corn is Green* in 1962. The Tyneside Theatre Company presented a wide and varied programme of plays old and new until the theatre was demolished in a road-widening scheme. It was replaced in 1970 by the present Playhouse, first known as the University Theatre, which is owned by the University of Newcastle. Seating 450 in a single steeply rising tier, it has a proscenium arch which deliberately fails to conceal the backstage workings. The Tyneside Theatre Company continued to occupy it until 1977, when financial troubles led to its being used for touring companies. A year later a new company, the Tynewear, supported by the local authority and the ARTS COUNCIL, took possession, and the theatre's name was changed. The company has presented a number of interesting productions, including BRECHT'S *The Resistible Rise of Arturo Ui*, *The Merchant of Venice*, the musical *Cabaret*, and *And a Nightingale Sang*... by C. P. TAYLOR. There is a Theatre In Education scheme. Behind the theatre is the Gulbenkian Studio, with movable seating for 120 to 200, which is run by the university. It is used by touring companies and by the ROYAL SHAKESPEARE COMPANY on its annual visit to the NEWCASTLE Theatre Royal.

NEWCASTLE-UPON-TYNE. There is little evidence of theatrical activity in this area much before the late 17th century, when actors are known to have played in rooms near the Quayside in the then centre of the town. They were compelled to act secretly, however, on account of the hostility of the Puritan element among the citizens. In the 18th century the County House of Northumberland, subsequently the city's courthouse, became a centre for dramatic activity until in about 1747 it was replaced by the Turk's Head Long Room. The first permanent theatre was built in Mosely Stteet in 1788, and having a royal patent was known immediately as the Theatre Royal. It was demolished in the 1830s to make way for Grey Street, in which a new Theatre Royal was built and still stands. It opened in 1837 with *The Merchant of Venice* and Rodwell's *The Young Widow*. Reconstructed and enlarged in 1895, it was extensively damaged by fire in 1899, the interior being completely burnt out, though the external structure was untouched. Rebuilt and enlarged by the addition of properties from the adjoining Shakespeare Street, the theatre reopened yet again in 1901. It housed touring companies until 1973, when it was bought by the Newcastle City Council. It is now managed by a Trust, and continues to be one of the country's major touring centres. Seating 1,400, it has since 1977 provided a regional base for the ROYAL SHAKESPEARE COMPANY, which presents seasons there lasting several weeks in the early part of the year.

NEW CHATHAM THEATRE, New York, see CHATHAM THEATRE (2).

NEW CHELSEA THEATRE, London, see ROYAL COURT THEATRE (1).

NEW COMEDY, term used to describe the last period of ancient Greek comic drama, which in the later 4th and 3rd centuries BC developed from the transitional MIDDLE COMEDY. It later became the model and the quarry for Roman comedy, and so influenced later European comic dramatists, particularly MOLIÈRE. Its finest exponent was MENANDER, who was immensely popular in antiquity but little more than a name in modern European literature until one of his complete works and considerable portions of several other plays were discovered among papyri from Egypt. New comedy was pure comedy of manners. It used stock characters—the testy old man, the boastful soldier, the whining parasite, the resourceful slave—and conventional turns of plot (the foundling was a constant figure); but all these are treated, by Menander in particular, with a delicacy of feeling and observation which make a drama of great charm. The CHORUS survives from the earlier type of comedy, but has nothing to do with the plot; Menander often makes it a band of tipsy singing revellers.

NEW ENGLISH OPERA HOUSE, London, see ROYALTY THEATRE (2).

NEWES, TILLY, see WEDEKIND, FRANK.

NEWINGTON BUTTS THEATRE, London. There are many references to performances at Newington in HENSLOWE's diary, but it is still uncertain whether they were given in an inn-yard or in an open-air enclosure. A theatre may have been built in about 1576, but the first reference to plays being given there is found in a letter from the Privy Council desiring that the Surrey justices should prohibit performances at Newington for fear of spreading the plague. A similar letter was sent to the Lord Mayor of London in 1586. In 1594 Henslowe recorded ten days of performances given by the CHAMBERLAIN'S MEN in association with the ADMIRAL'S MEN, among them MARLOWE's *The Jew of Malta* and Shakespeare's *Hamlet*. Thereafter the theatre fell into disuse, but Howes, in his continuation to Stow's *Annals* in 1631, includes in his list of London theatres 'one in former times at Newington Butts'. The accepted site of this is the present Elephant and Castle.

NEWLEY, ANTHONY, see MUSICAL COMEDY.

NEW LONDON THEATRE, London, in Drury Lane, designed by Sean KENNY and erected on the site of the WINTER GARDEN THEATRE, is part of a complex of buildings which includes a multi-storey car park and a restaurant. The auditorium, reached by lifts from the car park and by an escalator from the foyer, is on two levels, one-third of the lower level comprising a 60-ft revolve which holds the stage, orchestra pit, and first rows of the stalls. The stage can thus be used for either proscenium or open-stage productions, 206 of the total of 911 seats moving with the stage to complete the transformation. The auditorium can also be used as a conference hall during the day. The theatre opened on 11 Jan. 1973 with Peter USTINOV in his own play *The Unknown Soldier and His Wife*, which was followed by *Grease*, a rock-and-roll musical. It proved difficult, however, to find suitable product for the theatre, and from 1977 until 1980 it was used as a television studio. In 1981 it returned to theatrical use with the enormously successful musical *Cats*, which ingeniously exploited the full potential of the theatre. To avoid confusion with the newcomer, the former NEW THEATRE in St Martin's Lane changed its name to the Albery.

NEW NATIONAL THEATRE, New York, see CHATHAM THEATRE (2).

NEW OLYMPIC THEATRE, New York, see OLYMPIC THEATRE (3).

NEW PARK THEATRE, New York. (1) At 932 Broadway, between 21st and 22nd Streets. This opened on 13 Apr. 1874 with FECHTER in the last new part he was to create, in his own adaptation of a French play. The theatre had been intended for BOUCICAULT, but after lengthy litigation its real history began on 16 Sept. 1874, when John T. RAYMOND appeared there as Colonel Sellers in a dramatization of Mark Twain's *The Gilded Age*. Later successes were Woolf's *The Mighty Dollar* with the FLORENCES in 1875, and G. F. ROWE in his own play *Brass* in 1876. Henry E. ABBEY took over the New Park late in 1876, and started it on a prosperous career which lasted until, on 30 Oct. 1882, the day on which Lillie LANGTRY was to have made her New York début there, the theatre was totally destroyed by fire and never rebuilt.

(2) At the north-east corner of Broadway and 35th Street. Built on the site of the old Aquarium, this opened on 15 Oct. 1883. It was furnished with a good deal of material bought when BOOTH'S THEATRE was demolished, and began with a series of musical plays. It was then occupied for a short time by BELASCO, who imported mainly melodrama. On 11 Aug. 1884 little Minnie Maddern, later Mrs FISKE, was seen in Henry P. Taylor's *Caprice*, specially written for her, and the theatre then became a museum, occasionally housing itinerant companies and light opera.

NEW ROYALTY THEATRE, London, see ROYALTY THEATRE (2).

NEW THEATRE, London, in St Martin's Lane, built for Charles WYNDHAM who opened it on 12 Mar. 1903 with a revival of PARKER and Carson's *Rosemary*, after which it settled down to a consistently successful career. From 1905 to 1913 Fred TERRY and Julia NEILSON occupied it annually for a six-month season, and many of their most successful plays were seen there, including Baroness Orczy's *The Scarlet Pimpernel* in 1905. The theatre also housed an annual revival of BARRIE'S *Peter Pan* for several years. Among outstanding productions have been a dramatization of Louisa M. Alcott's *Little Women* (1919), in which Katharine CORNELL made her only London appearance; A. A. Milne's *Mr Pim Passes By* (1920); and, in 1924, SHAW'S *Saint Joan* with Sybil THORNDIKE as Joan of Arc. A year later came the long run of Margaret Kennedy's *The Constant Nymph*, which saw the first appearance of John GIELGUD at a theatre which was later to play an important part in his career, beginning in 1933, when he appeared there in Gordon Daviot's *Richard of Bordeaux*, followed by *Hamlet* (1934), *Romeo and Juliet*, and OBEY'S *Noah* (both 1935). Among later productions at this theatre were *The Taming of the Shrew* (1937), O'NEILL'S *Mourning Becomes Electra* (1938), and PRIESTLEY'S *Johnson over Jordan* (1939), with Ralph RICHARDSON. After the bombing of the OLD VIC and SADLER'S WELLS, the New became the London headquarters of both companies, opening on 14 Jan. 1941. Sadler's Wells withdrew in 1944 and the Old Vic in 1950, in which year T. S. ELIOT'S *The Cocktail Party* began a successful run. The theatre has since housed a series of excellent plays, including Dylan Thomas's *Under Milk Wood* (1956); *Summer of the Seventeenth Doll* (1957) by the Australian playwright Ray LAWLER; and Wolf Mankowitz's *Make Me an Offer* (1959) from THEATRE WORKSHOP. In 1960 Lionel BART'S *Oliver!*, a musical based on DICKENS'S *Oliver Twist*, began a run of several years, and on 12 Dec. 1968 HOCHHUTH'S controversial play *Soldiers* began a short run. It was followed by a popular musical, *Anne of Green Gables*. On 1 Jan. 1973, partly to avoid confusion with the NEW LONDON THEATRE, the New was renamed the Albery after its former director Bronson ALBERY. It housed the successful rock musical *Joseph and the Amazing Technicolor Dreamcoat* (1973) and successful revivals of MAUGHAM'S *The Constant Wife* (also 1973), Shaw's *Pygmalion*, Barrie's *What Every Woman Knows* (both 1974), and TURGENEV'S *A Month in the Country* (1975). The NATIONAL THEATRE'S production of Peter SHAFFER'S *Equus* was transferred to the Albery in 1976, and there were revivals there of Shaw's *Candida* (1977) and the musicals *Oliver!* (also 1977), which ran for almost three years, and *Pal Joey* (1980), which transferred from a FRINGE THEATRE, the Half Moon.

NEW THEATRE, New York, see CENTURY THEATRE (1), NASSAU STREET THEATRE, and PARK THEATRE (1).

NEW THEATRE COMIQUE, New York, see NEW YORK THEATRE (1).

NEWTON, JOHN, see ROWLEY, WILLIAM.

NEW VICTORIA PALACE, London, see OLD VIC.

NEW YORK CITY CENTER, see CITY CENTER OF MUSIC AND DRAMA.

NEW YORKER THEATRE, New York, at 254 West 54th Street, between Broadway and 8th Avenue. As the Gallo, this opened on 7 Nov. 1927 with opera, followed by Margaret ANGLIN in EURIPIDES' *Electra* and the ABBEY THEATRE company from Dublin in O'CASEY'S *Juno and the Paycock*. On 12 May 1930 the name was changed to the New Yorker, and the theatre reopened with IBSEN'S rarely seen *The Vikings*, in which realistic scenery was replaced by a background of colour and light in moving patterns. In 1932 a fine Spanish company under Fernando Díaz de Mendoza and María GUERRERO occupied the theatre for five weeks in an impressive repertory which ranged from Lope de VEGA to the ÁLVAREZ QUINTERO brothers. In 1933, as the Casino de Paree, the building became a music-hall and home of MUSICAL COMEDY, but on 21 Mar. 1939, after 18 months as the Federal Music Theatre, it again became the New Yorker with a production of *The Swing*

Mikado, based on GILBERT and Sullivan's operetta. In 1943 it was converted into a studio for radio and television shows.

NEW YORK SHAKESPEARE FESTIVAL, see PAPP, JOSEPH.

NEW YORK THEATRE, New York. (1) At 724–8 Broadway. Originally the church of the Messiah, this had been used for concerts and lectures before on 23 Dec. 1865, Lucy Rushton (1844–?), an actress of little ability but great personal charm, opened it in her own name, appearing with the veteran Charles WALCOT in SHERIDAN's *The School for Scandal*, BULWER-LYTTON's *The Lady of Lyons*, and *As You Like It*. On 3 Sept. 1866, renamed the New York, it opened under Mark Smith and Lewis Baker. Augustin DALY took his company there after his first FIFTH AVENUE THEATRE was burnt down, remaining for the whole of 1873. After his departure the theatre was known as FOX's Broadway and became a home of variety as the Globe, one of its more enduring names. The VAUDEVILLE duo HARRIGAN and Hart named it their New Theatre Comique in 1881, and extensively refurbished it: but it was burnt down in 1884 and not rebuilt.

(2) On Broadway between 44th and 45th Streets. This was originally the Olympia Music Hall, which opened on 17 Dec. 1895, with Yvette GUILBERT, as part of Oscar Hammerstein's OLYMPIA complex, in which the Lyric Theatre, later the CRITERION, was also included. Auctioned only three years later, it reopened on 24 Apr. 1899 as the New York Theatre, and among the many successes staged there were a revival of Alice Hegan Rice's *Mrs Wiggs of the Cabbage Patch* (1906), several of George M. COHAN's plays, and the musical comedy *Naughty Marietta* (1910). The theatre was later used for vaudeville and films and was finally demolished in 1935.

For the New York Theatre, Bowery, see BOWERY THEATRE (1).

NEW ZEALAND. The first recorded dramatic performance in New Zealand took place in Auckland on Christmas Eve 1841, in a room in a hotel on the waterfront named for the occasion the Albert Theatre, only two years after the Treaty of Waitangi which opened the way for the colonization of New Zealand by the British. The first theatre to be built, the Royal Victoria, was erected in Wellington in 1843, and it was not long before theatre buildings arose in all the principal towns. The discovery of gold in Otago in 1861 helped to put the country firmly on the map of theatrical touring companies from Australia, England, and America, and for the next 70 years tentative efforts at local productions were overshadowed by a constant stream of entertainment from outside. Initially the visitors appeared in Shakespeare, melodrama, GILBERT and Sullivan, musical comedy, and the popular 'well-made' plays of the period; but by the 1890s new plays by IBSEN, PINERO, and WILDE were also being brought; the earliest New Zealand plays, like George Leitch's *The Land of the Moa* (1895), were modelled on importations from overseas. Eventually the rise of the cinema, the First World War, and the subsequent period of depression practically put an end to the era of large-scale touring companies, but by the early 1930s a flourishing AMATEUR THEATRE movement was firmly established in nearly every city, town, and rural community throughout the country. It provided the soil in which a local professional theatre was to grow and flourish, and also continued to play a significant role in its own right up to the present time. From the beginning the larger societies in Auckland, Wellington, Christchurch, and Dunedin were strongly linked with the British Theatre Association (B.T.A.), then known as the British Drama League (B.D.L.), which in 1971 combined with the New Zealand Drama Council to form the New Zealand Theatre Federation. Although the B.D.L. encouraged the writing of one-act plays and sponsored one-act play festivals, most of the plays produced in New Zealand in the 1920s and 1930s, by such authors as Merton [Horace Emerton] Hodge (1904–58), whose light comedy *The Wind and the Rain* (1933) was successfully produced in London, or Alan Mulgan, J.A.S. Coppard, and Eric Bradwell, were full-length, and based on imported patterns. Even after the Second World War the most interesting New Zealand writing came not from dramatists nurtured by the B.D.L. but from poets and other non-dramatic writers like Allen Curnow, R.A.K. Mason, Frank Sargeson, James K. Baxter, and the expatriate Douglas STEWART. Assistance and production came mostly from a tiny group of adventurous small theatres: Wellington's Unity, Auckland's New Independent, Christchurch's Elmwood, Patric Carey's Globe in Dunedin, and several University drama societies.

Post-war tours, such as that of the OLD VIC company led by Laurence OLIVIER and Vivien LEIGH in 1948, revived efforts to found a professional national theatre which would be able to compete on equal terms with distinguished visitors. Robert Kerridge, a cinema magnate, had tried the previous year to establish a 'New Zealand Theatre' in Auckland, with a company of mainly English actors called the West End Players, but low standards and the hostility of the amateur theatres killed it within a year. A more modest operation known as the COMMUNITY ARTS SERVICE or C.A.S. was more successful, and lasted from 1947 to 1962; but despite its potential and its quality it never achieved national status, nor was Dame Ngaio Marsh's (1899–1982) British Commonwealth Theatre Company able to survive beyond its first season in 1951. However, in 1953 Richard and Edith Campion returned from training at the Old Vic Theatre School and founded the NEW ZEALAND PLAYERS, who until their sudden financial collapse in 1960 did excellent work up and down the country. In the same year, but too late to save the Players, the Arts Advisory Council of New Zealand was established, followed in 1964 by the Queen Elizabeth II Arts Council, and for the first time Government subsidies were made available to the theatre. The new régime was intended to encourage regional institutions, and consequently the Southern Comedy Players, who had been touring the South Island since 1957 with light comedy, took on a new role as a professional theatre in Dunedin under the Southern Theatre Trust; a Canterbury Theatre Trust on the same lines was also established in Christchurch, but both these collapsed in the late 1960s. The only successful venture on this model was that promoted by the Auckland Theatre Trust, which housed a professional and subsidized company in the MERCURY THEATRE that still flourishes. The history of the DOWNSTAGE THEATRE in Wellington, founded four years before the Mercury, is perhaps more typical of professional development in New Zealand since the 1960s. Established in 1964, its slow but steady progress enabled it to avoid the fate which had overtaken so many ventures before it, and over a period of ten years it rose from ramshackle semi-professionalism to become a highly professional ensemble working in the versatile purpose-built Hannah Playhouse, which seats 200 people, and saw in 1974 the first full season of New Zealand plays, including a one-man show by Bruce Mason, Robert Lord's *Well-Hung*, Brian McNeill's *The Two Tigers*, and Joseph Musaphia's *Victims*.

In the 1970s companies in other New Zealand centres, operating in small theatres seating usually no more than 250 people, followed the Downstage model of gradual growth through a period of semi-professionalism. Among them were the Court Theatre in Christchurch, founded in 1971, which included among its early productions the lively and important musical documentary *O! Temperance!* (1972) by its co-founder Mervyn Thompson, and Centrepoint Theatre in Palmerston North and the Fortune Theatre Company in Dunedin, which followed a similar pattern. All three are now fully professional regional theatres assisted by Q.E.II subsidies. Together with the Mercury Theatre and Auckland's newer Theatre Corporate, formed by a returned expatriate, Raymond Hawthorne, they comprise the Association of Community Theatres (A.C.T.), of which three original members did not survive—the Central in Auckland, the Four Seasons in Wanganui, and the Gateway in Tauranga.

Alternative theatres, as elsewhere, have come and gone, but several groups from the 1970s deserve mention, among them Amamus, which, strongly influenced by GROTOWSKI, moved from political documentary to exploratory theatre seeking to define a myth of national identity; Theatre Action, which injected the techniques of Le Coq's Paris School into local theatre, and the Red Mole, which carried them into cabaret spectaculars; Heartache and Sorrow, formed by expatriates in Amsterdam and Great Britain, which won an important award at the 1979 EDINBURGH FESTIVAL; and the radical Maori group, Maranga Mai ('Wake Up!'), which in 1980 provoked sharp political controversy in its own country.

One professional theatre which operates outside the subsidized A.C.T. framework is Circa Theatre, established in Wellington in 1976 by a group of actors formerly with the Downstage Theatre company, to act as a co-operatively run alternative for performers whose days were mostly committed to radio and television. Its first major success was Roger Hall's Civil Service comedy *Glide Time* (1976), directed by Anthony Taylor (later artistic director of Downstage), which drew an unusually large audience. Hall's continuing success has been phenomenal; his *Middle Age Spread* (1977) began a successful run at the LYRIC THEATRE in London in 1979, starring Richard Briers and Paul Eddington. Hall, like Bruce Mason and Joseph Musaphia before him, received great assistance and encouragement from the radio drama section of the then N.Z.B.C., for in the 1960s the radio was almost the only outlet for local dramatic talent. With the new mood of cultural nationalism, influenced in particular by the new wave of Australian playwriting in the 1970s, Playmarket was established in 1973 as an agency to advise on scripts and promote new works. It organizes regular workshops of new plays in connection with the professional theatres, publishes scripts and a bulletin, *Act: Theatre in New Zealand*, and in 1980 held the first national Playwrights' Workshop. One of the plays to emerge from this, Greg McGee's *Foreskin's Lament*, was immediately hailed as a milestone in New Zealand's dramatic writing, and received its first production at Theatre Corporate the same year.

Up to the 1970s training for the stage had been largely in the hands of committed directors, such as Ngaio Marsh and her influential Shakespeare productions at Canterbury University in the 1940s and later, and Nola Millar in Wellington, from whose New Theatre School the Q.E.II-sponsored New Zealand Drama School was formed. This has now reached a high standard, offering since 1974, under the direction of George Webby, a full-time two-year actors' training course. Various other theatres also run training and apprenticeship schemes, usually in conjunction with the Drama School, and several New Zealand universities now offer courses in drama and theatre arts.

NEW ZEALAND PLAYERS, company formed in 1953 by Richard and Edith Campion, which gave its first production at the Wellington Opera House. It then went on tour two or three times a year, and did excellent work, but remained fundamentally a travelling troupe with no permanent home. Over the next seven years it presented a representative selection of plays from overseas in all parts of the country, with ANOUILH's *Ring Round the Moon*, USTINOV's *The Love of Four Colonels*, FRY's *The Lady's Not For Burning*, with Keith MICHELL and Barbara JEFFORD, specially brought out from England, SHAW's *Saint Joan*, and Willis HALL's *The Long and the Short and the Tall*. The company also appeared in occasional plays by New Zealanders, including Douglas STEWART's *Ned Kelly*, and gave a workshop production of Bruce Mason's *The Pohutukawa Tree*, the first important modern New Zealand play. In spite of its artistic success, however, the constant touring proved too costly, particularly as audiences in general remained small, and the company finally disbanded in 1960, ironically enough while playing to full houses in Napier in MILLER's *A View From the Bridge*.

NEZVAL, VÍTĚSLAV, see CZECHOSLOVAKIA.

NIBLO'S GARDEN, New York, on the north-east corner of Broadway and Prince Street, a summer resort opened by William Niblo on the site of the Columbia Garden. Here, on 4 July 1827, he opened the small Sans Souci Theatre which was used, in the summer only, for concerts and also for plays, particularly after the destruction by fire in 1828 of the BOWERY (to whose displaced company the theatre was leased). Rebuilt in 1829 and renamed Niblo's Garden, the theatre prospered until it was burnt down on 18 Sept. 1846. It was not rebuilt until 1849, when on 30 July a new theatre was built as part of (and entered through) the Metropolitan Hotel. It was improved and enlarged in 1853, and two years later saw the last appearance in New York of RACHEL. On 12 Sept. 1866 came the first performance of *The Black Crook*, a fantastic mixture of drama and spectacle, with elaborate transformation scenes and a scantily-clad *corps de ballet*, which ran for 475 performances and was followed by a similar spectacle, *The White Fawn*, which was less successful. After a production on 16 Nov. 1868 of BOUCICAULT's melodrama, *After Dark; or, London by Night*, the theatre was given over to melodrama and spectacle, and such popular performers as CHANFRAU in Spencer and Tayleure's *Kit the Arkansas Traveller* and Lotta (Charlotte CRABTREE) in a version of the younger DUMAS's *La Dame aux camélias* entitled *Heartsease*, until on 6 May 1872 it was again burnt down. Rebuilt, it opened on 30 Nov., but its great days were over, and it was too far downtown for a front-rank theatre. It served for some time as a home for visiting companies and was finally demolished in 1895.

NICCOLINI, GIAMBATTISTA (1782–1861), Italian dramatist, significant for the popularity enjoyed by his robustly pamphleteering tragedies, for he saw drama as a vehicle for politics and patriotism: his plays *Giovanni da Procida* (1817), *Lodovico Sforza* (1834), and *Filippo Strozzi* (1847) are indistinguishable onslaughts on despotism and exaltations of Italian nationalism. *Antonio Foscarini* (1823) combines drama in the high romantic style with a conflict between duty and love, while *Nabucco* (1815), overtly following ALFIERI in its portrait of the struggle between absolute power and the ideal of freedom, is an allegory on the life of Napoleon. Niccolini's finest work for the stage—a dramatic poem rather than a play—is *Arnaldo da Brescia* (1843), a vigorous assault on the temporal power of the Pope and the corruption of the clergy.

NICHOLS, MIKE [Michael Igor Peschowsky] (1931–), American actor, director, and producer, born in Berlin, who for some years played a double act in night clubs with Elaine May, most of the material for which they wrote themselves, transporting it to the JOHN GOLDEN THEATRE, New York, in 1960. Three years later Nichols directed Neil SIMON's *Barefoot In the Park*, which was followed by three more of Simon's plays—*The Odd Couple* (1965), *Plaza Suite* (1968), and *The Prisoner Of Second Avenue* (1971)—and other new plays such as Ann Jellicoe's *The Knack*, Murray Schisgal's *Luv* (both 1964), David RABE's *Streamers*, and Trevor Griffiths's *Comedians* (both 1976). He also directed revivals of Lillian HELLMAN's *The Little Foxes* (1967) and CHEKHOV's *Uncle Vanya* (1973), and co-adapted the latter. In 1977 he produced the musical *Annie* and D. L. Coburn's *The Gin Game*, directing the latter both in New York and, in 1979, in London. He is also a successful film director.

NICHOLS, PETER RICHARD (1927–), English playwright, who had written a number of television plays before he had an unexpected success in the theatre with *A Day in the Death of Joe Egg* (1967; N.Y., 1968), transferred from the CITIZENS' THEATRE, Glasgow. This portrayal of the stresses imposed on parents by the care of a spastic child was followed by *The National Health* (1969; N.Y., 1974), set in a men's ward for incurables and including fantasy scenes in the style of a television serial; *Forget-Me-Not Lane* (1971), in which the action switches between past and present as the middle-aged narrator looks back on his adolescence in the 1940s; and the more conventional *Chez Nous* (1974; N.Y., 1977), with Albert FINNEY, Geraldine MCEWAN, and Denholm ELLIOTT in London. *The Freeway* (also 1974), like *The National Health*, was produced by the NATIONAL THEATRE, and *Privates on Parade* (1977), a musical about an army concert party, by the ROYAL SHAKESPEARE COMPANY. In *Born In the Gardens* (1980) Beryl Reid played an eccentric lady, newly widowed, whose middle-aged son still lives at home. Their apparent contentment is sharply contrasted with the misery of her two other, more ambitious, children, who have returned for their father's funeral. One of Nichols's most highly praised plays, *Passion Play*, was produced by the R.S.C. in 1981. A complex play about adultery, it shows great technical virtuosity in bringing on stage the *alter egos* of the two main characters.

NICKINSON, ISABELLA, see WALCOT, CHARLES.

NICKINSON, JOHN (1808–64), actor-manager, born in London, who played an important part in the development of the theatre in Canada, particularly in Toronto. Joining the army at 15, he was sent to Canada, where both in Quebec and in Montreal he acted with the Garrison Amateurs, appearing at Montreal's Theatre Royal in 1833. Three years later he left the army and became a professional actor, specializing in dialect comedy and playing such parts as Havresack in BOUCICAULT's *The Old Guard*, Pickwick in a dramatization of DICKENS's novel, and Sir Peter Teazle in SHERIDAN's *The School for Scandal*. He made his first appearance in New York in 1837 and for some years acted there in the winter, spending the summer on tour in Canada. He was for six seasons with MITCHELL at the OLYMPIC THEATRE in New York, and when it closed in 1850 he formed his own company, with which he toured the Great Lakes, becoming manager of the Royal Lyceum Theatre in Toronto in 1853. He remained there until 1858, when he was succeeded by his son-in-law Owen Marlowe, and returned to America where he died. In his company were his four daughters and a son. His eldest girl Charlotte (1832–1910), as Mrs Morrison, ran the stock company at Toronto's Grand Opera House most successfully until it burnt down in 1879.

NICOLET, JEAN-BAPTISTE (*c.* 1728–96), French acrobat and entertainer, son of a puppet-master, who played at the FAIRS of St-Germain and St-Laurent. In 1760 he had a booth on the BOULEVARD DU TEMPLE, which he soon transformed into a small permanent theatre, with a good troupe of acrobats and animal turns, notably monkeys. Nicolet himself played young lovers and HARLEQUINS when he replaced his puppets by living actors, and his theatre flourished in spite of the opposition of the COMÉDIE-FRANÇAISE. In 1772 he was summoned to Court by Louis XV, who allowed him to call his theatre the Spectacle des Grands Danseurs du Roi, a title which it retained until 1795, when it became the Théâtre de la GAÎTÉ. The freedom of the theatres under the Revolution allowed Nicolet to play the repertory of the Comédie-Française, which he did, choosing for preference the lighter pieces, until in 1795 he retired and his associate Ribié took over.

NICOLL, ALLARDYCE (1894–1976), Scottish theatre historian, successively Professor of English Language and Literature in London and Birmingham Universities and at one time head of the Department of Drama at YALE UNIVERSITY where he began a vast and comprehensive file of photographs of theatrical material from all over Europe which is constantly being enriched, forming a deposit of valuable material for the theatre research worker. He was the author of a number of useful and well-illustrated books on specialized aspects of the theatre, including a series of nine volumes begun in 1923, covering the history of the English theatre from the Restoration, each period having an invaluable hand-list of plays arranged under dramatists. Nicoll who for many years, as Director of the Shakespeare Institute, convened a conference of Shakespearian experts at STRATFORD-UPON-AVON during the summer, was also an excellent lecturer, and the founder of *Shakespeare Survey*, which he edited from 1948 to 1965. He was President of the SOCIETY FOR THEATRE RESEARCH from 1958 until his death.

NICOSTRATUS, see GREECE, ANCIENT.

NIC SHIUBHLAIGH, MAIRE [Maire Price, *née* Marie Walker] (?-1958), Irish actress, one of the founder players of the IRISH NATIONAL DRAMATIC SOCIETY, with which she appeared as Delia Cahel in YEATS's *Cathleen ni Houlihan* (1902), going with the company to the ABBEY THEATRE in 1904, where she played leading parts with the sisters Sarah ALLGOOD and Maire O'NEILL. She was at her best in tragic roles, among them Nora Burke in SYNGE's *In the Shadow of the Glen* (1903) and Moll Woods in Seamus O'KELLY's *The Shuiler's Child* (1910). She left the Abbey Theatre in 1905, but returned several times as a guest artist, and made her last appearance on the Dublin stage in 1948 in *The Shuiler's Child*. In 1950, in collaboration with her nephew Edward Kenny, she published a book about the renaissance of the Irish theatre entitled *The Splendid Years*.

NIEMCEWICZ, JULIAN URSYN, see POLAND.

NIGGER MINSTRELS, see MINSTREL SHOW.

NIGHT SCENES, see WEAVER, JOHN.

NILSEN, HANS JACOB (1897–1957), Norwegian actor, director, and theatre manager. He began his career in the 1920s as an actor, joining the company at the NATIONALTHEATRET in 1928. After 1933, however, he worked mainly as a director, first at the NATIONALE SCENE in Bergen from 1934 to 1939, then at the NORSKE THEATRET in Oslo from 1946 to 1950, and finally at the Folketeatret from 1952 to 1955. Among his productions the most noteworthy were those of HOLBERG's plays, and a startlingly 'anti-romantic' production of IBSEN's *Peer Gynt* in 1948, in which he also played the title role.

NIMROD THEATRE, Sydney, see AUSTRALIA.

NINA VANCE ALLEY THEATRE, Houston, Texas, named after its founder on her death in

1980. Its first home, in 1947, was a rented dance studio seating 87, which took its name, Alley Theatre, from the narrow passage leading up to it. Forced to move for safety reasons in 1949, the company then occupied a fan-manufacturing plant converted into an arena theatre, its first production being Lillian HELLMAN's *The Children's Hour*. By 1954 the resident company had become fully professional, being one of the first three such companies in the United States, the others being the ARENA STAGE, Washington, and the CLEVELAND PLAY HOUSE. In 1968 it moved into a permanent home, a striking building whose nine towers give it the appearance of a medieval castle. It has two auditoriums, the Large Stage seating 798 and having a thrust stage which can provide a variety of playing areas, no seat being more than 58 ft from the stage; the smaller Arena Stage, seating 296, based on the company's first home, and being unique of its kind in that the actors can enter without passing through the audience. In a season lasting from Oct. to June the theatre presents a balanced repertoire of world-wide drama, including original works and classic revivals, plays for children, lunchtime theatre, and readings of new plays. The building also houses the Alley Merry-Go-Round, America's largest theatre school for young people.

NŌ, see JAPAN.

NOAH, MORDECAI MANUEL (1785–1851), American playwright, whose first play, *The Wandering Boys* (1812), was a translation of PIXÉRÉCOURT's *Le Pèlerin blanc* (1801). First produced in Charleston, this was later seen at COVENT GARDEN with alterations by John Kerr, and in its amended form returned to New York, where it remained popular for many years. Noah's later plays were produced at the PARK THEATRE (1); it was after the third night of *The Siege of Tripoli* on 24 May 1820 that the theatre was destroyed by fire. In a later play, *The Grecian Captive* (1822), the hero and heroine made their entrances on an elephant and a camel respectively, a spectacular device due, no doubt, to the fertile brain of the manager Stephen PRICE. Noah's plays are simply written, with a good deal of action and sustained interest, and with the aid of lavish scenery, transparencies, and illuminations they held the stage for many years. He led an active life in politics and journalism, and in the prefaces to the printed editions of his plays gives an amusing account of his experiences in the theatre and of the difficulties of the native American playwright in competition with the established English drama.

NOBEL PRIZE FOR LITERATURE, has been awarded in recognition of their dramatic writings to the following: BECKETT, Samuel, 1969; BENAVENTE, Jacinto, 1922; BJØRNSON, Bjørnstjerne, 1903; ECHEGARAY, José, 1904; HAUPTMANN, Gerhart, 1912; MAETERLINCK, Maurice, 1911; O'NEILL, Eugene, 1936; PIRANDELLO, Luigi, 1934; SARTRE, Jean-Paul (who declined it), 1964; SHAW, George Bernard, 1925.

NOBILI, LAUDIVIO DE', see ITALY.

NOISES OFF, see SOUND EFFECTS.

NOKES, JAMES (?–1696), English actor, a member of DAVENANT's original company. He was a fine comedian, of whom Colley CIBBER in his *Apology* has left a masterly pen-portrait, and usually played foolish old husbands and clumsy fops, as well as a few ridiculous old ladies. His best part among the many that he created in Restoration drama seems to have been the title role in DRYDEN's comedy *Sir Martin Mar-all* (1667).

NOLI, FAN STYLIAN, see ALBANIA.

NORA BAYES THEATRE, New York, see FORTY-FOURTH STREET THEATRE.

NORSKE THEATRET, Det, a playhouse in Oslo (then Christiania) which opened on 2 Jan. 1913 in Bøndernes Hus. Its prime objective was to provide a stage for the performance of plays in Nynorsk, the officially standardized Norwegian language, to offset the Dano-Norwegian influence of IBSEN and BJØRNSON. It also aimed to stimulate Nynorsk dramatists into providing a continuous supply of new plays in the official language. Although the theatre lives on as a lively component of the Oslo theatrical scene, and has mounted many significant productions, among them an 'anti-romantic' version of Ibsen's *Peer Gynt* directed in 1948 by Hans Jacob NILSEN, the Nynorsk contribution to Norwegian drama remains meagre.

NORTHCOTT THEATRE, Exeter, Devon, named after its chief benefactor, G. V. Northcott, a Devon business man. It was erected in 1967 on a site donated, together with a large sum of money, by the University of Exeter, within whose precincts it stands. It has a proscenium stage and the auditorium, fan-shaped and steeply raked, seats 433. The theatre is, however, easily adaptable to other forms, particularly THEATRE-IN-THE ROUND, which involves setting seating on the main stage and using the apron stage as the playing area. The seating is then 580. Plays are presented for an average run of three weeks, and include a judicious mixture of old and new, with the first production of BOND's *Bingo* and *Halfway up the Stairs*, a play about Christopher Robin by Ian Fell and Christopher Robin Milne. The theatre's productions of two musicals by SONDHEIM, *A Little Night Music* and *Company*, toured nationwide. The building is also used by amateur and visiting companies and presents films and exhibitions, as well as acting as a regional repertory theatre of a very high standard.

NORTON, THOMAS (1532–84), a member of the Inner Temple, who in collaboration with a fellow-student, Thomas Sackville (1536–1608), later the first Earl of Dorset and Lord Treasurer under Elizabeth I and James I, wrote *Gorboduc, or Ferrex and Porrex*, the first surviving example of a regular five-act tragedy in the style of SENECA in English dramatic literature. Apparently Norton wrote the first three acts and Sackville the last two. The play was performed before Elizabeth I on New Year's Day 1561/2 during an entertainment in the Inner Temple Hall; its theme resembles that of *King Lear*, with Gorboduc, king of Britain, dividing his kingdom between his two sons, who quarrel over it and are both killed. Norton does not appear to have written anything else, but Sackville contributed to the second edition of *A Mirror for Magistrates* (1563) the Induction and 'The Complaint of Buckingham', the only contributions having any literary merit.

NORWAY. There is some evidence that Norway had a share in the Latin SCHOOL DRAMA of the late 16th century, but the first recorded performance of a play in the vernacular was in 1562, when pupils in Bergen put on *Adam's Fall* in the Cathedral grounds, directed by Absalon Pederssson Beyer. In general, however, the drama of Norway is bound up with that of Denmark until their separation in 1814, and indeed for some time afterwards: thus, HOLBERG was Norwegian-born but made his career in Denmark; and the writers of Det Norske Selskab (The Norwegian Society, established in 1772)—a group of Norwegians resident in Copenhagen—included two dramatists, Johan WESSEL and Nordahl Brun (1745–1816), author of *Zarine* (1772), a tragedy in the French manner, among their number. There was, of course, no Court theatre, since Norway had no resident king.

From the beginning of the 19th century serious interest in theatre began to show itself, mainly in the formation of many amateur societies: first in Christiania (later Oslo) (1780–1844), in which Bernt Anker and Envold de Falsen were the moving spirits; and later in Bergen, Trondheim, Kristiansand, Stavanger, and elsewhere. A modest theatre was built in Bergen in 1799, and another in Christiania in 1802.

In 1827 Christiania acquired its first established theatre with a company of professional actors under the direction of a Swede, Johan Peter Strömberg. To begin with the actors were Norwegian and Norwegian plays—by Holberg, Wessel, and Henrik Bjerregaard (1792–1842), author of the successful musical play *Fjeldeventyret* (*The Adventure on the Mountain*, 1824)—were among those performed; but within two years, Danish players had completely taken over. The building was burnt down in 1835 and a new theatre, the KRISTIANIA THEATER, was erected on another site in 1837. This remained Norway's premier theatre until it closed in 1899.

The first attempt to establish a genuinely Norwegian theatrical tradition was made by the poet Henrik WERGELAND, who wrote a number of plays: his play *Campbellerne* (*The Campbells*) was deliberately disrupted at its first performance in the Kristiania Theater on 28 Jan. 1838 as part of the struggle between the 'Danish' and the 'Norwegian' factions in the capital. In 1850 a theatre with Norwegian actors and directors, known as Det Norske Teatret (later Den NATIONALE SCENE), was founded in Bergen by Ole BULL; and in 1852 a Norwegian theatre (modestly calling itself at first a 'dramatic school') began operating in Christiania. The Bergen theatre was forced by financial difficulties to close down in 1863, and did not reopen until 1876. The Christiania Norwegian Theatre lasted as an independent institution only until 1863, when it was merged with the more Danishly inclined Kristiania Theater.

Both Henrik IBSEN and Bjørnstjerne BJØRNSON held appointments at these Bergen and Christiania theatres, and there they learned their craft. Between them they revolutionized Norwegian thought, literature, and language, and carried the dramatic reputation of the country ahead of that of any in Europe. Bjørnson's early historical plays and Ibsen's great poetic dramas *Brand* and *Peer Gynt* stirred the national imagination. Bjørnson's realistic and contemporary analyses and condemnation of the errors of society in part preceded Ibsen's similar but more profound studies of social evils. Ibsen's dramas went on to astound Europe, but both gave salutary shocks to the complacency of the society they condemned.

The effect on some of the more practical aspects of Norwegian theatre was also profound: standards of acting and production greatly improved; talented players of the early years, like Laura GUNDERSEN, were succeeded by actors and actresses of outstanding distinction, the most illustrious being Johanne DYBWAD; a more rigorous realism replaced the earlier tendency to romanticize and sentimentalize. The stage (to use SCHILLER's earlier phrase) was transformed into a 'moral institution', an instrument of social change, and it became the practice of writers who by temperament and talent were novelists rather than playwrights —men like Alexander KIELLAND, and later Knut HAMSUN and Hans KINCK—to attempt something in the dramatic mode. By the time Ibsen died in 1906 Norwegian drama was firmly established among the world's great dramatic literatures.

With the building of the present NATIONALTHEATRET in Christiania in 1899 and the establishment of a company there under the direction of Bjørn BJØRNSON, son of the dramatist, the theatre as an institution was firmly established in Norway. In 1913 the new NORSKE THEATRET, intended for the performance of plays in the officially standardized language Nynorsk, was added to the metropolitan theatres; and the Nye Teater, at which Lillebil IBSEN was leading lady, opened in 1929. In this immediate post-Ibsen period, the level of achievement in dramatic literature perhaps inevitably fell away. Gunnar HEIBERG's best work was already behind him by the turn of the century and the talents of Helge KROG and Tore ØRJASAETER were modest ones. Nordahl GRIEG was only beginning to reveal his potential dramatic power when he was killed in the Second World War.

Post-war developments were, however, considerable. The Riksteatret, a state touring company, was formed in 1948 under the direction of Frits von der Lippe, to perform in local theatres and halls and also occasionally abroad. The Folketeatret (People's Theatre) had hopes of attracting a new and largely working-class audience when it opened in 1952; but it folded in 1959 and was merged with Det Nye Teater to form the OSLO NYE TEATER, whilst its buildings were taken over by a Norwegian opera company. The Centralteatret closed in the same year.

It seemed impossible for the theatre to survive without financial support, and a full reorganization took place; since 1962 the Nationaltheatret, the Oslo Nye Teater, and the Norske Theatret in Oslo have all been subsidized, as have also the Riksteatret and the provincial theatres in Bergen, Trondheim, and Stavanger. Much excitement followed the range of new appointments made in 1967. Arild Brinchmann became head of the Nationaltheatret, Toralv MAURSTAD of the Oslo Nye Teater, and Knut Thomassen of the theatre in Bergen. Outstanding actors of the post-war period have included Wenche FOSS, Claes GILL, and Liv STRØMSTED.

By contrast, the post-war years were a relatively lean period for Norwegian dramatic literature: among those active were Finn Havrevold (1905–), Axel Kielland (1907–63), Odd Eidem (1913–), and Helga Hagerup (1933–). The 1960s saw some experimental work of interest from Jens Bjørneboe, Finn Carling, and Georg Johannesen; but the excitement caused by the opening of the Odinteatret in Oslo in 1964 was dampened by the company's move to Denmark in 1966.

NORWICH, Norfolk, best known today for its amateur MADDERMARKET THEATRE, was the headquarters of the Norwich CIRCUIT, which comprised Ipswich, Colchester, Yarmouth, and other towns. Its first theatre building was erected in 1757 by Thomas Ivory on Assembly Plain, but before that various inns in the city had served as playing-places, among them the White Swan which, although it was last used for entertainments in 1771, was not demolished until 1961. Ivory, who was succeeded by William Wilkins, obtained a patent for his theatre in 1768. It then became the Theatre Royal, a name retained by the new theatre built in 1826 on a site adjoining the old one. The Norwich circuit broke up in about 1852, and the STOCK COMPANY was replaced by touring companies. Enlarged in 1931, the Theatre Royal was burned down in 1934, but reopened a year later. It had a struggle to survive, and was for a time a cinema. In 1967 it was bought by the local authority, and since 1972 a careful selection of touring attractions designed to appeal to all tastes, from nude shows to the NATIONAL THEATRE company, has brought it great prosperity. It is run by a Trust, and seats 1,274.

NORWID, CYPRIAN KAMIL (1821–83), Polish romantic poet and playwright, who spent most of his life in exile in Paris where he died in poverty. He wrote between 1862 and 1881 over 20 deeply religious and historical poetic plays, only three of which were published in his lifetime, and he remained virtually unknown until, under the impetus of the Young Poland movement at the turn of the century, some of his plays were staged in the early 1900s. Even then his genius was not fully recognized until 1970, when Hanuszkiewicz directed an anthology of his works at the National Theatre in Warsaw under the general title of *Norwid*. One of his most interesting plays, *Behind the Scenes*, written in the 1860s and built round the device of a play-within-a-play, was given its first production in 1971 at the Słowacki Theatre in Cracow under the direction of Kazimierz Braun.

NORWORTH, JACK (1879–1959), American comedian, who first appeared in 'blackface', and then went into VAUDEVILLE, where with his second wife, the singer Nora Bayes (1880–1928) whom he married in 1907, he set up a double act which was successful in New York and London. They were divorced in 1913; Norworth returned to London a year later and from 1914 to 1918 appeared successfully in a number of REVUES. Back in America he opened a small, intimate theatre under his own name, which later became the BELMONT, but this was not a success and he soon left. He is best remembered as the singer of such tongue-twisters as 'Sister Susie' and 'Which Switch is the Switch, Miss, for Ipswich?' Among his own compositions the most popular was 'Shine On, Harvest Moon'.

NOTTARA, CONSTANTIN, see ROMANIA.

NOTTINGHAM. The first Theatre Royal in this town was built in about 1760 by James Augustus Whitley (*c.* 1724–81), an Irish actor-manager who ran the extensive Midland CIRCUIT which included Worcester, Wolverhampton, Derby, Retford, and Stamford as well as Nottingham. Rebuilt in the late 1770s, it began to decline during the first half of the 19th century, audiences becoming less fashionable and somewhat scanty, and the building itself being considered shabby and out of date. It always had a responsible management, however, and London stars continued to appear there until 1865, when a new theatre was built. The old one was converted into a MUSIC-HALL, becoming a restaurant in 1883 and a lace warehouse in 1901; bombed in 1941, it was eventually demolished. The second Theatre Royal, seating 1,100, was built by two lace dressers, John and William Lambert. It had a splendid interior and a portico with six giant Corinthian columns. It opened with SHERIDAN's *The School for Scandal*, a prologue being spoken by Madge KENDAL. Like its predecessor it attracted many of the outstanding players of the day—in 1882 both Lillie LANGTRY and Sarah BERNHARDT appeared there—and it became famous for its Christmas pantomimes. It was remodelled in 1897, and, with a seating capacity of 1,997, reopened with Augustin DALY's company in *The Taming of the Shrew*. At the same time a variety theatre, the Empire, was built alongside it, which survived until 1958 and was demolished in 1969. The Theatre Royal continued to house leading touring companies, Henry IRVING and Anna Pavlova both appearing there. By the 1960s, however, it had become very dilapidated, and some major touring companies refused to visit it because of the poor accommodation backstage. After existing for some years under the threat of closure, it was taken over by the Nottingham City Council and underwent a complete transformation, which provided it with all modern facilities while preserving its Victorian character. It reopened in 1978 with a seating capacity of 1,138, and is one of the country's leading touring theatres.

NOTTINGHAM PLAYHOUSE. This theatre, run by a Trust since 1948, was originally a converted cinema seating 467. André Van Gyseghem was its first director, and Val May was its director from 1957 to 1961, being succeeded by Frank DUNLOP, John NEVILLE, and other energetic managers. The new Playhouse, one of the first modern regional theatres, opened in 1963, funded by private contributions and Nottingham City Council. Its cylindrical auditorium gives a feeling of intimacy and seats 500 in the stalls and 250 in the balcony. It has a large and well equipped stage on which elaborate productions can be mounted. Outstanding presentations have included BRECHT's *The Resistible Rise Of Arturo Ui*, SHAW's *Widowers' Houses*, *Hamlet* with Alan BATES, and *King Lear* with Michael HORDERN, all of which were seen in London. Notable among the numerous premières have been Peter Barnes's *The Ruling Class* (1968) and his *Lulu* (1970), adapted from WEDEKIND, Christopher FRY's *A Yard Of Sun* (also 1970), Howard BRENTON's *The Churchill Play* (1974), and Trevor Griffiths's *Comedians* (1975). In the absence of a studio theatre productions likely to attract smaller audiences are given on Monday evenings. A Theatre in Education scheme is run by the Roundabout company.

NOUVEAU MONDE, Théâtre du, French-Canadian company founded in Montreal in 1951 by a group of actors, under Jean GASCON, who had been associated with Father Émile Legault's Compagnons de Saint-Laurent. They opened at the Salle Gésu on 9 Oct. with *L'Avare* by MOLIÈRE, the author of most of their early successes. A trio of his farces, played in the style of the COMMEDIA DELL'ARTE, was an outstanding success at the THÉÂTRE DES NATIONS in Paris in 1955, and brought the new company an international reputation. In the following year Gascon and several of his companions joined the company at the STRATFORD (ONTARIO) FESTIVAL theatre to play the French roles in *Henry V*, in a production that was later seen at the EDINBURGH FESTIVAL. Although for over 20 years the group had no permanent home, it evolved into a civic repertory theatre with high standards, presenting an international repertoire varied with a new Canadian play each season, one of its most successful authors being Marcel DUBÉ, whose *Le Temps des lilas* (1958) was taken on tour in 1958–9 together with Molière's *Le Malade imaginaire*. The company suffered a severe setback when in 1963 fire destroyed their offices, records, and rehearsal space, and again in 1966 when the Orpheum cinema, which they had rented in 1957, was scheduled for demolition. Under a new director Jean-Louis Roux, one of the original members of the company, who took over when Gascon resigned in 1966, they moved in 1967 to the Port-Royal in the new Place des Arts complex. Ill at ease, however, in these lavish surroundings, they were glad to find a permanent home in 1972 at the Théâtre de la Comédie-Canadienne, left vacant by the departure of Gratien GÉLINAS and his company. An immediate innovation there was the introduction of Lunch-Time Theatre (*Théâtre-midi*), and among the new plays produced were *Marche, Laura Secord* (1975), a musical revue which parodied the heroine of the War of 1812; *Marathon* (1976) by Claude Confortes, which helped Montreal to celebrate the Olympic Games held that year in the city; and the controversial *Les Fées ont soif* (*The Fairies are Thirsty*, 1978) by Denise Boucher, an attack on male chauvinism and the restraints imposed on women by the Catholic Church which provoked accusations of blasphemy and resulted in subsidies being withheld by the civic authorities. The company toured abroad in 1965 and 1971 and from 1965 to 1975 also supported Les Jeunes Comédiens, a touring company of young actors which began as a graduation class from the National Theatre School of Canada.

NOUVEAUTÉS, Théâtre des, Paris, a name borne by four successive playhouses. The first, built in 1827 in the Passage Feydeau near the Bourse, caught fire during the 1830 revolution, and although it survived it was closed down after the Government had taken exception to the production there of a Bonapartist play. The second, built in 1866 on the Boulevard St Martin, was burnt down in the same year. The third, which opened in 1878 on the Boulevard des Italiens, lasted long enough for FEYDEAU to make his name there. The present Théâtre des Nouveautés, on the Boulevard Poissonnière, dates from 1920 and has specialized in light comedies and musicals, plays by

Tristan BERNARD, COURTELINE and André Roussin being seen there.

NOVELLI, ERMETE (1851–1919), Italian actor, who made his first appearance at the age of 18 and though not at first well received became one of the outstanding actors of the Italian stage in his day. A large man, weighing some 18 stone, he was nevertheless light on his feet and quick in action, with fine, expressive features and a mobile countenance. At first he played exclusively in comedy, but it was finally in tragedy that he made his reputation, and he toured extensively in such Shakespearian parts as Othello, Lear, Shylock in *The Merchant of Venice*, Macbeth, and Hamlet. In the last-named he is said to have played the death-scene with brutal realism—indeed, from contemporary accounts his acting appears to have been exceedingly forceful and melodramatic—and in the scene with the Ghost he seemed to impart to the audience the certainty of a visitation from another world. Another part in which he excelled was the title role in a translation of Aicard's *Le Père Lebonnard*, a play first seen at ANTOINE's Théâtre-Libre. In 1900 Novelli attempted to found a permanent theatre in Rome, La Casa di GOLDONI, but the enterprise failed through lack of public support and had to be abandoned.

NOVELLO [Davies], (David) IVOR (1893–1951), actor-manager, dramatist, and composer, and one of the most consistently successful men of his day in the London theatre. He was the son of musical parents, his mother, Clara Novello Davies (*née* Davies), being a choral conductor who owed her Christian names to her father's admiration for the great singer, to whom she was not related and who was not, as is sometimes stated, her godmother. From his earliest days he showed facility and tunefulness in the composition of light music, and during the First World War was responsible for part of the score of several successful MUSICAL COMEDIES. He also wrote the popular song 'Keep the Home Fires Burning'. He made his first appearance on the stage in Sacha GUITRY's *Deburau* (1921), and three years later appeared in his own play *The Rat*, written in collaboration with Constance COLLIER. He subsequently wrote more than 20 comedies and musical plays, composing also the scores for the latter and appearing in most of the productions himself. Among his straight plays were *The Truth Game* (1928), *Symphony in Two Flats* (1929), *I Lived With You* (1932), *Fresh Fields* (1933), *Full House* (1935), and *We Proudly Present* (1947). He was the author, composer, and leading man of four successive musicals at DRURY LANE: *Glamorous Night* (1935), *Careless Rapture* (1936), *Crest of the Wave* (1937), and—after he had appeared in a spectacular revival of *Henry V* in 1938 at the same theatre—*The Dancing Years* (1939). The last was revived at the ADELPHI THEATRE in 1942 and was one of the big successes of the Second World War. Novello also wrote, composed, and played in another great success, *Perchance to Dream* (1945). He was appearing in his own musical *King's Rhapsody* (1949) at the time of his death. During its run he had written and composed *Gay's The Word* (1951), which had a long run starring Cicely COURTNEIDGE.

NOVELTY THEATRE, London, see KINGSWAY THEATRE.

NOVIUS, see FABULA 1: ATELLANA.

N.T.D., see O'NEILL, EUGENE.

NUNN, TREVOR ROBERT (1940–), English director. He had had some experience of amateur theatre in Ipswich, his birthplace, before going to CAMBRIDGE, where he acted in and directed several plays for the Marlowe Society. In 1962 he went to the BELGRADE THEATRE in Coventry as a trainee director, remaining there until in 1965 he joined the ROYAL SHAKESPEARE COMPANY, where his first production, at the ALDWYCH, was *The Thwarting of Baron Bolligrew*, a children's play by Robert BOLT. A year later he directed the first revival for 300 years of TOURNEUR's *The Revenger's Tragedy*, and in 1967 he was responsible for excellent revivals of VANBRUGH's *The Relapse* and of *The Taming of the Shrew*. In 1968 he was appointed the company's Artistic Director, directing in the same year productions of *Much Ado About Nothing* and *King Lear*. Among his other notable productions have been *The Romans* (1972), which included *Coriolanus, Julius Caesar, Antony and Cleopatra*, and *Titus Andronicus*; his own adaptation of IBSEN's *Hedda Gabler* (1975); a musical version of *The Comedy Of Errors*, for which he wrote the lyrics, and *Macbeth*, with Ian MCKELLEN and Judi DENCH (both 1976). In 1978 he became Chief Executive and joint Artistic Director with Terry HANDS, since when his productions have included a revival of KAUFMAN and HART's *Once in a Lifetime* in 1979, and *The Life and Adventures of Nicholas Nickleby*, based by David Edgar on DICKENS's novel, in 1980; the latter was revived in 1981 and presented in New York. In 1981 he also directed the successful musical *Cats*, based on T. S. ELIOT's *Old Possum's Book Of Practical Cats*—his first freelance production since becoming the R.S.C.'s Artistic Director—and a highly praised production of *All 's Well That Ends Well*. He is married to Janet SUZMAN.

NURSERIES, the name given during the Restoration period to training schools for young actors, the best known being that set up by KILLIGREW in Hatton Garden in about 1662. This moved in 1668 to the VERE STREET THEATRE, where it flourished until 1671, when DAVENANT's widow opened a new Nursery in the Barbican. This was still in use in 1682, as it is referred to in DRYDEN's poem *MacFlecknoe*, published in that year. A third Nursery is known to have opened briefly on Bun Hill, in Finsbury Fields, some time during 1671, and Nursery companies occasionally performed in borrowed playhouses. Little is known of their work, and although they must have provided at least a minimal training for young players, Joseph HAINES is the only actor of note known to have emerged from such a background.

NUŠIČ, BRANISLAV, see YUGOSLAVIA.

NYE TEATER, see OSLO NYE TEATER.

O

OAKLEY, ANNIE, see CODY, W. F.

OBALDIA, RENÉ DE (1918–), French writer, born in Hong Kong of a Panamanian father and a French mother. He was educated, and is now domiciled (as a French citizen), in Paris, and was already well known as a poet and novelist before his first play *Génousie* (1960), an idiosyncratic murder-mystery, was produced by Jean VILAR for the THÉÂTRE NATIONAL POPULAIRE. It was followed by *Le Satyre de La Villette* (1963), an ironic fantasy on the kind of situation explored by Nabokov in his novel *Lolita* which caused something of a scandal when first produced, and *Du Vent dans les branches de Sassafras* (1965), probably Obaldia's best known play and the first 'spoof' Western. It was produced in English at the PITLOCHRY FESTIVAL THEATRE in 1979, where Obaldia's *Monsieur Kelbs et Rozalie* (1975) was seen in 1980. With this, and such plays as *Le Cosmonaute agricole* (also 1965) and *Deux Femmes pour un fantôme* (1971), Obaldia, who had hitherto been regarded as an AVANT-GARDE writer in the style of IONESCO and BECKETT, seems to be aiming at a wider audience, who appreciate his witty but sympathetic presentation of the dilemmas of modern society.

OBERAMMERGAU, see PASSION PLAY.

OBEY, ANDRÉ (1892–1975), French dramatist, whose plays *Noé, Le Viol de Lucrèce*, and *La Bataille de la Marne* were all produced in 1931–2 by SAINT-DENIS for the COMPAGNIE DES QUINZE. The first, produced in London in 1935 with GIELGUD as Noah, was remarkable for the liveliness of its beasts; the second, which made use of a modified Greek chorus, was one of the sources of the libretto of Britten's opera *The Rape of Lucrece* (1946). Continuing to employ a distanced technique, often involving the use of a narrator and heightened language, Obey later wrote *Le Trompeur de Séville* (1937), on the theme of DON JUAN, and *L'Homme de cendre* (1950); also, on Biblical themes, *Lazare* (1952) and *Plus de miracles pour Noël* (1957), seen in London in 1965 as *Frost at Midnight*.

OBRAZTSOV, SERGEI VLADIMIROVICH, see PUPPET.

O'CASEY, SEAN [John Casey] (1880–1964), Irish dramatist, whose best plays, set in the slums of Dublin, show how intimately he knew the people of whom he wrote and the events of 1915–22 from which he drew his material. His treatment of his themes is closely related to that of the Irish realists before him—grim, clear-cut, and satirical; but in his work the comedy of satire points directly to tragic implications. His first play, produced at the ABBEY THEATRE in Dublin, was *The Shadow of a Gunman* (1923; London, 1927; N.Y., 1932), a melodramatic story of the war in 1920 and its effects on the lives of a group of people in a Dublin tenement house. This anticipates, in its subject and setting and in some of its objective commentary, the finer plays that followed; the men who talk, live, and die for an idea are contrasted with the women who live and die for actualities. O'Casey's next play, *Juno and the Paycock* (1924; London, 1925; N.Y., 1926), a moving, realistic tragedy, set in 1922 with much the same background, was popular with both English and Irish audiences, though for different reasons, as was *The Plough and the Stars* (Dublin and London, 1926; N.Y., 1927), a play on the Easter Rising of 1916 which caused a riot in the Abbey Theatre when it was first produced there. The consequent refusal of the Abbey to produce his next play, *The Silver Tassie* (London and N.Y., 1929), which was not seen in Dublin until 1935, led O'Casey to leave Ireland and settle in England. His next play, *Within the Gates* (London and N.Y., 1934), was set in London, and gave further evidence of the influence on his work of EXPRESSIONISM, already apparent in *The Silver Tassie*. Although remarkable in many ways, it is doubtful whether the extended use of stylization and symbolism helped O'Casey to master his material, nor was he particularly successful with his Cockney characters. His next play, *The Star Turns Red* (1940), was first seen at the UNITY THEATRE in London, but in 1943 O'Casey had his first Irish premières for 17 years with *Purple Dust* (N.Y., 1955; London, 1962) and *Red Roses For Me* (London, 1946; N.Y., 1955). In these, and again in *Oak Leaves and Lavender*, seen in London in 1947, he again used symbolist and expressionistic devices to reinforce ideals expressed by their Marxist heroes. *Cock-a-Doodle Dandy*, first seen at the People's Theatre, Newcastle-on-Tyne, in 1949, was produced in New York in 1958 and at the ROYAL COURT THEATRE in London in 1959, stimulated a new interest in O'Casey's later plays. In 1964 the Abbey Players were seen in London in the WORLD THEATRE SEASON at the ALDWYCH in *Juno and the Paycock* and *The Plough and the Stars*. In his last plays, *The*

Bishop's Bonfire (1955; London, 1961), *The Drums of Father Ned* (1958), *Behind the Green Curtains* (1961), and *Figuro in the Night* (also 1961; N.Y., 1963), O'Casey contrasts the repressive forces of the clergy and the moneyed classes in modern Ireland with the yearnings of Irish youth for artistic, sexual, and political freedom.

Two of O'Casey's early one-act plays produced at the Abbey, *Kathleen Listens In* (1923) and *Nannie's Night Out* (1924), were published in 1963 in a volume entitled *Feathers from the Green Crow*, and O'Casey himself rescued five others, all Irish farces—*The End of the Beginning, A Pound on Demand, Hall of Healing, Bedtime Story*, and *Time to Go*—and added them to his collected plays, published in 1951. Many episodes in his plays are drawn from the personal experiences recorded in his six volumes of autobiography, published in one volume in 1956 as *Mirror in My House*. Most of his dramatic criticisms, collected in *The Flying Wasp* (1937), *The Green Crow* (1957), *Under a Coloured Cap* (1963), and the posthumous *Blasts and Benedictions* (1967) edited by Ronald Ayling, vigorously attack the fourth-wall convention and the commercialism of the English theatre.

OCTAGON THEATRE, Bolton, the first fully professional flexible theatre to be erected in Britain, its octagonal shape being considered the most satisfactory for such a building. It can be adapted for open end, thrust stage, or in-the-round productions, seating 350, 324, and 420 respectively. The money required was raised mainly by public subscription, though the site was donated by Bolton Corporation. The theatre opened in 1967, only 18 months after the campaign to build it began, a remarkable achievement for a town with no great theatrical tradition, and which had been without a theatre since 1961. The company presents classics and good modern plays, its first production being the world première of Bill Naughton's *Annie and Fanny*. For several years three Shakespeare plays have been staged every autumn, and it is hoped ultimately to present all of them. The theatre enjoys a good measure of public support, and every effort is made to create a friendly atmosphere, the building being used for many activities—concerts, debates, recitals, fashion shows—apart from plays. The original Theatre in Education scheme has now been replaced by a Theatre Appreciation Programme for children with its own building and company of actors.

O'DEA, JIMMY [James Augustine] (1899–1965), Irish comic actor, who in 1920 appeared at the old Empire Theatre in Dublin. A few years later he took the role of Buttons in the QUEEN'S THEATRE annual Christmas pantomime, and from then became well known as a mimic and revue artist of great verve and energy, particularly in the character of a sharp-tongued old reprobate, 'Biddy Mulligan'. For many years the diminutive O'Dea appeared regularly each year at the GAIETY THEATRE in a Christmas pantomime and in a special show for the Dublin Horse Show Week, working with the comic actress Maureen Potter. He demonstrated his versatility when in 1940 MACLIAMMÓIR and EDWARDS persuaded him to play Bottom in *A Midsummer Night's Dream* at the GATE THEATRE.

ODELL'S THEATRE, London, see GOODMAN'S FIELDS THEATRE (3).

ODÉON, Théâtre National de l', Paris, the second theatre of France, ranking next to the COMÉDIE-FRANÇAISE, which in 1781 occupied the first building to be erected on the present site of the Odéon. Following the vicissitudes of the early Revolutionary period, the theatre acquired its present name in 1795. Rebuilt in 1816, and again after its destruction by fire two years later, it was managed by PICARD from 1816 to 1821; but it was only after Harel took over in 1829 that light comedies and operettas gave way to a classical and contemporary repertoire. André ANTOINE was its director from 1906 to 1916, GÉMIER from 1921 to 1928. During the reorganization of the Parisian theatres under André Malraux in 1959 the Odéon was removed from the control of the Comédie-Française, and renamed the Théâtre de France, under the direction of Jean-Louis BARRAULT, who presented there a number of important plays, including GENET's *Les Paravents* (1966), which led to rioting in the theatre. Barrault also created a small studio for experimental work, and provided a home for the THÉÂTRE DES NATIONS. After the demonstrations of May 1968, for which the Odéon provided a focal point, Barrault was dismissed and the theatre reverted to the control of the Comédie-Française, and its old name, in 1971.

ODETS, CLIFFORD (1906–63), American dramatist, considered one of the most gifted of those who developed a theatre of social protest during the 1930s. Born in Philadelphia but growing up in New York, he tried his hand at writing and became an actor after graduation from high school. In 1935 he attracted attention with a long one-act play about a taxi-drivers' strike, *Waiting for Lefty*, first produced by the GROUP THEATRE, of which he was a member, at the Civic Repertory Theatre. For a Broadway production he added to it another multi-scened one-acter, *Till the Day I Die*, a drama of the anti-Nazi underground movement in Germany. Almost simultaneously, the Group Theatre produced an earlier full-length play, *Awake and Sing*, which Odets re-wrote for the occasion. It portrayed the rejection by Jewish youths in the Bronx of the insidious materialism engendered by their mother's stranglehold on the family. Notable for its realism, contrapuntal technique, and mingling of humour with explosive passion, it brought its author instant recognition as the most promising new American playwright. Although his next play, *Paradise Lost* (also 1935), which focused on the bewilderment of the lower middle-class and the false ideals instilled in the success-orientated young, met with a tepid reception, Odets retrieved his reputation with *Golden Boy* (1937), the story of a sensitive Italian youth's deterioration after economic pressures turn him from music to professional boxing. *Rocket to the Moon* (1938), however, failed as a social parable in spite of excellent characterization and considerable pathos. *Night Music* (1940), an extravaganza depicting the struggles of disorientated youth, also failed, and *Clash by Night* (1941), in which Odets attempted to create a political allegory out of the personal humiliations of an unemployed labourer, emerged as a heavy-handed domestic triangle. He worked in films for some years, giving rise to an abrasive indictment of Hollywood, *The Big Knife* (1949), which was followed by *The Country Girl* (1950), a sympathetic handling of the story of a drunken actor and his loyal wife which was seen in London in 1952 as *Winter Journey*. Odets's last play, *The Flowering Peach* (1954), was a retelling of the Biblical story of Noah as a nostalgic Bronx Jewish fable. Though it ran for some months in New York, Odets's reputation rests on his plays of the 1930s, with their sympathetic, humorous, yet merciless exposure of middle-class ideals and strategies for survival and power; an unproduced play from this period, the 1937 drama *The Silent Partner*, received a posthumous off-off-Broadway production in 1972.

OEHLENSCHLÄGER, ADAM GOTTLOB (1779–1850), Danish poet and dramatist, who as a young man first came into contact with the theatre in Copenhagen as an actor. His attachments were to KOTZEBUE and his German contemporaries, to Shakespeare, and to the Danish poet Johannes EWALD, to whom he was in some ways a successor. He also became deeply interested in the myths and history of Scandinavia, the saga-play in his hands achieving a power which made him the greatest influence in Danish drama after HOLBERG. Some of his finest plays belong to the earlier stage of his career: *Sanct-Hans Aften Spil* (1802), which in 1913 was translated into English by G. Borrow as *Midsummer Night's Play*; *Aladdin* (1804), a fairy-tale play; and the two 'northern' tragedies, *Hakon Jarl hin Rige* and *Baldur hin Gode* (both 1807). *Axel og Valborg* (1810) is a sentimental piece, popular in its day, but with little appeal now. In 1810 Oehlenschläger was appointed Professor of Aesthetics in Copenhagen. The many plays of his maturity—he wrote over 30 between 1802 and 1845—show little distinction.

OENSLAGER, DONALD MITCHELL (1902–75), American scene designer, who worked under George Pierce BAKER at HARVARD. In 1923 he went to Europe to study scenic production there, and on his return to America worked with the PROVINCETOWN PLAYERS and the Greenwich Village Theatre. With Robert Edmond JONES, Jo MIELZINER, and Lee SIMONSON, he may be said to have contributed to the creation of a new age of stagecraft in the United States, and he left a permanent mark on the contemporary theatre there. He designed sets for a number of major drama productions (as well as operas and ballets), among them STEINBECK's *Of Mice and Men* (1937); SHAW's *The Doctor's Dilemma* in 1941 for Katharine CORNELL, *Pygmalion* in 1945 for Cedric HARDWICKE, and *Major Barbara* in 1956 for Charles LAUGHTON; also for IBSEN's *Peer Gynt* in 1951, *Coriolanus* in 1954, Spigelgass's *A Majority of One* (1959), *A Call on Kuprin* (1961) by Jerome Lawrence and Robert E. Lee and Henry Denker's *A Far Country* (also 1961) and *A Case of Libel* (1963). He was active up to the time of his death, and was for many years Professor of Scenic Design in the YALE Department of Drama.

OEUVRE, Théâtre de l', Paris, see LUGNÉ-POË.

OFF STAGE, see STAGE DIRECTIONS.

O'FLAHERTY, Mrs, see WINSTANLEY, ELIZA.

OGDEN, R. D'ORSEY (*fl.* 1861–65), British-born American actor and playwright, who at an early age emigrated with his family to the United States. By 1861 he was a popular entertainer in Richmond, Va., where a year later he took over the management of the theatre. In spite of the Civil War, the occupation of Richmond by Union forces, and adverse pressure and hostility from the press and the Church, Ogden managed to keep it open, his company, billed as the Confederate Theatre, presenting a wide range of native and foreign plays as well as Ogden's own farces, which were well received. His most noteworthy theatrical accomplishment, however, was the reconstruction and reopening of the RICHMOND THEATRE on 9 Feb. 1863, only 13 months after it had been totally destroyed by fire.

OHEL (Tent) **THEATRE,** the theatrical company of the Jewish Labour Federation, founded in Palestine in 1925 by Moshe Halevy (1895–1974), a former member of HABIMAH. The actors were at first encouraged to divide their time between the stage and their daily work in field and factory, so as not to sever their ties with the working class, but this proved impractical and Halevy soon found that it was necessary for them to devote their whole time to acting if they were to reach professional standards. The company's first public performance, given in Tel-Aviv on 24 May 1926, consisted of dramatizations of stories by Isaac Leib PERETZ. These, and Biblical plays, of which Stefan ZWEIG's *Jeremiah* (1929) was one, proved immensely popular with the kibbutz members who made up most of Ohel's audiences as the troupe travelled throughout the country, and were also well received in Europe on tour. By 1958, however, the theatre had lost sight of its original purpose and, its standards having declined, the Labour Federation withdrew its support. From then on Ohel led a precarious existence and in spite of occasional successes, such as revivals of Hašek's *The Good Soldier Schweik* and Ephraim Kishon's *The Marriage Contract* (1961), the company was eventually forced to disband in 1969.

OHIO ROSCIUS, see ALDRICH. LOUIS.

O'KEEFFE, JOHN (1747–1833), Irish-born dramatist, who was for about 12 years an actor in Dublin where his first plays, including *The Shamrock; or, St Patrick's Day* (1777), were produced. In his twenties his sight began to fail, which stopped his acting but did not prevent him from writing a vast number of plays, mainly farces and light operas, the latter containing many well known songs. The first of his plays to be seen in London was the farce *Tony Lumpkin in Town* (1778), produced at the HAYMARKET THEATRE as were his next three 'operatic farces'. The most successful of his later plays was *The Poor Soldier* (1783), which had a great vogue in the United States, as did the comedy *Wild Oats; or, the Strolling Gentleman* (1791), first seen at COVENT GARDEN. This was successfully revived in 1976 by the ROYAL SHAKESPEARE COMPANY, with Alan HOWARD in the leading role. One of O'Keeffe's farces, *The Little Hunch-Back; or, a Frolic in Bagdad* (1789), was included in the repertory of the TOY THEATRE. He published a volume of *Recollections* in 1826.

O'KELLY, SEUMAS (*c*. 1878–1918), Irish dramatist, poet, novelist, short story writer, and journalist, whose first plays, both in one act, *The Matchmakers* (1907) and *The Flame on the Hearth* (1908, later rewritten and published in 1922 as *The Stranger*), were produced in Dublin by an amateur company. He first came into prominence with *The Shuiler's Child* (1909), an IBSEN-inspired tragedy in two acts concerning the vagrant Moll Woods' rediscovery of her abandoned child. Staged professionally by the ABBEY THEATRE in 1910, this play was successfully revived in Dublin in 1948, with Maire NIC SHIUBHLAIGH once again in the leading role. Other plays by O'Kelly include *The Bribe* and *Lustre* (both 1913), the second written in collaboration with Count Casimir de Markiewicz; *Driftwood* (1915), first produced by Miss HORNIMAN's company at the GAIETY THEATRE, Manchester; *The Parnellite*, more propagandist in purpose, and *Meadowsweet* (both 1917).

OKHLOPKOV, NIKOLAI PAVLOVICH (1900–67), Soviet actor and director, whose first production, in 1921, was a May-Day spectacle in the central square of Irkutsk (his birthplace), of which he was author, director, and chief actor. In it can be seen the beginnings of that original style of production which, after four years with MEYERHOLD, from 1923 to 1927, he was able to develop more fully on his appointment in 1930 as artistic director of the REALISTIC THEATRE. There he produced a number of new plays, among them an adaptation of GORKY's *Mother* (1933) and POGODIN's *Aristocrats* (1934), for each of which he set up a new stage, or set of stages, drawing the audience into the development of the action. His work was necessarily experimental and its limited appeal led to the merging of his company with TAÏROV's in 1938. Okhlopkov then worked for a while at the VAKHTANGOV THEATRE, where he directed a notable production of ROSTAND's *Cyrano de Bergerac*, and in 1943 became director of the Theatre of the Revolution (later the MAYAKOVSKY THEATRE), where his revivals of OSTROVSKY's *Fear* in 1953 and of Shakespeare's *Hamlet* in 1954 were much admired. He also continued his policy of producing new Soviet plays, among them A. P. Shteyn's *Hotel Astoria* (1956), ARBUZOV's *Endless Distance* (1958), and Pogodin's *The Little Student* (1959).

OLD BOWERY THEATRE, New York, see BOWERY THEATRE (1).

OLD COMEDY, term used to distinguish the early comedy of Ancient Greece—specifically of Athens in its prime—from its subsequent development into MIDDLE COMEDY and NEW COMEDY. It had a much stricter and more complex form than Attic TRAGEDY, though it too was a blend of the choral and the histrionic. It seems clear that comedy had a dual origin, the histrionic scenes having an obvious affinity with such things as the Sicilian MIME and the burlesque performers whose MASKS (or representations of them) have been dug up in great numbers in Sparta, while the formal parts must have been developed out of some ritual performance by a CHORUS. The name 'comedy' is derived from *komos* and *ode*, meaning a 'revel-song'. One form of revel was associated with fertility rites: it was a mixture of singing, dancing, scurrilous jesting against bystanders, and ribaldry; ARISTOTLE derives comedy from this, and certainly comedy contained all these elements, including the use of the phallus, the symbol of fertility. Another form of *komos*, well represented on vase-paintings, was the Masquerade, in which revellers disguised themselves as animals or birds. Since the comic chorus was often of this type (as in the *Wasps, Birds*, and *Frogs* of ARISTOPHANES) the influence of this kind of revel on comedy seems clear enough.

Old Comedy (represented now only by Aristophanes) exhibits, with variations, a complex and elaborate form. There is, on the one hand, a succession of scenes in which form is strict; the chorus enters, there is a dispute between the chorus and an actor or between two actors each supported by a semi-chorus; then follows a formal debate—the *agon*—and finally an address made by the chorus direct to the audience—the *parabasis*, or 'coming-forward'. All this, but especially the *parabasis*, tends to be elaborately symmetrical in structure, and the contest is usually composed in metres other than the iambic trimeter of ordinary dramatic dialogue. Besides this element, essentially choral in form, there is another, consisting of scenes for actors alone which precede the entry of the chorus and follow the *parabasis*. These are written in iambic trimeters and show no trace of symmetrical structure.

Aristotle says that it was not until later times that comedy was taken seriously enough to be admitted to the civic DIONYSIA. The first contest seems to have taken place in 486 BC, some 50 years after the first tragic contest. Among the early masters were Chionides, MAGNES, and CRATES, the last-named being the first to introduce elements of plot and a unifying theme. Until that time comedy had probably retained the style and atmosphere of the original revel, with its sudden and disconnected attacks on members of the audience. Later masters of the form, contemporary with Aristophanes, were AMEIPSIAS, CRATINUS, and EUPOLIS. Old Comedy is the most local form of drama that has ever reached literary rank. It was a sort of national lampoon, in which anything prominent in the life of the city, whether persons or ideas, was unsparingly ridiculed—a unique mixture of fantasy, criticism, wit, burlesque, obscenity, parody, invective, and exquisite lyricism. The atmosphere of the whole is well suggested by the story that during the performance of Aristophanes' *Clouds* Socrates rose from his seat to give the audience an opportunity of comparing the mask of the stage Socrates with the man himself.

Instead of plot, Aristophanes uses a single fantastic situation, which is quickly developed and then exploited in a series of loosely-connected scenes. The most remarkable feature is the *parabasis*, for in it the dramatist drops all dramatic illusion, suspends the plot, and speaks directly to the audience on matters which are quite unconnected with the play and sometimes purely personal. Thus, in the *parabasis* of the *Knights*, Aristophanes gives a brilliant account of his predecessors and of his rival Cratinus. Aristophanes' last two plays, the *Ecclesiazousae* (*Women in Parliament*, 392 BC) and *Plutus* (388 BC), mark the transition to Middle Comedy. By its nature, Old Comedy was well suited to exploit the tensions generated by the Peloponnesian War, but not to survive the exhaustion of defeat.

In the late 18th and 19th centuries AD the term Old Comedy was used in England, and particularly in North America, to denote the repertory of comedies from Shakespeare to SHERIDAN.

OLD DRURY, see CHESTNUT STREET THEATRE, PHILADELPHIA; DRURY LANE THEATRE, LONDON; and PARK THEATRE, NEW YORK (1).

OLDFIELD, ANNE [Nance] (1683–1730), English actress, who succeeded Mrs BRACEGIRDLE as one of the leading players at DRURY LANE. She first came into prominence when Colley CIBBER, struck by her playing of Leonora in a revival of CROWNE's *Sir Courtley Nice*, cast her as Lady Betty Modish in his own play *The Careless Husband* (1704). From then on her career was one of unbroken triumph. Lovely in face and figure, she had an exceptionally beautiful voice and clear diction, and VOLTAIRE said of her that she was the only English actress whose speech he could follow without difficulty. She was particularly admired as Silvia in *The Recruiting Officer* (1706) and as Mrs Sullen in *The Beaux' Stratagem* (1707), both by FARQUHAR, who first encouraged her to go on the stage, but she was also considered outstanding in tragedy, playing with majesty and power such parts as Andromache in PHILIPS's *The Distrest Mother* (1712) and the title role in ROWE's *Jane Shore* (1714). She much preferred comedy, however, and in the year of her retirement gave a highly acclaimed performance as Lady Townly in *The Provoked Husband* (1728), based by Cibber on *A Journey to London*, an unfinished play by VANBRUGH. She made her last appearance in FIELDING's first play, *Love in Several Masques* (also 1728), and after her death two years later was buried in Westminster Abbey, near CONGREVE, but was not allowed a monument over her grave, because she had had two illegitimate sons.

OLD GLOBE THEATRE, San Diego, see SHAKESPEARE FESTIVALS.

OLDHAM, DEREK, see MUSICAL COMEDY.

OLD MAN, WOMAN, see STOCK COMPANY.

OLDMIXON, Mrs [*née* Georgina Sidus] (?–1836), English singer and actress, daughter of an Oxford clergyman, who as Miss George became an established favourite at the HAYMARKET THEATRE, where she appeared in the younger COLMAN's *Inkle and Yarico* (1787) when it was first produced. In 1793 she went to the United States and became a member of WIGNELL's company at the CHESTNUT STREET THEATRE in Philadelphia, where she took the lead in light operas, and also appeared in straight plays. In 1798 she joined DUNLAP's company at the newly-built PARK THEATRE in New York, playing Mrs Candour in SHERIDAN's *The School for Scandal* and Ophelia in *Hamlet*, and appearing again in *Inkle and Yarico*. For some years she was away from the stage, during which time her seven children were born; but in 1806 she reappeared under COOPER, playing the Nurse in *Romeo and Juliet* and other elderly parts. She retired in 1813, opening a seminary for young ladies in a suburb of Philadelphia and making occasional appearances on the concert platform.

'OLD MO', London, see MIDDLESEX MUSIC-HALL.

OLD TOTE THEATRE, Sydney, company founded in 1963, its name being derived from its first premises, the refurbished tote box of a race-track. Its opening production was CHEKHOV's *The Cherry Orchard*, and it later presented more classical plays than any other comparable Australian company, featuring top local actors and directors and sometimes an overseas guest, as when Tyrone GUTHRIE directed SOPHOCLES' *Oedipus the King* in 1970. In its first ten years the Old Tote presented 18 Australian plays, 12 of them being first performances, and it was the first drama company to perform in the SYDNEY OPERA HOUSE, appearing in *Richard II,* BRECHT's *The Threepenny Opera*, and David WILLIAMSON's *What If You Died Tomorrow?* The last of these productions was seen in London in Sept. 1974. The company depended on substantial subsidies from the Australia Council and the Government of New South Wales; when these were withdrawn in 1978 the Old Tote Theatre was replaced by the Sydney Theatre Company.

OLD VIC THEATRE, London, in the Cut, off the Waterloo Road. This was first called the Royal Coburg, after the husband of Princess Charlotte, its foundation stone and other materials coming from the old Savoy Palace, which had just been demolished to make way for the entrance to Waterloo Bridge. After many delays, due mainly to shortage of money, the theatre opened on 11 May 1818 with a melodramatic spectacle by William BARRYMORE, *Trial By Battle; or, Heaven Defend the Right*, based on a recent notorious murder trial. The journey across the river was too hazardous for a fashionable audience, but a series of melodramas of the most sensational kind soon attracted a large local audience, particularly as the plays were well staged and well known actors, immortalized in the TOY THEATRE prints, appeared in them. The interior too was handsomely decorated, one of the most interesting features being a 63-piece looking-glass curtain installed in 1824 which reflected the entire auditorium. Unfortunately its weight put too great a strain on the roof and it had to be removed. In 1833 the theatre was redecorated and reopened on 1 July with a revival of JERROLD's *Black-Ey'd Susan*, being renamed the Royal Victoria Theatre in honour of Princess (later Queen) Victoria. It soon became affectionately known as the Old Vic, but its standards noticeably declined and it finally sank to the level of a Penny GAFF. The audience was very rough, and in 1858 16 people lost their lives in a panic due to a false alarm of fire. In 1871, after a short period as a MUSIC-HALL under J. A. Cave, the theatre closed. It was sold by auction and renamed the New Victoria Palace. This closed early in 1880, and the building was then bought by Emma Cons, a social worker and the first woman member of the London County Council, with the intention of turning it into a temperance music-hall. Renamed the Royal Victoria Hall and Coffee Tavern, it opened on 26 Dec. 1880 under the management of William POEL, who remained until 1883. Intended as a place of inexpensive family entertainment, it prospered in spite of initial misgivings, much helped by the support of a Bristol M.P., Samuel Morley, whose efforts to promote the arts in South London included the founding of Morley College which for a time occupied part of the building. In 1900 the first full length opera, Balfe's *The Bohemian Girl*, was given, and scenes from Shakespeare were introduced into the concert programmes. In 1912 Emma Cons's niece Lilian BAYLIS, who had been assisting her aunt since 1898, took over and two years later engaged Rosina Filippi to direct the first full Shakespeare season. It was successful enough to warrant further productions of his plays, in spite of the outbreak of war later that year, and in the next nine years all the plays in the First Folio were performed, the completion of the project, with a staging of *Troilus and Cressida* on 5 Nov. 1923, coinciding with the tercentenary of its first publication. Ben GREET was in charge of productions from 1915, during which time Sybil THORNDIKE played most of Shakespeare's heroines and some of his heroes, including Prince Hal. After the war the theatre was found to be in urgent need of repair, and in 1927 the L.C.C. insisted on its closure for urgent and essential alterations which were paid for by Sir George Dance, a noted benefactor of the Old Vic. The company moved to the LYRIC THEATRE, Hammersmith, and returned on 14 Feb. 1928 with a performance of *Romeo and Juliet*. In Jan. 1931 Lilian Baylis opened the renovated SADLER'S WELLS THEATRE, and mixed seasons of opera, ballet (under the direction of Ninette de Valois), and drama alternated between the two theatres. This proved impractical, and two years later it was decided to use Sadler's Wells for opera and ballet and the Old Vic for drama. A succession of fine actors, among them John GIELGUD, Laurence OLIVIER, and Ralph RICHARDSON, made the name of the Old Vic famous far beyond its own territory, and its success was only temporarily checked by the death in 1937 of Lilian Baylis during a run of *Macbeth*. In common with all the London theatres, the Old Vic closed on the outbreak of war in 1939; it reopened a year later, but was badly damaged by bombing in May 1941 and closed again, the company moving to the NEW THEATRE. Repaired and redecorated it reopened on 14 Nov. 1950 with *Twelfth Night*. Between 1953 and 1958 a 'five-year' plan under Michael Benthall resulted in a second presentation of the 37 plays in the First Folio, opening with Richard BURTON in *Hamlet* and ending with Gielgud, Edith EVANS, and Harry ANDREWS as Wolsey, Katharine, and the King in *Henry VIII*. The Old Vic had from early times presented some plays by dramatists other than Shakespeare, and two interesting works produced in 1962 were IBSEN's *Peer Gynt* and GUTHRIE's modernized version of JONSON's *The Alchemist*.

Later that year it was decided that the theatre should house the NATIONAL THEATRE's company under Laurence Olivier pending the erection of their own theatre, and in June 1963, after a final performance of *Measure for Measure*, the theatre closed. It reopened on 22 Oct. (after extensive alterations) with the National Theatre company in *Hamlet*, starring Peter O'TOOLE, the other actors being mainly from the CHICHESTER FESTIVAL THEATRE company, which was also under the direction of Olivier. G. B. SHAW's *Saint Joan* and CHEKHOV's *Uncle Vanya*, both transferred from Chichester, followed, the latter proving one of the company's most successful achievements in ensemble acting. The first notable revival was FARQUHAR's *The Recruiting Officer*, and the first foreign play to be attempted by the new company was Max FRISCH's *Andorra*. Shakespeare's Quatercentenary was celebrated in 1964 by Olivier in *Othello*, possibly the most controversial *tour de force* of his career. The company's first world première was Peter SHAFFER's *The Royal Hunt of the Sun*, in a virtuoso production by John DEXTER which established Robert Stephens and Colin BLAKELY as stars. In the seasons that followed the company's greatest successes were with comedy, particularly with FEYDEAU's *A Flea in Her Ear* (1965), which was to remain in the repertory for some time. In 1967 Tom STOPPARD's *Rosencrantz and Guildenstern are Dead* introduced a new playwright of considerable talent, while the classics were prominent with an interesting all-male *As You Like It* and MOLIÈRE's *Tartuffe*, with Gielgud, who in 1968 appeared in a modern dress version of SENECA's *Oedipus* directed by Peter BROOK. The abortive attempt by the company's literary manager Kenneth TYNAN to introduce HOCHHUTH's controversial play *Soldiers* into the repertory, and the extension in 1970 of the company's activities to West End theatres—the CAMBRIDGE and the NEW—while the Old Vic was let to visiting companies, attracted some adverse criticism in spite of the excellence of some of the plays produced during this time. The return to the Old Vic was generally welcomed, SCOFIELD giving an excellent comic performance in ZUCKMAYER's *The Captain of Kopenick* in 1971, and old and new plays being equally successful, with Stoppard's *Jumpers* and a revival of O'NEILL's *Long Day's Journey into Night* in 1972 and Molière's *The Misanthrope* and Shaffer's *Equus* in 1973. The last play to be staged under Olivier's management was Trevor Griffiths's *The Party*, after which Peter HALL took over, and in 1974 staged a varied repertory in anticipation of the move to the National Theatre, ending with PINTER's *No Man's Land* (1975). The last production before the move took place on 28 Feb. 1976 was a special charity performance of *Tribute to the Lady*, a celebration of the life and work of Lilian Baylis. A year later the theatre, which was being used by the Old Vic Youth Theatre (which retained its headquarters there until 1982), became the London home of the PROSPECT THEATRE COMPANY. In 1980 it staged under its artistic director Timothy West a production of *Macbeth* with Peter O'Toole which, although mauled by the critics, was a financial success. A year later the company was forced to disband, and the theatre was bought by the Canadian entrepreneur Ed Mirvish, handsomely refurbished, and reopened in Nov. 1983.

OLIMPICO, Teatro, Vicenza, an outstanding example of academic Italian Renaissance theatre architecture. Designed by PALLADIO and completed after his death by his pupil Scamozzi, it opened on 3 Mar. 1585 with a production of SOPHOCLES' *Oedipus the King*, in a new translation by Giustiniani directed by Ingegneri. The building, which still stands, was based strictly on the contemporary neo-classical idea of a Greek theatre, formulated by VITRUVIUS, and in spite of its splendour it had little or no influence on the development of later theatre buildings in Italy.

OLIO, see BURLESQUE, AMERICAN.

OLIVIER, LAURENCE KERR (1907–), English actor, director, and manager, knighted in 1947, created a life peer in 1970 (as Baron Olivier of Brighton) and C. H. in 1981. He made his first appearance on the stage in 1922, playing Katharina in a schoolboy production of *The Taming of the Shrew*, and his professional début in London in 1924, playing a small part in Alice Law's *Byron*. From 1926 to 1928 he was with the BIRMINGHAM REPERTORY company, playing a wide variety of parts which included Richard Coaker in PHILLPOTTS's *The Farmer's Wife*, the Young Man in Elmer RICE's *The Adding Machine*, and Malcolm in Barry JACKSON's modern-dress production of *Macbeth*. He was the first to play Captain Stanhope in SHERRIFF's *Journey's End* when it was tried out in 1928, and in 1929 he made his New York début in Frank Vosper's *Murder on the Second Floor*. He had already attracted attention in COWARD's *Private Lives* (1930; N.Y., 1931) and Edna Ferber and George S. KAUFMAN's *Theatre Royal* (1934) when he played his first major Shakespearian roles, alternating Romeo and Mercutio with GIELGUD at the NEW THEATRE in 1935. During a subsequent season at the OLD VIC, where he played Hamlet in its entirety (being later seen in the part at Elsinore), Henry V, Macbeth, and Sir Toby Belch in *Twelfth Night*, he emerged as an actor of the front rank, returning to the Vic in 1938 to play Iago to the Othello of Ralph RICHARDSON and the title role in *Coriolanus*. After appearing in New York in S. N. BEHRMAN's *No Time for Comedy* (1939) and as Romeo in 1940, he spent four years with the Fleet Air Arm, rejoining the Old Vic company in 1944 and remaining with it until 1949. He gave some remarkable performances, notably in a double bill, seen in London in 1945 and New York in 1946, in which he appeared in the title role of SOPHOCLES' *Oedipus* *the King* and as Mr Puff in SHERIDAN's *The Critic*; his other roles included King Lear (1946) and Richard III (1949). In 1950 he was seen in FRY's *Venus Observed*, written specially for him, and in 1951, in London and New York, he starred with his second wife Vivien LEIGH in *Antony and Cleopatra* and SHAW's *Caesar and Cleopatra*, playing Antony and Caesar. He was then seen in RATTIGAN's *The Sleeping Prince* (1953), and spent a season at the SHAKESPEARE MEMORIAL THEATRE best remembered for his portrayal of the title role in *Titus Andronicus* (1955), with Vivien Leigh as Lavinia. He scored his first outstanding success in a modern role as the second-rate music-hall comedian Archie Rice in John OSBORNE's *The Entertainer* (1957; N.Y., 1958), playing opposite Joan PLOWRIGHT, later to become his third wife. He played Berenger in IONESCO's *Rhinoceros* (1960) in London, and in New York played first the title role (1960) and then Henry II (1961) in ANOUILH's *Becket*. Later the same year he was appointed the first director of the CHICHESTER FESTIVAL THEATRE. His starring role in David Turner's *Semi-Detached* (1962) was his last in the commercial theatre; a year later he became the first Director of the state-subsidized NATIONAL THEATRE company. He directed the opening production of *Hamlet* at the Old Vic, and appeared as Astrov in CHEKHOV's *Uncle Vanya* (which he also directed) and Brazen in FARQUHAR's *The Recruiting Officer* during the first season. In 1964 he was a controversial but memorable Othello. A year later he handed Chichester over to John CLEMENTS and concentrated all his attention on the National Theatre, where he was active as both actor and director. Among his important roles were Tattle in CONGREVE's *Love for Love* in 1965, Edgar in STRINDBERG's *The Dance of Death* in 1967, Shylock in a Victorian production of *The Merchant of Venice* in 1970, and James Tyrone in O'NEILL's *Long Day's Journey Into Night* in 1971. In 1973 he was succeeded by Peter HALL. Blessed with a strong interpretative intelligence, he is commonly regarded as the supreme actor of his generation.

OLIVIER THEATRE, the largest of the three theatres housed in the NATIONAL THEATRE complex in London, the others being the COTTESLOE and the LYTTELTON, and the second to be opened. Named after Laurence OLIVIER, the first Director of the National Theatre, it seats 1,165 in two stepped tiers and has an open stage 56 ft wide and 72 ft deep at the centre with a revolve 40 ft in diameter. It is used for plays which require larger casts or more spectacular effects than can be accommodated in the Lyttelton. The Olivier opened on 25 Oct. 1976 with MARLOWE's *Tamburlaine the Great*, with Albert FINNEY in the title role. Other notable productions have included WYCHERLEY's *The Country Wife* (1977), with Finney, O'CASEY's *The Plough and the Stars* (also 1977), with Cyril CUSACK, CONGREVE's *The Double Dealer* (1978), with Dorothy TUTIN and Michael BRYANT, and IBSEN's *The Wild Duck* (1979), with Bryant. Modern plays seen there include BOND's *The Woman* (1978), SHAFFER's *Amadeus* (1979), with Paul SCOFIELD, and Howard BRENTON's controversial *The Romans in Britain* (1980). The Olivier and Cottesloe Theatres were in 1980 placed under the joint direction of Peter GILL and Peter HALL.

OLMO, JACOB, see JEWISH DRAMA.

OLYMPIA, New York, on the east side of Broadway, between 44th and 45th Streets, a complex built for Oscar Hammerstein in 1895. It comprised a music-hall, later known as the NEW YORK THEATRE (2), a concert hall, and a legitimate theatre, the Lyric, which after the complex had been auctioned off in 1898 was reopened as the CRITERION.

OLYMPIA MUSIC-HALL, London, see STANDARD THEATRE.

OLYMPIA THEATRE, Dublin, one of the oldest surviving theatres in Dublin, situated off Dame Street. A music-hall opened on the site in 1855, which became Dan Lowrey's Star of Erin Music Hall in 1879. All the famous English performers of the time were seen here, and in spite of some harassment from the GAIETY THEATRE over the use of 'dramatic' material, Lowrey was enormously successful as a manager. In 1897 he completely rebuilt the hall, providing substantially the building that survives today. It was known as the Empire Palace of Varieties until 1922, when it was renamed the Olympia. For many years, legitimate drama has been part of the repertoire, especially after 1940, when the Second World War prevented the engagement of visiting performers and Irish plays were performed regularly. From 1951 to 1964 the theatre had its own company, but also saw the return of English touring companies. In 1964 it was sold, but was leased by a group of theatre managers dedicated to preserving it, and up to 1974 it successfully blended traditional variety with legitimate theatre, often in association with the DUBLIN THEATRE FESTIVAL. On 5 Nov. 1974, while the musical *West Side Story* was in rehearsal, a serious structural collapse closed the theatre, but after generous financial support from both public and private sources it reopened on 14 Mar. 1977 with Willy Russell's musical about the Beatles *John, Paul, George, Ringo . . . and Bert*. Brian FRIEL's *Translations* had a notable production there in 1980.

OLYMPIC THEATRE, London, in Wych Street, Strand. This was erected by Philip Astley while ASTLEY'S AMPHITHEATRE was being built, and was constructed mostly of timber from the French warship *Ville de Paris*, with a little brickwork and a tin roof, in the shape of a tent. It opened as the Olympic Pavilion on 18 Sept. 1806, housing circus acts and performances of horsemanship in an arena. It was not a success, and in 1813 it was bought by R. W. ELLISTON, who changed its name to the Little Drury Lane Theatre. This was objected to by DRURY LANE as infringing its patent, and Elliston's licence was withdrawn. However, by the end of the year he had obtained a new licence for BURLETTA only, and in December he reopened the theatre as the Olympic. It did well for the next five years with a mixed programme of pantomime, ballet, farce, and melodrama, and was then reconstructed and reopened with an excellent company which attracted a fashionable audience and made so much profit that Elliston was able to purchase the patent of Drury Lane on the proceeds. He leased the Olympic to a series of lessees, the majority of whom went bankrupt as did Elliston himself. The theatre was put up for sale and bought by John Scott, owner of the ADELPHI, who used it for melodramas until at the end of 1830 Mme VESTRIS leased it from him and opened on 3 Jan. 1831 with a programme which included PLANCHÉ's *Olympic Revels*. Her policy of low prices and beautifully staged light entertainment made the Olympic a success. In 1835 Charles J. MATHEWS made his first appearance there and three years later he and Mme Vestris were married, leaving the Olympic in 1839 to go to COVENT GARDEN. The theatre then led a precarious life until it was burned down on 29 Mar. 1849. Rebuilt by William Watts to seat 1,750 people, it reopened on Boxing Day, but had to close hurriedly the following March, during a run of A. C. MOWAT's *Fashion; or, Life in New York*, when Watts was arrested on charges of defalcation and forgery. It reopened in August, when William FARREN took over, with Frederick ROBSON as his star and later co-manager. It was under him that Tom TAYLOR's *The Ticket-of-Leave Man* had a successful run in 1863. Robson died prematurely the following year, and was succeeded by Horace Wigan, who introduced a series of new plays with Henry Neville and Kate TERRY. Neville himself became manager of the theatre for six years, during which period Wilkie COLLINS's *The Moonstone* (1877) was first produced. Charles WYNDHAM appeared in 1880 in Bronson HOWARD's *Brighton* with great success, and in 1883 Geneviève WARD was seen in her own productions of *Meg Merrilees* (based on Walter SCOTT's *Guy Mannering*), *Medea*, translated from the French of Legouvé, and Herman Merivale's *Forget-Me-Not*. A succession of managements came and went until the theatre closed in 1889, and after reconstruction opened under Wilson BARRETT two years later. It was never again successful, and closed for the last time on 17 Nov. 1897, standing derelict until it was demolished in 1904 during the reconstruction of the Aldwych.

OLYMPIC THEATRE, New York. (1) At 444 (later 442) Broadway, between Howard and Grand Streets. This handsome theatre opened on 13 Sept. 1837 with a mixed bill, but was not a success, and passed through many hands before on 9 Dec. 1839 William MITCHELL opened it as a home of light entertainment. As Mitchell's Olympic it flourished for over ten years with burletta, burlesque, and extravaganza, and survived even the depression of 1842–3 which proved fatal to many other enterprises. It was, incidentally, the first theatre in New York to play a weekly matinée. Among its outstanding successes were *Hamlet Travestie* by John Poole, and burlesques of *Richard III*, of BOUCICAULT's *London Assurance*, and of the opera *The Bohemian Girl* as *The Bohea-Man's Girl*. The season of 1847–8 saw PLANCHÉ's *The Pride of the Market*, and Frank CHANFRAU as Mose, his famous fireman character, in Benjamin BAKER's *A Glance at New York in 1848*. A year later the theatre was redecorated; but Mitchell was failing in health and took the easy way of importing foreign stars. This proved disastrous, and on 9 Mar. 1850 the Olympic closed abruptly. After a short spell under William BURTON it housed plays in German and finally closed on 25 June 1851. It was burnt down in 1854.

(2) At 622–4 Broadway, above Houston Street. This opened on 18 Nov. 1856 with Laura KEENE as Rosalind in *As You Like It*, and as Laura Keene's Varieties was the first American theatre to have a woman manager. It flourished until 1863 when Laura Keene left, and reopened on 8 Oct. of that year as the Olympic under Mrs John WOOD, who sponsored the first appearance in New York of Mrs G. H. GILBERT. Her last season was memorable for the début of G. F. ROWE, who appeared as Mr Micawber and Silas Wegg in his own adaptations of DICKENS's *David Copperfield* and *Our Mutual Friend* respectively. Later productions at this theatre were *A Midsummer Night's Dream* in 1867, with G. L. FOX as Bottom and a panorama of London painted by TELBIN, and Fox's famous pantomime *Humpty-Dumpty*, which opened on 10 Mar. 1868 and ran for 483 performances. In 1872 the theatre became a home of VARIETY. It finally closed on 17 Apr. 1880 and was demolished, shops being built on the site.

(3) The New Olympic Theatre, at 585 Broadway, was opened in 1856 by Chanfrau, who intended to revive the mixed bills of Mitchell's Olympic. He was unsuccessful, and after a few weeks the theatre was taken over by a MINSTREL SHOW, Buckley's Serenaders, and became Buckley's Olympic. A later manager was Tony PASTOR, from 1875 to 1881.

The ANTHONY STREET THEATRE was known as the Olympic in 1812; there was a circus known as the Olympic Arena in 1858, a short-lived Olympic on 8th Avenue in 1860, and an Olympic Music-Hall at 600 Broadway on the site of the old Alhambra, which flourished from 1860 to 1861; and WALLACK'S THEATRE (1) was renamed the Olympic in 1862 under Fox.

OMBRES CHINOISES, see SHADOW-SHOW.

O'NEAL, FREDERICK, see AMERICAN NEGRO THEATRE.

O'NEIL, NANCE, see RANKIN, ARTHUR MCKEE.

O'NEILL, ELIZA (1791–1872), Irish actress, who made her first appearance on the stage in her birthplace, Drogheda, where her father was manager of the local theatre. Going to Belfast and Dublin, she soon made a name for herself and in 1814 was engaged for COVENT GARDEN. Her first appearance as Juliet in *Romeo and Juliet* was overwhelmingly successful, as was her Lady Teazle in SHERIDAN's *The School for Scandal*, and for five years she had a career of unbroken triumph, being considered a worthy successor to Mrs SIDDONS, with less nobility, perhaps, but more sweetness and charm. On 13 July 1819 she made her last appearance on the stage as Mrs Haller in KOTZEBUE's *The Stranger*, and then retired on her marriage to Mr (later Sir William) Becher.

O'NEILL, EUGENE GLADSTONE (1888–1953), American playwright, born in New York City, son of the actor James O'NEILL. His education was fragmentary, including a year at Princeton, after which he signed on as a seaman on several voyages to South America, South Africa, and elsewhere. He was working as a reporter on a newspaper in New London, Conn., when his health broke down, and during six months spent in a sanatorium he began writing his first play, *The Web*. In 1914–15 he studied under George Pierce BAKER at HARVARD, and in 1916 became connected with the PROVINCETOWN PLAYERS who, with the Greenwich Village Theatre, first presented many of his early plays. His first full-length play *Beyond the Horizon* (1920), produced on Broadway, was a starkly effective study of character set in rural New England. Awarded a PULITZER PRIZE, it established him as a playwright of genuine talent and considerable skill, and was followed almost immediately by productions of the one-act *Exorcism*; *Diff'rent*, a grim bit of dramatic irony in two acts; and *The Emperor Jones* (seen in London in 1925), which uses a powerful Negro as its central figure to represent the violent urges of human nature: Paul ROBESON gave an electrifying performance in the title role. *Anna Christie* (1921; London, 1923) tells the story of a prostitute who is, presumably, 'purified' by the love of a man; its popularity was based largely on the romantic and external theatrical qualities of the acting and production. In quick succession other O'Neill plays were brought to the stage—*Gold* (1921), *The Straw* (also 1921), drawing on his experiences in a sanatorium and an expression of the indomitable human spirit, and *The First Man* (1922)—each a failure with the public, yet each revealing new aspects of the author's preoccupations. *The Hairy Ape* (1922) stemmed, according to the author, from *The Emperor Jones* rather than from the work of the European Expressionists, which it in many ways resembles. It is a hymn to the inarticulate American worker, debased and exploited by the society that benefits from his toil. In 1924 three new plays were produced—*Welded, All God's Chillun Got Wings*, and *Desire Under the Elms*. The first, a compact and rather bloodless study in marriage, was a quick failure; the second (seen in London in 1926), dealing with racial intermarriage, verged on the sentimental, but was threatened with demonstrations by racialist factions; the last (seen in London in 1931) showed a new maturity, using a powerful tale of sexual passion, incest, and infanticide to comment on contemporary American society. *The Fountain* (1925), a romantic, pseudo-historical play about Ponce de Leon and his quest for the Fountain of Youth, was short-lived. *The Great God Brown* (1926) remains one of the most tortuous and complicated of O'Neill's plays. Making use of elaborate masks, it studies the conflict between man's material and spiritual needs. Though it had some success in the theatre, it is more interesting and significant as showing the author's dissatisfaction with surface realism as a means of revealing character. *Marco Millions* (1928; London, 1938) was more serene, pleasantly ironic, and full of comedy and romantic colour, even though it too is a bitter satire on the aggressive business man who has lost touch with beauty and the eternal verities. *Strange Interlude* (1928; London, 1931), a play in nine acts, almost twice the length of an ordinary stage piece, is a work of extraordinary power which, with the copious use of asides and soliloquies, seeks to expose the motives of human character. In this play and in others that followed, O'Neill seemed to be clarifying his ideas on the soul of man and the destiny of the human race to the detriment of his art as a playwright. *Lazarus Laughed* (1928), first produced by the THEATRE GUILD, is the author's most pronounced philosophical affirmation of his belief in humanity; a munificent spectacle set in ancient Rome, it tells the story of the resurrection of Lazarus and his ultimate triumph over death. 'Fear is no more!' runs the refrain, 'Death is dead!' In 1929 came *Dynamo*, planned as the first part of an uncompleted trilogy on man's efforts to find a lasting faith. It was unsuccessful, but is remembered for its exciting set design by Lee SIMONSON. Another trilogy, *Mourning Becomes Electra* (1931; London, 1937), is in many respects O'Neill's most successful work. It transposes the events of AESCHYLUS' *Oresteia* to a Puritanical family in New England, replacing the old acceptance of fate with a modern doctrine of psychological causation. A nostalgic comedy, *Ah, Wilderness!* (1933), and a somewhat barren and over-intellectualized play about faith, *Days Without End* (1934; London, 1943), followed. Then for twelve years O'Neill retired from the theatre; he did an enormous amount of writing, including several plays that are parts of a series of nine interrelated plays, but refused to allow any of them to be staged. *The Iceman Cometh* (1946; London, 1958), a vast play about a group of bar-room derelicts and their pipe-dreams, enjoyed a long run in New York and aroused a great deal of critical comment. Like *A Moon for the Misbegotten* (1947; London, 1960), it is partly expository drama and partly a disquisition on faith.

Some of O'Neill's plays were produced posthumously. *Long Day's Journey Into Night*, written in 1941, was first seen in Stockholm on 2 Feb. 1956; it was produced in New York the following November and in London in 1958. A largely autobiographical study, set during a single day, of a miserly actor father, his morphine-addicted wife, and their two sons, it has been several times revived, notably at the NATIONAL THEATRE in 1971, with OLIVIER. *A Touch of the Poet*, written in 1940, was produced in Stockholm in 1957 and New York in 1958; it examines the painful conflict of an immigrant father and his American-born son. *More Stately Mansions*, written in 1938, was produced in Stockholm in Nov. 1962, in a translation which cut the playing time from ten hours to five; the play was seen in New York in 1967 and at GREENWICH in 1974.

O'Neill was awarded the NOBEL PRIZE for Literature in 1936, and in 1959 the Coronet Theatre in New York was renamed the EUGENE O'NEILL in his honour. He was an introspective and troubled man, continually dogged by illness; all his talent and energy went into his work for the theatre, through which he attempted to examine the soul of modern man. Though his language is often clumsy, and some of his plays become melodramatic and absurd, he is a major dramatist whose best plays, offering strong parts for actors, continue to be revived in many countries.

The Eugene O'Neill Theatre Center, Waterford, Connecticut, founded in 1963, provides premises and resources for a variety of programmes and organizations, including the annual National Playwrights Conference, inaugurated in 1965, at which writers can exchange ideas with directors and actors; the National Critics Institute has since 1968 run concurrently with the Conference. The Center also provides a home for the National Theatre of the Deaf (N.T.D.), founded in 1967, a permanent company of deaf actors performing mainly for hearing audiences; the Little Theatre of the Deaf (L.T.D.) was formed a year later and concentrates on visual theatre for smaller audiences and superior children's theatre.

O'NEILL, JAMES (1847–1920), Irish-born American actor, once considered the rival and successor of Edwin BOOTH. Taken to America as a small child, he made his stage début in Cincinnati in 1867, playing a small part in BOUCICAULT's *The Colleen Bawn.* In 1871 he joined the company of James V. McVicker in Chicago, and five years later was at the UNION SQUARE THEATRE in New York. He then went to San Francisco, where among other parts he played Christ in a Passion Play directed by David BELASCO at the Grand Opera House. He married in 1875 Ella [Mary Ellen] Quinlan, by whom he had two sons, the younger being the playwright Eugene O'NEILL. James O'Neill was considered a good Shakespearian actor, but romantic swashbuckling roles suited him best, and in 1883 he found the perfect part for himself in Edmond Dantès, the hero of the elder DUMAS's *The Count of Monte Cristo*. This became so popular with the American public that for the next 30 years O'Neill played almost nothing else, being seen in it at least 6,000 times. The resultant frustration told heavily on him and he took to drink, his gradual decline, and that of his long-suffering wife, being tragically depicted in Eugene's autobiographical play *Long Day's Journey Into Night* (Stockholm 1956; N.Y., 1956; London, 1958).

O'NEILL, MAIRE [Molly Allgood] (1887–1952), Irish actress, sister of Sara ALLGOOD, and with her a member of the ABBEY THEATRE company, where she created the part of Pegeen Mike in SYNGE's *The Playboy of the Western World* (1907) and the title role in *Deirdre of the Sorrows* (1910), written for her by Synge who was at that time her fiancé. She remained at the Abbey until 1911, returning occasionally to play there until 1917. She appeared subsequently at the Liverpool Repertory Theatre (see LIVERPOOL PLAYHOUSE), and during a long and distinguished career was seen in London and New York, often in Irish plays, and especially those of Sean O'CASEY, with her sister and her second husband Arthur SINCLAIR, whom she married in 1926.

ON STAGE, see STAGE DIRECTIONS.

O.P. (Opposite Prompt), see STAGE DIRECTIONS.

OPEN AIR THEATRE, London, in Queen Mary's Garden, Regents Park. This was founded by Sydney Carroll, and opened on the afternoon of 5 June 1933 with a production of *Twelfth Night* directed by Robert ATKINS, who ran the theatre for many years. The success of the venture was encouraging enough to persuade Carroll to make the theatre a permanent fixture of the London summer season. A marquee was later erected for use in wet weather. In 1961 David Conville took over from Atkins, and in 1975 the site was reconstructed.

OPEN SPACE THEATRE, London, see MAROWITZ, CHARLES.

OPEN STAGE, a raised platform built against one wall of the auditorium, with the audience on three sides. This was sometimes known in England as an 'arena stage', a term reserved in the United States for THEATRE-IN-THE-ROUND. The open stage derives basically from the Elizabethan platform STAGE, and is in use today in a number of new theatres, including the CHICHESTER FESTIVAL THEATRE and the STRATFORD (ONTARIO) FESTIVAL theatre. It was also used for productions at the Assembly Hall in the early days of the EDINBURGH FESTIVAL. Most of the new all-purpose theatres being built today, which are suitable for FLEXIBLE STAGING, make provision for open-stage productions, which, like those of theatre-in-the-round, call for different techniques in acting, staging, setting, and lighting from those used on proscenium-arch stages.

OPEN THEATER, group established in New York in 1963 by Joseph Chaikin, a former member of The LIVING THEATRE, with the intention of exploring by means of COLLECTIVE CREATION the unique powers of live theatre —which, according to Chaikin's book *The Presence of the Actor* (1972), is based upon the encounter between live performers and spectators—and to concentrate on abstraction and illusion in contrast to the realistic and psychological method of STANISLAVSKY typical of the modern conventional theatre. The group's first performances were exercises and improvisations developed into short productions by Jean-Claude VAN ITALLIE and Megan TERRY. These made frequent use of 'transformation', whereby one object or character becomes another, as, for instance, in a piece where the performers were first parts of an aeroplane as it took off, and were then 'transformed' into the people on the plane. The first long play to be developed collectively was *Viet Rock* (1966), a collage of transformations on the war in Vietnam which was created in a workshop conducted by Megan Terry and produced under the aegis of LA MAMA EXPERIMENTAL THEATRE CLUB. Later productions were *The Serpent* (1968), based on the Old Testament and including images from the assassinations of President Kennedy and Martin Luther King; *Terminal* (1969), a study of human mortality and responses to death; and *Mutation Show* (1971) which focused upon human adaptation. The last production created by the group before disbanding was *Nightwalk* (1973), an investigation into levels of sleep.

OPERA COMIQUE, London, at the junction of the Strand and Holywell Street. This stood back to back with the old GLOBE THEATRE in Newcastle Street, the pair being known as the Rickety Twins, having been hastily erected in the hope of compensation—which was not forthcoming—when the area was rebuilt. It opened on 29 Oct. 1870 with a French company under DÉJAZET in SARDOU's *Les Prés St Gervais*. In May 1871 actors from the COMÉDIE-FRANÇAISE under GÔT, driven from home by the Franco-Prussian War, made their first appearance outside Paris. The theatre was then used mainly by visiting foreigners—including RISTORI in 1873—with long periods of inactivity, until in Nov. 1877 D'Oyly Carte took over and produced GILBERT and Sullivan's *The Sorcerer* (1877), *H.M.S. Pinafore* (1878), *The Pirates of Penzance* (1880), and *Patience* (1881), the last transferring to the SAVOY THEATRE later in the year. In 1884 the Opera Comique closed for redecoration, reopening on 6 Apr. 1885, but with little success except for Herman's *The Fay o'Fire* in Nov., in which Marie TEMPEST made her second appearance in London. For some years it was used for special performances and try-outs, being partly rebuilt in 1895, and at the end of 1899 it closed for good, being demolished in 1902.

OPÉRA-COMIQUE, Paris, see COMÉDIE-ITALIENNE.

O.P. RIOTS, see COVENT GARDEN.

ORANGE MOLL, see MEGGS, MRS.

ORANGE STREET THEATRE, Dublin, see IRELAND.

ORCHESTRA, from the Greek word for 'a dancing place'. In ancient Greek theatres the name was given to the circular area where the CHORUS performed its songs and dances, as opposed to the smaller raised stage or *logeion* used by individual actors. In Roman theatres the orchestra was a semi-circular space in front of the stage, with seating reserved for senators and distinguished visitors. Towards the end of the 17th century the term was revived to describe the area in front of the stage where the musicians sat, and was soon extended to the players themselves. Orchestra stalls, in a proscenium theatre, are the front rows of seats on a level with the orchestra pit.

OREGON SHAKESPEAREAN FESTIVAL, see SHAKESPEARE FESTIVALS.

ØRJASÆTER, TORE (1886–1968), Norwegian poet and dramatist, whose work for the stage began in 1917 with a play about the famous 19th-century reindeer-hunter Jo Gjende. He developed the theme of self-awareness in *Anne på torp* (*Anne on the Farm*, 1930), but turned to EXPRESSIONISM in *Christophoros* (*St Christopher*, 1947) and in his treatment of an individual's atonement for collaboration with the Germans during the Occupation of Norway during the Second World War in *Den lange bryllaupsreisa* (*The Long Honeymoon*, 1949).

ORLENEV, PAVEL NIKOLAYEVICH (1869–1932), Russian actor, who made his début in 1886, and from 1898 to 1902 was at the Suvorinsky Theatre in St Petersburg, where on 12 Oct. 1898 he played the title role in A. K. TOLSTOY's *Tsar Feodor Ivanovich*, the play with which the MOSCOW ART THEATRE opened two days later. In 1904 Orlenev, with Alla NAZIMOVA, took a Russian company to Europe and in 1905 they were in America, where their new naturalistic style of acting and excellent ensemble playing aroused great interest. It was during this tour that CHEKHOV's *The Seagull* was first seen in the U.S.A. Orlenev returned to Russia, leaving Nazimova to pursue her successful American career, and after the Revolution toured the provinces, acting in factories and on collective farms. Nervous and undisciplined in private life, he was capable on stage of remarkable control and intense emotion in his favourite roles, among which were Raskolnikov in DOSTOEVSKY's *Crime and Punishment*, Oswald in IBSEN's *Ghosts*, and the title role in his *Brand*.

ORLOV, DMITRI NIKOLAYEVICH (1892–1955), Soviet comic actor, who enjoyed great success in the years just after the Revolution while working with MEYERHOLD, who, he said: 'Gave me eyes to see the tasks facing a worker in the revolutionary theatre.' Possessed of extreme exuberance and vivacity, he had also superb technical control of his voice, and later achieved great personal popularity with his performances as Stepan in POGODIN's *Poem about an Axe* (1931) and Globa in K. M. SIMONOV's *The Russian People* (1942).

ORRERY, Lord [Roger Boyle, first Earl of Orrery] (1621–79), Restoration nobleman and man of letters, author of some forgotten comedies and of *Mustapha* (1665) and *The Black Prince* (1667), which DRYDEN, generously though perhaps not accurately, hailed as the first examples of HEROIC DRAMA. The printed versions of these and Orrery's other plays are useful on account of their detailed stage directions.

ORTON, JOE (1933–67), English dramatist, whose first play, *Entertaining Mr Sloane* (1964; N.Y., 1965), shocked and amused its audience by the contrast between its prim-and-proper dialogue and the violence and outrageousness of its action. The same contrast was apparent in *Loot* (1966; N.Y., 1968), a satire on police corruption and the conventions of detective fiction. In 1967 an earlier play *The Ruffian on the Stair*, written for radio, was staged at the ROYAL COURT THEATRE in a double bill called *Crimes of Passion* with *The Erpingham Camp*, previously televised; the same bill was seen in New York in 1969. Orton was brutally murdered by his homosexual partner who then committed suicide—an episode as violent and bizarre as any in his plays. His last full-length work, *What the Butler Saw*, a parody of conventional farce, was produced in 1969 (N.Y., 1970), with Ralph RICHARDSON in the lead in London, and had a mixed reception. A one-act play, *Funeral Games*, was produced by the Basement Theatre at the end of 1970.

OSBORNE, JOHN JAMES (1929–), English dramatist who was for some years an actor, making his first appearance in 1948 and his London début in 1956. Although he had already had two plays produced in the provinces, he first came into prominence as a playwright when *Look Back in Anger* was produced at the ROYAL COURT THEATRE. It was the ENGLISH STAGE COMPANY's first outstanding success, and the date of the first night, 8 May 1956, is considered a landmark in the modern theatre; it was seen in New York in 1957. With its 'angry young man' hero, rude, eloquent, and working-class, it marked a radical departure from the traditional West End play. Osborne's next play, also seen at the Royal Court where virtually all his plays have been produced, was *The Entertainer* (1957; N.Y., 1958), in which Laurence OLIVIER gave an outstanding performance as the seedy music-hall artist Archie Rice. *Epitaph for George Dillon* (London and N.Y., 1958), written in collaboration with Anthony Creighton, and a musical, *The World of Paul Slickey* (1959), were less successful, but *Luther* (1961; N.Y., 1963), with Albert FINNEY in the name part, was another major achievement. After a double bill at the Royal Court, *The Blood of the Bambergs* and *Under Plain Cover* (1962) (billed as *Plays for England*), came *Inadmissible Evidence* (1964; N.Y., 1965), one of his best plays, in which Nicol Williamson gave a fine performance as a solicitor under stress. A year later *A Patriot for Me* was refused a licence by the Lord Chamberlain because it dealt with homosexuality, and it was therefore staged privately for members of the English Stage Society; it was produced in New York in 1969. In 1966 Osborne adapted Lope de VEGA's *La fianza satisfecha* as *A Bond Honoured* for the NATIONAL THEATRE, which presented it as part of a double bill. Two new plays in 1968, *Time Present* and *The Hotel in Amsterdam*, failed to enhance his reputation, though they provided excellent roles for Jill BENNETT, his fourth wife, as a bitchy actress, and Paul SCOFIELD respectively. They were followed by *West of Suez* (1971), with Ralph RICHARDSON, and *A Sense of Detachment* (1972), which has no plot and makes no attempt at realism, depending on the reaction between the actors on the stage and two actors pretending to be members of the audience. Osborne's adaptation of IBSEN's *Hedda Gabler* was also produced in 1972. *The End of Me Old Cigar* (1975) was seen at GREENWICH and *Watch It Come Down* (1976) at the NATIONAL THEATRE, but neither was enthusiastically received. His autobiography, *A Better Class of Person*, was published in 1981.

His second wife was the actress Mary Eileen Ure (1933–75), who played the feminine lead in *Look Back in Anger* in both London and New York. She was also seen in Arthur MILLER's *The Crucible* and *A View From the Bridge* (both 1956), at the SHAKESPEARE MEMORIAL THEATRE in 1959, and in New York in GIRAUDOUX's *Duel of Angels* (1960) and PINTER's *Old Times* (1971).

OSCARSSON, PER (1927–), Swedish actor, who from 1947 to 1952 was with the KUNGLIGA DRAMATISKA TEATERN in Stockholm, and then moved to the city theatre of Gothenburg. He is a tense and sensitive player, who has given memorable interpretations of Hamlet, STRINDBERG's Master Olof, Orin in O'NEILL's *Mourning Becomes Electra*, and Marchbanks in SHAW's *Candida*.

OSLO NYE TEATER (Oslo New Theatre), playhouse which opened on 26 Feb. 1929 after a successful fund-raising campaign by the two authors Johan Bojer and Peter Egge. Det Nye Teater (as it was first called) adopted a policy of paying particular attention to contemporary Norwegian drama, but this proved too ambitious an undertaking, and the repertoire (particularly under Gyda Christensen's direction, and with Lillebil IBSEN emerging as the company's chief actress) became more general. In 1947 the newly-appointed director Axel Otto Normann recalled the theatre to its initial more serious intentions, and the repertoire then contained a balance of classic and experimental theatre, interspersed with lighter pieces, opera, and operetta. In 1959 Det Nye Teater was merged with the Folketeatret (People's Theatre) under its present name, with Normann continuing as director. Once again, however, financial difficulties necessitated early reorganization and in 1967 the municipality of Oslo took over financial responsibility for running the theatre; opera was eliminated, the company was reduced in numbers, and a new director, Toralv MAURSTAD, appointed. With these changes came a new stability and vigour.

ÖSTERBERG, VALDEMAR, see DENMARK.

OSTERWA, JULIUSZ, see POLAND.

OSTLER, WILLIAM (?–1614), boy-actor who appeared with the Children of the Chapel Royal in JONSON's *The Poetaster* (1601). As an adult actor he joined the KING'S MEN, with whom he played in *The Alchemist* (1610), also by Jonson, and was the first to play Antonio in WEBSTER's *The Duchess of Malfi* (1614). He married the daughter of HEMINGE and had shares in both the GLOBE THEATRE and BLACKFRIARS, which his father-in-law tried to acquire from his widow after his early death.

OSTROVSKY, ALEXANDER NIKOLAYEVICH (1823–86), Russian dramatist, many of whose numerous plays remain in the Soviet repertory. He first attracted attention with *The Bankrupt* (1849), a study of corruption in the Moscow merchant class which cost him his job as a civil servant and condemned him to a life of constant struggle and near poverty. Banned from the stage for 13 years, it circulated freely in manuscript and was eventually staged as *It's All in the Family*. It was followed by a number of historical plays, a fairy-tale play, *The Snow Maiden* (1873) which provided the basis for an opera by Rimsky-Korsakov, and the series of realistic contemporary satires for which Ostrovsky is best known. He is difficult to translate into English, owing to the richness of his language and his use of local colour, but three of his satires—*Even a Wise Man Stumbles* (1868), also known as *Enough Stupidity in Every Wise Man*; *Easy Money* (1870), based on *The Taming of the Shrew*; and *Wolves and Sheep* (1875)—were published in 1944 in translations by David Magarshack. Another satire, *The Forest* (1871), about two strolling players in provincial Russia, was given its British première by the ROYAL SHAKESPEARE COMPANY in 1981, with Alan HOWARD. Of the plays dealing with the position of women in Russian society which he wrote in his later years the best is usually acknowledged to be *The Poor* (or *Dowerless*) *Bride* (1879). Outside Russia his best known play is a domestic tragedy, *The Storm* (1860). It was seen in translation in New York in 1900, and again in 1919 and 1962. In London it first appeared in 1929, and was revived at the OLD VIC by the NATIONAL THEATRE in 1966. Most of Ostrovsky's plays were first produced at the MALY THEATRE, Moscow, sometimes known as the House of Ostrovsky, where he found a friend and ideal interpreter in the actor Prov SADOVSKY. He was deeply concerned with the position of Russian actors, and served as President of the Society of Russian Dramatists and Operatic Composers from its foundation in 1870 until his death.

OSTUSHEV, A., see MALY THEATRE, MOSCOW.

OTHER PLACE, The, the ROYAL SHAKESPEARE COMPANY's studio theatre in Stratford-upon-Avon, which seats 140 and presents plays in repertory, opened in 1974. Its programmes, which consist of both new plays and classic works, have included some notable small-scale productions of Shakespeare, among them *Hamlet* in 1975, *Macbeth* in 1976 (directed by Trevor NUNN), *Pericles* in 1979, and *Timon of Athens* in 1980. Revivals of rarely-performed plays have included FORD's *Perkin Warbeck* in 1975 and *'Tis Pity She's a Whore* in 1977; BEAUMONT and FLETCHER's *A Maid's Tragedy* in 1980; and FARQUHAR's *The Twin Rivals* and Ford, DEKKER, and ROWLEY's *The Witch of Edmonton* in 1981. Among the new plays which have had their first production at The Other Place have been David Edgar's *Destiny* (1976), Pam Gems's *Queen Christina* (1977) and her musical *Piaf* (1978), and David Rudkin's *The Judgment on Hippolytus* (1978) and *Hansel and Gretel* (1980).

OTHER STAGE, New York, see PUBLIC THEATRE.

O'TOOLE, PETER (1932–), Irish-born actor, who was with the BRISTOL OLD VIC from 1955 to 1958, and made his first appearance in London in 1956 with the company, in SHAW's *Major Barbara*. He built up a formidable reputation in Bristol, particularly with his Jimmy Porter in OSBORNE's *Look Back in Anger* in 1957 and John Tanner in Shaw's *Man and Superman* in 1958. He made his first outstanding success in London as the earthy, cynical trouble-maker Private Bamforth in Willis HALL's *The Long and the Short and the Tall* (1959). In 1960 he was with the SHAKESPEARE MEMORIAL THEATRE company, playing Petruchio in *The Tam-*

ing of the Shrew and giving an electrifying performance as Shylock in *The Merchant of Venice*. He then moved into films and has made only intermittent returns to the stage, playing the title role in BRECHT's *Baal* and Hamlet in the NATIONAL THEATRE company's inaugural production in 1963, and the Yorkshire novelist-hero of David MERCER's *Ride a Cock Horse* in 1965. A year later he was in Dublin to play Jack Boyle in O'CASEY's *Juno and the Paycock*, and in 1969 he appeared at the ABBEY THEATRE there as Vladimir in BECKETT's *Waiting for Godot*. He returned to the Bristol Old Vic in 1973 to play the title role in CHEKHOV's *Uncle Vanya* and King Magnus in Shaw's *The Apple Cart*, being seen in *Uncle Vanya* in Chicago in 1978 as well as in COWARD's *Present Laughter*. In 1980 he returned to the London stage to give a controversial and highly criticized performance as Macbeth for the Old Vic Company (see PROSPECT THEATRE COMPANY).

OTTO, TEO (1904–68), German stage designer, who worked for a number of leading directors in Germany up to 1933, when he emigrated to Switzerland, where he was responsible for the sets for the first productions, in Zurich, of BRECHT's *Mutter Courage und ihre Kinder* (1941) and *Der gute Mensch von Sezuan* (1943), also designing the sets for the latter when it was produced at the ROYAL COURT THEATRE in London as *The Good Woman of Setzuan* in 1956. Among his later designs were those for the first productions of FRISCH's *Graf Öderland* (1951), *Don Juan, oder die Liebe zur Geometrie* (1953), *Andorra* (1961), and *Biografie* (1968), all in Zurich. He was responsible for the first productions, in Zurich, of BRECHT's *Mutter Courage und ihre Kinder* (1941) and *Der gute Mensch von Sezuan* (1943), also designing the sets for the latter when it was produced at the 1960. One of his most famous settings was that for GRÜNDGENS's revival of GOETHE's *Faust*, Parts I and II, at Hamburg in 1957/8. He had a fine sense of visual effect, coupled with an understanding of dramatic structure and the magic of the stage. He was the author of a 'diary in pictures', *Nie wieder* (1949), and of *Mèine Szene* (1965).

OTWAY, THOMAS (1652–85), English dramatist, who in 1670 made his first appearance on the stage in Aphra BEHN's *The Forced Marriage* with such conspicuous lack of success that he never acted again, but turned instead to playwriting. *Alcibiades* (1675) gave Mrs BARRY, with whom Otway was madly but hopelessly in love, her first outstanding role; it was followed in 1676 by *Don Carlos*, a tragedy in rhymed verse; by a new version of RACINE's *Bérénice* (1670) as *Titus and Berenice* (1676); and by what was to prove Otway's most successful play during his lifetime, a farce entitled *The Cheats of Scapin* (based on MOLIÈRE), given as an afterpiece to the Racine play. After another comedy, *Friendship in Fashion* (1678), and a tragedy, *The History and Fall of Caius Marius* (1679), Otway produced his two finest works, *The Orphan; or, the Unhappy Marriage* (1680) and *Venice Preserv'd; or, a Plot Discovered* (1682), in both of which Mrs Barry, the inspiration behind all Otway's work, played the heroines, Monimia and Belvidera, opposite BETTERTON. Both plays held the stage during the 18th century; *The Orphan* was last seen in London in 1925, after a century's neglect, and *Venice Preserv'd*, which continued to be revived during the 19th century, was given a superb production at the LYRIC THEATRE, Hammersmith, in 1953 with GIELGUD as Jaffier and SCOFIELD as Pierre. Otway, who has been called 'an Elizabethan born out of his time', and whose best work certainly shows greater depth and sincerity, coupled with fine writing, than that of any other dramatist of the period, wrote only two other plays, *The Soldier's Fortune* (also 1680), which was revived at the ROYAL COURT THEATRE in 1967 with unexpected success, and its sequel, *The Atheist; or, the Second Part of the Soldier's Fortune* (1683).

OUDS, see OXFORD.

OUEST, Comédie de l', see DÉCENTRALISATION DRAMATIQUE.

OUSPENSKAYA, MARIA, see AMERICAN LABORATORY THEATRE.

OUVILLE, ANTOINE LE METEL D' (*c*. 1590–1656/77), French dramatist, the profligate and licentious elder brother of BOISROBERT. Having lived for some time in Spain, he was the first to introduce the plays of CALDERÓN to the French stage with his adaptation of *La dama duende* as *L'Esprit folet* (*The Hobgoblin*, 1638–9). Ouville's *Les Trahisons d'Arbiran* (1638) may have been the source of part of MOLIÈRE's *Tartuffe* (1664).

OVERSKOU, THOMAS, see DENMARK.

OWENS, JOHN EDMOND (1823–86), London-born American actor, best known as an eccentric comedian and an outstanding interpreter of Yankee characters. His most famous part was that of Solon Shingle in Joseph S. JONES's *The People's Lawyer*, in which he toured successfully throughout the United States and also appeared in London in 1865, where he was seen by DICKENS. He was also good as Dr Pangloss in the younger COLMAN's *The Heir-at-Law*, as Caleb Plummer in *Dot*, BOUCICAULT's dramatization of Dickens's *The Cricket on the Hearth*, in the title role of John Poole's *Paul Pry*, and as Mr Toodles in *The Toodles*, adapted by W. E. Burton from R. J. Raymond's *The Farmer's Daughter of the Severnside*. In 1876 he appeared in New York as Perkyn Middlewick in H. J. BYRON's *Our Boys*, and later joined the stock company at MADISON SQUARE, where he played with Annie Russell. He retired in 1885.

OXBERRY, WILLIAM (1784–1824), English actor, editor, and publisher, the son of an auctioneer, apprenticed in youth to a printer. He had, however, always wished to go on the stage, and at 18 years of age broke his indentures and became a professional, having made several appearances as an amateur. He first appeared at Watford as Antonio in *The Merchant of Venice*, and after touring the provinces made an unsuccessful appearance at COVENT GARDEN, being harshly criticized by *The Monthly Mirror*, a paper which he was later to edit. Disillusioned, he withdrew to Scotland, where he toured, mainly in tragedy, returning to London eventually to play small parts at the LYCEUM and DRURY LANE. Though never outstanding, he became a useful comic actor, and was the author of a farce, *The Actress of All Work; or, My Country Cousin* (1819), probably inspired by the younger COLMAN's *The Actor of All Work* written two years earlier for Charles MATHEWS. This showed the same actress in six different parts, and later proved an excellent vehicle for the child prodigy Clara FISHER. Oxberry was also a publisher and was responsible for a number of volumes of theatrical interest, giving his name to *Oxberry's Dramatic Biography* (1825–6) published posthumously by his widow Catherine, probably with the assistance of her second husband Leman Rede.

OXBERRY, WILLIAM HENRY (1808–52), English actor, who, like his father (above), dabbled extensively in theatrical journalism, editing *Oxberry's Weekly Budget* (1843–4) in which several MELODRAMAS otherwise unknown were printed. He was at one time credited with the authorship of his father's farce *The Actress of All Work* (1819), produced when he was only 11 years of age. He was, however, the author of several plays, mainly from the French, the most successful being *Matteo Falcone; or, the Brigand and His Son* (1836), which was included in the repertory of the TOY THEATRE. Unlike his father, he was a quaint little man, a lively actor and dancer, unsuited to tragedy but good in burlesque, though he is said never to have known his part on first nights.

OXFORD, the home of England's oldest university, probably had performances of LITURGICAL DRAMA in medieval times, but the first reference to a play there dates from 1490 when *St Katherine* was performed at Magdalen College. In the 1540s several colleges acted plays by Nicholas Grimald, including *Troilus*, based on Chaucer, as was the *Palaemon and Arcite* by Richard Edwardes with which Christ Church entertained Elizabeth I in 1566. In 1567 a comedy with the intriguing title of *Wylie Beguylie* was produced at Merton but, as with so many plays of this period, it has not survived, though its title is echoed in the anonymous *Wily Beguiled* of 1602.

There seems to have been much less acting in Oxford than at CAMBRIDGE; the two colleges mainly concerned, in the surviving records, are Christ Church and St John's. At the first in the 1580s and early 1590s a number of plays in Latin, mostly tragedies, by Nicholas Gager were seen, and in Aug. 1605 James I attended the performance of a PASTORAL, his consort alone being present a few days later at Samuel DANIEL's *The Queen's Arcadia*, also a pastoral. St John's, whose first recorded play in about 1588 was Richard Lateware's Latin tragedy *Philotas*, saw another tragedy, *Labyrinthus*, a Latin translation of the Italian *La Cintia*, in 1603, and at Christmas 1607–8 embarked on an elaborate entertainment, *The Christmas Prince*, which included at least six shows of different kinds. The last play recorded at Oxford before the Commonwealth, an anonymous comedy entitled *Grobiana's Nuptials* (*c*. 1641), was probably produced at St John's.

All these plays were in the nature of academic exercises, somewhat in the style of the JESUIT DRAMA on the Continent, and there seems to have been very little acting outside the university, though STRANGE'S MEN were seen in an Oxford inn-yard in 1590–1. After the Restoration the town was visited by the KING'S MEN, who played in an adapted tennis-court in 1680, and by BETTERTON's company in 1703. The 18th century saw little beyond private theatricals and occasional quasi-official performances at Commemoration. It was the foundation of the Oxford University Dramatic Society (OUDS) in 1885, by a group of undergraduates which included Arthur BOURCHIER, later an important actor-manager, and Cosmo Gordon Lang, the future Archbishop of Canterbury and a cousin of Matheson LANG, that again made acting an extra-curricular activity in the university. The first production, given in the Town Hall, was *Henry IV, Part I*, in which Bourchier played Hotspur and Lang spoke a Prologue. Production was by Alan Mackinnon, who continued to direct until 1895, most of the plays produced being comedies by Shakespeare, given in a college garden in summer and indoors in winter.

During the First World War the OUDS was disbanded, but it started up again in 1919, the first production being Thomas Hardy's epic poem *The Dynasts*, the author himself being in the audience. The society then continued the former plan of indoor and outdoor productions, with occasional plays by dramatists other than Shakespeare—FLECKER's *Hassan* was given in 1931, with Peggy ASHCROFT as Pervaneh, since women were still not admitted to membership—until on the outbreak of the Second World War it was again suspended. A new society known as Friends of the OUDS was, however, formed by Nevill Henry Kendal Aylmer Coghill (1899–1980), English Tutor at Exeter College and later Merton Professor of English Literature, together with other senior members of the university, and from 1940 to 1946 twelve productions were given, six of them directed by Coghill. Under him the Friends not only maintained a high standard but also succeeded in paying off outstanding debts and laying by funds for the future. Their last production was IBSEN's *The Pretenders*. The society was then re-formed under Glynne Wickham, later Professor of Drama at Bristol University, women were admitted to full membership, which meant that professional actresses were no longer engaged for the female leads, and in June 1947 *Love's Labour's Lost* was given in the garden of Merton College. In May 1948 *The Masque of Hope* by Coghill, directed by Wickham, was acted before Princess Elizabeth (later Elizabeth II) in University College. In 1950 the society was forced for financial reasons to give up its premises, but it now has a club room and office in the Playhouse (below), enabling it to enlarge its activities considerably. It has organized tours of British and French university towns, helped to sponsor the new Oxford Festival of Theatre founded in 1979, and apart from its major productions, which continue to reach a high standard, produces several smaller shows each term. The archives of the OUDS have been deposited at the Bodleian Library.

In Feb. 1936 Coghill founded a new society, the Oxford University Experimental Theatre Club (E.T.C.), which unlike the OUDS was designed to leave everything connected with the production of a play in the hands of undergraduates. It also chose to do plays which did not come within the scope of the older society—neglected classics, new English or foreign plays, particularly of the AVANT-GARDE, and, when possible, plays by undergraduates. The first production was DRYDEN's *All For Love* in 1936; later productions have included plays by COCTEAU, PIRANDELLO, Ben JONSON, ARISTOPHANES, and a revival of the anonymous MORALITY PLAY *The Castell of Perseverance*. The E.T.C. has in recent years endeavoured to keep up with the major trends in experimental theatre, even becoming involved in COMMUNITY and street theatre, and concentrating mainly on modern works. Among its productions have been David HARE's *Teeth 'n' Smiles*, given in St Hilda's College in 1980.

One noteworthy activity is the acting of plays in Greek. This was begun in 1880, when the *Agamemnon* of AESCHYLUS was performed under the aegis of the Master of Balliol, Dr Jowett. Later productions included EURIPIDES' *Alcestis* in 1887, Aristophanes' *Frogs* in 1892, and his *Knights* in 1897. The tradition continued into the 1930s, and after a long gap was revived by the Oxford University Classical Society in 1967. Many college dramatic societies have also produced Greek plays in translation.

Although Oxford has no UNIVERSITY DEPARTMENT OF DRAMA, and the theatre has no place in the regular curriculum, it has a university theatre, the OXFORD PLAYHOUSE. Also in the town is the Apollo Theatre, formerly the New, which opened in 1934 and is used by touring companies. There is a project to build a small Samuel BECKETT Theatre, but it has not yet materialized.

OXFORD MUSIC-HALL, London, at the corner of Oxford Street and Tottenham Court Road. One of London's best known MUSIC-HALLS, it opened under the management of Charles MORTON on 26 Mar. 1861, having on its act-drop and on the cover of its programmes a view of Magdalen Tower, Oxford. It had a successful career, both George ROBEY and Harry Tate making their London débuts there, and was rebuilt in 1869, 1873, and 1893. In 1917 it was taken over by C. B. COCHRAN, who renamed it the Oxford Theatre and presented there an extravaganza, *The Better 'Ole* by Bruce Bairnsfather, the famous cartoonist and war correspondent, which ran for 811 performances. In 1920 the old hall was fully converted into a theatre, and opened on 17 Jan. 1921 with a spectacular long-running REVUE, *The League of Notions*. During 1922 it was used briefly for films, but returned to live theatre with plays starring Sacha GUITRY and Yvonne PRINTEMPS and a visit from Eleonora DUSE. In 1924 the OLD VIC company made its first West End appearance at the Oxford, with productions by Robert ATKINS of *As You Like It, The Taming of the Shrew, Twelfth Night*, and *Hamlet*. A series of unsuccessful ventures followed, and the theatre closed in May 1926. It was demolished in the following year, a Lyons Corner House being built on the site.

OXFORD PLAYHOUSE, in Beaumont Street, the last theatre to be built in Britain before the Second World War. It replaced an earlier and very inconvenient theatre in the Woodstock Road, known as the Red Barn and formerly a Big Game Museum. It was adapted by J. B. FAGAN, who from 1923 to 1928 intermittently gave seasons there of good British and Continental plays, including CHEKHOV and IBSEN, his company including at various times, in the early stages of their careers, Tyrone GUTHRIE, Flora ROBSON, and John GIELGUD. The venture did not receive the support it deserved, and after Fagan left the theatre stood empty for two years until it was reopened in 1930 with a company under Stanford Holme and his wife Thea which produced mainly modern comedies by, among others, Ben TRAVERS, Noël COWARD, and Oscar WILDE. This company was later run by Eric Dance, who in 1938, after a public appeal for funds, designed and built the present Playhouse, opening it with Fagan's comedy *And So To Bed*. It flourished during the war years, but afterwards audiences dwindled and it was not until 1956, after being closed for some months, that the theatre was firmly established, with the support of the ARTS COUNCIL and financial assistance from Richard BURTON. It reopened on 1 Oct. with GIRAUDOUX's *Electra* produced by the Meadow Players under the direction of Frank Hauser, the company remaining in occupation for 17 years and establishing a high reputation, many West End actors appearing with it as guest stars for special occasions. In 1961 the university bought the remaining lease of the theatre, and two years later it was redecorated and enlarged, the seating capacity being increased to its present 700. Richard Burton and his then wife Elizabeth Taylor appeared there in 1966 with the OXFORD University Dramatic Society in a production of MARLOWE's *Dr Faustus* to raise money for an extension, now known as the Burton Rooms. In all these enterprises a leading part was played by the Oxford don Nevill Coghill, who until his retirement was Chairman of the theatre's governing body. Because of financial difficulties, the Meadow Players ceased operations in 1973, and were succeeded a year later by Anvil Productions (the Oxford Playhouse Company), which has presented an enterprising programme, including several world premières and new translations of foreign plays. The theatre is also used by touring companies and by the main university dramatic societies.

OXFORD'S MEN. The first mention of a company of players under this name occurs as early as 1492, and in 1547 the 'players of the Earl of Oxford' (who were to be disbanded in 1562 on the death of the 16th Earl) caused a scandal by acting in Southwark while a dirge was being sung in St Saviour's for Henry VIII who had recently died. This company played mainly in the provinces. A new company under the patronage of the 17th Earl, Edward de Vere, who was himself something of a playwright, was formed in 1580 in conjunction with the players who had been for a short time under the patronage of the Earl of Warwick, and played at the THEATRE. The actors got into trouble for brawling and were banished to the provinces until 1584, when the boys belonging to the company appeared with other BOY COMPANIES at the first BLACKFRIARS, in two of LYLY's plays. They were also seen at Court in a lost play, *Agamemnon and Ulysses*, probably written for them by their patron. The company was disbanded in 1602, after some final appearances at the Boar's Head in Middlesex, where they may have been joined by the Earl of Worcester's Men.

OYONO-MBIA, GUILLAUME (*c.* 1940–), African dramatist, born in the French-speaking part of Cameroon. He is the author of several successful comedies in which he has adapted the conventions of French classical comedy as exemplified by MOLIÈRE to African situations, as in *Trois Prétendants, un mari* (*Three Suitors, One Husband*, 1964), in which a greedy father, anxious to make money out of his daughter's marriage, is outwitted by her when she contrives to marry the man she loves. A later play, *Notre Fille ne se mariera pas* (*Our Daughter Will Never Get Married*, 1971), makes fun of the affectations of Europeanized Africans and of misunderstandings between town and country dwellers.

OZEROV, VLADISLAV ALEXANDROVICH, see RUSSIA AND SOVIET UNION.

P

PACE-EGG PLAY, see MUMMERS' PLAY.

PACUVIUS, MARCUS (*c.* 220–130 BC), one of the leading dramatists of Rome, and the first to specialize in tragedy. The titles of 12 of these have survived, together with that of one FABULA *praetexta*, and fragments amounting to about 400 lines. Pacuvius, who was the nephew of ENNIUS, appears to have liked pathetic scenes, philosophical discussions, and complicated plots, and his language, though sometimes awkward and obscure, shows pictorial power: he was a painter as well as a poet. Many of his plays remained popular down to the end of the Republic, the most famous scene in Roman tragedy being the opening of his *Iliona*, where the ghost of a murdered boy rises to beg his mother for burial. Another favourite episode, in the *Chryses*, shows Orestes and Pylades brought captive before Thoas, each claiming to be Orestes who alone is to be punished. Cicero tells us that this generous rivalry between the two friends brought the audience to their feet, applauding. Cicero also maintains that in his *Niptra* Pacuvius, in the final scene, improved on his model SOPHOCLES. Whereas Sophocles allowed the mortally wounded hero to express his agony, Pacuvius makes him die with stoic self-control.

PAGE, GERALDINE (1924–), American actress, who made her début in Chicago in 1940 and in New York in 1945. She then appeared for several years with stock companies and did not come into prominence until 1952, when her performance as Alma Winemiller in a New York revival of Tennessee WILLIAMS's *Summer and Smoke* was widely acclaimed. In 1954 she was seen as Marcelline in Ruth and Augustus Goetz's adaptation of Gide's novel *The Immoralist*, and later in the same year she scored a big success as Lizzie Curry, the plain spinster in N. Richard Nash's *The Rainmaker*, making her London début in this role in 1956. She successfully succeeded Margaret LEIGHTON in the New York production of RATTIGAN's double bill *Separate Tables* (1957), and in 1959 created the role of the Princess in another play by Williams, *Sweet Bird of Youth*. A revival of O'NEILL's *Strange Interlude* in 1963 in which she played Nina Leeds was followed in 1964 by a production of CHEKHOV's *Three Sisters* in which she played first Olga and then Masha. She was later seen in such plays as SHAFFER's double bill *White Lies* and *Black Comedy* (1967), Jerome Kilty's *Look Away* (1973), AYCKBOURN's *Absurd Person Singular* (1974), and a triple bill of STRINDBERG's *Creditors*, *The Stronger*, and *Miss Julie* (1977). In 1980 she starred in *Clothes For a Summer Hotel*, Williams's play about Scott and Zelda Fitzgerald. One of America's best actresses, she has a wide range and great sensitivity.

PAGEANT, a word which has over the centuries completely changed its meaning, though retaining always its connection with the idea of 'pageantry'. In medieval times it was used variously to describe the 'carts', each one manned and decorated by a different guild, which carried the notables of the city in procession on civic occasions; the last vestige of this is the procession on Lord Mayor's Day, dating from the 13th century, which still takes place in London early in November. 'Pageant' was also the word applied to those fixed points within a city's walls at which an entertainment, ranging from a short speech to a long dramatic debate, could be given to welcome the visit of a Royal or noble personage or an important visitor from overseas. Such places were the market cross with its steps, any raised permanent structure such as a water conduit, and the space above and around the city gates, often augmented by the use of scaffolding. By extension it was then applied to the 'pageant wagon', a wheeled vehicle on which a scene from a religious play could be performed before an audience assembled at a fixed point before being drawn away to perform the same scene at another pre-arranged station, usually a crossroads, a market square, or other open space. These 'pageants' or perambulating stages, which were very popular in England during the period covered by the LITURGICAL DRAMA, were usually two-storied, the lower story serving being curtained off to serve as a dressing-room for the actors, though it could on occasion represent Hell.

A further use of the word 'pageant' was to describe the elaborate structures of wood and painted canvas which were used in the Tudor MASQUES. These were often ingenious machines, fixed or movable, designed and built at considerable expense for the purposes of a single night's entertainment.

Early in the 20th century there arose a passion for elaborate and partly processional open-air shows celebrating the history or legends of a particular town, and these were referred to as 'pageants'. The first was produced by Louis N.

PARKER at Sherborne in 1905, and it was followed by many others, at Warwick, Dover, York, Oxford, and other historic towns. The performers were usually amateurs, townspeople and schoolchildren, directed by a professional who was responsible for the songs, dances, and short interludes of dialogue which made up the whole, including sometimes re-enactments of medieval joustings and tournaments. The performances usually took place in a large park, or in the countryside beyond the city boundaries, and it was usual for the properties and costumes to be made by the local population.

From the countryside the pageant soon spread to the theatre, and during the First World War several patriotic pageants were produced in London. In 1918 DRURY LANE celebrated its own history in *The Pageant of Drury Lane*, again devised and directed by Parker.

PAGEANT LANTERN, see LIGHTING.

PALACE THEATRE, London, in Cambridge Circus, Shaftesbury Avenue. This theatre, with its four-tier auditorium seating 1,697 (reduced in 1908 to 1,462), opened on 31 Jan. 1891 as the Royal English Opera House. A year later, as the Palace Theatre of Varieties, it became a successful MUSIC-HALL under Charles MORTON, who was succeeded in 1904 by Alfred Butt. It took its present name in 1911, in which year the first Royal Command Variety Show was seen there, as was GROCK on his first visit to London. It continued to present variety bills and REVUES, the first being *The Passing Show* (1914) with Elsie Janis. In 1924 the Co-Optimists occupied the theatre, and in the following year the MUSICAL *No, No, Nanette!* began its long run, followed by a number of other successful musicals, including *The Girl Friend* (1927) and *Frederica* (1930). After some straight plays in the 1930s the theatre reverted to revue with a series of COCHRAN shows, and after appearing in *Under Your Hat* (1938) Cicely COURTNEIDGE and Jack Hulbert remained in the theatre for much of the Second World War in *Full Swing* (1942) and *Something in the Air* (1943). Among later successes were *Gay Rosalinda* (1945) and *Song of Norway* (1946); and in 1949 the long run of NOVELLO's *King's Rhapsody* began. During the 1950s the Palace was used by a number of foreign companies—the THÉÂTRE NATIONAL POPULAIRE, the RENAUD-BARRAULT, the BERLINER ENSEMBLE—brought to London on the initiative of Peter DAUBENY; Laurence OLIVIER was also seen there in John OSBORNE's *The Entertainer* (1957). On 24 Mar. 1960 the Rodgers and Hammerstein musical *Flower Drum Song* opened and was followed on 18 May 1961 by their *The Sound of Music*, which ran for six years. In 1968 Judi DENCH made a great success in *Cabaret*, and in 1970 Danny LA RUE began a long season in drag. He was followed by the rock musical *Jesus Christ Superstar*, which opened on 9 Aug. 1972 and ran for eight years. It was succeeded in 1980 by a revival of Rodgers and Hammerstein's *Oklahoma!*

PALACE THEATRE, Manchester, a large theatre built for the use of touring companies in 1891. It was visited by many famous stars of the entertainment world, particularly in MUSIC-HALL, VARIETY, and MUSICAL COMEDY, opera, ballet, and straight plays normally being left to the Opera House, built in 1912. When the decline in touring, combined with the rise of television, made it apparent that the city could support only one large theatre, the choice fell on the Palace because, unlike the Opera House (which is now used for bingo), it had room for expansion. In 1978 it was taken over by a Trust and redeveloped, the stage in particular being enlarged so as to take the full ballet and opera companies from COVENT GARDEN. Seating 2,000, it reopened in Mar. 1981 with the rock musical *Jesus Christ Superstar*, followed by a season of opera, the first to be given in the United Kingdom outside London by the Covent Garden company for 17 years. The Palace will present productions subsidized by the ARTS COUNCIL for 20 weeks annually, the rest of the year being devoted to major musicals, plays, variety, concerts, and a Christmas PANTOMIME.

PALACE THEATRE, New York, at 1564 Broadway, between 46th and 47th Streets. This epitome of the big American VAUDEVILLE theatre was built by the Chicago-based impresario Martin Beck. With a seating capacity of 1,358, it opened in Mar. 1913, but at first, partly because of its uptown location, it was not very successful. A visit by Sarah BERNHARDT brought it into the limelight, and it was acquired by the Keith-Albee syndicate. The outstanding stars of vaudeville made their appearance at the Palace and it became the mecca for entertainers from all over the world; but with the decline in popularity of vaudeville and the end of 'two-a-day' performances in 1932 the theatre had to fight for survival. There were combination bills of films and vaudeville, appearances by such stars as Judy Garland and Danny Kaye, and even 'four-a-day' programmes, but in the end the building became a cinema. It took on a new lease of life, however, after its acquisition in 1965 by the Nederlanders, when it became the home of big musicals and also provided offices for many theatrical organizations. Famous musicals staged there have been *Sweet Charity* (1966), *George M!* (1968), *Applause* (1970), and *Lorelei* (1974). There was a break in 1974 when the ROYAL SHAKESPEARE COMPANY appeared at the Palace in a revival of BOUCICAULT's *London Assurance*. In 1979 there was a revival of *Oklahoma!*, and 1981 saw the launching of a new musical *Woman of the Year* based on a film which had starred Spencer Tracy and Katharine HEPBURN.

PALACE THEATRE, Watford, Hertfordshire. Built as a music-hall in 1908, this theatre began to stage plays during the First World War, becoming a repertory theatre in 1932. In 1964 the local council took it over and established it as a CIVIC THEATRE run by a Trust. Attractive and intimate, it seats 467 plus 200 in the gallery. There is no resident company, but well known guest stars often appear in individual productions. The theatre has acquired a reputation for enterprising and ambitious programmes, giving the British premières of Tennessee WILLIAMS's *Sweet Bird of Youth* (1968), Hugh LEONARD's *Summer*, and Lillian HELLMAN's *The Autumn Garden* (both 1979). Other notable productions include PINTER's *The Homecoming* (1969), directed by and starring the author, John OSBORNE's *A Patriot For Me* (1973), the British première of Simon GRAY's *Molly* (1977), later seen in the West End, and the world professional première of James SAUNDERS's *The Girl In Melanie Klein* (1980) with Frank FINLAY. There is a Theatre in Education scheme.

PALAIS-ROYAL, Théâtre du, Paris, originally a small private playhouse in the home of Cardinal RICHELIEU, which was rebuilt at great expense in the last years of his life, with superb interior decorations and all the newest stage machinery. It was long and narrow, the floor rising in a series of shallow steps with two balconies on each side, and held about 600 people. It was formally inaugurated on 14 Jan. 1641 with a spectacular performance of DESMARETS's *Mirame* in the presence of Louis XIII and his Court. After Richelieu's death in the following year the theatre became the property of the King, and was used intermittently for Court entertainments until 1660, when it was given to MOLIÈRE in place of the demolished PETIT-BOURBON. It remained in use until 1670, when it was rebuilt and enlarged and equipped with the new machinery necessary for productions of opera and spectacular musical plays. It reopened with Molière's MACHINE PLAY *Psyché* (1671) and Molière played there in *Le Malade imaginaire* on the night of his death, 17 Feb. 1673. Lully, who held the monopoly of music in Paris, immediately claimed the theatre to house his new Academy of Music, and it was called by that name until it was burnt down in 1763. Rebuilt, it was again destroyed by fire in 1781. In 1784 the whole area occupied by the Palais-Royal was reconstituted by its owner, the Duc de Chartres (later Philippe-Égalité), cousin of Louis XVI, and several theatres were built there, most of which at some time called themselves the Palais-Royal. One, which opened in 1790 as the Variétés-Amusantes, was later occupied by TALMA and his pro-revolutionary comrades as the Théâtre de la République. It later became the COMÉDIE-FRANÇAISE. Another Palais-Royal, which opened in 1831 after having had a variety of other names, specialized in VAUDEVILLE and FARCE, and among other plays witnessed the first nights of LABICHE's famous comedy *Un Chapeau de paille d'Italie* (1851) and of many farces by Tristan BERNARD. In England towards the end of the 19th century the term 'Palais-Royal farce' was used for such broadly suggestive adaptations from the French as *The Pink Dominos* (1877) and FEYDEAU's *The Girl from Maxim's* (1902). The fortunes of the theatre declined somewhat during the 20th century, until in 1958 it was restored to house BARRAULT's company on its return from an extended tour abroad. It is one of the largest and best preserved 19th-century playhouses in Paris, and is now used for the presentation of light comedies and BOULEVARD PLAYS.

PALAPRAT, JEAN DE BIGOT (1650–1721), French dramatist, friend of MOLIÈRE, who collaborated with BRUEYS in several plays for the COMÉDIE-FRANÇAISE in which the actor J. B. RAISIN, a friend of both authors, played the leading parts. They were successful at the time but soon forgotten, as were the plays which Palaprat wrote, on his own or with BIANCOLELLI, for the COMÉDIE-ITALIENNE.

PALÁRIK, JAN, see CZECHOSLOVAKIA.

PALITZSCH, PETER (1918–), German director, who in the early 1950s was one of BRECHT's associates at the BERLINER ENSEMBLE, where, with WEKWERTH, he directed SYNGE's *The Playboy of the Western World* in 1956, VISHNEVSKY's *The Optimistic Tragedy* in 1958, the first production of Brecht's *Der aufhaltsame Aufstieg des Arturo Ui* (1959), and H. Baierl's *Frau Flinz* (1960). In 1962 he moved to the Federal Republic, where he again directed many of Brecht's plays, as well as WALSER's *Überlebensgross Herr Krott* (1963) and *Der schwarze Schwann* (1964), DORST's *Toller* (1968), and a production based on BARTON and HALL's Shakespeare cycle *The Wars of the Roses* (1967). He was artistic director at Stuttgart from 1967 to 1972, and at Frankfurt from 1972 to 1980.

PALLADIO, ANDREA (1518–80), Italian architect. Born Andrea di Pietro, the name by which he is now known (taken from Pallas Athene) was bestowed on him by his patron TRISSINO in recognition of his genius. It has in turn given its name to the Palladian style of architecture, based on the principles of classical antiquity as Palladio, under the influence of VITRUVIUS, interpreted them in his buildings and codified them in his *Quattro libri dell'architettura*, published in Venice in 1570. This work, translated into English by Inigo JONES who was a pupil and admirer of Palladio, brought his ideas to the notice of English architects, influencing both theatre and public buildings throughout the land. One of Palladio's greatest though least fruitful achievements was the Teatro OLIMPICO, which was finished after his death by his pupil Vincenzo Scamozzi (1552–1616).

PALLADIUM, see LONDON PALLADIUM.

PALLENBERG, MAX (1877–1934), Austrian actor, who worked for some time with REINHARDT. A versatile and subtle comedian much given to extemporization, he was excellent in broad comedy but could also play tragi-comic and serious parts, as was proved by his performance as the Barker in MOLNÁR's *Liliom*, and as Schweik in PISCATOR's production of *Die Abenteuer des braven Soldaten Schweik* (based on a novel by the Czech author Hašek) in 1928.

PALLIATA, see FABULA (2).

PALMER, ALBERT MARSHMAN (1838–1905), American theatre manager, who controlled a number of New York theatres, among them the MADISON SQUARE, the UNION SQUARE, and WALLACK'S (1), to which he gave his own name. Well educated, and a man of refined tastes, he sought to rival DALY and Wallack, and built up in each of his theatres a good company. He did not, like Daly, create stars, but chose his actors wisely and set many on the road to fame. In his early years he followed Wallack's plan of importing plays from Europe, but later he encouraged new American playwrights, being responsible, among other productions, for the premières of Clyde FITCH's *Beau Brummell* (1890) and Augustus THOMAS's *Alabama* (1891). He retired in 1896 to become manager for Richard MANSFIELD, who had made his first appearance under him at the Union Square Theatre. He did a great deal for the American stage, and his influence on staging, backstage conditions, and the fostering of native talent was extremely beneficial.

PALMER, JOHN. There were two 18th-century actors of this name. The first, known as 'Gentleman' Palmer (1728–68), married the actress-daughter of Mrs PRITCHARD, and made something of a reputation as a player of small comic parts such as Brush in the elder COLMAN and GARRICK's *The Clandestine Marriage* (1766). He was a very vain man, proud of the looks and elegant bearing which earned him his nickname, and would no doubt have hated to be confused, as he often has been, with John Palmer (1742–98), son of the pit door-keeper at DRURY LANE. This younger John made his first appearance at the HAYMARKET THEATRE under FOOTE, and was engaged by Garrick in 1767 for Drury Lane, where he was the first to play Joseph Surface in SHERIDAN's *The School for Scandal* (1777). He was also good as Falstaff, as Sir Toby Belch in *Twelfth Night*, and in such 'impudent' parts as Captain Absolute in Sheridan's *The Rivals*, Young Wilding in Garrick's adaptation of SHIRLEY's *The Gamester*, and Dick Amlet in VANBRUGH's *The Confederacy*. Tragedy was beyond him, and he rarely attempted it. In 1787 he opened the ROYALTY THEATRE in Wellclose Square, but as he had no licence he was summonsed by the PATENT THEATRES and forced to close. He died on the stage, while acting in KOTZEBUE's *The Stranger* at Liverpool.

PALMER, LILLI, see HARRISON, REX.

PALMER'S THEATRE, New York, see WALLACK'S THEATRE (1).

PALSGRAVE'S MEN, see ADMIRAL'S MEN.

PANOPTICON, Royal, London, see ALHAMBRA.

PANTALONE, one of the stock characters, or 'masks', of the COMMEDIA DELL'ARTE, from which is derived the Pantaloon of the English HARLEQUINADE. In both capacities he retained his place in the scenario as the father or guardian of the young girl, opposing her marriage to the young lover because he wished (as her father) to marry her to one of his friends, or (as her guardian) to marry her himself. Invariably represented as an elderly Venetian, he wore a long black coat over a red suit, with Turkish slippers, a long pointed beard, and a skullcap. By turns avaricious, suspicious, amorous, and gullible, he was the butt of the ZANNI, who finally outwitted him and brought the young couple together. The character came into the English theatre through the visits of Italian comedians to London, and in Elizabethan times 'pantaloon' became a generic term for any old man, as in Shakespeare's reference to the sixth age of man (*As You Like It*, II.vii): 'the lean and slipper'd pantaloon With spectacles on nose and pouch on side'; but with the development of WEAVER's 'Italian Night Scenes' the name was again limited to a specific role. In the harlequinade he was the father or guardian of COLUMBINE, who, with the help of CLOWN, finally succeeded in eloping with HARLEQUIN. English Pantaloons soon discarded the long coat and the skullcap, and James Barnes, one of the earliest and most famous Pantaloons of the early 19th century, played the part in short striped knee-breeches, a matching jacket with a short cape, and a fringe of beard with a stiff pigtail sticking up behind.

PANTOMIME, a word which has drastically changed its meaning over the years. Derived from the Latin PANTOMIMUS, it originally meant in English 'a player of several different parts', but a misunderstanding of the art of the Roman exponent of pantomime, fostered by the entertainments put on by the Duchesse du Maine at Sceaux as *ballets-pantomimes* in the belief that she was reviving an ancient art, led to its adoption as a description of a story told in dancing only, with no words. Thus *The Loves of Mars and Venus* at DRURY LANE in Mar. 1717 was billed as a 'new dramatic entertainment after the manner of the ancient pantomime'. A confusion in the public mind between such ballets and the story-telling dances of the HARLEQUINADE led to the adoption of the term 'pantomime' for this offshoot of the COMMEDIA DELL' ARTE, which became so popular in England when performed by John RICH and others that it was eventually lengthened to provide a whole evening's entertainment. To lessen the burden on the dancers the harlequinade was introduced, first by a classical fable which gave some Immortal the opportunity of handing HARLEQUIN his magic wand, and later by a story from the chapbooks of the day which served the same purpose. In the 19th century this opening scene was elaborated (and incidentally gave rise to the fashionable BURLESQUE and EXTRAVAGANZA), and actresses, who had already proved how attractive they could be in BREECHES PARTS, were cast as the young heroes of the fairy-tales which began to take the place of the classical and other stories, the so-called PRINCIPAL BOYS. At the same time the comic elderly characters, who had for some time been played by men, became the prerogative of the knockabout comedian, and were henceforth known as DAMES. The success of the fairy-tale openings—*Cinderella*, *Babes in the Wood*, *Aladdin*, *Red Riding Hood*—caused them to be spun out for so long that the harlequinade was relegated to a short scene at the end, and although it lingered on in some theatres, particularly in the provinces, as a vestigial appendix for some time, it eventually disappeared altogether. Pantomimes in their new and entirely English form were from very early on associated with Christmas, and theatres all over the country vied with each other to produce more colourful, more spectacular, and funnier shows than their rivals. In London both Drury Lane and COVENT GARDEN produced splendid pantomimes, and those of the BRITANNIA at Hoxton also became famous. Most of them opened on Boxing Day (26 Dec.) and ran till March, and some actors played in nothing else. With the importation of speciality acts from the MUSIC-HALLS, which held up the action but delighted the audiences, the show became such a hotchpotch of incongruous elements—slapstick, romance, topical songs, male and female impersonation, acrobatics, splendid settings and costumes, precision and ballet dancing, trick scenery and transformation scenes—that for a time the phrase 'a proper pantomime' was used outside the theatre in colloquial English to signify 'a state of confusion'.

The word *pantomime* was also used in France for the wordless PIERROT plays of DEBURAU until the genre disappeared with the death in 1930 of Séverin. Its most famous production was *L'Enfant prodigue (The Prodigal Son)*, widely performed during the 1890s with a girl as Pierrot. Another meaning was also given to the word during the craze for MELODRAMA, when it signified the use of dumb show to convey ideas wordlessly. In this sense it is an important element in all acting and dancing, particularly ballet. Modern performers in dumb show, in order to distinguish their art from the popular idea of pantomime, describe it as MIME.

PANTOMIMUS, the name given to a performer popular in Imperial Rome, who by movement and gesture alone represented the different characters in a short scene based on classical history or mythology which was sung in Greek by the chorus accompanied by musicians—usually flutes, pipes, cymbals, and trumpets. The *pantomimus* wore the costume of the tragic actor—a long cloak and a silken tunic—and a mask with no mouthpiece, changing it when necessary. As many as five masks could be used in one scene. The most famous *pantomimi* were BATHYLLUS of Alexandria, PYLADES of Cilicia, and PARIS, who was put to death by NERO out of professional jealousy. The art of the *pantomimus* was considered by St Augustine more dangerous to morals than the Roman circus, since it dealt exclusively with guilty passions and by its beauty and seductiveness had a disastrous effect on female spectators, although unlike its rival the Roman MIME it was never coarse or vulgar.

PAPER BOYS' PLAY, see MUMMERS' PLAY.

PAPP [Papirofsky], JOSEPH (1921–), American theatre director, who had had over 10 years' theatrical experience, mostly backstage, before founding the New York Shakespeare Festival in 1954. His first productions were given in the Emanuel Presbyterian Church at 729 East 6th Street, most of the actors giving their services. Two years later Papp presented *Julius Caesar* and *The Taming of the Shrew* in the East River Park amphitheatre, using a portable stage mounted on a truck, but in 1957 the City of New York offered him the use of a site in Central Park, where *Romeo and Juliet*, *Macbeth*, and *The Two Gentlemen of Verona* were seen. The company occupied various open-air sites in the Park up to 1962, when a permanent home, the Delacorte Theatre, also in the open air, was built there for it. Financed by public and private donations, it seats 1,936, and as with all Papp's other ventures in the Park admission is free. It opened with a production, directed by Papp himself, of *The Merchant of Venice*, and achieved an enviable reputation, with the appearance of stars such as James Earl JONES, Julie HARRIS, and Colleen DEWHURST. Among its productions a musical version of *The Two Gentlemen of Verona* in 1971 was particularly successful, its first appearance being followed by a long run at the ST JAMES THEATRE. Meanwhile Papp had founded the PUBLIC THEATRE in 1967, and in 1973 he also became director of the VIVIAN BEAUMONT and the Mitzi E. Newhouse theatres in the Lincoln Center. A financial crisis in 1977 forced him to relinquish the management of the last two theatres, though he still retained control of his earlier projects. In 1980 another crisis compelled the Delacorte to stage only one production, a modernized version of GILBERT and Sullivan's *The Pirates of Penzance*. Fortunately it was an outstanding success, later having a long run at the URIS THEATRE, and Papp's difficulties were finally resolved by the award of a permanent subsidy from the City of New York. In 1981 there was once again a Shakespeare season at the Delacorte, with *Henry IV, Part I* and *The Tempest*.

PAPPUS, probably the most popular of the stock characters of the early Roman FABULA *atellana*, judging by the number of plays about him, among them *Pappus the Farmer*, *The Bride of Pappus*, *Pappus Defeated at the Poll*. All the information available about him shows that he was an old, rather foolish man, probably bald, with a straggling beard, who is constantly tricked by younger men and fails in everything he sets out to do, unlucky as an election candidate, as a lover, or as a husband.

PARADE, in the French theatre, the short sketch acted by fairground actors outside their booth in order to induce the spectators to pay their entrance-fees to see the play given inside. A volume of plots of *parades* acted on the first-floor balcony of the Théâtre de la Foire Saint-Germain (see FAIRS), destroyed in 1756, shows the affinity of the genre with the scenarios of the Italian COMMEDIA DELL'ARTE, but there is also evidence of direct survival from French medieval FARCE. The *parade* died out with the disappearance of the old theatres of the fairs in the mid-18th century but was revived by BOBÈCHE and Galimafré on the post-Revolution BOULEVARD DU TEMPLE. A vogue for this essentially popular form among the more sophisticated audiences of the private theatres fashionable in the second half of the 18th century was catered for by a number of dramatists of the time, among them BEAUMARCHAIS.

PARADISO, an elaborate piece of machinery invented by Filippo Brunelleschi (1377–1446), for the church of San Felice in Florence, where every year a representation of the Annunciation, or visit of the angel Gabriel to the Virgin Mary to foretell the birth of Christ, took place at the appropriate time. Choirboys, representing cherubim, were placed in a copper dome which was suspended from the roof of the church. This was then lowered by crane to a platform, and from among them emerged the actor who was playing Gabriel. When he had finished his part, he re-entered the dome and returned to heaven. This device was the forerunner of many similar mobile chariots, which were improved by such additions as the clouds of cotton wool with which Francesco d'Angelo (1447–88) masked the machinery needed for Christ's Ascension. Later, painted canvas, mounted on battens, took the place of cotton wool, but the basic principle of Brunelleschi's device remained in use for the transport of any supernatural being until almost the end of the 18th century.

PARALLEL, see ROSTRUM.

PARIGI, GIULIO and ALFONSO, see SCENERY.

PARIS. There were two popular Roman pantomime dancers (see PANTOMIMUS) of this name. One was executed under NERO (AD 67) because, it is said, Nero was jealous of his art, the other under Domitian (AD 83). For the latter the poet Statius wrote the libretto of a pantomime *Agave*, now lost.

PARIS, Théâtre de, Paris. Opening as the Nouveau Théâtre in 1891, this playhouse came under the direction of RÉJANE for 13 years from 1906, and assumed its present name when Léon Volterra took over in 1919. Among the plays he presented were those of Bataille, VERNEUIL, and Sacha GUITRY, and Marcel Pagnol's Marseilles trilogy *Marius* (1929), *Fanny* (1931), and *César* (1936). Volterra retired in 1948, and in 1953 Elvire Popesco, the wife of Verneuil, became director of the theatre in partnership with Hubert de Malet, renovating the building, and creating the intimate Théâtre Moderne upstairs in 1958.

PARKER, LOUIS NAPOLEON (1852–1944), English dramatist and pageant-master, who was for a time a music master at Sherborne School in Dorset. The success of his early plays enabled him to resign in 1892 and go to London, where he devoted the rest of his long life to the theatre. Among his numerous works the most successful were *Rosemary* (1896), written in collaboration with Murray Carson, who worked with him on a number of other plays, *Pomander Walk* and *Disraeli* (both 1911), the latter providing a fine part for George ARLISS, in which he had a long run in New York. It was seen in London in 1916 with Dennis Eadie in the title role. Parker was much in demand as a producer of the civic PAGEANTS so popular in Edwardian England, and directed those at Sherborne in 1905, at Warwick in 1906, at Dover in 1908, and at York in 1909. During the First World War he was responsible for several patriotic pageants in London, and in 1918 devised *The Pageant of Drury Lane*.

PARK LANE THEATRE, New York, see DALY'S SIXTY-THIRD STREET THEATRE.

PARK THEATRE, Glasgow, see CURTAIN THEATRE, GLASGOW; PITLOCHRY FESTIVAL THEATRE.

PARK THEATRE, London, see ALEXANDRA THEATRE.

PARK THEATRE, New York. (1) At 21–5 Park Row, the first important theatre of the United States, known as the 'Old Drury' of America. Built and managed by HALLAM and HODGKINSON, originally as the New Theatre, though its name was soon changed, it opened on 29 Jan. 1798 with *As You Like It*, followed on 31 Jan. by SHERIDAN's *The School for Scandal*. Prosperity came for a time with the engagement of Thomas Abthorpe COOPER, then on the threshold of a brilliant career, but in spite of a good company, and a series of popular plays, dissension among the managers, of whom the playwright DUNLAP was also one, prevented the theatre from prospering until in 1808 Stephen PRICE took over. Under him the first indigenous American drama, *The Indian Princess; or, La Belle Sauvage* by J. N. BARKER, was given on 14 June 1809 with a success which was later repeated in London at DRURY LANE. Price also engaged the young American actor John Howard PAYNE, whose first play had been given at the Park when he was 14, and added the parents of Edgar Allan Poe to his resident company. Unfortunately he also inaugurated the policy of importing foreign stars which eventually led to the decline of the STOCK COMPANY, the first being G. F. COOKE. Because of the war with England from 1812 to 1815 and the consequent difficulty in importing stars, it was not until 1818 that the stock system was finally abandoned and the company retained purely to support visiting stars. One of the greatest of them, Edmund KEAN, was booked to appear at the Park when on 1 Sept. 1820 it burnt down, and the company moved to the old ANTHONY STREET THEATRE, where Kean made his American début on 29 Nov. 1820 as Richard III. The Park meanwhile was rebuilt, and reopened on 1 Sept. 1821 for the era of its greatest prosperity and influence, practically every player of any importance appearing there. This period saw also the emergence of the American playwright, and the theatre no longer had to depend mainly on imported plays. Although the success in 1824 of the newly opened CHATHAM THEATRE (1) led to the introduction of vast spectacular shows like W.T. Moncrieff's *The Cataract of the Ganges*, the constant succession of guest artists helped to keep Shakespeare and other classic authors in the repertory. In the autumn of 1827 the Park was first lit by gas; two years later, during a bad financial panic, it closed for a time and was given over to masked balls and popular lectures, but it was redecorated and reopened in 1834. Three years later it had to contend with the rivalry of J.W.WALLACK at the NATIONAL THEATRE; it was then that the Park saw the first appearance of Jean Davenport LANDER, aged 11, believed to be the original of the Infant Phenomenon in DICKENS's *Nicholas Nickleby*, who was seen in a range of parts which included Sir Peter Teazle in *The School for Scandal*, Young Norval in HOME's *Douglas*, Shylock in *The Merchant of Venice*, and Richard III.

The Park was by now becoming outmoded, clinging to the old system of frequent changes of bill with the attendant evils of under-rehearsal and over-fatigue. It no longer attracted the best acting talent, and when Price, who had managed it for over 30 years, died suddenly on 20 Jan. 1840, it was found to be in a bad financial state. In June 1841 the building and its contents were put up for sale to pay the rent, and the company moved to the FRANKLIN THEATRE, the Park being used as a circus. In spite of an occasional flicker of its former brilliance, such as the three-week run of BOUCICAULT's first play *London Assurance* in Oct. 1841 and a visit by MACREADY in 1843, it became more and more of a liability. In 1848 it was taken over by HAMBLIN, who ran it in conjunction with the BOWERY (1); but in spite of extensive rebuilding he could not restore its prosperity, and when on 16 Dec. 1848 it was once again destroyed by fire it was not rebuilt.

(2) The second Park Theatre was the first professional playhouse to be built in Brooklyn, after nearly 40 years of unsuccessful tentatives. It opened on 14 Sept. 1863 under F. B. CONWAY, who remained there until 1871 and established the theatre firmly in the affections of its audience, maintaining a good company and frequently importing stars for short seasons. After his departure it languished for a while, and in spite of appearances by a succession of stars in 1873–4, it was eventually given over to visits from touring companies. At the beginning of 1876 it fell a victim to the current craze for VARIETY and BURLESQUE, but after the burning of the BROOKLYN THEATRE on 5 Dec. it regained the local monopoly of legitimate drama. The last theatre in New York to have a stock company supporting visiting stars, it was burnt down in Nov. 1908 and not rebuilt.

From 1889 to 1894 the HERALD SQUARE THEATRE was known as the Park, as was the MAJESTIC THEATRE for some years from 1911.

PARSELLE, JOHN, see TAPPING, FLORENCE.

PASADENA PLAYHOUSE, Pasadena, California, one of the first community theatres in the United States, founded in 1918, when the Pasadena Community Playhouse Association was organized on an amateur basis. The group opened its own playhouse, seating 820, on 18 May 1925, and it was followed in 1928 by a College of Theatre Arts, originally 'School of the Theatre'. The Playhouse, which became an important training ground for stage and screen actors, operated four stages, a touring company, and radio and television studios. All the plays of Shakespeare were performed, and the first performances of many new plays were given. The Playhouse's high reputation could not, however, prevent its closure in 1970 for financial reasons.

PASHENNAYA, VERA NIKOLAYEVNA (1887–1962), Soviet actress, who joined the MALY THEATRE company in 1905 and remained there until 1922. She later went to the MOSCOW ART THEATRE, and in 1941 was appointed to the Chair of Acting in the Shchepkin Theatre School. She appeared in a wide range of parts, including SCHILLER's Maria Stuart, Emilia in *Othello*, and Tanya in TOLSTOY's *Fruits of Enlightenment*. Her conception of Vassya Zheleznova in GORKY's play of that name was characterized by psychological insight, and all her roles gave proof of dramatic force and inner concentration. In 1955 she celebrated her jubilee as an actress and a year later she gave a memorable performance of Praskovya Sharabai in SOFRONOV's *Money* at the Maly. In 1962 she was equally successful in the same author's *Honesty*.

PASO, term applied in 16th-century Spain to a short comic scene which later developed into the ENTREMÉS. It depended for its appeal on a simple plot, quick-fire dialogue and repartee, and the use of a few well known types from the COMMEDIA DELL'ARTE. One of the chief characters was the BOBO or rustic clown, who became the GRACIOSO of the entremés. The best known writer of *pasos* was Lope de RUEDA.

PASO, ALFONSO (1926–), Spanish playwright, author of over 100 plays, who began his career by collaborating with his father Antonio Paso (1870–1958), also a prolific minor playwright, in a series of light comedies. He was the co-founder with SASTRE of the experimental Arte Nuevo group, but soon abandoned its innovatory work to become Spain's most commercially successful writer for the stage, producing as many as five new plays in a season, mostly light comedies and comedy-thrillers.

PASQUATI, GIULIO, COMMEDIA DELL'ARTE actor active in the second half of the 16th century. An early and well-known interpreter of the mask of PANTALONE, he was associated with the GELOSI, and it was apparently his excellent acting that resulted in the company's being invited to act before the French Court in 1577.

PASQUINO, PASQUIN, one of the minor ZANNI or servant roles of the COMMEDIA DELL'ARTE. The name was adopted towards the end of the 16th century and passed into French comedy as Pasquin, being the name of the valet in the plays of DESTOUCHES. In the French 17th-century theatre the expression 'the Pasquin of the company' designated the actor who played the satiric roles in REGNARD and DUFRESNY. In the 18th century the word 'pasquin' or 'pasquinade' was applied in England to a lampoon, squib, or satiric piece, often political. Henry FIELDING used it as the title of a production at the HAYMARKET in 1736 and often as a journalistic pseudonym.

PASS DOOR, a fireproof door placed in an inconspicuous part of the proscenium wall, leading from the auditorium to the side of the stage and so backstage. It is usually used only by those connected with the theatre, ordinary members of the audience penetrating behind the scenes by means of the STAGE DOOR which opens on to the street.

PASSION PLAY, a medieval religious drama in the vernacular, which dealt with the events from the Last Supper to the Crucifixion, unlike the MYSTERY PLAY which presented in cyclic form Bible history from the Creation to the Second Coming. The establishment of the feast of Corpus Christi (the Thursday after Trinity Sunday) in 1313 gave a great impetus to the enactment of Passion plays in large cities throughout Europe, and also led to a tradition of open-air productions of the Good Friday story in many small towns and villages, all the actors being drawn from the local community. Most of these died out during the 15th century, but, helped by the Catholic counter-Reformation in the following century, a few were revived in Switzerland, Austria, and Germany. The only one to have become well known, among the few that still survive, is that given decennially since 1634 at Oberammergau in Bavaria. A fusion of two Augsburg plays, this was first performed during a visitation of plague and remains entirely amateur, the villagers dividing the parts among themselves and, as in earlier times, being responsible also for the production, music, costumes, and scenery. There have been several revisions in recent times of the Oberammergau text, caused mainly by the desire to muffle its anti-Semitism, and it still remains a subject of controversy.

PASTOR, TONY [Antonio] (1837–1908), American VAUDEVILLE performer and manager. He made his first appearance as a child with Barnum, and later travelled with circuses and MINSTREL SHOWS as a singing clown, ring-master, or 'blackface' ballad-singer. In 1861 he made his début in VARIETY, which had sunk to a low level of vulgarity. Determined to clean it up, he opened a theatre at 199–200 Broadway in 1865, moving ten years later to Buckley's Minstrel Hall, later the OLYMPIC THEATRE (3), which he renamed Tony Pastor's Opera House. He banned drinking and smoking and discouraged the more vulgar acts, appearing always in his own shows—he had a rich tenor voice well suited to the popular ballads of the day, of which he himself claimed to have composed over 2,000. In 1881 he presented the first performance of what was later called Vaudeville at his newly acquired Fourteenth Street Theatre, which he ran until his death, many famous stars including WEBER AND FIELDS and Lillian Russell, appearing there.

PASTORAL, dramatic form which evolved in Italy from pastoral poetry by way of the dramatic eclogue or shepherds' play. There were pastoral elements in early plays, notably in POLIZIANO'S *Orfeo* (1472) and BECCARI'S *Il Sacrificio* (1554); but the first true pastoral was TASSO'S *Aminta* (1573), which was widely imitated. Another important work was GUARINI'S *Il pastor fido* (1596–8), which was seen in London in 1602, and had a great influence, particularly on LYLY and Samuel DANIEL. The best of the many reworkings of inherently repetitive themes, and the representative pastoral of the next generation, was BONARELLI'S *Filli di Sciro* (1607). The first English pastoral was the charming and poetic *The Faithful Shepherdess* (1608) by FLETCHER. The genre never became acclimatized in England, though it had some influence on the MASQUE and on Shakespeare, perhaps because being purely artificial it cannot flourish alongside romantic drama. It was more at home in France, where it inspired a number of dramatists, among them HARDY, RACAN, and above all MAIRET, whose *Sylvie* (1626) is generally regarded as the finest French pastoral, and whose *Silvanire* (1630) served as a vehicle for the introduction of the UNITIES into French drama. Although the vogue for the pastoral was short-lived, its introspective lovers, who analyse their feelings at much length, stimulated the type of comedy written in France in the 1630s by such authors as CORNEILLE.

'PATAPHYSICS, see JARRY, ALFRED.

PATENT THEATRES, the THEATRES ROYAL, DRURY LANE and COVENT GARDEN, which operate under Letters Patent, or Charters, given by Charles II in 1662 to Thomas KILLIGREW for Drury Lane and Sir William DAVENANT for LINCOLN'S INN FIELDS, whence it descended in 1732 via DORSET GARDEN to Covent Garden. These were for the establishment of two companies to be known as 'the King's Servants' and 'the Duke of York's Servants' respectively. The company at Drury Lane was technically part of the Royal Household, and some members of it were sworn in as Grooms of the Chamber, being given an allowance of scarlet cloth and gold lace for their liveries. The charters, which in the course of the next 200 years changed hands many times fluctuating sharply in value, are still in existence and form an integral part of the leases of the theatres. They rendered the two theatres independent of the LORD CHAMBERLAIN as licenser of theatre buildings, though until the abolition of the theatre CENSORSHIP in 1968 they still remained accountable to him for the plays which they put on. The monopoly in 'serious' or 'legitimate' acting established by Charles II and reinforced by the Theatres Act of 1737 was finally broken in 1843.

PATERNOSTER PLAY, see LITURGICAL DRAMA.

PATTERSON, TOM, see STRATFORD (ONTARIO) FESTIVAL.

PAVILION, The, New York, see ANTHONY STREET THEATRE and CHATHAM THEATRE (1).

PAVILION MUSIC-HALL, London, see LONDON PAVILION.

PAVILION THEATRE, London, in the Whitechapel Road. The first theatre on this site opened on 10 Nov. 1828, and a year later saw several appearances by Fanny Clifton, later Mrs STIRLING. The building was destroyed by fire on 13 Feb. 1856, but was rebuilt with a capacity of 3,500. Under the management of Morris Abrahams, who became its manager in 1871, it catered with great success for the large Jewish population of the neighbourhood and was rebuilt in 1874, its three-tier auditorium then holding about 2,500. It was altered again in 1894 for Isaac Cohen, who presented a number of successful plays, including several by Sholom ALEICHEM, and also added an annual Christmas PANTOMIME to the theatre's repertory. It closed in Feb. 1933, and stood empty for many years, being severely damaged by bombing during the Second World War and finally demolished in 1961.

PAVY, SALATHIEL, see BOY COMPANIES.

PAXINOU [*née* Konstantopoulou], KATINA (1900–73), Greek actress, who in 1940 married as her second husband Alexis MINOTIS, under whose direction she frequently acted. Trained as a singer, she made her début as an actress in 1924, and was first seen in New York in 1930, one of her roles being Clytemnestra in SOPHOCLES' *Electra*. In 1932 she joined the Greek National Theatre company, playing Clytemnestra again in AESCHYLUS' *Agamemnon* and becoming the company's leading lady. She translated and directed several English-language plays, among them Eugene O'NEILL'S *Anna Christie*, in which she played the title role, in 1932. One of her finest parts was Mrs Alving in IBSEN'S *Ghosts*, in which she performed annually for six years beginning in 1934. She was also much admired as Phaedra in EURIPIDES' *Hippolytus*, Lady Windermere in WILDE'S *Lady Windermere's Fan*, both in 1937, Goneril in *King Lear*, and Mrs Chevely in Wilde's *An Ideal Husband*, both in 1938. In 1939 she made her first appearance in London as Sophocles' Electra, followed by Gertrude in *Hamlet*, in which Minotis played the title role. In the following year she gave her first performance in English, also in London, as Mrs Alving, and in 1942 played Ibsen's Hedda Gabler in English in New York. After some years in films she returned to the stage in New York in 1951 as Bernarda in GARCÍA LORCA'S *The House of Bernarda Alba*, a part in which she was seen in Greece in 1954, appearing there in 1957 in FRY'S *The Dark is Light Enough*. She was seen again in London in the 1966 WORLD THEATRE SEASON, playing Jocasta in Sophocles' *Oedipus Rex* to the Oedipus of

Minotis, and the title role in Euripides' *Hecuba* to his Talthybius.

PAYNE, BEN IDEN (1881–1976), English actor and director, best remembered for his work in connection with Shakespeare. He was with BENSON's company in 1899 and later helped Miss HORNIMAN to establish her repertory theatre in Manchester, appearing in many of the plays himself, and acting as her general manager from 1907 to 1911. In 1913 he went to the United States, where he worked in the Philadelphia and Chicago Little Theatres, and from 1919 to 1934 was visiting Professor of Drama at the Carnegie Institute of Technology. He then returned to England to become director of the SHAKESPEARE MEMORIAL THEATRE at Stratford-upon-Avon in succession to BRIDGES-ADAMS. After guiding the fortunes of the theatre through the difficult first years of the Second World War, he left in 1943 to lecture in America for the Ministry of Information, and remained there to become visiting professor at several universities, finally settling at the University of Texas in 1946. In 1950 he inaugurated a summer SHAKESPEARE FESTIVAL at Balboa Park, San Diego. By his first wife, the actress Mona Limerick [Mary Charlotte Louise Gadney], he had a daughter, Rosalind, also an actress, who married Donald WOLFIT.

PAYNE, JOHN HOWARD (1791–1852), American actor and dramatist, who as a precocious 14-year-old had his first play *Julia; or, the Wanderer* produced in New York. Three years later he went on the stage and made a great reputation, touring the larger American cities in such parts as Hamlet, Romeo, Young Norval in HOME's *Douglas*, and Rolla in KOTZEBUE's *Pizarro*. In 1811 he appeared at the CHESTNUT STREET THEATRE, Philadelphia, as Frederick in his own version of Kotzebue's *Das Kind der Liebe*, which, as *Lovers' Vows,* was already well known in English translations by Mrs INCHBALD and Benjamin Thompson. Two years later he sailed for England, and appeared at DRURY LANE with great success. A visit to Paris brought him the friendship of TALMA and the freedom of the COMÉDIE-FRANÇAISE, and for many years he was engaged in the translation and adaptation of current French successes for the English and American stages. Of his own plays—he has been credited with some 50 or 60—the best was *Brutus; or, the Fall of Tarquin* (1818), in which Edmund KEAN was seen at Drury Lane; he adapted several MELODRAMAS by PIXÉRÉCOURT, of which the most successful was *Therese; or, the Orphan of Geneva* (1821), which provided him with the money to pay off the debts incurred by his unsuccessful attempt to manage SADLER'S WELLS THEATRE. With Washington IRVING, a lifelong friend, he also adapted Duval's *La Jeunesse de Henry V* (1806) as *Charles II; or, the Merry Monarch* (1824). In 1842 he was appointed American Consul at Tunis, where he died. He is now best remembered for the lyric of 'Home, Sweet Home', part of the libretto which he wrote for Henry Bishop's opera *Clari, the Maid of Milan*, first performed at COVENT GARDEN in 1823.

PEACOCK THEATRE, Dublin, see ABBEY THEATRE.

PEDROLINO, one of the comic ZANNI or servant masks of the COMMEDIA DELL'ARTE, of which Giovanni PELLESINI was the best known exponent. The character had in it something of PULCINELLA, but is chiefly important for its later transformation into the French PIERROT.

PEELE, GEORGE (*c.*1558–*c.*1597), English dramatist, of good family and well educated—a translation by him of EURIPIDES' *Iphigenia* was acted in Oxford, probably at Christ Church, in about 1588—but a shiftless, dissolute man, the boon companion of GREENE, NASHE, and MARLOWE, who lived by his wits and by such sums as he could earn by playwriting and play-acting. He was the author of several civic pageants and of a classical comedy *The Arraignment of Paris*, acted at Court by one of the BOY COMPANIES probably in Dec. 1581; of *David and Bethsabe* (*c.* 1587); and of a chronicle play *Edward I* (1591), which has survived only in a mutilated form. His best and best known work *The Old Wives' Tale*, probably written and produced in about 1590 and published in 1595, is a mixture of high romance and English folk-tale, long dismissed by critics as negligible and pretentious nonsense. It is now recognized as a landmark in the development of English comedy, bringing a new and more subtle strain of humour into the farce of earlier days. Peele, who wrote some charming lyrics, may have been the author of the play on which Shakespeare based his *King John*, and may have written part of the first act of *Titus Andronicus*. Several other plays, some in collaboration with Greene, have also been tentatively assigned to him, including *Locrine* and *Jack Straw* (both 1591).

PEKING OPERA, see CHINA.

PÉLISSIER, HARRY, see COMPTON, FAY, and REVUE.

PELLESINI, GIOVANNI (*c.*1526–1612), actor of the COMMEDIA DELL'ARTE, who by 1576 had a company of his own but shortly afterwards joined the CONFIDENTI, becoming its director and the husband of its leading lady Vittoria Piissimi (*fl.* 1575–94), whom he may have married about 1582. He played the ZANNI role PEDROLINO, from which the later character of PIERROT was derived.

PELLICO, SILVIO (1789–1854), Italian writer and patriot, best known for his autobiography *Le mie prigioni* (*My Prisons*, 1832), a moving account of his ten-year imprisonment for political activities. He was also the author of several plays, mainly tragedies modelled on those of ALFIERI and inspired by the ideals of the Risorgimento. The most successful in its day was *Francesca da Rimini* (1815), in which the actor Luigi DOMENICONI created the part of Paolo. Francesca became a favourite part with many later actresses, notably Adelaide RISTORI, but the play was eventually eclipsed by the more richly lyrical treatment accorded it by D'ANNUNZIO in his version of the story, first seen in 1901.

PELLIO, TITUS PUBLILIUS, actor in Rome who over a long period directed and acted in the plays of PLAUTUS. He is referred to in the latter's *Menaechmi* as being already famous for his acting, and according to a note made in about 200 BC he directed and acted in the *Stichus*, also by Plautus. An allusion in the *Bacchides*, which was acted at some time after 186 BC, is usually taken to mean that he was appearing as Pistoclerus in that play, as well as having at some time earlier played in the *Epidicus*.

PEMÁN Y PEMARTÍN, JOSÉ MARÍA (1898–1981), Spanish dramatist, of strong conservative views, whose plays endeavour to reaffirm traditional Catholic and monarchist values. Typical of his historical verse dramas is *Cuando las Cortes de Cádiz* (*At the Time of the Cadiz Parliament*, 1934). Among his later works one of the most interesting is *Callados como muertos* (*Silent as the Dead*, 1952). He also made a number of translations and adaptations, among them a Spanish version of SOPHOCLES' *Oedipus the King*, performed in the ruins of Sagunto.

PEMBROKE'S MEN, company under the patronage of the 1st Earl of Pembroke, first mentioned in 1592. Among the play-books which the actors sold to the booksellers a year later were MARLOWE's *Edward II* and two anonymous plays, *The Taming of a Shrew* and *The True Tragedy of Richard Duke of York*. Shakespeare's *The Taming of the Shrew* and *Richard III* may have been revisions of these last two. Pembroke's Men are also named together with SUSSEX'S MEN on the title-page of *Titus Andronicus*, which Shakespeare refashioned from *Titus and Vespasian*. The 2nd and 3rd parts of his *Henry VI* were also in their repertory, the first part having been written for STRANGE'S MEN. Shakespeare's connection with Pembroke's Men ceased when he joined the CHAMBERLAIN'S MEN on their foundation in 1594, and the company underwent an eclipse until 1597 when they leased the newly-built SWAN THEATRE. They were soon in trouble over their production of NASHE's and Ben JONSON's satire *The Isle of Dogs*. This caused the group to disband, and some of the leading actors joined the ADMIRAL'S MEN at the ROSE THEATRE under HENSLOWE. The others went into the provinces and may have made an unsuccessful attempt to return to London in about 1600. No further record of them is known, but they may have joined a short-lived company belonging to the Earl of Worcester which formed the basis of QUEEN ANNE'S MEN, founded in 1603.

PENCIULESCU, RADU, see ROMANIA.

PENÇO DE LA VEGA, JOSEPH, see JEWISH DRAMA.

PENKETHMAN [Pinkethman], WILLIAM (?–1725), English comedian, whose early years are obscure, though he is known to have managed a theatre in Richmond and to have had a booth at Bartholomew Fair. He also toured the countryside with peepshows and scenic automata introduced from the Continent. He may have played small comic parts under BETTERTON in about 1682, but he first came into prominence when he played a masked and speaking HARLEQUIN in Aphra BEHN's *The Emperor of the Moon* (1687), and he probably took the part of Stitchem, a tailor, in SHADWELL's *The Volunteers; or, the Stock Jobbers* (1692). He was a member of the company at DRURY LANE under Colley CIBBER, who wrote for him the parts of Don Lewis in *Love Makes a Man; or, the Fop's Fortune* (1701) and Trappanti in *She Would and She Would Not; or, the Kind Imposter* (1702). He continued acting until a year before his death.

PENLEY, WILLIAM SYDNEY (1852–1912), English actor-manager, who made his first appearance on the stage at the ROYAL COURT THEATRE in Dec. 1871 in farce, and then toured in light and comic opera. He was for some years at the STRAND THEATRE, playing burlesque under Mrs Swanborough. The first outstanding success of his career came when he succeeded TREE in the title role of HAWTREY's *The Private Secretary* (1884), a part with which he became so identified that he is usually believed to have been the first to play it. He was also closely connected with Brandon THOMAS's *Charley's Aunt* (1892), playing Lord Fancourt Babberley during its run of 1,466 performances, a record for the period. In 1900 he opened the former Novelty Theatre (later the KINGSWAY) as the Great Queen Street Theatre and appeared there in revivals of his most successful parts, retiring a year later. Much of his success as a comedian lay in his dry humour, his serious, rather pathetic, face, and the solemnity of his voice and manner contrasted with the farcical lines of his part.

PENN, WILLIAM, see BOY COMPANIES.

PENNY GAFF, see GAFF.

PENNY PLAIN, TWOPENCE COLOURED, see TOY THEATRE.

PENTHOUSE THEATRE, Seattle, see THEATRE-IN-THE-ROUND.

PEOPLE SHOW, experimental theatre group formed in London in 1966 by Mark Long and others, which took its name from its first productions, written by Jeff Nuttall, who in its early years provided the group with partial scripts serving as themes to be developed in performance, following the methods of COLLECTIVE CREATION. The members worked without writer or director in an effort to develop individual creativity and variety within their productions. The early productions were planned specifically for the environment in which they were to take place, while later productions were made to tour. Each new production is improvised within a framework to which each performer contributes his own images, character, and materials. By 1980 the group had created more than 80 productions and had toured widely throughout Western Europe and the U.S.A.

PEOPLE'S NATIONAL THEATRE, London, see PRICE, NANCY.

PEPPER'S GHOST, device by which a ghost can be made to appear on stage, side by side with an actual person. Based, roughly, on the principle that a sheet of glass can be both reflective and transparent, it was perfected and patented, after several earlier attempts, by 'Professor' J. H. Pepper, a director of the Royal Polytechnic Institution in London, a place of popular scientific entertainment in the second half of the 19th century. It was first shown privately at the Polytechnic on 24 Dec. 1862, and then exhibited publicly with great success, Charles DICKENS using it in connection with readings of *The Haunted Man*. It was first used in the theatre—it had already appeared at several MUSIC-HALLS—on 6 Apr. 1863 at the BRITANNIA, and several plays given there were specially written for it. It was never widely adopted on the stage, perhaps because it was difficult to manipulate, and also impossible for the ghost to speak. This made it useless for such plays as BOUCICAULT'S *The Corsican Brothers*, where the Ghost has to speak in the last act. The device did, however, enjoy a semi-dramatic vogue in the Ghost shows based on popular melodramas which toured the FAIRS until the early 20th century.

PEPYS, SAMUEL (1633–1703), English naval administrator under Charles II and James II, whose *Diary* contains a good deal of information on the London theatre in the early days of the Restoration. In it he recorded the plays he saw and the actors in them, noted his opinion of them, and related stray items of backstage gossip retailed to him by his friends in the theatre, of whom one of the closest was Mary KNEPP. To him we owe many illuminating glimpses of the green room, of the theatre under reconstruction, and of the rowdy talkative audiences of his day.

PERCH PLATFORM, see LIGHTING.

PERCIVAL, SUSANNA, see MOUNTFORT, WILLIAM.

PERCY, ESMÉ SAVILLE (1887–1957), English actor, particularly admired for his work in Shakespeare and SHAW. He studied for the stage under Sarah BERNHARDT, and made his first appearance in BENSON'S company in 1904, later touring with him in a number of leading Shakespearian roles. In 1908 he was with GRANVILLE-

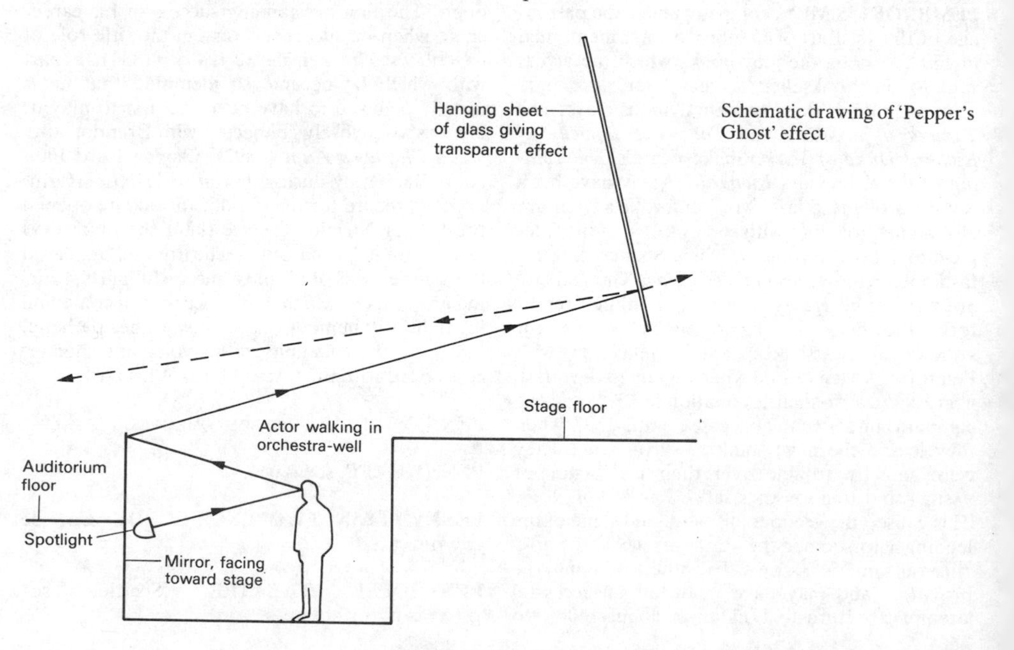

Schematic drawing of 'Pepper's Ghost' effect

BARKER at the ROYAL COURT THEATRE, and then joined Miss HORNIMAN's repertory company in Manchester. During and after the First World War he was in charge of entertainment for the troops, producing over 140 plays. He returned to London in 1923, and remained active in the theatre until his sudden death, which occurred shortly before the first night of Nigel Dennis's *The Making of Moo* (1957), in which he was to have appeared. He was a recognized authority on Shaw and from 1924 to 1928, with the Charles Macdona Players, appeared in a wide variety of Shavian parts, including the male leads in *Man and Superman, Androcles and the Lion, The Doctor's Dilemma, Pygmalion*, and *The Apple Cart*. He also directed the Hell scene (Act III, sc. 2) from *Man and Superman* as a separate production—*Don Juan in Hell*—at the LITTLE THEATRE in 1928, playing Don Juan himself. He did not, however, confine himself to Shaw, but was an actor of great versatility, playing Hamlet, Peter Quince in *A Midsummer Night's Dream*, Gaev in CHEKHOV's *The Cherry Orchard*, and Hotspur in *Henry IV, Part 1* with equal effectiveness. Among his later parts were Humpty Dumpty in Nancy PRICE's adaptation of Lewis Carroll's *Alice Through the Looking-Glass* in 1947, and Matthew Skipps in FRY's *The Lady's Not For Burning* (1949), of which he was co-director.

PEREGRINUS, see LITURGICAL DRAMA.

PERETZ, ISAAC LEIB (1852–1915), Jewish writer and lawyer. Born in Poland, he became secretary to the Warsaw Jewish community, and in 1876 published his first work, a volume of poems in Hebrew. Later, however, he turned to Yiddish as his medium. Similarly, in his philosophy—there is a definite struggle of ideas in Peretz—he changed from the rationalism of the Haskala (or Enlightenment) Movement to the symbolism and mysticism of Hassidism, which henceforth permeated his whole outlook. Some of his short stories dealing with life in Hassidic villages were dramatized and produced by OHEL on its first public appearance in Tel-Aviv in 1926. His dramatic poem *Night in the Old Market* was dramatized by GRANOVSKY and produced in 1925 at the MOSCOW STATE JEWISH THEATRE.

PERÉZ, COSMÉ (*c.* 1585–1673), Spanish comic actor, and a celebrated GRACIOSO, best remembered for his portrayal of Juan Rana, a character invented by CERVANTES who appears in a number of ENTREMÉSES, particularly those of QUIÑONES DE BENAVENTE, where he is represented as a rustic lawyer or doctor addicted to most of the seven deadly sins, above all, avarice and gluttony. Lady Anne Fanshawe, wife of the British Ambassador in Madrid, records in her memoirs Peréz's appearance in an entertainment at the Spanish Court in 1665, when he was said to be about 80 years of age.

PÉREZ GALDÓS, BENITO (1843–1920), the greatest Spanish novelist of the 19th century, whose works stand comparison with those of BALZAC or DICKENS (whose *Pickwick Papers* he translated from French into Spanish). In the 1890s he turned to the theatre, which had fascinated him in his youth, writing original plays and also dramatizing several of his own novels, among them *Realidad* (*Reality*) in 1892, *La loca de la casa* (*The Madwoman of the House*) in 1893, *Doña Perfecta* in 1896, and in 1904 *El abuelo* (*The Grandfather*), which, like IBSEN's *Ghosts* (1881), deals with the problems of heredity. Among his other plays *Electra* (1901) was considered anti-clerical and therefore highly controversial. Although the dialogue in his novels is vivid and revealing his plays tend to be unconvincing and stilted, for he lacked a sure stage-sense; they are none the less interesting in showing the influence on the contemporary Spanish stage of Ibsen and ZOLA.

PERFORMANCE GROUP, New York company founded in 1967 by Richard SCHECHNER to experiment with environmental theatre, group processes, training techniques, and performer-audience relationships. Most of its productions were developed by COLLECTIVE CREATION. The first, *Dionysus in 69* (1968), was a textual montage, loosely based on EURIPIDES' *Bacchae*, improvised by the performers under workshop conditions and providing in performance several opportunities for participation by the spectators. *Makbeth* (1969), a restructuring of Shakespeare's play, made the audience into members of the Scottish Court and spectators of the banquet at which Duncan became food for his children Cawdor, Malcolm, Macduff, and Banquo. Similarly, in a production of BRECHT's *Mother Courage* (1975), supper was served to the audience as part of the play, and two scenes were played in the street, watched by the spectators through an open door. In 1970 Schechner withdrew, and the group then reformed as the Wooster Group, led by Spalding Gray and Elizabeth LeCompte, authors of an experimental trilogy, *Three Places in Rhode Island* (1977), of which the centre piece, *Rumstick Road*, was revived in 1980 as a separate entity.

PERIAKTOI, scenic devices used in the Roman, and perhaps earlier in the Hellenistic, theatre. According to VITRUVIUS, they were triangular prisms set on each side of the stage which could be revolved on their axes to indicate a change of scene, each of their three surfaces bearing an indication of a locality, as waves for the sea, ships for a harbour, trees for a wood. The publication of Vitruvius's treatise in 1511 led to the adoption and improvement of the PERIAKTOI by Renaissance theatre architects, particularly by Bastiano Da San Gallo (1481–1551), who increased their size and their number, placing several one behind the other on each side of the stage, and providing removable painted canvas panels for each of the three faces of the prism, so as to make possible a greater number of variations in the scenery. These improved *periaktoi* were later known as *telari*.

PERKINS, (James Ridley) OSGOOD (1892–1937), American actor, who, after seeing active service in the First World War, made his first appearance on the stage in KAUFMAN and CONNELLY's *Beggar on Horseback* (1924). His performance as Walter Burns, the archetypal cynical newspaper editor in Hecht and MacArthur's *The Front Page* (1928), confirmed his reputation as a skilled comic actor, noted for his thin, expressive face and mobile hands; but he was also a sympathetic Astrov in CHEKHOV's *Uncle Vanya* in 1930. He played Sganarelle in MOLIÈRE's *The School for Husbands* (1933), supported the LUNTS in Noël COWARD's *Point Valaine* (1935), and was in S. N. BEHRMAN's *End of Summer* (1936). He was to have been the leading man in Rachel CROTHERS's *Susan and God*, but died of a heart attack soon after its first performance in Washington.

PERMANENT SETTING, see DETAIL SCENERY and FULL SCENERY.

PERUZZI, BALDASSARE (1481–1537), Italian scene designer, who was the first to apply the science of perspective to theatrical scenery. His flat work was as convincing as his BUILT STUFF. He was responsible for the scenery of Cardinal Bibbiena's *Calandria* when it was given in Rome in 1507, the year after its first production in Urbino. He had a great influence on the development of scenery in Italy, one of his oustanding pupils being SERLIO.

PERZYŃSKI, WŁODZIMIERZ, see POLAND.

PETIT-BOURBON, Salle du, Paris, the first Court Theatre of France, in the gallery of the palace of the Dukes of Bourbon. A long, finely proportioned room, with a stage at one end, it had formerly been used for balls and ballets, and the first recorded professional company to play in it was the COMMEDIA DELL'ARTE troupe the GELOSI in May 1577. It was often used by visiting Italian companies thereafter, and in 1604 the famous Isabella ANDREINI made her last appearance there. In 1645 Mazarin invited the great Italian scene-painter and machinist TORELLI to supervise the production of opera in the theatre, and in 1658, when it was again in the possession of a *commedia dell'arte* troupe under FIORILLO, MOLIÈRE's company, fresh from the provinces, was allowed to share it with them. For this privilege Molière paid a heavy rent and was allotted the less profitable days—Mondays, Wednesdays, Thursdays, and Saturdays—the Italians keeping the more lucrative Tuesdays and Sundays for themselves. He opened on 2 Nov. 1658 with five plays by CORNEILLE in quick succession, and not until the end of the month did he put on one of his own farces, *L'Étourdi*, followed by *Le Dépit amoureux*. The Petit-Bourbon also saw the first nights of *Les Précieuses ridicules* (1659) and *Sganarelle* (1660), before it was suddenly scheduled for demolition, and the company, in the full tide of their success—the Italians having already left—found itself homeless. Louis XIV gave them Richelieu's disused theatre in the PALAIS-ROYAL and the Petit-Bourbon disappeared, Molière taking the boxes and fittings with him and Torelli's scenery and machinery being given to VIGARANI.

PETITCLAIR, PIERRE, see CANADA.

PETRI, OLAUS (1493–1552), Swedish humanist, translator of the Bible and from 1531 to 1533 Chancellor to Gustavus Vasa, king of Sweden from 1523 to 1560. Petri is usually credited with the authorship of the first Swedish play in the vernacular, *Tobiae Comedia* published in 1550, and was the subject of STRINDBERG's first important play, *Mäster Olof* (1872).

PETRUSCU, CAMIL, see ROMANIA.

PHELPS, SAMUEL (1804–78), English actor and manager, who toured the provinces for several years, making a name for himself as a tragedian, and in the summer of 1837 made his first appearances in London at the HAYMARKET THEATRE under Ben WEBSTER, playing Hamlet, Richard III, Othello, and Shylock in *The Merchant of Venice* with considerable success. In the autumn of that year he joined the company at DRURY LANE, where he appeared with MACREADY in a number of plays by Shakespeare and also created the roles of Captain Channel in a then very popular play by Douglas JERROLD, *The Prisoner of War* (1842), and Lord Trensham in BROWNING's *A Blot in the 'Scutcheon* (1843). In 1843 he took advantage of the abolition of the PATENT THEATRES' monopoly in serious drama to assume the management of SADLER'S WELLS THEATRE, which he made for the first time a permanent home of Shakespeare, though other plays also figured in his repertory. With his fine and imaginative productions he did much to redeem the English stage from the triviality into which it had fallen, though he needed all his fortitude and obstinacy to maintain a consistently high standard in the face of much opposition. The plays he directed during his long tenancy of the theatre were remarkable for their scenic beauty, but he never succumbed to the prevailing desire for mere exhibition and pageantry, always making sure that the author's text was paramount. By the time he retired in 1862 he had produced all Shakespeare's plays except *Henry VI, Titus Andronicus, Troilus and Cressida*, and *Richard II*, the most successful productions being *Macbeth* in 1844 and *Antony and Cleopatra* in 1849. In 1854 *Pericles* was seen in its original form for the first time since the Restoration. He appeared in most of the plays himself, his best parts being considered Lear and Othello. He was not suited to romantic comedy, his manner being cold, with a somewhat harsh voice; he was, however, good in pathetic parts, and in spite of being mainly a tragedian gave an excellent comic performance as Bottom in *A Midsummer Night's Dream*. He was also much admired as Sir Pertinax McSycophant in MACKLIN's *The Man of the World*. After leaving Sadler's Wells he appeared in London and the provinces in Shakespeare and in dramatizations of SCOTT's novels. He remained on the stage until almost the end of his life, making his last appearance as Wolsey in *Henry VIII* on 31 Mar. 1878 at the Aquarium (later the IMPERIAL) Theatre.

PHERECRATES, ancient Greek writer of OLD COMEDY, active slightly earlier than ARISTOPHANES. He probably won his first victory in 437 BC. In one of his plays, the *Savages* referred to by PLATO in the *Protagoras*, he poked fun at the idea of the 'noble savage', and suggested that even certain contemporary Athenians were preferable to him. Numerous scattered fragments of his work survive.

PHILADELPHIA, Pennsylvania, one of the first American cities to have theatrical entertainment. In 1749 KEAN and Murray's company acted ADDISON's *Cato* in a converted warehouse known as Plumsted's Playhouse, from the name of its owner. The same building, which was subsequently used for other purposes but was not demolished until 1849, housed the elder Lewis HALLAM's company in 1754, two years after its arrival from England. David DOUGLASS brought his AMERICAN COMPANY to Philadelphia in 1759, performing in a temporary theatre erected on Society Hill, and returned in 1766 to build the SOUTHWARK THEATRE, considered by some to be the first permanent theatre building in the United States. Later theatres were the CHESTNUT STREET, which opened in 1793, the WALNUT STREET, which became a theatre in 1811 and is still in use, and the Arch Street, opened by William B. WOOD in 1828. In the 1830s, after the intense rivalry between the three theatres had led in 1829 to a financial crisis, Philadelphia's undoubted theatrical supremacy passed to New York. The Arch Street Theatre, however, was to have its greatest artistic and financial success from 1860 to 1892 under the able management of Mrs John DREW. The town, like Boston, cannot now be considered of major theatrical importance in spite of supporting a symphony orchestra and a ballet company. The Philadelphia Drama Guild, however, has done valuable work since its foundation in 1956, and from 1971 to 1980 was housed in the old Walnut Street Theatre. It then took up residence in a new theatre of considerable dimensions in the Annenberg Center at the University of Pennsylvania.

PHILANDRE, see FILANDRE.

PHILEMON (*c*. 361–263 BC), Greek poet of New Comedy, who was considered the equal of MENANDER. He was freely imitated by Roman writers of comedy, the *Mercator, Trinummus*, and *Mostellaria* of PLAUTUS being adaptations of three of his plays: he reputedly wrote about 100, but only fragments survive.

PHILIPE, GÉRARD (1922–59), French actor, who studied at the Paris Conservatoire, and made his début in 1943 in GIRAUDOUX's *Sodome et Gomorrhe*, first attracting attention in 1945 in the title role of CAMUS's *Caligula*. In 1951, by which time he had made an outstanding reputation both on stage and screen, he joined the THÉÂTRE NATIONAL POPULAIRE under VILAR and gave a superb performance as the hero of CORNEILLE's *Le Cid* (he was buried in the costume he wore for this part). He continued to act for the T.N.P. until his sudden death, and his fame and popularity did much to attract a young audience. Among the plays in which he appeared in Paris, at the AVIGNON FESTIVAL, and on tour in Russia, the U.S.A., and Canada, were KLEIST's *Prinz Friedrich von Homburg* (1951), MUSSET's *Lorenzaccio* (1952), HUGO's *Ruy Blas*, and Shakespeare's *Richard II* (both 1954). In 1958 and 1959 he appeared in *Les Caprices de Marianne* and *On ne badine pas avec l'amour*, both by Musset. He also played Eilif in the first French production of BRECHT's *Mother Courage* (1951).

PHILIPPIDES, writer of ancient Greek NEW COMEDY, who was active in the later 4th and early 3rd century BC. Well known to his contemporaries in Athens as a politically active citizen, he is said to have died of joy in old age on winning a dramatic contest.

PHILIPS, AMBROSE (1674–1749), English dramatist, son of a draper in Shrewsbury, well-educated, and a member of ADDISON's circle. His chief claim to fame is that in *The Distrest Mother* (1712), an adaptation of RACINE's *Andromaque* (1667), he wrote one of the best pseudo-classical tragedies in English, second only to Addison's *Cato* (1713). Henry FIELDING parodied it in *The Covent Garden Tragedy* (1732), proving that it still held the stage in his day. Philips, who was nicknamed Namby-Pamby by Swift for his poor verses, wrote two other undistinguished tragedies, *The Briton* (1722) and *Humfrey, Duke of Gloucester* (1723).

PHILLIPIN, see VILLIERS, CLAUDE DE.

PHILLIPS, AUGUSTINE (?–1605), Elizabethan actor, who was with STRANGE'S MEN, playing with them in TARLETON's *Seven Deadly Sins* (1590–1), and later with the ADMIRAL'S MEN. He joined the CHAMBERLAIN'S MEN on their foundation in 1594, and remained with them until his death. He was a friend of Shakespeare, with whom he acted in JONSON's *Every Man In His Humour* (1598) and *Sejanus* (1603), and is listed among the players in Shakespeare's plays. He was one of the five original shareholders in the GLOBE THEATRE, and on his death left Shakespeare a 30*s*. gold piece.

PHILLIPS, ROBIN (1942–), English director, who was first an actor, making his début as Mr Puff in SHERIDAN's *The Critic* at the BRISTOL OLD VIC in 1959. He became an associate director there a year later, and acted at the CHICHESTER FESTIVAL THEATRE in 1962 and OXFORD PLAYHOUSE in 1964 before becoming an assistant director with the ROYAL SHAKESPEARE COMPANY in 1965. From 1967 to 1968 he was associate director at the NORTHCOTT THEATRE in Exeter, but he returned to the R.S.C. to direct ALBEE's *Tiny Alice* (1970). Later in the same year he directed Ronald MILLAR's *Abelard and Heloise* at WYNDHAM's, seen a year later in New York. He then returned to Chichester, directing ANOUILH's *Dear Antoine* in 1971 and FRY's *The Lady's Not For Burning* in 1972, in which year he played Louis Dubedat in SHAW's *The Doctor's Dilemma* there. After a season as artistic director at the GREENWICH THEATRE in 1973 he took up the same position in 1974 at the STRATFORD (ONTARIO) FESTIVAL, which he ran with outstanding success until 1980. Among his notable productions were CONGREVE's *The Way of the World* (1976), *As You Like It* (1977), *Macbeth*, and COWARD's *Private Lives* (both 1978), all with Maggie SMITH, and *King Lear* (1979) with Peter USTINOV. His production of Edna O'Brien's *Virginia* (1980), with Maggie Smith as Virginia Woolf, was seen in London in 1981.

PHILLIPS, STEPHEN (1864–1915), English poet and dramatist, who for a short time was a member of the Shakespearian company of his cousin Frank BENSON. He had had two plays, *Herod* (1900) and *Ulysses* (1902), produced by TREE at HER MAJESTY'S THEATRE when, later in 1902, his poetic tragedy *Paolo and Francesca* was produced by George ALEXANDER at the ST JAMES'S THEATRE, with the young Henry AINLEY as Paolo. Its great success was believed to herald a new era of poetry in the English theatre, but his next play, *Nero* (1906), also seen at Her Majesty's, showed such a falling-off that, like the rest of Phillips's work, it was soon forgotten.

PHILLIP STREET THEATRE, Sydney, Australian playhouse notable for staging a series of revues, of which the first, *Top of the Bill*, opened in 1954. The material was wholly Australian and local talent was freely used. The series continued until in 1961 the rebuilding of the Phillip Street Theatre caused the company to transfer to the Phillip Theatre, where *A Bex, a Cup of Tea, and a Good Lie Down* ran for over a year and then went on tour. The management gradually changed its policy, and plays and one-man shows were presented with imported players, but without success, and in 1974 the building became a cinema.

PHILLPOTTS, EDEN (1862–1960), English dramatist and novelist, author of a number of light comedies of English rural life, of which the most successful was *The Farmer's Wife*. First seen at the BIRMINGHAM REPERTORY THEATRE in 1916, it was revived in 1924 with Cedric HARDWICKE as Churdles Ash. Transferred to the ROYAL COURT THEATRE, London, it had a long run and has been revived several times with success. Also produced at the Birmingham Repertory Theatre and in London were *Devonshire Cream* (1924), *Jane's Legacy* (1925), and *Yellow Sands* (1926), in the last of which Phillpotts collaborated with his daughter (Mary) Adelaide Eden (1896–), also a novelist and dramatist.

PHILOCLES, Greek tragic dramatist, the nephew of AESCHYLUS. One of his plays defeated SOPHOCLES' *Oedipus the King*, but unfortunately nothing of his work has survived.

PHLYAX, a form of ancient Greek MIME play, or FARCE, which bridged the gap between Athenian and Roman comedy. It was probably the model for the FABULA *atellana*, and, through its chief exponent Rhinthon of Tarentum, for the *fabula rhinthonica*. Much of it was improvised, and consisted of burlesques of earlier plays interspersed with scenes of daily life played by actors in ludicrously padded costumes, each male character having also a gigantic phallus. Our knowledge of the *phlyakes* derives mainly from vase paintings which portray the characters and settings of the 4th century BC. The form of stage depicted is important, for from it may have developed the salient forms of the Roman theatre, which differed markedly from the Greek. The most primitive type consisted of roughly-hewn posts supporting a wooden platform. Later the posts appear to have been joined by panels of wood with ornamental patterns, while later still the structure, though still not permanent, had a background for the actors which approximated to the Roman *scaenae frons*, with a practicable door and windows which were used in the course of the play.

PHOENIX ARTS CENTRE, Leicester, originally the Phoenix Theatre, was built in 1963, after the town had been without a theatre for seven years. It was originally envisaged that the site on which it was erected would be available for only five years, on which assumption a second theatre, the HAYMARKET, was built some years later, both theatres being run by the Leicester Theatre Trust. However, the Phoenix, an open-stage theatre with no proscenium arch, seating 274, which opened with Thornton WILDER's *The Matchmaker*, proved so successful that it continued to function. It became a young people's theatre, not only presenting plays but also financing Flying Phoenix, which took plays to schools and ran children's workshops attended every week by up to 100 youngsters. In 1979 the theatre severed its connection with the Trust and changed its name to indicate that its activities now embraced dance, music, films, and the visual arts as well as drama. It aims to attract all age groups, and Flying Phoenix has been replaced by Phoenix Roadworks, a COMMUNITY THEATRE project.

PHOENIX SOCIETY, London, group founded in 1919 under the auspices of the STAGE SOCIETY to present plays by early English dramatists, most of which, apart from those by Shakespeare, seemed to have fallen out of the repertory in spite of the work of the MERMAID SOCIETY in the early 1900s. The Stage Society began the work of revival in 1915, and continued to produce every year one Restoration comedy, by FARQUHAR, CONGREVE, or VANBRUGH, until 1919. The Phoenix Society was then constituted, and in the 6 years of its existence, up to 1925, staged 26 plays, by, among others, MARLOWE, JONSON, BEAUMONT and FLETCHER, HEYWOOD, FORD, DRYDEN, OTWAY, and WYCHERLEY. From the beginning, enthusiastic support was given to the Phoenix Society by actors and actresses, many of them already well established; two permanent adaptable sets were designed by Norman WILKINSON; and all but two of the productions (for which Edith CRAIG was responsible) were directed by Allan WADE. In 1923 a brilliant performance of Fletcher's PASTORAL The *Faithful Shepherdess* was given with elaborate scenes and dresses and music arranged and conducted by Sir Thomas Beecham. The influence of these performances helped considerably to combat the indifference, and in some cases hostility, once shown to early English drama, several of the plays staged by the Phoenix Society having since been frequently and successfully revived on the public stage.

PHOENIX THEATRE, London, on the corner of Charing Cross Road and Phoenix Street, from which it takes its name. A three-tier house with a capacity of 1,012, it opened on 24 Sept. 1930 with COWARD's *Private Lives*. Later successes by Coward were *To-night at 8.30* (1936) and *Quadrille* (1952). In its early years the Phoenix had a chequered history, but it later housed a number of important plays, including SAINT-DENIS's productions of BULGAKOV's *The White Guard* and of *Twelfth Night* (1938); GIELGUD's revivals of CONGREVE's *Love for Love* (1943) and of *The Winter's Tale* (1951) and *Much Ado About Nothing* (1952); revivals of VANBRUGH's *The Relapse* (1948), from the LYRIC, Hammersmith, and FARQUHAR's *The Beaux' Stratagem* (1949); and *Hamlet* (1955) with Paul SCOFIELD. Notable modern plays included WILDER's *The Skin Of Our Teeth* (1945) with Vivien LEIGH; RATTIGAN's *Playbill* (1948) and *The Sleeping Prince* (1953); VAN DRUTEN's *Bell, Book and Candle* (1954) with Rex HARRISON; and Lesley Storm's *Roar Like a Dove* (1957). OSBORNE's *Luther* with Albert FINNEY transferred from the ROYAL COURT in 1961. In 1963 BRECHT's *Baal*, with Peter O'TOOLE, had a short run, and in 1965 Gielgud appeared in his own version of CHEKHOV's *Ivanov*. In 1968 a musical dramatization of four of Chaucer's *Canterbury Tales* started a long run, finally closing in 1973. The musical *Godspell* was transferred from WYNDHAM's in June 1975, and in 1978 came Royce Ryton's comedy *The Unvarnished Truth* and Tom STOPPARD's play about press freedom, *Night and Day*.

PHOENIX THEATRE, London, in Drury Lane, see COCKPIT.

PHOENIX THEATRE, New York, on 2nd Avenue at 12th Street, formerly the Yiddish Art Theatre, run by Maurice SCHWARTZ. Seating 1,100, and intended for the production of non-commercial plays, it opened as the Phoenix on 1 Dec. 1953 with Sidney HOWARD's posthumous *Madam, Will You Walk?* The repertory was mainly European, and in the first seasons plays presented included *Coriolanus*, CHEKHOV's *The Seagull*, SHAW's *Saint Joan* with Siobhán MCKENNA, IBSEN's *The Master Builder*, STRINDBERG's *Miss Julie*, and TURGENEV's *A Month in the Country*. After the company withdrew the theatre became the Casino East in 1961, opening with a Yiddish revue, and after conversion into a burlesque house in 1965 it was renamed the Eden in 1969, presenting Kenneth TYNAN's *Oh, Calcutta!* In 1972 the long-running musical *Grease* began its career there, and several Yiddish musicals were also seen. The name was changed again in 1977 to the Entermedia, and the building was renovated to become a 1,143-seater entertainment complex for plays, concerts, and other functions. Meanwhile the Phoenix company had moved to the East 74th Street Theatre, seating 204, renamed the Phoenix 74th Street Theatre. It reopened on 18 Sept. 1961, achieving a critical success with Conway's *Who'll Save the Plowboy?* and both critical and commercial success in the following year with KOPIT's *Oh Dad, Poor Dad, Mamma's Hung You in the Closet and I'm Feelin' So Sad*. After three seasons in 74th Street the company moved to the LYCEUM THEATRE until 1969, and then produced plays at various theatres on and off Broadway until in 1972 the New Phoenix Repertory Company was formed, playing for the next three seasons at several university theatres and, for limited engagements, on Broadway. In 1975 it moved to the off-Broadway Playhouse Theatre at 359 West 48th Street, and in the following year to the new 250-seat Marymount Manhattan Theatre, where, as the Phoenix Theatre, it presented the work of new playwrights. From 1968 the East 74th Street Theatre was renamed the Eastside Playhouse, opening with a double bill by Harold PINTER, *Tea Party* and *The Basement*. Later productions were the rock musical *The Last Sweet Days of Isaac*

(1970) and Mark Medoff's *When You Comin' Back, Red Ryder?* (1973), the latter having been produced originally by the CIRCLE REPERTORY COMPANY. Since 1975 it has been used for light opera.

PHRYNICHUS, (1) Greek tragic poet, active slightly earlier than AESCHYLUS, who won victories in 512 and 476 BC. He was fined for writing a play on the capture of Miletus, based on a recent event very painful to Athens, and it was decreed that the play should never be revived. He was also the author of another historical play, the *Phoenician Women,* on the defeat of Xerxes. In his drama the CHORUS was more prominent and the actor less so than in the plays of Aeschylus. More than 50 years later ARISTOPHANES refers to Phrynichus' 'sweet lyrics' as being still very popular.

(2) Ancient Greek writer of OLD COMEDY, contemporary with Aristophanes, who in the *Frogs* ridiculed Phrynichus' use of outworn comic devices. Of the ten plays by him whose titles are known numerous fragments survive.

PIAF, EDITH [Giovanna Gassion] (1915–63), French singer and entertainer, who had a hard and unhappy childhood, supporting herself at an early age by singing in the streets. Once launched into cabaret and music-halls, however, she rapidly became a popular favourite and something of a cult among a group of influential critics. Although she never appeared in England, her recordings sold there by the million, and the personal tragedy of her life was well known—her first lover (and manager) was murdered, she made several unhappy marriages, and had continually to face the problems of alcoholism, drug addiction, and ill-health. Her massive yet superbly controlled voice, emerging from a tiny, shabbily dressed body, contrived to be at the same time strident and intensely moving. Her style was deceptively simple and nostalgic and highly personal. Although she was primarily a singer—she made extensive tours of Europe and North America with Les Compagnons de la Chanson, for instance—she also appeared in a number of films, and was seen on stage in *Le Bel Indifférent* (1941), specially written for her by COCTEAU, who was devoted to her and died on the same day as she did. She also appeared in ACHARD's *La P'tite Lili* (1951). Among the songs she made famous were 'Je ne regrette rien', 'Mon légionnaire', 'La vie en rose', of which she wrote both the words and the music, 'Le voyage du pauvre nègre', and 'Pour deux sous d'amour'. In 1958 she published her autobiography, *Au Bal de la Chance*, translated into English in 1965 as *The Wheel of Fortune*. She was the subject of a play by Pam Gems, *Piaf* (1980), presented by the ROYAL SHAKESPEARE COMPANY with Jane Lapotaire.

PICARD, LOUIS-BENOÎT (1769–1828), French actor-manager, and one of the few successful dramatists of the Napoleonic era, who with unquenchable gaiety flourished equally under the Revolution, the Empire, and the Restoration. He had the knack of amusing the public without touching on controversial subjects, which made him invaluable to Napoleon, and his caustic humour took as its target the newly rich and newly risen. He excelled in the portrayal of bourgeois and provincial interiors, as is evident in *Médiocre et rampant* (1797)—the title comes from a speech by BEAUMARCHAIS's Figaro; but more amusing is the lighthearted *Le Collatéral, ou la Diligence à Joigny* (1799), which with *La Vieille Tante* (1811) and *Les Deux Philibert* (1816) is among the best of his comedies. He was the originator of a new type of light satiric play with music, typified by *La Petite Ville* (1801) which, with music by Lehár, was produced in Vienna as *Die lustige Witwe* (1905) and two years later, as *The Merry Widow*, captivated the English-speaking world. The published texts of Picard's plays give careful directions for settings and costume, in which he aimed above all at pictorial effect, and he can be credited with the creation of one new comic character, the valet Deschamps. He was from 1816 to 1821 manager of the ODÉON, having given up acting in 1807 to qualify for the award of the Légion d'Honneur, which even Napoleon did not dare to give to an actor.

PICCADILLY THEATRE, London, in Denman Street. This theatre, with its three-tier auditorium seating 1,193, opened on 27 Apr. 1928 with a musical play *Blue Eyes* and then became a cinema. It returned to live theatre in Nov. 1929 with a revival of the operetta *The Student Prince*, and in 1933 housed BRIDIE's *The Sleeping Clergyman*. It was then used for the transfer of long runs at reduced prices. After being closed for some time at the beginning of the Second World War, it reopened with COWARD's *Blithe Spirit* (1941), and among later successes were GIELGUD in *Macbeth* (1942) and the American musical *Panama Hattie* (1943). The building was then damaged by flying bombs and had to be closed. It did not reopen until 1945, with Agatha CHRISTIE's *Appointment with Death*. Successful productions were WERFEL's *Jacobowsky and the Colonel* (1945), John VAN DRUTEN's *The Voice of the Turtle* (1947), USTINOV's *Romanoff and Juliet* (1956), and Benn W. Levy's *The Rape of the Belt* (1957). Edward ALBEE's *Who's Afraid of Virginia Woolf?* (1964) began its long run here, and Robert BOLT's *Vivat! Vivat Regina!* transferred from the CHICHESTER FESTIVAL THEATRE in 1970. In 1973 the musical *Gypsy* had a good run, as did a revival of the musical *Very Good, Eddie!* in 1976. The theatre later provided a West End home for several of the ROYAL SHAKESPEARE COMPANY's productions—O'KEEFFE's *Wild Oats* (1977), Peter NICHOLS's *Privates On Parade* (1978), Pam Gems's *Piaf* and KAUFMAN and HART's *Once In a Lifetime,* which ran in repertory (1980), and Willy Russell's enormously successful *Educating Rita* (also 1980).

PICCOLO TEATRO DELLA CITTÀ DI MILANO, the first *teatro stabile*, or permanent theatre company, to be set up in Italy. Founded in 1947 by Giorgio Strehler (1921–) and the actor-director Paolo Grassi (1919–81) with a municipal grant, it later received a State subsidy, and became the model for similar ventures in Rome, Turin, and elsewhere. It opened on 14 May with GORKY's *The Lower Depths*, the cast including the distinguished actor Marcello Moretti (1910–61), who in the same season played the title role in Strehler's production of GOLDONI's *Il servitore di due padroni*. The company quickly achieved an excellent reputation, its audiences including many people who had not formerly been in the habit of going to the theatre. Among Strehler's outstanding achievements was the introduction to Italian audiences of the plays of BRECHT, his Brechtian interests being also reflected in the work of his designers Gianni Ratto and Luciano Damiani. In 1967 the Piccolo Teatro was seen in London during the WORLD THEATRE SEASON, again in Goldoni's comedy. A year later Strehler resigned over the municipality's failure to provide a new theatre for the company, but he returned as sole director in 1972. The Piccolo Teatro's repertoire has encompassed revivals of the classics, contemporary foreign works, and new Italian plays, and the company has visited nearly 30 foreign countries.

PICKELHERING, see REYNOLDS, ROBERT.

PIERROT, stock character in the French and English theatres, derived from the COMMEDIA DELL'ARTE mask PEDROLINO. The transformation is usually attributed to an Italian actor named Giuseppe Giaratone or Giratoni, who joined the COMÉDIE-ITALIENNE in Paris in about 1665. He accentuated the character's simplicity and awkwardness, an important feature of his later manifestations, and dressed him in the familiar costume, a loose white garment with long sleeves, ruff, and large hat whose soft brim flapped round his whitened face. This, with some modifications, has remained his distinguishing garb ever since, but his character was fundamentally altered by DEBURAU, who for some 20 years played Pierrot at the FUNAMBULES. He was succeeded in the part by his son and later by Paul Legrand, who made the character less amusing and more sentimental. This was later developed by a host of imitators until the robust country bumpkin of early days became a lackadaisical, love-sick youth, pining away through unrequited love and much addicted to singing mournful ballads under a full moon. Meanwhile the Pierrot of Deburau had become well known in London, where in 1891 he was seen as the hero of a wordless play *L'Enfant prodigue* (*The Prodigal Son*), produced in Paris in 1889. The character, which had been ousted from the HARLEQUINADE by the English CLOWN, still occasionally appeared in PANTOMIME and REVUE, and regained much of his old vigour when he was incorporated into the new PIERROT TROUPES.

PIERROT TROUPES, form of entertainment which, like the Christmas PANTOMIME, appears to be purely English. It took the solitary, lovelorn PIERROT of DEBURAU's successors, and multiplied him by the score, turning him into a jolly, gregarious entertainer, a versatile member of an organized concert party. Dressed in a modified Pierrot costume—short white frilly frock for the girls, or Pierrettes, loose white (or black) suits for the men, with tall dunces' caps, the whole enlivened by brightly coloured (or black on white) buttons, ruffs, ruffles, and pompons—the various members of the troupe sang, danced, juggled, told funny stories, and engaged in humorous backchat. After an initial appearance at Henley Regatta, the Pierrots spread all over England, even ousting the black-faced MINSTREL SHOWS from the beaches and pier-pavilions of the seaside towns, and eventually reaching London with Pélissier's *Follies*, the apotheosis of the genre, in the 1900s. The Pierrot troupes were later replaced by the more sophisticated agglomeration of turns known as REVUE, but lingered on in summer resorts, and a successful revival of the old Pierrot show, staged by the Co-Optimists under Davy Burnaby (1881–1949), had an unexpected success during several London seasons in the 1920s.

PIKE'S OPERA HOUSE, New York, see GRAND OPERA HOUSE.

PIKE THEATRE CLUB, Dublin. Founded in 1953 by Alan Simpson, this theatre opened on 15 Sept. 1953 with G. K. Chesterton's *The Surprise*. A year later, on 19 Nov. 1954, came the première of BEHAN's *The Quare Fellow,* a considerable feat on such a tiny stage. The following year saw the English-language première of BECKETT's *Waiting for Godot* (28 Oct. 1955), which ran for six months and then toured Ireland, IONESCO's *The Bald Prima Donna* was staged in 1956, and there were also productions of plays by BETTI, FABBRI, and SARTRE. During the first DUBLIN THEATRE FESTIVAL in 1957 Simpson was arrested for staging Tennessee WILLIAMS's *The Rose Tattoo*, which was considered obscene; he won his case, a victory in the Irish struggle against CENSORSHIP. An important part of the Pike's reputation derived from its late-night revues, or 'Follies,' an innovation in Dublin. Financial losses eventually forced the Pike to stage productions at Dublin's major theatres, Dominic Behan's *Posterity Be Damned* (1959) playing at the GAIETY, and a rock musical, *The Scatterin'* (1960), at the OLYMPIA. The success of these shows (both transferred to London) indicated that the Pike had outgrown the little theatre movement, and the last play staged under Simpson's direction was *God's Child* (1959) by J. B. McGowan. Thereafter, the theatre was let to other groups, and later sold.

PILGRIM PLAYERS, see BROWNE, E. MARTIN, and JACKSON, BARRY.

PINERO [Pinheiro], ARTHUR WING (1855–1934), English dramatist, knighted in 1909. He was an actor for ten years, making his first appearance on the stage at the Theatre Royal, Edinburgh, on 22 June 1874, and used his experience to forward his playwriting, always his main objective. On 6 Oct. 1877 he had his first play, *£200 a Year*, produced at the GLOBE THEATRE. Many minor pieces followed, including *The Money Spinner* (1881) at the ST JAMES's. Popularity came with *The Magistrate* (1885), the first of the ROYAL COURT THEATRE farces which became all the rage. It was followed by *The Schoolmistress* (1886), *Dandy Dick* (1887), *The Cabinet Minister* (1890), and *The Amazons* (1893). The first three have been successfully revived many times and still rank as some of the best farces in the English language. In 1888 a frankly sentimental play, *Sweet Lavender*, confirmed Pinero's pre-eminence in the contemporary theatre, and with *The Profligate* (1889) he made his first venture into the 'theatre of ideas', following it at the end of May 1893 with *The Second Mrs Tanqueray*, a 'problem play' in which Mrs Patrick CAMPBELL startled the town. In a theatre long given over to FARCE, BURLESQUE, and MELODRAMA it appeared revolutionary—a 'serious' English play which made money. During the next 30 years Pinero was regularly productive. *The Notorious Mrs Ebbsmith* (1895) was Paula Tanqueray's successor; *Trelawny of the 'Wells'* (1898) was a theatrical romp which has been revived regularly, always with success; *The Gay Lord Quex* (1899) was a brilliant piece of theatricalism, containing a third act which is perhaps the author's masterpiece of contrivance. With the turn of the century came a long succession of serious plays —*Iris* (1901), *Letty* (1903), *His House in Order* (1906), and *Mid-Channel* (1909). They were progressively less successful, for Pinero had written himself out, and his last play, *A Cold June* (1932), was a pathetic failure. His reputation long rested on his farces and on *Trelawny of the 'Wells'*, whose title role presented an irresistible challenge to spirited young actresses.

PIN-SPOTTING, see LIGHTING.

PINTER [Da Pinta], HAROLD (1930–), English dramatist, who for some years acted in the provinces under the name of David Baron. His first plays, the one-act *The Room* (1957), seen at Bristol University, and the full-length *The Birthday Party* (1958), which had a short run in London, roused little interest; the latter was produced in New York in 1967. In 1960 there was a professional production of *The Room* in a double bill with *The Dumb-Waiter*. He came into prominence with *The Caretaker* (1960; N.Y., 1961), an archetypal Pinter play in which the three characters, one of them a tramp, demonstrate the difficulties of communication between human beings and their consequent isolation. Next came five one-act plays—*A Night Out, A Slight Ache* (both 1961), *The Collection* (1962), produced by the ROYAL SHAKESPEARE COMPANY at the ALDWYCH, and, in a double bill, *The Lover* and *The Dwarfs* (1963), both of which Pinter himself directed. His next full-length play, *The Homecoming* (1965; N.Y., 1967), was also seen at the Aldwych, as were another double bill, *Landscape* and *Silence* (1969; N.Y., 1970) and the full-length *Old Times* (London and N.Y., 1971). A double bill of *Tea Party* and *The Basement* was seen in New York in 1968 and London in 1970. *No Man's Land* (1975; N.Y., 1976), seen at the NATIONAL THEATRE, provided excellent parts for John GIELGUD and Ralph RICHARDSON. *Betrayal* (1978; N.Y., 1980), also at the National Theatre, was written in a more naturalistic style than most of Pinter's plays. In 1980 *The Hothouse*, written over 20 years earlier, was also presented in London. He has directed a number of plays by other writers, including James JOYCE's only play *Exiles* (1970) and several works by Simon GRAY.

Pinter is probably the most influential of modern English playwrights, his highly individual style, when attempted by others, being labelled 'Pinteresque'. The behaviour of his characters is observed with an objective and impartial eye, but their motives often remain obscure, their backgrounds indefinite, and their fate at the end of the play indeterminate. His ability to keep his delicate structures of uncertainty and ambiguity from collapsing rests on his consummate stagecraft; his scripts are as carefully balanced, with pauses and silences exactly indicated, as musical scores. The extreme linguistic realism of his early plays, embodying the solecisms, repetitions, and illogicalities of ordinary speech, has now been replaced by language controlled to the point of being highly stylized.

Many of the leading feminine roles in his plays were played by his first wife Vivien Merchant (1929–82), who also played Lady Macbeth for the Royal Shakespeare Company in 1967 and was in David MERCER's *Flint* and Joyce's *Exiles* (both 1970).

PINTILIE, LUCIAN, see ROMANIA.

PIPE, see LIGHTING.

PIPE BATTEN, see BARREL.

PIP SIMMONS THEATRE GROUP, experimental touring company, based on London, which took its name from its founder and director. Beginning in 1968, it developed a style of production which combined pop music, composed and performed by the members themselves, acrobatic movement, and narration, as well as dialogue. Its original scripts, by Pip Simmons, used existing works as starting points, and were developed in part during rehearsals by the methods of COLLECTIVE CREATION. The first production, in 1969, was a version of Chaucer's *The Pardoner's Tale*; this was followed in 1970 by *Superman*, which presented the comic-strip hero as a super-anarchist, and in 1971 by *Do It!*, based on a book by the American radical of the 1960s Jerry Rubin, and *Alice*, based on Lewis Carroll's books. The *George Jackson's Black and White Minstrel Show* in 1972 took its name from a black American who died in prison and the 19th-century theatrical form in which white Americans parodied their slaves. The group disbanded in 1973, but a year later Pip Simmons formed a new group, which in 1975 toured Western Europe with *An die Musik*. This marked the 30th anniversary of the ending of the Second World War by re-creating a concert given by Jewish prisoners in a Nazi concentration camp. Other works created by the newly formed group include a musical based on DOSTOEVSKY's short story *The Dream of a Ridiculous Man* in 1975 and *The Masque of the Red Death*, based on a short story by Edgar Allan Poe, in 1977.

PIRANDELLO, LUIGI (1867–1936), Italian dramatist, awarded the NOBEL PRIZE for Literature in 1934. Although well known as a novelist and critic before he turned to the theatre, it is as the playwright who brought Italy, after a period of stagnation, back into the mainstream of European drama that he is chiefly remembered. The success of his plays meant that the Italian straight theatre was once again able to challenge the supremacy of opera, and the diffusion of his work throughout Europe and America proved beneficial everywhere. His early plays were dramatizations of some of his own short stories, and it was not until the production of his best known work *Sei personaggi in cerca d'autore* in 1921 that he became internationally famous. As *Six Characters in Search of an Author*, this was produced in London and in New York in 1922, and has several times been revived, a production at the MAY FAIR THEATRE in London in 1963 achieving a run of 288 performances. The play, which is central to an understanding of Pirandello's work, concerns a group of characters from another play who invade the stage during a rehearsal and become the arbiters of their own destiny. It forms part of a trilogy of which the other sections, less well known in English, are *Ciascuno a suo modo* (*Each in His Own Way*, 1924), in which Pirandello argues that art may have a truth superior to that proposed by life and may therefore dictate a more appropriate pattern of conduct, and *Questa sera si recita a soggetto* (*To-night We Improvise*, 1929), which postulates the dominance of the actor by the part he is playing. In these, as in all his plays, particularly in his finest tragedy *Enrico IV* (1922), seen in London in 1925 and in New York in 1947 as *Henry IV*, Pirandello was concerned with the futility of human endeavour and the impossibility of establishing an integrated, objective personality for any human being. With consummate ability he pushed forward the frontiers of drama by bringing to the stage the mental and psychological preoccupations of his own day and becoming the chronicler of an age of decay and fragmentation. Among his other plays, those which have been successfully produced in English are *Così è (si vi pare)* (*Right You Are, If You Think You Are*, 1917); *Il giuoco delle parti* (*The Rules of the Game*, 1918); *Lazzaro* (*Lazarus*, 1929); and *Come tu mi vuoi* (*As You Desire Me*, 1930). Of the one-act plays, the best known is probably *L'uomo dal fiore in bocca* (1923), which as *The Man With a Flower in His Mouth* (the flower being an inoperable cancer) was produced in London in 1926. In 1969 *Quando si è qualcuno* (1933), was seen at the Theatre Royal, York, as *When One is Somebody*, in a translation by Marta Abba, who was for many years Pirandello's leading lady; and in 1970 *Liolà* (1916) was seen in London during the WORLD THEATRE season in a production by the Catania Stabile Theatre from Sicily. It also provided the libretto for Mulè's opera of the same title, first heard in 1935, a year after the first performance of Malipiero's opera based on Pirandello's *La Favola del figlio cambiato* (1933). Pirandello had an unhappy life, his wife, whom he married in 1894, becoming mentally ill in 1904. For 15 years he cared for her at home, finding his only consolation in his work. After she had been consigned to a mental home—she died in 1959—he was able to devote more of his time to the theatre and in 1925 established his own company in Rome, proving himself an excellent director with an acute awareness of the technical problems of stagecraft. He toured widely in Europe and America with this company, and became well known as an actor, particularly in his own plays.

PIRON, ALEXIS (1689–1773), French dramatist, who began by writing farces for the fairground theatres (see FAIRS), overcoming the difficulty of not employing more than one speaking actor (a prohibition laid on the unofficial theatres by a law of 1718) by a series of monologues, of which the first was *Arlequin Deucalion* (1722). Encouraged by its success, he sent a comedy, *L'École des pères*, to the COMÉDIE-FRANÇAISE, where it was produced in 1728. It represents an interesting mixture of old and new in the theatre of the time, for it stands on the threshold of the COMÉDIE LARMOYANTE which was shortly to be introduced by LA CHAUSSÉE, but its author still evidently believes that comedy should amuse first and only incidentally instruct. Piron's best work (and one of the outstanding comedies of the 18th century) was *La Métromanie* (1738), which makes gentle fun of bad poets who insist on reading their effusions to their reluctant friends. Of his tragedies the best was *Gustave Wasa* (1734), which remained in the repertory of the Comédie-Française for many years. Piron suffered all his life from some scurrilous verses that he wrote in his youth; they turned up at inopportune moments and finally cost him his chance of a seat in the French Academy.

PISANI, UGOLINO, see ITALY.

PISCATOR, ERWIN FRIEDRICH MAX (1893–1966), German director, who during the 1920s worked in BERLIN where he devised a form of EPIC THEATRE, intended to reinforce the impact of his strong pacifist and communist convictions, which encompassed the whole of society in its political and economic complexity and greatly influenced the later work of BRECHT, one of his co-workers. He anticipated the later trend away from a completed playscript, and was one of the first directors to use film-strips and animated cartoons in conjunction with live actors. Among his productions at this time were GORKY's *The Lower Depths* and a modern-dress version of SCHILLER's *Die Räuber* in 1926; *Gewitter über Gottland* (1927) by Elm Welk, which included film projections of the Russian Revolution, cost him his job at the VOLKSBÜHNE. He then opened his own theatre, where he produced TOLLER's *Hoppla, wir leben!* in 1927, in a revolving multi-level set with seven or eight acting areas and surfaces for film and slide projections; Alexei TOLSTOY's *Rasputin*, also 1927, staged in a huge revolving steel hemisphere symbolizing the earth; and *Die Abenteuer des braven Soldaten Schweik*, adapted by Brecht and others from the novel by Hašek, in 1928, with two steam-driven treadmills installed side by side across the stage so that Schweik could literally march from episode to episode while his environment trundled towards him. This was Piscator's most successful production, and technically his most ambitious, but so expensive that the theatre could only survive the beginning of its second season, which saw the production of Mehring's *Der Kaufmann von Berlin* (1929) before it closed. After some desultory freelance work, Piscator left Germany in 1933, spent some time in Paris, and reached New York in 1939. There he directed a Dramatic Workshop, where he mounted a large number of teaching productions including his own adaptation of TOLSTOY's *War and Peace* (1942). Working with him were John GASSNER and Mordecai Gorelik, and among his students were Judith Malina, Arthur MILLER, and Tennessee WILLIAMS. He returned to Germany in 1951, where he again produced *War and Peace*, and in 1962 became director of the West Berlin Freie Volksbühne, where his first production was HAUPTMANN's *Die Atriden-Tetralogie*, followed by a spate of new 'documentary' dramas—HOCHHUTH's *Der Stellvertreter* (1963), KIPPHARDT's *In der Sache J. Robert Oppenheimer* (1964), and WEISS's *Die Ermittlung* (1965). Piscator died suddenly during the rehearsals of Hochhuth's *Soldaten*. In England, where his production of *War and Peace* at the BRISTOL OLD VIC in 1962 gave British audiences their only chance of seeing his work, albeit at second-hand, his influence was apparent in the work of Joan LITTLEWOOD at THEATRE WORKSHOP, and in the U.S.A. in Malina and Beck's LIVING THEATRE. He was the author of *Das politische Theater* (1929, rev. 1964) and *Schriften zum Theater* (1968).

PISISTRATUS, tyrant of Athens, on and off, from 560 BC until his death in 528 BC. He did much for the economic and cultural development of the city, notably in reorganizing on a grand scale the DIONYSIA and instituting contests in DITHYRAMB and in TRAGEDY, this being the first official recognition of the latter form.

PISTOIA, Il, see SACRA RAPPRESENTAZIONE.

PIT, the ground floor of the theatre auditorium, generally excavated below ground level. In the early playhouses the stage and lower boxes were approximately at ground level, and the whole space sunk between these was called the pit, from the Elizabethan cockpit, used for cock-fighting. In the early 19th century the lower boxes were replaced by a raised circle, with the pit extending underneath; shortly after, the old rows of pit seats near the orchestra were replaced by the higher-priced stalls, and the name 'pit' was applied only to the more distant rows. Most modern theatres have no pit.

PITEL, LOUISE, see BEAUBOUR.

PITLOCHRY FESTIVAL THEATRE, in Perthshire, Scotland. The idea of building a 'theatre in the hills' for the presentation of a festival of plays in repertory came to John Stewart, founder of the Glasgow Park Theatre (itself a successor of the CURTAIN THEATRE) during a visit to the MALVERN FESTIVAL before the Second World War; but it was not until 19 May 1951 that the first theatre at Pitlochry opened. This was a large marquee with a fan-shaped auditorium and a very wide proscenium opening, as well as excellent front-of-house amenities. This tent theatre was so well designed that its features were retained in the more permanent structure built in 1953. Generally speaking, Pitlochry presents six plays annually, the season running from April to the beginning of October, with at least one Scots play and one new play in the repertory. Since Stewart's death in 1957 the artistic director of the Festival has been Kenneth Ireland, who has added concerts, art exhibitions, and lectures to the original programme. Financial difficulties which came to a head in 1969 were overcome by a successful appeal for public support and in 1978 work was begun on the site of a new theatre situated beside the River Tummell. At the end of the 1980 season the existing theatre closed, and the new theatre opened on 19 May 1981.

PITOËFF, GEORGES (1887–1939), Russian-born French actor, who for two years directed his own amateur company in St Petersburg, where he was in contact with STANISLAVSKY and MEYERHOLD. From 1915 to 1921 he lived in Geneva, and continued to produce, coming under the influence of the theorist and teacher of movement Jaques-Dalcroze. He then settled in Paris and with his wife Ludmilla (below) appeared in several theatres, including that of COPEAU, before in 1924 he took a company to the Théâtre des Arts (now the HÉBERTOT), and from there to the MATHURINS, where much of his best work was done. Like Copeau, he believed that the French theatre was suffering from bankruptcy, both of ideas and of imagination. With very little money, but a great deal of ingenuity and hard work, he attempted to remedy this state of affairs by giving audiences in Paris the chance of seeing the best work of foreign dramatists, among them Shakespeare, SHAW, CHEKHOV, and PIRANDELLO, as well as new plays by such innovatory French writers as CLAUDEL, COCTEAU, and ANOUILH. The value of his work lay not only in the plays he presented, but in the subtle and entirely personal interpretation he gave them, often centred on some brilliantly simple piece of scenography or decorative technique. He was himself an excellent actor, and a complete man of the theatre, adapting, translating, directing, and acting at one and the same time.

PITOËFF, LUDMILLA (1896–1951), wife of the above, an excellent actress, who appeared in many of her husband's productions, and after his death continued to direct the company, taking it on an extended tour of America and Canada. Among the many parts she played were Nora in IBSEN's *A Doll's House*, Marthe in CLAUDEL's *L'Échange*, and the title role in GOLDONI's *La locandiera*. In 1930 she appeared in London, playing SHAW's St Joan in French and the title role in the younger DUMAS's *La Dame aux camélias*.

PITOËFF, SACHA (1920–), French actor, son of the above, who inherited a great deal of his parents' talent, both as actor and director. He was a member of BARSACQ's Compagnie des Quatre Saisons, and then became stage manager of his mother's touring company, and of theatres in Lausanne and Geneva. Returning to Paris after the Second World War, he acted in various plays, founding his own company in 1949; he repeated some of his father's more famous productions (particularly of CHEKHOV and PIRANDELLO) and continued the tradition of translation and search for new authors, staging, for example, *nō* plays from JAPAN and MIDDLETON's *The Changeling*. As an actor, he conveys a strange, fierce spirituality with his gangling, nervous body and unusually expressive face, and is at his best in self-contained, solitary roles.

PITT, CHARLES and THOMAS, see DIBDIN, THOMAS JOHN.

PITTENEC, see LESAGE.

PIXÉRÉCOURT, (René-Charles) GUILBERT DE (1773–1844), French dramatist, who alone or in collaboration wrote nearly 100 plays and for 30 years provided the staple fare of the secondary theatres. His early years were eventful, and he led an unsettled life until in 1797 one of his many early plays, *Les Petits Auvergnats*, was put on at the AMBIGU. It was successful enough to persuade him to devote the rest of his life to the theatre, and in 1798 he produced the first of the long series of plays by which he is chiefly remembered and for which he himself coined the name MELODRAMA, with *Victor, ou l'Enfant de la forêt*. The most successful of these, which earned him the nickname of 'the CORNEILLE of the boulevards', was *Coelina, ou l'Enfant du mystère* (1800), which, in a translation by Thomas HOLCROFT as *A Tale of Mystery* (1802), was the first melodrama seen on the English stage. It was also translated into German and Dutch. The success of his plays made Pixérécourt a wealthy man, but he was ruined when the Théâtre de la GAÎTÉ, of which he was a director and where many of his most successful plays were produced, was burned down in 1835. This disaster, joined to the effects of a serious illness, ended his career, and he retired to Nancy to die a lingering death. He was an odd, tormented creature, who seems to typify in his work the mixture of ferocity and idealism of the French Revolution, when blood and tears were shed with equal facility. He had an appreciation of good literature and was well aware of the literary shortcomings of his own plays, but he believed sincerely in their importance as a vehicle for moral teaching. Although he wrote quickly, he took great trouble over his productions and settings, often inventing new machinery and spectacular effects, for, as he himself said, he wrote for 'those who cannot read'—a large, enthusiastic, and unlettered audience whose counterpart is catered for today by certain kinds of film and television. His work shows the extent of German influence on the French theatre of his time, and in its turn influenced the work of the Romantic playwrights: HUGO's best plays are in essence melodramas, redeemed from triviality by felicitous language and fine lyric poetry. Pixérécourt also had an immense influence in England, where most of his plays were seen soon after they appeared, their main characteristics being preserved in the contemporary TOY THEATRE. He himself spelt his name Pixerécourt, in accordance with the original pronunciation, but custom now enforces the use of the present spelling, Pixérécourt.

PLACE, The, see ROYAL SHAKESPEARE COMPANY.

PLACIDE, HENRY (1799–1870), American actor, elder son of a French emigrant rope-dancer, Alexandre (?–1813), who married an English actress by whom he had a large family. Henry made his first appearance on the stage at the age of nine, and in 1814 was at the ANTHONY STREET

THEATRE in New York. He is believed to be the 'Mr Placide' who is mentioned as playing in Halifax in 1816 and in St John, New Brunswick, a year later. He was back in New York in 1823 when he appeared in the younger COLMAN's *The Heir at Law* at the PARK THEATRE, where he remained for many years, occasionally going on tour and in 1841 visiting London. He was later with BURTON at his theatre in Chambers Street, New York. Like John GILBERT, he represented the best traditions of acting in polished comedy, Sir Peter Teazle in SHERIDAN's *The School for Scandal* being accounted one of his best parts. His elder sister Caroline (1789–1881), who married as her second husband the Canadian actor William Rufus Blake (1805–63), and his younger sisters Jane (1804–35) and Eliza (?–1874) were all on the stage, and his younger brother Thomas (1808–77), who in 1826 was in Montreal playing Roderigo to Edmund KEAN's Othello, was for some years manager of the Park Theatre, N.Y.

PLANCHÉ, JAMES ROBINSON (1795–1880), English dramatist of Huguenot descent, a prolific writer of BURLESQUES, EXTRAVAGANZAS, and PANTOMIMES, most of which were seen in the smaller theatres of London, particularly the ADELPHI and the OLYMPIC, though occasionally a melodrama was produced at one of the PATENT THEATRES. The best known of Planché's works, *The Vampire; or, the Bride of the Isles* (1820), introduced to the English stage the Vamp(ire) TRAP, and for Madame VESTRIS at the LYCEUM he wrote the spectacular extravaganza *The Island of Jewels* (1849) for which the painter Beverley produced some remarkable scenic effects. Having had his first play, a burlesque entitled *Amoroso, King of Little Britain*, produced in 1818, Planché ended his career with a pantomime, *King Christmas* (1871), and a spectacle play for Covent Garden, *Babil and Bijou; or, the Lost Regalia* (1872), written in collaboration with Dion BOUCICAULT. Although Planché's work was enormously successful in his own day it has no literary merit and divorced from its music and spectacular effects is quite unreadable. It depended largely on its staging, acting, and topicality, and taken as a whole provides an excellent picture of the London stage over more than 50 years. According to the theatre historian Sir St Vincent Troubridge (see *Notes and Queries*, vols. 180, 181) many of GILBERT's librettos for the Savoy operas were based on or suggested by the texts of Planché's extravaganzas. Planché was also a musician, directing the music at VAUXHALL Gardens during the season of 1826–7, and a serious student of art, his *History of British Costume* (1834) long remaining a standard work. In 1823 he designed and supervised the costuming of Charles KEAN's production of *King John*, one of the first to approximate to historical accuracy. An unauthorized production of one of his own plays led him to press for the reform of the laws governing dramatic COPYRIGHT, and it was largely due to his efforts that new legislation was passed, giving greater protection to British dramatists.

PLANCHON, ROGER (1931–), French director, actor, and dramatist, whose first production, *Bottines, collets montés*, a burlesque in the style of the GRENIER-HUSSENOT company which he mounted with his own amateur group, won a prize in a local competition in Lyon in 1950. The company then turned professional and, living more or less as a community, built their own 100-seat theatre, the Théâtre de la Comédie de Lyon, which opened in 1952. By 1957 they had won a considerable reputation as an experimental theatre, staging, among other plays, ADAMOV's *Professor Taranne* and *Paolo Paoli* (both 1952). Planchon was at this time consciously conducting investigations into various forms of stagecraft borrowed from every possible source, including the Elizabethan theatre and American gangster films. Like many of his contemporaries he was strongly influenced by VILAR, but after a meeting with BRECHT in 1954 became the leading director of his plays in France with translations of *Der gute Mensch von Sezuan* in 1954, *Furcht und Elend des Dritten Reiches* in 1956, and *Schweyk im Zweiten Weltkrieg* in 1962. His preoccupation with EPIC THEATRE had early on led him to seek a larger theatre, and in 1957, at the invitation of Villeurbanne, an industrial satellite-town of Lyon, he took his company to the 1,300-seat Théâtre de la Cité. There he addressed himself to factory workers rather than to sophisticated theatregoers and by conducting what was to become a classic example of a campaign of popularization through meetings, publications, exhibitions, and door-to-door salesmanship, succeeded in creating an entirely new audience. After a resoundingly successful visit to Paris in 1961, the company was awarded a government subsidy, and thus became the first national theatre in the French provinces, inheriting in 1972 the name THÉÂTRE NATIONAL POPULAIRE after the closing of the Palais de CHAILLOT. Planchon himself was a brilliant actor before he took on so many responsibilities, and as d'Artagnan in his own adaptation of the elder DUMAS's *Les Trois Mousquetaires* was seen in 1960 in London and at the EDINBURGH FESTIVAL. As a director he abandoned the traditional interpretations of French classics standardized by the COMÉDIE-FRANÇAISE, and in his productions of MARIVAUX's *La Seconde Surprise de l'amour* and MOLIÈRE's *George Dandin* in 1959 carried the biting social satire underlying the buffoonery and *marivaudage* as far as overt references to a Marxist view of the class struggle. His iconoclasm culminated during the liberating frenzy of 1968 in a *Mise en pièces et contestation de Cid*, which, basing itself on CORNEILLE's revered masterpiece, attacked the very foundations of French classical drama. He later aplyied the same methods to Shakespeare's *Henry IV, Parts 1 and 2*. His other notable productions include GOGOL's *Dead Souls*, adapted by Adamov, in 1959; MARLOWE's *Edward II*, in his own adaptation, in 1960; and the plays of GATTI, whom he helped to discover. In 1968 he committed himself wholeheartedly to an anti-Establishment declaration by the managers of most of France's subsidized playhouses, and thereafter played a less prominent role at Villeurbanne, laying more stress on the collective nature of the company and on the contributions of his associates, particularly the scenographer ALLIO and the director Jacques Rosner. He also began to write plays himself, in a style curiously closer to NATURALISM than to the critical REALISM of his former master Brecht; among them are *La Remise* (1962), *Bleus, blancs, rouges* (1967), *Dans le vent* (1968), *L'Infâme* (1970), *Le Cochon noir* (1974), and *Gilles de Rais* (1976).

PLATFORM, see ROSTRUM.

PLATFORM STAGE, see FORESTAGE.

PLATO (427–348 BC), Greek philosopher, of an aristocratic Athenian family. In his youth he composed tragedies and other forms of poetry, but on coming under the influence of Socrates and his rigorous intellectualism he burnt his plays and devoted his life to philosophy and mathematics.

The philosophical work he published was put into a dialogue form which was a development of the ancient MIME. In the more abstruse work, and in most of the *Republic*, the dialogue is only nominal; elsewhere it is consistently dramatic, with occasional passages of astonishing vividness and power. The character-sketches, particularly that of ION, or the opening scenes of the *Protagoras* are good examples of Plato's dramatic skill; his mastery of ironic comedy is shown by his picture of the sophists in the *Euthydemus*; of tragedy by the scene of Socrates' death in the *Phaedo*.

Plato's theories of literature and drama have had an immense influence. His 'inspirational' theory of poetry is the direct source of the idea of the *furor poeticus*—'the poet's eye in a fine frenzy rolling'—through a 16th-century translation of *Ion* which greatly influenced the French poets of the time. In other dialogues Plato is much less sympathetic to literature; his idea that a conception of 'the Good' can be reached only through the intellectual process of dialectic led him, apparently, to mistrust 'inspiration', and in the *Republic* and elsewhere he makes it clear that he would admit poetry into his ideal society only under a paralysing censorship. His main objections to drama were that it debilitates the community by appealing to the emotions not to reason, that it has an unfortunate influence on the more ignorant among the spectators, and that it propagates blasphemous and impossible ideas about the gods by keeping alive the stories of their immorality and internecine strife. These criticisms were important at the time for the reply which they drew from ARISTOTLE, but in stating them Plato was merely reproducing, though with a more intense conviction, the prevailing Greek idea that all art should be functional, and that drama in particular, as the community's most public form of expression, should serve the purposes of the State —an attitude already conveyed, though in comic form, by ARISTOPHANES in the *Frogs*.

PLATO COMICUS, Athenian poet contemporary with ARISTOPHANES, who gained several victories. Only fragments of his work survive. He was given the name Comicus posthumously to distinguish him from the philosopher (above).

PLATT, the Elizabethan term for an outline, or 'plot', of a play, giving the main points of the action, the division into acts, and the actors' entrances and exits. Posted up behind the scenes during the performance, it was intended to help in the organization of calls and properties, and was not, like the scenario of the COMMEDIA DELL'ARTE, a basis for improvisation but a purely utilitarian device. A few specimens have survived among HENSLOWE's papers.

PLAUTUS, TITUS MACCUS [Maccius] (*c.* 254–184 BC), Roman dramatist, of whose life and background very little is known. By the end of the 1st century BC he was credited with no fewer than 150 plays of which 20 and one fragment are believed to be authentic, though the *Mercator* and the *Asinaria* may be the work of another Titus Maccus. They are all free renderings of Greek OLD COMEDY—how free, a comparison between the newly discovered extensive fragment of MENANDER's *Dis exapaton* and Plautus's *Bacchides*, based on it, will quickly show—with complicated plots, strongly marked characters, and scenes of love-making, revelry, trickery, and debauchery, with a profusion of songs, jokes, puns, and topical allusions. Although Roman dramatists were warned against political and personal satire there seems to have been no ban on indecency, and several of the plays, among them the *Pseudolus* and the *Truculentus*, are set in brothels; but in the preface to the *Captivi* Plautus could boast of its high moral tone and, although he sometimes carried farce to outrageous lengths, when the honour of a respectable woman was in question he knew how to draw the line. The earliest of his plays to which a date can be assigned are the MILES GLORIOSUS and the *Cistellaria*, both probably 204 BC. It is known that the *Stichus* was first produced in 200 BC and the *Pseudolus* was written for the Festival of 191 BC. According to Cicero, the last production of a new play by Plautus was in 184 BC, the year of his death. Among the best known are the *Menaechmi*, the source of Shakespeare's *The Comedy of Errors*, and the *Amphitruo*, which deal with mistaken identities; the *Aulularia*, the source of MOLIÈRE's *L'Avare* which portrays a poor old man crazed by the discovery of a buried treasure; the *Mostellaria*, or 'ghost story', which displays the endless fertility of invention displayed by a slave in the service of his master; the *Captivi*, in which another slave, by his devotion and courage, enables his master to escape from captivity; and the *Rudens*, which deals with storm and shipwreck, a treasure recovered from the sea, and a long-lost daughter restored to her parents. Considering the limits imposed upon the FABULA *palliata*, the variety of plot found in Plautus is considerable; and if he lacks the subtle effects of his contemporary TERENCE, he offers instead a flow of wit and a vigour of language which go a long way towards explaining his supreme popularity on the Roman stage. Nor must the importance of music in his plays be overlooked. Although all the scores are lost, the texts which remain were probably to a large extent chanted or sung in a variety of metres to the accompaniment of a *tibia*, or pipe. Within a few years of his death he had become a classic, his plays being constantly revived, and even when they were no longer acted they were still read. Most of them have been translated, several acted by school or college societies, and the successful American musical *A Funny Thing Happened on the Way to the Forum* (1962; London, 1963) owed its existence to several of Plautus' plays combined.

PLAY, generic term applied to any work written to be acted, and covering such more limiting terms as comedy, drama, farce, or tragedy. It may designate back-chat between two mountebanks in the market-place or a full-length work given in a special building—a theatre—with a cast of highly trained professional actors aided by all the appurtenances of lighting, costuming, and production. The one essential is that it should be entirely or mainly spoken; if it has no dialogue it is a MIME; if danced, a ballet; if entirely sung, an opera. Hybrid forms are the BALLAD OPERA, BURLETTA, and MUSICAL COMEDY. (For the use of music as an adjunct to a play, see INCIDENTAL MUSIC.)

A play can be read, but only fulfils its original intention when acted. The text may therefore be regarded as an inert body of words to which the director, player, and audience must contribute to bring it to life. Although the fundamental principles of drama remain constant—action, conflict, unity of purpose, resolution—the form of a play may conform to certain conventions—five-act, three-act, unity of time and place, separation (or alternatively fusion) of tragedy and comedy—which vary from age to age, and even from country to country. The form, however, comes first, the rules afterwards, even with ARISTOTLE. In the same way, although the art of acting is to some extent dependent on the type of play in favour at the moment, it has an independent life of its own, and at certain points in the history of the theatre there may be conflict between the text and its interpreters. A good play may fail in its own day and only be appreciated in revival. Plays written to be read—CLOSET DRAMA—remain outside the main stream of the theatre, though this theory is refuted by the success of the plays of Alfred de MUSSET on the stage many years after they were written and by the influence on European drama of the tragedies of SENECA, which were probably not acted but read aloud. The POETIC DRAMA of the 19th century, which from a purely literary point of view contains many fine things, has not proved successful on the stage. It seems likely that an inherent lack of dramatic impulse allied to too great a weight of pure poetry will always hinder its immediate effect on an audience. The author alone cannot produce a play in the full sense of the word but must have the co-operation of many other people. The earlier name for a dramatist, playwright, by its affinity with such words as wheelwright, reveals this clearly, and makes the author a fellow worker in the theatre with actor, director, designer, stage carpenter, and so on. Some of the finest plays have been written by men experiencing all the advantages and disadvantages of actual daily participation in the work of the theatre, as with MOLIÈRE and Shakespeare.

PLAYERS CLUB, New York, founded in 1888 on the lines of the GARRICK CLUB in London. Its first president was Edwin BOOTH, who bought, donated, and endowed a house for it in Gramercy Park. He kept a suite of rooms there for the rest of his life, and like his successors, Joseph JEFFERSON and John DREW, died in office. The club has a large collection of theatrical relics, including jewellery and weapons owned by famous actors, paintings of American and foreign players, death-masks, and a fine library, which was opened to theatre research workers in 1957 as a memorial to Walter HAMPDEN, the club's fourth president, who served 27 years before his retirement in 1955. He was succeeded by Howard LINDSAY, who retired in 1965, being followed by Dennis KING, by Alfred Drake in 1970, and by Roland Winters in 1978. Ladies were formerly admitted to the premises only for an afternoon reception on Shakespeare's birthday, but since 1946 have been invited to four annual Open House nights featuring entertainments and refreshments.

PLAYERS' CLUB, Toronto, see HART HOUSE THEATRE.

PLAYERS' THEATRE, London, membership club which began life as Playroom Six, on the first floor of 6 New Compton Street. It opened in June 1927, and five months later Peggy ASHCROFT made her London début there as Bessie in *One Day More*, based on a story by Conrad. When in Nov. 1929 the company moved to the ground floor of the building, the name was changed to the Players' Theatre. A further move in Apr. 1936 took the players to premises in King Street, Covent Garden, once occupied by EVANS'S (late Joy's) 'song-and-supper-rooms'. The new venture was not a success, but in Dec. 1937 Peter Ridgeway reopened the premises as the New Players', with an entertainment in the style of the old 'song-and-supper' evenings; Harold Scott was the director of the first programme of what became famous as *Ridgeway's Late Joys*. The same mixture of songs and sketches continued to fill the bill after Ridgeway's death in 1938, and in 1940 the company, under Leonard Sachs, moved to 13 Albemarle Street. In 1945 they were able to acquire part of the premises formerly occupied by GATTI's-Under-the-Arches, opening there in 1946. Their increasing popularity became universal with the production in 1953 of Sandy Wilson's musical pastiche *The Boy Friend*, and the theatre continues in the same premises to present 'Victorian entertainment', including an annual PANTOMIME, to a small but devoted membership.

PLAYFAIR, NIGEL (1874–1934), English actor-manager and director, knighted in 1928. He was intended for the law, but in 1902 made his first appearance on the stage and later toured with BENSON's company. Two years later he played Hodson in the first production of SHAW's *John Bull's Other Island*, and in 1911 appeared as Flawner Bannal in the first production of the same author's *Fanny's First Play*. In 1914 he was seen as Bottom in GRANVILLE-BARKER's production of *A Midsummer Night's Dream*. He took over the LYRIC THEATRE, Hammersmith, in 1918, and made it one of the most popular and stimulating centres of theatrical activity in London, producing there DRINKWATER's *Abraham Lincoln* (1919); in 1920 GAY's *The Beggar's Opera* which ran for nearly 1,500 performances; CONGREVE's *The Way of the World* (1924), with Edith EVANS; BICKERSTAFFE's *Lionel and Clarissa* (1925), its first revival since the original production in 1768; *Riverside Nights* (1926), an intimate revue by A. P. Herbert and others; FARQUHAR's *The Beaux' Stratagem* (1927), again with Edith Evans; *When Crummles Played* (also 1927), a burlesque of LILLO's *The London Merchant* (1731) set against the theatrical background of DICKENS's *Nicholas Nickleby*; a stylized black-and-white revival of WILDE's *The Importance of Being Earnest* (1930), with John GIELGUD as John Worthing; and numerous other plays, old and new, in many of which he himself appeared. He was also responsible for the production of the ČAPEKS' *The Insect Play* (1923) at the REGENT THEATRE, being part-author of the English translation, and for productions at many other London theatres. He remained at the Lyric Theatre until 1932, his final production being A. P. Herbert's *Derby Day*, with music by Alfred Reynolds.

PLAYHOUSE, The, London, in Northumberland Avenue. As the Royal Avenue Theatre, holding about 1,500 people in four tiers, this opened on 11 Mar. 1882 with Offenbach's *Madame Favart*, the first of a series of light operas presented by various managements. In 1890 George ALEXANDER began his career as an actor-manager with Hamilton Aidé's *Dr Bill*, and a year later Henry Arthur JONES's *The Crusaders* was a success. The most important event of these early years was the production of SHAW's *Arms and the Man* (1894), in a season managed by Florence FARR and financed by Miss HORNIMAN. Mrs Patrick CAMPBELL and FORBES-ROBERTSON were at the Avenue in 1899, Charles HAWTREY followed with Ganthony's successful *A Message from Mars* late in the same year, and in 1904 came Somerset MAUGHAM's first play *A Man of Honour*. At the beginning of 1905 Cyril MAUDE took over the theatre, and started to rebuild it, but on 5 Dec., when the work was almost completed, part of Charing Cross station collapsed on it, killing six people and injuring twenty. Maude received £20,000 compensation and started building again. The new theatre, named the Playhouse, its three-tier auditorium having a seating capacity of 679, opened on 28 Jan. 1907 with Clyde FITCH's *Toddles*. With his wife Winifred EMERY as his leading lady, Maude remained until Sept. 1915, producing and playing in a number of successful plays both old and new. In Jan. 1916 Gladys COOPER began her long association with the Playhouse when she appeared as Emily Delmar in H. M. HARWOOD's *Please Help Emily*. With Frank Curzon until 1928, and then in management by herself, she produced and played in many plays, including Maugham's *Home and Beauty* (1919) and a revival in 1922 of PINERO's *The Second Mrs Tanqueray* in which she made a great success as Paula. In 1927 she appeared as Leslie Crosbie in *The Letter*, the third of Maugham's plays to be staged at this theatre and two years later played Stella Tabret in the fourth (and last) to be seen there, *The Sacred Flame*. Her final production was Keith Winter's *The Rats of Norway* in which Laurence OLIVIER also appeared. After her departure the theatre had no settled policy until in 1938–9 Nancy PRICE took it over as a base for her People's National Theatre, as the Playhouse Theatre. It then closed for a time, to reopen in 1942 under its former name, with a revival of Maugham's *Home and Beauty*. In 1943 a season by the OLD VIC company, bombed out of their own theatre, brought a revival of DRINKWATER's *Abraham Lincoln*, staged by Tyrone GUTHRIE, a translation of a new Soviet play, *The Russians* by K. SIMONOV, and an early play by Peter USTINOV *Blow Your Own Trumpet*. The last memorable production at this theatre was a dramatization of Agatha CHRISTIE's *Murder at the Vicarage* in 1949. Two years later it became a B.B.C. studio and housed a number of popular radio programmes until in 1975 it was given up.

PLAYHOUSE THEATRE, New York, at 137 West 48th Street, between Broadway and the Avenue of the Americas. Built by William A. BRADY, this opened on 15 Apr. 1911, and had its first success with BROADHURST's *Bought and Paid For* the same year. In 1915–16 Brady's wife Grace GEORGE appeared in a repertory which included H. A. JONES's *The Liars*, the first American performance of SHAW's *Major Barbara*, and a revival of his *Captain Brassbound's Conversion*. Later successful new plays were *For the Defence* (1919) by Elmer RICE and *The Show-Off* (1924) by George KELLY. The next important productions were Robert SHERWOOD's *The Road to Rome* (1927), and Rice's *Street Scene* (1929), which won him a PULITZER PRIZE. After a somewhat blank period the theatre had a further success with ABBOTT and Holm's *Three Men on a Horse* (1935), followed by Mark Reed's *Yes, My Darling Daughter* (1937) and by several successful revivals, including MAUGHAM's *The Circle* and Sutton Vane's *Outward Bound*, both in 1938. In 1945 Tennessee WILLIAMS's *The Glass Menagerie*, in which Laurette TAYLOR made her last appearance, began a long run, and in 1959 Gibson's *The Miracle Worker*, with Anne Bancroft, was highly successful. The theatre was demolished in 1968.

Another Playhouse Theatre is situated off-Broadway at 359 West 48th Street. Originally a Presbyterian church, it opened in 1970 as a two-theatre structure with the 499-seat Playhouse Theatre on the upper floor and the 200-seat Bijou Theatre on the lower.

PLAYROOM SIX, see PLAYERS' THEATRE.

PLAYWRIGHTS' THEATRE, New York, see PROVINCETOWN PLAYERS.

PLINGE, WALTER, name used on English playbills to conceal a doubling of parts, particularly in a Shakespeare play, the American equivalent being George SPELVIN. There are two versions of the origin of the name. It may have been that of the landlord of a public house near the stage door of the LYCEUM in about 1900 who was popular with BENSON's company, or it may have been a fictitious name for a convivial acquaintance, the announcement by one of the company that Mr Plinge was waiting being an invitation to go for a drink. Oscar ASCHE may have been the first to use it on a playbill. It still makes an occasional appearance in London and provincial cast lists. The name was never used, as Spelvin's was, for dolls or animal actors.

PLOUGH [or PLOW] **MONDAY**. One of the main English FOLK FESTIVALS, on the first Monday after Twelfth Night (6 Jan.). A survival of an ancient ritual of the agricultural year, which includes the blessing and procession of the plough through the village, and a collection or *Quête* in the style of the MUMMERS' PLAY. There is some evidence that the Wooing Ceremony form of the Mummers' Play was at one time connected with the Plough Monday procession and may indeed have grown out of it.

PLOWRIGHT, JOAN ANNE (1929–), English actress, the third wife of Laurence OLIVIER. She made her first appearance on the stage in 1951, and in 1956 joined the ENGLISH STAGE COMPANY, making an outstanding success at the ROYAL COURT THEATRE as Margery Pinchwife in WYCHERLEY's *The Country Wife*. She was also much admired as the Old Woman in IONESCO's *The Chairs* (1957), in which she made her first appearance in New York a year later; as Beatie, the working-class girl who discovers culture, in WESKER's *Roots* (1959), in which she had been seen earlier in the year at the BELGRADE THEATRE in Coventry; and as Daisy in Ionesco's *Rhinoceros* (1960), in which year she also played Josephine in Shelagh Delaney's *A Taste of Honey* in New York. During the first CHICHESTER FESTIVAL in 1962 she gave a most interesting performance as Sonya in CHEKHOV's *Uncle Vanya*, and then, with her husband, joined the NATIONAL THEATRE COMPANY at the OLD VIC on its inception, playing Sonya again and also the title role in SHAW's *Saint Joan* (1963), Maggie in BRIGHOUSE's *Hobson's Choice*, and Hilde in IBSEN's *The Master Builder* (both 1964). Among her other parts with the National have been Beatrice in ZEFFIRELLI's production of *Much Ado About Nothing* (1967), Portia in *The Merchant of Venice* (1970), and Rosa, the harassed Neapolitan mama in DE FILIPPO's *Saturday, Sunday, Monday* (1973), which also had a long West End run. At the LYRIC THEATRE in 1975 she appeared in repertory as Arkadina in Chekhov's *The Seagull* and Alma in Ben TRAVERS's *The Bed Before Yesterday*. In 1977 she had another long run in De Filippo's *Filumena*, playing the same role briefly in New York in 1980. Later in 1980 she returned to the West End, playing a working-class old lady in Alan BENNETT's short-lived *Enjoy*. Her forte is an earthy regional realism that places her characters in a precise social context with an emphasis on down-to-earth indomitability.

PLUCHEK, VALENTIN NIKOLAYEVICH (1909–), Soviet director, who studied under MEYERHOLD during the 1920s, the latter's influence being much in evidence in his later productions of BEAUMARCHAIS's *Le Mariage de Figaro* and GOGOL's *The Government Inspector*. While working as an actor he also organized a theatre group drawn from young workers in the electrical industry, and during the Second World War he was responsible, together with ARBUZOV, for a studio theatre in Moscow which took productions to units at the front line. In 1957 he became artistic director of the Moscow Theatre of SATIRE, where his great successes included revivals of MAYAKOVSKY's plays and Russian versions of such European plays as SARTRE's *Nekrassov*, SHAW's *Heartbreak House*, FRISCH's *Don Juan, oder Die Liebe zur Geometrie*, and MOLIÈRE's *Les Fourberies de Scapin*.

PLUMMER, (Arthur) CHRISTOPHER ORME (1929–), gifted and versatile Canadian actor, with a flair for irony and flamboyant comedy. He gained his early experience with the Canadian Repertory Company in Ottawa and made his New York début in 1954. For several years he appeared mainly in Shakespeare, being seen as Mark Antony in *Julius Caesar* and Ferdinand in *The Tempest* in the opening season at Stratford, Conn., in 1955 (see AMERICAN SHAKESPEARE THEATRE). In 1956 he played Henry V at the STRATFORD (ONTARIO) FESTIVAL, where his other roles included Aguecheek in *Twelfth Night* and Hamlet the following year, Leontes in *The Winter's Tale* (1958), the title roles in *Macbeth* and ROSTAND's *Cyrano de Bergerac* (1962), and Antony in *Antony and Cleopatra* (1967). In 1961 he was seen at the ROYAL SHAKESPEARE THEATRE as Benedick in *Much Ado About Nothing* and as Richard III. He also appeared in a number of modern plays in New York, including FRY's *The Dark is Light Enough* (1955), MacLeish's *J.B.* (1958), and SHAFFER's *The Royal Hunt of the Sun* (1965). In London he played Henry II in ANOUILH's *Becket* (1961) and was in revivals of GIRAUDOUX's *Amphitryon 38* and BÜCHNER's *Danton's Death* for the NATIONAL THEATRE company in 1971. He portrayed Anton CHEKHOV in Neil SIMON's *The Good Doctor* in New York in 1973, and in 1981 returned to the American Shakespeare Theatre to play Iago in *Othello* and the title role and Chorus in *Henry V*.

PLYMOUTH THEATRE, New York, at 236 West 45th Street. This 1,063-seat theatre opened on 10 Oct. 1917, and a year later saw the first production in New York in English of IBSEN's *The Wild Duck*, starring NAZIMOVA, who was also seen in his *Hedda Gabler* and *A Doll's House*. Later in the year John BARRYMORE appeared in a dramatization of TOLSTOY's *Redemption*, renamed *The Living Corpse*, and in 1919 he was seen with his brother Lionel BARRYMORE in BENELLI's *The Jest*. Among later successes were *What Price Glory?* (1924), a realistic portrayal of war by Maxwell ANDERSON and Laurence Stallings which had a long run; *Burlesque* (1927) by Arthur Hopkins and D. G. M. Walters; Elmer RICE's *Counsellor-at-Law* (1931); Robert SHERWOOD's adaptation of Deval's *Tovarich* (1936) and Sherwood's own *Abe Lincoln in Illinois* (1938), awarded a PULITZER PRIZE. Other notable productions were *Lute Song* (1946), adapted by Sidney HOWARD from a Chinese play; SHAW's *Don Juan in Hell* (1952), from *Man and Superman*; and GIRAUDOUX's *Tiger at the Gates* (1955). In the 1960s came WESKER's *Chips with Everything* (1963), Alec GUINNESS in Sidney Michaels's *Dylan* (1964); and Neil SIMON's *The Star-Spangled Girl* (1966) and *Plaza Suite* (1969); while the 1970s brought Peter SHAFFER's *Equus* (1974) and Simon GRAY's *Otherwise Engaged* (1977). In 1981 the Plymouth staged the ROYAL SHAKESPEARE COMPANY's productions of *Piaf* by Pam Gems and a marathon adaptation of DICKENS's *Nicholas Nickleby*.

POCOCK, ISAAC (1782–1835), English dramatist, a prolific writer of popular works which in a more literary age would have been relegated to the minor theatres, but which, in the general poverty of playwriting at the time, were accorded productions at COVENT GARDEN, DRURY LANE, or at least at the HAYMARKET THEATRE. He is mainly remembered today for one of his melodramas, *The Miller and His Men* (1813). This, with its romantic scenery, strong situation, and final blowing up and burning down of the mill, was one of the most popular products of the TOY THEATRE, and is occasionally revived in cardboard even now. He was also responsible for one of the many versions of the story of the thieving magpie, which as *The Magpie, or the Maid?* (1815) was another Toy Theatre favourite. Among his other plays the farce *Hit or Miss* (1810) provided the elder Charles MATHEWS with an excellent part as Dick Cypher, and his adaptation of SCOTT's *Rob Roy* (1818) first brought MACREADY into prominence.

POEL [Pole], WILLIAM (1852–1934), English actor and director, who altered his name in deference to his father's dislike of his chosen profession. He made his first appearance on the stage in 1876, and from 1881 to 1883 was manager of the OLD VIC under Emma Cons. His productions for the Shakespeare Reading Society led him in 1894 to found the Elizabethan Stage Society, which was to have an enormous influence on the staging and production of Shakespeare in the first half of the 20th century. On a stage modelled in accordance with his ideas of an Elizabethan stage, with the minimum of scenery, and with music by the Dolmetsch family, Poel produced in a variety of halls and courtyards a number of Elizabethan plays by Shakespeare, MARLOWE, JONSON, BEAUMONT and FLETCHER, MIDDLETON, ROWLEY, and FORD, beginning with *Twelfth Night* in 1895. He also staged the Dutch medieval MORALITY PLAY *Everyman* for the first time for 400 years. His last production for the Elizabethan Stage Society was *Romeo and Juliet* in 1905. Financially the venture had not been a success, but artistically it vindicated Poel's theories, and it undoubtedly stimulated other directors to experiment with simple settings and so free Shakespeare from the cumbersome trappings of the late 19th century. Poel continued to work in the theatre until his death, and was responsible for revivals, under various auspices, of the old improvised *Hamlet* play of the ENGLISH COMEDIANS, *Fratricide Punished*, first seen in England at the OXFORD PLAYHOUSE in 1924; of the anonymous *Arden of Feversham* (1925); and of PEELE's *David and Bethsabe* (1932) for the first time since 1599.

POELZIG, HANS (1869–1936), German architect, who converted the Zirkus Schumann in BERLIN into the Grosses Schauspielhaus for Max REINHARDT. Intended to achieve the maximum unity of stage and auditorium, it had seating for over 3,000 spectators on three sides of a U-shaped forestage made up of three movable segments. This could be shut off by a movable wall from a main stage 30 metres deep and 20 metres wide, flanked by banks of steps, and having an 18-metre revolve. Behind this was a hugh domed CYCLORAMA illuminated from the flies by banks of spotlights in five colours, and pierced with stars which were echoed in the light-tipped stalactites decorating the dome above the auditorium. This early attempt to get away from the proscenium arch was a seminal design for later arena theatres. The Grosses Schauspielhaus, which opened in 1919 with Reinhardt's production of AESCHYLUS' *Oresteia*, was later modified and renamed the Friedrichsstadt-Palast.

POETIC DRAMA, a term applied to plays written in verse or in a heightened, 'poetic' form of prose, which in the 19th and 20th centuries constituted an attempt to restore the medium of poetry to the stage. In earlier times all plays throughout Europe were in verse, and tragedy continued to be so written long after prose had become the accepted medium for comedy. Shakespeare interpolated comic scenes in prose into his great poetic plays; and by the time of DRYDEN, prose comedies existed side by side with tragedies in verse. As the theatre increasingly attracted a mass audience, prose (with a greater or lesser approximation to everyday speech) became the accepted mode of expression for all plays. Poets who wrote works in dramatic form rarely intended them for the stage; attempts by such distinguished performers as Edmund KEAN, MACREADY, and IRVING to present plays by even major poets usually failed because of the dramatists' lack of stagecraft. Lord BYRON's *Werner* (1830), BROWNING's *A Blot in the 'Scutcheon* (1843), SHELLEY's *The Cenci* (1886), and TENNYSON's *Becket* (1893), had some success, but on the whole the public preferred the rhetorical dramas of Sheridan KNOWLES and BULWER-LYTTON. The poet John Davidson (1857–1909) wrote a number of plays which were never performed, and Thomas Hardy's epic drama *The Dynasts* (pub. 1904–8) has never been staged in its entirety. Stephen PHILLIPS briefly brought blank verse back into the commercial theatre in London with *Herod* (1900), *Ulysses,* and especially *Paolo and Francesca* (both 1902); but he was the last of the poetic dramatists in the tradition of the 19th century.

About the turn of the century, poetic drama, under such diverse influences as the *nō* plays of JAPAN, IBSEN (whose 'realism' is fundamentally that of a poet), and the writings of the French symbolist poets, became more assured, and its authors were encouraged to think in terms of a theatre of their own. The leaders of the Irish literary revival, YEATS, SYNGE, and Lady GREGORY, produced plays of great poetic beauty combined with sound dramatic structure. They did not, however, establish the poetic theatre that had been hoped for, though Yeats's integrity as a poet and dramatist raised the standards of poetic drama and deeply influenced his Irish and English contemporaries. Synge's plays, though written in prose, are the work of a true poet, and so are those of O'CASEY.

Among the English poetic dramas of the early 20th century John MASEFIELD's *The Tragedy of Nan* (1908), in poetic prose, John DRINKWATER's *Rebellion* (1914), Gordon BOTTOMLEY's *Gruach*, and FLECKER's *Hassan* (both 1923) were the most important. In the 1930s a number of poets broke away from the traditional verse play, employing free verse and the concepts of modern symbolism. T. S. ELIOT's *Murder in the Cathedral* (1935) was notable for its fine poetry, and has frequently been revived, while *The Dog Beneath the Skin* (1936) and *The Ascent of F.6* (1937) by W. H. Auden and Christopher Isherwood were valued in their day for their wit, satire, and social criticism.

The first important American poetic dramatist was William Vaughn MOODY; his work and teachings influenced, among others, Josephine Preston Peabody and Percy Mackaye. Maxwell ANDERSON carried on a long and valiant fight to establish poetic drama on the American stage, but his poetry was not of the first rank and his dramatic construction often left something to be desired. Other American poets have written plays, for the most part characterized by great individuality and a considerable degree of experiment, but only a few have been produced and even those which were have not been revived. On the whole the production of poetic drama in the 20th-century theatre has had to depend on university theatres, drama schools, and groups specifically formed to present them. In England in the 1930s, a revival of RELIGIOUS DRAMA led to the writing and staging of a number of plays in verse at Canterbury and other cathedrals, including *The Zeal of Thy House* (1937) by Dorothy L. Sayers, which later had a London run, and *Christ's Comet* (1938) by Christopher Hassall. After the Second World War the MERCURY THEATRE, where *Murder in the Cathedral* and *The Ascent of F.6* both had their first London productions, was for a time exclusively devoted to the production of poetic drama, among its successful presentations being Ronald Duncan's *This Way to the Tomb*, Anne Ridler's *The Shadow Factory* (both 1945), and Christopher FRY's *A Phoenix Too Frequent* (1946), and poetic drama returned to the commercial theatre again with the production of new plays by Christopher Fry and T. S. Eliot, while Dylan Thomas's *Under Milk Wood* (1953), written for radio, had many stage readings and performances, but these were isolated phenomena.

After the collapse of the FEDERAL THEATRE PROJECT in the United States in 1939, the Poets' Theatre, founded in 1951 in Cambridge, Mass., became one of the most important agencies for the commissioning and production of poetic drama. Richard Eberhart, its founder and first president, wrote *The Apparition* (1951) and *The Visionary Farm* (1952), both well received; Archibald MacLeish's poetic *J.B.* (1958), a modern commentary on the story of Job, had a long run on Broadway. During the 1950s and 1960s adaptations in English of poetic plays by ANOUILH and GIRAUDOUX, and later those of BECKETT and IONESCO, which combine poetry of a high order with surrealism and the world of dreams, were commercially successful in London and New York, but poetic drama has during the past decades become increasingly identified with experimental work, and the theatre as a whole remains firmly committed to prose.

POGODIN [Stukalov], NIKOLAI FEDOROVICH (1900–62), Soviet dramatist, one of the outstanding figures in the Russian theatre of his time. Originally a journalist, his first play *Tempo* (1930) was a documentary on building produced at the VAKHTANGOV THEATRE, which also staged his best known work *Aristocrats* (1934), dealing with the regeneration of a number of petty criminals engaged on the construction of a canal from the Baltic to the White Sea. Both these were published in English translations in *Six Soviet Plays* (1934), and *Aristocrats* was produced in London at UNITY THEATRE in 1937. Of his other plays the most successful was *Poem About an Axe* (1931), produced by POPOV, as was *My Friend* (1932). Two war plays, *Moscow Nights* (1942) and *The Ferryboat Girl* (1943), were also well received, as was a trilogy based on the life of Lenin—*The Man With the Gun* (1937), *Kremlin Chimes* (1942), and *The Third Pathétique* (1958). A revised version of *Kremlin Chimes* was produced in 1956 by the MOSCOW ART THEATRE, who in the WORLD THEATRE SEASON of 1964 were seen in London in the second play of the trilogy.

POISSON, FRANÇOIS ARNOULD (1696–1753), French actor, grandson of BELLEROCHE, whose stutter he inherited. He was intended for the army but ran away to join a company of strolling players and later became a member of the COMÉDIE-FRANÇAISE, against the wishes of his family. He was excellent as the valet in comedy, and in heavy, humorous parts, his most admired role being Lafleur in DESTOUCHES's *Le Glorieux* (1732).

POISSON, PAUL (1658–1735), French actor, son of BELLEROCHE, who in 1686 joined the COMÉDIE-FRANÇAISE, playing many of his father's old parts, including CRISPIN, with the same stutter, which he inherited. He remained with the company, except for a short break between 1711 and 1715, until 1724. His wife, Marie-Angélique de l'École (1657–1756), was the step-daughter of MOLIÈRE's actor DU CROISY, and an excellent actress who remained with the Comédie-Française from its foundation until she retired in 1694. In her old age she wrote articles on Molière and his company for the *Mercure de France*. She was the mother of François (above) and of Philippe (1682–1743), who was on the stage for a short time but disliked it, retiring in 1722 to write plays of which the best was *L'Impromptu de campagne* (1733). His sister Madeleine-Angélique (1684–1770) was also a dramatist, author of four tragedies of which *Habis* (1714) was the most successful.

POLAND. The origins of the theatre in Poland, as in all European countries, must be looked for in the liturgy of the Christian church, which gave rise to LITURGICAL DRAMA and eventually to religious drama in the vernacular. Two plays of this type which have been revived on the modern stage by the director Kazimierz DEJMEK are *The Life of Joseph* by Mikołaj Rej (1505–69), revived in 1958, and *The Glorious Resurrection of Our Lord* (*c.* 1575) by Nicolas of Wilkowiecka, seen in London in the WORLD THEATRE SEASON of 1967. Dejmek's version of the latter includes scenes from other Polish medieval texts and comic interludes of low life which testify to the continued tradition of folk-plays in the vernacular which continued into the 17th century with such examples as *A Peasant Made King* (1637) by Piotr Baryka. Long before this the Renaissance had brought to Poland a knowledge of classical Roman drama, and it was the poet Jan Kochanowski who in 1578 produced the first Polish secular play of any importance, *The Dismissal of the Grecian Envoys*, a rhetorical drama in the style of SENECA. But this remained something of an anomaly, and for a time plays continued to be written in Latin, as in JESUIT DRAMA, or translated from Latin authors, the best example being Piotr Ciekliński's version of PLAUTUS's *Trinummus* published in 1597.

The scattered villages and small towns of the Polish countryside favoured for some time the development of itinerant rather than permanently settled companies; John Green, for instance was in Poland with his ENGLISH COMEDIANS in 1616, and perhaps even earlier. It was probably the Italian COMMEDIA DELL'ARTE travelling troupes which had the greatest influence on the development of Polish comedy; their productions, fused with the native Shrovetide entertainments, produced a unique form of village farce, much of which was destined to perish in the upheavals of the 17th century, though some 20 texts have fortunately survived.

The Court gave Poland its first permanent theatre building in 1637; but as in so many other Court Theatres of the time, this one in Warsaw was devoted to opera and used mainly by visiting French, German, and Italian companies. It was not until 1765 that a National Theatre, to be used solely for plays in Polish, opened in the capital, and even then the plays produced there were all translations or adaptations of foreign authors, the first production being *The Intruders*, based on MOLIÈRE's *Les Facheux* by Józef Bielawski (1739–1809). Prominent figures in the cultural life of the country saw that if a Polish theatre were to flourish the greatest need was for good Polish playwrights; one of the most influential during the late

18th century was the cousin of the reigning king, Prince Adam Kazimierz Crzartoryski, whose comedy *A Marriageable Miss* (1771), based on GARRICK's *Miss in her Teens*, was published with an introduction setting out the principles of playwriting.

In 1783, when the National Theatre was taken over by Poland's first professional director, Wojciech BOGUSŁAWSKI (1757–1829), Poles could at last look forward to a repertory of original plays by native playwrights interpreted by Polish-trained actors. It was necessarily a slow business, and apart from the satiric comedies of Franciszek Zabłocki (1754–1821), which still drew their inspiration mainly from France; the political comedy *The Envoy's Return* (1791) by Julian Ursyn Niemcewicz (1757–1841); Bogusławski's own plays, particularly *Cracovians and Mountaineers* (1794); and the neo-classical verse-tragedy, *Barbara Radziwiłłówna* (1817) by Alojzy Feliński (1771–1820), which was also the last of its kind, there was little of moment to record in Polish drama. On the other hand, the Polish theatre advanced rapidly. A training school for actors was set up in connection with the National Theatre by Bogusławski, who was himself the first Polish actor to play Hamlet, in his own version of the play staged in 1797, and he also founded theatres in Poznań, Kalisz, Danzig, Białystok, Vilna, and Lwów. Several of these later closed, but Bogusławski's dedication and persistence established for Polish theatre a firm basis from which it profited in due course.

It was Poland's misfortune that the finest plays of the 19th century were destined to be written in exile, and rarely, if ever, staged during the lifetime of their authors. The greatest of these was Adam MICKIEWICZ, whose early poems heralded the advent of Romanticism into Polish literature. Published as part of his poetic output, but later adapted and staged, was his vast four-part drama *Forefathers' Eve*, which was known and venerated as a masterpiece of Polish literature long before its first production in Cracow in 1901. Other dramatists who shared Mickiewicz's fate were Juliusz SŁOWACKI, whose finest tragedy, *Kordian* (written in 1832, and much influenced by *Macbeth*) was not staged until 1899; Zygmunt KRASIŃSKI, author of *Irydon* and *The Undivine Comedy*, both posthumously produced; and Cyprian Kamil NORWID, who remained almost unknown until about 1908. The only outstanding dramatist of the time to remain in Warsaw and see his plays on the stage was Aleksander FREDRO, author of a number of sparkling comedies which retain their place in the repertory. They derived from the submerged tradition of peasant farce combined with the long apprenticeship of Polish writers to the art of sophisticated French comedy, and their influence was strongly felt by the later writers of comedy, Józef Bliziński (1827–93) and Michał Bałucki (1837–1901). Though Bliziński also wrote, in such plays as *The Wrecks* (1877), problem plays dealing with the moral perplexities of his age, Bałucki's work was all of a piece, dealing only with the humorous side of middle-class life in such comedies as *Big Shots* (1881) and *The Bachelors' Club* (1890), both clearly indebted to Fredro's *Ladies and Hussars* (1825). Meanwhile social problems were being ventilated by such writers as Józef Narymski (1839–72), much in the style of AUGIER, and Gabriela Zapolska [Korwin Piotrowska] (1857–1921), a Polish actress who worked in ANTOINE's theatre in Paris, and became an advocate of NATURALISM in such plays as *The Morality of Mrs Dulska* (1907). Another Polish actress who won international fame was Helena Modrzejewska, better known as MODJESKA.

Among the writers of the Young Poland movement one of the most important in the theatre was Stanisław WYSPIAŃSKI, author of a number of poetic dramas based on national themes, and a director whose innovations in stagecraft had a great influence on his successors, notably Leon SCHILLER. Other Young Poland dramatists who helped to make the period between 1890 and 1914 a brilliant one for Polish drama, first in Cracow under Tadeusz Pawlikowski and later in Warsaw under Leon Schiller, were Stanisław Przybyszewski (1868–1927), most of whose plays were written with DUSE in mind; Łucjan Rydel (1870–1918), author of the charming folk-fantasy *The Enchanted Circle* (1899); Tadeusz Rittner (1873–1921), doomed to spend his life in exile but setting his plays in his native Galicia; Jan August Kisielewski (1876–1918), portrayer of city life in Cracow; and Włodzimierz Perzyński (1878–1930), whose gift for dialogue, and unusual grasp of dramatic technique—unusual for those days—was shown in such satiric comedies as *Prodigal Sister* (1904), which is still popular.

With a few exceptions the period between the wars was less fruitful dramatically, though materially the theatres flourished. In Warsaw the Polski Teatr, which had opened in 1913 under Arnold Szyfman [Arnold Stanisław Zygmunt Schiffman] (1882–1967), saw the world première of SHAW's *The Apple Cart* (1929), as well as the European premières of his *Too True to be Good* in 1932 and *The Simpleton of the Unexpected Isles* in 1935; the Reduta Teatr, founded in 1919 by Juliusz Osterwa (1885–1947), became known for its productions of new avant-garde plays but also revived a number of Polish classics, among them Zabłocki's *The Dandy's Courtship*, first seen in 1781; the Ateneum, directed by the distinguished actor Stefan Jaracz (1883–1947), and the Bogusławski, founded and directed from 1924 to 1926 by Leon Schiller, also did good work, as did the Jewish Art Theatre under Ida KAMIŃSKA. Another remarkable producer was Wilam Horzyca (1880–1959), and among the outstanding actors were Kazimierz Junosza-Stepowski (1882–1943), who played King Magnus in *The Apple Cart*, and the great doyen of the Polish stage, Ludwig Solski [Sosnowski] (1885–1954) who retained his powers until the end of his long life and was at his best in comedy, where his wit, dexterity, and perfect timing had full play. There was thus no lack of theatre buildings and good actors to play in them, but with a few exceptions there was very little original work from the many playwrights of the day. Some are best remembered for one play—Bruno Winawer (1883–1944), for instance, whose *Book of Job* (1921) was translated into English by Joseph Conrad, or Jarosław Iwaszkiewicz (1894–1980), whose *Summer in Nohant* (1936) was seen in London in 1946. The veteran novelist Stefan Żeromski (1864–1925), who took to playwriting late in life, scored a success with a play on Polish life in the 1880s, *My Little Quail Has Flown Away* (1924), as did Adolf Nowaczyński (1876–1944) with his comedy *War for War* (1928), and Maria Morozowicz-Szczepkowska (1889–1968) with *Dr Monica* (1933), seen in New York in 1934 with NAZIMOVA in the title role. These and many more kept up a general level of competence; but the most important writer of the time, and one destined later to have a great influence, not only in Poland but throughout Europe, Stanisław Ignacy WITKIEWICZ, remained practically unknown until the 1950s. His plays, which are now hailed as forerunners of the Theatre of the ABSURD, of the Teatro GROTTESCO, and even of EXISTENTIALISM, were performed only in small experimental theatres and attracted little attention. His contemporary Jerzy Szaniawski (1886–1970), though his early plays—*The Bird* (1923), *The Lawyer and the Roses* (1930), *The Girl From the Forest* (1939)—were staged, had to wait until after the failure of SOCIALIST REALISM in 1958 before he was accorded full recognition. Then, with such plays as *The Lady Archer* (1959) and *Nine Years* (1960), he became one of Poland's most popular playwrights.

The Second World War brought almost total destruction to the Polish theatre. Many buildings were destroyed by bombing, many prominent theatre men—actors, dramatists, directors—died or were driven out; but as soon as the war was over a remarkable resurgence led to an unprecedented flowering of new talent and a widespread increase in the number of theatre buildings and their audiences, until by 1976 there were at least 100 playhouses in use throughout the country. They included not only the older theatres, rebuilt and refurbished, but a number of new ones, among them the Contemporary Theatre in Warsaw, founded in 1949 by Erwin AXER and later run as the New Theatre by Kazimierz Dejmek; the People's Theatre in the new town of Nowa Huta, founded in 1955 by the actress Krystyna SKUSZANKA and her husband Jerzy Krasowski and later run by Józef SZAJNA, who had been with it from the beginning; the Cricot 2, an experimental theatre founded in 1956 in Cracow by the director and designer Tadeusz KANTOR; the Popular Theatre in Warsaw, founded in 1963 by Adam Hanuszkiewicz, who in 1966 took his company in Wyspiański's *The Wedding* to the WORLD THEATRE SEASON in London; and the Theatre Laboratory in Wrocław, founded in 1965 by Jerzy GROTOWSKI, probably one of the most influential directors of the present day. A director of the post-war period of whom great things were expected was Konrad ŚWINARSKI, who died prematurely in an air crash after a short but spectacular career. Among those still active are Janusz Warmiński, who in 1960 took over the rebuilt Ateneum, where he specializes in translations of English and American plays and Polish classics; Maciej Prus, who in 1979 revived Mickiewicz's *November Night* with Gustaw Holoubek in the lead; Andrzej Wajda, whose production of the same play at Cracow in 1974 was seen in the World Theatre Season a year later; Jerzy Jarocki, who has been responsible for some fine Shakespeare productions but has also directed a number of modern plays; and Zygmunt Hübner, who in 1978 staged his own adaptation of Moczarski's novel *Conversations with a Hangman*. Among the outstanding players of the day are Tadeusz Łomnicki, who in 1975 played The Host (Rydel) in Hanuszkiewicz's revival of Wyspiański's *The Wedding*; Irena Eichlerówna Eichler, who returned from exile after the Second World War to lend her resonant voice and commanding presence to leading roles in tragedy; and Halina Mikołajska, who played Ethel in KRUCZKOWSKI's *Julius and Ethel* (1954) at the Polski Teatr, and in 1970 was seen as the Mother in Witkiewicz's play of that name, revived by Axer.

A Polish playwright of the postwar era who, like so many before him, spent most of his life in exile, was Witold GOMBROWICZ, whose plays, mostly produced abroad, were not seen in Poland until after 1957; but the best known outside his own country today is undoubtedly Sławomir MROŻEK, author of the internationally famous *Tango* (1965). Others who should be mentioned are Tadeusz Rozewicz, author of the surrealist *Card Index* (1960), the satiric *Laocoon Group* (1962), and, more recently, of *On All Fours* (1972), the tragi-comedy of a writer, *White Wedding* (1975), and *Dead and Buried* (1979); and Ernest Bryll, whose controversial *The November Affair* (1968), directed by Skuszanka, upset the public, but is regularly revived through Poland. Among his later plays *Painted on Grass* (1970), a musical featuring the national folk-hero Janosik, has been equally popular.

The Polish theatre, which has always been noted for its readiness to experiment, will no doubt continue to do so as new playwrights and directors come to the fore, with new and well-trained players to work for them. Exciting work is being done in the Stary Theatre in Cracow, in Łódź, in Wrocław, with its three complementary theatres, in Gdańsk, Katowice, Lublin, and elsewhere. In Warsaw the established theatres continued to do good work, and were joined in 1975 by the rebuilt and ultra-modern Powszechny. Theatre criticism and journalism are lively and alert, both in the regular press and in a number of specialized journals, while the monthly bulletin of the Polish Centre of the International Theatre Institute keeps theatre-lovers abroad up to date with the latest developments. Two influential writers on the theatre were the writer and translator Tadeusz Żeleński (known as Boy) (1895–1941), who died at the hands of the Nazis, and the controversial but stimulating Jan KOTT, who emigrated in 1968 and now teaches in America.

POLEVOY, NICHOLAS ALEKSEYEVICH, see RUSSIA AND SOVIET UNION.

POLICHINELLE, French character, particularly prominent in PUPPET shows, which derived from the COMMEDIA DELL'ARTE mask PULCINELLA. Like his prototype, he was humpbacked with a big hooked nose, but not at all doltish, being regarded by many as the epitome of quick wit and sardonic Gallic humour. He flourished particularly in Lyon and the surrounding district, and in 1660 was imported into England in the wake of Charles II's entourage, where as Punchinello he took fashionable society by storm and was transmogrified into the very English figure of PUNCH.

POLISH STATE JEWISH THEATRE, see KAMIŃSKA, IDA.

POLITIS, FOTOS (1890–1934), the first director of the Greek National Theatre, founded in 1930. In the four years before his death he laid the foundations of the company's excellent ensemble playing, imbuing all the actors who worked under him with his own sense of devotion and self-abnegation in the service of the theatre. Although he was open to all the influences of the modern theatre and his productions were often experimental, he was convinced that the strength of the Greek theatre lay in its past, and he always insisted on the importance of maintaining the classical repertory, a task carried on by his successors MINOTIS and RONDIRIS.

POLIZIANO, ANGELO (1454–94), Italian scholar, sometimes known as Politian. Born Angelo Ambrogini, he dropped his surname after the death of his father and assumed a new name derived indirectly from his birthplace, Montepulciano in Tuscany. As a precocious youth he became the friend and protégé of Lorenzo de' MEDICI (the Magnificent), and ultimately tutor to his two sons. He was the author of the first important play to be written in the vernacular, the *Favola d'Orfeo* (1472), performed in Mantua to celebrate the betrothal of Clara Gonzaga. With no act divisions and a MULTIPLE SET, as in the SACRA RAPPRESENTAZIONE, it nevertheless takes its subject not from the Bible but from classical mythology and the prologue, formerly spoken by an angel, is given to the Roman god Mercury. It thus forms an interesting link between the old and new drama, and points forward to the imminent development of new forms of secular theatre, particularly the PASTORAL.

POLOTSKY, SIMEON [Samuil Yemelyanovich Petrovsky-Sitnianovich] (1629–80), Russian churchman, poet, and dramatist, who was the author of two plays, the *Comedy-Parable of the Prodigal Son* (published in 1685) and *Nebuchadnezzar the King, the Golden Calf, and the Three Youths Unconsumed in the Fiery Furnace*. In the first he inveighs against the gilded youth of Russia who disgrace their families through promiscuity and inebriation; in the second a wise and just Tsar puts an end to the tyranny from which his people have been suffering. For the first time full use seems to have been made in their production of incidental music, choral singing, dancing, lighting effects, and transformation scenes, to provide sumptuous theatrical representations. Both productions enjoyed some success at Court, and their historical realism and vitality laid the foundation of SCHOOL DRAMA in the ecclesiastical and educational institutions of Russia, as it developed in the latter years of the 17th century.

POLUS, the most famous Greek actor of the late fifth and early fourth century BC. The recipient of high civic honours, he continued acting to an advanced age. Later writers preserved the story that, when playing the title role in SOPHOCLES' *Electra*, he achieved a particularly pathetic effect by substituting for a property urn supposedly containing the ashes of Orestes a real urn holding the remains of his own son who had just died.

POMPONIUS, LUCIUS, see FABULA 1: ATELLANA.

PONSARD, FRANÇOIS (1814–67), French dramatist, who was first attracted to the theatre by seeing RACHEL act in Lyon, his tragedies demonstrating the same reaction against the excesses of Romantic drama as was evident in the sobriety of her acting. The success of his first play, *Lucrèce* (1843), coincided with the failure of HUGO's *Les Burgraves*, and owed its success partly to the excellent acting of BOCAGE and DORVAL but also to the indulgence of the public towards a somewhat mediocre play which nevertheless represented a return to common sense and classical restraint. Ponsard wrote a number of other plays, among them a social drama, *L'Honneur et l'argent* (1853), and a play set in the period of the Directoire, *Le Lion amoureux* (1866), which remained popular for many years; his finest work was probably the historical drama *Charlotte Corday* (1850), a successful blending of Romantic freedom with classical tradition. Rachel, who had refused to play the title role in this (ably interpreted by her cousin Mlle Judith), was consoled for its success by a charming one-act idyll based on one of Horace's odes, *Horace et Lydie* (also 1850), which Ponsard wrote specially for her.

PONTI, DIANA DA, actress of the COMMEDIA DELL'ARTE, known as Lavinia and first heard of in 1582, playing young lovers' parts. After appearing with the CONFIDENTI and possibly the GELOSI, she formed a company of her own, which may have been the DESIOSI. She is known to have appeared in Lyon in 1601 and is last heard of in 1605 belonging to, and probably leading, a company under the patronage of the Duke of Mirandola.

POPE, JANE (1742–1818), English actress, who was on the stage as a child, making her adult début on 27 Sept. 1759 at DRURY LANE as Corinna in VANBRUGH'S *The Confederacy*. She was immediately successful and later took over from Kitty CLIVE in such parts as hoydens, chambermaids, and pert young ladies. In 1765 she played Beatrice to GARRICK'S Benedick in *Much Ado About Nothing*, and she was in the original production of Garrick and the elder COLMAN'S *The Clandestine Marriage* (1766); she was also the first to play Mrs Candour in SHERIDAN'S *The School for Scandal* (1777) and Tilburina in his *The Critic* (1779). She only relinquished the youthful characters in which she excelled when forced to do so by age and obesity and then proved herself equally good in elderly duenna parts, retiring in May 1808. She has sometimes been confused with a Mrs Elizabeth Pope (formerly Miss Younge) who was playing Portia when MACKLIN made his last attempt at Shylock in *The Merchant of Venice* at COVENT GARDEN in 1789.

POPE, THOMAS (?–1604), one of the original actors in Shakespeare's plays, and an original SHARER in the GLOBE and CURTAIN THEATRES. He is first heard of as being one of the players taken by the Earl of Leicester to Germany and Holland in 1586–7 in company with KEMPE. He joined the CHAMBERLAIN'S MEN on their formation in 1594, and is usually referred to as a 'clown'. He may have succeeded to some of the parts originally played by Kempe, who left the company in 1600.

POPOV, ALEXEI DMITREVICH (1892–1961), Soviet director, who was at the MOSCOW ART THEATRE from 1912 to 1918 and in 1923 was at the VAKHTANGOV THEATRE, where he played an important part in the development of the company. In 1931 he was invited to direct the Theatre of the Revolution, later the MAYAKOVSKY THEATRE, where he was responsible for the production of POGODIN'S *Poem About an Axe* (1931) and *My Friend* (1932), and a version of *Romeo and Juliet* (1936) which remained in the repertory for many years. In 1936 he went to the RED ARMY THEATRE in Moscow, where two years later he directed a production of *The Taming of the Shrew* which made theatre history, as well as a spectacular version of *A Midsummer Night's Dream* in 1940, both designed by N. A. SHIFRIN. During the war he was evacuated with his company, but continued to produce new plays, in particular a brilliant documentary on the defence of Stalingrad. Back in Moscow he made his theatre, now known as the Central Theatre of the Soviet Army, one of the most exciting in the city, winning particular acclaim for his revival in 1951 of GOGOL'S *The Government Inspector* and in 1956 of the rewritten version of Pogodin's *Kremlin Chimes*. A tall, strikingly handsome man, with the courtly manners of an earlier generation, he visited London in 1955 to assist at the foundation of the INTERNATIONAL FEDERATION FOR THEATRE RESEARCH. As a director he was the opposite of OKHLOPKOV in that he first considered all the details of a production and then combined them into an integrated whole, whereas Okhlopkov looked at the complete production first and then turned to consider the details.

POPOV, ANDREI ALEXEYEVICH (1918–), Soviet actor and director, whose career has to some extent been overshadowed by that of his father Alexei (above), whom he succeeded as Artistic Director of the RED ARMY THEATRE in 1963. He established his reputation earlier with brilliant performances in such contrasting roles as Petruchio in *The Taming of the Shrew* in 1956, Khlestakov in GOGOL'S *The Government Inspector* in 1951, and the Tsar in A. K. TOLSTOY'S *The Death of Ivan the Terrible* in 1960. As director of the Red Army Theatre he maintained the traditional association with the armed forces, frequently staging plays with a military theme.

POPOVIČ, JOVAN STERIJA, see YUGOSLAVIA.

PORTA, GIAMBATTISTA DELLA, see DELLA PORTA.

PORTABLE THEATRES, see BOOTHS.

PORTAL OPENING, see FALSE PROS(CENIUM).

PORTER, ERIC RICHARD (1928–), English actor, who made his first appearance in 1945 at the SHAKESPEARE MEMORIAL THEATRE and after touring with WOLFIT and two seasons with the BIRMINGHAM REPERTORY COMPANY, joined GIELGUD'S company at the LYRIC THEATRE, Hammersmith. He was for a time with the BRISTOL OLD VIC, where he played Becket in a revival of T. S. ELIOT'S *Murder in the Cathedral* and Father James Browne in Graham GREENE'S *The Living Room*, as well as a number of classical parts, including the title roles in CHEKHOV'S *Uncle Vanya* and JONSON'S *Volpone*. In 1957 he toured England with the LUNTS in DÜRRENMATT'S *Time and Again*, in which in 1958 he made his début in New York, where it was renamed *The Visit*. Back in England, he was seen at the ROYAL COURT THEATRE in 1959 as Rosmer in IBSEN'S *Rosmersholm* and then joined the ROYAL SHAKESPEARE COMPANY, where his roles included a fine Ulysses in *Troilus and Cressida* (1960), the title roles in ANOUILH'S *Becket* (1961) and in *Macbeth* (1962), Barabas in MARLOWE'S *The Jew of Malta* in 1965 (in conjunction with Shakespeare's Shylock), and Ossip in GOGOL's *The Government Inspector* (1966). In 1968 he was seen as King Lear and also as Marlowe's Doctor Faustus, playing the latter role on tour in the U.S.A. a year later. At Christmas 1971 he doubled the parts of Mr Darling and Captain Hook in BARRIE's *Peter Pan*, and in 1976 he played Malvolio in *Twelfth Night*, the opening production at the ST GEORGE'S THEATRE.

PORTER, HAL (1911–), Australian novelist who has also written several plays, notably *The Tower* and *Eden House*, both produced at the ST MARTIN'S THEATRE in Melbourne, where the latter had an outstanding success in 1961 with Bettina Welch in the lead. Another play, set in Japan and originally entitled *Toda-San*, was seen at the ROYAL COURT THEATRE in London in 1965.

PORTER, MARY ANN (?–1765), English actress, who was on the stage as a child. A pupil of BETTERTON, she made her first appearance in 1699 at LINCOLN'S INN FIELDS THEATRE and understudied Mrs BARRY, later succeeding her in tragic parts for which she was ideally equipped, being tall and well-formed, with a deep voice which rendered her unsuitable for comedy; but passion, grief, and tenderness were all equally within her range, and she was particularly admired as Belvidera in revivals of OTWAY's *Venice Preserv'd*. Among her original parts were Hermione in PHILIPS's *The Distrest Mother* (1712) and Alicia in ROWE's *Jane Shore* (1714). She made her last appearance at COVENT GARDEN in 1742. In private life she was a quiet, respectable person, and enjoyed a long period of retirement in the company of her many friends.

PORTER'S HALL, London, playhouse erected by Philip Rosseter in 1615 in the precincts of Blackfriars, near Puddle Wharf. Although it was licensed by the authorities the residents objected to it and work on it was stopped, but not before it was sufficiently advanced for plays to be given. It was used by the LADY ELIZABETH'S MEN, PRINCE CHARLES'S MEN, and perhaps by the Children of the Revels, for whom it had been intended. In 1617 it was again suppressed and was apparently not used after 1618.

PORTE-SAINT-MARTIN, Théâtre de la, Paris, celebrated playhouse, built in 1782 to replace the Opéra, which had been burnt down. The opera company remained there until 1794, and after they left the building was apparently not used as a theatre again until 1802, when one of the first plays to be presented was a melodrama by PIXÉRÉCOURT, *L'Homme à trois visages, ou le Proscrit*. Several of his later works were also first seen there, and the theatre became noted for its strong dramas and spectacular shows. In 1822 an English company appeared unsuccessfully in *Othello* and in 1827 FRÉDÉRICK, playing for the first time with Mlle DORVAL whose career was to be linked spectacularly with the Porte-Saint-Martin, was seen in the famous melodrama *Trente ans, ou la Vie d'un joueur*, later to be equally successful in London in JERROLD's adaptation as *Fifteen Years of a Drunkard's Life* (1828). The great days of the theatre were in the 1830s, when it saw the first night of the elder DUMAS's *Antony* and *Le Tour de Nesle* and HUGO's *Marion Delorme* and *Lucrèce Borgia*, but with the decline of Romantic drama the fortunes of the theatre declined also and in 1840 it closed after the banning of BALZAC's *Vautrin*. When it reopened it had no settled policy, but continued to present revivals and commonplace and lachrymose melodramas like Dennery's *Marie-Jeanne, ou la Femme du peuple* (1846), in which Mme Dorval made her last appearance on stage. It was burnt down in the rioting of 1870 and rebuilt on the original plans, but somewhat smaller, and in 1873 became one of a chain of suburban theatres run by Henry Larochelle. It had a further moment of glory in the 1880s when it was acquired by Sarah BERNHARDT, who had appeared there 18 years earlier in the fairy-tale play *La Biche au bois* and now returned in a revival of MEILHAC and Halévy's *Frou-Frou*, and in 1898 the record run of ROSTAND's *Cyrano de Bergerac* again made the theatre one of the most popular in Paris. Because of its great size it was unable to compete later with the cinema, and from 1936 to 1978 it was devoted almost entirely to musical comedy. Marcel MARCEAU then took it over as a base for his École de mimodrame.

PORTMAN, ERIC (1903–69), English actor, who made his first appearance in 1924, playing a number of Shakespearian parts on tour with Robert Courtneidge and later at the OLD VIC THEATRE, where in 1928 he was much admired as Romeo to the Juliet of Jean FORBES-ROBERTSON. In the following year he was seen as Stephen Undershaft in a revival of SHAW's *Major Barbara* at WYNDHAM'S THEATRE; he also appeared as Eben in O'NEILL's *Desire Under the Elms* (1931) when it was first produced in London and as Byron in the initial production of Catherine Turney's *Bitter Harvest* (1936). He made his first appearance in the United States in 1937 in the New York production of a dramatization of Flaubert's *Madame Bovary*, and returned to London the following year to appear in PRIESTLEY's *I Have Been Here Before*. In later years he was outstandingly successful in RATTIGAN's double bill of *The Browning Version* (in which he played a repressed schoolmaster) and *Harlequinade* (1948), and gave fine performances as the Governor in the Christies' *His Excellency* (1950) and as Father James Browne in Graham GREENE's *The Living Room* (1953). He was seen later in another double bill by Rattigan, *Separate Tables* (1954), in which he also appeared successfully in New York in 1956, following it on Broadway with several important roles, notably Cornelius Melody in O'Neill's *A Touch of the Poet*

(1958) and Cherry in BOLT's *Flowering Cherry* (1959). He returned to England in 1964, and made his last appearance on the stage in a revival of GALSWORTHY's *Justice* at the ST MARTIN'S THEATRE in 1968.

PORTMAN THEATRE, London, see WEST LONDON THEATRE.

PORTO-RICHE, GEORGES DE (1849–1930), French dramatist, whose first play, a one-act comedy entitled *La Chance de Françoise*, was staged by ANTOINE at the THÉÂTRE-LIBRE in 1889. His reputation mainly rest on his *Amoureuse* (1891), a psychological analysis of the 'eternal triangle' in which RÉJANE gave an outstanding performance, which set a fashion for mildly erotic plays on sexual themes to which two later plays, *Le Passé* (1897) and *Le Vieil Homme* (1911), further contributed. Porto-Riche also made an adaptation of SHADWELL's *The Libertine*. He became a member of the French Academy in 1923.

PORTUGAL. Traces of medieval drama are tantalizingly scarce in Portugal, and the Latin LITURGICAL DRAMA is represented only by a 14th-century Christmas shepherds' ceremony for performance at Lauds. In the vernacular nothing remains. Most scholars conclude that religious drama never really took root in Portugal until the 16th century, though there were, of course, symbolic representations—for example, the placing of the reserved sacrament or of a cross in a sepulchre on Good Friday—and in the 1470s the Archbishop of Oporto specifically allowed 'some good and devout representation'.

From the 13th century there was a series of attempts, presumably unsuccessful, to prevent clergy from attending profane performances, to keep lewd *jogos* (games) outside the church precincts, and to prevent the use of ecclesiastical vestments for comic purposes. It is certain that there were burlesque sermons and parodic ceremonies, and likely that some rudimentary form of farce grew up. The performers would normally have been the *mimi* and *joculatores*, or wandering entertainers, but there are indications that priests also took part from time to time. The earliest performers known by name are Bonamis and Acompaniado, referred to in 1193 as actors of *arremedilhos*, a form of low mimicry.

These popular entertainments appealed to all levels of society, but later had to compete in Court circles with *momos* (mummers). In the elaborate mummings of Christmas 1500, classical, chivalresque, and exotic scenes were represented in pageantry with the help of rich costume and complex mechanisms. There was little dramatic action and no dialogue, but the mummers' technique can be traced in some of the plays of Gil VICENTE and the mechanism evolved was fruitfully used in later allegorical drama.

Another form capable of dramatic development was the dialogue poem, one of the most persistent and widespread of medieval European genres. The Galician-Portuguese *cancioneiros* (songbooks) contain numerous poems of the 13th and 14th centuries in which the dialogue is vigorous and often scurrilous, and this tradition is continued and expanded in the vast *Cancioneiro Geral* (1516). Here a few poets approach the ill-defined frontier that separates dialogue poems from true drama and one, Anrique da MOTA, seems to have crossed it.

The presence of shepherds in the festivities of Christmas 1500 suggests that the Spanish influence of Juan del ENCINA's religious eclogues was beginning to be felt, and this may well have been connected with King Manuel's marriage to the Spanish princess María. In June 1502 the Queen gave birth to a son, and among the celebrations was a monologue by Gil Vicente, the *Auto da Visitação* (*The Play of the Visitation*—when the Virgin Mary visited Elizabeth, the mother of John the Baptist, celebrated on 2 July). This work, Vicente's first, is in a type of Spanish heavily tinged by *sayagués*, the rustic dialect of the Salamanca region used by Encina and Lucas FERNÁNDEZ for the speech of their shepherds. Thus, both in genre and in language, Vicente began under the tutelage of Encina. He was, however, always a deviser of Court entertainments rather than a professional dramatist, and his later works merely accentuated this fact, moving away from the kind of drama that could take root outside the Court, and when the Court lost interest, fairly soon after Vicente's death in 1537, the path of his successors was hard.

Succeeding dramatists fell into two main groups, those who tried to follow the paths indicated by Vicente, some of whose works were assembled in a major collection published in 1587, and those who turned to classical and Renaissance models. The earliest figure of any importance in the former group is Afonso Álvares, a mulatto who was probably the originator of the hagiographic drama in Portugal. Álvares, commissioned by the Church, combined stories from the 'Golden Legend' of Jacopo de' Varazze (1230–98), with comic elements taken from Vicente. His *Auto de Santo António* was performed in 1531, and three more of his plays survive. The blind Baltasar Dias, living in Madeira, wrote plays of this and other types: though of poor literary quality, his work won and retained great popularity, especially by circulation in chapbooks. António Ribeiro Chiado (?–1591) wrote sketches of Lisbon life, usually dispensing with a plot but with a good eye for character types and a good ear for dialogue. António PRESTES was more highly regarded by his contemporaries. The poet Luís de CAMÕES wrote three plays of some interest in the mid-16th century; and the work of Francisco da COSTA remains of historical interest for its performance by Christian slaves in Morocco. In the closing years of the century Simão

Machado composed two talented and original plays, successful modifications of the Vicentine tradition: a historical drama, *O Cerco de Diu*, and a pastoral comedy, *Comédia da Pastora Alfea*.

In conscious opposition to the work of these dramatists was the classical tradition founded by Francisco de Sá de MIRANDA, whose plays were on the whole unsuccessful—for where Vicente's aim was performance, Miranda's was literary correctness. Greater, but still limited, success was attained by António FERREIRA, author of the greatest Portuguese tragedy *Castro*—before which there had been only one translation of Sophocles, a fragmentary tragedy by Miranda, and Latin plays written and performed in the College of Arts at Coimbra and the Jesuit College at Evora. This neo-Latin JESUIT DRAMA, whose best writer was Father Luís da Cruz (?–1604), had very little appeal outside the universities and perhaps the Court, but it and the Latin plays of George Buchanan (who came to Coimbra as professor in 1548) were put to good use by Ferreira. A third writer usually (though inaccurately) described as a classical dramatist was Jorge Ferreira de VASCONCELOS.

During the second half of the 16th century dramatic development was restricted by other forces, notably the activity of the Inquisition, whose proscriptive Indexes became increasingly comprehensive: the 1562 edition of Vicente's works escaped unscathed, but the second edition of 1586 was heavily censored, and the total loss of some chap-book plays may be due to this censorship. Inquisitorial activities and episcopal prohibitions also extended to theatrical performances, and a further hampering factor was loss of interest by the Court before the drama had found firm support elsewhere. On the other hand, the number of outlets grew: by the last quarter of the century the plays of Vicente's followers were being performed in the houses of the aristocracy and, if the subject was suitable, in churches and monasteries, while the universities and sometimes the aristocracy saw performances of classical drama and chap-books circulated widely. In 1580 Philip II of Spain became King of Portugal as well—a dual monarchy which lasted until 1640–and Portuguese literature quickly declined to a subordinate position. Some Portuguese dramatists wrote only in Spanish, though it has also been suggested that some Spanish dramatists wrote with half an eye on a Portuguese audience.

The Spanish connection was probably responsible for the beginning of the commercial theatre. In 1588 King Philip forbade the performance of plays in Lisbon except with the permission of the Hospital de Todos os Santos, which was to receive a share of the profits, a monopoly which lasted until 1743. Plays were performed in *pátios de comédias*, the equivalent of the *corrales* of Madrid, originally mere open courtyards, later covered and made more comfortable. The oldest, the Pátio do Poço Borratém, was already in existence in 1588, and the most famous, the Pátio das Arcas, was established in 1594. Other *pátios* were founded in the 17th century, and one of these, the Pátio dos Condes, attained great prosperity in the 18th century.

Portugal in the 17th century thus presents the paradox of flourishing theatrical performances and a moribund indigenous drama, the only noteworthy writer in Portuguese being Francisco Manuel de MELO. A vigorous tradition was to be found only in the Jesuit colleges, where elaborate spectacle was used for didactic ends, one of the most lavish productions being staged at the Colégio de Santo Antão in 1619 to celebrate the King's entry into Lisbon. By 1700 the neo-Latin drama had died out, and there remained only the commercial theatre, where the dominant forms were Spanish plays and popular comedies of little merit. These were challenged early in the 18th century by the Italian opera and by a new vogue for puppets (*bonecos* or *bonifrates*) which often parodied the opera, and from this improbable combination arose the work of António José da SILVA.

The influence of the Spanish drama continued to wane, and after controversy in the 1740s that of France established a theoretical superiority. French plays were often performed, though the Portuguese drama scarcely benefited from this, and in the second half of the century the Arcadian movement for a return to the PASTORAL tradition made an effort to work out a coherent dramatic theory. The leading figure here was Pedro António Correia Garção (1724–72), a good poet and critic, but attempts to write neo-classical plays in accordance with Arcadian theories by such dramatists as Manuel de Figueiredo (1725–1801) were notably unsuccessful.

João Baptista de Almeida GARRETT, the leading figure of Portuguese Romanticism, founded a National Theatre in Lisbon after his appointment as Inspector-General of Theatres in 1836, but he could not found a tradition. His immediate followers such as José da Silva Mendes LEAL, were minor writers who absorbed only the superficial elements of his work. A long-term result of his influence was that for over a century most of the leading Portuguese writers felt impelled to write at least one play, but most of these were mediocre or downright bad, and even the good ones had little success on the stage. The theatres of later 19th-century Lisbon were thus dominated by the well-intentioned social dramas of Gomes de Amorim (1827–91), and the well-made historical plays and polite comedies of João da Câmara (1852–1908), a writer who could also be an exponent of REALISM in *Os Velhos* (*The Old Ones*, 1893) and employ Maeterlinckian SYMBOLISM in *Meia Noite* (*Midnight*, 1900). Realistic scenes from modern life were also to be found in the work of Marcelino Mesquita (1856–1919), and became the prevailing fashion with the foundation of the Teatre Livre in

1904 and the success of Ramada Curto (1886–1961), Alfredo Cortês (1880–1946), and Raul BRANDÃO.

The censorship imposed in the early 1930s suppressed morally and politically dangerous plays and for a long time isolated Portugal from modern European developments in the theatre. The first signs of new life came in the late 1940s, with a number of plays aimed chiefly at the reader, while interest in theatrical technique was stimulated by the Teatro Estúdio do Salitre, founded in 1946. These two currents converged in the following 15 years, giving Portugal at least the promise of a vigorous modern drama. While the commercial theatres of Lisbon on the whole continued to present safe mediocrity to dwindling audiences, the Teatro Experimental in Oporto, founded in 1953, became an invaluable outlet for new drama.

Plays came from such established writers as Miguel Torga [Alfredo Rocha] (1907–), author of a minor classic *Mar* (*The Sea*, 1941), and from a group of younger dramatists, some of whose plays were staged in Oporto (the censor sometimes intervening), some abroad, and a few in the commercial theatre. Among the most successful were *O Vagabundo das Mãos de Ouro* (1961) by Romeu Correia (1917–) and *Felizmente há Luar* (1961) by Luís de Sttau Monteiro (1926–), the latter showing signs of BRECHT's influence and frowned on by the Salazar régime. The two leading figures were, however, Bernardo Santareno (1924–), a moralist showing equal concern for ultimate problems and for the lives of ordinary people in *A Promessa* (*The Promise*, 1957), *O Lugrè* (*The Lugger*, 1959), and *Anunciação* (*The Annunciation* 1962); and Luís Francisco Rebelo (1924–), who turned from EXPRESSIONISM to the social protest of plays like *O Dia Seguinte* (1953) and *Os Pássaros de Asas Cortadas* (1959), and who also emerged as an outstanding dramatic critic.

Following the death of the dictator Salazar in 1968 and the revolution of 1974, Portugal faced the prospect of a theatrical renewal but, in the aftermath of cultural (as of political) ferment, new directions were for some time difficult to discern. The strong physical presence of the theatre in Lisbon, however, given over though it had previously been to lightweight fare, offered considerable scope to newly emergent writers and companies.

POSSART, ERNST VON (1841–1921), German actor and theatre manager, who made his first appearance on the stage at Breslau in 1860, where he played Iago in *Othello*. His later roles included Shylock in *The Merchant of Venice*, Franz Moor in SCHILLER's *Die Räuber*, Mephistopheles in GOETHE's *Faust*, Carlos in his *Clavigo*, and the title role in LESSING's *Nathan der Weise*. In 1864 Possart was in MUNICH, where he became manager of the theatre in 1875. A fine-looking man, with an alert, intelligent face and flashing eyes, he was noted for the beauty of his voice and the dignity and ease of his movements. He was seen in New York in 1910, with much success.

POSSET, JEAN, see ENGLISH COMEDIANS.

POTIER DES CAILLETIÈRES, CHARLES-GABRIEL (1774–1838), French actor, considered by some the greatest comedian of his day. After some years in the provinces, he made his début in Paris in 1809 at the Théâtre des VARIÉTÉS, and once he had established his reputation he never lost his hold on the affections of the public, spending his last years at the PALAIS-ROYAL, where he had first appeared in 1831. His one fault was a weak voice, but this was offset by the subtlety and vivacity of his acting, which conveyed his meaning without the need for words.

POTTER, MAUREEN, see O'DEA, JIMMY.

POUGET, MADELEINE DE, see BEAUCHÂTEAU.

POULSEN, EMIL (1842–1911) and OLAF (1849–1923), Danish actors, who made their first appearance on the stage on the same day in 1867, Emil playing the title role in HOLBERG's *Erasmus Montanus* while Olaf played Jacob. Both were excellent actors, best remembered for their appearances in Holberg, IBSEN, and Shakespeare, Olaf being outstanding in the comic roles of Bottom in *A Midsummer Night's Dream* and Falstaff in *The Merry Wives of Windsor* and *Henry IV*. Emil's elder son Adam (1879–1969) was for some years director of the DAGMARTEATRET, and his younger, Johannes (1881–1938), was with the KONGELIGE TEATER from 1909 until his death.

POWELL, GEORGE (1668–1714), English actor, who created many gallant and tragic roles in the drama of his day. He was the first Bellamour in CONGREVE's *The Old Bachelor* (1693) and the first to play Lothario in ROWE's *The Fair Penitent* (1703). He might have risen to great heights, but drink and brawling kept him back and he was soon supplanted by WILKS. He caused many disturbances in the theatre, being usually imperfect in his part and drunk on the stage, where, it is said, he made such violent love that VANBRUGH grew nervous for the actress. In spite of his debaucheries Christopher RICH tolerated him because, in his capacity as director of rehearsals, he kept the actors content on low salaries. He was the author of two tragedies given at DRURY LANE and of three comedies, of which the most successful was *The Imposture Defeated: or, a Trick to Cheat the Devil* (1697).

POWELL, MARTIN, see PUPPET.

POWER, NELLIE, see MUSIC-HALL.

POWER, TYRONE (1797–1841), Irish actor, who first appeared on the stage in 1815, after serving in both the army and the navy. For some years he played light comedy in the minor theatres, and made little impression until in 1826 he took to specializing in such 'Irish' roles as Sir Lucius O'Trigger in SHERIDAN's *The Rivals* and Dennis Brulgruddery in the younger COLMAN's *John Bull; or, the Englishman's Fireside*. He also appeared in a number of his own plays, among them *St Patrick's Eve* (1832), *Paddy Cary, the Boy of Clogheen* (1833), and *O'Flannigan and the Fairies* (1836). A handsome, high-spirited man with a rich brogue, he was a firm favourite in the United States, and was returning from his third visit there when he was drowned in the sinking of the S.S. *President*. Of his eight children, Maurice was on the stage, as were his grandson and great-grandson (below). His grand-daughter Norah was the mother of Tyrone GUTHRIE.

POWER, TYRONE (1914–58), great-grandson of the above, his father (Frederick) Tyrone Edmond (1869–1931), a well known actor in New York mainly in Shakespearian parts, being the son of the first Tyrone's son Harold. The third Tyrone was best known as a film star, but he was for some years on the American stage, making his first appearance in New York in 1931 as a page in *Hamlet*. Among his later parts were Fred in Edward SHELDON's *Romance* (1934) with Leontovich; Benvolio in *Romeo and Juliet* in 1935; de Poulengy in a revival of SHAW's *Saint Joan* in 1936 with Katharine CORNELL; and Gettner in the New York production of Christopher FRY's *The Dark is Light Enough* (1955). In 1950 he was seen in London in the title role of Heegan and Logan's *Mister Roberts*.

PRADON, NICOLAS [Jacques] (1644–98), French dramatist, remembered only because his *Phèdre et Hippolyte*, performed two days after RACINE's *Phèdre* in Jan. 1677, was so praised by the latter's enemies that it was at first adjudged the better play, a verdict which time has reversed. Even Pradon's first play, *Pyrame et Thisbé* (1674), was only successful because of the cabal against Racine. Whatever posthumous fame he has achieved he owes to the greatness of the men he roused against him, including the redoubtable BOILEAU.

PRAETEXTA, see FABULA: 3.

PRAGA, MARCO (1862–1929), Italian dramatist and novelist, son of the poet and painter Emilio Praga, and an influential exponent of VERISMO. His most important plays were *Mater dolorosa* (1888), based on the celebrated novel of the same name by his friend Girolamo ROVETTA; *Le vergini* (*The Virgins*, 1889); *La moglie ideale* (*The Ideal Wife*, 1890), in which DUSE starred, as she did in *Il divorzio* (*The Divorce*, 1915); *L'erede* (*The Heir*, 1893); *La crisi* (*The Crisis*, 1904); and *La porta chiusa* (*The Closed Door*, 1913). In all these plays, which were much influenced by contemporary French authors and demonstrate keen interest in psychological and sociological problems and intense social commitment, he attacks the conventions of the bourgeois world of his time in Milan, his birthplace. For a short period he was one of the directors of the Stabile Milanese, a theatrical company founded in 1915, and he was also the author of ten volumes, published between 1919 and 1928, of theatrical criticisms originally written for the *Illustrazione Italiana*. Although he supported the early works of PIRANDELLO, he rejected his later plays as he did all new developments in the theatre, devoting his time to writing popular but facile works on the theme of the 'eternal triangle'.

PRATINAS OF PHLIUS (*c*. 540–470 BC), Greek dramatist, credited with the introduction into Athenian festivals of the SATYR-DRAMA, of which he wrote about 32. He also wrote some 18 tragedies which rivalled those of AESCHYLUS.

PRAY, MALVINA, see FLORENCE, W. J.; MARIA, see WILLIAMS, BARNEY.

PREHAUSER, GOTTFRIED (1699–1769), Austrian actor, who spent many years in a travelling company before being invited by STRANITZKY to VIENNA in 1725 to take over the part of HANSWURST, which he had already played at Salzburg in 1720. In his hands the character, and the plays in which he appeared, underwent subtle changes, becoming noticeably more Viennese and bourgeois in background and situation. With the death of Prehauser and the decline of improvised farce in the late 18th century, Hanswurst disappeared from the stage.

PRESTES, ANTÔNIO (?–1587), Portuguese dramatist, said by Manuel de MELO in the 17th century to have been the equal of Gil VICENTE. He has however been condemned by most modern critics for his structural weakness and his obscurity of language. Seven of his plays are extant: six domestic comedies with a legalistic bias (he was himself a judicial functionary in Santarem) but some amusing scenes, and an ambitious morality, the *Auto de Ave Maria*.

PRESTON, THOMAS, (*fl*. 16th cent.), English dramatist, author of a popular tragi-comedy, *Cambyses King of Persia* (*c*. 1569), written with bombastic eloquence, thus giving rise to Falstaff's remark in Shakespeare's *Henry IV, Part I* that he would speak passionately in 'King Cambyses' vein'. This play marks the transition from the medieval MORALITY PLAY to the Elizabethan historical drama. Its author, who may also have written *Sir Clyomon and Sir Clamydes* (*c*. 1570), was evidently not the Thomas Preston (*c*. 1537–98) who in 1592, as Master of Trinity Hall, petitioned for the banning of plays in CAMBRIDGE.

PRÉVILLE [Pierre-Louis Dubus] (1721–99), French actor, who was in a provincial company before joining the COMÉDIE-FRANÇAISE in 1753, where, in roles played previously by F. A. POISSON, he proved himself the finest comedian the company had had since J. B. RAISIN. He was also excellent in MARIVAUX's plays, ably partnered by Mlle DANGERVILLE, created the role of Figaro in BEAUMARCHAIS's *Le Barbier de Séville* (1775), and enjoyed a great personal success as the six characters in one in a revival of BOURSAULT's *Le Mercure galant*. He retired in 1786, returning to the theatre for a short time in 1791.

PRICE, MAIRE, see NIC SHIUBHLAIGH, MAIRE.

PRICE, (Lilian) NANCY BACHE (1880–1970), English actress and theatre manager, who made her first appearance on the stage in BENSON's company in 1899, being seen in London as Olivia in *Twelfth Night* a year later. She made her first success as Calypso in Stephen PHILLIPS's *Ulysses* (1902), and then appeared as Rosa Dartle in *Em'ly*, a dramatization of DICKENS's *David Copperfield*, and as Hilda Gunning in PINERO's *Letty* (both 1903). She continued to appear in a wide range of parts, but is now chiefly remembered as the founder and guiding spirit of the People's National Theatre in London. She began this venture in 1930 with a revival at the FORTUNE THEATRE of Anstey's *The Man from Blankley's*, and during the next few years was responsible for the production of over 50 plays, ranging from EURIPIDES to PIRANDELLO, and including Susan GLASPELL's *Alison's House* (1932), *Lady Precious Stream* (1934), a Chinese play by S. I. Hsiung, and Mazo de la Roche's *Whiteoaks* (1936), in which she played for two years the part of old Adeline Whiteoaks. These were all produced at the LITTLE THEATRE, which she made her headquarters, though some of her productions were seen elsewhere. During the Second World War she toured as Madame Popinot in Seymour HICKS and Ashley DUKES's *Vintage Wine* and was in Liverpool with the OLD VIC company, but returned to London in 1948, making her last appearance in Eden PHILLPOTTS's *The Orange Orchard* (1950).

PRICE, STEPHEN (1783–1840), American theatre manager, who in 1808 bought a share in the management of the PARK THEATRE, New York, and gradually assumed complete control. He inaugurated the pernicious policy of importing famous European actors, beginning with G. F. COOKE in 1810–11, which by 1840 had wrecked the old resident companies of the larger American towns. He spent a good deal of time in London between 1820 and 1839, engaging English and Continental actors and singers for America, and demonstrating his love of spectacular and freakish effects—real horses and tigers on stage, for example—during his tenancy of DRURY LANE from 1826 to 1830. Price, whose whole theatrical career was closely bound up with that of the Park Theatre, died opportunely just as its fortunes were beginning to decline: he was the first notable American manager who was not also an actor or playwright.

PRIESTLEY, J(ohn) B(oynton) (1894–), English dramatist, novelist, and critic, awarded the Order of Merit in 1977. His first play, a dramatization of his own best-selling novel *The Good Companions* (London and N.Y., 1931) undertaken in collaboration with Edward KNOBLOCK, was followed by *Dangerous Corner* (London and N.Y., 1932), an ingenious play in which a chance remark at a dinner party produces a chain of revelations which lead eventually to a suicide: but at the critical moment the play returns to its beginnings and the words pass unnoticed. After *Laburnum Grove* (1933; N.Y., 1935) and *Eden End* (1934; N.Y., 1935), the latter mingling gentle melancholy and rich humour in a way Priestley never again achieved, came two excellent plays influenced by Dunne's *An Experiment With Time*, *Time and the Conways* and *I Have Been Here Before* (both 1937; N.Y., 1938), the former being particularly effective, with its second act set 20 years later than the first and third. These 'time-plays' were followed by a rollicking farce, *When We Are Married* (1938; N.Y., 1939), which concerns three Yorkshire couples who find after many years that their marriages are not legal. In his next two plays, *Music At Night* (also 1938) and *Johnson Over Jordan* (1939), in which Ralph RICHARDSON gave a fine performance, Priestley sought to give modern drama a new depth, but the technical means he employed were not to the taste of the public, though *They Came To a City* (1943), an earnest left-wing political tract, proved surprisingly popular in the West End. Another play in the style of *Dangerous Corner*, *An Inspector Calls* (1946; N.Y., 1947), in which Richardson again appeared, was followed in 1947 (N.Y., 1948) by one of Priestley's best plays, *The Linden Tree*, in which Lewis CASSON and Sybil THORNDIKE played an academic and his wife confronted at a family reunion by the contrasting ideologies of their three adult children. His later plays, such as *Home Is Tomorrow* (1948) and *Summer Day's Dream* (1949), proved less memorable. *Dragon's Mouth* (1952; N.Y., 1955) and *The White Countess* (1954) were written in collaboration with his third wife, the archaeologist Jacquetta Hawkes, while his last plays included *Mr Kettle and Mrs Moon* (1955), *The Glass Cage* (1957), and a dramatization of Iris Murdoch's novel *A Severed Head* (1963; N.Y., 1964) in collaboration with the author. Priestley acted as chairman of theatre conferences in Paris in 1947 and Prague in 1948, and of the British Theatre Conference in 1948; he was also the first President of the INTERNATIONAL THEATRE INSTITUTE.

PRIME MINISTER OF MIRTH, see ROBEY, GEORGE.

PRIMROSE, GEORGE, see DOCKSTADER, LEW.

PRINCE, HAL, see MUSICAL COMEDY and SONDHEIM, STEPHEN.

PRINCE CHARLES'S MEN, company usually known as the Prince's Men formed by ALLEYN in 1616, on the death of his father-in-law HENSLOWE, under the patronage of the Prince of Wales (later Charles I). They were seen at several theatres including PORTER'S HALL and the HOPE before settling in 1619 at the COCKPIT, where they remained until 1622. They were then ousted by the newly-formed LADY ELIZABETH'S MEN and went first to the CURTAIN and then to the RED BULL. On the accession of Charles I in 1625 they were disbanded, many of the leading actors joining the KING'S MEN, but in 1631 a new company, under the same name and the patronage of the young Prince Charles (later Charles II) was formed to play at SALISBURY COURT, taking in some of the actors from the FORTUNE, where the new company itself played from 1640 to the closing of the theatres in 1642.

PRINCE EDWARD THEATRE, London, in Old Compton Street, Soho. It opened on 3 Apr. 1930 with a musical, *Rio Rita*, which had a short run. Except for a play on the adventures of Sexton Blake, in which Arthur Wontner played the title role, the theatre was used mainly for musicals and revues, of which the best were *Nippy* (also 1930) and *Fanfare* (1932). In 1935, after the run of a Christmas PANTOMIME, *Aladdin*, it closed, to reopen on 2 Apr. 1936 as the London Casino, a cabaret-restaurant featuring a spectacular stage show. On the outbreak of war in 1939 it became the Queensberry All-Services Club, reopening as a theatre on 20 Dec. 1946 with another pantomime, *Mother Goose*, followed by a revival of Ivor NOVELLO's *The Dancing Years*. Three editions of Robert Nesbitt's revue *Latin Quarter*, beginning in 1949, kept the theatre open, but in 1954 it became a cinema, except for a lavish *Cinderella* in 1974. In 1978 it again reverted to use as a theatre under its original name with the musical *Evita* by Tim Rice and Andrew Lloyd Webber, which ran into the 1980s.

PRINCE HENRY'S MEN, see ADMIRAL'S MEN.

PRINCE OF WALES' THEATRE, London, in Coventry Street. (The name is also found on playbills and programmes as Prince of Wales and Prince of Wales's.) It opened as the Prince's Theatre on 18 Jan. 1884, and on 3 Mar. a free adaptation of IBSEN's *A Doll's House* (as *Breaking a Butterfly*) was produced, but aroused little interest. It was followed by *The Private Secretary*, a German play adapted by HAWTREY, which with TREE in the role of the Revd Robert Spalding was not at first a success but when transferred to the GLOBE had a long run. The first successful production at the Prince of Wales, as it was renamed in 1886, was the wordless play *L'Enfant prodigue* (1891), superbly mimed, with Jane May as the prodigal son, which served to introduce PIERROT, in the person of Zanfretti, to London in something other than PANTOMIME. *In Town*, often considered the first English MUSICAL COMEDY, was presented here by George EDWARDES in 1892, and was followed by the equally successful *A Gaiety Girl* (1893). On 2 Mar. 1895 Basil Hood's *Gentleman Joe, the Hansom Cabby* began a long run, with Arthur Roberts in the title role, and the theatre then moved over to straight plays with FORBES-ROBERTSON and Mrs Patrick CAMPBELL in MAETERLINCK's *Pelléas and Mélisande* and MARTIN-HARVEY in Wills's adaptation of DICKENS's *A Tale of Two Cities* as *The Only Way*, transferred from the LYCEUM. Marie TEMPEST, forsaking musical comedy, played the title roles in *English Nell* (1900), based on Simon Dale's novel about Nell GWYNN, and *Peg Woffington*, a dramatization of Charles READE's novel, and Becky Sharp in a dramatization of Thackeray's *Vanity Fair* (both 1901). The theatre returned to musical comedy between 1903 and 1910, including *Miss Hook of Holland* (1907) and *The King of Cadonia* (1908). For the next two decades the theatre housed musicals such as *Yes, Uncle!* (1917), plays such as Avery Hopwood's farce *Fair and Warmer* (1918) and Ivor NOVELLO's *The Rat* (1924), written in collaboration with Constance COLLIER, and revues including *A to Z* (1921), *Co-Optimists* (1923), and *Charlot's Revue* (1924). During much of the 1930s the theatre was given over to non-stop REVUE, and when *Encore les Dames* closed in 1937 the building was demolished. A new theatre under the old name, seating 1,139 in two tiers, opened on 27 Oct. 1937 with *Les Folies de Paris et Londres*. In 1943 Sid Field made his London début here in *Strike a New Note*, and he returned to star in *Piccadilly Hayride* (1946). Three years later Mary Chase's comedy about an imaginary rabbit, *Harvey*, had a long run, as did Paul Osborn's *The World of Susie Wong* (1959) and Neil SIMON's *Come Blow Your Horn* (1962), while *Funny Girl* (1966) with Barbra Streisand, *Sweet Charity* (1967), and *Promises, Promises* (1969) brought to London three successful Broadway musicals. Bertolt BRECHT and Kurt Weill's *The Threepenny Opera* was revived in 1972 with Joe Melia as Macheath and Vanessa REDGRAVE as Polly, and in the following year a variety spectacular starring Danny LA RUE had a great success. Later long-running shows were Bernard Slade's *Same Time, Next Year* (1976) with Michael Crawford, the musical *I Love My Wife* (1977), and AYCKBOURN's *Bedroom Farce* (1978) transferred from the NATIONAL THEATRE. (See also SCALA THEATRE.)

PRINCESS'S THEATRE, Glasgow, see CITIZENS' THEATRE.

PRINCESS'S THEATRE, London. In 1836 a building in Oxford Street, known as the Queen's Bazaar, was converted into a theatre, and opened on 30 Sept. 1840 as the Princess's, named after

Queen Victoria before her accession. After an unsuccessful series of promenade concerts, the theatre underwent further alteration and reopened on 26 Dec. 1842 with Bellini's *La Sonnambula*. Other operas and light dramatic pieces followed, but it was not until the Theatres Act of 1843 allowed the staging of plays that the new theatre had any success. In 1845 the American actor Edwin FORREST, with Charlotte CUSHMAN, made his London début, and was followed by MACREADY in a season of his most popular parts. In 1850 Charles KEAN brought a company to the Princess's to play Shakespeare and adaptations of French plays suited to the taste of the time, among them in 1851 Oxenford's version of *Pauline*, which was seen by Queen Victoria, who also attended the first night of *The Winter's Tale* in 1856 in which the nine-year-old Ellen TERRY played Mamilius. Kean had eight years of almost unbroken success before leaving, making his last appearance as Wolsey in *Henry VIII*, the first play to be lit by limelight. Augustus HARRIS then took over, engaging IRVING as his leading man and in 1860 bringing FECHTER to London, where his portrayal of Hamlet caused a sensation. The theatre later became famous for its productions of MELODRAMA, among them BOUCICAULT's *The Streets of London* (1864), which included a thrilling fire scene, and *Arrah-na-Pogue* (1865). Charles READE's *It's Never too Late to Mend* (also 1865) began with a riotous first night when the audience objected to the savagery of one of the scenes, but achieved a run of 148 performances. Several managers, among them Ben WEBSTER and Chatterton, tried to revive the glories of Kean's epoch, alternating Shakespeare and revivals of serious plays, but were forced to revert to melodrama. In 1875 Joseph JEFFERSON revived Boucicault's adaptation of *Rip Van Winkle* with great success, and in 1879 Charles WARNER startled the audience with his performance as Coupeau in Reade's *Drink*, based on ZOLA's *L'Assommoir*. On 19 May 1880 the theatre closed and was rebuilt, reopening as the Royal Princess's on 6 Nov. with Edwin BOOTH as Hamlet. A year later Wilson BARRETT took over, achieving his first success with Sims's *The Lights of London* (1881) and on 16 Nov. 1882 came the first night of JONES and Herman's *The Silver King*, which ran for a year. After Barrett left in 1886 the theatre's prosperity declined, and shortly after the last production, Kremer's *The Fatal Wedding* in 1902, it was leased by an American syndicate. Difficulties over the lease, however, and the need for extensive alterations, led to the building becoming a warehouse, and it was finally demolished in June 1931.

PRINCESS THEATRE, Melbourne. This opened in 1886 with Nellie STEWART as Yum-Yum in *The Mikado*. Previously a wooden structure known as 'the old barn' had occupied the site for some 40 years, but, rebuilt by J. C. WILLIAMSON, it was considered one of the most beautiful and up-to-date theatres in the southern hemisphere, having an Italian Renaissance exterior, French interior decorations, a sumptuous GREEN ROOM, a sliding roof, miniature waterfalls on each side of the proscenium, and even electricity. It has housed all forms of entertainment, and even has its own ghost—the English baritone Federici [Frederick Baker], who suffered a fatal heart attack immediately after singing Mephistopheles in Gounod's *Faust*.

PRINCESS THEATRE, New York, at 104 West 39th Street, one of the city's smallest and most perfect playhouses, seating 299. It opened on 14 Mar. 1913 with the Princess Players in four one-act plays, its first important production being a translation of BRIEUX's *Maternité* in 1915. In 1920–1 the PROVINCETOWN PLAYERS appeared in several productions, including O'NEILL's *The Emperor Jones*, and in 1922 PIRANDELLO's *Six Characters in Search of an Author* had its first production in New York. Maxwell ANDERSON's first play *The White Desert* was also presented in 1922, but was a failure. In 1928 the theatre was renamed the Lucille La Verne, but after two productions reverted to its original name only to change it to the Assembly in 1929. It then became a cinema, except for a short period in 1937, when (after it had become the recreation centre of the International Ladies' Garment Workers Union) it reopened as the Labor Stage with a topical revue, *Pins and Needles*, which ran into three editions with 1,108 performances. In 1944 it became Theatre Workshop, but from 1947, under various names, was a cinema until it was demolished in June 1955.

PRINCE'S THEATRE, Bristol, see BRISTOL.

PRINCE'S THEATRE, London, Coventry Street, see PRINCE OF WALES' THEATRE; Shaftesbury Avenue, see SHAFTESBURY THEATRE; King Street, Piccadilly, see ST JAMES'S THEATRE.

PRINCIPAL BOY, the chief character in the English PANTOMIME—Aladdin, Dick Whittington, Robinson Crusoe, Idle Jack in *Jack and the Beanstalk*, Prince Charming in *Cinderella*—traditionally played by a woman. If, as has been suggested, the pantomime is a survival of the Roman feast of Saturnalia, which took place approximately at what is now the Christmas season, this custom may have originated from the exchange of costume between the sexes which was one of its main features, an example of topsy-turvydom well in the spirit of the modern pantomime, in which the DAME is played by a man. It was certainly influenced by the playing of boys' parts by women singers in 18th-century opera and by the widespread popularity of BREECHES PARTS on the English stage. The *travesti* fashion of the 19th century in BURLESQUE and EXTRAVAGANZA standardized the blonde wig, short tunic, fleshings (flesh-coloured tights), and high heels of the typical Principal Boy, who was firmly established by the

1880s when Augustus HARRIS, at DRURY LANE, started the fashion for opulent curves characteristic of such outstanding players of the part as Harriet Vernon, Nellie STEWART, and Queenie Leighton. In the 20th century the tradition was carried on, though with less emphasis on curves, by Phyllis Neilson-Terry, Madge Elliott, Jill Esmond, Fay COMPTON, and Dorothy Ward. As early as 1938 the role began to be played by actors, but this was not favourably received until from the 1950s male pop singers with a following among family audiences took it over in the few pantomimes still being presented.

PRINTEMPS, YVONNE (1895–1977), French actress and singer, who made her first appearance in Paris in revue in June 1908. She was for some time at the FOLIES-BERGÈRE, and also appeared at the PALAIS-ROYAL. In 1916 she joined the company of Sacha GUITRY whom she married in 1919, appearing with him in a number of plays including his own *Nono* in which she was first seen in London in 1920. She played in English for the first time in COWARD's *Conversation Piece* (1934), and with her second husband Pierre FRESNAY was also seen in London in Ben TRAVERS's *O Mistress Mine* (1936). She made her first appearance in New York in 1926 in Sacha Guitry's *Mozart*, playing the title role, one of her most charming parts, and in 1937 took over the Théâtre de la Michodière, where she appeared in a succession of musical plays during the next 20 years.

PRITCHARD [*née* Vaughan], HANNAH (1711–68), English actress, wife of the actor William Pritchard (1707–63), who was for some years Treasurer at DRURY LANE. She had already had some experience in small parts at the HAYMARKET and elsewhere when in 1733 she made her first appearance at Drury Lane, where she remained for some years, spending the summer seasons from 1741 to 1747 at the Jacobs Wells Theatre in BRISTOL, and being from 1743 to 1747 under RICH at COVENT GARDEN. She quickly achieved an enviable reputation in light comedy, including in her extensive repertory a number of leading parts which she continued to play throughout her career, among them Lady Brumpton in STEELE's *The Funeral* and Lady Townly in Colley CIBBER's *The Provok'd Husband*. She was also much admired as Rosalind in *As You Like It* and Beatrice in *Much Ado About Nothing*, and played Nerissa to the Portia of Kitty CLIVE in MACKLIN's epoch-making production of *The Merchant of Venice* in 1741. In 1748 GARRICK, to whose Chamont in OTWAY's *The Orphan* she had played Monimia in 1742, invited her to join his new company at Drury Lane, and she remained as one of his leading ladies for 21 years, retiring only a few months before her death. Under him she continued her outstanding career in comedy, one of her creations being Mrs Oakly in the elder COLMAN's *The Jealous Wife* (1761); but she widened her range to include tragedy, giving more prominence than usual to Gertrude in *Hamlet* and Queen Katharine in *Henry VIII*. She was also the first and only interpreter of the heroine in Dr JOHNSON's *Irene* (1749); but the part in which she excelled all her contemporaries was Lady Macbeth, which she first played with Garrick in 1748, and in spite of the increasing obesity which troubled her later years chose for her farewell to the stage on 25 Apr. 1768, after which Garrick never played Macbeth again.

Mrs Pritchard's two brothers, Henry (1713–79) and William (1715–63) Vaughan, were both on the stage, playing mainly low comedy, and her daughter Hannah Mary (1739–81), who married the actor John 'Gentleman' PALMER, became a member of Garrick's company in 1756, making her début as Juliet to her mother's Lady Capulet. She retired on the death of her husband in 1768.

PRIVATE THEATRES IN ENGLAND. The vogue for private theatricals in English high society, which reached its peak from 1770 to 1790, resulted in the building of several private theatres. Earlier, dilettanti productions had taken place in specially adapted rooms, an example of which can be seen in Hogarth's painting of a children's performance of DRYDEN's *The Indian Emperor* in the house of John Conduit, Master of the Mint, in 1731. In 1740, in an open-air theatre in Cliveden which is still extant, Arne's 'Rule, Britannia' was first sung during a performance of Thomson's masque *Alfred*, given before Frederick, Prince of Wales, who was then living there.

The first specially erected of the indoor private theatres, and one which could vie with such Continental Court theatres as CELLE, ČESKY KRUMLOV, DROTTNINGHOLM, or GRIPSHOLM, was erected at Wargrave in 1789 for the Earl of Barrymore (1760–93). Rectangular in shape, it held 400 spectators, and contained two tiers of boxes and two stage boxes unusually placed over the orchestra well. It was provided with a series of workrooms and an adjoining salon for refreshments. After ruining its owner, who also had a small puppet theatre in Savile Row, London, it was demolished in 1792.

In 1793 the Lady Elizabeth Craven, Margravine of Anspach (1750–1828), built a theatre for the production of her own plays at Brandenburgh House, Hammersmith, which remained in use until 1804. It was in castellated style, and is said to have resembled the Bastille rather than a Temple of the Muses. An engraving of the interior shows a large central box and a parterre raised on a shallow platform after the Continental model.

A theatre in a simpler style was converted from a kitchen for Sir Watkins Williams Wynn at Wynnstay, and was used for annual performances from 1771 to 1789. Other private theatres were the Duke of Marlborough's at Blenheim Palace in Oxfordshire, converted from a greenhouse and in use from 1787 to 1789, and the Duke of Richmond's at Richmond House, London, constructed out of two rooms by Wyatt in 1787 to hold about 150 spectators. The well known scene-painter Thomas Greenwood was employed to provide scenery for these two theatres; Inigo Richards worked at Wynnstay and Malton at Brandenburgh. The wardrobe book of Wynnstay and the sale catalogue of Wargrave give ample evidence of the sumptuous décors provided for performances at these theatres, where professionals frequently acted with or coached the amateurs.

The Duke of Devonshire's private theatre at Chatsworth, Derbyshire, dating from about 1830, is the oldest still in existence. Another theatre of that period was at Burton Constable, in Yorkshire, which functioned from 1830 to 1850. Later Victorian private theatres include Charles DICKENS's at Tavistock House, in London; a small one attached to Campden House in Kensington in the 1860s; one at Capethorne Hall, Cheshire, in use in 1870; the artist Herkomer's at Bushey, Hertfordshire; and the singer Patti's at 'Craig y Nos, Wales, still extant. In the 20th century Lord Bessborough gave annual productions for some years at his theatre at Stansted, Essex, and Lord Faringdon built a theatre, still in use, at Buscot Park, Berkshire. The late John Christie's opera house at Glyndebourne, Sussex, and Nicholas Sekers's theatre at Rosehill, Whitehaven, Cumbria, though founded by private individuals, both employ professional companies, are open to the general public, and charge for admission, and are thus not strictly to be classed as private theatres.

PROCTOR, FREDERICK FRANCIS, see VAUDEVILLE, AMERICAN.

PRODUCER, American term for the man responsible for the financial side of play-production, for the buying of the play, the renting of the theatre, the engagement of actors and staff, and the handling of the receipts. In England most of these functions are assumed by the manager. Formerly the person who was responsible for the actual staging of the play and the conduct of rehearsals was known in England as the producer, but this has now been universally replaced by the American term DIRECTOR.

PROFILE BOARD, see FLAT.

PROFILE SPOT, see LIGHTING.

PROJECTOR, see LIGHTING.

PROLETKULT THEATRE, Moscow, see TRADES UNIONS THEATRE.

PROLOGUE, from the Greek *prologos*, originally applied to the speech or dialogue which preceded the first entry of the CHORUS in a Greek play, and by extension to an introductory poem or speech which explained or commented on the action which was to follow. It was first used by EURIPIDES, and was later employed by the Elizabethan dramatists, who called it the Chorus. Together with the Epilogue (from the Latin *epilogus*, peroration), which closed the action, the prologue was extensively used during the Restoration period, providing incidentally a good deal of information on the contemporary theatre, and survived well into the 18th century. It disappeared with the crowded bills of the 19th century and is now used only on special occasions. At their best the prologue and epilogue were witty and sometimes scurrilous commentaries on the politics and social conditions of the day, as they still are in the WESTMINSTER PLAY; and they were often written by outstanding men of the theatre, of whom DRYDEN and GARRICK were the greatest, and spoken by the finest actors of the time.

PROMENADE, see THEATRE-IN-THE-ROUND.

PROMENADE PRODUCTIONS, see BRYDEN, BILL, and COTTESLOE THEATRE.

PROMPT CORNER, in the British theatre, a desk against the inner side of the proscenium wall on the stage (i.e., the actor's) left, where the PROMPTER installs his Prompt Book (a copy of the play, generally interleaved, and carrying the full directions and warnings necessary to the management of the production), and where a board of switches for signals, communicating to various parts of the theatre, is generally situated. Among the signals is the Bar Bell to warn patrons in the bars and foyers of the approaching end of an interval. On the Continent, and in opera houses generally, the Prompter's Box is usually sunk beneath a coved hood in the centre of the footlights.

PROMPTER, the person responsible for supplying the necessary 'next line' to an actor who has forgotten his part—a function assumed in the Elizabethan theatre by the BOOK-HOLDER. The prompter attends all rehearsals, and should note in his copy of the play all alterations in the script and, if so instructed by the STAGE-MANAGER, all the moves, which should also be in the stage-manager's copy. Prompt copies of early plays are exceedingly valuable documents, as they often contain lists of actors' names and sometimes provide the only clue to the author.

PROMPT SIDE, see STAGE DIRECTIONS.

PROPHET PLAY, see LITURGICAL DRAMA.

PROPS, the usual term for stage properties. It covers anything essential to the action of the play which does not come under the heading of COSTUME, SCENERY, or furniture. Hand-props are those which an actor handles—letters, documents, revolvers, newspapers, knitting, snuff boxes, and so on. These are given to him as he goes on stage, and taken from him as he comes off, and are not his personal responsibility. Other props—stuffed birds, food in general, dinner-plates, telephones—are placed on stage by the property man, who is responsible for all props under the direction of the STAGE-MANAGER. He has for storage a property room backstage, from which he is expected to produce at a moment's notice anything that may be required. He must also prevent the removal from it of oddments by members of the company.

PROSCENIUM, a word which is classical in origin, and in the later Greek theatre meant the area in front of the stage. It was used in Renaissance Italy for the draperies surrounding the stage picture, introduced into England by Inigo JONES with his scenery for Court MASQUES, and in its modern meaning is applied to the permanent or semi-permanent wall dividing the AUDITORIUM from the STAGE. It has hanging in it the front CURTAIN, and can be made smaller by the use of a FALSE PROS(CENIUM). In its heyday it was a feature of considerable architectural complexity, forming an essential link between the auditorium and the scene, but it is now generally regarded as a hindrance, particularly in the production of Elizabethan plays and modern EPIC THEATRE. The tendency is to abolish it wherever possible in favour of the OPEN STAGE or THEATRE-IN-THE-ROUND, but one obstacle which confronts those who wish to experiment with these new forms in the older London theatres is the presence within the proscenium arch of the Safety Curtain demanded by the G.L.C. Fire Regulations.

PROSCENIUM BORDER, PELMET, see BORDER.

PROSCENIUM DOORS, or Doors of Entrance, a permanent feature of the English Restoration playhouse. Set on each side of the FORESTAGE, they had practicable knockers and bells, and provided the usual means of exit and entrance for the actor. Leaving by one door and returning by another, he was presumed to be in another room even though the WINGS and BACKCLOTH remained unchanged. The number of doors varied from four to six not only in different theatres but also in the same theatre at different periods. By the early 18th century they were reduced to one on each side, and by the beginning of the 19th they were used only by the actor 'taking a bow' after the play. They were then known as Call Doors.

PROSPECT THEATRE COMPANY, founded, as Prospect Productions, in Oxford in 1961. After three successful summer seasons it became a touring company based on the Arts Theatre, CAMBRIDGE, with Toby [Sholto David Maurice] Robertson (1928–) as its Artistic Director. It soon achieved an outstanding reputation for its high standard of acting and ensemble playing and its adequate but inexpensive staging of classic plays with brilliant costumes but minimal sets, and from 1967 it had a unique association with the EDINBURGH FESTIVAL, where it presented numerous productions. After 1969, when the association with Cambridge ended, the lack of a base large enough for its extended activities was a severe handicap, until in 1977 a highly successful season at the OLD VIC THEATRE, recently vacated by the NATIONAL THEATRE company, led to its establishment there, and a change of name two years later to the Old Vic Company. Among its notable productions were *Richard II* in 1968 and MARLOWE'S *Edward II* in 1969, with Ian MCKELLEN playing the title role in both: TURGENEV'S *A Month in the Country* in 1975 with Dorothy TUTIN; *Antony and Cleopatra* with Alec MCCOWEN and Dorothy Tutin and SHAW'S *Saint Joan* with Eileen ATKINS, both in 1977. A production of *Hamlet* with Derek JACOBI a year later was taken on an extensive overseas tour, being seen at Elsinore, Denmark, and in China, marking the first visit by a British company to the People's Republic. In 1980 Toby Robertson was succeeded by Timothy Lancaster West (1934–), who had first joined the company in 1966, when he played Prospero in *The Tempest* and Dr JOHNSON in *Madam, Said Dr Johnson*, based by Robertson and Bill Dufton on Boswell, and later revised as *Boswell's Life of Johnson* (1970); his many later roles included King Lear (1971) and Holofernes in *Love's Labour's Lost* (1972). West's first season as Artistic Director included a much criticized but commercially successful production of *Macbeth* with Peter O'TOOLE and *The Merchant of Venice* with West himself as Shylock, and the company appeared to be prospering when at the end of the year the loss of its ARTS COUNCIL grant was announced, and it had to be disbanded.

PROTAGONIST. In ancient Greek drama the tragic poet was originally allowed only one actor and a CHORUS. By the 5th century BC. when three actors were allowed, each playing, if necessary, more than one part, they were named the Protagonist, Deuteragonist, and Tritagonist (from AGON, contest). This did not mean, as far as we can tell, that the Protagonist was the best actor of the three. He may have been allotted the best role, but in the plays of EURIPIDES and SOPHOCLES there are usually at least two major roles of equal importance. It may merely have been that he was the first to speak. There is certainly no indication that the Tritagonist was a third-rate or small-part player. All three probably enjoyed equal status and consideration.

PROVINCETOWN PLAYERS [and **PLAYHOUSE**], a group of American actors and playwrights founded in 1916 by Susan GLASPELL and others, whose ardent experimentalism gave Eugene O'NEILL the opportunities he needed at a vital stage in his career. Their first season, at the Wharf Theatre, Providence, R.I., a converted fishing shack, included the first of his plays to be staged, the one-act *Bound East for Cardiff*, as well as new works by Susan Glaspell, Edna St Vincent Millay, Laurence LANGNER, and others. Later in the year the group moved to the Playwrights' Theatre in Greenwich Village, New York (although they continued to make the Wharf Theatre their summer headquarters until 1921), leaving it in 1918 for the Provincetown Playhouse, a converted stable a few doors away. They ceased operations in 1921, but three years later the Playhouse reopened under the management of Kenneth MACGOWAN, Robert Edmond JONES, and O'Neill himself, working in conjunction with the Greenwich Village Theatre, which saw the first production of O'Neill's *Desire Under the Elms* (1924). After Macgowan's departure it functioned for a further year as the Irish Theatre, but then closed and was demolished in 1930. At the Provincetown Playhouse a number of contemporary European plays were staged, including HASENCLEVER's *Jenseits* (1920) as *Beyond* in 1925; classical revivals included CONGREVE's *Love for Love* in the same year. Some of the original members of the Provincetown Players continued to work at the Playhouse, but in 1929, after an unsuccessful move to the GARRICK THEATRE on Broadway, the group disbanded. The Playhouse has since been leased by various companies, and productions have included light opera, children's theatre, and showcase presentations. Edward ALBEE's *The Zoo Story* and BECKETT's *Krapp's Last Tape* had their American premières there in a double bill in 1960.

PRZYBYSZEWSKI, STANISŁAW, see POLAND.

PUBLIC THEATRE, New York, at 425 Lafayette Street, was converted by Joseph PAPP from the old Astor Library in 1967. Its auditoriums include the three-quarters arena Anspacher, seating 275, which opened on 17 Oct. 1967 with the rock musical *Hair*, later to have a long run at the BILTMORE; the flexible Other Stage, with seating for 108, which opened on 2 Nov. 1968 as a workshop for new playwrights, its first production being Robert Nichols's *The Expressway*; the Newman, a proscenium-arch theatre seating 300, which opened on 4 Oct. 1970 with Dennis J. Reardon's *The Happiness Cage*; and Martinson Hall, a flexible theatre seating 191 which opened on 7 Mar. 1971 with a musical play *Blood*. Notable among the theatre's productions are the following, all of which were transferred elsewhere: Charles Gordone's *No Place to be Somebody* (1969), the first off-Broadway play to win a PULITZER PRIZE; David RABE's *Sticks and Bones* (1971); Jason Miller's *That Championship Season* (1972); the musical *A Chorus Line* (1975); *For Colored Girls Who Have Considered Suicide: When the Rainbow is Enuf* (1976), based on the poems and prose works of Ntozake Shange; *Runaways*, a musical about children who have run away from home; and the feminist musical *I'm Getting My Act Together and Taking it on the Road* (both 1978). The theatre's future has been assured by the granting of a permanent subsidy from the City of New York.

PUBLILIUS SYRUS, a writer of MIMES in which he himself appeared. Originally a slave, he appears to have served his apprenticeship to the stage in provincial Italian towns. Manumitted, and probably brought to Rome in about 46 BC, he challenged and vanquished all other mime-players including LABERIUS. None of his mimes has survived, but he was said to be also the author of an extant anthology of Stoic maxims, presumably from his lost plays, which had a wide circulation and was used as a school book

PUDDLE WHARF THEATRE, London, see PORTER'S HALL.

PULCINELLA, one of the ZANNI or comic servant masks of the COMMEDIA DELL'ARTE, a humpbacked, doltish fellow who may have inherited some of the characteristics of the MACCUS of Roman comedy. In his Italian form he originated from Naples. Little is known of his roles or of his original dress, but he is important as the ancestor of the French POLICHINELLE and the English PUNCH. Some Italian critics believe that the popularity of Pulcinella and the disproportionate attention paid to his buffoonery was one of the main reasons for the decline of the *commedia dell'arte*.

PULITZER PRIZE, founded in 1918, is awarded annually (unless the judges decide to withhold it) for the best play, preferably one dealing with American life, produced in New York during the preceding twelve months. It has been awarded to the following playwrights:

ABBOTT, George, and Jerome Weidman, 1960 (*Fiorillo!*); AKINS, Zoë, 1935 (*The Old Maid*); ALBEE, Edward, 1967 (*A Delicate Balance*), 1975 (*Seascape*); ANDERSON, Maxwell, 1933 (*Both Your Houses*); Bennett, Michael, *et al.*, 1976 (*A Chorus Line*); Chase, Mary, 1945 (*Harvey*); Coburn, D. L., 1978 (*The Gin Game*); CONNELLY, Marc, 1930 (*The Green Pastures*); Cristofer, Michael, 1977 (*The Shadow Box*); DAVIS, Owen, 1923 (*Icebound*); Frings, Ketti, 1958 (*Look Homeward, Angel*); Gale, Zona, 1921 (*Miss Lulu Bett*); Gilroy, Frank D., 1965 (*The Subject was Roses*); GLASPELL, Susan, 1931 (*Alison's House*); Gordone, Charles, 1970 (*No Place to be Somebody*); GREEN, Paul, 1927 (*In Abraham's Bosom*); Hackett, Albert, and Frances Goodrich, 1956 (*The Diary of Anne Frank*); Hammerstein, Oscar, *et al.*, 1950 (*South Pacific*); HOWARD, Sidney, 1925 (*They Knew What They Wanted*); Hughes, Hatcher, 1924 (*Hell-Bent for Heaven*); INGE, William, 1953 (*Picnic*); KAUFMAN, George, and Morrie Ryskind, 1932 (*Of Thee I Sing*), and Moss HART, 1937 (*You Can't Take It With You*); KELLY, George, 1926 (*Craig's Wife*); KINGSLEY, Sidney, 1934 (*Men in White*); Kramm, Joseph, 1952 (*The Shrike*); LINDSAY, Howard, and Russel Crouse, 1946 (*State of *the Union*); Loesser, Frank, and Abe Burrows, 1962 (*How To Succeed in Business Without Really Trying*); MacLeish, Archibald, 1959 (*J.B.*); MILLER, Arthur, 1949 (*Death of a Salesman*); Miller, Jason, 1973 (*That Championship Season*); Mosel, Tad, 1961 (*All the Way Home*); O'NEILL, Eugene, 1920 (*Beyond the Horizon*), 1922 (*Anna Christie*), 1928 (*Strange Interlude*), 1957 (*Long Day's Journey Into Night*); Patrick, John, 1954 (*The Teahouse of the August Moon*); RICE, Elmer, 1929 (*Street Scene*); Sackler, Howard, 1969 (*The Great White Hope*); SAROYAN, William, 1949 (declined) (*The Time of Your Life*); SHEPARD, Sam, 1979 (*Buried Child*); SHERWOOD, Robert, 1936 (*Idiot's Delight*), 1939 (*Abe Lincoln in Illinois*), 1941 (*There Shall Be No Night*); WILDER, Thornton, 1938 (*Our Town*), 1943 (*The Skin of Our Teeth*); Williams, Jesse L., 1918 (*Why Marry?*); WILLIAMS, Tennessee, 1948 (*A Streetcar Named Desire*), 1955 (*Cat on a Hot Tin Roof*); Wilson, Lanford, 1980 (*Talley's Folly*); Zindel, Paul, 1971 (*The Effect of Gamma Rays on Man-in-the-Moon Marigolds*).

PUNCH AND JUDY, English PUPPET-show, presented on the miniature stage of a tall collapsible BOOTH traditionally covered with striped canvas. It was once a familiar sight in the streets of large cities and can still be seen occasionally in seaside towns. Punch, the chief character, with his humped back and hooked nose, evolved from the PULCINELLA of the COMMEDIA DELL'ARTE, and first appeared in London as part of the Italian marionette shows popular after the Restoration, as PEPYS noted in his diary for 9 May 1662. While retaining the physical peculiarities of his Italian prototype, Punch soon became the ubiquitous English buffoon of every puppet-play of the period, equally at home with Adam and Eve, Noah, or Dick Whittington, taking over many of the characteristics of the old VICE of the medieval MYSTERY PLAY. When in the early years of the 18th century fashionable London grew tired of his antics, he migrated to the country fairs, took a wife (first called Joan, later Judy), and adopted the familiar high-pitched voice produced by introducing a 'swazzle' or squeaker into the mouth of the showman who spoke for him. Towards the end of the century he went into eclipse, but emerged again in the 19th century as a hand- or glove-puppet, a reversion to the style of the early English puppet-show which had temporarily been ousted by the Italian stringed marionettes. The change proved economically worthwhile, for one man could carry the portable booth on his back and present all the characters with his own two hands, with a mate (or wife) to 'bottle', or collect pennies from the audience. In the more or less standardized version of the play, which dates from about 1800, Punch, on the manipulator's right hand, remains on stage all the time, while the left hand provides a series of characters—baby, wife, priest, doctor, policeman, hangman—for him to nag, beat, and finally kill, until he is eaten by a crocodile, carried off by the Devil, or allowed to remain in solitary triumph, his only companion being his faithful TOBY—a live dog, usually a terrier, who sits on the ledge of the booth during the entire performance. When in 1962 Punch's 300th anniversary was celebrated by a service in St Paul's Church, Covent Garden, some 50 Punch-and-Judy showmen attended with their dogs.

PUNCH AND JUDY THEATRE, New York, at 153 West 49th Street, between 7th Avenue and the Avenue of the Americas. This delightful playhouse, seating 300, opened on 10 Nov. 1914 with Harold CHAPIN'S *The Marriage of Columbine*; one of its first successes was a dramatization of R. L. Stevenson's *Treasure Island* (1915). In 1926, renamed the Charles Hopkins after the actor-director who built and owned it, it reopened with Karel ČAPEK'S *The Makropulos Secret*, and a record run was set up for the theatre by the 294 performances of Benn W. Levy's *Mrs Moonlight* (1930), seen two years earlier in London. In 1933 the building became a cinema.

PUNCH'S PLAYHOUSE, London, see STRAND THEATRE (1).

PUPPET, an inanimate figure controlled by human agency, which can be larger than life or only a few inches high. It is probably as old as the theatre itself, and it is possible that many of the wonder-working idols of pagan times were in effect immense puppets controlled by their officiating priests; but in its modern sense a puppet is a semblance of a creature—man, bird, beast, fish—given movement and the appearance of life by direct human assistance.

There are several different kinds of puppets, among them the Hand- or Glove-Puppet, the Rod-Puppet, the Marionette, all of which are rounded figures, and the flat puppets of the SHADOW-SHOW and the TOY THEATRE. Because of the popularity of the PUNCH AND JUDY show the hand-puppet is the best known in England. It has a firm head and hands, made of wood or papier mâché, and a loose open costume, into which the puppeteer puts his hand, inserting his first finger into the head and his second finger and thumb each in a hand; he then stands behind a screen or booth and holds the puppets above it. The disadvantage of hand-puppets is that their gestures are limited to the movements of the manipulator's fingers, and one performer cannot handle more than two characters at once; but they are simple and quick to make and easy to carry about. The successful hand-puppet play—or MOTION, as it was called in England—concentrates on broad, simple effects, humorous dialogue, and knockabout comedy. Many of the popular national puppet characters are hand-puppets, carried across Europe by wandering showmen. Apart from the English Punch there is the French GUIGNOL, a generous, bibulous, and witty Lyonnais silk-weaver; the German KASPERL, a slyly astute peasant; the Russian Petrushka, immortalized in Fokine's ballet; and the Italian PULCINELLA, the father of them all. There are hand-puppets in China not very dissimilar from the European types; in INDIA, where they were once very common, they survive mainly in Kerala.

An extension of the hand-puppet, less well known in Western Europe but still to be found in India, is the Rod-Puppet, a full-length rounded figure supported and controlled from below. Its movements are comparatively slow and limited, but the control is absolute, and broad gestures of rare beauty with the arms can be obtained. The most famous and beautiful rod-puppets are found in Java. In Europe the only native tradition of rod-puppets is found in the Rhineland, but the famous Russian puppeteer Sergei Obraztsov used rod-puppets. Under his direction the State Central Puppet Theatre, using thousands of different puppet characters, mounted spectacular productions for both adults and children. Some striking effects with rod-puppets were achieved in Vienna by Richard Teschner, whose stage was seen through a convex lens which enlarged the figures, lending them an aura of enchantment and mystery. The highly-polished nature of his performances was achieved not only by the meticulousness of his staging, but also by the insertion of string-controls inside the rods. The simplest form of rod-puppet is the FOOL's *marot* or bauble, a replica of his own head with its cap and bells, fastened to a stick. The modern Bread-and-Puppet Theatre under Peter Schumann in New York used giant puppets manipulated both by internal and external operators with rods sometimes as long as thirty feet.

The most elaborate form of puppet is probably the string puppet or Marionette. Originally controlled from above by rods or wires running to the centre of the head and to each limb, as in the 'Orlando Furioso' Sicilian puppets used in productions of ARIOSTO's poem, from which they derive their name, they became between 1770 and 1870 entirely manipulated by strings, thus allowing far more flexibility in limb and head movements. A standard marionette has a string to each leg and arm, two to the head, one to each shoulder (which take the weight of the body), and one to the back, that is, nine strings (actually fine threads) in all. An elaborate figure can have two or three times this number. All the strings are gathered together on a wooden 'crutch' or control, held in one hand by the manipulator, while with the other he plucks at whatever strings are required. The figures vary in size from 12 to 18 inches for home use up to 2 or 3 feet for public performances. The Bunraku puppets of JAPAN, seen in London in 1968 during the WORLD THEATRE SEASON, are about two-thirds life-size. They are sometimes strung like marionettes, but more often manipulated by as many as three operators to each figure, working in full view of the audience, and controlling their charges by means of wires and levers in their backs. Indian string puppets, now mostly used for the Tamil 'dance of the dolls' are manipulated somewhat differently.

The Italian Fantoccini puppets, who appeared with much success in London at the Restoration, and the Puppet Theatre in the Piazza in Covent Garden run by the puppeteer Martin Powell between 1710 and 1713, were all marionettes, as were those used by Samuel FOOTE for satirical purposes in 1733, and by Charles DIBDIN, who erected a puppet theatre at Exeter 'Change in 1775. The fortunes of the marionette then waned, but there was a revival of interest in the early 20th century, fostered by Gordon CRAIG with his emphasis on the actor's role as an 'Übermarionette'. This led to an artistic flowering which bore fruit in the work of the Hogarth Puppets; John Wright's marionettes, at the Little Angel Theatre in Islington; the puppet theatre attached to the Midlands Art Centre in Birmingham; and the Caricature Theatre based on Cardiff. But in spite of the foundation in 1925 of a British Puppet and Model Theatre Guild, which publishes a quarterly journal and holds an exhibition every year, puppets in England have a limited appeal and are often thought of only as educational. In the United States, where there was no tradition of puppetry, they were also slow to establish themselves, and still attract only a minority audience. There is, however, a marionette theatre in Greenwich Village, run by Bil Baird, and the Detroit Institute of Art, to which the McPharlin family presented in 1952 a magnificent collection of puppets, has a theatre in which shows are given for children and adults by visiting puppeteers. The true home of the puppet-theatre is the Far East and Eastern Europe, where it covers everything from elementary education in backward areas to sophisticated cabaret shows in the big cities. The history of puppetry is complicated and highly specialized, and there is a vast literature devoted to the subject.

PURDY'S (NEW) NATIONAL THEATRE, New York, see CHATHAM THEATRE (2).

PURIM PLAYS, associated with the Jewish Festival of Purim on the 14th Adar (roughly the middle of March), appear to have originated in France and Germany as early as the 14th century, mainly as extemporized entertainments centring on the Old Testament story of Esther and Haman. Under the influence of the masquerades and mumming of the Italian Carnival, they developed into plays featuring racy dialogue, with interposed songs and dances, which widened their scope to include other Old Testament figures such as Joseph and his brethren, David and Goliath, Moses and Aaron. Mostly in one act, they featured comic rabbis, apothecaries, midwives, and devils, the whole ending with a final chorus foretelling Israel's salvation. The religious authorities, who had at first opposed the acting of Purim plays, finally bowed to public demand and tolerated them as long as they did not overstep the bounds of decency. Given mainly in the countryside, in a barn, a workshop, a stable, or in the house of some kindly disposed rich man, they spread rapidly across Europe to the east, and remained popular in some remote districts until the end of the 19th century. They had a considerable influence on the development of JEWISH DRAMA, and took on a literary form in the 17th century. Under the influence of the COMMEDIA DELL'ARTE Goliath or Haman assumed some of the characteristics of the CAPITANO, Abraham or Jacob those of PANTALONE, the Devil those of HARLEQUIN. The ENGLISH COMEDIANS had in their repertory a version of *Esther and Haman*, printed in German in 1640: if there was an original version in English it is lost. The first original German Purim play is an *Ahasuerus* acted at Frankfurt-am-Main in 1708. In the same city in 1712 a play on the sale of Joseph by his brothers by Beerman of Limburg proved so popular that it was printed, and produced as late as 1858 at Minsk. It appears to have been very spectacular, with fire, thunder, and other wonders. A point of interest is the appearance in it of the Pickelhering of the English Comedians, who thereafter becomes the Purim clown. Another Purim play by Beerman, dealing with David and Goliath, was so popular that Christians were forbidden by their priests to attend it.

The Purim players were constantly in trouble with the authorities for the accretion of vulgar and bawdy ornamentations which gathered round the original plots, and in 1720 the reaction set in. A new play on Esther, shorn of much of the extraneous material, was produced in Prague, and proved so popular that a 3rd reprint of the text was called for in Amsterdam as late as 1774. In Germany the Haskala (Enlightenment) groups continued the fight against vulgarity and improved a number of accepted texts by relegating the comic figures to a subplot, and stressing the religious and educational elements. By the early 19th century Purim plays were covering the whole range of drama, but the founding by GOLDFADEN of a permanent Yiddish theatre which embodied in its productions much of the spirit and method of the Purim plays finally led to their disappearance.

PURITAN INTERREGNUM, the name given in English history to the period of Puritan domination in the mid-17th century. Opposition to the building of playhouses and the performing of plays by professional companies had been growing steadily among the Puritans since the beginning of the century, and this culminated in 1642 in a Parliamentary Ordinance which led to the closing of theatres throughout the country and the dispersal of the acting companies. Under Charles I the London theatre, its actors and its dramatists, had become increasingly attached to the Royalist cause, and the assembling of audiences in large playhouses had provided excellent opportunities for subversive activities. The Puritans maintained that they had no objection in principle to drama; under the Commonwealth plays continued to be acted in schools, with the approval of Cromwell himself, and even in private houses, and in 1656 DAVENANT was allowed to produce publicly at Rutland House his 'entertainment with music' *The Siege of Rhodes* now regarded as the first English opera. But the Puritans held that there were sound political and social, as well as religious, reasons for the banning of stage-plays, which, as the Ordinance said, did not agree 'with private calamities . . . nor with the seasons of humiliation'. As a result, for 18 years actors were deprived of their livelihood and their theatres stood derelict, many of them never to be used again. Some actors joined the army, some, like Andrew CANE, originally a goldsmith, returned to their old trades. Only the boldest, or the most desperate, tried to evade the ban. Evidence of surreptitious performances is given by court judgements imposing fines or terms of imprisonment on actors found playing, usually in the smaller theatres like the COCKPIT, or the RED BULL. The worst consequence of the closing of the theatres was perhaps that ordinary people lost the habit of playgoing and it took a long time to win them back again: even now the bulk of the population has not been entirely won over.

PUSHKIN, ALEXANDER SERGEIVICH (1799–1837), Russian poet, who at the age of eight reputedly wrote little plays in French, which he acted with his sister. At 15 he included among his favourite writers the French dramatists MOLIÈRE, RACINE, and VOLTAIRE, the Russians Ozerov and FONVIZIN, and above all Shakespeare, under whose influence he started work on his great drama *Boris Godunov*. He had already realized that a truly Russian drama could only be created by returning to Russian themes and Russian folklore, and by making the Russian language a literary instrument worthy to rank with the French and German languages for which it had so often been discarded in its own country. His association with the Decembrist revolutionaries shows that he realized also the social implications of his search for the soul of Russia. It is evident from his letters and other sources that Pushkin contemplated writing a series of dramatic works of which *Boris Godunov* alone was completed. It is notable as being the first Russian tragedy on a political theme—the relationship between a tyrant and his people—which, though set back in time, was actually a burning contemporary problem; and it does not rely on a love-intrigue. In other respects, too, it was revolutionary; it was broken up into scenes and episodes, it mingled poetry with prose, and it made use of colloquial Russian speech. It was not published until six years after its completion in 1825, owing to trouble with the censorship, and was not seen on the stage for nearly 50 years, being given its first production in 1870. Four years later it was used for the libretto of an opera by Mussorgsky, in which form it is best known today. Just before his death in a duel Pushkin completed a series of one-act tragedies, one dealing with DON JUAN, one with the rivalry of Mozart and Salieri, and a third with a miser who owes something to Harpagon in Molière's *L'Avare*, but more to Shylock in *The Merchant of Venice*. With some unfinished scenes taken from Russian folklore, these made up the total of Pushkin's work for the theatre. Yet though he is primarily remembered as a poet, the Russian theatre owes him a great debt, since it was he who first made Russian a literary language. There are theatres named after him in Leningrad and Moscow (below), and his works are quoted by Russians as much as Shakespeare's are by the English.

PUSHKIN THEATRE, Leningrad. This theatre, renamed in 1937 in honour of Russia's greatest poet, was founded in what was then St Petersburg as the Alexandrinsky. It opened in 1824, the same year as the MALY THEATRE in Moscow. It had a fine leading actor in KARATYGIN, but no dramatists of the calibre of GOGOL and OSTROVSKY, whose plays were then being staged at the Maly, and never developed a settled policy. For many years its programmes consisted of opera and ballet, and later the patriotic melodramas of Polevoy and Kukolik, and it was not until the end of the 19th century that the first stirrings of REALISM were felt with the production of such plays as STRINDBERG's *The Father*, in which the actor Mamont Dalsky (1865–1918) scored a personal triumph. The first production of CHEKHOV's *The Seagull* in 1896 was a complete failure, the company's old-fashioned technique being inadequate to the task of conveying the subtlety of the author's characterization. Just before the October Revolution MEYERHOLD was working at the Alexandrinsky, his last production there being a revival of LERMONTOV's *Masquerade*. Under the guidance of LUNACHARSKY the theatre weathered the storms of the early 1920s, and by 1924 was ready to include Soviet plays in its repertory, one of the directors at this time being RADLOV. In 1937 Meyerhold returned to produce *Masquerade* again, and during the Second World War the company went on tour, returning to Leningrad in 1944. Interesting landmarks during its later history were a successful production of *The Seagull* by Leonid Vivyen in 1954, in which year he also staged *Hamlet*, and the 1955 revival by TOVSTONOGOV of VISHNEVSKY's *The Optimistic Tragedy*. More recent productions have included Ostrovsky's *The Last Sacrifice* and Shteyn's *Night Without Stars* (both 1975). The theatre has traditionally been a stronghold of fine acting, and although the death of CHERKASSOV in 1966 and of several other leading actors in the early 1970s weakened the company's strength for a time, enough good young actors were left to sustain its high standards.

PUSHKIN THEATRE, Moscow. This theatre, named in honour of the great Russian poet (above), opened in 1951 under Vasily Vasilyevich Vanin (1898–1951), who shortly before his death staged there a revival of *Krechinsky's Wedding* by SUKHOVO-KOBYLIN. He was succeeded as director by Babochkin, and from 1953 to 1961 by TUMANOV. Among the interesting productions seen at this theatre were a revival of OSTROVSKY's *At a Busy Place* in 1952; a new translation of WILDE's *The Importance of Being Earnest* in 1957, directed by Petrov; a new play by Çasona, *The Trees Die Standing* (1958); and a dramatization of Goncharov's *Oblomov* (1969).

PYLADES, Greek who in about 20 BC came to Rome under Augustus, and introduced to Roman audiences the 'Italian dance' or pantomime, from which he took his designation of PANTOMIMUS. He excelled in tragic roles, leaving comedy to his contemporary BATHYLLUS, and his popularity contributed in no small measure to the decline of tragedy on the Roman stage. He was the first *pantomimus* to be accompanied during his silent miming by a choir or orchestra instead of a single singer.

Q

QUAGLIO, family of scenic artists extending over several generations, of whom the first were the brothers Lorenzo (1730–1804), important in the history of neo-classical design, and Giuseppe (1747–1828). Of Italian origin, they worked mainly in foreign Courts, and in the late 18th century established themselves in MUNICH, where their sons, grandsons, and great-grandsons were connected with the Court Theatre. A Quaglio was also working at the Berlin Court Theatre as late as 1891.

QUARTERMAINE, LEON (1876–1967), English actor, who made his début in Sheffield in 1894, and was first seen in London in 1901 with FORBES-ROBERTSON, with whom he went to America in 1903. In 1913 he was at the ST JAMES'S THEATRE with GRANVILLE-BARKER and Lillah MCCARTHY, where he created the part of the Emperor in SHAW's *Androcles and the Lion* (1913) and appeared in revivals of IBSEN's *The Wild Duck* and GALSWORTHY's *The Silver Box*. He was a fine Shakespearian actor, and in later years his Banquo in *Macbeth*, John of Gaunt in *Richard II*, Buckingham in *Henry VIII*, and Cymbeline were memorable. He also gave a fine performance as the Nobleman's Man in Ashley DUKES's *The Man With a Load of Mischief* (1924). His first wife was the actress Madge Titheradge (1887–1961) and his second was Fay COMPTON. His brother Charles (1877–1958) was also a distinguished actor, whose career was mainly in films.

QUAYLE, (John) ANTHONY (1913–   ), English actor and director. He made his first appearance on the stage in London in 1931, and soon gave proof of solid qualities, notably during several seasons with the OLD VIC, where he played a wide variety of parts including John Tanner in SHAW's *Man and Superman* (1938). He was first seen in New York in 1936 in WYCHERLEY's *The Country Wife*. After six years in the army during the Second World War he returned to the theatre to play Jack Absolute in SHERIDAN's *The Rivals* (1945) and direct a dramatization of DOSTOEVSKY's *Crime and Punishment* (1946) starring John GIELGUD and Edith EVANS. In 1948 he succeeded Barry JACKSON as director of the SHAKESPEARE MEMORIAL THEATRE in Stratford-upon-Avon, where he directed a number of plays and also appeared as Petruchio in *The Taming of the Shrew* in 1948, Falstaff in both parts of *Henry IV* in 1951 and *The Merry Wives of Windsor* in 1955, Coriolanus in 1952, and Othello in 1954. In 1956 he played the title role in MARLOWE's *Tamburlaine the Great* in New York, and shortly afterwards left Stratford to appear in London in a succession of non-classical parts which included Eddie in Arthur MILLER's *A View From the Bridge* (1956), James Tyrone in O'NEILL's *Long Day's Journey Into Night* (1958), Marcel Blanchard in FEYDEAU's *Look After Lulu* (1959), and Cesareo Grimaldi in BILLETDOUX's *Chin-Chin* (1960). After appearing as Sir Charles Dilke in Bradley-Dyne's *The Right Honourable Gentleman* (1964) he was seen in New York in the title role of BRECHT's *Galileo* and in USTINOV's *Half-way Up the Tree* (both 1967), and in 1970 he returned there as Andrew Wyke in Anthony Shaffer's *Sleuth*, having previously played the part in London. In 1970 he also directed DOSTOEVSKY's *The Idiot* for the NATIONAL THEATRE. In 1976 he partnered Peggy ASHCROFT in ARBUZOV's two-character play *Old World*, repeating the role opposite Mary MARTIN in New York in 1978, where the play was called *Do You Turn Somersaults?* In the same year he directed and appeared in *The Rivals* and also played Lear for the PROSPECT THEATRE COMPANY at the Old Vic.

QUEEN ANNE'S MEN, company, usually known as the Queen's Men, formed on the accession in 1603 of James I from a combination of WORCESTER's and OXFORD'S MEN. It included Christopher BEESTON, who later became its manager, the playwright Thomas HEYWOOD, and Richard Perkins, and was under the direct patronage of James's wife. It had a successful career playing at the CURTAIN and at the RED BULL until 1616, when after internal dissensions Beeston took some of the company to his new theatre the COCKPIT. The venture was not a success, and on the death of Queen Anne in 1619 the company disbanded, some of the actors going back to the Red Bull or to the provinces until 1625, when several, including Beeston, joined the newly formed QUEEN HENRIETTA'S MEN.

QUEEN ELIZABETH'S MEN, Elizabethan company, usually known as the Queen's Men (as, somewhat confusingly, were also the later QUEEN ANNE's and QUEEN HENRIETTA'S MEN). Formed in 1583, it included the jester Richard TARLETON, whose death in 1588 was a great blow. The company played at several of the INNS USED AS THEATRES, notably the Bull, the Bell, and the Bel Savage and also at the THEATRE, the CURTAIN, and the ROSE up to 1594, when it was disbanded and replaced by the ADMIRAL'S MEN.

QUEEN HENRIETTA'S MEN, company of players, usually known as the Queen's Men, formed on the accession of Charles I in 1625 and placed under the direct patronage of his wife. It was headed by Christopher BEESTON, and included several other members of the disbanded QUEEN ANNE'S MEN, to which it was in some ways a successor. Among the successful plays given by this company at their theatre the COCKPIT were about 20 which their official dramatist James SHIRLEY wrote for them, either alone or in collaboration, between 1625 and 1636. They also appeared at Court in Thomas HEYWOOD's masque *Love's Mistress* (1634), for which Inigo JONES designed some excellent scenery. When plague closed the theatres in 1636 the company was disbanded, and replaced at the Cockpit by Beeston's Boys, but a new company under the old name was formed in 1637 and played successfully at SALISBURY COURT until the closing of the theatres in 1642.

QUEEN'S THEATRE, Dublin, one of the most popular theatres in the city, used mainly for MELODRAMA. The first building on the site, the New Olympic Circus, opened in 1823, and was succeeded by the Adelphi Theatre in 1829. This was forced to close for want of a licence, but reopened as the Queen's Royal Theatre on 5 Oct. 1844. It was a large building, seating almost 2,000, and was probably the last theatre to devote the whole of its floor space to the PIT. A STOCK COMPANY was employed, and there was a frequent change of bill. Frederick ROBSON made his Irish début here in Oct. 1850 and Henry IRVING in March 1860. In 1887 the Queen's entered on its most famous period with Irish melodramas on patriotic themes, which lasted in popularity into the new century and in which the actor F. J. MCCORMICK began his career. In 1907 the theatre closed, but it was rebuilt in 1909 and carried on with melodramas and pantomimes. In 1928 came the revue acts of Jimmy O'DEA, soon to become legendary. Ironically, the company that killed the popular idea of the 'stage Irishman', the ABBEY THEATRE players, was the last to play at the Queen's. They leased the theatre in 1951 after fire had destroyed their own building and remained there until the opening of the new Abbey Theatre in 1966, after which the Queen's was closed. It was demolished in 1969.

QUEEN'S THEATRE, London. (1) In Long Acre. This building, erected in 1849, was known originally as St Martin's Hall, and was used on a number of occasions by Charles DICKENS reading from his own works. In 1867 it was converted into the second largest theatre in London, and opened on 24 Oct. with Labouchère as manager, his future wife Henrietta HODSON being a member of the company. The first production, an adaptation of Charles READE's novel *White Lies* as *The Double Marriage*, failed in spite of a strong cast, and the theatre's first success was H. J. BYRON's *Dearer than Life*, which opened on 8 Jan. 1868. In Dec. of the same year Ellen TERRY and Henry IRVING appeared together for the first time, in *Katharine and Petruchio*, GARRICK's 'boiled-down version' (as Ellen Terry called it) of *The Taming of the Shrew*. A revival of Tom TAYLOR's *A Fool's Revenge* in 1869 was followed by several new plays from his pen, including *'Twixt Axe and Crown; or, the Lady Elizabeth* (1870), *Joan of Arc* (1871), and *Dead or Alive* (1872); but the chief event in the theatre's history was the return of Ellen Terry to the stage, after six years in retirement, to take over the part of Philippa in Reade's *The Wandering Heir* (1873) from Mrs John WOOD. After that the Queen's had little success. It changed its name to the National, reopening on 27 Oct. 1877 with Farnie and Reece's drama *Russia; or, the Exiles of Siberia*, and finally closed in 1879.

(2) In Shaftesbury Avenue. It is the sister theatre of the present GLOBE THEATRE, which it adjoins, the auditoriums and stages of the two playhouses being separated only by a party wall. With a seating capacity of 1,160 in three tiers it opened on 8 Oct. 1907, and had its first success a year later with a musical *The Belle of Brittany*. Fashionable tango-teas were held in 1913, and in 1914 the theatre had another success with Glass and KLEIN's *Potash and Perlmutter*. Among later productions were *Bluebeard's Eighth Wife* (1922) by Arthur Wimperis, FAGAN's *And So To Bed* (1926) with Yvonne ARNAUD, and the MALVERN FESTIVAL production of SHAW's *The Apple Cart* (1929) with Edith EVANS and Cedric HARDWICKE. In the following year John GIELGUD repeated his OLD VIC triumph as Hamlet, and in the 1930s successful productions included Besier's *The Barretts of Wimpole Street* (1930); *Evensong* (1932), based by Edward KNOBLOCK on Beverley Nichols's novel; and Robert MORLEY's *Short Story* (1935), which had Margaret RUTHERFORD, Marie TEMPEST, Sybil THORNDIKE, Rex HARRISON, and A. E. MATTHEWS in the cast. In 1937–8 Gielgud returned to the Queen's with a season which included SHERIDAN's *The School for Scandal*, CHEKHOV's *Three Sisters, Richard II*, and *The Merchant of Venice*. He was also in Dodie SMITH's *Dear Octopus* (1938), which was still running when the theatres closed on the outbreak of the Second World War in 1939. The Queen's had reopened and was occupied by Daphne Du Maurier's *Rebecca* when on 24 Sept. 1940 it was badly damaged by bombs. It did not reopen until 1959, after the complete restoration of the front-of-house, the first production being Gielgud's solo recital *The Ages of Man*. Notable productions included Michael REDGRAVE in *The Aspern Papers* (also 1959), which he himself adapted from Henry JAMES's novel, and in BOLT's *The Tiger and the Horse* (1960), and Anthony Newley in the musical *Stop the World—I Want to Get Off* (1961), which he both wrote and starred

in. In 1963 Maggie SMITH gave a fine performance in Jean Kerr's *Mary, Mary*, and in 1964 there was a revival of Chekhov's *The Seagull* in a new version by Ann Jellicoe, with Peggy ASHCROFT and Vanessa REDGRAVE. Noël COWARD made his last appearance on the stage in his *Suite in Three Keys* (1966), and Peter USTINOV's *Halfway Up the Tree* (1967) and Barry England's *Conduct Unbecoming* (1969) both had long runs. Alan BENNETT's *Getting On* was seen in 1971, and in 1972 Maggie Smith returned in a revival of Coward's *Private Lives*, which was still running when Coward died a year later. The NATIONAL THEATRE production of DE FILIPPO's *Saturday, Sunday, Monday* moved here in 1974, and in 1975 Alan BATES was seen in Simon GRAY's *Otherwise Engaged*. In 1977 Alan AYCKBOURN's *Just Between Ourselves* had a comparatively short run, but Alan Bennett's *The Old Country*, starring Alec GUINNESS, was more successful. Tom COURTENAY gave a brilliant performance in 1980 in the title role of Ronald Harwood's *The Dresser*, originally seen at the ROYAL EXCHANGE THEATRE, Manchester, and in the following year another play by Simon Gray, *Quartermaine's Terms*, received high praise. (See also DORSET GARDEN THEATRE, HER MAJESTY'S THEATRE, and SCALA THEATRE.)

QUEM QUAERITIS? (Whom seek ye?), the first line of (and so the short name for) the earliest of the Tropes introduced into the Mass of the medieval Christian Church, in which LITURGICAL DRAMA had its origins. It followed closely the Gospel account of the meeting between the angel at Christ's tomb and the three Marys on Easter morning. Introduced during the 10th century at the Benedictine Abbey of St Gallen in Switzerland and established in England after the Synod at Winchester in *c*. 960, it became, some time at the beginning of the 12th century, detached from its place at the beginning of the Easter Mass, either as an Introit trope or as a processional chant, and reappeared as a separate scene at Matins on Easter morning, with four male singers to act the parts of the women and the angel and with a building, temporary or permanent, to represent the empty tomb, in which a cross was laid on Good Friday and removed early on Sunday morning. Anthems were added to the original trope, then proses, and finally metrical hymns, the most important being the *Victimae paschali* written by Wipo of St Gallen in 1125–50 which found its way into the scenario during the 13th century, forming part of the second scene which had by then been added, showing the Marys announcing the Resurrection to the congregation. New voices were added to represent the disciples, and a third scene was introduced in which the apostles John and Peter were shown going into the sepulchre. In most churches this brought the drama to a close, but in a few a fourth scene was added, with the dialogue between the Risen Christ and Mary Magdalene. An important step in dramatic evolution was an interpolation (probably in the 13th or 14th century) of a scene with no scriptural basis—the dialogue between the Marys and the spice-sellers. But it is possible that these characters, who introduced the first hint of secularized drama into the Bible narrative, were a counter-influence on the later liturgical play from the plays in the vernacular which had by this time developed from the original Latin play.

QUESTORS THEATRE, London, in Mattock Lane, Ealing, the home of one of Britain's leading amateur groups founded in 1929 by Alfred Emmet, which in 1933 adapted a disused chapel for its productions. After the Second World War the company, still under the enthusiastic leadership of Emmet, raised funds to construct a sophisticated flexible theatre, adaptable for proscenium, forestage, thrust stage, in-the-round, and other types of open-stage productions and seating between 325 and 450. Rehearsal rooms, a lecture room, wardrobe, workshop, scenery store, and administrative offices are also housed in the complex. The building proceeded in stages as money became available, the theatre itself being opened by the Queen Mother in 1964. A professionally staffed part-time training course for young actors has been run since 1947 and young people's groups, from the age of five upwards, meet weekly at the theatre.

The Questors was one of the founders of the Little Theatre Guild of Great Britain (1946) and participated in the founding of the International Amateur Theatre Association (1952). (See AMATEUR THEATRE.) From 1960 to 1977 it held an annual Festival of New Plays: one of its first successes was *Next Time I'll Sing to You* (1962) by James SAUNDERS, and other plays by him were given at subsequent festivals. Among foreign plays first introduced to English audiences by the Questors have been works by Marcel Aymé, James JOYCE, William SAROYAN, Jacques AUDIBERTI, Friedrich DÜRRENMATT, Günter Grass, and Michel TREMBLAY. International Theatre Weeks have been held every two or three years since 1969 with participation by outstanding companies from Canada, the U.S.A., and Europe.

Emmet retired from the directorship of the theatre in 1969 but remained closely involved in its work; he was succeeded by a directorate of four.

QUÊTE, see MUMMERS' PLAY.

QUICK, JOHN (1748–1831), English actor, who as a boy of 14 joined a provincial company, and in 1767 was engaged by Samuel FOOTE for the HAYMARKET THEATRE. His good work there caused him to be taken into the company at COVENT GARDEN, where he remained for the rest of his career except for occasional visits to the provinces and a brief managership of the King Street theatre in BRISTOL. He created the part of Postboy in the elder COLMAN's *The Oxonian in Town* (1767) and was in the first production of GOLDSMITH's *The Good-Natur'd Man* (1768), also playing Tony Lumpkin in his *She Stoops to Conquer* (1773) and Bob Acres in SHERIDAN's *The Rivals* (1775). He had a vast range of comic parts, from Polonius in *Hamlet* (a part which for a long time remained the property of the leading comedian in the company) and Shallow in *Henry IV, Part II* to the Clown in PANTOMIME and numerous rustics and comic servants in forgotten comedies. He hankered after tragedy, and for his benefit in 1790 chose to appear as Richard III, only to be laughed off the stage. After that he remained faithful to comedy until his retirement in 1798, from which he occasionally emerged, being last seen at a benefit performance at the Haymarket as Don Felix in Mrs CENTLIVRE's *The Wonder, a Woman Keeps a Secret*. A small, impetuous man, with a chubby smiling face, Quick was the favourite actor of George III. His portrait by Zoffany hangs in the GARRICK CLUB in London.

QUICK-CHANGE ROOM, a small, closed recess opening off the stage, where players can change their clothes when a brief absence from the stage does not allow them to return to their dressing-rooms.

QUILLEY, DENIS CLIFFORD (1927–), English actor, who appears both in straight plays and in MUSICAL COMEDY. He made his first appearance at the BIRMINGHAM REPERTORY THEATRE in 1945, his roles including Lyngstrand in IBSEN's *The Lady From the Sea*, and was first seen in London in 1950, when he took over the part of Richard in FRY's *The Lady's Not For Burning* from Richard BURTON. After appearing in the REVUE *Airs on a Shoestring* (1953) he played his first leading role in the West End, Geoff Morris in the musical *Wild Thyme* (1955), and was in the long-running musical *Grab Me a Gondola* (1956). In 1959 he had the title role in *Candide*, Bernstein's musical based on VOLTAIRE, and in 1960 he took over from Keith MICHELL in the musical *Irma La Douce*, in which he made his New York début a year later. Back in England he was seen in two more musicals, *The Boys From Syracuse* (1963), based on *The Comedy of Errors*, and *High Spirits* (1964), based on COWARD's *Blithe Spirit*. Although he was seen at NOTTINGHAM PLAYHOUSE in 1969 as Macbeth and as Archie Rice in OSBORNE's *The Entertainer*, his career as a straight actor received its biggest impetus when he joined the NATIONAL THEATRE company in 1971. He remained for over five years, playing a wide variety of parts, among them Jamie in O'NEILL's *Long Day's Journey into Night* in 1971, Hildy Johnson in Hecht and MacArthur's *The Front Page* in 1972, Caliban in *The Tempest* in 1974, and Claudius in *Hamlet* in 1975. In 1977 he went to the ROYAL SHAKESPEARE COMPANY to appear in Peter NICHOLS's musical *Privates on Parade*, and was also seen at the Albery (see NEW THEATRE) as Morell in SHAW's *Candida*. In 1980 he starred in the London production of SONDHEIM's musical *Sweeney Todd*, playing opposite Sheila HANCOCK.

QUIN, JAMES (1693–1766), English actor, the last of the declamatory school whose supremacy was successfully challenged by GARRICK. He made his first appearance at the Smock Alley Theatre in Dublin in 1712 and two years later was playing small parts at DRURY LANE, where he made an unexpected success as Bajazet in ROWE's *Tamerlane* when the actor billed to play the part was taken ill. In 1718 he went to LINCOLN'S INN FIELDS THEATRE where he remained for 14 years, appearing first as Hotspur in *Henry IV, Part I*, and then in a range of Shakespearian parts which included Othello, Lear, Falstaff, the Ghost in *Hamlet*, and Buckingham in *Richard III*. He then went to COVENT GARDEN and finally returned to Drury Lane, retiring in 1751 to live in Bath. A man of great gifts but very little education, he was obstinate and quarrelsome and a great stickler for tradition, refusing to the end to alter one detail of his original costumes. A portrait of him as the hero of Thomson's *Coriolanus* (1748) shows him equipped with plumes, peruke, and full spreading short skirt in the style of the later 17th century. There is a description of his acting in Smollett's *Humphry Clinker* (1771).

QUINAULT, see DUFRESNE.

QUINAULT, PHILIPPE (1635–88), French dramatist and librettist, who became valet to TRISTAN L'HERMITE, under whose name his first play *Les Rivales* (1653) was accepted by the actors at the Hôtel de BOURGOGNE. When they discovered the truth they were reluctant to pay the fee agreed upon, and the resultant negotiations, which resulted in the author being given a fixed share in the takings from each performance, has been cited as the origin of the ROYALTY system, though the point is still disputable. The last of Quinault's plays to profit from the advice of Tristan, shortly before his death, was *La Comédie sans comédie* (*A Play with No Plot*, 1655), which consisted of four scenes in different genres designed to show off the talents of the actors at the MARAIS. By 1666 Quinault had written a number of plays, of which the most successful were the tragedy *Astrate roi de Tyr* (1664) and the comedy *La Mère coquette* (1666). He then married a wealthy young widow who despised the theatre and made him give it up; but in 1668, being well known for his facility in light verse, he was persuaded to write the libretto for the first of Lully's French operas *La Grotte de Versailles*. He continued to write Lully's librettos until the latter's death in 1687, and such was the purifying power of music that his wife made no objections. He also collaborated with MOLIÈRE and CORNEILLE in the lyrics for *Psyché* (1671) and contributed to *Les Fêtes de l'Amour et de Bacchus* (1672) in which Molière also had a hand. Quinault, who was elected a member of the

French Academy in 1670, was a modest and kindly man, and he had the good fortune to occupy, with Thomas CORNEILLE, the interval between the great days of Pierre Corneille and the rise of RACINE—who incidentally thoroughly disliked him. His verse was harmonious and much admired by Louis XIV, who with royal condescension would often suggest subjects for his pen; but he could never win the approval of BOILEAU, who considered him insipid and sentimental.

QUIÑONES DE BENAVENTE, LUIS (*c.* 1593–1651), Spanish dramatist, friend of Lope de VEGA, and author of a number of early short comedies. He contributed largely to the development of the LOA and ENTREMÉS, and for the latter genre he took from CERVANTES' *Los alcaldes de Dagazo* the character of Pedro Rana, renamed him Juan, and introduced him into many of his comedies. He stopped writing for the stage in 1640.

QUINTERO, JOSÉ BENJAMIN (1924–), American director, a co-founder of the off-Broadway group at the CIRCLE-IN-THE-SQUARE. His productions there include an excellent revival of O'NEILL's *The Iceman Cometh* in 1956, and in the same year he directed the American première of O'Neill's *Long Day's Journey into Night* at the HELEN HAYES THEATRE. In 1958, at the Festival of Two Worlds in Spoleto, Italy, he directed the same author's *A Moon for the Misbegotten*. Among his later productions at the Circle-in-the-Square was Brendan BEHAN's *The Hostage* in 1958. Quintero continued to concentrate on plays by O'Neill, including *Marco Millions*, for the Lincoln Center Repertory Company at the WASHINGTON SQUARE THEATRE in 1964, *A Moon for the Misbegotten* in Oslo, Norway, in 1975, and *Anna Christie* in New York in 1977. The last two starred the Norwegian actress Liv Ullmann, whom he also directed in COCTEAU's *The Human Voice* in Melbourne, Australia, in 1978 and off-Broadway in 1979.

QUINTERO, SERAFÍN and JOAQUÍN ÁLVAREZ, see ÁLVAREZ QUINTERO.

Q THEATRE, London, on the Brentford side of Kew Bridge, a 500-seat theatre run by Jack de Leon. It opened on 26 Dec. 1924 in a converted hall, with a revival of Gertrude Jennings's comedy *The Young Person in Pink*. Over 1,000 plays were presented here, half of which were new; of these over 100 were transferred to the West End, Frederick Knott's *Dial M for Murder* being probably the best remembered. Terence RATTIGAN's first play *First Episode* (1934) was tried out at the Q before being transferred to the COMEDY, with Max ADRIAN, who had made his first professional appearance at the Q, in the leading role. Other well known actors who made their débuts at this theatre were Anthony QUAYLE in *Robin Hood* (1931) and Dirk Bogarde in a revival of PRIESTLEY's *When We Are Married* in 1939. In 1955 the Middlesex County Council refused to renew the theatre's licence unless it was rebuilt. The campaign to raise the necessary funds ended with Jack de Leon's death in Nov. 1956, and the building was demolished in May 1958.

R

RABE, DAVID (1940–   ), American dramatist, whose importance was immediately recognized when his first two plays, *The Basic Training Of Pavlo Hummel* and *Sticks and Bones*, were produced in New York at the PUBLIC THEATRE in 1971. Both related to the war in Vietnam, where he had served as a draftee. *Pavlo Hummel* concerns a neurotic and unhappy teenager who after undergoing his basic military training—used as a symbol of the ordinary man's inability to control his own life—is killed in Vietnam after a quarrel with another soldier. In *Sticks and Bones*, produced later in New York but written earlier, a man blinded in Vietnam returns to his parents, Ozzie and Harriet, whose behaviour resembles that of the characters in a popular television show whose names they bear. It was produced in London in 1973, and in the same year *The Orphan*, loosely based on the *Oresteia* of AESCHYLUS and described by the author as the third play of his Vietnam trilogy, was seen in New York, but was not well received. Later plays are *Boom Boom Room* (also 1973, revised in the following year as *In the Boom Boom Room*, under which title it was produced in London in 1976) and *Streamers* (1976; London, 1978), a powerful play set in an army barracks in Virginia.

RABÉMANANJARA, JACQUES (1913–   ), the best known playwright of French-speaking Africa before 1965, a native of Madagascar (now the Malagasy Republic). His plays, which have rarely been performed since they read better than they act, have historical themes and are written in a mixture of poetry and poetic prose. They include *Les Boutriers de l'aurore* (*Ships of Dawn*, pub. 1957), *Agapes des Dieux* (*Feasts of the Gods*, pub. 1962), perhaps the most moving and successful of his works, and *Les Dieux Malgaches* (*The Gods of Madagascar*, pub. 1964). Although Rabémananjara seems not to have been influenced by Polynesian culture to which the culture of Madagascar is so strongly allied, in general his attempt to combine modern French dialogue with historical and mythical characters in the manner of COCTEAU lacks the irony which sustains the device in the latter's work.

RABINOVICH, ISAAC MOISEIVICH, see MOSCOW STATE JEWISH THEATRE.

RACAN, HONORAT DE BUEIL, Seigneur de (1589–1670), French poet and dramatist, who with THÉOPHILE DE VIAU marks the entry into French dramatic literature of the nobleman and courtier. As a page at Court Racan had much enjoyed the plays of Alexandre HARDY, but his own *Bergeries* (1620), which was to play an important part in the development of the PASTORAL in France, showed the influence of TASSO's *Aminta* (1573) and of GUARINI's *Il pastor fido* (1598) rather than of Hardy's ranting tragi-comedies. Written in accordance with the UNITIES, it was also influenced by Malherbe's attempts to reform and purify the French language, and contains some notable lyric passages.

RACHEL [Élisa Félix] (1820–58), child of a poor Jewish family, and one of the greatest tragediennes of the French stage. Befriended by the impresario Choron, who found her singing in the streets, the 13-year-old girl was sent to Saint-Aulaire's drama school in the old Théâtre Molière. Further study at the Conservatoire was cut short by her father, who was anxious to make money out of her, and in 1837 she was taken into the company of the GYMNASE, where she attracted the attention of Jules Janin, critic of the *Journal des Débats*. Encouraged by him, and coached by SAMSON, she entered the COMÉDIE-FRANÇAISE in 1838, appearing as Camille in CORNEILLE's *Horace* with some success. Again encouraged by Janin, she appeared in other plays by Corneille which had been almost entirely neglected since the death of TALMA and in the works of RACINE, whose *Phèdre* was to be her greatest part. She also appeared in some modern plays, including a revival of *Marie Stuart*, by Lebrun, and the first production of *Adrienne Lecouvreur* (1849), by SCRIBE and Legouvé. But it was in the great classical roles that she appeared mainly on tour, either in the French provinces, in Europe, going as far as Russia, in London, where she first appeared in 1841 with outstanding success, or in America, where, on her one visit in 1855, she finally aggravated her tubercular condition, the result of early hardships and later overwork combined with a feverish succession of amorous intrigues. In spite of superb acting her visit to the United States was a failure financially, mainly on account of the language barrier. An account of her last tragic journey, written by a member of her company, Léon Beauvallet, was later published, and in 1968 was issued in an English translation by Collin Clair. Her four sisters, Sarah [Sophie] (1819–77), Lia [Adelaide] (1828–72), Rebecca [Rachel] (1829–

54), and Dinah [Mélanie Émilie] (1836–1909), were all on the stage, as was her brother Raphaël (1825–72).

RACINE, JEAN (1639–99), French playwright and poet, one of the greatest tragic dramatists in the history of the theatre. Orphaned at the age of four, he was brought up by his grandparents and his aunt Agnès, later Abbess of Port-Royal, where Racine was educated after a few years at the Collège de Beauvais. He was an excellent scholar, though somewhat undisciplined, an enthusiastic admirer of the Greek dramatists, and at 19 already a good poet. He soon escaped from the restraining influence of Port-Royal and the Jansenists and led a free, though not particularly dissipated, life. He was quickly accepted in literary circles, where he made the acquaintance of MOLIÈRE, who in 1664 staged his first play *La Thébaïde, ou les Frères ennemis* at the PALAIS-ROYAL. It was successful enough for Molière to accept his second play *Alexandre le Grand* (1665), but a fortnight after its production Racine, who was not a man to be grateful for favours, allowed the actors at the Hôtel de BOURGOGNE to stage it in direct competition with Molière. His excuse was that he was not satisfied with the Palais-Royal production, and it was true that the Hôtel de Bourgogne had a higher reputation in tragedy; but it is possible that feminine intrigue played a part in the affairs, as Mlle DU PARC, a fine tragic actress and Racine's mistress at the time, left Molière's company in order to play the lead in Racine's next play *Andromaque* (1667) at the Hôtel de Bourgogne. After this double betrayal Molière never spoke to Racine again and Racine did not long enjoy his triumph as Mlle du Parc died suddenly the following year. There was some unkind gossip about the affair at the time, and in 1680 the infamous poisoner Catherine Voisin alleged at her trial that Racine had removed Mlle du Parc to make way for Mlle CHAMPMESLÉ, who had come from the MARAIS to play Hermione to Du Parc's Andromaque. She certainly became Racine's mistress and was closely associated with his later work, creating among others the roles of Phèdre and Bérénice.

With the production of *Andromaque* Racine first achieved recognition as an outstanding dramatist, rival of the ageing CORNEILLE and in some ways superior to him. It was followed by Racine's only comedy *Les Plaideurs* (*The Litigants*, 1668), based partly on ARISTOPHANES' *Wasps*. It was intended for the Italian actors at the Palais-Royal, but on the departure of Tiberio FIORILLO (Scaramouche) from Paris it was transferred to the Hôtel de Bourgogne, where after a slow start it was extremely successful and frequently revived. It remained in the repertory, and in 1920 stood sixth in the list of plays most frequently given at the COMÉDIE-FRANÇAISE. Sadly, Racine wrote no more comedies. His next tragedy, *Britannicus* (1669), was not a success, in spite of its exquisite poetry, but according to BOILEAU the portrait of Nero deterred Louis XIV from featuring himself further in Court ballets and entertainments. With his next play Racine found himself at odds with Corneille. Either by accident or design they were both working on the same subject, and the production of Racine's *Bérénice* (1670) took place only a week before Molière produced Corneille's *Tite et Bérénice*. Racine's version was more generally admired, and he followed it with two more tragedies, *Bajazet* (1672) and *Mithridate* (1673), both historical and oriental subjects treated in a contemporary style very different from Corneille's. Racine was often criticized for abandoning the latter's heroic mood and portraying great rulers overcome by degrading passions, but he was now firmly established as the leading tragic dramatist of his day. For his next play he returned to Greek tragedy with *Iphigénie* (1674), which was a brilliant success, but with *Phèdre* (1677), his greatest play, his career as a dramatist came to an abrupt end. It was known that he was writing on the subject, and some of his many enemies—for he was not a particularly likeable man—persuaded PRADON, a mediocre playwright of the day, to compose a tragedy on the same theme. The two plays were produced within two days of each other, and Pradon's supporters made sure that his play was for a time the more successful. Racine's chagrin at this setback may have been one of the reasons which prompted him to retire from the theatre. Another, more tangible one, was his appointment as historiographer-royal, a Court appointment incompatible with an active career in the theatre. He married, had seven children, made up his quarrel with Port-Royal, and turned his great gifts to the studying and recording of contemporary French history. It was not until 1689 that, at the request of Mme de MAINTENON, he wrote the tender and poetic *Esther* for performance by the young ladies at her school in Saint-Cyr, following it in 1691 with *Athalie*. Both were successful, though the latter was given with less pomp and ceremony than the first; but at Racine's own request neither was seen in public. It was not until 1716 that *Athalie* was performed at the Comédie-Française, and recognized as one of Racine's dramatic masterpieces. It has frequently been revived. *Esther*, on the other hand, was unsuccessful when produced there in 1721, being more of a sacred oratorio than a tragedy, but the chief role has tempted many famous actresses including RACHEL. *Phèdre* is regarded as the summit of Racine's achievement, and the role of the heroine, which has been interpreted by such diverse actresses as CLAIRON, Rachel, and BERNHARDT, has a prestige as great as that of Hamlet on the English stage. The conventions of his time, which forbade excess in any form, make the vitality and passions of Racine's characters—particularly his women—seem all the greater. It may be said that the rules he chose to observe served only to enhance his genius, while his great poetic gifts saved his plays from monotony or rigidity.

Up to the end of the 17th century only two of Racine's plays had been seen on the English stage —*Andromaque*, translated by CROWNE, in 1674, and *Bérénice*, translated by OTWAY, in 1676. In 1707 Edmund Smith made the first translation of *Phèdre*, which as *Phaedra and Hippolitus* was seen at the HAYMARKET, and in 1712 Ambrose PHILIPS's very free adaptation of *Andromaque* as *The Distrest Mother* was so successful at DRURY LANE that it may have encouraged the actors there to put on ADDISON's *Cato* (1713), a play which more than any other shows the short-lived influence on the English stage of French neo-classical tragedy. Like Corneille, however, Racine has never been popular in England, and many translations of his plays, even modern ones, are intended for reading rather than acting. Exceptions are Charles Johnson's *The Victim* (1713), based on *Iphigénie*, and *The Sultaness* (1716), based on *Bajazet*, and, in modern times, a version of *Phèdre* by the American poet Robert Lowell and one by Margaret Rawlings in which she herself appeared in London in 1958. Four of Racine's plays were in the repertory of a French company which appeared at the Haymarket in Dec. 1721, and since then most visiting companies have included at least one in their programme—in May 1953 the Comédie-Française was seen in *Britannicus*—but again the only one which is at all familiar to London audiences is *Phèdre*, which was in the 1721 programme and has since been brought to London by Rachel, Bernhardt, Edwige FEUILLÈRE (in 1957), and Marie BELL (in 1960). In 1930 it was played in French by Sybil THORNDIKE.

RADA, see SCHOOLS OF DRAMA.

RADIO CITY MUSIC-HALL, New York, in the Rockefeller Center. This, the largest indoor theatre in the world, seating 6,200, was planned as part of the 24-acre Radio City, opening on 27 Dec. 1932 with a staff which included Robert Edmond JONES as designer and Martha Graham as choreographer. Its companion house, the RKO Roxy cinema, opened on 29 Dec. The opening production was not a success and the Music-Hall closed, to reopen a month later as a combined film and VARIETY theatre, which it has remained ever since. The variety shows, presented four times daily, are built around the Rockettes, a resident *corps de ballet*, and a permanent symphony orchestra, and have a central theme, as in the holiday extravaganzas *The Nativity* and *The Glory of Easter*. The stage, with a proscenium height of 60 ft and a width of 100 ft, has a three-ton contour curtain which is motor-controlled to arrange the folds in a variety of patterns. The full stage, 66 ft deep by 140 ft wide, is equipped with a rain and steam curtain and a circular motorized turntable with elevator sections which can be raised and lowered 40 ft, making the theatre eminently suitable for spectacular effects. The auditorium and public areas were restored to their original art deco glory when the building was designated by the Landmark Preservation Commission of the City of New York in 1979.

RADIUS, see RELIGIOUS DRAMA.

RADLOV, SERGEI ERNESTOVICH (1892–1958), Soviet actor and director, whose work was done mainly in Leningrad, where he started his theatrical career under MEYERHOLD and later opened his own theatre. He did some important work in expressionist forms of staging during the 1920s and pioneered the staging of modern operas by Berg and Prokofiev; but although his outstanding productions included OSTROVSKY's *The Dowerless Bride* and IBSEN's *Ghosts*, it was as a director of Shakespeare that he was pre-eminent, with productions of *Romeo and Juliet*, *Hamlet*, and *Othello*, the last being seen also at the Moscow MALY. His finest work was his production of *King Lear* for the MOSCOW STATE JEWISH THEATRE, with MIKHOELS in the title role.

RAILS, see FLAT.

RAIMUND, FERDINAND [Jakob Raimann] (1790–1836), Austrian playwright and actor, who, though secretly preferring tragedy, was a popular farce-player at the Josefstädter and Leopoldstädter theatres in VIENNA from 1813 to 1823. Although considerd the greatest comic actor of his day, he decided that his unique combination of gifts needed a vehicle specially created to display them and so began to write his own plays. The first *Der Barometermacher auf der Zauberinsel* (*The Barometer-Maker on the Magic Island*, 1823) was a great success, and was followed by *Das Mädchen aus der Feenwelt, oder der Bauer als Millionär* (*The Girl from Fairyland, or the Millionaire Farmer*, 1826), which preaches, with the help of magical forces and a whole host of allegorical personages, the doctrine of contentment on small means. Perhaps Raimund's greatest play is *Der Alpenkönig und der Menschenfeind* (*The King of the Alps and the Misanthrope*, 1828), in which a kindly mountain spirit cures a misanthropist by assuming his shape and character, while the misanthrope, disguised as his own brother-in-law, has to watch the havoc caused by his suspicions and ill-will. However, all ends happily. Encouraged by his success, Raimund set about educating himself by studying Shakespeare and—typically for a Viennese playwright—CALDERÓN. The result was a number of plays which bordered too consciously on the mystical and were out of tune with the demands of his audiences and with his own essentially unreflective genius. It was not until he returned to his earlier style in *Der Verschwender*

(*The Prodigal*, 1834) that he was again successful; but his last years were overshadowed by the rising popularity of NESTROY and he finally commited suicide.

RAIN BOX, see SOUND EFFECTS.

RAISIN, JEAN-BAPTISTE (1655–93), French actor who, with his sister Cathérine (1649–1701) and brother Jacques (1653–1703), was a member of a children's company established in Paris by his father under the patronage of the Dauphin. Among the other children were Michel BARON and Jean de VILLIERS; the latter later married Cathérine. As an adult actor Raisin, who was known as 'little MOLIÈRE' from his excellent acting in comedy, was at the Hôtel de BOURGOGNE with his wife Françoise Pitel de Longchamp (1661–1721), known as Fanchon. They both became members of the COMÉDIE-FRANÇAISE on its foundation in 1680, but it was not until after the death of ROSIMOND in 1686 that Raisin was given leading parts. His brother Jacques joined him at the Comédie-Française in 1684. A tall, thin, somewhat solemn man, he was at his best in tragedy.

RAKE, the slope of the stage floor from the back wall to the footlights. This had once a definite purpose in that it aided the illusion of scenes painted in perspective. With the passing of such scenes it ceased to have any practical purpose, since, contrary to traditional belief, it did not give the dancer a better basis for a leap, nor did it make the actors up-stage more visible to the audience. (This is better done by grouping them on ROSTRUMS at least a foot higher than the group in front.) The raked stage was limited in practice to a slope of 4% and was often as small as 1½%. It had serious disadvantages in the setting of scenery, as pieces set diagonally could not join neatly to squarely vertical neighbours; also the side-flats of a BOX-SET needed fox wedges under the base of each to compensate for the slope, or else they had to be built out of true with sloping bottom rails and ceased to be interchangeable. Moreover, any setting of pieces on a BOAT TRUCK became dangerous, since the truck might run off on its own down the incline. The modern practice is to have a flat stage floor, and to rake the floor of the AUDITORIUM, although a temporary rake may be used to achieve a particular effect by laying plywood sheets over joist-frames laid parallel to the front of the stage and increasing in height up-stage.

RAKESH, MOHAN, see INDIA.

RÁKOSI, JENÖ, see HUNGARY.

RAMSAY, ALLAN (1686–1758), Scottish poet, whose *Tea-Table Miscellany*, published in four volumes between 1725 and 1740, was the first considerable collection of the words of Scottish songs. Ramsay was also the author of *The Gentle Shepherd*, a PASTORAL published in 1725 which, under the influence of the success in London of GAY's *The Beggar's Opera* in 1728, was fitted to Scotttish airs and turned into a BALLAD OPERA for performance by the boys of Haddington Grammar School in the Taylor's Hall in Edinburgh in 1729. In 1730, with the original dialogue turned into English by Theophilus CIBBER, it was produced at DRURY LANE as *Patie and Peggy; or, the Fair Foundling*. In its original form it was revived for the EDINBURGH FESTIVAL of 1949 under the direction of Tyrone GUTHRIE. Ramsay, encouraged by the success of *The Gentle Shepherd*, ventured to open a theatre in Edinburgh in 1736, but it soon closed down under the Licensing Act of 1737.

RANCH, HIERONYMUS JUSTESEN, see DENMARK.

RANDALL, HARRY, see LENO, DAN, and MUSIC-HALL.

RANKIN, ARTHUR McKEE (1841–1914), actor-manager, born in Sandwich, Ontario, who appeared as an amateur under the name of George Henley in Rochester, N.Y., in 1861. Four years later, under his own name, he joined the company at the Arch Street Theatre, PHILADELPHIA, where he remained for four years, playing a wide variety of parts. He then married an actress, Kitty Blanchard (1847–1911), and with her starred at the UNION SQUARE THEATRE, New York, and also toured widely, in a series of somewhat crude melodramas. The most successful play in his repertory was Joaquín Miller's *The Danites* (1877), in which he appeared in London in 1880. He also played with some success the French-Canadian lead in *The Canuck* (1890), being partly responsible for its dramatization from the novel *Border Canucks*. He was for some years manager of the Third Avenue Theatre, New York, and after separating from his wife toured and again visited London in 1902 with Nance O'Neil [Gertrude Lamson] (1874–1965) as his leading lady, a notable actress whose famous roles included Magda in SUDERMANN's play, Leah in Augustin DALY's *Leah the Forsaken*, Marguerite Gautier in *Camille* by DUMAS *fils*, and Juliet in *Romeo and Juliet*. His three daughters by his first wife all married well-known actors, Gladys (1873–1914) becoming the wife of Sidney Drew, Phyllis (1874–1934) of Harry, son of E. L. DAVENPORT, and Doris (1880–1946) of Lionel BARRYMORE.

RAPIER PLAYERS, see LITTLE THEATRE, BRISTOL.

RASTELL, JOHN (?–1536), brother-in-law of Sir Thomas More, and father-in-law, through the marriage of his daughter Elizabeth, of John HEYWOOD, some of whose INTERLUDES were printed by Rastell's son William. John Rastell is believed to have been the author of *Calisto and Meliboea*, an adaptation of part of de ROJAS's *Celestina*, and of *The Dialogue of Gentleness and Nobility*, both of which were acted in his own garden at Finsbury in about 1527, and printed by him in the same year. He may also have been the author of an earlier interlude entitled *The Play of the Four Elements* (*c.* 1517).

RATTIGAN, TERENCE MERVYN (1911–77), English playwright, knighted in 1971, author of a number of well constructed and theatrically effective works. His first play, *First Episode* (1933; N.Y., 1934), written in collaboration with Philip Heimann, was followed by an immensely successful light comedy, *French Without Tears* (1936; N.Y., 1937), and by *After the Dance* (1939). He collaborated with Anthony Maurice in *Follow My Leader* (1940) and with Hector Bolitho in *Grey Farm* (N.Y., 1940), and then achieved a further success with *Flare Path* (London and N.Y., 1942), a topical war play with an R.A.F. background. Next came two more comedies, *While the Sun Shines* (1943; N.Y., 1944) and *Love In Idleness* (1944), the latter providing an excellent vehicle for the LUNTS, who played it in London and in 1946 in New York as *O Mistress Mine*. Rattigan, hitherto considered no more than an astute purveyor of light entertainment, now began to show signs of a more serious purpose with *The Winslow Boy* (1946; N.Y., 1947), based on the true story of a father's fight to clear his young son of a charge of petty theft. *Playbill* (1948; N.Y., 1949) consisted of two short plays, *The Browning Version* and *Harlequinade*, the former an excellent study of a repressed schoolmaster, played by Eric PORTMAN, and his ill-matched wife. *Adventure Story* (1949), with Paul SCOFIELD as Alexander the Great, proved an interesting failure, and with *Who Is Sylvia?* (1950) Rattigan returned to his former vein of light comedy. His next play, however, *The Deep Blue Sea* (London and N.Y., 1952), was a deeply emotional study of a judge's wife (played successively by Peggy ASHCROFT, Celia JOHNSON, and Googie WITHERS) who falls in love with a feckless, drunken, ex-R.A.F. fighter pilot and twice attempts suicide. An excursion into Ruritanian romance followed with *The Sleeping Prince* (1953; N.Y., 1956), in which Laurence OLIVIER and Vivien LEIGH appeared. In 1953 Rattigan also published his collected plays in 2 vols., each with a long introduction by the author in which he first used the term 'Aunt Edna' to indicate the ordinary unsophisticated playgoer who has no use for experimental, avant-garde plays. It has since proved a useful term for drama critics and an Aunt Sally for the progressives. Rattigan's next play, *Separate Tables* (1954; N.Y., 1956), was a double bill (*The Window Table* and *Table Number Seven*), which portrayed with compassion the problems of a group of characters in a Bournemouth hotel, Eric Portman and Margaret LEIGHTON giving excellent performances. It was followed by *Variations on a Theme* (1958), again with Margaret Leighton; *Ross* (1960; N.Y., 1961) with Alec GUINNESS as T. E. Lawrence; *Joie de Vivre* (also 1960), a musical version of *French Without Tears* which lasted only four nights; and *Man and Boy* (London and N.Y., 1963). After an unusually long interval came *A Bequest to the Nation* (1970), on Nelson and Lady Hamilton; another double bill, *In Praise of Love* (*Before Dawn* and *After Lydia*) (1973; a 'full version' of *After Lydia* was produced in New York as *In Praise of Love* in 1974); and *Cause Célèbre* (1977), based on a real-life murder case, which was still running when Rattigan died. *Playbill* was revived at the NATIONAL THEATRE in 1980.

RAUCOURT, MLLE [Françoise-Marie-Antoinette-Josèphe Saucerotte] (1756–1815), French actress, who first appeared at the COMÉDIE-FRANÇAISE in 1771 with great success. She was excellent in stern, tragic parts, for which her queenly figure and deep voice were eminently suitable. Unfortunately she led a wild life, got into debt, lost the affection of her audience and fled from Paris. Recalled by order of Marie-Antoinette, she was with some difficulty reinstated and immediately found herself in conflict with the all-powerful Mme VESTRIS. She also played a leading part in the troubles of the Revolutionary period, when she opposed TALMA's secession from the company, and was herself imprisoned with a number of her comrades.

RAVENSCROFT, EDWARD (*fl.* 1671–97), English dramatist, whose career as a playwright started in 1671 with an adaptation of MOLIÈRE's *Le Bourgeois Gentilhomme* as *Mamamouchi; or, the Citizen Turned Gentleman*. Another adaptation of Molière, *The Careless Lovers* (1673), based on *Monsieur de Pourgeaugnac*, was followed by an original comedy, *The Wrangling Lovers; or, the Invisible Mistress* (1676). An adaptation of Molière's *Les Fourberies de Scapin* was forestalled by OTWAY's version, and had to be postponed until 1677, in which year there also appeared *King Edgar and Aldreda*, a tragi-comedy, and *The English Lawyer*, an English version of George RUGGLE's *Ignoramus*. Ravenscroft's best work, however, was a farce, the outrageous *The London Cuckolds* (1681), which it became the tradition to produce at both PATENT THEATRES on Lord Mayor's Day until GARRICK stopped it at DRURY LANE in 1751 and COVENT GARDEN dropped it three years later. It was revived in 1782 for the benefit of John QUICK, and then disappeared until 1979, when it was revived at the ROYAL COURT THEATRE. Ravenscroft's later plays included *Dame Dobson; or, the Cunning Woman* (1683); an 'improved' version of *Titus Andronicus* subtitled *The Rape of Lavinia* (1686); two comedies, *The Canterbury Guests; or, a Bargain Broken* (1694) and *The Anatomist; or, the Sham Doctor* (1696), again based on Molière; and his last play, the tragedy *The Italian Husband* (1697).

RAYMOND, JOHN T. [John O'Brien] (1836–87), American actor, who made his first appearance in 1853 and was immediately recognized as an excellent comedian. He toured and worked in stock companies for some years, and in 1861 joined Laura KEENE's company, taking over from Joseph JEFFERSON the part of Asa Trenchard in Tom TAYLOR's *Our American Cousin*, in which he was also seen in London in 1867. Among his other parts were Tony Lumpkin in GOLDSMITH's *She Stoops to Conquer*, and Crabtree in SHERIDAN's *The School for Scandal*, but he was best known for his playing of Colonel Mulberry Sellers in Mark Twain's *The Gilded Age* (1874) and Ichabod Crane in *Wolfert's Roost* (1879), a dramatization of Washington IRVING's *The Legend of Sleepy Hollow*. An able and energetic actor who remained on the stage until his death, he had a long, imperturbable face and a slow seriousness of manner which made his comedy all the more appealing. He was popular with the public and with his fellow actors, and once established as a star appeared mainly in plays by and about Americans, except for PINERO's *The Magistrate* in which he was inimitable as Posket.

RAYNAL, PAUL (1885–1971). French dramatist, who attempted to follow in the steps of CORNEILLE in writing heroic plays on contemporary tragedies within appropriately formalized conventions. This led occasionally to the use of inflated rhetoric and a sense of disproportion between style and content, but sometimes he succeeded admirably in obtaining the effect he wanted. The best known of his plays, and the only one to have been seen in English, is *Le Tombeau sous l'Arc de Triomphe* (1924), translated by C. Day Lewis as *The Unknown Warrior* (1928), which depicts with great insight the feelings of a common soldier on leave. Raynal's later plays include *La Francerie* (1933), showing the opposition between two nations at war and two ways of life, and *Le Matériel humain* (1947), dealing with the irreconcilable opposition between the individual and the 'system', in which he achieved greater complexity by abandoning his previous allegiance to the neo-classical UNITIES.

READE, CHARLES (1814–84), English novelist, who was also the author of a number of plays, of which the best known are *Masks and Faces* (1852), dealing with GARRICK and Peg WOFFINGTON and written in collaboration with Tom TAYLOR: *The Courier of Lyons* (1854), which as *The Lyons Mail* (1877) provided a marvellous vehicle for IRVING in the dual roles of Dubosc and Lesurques, parts played later by MARTIN-HARVEY; and *It's Never Too Late To Mend* (1864), based on his own novel. He also dramatized TENNYSON's *Dora* (1867), and in 1874 persuaded Ellen TERRY to return to the theatre after her liaison with GODWIN to take over the part of Philippa Chester in his play *The Wandering Heir* (1873), first produced with Mrs John WOOD in the part. Among his later plays the best was a version of ZOLA's novel *L'Assommoir* (1877) as *Drink* (1879), which he wrote in collaboration with Charles WARNER who played Coupeau. Reade was essentially a novelist, and his best work for the theatre was done in collaboration with more theatrically-minded men, or based on existing foreign plays.

REALISM, a movement in the theatre at the end of the 19th century which replaced the WELL-MADE PLAY and the declamatory acting of the period by dramas which approximated in speech and situation to the social and domestic problems of everyday life, played by actors who spoke and moved naturally against scenery which reproduced with fidelity the usual surroundings of the people they represented. The movement began with IBSEN and spread rapidly across Europe, upsetting the established theatre and demanding a new type of actor to interpret the new plays. This was achieved by the system of STANISLAVSKY and by the later advocates of NATURALISM, which was the logical outcome of realism.

REALISTIC THEATRE, Moscow, a small theatre, also known as the Krasnya Presnya Theatre from the district in which it was situated. It opened in 1921 and was originally one of the studios attached to the MOSCOW ART THEATRE, on whose methods its early productions were based. A decisive change in its history came with the appointment in 1930 of OKHLOPKOV as its Artistic Director. His style of production, particularly for new plays, among which the most important was POGODIN's *Aristocrats*, involved the virtual reconstruction of the acting area for each play, often with several stages used simultaneously or in quick succession, while actors and audience mingled freely, the latter sometimes being called on to take part in the action of the play. This experimental technique, while interesting and valuable, had necessarily a limited appeal, and in 1938 an unsuccessful attempt was made to merge the company with that of TAÏROV at the KAMERNY THEATRE. Okhlopkov then went to the Theatre of the Revolution (later the MAYAKOVSKY THEATRE) and the Realistic was closed.

REANEY, JAMES CRERAR (1926–), Canadian poet and playwright. His first play *The Killdeer* (1959, revised 1968), a flawed but haunting work set in rural Ontario, unravelled a skein of emotional entanglements through a macabre murder mystery. *The Easter Egg* (1962) had a similar theme of stunted childhood and spiritual rebirth; it was followed by *The Sun and the Moon* (1965) (originally *The Rules of Joy*, 1958), and *The Three Desks* (1966), early works which were awkwardly realistic and even melodramatic in form. Other plays, including a series written for children, *One-Man Masque* (1959), the marionette play *Apple Butter* (1965), *Listen to the Wind*

(1966), *Ignoramus*, and *Geography Match* (1967), experimented with the repetition of words and the playing of games, elements of improvisation that showed Reaney moving towards a new style. *Colours in the Dark* (1967) was written for performance at the STRATFORD (ONTARIO) FESTIVAL, as part of the celebration for Canada's Centennial year. Kaleidoscopically combining scenes from the author's childhood in Ontario, the history of Canada, and even the Biblical account of the Creation, this delightful play has been several times revived. *Sticks and Stones* (1973), *St Nicholas Hotel* (1974), and *Handcuffs* (1975), together form a moving and detailed account of the events leading up to the massacre in 1880 of a legendary Irish family, the Donnellys, who emigrated to Southwestern Ontario. The trilogy, presented at the Tarragon Theatre in Toronto, was created by Reaney in collaboration with the actors and the director, Keith Turnbull, and gives the feeling of an improvised but organic work of COLLECTIVE CREATION, rich in mood and poetry, but employing the sparse inventive staging techniques of Peking Opera.

REBELO, LUÍS FRANCISCO, see PORTUGAL.

REBHUN, PAUL (*c.* 1500–46), German dramatist, studied in Wittemberg and wrote his *Suzanna* (1535) with direct encouragement from Luther. This Biblical play, with its MULTIPLE SETTING and division into five acts, shows an interesting mingling of medieval and Renaissance theatre. Each act is followed by a lyrical CHORUS, and within the body of the play Rebhun used lines of differing lengths and rhythms.

RECITAL THEATRE, New York, see DALY'S SIXTY-THIRD STREET THEATRE.

RED ARMY THEATRE, Moscow, the first and better known name of the Central Theatre of the Soviet Army. This was founded in 1919, its company being drawn from groups of actors who had for some time been entertaining army audiences, and was intended for the performance of plays about, or of interest to, the men of the Red Army. Its first director was ZAVADSKY, but it was Alexei POPOV, who directed it from 1937 until his death in 1961, who widened its scope and made it one of the outstanding theatres of the Soviet Union. Under him the company moved in 1940 from its original building into a specially designed and admirably equipped theatre shaped like a five-pointed star and capable of seating an audience of 2,000. One of Popov's first productions, *The Taming of the Shrew* in 1938, was an immense success and remained in the theatre's repertory for many years; but he did not neglect new Soviet plays, reviving VISHNEVSKY'S *The Unforgettable 1919* in 1952, soon after the author's death, and giving the first production of Shteyn's *A Game Without Rules* (1962). In 1963 Popov's son Andrei POPOV became Artistic Director of the theatre, which now includes a number of foreign plays in its repertory. For the 30th anniversary of the end of the Second World War it produced in 1974–5 an adaptation of Kurt Vonnegut's novel *Slaughterhouse Five* as *The Wanderings of Billy Pilgrim*.

RED BULL THEATRE, London, in Upper Street, Clerkenwell. This was originally an inn in whose yard plays were occasionally given, and was adapted as a permanent theatre by Aaron Holland in about 1605. A square structure with a tiring house adjoining the stage and galleries round, it was occupied by QUEEN ANNE'S MEN until 1617 and then by other companies. It was renovated and partly rebuilt in 1625 and may have been roofed in for use in bad weather. Contemporary dramatists referred to it in slighting terms, and it appears to have been what was later known as a GAFF, or Blood Tub, specializing in strong drama with plenty of devils and red fire. Thomas KILLIGREW served his apprenticeship there, appearing as one of the crowd which was recruited from the boys in the audience. When the theatres were closed during the Commonwealth, surreptitious shows and puppet-plays were sometimes given at the Red Bull, and at the Restoration a company under Michael MOHUN appeared there. Killigrew also reappeared there before he went to the VERE STREET THEATRE, taking some of the best actors in the company with him. On 23 Mar. 1661 PEPYS saw those who remained behind in a poor production of ROWLEY'S *All's Lost by Lust*. The theatre fell into disuse soon after; it was still standing in 1663 but by 1665 it had vanished.

REDERYKERS, see CHAMBERS OF RHETORIC and NETHERLANDS.

REDFORD, JOHN, see ENGLAND.

REDGRAVE, MICHAEL SCUDAMORE (1908–), English actor, knighted in 1959. Formerly a schoolmaster, he made his first appearance on the professional stage with the LIVERPOOL PLAYHOUSE company, where he remained from 1934 to 1936, playing a wide variety of parts and marrying a fellow member of the company, Rachel Kempson (1910–). In 1936 they were together at the OLD VIC, where Redgrave made his first London appearance as Ferdinand in *Love's Labour's Lost*, his other roles including Mr Horner in WYCHERLEY'S *The Country Wife*, in which he displayed a gift for comedy which has been too rarely exploited. He joined GIELGUD'S repertory season at the QUEEN'S THEATRE in 1937, playing Bolingbroke in *Richard II*, Charles Surface in SHERIDAN'S *The School for Scandal*, and Tusenbach in CHEKHOV'S *Three Sisters*. He was then seen at the PHOENIX THEATRE as Alexei Turbin in BULGAKOV'S *The White Guard* and as a richly comic Sir Andrew Aguecheek in *Twelfth Night* (both 1938). His first notable appearance in a modern play was as Harry in T. S. ELIOT'S *The*

Family Reunion (1939), after which in 1940 came an interesting Macheath in GAY's *The Beggar's Opera* and Charleston in Robert ARDREY's *Thunder Rock*. After serving in the Royal Navy he returned to the theatre in 1943, when he gave a thoughtful performance as Rakitin in TURGENEV's *A Month in the Country*, followed by the title role in Thomas Job's *Uncle Harry* (1944) and Stjerbinsky in S. N. BEHRMAN's adaptation of WERFEL's *Jacobowsky and the Colonel* (1945). His first outstanding role in Shakespeare was Macbeth (1947), in which he made his New York début in 1948; he then played the Captain in STRINDBERG's *The Father* (also 1948), subsequently rejoining the Old Vic company at the NEW THEATRE, where in 1950 he played Hamlet for the first time. In 1951 he was at the SHAKESPEARE MEMORIAL THEATRE, playing Richard II and Prospero in *The Tempest*, and he appeared in London a year later in ODETS's *Winter Journey* (known in America as *The Country Girl*) as a drunken actor dependent on his wife's support. Returning to Stratford in 1953, he played three more major Shakespearian roles—Shylock, King Lear, and Antony—and then deserted the classics for some years, being seen in GIRAUDOUX's *Tiger at the Gates* (London and N.Y., 1955), RATTIGAN's *The Sleeping Prince* (N.Y., 1956), and, in London, N. C. HUNTER's *A Touch of the Sun* (1958), in which his elder daughter Vanessa (below) also appeared. In 1958 he also played Hamlet and Benedick in *Much Ado About Nothing* at Stratford. A year later he was seen in his own adaptation of Henry JAMES's *The Aspern Papers*, followed by Robert BOLT's *The Tiger and the Horse* (1960) and Graham GREENE's *The Complaisant Lover* (N.Y., 1961). At the first CHICHESTER FESTIVAL in 1962 he gave an outstanding performance as Chekhov's Uncle Vanya, repeating it the following year during the NATIONAL THEATRE's first season at the Old Vic, where he was also seen as Claudius in the inaugural production of *Hamlet* in 1963, as a majestic and magisterial Hobson in Harold BRIGHOUSE's *Hobson's Choice*, and as a neurotic, foolishly heroic Solness in IBSEN's *The Master Builder* (both 1964). He took over from Alec GUINNESS in John MORTIMER's *A Voyage Round My Father* in 1972, and was in Simon GRAY's *Close of Play* at the National Theatre in 1979. A fine actor, with a good presence and a superb speaking voice, he has been particularly successful in the portrayal of men of intellect and sensibility flawed by emotional turbulence. He has also directed many plays, and is the author of two books on acting, *Mask or Face* (1958) and *The Actor's Ways and Means* (new ed. 1979).

His two younger children, Corin (1939–) and Lynn (1943–), are also on the stage, the latter being well known in the U.S.A. where she now resides.

REDGRAVE, VANESSA (1937–), English actress. She made her first appearance on the stage at Frinton in 1957, and was first seen in London in N. C. HUNTER's *A Touch of the Sun* (1958), in which her father, Michael REDGRAVE, also appeared. She then played Sarah Undershaft in SHAW's *Major Barbara*, and joined the SHAKESPEARE MEMORIAL THEATRE company to play Helena in *A Midsummer Night's Dream* and Valeria in *Coriolanus* (both 1959). A year later she appeared, again with her father, in Robert BOLT's *The Tiger and the Horse*, and back in Stratford in 1961 and 1962 she was much admired as Rosalind in *As You Like It*, Katharina in *The Taming of the Shrew*, and Imogen in *Cymbeline*. In 1964 she was seen in London as a touchingly vulnerable Nina in CHEKHOV's *The Seagull*, and in 1966 scored a great personal triumph as the embattled Scottish spinster schoolmistress in an adaptation of Muriel Spark's novel *The Prime of Miss Jean Brodie*. Her involvement in Marxist politics and in filming somewhat curtailed her theatrical activities, but she was seen briefly in 1972 as Polly Peachum in BRECHT's *The Threepenny Opera* and Viola in *Twelfth Night*, and in 1973 played Cleopatra in *Antony and Cleopatra* at the Bankside Globe (see WANAMAKER, SAM). She returned to the West End later in the same year as Gilda in COWARD's *Design for Living*, and three years later gave a highly praised performance as Ellida in IBSEN's *The Lady from the Sea* in New York, repeating the role in London in 1979. She was formerly the wife of the director Tony RICHARDSON. An excellent actress, she has a great talent for conveying ecstasy and excitement.

RED LION INN, London, see INNS USED AS THEATRES.

REFLECTOR, see LIGHTING.

REGENCY THEATRE, London, see SCALA THEATRE.

REGENT THEATRE, London, in the Euston Road. This theatre, seating 1,310 persons in three tiers, opened on 26 Dec. 1900 as the Euston Music-Hall, and became a playhouse in 1922, opening on 11 Sept. with *Body and Soul* by Arnold BENNETT. It had a short but interesting life, being occupied soon after its opening by Rutland Boughton's 'music-drama' *The Immortal Hour*, with Gwen FFRANGCON-DAVIES. This was directed by Barry JACKSON, whose own play for children *The Christmas Party* opened on 20 Dec. In 1923 *The Insect Play* by J. and K. ČAPEK had a short run, with GIELGUD playing Felix, and a year later Gielgud and Gwen Ffrangcon-Davies starred together in *Romeo and Juliet*, Gielgud being then 19 years old. The theatre was used extensively by play-producing societies like the Fellowship of Players, who staged a number of Shakespeare's plays for short runs; the PHOENIX SOCIETY, which revived Elizabethan and Restoration comedies; the Pioneer Players, whose production of Susan GLASPELL's *The Verge* (1925) was directed by Edith

CRAIG; the Repertory Players, who in 1926 produced O'NEILL's *Beyond the Horizon*; the STAGE SOCIETY, with James JOYCE's *Exiles* (also 1926); and the Three Hundred Club, with D. H. LAWRENCE's *David* (1927). During a short repertory season in 1925 the Macdona Players were responsible for the first London production of SHAW's *Mrs Warren's Profession*, and for *Man and Superman* in its entirety for the first time, with Esmé PERCY as Tanner. The last play to be presented at this theatre was a Jewish comedy by Izak Goller, *Cohen and Son*, which opened on 3 Oct. 1932. When it closed, the building became a cinema.

REGIONAL THEATRES, in the USA, see UNITED STATES OF AMERICA.

REGNARD, JEAN-FRANÇOIS (1655–1709). French dramatist, who wrote for the COMÉDIE-ITALIENNE from 1688 to 1696, often in collaboration with DUFRESNY, and from 1694 to 1708 for the COMÉDIE-FRANÇAISE also, ranking second to MOLIÈRE as a writer of comedy. His first successful production at the Comédie-Française was *Attendez-moi sous l'orme* (1694); of those that followed the best were *Le Joueur* (1696) and *Le Légataire universel* (1708). The latter, partly based on the traditional Italian story of Gianni Schicchi, remained in the repertory of the Comédie-Française until the early 20th century.

Though Regnard inevitably suffers by comparison with his illustrious predecessor, he had a considerable talent for devising comic situations and for writing witty dialogue, and it is interesting to trace in his plays the gradual emergence as the central figure of the valet, forerunner of BEAUMARCHAIS's Figaro. His greatest fault is the weakness of his character development, which anticipates the main fault of the later moral DRAME.

REHAN [Crehan], ADA (1860–1916), Irish-born American actress, who was taken to the United States as a child of five and at 14 appeared on the stage for the first time in New Jersey, joining Mrs John DREW's stock company at the Arch Street Theatre in PHILADELPHIA a year later. There, by a printer's error, she was billed as Rehan (for Crehan), a name which she retained and made famous. She worked with several stock companies, playing opposite famous actors of the day, until in 1879 she was engaged for his New York company by Augustin DALY, with whom she remained until his death. She soon became one of the most popular actresses of the day in New York and in London, where she made her first appearance in 1884 at the CHARING CROSS THEATRE (then Toole's). In 1891 she laid the foundation stone of Daly's own theatre in London, where she was seen in a wide range of parts, including one always connected with her, Katharina in *The Taming of the Shrew*. She had first played this in New York in 1887, when the Induction to the play was given for the first time there. Other parts in which she was much admired were Lady Teazle in SHERIDAN's *The School for Scandal* and Rosalind in *As You Like It*. She was essentially a comédienne, and only that side of her art was developed by Daly; unfortunately the turn of the century demanded a new style of acting, and after Daly's death in 1899 she found herself, while still young, somewhat outmoded. She continued on her own, presenting plays from her former repertory, but with dwindling success, in spite of her attractive personality, and she made her last public appearance in May 1905.

REHEARSAL, a session during which the DIRECTOR, his cast, and his technical staff work on a play, preparing it for presentation. We know nothing of rehearsals in the classical theatre, and little of those in the medieval, though players, particularly women unable to read, must have been taught their parts orally, and great skill was obviously needed to direct the large crowds demanded by the LITURGICAL DRAMA. Companies in later times were no doubt rehearsed by their leader, who was also the chief actor and often, as with MOLIÈRE, the author of the play. A glimpse of Elizabethan actors in rehearsal is given in *A Midsummer Night's Dream*. In the days of the STOCK COMPANY, in England and elsewhere, there were very few rehearsals. New plays were merely gone through to check cues and settle entrances and exits. A newcomer in an old play was left to learn his way about by trial and error, while a visiting star walked through his lines and left the company to adapt itself to his acting during the actual performance. New managements constantly began their reforms with an endeavour to institute regular rehearsals, and some degree of co-ordination was finally achieved by the STAGE-MANAGER, who eventually became the all-important stage DIRECTOR of modern times. (See also AUDITION and IMPROVISATION.)

REID, (Daphne) KATE (1930–   ), English-born Canadian actress, who trained for the theatre in New York and Toronto, making her first appearances as a student at HART HOUSE THEATRE in 1948. She then worked as a professional in summer stock in Canada, making her London début in 1958 as Catherine Ashland in Chetham-Strode's *The Stepmother*. In 1959 she joined the company at the STRATFORD (ONTARIO) FESTIVAL theatre, appearing there during several seasons in a number of Shakespeare's plays, notably as Lady Macbeth and Katharina in *The Taming of the Shrew* in 1962; in 1965 she also appeared there as Madame Ranevskaya in CHEKHOV's *The Cherry Orchard*. She made her début in New York in 1962, playing Martha in ALBEE's *Who's Afraid of Virginia Woolf?* at matinées, and she played Caitlin Thomas opposite Alec GUINNESS in Sidney Michaels's *Dylan* (1964). Two years later she starred in Tennessee WILLIAMS's double bill *Slapstick Tragedy*, and in 1968 she was in Arthur

MILLER's *The Price*, repeating her role in London the following year. In 1969 she was also seen at Stratford, Conn. (see AMERICAN SHAKESPEARE THEATRE), in Chekhov's *Three Sisters* and as Gertrude in *Hamlet*, returning in 1974 to play the Nurse in *Romeo and Juliet* and Big Mama in Tennessee Williams's *Cat on a Hot Tin Roof*, and moving with the latter to New York, where earlier in the year she had appeared in Brian FRIEL's *The Freedom of the City*. At the SHAW FESTIVAL in Ontario in 1976 she played in *Mrs Warren's Profession* and *The Apple Cart*, and in 1979 she returned to New York as Henny in John GUARE's *Bosoms and Neglect*, repeating the role at Stratford (Ontario) in 1980 as well as appearing in *Twelfth Night* and D. L. Coburn's *The Gin Game*.

REIMERS, ARNOLDUS (1844–99), Norwegian actor, brother of Sofie REIMERS, and married to Johanne JUELL-REIMERS. He spent his entire career, from his début in 1867 until his death, with the KRISTIANIA THEATER. His finest roles were in IBSEN, particularly Dr Stockmann in *An Enemy of the People* and Hjalmar Ekdal in *The Wild Duck*.

REIMERS, SOFIE (1853–1932), Norwegian actress, sister of the above, who began her career in Bergen at the NATIONALE SCENE. Two years later she moved to the KRISTIANIA THEATER, transferring to the NATIONALTHEATRET on its opening in 1899, and remaining there until she died. She is best remembered for her roles in broad comedy, especially in the plays of HOLBERG and WESSEL. In 1918 she published a volume of memoirs.

REINHARDT [Goldmann], MAX (1873–1943), Austrian actor, director, and impresario, who dominated the stage in BERLIN between 1905 and 1918, and remained an influential figure in the German-speaking theatre until he left for the United States in 1938. He regarded the script of a play as a score to be interpreted, and his integrated productions, with their careful harmonization of voice, movement, music, and setting, established for many years the pre-eminence of the director, more especially in the works of SYMBOLISM and Impressionism which superseded the earlier NATURALISM of BRAHM and was, after the First World War, itself superseded by the EXPRESSIONISM of JESSNER and PISCATOR.

As a young man Reinhardt appeared in VIENNA and SALZBURG, and in 1894 joined the company at the DEUTSCHES THEATER in Berlin, where under the tuition of Brahm he became an excellent actor, particularly in the portrayal of old men such as Probst in IBSEN's *Brand* and King Arkel in MAETERLINCK's *Pelléas et Mélisande*. Some experience of directing plays by STRINDBERG, WEDEKIND, and WILDE at the Kleines Theater in 1902, and above all the success of his production in 1903 of GORKY's *The Lower Depths*, in which he played Luka, led him to abandon acting and give himself up entirely to directing. At the Neues Theater am Schiffbauerdamm he produced HOFMANNSTHAL's *Elektra* (1903) against a primitive Mycaenean façade, the action taking place at night by the light of flickering torches. He went on to direct LESSING's *Minna von Barnhelm* and SCHILLER's *Kabale und Liebe*, and in 1905 the first of the 12 versions of *A Midsummer Night's Dream* for which he was eventually to be responsible. For this he used Lautenschläger's revolving stage, which eliminated the cumbersome scene changes of naturalism and allowed the scenes to flow one into another with a partial turn of the revolve, on which his designer Karl WALSER built for him a forest in realistic detail, with moss, trees trailing foliage, and a pond lit from beneath on which the fairies danced by moonlight. A production of *King Lear* in 1908 was, in contrast, set in an angular, archaic palace, decorated with a barbaric chevron design and furnished with massive chairs.

By this time Reinhardt had succeeded Brahm as director of the Deutsches Theater and had bought the theatre outright, erecting next door to it the Kammerspiele for intimate productions, and for some years he directed several plays each season at both theatres. At the Deutsches Theater there was always a play by Shakespeare, and among the German classics which he revived were Schiller's *Die Räuber* in 1908 and *Don Carlos* in 1909; and GOETHE's *Clavigo* in 1908 and both parts of *Faust*, Part I in 1908, Part II in 1911. Meanwhile at the Kammerspiele he concentrated on modern plays, including Ibsen's *Ghosts*, the first production of Wedekind's *Frühlings Erwachen* (both 1906), and the chamber plays of Strindberg. From 1915 to 1920 he sponsored at the Berlin VOLKSBÜHNE matinée performances of new plays by young authors (*Das junges Deutschland*), where he himself directed SORGE's *Der Bettler* (1917) and GOERING's *Seeschlacht* (1918). Elsewhere he was exploring the potential of vast acting areas, and no stage was too big for him; in 1910 he produced SOPHOCLES' *Oedipus the King* in the Zirkus Schumann in Berlin; in 1911 he directed Vollmöller's *The Miracle* in London, converting Olympia into a vast flamboyant Gothic cathedral embracing both actors and audience; and in 1919 he opened POELZIG's conversion of the Zirkus Schumann, the Grosses Schauspielhaus, with AESCHYLUS' *Oresteia* and Romain ROLLAND's *Danton*, in which, as in previous productions, his superb handling of crowd scenes was seen at its best. Since 1917 he had also been involved in the founding and running of the Salzburg Festival, where in 1920 he directed Hofmannsthal's morality play *Jedermann* in an open space in front of the Cathedral, following it in 1922 with the same author's *Das Salzburger grosse Welttheater* in the Kollegienkirche and in 1933 with Goethe's *Faust* in the Felsenreitschule. In 1924 he returned to Vienna and took over the Theater in der Josefstadt, and there and at the Komödie am Kurfürstendamm in Berlin he directed nearly 30 plays, paying particular attention to the schooling and directing of the actor, who was now his main concern. His repertory in both theatres was predominantly modern, and included not only SHAW and PIRANDELLO, but COCTEAU and MOLNÁR. In 1938 he left Austria for the United States, married Helene Thimig, who with Gertrude EYSOLDT had been one of his leading ladies for many years, and settled in Hollywood where he died.

RÉJANE [Gabrielle-Charlotte Réju] (1857–1920), French actress, who made her first appearance in 1875 at the Théâtre du VAUDEVILLE. She was soon recognized as a leading player of comedy and appeared at many Parisian theatres. She was also a frequent visitor to London, making her first appearance there in 1894; SHAW, in *Our Theatres in the Nineties*, while despising her choice of play, much admired her acting. Few of her parts were memorable, with the exception of the title role in SARDOU's *Madame Sans-Gêne*, in which she was seen in New York in 1895, and she never ventured on the classics, but in her own line of light comedy she was unapproachable. She opened her own theatre in Paris in 1906 and retired in 1915.

RELIGIOUS DRAMA, in England and the United States. With the disappearance of the medieval LITURGICAL DRAMA, religion as a subject for plays was, with rare exceptions, replaced on the English stage by classical history and mythology and plots drawn from contemporary life. It came back tentatively in the late 19th century, when professional dramatists like Henry Arthur JONES, with *Saints and Sinners* (1884), *The Tempter* (1893), and *Michael and his Lost Angel* (1896), began to include religion and the clergy in serious plays of contemporary life. The best known play of this kind is probably Jerome K. JEROME's *The Passing of the Third Floor Back* (1908). Meanwhile the Biblical 'spectacular', of which Wilson BARRETT's *The Sign of the Cross* (1895) is typical, was also popular until taken over by the cinema. In all these productions the emotional and not the intellectual content of the religious story was uppermost. The latter element was first introduced into modern drama by G. B. SHAW with *Androcles and the Lion* (1913; N.Y., 1915) and *Saint Joan* (1923; London, 1924), followed later by James BRIDIE, who made modern comedies out of Bible stories (*Tobias and the Angel*, 1930; *Jonah and the Whale*, 1932) and presented the conflict of good and evil in *Mr Bolfry* (1943). In 1935, a religious play from France, André OBEY's *Noah*, had a considerable success in the West End.

The revival of a truly religious drama, however, came partly through the interest aroused by revivals of the old MORALITY PLAY *Everyman*, staged from 1902 onwards by William POEL, and partly through the efforts of George Bell, who as Dean of Canterbury in 1928 commissioned for performance in the Cathedral MASEFIELD's *The Coming of Christ*. Bell became Bishop of Chichester a year later, and not only appointed E. Martin BROWNE the first Director of Religious Drama for his diocese but also became the first President of the newly-founded Religious Drama Society of Great Britain, now known as Radius. The annual Canterbury Festival continued to produce notable religious plays, and in 1935 T. S. ELIOT's *Murder in the Cathedral* on the assassination of Becket, directed by Martin Browne, moved to London and had an immense success in the commercial theatre. In the period after the Second World War other religious plays by Eliot, as well as by Graham GREENE and Christopher FRY, enjoyed a similar success with West End audiences. Fry's *A Sleep of Prisoners* was commissioned by the Religious Drama Society for the 1951 Festival of Britain, and for the same festival the medieval York MYSTERY PLAY had its first full-scale revival under Browne. It has since been revived triennially, and other cycles—the Chester, Wakefield, and Lincoln among them—have also been seen in cut and modernized versions, both in their own towns and in London. In 1980 the NATIONAL THEATRE staged a highly praised adaptation of several sections of Mystery Plays, telling the Bible story from the Creation to the Nativity; but religious subjects, from the late 1960s onwards, reached the commercial stage only through the rock MUSICALS *Joseph and his Amazing Technicolor Dreamcoat* (1968), *Godspell* (1971), and *Jesus Christ Superstar* (1972). Drama became part of the regular work of the new Coventry Cathedral on its dedication in 1962, with the appointment of a Director of Drama, the only one now permanently attached to a religious foundation, though churches of all denominations produce religious plays under the aegis of Radius, which acts as a consultant body, having a reading panel for plays and sponsoring publication of a series of plays on religious themes. It is allied with Sesame, a movement which encourages the use of drama therapy for the physically and mentally handicapped.

In the United States, where religious plays were being written in the 1930s by Fred Eastman in Chicago, a rapid and widespread development took place in the 1950s, when a Program in Religious Drama, with E. Martin Browne as Visiting Professor, was established at the Union Theological Seminary, New York. Many of the most significant European religious plays were introduced to America at the Union Theological Seminary, or at similar institutions elsewhere which followd its example as far as their resources permitted, while many parishes used trained directors to put on productions of religious plays which were later toured throughout the surrounding areas. In 1961 Robert E. Seaver, who succeeded Martin Browne at the Union Theological Seminary, staged a musical, *For Heaven's Sake*, the first of several productions which made use of the American talent for spectacle and music. The Rockefeller Foundation, which originally sponsored the Program in Religious Drama, also made possible the first international exchanges on the subject, when at Oxford in 1955 and at Royaumont, near Paris, in 1960 delegates from some 20 European and Far Eastern countries assembled to discuss and watch performances of religious plays in many styles and languages. Particularly interesting were the accounts of a liturgical movement in Sweden; of the formation of an active religious drama society in Greece; and of the Christian use of drama stemming from native traditions in India, Japan, Thailand, and Uganda.

RENAISSANCE, Théâtre de la, Paris. The first theatre of this name opened in 1826 as the Salle Ventadour, but was taken over and renamed ten years later by Anténor Joly, who produced there plays by HUGO and the elder DUMAS. It had a brief moment of glory in 1838, when FRÉDÉRICK made an outstanding success in the title role of Hugo's *Ruy Blas*, but its fortunes thereafter declined and it closed in 1841.

A second Renaissance opened in 1873 on a site cleared by the burning down of the Théâtre de la PORTE-SAINT-MARTIN two years previously. It began with strong drama, including ZOLA's *Thérèse Raquin* (1873), but soon changed to lighter works by LABICHE and FEYDEAU, though in 1885 BECQUE's *La Parisienne* had a great success. In the 1890s it came under the management of Sarah BERNHARDT, her productions, in all of which she appeared, including ROSTAND's *La Princesse lointaine* (1895), a translation of SUDERMANN's *Heimat* as *Magda* (in which she also appeared in London), and MUSSET's *Lorenzaccio* (written in 1834, but not previously produced), together with revivals of RACINE's *Phèdre* and the younger DUMAS's *La Dame aux camélias*. When she left her place was taken by GÉMIER and later by Lucien GUITRY. The theatre's prestige declined with the coming of the cinema, but it saw successful revivals of plays by Sacha GUITRY and Feydeau, and in 1956 it was taken over by the actress Vera Korene, who restored it and brought to it a more contemporary repertoire, including plays by SARTRE and ALBEE.

RENAUD, MADELEINE-LUCIE (1900–), French actress, who in 1923 made her début at the COMÉDIE-FRANÇAISE and soon became known as an outstanding interpreter of classical comedy, particularly in the plays of MOLIÈRE and MARIVAUX. In 1940 she married as her second husband J.-L. BARRAULT, who had just joined the company, and appeared in several of his productions, including in 1943 CLAUDEL's *Le Soulier de satin*. In 1946 they left the Comèdie-Française together to establish the Compagnie Renaud-Barrault, with which she has appeared in a wide range of parts, adding to her former repertory plays by Lope de VEGA, CHEKHOV, FEYDEAU, ANOUILH, SALACROU, as well as such later authors as FRY and BECKETT, in whose *Oh! les beaux jours* she was seen in London in 1965. In 1973, in a theatre at the Quai d'Orsay, she appeared as Maude in Colin Higgins's *Harold et Maude*. She was in Marguerite DURAS's *L'Eden cinéma* (1977), having appeared earlier in the same author's *Des Journées entières dans les arbres*, and in 1979 she played a woman suffering from the effects of a stroke in KOPIT's *Wings*. The beauty of her voice and person gave her an initial advantage to which years of hard work and technical mastery added embellishments, but it is above all the fine, mature intelligence which governs all her work which made her outstanding. It would be impossible to apportion between her and her husband the part of each in their company's success, since, as Barrault himself said, their gifts are complementary; as was their training, she having achieved eminence by the direct classical route, he by a process of trial and error.

REPERTORY, REPERTOIRE, the collection of parts played by an actor or actress, or, more usually, the plays in active production at a theatre in any one season, or which can be put on at short notice, each taking its turn in a constantly changing programme. This system was at one time common to all theatres, and is still in use on the Continent, but in the commercial theatres of London and New York it has been superseded by the continuous run of one play, which may last a year or more. An effort was made in the early 20th century to reintroduce the true repertory system into England with the establishment of REPERTORY THEATRES, but it was not successful, and today only the ROYAL SHAKESPEARE COMPANY and the NATIONAL THEATRE company adhere to it, though it is adopted by seasonal theatres such as the CHICHESTER FESTIVAL THEATRE.

REPERTORY THEATRES. After the disappearance in the late 19th century of the old STOCK COMPANIES in England and Scotland, a movement was set on foot to bring back into the theatre the 'true' REPERTORY system as practised on the Continent. It was begun by such pioneers as Frank BENSON at Stratford-upon-Avon, J. T. GREIN, Lena ASHWELL, Gertrude KINGSTON, and Charles FROHMAN in London, and Alfred Wareing in GLASGOW. The strongest impetus to its development outside London, however, was given by Miss HORNIMAN in Manchester. Attempts to re-create 'true' repertory met with little success, theatre staffs and audiences having become accustomed to the continuous run of one play at a time, and the term 'repertory theatre' has now come to signify merely that a theatre puts on its own productions for a series of short runs instead of housing TOURING COMPANIES.

The oldest surviving repertory theatre in Britain is the LIVERPOOL PLAYHOUSE, where repertory was introduced in 1911. Two years later came the BIRMINGHAM REPERTORY THEATRE. The success of these two theatres led to others being founded, mostly presenting twice-nightly productions for a week at a time, gradually changing to once-nightly. By the early 1950s there were over 100 repertory companies, but the general standard was not high. It was difficult to be both adventurous and successful, and most managements settled for lightweight West-End successes with the occasional new but uncontroversial, often local, play. The spread of television in the 1950s destroyed many of the weakest companies, and by 1960 only 44 were left. At its best the old repertory theatre was an excellent training ground for young actors. A weekly change of bill, with the strain of rehearsing one play while acting in another and learning lines for a third, and the added tension of frequent first nights, could do harm, and most actors remained in repertory only for a year or two. Yet it made for versatility and resourcefulness, was excellent for training the memory, and soon gave beginners self-confidence; many excellent players graduated from weekly rep.

A new era was ushered in by the opening of the BELGRADE THEATRE in Coventry in 1958, in which the CIVIC THEATRE, founded and run in close co-operation with the local authority, was to predominate. Once the novelty of television had worn off, audiences began to drift back, augmented by new audiences introduced to drama by television, and many found themselves confronted by fine new subsidized theatres, now tending to be called 'regional' rather than 'repertory'. They had bars and restaurants, spacious foyers which housed art and other exhibitions, and they often incorporated a small studio theatre for experimental purposes. More money, better working conditions, and longer runs (three or four weeks is now usual) led to a rise in standards, and the best repertory theatres can challenge comparison with the West End, to which some of their productions have been transferred. Financial and other problems still exist, but on the whole the new theatres are community assets. The old bond between the local theatre and its audience, impossible in the changing world of the commercial West End, is being forged anew, though performers are more transient now that casting is normally done for individual plays rather than the season. The number of repertory theatres has greatly increased since 1960, and now exceeds the number used by touring companies. Among them, apart from those already mentioned, are the BRISTOL OLD VIC; the CRUCIBLE, Sheffield; the EVERYMAN, Liverpool; the HAYMARKET, Leicester; LEEDS Playhouse; the LIBRARY, Manchester; the Mercury, Colchester; NEWCASTLE PLAYHOUSE; the NORTHCOTT, Exeter; NOTTINGHAM PLAYHOUSE; the OCTAGON, Bolton; OXFORD PLAYHOUSE; the PALACE, Watford; the ROYAL EXCHANGE, Manchester; the STEPHEN JOSEPH THEATRE-IN-THE-ROUND, Scarborough; the VICTORIA, Stoke-on-Trent; the Theatre Royal, WINDSOR (which is privately owned and not subsidized); and the Theatre Royal, YORK. Around London are the GREENWICH THEATRE and the LYRIC THEATRE, Hammersmith; in London the ROYAL COURT THEATRE. 'True' repertory is maintained by the ROYAL SHAKESPEARE COMPANY and the NATIONAL THEATRE company.

REPUBLIC THEATRE, New York, at 207 West 42nd Street, between 7th and 8th Avenues. Built by the first Oscar Hammerstein, this opened on 27 Sept. 1900 with James A. HERNE in his own play *Sag Harbor*, in which the young Lionel BARRYMORE made his first appearance in New York. On 19 Sept. 1902 the theatre was taken over by David BELASCO, who named it after himself and produced there several of his own plays, including *The Darling of the Gods* (1902), *Sweet Kitty Bellairs* (1903), *Adrea*, and *The Girl of the Golden West* (both 1905). He also produced KLEIN's *The Music Master* (1904), in which WARFIELD made a great success, and Tully's *The Rose of the Rancho* (1906), with Frances Starr. When in 1910 Belasco's second theatre, the Stuyvesant, was renamed the BELASCO, the Republic reverted to its original name. The last play to be seen there, before it became a BURLESQUE house in 1931 and then a cinema named the Victory in 1942, was John Huston's *Frankie and Johnny* (1930).

RÉPUBLIQUE, Théâtre de la, see COMÉDIE-FRANÇAISE.

RÉTORÉ, GUY, see EST PARISIEN.

RETURNS, see BOX-SET.

REUCKER, ALFRED, see ZÜRICH.

REUMERT, POUL (1883–1968), Danish actor who, beginning his career in Copenhagen in 1902 in the city's minor theatres, worked from 1911 mostly with the KONGELIGE TEATER. He became Denmark's leading actor of his generation, renowned particularly for his playing of MOLIÈRE and HOLBERG, as well as of STRINDBERG and other contemporary dramatists.

REUTER, CHRISTIAN (1665–*c*. 1712), German dramatist, who while still a student wrote two comedies satirizing bourgeois attitudes, *L'Honnête Femme, oder Die ehrliche Frau zu Plissine* (1695) and *La Maladie et la mort de l'honnête femme, das ist: Der ehrlichen Frau Schlampampe Krankheit und Tod* (1696). These gave offence to his landlady, with whom he was at loggerheads, and he found himself in prison, where he wrote yet another diatribe directed at her—*Letztes Denck- und Ehren-Mahl der weyland gewesenen Ehrlichen Frau Schlampampe* (1697). He later mounted an equally forceful attack on aristocratic absurdities in *Graf Ehrenfried* (1700), in which the young hero wastes his substance in riotous living.

REVEALS, see BOX-SET.

REVELS OFFICE, see MASTER OF THE REVELS.

REVENGE TRAGEDY, the name given to those Elizabethan plays, of which KYD's *The Spanish*

Tragedy (*c*. 1585–9) was the first, dealing with blood deeds demanding retribution. Their sublimity could easily turn to MELODRAMA; indeed, in a cruder form, the revenge motif underlay many of the famous melodramas of the 19th century. Among Shakespeare's plays, *Titus Andronicus* (*c*. 1592) may be considered the lowest form of the Revenge Tragedy and *Hamlet* (*c*. 1600–1) its finest flowering. Under the same heading come such plays as CHAPMAN's *Bussy d'Ambois* (*c*. 1604), TOURNEUR's *The Revenger's Tragedy* (*c*. 1606) and *The Atheist's Tragedy* (*c*. 1611), John WEBSTER's *The White Devil* (1612) and *The Duchess of Malfi* (1614), and MIDDLETON's *The Changeling* (1622).

REVERBERATOR, see LIGHTING.

REVESBY PLAY, see MUMMERS' PLAY.

REVOLUTION, Theatre of the, Moscow, see MAYAKOVSKY THEATRE.

REVOLVING STAGE, a scenic device which originated in the *kabuki* theatre in JAPAN in the 17th century, and was brought into Europe by Lautenschläger, who in 1896 installed one in a theatre in Munich. The advantage of it was that three or even more scenes could be set in advance on the revolve and presented to the audience in turn. The scenery itself could be solid BUILT STUFF, but considerable ingenuity was needed to fit the pieces into the various segments of a circle, and to overcome this it is now usual to combine a revolving section with laterally and vertically moving stage machinery. The LYTTELTON and OLIVIER stages at the NATIONAL THEATRE in London incorporate a disc and a drum revolve respectively.

REVOLVING WINGS, see BOOK WINGS.

REVUE, French word meaning 'survey', used for an entertainment consisting of a number of short items—songs, dances, sketches, monologues—which are not normally related. Unlike the English MUSIC-HALL and American VAUDEVILLE, in which a succession of performers appeared, revue players reappear in various numbers throughout the programme; the material is usually, though not necessarily, topical. In France the first revues in the modern sense were seen in the 1820s, but it was not until the end of the 19th century that they spread to England and America. The first to be staged in London, at the ROYAL COURT THEATRE, was *Under the Clock* (1893), by Seymour HICKS and Charles Brookfield, but it was only part of an evening's entertainment, not an independent show. The first American revue, described as a 'review', was *The Passing Show* (1894), by George W. Lederer (1906–82) and Sydney Rosenfeld. *Pot-Pourri* (1899) was the first English revue to be so described on the playbill. A specialized type of revue that became widely popular at holiday resorts was that presented by the characteristically English concert-party, often in a tent on the beach or in a small theatre at the end of a pier. One of the first to devise such entertainments was Harry Gabriel Pélissier (1874–1913), the first husband of Fay COMPTON, whose company wore black-and-white PIERROT costumes against a setting of black-and-white curtains, Pélissier himself acting as compère and writing much of the material. The London success of the *Pélissier Follies* at the APOLLO THEATRE from 1908 to 1912 finally established the popularity of modern revue with West End audiences, and the pierrot costume, in various colour combinations, was widely adopted. One of the most famous concert-parties of this type was *The Co-Optimists*, who, under Davy Burnaby, appeared in London during several seasons in the 1920s.

In America ZIEGFELD's *Follies*, an entertainment somewhat similar to Pélissier's, which ran through 25 editions and established revue as an important feature of the New York theatrical scene, was first seen in 1907. The star of the series from 1910 onwards was Fanny Brice [Fannie Borach] (1891–1951), a singer and comedienne whom Ziegfeld discovered doing impersonations in BURLESQUE. She later married Billy Rose and appeared in his *Crazy Quilt* (1931). (She was the subject of a Broadway musical, Styne's *Funny Girl*, in 1964.) The edition of 1910 also introduced another outstanding newcomer Bert WILLIAMS, and in two later editions the musical-comedy star Marilyn Miller also appeared. Another entertainer, later famous in films, Al Jolson [Asa Yoelson] (1883–1950), appeared at the newly built WINTER GARDEN THEATRE in the revue *La Belle Paree* (1911), and so began a long association with that playhouse during which he starred there in many musical shows, including *Whirl of Society* (1912) and *Dancing Around* (1914). To fill a gap during Jolson's holiday, and provide a counterblast to the successful Ziegfeld shows, the Shuberts presented in 1912 the first edition of their revue *The Passing Show* (not to be confused with the earlier one of the same name), which became another annual event, running to 12 editions.

Just before the First World War revue was at the height of its popularity in London, and was seen at the ALHAMBRA, the EMPIRE, and the London HIPPODROME, this last housing from 1912 to 1925 the revues of Albert Pierre de Courville (1887–1960). Much of the music, and many of the artists, came from America, de Courville's *Hullo Ragtime!* (1912), being a typical example. In 1914 Alfred Butt (1878–1962) inaugurated a series of revues at the PALACE THEATRE, the first of which, also called *The Passing Show*, introduced to London the American star Elsie Janis [Bierbower] (1889–1956), who, originally on the stage as a child, as 'Little Elsie', then went into musical comedy, and eventually became famous in such revues as *The Century Girl* (1916), *Hullo! America* (1918), *Elsie Janis and her Gang* (1919 and

1922), and *Puzzles of 1925*. She also appeared in her own play *A Star for a Night* (1911). She made her last appearance in London in 1929 and in New York in 1939.

A new type of 'intimate' revue, which relied more on witty dialogue than on dress and dancing, was essayed by C. B. COCHRAN with *Odds and Ends* (1914) at the AMBASSADORS, but his best known revues were those at the LONDON PAVILION from 1918 to 1931, which included three by Noël COWARD—*On With the Dance* (1925), *This Year of Grace* (1928), seen in New York the same year, and *Cochran's 1931 Revue*—*Blackbirds* (1926), with Florence Mills (1895–1927), Rodgers and Hart's *One Dam Thing After Another* (1927), and Cole Porter's *Wake Up and Dream* (1929), which starred Jessie MATTHEWS and, in New York in the same year, Jack BUCHANAN. During roughly the same period André Charlot (1882–1956), who came from Paris in 1912 and was associated with the pre-war productions of revue at the Alhambra, was putting on such shows as *A to Z* (1921) with Jack Buchanan; Coward's *London Calling* (1923) with Gertrude LAWRENCE; and, most notably, *Charlot's Revue* (1924 and 1925), again with Gertrude Lawrence and Jack Buchanan, joined in this instance by Beatrice LILLIE, which was popular in both London and New York. The outstanding name in revue in New York at the time, however, was George White [Weitz] (1890–1968), whose *Scandals of 1919* was successful enough to warrant 12 further editions. Also in 1919 came the first of the *Greenwich Village Follies*, of which seven more editions were to appear during the 1920s. The first was presented by the Canadian John Murray Anderson (1886–1954), known as 'the king of revue', who was involved as director, lyricist, or author with several of the later editions. He produced and staged 34 musical comedies and revues, including three of the later editions of the *Ziegfeld Follies*, and at the end of his career put on *John Murray Anderson's Almanac* (1953) with Hermione Gingold (1897–). Ed Wynn [Edwin Leopold] (1886–1966), one of the greatest American comedians of his day, had his first starring role in revue in the *Ed Wynn Carnival* (1920), having already spent 10 years in vaudeville and appeared in the *Ziegfeld Follies* and the Shubert revues. He presented on stage a personality of staggering ineptitude, with a lisp, fluttering hands, and outrageous costumes. He later appeared in such revues as *The Grab Bag* (1924), *The Laugh Parade* (1931), and *Boys and Girls Together* (1940), as well as in musical comedies, usually tailored to suit his personality, among them *Simple Simon* (1930) and *Hooray for What!* (1937). Another major star of American revue was Eddie Cantor [Isidore Itzkowitz or Iskowitz] (1892–1964), an eye-rolling singer and comedian who often appeared in blackface. He was in several editions of the *Ziegfeld Follies* as well as in other revues such as *Make It Snappy* (1922) and musical comedies—*Kid Boots* (1923), seen in London in 1926 with Leslie HENSON, and *Whoopee* (1928).

During the 1920s several more revue series began in New York. In 1923 Earl Carroll (1893–1948) presented the first of 11 editions of his *Vanities*; during his career he also presented two *Sketch Book* revues as well as some 60 legitimate plays of which the best known was Leon Gordon's *White Cargo* (1923; London, 1924). W. C. FIELDS forsook vaudeville, in which he had starred for many years, to appear in several editions of the *Ziegfeld Follies* between 1915 and 1925, and also in *George White's Scandals* and the *Earl Carroll Vanities*. Other series were the Shuberts' *Artists and Models*, which went into six editions from 1923 and emulated the nudity of the *Vanities*, and the *Grand Street Follies*, a more intimate type of revue which specialized in satires on current Broadway shows and went into six editions beginning in 1924. Among other successful revues were *The Band Wagon* (1931), whose score by Arthur Schwartz and Howard Dietz was perhaps the best ever written for a revue and introduced such standards as 'Dancing in the Dark', and *As Thousands Cheer* (1933), with music by Irving Berlin, in which Marilyn Miller made her last appearance on the stage. The latter also starred Ethel Waters (1900–77), a well-known Negro singer who also achieved success as an actress in the DuBose Heywards' *Mamba's Daughters* (1939) and Carson McCullers's *The Member of the Wedding* (1950).

Although long-running series of revues under the same title, on the American model, were not seen in London, the genre nevertheless flourished. Apart from the Cochran and Charlot revues, Jack Hulbert and Cicely COURTNEIDGE co-starred in several, while Clarkson [Arthur] Rose (1890–1968) presented every summer his famous seaside revue *Twinkle*, most of which he wrote himself, combining it with appearances in PANTOMIME at Christmas. Continuous revue, featuring statuesque female nudes, which ran from 2 p.m. to midnight, was introduced at the WINDMILL THEATRE in 1932, and the PRINCE OF WALES' THEATRE also housed continuous revue in the 1930s. During the same decade intimate revue featured topical satire of the type provided on the Continent in the inter-war years by cabaret in nightclubs and bars. From 1934 to 1939 Norman MARSHALL staged an intimate and witty revue annually at the GATE THEATRE. Intimate revue also flourished at the LITTLE THEATRE with *Nine Sharp* (1938) and *The Little Revue* (1939), written by Herbert Farjeon (1887–1945), the major revue librettist of the period. Grandson of the American actor Joseph JEFFERSON, he was also a dramatic critic and author, and with his sister Eleanor was responsible for the musical plays *The Two Bouquets* (1936), *An Elephant in Arcady* (1938), and *The Glass Slipper* (1944). Both the Farjeon revues featured the Australian-born comedian Cyril Ritchard (1898–1977), who had settled in England and was later seen in Coward's revue *Sigh No More* (1945), as well as in a number of straight roles, among them Tattle in CONGREVE's *Love for Love* in New York in 1947 and Sir Novelty Fashion in VANBRUGH's *The Relapse* in London in 1948 (N.Y., 1950). He eventually moved to America, where he directed a number of plays, had a long run as Biddeford Poole in *The Pleasure of His Company* (1958) by Cornelia Otis Skinner and Samuel Taylor, and played Bottom, Oberon, and Pyramus in *A Midsummer Night's Dream* (1967) at Stratford, Conn. (see AMERICAN SHAKESPEARE THEATRE), which he also directed.

The vogue for revue continued during and after the Second World War, George Black (1890–1943) presenting in London a series beginning with *Apple Sauce* (1940), and his sons George and Alfred continuing the tradition with such productions as *Piccadilly Hayride* (1946) and *Take It From Us* (1950). Also active in revue was Robert Nesbitt, whose productions included *Strike a New Note* (1943), *Fine Feathers* (1945), and three *Latin Quarter* revues (in 1949, 1950, 1951), the nearest approach in London to a series in the style of the *Ziegfeld Follies*. On a more intimate scale, the biggest success of the 1940s was the *Sweet and Low* series with Hermione Gingold, for which the books were written by Alan Melville (1910–). He was connected with many other revues, including *A La Carte* (1948) with Hermione BADDELEY, *At the Lyric* (1953), in which she was joined by Dora Bryan (1924–), and *Six of One*, also with Dora Bryan. Melville was also the author of several plays, among them *Castle in the Air* (1949), *Dear Charles* (1952), and *Simon and Laura* (1956). Other outstanding revues of the postwar period included *Tuppence Coloured* (1947) and *Penny Plain* (1951), in both of which Joyce GRENFELL appeared, *Oranges and Lemons* (1948), and *Airs on a Shoestring* (1953), Max ADRIAN starring in all four. Dora Bryan was seen in *The Lyric Revue* (1951), *The Globe Revue* (1952), and the revue *Living for Pleasure* (1958), as well as in MUSICAL COMEDY and plays such as SHAW's *Too True to be Good* (1965) and Ronald MILLAR's *They Don't Grow on Trees* (1968), in which she played nine parts. Revue also became a regular feature of the little theatres of the period, such as the Boltons, with several editions of *The Boltons Revue*, and the New Lindsey, with *Ring in the New* (1951) and *Lighting a Torch* (1952).

A notable London revue was that written by the South-African born choreographer John Cranko, *Cranks* (1955), which introduced contemporary techniques of presentation, together with new subject-matter and new performers, innovations also apparent in Bamber Gascoigne's *Share My Lettuce* (1957), in which Maggie SMITH made her London dèbut. Also in the cast was Kenneth Williams (1926–), an actor well known for his 'camp' style and wide range of silly and affected voices. He was later seen in the revues *Pieces of Eight* (1959) and *One Over the Eight* (1961), and in such diverse straight parts as the Dauphin in SHAW's *Saint Joan* (1954), Julian in Peter SHAFFER's *The Public Eye* (1962), and the title role in Robert BOLT's *Gentle Jack* (1963). One revue which stood somewhat apart from the general trend was *At the Drop of a Hat* (1956), a two-man entertainment by Michael Henry Flanders (1922–75), confined to a wheel-chair by poliomyelitis, and Donald Ibrahim Swann. It ran for two years in the West End and then toured all over the world. A second version, *At the Drop of Another Hat* (1963), was also successful, both shows being seen in New York in 1959 and 1966 respectively. Flanders, who had already contributed to such revues as *Penny Plain* and *Airs on a Shoestring*, wrote the words for both, and Swann the music.

In 1961 *Beyond the Fringe*, a brilliantly clever satirical revue which originated with the CAMBRIDGE Footlights Club and had already been seen at the EDINBURGH FESTIVAL the previous year, arrived in London and after a long run there opened in New York in 1962 with equal success. It was the joint work of Alan BENNETT, Jonathan MILLER, Peter Cook, and Dudley Moore, and seemed destined to inaugurate a new era in satirical revue. Its success was, however, difficult to follow in the theatre, and its vein of satire continued on television with such programmes as *That Was the Week That Was* in Britain and Rowan and Martin's *Laugh-In* in America, which kept abreast of current events in a way which would be virtually impossible for a theatrical company. Topicality in theatrical revues became increasingly rare outside the FRINGE THEATRE, and since on economic grounds spectacular revues such as those of Ziegfeld were also impossible, the genre could survive only through specialization.

One such specialization was nudity. Though brief female nudity had long been permissible in the theatre, *Oh, Calcutta!* (N.Y., 1969; London, 1970), devised by the drama critic Kenneth TYNAN, in which both men and women appeared naked for much of the show, was something entirely new, and curious audiences kept it running for years on both sides of the Atlantic. The all-black revue *Bubbling Brown Sugar* (N.Y., 1976; London, 1977) evoked the atmosphere of bygone Harlem, and there were entire evenings of the work of a particular composer—*Jacques Brel is Alive and Well and Living in Paris* (N.Y. and London, 1968), *Oh, Coward!* (N.Y., 1972; London, 1975), *Side by Side by Sondheim* (London, 1976; N.Y., 1977), and *Ain't Misbehavin'* (N.Y., 1978; London, 1979), devoted to the pianist Fats Waller—which broke new ground, opening up a rich vein of popular and hitherto untapped material.

REY DE ARTIEDA, ANDRÉS (1549–1613), Spanish dramatist, whose four-act tragedy *Los amantes* (*The Lovers*, 1581) marked the first attempt at the serious depiction of tragic love on the Spanish stage, treating it as a universal rather than a personal problem. It is also the first Spanish dramatization of the story of the lovers of Teruel, based on Boccaccio's tale of Girolamo and Silvestra, which was later treated by TIRSO DE MOLINA and most successfully by Juan Eugenio HARTZENBUSCH.

REYNOLDS, FREDERICK (1764–1841), English dramatist, whose first plays, *Werter* (based on GOETHE's *Die Leiden des jungen Werthers*) and *Eloïsa* (based on ROUSSEAU's *La Nouvelle Héloïse*), were tragedies seen at COVENT GARDEN in 1786. He then wrote for the same theatre a number of comedies, of which the most successful was *How to Grow Rich* (1793). Reynolds's most notorious play, however, was the DOG DRAMA *The Caravan; or, the Driver and his Dog* (1803), which saved DRURY LANE from financial disaster when it was produced there, mainly because of the appeal of a real dog, Carlos, which dived into a tank of real water to rescue a child from drowning. Later Reynolds wrote a number of ephemeral but highly successful MELODRAMAS, mostly seen at Covent Garden, and after providing librettos for several comic operas based on Shakespeare's plays, turned his attention to PANTOMIME. His *Harlequin and Old Gammer Gurton; or, the Lost Needle* (1836) was produced at Drury Lane and his last work *Harlequin and the Enchanted Fish; or, the Genii of the Brazen Bottle* (1840) at the ADELPHI THEATRE, both on Boxing Day (Dec. 26).

REYNOLDS, LYDIA, see LENO, DAN.

REYNOLDS, ROBERT, English actor, who was with QUEEN ANNE'S MEN soon after their foundation in 1603, but in about 1616 joined the ENGLISH COMEDIANS under Robert BROWNE, eventually succeeding John Green, Browne's successor, as leader of the company and remaining active on the Continent until about 1640. He was the originator of Pickelhering, a German-speaking clown somewhat in the style of John Spencer's earlier Hans Stockfisch, with a bizarre costume and grotesque make-up. He pattered in bad German with a comically repetitive stutter, and performed acrobatic antics which, in an amusing combination of stupidity and presumption, brought him into conflict with the other members of the cast. The character contributed materially to the creation of the purely German clown HANSWURST.

RHINTHON OF TARENTUM, see FABULA 4: RHINTHONICA.

RHODES, JOHN (*c*. 1606–?), London bookseller, said by Downes in *Roscius Anglicanus* (1709) to have been connected before the Commonwealth with BLACKFRIARS THEATRE, probably as wardrobe-keeper or prompter. At the Restoration he obtained a licence to reopen the COCKPIT—of which he had become Keeper in 1644—with a small company of players, among whom were his young apprentices Thomas BETTERTON and Edward KYNASTON. His licence was rendered null by the patent granted to KILLIGREW and DAVENANT, and his actors were taken over by them, Betterton becoming leading man in Davenant's company, and later the leading actor of his day. There was also a John Rhodes, presumably a different man, who was part-owner of the FORTUNE playhouse in 1637.

RICCOBONI, LUIGI (1676–1753), actor of the COMMEDIA DELL'ARTE, known as Lélio. The son of Antonio, who played PANTALONE in London in 1678/9, Luigi was a fine actor who ran his own company in Italy for many years, and in 1716 returned with an Italian company to Paris, where the COMÉDIE-ITALIENNE had been closed for nearly 20 years. His attempts to re-establish the old improvised comedy, which was already dying out in Italy, met with mixed fortunes and he relied increasingly on written texts, which hastened the complete absorption of the Italians into the French theatre. Riccoboni visited London in 1727 and 1728, and was the author of several books on the theatre in French and Italian, one of which was translated into English in 1741 as *An Historical and Critical Account of the Theatres in Europe*. He was twice married, his first wife being the actress Gabriella Gardellini, known as Argentina, and his second Elena Balletti (1686–1771), known as Flaminia. His son, Antoine-Françoise (1707–72), was also an actor, known as Lélio *fils*.

RICE [Reizenstein], ELMER (1892–1967), American lawyer and dramatist, whose first play, *On Trial* (1914), was the first on the American stage to employ the flashback technique of the cinema. His first major contribution to the theatre was the expressionistic fantasy *The Adding Machine* (1923), which satirized the growing regimentation of modern man in the machine age through the life and death of the arid bookkeeper, Mr Zero. *Street Scene* (1929), which followed, was awarded a PULITZER PRIZE for its realistic chronicle of life in the slums. The author later adapted it as the libretto of an opera with music by Kurt Weill, first performed in 1947. *Counsellor-at-Law* (1931) drew an equally realistic picture of the legal profession. The depression of the 1930s inspired *We, the People* (1933), the Reichstag trial was paralleled in *Judgement Day*, and conflicting American and Soviet ideologies formed the subject of *Between Two Worlds* (both 1934). When these plays failed on Broadway Rice retired from the theatre, but returned two years later to help found and run the Playwrights' Producing Company. His later plays included *American Landscape* (1938), *Two on an Island*, *Flight to the West* (both 1940), the latter a fervent denunciation of Nazism, and *A New Life* (1942). He recaptured something of the success of his early plays with the fantasy *Dream Girl* (1945) and presented a modern psychoanalytical variation on the Hamlet theme in *Cue For Passion* (1958), in which Diana WYNYARD played the Gertrude-like character Grace Nicholson. In 1964 Rice published an autobiography, *Minority Report*.

RICE, JOHN (*c.* 1596–?), Elizabethan boy-actor, apprenticed to HEMINGE, who in 1607 lent him to the Merchant Taylors' company to deliver a speech before James I. Three years later he appeared with Richard BURBAGE as a nymph in a water-pageant. He was with the KING'S MEN in 1614, when he played the Marquis of Pescara in the first production of WEBSTER's *The Duchess of Malfi*, and is in the actor-list of those who appeared in Shakespeare's plays. He probably became a full member of the company in 1619, perhaps in succession to Nathan FIELD, and may have taken Holy Orders, as there was a 'John Rice clerk of St Saviour's in Southwark' to whom Heminge in his will of 1630 left 20*s*.

RICE, THOMAS DARTMOUTH (1808–60), American VAUDEVILLE performer and Negro impersonator, known as 'Jim Crow' from the refrain of his most famous song-and-dance act, which he first performed in 1828 in Kentucky. It caught the public fancy and was soon being performed all over the United States. The 'trucking dance' with which Rice accompanied the song was an important element in its success, and his decision to perform it in 'blackface' led directly to the craze for MINSTREL SHOWS, though Rice himself never became part of a troupe, preferring to work alone. In 1833 he visited Washington, where the four-year-old Joseph JEFFERSON the third, later one of America's finest actors, appeared with him, being tumbled out of a sack at the conclusion of Rice's turn to mimic his song and dance. Rice himself was seen as Jim Crow in London at the SURREY THEATRE in 1836, accompanied by a full-scale publicity campaign featuring 'Jim Crow' pipes, hats, and cartoons. The name later became a term of reproach applied to subservient Negroes, and was also applied to the racial segregation laws in the American South.

RICE, TIM, see MUSICAL COMEDY.

RICH, CHRISTOPHER (?–1714), English lawyer who in 1689 bought a share of the DRURY LANE patent, and by 1693 had achieved complete control of the theatre. He soon became known as a tyrant, a twister, and a mean man, and under his management the affairs of the theatre and its actors went from bad to worse. Salaries were cut, expenses pared to the minimum, and Rich was constantly involved in lawsuits. In the end BETTERTON, with most of the abler actors, broke away and formed a rival company, leaving Rich to carry on with a mediocre group of players. He was finally forced out and took over the deserted theatre in LINCOLN'S INN FIELDS, but died before it was ready for occupation, leaving it to his son John (below).

RICH, JOHN (*c.* 1692–1761), English actor and theatre manager, son of Christopher RICH, from whom he inherited the old LINCOLN'S INN FIELDS THEATRE, where he was responsible in 1728 for the production of GAY's *The Beggar's Opera*. With the profits from this highly successful venture he built the first COVENT GARDEN theatre, transferring to it the patent granted by Charles II to William DAVENANT, and opening it in 1732 with a company headed by James QUIN. Though almost illiterate John Rich was an excellent actor in dumb-show and, developing the ideas of WEAVER, he popularized PANTOMIME in England, playing HARLEQUIN himself under the name of Lun. He produced a pantomime annually from 1717 to 1760, his own masterpiece in acting being 'Harlequin Hatched from an Egg by the Sun', which he performed in *Harlequin Sorcerer* at Tottenham Court Fair in 1741.

RICHARDS, ELIZABETH REBECCA, see EDWIN, JOHN, the younger.

RICHARDSON, IAN (1934–), Scottish actor, born in Edinburgh, who studied for the stage in Glasgow, and in 1958–9 was at the BIRMINGHAM REPERTORY THEATRE. In 1960 he joined what was to become in 1961 the ROYAL SHAKESPEARE COMPANY, appearing in Stratford in 1960 as Arragon in *The Merchant of Venice* and Aguecheek in *Twelfth Night*, and making his first appearance in London as Count Malatesti in WEBSTER's *The Duchess of Malfi*. Among his later parts were Antipholus of Ephesus in *The Comedy of Errors* in 1962 and Edmund in *King Lear* in 1964, in which he toured Europe and the U.S.S.R. and made his first appearance in New York. In London, also in 1964, he was seen as the Herald in WEISS's *Marat/Sade*, in which a year later he was to play Marat both in London and in New York. Later in 1964 he had given an excellent performance as Ford in *The Merry Wives of Windsor*, whipping himself into a tumult of sexual jealousy, and in 1965 he was seen in BRECHT's *Puntila*. With a fine, resonant voice, great bodily plasticity, and a powerful sense of controlled emotion, he was admirable as Vendice in TOURNEUR's *The Revenger's Tragedy* in 1966, in the title roles of *Coriolanus* in 1967 and *Pericles, Prince of Tyre* in 1969, and particularly as Prospero in *The Tempest* in 1970. He temporarily left the company to appear in the musical *Trelawny* (1972), based on PINERO's *Trelawny of the 'Wells'*, but returned to alternate the roles of Richard II and Bolingbroke in *Richard II* with Richard Pasco both in London in 1973 and in New York a year later; later in 1974 he played in GORKY's *Summerfolk* and WEDEKIND's *The Marquis of Keith* in London. After playing

Theatre-in-the-round enjoyed a considerable vogue from the early 1960s. The Victoria Theatre, Stoke-on-Trent (*above*), and the Arena Stage, Washington (*left*), exercised a noticeable influence on theatrical production in their respective countries

Design by Mario Chiari for Pirandello's *Enrico IV* (*above*), produced by the Piccolo Teatro della Città di Milano in 1961

The Royal Hunt of the Sun (*right*) by Peter Shaffer, directed by John Dexter for the National Theatre, London, in 1964.

Strindberg, *The Dance of Death* (*left*); Jean Gascon and Denise Pelletier in Gascon's production for the Théâtre du Nouveau Monde, Montreal, 1964.

Serjeant Musgrave's Dance (*below*) by John Arden at the Royal Court Theatre, 1965. The play instantly attained the status of a modern classic.

Yvonne by Witold Gombrowicz, directed for the Royal Dramatic Theatre, Stockholm, by Alf Sjöberg, 1965.

Saved by Edward Bond, Royal Court Theatre, 1965. The play's brutality caused an outcry.

The Asolo State Theatre, an 18th-century Italian playhouse rebuilt in Sarasota, Florida.

The Chichester Festival Theatre, 1966; *The Clandestine Marriage* by Colman and Garrick. The thrust stage was inspired by the design of the Stratford (Ontario) Festival Theatre (plate 72).

Madeleine Renaud in Genet's *The Screens* (*right*), Théâtre de l'Odéon, Paris, 1966, directed by Roger Blin.

The Living Theatre's production of *The Brig* (*below*) by Kenneth Brown, at the Théâtre de l'Odéon, Paris, during the Théâtre des Nations festival of 1966.

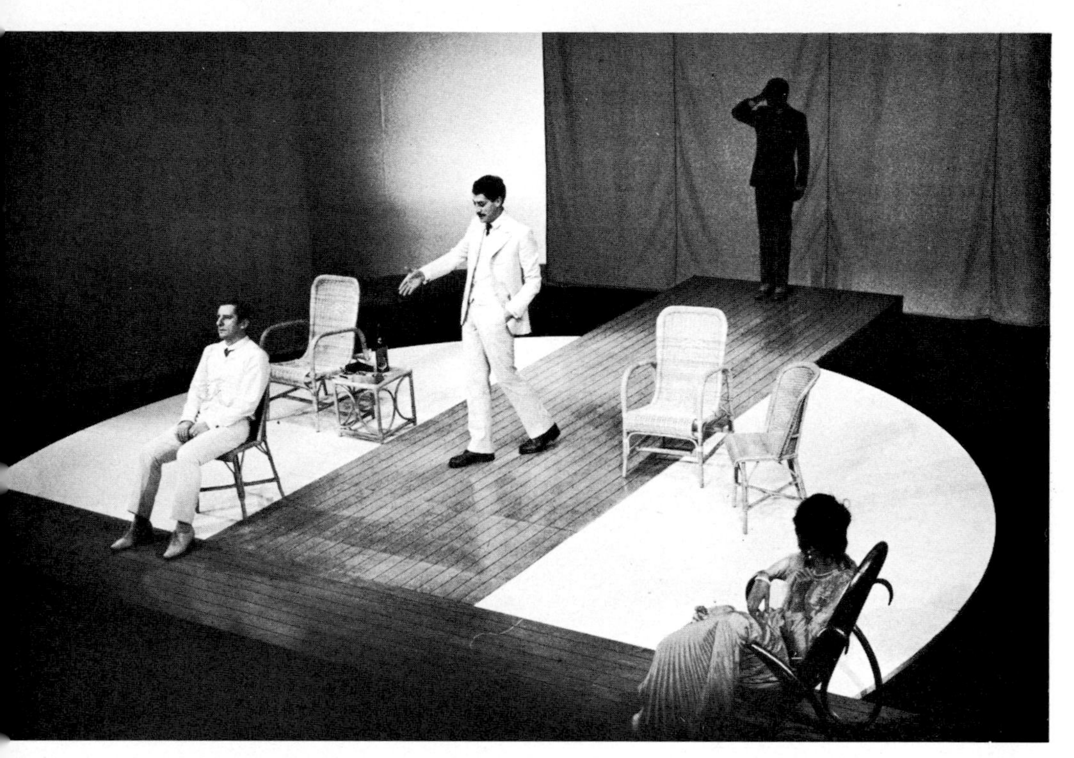

Claudel's *Partage de midi*, revived by the Comédie-Française in 1975, directed by Antoine Vitez.

Marlowe's *Tamburlaine*, the opening production in the Olivier Theatre, 1976, directed by Peter Hall; Albert Finney in the title role.

The Guthrie Theatre, Minneapolis, with its seven-sided thrust stage; Noël Coward's *Design for Living*, produced in 1977.

Stage and auditorium of the Stratford (Ontario) Festival Theatre.

The Punch and Judy show can not be traced in England beyond the 17th century, but comparison of the detail from a French 14th-century illuminated manuscript (*left*) with the modern English seaside scene (*right*) shows that the form must have very ancient roots.

Juvenile Drama: characters from a toy theatre sheet of a popular nautical drama.

Typical 19th-century melodrama: Kean in Boucicault's *The Corsican Brothers*, 1851 (*above*), Augustin Daly's *Under the Lamplight*, 1867 (*below*).

The great period of English pantomime: *Puss in Boots*, 1887 (*above*), and *Sleeping Beauty*, 1900 (*below*), both at Drury Lane.

French farce: *La Ronde du commissaire* by Meilhac and Gille, at the Gymnase, 1884, showing the form's dependence on complicated activity and a solid practicable set with a number of doors for abrupt appearances and exits.

English farce: Ben Travers's *Thark*, 1927, with the team that made 'Aldwych farce' a genre in its own right, including Mary Brough, Ralph Lynn, Robertson Hare, and Tom Walls.

English music-hall: Dan Leno (*above left*), Marie Lloyd (*above right*), Harry Tate (*below left*), George Robey (*below right*). Robey was unusual among music-hall artists in establishing a reputation in the straight theatre.

Rose-Marie, Drury Lane, 1925. Rudolf Friml's music spanned the transition between Viennese operetta and the Broadway musical of the 1930s.

Clowns in Clover, 1927, a musical revue drawing on the tradition of the seaside pierrot show.

Ever Green, Adelphi Theatre, 1930; Jessie Matthews and Sonnie Hale supported by C.B. Cochran's 'Young Ladies'.

Helen, 1932, another Cochran muscial, was directed by Reinhardt and designed by Oliver Messel.

Oklahoma, 1947, one of the most influential musicals of all time with its combination of vigour and rusticity.

My Fair Lady, 1958, was derived from Shaw's *Pygmalion*. This Ascot scene, which does not occur in the play, was set during a period of Court mourning to justify Cecil Beaton's witty use of only black, white, and grey for the costume designs.

the title role in *Richard III* in 1975 he returned to New York in 1976 to play Higgins in a revival of the musical *My Fair Lady*.

RICHARDSON, RALPH DAVID (1902–84), English actor, knighted in 1947, who made his first professional appearance at Lowestoft in 1921, playing Lorenzo in *The Merchant of Venice*. After touring the provinces for some time, he joined the BIRMINGHAM REPERTORY company in 1926, making his London début later the same year as the Stranger in SOPHOCLES' *Oedipus at Colonus*. He played in a number of London productions before going in 1930 to the OLD VIC, where his reputation was chiefly made, his roles including Caliban in *The Tempest*. On the reopening of SADLER'S WELLS in Jan. 1931 he played Sir Toby Belch in *Twelfth Night*, and in the Vic-Wells company's 1931–2 season his roles included Petruchio in *The Taming of the Shrew*, Bottom in *A Midsummer Night's Dream*, and Henry V. After leaving the Old Vic he was seen in several important modern plays, including MAUGHAM's *For Services Rendered* (1932) and *Sheppey* (1933) and PRIESTLEY's *Eden End* (1934) and *Cornelius* (1935). His New York début, as Chorus and Mercutio in *Romeo and Juliet* at the end of 1935, was followed back in London by the title roles in Barré Lyndon's *The Amazing Dr Clitterhouse* (1936), *Othello* at the Old Vic (1938), and Priestley's *Johnson Over Jordan* (1939), an outstanding performance. He then served in the Fleet Air Arm until 1944, when he returned to the Old Vic, remaining for three seasons and playing, among other parts, the title roles in IBSEN's *Peer Gynt*, CHEKHOV's *Uncle Vanya*, and ROSTAND's *Cyrano de Bergerac*, Falstaff in *Henry IV*, and the Inspector in Priestley's *An Inspector Calls*. In 1949 came another of his much-acclaimed roles, Dr Sloper in Ruth and Augustus Goetz's *The Heiress*, based on Henry JAMES's novel *Washington Square*. After playing David Preston in SHERRIFF's *Home at Seven* (1950) and Vershinin in an all-star production of Chekhov's *Three Sisters* in 1951 he went to the SHAKESPEARE MEMORIAL THEATRE to play Prospero in *The Tempest* and the title role in JONSON's *Volpone* (both 1952), and then returned to the West End, where he was seen in Sherriff's *The White Carnation* and N. C. HUNTER's *A Day By the Sea* (both 1953). After a tour in Australia he returned to the Old Vic in 1956 to give a fine performance in the title role of *Timon of Athens*, and was then seen in New York in ANOUILH's *The Waltz of the Toreadors* (1957). Later in the year he created in London another notable role, the self-deluding Cherry in Robert BOLT's *Flowering Cherry*, which was followed by Graham GREENE's *The Complaisant Lover* (1959). Sir Peter Teazle in SHERIDAN's *The School for Scandal* (1962; N.Y., 1963) preceded an appearance in PIRANDELLO's *Six Characters in Search of an Author* (1963), the first play to be produced at the MAY FAIR THEATRE, and in another play by Greene, *Carving a Statue* (1964), at the HAYMARKET, where he was also seen as William the Waiter in SHAW's *You Never Can Tell*, Sir Antony Absolute in Sheridan's *The Rivals* (both 1966), and Shylock in *The Merchant of Venice* (1967). He then starred in another series of modern plays: Joe ORTON's controversial black farce *What the Butler Saw* (1969), David STOREY's *Home* (London and N.Y., 1970), OSBORNE's *West of Suez* (1971), in which he gave a superb performance as an elderly writer turned television pundit, and William Douglas HOME's *Lloyd George Knew My Father* (1972). Joining the NATIONAL THEATRE company, he played the title role in Ibsen's *John Gabriel Borkman* and Hirst in PINTER's *No Man's Land* (both 1975), going with the Pinter play to New York in 1976, and he returned to London to give an outstanding performance in a part specially written for him, the Author in William Douglas Home's *The Kingfisher* (1977). He later returned to the National Theatre in several productions, including the starring role in David Storey's *Early Days* (1980). The range and variety of his work made him difficult to classify as an actor. Though not by nature a tragedian he could rivet the attention of the audience in a tragic or sinister role, was outstanding in pathos, and in comedy gave free rein to his own eccentric and amusing personality.

RICHARDSON, TONY [Cecil Antonio] (1928–), English director, who in 1955 joined the company at the ROYAL COURT THEATRE, where among other new plays he directed John OSBORNE's *Look Back in Anger* (1956) and *The Entertainer* (1957), both of which he subsequently directed in New York. Other outstanding productions for the Royal Court were IONESCO's *The Chairs* and *The Lesson* (1958) in a double bill. He then went to the SHAKESPEARE MEMORIAL THEATRE, where he directed *Pericles* (also 1958) and *Othello* (1959) with Paul ROBESON in the title role. He returned to the Royal Court to direct FEYDEAU's *Look After Lulu* (1959), Osborne's *Luther* (1961), and *A Midsummer Night's Dream* (1962), and in 1962 he also directed David Turner's *Semi-Detached* at the SAVILLE THEATRE. After directing *Luther* in New York in 1963 he remained on Broadway, being responsible for productions of BRECHT's *Arturo Ui* (also 1963) and Tennessee WILLIAMS's *The Milk Train Doesn't Stop Here Anymore* (1964). Back in London he directed, also in 1964, Ann Jellicoe's new version of CHEKHOV's *The Seagull*, with Vanessa REDGRAVE (to whom he was then married) as Nina, Peggy ASHCROFT as Madame Arkadina, and Paul ROGERS as Sorin. Later in the same year he was responsible for the first English production of Brecht's *Saint Joan of the Stockyards*. In 1969 he directed Nicol Williamson in *Hamlet*, a production later seen in New York and on tour in the U.S.A., and in 1972 he was responsible for productions of Brecht and Weill's

The Threepenny Opera, again with Vanessa Redgrave, and John MORTIMER's *I, Claudius*, based on Robert Graves's novels. He again directed Vanessa in *Antony and Cleopatra* at Sam WANAMAKER'S Bankside Globe Theatre in 1973, and in IBSEN's *The Lady From the Sea* in New York in 1976. He now lives in Los Angeles, where in 1979 he directed *As You Like It*.

RICHELIEU, ARMAND-JEAN DU PLESSIS DE, Cardinal (1585–1642), French statesman, for many years the virtual ruler of France. He did a great deal for the theatre, and by his patronage of MONTDORY helped to establish a permanent professional playhouse in Paris and to raise the status of the actor. He had strong leanings towards dramatic authorship, and set up a company of five—CORNEILLE, ROTROU, BOISROBERT, Colletet, and Claude de l'Étoile—to write plays under his direction. They were not very successful, and Corneille resigned after a disagreement over his share of the plot. Richelieu built a very well-equipped theatre in his palace which was first used on 14 Jan. 1641 and later, as the PALAIS-ROYAL, became famous under MOLIÈRE.

RICHEPIN, JEAN (1849–1926), French poet and dramatist, best remembered for his poems about tramps and vagabonds (*les chansons des gueux*). His plays, which formed an important part of his work but are now forgotten, were all given at the COMÉDIE-FRANÇAISE except for *Le Chemineau* (*The Tramp*, 1897), which the author took to the ODÉON after the actors at the Comédie-Française had asked for alterations in the script. It was an immense success, rivalling the popularity of ROSTAND's *Cyrano de Bergerac* of the following year. Other plays successful in their day were *Le Filibustier* (1888) and *Par le glaive* (1892).

RICHMOND, Surrey, small town on the River Thames, not far from London, which became a fashionable resort after the opening of Richmond Wells in 1696. The first recorded theatrical performance was given in 1714, when the Duke of Southampton's Servants presented an adaptation of *The Virgin Martyr* (1620) by DEKKER and MASSINGER, retitled *Injured Virtue*. Four years later PENKETHMAN set up a temporary theatre building on Richmond Hill which was regularly visited by the Prince of Wales, then residing at Richmond Lodge. The venture was a success and on 6 June 1719 Penkethman opened a permanent theatre in a converted stable with a production of the younger KILLIGREW's new comedy *Chit-Chat*, and continued to run it successfully until his death in 1725. In 1730 Thomas Chapman, who had been a member of Penkethman's company, built a new theatre which opened with a season by the company from the LINCOLN'S INN FIELDS Theatre and survived until the end of 1767, Ned SHUTER being a regular member of the company from 1744. Competition from a new theatre on Richmond Green, which opened in 1765, forced it to close.

The new theatre, built for James Love, a minor actor from DRURY LANE, had a three-tier auditorium and a proscenium opening 24 ft wide. Its first production was BICKERSTAFFE's *Love in a Village*, first seen in 1762. For some years the theatre provided summer employment for actors from the PATENT THEATRES, but once the novelty had worn off the audiences dwindled, though Dorothy JORDAN was always well received, as was the talented amateur the Earl of Barrymore. In 1831 Edmund KEAN took over the theatre, which was then known as the King's, as manager and leading actor, living for some time in the house attached to it. After his death in 1833 William Sidney and his wife took over with a mixed programme of MELODRAMA and FARCE and the theatre finally closed in 1884.

The present Richmond Theatre, which holds 875, opened on Richmond Green on 18 Sept. 1899 with a production of *As You Like It*, Ben GREET playing Touchstone and Dorothea Baird Rosalind. It has housed touring companies and also functioned as a VARIETY theatre (being known for a time as the Richmond Hippodrome and Theatre), but for much of its history has been a REPERTORY THEATRE. Since 1973, however, it has staged touring productions, plus an annual PANTOMIME, and it is now the most popular stopping-place for plays on pre-London runs. The building is preserved as being of architectural merit.

RICHMOND, Yorkshire, see GEORGIAN THEATRE.

RICHMOND HILL THEATRE, New York, at the south-east corner of Varick and Charlton Streets. Opened as a summer resort in 1822, this began its life as a theatre on 14 Nov. 1831 with HOLCROFT's *The Road to Ruin*. It had a good company, led by Mrs DUFF, who played there for two seasons, mainly in revivals. The theatre had the temerity to stage Sheridan KNOWLES's *The Hunchback* on the same night—18 June 1832—as the famous PARK THEATRE, but was forced to close owing to an outbreak of plague which killed one of the company, an actor named Woodhill. When the theatre reopened it housed a season of Italian opera under the sponsorship of Lorenzo Da Ponte, Mozart's librettist, and was later managed by Mrs Hamblin, the famous actor James E. MURDOCH making his first appearance in New York there. Later known as the Tivoli Gardens, it reverted to its original name, housing circus and variety shows from 1845 to 1848, and was demolished in 1849.

RICHMOND THEATRE, Richmond, Virginia. The first theatre in Richmond was built in the 18th century at the intersection of Twelfth and Broad Streets, and was destroyed by fire on 26 Dec. 1811 with the loss of 72 lives. A new theatre, the Marshall, opened in 1818 and saw the first appearance in America of the elder BOOTH, in 1821, and Edwin FORREST, in 1841; Joseph JEFFERSON the third managed it from 1854 to 1856, and in 1859 John Wilkes BOOTH, the assassin of President Lincoln, was a member of its stock company. The theatre burned down on 1 Jan. 1862 and was replaced in 1863 by a lavishly decorated new theatre which became known as the New Richmond. It was destined to become the leading Confederate theatre, all the outstanding theatre personalities of the South appearing there. It retained much of its glamour in the post-war years, but was finally demolished in 1896.

RICINIATA, see FABULA: 5.

RICKARDS, HARRY, see AUSTRALIA.

RIFBJERG, KLAUS (1931–), Danish lyric poet and dramatist whose first plays—*Hva skal vi lave?* (*What Shall We Make?*, 1963), *Diskret Ophold* (*Discreet Stay*, 1964), and *Udviklinger* (*Developments*, 1965)—mounted a satirical attack on the conventions of both the stage and society. *Hvad en mand har brug far* (*What a Man Needs*, 1966) traces the self-examination of a materially successful man, conducted through a series of encounters in an empty theatre. Also reminiscent of PIRANDELLO, *Voks* (*Wax*, 1968) is an identity game played by characters incapable of achieving individuality; *År* (*A Year*, 1970) deals more naturalistically with the Danish response to the German Occupation.

RIGG, DIANA (1938–), English actress, who had spent some time in provincial repertory companies before joining the SHAKESPEARE MEMORIAL THEATRE company in 1959. She made her first appearance in London with the company, renamed the ROYAL SHAKESPEARE, in 1961 in GIRAUDOUX's *Ondine*, and remained with it until 1964, making her first appearance in New York in that year as Adriana in *The Comedy of Errors* and Cordelia in *King Lear*. She returned to the R.S.C. in 1966 to play Viola in *Twelfth Night*. After starring opposite Keith MICHELL in Ronald MILLAR's *Abelard and Heloise* in London (1970; N.Y., 1971) she joined the NATIONAL THEATRE company, where she displayed a gift for comedy in handling the verbal pyrotechnics of STOPPARD's *Jumpers* (1972) and showed her versatility as a striking Lady Macbeth (also 1972) and an enchantingly frivolous Célimène in MOLIÈRE's *The Misanthrope* (1973; N.Y., 1975). In 1974 she played Eliza Doolittle to Alec MCCOWEN's Professor Higgins in a revival of SHAW's *Pygmalion*, and she then returned to the National Theatre in *Phaedra Britannica* (1975), a transposition of RACINE's *Phèdre* to 19th-century British India. After appearing again for the National Theatre in MOLNÁR's *The Guardsman* in 1978, she was seen in another play by Stoppard, *Night and Day* (also 1978), in which she played the intelligent, bored wife of a mining engineer in an African state.

RING, The, London, a large octagonal structure, originally a chapel, which stood in Blackfriars Road, London, about 500 yards from Blackfriars Bridge. Built in 1783, it later became a well-known boxing ring, and sprang into temporary theatrical fame when Robert ATKINS used it for some of the earliest THEATRE-IN-THE-ROUND productions in England, producing there on 29 Nov. 1936 *Henry V*, on 17 Jan. 1937 *Much Ado About Nothing*, and on 14 March 1937 *The Merry Wives of Windsor* with Irene and Violet VANBRUGH as Mistress Page and Mistress Ford. It then sank back into obscurity and was demolished some time after the Second World War. It has sometimes been confused with the nearby ROTUNDA.

RISE-AND-SINK, a method of effecting a TRANSFORMATION SCENE made possible by the provision of space above the stage with a low GRID. Though not as lofty as the later FLIES, this would take the depth of half a backcloth, and a scene could therefore be changed quickly by causing the upper half to ascend and the lower half, possibly framed out and with a profiled upper edge, to descend into the CELLAR on SLOTES.

RISTORI, ADELAIDE (1822–1906), Italian actress, who made an international reputation, particularly in tragedy. On the stage as a child, she was only 14 when she gave an outstanding interpretation of the title role in PELLICO's *Francesca da Rimini* and at 18 she played Mary Queen of Scots in SCHILLER's *Maria Stuart*, a part with which she was later closely associated. She then joined the COMPAGNIA REALE SARDA and at the time of her marriage in 1847 was the leading lady of Domeniconi's company where she played opposite SALVINI. After a brief retirement she rejoined the Reale Sarda and in 1855 went with Ernesto ROSSI to Paris, where she soon became a serious rival to RACHEL in tragedy, though she was also seen in comedies by GOLDONI. She visited London and toured the main provincial towns of Great Britain a number of times between 1856 and 1882, on her last visit playing Lady Macbeth in English at DRURY LANE to a mixed reception. She made the first of four highly successful tours of the United States in 1866 and retired from the stage in 1885. A stately woman, of strong physique and commanding presence, she sometimes wasted her talents on popular romantic melodrama, but could encompass with equal brilliance parts as diverse as EURIPIDES' Medea and Goldoni's Mirandolina in *La locandiera*. The critic G. H. LEWES, who saw her in her prime, thought that although beautiful, graceful, and possessed of a most musical voice, she lacked the natural passion that would have made her an actress of genius; and Mrs KENDAL, in a characteristic remark, said of her that she was a greater actress than BERNHARDT because she had no sex appeal. In 1888 she published her memoirs, which provide an interesting account of her life and a penetrating study of her approach to her art.

RITCHARD, CYRIL, see REVUE.

RITTERDRAMA, offshoot of the STURM UND DRANG drama, in which the valour of medieval knights was displayed in scenes of battle, jousting, and pageantry, often with a marked vein of Bavarian local patriotism. Written in prose and irregular in form, this 'feudal drama', had as its theme strong passions and contempt for the conventions and fostered the taste for romantic and medieval settings kindled by GOETHE's *Götz von Berlichingen* (1773) and KLINGER's *Otto* (1774). Indeed, the acquisition by theatre companies of wardrobes of medieval costumes for the *Ritterdrama* led to a new awareness of 'picturesqueness' and historical realism which was carried over into the staging of melodrama. Among the authors of such plays were Josef August von Törring (1753–1826), Bavarian Minister of State, with *Kasper der Thorringer* (pub. 1785) and *Agnes Bernauerin* (1780), and Joseph Marius Babo (1756–1822), for some time director of the Court theatre in MUNICH, with *Otto von Wittelsbach* (1782). Reactionary influences caused the *Ritterdrama* to be banned from the Munich stage, but its vogue continued elsewhere, notably in Austria, where Karl Friedrich Hensler (1761–1825) fused this type of drama with the native operatic fairy-tale in *Das Donauweibchen* (1797).

RITTNER, TADEUSZ, see POLAND.

RITZ THEATRE, New York, at 225 West 48th Street, between Broadway and 8th Avenue. This opened on 21 Mar. 1921 with DRINKWATER's *Mary Stuart*, which failed, as did his *Robert E. Lee* (1923). Success came in 1924 with the production of Sutton Vane's *Outward Bound* and GALSWORTHY's *Old English*, with George ARLISS as Sylvanus Heythorp, and in the following year Ashley DUKES's *The Man with a Load of Mischief*, with Ruth Chatterton and Robert LORAINE, had a short run. A further series of failures was broken in 1927 by the long run of John McGowan's *Excess Baggage*, a comedy on the heartbreaks of vaudeville, while in the autumn of 1932 Ruth DRAPER gave a three-week season of monologues. In 1937 the theatre was taken over by the FEDERAL THEATRE PROJECT, which presented there Arthur Arent's LIVING NEWSPAPER *Power*. A year later T. S. ELIOT's *Murder in the Cathedral* had a short run, and after a Federal Theatre production of Collodi's *Pinocchio* in 1939 the theatre was taken over for radio and television. Unsuccessful attempts to reopen it were made in 1970 and 1972, and in 1973 it was taken over by the Robert F. Kennedy Theatre for Children, which remained there until 1976 when a financial crisis caused it to leave, since when the theatre has remained closed.

RIVAS, Duke of, see SAAVEDRA, ÁNGEL.

RIVERSIDE STUDIOS, Hammersmith, London, arts centre housing a theatre, concerts, films, dance programmes, and exhibitions. Originally a foundry, it was converted into film studios between the wars, and was at one time the largest television centre in Europe. When the B.B.C. vacated the premises in 1974 the lease was acquired by the Hammersmith Borough Council which in 1975 formed an independent trust to administer the building. The centre opened on a part-time basis in 1976 and full time in 1978. The theatre accommodates visiting companies, and in 1978 played host to the Joint Stock Theatre Group in a dramatization of Robert Tressell's *The Ragged Trousered Philanthropists,* directed by William GASKILL, and the Catalan troupe La Claca in Joán Miró's *Mori el merma.* Its own productions were highly praised, notably CHEKHOV's *The Cherry Orchard,* MIDDLETON and ROWLEY's *The Changeling* (1978), *Measure for Measure* (1979) with Helen MIRREN as Isabella, and *Julius Caesar* (1980), all staged by Peter GILL, the centre's first director. The theatre was taken over by United British Actors, a company formed by Albert FINNEY, Glenda JACKSON, Harold PINTER, Diana RIGG, and Maggie SMITH, their first production being *The Biko Inquest* (1984).

RIX, BRIAN NORMAN ROGER (1924–), English actor and manager, who made his first appearance on the stage in 1942 and a year later was seen in London with Donald WOLFIT's company in *Twelfth Night*. From 1944 to 1947 he served with the R.A.F., and on demobilization formed his own repertory company at Ilkley, in Yorkshire, in 1948 and a second company at Margate in 1949. He had his first outstanding success with Colin Morris's farce *Reluctant Heroes* (1950), which he presented on tour and then took to the WHITEHALL THEATRE in London, where it ran for nearly four years with himself as Gregory. It was succeeded by four more highly successful farces which he presented at the same theatre, Rix appearing in them all himself: John Chapman's *Dry Rot* (1954) and *Simple Spymen* (1958); Ray Cooney and Tony Hilton's *One for the Pot* (1961); and Ray Cooney's *Chase Me, Comrade* (1964). The combined runs of these easily exceeded the record previously held by the ALDWYCH THEATRE of 10 years' continuous presentation of farce by one manager in the same theatre. In 1967 Rix moved to the GARRICK, where he presented and starred in Ray Cooney and Tony Hilton's *Stand By Your Bedouin*, Anthony Marriott and Alistair Foot's *Uproar in the House*, and Harold Brooke and Kay

Bannerman's *Let Sleeping Wives Lie*, the last continuing to run until 1969. Further farces presented and acted in by Rix were *She's Done It Again!* (also 1969) and *Don't Just Lie There, Say Something* (1971), both by Michael Pertwee, at the Garrick, and Pertwee's *A Bit Between the Teeth* (1974) at the CAMBRIDGE THEATRE. In 1976 he returned to the Whitehall to co-present and star in *Fringe Benefits* by Peter Yeldham and Donald Churchill. He left the theatre in 1980 to work full time with a charity for the mentally handicapped. His autobiography, *My Farce from My Elbow*, was published in 1975.

ROBARDS, JASON (1922–), American actor, who made his stage début in 1947 but spent some years in obscurity, including periods as assistant stage manager, and was still virtually unknown when in 1956 he gained enormous acclaim for his performance as Hickey in O'NEILL's *The Iceman Cometh* at the CIRCLE-IN-THE-SQUARE. He followed it in the same year with a fine portrayal of the alcoholic elder son James in the same author's *Long Day's Journey Into Night*, and sealed his reputation as an interpreter of O'Neill with four more roles—Erie Smith in *Hughie* in 1964, James again, 10 years later, in *A Moon for the Misbegotten* in 1973, James Tyrone, father of the character he originally played, in another production of *Long Day's Journey Into Night* in 1976, and Cornelius Melody in *A Touch Of the Poet* in 1977. He was at the STRATFORD (ONTARIO) FESTIVAL in 1958, returning to New York to play Manley Holliday in Budd Schulberg's *The Disenchanted*. In 1960 he played Julian Berniers in Lillian HELLMAN's *Toys in the Attic*. Later roles have included Murray Burns in Herb Gardner's long-running *A Thousand Clowns* (1962), Quentin in Arthur MILLER's *After the Fall* (1964), the Vicar of St Peter's in John WHITING's *The Devils* (1965), and Frank Elgin in a revival of ODETS's *The Country Girl* in 1972.

ROBERTS, ARTHUR, see MUSIC-HALL.

ROBERTS, RACHEL, see HARRISON, REX.

ROBERT S. MARX THEATRE, see CINCINNATI PLAYHOUSE IN THE PARK.

ROBERTSON, AGNES KELLY (1833–1916), actress, born in Scotland, where she appeared as a child, making her adult début in London in 1851 with Charles KEAN. Two years later she became the common-law wife of Dion BOUCICAULT and went with him to America, playing the heroines in many of his plays. She was particularly admired in the title role of *Jessie Brown; or, the Relief of Lucknow* (1858) and as Jeanie Deans in *The Trial of Effie Deans* (1860), based on SCOTT's novel *The Heart of Midlothian*. In 1888, wishing to marry a young actress, Boucicault publicly repudiated her, saying they had never been legally married. Their four children were all on the stage, the best known being 'Dot' and Nina BOUCICAULT; Aubrey died young and Eva retired on marriage. Agnes continued to act under her married name, making her last appearance in London in 1896 as Mrs Cregan in a revival of Boucicault's *The Colleen Bawn*, in which she had played the heroine, Eily O'Connor, on its first production in 1860.

ROBERTSON, MADGE, see KENDAL, WILLIAM.

ROBERTSON, T(homas) W(illiam) (1829–71), English dramatist, eldest of the 22 children of an actor. Several of his brothers and sisters were on the stage, the most famous being the youngest girl Madge who became Mrs KENDAL. Robertson himself acted as a child, and later appeared as an adult in LINCOLN, where the Robertson family had for many years been in control of the theatres on the Lincoln CIRCUIT. There he made himself generally useful, painting scenery, writing songs and adapting plays, and playing small parts. He was, in fact, trained in the old school which he was later to destroy, a process which can be studied, with reservations, in PINERO's *Trelawny of the 'Wells'* (1898). Yet his earliest plays were in no way remarkable. He wrote them quickly and sold them cheaply to Lacy, the theatrical publisher. The first of them, *The Chevalier de St George* (1845), was produced at the PRINCESS'S THEATRE, and although they were all moderately successful, it was not until the production of *David Garrick* (1864) at the HAYMARKET THEATRE that he came to the attention of the public. Though built up very largely on the old formulae and abounding in type characters, the printed copy of this play, with its elaborate directions for realistic scenery and costume and its wealth of stage directions, is a definite pointer in the direction which Robertson was to take almost immediately, with such plays as *Society* (1865), *Ours* (1866), *Caste* (1867), *Play* (1868), and *School* (1869). Their monosyllabic titles alone come as a refreshing change from those of earlier, and even some contemporary, plays. They were all seen at the Prince of Wales, later the SCALA THEATRE, where they established the reputation not only of the author but also of the newly formed BANCROFT management. With this series Robertson founded what has been called the 'cup-and-saucer drama' —the drama of the realistic, contemporary, domestic interior. His rooms were recognizable, his dialogue credible; his plots, though they now seem somewhat artificial, were true to his time, embodying serious social content, and an immense advance on anything that had gone before. *Caste* in particular still holds the stage, and some of the others would revive well. Robertson, a convivial creature with a brilliant flow of conversation, was active in the production of his own plays, fulfilling some of the functions of the modern DIRECTOR. He enjoyed a few years of fame and adulation before dying at the height of his success, leaving a permanent mark on the theatre of his time and foreshadowing the work of many modern dramatists.

ROBERTSON, TOBY, see PROSPECT THEATRE COMPANY.

ROBESON, PAUL BUSTILL (1898–1976), American Negro actor and singer, who made his first appearance on the stage in 1921. He created a sensation when he appeared for the PROVINCETOWN PLAYERS in 1924 as Jim Harris in O'NEILL'S *All God's Chillun Got Wings*, and even more by his playing of Brutus Jones in the same author's *The Emperor Jones*, in which he was first seen in London a year later and in Germany in 1930. His singing of 'Ole Man River' in the London production of the musical *Show Boat* (1928) first revealed the haunting quality of his superb bass voice, and led him to devote much of his time to touring in recitals of Negro spirituals. He was, however, seen again in London in the title role of *Othello* (1930), which when he appeared in it in New York in 1943 achieved the longest run of a Shakespeare play on Broadway up to that time. He also played the part at Stratford-upon-Avon in 1959. Another of his outstanding roles was 'Yank' in O'Neill's *The Hairy Ape* (1931). During his active career Robeson did much to further the interests of the Negro people, and it was unfortunate that his visit to Russia in 1963 as an avowed Communist aroused so much controversy in the United States and elsewhere, as did the plain statement of his beliefs published in 1958 as *Here I Stand*. A biography by his wife, who shared in his opprobrium, shows him to have been highly gifted, sincere, and courageous, and he was certainly one of the best known Negro artists of his time.

ROBEY [Wade], GEORGE EDWARD (1869–1954), English actor and MUSIC-HALL comedian, knighted in 1954. He made his first appearance at the OXFORD MUSIC-HALL in 1891, at a trial matinée, and was so successful that he was at once engaged to appear again ten days later. He was soon playing at all the leading London and provincial halls, earning for himself the nickname of 'Prime Minister of Mirth'. Originally a singer of comic songs, he was equally admired in a series of humorous sketches featuring Shakespeare, Charles II, Henry VIII, The Caretaker, The Gladiator, and other historic and imaginary characters, and he made a great success in the REVUE *The Bing Boys Are Here* (1916), in which he played Lucius Bing, as he did in the sequel, *The Bing Boys on Broadway* (1918). He was himself the author of a revue, *Bits and Pieces* (1927), in which he appeared with great success, but it is for his music-hall turns that he is chiefly remembered. His humour was robust, and on stage he seemed to consist almost entirely of a bowler hat and two enormous black eyebrows. He showed his versatility by appearing as Dame Trot in the PANTOMIME *Jack and the Beanstalk* (1921); Menelaus in *Helen!* (1932), a new English version of Offenbach's *La Belle Hélène*; and Falstaff in *Henry IV, Part I* at HER (then His) MAJESTY'S THEATRE in 1935, being the first music-hall star to appear in Shakespeare.

ROBIN HOOD, legendary English hero, whose name first appears in *Piers Plowman* (1377). He typifies the chivalrous outlaw, champion of the poor against the tyranny of the rich. It is impossible to identify him with any historical personage, though Anthony MUNDAY made him the exiled Earl of Huntingdon. Since he is always dressed in green he may be a survival of the Wood-man, or Jack-in-the-Green, of the early pagan FOLK FESTIVALS, or he may have been imported by minstrels from France as in ADAM DE LA HALLE'S *Le Jeu de Robin et Marion*. By the end of the 15th century he and his familiar retinue of Maid Marian, Little John, Friar Tuck, and the Merry Men, with their HOBBY HORSE and MORRIS DANCE, were inseparable from the May-Day revels, and the protagonists of many rustic dramas. These, however, cannot be considered folk plays, as were the MUMMERS' PLAY and the PLOUGH MONDAY PLAY, since they were usually written by minstrels. The May-Day festivities found their way to Court, where they became mixed up with allegory and pseudo-classicism: Henry VIII, in particular, enjoyed many splendid Mayings, including one in which he was entertained by Robin Hood to venison in a bower. After that their popularity waned, and they were finally suppressed by the Puritans. The story of Robin Hood and his Merry Men became a favourite subject for 19th- and 20th-century Christmas PANTOMIME.

ROBINS, ELIZABETH (1862–1952), American actress, who made her first appearance on the stage with the Boston Museum stock company in 1885. She later toured with Edwin BOOTH and Lawrence BARRETT, and appeared in the elder DUMAS'S *The Count of Monte Cristo* with James O'NEILL. In 1889 she made her first visit to London, where the greater part of her professional life was passed, and became prominently identified with the introduction of IBSEN to the London stage, playing Martha Bernick in *The Pillars of Society* (1889), Mrs Linden in *A Doll's House*, the title role in *Hedda Gabler* (both 1891), being particularly fine in the latter part, Hilda in *The Master Builder*, Rebecca West in *Rosmersholm*, Agnes in *Brand* (all 1893), Astra Allmers in *Little Eyolf* (1896), and Ella Rentheim in *John Gabriel Borkman* (1896). She held the stage rights of most of these plays, and was responsible, sometimes in conjunction with such advanced groups as GREIN'S Independent Theatre, for their initial productions. She was also seen in a wide variety of other parts, notably the title role in ECHEGARAY'S *Mariana* (1897), after which she retired, apart from a brief return to the stage as Lucrezia in Stephen PHILLIPS'S *Paolo and Francesca* (1902), and devoted herself to writing novels.

ROBINSON, BILL, see VAUDEVILLE, AMERICAN.

ROBINSON, (Esmé Stuart) LENNOX (1886–1958). Irish dramatist, actor, director, and critic, some of whose plays are well known in London and New York. His first play, *The Clancy Name* (1908), was staged at the ABBEY THEATRE, Dublin, with which he remained connected until his death. In his early plays—*The Cross Roads* (1909), *Harvest* (1910), *Patriots* (1912), *The Dreamers* (1913)—he treated political and patriotic themes as matter for tragedy, with no weakening into sentiment, but his comedy *The White-Headed Boy* (1916; London, 1920; N.Y., 1934) first made him known outside his own country. In this, and in *Crabbed Youth and Age* (1922; N.Y., 1932), his skill as a structural craftsman and as a creator of character in the style of the comedy of manners never faltered. In *The Big House* (1926; N.Y., 1933; London, 1934), Robinson became the first Irishman to write a play on the changing order of Ireland's civilization, but in *The Far-Off Hills* (1928; London, 1929; N.Y., 1932) and *Church Street* (Dublin and N.Y., 1934) he returned to comedy, in which his best work has been done, achieving in the latter play a tragic-comedy or 'mingled drama' whose satire was genial and ironies tragic. His later works included *Give a Dog . . .* (also 1928; London, 1929); *Ever the Twain* (1929); *All's Over Then* (1932; London, 1934); *Is Life Worth Living?* (also known as *Drama at Inish*, Dublin, London, and N.Y., 1933); *When Lovely Woman* (1936); *Killycreggs in Twilight* (1937; London, 1940); *Bird's Nest* (1938); *Forget Me Not* (1941); and *The Lucky Finger* (1948).

ROBINSON [*née* Darby], MARY (1758–1800), English actress, who from her fine playing of Perdita in *The Winter's Tale* in Nov. 1779 was thereafter known by that name. The spoilt child of a spendthrift father, she made an unhappy marriage at 16 with a dissolute young man, with whom she shortly afterwards went to prison for debt. Freed, she was coached for her first appearance (as Juliet at DRURY LANE in 1776) by GARRICK, the success of her début being already assured by her reputation for beauty and profligacy. Some critics saw in her the makings of a fine actress, better suited to tragedy than to the light, girlish parts she usually played, but after making her last appearance on 31 May 1780 as Eliza in Lady Craven's *The Miniature Picture*, she left the stage to become the mistress of the Prince Regent. The affair was short-lived, and she would probably have returned to the theatre had not a severe attack of rheumatic fever left her at the age of 24 too helpless to do so. She spent the rest of her life wandering from one spa to another and supporting herself by writing novels and poems, now forgotten.

ROBINSON, RICHARD (?–1648), English actor, who appears in the actor-list of Shakespeare's plays. He was probably a boy player at the second BLACKFRIARS THEATRE, and was certainly one of the KING'S MEN from 1611 onwards. As a young lad he played women's parts, and was much praised by Ben JONSON. He witnessed the will of Richard BURBAGE and may have married his widow, and in 1647 he was one of those who signed the dedication in the volume of collected plays ascribed jointly to BEAUMONT and FLETCHER.

ROBSON, FLORA (1902–84), English actress appointed DBE in 1960. She made her first appearance on the stage in Clemence DANE's *Will Shakespeare* (1921) and then toured with Ben GREET and was at the OXFORD PLAYHOUSE under J. B. FAGAN. After leaving the theatre for several years she returned in 1929 and was seen in London in Eugene O'NEILL's *Desire Under the Elms* (1931), her performance as Abbie Putnam, followed by appearances in BRIDIE's *The Anatomist* (also 1931), PRIESTLEY's *Dangerous Corner*, and MAUGHAM's *For Services Rendered* (both 1932), bringing her to the head of her profession. In 1933 she joined the company at the OLD VIC, playing a wide variety of parts and showing in such roles as Gwendolen Fairfax in WILDE's *The Importance of Being Earnest* and Mrs Foresight in CONGREVE's *Love for Love* an unsuspected talent for high comedy. During the next few years she appeared in the West End in a number of modern works, playing the title roles in Bridie's *Mary Read* (1934), Wilfrid Grantham's *Mary Tudor* (1935), and O'Neill's *Anna Christie* (1937). She went to the U.S.A. in 1938, making her New York début in 1940 in Edward Percy and Reginald Denham's *Ladies in Retirement*, and appearing in Hollywood films. On returning to London she was seen as Thérèse in *Guilty* (1944), a dramatization of ZOLA's novel *Thérèse Raquin*, following it with John Perry's *A Man About the House* (1945) and James Parish's *Message for Margaret* (1946). In 1948, after playing Lady Macbeth in New York, she again displayed her talent for comedy as Lady Cicely Waynflete in SHAW's *Captain Brassbound's Conversion*, and then achieved a great success in Lesley Storm's *Black Chiffon* (1949; N.Y., 1950), bringing to the role of a middle-class kleptomaniac the claustrophobic effect of controlled nervous tension in which she excelled. She played Paulina in GIELGUD's production of *The Winter's Tale* in 1951, and among her later parts were the governess in *The Innocents* (1952), based on Henry JAMES's *The Turn of the Screw*; Janet in Hugh Mills's *The House By the Lake* (1956), which had a long run; Mrs Alving in IBSEN's *Ghosts* (1958) at the Old Vic; and Miss Tina in Michael REDGRAVE's adaptation of Henry James's *The Aspern Papers* (1959). She was seen in revivals of Ibsen's *John Gabriel Borkman* (1963), Wilde's *The Importance of Being Earnest* (this time as Miss Prism), and ANOUILH's *Ring Round the Moon* (both 1968). Her last West End appearance was in 1969 in a revival of Rodney

Ackland's *The Old Ladies*, based on a novel by Hugh Walpole.

ROBSON, FREDERICK [Thomas Robson Brownhill] (1821–64), English actor who first made his name as a singer of comic songs, appearing in 1844 at the GRECIAN THEATRE; he may have been seen there in his later famous character of Jem Baggs in a revival of Mayhew's *The Wandering Minstrel*, which had been produced originally at the Fitzroy Theatre (later the SCALA) in 1834. After enjoying great local popularity, Robson visited Dublin in 1850 and on his return joined the company at the OLYMPIC THEATRE, famous for its BURLESQUES. It was there that he probably first sang in *The Wandering Minstrel* the popular ballad 'Villikins and his Dinah' by E. L. Blanchard, which later figured prominently in the repertory of Sam COWELL. Short and ugly, and a heavy drinker, he was nevertheless a powerful actor of great charm, affectionately known as 'the great little Robson'. A number of burlesques were written specially for him, the best of which was PLANCHÉ's *The Yellow Dwarf* (1854), which exploited to the full his genius for blending the comic with the macabre. He also appeared with much success in the title role of Palgrave Simpson's *Daddy Hardacre* (1857) and as Sampson Burr in Oxenford's drama *The Porter's Knot* (1858).

ROBSON, STUART [Henry Robson Stuart] (1836–1903), American comedian who first appeared on the stage as a boy of 16 and after ten years in various stock companies became principal comedian in Laura KEENE's company. He also spent some years with Mrs John DREW at the Arch Street Theatre in PHILADELPHIA and was for a time with the younger William WARREN at the Boston Museum. In 1873 he was seen in London, and shortly afterwards began a long association with the comedian W. H. CRANE. They appeared together as the two Dromios in *The Comedy of Errors* in 1877 and parted amicably in 1889, after the outstanding success of Bronson HOWARD's *The Henrietta* (1887) which was specially written for them. Robson then appeared successfully in several new plays, dying suddenly on tour shortly after celebrating his stage jubilee.

ROD-PUPPET, see PUPPET.

RODRÍGUEZ BUDED, RICARDO (1926–), Spanish dramatist, who helped to introduce social realism in the bourgeois comfort of contemporary Spanish drama, notably in *La madriguera* (*The Warren*, 1960), set in a working-class lodging house, whose inmates, deprived of decent housing and privacy, become criminals. In *El charlatán* (1962) an element of farce is added to the portrayal of a girl cruelly exploited by her own parents.

ROGERS, GINGER, see MUSICAL COMEDY.

ROGERS, PAUL (1917–), English actor, who has made his reputation mainly in classical parts and serious modern plays. He was first seen, in London, in 1938, went to the SHAKESPEARE MEMORIAL THEATRE in 1939, and after serving in the Royal Navy returned to the theatre in 1946 with the Colchester repertory company, moving to the BRISTOL OLD VIC a year later. From 1949 to 1960 he appeared regularly with the OLD VIC company, leaving it to appear in T. S. ELIOT's *The Confidential Clerk* (1953) and *The Elder Statesman* (1958). He made his New York début in 1956 as John of Gaunt in *Richard II*, and during his time with the Old Vic played many leading roles in Shakespeare, including Shylock in *The Merchant of Venice*, Henry VIII (both 1953), Macbeth (1954 and 1956), and King Lear (1958). After leaving the company he was seen in the commercial theatre in 1961 in the London production of Archibald MacLeish's *J. B.*, followed by USTINOV's *Photo Finish* (1962; N.Y., 1963) and CHEKHOV's *The Seagull* (1964). A year later he became a member of the ROYAL SHAKESPEARE COMPANY, where he gave brilliant performances as Max in PINTER's *The Homecoming* (1965; N.Y., 1967) and as the muffin-faced mayor in GOGOL's *The Government Inspector* (1966). He returned to the West End to give a display of virtuosity in three roles in Neil SIMON's *Plaza Suite* (1969), and in 1970 took over the part of the tweedy, fiction-writing bully in Anthony Shaffer's *Sleuth*, which he also played in New York in 1971. He joined the NATIONAL THEATRE company in 1974, appearing in SHAW's *Heartbreak House* in 1975 and JONSON's *Volpone* and GRANVILLE-BARKER's *The Madras House* in 1977, and returning to the R.S.C. to play in Granville-Barker's *The Marrying of Ann Leete* in 1975 and GORKY's *The Zykovs* in 1976.

ROJAS, FERNANDO DE (*c.* 1465–1541), Spanish novelist, now known to be the author of *La comedia de Calisto y Melibea*, better known as the CELESTINA. De Rojas was the child of Jewish parents forcibly converted to Christianity, and it is thought that his heroine Melibea was in the same case, and so prevented from marrying the young nobleman Calisto. Their clandestine love-affair leads to their tragic deaths. Certainly de Rojas was at pains to conceal his authorship of the book, of which the first surviving edition dates from 1499, by presenting it as the unfinished work of an earlier writer which he merely transposed and completed. In 1502 he published a new edition in which the original 16 'acts' were extended to 21, introducing a new character, the MILES GLORIOSUS Centurio. The most important character in the book is the old bawd Celestina, around whom revolve the amusing and licentious scenes of low life which give the book much of its appeal. Its popularity is attested by at least 60 reprints in the 16th century alone. It was translated into English in 1631 by James Mabbe as *The Spanish Bawd*. Two modern translations are those of P. Hartnoll (1959) and J. Cohen (1964). The best modern Spanish edition is that of Criado de Val and G. D. Trotter (3rd ed., 1970).

ROJAS VILLANDRANDO, AGUSTÍN DE (*c.* 1572–1625), Spanish actor and dramatist, who was for a short time a strolling player, and wrote entertainingly of his experiences in a picaresque novel *El viaje entretenido* (*The Amusing Journey*), published in 1603. In addition to a number of his own short playlets, or *loas*, this contains also a good deal of information on the early days of the Spanish theatre, and was not without influence on other European writers, notably SCARRON (in *Le Roman comique*, 1651) and GOETHE (in *Wilhelm Meister*, 1821).

ROJAS ZORRILLA, FRANCISCO DE (1607–48), Spanish dramatist, friend of CALDERÓN, who in his comparatively short working life produced an astonishing number of plays. The best-known is *Del rey abajo, ninguno* (*None But the King*, 1650), in which Garcia del Castañar, a nobleman living quietly in the country, suspects the king of having seduced his wife. He can do nothing about it, except plan to kill his wife to avenge his honour, but when he discovers that the seducer is merely a courtier he kills him instead. Rojas was also the first to write a true COMEDIA *a figurón* in *Entre bobos anda el juego* (*The Sport of Fools*, pub. 1645). His plays were well known in France, where they influenced among others LESAGE and Thomas CORNEILLE.

ROLLAND, ROMAIN EDMÉ PAUL ÉMILE (1866–1944), French writer, awarded the NOBEL PRIZE for Literature in 1915. He is best known for his 10-volume *roman-fleuve Jean Christophe* (1906–12), and he was a theorist of the theatre rather than a working playwright, his *Le Théâtre du peuple* (1903) having a considerable influence on GÉMIER. Two of his plays on the French Revolution, *Danton* (1900) and *Le 14 juillet* (1902), had some success, however, and the former was given a brilliant production by REINHARDT in Berlin in 1919. *Robespierre*, written in 1938, also had its first production in Germany, in Leipzig in 1952.

ROLL CEILING, see CEILING-CLOTH.

ROLLENHAGEN, GEORG (1542–1609), German pedagogue and Lutheran pastor, author of three plays on Biblical themes for performance in schools—*Des Ertzvaters Abraham Leben und Glauben* (1569), *Tobias* (1576), and *Vom reichen Manne und armen Lazaro* (1590). His son Gabriel Rollenhagen (1583–1619), a lawyer, dramatized the story of Euryalus and Lucretia, as told by Pope Pius II in his *Historia de duobus amantibus* (1444), as *Amantes amentes* (1609), mainly in Latin with comic scenes in Low German.

ROLLER, ANDREI ADAMOVICH [Andreas Leongard] (1805–91), Russian theatre designer. After studying in Vienna he worked as a stage mechanic and designer in Austria, France, and England, and from 1834 to 1879 was chief stage designer and machinist in the Imperial theatres in St Petersburg. He created a number of striking décors in the romantic tradition, devising splendid transformation scenes and ingenious tableaux, and experimenting with the use of light for dramatic emphasis—shafts of moonlight through a window or sunlight filtering through chinks in a vaulted interior. He played an active part in the reconstruction of the scenic apparatus for the theatre at the Tsarskoye Selo Hermitage, as well as the Bolshoi and Hermitage theatres in St Petersburg, and was responsible for much of the restoration of the Winter Palace after its destruction in a disastrous fire. His work had a great influence on the development of scenic design in Russia.

ROLL-OUT, see TRICKWORK.

ROMAINS, JULES [Louis-Henri-Jean Farigoule] (1885–1972), French poet, novelist, and dramatist. His first play, *L'Armée dans la ville*, was produced by ANTOINE in 1911, but it was not until after the First World War that he began his close association with the theatre through his friendship with COCTEAU and COPEAU. He worked for a time at the VIEUX-COLOMBIER, where his *Cromedeyre-le-Vieil*, a mythic portrayal of a mode of communal existence and experience, was produced in 1920. Louis JOUVET produced and played in the three farces which followed—*M. Le Trouhadec saisi par la débauche* (1922), *Knock, ou le Triomphe de la médecine* (1923), and *Le Mariage de M. Le Trouhadec* (1925). *Knock*, in which Jouvet played the quack doctor, adept at manipulating human credulity, had an immense success and was frequently revived. It was translated into English by GRANVILLE-BARKER and seen with equal success in London in 1926 and New York in 1928. The Trouhadec plays, which concern the improbable adventures of a professor of geography, the embodiment of self-perpetuating (and self-deceiving) pedantry and bureaucracy, were also well received. Romains's later plays include *Jean le Maufranc* (1926), which was initially unsuccessful but rewritten as *Musse, ou l'École de l'hypocrisie* (1930) offered DULLIN a fine part; *Le Dictateur* (also 1926), which had its greatest success outside France; an excellent adaptation of JONSON's *Volpone* (1928); *Boën, ou la Possession des Biens* (1930), in which GÉMIER made one of his last appearances; and *Donogoo* (1931), a stage version of a film scenario of 1920 in which the Le Trouhadec saga had its beginnings. Romains then concentrated on his 27-volume panoramic novel *Les Hommes de bonne volonté* and, except for some one-act plays and *L'An Mil*, produced by Dullin in 1947, wrote no more for the theatre. A revival of *Donogoo* at the COMÉDIE-FRANÇAISE in 1951 proved a great success, as did *M. Le Trouhadec saisi par la débauche* and the one-act *Amédée et les messieurs en rang* in 1956. *Volpone* was also successfully revived by BARRAULT at the MARIGNY in 1955, with himself as Mosca and LEDOUX as Volpone.

ROMANELLI, SAMUEL, see JEWISH DRAMA.

ROMANIA. In the Middle Ages primitive ritual dances and FOLK FESTIVALS throughout the Balkan peninsula came under the influence of the LITURGICAL DRAMA spreading from Russia and Poland. By the end of the 17th century puppet-plays from the Ukraine and the productions of the strolling players who from time to time visited Jassy and Bucharest had resulted in the development of a rudimentary native drama, which existed side by side with foreign plays imported by visiting German, French, Italian, and later Russian companies, some of which settled permanently in the larger towns. The modern Romanian theatre is therefore comparatively young, the first play in Romanian, as far as can be ascertained, being a translation by Gheorghe Asachi (1788–1869) of a German pastoral, *Myrtle and Chloe*, acted in Jassy by amateurs at Christmas 1816. It was not until the 1830s and 1840s that a distinctive school of Romanian dramatists began to appear, with the plays of Costache Bălăcescu, Cezar Bolliac, and Costache Faca (1800–45), whose comedy *Frantuzitele* (1833) preserves in its dialogue the curious half-French half-Romanian argot spoken by the Francophiles of the day.

The first truly Romanian theatre opened at Jassy in 1840, under the direction, amongst others, of the dramatist Vasile Alecsandri (1811–90), whose first play *Iórgu de la Sadagúra*, a comedy which depicts the conflict between the older generation and the new westernized youngsters, was seen at the theatre four years later. Of Alecsandri's other plays, very popular in their time, several have recently been revived—*The Tyrannic Prince* (1879) in 1945, for the reopening of the theatre, gutted during the Second World War; *Ovid* (1885) in 1957; and a musical version of a comedy centring on a Romanian 'Mrs Malaprop', *Madame Chirita* in 1970. A younger contemporary of Alecsandri was Bogdan Hasdéu (1838–1907), author of a romantic gipsy drama *Rasvan and Vidra* (1867) and of Romania's first successful historical tragedy *Domnita Rosanda* (1868).

Meanwhile in Bucharest the playwright Ion Eliáde Răduléscu (1802–72) had founded a society for the production of plays in translation which in 1833 took on the task of training young actors and the fostering of a national interest in the drama. Its emphasis on social and revolutionary plays led to its suppression in 1837, but not before it had paved the way for the establishment of a permanent theatre. This opened on 31 Dec. 1852 and became the National Theatre two years later. Its first artistic director was Mattei Millo (1814–96), actor and playwright, most of his comedies being satires on the hypocrisy of contemporary society. It was due to his efforts that Romanian acting and direction were placed on a firm professional footing, and he gave as much attention to the visual aspects of play production as to the interpretation of the text. The National Theatre is now named after Ion Luca Caragiale (1852–1912), son and grandson of actors and Romania's best known playwright. His works, which all had their first performances at the National Theatre, include a comedy of marital jealousy, *Stormy Night* (1879); a satire on local government, *The Lost Letter* (1884); the light-hearted *Carnival Scenes* (1885); a tragedy, *False Witness* (1890), used in 1928 for the libretto of an opera by Dragoi; and his last play, *Monsieur Leonida Faces the Reaction* (1912), revived in 1974 for the inauguration of the National Theatre's second stage.

With the turn of the century indigenous drama seemed to suffer something of a setback, the only play of note being *Manesse* (1900) by Ronetti Roman (1854–1908), a powerful and controversial work set in a Moldavian village at a time of conflict between Jews and Christians. Yet in spite of the dearth of good new plays the theatre as an institution flourished materially under a series of able directors, one of whom, Ion Ghica, presented for the first time a number of Shakespeare's plays in translation. Among the actors of the time were Mihail Pascaly, an excellent Hamlet, as was the later Grigore Manolescu, who also played Romeo to the Juliet of Aristiza Romanescu, best remembered for her Ophelia. Outstanding also was Constantin Nottara (1859–1935), after whom the Army Theatre in Bucharest was later renamed.

Under the directorships of Alecsándru Davilá (1860–1929), who studied in Paris before the First World War, and Pompiliu Eliáde (1869–1914) the new stagecraft of ANTOINE and STANISLAVSKY and the REALISM of IBSEN and SUDERMANN came to the fore, side by side with the more conventional romantic historical dramas of Barbu Stefanescu (1858–1918), who wrote under the name of Delavrancea and whose trilogy *Sunset* (1909), *Storm*, and *Morning Star* (both 1910) on the history of 15th-century Moldavia, has often been revived. Another interesting dramatist of the period was George Diamondi (1867–1917), whose realistic plays, among them *The Beasts* (1910) and *The Call of the Wood* (1913), remained in the repertory for some time.

The period during and immediately after the First World War was characterized by the influence of the symbolists, particularly MAETERLINCK, on Romanian dramatists, among them Zaharia Bârsán (1878–1948), at one time director of the National Theatre at Cluj in Transylvania and author of the popular lyric drama *Red Roses* (1915); Ion Minulescu (1881–1959), with *The Cranes Are Leaving* (1920); the prolific Victor Eftímiu (1889–

1972), who also directed the theatre at Cluj and endeavoured to combine SYMBOLISM with elements of folk-lore in such works as *The Black Cock* (1920) and *The Man Who Saw Death* (1928); and Lucian Blaga (1895–1961), with his 'pagan mystery' *Zamolxe* (1922). On the whole the plays of the period between the two world wars were by writers who were not primarily dramatists, and few of them have been revived. Exception must be made, however, for the novelists Camil Petruscu (1894–1957), author of *Witches' Games* (1919) and the comedy *Mitică Popescu* (1926), and George Mihail Zamfirescu (1895–1939), whose first play *Miss Anastasia* (1928), a tragi-comedy of working-class life, was produced by the company founded in 1914 by the great actress Lucia Sturdza Bulandra (1873–1961) with her husband Tony Bulandra and run by them until his death in 1943. Among the other actresses of this period were Agepsina Macri (the wife of Eftímiu) (1890–1961), who excelled in tragedy and was much admired as Lady Macbeth, and the comédienne Maria Filotti (1891–1963).

A dramatist whose work covers the period of transition was Mihail Sebastian (1907–45), whose death in a car crash deprived Romania of one of her best young dramatists. His three satirical comedies *Holiday Games* (1938), *The Nameless Star* (1944), and *Stop Press* (1948) are popular throughout Eastern Europe. In the troubled period which followed 1945 several of his contemporaries, notably IONESCO, who made his home in Paris, left Romania and settled elsewhere. Among those who stayed, Petruscu remained faithful to the historical play with *Bălcescu* (1949), dealing with the exploits of a hero of the 1848 Revolution, but younger writers turned to more modern social and political themes acceptable to the new communist régime. One of the most prolific was Aurel Baranga (1913–79), writer of satiric and farcical comedies, one of which, *The Angry Lamb* (1954), a typical bureaucratic burlesque, starred the leading actor Radu Beligan (1918–), who appeared also in *Journalists* (1956) and *Transplanting the Unknown Heart* (1969), by Al(exandru) Mirodan (1927–). The second of these was first produced at the Bulandra (formaly the Municipal) Theatre, named after the actress, who had become its manager after her husband's death. She appeared there in many outstanding roles, among them Valeria Zapan in Baranga's *Arc de Triomphe* (1954) and Professor Dinescu in *The Crumbling Citadel* (1955) by another popular post-war dramatist Horia Lovinescu (1917–), author of a sequel to CHEKHOV's *Three Sisters* entitled *The Boga Sisters* (1957). A woman of immense vitality and versatility, who a few days before her death was appearing as the Mother in Tennessee WILLIAMS's *The Glass Menagerie*, Bulandra was much helped in her work at the Municipal Theatre by the actor and designer Liviu Ciulei (1923–), who became the theatre's artistic director on her death in 1961. He directed a number of new plays, but also continued to act, playing Edgar in DÜRRENMATT's *Play Strindberg* in 1972 and directing and acting in a revival of Caragiale's *The Lost Letter* in 1973. Ciulei later moved to the United States, becoming Artistic Director of the GUTHRIE THEATRE, Minneapolis, in 1980.

A playwright whose popularity equals that of Baranga and Lovinescu is Paul Everac [Petre Constantinescu], who had his first four plays produced in 1957 and has continued to write mainly on industrial themes, his *Open Windows* (1959) being about metal workers and *The Invisible Courier* (1964) about worker responsibility. But much of the best work in the Romanian theatre today is found in the directing of foreign classics or new plays from abroad. Lucian Giurchescu helped to popularize BRECHT with productions ranging from *Herr Puntila und sein Knecht Matti* in 1961 to *Mutter Courage und ihre Kinder* in 1972; Radu Penciulescu was responsible for Ionesco's *Rhinoceros* in 1963, which with Beligan as Berenger toured widely in Europe; he also directed interesting productions of Shakespeare's *Richard II* in 1966 and *King Lear* in 1970, and at the Bulandra in 1972 put on HOCHHUTH's *Der Stellvertreter*. Lucian Pintilie directed FRISCH's *Biedermann und die Brandstifter* in 1964 and staged a revival of Caragiale's *Carnival Scenes* which was seen at the EDINBURGH FESTIVAL in 1971. A spectacular producer of satirical plays is Dinu Cernescu, who was responsible for remarkable revivals of Baranga's *The Angry Lamb* and Alexsandri's *Ovid* in 1962 and of GHELDERODE's *Escorial* in 1968; he also produced in 1971 Shakespeare's *Measure for Measure* in accordance with the ideas of Jan KOTT. Andrei Serban (1943–), a pupil of Peter BROOK, went to New York in 1970 to work at LA MAMA, and later achieved a high reputation with such productions as Chekhov's *The Cherry Orchard* (1976) and the musical *The Umbrellas of Cherbourg* (1979), derived from a film by the French director Jacques Demy, with which he also made his London début in 1980. David Esrig, like Pintilie and Ciulei, incurred the disapproval of the authorities; he was forced to work in exile for some time, had a great success at the 1965 THÉÂTRE DES NATIONS Festival with a highly fanciful production of *Troilus and Cressida*, and has since returned to Bucharest to direct the work of new dramatists; the end of the 20th century may yet see a fine flowering of Romanian drama.

The number of theatres and the size of their audiences continue to increase. Under the monarchy theatres were found only in Bucharest, Jassy, Craiova, and Cluj. In 1976 there were 46 State theatres, and in Bucharest alone the National Theatre is supplemented by 16 others, including the Nottara, the Mic, and the Bulandra. There are also in Romania six Hungarian theatres, two German and two Yiddish, as well as a number of workers and children's theatres and many flourishing puppet-theatres.

ROMASHOV, BORIS SERGEIVICH (1895–1958), Soviet dramatist, whose first play, *Meringue Pie* (1925), was a satire on bourgeois elements in Soviet society. Among his later plays, which dealt mainly with episodes of the Civil War, were *The End of Krivorilsky* (1927), *The Fiery Bridge* (1929), and *Fighters* (1934). This last, perhaps his most important work, dealt with the Red Army in peacetime, and the clash between private and public interests among the officers. In 1942 a play dealing with the defence of Moscow, *Shine, Stars!*, was put on at the Sverdlovsk Theatre less than a year after the events it dealt with took place and proved immensely popular. In 1947 *A Great Force*, which depicted the conflict between conservative and progressive scientists, was seen at the MALY THEATRE.

ROME. Roman drama was produced by a succession of writers who adapted Greek originals to the native taste for rhetoric, spectacle, and sensationalism, buffoonery, homely wit, and biting repartee. Comedy almost certainly arose from a blending of the primitive *fescennina iocatio*, or bawdy extempore joking by clowns at harvest festivals and marriage ceremonies, with the performances of masked dancers and musicians who came from Etruria. From Etruria also came such theatrical terms as *histrio* for actor and *persona* for the person indicated by the mask. By the 3rd century BC Roman audiences would have been familiar with the secular entertainments of other Italian peoples—the *phylax* comedies and FABULA *rhinthonica* of Tarentum, for instance, and the Oscan FABULA *atellana*, both of which were not unlike the comedy of mainland Greece. Although MIME may well have been international from its inception, it was not until the beginning of the 3rd century that the spread of Greek drama under the auspices of the ARTISTS OF DIONYSUS led to the emergence of an organized Roman theatre, particularly after 250 BC, when plays were produced at the *ludi scaenici*, or public festivals, as part of the general celebrations. The advance to true drama was made by Livius ANDRONICUS who was the first to produce at a public festival a FABULA *palliata*, or Greek play in translation. From then until the end of the Republic adaptations of Greek tragedies and comedies predominated in the Roman theatre, together with a very few examples of the FABULA *praetexta*, on a historical Roman theme, and a considerable number of the new-style FABULA *togata*, or play on everyday Roman life. Among Roman dramatists there was no invention of plot or character on a grand scale. They were mainly concerned with supplying a non-reading public with live entertainment which blended a well known story with topical allusions and a plentiful supply of comic business. Only in TERENCE is it possible to discern a conscious artistic impulse to improve on his Greek model. Other dramatists following Andronicus, among them CAECILIUS STATIUS, NAEVIUS, and PLAUTUS, were content to take what was offered them by the Greeks and concerned themselves only with adapting it to the demands of a Roman audience, with the proviso that since dramatic performances were provided free by the authorities nothing should be said or indicated which might reflect on them or on their management of public affairs: political and personal allusions were therefore banned. With moral and religious matters the censorship was less concerned: Plautus's *Amphitruo* shows the gods Jupiter and Mercury as heartless deceivers, and it is precisely those plays of Plautus and Terence which are known to have been particularly popular—among them the *Casina* and the *Eunuchus*—which are nowadays considered the most improper. But Plautus's boast in the prologue and epilogue to the *Captivi* that this play is of a higher moral tone than most Greek comedies suggests that public taste might itself impose limits on the dramatist's freedom; in general the extant Latin comedies are fairly free from indecency.

At first Rome had no permanent theatre. The simple wooden buildings required for a production, the stage and behind it the scene-building, could be put up as and when required. The stage itself represented a street leading from the centre of the town to the harbour on one side and the open country on the other; or it might represent the space in front of a palace or a row of houses, or the courtyard of one or more buildings to which, in all cases, access was gained by three doors in the back wall. Interior scenes and changes of setting within a play were unknown at this time, and the absence of a front curtain made it necessary for every play written within the period of the extant comedies to begin and end with an empty stage. The characters in comedy were types rather than individuals—old gentlemen, usually miserly and domineering, young gentlemen, usually extravagant and spineless, jealous wives, treacherous pimps, intriguing slaves, boastful captains—and custom prescribed the appropriate costume, wig, and mask for each type. Since respectable women were not supposed to appear in public, the unmarried heroines of Greek NEW COMEDY and its Latin derivatives were somewhat less than respectable—courtesans, or maidens separated by some mishap from their parents and brought up in humble circumstances. If marriage was to be the outcome of the hero's wooing, then it was necessary to show in the course of the play that the girl he wished to marry was not only chaste, but the long-lost child of respectable parents. It is hardly surprising that in general the plots of Roman comedies turn on intrigues, attempts to raise money, and swindling and deception of many kinds.

The use of costumes and MASKS enabled small companies of five or six actors to perform almost any play by doubling of parts. The leading actor was probably also the director. Such actors, like Ambivius TURPIO or Cicero's contemporaries AESOPUS and ROSCIUS GALLUS, could rise to fame and fortune, since the theatrical profession was not yet regarded as degrading in itself. There is no mention of slaves acting until the day when Roscius, near the end of his career, trained a slave named Panurgus to appear on stage. Revivals of old plays were rare, since audiences would naturally be attracted to a new work. This would be bought from the author by a manager authorized by the festival authorities to hire the necessary actors and costumes and settle all business matters, his profit arising from what he could save on the fixed sum allotted to him. Most of the present knowledge of Roman drama comes from such manuscripts as have survived, sometimes with later alterations and additions, and the comments and lists of later writers concerned with drama and literature. The best hope of survival, even in fragments, was probably as a school text for later generations. The records are confused, and in later times there was always the temptation to fill in the biographical details of a dramatist's life by inference or imagination. The safest basis for the study of Latin drama is the text of the plays and the prologues.

Though tragedy seems to have attracted fewer writers than comedy, the popularity and influence of the leading tragic writers appear to have been great. In general they selected as models the more melodramatic of their Greek originals. Exciting plots, flamboyant characters, gruesome scenes, violent rhetoric, were apparently more attractive to Roman crowds than the poetic qualities for which Greek tragedies are admired today. In substance the plays underwent little alteration, the chief difference between the Greek original and its Latin adaptation being the development of rhetoric at the expense of truth and naturalness. In comedy, on the other hand, the free-and-easy methods of the early writers seems to have been succeeded by greater fidelity to the originals; but that something was thereby lost seems to be indicated by the rapid development at the hands of AFRANIUS, ATTA, and TITINIUS of the *fabula togata*.

With the end of the 2nd century BC a fundamental change came over the Roman theatre. Sextus Turpilius, the last writer of comedy for the stage, died in 104; ACCIUS, the last writer of tragedy, in 85. Few new plays were being written, and those mostly not for production. Theatres, which had for some time been permanent buildings, the first, in stone, dating from 55 BC, became more and more elaborate as the level of entertainment presented in them sank very low. Rustic farce (*fabula atellana*) and mime predominated; such plays as were staged were marred by tasteless extravagance. The introduction of a curtain led to elaborate scenic effects and quick changes of settings, to the detriment of the spoken text. A divorce thus set in between the drama and the stage, Cicero, for instance, thinking of plays as something to be read, while he regarded theatrical performances as only one, and not the best, of the various forms of popular entertainment which included gladiatorial and animal combats. It is true that under the Empire many plays were written, and many fine theatre buildings were erected in Europe, Asia, and Africa, the ruins of which still stand on such sites as Orange, Aspendus, Sabratha, and Leptis Magna; but the plays and the theatres had very little to do with each other. Popular taste was all for mime and PANTOMIME, and the stage was given over to these and other forms of light entertainment, while the writing of plays became the prerogative of educated literary men. The tragedies ascribed to SENECA and such works as the anonymous *Octavia* were CLOSET DRAMAS, intended to be read among friends. Though full of clever rhetoric, they were not written with the limitations of the stage in mind, and were probably never performed. Yet as the only examples of Latin tragedy to reach the modern world they were destined to exercise an immense influence on the playwrights of the Renaissance.

It was not only the degeneration of public taste which led to the disintegration of the serious theatre in Imperial Rome. There was the added problem of the constant hostility of the Christian Church. No Christian could be an actor, under pain of excommunication, and all priests and devout persons refrained from attendance at theatrical performances of any kind. Tertullian in *De Spectaculis*, probably written at the beginning of the 3rd century, urged Christians to look for spectacle to the services of the Church, not forseeing that with the rise of LITURGICAL DRAMA he would be taken literally. The Code of Theodosius, AD 434, forbade all public performances on Sunday, and in the 6th century the theatres in Europe finally closed, the much-harassed actors being forced to rely on private hospitality or take to the road. In Constantinople, however, the rules were somewhat relaxed, possibly under the influence of the Empress Theodora, who had herself (according to contemporary gossip) been an actress, and the Byzantine government still considered the provision of actors and public performances as being among its duties until the organized theatre in the East perished in the Saracen invasions of the 7th and 8th centuries. In the West, in spite of a continued interest in drama as literature, an interest which St Augustine in *Civitate Dei* (*c.* 420) upholds as a means of education, the theatre in the East perished in the Saracen invabarbarians, the last reference to it dating from 533 AD. From then until the 10th century there was nothing but an undercurrent of itinerant entertainers—mimes, acrobats, jugglers, bearleaders, jongleurs, and minstrels—who kept alive some of the traditions of the classical theatre, while the Church quietly absorbed such pagan rites as the FOLK PLAY and the MUMMERS' PLAY into its own ritual, unconciously preparing the way for the revival of the theatre it had tried to suppress.

RONCONI, LUCA, see ITALY.

RONDIRIS, DIMITRIOS (1899–1981), Greek director, who made his first appearance on the stage in 1919, playing Florizel in a translation of *The Winter's Tale*. During the next 10 years he was seen in a number of Greek companies, and then went to Vienna, where for three years he studied the history of art and attended Max REINHARDT's seminars. On his return to Athens he was appointed a director of the Greek National Theatre and in 1936 inaugurated the first modern festival of Greek tragedy with a production of SOPHOCLES' *Electra* in the Theatre of Herod Atticus in Athens. This festival has now become an annual event, as has the festival of Greek plays at EPIDAUROS which he also organized. In 1939 he directed a production of *Hamlet*, with Alexis MINOTIS in the title role, which was seen in London. His production of *Electra* was seen in New York in 1952, with Katina PAXINOU in the title role. Five years later he founded the Piraikon Theatre in the Piraeus, where for some years before his retirement he continued staging classical tragedies in a modern style, taking his company on tour as far afield as Russia and Israel.

ROPE HOUSE, see HAND WORKING.

ROSCIUS GALLUS, QUINTUS (*c*. 120–62 BC), Roman actor, the most famous of his day. He was a friend of Cicero, who delivered on his behalf during a lawsuit the speech *Pro Roscio Comoedo*, and though of middle-class origins was raised to equestrian rank by the dictator Sulla. He appeared in the plays of PLAUTUS—one of his best parts being Pseudolus— and TERENCE, as well as of lesser dramatists, and by his example did much to raise the status of actors in the Roman world. Much of his success was due to careful study of his parts; he thought out and rehearsed every gesture before using it on stage, and his name became synonymous with all that was best in acting. Shakespeare referred to him in *Hamlet*, and many outstanding actors were called after him.

(For the African Roscius, see ALDRIDGE, IRA; for the Dublin, or Hibernian, Roscius, see BROOKE, G. V.; for the Ohio Roscius, see ALDRICH, LOUIS; for the Scottish Roscius, see JOHNSTON, H. E.; for the Young American Roscius, see COWELL, SAM; for the Young Roscius, see BETTY, WILLIAM.)

ROSE, BILLY, see BILLY ROSE THEATRE and REVUE.

ROSE, CLARKSON, see REVUE.

ROSENPLÜT, HANS (*c*. 1400–*c*. 1470), German poet and dramatist, known also as *der Schnepperer* ('chatterbox'), and the first author of Carnival Plays (see FASTNACHTSSPIEL) to be known by name. Of those attributed to him, all of which were performed in his native city Nuremberg, the best known is *Des Türken Vasnachtspil* (1456), in which he vents his spleen against the nobility—a characteristic of the townsmen of his time—by comparing the orderly government of Turkey to the oppression and exploitation current in European communities.

ROSE THEATRE, London, in Rose Alley, off Southwark Bridge Road. This theatre, 94 ft square, constructed of wood and plaster on a brick foundation, octagonal in shape, and partly thatched, opened in the autumn of 1587. It was built by a carpenter, James Grigges, for Philip HENSLOWE and his partner John Cholmley, and stood in an area known as the Liberty of the Clink on the site of a former rose garden, about halfway between the later GLOBE and HOPE theatres. It is not known which company first played there, but in 1592, after extensive repairs had been done and alterations made, STRANGE'S MEN were there and from 1594 the ADMIRAL'S MEN, under Henslowe's son-in-law ALLEYN, used it almost exclusively until they moved to the FORTUNE in 1600. It is probable that Shakespeare's *Henry VI* was first given at the Rose, which according to Henslowe's accounts was a highly profitable venture. After the departure of Alleyn, several companies used it, notably PEMBROKE'S MEN and Worcester's Men, the latter (as QUEEN ANNE'S MEN) being still in residence in 1604, just before Henslowe's lease expired. The building may then have been used for some kind of amusement, but not for plays, and it was pulled down the following year.

ROSIMOND [Claude la Roze] (*c*. 1640–86), French actor who was playing at the MARAIS when he was invited to join the company at the PALAIS-ROYAL after the death of MOLIÈRE, playing as his first part Molière's role in *Le Malade imaginaire*. He was already known as a dramatist, and it may have been his double reputation that induced LA GRANGE and his companions to look on him as a successor to Molière. If so, he proved a disappointment, as he produced only one more play, which was not a success. One of his comedies on the subject of DON JUAN was the source for SHADWELL'S *The Libertine* (1675).

ROSS, W. G. (?–*c*. 1876), Scottish comedian, who was a small-part actor and singer when in the 1840s he became famous for his singing of one song, 'Sam Hall'. Based on a traditional street-ballad, this was a long soliloquy by a foul-mouthed murderer spending his last night in a condemned cell, an unpleasant piece of realism with a reiterated refrain of 'Damn your eyes!', which Ross delivered sitting astride an old kitchen chair in shabby clothes and battered hat, a blackened clay pipe between his teeth which he removed only to spit. It became so popular that the Cyder Cellars and the Coal Hole were crowded to suffocation whenever Ross was billed to sing it, but its oaths and murderous intensity as delivered by Ross proved too much for EVANS'S and it was never heard there. After a time the novelty of the song wore off, and Ross could never find another to equal it. He drifted back to the stage—he does not seem to have appeared at any of the new music-halls—and died while a member of the GAIETY chorus. When 'Sam Hall' was revived by Philip Godfrey at the PLAYERS' THEATRE in 1937, it was found to have lost none of its power of holding and terrifying an audience.

ROSSETER, PHILIP, see BOY COMPANIES and PORTER'S HALL.

ROSSI, ERNESTO FORTUNATO GIOVANNI MARIA (1827–96), Italian actor, who excelled in tragedy. As a young boy he acted with a children's company which played in private houses and he made his professional début at the age of 18 in his birthplace, Livorno (Leghorn). He later joined the company of Gustavo MODENA, and in 1852 played opposite Adelaide RISTORI, with whom he went to Paris in 1855. Although his Othello was later eclipsed by that of SALVINI, he was the first Italian actor to play the part, in Milan in 1856, in a translation by Carcano; he also played Hamlet, in an early version of the play by Rusconi, and was considered good as Lear, though not in England or America where his interpretations of Shakespeare did not meet with the approval of English-speaking audiences. He travelled widely, being much admired in Paris and in Germany, and died while returning from a successful season in St Petersburg. He published two volumes of theatrical memoirs, in 1886 and 1888, and also translated *Julius Caesar* into Italian.

ROSSO DI SAN SECONDO, PIER MARIA (1887–1956), Italian novelist and dramatist, the most lyrical of the writers of the GROTTESCO school, and perhaps the most subtle. Influenced by CHIARELLI and PIRANDELLO, his constant theme is the conflict between the dream and the reality, between the illusion of freedom and the conventions of modern society, and although his plays are based on concrete situations, they have the quality of symbolic fantasies. His best known play is *Marionette, che passione!* (1917), in which the characters are puppets, differing only in their dress and each one typifying a different facet of despair. Though a Sicilian Rosso was cosmopolitan, and participated in the 20th century's preoccupation with the legend of Persephone: much of his work, which includes *La bella addormentata* (*The Sleeping Beauty*, 1919), *Una còsa di carne* (*An Affair of the Flesh*, 1924), and *La scala* (*The Ladder*, 1926), his greatest success commercially, can be seen as a series of new and subtle versions of the myth, culminating in *Il ratto di Proserpina* (*The Rape of Persephone*, 1954, but written in 1933). Rosso helped to bring modern Italian tragedy to maturity, substituting lucid poetry for rhetoric, the dance of life for the dance of death, and deep compassion for the obtuseness of self-pity.

ROSTAND, EDMOND EUGÈNE ALEXIS (1868–1918), French romantic dramatist, whose colourful poetic plays came as a relief after the drab realities of the naturalistic school. His first play *Les Romanesques* (1894), a delicious satire on young lovers, was followed by the more serious but still tender and lyrical *La Princesse lointaine* (1895). A Biblical play, *La Samaritaine* (*The Woman of Samaria*, 1897), was less successful, but in *Cyrano de Bergerac* (1898), written for COQUELIN *aîné*, he achieved a marvellous fusion of romantic bravura, lyric love, and theatrical craftsmanship, and its success was overwhelming. It became a perennial favourite, not only in France but in England and America, where something of its quality was apparent even through a pedestrian translation. *L'Aiglon* (*The Eaglet*, 1900), in which BERNHARDT played the ill-fated son of Napoleon, had less vigour, but appealed by its pathetic evocation of fallen grandeur and the frank sentimentality of its theme. Rostand's last play was *Chantecler* (1910), which is by some critics accounted his best, as it is certainly his most profound, work. The verse is masterly and the allegory unfolds effortlessly on two planes of consciousness, the beast's and the man's. *La Dernière Nuit de Don Juan* was left unfinished, but indicates how much Rostand's undoubted talent might have matured.

A free adaptation of *Les Romanesques* as *The Fantasticks*, by Tom Jones and Harvey Schmidt, opened at the Sullivan Street Playhouse, New York, in 1960 and was still running 20 years later.

ROSTRUM, any platform, from a small dais for a throne to a large battlement on which actors can assemble. Known in America as a Parallel, it is usually made with a removable top and hinged side-frames, to fold flat for packing, though permanent 'stock' or Rigid Rostrums are also used. These are made of 2 × 4 in. or 1 × 6 in. fir or pine framing with a fixed plywood top, and legged up to any desired height with 2 × 4 in. fir legs, thus offering a choice of heights that the collapsible rostrum lacks. Recently a variety of metals have been used for rostrum understructures, and experiments have been made with Skin-Tension Platforms, consisting of polyurethane or polystyrene foam sandwiched between two sheets of fibre-glass, forming a very strong combination. A Rostrum is usually approached by steps or a ramp, and quitted off-stage by 'lead-off' steps. When necessary a canvas-covered FLAT, known as a rostrum-front, is placed to hide the front of the platform from the audience.

ROSWITHA, see HROSWITHA.

ROTIMI, OLA [Emmanuel Gladstone Olawole] (1938–   ), Nigerian playwright, born of Yoruba and Ijaw parents. His plays, written in English, reflect his deep concern with the history, culture, and political problems of his people. *To Stir the God of Iron* (1963) was first produced at Boston University, where he was studying Theatre Arts; at Yale, where he also studied, *Our Husband Has Gone Mad Again* (1965) received the major 'Play of the Year' award. In 1966 Rotimi took up a research fellowship at the University of Ife, where he directed his own adaptation of SOPHOCLES' *Oedipus Rex* as *The Gods Are Not to Blame*. This play was also seen in London in 1978. Among his other plays *Kurunmi* (1969) and *Ovonramwen Nogbaisi* (1971) are historical, while *If* (1973) was prompted by the aftermath of the Nigerian civil war. In 1977 Rotimi became head of the Creative Arts Centre at the University of Port Harcourt in Nigeria.

ROTROU, JEAN DE (1609–50). French dramatist, next to CORNEILLE the most important of his day. He was only 19 when he had two plays produced at the Hôtel de BOURGOGNE, where he may have succeeded HARDY as official dramatist to the troupe. His popularity may be gauged from the fact that he had four plays produced in Paris in 1636. More than 30 of his works survive, some of them the best extant examples of the TRAGI-COMEDY of the time, though he was also instrumental with MAIRET and Corneille in establishing neo-classical tragedy, of which his *Hercule mourant* (1634) is an early example. Rotrou, like Corneille, was interested in Spanish literature, and translated one of Lope de VEGA's plays as *La Bague de l'oubli* (1629), the first extant French play to be based on a Spanish source and the first notable French comedy, as distinct from farce. He was also the author of one of the many versions of the story of Amphitryon, *Les Sosies* (1637), which is considered one of his best plays. Of his later works a tragedy, *Cosroës* (1649), remained in the repertory until the early 18th century, while *Venceslas* (1647), a tragedy based on a play by ROJAS DE ZORRILLA, was still being played up to 1857, the role of Ladislas providing a memorable vehicle for such actors as BARON, LEKAIN, and TALMA; its first interpreter was probably MONTFLEURY. In many ways the most appealing of Rotrou's later plays, and a masterpiece of baroque tragedy, is *Le Véritable Saint-Genest* (1645). Again based on a play by Lope de Vega, this portrays the conversion of the actor Genest while playing the part of the martyr Saint Adrian, with the result that he is himself taken away to suffer martyrdom. Rotrou, a man of great charm and nobility of character, held important municipal offices in his native town of Dreux, and died there during a plague, having refused to abandon his official post and seek shelter elsewhere.

ROTUNDA, The, London, a hall in Blackfriars Road, at the corner of Stamford Street, sometimes confused with the RING, which was on the opposite side of the road about 500 yards away. It opened in 1790 as a museum, but was used for musical and dramatic entertainments as early as 1829 and from 1833 to 1838 was known as the Globe Theatre. It became the Rotunda under the management of the composer John Blewitt, who called it a 'musick hall', in the old sense. Although it held only 150 people, it enjoyed much local popularity until, having changed its name again to the new-style Britannia Music-Hall, it lost its licence in 1886, after an illegal cockfight. It then became a warehouse and was finally demolished in 1945.

ROUEN, France, a town second only to Paris as a dramatic centre in the late Middle Ages. Most known types of LITURGICAL DRAMA were acted there under the sponsorship of the Cathedral, and some texts survive in several versions. The Rouen *Officium Pastorum* (*Office of the Shepherds*) and the *Festum Asinorum* (*Feast of the Ass*, also known as the *Procession of the Prophets*) represent the highest development of these plays; the latter contains 28 principal prophets and a chorus of six Jews and six Gentiles. The first record in France of a vernacular MYSTERY PLAY comes from Rouen, in 1374; there are also records of Passion and Miracle Plays, usually performed at Whitsun, from 1410 onwards; of Nativity Plays from 1474; and of MORALITY PLAYS in the 16th century, in which century religious plays were finally suppressed because of the admixture of burlesque and abuse. In later dramatic history Rouen is noteworthy as being the birthplace of CORNEILLE, who probably saw there the plays of Alexandre HARDY, and wrote his first play *Mélite* for a strolling company under MONTDORY which took the play back to Paris to be produced in 1630.

ROUND HOUSE, The, London, in Chalk Farm Road, Camden Town. This disused locomotive shed, dating from the mid-19th century, was taken over in 1961 by Arnold WESKER to serve as the home of his Centre 42, which aimed to make the arts more widely available, primarily through trade union involvement. The venture was not a success, but the theatre has since housed some highly interesting productions 'in the round' including in 1968 an experimental production of *The Tempest* by Peter BROOK. Later in the same year a play about Nelson, *The Hero Rises Up*, by John ARDEN and his wife, had a short run, and 1969 saw the production of Kafka's *Metamorphosis* and *In the Penal Colony* and the LIVING THEATRE in *Frankenstein, Paradise Now*, and *Mysteries*. In 1970 the Meadow Players from Oxford appeared in GENET's *The Blacks*, and the erotic revue *Oh! Calcutta* had a short run before moving on to the West End. An English version of BARRAULT's *Rabelais* (seen in French at the OLD VIC in 1969) ran for a seven-week season from 8 Mar. 1971, and on 17 Nov. the American musical *Godspell* opened, transferring later to WYNDHAM's. The OPEN THEATER from New York appeared in 1973, as did the Grand Magic Circus. The theatre continued to be used for productions that might not otherwise find a home in London, including BRENTON's *Epsom Downs*, presented by Joint Stock, Tennessee WILLIAMS's *The Red Devil Battery Sign* (both 1977), and David RABE's *Streamers* (1978). After a complete reconstruction in 1979 the theatre reopened with Ronald Harwood's adaptation of Evelyn Waugh's novel *The Ordeal of Gilbert Pinfold* with Michael HORDERN, originally seen at the ROYAL EXCHANGE THEATRE, Manchester; other productions from the same theatre include IBSEN's *The Lady From the Sea* (also 1979) with Vanessa REDGRAVE and, in 1981, WEBSTER's *The Duchess of Malfi* with Helen MIRREN and MOLIÈRE's *The Misanthrope* with Tom COURTENAY. The theatre has also staged AYCKBOURN's *Season's Greetings* (1980) and *Suburban Strains* (1981). The production by the Rustaveli company from GEORGIA of a Georgian version of *Richard III* in 1980 made a great impression.

ROUSSEAU, JEAN-JACQUES (1712–78), French philosopher and man of letters, who had some facility in music and in 1752 produced at Court a successful light opera, *Le Devin du village* (*The Village Soothsayer*), which contained many simple and charming songs. Paradoxically—though it may be accounted for by his quarrel with DIDEROT—in his *Lettre à d'Alembert contre les spectacles* (1758) he opposed the 18th-century view that the stage could be used for political and moral teaching, and declared that, being intended only for amusement, it was harmful and useless and should be suppressed. The letter, written as a protest against d'Alembert's suggestion that a playhouse should be built in Geneva, may also have been meant as an attack on VOLTAIRE, whose passion for the theatre was well known. It contained a virulent attack on the comedies of MOLIÈRE, particularly *Le Misanthrope*. Though no dramatist, Rousseau is important for the influence of his ideas on the drama, as well as on French and European literature in general.

ROVETTA, GEROLAMO (1851–1910), Italian novelist and playwright of the school of VERISMO, whose prolific output included several historical dramas of which the most successful was *Romanticismo* (1901), dealing with the heroism and suffering of Manzini's followers in Italy's struggle for independence. Once considered a supreme naturalist, Rovetta is now adjudged to have sacrificed his admirable capacity for clear narrative by cluttering it up with unnecessary details. Typical of his other works, which began in 1870 with *In sogno* (*In Dreams*) and *Colera cieca* (*Blind Rage*) and ended in 1909 with *Molière e sua moglie* (*Molière and His Wife*), are *La trilogia di Dorina* (1889), in which a marquis who demands that his mistress should be pure is driven by jealousy to marry her once her virtue is gone; *I dishonesti* (1892), which asserts that the shattering of a man's belief in his wife's chastity may turn him into a thief; and *Il poeta* (1897), an interesting *fin-de-siècle* study of unmitigated evil.

ROWE, GEORGE FAWCETT (1834–89), American actor, who made his first appearance in New York in 1866 and for many years was a popular player of young lovers, for which his fair, handsome, boyish face and elegant figure were eminently suited. In 1872 he made a great success as Digby Grant—played in London two years previously by IRVING—in James ALBERY's *Two Roses*. In later years he was much admired as Micawber in *Little Em'ly* (a dramatization of DICKENS's *David Copperfield*), as Waifton Stray in his own play *Brass*, and as Hawkeye in *Leatherstocking*, his own dramatization of Fenimore Cooper's *The Last of the Mohicans*. He also translated and adapted *Sphinx* (1875), by Octave Feuillet, in which Clara MORRIS gave a powerful performance.

ROWE, NICHOLAS (1674–1718), English dramatist, one of the few of the Augustan age to display any real dramatic power. Of his seven tragedies the early ones, *The Ambitious Step-Mother* (1700) and *Tamerlane* (1701), in which BETTERTON was outstanding in the title role, were written in a somewhat frigid neo-classical style, but his later masterpieces, which include *The Fair Penitent* (1703), based on MASSINGER's *The Fatal Dowry*, and *The Tragedy of Jane Shore* (1714), have genuinely moving and poetic passages, and both were frequently revived, Mrs SIDDONS being particularly admired in the leading roles first played by Mrs BARRY and Anne OLDFIELD respectively. His other tragedies include *Ulysses* (1705) and *Lady Jane Grey* (1715); he also wrote an unsuccessful comedy *The Biter* (1704). He published in 1709 an edition of Shakespeare, adding stage directions and act- and scene-divisions, and working to make the text less corrupt. He became Poet Laureate in 1715.

ROWLEY, SAMUEL (*c*. 1575–1624), English actor and dramatist, who is credited with having had a hand in a number of lost plays which preceded, and possibly provided material for, some of Shakespeare's, including *The Taming of a Shrew* (*c*. 1589). His only known extant play, a chronicle drama on the life of Henry VIII, *When You See Me, You Know Me* (1603), was performed by the ADMIRAL'S MEN, whom he joined in 1597. They also appeared in several of his lost plays. He is believed to have altered MARLOWE's *Doctor Faustus* for HENSLOWE, mainly by adding some comic passages featuring the Devil for a revival in 1602.

ROWLEY, WILLIAM (*c*. 1585–*c*. 1637), English actor and dramatist, who in 1619 played the fat clown Plumporridge, in the Inner Temple MASQUE, as a foil to the thin clown of John Newton (?–1625), who represented 'A Fasting Day'. He also played the clown in his own play *All's Lost by*

Lust (1622) and the Fat Bishop in *A Game at Chess* (1624) by MIDDLETON, with whom Rowley collaborated in several plays, among them *A Fair Quarrel* (1617), *The World Tossed at Tennis* (1620), and their best known work *The Changeling* (1622). He also worked with DEKKER and FORD on *The Witch of Edmonton* (1621), with FLETCHER on *The Maid in the Mill* (1623), and with WEBSTER on *A Cure for a Cuckold* (1625).

ROYAARDS, WILLEM, see NETHERLANDS.

ROYAL ACADEMY OF DRAMATIC ART, see SCHOOLS OF DRAMA.

ROYAL ADELAIDE THEATRE, London, see GATTI'S.

ROYAL ADELPHI THEATRE, London, see ADELPHI THEATRE.

ROYAL ALFRED THEATRE, London, see WEST LONDON THEATRE.

ROYAL AMPHITHEATRE, London, see ASTLEY'S AMPHITHEATRE.

ROYAL AQUARIUM THEATRE, London, see IMPERIAL THEATRE.

ROYAL ARCTIC THEATRE, Canada, an unusual form of entertainment which flourished intermittently in the 19th century. During the Royal Navy's search for a northwest passage to the Pacific British ships had to winter in the Arctic circle and to keep up morale plays were regularly performed, either in the ships that lay frozen in the ice or in snow-houses built alongside. Although known by various names—North Georgia; Queen's Arctic; Theatre Royal, Melville Island; Royal Alexandra, Discovery Bay—the term Arctic Theatre has come to be used as an overall title for the whole enterprise. There were three main periods of activity, the first in 1819, when on 5 Nov. GARRICK'S *Miss in her Teens* is known to have been acted on the quarter-deck of H.M.S. *Hecla*, followed at some unknown date by an original play, of which Lieutenant (later Admiral) W. E. Parry was co-author, optimistically entitled *The North-West Passage, or Voyage Finished*. This tradition of play-acting was revived between 1850 and 1853, in the wake of the activity which followed the disappearance of Sir John Franklin. A series of theatrical 'seasons' was advertised in playbills and in 'The Illustrated Arctic News' published on board such ships as H.M.S, *Assistance, Resolute*, and *Intrepid*; among the plays given were PLANCHÉ'S *Charles XII*, W. B. Rhodes's burlesque *Bombastes Furioso*, and a comedy by an unknown author called *The Scapegrace*. Appropriately, an original HARLEQUINADE performed in 1851 was entitled *Zero, or Harlequin Light*, and involved the defeat of a polar bear and the spirits of winter by Harlequin and Columbine, disguised as Sun and Daylight. It was reported that one stage made of ice was so engulfed in clouds of vapour caused by the breath of the audience that the actors were virtually invisible. It was in such conditions that what was probably the most northerly production of Shakespeare ever given was seen, when on 21 Dec. 1852 the first act of *Hamlet* was performed, with, improbably, the roles of Hamlet and Ophelia doubled by the same actor.

The Arctic Theatre came to an end in 1875–6, when British ships were attempting to reach the North Pole. A PANTOMIME, *Aladdin and the Wonderful Scamp*, perhaps based on E. L. Blanchard's DRURY LANE version of 1874, was performed below decks on H.M.S *Alert* on 23 Dec. 1875, and the officers and crew of H.M.S. *Discovery*, anchored further south, alternated productions every two weeks in a snow-theatre on land. From the illustrations and accounts which have survived it appears that the scenery, props, and costumes were often skilfully made and extremely elaborate. Female roles were, of course, played by men.

ROYAL ARTILLERY THEATRE, London, at Woolwich. This was built by public subscription, with the help of the War Office, to take the place of a theatre constructed in the old garrison church in 1863 and used for amateur dramatics. The new theatre was fully professional and from 1909 to 1940 was controlled by Mrs Agnes Littler who, with her husband, was injured when a bomb dropped outside the building in Mar. 1918. Among the famous actors who appeared here were Lillie LANGTRY, Violet VANBRUGH, George ROBEY, and Albert CHEVALIER. During the Second World War the theatre, which was used by various Service units for entertaining the troops, was again damaged by bombing, and, attempts to reopen it having failed, it closed down finally in 1956.

ROYAL AVENUE THEATRE, London, see PLAYHOUSE.

ROYAL BRUNSWICK THEATRE, London, see ROYALTY THEATRE (1).

ROYAL CIRCUS, London, see SURREY THEATRE.

ROYAL COBURG THEATRE, London, see OLD VIC THEATRE.

ROYAL COURT THEATRE, London. (1) In Lower George Street, Chelsea. A badly adapted Nonconformist chapel, it opened as the New Chelsea on 16 Apr. 1870 and had no success at all, even when it changed its name to the Belgravia, until Maria Litton took it over, and after reconstruction and redecoration opened it as the Royal Court on 25 Jan. 1871 with *Randall's Thumb* by W. S. GILBERT, who provided further successes with *The Wedding March* and *The Happy Land* (both 1873). In the latter he burlesqued contemporary politicians so mercilessly that the Lord Chamberlain intervened, and the actors' make-up had to be altered. In 1875 HARE took over with a good company led by the KENDALS, and produced a number of successful plays. He was followed in 1879 by Wilson BARRETT, and with him as her leading man MODJESKA made her first appearance in London a year later. In 1885 a company which included Marion TERRY, Mrs John WOOD, and Brandon THOMAS inaugurated a successful series of farces by PINERO, beginning with *The Magistrate* and followed by *The Schoolmistress* (1886) and *Dandy Dick* (1887). The theatre finally closed on 22 July 1887 to make way for street improvements, and was completely demolished.

(2) On the east side of Sloane Square. This theatre was built to replace the above, and had a three-tier auditorium with a seating capacity of 642. It opened on 24 Sept. 1888 with GRUNDY'S *Mamma!*, and was for a time less successful than the old theatre, in spite of the popularity of Pinero's new farce *The Cabinet Minister* (1890). It was his record-breaking *Trelawny of the 'Wells'* (1898) that first brought the Royal Court back into the limelight, its next outstanding success being Charles Hannon's *A Cigarette-Maker's Romance* (1901), starring MARTIN-HARVEY. In 1904 J. E. VEDRENNE and Harley GRANVILLE-BARKER took over the theatre and produced there a remarkable series of plays, both new and old, which ranged from Shakespeare to SHAW, GALSWORTHY, Barker himself, and St John HANKIN. After Barker and Vedrenne had moved to the SAVOY, Somerset MAUGHAM'S *Lady Frederick* (1907) filled the house to capacity; but then its fortunes declined until after the First World War, when J. B. FAGAN took over, one of his first productions being Shaw's *Heartbreak House* (1921). In 1924 Barry JACKSON brought his BIRMINGHAM REPERTORY THEATRE company to the Court, opening with the five parts of Shaw's *Back to Methuselah*, followed by Eden PHILLPOTTS's long-running comedy *The Farmer's Wife*. In 1928 Jackson returned with controversial productions of *Macbeth* and *The Taming of the Shrew*, both in modern dress, and in 1932, after three seasons of Shaw's plays by the Macdona Players, the theatre became a cinema. It was badly damaged in Nov. 1940 when Sloane Square underground station was bombed, but after extensive renovation it reopened as a theatre in 1952, the only productions of note being Frank Baker's *Miss Hargreaves*, starring Margaret RUTHERFORD, the revue *Airs on a Shoestring* (1954), and BRECHT'S *The Threepenny Opera* (1956), until in April of that year it was taken over by the ENGLISH STAGE COMPANY under George DEVINE. A policy of presenting new playwrights was justified by the success of John OSBORNE'S *Look Back in Anger*, followed in 1958 by Arnold WESKER'S *Chicken Soup with Barley*, transferred from the BELGRADE THEATRE, Coventry. In the following year the remaining plays of the trilogy, *Roots* and *I'm Talking About Jerusalem*, consolidated Wesker's reputation as a playwright. In 1957 came Osborne's *The Entertainer*, with Laurence OLIVIER, and in 1959 *Serjeant Musgrave's Dance* by John ARDEN, who, like Osborne and Wesker, remained closely associated with this theatre. Before the building closed in Feb. 1964 for reconstruction it had seen a number of important productions, including N. F. Simpson's *One-Way Pendulum* (1959) and IONESCO'S *Rhinoceros* (1960). After the reopening came two plays by Osborne, *Inadmissible Evidence* (1964) and *A Patriot for Me* (1965), and in 1965, following the death of Devine, William GASKILL took over. Under him David STOREY'S first play *The Restoration of Arnold Middleton* was seen in 1967, followed by Osborne's *Time Remembered* and *The Hotel in Amsterdam* (both 1968), the latter with Paul SCOFIELD. In 1969 the theatre celebrated the end of stage CENSORSHIP by producing three plays by Edward BOND, and in the same year Storey's *In Celebration* and *The Contractor* were well received. An outstanding success was also scored by his *Home*, which, with John GIELGUD and Ralph RICHARDSON, went on to the APOLLO and eventually to New York. Gielgud was seen again in Charles WOOD'S *Veterans* (1972), and Alan BATES, who had been in the original production of *Look Back In Anger* 18 years previously, returned in 1974 to star in Storey's *Life Class*. Howard BARKER'S *Stripwell* (1975) received a major production, and David HARE'S *Teeth 'n' Smiles* (also 1975) and Mary O'Malley's *Once a Catholic* (1977) both transferred to WYNDHAM'S, the latter with enormous success. In 1977 the theatre, which had been under the artistic control of Oscar Lewenstein since 1972, was taken over by Stuart Burge, who was succeeded in 1979 by Max Stafford-Clark. Notable later productions have included Martin Sherman's *Bent* (1979), with Ian MCKELLEN, which transferred to the CRITERION THEATRE, and *Hamlet* (1980) with Jonathan Pryce. The theatre has also played host to such visiting companies as the Paper Bag Players in 1967, the Bread and Puppet Theatre in 1969, the LA MAMA company in 1970, and the PEOPLE SHOW in 1973, and there have been several visits from the TRAVERSE THEATRE company and the PIP SIMMONS THEATRE GROUP. In 1969 the Royal Court added to its amenities the Theatre Upstairs, an adaptable theatre housed in a former rehearsal room and intended for experimental and low-budget works. Its productions have included several of Sam SHEPARD'S plays.

ROYAL DRAMATIC THEATRE, Stockholm, see KUNGLIGA DRAMATISKA TEATERN.

ROYAL ENGLISH OPERA HOUSE, London, see PALACE THEATRE.

ROYALE THEATRE, New York, at 242 West 45th Street, between Broadway and 5th Avenue.

Built by the Chanins, with a seating capacity of 1,059, it opened on 11 Jan. 1927 with a MUSICAL COMEDY, and later housed Winthrop AMES's productions of GILBERT and Sullivan. Mae WEST was seen here in her own play *Diamond Lil* (1928), which ran for nearly a year, and the THEATRE GUILD put on Maxwell ANDERSON's *Both Your Houses* (1933), which was awarded a PULITZER PRIZE. The controversial *They Shall Not Die* (1934), based by John Wexley on the Scottsboro case, had a distinguished cast and settings by Lee SIMONSON. In the autumn of the same year the theatre was renamed the John Golden, presenting a series of moderately successful comedies, but from 1936 to 1940 it was used for broadcasting. It returned to drama under its old name on 19 Dec. 1940, successful productions including FRY's *The Lady's Not for Burning* (1950), Thornton WILDER's *The Matchmaker* (1955), a REVUE entitled *La Plume de ma tante* (1958), Tennessee WILLIAMS's *The Night of the Iguana* (1961), and Abe Burrows's *Cactus Flower* (1965). Robert Shaw's *The Man in the Glass Booth* (1968) and Robert Marasco's *Child's Play* (1970) had good runs, but the biggest hit in the theatre's history was the musical *Grease*, which moved there in 1972 and ran until 1980.

ROYAL EXCHANGE THEATRE, Manchester. In 1968 the 69 Theatre Company, an offshoot of the 59 Theatre Company which had successfully played for a season at the LYRIC THEATRE, Hammersmith, took possession of the university theatre in MANCHESTER and achieved a great success, seven of its 21 productions being transferred to London. The company, which included actors, directors, and designers trained under Michel SAINT-DENIS at the OLD VIC Theatre School, was looking for a permanent home for the development of its ideas and professional standards, and decided to stay in Manchester. In 1972 it leased the Royal Exchange, formerly used for cotton trading, within which a highly successful temporary theatre was erected in 1973. It was dismantled shortly after, and the company mounted four productions in Manchester Cathedral while waiting for its new theatre to be ready. This theatre, which opened in 1976, is a module enclosed in clear glass and suspended from four of the pillars in the hall of the Exchange. It is a THEATRE-IN-THE-ROUND, based on a seven-sided figure, and no seat is more than about 30 ft from the stage. The ground floor seats 400 and the two balconies 150 each. The auditorium can be converted for thrust-stage productions, and the foyer, one of the largest in the world, consists of the whole of the Royal Exchange hall outside the module. The venture was made possible by grants from two local authorities, the ARTS COUNCIL, and public subscription. The company has retained its high reputation in its new home and attracted a large number of outstanding players, including Albert FINNEY in CHEKHOV's *Uncle Vanya* and COWARD's *Present Laughter* (both 1977). Vanessa REDGRAVE in IBSEN's *The Lady from the Sea* (1978), and Helen MIRREN in WEBSTER's *The Duchess of Malfi* (1980). The last two were transferred to London, as were three plays by Ronald Harwood (1934–   ) which had their world premières at Manchester—*The Ordeal of Gilbert Pinfold* (1977), based on Evelyn Waugh's novel, with Michael HORDERN; *A Family* (1978) with Paul SCOFIELD; and *The Dresser* (1980) with Tom COURTENAY, which was also seen in New York. The theatre is used for many other activities apart from plays, including lectures, puppet-shows for children, drama workshops, and concerts.

The co-founder and joint Artistic Director of the 69 Theatre Company, Michael Elliott (1931–   ), became one of the Resident Artistic Directors. His first London stage production was Ibsen's *Brand* in 1959 for the 59 Theatre Company. He directed the ROYAL SHAKESPEARE COMPANY's famous production of *As You Like It* in 1961–2 and was Artistic Director at the Old Vic in 1962–3, the last season before it housed the NATIONAL THEATRE company. In Manchester he has directed such productions as *The Tempest* (1969), *Uncle Vanya*, *The Ordeal of Gilbert Pinfold*, an adaptation of DOSTOEVSKY's *Crime and Punishment* (1978), and *The Dresser*. He has twice directed Ibsen's *Peer Gynt*, at the Old Vic in 1962 and in Manchester in 1970.

ROYAL GROVE, London, see ASTLEY'S AMPHITHEATRE.

ROYAL (HOLBORN) MUSIC-HALL, London, see HOLBORN EMPIRE.

ROYAL OPERA HOUSE, London, see COVENT GARDEN.

ROYAL SHAKESPEARE COMPANY. In 1961 the company playing at the SHAKESPEARE MEMORIAL THEATRE was reorganized and given its present name, the theatre being at the same time renamed the ROYAL SHAKESPEARE. Its first director was Peter HALL, under whom the company embarked on an ambitious programme which included the maintenance of a permanent base in London at the ALDWYCH THEATRE, where the accent was on modern plays and non-Shakespearian classics, which were also seen during a six-month season at the ARTS THEATRE in 1962. The plays of Shakespeare remained the preoccupation of the Stratford-based company. Most of the Stratford productions were later transferred to London, and R.S.C. companies toured all over the world. In 1970, two years after Hall had been succeeded by Trevor NUNN, with the title of Artistic Director, a Stratford company became the first British theatrical ensemble to visit Japan, appearing there with great success in *The Winter's Tale* and *The Merry Wives of Windsor*. In 1965 the company had established Theatregoround, a travelling troupe which with more flexible staging and smaller casts than usual was able to perform in schools, factories, and church halls. For a time it was mainly through this organization that the work of the R.S.C. was seen in the provinces, but unfortunately the project had to be abandoned in 1971. Its basic aims were, however, served by The Place in London, a small theatre seating 330, where for three seasons, in 1971, 1973, and 1974, plays were staged which could not have been risked at the Aldwych, among them *Subject to Fits*, based on DOSTOEVSKY's *The Idiot*, STRINDBERG's *Miss Julie*, and Athol FUGARD's *Hello and Goodbye*. The Place was superseded in 1974 by the OTHER PLACE in Stratford. In the mid-1970s the R.S.C. faced a financial crisis and the prospect of increasing competition from the more heavily subsidized NATIONAL THEATRE, which was about to move into its permanent home. Nevertheless the years from 1975 to 1977 were the most successful in the company's history, with excellent work being done in both its large and small theatres, the opening of the WAREHOUSE in 1977, and, in the same year, a season in NEWCASTLE-UPON-TYNE which proved so popular that it became an annual event. The administrative side of the Stratford organization was strengthened by the appointment in 1978 of Terry HANDS as Joint Artistic Director with Nunn, and in 1979–80 20 new productions were added to the repertoire. The R.S.C. is particularly admired for its high standard of ensemble playing, unusual in England. Financial problems still remain, however, and it is hoped that the long-delayed move of the company's London base to the Barbican arts complex in the City of London (finally achieved in 1982), with its attendant subsidies from the City, will help to resolve them. (See also BARTON, JOHN.)

ROYAL SHAKESPEARE THEATRE, the name given to the SHAKESPEARE MEMORIAL THEATRE, Stratford-upon-Avon, on the formation of the ROYAL SHAKESPEARE COMPANY in 1961. The first season included Vanessa REDGRAVE's enchanting Rosalind in *As You Like It* and ZEFFIRELLI's disastrous *Othello*, while 1962 provided outstanding productions of *The Comedy of Errors* and *King Lear*, the latter directed by Peter BROOK with Paul SCOFIELD in the title role. In 1963 the theatre mounted John BARTON's three-play adaptation of the *Henry VI* trilogy with *Richard III* as *The Wars of the Roses*, followed a year later by *Richard II* and *Henry IV, Parts I and II*. In 1965 there was David Warner's controversial Hamlet, in 1968 Eric PORTER's King Lear, and in 1970 another production by Peter Brook, an idiosyncratic reading of *A Midsummer Night's Dream*. Four Roman plays, *Coriolanus, Julius Caesar, Antony and Cleopatra*, and *Titus Andronicus*, under the joint title of *The Romans*, were seen in 1972, and in 1974 the season, which had previously begun in March and ended in December, was extended into January, an innovation which proved so successful that it was adopted permanently. A highly acclaimed production of *Henry V* in 1975, with Alan HOWARD, established a record by playing for over 100 performances at Stratford during the year; it was revived three times and widely seen in Britain and Europe, as well as in New York. In 1976 there was a popular musical version of *The Comedy of Errors*, and a year later the first production since Shakespeare's lifetime of the unadapted texts of the *Henry VI* trilogy. Alan Howard appeared again in 1980 in *Richard II* and *Richard III*, both directed by Terry HANDS.

ROYAL SOHO THEATRE, London, see ROYALTY THEATRE (2).

ROYAL STANDARD MUSIC-HALL, Pimlico, see VICTORIA PALACE.

ROYAL THEATRE, Copenhagen, see KONGELIGE TEATER.

ROYALTY, in the modern commercial theatre the payment made to a dramatist—usually a fixed percentage of the takings—every time his play is performed. Amateur societies generally pay a fixed fee for each night or group of nights of a particular production. The payment of royalties is of comparatively recent origin. In Elizabethan times plays were either bought outright by a manager, as was done by HENSLOWE, or formed part of the stock-in-trade of the company to which the actor-author belonged; this was the system under which Shakespeare worked. While a play remained in manuscript no other manager or company could perform it, but once it was printed anyone could stage it on payment of a small fee. This probably helps to account for the number of early plays which have failed to survive, since they were never printed. Some authors, particularly if a play had been successful, would go to the trouble of recovering the original prompt-copy for the sake of the small fee they might receive when it was printed. The first to do this systematically was Ben JONSON, whose collected plays and MASQUES were printed in 1616, and it is to his pioneer work in this field that we owe the publication of the plays of Shakespeare by CONDELL and HEMINGE in the First Folio of 1623. From then onwards dramatists began to bear in mind the importance of a reading public which was to increase considerably, particularly when the rise of Puritanism made it safer to read plays than to see them. As early as 1602 the custom arose of giving the author the profits of the third, sixth, and ninth performances of a play, the so-called 'author's nights' (see BENEFIT), but with few theatres, a potentially small audience, and a constant change of bill, many plays failed to achieve even a third night. Under Restoration conditions dramatists continued to receive little monetary reward for their work; most of them, being men of substance, had little need of it and the others managed as best they could, sometimes with the help of a patron. Conditions were little better in the 18th century. One of the most successful plays of the time, *The Beggar's Opera* (1728), brought GAY less than £700, which was considered enough to make him rich. At the end of the century a return was made to the old practice of buying plays outright. Thomas MORTON is said to have received £1,000 for one of his comedies, and Mrs INCHBALD £800 for *Wives as They Were and Maids as They Are* (1797). Unfortunately in the early 19th century the prestige, and consequently the monetary value, of dramatic works declined sharply. Farces and musical pieces earned more than straight plays, and even they were priced very low: it was possible to buy a BURLETTA outright for two guineas. Consequently the theatre was inundated with a stream of thefts, plagiarisms, and adaptations of French and German plays turned out quickly by poorly paid hacks.

The first movement to secure proper recognition of dramatic authorship in England was made by PLANCHÉ, and it was mainly owing to his efforts, and those of BULWER-LYTTON, that adequate COPYRIGHT protection was established for plays written after 1833, though the law was sometimes difficult to enforce. The first dramatic copyright law in the U.S.A. was not passed until 1856. In France, where BEAUMARCHAIS had made great efforts to establish a royalty system earlier, SCRIBE secured for himself the payment of a royalty after the success of *Le Solliciteur* (1817), and by 1832 a payment to the author of 12 per cent of the nightly receipts was usual. One of the first to profit by the new system in England was BOUCICAULT. For his highly successful *London Assurance* in 1841 he received only £300, whereas in 1860 he netted £10,000 for *The Colleen Bawn*, which seems to have been the first play in London to be paid for on the royalty system. Other authors were quick to follow Boucicault's example, and by degrees the standard rate of royalty became 5–10 per cent, rising perhaps to 20 per cent for established authors. During the early part of the 20th century, with the theatre largely dependent upon successful West End runs and subsequent provincial tours, dramatists tended to make either a great deal of money or very little. Since the Second World War, however, there have been increasing opportunities for them to supplement their incomes by writing for the radio, the cinema, and television. Limited subsidies have also been secured through the ARTS COUNCIL, either indirectly, because of the increase in the number of companies whose grants enable them to take risks with new works, or, less often, through bursary schemes, personal grants, and the attachment of resident dramatists to particular theatres. (See also SHARING SYSTEM.)

ROYALTY THEATRE, London. (1) In Well Street, Wellclose Square. In 1787 John PALMER and John BANNISTER from DRURY LANE opened a theatre for which they had obtained a licence from the Governor of the Tower of London, in whose precincts the building stood, and the magistrates of the Tower Hamlets. Under the provisions of the Licensing Act of 1737 this made it technically an 'unlicensed' theatre, and when it opened on 20 June with *As You Like It* and GARRICK's *Miss in Her Teens* the PATENT THEATRES objected; Palmer was arrested and the theatre closed. Later it reopened with BURLESQUE and PANTOMIME only but led a precarious existence, though several managers including the father of William MACREADY tried to improve its fortunes. In 1813 another Palmer took it over and changed its name to the East London, but it was still unsuccessful and in 1826 it was burned down. Rebuilt as the Royal Brunswick, it became the shortest-lived playhouse in London's theatre history. Opening on 25 Feb. 1828, it collapsed three days later owing to the weight of the roof, which the walls were not strong enough to support. A rehearsal of SCOTT's *Guy Mannering* was in progress at the time; 15 people were killed and 20 injured.

(2) In Dean Street, Soho. This small theatre was built for Fanny KELLY, and was used by her in conjunction with a school of acting. It opened on 25 Mar. 1840, but the newly installed stage machinery, worked by a horse, proved so noisy that it had to be removed. The theatre was not a success, and closed in 1849, reopening on 30 Jan. 1850 as the Royal Soho. After a further period as the New English Opera House it again became a school of acting, and on 12 Nov. 1861 reopened as the New Royalty with a revival of a MELODRAMA by Almar, *Attar Gull; or, the Serpent of the Jungle*, in which the 13-year-old Ellen TERRY played Clementine. On 25 Mar. 1875 a one-act musical farce *Trial by Jury*, put on as a stop-gap, brought GILBERT and Sullivan together for the first time. In 1882 the theatre was partially reconstructed, reopening on 23 Apr. 1883. The Independent Stage Society produced IBSEN's *Ghosts* there in 1891, and Bernard SHAW's first play, *Widowers' Houses*, in 1892. At the end of that year Brandon THOMAS's farce *Charley's Aunt* opened on 21 Dec. before moving to the GLOBE THEATRE a month later. Ibsen's *A Doll's House* (1893) and *The Wild Duck* (1894) maintained the theatre's reputation as a home of modern drama, and between these two productions William POEL mounted his seminal 'Elizabethan' production of *Measure for Measure*. The Royalty was reconstructed again in 1895, in 1906, and in 1911, and between these dates a considerable number of well known managements came and went. In 1900 Mrs Patrick CAMPBELL revived SUDERMANN's *Magda* and MAETERLINCK's *Pelléas and Mélisande*. She also gave the first English production of BJØRNSON's *Beyond Human Power*. Under VEDRENNE, who took over in 1911, came GALSWORTHY's *The Pigeon* and BENNETT and KNOBLOCK's *Milestones* (both 1912). In 1919 came the first performance of MAUGHAM's *Caesar's Wife*,

with Fay COMPTON, and two years later *The Co-Optimists*, with Davy Burnaby, Laddie Cliff, and Phyllis Monkman, began a successful run. In 1924 Noël COWARD's *The Vortex* was transferred to the Royalty from the EVERYMAN THEATRE, and in 1925 Sean O'CASEY had his first London success with *Juno and the Paycock*. John DRINKWATER's *Bird in Hand* (1928) had a long run, and two comedies, *While Parents Sleep* (1932) by Anthony Kimmins and BRIDIE's *Storm in a Teacup* (1936), opened here before moving to the GARRICK and the HAYMARKET respectively. On 22 Sept. 1937 PRIESTLEY's *I Have Been Here Before* opened and gave the theatre its last success before it closed on 25 Nov. 1938. It was badly damaged by bombing during the Second World War, and was demolished in 1955.

(3) In Portugal Street, Kingsway. Built on the site of the STOLL THEATRE, this theatre has a two-tier auditorium holding 997. It opened on 23 June 1960 with DÜRRENMATT's *The Visit* starring the LUNTS. After the transfer to WYNDHAM's of William Gibson's *The Miracle Worker*, which had opened on 9 Mar. 1961, it became a cinema, but it reopened as a live theatre on 2 Apr. 1970, the erotic revue *Oh! Calcutta* being transferred there from the ROUND HOUSE. It occupied the theatre until Jan. 1974 before moving on to the DUCHESS. In 1977 the Royalty housed the successful musical *Bubbling Brown Sugar*, but it later had little success, and in 1981 became a television studio.

ROYAL VICTORIA HALL AND COFFEE TAVERN, see OLD VIC THEATRE.

ROZEWICZ, TADEUSZ, see POLAND.

ROZOV, VICTOR SERGEYEVICH (1913–), Soviet dramatist, who in 1938 became an actor and director at the MAYAKOVSKY THEATRE in Moscow. He began writing plays after the Second World War, mainly on the problems confronting young people, and soon became one of the most prolific and popular playwrights of the U.S.S.R. His works have also found a responsive public abroad, especially in socialist countries. One of his best known plays, *Alive For Ever*, was the opening production in 1957 of the group that became the SOVREMENNIK THEATRE, and provided the scenario for the film *The Cranes Are Flying*. It was translated into English by F. D. Reeve and published in a volume of *Contemporary Russian Drama* (1968), with an introduction by Rozov himself. Among his other plays are *Unequal Combat* (1960), *On the Day of the Wedding* (1964), and *In Search of Happiness* (1957); *Brother Alyosha* (1972) was an adaptation of DOSTOEVSKY's novel *The Brothers Karamazov*.

RUCELLAI, GIOVANNI (1475–1525), Italian dramatist, whose *Oreste* (1514) and *Rosmunda* (1516) are tragedies constructed on the Greek classical model, as laid down by TRISSINO. The first, on a classical theme, owes something to the *Iphigenia in Tauris* of EURIPIDES, but in the second Rucellai followed the new fashion of taking his subject from the recent past.

RUEDA, LOPE DE (*c*. 1505–65), Spain's first actor-manager and popular dramatist, whose plays were performed by his own company in squares and courtyards throughout the country, as well as in palaces and great houses. He was much admired by CERVANTES, particularly for his playing of comic rascals and fools, and his reputation led the great COMMEDIA DELL'ARTE actor GANESSA to visit Spain with his troupe in 1574. Rueda's dialogue, mainly in prose, is natural, easy, and idiomatic, with a strong sense of the ridiculous and a happy satirizing of the manners of his day. He was the originator of the PASO, or comic interlude, of which the best is *Las aceitunas* (*The Olives*). Two of his plays are based on Italian originals which were also used by Shakespeare—*Eufemia*, like *Cymbeline*, deriving from Boccaccio, and *Los engañados*, like *Twelfth Night*, from the anonymous *Gl'Ingannti* (1531).

RUF, JACOB, see SWITZERLAND.

RUGGERI, RUGGERO (1871–1953), Italian actor, who made his first appearance in 1888 and from 1891 to 1899 was with Novelli's company, where he laid the foundations of his future career, retaining until the end the taste he then acquired for strong dramatic and romantic parts. He was a fine if somewhat old-fashioned Hamlet and an excellent Iago in *Othello*, and in 1904 he scored a great success as Aligi in D'ANNUNZIO's *La figlia di Jorio*. He was for some years leading man in the company run by PIRANDELLO, in whose *Enrico IV* he appeared in London shortly before his death and in whose *Sei personaggi in cerca d'autore* he was much admired. He played opposite most of the outstanding actresses of his day, particularly Emma GRAMATICA. In his later years he was hostile to modern trends in the theatre, disliking the plays of BRECHT and considering T. S. ELIOT the best contemporary dramatist.

RUGGLE, GEORGE (1575–1622), scholar of Clare Hall, Cambridge, who is believed to have written two comedies performed by his fellow students in about 1600. In 1615 his *Ignoramus*, a satire on lawyers in Latin with some English, was performed before James I, who was so delighted with it that he returned to Cambridge a week later to see it again. It was largely based on DELLA PORTA's *La Trappolaria* (1596), but the chief part is a satirical portrait of the then Cambridge Recorder, Francis Brackyn. An English translation of Ruggle's play, as *Ignoramus; or, the Academical Lawyer*, by Ferdinando Parkhurst, was acted at Court in 1662 before Charles II. Edward RAVENS-

CROFT also made an adaptation of it, performed at DRURY LANE in 1677 as *The English Lawyer*.

RUIZ DE ALARCÓN Y MENDOZA, JUAN (1580–1639), Spanish dramatist of the Golden Age, a hunchback embittered by the ridicule levelled at his deformity, and against which he reacted by making the admirable hero of *Las paredes oyen* (*Walls Have Ears*, 1617) similarly deformed. Alarcón, who was born in Mexico City, went to Spain in 1600 and there attained high political office. His plays are satirical, well-constructed, and contain some excellent character drawing, though he lacked the verbal brilliance and fertile imagination of his rival and enemy Lope de VEGA. They were more popular outside his own country, especially in France, where CORNEILLE adapted *La verdad sospechosa* (*Truth Itself Suspect*) as *Le Menteur* (1643). This in its turn led to STEELE's *The Lying Lover; or, the Ladies' Friendship* (1703) on which FOOTE based *The Liar* (1762), and GOLDONI too was indebted to Alarcón's play for his own *Il bugiardo* (1750). Another of Alarcón's plays, *El examen de maridos* (pub. 1630), originally attributed to Lope de Vega, was based on the same Italian source as Shakespeare's *The Merchant of Venice*.

RUN, the total number of consecutive performances of one play, usually at one theatre. When audiences were small and mainly local all theatres worked on the REPERTORY system and changed their bill practically every night. A play which was successful on its first appearance would be retained in the repertory; an unsuccessful one might be given again once or twice and then dropped. In the 19th century, with the vast increase in the number of potential theatregoers and the improvement in transport, plays began to remain in a theatre for as long as the actor-managers were willing to appear in them, and later for as long as the audiences would pay to see them, with the result that one successful production could tie up a theatre and a number of actors over a period of years. In London Agatha CHRISTIE's *The Mousetrap* opened on 25 Nov. 1952 and was still running 30 years later, having moved from the AMBASSADORS to the ST MARTIN'S. The WHITEHALL, with a series of successful farces from 1954 to 1966, had only four plays in 12 years; and five American musicals, from *Oklahoma!* in 1947 to *My Fair Lady* in 1958, accounted for 16 years in the life of DRURY LANE.

In New York the musical *The Fantasticks*, which opened at the Sullivan Street Playhouse on 3 May 1960, was still running in the 1980s, and the musicals *Fiddler on the Roof* and *Grease* each ran for eight years.

RUNDHORIZONT, see CYCLORAMA and LIGHTING.

RURAL DIONYSIA, ancient Greek festivals held in honour of the god DIONYSUS. These preceded the Athenian festivals—the LENAEA and the City DIONYSIA—and were held in December in each Attic township, or *deme*. Little is known of the scope and local organization of such festivals, though dramatic contests certainly figured in them and original plays, perhaps by apprentice or little known poets, were produced. Later on revivals of the plays which had been presented in Athens were also seen at the Rural Dionysia.

RUSHTON, LUCY, see NEW YORK THEATRE (1).

RUSIÑOL I PRATS, SANTIAGO (1861–1931), Spanish novelist, painter, and dramatist, who with Àngel GUIMERÀ was at the centre of the Catalan revival of letters at the end of the 19th century usually referred to as the *modernista* movement. His finest plays are *L'alegria que passa* (*Happiness which passes*, 1891) and *L'auca del senyor Esteve* (*Mr Esteve's Broadsheet*, 1917).

RUSSELL, FRED, see MUSIC-HALL.

RUSSELL, GEORGE WILLIAM, see AE.

RUSSELL, LILLIAN, see BURLESQUE, AMERICAN, and VAUDEVILLE, AMERICAN.

RUSSELL, SOL SMITH (1848–1902), American actor, nephew of the pioneer manager Sol SMITH, who made his début at the Defiance Theatre in Cairo, Illinois, at the age of 14. He became a popular comedian, was for a time in DALY's company, and also toured America with a company of his own, one of his most successful productions being E. E. Kidder's *A Poor Relation*. He was also good in *A Bachelor's Romance* by Martha Morton, but his finest part was probably Bob Acres in SHERIDAN's *The Rivals*.

RUSSELL, WILLY, see EVERYMAN THEATRE, LIVERPOOL, and WAREHOUSE.

RUSSELL STREET THEATRE, Melbourne, see MELBOURNE THEATRE COMPANY.

RUSSIA AND SOVIET UNION. Compared with most of Europe the Russian theatre had a late and difficult beginning, owing originally to the hostility of the Church and later to the iron hand of absolutism, the illiteracy of the bulk of the population, and the neglect of the Russian language among the nobility. As far back as the 11th century there were wandering troupes—the *skomorokhak*—which performed at fairs, weddings, holiday gatherings of all kinds, and at carnivals and such religious festivals as Christmas and Easter. These troupes included a variety of entertainers—clowns, buffoons, puppet-masters, jugglers, acrobats, ballad-singers—often accompanied by performing animals, particularly bears. Varying their programmes to suit their audiences, they danced, sang, tumbled, played musical instruments, and declaimed folk-ballads and folk-tales which often served as vehicles for indirect criticism of current abuses.

These nomadic entertainers never received from Church or State the support which might have enabled them to develop into a national theatre; indeed, in the 14th century the indifference of the authorities turned to intolerance and they were driven out of Muscovy (the then comparatively small area actually ruled from Moscow). They eventually settled in villages across the border to the north where their descendants kept alive by oral tradition, with all the embellishments and falsifications which this entailed, a popular form of entertainment consisting mainly of mime-plays and farces in which the puppet-theatre played a large part. In the 16th and 17th centuries the strolling players again took to the road and became a familiar sight at fairs and village festivals.

There is little information available regarding Tsarist amusements at this time. It seems likely that during the period of reaction under Boris Godunov (1598–1605) there were none, while the 'time of the troubles' after his death left little leisure for play-acting. With the establishment of the first Romanov, however, in 1613, a 'house of amusement' was built by royal command. The chief actors at this first Russian theatre were probably foreigners, perhaps German; but it is significant that at the wedding celebrations of the Tsar in 1626 the traditional church choir was replaced by a company of Russian strolling players, and this example was followed by the leading nobles of the Court. The actors' triumph was short-lived, for in 1648 a decree of the Tsar Alexis (1645–76) forbade all types of worldly amusement under threat of severe punishment and the strolling players disappeared once more, though again only temporarily, from the scene. Meanwhile, away from the Court, SCHOOL DRAMA was coming into being. After the 'time of the troubles' it was necessary for the Russian Church to defend itself, particularly against the Jesuits, and for this education and training in theology were essential. In 1633 the first Russian theological academy was founded on the lines of existing Jesuit schools in Europe, and Russian churchmen duly followed the example of JESUIT DRAMA in using the theatre as a means of instruction and propaganda. So began the development of written plays on Biblical and historical themes, many of them adapted from the Polish, but with interpolated scenes of daily life, a development assisted by the works of St DMITRI OF ROSTOV. As the theological students themselves took over the responsibility for dramatic entertainments, farce, always a potent social weapon, crept into the repertory, and in response to an increasing demand new plays by Russian authors made their appearance, one of the earliest known writers of such plays being Simeon POLOTSKY. The Tsar, reversing his former policy towards the theatre under the influence of his journeys in Poland and his Anglophile second wife Natalia Naryshkina, now sought to bring to his surroundings some of the brilliance of the contemporary European Courts, and the performance of a play was planned as part of the celebrations for the birth of an heir (later Peter the Great) in 1672. For actors the Court officials looked to the free German quarter of Moscow, where plays had not been banned, and from this source the authorities procured the services of a priest, Johann Gottfried GREGORY, who in 1673 was put in charge of a theatre school for the training of Russian actors which continued to flourish after his death. However, on the death of Alexis in 1676 it was closed, and when under Peter the Great (1689–1725) a more liberal atmosphere again prevailed, all the work had to be started afresh. Peter, who intended to use drama more for political purposes than for amusement, built a theatre for a new company under Johann Kunst, 'an eminent master of theatrical science' recruited from Danzig, in the Red Square, Moscow—the first public theatre in Russia—and issued a decree ordering everyone, particularly foreigners, to attend. The price of admission was the equivalent of a few pence, and according to contemporary records audiences averaged 124, though during the summer seasons it sometimes rose to 400. The death of Kunst less than a year after his arrival caused this venture to be abandoned in 1707, and drama, together with spectacular ballet and opera, became once more the prerogative of Court circles. The Tsar's sister had her own private playhouse, for which she herself wrote a number of plays. Some of these, based on old Russian legends, were later included in the repertory of the strolling players and of the student groups which gave amateur performances at holiday times. Meanwhile School Drama continued to flourish, and there are records of several plays which dealt allegorically with the political events of the day, mainly for the purpose of eulogizing Peter the Great and aiding his reforms. It even seemed for a time as if a national theatre might arise out of all this activity, particularly with the works of Feofan Prokopovich; but although a new spirit began to be apparent in School Drama when it introduced into its productions more scenes of comedy and more peasant characters, the dead weight of authority, the popularity of foreign actors, and the too-rigidly moral and educational purpose behind its inception, combined to stifle its aspirations. After a last splendid offering of panegyric plays for the coronation of the Empress Elizaveta Petrovna in 1742, Russian School Drama disappeared, but not without having made a contribution to the technique of writing and staging plays important in the light of future developments.

The impetus towards a national theatre was given by the plays of SUMAROKOV, a member of the Society of Lovers of Russian Literature, founded in connection with the Cadet College opened in 1732 for the education of the sons of the nobility. Though intended as a military academy, its curriculum included some training in cultural subjects, such as poetics, declamation, deportment, music, and dancing. Its students were thus admirably fitted to assist in Court festivities, and acquitted themselves well in Sumarokov's *Khorev* (1749) and other plays when summoned to play before the Empress. But neither they, nor the Court singers whom the Empress sent to the College for the express purpose of having them trained as actors, could sustain the burden of constant acting, and for most of the time the Court was forced to rely on the French players who had taken over from Carolina NEUBER (who had herself taken the place in 1740 of a COMMEDIA DELL'ARTE company) and who were to remain in Moscow for some 15 years. However, there was in Yaroslavl at this time an amateur Russian company formed in 1750 under the brothers Feodor and Gregori VOLKOV, which became so well known for its excellent performances in adapted School Dramas that in 1752 it was summoned to appear before the Empress. This it did with such success that the Volkov brothers, the leading actor DMITREVSKY, and several other members of the company were sent to the Cadet College for training. They appeared before the Court again in 1755 in Sumarokov's *Sinav and Truvor* and by 1757 were in receipt of a small subsidy, negligible compared with that given to the French players but at least a sign of official recognition. The cadets now ceased to appear in plays except for their own amusement and acting in Russian was left to the newly established professional company, which now included actresses, among them a fine player of tragic heroines, Tatiana Mikhailovna Troepolskaya (?–1774), who was particularly admired in the plays of Sumarokov. The company's repertory was necessarily restricted, relying on adaptations and imitations of French COMÉDIE-LARMOYANTE and German BÜRGERLICHES TRAUERSPIEL, by such authors as Lukin and Veryovkin, though some dramatists, among them Kheraskov and Maykov, continued to write in the neo-classical manner favoured by Sumarokov. There were also a number of comedies and light operas, many based on idealized scenes of village life, among them those of the Empress Catherine the Great (1729–96) herself. The only dramatist in this lighter vein to be remembered today is FONVIZIN, whose play *The Minor* (1782) remains in the repertory.

The emergence of a distinctively Russian acting tradition owed less than the development of playwriting to the foreign plays, and stemmed from the 'serf' theatres which developed towards the end of the 18th century, when it became the custom for the landed gentry to provide entertainment for relations and friends who visited their country estates (as Konstantin tries to do in CHEKHOV's *The Seagull*). Many of the best actors in these localized companies were serfs, and it was from among them that some of the finest actors of the rapidly expanding Russian professional theatre were eventually drawn. In the early years of the 19th century the only Russian dramatist of any importance was Vladislav Alexandrovich Ozerov (1770–1816), who wrote frigid historical plays much influenced by French neo-classicism; Fyodorov and Ilyin were among the best of those writing in a popular sentimental vein. Modern problems and the details of contemporary life had as yet no place on the official Russian stage, which was completely under the control of the Secret Police, even to the choice of play and the allotting of roles.

There were at this time two Imperial theatres in St Petersburg—the Bolshoi (large), used mainly for opera and ballet, and the Maly (little), used for plays, not to be confused with the Moscow theatres of the same names which still flourish. The Bolshoi in Moscow was burnt down at the time of the Napoleonic invasion in 1812, rebuilt in 1824, and subsequently much altered after a disastrous fire in 1853. Like the St Petersburg Bolshoi it specializes in opera and ballet, plays being the usual offerings of the Moscow MALY.

Two popular but otherwise unimportant dramatists who emerged in the first half of the 19th century were Nicholas Alekseyevich Polevoy (1796–1846) and Nestor Vasilyevich Kukolnik (1809–68). The former was a merchant of Siberia who became a literary critic, dared to criticize the plays of Kukolnik, and had his paper suppressed as a result. He then turned his attention to writing plays, and the first of his many patriotic dramas was produced in 1837. Kukolnik, who began his playwriting career in 1833 with an innocuous play on the life of the Italian poet Tasso, had his first taste of success with a stirring patriotic drama dealing with the accession of the first Romanov. Fired by its enthusiastic reception, he continued to turn out a series of dramas which gave the chief characters plenty of opportunities for asserting their loyalty to the existing régime. Apart from the works of these two dramatists, which earned the approbation of Nicholas I (1825–55), the staple fare was melodrama, both home-made and imported, tolerated, it was said later, because it provided 'an emotional lightning-conductor which grounded the energy of social protest'. In spite of the Tsar's efforts to suppress unfavourable criticism, such men as the redoubtable Vissarion Grigorievich Belinsky (1811–48) savagely attacked such contemporary writings. He included the actors in his censure, though by now outstanding Russian players were beginning to appear. The traditions begun by the Volkovs and Dmitrevsky were worthily upheld, first by the great actor SHCHEPKIN, the man who first raised the status of the actor in Russia from serfdom to freedom, and then by two men of very different temperaments—MOCHALOV, an emotional, romantic actor who appeared mainly at the Maly in Moscow, and KARATYGIN, an ideal exponent of neo-classical tragedy and the leader of the company at the Alexandrinsky (now the PUSHKIN) Theatre founded in St Petersburg in 1824. Other notable players included SEMENOVA, SHUSHERIN, and SOSNITSKI.

By this time the serf-theatres were in decline, the loss of their fortunes causing many of their former patrons to transport their private companies to the nearest town and open public theatres there with the idea of making money. Other companies were taken over by rich merchants as commercial propositions. Their standards were low, and since all came under the jurisdiction of the Secret Police they could produce only plays permitted by the authorities. Apart from melodrama and frigid neo-classical tragedy, the chief theatrical fare of the time was the increasingly popular VAUDEVILLE, imported from France in about 1840 and tolerated by the authorities since it was devoid of all suspicion of social criticism and relied mainly on singing, dancing, and witty but inoffensive dialogue. The reputed founder of the genre in Russia was Alexander Alexandrovich Shakhovsky (1777–1846), whose first vaudeville was seen at the Hermitage Theatre in 1795; but more eminent writers, even the great PUSHKIN himself, did not hesitate to write *en vaudeville*. A notable exception was GOGOL, who never forgave the actor at the Alexandrinsky Theatre who played Khlestakov in *The Government Inspector* as a vaudeville character.

It would seem impossible for a free national drama to flourish in such an atmosphere, yet the 19th century, from the defeat of Napoleon in 1812 onwards, saw the emergence of the great dramatists of Imperial Russia—Pushkin, Gogol, Chekhov, TURGENEV, TOLSTOY, and GORKY. The Napoleonic invasion had raised the Russian people to unprecedented heights of patriotism, and proved a vital inspiration to many writers. Even the nobility woke up to the emergence of a national consciousness and Pushkin, the first great writer to use Russian as a literary medium, was universally acclaimed; his works, important in themselves, also provided a storehouse of plots on which his contemporaries and successors drew for themes. Both he and his contemporary LERMONTOV represented the educated lesser nobility who had suffered under the oppression of the Court and the higher bureaucracy, and were in sympathy with the aspirations of the people. Their plays—tragedies in verse—were implicit criticisms of the aristocracy, which GRIBOYEDOV was in the same period satirizing in his comedies. For the first time Russian dramatists were writing about living people, and were concerned to present the truth about the existing order. Although they all died young, they had prepared the way for Gogol and his actor friend Shchepkin to effect the reform of the Russian stage, though matters were delayed by the period of reaction which followed the Decembrist revolt of 1825 and only ended with the accession of Alexander II (1855–81), on the whole the most liberal of the Tsars. Under him the nobility began to give place to an emerging bourgeoisie, from which arose an intelligentsia (many of them recruited from the ranks of the lesser nobility) supported by the wealthy merchants, who were for the most part of humble origin and felt themselves despised by the more reactionary among the aristocracy. The changes in social life which this entailed were reflected in the theatre in the works of OSTROVSKY, whose first play, produced in 1853, startled the audience by its simple and photographically exact presentation. Thus, for the first time a heroine appeared in a cotton frock, with naturally smooth hair, instead of the hitherto obligatory silk dress and French coiffure. Shchepkin, the moving spirit behind these reforms, was helped by some of the younger actors, notably MARTYNOV and Prov SADOVSKY. In spite of the advances that were made and an occasional excellent play such as Turgenev's *A Month in the Country*, first produced in 1861 many years after it was written, the Russian theatre still had a long way to go before it could deal adequately with the works of one of Russia's greatest dramatists, Anton Chekhov. One important event was the visit of the MEININGER COMPANY to Russia in the 1880s, another the influence of Leo Tolstoy, whose finest plays were contemporary with Chekhov's early works. The failure of the Alexandrinsky Theatre to come to terms with *The Seagull* in 1896 made plain the incompetence of the old Imperial theatre and showed the need for a new approach to acting and production based on truth and REALISM. Luckily this was to come almost at once. The ending of the Imperial theatres' monopoly in Moscow, which had already resulted in the founding of the KORSH THEATRE, in 1882, was followed by the establishment in 1889 under STANISLAVSKY and NEMIROVICH-DANCHENKO of the MOSCOW ART THEATRE, where Chekhov and other writers of the European naturalistic movement found their ideal interpreters. Important actors with the company in its formative years included Olga KNIPPER, who later married Chekhov, Maria GERMANOVA, MOSKVIN, KACHALOV, and LEONIDOV; at the Moscow Maly Maria YERMOLOVA was the leading tragic actress, while Pavel ORLENEV and Alla NAZIMOVA made their reputations in St Petersburg.

Although the dramatists Leonid ANDREYEV, Alexandr BLOK, and Nikolai EVREINOV were producing good and important work at this time, the last outstanding dramatist to appear before the Revolution was Maxim Gorky, who, like the Moscow Art Theatre, helped to form a bridge between the old and the new, carrying into the Soviet theatre the fruits of many years of labour and reform. Prophet of the Revolution and one of its builders, Gorky represented the new element in the theatre as in the nation—the proletariat. The Soviet régime which came to power following the Revolution of October 1917 had the intelligence to take over the best of what it inherited and immensely fostered its expansion. The mere physical development of the theatre under Soviet rule was enormous. Before the outbreak of the Second World War there were nearly 800 professional theatres in the U.S.S.R., of which 410 staged plays in non-Russian languages.

The enlightened outlook of the first Commissar for Education, A. V. LUNACHARSKY, ensured the survival, with full government support, of the Moscow Art Theatre, as well as the Maly and the newly formed KAMERNY THEATRE under the innovatory director TAÏROV, while in Petrograd (formerly St Petersburg and later to become Leningrad) Theatres for the People, an experiment in popular theatre which dated from the turn of the century, continued to flourish. Two of the studios of the Moscow Art Theatre achieved independence in 1921 as the REALISTIC THEATRE and the VAKHTANGOV THEATRE. Already in 1918 the Organization for Cultural Enlightenment had opened the First Workers' Theatre in Moscow, which as the Proletkult, now the TRADES UNIONS THEATRE, was notable for housing the early productions of Eisenstein. Other important Soviet theatres in Moscow are the RED ARMY THEATRE, opened in 1919; the Sohn (briefly the MEYERHOLD theatre, after that director had established his workshop there) in 1920; the Theatre of the Revolution (later the MAYAKOVSKY THEATRE), where A. D. POPOV worked, in 1922; the MOSSOVIET (as the Trades Unions Theatre) in 1923; the Theatre of SATIRE in 1924; the Theatre of Working Youth (later the LENKOM) in 1927; and the YERMOLOVA THEATRE in 1930. The MOSCOW STATE JEWISH THEATRE under GRANOVSKY also moved to the capital after its formation in Leningrad in 1919, while other theatres established in Leningrad included the Grand GORKY THEATRE in 1919, a theatre for children, the Theatre of Young Spectators, in 1921, and the Theatre of Comedy, which housed some of the outstanding productions of AKIMOV, in 1929.

The first decade after the Revolution saw also a flowering of new playwriting talent, more impressive for quantity than for quality. Early Soviet dramatists included FAIKO, TRETYAKOV, BILL-BELOTSERKOVSKY, TRENEV, ROMASHOV, BULGAKOV, and IVANOV, whose *Armoured Train 14–69* (1927) was an outstanding example of the naturalistic treatment of recent Russian history prevalent in dramatic writings of the time. The work of MAYAKOVSKY—whose *Mystery-Bouffe* (1918) is usually considered the first Soviet play—represented a new vein of formal experiment with which Meyerhold and his disciples, among them RADLOV, remained associated. Directors such as DIKIE and GORCHAKOV continued to follow the teachings of Stanislavsky, while Michael CHEKHOV and A. D. Popov worked in the style developed by VAKHTANGOV. The director OKHLOPKOV was particularly associated with the new art of staging mass spectacles for which LYUBIMOV-LANSKOY was also noted. KURBAS and MARDZHANOV were important for their work outside the main centres, Kurbas in the Ukraine and Mardzhanov in Georgia. Unfortunately the post-Revolutionary spate of activity was all too brief. In 1932 the call by the new cultural overlord Zhdanov for SOCIALIST REALISM effectively ended the explorations begun during the previous decade into new ways of bringing the theatre to a socialist society, whether by re-interpretations of Russian and foreign classics or by CONSTRUCTIVISM and FORMALISM, the latter in particular becoming a term of abuse during the 1930s and 1940s for any work which deviated from the requirements of officialdom. The period of Stalinist rule thus witnessed a continuing expansion of the theatre materially but an all-pervasive control over its tone and content. The most successful dramatists were those who like KATAYEV and ARBUZOV concentrated on lightweight treatment of personal relationships; or like POGODIN wrote patriotic dramas in praise of Lenin; or like V. M. Kirshon (1902–38) treated aspects of the first five-year plan. Others who were active during this period included AFINOGENOV, LEONOV, and, in adaptations of his novels, N. Y. Virta. In *The Optimistic Tragedy* (1933) VISHNEVSKY successfully dramatized one of the formal paradoxes of Socialist Realism, as implied in his title.

The Second World War provided much more acceptable material for the theatres of the day, and some notable plays emerged, among them those of KORNEICHUK, SIMONOV, ROZOV, and SOFRONOV. The cooling of Soviet relations with the West in the post-war period also produced, under the continuing tutelage of Zhdanov, a spate of negatively patriotic plays and earnest works on industrial or agricultural problems. Only after the death of Stalin in 1953 did relative freedom return to the theatre, and the influence of Meyerhold again began to be felt, even before his complete rehabilitation, notably in the work of the veteran Okhlopkov. The system of allocating plays to theatres through a centralized Arts Council was abandoned in 1954, and dramatists who made good use of the new political mood included A. P. Shteyn and VOLODIN. Such new directors as YEFREMOV at the recently established SOVREMENNIK THEATRE and TOVSTONOGOV at the Grand Gorky Theatre led the way in introducing new foreign dramatists to Soviet audiences, while in 1964 the new Theatre of Drama and Comedy, familiarly known as the TAGANKA THEATRE, opened under LYUBIMOV. The Theatre of Satire, directed from 1957 by PLUCHEK, and the Mayakovsky, from 1967 by Goncharov, gave rise to a degree of fruitful conflict between those more experimental directors who took up where Meyerhold had been forced to leave off and those who still followed the naturalistic tradition of Stanislavsky. The enormous theatrical output of the U.S.S.R. and, paradoxically, the professional security of workers in the theatre, has meant that much of the material presented, particularly outside the main centres, continues to be at best mediocre, but firm foundations have been laid for the future broadening of the theatre's range, and for its extension beyond the boundaries of the great cities.

Russia is of course only one of the Soviet Socialist Republics, and as early as 1933 the Institute of Theatre Art founded by Lunacharsky was extending its training facilities to actors of the non-

Russian-speaking republics of the Union. The groups founded then were known as National Theatre Workshops, and the first fully trained graduates came from them in 1936. Five years later four National Studios were established—in Turkmenian, Tadzhik, Kirghiz, and Adygei. For the first time audiences in these regions were able to attend plays by such authors as Shakespeare, MOLIÈRE, GOLDONI, and BEAUMARCHAIS, as well as Russian plays, in their own languages. Now most of the nationalities which make up the Soviet Union have a studio for training actors in Moscow or Leningrad, and some have set up similar training schools in their own capitals. Among the National Republics of the Soviet Union a number have an interesting theatrical history, and continue to develop their own theatrical culture, notably Byelorussia, GEORGIA, UZBEKISTAN, and even the remote Arctic region of Yakutia.

RUTEBEUF, 13th-century Parisian poet and JONGLEUR, of whose life little is known, in spite of the bitter and witty poems on his poverty and marriage which have survived. His work shows great versatility, lyrical power, and satirical force, his satire being directed against almost everyone. His short play *Le Miracle de Théophile*, dramatizes a FAUST-like legend of a priest who sells his soul to the devil in return for ecclesiastical preferment, but repents and is saved by the intervention of the Virgin Mary. It is the first surviving MIRACLE PLAY of its kind, and besides its formal beauty and the intricate imagery of Théophile's dramatic monologues, it is remarkable for the figure of a Jewish-Mohammedan wizard who is able to conjure up the devil by speaking stage gibberish. It was revived in a modern adaptation in 1930 by Gustave COHEN for production by his students at the Sorbonne, who took from it their name of Les Théophiliens.

RUTHERFORD, MARGARET (1892–1972), English actress, appointed DBE in 1967, who was late in starting her stage career, being in her early thirties when a small legacy enabled her to give up teaching elocution and the piano. She made her first appearance in pantomime at the OLD VIC in 1925, under Robert ATKINS, subsequently playing several small parts without much success. The turning-point in her career came when she played the formidable and extremely eccentric Bijou Furze in *Spring Meeting* (1938) by M. J. Farrell and John Perry. She was immediately recognized as an outstanding comédienne, and in the following year played Miss Prism in WILDE's *The Importance of Being Earnest*. After a serious role as Mrs Danvers, the housekeeper in Daphne Du Maurier's *Rebecca* (1940), she achieved a great success as the medium Madame Arcati in COWARD's *Blithe Spirit* (1941). She then made her New York début as Lady Bracknell in *The Importance of Being Earnest* (1947), and returned to London to appear as Miss Whitchurch in John Dighton's farce *The Happiest Days of Your Life* (1948); Madame Desmortes (in a bath-chair) in ANOUILH's *Ring Round the Moon* (1950); and Lady Wishfort in CONGREVE's *The Way of the World* in 1953, in which she gave a masterly impression of an 'old peeled wall'. She was then seen in Anouilh's *Time Remembered* (1954) and, after an Australian tour, in Rodney Ackland's adaptation of John Vari's *Farewell, Farewell, Eugene* (1959; N.Y., 1960). She played Bijou again in *Dazzling Prospect* (1961), the sequel to *Spring Meeting*, and was then seen as two of SHERIDAN's ladies, Mrs Candour in *The School for Scandal* in 1962 and Mrs Malaprop in *The Rivals* (her last appearance on the stage) in 1966. At her best in comedy, she could convey pathos and also inject a certain sinister element into some of her personifications of unpredictable old ladies.

RUZZANTE, Il, see BEOLCO, ANGELO.

RYDEL, ŁUCJAN, see POLAND.

RYGA, GEORGE (1932–), Canadian dramatist, author of a number of 'ballad-plays' which incorporate elements of folk-song and choreography. The best known, *The Ecstasy of Rita Joe* (1967), a tragedy dealing with the death of a local Indian girl, was first seen at the Vancouver Playhouse, and has become something of a Canadian classic. Ryga's other plays include *Indian* (1964), based on an earlier television play; *Nothing But a Man* (1966), on the subject of the Spanish playwright GARCÍA LORCA; *Grass and Wild Strawberries* (1968), described as 'a multi-media rock musical which attempts to bridge the generation gap between the drug culture and the establishment'; and *Captives of the Faceless Drummer* (1971), an exploration of urban violence which leads to the kidnapping of a diplomat. The withdrawal of this last play by the directors of the Vancouver Playhouse after it had aroused a considerable amount of controversy highlighted Canada's need for 'alternative' experimental theatres and materially assisted the development of new Canadian playwrighting during the 1970s. Ryga's later plays, *Sunrise on Sarah* (1972) and *Portrait of Angelica* (1973), both first seen at the Banff School of Fine Arts, seem to have aroused less interest.

RYLANDS, GEORGE, see CAMBRIDGE.

S

SAAVEDRA REMÍREZ DE BAQUEDANO, ÁNGEL, Duke of Rivas (1791–1865), Spanish poet and playwright, who, under the influence of Victor HUGO, abandoned his former neo-classical form and wrote the first, and possibly the finest, Spanish Romantic drama, *Don Álvaro, o la fuerza del sino* (1835). Apart from its emotional range, it is noteworthy for the author's use of local colour, and his detailed descriptions of the stage sets. It later provided Piave with the libretto of Verdi's opera *La Forza del destino* (1862). It is the only one of Rivas's plays to remain in the repertory; his later plays were costume dramas in the style of DUMAS *père*.

SABBATTINI, NICOLA, see LIGHTING.

SACCHI, ANTONIO (1708–88), Italian actor of the COMMEDIA DELL'ARTE, who played the part of Truffaldino. A player of great skill and intelligence, he had his own company, with which he alternated improvised and written comedy. He was much admired by GOZZI and GOLDONI, who both wrote plays for him, and he was the first to play the title role in the latter's *Il servitore di due padrone* (*The Servant of Two Masters*, 1746). Although mainly active in Venice, he toured widely in France, Bavaria, and Portugal, and in 1742 was in Russia.

SACHS, HANS (1494–1576), German dramatist, a cobbler by trade, best known as the hero of Wagner's opera *Die Meistersinger von Nürnberg* (1868). From about 1518 onwards he poured out a stream of plays on subjects taken from the Bible, folk legend, folk-lore, history, and the classics, all treated in very much the same way for the purpose of moral betterment—he was one of the first writers of his day to support Martin Luther. Many of them are tragedies, some comedies—that is, plays with happy endings, though not necessarily humorous; but his fame rests chiefly on his Carnival Plays (see FASTNACHTSSPIEL), of which he wrote about 200, notable for their vivid folk-pictures of daily life and their homespun humour. The dialogue is unforced and the rough-and-ready verse not unpleasing, while the horseplay inseparable from Shrovetide is kept under control. Sachs trained his actors and directed his plays himself, using for his productions a disused church which thus became Germany's first theatre.

SACKLER, HARRY, see JEWISH DRAMA.

SACKLER, HOWARD, see UNITED STATES OF AMERICA.

SACKVILLE, THOMAS, 1st Earl of Dorset, see NORTON, THOMAS.

SACKVILLE, THOMAS, see ENGLISH COMEDIANS and WOLFENBÜTTEL.

SACRA RAPPRESENTAZIONE (pl. *sacre rappresentazioni*), the religious play of 15th-century Italy, which, like the Spanish AUTO SACRAMENTAL and the English and French MYSTERY PLAY, developed from the LITURGICAL DRAMA, but with the addition of the LAUDA. It was chiefly the product of Florence, where it was produced not, as in England, by craft guilds or, as in France, by specially organized adult lay bodies, but by groups of young people from religious or educational centres, who enacted scenes from the Bible or Church history either indoors, in a hall, or in the open square or marketplace, in a MULTIPLE SETTING. One play consisting of a cycle of episodes from the Old and New Testaments required no fewer than 22 booths or *edifizi*, indicating different localities; and a play on the life of Saint Olive, based on a popular legend, took two days to perform. As elsewhere, the over-elaboration of comic interpolations and the gradual introduction of secular and often impious elements led to the decline of the *sacra rappresentazione*, which in the end could not hold its own against the pressure of the newly discovered classical forms; but the traditions which had governed its production, particularly its free-flowing style of composition and its mingling of comedy and tragedy, were still strongly felt even by those late 15th-century dramatists whose plays were being written under the influence of the tragedies of SENECA, as can be seen in the *Timon* (1497) and *Sofonisba* (1502) of Galleotto del Carretto (?-*c.* 1530) and the *Filostrato e Panfile* (1499) of Antonio Caminelli (1436–1502), known as 'Il Pistoia'.

SADDLE-IRONS, see BOX-SET.

SÁ DE MIRANDA, FRANCISCO DE, see MIRANDA, F. SÁ DE.

SADLER'S WELLS THEATRE, London, in Rosebery Avenue, Finsbury. In 1683 Thomas Sadler discovered a medicinal spring in his garden and established there a popular pleasure-garden which became known as Sadler's Wells. Two years later he built a wooden music-room to house concerts, which in 1699 was run by a Mr Miles who renamed it Miles's Musick House. Miles died in 1724, and although the Wells continued to function it rather lost its reputation until in 1746 a local builder, Thomas Rosoman, took it over and restored its fortunes. In 1753 he engaged a regular resident company of actors, and the old 'musick house' became a theatre. A version of *The Tempest*—probably DRYDEN's—was performed there in 1764, and in the following year Rosoman demolished the old wooden building, replacing it with a stone one. This opened with a mixed bill on 8 Apr. and Rosoman continued to run it successfully until his retirement in 1772. Under Tom KING of DRURY LANE who succeeded him, the theatre prospered, attracting fashionable audiences. In 1781 GRIMALDI made his first appearance there at the age of three, and in 1801 Edmund KEAN, as 'Master Carey', aged about 13, was also in the company. In 1804 Charles DIBDIN, who had been associated with the Wells for some time and took over when King died, yielded to the fashionable craze for AQUATIC DRAMA and installed a large tank on stage, filled with water from the New River, for the production of mimic sea battles. As the Aquatic Theatre, the renovated Wells opened with *The Siege of Gibraltar*, complete with naval bombardment, but soon returned to its former name. A false alarm of fire in 1807 caused a panic which resulted in the death of 20 people. For some years afterwards the theatre was in low water, but recovered, and on 17 Mar. 1828 Grimaldi, who had been closely associated with it for many years, returned to make his farewell appearance, Tom MATTHEWS succeeding him as Clown. Following the breaking of the monopoly of the PATENT THEATRES in 1843, the theatre was let to Samuel PHELPS, who on 27 May 1844 inaugurated with *Macbeth* a long series of productions which led him, before his retirement in 1862, to stage no fewer than 34 plays by Shakespeare. After this the Wells reverted to mixed popular entertainment, being used as a skating rink and for prize-fighting before being closed in 1878 as a dangerous structure. A year later BATEMAN's widow, who had taken over the LYCEUM on her husband's death, moved to Sadler's Wells, had the interior reconstructed, and reopened it on 9 Oct. with her daughter Isabel as the leading lady of a good company. However, attempts to revive the theatre's reputation were only partially successful, and Mrs Bateman died in 1881 heavily in debt. The theatre then became a local home for MELODRAMA, and in 1893 a music-hall, before finally closing down in 1906. Twenty-one years later Lilian BAYLIS took over the derelict building and erected a new theatre intended as a North London counterpart to the OLD VIC in South London. This opened on 6 Jan. 1931 with *Twelfth Night*, in which GIELGUD played Malvolio. It had a seating capacity of 1,650 (later reduced to 1,499) in three tiers, and a proscenium width and stage depth of 30 ft each. The original policy of alternating productions between Sadler's Wells and the Old Vic proving uneconomic, it was decided in 1934 to make the latter the permanent home of drama while the Wells became the home of the opera and ballet companies. It closed in 1940 and a year later suffered damage by enemy action, not reopening until 7 June 1945. After the departure of the ballet company to COVENT GARDEN the opera company carried on alone, though the theatre was sometimes used by visiting companies, including in 1958 the MOSCOW ART THEATRE who were seen in three plays by CHEKHOV—*The Cherry Orchard* on 15 May, *Three Sisters* on 16 May, *Uncle Vanya* on 20 May—and a new play by Leonid Rakhmanov, *The Troubled Past*, on 21 May. In 1968 the Sadler's Wells opera company moved to the COLISEUM, becoming the English National Opera, and the theatre has since been used by a number of distinguished visiting companies, both from Britain and overseas.

SADOVSKY [Yermilov], PROV MIKHAILOVICH (1818–72), Russian actor, who first appeared on the provincial stage in Tula at the age of 14 and in 1839 joined the company at the MALY THEATRE in Moscow, coming to the fore in OSTROVSKY's plays, of whose comic roles he proved to be the ideal interpreter. Unlike KARATYGIN and MOCHALOV, he followed SHCHEPKIN in a realistic approach to his art and was largely responsible for keeping Ostrovsky's plays in the repertory until they won public acceptance.

SAFETY CURTAIN, see CURTAIN.

SAINETE, name given in 17th-century Spain to a short comic scene played between the acts of a serious full-length play. Originally it resembled the PASO and the ENTREMÉS, but later differed from them by its greater use of music, its diversity of scenes and characters from city life, and a hint of social criticism in its racy dialogue. The great writer of *sainetes* in the 18th century was Ramón de la CRUZ, but the genre lives on, and can be found in the works of such modern writers as Carlos Arniches and the ÁLVAREZ QUINTERO brothers. With the one-act ZARZUELA, the *sainete* forms part of the GENERO CHICO.

ST ALBANS, Hertfordshire, see VERULAMIUM.

SAINT-DENIS, MICHEL JACQUES (1897–1971), French actor and director, who began his career under his uncle Jacques COPEAU at the VIEUX-COLOMBIER. In 1930 he founded the Compagnie des Quinze, for which he directed *Noé, Le Viol de Lucrèce*, and *La Bataille de la Marne*, all by André OBEY, as well as a number of other plays, in all of which he appeared himself. The company achieved a great reputation, but was eventually forced to disband, and Saint-Denis settled in London, directing Obey's *Noé* in translation, with John GIELGUD as Noah, in 1935 and the Elizabethan tragi-comedy *The Witch of Edmonton* at the OLD VIC in 1936. He then founded the London Theatre Studio for the training of young actors. It had already made several interesting contributions to the English theatre, one of its graduates being Peter USTINOV, when the outbreak of the Second World War caused it to close down. After working with the French section of the B.B.C. during the war, as Jacques Duchesne, Saint-Denis returned to the theatre, becoming head of the short-lived drama school at the Old Vic. On its demise he returned to France to found and run the Centre Dramatique de l'Est, based on Strasbourg. In 1957 he was appointed artistic adviser to the VIVIAN BEAUMONT THEATRE project in New York in connection with the Lincoln Center, and in 1962 became general artistic adviser to the ROYAL SHAKESPEARE COMPANY, for whom he directed at the ALDWYCH THEATRE in 1965 BRECHT's *Squire Puntila and his Servant Matti*. As a frequent adjudicator of the Dominion Drama Festival in CANADA, he also helped in the foundation of the National Theatre School of Canada in Montreal in 1960, advising particularly on its basic curriculum and method of training.

SAINTE-BEUVE, CHARLES-AUGUSTIN (1804–69), one of the best of French critics, whose Monday articles, contributed regularly to a succession of journals—the *Constitutionnel*, the *Moniteur*, and the *Temps*—provide in their collected form as *Causeries du lundi, Nouveaux lundis, Premiers lundis* a panorama of the French theatre during his lifetime. He aspired to write plays and novels, but finally gave up creative work to become, as he himself said, 'merely a spectator, an analyst'. He was in his youth a friend of Victor HUGO, whose wife Adèle became his mistress, and of the actor MOLÉ, and little that went on in the theatre escaped him. Besides his Monday articles he was responsible for many books of theatrical and literary criticism.

SAINT-ÉTIENNE, Comédie de, see DÉCENTRALISATION DRAMATIQUE.

ST GEORGE, the patron saint of England, slayer of the Dragon, and usually the central figure in the English MUMMERS' PLAY. After the accession of George I the name is often found as King George. It seems likely that St George found his way into the Mummers' play by way of Richard Johnson's *Famous History of the Seven Champions of Christendom*, which appeared in 1596. He usually introduces himself as follows:

> In come I, Saint George,
> The man of courage bold...
> I fought the fiery dragon
> And drove him to the slaughter
> And by these means I won
> The King of Egypt's daughter.

The Haghios Gheorgios Plays of Thrace and Thessaly show striking similarities with the Mummers' Play and seem to link it to the spring FOLK FESTIVALS of primitive agricultural communities.

ST GEORGE'S THEATRE, London, in Islington. This was founded by the actor George Murcell in 1976 for the production of plays by Shakespeare only. The circular plan of the interior, built within a converted church, is based on that of the Elizabethan GLOBE THEATRE, and the stage has a permanent setting freshly designed annually for each season. As in Elizabethan times, music plays an integral part in the production, and the theatre has a group of musicians who have specialized in the music of Shakespeare's day. Alterations to the text are kept to a minimum. Current examination syllabuses are borne in mind when choosing the plays for the season, and special matinées are given for schools and colleges, together with morning 'workshops' on the plays. In the opening performance on 23 Apr. 1976 Eric PORTER played Malvolio in *Twelfth Night*, and in the same season Alan BADEL was seen as Richard III. Plays are performed in repertory.

SAINTHILL, LOUDON (1919–69), theatre designer, born in Tasmania, who worked for some years in Australia before moving to England. In 1951 he achieved an outstanding success with his designs for *The Tempest* at the SHAKESPEARE MEMORIAL THEATRE, where he was subsequently responsible for the décor of *Pericles* in 1958 and *Othello* in 1959. In London he worked for the OLD VIC THEATRE, and also designed the sets for such diverse productions as Errol Johns's *Moon on a Rainbow Shawl* (1958), Tennessee WILLIAMS'S *Orpheus Descending* (1959), *Half-a-Sixpence* (the musical version of H. G. Wells's *Kipps*), and *The Wings of the Dove* (based on Henry JAMES's novel) (both 1963). The year before his death he designed the costumes for a musical (N.Y., 1969) based on Chaucer's *Canterbury Tales*. He was also active as a designer in ballet, opera, and pantomime.

ST JAMES'S THEATRE, London, in King Street, Piccadilly. Designed by Samuel BEAZLEY, this theatre was built for John Braham, who opened it with a seasonal BURLETTA licence on 14 Dec. 1835. His venture was a failure, but among the plays he staged were the first dramatic works of Charles DICKENS, the two-act farce *The Strange Gentleman* and a ballad-opera *The Village Coquettes* (both 1836). There was also an adaptation by another hand of *Oliver Twist* which closed after one night. Alfred Bunn took over, but the only things that made money were wild beast shows and, in 1840, a visiting German company. The theatre was renamed the Prince's in honour of Queen Victoria's consort, but on 7 Feb. 1842 John Mitchell took over, gave the theatre back its original name, and established it as a home of French drama, RACHEL appearing there in 1846, 1850, 1853, and 1855. Between these visits the theatre was filled by a variety of entertainments, including conjuring tricks, dramatic readings, and shows by the Ethiopian Serenaders. In 1851 Hezekiah BATEMAN brought his young daughters Ellen and Kate to the theatre to appear in scenes from Shakespeare's tragedies. A succession of undistinguished managements came and went. Louisa Herbert opened on 23 Feb. 1863 in *Lady Audley's Secret*, adapted from Miss Braddon's famous novel; in Nov. 1866 IRVING enjoyed his first London success as Rawdon Scudamore in BOUCICAULT's *Hunted Down*; and a month later W. S. GILBERT's first burlesque, *Dulcamara; or, the Little Duck and the Great Quack*, was produced. In 1869 Mrs John WOOD took the theatre and had it partly reconstructed, remaining in nominal control until 1876 when she left for America. John HARE and W. H. KENDAL took over in 1879 and again reconstructed the building as a four-tier theatre with a seating capacity of 1,200, reopening it on 4 Oct. They enjoyed a steady success and established a tradition of good stagecraft, much attention being paid to scenery, furniture, and costumes. Among the productions by their excellent company were several early plays by PINERO, including *The Money Spinner* and *The Squire* (both 1881), and they also engaged the young George ALEXANDER who later became manager of the theatre, inaugurating its most brilliant period on 31 Jan. 1891 with the transfer of R. C. CARTON's *Sunlight and Shadow*. Among his outstanding productions were WILDE's *Lady Windermere's Fan* (1892) and *The Importance of Being Earnest* (1895), Pinero's *The Second Mrs Tanqueray* (1893), and an adaptation of Anthony Hope's *The Prisoner of Zenda* (1896). In 1899 the theatre closed for seven months for the reconstruction of the interior, to reopen on 1 Feb. 1900 with Hope's *Rupert of Hentzau*. Later productions included Stephen PHILLIPS's *Paolo and Francesca*, which introduced the young Henry AINLEY to London, W. M. Forster's *Old Heidelberg* (1903), Pinero's *His House in Order* (1906), and JEROME's *The Passing of the Third Floor Back* (1908), with FORBES-ROBERTSON. The last play in which Alexander appeared was Louis N. PARKER's *The Aristocrat* (1917). He died a year later, and Gilbert Miller took over the theatre, remaining in control until it was demolished. During the next few years many famous players appeared at the St James's, including Sybil THORNDIKE and Lewis CASSON, Edith EVANS, and Noël COWARD, but there were few memorable plays until 1923, when George ARLISS starred in William ARCHER's long-running thriller *The Green Goddess*. In 1925 Gladys COOPER and Gerald DU MAURIER also had a great success with LONSDALE's *The Last of Mrs Cheyney*. The 1930s saw A. A. Milne's *Michael and Mary* (1930), *The Late Christopher Bean* (1933), adapted by Emlyn WILLIAMS from René Fanchois's *Prenez-garde à la peinture*, an excellent adaptation of Jane Austen's *Pride and Prejudice* (1936), with scenery by Rex Whistler, and Clifford ODETS's *Golden Boy* (1938). The biggest success of the war years was TURGENEV's *A Month in the Country*, with Michael REDGRAVE as Rakitin. The run of Agatha CHRISTIE's *Ten Little Niggers* was interrupted when the roof was damaged by bombing in Feb. 1944, and the play was transferred to the CAMBRIDGE THEATRE. Repairs were eventually undertaken, and in 1949 Paul SCOFIELD had his first leading role in the West End as Alexander the Great in RATTIGAN's *Adventure Story*, when the actor was more successful than the play. In Jan. 1950 Laurence OLIVIER and Vivien LEIGH took over the theatre, opening with FRY's *Venus Observed* and scoring a great success alternating SHAW's *Caesar and Cleopatra* with Shakespeare's *Antony and Cleopatra* during 1951, for the Festival of Britain. In the same year the St James's returned to an old tradition when it housed the distinguished company of Madeleine RENAUD and Jean-Louis BARRAULT in a season which included SALACROU's *Les Nuits de la colère*, CLAUDEL's *Partage de midi*, with Edwige FEUILLÈRE, and MARIVAUX'S *Les Fausses Confidences*, with Madeleine Renaud. The only productions of note after this year were Odets's *Winter Journey* (1952) and Rattigan's *Separate Tables* (1954), which gave the theatre its longest run—726 performances—with Margaret LEIGHTON and Eric PORTMAN playing the leading parts in both parts of the play. The theatre closed on 27 July 1957, and a campaign was immediately got under way to preserve it, led by Vivien Leigh. This was unsuccessful, and the building was pulled down and replaced by a block of offices.

ST JAMES THEATRE, New York, at 246 West 44th Street, between Broadway and 8th Avenue. This large theatre, seating 1,583, opened as Erlanger's on 26 Sept. 1927 for spectacular and musical shows, and received its present name on 7 Dec. 1932. Among its early productions were revivals of GOLDSMITH's *She Stoops to Conquer* and SARDOU's *Diplomacy* (both 1928). In 1929 Mrs FISKE made one of her last stage appearances in Fred Ballard's *Ladies of the Jury*, which was followed a year later by Lion FEUCHTWANGER's *Jew Süss*, dramatized by Ashley DUKES. The theatre then housed light opera and musical shows, returning to straight drama with a dramatization of James Hilton's *Lost Horizon* in 1934. In 1937 Margaret WEBSTER's highly acclaimed production of *Richard II* starred Maurice EVANS, who also appeared in her production of *Hamlet* in 1938 and in *Henry IV, Part 1*, in which he played Falstaff, in 1939. A year later he was seen with Helen HAYES in *Twelfth Night*, produced under the auspices of the THEATRE GUILD, and in 1941 Canada LEE gave a fine performance in

Paul GREEN's *Native Son*. The theatre was then occupied for several years by *Oklahoma!* (1943), a musical based on Lynn Riggs's play *Green Grow the Lilacs* (1931). Other highly successful productions were the musical *The King and I* (1951), in which Gertrude LAWRENCE made her last appearance on stage; ANOUILH's *Becket* (1960); and OSBORNE's *Luther* (1963), in which Albert FINNEY made a successful Broadway début. This last moved to the LUNT-FONTANNE in 1964 to make way for *Hello, Dolly!*, a musical based on Thornton WILDER's *The Matchmaker*. Later hits were PAPP's musical version of *Two Gentlemen of Verona* in 1971; the London NATIONAL THEATRE's production of MOLIÈRE's *The Misanthrope* in 1975; and two notable new musicals, *On the Twentieth Century* (1978) and *Barnum* (1980).

ST MARTIN'S HALL, London, see QUEEN'S THEATRE (1).

ST MARTIN'S THEATRE, London, in West Street, St Martin's Lane, a small theatre with a seating capacity of 550 which opened on 23 Nov. 1916 with C. B. COCHRAN's production of a 'comedy with music' *Houp La!* It was followed by the first public performance on 17 Mar. 1917 of BRIEUX's *Damaged Goods*, which created something of a sensation. Later in the same year Seymour HICKS appeared in his own play *Sleeping Partners* and in 1920 Alec L. Rea and Basil DEAN took over, producing a number of notable new plays, among them GALSWORTHY's *The Skin Game* (1920) and *Loyalties* (1922), Clemence DANE's *A Bill of Divorcement* (1921) with Meggie ALBANESI, Charles McEvoy's *The Likes of Her* (1923) with Hermione BADDELEY, and Frederick LONSDALE's *Spring Cleaning* (1925). Dean then withdrew, but Rea continued in management until 1937 with such productions as Arnold Ridley's *The Ghost Train* (1925), Reginald Berkeley's *The White Chateau* (1927), considered by some critics the best play yet written about the First World War, Rodney Ackland's *Strange Orchestra* (1932), the first modern play to be directed by John GIELGUD, and, later the same year, Merton Hodge's *The Wind and the Rain*, which moved on elsewhere to finish its long run. J. B. PRIESTLEY's *When We Are Married* (1938) was produced under the management of J. W. Pemberton. During the Second World War, and for some years afterwards, the theatre had no settled policy, and plays succeeded one another quickly, only Kenneth Horne's *Love in a Mist* (1941), Rose Franken's *Claudia* (1942), and Edward Percy's *The Shop at Sly Corner* (1945) achieving good runs. Later productions included the revue *Penny Plain* (1951) with Joyce GRENFELL and two plays by Hugh and Margaret Williams, *Plaintiff in a Pretty Hat* (1957) and *The Grass is Greener* (1958). The 1960s saw a revival of SYNGE's *The Playboy of the Western World* (1960) with Siobhán MCKENNA, Hugh LEONARD's *Stephen D* (1963), based on James JOYCE's *Portrait of the Artist as a Young Man*, Donald Howarth's *A Lily in Little India* (1966), and Robert Shaw's *The Man in the Glass Booth* (1967). The new decade began with a resounding success when Anthony Shaffer's thriller *Sleuth* opened on 12 Feb. 1970 with Anthony QUAYLE. It moved to the GARRICK three years later, and several short runs followed before the theatre was taken over on 25 Mar. 1974 by Agatha CHRISTIE's *The Mousetrap*, then in its 22nd year, which was still running in the 1980s.

ST MARTIN'S THEATRE, Melbourne. This was founded in 1931 as the Melbourne Little Theatre, with an amateur company which performed first in a park kiosk and then a church hall, usually presenting Australian plays, some written by local authors, notably Hal PORTER. In 1956 the company opened its own theatre, becoming semi-professional with three full-time producers, and in 1962 it changed to its present name and became fully professional. In spite of government subsidies the theatre was, however, losing its hold on its audiences when in 1972 a new director was appointed, who staged some experimental productions; finally, in September 1973, the theatre was leased to the already successful MELBOURNE THEATRE COMPANY.

SAINT PLAY, in early religious drama a play based on the life and legends of a Christian saint or martyr. One of the earliest extant plays, in a form similar to LITURGICAL DRAMA but not connected with the liturgy, is a short work by Hilarius, an English 'wandering scholar' resident in France, dramatizing one of the miracles performed by St Nicholas. Another deals with an episode in the life of St Paul. As the patron saint of children St Nicholas became a favourite subject of plays intended to be performed at or near Christmas—his feast being on 6 Dec.—by choirboys and schoolboys, and perhaps linked with the tradition of the BOY BISHOP. The celebration of the name-day of other saints chosen as patrons of churches, colleges, and guilds offered ample opportunity for the dramatization of episodes in their lives which employed, instructed, and entertained both the local clergy and the laity. In France similar plays, and those specifically devoted to incidents in the life of the Virgin Mary, were known as MIRACLE PLAYS.

SALACROU, ARMAND (1899–), French dramatist, the finest but least characteristic of whose 30-odd plays is *Les Nuits de la colère*, which deals in a chronologically free-flowing form with the German occupation of France during the Second World War. Superbly acted by the RENAUD-BARRAULT company at the MARIGNY, it opened on Dec. 12, 1946, and was seen in London during the company's visit in 1951. It was produced in New York in 1947 in an English translation as *Nights of Wrath* and broadcast by the

B.B.C. as *Men of Wrath*. Unlike his younger contemporaries ANOUILH and SARTRE, Salacrou has made little impact upon the English-speaking theatre, and although a number of his plays have been given in the original in London and elsewhere, mainly by student groups, the only others to have been performed in translation appear to be *L'Inconnue d'Arras* (1935), seen in 1948 and 1954 as *The Unknown Woman of Arras*, in which a man's entire life is relived in the instant of his suicide; the farcical *Histoire de rire* (1939), produced at the ARTS THEATRE in 1957 as *No Laughing Matter*; and *L'Archipel Lenoir* (1947), which as *The Honour of the Family* was produced at the MADDERMARKET, Norwich, in 1958 and as *Never Say Die* in London in 1966. Salacrou's other plays include two early experiments in Surrealism, *Tour à terre* and *Le Pont de l'Europe* (both 1925); his first success, the comedy *Atlas-Hotel* (1931); *Une Femme libre* (1934), a study of differing kinds of male domination; the melodramatic *Un Homme comme les autres* (1936); a historical drama, *La Terre est ronde* (1938), set in Florence in the time of Savonarola; *Les Fiancés du Havre* (1944), a satirical day-dream first staged at the COMÉDIE-FRANÇAISE; *Une Femme trop honnête* (1956); and *Comme les Chardons* (1964).

SALISBURY COURT THEATRE, London, the last playhouse to be built in London before the Civil War. A private, roofed theatre, built of brick, it was erected in 1629 by Richard Gunnell and William Blagrove on part of the site of Dorset House, between Fleet Street and the Thames, the area being 140 ft by 42 ft, and cost £1,000. It was used by the KING'S MEN from 1629 to 1631, one of their productions being FORD's *The Broken Heart*, by PRINCE CHARLES'S MEN from 1631 to 1635, and by QUEEN HENRIETTA'S MEN from 1637 to 1642, when the theatres closed. During the Commonwealth surreptitious performances were given there, but the interior fittings were destroyed by soldiers in Mar. 1649, during a raid. William BEESTON restored it in 1660, and RHODES's company played there, as did DAVENANT's before he went to his LINCOLN'S INN FIELDS THEATRE. In 1661 George JOLLY was in possession and Beeston himself ran a company there from 1663 to 1664. The building was burnt down in the Great Fire of London in 1666. It was sometimes confused with the WHITEFRIARS THEATRE, which it replaced.

SALLE DES MACHINES, Paris, large, lavish and well-equipped theatre in the Tuileries, built by VIGARANI in 1660 to house the spectacular shows given in honour of the marriage of Louis XIV. It continued in use for many years for Court entertainment, and was later under the control of the artist and scenic designer Jean Bérain. It was, however, under SERVANDONY that it reached the height of its splendour, many magnificent spectacles being given there with his designs and machinery, some of them being wordless displays of lighting and décor well suited to a house whose acoustics were reputedly abominable. For a few years after 1770 it was the home of the COMÉDIE-FRANÇAISE.

SALLES, AURÉLIA DE (*fl.* late 16th century), French actress, mentioned in a document of 1581 as being a member of a company which had arrived in Mâcon to 'act plays after the manner of the ancients'. As the first French actress to be known by name she thus easily antedates Marie VENIER for whom this distinction was formerly claimed.

SALOM, JAIME (1925–), Spanish dramatist, by profession a doctor. His early plays were written in the light-hearted style prevalent at the time, but he later moved to more serious themes, as in *La casa de las Chivas* (1968), a play dealing with the Spanish Civil War, and *Los delfines* (*The Dauphins*, 1969), a contemporary study of society in transition. His later work includes *Historias intimas del paraíso* (1978) and *El corto vuello del gallo* (1980).

SALTICA, see FABULA: 6.

SALTIKOV-SHCHEDRIN, MIKHAIL EVGRAPOVICH (1826–89), Russian satirist, author of one outstanding play, *The Death of Pazukhin*, published during his lifetime, but not performed until 1901. It was revived by the MOSCOW ART THEATRE in 1914, and staged by them in New York in 1924. Another play, *Shadows*, was found among his papers and first produced in 1914. Both reveal the rottenness and corruption of the Tsarist society of the time, a theme to which Saltikov-Shchedrin returned again and again in his other writings. Though his so-called 'Dramatic Essays', written in dialogue form, were not intended for the stage, several were subsequently dramatized.

SALVINI, TOMMASO (1829–1915), Italian actor, who made an international reputation, particularly in the tragic heroes of Shakespeare. He began his career at the age of 14 in a company run by Gustavo MODENA, who taught him the rudiments of his art, and five years later made an outstanding success in the title role of ALFIERI's *Oreste*, playing opposite RISTORI in a company run by Domeniconi. His political activities caused him to leave Italy for a time and on his return he joined Dondini's company, playing Hamlet and Othello with Clementina Cazzola, with whom he formed a close attachment that lasted until her death in 1868. It was in 1861 that he added to his repertory one of his finest modern parts, that of the ex-convict Corrado in GIACOMETTI's *La morte civile*. His international career began effectively in 1869, when he toured Spain and Portugal; in 1871 he progressed in triumph through the major cities of Latin America, making his first appearance in the United States on 16 Sept. 1873. He returned there on several occasions, and in 1886 played the Ghost and Othello to the Hamlet and Iago of Edwin BOOTH. He also added Coriolanus to his repertory. He was first seen in London in 1875, as Othello at DRURY LANE, creating a sensation by depicting the savage beneath the veneer of civilization. His Hamlet inevitably invited comparison with IRVING's but was well received. On a visit to Scotland in 1876 he was seen in *Macbeth*, which he considered Shakespeare's finest play; this was not seen in London until 1884, when he also played Lear to a British audience for the first time. On a visit to Russia in 1891 he was seen by STANISLAVSKY, who left a description of his Othello in *My Life in Art* and commented approvingly on his habit of arriving at the theatre three hours before a performance in order to work himself into his rôle. Possessed of a powerful physique, graceful in movement and gesture, and above all equipped with a superbly flexible and musical voice, he studied his roles long and carefully, and his interpretations were rich in imaginative insight. On his first visit to London he married an Englishwoman, Carlotta Sharpe, who died three years later. In 1899 he married Genevieve Bearman, an American, and retired a year later, returning briefly to the stage in 1902 to take part in the celebrations in Rome in honour of Ristori's 80th birthday. He also published a volume of reminiscences, part of which appeared in an English translation in 1893.

SALZBURG, town in Austria, where there was already a lively tradition of baroque JESUIT DRAMA in the 17th century, with specially-written plays staged with great splendour, and also of early Italian opera. The town's fame today rests on its Festival, planned in 1917 by Hugo von HOFMANNSTHAL, Max REINHARDT, Richard Strauss, and Alfred Roller. It opened in 1920 with an open-air production in front of the Cathedral of Hofmannsthal's *Jedermann* (*Everyman*), directed by Reinhardt, who dominated the Festival until 1938. This was an adaptation of an old MORALITY PLAY, a lavish and exciting spectacle which has been constantly revived. In 1922 an adaptation of CALDERÓN's *El gran teatro del mundo*, again by Hofmannsthal, was seen in the Collegiate Church, and a series of revivals of European classics was inaugurated with a translation of MOLIÈRE's *Le Malade imaginaire*, starring Max PALLENBERG. In 1926 operas, mainly by Mozart, were added to the Festival's productions, in the new Festival Theatre designed by Holzmeister, and in 1933 Reinhardt directed a historic production of GOETHE's *Faust* in the old Riding School, in a set consisting of a number of rooms on different levels cut in the cliff behind the stage. When the Festival was resumed in 1945, after the Second World War, music and not drama formed its main attraction, operas being given in the new theatre, again designed by Holzmeister, which opened in 1961 with an exceptional performance of *Faust*, directed by Leopold Lindtberg. The Salzburg Festival is now firmly established in the annual calendar of international cultural events.

ŠAMBERK, FRANTIŠEK FERDINAND, see CZECHOSLOVAKIA.

SAM H. HARRIS THEATRE, New York, at 226 West 42nd Street, between Broadway and 8th Avenue. This opened as a cinema on 7 May 1914 but as the Candler Theatre was sometimes used for plays, Elmer RICE's *On Trial* being seen there in 1914 and GALSWORTHY's *Justice* in 1916. Later that year it was renamed the Cohan and Harris Theatre, receiving its present name on 21 Feb. 1921, when COHAN relinquished his share in the management. In 1922 theatrical history was made when John BARRYMORE played Hamlet 101 times, thus breaking by one performance the record set up by Edwin BOOTH. The building again became a cinema in 1932.

SAMSON, JOSEPH-ISIDORE (1793–1871), French actor, who entered the Conservatoire at 16, and subsequently spent several years in the provinces. In 1819 he appeared in Paris at the opening of the new ODÉON theatre, and made such a good impression that he was retained as leading man until 1826, when the COMÉDIE-FRANÇAISE claimed him. Apart from a few years at the PALAIS-ROYAL, he remained there until his retirement in 1863. A handsome man, with a fine profile and a mass of curly hair, he was accounted a good actor, and also one of the finest teachers of acting at the Conservatoire where he remained on the staff until his death; and it is as the teacher of RACHEL that he is now chiefly remembered. By instructing her in the classical tradition, which he himself had inherited from TALMA, he contributed not a little to the revival of French tragedy with which Rachel was associated.

SAM S. SHUBERT THEATRE, New York, at 225 West 44th Street, usually known as the Shubert, opened on 2 Oct. 1913 under the control of the managers Lee (1875–1953) and Jacob J. (1880–1964) Shubert, who named it after their dead brother (1876–1905). The opening production was *Hamlet* starring FORBES-ROBERTSON, the first American play to be seen there being Percy Mackaye's *A Thousand Years Ago* (1914). The theatre, which seats 1,469, was intended mainly for musicals, but among its straight successes were a dramatization of Harold Frederic's novel *The Copperhead* (1918), which brought fame to Lionel BARRYMORE; FAGAN's *And So To Bed* (1927), with Yvonne ARNAUD as Mistress Pepys; and Sinclair Lewis's *Dodsworth* (1934). Elisabeth BERGNER made her Broadway début at this theatre in Margaret Kennedy's *Escape Me Never* (1935), under the auspices of the THEATRE GUILD, which also sponsored the LUNTS in SHERWOOD's *Idiot's Delight* (1936) and GIRAUDOUX's *Amphitryon 38* (1937), adapted by S. N. BEHRMAN. A series of successful musicals began with *Bloomer Girl* (1944), which ran for 18 months, and included *Paint Your Wagon* (1951) by Lerner and Loewe; *Stop the World, I Want to Get Off* (1962), starring Anthony Newley, part-author of the show with Leslie Bricusse; and the same authors' *The Roar of the Greasepaint, the Smell of the Crowd* (1965). In 1966 GIELGUD and Vivien LEIGH starred in CHEKHOV's *Ivanov*, and later the theatre housed the musicals *Promises, Promises* (1968), SONDHEIM's *A Little Night Music* (1973), and *A Chorus Line* (1975), which moved from the PUBLIC THEATRE and ran into the 1980s.

SAND-CLOTH, see CLOTH.

SAN DIEGO NATIONAL SHAKESPEARE FESTIVAL, see SHAKESPEARE FESTIVALS.

SAN FRANCISCO MIME TROUPE, American political theatre troupe formed in 1959 by R. G. Davis. Their early performances were adaptations of COMMEDIA DELL'ARTE scenarios, which they performed free in the parks of San Francisco for spectators who would not normally go to the theatre. In 1970 Davis was replaced by a collective organization, since when the focus of each piece, put together by COLLECTIVE CREATION, is a political problem selected by the group and then written by one or more members after thorough study. The group has adapted to its own needs a variety of forms, including the MINSTREL SHOW, VAUDEVILLE, MELODRAMA, and parodies of spy films and television detective series. The style of playing is characterized by broad, comic action, with musical accompaniment by the Guerrilla Band, and incorporates juggling, singing, and disguises. The actors live co-operatively in small groups on a minimal wage, their income being derived from the collections made after performances in the parks, fees paid by colleges and institutions, and by the profit on their annual tours in the U.S.A. and Europe.

SANGER'S GRAND NATIONAL AMPHITHEATRE, London, see ASTLEY'S AMPHITHEATRE.

SAN PAREIL, London, see ADELPHI THEATRE.

SANS SOUCI, The, London. On 31 Oct. 1791 Charles DIBDIN opened a small theatre under this name in the Strand. It closed in March 1796, and he then leased a new building at the corner of Leicester Place, Leicester Square, which he opened, again as the Sans Souci, on 8 Oct. 1796. There he appeared in his one-man 'Table Entertainments', of which he was author, composer, narrator, singer, and accompanist, until 1804, when he sold it. It was used occasionally by other companies, Edmund KEAN as a boy, billed as 'Master Carey', giving acrobatic performances there. Although described as an 'elegant little theatre', it was only suitable for small, usually amateur, productions, and in 1832 was given over to VAUDEVILLE. In 1834 it was used by a French company, and after that became a warehouse and a hotel, being demolished in 1898.

SANS SOUCI, New York, see NIBLO'S GARDEN.

SANTARENO, BERNARDO, see PORTUGAL.

SARACEN'S HEAD INN, London, see INNS USED AS THEATRES.

SARAH-BERNHARDT, Théâtre, see BERNHARDT.

SARAT, AGNAN (?–1613), French actor, who in 1578 took a company to Paris and leased the theatre of the Hôtel de BOURGOGNE from the CONFRÉRIE DE LA PASSION. After a short stay he disappeared again into the provinces, but in 1600 returned to Paris as chief comedian in the company of VALLERAN-LECOMTE, with whom he remained until his death. He is pictured in several of his comic roles in the *Recueil de Fossart*, where his ugly, squashed nose is much in evidence. Like some of his fellow comedians, he always put flour on his face when playing in farce.

SARDOU, VICTORIEN (1831–1908), French dramatist, one of the most consistently successful of his day. Like SCRIBE, whose successor he was, he wrote copiously on many subjects in many styles, with expert craftsmanship and superficial brilliance. His first outstanding success was a comedy, *Les Pattes de mouche* (1860), seen in London in 1861 as *A Scrap of Paper*. Of his historical plays, the best known is *Madame Sans-Gêne* (1893). The part of the Duchess of Dantzig, a washerwoman whose husband became one of Napoleon's marshals, was created by RÉJANE and played in an English adaptation by Ellen TERRY in 1897. The best of the romantic melodramas were *Fédora* (1882) and *La Tosca* (1887), the latter providing the libretto for Puccini's opera *Tosca* (1900). Both plays were written for BERNHARDT, who revived them many times, and appeared also in a number of Sardou's other plays. Two social dramas which are typical of their kind were *Dora* (1877) and *Divorçons* (1880). The former, in a translation by Clement SCOTT as *Diplomacy*, had a great success in London in 1878; the latter, under its original title, was seen at the DUKE OF YORK'S THEATRE in 1907. Sardou, who was always ready to exploit whatever dramatic form seemed assured of popular success, was a favourite target of contemporary critics, and SHAW, who disliked everything he stood for, coined the word 'sardoodledom' to epitomize his WELL-MADE PLAYS.

SARMENT, JEAN (1897–1976), French dramatist, novelist, and actor, akin to ANOUILH in his ambiguous attitude towards the role of illusion in human affairs, notably in the contrast between adolescent aspirations and adult actualities, a contrast summed up in the title of an early play *Je suis trop grand pour moi* (1924). Although circumstances forced him to write BOULEVARD PLAYS he did succeed in creating stage worlds of persuasive cynicism. One of the best of his 20 or so plays, *Bobard* (1930), illustrates the idea that illusion is self-sufficient so long as its object remains unfulfilled, while *Le Discours des prix* (1934) recommends an acutely conscious outward conformism as the price to be paid for privacy and freedom. One of his early plays *Le Pêcheur d'Ombres* (1921) was seen in London in 1934 as *The Fisher of Shadows*.

SAROYAN, WILLIAM (1908–81), American novelist and dramatist, born in Fresno, California, of Armenian parents. His first play, a long one-acter produced under the auspices of the THEATRE GUILD, was *My Heart's in the Highlands* (1939), which dealt with a poet's struggle to maintain his integrity in a materialistic world. Although not completely successful it was much admired by, among others, George Jean NATHAN. Saroyan's next play, *The Time of Your Life* (also 1939), a fresh and invigorating work about a group of characters in a saloon, was awarded a PULITZER PRIZE. Not as successful, but still interesting, were *Love's Old Sweet Song* (1940) and *The Beautiful People* (1941). A less optimistic note was struck with *Hello, Out There* (1942), a one-acter depicting the lynching of an innocent tramp, and *Get Away, Old Man* (1943), which dealt with a young writer's conflict with a ruthless Hollywood mogul; but after the comparative failure of several plays which were not seen on Broadway, Saroyan recaptured some of his early mood and success with *The Cave Dwellers* (1957), about some decrepit performers living in an abandoned theatre. *Sam, the Highest Jumper of Them All* (1960) was created with and for Joan LITTLEWOOD'S THEATRE WORKSHOP in London, but he then wrote little further for the stage, his last work being *The Rebirth Celebration of the Human Race at Artie Zabala's Off-Broadway Theater* in 1975. His plays have a strong vein of fantasy, and avoid conventional dramatic structure.

SARRAZIN, MAURICE (1925–72), French theatre director, who in 1945 founded an amateur group called Le Grenier de Toulouse which a year later won first prize in a competition held in Paris. In Jan. 1949 the company became one of the State-subsidized dramatic centres set up under the policy of DÉCENTRALISATION DRAMATIQUE, covering an area from the Atlantic at Bordeaux to the banks of the Rhône. It was staffed by amateurs until 1950, and in 1965 celebrated its 20th anniversary by taking over a theatre built for it by the municipality of Toulouse which was named in honour of Daniel SORANO. The company, which first came into prominence with a performance of *The Taming of the Shrew* (as *La Mégère apprivoisée*) which was successfully transported to Paris, was run by Sarrazin until his death. It offered its audiences a cleverly mixed programme of classics and translations together with plays by new authors, notably GATTI.

SARTHOU, JACQUES (1920–), French actor and director, who in 1945 appeared in a production by Jean VILAR of T. S. ELIOT's *Murder in the Cathedral* at the VIEUX-COLOMBIER. In 1952 he formed a company to bring theatre to the working-class suburbs of Paris, based on the Théâtre de l'Île-de-France. This also performed in summer in the provinces, and sponsored several schools of dramatic art in such suburbs as Colombes and Noisy-le-Sec. In 1961 a festival was held in Kremlin-Bicêtre, a predominantly working-class south-eastern suburb of Paris. Among the plays performed were GIRAUDOUX's *Supplement au voyage de Cook*, Victor HUGO's *Marie Tudor*, and a triple bill of MOLIÈRE, MUSSET, and MARIVAUX, all directed by Jean Puyberneau.

SARTRE, JEAN-PAUL-CHARLES-AYMARD (1905–80), French dramatist and philosopher, probably the best known of the post-war French playwrights outside his own country. His philosophy of existentialism—the responsibility of each man for his own acts and for the consequences of those acts—is implicit in his dramatic works. His first play, *Les Mouches* (1942), was a modern interpretation of the story of Orestes; as *The Flies* it was seen in New York in 1947 and in London in 1951. *Huis-Clos* (1944), which followed, was produced in London as *Vicious Circle* and in New York as *No Exit*, both in 1946, the year in which *Morts sans sépultures* and *La Putain respectueuse* were first seen in Paris. As *Men Without Shadows* and *The Respectable Prostitute* these were seen in London in 1947. The second was produced in America in 1948 as *The Respectful Prostitute*, a more appropriate title, since the prostitute, Lizzie, betrays her coloured lover out of a craven respect for the dictates of society. Sartre's plays have often been given different titles in translation; *Morts sans sépultures* was seen in New York in 1948 as *The Victors; Les Mains sales* (1948) was produced in London as *Crime Passionel* and in New York as *Red Gloves*, both in 1948. Sartre's next play, *Le Diable et le Bon Dieu*, based on the same story as GOETHE's *Götz von Berlichingen*, has been variously translated as *The Devil and the Good Lord* or *Lucifer and the Lord*. In 1956 UNITY THEATRE gave the first English production of *Nekrassov* (1955), later seen at the ROYAL COURT, where in 1961 *Les Séquestrés d'Altona* was produced as *Altona*. This was not staged in New York until 1965, when it was produced at the VIVIAN BEAUMONT THEATRE as *The Condemned of Altona*. In 1969 one of Sartre's many film scripts, *L'Engrenage*, was successfully adapted for the theatre, and in 1970 his adaptation of the elder DUMAS's *Kean* (1836), first seen in Paris in 1953 with Pierre BRASSEUR as Kean and used as the basis for an

American musical in 1961, was produced in London with Alan BADEL in the title role.

SASTRE, ALFONSO (1926–   ), Spanish dramatist, many of whose plays have not been performed in Spain because of their pacifism and social criticism, though they have been published and some have been translated into English. One of the most interesting is *Escuadra hacia la muerte* (*The Condemned Squad*), set in a bunker during the next world war. Although this was banned from the public stage, it was performed in 1953 by the Teatro Popular Universitario, which Sastre helped to organize. He was also co-founder with Alfonso PASO of the experimental Arte Nuevo group. Among his other plays are: *Muerte en el barrio* (*Death in the Neighbourhood*, 1955); *El cuervo* (*The Raven*, 1957); *La cornada* (*The Death Thrust*, 1957); *En la red* (1960), on the independence movement in Algeria; and *Guillermo Tell tiene los ojos tristes* (*Sad Are the Eyes of William Tell*, pub. 1966), a tragic version of the Swiss legend. He later turned to writing novels, though his *Ahola no es de leil* (*It's No Laughing Matter*) was staged in Spain in 1979.

SATCHELL, ELIZABETH, see KEMBLE, STEPHEN.

SATIRE, Theatre of, Moscow. This theatre, which opened in Oct. 1924, was intended for the production of revues on domestic and political topics, but soon began staging satirical comedies such as KATAYEV's *Squaring the Circle* (1926). This policy was maintained under GORCHAKOV, though its tone was muted during the last years of Stalin's rule; after his death the theatre flourished again and there were revivals of MAYAKOVSKY's three major plays. Under the directorship of PLUCHEK, appointed in 1957, it mounted a series of successful productions, outstanding among which were SHAW's *Heartbreak House*, GOGOL's *The Government Inspector*, and in 1974 BEAUMARCHAIS's *The Marriage of Figaro*. During the late 1960s the company, originally housed on Mayakovsky Square, moved into a magnificent new theatre next door to the Tchaikovsky Concert Hall.

SATYR-DRAMA, burlesque plays which followed, and served as ribald comments on, the statutory tragic TRILOGY in the annual dramatic contest instituted by PISISTRATUS in connection with the DIONYSIA. Pratinus of Phlius (*fl. c.* 496 BC) is said to have introduced the satyr-drama to the Athenian stage, and ARION was the first poet to give metrical form to it. In the satyr-plays a heroic figure, sometimes the chief character of the preceding trilogy but very often Hercules, was shown in a farcical situation, always with a chorus of Sileni, or satyrs. These were the legendary companions of DIONYSUS, and were portrayed as being half-human, half-animal, with the ears and tail of a horse. The characteristics of satyr-drama were swift action, vigorous dancing, boisterous fun, and much indecency in speech and gesture. Although ARISTOTLE said that Greek tragedy 'developed from the satyr-play', the connection between them is not clear and must date from long before the time of the first official festival in 534 BC. Only one satyr-play has survived in its entirety, the *Cyclops* of EURIPIDES, though there are also fragments of an *Ichneutae* by SOPHOCLES. The popularity of the satyr-play declined during the second half of the 5th century, and in 438 BC Euripides' tragicomedy *Alcestis* took its place after the statutory trilogy, to which it was connected only by its general treatment and by the appearance of a 'genial' Hercules.

There is no connection between satyr-drama and modern satire, or between satyr-drama and any form of Greek comedy.

SAUNDERS, JAMES (1925–   ), English dramatist, who after writing a number of one-act plays vaguely reminiscent of IONESCO, among them *Alas, Poor Fred* and *The Ark* (both 1959), created a sensation with *Next Time I'll Sing to You*, based on the life and death of the Great Canfield hermit Alexander James Mason. It was first seen at the QUESTORS THEATRE in 1962, and in a revised form was successfully produced in the West End and in New York a year later. More one-act plays followed, and in 1964 came another unusual full-length play, *A Scent of Flowers*, dealing with the period immediately following a young girl's suicide, the girl herself appearing as the chief character; it was produced at the DUKE OF YORK'S THEATRE, and was also seen in New York in 1969. In 1967, in collaboration with the author, Saunders dramatized for the BRISTOL OLD VIC Iris Murdoch's novel *The Italian Girl*, which later had a long run at WYNDHAM'S THEATRE. It was followed by *The Borage Pigeon Affair* and *The Travails of Sancho Panza* (both 1969), the latter a children's play specially written for production by the NATIONAL THEATRE company, and by the one-act *After Liverpool* (1970). Among later plays, most of them produced either at the Questors (where about 20 of Saunders's plays have been seen) or at the Orange Tree Theatre, Richmond, Surrey, are the one-act *Games* and *Savoury Meringue* (both 1971); *Hans Kohlhaas* (1972), based on a play by KLEIST; the one-act *Bye Bye Blues* (1973); and the full-length *The Island* (1975). The highly successful *Bodies* (1977), about two couples who meet again after changing partners ten years earlier, had a long run at the AMBASSADORS THEATRE in 1979. Later came the one-act *Birdsong* (1979); *Random Moments in a May Garden* (1980); and *The Girl in Melanie Klein* (also 1980), set in a private asylum and based on a novel by Ronald Harwood, which had its professional première at the PALACE THEATRE, Watford, with Frank FINLAY.

SAUNDERSON, MARY, see BETTERTON, THOMAS.

SAURIN, BERNARD-JOSEPH (1706–81), French lawyer who at the age of 40, on receiving a pension from a wealthy friend, retired to devote his time to literature. Two of his plays were based on English works—a tragedy, *Blanche et Guiscard* (1763), on Thomson's *Tancred and Sigismunda* (1745), and a *drame bourgeois, Beverleï* (1768), on Edward MOORE's *The Gamester* (1753). The former was not as successful as the actors had hoped, according to Mlle CLAIRON who played in it, but the latter remained in the repertory for many years. Saurin probably took it from a translation by DIDEROT, as there is no evidence that he knew English, but he omitted most of the melodrama and concentrated on the pathetic situation of the gambler's family, particularly Tomi, his infant son. Of Saurin's other plays, the most successful were a comedy entitled *Les Moeurs du temps* (1759) and a tragedy, *Spartacus* (1760).

SAVILLE THEATRE, London, in Shaftesbury Avenue. This opened on 8 Oct. 1931 with a series of musical plays, including a musical version of Walter HACKETT's *Ambrose Applejohn's Adventure* entitled *He Wanted Adventure* (1933), with Bobby Howes in the lead. *Jill Darling!*, starring Frances Day, was a success in 1934, and the theatre continued to house musical comedy and revue until the end of 1938, when it went over to straight plays with SHAW's *Geneva*. In the following year, PRIESTLEY's *Johnson Over Jordan* transferred from the NEW THEATRE. The Saville, which held 1,250 persons on three tiers, was damaged by enemy action in 1940, but was hastily repaired and continued in use. In 1951 Ivor NOVELLO's last work *Gay's the Word* was seen, with Cicely COURTNEIDGE in the lead, and was followed by another musical, *Love from Judy*, which ran for 594 performances. John CLEMENTS took over in 1955 and inaugurated a fine series of revivals which continued for two years and included IBSEN's *The Wild Duck*, with Emlyn WILLIAMS; CHEKHOV's *The Seagull*; and CONGREVE's *The Way of the World*, with Clements himself as Mirabel, his wife Kay Hammond as Millamant, and Margaret RUTHERFORD as Lady Wishfort. Later productions of note were N. C. HUNTER's *A Touch of the Sun* (1958), with Michael REDGRAVE and his daughter Vanessa, the musical *Expresso Bongo*, also 1958, with Paul SCOFIELD, and Anthony Kimmins's *The Amorous Prawn* (1959), with Evelyn LAYE. The last ten years of the theatre's life were dogged by uncertainty, but among the interesting productions were USTINOV's *Photo Finish* and David Turner's *Semi-Detached* with Laurence OLIVIER (both 1962); a musical version of DICKENS's *Pickwick Papers*, with Harry Secombe as Mr Pickwick (1963), which ran for 694 performances; Teichmann and KAUFMAN's *The Solid Gold Cadillac* (1965), with Margaret Rutherford; and, also in 1965, Arnold WESKER's two-person play *The Four Seasons*, with Alan BATES and Diane Cilento. In 1967 the theatre was occupied by dance companies, including Martha Graham's, and in 1968 there were revivals of Novello's *The Dancing Years* and of *Lady, Be Good!* Danny LA RUE starred in *Queen Passionella and the Sleeping Beauty* at Christmas, and the last notable performance, before the theatre closed in 1970 and was converted into twin cinemas, was by Leonard Rossiter in the title role of BRECHT's *The Resistible Rise of Arturo Ui* (1969).

SAVOY THEATRE, London, in Beaufort Buildings, in the Strand. Built to provide a permanent new home for the operettas of GILBERT and Sullivan, this theatre had a seating capacity of about 1,000 in four tiers, and with its delicate colouring and new electric LIGHTING was considered revolutionary. It opened on 10 Oct. 1881 with *Patience*, transferred from the OPERA COMIQUE, and gave its name to the whole series of Gilbert and Sullivan collaborations, most of which were seen there, as were *Merrie England* (1901) and *The Princess of Kensington* (1903), both with music by Edward German. In 1907 GRANVILLE-BARKER and J. E. VEDRENNE, fresh from their successful seasons at the ROYAL COURT, staged the first London production of SHAW's *Caesar and Cleopatra*, and from 1912 to 1914 a number of Shakespeare revivals. It was at the Savoy in 1911 that the popular children's play *Where the Rainbow Ends* was seen for the first time. Later productions included COWARD's *The Young Idea* (1923), VAN DRUTEN's *Young Woodley* (1928), and SHERRIFF's *Journey's End* (1929). The building was then closed for reconstruction. The new theatre, seating 1,121 in three tiers, opened on 21 Oct. 1929 with a revival of *The Gondoliers*. During the next decade it was used mainly for transfers, but in 1941 *The Man Who Came to Dinner*, by George S. KAUFMAN and Moss HART, began a long run, as did two American comedies based on volumes of short stories, Rose McKenney's *My Sister Eileen* (1943) and Clarence Day's *Life with Father* (1947), COWARD's *Relative Values* (1951) with Gladys COOPER, and Agatha CHRISTIE's *Spider's Web* (1954). Later successes included Alec Coppel's *The Gazebo* (1960), Coward's musical *Sail Away!* (1962), Ronald MILLAR's *The Masters* (1963), based on a novel by C. P. Snow, and *The Secretary Bird* (1968) by William Douglas HOME, who had a further success in 1972 with *Lloyd George Knew My Father* starring Peggy ASHCROFT and Ralph RICHARDSON. Agatha Christie's *Murder at the Vicarage* began a long run in 1975, transferring to the FORTUNE THEATRE in the following year. Robert MORLEY starred in a revival of Ben TRAVERS's *Banana Ridge* in 1976, the ROYAL SHAKESPEARE COMPANY's production of Shaw's *Man and Superman* was seen in 1977, and in 1979 a revival of Ray Cooney and John Chapman's *Not Now, Darling* was very popular.

SCALA, FLAMINIO (*fl.* 1600–21), actor of the COMMEDIA DELL'ARTE, known as Flavio. He may have been an early director of the GELOSI, but by 1610 was certainly associated with the second CON-

FIDENTI troupe, possibly more as author and manager than as actor. Of noble birth, he was unusually well educated for the time, and left a number of informative letters. He was also the author of *Il teatro delle favole rappresentative* (1611), an early collection of scenarios which in 1967 was edited and translated into English by Salerno.

SCALA THEATRE, London, in Tottenham Street, Tottenham Court Road. This opened as the King's Concert Rooms in 1772 and in 1802, as the Cognoscenti Theatre, became the headquarters of a private theatrical club The Pic-nics, which was successful enough to attract the hostility of the PATENT THEATRES. Closed in 1808, it reopened as the Tottenham Street Theatre in 1810, and in Dec. 1814 was sold to the father of the scene painter William BEVERLEY. He renamed it the Regency Theatre, but it had little success, and in 1820 was reopened by Brunton as the West London, his daughter Elizabeth, who later married the actor Frederick YATES, starring in many of his productions. It was constantly in trouble with the Patent Theatres and was closed for several years, reopening in 1831 as the Queen's or alternatively the Fitzroy, the names being interchangeable. It then became a home for lurid melodrama, and was nicknamed the 'Dust Hole'. In 1865, taken over by Marie Wilton, it was completely redecorated, renamed (by royal permission) the Prince of Wales, and reopened in the presence of the future Edward VII with immediate success. Marie Wilton's leading man was Squire BANCROFT, whom she later married, and together they built up an excellent company, including Ellen TERRY, Mr and Mrs KENDAL, and John HARE. Under their management the epoch-making 'cup-and-saucer' dramas of T. W. ROBERTSON were first produced, beginning with *Society* in 1865. *Caste*, the only one to have been revived in recent times, was seen in 1867. Other important productions were *Masks and Faces* (1875) by Tom TAYLOR and Charles READE and SARDOU's *Diplomacy* (1878). By 1880, when the Bancrofts left to go to the HAYMARKET, the despised 'Dust Hole' had become a fashionable theatre, but in 1882 it was condemned as structurally unsound and closed for repairs. Owing to a long-drawn-out dispute it was not used again except as a Salvation Army hostel. In 1903 it was demolished, only the original portico remaining to serve as the stage-door entrance of a new theatre which seated 1,193 persons in a three-tier auditorium. Renamed the Scala, it opened on 23 Sept. 1905 under the management of FORBES-ROBERTSON, but the venture was not a success and the theatre often stood empty or was used for films, puppet-shows, and amateur productions. In 1926 Ralph RICHARDSON made his first London appearance at the Scala for the Greek Play Society as the Stranger in SOPHOCLES' *Oedipus at Colonus*. Various touring companies used the theatre, among them Donald WOLFIT's and the D'Oyly Carte in GILBERT and Sullivan, and from 1945 there was an annual Christmas revival of BARRIE's *Peter Pan*, until in 1969 the theatre closed, being demolished in 1972.

SCAMMACCA, ORTENSIO (1562–1648), Italian dramatist, author of nearly 50 plays on sacred or moral themes, ostensibly in the didactic tradition of JESUIT DRAMA, but containing a good deal of sensational matter eked out with love intrigues and interlarded with piety. The religious element is, however, preponderant, and even in plays drawn from classical sources angels and devils make their appearance, while women, unusually, are given important parts.

SCAMOZZI, VINCENZO, see PALLADIO and THEATRE BUILDINGS.

SCAPINO, SCAPIN, one of the ZANNI or servant roles of the COMMEDIA DELL'ARTE. Like BRIGHELLA Scapino was crafty and unprincipled, and when in danger lived up to his name, which means 'to run off' or 'to escape'. The first actor to play him was Francesco GABRIELLI, and through him and later actors of the part the character passed into French comedy to end up as the quick-witted and unscrupulous valet of MOLIÈRE's *Les Fourberies de Scapin* (1671).

SCARAMUCCIA, literally, 'a little skirmisher', a character of the COMMEDIA DELL'ARTE generally classed with the ZANNI roles, but considered by some to have approximated more to the blustering braggart CAPITANO. The actor most closely associated with the part, though he may not actually have created it, was Tiberio FIORILLO, who, abandoning the use of a mask, played the role with white powdered face, small beard, and long moustache, in a costume predominantly black but set off by a white ruff. In France he became Scaramouche.

SCARBOROUGH, see STEPHEN JOSEPH THEATRE IN THE ROUND.

SCARRON, PAUL (1610–60), French dramatist and novelist, who was crippled by rheumatism at the age of 30 and forced to rely on his pen for a livelihood. He wrote a number of witty though slightly scabrous farces, of which the first two, *Jodelet, ou le Maître-valet* (1643) and *Jodelet souffleté* (1645), were produced at the MARAIS with the comedian JODELET himself in the title roles. In 1652 Scarron married the beautiful but penniless orphan Françoise d'Aubigné, who as Mme de MAINTENON was to become the second wife of Louis XIV and the virtual ruler of France. Meanwhile Scarron continued to write for the theatre. The best example of his burlesque comedy, in which he obtained his comic effects by ingenious word-play and by the incongruity between subject-matter and style, is *Don Japhet d'Armenié* (1647), which was frequently revived in later years by MOLIÈRE. *L'Écolier de Salamanque*

(1654), unsuccessfully plagiarized by BOISROBERT, is notable for the character of the valet CRISPIN, long played by successive members of the POISSON family. Scarron, whose interest in Spanish literature had led him to translate a number of Spanish plays, modernizing them and adding much material of his own, may have taken from the novelist Agustín de ROJAS the idea of his most important work *Le Roman comique* (1651), a novel which depicts the adventures of an itinerant provincial theatre company. It has considerable documentary value, and the actor-manager Léandre is probably based on FILANDRE.

SCENARIO, the skeleton plot of the COMMEDIA DELL'ARTE play, the term replacing some time in the early 18th century the older word *soggetto*. These are not such synopses as might be drawn up by an author for his written drama nor are they identical with the Elizabethan PLATT; they were theatrical documents prepared for the use of a professional company by its leader, or by an enthusiastic amateur admirer of the *commedia dell'arte* style, and consisted of a scene-by-scene résumé of the action, together with notes on locality and special effects. Their informal elasticity allowed the insertions of extraneous business (the BURLA or LAZZO) at the discretion—or according to the ability—of the actors. The term, whose plural in English is now scenarios, is commonly used today for the script of a film or the synopsis of a musical play.

SCENE OF RELEAVE (RELIEF), see MASQUE and SCENERY.

SCENE ROOM, see GREEN ROOM.

SCENERY, term covering everything used on stage to represent the place in which an action is performed, including hangings, cut-outs, painted FLATS, BOX-SETS, BUILT STUFF, etc., but not usually movable furniture and PROPS.

Scenery is a comparatively recent innovation in the history of the theatre. Greek plays were acted against a stage wall, which by Roman times had become a grandiose architectural façade, the *scenae frons*, and the use of stage machinery is indicated by mentions of the EKKYLEMA and the MECHANE, but except for the PERIAKTOI of Hellenistic and Roman times the classical stage had no scenery as we know it. In the early medieval period the interior of a church provided the setting for LITURGICAL DRAMA, although theatrical effects were provided, as in the PARADISO used in the church of San Felice in Florence. As the drama moved out of doors the MULTIPLE SETTING began to evolve, with 'mansions' representing different localities placed around an open *platea*, or acting area, the outside wall of the church forming a backdrop; this style was to have a long-lasting influence on the theatres of France. In England, by contrast, MYSTERY PLAYS were presented on mobile wagons, scene by scene, and the public playhouses of the 16th–17th centuries owed much to their portable precursors. Theatres such as Shakespeare's GLOBE were probably gaily decorated, but apart from movable properties on the platform stage and in the 'discovery space' (see INNER STAGE) scenery in the modern sense was not used. It is first found in the courts of Renaissance Italy where entertainments were presented with all possible splendour. Perspective painting, developed in the mid-15th century, was used to make enclosed spaces seem larger; these principles were applied to theatrical illusion by PERUZZI, who was also influenced by the newly discovered architectural works by VITRUVIUS. Peruzzi's pupil SERLIO published in 1545 descriptions of perspective stage settings for tragic, comic, and satyric plays, basically consisting of a pair of side scenes receding symmetrically at right angles to the front of the stage. In the second half of the century BUONTALENTI was using a painted backcloth with *telari*, three-sided prisms in imitation of the classical *periaktoi*, forming side-wings. His innovations and those of his pupils (notably Giulio (1590–1636) and Alfonso (?–1656) Parigi, father and son) spread all over Europe and were introduced into England by Inigo JONES for the elaborate Court MASQUES of 1600–40.

The scenic system of the masque—the origin of modern scenery—used, in its final form, a decorative PROSCENIUM arch behind which sets of WINGS, called side scenes or side shutters, framed the back scene on either side. This was the Flat Scene, from which the modern term 'flat' is derived. The back scene consisted of pairs of painted shutters, from two to four in number, centrally divided and sliding in GROOVES placed about half-way upstage. The wings, too, could be in groups sliding in grooves, so that they could be changed according to the changes of the back scenes. In some masques the shutters were further divided horizontally: the upper pair would show a mountain-top, for example, while the lower ones disclosed a cave within the mountain. At specially dramatic moments all the shutters were withdrawn to display the Scene of Relief, a three-dimensional set piece placed before a BACKCLOTH and often framed by a narrow pair of wings having their own grooves immediately upstage of the last pair of shutters. It seems likely that scenes of relief were constructed in a manner similar to the Set Scenes of a modern PANTOMIME with their successive profiled GROUNDROWS.

These private performances had no influence on the English public theatres, which still used the bare apron stage derived from the medieval tradition and where development along Continental lines was prevented by the outbreak of the Civil War and the proscription of playhouses until the monarchy was restored in 1660. In 17th-century Italy the widespread passion for opera gave stimulus to the work of scene designers, including the Burnacinis, Giovanni (?–1656) and his son Lodovico Ottavio (1636–1707), in Venice; the Mauro family of five brothers, who worked in Venice, Vienna, and Dresden; two generations of the Galliari family in north Italy and Berlin; the great sculptor BERNINI; Filippo JUVARRA; and above all the BIBIENAS, whose baroque architectural designs were found in almost every capital city of the time. The diagonal perspective first used by the Bibienas, and the growing popularity of landscape painting, gradually changed the whole character of theatre decoration. In Paris, where MAHELOT had begun to oust the old-fashioned multiple setting from the Hôtel de BOURGOGNE, TORELLI (a pupil of ALEOTTI) created a rage for MACHINE PLAYS with his inventions, including the CARRIAGE-AND-FRAME method of rapid scene-changing. French taste was moving towards lightness, and fantasy progressing into neo-classicism, as can be seen in the succession of the VIGARANIS, Bérain, and SERVANDONY at the SALLE DES MACHINES; the QUAGLIO family was to continue the trend in Germany, as did GONZAGA in Russia. The increasing elaboration of the JESUIT DRAMA both echoed and influenced developments all over Europe.

Meanwhile in England the Restoration playhouse set the basic scenic style for over a century: an apron stage, with up to three pairs of PROSCENIUM DOORS opening on to it, drew on the medieval tradition, while scenes composed of sets of wings and backcloths were a direct inheritance from the Flat Scene of the masque. The earliest English scene painter known by name is Robert Streater or Streeter (1624–80), whose 'perspectives' for DRYDEN's *The Conquest of Granada* (1670–1) are noted in John Evelyn's diary (9 Feb. 1671); skilled scene painters working at the DORSET GARDEN THEATRE and at DRURY LANE include AGGAS, FULLER, and WEBB; but England was untouched by the aesthetic movements of France and Italy, and scene design changed little until GARRICK brought Philip de LOUTHERBOURG to Drury Lane. Gradually the old architectural setting was abandoned in favour of romantic landscapes, with transparencies and elaborate cut-outs helping to create an attractive stage picture which was still in use 100 years later and lasted even longer in pantomime.

The 19th century, dominated by William GRIEVE and his family and by the histrionic talent of Clarkson STANFIELD, saw also an enthusiasm for neo-Gothic design and a growing insistence on painstaking architectural detail. This passion for authenticity had begun with the designs of CAPON for KEMBLE's Shakespeare revivals at Drury Lane, 1794–1802, and at COVENT GARDEN in the 1810s, and was continued by KEAN in the 1850s at the PRINCESS's, where one of his artists was William TELBIN. This romantic historicism continued with Hawes CRAVEN, working for IRVING at the LYCEUM, and reached its peak of elaboration with Beerbohm TREE's productions of Shakespeare; the popular easel painter Lawrence Alma-Tadema (1836–1912) designed settings for both Irving and Tree. The style was popular all over Europe and was exemplified in Tsarist Russia by the work of Andrei ROLLER. This was the great era of stage illusion, of TRAPS, GAUZES, and TRANSFORMATION SCENES, and of *trompe-l'oeil* scene painting, with every detail painted on stretched canvas—doors, windows, draperies, even furniture, as well as outdoor vistas: William BEVERLEY was one of the period's most successful practitioners. For spectacular pieces such as pantomimes, cut-cloths developed to such excess that the stage picture resembled a lacy valentine. As drama began to follow the rise of realism in literature and painting, these theatrical conventions became unacceptable. The box-set, in use since the 1830s, became the normal means of presenting interior scenes; still constructed of flats lashed together, it was now supplied with real furniture and accessories and practicable doors and windows. When André ANTOINE founded the THÉÂTRE LIBRE in 1887 he insisted on complete verisimilitude for his productions of naturalistic contemporary plays (even real food was used, and real fountains played, on stage); but he used the scenery and properties to reinforce the mood of a play in a totally new manner.

The French Symbolists led by the poet Paul FORT, attacked the methods of the Théâtre Libre for ignoring imagination and fantasy in the search for the exact. Fort founded the Théâtre Mixte which, after two performances, became the Théâtre d'Art, and in his manifesto enunciated many of the principles which were later adopted by the modernist school. Scenery was to be simplified, evocative rather than descriptive; there was to be frank stylization, complete harmony between scenery and costume, and the absolute abandonment of the perspective backcloth. Among the painters who painted decorations for the Théâtre d'Art were Vuillard, Bonnard, Maurice Denis, Odilon Redon, and K. X. Roussel. Norman WILKINSON and Charles Ricketts, working for GRANVILLE-BARKER, were to be the main followers in England of the symbolist approach.

The works of some new dramatists, especially MAETERLINCK, were sufficiently imaginative to give scope to the new method. In 1892, when the Théâtre d'Art had become the Théâtre de l'Œuvre, LUGNÉ-POË (who was to start Georges WAHKEVITCH on a distinguished career in theatrical design) collaborated with Camille Mauclair and Édouard Vuillard to present *Pelléas et Mélisande* at the Bouffes-Parisiens; STANISLAVSKY was present at this performance, and afterwards admitted how much he owed to the experimental work which was being carried out in Paris towards the end of the century. The MOSCOW ART THEATRE, founded by Stanislavsky in 1898 with V. A. SIMOV as a regular scene designer, adopted Antoine's naturalism and the realistic effects of the MEININGER COMPANY, which had visited Moscow in 1885 and 1890: this low-key manner was in harmony with Stanislavsky's aim of presenting actual conditions of life through drama.

In 1899 Adolphe APPIA published his epoch-making work on the reform of staging, *Die Musik und die Inscenierung*, which stressed in particular the illogicality of placing three-dimensional actors against flat scenery. His proposals for solid settings of extreme simplicity, lit so as to emphasize instead of flattening the human form, were to have an immense influence on 20th-century stage décor. Electric lighting, applied in the theatre from the 1890s, made Appia's ideas practicable and opened the way to other developments in stage design. One of the most widespread was the CYCLORAMA, which evolved from Mariano Fortuny's *Kuppelhorizont* or sky-dome (see LIGHTING). This solid, curved rear wall could, with shadowless lighting, represent indefinite open space; it was later used for projected clouds and various effects of light. To a great extent the cyclorama made the painted back-cloth obsolescent, although scene-painting continued to be practised and achieved a late flowering in the designs executed for the Diaghilev Ballet by Léon Bakst and Alexandre Benois and by French Cubist painters including Picasso and Derain.

The trend towards greater simplicity continued, however. Gordon CRAIG (the son of a distinguished scenic designer, E. W. GODWIN) was more of a theorist than a practitioner and the most influential of Appia's successors. He evolved a system of large screens, with a few movable features such as flights of steps, to build up an imaginative stage picture with no concessions to realism: his production of *Hamlet* at the Moscow Art Theatre in 1912 is still controversial. He also urged a theatre completely created by a single man—author, director, designer, costumier—with actors (*Übermarionette*) under his dictatorial control. However extreme, this theory was in tune with the increasing importance of the theatre director during the 20th century. Max REINHARDT came near to filling the role proposed by Craig. His designers, notably Karl WALSER, Ernst STERN, and Oskar Strnad, provided scenery as eclectic as his choice of play: he used semi-permanent settings, COMPOSITE SETTINGS, proscenium, apron, and arena stages; mounted productions in theatres, circuses, exhibition halls, ballrooms; and in his historic production of HOFMANNSTHAL's *Jedermann* (1911) the streets of Salzburg became his stage and the façade of its cathedral his back-drop. In the years 1910–33 Reinhardt dominated the stage of central Europe with his grand theatrical enterprises; his more intimate productions showed great subtlety and individuality as well as an awareness of current trends in the visual arts. David BELASCO, in contrast, opted for a kind of spectacular naturalism for the romantic dramas produced at his own theatre in New York, 1907–31. He was adventurous in his use of lighting, and took advantage of current developments in stage machinery including the REVOLVING STAGE, which Karl Lautenschläger (1843–1906) had introduced at Munich in 1896. This, like wagon stages, sliding stages, and the various rise and fall platforms pioneered by the ASPHALEIAN SYSTEM and further developed by Adolf Linnebach (1876–1963) at Dresden, attained considerable refinement during the first 30 years of the century; but elaborate machinery, intended to aid frequent scene-changing, went against the main line of artistic progress and was soon limited to spectacular musical productions.

In the years immediately following the First World War, the European stage saw a brief flowering of EXPRESSIONISM, taking the form on the one hand of an extreme simplification of scenery—Leopold JESSNER and his designer Pirchan in Berlin using bare flights of steps connecting different acting levels, for example—and on the other the distortion of inanimate objects to reflect the moods of a play. In the Soviet Union EVREINOV's theory of 'monodrama' took up the same theme; the Dutch director Herman Rosse experimented with animated backgrounds and film projection; and KOMISARJEVSKY mounted productions of Shakespeare in expressionist settings at Stratford-upon-Avon during the 1920s and 1930s. Expressionism appeared on the American stage: Robert Edmond JONES, in his designs for *Macbeth* in 1921, showed lop-sided cardboard arches that emphasized the insecurity of the hero's moods and fortunes. Other American scene designers who came to the fore at this time included Norman BEL GEDDES and Lee SIMONSON.

England remained for the most part indifferent to Continental and American developments, Lovat FRASER, the only artist who might have inaugurated a movement of far-reaching significance, dying young after producing his admirably simple semi-permanent set for a revival of GAY's *The Beggar's Opera* in 1920. In general England remained faithful to realism and the box-set, and even the best scene designers, like Charles Ricketts, found little scope for their talents. In France the best theatre directors of the post-war age—COPEAU, JOUVET, DULLIN—still strove to get away from the whole tradition of the painted scene. Copeau indeed dispensed with scenery entirely, and the sets of Dullin were based on Craig's idea of movable screens. In Russia the Revolution had swept away the conventional forms of theatre décor, reducing the set to bare scaffolding or the sparse clean lines of metal machinery. Décor became frankly symbolic, simplified to the point of abstraction in the then dominant mode of Constructivism. The movement's chief exponents in Soviet theatre, Vsevolod MEYERHOLD and Alexander TAÏROV, adoped a frank theatricality, rejecting all kinds of realism and taking the action among the spectators, who were in turn drawn into the action. Even the Moscow Art Theatre was affected by the artistic ferment of the 1920s; but official reaction against formalism in all the arts put an end to such experiments and from the early thirties SOCIALIST REALISM became the only acceptable manner, in the theatre as elsewhere, for more than two decades. This meant, in essence, soberly naturalistic décors, although historical pieces provided the chance to present sumptuous architectural settings. N. A. SHIFRIN did good work within the approved modes, and a few designers such as AKIMOV and P. V. WILLIAMS strove to sustain the modernist spirit. Constructivism found an English disciple in Terence GRAY at the Festival Theatre in Cambridge, and PISCATOR became its chief practitioner in Germany, supported by members of the Bauhaus, including Walter Gropius and Laszlo Moholy-Nagy.

Throughout Europe, the period of fundamental innovation ended in about 1930, the growing threat of war inhibiting the expansion of new ideas; the English theatre now began to catch up with Continental developments, helped by an influx of émigré artists and designers. The United States, too, received an injection of European talent, notable names including Christian BÉRARD, Eugene Berman, and Pavel Tchelichew, to join the growing number of American designers making their reputations, such as Boris ARONSON, Jo MIELZINER, and Donald OENSLAGER.

The popular revues of C. B. COCHRAN used designers of the calibre of Rex Whistler and Oliver MESSEL, whose witty, stylized settings owed much to earlier European experiments. The commercial theatre of the pre-war decade was marked by elegance and decorativeness, taken up in the years immediately after the Second World War in a revival of opulent romanticism: Messel, Leslie HURRY, and Loudon SAINTHILL had particular success in this vein. But the preoccupations of the playwrights of the fifties—OSBORNE, WESKER, BEHAN in Britain, Arthur MILLER, Tennessee WILLIAMS, Lillian HELLMAN in America—forced designers into a visual austerity that extended into the classical repertory as post-war euphoria declined. Monochrome designs, sometimes with minimal colour accents, came into vogue, perhaps in reaction to the pictorial splendour of musicals, which reached a peak in Cecil BEATON's scenery and costumes for *My Fair Lady* (1956). The increasing use of THEATRE-IN-THE-ROUND and FLEXIBLE STAGING placed new difficulties in the way of designers, although some, including Tanya MOISEIWITSCH and Sean KENNY, welcomed the challenge; scene-painting in particular languished. Materials produced by new industries almost ousted the wood, canvas, and papier-mâché of traditional scenery: the vast range of plastics, especially, provided novel textural effects and made possible the building of structures at once massive and lightweight. Modern metal alloys, fibre glass (GRP or glass reinforced plastic), fibre board, ply wood, and newly developed adhesives, all helped to widen the designer's range, although audiences tended to resist their more brutal evocations of contemporary artistic trends. Projected scenery had been used by Piscator as early as 1924 and is claimed to have been first used in England in connection with the STAGE SOCIETY's producton of STRINDBERG's *The Road to Damascus* at the WESTMINSTER THEATRE on 2 May 1937; the technique benefited from post-war developments in optical technology, and was used with particular success by Josef SVOBODA, who took up the European tradition (inherited from the Symbolists and the Constructivists by way of the Epic Theatre of BRECHT and Piscator) to achieve a heightened realism that proved applicable to a wide range of subjects. Although the naturalistic box-set is still used for single-setting plays, the great range of technical choice has tended to be applied to a deliberate anti-illusionism, with stage mechanisms and lighting equipment exposed to view. Within this general consensus styles of interpretation vary widely, from the multi-media documentary mode applied by John BURY to Joan LITTLEWOOD's production of *Oh! What a Lovely War* (1963) to the bold abstractionism of Ralph KOLTAI's designs for Shakespeare at the NATIONAL THEATRE. (See also COSTUME).

SCHECHNER, RICHARD (1934-), American critic and director, who became Professor of Drama at New York University in 1967, when he founded the PERFORMANCE GROUP to put into practice the ideas of ARTAUD and GROTOWSKI, of which he was a leading advocate in the United States, having edited since 1962 the *Tulane Drama Review*, making it the leading international journal of radical theatre. The best known of the many productions of the Performance Group under his direction was *Dionysus in '69*, a modern version of EURIPIDES' *Bacchae* involving audience participation. Schechner withdrew from the leadership of the Group in 1970, and has since turned to the study of Shamanism and the relationship of primitive ritual to theatre. He is the author of several books on modern theatre problems; *Environmental Theater* (1973) in particular, though marred by political bias and a certain subjectivity, displays at its best his capacity for brilliant insight and cogent argument.

SCHELANDRE, JEAN DE (*c.* 1585–1635), French dramatist, who passed some of his life in England. His tragedy *Tyre et Sidon* (1608) is of some historical importance. Written in the manner of the early tragedies of Alexandre HARDY, it was recast 20 years later as a TRAGI-COMEDY, and published in 1628 with an important preface by François Ogier in which he defended the free, irregular drama which was already under attack from the theorists and was to give way a few years later to TRAGEDY based on the UNITIES and a rigid neo-classical aesthetic.

SCHICKSALSTRAGÖDIE, the name given to the 'fate drama' of early 19th-century Germany inaugurated by WERNER's *Der vierundzwanzigste Februar* (1809), in which a malignant fate dogs the footsteps of the chief character, driving him by a chain of fortuitous circumstances to commit a horrible crime, often the unwitting murder of a son by his own father. The genre was further exploited by Adolf Müllner in *Der neunundzwanzigste Februar* (1812) and *Die Schuld* (*Guilt*, 1813) and by the young GRILLPARZER in his play *Die Ahnfrau* (1817).

SCHILDKRAUT, RUDOLF, see JEWISH DRAMA.

SCHILLER, (Johann Christoph) FRIEDRICH VON (1759–1805), German poet and dramatist. He was only 22 when his first play *Die Räuber* (*The Robbers*) was accepted by DAHLBERG, who produced it at MANNHEIM in 1782. It was an immediate success, particularly with the younger generation, and has been constantly revived; it was given a memorable political production by PISCATOR in 1918. In 1783 Schiller was appointed official dramatist to the Mannheim theatre, writing for it *Fiesco* (1783) and *Kabale und Liebe* (1784). Schiller, who was absent without leave from his duties as an army doctor and living under an assumed name, was heavily in debt; he was befriended by Charlotte van Kalb, whose influence is apparent in his first historical tragedy, *Don Carlos* (1789). He married soon after, eking out a miserable existence at Jena as a teacher of history and publishing two historical works, one on the Netherlands, one on the Thirty Years War. The research undertaken for the latter provided him with the material for his great dramatic trilogy *Wallenstein*, which was completed in 1799 and translated into English by COLERIDGE a year later. Schiller's last years, before his early death from tuberculosis, were spent in WEIMAR, where he enjoyed the friendship and collaboration of GOETHE who produced some of his best works, notably *Maria Stuart* (1800), *Die Jungfrau von Orleans* (1801), *Die Braut von Messina* (1803), notable for the lyric beauty of its choruses, and his last play *Wilhelm Tell* (1804); a revival of this by JESSNER in 1919 was one of the first manifestations of EXPRESSIONISM.

All Schiller's plays were translated into English, at first for reading rather than for the stage. The most influential was *Die Räuber*, which reinforced the STURM UND DRANG movement unleashed by Goethe's *Götz von Berlichingen* (1773). As *The Red-Cross Knights*, in an adaptation by Holman, it was seen at the HAYMARKET in 1799 and, as *The Robbers*, at DRURY LANE in 1851, a year after PLANCHÉ's adaptation of *Fiesco*, and as *The Highwayman* at the ROUND HOUSE in 1974. *Kabale und Liebe*, as *The Harper's Daughter*, was seen at COVENT GARDEN in 1803 and, as *Power and Principle*, at the STRAND in 1850. It was also seen in German in London during the WORLD THEATRE SEASON of 1964. *Maria Stuart*, which brings together Elizabeth I and Mary Queen of Scots, who in real life never met, was performed at the OLD VIC in 1958 and 1960 in a translation by Stephen Spender. Earlier translations were staged at Covent Garden in 1819 and the ROYAL COURT in 1880. In 1974 a translation of *Don Carlos* was given at the Everyman Theatre in Cheltenham.

SCHILLER [de Schildenfeld], LEON (1887–1954), Polish director and designer, an innovator in the style of WYSPIAŃSKI, who, with CRAIG and STANISLAVSKY, constituted the main influences on his work. An exponent of 'total theatre', he was also a champion of both romantic and realistic drama, and became one of the outstanding figures in Polish theatre history. After the First World War he worked at the Polski Teatr in Warsaw, and at Osterwa's Reduta Theatre before founding his own Bogusławski Theatre in 1924, where he directed, among other classics, a revival of KRASIŃSKI's *The Undivine Comedy* in 1926, with Irena Solska as his leading lady. His political affiliations with the Left, and a controversial production of BRECHT's *Die Dreigroschenoper*, led to the closing of his theatre in 1930, and he went to Lwów, where his outstanding productions were of TRETYAKOV's *Roar, China!* and MICKIEWICZ's *Forefathers' Eve*; he also directed a production of the latter in Bulgarian in Sofia in 1937. Interned at Auschwitz, he was released in 1941 and, since the Germans had closed all the Polish theatres, set to work secretly to produce, in a convent near Warsaw, some of the old Polish liturgical dramas. In 1946 he became director of the theatre in Łódź, where his last important production was *The Tempest* in 1947. After his death a collection of his writings on the theatre was published under the title of *Teatr Ogromny*.

SCHIMMEL, HENRIK JAN, see NETHERLANDS.

SCHLEGEL, AUGUST WILHELM VON (1767–1845), German dramatist and writer, nephew of Johann von SCHLEGEL. He wrote 16 plays, now forgotten, and is mainly remembered as a sensitive critic of enormous range. A series of lectures, given in VIENNA and published between 1809 and 1811 as *Über dramatische Kunst und Literatur*, marks the first attempt at a history of the development of world drama, and with his friend TIECK he translated 17 of Shakespeare's plays. These versions, in spite of their romanticism, are still the ones most often performed in Germany. He also translated Dante, and some of the plays of CERVANTES and CALDERÓN.

SCHLEGEL, JOHANN ELIAS VON (1719–49), German dramatist, whose comedies and historical tragedies such as *Hermann* (1743) were strongly influenced by GOTTSCHED. He later became a strong champion of Shakespeare, as witness his essay *Vergleichung Shakespeares und Andreas Gryphius* (1741), in which Shakespeare's *Julius Caesar* is compared favourably with GRYPHIUS's work on the same subject. In 1743 he was given a diplomatic appointment in Copenhagen, where he wrote a treatise on the Danish theatre which was influential in subsequent attempts to found a

German National Theatre. His objective valuation of the drama of different nations and his views on the relation of art and nature were well in advance of his time, but he was destined to be eclipsed by LESSING's more trenchant mind and greater forcefulness of expression.

SCHMELTZL, WOLFGANG (*c.*1505–*c.*1557), Viennese schoolmaster who between 1540 and 1551 wrote a number of plays which were in complete contrast to the humanistic works of CELTIS and WATT. Under the influence of Paul REBHUN he turned to Biblical subjects—*Der verlorene Sohn* (*The Prodigal Son*, 1540), *Judith* (1542), *Die Hochzeit zu Cana* (*The Marriage Feast at Cana*, 1543), *David und Goliath* (1545), *Samuel und Saul* (1551)—writing in the vernacular, and introducing into his plays elements of popular theatre.

SCHNITZLER, ARTHUR (1862–1931), Austrian dramatist, a doctor by profession, who brought to his plays something of the dispassionate attitude of the consulting-room. His first work for the theatre was *Anatol* (1893), a series of sketches depicting the adventures of a young Viennese philanderer. This was followed by *Liebelei* (1896), in which a working-class girl kills herself on learning of the death of the young aristocrat who had been merely trifling with her affections. *Der grüne Kakadu* (*The Green Cockatoo*, 1899) is a one-act play on an incident of the French Revolution in which he handles with a sure touch the change from irresponsible make-believe to grim reality. *Reigen* (*The Round Dance*) was written in 1896–7 and Schnitzler always asserted that it was intended for private circulation only (though it was published in 1903). It was performed in Magyar in Budapest in 1912 and in the original German in Berlin and Vienna in 1921; the riots in both theatres were apparently mounted by right-wing anti-Semitic groups, and the case brought against this linked sequence of ten loveless sexual encounters on the grounds of obscenity was unsuccessful. Schnitzler however forbade all performances and it became well known only through a French film, *La Ronde* (1950). Immediately on the expiry of COPYRIGHT the play was given a number of productions, notably by the ROYAL SHAKESPEARE COMPANY in London in 1982. Among Schnitzler's later plays are *Der einsame Weg* (*The Lonely Way*, 1904), a sensitive play of delicate half-lights; *Das weite Land* (1911), seen in London at the NATIONAL THEATRE in 1979 as *Undiscovered Country*, in an adaptation by Tom STOPPARD; *Der Ruf des Lebens* (*The Call of Life*, 1906); and *Professor Bernhardi* (1912), Schnitzler's only contribution to the problem-play, in which he views from all angles the repercussions of an anti-Semitic incident in a Viennese hospital. Schnitzler's world vanished in the First World War, and the plays he wrote after 1918—*Komödie der Verführung* (*A Comedy of Seduction*, 1924) and *Der Gang zum Weiher* (*The Walk to the Lake*, 1925) are little more than nostalgic echoes of the past.

SCHÖNEMANN, JOHANN FRIEDRICH (1704–82), German actor, originally a harlequin in a travelling troupe. He later joined the company of Carolina NEUBER, and in 1740 formed his own troupe, in which were Sophia SCHRÖDER, EKHOF, and ACKERMANN, all destined to play an important part in the development of the German theatre. Schönemann, who was good in comedy but less successful in tragedy, to which he brought the pompous declamatory style evolved by GOTTSCHED, was at first an able and astute manager, but he later ruined himself by his hobby of horse-dealing and abandoned the company to Ekhof.

SCHÖNHERR, KARL (1867–1943), Austrian dramatist, whose powerful and realistic dialect dramas of Tyrolean peasant life won him widespread recognition. Writing under the influence of ANZENGRÜBER but with a closer affinity to NATURALISM, his themes are the peasant's clinging to the soil, as in *Erde* (*Earth*, 1907), his bewilderment in the face of religious conflict, as in *Glaube und Heimat* (*Faith and Homeland*, 1910), and his defence of the Tyrol against Napoleon's armies, as in *Volk in Not* (*People in Need*, 1915). A doctor himself, Schönherr wrote about the medical profession in *Der Kampf* (*The Struggle*, 1920) and the psychology of pretence in *Der Komödiant* (*The Comedian*, 1924). A sound instinct for the theatre which did not hesitate to use the resources of melodrama where necessary was the basic element in his success.

SCHOOL DRAMA, a term applied to the academic, educational plays which appeared in all European countries during the Renaissance. Written under the influence of the humanists by scholars for performance by schoolboys, they were originally in Latin, a tradition which persisted longest in JESUIT DRAMA. Elsewhere they tended to slip quickly into the vernacular, and had some influence on the development of the non-academic, popular, and later professional drama.

There was a good deal of dramatic activity in English schools and colleges in the first half of the 16th century, and the first two regular English comedies—*Ralph Roister Doister* by Nicholas UDALL and *Gammer Gurton's Needle* probably by William STEVENSON—were given at Eton (or perhaps Westminster) and Christ's College, Cambridge, respectively. The tradition of an annual school production in English has survived in many schools, but the only one in Latin today is the WESTMINSTER PLAY.

SCHOOLS OF DRAMA. Until the present century entry into the theatrical profession was a haphazard affair, the beginner usually joining an established company in the provinces and learning the job through trial and error. Sometimes study with a well known actor, or experience with an amateur group, might prove a help in gaining a foothold on the stage. Today, though some actors still join the profession without any formal training, most of them go through a three-year course at a recognized drama school, of which there are some 30 in Great Britain, or study in a UNIVERSITY DEPARTMENT OF DRAMA. The leading London schools include the London Academy of Music and Dramatic Art (LAMDA), founded in 1861 under the auspices of the Academy of Music by Dr T. H. Yorke-Trotter, and now housed in Tower House, Cromwell Road, with a small theatre nearby in Logan Place (where in 1964 Peter BROOK and Charles MAROWITZ gave the first English season of the Theatre of CRUELTY); the MOUNTVIEW THEATRE SCHOOL; the Royal Academy of Dramatic Art (RADA), founded in 1904 at HER (then his) MAJESTY'S THEATRE by Beerbohm TREE, and run for many years on its present site in Gower Street by Sir Kenneth Barnes (1878–1957), under whom its theatre, the Vanbrugh, named in honour of his sisters Irene and Violet VANBRUGH, was built to replace an earlier one, destroyed by bombing in Apr. 1941; and the Central School of Speech and Drama, founded in 1906 in the Albert Hall by the actress Elsie Fogerty (1866–1945), mainly for the teaching of poetic speech, and now housed in the EMBASSY THEATRE. In SCOTLAND the most important school of drama is at the Royal Scottish Academy of Music and Drama, founded in 1950, mainly through the efforts of James BRIDIE.

Apart from the recognized schools of drama, for which the local authorities will give grants to assist promising students, there are a number of stage schools for children, of which the best known was founded by Italia CONTI. These are privately operated, and give stage training as well as the elements of a standard education. Of recent years many Polytechnics and Colleges of Education have introduced drama courses into their curriculum, but in most cases they lack the expertise and facilities of the specialized schools and only help to swell the ranks of an already overcrowded profession. Training at any type of drama school does not of itself entitle a person to membership of BRITISH ACTORS' EQUITY, a prerequisite of professional employment, which has its own methods of selecting its members.

In the U.S.A. most actor training has now fallen within the sphere of the universities, sometimes, as at YALE, in conjunction with fully professional theatre activities. An early and notable school was the American Academy of Dramatic Arts, New York, which was founded in 1884 by Franklin Haven Sargent as the Lyceum Theatre School of Acting and received charters from the University of the State of New York in 1899 and 1952. It offers a short Junior Course, followed by a Senior Course restricted to those who have successfully completed the Junior Course. The GOODMAN MEMORIAL THEATRE in Chicago combined, notably between the wars, a resident professional company with a student training programme, an example followed by many other theatres at various times.

The leading and longest established French school is the Paris Conservatoire, which began in 1786 as the École de Déclamation, taking its present name in 1793. Among its first pupils was TALMA. A later pupil was SAMSON, who returned to become one of its finest teachers. The school was reorganized after the 'events' of May 1968. The oldest and most important training establishment in the Soviet Union is the LUNACHARSKY State Institute of Theatre Art (GITIS), founded on 22 Sept. 1878. Originally both a music and drama school, it became a Conservatoire in 1886, and counted among its teachers NEMIROVICH-DANCHENKO, whose pupils MOSKVIN and Olga KNIPPER became members of the first MOSCOW ART THEATRE company. In 1934, after many changes of name and status, it was given the name of the Soviet Union's first Minister for Education. Its pupils are drawn from some 40 different nationalities, half of them being external students already working in the theatre. In 1958 the Institute opened its own theatre building for practical work.

SCHOUWBURG, Amsterdam, see CAMPEN, JACOB VAN.

SCHREYVOGEL, JOSEF (1768–1832), Austrian theatre director, who in 1814 was appointed head of the three major theatres in VIENNA, the Burgtheater, the Kärntnertor, and the Theater an der Wien. Well-read and a great traveller, he endeavoured to further the literary reforms inaugurated by his predecessor SONNENFELS and was careful to provide a balanced programme of plays, old and new, German and foreign, particularly at the Burgtheater, which under him became one of the outstanding theatres of Europe. One of his greatest achievements was the discovery and fostering of the genius of GRILLPARZER.

SCHRÖDER, FRIEDRICH LUDWIG (1744–1816), German actor, the son of Sophia SCHRÖDER. He played as a child in the travelling company of his stepfather ACKERMANN but in 1756, on the outbreak of war, became separated from it and for the next few years lived by his wits, becoming in the process an expert acrobat and rope-dancer. He eventually rejoined the company in Switzerland and resumed his acting career, but considered it very inferior to that of a tumbler. It was the advent of EKHOF, then at the height of his powers, that made him realize what acting could be. During the next few years he studied and practised his art to such good effect that he gradually took over most of Ekhof's parts, leaving the older man no option but to withdraw; at the new National Theatre in HAMBURG he was much admired, particularly as the quick-witted, light-heeled valets of French comedy. Following Ackermann's death in 1771, he assumed artistic control of the company, his mother retaining financial control, a situation which eventually caused trouble. At the head of an admirable company, which included his half-sisters Dorothea and Charlotte Ackermann, he was at the forefront of the new movement in Germany, responsible for the production of several important new plays, including *Emilia Galotti* (1772) by LESSING and *Götz von Berlichingen* (1773) by GOETHE. His chief enterprise was the introduction of Shakespeare in action to young Germans who had previously met him only on the printed page. The adaptations of Shakespeare's tragedies, in which Romeo, Juliet, Cordelia, Ophelia, and even Hamlet survived, were made by Schröder himself; he probably knew how much of the original his audience would take. He began with *Hamlet* in 1776, in which BROCKMANN played the title role and Schröder the Ghost (he later played Laertes, the Gravedigger, and Hamlet), and by 1780 11 of Shakespeare's plays had been performed, of which *Othello* was a failure and *King Lear* an outstanding success. To offset all this pioneer work Schröder continued to give his audiences a more conservative repertory of now-forgotten plays and monodramas, ballets and light musical pieces, which enhanced the prosperity and reputation of his actors. The first glorious phase of his Hamburg management ended when, tiring at last of the constant friction with his mother over money matters, he went in 1782 with his wife Christine Hart (?–1829), as guest-artist to the Burgtheater in VIENNA, where he remained for four years. Though this experience added little to his own development, he exercised a salutary influence on his fellow actors, modifying their pompous ranting in tragedy and old-fashioned fooling in comedy, and may be said to have laid the foundations of the subtle ensemble playing which was later a distinguishing feature of the Burgtheater's productions. He enjoyed his years in Vienna, but in 1786 had once more to take over the company which his mother, at the age of 72, could no longer control. He again gave it an important position in German theatrical life, but the fire and enthusiasm of the earlier period were lacking and the energy that should have gone to the production of the last great plays of SCHILLER and GOETHE was dissipated on the trivialities of IFFLAND and KOTZEBUE. Nevertheless, once given a free hand financially, Schröder prospered, and in 1798 he was able to buy and retire to a country estate at Holstein, where he died. He and his sister Charlotte were depicted in Goethe's novel *Wilhelm Meisters Lehrjahre* (1795) as Serlo and Aurelie.

SCHRÖDER [*née* Biereichel], SOPHIA CARLOTTA (1714–92), German actress, who was persuaded by the actor EKHOF to leave her drunken and unsatisfactory husband and join him in the company of SCHÖNEMANN. She made her début at Luneburg as Monime in an adaptation of RACINE's *Mithridate* on 15 Jan. 1740, and was immediatelv successful. When ACKERMANN left Schönemann to form his own company, she went with him as his leading lady, but shortly after returned to her husband. Their reconciliation resulted in the birth of one of Germany's greatest actors, Friedrich SCHRÖDER, but the husband was unable to overcome his intemperance and Sophia returned to the stage, marrying Ackermann after her husband's death and bearing him two daughters who became actresses with the company. She toured with him indefatigably, and after his death in 1771 retained a tight hold on the company, particularly as regards finance, finally driving her son, who was its leading actor and artistic manager, to leave for VIENNA. He returned in 1786, by which time she no longer had the strength to refuse his demand for a free hand, and spent the last six years of her life in unwilling retirement.

SCHRÖDER, SOPHIE (1781–1868), Austrian actress, who played leading parts at the VIENNA Burgtheater under the artistic direction of SCHREYVOGEL. She created the part of Bera in GRILLPARZER's *Die Ahnfrau* (1817) and of Medea in his *Das goldene Vliess* (1821), and it was said that her playing of the title role in his *Sappho* (1818) established the 'noble simplicity' which remained the hallmark of the Burgtheater's style of acting for many years. She was much admired in the plays of GOETHE, SCHILLER, KLEIST, and particularly Shakespeare, her Lady Macbeth in 1821 eschewing violent outbursts and impressing the spectators with her tenacity of purpose and unshakeable resolution: the contemporary critic Heinrich Laube speaks of her 'ideal form' enlivened by a passionate temperament, her grace of movement and gesture, and the moving purity of her diction.

SCHUCH, FRANZ (*c.* 1716–64), Austrian harlequin-player, leader of a company of travelling players to which EKHOF belonged for a short time after he left SCHÖNEMANN. Johann BRANDES had his first stable engagement with Schuch and described him in his memoirs as a fine comedian and improviser, homely without descending to triviality, humorous without vulgarity; LESSING also thought highly of him. After his death the company was eventually taken over by DÖBBELIN, who had previously been one of its members, and established by him in BERLIN, where it prospered.

SCHUSTER, IGNAZ (1779–1835), Austrian actor, one of the most popular comic figures in the Viennese popular theatre. After a period with Karl MARINELLI's Baden company, he joined the Leopoldstädter company in VIENNA in 1801 and remained with it until his death. From 1805 onwards he appeared in local *Singspiele*, particularly in Joachim Perinet's adaptations of works by Philipp

HAFNER. Schuster, who was small and deformed, did not achieve fame until 1813 when he created the part of STABERL the umbrella-maker in BÄUERLE's *Die Bürger in Wien*. Thereafter he was the hero of innumerable *Staberliaden*, many of them written by Karl CARL, and gradually replaced Anton HASENHUT in popular favour.

SCHWARTZ, EVGENYI LVOVICH, see SHWARTZ, E. L.

SCHWARTZ, MAURICE (1889–1960), Jewish actor and director. Born in the Ukraine, he went to the United States as a child, and later appeared in Yiddish plays in several towns before joining the company of David Kessler in New York. In 1918 he was at the Irving Place Theatre, where in 1919 he directed a performance of a play by Peretz HIRSCHBEIN. This proved a turning-point in his career; but his greatest achievement was his discovery of the work of Sholom ALEICHEM, the quintessence of Jewish folk humour and characterization. Schwartz also introduced to the stage the works of a number of other Jewish writers, including Halper LEIVICK. In 1924 he undertook a tour of Europe which proved successful, and he returned to New York to open a theatre on Broadway for the production of European classics in Yiddish. This, however, failed, and in 1926 he opened a Yiddish Art Theatre on 2nd Avenue, the traditional home of New York Yiddish drama. He later toured South America and also visited Palestine where he worked with OHEL. His repertory was extensive, comprising some 150 roles, and under his influence a number of Yiddish Art Theatres were founded in New York; but the widespread adoption of English by Yiddish families, and a slackening of Jewish immigration to the States, caused a decline in their audiences. In1959 Schwartz went to Israel hoping to establish a Yiddish theatre there, but died after producing only one play, Singer's *Yoshe Kalb*.

SCIENCE FICTION THEATRE OF LIVERPOOL, see EVERYMAN THEATRE, LIVERPOOL.

SCIOPTICON, see LIGHTING.

SCISSOR CROSS, see STAGE DIRECTIONS.

SCISSOR STAGE, see BOAT TRUCK.

SCOFIELD, (David) PAUL (1922–), English actor. He made his first appearance on the stage in 1940, and had played a number of parts outside London when he first came into prominence with his portrayal of the Bastard in *King John* at the BIRMINGHAM REPERTORY THEATRE in 1945. He then went to the SHAKESPEARE MEMORIAL THEATRE for two seasons, his wide range of parts including Mephistophilis in MARLOWE's *Doctor Faustus* and the title roles in *Henry V* and *Pericles*, and he was also seen in London in 1946 as Tegeus-Chromis in FRY's *A Phoenix Too Frequent*. His Young Fashion in VANBRUGH's *The Relapse* in 1948 was followed by a return to Stratford, where he played Hamlet and Troilus in *Troilus and Cressida*. In 1949 he appeared at the ST JAMES'S THEATRE as Alexander the Great in RATTIGAN's *Adventure Story* and Konstantin in CHEKHOV's *The Seagull*, and a year later he was seen as the twin brothers Hugo and Frederick in ANOUILH's *Ring Round the Moon*. At the EDINBURGH FESTIVAL of 1952 he played Philip Sturgess in Charles MORGAN's *The River Line*, repeating his performance in London, and in 1953 he gave a fine performance as Pierre to GIELGUD's Jaffier in OTWAY's *Venice Preserv'd*, in the same season at the LYRIC THEATRE, Hammersmith, also offering his Richard II and Witwoud in CONGREVE's *The Way of the World*. After appearing in two modern plays, Wynyard Browne's *A Question of Fact* (also 1953) and Anouilh's *Time Remembered* (1954), he went to Russia in 1955 to play Hamlet at the MOSCOW ART THEATRE with the first English company to appear in Russia since the Revolution. The production was directed by Peter BROOK, who also directed him as the drunken priest in *The Power and the Glory*, based on Graham GREENE's novel by Denis Cannan and Pierre Bost, and as Harry in a revival of T. S. ELIOT's *The Family Reunion* (both 1956). In 1958 he made his début in a musical, *Expresso Bongo*, and after Greene's *The Complaisant Lover* (1959) was seen in probably his most famous role, Sir Thomas More in Robert BOLT's *A Man For All Seasons* (1960), in which he made his first appearance on Broadway a year later. Peter Brook then directed him in *King Lear* in 1962 in a production which went from Stratford to London, several European countries, including Russia again, and New York. Later roles included Timon in *Timon of Athens* at Stratford in 1965, Khlestakov in GOGOL's *The Government Inspector* and the homosexual barber in Charles Dyer's *Staircase* at the ALDWYCH in 1966, and Macbeth at Stratford in 1967. He was then seen in OSBORNE's *The Hotel in Amsterdam* (1968), and gave a superb performance in the title role of Chekhov's *Uncle Vanya* at the ROYAL COURT in 1970. He was first seen with the NATIONAL THEATRE company in 1971 in ZUCKMAYER's *The Captain of Köpenick* and PIRANDELLO's *The Rules of the Game*, and he then took the leading role in Christopher HAMPTON's *Savages* (1973), following it with Prospero in *The Tempest* in 1974 and Athol FUGARD's *Dimetos* (1976). He returned to the National Theatre in 1977, playing in JONSON's *Volpone* and GRANVILLE-BARKER's *The Madras House*; in 1979 he appeared there in Peter SHAFFER's *Amadeus* as Salieri, the bitterly envious contemporary of Mozart, and in 1980 he played Othello there. He has a particular gift for conveying moral worth without seeming priggish; but few modern actors have been seen in such a wide variety of outstanding roles.

SCÔPS, see ENGLAND.

SCOTLAND. The seemingly belated appearance of an indigenous Scottish drama was really a revival after early arrested growth, for in the 15th and 16th centuries MYSTERY and MORALITY PLAYS flourished there as elsewhere, though the only example to survive is Sir David LYNDSAY's *Ane Pleasant Satyre of the Thrie Estaitis* (1552); and although it seemed at one time as if the Scottish Renaissance was tending towards the theatre, again only one play of the period survives, the anonymous comedy *Philotus*. The country's turbulent history, and the natural humour of the people, would have seemed to offer a fitting soil for drama, and dramatic imagination shone richly in the great ballads; but a Scots parallel to the Elizabethan stage did not develop. On the contrary, legislation against the ROBIN HOOD folk-plays began in 1555, and in 1575 the General Assembly of the Church of Scotland put an end to LITURGICAL DRAMA by banning 'clerk-plays or comedies based on the canonical scriptures'. The subsequent removal of the Court to London in 1603 and the civil and religious strife of the 18th century effectively stifled the growth of an indigenous Scottish drama. Some activity continued, however, in Edinburgh, including the first known production in Scotland of Shakespeare's *Macbeth* in 1672. A company of actors was established there in 1715, and another, managed by Tony ASTON, was active from 1725 to 1728. In 1733 the poet Allan RAMSAY, author of *The Gentle Shepherd*, a pastoral performed as a BALLAD OPERA in 1729, founded the Edinburgh Players, and in 1736 opened a theatre in Carubber's Close; after six months it was closed down under the provisions of the Licensing Act of 1737. But from 1741 onwards there was a theatrical season every year, the law being evaded by charging not for the play but for a concert which preceded it. The Concert Hall in the Canongate, which opened in 1747, saw the first production of John HOME's tragedy *Douglas* (1756), under the management of Dudley DIGGES who played the hero, Young Norval, and in 1767 it became Edinburgh's first Theatre Royal. Two years later it was replaced by a new theatre at the east end of Princes Street, which from 1791 to 1800 was managed by Stephen KEMBLE.

In Glasgow the first permanent playhouse was built in 1753. There was a Theatre Royal in Queen Street in 1805, a fine building with excellent new machinery. When it burned down in 1829 the name was taken over by the Caledonian Theatre in Dunlop Street. Dumfries had its first theatre in 1792, based on the design of the Theatre Royal in BRISTOL; Robert Burns was an active supporter of the venture. In 1811 this became the Theatre Royal, Dumfries. The first theatre in Dundee, the Royal, lasted only from 1800 to 1810, when a more salubrious building, also called the Royal, opened and survived until 1885, housing a succession of Scottish and English companies. All these cities had STOCK COMPANIES playing Shakespeare and the standard classical repertory. There were also regular visits by leading actors from London, including Edmund and Charles KEAN and Mrs SIDDONS; and it was even possible for an actor like Charles MCKAY to pursue a distinguished acting career entirely in Scotland. From the turn of the century many of the melodramas and romances had a Scottish flavour, and a popular 'grand Scottish National Pantomime' entitled *Oscar and Malvina* (1791) drew its inspiration from Macpherson's *Ossian*. The local interest indicated by such titles as *Sawney Bean's Cave* or *The Falls of Clyde*, both produced in the early years of the 19th century, was further reinforced by the success of the novels of Sir Walter SCOTT. Dramatic versions of these were popular throughout the 19th century, the first to be seen on the Scottish stage, in Edinburgh and Glasgow, being *Guy Mannering* in 1817, only two years after its publication. Later in the century plays like *Deacon Brodie* (1880) by R. L. Stevenson and W. E. Henley and the early plays of J. M. BARRIE heralded the distinctively Scottish drama that was to arise in the 20th century.

The latter half of the 19th century saw the stock companies replaced by touring companies, mainly from England, but Scotland made a fresh contribution to the theatre in the years 1909 to 1914, when the REPERTORY THEATRE movement made its way north of the border and led to the founding of the GLASGOW REPERTORY THEATRE, which during its five seasons brought together some of the most brilliant actors and actresses of a generation, and produced plays by outstanding English and Continental dramatists, as well as a few Scottish plays. Unfortunately the outbreak of war in 1914 caused it to close down. What remained of its funds was used to launch the movement that in 1921 produced the SCOTTISH NATIONAL PLAYERS, who in the 1930s were succeeded by a new group of Glasgow amateurs, the CURTAIN THEATRE company. This became the main vehicle for the production of new plays in the vernacular, among them works by the Irish dramatist Paul Vincent CARROLL, then resident in Glasgow. Meanwhile the Perth Theatre, which had housed touring companies since 1900, went over to repertory in 1935 under the management of Marjorie Dance and David Steuart, and in 1939 presented a Scottish Theatre Festival for which James BRIDIE wrote *The Golden Legend of Shults*; and in Dundee a repertory theatre, with a non-profit-making company formed with municipal and commercial support, opened in 1939 with FLECKER's *Hassan*.

The work of the Curtain Theatre for the creation of a Scottish drama was taken up in Glasgow in 1941 by UNITY THEATRE, followed in 1943 by the CITIZENS' THEATRE and in Edinburgh by the GATEWAY THEATRE, which opened in 1946. The first EDINBURGH FESTIVAL in 1947 presented no Scottish drama, but at the second a revival of Lyndsay's *Ane Pleasant Satyre of the Thrie Estaitis* proved a revelation of the riches of Scottish acting and of the Scots tongue. A fifth Scottish repertory company was established in 1951 with the foundation of the PITLOCHRY FESTIVAL THEATRE. In 1963 the TRAVERSE THEATRE CLUB opened its small theatre in the Royal Mile in Edinburgh, and soon became internationally known for its policy of presenting new plays. The Gateway Theatre gave its final performance at the Edinburgh Festival of 1965, handing over to the Edinburgh Civic Theatre Company at the LYCEUM.

Two small theatres outside the urban centres which have achieved much success are the Mull Little Theatre, seating 35 in a converted byre, which opened in 1967 and is the smallest theatre in Great Britain, and the Byre Theatre at St Andrews, which also began in a converted cowshed, moving in 1970 to a new building seating 128.

Much of the post-war development is to be credited to the Scottish Arts Council, which makes grants to the repertory theatres to help with their running costs and supports creative policies through new drama schemes. It also arranges tours to centres which do not have repertory companies. Local government has also provided substantial financial support, but following upon the 1974–5 reorganisation it has found itself, in common with the Arts Council, short of funds. This means that new theatres scheduled for several cities, including Edinburgh's long-promised opera house and theatre complex, may not now be built. Much progress has, however, been made by local authorities, working in partnership with the Scottish Arts Council, in the establishment of arts centres, such as the MacRobert Centre in Stirling, most of which have theatrical facilities. A somewhat different venture was the establishment of a Scottish branch of the experimental COLLECTIVE CREATION group, the 7:84 Theatre Company, which scored a great success in 1973 with John McGrath's *The Cheviot, the Stag, and the Black, Black Oil*. The company tours its productions, which have a clear left-wing commitment and are largely scripted by McGrath, throughout the country and to the islands, making use in the absence of a theatre of any accommodation it can find to bring drama to 'non-theatre' people.

In 1950 the training of Scottish actors was given a great stimulus when (with James Bridie as prime mover) a College of Dramatic Art was established within the Royal Scottish Academy of Music in Glasgow. The amenities of the College include the Athenaeum Theatre and, since 1962, a closed-circuit television studio. The importance which the teaching of drama had achieved was recognized in 1968 by the new name of Royal Scottish Academy of Music and Drama. The University of Glasgow, which had from the outset undertaken a share of the teaching, further extended the provision of dramatic education by the establishment in 1966 of a Department of Drama.

Whereas James Bridie had to turn to the resources, artistic and financial, of the English theatre to achieve his full development as a dramatist, such new dramatists as Robert KEMP and Robert MCLELLAN have had the alternative of opting to write for the Scottish theatre, even if this has meant smaller audiences. Their work illustrates another choice which confronts the Scottish playwright, that between English and Scots, or 'Lallans'. (There is also an interesting Gaelic drama of modern origin, though the audience for it is relatively small.) Kemp, using Lallans, has achieved far better translations of MOLIÈRE than would be possible in contemporary English, and a similar claim can be made for Douglas Young's translations of ARISTOPHANES.

It may be argued that too high a proportion of post-war Scots plays are set in the past, even when that past is as dramatic as Scotland's; and that the riches of Lallans have lured the successors of Bridie and the Unity playwrights away from contemporary life. Among the most important of the writers who since 1960 have tried to correct the balance are Stewart Conn, Cecil TAYLOR, and Stanley Eveling, who, although an Englishman, has had much of his work presented by the Traverse Theatre, as indeed has Taylor. Conn, in *The Burning* (1971), has also written a historical play about the persecution of witches during the reign of James VI, and Bill BRYDEN, in *Willie Rough* (1972), has depicted more recent urban history. Standing rather apart from these writers is Tom Gallacher, in whose work there is an emphasis on more overtly literary subjects than is usually found in the Scottish theatre. At his best, in *Mr Joyce is Leaving Paris* (1971) or *Schellenbrack* (1973), for instance, Gallacher combines skilled craftsmanship, subtlety of perception, and a philosopher's delight in the artifices of the theatre.

The amateur movement in Scotland was strengthened in 1926 by the founding of the Scottish Community Drama Association on the lines of the British Drama League (now the British Theatre Association), which since 1932 has run a National Festival of Community Drama. This was at first limited to one-act plays, but since 1945 has paid more attention to full-length works, as well as to works by Scottish dramatists. The present Byre Theatre at St Andrews, now professional, began as an amateur group in 1933. Important among the amateur Little Theatre groups now is the Dundee Dramatic Society, founded in 1924, which acquired its own theatre in 1936. From 1939 to 1943 the Rutherglen Repertory Theatre also had a successful record.

In 1936 the Glasgow Jewish Theatre Institute Players were formed, opening their own theatre in 1938. During the Second World War some members joined the Unity Theatre, but in 1944 the group was reformed. Groups well housed in Art Centres include the Dumbarton People's Theatre and the Greenock Players.

The decline in the number of amateur groups since the war has been accompanied by a gradual rise in standards and a more open attitude to experimental work. Since the late 1960s there has also been a noticeable growth in audience support. There is an increasing number of little theatres, like the Cottage Theatre in Cumbernauld and the Harbour Arts Centre in Irvine, which have been constructed in arts centres and are used both by local amateurs and visiting professionals. Small professional companies have been established in association with both these centres, at Irvine in 1974 and at Cumbernauld in 1980.

SCOTT, CLEMENT WILLIAM (1841–1904), English dramatist and dramatic critic, who for nearly 30 years from 1872 reviewed plays for the *Daily Telegraph*, putting up a determined resistance to the new drama as typified by IBSEN. His attack on *Ghosts* (1891) as 'a wretched, deplorable, loathsome history' was the outcome of his obstinate refusal to consider anything outside his own range of conventional morality. He was for many years editor of *The Theatre*, which he founded in 1877. After he withdrew in 1890 it distressed him by supporting his rival William ARCHER in his campaign in favour of modern drama. Scott was the author of a number of plays based on French originals and usually written in collaboration. The only one to survive is *Diplomacy* (1878), based by Scott and B. C. Stephenson on SARDOU's *Dora*, which was last revived in London in 1933.

SCOTT, Sir WALTER (1771–1832), Scottish novelist, whose works, like those of Charles DICKENS, were quickly and widely adapted for the stage. The first to be used were apparently *The Lady of the Lake*, in a version by Thomas DIBDIN at the SURREY THEATRE, and *Marmion*, as *The Spectre Knight* by Charles DIBDIN at SADLER'S WELLS, both in 1810. *Marmion* was also dramatized by the American playwright James BARKER in 1812, a version which held the stage for over 40 years. Between 1811 and 1843 more than half a dozen versions of *The Lady of the Lake* were produced, some of the later ones showing clearly the influence of Rossini's opera *La donna del lago* (1819); two versions bore the alternative title of *The Knight of Snowdoun*. Scott himself helped Daniel TERRY to prepare a stage version of *Guy Mannering* for COVENT GARDEN in 1816; this was used by SCRIBE in 1825 for part of his musical play *La Dame blanche* and so returned to England as *The White Lady* (1826) and *The White Maid* (1827). One of the most popular novels in the theatre was *Rob Roy*, which was seen in Edinburgh, where many first performances of Scott took place, in Jan. 1818, three versions being seen in London the same year. It was also done under the title of *Gregarach* (1821) and *Roy's Wife* (1825). In one form or another it held the stage in Scotland up to modern times and its success intensified the speed with which other novels were adapted as they appeared. *The Heart of Midlothian* was particularly popular, four versions of it appearing in 1819, a year which saw also the first two dramatizations of *The Bride of Lammermoor*, also seen as *The Mermaiden's Well* (1828) and later providing Donizetti with the subject of *Lucia di Lammermoor* (1835). Also in 1819 came the first version of *The Legend of Montrose*, as *Montrose*, by the indefatigable Thomas Dibdin; it was seen the following year as *The Children of the Mist*. *Ivanhoe* was also to prove popular, with no less than seven versions in 1820, as was *Kenilworth*, with seven different versions in 1821, one of them entitled *Elizabeth and Essex*; it was later to appear as *Tilbury Fort* (1829) and *The Earl of Leicester* (1843). *Waverley*, Scott's first novel published in 1814, was not dramatized until 1822, when it was seen in Perth, and two years later in London. Three versions of *Peveril of the Peak* and two of *Redgauntlet* were in circulation between 1823 and 1826; three versions of *Quentin Durward* were seen in 1823; and three of *Woodstock*, one of them at the Surrey by Charles Dibdin, in 1826, the year which saw *The Talisman*, as *The Knights of the Cross*, at DRURY LANE. *The Fair Maid of Perth* was seen in 1828, *The Highland Widow as Dougal the Piper* in 1836 and as *Military Punishment* in 1846.

By the mid-19th century enthusiasm for Scott was beginning to fade and Dickens was in the ascendant, though it is noteworthy that eight new versions of *Ivanhoe* appeared between 1859 and 1896 and it was well enough known to be burlesqued, as many others of Scott's novels had been, in 1862 by H. J. BYRON. *The Heart of Midlothian* was also used again several times, notably by Dion BOUCICAULT in 1860 for *The Trial of Effie Deans*, first seen at Laura KEENE's theatre in New York and three years later in London. Unlike Dickens, Scott has not survived into the 20th century, either on stage or as material for the reading public. The dramatization of his novels and poems undoubtedly had a strong influence on the development and general popularity of MELODRAMA, but though he was, again like Dickens, the author of several plays, they were of no importance and none has survived; his first venture into drama was with a translation in 1799 of GOETHE's *Götz von Berlichingen mit der eisernen Hand*, which does not appear to have been performed. A tragedy *The House of Aspen* (1829) was seen at the Surrey, and another, *Auchindrane* (1830), in Edinburgh. Neither has been revived.

SCOTTISH COMEDIANS, see LAUDER, HARRY.

SCOTTISH COMMUNITY DRAMA ASSOCIATION (S.C.D.A.), see SCOTLAND.

SCOTTISH NATIONAL PLAYERS, company which made its first appearance in Jan. 1921 in a Glasgow hall under the auspices of the St Andrew Society of Glasgow, presenting three new Scottish one-act plays. By the end of that year a further five one-act plays and two full-length ones had been produced. The company, which became independent in 1922, was wound up in 1934, two years after a proposal to turn professional had been rejected, but the Players continued independently, their last production (after a war-time interruption) being in 1947. A notable feature of their work was the summer camping tours, which brought drama to relatively remote parts of the country. Although the movement did not succeed in its aim of creating a Scottish National Theatre similar to the ABBEY THEATRE, Dublin, it did call into being a body of plays which sought to reflect the variety of Scottish life, with the emphasis on the rural rather than the urban scene; among the 30 or so authors who wrote for the company were Gordon BOTTOMLEY, John BRANDANE, James BRIDIE, Joe CORRIE, and Robert KEMP. During the years 1921–31 the Players produced 62 new plays, their director in 1926 and 1927 being Tyrone GUTHRIE.

SCOTTISH ROSCIUS, see JOHNSTON, H. E.

SCRIBE, (Augustin) EUGÈNE (1791–1861), French dramatist, originator and exploiter of the WELL-MADE PLAY. A prolific writer, he was responsible, alone or in collaboration, for more than 400 works, comprising tragedies, comedies, vaudevilles, and librettos for light opera. His early plays were failures and it was not until 1815 that he achieved fame with *Une Nuit de la Garde Nationale*. Even more successful was *Un Verre d'eau* (1850), translated into English by W. E. Suter as *A Glass of Water; or, Great Events from Trifling Causes Spring* (1863) and by GRUNDY as *The Queen's Favourite* (1883). The most successful of Scribe's plays, however, and the only one now remembered, was *Adrienne Lecouvreur* (1849), written in collaboration with Legouvé, who in 1884 wrote Scribe's biography. The play, though historically incorrect, provided a fine part for RACHEL and later for BERNHARDT. In translation it was played by RISTORI, MODJESKA, and Helen FAUCIT, among others. Scribe's plays, skilfully constructed with the utmost neatness, economy, and banality, came as a relief to a middle-class audience surfeited with the incoherence of the Revolution and the excess of the Romantics. Though his collaborators contributed much to the common stock, the stagecraft was his alone; with no depth or delicacy of perception, he had an uncanny flair for knowing what the public wanted, and how to give it to them with the maximum dramatic effect. Although he was to some extent unfairly blamed for all the shortcomings of the dramatists who succeeded him, he had an immense influence on them, particularly on LABICHE and SARDOU, and there is no doubt that his much-sought-after librettos, written for such musicians as Meyerbeer, Offenbach, and even Verdi (*Les Vêpres Siciliennes*, 1855), helped to make French romantic opera a model of theatrical effectiveness.

SCRIM DROP, see BACKCLOTH and CLOTH.

SCRUTO, an important element of TRANSFORMATION SCENES and TRAPS. It consists of a number of narrow strips of thin wood hinged together or attached side by side on a canvas backing, so as to form a continuous flexible sheet like the cover of a roll-top desk. Upon this sheet a subject can be painted, and the whole attached to lines so as to roll down or up on cue, replacing one scene by another. For traps, a roll of scruto formed a cover across the opening which could be quickly removed when necessary by rolling or sliding it aside.

SCUDÉRY, GEORGES DE (1601–67), French dramatist, brother of the celebrated Madeleine de Scudéry, in whose novels he had a hand. He wrote in all 16 plays, of which the first, *Ligdamon et Licias, ou la Ressemblance* (1630), which turns on the unlikely coincidence of two young men, not related, being as alike as identical twins, shows already the lively imagination, love of excessive rhetoric, and predilection for the unusual which was to be apparent in all his later works. The best of these were the tragedy *La Mort de César* (1635); the tragi-comedy *Le Trompeur puni* (1631), which FLORIDOR's company played at the COCKPIT in London in 1635; and the comedy *La Comédie des comédiens* (1635), which portrays on stage the company of which MONTDORY was the leader, several parts being given the names of the actors playing them. Scudéry, who had been resentful at not being one of the five dramatists chosen to write RICHELIEU's plays for him, was consoled a little by his election as a founder-member of the French Academy, an honour he had done little to deserve. He took an active part in the literary quarrel over CORNEILLE's *Le Cid* (1637) and had the pleasure of finding his own tragi-comedy *L'Amour tyrannique* (1638) praised above that of Corneille by Richelieu: but it is now forgotten.

SEA ROWS, see GROUNDROW.

SEATTLE REPERTORY THEATRE, Seattle, Washington, seating 895, is housed in the Seattle Centre Play House, a handsome building erected at the time of the World's Fair in 1963; it opened with *King Lear*, Max FRISCH's *The Firebugs*, Christopher FRY's *The Lady's Not For Burning*, Arthur MILLER's *Death of a Salesman*, and ARDREY's *Shadow of Heroes*. Six major productions are presented annually in a season which runs from Oct. to May, and cover a wide field, though since a financial crisis in 1970 there has been less emphasis on the classics. A typical season in the early 1980s included *The Two Gentlemen of Verona*, SHAW's *Major Barbara*, ODETS's *Awake and Sing*, and Christopher HAMPTON's *Savages*. The company also tours extensively and runs a mobile children's theatre. It works in close collaboration with the 423-seat A Contemporary Theatre, also in Seattle, which was founded in 1965 as a summer attraction and now has a season running from May to Dec. during which it aims to 'produce the important plays of our time and other alternative theatre experiences'.

SEBASTIAN, MIHAIL, see ROMANIA.

SEDAINE, MICHEL-JEAN (1719–97), French dramatist, forced by the death of his father to leave school and become a manual labourer; like Ben JONSON in similar circumstances, he soon made good the deficiencies in his education, with the help of good friends, among them the painter David. Finding himself endowed with a gift for amusing lyric verse, he began writing librettos for light operas, one of the best being for *Le Roi et le fermier* (1762), with music by Monsigny. This was probably based on DODSLEY's *The King and the Miller of Mansfield* (1737) by way of COLLÉ's adaptation *La Partie de chasse d'Henri IV*, which was published and acted in the provinces some years before its production at the COMÉDIE-FRANÇAISE in 1774. The play for which Sedaine is best remembered is *Le Philosophe sans le savoir* (1765), a domestic drama written under the influence of DIDEROT which interpreted the latter's dramatic theory much better than he was able to do in his own plays. In it Sedaine achieved a successful combination of realism and sentiment, and his picture of a bourgeois family interior is both charming and convincing. It remained in the repertory of the Comédie-Française until well into the 20th century, the only play of its type to do so.

SEDLEY, Sir CHARLES (*c.* 1639–1701), Restoration dramatist, wit, and man of letters, friend of Rochester and ETHEREGE. He wrote several plays, of which the best are *The Mulberry-Garden* (1668), a comedy of contemporary manners which owes something to MOLIÈRE's *L'École des maris* (1661) and something to Etherege's *Comical Revenge* (1664), and the lively but licentious *Bellamira; or, the Mistress* (1687), based on the *Eunuchus* of TERENCE. Sedley was also the author of a dull tragedy on the subject of Antony and Cleopatra, written in imitation of DRYDEN's heroic drama *All For Love* (1677).

SEDLEY-SMITH, WILLIAM HENRY (1806–72), American actor, stage manager, and dramatist, who in his day was considered one of the best light comedians in the United States. Born in Wales, he left home at 14 to become an actor, adding Sedley to his own name of Smith, and retaining it for the rest of his life. After touring the English provinces for some years he went to America in 1827, making his first appearance there at the WALNUT STREET THEATRE, Philadelphia, as Jeremy Diddler in Kenney's farce *Raising the Wind*. For many years he toured the larger American cities, and in 1840 was in New York supporting the elder Junius Brutus BOOTH. He was stage manager of the Boston Museum from 1843 to 1860, making his last appearance in Boston in 1869 and then going to San Francisco, where he became manager of the California Theatre until his death, working for a time with the young David BELASCO. He is chiefly remembered today for his famous temperance drama *The Drunkard; or, the Fallen Saved* (1844), a melodramatic tract which achieved an astonishing success. After its first performance at the Boston Museum it was revived by Barnum at his AMERICAN MUSEUM in New York in 1850, and became the first American play to run for nearly 200 performances. Frequently revived, it was put on again in Los Angeles in 1933 (at about the time prohibition ended) and ran for 26 years, notching up 9,477 performances.

SEGELCKE, TORE (1901–79), Norwegian actress, one of the leading Scandinavian players of her generation. After making her début at Det NORSKE THEATRET in Oslo in 1921, she went to Bergen in 1924, staying there for four years before returning to Oslo to the NATIONALTHEATRET, where she remained. Her IBSEN roles included Aase in *Peer Gynt*, Mrs Alving in *Ghosts*, Hilde Wangel in *The Master Builder*, and, supremely, Nora in *A Doll's House*, a performance which won her international fame. She also appeared in plays by Shakespeare, BJØRNSON, O'NEILL, and ANOUILH.

SEGMENT STAGE, see BOAT TRUCK.

SELWYN THEATRE, New York, at 229 West 42nd Street, between 7th and 8th Avenues. This opened on 3 Oct. 1918 and had a somewhat undistinguished career, its first successful run being scored in 1923 by Charles Ruggles in the English musical farce *Mr Battling Butler* (produced in London the previous year as *Battling Butler*). One of the most successful plays staged there was *The Royal Family* (1927), based on the personalities of the DREW and BARRYMORE families by Edna Ferber and George S. KAUFMAN, which was seen in London in 1934 as *Theatre Royal*. In Dec. 1929 *Wake Up and Dream*, with music by Cole Porter, starred Jessie MATTHEWS and Jack BUCHANAN in company with Tilly Losch. The last outstanding event in the history of the theatre was a series of six matinées of SOPHOCLES' *Electra*, staged there in 1932 with Blanche YURKA and Mrs Patrick CAMPBELL. It became a cinema in the early 1930s.

SEMENOVA, EKATERINA SEMENOVNA (1786–1849), Russian actress, who studied under DMITREVSKY, made her first appearance in 1803, and came to the fore in the tragedies of OZEROV. Later she played the heroines of Shakespeare, RACINE (especially Phèdre), and SCHILLER, making a great impression on her contemporaries by her beauty and lovely contralto voice. According to PUSHKIN, who dedicated some of his poems to her, she had no peer on the Russian stage, though her acting lacked continuity and was marred by gusts of emotion.

SEMPER, GOTTFRIED, see THEATRE BUILDINGS.

SENECA, LUCIUS ANNAEUS (*c.* 4 BC–AD 65), Roman dramatist, philosopher, satirist, and statesman, the tutor, and later the victim, of NERO. Nine tragedies adapted from the Greek are attributed to him—the *Hercules Furens, Medea, Phaedra* (or *Hippolytus*), and *Troades* (all possibly based on EURIPIDES), the *Agamemnon* (on AESCHYLUS), the *Oedipus, Phoenissae* (or *Thebais*), and *Hercules Oetaeus* (on SOPHOCLES), and the *Thyestes* (on an unknown original). *Octavia*, based on the life of Nero's unhappy wife, was formerly attributed to Seneca, but is now considered not to be by him, though its author is still unknown.

As the only extant dramas from the Roman empire Seneca's tragedies are important historically, and their influence on the development of drama in modern times has been profound, in spite of the fact that they were CLOSET DRAMAS, that is, plays written to be read aloud and not acted. In fact there are scenes which could not be staged, nor do the actors enter and leave as in a stage play. Characters speak, then relapse into silence, and it is not clear whether they are still present; objects which in a production would be visible to the actors are ignored; character drawing is stereotyped—Hercules will always be heroic, Ulysses always crafty. The *recitatio*, or recited drama, aims at immediate, often verbal, effects. Plot and character are subordinated to the necessity to startle and astonish the listeners by novel excesses in emotion or expression. Yet, in spite of the fact that Seneca's alterations of his Greek models are usually for the worse, it would be unfair to deny the dramatic power of many of his scenes or the beauty of some of his choral passages, which offset the atmosphere of gloom and horror, brutality and treachery, which pervades these plays and reflects that of contemporary life. When Seneca writes of the intrigues of countries, the instability of princes, the crimes of tyrants, the courage of men in peril of death, there is something more than literary artifice and imagination: he had experienced it all. Perhaps that is why, for the Renaissance, Seneca was the model writer of tragedy. His Latin was easily understood, his plays were divided into the five acts demanded by Horace, their plots, however melodramatic, were universally intelligible, and even his rant and rhetoric appealed to the taste of the time. His line-by-line exchange of dialogue, his chorus, his tyrants, ghosts, and witches, his corpse-strewn stage, all reappear in Elizabethan drama, and even if such effects had already been used by dramatists before Shakespeare, they were reinforced by the reading, and possibly the acting, of translations of the tragedies as early as the 1550s, long before the publication of 'Tenne Tragedies' in 1581. Their influence is already apparent in the earliest English tragedy, *Gorboduc* (1562) by NORTON and SACKVILLE; in GASCOIGNE's *Jocasta* (1566); in Shakespeare's early plays, particularly *Richard III* (*c.* 1593) (which shows what splendid results can be achieved when Senecan material is used by a man of genius) and *Titus Andronicus* (*c.* 1594); and in JONSON's two tragedies *Sejanus* (1603) and *Catiline* (1611).

ŠENOA, AUGUST, see YUGOSLAVIA.

SERBAN, ANDREL, see ROMANIA.

SERLIO, SEBASTIANO (1475–1554), Italian painter and architect, a pupil of PERUZZI and through him well acquainted with the works of VITRUVIUS. He published a treatise on architecture, largely based on Peruzzi's notes and drawings, of which the second part, dealing with perspective in the theatre, appeared in 1545. It was published in an English translation as *The Second Book of Architecture* in 1611. Writing with temporary theatres set up in princely or ducal banqueting-halls in mind, Serlio described and illustrated three basic permanent sets, the tragic, the comic, and the satyric. These, with their symmetrical arrangement of houses or trees in perspective on either side of a central avenue, had an immense influence on scene design everywhere. They survived the introduction of the *scena d'angolo,* or diagonal perspective, by the BIBIENAS, and traces of them can still be seen in the scenery of 19th-century MELODRAMA. Serlio was also a pioneer of LIGHTING, one section of his book dealing with the general illumination of the stage and theatre and the imitation of such natural phenomena as sunshine and moonlight.

SERREAU, JEAN-MARIE (1915–73), French actor and director, who in the 1950s and 1960s exercised a great influence on the French stage with his productions of plays by new authors. His style was sober and intelligent, based on careful use of the individual actor's talents, with plain but imaginative stage designs. He directed a number of productions of BRECHT's plays before they were generally known in France, and championed the Theatre of the ABSURD, bringing a company to London in 1963 with two of IONESCO's plays. From 1952 to 1955 he made the small Left-Bank Théâtre de Babylone a focal point for the AVANT-GARDE, and from 1972 to 1973 managed the Théâtre de la Tempête, which joined the Théâtre du SOLEIL and other experimental groups in the Cartoucherie de Vincennes. He also staged plays from the Third World written in French, including those of CÉSAIRE and YACINE, for which he assembled casts that could add non-European rhythms and gestures to the usual European dramatic conventions, and took Brecht's plays to Africa with black actors.

SERVANDONY, JEAN-NICOLAS (1695–1766), French scenic artist, born in Lyons, who in an effort to appear fashionably Italian changed the spelling of his name to Servandoni. After studying in Italy, he worked there and in other European countries, and was one of the first to adopt the neo-classic style in reaction against the universally popular baroque. He then settled in Paris, where he assumed control of the SALLE DES MACHINES. The influence of the work he did there in a series of spectacular productions was soon apparent in Germany and even in Italy. Later he worked in Dresden and Vienna, and in 1749 was in London, where he married.

SESAME, see RELIGIOUS DRAMA.

SET, the surroundings, visible to the audience, in which a play develops. Originally the phrase was 'set scene'—that is, an arrangement of painted and built components prepared or 'set up' in advance and revealed by the opening of a front scene, as opposed to a 'flat scene', where the FLATS slid on and off stage in full view of the audience. An alternation of set and flat scenes was common in the English theatre until almost the end of the 19th century, and with the elaboration of BUILT STUFF in Victorian times, a specially written front scene, the CARPENTER'S SCENE, was often provided to allow time for its erection. The word 'set' now covers everything arranged on the stage, ranging from the simplicity of a CURTAIN SET to the detailed naturalism of a BOX-SET, and its derivative, 'setting', has become the general term for the whole theatrical art of designing and staging the scenery of a play.

SET PIECE, a FLAT cut to the silhouette of, for example, a house or a fountain; also the name given to the solid three-dimensional elements of full scenery, also known as BUILT STUFF.

SETTLE, ELKANAH (1648–1724), Restoration dramatist, who began his theatrical career by staging DROLLS at the London fairs. He then turned to playwriting, his first play, *Cambyses, King of Persia* (1671), being put on at LINCOLN'S INN FIELDS THEATRE. A second, *The Empress of Morocco* (also 1671), was a HEROIC DRAMA first presented at Court and revived in 1673 at DORSET GARDEN with BETTERTON in the leading role. It was the first English play to be published with scenic illustrations, which have provided valuable evidence on the theatre of the time. It was also successful enough to be parodied in a farce by Duffett produced at DRURY LANE towards the end of 1673. Settle wrote a number of other plays, mainly tragedies, now forgotten but successful enough in their day to enrage DRYDEN, who considered his own popularity at Court endangered by that of Settle and satirized him as Doeg in *Absalom and Achitophel*. Towards the end of his life Settle returned to Bartholomew Fair, writing and acting in drolls at Mrs Minn's (or Myn's) booth, where he is also recorded as having played a 'dragon in green leather of his own invention'.

SET WATERS, see GROUNDROW.

SEYLER, ABEL (1730–1801), German theatre manager who was in ACKERMANN'S company at HAMBURG, where he married the actress Sophie HENSEL, a malicious, hot-tempered woman who made life so intolerable for him when he took over the management of the company that he decided to leave Hamburg and tour with a new company. This was successful for a while, particularly when EKHOF joined it and Sophie decided to retire to Vienna, but she soon returned and forced Seyler on to the road again. In 1779 Seyler, who had done good work at Gotha under IFFLAND, was asked by DALBERG to assist in the management of the new theatre in MANNHEIM, but again his wife's ungovernable temper caused their dismissal and they resumed their travels. After his wife's death in 1789, Seyler retired to SCHRÖDER'S hospitable house at Holstein.

SEYLER, ATHENE (1889–   ), English actress, outstanding in comedy. After training at the Royal Academy of Dramatic Art, where she won the Gold Medal in 1908, she appeared in the West End from 1909 in a wide variety of mainly modern roles before making her first venture into Restoration comedy, at which she was to excel, playing Cynthia in a revival of CONGREVE'S *The Double Dealer* in 1916. It was followed by Melantha in DRYDEN'S *Marriage à la Mode* in 1920, Mrs Frail in Congreve's *Love for Love* in 1921, and Lady Fidget in WYCHERLEY'S *The Country Wife* in 1924. In 1920 she also made her first appearance in Shakespeare, playing Rosalind in *As You Like It*. Her later roles in his plays included Titania and Hermia in *A Midsummer Night's Dream*, Beatrice in *Much Ado About Nothing*, Emilia in *Othello*, and the Nurse in *Romeo and Juliet*. She was also seen in plays by Oscar WILDE (*The Importance of Being Earnest*, as Lady Bracknell and Miss Prism, *Lady Windermere's Fan*, and *A Woman of No Importance*); SHERIDAN (Mrs Candour in *The School for Scandal* and Mrs Malaprop in *The Rivals*); SHAW (*Candida*, Mrs Higgins in *Pygmalion*, and Mrs Mopply in *Too True to be Good*); and CHEKHOV (Madame Ranevskaya in *The Cherry Orchard*). Outstanding modern plays in which she appeared include Lillian HELLMAN'S *Watch on the Rhine* (1942), Mary Chase's *Harvey* (1949), RATTIGAN'S *Who Is Sylvia?* (1950), and John VAN DRUTEN'S *Bell, Book and Candle* (1954). In 1958 she was seen in Peter Coke's long-running *Breath of Spring*, and in 1962 she played an Old Bawd in John FLETCHER'S *The Chances* at the first CHICHESTER FESTIVAL with unimpaired wit and vitality. She was last seen on the stage in 1966, as Martha Brewster in a revival of Joseph Kesselring's *Arsenic and Old Lace*. In 1950 she was elected

President of the Council of RADA, the first former pupil to receive this appointment. Her *The Craft of Comedy* (1944), written in collaboration with Stephen Haggard, is an essential handbook for all aspiring actors.

She married as her second husband the actor Nicholas James Hannen (1881–1972), who appeared with her in a number of productions, playing Oberon to her Titania and Benedick to her Beatrice. His first London success was as Nelson in GRANVILLE-BARKER's dramatization of Hardy's *The Dynasts* (1914), and his many roles ranged from Greek tragedy (Menelaus in EURIPIDES' *The Trojan Women*) to modern comedy (Samson Raphaelson's *Accent on Youth*). He made a big impression in Allan MONKHOUSE's *The Conquering Hero* (1924), which was followed by Philip Madras in Granville-Barker's *The Madras House* (1925). He was with the OLD VIC company on several occasions, including 1944–7, when his roles included the title role in *Henry IV, Parts I and II*.

SEYMOUR, WILLIAM GORMAN (1855–1933), American actor, son of an actress, Lydia Eliza Griffith (1832–97), and her husband James Cunningham (1823–64), who adopted the name Seymour when he left Belfast to go on the stage and became well known in America from 1840 to 1864 for his playing of stage Irishmen. William Seymour made his first appearance on the stage at the age of 2, and at 7 played the Duke of York to Lawrence BARRETT's Richard III. He continued to play children's parts with most of the stars of the day, and in 1869, at BOOTH'S THEATRE, played François to Edwin BOOTH's Richelieu in BULWER-LYTTON's play of that name, and the Player Queen to his Hamlet. He also played Henrick with Joseph JEFFERSON in a long run of Washington IRVING's *Rip Van Winkle*, and in 1871 joined the stock company at the Globe Theatre, Boston, appearing with Edwin FORREST at his farewell performance. From 1875 until his retirement in 1927 he continued to act, and was also manager of a number of theatres, being general stage director for Charles FROHMAN, and at the EMPIRE THEATRE, New York, from 1901 to 1919. He married May, the daughter of E. L. DAVENPORT, three of their children going on the stage. After his death his library was presented to Princeton University, where it formed the nucleus of a William Seymour Theatre Collection.

SEYRIG, DELPHINE (1932–   ), Lebanese-born French actress, who became well known in the early 1960s through Resnais's film *L'Année dernière à Marienbad* and her performance as Nina in CHEKHOV's *The Seagull*. She has since become one of the leading actresses of Paris. The mysterious resonance and slightly foreign intonation of her voice, and the supple exoticism of her gestures, as well as her extremely sophisticated presence on stage, made her much sought after by directors for such various plays as PIRANDELLO's *Enrico IV* and *Non si sa come*, TURGENEV's *A Month in the Country*, ARRABAL's *Le Jardin des délices* (1969), and HANDKE's *Der Ritt über den Bodensee* (1973). She has also appeared in British plays by PINTER (*The Collection* and *The Lover*, 1965; *Old Times*, 1971), James SAUNDERS (*Next Time I'll Sing to You*, 1966), and STOPPARD (*Rosencrantz and Guildenstern Are Dead*, 1967).

SHADOW-SHOW, form of puppetry in which flat, jointed figures are passed between a translucent screen and lighted candles or, nowadays, electric light bulbs, so that the audience, seated in front of the screen, sees only their shadows. It originated in the Far East, particularly in Java and INDIA, and in an increasingly crude form spread to TURKEY and so to GREECE, where it gave rise to plays centred on the comic character Karaghiozis (in Turkish, Karagöz) which can still be seen in a rudimentary form in some Greek villages and in the back streets of Athens. As 'les Ombres Chinoises', shadow-shows were popular in Paris for about 100 years. In 1774 Dominique Séraphin opened a theatre devoted to them in Versailles, moving in 1784 to the Palais Royal, where his nephew continued his work until 1859. It was Séraphin who first introduced to Paris the classic shadow-play *The Broken Bridge*, in which a frustrated traveller indulges in an impassioned but silent argument with a workman on the other side of the river. This was well known in the streets of London, where, as the Galanty Show, shadow-plays continued to be given up to the end of the 19th century, usually in PUNCH AND JUDY booths with a thin sheet stretched across the opening and candles behind. There was a literary revival of the shadow-show at the Chat Noir in Paris in the 1880s, and in the 1930s Lotte Reiniger employed the technique of the shadow-show for her animated films. Her puppets were made of tin, as were those used at the Chat Noir and in the English Galanty Show, but in Java and Bali, where the shadow-play survives in its traditional form, they are cut from leather. Manipulation is by bamboo rods or concealed wires running up the centre of the figure and operated from below the screen, except in Turkey and Greece, where the rod is held at right angles to it, and fastened to the flat figure in the centre of the back.

SHADWELL, THOMAS (*c.* 1642–92), Restoration dramatist, whose first play *The Sullen Lovers; or, the Impertinents* (1668) was based on MOLIÈRE's *Les Fâcheux* (1661). The comedies which followed it seem, however, more indebted to JONSON, whom Shadwell much admired. The best known of them is *Epsom Wells* (1672), which had an unexpected success when revived in 1969 at the Thorndike Theatre in Leatherhead. Shadwell has been much criticized for his adaptation of Shakespeare's *The Tempest* as an opera, *The Enchanted Island* (1674), in which, following the examples of DAVE-

NANT and DRYDEN, everything was subordinated to the stage machinery and scenery. His own opera *Psyche* (1675) was staged with equal extravagance, the scenery alone costing DORSET GARDEN £800. Shadwell then returned to writing comedy with *The Libertine* (also 1675) and *The Virtuoso* (1676) before rewriting *Timon of Athens* as *The Man-Hater* (1678), and in his last years produced two of his best comedies, *The Squire of Alsatia* (1688) and *Bury Fair* (1689), which give interesting though somewhat scurrilous pictures of contemporary manners. His last play, *The Volunteers; or, the Stock Jobbers*, was produced posthumously. It was ironic that Shadwell, who was mercilessly satirized by Dryden in *MacFlecknoe; or, a Satire upon the True-Blue Protestant Poet T. S.* (1682), should have succeeded him on political grounds as Poet Laureate in 1688.

SHAFFER, PETER LEVIN (1926–), English dramatist, whose first play *Five Finger Exercise* (1958), a naturalistic study of family tensions directed by GIELGUD, had a great success in London and was seen a year later in New York. It was followed by a double bill, *The Private Ear* and *The Public Eye* (1962; N.Y., 1963), with Maggie SMITH in London, which also did well, and *The Merry Roosters Panto* (1963), written for Joan LITTLEWOOD. Shaffer's next play, *The Royal Hunt of the Sun* (1964), an epic tragedy dealing in a symbolic manner with the murder of the Aztec king Atahualpa by Pizarro, was first performed at the CHICHESTER FESTIVAL and then seen with great success in London and New York, as was the one-act *Black Comedy* (1965), an ingenious farce in which most of the action is supposed to take place in the dark, although the stage is lit. In New York, and later in London, the latter was joined in a double bill with another one-acter, *White Lies* (N.Y., 1967; London, as *The White Liars*, 1968). It was followed by a full-length play, *The Battle of Shrivings* (1970), in which an elderly pacifist reminiscent of Bertrand Russell (played by Gielgud) was shown in conflict with an anti-liberal poet. Two later plays were *Equus* (1973; N.Y., 1974), dealing with a stable-boy who blinds horses and his subsequent treatment by a psychiatrist, and *Amadeus* (1979; N.Y., 1980), portraying the bitterness aroused in the composer Salieri (played by Paul SCOFIELD followed by Frank FINLAY in London, and Ian MCKELLEN followed by John WOOD in New York) by the success of Mozart.

Shaffer's twin brother Anthony achieved an unexpected success with his first play *Sleuth* (London and N.Y., 1970) which his later plays, including *Murderer* (1975), have not equalled.

SHAFTESBURY THEATRE, London. (1) In Shaftesbury Avenue. This four-tiered theatre, with a seating capacity of 1,196, was the first to be built in the new Shaftesbury Avenue, and opened on 20 Oct. 1888 with FORBES-ROBERTSON as Orlando in *As You Like It*, which was not a success; but E. S. WILLARD did well in Henry Arthur JONES's *The Middleman* (1889) and *Judah* (1890). In 1898 Hugh Morton's musical *The Belle of New York*, which introduced Edna May to London, had a long run. The first Negro musical *In Dahomey*, starring Bert WILLIAMS, was seen at this theatre in 1903, and ran for 251 performances. Seasons of GRAND GUIGNOL and revivals followed, and in 1909 Cicely COURTNEIDGE made her London début in Lionel Monckton's *The Arcadians* under the management of her father Robert Courtneidge. In 1921 Clemence DANE's *Will Shakespeare* was well received, and the theatre scored an immense success with *Tons of Money* (1922) by Will EVANS and Valentine, which brought together Ralph LYNN, Robertson HARE, Tom WALLS, and Mary BROUGH, so laying the foundation of the future ALDWYCH farces. The next year saw the London début of Fred Astaire and his sister Adèle in the musical farce *Stop Flirting*, but future productions were less successful; the last was a revival of Oscar Straus's *The Chocolate Soldier* (1940), and on the night of 16–17 April 1941 the theatre was destroyed by bomsb.

(2) At the Holborn end of Shaftesbury Avenue. This theatre, with a seating capacity of 1,300 in three tiers, was built by the MELVILLE brothers to house their own MELODRAMAS. It opened on 26 Dec. 1911 as the New Prince's Theatre, but the 'New' was soon dropped. After 1916 it had no settled policy, and its productions ranged from straight plays to ballet, pantomime, and opera. There were also revivals of GILBERT and Sullivan during the 1920s, and distinguished foreign visitors included Sarah BERNHARDT in 1921, on her last visit to London; Diaghilev's Ballets Russes in 1921 and 1927; and Sacha GUITRY with Yvonne PRINTEMPS in 1922. In 1924 DARLINGTON'S *Alf's Button* had an unexpected success. Sybil THORNDIKE and Henry AINLEY were seen in *Macbeth* in 1926, and in 1927 George ROBEY appeared in the revue *Bits and Pieces*. A year later a musical comedy *Funny Face*, with music by George Gershwin, began a long run. From 1929 to 1946 the theatre was intermittently managed by Firth Shephard, and under him two dramatizations by Ian HAY of stories by Edgar WALLACE, *The Frog* (1936) and *The Gusher* (1937), had long runs, as did two musicals devised by Douglas Furber, *Wild Oats* (1938) and *Sitting Pretty* (1939). The theatre was badly blasted in 1940–1 but managed to stay open, and for a time housed the SADLER'S WELLS ballet and opera companies. After the war the main successes were again the Gilbert and Sullivan seasons, but there were long runs of *His Excellency* (1950) by Dorothy and Campbell Christie and of two American musicals, *Pal Joey* (1954) and *Wonderful Town* (1955). In 1962 the theatre closed for renovation, and it reopened under its present name on 28 Mar. 1963 with another American musical *How To Succeed in Business Without Really Trying*. In 1966 a farce by Philip King and Falkland Cary, *Big Bad Mouse*, began a

long run, and it was followed by the epoch-making American rock musical *Hair* (1968), which just failed to complete its 2,000th performance in 1973 owing to the collapse of the auditorium ceiling on 20 July. The theatre reopened at the end of 1974 with a revival of *West Side Story*. It was followed by a series of short runs which left the theatre in the doldrums until in 1980 the musical *They're Playing Our Song*, written by Neil SIMON, scored a big success.

SHAKESPEARE. 1. LIFE AND WORKS, William Shakespeare (1564–1616), first son and third child of John Shakespeare, a glover yeoman of Stratford-upon-Avon, and his wife Mary Arden, a minor heiress of Wilmcote, was christened on 26 Apr. 1564; tradition asserts that his birthday was 23 Apr., St George's Day. John Shakespeare became an alderman in 1565 and in 1568 bailiff (that is, mayor). His position would have qualified William to attend the grammar school of his native town, but its archives have not survived and the first record of his activities is that of his marriage, evidently a hasty one, to a lady whom extant documents almost certainly (but not positively) identify as Anne Hathaway of Shottery. The wedding took place about the end of Nov. 1582; a daughter, Susanna, was born in May 1583, and twins followed in 1585. Another gap in our knowledge extends from this time until 1592, when a pamphlet written by the dying Robert GREENE shows Shakespeare evidently well established in London as actor and dramatist. There is a story that he left Stratford because of a deer-stealing escapade at Charlecote; another tale asserts that he served as a schoolmaster in the country: whatever his activities, during those years he must have added to his education and gained experience of life in circles higher than his own domestic surroundings. From this time contemporary documents yield us more information. In 1593 and 1594 he dedicated his poems *Venus and Adonis* and *The Rape of Lucrece* to the Earl of Southampton, in terms that suggest familiarity, and by the beginning of 1595 he had evidently become a SHARER in the company known as the CHAMBERLAIN'S MEN. Evidence of his rise in the world appears in his father's successful application in 1596 to the Heralds' College for a coat of arms, and by the poet's purchase in 1597 of the large house known as New Place in Stratford. Other documents show him, in 1596, associated with Francis Langley, builder of the SWAN THEATRE, in a complicated quarrel involving William Gardiner, a Justice of the Peace in London, and at the same time concerned with Stratford interests (the only extant letter addressed to him is from his fellow townsman Richard Quiney in 1598). About that time he was resident at St Helens, Bishopsgate, but by 1599 he had moved to the Liberty of the Clink, and in 1604 was a lodger in the house of Christopher Mountjoy in Cripplegate. In 1610 he seems to have taken up residence at New Place, although his presence in London during the summer of 1612, the spring of 1613, and the winter of 1614, together with his purchase of some Blackfriars property in 1613, demonstrates his continued association with the metropolis. His will, signed on 25 Mar. 1616, preceded his death (23 Apr.) by about a month; tradition says he died after a too convivial evening with DRAYTON and JONSON. He was buried in the chancel of Stratford church.

There are other documents, but these mainly concern business transactions at Stratford, and do not add much to our knowledge of the man. The record of his life can, however, be expanded both by relating his literary work to those meagre facts and by cautiously making use of traditional material. The prefatory matter to the First Folio shows how highly he was esteemed by his fellow actors and by his great contemporary Ben Jonson: tradition speaks, with probable truth, of royal esteem as well. There must be a personal story behind the sonnets, and with that story the puzzling narrative of *Willobie his Avisa* (1594) is likely to be connected. It has been recorded that he 'died a Papist', and this suggestion is not without some vague corroborative evidence.

Bare factual information, however, and legends regarding his adventures fade into insignificance beside the extraordinary range of his poetic achievement. He came just at the right moment to make full and fresh use of the teeming drama of his time, finding a novel and flexible stage apt for his purposes and an eager audience representative of all classes in the community to encourage and inspire. The man and the time were in harmonious conjunction. Starting to write probably about 1590, he contributed at least 36 plays to the theatre. Of these, 16 were printed in quarto during his lifetime, but, apart from the fact that some are obviously bad texts, surreptitiously obtained, the publishing conditions of the age make it probable that he himself did not read the proofs. In 1623 HEMINGE and CONDELL of the KING'S MEN, as the Chamberlain's Men were called after 1603, issued the entire body of his dramatic work in folio form; this volume presents the only texts of another 20 plays, and is probably the most important single volume in the entire history of literature. Arranging the contents under the headings of Comedies, Histories, and Tragedies, the editors of the Folio give no indication of the dates of composition of the separate items, but from a careful scrutiny of such external evidence as exists, and from 'internal' tests (the quality of the blank verse, use of prose and rhyme, etc.), most scholars are agreed, at least in general terms, concerning their chronology.

Shakespeare probably started his career by writing, unaided or in collaboration, a historical tetralogy consisting of the three parts of *Henry VI* and *Richard III*. Parts II and III of *Henry VI* seem to have come first in 1591, Part I in 1592, and *Richard III* in 1593. Richard BURBAGE won fame in the role of Richard, and the play was still being presented in 1633 as a record of a Court performance shows. The success of these plays no doubt encouraged Shakespeare to write *King John*, based on an older two-part dramatization of that monarch's reign. This tragedy stands alone, but shortly afterwards, about 1595, another historical tetralogy was started with *Richard II*, which, during the Essex conspiracy in 1601, won notoriety because of its abdication scene; it was still in the repertory of the GLOBE THEATRE in 1631, and from the year 1607 comes an interesting record of its popularity, when it was produced on the high seas by sailors in the fleet of William Keeling. The two parts of *Henry IV* (probably about 1597 or 1598) carry on the story of Henry Bolingbroke and introduce a richly contrasting comic element with the character of Falstaff (originally named after the historical Sir John Oldcastle), while the general theme is rounded off with *Henry V* (probably 1599). With this play Shakespeare closed his career as a writer of histories, save for the late *Henry VIII*, produced in the summer of 1613, in which he is believed by many scholars to have collaborated with the young FLETCHER.

Early in his career he started to try his hand at comedy, experimenting with the courtly, satirical *Love's Labour's Lost*, *The Comedy of Errors* in the style of PLAUTUS, and the more robust *The Taming of the Shrew*. These plays convey the impression of a young dramatist unsure of his orientation, yet all are skilful and succeeded in holding the stage. The first was revived in 1605 and was still being played in 1631; the second, which is known to have been acted at Gray's Inn in 1594, was revived in 1604; a quarto of the third, issued in 1631, indicates that it was still in the playhouse repertory, and a Court performance is recorded in 1633. *The Two Gentlemen of Verona* shows, perhaps, even less assurance, although in its mingling of the comedy of humour and romance it is a more obvious forerunner of the comedies written between 1595 and 1599, *A Midsummer Night's Dream*, *Much Ado About Nothing*, *As You Like It*, and *Twelfth Night*. Little is known concerning the early stage-history of these rich and lyrical plays. *A Midsummer Night's Dream* was given a Court performance in 1604; *Twelfth Night* was acted at the Middle Temple in 1602, was given at Court in 1618 and 1623, and was still popular in 1640. Some time following the production of *Henry IV* comes *The Merry Wives of Windsor*, an aberration in this series; probably tradition is right in saying it was written at the command of Queen Elizabeth, who wanted to see Falstaff in love. It was revived at Court in 1604 and 1638. In *The Merchant of Venice* a break in the almost perfect balance observable in the other comedies is patent, and this leads to a couple of so-called 'dark comedies'—*All's Well that Ends Well* and *Measure for Measure* (given at Court in 1604), in which the romantic material is strained almost to breaking. With these may be associated the cynically bitter *Troilus and Cressida*, which possibly was acted not on the public stage but privately.

These were composed at the same time as Shakespeare was reaching towards the deepest expression of tragic concepts. Already at the very beginning of his career he wrote (possibly in collaboration) *Titus Andronicus*, a play which despite or because of its bloodiness remained popular. Again, in the midst of his lyrical comedies he made a second attempt at tragedy in *Romeo and Juliet* (about 1595), evidently a popular play although there are no records of specific early performances. Then came the great series of tragedies and Roman plays. *Julius Caesar*, which was seen by a visitor to London in 1599, was probably the first, but *Hamlet* must have come very soon after. *Othello* may have been new when it was presented at Court in 1604; it was evidently very popular. Performances are recorded in 1610, 1629, 1635, and there were Court productions in 1612 and 1636. *King Lear* must have followed not many months later; it appeared at Court in 1606, and about the same time came *Macbeth*, which, tied in theme with *Julius Caesar*, clearly addresses itself to a Jacobean Court. The classical subject-matter of *Julius Caesar* is paralleled in *Antony and Cleopatra*, in *Coriolanus*, and in *Timon of Athens*, which seems to have survived only in a draft (all probably about 1607 or 1608).

In *King Lear* Shakespeare had turned to ancient British history, and the atmosphere of this play is reproduced, albeit with a changed tone, in *Cymbeline*. Seen by Simon Forman in 1611, it was written probably about 1610; there was a revival at Court in 1634. Another play seen in 1611 by Forman, *The Winter's Tale*, is similar in spirit, darker than the early comedies of humour and including incidents reminiscent of the tragedies, yet ending with solemn happiness. Evidently popular, it had Court productions in 1611, 1613, 1618, 1619, 1624, and 1634. *The Tempest*, gravest and serenest of all the dramas, was presented at Court in 1611 and in the following year.

To Shakespeare have been attributed, in whole or in part, several other dramas. His hand in *The Two Noble Kinsmen*, which was printed in 1634 as by him and Fletcher, and in *Pericles*, printed as his in 1609 and added to the Third Folio of 1664, has generally been accepted. There may be some pages of his own writing in the manuscript of *Sir Thomas More*, dating probably from the mid-1590s. Less likely, although still possible, is his participation in *Edward III*, printed in 1596.

Of his non-dramatic work, *Venus and Adonis* was carefully printed in 1593, and *The Rape of Lucrece* in 1594. The *Sonnets* were first printed in 1609, with a publisher's dedication to 'Mr W. H.' over which a lively, acrimonious, but inconclusive controversy has raged for many years. Some of his work appeared in *The Passionate Pilgrim* (1599) and in *Love's Martyr* (1601).

Although it has been conjectured that a few of Shakespeare's plays were originally written for private or for Court performance, his strength rests in the fact that he was essentially a 'public' dramatist, addressing himself to the demands of a widely representative audience eager to listen to rich poetic utterance, keenly interested in human character, and apt to welcome both the delicacy of romantic comedy and the rigours of tragedy. Little is known concerning the contemporary production of his dramas, but sufficient information concerning the Elizabethan stage is extant to give us a general impression of the methods used in their presentation. The absence of scenery helps to explain the strong emphasis on poetic utterance; for the Elizabethans the visual appeal lay in the rich costumes worn by the actors, which threw stress on the persons speaking rather than on dumb, inanimate properties. Here was an opportunity for the dramatist to reveal an essential reality that has kept his plays vivid over three centuries. Again and again commentators have spoken of Shakespeare as a 'child of nature', or as one who vied with nature in his creative power, and possibly there is no other author who so harmoniously and with such ease reveals the basic power of the great playwright—the power at once to create the characters of his imagination and to enter into them. Transcendence and immanence are the two bases of his genius, and it is the presence of these qualities that explains the prime paradox—that Shakespeare himself belongs in a sphere beyond our grasp even while we have the impression of knowing him intimately; it likewise explains that peculiar irony which is the characteristic feature of both his comedy and his tragedy. This quality is, of course, dependent upon the nature of his own literary genius, yet one cannot too firmly assert that for its expression in these plays the Elizabethan theatre and the contemporary audience were definitely responsible.

2. PRODUCTION IN ENGLISH. Although some of Shakespeare's plays remained continuously in the repertory of the English theatre, from the Restoration until the end of the 19th century few people had an opportunity of seeing them in their original form. For this the change in theatre building and theatrical technique was partly responsible, but the main onus lay on those who, while professing their admiration for Shakespeare, deliberately altered his texts to make them conform to the requirements of a new age. During the Commonwealth his comedies were pillaged to provide short entertainments or DROLLS, like that of 'Bottom the Weaver', taken from *A Midsummer Night's Dream*. There was every excuse for this, in the precarious state of the theatre at that time, since a full-length production would have had little chance of survival. But there was no excuse for later remodellings except that, to a small sophisticated audience much under the influence of French classical tragedy, Shakespeare was a barbarian, whose occasional poetic beauties entitled him to some consideration, but whose work stood in need of purification and revision.

It was in this spirit that the Restoration and early 18th-century theatre approached Shakespeare. *Macbeth* was revised and embellished with singing and dancing by DAVENANT; his 'singing witches' were to last until 1847. *Romeo and Juliet* was sometimes played with a happy ending. Neither play was revived in its proper form until 1744. *The Tempest*, adapted by Davenant and DRYDEN, was then made into an opera by SHADWELL. *A Midsummer Night's Dream* was combined with masque-like episodes to music by Henry Purcell to become *The Fairy Queen*, and LACY made a new version of *The Taming of the Shrew*. In 1681 Nahum TATE rewrote *King Lear*, omitting the Fool (who was not seen again until 1838), sending Lear, Gloucester, and Kent into peaceful retirement, and keeping Cordelia alive to marry her lover Edgar. He also tackled *Richard II* and *Coriolanus*, but with less success. An adaptation which survived even longer than Tate's *Lear* was the *Richard III* of Colley CIBBER. First given in 1700, and containing passages from *Henry IV*, *Henry V*, *Henry VI*, and *Richard II*, as well as a good deal of Cibber's own invention, it proved immensely popular and provided an excellent part for a tragic actor. Also popular were versions of other plays by Cibber, Shadwell, and later GARRICK. This great actor, though in some ways he tried to prune the excrescences of the Restoration texts and gave the first recorded performances of *Antony and Cleopatra*, retained Cibber's *Richard III*, gave Macbeth a dying speech, and caused Juliet to awake before the death of Romeo, giving them a touching final conversation. He even made short versions of four of the comedies, and his *Katharine and Petruchio* (1756) remained popular until well into the 19th century.

Garrick's tampering with *Hamlet*, including his omission of the Grave-diggers, marks the end of this phase in the treatment of Shakespeare's texts. The tide had already begun to turn slowly in favour of the originals, and in 1741 MACKLIN had rescued Shylock, in the so-called *Jew of Venice*, from the hands of the low comedian and by his interpretation of the part is said to have caused Pope to exclaim: 'This is the Jew, That Shakespeare drew.' John Philip KEMBLE still thought it necessary to cut and revise each play he produced. He did, however, try to follow the lead, again given by Macklin, who in 1773 had dressed Macbeth in Highland costume, in reforming the costuming of Shakespeare's plays. His Othello was still a scarlet-coated general, his Richard III wore silk knee-breeches, and Lear defied the storm in a flowered dressing-gown. But, helped by his sister Sarah SIDDONS, who was the first to discard the hoops, flounces, and enormous headgear of earlier tragic heroines, he made an effort to combine picturesqueness with accuracy, and his Coriolanus wore what in contemporary thought approximated to a Roman costume. The *Examiner* scoffed at his innovations, but they bore fruit in Charles KEMBLE's *King John*, staged in Mar. 1823 with historically accurate costumes designed by

J. R. PLANCHÉ. The campaign against the radical adaptations was almost won by this time, and Edmund KEAN had taken a step in the right direction on 10 Feb. 1826 when he restored the original ending to *King Lear*. Helped by the research of scholars, and by the criticisms of such men as COLERIDGE and HAZLITT, the original plays were gradually emerging from the accretions of more than a century's rewriting. Madame VESTRIS and the younger MATHEWS revived *Love's Labour's Lost* and *A Midsummer Night's Dream* in 1839–40 with the original text, and the freedom of the theatres in 1843 enabled Samuel PHELPS to embark on his fine series of productions at SADLER'S WELLS from 1844 to 1862, while Charles KEAN staged his equally remarkable Shakespeare seasons at the PRINCESS'S.

Shakespeare was now presented in a reasonably correct form, but he suffered a new distortion at the hands of his admirers by their emphasis on detail, scenery, and pageantry. The archaeological correctness—which apparently led also to dullness—of Charles Kemble heralded the magnificent spectacles of MACREADY at COVENT GARDEN from 1837 to 1843. He should be credited with having restored the Fool to Lear, though played by a woman, and with having revived *The Tempest* without Dryden's interpolations, which included a male counterpart of Miranda. Phelps, who in 1845 restored a male Fool, and in 1847 put on *Macbeth* without the singing witches, as well as reviving *The Winter's Tale*, was more concerned with the text than with the scenery, which was pleasantly sober and unobtrusive. Elsewhere the newly restored text was in danger of disappearing under the elaboration of detail, while the action of the play, designed for an untrammelled stage, was constantly rearranged and held up because of the necessity for elaborate scene changes. Still in this tradition were Henry IRVING'S productions at the LYCEUM (1878–1902) and Beerbohm TREE'S at the HAYMARKET (1887–97) and later at HER MAJESTY'S.

The publication in 1888 of de Witt's drawing of the SWAN THEATRE encouraged attempts to reproduce not only the text of Shakespeare's plays but also the physical conditions in which they were first seen. The main interest had already switched from the problem of the text to the problem of interpretation. This became even more important as the director gained the upper hand in the theatre. To this was added the problem of providing a building suitable for Shakespeare, an approximation to the original Elizabethan stage. This was first tackled by William POEL, with his early productions of Shakespeare and the later founding of the Elizabethan Stage Society, and by Nugent Monck with the MADDERMARKET at Norwich. Robert ATKINS'S productions at the RING, Blackfriars, made an effort to solve the difficulties by presenting Shakespeare-in-the-round. After the Second World War the stages at the SHAKESPEARE MEMORIAL THEATRE, Stratford-upon-Avon, and the OLD VIC crept out beyond the proscenium arch, which Stratford finally abolished altogether. These and other tentatives, including the MERMAID THEATRE in London and the open stage at the STRATFORD (ONTARIO) FESTIVAL theatre showed that Shakespeare cannot adequately be presented in the proscenium-arch theatre. The form of stage best suited to his plays is still a matter of dispute, and in view of the divergent opinions on the details of his original stage it seems as if it will not soon be settled.

Even in the proscenium-arch theatre, however, efforts have been made to present the plays more simply and coherently, allowing the action to flow unchecked for as long as possible. One of the landmarks in the history of Shakespearian production was undoubtedly GRANVILLE-BARKER'S first season at the SAVOY in 1912; another was the introduction in 1914 of the first season of Shakespeare at the Old Vic, where the pioneer work of Lilian BAYLIS did more for Shakespeare *vis-à-vis* the general public than the lavish but intermittent spectacles of the West End theatres. Experiments in presenting the plays have been frequent, from Barry JACKSON'S modern-dress *Hamlet* to the fantastications of KOMISARJEVSKY, the elaborations of REINHARDT, and the challenges to tradition of Peter BROOK. There have been a *Lear* with Japanese décor, a *Hamlet* set in Victorian times, a *Romeo and Juliet* in modern Italian style by ZEFFIRELLI. But, with all its divergencies and aberrations, the main trend since 1900 has been the simplification of the background, by the use of a permanent set, a bare stage, or symbolic settings, and a consequent insistence on the importance of the text, the free flow of the verse, and the unhampered action of the plot. These ideals have been discernible behind most of the productions of the ROYAL SHAKESPEARE COMPANY, diverse in style though these have been. A happy combination of scholarly research and theatrical experience seems the best method of dealing with the many problems that inevitably arise in producing plays written in haste, printed without the author's supervision, and designed for a vanished playhouse.

In America the first productions of Shakespeare were given by visiting English actors, and the situation was therefore much the same as in contemporary England. For instance, the first recorded play, *Richard III*, acted in New York City on 5 Mar. 1750 by Thomas Kean and William Murray's company at their NASSAU STREET THEATRE, was in Colley Cibber's version. Little information is available on the texts of subsequent productions, but it is reasonable to suppose that they were substantially those current in the English theatre at the time. Although there was not, as in Europe, a language barrier or a preconceived notion of classical writing to be overcome, there were moral difficulties. Even Shakespeare's reputation was not always sufficient to overcome deep-rooted prejudices against play-going, and *Othello* was first introduced to Boston as 'a Moral

Dialogue against the Sin of Jealousy'. Probably more people in the 18th century were reading Shakespeare as a poet than seeing him as a playwright. Although the Eastern cities had opportunities of seeing the full-length plays in theatres modelled on contemporary English lines, in the West it was the lecturers, elocutionists, entertainers, and showboat companies who first popularized Shakespeare, in isolated scenes and speeches. It may be said that from the earliest times Shakespeare played a large part in the emergent culture of the pioneer peoples.

After the first 50 years it is difficult to disentangle the imported productions of Shakespeare from those of the young but vigorous American theatre. Thomas Abthorpe COOPER and Henry WALLACK, who both appeared in Shakespeare early in their respective careers, were typical of the new generation of actors who, though born in England, spent the latter and greater part of their working lives in America. The honour of being the first native-born actor to play leading roles in Shakespeare must probably go to John Howard PAYNE, who as a youth of about 17 played Hamlet and Romeo in 1809, but the greatest was undoubtedly Edwin BOOTH, whose Hamlet in 1864 was generally admired. The 19th-century personality cult of the 'star' actor and the insistence on elaborate trappings was as prevalent in America as in England, and was reinforced by the many tours undertaken by such London companies as Charles Kean's and later Irving's.

The 20th century saw, as in England, the gradual liberation of the play from overdecoration, with the simplified settings of Robert Edmond JONES and Norman BEL GEDDES, and the new approach to a purified text. This was reinforced by a phenomenon peculiar to America—the invasion of the world of the theatre by the universities, which did not take place in England until much later. The proliferation of UNIVERSITY DEPARTMENTS OF DRAMA whose syllabuses lead to a degree in drama has not been without its dangers, but it has led to much scholarly work on the problems of Shakespeare and the Elizabethan theatre in general, and to a wider spread of interest in and productions of Shakespeare than would have been the case otherwise. This is particularly fortunate since the commercial theatre on Broadway has not on the whole been enthusiastic about Shakespeare's plays. With no tradition of Shakespearian acting, with the same handicap as in London of unsuitable theatres, and with no pressing demand from the audience, managers have found them expensive to stage and uncertain in their box-office returns. There have been exceptions, but these have mostly been due to the efforts of an individual—John BARRYMORE, Eva LE GALLIENNE, Orson WELLES, Maurice EVANS, Margaret WEBSTER, Joseph PAPP. It is therefore the universities who have had to bear the main burden of keeping Shakespeare alive on the American stage, either by incorporating his plays in the repertory of a community theatre, or by organizing festivals devoted solely to his works. The establishment of the AMERICAN SHAKESPEARE THEATRE at Stratford, Conn., has done much to consolidate the position of Shakespeare in the American theatre, where he is often 'more honoured in the breach than the observance'. The main dangers of academic Shakespeare are pedantry in the presentation and immaturity in the actors, but a healthy spirit of experiment may do much to redress the balance. It certainly seems as if the future of Shakespeare in the United States lies rather with the community and the off-Broadway theatres than with the commercial theatre.

SHAKESPEARE FESTIVALS. The main festivals of Shakespeare's plays by professional companies are given in the three Stratfords—in England at the ROYAL SHAKESPEARE THEATRE, in Canada at the STRATFORD (ONTARIO) FESTIVAL theatre, and in the United States at the AMERICAN SHAKESPEARE THEATRE. Professional open-air festivals are also held in Central Park, New York City, under the auspices of Joseph PAPP, and in Regent's Park, London, where performances were given from 1900 onwards by Ben GREET and his Woodland Players, a permanent OPEN-AIR THEATRE being established there in 1933 and run for many years by Robert ATKINS. London's ST GEORGE'S THEATRE also provides all-the-year-round Shakespeare productions, and an attempt is being made by Sam WANAMAKER to gather funds for the building of a new GLOBE THEATRE near the site of the original one.

The oldest Shakespeare festival in America, founded in 1935, is that held out of doors during the summer at Ashland, Oregon. Plays were first given by students in a roofless structure, which was damaged by fire in 1940. The festival was then suspended until 1947, when a new stage was built and professional actors were engaged. An Elizabethan-type theatre, the Old Globe, was designed by Thomas Wood Stevens for the 1935–6 California Pacific International Exposition, and later transported to Balboa Park, San Diego, where a summer Shakespeare Festival was inaugurated by Ben Iden PAYNE in 1950. The first actors were local amateurs, but in 1954 student actors and technicians from colleges and drama schools throughout the country were enrolled, and since 1958 the major roles have been played by professionals. The theatre, originally roofless like its predecessor, was later roofed over. Since it was damaged by fire in 1978 plays have been performed on a Festival Stage adjacent to the Old Globe, but a 350-seat Elizabethan theatre is now being built to replace it. The Great Lakes Shakespeare Festival at Lakewood, Ohio, founded in 1962, stages plays by Shakespeare and others indoors in the summer. There are about 30 Shakespeare festivals in the United States.

A Danish festival, devoted entirely to productions of *Hamlet*, has been held in the courtyard of

Kronborg Castle at Elsinore. It was inaugurated in 1937, with a performance by Laurence OLIVIER and the OLD VIC company. In 1938 there was a German company headed by GRÜNDGENS, and in 1939 another English company under John GIELGUD. After the Second World War efforts were made to revive this festival, and Swedish, Norwegian, Finnish. American, and Irish Hamlets were seen, with Michael REDGRAVE heading a third English company in 1950. After 1954, however, there were no further productions until the PROSPECT THEATRE COMPANY's visit in 1978, with Derek JACOBI in the title role.

There are a number of amateur festivals in England, of which the most important is probably that held in London in the week nearest to Shakespeare's birthday (23 Apr.), which includes a production in the yard of the George Inn, Southwark, a site which more nearly than any other appears to resemble the innyards in which London companies appeared before the erection by James BURBAGE of the THEATRE in 1576.

SHAKESPEARE IN TRANSLATION. There is probably no country in the world where some, at least, of Shakespeare's plays are not known in translation, and it would be impossible to list all the versions that have been made. Many of them were intended for reading only, and although the ENGLISH COMEDIANS, travelling on the Continent in the first quarter of the 17th century, included in their repertory cut versions of such plays as *A Midsummer Night's Dream, Romeo and Juliet, Hamlet, King Lear*, and perhaps *The Merchant of Venice* and *The Tempest,* they were played in English, and Shakespeare remained virtually unknown outside his own country until well into the 18th century. It was VOLTAIRE, in his *Lettres philosophiques* (1734), who first drew attention to him, but even then early versions of some of the plays did little to help his reputation. Among them were a frigid adaptation in verse of *Julius Caesar* by the German Ambassador to London, Baron C. W. von Borck, published in 1741; a prose summary of some of the plays in French by LA PLACE in 1744, with some of the more important poetic scenes put into dull verse; and bowdlerized versions of some plays in German prose by Christoph Wieland, 1762–6. The first translations to be widely read were the French prose versions by LETOURNEUR, 1776–82. They were the only ones known in Italy until the publication in 1819–22 of Italian versions by Michele Leoni, and together with those of La Place were read by MANZONI, the first Italian writer to be influenced by Shakespeare. Letourneur's versions also provided DUCIS, who knew no English, with the basis of some of the first French stage versions, used by MOLÉ. The latter were tailored to fit the UNITIES, so that Desdemona, for instance, in *Othello*, was wooed, wedded, and murdered in the space of a day, while *Hamlet, Romeo and Juliet*, and *King Lear* were provided with conventionally happy endings.

In Germany, attention was drawn to Shakespeare by LESSING's *Hamburgische Dramaturgie* (1767–9), but it was the great actor F. L. SCHRÖDER who first introduced him to German theatre audiences, with a production of *Hamlet* at Hamburg in 1776. Though considerably adapted, the play still gave the audience a shock, but when they had recovered from the Ghost they broke into applause. This encouraged Schröder to stage *Othello*, which failed, *The Merchant of Venice, Measure for Measure*, which had a cool reception, and finally, in 1778, *King Lear*, an unqualified success. New translations were made by J. J. Eschenburg (the Mannheim Shakespeare, 1775–82) and by A. W. von SCHLEGEL, whose versions of 17 of the plays were closer to the originals, though still very much bowdlerized. They were later added to by TIECK and others, and republished with revisions and additions by F. Gundolf in 1908. Good acting versions of the most important plays were provided between 1869 and 1871 by the actor Emil DEVRIENT, and once established on the German stage Shakespeare flourished until he was finally claimed as '*unser*' (our) Shakespeare, and his genius was thought akin to that of the Teutons. In the 20th century versions have come from both East and West Germany, Anselm Schlösser's in 1966 being best known in the former, and Rudolf Schaller's in 1961 in the latter. Bertolt BRECHT also made a version of *Coriolanus* which, ironically, has been translated back into English.

In France it was Romanticism that finally swept Shakespeare into favour, and the visit of an English company under Charles KEMBLE to Paris in 1827 set the seal on his growing reputation. New translations by Laroche were published in 1839, with an introduction by the elder DUMAS, and between 1856 and 1867 translations appeared of the complete works by François Victor-Hugo, son of Victor HUGO, whose book *William Shakespeare* (1864) was intended as a preface to his son's work. These translations are now the ones most commonly used, though some useful stage versions have been issued by the THÉÂTRE NATIONAL POPULAIRE, and in the 1940s–50s Jean-Louis BARRAULT made a great success in a new version of *Hamlet* by André GIDE.

In Italy Shakespeare established himself more slowly. Carlo Rusconi's *Hamlet* did not appear until 1839, and Giulio Carcano's standard translations in blank verse were not begun until 1843, being completed in 1882. Translations by C. V. Lodovici, Giorgio Melchiori, and Mario Praz appeared in the mid-1960s. The first Italian actor to succeed in Shakespeare was Ernesto ROSSI in 1856 as Othello, which was also the finest role of Tommaso SALVINI. Their success in Shakespeare's other tragic heroes, and the acclaim which greeted Adelaide RISTORI's Lady Macbeth, seem to show that in spite of the success of ZACCONI and NOVELLI as Petruchio in *The Taming of the Shrew*, and of *A Midsummer Night's Dream* with Mendelssohn's music, it is the tragic aspect of Shakespeare's genius which appeals to the Italians rather than his comic spirit.

In Spain, little was heard of Shakespeare until well into the 19th century, nor have any of the great actors there popularized him on the stage. The first Spanish translations were based on Ducis. In the 1870s Jaime (James) Clark and Guillermo (William) Macpherson, Englishmen resident in Spain, undertook to make new translations from the English originals, but neither of them completed the task. The only complete versions so far are those of Luis Astrana Marín in Spanish, begun in 1920 and finished in 1950, and Josep María de Sagarra in Catalan. Spanish translations are used in South America, except in Brazil, where Portuguese versions, mainly from the French, are current. In 1948 the University of Coimbra began publication of a complete series of Portuguese translations, but it was never finished. Another series, by L. de Sousa Rebelo and others, was begun in 1960, but progress on it is slow, and Shakespeare does not seem to have become popular on the Portuguese stage.

The first Greek translator was Demetrios Bikelas, whose versions, made between 1876 and 1884, were acted in Athens. The Ikaros edition, begun in 1949 under the auspices of the British Council in Athens, with translations by B. Rotas and K. Karthiou, has now been completed.

The first translations in central Europe were made from German versions, but from the middle of the 19th century onwards new versions direct from the originals were made, often by distinguished poets. In Hungary a complete translation of all the plays was made as early as 1864 by a group of poets and writers under the auspices of the Kisfaludy Society, though some of the plays had already been translated from the English by Mihály Vörösmarty. In Poland, where the great actress MODJESKA appeared in many Shakespearian parts before playing them in English in England and the U.S.A., the first direct translation was of *Hamlet* in 1840. New versions by a group of three Polish poets appeared in 1911–13, and a six-volume edition of Serbian translations, mostly by Borivoge Nedić, which included also a Croatian translation of the poems, was published in 1978. The first Macedonian translations date from 1949. The Yugoslav theatre director Branko Gavella made some new versions which he directed and played in himself, the playwright Ivan Cankar made a number of Slovene versions, while versions in Czech were made by Vrchlický in the early years of the 20th century, to which have now been added the new acting versions of Emil Saudek. Fan Noli, a noted politician who died in 1965, translated a number of plays into Albanian.

There are excellent versions of most of Shakespeare's plays in Norwegian by H. Rytter, published in 1932. The first Swedish versions were made in 1847–51 by Carl August Hagberg, and Per HALLSTRÖM's magnificent translations appeared in the 1920s–30s. A three-volume set of translations by A. Ohlmarks was published in 1962–4. In the early 19th century in Denmark the actor Peter Foersom appeared in translations which he had made himself, and in 1873 Edvard Lembcke published Danish versions of all the plays except *Titus Andronicus* and *Pericles*; a revised edition of his work appeared in 12 volumes in 1975–8. In Finland Shakespeare is played in both the official languages, Finnish and Swedish, and his plays have figured in the repertory of the Iceland National Theatre, founded in 1950.

In Belgium, which also has a double-language theatre, French and Flemish versions are found, the first play in Flemish being *Romeo and Juliet*, produced in 1884 and followed by *Hamlet* in the same year. In 1929 a new version of *The Merry Wives of Windsor* was staged; but in general the translations used in the Flemish theatre are those of the Dutch author, L. A. J. Burgersdijk, first published in 1885, and revised and modernized in the 1940s by F. de Backer and G. A. Dudok. These are the versions used also in the Netherlands, as are the Dutch versions of A. S. Kok.

In Russia adaptations of Shakespeare seen in the 18th century bore little resemblance to the originals. Among them was a version of *The Merry Wives of Windsor* entitled *What It Is Like To Have Linen in a Basket* (1786) by the Russian empress Catherine the Great. Later versions were usually made from the French or German translations current at the time, the first to be staged being *Richard III* (1833). Later translations were made directly from the originals. In Soviet Russia many of Shakespeare's plays are in the permanent repertory of the main theatres, the most popular being *Hamlet*, and translations are available in many of the languages of the U.S.S.R.

Shakespeare has also appeared on the Jewish stage, in Hebrew as early as 1874 with a version of *Othello*, and in Yiddish in 1880 with GOLDFADEN's version of *Romeo and Juliet* as *The Two Kune Lemels*. None of the plays has been performed in Ladino, but two, *The Comedy of Errors* and *Romeo and Juliet*, formed the basis of novels written in that language. In India, where Shakespeare was used for the teaching of English, and some early performances were given by students, productions by Europeans for Europeans were formerly staged, but have now been abandoned. The first versions in the vernacular may have been *The Comedy of Errors* in Marathi in 1843 and *The Taming of the Shrew* in Gujarati in 1850. There are also versions of some of the plays in Tamil, Bengali, and Kannada. A version of *Othello* in Marathi, *Jhumjararava* by Govind Ballal Dewal, in which he himself appeared in the name part, was very popular. The hero became a South Indian Maratha adventurer, and the heroine a fair high-class Hindu maiden.

In Japan in the early 1900s Oto Kawakami, the founder and leader of a new school of acting, appeared with his wife Sada Yacco in *Hamlet* and *Othello*, with the scene set in Japan and the actors wearing contemporary Japanese costume. Kawakami's productions were seen in London and the United States, where they aroused much interest. The most important translations into Japanese, however, are those of Dr Tsubouchi, completed in 1928. In 1974 a periodical called *Shakespeare Translation* was founded by Professor Jiri Ozu. An 11-volume set of translations by various scholars appeared in China in 1978.

In Africa the outstanding translations are those of *Julius Caesar* and *The Merchant of Venice* into Swahili, by Julius Nyerere. The best of several Zulu translations were those of *Macbeth*, and there are several Xhosa translations by B. B. Mdledle, which appeared in 1956–60.

The many immigrant groups in the U.S.A. were from early in the 19th century able to attend performances in their own languages, given either by their own amateur groups or by visiting stars from Europe.

SHAKESPEARE MEMORIAL THEATRE, Stratford-upon-Avon, Warwickshire. This theatre, devoted to the production of plays by Shakespeare, stood on a riverside site in his birthplace, donated by Charles Edward Flower, member of a local family of brewers. A bright-red brick building in a pseudo-Gothic style, it opened in 1879 on 23 Apr. (Shakespeare's birthday), and attracted a good deal of adverse criticism on account of its gabled and turreted exterior, bare interior, and inadequate stage. It was, however, destined to house many fine productions with outstanding actors during the annual festival of Shakespeare's plays, which from 1886 to 1919 were directed mainly by Frank BENSON, and afterwards by W. BRIDGES-ADAMS, and even to gain the affection of some of those who visited it regularly, until on 6 Mar. 1926 it was destroyed by fire, leaving the library and picture gallery, added in 1883, still standing, though badly damaged. The company moved to a local converted cinema while plans were put in hand for a new theatre, on the same site but with an extension into the adjoining Bancroft gardens. The shell of the old theatre was converted into a conference hall, now used for rehearsals. Much of the money needed to build a new theatre came from the United States of America, and the moving spirit of the appeal was again a Flower—Sir Archibald.

The new building, designed by Elizabeth Scott, grandniece of the architect Sir Gilbert Scott, opened on 23 Apr. 1932. It was purely functional both inside and out, with high windowless walls, a fan-shaped auditorium seating about 1,500, and a wide stage. Again it caused widespread controversy; just as the first theatre had been dubbed 'a wedding cake', so the second was dismissed as 'a factory' or 'a tomb'. The actors suffered from cramped conditions backstage and from the distancing effect on their performances of the large orchestra pit.

Two years later Bridges-Adams retired, after extending the annual season from three or four weeks to five months and inviting KOMISARJEVSKY to direct several plays, including a controversial production of *Macbeth* with aluminium screens and vaguely modern uniforms. He returned under Bridges-Adams's successor Ben Iden PAYNE, who introduced dramatists other than Shakespeare into the programme—JONSON for the tercentenary of his death in 1937, GOLDSMITH in 1940, SHERIDAN in 1941. This policy continued until 1946, since when Shakespeare has reigned virtually supreme. A full programme was maintained during the Second World War, Payne being succeeded by Milton Rosmer in 1943 and Robert ATKINS in 1944; under the latter the forestage was carried out over the orchestra pit, with a welcome gain in contact between actors and audience. In 1945 Barry JACKSON took over and initiated a number of reforms, including the spacing out of first nights over the whole season instead of crowding them all into the first fortnight and the appointment of a different director for each play instead of a resident director for the season. Improvements were made both in the auditorium and backstage, including the enlargement and refitting of the workshops. Jackson also invited promising youngsters to join the company, including Paul SCOFIELD and Peter BROOK, and in 1948 Robert HELPMANN appeared as King John, Shylock, and Hamlet. In the autumn of that year Anthony QUAYLE took over as director, and under him a number of leading players, including Peggy ASHCROFT, John GIELGUD, Diana WYNYARD, and Michael REDGRAVE, appeared in a series of brilliant productions. The front curtain was removed, thus further integrating stage and auditorium. Glen BYAM SHAW succeeded Quayle in 1956, after being co-director for some years. Overseas touring, which began with tentative visits to North America and Australia before 1939, increased after the Second World War, and there were visits to Moscow in 1955 and Leningrad in 1958.

The formation of the ROYAL SHAKESPEARE COMPANY in 1961, with Peter HALL as director, began a new era, with the company appearing not only at Stratford, where the theatre was renamed the ROYAL SHAKESPEARE THEATRE, but also in London, at the ALDWYCH THEATRE.

SHAKHOVSKY, ALEXANDER ALEXANDROVICH, see RUSSIA AND SOVIET UNION.

SHANK, JOHN (?–1636), English actor, who appears in the actor-list of Shakespeare's plays, and whose name is found in many different spellings. He apparently began his career with QUEEN ANNE'S MEN, though no proof of this has yet been found. He is known to have been with the KING'S MEN, joining them either shortly before, or at the time of, the death in *c.* 1611 of ARMIN, whose position as chief clown he inherited. He was a comedian, well thought of as a singer and dancer of JIGS, and appears to have been very popular with his audience; though the written lines of his roles are often few, it seems that he was allowed considerable licence in gagging. From the number of boy-apprentices with whom his name is connected, and from the fact that so many young men are recorded in the parish burial register as having died at his house in Cripplegate, it has been inferred that he undertook the training of apprentices who lodged with him. His son John, also an actor, who played at the FORTUNE THEATRE, was a dissolute young man who was court-martialled for cowardice while fighting for the Parliamentarians during the Civil War.

SHARERS, the name given in the Elizabethan theatre to those members of a company who owned part of the wardrobe and playbooks, as distinct from the apprentice, who was not paid, or the hired man, who was on a fixed wage. Those who also owned shares in the actual building were known as HOUSEKEEPERS.

SHARING SYSTEM. All the English provincial companies in the first half of the 18th century worked on shares, though somewhat differently from the Elizabethans. After the expenses of the night had been paid the profits, even what remained of the candles, were shared among all the members of the company equally, except that the manager took four extra parts known as dead shares for his expenses in connection with the scenery, wardrobe, and so on. Every night an account was made up and the money paid out, the younger and less experienced members getting the same as the leading actors. Although this 'commonwealth' system was good in many ways it was open to abuse by unscrupulous managers, for the accumulation of important bills which the management owed, known as the stock debt, sometimes figured among the expenses long after it had been paid off. Such abuses led to the system falling out of favour in the mid-century and being replaced by the payment of salaries, which varied in proportion to the importance of the player and the affluence of the manager but were always very low in comparison to London salaries. The actors often depended more on their BENEFIT performances, of which they had at least one every year, than on their shares or their salaries.

SHAW, GEORGE BERNARD (1856–1950), Irish dramatist, critic, and social reformer, born in Dublin. His mother, a singer and teacher of singing, early imbued him with a knowledge and love of music which qualified him, after migrating to London in 1876, to become music critic of the *Star* (under the pseudonym of Corno di Bassetto) from 1888 to 1890 and of the *World* from 1890 to 1894. Admirable though his music articles were, their quality was undoubtedly surpassed by that of the dramatic criticism he contributed to the *Saturday Review*—where his successor was Max BEERBOHM —between Jan. 1895 and Dec. 1898. Before writing for the *Star*, Shaw had done book reviewing for the *Pall Mall Gazette* and art criticism for the *World*, and, subsidized by his mother, had written five unsuccessful novels which were later included in his collected works. His interest in social and political reform had also led him in 1884 to join the Fabian Society, on whose behalf he soon became a fluent and effective speaker. Though he had no particular love for the theatre of his own time, he was not slow to recognize its value as a platform which could be used to further the interests of the causes he had espoused. This, and his admiration for the 'new drama' of IBSEN, which in translations by Shaw's friend William ARCHER was becoming known in London, led him to the writing of plays. The production on 9 Dec. 1892, under the auspices of GREIN'S Independent Theatre Club, of *Widowers' Houses* (begun in 1885) inaugurated his long career as the foremost dramatist of his day. This was a private production for Club members, as were the productions in 1902 and 1905 of *Mrs Warren's Profession* and *The Philanderer* (both written in 1893) respectively. The first of Shaw's plays to be presented publicly was *Arms and the Man*, produced at the Avenue (later PLAYHOUSE) Theatre on 21 Apr. 1894 as part of a repertory season run by Florence FARR, who had played in the first production of *Widowers' Houses*. The others ran into trouble with the censor and, taken in conjunction with Shaw's lectures and writings on behalf of the Fabians, caused him to be regarded in many quarters as a subversive influence. Deliberately disregarding the current conventions of the WELL-MADE PLAY, he set out to appeal to the intellect and not the emotions of his audiences, and introduced on stage subjects previously confined to the law-courts, the church pulpit, or the political platform—slum landlordism, prostitution, war, religion, family quarrels, health, economics, the position of women. Thought, not action, was the mainspring of the Shavian play; but, as audiences were eventually to realize, it was thought seasoned by wit and enlivened by eloquence, which in the long run provided more rewarding entertainment than the theatre had been able to offer for a very long time. The popularity of Shaw's plays dates from the seasons of 1904–7 at the ROYAL COURT THEATRE, run by GRANVILLE-BARKER and J. E. VEDRENNE, when ten were performed in repertory—*Candida, John Bull's Other Island, How He Lied to Her Husband, You Never Can Tell* (first publicly produced at the STRAND in 1900), *Man and Superman* (without Act III, produced separately in 1907 as *Don Juan in Hell*), *Major Barbara, The Doctor's Dilemma, Captain Brassbound's Conversion*, with Ellen TERRY, *The Philanderer*, and the one-act *The Man of Destiny*. When Vedrenne and Barker moved to the SAVOY THEATRE at the end of 1907, they produced there *The Devil's Disciple*, first seen in New York with Richard MANSFIELD as early as 1897, and *Caesar and Cleopatra*, first performed in 1906 in German in Berlin under Max

REINHARDT and, in English, in New York the same year.

The chronology of Shaw's plays is complicated by the fact that many of them had COPYRIGHT, private, amateur, or foreign professional productions before being seen in London. His most important works after *Misalliance*, which failed when presented by Charles FROHMAN at the DUKE OF YORK'S THEATRE in London in 1910 and had an unexpected success in a New York revival in 1953, are *Fanny's First Play* (1911; N.Y., 1912), Shaw's first commercial success, in which he satirized the contemporary London dramatic critics William Archer (Gunn), A. B. WALKLEY (Trotter), and E. A. Baughan (Vaughan); *Androcles and the Lion* (1913; N.Y., 1915); *Pygmalion* (1914) first seen in Vienna in 1913 with Lilli Marburg as Eliza, the part played in London by Mrs Patrick CAMPBELL, and the source of the overwhelmingly successful American musical *My Fair Lady* (N.Y., 1956; London, 1958); *Heartbreak House* (N.Y., 1920; London, 1921), a 'fantasia' in the Russian (i.e., Chekhovian) manner first produced by the THEATRE GUILD, as was the 'metabiological Pentateuch' *Back to Methuselah* (1922), considered by Shaw his masterpiece, which was first seen in London in 1924 in a production brought from the BIRMINGHAM REPERTORY THEATRE by Barry JACKSON; *Saint Joan* (N.Y., 1923), Shaw's finest play in the opinion of most critics, seen in London in 1924 with Sybil THORNDIKE as an unforgettable Joan, the part played in New York by Winifred Renihan; *The Apple Cart* (1929), brought to London from the MALVERN FESTIVAL, as were *Too True to Be Good* (1932), *Geneva* (1938), and Shaw's last memorable full-length play, the delightful *In Good King Charles's Golden Days* (1939). Subtitled by Shaw 'a true history that never happened', this brings together Charles II, Isaac Newton, George Fox, Nell GWYNN, the Duchess of Cleveland, and the Duchess of Portsmouth. Shaw wrote a number of one-act pieces, often for specific occasions and seldom revived, of which the best are probably *The Dark Lady of the Sonnets* (1910), the first of Shaw's plays to be given at the OLD VIC as a curtain-raiser to the 1930 production of *Androcles and the Lion*, and *Great Catherine* (1913), written for Gertrude KINGSTON, which provided the libretto for an opera by the Dutch composer Ignace Lilien, first performed at the Wiesbaden Festival in 1932.

It would be impossible to list all the revivals and translations of Shaw's major plays. *Pygmalion, Major Barbara*, and *Saint Joan* are the most popular, followed by *Candida*, *Man and Superman*, *Arms and the Man, The Doctor's Dilemma*, and *You Never Can Tell. Heartbreak House* is also held in high esteem, but perhaps more by the critics than by the general public. Many of the first productions, particularly of the earlier plays, were directed by Shaw himself, who earned the respect and admiration of his actors and might, had occasion offered, have been a successful actor himself. He also supervised the printing of his plays with meticulous attention to layout and typography, his detailed stage directions forming a running commentary which helps the reader to visualize the scene, and his prefaces to the published plays enlarge on their arguments. His letters to Mrs Patrick Campbell, published in 1952, provided a basis for *Dear Liar* (1960), a duologue by the American actor Jerome Kilty, performed in New York by Katharine CORNELL and Brian Aherne and in London by Kilty himself and his wife Cavada Humphrey.

SHAW, GLEN BYAM, see BYAM SHAW.

SHAW FESTIVAL, at Niagara-on-the-Lake, Ontario, Canada, an annual event inaugurated in 1962 by the playwright and impresario Brian Doherty (1906–74). Originally amateur, it became professional in 1965, and although intended for the production of plays by Shaw only—the opening productions being *Candida* and the 'Don Juan in Hell' scene from *Man and Superman*—it now presents also plays from Europe by his contemporaries and new works by Canadian dramatists. A new proscenium theatre seating 820 opened in 1973, and the season, which at first lasted for eight nights only and in the 1960s for 3–6 weeks, was extended from 1 May to mid-October. The company is predominantly Canadian, but visiting guest actors have included Jessica TANDY, Micheál MACLIAMMÓIR, and Ian RICHARDSON.

SHAW THEATRE, London, see NATIONAL YOUTH THEATRE.

SHCHEPKIN, MIKHAIL SEMENOVICH (1788–1863), Russian actor, son of a serf, who in his youth played small parts in private amateur theatricals in Kursk, his birthplace, and also appeared at the Kursk theatre during several summer seasons. In 1808 he took over a number of leading comic roles there, which he continued to play until the company was disbanded in 1816. A movement was then set on foot by Prince Repin, one of his admirers and patrons, to purchase his freedom, and in 1822 he was invited to join the company at the MALY THEATRE in Moscow, where he made his début in Nov. 1822 in Zagoskin's comedy *Gospodin Bogatonov; or, a Provincial in the Capital*. From 1825 to 1828 he was in St Petersburg, but it was in Moscow that he did his best work, particularly during the 1840s, when GRIBOYEDOV's *Wit Works Woe* and GOGOL's *The Government Inspector* provided him with two of his best parts. He was also noted for his work in a number of minor comic characters in MOLIÈRE and Shakespeare, though sometimes, particularly in his later years, he was criticized for too boisterous buffoonery. Nevertheless his influence on the Russian stage was farreaching, and he was recognized by STANISLAVSKY as the founder of the tradition of realistic acting on the Russian stage, his synthesis of

MOCHALOV's passion and KARATYGIN's technical perfection creating a style well suited to the plays of OSTROVSKY, CHEKHOV, and GORKY. Encouraged by PUSHKIN, he published his autobiographical notebooks, which are of great artistic and theatrical interest.

SHCHUKIN, BORIS VASILIEVICH (1894–1939), Soviet actor, whose early death robbed the Russian stage of one of its outstanding figures. He was for 20 years at the VAKHTANGOV THEATRE, where he played leading roles. In his last years he gave some exceptionally fine performances, notably in the title role of GORKY's *Yegor Bulychov and Others*, in AFINOGENOV's *Distant Point*, and as Lenin, whom he was the first actor to impersonate, in POGODIN's *The Man with the Gun* (1937).

SHEFFIELD, see CRUCIBLE THEATRE.

SHELDON, EDWARD BREWSTER (1886–1946), American dramatist, who trained under George Pierce BAKER and had his first play *Salvation Nell* (1908) produced when he was only 22, with Mrs FISKE as the heroine. It was a great success and Sheldon was immediately hailed as the rising hope of the new American school of realistic dramatists, a position he appeared to consolidate with *The Nigger* (1909), a courageous handling of the Negro problem, and *The Boss* (1911), a study of modern industrial conditions. But all his serious work was overshadowed by the immense popular success of his romantic play *Romance* (1913), which with Doris Keane (1881–1945) as an Italian opera singer had long runs in London and New York. It made an immense emotional appeal to audiences all over the world, and was translated into French and other languages. Handicapped by serious illness, Sheldon did most of his later work in collaboration, translating and adapting a number of popular successes, one of the best being a version of BENELLI's *La cena delle beffe* (1909) as *The Jest* for John and Lionel BARRYMORE in 1919.

SHELLEY, PERCY BYSSHE (1792–1822), English Romantic poet, and author of several plays in verse, of which the best known is *The Cenci.* Published in 1819, this was first acted by the Shelley Society in 1886, with Alma MURRAY as an outstanding Beatrice. It has been several times revived, notably in 1922 and 1926 with Sybil THORNDIKE, and in 1959 at the OLD VIC with Barbara JEFFORD. Though pure poetry, it is poor drama, being confused in action and in style somewhat too indebted to Shakespeare.

SHEPARD, SAM [Samuel Shepard Rogers] (1943–), prolific American dramatist, whose plays reflect, with striking visual imagery, the violent conflict between the dreams of the early American pioneers and modern industrialization. His early one-act plays were performed off-off-Broadway, beginning with *Cowboys* (1964) and including *Chicago* (N.Y., 1965; London, 1976), *Icarus's Mother* (N.Y., 1965; London, 1970), *Red Cross* (N.Y., 1966; London, 1970), and *Melodrama Play* (N.Y. and London, 1967). His first full-length plays were *La Turista* (N.Y., 1966; London, 1969) and *Operation Sidewinder* (1970), in which a computer shaped like a sidewinder rattlesnake becomes a real snake. In the early 1970s he lived for some years in London, where his work included one of his best plays *The Tooth of Crime* (London, 1972; N.Y., 1973) about an ageing rock star who is displaced by a rival and commits suicide. *The Curse of the Starving Class* (London, 1977; N.Y., 1978) portrays the selfishness and indifference within a poor farming family in California; while the PULITZER PRIZE-winning *Buried Child* (N.Y., 1978; London, 1980), also set on a farm, depicts a family in an even more advanced state of disintegration, though the settings and character development are somewhat more realistic than in his previous plays. *Seduced* (N.Y., 1979; London, 1980) features a dying recluse based on Howard Hughes. It was followed by *True West* (N.Y., 1980; London, at the COTTESLOE, 1982), about rival Californian brothers, one a successful Hollywood script writer and the other a petty crook.

SHERIDAN, MARK, see MUSIC-HALL.

SHERIDAN, RICHARD BRINSLEY (1751–1816), Dublin-born dramatist, theatre manager, and politician, son of an actor, Thomas Sheridan (1721–88), and a writer, Frances [*née* Chamberlaine] (1724–66), whose plays *The Discovery* and *The Dupe* (both 1763) were produced by GARRICK at DRURY LANE. Sheridan was educated at Harrow and intended for the law. At 21, however, he made a romantic marriage with the singer Elizabeth Ann Linley (1754–92) and settled in London, where his first play *The Rivals* was produced at COVENT GARDEN in 1775, followed later the same year by a farce, *St Patrick's Day; or, the Scheming Lieutenant*, and a comic opera, *The Duenna*. A year later he bought Garrick's share in the Drury Lane patent, and his later plays were all produced there. They include *A Trip to Scarborough* (based on VANBRUGH's *The Relapse*) and *The School for Scandal* (both 1777). The latter, in which Mrs ABINGTON played Lady Teazle, is a masterpiece of the English comedy of manners, with all the wit but none of the licentiousness of the Restoration comedy from which it derives. Like *The Rivals* it became immensely popular, and has been constantly revived. Sheridan's next play, *The Critic: or, a Tragedy Rehears'd* (1779), was the best of many burlesques stemming from BUCKINGHAM's *The Rehearsal* (1671); it too has been revived many times, notably in 1945 by the OLD VIC company with Laurence OLIVIER as Mr Puff. Its original production exploited to the full the popular taste for spectacle and PANTOMIME, with remarkable scenic effects by de LOUTHERBOURG and lavish costumes. In 1780 Sheridan became a Member of Parliament, and thereafter wrote little for the stage, though he figures as part-author of two spectacular entertainments given at Drury Lane in 1794 and 1797, and wrote a pantomime, *Robinson Crusoe: or, Harlequin Friday* (1781), in which Delpini arranged the dumbshow, playing Crusoe to the Friday of Joseph GRIMALDI's father. In 1794 he rebuilt Drury Lane, and it was there that his last play, *Pizarro* (1799), an adaptation of KOTZEBUE's *Die Spanier in Peru*, was given. The building was destroyed by fire in 1809, and Sheridan, always in financial difficulties, and constantly at loggerheads with the managers of the smaller London playhouses, finally withdrew from the theatre.

SHERIDAN SQUARE PLAYHOUSE, see CIRCLE REPERTORY COMPANY.

SHERMAN THEATRE, Cardiff, see WALES.

SHERRIFF, R(obert) C(edric) (1896–1975), English dramatist and novelist, who became widely known for his realistic and moving play *Journey's End* (1928), the first to deal successfully in the theatre with the First World War. Originally produced on a Sunday evening by the STAGE SOCIETY, it was taken into the commercial theatre by Maurice BROWNE, who played Lieutenant Raleigh, the part of Captain Stanhope, first played by Laurence OLIVIER, being taken over by Colin Clive; also in the all-male cast was Robert SPEAIGHT. Dealing with the reactions of a small group of men in a dug-out just before an attack, it made an immediate impact, and was subsequently translated and played all over the world, being first seen in New York in 1929. Among Sherriff's other plays are *Badger's Green* (1930), on village cricket; *St Helena* (1935; N.Y., 1936), written in collaboration with Jeanne de Casalis and dealing with the last years of Napoleon; *Miss Mabel* (1948), in which Mary Jerrold played the title role; and two plays in which Ralph RICHARDSON gave impressive performances—*Home at Seven* (1950), a study of amnesia, and *The White Carnation* (1953). *The Long Sunset* (1955) was set towards the end of the Roman occupation of Britain; and *The Telescope* (1957) later provided the basis for a musical by Peter Powell, *Johnny the Priest*, seen in 1960, in which year Sherriff's last play *A Shred of Evidence* was also produced, with Paul ROGERS as Richard Medway. His autobiography, *No Leading Lady*, was published in 1968.

SHERWOOD, ROBERT EMMET (1896–1955), American dramatist, who scored a success with his first play *The Road to Rome* (1927; London, 1928), a satirical treatment of Hannibal's march across the Alps which deflated military glory. It was followed by *The Love Nest* (also 1927), based on a short story by Ring Lardner, and *The Queen's Husband* (1928; London, 1931), which drew an amusing portrait of a henpecked king. *Waterloo Bridge* and *This is New York* (both 1930) were failures, but with his next play, *Reunion in Vienna* (1931; London, 1934), brilliantly interpreted by the LUNTS, Sherwood again achieved success. With *The Petrified Forest* (1935; London, 1942), he began to take cognizance of the rapidly deteriorating world situation, and although there was a light-hearted interval with *Tovarich* (1935; N. Y., 1936), based on a play by Jacques Deval, in the PULITZER PRIZE-winner *Idiot's Delight* (1936; London, 1938) his ironic pessimism grew darker; in it he foretold a Second World War (he had fought in the first one), and the intellectual bankruptcy of Western civilization. *Abe Lincoln in Illinois* (1938) showed Lincoln as a man of peace who entered the political arena reluctantly, and paralleled contemporary political struggles with those of his day. *There Shall Be No Night* (1940; London, 1943), written in response to the invasion of Finland, showed a pacifist scientist choosing war as preferable to slavery. After a few years of political activity, during which he wrote no new plays, Sherwood returned to the theatre with *The Rugged Path* (1945). His last play, *Small War on Murray Hill*, a mildly romantic comedy about the American Revolution, was produced posthumously in 1957.

SHIELDS, ARTHUR (1900–), Irish actor, who was a member of the ABBEY THEATRE company from 1914 to 1937, playing in seven plays by Sean O'CASEY. He also created many roles in the plays of Lady GREGORY, George SHIELS, and Lennox ROBINSON, perhaps his most famous part being Denis in Robinson's *The White-Headed Boy* (1916), which he first played at the AMBASSADORS THEATRE in London in 1920 and appeared in over 1,000 times in England, Australia, and the U.S.A. After 1937 he remained in America, directing and managing a company which frequently included his brother Barry FITZGERALD, and in 1941 he turned almost exclusively to films.

SHIELDS, ELLA, see MUSIC-HALL.

SHIELS [Morshiel], GEORGE (1881–1949), Irish dramatist, who drew upon a wide knowledge of character and social relations in both parts of Ireland and in America for his realistic comedies. He made his name with the one-act *Bedmates* (1921), his later sympathetic satires of contemporary life including *Paul Twyning* (1922; London, 1933), *Professor Tim* (1925; London, 1927), and *The New Gossoon* (1930), which was seen in London in 1931, the title being subsequently altered to *The Girl on the Pillion*. Shiels's serious plays, such as *The Passing Day* (1936; London, 1951), *The Rugged Path* (1940), and *The Summit* (1941), showed that his dramatic gifts were not limited to comedy. *The Rugged Path*, in particular, had an unprecedentedly long run at the ABBEY THEATRE, where 22 of his plays were produced, and was also seen in England, at Chepstow, in 1953.

SHIFRIN, NISSON ABRAMOVICH (1892–1959), Soviet designer, whose finest work was done for productions of plays by Shakespeare, notably for POPOV's production at the RED ARMY THEATRE in 1938 of *The Taming of the Shrew*, where he combined tapestries with solid furnishings to produce a successful synthesis of Elizabethan and Renaissance settings. In 1940 he designed some lovely settings for the same director's production of *A Midsummer Night's Dream*, and in 1957 his designs for *The Merry Wives of Windsor*, based on the simple lines and rich colouring of Russian folk-art, were much admired. Among his designs for modern Soviet plays the most successful were probably those for a documentary dealing with the siege of Stalingrad, based on sketches made on the spot, and an evocation of the endless steppes, fringed by leafless branches under a clear sky, made for a dramatization of Sholokhov's *Virgin Soil Upturned*, also in 1957.

SHILLING THEATRE, Fulham, see GRAND THEATRE (2).

SHIRLEY, JAMES (1596–1666), the leading dramatist of London when the Puritans shut the playhouses in 1642. He wrote about 40 plays, most of which have survived in print though not on the stage. These include tragedies like *The Maid's Revenge* (1626); *The Traitor* (1631), Shirley's most powerful play, a REVENGE TRAGEDY into which he imported a MASQUE of the Lusts and Furies; *Love's Cruelty* (1631); and *The Cardinal* (1641). His best work, however, is found in his comedies, which provide a link between those of Ben JONSON and the Restoration playwrights. The most successful were *The Witty Fair One* (1628), *Hyde Park* (1632), *The Gamester* (1633), which was later adapted by GARRICK, *The Lady of Pleasure* (1635), and *The Sisters* (1642). A prompt-book of this last, dating from the early years of the Restoration, supplies some interesting stage directions, and is now in the library of Sion College. Shirley survived the Commonwealth and was popular in the early days of the Restoration, no less than eight of his plays being revived, including *The Cardinal,* which PEPYS saw in 1667, the author having died of exposure during the Great Fire of London.

SHKRELI, AZAM, see ALBANIA.

SHOREDITCH EMPIRE, London, see MUSIC-HALL.

SHOWBOAT, the name given to the floating theatres of the great North American rivers of the West, particularly on the Mississippi and the Ohio, which represented an early and most successful attempt to bring drama to the pioneer settlements. It is not known who first built a showboat or at what date. The first record of actors travelling by boat dates from 1817, when Noah LUDLOW took a company of players along the Cumberland river to the Mississippi in his 'Noah's Ark'. But they acted on land, and it was apparently William CHAPMAN, formerly an actor in London and New York, who first commissioned the building of a true Showboat, the 'Floating Palace', described in 1831 as a flat boat propelled downstream by the current and, when necessary, poles and sweeps, having on it a structure labelled Theatre, with a ridge roof above which projected a staff flying a flag. The interior was long and narrow, with a shallow stage at one end and benches across the width of the boat, the whole being lighted by candles. Here the Chapmans with their five children played one-night stands along the rivers wherever enough people could be found to provide an audience. The entrance fee was about 50 cents, and the staple fare strong melodrama or fairy-tale plays, ranging from KOTZEBUE's *The Stranger* to *Cinderella*. Starting in the autumn from as far upstream as possible, usually Pittsburgh, the showboat made its way downstream to New Orleans, where it was abandoned, the company returning to their starting-point to descend the river in a new boat. Later, steam-tugs were used to take the showboat, renamed the 'Steamboat Theatre', back to its point of departure, some managers even owning their own tugs. By the time Chapman died in 1839 there were a great number of showboats on the rivers, but his widow, with her two sons Henry and George, continued to operate under the name of 'Chapman's Floating Palace' until 1847, when they sold the boat to Sol SMITH, who lost it a year later in a collision. Another showboat captain of these early days was Henry Butler, an old theatre manager who took a combined museum and playhouse up and down the Erie Canal for many years, showing stuffed animals and waxworks by day, and at night producing NAUTICAL DRAMAS like JERROLD's *Black-Ey'd Susan*, with a sailor-turned-actor, Jack Turner, in the hero's role. A more elaborate boat was the 'Floating Circus Palace' owned by G. R. Spaulding and Charles J. Rogers. Built in Cincinnati in 1851, it was intended for spectacular equestrian shows, having living quarters, dressing rooms, stables, a Museum, and a ring capable of holding 45 horses. The steamboat which accompanied it, the *James Raymond*, had a theatre which was used for straight dramatic performances and on which the first calliope in a steamboat was installed in 1858. The history of Spaulding's enterprise coincides with the heyday of the showboats, which increased in numbers and popularity until the Civil War in 1861 drove them off the rivers. Some of them, including Spaulding's, were commandeered by the Confederates. When the war ended there were no showboats in operation at all, and although a few crept back afterwards it was not until 1878, when a Captain A. B. French (?–1902) took his *New Sensation*, the first of five successive boats, along the Mississippi that showboats again became familiar sights.

French had to live down a good deal of prejudice, but the high moral tone of his productions and the good behaviour of his small company soon made his productions popular. At one time French and his wife, the first woman to hold a pilot's licence and master's papers on the Mississippi, ran two showboats, piloting one each. Like other showboat captains of the time, they usually avoided the cities and larger towns, not wishing to risk comparison with the theatres springing up everywhere. A formidable rival to the Frenches was Captain E. A. Price, owner of the *Water Queen*, built in 1885. This had a stage 19 ft wide, lit by oil, a good stock of scenery, a company of about 50, and, like all showboats, a steam calliope. It was later a floating dance-hall in Tennessee, and in 1935 was used in the filming of Edna Ferber's *Showboat*, which, successful as novel, play, and film, was the first work to popularize the story of a floating theatre. Another river-boat which figures in a film was the sidewheeler *Kate Adams*, built in 1898, which in 1926 appeared as *La Belle Revere* in *Uncle Tom's Cabin*. She was later burnt to the waterline. Another well known manager was Captain E. E. Eisenbarth, owner of the first boat to bear the name *Cotton Blossom*. This was capable of seating a large audience, and on a stage 20 ft by 11 ft presented a three-hour entertainment of straight solid drama, usually popular melodrama. *The Cotton Blossom* was one of the first showboats to be lit by electricity. The name persisted, and a later *Cotton Blossom* was owned by a Captain Otto Hitner. In 1907 Captain Billy Bryant, author of *Children of Ol' Man River* (1936), began his career as a showboat actor when his father launched *The Princess* with a programme featuring himself and his family. By 1918 the Bryants were able to build their own boats, on which they gave successful revivals of many good old melodramas, the most popular being *Ten Nights in a Bar Room*, one of the many dramatizations of a temperance novel by William Pratt. Vaudeville, interspersed with songs and magic-lantern shows, filled up the intervals, while Captain Billy's own speeches before and after the show became famous. A typical one was printed in the *New York Times* for 12 Oct. 1930. Other well known showboat personalities were the Menkes, four brothers who in 1917 bought French's fifth and last *New Sensation*, built in 1901, from Price, who had purchased it from French's widow in 1902. They kept the old name and in 1922–3 took the boat on a trip which lasted a year and covered 5,000 miles. The plays given in her, as in the other boats owned by the Menkes—*Golden Rod, Hollywood, Wonderland, Sunny South*, and *Floating Hippodrome*—were still mainly melodramas and typical Victorian entertainments, though in the late 1920s they took to presenting musical comedies. In the years before the slump of 1929 a number of new managements made their appearance on the water—the *Majestic* under Nico and Reynolds, the *America*, the *River Maid*, another *Princess*, a new *Water Queen* under Captain Roy Hyatt—mostly offering melodrama and variety. But they were hard hit by the economic depression, and for some time the Menkes' *Golden Rod* and Bryant's *Showboat* were the only ones still functioning. The last to be built was the *Dixie Queen*, launched by Al Cooper in 1939, which was converted into an excursion boat in 1943. In his book on the Mississippi, *Big River to Cross*, the author Ben Lucien Burman was able to mention five by name as still working, as well as many other smaller and less well known ones on distant waters.

Found nowhere but in the United States, the showboat represents a survival of the colourful pioneering days of the Golden West.

SHOW PORTAL, see FALSE PROS(CENIUM).

SHTRAUKH, MAXIM MAXIMOVICH (1900–74), Soviet actor and director, who is 1921 was at the Proletkult (later the TRADES UNIONS) Theatre, where he established a reputation as a brilliant performer in 'eccentric' roles. From 1929 to 1931 he was with MEYERHOLD, and later at the MAYAKOVSKY THEATRE, where he appeared as Lenin in KORNEICHUK'S *Truth* (1937), subsequently, mainly through films, becoming the best known interpreter of this role. He was seen in VISHNEVSKY'S *The Unforgettable 1919* in 1949, and a year later joined the company at the MALY THEATRE. In 1959 he returned to the Mayakovsky, remaining there until he died. Some of his later appearances were in plays by POGODIN and Shteyn, and he also directed a number of plays including some by BRECHT.

SHUBERT THEATRE, New York, see SAM S. SHUBERT THEATRE.

SHUMSKY, YAKOV DANILOVICH (?–1812), Russian comic actor, who was in the company of amateurs run by F. G. VOLKOV in Yaroslav. Sent to St Petersburg with his fellow actors to play before the Court in 1752, he was one of the players who, with DMITREVSKY, formed the first professional Russian theatrical company four years later. He was renowned for his playing of eccentric old women (still enacted, as in France, by men), his most celebrated role being that of the nurse in FONVIZIN'S comedy *The Minor* (1782). He also excelled in the fops and comic servants of HOLBERG and MOLIÈRE. He retired in 1785.

SHUSHERIN, YAKOV EMELYANOVICH (1753–1813), Russian actor, who joined the Moscow troupe of M. E. MEDDOCKS about 1770. Later he studied for a time under DMITREVSKY and in 1785 had a great success as Iarbas in *Dido* by Y. B. Knyazhin. He went to St Petersburg in 1786, where he played the title role in SUMAROKOV'S *Dmitri the Impostor* and Count Appiani in LESSING'S *Emilia Galotti*, two of his best roles, becoming a great favourite with the Empress Catherine the Great after appearing successfully in her comedy *The Discontented Family* and her chronicle play *The Early Rule of Oleg*. He survived the transition from the classical to the sentimental style and was as popular in the fashionable COMÉDIE LARMOYANTE as in *King Lear*, Ozerov's neo-classical *Oedipus in Athens*, and SOPHOCLES' *Philoctetes*.

SHUTER, NED [Edward] (1728–76), English actor, who made his first appearance on the stage at Richmond in 1744 and two years later was playing small parts in GARRICK's company at DRURY LANE. He later moved to COVENT GARDEN, where he created a number of parts, including Justice Woodcock in BICKERSTAFFE's *Love in a Village* (1762), Mr Hardcastle in GOLDSMITH's *She Stoops to Conquer* (1773), and Sir Anthony Absolute in SHERIDAN's *The Rivals* (1775). Shuter, whose portrait was painted by Zoffany, made his last appearance at the HAYMARKET on 23 Sept., only five weeks before his death, playing Scrub in FARQUHAR's *The Beaux' Stratagem*. He was especially admired for the mobility and expressiveness of his features and for his extempore fooling.

SHWARTZ [SCHWARTZ, SHVARTS], EVGENYI LVOVICH (1896–1958), Soviet dramatist, who in 1925 abandoned acting to concentrate on writing plays for young people. Several of the most successful were based on fairy-tales—*The Naked King* (1934) (based on Andersen's *The Emperor's New Clothes*), *Red Riding Hood* (1937), *The Snow Queen* (1938)—but his best known work, *The Dragon*, was entirely original. First seen in Leningrad in 1944, it was produced in an English translation at the ROYAL COURT THEATRE in 1967, and proved to be an outspoke and hilarious debunking of political tyranny, with a daring hero, a beautiful heroine, a talking cat, villains, clowns, and a three-headed dragon. Among his other plays, *The Shadow* (1940) was directed by AKIMOV, who was also responsible for the production of his last play, a satirical fairy-tale entitled *An Ordinary Miracle* (1956).

SIDDONS [*née* Kemble], SARAH (1755–1831), English actress. The eldest of the 12 children of Roger Kemble, an actor-manager in the provinces, she spent her childhood travelling with his company, and at the age of 18 married William Siddons (1744–1808), also a member of the company. They played together in the provinces, returning there in 1775 after Mrs Siddons had made a first and unsuccessful appearance at DRURY LANE under GARRICK, and were seen in YORK with Tate WILKINSON and BATH with John PALMER. A second appearance by Mrs Siddons in London in 1782 was more successful, and she was soon acclaimed as a tragic actress without equal, a position she maintained until the end of her career. She began, however, at the zenith of her powers, and unlike her brother John Philip KEMBLE did not improve with age. Among her early parts were Isabella in SOUTHERNE's *The Fatal Marriage*, Belvidera in OTWAY's *Venice Preserv'd*, and the title role in ROWE's *Jane Shore*. Later she proved outstanding as Constance in *King John*, Zara in CONGREVE's *The Mourning Bride*, and above all Lady Macbeth, the part in which she made her farewell appearance on 29 June 1812. She returned in 1819 to play Lady Randolph in HOME's *Douglas* for the BENEFIT of her younger brother Charles KEMBLE, but was only the shadow of her former self. In her heyday critics were unanimous in their praise of her beauty, tenderness, and nobility. A superbly built and extremely dignified woman, with a rich, resonant voice and great amplitude of gesture, she wisely refused to appear in comedy. She was not much liked by her fellow actors and had a dislike of publicity which led her to rebuff her admirers, but her intelligence and good judgement made her the friend of such men as Dr JOHNSON and Horace Walpole. Painters such as Reynolds, Lawrence, and Gainsborough delighted in painting her, the first immortalizing her beauty in 1784 in his portrait 'The Tragic Muse'. Towards the end of her career she became somewhat stout, and her acting was considered old-fashioned. She was also extremely prudish and refused to wear man's attire when playing Rosalind in *As You Like It* in 1785, appearing in a costume which was neither that of a man nor a woman and extremely unbecoming. In any case she was poor in the part and never appeared in it again. Yet her brother John Philip once referred to her as 'one of the best comic singers of the day'; as there is no record of her ever having appeared before the public in that capacity one can only surmise that she unbent in private. She had seven children, four girls dying in infancy. Her son Henry (1775–1815), who married the sister-in-law of Sam COWELL, was for a long time connected with the Edinburgh Theatre, but was accounted a poor actor.

SIERRA, GREGORIO MARTÍNEZ, see MARTÍNEZ SIERRA.

SIGURJÓNSSON, JÓHAN (1880–1919), the first fully professional Icelandic dramatist, although he wrote almost exclusively in Danish. *Bjærg-Ejvind og hans Hustru* (*Eyvind of the Hills and His Wife*, 1911) was a tragedy of misplaced love, and *Ønsket* (*Loft's Wish*, 1915) told of a man who, like FAUST, attempted to harness the powers of evil.

SILENCE, Theatre of, term used for plays which, like those of MAETERLINCK, who is regarded as the founder of the genre, and particularly of J.-J. BERNARD, are important as much for what they omit from their dialogue as for what they actually say—a theatre, in fact, of pregnant pauses, during which the imagination of the audience supplies the missing ingredient, which is not only unexpressed but perhaps cannot be expressed in words. Hence the French term for this type of play, *le théâtre de l'inexprimé*. It is also known as the theatre of *l'école intimiste*.

SILICON CONTROLLED RECTIFIER (S.C.R.), see LIGHTING.

SILL-IRONS, see BOX-SET.

SILVA, ANTÓNIO JOSÉ DA (1705–39), Brazilian-born Portuguese playwright, one of the most popular of his day in Lisbon, who as a practising Jew was put to death by the Inquisition. His plays, produced at the Teatro do Bairro Alto (where the first Portuguese company of professional actors was established), are best described as operatic comedies; most of them, including the *Vida do Grande D. Quixote* (1733), seem to have been written for puppets, though the *Guerras do Alecrim e da Mangerona* (1737), probably his best work, had human actors. His structure is weak, but there is excellent comedy of situation and language in his plays, often provided by the GRACIOSO or comic servant, taken over from Spanish theatre, and some vigorous social satire.

SILVA, NINA DE, see MARTIN-HARVEY, JOHN.

SILVIA [Zanetta-Rosa-Giovanna Benozzi] (*c*. 1701–58), Italian actress, who was in the company taken by RICCOBONI to Paris in 1716. Although engaged for secondary roles, Silvia quickly became the leading lady of the COMÉDIE-ITALIENNE, unsurpassed in the playing of MARIVAUX's heroines, from *Arlequin poli par l'amour* (1720) to *Le Jeu de l'amour et du hasard* (1730).

SIM, ALASTAIR (1900–76), outstanding Scottish actor, who could combine a roguish conspiratorial glee with prim old-maidish fastidiousness. He made his first appearance on the stage in 1930, and in 1932–3 was at the OLD VIC, appearing later in Hubert Griffith's *Youth at the Helm* (1934) and Edgar WALLACE's *The Squeaker* (1937). In 1941 he first played Captain Hook in BARRIE's *Peter Pan*, a role which he was to repeat many times; but he first made a decided impact on the theatrical scene with his Mr McCrimmon in James BRIDIE's *Mr Bolfry* (1943), the first of a series of roles in plays (which he also directed) by the same author: *It Depends What You Mean* (1944), *The Forrigan Reel* (1945), *Dr Angelus* (1947), *The Anatomist* (1948), and *Mr Gillie* (1950). After playing the title role in a revival of *Mr Bolfry* in 1956 and the Emperor in William Golding's *The Brass Butterfly* (1958), he was seen again at the Old Vic in 1962 as a somewhat improbable Prospero in *The Tempest*. He found himself on more familiar ground as Colonel Tallboys in SHAW's *Too True To Be Good* in 1965, and as a singularly decrepit Lord Ogleby in COLMAN the elder's *The Clandestine Marriage* at the CHICHESTER FESTIVAL in 1966, where in 1969 he gave a masterly performance as Mr Posket in PINERO's *The Magistrate*, seen in London the same year. It was followed by an appearance in William Douglas HOME's *The Jockey Club Stakes* (1970), and by another excellent Pinero performance as the sporting Dean in *Dandy Dick*, seen in Chichester and London in 1973. He made his last appearance on the stage as Lord Ogleby in London in 1975.

SIMON, (Marvin) NEIL (1927–), American dramatist, a witty, prolific, and accurate observer of American marital and family relationships who once had four plays running simultaneously on Broadway. He began by writing sketches for REVUE, his first show for Broadway being *Catch a Star!* (1955); he collaborated on the sketches with his brother Daniel, with whom he also wrote his first play *Come Blow Your Horn* (1961; London, 1962). After the book of the musical *Little Me* (1962; London, 1964) came a string of successful plays: *Barefoot In the Park* (1963; London, 1965); *The Odd Couple* (1965; London, 1966), about two heterosexual males—one fussy, one not—sharing accommodation; *The Star-Spangled Girl* (1966); and *Plaza Suite* (1968; London, 1969), three one-act plays set in the same hotel suite. He also wrote the book for two more musicals, *Sweet Charity* (1966; London, 1967) based on Fellini's film *Nights of Cabiria* and *Promises, Promises* (1968; London, 1969) based on Billy Wilder's film *The Apartment*. His last play of the 1960s was *The Last Of the Red Hot Lovers* (1969; London, 1979) in which the three acts are concerned with a man's three unsuccessful attempts to commit adultery. Simon's success continued into the 1970s with *The Gingerbread Lady* (1970; London, 1974), a more serious play about an alcoholic cabaret star; *The Prisoner Of Second Avenue* (1971), about the stresses of redundancy and New York apartment life; and *The Sunshine Boys* (1972; London, 1975). He broke new ground in 1973 with *The Good Doctor*, a collection of sketches based on CHEKHOV's stories, and again in 1974 with *God's Favorite*, based on the Book of Job. Neither these works nor *The Trouble With People . . . and Other Things*, which he wrote in the same year with Daniel, were successful, but his popularity returned with *California Suite* (1976), constructed on the same principle as *Plaza Suite*, which was followed by the semi-autobiographical *Chapter Two* (1977; London, 1981), about a widower's second marriage, the book of the musical *They're Playing Our Song* (1978; London, 1980), and *I Ought To Be In Pictures* (1980).

SIMONOV, KONSTANTIN MIKHAILOVICH (1915–79), Soviet dramatist, whose first play, *The Russian People*, dealing with the impact of war on a group of civilians and soldiers near the front line, was produced in 1942 and frequently revived. It was seen in London in 1943, as *The Russians*, in a production by the OLD VIC company. It was followed by the prize-winning *A Fellow from Our Town* (also 1942) and *Wait for Me* (1943). This was played all over the U.S.S.R. and established the author's reputation, which was further enhanced by the success of *The Russian Question* (1947), *Friends and Enemies* (1949), *A Foreign Shadow* (1950), the comedy *A Good Name* (1953), *The Story of a Love* (1954), and *The Fourth* (1961).

SIMONOV, REUBEN NIKOLAIVICH (1899–1968), Soviet actor and director, who in 1920 joined the MOSCOW ART THEATRE's Third Studio (later the VAKHTANGOV THEATRE). He was an excellent actor, his best parts being Cyrano de Bergerac in ROSTAND's play of that name and Kostya Kapitan in POGODIN's *Aristocrats*; but after a brilliant performance by students, whom he himself had trained, in the dramatization of Sholokhov's *Virgin Soil Upturned* in 1931, he devoted himself almost entirely to production, becoming chief director of the Vakhtangov from 1939 until his death, when he was succeeded by his son Yevgenyi Reubenovich (1925–). Among the elder Simonov's outstanding productions were Pogodin's *The Man with the Gun* (1937), KORNEICHUK's *The Front* (1943), and his own adaptation of GORKY's *Foma Gordeyev* (1956).

SIMONSON, LEE (1888–1967), American designer, whose first work for the stage was done in connection with the WASHINGTON SQUARE PLAYERS. He later became one of the founders and directors of the THEATRE GUILD, for which many of his finest sets were done, among them the décor of MASEFIELD's *The Faithful* (1919), with a Japanese setting, TOLSTOY's *The Power of Darkness*, STRINDBERG's *The Dance of Death* (both 1920), MOLNÁR's *Liliom* (1921), TOLLER's *From Morn To Midnight* (1922), IBSEN's *Peer Gynt*, Elmer RICE's *The Adding Machine* (both 1923), O'NEILL's *Marco Millions* (1928) and *Dynamo* (1929)—for which he did one of his best designs, a constructivist set based on the interior of a power house—Robert SHERWOOD's *Idiot's Delight* (1936), and Maxwell ANDERSON's *The Masque of Kings* (1937). He also did the settings for the Theatre Guild's world premières of SHAW's *Heartbreak House* (1920), *Back to Methuselah* (1922), and *The Simpleton of the Unexpected Isles* (1935). For other managements he designed mainly for new American plays, among them Sherwood's *The Road to Rome* (1927), O'Neill's *Days Without End* (1934), and Maxwell Anderson's *Joan of Lorraine* (1946). He was for many years a director of the AMERICAN NATIONAL THEATRE AND ACADEMY, and of the Museum of Costume Art in New York, and directed the International Exhibition of Theatre Art held in New York in 1934. He also wrote an autobiography, *Part of a Lifetime* (1943), and two books on scenic design.

SIMOV, VICTOR ANDREYEVICH (1858–1935), Russian stage designer, who joined the MOSCOW ART THEATRE on its foundation in 1898, and except for a short interval after the First World War, when he worked as a freelance, spent the rest of his life there. He was responsible for the settings of the theatre's first production, Alexei TOLSTOY's *Tsar Feodor Ivanovich*, and of CHEKHOV's plays as they joined the repertory, and among his other pre-war designs were those for GORKY's *The Lower Depths* (1902), Shakespeare's *Julius Caesar* (1903), and Leo TOLSTOY's *The Living Corpse* (1911). After his return to the Moscow Art Theatre in 1925 he designed the settings for, among other productions, IVANOV's *Armoured Train 14–69* (1927) and GOGOL's *Dead Souls* (1932).

SIMPSON, ALAN, see PIKE THEATRE CLUB, DUBLIN.

SIMPSON, EDMUND SHAW (1784–1848), American actor and manager, who was playing in Dublin when COOPER and PRICE engaged him for the PARK THEATRE, New York. He made his first appearance there in Oct. 1809, and remained for 38 years, playing Richmond to the Richard III of KEAN, COOKE, and the elder BOOTH, and remaining a prime favourite with the audience until 1833, when he retired from the stage because of lameness caused by an accident. He had been appointed acting manager of the theatre in 1812, and Price's partner in 1815, and had kept the theatre running well, in spite of the disastrous fire of 1820, mainly by importing English stars and encouraging the fashion for Italian opera. He continued in management after his retirement from acting, and in 1837, by which time the prestige of the Park was on the wane, he became sole lessee. In 1848 he surrendered it to HAMBLIN, manager of the BOWERY, against an annuity, and died almost immediately.

SIMULTANEOUS-SCENE SETTING, see COMPOSITE SETTING.

SINCLAIR, ARTHUR [Francis Quinton McDonnell] (1883–1951), Irish actor, second husband of Maire O'NEILL, who made his first appearance on the stage in YEATS's *On Baile's Strand* (1904) with the IRISH NATIONAL DRAMATIC SOCIETY and was with the ABBEY THEATRE until 1916, when he led a mass resignation of the players over St John ERVINE's management of the theatre. He then formed his own company and toured Ireland and England, subsequently appearing in variety theatres in Irish sketches. He built up a great reputation as an Irish comedian, among his finest parts being Old Mahon in SYNGE's *The Playboy of the Western World* and John Duffy in Lennox ROBINSON's *The White-Headed Boy*. He was much admired in

O'CASEY's plays, especially as Captain Boyle in *Juno and the Paycock*, Fluther Good in *The Plough and the Stars,* and Seumas Shields in *The Shadow of a Gunman*.

SINDEN, DONALD ALFRED (1923–), English actor with a rich, resonant voice, equally at home in the classics and in light modern comedy, in which he toured from 1941 to 1945 entertaining the Forces. He joined the SHAKESPEARE MEMORIAL THEATRE company in 1946, and was later with the BRISTOL OLD VIC, returning to it in 1950 after a long run in *The Heiress* (1949), adapted from Henry JAMES's novel *Washington Square*. From 1952 to 1957 he was in films, but he returned to the stage in 1957 in *Odd Man In* by Claude Magnier, adapted by Robin Maugham. During the next few years his roles were unremarkable except for his doubling of Captain Hook and Mr Darling in BARRIE's *Peter Pan* in 1960 and the title role in Archibald MacLeish's *J.B.* in 1961; but in 1963 he joined the ROYAL SHAKESPEARE COMPANY, where he played Richard Plantagenet in *The Wars of the Roses*, John BARTON's adaptation of *Henry VI* and *Richard III*, and was also seen in Henry LIVINGS's *Eh?* (1964). For some years he alternated between the R.S.C. and modern comedy, among the productions in which he appeared being Terence Frisby's *There's a Girl in My Soup* (1966); VANBRUGH's *The Relapse* in 1967, in which he gave a brilliant performance as Lord Foppington; Ray Cooney and John Chapman's *Not Now, Darling* (1968); Malvolio in *Twelfth Night* and the title role in *Henry VIII* (both 1969). In 1970 he played Sir William Harcourt Courtly in the R.S.C.'s revival of BOUCICAULT's *London Assurance*, seen at the NEW THEATRE in 1972 and in New York in 1974, and he also appeared in RATTIGAN's double bill *In Praise of Love* (1973) at the DUCHESS THEATRE and as Dr Stockman in IBSEN's *An Enemy of the People* at the CHICHESTER FESTIVAL THEATRE in 1975. Later appearances have only served to confirm his versatility, ranging from Lear, Benedick (both 1976), and Othello (1979) to farces by Alan BENNETT (*Habeas Corpus*, N.Y., 1975) and John Chapman and Anthony Marriott (*Shut Your Eyes and Think of England*, 1977).

SINGSPIELE, see BALLAD OPERA.

SIPARIUM, see CURTAIN.

SIRCAR, BADAL, see INDIA.

SIXTY-NINE THEATRE COMPANY, see ROYAL EXCHANGE THEATRE, MANCHESTER.

SJÖBERG, ALF (1903–80), Swedish director and manager attached to the KUNGLIGA DRAMATISKA TEATERN from 1925, first as an actor and from 1929 as director. During his long career, his imaginative direction and sensitive psychological interpretations left a deep mark on the Swedish theatre. His major productions included *As You Like It* (1938), *The Merchant of Venice* (1944), *Twelfth Night* (1946), *Richard III* (1947), and *Hamlet* (1960), as well as BRECHT (whom he introduced to Sweden), IBSEN (*Brand*, 1950; *Rosmersholm*, 1959), GARCÍA LORCA (*Blood Wedding*, 1944), SARTRE (*Les Mouches*, 1945), and Arthur MILLER (*Death of a Salesman*, 1949). His films, notably *Frenzy* (1944) and adaptations of STRINDBERG's *Miss Julie* (1951) and *The Father* (1969), made him internationally famous.

SKELTON, JOHN (*c.* 1460–1529), English poet and satirist, tutor to the future Henry VIII. He is believed to have written an interlude played before the Court at Woodstock, a comedy, and three MORALITY PLAYS; the only one to survive is *Magnyfycence*, acted probably between 1515 and 1523. First printed in 1530 by John RASTELL, it was reprinted by the Early English Text Society in 1906. In this play Magnyfycence, a benevolent ruler, is corrupted by bad counsellors (Folly, Mischief, etc.), but restored by good ones (Good Hope, Perseverance, etc.).

SKINNER, OTIS (1858–1942), American actor, who made his first appearance on the stage in Philadelphia in 1877 and was seen in New York, at NIBLO'S GARDEN, in 1879. After some years with Edwin BOOTH and Lawrence BARRETT he joined Augustin DALY and with him made his first appearance in London, as Romeo. For two years he toured with MODJESKA, and later played Young Absolute in SHERIDAN's *The School for Scandal* to the Sir Anthony of Joseph JEFFERSON. Among his later successes were *Your Humble Servant* (1910) and *Mr Antonio* (1916), both specially written for him by Booth TARKINGTON, but he made his greatest and most lasting success as Hajj in KNOBLOCK's *Kismet* (1911). In 1926 he appeared as Falstaff in *Henry IV, Part I*, and in 1928 played the same character in *The Merry Wives of Windsor*. Among his last appearances were Shylock in *The Merchant of Venice* in 1931 and Thersites in *Troilus and Cressida* in 1933. His daughter, Cornelia Otis Skinner (1902–79), first appeared with her father's company in 1921, and became celebrated in America as a *diseuse*, in which capacity she was seen in London in 1929 in *The Wives of Henry VIII* and other sketches. She was also well known as a novelist and writer and was co-author with Samuel Taylor of *The Pleasure of His Company* (1958) in which she also appeared.

SKUSZANKA, KRYSTYNA (1924–), Polish director, who first worked at Opole, and from 1955 to 1963 managed the People's Theatre in the new industrial town of Nowa Huta, with her husband Jerzy Krasowski and the director-designer Józef SZAJNA. Her production, with her husband, of GOZZI's *Turandot*, seen at the THÉÂTRE DES NATIONS in Paris in 1956, gained her an international reputation, as did her production of *The Tempest* at Nowa Huta in 1959. She had also given the first Polish production of SHWARTZ's *The Dragon* and staged an ambitious revival of MICKIEWICZ's *Forefathers' Eve* before being transferred from Nowa Huta to the Polski Teatr in Warsaw in 1963. She then worked at the Wrocław Theatre, where her productions revealed an increasing mastery of her medium and a more mellow style, and in 1972, accompanied by her husband, she took over the direction of the Słowacki Theatre in Cracow.

SKY BORDER, see BORDER.

SKY-CLOTH, SKY DROP, see CLOTH.

SKY-DOME, see LIGHTING.

SLAPSTICK, the American term for the bat, or flexible divided lath, used by a HARLEQUIN. With it he gave the signal for the TRANSFORMATION SCENE and other changes of scenery in the HARLEQUINADE and belaboured the backsides of his enemies, making a great noise under pretence of striking heavy blows. The word became current in England in about 1900, and was later applied to the boisterous comedy of the CLOWN and the low comedian, particularly to the knockabout farces and sketches of the MUSIC-HALL.

SLIPS, the name given in the late 18th and early 19th centuries to designate the ends or near-stage extremities of the upper tiers of seats in the theatre. It is still in use at COVENT GARDEN.

SLIP STAGE, see BOAT TRUCK.

SLOMAN, CHARLES (1808–70), a performer in the 'song-and-supper rooms' which preceded the early MUSIC-HALLS, and the original of Young Nadab in Thackeray's novel *The Newcomes*. He was best known for his extempore doggerel verses on subjects suggested by members of the audience or on the dress and appearance of those seated in the hall; but he was also a writer of songs, both comic and serious, for himself and his contemporaries, in particular Sam COWELL. He was at the ROTUNDA in 1829, and probably appeared at EVANS's, being at the height of his fame in the 1840s. He later fell on hard times, and one of his last engagements was as CHAIRMAN of the MIDDLESEX MUSIC-HALL, a position he had previously filled when it was the Mogul Tavern or 'Old Mo'. He died soon after, a pauper, in the Strand workhouse.

SLOTE or **SLOAT** (Hoist in America, *Cassette* in France), a form of TRAP for raising a long narrow piece of scenery—a GROUNDROW, for example—from below to stage level, or for carrying an actor up or down from the stage. It appears that the slote was already standard equipment at DRURY LANE as early as 1843, and in *The Orange Girl*, by H. T. Leslie and N. Rowe, produced at the SURREY THEATRE in 1864, there are elaborate stage-directions for the conveyance of the heroine by a slote from a height downwards to a trap. Since the slote is also referred to as a 'slide' one of its variants may have been the Corsican Trap (see TRAPS). To allow free passage to the slote the floor-board above it was divided vertically, each half dropping and being pulled to the side of the stage by ropes, leaving an opening in the stage floor. On occasion a CLOTH might be attached to the base of a groundrow and be drawn up from its rolled position in a Slote Box below by means of lines from the GRID. It would thus unroll as the groundrow rose in the air, for instance during a TRANSFORMATION SCENE. With two slotes set in each of three sections of the stage, the elements of six separate scenes could be raised from below to stage level.

SŁOWACKI, JULIUSZ (1809–49), Polish romantic poet and playwright, who was exiled in 1831 and thereafter lived mainly in Paris. He wrote over 20 plays, of which several remained unfinished. The only one to be staged during his lifetime was *Mazeppa*, in a Hungarian version seen in Budapest in 1847. The Polish original was seen in Cracow soon after Słowacki's death, but it was not until 1899 that his finest play, *Kordian*, written in 1833, was finally produced. This, like the unfinished *Horsztyński* (written in 1835), shows very clearly the influence of Shakespeare, who first became known to Słowacki in 1831 when on a visit to London he saw Edmund KEAN in *Richard III*. All his plays had been seen on the Polish stage by 1905, but it was not until Leon SCHILLER's rediscovery of the Romantic playwrights that Słowacki's plays became part of the permanent repertory. The revival of *Kordian* by AXER in 1954, with Tadeusz Łomnicki in the title role, marked the end of SOCIALIST REALISM in the Polish theatre, and led to further productions of such works as *Beniowski* in 1971 and *Balladyna* in 1974, both directed by Hanuszkewicz. Other important plays by Słowacki are *Lilla Weneda*, based on an early legend, and two late tragedies, *The Silver Dream of Salomé* and *Samuel Zborowski*. His only comedy, a satire on the excesses of Romanticism entitled *Fantazy*, was written in about 1841 and first performed in 1867. His translation of CALDERÓN's *Il principe constante* was produced by GROTOWSKI's Laboratory Theatre and was seen in London in 1969.

SLY, WILLIAM (?–1608), English actor, who appears in the actor-list of Shakespeare's plays. He was connected with the theatre from about 1590, when his name appears in the cast of *Seven Deadly Sins, Part II*, a lost play, probably by TARLETON, of which the PLATT is preserved among HENSLOWE's papers at Dulwich College, where there is also a portrait of Sly by an unknown artist. He joined the CHAMBERLAIN'S MEN on their formation in 1594, appearing in four plays by Ben JONSON.

Every Man in His Humour (1598), *Every Man Out of His Humour* (1599), *Sejanus* (1603), and *Volpone* (1606). Though he was not one of the original shareholders of the GLOBE THEATRE, he became one at some time, since he mentions it in his will. He also had a seventh share in the BLACKFRIARS, later taken over by Richard BURBAGE.

SMIRNOV, BORIS ALEXANDROVICH (1908–), Soviet actor, who rose to prominence in Leningrad in the 1930s and 1940s as a fine performer of classical roles. In 1955 he joined the MOSCOW ART THEATRE, and a year later appeared as Lenin in the second (revised) version of POGODIN's *Kremlin Chimes*, being seen in the part in London during the WORLD THEATRE SEASON of 1964. In 1959 he again played Lenin in *The Third Pathétique*, the final play in Pogodin's trilogy which began in 1937 with *The Man With the Gun*.

SMITH, ALBERT (1816–60), an interesting but forgotten figure of literary London, who dramatized several of DICKENS's novels for the stage and also produced some ephemeral plays and novels of his own. His chief enterprise was a series of one-man entertainments, of which the first was *The Overland Mail* (1850). This was an amusing account of a recent trip to India, interspersed with topical songs and stories and illustrated by scenery specially painted for the occasion by William BEVERLEY. It proved such a success that Smith followed it with *The Ascent of Mont Blanc*, given at the Egyptian Hall, Piccadilly, again with scenery by Beverley, and then by a similar programme on China. He married in 1859 Mary, the actress-daughter of Robert KEELEY, and died at the height of his popularity. His simple entertainment, whose charm lay as much in the spontaneity and wit of its presentation as in the actual material, was frequently patronized by Queen Victoria and the royal children.

SMITH, DODIE [Dorothy] GLADYS (1896–), English playwright and novelist, who until 1935 wrote under the pseudonym of C. L. Anthony. She trained as an actress, but after some years spent in touring left the stage to become buyer for a commercial firm, until the success of her play *Autumn Crocus* (1931), which starred Fay COMPTON and Jessica TANDY, enabled her to devote all her time to writing. Two moderately successful plays, *Service* (1932) and *Touch Wood* (1934), were followed by the long-running *Call It A Day* (1935), with Fay Compton and Marie LÖHR, about a day in the life of a prosperous English middle-class family, and *Bonnet Over the Windmill* (1937). Her best known play is *Dear Octopus* (1938). With a cast led by Marie TEMPEST as a matriarch presiding over a reunion of her family—the 'octopus' of the title—and John GIELGUD as her eldest son, it was still running in Sept. 1939 when the Second World War broke out and the theatres closed; it was revived in 1940 with a slightly different cast. Although Dodie Smith continued to write for the theatre she never again achieved such a success, though *Dear Octopus* was revived in 1967. Among her later plays were *Letter From Paris* (1952) and *I Capture the Castle* (1954), based on one of her own novels. She has written two volumes of autobiography, *Look Back With Love* (1974) and *Look Back With Mixed Feelings* (1978).

SMITH, EDWARD TYRRELL (1804–77), English theatre manager, the first to inaugurate the 'morning performance' which under HOLLINGSHEAD became the modern afternoon matinée. He was probably the most reckless theatrical speculator of his day, having an interest in many places of entertainment, from music-halls and minor houses like the WEST LONDON and SURREY theatres up to DRURY LANE itself, which he rented in 1852 and opened with a dramatization of Harriet Beecher Stowe's *Uncle Tom's Cabin* and one of BLANCHARD's pantomimes. He would sponsor anything that he thought would make money—RACHEL, Gustavus BROOKE, the younger MATHEWS, Shakespeare, opera, circuses, Chinese conjurers, a Human Fly that crawled on the ceiling—and in the process he ruined himself. Yet he was in his own day a noted character and in spite of his follies made and kept friends everywhere.

SMITH, MAGGIE [Margaret] NATALIE (1934–), English actress, with a flair for high comedy but also outstanding in dramatic roles. She made her first appearance on the stage in OXFORD in 1952, playing Viola in an OUDS production of *Twelfth Night*, and after training at the OXFORD PLAYHOUSE Drama School went to New York to appear in the REVUE *New Faces of '56*. Back in London she made an outstanding success as the leading comédienne in another revue, *Share My Lettuce* (1957), and then went to the OLD VIC, where she gave an admirable performance as Lady Plyant in CONGREVE's *The Double Dealer* in 1959 and played Maggie Wylie in BARRIE's *What Every Woman Knows* in 1960, in which year she took over from Joan PLOWRIGHT the part of Daisy in IONESCO's *Rhinoceros*. Her growing reputation was enhanced by performances in ANOUILH's *The Rehearsal* (1961), Peter SHAFFER's double bill *The Private Ear* and *The Public Eye* (1962), and Jean Kerr's *Mary, Mary* (1963). She then joined the NATIONAL THEATRE company at the Old Vic, giving an outstanding performance as Silvia in FARQUHAR's *The Recruiting Officer* and a year later showing unsuspected depths of passion as Desdemona to the Othello of Laurence OLIVIER. Among her later parts with this company were Hilde Wangel in IBSEN's *The Master Builder*, Myra in a revival of COWARD's *Hay Fever* (both 1964), Beatrice in *Much Ado About Nothing* to the Benedick of her first husband Robert Stephens, and the title role in STRINDBERG's *Miss Julie* (both 1965). In 1970 she gave excellent performances as Mrs

Sullen in Farquhar's *The Beaux' Stratagem* and Hedda Gabler in Ingmar BERGMAN's production of Ibsen's play, in which she carried a strong charge of sexual frustration. On leaving the National Theatre she played a much-acclaimed Amanda, again opposite Robert Stephens, in Coward's *Private Lives* (1972; N.Y., 1975). She then spent several seasons at the STRATFORD (ONTARIO) FESTIVAL, where she was seen in many important roles including Cleopatra in *Antony and Cleopatra*, Millamant in Congreve's *The Way of the World*, and Masha in CHEKHOV's *Three Sisters* (all 1976), and Lady Macbeth (1978). In 1979 she took over from Diana RIGG in the London production of Tom STOPPARD's *Night and Day* and then played the role in New York, after which she returned to Stratford, Ontario, where she starred as Virginia Woolf in Edna O'Brien's *Virginia* in 1980, being seen in the same play in London in 1981.

She married as her second husband the dramatist Beverley Cross (1931–), best known for his adaptation of Marc Camoletti's *Boeing-Boeing* (1962), which had a long run.

SMITH, RICHARD PENN (1799–1854), American dramatist, born in Philadelphia, a lawyer by profession, and the author of some 20 plays, of which 15 were acted. They are of all types, ranging from farce to romantic tragedy, and represent the transition in the American theatre from the play imported or inspired by Europe to the true native production of later years. Most of Smith's comedies were adaptations from the French, while his romantic plays were based mainly on incidents from American history. What is believed to have been his best work, a tragedy entitled *Caius Marius*, has not survived. It was produced in 1831 by Edwin FORREST with himself in the title role, and proved extremely successful. It was possibly Forrest's aversion to the printing of plays in which he appeared that caused it to be lost. Another interesting play, also lost, was *The Actress of Padua* (1836), which was based on Victor HUGO's *Angelo*, and marks the first appearance in American theatrical history of French Romanticism. It was revived by Charlotte CUSHMAN in the early 1850s, and was seen in New York as late as 1873, probably with some alterations by John BROUGHAM, to whom it has been attributed.

SMITH, SOL(omon) FRANKLIN (1801–69), a pioneer of the American theatre on the frontier. After a hard childhood he ran away from home to become an actor, travelling with the DRAKES and other itinerant companies. In 1823 he had formed a company of his own, but it was not a success and in 1827 he and his wife joined the company of J. H. CALDWELL, with whom they visited St Louis and other Mississippi towns. In 1835 Smith went into partnership with Noah LUDLOW, and together they prospered, dominating theatrical activity in the South and West and building in St Louis the first permanent theatre west of the Mississippi. Smith, who later appeared as a guest artist at the PARK THEATRE in New York under the management of SIMPSON and also in Philadelphia under WEMYSS, was an excellent low comedian, excelling in such parts as Mawworm in BICKERSTAFFE's *The Hypocrite*. His son Marc(us) (1829–74) was also on the stage, specializing in Old Comedy with Lester WALLACK and Mrs John WOOD, and being well received in London. He made his last appearance at the UNION SQUARE THEATRE in Hart Jackson's adaptation of a French play, *One Hundred Years Old*, in 1873.

SMITH, WILLIAM (?–1696), English actor, a tall, handsome man with a well-modulated voice, who was a friend of BETTERTON and a member of his company at DORSET GARDEN, where he played a number of important new roles, including Sir Fopling Flutter in ETHEREGE's *The Man of Mode* (1676) and Chamont and Pierre in OTWAY's *The Orphan* (1680) and *Venice Preserv'd* (1682) respectively. Driven from the stage by the animosity of a group of young noblemen whom he was presumed to have offended, he retired to live quietly on his considerable income, but later returned to the theatre and died while playing in *Cyrus the Great* (1696), an adaptation by Banks of SCUDÉRY's *Le Grand Cyrus*.

SMITH, WILLIAM (1730–1819), English actor, known as 'Gentleman' Smith on account of his elegant figure, fine manners, and handsome face. Coached by Spranger BARRY, he made his first appearance at COVENT GARDEN on 8 Jan. 1753. He remained there until 1774, and then went to DRURY LANE, where he created the part of Charles Surface in SHERIDAN's *The School for Scandal* (1777). He also played Macbeth to Mrs SIDDONS's Lady Macbeth when she first appeared in the part in 1785, and alternated Hamlet and Richard III with GARRICK, whom he greatly admired, though his own style of acting approximated more to that of QUIN. He was playing most of the big tragic roles when John Philip KEMBLE first went to Drury Lane in 1783, and continued to appear in them until his retirement in 1788. In an age which expected its actors to turn their hands to anything from tragedy to pantomime—even Garrick is said to have once played HARLEQUIN—Smith's proudest boast was that he had never blacked his face, never played in a farce, and never ascended through a trap-door. He would also never consent to appear at the theatre on a Monday during the hunting season, as he was a zealous rider to hounds.

SMITH, WILLIAM HENRY SEDLEY-, see SEDLEY-SMITH, W. H.

SMITHFIELD FAIR, see FAIRS.

SMITHSON, HARRIET [Henrietta] CONSTANCE (1800–54), English actress of Irish descent, who made her first appearance at DRURY LANE in 1818 as Letitia Hardy in Hannah COWLEY's *The Belle's Stratagem* and then played Lady Anne and Desdemona to the Richard III and Othello of Edmund KEAN. She spent some time at COVENT GARDEN and the HAYMARKET THEATRE and then returned to Drury Lane to create the part of the Countess Wilhelm in John Howard PAYNE's adaptation of PIXÉRÉCOURT's *Valentine* as *Adeline; or, the Victim of Seduction* (1822). In 1827 she went with Charles KEMBLE to Paris, playing Ophelia to his Hamlet, and returned there a year later with MACREADY, as Desdemona in *Othello*. To the amazement of the London critics, who had not thought highly of her acting, she was received on both occasions with acclamation, and aroused enormous enthusiasm among the young Romantics of the day, the critic Janin declaring that she had revealed Shakespeare to France and made his plays the prerogative of the tragic actress, thus forestalling RACHEL, who many years later mentioned her as 'a poor woman to whom I am much indebted'. Her fame was short-lived, as she retired in 1833 to make an ill-judged and unhappy marriage with the composer Berlioz, who had fallen madly in love with her the first time he saw her on stage and after a period of doubt and despair again fell temporarily under her spell when they finally met off stage in 1832.

SMOČEK, LADISLAV, see CZECHOSLOVAKIA.

SMOCK ALLEY THEATRE, Dublin, see IRELAND.

SMOKTUNOVSKY, INNOKENTI MIKHAILOVICH (1925–), Soviet actor, who served a long apprenticeship in the provinces before joining the Leningrad Grand GORKY THEATRE company in 1957. He remained there until 1960, scoring a great success by his playing of Prince Myshkin in a dramatization by TOVSTONOGOV of DOSTOEVSKY's *The Idiot* in 1957; the production was seen during the 1966 WORLD THEATRE SEASON in London. After some years in films, during which he became internationally known for his Hamlet, Smoktunovsky returned to the theatre in 1973, appearing at the MALY THEATRE in the title role of A. K. TOLSTOY's *Tsar Feodor Ivanovich*.

SNOW, SOPHIA (1745–86), English actress, daughter of the famous trumpeter Valentine Snow. She first appeared on the stage in 1764 at DRURY LANE and married a member of the company, Robert BADDELEY. Unfortunately she was dissipated and extravagant and he soon left her, though he later fought a duel on her behalf with the brother of David GARRICK. A strikingly beautiful woman, she excelled in such parts as Ophelia, Desdemona, and Imogen (in *Cymbeline*), and was painted by Zoffany as Fanny in the elder COLMAN and GARRICK's *The Clandestine Marriage* (1766). She was last seen in London in 1781 and then played in the provinces, mainly at York and Edinburgh, until her death.

SOCIALIST REALISM, term applied, in all artistic fields, to the sober style that succeeded the wave of experiment following the Russian Revolution. As expounded by LUNACHARSKY, it was intended to make the theatre an instrument for the education of the masses in Communism. Apparently the term was first used in 1932, during a period of protest against the FORMALISM of such directors as MEYERHOLD and TAÏROV, whose work at the time was considered too abstract for the new audiences, and useless as social propaganda. While admitting that a production must not be untrue, either to present-day facts or to the facts of the past, it nevertheless entailed the depiction of the truth in terms that a worker-audience could understand, and the interpretation of the classics in the light of present-day trends. Everything in the theatre, even the writing of new plays, was therefore bound up with the approach to, or the exposition of, the upheavals which led to the Revolution of 1917. Although the criteria of Socialist Realism, of which GORKY (with *Mother* and *Enemies*) is considered the founder and MAYAKOVSKY the first brilliant exponent, have shifted with changing conditions over the years its basic policy on the problems of theatrical creation and interpretation apparently remains the same.

SOCIÉTÉS JOYEUSES, see FRANCE.

SOCIETY FOR THEATRE RESEARCH, founded in 1948 after a meeting held in the OLD VIC THEATRE on 15 June which brought together all those interested in matters concerning theatre history and the preservation of ephemeral and other material relating to the theatre. The society, whose first president was Mrs Gabrielle ENTHOVEN, arranges a winter programme of lectures, publishes a journal, *Theatre Notebook*, three times a year and an annual publication, as well as occasional pamphlets. It encourages the preservation of theatre buildings, of source material on the theatre, and of photographic records of all kinds pertaining to the theatre. It was instrumental in founding the British THEATRE MUSEUM, to which it presented its library, and in the summer of 1955 held in London the first international conference on theatre history which resulted in the foundation of the INTERNATIONAL FEDERATION FOR THEATRE RESEARCH. It administers an annual prize for diction, awarded to a drama student in memory of William POEL, and an annual Edward Gordon CRAIG Memorial Lecture.

In 1956 an American Society for Theatre Research was founded by Professor A. M. Nagler and Alan Downer. Unlike most national theatre societies, it is concerned not only with American theatre but with all aspects of theatre studies which may interest its members. It holds an annual meeting in New York, and publishes an occasional newsletter and an annual volume of essays, *Theatre Survey*.

SOCK, see BUSKER.

SOFRONOV, ANATOL VLADIMIROVICH (1911–), Soviet dramatist, who was born in Minsk and began his literary career while working in a factory. His plays include *In One Town* (1946), *The Moscow Character* (1948), *Beketov's Career* (1949), *Impossible to Live Otherwise* (1952), and *A Man in Retirement* (1957). These all deal with topical social problems in the daily working life of the Soviet people. *Honesty* (1962), in which Vera PASHENNAYA gave a fine performance as the heroine Praskovya Ivanovna, is concerned with the problems of co-operative farming.

SOGGETTO, Commedia a, see COMMEDIA DELL'ARTE.

SOLDENE, EMILY, see MUSIC-HALL.

SOLEIL, Théâtre du, French workers' cooperative drama group, formed in Paris in 1964 under the direction of Ariane Mnouchkine. Their first productions, which were not particularly successful, established their method of work—months of rehearsal, participation of all the actors on the technical side, and experiments with various levels of illusion and acting conventions. The company became famous in 1967 with a production of WESKER's *The Kitchen* in the disused Cirque Medrano, and in 1969 produced, by COLLECTIVE CREATION, a play of their own entitled *Les Clowns*. In 1972, being homeless, they were offered cheaply, by the town council of Paris, the lease of a shed in the abandoned Cartoucherie de Vincennes, a former army training ground and ammunition dump where in 1975 they staged *L'Âge d'or*. Their presence attracted a large audience of supporters of popular theatre, and adjacent sheds soon housed other experimental companies, among them SERREAU's Théâtre de la Tempête; a political group, the Théâtre de l'Aquarium; and the Théâtre Bulle, a Latin-quarter group working on popular and children's theatre.

SOLSKI, LUDWIG, see POLAND.

SOMĪ, LEONE DI, see LIGHTING.

SØNDERBY, KNUT (1909–66), Danish novelist and dramatist, whose early play *En kvinde er overflødig* (*A Woman is Superfluous,* 1936) was a dramatization of one of his own novels. His best known play is *Kvindernes Opror* (*The Woman's Uprising*, 1955), a modern version of ARISTOPHANES' *Lysistrata*.

SONDHEIM, STEPHEN JOSHUA (1930–), American composer and lyricist, whose first music for the theatre was the incidental music for N. Richard Nash's *Girls of Summer* (1956). He then wrote the lyrics for two major musicals, *West Side Story* (1957; London, 1958) and *Gypsy* (1959; London, 1973). The first shows for which he wrote both music and lyrics were *A Funny Thing Happened on the Way to the Forum* (1962; London, 1963), drawn from the comedies of PLAUTUS, which starred Zero MOSTEL in New York and Frankie Howerd in London, and the short-lived *Anyone Can Whistle* (1964) with Angela Lansbury. He wrote only the lyrics for *Do I Hear a Waltz?* (1965), but thereafter he was responsible for both words and music. In the 1970s he firmly established his pre-eminence with a series of brilliant musicals which moved progressively away, in form and content, from traditional concepts of the genre, the music and the densely packed lyrics being ever more closely integrated with the text. *Company* (1970; London, 1972), about the problems of marriage and bachelorhood, was followed by *Follies* (1971), in which a group of former artists from the Follies (see REVUE) meet again in a partly demolished theatre. *A Little Night Music* (1973; London, 1975), which contains some of Sondheim's most hauntingly beautiful music, was based on Ingmar BERGMAN's film *Smiles of a Summer Night*. *Pacific Overtures* (1976) looked from the Japanese point of view at the opening up of Japan to American trade in the 19th century, the music making use of Oriental instruments. *Sweeney Todd* (1979; London, 1980) has more singing than spoken dialogue, and could be considered virtually an opera. All his work from *Company* onwards was directed by Hal Prince.

'SONG-AND-SUPPER ROOMS', see EVANS'S.

SONNENFELS, JOSEF VON (1733–1817), Austrian director, a fervent advocate of the reforms of GOTTSCHED, which he supported in his publication *Briefe über die Wienerische Schaubühne* (1767), urging the abolition of the typical Viennese impromptu burlesque featuring HANSWURST in favour of serious plays modelled on classical French and German drama. In 1776 he was appointed director of the Burgtheater in VIENNA by Joseph II, who hoped to use drama as an educational force in his schemes for the reform of Austria. With this end in view, he banished the popular theatre to the suburbs (where it flourished) and decreed that the Burgtheater should become a national theatre for the production of serious and improving plays by dramatists like AYRENHOFF. The reforms begun by Sonnenfels were consolidated by his successor SCHREYVOGEL, who was more successful in blending the new drama with the old.

SOPHOCLES (496–406 BC), Greek tragic dramatist, born of good family at Colonus near Athens. As a boy he was celebrated for the beauty of his voice and figure, and took part in a boys' dance which celebrated the victory of Salamis in 480 BC.

(Athenian tradition linked its three great tragic poets to this battle; AESCHYLUS fought in it, and EURIPIDES was said, inaccurately, to have been born while it was in progress.) Sophocles is said to have written over 100 plays, of which seven tragedies are extant, as well as substantial parts of a SATYR-DRAMA, the *Ichneutae* (*The Trackers*), dealing with the theft by Hermes of Apollo's cattle. He won 18 victories, the first—over Aeschylus—in 468, as well as holding important civic and military offices, and seems always to have enjoyed the respect and esteem of his fellow-Athenians. The extant plays are: *Ajax* and *Antigone* (perhaps written closely together, *c.* 442–441), the *Trachiniae, Oedipus the King* (*c.* 429), *Electra, Philoctetes* (409), and *Oedipus at Colonus* (written at the end of his life, and produced posthumously by his son).

Sophocles' life embraced the most vital and crucial period of Athenian history, from the defeat of the foreign menace in the Persian wars, through the subsequent economic and cultural expansion, to the years of decline; he died just before the final defeat of Athens by Sparta in the Peloponnesian War. Although his plays were performed side by side with those of Euripides they reveal a serenity which his younger contemporary lacked—a serenity that comes from triumph over suffering, not avoidance of it. Few moments in drama are more poignant than Sophocles' tragic climaxes. His language, compared to that of Aeschylus, is clearer and more incisive; his characters are more fully rounded; but he inherited his predecessor's concern with questions of moral law, though he set those questions in a framework with which his audience could more immediately identify. His technique is often to isolate powerful, resourceful individuals against a background of crisis—*Oedipus the King* is set in a city ravaged by plague, *Antigone* in the same city decimated by war; the chief character in *Philoctetes* is a desperate castaway on a desert island—and to show their response, under pressure, to the various demands upon them.

Writing in an age when many hailed expediency as the only guiding principle, he constantly reaffirmed the necessity to respond to a higher moral imperative; though he is always ready to pay tribute to those purely human attributes by which man distinguishes himself from the rest of creation, contriving to hold a balance between the old religion and the new morality. His final play exemplifies the serenity which pervades his work—Oedipus, after a life of torment, goes at last to a tranquil rest—but also foreshadows the closing of an age; although great spirits like Oedipus may continue to live in legend and tradition, the living world is left to lesser men.

The analysis of tragedy made by ARISTOTLE in his *Poetics* is based in the main on Sophoclean drama, which he regarded as the mature form of tragedy, therefore neglecting the earlier Aeschylean form.

SOPHRON OF SYRACUSE (*fl.* late 5th cent BC), Greek writer of MIME plays. His work is believed to have influenced the semi-dramatic form of the dialogues of PLATO and also the style of THEOCRITUS, but nothing of it has survived.

SORANO, DANIEL (1920–62), French actor, one of the finest of his day, who began his career with SARRAZIN's Grenier de Toulouse, which in 1965 celebrated its 20th anniversary by taking possession of a new theatre named after him, and with which he was seen on a visit to Paris, where he made a great impression as Scapin in MOLIÈRE's *Les Fourberies de Scapin*. He also played Biondello in their excellent production of *The Taming of the Shrew* as *La Mégère apprivoisée.* In 1952 Sorano joined the company of the THÉÂTRE NATIONAL POPULAIRE, where his most notable roles, both classic and modern, were Sganarelle in *Don Juan*, Mascarille in *L'Étourdi*, and Argan in *Le Malade imaginaire*, all by Molière; Arlequin in MARIVAUX's *Le Triomphe de l'amour*; Figaro in BEAUMARCHAIS's *Le Mariage de Figaro*, Don César de Bazan in HUGO's *Ruy Blas*, and the Chaplain in BRECHT's *Mutter Courage und ihre Kinder.* He was an excellent interpreter of Shakespeare and other Elizabethan dramatists, appearing in 1961 in a French version of FORD's *'Tis Pity She's a Whore* under the direction of VISCONTI, and being particularly admired as Richard III and as the Porter in *Macbeth*. His last role was Shylock, with BARRAULT's company, a powerful and sober interpretation.

SORGE, REINHARD JOHANNES (1892–1916), German poet, who began as a disciple of Nietzsche and fell, a devout Catholic, in the First World War. His most important play *Der Bettler* (*The Beggar*), written in 1912 but not performed until 1917 in a production by REINHARDT, was a drama of social protest which foreshadowed the revolt of youth against the older generation and the striving for a higher spiritual orientation, two of the most insistent themes of EXPRESSIONISM. The structure of the play is determined by the subject matter: the main character's inner development provides the sole link of a series of loosely connected episodes. The script specifies the use of sophisticated spot lighting, then in its infancy, to effect changes within COMPOSITE SETTINGS. The plays he wrote after his conversion to Catholicism are on mystical and religious themes—*König David* (pub. 1916) and *Der Sieg des Christos* (pub. 1924).

SORMA [Zaremba], AGNES (1865–1927), German actress, who in 1883 was engaged for the newly founded DEUTSCHES THEATER in BERLIN, where she soon became popular in young girls' parts. Some of her first mature successes were scored in revivals of GRILLPARZER's works, particularly *Weh dem, der lügt!* in which she played opposite Joseph KAINZ. She was also seen as Juliet in

Romeo and Juliet, Ophelia in *Hamlet*, and Desdemona in *Othello*, and later as Nora in IBSEN's *A Doll's House*, a part she continued to play for many years, notably on an extended tour of Europe and on her first visit to New York in 1897. She was a distinguished interpreter of the heroines of SUDERMANN and HAUPTMANN and, in a lighter vein, of the Hostess in GOLDONI's *La locandiera*. From 1904 to 1908 she worked under Max REINHARDT.

SOSNITSKY, IVAN IVANOVICH (1794–1871), Russian actor, who studied under DMITREVSKY and soon became known for his delicate and elegant, though somewhat mannered, playing of worldly young men. In later years he tended towards a more realistic interpretation of his parts, and in the 1830s added a vein of satire to his art. One of his favourite roles was that of Repetilov in GRIBOYEDOV's *Wit Works Woe*, which he played on its first production at the Bolshoi Theatre in St Petersburg in 1831 and continued to play until his death. He was also highly successful in the comedies of MOLIÈRE.

SOTHERN, EDWARD ASKEW (1826–81), English actor, who first appeared in the provinces under the name of Douglas Stewart and then went to America, where he played unsuccessfully in Boston, being considered by a contemporary critic 'under-taught and over-praised'. After some years on tour in North America he joined Lester WALLACK as Sothern, by which name he was known thereafter, and in 1858 was with Laura KEENE, making an immense success in Tom TAYLOR's *Our American Cousin* as Lord Dundreary, a part ever after associated with him. Joseph JEFFERSON, who played Asa Trenchard in the same production, describes in his autobiography how Sothern, who was at first dismayed by the few lines offered him, began to introduce 'extravagant business' into his part. By the end of the month he was the equal of any other character; by the end of the run he was the whole play. He was equally successful in London in 1861, where long side-whiskers, as he wore them, became fashionable as 'dundrearies'. The play became practically a series of monologues, and several other sketches were written round Sothern's creation. Another of his great parts was the title role in *Brother Sam* (1865), which he wrote in collaboration with BUCKSTONE and Oxenford, some people preferring it to Dundreary. He also appeared in T. W. ROBERTSON's *David Garrick* (1864), in which he failed in the love scene, and in H. J. BYRON's comedy of an old actor, *A Crushed Tragedian* (1878), which after a poor reception in London was a great success in New York. Sothern was essentially an eccentric comedian, and it was in that line of business that he did his best work.

SOTHERN, EDWARD HUGH (1859–1933), one of three actor-sons of Edward Askew SOTHERN, who inherited much of his father's charm and talent. Educated in England, he intended to become a painter but took to the stage, becoming in 1904 Daniel FROHMAN's leading man at the LYCEUM THEATRE in New York and remaining there until 1907. A light comedian and an excellent romantic hero in such popular plays as Hope's *The Prisoner of Zenda* and Justin McCarthy's *If I Were King*, he later formed his own company for the production of plays by Shakespeare with his second wife Julia MARLOWE, whom he married in 1911 and took with him on his return to England after an absence of 25 years. Their joint appearance in such plays as *Romeo and Juliet* were very well received. On Julia's retirement in 1924, after an accident, she and Sothern presented the scenery, costumes, and properties for ten of his Shakespeare productions to the SHAKESPEARE MEMORIAL THEATRE. He continued to act intermittently until 1927 and devoted much of his later years to public readings and lectures. A small but dignified man, with a handsome, sensitive face, he was the ideal romantic hero of the late 19th century, and although by hard work he achieved some success in tragedy he was at his best in high comedy, his curious combination of gifts making him an excellent Malvolio in *Twelfth Night*. He also revived on several occasions his father's old part of Lord Dundreary in Tom TAYLOR's *Our American Cousin*.

SOTIE, the topical and satirical play of medieval France, whose best known author is Pierre GRINGORE. The *sotie* was not a farce, though the two forms had elements in common, and it was often inspired by political or religious intrigue. It was intended for amusement only and had no moral purpose, though it often served as a prelude to a MYSTERY or MORALITY PLAY. The actors, or *sots* (fools), wore the traditional fool's costume—dunce's cap, short jacket, tights, and bells on their legs. *Soties* were acted not only by amateur companies of students and law clerks but by semi-professional and more or less permanent companies, somewhat in the tradition of the COMMEDIA DELL'ARTE, each with its own repertory. There are a number of extant texts, of which the *Recueil Trepperel* is the most representative.

SOUL-CAKERS' (or **SOULERS'** or **CHESHIRE SOULING**) **PLAY,** see MUMMERS' PLAY.

SOUND, see ACOUSTICS AND SOUND.

SOUND EFFECTS. Until the introduction of disc recordings into the theatre in the 1950s, all sound effects were produced live offstage, many by very old methods that still survive unchanged. It is widely held that the spontaneity of 'live' effects is superior to the recorded reproduction of reality and some effects, such as weather and bells, are better when produced live. Some examples of live sound effects are:

Wind, produced by a wind machine, a hand-

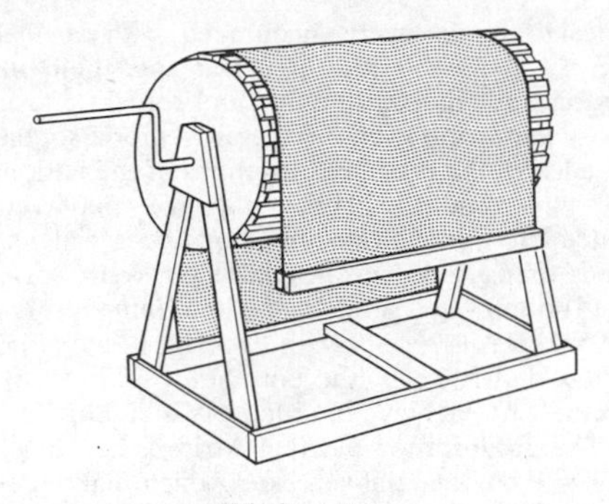

Wind machine

driven drum on to which is mounted a series of strips which rub against a length of sail canvas draped over the drum and fixed at one end. This is particularly effective when operated sensitively, but the machine needs regular maintenance.

Rain, surf, and *hail*, usually represented by dried peas or small lead shot being shaken in a shallow box, usually containing some fixed obstruction, such as nails. A cylinder, which is easier to operate by hand, can be used instead of a box, and the substitution of larger shot or marbles produces an effective hailstorm.

Thunder is still produced most effectively by means of a thunder-sheet. This is an iron sheet, as large as practicable, at least 6′ by 2′6″, suspended in the flies and vibrated by jerking a handle at the bottom. Less resonant thunder can be produced in the same way by using large sheets of plywood. The thunder-sheet replaced the 18th-century thunder-run, a series of wooden troughs built into the theatre's back wall, each separated by a gap so that cannon balls rumbled down and fell with a considerable crash. This method lost favour because once the cannon balls had been released there was no way of controlling the effect. A thunder-run still survives at the Theatre Royal, Bristol (see BRISTOL OLD VIC), though it is seldom used.

Door slams are only effectively produced by slamming an actual door. Should one not be available offstage, then a small door is used, 3′ by 2′ by 1″, fixed to a portable base and heavily constructed with all appropriate locks, etc.

Horses' hooves are still simulated most effectively by coconut shells, although much skill is needed in the operator. The shells can be used against each other, or against slate or carpet, to simulate cobbles, tarmac, or turf.

Explosions and gunfire. Individual explosions are best obtained by maroons placed in appropriate bombtanks and fired electrically. For shots, other than those obtained by striking wood against leather, a starting pistol should be used. It should be noted that there are strict regulations governing explosions and the use of firearms in the theatre. Recordings are the best way to convey the sound of continuous bombardment.

Bells. Doorbells and telephones best utilize their real components. It is desirable for actors to be able to press the doorbells themselves, and also that the operation of lifting the telephone handset cuts off the bell, although it is actuated from offstage. For chimes, a set of tubular bells is preferable to recordings.

Breaking glass is simulated by a quantity of broken glass and china flung from one bucket into another.

Effects recorded on 78rpm discs persisted well into the 1960s, usually alongside live effects. A theatre would typically have two independent turntables and amplifiers, as an insurance against break-downs, each with switch or fader access to several loudspeakers, which could be positioned according to the apparent sources of sound effects, and a graduated pickup arm with twin scales which allowed the needle to pinpoint a particular groove, often marked in chinagraph pencil. The records were generally hired from specialist companies.

Tape decks are also installed in pairs, for back-up purposes but also because one may be taken up by tapes of house-music or by loops—small continuous tapes of repetitive sounds like birdsong, which are placed directly on the capstan drive, not wound on reels. Other effects are combined on a conventional tape in the order in which they will be used. Each one is preceded by a length of contrasting leader, on which the cue number and description of the next effect can be written, and followed by a metal foil or translucent strip. These devices stop the tape (by mechanical or photoelectric means respectively), and if the succeeding leader is correctly measured the sound effect needed next will be precisely positioned in relation to the playback head. Thus the operator need only concern himself with adjustments to the volume and balance of each effect and with the all-important job of starting each one on cue. Many sound effects are now recorded on cartridge machines similar to those used for advertising 'jingles' in commercial broadcasting.

SOUTAR, ROBERT, see FARREN, NELLIE.

SOUTH AFRICA. The first theatrical entertainments in South Africa were given in Cape Town in the late 18th century by French, and later British, troops stationed there. The first playhouse, the African Theatre, opened in Hottentot Square in 1801, and was used by English, Dutch, French, and German amateurs, and also by English companies on tour, until in 1839 it was closed during an upsurge of puritanism. In 1843 two wine stores were converted into little theatres, and in 1855 Sefton Parry, an English theatrical manager, opened the Drawing-Room Theatre, where he presented performances by semi-professional companies.

The Diamond Rush of the 1860s brought with it a number of theatrical adventures, notably

Captain Disney Roebuck (*c.* 1821–85), who from 1873 onwards provided Cape Town with regular entertainment; with the Gold Rush of the 1880s came real prosperity, a much inflated population, and considerable theatrical activity. Excellent theatres were built in Cape Town—the Opera House in 1893, the Tivoli in 1903—Johannesburg—the Standard in 1891, the Globe, later the Empire, in 1892, the Gaiety in 1893, His Majesty's in 1903—and also in Pretoria, Durban, Port Elizabeth, Bloemfontein, Kimberley, Grahamstown, and many smaller towns. Almost every great player of the time came from Europe and America to play in them. The central figure during this period was Leonard Rayne (1869–1925), who was lessee of many of the theatres and as well as importing overseas artists had his own company, in which he played leading roles partnered by his wife Freda Godfrey. After his death the South African theatre, in common with the theatre everywhere, suffered a decline, many of the old buildings being turned into cinemas or torn down. Fortunately this decline in professional theatre was accompanied by the rise of the amateur movement and by the emergence of the Afrikaans professional theatre under Paul de Groot (1882–1942), who in 1926 led the first company to tour in plays in Afrikaans. He was followed by many others, including André Huguenet, the Hanekoms, Wena Naude, Pikkie Uys, and Anton Ackermann, who toured the country during the 1930s and 1940s with European plays in translation and simple, somewhat melodramatic and sentimental works by the many folk-dramatists who arose to satisfy their needs. Unfortunately the emergence of an Afrikaans dramatic repertory was not matched by a similar one in English, the English amateur companies continuing to stage only popular successes from London and New York, often with considerable personal success for the actors. The two leading groups, the Cape Town Repertory Theatre Society founded in 1921 and the Johannesburg Rep founded in 1927, were, however, able to build their own theatres, while in Cape Town the Drama Department of the University presented in its Little Theatre, which opened in 1931, an international repertory based mainly on the classics.

The period covered by the Second World War saw a brilliant renaissance of the English-speaking professional theatre when a company formed by Gwen FFRANGCON-DAVIES and Marda Vanne gave between 1941 and 1946 seasons of distinction in many of the larger towns. After the war their standard was maintained by other companies, notably those of Brian Brooke, who established a repertory theatre in Cape Town in 1946 and in Johannesburg in 1955. In 1947 the National Theatre Organization was instituted, with headquarters in Pretoria under the direction of P. P. Breytenbach. It evolved into four Councils of Performing Arts, one for each province, to manage regional touring companies playing in both English and Afrikaans as well as to co-ordinate other theatrical activities. Local dramatists were encouraged, with particular effect on those writing in Afrikaans; their plays epitomized, mostly in naïve and romantic terms, Afrikaner aspirations, becoming more sophisticated with the post-1948 triumphs of the nationalist government. English-language playwrights tended to concentrate more on the racial questions that are a central feature of South African life, notably Lewis Sowden with *Kimberley Train* (1958), Basil Warner with *Try For White* (1959), and Athol FUGARD, the only South African-born dramatist to achieve international standing.

The urbanization of tribal Africans has resulted in some commercial successes which make use of traditional music and dance; examples include *King Kong* (1958), based on the life of an African boxer, which was seen in London, and *Ipi Tombi* (1974), which began a long run in London in 1975. Straight plays from non-white authors include *The Festival* (1970) by H. W. D. Manson; *Sikhalo* (*c.* 1972) by Gibson Kente; *Phiri* (*The Wolf*, also 1972), an adaptation by the Phoenix Players of Ben JONSON's *Volpone* set in Soweto; and Welcome Msomi's *Umbatha*, a Zulu version of *Macbeth*, seen in London during the WORLD THEATRE SEASONS of 1972 and 1973. Most activity in this area, however, is among small non-white or precariously multi-racial groups working with limited facilities outside the mainstream commercial theatre.

South Africa is well provided with theatre buildings of a high standard, few towns of any size being without at least one. Johannesburg, with a population of around 2 million, has 10; Cape Town, with a population of about 1 million, has 4. Since leaving the British Commonwealth in 1961 the country has suffered from a certain degree of cultural isolation and, especially since the mid-1970s, from international dislike of apartheid. Standing Council Instructions formulated by BRITISH ACTORS' EQUITY, for instance, effectively discourage its members from accepting engagements in South Africa, while strict CENSORSHIP there of both the political and moral content of plays inevitably hinders artistic development.

SOUTH AUSTRALIAN THEATRE COMPANY, Adelaide, see AUSTRALIA.

SOUTHERN COMEDY PLAYERS, Dunedin, see NEW ZEALAND.

SOUTHERNE, THOMAS (1660–1746), English dramatist, a friend of DRYDEN, for whose plays he wrote a number of prologues and epilogues, and of Aphra BEHN, two of his plays, *The Fatal Marriage; or, the Innocent Adultery* (1694) and *Oroonoko* (1695), being based on her novels. His tragedies, which include his first play *The Loyal Brother; or, the Persian Prince* (1682), *The Fate of Capua* (1700), and *The Spartan Dame* (1719), show a mingling of heroic and sentimental drama which had some influence on the development of

18th-century tragedy as exemplified by Nicholas ROWE. He was also the author of three comedies which enjoyed some success when first produced: *Sir Anthony Love; or, the Rambling Lady* (1690), *The Wives' Excuse; or, Cuckolds Make Themselves* (1691), and *The Maid's Last Prayer; or, Any, Rather than Fail* (1693). They contain some witty scenes, but are weak in construction and overloaded with detail and have not been revived. It was on Southerne's recommendation that Colley CIBBER's first play *Love's Last Shift; or, the Fool in Fashion* was produced at DRURY LANE in 1696.

SOUTH LONDON MUSIC-HALL, see MUSIC-HALL.

SOUTH STREET THEATRE, Philadelphia, see SOUTHWARK THEATRE.

SOUTHWARK FAIR, see FAIRS.

SOUTHWARK THEATRE, the first permanent playhouse to be erected in Philadelphia, and possibly the first in America. A rough brick and wood structure painted red, its stage lit by oil lamps, it was built in 1766 by David DOUGLASS, manager of the AMERICAN COMPANY, and opened on 12 Nov. with VANBRUGH's *The Provoked Wife* and Isaac BICKERSTAFFE's *Thomas and Sally*. Early in 1767 it saw the production for one night of GODFREY's *The Prince of Parthia*, the first American play to be staged professionally. During the War of Independence the building, which from its position was sometimes known as the South Street Theatre, was closed, but after the departure of the British it reopened for a short time in the autumn of 1778. In 1784 the younger HALLAM and John HENRY, who had assumed the management of the American Company, stopped in Philadelphia on their way back to New York from the West Indies, and reopened the theatre for a while. It continued to be used for plays after their departure until in 1821 it was partly destroyed by fire. Rebuilt and used as a distillery, it was finally demolished in 1912.

SOVIET ARMY, Central Theatre of the, Moscow, see RED ARMY THEATRE.

SOVIET UNION, see RUSSIA AND SOVIET UNION.

SOVREMENNIK THEATRE, Moscow, founded in 1958, after a group of young actors from various other playhouses in the city had come together under YEFREMOV to stage the first production of ROZOV's *Alive for Ever* (1957). The success which greeted their venture led them to remain together, and the building in which they had appeared, lent to them by the MOSCOW ART THEATRE, was given the name Sovremennik or Contemporary. A strong humanist and lyrical element was characteristic of the company's best work in its early years, as was shown in their productions of plays by Volodin and Zorin, as well as Rozov, and in their treatment of Shatrov's *The Bolsheviks* in 1967. They were responsible for the first Soviet production of OSBORNE's *Look Back in Anger*, and in 1962 presented an extremely popular interpretation of William Gibson's *Two for the Seesaw*; ALBEE's *The Ballad of the Sad Café* was staged in 1967. When Yefremov moved to the Moscow Art Theatre in 1971 the company came under the control of its leading players, Oleg Tabakov and Galina Volchok. In 1975 Peter James, then in charge of the CRUCIBLE THEATRE, Sheffield, was invited to direct *Twelfth Night*, the Sovremennik's first Shakespeare production.

SOYA [Carl Erik Soya-Jensen] (1896–), Danish dramatist and novelist, whose earliest plays, *Parasitterne* (*The Parasites*, 1929) and *Den leende Jomfru* (*The Smiling Virgin*, 1930), are naturalistic in style and cynical in tone. *Hvem er jeg?* (*Who am I?*, 1931) is coloured by Freudian symbolism. His plays of the 1940s—*Brudstykker af et Mønster* (*Fragments of a Pattern*, 1940), *To Traade* (*Two Threads*, 1943), *30 års Henstand* (*30 Years' Respite*, 1945), and the farce *Frit Valg* (*Free Choice*, 1948), together designated by the author as his 'tetralogy'—deal generally with problems of human destiny and individual guilt.

SOYINKA, WOLE [Akinwande Oluwole] (1934–), Nigerian writer, the first to become known for his dramatic works outside his own country. Born at Abeokuta in Western Nigeria, he attended University College, Ibadan, before taking a degree in English at Leeds University, where he wrote *The House of Banegegi* (1956), *The Lion and the Jewel* (1957), and *The Swamp Dwellers* (1957). In 1958 he joined a writers' group conducted by George DEVINE and William GASKILL at the ROYAL COURT THEATRE in London, where his one-act play *The Invention* (1959) was given a Sunday night performance. He returned to the University of Ibadan in 1960 as a research fellow in drama, organized the 1960 Masks Company, and wrote *The Dance of the Forests* for the Nigerian independence celebrations. He then produced two satirical revues, *The Republican* (1964) and *Before the Blackout* (1965), and also in 1965 had a play, *The Road*, produced at the Commonwealth Arts Festival in London. This was followed by *Kongi's Harvest* (1966), produced at the Dakar World Festival of Negro Arts and, in 1968, in New York. Active in politics, he was arrested in 1965, accused of substituting a tape recording of his own voice for a broadcast election address by the Western Premier Akintola. His attempts to negotiate a truce in the Biafran War in 1967 again led to his arrest and to two years in jail. He returned to the School of Drama at Ibadan, of which he had been appointed director in 1967, in 1969, but resigned in 1972 after disturbances on the campus. His next play, *Madmen and Specialists* (1971), expressed his horror of war and destruction.

In 1973 the NATIONAL THEATRE in London commissioned his adaptation of EURIPIDES' *Bacchae*, and he edited the magazine *Ch'Indaba* (formerly *Transition*) until he returned home in 1975 to occupy the Chair of Comparative Literature at the University of Ife. A play written in the same year, *Death and the King's Horseman*, deals with an incident in Yoruba history. He became Head of the university's Department of Drama and president of the Nigeria Centre of the INTERNATIONAL THEATRE INSTITUTE.

Soyinka, who fully deserves his international reputation, is a sophisticated craftsman, with a fine command of the English language and an innovative approach to stagecraft. He is essentially a political satirist, and his work has become progressively more powerful and demanding.

SPAIN, as a favoured province of the Roman Empire, possessed many theatres and circuses; and after the fall of imperial power the repeated condemnations by Church Councils of charioteer (*auriga*) and actor (*pantomimus*) suggest that they continued to flourish for many years. Nevertheless, the most significant evidence of dramatic activity during the early Christian era concerns the pagan rites and festivals, many containing dramatic elements, which also earned the condemnation of the ecclesiastical authorities.

During the first centuries AD the Church stood firm against all theatrical representation, but to little avail. An attempt was made to convert pagan song and ritual dance to Christian purposes by the introduction of mime and dance into the ceremonies of the Church, but by 627 the third Council of Toledo was condemning the use of dances and lewd songs in the churches, to the prejudice of the divine nature of the services.

In the south of Spain the pagan spectacles were ended, as far as we can judge, by the invasion of the Moors in 711. In the north-east the growth of the power of the Church, and the added wealth and attraction of its festivals, gradually emasculated the old traditions until all that survived was the lighting of ceremonial fires, the burning of precious objects, and rustic rites and dances whose origins were unknown to those who performed them.

The singing of Tropes is said to have made its appearance in the north-east in the early part of the 9th century; this deliberate amplification of a passage in the authorized liturgy text gradually became detached from its context and eventually developed into authentic LITURGICAL DRAMA. In Catalonia the Latin Easter play was widespread, and many examples of the *Peregrinus* and *Planctus* have been identified.

At Christmas and Whitsun lavish spectacles were staged in the cathedral of Valencia. Whilst actors wearing masks performed and sang on scaffolds set on the floor of the cathedral, mechanical doves flew down from the vaulting, heavens constructed in the roof opened, and platforms descended in the shape of clouds or globes bearing angels, saints, and statues of the Virgin Mary—whose Assumption became a peculiarly Spanish theme, of which various versions are known, the most famous being the still extant performance at Elche. This play, which lasts two days (14 and 15 Aug.) and utilizes three separate *aracelis*, or cloud-machines, represents a 16th-century crystallization of a play which is probably much earlier in origin, a theatrical performance which is today unique.

From the 9th to the 11th centuries much of Castile remained under Moorish domination and was thus relatively untouched by the developments in the north-east; for this reason, little evidence has survived of the liturgical play in Castile. Nevertheless, it was in Toledo that the fragmentary *Auto de los Reyes Magos* appeared, and it has been argued that this play may be evidence of the introduction from France of such works, already in a vernacular state, following the replacement of the old Hispanic liturgy (or Mozarabic rite) by the Roman towards the end of the 11th century.

It is to the foregoing types of performance that Alfonso X (1252–84) mainly refers in the oft-quoted passage from the laws of the *Siete partidas*: in churches under the direct control of archbishops, bishops, or their representatives, but not in small towns or villages, the Nativity and Easter plays may be performed, to present the scriptural happenings as a physical reality and turn the hearts of men to devotion and good works. But the clergy are to refrain from bringing their cloth into disrepute, and the laity are prohibited from wearing religious habits for the same reason.

A struggle was thus taking place between the desire to save the churches from irreligious pollution and the urge to instruct and enlighten their congregations, appealing to them through their dramatic sensibilities. The establishment of the Corpus Christi procession in Spain in the first half of the 14th century allowed these two aims to be pursued simultaneously: priest and layman could exercise their dramatic ingenuity to the full, but their efforts were excluded for the most part from the churches. The form taken by the procession was influenced by the *juglares*, or wandering players of high and low degree, and many other existing dramatic forms—processional floats, counterfeit figures, devils, mock savages, and mimic battles—were incorporated into this, the chief procession of the Church, a statement of faith, ostentatious and magnificent, parading in Spain not only before Christians but before Jews and Moors. In the early references to the procession it is this feeling of religious fervour that predominates, but gradually its place was taken by a more worldly and satirical attitude; the spectacular began to appeal for its own sake. The Church, however, was alive to the danger, and the procession was thoroughly reformed by the provisions of the Council of Trent, which in 1551 declared that the Corpus Christi procession should proceed from a cathedral and that the town council should be in charge of the entire festivity.

The Court officials responsible for regal displays and courtly entertainments soon saw the possibilities of adapting this religious pomp for their own purposes. Elements of the Corpus Christi procession, notably the counterfeit animals, were borrowed and used to grace coronations; the heaven filled with stars and the clouds which descended from it were used to greet royal visitors at a city gate. Counterfeit figures filled with fireworks gradually developed into representations in their own right. It was through these entertainments, in all probability, that the people of culturally isolated Spain enjoyed their closest contact with the Renaissance of classical learning, as Perseus and Andromeda, Hercules and the Titans, sailed across the night sky in every great city of Spain, fought out pitched battles in the principal squares, or graced river pageants and water spectacles.

The same period saw the tourney develop from a display of prowess at arms to a semi-dramatic entertainment. The introduction of firearms had largely defeated its original purpose, and gradually the spectacular element increased in importance, display coming to be valued for its own sake. As the chivalresque novel increased in popularity during the 16th century, it became usual to introduce some features of this literary genre into the tourney. Often the joust was in earnest and the novelesque element was used as a backcloth, the participants being rewarded according to their deserts by a human being impersonating a figure from a chivalresque novel. Again, the challenge of a real tourney might echo literary language. Occasionally whole episodes from a book of chivalry were consciously enacted in what may best be described as a pageant play. These entertainments, together with the ENTREMÉS (French *entremets*) of the court banquet, mummings (*momos*), masquings, processions, and masquerades, formed the principal diversions of the Courts of Spain during the period which saw the birth of the Court theatre.

This came into being at the end of the 15th century, and reflected both the religious performances of its day and the secular inspiration of the Renaissance. Gómez MANRIQUE combined courtly mumming and the Nativity play in his *Representación del nacimiento*. Lucas FERNÁNDEZ and Juan del ENCINA developed elements of the liturgical plays in their Christmas and Easter works: and Encina's secular plays most clearly show Renaissance influence, extolling the power of love and contrasting unfavourably life at Court with the simple pleasures of the country. Gil VICENTE, writing for a Castilian Queen at the Court of Portugal and equally at home in Castilian and Portuguese, offered a wide range of subjects: his simpler religious plays are very similar in subject matter to those of Encina, but to others he added an element of pageantry. Indeed, in the plays of Encina and Vicente we find already well developed two of the three main themes of the 17th century Court theatre: the pastoral and the chivalresque. The later 16th century was to add the third, the mythological.

The year 1500 saw also the first extant edition of *La Celestina*, a prose work largely by Fernando de ROJAS which, although it cannot be classed as a play, nevertheless exerted considerable influence on the development of the drama, particularly through the comparative realism and raciness of its prose dialogue.

In the prologue to his collected works TORRES NAHARRO distinguishes between the *comedia a noticia* and the *comedia a fantasía*—that is, between realistic and novelesque plays. His works themselves are Italianate, a feature also of the plays of Lope de RUEDA. Rueda's fame rests more securely on the PASOS, the short farcical interludes which exploit simple situations and which still have considerable vitality. He appears to have come early into contact with Italian COMMEDIA DELL'ARTE players performing in Spain, and the influence of these troupes, particularly that of GANASSA, on the development of the Spanish theatre was considerable. Ganassa was responsible for structural changes in the early commercial theatres, and it may well be that the typical composition of the Spanish commercial troupe of the late 16th and 17th centuries owed much to the appearance of these polished professionals in Spain at the precise moment when commercial theatres were being established throughout the peninsula.

Whilst the neo-classical dramatists of Spain cannot be ignored by literary historians, their works had little impact on the development of the commercial theatre, and a far more potent influence is to be found in the Senecan tragedies of Juan de la CUEVA. Other dramatists of the latter part of the 16th century are Juan de TIMONEDA, Andrés REY DE ARTIEDA, and Cristóbal de VIRUÉS; and ten plays have survived from the pen of Míguel de CERVANTES SAAVEDRA, together with eight interludes.

The dramatic careers of Cueva and Cervantes saw the establishment of the first commercial theatres of the great towns of Spain, notably in Madrid. To these courtyard theatres, or *corrales*, came professional troupes, the actors bound by yearly contracts to the manager (*autor de comedias*), who himself contracted with the lessee of the theatre (*arrendador*) or with the municipality to give a stated number of performances. These contracts always stress the need for novelty in the repertoire, the audience's appetite for new plays being insatiable. The manager usually bought his plays outright from the dramatist, who then had no further financial interest in them. In all the great cities the audience consisted largely of tradespeople, servants, soldiers, the bourgeoisie, the professional classes, and the nobility. The actors certainly played also in small towns and villages, but the plays were written for an urban audience.

Since the common people paid for the plays, said Lope de VEGA CARPIO, it was only fitting that the dramatist should seek to please them. This, however, does not mean that the works of Lope himself, despite his enormous output, are trivial, but rather that the values which they stress are those commonly held in his own day. He writes in freely flowing verse, adapting his verse form to the subject matter, avoiding irrelevancies, and building up to a tense climax. His plays are linear works, with clear-cut divisions between main and sub-plots.

Among Lope's contemporaries are Guillén de CASTRO Y BELLVÍS, Antonio MIRA DE AMESCUA, Luis VÉLEZ DE GUEVARA, and TIRSO DE MOLINA. Francisco de ROJAS ZORRILLA belongs to the next generation of dramatists, as do Juan RUIZ DE ALARCÓN and Agustín MORETO. All these dramatists wrote copiously, and it is not surprising that many of their productions are refashionings of previous plays (*refundiciones*)—a practice which was, of course, recognized, and carried with it no suggestion of plagiarism. Each generation of dramatists rewrote the audience's favourite plays to suit the changing taste of the time.

Many of these playwrights also turned their attention to the minor theatrical forms, among them the *entremés* and the LOA. Foremost among the *entremesistas* was Luis QUIÑONES DE BENAVENTE. The minor entertainers—acrobats, jugglers, and puppet-players—had a monopoly of the public theatres during Lent, when the performance of plays was prohibited.

Pedro CALDERÓN DE LA BARCA, the greatest dramatist of 17th-century Spain, was not as prolific as his predecessor, Lope de Vega, but nevertheless some 200 of his dramatic pieces have survived. His first plays were written in the early 1620s, some 40 years after Lope de Vega began to write for the Spanish stage, but his most famous COMEDIAS date from the period 1625–40; during these years he also wrote AUTOS SACRAMENTALES and in the late 1630s Court plays destined for the newly built palace of Buen Retiro. He wrote little during the period 1644–51, when the commercial theatres were closed, partly as a result of Court mourning and partly because of the attacks of reformers. Calderón was a master craftsman, equally at home with the mythological play of spectacle performed on the ornamental waters of Buen Retiro or in the Coliseo; with plays which presented religious or moral themes; with ingenious and complicated comedies; with works which develop and comment upon the theme of honour; or with the elaborate machinery of the *auto sacramental*, in which his genius is particularly evident. The *auto sacramental* of this period was, to use Calderón's own terms, a sermon in verse, theology put upon the stage, and it is only when the plays have been visually recreated in performance that their import can be clearly seen and their didactic purpose fully realized.

The Court theatres of 17th-century Spain employed elaborate machinery. No permanent theatre was available until the late 1630s, when the Coliseo in the palace of Buen Retiro came into use. Before that, plays or entertainments had been presented by commercial troupes in the private apartments of the king and queen in the Alcázar, or in its magnificently decorated Salón Dorado. The elaborate Court plays combined elements taken from the spectacles of Italian Courts with contemporary staging techniques: the use of perspective in scene design, concealed lighting and orchestra, and transformations which took place before the eyes of the audience. They offered a type of spectacle distinct from that of the *comedia*, the latter appealing by reason of the ingenuity and novelty of its plot, the former by reason of the lavishness of its spectacular production. It can indeed be argued that the relative growth of importance of the Court theatre in the last decades of the 17th century, which also saw the rise of the ZARZUELA, a short play with music on a mythological theme, was one of the causes of the decline of the commercial stage. Such dramatists as Francisco Antonio de BANCES CANDAMO wrote elaborate Court plays with political implications, unsuitable both as regards subject matter and staging techniques for performance in the public theatres. From 1680 onwards the commercial theatre rapidly declined, and, after the death of Calderón, for instance, no new *autos sacramentales* of any literary distinction were written.

The 18th century opened inauspiciously for the Spanish theatre. The new dynasty imported Italian players and broke the long monopoly of the two *corrales de comedias* of the capital, while the economic effects of the War of the Spanish Succession were disastrous. The typical Court entertainment of the period was the Italian opera, while up to the middle of the century the stock-in-trade of the commercial theatres continued to be rewritings of 17th-century *comedias*. Representative dramatists of the period are Antonio de Zamora (*c*. 1664–1728), author of a realistic play on DON JUAN, and José de Cañizares (1676–1750), several of whose plays were based on Lope de Vega.

French influence on the Spanish theatre was noticeable in the last decades of the 17th century, and the accession of a Bourbon king naturally stimulated interest in French literature, although the immediate effect on the Madrid stage was to encourage Italian opera. Neo-classical theories began to be expounded, though the era of neo-classical influence cannot be said to have begun in earnest until 1737, the year of the publication of the *Poética* of Ignacio de LUZÁN. Early attempts to write in the vein were, however, unsuccessful, as the lifeless comedies and dull tragedies of Nicolás Fernández de MORATÍN demonstrate. One of the few effective neo-classical tragedies in Spain is the *Raquel* (1772) of Vicente GARCÍA DE LA HUERTA, but the finest neo-classical dramatist was undoubtedly Leandro Fernández de MORATÍN. The triumph of neo-classicism is usually taken to be the banning in 1765, after considerable controversy, of the *autos sacramentales*; and it is clear that the *autos* were out of tune with the new age, both technically and in the treatment of their subject matter.

Neo-classical plays were not all that the 18th-century Spanish theatre had to offer. Rewritings of old plays still appeared on the boards, together with such new genres as the *melólogo*, or dramatic monologue set to music, and the *escena muda*, or dumb ballet. The *zarzuela*, too, developed in the hands of Ramón de la CRUZ CANO Y OLMEDILLA into a popular musical of contemporary life; but it could not survive competition from the improved Italian opera in the last years of the century, and Cruz is better known for his SAINETES, racy, lively, satirical sketches which mark the trend towards a more realistic portrayal of life which was to culminate in the REALISM and NATURALISM of the second half of the 19th century.

Neo-classicism never recovered from the invasion of Spain during the Napoleonic wars. If Luzán's *Poética* of 1737 can be said to mark the beginning of its influence, the publication in 1827 of another *Poética*, that of Francisco MARTÍNEZ DE LA ROSA, may be said to signal its close. Martínez de la Rosa, one of the last writers to defend neo-classicism, was already much influenced by Romanticism—essentially an alien growth, nurtured in France and England by the Liberal exiles of the 1820s and brought back to Spain with their return to power in 1835. The Romantic plays of Mariano José de LARRA, Juan Eugenio HARTZENBUSCH, and Antonio GARCÍA GUTIÉRREZ have not survived in performance, the only play of the period still to be found in the repertoire of the modern theatre being the *Don Juan Tenorio* (1844) of José ZORRILLA Y MORAL, still played each year on All Saints Day—a version in which the hero is sentimentalized and where his defiance of God is finally overcome. At the end of the play, Don Juan's soul is received into Heaven—an apotheosis which symbolizes the fate of Romanticism in Spain.

Almost as significant as the Romantic plays of the 1830s were the bourgeois comedies of BRETÓN DE LOS HERREROS. In the second half of the 19th century the bourgeois drama predominated, at the hands of Adelardo LÓPE DE AYALA, Manuel TAMAYO Y BAUS, and José ECHEGARAY. The finest realistic dramatist was Benito PÉREZ GALDÓS, better known as the leading novelist of his day. Also important for their social themes are the thesis plays of Joaquín Dicenta Benedicta, among them *Juan José* and *Daniel* (both 1906).

By 1894 the realistic drama was beginning to give way to new themes. Jacinto BENAVENTE Y MARTÍNEZ, a playwright who suggests that will and energy can overcome reality—reflecting in this the nonrational philosophies of his times—staged his first play *El nido ajeno* (*The Alien Nest*) in that year; and Ricardo de la VEGA scored a lasting success with *La verbena de la Paloma* (*The Festival of La Paloma*), a modern *zarzuela*. Typical also are the short pieces of Carlos Arniches, putting the people of Madrid upon the stage in the GENERO CHICO, a kind of sketch which was the lineal descendant of the *entremés*.

The last years of the century saw also a revival of the Catalan drama, at the hands of Santiago RUSIÑOL and Ángel GUIMERÁ, while Adría GUAL broke with rhetorical drama, experimented with symbolism, and, in his own Teatro Intím, set new artistic standards for scene design, hitherto neglected in Spain. The influence of Gual was not immediately felt in Madrid, and it was not until the second decade of the new century that the young designers he had shaped and trained triumphed in the capital. The earlier years saw various reactions in Madrid: the ÁLVAREZ QUINTERO brothers, Jacinto GRAU, and Gregorio MARTÍNEZ SIERRA can all, in their different ways, be seen as attempting new themes and making fuller use of stage resources.

A significant indication of the revolt from realism was the revival of interest in puppetry, a popular theatrical form still alive in Spain. The artists who frequented Els Quatre Gats, the Barcelona café of the turn of the century, were entertained by popular puppets and an artistic shadow-show, the latter a reminiscence of the shadow theatres of the cafés of Montmartre, notably Le Chat Noir. Such writers as Santiago, Rusiñol, and Benavente turned for inspiration to the puppet theatre. Benavente and Ramón María del VALLE-INCLÁN both wrote highly sophisticated plays for children's theatres, and Valle-Inclán made excellent use of the puppet metaphor in *La marquesa Rosalinda* (1913), a graceful satire of his own exquisite style.

The new theatre was to see its culmination in the years 1917–25, when Martínez Sierra took over the management of the Teatro Eslava. Although the Eslava was small and not particularly well equipped, Sierra presented a large range of extremely varied works—classical and modern, indigenous and foreign, poetic theatre, children's plays, pantomime, and *autos*—and introduced contemporary techniques in lighting, acting, and scene-painting. During his management of the Eslava the Spanish theatre was finally Europeanized. Cipriano Rivas Cherif, a pupil of Gordon CRAIG, was, like Sierra, an advocate of the *reteatralización* of the theatre. His original company, El Caracol founded in 1928, presented the first surrealist plays seen on the Spanish stage. In the same year he was appointed manager of the Teatro Español, and in collaboration with the director of the company and its leading actress Margarita Xirgu set up an influential theatre study group, the Teatro Escuela de Arte.

In the early 1920s Valle-Inclán turned from gentle satire to savage caricature. His *esperpentos* are a deliberate defamation of Spanish society and traditional values, an attack on humbug and sham. In the political atmosphere of the time such plays could not be publicly performed, and the important private theatres and theatre clubs of the period had often a political as well as an artistic purpose. El Mirlo Blanco, the theatre of the brothers Baroja, staged plays by Valle-Inclán; El Club Anfistora played both Lope de Vega and the outstanding dramatist of the modern Spanish theatre, Federico GARCÍA LORCA.

With the coming into power of the Third Republic in 1931, various *teatros proletarios* were formed with the intention of taking drama to the people. The Republic also subsidized two travelling theatres, La Barraca, under the direction of Lorca and Eduardo Ugarte, which presented mainly classical plays; and El Teatro del Pueblo, under the dramatist Alejandro Casona [Rodríguez Álvarez] (1900–65), author of an excellent play on the tragic love story of Inés de Castro.

The Civil War of 1936–9 brought to a sudden stop a resurgence in the Spanish theatre of solid achievement and high promise. Censorship was rigid, and the atmosphere of the years immediately after the war was not conducive to a rebirth of artistic endeavour. The dramatists who enjoyed the greatest success had all begun their careers before 1936, like José Maria PEMÁN Y PEMARTÍN and above all Benavente. It was not until the mid-1940s that new dramatists such as Antonio BUERO VALLEJO and Alfonso SASTRE began to emerge.

Most of the new work was necessarily non-political, and so tended to be comic or escapist in emphasis. Often more interesting experiments were to be found in the theatre clubs and little theatres—*teatros minoritarios*—and Sastre and Alfonso PASO founded the experimental Arte Nuevo group. Originating in student activities, the Teatro Español Universitario, or Spanish University Theatre, spread throughout the country, and introduced the work of contemporary foreign dramatists well in advance of the professional theatres. The T.E.U. of Murcia was under the direction of González Vergel and Ángel F. Montesinos, the Teatro Popular Universitario of Madrid under Sastre and Pérez Puig, and the T.E.U. of Barcelona under Antonio Chic. Other notable work was done in Barcelona—this at a time when the city lacked its own, third National Theatre—by the Theatre School of Adría Gual, under the direction of Ricardo Salvat; and in Madrid by the Dido company, directed by Josefina Sánchez Pedreño.

The censorship exercised by the church was somewhat relaxed in 1963, and the work of Lorca, Casona, and Valle-Inclán began to return to the stage. Sastre exploited social themes, while dramatists like RODRIGUEZ BUDED treated political and cultural subjects with a new explicitness. More allusive and tangential—and so more widely performed—is the work of José-María Bellido, whose *Football* (1963) allegorized world power in terms of the rivalries of a prolonged team game. In spite of the creation of the National Experimental Theatre in Madrid in 1965, most new Spanish work for the theatre has remained conservative in form as in content, and the commercial theatre continues to depend on foreign successes.

The best known Spanish dramatist now at work, Fernando ARRABAL, is an exile writing in French, while the best known Spanish-speaking director is the Argentinian Victor GARCIA. His productions for Núria ESPERT I ROMERO's company in Barcelona of GENET's *Les Bonnes* in 1969 and of Lorca's *Yerma* in 1971, seen in London during the WORLD THEATRE SEASONS of 1971 and 1972, are the chief international successes to have been achieved by the Spanish theatre since the Civil War.

SPEAIGHT, ROBERT WILLIAM (1904–76), English actor and author, who played with the OUDS while at OXFORD and made his professional début at the LIVERPOOL PLAYHOUSE in 1926. A year later he was in London, and before joining the OLD VIC company in 1931, where his roles included King John and Hamlet, had been seen in a number of plays, including SHERRIFF's *Journey's End* (1928) and Oscar WILDE's *Salome* (1931). He came into prominence when he appeared in the Chapter House of Canterbury Cathedral as Becket in T. S. ELIOT's *Murder in the Cathedral* (1935), playing the same part later in London and on tour and in 1938 in New York, where he was also seen as Chorus in *Five Kings* (1939), an adaptation by Orson WELLES of Shakespeare's history plays. Back in London he was in Ronald Duncan's *This Way to the Tomb* (1946) and made his last appearance in the West End in Charles MORGAN's *The Burning Glass* (1954). He also acted in plays and pageants connected with religious foundations, notably at Peterborough Cathedral as the Chronicler in James Kirkup's *Upon This Rock* (1955). At the EDINBURGH FESTIVAL he played King Agis in Thornton WILDER's *A Life in the Sun* (1955), Don Pedro de Miura in HOCHWÄLDER's *The Strong Are Lonely* (1956), and Dom Diogo in Jonathan Griffin's *The Hidden King* (1957), and in 1960 he played Becket in Sydney, Australia, where he was also seen in 1962 as Sir Thomas More in BOLT's *A Man for All Seasons*. His books include *Acting* (1939); *Drama Since 1939* (1947); *Nature in Shakespearian Tragedy* (1955); *The Christian Theatre* (1960); *Shakespeare on the Stage* (1973); and *Shakespeare: the Man and His Achievement* (1977). A book of memoirs, *The Property Basket*, was published in 1970.

SPECTACLE THEATRES, name given to the early ornate playhouses of which only Italy has been able to preserve any 16th- and 17th-century examples. These are PALLADIO'S TEATRO OLIMPICO in Vicenza, Scamozzi's Court Theatre at Sabbioneta, and ALEOTTI's Teatro Farnese at Parma. All these theatres, and many like them which no longer exist, were sumptuously decorated and had finely equipped stages, with machinery capable of dealing with the most elaborate settings. (See THEATRE BUILDINGS.)

SPECTATORY, see AUDITORIUM.

SPELVIN, GEORGE, fictitious stage name, the American equivalent of Walter PLINGE, used to cover doubling. It is first found in New York in 1886 in the cast-list of Charles A. Gardiner's *Karl the Peddler*, and in a comedy entitled *Hoss and Hoss* (1895), by William Collier, Sr., and Charles Reed, is even given credit for supplying some of the gags. It is estimated that George Spelvin or his relatives (several variations of the Christian name have been used) have figured in more than 10,000 Broadway performances since George's début. The name has also been applied, theatrically, to dead bodies, dolls substituting for babes in arms and animal actors. The American magazine *Theatre Arts* also sporadically featured a critic of critics who masqueraded under Spelvin's name.

SPENCER, JOHN, see ENGLISH COMEDIANS.

SPERONI, SPERONE (1500–88), Italian dramatist, whose *Canace* (1543) is considered the bloodiest of the many bloodthirsty tragedies of this period. Dealing with incestuous love and ending with a pile of corpses, it caused fierce controversy when it was first performed at Padua.

SPERR, MARTIN (1944–), German dramatist and director, whose plays are essentially portrayals of life in the villages and small towns of his native Bavaria. *Jagdszenen aus Niederbayern* (*Hunting Scenes from Lower Bavaria*) was first seen at Bremen in 1966; *Landshuter Erzählungen* (*Tales from Landshut*, 1967), a series of 17 scenes which explore the struggle of two competing businesses for local supremacy, was produced at the MUNICH Kammerspiele; and *Koralle Meier* (1970), which shows a pre-war Bavarian town coming to terms with its neighbouring concentration camp, was first seen in Stuttgart.

SPIELTREPPE, see JESSNER, LEOPOLD.

SPOT, SPOTLIGHT, see LIGHTING.

SPOT BAR, see BARREL.

ŠRÁMEK, FRÁŇA, see CZECHOSLOVAKIA.

STABERL, Austrian stock comic character, the umbrella-maker Chrysostomos Staberl, played by Ignaz SCHUSTER in BÄUERLE's *Bürger von Wien* (1813) with such success that he became a household word and appeared in some 25 sequels by various hands. He marks the transition from such imported clowns as HANSWURST and KASPERL to native comedy, and was instantly recognizable as an archetypal Viennese working-class character, genial and shrewd, but something of a grouser and layabout.

STABILE, Teatro, see TEATRO STABILE.

STADT THEATRE, New York, see BOWERY THEATRE.

STAGE, the space in which the actors appear before the public. In its simplest form this is a cleared area with the audience sitting or standing all round, indoors or out, with or without a raised platform. At its most complicated, as in the picture-frame or PROSCENIUM-arch theatre, it is an elaborate structure, having a stage floor with a RAKE, surmounted by FLIES and a GRID, with a cellar underneath to house the machinery needed for working the TRAPS and for mechanical scene-changing. In this case the BOARDS will be removable to permit the passage of actors and props from the cellar. An intermediate form of stage, found in the Elizabethan playhouse, is a high platform built against a permanent wall structure with a cellar below. This may or may not have an INNER STAGE for the provision of simple scenic effects. The part of the stage floor which is in front of the proscenium arch is known as the FORESTAGE or apron stage.

The word 'stage' is also used of the whole ensemble of acting and theatre production, excluding the texts of the plays, which belong to dramatic literature. To be 'on the stage' is to be an actor; to be 'on stage' is to appear in the scene then being played; 'off-stage' refers to a position close to the stage where the actor is invisible, though not inaudible, to the audience; BACK STAGE includes the area around and behind the stage used by theatre workers, who enter the building through the STAGE DOOR.

STAGE BRACE, see FLAT.

STAGE-CLOTH, see CLOTH.

STAGE DIRECTIONS, notes added to the script of a play to convey information about its performance not already explicit in the dialogue. Generally speaking, they are concerned with the actors' movements and the scenery or stage effects.

Stage directions concerning the actors' movements are, in the English theatre, based on two important peculiarities: they are all relative to the position of an actor facing the audience—right and left are therefore reversed from the spectators' point of view—and they all date from the time when the stage was raked, or sloped upwards towards the back. Thus, movement towards the audience is said to be 'down stage', movement away from the audience 'up stage'. (When one actor moves upstage of another, the latter must turn away from the audience to face him; hence the expression 'to upstage', meaning to focus attention on oneself at the expense of someone else.)

The stage is further divided into nine main zones, three upstage (at the back), 'up left', 'up centre', and 'up right'; three downstage (at the front), 'down left', 'down centre', and 'down right'; and three across the middle, 'left', 'centre', and 'right'. Three further terms, relating to the back wall, are 'left centre back', 'centre back', and 'right centre back'. Further sub-divisions relating to the central area are indicated by 'up left centre', 'down left centre', 'up right centre', and 'down right centre'. Initials are generally used when writing all the above. Lateral movements to or from a centre line are qualified as 'off' and 'on'. 'To go off' is to leave the stage; 'to go off a little' is to withdraw to one side while remaining in view of the audience.

Stage directions may also indicate a change of position on stage, as 'cross', meaning 'go across the stage'. A 'scissor cross' is the simultaneous crossing of two characters from opposite directions and unless done intentionally, usually for a humorous effect, is regarded as an ugly movement, denoting clumsy technique. Movement round an object on stage is expressed as 'above' or 'below' that object, and not as 'in front' or 'behind'. This, too, dates back to the steeply raked stage of earlier times, and is also useful in avoiding confusion between the front or back of the object and the front or back of the stage.

The simplest examples of stage directions relating to the actors' movements are such single words as 'enters' or 'turns', to which may be added the Latin word *exit* ('he goes out'), now inflected as a normal English verb: 'you exit, he exits, they exit'. The plural form *exeunt* is obsolete except in the conventional phrase *exeunt omnes* ('all go out'). The antithesis *manet* ('he remains') has disappeared, but as early as 1698 it was used to indicate that a character remained on stage at the end of a scene to take part in the next scene, even if the scenery had to be changed (as was done in full view of the audience), thus ensuring that the action of the play carried on without a break.

In printed copies of old plays, particularly MELODRAMAS, such terms as R.U.E. ('right upper entrance') and L.2.E. ('left second entrance') are found. These relate to the time when the side-walls of the stage were composed of separate WINGS, the entrance being the passage between one wing and its neighbour. An earlier variant of this terminology (found as far back as 1748) was to indicate an entrance as 'in the second (or third) GROOVES', signifying that the actor entered behind the wing running in the second (or third) groove in the stage floor.

Two other terms, which apply only to the English-speaking theatre, are 'prompt side' (on the stage left), which houses the PROMPT CORNER, and 'opposite prompt' (on the stage right), usually spoken and written as O.P.

STAGE DOOR. Situated at the back or side of the theatre, this provides the usual means of access to the area backstage for actors and stage-hands. Immediately inside it is the cubicle of the stage-door keeper (formerly known as the hall keeper) who checks the arrival and departure of staff, prevents the entry of unauthorized persons, and transmits messages. Nearby is the Call Board, on which schedules of rehearsals and other items of information needed by the actors are posted, including the 'Notice' informing them of the end of the play's run. The stage door is never used by the audience, who enter by the main front and subsidiary side doors, but in Elizabethan times the stage (or tiring-house) door was used by those members of the audience who had seats on the stage.

STAGE FLOOR, see BOARDS.

STAGE-KEEPER, a functionary of the Elizabethan theatre who was responsible for the sweeping and clearing of the acting space, and probably for other odd jobs.

STAGE LIGHTING, see LIGHTING.

STAGE-MANAGER, the member of a theatre staff responsible for the overall management of everything connected with the stage and backstage. In the days before the pre-eminence of the DIRECTOR, the stage-manager also conducted rehearsals, which were often perfunctory and were called usually for the instruction of new members of the company and to co-ordinate the movements of the resident actors with those of a guest-star. In the 19th century men like BELASCO, though called stage-managers, occupied virtually the position held by the director today, often combining it with the office of PROMPTER.

STAGE PROPS, see PROPS.

STAGE RAKE, see RAKE.

STAGE SETTING, a term sometimes used to define the arrangement of a stage with curtains or drapery of some kind in such a way as to provide a general background suitable for any play, as distinct from SCENERY, which is suitable only for one play, or part of a play. It is also known as a CURTAIN SETTING, and is one of the forms of a stage SET.

STAGE SOCIETY, The Incorporated, London, organization founded in 1899 on the initiative of Frederick Whelen, after the demise of GREIN's Independent Theatre, to produce plays of artistic merit which were not likely to be performed in the commercial theatre. In order to make use of professional actors, performances were given in selected theatres on Sunday nights, when they were normally closed. This led to a police raid on 26 Nov. 1899 on the ROYALTY THEATRE, where the society was giving its first production, SHAW's *You Never Can Tell*. Whelen argued, successfully, that the theatre was being used as a private place and was therefore not subject to the ban on Sunday opening. Among the plays produced later at similar Sunday performances was Shaw's *Mrs Warren's Profession* on 5 Jan. 1902, which firmly established the society's right to perform plays which had been refused a licence by the LORD CHAMBERLAIN. Other unlicensed plays were MAETERLINCK's *Monna Vanna* (also 1902), TOLSTOY's *The Power of Darkness* (1904), GRANVILLE-BARKER's *Waste* (1907), PIRANDELLO's *Six Characters in Search of an Author* (1922), and James JOYCE's *Exiles* (1926). These were all seen later in the commercial theatre, where the most successful of the society's productions, R. C. SHERRIFF's *Journey's End* (1928), was seen only a few weeks after its original production.

The Stage Society functioned for 40 years, during which time it staged over 200 plays, many of them first performances of American and foreign plays in England. During the First World War it experimented with the revival of classic plays (in the absence of suitable modern material) and so aroused an interest in Jacobean and Restoration plays which was to lead to the establishment of the PHOENIX SOCIETY and be a marked feature of the 1920s. In 1926 in face of rising costs and a declining membership, which had reached its peak just before the war, the society merged with Phyllis Whitworth's Three Hundred Club, whose productions since its foundation in 1923 had included *A Comedy of Good and Evil* (1924), by Richard Hughes, J. R. Ackerley's *The Prisoners of War* (1925), and J. E. FLECKER's *Don Juan* (1926). Important productions after the merger, which lasted until 1931, were *The Widowing of Mrs Holroyd* (1926) and *David* (1927), both by D. H. LAWRENCE, and John VAN DRUTEN's *Young Woodley* (1928) and *After All* (1929). The society's last production was GARCÍA LORCA's *Bodas de sangre* as *Marriage of Blood* (later known as *Blood Wedding*) on 19 Mar. 1939 at the SAVOY THEATRE. The Second World War created conditions in which the society could not survive, and an attempt to revive it after the war was not successful.

Besides its work in producing interesting and important plays which might not otherwise have been seen in London, the Stage Society gave Granville-Barker his early opportunities as a director, which may be said to have led directly to the VEDRENNE-Barker seasons at the ROYAL COURT THEATRE, and also provided excellent acting opportunities for many rising young players, among them Peggy ASHCROFT, Edith EVANS, and Laurence OLIVIER, and for a number of directors, including Norman MARSHALL, Michel SAINT-DENIS, and Allan WADE. Although the settings of its productions were not normally elaborate, among those who designed scenery for it was Theodore KOMISARJEVSKY, and it has been claimed that the first use of projected scenery in the English theatre was in connection with the Stage Society's production of STRINDBERG's *The Road to Damascus* at the WESTMINSTER THEATRE on 2 May 1937. The society was also in the vanguard of the long struggle to abolish theatrical CENSORSHIP, which had been one of the causes of its establishment.

STAGG, CHARLES and MARY, see WILLIAMSBURG, VIRGINIA.

STAGNELIUS, ERIK JOHAN (1793–1823), Swedish poet and dramatist, who brought an essentially lyric talent to his dramatic work. His style, though highly wrought, is lucid, his imagery is ornate and richly textured, his preoccupations are at once sensual and ascetic. *Martyrerna* (*The Martyrs*, 1821) and *Bacchanterna* (*The Bacchae*, 1822) are both verse dramas. *Thorsten Fiskare, Sigurd Ring*, and *Wisbur* were published posthumously.

STALLINGS, LAWRENCE, see ANDERSON, MAXWELL.

STALLS, the individual seats between the stage front and the PIT, those nearest the stage being known as Orchestra Stalls. They first appeared in the 1830s to 1840s, after the raising of the first circle had allowed the pit to extend further back. With the exception of BOXES they are the most expensive seats in the theatre. They were at one time called by their French name of *fauteuils*. In some theatres the term Balcony Stalls was applied to the front rows of the Dress Circle.

STANDARD THEATRE, London, in Shoreditch. This was originally a pleasure-garden attached to the Royal Standard public-house, which was already in use in 1837. The first substantial building on the site, which held over 3,000 people, had a circus-ring which could be boarded over for concerts and dramatic performances. Burnt down on 21 Oct. 1866, it was rebuilt on a larger scale, and reopened in Dec. 1867 as the Standard Theatre under the management of John Douglass, who remained there until his death in 1888. He maintained a good stock company, and his pantomimes rivalled those of the BRITANNIA in Hoxton and even those of DRURY LANE. It was the first of the north-eastern suburban theatres to attract visiting stars from the West End, Henry IRVING being seen there in 1869. In 1876 it was again rebuilt, with a seating capacity of 2,463 in four tiers, and reopened as the New National Standard Theatre, but was always known by its old name. In 1889 it was taken over by the MELVILLE brothers, who ran it successfully until 1907, after which, as the Olympia, Shoreditch, it became a MUSIC-HALL, and in 1926 a cinema. It was badly damaged by bombs in 1940 and demolished soon after.

STANDARD THEATRE, New York, on the west side of Broadway, between 32nd and 33rd

Streets. This opened as the Eagle on 18 Oct. 1875, and had a prosperous career as a home of variety and light entertainment. Renamed the Standard, it witnessed on 15 Jan. 1879 the first production in New York of GILBERT and Sullivan's *H.M.S. Pinafore*, which ran for 175 nights. An unsuccessful season in 1880 had for its only bright spot Annie Pixley in the title role of Bret Harte's *M'liss; An Idyll of Red Mountain* and shortly afterwards the theatre changed hands at a very low price. *Patience*, beginning on 22 Sept. 1881, ran through the whole season, as did *Iolanthe* in the following year. On 14 Dec. 1883 the theatre was destroyed by fire, but it was rebuilt and reopened on 23 Dec. 1884 as a home of light opera. In 1896, renamed the Manhattan, it opened as a playhouse under William A. BRADY with *What Happened to Jones* by George BROADHURST, and from 1901 to 1906 housed a company run by Mrs FISKE. It was demolished in 1909 to make way for Gimbel's department store.

STANFIELD, CLARKSON (1793–1867), English artist, who was for a few years a scene painter, by many considered the most gifted to work in the London theatre. His first scenes were painted for the ROYALTY THEATRE in Wellclose Square; in 1831 he was at the Theatre Royal in Edinburgh, and he then returned to London to become scenic director at the Coburg Theatre (later the OLD VIC) and eventually at DRURY LANE. By 1834 he had achieved sufficient eminence as an artist to give up scene painting as a profession, though he occasionally did some work for his friends, assisting MACREADY with scenery for his PANTOMIMES in 1837 and 1842 and painting a backdrop in 1857 for DICKENS's production at his private theatre in Tavistock House of Wilkie COLLINS's *The Frozen Deep*. His last theatrical work was a drop scene for the ADELPHI THEATRE, which he painted for Ben WEBSTER in 1858.

STANISLAVSKY [Alexeyev], KONSTANTIN SERGEIVICH (1863–1938), Russian actor, director, and teacher of acting. He had already had some experience of acting with amateur groups, and had directed in 1891, for the first time on the Russian stage, *The Fruits of Enlightenment* by Leo TOLSTOY as well as a dramatization of DOSTOEVSKY's *Sela Stepanchikov*, when in 1898, in partnership with NEMIROVICH-DANCHENKO, he founded the MOSCOW ART THEATRE, which opened a new epoch in Russian and indeed in world theatre. In its first productions (Alexei TOLSTOY's *Tsar Feodor Ivanovich*, 1898; OSTROVSKY's *The Snow Maiden*, 1900; Leo Tolstoy's *The Power of Darkness*, 1902; Shakespeare's *Julius Caesar*, 1903) Stanislavsky put into practice the theories he had formulated under the influence of the MEININGER COMPANY and out of his own experience. Rejecting the current declamatory style of acting, with unrealistic costumes and scenery and stereotyped casting, he sought for a simplicity and truth which would give the complete illusion of reality. He trained his actors in a new way of acting, basing his methods on the psychological development of character and the drawing-out of latent powers of self-expression by precept and long practice, a process he described in *My Life in Art* (1924), *An Actor Prepares* (1936), *Building a Character* (1950), and the composite *Creating a Role* (1961). On these a whole system of actor-training has been built up, particularly by devoted Stanislavsky adherents in the United States where his system was elaborated into the METHOD. Among Stanislavsky's greatest achievements were his productions of the plays of CHEKHOV—*The Seagull* in 1898, *Uncle Vanya* in 1899, *Three Sisters* in 1901, and *The Cherry Orchard* in 1904—which he produced as lyric dramas, underlining the emotional elements with music and showing how Chekhov's apparently passive dialogue demands great subtlety and a psychologically-orientated internal development of the role, with great simplicity of external expression (an approach which Chekhov felt diminished the essential comedy in his work). During the years of upheaval leading to the uprising of 1905 Stanislavsky produced the plays of that 'stormy petrel of the Revolution' Maxim GORKY, including *The Lower Depths* (1902). In the years of reaction (1905–16) he turned to SYMBOLISM with MAETERLINCK and ANDREYEV in aestheticized, stylized productions. Under the Soviet régime, after an initial period of adjustment, he continued his work, but in later years gave up acting, owing to ill health, to concentrate on production and teaching. Among his own most notable roles were Astrov in *Uncle Vanya*, Vershinin in *Three Sisters*, Gaev in *The Cherry Orchard*, Dr Stockmann in IBSEN's *An Enemy of the People*, and Rakitin in TURGENEV's *A Month in the Country*.

STAPLETON, (Lois) MAUREEN (1925–), American actress, who made her New York début in SYNGE's *The Playboy of the Western World* in 1946, but achieved stardom as the tempestuous Serafina in Tennessee WILLIAMS's *The Rose Tattoo* in 1951. After roles such as Anne in *Richard III* 1953 and Masha in CHEKHOV's *The Seagull* in 1954 she starred in two more plays by Williams—as Flora in *Twenty-Seven Wagons Full of Cotton* (1955) and Lady Torrance in *Orpheus Descending* (1957). She then played Ida in S. N. BEHRMAN's *The Cold Wind and the Warm* (1958) and Carrie in Lillian HELLMAN's *Toys In the Attic* (1960), and in 1965 was seen in another Williams role, that of Amanda Wingfield in a revival of *The Glass Menagerie*, while in 1966 she played Serafina again. She was seen to advantage in two plays by Neil SIMON, playing three separate roles in *Plaza Suite* (1968) and the alcoholic Evy Meara in *The Gingerbread Lady* (1970); in 1972 she appeared as an effective Georgie Elgin opposite Jason ROBARDS in a revival of ODETS's *The Country Girl*. In 1975 she again appeared as Amanda, and in 1981 she played Birdie Hubbard to Elizabeth Taylor's Regina in a revival of Lillian Hellman's *The Little Foxes*. She is a METHOD actress of great power and subtlety, excelling in character roles.

STAR THEATRE, New York, at the north-east corner of Broadway and 13th Street. Though opened by the elder James WALLACK under his own name on 26 Sept. 1861, he himself never appeared there, and after his death in 1864 it was managed by his son Lester WALLACK, who left it in 1881 to open his own theatre on Broadway. The old theatre then housed plays in German until on 4 Jan. 1882 it reopened as the Star. It was here, on 29 Oct. 1883, that IRVING, under the management of H. E. ABBEY, with Ellen TERRY and his LYCEUM company, made his first appearance in New York. He confined himself to modern plays, not wishing to challenge Edwin BOOTH in Shakespeare, and his staging, lighting, and acting proved a revelation to New York audiences. Later visitors to the Star were Booth himself, John MCCULLOUGH on his farewell visit to New York in 1884, Mary ANDERSON in 1885, returning after two years in London, MODJESKA in Maurice BARRYMORE's *Nadjezda* (1886), BERNHARDT in her most popular parts, and Wilson BARRETT making his American début, also in 1886. The theatre was demolished in 1901.

STAR TRAP, see TRAPS.

STEAD, JAMES HENRY, see MUSIC-HALL.

STEELE, Sir RICHARD (1672–1729), English soldier, politician, essayist, pamphleteer, and incidentally dramatist, in which capacity he was one of the first to temper the licentiousness of Restoration drama with sentimental moralizing. His three early comedies—*The Funeral; or, Grief à-la-mode* (1701), *The Lying Lover; or, the Ladies' Friendship* (1703), and *The Tender Husband; or, the Accomplish'd Fools* (1705)—had only a moderate success, and he turned his attention to founding and editing, with ADDISON, *The Tatler* (1709–11) and *The Spectator* (1711–12), and also the first English theatrical periodical, *The Theatre*, which appeared twice a week from 1719 to 1720. It was not until 1722 that he produced his last and most important play, *The Conscious Lovers*, a sentimentalized adaptation of TERENCE's *Andria*, marked throughout by a high moral tone and given at DRURY LANE under Colley CIBBER's direction with an excellent cast which included Barton BOOTH, Robert WILKS, and Anne OLDFIELD. It was a great success, and was immediately translated into German and French, exercising an immense influence on the current European drift towards COMÉDIE LARMOYANTE.

STEELE, TOMMY, see MUSICAL COMEDY.

STEIN, PETER, see BERLIN and GERMANY.

STEINBECK, JOHN ERNST (1902–68), American novelist, awarded the Nobel Prize for Literature in 1962. He adapted for the stage two of his own novels, which were already dramatic both in structure and dialogue—*Of Mice and Men* (1937; London, 1939), a realistic picture of itinerant farm labour and of a feeble-minded farmhand, and *The Moon Is Down* (1942; London, 1943), which portrays the occupation by the Germans of a peaceful town, and its resistance movement. This was more favourably received in Europe than in America. Dramatized by other hands were *Tortilla Flat* in 1938 and *Burning Bright* in 1950. In 1955 Rodgers and Hammerstein based a musical, *Pipe Dream*, on Steinbeck's novel *Sweet Thursday*.

STEPHEN JOSEPH THEATRE IN THE ROUND, Scarborough, North Yorkshire. Stephen JOSEPH set up his first experimental THEATRE-IN-THE-ROUND in the public library of this seaside resort for a summer season in 1955. His aim was to create an ambience in which young playwrights could work closely with a small group of actors. The season became an annual event, and Alan AYCKBOURN's first play was produced there in 1959 as was James SAUNDERS's *Alas, Poor Fred*, his first play to be produced professionally. The theatre survived Joseph's death in 1967 through the perseverance of its manager, who invited Ayckbourn to become Director of Productions in 1970. In 1970 the company moved to new premises on the ground floor of a school and the theatre took its present name. The new auditorium seats 300 and there is a studio theatre; the company now operates throughout the year. Most of Ayckbourn's plays have been launched from Scarborough and many other writers have benefited from the policy of presenting new (often first) plays. The company also tours in Britain and abroad.

STEREOPTICON, see LIGHTING.

STERN, ERNST (1876–1954), scene designer, born in Bucharest, who worked for many years in Germany, making his début in 1904 with designs for Emanuel von Bodman's *Die heimliche Krone*. From 1906 to 1921 he worked for Max REINHARDT, solving the problems set by the director's predilection for the revolving stage, and the consequent need to think in terms of ground-plans and models rather than pictures. His sets for Reinhardt's Shakespearian productions included *Twelfth Night* in 1907, *Hamlet* in 1909, and *A Midsummer Night's Dream* in 1913; of his other designs, the most important were probably those for GOETHE's *Faust II*, AESCHYLUS's *Oresteia*, Vollmöller's *The Miracle* at Olympia in London (all 1911), and IBSEN's *John Gabriel Borkman* in 1917. From the late 1920s onwards he worked mainly in London, designing the sets for COWARD's *Bitter Sweet* (1929), another musical, *Ever Green* (1930), and for *White Horse Inn* (1931), reproducing the designs which he had made for the version produced earlier in Germany. Between 1943 and 1945 he also designed a number of sets for Shakespeare productions by Donald WOLFIT, and he was for many years active as a film designer in London and Hollywood.

STERNHEIM, CARL (1878–1942), German dramatist, at the height of his popularity in the 1920s. Banned under the Nazis, the best of his plays have since been revived many times, and he is considered one of the few German playwrights of this century to have excelled in comedy. When first produced, his clipped dialogue and grotesque situations seemed to show his affinity with EXPRESSIONISM, but in revival he appears more closely related to REALISM. Several of his bitter anti-bourgeois satires follow the fortunes of one family in its rise to social eminence under the collective title of *Aus dem bürgerlichen Heldenleben* (*Scenes from the Heroic Life of the Middle Classes*). The first of these was *Die Hose* (1911); produced by REINHARDT in BERLIN, this caused such a scandal because of its 'indelicate subject'—the loss of a lady's knickers in embarrassing circumstances—that it was banned, and its author left Germany to reside permanently in Brussels. In a translation by Eric BENTLEY, it was seen in London in 1963 as *The Knickers*. It was followed by *Die Kassette* (*The Money Box*, 1911) and by *Bürger Schippel* (1913), which shows the chief character—a marvellous part for a comic actor, played in London by Harry Secombe in an adaptation by C. P. TAYLOR in 1975—rising to middle-class status. *Der Snob* (1914) deals with the next generation's adaptation to high society, and was the first play to be revived in Berlin in 1945. It was followed by *Der Kandidat* (1915), *Tabula rasa* (1919), and *Das Fossil* (1925). Among Sternheim's other plays *Die Marquise von Arcis* (1919) was adapted by Ashley DUKES in 1935 as *The Mask of Virtue*, in which Vivien LEIGH made her first outstanding success in London.

STEVENSON, WILLIAM (?–1575), a Fellow of Christ's College, Cambridge, believed to have been the author of *Gammer Gurton's Needle*, a play which, with *Ralph Roister Doister* by UDALL, stands at the beginning of English comedy. No definite date can be assigned for its performance but it was probably given in Cambridge between 1552 and 1563. It was printed in 1575, the year of Stevenson's death, and may be identical with a play referred to as *Diccon the Bedlam*, the name of the chief character in *Gammer Gurton's Needle*. This play has also been attributed, with little likelihood, to Dr John Still, Bishop of Bath and Wells in 1593, and to Dr John Bridges, who is mentioned as its author in the *Martin Marprelate* tracts. Although structurally it conforms to the classic type, its material is native English, and one of its characters, Hodge, has given his name to the conventional English farm labourer.

STEWART, DOUGLAS, see SOTHERN, EDWARD ASKEW.

STEWART, DOUGLAS (1911–), New Zealand-born poet, dramatist, and critic, a fine writer of verse plays, most of them first produced in Australia where he now resides. They include *Fire on the Snow* (1941), *The Golden Lover* (1943), *Ned Kelly* (1945), *Shipwreck* (1947), and *Fisher's Ghost* (1961). Some of them were originally written for radio. A revival of *Ned Kelly* was put on in Australia in 1956 by the ELIZABETHAN THEATRE TRUST with Leo McKern in the title role.

STEWART, NELLIE (1858–1931), Australian actress, a descendant of Richard YATES, who made her début in Melbourne at the age of five with Charles KEAN in KOTZEBUE's *The Stranger*. Her early popularity was due largely to her dynamic stage presence and fine singing voice, and in 13 years she sang leading roles in 35 different comic operas. By 1902 she had lost her singing voice, but then appeared in what was to be her greatest success, KESTER's *Sweet Nell of Old Drury*, which was constantly revived; she also scored a success in SHELDON's *Romance*. In later years she played Romeo in the balcony scene of *Romeo and Juliet* to the Juliet of her daughter Nancy Stewart (1892–1973). Her marriage in 1884 to Richard Goldsbrough Row was unsuccessful, but she later formed a business and personal relationship with the impresario George Musgrove (1854–1916) which lasted until his death.

STICHOMYTHIA, a type of dialogue employed occasionally in classical Greek verse drama, in which two characters speak single lines alternately during passages of great emotional tension. The device, which has been likened to alternate strokes of hammers on the anvil, was used by Shakespeare, notably in *Richard III* (see I, ii, ll. 193–203; IV, iv, ll. 344–70). Echoes of it can be found in modern prose plays, particularly in the clipped dialogue of the 1920s, as in some of COWARD's early works, and it is found even in BECKETT's *Waiting for Godot* (1955), where, though outwardly comic, it engenders a mood of almost hysterical excitement. To be effective, the device must be used sparingly. It is sometimes referred to as 'cat-and-mouse', 'cut-and-parry', or 'cut-and-thrust' dialogue.

STIERNHIELM, GEORG (1598–1672), Swedish poet and polymath, author of a number of imitations of contemporary French MASQUES, such as *Then fångne Cupido* (*Cupid Captured*, 1649) and *Parnassus Triumphans* (1651). These were much patronized by the Court, and Stiernhielm is often called 'the father of the Swedish art of poetry'.

STILES, see FLAT.

STILL, Dr JOHN, see STEVENSON, WILLIAM.

STIRLING, FANNY [*née* Mary Anne Kehl] (1813–95), English actress, who made her first appearance in London in 1829 as Fanny Clifton. She married an actor-manager and dramatist, Edward Stirling [Lambert] (1809–94), most of whose plays were dramatizations of novels by DICKENS. She appeared under his management at the ADELPHI, and though not good in tragedy made a success in soubrette and comedy roles, one of her best parts being Peg Woffington in Tom TAYLOR's *Masks and Faces* (1852). Though her acting was later considered old-fashioned, she was recognized as the last great exponent of the grand style in comedy, particularly in such parts as the Nurse in *Romeo and Juliet* and Mrs Malaprop and Mrs Candour in SHERIDAN's *The Rivals* and *The School for Scandal*. She retired from the stage in 1870, but gave recitals and taught elocution. On her husband's death she married again, but died a year later.

STOCK COMPANY, the name applied in the 1850s to a permanent troupe of actors attached to one theatre or group of theatres and operating on a true repertory system, with a nightly change of bill, to distinguish it from the newly emergent TOURING COMPANIES. The system was, however, in use long before, and the term could have been applied to the companies at COVENT GARDEN and DRURY LANE from the Restoration onwards, to the 18th-century CIRCUIT companies in the English provinces and the resident companies of the early 19th century in the larger towns of the U.S.A. In both countries the stock company found itself threatened by the establishment of the long run and the touring company, which in the 1880s, helped by cheap and easy railway transport, finally triumphed. The stock company then ceased to exist, the last in London being that of Henry IRVING at the LYCEUM THEATRE. It had been an excellent training-ground for young actors, combining a variety of stage experience with some element of security, functions that were to some extent taken over by the REPERTORY THEATRES.

Each player in the old stock company undertook some special 'line of business', though usually ready to play something else when required. The recognized leader of the troupe was the Tragedian, who played Hamlet and Macbeth and might appear also in serious comedy. The Juvenile Tragedian played Laertes or Macduff, combining such roles with light comedy. The Juvenile Leads played the young lovers and the youthful heroes and heroines. The Old Man and the Old Woman appeared in such parts as Sir Anthony Absolute in SHERIDAN's *The Rivals* and the Nurse in *Romeo and Juliet*, while the Heavy Father (or Heavy Lead) played tyrants in tragedy and from the 1830s onwards villains in MELODRAMA. The Heavy Woman played Lady Macbeth or Emilia in *Othello*. The Low Comedian, who ranked next in importance to the Tragedian, played leading comic parts of a broad, farcical, or clownish type, together with minor roles in tragedy, and the Walking Lady and Gentleman played the secondary parts, such as Careless in Sheridan's *The School for Scandal*. They were usually beginners, and were poorly paid. General Utility, or simply Utility, also poorly paid, played minor roles in every type of play; the Supernumerary, or Super, was engaged merely to walk on, had nothing to say, and was not paid at all. The term Super is still used in the theatre for members of a crowd; a small, individualized part with few or no lines is now called a Walk-on; the term Juvenile Lead persists only in an ironic or pejorative sense; but Heavy Father has become part of the language in general. Apart from the above, the company might contain such specialized players as the First Singer, the First Dancer, the Countryman, or the Singing Chambermaid.

STOCKFISCH, HANS, see ENGLISH COMEDIANS.

STODOLA, IVAN, see CZECHOSLOVAKIA.

STOKE-ON-TRENT, see VICTORIA THEATRE, STOKE-ON-TRENT.

STOLL THEATRE, London, in Kingsway. This theatre, which held 2,430 persons on four tiers, was built for Oscar Hammerstein I who, intending it as a rival to COVENT GARDEN, opened it as the London Opera House on 13 Nov. 1911. The venture proved a failure and Hammerstein gave up in July 1912. Revue and variety were then staged there under a succession of managers, including C. B. COCHRAN, until on 15 May 1916 it was taken over by Oswald Stoll, who already controlled the COLISEUM and a chain of variety theatres. He opened it a year later as a cinema under the name of the Stoll Picture Theatre. It returned to use as a live theatre, still as the Stoll, in 1941, being used for revivals of popular musicals. In 1947 the huge stage was used for ice shows, and in 1951 the theatre became the headquarters of the Festival Ballet. Gershwin's *Porgy and Bess* had its first London production there in 1952, and two years later Ingrid Bergman made her first appearance in London in *Joan of Arc at the Stake* by CLAUDEL and Arthur Honegger. A musical version of KNOBLOCK's *Kismet* opened on 20 Apr. 1955, and with 648 performances gave the theatre its longest run to date. On 1 July 1957 Laurence OLIVIER and Vivien LEIGH began a five-week season of *Titus Andronicus*, directed by Peter BROOK, and this proved to be the last play seen at the Stoll, which closed on 4 Aug., the new ROYALTY THEATRE in Portugal Street being built on part of the site.

STONE, JOHN AUGUSTUS (1801–34), American dramatist, author of *Metamora, or the Last of the Wampanoags* (1829), in which FORREST, who had awarded it first prize in a competition for a play based on American history, played the title role. It was a great success, being revived as late as 1887. After playing in it for some years Forrest commissioned BIRD to rewrite it, but as no complete manuscript of either version exists, it is not possible to say what alterations Bird made. Stone, who was a mediocre actor, wrote and appeared in a number of romantic historical plays which have not survived. Disappointed as actor and author, he drowned himself in the Schuylkill River at Philadelphia.

STOPPARD [Straussler], TOM (1937–   ), Czech-born English dramatist, who first came into prominence with the production by the NATIONAL THEATRE company of *Rosencrantz and Guildenstern Are Dead* (1967), a play in the style of the Theatre of the ABSURD in which two minor characters from *Hamlet* are shown as having no existence outside Shakespeare's play, and incapable of making any independent decision except choosing to die. Its unexpected success, both in England and abroad—it began a long run in New York in the same year—led to the production in London of a play written earlier, *Enter a Free Man* (1968; N.Y., 1974), first seen in Hamburg in 1964. Stoppard's next play, *The Real Inspector Hound* (also 1968; N.Y., 1972), was a one-acter about two dramatic critics who get drawn into the action of a play. It was followed by two more one-act plays, *After Magritte* (1970; N.Y., 1972) and *Dogg's Our Pet* (1972). In *Jumpers* (also 1972; N.Y., 1974) a Professor of Moral Philosophy defends intuitive values against contemporary rationalism; while *Travesties* (1974; N.Y., 1975) was a lively debate, involving James Joyce and Lenin, on the justification for art. Produced in London by the National Theatre company and the ROYAL SHAKESPEARE COMPANY respectively, they provided supreme examples of Stoppard's astonishing verbal dexterity. A double bill, *Dirty Linen* and *New Found Land* (1976; N.Y., 1977), ran for four years at the ARTS THEATRE, and after a 'music-drama', *Every Good Boy Deserves Favour* (1977; N.Y., 1979), Stoppard had another success with *Night and Day* (1978; N.Y., 1979), his most naturalistic play to date, a debate on the freedom of the press set in a fictitious African country; it starred Diana RIGG in London and Maggie SMITH in New York. Another double bill, *Dogg's Hamlet* and *Cahoot's Macbeth* (London and N.Y., 1979), was followed by adaptations of SCHNITZLER's *Das weite Land* as *Undiscovered Country* (also 1979) and of NESTROY's *Einen Jux will er sich machen* as *On the Razzle* (1981), both produced at the National Theatre.

STORER, MARIA (*c.* 1760–95), American actress, who was first on the stage as a child, and later became a member of the AMERICAN COMPANY, where her imperious temper caused much trouble. She retired from the theatre at the same time as her husband John HENRY, and went mad after his sudden death on board ship, surviving him by only a year. Her sisters Ann or Nancy Storer, later Mrs Hogg (1749–1816), and Fanny, later Mrs Mechtler, were also actresses.

STOREY, DAVID MALCOLM (1933–   ), English novelist and playwright, who was already well known through his books, particularly *This Sporting Life* (pub. 1960), when his first play, *The Restoration of Arnold Middleton* (1967), was staged at the ROYAL COURT THEATRE, where most of his later plays were seen. It was followed by *In Celebration* (1969), centring on a 40th wedding anniversary and starring Alan BATES, and the extremely successful *The Contractor* (1970; N.Y., 1973), which dealt with the erection and dismantling of a marquee for a wedding reception. Both these plays were directed by Lindsay ANDERSON, who was also responsible for the production of several of Storey's later works. His next plays were *Home* (London and N.Y., 1970), an impressionistic play in which John GIELGUD and Ralph RICHARDSON gave fine performances as two inmates of a mental home, and *The Changing Room* (1971; N.Y., 1973), which dealt with a Rugby football team. After a symbolic drama entitled *Cromwell* and *The Farm* (both 1973), seen in New York in 1978 and 1976 respectively, came *Life Class* (1974; N.Y., 1975), again with Alan Bates in London, in which an art class provides a background for a discussion of life and art; *Mother's Day* (1976); *Sisters* (1978), produced at the ROYAL EXCHANGE THEATRE, Manchester; and, at the NATIONAL THEATRE, *Early Days* (1980), in which Ralph Richardson played an elderly politician reviewing the course of his life.

STORM AND STRESS, see STURM UND DRANG.

STRAND ELECTRIC, STRAND LIGHT CONSOLE, see LIGHTING.

STRAND MUSICK-HALL, see GAIETY THEATRE.

STRAND THEATRE, London. (1) At 168–9 Strand. In 1831 a Yorkshire comedian named Benjamin Rayner acquired a building near Somerset House which from 1803 to 1828 had been used to house panoramas and converted it into an unlicensed theatre; it opened on 25 Jan. 1832 as Rayner's New Strand Subscription Theatre. At this time the last battles between the unlicensed houses and the PATENT THEATRES, which were to result in the passing of the Theatres Act of 1843, were still being waged, and the opening attraction was a BURLESQUE on the current situation entitled *Professionals Puzzled; or, Struggles at Starting*. Rayner also revived one of his best known vehicles, a melodrama entitled *The Miller's Maid*; but the enterprise was not a success and in Nov. the theatre closed. It reopened in 1833, when Fanny KELLY presented her solo entertainment with great success and started the dramatic school which later moved to the ROYALTY THEATRE; later it closed again, but in 1836 Douglas JERROLD reopened it, adding a gallery to the auditorium. He enjoyed some success with dramatized versions of novels by Charles DICKENS, *The Pickwick Papers* being retitled *Sam Weller*. After a succession of managers William FARREN took over in 1848, starring Mrs STIRLING as Olivia in an adaptation of GOLDSMITH's novel *The Vicar of Wakefield*. He was succeeded in 1850 by Copeland from Liverpool, who renamed the theatre Punch's Playhouse but could not make a success of it. In 1858 W. G. Swanborough took over, had the theatre refurbished, and presented his daughter Ada in a season of burlesques by H. J. BYRON, in one of which, *The Maid and the Magpie; or, the Fatal Spoon*, the part of Pippo was played by Marie Wilton, the future Lady BANCROFT. The unexpected success of these productions enabled the Swanboroughs to have the theatre largely reconstructed in 1865, and they remained there until 1872. New safety regulations, however, forced the theatre to close in 1882, and it was again largely rebuilt, reopening on 18 Nov. under J. S. Clarke. A dramatization of F. Anstey's *Vice-Versa* ran for six months in 1883, and Edward COMPTON and his wife Virginia Bateman ran a successful season of Old Comedy. In 1884 a revival of H. J. Byron's *Our Boys* had a long run, and four years later Willie EDOUIN staged some other successful revivals. In 1901 a musical play, Dance's *The Chinese Honeymoon*, introduced Lily Elsie to London, and had a long run before the theatre finally closed on 13 May 1905. It was demolished later in the same year, the site now being occupied by the Aldwych underground station.

(2) In the Aldwych. This theatre was built for the American impresarios Sam and Lee Shubert, its exterior being identical to that of the ALDWYCH THEATRE at the other end of the block. It has a four-tier auditorium seating 1,084 and opened on 22 May 1905 as the Waldorf. After the destruction of his own theatre in 1905 Cyril MAUDE made it his headquarters, remaining until 1907, when Julia MARLOWE and E. H. SOTHERN appeared in a series of plays which included KESTER's *When Knighthood was in Flower*. In 1909 the name of the theatre was changed to the Strand, but it had little success, and in 1911 it became the Whitney, after its American manager, reverting to the Strand in 1913, when he left, and scoring a success at last with Matheson LANG in *Mr Wu* by Harry Vernon and Harold Owen. In 1915 the building, which was then occupied by Fred TERRY and his wife Julia NEILSON, was slightly damaged during a Zeppelin raid. Arthur BOURCHIER took over in 1919 and successfully produced A. E. W. Mason's *At the Villa Rose* (1920), Ian HAY's *A Safety Match* (1921), and a dramatization of R. L. Stevenson's *Treasure Island* (1922). It was at the Strand that *Anna Christie* by Eugene O'NEILL was first seen in London in 1923, and another successful American play was Abbott and Dunning's *Broadway* (1926). Later successes included a farce by Austin Melford, *It's a Boy* (1930), *1066 and All That* (1935), based on the book by Sellar and Yeatman, Vernon Sylvaine's *Aren't Men Beasts!* (1936), and Ben TRAVERS's *Banana Ridge* (1938). In 1940, at the height of the blitz, Donald WOLFIT gave midday productions of Shakespeare. During one of them the building was badly blasted; the dressing-rooms were damaged and the actors had to scramble over débris to reach the stage, but the performance continued, though costumes had to be dug out of the ruins and hastily brushed down. The theatre was soon repaired, and had an outstanding success with Kesselring's *Arsenic and Old Lace* (1942). Post-war successes included Vernon Sylvaine's farce *Will Any Gentleman?* (1950); *Sailor, Beware!* (1955), a farce by Philip King and Falkland Cary which established Peggy Mount as a star; the revue *For Adults Only* (1958); and two plays by Ronald MILLAR based on novels by C. P. Snow, *The Affair* (1961) and *The New Men* (1962). An American musical *A Funny Thing Happened on the Way to the Forum* (1963) with Frankie Howerd and Robertson HARE was followed by the thriller *Wait Until Dark* (1966) by Frederick Knott. In 1968 *Not Now, Darling* by Ray Cooney and John Chapman began a long run, and a revival of PRIESTLEY's *When We Are Married* did well in 1970. The comedy *No Sex Please —We're British* by Anthony Marriott and Alistair Foot opened on 3 June 1971 and despite poor reviews ran there until moved to the GARRICK THEATRE in 1982.

STRANGE'S MEN, company which after playing in the provinces first appeared at Court in 1582, at the same time as the company of Lord Strange's father, the 4th Earl of Derby, a duplication which has caused some confusion in the records. This company, which appears to have amalgamated intermittently with the ADMIRAL'S MEN, was at the THEATRE in 1590–1 and may have been the first to employ Shakespeare, either as actor and playwright or as playwright only. They were also at the ROSE under HENSLOWE in 1592–3, when they may have produced the first part of Shakespeare's *Henry VI* and perhaps *The Comedy of Errors*. They also appeared in the lost *Titus and Vespasian* (1592), which may have been an early draft of *Titus Andronicus* (1594), and in *A Jealous Comedy* (1593), a possible source of *The Merry Wives of Windsor* (1600). The company's repertory included the anonymous *A Knack to Know a Knave*, GREENE's *Orlando Furioso* and *Friar Bacon and Friar Bungay*, and MARLOWE's *The Jew of Malta*, as well as a number of plays now lost. They separated themselves from the Admiral's Men in 1594, and on the death of their patron later in that year went into the provinces, some of their actors remaining in London to join the newly-formed CHAMBERLAIN'S MEN.

STRANITZKY, JOSEPH ANTON (1676–1726), Austrian actor, originator of the comic figure HANSWURST. He began his theatrical career with a travelling puppet-show and in about 1705 arrived in VIENNA, where he headed a company of 'German Comedians' modelled on the lines of the ENGLISH COMEDIANS. They played at the Ballhaus from 1707 and in 1711 he took over the newly-built Kärntnertortheater, making it the first permanent home of German-language comedy. For this company Stranitzky adapted old plays and opera librettos, making Hanswurst (a role which he bequeathed to PREHAUSER) the central comic character.

STRASBERG, LEE (1901–82), American theatre director, who studied at the AMERICAN LABORATORY THEATRE and had had some acting experience with the THEATRE GUILD before, in 1931, he helped to found its offshoot the GROUP THEATRE. With Cheryl CRAWFORD he directed its first production, Paul GREEN's *The House of Connelly*, and later was responsible for a number of other productions, including Maxwell ANDERSON's *Night Over Taos* (1932), Sidney KINGSLEY's *Men in White* (1933), and Paul Green's *Johnny Johnson* (1936). When the Group Theatre ceased production he directed such plays as ODETS's *Clash by Night* (1941) and *The Big Knife* (1949) for other managements. A firm believer in the METHOD for the formation of actors, he became in 1950 a director of the Actors' Studio, where in 1965 he was responsible for the production of CHEKHOV's *Three Sisters* which was seen during the WORLD THEATRE season in London. He also directed for the Studio the première of Odets's *The Silent Partner* (1972), Paul Zindel's *The Effect of Gamma Rays on Man-in-the-Moon Marigolds*, and O'NEILL's *Long Day's Journey into Night* (both 1973).

STRASHIMIROV, ANTON, see BULGARIA.

STRATFORD (ONTARIO) FESTIVAL, Canada, annual drama and music festival which lasts from mid-May to the end of October. It was founded for the production of plays by Shakespeare, along the lines of the SHAKESPEARE MEMORIAL THEATRE festival at Stratford-upon-Avon, on the initiative of Tom Patterson, a citizen of Stratford, Ontario. Its first theatre was an immense tent housing an approximation to an Elizabethan open stage, thus fulfilling a long-standing dream of Tyrone GUTHRIE, its first director, and his designer Tanya MOISEIWITSCH, both from England. The opening production on 13 July 1953 was *Richard III* with Alec GUINNESS in the title role, followed by *All's Well That Ends Well*. Concerts, opera, and a film festival were later added to the programme of plays, and in 1957 a permanent theatre, designed by Robert Fairfield, was erected, which retained the spirit of the tent and the original stage. The conical roof, locked by 34 girders which meet at the centre like the spokes of a wheel, and the cantilevered balcony, are without visible means of support. The auditorium sweeps in a full semicircle round the stage, and although the building seats 2,262 people no seat is more than 65 feet from the stage. This makes for an Elizabethan intimacy between players and audience which contributes to the theatre's excitement. The stage, lit from above, is designed to facilitate the swift succession of scenes which is the essence of modern interpretations of Shakespeare; it has a deep, practicable TRAP, and is accessible from two tunnels located under the audience. In 1962 the number of pillars supporting the triangular stage balcony was reduced, and the width of the back wall extended, with new entrances and staircases. Further modifications in 1975 allow the pillared balcony to be revolved or removed completely to provide a more versatile backdrop.

In 1956 Guthrie was succeeded as director by Michael Langham, whose productions were perhaps less theatrically flamboyant and sculptural, but nevertheless imbued with great delicacy in the comedies and a fine sense of irony in the histories and tragedies. Langham continued the general policy of maintaining a permanent Canadian company with visiting stars, though it was not long before the stars themselves were Canadian. In Sept. 1956 the company appeared at the EDINBURGH FESTIVAL in a distinctively Canadian production of *Henry V* in which the French king was played by Gratien GÉLINAS, with actors from the Théâtre du NOUVEAU MONDE as his courtiers, Henry V being played by the Canadian actor Christopher PLUMMER, supported by an English-language cast. A year later Plummer inaugurated the new theatre building with his performance of Hamlet. Longstanding Canadian actors at Stratford have included William HUTT, Martha Henry, Douglas Rain, and Kate REID. Other international tours undertaken by the company, which had already been seen in New York in MARLOWE's *Tamburlaine the Great* in Jan. 1956, have included a visit to the eastern United States with *The Two Gentlemen of Verona* and a Canadian adaptation of KLEIST's *The Broken Jug* in 1958; visits to New York in 1960 and 1961 with GILBERT and Sullivan's *H.M.S. Pinafore* and *The Pirates of Penzance*, also seen in London in Feb. 1962; a trip to the CHICHESTER FESTIVAL THEATRE in Apr. 1964, for the Shakespeare Quatercentenary celebrations, with *Timon of Athens, Love's Labour's Lost*, and MOLIÈRE's *Le Bourgeois gentilhomme*; a tour of Europe as far as Russia in the spring of 1973 with *King Lear* and *The Taming of the Shrew*; and a tour of Australia in 1974 with Molière's *Le Malade imaginaire*. There were also extensive cross-Canada tours in 1967 and 1975, as well as a series of out-of-Stratford openings between 1968 and 1972.

In 1963 a second theatre, the Avon, a proscenium house seating 1,100 persons, was opened, and in 1971 an experimental Third Stage was added, primarily for the production of new Canadian plays.

Two Canadian directors, Jean GASCON and John

HIRSCH, succeeded Langham in 1967, and after Hirsch resigned in 1969 Gascon remained as sole director until 1974. The immediate impact was an expansion in repertoire to include Jacobean authors and Molière, as well as European classics that ranged from the elder DUMAS's *The Three Musketeers* to Samuel BECKETT's *Waiting for Godot*. Gascon also introduced some of Shakespeare's less well known plays, among them *Cymbeline* and *Pericles, Prince of Tyre*. With *Titus Andronicus* in 1978, the canon was complete. Gascon's COMMEDIA DELL'ARTE gifts were apparent in the comedies, and he gave the serious dramas a baroque and glowing strength. He was succeeded in 1975 by an English director, Robin PHILLIPS, amid a storm of nationalistic controversy, somewhat muted by his surrounding himself with a Canadian advisory committee and by the success of his many innovations. He pioneered a Young Company, giving stardom to Canadians such as Richard Monette and Alan Scarfe, and he invited many Canadian directors to share productions with him, Peter Moss perhaps achieving the most notable success. He also experimented with double and triple casting of certain leading roles, and most astutely utilized both the Avon Theatre and the Third Stage for Shakespeare productions, hitherto reserved for the main festival stage. Among his other innovations, he gave the festival a more truly festive character by opening several seasons with a casual 'Gala' or 'Revel'. His own productions were cool and subtle, crisply realistic and well-spoken, and made visually distinctive by Daphne Dare's somewhat monochromatic designs. Although stars had been imported on occasion over the years, among them Irene WORTH, Paul SCOFIELD, and Julie HARRIS, with Phillips the idea became central. In addition to the Canadian actor Hume CRONYN and his wife Jessica TANDY, Phillips essentially built four of his six seasons around Maggie SMITH and Brian Bedford, and in 1979 and 1980 Peter USTINOV was invited to play King Lear. Unfortunately, by Nov. 1980 financial problems brought on by overexpansion and the uncertainty of the nature of Phillips's association with the festival meant that the Board of Governors accepted Phillips's resignation, leaving the leadership of the festival in some jeopardy; but at the end of the year John Hirsch was appointed sole director on a three-year contract.

Only one Canadian play, *The Canvas Barricade* (1963) by Donald Jack, has been shown on the festival stage, but a number have had major productions at the Avon, among them Michael Bawtree's *The Last of the Tsars* (1966); James REANEY's *Colours in the Dark* (1967); Tom Hendry's *The Satyricon* (1969); Larry Fineberg's *Eve* (1976); Sheldon Rosen's *Ned and Jack* (1978–9); and *Foxfire* (1980) by Susan Cooper and Hume Cronyn. *Mark* (1972) by Betty Jane Wylie; *The Collected Works of Billy the Kid* (1973) by Michael Ondaatje; *Walsh* (1974) by Sharon Pollock; *Fellowship* (1975) by Michael Tait; *Medea* by Larry Fineberg and *Star-Gazing* (both 1978) by Tom Cone; and *Victoria* (1979) by Steve Petch, have all been seen on the Third Stage, as have two plays for children, *Pinocchio* (1972) and *Inook and the Sun* (1973).

STRATFORD-UPON-AVON, Warwickshire, the birthplace of Shakespeare, and the scene of an annual festival of his plays. The first festival, organized by GARRICK in 1769, was chiefly remarkable for the fact that none of Shakespeare's plays was performed. After that there were only sporadic performances by strolling companies, in spite of the founding of a Shakespeare Club in 1824 and the building of a theatre on part of the garden on the site of New Place, Shakespeare's last home, in 1827. It was demolished in 1872, the last play to be performed there being *Hamlet*. In Apr. 1864, on the initiative of the then mayor, Edward Fordham Flower, member of a local family of brewers, actors from London were invited to appear in a Grand Pavilion, specially built for the occasion, to celebrate the 300th anniversary of Shakespeare's birth, the plays performed being *Twelfth Night, The Comedy of Errors, Romeo and Juliet, As You Like It, Othello*, and *Much Ado About Nothing* Some years later Flower's son, Charles Edward Flower, donated the site and provided most of the money for the building of the SHAKESPEARE MEMORIAL THEATRE, which later became the ROYAL SHAKESPEARE THEATRE.

Apart from this theatre, with its library and picture gallery, Stratford-upon-Avon has also a library and conference centre attached to Shakespeare's birthplace, and an Institute of Shakespeare Studies sponsored by the University of Birmingham, which grew out of the lectures arranged by Professor Allardyce NICOLL for the BRITISH COUNCIL at Mason Croft, the former home of the novelist Marie Corelli; it provides a centre for research work all the year round and a meeting-place for Shakespeare scholars from all over the world during the summer conference. Hall Croft, the home of Shakespeare's daughter Susanna after her marriage in 1607 to Dr John Hall, also provides a useful centre of activity during the Shakespeare season.

STRATTIS (*fl.* early 4th cent. BC), Greek writer of OLD COMEDY, whose works, of which only fragments of 17 known titles survive, are highly satirical at the expense of his contemporaries, especially the tragic dramatists.

STRATTON, EUGENE, see MUSIC-HALL.

STREATER (or STREETER), ROBERT, see SCENERY.

STREET THEATRE, see COMMUNITY THEATRE.

STREHLER, GIORGIO, see PICCOLO TEATRO DELLA CITTÀ DI MILANO.

STREISAND, BARBRA, see MUSICAL COMEDY.

STRINDBERG, (Johann) AUGUST (1849–1912), Swedish dramatist who also wrote prolifically in the fields of fiction, criticism, and social commentary. After an unhappy childhood, his university career was cut short for financial reasons, and these early miseries set the tone both of his later life and of his writings. His first major play, *Mäster Olof* (1872), was a study of the humanist Olaus PETRI; a prose work, it was later rewritten in verse. In 1877 Strindberg married the actress Siri von Essen; the first of his three disastrous experiences of marriage, it ended in divorce in 1891. Discouraged by the lack of success of his early dramas, he spent the greater part of the years 1883 to 1896 abroad, though he had to return to Sweden during this time to stand trial for blasphemy in his collection of stories *Giftas* (*Married Life*); he was acquitted. Misogyny and despair mark the plays of this period, notably *Fadren* (*The Father*, 1887), *Fröken Julie* (*Miss Julie*), and *Fordringsägare* (*The Creditors*), both 1888; and the one-act *Den Starkare* (*The Stronger*, 1889). In them, as well as in the important Preface to *Fröken Julie*, Strindberg denounces the moral corruption of human nature, and concentrates on sin, crime, abnormality, and the pitiless battle between the sexes as he saw it. In 1895–7, following his divorce from his second wife Frida Uhl, Strindberg suffered a severe mental breakdown, his 'Inferno crisis'; he became interested in alchemy and occultism and, abandoning his previous allegiance to the philosophy of Nietzsche, came under the influence of Swedenborg. The NATURALISM of his earliest dramatic works developed into a highly-charged REALISM (he preferred the term 'neo-naturalism') which has been seen as an anticipation of SARTRE. In this mode he explored his conception of spiritual reality in *Advent* (1898); in five plays completed in 1899—the historical dramas *Folkungsagan* (*The Saga of the Folkungs*), *Gustaf Vasa*, *Erik XIV*, and *Gustaf Adolf*, and in *Brott och Brott* (*Crimes and Crimes*); in *Påsk* (*Easter*, 1900); and supremely in *Dodsdansen I–II* (*The Dance of Death*, 1901). His work thereafter became progressively more experimental, symbolic, and dream-like with *Till Damaskus I–II* (1898–1904), and (in 1901) with *Kronbruden* (*The Crown Bride*, seen in New York as *The Bridal Crown*), *Svanehvit* (*Swanwhite*), and *Ett drömspel* (*A Dream Play*), all of them containing elements in which the major movements of 20th-century drama are rooted, notably EXPRESSIONISM, Surrealism, and the Theatre of the ABSURD. Also in 1901 Strindberg married his third wife, the actress Harriet Bosse; they were divorced in 1904.

Strindberg was always closely concerned with the practical presentation of his plays, and in 1907, with the director August Falcke, he established the INTIMA TEATERN in Stockholm, writing his last group of plays—*Spöksonaten* (*The Ghost Sonata*), *Oväder* (*The Storm*), *Brända tomten* (*The Burnt House*), and *Pelikanen* (*The Pelican*)—for performance there. His last drama, *Stora Landsvägen* (*The Great Highway*, 1909), contains many autobiographical elements.

The names of Strindberg and IBSEN are often coupled, and their Scandinavian background, their contemporaneity, and their examinations of the modern family have encouraged this view. Strindberg's vision must, however, be judged as uniquely subjective and pessimistic, and the wide-ranging experimentalism of his work is attributable to the unselfconsciousness of his inward gaze. His historical dramas have assumed lasting importance in Sweden, but although many of his plays have been translated and acted in English, only a few are regularly performed in English-speaking countries. The best known are probably *The Father* (first seen in New York in 1912 and in London in 1927) and *Miss Julie* (produced privately in London in 1927 by the STAGE SOCIETY and publicly in 1935; not seen in New York until 1956). In America the PROVINCETOWN PLAYERS first performed *The Ghost Sonata* in 1924 (as *The Spook Sonata*, under which title it was seen in London in 1927) and *A Dream Play* in 1926 (seen in London in 1933), and both productions had a great influence on the development of O'NEILL. Among other productions were *Advent* (London, 1921); *The Creditors* (N.Y. 1922; London, 1952); *Easter* (N.Y., 1926; London, 1928); *The Dance of Death* (N.Y., 1923; London, 1928—this was given a superb revival in 1967 by the NATIONAL THEATRE at the OLD VIC, with Laurence OLIVIER as the Captain); *The Road to Damascus* (London, 1937; as *To Damascus*, N.Y., 1961); *The Stronger* (N.Y., 1937); *There Are Crimes and Crimes* (London, 1946).

STRIP LIGHT, see BATTEN.

STRIP-TEASE, see BURLESQUE, AMERICAN.

STRITCH, ELAINE, see MUSICAL COMEDY.

STROBE LIGHT, see LIGHTING.

STRØMSTED, LIV (1922–), Norwegian actress who spent most of her career (apart from a spell with the Studioteatret between 1945 and 1949) with the NATIONALTHEATRET in Oslo, which she left in 1962. She revealed a versatile talent, capable of ranging from Shakespeare's Juliet in 1952 to BRECHT's Shen Te in *The Good Woman of Setzuan* in 1958. Among her best remembered roles in IBSEN are Hilde Wangel in *The Master Builder* in 1950, the Woman in Green in *Peer Gynt* in 1955, Svanhild in *Love's Comedy* in 1956, and Nora in *A Doll's House* in 1957.

STROUPEŽNICKÝ, LADISLAV, see CZECHOSLOVAKIA.

STUDIO ARENA THEATRE, Buffalo, New York State, a non-profit-making organization which serves the 1.5 million people of the western part of the state. It grew out of a community theatre group founded in 1927, and since 1965, when it took over an adapted nightclub, has been fully professional. In 1978 it moved across the street to a new 637-seat thrust-stage theatre converted from a BURLESQUE house. A subscription series of seven plays is offered in a season which runs from Sept. to May, productions including a wide variety of old and new plays and musicals. There is a decided emphasis, however, on contemporary theatre, and over 20 world or American premières have taken place, including ALBEE's *Box Mao Box* (1968), Truman Capote's *Other Voices, Other Rooms* (1973), Howard Sackler's *Semmelweiss* (1977), and Paul Giovanni's *The Crucifer of Blood* (1978). The theatre school, founded in 1927, offers an integrated training programme and plays for children.

STURDZA-BULANDRA, LUCIA, see ROMANIA.

STURM UND DRANG (Storm and Stress), the name given to one aspect of 18th-century German Romanticism which carried to excess the doctrine of the rights of man and ROUSSEAU's plea for a return to nature. It took its name from the title of a play by KLINGER, produced in 1776, and was much influenced by Shakespeare. Among the themes which frequently recur in the plays which characterize the movement are the tragedy of the unmarried mother executed for infanticide while her seducer goes free, treated by GOETHE and Klinger; the conflict between hostile brothers in love with the same woman, as in the *Julius von Tarent* (1776) of LEISEWITZ and SCHILLER's *Die Räuber* (1782); and the overmastering power of love, hurling even honourable natures into crime, as in *Golo und Genoveva* (*c.* 1780) by 'Maler' Müller. The movement had repercussions all over Europe, and its influence can clearly be seen in early 19th-century English MELODRAMA. A native German offshoot of it was the RITTERDRAMA.

STUYVESANT THEATRE, New York, see BELASCO THEATRE (2).

ŠUBRT, FRANTIŠEK ADOLF, see CZECHOSLOVAKIA.

SUDAKOV, ILYA YAKOVLEIVICH (1890–1969), Soviet actor and director, who trained at the MOSCOW ART THEATRE where, under STANISLAVSKY's supervision, he directed BULGAKOV's *Days of the Turbins* (1926) and IVANOV's *Armoured Train 14–69* (1927) and together with NEMIROVICH-DANCHENKO staged Ivanov's *Blockade* (1929), AFINOGENOV's *Fear* (1935), and TRENEV's *Lyubov Yarovaya* (1936). In 1937 he became Artistic Director of the MALY THEATRE, where some of his best productions were GUTZKOW's *Uriel Acosta* (1940), *Invasion* (1942) by LEONOV, and *Front* (1943) by KORNEICHUK. Among his post-war productions were revivals of CHEKHOV's *Uncle Vanya* (1947), Afinogenov's *Mother of Her Children* (1954), and Alexei TOLSTOY's *The Road to Calvary* (1957). In 1959 he became director of the Gogol Theatre (formerly the Moscow Transport Theatre), where he remained until his death.

SUDERMANN, HERMANN (1857–1928), German novelist and dramatist, whose first play, *Die Ehre* (*Honour*), was produced with great success in 1889. He became the main exponent of the new theatre of REALISM in Germany, much influenced by IBSEN, as can be seen in *Das Glück im Winkel* (*Happiness in a Quiet Corner*, 1896) and *Johannsfeuer* (*The St John's Eve Fire*, 1900). The play by which he is chiefly remembered is *Heimat* (1893), a work which provided a popular melodramatic vehicle for many famous actresses, including DUSE and BERNHARDT, who both played it in London at the same time in 1895. As *Magda*, in a translation by Louis N. PARKER, it was first seen in English in 1896 at the LYCEUM THEATRE with Mrs Patrick CAMPBELL in the title role, a part played in 1923 by Gladys COOPER and in 1930 by Gwen FFRANGCON-DAVIES. It was produced in New York in 1904 and revived in 1926.

SUD-EST, Centre dramatique du, see DÉCENTRALISATION DRAMATIQUE.

SUGG, CATHARINE LEE, see HACKETT, JAMES HENRY.

SUKHOVO-KOBYLIN, ALEXANDER VASILIEVICH (1817–1903), Russian dramatist, whose whole life was overshadowed by the death of his mistress, whom he was suspected of having murdered. He had already started to write a play, *Krechinsky's Wedding*, before this tragic event, and while in jail he finished it. It was finally staged at the Moscow MALY THEATRE in 1855. Its subject, like that of his two later plays—*The Case* (written in 1857 but not performed until 1881) and *Tarelkin's Death* (written in 1868 and staged in 1900)—was the decay of the patriarchal life of old Russia and the growing power of a corrupt bureaucracy. Although Sukhovo-Kobylin always asserted that he was not a revolutionary, he was regarded as a dangerous man and his plays were banned by the censor. He was a great friend and admirer of GOGOL, whose influence is seen in the character of Krechinsky, and was also much influenced by SALTIKOV-SHCHEDRIN. Worn out by his struggles with the censorship, he finally gave up the theatre; in 1857 he settled permanently in France.

SULLIVAN, Sir ARTHUR, see GILBERT, WILLIAM SCHWENCK.

SULLIVAN, (Thomas) BARRY (1821–91), Irish actor, who gained his experience in Ireland, Scotland (where he managed a theatre in Aberdeen for three years), and the English provinces. He made his first appearance in London in 1853 as Hamlet under BUCKSTONE at the HAYMARKET, and then spent a season at SADLER'S WELLS with PHELPS. In the autumn of 1858 he went to New York and then toured the States, being well received in San Francisco, and made a long visit to Australia, returning to England to play Benedick in *Much Ado About Nothing* to the Beatrice of Helen FAUCIT in the inaugural performance at the SHAKESPEARE MEMORIAL THEATRE in Stratford-upon-Avon in 1879. He made his last appearance on the stage in 1887 as Richard III, in Liverpool. He was never a first-class actor, but his vigour and forcible delivery in tragedy—he was unsuited to comedy or romance, which he seldom played—made him a success with less sophisticated audiences, for whom he kept alive the old traditions of Shakespearian acting.

SULZER, JOHANN GEORG, see SWITZERLAND.

SUMAROKOV, ALEXEI PETROVICH (1718–77), Russian dramatist, who wrote in the neo-classical style imported from Germany and France, but took his subjects from Russian history and in his dialogue endeavoured to refine and purify the Russian language, writing with great intensity and economy. His plays, of which the first, *Khorev*, was produced in 1749, were acted by students from the Cadet College for sons of the nobility, of which Sumarokov had been one of the first pupils on its foundation in 1732. It was largely owing to his efforts that the first professional Russian company was formed, and in 1756 he was appointed head of the Russian (as distinct from the Italian and French) Theatre in St Petersburg, remaining there until 1761, when his liberal outspokenness caused his plays to be banned and he himself to be dismissed from office.

SUMNER, JOHN (1924–), English-born theatre director, who in 1959 became administrator of the MELBOURNE THEATRE COMPANY, having been manager of its parent body the Union Theatre Repertory Company which he founded in 1953. He directed the original production of Ray LAWLER'S *Summer of the Seventeenth Doll* (1955), and also the London and New York productions in 1957 and 1958. He was briefly general manager of the Elizabethan Theatre in Sydney, and has continued to direct three or four productions for the Melbourne Theatre Company each year, among them Lawler's play on Captain Bligh, *The Man Who Shot the Albatross* (1971), and the first two plays of the *Summer of the Seventeenth Doll* trilogy, *Kid Stakes* (1975) and *Other Times* (1976), the three being played together in repertory in 1977 as *The Doll Trilogy*.

SUN MUSIC-HALL, London, see MUSIC-HALL.

SUPER(NUMERARY), see STOCK COMPANY.

SURREY MUSIC-HALL, SURREY GARDEN MUSIC-HALL, see WINCHESTER MUSIC-HALL.

SURREY THEATRE, London, in Blackfriars Road, Lambeth. This stood on the site of the Royal Circus, which opened on 4 Nov. 1782 and continued in use until 1810, although it had a troubled existence, being burnt down in 1799 and 1805. Rebuilt in 1806, it was converted into a theatre by Robert ELLISTON, who gave it the name by which it was thereafter known. To avoid trouble with the PATENT THEATRES, he put a ballet into every production, including *Macbeth, Hamlet*, and FARQUHAR'S *The Beaux' Stratagem*. Elliston left in 1814, and the Surrey became a circus again until Thomas DIBDIN reopened it as a theatre in 1816, but with little success. Not until Elliston returned did its fortunes change, with the production on 8 June 1829 of Douglas JERROLD'S *Black-Ey'd Susan*, which with T. P. COOKE as William, the nautical hero, had a long run. Elliston himself made his last appearance at this theatre on 24 June 1831, twelve days before he died. Osbaldiston then took over, and among other plays produced Edward FITZBALL'S *Jonathan Bradford; or, the Murder at the Roadside Inn*, which ran for 260 nights, but it was Richard Shepherd (who succeeded Alfred Bunn in 1848 and remained at the theatre until 1869) who established its reputation for rough-and-tumble TRANSPONTINE MELODRAMA. On 30 Jan. 1865 the theatre was burnt down, but a new theatre, seating 2,161 people in four tiers, opened on 26 Dec. 1865. Little of note took place until 1881, when George CONQUEST took over, staging sensational dramas, many of them written by himself, which proved extremely popular, and each Christmas an excellent PANTOMIME. The Surrey prospered until his death in 1901, but thereafter went rapidly downhill until in 1920 it became a cinema. It finally closed in 1924 and the building was demolished in 1934.

SUSIE, see TOBY (in the United States).

SUSSEX'S MEN, company of players founded in about 1569 by the 3rd Earl of Sussex. They were seen at Court in 1572, but then played chiefly in the provinces until in 1593–4 they appeared for a six-week season at the ROSE under the management of HENSLOWE. The one new play in their repertory was *Titus Andronicus*, which Shakespeare may have refashioned for them from an earlier play, *Titus and Vespasian*, in which PEMBROKE'S MEN had appeared in 1592. Having lost their patron, the 4th Earl who died in 1593, they came under his son but were disbanded soon after they left the Rose.

SUTHERLAND, EFUA (*c.* 1930–), one of the founders of the modern Ghanaian theatre. She started the Experimental Theatre Players in 1958, and in 1960 opened the Ghana Drama Studio in Accra, an open courtyard with a covered stage, the first attempt to design a specifically African theatre building. She joined the staff of the School of Music and Drama at the University of Accra at Legon in 1962, but continued to work in the Drama Studio and on projects designed to take drama to people in rural areas.

Mrs Sutherland's example and teaching, and that of her colleague Joe de Graft, inspired a whole generation of Legon graduates. Her own plays blend European and African forms and material; *Edufa* (1957) is an African version of EURIPIDES' *Alcestis*; an exhortation to national reawakening is contained in *Foriwa* (1967); *The Marriage of Anansewa* (1975) draws on a Ghanaian folk-tale with its popular hero Ananse the cunning spider.

SUZMAN, JANET (1939–), South-African born actress who has made an outstanding reputation for herself in England, mainly with the ROYAL SHAKESPEARE COMPANY, which she joined in 1962 after making her first appearance on the stage earlier the same year in Ipswich in Keith Waterhouse and Willis HALL's *Billy Liar*. She was first seen in London in *The Comedy of Errors*, and came into prominence with her embattled Joan La Pucelle in *Henry VI* in 1963. A modern role in PINTER's *The Birthday Party* (1964) was followed by an impressive gallery of Shakespeare's heroines—Portia in *The Merchant of Venice* and Ophelia in *Hamlet* (both 1965), Katharina in *The Taming of the Shrew* (1967), Rosalind in *As You Like It*, and Beatrice in *Much Ado About Nothing* (both 1968), her Katharina and Beatrice both being seen also in the U.S.A. In 1972 she scored a striking success as Cleopatra in *Antony and Cleopatra*, endowing her with vocal strength, voluptuousness, and a mettlesome intellect; in the same year she also played Lavinia in *Titus Andronicus*. She gave a sensitive performance as a dilapidated South African whore in Athol FUGARD's *Hello and Goodbye* in 1973. Outside the R.S.C., she played Masha in Jonathan MILLER's production of CHEKHOV's *Three Sisters* in 1976, and a year later was seen in BRECHT's *The Good Woman of Setzuan* at the ROYAL COURT and in IBSEN's *Hedda Gabler*. She returned to the R.S.C. in 1980 to appear as Clytemnestra and Helen in *The Greeks*, based mainly on EURIPIDES' Trojan plays. She is married to Trevor NUNN.

SVOBODA, JOSEF (1920–), Czech scene designer with an international reputation, whose work has had an immense influence throughout Europe and in America. He designs sets for playhouses and opera houses all over the world while remaining chief designer at the Czech National Theatre, a post which he has held since 1948. In Prague some of his best work was done for GOGOL's *The Government Inspector* in 1948, *Hamlet* in 1959, CHEKHOV's *The Seagull* in 1960, and SOPHOCLES' *Oedipus Rex* in 1963. He also designed the production of the ČAPEKS' *The Insect Play* which the Czech National Theatre took to London in the WORLD THEATRE SEASON of 1966, as a result of which he was commissioned by the NATIONAL THEATRE to design sets for productions at the OLD VIC of OSTROVSKY's *The Storm* (1966) and Chekhov's *Three Sisters* (1967). He returned to the Old Vic in 1970 to do the designs for Simon GRAY's adaptation of DOSTOEVSKY's *The Idiot*. His work is notable for its originality and for its technical ingenuity, since, more than any other modern stage designer, he makes full use of machinery, lighting, and electronic devices. He is also well known for his Laterna Magica, a REVUE-type entertainment combining live action with projected images which arose out of the work which he did for the Brussels World's Fair in 1958. It is now regularly presented in Prague in association with the Czech National Theatre.

SWANN, DONALD, see REVUE.

SWANSTON, ELIARD [EYLLAERDT, HILLIARD] (?–1651), a prominent member of the KING'S MEN from 1624 to the closing of the theatres in 1642. He appears to have been on the stage for at least two years before joining the company, and was not only one of its leading actors, but also important on the business side, together with LOWIN and TAYLOR. He played a variety of roles, many of them in revivals, including Shakespeare's Othello and Richard III and CHAPMAN's Bussy d'Ambois. During the Civil War he became a Parliamentarian, in contrast to most of the other actors, who remained staunchly Royalist, and earned his living as a jeweller.

SWAN THEATRE, London, on Bankside in Southwark, in Paris Garden. The fourth theatre to be built in London, it was a circular building constructed for Frances Langley in about 1595. Although it did not have a very exciting history, a good deal is known about it, as Johannes de Witt, a Dutch visitor to London in 1596, sent a sketch of its interior, probably made from memory, to a friend in Utrecht, Arend von Buchel, who copied it into his commonplace book. This is the only known drawing of an Elizabethan theatre interior and presents many interesting and puzzling features. According to de Witt's account of it, it held about 3,000 persons, and was of wood on a brick foundation with flint and mortar work between wooden pillars painted to resemble marble. It served HENSLOWE as a model for the HOPE, and is frequently mentioned in his diary, but it had no regular company, and was used as often for sports and fencing as for plays. In 1597 trouble arose over the presentation by PEMBROKE'S MEN of *The*

Isle of Dogs, a 'seditious' comedy by JONSON and NASHE, which landed Jonson and others in prison. The company was broken up, and the leading actors joined the ADMIRAL'S MEN at the ROSE. The Swan was then used intermittently, and was the scene of a curious incident in 1602, when one Richard Venner announced that on 6 Nov. a spectacular satire on recent history entitled *England's Joy* would be given by 'gifted amateurs'. After taking the money at the door, Venner decamped (or, according to some, was arrested), and the infuriated audience wrecked the interior of the playhouse. It was still used occasionally for plays, the last to be seen there being MIDDLETON's *A Chaste Maid in Cheapside* (1611), performed by the LADY ELIZABETH'S MEN. When the Hope Theatre opened in 1614 the Swan fell into disuse, though it was still being used in 1620 for prize-fights. The last mention of it is in *Holland's Leaguer*, a pamphlet issued in 1632, which says it 'was now fallen to decay, and like a dying *Swanne*, hanging downe her head, seemed to sing her owne dierge'.

SWAZZLE, see PUNCH AND JUDY.

SWEDEN. The presence of the early Latin SCHOOL DRAMA, or Reformation Bible-play, in the mid-16th century suggests that Sweden already had some dramatic tradition and had contributed to the LITURGICAL DRAMA common to many European literatures, but the documented history of Swedish drama begins only with the vernacular *Tobiae Comedia* (pub. 1550), generally attributed to Olaus PETRI. To the School Drama proper belong such plays as *Josephi Historia* (pub. 1601) and *Dawidhs Historia* (pub. 1604). Soon native comedy became mingled with the religious material as well as with the imitations of classical drama which also appeared. In *Judas Redivivus* (1614) Jacobus Rondeletius produced a 'Christian tragicomedy' with much dramatic power, especially in its comic parts. Johannes MESSENIUS and his followers began to draw their material from Swedish saga and history instead of from Biblical subjects. In a similar tradition are Andreas Johannis Prytz with *Olof Skottkonung* (1620) and other chronicle plays, all with a religious purpose, and Nikolaus Holgeri Catonius with *Troijenborg* (1632), sometimes considered the best play of the period. Urban Hiärne (1641–1724) belatedly carried the School Drama to its climax with the famous and popular *Rosimunda* (1665), but from the middle of the 17th century the native drama was left to university circles, while the Court patronized the new French MASQUE and its imitations in Swedish by Georg STIERNHIELM. By 1690 Swedish actors were established in Stockholm, but the classical French drama still predominated, carrying into the 18th century the traditions of 17th-century France with heavy tragedies based on VOLTAIRE and comedies, with rather more vitality, after MOLIÈRE.

Olof DALIN, the most eminent writer of mid-18th century Sweden, occasionally wrote plays, while genuine comedy, sometimes owing much to HOLBERG, was contributed by Carl GYLLENBORG and by R. G. Modée (1698–1752), who presented bourgeois material in classical form. Erik Wrangel (1686–1765) produced two tragedies and a comedy between 1739 and 1748.

A Royal Swedish Theatre was opened in 1737, but performances by Swedish nationals were still rare. A Royal Swedish Opera was founded in 1773, and the Royal Dramatic Theatre (KUNGLIGA DRAMATISKA TEATERN) in 1788, with support from King GUSTAV III. The King also gathered around him a group of playwrights: Carl Israel Hallman (1732–1800); Johan Henrik Kellgren (1751–95), who between 1780 and 1788 had a share in a number of historical plays; Carl Gustaf Leopold (1756–1829), with themes from Scandinavian legends; Gudmund Jöran Adlerbeth (1751–1818) at the turn of the century, with imitations of RACINE and Voltaire; Olof Kexél (1748–96), with comedies of French and English derivation; and Carl Envallsson (1756–1806), who parodied classical form with *Iphigenie den andra* (*The Other Iphigenia*, 1800). In content, theme, and form, however, the French influence still held sway.

Swedish Romantic poets, unlike their English contemporaries, had every encouragement to write for the theatre, but there is little drama worthy of note in early 19th-century Sweden. Thus the dramatic works of Atterbom and STAGNELIUS are of far less importance than their poetry. Later the historical dramas of Bernhard von Beskow (1796–1868) began to show the influence of SCHILLER. In mid-century, August Theodore Blanche (1811–68) wrote comedies in the style of J. L. HEIBERG; history plays by Johan Børjessen (1778–1866) were influenced by Shakespeare, then becoming available in K. A. Hagberg's translation; Fredrik August Dahlgren (1816–95) was also influenced by Shakespeare, and by CALDERÓN: his peasant play *Vermländingarne* (*The People of Vermland*, 1846) won lasting popularity. By the later years of the century, the works of IBSEN and BJØRNSON were making their mark in Sweden; their effect was at first modified and then reinforced by the work of STRINDBERG in the 1880s. From this period Alfhild Agrell (1849–1923) is remembered chiefly for *Räddad* (1882); Charlotte Leffler (1849–92) as a disciple of Ibsen; and Ernest Ahlgren [Victoria Benedictsson] (1849–88) for one play, *Final* (1885).

The importance of Strindberg in Swedish as well as in world drama would be hard to overrate. His themes sometimes provoked hostility as great as that provoked by Ibsen's in Norway; but in his work Swedish drama took on a new vitality and originality in both matter and form. The realistic plays of the first half of his career and the impressionist dramas of the second half are best known abroad, but in Sweden his treatment of themes from the nation's history also assumed lasting importance. Perhaps inevitably, his work, once accepted, was followed by a host of imitators.

Swedish drama from Strindberg's death to the end of the First World War is best represented by the earlier plays of Per HALLSTRÖM, the later work of Tor HEDBERG, and individual plays such as *Elna Hall* (1917) by Ernst Didring (1868–1931). The most notable dramatists of the inter-war years were Hjalmar BERGMAN and Pär LAGERKVIST, both closely linked to Strindberg by theme, mood, and constant experiment in form and technique.

From as early as the 1880s important innovations in theatrical presentation were emerging alongside the new naturalistic literature, nourished particularly by the dramas of Ibsen, Bjørnson, and Strindberg. The stress was on freedom of movement on the stage, on a more natural mode of speech, and on realistically convincing sets. After the turn of the century attention was directed above all to the intimate portrayal of psychological subtleties. In the early years of the century Max REINHARDT was one of the dominant influences, clearly seen in the work of directors such as Olof MOLANDER. Among the leading Swedish players of this century up to the Second World War have been Ander de Wahl, Ivan HEDQVIST, Gösta EKMAN, Lars Hansson, Harriet Bosse (Strindberg's third wife), Gerda LUNDEQUIST, Pauline Brunius, Bengt Ekerot, Tore Teje, and Inga Tidblad; among directors, Per Lindberg, Olof Molander, and Alf SJÖBERG were the outstanding names. In the post-war years the Kungliga Dramatiska Teatern in Stockholm, under the direction of Karl Ragnar Gierow, won acclaim for its posthumous first performances of plays by O'NEILL; and the municipal theatres of Gothenburg and Malmö also attracted international interest. Players of note have included Anders EK, Gun Wållgren, Per OSCARSSON, and Ulf Palme; as directors Ingmar BERGMAN and Per Axel Branner have done important work; and Alf Sjöberg's last years added enormously to his reputation.

The novelist Vilhelm MOBERG wrote extremely popular peasant dramas during and after the Second World War; in the 1950s Sara Lidman, Werner Aspenström, and Stig DAGERMAN emerged as dramatists of importance, with the novelist Harry MARTINSON achieving a more idiosyncratic success. During the 1960s Lars Forsell and Sandro Key-Åberg were new writers of note, while among the many new experimental theatre groups, Narren Teatern was outstanding, both in initial impact and, more unusually, staying power.

Swedish theatre enjoys substantial support from public funds and from State lotteries. Among the beneficiaries are not only the Kungliga Dramatiska Teatern (with its associated experimental theatre) but also the Stockholm Municipal Theatre (built in 1960), two theatres in Gothenburg, and others in Uppsala, Hälsingborg, Linköping-Norrköping, and Malmö. Commercial theatres in Stockholm include the Svenska Dramatiska Teatern, burnt down in 1925 and rebuilt, and the INTIMA TEATERN, which takes its name from the theatre founded in 1907 by Strindberg and Falcke. There is also a National Touring Theatre, founded in 1930 and merged with another itinerant company in 1966, to take plays to the less well-served provinces. A drama school has existed in Stockholm since 1787, attached to the Kungliga Dramatiska Teatern until it became independent in the 1960s, and there are also academies of drama in Gothenburg and Malmö.

ŚWINARSKI, KONRAD (1930–75), Polish director, who first trained as an artist and designer, and worked with the BERLINER ENSEMBLE. He then turned to direction, and in 1963 was responsible for the world première of WEISS's *Marat/Sade* at the Schillertheater in Berlin. He later went to the Stary Theatre in Cracow, working under Zygmunt Hübner until 1969, when he took over the management of the theatre himself. He had already caused some controversy with a strikingly revolutionary version of KRASIŃSKI's *The Undivine Comedy* in 1965, following it with a revival of SŁOWACKI's *Fantasy*, two Shakespeare plays, and three plays by WYSPIAŃSKI. In 1971 he was involved in the production of a new play, *Farewell Judas* by Ireneusz Iredyński. His mobile arena production of MICKIEWICZ's *Forefathers' Eve* in Southwark Cathedral during the London WORLD THEATRE SEASON of 1975 enabled English playgoers to see something of this internationally-famous producer's work. Great things were expected of him, and he had already completed the staging of a revival of MAYAKOVSKY's *The Bed Bug* at the National Theatre in Warsaw when he was killed in an air crash, leaving his eagerly awaited production of *Hamlet*, intended for Cracow, only half finished. His early death was a great loss to the Polish theatre.

SWINBOURNE [*née* Vandenhoff], CHARLOTTE ELIZABETH (1818–60), English actress, sister of George VANDENHOFF. A small, fair woman of a gentle disposition, she was an excellent player of parts requiring delicacy and pathos. She made her first appearance in 1836 as Juliet, and among her later parts were Cordelia in *King Lear*, Julia in Sheridan KNOWLES's *The Hunchback*, and Pauline in BULWER-LYTTON's *The Lady of Lyons*. She was the first to play Lydia in Knowles's *The Love Chase* (1837) and Parthenia in Mrs Lovell's *Ingomar* (1851), and was much admired in the title roles of SOPHOCLES' *Antigone* in 1845 and EURIPIDES' *Alcestis* in 1855.

SWINEY, OWEN (*c.* 1675–1754), Irish actor and manager, who in about 1700 was at DRURY LANE, where for a time he acted as right-hand man to Christopher RICH. In 1705 he joined with Colley CIBBER and others in an attempt to break Rich's stranglehold over his actors by leasing VANBRUGH's old theatre in the HAYMARKET, where he produced FARQUHAR's *The Beaux' Stratagem* with Anne OLDFIELD as Mrs Sullen. This brought him into conflict with the licensing laws and he was forbidden to put on plays. He was for a time able to stave off ruin by pandering to the popular taste for opera, but after an unsuccessful attempt to run Drury Lane in partnership with WILKS, DOGGETT, and Cibber, he evaded his creditors by going to Venice in 1710. On his return to London in 1730 he called himself MacSwiney, by which name he is sometimes known. He became the friend and patron of Peg WOFFINGTON, to whom he imparted the traditions of Anne Oldfield, and on his death left her all his property.

SWITCHBOARD, see LIGHTING.

SWITZERLAND. The masked ceremonies and other mime plays still found in almost every part of Switzerland derive from the original drama of the Celts and ancient Germans. The primitive plays, which have survived in the district of Berne, in the German-speaking section of Valais, and in the Rhaeto-Romanic Grisons, go back to the even older magic rituals of the hunting play, while the processions of the Magi and of St Nicholas betray a heathen origin in a Christian setting (see FOLK FESTIVALS).

It is, however, from the Roman occupation that the first physical traces of the Swiss theatre date. Under Tiberius (AD 14–37) the first Roman theatre was built at Augusta Raurica (Kaiseraugst bei Basel), and the first amphitheatre at Vindonissa (Windisch bei Brugg). There were also amphitheatres at Octodurus (Martigny), at the foot of the Great St Bernard, at Aventicum (Avenches), the capital of Helvetia, and at the Roman settlement on the narrow peninsula near Berne, the arena of which was discovered in 1960. A new theatre was built about AD 150 in Basle on the site of the first theatre, and its ruins are today among the most imposing of those north of the Alps. As was also the case with the contemporary theatre at Aventicum, the theatre formed a single entity with the temple, and comprised a vast festival area.

Illustrated documents in the Benedictine Abbey of St Gallen, which date from the 10th and 11th centuries, and the semi-dramatic *Ordinarium Sedunense* of the church of St Valeria in Sion, composed in the 13th century, are extant as evidence of the origins of Christian LITURGICAL DRAMA in Switzerland. The use of mimicry and gesticulation can be inferred from an Easter celebration in the monastery of Einsiedeln during the 12th century, where the progress of the Apostles to the Sepulchre already possesses certain comic elements. An Epiphany celebration of the same period displays even more elaborate features.

It seems likely that the first MYSTERY PLAYS in the vernacular also originated in Switzerland, in the Easter play at Muri, first recorded about 1250. This is the only contribution to religious drama in the Germanic tongue made by the aristocratic poetry of chivalry, and was performed in a MULTIPLE SETTING in front of the church. The so-called Christmas play of St Gallen—also remarkable for its solemn and aristocratic character—dates from about 1300, and possibly belongs also to Muri. The oldest known Swiss Mystery Play in French is a comprehensive Epiphany play performed in Neuchâtel, which has remarkably poetic pastoral scenes, while a *Mystère de Saint Bernard de Menthon*, first written down in 1453, shows affinities with the dialect and customs of the Valais.

Possibly the most important Mystery Play in Switzerland before the Reformation was the Rheinau play of the Last Judgement, printed in 1487. It is one of the best texts in the German language, and obviously influenced similar plays at Donaueschingen, Copenhagen, and Berlin. The PASSION PLAYS of Donaueschingen and Villingen, however, derive from an early Lucerne play, as does the later famous Lucerne Passion Play, which in the last years of the 16th century assumed great importance under the direction of Renward Cysat. The large number of extant documents relating to his productions rank among the most important for the history of the later medieval theatre, and from them it has been possible to reconstruct the original stage setting.

Mystery plays received a fresh impulse during the 16th century from other Catholic areas—for example from Freiburg, where the Epiphany plays, dating from 1433, were produced with ever-increasing splendour, and continued up to the end of the 18th century, unlike the Lucerne play, which was last performed in 1616. Even in places like Basle, Berne, Biel, Geneva, Lausanne, and ZÜRICH, which had gone over to the Reformed faith, there were still traces of the Mystery Plays.

Carnival plays, too, attained their peak of activity in the 16th century. Pamphilius GENGENBACH, a printer in Basle, was the author of the moralizing FASTNACHTSSPIEL *Die zehn Alter der Welt* (1515), the first drama of any importance in the German language in the 16th century and the first to appear in print. The Carnival play was also used in the service of the new learning even before the Reformation, as is shown, for instance, by the works of Niklaus MANUEL of Berne, the most important Swiss dramatist of the 16th century.

In 1512 the first known *Tellspiel* (based on the exploits of William Tell, the legendary hero of 14th-century Switzerland) was performed in Altdorf. In 1545 a new version was made by the prolific dramatist Jacob Ruf and it has been constantly performed ever since, while the original subject of the legend has passed into world literature. Finally, the SCHOOL DRAMA of the humanists flourished throughout Switzerland. Even Ulrich Zwingli (1485–1531), the great reformer, wrote incidental music for a production of ARISTOPHANES' *Plutus*, while Calvin's disciple Théodore de Bèze (1519–1605) wrote, among other things, a play, *Abraham sacrifiant* (1550).

Unfortunately the new puritanical slant given to Protestantism in Geneva and Zürich, which found expression in the sumptuary laws of Geneva,

1617, and in the condemnatory tract *Bedenken von Comoedien* (1624) by Johann Jakob Breitinger, head of the church in Zürich, put a virtual end to theatrical activity; though some School Drama survived in reformed Berne and the Cathedral there was sometimes used for performances, for instance in 1692, when an allegorical play which inveighed against the persecution of the Huguenots by Louis XIV led to a protest from the French envoy.

Shortly after the middle of the 17th century there was a manifestation of folk drama in the reformed city of St Gallen, stimulated by the dramatic pageants of the Catholic monastery there, in which Josua Wetter, writer of popular plays and much under the influence of GRYPHIUS, was particularly prominent. In Catholic Switzerland the JESUIT DRAMA of the Order's schools in Lucerne, Freiburg, Solothurn, Brigue, Porrentruy, and Sion also influenced the growth of native folk drama, mainly through the work of their former pupils. This reached its height not only in the religious plays of the pilgrims at Einsiedeln, but also in the patriotic plays like those given in the small town of Zug, where in 1672 the *Eidgenössisches Contrafeth der Jungfrauen Helvetiae*, which provided a cross-section of Swiss history, was given on an open-air stage with five acting areas. With its baroque overtones, this two-day play shows some affinity with the English drama of Shakespeare's time.

All local drama was soon swept aside by the influx of foreign travelling companies which took place from the end of the 16th century. More than 230 such troupes are known by name up to 1800, mainly of German, Austrian, and French origin, but also Italian and English. Their repertory consisted of plays, ballad operas, pantomimes, puppet-plays, and shadow-shows. They pitched their booths in the open air or set up stages in warehouses and tennis-courts. A permanent stage was at their disposal in Baden from 1673, in Lucerne from 1741, and in Solothurn from 1755.

During the Enlightenment society in the larger towns delighted in theatrical performances; VOLTAIRE himself roused the enthusiasm of French Switzerland for the theatre, first at Les Délices, outside Geneva, and then at Mon Repos, between Lausanne and Ouchy. The first performance of his *Tancred*, played by a company composed for the most part of members of Geneva families, took place in 1760 in Tournay near Geneva. Geneva also provided the COMÉDIE-FRANÇAISE with its first Swiss professional actor in the person of Aufresne [Jean Rivaz] (1728–1804), and built its first permanent theatre for foreign visitors in 1783.

In German Switzerland the company of Konrad ACKERMANN, which operated with great success in 1758–60, was the principal source of inspiration. The architect Niklaus Sprüngli, a pupil of Blondel and SERVANDONY, built for the Grande Société in Berne an 'Hôtel de Musique', one of the finest of the old Swiss theatres, though it was not used for plays until 1798. Abel SEYLER, who was born in Liestal (Basle), not only helped to found the first German National Theatre in HAMBURG, but later became director of the Court theatres of Gotha, Hanover, and WEIMAR.

The Enlightenment also led to a new appraisal of Swiss folk drama, in so far as it was not expressly religious in character. The so-called Helvetic movement, which sought to achieve a comprehensive political and cultural revival in Switzerland, rightly regarded the theatre as one of the best instruments for its purpose, and as early as 1758 Jean-Jacques ROUSSEAU in Geneva was sponsoring a national open-air theatre based on classical models. Johann Georg Sulzer, of Zürich, inaugurated Swiss dramatic criticism with his *Allgemeinen Theorie der Schönen Künste* (1771–4), in which, for the national festival plays, he argued in favour of sweeping movements of chorus and ballet on an enormous open-air stage. There was also, towards the end of the 18th century, a revival of the *Tellspiel* in German Switzerland, while in French Switzerland the traditional wine-harvest festival in Vevey developed into a 'Play of the Four Seasons' with mime, ballet, and music; this was produced for the first time in 1797 in the market-square, and was revived five times in the 19th century, with increasing elaboration. It continues to be given at intervals.

Although in the 18th century there had already been an increasing demand for the establishment of a Swiss National Theatre with a permanent professional company of Swiss actors, and the 19th century brought a certain measure of consolidation into the Swiss theatre, the acting companies still consisted mainly of foreigners, with Germans and French predominating up to 1833. Regular theatre seasons in Berne date from 1800, in St Gallen from 1805. Lausanne had its first permanent theatre in 1804, Basle and Zürich in 1834, Biel in 1842, and Chur in 1861. New theatres were erected in Lucerne in 1839, in St Gallen in 1857, in Basle in 1873, in Geneva in 1879, and in Zürich in 1891.

The municipal theatre in Berne, which had been spasmodically supported by the authorities since the early years of the 19th century, began to receive a regular subsidy in 1837; Zürich in 1856, Basle in 1859. The repertory, which included operas as well as plays and the operettas fashionable in Paris and Vienna, was more or less in line with that of the smaller theatres of Germany, France, and Austria, though foreign producers put on more classical works than they would probably have done at home, and they also responded occasionally to the demands of the local audience for a Swiss play. As there was no censorship, works which were forbidden in neighbouring countries could be staged, and the first public performance of IBSEN's *Ghosts* (1881), for example, took place in the municipal theatre in Berne.

While this international activity was taking place in the indoor theatres, the characteristic

Swiss folk drama was being performed in the open air, all classes participating under the direction of schoolmasters and poets, physicians and clergymen. The favourite play was SCHILLER's *Wilhelm Tell*, and sometimes the whole landscape would be incorporated into the production, players and audience alike moving from one stage to another, as at the final performance in 1882 of the Passion Play at Lumbrein in Grisons. Festival plays performed by thousands of players and singers on vast stages before audiences of tens of thousands made a spectacle unique in Europe. In 1886 a three-tier stage of steps and platforms was erected on the battlefield of Sempach for the performance of the dramatic cantata *Siegesfeier der Freiheit*. This started a long series of historical festival plays, whose progressive advance in staging attracted attention far beyond Switzerland. It was to them that the outstanding theatrical personality of Switzerland, Adolphe APPIA, owed much of his inspiration.

With the 20th century the theatres of Switzerland became important centres of European culture. Oskar EBERLE achieved an international reputation in the 1930s with his AVANT-GARDE productions of traditional works. In French Switzerland Émile Jaques-Dalcroze, who worked with Appia, achieved recognition as a director, notably for his system of eurhythmics, as did René Morax, who with his brother Jean directed the Festival play at Vevey in 1905, for which he wrote a new libretto. He also built in 1908 at Mezières, near Lausanne, a festival theatre on new architectural and scenic lines, the Théâtre du Jorat, which became the centre of theatrical activity in western Switzerland.

It was, however, the influx of political and cultural refugees from the totalitarian states which from the late 1930s gave the Swiss theatre an outstanding international reputation. One of the first results was the foundation of the famous art theatre Cabaret Cornichon in 1933, and it was responsible also for the broadening of the repertory, which was most apparent in the Zürich Schauspielhaus, directed after 1938 by Oskar Wälterlin, successor of the famous Alfred Reucker (see ZÜRICH). Before the end of the Second World War the Swiss-German theatre was the only one in the German-language zone which could make its voice heard, and even when the war was over the Zürich Schauspielhaus, where BRECHT's *Mutter Courage und ihre Kinder* had its first production in 1941, retained its position as one of the principal playhouses in Europe. Fritz HOCHWÄLDER, who went into exile in Zürich in 1938, remained to become a Swiss dramatist by adoption.

Avant-garde theatres also opened in Berne (the Ateliertheater), in Basle (the Komoedie), in Lausanne (the Théâtre des Faux-Nez), and in Geneva (the Théâtre de Carouge and later the Theater an der Winkelweise). A new generation of scenic artists came into being as a result, and made a decided international impact, while notable writers to emerge included Max FRISCH and Friedrich DÜRRENMATT, who became well known in English translations. The spate of post-war activity has been consolidated rather than developed in new directions, and while the Schauspielhaus continued to do excellent work, notably under Peter Löffler from 1964 to 1970, a flourishing physical theatre has produced little new writing of note.

SWORD PLAY, see MUMMERS' PLAY.

SYDNEY OPERA HOUSE, an unusual and controversial building, designed by Jorn Utzon, which took nearly 15 years to erect. About 8,000 people helped in the construction, the largest number on site at any one time being 1,500. The building opened in 1973 with a permanent staff of about 350. It has four auditoriums, one of which —the Drama Theatre, seating about 500—is devoted to plays. It was first occupied by the OLD TOTE THEATRE company, which in Jan. 1979 was replaced by a specially selected Sydney Theatre Company. During 1979 the Drama Theatre was occupied by visiting companies, the Nimrod Theatre company, for instance, being seen in Dec. in GOLDONI's *The Venetian Twins*, but on 1 Jan. 1980 the new permanent company took over with Darrell's *The Sunny South* followed by Simon GRAY's *Close of Play*.

SYMBOLISM. Symbols have been used on the stage since the earliest times. Much of Elizabethan 'stage furniture' was symbolic, a throne standing for a Court, a tent for a battlefield, a tree for a forest. Symbolic elements are found in CHEKHOV and in the later plays of IBSEN and STRINDBERG. But Symbolism as a conscious art-form, conceived as a reaction against REALISM, came into the theatre with MAETERLINCK, writing under the influence of Mallarmé and Verlaine. His characters have no personality of their own, but are symbols of the poet's inner life. This aspect was intensified in the early plays of YEATS. Other dramatists to come under the influence of Symbolism include ANDREYEV and EVREINOV in Russia, Hugo von HOFMANNSTHAL and the later HAUPTMANN (with *Die versunkene Glocke*) in Germany, SYNGE (*The Well of the Saints*) and O'CASEY (*Within the Gates*) in Ireland, and O'NEILL in the United States.

SYMONDSBURY PLAY, see MUMMERS' PLAY.

SYMPHONIC PLAYS, see GREEN, PAUL.

SYNCHRONOUS WINCH SYSTEM, see GRID.

SYNDICATE, The, see THEATRICAL SYNDICATE.

SYNGE, (Edmund) JOHN MILLINGTON (1871–1909), Irish dramatist, with YEATS a leading figure in the Irish dramatic movement. Although he died prematurely, his six completed plays establish him as the greatest of modern Irish dramatists. His control of structure, whether in comedy or tragedy, is assured; his revelation of the characters and thought-processes of a subtle and imaginative peasantry is penetrating; his language, and especially his imagery, is rich, live, and essentially poetic. An early play, *When the Moon Has Set*, apparently written in 1901 and rejected for production by Yeats and Lady GREGORY, was found among his papers after his death and published with his other works in 1962–8; but the first of his plays to be performed, by the IRISH NATIONAL DRAMATIC SOCIETY, was *In the Shadow of the Glen* (1903), the first of a series of grave, original studies of Irish thought and character which drew upon the author the hostility of some of his early audiences. It was first seen in London in 1904, and was followed by *Riders to the Sea*, produced by the ABBEY THEATRE in the same year, a one-act tragedy whose brevity and intensity make it one of the best of modern short plays. It was used as the libretto for an opera by Vaughan Williams in 1937, having been first seen in London in 1904, in a double bill with *In the Shadow of the Glen*, and New York in 1920. A third play, *The Tinker's Wedding*, begun as early as 1902 but revised several times before its publication in 1908, was given its first production in London in 1909, being considered 'too dangerous' for an Abbey audience; but its comedy, drawn from the life of the Irish roads, is richer and more jovial than any other that Synge wrote. It was seen in New York, with *Riders to the Sea* and *In the Shadow of the Glen*, in 1957. *The Well of the Saints* (Dublin and London, 1905; N.Y., 1931) is another comedy in which poetic beauty is mingled with an underlying irony that is potentially tragic. The climax of Synge's achievements in the theatre is the comedy of bitter, ironic, yet imaginative realism, *The Playboy of the Western World* (1907). The unsparing though sympathetic portraiture in this play caused riots in the Abbey Theatre on its first production, and disturbances led by certain Irish patriots when it was first produced in New York in 1911. It had its first London production in 1907. It has long been accepted as Synge's finest work; his power is here seen at its fullest, as it could not be in the unfinished and unrevised *Deirdre of the Sorrows*, Synge's last play, in which he turned back, as Yeats and AE had done before him, to the ancient legends of Ireland. This was first produced posthumously at the Abbey in 1910 and was seen in London in the same year.

SYRUS, PUBLILIUS, see PUBLILIUS SYRUS.

SZAJNA, JÓZEF (1922–), Polish director and scene designer. He was with the People's Theatre in Nowa Huta from its foundation in 1955, becoming its director from 1963 to 1966, when he left for Warsaw. Since 1971 he has been a director and designer at the Studio Theatre there. His memories of Auschwitz colour all his work, particularly his concentration-camp settings for *The Deserted Field* (1964) by Tadeusz Houj, which he staged at Nowa Huta, and those for the revival of WYSPIAŃSKI's *Akropolis* by GROTOWSKI in 1966, and is still apparent in his later productions, in which the unusual costumes of the actors are integrated with powerfully evocative sculptural décors, as in GOETHE's *Faust* (1971), *Dante*, drawn from the *Divine Comedy* and first seen in Florence in 1974, and *Cervantes* (1976), based on themes from the life of the author of *Don Quixote*. Szajna's productions have been seen in many theatres outside Poland, and in 1970 he was in England, where he staged *Macbeth* at Sheffield.

SZANIAWSKI, JERZY, see POLAND.

SZIGLIGETI, EDE, see HUNGARY.

SZYFMAN, ARNOLD, see POLAND.

T

TABARIN [Antoine Girard] (?–1626), French actor, who in about 1618 set up a booth on the Pont Neuf with his brother, the quack-doctor Mondor. Here, and in the Place Dauphine, Tabarin would mount a trestle platform and, donning his famous, floppy, and polymorphic hat, would put the holiday crowd into a good humour before Mondor began the serious business of the day, selling his nostrums and boluses. Most of his material Tabarin wrote himself, or rather sketched out in the style of the COMMEDIA DELL'ARTE scenarios; and when a person unknown brought out in 1622 *Le Recueil général des rencontres, questions, demandes, et autres oeuvres tabariniques*, he himself published in the same year *L'Inventaire universel des oeuvres de Tabarin*, containing his farces, puns, jokes, and monologues. There is reason to believe that he was much influenced by the Italian actors who so often played in Paris, and from them he took the 'sack-beating' scene which MOLIÈRE later borrowed for *Les Fourberies de Scapin* (1671). Tabarin himself never appeared in a theatre, but he was remembered long after his death, and his name passed into everyday speech (*faire le tabarin* = play the fool).

TABERNARIA, see FABULA 7: TOGATA.

TAB(LEAU) CURTAIN, TABS, see CURTAIN.

TADEMA, Sir LAWRENCE ALMA-, see SCENERY.

TAGANKA THEATRE, Moscow. This theatre, whose official name is the Theatre of Drama and Comedy, is situated on Taganka Square, at some distance from the centre of the capital, and seats about 600 people. It was founded in 1946 and during the next 18 years presented more than 50 Soviet plays. In 1964 a group of graduates from the SHCHUKIN school of drama led by LYUBIMOV took it over and made it a centre for AVANT-GARDE work, presenting a wide range of innovatory productions including BRECHT's *The Good Woman of Setzuan* in 1964, an adaptation of John Reed's book *Ten Days That Shook the World* in 1967, Peter WEISS's *How Mr Mockinpott was Relieved of his Suffering*, and GORKY's *Mother* in 1973, as well as MOLIÈRE's *Tartuffe* in 1969. There was a modern-dress *Hamlet* in 1974 and among the theatre's other interesting experiments have been 'montage' versions of poems by, among others, MAYAKOVSKY and PUSHKIN.

TAGORE, RABINDRANATH (1861–1941), Indian poet, philosopher, and playwright, who in 1913 was awarded the Nobel Prize for Literature; he was knighted in 1915. In his numerous plays he successfully combined elements of Western drama, of which he had a wide knowledge, with classical Sanskrit drama and popular Bengali folk-drama (see INDIA). He wrote mainly in Bengali, his native language, often translating his plays into English himself. They were first performed either in Calcutta or at the school on his own estate, Santiniketan ('the house of peace'), and he frequently appeared in or directed them. The earlier ones were usually in verse or poetic prose, with interludes of colloquial speech, and were designed to be played on a bare stage with few, but highly symbolic, props. The best known are *Sacrifice* (*Visarjana*, 1890), *The King of the Dark Chamber* (*Raja*, 1910), *The Post Office* (*Dakaghan*, 1913), *Chitra* (1914), and *Red Oleanders* (*Rakta-Karabi*, 1924). Several of them, particularly *The Post Office*, have been translated and performed in many languages. In later years Tagore composed a number of dance-dramas in the Indian tradition, among them *Chitrangada* (1936), based on the earlier *Chitra*, and *Shyama* (1939), from a romantic Buddhist legend.

TAILS, see BORDER.

TAÏROV, ALEXANDER YAKOVLEVICH (1885–1950), Soviet director, who spent some time as an actor in the provinces and with MEYERHOLD in St Petersburg before joining Gaideburov's Mobile Theatre in 1908. Five years later, while working for MARDZHANOV, he met his future wife Alisa KOONEN, and together they opened the KAMERNY THEATRE in 1914. The first productions there, which included WILDE's *Salome*, showed clearly Taïrov's preoccupation with new spatial possibilities in staging, the varying levels, contrasting colours and scenic shapes, and beautiful costuming, allied with musically-accompanied rhythmic movement, serving to emphasize dynamic, often contrasting, tensions within a single dramatic action. The cubist-influenced designs of Alexandra Exter and Alexander Vesnin, his scenic artists, perfectly matched Taïrov's intentions. In common with Meyerhold he was antipathetic to NATURALISM but, unlike him, held the importance of the actor to be fundamental, demanding from his company all-round ability, including virtuoso pantomimic and acrobatic techniques, but allowing for the development of individual approaches to the text. The position of his wife within the company helped to emphasize the place of the actor in his productions, and almost every one of them became as much a medium for her remarkable talents as for his own experiments. The 1917 Revolution led to the abandonment of Taïrov's aesthetic excesses and to a sharpening of his social consciousness, as reflected in his staging of RACINE's *Phèdre* in 1922 and in his productions, much influenced by EXPRESSIONISM, of OSTROVSKY's *The Storm* in 1924 and of three plays by O'NEILL —*The Hairy Ape* and *Desire Under the Elms* in 1926 and *All God's Chillun Got Wings* in 1929. In the late 1920s Taïrov was severely criticized for his FORMALISM and for his choice of plays, which included BULGAKOV's *The Red Island*. He replied to his critics by staging in 1934 VISHNEVSKY's *An Optimistic Tragedy*, now considered one of the finest products of SOCIALIST REALISM. An attempt in 1938 to merge Taïrov's company with that of his rival OKHLOPKOV failed, owing to their fundamental incompatibility, and Taïrov celebrated its failure by staging one of his best productions, a dramatization of Flaubert's *Madame Bovary* with Alisa Koonen as Emma. Two Russian plays presented in 1947, Jacobson's *Life in the Citadel* and a revival of Ostrovsky's *Guiltless, Guilty*, were much admired, and although Taïrov's theatre was put by the authorities under the control of a committee he was not dismissed, as Meyerhold had been in similar circumstances, but continued to work there until a year before his death.

TALFOURD, Sir THOMAS NOON (1795–1854), English lawyer and man of letters, author of the tragedies *Ion* (1836), *The Athenian Captive* (1838), and *Glencoe; or, the Fate of the Macdonalds* (1840). Modelled on French tragedy, with careful observance of the UNITIES and a somewhat uninspired flow of blank verse, they were produced with MACREADY in the leading roles, *Ion*, which was several times revived up to 1850, at COVENT GARDEN, the others less successfully at the HAYMARKET THEATRE. Talfourd wrote a good deal of dramatic criticism, and was the literary executor of LAMB, whose works he edited after the latter's death, as he did the posthumous publications of HAZLITT. Becoming a member of Parliament, he supported the rights of authorship in the agitation over the COPYRIGHT Act of 1833, which was known by his name. He was an intimate friend of DICKENS and BULWER-LYTTON the first dedicating to him *The Pickwick Papers* (1837) and the second *The Lady of Lyons* (1838).

TALK OF THE TOWN, London, see HIPPODROME.

TALMA, FRANÇOIS-JOSEPH (1763–1826), French actor, brought up in England, where he was just about to join a London company when his father sent him back to Paris. There he entered the newly founded École de Déclamation, shortly to become the Conservatoire, and after tuition from MOLÉ and others made his début at the COMÉDIE-FRANÇAISE on 21 Nov. 1787 in VOLTAIRE's *Mahomet*. Although handsome, with a fine presence and a resonant voice which made him an excellent speaker of verse, he had played only small parts when on 4 Nov. 1789 he appeared in the title role of CHÉNIER's *Charles IX*, which all the older actors had refused because of its political implications. Talma declaimed the revolutionary speeches with such fervour that the theatre was in an uproar, and was eventually closed. Supported by some of the younger members of the company, he moved to the Théâtre de la Révolution (the present Comédie-Française), where he appeared in CORNEILLE's *Le Cid* and other classical revivals, as well as in some of Shakespeare's tragic parts in translations by DUCIS. In 1799, when Napoleon, who had met and become friends with Talma at Mlle MONTANSIER's *salon*, reconstituted the Comédie-Française, drawing up a firm code for its administration, Talma rejoined the company and put in hand many reforms, notably in the costuming of plays. He was the first French actor to play Roman parts in a toga instead of contemporary dress, and also reformed theatrical speech, suppressing the exaggerations of the declamatory style and allowing the sense rather than the metre to dictate the pauses. One of his great successes at this time was in a revival of LA FOSSE's *Manlius Capitolinus* in 1805, two years after Napoleon had taken him to Erfurt, where he acted before an audience which included five crowned heads. In 1817 he visited London, appearing with Mlle GEORGE at COVENT GARDEN in extracts from his best parts, and he attended John Philip KEMBLE's farewell performance and banquet. He remained on the stage until shortly before his death with no lessening of his great powers, and made his last appearance on 3 June 1826, only four months before his death, in a poor play by Delaville, *Charles VI*. He was for a time the lover of Pauline Bonaparte, and in 1801 divorced his first wife in order to marry his mistress, an actress in his company, Caroline Vanhove (1771–1860).

TAMAYO Y BAUS, MANUEL (1829–98), Spanish dramatist, who with LÓPEZ DE AYALA represents the transition period from Romanticism to Realism. His early plays were first performed by his parents, who were both on the stage, and were mainly translations from the French and German. But in 1855 a prose play on the life of the mad Doña Juana de Castile (sister of Katharine of Aragon) entitled *La locura de amor* (*The Insanity of Love*) had some success, as had *La bola de nieve* (*The Snowball*, 1856), which again dealt with the theme of jealousy but in a modern setting. Among his later plays, the best of those based on modern domestic and social problems was *Lo positivo* (*Materialism*, 1862), which deals with the conflict between sentiment and interest. His finest play, and the only one now remembered, was *Un drama* *nuevo* (*A New Play*, 1867), written under the pseudonym Joaquín Estebánez. It is set in Elizabethan times, with Shakespeare as one of the characters. An actor who is playing Yorick (the 'king's jester' referred to in *Hamlet*), kills on stage during a mock fight the young actor with whom he suspects his wife is in love, egged on by the Iago-like character called Walton.

TANDY, JESSICA (1909–), American actress, born in London, who began her stage career at the BIRMINGHAM REPERTORY THEATRE in 1928, being seen briefly in London a year later before going to New York in 1930 to play Toni Rakonitz in G. B. Stern's *The Matriarch*. On her return to London she appeared in a wide variety of plays, including Dodie SMITH's *Autumn Crocus* (1931), *Hamlet* in 1934 as Ophelia opposite John GIELGUD, and RATTIGAN's *French Without Tears* (1936). After a season with the OLD VIC company in *Twelfth Night* and *Henry V* in 1937 she returned to New York to play Kay in the Broadway production of J. B. PRIESTLEY's *Time and the Conways* in 1938, and in the 1940s settled permanently in the United States. The role of Blanche du Bois in Tennessee WILLIAMS's *A Streetcar Named Desire* (1947), which she played on Broadway for over two years, established her as one of America's leading actresses.

Her first husband was the actor Jack Hawkins; with her second husband Hume CRONYN she formed a notable stage partnership. Their early plays together included Jan de Hartog's two-character study of a marriage *The Fourposter* (1951) and N. C. HUNTER's *A Day By the Sea* (1955). She appeared without her husband in Peter SHAFFER's *Five Finger Exercise* (1959) and at the AMERICAN SHAKESPEARE THEATRE in 1961, where she played Lady Macbeth; they were reunited in the London production of Hugh Wheeler's *Big Fish, Little Fish* (1962). They were together in the opening season at the GUTHRIE THEATRE, Minneapolis, in 1963, when she played Gertrude in *Hamlet*, Olga in CHEKHOV's *Three Sisters*, and Linda Loman in Arthur MILLER's *Death of a Salesman*. They returned to the Guthrie in 1965 and were also seen in the New York productions of DÜRRENMATT's *The Physicists* (1964) and ALBEE's *A Delicate Balance* (1966). She starred independently in Tennessee Williams's *Camino Real* (1970) and Albee's *All Over* (1971), and continued her partnership with Cronyn in BECKETT's *Happy Days* (1972) and the double bill *Noël Coward in Two Keys* (1974). They were both at the STRATFORD (ONTARIO) FESTIVAL in 1976, where she played Lady Wishfort in CONGREVE's *The Way of the World* and Hipployta/Titania in *A Midsummer Night's Dream*. In New York in 1977 and London in 1979 they starred in another two-character play, D. L. Coburn's *The Gin Game*, set in an old people's home. Again at Stratford (Ontario) she played Mary Tyrone in O'NEILL's *Long Day's Journey into Night* in 1980.

TANGUAY, EVA, see VAUDEVILLE, AMERICAN.

TANVIR, HABIB, see INDIA.

TAPPING [*née* Cowell], FLORENCE (1852–1926), English actress, daughter of Sam COWELL, and best known under her second married name as the wife of the actor and stage manager Alfred B. Tapping. She was on the stage as a child, and in her teens toured with Charles WYNDHAM in America. She also toured India and Australia with Marie de Grey, and on her return played in Old Comedy with Kate VAUGHAN and Ben GREET. She was for some years with the KENDALS, and also in Miss HORNIMAN's repertory company at the GAIETY THEATRE, Manchester, rejoining it in London in 1916 to play Mrs Jeffcote in a revival of HOUGHTON's *Hindle Wakes*. She celebrated her stage jubilee in 1914, while appearing in Manchester. By her first husband, the actor John Parselle (1820–85), she had a daughter, Sydney, who became an actress under her great-grandmother's name of FAIRBROTHER.

TARKINGTON, (Newton) BOOTH (1869–1946), American novelist, and author of a number of plays, of which the best known is the romantic costume drama *Monsieur Beaucaire* (1901). Based in collaboration on his novel of the same title, this was spoilt artistically by the substitution of a conventionally happy ending for the ironic ending implicit in the book. It was nevertheless very successful in America, with Richard MANSFIELD in the name part, and also in London, where it later provided the libretto for a musical play with a score by Messager. Among Tarkington's other plays were two for Otis SKINNER, *Your Humble Servant* (1910) and *Mr Antonio* (1916); a charming comedy of youth entitled *Clarence* (1919); and a social drama on the theme of snobbery, *Tweedles* (1923), which failed in production.

TARLETON [Tarlton], RICHARD (?–1588), the most famous of Elizabethan clowns, probably the original of Yorick in Shakespeare's *Hamlet*, and a favourite of Elizabeth I until he angered her by a joke at the expense of the Earl of Leicester. A drawing of him in a manuscript now in the British Museum, reproduced in *Tarleton's Jests* (a posthumous work), shows that he was short and broad, with a large, flat face, curly hair, a wavy moustache, and a starveling beard. Tarleton himself said that he had a flat nose and a squint, a peculiarity well brought out in a second portrait of him discovered in 1920. His usual dress was a russet suit and buttoned cap, with short boots strapped at the ankle, as commonly worn by rustics at the time. A money-bag hangs from a belt at his waist, and he is shown playing on a tabor and pipe. He was one of QUEEN ELIZABETH'S MEN, and though none of his parts is definitely known, much of his clowning is believed to have been extempore. Both MARLOWE and Shakespeare may have had him in mind when they railed at 'clownage'—Marlowe in the prologue to *Tamburlaine* and Shakespeare in Hamlet's advice to the players, and it may have been the latter's desire to confine Tarleton's gagging within reasonable limits that led him to write such richly comic parts as Launce and Speed in *The Two Gentlemen of Verona*, Bottom in *A Midsummer Night's Dream*, Dogberry in *Much Ado About Nothing*, and the First Gravedigger in *Hamlet*. There can be no doubt that the genius of Tarleton was responsible for much of the mingling of tragedy and comedy in early English plays, but his greatest achievements lay in the JIG; the music for some of these has survived, but the only known libretto, *Tarltons Jigge of a horse loade of Fooles* (*c.* 1579), is now considered to be one of COLLIER's forgeries. Tarleton is, however, known to have written for the Queen's Men a composite play, now lost, entitled *The Seven Deadly Sins*, the first part containing five short plays, the second three. The outline, or PLATT, of the second part has been preserved in manuscript. A number of books published under Tarleton's name after his death are probably spurious, but his enduring popularity may be judged by the number of taverns named after him; one, The Tabour and Pipe Man, with a sign-board taken from the frontispiece to *Tarleton's Jests*, still stood in the Borough in London 200 years after his death, while the action of William Percy's *Cuckqueans and Cuckolds Errant* (1601) is said to take place in the Tarlton Inn, Colchester. Tarleton himself for some time ran an eating-house in the City, in Paternoster Row.

TASMANIAN THEATRE COMPANY, see HOBART.

TASSO, TORQUATO (1544–95), Italian poet and playwright, whose *L'Aminta* (1573) was the first true PASTORAL and the pattern for many that followed in Italy and France. It was also the source of Berowne's speech in *Love's Labour's Lost* IV. iii: 'From women's eyes this doctrine I derive.' This tale of rustic life—rustic in an artificial sense, since it deals with the loves of idealized shepherds and shepherdesses—was first translated into English in 1591 as *Phillis and Amyntas*, and in the following century at least four separate translations were made. The latest version is apparently that prepared by Leigh HUNT in 1820. Tasso, who is best remembered for his great epic poem *Gerusalemme liberata* (1581), which may have influenced the writing of *Cymbeline*, was also the author of *Torrismondo* (1587), which shows an early mingling of tragedy and romance. Though classic in form, it deals in romantic fashion with the love of King Torrismondo for Rosmonda, who turns out to be his sister. Finding himself guilty of incest in marrying her, he commits suicide. Tasso (himself the subject of a fine poetic play by GOETHE, though the story on which it is based is now discredited) wrote only one other work for the stage—the posthumously published comedy *Intrighi d'amore*.

TATE, HARRY, see MUSIC-HALL.

TATE, NAHUM (1652–1715), a poor poet and worse playwright, who collaborated with DRYDEN in the second part of *Absalom and Achitophel* (1682) and with Brady in a metrical version of the Psalms, published in 1696, which long remained popular in the Church of England. In 1692 he succeeded Thomas SHADWELL as Poet Laureate, and he was pilloried by Pope in *The Dunciad* (1728). Otherwise he is mainly remembered for his extremely odd versions of some of the plays of Shakespeare. His *The History of King Richard the Second* (1680)—published, after an unsuccessful production, as *The Sicilian Usurper*—was intended to blacken the character of Bolingbroke and render Richard II wholly sympathetic. In *The History of King Lear* (1681) the character of the Fool is omitted, and Cordelia survives to marry Edgar. This remained the standard acting text (adapted again by the elder COLMAN in 1768) and was not replaced by the original version until MACREADY's production of 1838. *Coriolanus*, as *The Ingratitude of a Common-Wealth; or, the Fall of Caius Martius Coriolanus* (also 1681), was not so badly mangled, except in the last act, which incorporates the worst features of *Titus Andronicus*, probably in a bid for popular success which failed.

TAVISTOCK HOUSE THEATRE, London, see DICKENS, CHARLES.

TAVISTOCK REPERTORY COMPANY, see TOWER THEATRE, LONDON.

TAYLOR, C(ecil) P(hilip) (1929–81), Scottish dramatist who settled in Northumberland, and was formerly an electrician. A prolific writer who wrote well over 60 stage plays in 20 years, as well as many for television and radio, he had a strong commitment to regional theatre. His first play *Aa Went to Blaydon Races* was produced in Newcastle, and the TRAVERSE THEATRE, Edinburgh, gave the first performances of *Happy Days Are Here Again* and *Of Hope and Glory* (both 1965), *Allergy* (1966), *Lies About Vietnam* and *Truth About Sarajevo* (both 1969), *Passion Play* (1971), *The Black and White Minstrels* (1972), with Alan HOWARD, *Next Year in Tel Aviv* and *Columba* (both 1973), *Schippel* (1974), adapted from STERNHEIM, *Gynt!* (1975), adapted from IBSEN, and the two-part *Walter* (1977). Although Taylor's primary concern in political, political ideas do not dominate his characters, but rather act through them: thus in *Allergy*, a hilarious one-acter, the failures of the left are presented through the all-too-human weaknesses of the bearers of the revolutionary torch, while in *Bread and Butter* (1966) the defeats and compromises the left has suffered in the recent past are seen through the changing circumstances of two ordinary Glasgow-Jewish couples. Taylor's most highly praised play was his penultimate one, *Good*, produced by the ROYAL SHAKESPEARE COMPANY at the WAREHOUSE in 1981 and commercially in the West End in 1982. Tracing the self-deception and moral inertia which enable a liberal German professor, played by Howard, to work in Auschwitz, it embodies Taylor's constant theme of the need for absolute honesty. Apart from *Good*, only *Schippel* (as *The Plumber's Progress*, 1975) and *And a Nightingale Sang ...* (1979) have had mainstream London productions, though *Bandits* was seen at the Warehouse in 1977 and *The Black and White Minstrels* (1974), again with Howard, and *Goldberg* (1976) were produced at the HAMPSTEAD THEATRE CLUB. The NEWCASTLE PLAYHOUSE's posthumous production in 1982 of his last play, *Bring Me Sunshine, Bring Me Smiles*, was also seen in London. Of all modern Scottish dramatists, Taylor most successfully learned the lesson of BRIDIE's career, that an honest presentation of what one knows best can be of more than local significance.

TAYLOR, JOSEPH (*c.* 1585–1652), English actor, who joined the KING'S MEN in 1619, by which time he was already well known, and took over many of Richard BURBAGE's parts. He also appeared as the handsome young lovers or dashing villains of BEAUMONT and FLETCHER. It is unlikely that, as was once thought, he was coached by Shakespeare himself in the part of Hamlet, but he may well have seen Burbage act it, and he is certainly said to have appeared to advantage in the part. Other roles in which he gave excellent performances in revivals appear to have been Iago in *Othello*, Truewit and Face in Ben JONSON's *Epicoene* and *The Alchemist*, and Ferdinand in WEBSTER's *The Duchess of Malfi*. With LOWIN he became one of the chief business managers of the King's Men after the deaths of CONDELL and HEMINGE.

TAYLOR [*née* Cooney], LAURETTE (1884–1946), American actress, who had already had a long and distinguished career, being on the stage from childhood, when in 1912 she appeared at the CORT THEATRE, New York, as Peg in *Peg o' My Heart*, by her second husband, John Hartley MANNERS. She also played the part in London in 1914, with immense success, and it was always thereafter associated with her. After Manners's death in 1928 she retired from the stage for a time, but returned to continue her successful career, reappearing in New York in 1938 as Mrs Midgit in a revival of Sutton Vane's *Outward Bound*. In 1945 she gave an outstanding performance as the mother in Tennessee WILLIAMS's *The Glass Menagerie*, which ran for over a year.

TAYLOR, TOM (1817–80), English dramatist, and editor of *Punch*, whose first play, *A Trip to Kissingen* (1844), was produced by the KEELEYS at the LYCEUM. A prolific writer, he continued his output of plays until about two years before his death. His playwriting was combined with other activities: he was successively Professor of English at London University and, for more than 20 years, a civil servant in the Health Department. The best known of his works are probably *To Parents and Guardians* (1846); *Masks and Faces* (1852), a comedy on the life of Peg WOFFINGTON written in collaboration with Charles READE and frequently revived; *Still Waters Run Deep* (1855), a play based on a French novel and remarkable in its time for its frank discussion of sex; *Our American Cousin* (1858), first produced in New York and noteworthy because of the appearance in it of E. A. SOTHERN as Lord Dundreary, a part which he enlarged until it practically swamped the play; *The Overland Route* (1860); *The Ticket-of-Leave Man* (1863), a melodrama on a contemporary theme of low life which had much influence on such later works as H. A. JONES and Herman's *The Silver King*; and two plays written in collaboration, *New Men and Old Acres* (1869) and *Arkwright's Wife* (1873). Taylor was himself an enthusiastic amateur actor, playing at DICKENS's private theatre in Tavistock House and being one of the leading members of the Canterbury Old Stagers. He had little originality, borrowing his material freely from many sources, but his excellent stagecraft and skilful handling of contemporary themes make him interesting as a forerunner of T. W. ROBERTSON.

TCHEHOV, TCHEKHOV, see CHEKHOV.

TEARLE, GODFREY SEYMOUR (1884–1953), English actor, knighted in 1951. He made his first appearance at the age of nine in the company run by his father Osmond TEARLE and rejoined it six years later, remaining until his father's death. In spite of his fine voice and natural authority he was slow to achieve recognition, but he was an ideal romantic hero and in 1920 scored a popular success in the dramatization of Robert Hichens's novel *The Garden of Allah*. A year later he played Othello at the ROYAL COURT THEATRE, but for the rest of the 1920s he appeared mostly in conventional comedies. His Hamlet at the HAYMARKET THEATRE in 1931 was not a great success; his finest part was Edward Ferrars in Charles MORGAN's *The Flashing Stream* (1938; N.Y., 1939), though he was also much admired as Maddoc Thomas in Emlyn WILLIAMS's *The Light of Heart* (1940). He achieved belated recognition as a Shakespearian actor with his Antony to the Cleopatra of Edith EVANS in 1946, a role he repeated in New York the following year, and with Othello and Macbeth at the SHAKESPEARE MEMORIAL THEATRE in the 1948–9 season. Other later roles included Hilary Jesson in PINERO's *His House in Order* in 1951 and the title role in Raymond MASSEY's *The Hanging Judge* (1952).

TEARLE, (George) OSMOND (1852–1901), English actor, who made his début in Liverpool in 1869 and two years later was seen in Warrington as Hamlet, a part he subsequently played many times. After six years in the provinces he appeared in London and soon after formed his own company, which he took on tour. In 1880 he went to America where he joined the stock company at Lester WALLACK's theatre in New York, making his first appearance there as Jaques in *As You Like It*. He alternated for many years between England and New York, and in 1888 organized a Shakespeare company which appeared with much success at STRATFORD-UPON-AVON and provided an invaluable training-ground for young actors. He was himself a fine Shakespearian actor, combining excellent elocution with a natural elegance and dignity. He made his last appearance in Carlisle, dying a week later. By his second wife, the actress Minnie [Marianne] Levy Conway (1854–96) daughter of F. B. CONWAY, he was the father of Godfrey TEARLE.

TEASER, see BORDER and FALSE PROS(CENIUM).

TEATRO CAMPESINO, CHICANO, DE LA ESPERANZA, DE LA GENTE, DE LOS BARRIOS, DEL PIOJO, see CHICANO THEATRE.

TEATRO ESLAVA, Madrid, see MARTÍNEZ SIERRA, GREGORIO.

TEATRO FARNESE, see ALEOTTI and THEATRE BUILDINGS.

TEATRO INTÍM, Barcelona, see GUAL, ADRÍA.

TEATRO POPOLARE ITALIANO, see GASSMANN, VITTORIO.

TEATRO POR HORAS, see GÉNERO CHICO.

TEATRO SPERIMENTALE DEGLI INDIPENDENTI, see INDIPENDENTI, TEATRO SPERIMENTALE DEGLI.

TEATRO STABILE (pl. *teatri stabili*), the name given to the permanent professional companies established in the major cities of Italy since the Second World War. Although attempts had been made earlier—as with the COMPAGNIA REALE SARDA—to set up stable troupes in some of the larger towns, these had proved short-lived, whereas some of the *teatri stabili* have survived long enough to become an important part of the Italian theatrical scene, among them the PICCOLO TEATRO DELLA CITTÀ DI MILANO, and those in Genoa, Rome, and Turin.

TEIRLINCK, HERMAN, see BELGIUM.

TELARI, see PERIAKTOI.

TELBIN, WILLIAM (1813–73), English scene painter, who in 1840 was working for MACREADY at DRURY LANE. He had previously been employed by several provincial theatres, and was later at COVENT GARDEN and the LYCEUM. Two of his sisters were on the stage in New York. His son, William Lewis Telbin (1846–1931), was for many years a scene designer at the Theatre Royal in Manchester. Later he worked in London, and in 1902, after a visit to Italy, he designed the settings and costumes for ALEXANDER's production of Stephen PHILLIPS's *Paolo and Francesca* at the ST JAMES'S THEATRE. He also worked for Henry IRVING at the Lyceum, collaborating with Hawes CRAVEN on TENNYSON's *The Cup* (1881) as well as on *Romeo and Juliet* in 1882 and Wills's adaptation of GOETHE's *Faust* in 1885, and creating a splendid cathedral scene for *Much Ado About Nothing*, also in 1882.

TÉLLEZ, FR. GABRIEL, see TIRSO DE MOLINA.

TEMPEST, MARIE [Mary Susan Etherington] (1864–1942), English actress, appointed DBE in 1937. Trained as a singer, she was seen in operetta and musical comedy and in 1890 went to New York, later touring Canada and the United States in light opera. Returning to London in 1895 she appeared under George EDWARDES at DALY'S THEATRE, being seen in *The Geisha* (1896). *San Toy* (1899), and other musical comedies. She then deserted the musical stage and from the time of her successful appearance as Nell GWYNN in Anthony Hope's *English Nell* (1900) appeared only in comedy. She was much admired in two plays by her second husband Cosmo Gordon-Lennox (1869–1921), *Becky Sharp* (1901), based on Thackeray's *Vanity Fair*, and *The Marriage of Kitty* (1902), which she subsequently revived many times. She toured America and Australia from 1914 to 1922, going also to India and South Africa, and venturing as far as the Philippine Islands with a varied repertory of light comedies. On her return to London she became noted for playing charming and elegant middle-aged women—Judith Bliss in Noël COWARD's *Hay Fever* (1925), the title role in St John ERVINE's *The First Mrs Fraser* (1929), and Fanny Cavendish in *Theatre Royal* (1934) by Edna Ferber and George S. KAUFMAN. In 1935 she celebrated her stage jubilee with a matinée at DRURY LANE, but she continued to act, being seen as Georgina Leigh in Robert MORLEY's *Short Story* (1935) with her third husband, W. Graham Browne (1870–1937), who had appeared with her in many previous productions, some of which he directed. She made her last appearance as Dora Randolph in Dodie SMITH's *Dear Octopus* (1938).

TEMPLETON, FAY, see VAUDEVILLE, AMERICAN.

TENDULKAR, VIJAY, see INDIA.

TENNIS-COURT THEATRES. Both in Paris and in London in the mid-17th century, tennis courts were converted into theatres. Among the most famous were the ILLUSTRE-THÉÂTRE (1644), where MOLIÈRE first acted in Paris, and, in London, KILLIGREW'S VERE STREET THEATRE (1660) and DAVENANT'S LINCOLN'S INN FIELDS THEATRE (1661). They were used only for a short time, and their conversion into theatres must have entailed no more alteration than could have been easily removed to allow them to revert to their original use. The rectangular shape of the tennis-court auditorium, which approximated more to the private than the public Elizabethan theatre auditorium, had an important influence on the development of the French playhouse, and may also have influenced the eventual shape of the English Restoration theatre, just as the boxes round the pit may have developed partly from the 'pent-house' or covered way which ran along one side and one end of the court itself.

TENNYSON, ALFRED, Lord (1809–92), English poet, who succeeded Wordsworth as Poet Laureate. Though not very much in touch with the contemporary stage, he nevertheless contributed to the poetic drama of his day, his first play *Queen Mary* (1876), a frigid tragedy in blank verse on Elizabethan lines, *The Cup* (1881), and *Becket* (1893), being produced at the LYCEUM by Henry IRVING, whose Becket was considered by many the finest achievement of his career. Of Tennyson's other plays in verse *The Falcon*, based on an episode in the *Decameron*, was produced at the ST JAMES'S THEATRE by the KENDALS in 1879; *The Promise of May* (1882), a drama of modern village life, was unsuccessful at the GLOBE THEATRE; and *The Foresters* (1892), with music by Sullivan, was seen at the Lyceum. *Harold* was published in Tennyson's lifetime but not performed until Apr. 1928, when it was given by the BIRMINGHAM REPERTORY company at the ROYAL COURT THEATRE, London, with Laurence OLIVIER as King Harold. Like many other writers of his day, Tennyson failed to amalgamate fine poetry and good theatre. None of his plays has been revived, and the success of *Becket* was mainly due to the beauty and compelling power of Irving's interpretation.

TERENCE [Publius Terentius Afer] (*c.* 190–159 BC), Roman dramatist, a freed slave, probably of African parentage. Six of his plays in the FABULA *palliata* form are extant—the *Andria* (*The Girl From Andros,* 166 BC), the *Hecyra* (*The Mother-in-Law*, 165), the *Heauton Timorumenos* (*The Self-Punisher*, 163), the *Eunuchus*, the *Phormio* (both 161), and the *Adelphi* (*The Brothers*, 160). Although they are all based on Greek comedies (four of them on plays by MENANDER), their originality is not at all comparable with that of PLAUTUS. Where Plautus is topical, rough, farcical, and anti-realistic, Terence's plays are distinguished by their urbanity, elegance, and smooth construction. In all there are contrasts of character—in the *Adelphi*, for instance, between the strict father and the genial uncle, and the two brothers, one rash, one timid. In the *Andria* there is a situation not found in Greek comedy—a young man of good family in love with a young lady of his own station, an episode which was only possible because of the greater freedom of women in Rome than in Athens. The background of all the plays is neither Greek nor Roman but is independent of time and place, and perhaps because of this Terence's works later had a universal appeal. In his own day, though he achieved some measure of success in the theatre, his audiences were not uncritical, and the *Hecyra* twice failed in production. Even Julius Caesar found Terence lacking in comic force, though many of his pithy sayings were quoted by contemporary writers. But his interest in humanity, summed up in the famous remark by a character in the *Heauton Timorumenos*: 'I am a man; and all human affairs concern me', gave his work an abiding appeal. In the schools of the Middle Ages his plays were not only read but acted; in the 10th century the Abbess HROSWITHA even made adaptations of them for her nuns at Gandersheim. With the coming of the Renaissance they were translated into several languages; their influence spread to France, where it reached as far as MOLIÈRE, and to England, where it can be traced from the first English comedy, UDALL'S *Ralph Roister Doister* (*c.* 1553), through LYLY and Shakespeare to STEELE, whose *The Conscious Lovers* (1727) is an adaptation of the *Andria*. Several of the manuscripts and early printed editions of Terence have illustrations which provide useful information about the staging of Renaissance plays, notably the TERENCE-STAGE.

TERENCE-STAGE, name given to the setting shown in Renaissance editions of the plays of TERENCE. In the Trechsel edition of 1493 a woodcut shows a two-storey structure with arches (*fornices*) below and an auditorium (*theatrum*) above. Three tiers of spectators are seated in front of a stage-wall (*proscenium*) divided into sections by columns, with curtains hanging between them. There is a large forestage on which a musician is seated playing a wind instrument. In a box on the side wall between the spectators and the stage are two officials (*ædiles*). The appearance of this stage is vouched for by another woodcut in the same volume, showing a scene from the *Eunuchus* with four curtained arches, each labelled with the name of a character in the play. Although this is reminiscent of the 'houses' of the LITURGICAL DRAMA the pillared façade is more like the *scaenae frons* of the late classical theatre building. There were evidently alternative forms of the *Eunuchus* stage, as a woodcut illustrating the *Heauton Timorumenos* shows the four entrances forming a three-sided structure jutting forward on to the stage, while another, for the *Adelphi*, shows five entrances, the central one also jutting forward. It is possible that some such arrangement formed the much-disputed INNER STAGE of the Elizabethan theatre. Although these illustrations are of academic performances—the spectators in the *theatrum* might equally well be expecting a lecture—this form of stage was later adapted for a less restricted audience, and resulted in the theatre built by Scamozzi at Sabbionetta in about 1588 and the Teatro Farnese built by ALEOTTI at Parma in about 1619. From these, based on an open loggia and a semi-circular amphitheatre, came the horseshoe-shaped auditorium and the proscenium-arch stage with its elaborate curtain, typical of later stage buildings all over the world.

TERPSICHORE, see MUSES.

TERRISS, ELLALINE (1871–1971), English actress, daughter of William TERRISS. Born in the Falkland Islands while her father was sheep-farming there, she returned to England with her parents at an early age, and in 1888 made her first appearance on the stage at the HAYMARKET THEATRE under TREE. She was also with WYNDHAM at the CRITERION THEATRE for three years. She made a great success in the title role of *Bluebell in Fairyland* (1901), a children's play by her husband Seymour HICKS which was revived annually at Christmas for many years, and also appeared with him in a number of his other plays, including *The Beauty of Bath* (1906) and *The Gay Gordons* (1907), though one of her best performances was given somewhat earlier as Phoebe Throssel in BARRIE's *Quality Street* (1902). She accompanied Hicks on tour, both in straight plays and in MUSIC-HALL sketches, and went with him to France in the First World War to play to the troops, afterwards travelling with him in Australia and Canada. She left the stage in 1929, after appearing in a revival of Hicks's *The Man in Dress Clothes*, but returned briefly to play Mrs Thornton in his *The Miracle Man* (1935).

TERRISS [Lewin], WILLIAM CHARLES JAMES (1847–97), English actor, formerly a sailor, known affectionately as 'Breezy Bill' or 'No. 1, Adelphi Terriss', since his best work was done at the ADELPHI THEATRE. He first appeared on the stage, unsuccessfully, in Birmingham in 1867, and then emigrated with his wife, the actress Amy [*née* Fellowes] (?–1898), to the Falkland Islands, where his daughter Ellaline TERRISS was born. Returning to England in 1873, he again entered the theatre, playing Doricourt in a revival of Hannah COWLEY's *The Belle's Stratagem*, and this time was successful. One of his first outstanding appearances, in 1875, was as Nicholas Nickleby in Halliday's dramatization of DICKENS's novel; but he finally made his name in *Olivia* (1878), based by William Wills on GOLDSMITH's novel *The Vicar of Wakefield*, in which he played Squire Thornhill to Ellen TERRY's Olivia. A year later he played Romeo to the Juliet of Adelaide NEILSON at the HAYMARKET THEATRE, and in 1880 joined Henry IRVING's company at the LYCEUM, where he remained for five years. He is best remembered, however, as the hero of a series of famous MELODRAMAS staged at the Adelphi, among them Sims's *Harbour Lights* (1885), BELASCO's *The Girl I Left Behind Me*, and Seymour HICKS's *One of the Best* (both 1895). The success of such plays owed much to his handsome debonair presence and vigorous acting. He was one of the most popular actors of his day, and his assassination at the stage door of the Adelphi on 16 Dec. 1897, by an actor who had allowed a grudge against Terriss to unhinge his mind, was felt as a personal loss by his many admirers. The Rotherhithe Hippodrome (1908) was originally named after him when it opened on 16 Oct. 1899.

TERRY, BENJAMIN (1818–96), English actor, son of an innkeeper at Portsmouth, who with his wife Sarah Ballard (1817–92), daughter of a local builder, known on the stage as Miss Yerret, toured extensively in the provinces. He later played small parts with Charles KEAN at the PRINCESS'S THEATRE in London, where his wife assisted in the wardrobe and two of his daughters were child-actors with the company. He had eleven children in all, of whom five went on the stage—Ellen, Florence, Fred, Kate, and Marion (see below)—while two other sons, George (1850–1928) and Charles (1857–1933), were connected with theatre management. Charles was the father of Minnie (1882–1964) and Beatrice (1890–), who were both on the stage, Minnie later marrying the actor Edmund Gwenn and appearing mainly in London, while Beatrice spent the latter part of her career in the U.S.A.

TERRY, DANIEL (1789–1829), English actor, friend of Sir Walter SCOTT, several of whose novels he dramatized for COVENT GARDEN. After some experience in the provinces he made his first appearance in London in 1812 at the HAYMARKET, playing the fop Lord Ogleby in COLMAN and GARRICK's *The Clandestine Marriage*. From 1813 to 1822 he was at Covent Garden, and then went to DRURY LANE. A good-looking man with a fine voice and an alert, sensitive face, he was at his best in character parts, particularly those of old men, or in strong emotional drama. In 1825, in partnership with Frederick YATES, he took over the ADELPHI THEATRE, but the venture was not a success, and he soon retired, dying shortly afterwards. As far as is known he was not related to the family of Benjamin TERRY, being of Scottish extraction.

TERRY, EDWARD O'CONNOR (1844–1912), English actor and manager, not, as far as is known, connected with the family of Benjamin TERRY. After some years in the provinces, when he was briefly in the same company as the young Henry IRVING and also played such Shakespearian parts as Touchstone in *As You Like It* and Dogberry in *Much Ado About Nothing* under Charles Calvert in Manchester, he made his first appearance in London in 1867, at the SURREY THEATRE, and then played in BURLESQUE and light comedy for several years with Mrs Swanborough at the STRAND THEATRE. In 1876 he was engaged by HOLLINGSHEAD for the GAIETY THEATRE, where he became one of the famous 'quartet' with Nellie FARREN, Kate VAUGHAN, and Edward Royce. In 1887 he opened a theatre under his own name in the Strand, on the site of a famous 'song-and-supper room', the Coal Hole, the first production, on 17 Oct. being his own adaptation of a German play as *The Churchwarden*. A year later he scored his first outstanding success as Dick Phenyl in *Sweet Lavender* by PINERO, whose *The Times* (1891) was also produced at Terry's. In 1902 Terry appeared for the first time in musical comedy with *My Pretty Maid*, followed in 1903 by *My Lady Molly*, which had a long run. After this his only commercial success was *Mrs Wiggs of the Cabbage Patch* (1907), dramatized by Alice Hegan Rice from her book of the same name. In 1910 he gave up his theatre, which then became a cinema and was demolished in 1923, and toured extensively in Australia, South Africa, and the U.S.A. Though not a good straight actor, Terry was an excellent 'eccentric comedian' and a careful and conscientious manager. He married as his second wife the widow of Sir Augustus HARRIS.

TERRY, ELLEN ALICE (1847–1928), English actress, the second daughter of Benjamin TERRY, appointed DBE in 1925. She made her first appearance on the stage at the age of nine, playing Mamillius in *The Winter's Tale* in Charles KEAN's company at the PRINCESS'S THEATRE. She remained with the Keans until their retirement in 1859, and in the summer of that and succeeding years toured with her elder sister Kate TERRY in *A Drawing-Room Entertainment*, in which they played together in short sketches. In 1861 Ellen joined the company at the HAYMARKET THEATRE, leaving in 1864 to marry the painter G. F. Watts, an ill-judged union with a man twice her age which soon came to an end. She returned to the theatre for a time, leaving it again shortly afterwards to live with Edward GODWIN, by whom she had two children Edith and Edward Gordon CRAIG. When, on the insistence of Charles READE, she reappeared on the London stage in 1874 as Philippa in his drama *The Wandering Heir*, taking over the part from Mrs John WOOD, she was as brilliant as ever, and her long absence from the theatre seemed only to have increased the excellence of her acting. After playing for a year with the BANCROFTS, she went to the ROYAL COURT THEATRE under HARE, playing for him one of her most successful parts, the title role in *Olivia* (1878), Wills's adaptation of GOLDSMITH's novel *The Vicar of Wakefield*. During the run of this play she married her second husband, the actor Charles Kelly [Wardell] (1839–85). In 1878 Henry IRVING, who had recently begun his tenancy of the LYCEUM, engaged Ellen Terry as his leading lady, and so inaugurated a partnership which was to become one of the outstanding features of the London theatrical scene for the next 25 years. She appeared with him in a wide variety of parts, including a good deal of Shakespeare—notably Ophelia, Beatrice, Desdemona, Juliet, Viola, Lady Macbeth, and Imogen—in revivals of contemporary plays—BULWER-LYTTON's *The Lady of Lyons* and Selby's *Robert Macaire*—and in a few plays specially written for him—Wills's *Charles I* (1879) and *Faust* (1885), Merivale's *Ravenswood* (1890), based on SCOTT's *The Bride of Lammermoor*, TENNYSON's *The Cup* (1881) and *Becket* (1893), and Comyns Carr's *King Arthur*. After leaving the Lyceum Ellen Terry became manager of the IMPERIAL THEATRE, where in 1903 she appeared in *Much Ado About Nothing* and IBSEN's *The Vikings*, being seen in the same year in HEIJERMANS's *The Good Hope* (for the STAGE SOCIETY), and two years later in BARRIE's *Alice Sit-By-The-Fire*. In 1906 she celebrated her stage jubilee with a mammoth matinée at Drury Lane at which 22 members of the Terry family assisted. She was at the same time appearing as Lady Cicely Waynflete, a part specially written for her by SHAW in *Captain Brassbound's Conversion*. Playing opposite her was a young American actor, James Carew [Usselman] (1876–1938), whom she married as her third husband in 1907. Although the marriage lasted only a short time, she seldom acted afterwards, but toured America and Australia, giving readings of and lectures on Shakespeare. Four of the lectures were published in 1931, proving once again how excellent was her critical faculty and how masterly her handling of the written as well as the spoken word. These qualities had already been apparent in ther autobiography, published in 1908, and in her correspondence with Bernard Shaw, also published in 1931. Throughout her career she was an inspiration to those who played with her. She was not at her best in tragedy, though some critics admired her Lady Macbeth, the role in which she was painted by Sargent, and she unfortunately never played Rosalind in *As You Like It* which seemed, above all other parts, to have been written for her, but to a hundred other roles she imparted a freshness and vitality which was never forgotten by those who saw her.

TERRY, FLORENCE (1855–96), English actress, the fourth daughter of Benjamin TERRY. On the stage as a child, she later understudied her sister Marion TERRY, and then went to the LYCEUM to play Nerissa in *The Merchant of Venice* to the Portia of her sister Ellen TERRY. She also played two of the latter's parts—Olivia in GOLDSMITH's

The Vicar of Wakefield and Lilian Vavasour in Tom TAYLOR's *New Men and Old Acres*—on tour. A young actress of great promise, she left the stage in 1882 to marry a lawyer, William Morris.

TERRY, FRED (1864–1932), English actor, the youngest child of Benjamin TERRY. He made his first appearance on the stage at the HAYMARKET THEATRE under the BANCROFTS, and four years later, at the LYCEUM, played Sebastian in *Twelfth Night* to the Viola of his elder sister Ellen TERRY. A handsome, romantic actor, he became best known for his performance as Sir Percy Blakeney in Baroness Orczy's *The Scarlet Pimpernel*, which he produced under his own management in 1905 and frequently revived in London and on tour. He was also much admired as Charles II in *Sweet Nell of Old Drury* (1900) and as Sir John Manners in *Dorothy o' the Hall* (1906), both by Paul KESTER, and in the title roles of *Matt o' Merrymount* (1908), by Beulah Dix and Mrs Sutherland, and *Henry of Navarre* (1909), by William Devereux. In these and many other productions he was partnered by his wife Julia NEILSON until his retirement in 1929. His son Dennis (1895–1932) and daughter Phyllis (1892–1977) were both on the stage.

TERRY, KATE (1844–1924), English actress, the eldest daughter of Benjamin TERRY. Trained for the stage by her father, she made her first appearance in London as Prince Arthur in Charles KEAN's production of *King John* at the PRINCESS'S THEATRE in 1852, and remained with the Keans until 1859, playing Ariel in *The Tempest* in 1857 and Cordelia in *King Lear* in 1858. She then joined the Bristol stock company, but returned to London in 1861 to play Ophelia to the Hamlet of Charles FECHTER. She appeared in several plays by Tom TAYLOR, and seemed to be heading for a brilliant career when in 1867 she left the stage to marry Arthur Lewis, by whom she had four daughters, the eldest, also Kate, becoming the mother of John GIELGUD, and the youngest, Mabel TERRY-LEWIS, an outstanding actress. In 1898 Kate Lewis, who at the height of her career had been considered by some critics to be a better actress than her more famous sister Ellen TERRY, returned to the stage to play opposite HARE in Stuart Ogilvie's *The Master*, but the venture was not a success, and she never again appeared on stage.

TERRY, MARION (1852–1930), English actress, the third daughter of Benjamin TERRY. She made her first appearance on the stage in 1873, and three years later scored a success as Dorothy in W. S. GILBERT's *Dan'l Druce, Blacksmith*. In 1879 she was engaged by the BANCROFTS for the PRINCE OF WALES' THEATRE, where she made her first appearance as Mabel in a new play by James ALBERY entitled *Duty*. An able and attractive actress, she was later with ALEXANDER at the ST JAMES'S THEATRE where she created the part of Mrs Erlynne in WILDE's *Lady Windermere's Fan* (1892); with FORBES-ROBERTSON at the LYCEUM, where she played Andrie Lesden in Henry Arthur JONES's *Michael and his Lost Angel* (1896); and at the VAUDEVILLE, where she appeared as Susan Throssel in the first production of BARRIE's *Quality Street* (1902). She continued to play elegant aristocratic parts until increasing arthritis forced her to retire, her last part being the Principessa della Cercola in Somerset MAUGHAM's *Our Betters* (1923).

TERRY, MEGAN [Marguerite Duffy] (1932–), American dramatist, and a founder member of the OPEN THEATER, the New York Theatre Strategy, and the Women's Theatre Council. Her early plays, beginning with *Beach Grass* (1955), were produced in her native city of Seattle, her first New York production being *Ex-Miss Copper Queen On a Set Of Pills* (1963). An off-off-Broadway writer, needing the flexibility and freedom which that implies, she has throughout her career been an outspoken critic of modern American society, her plays constituting a series of terse exposés of what she regards as the decay of human sensibility in contemporary life. They include *Keep Tightly Closed In a Cool Dry Place* and *Calm Down Mother*, performed in a double bill in 1965, and produced in London in 1968 and 1969 respectively; *Viet Rock* (1966), a rock musical about the Vietnam war, first produced by LA MAMA EXPERIMENTAL CLUB and later the same year by Robert Brustein at YALE; *Approaching Simone* (1970), a study of the French philosopher Simone Weil; and *The Tommy Allen Show* (1971) about a television star. *Hothouse* (1974), originally written 20 years earlier, showed the relationships of three generations of women in an Irish-American family. After the Open Theater disbanded in 1973 she became playwright in residence at the Magic Theatre in Omaha, where most of her subsequent plays have been premièred, including *Babes In the Bighouse* (1974) and *Goona Goona* (1979).

TERRY-LEWIS, MABEL GWYNEDD (1872–1957), English actress, daughter of Kate TERRY. She made her first appearance in 1895 with John HARE in GRUNDY's *A Pair of Spectacles*, and in 1898 was seen with her mother, who had been absent from the stage for 22 years, in G. Stuart Ogilvie's *The Master*. She then appeared in revivals of several of T. W. ROBERTSON's plays, played Muriel Eden in PINERO's *The Gay Lord Quex* (1899), and was seen as Gloria Clandon in the first public performance of SHAW's *You Never Can Tell* (1900). Among her later parts were Madeleine Orchard in MARTIN-HARVEY's production of *After All* (1902), based by Wills and Langbridge on BULWER-LYTTON's novel *Eugene Aram*, and Isabel Kirke in H. H. Davies's *Mrs Gorringe's Necklace* (1903). She retired on her marriage in 1904, and did not return to the stage until after the death of her husband, when she was seen in H. M. HARWOOD's *The Grain of Mustard Seed* (1920). She continued her successful career, playing in a number of revivals, and creating some 30 parts, mainly in plays now forgotten. Among the best of her elegant and aristocratic characters were Dona Filomena in the ÁLVAREZ QUINTEROS' *A Hundred Years Old* (1928), Lady Bracknell in WILDE's *The Importance of Being Earnest*, which she played in 1930, and her last part, Lady Damaris in Pryce's *Frolic Wind* (1935). She was extremely popular in New York, where she appeared several times from 1923 onwards.

TERSON, PETER (1932–), English dramatist, whose first play to be produced, *A Night to Make the Angels Weep* (1964), was performed at the VICTORIA THEATRE, Stoke-on-Trent, where Terson was resident dramatist for 18 months. It was followed by *The Mighty Reservoy* (also 1964), *All Honour Mr Todd*, and *I'm in Charge of These Ruins* (both 1966). Other works for Stoke were *Mooney and his Caravans* (1967) and a dramatization of *Clayhanger* staged in honour of Arnold BENNETT's centenary (also in 1967), in which Terson collaborated with Joyce Cheeseman. In the same year he began a long association with the NATIONAL YOUTH THEATRE with *Zigger-Zagger*, his best known play, about a football fan. From then onwards he wrote a play for the N.Y.T. almost every year, all of them relating to young people and their problems. They include *The Apprentices* (1968), *Fuzz!* (1969), *Good Lads at Heart* (1971; staged by the N.Y.T. in New York in 1979), and *Geordie's March* (1973). Among later subjects have been life in a hotel (*The Bread and Butter Trade*, 1976), relationships between parents and children (*Family Ties*, 1977), an unemployed youth who joins the National Front (*England My Own*, 1978), and a young soldier who tries to get out of the army (*The Ticket*, 1980). Terson has continued to write plays for the Victoria Theatre, among them a play for children, *The Adventures of Gervaise Becket, or the Man Who Changed Places* (1969), the semi-documentary *The 1861 Whitby Lifeboat Disaster* (1970), *But Fred, Freud is Dead* and an adaptation of Herman Melville's *Moby Dick* (both 1972), and *The Pied Piper* (1980), another play for children. He has also written plays for other theatres. Terson has been widely praised for his honesty and directness, and has even been compared to CHEKHOV. His plays for the N.Y.T. are thematic and extensively reworked in rehearsal, whereas his plays for the Victoria Theatre and elsewhere are more personal and subject only to the normal rehearsal adjustments. He works at great speed, writing a play in two weeks and sometimes in as little as three days.

TETRALOGY. In the ancient Greek DIONYSIA each dramatic poet was originally supposed to submit for production and competition four plays on the same theme, character, or family, consisting of a trilogy of tragic plays followed by a burlesque of the main theme in a SATYR-DRAMA. This rule was followed by AESCHYLUS, whose *Oresteia* is the only true trilogy to survive, though it is reasonable to suppose that the satyr-drama which followed the *Agamemnon, Libation Bearers*, and *Eumenides* dealt also with the House of Atreus; also that the surviving play *Prometheus Bound* was the first of a trilogy on Prometheus. Later dramatists ignored the unity of subject, and submitted 'pseudo-tetralogies', in which all four plays dealt with different subjects.

THADDÄDL, the last Viennese comic figure in the tradition of HANSWURST, created by the Austrian actor Anton HASENHUT. Based on Taddeo, a German importation into the COMMEDIA DELL' ARTE, Thaddädl, as developed by Hasenhut, was a clumsy youth with a falsetto voice, a perpetually infatuated, idiotically infantile booby. Although much admired by GRILLPARZER, among others, the character never achieved universal popularity, and was soon forgotten. The role at its best is found in Kringsterner's *Der Zwirnhändler* (*The String Merchant*, 1801).

THALIA, see MUSES.

THALIA THEATRE, New York, see BOWERY THEATRE (1).

THEATR CLWYD, Mold, see WALES.

THEATER AN DER WIEN, see VIENNA.

THEATRE, The, London, the first playhouse to be built in London. It stood outside the jurisdiction of the civic authorities, who were hostile to the project, on the site now occupied by Holywell Lane, Shoreditch, and was erected by James BURBAGE with the reluctant financial assistance of his brother-in-law John Brayne, whom he took into partnership. A circular wooden building, with some ironwork, it was roofless, and the cost of admission was one penny for standing room in the area around the stage. A further penny was charged for admission to the three galleries, and yet a third would procure a stool, or what was described as 'a quiet standing', probably in one of the compartments into which one gallery—it is not known which—was divided. The theatre opened in the autumn of 1576, probably with LEICESTER'S MEN (later known as OXFORD'S MEN) in residence. They remained there until the 1580s, though the stage was sometimes used by other troupes, but by 1590 the ADMIRAL'S MEN were established there, and remained until a quarrel with Burbage's son Richard BURBAGE caused most of the actors to move without him to HENSLOWE'S ROSE THEATRE. In 1594 the young Burbage became the leading actor of the newly formed CHAMBERLAIN'S MEN, with which Shakespeare was to be intimately associated, and they remained in the Theatre until 1596, when *Hamlet* was probably acted there for the first time. Later in the year difficulties with the ground landlord, Giles Alleyn, caused the company to move to the CURTAIN. In 1597 the lease of the property, which Richard had inherited jointly with his brother Cuthbert on their father's death in Feb. of that year, ran out and taking advantage of a clause in it, the brothers instructed the carpenter Peter Street to dismantle the building and carry the timber to a site across the Thames and on to Bankside, where it was used to build the first GLOBE THEATRE. The Theatre, which had been temporarily out of use as a result of the trouble over *The Isle of Dogs* by Ben JONSON and others, which closed all London playhouses for a time, never reopened, and in 1598 was already being referred to as unoccupied.

THÉÂTRE ANTOINE, Paris. Built in 1866 on what is now the Boulevard de Strasbourg as the Théâtre des Menus-Plaisirs, this playhouse changed its name several times in the next 30 years, taking its present name in 1896 from the director-manager ANTOINE. Its most brilliant periods were under Antoine from 1896 to 1906, under Firmin GÉMIER from 1906—they both went from it to manage the ODÉON—and under Simone Berriau from 1943 onwards, during which time it was an important focus of EXISTENTIALISM.

THEATRE BEHIND THE GATE, Prague, see KREJČA, OTOMAR.

THEATRE BUILDINGS. The provision of permanent roofed buildings specially erected for performances of plays came comparatively late in theatrical history. Greek open-air theatres evolved from the ritual DITHYRAMB performed round the altar of DIONYSUS, which took place in front of the temple, and later on a site cut out of a neighbouring hillside. This provided a natural auditorium of rising tiers of seats which extended a little more than halfway round a circular ORCHESTRA, or playing-place, backed by a low stage with a stage-wall (*skénê*) behind. This formed one wall of the dressing-rooms and storage rooms behind and was pierced by doors through which the actors came on stage. It also housed the machinery which worked the crane by which the god (the DEUS EX MACHINA) finally appeared from heaven to resolve the complications of the plot. All that is known of Greek theatres in the 5th century BC, the age of the great classical tragedies and comedies, has had to be inferred from the ruins of those that still exist, many of which have been subsequently altered, rebuilt, and finally abandoned. Some, like those at EPIDAUROS, have been refurbished and are used for annual festivals of Greek classical plays. The only things which seem certain about the early theatres are that the audience sat first on wooden benches and then on stone; that the CHORUS occupied the circular orchestra; that a raised stage was provided for the actors, of which, in classical times, there were never more than three; and that the acoustics of these early theatres were perfect, as can be verified by anyone visiting them today. In the great Hellenistic theatres which replaced the simpler ones of early times, the stage was raised, often to a height of several feet, and the stage-wall became more elaborate. Columns supported a stage-roof, and the ramps which led up to the stage were built over a colonnade or *proskênion* (see PROSCENIUM).

The Roman theatre, unlike the Greek, was built on the flat. The early ones were of wood and have disappeared. Later ones, in stone, still exist. The much diminished orchestra is little more than a semi-circle terminated by a most elaborate stage-wall (*scaenae frons*) often three storeys high, in front of which was the stage, usually about five feet above ground level. This was separated from the auditorium by a curtain (*aulaeum*) which descended into a trough. The exterior of the theatre, which rose in a series of colonnades to a great height, was solidly constructed, and, in the case of AMPHITHEATRES used for chariot-races and gladiatorial combats, which had a circular arena, was also completely circular.

The destruction of the Roman Empire saw the collapse of organized theatre. When it was reborn in LITURGICAL DRAMA plays were first given in churches, sometimes with considerable elaboration (see, for example, PARADISO), later in the open air either in front of the church, which provided a stage wall, or on platforms erected in the market-place. In England Biblical plays were often acted on PAGEANTS, while the CORNISH ROUNDS represent a rare form of permanent open-air theatre. The Renaissance, which was in full flower in Italy while other countries still clung to their medieval traditions, brought about a great change in the design of theatres. For the first time plays were produced indoors, often on temporary stages in a hall or palace. This continued to be common practice through the 16th century: SERLIO set up a temporary theatre in a courtyard for the Academicians of Vicenza in the 1530s and another was erected in 1560 by Bartolomeo Neroni (*c.* 1500–71/3) in the Palace of the Senate in Siena. Illustrations in late 15th-century editions of TERENCE's plays show the Renaissance stages on which they were acted. They combined elements of medieval staging with what had been learned of classical staging through the newly discovered works of VITRUVIUS, and provided models which, in various combinations, developed into the theatres we know today (see TERENCE-STAGE).

The main innovation in 16th-century Italy was the proscenium arch, which framed the elaborate stage picture provided for a courtly entertainment. The rise of opera and ballet in Italy led to the evolution of the horseshoe-shaped auditorium characteristic of opera houses all over the world and typified by the 1589 theatre at Sabbionetta built by Vincenzo Scamozzi (1552–1616) and by ALEOTTI's 1619 Teatro Farnese at Parma, while the academic tradition of the classical play under the influence of Vitruvius culminated in PALLADIO's great Teatro OLIMPICO at Vicenza, with its superb *scaenae frons*, which was first used in 1585 after it had been completed by Scamozzi. During the 16th century new theatres were being built all over Europe. At first each country had its own style. The early French theatres, like the Hôtel de BOURGOGNE (1548), were long and narrow, with a space in front of the high stage (originally intended for the ball which followed the performance), rising tiers of seats, and galleries at the side. Many French theatres up to the time of MOLIÈRE were adapted TENNIS COURTS, but the Court theatres followed the Italian pattern, with a centrally-placed dais—later a Royal Box—to accommodate the King and Queen. The VIGARANIS' 1660 SALLE DES MACHINES showed the influence of Palladio. The first theatre in Amsterdam, CAMPEN's Schouwburg of 1638, was directly derived from the Terence-stage; within 30 years it had been rebuilt in the Italian style. In Spain the early theatres followed the pattern of the open-air stages erected in courtyards (*corrales*) and in public squares, with a stage raised on scaffolding and spectators at windows and on balconies all round somewhat similar to the unroofed playhouses of Elizabethan England, typified by the THEATRE, the FORTUNE, the ROSE, and the GLOBE, which had similar roots in the inn-yards used by travelling players (see INNS USED AS THEATRES). These public playhouses persisted alongside private indoor theatres such as the BLACKFRIARS THEATRE until they were all proscribed in 1642.

Italian architects and designers, notably the BIBIENAS and the Burnacinis, dominated the building and decoration of theatres all over the Continent during the 17th century; and when theatrical activity was resumed in London after the Restoration the new playhouses were modelled on the European pattern. The English theatres retained some idiosyncrasies, however: the horseshoe plan and proscenium stage of the Italian opera house were combined with the large FORESTAGE inherited from the Elizabethan playhouse, with PROSCENIUM DOORS opening on to it. In addition, BOXES tended to be sited on the side walls only, with galleries above them and at the back of the auditorium. Restoration theatres incorporating these features included the DORSET GARDEN (1671) and the rebuilt DRURY LANE (1674), the style continuing well into the next century with LINCOLN'S INN FIELDS (1714), remodelled by RICH, and his new COVENT GARDEN (1732). While theatres on the Continent became progressively more elaborately decorative, the English clung to a quieter style, avoiding the excesses of both the rococo and the baroque: three existing theatres of 1766, the Theatre Royal, Bristol (see BRISTOL OLD VIC), ČESKÝ KRUMLOV, and the DROTTNINGHOLM THEATRE, demonstrate the contrast. But theatres in general confined their splendour to the auditorium until the second half of the 18th century: circulation areas were limited, façades were comparatively modest, and they rarely appeared in dominating positions. The provision of grand staircases, foyers, and porticos began with the opera houses of Germany and, later, Italy, to be taken up by legitimate theatres only in the 19th century.

New building techniques, especially the use of cast and wrought iron framing, and the rise of a prosperous middle class, stimulated a boom in theatre building all over the world after 1800. Most were modelled on the opera houses of Europe, epitomizing the pomp and grandeur copied from earlier social institutions; horseshoe-shaped auditoriums with proscenium-arch stages became universal, and while theatres in English-speaking countries still preferred a succession of galleries to the tiers of small boxes favoured on the Continent, the characteristic forestage and proscenium doors of the Georgian theatre gradually disappeared. Foyers and staircases serving the more expensive parts of the auditorium received ever-increasing attention, exteriors became handsome focal points of city streets, and the decorative convention settled into a codified baroque—the gilt and plush now regarded as an essential part of theatrical tradition.

Richard Wagner (1813–83) instituted the first reaction against what he regarded as the florid frivolity of the conventional opera house. His Festspielhaus, which opened in Bayreuth in 1876, was based on a scheme he had planned with the architect Gottfried Semper (1803–79) for a site in Munich. In it the audience was seated in a single steeply-raked wedge, eliminating the hierarchy of pit, boxes, and galleries, the intrusive columns needed to support the overhanging upper floors, and the accretion of decorative detailing these structural elements had always attracted. The perfect sight-lines concentrated every eye on the stage picture, a blow against the convention that the audience itself constituted a part of the theatrical spectacle, and the acoustics were excellent for Wagner's large-scale music dramas. The wedge of seats was placed in a rectangular auditorium: in later applications of Wagner's ideas the walls were splayed out with unfortunate effects on acoustic quality, as in Otto March's Volkstheater at Worms (1888); and the problem was emphasized when Wagnerian theatres such as the Prinzregenten, Munich (1901), and the Schiller, Berlin (1906), both designed by Max Littmann (1862–1931), were used for dramatic rather than operatic performances. With the Künstlertheater, Munich (1908), Littmann discarded the wedge plan while retaining the flat ceiling, the steeply-raked amphitheatre confined between the parallel walls of a rectangle, and the austere manner of decoration; this auditorium was to become a model for art theatres and university theatres all over the world. The Künstlertheater's stage was less satisfactory, little more than a narrow platform behind a proscenium arch with no flying space (see FLIES), a reaction against the enormous, increasingly mechanized stages being built all over Europe. By 1912, however, Littmann had designed the Hoftheater, Stuttgart, typical of German civic theatres of the period, with elaborate stages, immense workshop and storage areas, and spacious foyers, circulation spaces, and restaurants. Something of Wagner's democratic seating was apparent in the deep amphitheatral balconies which continental theatres had begun to absorb from the English-speaking world; as important was the application of the cantilever to building technology: overhanging tiers could now dispense with the supporting pillars which had interrupted the view of the stage, and increasing simplicity leading often to starkness was the result.

Germany led the world in theatre design up to the First World War, theatres in Britain and the United States remaining unaffected by modernist designs and building techniques until the 1920s. Then a boom in cinema architecture and a passion for art deco affected both the modernization of old theatres and the conception of new ones. In London the new puritanism resulted in the DUCHESS (1929), the CAMBRIDGE (1930), the SAVILLE (1931); in the United States, in the Pasadena Playhouse (1925), Joseph URBAN's ZIEGFELD, New York (1927), the Pickwick, Park Ridge (1928), the Wilson, Detroit (1928), and RADIO CITY MUSIC HALL (1932). Theatre-centres like Stuttgart had a great influence in the United States, resulting in, for example, the CLEVELAND PLAY HOUSE (1927), the Iowa University Theatre (1936), and the Hopkins Art Center of Dartmouth College (1962). The Chicago Opera House complex (1929) brought something of the German civic conception into the commercial scene. The Opera House itself, seating 3,471, was a tamed Bayreuth, with a flattened rake and a wedge auditorium curved slightly in the manner of the cinema, and balconies subdivided into compartments. The SHAKESPEARE MEMORIAL THEATRE at Stratford-upon-Avon (1932) and the Théâtre du Palais de Chaillot (1937) in Paris were in the same idiom. Concrete and steel had become the basis of the new aesthetic; boxes and encircling galleries were ousted in favour of balconies which faced squarely on to the stage, as in the super-cinemas, with an inevitable loss of intimacy and atmosphere. With the whole audience seated facing the stage in steeply rising tiers, the treatment of large expanses of ceiling presented problems both structurally and acoustically: in Chicago it was stepped up, while at Stratford it flowed in an unbroken line from a low proscenium opening to a very high back balcony.

The separation of performers and witnesses, initiated at Bayreuth and strengthened by the copying of cinema design, suited the styles of commercial theatre prevalent between the wars—social comedy and drama, spectacular musicals and revues. Innovatory directors looked for a more truly dramatic form, notably COPEAU at the VIEUX-COLOMBIER (1919) and Terence GRAY at the Festival Theatre, Cambridge (converted 1926), who both banished the proscenium arch in search of closer audience involvement. Even more radical ideas were developed by REINHARDT, for whom Hans POELZIG converted a circus into the Grosses Schauspielhaus, Berlin (1919; later the Friedrichstadt-Palast), seating 3,000 spectators in a steeply ascending amphitheatre round three sides of a great horseshoe forestage backed by a wide, low proscenium opening inconspicuously treated. Reinhardt's concept of Theatre for the Masses survived chiefly in the Soviet Union where a number of vast indoor amphitheatres were built, including those at Rostov-on-Don (1936), Moscow (RED ARMY THEATRE, 1940), Minsk (1941), and Tashkent (1948). The next step was inevitably towards the complete arena, as visualized in 1926 by Walter Gropius. His unrealized Totaltheater, intended for PISCATOR, was oval in plan with a steep, 2,000-seat auditorium wrapped around a forestage backed by a proscenium stage. The forestage was to rotate into the centre of the auditorium with subsidiary walkway stages encircling the audience, scenery being projected on to screens mounted between the pillars supporting the domed ceiling, itself a screen for a battery of built-in projectors.

The Grosses Schauspielhaus and the Totaltheater embodied ideas that were a crucial influence on later experiments with THEATRE-IN-THE-ROUND and FLEXIBLE STAGING, as were Norman BEL GEDDES's schemes of the 1920s and 1930s. The Malmö Stadsteater (1944), conceived in the early 1930s with Poelzig on the judging committee, has an adaptable amphitheatral auditorium seating from 400 to 1,700, a deep U-shaped apron stage, and a 66-foot wide proscenium opening. Because financial considerations demand that the greatest possible number of seats be sold, the smaller forms of the auditorium are rarely used; and, although the backstage facilities, circulation spaces, and architectural standard are of the highest order, the flexible stage is rarely ideal for any particular production. A more fertile line of development was taken up by drama departments in American colleges and universities, where small theatres-in-the-round proved both economical and artistically exciting. Notable examples of this trend include the University of Miami's theatre (1950); Teatro S Erasmo, Milan (1953); Théâtre en Rond de Paris (1954); CIRCLE-IN-THE-SQUARE, New York (1960 and 1972); ARENA STAGE, Washington (1961); VICTORIA THEATRE, Stoke-on-Trent (1962); STEPHEN JOSEPH THEATRE, Scarborough (1976); ROYAL EXCHANGE THEATRE, Manchester (1976). A more dominant theme has been the open or thrust stage, with the audience seated around the deep apron area in an arc of up to 180° and with no proscenium wall. Major examples of this kind, some having a degree of flexibility, are: STRATFORD (ONTARIO) FESTIVAL theatre (1953, rebuilt 1957); CHICHESTER FESTIVAL THEATRE (1962); GUTHRIE THEATRE, Minneapolis (1963); NINA VANCE ALLEY THEATRE, Houston (1968); LEEDS Playhouse (1970); CRUCIBLE, Sheffield (1971). In contrast to the amphitheatral seating found in post-1950 theatres-in-the-round and thrust stage theatres, a number of post-war German theatres returned to galleried auditoriums though in a modern context; the best of them —the Thaliatheater, Hamburg (1950), the Landstheater, Hanover (1950), the Stadttheater, Kiel (1953)—were designed by Werner Kallmorgen. But the boom in theatre building in Germany during the 1950s and 1960s encompassed all the contemporary idioms, the characteristic common to all being the provision of liberal backstage, workshop, and front-of-house amenities.

A development of the 1970s has been the small 'workshop' theatre, either added to an existing theatre (though not necessarily in the same building) or incorporated in a new theatre complex. This is in part an acknowledgement of the impossibility of designing a single stage and auditorium that will suit every kind of play and style of production. In general the 'workshop' will be very small, with a flexible stage that incorporates a theatre-in-the-round option, and with a comparatively simple, sometimes spartan, standard of comfort. Britain's NATIONAL THEATRE (1976) incorporates such a workshop in the COTTESLOE THEATRE: with the proscenium-stage, balconied LYTTELTON THEATRE and the thrust-stage, amphitheatral OLIVIER THEATRE, the National could be seen as presenting a cross-section of 20th-century theatre styles as they have evolved up to the 1980s.

THÉÂTRE COMIQUE, New York, see HARRIGAN, NED.

THEATRE CORPORATE, Auckland, see NEW ZEALAND.

THÉÂTRE D'ART, Paris, see FORT, PAUL, and LUGNÉ-POË.

THÉÂTRE DE L'ATELIER, Paris, see ATELIER.

THÉÂTRE DE L'HÔTEL D'ARGENT, see ARGENT.

THÉÂTRE DE L'HÔTEL DE BOURGOGNE, see BOURGOGNE.

THÉÂTRE DE L'INEXPRIMÉ, see SILENCE, THEATRE OF.

THÉÂTRE DE L'ŒUVRE, Paris, see LUGNÉ-POË.

THEATRE DE LYS, New York, at 121 Christopher Street in Greenwich Village. This off-Broadway playhouse, run by Lucille Lortel and formerly a cinema, opened on 27 Oct. 1952 and housed from 1954 to 1961 Marc Blitzstein's English adaptation of BRECHT's *Die Dreigroschenoper* as *The Threepenny Opera*. It was followed by George Tabori's compilation of Brechtian items known as *Brecht on Brecht* (1962), which also had a long run; John ARDEN's *Serjeant Musgrave's Dance* (1966); and *The Deer Park* (1967), adapted by Norman Mailer from his own novel. Notable productions of the 1970s included the revue *Berlin to Broadway with Kurt Weill* (1972); Michael Weller's *Moonchildren* (1973); David Mamet's *A Life in the Theatre* (1977); and Sam SHEPARD's *Buried Child* (1978). One of the outstanding features of this theatre was its Matinée Series, run for the AMERICAN NATIONAL THEATRE AND ACADEMY by Lucille Lortel, during which established actors were able to experiment with new forms and techniques; the series has been discontinued since 1975.

THÉÂTRE DES NATIONS, international festival of drama under the auspices of UNESCO, which began in 1954 when a number of foreign companies were invited to perform in Paris at the Théâtre Sarah-Bernhardt, now known as the Théâtre de la Ville. The festival was officially established in 1955, and from 1957 to 1965 a two-month season of foreign plays was organized annually by A. M. Julien, manager of the Théâtre Sarah-Bernhardt, with State and municipal support. During this period some 50 countries were represented by about 150 companies with 180 plays and 150 other entertainments, and the festival played an important part in bringing groups such as the BERLINER ENSEMBLE, the ROYAL SHAKESPEARE COMPANY, the PICCOLO TEATRO DELLA CITTÀ DI MILANO, the American LIVING THEATRE, and the Peking Opera of CHINA to the attention of Parisians, and indeed to European audiences in general, since the prestige and publicity of their Paris season brought the companies invitations to tour to other festivals. In 1966 BARRAULT took over the organization of the festival and transferred it to the ODÉON, which he was then managing. He was in the process of giving it a firmer international basis when the 1968 riots caused his own dismissal from State employment. After that the Théâtre des Nations led a sporadic and vestigial existence, until in 1973 the INTERNATIONAL THEATRE INSTITUTE, which had been instrumental in founding it, agreed to its becoming a peripatetic event, and Warsaw was chosen for the first reunion in 1975.

THÉÂTRE DU VIEUX-COLOMBIER, see VIEUX-COLOMBIER, THÉÂTRE DU.

THÉÂTRE-FRANÇAIS, Paris, see COMÉDIE-FRANÇAISE.

THÉÂTRE FRANÇAIS, New York, see FOURTEENTH STREET THEATRE.

THEATREGOROUND, see ROYAL SHAKESPEARE COMPANY.

THEATRE GUILD, New York, theatre production company founded in 1919 as a society presenting non-commercial plays for a subscription audience. Its main precursor was the WASHINGTON

Medieval mummers wearing animal masks.

A character from Terence's adaptation of the *Eunuchus* of Plautus, *c*.1500.

Dress of a contemporary Roman prostitute, from Veccello's costume book, 1598 edition.

Dress for a masquer at the marriage of Lord Hay, 1606/7.

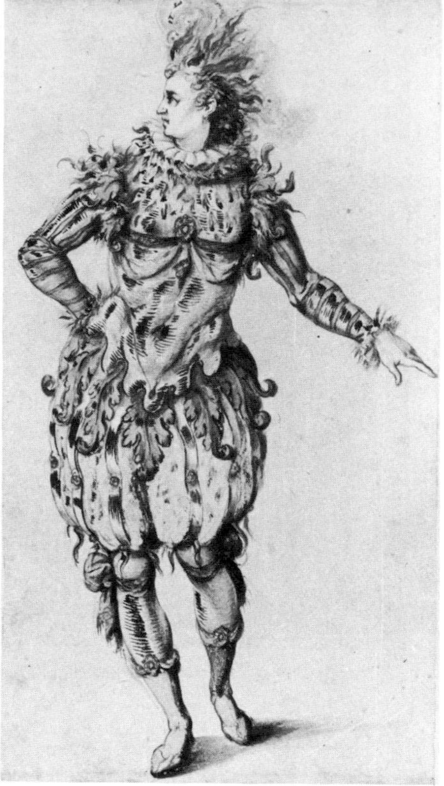

Design by Inigo Jones for a masquer's costume, 1613.

Design by the elder Bérain, probably for Hermione in Lully's opera *Cadmus et Hermione*, 1673.

Design by Burnacini for a masquer's costume, late 17th century.

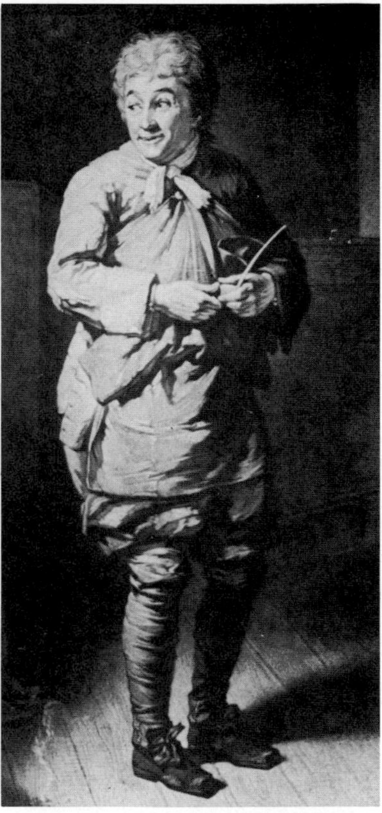

David Garrick as Hamlet in contemporary dress and as Abel Drugger in Ben Jonson's *The Alchemist,* (*c.*1750), in a costume expressing character.

Elizabeth Farren as Hermione in *The Winter's Tale*, Drury Lane, 1788.

Talma as Nero in Racine's *Britannicus*, painted by Delacroix. He was the first French actor to play Roman parts in a toga instead of contemporary dress.

Edmund Kean as Othello, *c.*1800. The essentials of the costume, representing the exotic hero-figure, are accurate, although the decoration has been elaborated to accommodate the trimmings for this tinsel picture.

Charles Kemble as Charles Surface in Sheridan's *The School for Scandal*, 1821. He wears the dress of an elegant contemporary gentleman.

Planché's design for Constance in Kemble's production of *King John*, 1823, showing the painstaking 'archeological' detail espoused at Covent Garden under Kemble's management.

Henry Irving as Mathias in *The Bells*, 1883.

ırah Bernhardt in the title role of Catulle Mendès' *Medée* at the Théâtre de la Renaissance in the 1890s.

Beerbohm Tree as Falstaff in *Henry IV, Part 1*, 1896.

Johnston Forbes-Robertson and Mrs Patrick Campbell in the tomb scene from *Romeo and Juliet*, Lyceum Theatre, 1895.

A severely historical presentation of *Julius Caesar* at the Moscow Art Theatre, 1903 (*left*), contrasts with a Constructivist costume design for *Romeo and Juliet* by Alexandra Exter, 1921 (*right*).

Granville-Barker's Shakespeare productions at the Savoy Theatre featured work by innovative young designers: (*left*) Leontes in *The Winter's Tale*, 1912, by Charles Ricketts; (*right*) Snout (as Wall) in *A Midsummer Night's Dream*, 1914, by Norman Wilkinson of Four Oaks.

Costume design by Claud Lovat Fraser (*above left*) for *The Captain* by Beaumont and Fletcher, 1912.

Costume design by Michael Weight (*above right*) for the famous 'black-and-white' production of Wilde's *The Importance of Being Earnest* at the Lyric Theatre, Hammersmith, 1930.

Shaw's *Back to Methuselah* (*left*), the Birmingham Repertory Company's production at the Royal Court Theatre, 1924. Paul Shelving's designs echoed the influence of Diaghilev's Russian Ballet, especially in the use of naïve forms and pure colour.

Tyrone Guthrie's production of *Measure for Measure* at the Old Vic, 1933, with costume designs by John Armstrong.

A Midsummer Night's Dream at the Open Air Theatre, Regent's Park, 1934.

Two costumes designed by Roger Furse for the Old Vic: King Lear (John Gielgud), 1940, and Falstaff (Ralph Richardson), 1945.

Richard Burton as Henry V, Shakespeare Memorial Theatre, 1951; romantic costumes by Tanya Moiseiwitsch were set off by her stark permanent staging.

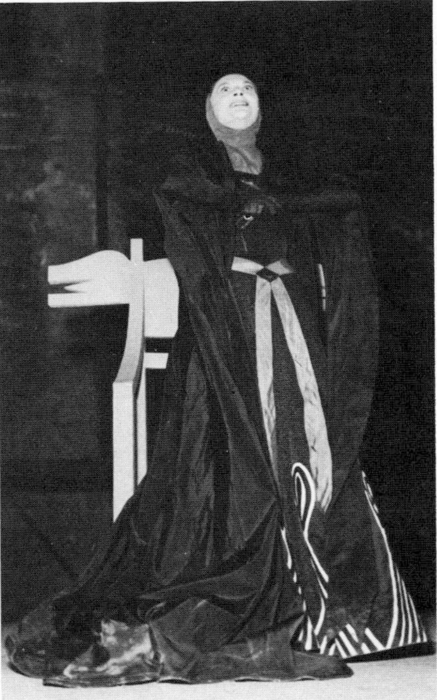

Lady Macbeth, played by Maria Casarès at the Avignon Festival, 1954.

Titus Andronicus, Shakespeare Memorial Theatre, 1955, directed and designed by Peter Brook.

Cymbeline, Shakespeare Memorial Theatre, 1957, directed by Peter Hall, costumes by Lila de Nobili.

A Midsummer Night's Dream, directed by John Hirsch for the Stratford (Ontario) Festival, 1968; Martha Henry as Titania, Douglas Rain as Bottom.

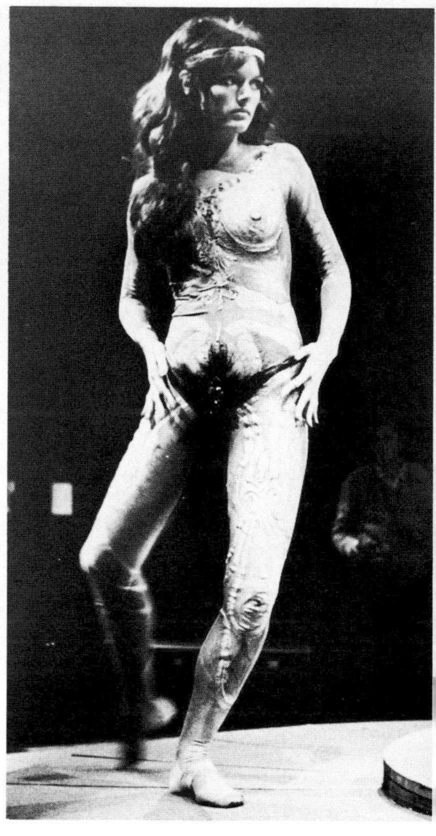

Jean-Louis Barrault's *Rabelais*, Elysée-Montmartre Theatre, 1968, costume by Matias.

Henry V, Royal Shakespeare Company, 1975, costumes by Farrah.

Equus by Peter Shaffer, staged by the Pantomima Theatre of Wroctaw, 1978.

A children's fairy-tale play *Kokori*, Theater der Freundschaft, Berlin, 1979.

Goldoni's *The Good-Humoured Ladies*, Citizens' Theatre, Glasgow, 1979, designed by Sue Blane after Bakst.

String puppets of Rama and Sita (*above*), from a performance of the *Rāmāyaṇa*. Indian puppet theatre retains many elements of religious ritual.

All-male Kathakali performance of the *Mahābhārata* (*right*). Kathakali is also performed by puppets, who wear the same traditional costumes as the human dancers.

Yakshagan, Karnataka. The stylized, brilliantly coloured make-up and fantastic costumes are established by tradition

The three witches in a Hindi translation of *Macbeth*, 1979, produced by the National School of Drama company, New Delhi.

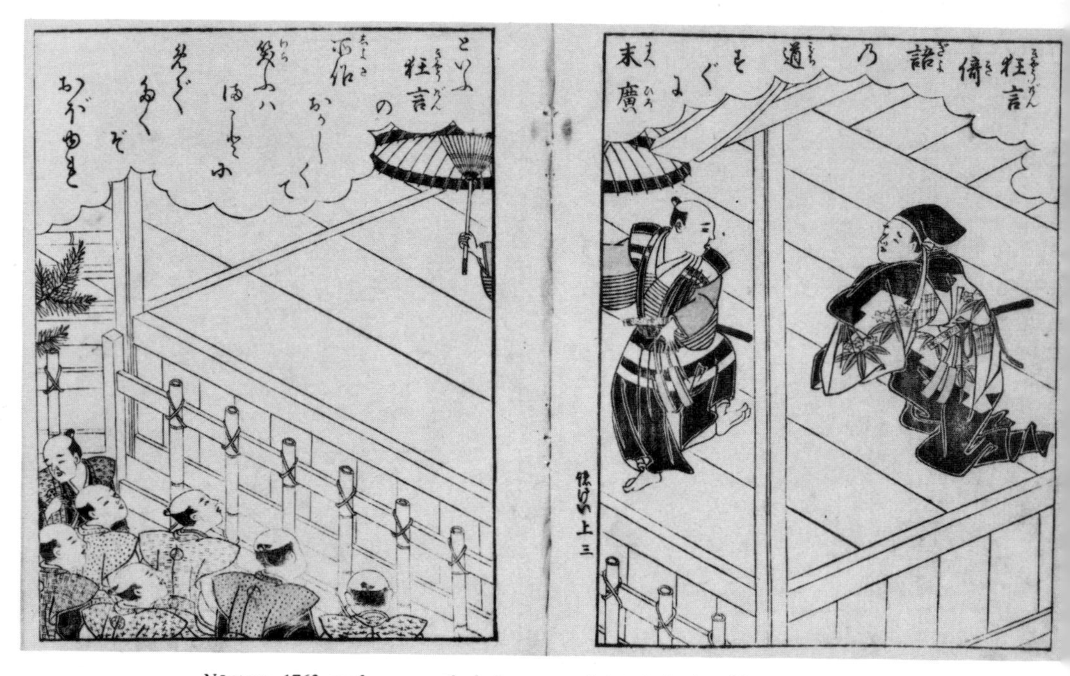

Nō stage, 1763, performance of a *kyōgen*, a comic interlude played between *nō* dramas.

Kabuki stage, 1798. Many of the traditions of Kabuki have persisted unchanged to the present day.

Kantan. The leading actor in his stylized canopied bed performs one of the short dances that punctuate the action.

Ichikawa Ennosuke III in 'Renjishi', a Kabuki dance first performed in 1871, in which he portrays a 'shishi' lion teaching his cub.

Traditional open-air theatre, from a scroll-painting of the Ching dynasty.

Raid on the White Tiger Regiment, 1967, one of the five dramas acceptable during the Cultural Revolution.

SQUARE PLAYERS. Particularly in its early years, the Guild had a great influence on the American theatre with its productions of new or little-known works by American and foreign authors, and many distinguished players appeared under its aegis. It occupied the GARRICK THEATRE, where the first production, on 19 Apr. 1919, was BENAVENTE's *The Bonds of Interest*, followed on 13 May by St John ERVINE's *John Ferguson*, a striking success. On 10 Nov. 1920 SHAW's *Heartbreak House* opened; the Theatre Guild was to become a major force in bringing Shaw's work before New York audiences. Other Guild productions at the Garrick included A. A. Milne's *Mr Pim Passes By*, MOLNÁR's *Liliom* (both 1921), ANDREYEV's *He Who Gets Slapped*, ČAPEK's *R.U.R.* (both 1922), IBSEN's *Peer Gynt*, RICE's *The Adding Machine* (both 1923), and Molnár's *The Guardsman* (1924) starring the LUNTS. The society opened its own Guild Theatre on 13 Apr. 1925 with a production of *Caesar and Cleopatra*, with Helen HAYES. Other plays by Shaw presented here include *Arms and the Man* (1925), *Pygmalion* (1926), *The Doctor's Dilemma* (1927), *Major Barbara* (1928), *The Apple Cart* (1930), *Getting Married* (1931), and *Too True to be Good* (1932). Other authors whose work was presented by the Guild during these years include WERFEL, whose *Goat Song* in 1926 aroused controversy; Jacques COPEAU, as adapter of DOSTOEVSKY's *The Brothers Karamazov* in 1927; O'NEILL, MARLOWE, JONSON, and TURGENEV—NAZIMOVA appeared in *A Month in the Country* in 1929. Many productions were mounted at other New York theatres and several toured outside the city.

The non-commercial aims of the Theatre Guild were vulnerable during the financial recession of the thirties and early forties, but its fortunes were secured by the production of Rodgers and Hammerstein's musical *Oklahoma!*, presented at the ST JAMES THEATRE on 31 Mar. 1943. (This was based on a straight play, *Green Grow the Lilacs* by Lynn Riggs, which had been produced by the Guild in 1931.) The Guild Theatre was taken over by the AMERICAN NATIONAL THEATRE AND ACADEMY in 1950 (see ANTA THEATRE); but the Guild itself continued to mount a notable list of new plays, revivals, and musicals within the commercial system, though its producing activities have fallen off considerably in recent years.

THÉÂTRE HISTORIQUE, Paris, see DUMAS PÈRE.

THEATRE-IN-THE-ROUND, a form of play presentation in which the audience is seated all round the acting area. One of the earliest forms of theatre, it was probably used for open-air performances, street theatres, and such rustic sports as the Mayday games and the Christmas MUMMERS' PLAY, and was revived in the 20th century by those who rebelled against the so-called 'tyranny' of the PROSCENIUM arch.

Modern theatre-in-the-round first came into prominence in the Soviet Union, where in the 1930s OKHLOPKOV in his REALISTIC THEATRE produced a number of plays on stages with the audience on all sides. At the same time, in England, Robert ATKINS was producing Shakespeare in the RING in Blackfriars, an interesting experiment which seemed to have no immediate impact. It was in America, where the idea had already been mooted by Robert Edmond JONES in 1920, that the first theatre-in-the-round was erected, by Glenn Hughes at the University of Washington in Seattle in 1940. His Penthouse Theatre had an elliptical acting area and auditorium contained within a circular foyer, the space between them on two sides being used for prop rooms and a lighting control booth. In the rapidly expanding world of American university drama, theatre-in-the-round flourished. Outside the universities, the most important exponent of the new method was Margo JONES, working in Dallas. In recent years such theatres as CIRCLE-IN-THE-SQUARE have made this type of production familiar to off-Broadway playgoers, and in 1961 Harry Weese designed the ARENA STAGE in Washington, D.C., which opened with BRECHT's *The Caucasian Chalk Circle*.

In England Stephen JOSEPH in Stoke-on-Trent and Scarborough proved how successful the form can be, even with plays written for a proscenium stage. Arena-type productions were also featured at the interesting but short-lived Pembroke Theatre in Croydon. The VICTORIA THEATRE, Stoke-on-Trent, the STEPHEN JOSEPH THEATRE, Scarborough, and the ROYAL EXCHANGE THEATRE, Manchester are permanent professional theatres-in-the-round. A number of new theatres, both professional and amateur, have been built with adaptable stages; theatre-in-the-round can be used in such buildings for those plays that lend themselves to the method, as well as for plays specially written for it. Inherent technical problems include special demands on lighting and set designers, as well as the need for frequent movement by the actors in order to give all sections of the audience a view of each character's face.

A fertile development of theatre-in-the-round is the 'promenade' production, of which examples have been seen at the ROUND HOUSE and at the COTTESLOE THEATRE, in which the players move around the auditorium for the different episodes in the play, with the standing audience clustering about them.

THÉÂTRE-ITALIEN, Paris, see COMÉDIE-ITALIENNE.

THÉÂTRE LIBRE, Paris, theatre club, which occupied successively three different buildings, founded in 1887 by André ANTOINE for the production of plays by new naturalistic French and foreign playwrights. It opened with a programme of four one-act plays—*Jacques Damour* by Léon Henrique, based on a story by ZOLA; *Mademoiselle Pomme* by Duranty and Paul Alexis; *Le Sous-*

Préfet by Arthur Byl; and *La Cocarde* by Jules Vidal—and before it finally closed in April 1896, mainly because of financial difficulties, it had staged a total of 184 plays, among them those of such outstanding dramatists as BECQUE, BRIEUX, HAUPTMANN, IBSEN, and STRINDBERG. Its innovations in playwriting, direction, and acting had a great influence on the contemporary French theatre, and its settings, which were scrupulously exact reproductions of real life, helped to liberate the stage from the artificial prettiness of an earlier epoch.

THEATRE MASQUE, New York, see JOHN GOLDEN THEATRE (2).

THEATRE MUSEUM, London. In 1957, on the initiative of the SOCIETY FOR THEATRE RESEARCH, a British Theatre Museum Association was formed with Laurence IRVING as chairman. The Theatre Museum officially came into being on 16 Sept. 1974, when the collections assembled under the Association's auspices were transferred to the Victoria and Albert Museum, by which the Theatre Museum is administered. The Association was dissolved in 1977. The museum will, it is hoped, open in converted premises in the Flower Market, Covent Garden, in 1984. Its most important collection is that of Gabrielle ENTHOVEN, established at the Victoria and Albert Museum in 1925; another important collection is that of Harry R. Beard, which includes programmes, playbills, and manuscripts relating to the history of the British stage, particularly in its musical aspects. Specialized deposits include the Guy Little collection of theatrical photographs, mainly *cartes-de-visite*; the M. W. Stone and the Hinkins collections of TOY THEATRE sheets and plays texts; the Nancy PRICE collection of prompt books; and the Melville collection of Edwardian MELODRAMAS and PANTOMIMES. The museum will also house the whole drama archive and collection of designs of the ARTS COUNCIL, and there will be material on ballet, opera, music-hall, puppetry, the circus, and 'pop' music.

THEATRE MUSIC, THEATRE ORCHESTRA, see INCIDENTAL MUSIC.

THÉÂTRE NATIONAL POPULAIRE, Paris. The first T.N.P. was founded by GÉMIER, who from 1920 to 1933 was successful in obtaining government grants to defray the occasional cost of inviting a variety of companies to play to worker-audiences at the Palais de CHAILLOT in the Trocadéro. He also organized provincial tours, pioneering the use of motorized transport for stage equipment. His work came to an end when in 1934 the Trocadéro was reconstructed to house the International Exhibition of 1937.

The second T.N.P. was directed by VILAR from 1951 to 1963 and by Georges WILSON from 1963 to 1972, and was again housed in the Palais de Chaillot, which since 1945 had been used as an assembly hall by the United Nations. The enterprise was able to support its own company, which developed from the activities of Vilar at the AVIGNON FESTIVAL. For the first five years productions were staged at Chaillot and then sent on tour, especially in the working-class suburbs of Paris; later, efforts were directed rather to attracting spectators from those areas through organized trips and audience associations. In 1963 the theatre had 350 associations and 33,000 subscribers on its books. An entirely new approach to public relations (including such innovations as postal reservations; a theatre journal, *Bref*; guided tours backstage and meetings with actors and technical staff; abolition of tipping in the auditorium; and the closing of doors at the advertised time) gave the T.N.P. enormous popularity in the 1950s, though even then there were criticisms of the low proportion of manual workers (between two and three per cent) in its audiences, and the preponderance of well known plays in its repertory. Certainly Vilar based his work mainly on the classics, both French and foreign, but he also produced a number of contemporary plays—BRECHT's *Mutter Courage und ihre Kinder* in 1953, JARRY's *Ubu-roi* in 1958, and Brecht again with *Der aufhaltsame Aufstieg des Arturo Ui* in 1960. He had to contend with many theatrical problems posed by the enormous auditorium (seating 1,800 people, with a stage 34 metres wide) and overcame them by adopting reforms still at that time considered experimental: sparse use of scenery, abolition of the footlights and proscenium arch, complex lighting effects, and the use of large acting areas defined by lifts and revolving platforms. Playing in many of the productions himself, and ably seconded by Gérard PHILIPE in such classic roles as CORNEILLE's Le Cid and Maria CASARÈS in the tragic heroines of the Romantics, Vilar established something of a 'T.N.P. style', which, unlike that of the COMÉDIE-FRANÇAISE, was based on economy of gesture, simplicity of speech, and an intimacy engendered by careful stage design and spotlighting. Under Wilson the repertory became more modern and international, but the loss of Vilar and Philipe, together with the development of other popular theatres in the Parisian suburbs and provinces under the policy of DÉCENTRALISATION DRAMATIQUE, led to a decline in popularity for the T.N.P. Whereas the average number of seats sold had remained at over 90 per cent until 1966, it dropped in 1970–1 to 37 per cent and in 1972 the theatre was closed by the government, its subsidy and title passing to the group controlled by PLANCHON in Lyon, which later moved to Villeurbanne.

THEATRE OF COMEDY, Leningrad, see AKIMOV; Moscow, see RUSSIA AND SOVIET UNION.

THEATRE OF CRUELTY, see CRUELTY, THEATRE OF.

THEATRE OF DRAMA AND COMEDY, Moscow, see TAGANKA THEATRE.

THEATRE OF FACT, see FACT, THEATRE OF.

THEATRE OF SATIRE, Moscow, see SATIRE, THEATRE OF.

THEATRE OF SILENCE, see SILENCE, THEATRE OF.

THEATRE OF THE ABSURD, see ABSURD, THEATRE OF THE.

THEATRE OF THE BALTIC FLEET, Leningrad, founded in 1934 under an able director, A.V. Pergament, from a group of actors conscripted into the navy. It gave performances on board ship and in Baltic ports, continuing to work during the whole of the Second World War in spite of suffering many casualties. It was in Leningrad during the siege which began in 1942. In 1943 VISHNEVSKY wrote for it *At the Walls of Leningrad* and, in collaboration with Kron and Azarov, a musical play, *Wide Spreads the Sea*, also dealing with the siege of Leningrad. It was given a permanent base at Fleet headquarters, most of its performances being intended for naval audiences, and plays on naval and military subjects predominate in its repertory.

THEATRE OF THE REVOLUTION, Moscow, see MAYAKOVSKY THEATRE.

THEATRE ROYAL, term applied to the two London theatres, COVENT GARDEN and DRURY LANE, which operate under Letters Patent granted by Charles II in 1662. These PATENT THEATRES originally had a monopoly of serious acting which lasted until 1843. The HAYMARKET THEATRE was given the courtesy title of Theatre Royal by virtue of a Patent for the summer months only obtained in 1766 by Samuel FOOTE from George III. During the 19th century a proliferation of Theatres Royal all over the country—none of them having any right to the title, since these theatres were licensed by the local magistrates and not directly by the Crown—caused the term to become nothing more than a generic title for a playhouse.

For individual theatres bearing the name see also ADELPHI THEATRE, LONDON; ASTLEY'S AMPHITHEATRE; BATH; BRIGHTON; BRISTOL OLD VIC; BURY ST EDMUNDS; EXETER; JOHN STREET THEATRE, NEW YORK; LINCOLN; MANCHESTER; NEWCASTLE-UPON-TYNE; NORWICH; NOTTINGHAM; WINDSOR; and YORK.

THEATRE ROYAL, Stratford East, London, in Gerry Raffles Square, named after the theatre's most famous manager, seats 470, and opened on 17 Dec. 1884 with BULWER-LYTTON'S *Richelieu*. It was used for a time for revivals of other old plays, the 13-year-old Ada Reeve appearing there with her father in Pettitt's *The Black Flag* in 1887. For many years it was run by the Fredericks family, who also controlled the Borough Theatre in Stratford High Street, used mainly by touring companies. The Royal had a somewhat undistinguished career, relying on melodrama and an annual pantomime. In 1902 it was badly damaged by fire; the stage was then enlarged and the auditorium redecorated. Between 1927 and 1933 the theatre was given over to twice-nightly variety, and from 1947 to 1949 it returned to straight plays under David Horne, who inaugurated his management on 24 Mar. with his own play *Damaris*, in which he played opposite Eileen Thorndike. Several other new plays, some of them by Horne, were tried out, but the venture was not a success, and on 16 Jan. 1950 Tod Slaughter's *Spring-Heeled Jack, the Terror of Epping Forest* heralded the return to more popular fare. The Christmas play that year was Stevenson's *Treasure Island*. In 1953 Joan LITTLEWOOD took over the theatre and made it the home of her THEATRE WORKSHOP until 1963. A new management opened on 18 Sept. 1964 with a translation of Max FRISCH'S *Graf Öderland* as *Edge of Reason*; but Theatre Workshop returned in 1967, and did not finally leave until the end of 1973. After the Workshop era the theatre went through a very difficult period: attendances were low and there was a danger that its subsidy from the ARTS COUNCIL would be reduced. In 1980, however, under a new manager and with a programme tailored to the needs of the area, there were signs of recovery. Productions included Howard BRENTON and Tony Howard's *A Short, Sharp Shock*, a satire on Mrs Thatcher's Government, and Henry LIVINGS'S *This Jockey Drives Late Nights*, and in 1981 the theatre had its biggest success since Joan Littlewood's departure with Nell Dunn's *Steaming*, which transferred to the West End for a long run. Although the theatre mainly presents new plays it stages the occasional classic, its production of BOUCICAULT'S *The Streets of London* in 1980 also moving to the West End. The theatre provides an entertainment nearly every Sunday, and is strongly involved in educational and community activities. The studio theatre, the Square Thing, opened in 1979 with *London Calling* by Tony Marchant (later seen on the main stage as *Thick as Thieves*). Situated in the theatre's car park and seating about 100, it presents mainly new work and caters especially for the neighbourhood's ethnic minorities.

THEATRES ACTS, see LICENSING OF THEATRES and LORD CHAMBERLAIN.

THÉÂTRES NATIONAUX, see DÉCENTRALISATION DRAMATIQUE.

THEATRE UNION, see LITTLEWOOD, JOAN.

THEATRE UPSTAIRS, see ROYAL COURT THEATRE.

THEATRE WALES, see WALES.

THEATRE WORKSHOP, company founded in Kendal in 1945 by a group of actors who were dissatisfied with the commercial theatre 'on artistic, social, and political grounds'. Joan LITTLEWOOD, who had worked with some of the members of the group earlier in the North of England, became artistic director and Gerald Raffles general manager. After seven years spent touring in England and Europe, the company took over the THEATRE ROYAL, Stratford, in London (Chaucer's Stratford-atte-Bowe), and after repairing and redecorating the building themselves, opened in Feb. 1953 with *Twelfth Night*. Their first new play, on 14 Nov. of the same year, was *The Travellers* by Joan Littlewood's husband Ewan McColl, a founder-member of the group. The company quickly made a name for itself, and was invited to represent Great Britain at the Paris THÉÂTRE DES NATIONS in 1955 with LILLO's version of *Arden of Feversham* and JONSON's *Volpone* and in 1956 with an adaptation by McColl of Hašek's *The Good Soldier Schweik*. They also visited Zürich, Belgrade, and Moscow, where they appeared in the MOSCOW ART THEATRE, filling it to capacity. Left-wing, and indeed almost Communist in their ideology, they sought always to revivify the English theatre by a fresh approach to established plays—two of their most interesting and controversial productions being Lope de VEGA CARPIO's *Fuenteovejuna* as *The Sheep Well* and Shakespeare's *Richard II* (both 1955)—or by commissioning working-class plays, many of which were subsequently transferred to West-End theatres, among them Brendan BEHAN's *The Quare Fellow* (1956) and *The Hostage* (1958), which was also seen on Broadway in 1960; Shelagh Delaney's *A Taste of Honey* (1958); Frank Norman and Lionel BART's *Fings Ain't Wot They Used T'Be* (1959); and Stephen Lewis's *Sparrers Can't Sing* (1960). Finding her resources drained by these transfers, and feeling also that commercial success was inimical to her ideals, Joan Littlewood left to work elsewhere, and the company carried on under Gerald Raffles, until in 1963 she returned to produce a 'musical entertainment' *Oh, What a Lovely War!*, by Charles Chilton and the members of the company. This also proved successful enough to be transferred to the West End, after which the company dispersed. They were back in residence, however, in 1967, when once more under Joan Littlewood's direction they presented a succession of plays—Barbara Garson's *MacBird!* (a reworking of Shakespeare's *Macbeth* first seen at the Village Gate in Bleecker Street, New York); *Mrs Wilson's Diary*; *The Marie Lloyd Story*; and *Intrigues and Amours* (based on VANBRUGH's *The Provoked Wife*). After directing Peter Rankin's *So You Want to be in Pictures?* (1973) Joan Littlewood went abroad; the group finally dispersed, and the Stratford theatre came under new management.

THEATRE WORKSHOP, New York, see PRINCESS THEATRE.

THEATR GWYNEDD, Bangor, see WALES.

THEATRICAL COMMONWEALTH, a group of actors under Thomas TWAITS who seceded from the PARK THEATRE, New York, in 1813, after disputes with the management, and set up for themselves in a converted circus on Broadway. Although they were none of them outstanding, they reputedly gave excellent productions of *As You Like It* and SHERIDAN's *The School for Scandal* and *The Rivals*. Emboldened by their success, the rebels even had the temerity to put on Frederick REYNOLDS's *The Virgin of the Sun* on the same night as the Park, and survived. They disbanded after the death of Mrs Twaits, a member of the company, at the end of 1813, several of them returning to the Park.

THEATRICAL SYNDICATE, association of American businessmen in the theatre, formed in 1896, which included the firm of Klaw and Erlanger, Al Hayman, Sam Nixon [Samuel F. Nirdlinger], J. Fred Zimmerman, and Charles FROHMAN. For about 16 years they controlled most of the theatres of New York and many of those in other big towns, and gradually they exerted a stranglehold over the whole entertainment life of the United States. The Syndicate's original intentions—to rationalize theatre organization and prevent exploitation—developed into a determined commercialism which suppressed competition and depressed aesthetic standards. It was powerful enough to harm those who opposed its monopoly, forcing Mrs FISKE to play in second-rate theatres on tour and Sarah BERNHARDT to appear in a tent. Both these players, with BELASCO and others, helped in the end to break it.

THEATR Y WERIN, Aberystwyth, see WALES.

THEOCRITUS (*fl.* 310–250 BC), Greek poet of the HELLENISTIC AGE, born in Syracuse. He is best known for his Idylls, among which several pastoral pieces in dialogue celebrating the old country life of Sicily are structurally similar to the MIME plays of HERODAS. One of the best known, the *Adoniazusae* or *Women at the Festival of Adonis*, said to have been inspired by a work of SOPHRON, is in fact pure mime. It was translated into English by Matthew ARNOLD.

THEODORUS, see GREECE, ANCIENT.

THEOGNIS, Greek tragic poet who, according to ARISTOPHANES, was so frigid that when his play was produced in Athens the rivers froze in Thrace. He became one of the thirty tyrants of Athens in 404 BC.

THÉOPHILE DE VIAU (1590–1626), French poet and dramatist, who like his contemporary RACAN marks the incursion into the French theatre of the young nobleman, but unlike him had a short and stormy life. As a Huguenot and a free-thinker, he was suspected of being part-author of *Le Parnasse satirique* (1622), a collection of licentious and blasphemous verse, and was first exiled and then condemned to be burned at the stake; he escaped, was again imprisoned, and died in exile a year after being released. His only play, *Pyrame et Thisbé*, a tragedy based on a episode in Ovid's *Metamorphoses* (the source of the mechanicals' play in *A Midsummer Night's Dream*), was published in 1623 and perhaps acted before that. It was certainly seen at Court in 1625, and was in the repertory of the Hôtel de BOURGOGNE in 1633–4, being one of the oldest plays mentioned in the Mémoire de MAHELOT. It is now recognized as a masterpiece of baroque tragedy, though its elaborate conceits caused it to be looked on unfavourably by BOILEAU and other neo-classical critics. Regular in construction, according to the interpretation of the UNITIES current at the time, it is an excellent illustration of the final period of the MULTIPLE STAGE, before this gave way to the single set characteristic of French classical drama.

THEORIC FUND, a grant of two obols distributed to the poorer citizens of Athens to enable them to pay for admission to the theatre during the Festivals of DIONYSUS. It was introduced in the time of Pericles, suppressed during the Peloponnesian War, revived in 394 BC, when it was raised to one drachma a head, and finally abolished after the defeat of the Athenians by Philip of Macedonia at the battle of Chaeronea in 338 BC.

THESIGER, ERNEST (1879–1961), English actor, who first studied art, but in 1909 made his stage début under George ALEXANDER and three years later was with TREE. In the First World War he joined the army, but was invalided out and returned to the stage in Walter W. Ellis's farce *A Little Bit of Fluff* (1915), which had a long run. An actor of style, wit, and sensibility, he was also extremely versatile, creating the parts of Cameron in BARRIE's *Mary Rose* (1920) and Mr Bly in GALSWORTHY's *Windows* (1922), and playing the Dauphin in the first London production of SHAW's *Saint Joan* (1924) with a brilliantly comic blend of pathos and arrogance. He was also seen in COWARD's revue *On With the Dance* (1925) and *Cochran's Revue* (1926), in which he partnered Douglas Byng in an amusing sketch about two old ladies preparing for bed. In more serious vein he appeared for the PHOENIX SOCIETY in revivals of several plays by MARLOWE, being seen as Ithamore in *The Jew of Malta* in 1922, Piers Gaveston in *Edward II* in 1923, and Mephistophilis in *Dr Faustus* in 1925. One of his more unusual creations was Master Crummles in *When Crummles Played* (1927), drawn from DICKENS's *Nicholas Nickleby*, at the LYRIC THEATRE, Hammersmith. He appeared at six MALVERN FESTIVALS in a wide variety of plays, including five by Shaw, among them the first productions of *Geneva* (1938) and *In Good King Charles's Golden Days* (1939), in which he appeared as Charles II. Among his later roles were John Cadmus in WHITING's *Marching Song* (1954) and Baron Santa Clara in Enid BAGNOLD's *The Last Joke* (1960).

THESPIS, Greek poet from Icaria in Attica, who is usually considered the founder of drama, since he was the first to use an actor in his plays in addition to the CHORUS and its leader. He won the prize at the first competitive DIONYSIA in Athens, *c.* 534 BC. Only the titles of his plays have survived and even these may not be authentic. Tradition has it that Thespis took his actors round in a cart, which formed their stage. In the 19th century the adjective Thespian came to be used of actors and acting in general, and often figured in the names of amateur companies, while 'the Thespian art' was journalese for the art of acting.

THIRD STAGE, see STRATFORD (ONTARIO) FESTIVAL.

THIRTY-NINTH STREET THEATRE, New York, at 119 West 39th Street. This opened on 18 Apr. 1910 with NAZIMOVA, after whom it was originally named by the Shuberts, in IBSEN's *Little Eyolf*, which ran for six weeks. In 1911, as the Thirty-Ninth Street Theatre, it reopened on 20 Oct. with Mason's *Green Stockings*, and early in 1912 housed *A Butterfly on the Wheel*, a drama by E. G. Hemmerde and Francis Neilson which starred the English actors Charles Quartermaine and Madge Titheradge. Later in the year Annie Russell appeared in a season of English classics, and in 1913 John BARRYMORE was seen in the Harvard Prize play, *Believe Me, Xantippe!* by Frederick Ballard. In 1919 Walter HAMPDEN appeared as Hamlet. The theatre was demolished in 1925.

THOMAS, AUGUSTUS (1857–1934), American dramatist, and one of the first to make use in his plays of American material. He succeeded BOUCICAULT as adapter of foreign plays at the MADISON SQUARE THEATRE under PALMER, but his first popular success, an original drama entitled *Alabama* (1891), enabled him to resign and devote all his time to his own work. Among his later plays were several others based on a definite locality—*In Mizzoura* (1893), *Arizona* (1899), *Colorado* (1901), and *Rio Grande* (1916). His most successful play was *The Copperhead* (1918), in which Lionel BARRYMORE made a hit as Milt Shanks. An interest in hypnotism and faith-healing was shown in *The Witching Hour* (1907), *Harvest Moon* (1909), and *As a Man Thinks* (1911), but on the whole Thomas's plays were not profound, and provided entertainment of a kind acceptable to his audiences. In 1922 he published his autobiography until the title *The Print of My Remembrance*.

THOMÁS, CORNELIA FRANCES, see JEFFERSON, JOSEPH.

THOMAS, (Walter) BRANDON (1856–1914), English actor, playwright, and writer of coon songs, which he sang in MUSIC-HALLS himself. After some amateur experience, he made his professional début on the stage with the KENDALS in 1879, remaining with them until 1885. He wrote about a dozen plays, but only one is now remembered, the farce *Charley's Aunt*, which has been seen in London a number of times since its first production at the ROYALTY THEATRE on 21 Dec. 1892, when it ran for four years with W. S. PENLEY in the title role. In later revivals Thomas, who played Colonel Chesney in the first production, took over the chief part himself. The play has been filmed, made into a musical, *Where's Charley?* (N.Y., 1948; London, 1958), which was successful in America but not in England, and has figured in the repertory of almost every amateur and provincial theatre, as well as being played all over the world in English and in innumerable translations. At one time it was running simultaneously in 48 theatres in 22 languages, including Afrikaans, Chinese, Esperanto, Gaelic, Russian, and Zulu.

THOMASHEFSKY, BORIS, see JEWISH DRAMA.

THOMASSIN, see VICENTINI, TOMMASO ANTONIO.

THOMPSON SHELTERHOUSE, see CINCINNATI PLAYHOUSE IN THE PARK.

THORNDIKE, (Agnes) SYBIL (1882–1976), English actress, appointed DBE in 1931 and CH in 1970. She began her long and distinguished career under Ben GREET in 1904, playing a wide variety of Shakespeare parts on tour in England and the U.S.A., and in 1908 joined Miss HORNIMAN's repertory company at the GAIETY THEATRE, Manchester, where in 1909 she married the actor and director Lewis CASSON, with whom she was associated in much of her later work. During the next five years she divided her time between London—where she was seen, among other parts, as Emma Huxtable in GRANVILLE-BARKER's *The Madras House* (1910) and Beatrice Farrar in HOUGHTON's *Hindle Wakes* (1912)—and Manchester, where she gave an excellent performance in the title role of St John ERVINE's *Jane Clegg* (1912). She also made her first appearance in New York during this time, as Emily Chapman in Somerset MAUGHAM's *Smith* (1910), opposite John DREW. From 1914 to 1918 she was at the London OLD VIC under Lilian BAYLIS, playing not only most of Shakespeare's young heroines but also, owing to the absence of so many young actors on war service, such parts as Prince Hal in *Henry IV*, Puck in *A Midsummer Night's Dream*, the Fool in *King Lear*, Ferdinand in *The Tempest*, and Launcelot Gobbo in *The Merchant of Venice*. On leaving the Old Vic she was seen as Synge de Coûfontaine in CLAUDEL's *The Hostage* and as Hecuba in Gilbert MURRAY's translation of EURIPIDES' *Trojan Women* (both 1919), as well as in a number of short-lived modern plays, and from 1920 to 1922 was in Grand Guignol seasons at the LITTLE THEATRE with her husband and brother. One of her finest performances was given in the title role of SHAW's *Saint Joan* (1924) in the first London production, and she played another Shaw heroine, Barbara Undershaft, in a revival of *Major Barbara* in 1929. She appeared in VAN DRUTEN's *The Distaff Side* in 1933, repeating the role a year later in New York, where she was also seen as Mrs Conway in PRIESTLEY's *Time and the Conways* in 1938. Later in the same year London audiences saw one of her most memorable performances, as the elderly schoolmistress Miss Moffat in Emlyn WILLIAMS's *The Corn Is Green*, in which she starred opposite the author. During the Second World War she toured tirelessly with the Old Vic company for ENSA to mining towns and villages, playing Shaw's Candida, Lady Macbeth, and Euripides' Medea. She appeared at the NEW THEATRE with the Old Vic company as Aase in IBSEN's *Peer Gynt* (1944) and Jocasta in SOPHOCLES' *Oedipus the King* (1945). In 1947 she and her husband were seen to advantage in one of Priestley's best plays, *The Linden Tree*, and in 1949 she began a long run in the comedy *Treasure Hunt* by M. J. Farrell and John Perry. She was then in two long-running plays by N. C. HUNTER, *Waters of the Moon* (1951) and *A Day By the Sea* (1953), and made three overseas tours before returning to London in a revival of T. S. ELIOT's *The Family Reunion* in 1956 and visiting New York in Graham GREENE's *The Potting Shed* in 1957. She and Lewis Casson celebrated their golden wedding in 1959 by appearing in *Eighty in the Shade*, specially written for them by Clemence DANE, and were seen together in COWARD's *Waiting in the Wings* (1960) and as the Nurse and 'Woffles' in CHEKHOV's *Uncle Vanya* at the first CHICHESTER FESTIVAL in 1962. She then starred in a new comedy by William Douglas HOME, *The Reluctant Peer* (1964), and in 1966 made her last appearance on the London stage in a revival of Kesselring's *Arsenic and Old Lace* with Casson and Athene SEYLER. Her last performance was in John Graham's *There Was an Old Woman* (1969) during the opening season of a theatre at Leatherhead, Surrey, named after her. She is the author of *Religion and the Stage* (1927). Her son John Casson published *Lewis and Sybil* in 1962.

Her brother (Arthur) Russell Thorndike (1885–1972) was an actor who also wrote the play

Dr Syn (1925), in which he played the title role, and a biography of his sister (1929). Their younger sister Eileen Thorndike (1891–1953) was also an accomplished actress.

THORNE, CHARLES ROBERT (1840–83), American actor, one of the five children of Ann Maria Mestayer (?–1881), member of a famous circus family, and her husband Charles Thorne (*c.* 1814–1893), an actor who first appeared in New York in 1829 and later managed theatres there, in San Francisco, and in Toronto, during the 1830s. The younger Charles appeared with his parents on tour as a child, and in 1860 was with Joseph JEFFERSON in New York, playing in his revival of Tom TAYLOR's *Our American Cousin.* He then went to Boston; but his career really began when in 1871 he joined the company at the UNION SQUARE THEATRE under PALMER, where he was for many years immensely popular as the dashing young heroes of melodrama. A good-looking, athletic and attractive person, not over-intelligent, he represented the ideal romantic hero of the time. In 1874 he was seen in London with some success. He made his last appearance on the stage in 1883 in BOUCICAULT's *The Corsican Brothers*, but was forced to retire from the cast through illness and died soon afterwards.

THORNE, THOMAS (1841–1918), English actor-manager, one of eight brothers and sisters who were all on the stage, the best-known being Sarah (1837–99), who took over the management of the theatre in Margate from her father, Richard Samuel Thorne, and ran it for many years. Thomas made his first appearances under his father, but in 1862 was in London, where from 1864 to 1870 he was a member of the company at the STRAND THEATRE. In partnership with David JAMES and Henry MONTAGUE he then opened the VAUDEVILLE THEATRE, where H. J. BYRON's comedy *Our Boys* (1875), with James as Perkyn Middlewick, ran for many years, making a fortune for all concerned in its production. From 1882 to 1892 Thorne was sole manager of the theatre, after which he toured in revivals of some of his former successes, including *Our Boys*, until his death.

THREE-FOLD, see FLAT.

THREE HUNDRED CLUB, see STAGE SOCIETY.

THROWLINE, see BOX-SET and FLAT.

THUNDER RUN, THUNDER SHEET, see SOUND EFFECTS.

THYRISTOR, see LIGHTING.

TIECK, (Johann) LUDWIG (1773–1854), German romantic poet and dramatist, whose early plays were satirical fairytales—*Ritter Blaubart* (*Bluebeard*, 1796), *Der gestiefelte Kater* (*Puss-in-Boots*, 1797), and *Die verkehrte Welt* (*The World Upside-down*, 1798). These were followed by a number of verse dramas—*Leben und Tod der heiligen Genoveva* (1799) and *Kaiser Oktavianus* (1804). In 1824 Tieck became director of the Court Theatre in DRESDEN, where he insisted on clear diction and simplified staging, though a performance of *A Midsummer Night's Dream* on a specially constructed Elizabethan stage, as Tieck imagined it to have been, remained an isolated experiment. He became an influential critic, and his description of work at Dresden, collected as *Dramaturgische Blätter* (1826), reveal him as a man of insight and good taste. His interest in the Elizabethan theatre led him to translate several plays by Ben JONSON, and with his daughter Dorothea he completed SCHLEGEL's translations of Shakespeare, whose reputation in Germany he did much to further.

TILLEY, VESTA [Matilda Alice Powles] (1864–1952), one of the outstanding stars of the British MUSIC-HALL, famous for her MALE IMPERSONATIONS. She made her first appearance at the age of four at the St George's Hall in Nottingham, where her father William Powles, known as Harry Ball, was CHAIRMAN, and soon after was seen in Birmingham, giving an impression of the popular contemporary tenor Sims Reeves. As The Great Little Tilley, she made her first appearance in London in 1878, at the Royal, Holborn, singing 'Near the Workhouse Door' in the character of 'Poor Jo'. It was then that she added Vesta to her pet name of Tilley, retaining it throughout the rest of her successful career. She was seen as PRINCIPAL BOY in a number of PANTOMIMES, including several at DRURY LANE, where she first appeared at Christmas 1892 as Captain Tra-la-la in BLANCHARD's *Sindbad.* But her true home was the music-hall, where her immaculately cut suits and elegant gestures provided a pattern followed everywhere by suburban and provincial young-men-about-town. She was not the first music-hall star to wear male attire, but she was undoubtedly the greatest, outshining even her excellent contemporaries Hetty King and Ella Shields. She was particularly admired for her impersonations of soldiers, and could reproduce to the life, with a faint edge of satire, the 'strut' of the guardsman or the young officer's 'glide'. Although she presented a completely convincing male image, she made no attempt to sing like a man, her voice, like her songs, many of which were written by her husband Walter (later Sir Walter) de Frece (1871–1935), whom she married in 1890, presenting the woman's view of the men she represented. Two of her most famous numbers were 'Following in Father's Footsteps' and 'Jolly Good Luck to the Girl who Loves a Soldier'.

TILNEY, Sir EDMUND, see MASTER OF THE REVELS.

TIMES SQUARE THEATRE, New York, at 219 West 42nd Street. This opened on 30 Sept. 1920 with Vajda's *Fata Morgana*, translated as *The Mirage*, which ran for six months. Several musical plays were followed by Channing Pollock's *The Fool* (1922), which was more successful outside New York than on Broadway, and MAETERLINCK's *Pelleas and Melisande* (1923) which, in spite of the performance of Jane COWL, had only a short run. *Charlot's Revue of 1924*, with a large cast of London favourites, was, however, a success, as was Anita Loos's *Gentlemen Prefer Blondes* (1926), while *The Front Page*, a play about journalism by Ben Hecht and Charles MacArthur, ran for 281 performances in the season 1928–9. In Jan. 1931 came a long run of COWARD's *Private Lives*, with the author, Gertrude LAWRENCE, Jill Esmond, and Laurence OLIVIER. The last play at this theatre before it became a cinema was Edward Roberts and Frank Cavett's *Forsaking All Others* (1933) which reintroduced Tallulah BANKHEAD to Broadway.

TIMONEDA, JOAN (*c.* 1520–*c.* 1583), Spanish dramatist, poet, novelist, and publisher, responsible for the publication, among many other works, of the plays of Alonso de la VEGA and Lope de VEGA CARPIO. Of his own plays, those published in 1559, two of which are based on PLAUTUS, link him with the neo-classical writers of his own day, but those which appeared in 1565 under the pseudonym Joan Diamonte (which may not be by him) are in the tradition of popular theatre, and look forward to the GRACIOSO of the later Spanish COMEDIAS. Timoneda was also the author of five AUTOS SACRAMENTALES, of which the best is *La oveja perdida* (*The Lost Sheep*).

TIPTEERERS' PLAY, see MUMMERS' PLAY.

TIREMAN, in the Elizabethan theatre the man in charge of the wardrobe, which was kept in the tiring-house, or dressing-room. He also saw to the provision of stools for those members of the audience who sat on the stage, and in the private roofed theatres he looked after the lights.

TIRSO DE MOLINA [Fray Gabriel Téllez] (*c.* 1571–1648), Spanish ecclesiastic, whose secular works include a number of plays, of which more than 80 are extant, though he claimed to have written 400. Their technique derives from that of his near contemporary Lope de VEGA CARPIO, whom he much admired, but it is modified in his case by his greater interest in the psychology of his characters. He was particularly good at drawing women at their wittiest and most intelligent in such comedies as *Don Gil de la calzas verdes* (*The Man in Green Breeches, c.* 1611), in which he employs his favourite comic device of women disguised as men, and *El vergonzoso en palacio* (*The Bashful Man at Court*, *c.* 1612). Among his historical plays, the best is undoubtedly *La prudencia en la mujer* (*Prudence in Woman*, *c.* 1622), in which he draws an excellent picture of the heroic Queen Maria and points a moral for the Spanish politicians of his own time. Tirso was also the author of four AUTOS SACRAMENTALES, and of a number of religious plays of which an excellent example is *El condenado por desconfiado* (*Damned for Lack of Faith, c.* 1624); but his masterpiece is *El burlador de Sevilla y convidado de piedra* (*The Trickster of Seville and the Stone Guest*, *c.* 1630), the first of many plays on the subject of DON JUAN.

TITHERADGE, DION, see REVUE.

TITINIUS (*fl.* 160 BC), Roman dramatist, contemporary with TERENCE, and with AFRANIUS and ATTA the only known writer of the FABULA *togata*, or comedy based on the daily life of Roman citizens. The titles of 15 of these are known, but only about 180 lines survive. His *Fullonia* opens with a quarrel between fullers and weavers, but appears to have dealt also with a wife who complains that her husband is squandering her dowry, and a philanderer who tries to escape an angry husband by throwing himself into the Tiber. Titinius himself confessed his debt to MENANDER in his playwriting, but on the scanty evidence available it is difficult to say how far this extended.

TIVOLI GARDENS, New York, see RICHMOND HILL THEATRE.

TIVOLI MUSIC-HALL, London, in the Strand. Erected on the site of a beer-hall, whose name it took, this three-tier building holding 1,500 people opened on 24 May 1890 as a theatre, but had no success until in 1893 it was taken over by Charles MORTON who made the 'Tiv', as it was called, one of London's most popular MUSIC-HALLS. Its shows lasted for anything up to four hours, and the bill often included as many as 25 turns, featuring all the great names of the music-hall. In the years immediately before the First World War its popularity began to decline and it closed on 7 Feb. 1914, being demolished and replaced by a cinema in 1923. This was demolished in 1957 and the site used for a department store.

TOBY, in England, the dog in the PUNCH AND JUDY show, a purely English accretion, having no connection with the COMMEDIA DELL'ARTE or the HARLEQUINADE. The name seems to have come into use with the introduction of a live dog, usually a small terrier, into the performance somewhere between 1820 and 1850. The name may have been used because the first dog to be so employed was already called Toby, or from association with the Biblical Tobias, a favourite subject for puppet-plays, in which Tobias and the Angel were usually accompanied by a dog. Wearing a ruff round his neck, Toby sits on the narrow ledge of the puppet-booth window, and takes no part in the action, unless Punch pets him or, alternatively, urges him to bite the other characters. After the show he goes round among the audience, with whom he is always a firm favourite, collecting pennies in a little bag which he holds in his mouth. When in 1962 some 50 showmen attended a service to celebrate the 300th anniversary of Punch's arrival in London, each one was accompanied by his Toby.

TOBY, in the United States, a stock character in folk theatre. Most provincial companies and travelling tent-shows in rural communities of the Mississippi Valley and the south-western states featured a Toby-comedian, and he was also to be found aboard the river SHOWBOATS. He represented the country bumpkin triumphant over evil (usually personified as 'city slickers'). With a freckled face and a blacked-out front tooth, he wore a rumpled red wig, battered hat, calico shirt, baggy jeans, and large, ill-fitting boots or shoes. He emerged in the 1900s, and Frederick R. Wilson, member of a touring tent-show company known as Horace Murphy's Comedians, was the first of a long line of actors to specialize in Toby roles. Plays were specially written for him, and Broadway successes altered to include Toby as the chief character. Toby-comedy included generous use of the topical 'ad-lib' and expert theatrical gymnastics—the pratfall, glides, splits, and rubber-legs. In revivals of Harriet Beecher Stowe's *Uncle Tom's Cabin* the Toby-comedian of the troupe would put on black-face and a gunnysack costume to impersonate Topsy. Toby had a female counterpart, the grand-daughter of the 'rough soubrette', known as Susie. The last Toby-show of any importance closed in Wapello, Iowa, on 1 Oct. 1962, after nearly 50 years under the same managers, Neil Schaffner and his wife, who played Toby and Susie.

TODOROV, PETKO, see BULGARIA.

TOGATA, see FABULA: 7.

TOGGLE, TOGGLE RAIL, see FLAT.

TOLLER, ERNST (1893–1939), German poet and dramatist, one of the best and most mature exponents of EXPRESSIONISM. His first play, *Die Wandlung* (*Transfiguration*, 1919), written during his imprisonment as a pacifist after being invalided out of the trenches in 1916, is a plea for tolerance and the abolition of war. It was followed by *Masse-Mensch* (1920) and *Die Maschinenstürmer* (1922), which, as *The Machine Wreckers*, in a translation by Ashley DUKES, was seen in London in 1923. It is less expressionist in technique than *Masse-Mensch*, which as *Man and the Masses* was produced in New York in 1924 by Lee SIMONSON for the THEATRE GUILD, and also less pessimistic, since Toller, in the person of his hero Jim Cobbett, foresees the day when the rebellious workers will be an organized and stable body of intelligent men. But Toller's later plays, of which *Hoppla, wir leben!* (1927), first produced by PISCATOR, was staged at the GATE THEATRE in London in 1929 as *Hoppla!*, became progressively less hopeful as he watched the decline of freedom in Germany, and in 1933 he left for England and the U.S.A., where he committed suicide on hearing of the outbreak of the Second World War. Among Toller's other plays were *Wunder in Amerika* (on Mary Baker Eddy) and *Feuer aus den Kesseln* (on a naval mutiny in 1917), both first produced in 1930. The former, as *Miracle in America*, was seen in London in 1934, the latter, as *Draw the Fires!*, in 1935. *Die blinde Göttin* (1932), on a miscarriage of justice, was translated in 1953 as *Blind Man's Buff*. Tankred DORST's play *Toller* (1968), incorporating scenes from *Masse-Mensch*, is an attack on his predecessor's theatrical attitude to revolutionary politics.

TOLSTOY, ALEXEI KONSTANTINOVICH (1817–75), Russian diplomat and poet, author of a fine historical trilogy, containing excellent crowd scenes and written with much semi-oriental imagery, in which he idealized old feudal Russia. All three plays—*The Death of Ivan the Terrible, Tsar Feodor Ivanovich*, and *Tsar Boris*—were written between 1866 and 1870, but were banned by the censor, who finally allowed the second to be put on as the opening production of the MOSCOW ART THEATRE in 1898, two days after it had been seen in St Petersburg. It has frequently been revived, and there have also been productions of the complete trilogy.

TOLSTOY, ALEXEI NIKOLAIVICH (1882–1945), Soviet novelist and dramatist, whose first play, written after the October Revolution, shows the influence of the uprising reaching out as far as the planet Mars. His later works include a historical drama on Peter the Great and a two-part play on Ivan the Terrible; the first part, dealing with Ivan's youth and marriage, was produced at the MALY THEATRE in 1943, the second, dealing with Ivan's struggles to unite Russia, at the MOSCOW ART THEATRE later the same year. Of his plays on modern themes the best is *The Road to Victory* (1939), dealing with an episode of the Revolution in which both Lenin and Stalin appear.

TOLSTOY, LEO NIKOLAIVICH, Count (1828–1910), Russian novelist, critic, and dramatist. Under the influence of TURGENEV and OSTROVSKY, he started some comedies in the 1850s, which, however, remained unfinished. It was not until 1886 that he once more turned to the theatre, and by then his whole philosophy of life had undergone a change. Under the influence of M. V. Lentovsky, director of one of the Moscow People's Theatres, he wrote *The Power of Darkness*, possibly the most forceful peasant play ever written. Its main outline was taken from a criminal case heard at Tula, but in Tolstoy's hands it became a stark naturalistic document which was for many years banned by the censor, and first acted abroad, in Paris in 1888 under ANTOINE, and in Berlin by the FREIE BUHNE in 1890. It was not seen in Russia until 1895, when it was staged both at the Alexandrinsky Theatre in St Petersburg (now the PUSHKIN THEATRE in Leningrad) and at the Moscow MALY. It was first seen in London in 1904, and in New York in 1920. Tolstoy's next play, a short comedy entitled *The First Distiller* (1887) which attacked alcoholism, was followed by *The Fruits of Enlightenment*, begun in 1886, a comedy which satirizes the parasitic life of the country gentry. Published in 1891, it was produced in the same year by STANISLAVSKY, and the following year was seen at the Maly, with little success. Even the actors at the MOSCOW ART THEATRE, where the play was revived some years later, could not at first tackle the peasant characters successfully, and many years of experiment and experience were needed before Tolstoy's famous drama could be adequately portrayed. It was first seen in London in 1928. Tolstoy's last two plays, published in 1912, were left unfinished at his death. *Redemption* (or *The Living Corpse*), an attack on the evils of contemporary Russian marriage laws, was produced by the Moscow Art Theatre in 1911; John BARRYMORE played the hero, Fedya, in New York in 1918, Donald WOLFIT in London in 1946. *The Light That Shines in Darkness*, in which the useless life of a wealthy family is contrasted with that of the poverty-stricken and overworked peasants, does not appear to have been staged. Three of Tolstoy's novels were dramatized and produced at the Moscow Art Theatre—*Resurrection, Anna Karenina* (seen in London and New York in 1907), and *War and Peace* (seen in London in 1943 and 1962).

TOMELTY, JOSEPH (1911–   ), Ulster actor, dramatist, and novelist, who was one of the original actors in the ULSTER GROUP THEATRE, serving as its general manager from 1942 to 1951. His first play, *Barnum Was Right* (based on an earlier radio play *The Beauty Competition*), was produced at the Empire Theatre, Belfast, and published in 1953 as *Mugs and Money*. Its sequel, *Right Again, Barnum* (1943), and a domestic tragicomedy, *The End House* (1944), were also produced by the Group Theatre, the latter play being seen at the ABBEY THEATRE in 1944, as was *All Souls' Night* (1948), a tragedy of Northern Irish fisher folk, in 1949. This was followed by *Is the Priest at Home?* (1954), a sympathetic portrayal of the varied responsibilities thrust by the community on its parish priest. Its sequel, *One Year at Marlfield* (1965), was first seen at the Grove Theatre in Belfast.

TOOLE, JOHN LAURENCE (1830–1906), English actor and theatre manager, who was for a short time, like GARRICK, clerk to a wine merchant. Success in amateur theatricals turned his thoughts to the stage, and he made his first appearance as a professional in Dublin in 1852. After further experience in the English provinces he established himself in London in 1856 as a low comedian, playing Fanfaronade in Charles Webb's *Belphegor the Mountebank*, in which Marie Wilton, later the wife of Squire BANCROFT, also made her London début. In 1857 Toole was with Henry IRVING, who remained a lifelong friend, and a year later, on the recommendation of DICKENS, another close friend, was engaged by Ben WEBSTER for the ADELPHI THEATRE, where he remained for nine years. Among his best parts were two of Dickens's characters—Bob Cratchit in *A Christmas Carol* (1859) and Caleb Plummer in *Dot* (1862), a dramatization of *The Cricket on the Hearth* by BOUCICAULT. In this latter part Toole combined humour with pathos in a way which showed how well he might have played serious parts; but the public preferred him in farce, and in 1869 he began a 5-year association with HOLLINGSHEAD at the GAIETY THEATRE, where he proved admirable in BURLESQUE. He made an appearance in New York, in 1874, but his humour was too cockneyfied for the Americans, and he never returned there. In 1879 he went into management at the CHARING CROSS THEATRE, to which he gave his own name three years later. His most important productions there were PINERO's early comedy *Imprudence* (1881) and BARRIE's first play *Walker, London* (1892). He retired in 1895.

TOP DROP, see BORDER.

TOPOL, CHAIM (1934–   ), Israeli actor, who had his first experience of acting while in the army, and afterwards founded a theatre of satire in Tel-Aviv which achieved widespread popularity. In 1961 he went to Haifa to help with the establishment of the Municipal Theatre there, acting as assistant to the director, Joseph Millo, and appearing with the company at the Venice Biennale as Azdak in BRECHT's *The Caucasian Chalk Circle*, the part in which he was seen at the CHICHESTER FESTIVAL THEATRE in 1969. He also played such parts as Petruchio in *The Taming of the Shrew*, Pat in Brendan BEHAN's *The Hostage*, John in IONESCO's *Rhinoceros*, and the Soldier in FRISCH's *Andorra*. In 1965 he was with the CAMERI THEATRE in Tel-Aviv, and he was already beginning to be well known in films when in 1967 he appeared in London as Tevye in the American musical *Fiddler on the Roof*, based on ALEICHEM's *Tevye the Milkman*, in which he scored an immense success.

TOPOL, JOSEF, see CZECHOSLOVAKIA.

TORCH THEATRE, Milford Haven, see WALES.

TORELLI, ACHILLE (1841–1922), Italian playwright, the first, with FERRARI, to introduce realistic social drama into the Italian theatre. His first successful play was *I mariti* (1867), a study of modern marriage as exemplified in the vicissitudes of four married couples. Of his later plays the most important are *La moglie* (1868) and *L'ultima convegno* (1898).

TORELLI, GIACOMO (1608–78), Italian machinist and scene-painter. A practical man of the theatre rather than an artist, he made many important innovations in the designing and setting of scenery, being the inventor of the CARRIAGE-AND-FRAME device for moving several sets of WINGS on and off stage simultaneously. He was a pupil of ALEOTTI, and may have worked with him on the building of the Teatro Farnese, which was not completed until 1628. In 1640 he was responsible for the building of the Teatro Novissimo in Venice, where the magical effects of his stage mechanisms won for him the nickname of 'the great magician'. Five years later he went to Paris and inaugurated the fashion for MACHINE PLAYS of which the finest was CORNEILLE's *Andromède* (1650). This was produced at MOLIÈRE's first theatre, the PETIT-BOURBON, which Torelli had completely refurbished backstage, painting new scenery and installing machinery of his own invention. It was all destroyed by his rival Gaspare VIGARANI in 1660, but the designs were preserved, and over a hundred years later the complete survey of contemporary theatre machinery given under 'Machines du Théâtre' in DIDEROT's *Encyclopédie* (in 1772) shows that it was almost all based on Torelli's ideas.

TORGA, MIGUEL, see PORTUGAL.

TORMENTOR, see FALSE PROS(CENIUM) and LIGHTING.

TORRES NAHARRO, BARTOLOMÉ DE (*c.* 1485–*c.* 1524), one of the leading Spanish dramatists of the 16th century, contemporary with ENCINA and Gil VICENTE. Little is known of his life, but from 1513 he was in Rome, where most of his plays appear to have been written. His collected works appeared in 1517 under the title of *Propalladia*. The important prologue distinguishes between two types of play, the *comedia a noticia* and the *comedia a fantasia*, terms which may be said to correspond to realistic and novelesque genres. Of the former type two examples have survived, *Soldadesca*, dealing with a braggart Spanish captain, and *Tinellaria*, depicting life in the kitchen of an Italian palace. The best known example of Torres Naharro's novelesque plays is *Himenea*, based on part of the CELESTINA, where the relationship between master and man and the theme of the conflict of love and honour both foreshadow the plays of Lope de VEGA CARPIO.

TÖRRING, JOSEF AUGUST VON, see RITTERDRAMA.

TORTORITI, GIUSEPPE (*fl.* late 17th–early 18th cent.), actor of the COMMEDIA DELL'ARTE. He assumed the part of Pasquariello and was with the Duke of Modena's company when it visited London in 1678–9. Later he was with the COMÉDIE-ITALIENNE, and on the death of FIORILLO in 1694 succeded to the part of SCARAMUCCIA. After the banishment of the Italian players from Paris in 1697 he toured with his own company, but failed to establish himself and is not heard of again.

TOTALTHEATER, see THEATRE BUILDINGS.

TOTTENHAM STREET THEATRE, London, see SCALA THEATRE.

TOULOUSE, Grenier de, see DÉCENTRALISATION DRAMATIQUE and SARRAZIN.

TOURING COMPANIES. After the demise in Britain of the local STOCK COMPANIES, the gap was partly filled by the formation of touring troupes based on London or some other large city; pre- and post-London tours were also made by the casts of West End productions. The development of touring was originally hastened by the cheap and easy transport provided by the railways, but today touring companies mainly travel by road, often with their own fleet of lorries and cars for scenery and personnel. Since many touring companies present opera, ballet, or large-scale musicals, they need larger theatre buildings than the REPERTORY THEATRES, which usually limit themselves to straight plays or modest musicals; on the other hand they need less storage and wardrobe space and fewer workshops. In the 1960s there were about 35 theatres graded as No. 1 for the use of touring companies, but soaring costs and the lack of suitable productions made it difficult to keep them all open throughout the year; the sites also proved increasingly valuable for development purposes, and their owners were often induced to sell them. After a number had been lost urgent action was taken to preserve those that remained through purchase by the local authority or by subsidy, and there is now a touring circuit of over a dozen well-equipped theatres which can accommodate the large subsidized opera, ballet, and drama companies, while a further 20 theatres can receive less spectacular productions. Subsidized productions, however, can occupy only a part of a theatre's programme, the rest of which must be supplied by commercial shows, such as appearances by pop groups and other performers and touring productions of West End hits and plays from the repertoire. Pre-London tours are becoming rare because of cost and because West End theatres increasingly take their productions direct from the subsidized and FRINGE THEATRE; post-London tours with the original casts are now also rare. Some touring is undertaken by regional companies such as the CAMBRIDGE Theatre Company, and fringe theatre groups tour studio and university theatres and other more makeshift venues.

TOURNEUR, CYRIL (1575–1626), English dramatist, of whom little is known, though he was connected with the Cecils, and may have been sent by them on secret missions abroad. He was certainly in Cadiz with Sir Edward Cecil in a secretarial capacity the year before his death. Of the three tragedies doubtfully ascribed to him, *The Nobleman* (1607) is lost, the manuscript having been destroyed by WARBURTON's cook. *The Revenger's Tragedy*, which has also been credited to MIDDLETON, was published anonymously in 1607 and probably acted a year or two previously by the KING'S MEN. It was revived under Tourneur's name by the ROYAL SHAKESPEARE COMPANY, at Stratford-upon-Avon in 1966 and at the ALDWYCH THEATRE in London three years later. *The Atheist's Tragedy; or, the Honest Man's Revenge* was probably written in 1606, since echoes of *King Lear* (1605) have been noted in it.

TOVSTONOGOV, GEORGYI ALEXANDROVICH (1915–), Soviet director, who began his theatrical career in 1931 in Tbilisi, the capital of his native Georgia. In 1949 he became director of the Lenkom Theatre in Leningrad, and in 1956 of the GORKY THEATRE in the same city, where one of his most successful productions was Shteyn's *The Ocean* (1961). He staged many other successful, though often controversial, productions both at this theatre and at Leningrad's PUSHKIN THEATRE, notably GRIBOYEDOV's *Wit Works Woe* in 1962, CHEKHOV's *Three Sisters* in 1964, and GORKY's *The Smug Citizens* in 1966. He has a very individual style; his dramatization of DOSTOEVSKY's *The Idiot* (1957), which was seen in London during the 1966 WORLD THEATRE SEASON with SMOKTUNOVSKY as Prince Myshkin, was prefaced by the showing of a vast facsimile of the original title-page of the novel, and ARBUZOV's *It Happened in Irkutsk* (1959) opened with the hero seated at the top of rising tiers of seats, striking chords on a piano: in 1969 Shakespeare's *Henry IV* was set in a fair-booth, and his 1972 production of GOGOL's *The Government Inspector* made Khlestakov a phantasmagorical figure which clearly owed something to MEYERHOLD's famous production of 1926. Tovstonogov's experiments may not always succeed, but his work is full of life and vigour, and shows a fresh approach not only to revivals of the classics but also to new Soviet plays, many of which he has directed in his own theatre. His *Profession of the Stage Director* was published in an English translation in 1972.

TOWER THEATRE, London, in Canonbury Place, Islington, the home of the Tavistock Repertory Company, an amateur group founded in 1932, which up to the outbreak of war in 1939 was housed at the Mary Ward Settlement, Tavistock Square, under Duncan Marks. It lapsed for a while, but in 1952 was reformed under F.O.M. Smith, and moved to Canonbury Tower, Islington, a well preserved 16th-century building in which Oliver GOLDSMITH once lived. The main premises were equipped as a club, while the adjacent hall was converted into a proscenium theatre seating about 160. As the Tower Theatre, this opened on 15 Feb. 1953, and is licensed for public performances. The company, which has a long list of successful productions to its credit, presents up to 20 plays a year, and has given the first performances in London of several outstanding plays, including Tennessee WILLIAMS's *The Milk Train Doesn't Stop Here Anymore* (1968), ARRABAL's *The Cemetery* (1969), and John WHITING's *The Gates of Summer* (1970).

TOY THEATRE, or Juvenile Drama, collections of theatrical material, popular in the 19th century, which consisted of drawings of actors, scenery, and properties in a successful contemporary play, suitable for cutting out and mounting on cardboard for a performance in which they were drawn on metal long-handled slides across a small model stage, while an unseen assistant recited an extremely condensed version of the text. The sheets which made up the complete set, usually eight to twelve in all, could be bought for a 'penny plain, twopence coloured', the colouring being done by hand in bold, vivid hues that are as fresh today as when first applied. These sheets, which may have originated in those sold by theatrical agencies in Paris for the benefit of provincial and foreign managements, were probably first intended, in England at any rate, as theatrical souvenirs. They capture with astonishing fidelity the theatre of GRIMALDI, KEAN, KEMBLE, LISTON, and VESTRIS, and the productions of ASTLEY'S, COVENT GARDEN, DRURY LANE, the OLYMPIC, and the SURREY in the early 19th century. The total repertory of some 300 plays includes melodramas such as POCOCK's *The Miller and His Men*, LILLO's *George Barnwell*, or BOUCICAULT's *The Corsican Brothers*, ballad operas like DIBDIN's *The Waterman*, contemporary versions of Shakespeare, and many long-forgotten pantomimes. At first considerable care was taken to reproduce the costumes, attitudes, and even the features of the actors, as well as the details of wings, backcloth, and scenic accessories.

The first English Toy Theatre sheets were issued in 1811 by William West, a stationer in Exeter Street, Strand, and the production and sale of what soon became a popular children's plaything was a thriving industry between 1815 and 1835, with some fifty publishers engaged in it, among them West, Jameson, Hodgson, Skelt, Green, Park, and Webb. As the demand for sheets grew, so the quality of the drawings and reproduction fell off, but trade continued brisk until the 1850s and beyond, particularly when the sheets were given away as supplements to boys' magazines. The old-style Juvenile Drama never quite disappeared, and as late as 1932 two shops in Hoxton, Webb's and Pollock's, still printed the sheets of the old plays from the original blocks. There are good collections of Toy Theatre sheets in several London museums, most of which will, it is hoped, be transferred to the THEATRE MUSEUM.

Similar toy theatres were popular on the Continent in the 19th century, particularly in Germany, Denmark, and Spain, where the characters moved in grooves rather than on slides, but the plays were usually specially written for children, and not, as in England, taken from the adult repertory.

TRABEATA, see FABULA: 8.

TRADES UNIONS THEATRE, Moscow. This theatre, originally known as the First Workers' Theatre, was one of several established during the early years of the Revolution which were affiliated to Proletkult (The Organization for the Cultural Enlightenment of the Proletariat), by which name it was also known in its early days. Its work was much influenced by the theories of Bogdanov relating to the independent development of a specifically proletarian culture, and it is chiefly memorable for the work done there between 1921 and 1923 by Sergei Eisenstein, especially his 'agit-montage' productions of OSTROVSKY's *Enough Stupidity in Every Wise Man* and TRETYAKOV's *Gas Masks*. After Eisenstein left to devote the rest of his life to the cinema, the theatre passed through many vicissitudes until in 1932 it was given a new name as well as a new director, Alexei DIKIE, whose first production was WOLF's *Sailors of Cattaro*. This was well received, but later productions were less successful and in 1936 the theatre finally closed.

TRAFALGAR SQUARE THEATRE, London, see DUKE OF YORK'S THEATRE.

TRAFALGAR THEATRE, New York, see BILLY ROSE THEATRE.

TRAGEDIAN, see STOCK COMPANY.

TRAGÉDIE-BOURGEOISE, see DRAME.

TRAGEDY, play dealing in an elevated, poetic style with events which depict man as the victim of destiny yet superior to it, both in grandeur and in misery. The word is of Greek origin and means 'goat-song', possibly because a goat was originally given as a prize for a play at the DIONYSIA, or festival of DIONYSUS. The classic Athenian tragedies of AESCHYLUS, EURIPIDES, and SOPHOCLES developed from the choral lyric, an art which reached its height among the Dorian peoples of the Peloponnese during the 6th century BC. The earliest plays began with the *parados*, or entrance of the CHORUS, which was soon preceded by a *prologos* for the actor or actors; an actor might be given a solo *monody* to sing before the chorus appeared. Each formal ode, or *stasimon*, for the chorus alternated with a dramatic scene, or *episode*; lyrical dialogue between an actor and the chorus was called a *kommos*; and all that followed the final stasimon was the *exodus*. The chorus sang, or chanted, in unison, but probably spoke through its leader. As nothing is known about the music and dancing of the chorus, and the music-rhythms of the odes cannot be translated into speech-rhythms, it is impossible to dogmatize about the original productions of the great texts which have come down to us, and all translation and revivals can only be approximations. It was the subject-matter of the plays which exercised the greatest influence on the drama of the future. Taken from the myths of gods and heroes, it retained a link with its religious origins by the beneficent intervention, usually at the end of the play, of a god—the DEUS EX MACHINA—who descended from above the stage by means of a crane or pulley. The Roman theatre produced excellent writers of comedy in PLAUTUS and TERENCE, but no tragedies for the stage have survived; those by SENECA, which had an immense influence on later European drama, were CLOSET PLAYS, written to be read and not acted.

Tragedy in Renaissance Italy, more under the direct influence of the Greeks than of Seneca, developed early, but did not produce any outstanding playwright until the 18th century, with ALFIERI. In France tragedy developed under the influence of Seneca, modified by the contemporary interpretation of ARISTOTLE which gave rise to the theory of the UNITIES of time, place, and action, though only the last was consistently observed by Greek dramatists, the unities of time and place being imposed on the play by the continuous presence of the chorus. The greatest exponents of French classical tragedy were CORNEILLE and RACINE, whose successors up to the end of the 18th century continued to employ their outward forms but without their inward excellence.

In England, where the influence of Seneca was paramount, MARLOWE and Shakespeare evolved a form of tragedy mingled with comedy which was *sui generis*. Because of its powerful appeal to English audiences, the English theatre remained impervious to the influence of French classical tragedy, even after the Restoration, when such plays as ADDISON's *Cato* (1713) brought the letter but not the spirit of Corneille and Racine briefly on the English stage. Spain, too, had her native tragedy, formulated by CALDERÓN, and the efforts of LUZÁN to import French tragedy failed, as did the attempts of GOTTSCHED and Carolina NEUBER in Germany. The German theatre later produced its own writers of tragedy in GOETHE and SCHILLER; but it was the melodramatic aspect of their tragedies which had the greatest appeal, and this, added to the influence of Shakespeare all over Europe at the end of the 18th century, produced the highly coloured MELODRAMA which in the 19th century replaced true tragedy everywhere. Meanwhile, in the 18th century, in the plays of LILLO, LESSING, and MERCIER, efforts had been made to apply the formula of classical tragedy to middle-class existence, resulting in 'domestic tragedy' or *tragédie bourgeoise*. It was not a success. Tragedy in the narrow theatrical sense demands a cast of heroes or demi-gods, an unfamiliar background—exotic, romantic, or imaginary—and a sense of detachment heightened by the use of verse or rhetorical prose. Even the plays of IBSEN and his successors, though often tragic in their implications, are dramas rather than tragedies in the Greek sense. In modern times efforts have again been made to tame tragedy and bring it within the family circle. But it is interesting to note that *Murder in the Cathedral* (1935) by T. S. ELIOT, which has as protagonists a king and an archbishop, was a success, unlike his *The Family Reunion* (1939) which, though based on a Greek myth, was firmly rooted in suburbia.

TRAGIC CARPET, a green baize stage CLOTH which by the end of the 17th century was invariably spread before the performance of a tragedy, to prevent the actors from soiling their costumes when collapsing on the dusty boards. It continued in use until well into the 19th century, and is often referred to in theatrical memoirs and letters. Earlier mentions of permanent green cloth coverings, once thought to refer to the tragic carpet, are now believed to refer to the benches in the auditorium.

TRAGI-COMEDY, term adopted from PLAUTUS' *Amphitryon* by the theorists of Italian Renaissance drama to indicate a play in which the elements of comedy and tragedy are freely mingled, and so applied to some of the works of such dramatists as GIRALDI and DOLCE. It was later adopted in France to classify the plays of GARNIER and Alexandre HARDY. It is not easy to define the genre in detail, but in general the plot should be original or drawn from modern romance (not Biblical, historical, or mythological sources, as in tragedy); the characters of noble, but not royal, rank; the development episodic, not subject to the unity of time but embracing the greater part of the protagonist's life; the incidents sensational, often horrific, sometimes presented on stage; and lastly, a happy ending—though this last feature, central to the practice of Giraldi and of French writers of tragi-comedy, tended not to be retained by Giraldi's successors. Indeed, in Italian drama after Giraldi, the genre lost its separate identity, and the name was frequently applied to horror tragedies like Dolce's *Marianna* (1565). In France, however, tragi-comedy played an important part in the development of drama up to the middle of the 17th century, as a rival (and in the period 1600–30 a very successful rival) to the more restrained forms of tragedy based on classical models. Hardy's career marks the pre-eminence of the freer, more theatrical tragi-comedy, and the 1630s saw keen competition between the partisans of the two genres, with tragedy finally triumphing in the 1640s. *Le Cid* (1637), CORNEILLE's masterpiece, was first called a tragi-comedy and not labelled 'tragedy' until 1648, by which time the rival term had lost its popularity.

TRAINING, Dramatic, see SCHOOLS OF DRAMA.

TRANSFORMATION SCENE, an important element in the English PANTOMIME. An instantaneous change of part of a scene, like a shop-window or a house-front, was usually done by the use of FALLING FLAPS, or hinged sections of canvas, painted on both sides, and kept in position by catches. When these were released, the flaps fell by their own weight, presenting their other side to the audience. A variant of this was the CHASSIS À DÉVELOPPEMENT. By the use of the earlier CARRIAGE-AND-FRAME or DRUM-AND-SHAFT methods, the whole scene could be changed, the BACKCLOTH and side WINGS being drawn off simultaneously to reveal new ones behind. For the swift changes needed for the pantomime a more spectacular method, made possible by the FLIES found in newer theatre buildings, was the RISE-AND-SINK, in which the upper part of the scene ascended into the GRID while the lower descended into the CELLAR. A quick change could also be achieved by the use of SCRUTO, thin strips of wood fastened to a canvas backing so as to form a continuous flexible sheet; by the Fan Effect, in which sectors of the back scene, pivoting centrally at the foot of the scene, sank sideways upon each other like collapsing fans, revealing a new scene behind; or by placing rolls of painted canvas like columns across the stage, with lines top and bottom, which when pulled drew the new scene across the old. Various TRAPS were used to enhance the illusion. With the advent of more sophisticated LIGHTING the usual method of effecting a transformation scene became the TRANSPARENCY, which when lit from the front was as opaque as canvas, but faded from sight when lit from behind.

TRANSPARENCY, a method of effecting a TRANSFORMATION SCENE, made possible in the second half of the 19th century by the development of sophisticated stage LIGHTING techniques. It is achieved by painting a scene, not upon canvas with opaque size-paint, but on linen, calico, loose-weave muslin, or sharkstooth gauze in transparent dyes. When lit from the front this appears as a normal painted CLOTH; but as the lights are brought up behind, the subject fades to reveal a more distant scene, or is supplemented by further painting on the back of the fabric. In this way the whole aspect of the scene is changed; a building may appear to be ravaged by fire, or a quiet country landscape become a battlefield. Any features on the original cloth which are required to remain in dark silhouette after the scene-change are executed in opaque paint.

TRANSPONTINE MELODRAMA, a term applied, usually in genial derision, to a type of crude and extravagantly sensational play staged in the mid-19th century in London theatres 'across the bridges' (i.e., on the south side of the Thames) such as the SURREY THEATRE and the OLD VIC. By extension the term was later attached to such MELODRAMAS wherever performed.

TRAPS, devices by which scenery and actors can be raised to stage level from below. They are now used mainly in PANTOMIME, but were formerly essential for the acrobatic TRICKWORK of the Dumb Ballet, for ghostly apparitions, and for the sudden metamorphosis of one character into another. Although the modern theatre makes very little use of traps, they can still be found in some older theatres, arranged in a traditional pattern and capable of being used for their original purpose. To accommodate them the joists of the stage floor are cut and specially framed, and the opening concealed in a variety of ways. Corner Traps, which are at each side of the stage just behind the front CURTAIN, and are used for conveying a single standing figure, are usually 2 ft square and are closed by a square of boards, battened out, or a piece of SCRUTO, which can be quickly rolled or slid aside. Nearer the centre of the stage is the Star Trap, which projects an actor on stage at great speed, and is used in pantomime for the arrival of the Demon King. It has a circular opening with triangular segments of wood fitted radially and hinged to the circumference of the opening with leather; as the actor shoots through, the segments part upwards, falling together immediately behind him. In a Bristle Trap the wood segments are replaced by bristles painted to match the stage floor; or a diaphragm of rubber can be used, with a slit across it. Almost in the centre of the stage is the large rectangular Grave Trap, which measures roughly 6 ft long and 3 ft wide. It has a platform below, which can be raised and lowered, and takes its name from its use for the graveyard scene in *Hamlet*, just as the less commonly found Cauldron Trap, usually square and placed further upstage, is named from the witches' scene in *Macbeth*. One of the most ingenious trap mechanisms was that devised for the apparition in BOUCICAULT's *The Corsican Brothers* (1852), which had to rise slowly out of the earth while gliding across the full width of the stage. This Corsican Trap, or Ghost Glide, consisted of a Bristle Trap set between two long sliders, the first of which (sometimes made of scruto) drew the trap across the stage while the second closed the aperture behind it. An inclined rail track below the stage carried a small BOAT TRUCK on which stood the actor playing the ghost. This rose from a point some 6 ft below the stage at the beginning of its travel to stage level at the end of it, and was drawn by a line working simultaneously with that drawing the trap door, so that the actor's body rose gradually through the Bristle Trap, whose aperture remained invisible to the audience. The device became so popular that in 1858 it was installed even in such a small playhouse as the Theatre Royal in Ipswich.

Two other traps were the SLOTE, which could be used for people or scenery, and the Vamp Trap, consisting of two spring-leaves in the backcloth or stage floor so arranged that the actor's body appeared to pass through a solid wall. It is said to have taken its name from PLANCHÉ's *The Vampire; or, the Bride of the Isles* (1820), in which it was first used.

For the swift substitution of one character for another a pair of traps, similar to the Corner Traps, was placed side by side, so geared that one rose as the other fell. When the actor on stage took his place on the raised trap, his partner stood ready on the lower trap, and at the crucial moment a diversion was caused. The traps were released, and in a second a different figure stepped from a part of the stage floor so near the original spot as to be indistinguishable from it. To complete the effect, the trapdoor of the second opening was arranged to flap over, or slide, into a new position covering the aperture left by the descent of the first trap, all as part of one concerted movement.

The mechanism of the traps which work in conjunction with the various openings described above varies in detail, but most of them consist of a rigidly built platform on strong legs, which slides up and down between four uprights, or corner posts. In a few small traps the platform slides between two posts only: these are grooved down the inner face to take two projections from the sides of the platform. In both cases ropes from the framework lead up to pulleys at the head of the uprights, and from there to some form of counterweight. In the larger and slow-moving traps the counterweight approximately balances the trap and its load, which is made to move by means of a hand winch acting on the DRUM-AND-SHAFT principle. For small, fast-moving traps the counterweight is very much heavier than the trap and its load, which has to be kept in position by a lever. When the lever is released, the counterweight plummets down and the trap rises sharply. To lessen the shock when the platform reaches the limit of its travel and has to be checked at stage level, the counterweight usually consists of a chain of iron balls, each about 9 ins. in diameter, hung one below another. The length of the chain is so adjusted that each ball in turn comes to rest on the floor until, when the summit of the ascent is nearly reached, all the balls are 'dead' and the trap travels for its last few inches practically on its own momentum, stopping far more gently than it would have done if the full force of the counterweight had been equally exerted throughout the whole of its run. (See also TRICKWORK.)

TRAVELLER, see CURTAIN.

TRAVERS, BEN (1886–1980), English dramatist, an outstanding writer of farce, who first came into prominence with *A Cuckoo in the Nest* (1925), which inaugurated a series of 'Aldwych farces', so called because they were all presented at the ALDWYCH THEATRE. Their casts included Mary BROUGH, Robertson HARE, Ralph LYNN, and Tom WALLS. The series continued with *Rookery Nook* (1926), *Thark* (1927), *Plunder* (1928), *A Cup of Kindness* (1929), *A Night Like This* (1930), *Turkey Time* (1931), *Dirty Work* (1932), and *A Bit of a Test* (1933). Travers also wrote a number of plays for other theatres, including a comedy, *O Mistress Mine* (1936), and the farces *Banana Ridge* (1939), *Spotted Dick* (1940), and *She Follows Me About* (1945), in all of which Robertson Hare again appeared, being joined in *Outrageous Fortune* (1947) and *Wild Horses* (1952) by Ralph Lynn. In the autumn of 1968 *Corker's End* was produced at the Yvonne Arnaud Theatre and in 1975 *The Bed Before Yesterday*, starring Joan PLOWRIGHT, began a long run in London. *Plunder* was revived at the NATIONAL THEATRE in 1978. Travers published two volumes of autobiography, *Vale of Laughter* (1957) and *A-sitting on a Gate* (1978).

TRAVERSE CURTAIN, see CURTAIN.

TRAVERSE THEATRE CLUB, Edinburgh, lively experimental theatre which opened in Jan. 1963, mainly owing to the efforts of its first chairman Jim Haynes, an American living and working in Edinburgh who later founded the ARTS LABORATORY in London. The original auditorium held 60 spectators seated on either side of the acting area, but in 1969 the company moved to new premises in the Grassmarket with a multi-purpose acting area and a seating capacity of up to 120. The Traverse has given the first productions in English of many foreign plays, including works by ARRABAL, JARRY, MROŻEK, and Peter WEISS. It has also had a particularly close relationship with the Scottish dramatists Stanley Eveling and C. P. TAYLOR, its Artistic Director Chris Parr, appointed in 1975, devoting considerable energy to continuing the tradition of developing Scottish talent. He was succeeded in 1981 by Peter Lichtenfels. The contributions of the Traverse to the EDINBURGH FESTIVAL have been remarkable in scope and size, and include the British première of BRECHT's *Happy End* (1964). In 1967 the theatre sponsored a visit by the LA MAMA troupe from New York, and it has also played host to FRINGE groups from France, Germany, Canada, the Netherlands, Nigeria, and the U.S.A., as well as dance and MIME troupes. Traverse productions have been seen in London, Amsterdam, and Boston, U.S.A.

TREE, ELLEN, see KEAN, CHARLES.

TREE, HERBERT DRAPER BEERBOHM (1853–1917), English actor-manager, knighted in 1907. The half-brother of Max BEERBOHM, he was working in the city office of his father, a grain merchant, when some successful appearances in amateur dramatics decided him to go on the stage. He made his professional début in 1878, appeared with Geneviève WARD as Prince Maleotti in Herman Merivale's *Forget-Me-Not* (1879), and was the first to play the Revd Robert Spalding in HAWTREY's *The Private Secretary* (1884), a part later closely associated with W. S. PENLEY. He made a great success later the same year as Paolo Macari in *Called Back*, based on a novel by Hugh Conway. Early in 1887 he became manager of the COMEDY THEATRE, where his most successful production was W. O. Tristram's comedy *The Red Lamp*, with which he inaugurated his management of the HAYMARKET THEATRE later the same year. Among his productions there were *The Merry Wives of Windsor* (1889), Henry Arthur JONES's *The Dancing Girl* (1891), in which he played the Duke of Guisebury to the Drusilla Ives of Julia NEILSON, *Hamlet* (1892), and Oscar WILDE's *A Woman of No Importance* (1893). The most successful of his productions was *Trilby* (1895), based by Paul Potter on George Du Maurier's novel. Trilby was played by Dorothea Baird, and Tree himself appeared as Svengali, a part which he revived many times. The success of his tenancy of the Haymarket enabled him to build HER MAJESTY'S THEATRE, which he opened with Gilbert Parker's *The Seats of the Mighty* (1897). There he carried on IRVING's tradition of lavishly spectacular productions of Shakespeare, staging between 1888 and 1914 18 of his plays with a magnificence much to the taste of the time, and achieving at least once, in *Richard II*, a remarkable synthesis of style and setting. He also produced a number of new plays, among them Stephen PHILLIPS's *Herod* (1900) and *Ulysses* (1902), and American importations like Clyde FITCH's *The Last of the Dandies* (1901) and BELASCO's *The Darling of the Gods* (1903). He was also responsible for the first production in English of SHAW's *Pygmalion* (1914), in which he played Higgins with Mrs Patrick CAMPBELL as Eliza Doolittle. The play was directed by Shaw, but in general Tree was his own director. A firm disciplinarian, and the founder of the Royal Academy of Dramatic Art, he was in essence a romantic actor, delighting in grandiose effects and in the representation of eccentric characters which allowed his imagination free play. He often chose parts outside his temperamental and physical range, and his tragic acting lacked fundamental force, but at his best he was outstanding. He also ran his company well, helped by his wife Helen Maud Holt (1863–1937), an excellent actress who was an active and intelligent partner in all her husband's enterprises. Her eldest daughter Viola (1844–1938) was on the stage, as was her grandson David.

TREE BORDER, see BORDER.

TREMAYNE, WILLIAM A., see CANADA.

TREMBLAY, MICHEL (1942–), French-Canadian dramatist who first attracted attention when his play, *Les Belles-soeurs* (1968), was put on by the Théâtre du Rideau Vert in Quebec. This rowdy comedy about two working-class sisters-in-law pioneered the use on stage of *joual* (from *cheval*), the Québecois *argot* that mixes broken French, swear-words, slang, and idiomatic English. This innovation caused considerable controversy, both in Quebec and in Paris when the play was performed there in 1973. Tremblay's later plays include *En pièces détachées* (1969), depicting the struggles of a young waitress in a cheap and tawdry world, variously translated as *Like Death Warmed Over, Montreal Smoked Meat*, and *Falling Apart*; *La Duchesse de Langeais* (1970), about an ageing homosexual; *A toi, pour toujours, ta Marie-Lou* (1971), a tragic study of a loveless family; and *Hosanna* (1973), about the humiliation of a transvestite. Tremblay, who has had several of his plays acted successfully in English translations but allows them to be done in Quebec only in the original versions, has pointed out the symbolic relationship between his themes and Quebec's search for a political and cultural identity.

TRENEV, KONSTANTIN ANDREIVICH (1884–1945), Soviet dramatist and short-story writer, whose first play, dealing with an 18th-century peasant insurrection, was staged by the MOSCOW ART THEATRE in 1925. In 1926 the MALY THEATRE produced the first version of his *Lyubov Yarovaya*, which, after intensive alterations, became one of the outstanding productions of the Moscow Art Theatre in 1937. The story of a school teacher in the Revolution, caught between love for her White Russian husband and loyalty to her ideals, it became popular all over the U.S.S.R. and is still acted today. Trenev was also the author of *On the Banks of the Neva* (1937), one of the first Soviet plays to bring Lenin on to the stage. His later plays include *Anna Luchinina* (1941), *Meeting Halfway* (1942), and *The General* (1944), in which the chief character is the famous Russian Field-Marshal Kutuzov (1745–1813).

TRETYAKOV, SERGEI MIKHAILOVICH (1892–1939), Soviet poet and dramatist, who worked at the Proletkult (later the TRADES UNIONS) theatre in 1923–4 with Sergei Eisenstein for whom he wrote two propaganda plays—*Are You Listening, Moscow?* and *Gas Masks*. His best known play, *Roar, China!*, was produced at MEYERHOLD's theatre in 1926, though not by Meyerhold himself. Although melodramatic and written avowedly for propaganda purposes, it is a vivid historical document, presenting with intense conviction a conflict between Chinese coolies and foreign (British) imperialists which had taken place only two years previously. It was seen in translation in New York in 1930, and also in England. During the 1930s, Tretyakov, whose later plays are negligible, was the first to champion BRECHT in the Soviet Union and translated several of his plays which were published in 1934 as *Epic Dramas*.

TREVELYAN [Tucker], HILDA (1880–1959), English actress, who as a child appeared in JONES and Herman's melodrama *The Silver King* (1889) but was not seen again until 1894, when she toured in musical comedy. She played Wendy in the first production of BARRIE's *Peter Pan* (1904) and in many subsequent revivals, and was closely associated with many other plays by him, including *Alice-Sit-By-the-Fire* (1905), *What Every Woman Knows* (1908), and *The Twelve-Pound Look* (1910), in which she toured the variety theatres. She was for a time manager with Edmund Gwenn of the VAUDEVILLE THEATRE, but the venture was not a success, and she returned to the stage to appear in *A Kiss for Cinderella* (1916), again by Barrie, and in revivals of his *The Admirable Crichton, The Little Minister, Quality Street*, and *Mary Rose*. She also appeared many times as Avonia Bunn in PINERO's *Trelawny of the 'Wells'* (1898), which she first played as understudy to Pattie Browne in the original production, but later made very much her own. Among her other parts were Mrs Wyatt in Michael Baringer's *Inquest* (1931) and Barbara Fane in Ian HAY's *The Housemaster* (1936), which had a long run. She retired in 1939.

TREW, CHRISTINE PATTI, see LONGFORD, 6TH EARL OF.

TRICKWORK, an outstanding feature of the 19th-century English theatre, particularly in PANTOMIME, which was much admired abroad. Such stage tricks as depend mainly upon the working of ingenious mechanisms are dealt with under TRANSFORMATION SCENE and TRAPS; but the apotheosis of the trick comes when the brilliant timing and acrobatic skill of the actor is combined with the sureness and expertise of the experienced stage-carpenter.

The humblest example is the Roll-Out, in which a flap of loose canvas is left at the bottom of a piece of scenery, through which a player can suddenly roll from behind, and leap to his feet on stage. The Leap, in fact, is the supreme test of the trick player. In essence no more than an acrobat's jumping through a trap in the scenery, as in the Vamp Trap, it can, by the skilful interplay of a group of highly trained actors and the clever multiplication and placing of different types of trap, become an entertainment fit to stand on its own. Such an entertainment was the Dumb Ballet, of which acts like *Fun in a Bakehouse* or *Ki Ko Kookeeree* were the supreme examples. There is still in existence, dating from about 1871, a diagrammatic plot of the latter, made to be forwarded each week to the stage-manager of the theatre in which the troupe would next be appearing. It shows a coloured sketch of the scene, with the hinged flaps of at least eight varieties of trap, each used in dazzling succession as the actors went through their evolutions in the Victorian backyard which formed their setting. The best known exponents of the Dumb Ballet were the HANLON-LEES. London audiences had a chance to see something resembling the old Trickwork (which is still employed in the COMMEDIA DELL'ARTE theatre in the Tivoli Gardens, Copenhagen) when Peppino DE FILIPPO brought his company to the WORLD THEATRE SEASON of 1964 in *Metamorphoses of a Wandering Minstrel.*

TRILOGY, see TETRALOGY.

TRINITY SQUARE REPERTORY COMPANY, Providence, Rhode Island, one of the best and most adventurous companies in the U.S.A., was founded in 1964 by a number of local citizens who played in a rented theatre in a church building, and became fully professional later the same year. In 1973 it moved to a permanent home in the Lederer Theatre Complex, a renovated cinema which contains two auditoriums. The downstairs theatre, seating 297, re-creates the intimate atmosphere of the original theatre; the upstairs, seating up to 500, is flexible, allowing for continual readjustment of audience and performing areas. The year-round season offers revivals of classic and contemporary plays, and the theatre is especially noted for its encouragement of living American playwrights such as Sam SHEPARD and John GUARE; many world premières have been staged. Trinity Summer Rep was initiated in 1978 in the downstairs theatre, and the same season marked the beginning of the Trinity Rep Conservatory, a training programme for actors, directors, and playwrights. Project Discovery, for high school students, was started in 1966.

TRISSINO, GIANGIORGIO (1478–1550), Italian diplomat and writer, whose tragedy *Sofonisba* (1524) shows the first direct contact during the Renaissance in Italy with classical Greek drama. Its material is taken from Livy, but handled strictly in accordance with the rules of Greek tragedy, and though in itself a somewhat lifeless piece, it is important as having inaugurated the style of rigid adherence to Greek models followed by later writers of Italian tragedies. Trissino was also the author of a comedy, *I simillimi* (1547), based on the *Menaechmi* of PLAUTUS.

TRISTAN L'HERMITE, FRANÇOIS (*c.* 1602–55), French dramatist and man of letters. He was poor and a compulsive gambler, and after a stormy youth settled in Paris to make a living by his pen. His first play, a tragedy entitled *La Mariane* (1636) on the subject of Herod's jealous love for his wife, was produced at the MARAIS with MONTDORY and Mlle Villiers (Cathérine RAISIN) in the leading roles. It had an unprecedented success, competing for public favour with *Le Cid*, and for some time Tristan was considered a serious rival to CORNEILLE. *La Mariane* remained in the repertory until 1704, Herod providing an excellent part for a passionate and fiery actor. Of Tristan's other plays, *La Mort de Sénèque* (1644) is one of the best of the Roman tragedies after those of Corneille and RACINE. *Le Parasite* (1654), a comedy, is interesting as an attempt to introduce into France the stock figure of the sponger from Latin comedy. Two of Tristan's plays were in the repertory of MOLIÈRE's short-lived ILLUSTRE-THÉÂTRE.

TRITAGONIST, see PROTAGONIST.

TRIVELLINO, see LOCATELLI, DOMENICO.

TROCADERO PALACE OF VARIETIES, London, in Great Windmill Street. Known familiarly as the 'Troc', this was originally an entertainment hall built on the site of a 1744 tennis-court. It opened in 1829 as the Subscription Theatre, and throughout its career under different names it functioned always as a subscription club, to avoid legal difficulties over the licensing laws. It was nevertheless closed by the authorities in 1835, and became derelict until in 1850 it reopened as the Argyll Rooms, combining a restaurant with entertainments of various kinds; from 1863 until it was again closed down in 1878 it was a popular haunt of the *demi-monde*. In 1882 it reopened as the Trocadero Palace of Varieties, seating 600 people. The venture was not a success, and the building found itself badly situated after the construction of Shaftesbury Avenue; in 1895 the catering company of J. Lyons and Co. took it over, extending it with a large restaurant fronting the new road, and converted the old music-hall into a grill room. C. B. COCHRAN was later invited to organize cabaret entertainments in the restaurant for late diners, which he did with great success, his troupes of 'Young Ladies' becoming immensely popular. After the Second World War attendances at the Trocadero declined, and in 1965 the building was acquired by Mecca, an entertainments firm, which converted the grill room into a bowling alley.

TROEPOLSKAYA, TATIANA MIKHAILOVNA, see RUSSIA AND SOVIET UNION.

TROPES, see LITURGICAL DRAMA.

TROUPES PERMANENTES, see DÉCENTRALISATION DRAMATIQUE.

TROUVÈRES, see JONGLEURS.

TRUCKS, see BOAT TRUCKS.

TRUFFALDINO, see SACCHI, ANTONIO.

TUCCIO, STEFANO (1540–97), playwright, born in Sicily, who was important in the development of JESUIT DRAMA. A member of the order, he wrote a number of plays for production by the pupils of the Jesuit college at Messina, one of which, *Juditha* (1564), based on the story of Judith and Holofernes, called for a cast of 26 men, 6 women, an angel, a demon, and a large chorus. Tuccio's most ambitious work was a trilogy on the life of Christ—*Christus Natus*, on the Nativity, *Christus Patiens*, on the Crucifixion, and *Christus Judex*, on the Second Coming. This was produced at Messina in 1569 and was one of the first Jesuit plays to be translated into the vernacular, an Italian version being seen at Bari in 1584. It was subsequently played in German at Olmütz in 1603 and in Polish at Warsaw in 1752.

TUKE, Sir SAMUEL (?–1674), a gentleman at the Court of Charles II, who, on the suggestion of the king, wrote *The Adventures of Five Hours*, a tragi-comedy adapted from CALDERÓN, which was produced at LINCOLN'S INN FIELDS THEATRE on 8 Jan. 1663. Although it was a great success, being highly praised by PEPYS, it remained Tuke's only play, the rest of his time being spent in attendance on Charles II, for whom he performed many secret missions.

TUMANOV [Tumanishvili], JOSEPH MIKHAILOVICH (1905–), Soviet director, born in Georgia, from 1928 to 1932 a pupil and then a colleague of STANISLAVSKY, on whom he lectured in London in 1963 on the occasion of the Stanislavsky centenary celebrations. He was director of a number of Moscow theatres, including, from 1953 to 1961, the PUSHKIN THEATRE, where his productions included a dramatization of DICKENS's *Little Dorrit* (1953) and Ewan McColl's *The Train Can Stop* (1954).

TUMBLER, TUMBLING, see DROP and FLAT.

TURGENEV, IVAN SERGEIVICH (1818–83), Russian novelist and dramatist, who in 1843, while studying at Berlin University, published his first play, a romantic swashbuckling drama set in Spain. His second play, a satirical comedy in the style of GOGOL entitled *Penniless; or, Scenes from the Life of a Young Nobleman*, was published in 1846. He later wrote several short plays, in the style of MUSSET's *Comédies et Proverbes*, among them *Where It's Thin, It Breaks*; *The Bachelor* (1849), written for SHCHEPKIN and later revived with equal success by MARTYNOV and KARATYGIN; and *The Boarder* (1850). In 1850 Turgenev wrote his dramatic masterpiece, *A Month in the Country* (originally entitled *The Student*): it was published in 1869, but not staged until 1872. It is important as the first psychological drama in the Russian theatre, and in it Turgenev proved himself the forerunner of CHEKHOV in that he shifts the dramatic action from external to internal conflict. Because of battles with the censorship, imprisonment, and exile, Turgenev then gave up the theatre and *A Month in the Country* is the only one of his plays to be known outside Russia. It was first seen in London in 1926 and several times revived, notably by the OLD VIC at the NEW THEATRE in 1949, with REDGRAVE as Rakitin, and by the NATIONAL THEATRE in 1981. In New York the play was first produced by the THEATRE GUILD in 1930, and there also has been revived several times.

TURKEY. 1. TRADITIONAL. Several influences combined to produce the Turkish theatre and dance-drama, among them the shamanistic rituals of the steppes north of the Great Wall of China, the earliest home of the Turks, and the culture of the Anatolian peninsula, the bridge between Asia and Europe which the Turks have occupied for so long and which has seen the passage of so many migrant civilizations.

The influence of Islam, the adopted religion of the Turks, has been largely negative, since it has traditionally frowned on theatrical art, except for the special case of the Shi'ite passion-play.

Turkey's imperial expansion westward from the 15th to the 19th centuries brought her into contact with the culture of the countries over which she ruled, and this doubtless contributed elements of organized entertainment to what is now recognized as the theatre and dance of Anatolian Turkey. In the absence of precise records, nothing much can be said of the Turkish theatre from the 12th to the 14th centuries, though no doubt the entertainers normally in attendance on medieval rulers would have been in evidence at the Court. An early description of a Turkish dramatic performance, in which an actor impersonated the Byzantine Emperor Alexius, lying on a sofa during an illness, with doctors and attendants bustling about, can be found in the epic prose-poem the *Alexiad*, an account of the period 1069–1118 written by the Emperor's eldest daughter Anna Comnena (1083–1148).

During the Ottoman period any important event, such as the birth of a prince, a royal marriage, a conquest, or the arrival of a foreign ambassador, provided the opportunity for public festivities, which sometimes lasted as long as 40 days and 40 nights. These festivities included not only processions, fireworks, equestrian games, and hunting, but also dancing, music, recitals of poetry, and performances by jugglers, mountebanks, and buffoons. Here, as in Renaissance Italy, the chief forms of festal display were processions and shows, which might include a magnificent parade of masked figures on foot or in chariots—animals, giants, dragons—artificial islands, or ships which seemed to float, moved by men concealed within them, spectacular mock fights between Turks and Christians, with the Turks always the victors. From the 15th century onwards

Tahtakale was a fairground and centre of entertainment in Istanbul. All in all, the early public festivities included almost everything that was later to be found in the circus, and each type of entertainer had his separate designation. It is probable that in the intervals between the great city festivals for which they were engaged, these performers would travel from one village to another giving shows where they could. All were organized into Guilds or Colleges, according to their activities, with different degrees of elaboration in the several localities according to their traditions and their ethnic groups. In Turkey, drama was brought before the public in three ways—by storytellers, by PUPPETS (either rounded marionettes or flat SHADOW-SHOW figures), and by live actors. Imitation (called *taklit*) and mimicry of dialectic peculiarities are common to all three forms, and the plot always revolves round certain stock characters, easily recognized by the audience because of their unvarying costume and their introductory music. There are, in fact, no true plays, in the modern sense of written scripts, but only scenarios giving a bare outline of plot and action. The storyteller, puppeteer, or actor memorized a number of stock phrases, usually in rhymed couplets, expressive of the colourful everyday life and language of their time. They relied very little on properties, and hardly at all on scenery. Women's parts were played by men. The plays, which were performed in public squares, in coffee-shops, in inn yards, or in private houses, had little or no consecutive action, depending for their comic effect on lively slapstick and monologues or dialogue involving puns, repartee, misunderstandings, crude practical jokes, *double entente*, and interpolated quips. Longer speeches were delivered according to clearly formulated rules of intonation. Performances were freely interspersed with songs and dances.

The simplest form of Turkish theatre is the dramatically presented story told by a single speaker, the Meddah (literally praise-giver), a clever mimic who can portray, by gesture and change of voice, more than one person. Apart from that, he relies mainly on two props, a cudgel and a kerchief wrapped round his neck.

The shadow-show (*hayali zıl*), performed indoors, has always been one of the great delights of the Turkish people. The chief characters of the Turkish shadow-show are Karagöz, who represents the ordinary Turk, of any trade or profession, and Hacıvat, who represents the false, pretentious, affected Turk. The greater part of the play consists of dialogues between them, but they are supported by a number of minor characters, who are simple caricatures of different races, professions, and human defects. There were in Turkey many kinds of puppets including *el kuklasi* (hand-puppets), *ipli kukla* (string-puppets), *iskemla kuklasi* (jigging puppets), and *dev kukla* (giant puppets or effigies). There was also a form of puppet show—*kukla*—which is often mistaken for the Karagöz play. The early *kukla* player, the *kuklabaz* or *suretbaz*, had an assistant who mingled with the audience and carried on a conversation with a puppet alone on the stage who addressed the public. The chief characters of this puppet play are *Ihtiyar* (the old man) and *Ibiş* (his servant) who correspond to Karagöz and Hacivat in the shadow show, though they are not so well defined. The plots of the *kukla* were borrowed from popular tales and legends and even from the plays of the live actors, *Orta Oyunu*.

The *Orta Oyunu*, or middle play, to which in characters, plots, and atmosphere the Karagöz bears a strong resemblance, with the substitution of puppets for live actors, was preceded by some rudiments of dramatic art in the impromptu farces, where animal mimicry played an important part (the deer was the principal character, or the camel), which can still be seen in parts of Anatolia. There were also performances in the open, acceptable to a street-corner audience, in which the onlookers took part, and in which a prearranged comic situation was worked out with considerable improvisation. The actors, impersonating such types as the watchman, the tax-collector, or the fortune-hunter, might play practical jokes on the shopkeepers or owners with the object of obtaining money from them.

As time went on these farces, whether they were called *Kol Oyunu* (company play) or *Meydan Oyunu* (public-square play) or even *Taklit Oyunu* (mimicry play) were all absorbed into the *Orta Oyunu* (middle play), which bears some resemblance to the COMMEDIA DELL'ARTE or the Roman MIME. The characters represent various ethnic groups and the actors imitate their dialects while depicting their occupations and professions in daily life. Their 'stage' is an open space round which the audience stands or sits (the raised platform sometimes appears, as the result of Western influence). In some cases a low railing marks off the actors' enclosure, but there is no curtain or dressing room, and the actors get ready and sit waiting with the orchestra for their cues in full view of the audience. Stage scenery consists of a large screen to represent a house and a smaller one to represent a shop. The imagination of the audience has to supply the rest. Near the stage area, beside the main entrance to it, sits the orchestra. Each actor has his own music and introduces himself with a short song or dance. The chief character is Pişekâr, who opens the play and rarely leaves the stage. He wears bright colours—red trousers, a yellow gown and a variegated nightcap—and carries a cudgel with which he sometimes hits the other actors. The secondary character is Kavuklu, 'the man with the large wadded hat'. He is the comic character and his dialogues with Pişekâr form the main part of the play. Among the supporting characters the most important is Zenne (the woman), played by a man. A subsidiary actor may play several parts, changing his costume with his character. Each play consists of two parts, one the dialogue between Pişekâr and Kavuklu, the other, the play proper, in which the actors impersonate the different types found in Turkey.

2. MODERN. The evolution of the modern Turkish theatre and of the adoption of Western drama can conveniently be divided into two periods. From the early 19th century to 1923 the Levantines brought the contemporary theatre to educated Turkish audiences and held it up as a model. Public performances began in about 1839, under Abdül Mecit I, by which time Istanbul had two amphitheatres, a theatre for Italian opera, and a theatre for Turkish improvised plays. Modern Turkish drama began in 1859, when Ibrahim Şinasi completed his comedy *Şair Evlenmesi*. In one act, it makes fun of the inconveniences and embarrassments which may arise from the conventionally 'arranged' marriage. There is, however, evidence that plays had been written earlier. After 1859 Turkish drama developed rapidly, with the help of Armenian actors, actresses, and managers. Among the important playwrights were Racaizade Ekrem, Namık Kemal, Ali Bey, Ahmet Mithat Efendi, Ahmet Vefik Paşa, and Ebüzziya Tevfik. Parallel with this development some of the actors from the *Orta Oyunu* theatre, seeking to evade the stranglehold of the theatre manager's patent which covered only the legitimate theatre 'with a prompter', started putting on plays 'without a prompter'. These were impromptu performances in which a general outline was filled out with improvised incidents taken from local events or newspaper reports, or even street gossip. This form of theatre was known as *Tulûat Tiyatrosu*. In 1913–14 the Mayor of Istanbul, Cemil Paşa (Topuzlu) invited ANTOINE, then director of the Paris ODÉON, to come and organize a permanent municipal theatre and drama school in the city. This is still in existence, and today runs four theatres and an opera house.

The establishment of the Republic of Turkey in 1923, and the reforms of 1925–8, began a new era in the cultural development of the country. Dramatic art received official support and Western theatre became an integral part of Turkish life, the state assuming full responsibility for the actors' training and professional career. A State Conservatory (Devlet Konservatuari) was established in Ankara in 1936, by Paul Hindemith, with a dramatic section for the training of actors, dancers, and singers under Carl Ebert. All training is free, and takes five years, after which the students become salaried members of the various State companies. Two-thirds of the plays produced are from world drama, the rest are by Turkish playwrights, who have been encouraged in their activities by the payment of substantial royalties on their plays, and the sole use of one State theatre. There is also a department of drama, with a degree course, in Ankara University. During the last few years Turkish playwriting has undergone great changes. Among the present playwrights of distinction are Turgut Özakman, Refik Erduran, Cahit Atay, Haldun Taner, Orhan Asena, Hidayet Sayın, and Güngör Dilmen.

TURKISH KNIGHT, see MUMMERS' PLAY.

TURLUPIN [Henri Legrand] (*c.* 1587–1637), French actor, who as Belleville played serious parts in tragedy, but excelled in the parts of roguish valets and is best remembered for his farce-playing at the Hôtel de BOURGOGNE in partnership with GAULTIER-GARGUILLE and GROS-GUILLAUME. He probably played at the Paris FAIRS before joining a professional company in about 1615. He figures as himself, with the rest of the company, in Gougenot's *La Comédie des comédiens* (1631).

TURNHAM'S MUSIC-HALL, London, see METROPOLITAN MUSIC-HALL.

TURPIO, AMBIVIUS (*fl.* 2nd cent. BC), Roman actor, who appeared in the first productions of all the plays of TERENCE, as well as popularizing those of CAECILIUS STATIUS.

TUTIN, DOROTHY (1930–), English actress, who made her first appearance as the young Princess Margaret in William Douglas HOME's *The Thistle and the Rose* (1949), and was then with the BRISTOL OLD VIC and the OLD VIC in London. She first attracted attention by her brilliant performance as Rose, a young Catholic orphan in love with a middle-aged man, in Graham GREENE's *The Living Room* (1953), following it with an equally outstanding performance as Sally Bowles in VAN DRUTEN's *I am a Camera* (1954). A year later she appeared with great success as St Joan in ANOUILH's *The Lark* and as Hedvig in IBSEN's *The Wild Duck*. In 1958 she went to Stratford-upon-Avon, where she was an outstanding Viola in *Twelfth Night* and also played Juliet in *Romeo and Juliet* and Ophelia in *Hamlet* (all 1958), Portia in *The Merchant of Venice*, and Cressida in *Troilus and Cressida* (both 1960). Remaining with the company when it became the ROYAL SHAKESPEARE COMPANY, she gave an absorbing study of sexual frustration as Sister Jeanne in John WHITING's *The Devils*, based on Aldous Huxley's *The Devils of Loudun*, added Desdemona in *Othello* to her gallery of Shakespearian heroines, and played Varya in CHEKHOV's *The Cherry Orchard* (all 1961). She made her first appearance in New York in John BARTON's anthology *The Hollow Crown* (1963). At the Bristol Old Vic and then in London she gave a touching account of the young Queen Victoria's resilience and humour in William Francis's *Portrait of a Queen* (1965), in which she was also seen in New York in 1968. She played Rosalind in *As You Like It* (1967) and Kate in PINTER's *Old Times* (1971) for the R.S.C. and then made the first of two appearances (1971 and 1972) as Peter Pan—to which her frail, elfin looks were admirably suited—in the play by

BARRIE, in whose *What Every Woman Knows* she was seen in 1974. After starring in TURGENEV's *A Month in the Country* at the CHICHESTER FESTIVAL in 1974 and in London for PROSPECT in 1975, she played Cleopatra opposite Alec MCCOWEN's Antony for Prospect in 1977. She then joined the NATIONAL THEATRE company, where her roles included Madame Ranevskaya in *The Cherry Orchard*, Lady Macbeth, and Lady Plyant in CONGREVE's *The Double Dealer* (all 1978), and Lady Fanciful in VANBRUGH's *The Provoked Wife* (1980). A sensitive actress of great charm and versatility, she suffers from the dearth of worthwhile parts for women in modern plays.

TWAITS, WILLIAM (?–1814), English-born American actor, a low comedian who was appearing under the elder MACREADY in Birmingham and Sheffield when WOOD engaged him for the CHESTNUT STREET THEATRE in Philadelphia. He played there opposite Joseph JEFFERSON, and in 1805 made his first appearance in New York at a benefit night for DUNLAP, who describes him as short, with stiff carroty hair, a mobile and expressive face, and a powerful asthmatic voice which he used with great comic effect. He was at his best in farce and broad comedy, and in accordance with the tradition of the day he appeared as Polonius in *Hamlet*. He was also seen as Richard III and, somewhat incongruously, as Mercutio in *Romeo and Juliet*. More suited to his peculiar talents were the parts of the First Grave-digger in *Hamlet*, Dogberry in *Much Ado About Nothing*, Launcelot Gobbo in *The Merchant of Venice*, and Goldfinch in HOLCROFT's *The Road to Ruin*, all of which he played under COOPER's management at the PARK THEATRE, New York. He married Elizabeth, one of the daughters of Mrs Anthony WESTRAY, and shortly before his death appeared at the newly opened ANTHONY STREET THEATRE, New York.

TWENTIETH CENTURY THEATRE, London, see BIJOU THEATRE.

TWO-FOLD, see FLAT.

TWOPENNY GAFF, see GAFF.

TYL, JOSEF KAJETÁN (1808–50), Czech playwright and actor, and virtually the founder of the modern Czech theatre. His first play, *The Fair* (1834), was a patriotic folk-comedy with music by F. Skroup; one of its songs, 'Where is my home?', has become the Czech national anthem. It was produced by Tyl's own company, the Kajetán, which he continued to direct until 1837. During the troubled years of 1845 to 1851 his plays reflected the revolutionary struggle in town and country, the fairy-tale atmosphere of *The Bagpiper of Strakonice* (1847)—the source of Weinberger's opera *Schwanda the Bagpiper* (1927)—contrasting strongly with the more realistically portrayed characters of *The Bloody Trial, or the Miners of Kutná Hora* (1848) and *A Stubborn Woman* (1849). All these plays have been revived in recent years, as well as the historical plays *Jan Hus* (1848), a verse-drama which presents Hus as a fighter for the national and social rights of the Czech nation, and *The Bloody Christening, or Drahomíra and her Sons* (1849).

TYLER, ROYALL (1757–1826), American dramatist, author of *The Contrast*, a light comedy in the style of SHERIDAN's *The School for Scandal* and the first indigenous play to be produced in America. Tyler was a friend of Thomas WIGNELL, the leading comedian of the AMERICAN COMPANY, and it was probably owing to his influence that the play was given by them at the JOHN STREET THEATRE on 16 Apr. 1787. In return for his help Tyler gave him the COPYRIGHT. It was a success and was several times revived, though when given in Boston, Tyler's birthplace, it had, like *Othello*, to be disguised as a 'Moral Lecture in Five Parts'. It was published in 1790, George Washington heading the list of subscribers, and in 1917 was revived by Otis SKINNER. Tyler wrote several other plays, but none was as successful as *The Contrast*.

TYNAN, KENNETH PEACOCK (1927–80), English dramatic critic, who worked for the *Observer* from 1954 to 1958 and again from 1960 to 1963, and for the *New Yorker* from 1958 to 1960. He was literary manager of the NATIONAL THEATRE from 1963 to 1969 (after which he became resident in the United States): as a strong opponent of CENSORSHIP he fought, and lost, a battle for the production of HOCHHUTH's controversial play *Soldiers* at the National, and he was one of the play's co-producers when it was finally staged at the NEW THEATRE in 1968. Equally controversial, but a good deal more successful, was the 'evening of elegant erotica' *Oh, Calcutta!* which he devised. It opened in New York in 1969 and in London a year later, having long runs in both cities. His books included three volumes of collected criticism—*He That Plays the King* (1950), *Curtains* (1961), and *Tynan Right and Left* (1968); *A View of the English Stage 1944–63* and *The Sound of Two Hands Clapping* (both 1975); and *Show People* (1980).

TYRONE GUTHRIE THEATRE, Minneapolis, see GUTHRIE THEATRE.

U

UDALL, NICHOLAS (1505–56), English scholar, headmaster in turn of Eton and Westminster, and the author of *Ralph Roister Doister*. Written while he was at Eton for performance by the boys in place of the usual Latin comedy, it was probably performed there between 1534 and 1541, though efforts have been made to connect it with Udall's headmastership at Westminster and date it 1552. It was not printed until about 1566–7. This comedy, the first play in English to deserve that name, is much influenced by TERENCE and PLAUTUS, and turns on the efforts of a vainglorious fool to win the heart and hand of a wealthy London widow. Although Udall is known to have written several other plays, as well as dialogues and pageants for Court festivals, where he was connected with the Revels Office under Mary Tudor, they are lost, or survive only in fragmentary form. He was once credited with the authorship of *Thersites*, an interlude acted at Court in 1537 which is now considered to be the work of John HEYWOOD.

ULSTER GROUP THEATRE, Belfast, intimate theatre sited in the Minor Hall of the Ulster Hall, Bedford Street. This was opened in 1932 by Griffith Knight as the Little Theatre and offered weekly repertory for five years; as the Playhouse it continued under Harold Norway until 1939. A year later the hall was leased as the Group Theatre by three amateur groups, initially to share theatre facilities while presenting independent production programmes: the Carrickfergus Players, the Northern Ireland Irish Players, and the Ulster Literary Theatre. The last-named, founded in 1904 in emulation of the ABBEY THEATRE, Dublin, had achieved a considerable reputation internationally and had mounted first productions of nearly 100 plays. Eventually the three became fully amalgamated and by 1950 the company was a professional one, though working on a minimal budget.

James BRIDIE, stationed in Ulster during the Second World War, was closely associated with the Group Theatre, and other playwrights whose work was successful there included St John ERVINE, George SHIELS, and Joseph TOMELTY. Important plays from England and America were also produced by a company that included at various times Tomelty, Colin BLAKELY, and J. G. DEVLIN, the major directors being Harold Goldblatt and R. H. MacCandless. The Group Theatre was seen in London, at the ARTS THEATRE, in 1953 and at the EDINBURGH FESTIVAL in 1958, with a bitterly sectarian play *The Bonefire* by Gerald McLernon. The production of another contentious play about the 'troubles', Sam Thompson's *Over the Bridge* (also 1958), was hastily withdrawn amid controversy; the arrival of television in Northern Ireland concurrently affected box-office takings, and the company came near to collapse. It was rescued by two popular local comedians, James Young and Jack Hudson, who as joint managing directors from 1960 to 1972 staged productions of enormously successful local comedies, many specially commissioned from Sam Cree. The theatre was then closed for some years owing to civil disturbances. In 1976 it was taken over by the Belfast City Council and thoroughly modernized and re-equipped, reopening in 1978 to accommodate amateur companies. (See also LYRIC PLAYERS, BELFAST, and BELFAST CIVIC ARTS THEATRE.)

UNAMUNO Y JUGO, MIGUEL DE (1846–1936), Spanish philosopher, essayist, novelist, and playwright. The best known of his plays is *Sombras de sueño* (*Dream Shadows*, 1930). He also dramatized his own novel *Abel Sánchez* (1917), a modern version of the story of Cain and Abel, as *El otro* (*The Other*, 1926), and in *El hermano Juan* (*Brother Juan*, 1929) made a typically personal contribution to the development of the DON JUAN theme, stressing the relationship between character and author in a manner reminiscent of PIRANDELLO. Of Unamuno's eleven plays, nine were staged, but they were not on the whole as successful in performance as the adaptation of one of his short stories, *Nada menos que todo un hombre* (*Nothing Less Than a Total Man*), made by Julio de Hoyos.

UNDERHILL, CAVE (*c.* 1634–*c.* 1710), English actor, esteemed one of the best comedians of his day, of whom Colley CIBBER, in his *Apology*, has left an excellent account. He was well equipped by nature for such parts as Clodpate in SHADWELL'S *Epsom Wells* (1672) and Lolpoop in his *The Squire of Alsatia* (1688), and was much admired as Sir Sampson Legend in CONGREVE'S *Love for Love* (1695) and as the Grave-digger in *Hamlet*, in which he made his last appearance shortly before he died.

UNDERWOOD, JOHN (*c.* 1590–1624), boy-actor at the BLACKFRIARS, where he appeared in Ben JONSON'S *Cynthia's Revels* (1600) and *The*

Poetaster (1601). As an adult actor he joined the KING'S MEN and from 1608 to his death played with them regularly, though his only known role is that of Delio in WEBSTER's *The Duchess of Malfi* (1614). It has been conjectured that he played juvenile leads, princes, gallants, and libertines. He owned shares in the BLACKFRIARS, GLOBE, and CURTAIN Theatres.

UNION OF SOVIET SOCIALIST REPUBLICS, see RUSSIA AND SOVIET UNION.

UNION SQUARE THEATRE, New York, on the south side of Union Square, between Broadway and 4th Avenue. This opened as a VARIETY hall on 11 Sept. 1871, with the VOKES family in their pantomime-spectacle *The Belles of the Kitchen* as one of the early attractions. On 1 June 1872 A. M. PALMER took over, and for ten years made the theatre one of the finest in New York, with an admirable stock company and visits from all the best players in the United States. *The Two Orphans* by John Oxenford, produced in Dec. 1874, made a fortune for Palmer and a star of Kate CLAXTON. Other outstanding successes were a dramatization of Charlotte Brontë's *Jane Eyre*, with Charlotte Thompson, and *Camille*, based on the younger DUMAS's *La Dame aux camélias*, with Clara MORRIS. W. H. GILLETTE made his first appearance in New York under Palmer, as did Richard MANSFIELD, in Feuillet's *A Parisian Romance* (1883), playing a small part which made him famous overnight. This was the last play to be put on under the management of Palmer, who in 1885 disbanded the stock company and went to the MADISON SQUARE THEATRE. The Union Square was then used by travelling companies until in Feb. 1888 it was burnt down. It was rebuilt, but never regained its former brilliance, and under various names was mostly devoted to continuous VAUDEVILLE. It later became a BURLESQUE house and then a cinema, and in 1936 it was demolished.

UNION THEATRE, New York, see CHATHAM THEATRE (2).

UNITED STATES OF AMERICA. Dramatic performances, medieval in spirit, were mounted in colonial America from the early 16th century. They were, directly or indirectly, sponsored by the Catholic Church: for example, a series of religious plays (*La Anunciación de la Natividad de San Juan Bautista hecha a su padre Zacarias, La Anunciación de Nuestra Señora, La Visitación de Nuestra Señora a Santa Isabel*, and *La Natividad de San Juan*) was produced in Mexico on Corpus Christi Day 1538; the following year there was a *Conquista de Rodas*, followed by a *Conquista de Jerusalén*. As early as 1598 there was a performance in New Mexico of 'an original play on a subject connected with the conquest of New Mexico'; professional Spanish actors were presenting dramas in the New World colonies by the beginning of the 17th century; and in 1606 a MASQUE was performed at Port Royal in Acadia (see CANADA).

The ancient religious dramas persisted into the 1830s, but it was from a secular, British source that a distinctively American theatre emerged. Attempts to establish an English-speaking community in the New World began in the first decade of the 17th century. For a century conditions of life were too harsh to allow such luxuries as play-going, although the drama was not unknown; even in Puritan New England Increase Mather, to his alarm, heard in 1687 'much discourse of beginning Stage Plays', but nothing seems to have been attempted beyond one or two amateur, and probably crude, productions. The first professional actor to appear before the colonists was Tony ASTON, who was apparently in 'Charles-Town', South Carolina, in 1703 and later in New York. Since he had no company with him, he presumably gave monologues and readings only. Between 1699 and 1702 a Richard Hunter, otherwise unknown, applied for a licence to act plays in New York. A decade later WILLIAMSBURG, Virginia, had a playhouse where Charles Stagg, with his wife Mary, presented a few dramas. Thereafter theatrical activity began to spring up in many localities. In the 1730s Charleston had its DOCK STREET THEATRE, where such Restoration tragedies as OTWAY's *The Orphan* and recent comedies such as FARQUHAR's *The Recruiting Officer* could occasionally be seen. Judging by their successors on the frontier, these early playhouses consisted simply of a wooden hut, with a platform at one end concealed by a frayed curtain; the scenery amounted to some paper WINGS, while dingy illumination came from a few smoky candles.

By the middle of the century villages were becoming towns, travel had become safer, and a small but active leisured society was emerging: the time was ripe for the drama to take root in America. A succession of shortlived theatres opened in New York—the NASSAU STREET in 1750, Hallam's Nassau Street in 1753, CRUGER'S WHARF in 1758, the CHAPEL STREET in 1761—but during the period from 1750 to 1792 the main developments took place in the South. In Philadelphia in 1749 Thomas KEAN and Walter Murray constructed a playhouse out of a warehouse; after presenting an interesting repertory including *Richard III*, they formed the 'Virginia Company of Comedians', playing at Williamsburg, Annapolis, and elsewhere in the area. In 1752 Williamsburg welcomed another company headed by the elder Lewis HALLAM—the first theatrical group specially dispatched from London to America. During the 1750s this company, later headed by David DOUGLASS, introduced many plays to American audiences, including *King Lear, Romeo and Juliet, Hamlet*, and *The Merchant of Venice*; they were active, too, in the building of new theatres, among them the SOUTHWARK, or South Street, in Philadelphia in 1766 and the JOHN STREET in New York in 1767. Even during the War of Independence theatrical activity was maintained: in 1777 the John Street Theatre, renamed the Theatre Royal, was used by British soldiers, and the following year ADDISON's *Cato* was performed at Valley Forge before a 'splendid audience' which included George Washington himself.

With independence some of the infant States tried unsuccessfully to prohibit the playhouse. Lewis HALLAM junior, in partnership with John HENRY, broadened the scope of the AMERICAN COMPANY which, with the comedian Thomas WIGNELL and the second Mrs Owen MORRIS, won considerable fame, was honoured by Washington, and maintained for several years a virtual monopoly in New York, Philadelphia, Baltimore, and Annapolis. It presented in 1787 Royall TYLER's *The Contrast*, the first native comedy written for professionals, and attracted some minor English actors to America. By the close of the century the rapidly growing cities could support more than a single company. Wignell, with a troupe that included Mrs OLDMIXON, Thomas JEFFERSON, James FENNELL, Mrs MERRY, and later Thomas COOPER, started performances in 1794 in Philadelphia at the CHESTNUT STREET THEATRE. Modelled on the Theatre Royal at BATH, this new playhouse was a milestone in American theatrical development. Soon Wignell's company extended its activities southward and was responsible for the erection of theatres at Baltimore in 1794 and Washington in 1800.

In New York the PARK THEATRE, which opened in 1798, marked another advance in theatre architecture. Under the successive managements of the playwright William DUNLAP and the actor Thomas Cooper, the Park company became consistently popular; many stage stars made their reputations there, including the eccentric comedian William TWAITS and the actor-playwright John Howard PAYNE. In 1809 it was taken over by Stephen PRICE, who established the star system at the expense of the old STOCK COMPANY, bringing sensational actors like George Frederick COOKE and Edmund KEAN before American audiences. Though Philadelphia struggled to maintain its lead, the Park Theatre finally established New York as the dominating force in American theatre.

As the influence of the South declined, the stage gained acceptance in the formerly puritanical North, where plays had had to be disguised as moral lectures and theatres as exhibition rooms; but Boston's 'New Exhibition Room' of 1792 was in fact a playhouse. The enthusiastic Joseph Harper introduced the stage to Boston, Providence, and Newport; the Federal Street Theatre, erected in Boston in 1794, was followed two years later by the Haymarket. Albany, which had seen Hallam's company in 1769, got its first permanent theatre, the Green Street Theatre, in 1812. The famous 'Boston Museum', which opened in 1841, was to prove of considerable significance, and the new Federal Street Theatre of 1854 had an auditorium seating 3,000 and an excellent stage. Meanwhile, as the American frontier was pushed south and west, the theatre followed. As early as 1791 a group of French actors had appeared in New Orleans and soon the city boasted a St Pierre Theatre, erected in 1807, and in 1808 and 1809 a St Philippe Theatre and an Orleans Theatre. In 1817–18 Noah LUDLOW brought his American Theatrical Commonwealth Company to this largely French-speaking city, and his success led to the building in 1822 of the American Theatre by James H. CALDWELL who achieved almost complete power in the district; he was one of the first in America to install gas LIGHTING in a theatre, and his St Charles Theatre in New Orleans, which opened in 1835, was then the most magnificent playhouse in the United States.

In 1837, in partnership with Sol SMITH, Ludlow, who had been active in St Louis in the 1820s, opened a large new theatre there. It remained the only one until 1851, when John Bates opened another; this was followed in 1852 by the Varieties and the People's, and in 1866 by the Olympic.

By 1835 Washington had its first adequate playhouse, the National Theatre, followed in 1862 by FORD's, where Lincoln was assassinated. By 1858 Indianapolis had the 1,500-seat Metropolitan, and in 1875 boasted a fine Opera House. Chicago had no permanent theatre until 1847, but McVicker's Theatre was built in the growing city in 1857, and others followed, both before and after the disastrous fire of 1871; the Auditorium, erected in 1889, was reputed to be the largest theatrical building in America at the time. In the far west, San Francisco experienced similar theatrical development during the 1850s. The Mormon community in Utah positively encouraged dramatic entertainment, both amateur and professional, and finally built an imposing theatre in Salt Lake City which opened in 1862. The Mississippi SHOWBOATS represented a colourful and unique medium for public performances.

The greater part of this rapid development took place in the half century which led up to the American Civil War. Theatrical activity continued during the war, and in the Confederate States became a vehicle for propaganda, especially in Richmond, Virginia. After 1870 the face of the United States changed rapidly owing to the spread of the railways: touring companies from the east became a regular feature of life in the western and southern states, and although a few major stars emerged from these areas, notably Julia DEAN and Lotta CRABTREE, New York was finally confirmed as the source of the nation's theatrical activity. There the Park Theatre, where Kean's rival Junius Brutus BOOTH had made his New York début in 1821, was confronted with its first serious competitor in 1824, when the CHATHAM GARDEN THEATRE presented a distinguished company that included Henry WALLACK and the first Joseph JEFFERSON. In 1826 the Lafayette Theatre and the first BOWERY THEATRE opened, and in 1833 New York acquired an Italian Opera House, taken over four years later as the National Theatre by James WALLACK, founding member of a theatrical dynasty that lasted 60 years. The period 1820–70 was one of great actors and managers. At the OLYMPIC THEATRE the comedian William MITCHELL put on burlesques, farces, and adaptations of DICKENS's novels; William Evans BURTON presented at his own theatre from 1848 some of the most ambitious productions ever seen in the United States; John BROUGHAM, already well known on the English stage, opened New York's first LYCEUM THEATRE in 1850. Starting in 1856, Laura KEENE's theatre saw many triumphs, including Tom TAYLOR's *Our American Cousin* (1858) with E. A. SOTHERN and Joseph JEFFERSON III in the cast. In 1869 BOOTH'S THEATRE, New York home of the great Edwin BOOTH, opened, as did the FIFTH AVENUE THEATRE which, under the management of Augustin DALY, was to set unprecedentedly high production standards.

W. C. MACREADY in 1826, Charles KEAN in 1830, and Charles KEMBLE in 1832 began the steady stream of important English actors who visited the United States, and by the middle of the 19th century the stages of New York and London had become virtually indistinguishable. Gradually, however, an individual style grew up among American players. James Henry HACKETT invented the stage Yankee; Edwin FORREST became the most representative American actor of his age; Charlotte CUSHMAN was the great American tragedienne of the first half of the century; and Mrs John DREW almost provided in herself a history of the American playhouse. Although English plays, classic and contemporary, formed the staple fare both in New York and on the frontier, a native drama was in the making.

The first known drama by a native American was *The Prince of Parthia* (1767) by Thomas GODFREY. Royall Tyler's *The Contrast* (1787) made use of local types, and slightly later William Dunlap became established as America's first professional dramatist. *The Indian Princess* (1808) by James Nelson BARKER introduced Red Indians into American drama and, as *Pocahontas; or, the Indian Princess*, was presented at DRURY LANE in 1820. Other notable American playwrights of the 19th century include John Howard PAYNE, Robert Montgomery BIRD, and George Henry BOKER. Most American dramas were fashioned after English models, but the less literary form of the MINSTREL SHOW, originally the creation of T. D. RICE, used a folk idiom, however debased, that was purely American.

Black American actors were rarely seen on the stage until after the Civil War; although an all-Negro troupe, the African Company, had played in Shakespeare in New York in 1821, the tragedian Ira ALDRIDGE achieved real success only in Europe. Negro roles were customarily played by white actors in blackface and were initially restricted to comic stereotypes echoing that of the stage Irishman; adaptations of Harriet Beecher Stowe's abolitionist novel *Uncle Tom's Cabin* (1852) added the submissive servant, the light-fingered imp, and the tragic mulatto to the range of stage Negro clichés. The heroine of Dion BOUCICAULT's *The Octoroon* (1859) is a striking example of the last-named type: the play was devised so as to give offence to neither North nor South, and was even provided with alternative endings to suit different audiences. Not until the turn of the century was there any consistent attempt to present the real experiences of black Americans in dramatic terms, and the success of Thomas Dixon's *The Clansman* (1906, best known in D. W. Griffith's film version of 1915, *The Birth of a Nation*) showed that white audiences still resisted the development. Attitudes to the indigenous American, or Red Indian, were even more dismissive. Early plays about frontier life attempted a Rousseauesque conception of the noble savage, especially as portrayed by Edwin Forrest in STONE's *Metamora; or, the Last of the Wampanoags* (1829), and popular melodramas throughout the 19th century found a convenient villain in the barbarous 'red devil'. Arthur KOPIT's *Indians* (1968) was the first stage work to consider the Indian experience in the context of American history, drawing parallels with the current war in South-East Asia.

The last 30 years of the 19th century saw a rush of theatre buildings in New York. The FOURTEENTH STREET THEATRE, rebuilt as the Lyceum in 1873, the Eagle (1875, later the STANDARD), the MADISON SQUARE (1879), WALLACK's (1882), another LYCEUM (1885), the BROADWAY (1888), Harrigan's (1890, renamed the GARRICK in 1895), Abbey's (later the KNICKERBOCKER), the EMPIRE, and the AMERICAN (1893), Oscar Hammerstein's OLYMPIA (1895), and many others, made New York's theatrical provision second to none. Great American actors like Robert MANTELL, William GILLETTE, Richard MANSFIELD, Mrs FISKE, Julia MARLOWE, and Maude ADAMS, appeared on the same stages as visiting giants such as Henry IRVING, COQUELIN, Sarah BERNHARDT, Ellen TERRY, and Mrs Patrick CAMPBELL.

As in other countries, directors and managers became dominant during this period. Daly engaged for his theatres in New York and London such actors of note as John DREW, Adelaide NEILSON, and Ethel BARRYMORE; his directing was sensitive and meticulous, and he broke new ground by introducing to New York audiences *Saratoga* (1870) by Bronson HOWARD, forerunner of the modern American dramatists. Daly's main rival, apart from Lester WALLACK, was Albert M. PALMER, first at the UNION SQUARE THEATRE, later at the Madison Square Theatre, for which elaborate machinery was devised by Steele MACKAYE. The remarkable advances in theatrical realism characteristic of this period were typified in the productions of David BELASCO; and in response to changing styles of presentation a genuinely American drama began to evolve in the work of such playwrights as Augustus THOMAS, Clyde FITCH, and James A. HERNE. The adventurousness of the late 19th century was, however, stifled by the formation of the first THEATRICAL SYNDICATE in 1896; in spite of resistance from independent managers like Belasco, combinations of business interests assumed control of the commercial theatres not only in New York but all over North America, setting up defined touring circuits catering for mass audiences.

Between 1900 and 1928 the number of theatres in active use in New York grew from 43 to 80: the most prestigious new arrivals included the REPUBLIC (1900), the MAJESTIC (1901), Daniel FROHMAN's new LYCEUM (1903), the Stuyvesant (later the BELASCO) and the GAIETY (1907), MAXINE ELLIOTT's (1908), the New (1909, later the CENTURY), the Globe (1910, later the LUNT-FONTANNE), the WINTER GARDEN and the CORT (1912), the BOOTH and the LONGACRE (1913), HENRY MILLER's (1918), the MUSIC BOX (1921), the EARL CARROLL (1922), the IMPERIAL (1923), and the Guild (1924, later the ANTA). The definition of Broadway as a synonym for highly professional, though insubstantial, theatrical entertainment was consolidated during these years, alongside the most exciting developments the American theatre had yet seen.

Reaction to the increasing commercialism of Broadway and the trusts came from professional actors and directors working outside the big management combines and from amateurs of the Little Theatre movement all over the United States. In New York Winthrop AMES made a brave attempt to re-establish repertory at the New Theatre in 1909; more lasting influence was exerted by the WASHINGTON SQUARE PLAYERS, formed in 1915 and later to evolve into the THEATRE GUILD, and by the PROVINCETOWN PLAYERS and the NEIGHBORHOOD PLAYHOUSE, both formed in 1916; all these groups brought classic plays and modern European work before audiences as well as fostering the development of new American playwrights. Foremost among them was Eugene O'NEILL, who probably did more than any of his predecessors to establish a native American drama on the stages of the world and paved the way for other dramatists who, however different their preoccupations, were also distinctively American. Most were concerned with contemporary reality, among them Sidney HOWARD, Robert SHERWOOD, Maxwell ANDERSON in his early plays, though Elmer RICE earned a reputation for vigorous experimentalism. A flowering of talent in scenic design supported the new writers, with the emergence of Boris ARONSON, Norman BEL GEDDES, Robert Edmond JONES, Lee SIMONSON, and others, whose work showed the influence of the innovatory ideas flourishing in post-war Europe.

Broadway in the 1930s discovered a succession of excellent comedy writers, some of whose plays have stood the test of time: S. N. BEHRMAN, Philip BARRY, George S. KAUFMAN, George ABBOTT, and John VAN DRUTEN. Like Europe, however, the United States was by now moving towards greater seriousness and an increase in political awareness in response to economic instability and the prospect of another war. Marc CONNELLY, Paul Eliot GREEN, Lillian HELLMAN, and John STEINBECK represent a fresh provocativeness in social drama, while the left-wing Theatre Union and GROUP THEATRE encouraged overtly political statements from a number of new playwrights led by Clifford ODETS. Signs of a nationwide dramatic movement emerged under the aegis of the FEDERAL THEATRE PROJECT: its abrupt dissolution in 1939 was a severe blow to the non-commercial American theatre.

During the Second World War Broadway flourished on a diet of comedy and musicals. Two plays of 1939, *Life with Father* by Howard LINDSAY and Russel Crouse and Kaufman and HART's *The Man Who Came to Dinner*, gained enduring popularity, while Joseph Kesselring's *Arsenic and Old Lace* (1941) became an instant classic among comic melodramas. Serious dramatists had fewer opportunities, and the only new names of any importance in this area were William SAROYAN and Thornton WILDER, although Lillian Hellman, Sidney KINGSLEY, and Elmer Rice continued to be active during the war years. The late 1940s were marked by the appearance of O'Neill's last plays after a silence of 12 years and by the emergence of Tennessee WILLIAMS and Arthur MILLER.

Outside New York the growth of regional, civic, and university theatres gathered pace throughout the 1930s and 1940s. Following the lead given by G. P. BAKER's '47 Workshop' which started at HARVARD in 1912, theatre arts courses became customary in universities and colleges, offering opportunities for lively experimental work outside the mainstream commercial theatre.

The New York theatre entered the 1950s in a depressed state. Production costs were climbing, and television was not only wooing the audience away from live entertainment but also taking over theatres for conversion to television studios. The political situation—the Cold War internationally and the investigations of the House Committee on UnAmerican Activities at home—deepened the general gloom. Serious drama was still dominated by the psychological plays of Tennessee Williams and by the strong social and moral concerns of Arthur Miller. Eugene O'Neill was posthumously represented on Broadway by his magnificent autobiographical drama *Long Day's Journey into Night* (1956); its success led to the production of three of his uncompleted plays also dealing with tortured family life. The major new dramatist of the 1950s was William INGE, who wrote a series of plays dealing sensitively with small-town life. Robert ANDERSON made a great impression with *Tea and Sympathy* (1953), but his subsequent plays were less successful. Paddy Chayefsky had some success with stage plays, particularly *The Tenth Man* (1959), but later moved to writing television and film scripts, and William Gibson (1914

–) wrote a lively, poignant two-character comedy, *Two for the Seesaw* (1958), followed by *The Miracle Worker* (1959), the story of Helen Keller and her teacher.

The 1950s also saw productions of new plays by Sidney Kingsley, John Van Druten, and Archibald MacLeish. Other notable works included Carson McCullers's stage adaptation of her novel *The Member of the Wedding* (1950); *Inherit the Wind* (1955), based by Jerome Lawrence and Robert E. Lee on the famous Scopes trial of 1925; the dramatization of *The Diary of Anne Frank* (1955) by Frances Goodrich and Albert Hackett; *Look Homeward, Angel* (1957) by Ketti Frings, based on Thomas Wolfe's autobiographical novel; and *A Raisin in the Sun* (1959), by Lorraine Hansberry, a young black playwright, who wrote about a struggling black middle-class family. Jean ANOUILH was well represented on the New York stage during the 1950s and GIRAUDOUX's *Tiger at the Gates*, transferred from London, was warmly welcomed in 1955. Not so well received was Samuel BECKETT's *Waiting for Godot*, in 1956, which confused and irritated Broadway audiences while alerting them to a new style of drama, the Theatre of the ABSURD.

American pre-eminence in MUSICAL COMEDY, established in the 1930s by Cole Porter, Jerome Kern, George and Ira Gershwin, and others, was reinforced in the 1950s by a succession of memorable musicals: *Guys and Dolls* and *Call Me Madam* (both 1950), *The King and I* (1951), *Wonderful Town* (1953), *My Fair Lady* and *The Most Happy Fella* (both 1956), *West Side Story* (1957), and *The Sound of Music* (1959), all of which had long runs in both New York and London.

In the mid-1950s off-Broadway theatre groups began to attract the attention of critics and the theatre-going public with new experimental plays and innovative productions of revivals. The PHOENIX THEATRE company, formed in 1953 by T. Edward Hambleton and Norris Houghton, produced a number of European classics. The LIVING THEATRE, founded in 1947 by Julian Beck and Judith Malina, had built up a reputation for innovative work in the United States and Europe; in 1959 they staged a chilling drama about drug addiction, Jack GELBER's *The Connection*. Marc Blitzstein's adaptation of BRECHT and Weill's *The Threepenny Opera* (1955) ran for over 2,600 performances, but the record holder is *The Fantasticks*, based by Tom Jones and Harvey Schmidt on ROSTAND's *Les Romanesques*, which opened in 1960 and was still playing in the 1980s.

Regional theatres were active in the 1950s, but had not yet achieved national prominence except for a few outstanding companies, several of them run by women. Margo JONES had founded the precursor of THEATRE-IN-THE-ROUND in Dallas, Texas, in 1945. Two years later the NINA VANCE ALLEY THEATRE opened in Houston, Texas, and in 1950 Zelda Fichandler founded the ARENA STAGE in Washington, DC. The Actors Workshop in San Francisco was formed by Jules Irving and Herbert Blau in 1952, flourishing until 1966 when Irving and Blau, with their company, were invited to take over the VIVIAN BEAUMONT THEATRE at the Lincoln Center, New York, unfortunately without success.

The prospects for Broadway in the early 1960s continued to be bleak. In June 1960 AMERICAN ACTORS' EQUITY called a 10-day strike, the first since the union was formed in 1919, and as the decade progressed newspaper and transport strikes also had a deleterious effect on the Broadway scene. To combat these problems producers began to stage low-priced previews in New York instead of going to the expense of out-of-town tryouts. The Theatre Development Fund was started in 1967 by private and public endowment to subsidize worthwhile plays in both the commercial and the non-profitmaking theatre and to attract new audiences by reduced-price tickets. The Fund has been very successful and has helped to start similar services in Boston, Los Angeles, Minneapolis-St Paul, and San Francisco.

One of the few bright spots in the theatrical scene was the appearance off-Broadway in 1960 of Edward ALBEE. His first full-length play to be produced on Broadway, *Who's Afraid of Virginia Woolf?* (1962), gained several awards and was followed by several more successes. Other serious works of the decade included *All the Way Home* (1960), adapted by Tad Mosel from James Agee's novel *A Death in the Family*, and Frank D. Gilroy's family drama *The Subject Was Roses* (1964). *The Great White Hope* (1968) by Howard Sackler (1929–82), based on the life of the famous black fighter Jack Johnson, came to Broadway from Washington's Arena Stage, signalling the emergence of regional theatre as a source of Broadway plays. Broadway also turned to London for nourishment, a practice that has increased until Broadway audiences sometimes see successful West End productions while they are still running in London. Among the imported productions of the 1960s were Shelagh Delaney's *A Taste of Honey* in 1960, IONESCO's *Rhinoceros* and Robert BOLT's *A Man for All Seasons* in 1961, Peter BROOK's production of Peter WEISS's *Marat/Sade* in 1965, Harold PINTER's *The Homecoming* and Tom STOPPARD's *Rosencrantz and Guildenstern Are Dead* in 1967, and Jay Allen's *The Prime of Miss Jean Brodie* (based on a novel by Muriel Spark) in 1968. In many cases this was New York's first glimpse of a new group of young British actors who were to have important transatlantic careers.

Neil SIMON, a new writer who established his pre-eminence in sharp contemporary comedy with *Barefoot in the Park* (1963) and *The Odd Couple* (1965), was also the author of books for musicals, such as *Sweet Charity* (1966) and *Promises, Promises* (1968). Among the other outstanding musicals of the 1960s were *Hello, Dolly!* (1964), *Cabaret*

(1966), and *Hair* (1968), the last of which startled the traditional Broadway audience with its ear-splitting sound levels and joyful display of nudity. It had first been seen off-Broadway, at Joseph PAPP'S PUBLIC THEATRE, and marked the emergence of another major tributary source for Broadway successes.

In some seasons off-Broadway theatres presented more new playwrights than appeared on Broadway and new repertory and experimental groups received increasing attention from the audience and the press. The Judson Poets' Theatre, founded in 1961 at Judson Memorial Church by the Revd Al Carmines and Robert Nichols, offered original and witty musical commentaries on controversial social issues; the LA MAMA EXPERIMENTAL THEATRE CLUB, created in 1962 by Ellen 'La Mama' Stewart, offered scope for AVANT-GARDE artists; and Joseph Chaikin's OPEN THEATER (1963) also presented avant-garde plays such as *The Serpent* (1968) by Jean-Claude VAN ITALLIE, the company's leading playwright. In Central Park Papp's New York Shakespeare Festival continued his policy, begun in 1956, of presenting Shakespeare's plays free on summer evenings. The AMERICAN PLACE THEATRE, formed by Wyn Handman and Sidney Lanier to find, develop, and stage new plays, opened in St Clement's Church in 1964, and in 1971 took over its own theatre in the Broadway area. In 1967 the Ford Foundation provided funds for the Negro Ensemble Company to offer a platform for works on racial themes and problems and to provide professional training for black artists; the company was instrumental in expanding the audience, both black and white, for such work. Other off-Broadway groups that emerged during the 1960s include the Roundabout Theatre in 1965; CSC Repertory in 1967; CIRCLE REPERTORY in 1969; and the Manhattan Theatre Club in 1970. All these groups developed a faithful subscription audience, although still needing subsidies and private donations to exist. In 1969 the Public Theatre presented *No Place to Be Somebody*, Charles Gordone's drama set in a Greenwich Village bar, the first off-Broadway play to win a PULITZER PRIZE.

The 1960s saw the birth of multi-theatre cultural centres throughout the United States, led in 1959 by the DALLAS Theater Center designed by Frank Lloyd Wright. In New York City in 1965 the Vivian Beaumont Repertory Theatre at the Lincoln Center for the Performing Arts joined the Metropolitan Opera, the New York Philharmonic, and the New York City Opera and Ballet Companies in a cultural complex built a few blocks north of the Broadway area. The Los Angeles Theatre Group/MARK TAPER FORUM, with Gordon Davidson as artistic director, began operations in 1966 and the John F. Kennedy Center in Washington, DC, in 1971. Resident professional companies were also building or expanding their facilities, establishing solid reputations, and acquiring new audiences: the McCarter Theatre Company in Princeton, New Jersey, in 1960; the GUTHRIE THEATRE in Minneapolis, Minnesota, the SEATTLE REPERTORY THEATRE, Washington, under Stuart Vaughan, and the Hartman Theatre Company in Stamford, Connecticut, all in 1963; the Free Southern Theatre, founded in 1963 to advance the cause of civil rights in the South, which moved from Jackson, Mississippi, to New Orleans in 1965; the ACTORS THEATRE OF LOUISVILLE, Kentucky, and the TRINITY SQUARE REPERTORY COMPANY, formed by Adrian Hall in Providence, Rhode Island, both of which began operations in 1964; the LONG WHARF THEATRE in New Haven, Connecticut, in 1965; William Ball's AMERICAN CONSERVATORY THEATRE in San Francisco in 1966; and the BOSTON Repertory Theatre in 1970.

In 1960 the New York State Council on the Arts was established to provide financial and technical aid to individuals and organizations in the State, proving so successful that when the National Endowment for the Arts was created by Congress in 1965 it was modelled on the Council in many significant respects. Both units gave excellent support to the arts throughout the 1970s, but severe budget cuts in 1981 were immediately reflected in the reduction or elimination of grants. In the private sector the Theatre Communications Group, founded in 1961 by the Ford Foundation, created a strong network to share information and expertise among non-commercial theatres throughout the country. At the privately endowed Eugene O'Neill Theatre Center in Waterford, Connecticut, an annual National Playwrights Conference was convened in 1965 and a National Critics Institute where new and established writers could exchange ideas opened in 1968. From 1967 the Center also sponsored the National Theatre of the Deaf, a touring company performing in sign as well as in spoken language.

Broadway's financial and artistic doldrums continued into the 1970s, in spite of a new source of funds from motion picture companies investing heavily in stage shows in order to secure future film properties. By the end of the decade the top ticket prices reached $30, although the Theatre Development Fund continued to offer reduced-price tickets for many worthwhile shows. Through special dispensations in local building regulations three new theatres were built in the Times Square area, the cavernous URIS THEATRE, with Joseph E. Levine's Circle-in-the-Square Uptown in its basement, in 1972, and the MINSKOFF THEATRE in 1973. Many of the best Broadway productions of this period originated off-Broadway, among them *Sticks and Bones* (1971), David RABE's play about a haunted Vietnam veteran, Jason Miller's drama of disillusionment *That Championship Season* (1972), and Ntozake Shange's collage of poetry and prose *For Colored Girls Who Have Considered Suicide: When the Rainbow is Enuf* (1976). All three originated at the Public Theatre. Other off-Broadway plays that moved to Broadway were *The River Niger* (1973), Douglas Turner Ward's drama about the destruction of a black family; Miguel Pinero's prison drama *Short Eyes* (1974); *The Elephant Man*, Bernard Pomerance's portrait of John Merrick (which had come to off-Broadway from London) in 1979; *Home* (1980) by Samm-Art Williams, a drama of the black experience in North and South; *Talley's Folly* and *The Fifth of July* (both 1980), two plays in a southern family chronicle by Lanford Wilson. (The third play in the sequence, *A Tale Told*, opened off-Broadway in 1981.) Among Broadway's transatlantic imports were Peter SHAFFER's *Equus* in 1974 and *Amadeus* in 1980; Hugh LEONARD's *Da* in 1978; *Whose Life Is It Anyway?*, a polemical drama by Brian Clark, in 1979; and Harold Pinter's *Betrayal* in 1980. Three very powerful plays arrived on Broadway from regional theatres: Michael Cristofer's *The Shadow Box* (1977), set in a hospice for terminally ill cancer patients, Arthur KOPIT's study of a stroke victim *Wings* (1979), and Mark Medoff's play about the world of the deaf *Children of a Lesser God* (1980).

While Broadway seemed preoccupied by illness, death, and the disintegration of the family, off-Broadway was examining the place of women in modern society. Wendy Wasserstein's *Uncommon Women and Others* (1977) and *Isn't It Romantic?* (1981) dealt with a group of women in college and afterwards; Marsha Norman's *Getting Out* (1978) concerned a young ex-convict trying to regain a foothold in society; and Beth Henley's *Crimes of the Heart* (1980) was a modern American version of CHEKHOV's *Three Sisters*. Gretchen Cryer and Nancy Ford presented a musical on the subject in *I'm Getting My Act Together and Taking It on the Road* (1978).

Musical comedy was dominated by Stephen SONDHEIM, who created a string of artistic and often commercial successes. Broadway seemed, however, to be turning from true musicals back to the old, much more economical, REVUE format—a *mélange* of songs by one or more composers: *Rodgers and Hart* (1975), *Side by Side by Sondheim* (1977), Bob Fosse's *Dancin'* (1978), *Perfectly Frank* (Loesser) (1980), *Duke Ellington's Sophisticated Ladies* (1981); or skits and songs as in *Sugar Babies* (1979), a revival of the old-fashioned BURLESQUE. Even Michael Bennett's highly acclaimed *A Chorus Line* (1975) was little more than a succession of audition pieces, albeit with touching emotional details.

Off-Broadway's success in spawning productions that went on to gain awards on Broadway, together with a more conservative economic and political atmosphere generally, brought a shift away from the experimental and avant-garde aims from which off-Broadway had emerged and a new movement, off-off-Broadway, began to fill the vacuum left by this 'upward mobility'. The Off-Off-Broadway Theatre Alliance (OOBA) was established in 1972 as a co-operative organization to provide administrative, public relations, and information services to approximately 65 member groups, while the Theatre Development Fund initiated a monthly calendar and directory of the various groups and their offerings. In 1978, as part of an effort to revitalize the Times Square area and in recognition of the growing importance of off-off-Broadway theatre, a row of former burlesque houses and massage parlours was converted into eight theatres for off-off-Broadway groups.

Regional theatre continued to expand throughout the 1970s: the Theatre Communications Group reported that in 1977 there were non-commercial professional theatres in 34 states plus Washington, DC, and that touring companies from these groups visited 13 more states. In this expansive atmosphere, various cities have begun to emulate New York in offering both commercial and experimental theatre, audiences throughout the country having developed to the point where they demand more than just a touring production of a Broadway hit; regional theatre is thus at last fulfilling its early promise as a vital element of the American theatre.

UNITI, company of the COMMEDIA DELL'ARTE which has caused much discussion, since it is possible that the name refers not to a distinct company, as did GELOSI and CONFIDENTI, for example, but to a combined troupe of actors from different companies formed for a special occasion. Because of this some critics consider the Uniti to be part of the history of the Gelosi, some of the Confidenti. It remains certain that a company under that name was directed by Drusiano MARTINELLI and his wife Angelica in 1584, and for some years afterwards. A company calling itself the Uniti, noted in 1614, may have been a temporary amalgamation. Among its members were Silvio FIORILLO and his son.

UNITIES, of time, place, and action, elements of drama introduced into French dramatic literature by Jean MAIRET. They demand that a play should consist of one action, represented as occurring in not more than 24 hours, and always in the same place. This doctrine was said to be based on the *Poetics* of ARISTOTLE, though in fact he insists only on the unity of action—and EURIPIDES disregarded even that—merely mentions the unity of time, and says nothing about the unity of place, though this was to a certain extent imposed on the Greek dramatist by the presence of the CHORUS. The observance of the unities, defined by BOILEAU in his *Art poétique* (1674), became an essential characteristic of French classical tragedy (though both CORNEILLE and RACINE ignored them when they wished) and found its way, with neo-classicism, into Spain and Italy. In England the influence of Shakespeare, who certainly had no regard for the unities of time and place and very little for that of action, was strong enough to counteract the efforts of Restoration writers of tragedy to force English plays into the French mould.

UNIT SETTING, see DETAIL SCENERY.

UNITY THEATRE, Glasgow, left-wing theatre group formed in 1941 by the war-time amalgamation of four amateur companies, the Transport, the Clarion, the Jewish Institute, and the Workers' Theatre. It rapidly achieved success, and in the summer of 1945 presented in London a highly praised production of GORKY's *The Lower Depths*. In the following year a full-time professional group was formed, which, like the SCOTTISH NATIONAL PLAYERS, toured throughout Scotland as well as playing in Glasgow. Its biggest success came with *The Gorbals Story* (1948), a study of life in an overcrowded slum, written by Robert MacLeish and produced by Robert Mitchell, a mixture of broad comedy and pathos reminiscent of the early plays of O'CASEY. This was revived many times, toured all over Britain, taken into the West End briefly, and eventually filmed. Financial insecurity and the lack of a permanent home led to Unity's eventual downfall, and by the early 1950s it had ceased to be effective, although a number of its actors continue to make a vital contribution to the Scottish theatre.

UNITY THEATRE, London, in Goldington Street, St Pancras, small theatre holding 200, having a proscenium stage with a width of 30 ft and a stage depth of 12 ft. An amateur club, it developed in the 1930s from a left-wing group which had acquired a converted church hall in Britannia Street, King's Cross, and opened it as the Unity Theatre on 19 Feb. 1936. The only noteworthy production there was Ben Bengal's *Plant in the Sun*, in which Paul ROBESON played the lead. Shortly after, the group moved to new premises, opening on 25 Nov. 1937, and a year later presented not only the first BRECHT play to be seen in London—*Señora Carrer's Rifles*—but also the first English LIVING NEWSPAPER, a documentary on a London bus-strike. Among other notable first productions at Unity, which at one time achieved professional status, have been ODETS's *Waiting for Lefty* (1936), O'CASEY's *The Star Turns Red* (1940), SARTRE's *Nekrassov* (1956), and ADAMOV's *Spring '71* (1962). A scheme was then put in hand to develop the theatre as a cultural centre for the Labour movement in London, and rebuilding had begun when in 1975 the premises were destroyed by fire.

UNIVERSITY DEPARTMENTS OF DRAMA. The first attempt to present theatre history and practice as an academic study leading to a university degree was made in the United States, when the Carnegie Institute of Technology, Pittsburgh, established in 1914 a department of drama offering a degree in theatre arts. More influential, however, was the establishment in 1925 of a postgraduate Department of Drama at YALE, headed by George Pierce BAKER. The movement grew so quickly that when in 1936 the American Educational Theatre Association (A.E.T.A.) was founded, it had 80 members. Now, as the American Theatre Association (A.T.A.) dealing with educational and non-commercial theatre nationwide, it covers 1,600 drama departments in United States colleges and universities.

In 1960 the McCarter Theatre at Princeton became the first professional theatre in the United States to be entirely under the control of a university. There was an expansion of university and college theatre building during the next 20 years, at the end of which virtually every educational institution with a theatre department had its own theatre, its facilities often comparing favourably with those of a Broadway or regional theatre. (See, for example, the KRANNERT CENTRE at the University of Illinois.) Some of the university theatres also function as civic theatres, serving the surrounding districts as well as the student body, often under a professional director. In some cases professional actors have been engaged to play in a university theatre, in addition to student groups.

In Britain the gradual acceptance of the theatre as a subject for academic study resulted in the establishment in 1946–7 of the first Department of Drama, at Bristol University, which works closely with the BRISTOL OLD VIC; it was followed after 13 years by the creation at Bristol of the first Chair of Drama at a British university. A Department of Drama was established in 1961–2 at MANCHESTER, with Hugh HUNT as its first Professor. Similar departments have now been set up at Hull, Birmingham, Glasgow, Exeter, Lancaster, and various other universities, and there are two within the University of London. Many of these universities also have Chairs of Drama. By the end of the 1970s almost 20 British universities, and a similar number of other institutions of higher education, offered drama within a B.A.degree, many more offering drama in a B.Ed.degree. Many universities now have a theatre open to the public, among them the University Theatre, Manchester, the NORTHCOTT at Exeter, OXFORD PLAYHOUSE, and the Sherman at Cardiff (see WALES).

There has also been a great expansion of university theatre studies throughout the world, particularly in Canada, France, Germany, and Italy. Theatre is studied at universities in the U.S.S.R. and other Communist countries and in places as diverse as India, Japan, Korea, and Nigeria.

UNIVERSITY WITS, name given to a group of somewhat dissolute Elizabethan playwrights educated at Oxford or Cambridge and therefore contemptuous of such 'uneducated' writers as Shakespeare and Ben JONSON. Among them were GREENE, NASHE, PEELE, and, the most important in the development of the English theatre, MARLOWE. A lesser member of the group was Thomas Lodge (*c.* 1557/8–1625), on whose pastoral romance *Rosalind* (1590) Shakespeare based *As You Like It*.

UNREASON, Abbot of, see MISRULE, ABBOT OF.

UNRUH, FRITZ VON (1885–1970), German dramatist. A cavalry officer, coming from an ancient military family, he resigned his commission in 1911 and began his writing career with the heroic, patriotic plays *Offiziere* (1911) and *Louis Ferdinand, Prinz von Preussen* (1913), which illustrated the tensions between the older and younger generations in their attitude not only to war but also to material, as opposed to spiritual, progress. The First World War brought about a change of heart, and von Unruh became an exponent of EXPRESSIONISM, creating visions of the 'new Man' in classical form. Returning to SCHILLER and KLEIST, he wrote *Ein Geschlecht* (*One Generation*, 1918), the first part of a planned trilogy of which only the second part, *Platz* (1920), followed. This drama of a family focuses on the matriarch and her children, exemplifying through the symbolic treatment of love and hate, incest and matricide, all the contemporary themes of pacifism and revolution raised to the level of myth.

UP STAGE, see STAGE DIRECTIONS.

URBAN, JOSEPH (1872–1933), architect and stage designer, born in Vienna, where he worked for many years before going to Boston in 1911–12 to design sets for the opera there. He was also in New York, where he designed settings for the *Ziegfeld Follies*, the Metropolitan, and James K. HACKETT's Shakespearian productions. He built the ZIEGFELD THEATRE in New York, with its egg-shaped interior, and introduced much of the new Continental stagecraft to American theatre-audiences, making use of broad masses of colour and novel lighting effects on costume and scenery.

URBÁNEK, FERKO, see CZECHOSLOVAKIA.

URE, MARY, see OSBORNE, JOHN.

URIS THEATRE, New York, at 51st Street, west of Broadway. This theatre, seating 1,900, was designed under new building laws which allow playhouses to be built within another type of structure. The CIRCLE-IN-THE-SQUARE theatre is in the basement of the same building. It opened on 19 Nov. 1972 with *Via Galactica*, a musical set in 2972, which ran for only a week. For some years it had a luckless career, though the musical *Seesaw* and the stage version of the film *Gigi* (both 1973) had moderate runs, and Gershwin's *Porgy and Bess* (1976) was performed with virtually the complete score for the first time. A revival of *The King and I* in 1977 had some success, SONDHEIM's *Sweeney Todd* (1979) had a *succès d'estime*, and in 1981 the theatre scored a big hit with a production by PAPP of GILBERT and Sullivan's *The Pirates of Penzance* which had originated at the New York Shakespeare Festival.

U.S.S.R., see RUSSIA AND SOVIET UNION.

USTINOV, PETER ALEXANDER (1921–), English actor, dramatist, and director. Trained under SAINT-DENIS at the London Theatre Studio, he made his first appearance in London in Aug. 1939 with the PLAYERS' THEATRE *Late Joys* in his own sketches, and his memorable impersonation of the ageing opera-singer Madame Liselotte Beethoven-Finck admirably displayed his great gift for mimicry. His first play *Fishing for Shadows* (1940), in which he himself appeared, was based on SARMENT's *Le Pêcheur d'ombres* (1921), and was followed in quick succession by *House of Regrets* (1942), *Blow Your Own Trumpet* (1943), *The Banbury Nose* (1944), and *The Tragedy of Good Intentions* (1945). Ustinov returned to the London stage in 1946, after serving in the army, to play Petrovitch in Rodney Ackland's adaptation of DOSTOEVSKY's *Crime and Punishment*. His next play *The Indifferent Shepherd* (1948), about a clergyman who has problems with both his beliefs and his marriage, starred Gladys COOPER; in the same year he appeared in his adaptation of Ingmar BERGMAN's *Hets* (*Frenzy*). In 1949 he directed and acted in Linklater's *Love in Albania*, after which he appeared only in his own plays for the next 30 years. His first popular success came with *The Love of Four Colonels* (1951; N.Y., 1953), a fantasy in which four officers from the occupying forces, American, British, French, and Russian, woo a Sleeping Beauty in a German castle; he played a leading role himself in London. After *The Moment of Truth* (also 1951) and *No Sign of the Dove* (1953) came another popular success, *Romanoff and Juliet* (1956; N.Y., 1957), in which he also appeared, about the romance between the son of the Russian ambassador and the daughter of the American ambassador in a Ruritanian country. This play and *The Love of Four Colonels* have a fantastic fairytale quality which nevertheless contains a hard core of commonsense and a shrewd perception of contemporary problems. His play *Photo Finish* (1962; N.Y., 1963) was memorable for his own performance as a bed-ridden 80-year-old confronted with himself at the ages of 20, 40, and 60. The comedy *Halfway Up the Tree* (1967) provided a good role for Robert MORLEY as General Fitzbuttress, a part played in New York in the same year by Anthony QUAYLE. Ustinov then played the Archbishop in his own play *The Unknown Soldier and His Wife* (1968) at the CHICHESTER FESTIVAL, repeating the role in London in 1973, in which year *R Loves J*, a musical based on *Romanoff and Juliet* for which he wrote the book, was also seen at Chichester. A year later he appeared in New York in his own play *Who's Who in Hell*, and in 1979 he was seen at the STRATFORD (ONTARIO) FESTIVAL as King Lear—his first Shakespearian role. He has also directed a number of plays, including several of his own. His autobiography *Dear Me* was published in 1977.

UTILITY, see STOCK COMPANY.

UZBEKISTAN, one of the constituent republics of the Soviet Union (see RUSSIA AND SOVIET UNION), which had from very early times a flourishing tradition of circus, PUPPETS, SHADOW-SHOWS and singers of folk-ballads, while the nomadic tribes had their own primitive musical instruments. To the rich heritage of song and dance has now been added the repertory of the Western theatre and of the great Russian masters, without, however, swamping the national characteristics. Particularly notable since the Revolution have been the names of Tamara Petrosiants and Kari Yakubov, and of the Khamza Theatre, one of the leading playhouses, which opened in Tashkent in 1920. It was for a time located in Samarkand, but in 1929 it returned to Tashkent, now the official capital, and was given the name of the Uzbek poet Khamza.

V

VADSTENA THEATRE, Sweden, private theatre situated in the middle of a row of early 19th-century houses in the small town of Vadstena. It was built in 1826 and the last professional production there was given in 1878. Its origin, architect, and original owner are unknown. The theatre is still in working order and is occasionally used for amateur performances.

VAKHTANGOV, EUGENE V. (1883–1922), Soviet actor and director, who joined the MOSCOW ART THEATRE under STANISLAVSKY, and soon proved himself not only a fine and sensitive player, but also an excellent organizer. In 1921 he was put in charge of a studio which almost immediately distinguished itself with a successful production of *Macbeth*, in which Vakhtangov played the title role, and under its dynamic leader began to break away from the parent theatre, being officially named the VAKHTANGOV THEATRE four years after its founder's death. During his short career Vakhtangov, who made the widest possible use of theatre forms and developed an unsurpassed use of the grotesque, sought to weld his company into an ensemble, his finest work, which he did not live to see, being a brilliant production of GOZZI's *Turandot*. This remained in the repertory of the theatre for many years, exactly as he had conceived it. He was also the founder, in 1918, of the HABIMAH THEATRE, which in the year of his death first gave its definitive performance of ANSKY's *The Dybbuk*.

VAKHTANGOV THEATRE, Moscow, MOSCOW ART THEATRE studio founded in 1921 under the direction of the Soviet director (above), who presented there two of his last and best productions —MAETERLINCK's *Le Miracle de Saint Antoine* in 1921 and GOZZI's *Turandot* in 1922. In 1926 it was officially given its present name, and many important Soviet directors worked there, among them MEYERHOLD, POPOV, ZAVADSKY, and AKIMOV, whose somewhat unorthodox production of *Hamlet* in 1932, with music by Shostakovich, caused controversy and drew the public's attention to the company. The first two parts of GORKY's uncompleted trilogy, *Yegor Bulychov and Others* (1932) and *Dostigayev and Others* (1933), were first seen at the Vakhtangov. Both OKHLOPKOV and SHCHUKIN worked there, the latter later making a great impression as Lenin in *The Man With the Gun* (1937) by POGODIN, several of whose plays were first seen at this theatre. After the Second World War the Vakhtangov again produced a number of new Soviet plays, including ARBUZOV's *City at Dawn* (1957), *It Happened In Irkutsk* (1959), and *Midnight* (1960). The high artistic standards of its founder, whose *Turandot* remains in the repertory, were sustained until his death in 1968 by Reuben SIMONOV, and maintained under his son Yevgenyi who succeeded him, his productions including *Antony and Cleopatra* in 1972 and *Little Tragedies*, based on PUSHKIN, in 1975.

VALANCE, see BORDER.

VALLE-INCLÁN, RAMÓN MARÍA DEL (1866–1936), Spanish novelist and poet, author also of a number of plays and of novels in dialogue. Typical of his writing before 1918 is *La cabeza del dragón* (*The Dragon's Head*, 1909), a sophisticated and mannered children's play with a satirical content. *Cuento de abril* (*April Story*, 1910) and *Voces de gesta* (*Epic Voices*, 1912) are verse plays, the latter dealing with the Carlist wars, one of his favourite themes. *El embrujado* (*The Bewitched*, 1913) is a sinister, tense play of vice and poverty, and *La marquese Rosalinda* (also 1913) shows once again the obverse side of Valle's writing, satirizing the exquisite. In the plays written after the First World War the satirical vein predominates: in *Divinas palabras* (*Divine Words*, 1920), a tragedy of poverty in primitive Galicia; in *Farsa y licencia de la reina castiza* (*The Farce and Licentiousness of the Noble Queen*, 1922), which censures the Court of Isabel II; in *Luces de Bohemia* (*Lights of Bohemia*, 1923), a biting attack on Madrid society at the turn of the century in which Valle attempts to define the *esperpento*, a term applied to his earlier plays *Los cuernos de don Friolera* (*The Horns of Don Friolera*, 1921), a bitter attack on military morals and the decadence of contemporary Spain, and *El hija del capitán* (*The Captain's Daughter*). The *esperpento*, which foreshadows the later Theatre of the ABSURD, holds up to ridicule the shams and evasions of contemporary society, warping and twisting the characters into parodies of heroic figures.

VALLERAN-LECOMTE (*fl.* 1590–*c.* 1615), French actor-manager, who is first heard of at Bordeaux in 1592. A year later he was asking permission in Frankfurt to present Biblical tragedies, claiming that he had already played these and the plays of JODELLE in Rouen and Strasbourg. Soon afterwards he had his own company, with Marie VENIER as his leading lady and the prolific Alexandre HARDY, often regarded as France's first professional playwright, as his salaried author. By the early years of the 17th century he was established intermittently at the Hôtel de BOURGOGNE as the tenant of the CONFRÉRIE DE LA PASSION, with occasional tours of the provinces and the Low Countries. The old actor Agnan SARAT, who more than 20 years before had brought his own company to Paris, was Valleran-Lecomte's chief comedian, and remained with him until his death. Valleran-Lecomte himself appears to have been well received in his early years as the young lover in COMMEDIA DELL'ARTE-type comedies, and was always acceptable in farce. When his company first played in Paris, according to Tallemant des Réaux in his *Historiettes*, he took the entrance money at the door himself, and he must have been one of the first managers to establish a school of acting for young players.

VALLI, ROMOLO (1925–80), Italian actor and theatre manager. He made his first appearance on the stage after the Second World War, and during the early 1950s worked under Giorgio Strehler at the PICCOLO TEATRO DELLA CITTÀ DI MILANO. He became well known for his portrayal of jovial, worldly characters, one of his best parts at this time being the Count d'Albafiorita in GOLDONI's *La locandiera* in a production by VISCONTI which was seen in Paris at the THÉÂTRE DES NATIONS festival of 1956. He was also much admired for his gloomy and stiff-necked Malvolio in *Twelfth Night*, directed by Giorgio de Lullo, with whom he founded in 1954 the Compagnia dei Giovani, later known as I Giovani del Teatro Elisio. It was with this company that Valli appeared with great success in London, being seen in FABBRI's *La Bugiarda* (*The Liar*) and PIRANDELLO's *Sei personaggi in cerca d'autore* (*Six Characters in Search of an Author*) in the WORLD THEATRE SEASON of 1965 and a year later in the latter's *Il giuoco delle parti* (*The Rules of the Game*). He was for a time the manager of the Spoleto Festival in Italy, retiring in 1978. He was killed in a car accident shortly after appearing in a new play, *Prima del silencio* by Giuseppe Patroni Griffi, at the Teatro Elisio.

VAMP TRAP, see TRAPS.

VANBRUGH [Barnes], IRENE (1872–1949), English actress, appointed DBE in 1941, sister of Violet VANBRUGH and wife of the younger Dion BOUCICAULT. She made her first appearance in London in 1888, and was later with ALEXANDER at the ST JAMES'S THEATRE, where she played Gwendolen Fairfax in the first production of WILDE's *The Importance of Being Earnest* (1895). She made her first outstanding success as Sophie Fullgarney in *The Gay Lord Quex* (1899) by PINERO, of whose heroines she became the leading exponent, having already played Ellean in *The Second Mrs Tanqueray* on tour in 1894 and Rose in *Trelawny of the 'Wells'* on its first production in 1898. She subsequently created the title role in *Letty* (1903) and Nina Jesson in *His House in Order* (1906). She was also much admired in BARRIE's plays, particularly *The Admirable Crichton* (1902), *Alice Sit-By-The-Fire* (1905), and *Rosalind* (1912). In 1907 she appeared in the first public performance of SHAW's *The Man of Destiny*, playing opposite the Napoleon of her husband. She was again seen in Shaw when she played Catherine of Braganza in *In Good King Charles's Golden Days* at the MALVERN FESTIVAL in 1939, having celebrated her stage jubilee the previous year with a matinée at His (now HER) MAJESTY's. The theatre at the Royal Academy of Dramatic Art, one of London's outstanding SCHOOLS OF DRAMA, was named after her and her sister by their brother Sir Kenneth Barnes, who was Principal of the school for many years.

VANBRUGH, Sir JOHN (1664–1726), English dramatist and architect, in which latter capacity he was responsible for the design of the first Queen's Theatre on the site of the present HER MAJESTY's. This was built to house a company led by BETTERTON, who appeared there in Vanbrugh's own play *The Confederacy* (1705), based on DANCOURT's *Les Bourgeoises à la mode* (1692) and often billed as *The City Wives' Confederacy*. Vanbrugh's best plays are undoubtedly *The Relapse; or, Virtue in Danger* (1696), a sequel to and parody of *Love's Last Shift* (also 1696) by CIBBER, and *The Provoked Wife* (1697). His last play, originally called *A Journey to London*, was left unfinished at his death. It was completed and produced by Cibber in 1728 as *The Provoked Husband. The Relapse* was later rewritten for a more prudish stage by SHERIDAN as *A Trip to Scarborough* (1777), but in its original form it had a long run in London in 1947–8, and was also revived by the ROYAL SHAKESPEARE COMPANY at the ALDWYCH in 1967.

VANBRUGH [Barnes], VIOLET AUGUSTA MARY (1867–1942), English actress, sister of Irene VANBRUGH, made her first appearance in London in 1886 and had a consistently successful career, celebrating her stage jubilee in 1937 by playing Mistress Ford in *The Merry Wives of Windsor*, a part which she first played in 1911 and frequently revived. She married Arthur BOURCHIER in 1894 and for some time appeared mainly under his management, but in 1910 was with TREE, giving a magnificent performance as Queen Katherine in *Henry VIII*, probably her finest part. She was also excellent as Lady Macbeth. Although her greatest successes were scored in Shakespeare she appeared in many modern plays, including several by Henry Arthur JONES, and was much admired as Lady Tonbridge in Gertrude Jennings's *The Young Person in Pink* (1920), as

Mrs Vexted in Millar's *Thunder in the Air* (1928), and as Princess Stephanie in Beverley Nichols and Edward KNOBLOCK's *Evensong* (1932).

VANCE, ALFRED [Alfred Peck Stevens] (1840–88), English actor, who deserted the stage for the MUSIC-HALLS and became the most versatile character-singer of his day. He had a repertory of some 20 songs, featuring such types as the heavy swell, the politician, the Irishman, or the yokel, which became notorious for their vulgarity and *double entendre*. These he modified very much, however, when he took his own VARIETY company, presenting family entertainment, on tour to summer holiday resorts. His most popular character was the flash Cockney crook 'The Chickaleary Cove', written in the style of his predecessor Sam COWELL but in the genuine cockney of the period, with no attempt to make it intelligible to outsiders. Known as 'The Great Vance', he also excelled in the singing of straightforward moral 'motto' songs like 'Act on the Square, Boys', which had a considerable vogue, was a great friend and rival of George LEYBOURNE—they were the *lions comiques* of the period—and when Leybourne at the CANTERBURY made a success of 'Champagne Charlie', Vance at the OXFORD replied with a song in praise of Veuve Cliquot; between them they went through the whole wine-list. Vance died on stage during a performance on Boxing Day at the Sun Music-Hall in Knightsbridge.

VANCE, NINA, see NINA VANCE ALLEY THEATRE.

VANČURA, ANTONÍN and VLADISLAV, see CZECHOSLOVAKIA.

VANDENHOFF, GEORGE (1813–85), British-born American actor, son of John Vandenhoff (1790–1861), a good actor who nevertheless failed to reach a commanding position. George first appeared at COVENT GARDEN in 1839, playing Mercutio in Madame VESTRIS's production of *Romeo and Juliet*. In 1842 he went to New York, and after making a successful début at the PARK THEATRE as Hamlet decided to remain there. He was later leading man at both the CHESTNUT STREET THEATRE in Philadelphia and at Palmo's Opera House, later the CHAMBERS STREET THEATRE, in New York. In 1845 he staged an English translation of SOPHOCLES' *Antigone*, with music by Mendelssohn, on a stage approximating to the contemporary idea of a Greek theatre. He returned to London in 1853, and after a brief appearance as Hamlet went into retirement, having little liking for the profession he had so successfully adopted. A tall, scholarly man, somewhat aloof, he was later called to the bar, but continued to give poetry readings, and in 1860 published his reminiscences as *Leaves from an Actor's Notebook*. He made a final appearance on the stage in 1878, playing Wolsey in *Henry VIII* to the Katherine of Genevieve WARD. He was the brother of Mrs SWINBOURNE.

VANDERBILT THEATRE, New York, at 148 West 48th Street, between the Avenue of the Americas and 7th Avenue. This modest but handsome playhouse opened on 7 Mar. 1918, and in 1921 had a success with the first production of O'NEILL's *Anna Christie*. Later productions were less successful, except for Owen Davis's *Lazybones* (1924), with George ABBOTT, and Hughes's *Mulatto* (1935). The last play of any importance at this theatre, which in 1939 was taken over for broadcasting, was an all-star revival of WILDE's *The Importance of Being Earnest*. The building was demolished in 1954.

VAN DRUTEN, JOHN (1901–57), dramatist, of Dutch extraction, born in London, but later an American citizen. He first came into prominence with *Young Woodley*, a slight but charming study of adolescence which was produced in New York in 1925. Unaccountably banned by the censor in England, it was first produced privately by the Three Hundred Club (see STAGE SOCIETY) in 1928, and when the ban was removed had a successful run at the SAVOY. Van Druten's later plays, which are mainly light comedies, include *After All* (1929; N.Y., 1931), a study of family relationships; *London Wall*, a comedy of office life, and *There's Always Juliet* (both 1931, the latter seen in New York in 1932); *The Distaff Side* (1933; N.Y. 1934), in which Sybil THORNDIKE gave a moving performance as the mother; and *Old Acquaintance* (1940; N.Y., 1941), a study of two women writers which was successful on both sides of the Atlantic. *The Voice of the Turtle* (1943), a war-time comedy which had a long run in New York, was coldly received in London in 1947, but Van Druten achieved a success in both capitals with *Bell, Book and Candle* (1950; London, 1954), an amusing comedy about witchcraft. In 1938 he published a volume of reminiscences, *The Way to the Present*. His adaptation of stories by Isherwood as *I am a Camera* (1951; London, 1954) was later used as the basis of a musical entitled *Cabaret* (1966; London, 1968).

VANIN, VASILY VASILYEVICH, see PUSHKIN THEATRE, MOSCOW.

VAN ITALLIE, JEAN-CLAUDE (1936–   ), Belgian-born American dramatist, whose early works, including *War* (1963; London, 1969), *The Hunter and the Bird* (1964), and *Where Is De Queen?* (1965), revealed him as an experimental writer of considerable imaginative powers. His first critical success was *America Hurrah!* (1966; London, 1967), a trilogy of short plays which exposed his vision of the dehumanizing nature of modern society in a series of striking metaphors. It was produced in both New York and London by the OPEN THEATER, and Van Itallie collaborated with the group to create *The Serpent* (Rome, 1968; N.Y., 1970), a confrontation of innocence and corruption which drew on familiar and sometimes painfully evocative images, and was the embodiment of Van Itallie's conviction that the theatre should rediscover something of its original function as a communal rite and spiritual exemplar. His later works include *Photographs: Mary and Howard* (Los Angeles, 1969), *Eat Cake* (Denver, 1971), *The King of the United States* (1972), revised as *Mystery Play* (1973), and *The Bag Lady* (1979). From 1972 to 1976 he was Playwright-in-Residence at the McCarter Theatre, Princeton, N.J., where his adaptation of CHEKHOV's *The Seagull* was seen in 1973; it was produced in New York in 1975, and his adaptation of Chekhov's *The Cherry Orchard* was produced at the VIVIAN BEAUMONT THEATRE, New York, in 1977, with Irene WORTH. He later adapted the same author's *Three Sisters* (1979) and *Uncle Vanya* (1980).

VARIÉTÉS, Théâtre des. The first playhouse of this name opened on the BOULEVARD DU TEMPLE in 1779, its company moving to several other theatres before being expelled by Napoleon from the present PALAIS-ROYAL. The name was then given to a theatre built in 1807, and saved from redevelopment in 1973 by a conservation order. Except for a partial reconstruction in 1823 and the replacement of the balcony boxes by seats in 1900, it remains in its original state, and is now mainly used for revues and popular musical shows. Among the plays seen there earlier were the elder DUMAS's *Kean* (1836), but its most successful period was between 1860 and 1900, when it housed a number of well known comic operas and VAUDEVILLES. It is described in ZOLA's novel *Nana*.

VARIETY. When in the late 19th century, the era of their greatest prosperity, the MUSIC-HALLS were rebuilt so as to abolish the individual supper-tables in favour of normal theatre seating, they lost much of the boisterous element, both on stage and among the audience, of the earlier days, as well as a good deal of the original 'free-and-easy' sparkle. They were then renamed 'Palaces of Variety' and the turns were billed as 'variety', a word already in use as the equivalent of 'music-hall', the former unbroken succession of single acts being replaced by the twice-nightly programme first instituted by Maurice de Frece at the Alhambra, Liverpool; ballets and spectacular shows were imported, as well as short plays from the legitimate stage, and the whole elaborate set-up was far removed from the simple robust humour of the old music-hall. The old name remained in use, however, side by side with the new, and the history of Variety will be found under MUSIC-HALL for England and VAUDEVILLE for the U.S.A.

VARIETY ARTISTS' FEDERATION, see BRITISH ACTORS' EQUITY ASSOCIATION.

VARIUS RUFUS, LUCIUS (*c.* 74–14 BC), Roman dramatist, friend of Virgil and Horace, and author of a lost tragedy, *Thyestes*, which was performed in 29 BC at the games in celebration of the battle of Actium two years earlier. The author is believed to have received a large sum of money as a reward from Augustus, who as Octavius had been the victor in the battle, and Quintilian said that in his opinion it was as good as any Greek tragedy.

VÁRKONYI, ZOLTÁN, see HUNGARY.

VARRO, MARCUS TERENTIUS (116–27 BC), Roman critic and poet, whose works on drama were very influential. His poetry included some CLOSET DRAMA, and he also produced DIDASCALIAE and works of dramatic and theatre history; all are now lost.

VASCONCELOS, JORGE FERREIRA DE (*c.* 1515–*c.* 1585), Portuguese writer in the classical tradition, whose *Eufrósina* (pub. 1555) is, like its source the CELESTINA, a novel in dialogue rather than a play, intended for reading aloud and not for acting. The same is true of Vasconcelos's other works in this form, *Ulissipo*, a portrayal of middle-class life in Lisbon, and *Aulegrafia*, in which urban values are satirically contrasted with those of courtly love.

VAUDEVILLE, French word, possibly a corruption of Vau (or Val) de Vire, meaning 'songs from the Valley of Vire' (in Normandy), where in the 15th century one Olivier Basselin composed satirical couplets, sung to popular airs, against the English invaders; or, alternatively, 'voix des villes' ('songs of the city streets'). In 1674 BOILEAU, in his *Art poétique*, used it in its present form, to describe a satirical, often political ballad. It acquired its later meaning of 'a play of a light or satiric nature, interspersed with songs' by way of the little theatres in the Paris FAIRS. Owing to the monopoly of the COMÉDIE-FRANÇAISE, plays in such theatres (sometimes no more than booths) could only be given in dumb-show, with interpolated choruses on well known tunes, often parodying the productions at the legitimate theatre. These *pièces en vaudevilles* were the staple fare of the OPÉRA-COMIQUE, and were written by many well known dramatists, including LESAGE and Favart. Their popularity paved the way for the immense vogue of light opera and operetta in mid-19th century Paris, with librettos written by such men as the prolific SCRIBE and his numerous collaborators. When this particular kind of light comedy lost its popularity, the use of the term was extended to sketches on the VARIETY stage, whence its present-day meaning in the United States of America (see below).

VAUDEVILLE, American, the name adopted in the U.S.A. in the late 19th century for the type of respectable family entertainment pioneered by Tony PASTOR in the 1860s, and later developed by

Benjamin Franklin Keith (1846–1914) and his partners, the elder Edward ALBEE and Frederick Francis Proctor (1851–1929). It replaced VARIETY, which was at a low ebb, catering mainly for the alcoholics and prostitutes who frequented the local beerhalls. After a long struggle to provide 'clean' entertainment in several New York theatres, Pastor finally settled in Fourteenth Street, where on 24 Oct. 1881 he presented the first programme of what was later to be known as vaudeville. It consisted of eight contrasting acts—comedy, acrobatics, singing, and dancing—the cast being headed by Ella Wesner, whose MALE IMPERSONATIONS featured several typical 'dandies' of the period and a number of English MUSIC-HALL songs. The theatre became popular with respectable family groups, including women and young girls, and the financial returns were gratifying. The transition from the lusty and often lewd acts of earlier times to the simpler and more refined bills of the turn of the century was welcomed also by the performers up and down the country, who enjoyed working in the spacious theatres, of which the Keith in Boston, built by Albee in 1894, was typical, with its comfortable dressing-rooms, new scenery, good furnishings, and rigorous management. The response of the new audiences led to a new type of theatrical presentation, and out of the new sketches, never lasting more than 20 minutes and written to a specific 'punch' formula, animal acts, with trainers in hunting or riding kit, and magic, musical, or spectacular numbers, came a new technique, which did away with the excessive slapstick of the earlier comics and presented true humour blended with pathos. One of the finest of these turns of the mid-1890s was that of James J. Morton (1862–1933) and Maude Ravel, in which Morton, with a cadaverous, expressionless countenance, made strenuous efforts to help Maude with her songs and failed every time, reducing the audience to helpless laughter.

The new conditions developed the 'headline' system, which led to the appearance in vaudeville of star names from the theatre—Sarah BERNHARDT, Lillie LANGTRY, Mrs Patrick CAMPBELL among others—in short scenes and sketches. They were engaged purely for their publicity value; but by this time audiences were creating their own vaudeville stars, the managers responding to public demand by giving top billing and increased salaries to performers with definite box-office appeal who, influenced also by such English music-hall stars as Marie LLOYD, Vesta TILLEY, George ROBEY, and Albert CHEVALIER, were inspired to create new acts, new *personae*, new twists of presentation. 'Nut' acts, or BURLESQUE, of which Collins and Hart, or Duffy and Sweeney, were incomparable examples, became very popular, and a whole generation of artistes, both black and white, grew up to grace the heyday of vaudeville, which lasted roughly from 1881 to the closing of Broadway's Palace Theatre as a 'two-a-day' in 1932, by which time films and radio had driven it into decline. Among them were some who deserve to be remembered—notably Lew DOCKSTADER, W. C. FIELDS, and Bert WILLIAMS; and Charlie Case (1858–1916), who pioneered the quiet, slily humorous monologue, delivered while fiddling with a short length of string, his only prop; or Joe Jackson (?–1942), whose bicycling tramp, in 'fright' wig and red nose, created by the look in his eyes, by his gestures, and by the furrowing of his quizzical brow, the puzzlement and poignancy that Case conveyed in words; or Bill Robinson (1878–1949), a famous tap-dancer and Negro comedian affectionately known as 'Bojangles', who was seen in the REVUE *Blackbirds* (1927) and, with outstanding success, in *The Hot Mikado* (1939). These were solo turns; but there were many fine groups and double acts, one of the best of the latter being that of WEBER AND FIELDS; later came the Howard brothers, Eugene (1880–1965) and Willie (1883–1949), who became vaudeville 'headliners' in 1912, and were later seen in the *Ziegfeld Follies* and *George White's Scandals*, Willie, who went on alone after Eugene retired, dying during the pre-Broadway tour of *Along Fifth Avenue*; and a double singing act just after the turn of the century was that of Jack NORWORTH and his second wife Nora Bayes (1880–1928). Other women singers who are still remembered are Lillian Russell [Helen Louise Leonard] (1861–1922), a beautiful woman with a vivid and flamboyant personality who first appeared under Pastor in 1881; Marie Dressler [Leila Koeber] (1871–1934), remembered for 'Heaven Will Protect the Working Girl', which she sang in a musical, *Tilly's Nightmare* (1894), who left the stage for films in 1914; Marie Cahill (1870–1933), who, like Elsie Janis (1889–1956), was equally well known in MUSICAL COMEDY; Eva Tanguay (1878–1947), singer of 'I Don't Care!', and Fay Templeton (1865–1939), who both had successful careers in the theatre before going into vaudeville, the first in 1907, the latter in 1916. .

Many of the big names of vaudeville survived to become even better known in films and on the radio, among them Ed Wynn (1886–1966), who started in vaudeville at 15 and later starred in *The Perfect Fool* (1921)—which became his nickname—and other shows which he wrote and directed himself; Eddie Cantor (1893–1964), known as 'Banjo-Eyes', who was first seen at the Clinton Music-Hall in New York in 1907, and later appeared in revue and musical comedy, scoring a great success as an eccentric comedian in *Kid Boots* (1923); and that other eccentric comedian, his exact contemporary, Jimmy Durante (1893–1980), whose big nose earned for him the sobriquet 'Schnozzle'; not to mention Bert LAHR, who spent many years in vaudeville and burlesque.

These are only a few of the practitioners who benefited from Tony Pastor's enterprise and helped to build up the glittering world of American vaudeville. Their counterparts still exist, providing the staple fare of popular television.

VAUDEVILLE, Théâtre du, Paris, playhouse which opened in 1792 in the rue de Chartres to house the actors forced to leave the COMÉDIE-ITALIENNE when its licence was renewed for musical plays only. The company, headed by Rozières, was frequently in trouble for the topical and political allusions in its productions. It eventually fell back on semi-historical pieces, based on anecdotes of heroic figures. Closed in 1838, it reopened two years later on the Place de le Bourse; closed again by town planners in 1869, it moved to the Chaussée d'Antin, where it operated until 1927 and then became a cinema. In its heyday this third theatre housed productions of comedies by LABICHE and others, as well as the social dramas of BRIEUX and IBSEN.

The Européen-Vaudeville on the same site is a music-hall which has no connection with the former theatre.

VAUDEVILLE THEATRE, London, in the Strand. The first theatre of this name was built in 1870 for three popular actors, David JAMES, Henry MONTAGUE, and Thomas THORNE. It held over 1,000 people, and opened on 16 Apr. 1870 with a comedy by Andrew Halliday, *For Love or Money*. It had its first success the following June with ALBERY'S *Two Roses*, which introduced Henry IRVING to London audiences, and in 1875 H. J. BYRON'S *Our Boys*, in which both James and Thorne appeared, began a four-year run. In 1884, with Thorne as sole manager, the theatre scored another success with Henry Arthur JONES'S *Saints and Sinners*, and with a series of productions starring Cyril MAUDE. The building then closed for reconstruction, the seating capacity being reduced to 740 accommodated on four tiers. It reopened on 13 Jan. 1891 with a comedy by Jerome K. JEROME called *Woodbarrow Farm*, and later in the same year came the first performances in England (at matinées only) of IBSEN'S *Rosmersholm* (23 Feb.) and *Hedda Gabler* (20 Apr.), both with Elizabeth ROBINS. In 1892 the GATTI brothers bought the theatre and from 1900 to 1906 Seymour HICKS and his wife Ellaline TERRISS appeared in a series of long runs under the direction of Charles FROHMAN, Hicks scoring one of his greatest successes in 1901 in the title role of *Scrooge*, an adaptation by BUCKSTONE of DICKENS'S *A Christmas Carol*. A Christmas entertainment devised by Hicks himself, *Bluebell in Fairyland*, proved popular and was revived annually for many years. Among later successes were BARRIE'S *Quality Street* (1902) and several musical comedies, in one of which, *The Belle of Mayfair* (1906), Gladys COOPER made her West End début. Charles HAWTREY then appeared in a series of comedies, and in 1915 the theatre became the home of CHARLOT'S revues, which continued with great success until 1925, when the theatre again closed for rebuilding, though retaining its original façade. It reopened on 23 Feb. 1926, continuing mainly with revues until 1937, its seating capacity having been reduced to 659 on three tiers. A period of mixed fortune followed until in 1938 Robert MORLEY'S *Goodness, How Sad!* proved successful. Wartime productions included Esther McCracken's *No Medals* (1944), which with Fay COMPTON as Martha Dacre ran for two years. In 1947 William Douglas HOME enjoyed two good runs with *Now Barabbas*... and *The Chiltern Hundreds*, the latter with A. E. MATTHEWS. The 1950s were dominated by the record-breaking musical by Dorothy Reynolds and Julian Slade, *Salad Days*, which ran from 5 Aug. 1954 to 27 June 1960. Later successes were Ronald MILLAR'S *The Bride Comes Back* (1960), WESKER'S *Chips with Everything* (1962), and Joyce Rayburn's *The Man Most Likely To*... (1968), which ran for two years. The theatre was then bought by Peter Saunders, and after extensive refurbishing it reopened on 24 June 1970, its first outstanding success being the farce *Move Over, Mrs Markham* by John Chapman and Ray Cooney. A series of short runs followed, and in 1974 Alan AYCKBOURN'S *Absurd Person Singular* was transferred to the theatre from the CRITERION. In 1977 Agatha CHRISTIE'S *A Murder is Announced*, which gave her three plays running in London simultaneously, began a long run. In 1979 a revival of T. S. ELIOT'S *The Family Reunion* transferred from the ROUND HOUSE, and Alan BATES starred in Simon GRAY'S *Stage Struck*.

VAUGHAN, HANNAH, HENRY, WILLIAM, see PRITCHARD, HANNAH.

VAUGHAN, KATE [Catherine Candelin] (*c.* 1852–1903), English actress, who made her début in the music-halls in 1870 and also appeared in BURLESQUE with her sister Susie [Susan Mary Charlotte Candelin] (1853–1950). In 1876 Kate became a member of the famous quartet, with Nellie FARREN, Edward TERRY, and Edward Royce, at the GAIETY THEATRE, where in 1883 she scored a success as Peggy in a revival of GARRICK'S *The Country Girl* (1766), based on WYCHERLEY'S *The Country Wife* (1673). This led to her appearance in a season under her own management in 1887, when she played Lydia Languish and Lady Teazle in SHERIDAN'S *The Rivals* and *The School for Scandal*, Kate Hardcastle in GOLDSMITH'S *She Stoops to Conquer*, and Peg WOFFINGTON in READE and TAYLOR'S *Masks and Faces*. She then toured in these plays in partnership with H. B. CONWAY.

VAUTHIER, JEAN (1910–), Belgian dramatist writing in French, whose first play, *L'Impromptu d'Arras*, was performed at the Arras Festival in 1951. This was followed by an adaptation of MACHIAVELLI'S *La Mandragora* as *La Nouvelle Mandragore*, which was produced by VILAR in 1952 at the THÉÂTRE NATIONAL POPULAIRE in Paris. Vauthier's important works deal mainly with different aspects of one character, a quarrelsome, cantankerous author who first appears in *Capitaine*

Bada (1952), where he is engaged in a constant struggle with his *alter ego*; one half of him wants to live happily with his wife Alice, the other is always in search of some distant and unattainable ideal. In Vauthier's most successful play, *Le Personnage Combattant* (1956) in which BARRAULT gave a fine performance, Bada returns to the hotel room he once lived in, and during a night-long battle with his own weaknesses tries to recapture his youthful idealism. In *Le Sang* (1970), he has become a theatre director who creates havoc among his company by making changes in his play at the last moment. Vauthier, who writes with a verbal fluency reminiscent of AUDIBERTI, though with more acidity, has attempted to heighten the language of passion by setting it in a musical framework with musical notation. Speeches tend to be solos, and actors are supposed to move like dancers in their efforts to dominate the world of material objects.

VAUXHALL, London, place of entertainment on the south bank of the Thames, originally known as Spring Gardens, Foxhall, which opened in 1660. During the 18th century it was used for concerts, glee-singing, fireworks, and occasional spectacular dramatic shows. It figures in many memoirs and novels of the period, including Fanny Burney's *Evelina* (1779). In the early 19th century it was known as the Royal Gardens, Vauxhall, and some of Bishop's operettas were staged there. In 1849 it was lit by gas-lamps. It was described by DICKENS, and was often in trouble owing to rioting and disorders, which finally led to its being closed on 25 July 1859, the site being built over, though the name persists.

There was a Vauxhall in New York in the early part of the 19th century (see below) and also one in Moscow (see MEDDOKS).

VAUXHALL GARDEN AND THEATRE, New York, at 4th Avenue and Astor Place. This opened in 1808 to replace the colonial Vauxhall on Greenwich Street; it had a small open-air theatre for summer shows which was burnt down almost immediately, but the gardens long remained a favourite resort, and several seasons of plays were given there in the 1840s, a new saloon theatre having been opened in 1838. This was demolished in 1855.

VAZOV, IVAN, see BULGARIA.

VEDRENNE, JOHN EUGENE (1867–1930), English theatre manager, best remembered for his seasons at the ROYAL COURT THEATRE from 1904 to 1907, and at the SAVOY later in 1907, in association with Harley GRANVILLE-BARKER. Together they presented a number of outstanding plays, including some of SHAW's for the first time. Vedrenne, who was originally in commerce, became a concert agent and business manager for several theatres, and was associated with such stars as Frank BENSON, Johnston FORBES-ROBERTSON, and Lewis WALLER.

VEGA, VENTURA DE LA [Buenaventura José Mariá Vega y Cárdenas] (1807–65), Spanish playwright, whose best known play is *El hombre del mundo* (*A Man of the World*, 1845), a bourgeois drama in verse in the style of MORATÍN. Among his other plays were an unsuccessful historical drama, *Don Fernando de Antequera*, a tragedy, *La muerte de Cesar* (1865), several comedies, and a number of translations from the French. His son Ricardo (1841–1910), who excelled in the GÉNERO CHICO, was the author of the popular ZARZUELA *La canción de la Lola* (*Lola's Song*, 1880), with music by Valverde and Chueca, whose success was only equalled by that of *La verbena de la Paloma* (*The Feast of La Paloma*, 1894), for which Bretón wrote the music.

VEGA CARPIO, LOPE FÉLIX DE (1562–1635), Spanish playwright, and the most prolific dramatist of all time; he himself claimed to have written 1,500 plays, though the actual number of extant texts is from 400 to 500, mostly written in neat, ingenious, and superbly lyrical verse. His first plays date from the period of the early open-air theatres of Madrid. When he began to write, the professional theatre in Spain was in its infancy, and he can rightly be considered the consolidator, if not the founder, of the commercial theatre in Spain. He played a large part in the development of the Spanish COMEDIA, and the formula which he evolved for its construction in the last decade of the 16th century remained largely unchanged for 100 years. His *Arte nuevo de hacer comedias* (*New Art of Writing Plays*, c. 1609) is an interesting, if ironical, exposition of his views on the art of the dramatist, written with one eye on classical precedent but based securely on practical experience.

Critics have endeavoured to cope with Lope's enormous output by dividing his plays into various categories, but it is really more important to stress the unity of his dramatic production. His world is that of a Spanish Catholic of his day—he took orders in 1614. Society is viewed in an idealized light, and its multiple social distinctions are reduced to three: king, nobles, and commoners, the last usually peasants. The king, or his accredited agent, is God's vice-regent, dispensing God's justice and maintaining the natural harmony intended by the Creator. Human behaviour is governed by three things: the four 'humours' so prominent in Elizabethan drama; the systematic conceptions of Catholic moral theology (basically Thomist); and a highly stylized code of honour. It is against this background that such plays as *Peribáñez y el Comendador de Ocaña* (c. 1608), *Fuenteovejuna* (*The Sheep-Well*, c. 1612), and *El mejor alcalde el rey* (*The King the Best Magistrate*, c. 1620) should be considered. *Fuenteovejuna* deals with the rising of a village against its brutal lord; shorn of its sub-plot—the rising of that same lord against his king—the play has been interpreted in terms of the class struggle. The two plots are, however, interdependent, and both are essential to the true understanding of the play. Here and elsewhere Lope reveals himself clearly as conservative rather than revolutionary: the brutal overlord is removed because he disrupts the harmony of society, failing in his duties to the king above him as well as to the commoners beneath him.

This and other historical dramas of Lope have suffered from an attempt to interpret them according to the ideas of a later age, as have those concerned with the stylized code of honour. Lope's handling of this latter theme is diverse. In the early play *Los comendadores de Córdoba*, the outraged husband, discovering that he is openly and publicly dishonoured, takes open, public, and bloody revenge. In the mature *El castigo sin venganza*, the duke secures his public position but revenges himself privately upon the bastard son and adulterous wife who have caused the loss of his honour. His public reputation is saved, since his revenge is secret, and with it the honour of the state, but this secret revenge entails the killing of the only two persons he has loved. The subtlety of the play consists in Lope's demonstration that this state of affairs is the direct consequence of the duke's own licentiousness. The theme of honour is more than a barbaric convention.

Lope also wrote for the entertainment of the Court, usually plays with a mythological or pastoral plot which made use of the most up-to-date machinery and scenic effects. *La selva sin amor* (*The Loveless Forest*, 1629) is a particularly interesting work, being partly set to music and thus a forerunner of the later ZARZUELA. Equally spectacular were his AUTOS SACRAMENTALES played on three carts, the centre being the main playing-space, and the outer carts bearing complicated machinery. They have not the lyrical beauty or the subtle symbolism of those by CALDERÓN, but they present their theme simply, forcibly, and dramatically, and show an assured grasp of technique.

VEIGEL, EVA MARIA, see GARRICK, DAVID.

VÉLEZ DE GUEVARA, LUIS (1579–1644), Spanish writer, whose novel *El diablo cojuelo* (1641) had a great influence on LESAGE's *Le Diable boîteux* (1707). Among his plays, which are in general fast-moving, exciting, and lyrical, *Reinar después de morir* (*Queen After Death*) is the best of those which deal with the fate of Inés de Castro. He wrote also a number of religious plays, AUTOS SACRAMENTALES, and ENTREMESES, and in *La serrana de la Vera* (*The Maid from the Mountains of La Vera*), built round the character of a tough, resourceful peasant woman, he produced an excellent COMEDIA which, in the opinion of some critics, is better than the play of the same name by Lope de VEGA CARPIO on which it is based. Vélez was the father of Juan (1611–75), author of an important early ZARZUELA, *Los celos hacen estrellas* (*c.* 1662), and of several *entremeses*.

VELTEN, JOHANNES (1640–92), German actor, who by 1669 had been a member of an itinerant troupe for some years. In 1678, after marrying an actress, he took over the company himself, and added to its usual repertory translations of some of MOLIÈRE's plays and, in 1686, *Der bestrafte Brüdermord* (*Fratricide Punished*), a version of *Hamlet* as played by the ENGLISH COMEDIANS. In this, as in all his adaptations, he tried to do away with improvization, and to make each play an artistic entity. After his death his company carried on under his widow. Under different names, and with many fluctuations of fortune, it held together until in 1771, after passing through the hands of Carolina NEUBER, SCHÖNEMANN, KOCH, and ACKERMANN, it came under the leadership of SCHRÖDER.

VENICE THEATRE, New York, see CENTURY THEATRE (2).

VENIER [Vernier], MARIE (*fl.* 1590–1627), the first Parisian actress to be known by name (but not the first in France, as was once claimed, a distinction which now goes to Aurélia de SALLES). She is chiefly remembered as the leading lady of VALLERAN-LECOMTE's troupe at the Hôtel de BOURGOGNE from 1607 to 1610. She was still alive in 1627. The daughter of an actor-manager, she married a member of her father's company, Laporte [Mathieu de Febvre, 1572–*c.* 1626/7], who also joined Valleran-Lecomte's company. In 1610 he and his wife got into trouble for acting at the Hôtel d'ARGENT, thus infringing the monopoly of the Hôtel de Bourgogne, and soon after this Laporte retired.

VENNE, LOTTIE (1852–1928), English actress, best known for her work in BURLESQUE at the STRAND THEATRE, where she played from 1874 to 1878. She then turned to straight comedy, though she continued intermittently to appear in musical comedy and farce. She was an excellent mimic, and in 1892 her imitations of Marion TERRY, who was appearing as Mrs Erlynne in WILDE's *Lady Windermere's Fan*, and of Lady TREE as Ophelia in the current production of *Hamlet* at HER MAJESTY's, ensured the success of Brookfield's burlesque *The Poet and the Puppets*. A fine-looking woman, high-spirited and good-humoured, she was sensible enough in her later years to play elderly roles in comedy, thus retaining the affection of her public and the respect of the critics.

VERBRUGGEN, Mrs J. B., see MOUNTFORT, WILLIAM.

VERE STREET THEATRE, London, in Clare Market. This was originally a TENNIS-COURT built in

1634 by Charles Gibbons. It was first used for plays after the closing of the theatres in 1642, the first recorded performance being a revival of KILLIGREW's *Claracilla* in 1653. In 1660 Killigrew converted it into a playhouse, opening on 8 Nov. with *Henry IV, Part 1*; in a production of *Othello* a month later the part of Desdemona was played by a woman, the first on the English stage. Killigrew remained in Vere Street until his new theatre in DRURY LANE was ready in 1663. After he left it was used as a fencing school, and from Apr. 1669 to May 1671 George JOLLY ran it as a NURSERY, supplying actors to both Killigrew and DAVENANT. From 1675 to 1682 it was a Nonconformist meeting-house; it then became a carpenter's shop and a slaughter-house before being destroyed by fire in 1809. The STOLL THEATRE was built on part of the site in 1911.

VERFREMDUNGSEFFEKT, term associated with the EPIC THEATRE of BRECHT, in which an effect of 'alienation' from the play is achieved by means of various techniques designed to destroy theatrical illusion and promote in the audience a critical attitude towards what is happening on stage. This is often done by placing what is familiar in an unfamiliar setting, a process which is particularly suited to the subtle use of language from common sources evident in such plays as Brecht's own *Der aufhaltsame Aufstieg des Arturo Ui* (1958) and *Die heilige Johanna des Schlachthöfe* (1959). Other distancing techniques include the use of placards, masks, songs which deliberately interrupt the flow of the action, documentary films, cartoons, visible lighting sources, and 'indicative' sets. The actor too has his part to play in alienation; he must stand slightly apart from the character he is portraying, narrating events almost as if he were merely an eye-witness of them.

VERGA, GIOVANNI (1840–1922), Italian playwright, born in Sicily, better known as a novelist and short-story writer. He was, however, the only completely successful writer of tragedy in the Italian theatre between ALFIERI and PIRANDELLO, employing VERISMO not by formula but by conviction, his portraits of Sicilian life being unflinching confrontations of grey desperation. *Cavalleria rusticana* (*Rustic Chivalry*, 1884), the first and most famous example of this movement, a dramatization of one of his own short stories which later provided the libretto for Mascagni's opera, is a sparse and swiftly moving example of Verga's power to fuse two interlinked tragedies, that of a community and that of a religious man, into a coherent whole. Comparable achievements are *La lupa* (*The She-Wolf*, 1896), a most perceptive study of female sexuality and man's rank fear of it, and *La caccia al lupo* (*The Wolf-Hunt*, 1901), where he succeeds in transmuting the melodrama of jealousy into the poignancy of inescapable aloneness. Only recently have critics come to discern the unequivocal poetry informing Verga's dramas, notably the overlooked excellence of *In portineria* (*In the Porter's Lodge*, 1885), a play set, untypically, outside Sicily, and *Dal tuo al mio* (*From Yours to Mine*, 1903), a terse and evocative analysis of the clash between two classes of society.

VERGERIO, PIER PAOLO (1370–1445), Italian dramatist, author of *Paulus*, a Latin comedy written in 1389 or 1390 while he was a student in Bologna. The play carries the sub-title *Ad Juvenum Mores Corrigendo* (*To Correct the Behaviour of Youth*), and its central character is a student torn between a dissolute life (represented by Herotes, the bad servant) and a life of study (represented by Stichus, the good servant). Paulus finally succumbs to Herotes. The play combines contemporary material with a classical structure, the principal influence being that of TERENCE, although its medium is rhythmic prose instead of verse.

VERISMO, Italian cultural movement of the later 19th century, having affinities with French NATURALISM. Subject to multiple critical interpretations, the term evades precise definition, but in general the movement constituted a reaction against the Italian academic and rhetorical tradition in favour of an objective exploration of the immediate problems of urban bourgeois and provincial society. Such an exploration was inextricably linked to an awareness of the political and social realities that came in the aftermath of the *Risorgimento*. One notable exponent of *verismo* in the theatre was Giuseppe GIACOSA, and later dramatists, among them PIRANDELLO and D'ANNUNZIO, were veristic in their early writings. But perhaps the richest achievements of *verismo* were in the field of regional literature, and are to be found in the work of writers like CAPUANA and notably (some would say exclusively) Giovanni VERGA.

VERKADE, EDUARD, see NETHERLANDS.

VERNEUIL, LOUIS [Louis-Collin Barbie de Bocage] (1893–1952), French actor and dramatist, who acted in and directed a number of his own BOULEVARD PLAYS, several of which were translated into English. BERNHARDT, whose grand-daughter was Verneuil's first wife, made her last appearance in London in his *Daniel* (1921). His best-known work was *Ma Cousine de Varsovie* (1923), successfully revived in 1955. In 1930 Verneuil went to the U.S., where he worked for some time in films, and wrote (in English) *Affairs of State* (N.Y., 1950; London, 1952). He returned in Sept. 1952 to Paris, where he committed suicide two months later.

VERSAILLES. Although there was a good deal of theatrical entertainment at the palace of Versailles under Louis XIV, there was no permanent theatre there and plays were given on temporary stages erected indoors or in the gardens. It was not until 1768 that Louis XV instructed his chief architect, Ange-Jacques Gabriel (1698–1782), to build a theatre in the north wing of the château. Oval in design, and not rectangular, as earlier French theatres were, it was built of wood, much of it painted to resemble marble. The stage, almost as large as that of the Paris Opéra, was well supplied with machinery, and the floor of the auditorium could be raised to stage level to form a large room for balls and banquets. Lighting was provided by crystal chandeliers. The theatre was first used on 16 May 1770 for a banquet in honour of the marriage of the future Louis XVI to Marie-Antoinette. The first plays to be given there were RACINE's *Athalie* on 23 May, with Mlle CLAIRON in the title role, and on 20 June VOLTAIRE's *Tancrède*. During the Revolution the theatre served as the meeting place of the Versailles branch of the Jacobins. When in 1837 Louis-Philippe made Versailles a museum of French military history, the opening ceremony was followed on 10 June by a gala performance of MOLIÈRE's *Le Misanthrope*. The theatre was then used occasionally for concerts and on 18 Aug. 1855 for a banquet in honour of Queen Victoria. In 1871 it was taken over by the Assembly, who met there during the Commune. A floor was laid over the pit, and everything above it was painted brown. This fortunately preserved the decorations below, and when in 1952 restoration began on the château, the theatre too was restored to its original colours of dark blue, pale blue, and gold. It was even found possible to replace the original material on the seats, made by the firm which had supplied it in 1768. The restoration was completed in time for an official visit by Queen Elizabeth II of England on 9 Apr. 1957, when a theatrical and musical entertainment was given. The theatre is still occasionally used for concerts, operas, and plays.

A theatre built by Mlle MONTANSIER in 1777 on a site near the palace remained in use until 1886.

VERULAMIUM. The Roman theatre in this 2nd-century city (now St Albans) was for a long time the only one known in Britain (traces of others have now been found at Canterbury and near Colchester). It was probably built between AD 140 and 150 and used mainly for sport, particularly cock-fighting. The ORCHESTRA, which was completely circular, with seating round two-thirds of it and a small stage-building in the remaining space, could have been used for MIMES and dancing, and the stage for small-scale entertainments. In AD 200 the stage was enlarged, possibly to allow for the positioning of a slot for a curtain, and at the end of the 3rd century, after a period of disuse, the theatre was rebuilt with modifications. The auditorium was extended over part of the orchestra, the floor levels were raised, and a triumphal arch was built spanning Watling Street. The building was finally abandoned at the end of the 4th century and the site used as a municipal rubbish-dump. It was rediscovered in 1847 and excavated in 1934 by Dr Kathleen Kenyon.

VESTRIS [Gourgaud], FRANÇOISE (1743–1804), French actress, wife of the Italian ballet-dancer Angelo Vestris who settled in Paris in 1747. With her brother DUGAZON and sister Marie-Marguerite Gourgaud, she was a member of the COMÉDIE-FRANÇAISE, where she was the pupil of LEKAIN, making her début in 1768. She created a number of tragic heroines, including the title roles in BELLOY's *Gabrielle de Vergi* (1772) and VOLTAIRE's *Irène* (1778), and Catherine de Médicis in Marie-Joseph CHÉNIER's *Charles IX* (1789). She much admired Chénier, and managed to secrete a copy of his *Timoléon* when he had been ordered by the censor to destroy it.

VESTRIS, Mme [*née* Lucia Elizabetta (or Lucy Elizabeth) Bartolozzi] (1797–1856), English actress, the wife of the French ballet-dancer Armand Vestris (1788–1825), who deserted her in 1820. As Mme Vestris, she had a distinguished career on the London stage. Although an excellent singer, she preferred to appear in light entertainment rather than grand opera, and was at her best in BURLESQUE, or in the fashionable ladies of high comedy. She made her first success in the title role of MONCRIEFF's *Giovanni in London* (1817), a burlesque of Mozart's *Don Giovanni*, and then played in Paris for several years with such success that she was able to return to London on her own terms, playing alternately at COVENT GARDEN and DRURY LANE. In 1830 she took over the OLYMPIC THEATRE, opening with *Olympic Revels*, by PLANCHÉ, who furnished her with a succession of farces and burlesques both at this theatre and later at the LYCEUM. During her tenancy of the Olympic she took into her company the younger Charles MATHEWS, marrying him in 1838, and the rest of her career ran parallel to his. She was an excellent manageress, effecting a number of reforms in theatre management and many improvements in scenery and effects. She was one of the first to use historically correct details in costume, anticipating the reforms of Charles KEAN, was responsible for the introduction of real properties instead of fakes, and in Nov. 1832 introduced the BOX-SET, complete with ceiling, on to the London stage.

VEZIN, HERMANN (1829–1910), English actor, born in the U.S.A., who moved to England in 1850 and remained there until his death. After working for a time in the provinces, he took over the SURREY THEATRE and appeared in a number of classic parts; PHELPS then engaged him for SADLER'S WELLS, and by 1861 he was recognized as an outstanding actor in both comedy and tragedy, his best parts being Macbeth, Othello, Jaques in *As You Like It*, and Sir Peter Teazle in SHERIDAN's *The School for Scandal*. He played opposite Ellen TERRY in *Olivia* (1878), Wills's dramatization of

GOLDSMITH's novel *The Vicar of Wakefield*, was Iago to the Othello of John MCCULLOUGH, and was for a time in IRVING's company at the LYCEUM. He made his last appearance under TREE in 1909, playing Rowley in *The School for Scandal*. A scholarly man, of small stature, with clear-cut features and a dignified bearing, he was remarkable for the beauty and clarity of his diction, and lacked only warmth and personal magnetism. He was the second husband of the actress Jane Thomson (below).

VEZIN [*née* Thomson], JANE ELIZABETH (1827–1902), Australian actress, on the stage as a child, who married as her first husband the American actor Charles YOUNG. In 1857 she arrived in England, where she became well known on the London stage as Mrs Hermann Vezin (see above). She made her first appearance at SADLER'S WELLS soon after her arrival, her charm, romantic fervour, and sweet voice winning her instant approbation, and from 1858 to 1875 she had few rivals as an exponent of poetic and Shakespearian drama.

VIAN, BORIS (1920–59), French novelist and dramatist, a disciple of JARRY, whose plays were violent satires on contemporary French society, especially in its more militaristic and socially oppressive aspects. His most important play, *Les Bâtisseurs d'Empire, ou le Schmürz* (1959), seen in London in 1962 as *The Empire Builders*, is a symbolist tragi-comedy in which a middle-class family lives through the gradual decline and disintegration of its ideals, the Schmürz, a figure of pervasive but ill-defined horror, being the embodiment of the shadowy fears of the bourgeoisie. Two other plays often revived are *L'Équarrissage pour tous* (*The Knacker's ABC*, 1950) and *Le Goûter des généraux*, which was not produced until 1963 and was seen in London in 1967 as *The Generals' Tea-Party*; both attack by implication, one in a knacker's yard, the other in a parody of a children's tea-party, the general futility of war.

VIAU, THÉOPHILE DE, see THÉOPHILE DE VIAU.

VICE, a character in the English MORALITY PLAY. Originally an attendant on the Devil, whom he helped in his attacks on mankind, he eventually became a figure of fun and a cynical and sardonic commentator on the action of the play. In *Twelfth Night* the Clown, who took over many of the Vice's attributes, refers to him as 'the old Vice . . . with dagger of lath'.

VICENTE, GIL (*c*. 1465–*c*.1536), Portuguese dramatist, and as Court poet from 1502 to 1536 primarily a deviser of courtly entertainments. He also wrote prolifically in a number of dramatic forms—eclogues, moralities, farces, romantic plays, and allegorical spectacles. Of his 44 extant works, 17 are in Portuguese, 11 in Spanish, and 16 use both languages, though all have Portuguese titles.

The eclogues, of which the *Auto pastoril castelhano* (1502) is a good example, are more highly developed than their prototype, the *Auto da Visitação*, and make good use of dialogue, but they are obviously dependent on ENCINA. Gradually, however, a new form grew out of them: the MORALITY PLAY, sharing some features with medieval European tradition, but owing most to Vicente's own powers of invention. The first step was taken in the *Auto da Fé* (1510). The *Auto dos Quatro Tempos* (*c*. 1511) was already half-way to the morality, and from it Vicente went on to his masterpieces in this form (1517–*c*. 1519)—the *Auto da Alma* and the trilogy of the *Barcas* (boats); *Barca do Inferno*, *Barca do Purgatório*, and *Barca do Glória*. These combined morality plays with outspoken social comment, possible only because of Vicente's privileged position at Court. Where the eclogues needed only actors, the *Barcas* trilogy called for fairly elaborate scenery, inspired by the mumming tradition and perhaps the Corpus Christi processions of Spain. Vicente is unquestionably the greatest religious dramatist of this period in the Peninsula, and his true successors are to be found in the Spanish writers of AUTOS SACRAMENTALES.

Whereas in the morality Vicente had to rely on a Spanish tradition which he then transformed, his farces seem to have had behind them not only the sketches of Anrique da MOTA but indigenous popular entertainments. He was thus able to find his feet at once, and the *Auto da Índia* (1509) has great technical assurance. In this and in later works—of which *Farsa de Inês Pereira* (1523) is among the best—Vicente combined knockabout humour and pungent social satire. He evolved a series of 'special languages' for different racial and social groups, his peasants learning to speak rustic Portuguese instead of *sayagués*. Here, as in the eclogues, the staging was simple: a dividing curtain would allow simultaneous scenes, and not much else would be needed. Vicente went on writing farces all his life, but from about 1520 his attention was mainly devoted to other forms. For a few years he experimented with romantic comedy, probably under the influence of the Spaniard Bartolomé de TORRES NAHARRO. *Comédia Rubena* (1521) shows the amatory misfortunes and final triumph of the heroine; *Dom Duardos* (1522) and *Amadis de Gaula* (1523?) put chivalresque fiction on the stage; and *Comédia do Viuvo* (1524) leans heavily on the figure of the disguised nobleman.

The morality and the romantic comedy were theatrically the most promising forms evolved by Vicente, but in the last 12 years of his life he turned chiefly to secular allegorical fantasies: a modern critic has compared Vicente's final technique to that of BRECHT. The characteristic note of the last period is a blend of allegory, lyricism, uninhibited satire, and lavish staging. Sometimes there is an organic plot, as in *Auto da Feira* (*c*. 1526–8), sometimes the work is held together only by the visually dazzling allegorical framework, as in the *Frágua do Amor* (1524). The splendour and complexity of the stage devices used in these works are a direct outcome of the 15th-century Court mummings.

VICENTINI, TOMMASO ANTONIO (1682–1739), Italian actor, born at Vicenza, who as a young man joined a travelling company, playing tragic heroes and when necessary (as in Rome, where women were not allowed on the stage) young heroines. At some point he turned to comedy and became a noted ARLECCHINO under the name of Thomassin. He and his wife Margarite Rusca were members of the COMMEDIA DELL'ARTE company which Lélio, the younger RICCOBONI, took to Paris in 1716, and he played Arlequin in *L'ingan no fortunato, ou L'Heureuse Surprise*, with which they opened their first season on 18 May at the PALAIS-ROYAL, before moving permanently to the Hôtel de BOURGOGNE. Thomassin scored an immediate triumph, and even when the Italian players began to lose their appeal he remained a great favourite with the Parisian public. He was so agile that it was said of him that he could turn a somersault holding a glass of water, without spilling a drop, and though renowned for the originality and vivacity of his jokes he could move an audience to tears as easily as to laughter. He appeared in the first productions of many of MARIVAUX's plays: *La Surprise de l'amour* (1722) and *Le Prince travesti* (1724) were probably written as much for him as for the company's leading lady SILVIA.

VICTORIA PALACE, London, in Victoria Street, opposite Victoria Station. This stands on the site of the Royal Standard Hotel, whose manager, John Moy, obtained a licence in 1840 for the presentation of entertainments. The room in which they were given became known as Moy's Music Hall. In 1863 the hall was renovated and reopened on 26 Dec. as the Royal Standard Music Hall, which was again rebuilt in 1886 and continued its successful career until in 1910 it was acquired by Sir Alfred Butt, who built on the site the present theatre. This opened on 6 Nov. 1911, its four-tier auditorium holding 1,565. From 1921 to 1929 and again in 1931 it housed an annual Christmas play, Frederick Bowyer's *The Windmill Man*, with Bert Coote in the lead as a mad gardener. Otherwise it continued to function as a music-hall, Gracie Fields appearing there in *The Show's the Thing* in 1929. Revue alternated with variety until in 1934 Walter Reynolds's patriotic play *Young England* opened on 10 Sept. Written in all seriousness, it was greeted with hilarity and drew vast audiences which came to cheer, jeer, and join in the dialogue. A year later Seymour HICKS took over and produced his own play *The Miracle Man*, following it with revivals of some of his former successes. Lupino LANE succeeded him in 1937, presenting on 16 Dec. a musical play *Me and My Girl*, which ran until the official closing of the theatres on the outbreak of the Second World War in 1939. From 1947 to 1962 the Victoria Palace was almost exclusively the home of the CRAZY GANG, each of their shows over this period achieving more than 800 performances. After the closure of the last production, *Young in Heart*, the popular B.B.C. television show *The Black and White Minstrels* took over and played live, opening on 25 May 1962, followed after a long run by *Magic of the Minstrels. Carry On, London!*, a stage version of the popular 'Carry On' film comedies, followed, and a new American musical, *Annie*, ran from May 1978 into 1981.

VICTORIA THEATRE, London, see OLD VIC THEATRE.

VICTORIA THEATRE, New York, see GAIETY THEATRE.

VICTORIA THEATRE, Stoke-on-Trent, Staffordshire. In 1962 Stephen JOSEPH, whose visits to neighbouring Newcastle-under-Lyme with his touring THEATRE-IN-THE-ROUND company had aroused great interest, converted a cinema in Stoke-on-Trent into Britain's first permanent, full-time theatre-in-the-round, seating 347 (389 from 1971). He was helped by Peter Cheeseman (1932–   ), who became the theatre's Artistic Director. The resident company presents ten productions each year, including two children's plays and at least one major classic and one major comedy. Over a third of the productions are original works, outstanding among which are the famous local musical documentaries, a form pioneered by this theatre and embodying Cheeseman's commitment to community involvement. Beginning with *The Jolly Potters* (1964), about 19th-century potworkers, they have included *The Staffordshire Rebels* (1965), on the local experience of the Civil War; *The Knotty* (1966), about the life and death of the local railway—the most popular show ever presented at the theatre; and *Plain Jos* (1980), on Josiah Wedgwood. The playwright most associated with the Victoria is Peter TERSON, whose first play to be staged, *A Night to Make the Angels Weep* (1964), was produced there. He was the theatre's first resident playwright and has continued to write for it. Alan AYCKBOURN was a founder member of the company and two of his plays, *Christmas v. Mastermind* (1962) and *Mr Whatnot* (1963), were given their first productions there. Although not a CIVIC THEATRE the Victoria works closely with the local authorities, from which it receives financial assistance, and does a great deal of work with the community, including schools, hospitals, old people's groups, and churches. Plans are on foot for the erection of a new theatre-in-the-round seating 700 on another site. The Victoria is run by a Trust and, like the EVERYMAN THEATRE, Liverpool, has a predominantly young audience.

VIENNA, capital up to 1918 of the former Austro-Hungarian empire, and until the mid-19th century the focal point of the German-speaking theatre. Its theatrical history begins at the end of the 15th century, when the humanists, in particular Konrad CELTIS, encouraged their students at the university to perform the works of PLAUTUS and TERENCE as well as new plays in Latin. By the mid-16th century plays in the vernacular, like those of Wolfgang SCHMELTZL on Biblical subjects, were incorporating elements of indigenous folk-comedy which were to be important in the later development of the theatre; but these had for a time to give way to the splendours of JESUIT DRAMA, whose lavish productions contributed greatly to the triumph of the baroque opera from Italy which was to supersede it. Meanwhile the influence of the ENGLISH COMEDIANS, who first appeared in the neighbourhood of Vienna in 1608, joined with indigenous drama to give rise to the HAUPT- UND STAATSAKTION, with its heroic background and its all-pervading HANSWURST, who with his companions under STRANITZKY occupied in 1711 the Kärntnertor, the first permanent theatre building in Vienna. There the ALTWIENER VOLKSTHEATER reigned supreme until in 1776 the Emperor Josef II, continuing the cultural reforms of his mother Maria Theresa, made the Burgtheater, built in 1741 in the Imperial palace for French and German companies, into the home of serious drama in German, banishing the light-hearted comedies with their songs and dances to the remote suburbs. His reforms were ably seconded by SONNENFELS, the Austrian exponent of 'regular drama' in the manner of GOTTSCHED, and this, together with the guest appearances of SCHRÖDER, whose impassioned, realistic acting was new to Viennese audiences, laid the foundations of the Burgtheater's future importance. Its pre-eminence was established by SCHREYVOGEL, who took over in 1814 and, acting on GOETHE's conception of 'world literature' as a cosmopolitan canon of major works with which a cultivated man ought to be familiar, built up a repertory which included not only the classical plays of Europe and the works of such recent German dramatists as Goethe and SCHILLER but also the first plays of the Austrian playwright GRILLPARZER and such lighter fare as KOTZEBUE and Raupach. Under him the last vestiges of French declamation vanished in favour of a polished but natural diction, the Viennese *parlando*, and he recruited from many parts of the German-speaking theatre a company of distinction, headed for some years by the actress Sophie SCHRÖDER. Meanwhile, contrary to all expectations, the old folk-comedy flourished in the suburbs, at the Leopoldstädter Theater (1781), the Josefstädter Theater (1788), and the Theater an der Wien (1801), the first manager of the last being Mozart's librettist Emanuel Schikaneder (1751–1812). Two comic characters typify this last phase of the old extempore farce—KASPERLE and THADDÄDL. Their brief but glorious reign gave way in the later 19th century to the unique mixture of farce, fairy-tale magic, and parody which culminated in the great but practically untranslatable plays of RAIMUND and NESTROY and finally disappeared in the vogue for Viennese operettas, which set out to conquer the world to the lilting rhythm of the waltz. The Burgtheater, after a period of decline following the departure of Schreyvogel, was revivified by Laube, who retained the European classics in the repertoire if they had a contemporary appeal, but also introduced the WELL-MADE PLAYS of France, partly because they were such admirable vehicles for the intimate casual style which had become the hallmark of the Burgtheater. Its world-famous acoustics had been reinforced by the introduction of the BOX-SET, which Laube pioneered in the German-speaking theatre. By the time of his retirement in 1867 pre-eminence had passed to BERLIN, with its openness to formal experiment, where an actor could be vital, dynamic, even grotesque, but not, as in Vienna, stylish and elegant. In 1888 the Burgtheater company moved to a new theatre which, destroyed in 1945 and rebuilt in 1955, still stands on the Ring facing the Town Hall. With the founding of the Austrian Republic in 1918, responsibility for the Burgtheater passed from the Court to the State. Heavily subsidized, it is now too unwieldy for one man to influence creatively, and its somewhat conservative repertory absorbs the theatre of the AVANT-GARDE only when it has become established. But its brilliant company, in beautifully set and costumed productions directed by outstanding men of the theatre, still makes it one of the great theatres of Europe. Of the other Vienna theatres, the Leopoldstädter was demolished in 1945; the Josefstädter now has a widely based repertory in which the tradition of fine acting in an intimate style, initiated by REINHARDT when he took over in 1925, still survives; the Theater an der Wien is mainly notable for its operettas and musical shows. The Volkstheater, which presents classical and modern plays with a political flavour, was reconstituted from the earlier Deutsches Volkstheater in 1948, attracting social-democratic support and the co-operation of the Trades Unions to secure its finances. There are also the Raimundtheater, built in 1893 for folk-plays; the Akademietheater, seating about 500 people, which was opened in 1922 by Max Paulsen; the Kammerspiele, the home of light comedy; and a number of experimental 'cellar' theatres. In summer the finely preserved rococo theatre in the Schönbrunn Palace, built in 1747, offers a season of plays mainly for visitors.

VIERTEL, BERTHOLD (1885–1955), Austrian poet, actor, and theatre director, who figures as Friedrich Bergmann in Christopher Isherwood's novel *Prater Violet* (1945). After working in Vienna and Dresden, he went to BERLIN in 1922, where he directed HEBBEL's *Judith* at the DEUTSCHES THEATER for REINHARDT and achieved a great success with a production of BRONNEN's *Vatermord* (*Patricide*). He then founded his own company, Die Truppe, but in 1927 left Germany, being in London from 1933 to 1937 and then in the U.S., where he worked in Hollywood and in 1945 produced in New York BRECHT's *Furcht und Elend des Drittes Reiches* as *The Private Life of the Master Race*. In 1949 he returned to Berlin, producing GORKY'S *Vassa Zheleznova* for the BERLINER ENSEMBLE, and in 1951 became a director at the VIENNA Burgtheater, where among his productions was Tennessee WILLIAMS's *A Streetcar Named Desire*.

VIEUX-COLOMBIER, New York, see GARRICK THEATRE.

VIEUX-COLOMBIER, Théâtre du, Paris, originally the Athenée, built in the early 19th century close to the dovecote of the former Abbey of St Germain. Until 1913 it was no more than a suburban playhouse for travelling companies; but it was then taken over by Jacques COPEAU as an experimental playhouse for the production of new and serious plays, in reaction against what he regarded as the frivolous and easy entertainment offered elsewhere. With the help of his stage manager Louis JOUVET, Copeau redesigned the auditorium. Originally an oblong with a stage at one end, it was adapted so as to do away with the curtain (for the first time in modern French stage history) and integrate the proscenium arch into a series of platforms and openings surrounding the audience on three sides. On a small scale—the theatre only held 200 spectators, and the stage mostly remained bare—Copeau tried to use features of the Elizabethan stage in an attempt to involve the audience in a total experience of theatre. He opened with a translation of HEYWOOD's *A Woman Killed with Kindness*, and from 1913 to 1914 and again from 1919 to 1926, when the theatre became a cinema, he presented memorable productions of Shakespeare, MOLIÈRE, GOLDONI, and MUSSET, and also introduced his predominantly intellectual audiences to COURTELINE, CLAUDEL, GHÉON, ROMAINS, and GIDE. In 1930 the Vieux-Colombier again became a theatre, when a group of Copeau's disciples, the COMPAGNIE DES QUINZE under the leadership of Michel SAINT-DENIS, re-established the tradition of intellectual experiment, notably with plays by André OBEY. After the departure of the Compagnie, several notable events took place at the theatre, including the production of SARTRE's *Huis Clos*, VILAR's production of T. S. ELIOT's *Murder in the Cathedral*, and first performances of plays by GARCÍA LORCA and Claudel; but successive managements found the theatre difficult to run successfully, and it finally closed in 1972.

VIGARANI, CARLO (?–1693), Italian scene designer and machinist, son of Gaspare (below). With his father he went to Paris in 1659, and worked with him on the construction of the SALLE DES MACHINES. After Gaspare's death he remained in Paris, and in 1664 was responsible for the mechanics of *Les Plaisirs de l'île enchantée*, an entertainment devised for Louis XIV at Versailles by MOLIÈRE and Lully. Carlo later became a naturalized Frenchman, lived in an apartment in the Louvre, and was frequently employed by Lully for his operatic spectacles.

VIGARANI, GASPARE (1586–1663), Italian machinist and scene designer, inventor of many stage effects. He was working in Modena when in 1659 he was summoned to Paris by Mazarin to supervise the entertainments being prepared for the approaching marriage of Louis XIV. For these he built the SALLE DES MACHINES, and claimed for it all the scenery and machinery made by TORELLI for the PETIT-BOURBON, which was being demolished. When he received it he burnt it all, thus destroying much of the best work of his famous rival, of whom he was extremely jealous.

VIGNY, ALFRED DE (1797–1863), French poet, novelist, and dramatist. A Romantic by temperament, he gave up a military career in 1827 and turned to literature, being for a time friendly with Victor HUGO and a frequenter of literary circles in Paris. His first play was a translation of *Othello*, as *Le More de Venise*, which was performed at the COMÉDIE-FRANÇAISE in 1829. This was followed by a historical drama, *La Maréchale d'Ancre*, produced at the ODÉON in 1831, and then by *Chatterton* (1835), Vigny's masterpiece. Produced at the Comédie-Française with Geffroy as the young English poet and Mlle DORVAL (who had been Vigny's mistress for some years) as Kitty Bell, this proved to be one of the great successes of the Romantic period. It also marked the end of Vigny's association with the theatre.

VILAR, JEAN (1912–71), French actor, manager, and director, who worked under DULLIN at the ATELIER for some time before the Second World War, and during the occupation of France by the Germans joined a band of young actors, La Roulette, which toured the provinces. Back in Paris in 1943 he directed STRINDBERG's *The Dance of Death* and MOLIÈRE's *Don Juan* and made a brief appearance in SYNGE's *The Well of the Saints* before taking over the 100-seat Théâtre du Poche. He first came into prominence in 1945 with his production, first at the VIEUX-COLOMBIER and later in front of the Abbey of Bec-Hellouin, of T.S. ELIOT's *Murder in the Cathedral*, in which he played Becket. As a result of this he was invited in the summer of 1947 to organize the open-air AVIGNON FESTIVAL, which he ran until his death. Its growing importance established him as the leading director of the French theatre, a position that was confirmed in 1951 when he became head of the THÉÂTRE NATIONAL POPULAIRE at the Palais de Chaillot. His productions, vast in scope and bold in conception as the great stage demanded, aroused both enthusiasm and controversy, and contributed to the simplification and de-romanticizing of scenery and to the increased attention paid to the quality of lighting and costume. Vilar, who was an excellent actor as well as an inspired director, appeared in many of his own productions, among them the title roles in *Macbeth* and *Richard II*, Molière's *Don Juan*, and PIRANDELLO's *Enrico IV*; he also played Harpagon in Molière's *L'Avare* and the Gangster in BRECHT's *Arturo Ui*, and in 1951 was seen as Heinrich in SARTRE's *Le Diable et le Bon Dieu*. In the same year he played SOPHOCLES' Oedipus for the RENAUD-BARRAULT company. Among the classics he revived for the T.N.P. were CORNEILLE's *Le Cid*, with the young Gérard PHILIPE in the title role, LESAGE's *Turcaret*, and HUGO's *Ruy Blas* and *Marie-Tudor*. He also introduced his audiences to KLEIST's *Prinz Friedrich von Homburg*, again with Philipe, GOLDONI's *I rusteghi*, BÜCHNER's *Dantons Tod*, CHEKHOV's *Platonov*, Brecht's *Mother Courage*, BOLT's *A Man for All Seasons*, and GATTI's first play, *Le Crapaud-Bufle*. A spare, ascetic-looking man, Vilar could on stage appear amazingly handsome, and his voice had great sonority and emotional overtones. His view of the theatre as an educational and spiritual force gave a certain gravity to all he did, most apparent in his playing of Becket. In 1963 he resigned from the T.N.P. in protest against the inadequate support given to it by the French government, and went to Italy, where he directed several operas. In spite of a long-standing love of music, he turned down an offer to direct the reorganization of the Paris opera houses, and engaged in freelance activities until his death.

VILDRAC [Messager], CHARLES (1882–1971), French poet, novelist, and dramatist, whose first full-length play *Le Paquebot Tenacity* had a great success when first produced by COPEAU at the VIEUX-COLOMBIER in 1920. As *The S. S. Tenacity* it was produced in London by the STAGE SOCIETY in the same year, and two years later was seen in New York. A study of two ex-soldiers, very different in character, waiting to emigrate to Canada, the play contrasts ambition and resignation, and appears to view both fatalistically. Vildrac's elliptical technique and consciously muted tone of voice were again employed in *Madame Béliard* (1925), a passionate triangle drama with a passionless woman at its apex, and in *La Brouille* (1930), in which motives are elucidated indirectly, through the efforts of third parties to bring about a reconciliation between two friends.

VILLE, Théâtre de la, Paris, see BERNHARDT, SARAH, and THÉÂTRE DES NATIONS.

VILLEURBANNE, near Lyon, see PLANCHON, ROGER.

VILLIERS, CLAUDE DESCHAMPS DE (1600–81), French actor, and the author of several plays, including farces and a version of the DON JUAN legend which may have had some influence on MOLIÈRE. He played in farce himself as Philippin, the name being often given to the character he was to portray. He was a member of a company formed in 1624 by LENOIR and MONTDORY, and with his second wife Marguerite Béguet (?–1670) accompanied them to Paris in 1630, remaining with them when they opened the MARAIS. After the retirement of Montdory they went to the Hôtel de BOURGOGNE. Both were excellent actors, the wife being best remembered as the first actress to play the part of Chimène in CORNEILLE's *Le Cid* (1637). The husband was caricatured in Molière's skit on the Hôtel de Bourgogne *L'Impromptu de Versailles* (1664). He retired on the death of his wife, who had left the stage four years earlier.

VILLIERS, GEORGE, see BUCKINGHAM, DUKE OF.

VILLIERS, JEAN DE (1648–1701), French actor, who as a child appeared with the RAISIN children in the Troupe du Dauphin. A royal edict removed him from the Dauphin's company in 1672 and sent him to join MOLIÈRE at the PALAIS-ROYAL. After a provincial tour he joined the company at the Hôtel de BOURGOGNE in 1679, returned to the provinces, and finally joined the COMÉDIE-FRANÇAISE in 1684. Better in comedy than in tragedy, he was much admired in the part of a ridiculous marquis. He married Cathérine Raisin in 1679; their son and daughter were both on the stage.

VILNA TROUPE, company founded at Vilna in Russia in 1916 by David Hermann (1876–1930) as the Union of Yiddish Dramatic Artists (called Fado, from the initials of the original name), with the intention of continuing Peretz HIRSCHBEIN's reform of the Yiddish stage. The first production was Sholom ASCH's *Landsleute*, which was immediately successful. The company moved in 1917 to Warsaw, where Hermann scored an immense success with his production, a month after the author's death, of ANSKY's *The Dybbuk*, in Yiddish. This was taken on tour to Berlin, London, and New York. Returning to Warsaw in 1924 the company found that they had lost their theatre, and at the end of the year they moved to Vienna. There they split up, one section going to America, where it joined forces with Maurice SCHWARTZ, the other, led by Hermann, going to Romania. This latter group returned to Warsaw in 1927 and remained there until Hermann's death. The repertory of the Vilna Troupe was at first fairly extensive and included a number of Yiddish classics, but later it became associated with and entirely dependent on a single play, *The Dybbuk*, a limitation which probably accounts for its decline.

VINCENNES, Cartoucherie de, Paris, see SOLEIL, THÉÂTRE DU.

VINCENT, Mrs J. R. [*née* Mary Ann Farley] (1818–87), American actress, born in England, where she joined a provincial company, married a fellow-actor, and went with him in 1846 to the National Theatre, Boston. He died shortly afterwards, but she continued to act, and in 1852 joined the company at the Boston Museum, where she remained for 35 years, first as leading comedienne, later playing duenna and old-lady parts. At her jubilee in 1885 she appeared in two of her finest roles, Mrs Hardcastle in GOLDSMITH's *She Stoops to Conquer* and Mrs Malaprop in SHERIDAN's *The Rivals*. The Vincent Memorial Hospital in Boston was founded in memory of her.

VINING, FANNY ELIZABETH, see DAVENPORT, EDWARD LOOMIS.

VIOLETTA (VIOLETTE, VILLETTI), Mlle, see GARRICK, DAVID.

VIRGINIA COMPANY OF COMEDIANS, see KEAN, THOMAS.

VIRGINIA MUSEUM THEATRE, Richmond, Virginia, one of Virginia's two state theatres, the other being the BARTER THEATRE. It was founded in 1955 under the auspices of the Virginia Museum, and is a handsome, lavishly appointed building seating 500, with flexible staging. It suffered during its early years from changes of policy and director, but since 1972, with professional actors augmented by visiting Broadway stars and some local talent, its eight-month season has become increasingly adventurous and well attended. It now has a Conservatory which operates a training programme.

VIRUÉS, CRISTÓBAL DE (*c.* 1550–*c.* 1614), Spanish dramatist, author of five neo-classical tragedies, or more correctly melodramas, written before 1590 but not published until 1609. These are: *Elisa Dido* (written strictly in accordance with the UNITIES), *Atila furioso, La gran Semíramis, La cruel Casandra*, and *La infeliz Marcela.*

VISCONTI, LUCHINO (1906–76), Italian director and scene designer, who as a young man went to Paris and became involved in film production. Although much of his early work was in films, he directed plays in Milan in 1937, and in 1945 was responsible for the introduction of plays by a number of European dramatists, particularly COCTEAU, into the repertory of the Teatro Eliseo in Rome, where in 1948 he staged Shakespeare's *As You Like It* in Italian as *Rosalinda*. In 1946, in which year he directed BEAUMARCHAIS's *Le Mariage de Figaro*, he joined the Paolo Stoppa-Rina Morelli company, for which he directed such modern works as Tennessee WILLIAMS's *The Glass Menagerie* in 1946, ANOUILH's *Eurydice* in 1947, MILLER's *Death of a Salesman* in 1951, and his *A View from the Bridge* in 1958. He was also responsible for productions of CHEKHOV's *Three Sisters* (1952) and *Uncle Vanya* (1955). In Paris in 1961 he staged a sumptuous version of FORD's *'Tis Pity She's a Whore*, as *Dommage qu'elle soit putain*, with Alain Delon and Romy Schneider. He worked with ZEFFIRELLI on *Troilus and Cressida* in Florence, and also on Tennessee Williams's *A Streetcar Named Desire*. His later work was mainly in films and in opera, but he returned to live theatre in 1973 to direct PINTER's *Old Times* in Rome, a production which was disowned by the author.

VISÉ, JEAN DONNEAU DE (1638–1710), French journalist, critic, and man of letters, founder in 1672 of the periodical *Le Mercure galant*. An article by him on MOLIÈRE's early comedies sparked off a pamphlet war on the subject of *L'École des femmes* (1662), and de Visé's one-act play *Zélinde* (1663), probably produced at the Hôtel de BOURGOGNE, was a reply to Molière's own *Critique de l'École des femmes*. A reply to Molière's *L'Impromptu de Versailles* (also 1663) is also attributed to de Visé. It was the most scurrilous of the attacks on Molière, and contained the first public references to his conjugal misfortunes. However, de Visé's next play, *La Mère coquette*, was acted by Molière's company only two years later, and in 1666 de Visé emerged as Molière's champion with a perceptive and very favourable report on *Le Misanthrope*. Among de Visé's other plays, some of which were written in collaboration with Thomas CORNEILLE, was the highly successful *La Devineresse* (*The Fortune-Teller*, 1679), which exploited contemporary interest in the trial of the notorious poisoner Cathérine Voisin (see also RACINE).

VISHNEVSKY, VSEVOLOD VITALEVICH (1900–51), Soviet dramatist, whose first successful play *The Optimistic Tragedy* (1932) was produced at the KAMERNY THEATRE by TAÏROV, whose wife Alisa KOONEN played the leading role of a woman commissar with the Red Fleet who is killed in battle during the early days of the Soviet régime. It was published in an English translation in *Four Soviet Plays* (1937). In 1943, after several less successful productions, Vishnevsky collaborated with Alexander Kron and Alexander Azarov in a musical, *Wide Spreads the Sea*, and in the same year wrote *At the Walls of Leningrad*. Both were well received, as was Vishnevsky's last play, *The Unforgettable Year 1919* (1949), of which Stalin is the hero. Both this and *The Optimistic Tragedy*, which was revived by TOVSTONOGOV in Leningrad in 1955, are still in the Soviet repertory.

VISOR, see MASK.

VITEZ, ANTOINE (1930–), French actor and director, probably the most intellectual and uncompromising of his generation as well as the most deeply marked by Communism and the theories of BRECHT. After working in Marseilles and at the Maison de la Culture in Caen in the 1960s (see DÉCENTRALISATION DRAMATIQUE), he became one of the moving spirits of the 'red belt' of popular playhouses in the industrial suburbs of Paris, taking productions to people in halls and schools, first at Nanterre, and then at Ivry, where he took over the Théâtre des Quartiers. His versions of SOPHOCLES' *Electra* (1971) and Brecht's *Mother Courage* (1973) established him as an original though perhaps over-ambitious director, able to handle actors well. As part of the modernization following the events of 1968 he was appointed director and teacher at the Conservatoire, chosen by the reformist party not only because of his expertise in MIME—a discipline traditionally neglected by the official body of French drama teachers—but because of his firm Marxist beliefs. From 1972 to 1974 he was joint Artistic Director of the Théâtre National de CHAILLOT, becoming sole director in 1981.

VITRAC, ROGER (1899–1952), French poet and playwright, one of the leaders of DADA, who in 1927 was associated with Antonin ARTAUD in the foundation of the Théâtre Alfred JARRY, where his first two plays were performed under Artaud's direction—*Les Mystères de l'amour* as part of the opening programme in 1927 and *Victor ou les Enfants au pouvoir* as its fourth production in 1928. *Les Mystères de l'amour*, which evokes the amorous and sadistic fantasies of a pair of lovers, is probably one of the most successful attempts to write a play on the surrealist principle of automatic writing and foreshadowed the Theatre of CRUELTY, just as *Victor*, with its monstrous characters and tragi-comic denouement, foreshadowed the Theatre of the ABSURD. It was successfully revived in Paris in 1946 and again in 1962, the second revival being directed by ANOUILH, who considers that Vitrac has had an important influence on his own work. It was seen in translation in London in 1964. Vitrac's other plays include *Le Peintre* (1930), *Le Coup de Trafalgar* (1934), *Le Camelot* (1936), the uncharacteristically tragic *Les Demoiselles du large* (1938), *Le Loup-garou* (*The Werewolf*, 1940) and *Le Sabre de mon père* (1951), a sexual extravaganza set in a fashionable nursing home.

VITRUVIUS POLLIO, MARCUS (*fl.* 70–15 BC), Roman author of a treatise in ten books, *De Architectura*, of which Book V deals with theatre construction, illustrated by diagrams. Discovered in manuscript at St Gallen in 1414, this was printed in 1484. The first edition with illustrations was published in 1511, and an Italian translation appeared in 1531. This work had a great influence on the building of Renaissance theatres, and from it the new generation of theatre designers took—though not always accurately—the idea of such devices as the PERIAKTOI and in general the proportions and acoustic properties of the later Hellenistic and Roman theatres (see THEATRE BUILDINGS).

VITTORIA, see PELLESINI, GIOVANNI.

VIVIAN BEAUMONT THEATRE, New York, at 150 West 65th Street. Designed by Eero Saarinen with the collaboration of Jo MIELZINER and seating 1,140, this theatre opened on 21 Oct. 1965. It forms part of the Lincoln Center for the Performing Arts, and housed the Lincoln Center Repertory Company from the WASHINGTON SQUARE THEATRE. Its opening production was BÜCHNER'S *Danton's Death*; later presentations included SARTRE'S *The Condemned of Altona*, BRECHT'S *The Caucasian Chalk Circle* (both 1966), Tennessee WILLIAMS'S *Camino Real*, and Sam SHEPARD'S *Operation Sidewinder* (both 1970). At the end of the 1972–3 season the theatre was taken over by Joseph PAPP, reopening with David RABE'S *Boom Boom Room*. Among later productions were STRINDBERG'S *The Dance of Death* in 1974, *Hamlet* in 1975, BRECHT and Weill's *The Threepenny Opera* in 1976, and CHEKHOV'S *The Cherry Orchard*, with Irene WORTH, in 1977. A smaller downstairs theatre, seating 299, in the same building, which opened in 1966 as the Forum, was also taken over by Papp in 1973 and renamed the Mitzi E. Newhouse Theatre; it staged plays by Shakespeare, opening with *Troilus and Cressida*, and new plays such as David Rabe's *Streamers* (1976).

In 1977 a financial deficit forced Papp to give up the theatres, and a directorate of distinguished theatrical figures, including Edward ALBEE, Robin PHILLIPS, and Woody Allen, was formed to guide them and participate actively in the productions. The Vivian Beaumont reopened in 1980 with a revival of Philip BARRY'S *The Philadelphia Story*, which was followed by *Macbeth* and Woody Allen's *The Floating Light Bulb*. The Mitzi E. Newhouse Theatre reopened in 1981 with a programme of one-act plays; two more programmes were planned, but were cancelled because of a disappointing critical and public response to the opening bill.

VLAAMSE VOLKSTONEEL, see BELGIUM.

VOJNOVIČ, IVO, see YUGOSLAVIA.

VOKES, FREDERICK MORTIMER (1846–88), JESSIE CATHERINE BIDDULPH (1851–84), VICTORIA (1853–94), and ROSINA (1854–94), family of English PANTOMIME players, children of a theatrical costumier, who first appeared together in Edinburgh in 1861, their number being augmented by Walter Fawdon (?–1904), who also took the name of Vokes. They were at the London LYCEUM in *Humpty-Dumpty* in 1865, and from 1869 to 1879 were the mainstay of the DRURY LANE pantomimes. They were also seen at the ADELPHI and elsewhere in two amusing sketches, *The Belles of the Kitchen* and *A Bunch of Berries*, making their first appearance in America in the latter in 1871. When Rosina left to marry the composer Cecil Clay her place was taken by Fred's wife Bella Moore, but the group finally broke up on the death of Jessie. Rosina came out of retirement in 1885 to take a light comedy and burlesque company to America and Canada, where she became very popular.

VOLKOV, FEODOR GRIGORYEVICH (1729–63), Russian actor, who with his brother Grigori was running an amateur theatre in Yaroslavl when in Jan. 1752 he was commanded to give a performance in St Petersburg before the Tsar and his Court. This led to his being sent with DMITREVSKY and other members of the company to the Cadet College for the sons of the nobility, where he was trained as an actor for the Court theatre and also given a good general education. He appeared at Court again in 1755 and in August the following year joined the first professional Russian company under SUMAROKOV, soon becoming his chief assistant and leading actor. Attached as they were to the party responsible for the overthrow of Peter III and the accession of Catherine the Great, Volkov and his brother were rewarded with Court offices and put in charge of the celebrations in honour of Catherine's coronation in Moscow. It was while directing a street masquerade there that Feodor caught cold and died.

VOLKSBÜHNE, or People's Theatre, an organization which originated in BERLIN in 1890 with the Freie Volksbühne, formed to bring good plays, preferably with a pronounced social content, to the working class at prices they could afford. In 1914 the organization opened its own theatre on the Bülowplatz, which soon became one of the leading Berlin playhouses, and by 1930 the movement had spread all over Germany, having over 300 branches with nearly half a million members, tickets being allocated by lot as they had been from the beginning. Dissolved by the Nazis in 1933, the movement was revived in 1948 and by the end of the 1950s had about 90 local centres in West Germany, the corresponding East German groups being absorbed into the State system which embraces all theatrical enterprises. From 1949 the Freie Volksbühne had its own theatre in West Berlin, the original theatre on the Bülowplatz being inaccessible in East Berlin, and in 1963 it moved into new premises under PISCATOR. The Volksbühne in East Berlin was revitalized under Benno BESSON's directorship from 1969 to 1979.

VOLTAIRE [François-Marie Arouet] (1694–1778), French man of letters, who took as his pseudonym an anagram of 'Arouet l(e) i(eune)'. Writer, philosopher, and historian, he is important in theatre history as a playwright and critic. He was passionately addicted to the theatre throughout his long life, befriending many actors and actresses, including LEKAIN, LA NOUE, and Adrienne LECOUVREUR; himself a keen actor, he built several private theatres (notably at Ferney, his last home) where he could indulge his taste for private theatricals. His contemporaries considered his tragedies as good as those of CORNEILLE and RACINE, but despite certain superficial innovations his plays for the most part conformed to the models established in the previous century and hardly any of them survived in the repertory even into the 19th century. However, when he was elected to the French Academy in 1746 his tragedies constituted his most substantial qualification, and when he paid his final visit to Paris at the age of 84 to attend the performance of his last tragedy, *Irène*, it was fittingly on the stage of the COMÉDIE-FRANÇAISE that he received the triumphal welcome which marked the apotheosis of his whole career.

His first play was a tragedy, *Oedipe* (1718), written in the Bastille, where he had been imprisoned as the author of a political lampoon. Its success, and that of other writings, brought him fame, social advancement, and a Court pension; he speculated to good effect and became a very wealthy man. Exiled in 1726 after a quarrel with the Marquis de Rohan, he went to London, where he remained until 1729. This visit was of the greatest importance to Voltaire's intellectual development (its general effect can be studied in his *Lettres philosophiques* of 1734), and to his career as a dramatist, for he learned English, frequented the London playhouses, and read Shakespeare and the Restoration dramatists in the original. He appreciated for a time, as far as was possible for any Frenchman of his day, brought up in the classical tradition, the works of Shakespeare, and some of the plays he wrote on his return to Paris —*Eriphyle, Zaïre* (both 1732, the latter, his masterpiece, based on *Othello*), *La Mort de César* (1735)—show traces of Shakespearian influence. But the differences far outweigh any similarities of detail that might exist, and it was not long before Voltaire's appreciation of Shakespeare gave way to harsh criticism. Bound, as were his contemporaries in the French theatre, by an entirely rationalistic approach to dramatic language, Voltaire wrote in verse, not poetry, and his plays show little creative imagination. Among his innovations in drama—and they are merely modifications of an existing formula, not fundamental changes—are the adoption in tragedy of subjects from French national history, alongside those from mythology or ancient history, and the use of the tragic form as a vehicle for the expression of controversial ideas. It is this latter feature which makes plays like *Alzire* (1736), *Mahomet ou le Fanatism* (1741), and *L'Orphelin de la Chine* (1755) still interesting to read today. They stand beside the philosophical tales and pamphlets as examples of Voltaire's unceasing attack on religious bigotry and intolerance, and on tyrannical oppression in all its forms. As a critic he left a substantial volume of writings on the theatre, the most important, apart from the prefaces to his own plays, being his *Commentaire sur Corneille* (1764); though he admired Corneille, his preference was for Racine. His relations with contemporary dramatists were not always harmonious, and he conducted a positive feud with CRÉBILLON. His comedy *L'Ecossaise* (1760) was a biting personal satire on his literary enemy Fréron, and he was critical both of MARIVAUX's idiosyncratic comedy and of the COMÉDIE LARMOYANTE established by LA CHAUSSÉE, describing the *drame bourgeois* which developed from it as 'a kind of tragedy for chambermaids'. This did not prevent him from writing a number of such DRAMES himself—*L'Enfant prodigue* (1736), for instance, and *Nanine* (1749), based on Richardson's novel *Pamela* (1740).

Voltaire was not a great dramatist, but his plays show a breadth of treatment and force of description hardly surpassed in his own day. He had the gift of attracting and interesting his audience, and if his plays are never revived now the fault lies in his facility, which led him to write too much and too carelessly, and in the fact that he lived in an age of transition and reflected its momentary preoccupations. By slackening the rigid form of tragedy to something more acceptable to the larger but less educated audience for which he was writing, he drove it a step further on the road to MELODRAMA, and to him, rather than to the Romantic dramatists, goes the honour of having first introduced local colour into the theatre. Many of his later works were marred by the introduction of philosophical propaganda, as in *Les Guèbres* which was never acted. The only play in which he achieved the impact of true tragedy was *Zaïre*, and there are innumerable tributes to the genuine emotion it aroused, both when it was first performed in 1732 with DUFRESNE and Mlle GAUSSIN in the leading roles and again later in the century, when Lekain took the role of Orosmane. When put on in London, at DRURY LANE, in Aaron HILL's translation as *Zara* (1736), it was the occasion of a memorable début by Susanna, wife of Theophilus CIBBER. It was also played with great success in Milan, and there were no less than seven translations of it into Italian in Voltaire's lifetime. One reason for actors and audiences alike to revere the memory of Voltaire is that he was instrumental in doing away with the intolerable nuisance of spectators on the stage, after which dramatists were able to use the larger stage area for the introduction of more spectacle and greater freedom of action, as Voltaire himself did in *Tancrède* (1760).

VONDEL, JOOST VAN DEN (1587–1679), Flemish dramatist, who began to write plays late in life, and produced some of his best work after 60. A humanist and a deeply religious man, he translated a number of classical plays, but his own works, while retaining the form and technique of Greek tragedy, are imbued with the Christian spirit. Many of his plays were on subjects taken from the Old Testament—Adam, Noah, Joseph, Saul, Samson—and these had a great influence on the Jewish drama which developed in Amsterdam during the first half of the 17th century, many of the Hebrew plays written at the time being based on his. Among his later plays was one on Mary Queen of Scots, *Maria Stuart* (1646); his masterpiece *Lucifer*, banned by the authorities after two performances, was published in 1654 and was known to Milton, on whose epic poem *Paradise Lost* (1667) it had some influence. Vondel also exercised a certain influence on the development of baroque drama in Germany in the 17th century, mainly through the translation and adaptation of some of his plays by GRYPHIUS.

VOS, JAN, see NETHERLANDS.

VOSKOVEC, JIŘÍ, see CZECHOSLOVAKIA.

VRCHLICKÝ, JAROSLAV, see CZECHOSLOVAKIA.

V.T.O., see ALL-RUSSIAN THEATRICAL SOCIETY.

VUOLIJOKI, HELLA (1888–1954), Finnish dramatist, credited by BRECHT with having suggested the basic idea of his play *Herr Puntila und sein Knecht Matti*. Important as a translator, she also wrote *Niskavuori* (1936–53), a cycle of plays following the social changes in a rural setting, and a tragedy, *Justina* (1937).

W

WADE, ALLAN (1881–1954), English actor, manager, and director, one of the founders of the PHOENIX SOCIETY and director of all but two of the 26 plays produced by it. He made his first appearance on the stage in 1904, and shortly afterwards was with Frank BENSON, leaving him in 1906 to become assistant to GRANVILLE-BARKER at the ROYAL COURT THEATRE, where he was instrumental in arranging the first visit of the company from the ABBEY THEATRE, Dublin, in 1909. Author of a *Bibliography of W. B. Yeats* (1908); he was also responsible for translations of GIRAUDOUX's *Intermezzo* and COCTEAU's *The Infernal Machine*, produced by the STAGE SOCIETY in 1934 and 1935 respectively.

WAGGON STAGE, see BOAT TRUCK.

WAGNER, RICHARD, see THEATRE BUILDINGS.

WAITS. Originally night watchmen in medieval palaces, castles, and walled towns, who indicated the hour by playing upon some musical instrument, these later became musicians in the service of a noble person and eventually of a town, functioning as a town band. It is not always easy to distinguish the early waits from the MINSTRELS, but it seems likely that the former were invariably resident, and had more formal musical training. From the early 16th century they are found performing any music called for by a play, and in Elizabethan London were probably hired by the theatres from the corporation when needed, to supplement the musicians among the actors. They varied in number from four to nine, and wore the livery of the noble house or town to which they were accredited. They were paid a nominal wage and received extra money from private individuals or theatres which hired them. They were not finally disbanded until the early 19th century.

WAKHEVITCH, GEORGES (1907–   ), stage designer, born in Odessa, whose career has been passed almost entirely in the French theatre. He has a wide knowledge of both the technical and theoretical aspects of design (he is himself responsible for all working plans, and oversees the construction of his sets), and does not hesitate to mingle built and painted scenery. His first work in the theatre (he began in the cinema) was for LUGNÉ-POË at the Théâtre de l'Œuvre, and he then designed for the Rideau Gris of Marseilles, where his *Macbeth* was particularly admired. He has since worked extensively for opera houses, and has been responsible for the décor of, among other plays, ANOUILH's *L'Invitation au Château* (1947) at the Théâtre de l'ATELIER, of the revival of ROMAINS's *Donogoo* (1951) and CLAUDEL's *L'Annonce faite à Marie* (1955) at the COMÉDIE-FRANÇAISE, and of Claudel's *L'Échange* for the 1951 production by BARRAULT.

WALCOT, CHARLES MELTON (1840–1921), American actor, who first appeared under the name of Brown to avoid confusion with his father, also Charles Melton Walcot (1816–68), a London-born architect who emigrated to the United States in 1839 and abandoned his career to go on the stage. They were, however, totally dissimilar, the father, an eccentric comedian and writer of BURLESQUES who was for many years connected with Mitchell's OLYMPIC, having in his later years the appearance of a venerable clergyman with side-whiskers, whereas the son sported a moustache and always wore the jovial expression of a genial English squire. He married in 1863 Isabella Nickinson (1847–1906), an accomplished actress who for many years played with him at the WALNUT STREET THEATRE in Philadelphia. They then joined Daniel FROHMAN's company in New York, and from 1887 onwards appeared in most of the LYCEUM successes, Walcot, playing comic or dignified elderly parts, helping to bridge the gap between the romantic actors of earlier years and the newer REALISM.

WALDIS, BURKARD (*c.* 1490–*c.* 1556), German dramatist, at one time a Franciscan monk, later the author of a virulent anti-Catholic play in support of the Reformation. Given at Riga in 1527, it was a retelling of the parable of the prodigal son (*Vom verlorenen Sohn*) showing him robbed of his fortune in a papist house at the instigation of the devil. In two acts, written in a Low German dialect, it is not without its lively moments, though action is somewhat subordinated to argument. It appears to have been performed on a stage similar to that required for the plays of NEIDHART VON REUENTAL, with two actions located on opposite sides of the available space; but they do not in this case proceed simultaneously.

WALDORF THEATRE, London, see STRAND THEATRE(2).

WALDORF THEATRE, New York, at 116 West 50th Street, between the Avenue of the Americas and 7th Avenue. This opened on 20 Oct. 1926 and had a brief, undistinguished life as a theatre. In 1930 came Leo Bilgakov's production of GORKY's *The Lower Depths* as *At the Bottom*, which had 72 performances. It was followed by eight performances of CHEKHOV's *The Seagull*, and a revival of Dreiser's *An American Tragedy* in 1931 ran for 17 weeks. The last production at this theatre in 1933 was PRIESTLEY's *Dangerous Corner*, after which the building became a cinema. It was converted into stores in 1937.

WALES has a strong musical tradition and its National Opera company enjoys an international reputation, but there is no long-standing theatrical tradition in the country and for a time Swansea was the only city to support a resident professional troupe. In 1962 the Welsh Committee of the ARTS COUNCIL established the Welsh Theatre Company, which was intended to be a permanent association performing plays throughout the country in both English and Welsh. The Welsh-language part of the company was established as a separate entity in Bangor in 1968 under the name of Cwmni Theatr Cymru, and continues to tour throughout Wales, but the English-speaking half, after merging with the Welsh National Opera in 1973, ceased operations in 1977.

Cardiff, which has the New Theatre, built in 1906 and now a CIVIC THEATRE seating 1,168 and housing touring productions, also has the Sherman Theatre, built in 1973 by University College with money provided by the Sherman Foundation. It has two auditoriums; the larger, seating 472, is a proscenium theatre that plays host to a wide range of professional drama companies, Welsh-language theatre being well represented. It is also used by ballet and opera companies and for concerts, films, lectures, recitals, and other activities. In 1981 Theatre Wales, a newly founded company employing Welsh actors in English-language plays, gave its first performances there, presenting ORTON's *What the Butler Saw* and PINTER's *No Man's Land*. Attached to the Sherman Theatre is the smaller Arena Theatre, a flexible space with an octagonal stage and seating for a maximum of 200. It is used by the smaller, more experimental professional companies and by amateurs, including the Sherman Arena Company which draws its actors from both the college and the community. There is also in Cardiff the Chapter Arts Centre, a mixed media centre with a theatre which provides a home for experimental groups such as Moving Being.

During the 1970s a number of theatres were built elsewhere in Wales, including the Torch Theatre at Milford Haven, which has a resident repertory company that also tours within the country; Theatr y Werin, a university arts centre at Aberystwyth which has a mixed-programme theatre; Theatr Gwynedd, a university-owned building in Bangor which provides a base for the Cwmni Theatr Cymru; and Theatr Clwyd, a civic theatre in Mold with a resident repertory company which also takes plays in English on tour.

WALKING GENTLEMAN, LADY, see STOCK COMPANY.

WALKLEY, ALFRED BINGHAM (1855–1926), English dramatic critic, whom SHAW satirised as Mr Trotter in *Fanny's First Play* (1911). He wrote for a number of papers before becoming dramatic critic of the *Times* in 1900, retaining this position until 1926. A cultured and conscientious writer who took himself and his duties very seriously, he wrote rather as a literary than a dramatic critic, devoting more space to the play than to the actors. His criticisms were collected in several volumes published between 1892 and 1925.

WALLACE, (Richard) EDGAR HORATIO (1875–1932), English journalist, novelist, and playwright. He was the first to make a speciality of detective drama, basing many of his plays on his own very popular books. Among the most successful of his plays were *The Ringer* (1926), *The Terror* (1927), *The Squeaker* (1928), *The Flying Squad* (1929), *On the Spot, Smoky Cell* (both 1930), and *The Case of the Frightened Lady* (1931). In all his works he showed unusual precision of detail, narrative skill, and inside knowledge of police methods and criminal psychology, the fruits of his apprenticeship as a crime reporter.

WALLACE, NELLIE [Eleanor] JANE (1870–1948), one of the great stars of the English MUSIC-HALL. Born in Glasgow, she made her first appearance as a clog-dancer in Birmingham in 1888. After touring the 'halls' as one of the Three Wallace Sisters, she was seen in the provinces in a series of straight comedies, but returned to the music-halls as a solo turn in the early 1890s, billed as 'the Essence of Eccentricity'. In her songs and sketches she was invariably the ever-hopeful spinster, seemingly unaware of her shabby clothes and her plain face, plastered with badly-applied make-up, surmounted by a battered hat, and with a moth-eaten feather boa—'Me Furs'—twisted round her throat. She would rattle on, equally unaware, it appeared, of the ambiguity of her remarks, until a sudden sideways leer proved the contrary. Her best-known song was 'I lost Georgie in Trafalgar Square/Lost 'im on me 'oneymoon, but I don't care', in which she was the epitome of the rapacious female on the rampage. A mistress of the grotesque, she was probably the only woman to make a success of the DAME in PANTOMIME, playing the Widow Twankey in *Aladdin*, the Cook in *Dick Whittington*, and Dame Durden in *Jack and the Beanstalk*. In 1935 she gave a superb performance as The Wicked Witch Carabosse in *The Sleeping*

Beauty, or, What a Witch, at the VAUDEVILLE THEATRE. She also appeared in REVUE, touring in *Love and Money* from 1928 to 1930 and in *The Queen of Clubs* in 1932. She made her last appearance on tour in *That'll Be the Day* in July 1945.

WALLACK, HENRY JOHN (1790–1870), London-born American actor, who with his parents, both actors, emigrated to the United States in 1818. He made his New York début at the ANTHONY STREET THEATRE and in 1824 was leading man at the CHATHAM THEATRE, which he took over two years later. After a visit to London, during which he was seen at COVENT GARDEN, he returned to New York in 1837 and worked as stage-manager for his brother, the elder James WALLACK, at the NATIONAL THEATRE while continuing to act. In 1847 he made a great success as Sir Peter Teazle in SHERIDAN's *The School for Scandal*. He then returned to England for a time to manage the Theatre Royal, MANCHESTER, and back in New York made his last appearance on the stage in 1858 as Falstaff in *The Merry Wives of Windsor*. Two of his sisters were on the London stage, one, as Mrs Stanley, being a popular member of the company at the Coburg Theatre (later the OLD VIC), the other, who married an actor named Pincott, becoming the mother of the actress Leonora, later well known as Mrs Alfred Wigan.

WALLACK, JAMES WILLIAM the elder (1791–1864), London-born actor who became equally well known in England and the United States, where he first appeared in 1818, having previously played leading roles at DRURY LANE. A romantic and tragic actor in the style of John Philip KEMBLE, he had a most successful season at the NATIONAL THEATRE in New York in 1837–8, with his elder brother Henry WALLACK as his stage-manager, and then went to NIBLO'S GARDEN and on tour. He made his last appearance in London in 1851, and a year later opened BROUGHAM's former theatre in New York as WALLACK'S LYCEUM. Elegantly redecorated, well equipped, and furnished with a good stock company in a repertory of Shakespeare and standard comedies, with some modern plays, it flourished for nine years, Wallack himself making his last appearance on the stage there in 1859. Two years later he opened the STAR THEATRE on Broadway, but his inaugural speech marked his last public appearance, and he retired, leaving the traditions he had established to be carried on by his son Lester WALLACK and his nephew, James WALLACK the younger.

WALLACK, JAMES WILLIAM the elder (1791–(1818–73), English-born American actor, son of Henry WALLACK, under whom he served his apprenticeship to the stage. In 1837 he joined the company of his uncle James WALLACK the elder at the NATIONAL THEATRE in New York, rising quickly from 'walking gentleman' to juvenile lead. A man of rugged physique, he was at his best in such parts as Macbeth, Othello, Iago, and Richard III. In 1865 he became a member of the stock company at the STAR THEATRE under his cousin, Lester WALLACK. He was excellent as Fagin in a dramatization of DICKENS's *Oliver Twist* in 1867 and in 1872 played Mathias in Lewis's *The Bells* at BOOTH'S THEATRE, most terrifyingly. He was not so successful in comedy, except for Jaques in *As You Like It* and Mercutio in *Romeo and Juliet*.

WALLACK, LESTER [John Johnstone] (1820–88), American actor, son of James WALLACK the elder, with whom he first appeared on the stage in the English provinces. In 1845 he was in MANCHESTER with Charlotte CUSHMAN and Helen FAUCIT, and soon after made his début in New York at the BROADWAY THEATRE. In 1850, as a member of BURTON's company at the CHAMBERS STREET THEATRE, he gave excellent performances as Aguecheek in *Twelfth Night* and Charles Surface in SHERIDAN's *The School for Scandal*, and then became stage-manager for his father at WALLACK'S LYCEUM, where he also played a wide range of comic and romantic parts. Under his management the STAR THEATRE, which the elder James Wallack opened in 1861 and immediately relinquished, flourished until 1881, when Lester moved to a new WALLACK'S THEATRE, remaining there until the year before his death. He concentrated on the production of plays by English dramatists, particularly T. W. ROBERTSON, and one of his finest parts was Elliot Grey in his own adaptation of a novel, *Rosedale* (1863), in which he appeared for many years. He was also good as Benedick in an 1867 production of *Much Ado About Nothing*, after which he staged no more of Shakespeare's plays until 1880, when he put on *As You Like It*. His memoirs were published in 1889.

WALLACK'S LYCEUM, New York, at 485 Broadway. As Brougham's Lyceum, this opened on 23 Dec. 1850 under John BROUGHAM, who put his trust in short burlesques and farces which had once been popular but were now outmoded. Consequently, in less than two years he was forced to give up his theatre, which came under the control of the elder James WALLACK, who on 25 Jan. 1852 started it on a successful career as Wallack's Lyceum. Elegantly redecorated, well equipped, and furnished with a good stock company which appeared in a repertory of plays by Shakespeare and standard comedies, with some modern plays, the theatre flourished for nine years, Wallack himself making his last appearance on stage there in 1859. It reopened on 2 Sept. 1861 as the Broadway Music-Hall, then became the Olympic under FOX, and was finally refurbished by George Wood and renamed the Broadway Theatre. The most notable event of its remaining years was the production of FLORENCE's pirated version of ROBERTSON's *Caste* in 1867. Thereafter the theatre went out of fashion, along with its neighbourhood, and was demolished in 1869.

WALLACK'S THEATRE. (1) At Broadway and 30th Street. This was opened by Lester WALLACK on 4 Jan. 1882 with a revival of SHERIDAN's *The School for Scandal*. Lillie LANGTRY, who was to have made her New York début at the second PARK THEATRE on 30 Oct. that year—the day it was burnt down—first appeared at Wallack's instead. The theatre was not a success, and in 1887 Lester transferred the lease to other hands, allowing them to retain the old name. The last stock season was given under Henry ABBEY in 1888, and the theatre was then leased by PALMER, who opened it under his own name on 8 Oct. of that year with a French company. It reverted to its original name on 7 Dec. 1896, and finally closed in 1915, after a season of plays directed by GRANVILLE-BARKER which included the first American production of SHAW's *Androcles and the Lion*. It was then demolished.

(2) A theatre at 254 West 42nd Street, between 7th and 8th Avenues, which in 1924 was named the Wallack after the famous theatrical family, opened originally on 5 Dec. 1904 as the Lew Fields Theatre. Two years later it was taken over by James K. HACKETT, who named it after himself and appeared there with his own company in a number of successful productions, including Anthony Hope's *The Prisoner of Zenda* in 1908. The same year saw the first production of SHELDON's *Salvation Nell*, with Mrs FISKE. In 1911 and 1920 the theatre changed its name with its managers, to the Harris and the Frazee respectively, and in 1921 housed a successful production of KAUFMAN and CONNELLY's *Dulcy*, with Lynn Fontanne. It became a cinema in 1931.

The STAR THEATRE was known as Wallack's from its opening in 1861 until 1882.

WALLER, EMMA (1820–99), British-born American actress, who emigrated to the United States with her husband, Daniel Wilmarth Waller (1824–82), and made her first appearance at the WALNUT STREET THEATRE in Philadelphia in Oct. 1857 as Ophelia to his Hamlet. She was seen in New York a year later, and from then until her retirement in 1878 played leading roles there and on tour throughout the country. Among her finest parts were Lady Macbeth; Queen Margaret in Colley CIBBER's version of *Richard III*, which she played with Edwin BOOTH; Meg Merrilies in a dramatization of SCOTT's *Guy Mannering*, in which she was seen many times; and Julia in Sheridan KNOWLES's *The Hunchback*. In the fashion of the time she also played Hamlet and Iago. After her retirement she continued to give public readings and taught elocution.

WALLER, LEWIS [William Waller Lewis] (1860–1915), English actor-manager, born in Spain of English parents, who made his first appearance as a professional at TOOLE's Theatre in 1883. A robust and dynamic actor, with a magnificent voice, he was at his best in costume parts and particularly in Shakespeare. His Brutus in *Julius Caesar*, Faulconbridge in *King John*, and above all his Henry V were memorable. In modern-dress comedy, such as WILDE's *An Ideal Husband* with which he opened his management at the HAYMARKET in 1895, he did not appear to such advantage. He was much admired as d'Artagnan in one of the many dramatizations of the elder DUMAS's *The Three Musketeers* (1898) and in the title role of *Monsieur Beaucaire* (1902), based by Booth TARKINGTON on his own novel. The latter role was perhaps the supreme example of Waller's talent. He visited the United States for the first time in 1911, where he was successful in the New York production of *The Garden of Allah*, adapted by Robert Hichens and Mary Anderson from the former's book. He was one of the outstanding romantic actors of his day, one of the first so-called 'matinée idols', but entirely without conceit, and the hysteria of his more fervid supporters caused him much embarrassment. He married an actress, Florence West (1862–1912), sister-in-law of the critic Clement SCOTT, who appeared with her husband in many of his outstanding successes, notably as Miladi in *The Three Musketeers*.

WALLS, TOM (1883–1949), English actor, director, and theatre manager. He made his first appearance in Glasgow in 1905, and subsequently toured North America, returning to London in 1907. After a successful career in MUSICAL COMEDY he went into management, his first venture, at the SHAFTESBURY THEATRE, being the immensely popular farce *Tons of Money* (1922) by Will EVANS and Valentine which, transferred to the ALDWYCH THEATRE a year later, inaugurated the long succession of 'Aldwych farces' by Ben TRAVERS, in all of which Walls appeared up to and including *Turkey Time* (1931). He played the role of the flashy opportunist, teamed with Ralph LYNN's 'silly ass' and the bewildered 'little man' of Robertson HARE. He was also manager of the FORTUNE THEATRE, opening with LONSDALE's *On Approval* (1927), which he directed himself. After some years in films, he reappeared on the London stage in 1938, continuing his successful career in light comedy and farce, and in 1939 took over the Alexandra Theatre in Stoke Newington, which he ran as a repertory theatre, mainly with transfers from the West End.

WALNUT STREET THEATRE, Philadelphia, the oldest playhouse in the U.S.A. Originally built as a circus in 1809, it was first used as a theatre two years later, competing with the CHESTNUT STREET THEATRE and, from 1828, with the Arch Street Theatre. After the financial crisis experienced by the theatres of PHILADELPHIA in 1829 the Walnut Street continued to operate with a good stock company which supported visiting guest players. Seating 1,052, it was remodelled in 1970, and from 1971 to 1980 provided a home for the Philadelphia Drama Guild.

WALSER, KARL (1877–1943), Swiss stage designer whose light touch and romantic vision were often made use of by REINHARDT in BERLIN before the First World War. The changing vistas of his grassy woodland glade, set on a revolving stage, for the first of Reinhardt's many productions of *A Midsummer Night's Dream*, at the Neues Theater in 1905, were a sensational innovation. He also designed the sets for revivals of NESTROY's *Einen Jux will er sich machen* in 1904 and KLEIST's *Käthchen von Heilbronn* in 1905, and for the first production, by Reinhardt at the Kammerspiele, of WEDEKIND's *Frühlings Erwachen* (1906).

WALTER, EUGENE (1874–1941), American dramatist, whose early plays, though somewhat melodramatic, seemed to point the way towards a more realistic and sober approach to social problems by contemporary American playwrights. The best of them were *Paid in Full* (1908) and *The Easiest Way* (1909). A later play, *Fine Feathers* (1913), was good also, but less successful. Walter failed to live up to his early promise, and his last plays are negligible. Among them were dramatizations of two popular novels by John Fox, Jr, *The Trail of the Lonesome Pine* (1912) and *The Little Shepherd of Kingdom Come* (1916).

WÄLTERLIN, OSKAR, see SWITZERLAND.

WANAMAKER, SAM (1919–), American actor and director, who studied for the stage at the GOODMAN THEATRE in Chicago, where in 1964 he returned to play Macbeth. He gained experience from 1936 to 1939 as actor and director with Chicago summer stock companies, and made his first appearance in New York in 1942. After serving with the U.S. armed forces he returned to the stage in 1946, playing Jimmy Masters in Maxwell ANDERSON's *Joan of Lorraine* with Ingrid Bergman. He made his first appearance in London in 1952 as Bernie Dodd in ODETS's *Winter Journey*, which he also directed, and then remained in England, directing and acting in a number of plays, including Odets's *The Big Knife* (1954) and N. Richard Nash's *The Rainmaker* (1956). In 1957 he was appointed Artistic Director of the New Shakespeare Theatre in Liverpool, and two years later he was with the company at Stratford-upon-Avon, playing Iago to the Othello of Paul ROBESON. He directed several plays in America in the early 1960s, but his work in the theatre has since been concentrated on his efforts to rebuild Shakespeare's GLOBE THEATRE on Bankside near the original site. In the early 1970s he founded a Trust for this purpose, and a temporary theatre tent opened in 1972 for a summer season which included a modern-dress production of *Hamlet* with Keith MICHELL. The second season, in which Vanessa REDGRAVE played Cleopatra in *Antony and Cleopatra*, came to an abrupt end when the tent collapsed in a storm. A second temporary theatre of tubular steel opened in 1975 with Charles MAROWITZ's version of *Hamlet*, and three summer seasons were given there. It is hoped to open a permanent theatre in the mid-1980s, and the site will contain a replica of the COCKPIT in Drury Lane as well as of the Globe. In 1972 Wanamaker also founded the nearby Bear Gardens Museum, a theatre museum covering Shakespeare's era.

WARBURTON, JOHN (1682–1759), English antiquarian and book collector, who had at one time in his possession some 60 Elizabethan and Jacobean plays in manuscript, many of them unique copies of works which had never been printed. Unfortunately, through his own carelessness and the ignorance of his servant Betsy Baker, who somehow got possession of them, all but three were, as Warburton himself says, 'unluckely burnd or put under Pye Bottoms'. He left a list of the titles of these lost plays, from which it appears that the chief sufferers from Betsy's depredations were Thomas DEKKER, John FORD, and Philip MASSINGER.

WARD [Buchanan], FANNIE (1872–1952), American actress, born in St Louis, who made her first appearance on the stage at the Broadway, New York, in 1890 as Cupid in J. Cheever Goodwin's *Pippino*. In 1894 she made her first appearance in London as Eva Tudor in Lionel Monckton's *The Shop Girl* at the Gaiety. She had only one line to say—'Watch my wink!'—specially written in for her by George EDWARDES, but her beauty and charm made an instant impression and she remained in London until 1906, appearing with undiminished success in a series of light and musical comedies. On her return to New York she was seen as Rita Forrest in J. Hartley MANNERS's *A Marriage of Reason* (1907), and in the same year she returned to London to play Nance in Channing Pollock's *In the Bishop's Carriage*, one of her most successful parts. From then onwards she pursued her career in London and New York, occasionally touring, and in 1927 and 1928 was seen in VARIETY at the London COLISEUM.

WARD, (Lucy) GENEVIÈVE TERESA (1838–1922), American-born actress, the first to be created DBE (in 1921). Originally an opera-singer (as Madame Ginevra Guerrabella), she lost her voice through illness and overwork, and turned to straight acting, making her début in 1873 in MANCHESTER as Lady Macbeth, a part which she later played in French at the PORTE-SAINT-MARTIN in Paris. Among her outstanding roles were Julia in Sheridan KNOWLES's *The Hunchback*, Portia in *The Merchant of Venice* (both 1874), SOPHOCLES' Antigone, Mrs Haller in KOTZEBUE's *The Stranger*, and Belvidera in OTWAY's *Venice Preserv'd* (all 1875). She was first seen in New York, her birthplace, in 1878, in the title role of W. G. Wills's *Jane Shore*. A year later, under her own management in London, she produced at the LYCEUM *Forget-Me-Not* by Herman Merivale and F. C. Grove, which proved such a success that she toured in it all over the world. Playing in the original production was the young Johnston FORBES-ROBERTSON. After her last visit to America in 1891 she joined Henry IRVING at the Lyceum to play Queen Eleanor in TENNYSON's *Becket* (1893) and Morgan Le Fay in Comyns Carr's *King Arthur* (1895). After 1900 she rarely appeared on the stage, but she was occasionally seen with Frank BENSON's company, particularly as Volumnia in *Coriolanus*, in which she made her last appearance in London at the OLD VIC THEATRE in Apr. 1920, and Queen Margaret in *Richard III*, in which she made her final appearance in the provinces on tour later the same year.

WARDE, FREDERICK BARKHAM (1851–1935), British-born American actor, who in 1867 joined a small touring company and after some years' experience in stock, playing with such actors as IRVING and Adelaide NEILSON, went to BOOTH'S THEATRE in New York, making an immediate success. He remained there for three years, and then went on tour until in 1881 he formed his own company, playing mainly in Shakespeare and in such old favourites as Sheridan KNOWLES's *Virginius*, BIRD's *The Gladiator*, and BULWER-LYTTON'S *The Lady of Lyons*, since he was unable to find any modern plays to his taste. He retired from the theatre in 1919, though he continued the lectures on Shakespeare and drama in general which he had begun in 1907. He was the author of *The Fools of Shakespeare* (1913) and of a volume of reminiscences, *Fifty Years of Make-Believe* (1920).

WAREHOUSE, The, the ROYAL SHAKESPEARE COMPANY's studio theatre in London, seating 200. It opened in 1977 with the intention of showing in London plays which had proved successful at the OTHER PLACE in Stratford-upon-Avon, and of encouraging young British playwrights. Among the latter have been Howard BARKER (*That Good Between Us*, 1977; *The Hang of the Gaol*, 1978; *The Loud Boy's Life*, 1980), Barrie Keeffe (*Frozen Assets*, 1978; *Bastard Angel*, 1980), and Howard BRENTON (*The Churchill Play* and *Sore Throats*, 1979; *Thirteenth Night*, 1981). Its most successful production was Willy Russell's *Educating Rita* (1980), which was transferred for a long run to the PICCADILLY THEATRE and was also seen in America. The R.S.C. left the Warehouse on the transfer of its London base from the ALDWYCH THEATRE to the Barbican in 1982.

WARFIELD, DAVID (1866–1951), American actor, who began as a programme seller and later an usher at the San Francisco theatre and in 1888 joined a travelling company, playing Melter Moss in Tom TAYLOR's *The Ticket-of-Leave Man*. This failed after a week, and he went into variety, appearing in New York in 1890 in vaudeville and musical comedy. He was Karl in the original production of the popular musical comedy *The Belle of New York* (1897) and later spent three years in a burlesque company. He was adept at presenting the New York East Side Jew of his day, and had already given proof of fine qualities in his acting when in 1901 BELASCO starred him in KLEIN's *The Auctioneer*. This was an instantaneous success and had a long run, but it was as the gentle, pathetic, self-sacrificing Anton von Barwig in *The Music Master* (1904), also by Klein, that Warfield set the seal on his growing reputation. He played nothing else, in New York and on tour, for three years. Among his later successes were Wes Bigelow in Belasco's *A Grand Army Man* (1907), and the title roles in *The Return of Peter Grimm* (1911), also by Belasco, and Wills's *Vanderdecken* (1915). He was also seen as Shylock in Belasco's 1922 production of *The Merchant of Venice*. He retired from the stage two years later.

WARMIŃSKI, JANUSZ, see POLAND.

WARNER [Lickfold], CHARLES (1846–1909), English actor, who made his first appearance as a boy of 15 at SADLER'S WELLS, where his father, James Lickfold, was a member of the company. Forced against his will to study architecture, he ran away from home and returned to the theatre under the name of Warner, which he afterwards retained. He first appeared in London in 1864, playing Paris in *Romeo and Juliet*. A year later he played Romeo, and Iago in *Othello*. He scored a great personal success as Steerforth in *Little Em'ly* (1869), an adaptation by Andrew Halliday of DICKENS's *David Copperfield*, and was the first to play Charles Middlewick in H.J. BYRON's *Our Boys* (1875); but though good in comedy, he was at his best in MELODRAMA, in which he played at the ADELPHI THEATRE for many years. His finest and most memorable part was Coupeau in *Drink* (1879), an adaptation of ZOLA's novel *L'Assommoir* in which he collaborated with Charles READE, first seen at the PRINCESS'S THEATRE. He made his last appearance in 1906 in *The Winter's Tale*, playing Leontes to the Hermione of Ellen TERRY at HER (then His) MAJESTY'S THEATRE, and then went to America where he committed suicide.

WARREN, WILLIAM the elder (1767–1832), British-born American actor, who made his early appearances in the English provinces, and in 1788 was with Tate WILKINSON, playing in support of Sarah SIDDONS. In 1796 he was invited by Thomas WIGNELL to join his company in PHILADELPHIA, and there, except for a number of short visits to New York, he spent the rest of his life. He succeeded Wignell as manager of the CHESTNUT STREET THEATRE, in partnership with William WOOD, and under them in 1820 Edwin FORREST made his first appearance, playing Young Norval in HOME's *Douglas*. Warren was a fine actor of the old school, among his best parts being Sir Toby Belch in *Twelfth Night*, Sir Anthony Absolute and Sir Peter Teazle in SHERIDAN's *The Rivals* and *The School for Scandal*, and Old Dornton in HOLCROFT's *The Road to Ruin*. He retired in 1829, making a final farewell appearance the year before his death. In 1806 he married as his second wife the actress Mrs MERRY, and after her death two years later married the sister-in-law of the first Joseph JEFFERSON, with whose father Thomas Jefferson he had appeared in England as a young man. There were six children of this marriage, all connected with the stage. The best known was William WARREN the younger; his brother became a theatre manager and, by the marriage of his daughter Sarah Isabel, the father-in-law of the third Joseph JEFFERSON; his four sisters all married actors or theatre managers.

WARREN. WILLIAM the younger (1812–88), American actor, son of the above, who spent most of his professional life at the Boston Museum. He made his first appearance at the Arch Street Theatre in PHILADELPHIA, shortly after his father's death, playing Young Norval in HOME's *Douglas*, and then toured for several years and was a member of several good stock companies. He was seldom seen in New York and only once, in 1845, in England. He began his long association with the Museum in 1847, and remained there until his retirement in 1883. In his early days he acted a wide variety of parts, but later specialized in comedy, being particularly admired as Touchstone in *As You Like It*, Polonius in *Hamlet*, and Bob Acres and Sir Peter Teazle in SHERIDAN's *The Rivals* and *The School for Scandal*. He appeared also in a number of new plays, which his excellent acting often redeemed from mediocrity, and in 1882 celebrated his stage jubilee by playing Dr Pangloss in the younger COLMAN's *The Heir-at-Law*. He was last seen on the stage shortly before his retirement as Old Eccles in T. W. ROBERTSON's *Caste*.

WARWICK'S MEN, see OXFORD'S MEN.

WASHINGTON, DC. The first theatre in the federal capital, known as the United States or National Theatre, was adapted from a building already in existence. Thomas WIGNELL and his company played there for some months in 1800, after which it became a post and patent office. The first purpose-built theatre was the Washington, which opened in 1804 and was later rebuilt and renamed the Washington City Assembly Rooms. A New Washington Theatre was opened in 1821 by a company under WARREN and WOOD, which used it during the summer months only. Joe COWELL appeared there in 1828, and by 1831 it was under the management of Francis C. WEMYSS. A second National Theatre, which opened in 1835, played an important part in the social life of the city, one innovation being the turning of the old pit into a 'parquette area', with comfortable seating, which made it more acceptable to the better-class citizens. There were several fires in the 19th century and the theatre had to be rebuilt several times. The present auditorium dates from 1922. FORD'S THEATRE, which opened in 1862, became famous as the scene of Abraham Lincoln's assassination; after a prolonged closure it is once again in use as a theatre. Among other present-day theatres in Washington are the ARENA STAGE, the Folger Theatre Group, founded in 1970 under the auspices of the Folger Shakespeare Library to present plays in a reconstruction of an Elizabethan playhouse, and the EISENHOWER THEATER.

WASHINGTON SQUARE PLAYERS, New York, a play-producing society founded in 1914 by a group which included Edward Goodman and Lawrence LANGNER. Their first programme, in Feb. 1915, was seen at the BANDBOX THEATRE, and consisted of a number of one-act plays, some of them specially written for the occasion. In 1916 they began to present full-length plays, among them CHEKHOV's *The Seagull*, IBSEN's *Ghosts*, and SHAW's *Mrs Warren's Profession*. In the same year they moved to the COMEDY THEATRE, where their productions included O'NEILL's one-act play *In the Zone* (1917) and Elmer RICE's *Home of the Free* (1918). Lee SIMONSON worked for the Washington Square Players, and Katharine CORNELL made her first professional appearances with them. They disbanded in 1918, but it was out of their work that the THEATRE GUILD evolved.

WASHINGTON SQUARE THEATRE, New York, at 40 West 4th Street, a shed-like corrugated-steel building designed by Jo MIELZINER, which was built as the temporary home of the Lincoln Center repertory company pending the completion of the VIVIAN BEAUMONT THEATRE. It opened on 23 Jan. 1964 with Arthur MILLER's *After the Fall* directed by Elia KAZAN, who, with Robert Whitehead, had been appointed director of the new company. Miller's play was followed by a revival of O'NEILL's *Marco Millions* and by a new play, *But For Whom Charlie*, by S. N. BEHRMAN. The venture was not a success, and Kazan and Whitehead withdrew, to be succeeded by Herbert Blau and Jules Irving from the San Francisco Actors' Workshop, who in 1965 moved with the company to its permanent home at the Lincoln Center. From Nov. 1965 to Mar. 1968 the Washington Square Theatre was occupied by a musical, *Man of La Mancha* based on CERVANTES' *Don Quixote*. It was demolished in 1968.

WATERHOUSE, KEITH, see HALL, WILLIS.

WATER ROWS, see GROUNDROW.

WATERS, ETHEL, see REVUE.

WATFORD, Herts., see PALACE THEATRE, WATFORD.

WATT, JOACHIM VON (1484–1551), Austrian humanist, a pupil of CELTIS, who provided in his *Gallus pugnans* (1514) the earliest examples of the Viennese *Posse* (farce). Cocks and hens appear before the court to argue out the morality of cock-fighting and the rights of the sexes. Humanist advocates represent them, capons act as arbitrators and advise each party to fulfil the functions assigned to it by nature, and a Viennese lad expresses the views of the dissatisfied audience when he declares that they should be consigned to the pot: in this he appears as a prototype of HANSWURST, with his cynical comments on the behaviour of his betters.

WEAVER [WEVER], JOHN (1673–1760), dancing-master at DRURY LANE, who in 1702 introduced from France on to the English stage short pieces in dumb show known as 'Italian Night Scenes'. These presented characters from the COMMEDIA DELL'ARTE in an English setting—usually an inn—where they played jokes, taken from the Italian scenarios, on typical English tradesmen and their families. One of the earliest featured 'a cooper, his wife, and his man'. Performed by a growing number of English dancers and popular up to 1723, these Night Scenes eventually gave rise to the English HARLEQUINADE, from which John RICH developed the specifically English form of PANTOMIME.

WEBB, JOHN (1611–72), English artist and scenic designer, a pupil of Inigo JONES. He was employed by DAVENANT to design and paint scenery for his productions, beginning with *The Siege of Rhodes* (1656), for which Webb prepared landscapes showing the general layout of the town and harbour, based probably on actual engravings of the time. In the epilogue to *The World in the Moon* (1697), the author, Elkanah SETTLE, emphasizes the local origin of Webb's elaborate scenery, in marked contrast to the French fashion of the moment, with a play on the scene-painter's name: ''Tis home-spun Cloth; All from an English Web.'

WEBER AND FIELDS, American comedy duo consisting of Joseph Weber (1867–1942) and Lew Fields (1867–1941), the sons of poor Jewish immigrants, who began acting as children in dime museums and beer gardens in and around New York and later appeared in 'burnt-cork' minstrelsy. They soon discarded this for a 'knockabout' act which combined slapstick clowning with the immigrant's difficulties with the English language, Fields, tall, thin, and tricky, appearing as Myer, and Weber, short, squat, and guileless, playing his stooge Mike. In 1885 they established their own company, writing BURLESQUES of the serious dramas of the day, and ten years later opened the former Broadway Music-Hall under their own names, remaining there until 1904, when their careers diverged, though they were seen together again in *Hokey-Pokey* (1912). Weber ceased to act in 1918, but continued to direct plays for another ten years, while Fields directed and appeared in a number of MUSICAL COMEDIES, remaining on the stage until 1929. He was the father of the dramatist Joseph FIELDS.

WEBSTER, BEN(jamin) NOTTINGHAM (1797–1882), English actor, manager, and dramatist, one of a large family of actors and musicians later connected by marriage with the LUPINOS. He was first a dancer, appearing as HARLEQUIN and PANTALOON in PANTOMIME in the English provinces and at DRURY LANE. He then became an actor, proving himself an excellent broad comedian, and was with Mme VESTRIS at the OLYMPIC THEATRE. In 1837 he became lessee of the HAYMARKET THEATRE, where he remained for 16 years, engaging all the best actors of the day and putting on many notable productions, in many of which he himself appeared. In 1844 he took over the ADELPHI THEATRE as well, and was associated there with Mme CELESTE and Dion BOUCICAULT, collaborating with the latter in two plays, *The Fox and the Goose; or, the Widow's Husband* and *Don Caesar de Bazan* (both 1844). Webster also made an adaptation of DICKENS's *The Cricket on the Hearth* in which he appeared at the Haymarket in 1846 as John Peerybingle with great success; but his finest part was Triplet in Tom TAYLOR and Charles READE's *Masks and Faces* (1852) in which he was seen both at the Adelphi and at the Haymarket. He opened the New Adelphi in 1859 and continued in management until 1874, when he retired. He retained all his faculties until the end. As a character actor he was unsurpassed in his own day, but he had little liking for farce, in spite of having written several successful farces himself, and he tempered his comedy with a somewhat grim humour. His grandson Ben Webster became a well known actor in London and New York and was the father of Margaret WEBSTER.

WEBSTER, JOHN (*c.* 1580–1634), English dramatist, whose fame rests almost entirely on two plays, *The White Devil* (1612) and *The Duchess of Malfi* (1614). Both are founded on Italian *novelle* and are passionate dramas of love and political intrigue in Renaissance Italy, compounded of crude horror and sublime poetry. In the latter respect Webster approached Shakespeare more nearly than any of his contemporaries, and both these plays have held the stage down to the present day, the most recent revival of *The White Devil* being that at the NATIONAL THEATRE in 1969 and of *The Duchess of Malfi* by the ROYAL SHAKESPEARE COMPANY in 1971 and the ROYAL EXCHANGE THEATRE, Manchester, in 1980. Both provide scope for great acting and fine settings, and in the category of poetic drama remained unsurpassed by any later work, except that of OTWAY, until a new conception of tragedy was introduced into European dramatic literature by IBSEN. Except for *The De-*

vil's Law Case (*When Women Go to Law*) (1620), a comedy given at the RED BULL, Webster's other plays, which are of little importance, were written in collaboration—*Westward Hoe* (1604) and *Northward Hoe* (1605) with DEKKER, *Appius and Virginia* (*c.* 1608) probably with Thomas HEYWOOD, and *A Cure for a Cuckold* (1625) with William ROWLEY. Little is known of Webster's life; he may have been an actor in his early years, but he was probably not the John Webster who appeared with the ENGLISH COMEDIANS in Germany under Robert BROWNE in 1596.

WEBSTER, MARGARET (1905–72), English actress and director, the daughter of Ben(jamin) Webster (1864–1947), himself the grandson of B. N. WEBSTER, and Dame May Whitty (1865–1948). They were together in IRVING's company at the LYCEUM from 1895 and later became very popular in the United States, Margaret being born in New York during one of their visits. She was on the stage for some time as a child, making her first appearance in 1917, and her adult début in the chorus of EURIPIDES' *Trojan Women* with Sybil THORNDIKE in 1924. She was a member of FAGAN's repertory company in OXFORD in 1927, toured with Ben GREET in Shakespeare, and in 1929 was at the OLD VIC. After playing a wide variety of parts in London she went to New York in 1936, and, while continuing to act, made an outstanding reputation as a director, particularly of Shakespeare's plays. Her production of *Othello* in 1943 with Paul ROBESON, in which she played Emilia, broke all records for a Shakespearian play on Broadway. With Eva LE GALLIENNE and Cheryl CRAWFORD she founded the American Repertory Company, playing Mrs Borkman in IBSEN's *John Gabriel Borkman* for it the same year, and the Cheshire Cat and the Red Queen in Lewis Carroll's *Alice in Wonderland* in 1947. She also organized a Shakespeare company which toured the United States in 1948–50 by bus and truck, and directed and appeared in *An Evening with Will Shakespeare* (1952) on tour. Among her other post-war productions were Eva Le Gallienne's translation of HOCHWÄLDER's *Das heilige Experiment* (1943) as *The Strong are Lonely* (N.Y., 1953; London, 1955), SHAW's *Back to Methuselah* in New York in 1958, and Noël COWARD's *Waiting in the Wings* (London, 1960). In 1964 she was seen in London in a solo performance based on the life and works of the Brontë sisters.

WEDEKIND, FRANK [Benjamin Franklin] (1864–1918), German dramatist and actor, who first appeared in cabaret, singing his own songs, and later acted in his own plays with his young wife Tilly [Mathilde Newes] (1886–1970). The first of these was *Die junge Welt* (written in 1889, prod. 1908); the best known is *Frühlings Erwachen*, written in 1891, which harks back to BÜCHNER and LENZ in its staccato structure and intensified realism, but looks forward to EXPRESSIONISM and SYMBOLISM in its graveyard and schoolroom scenes in which it analyses the situation of two 14-year-old lovers who pay with their lives for the moral dishonesty of their tyrannical parents. Produced by REINHARDT at the BERLIN Kammerspiele in 1906, its frank depiction of the results of sexual repression kept it off the English stage until 1963, when as *Spring Awakening* it was seen at the ROYAL COURT THEATRE, though the Sunday Theatre Club had given a private performance of it for members only as early as 1931 as *Spring's Awakening*. In 1974 a new translation by Edward BOND, produced by the NATIONAL THEATRE and seen also in New York in 1978, proved that though dated the play still retained its hold over an audience. Two of Wedekind's later plays, *Erdgeist* (written in 1893, staged in 1902) and *Die Büchse der Pandora* (written in 1894, staged 1905), served to reinforce his main thesis—that the repression of sexuality results in perversion and tragedy. The two plays, later staged together as *Lulu*, form the subject of an unfinished opera by Alban Berg and a classic silent film by Pabst, and in an adaptation by Peter Barnes were seen successfully in London in 1970; *Lulu* was also produced in New York in the same year. Among Wedekind's other plays, *Der Marquis von Keith* (1901) was revived by JESSNER in a famous expressionist production in 1920, and was seen at the Schillertheater in Berlin in 1963. It was not seen in London until 1974, and was produced in New York in 1979. Other plays which are sometimes revived are *König Nicolò, oder So ist das Leben* (1902) and *Schloss Wetterstein*, first produced in 1917 with Elisabeth BERGNER playing opposite the author. All Wedekind's plays, with their sex-ridden men, women, and children, their gentlemen crooks, and their grotesque yet vivid cranks, typify the feverish spirit of the years before 1914. In an atmosphere of relaxed morals and increasing juvenile delinquency they seem perhaps less shocking now than they did to their first audiences, but they remain valid statements of repressed and thwarted sexuality.

WEHR THEATRE, see MILWAUKEE REPERTORY THEATRE.

WEIGEL, HELENE (1900–71), Austrian-born actress and theatre manager, who began her career in Frankfurt playing such powerful low-life characters as Marie in BÜCHNER's *Woyzeck* and the title role in SCHÖNHERR's *Der Weibsteufel*. In 1923 she went to BERLIN, and under JESSNER played minor roles, in which she was admired by some critics, though others found her too strident and uncontrolled. She made her first big success as Klara in HEBBEL's *Maria Magdalene* in 1925, and after appearing as Grete in TOLLER's *Der deutsche Hinkemann* in 1927, the Widow Begbick in BRECHT's *Mann ist Mann* (also 1927), and the title role in his adaptation of GORKY's *The Mother* in 1932, was considered unrivalled as the exponent of working-class women. In 1928 she married

Brecht and accompanied him on his wanderings from 1933 to 1948, appearing during this time only in his *Die Gewehre der Frau Carrar* in Paris in 1937 and *Furcht und Elend des Dritten Reiches* in the following year. Returning to East Berlin with him in 1948, she helped to found and run the BERLINER ENSEMBLE, playing with the company in the title role of her husband's *Mutter Courage und ihre Kinder*, Natella in his *Die kaukasische Kreidekreis*, and Volumnia in his adaptation of *Coriolanus*. After Brecht died she took control of the company, taking it to London in 1956 and 1965. Among the parts which she played after her husband's death the most important was Frau Luckerniddle in his *Die heilige Johanna der Schlächthofe* in 1968. Her acting in later life had an effortless simplicity, and in *Mutter Courage* she illustrated to perfection Brecht's theory of acting (see VERFREMDUNGSEFFEKT), seeming to stand beside the figure without identifying with it.

WEIMAR, town in Thuringia, East Germany. The first theatre in the town was built in 1696 and used by visiting professionals, including EKHOF in 1772. Three years later this was destroyed and replaced by a temporary theatre in the palace of the Duchess Anna Amalia, used by professionals and amateurs. There from 1775 to 1783 GOETHE, with a group of courtiers, produced plays for royal occasions, constantly hampered by lack of money and the irresponsibility of his amateur company. He acted himself, both in comedy and tragedy, playing Orestes to Corona Schröter's Iphigenia in his own play *Iphigenie auf Tauris*. By 1781 the Court was beginning to lose interest in amateur theatricals, and three years later a new Court theatre opened with a resident professional company of which Goethe became artistic director in 1791. With the help of SCHILLER, who shared his vision of a theatre which combined dramatic appeal with a high literary content, he established a repertory which included plays by himself, as well as by Schiller, his collaborator from 1799 to 1805, LESSING, Shakespeare, CALDERÓN, and VOLTAIRE. Guest artists like SCHRÖDER and IFFLAND were imported to strengthen the resident company, and the theatre soon became famous for its productions and particularly for Goethe's fine handling of crowd scenes. After Schiller's death in 1805 Goethe continued to direct the theatre alone until 1817, by which time it had already begun to decline in popularity. The building, which had been renovated in 1798, was burnt down in 1826 and rebuilt, but it failed to regain its audience and had a chequered career until in 1848 it came under the direction of Liszt, who remained there until 1858, producing, as well as new German plays, two operas by Wagner, *Tannhäuser* and *Lohengrin* (its first performance). In 1861 Dingelstedt produced HEBBEL's *Die Nibelungen*, and in 1864 a season of Shakespeare's historical plays. After being damaged in the Second World War the theatre was rebuilt and reopened in 1948, since when it has continued to serve the town as a repertory theatre.

WEISE, CHRISTIAN (1642–1702), German dramatist, headmaster of a school at Zittau, where he wrote and produced a number of long plays which were purely academic. The chronicle plays and tragedies of his earlier days have little to recommend them, but the later comedies, in which Weise portrays the middle class on stage for the first time in the German theatre, foreshadow to some extent the work of LESSING a century later. His *Komödie der bösen Catharina* (1702) was based on the same plot as *The Taming of the Shrew*. He used a simple stage and, although he admitted the FOOL in the shape of Pickelhering into his plays, he demanded clear delivery in place of declamation and attempted individual characterization.

WEISS, PETER ULRICH (1916–82), German dramatist and novelist, resident after 1939 in Sweden. First known as a graphic artist and novelist, he gained international renown with his first play, *Die Verfolgung und Ermordung Jean Paul Marats dargestellt durch die Schauspielgruppe des Hospizes zu Charenton unter Anleitung des Herrn de Sade* (*The Persecution and Assassination of Marat as Performed by the Inmates of the Charenton Asylum under the Direction of the Marquis de Sade*, 1964). Known briefly as the *Marat/Sade*, it shows Sade and Marat in the asylum's bath-house, debating their contrasting philosophies of absolute individualism and unswerving dedication to social revolution, while the historical events leading up to Marat's assassination are acted out by the increasingly unruly lunatics. Peter BROOK produced it later the same year in London, in translation, with the ROYAL SHAKESPEARE COMPANY (N.Y., 1965). Its success led to the simultaneous production in 14 German theatres of Weiss's next play *Die Ermittlung* (1965), which as *The Investigation*, was given a rehearsed reading in London, again under Brook, and was produced in New York in 1966. One of a wave of 'documentary dramas' (see Theatre of FACT), it was based on the transcript of the 1964 Frankfurt War Crimes Trial, and attempts to apportion the blame for the Auschwitz atrocities. This, like the *Marat/Sade*, and *Der Turm* (*The Tower*, 1967), was published in English, as was *Die Versicherung* (*The Insurance*, 1969), a surrealist allegory in which men and beasts intermingle. An earlier play, *Gesang vom lusitanischen Popanz* (1967), an account of the uprising in Angola and its suppression by the Portuguese, was staged, as *Song of the Lusitanian Bogey*, by the Negro Ensemble Company in New York in 1968 and London in 1969. A further documentary on the war in Vietnam followed, its cumbersome title being shortened for practical purposes to *Viet Nam Diskurs* (*Vietnam Discourse*, 1968). After an excursion into more popular entertainment with *Wie dem*

Herrn Mockingpott das Leiden ausgetrieben wird (*How Mr Mockingpott was Relieved of his Sufferings*, also 1968), in which the chief characters are two clowns, Mockingpott and Wurst, Weiss reverted to documentaries with *Trotzki in Exil* (1970), which portrayed Trotsky not only as literally an exile from his own country but as an exile of the mind, his vision of the future being rejected by communists and capitalists alike, and *Hölderlin* (1971), presenting the poet as the archetypal revolutionary writer.

WEKWERTH, MANFRED (1929–), East German director, who had already had some experience with amateur companies when in 1951 he joined the BERLINER ENSEMBLE, where his first production was BRECHT's adaptation of GORKY's *Mother*. He directed plays with Brecht, including *Der kaukasische Kreidekreis* in 1954, and PALITZSCH, notably SYNGE's *The Playboy of the Western World* in 1956. He was also concerned in the production of KIPPHARDT's *In der Sache J. Robert Oppenheimer* in 1965, in which year his production of Brecht's adaptation of *Coriolanus* was seen at the OLD VIC during the visit there of the Berliner Ensemble. In 1971 he directed Shakespeare's original play for the NATIONAL THEATRE in London. A year later he was responsible for a striking production of *Richard III* at the DEUTSCHES THEATER in Berlin. He became manager of the Berliner Ensemble in 1977.

WELFARE STATE INTERNATIONAL, The, English experimental travelling troupe led by its founder, John Fox, and Boris Howarth. First formed in 1968, its method of COLLECTIVE CREATION combines the visual arts, theatre, music, myth-making, and magic. The members of the group call themselves Civic Magicians and Engineers of the Imagination, and believe that art is not a separate part of life, but intrinsic to it. They create 'ceremonies' in celebration of such events as the coming of winter or spring, the uniting of man and woman, and the naming of children. Other works centre on a hermaphrodite hero named Lancelot Quail, incorporating a variety of surrealistic events and images connected with his continuing search for beauty. One such adventure, *Beauty and the Beast* (1973), took place on a former municipal rubbish tip, where the group was then living in caravans. The entire site became an environmental sculpture constructed from discarded junk found on the dump. In another piece a giant temporary sculpture was constructed in Birmingham, using six up-ended car bodies sprayed with sand, a wooden tower with a radar scanner, and a river of 1,000 car mirrors. Such events have been created in various parts of England, and in France, the Netherlands, Poland, West Germany, and Australia.

WELLES, (George) ORSON (1915–), American actor and director, a forceful and idiosyncratic personality who has had a stormy and spectacular career in both the theatre and the cinema. He made his first professional appearance on stage at the GATE THEATRE, Dublin, in 1931, playing the Duke of Württemberg in FEUCHTWANGER's *Jew Süss*, and from 1933 to 1934 was with Katharine CORNELL, playing Mercutio and later Tybalt in *Romeo and Juliet*, Marchbanks in SHAW's *Candida*, and Octavius in Besier's *The Barretts of Wimpole Street*. Then followed a five-year collaboration with the producer John Houseman, first as an actor in Archibald MacLeish's *Panic* (1935) and then in the FEDERAL THEATRE PROJECT, for which he directed several controversial productions, among them the 'voodoo' *Macbeth* in 1936 and MARLOWE's *Dr Faustus* in 1937, in which he also appeared. Finally he partnered Houseman in the running of the Mercury Theatre, formerly the COMEDY. In 1939 he was seen as Falstaff in his own adaptation of Shakespeare's historical plays under the title *Five Kings*. He then turned his attention to the cinema, but returned to New York in 1941 to direct Paul GREEN's *Native Son* and in 1946 to direct and play in his own adaptation of Jules Verne's *Around the World in 80 Days*. He made his first appearance in London as Othello in 1951, and was seen there again in 1955 in Melville's *Moby Dick*, which he adapted, directed, and designed. In 1956 he directed and played the title role in *King Lear* in New York, and in Dublin in 1960 re-created the role of Falstaff in *Chimes at Midnight*, his own adaptation of *Henry IV* and *Henry V*. He also designed and directed the London production of IONESCO's *Rhinoceros* in the same year, and in 1962 his adaptation of *Moby Dick* was seen in New York. He then concentrated on films.

WELL-MADE PLAY, the English version of *une pièce bien faite*, a label applied in early 19th-century France, at first in a complimentary sense, to plays written by dramatists skilled above all in putting together a plot. It soon took on a pejorative meaning, and came to be used ironically of all plays in which the action develops artificially, according to the strict laws of logic and not to the unpredictable demands of human nature; and in which the plot, to which the characters are completely subordinated, is conceived in terms of exposition, knot, and denouement, with a series of contrived climaxes to create suspense. It is commonly used in France of the works of SCRIBE and SARDOU, and in England of such playwrights as ROBERTSON, JONES, PINERO, COWARD, and RATTIGAN. The well-made play was the chief target of ZOLA and other proponents of NATURALISM in France, and of SHAW's hostile criticism of productions in London from 1895 to 1898, collected together in *Our Theatres in the 'Nineties* (1931).

WEMYSS, FRANCIS COURTNEY (1797–1859), British-born American actor, who had already had a good deal of experience in the

English provinces when in 1821, while playing in London, he was engaged for the CHESTNUT STREET THEATRE in Philadelphia. He made his début there the following year with such actors as William WARREN the elder, William WOOD, Henry WALLACK, and the first Joseph JEFFERSON. In 1824 he made his first appearance in New York, and was playing Duncan to MACREADY'S Macbeth on the occasion of the ASTOR PLACE OPERA HOUSE riot in 1849. He was manager of theatres in several large American cities, including the Chestnut Street and Barnum's AMERICAN MUSEUM. A pleasant, cultured, and courtly man, though never of the first rank as actor or manager, he became extremely popular in America and was the author of a most entertaining autobiography, *Twenty-Six Years of the Life of an Actor and Manager*, published in 1847.

WERFEL, FRANZ (1890–1945), Austrian dramatist and novelist, who first became known as one of the early poets of EXPRESSIONISM. His first theatrical success was *Die Troerinnen* (1916), an adaptation of EURIPIDES' *Trojan Women* acclaimed as a disguised protest against war, but his main contribution to expressionist drama was the 'magic trilogy' *Der Spiegelmensch* (*The Mirror Man*, 1921), a modern version of the Faust-Mephistopheles theme, showing in symbolic images man in constant conflict with his *alter ego*. This was followed by *Bocksgesang* (also 1921), which, as *The Goat Song*, was produced in New York by the THEATRE GUILD in 1926. In it man's rebellion against the established order is symbolized by a monster, half-goat half-man, who leads a peasants' revolt in the 18th century. The revolt is crushed and the monster perishes, but not before he has passed on his heritage of cruelty and bestiality to his child by a young nobleman's wife. In his later plays Werfel turned to historical themes presented in a more realistic manner, his greatest theatrical success being *Juarez und Maximilian* (1924) on the tragedy of the Habsburg Emperor of Mexico. Two further historical plays proved less successful: *Paulus unter den Juden* (1926), which centres on the conflict between inspired prophecy and established religion among early Christians, and *Das Reich Gottes in Böhmen* (1930), on the Hussite movement in the author's homeland, Bohemia. Roused by the Nazi persecution of the Jews, Werfel then wrote a verse play, *Der Weg der Verheissung*, illustrating the tragic history of Judaism through the ages. It was staged in New York in 1936 in a spectacular production by REINHARDT. His last play, written after he had left Germany for the U.S.A., was, unexpectedly, a comedy, *Jacobowsky und der Oberst* (set in 1940, after the fall of France), in which a Jewish refugee contrives, by his superior intelligence and ingenuity, to smuggle an anti-Semitic Polish officer through the German lines to safety. Translated by S. N. BEHRMAN, this was successfully staged in New York in 1944 as *Jacobowsky and the Colonel* and was seen in London, with Michael REDGRAVE as the Colonel, in 1945.

WERGELAND, HENRIK ARNOLD (1808–45), outstanding Norwegian poet, author of a vast epic poem, *Skabelsen, Mennesket, og Messias* (*Creation, Man, and Messiah*, 1830), and an occasional playwright. His earliest dramatic works were farces, written under the name of Siful Sifadda. Among them were *Ah!* (1827), *Irreparabile tempus* (1828), *Phantasmer* (1829), and *Harlequin Virtuos* (1830). Other dramatic works on a larger scale, written to some extent under the influence of Shakespeare, included the tragedy *Sinclars Død* (*The Death of Sinclair*) and the comedy *Opium* (both 1828). In Wergeland's later plays there is less of this influence and more of his own fertile, fantastic, and sometimes apocalyptic imagination. *Den indiske Cholera* (*Indian Cholera*) and *Barnemordersken* (*The Child Murderess*, both 1835), the latter set in 13th-century France, are perhaps his most substantial achievements in a genre which was not altogether suited to his exuberant artistic temperament. The later works—*Campbellerne* (1837), *Hytten* (*The Cottage*, 1838), and the comedy *Søkadetterne iland* (*Sea Cadets Ashore*, 1839)—are of less importance.

WERICH, JAN, see CZECHOSLOVAKIA.

WERNER, (Friedrich Ludwig) ZACHARIAS (1768–1823), German dramatist, whose one-act play *Der vierundzwanzigste Februar* (1810), established the so-called SCHICKSALSTRAGÖDIE, or Fate Drama, in Germany. Werner was a poet and mystic who became a Roman Catholic priest in Vienna, where he was renowned for his preaching. His other plays, which include *Das Kreuz an der Ostsee* (1806), *Martin Luther* (1807), and *Kunigunde, die Heilige* (1815), bear a strong religious imprint, and mingle scenes of telling realism with weird fantasy. They were among the few German Romantic verse-dramas to achieve success on the contemporary stage.

WESKER, ARNOLD (1932–   ), English dramatist, who left school at 16 and had had a variety of jobs, including carpenter's and plumber's mate and farm labourer, before his first play *Chicken Soup with Barley* (1958) was produced at the BELGRADE THEATRE in Coventry and subsequently transferred to the ROYAL COURT in London. It formed the first part of a trilogy about a family of Jewish Communists, the later sections being *Roots* (1959; N.Y., 1961), in which Joan PLOWRIGHT scored an outstanding success in London, and *I'm Talking About Jerusalem* (1960). *The Kitchen*, based on Wesker's experiences as a pastry-cook, was first seen at the Royal Court as a 'production without décor' in 1959, and in a revised version was produced at the Belgrade two years later; it was also produced in New York in 1966. In 1961 Wesker accepted the directorship of Centre 42 (see ROUND HOUSE), which he retained until he disbanded it in 1970. His highly successful next play, an anti-Establishment study of R.A.F. conscripts, *Chips With Everything* (1962; N.Y., 1963), was followed by *The Four Seasons* (1965; N.Y., 1968), about the year two lovers spend together, and *Their Very Own and Golden City* (1966), previously entitled *Congress*. In 1970 Wesker himself directed the première of his play *The Friends* in Stockholm, and later in the same year it had a short run at the Round House. Later plays included *The Old Ones* (1972), *The Wedding Feast* (1974), and *The Merchant* (1976), a reworking of *The Merchant of Venice* in which Shylock receives more sympathetic handling. The last two were first seen in Stockholm, *The Merchant* receiving its English-speaking première in New York in 1977, directed by John DEXTER. A year later *Love Letters on Blue Paper* was produced at the NATIONAL THEATRE in London, and in 1980 he revised *One More Ride on the Merry-Go-Round*, written under a pseudonym in 1978. In 1981 *Caritas*, the story of a young anchoress who wishes to return to the world but is not released from her vows, received its world première at the National Theatre. He became chairman of the British Centre of the INTERNATIONAL THEATRE INSTITUTE in 1978.

WESSEL, JOHAN HERMAN (1742–85), Dano-Norwegian poet and dramatist, who was born in Norway but settled in Copenhagen. His reputation rests almost entirely on the mock-tragedy *Kierlighed uden Strømper* (*Love Without Stockings*, 1772), which with brilliant parody destroyed the French neo-classical tradition then in vogue in Denmark. Wessel was a comedy writer in the manner of HOLBERG; of his small remaining output *Lykken bedre end Forstanden* (*Luck Better Than Sense*, 1776) is perhaps the best known.

WEST, FLORENCE, see WALLER, LEWIS.

WEST, MAE (1892–1980), American actress and entertainer, and a significant figure in the history of stage CENSORSHIP. She was on the stage as a child, among her many juvenile roles being Little Willie in Mrs Henry Wood's *East Lynne*. She retired at the age of 11, returning as an adult actress in MUSICAL COMEDY, her first major role on Broadway being in *Winsome Widow* (1912). She was then seen in REVUE, and soon became renowned for her mastery of sexual innuendo expressed through the character of a curvaceous, Edwardian-style vamp. She first tangled with the censorship over *Sex* (1926), at DALY'S THEATRE, and was charged with corruption of morals, fined, and sent to Welfare Island for ten days. Two years later she played the title role in her own production of *Diamond Lil*, probably her best known vehicle. After some years in films she returned to Broadway in her own play *Catherine Was Great* (1944). In 1948 she was seen in London in *Diamond Lil*, which she then took on tour until 1951. She published her autobiography *Goodness Had Nothing To Do With It* in 1959, and was the author of several novels. A glamorous and extrovert personality, her fame can be judged by the fact that the R.A.F. in 1940 named their inflatable life-jacket a 'Mae West'.

WEST, TIMOTHY, see PROSPECT THEATRE COMPANY.

WESTCOTT, SEBASTIAN, see BOY COMPANIES.

WEST END PLAYERS, Auckland, see NEW ZEALAND.

WESTERN, (Pauline) LUCILLE (1843–77) and HELEN (1844–68), American actresses, daughters of a comedian and his actress wife who, by a second marriage after his early death, became Mrs Jane English, the name by which she is usually known. The children toured with their mother and step-father in a mixed entertainment which gave plenty of scope for their precocious talents in acting and dancing. Helen died before she could become well known as an adult actress but Lucille was noted for her playing of strong emotional parts in such plays as Mrs Henry Wood's *East Lynne*, the younger DUMAS's *Camille* (*The Lady of the Camellias*), TAYLOR and READE's *Masks and Faces*, and KOTZEBUE's *The Stranger*. One of her finest performances was given as Nancy in a dramatization of DICKENS's *Oliver Twist*. She died at the height of her success.

WEST LONDON THEATRE, in Church Street, off the Edgware Road. It opened in 1832 as an unlicensed theatre for crude melodrama and comic songs and was known as the Royal Pavilion West until 1835, when it was renamed the Portman. Two years later it closed, reopening on 13 Nov. 1837 as the Marylebone Theatre, with much the same mixture as before. In 1842 it was largely rebuilt, and under the management of John Douglass until 1847 it achieved a measure of prosperity with popular melodrama and pantomime, but after his retirement managers came and went without success. In 1868 the building was renovated and enlarged, and renamed the Royal Alfred, an effort being made to provide better entertainment; but in 1873 it reverted to its old name of the Marylebone and to melodrama, which continued to fill the bill when towards the end of the 19th century it was renamed the West London. It became a cinema in 1932, was damaged by enemy action in 1941, and was then used as a warehouse. While being demolished in 1962 it was totally destroyed by fire.

WEST LONDON THEATRE, Tottenham Court Road, see SCALA THEATRE.

WESTMINSTER PLAY, a production of a Latin play given annually at Westminster School,

London, the only survival of many similar productions given at such schools as Eton, Winchester, Shrewsbury, and St Paul's in the 16th and early 17th centuries, and even earlier at OXFORD and CAMBRIDGE. They probably originated in the Christmas ceremonies connected with the election of the medieval BOY BISHOP during the FEAST OF FOOLS. The acting of Latin plays at Westminster was introduced during the headmastership of Dr Alexander Nowell, which lasted from 1543 to 1555, and when the school was refounded by Elizabeth I in 1560 a clause was inserted in the Statutes enjoining the annual presentation of a Latin play by the 'Scholars in College', usually referred to in the records as the 'Children of the Grammar School'. Plays in English were occasionally given by the 'Town Boys' or 'Oppidans'; the choristers of the Abbey, who formed a separate school with their own Master and were known as the 'Children of Westminster', also played in English, often appearing at Court and forming one of the popular BOY COMPANIES, but it is the 'Scholars' play' only which has maintained a more or less unbroken tradition until the present day.

It was perhaps natural that the Latin play should almost invariably be chosen from among the works of PLAUTUS and TERENCE, but occasionally performances were given of plays by other authors. The most interesting of these was the *Sapienta Salamonis*, given in Jan. 1566 in the ancient and still existing College Hall, formerly the Abbot's Dining Hall. It was graced by the presence of the Princess Cecilia of Sweden and of Elizabeth I, who two years earlier had attended a performance of the *Miles Gloriosus*, when the exertions of the young actors were such that xijd. had to be expended 'for buttered beere for ye children being horse'.

The Westminster Play was given every Christmastide until the outbreak of the Civil War in 1642, and then, curiously enough, there is no further mention of it until 1704, although it is evident that by that time it was already regarded as a well-established annual event. The missing period covers most of the long headmastership, from 1638 to 1695, of the famous Dr Richard Busby (1606–95), who was, however, no enemy to acting. As an undergraduate at Oxford he appeared before Charles I and Queen Henrietta Maria at Christ Church in a production of William Cartwright's *The Royal Slave*, with such success that he momentarily thought of going on the stage. The Play may indeed have been revived at the Restoration, for James Russell, writing as a Westminster boy in about 1660–6, said that 'the boys are to act a play very shortly'; and it is also recorded that Barton BOOTH, who was at Westminster from 1689 to 1698, was first encouraged to act by Busby 'at the rehearsals of a Latin play acted at that School'. In 1704 the play was Plautus' *Amphitruo*, with a specially composed Prologue and Epilogue, and in that form, and from that date, the records are continuous. The traditional cycle of plays always included the *Adelphi, Andria*, and *Phormio*, but the fourth play varied. The *Eunuchus* was acted until 1860, when the *Trinummus* of Plautus took its place. In 1907, the *Eunuchus* was revived as the *Famulus*, but in 1926 it gave place to a revival of the *Rudens*, which also established itself in the cycle. After the Restoration the Play was transferred from the College Hall to the Old Dormitory of the King's Scholars in Dean's Yard. In 1730 it was again transferred, to the New Dormitory in Little Dean's Yard, which in May 1941 was gutted by an incendiary bomb. When the Play was revived in 1954 with a production of the *Phormio*, it was staged, as it still is, in the open air in Little Dean's Yard, around which the school is situated. The situation is ideal for a Latin play, as the audience faces Ashburnham House, whose two doors can be taken as the entrances to two houses. The actors in the Play are no longer the Queen's Scholars but the Classical Scholarship Form, who are also responsible for writing the Prologue, their Master providing the Epilogue, composed in elegiacs. Both still remain full of puns and topical allusions, as of old. In accordance with ancient custom the actors wear modern dress and stage equipment is kept to a minimum. A synopsis of the Play is provided for the benefit of the audience. Now that the production of the Westminster Play has become a biennial event, it has been felt that the old cycle of plays could be broken, and in recent years there have been performances of the *Mostellario*, the *Bacchites*, and in 1970 the *Captivi*. It has become much less a social event than in the past, when it was attended not only by Royalty, but by ambassadors, statesmen, judges, and bishops, and more of a genuine performance, aimed at the enjoyment and understanding of the plays for their own sake.

WESTMINSTER THEATRE, London. (1) Built by an undertaker named Gale, this opened under the management of T. D. Davenport, father of Jean LANDER, on 17 May 1832. During the four years of its existence the theatre was never able to obtain a licence and it eventually closed. The site was later occupied by part of the IMPERIAL THEATRE and subsequently by the Central Hall, Westminster.

(2) In Palace Street, near Victoria Station, opened on 7 Oct. 1931. It had originally been a chapel which in 1924 had been converted into a cinema, a new frontage being added. The two-tier auditorium seats 600. The crypt of the old building was converted into dressing-rooms and a bar, and the theatre opened with Henry AINLEY in BRIDIE's *The Anatomist*. Under the management of Anmer Hall a wide variety of English and foreign plays was staged, among them revivals of PIRANDELLO's *Six Characters in Search of an Author* in 1932, and

of GRANVILLE-BARKER's *Waste* in 1936. Several of Denis JOHNSTON's plays were first seen here, including *The Moon in the Yellow River* (1934), *The Old Lady Says No!* (1935), and *A Bride for the Unicorn* (1936), and O'NEILL's *Ah, Wilderness!* (also 1936) was followed by his *Mourning Becomes Electra* (1937) and *Marco Millions* (1938). *The Zeal of Thy House* by Dorothy L. Sayers was brought to the Westminster after its production in the Chapter House of Canterbury Cathedral, and T. S. ELIOT's *The Family Reunion* (1939) was seen under the auspices of the London Mask Theatre directed by J. B. PRIESTLEY and Michael MACOWAN. The Westminster was the first theatre to reopen after the compulsory closure in Sept. 1939, the production being Priestley's *Music at Night*. Paul SCOFIELD made his début there in 1940 in O'Neill's *Desire Under the Elms*, and other successful productions included AFINOGENOV's *Distant Point* (1941) and Bridie's *Mr Bolfry* (1943) and *It Depends What You Mean* (1944). *Message for Margaret* (1946) by James Parrish proved a personal triumph for Flora ROBSON, who returned in Lesley Storm's *Black Chiffon* (1949). The theatre had been bought in 1946 by the evangelical 'Oxford Group', which has since used it mainly for productions promoting its 'Moral Rearmament' movement such as Malcolm Muggeridge's *Sentenced to Life* (1978). Occasional returns to the commercial theatre have included revivals of COWARD's *Relative Values* (1973) and Gershwin's *Oh, Kay!* (1974).

WESTON, THOMAS (1737–76), English actor, whose father was head cook to George II. He joined a strolling company and, thinking himself a tragedian, played Richard III abominably. He found his true vocation in comedy, and after playing in booths at Bartholomew Fair he was engaged for small parts at the HAYMARKET THEATRE by Samuel FOOTE, who thought highly of his comic talent and wrote for him the part of Jerry Sneak in *The Mayor of Garret* (1763). Weston was subsequently at DRURY LANE, where he was considered even better than GARRICK in the part of Abel Drugger in JONSON's *The Alchemist*. The German critic Lichtenberg has left a fine description of his playing of Scrub in FARQUHAR's *The Beaux' Stratagem*. He was somewhat dissipated, constantly in debt, and finally died of drink.

WESTON'S MUSIC-HALL, London, see HOLBORN EMPIRE.

WESTRAY, Mrs ANTHONY (?–1836), British-born actress who in 1792 married as her second husband an actor named John Simpson and four years later went with him to America, where he died. Of the three daughters of her first marriage, who were all with her at the PARK THEATRE, New York, the eldest, Ellen, became Mrs John Darley, and after 20 years at the Park joined her sister Juliana, wife of William WOOD, at the CHESTNUT STREET THEATRE in Philadelphia, where she was extremely successful. The third sister, Elizabeth, married as her second husband the comedian William TWAITS, having previously been known on the stage as Mrs Villiers.

WETTER, JOSUA, see SWITZERLAND.

WHARF THEATRE, New York, see CRUGER'S WHARF THEATRE.

WHARF THEATRE, Provincetown, R.I., see PROVINCETOWN PLAYERS.

WHEATLEY, WILLIAM (1816–76), American actor, the son of Irish players who emigrated to the United States in 1803 and joined the company at the PARK THEATRE, where William, who made his first appearance at the age of ten, was for some years the chief player of boys' parts. He returned as an adult actor in 1834 and soon proved himself a hard worker, an excellent actor, and a good manager. From 1862 to 1868 he was at NIBLO'S GARDEN, where his greatest achievement was the production of the spectacular ballet-extravaganza *The Black Crook* (1866). Its success netted him a fortune, on which he retired.

WHELAN, ALBERT, see MUSIC-HALL.

WHITE, GEORGE, see REVUE.

WHITE, PATRICK VICTOR MARTINDALE (1913–), Australian novelist and playwright, Nobel Laureate 1973. Born in London of Australian parents, he was brought up partly in England and partly in Australia, where he settled after the Second World War, becoming internationally known after the publication of his novel *Voss* in 1957. While living in London in the 1930s he contributed a number of sketches and lyrics to the revues at the LITTLE and GATE theatres, and wrote several naturalistic plays which failed to attract attention. The first to be staged professionally was a comedy, *Return to Abyssinia* (1946), seen at the Boltons Theatre in London, but as a dramatist he is best known for four plays which caused much controversy when first produced—*The Ham Funeral* and *The Season at Sarsaparilla* (both 1961), first seen in Adelaide; *A Cheery Soul* (1962), from his own story of the same name, produced in Melbourne by John SUMNER, who had also the previous year directed a production in Melbourne of *The Season at Sarsaparilla* with Zoë CALDWELL as Nora Boyle; and *Night on Bald Mountain* (1964), which again had its première in Adelaide. A revival in 1976 of *The Season at Sarsaparilla* by the OLD TOTE THEATRE, Sydney, directed by Jim Sharman, was followed by a new play *Big Toys* (1977), also directed by Sharman for the Old Tote.

'WHITE-EYED KAFFIR' [Chirgwin, G. H.], see MUSIC-HALL.

WHITEFRIARS THEATRE, London, a 'private' roofed theatre in the refectory hall of the old Whitefriars monastery, which was situated in the area around the present Bouverie Street. The hall was adapted in about 1605 by Thomas Woodford and Michael Drayton, on the lines of the second BLACKFRIARS THEATRE, its dimensions being about 35 ft by 85 ft. It was used by the Children of the King's Revels from 1608 to 1609, by the Queen's Revels from 1609 to 1613, when the plays performed included FIELD's *A Woman is a Weathercock*, MARSTON's *The Insatiate Countess*, and Ben JONSON's *Epicœne, or the Silent Woman*, and from 1613 to 1614 by the LADY ELIZABETH'S MEN, who appeared there in Jonson's *Bartholomew Fair*. Its later history is obscure, but it was still in use in 1621, being replaced by the SALISBURY COURT THEATRE in 1629. The two were often confused, PEPYS in his diary for 1660 noting that he saw MASSINGER's *The Bondman* at Whitefriars, in mistake for Salisbury Court.

WHITEHALL THEATRE, London, in Whitehall, near Trafalgar Square. An intimate modern playhouse holding 628 persons on two tiers, it opened on 29 Sept. 1930 under the management of Walter HACKETT, its first production being his *The Way to Treat a Woman*, transferred from the DUKE OF YORK's. He and his wife Marian Lorne controlled the theatre until 1934, he writing the plays and she appearing in them. Among the most successful were *The Gay Adventure* (1931), *Road House* (1932), and *Afterwards* (1933). Subsequent successes at this theatre included Norman Ginsbury's *Viceroy Sarah*, St John ERVINE's *Anthony and Anna* (both 1935), Alec Coppel's *I Killed the Count* (1937), and Philip King's *Without the Prince* (1940). After the outbreak of war in 1939 the theatre housed non-stop revue with Phyllis Dixey, who remained there except for one short break until Dec. 1945. In Dec. of that year *Worm's Eye View* by R. F. Delderfield began a long run, returning after a short tour in May 1947 for a further 1,745 performances. In 1950 the first of a series of plays produced by Brian RIX that rivalled the popularity of the pre-war ALDWYCH farces, *Reluctant Heroes* by Colin Morris, opened on 12 Sept. and was followed by John Chapman's *Dry Rot* (1954) and *Simple Spymen* (1958), Ray Cooney and Tony Hilton's *One for the Pot* (1961), *Chase Me, Comrade!* (1964) by Ray Cooney, and *Come Spy With Me* (1966) by Bryan Blackburn. When Rix moved to the GARRICK THEATRE in 1967, the theatre had another success with *Uproar in the House* by Anthony Marriott and Alistair Foot, and in 1969 *Pyjama Tops*, a comedy by Paul Raymond based on a French farce *Moumou*, began a run of 5½ years and prompted a sequel in the same style, *Come Into My Bed* (1976) by André Launay. The Black African musical *Ipi Tombi* moved here from the CAMBRIDGE THEATRE in 1978 and ran until 1980.

WHITELAW, BILLIE (1932–   ), English actress, one of the best of her generation, who was sent by her mother for theatrical training at her local theatre to cure a stutter. She made her first appearance in London in FEYDEAU's *Hotel Paradiso* (1956) and was in the THEATRE WORKSHOP production of Alun Owen's *Progress to the Park* (1960), going with it to the West End in 1961. In 1962 she starred in Willis HALL and Keith Waterhouse's revue *England, Our England* and in 1964 joined the NATIONAL THEATRE company, with which she appeared in Samuel BECKETT's one-act *Play*, being seen also in the company's production of MARSTON's *The Dutch Courtesan* at the CHICHESTER FESTIVAL THEATRE where she also played Avonia Bunn in PINERO's *Trelawny of the 'Wells'* in 1965, having earlier given an excellent account of Maggie in BRIGHOUSE's *Hobson's Choice* at the OLD VIC. She joined the ROYAL SHAKESPEARE COMPANY to appear in David MERCER's *After Haggerty* (1971) and played Mouth in Beckett's *Not I* at the ROYAL COURT in 1973 and 1975. In 1975 she also gave an excellent performance as Lucy, the inefficient librarian in a newspaper office, in Michael FRAYN's *Alphabetical Order*. After Simon GRAY's short-lived *Molly* (1977) she returned to Beckett at the Royal Court in a revival of *Happy Days* in 1979. A year later, with the R.S.C. again, she played Andromache and Athena in *The Greeks*, a trilogy based mainly on EURIPIDES, and in 1981 she returned to star in *Passion Play*, Peter NICHOLS's study of adultery.

WHITING, JOHN (1917–63), English dramatist, who first came into prominence when his play *A Penny for a Song* (1951) was produced at the HAYMARKET THEATRE. A light-hearted comedy set in the time of the Napoleonic wars, it was well received by the critics but failed with the public. Also in 1951 Whiting won a playwriting competition to mark the Festival of Britain with *Saint's Day*, produced at the ARTS THEATRE; it was condemned by the critics for obscurity but publicly defended by theatre people of the stature of GIELGUD, GUTHRIE, and Peggy ASHCROFT who considered it intensely poetic and theatrically exciting. Whiting's other plays were *Marching Song* (1954), a darkly symbolic drama of conscience; *The Gates of Summer*, seen on tour in 1956 and in London at the TOWER THEATRE in 1970; and *The Devils* (1961), based on a historical work, *The Devils of Loudun* by Aldous Huxley, which was staged with great success by the ROYAL SHAKESPEARE COMPANY at the ALDWYCH THEATRE. In 1962 the R.S.C. presented a rewritten and much improved version of *A Penny for a Song*, which in 1967 was used as the libretto of an opera by Richard Rodney Bennett. Whiting also translated ANOUILH's *Madame de ...* and *Le Voyageur sans bagages*, as *The Traveller Without*

Luggage, both seen in 1959, and in 1965 the BRISTOL OLD VIC put on *Conditions of Agreement*, found among his papers after his death, which was also produced in New York in 1972. In 1965 an annual award of £1,000 was established, to be awarded in his memory to a promising young playwright, among the early recipients being Edward BOND, Tom STOPPARD, and Peter TERSON. Whiting is generally agreed to have been in advance of his time: he died just as his reputation in England was beginning to equal that he achieved on the Continent, especially in Germany.

WHITLEY, JAMES AUGUSTUS, see MANCHESTER and NOTTINGHAM.

WHITLOCK [*née* Elizabeth Kemble], Mrs CHARLES, see KEMBLE, JOHN PHILIP.

WHITNEY THEATRE, London, see STRAND THEATRE (2).

WHITTLE, CHARLES, see MUSIC-HALL.

WHITTY, MAY, see WEBSTER, MARGARET.

WIED, GUSTAV (1858–1914), Danish dramatist and satirist, whose so-called 'satyr plays' are marked by a deeply ironic and bitter wit. Among the best of them are the four one-acters published together in 1897 under the title of *Adel, Gejstlighed, Borger og Bonde* (*Nobility, Clergy, Burgher, and Peasant*). Also outstanding are the longer play *Danesmus* (*Dancing Mice*, 1905) and the comedy *Ranke Viljer* (1906). As *2 × 2 =5*, this last had a great success in Germany in 1908, and was seen in an English translation in New York in 1928 in a production at Eva LE GALLIENNE's Civic Repertory Theatre.

WIETH, MOGENS (1919–62), Danish actor, one of the outstanding Danish players of this century. He made his début in 1939 as a member of the KONGELIGE TEATER, Copenhagen, where he quickly established himself as a classical actor. Among the many important roles he played were Othello, IBSEN's Peer Gynt, the Narrator in Stravinsky's *The Soldier's Tale*, John Worthing in WILDE's *The Importance of Being Earnest*, and Higgins in the musical *My Fair Lady*. In 1948, during the celebrations held for the 200th anniversary of the Kongelige, he played ten leading parts in 18 days. In 1953 he played Helmer in English in a production of Ibsen's *A Doll's House* in London, and in 1962 was engaged to play Othello and Peer Gynt with the OLD VIC company. He had just started rehearsals when he died suddenly in London.

WIGNELL, THOMAS (1753–1803), American actor of English extraction, a cousin of the younger Lewis HALLAM, by whom he was persuaded in 1774 to join the AMERICAN COMPANY, of which he soon became leading man. In 1787 he was instrumental in arranging the production in New York of the first American comedy, Royall TYLER's *The Contrast*, and in 1789, in which year George Washington, who much admired him, attended his benefit night, he spoke the prologue to DUNLAP's *The Father; or, American Shandyism*, in which he played the comic doctor. Shortly afterwards he left for England, where he recruited a fine company to play at the newly opened CHESTNUT STREET THEATRE in Philadelphia. Among its members were James FENNELL, Mrs OLDMIXON, Mr and Mrs Owen MORRIS, with whom Wignell had acted in New York, and Mrs MERRY, whom he married seven weeks before his death. The company soon achieved an enviable reputation, and on a visit to New York in 1797 was considered superior even to the American Company. Wignell, however, took it back to Philadelphia, where he remained until his death.

WILDE, OSCAR FINGAL O'FLAHERTIE WILLS (1854–1900), Irish wit and writer, best known as a dramatist. Educated in Dublin, he then went to Oxford, where he won the Newdigate Prize for English Verse and, in spite of a reputation for idleness, took a First Class degree in Classics. By 1881 he was well enough known as a poet and aesthete to be satirized in GILBERT and Sullivan's operetta *Patience*, and a year later he went to New York, where his first plays, *Vera; or, the Nihilists* (1882) and *The Duchess of Padua* (1891) (as *Guido Ferrandi*), a blank-verse tragedy, were produced, without much success. It was with a succession of comedies, beginning with *Lady Windermere's Fan* (1892), produced at the ST JAMES'S THEATRE by George ALEXANDER, that he ultimately achieved fame, following up his initial success with *A Woman of No Importance* (1893) and *An Ideal Husband* (1895), both seen at the HAYMARKET THEATRE. His most characteristic play, *The Importance of Being Earnest* (also 1895), with Alexander as John Worthing, was seen at the St James's. In it Wilde discarded his former vein of sentiment, which he knew to be false and a concession to the taste of the times, and returned to the pure comedy of CONGREVE. Though all his comedies have been revived, *The Importance of Being Earnest* wears best, and has proved the most successful. Few comedies of the English stage have such wit, elegance, and theatrical dexterity. In the year of its first production Wilde, who was married with two young sons, was sentenced to two years' imprisonment with hard labour for homosexuality, and afterwards went to Paris, where, broken in health and fortune, he soon died. His last play, the poetic one-act *Salomé*, written in French, was banned by the censor in England, but was produced in Paris in 1896 by Sarah BERNHARDT, and later used as the libretto of an opera by Richard Strauss. It was first seen privately in London at the BIJOU THEATRE in 1905, and has since been revived occasionally. Wilde left an unfinished one-act play, *A Florentine*

Tragedy, which was completed by Sturge Moore and produced in London in 1906. His life was the subject of a play by Leslie and Sewell Stokes seen in London in 1936 with Robert MORLEY as Oscar Wilde, and in 1960 Michéal MACLIAMMÓIR toured in a one-man entertainment based on his work, *The Importance of Being Oscar*.

WILDER, THORNTON NIVEN (1897–1975), American novelist and dramatist. Among his earlier works for the theatre were *The Trumpet Shall Sound* (1927), a play about the American Civil War; four one-act plays (1931), one of which, *The Long Christmas Dinner*, provided the libretto for a short opera by Hindemith (1961); a translation of OBEY'S *Le Viol de Lucrèce*, as *Lucrece* (1932), for Katharine CORNELL; and a new version of IBSEN's *A Doll's House* (1937), for Ruth Gordon. His most important works for the theatre were, however, his two stimulating and provocative plays, *Our Town* (1938; London, 1946), a picture of a small American community, in which scenery was reduced to a minimum and the characters mimed instead of using props; and *The Skin of Our Teeth* (1942; London, 1945), a survey of man's hairsbreadth escape from disaster through the ages, both of which were awarded a PULITZER PRIZE. The latter play provided in Sabina an excellent part for Tallulah BANKHEAD in New York and Vivien LEIGH in London, where Laurence OLIVIER was responsible for its production. Perhaps Wilder's most popular work was his adaptation of one of NESTROY'S farces (*Einen Jux will er sich machen*, based on *A Day Well Spent* by John Oxenford). Originally produced as *The Merchant of Yonkers* (1938), it was rewritten and retitled *The Matchmaker*, under which title it was seen at the Edinburgh Festival in 1954, later that year in London, and in New York in 1955. As a musical entitled *Hello, Dolly!* it had a spectacular success (1964; London, 1965). *A Life in the Sun* (1955), on the legend of Alcestis, was commissioned by Tyrone GUTHRIE for the EDINBURGH FESTIVAL, and a trilogy of one-act plays, *Three Plays for Bleecker Street* (1962), was produced at the CIRCLE-IN-THE-SQUARE in New York.

WILDGANS, ANTON (1881–1932), Austrian poet and dramatist, who combined NATURALISM and EXPRESSIONISM in popular works such as the brief *In Ewigkeit Amen!* (1913), *Armut* (1914), his most successful play, and *Liebe* (1916). Of a projected Biblical trilogy only the first part, *Kain* (1920), was completed. Wildgans was director of the Burgtheater in VIENNA from 1921 to 1922 and again from 1930 to 1931, when he made heroic efforts to preserve its identity, which was fading under unimaginative ministerial control. He stopped many malpractices, but fell victim to intrigue and to his own impractical idealism and was dismissed.

WILKIE, ALLAN, see AUSTRALIA.

WILKINSON, NORMAN (1882–1934), English artist, designer of some outstanding settings and costumes for the theatre. He was usually referred to as 'Norman Wilkinson of Four Oaks', to distinguish him from a contemporary marine artist of the same name. His first worked for Charles FROHMAN at the DUKE OF YORK'S THEATRE in 1910, but came into prominence with his designs for GRANVILLE-BARKER's seasons at the SAVOY, his permanent set for *Twelfth Night* in 1912 being much admired. Particularly memorable were his designs for *A Midsummer Night's Dream* in 1914, with its gilded fairies and iridescent forest. He was later with Nigel PLAYFAIR at the LYRIC THEATRE, Hammersmith, worked for C. B. COCHRAN, and for the PHOENIX and STAGE SOCIETIES, and in 1932 designed *A Midsummer Night's Dream* for the SHAKESPEARE MEMORIAL THEATRE, of which he was a governor.

WILKINSON, TATE (1739–1803), English actor and provincial theatre manager, a man of good family whose passion for the theatre led him to adopt the stage as a profession. He was recommended to GARRICK who engaged him for DRURY LANE, where his imitations of contemporary players were much admired, particularly by Samuel FOOTE who took him to Dublin where he was a great success. He was less well received, however, in London on his return, and eventually went into the provinces. There he took over the YORK circuit, which included Hull and Leeds, and ran it for about 30 years with marked success, many actors and actresses, among them Mrs JORDAN, John Philip KEMBLE, and the elder Charles MATHEWS, gaining their first experience under him. As he grew older he became very eccentric and Foote, after a violent but short-lived quarrel, satirized him as Shift in *The Minor* (1760). In spite of his many foibles he was much loved and respected and did a great deal to keep the theatre alive in provincial towns in the North of England. He left an interesting account of his life in his *Memoirs* (1790) and *The Wandering Patentee* (1795).

WILKS, ROBERT (1665–1732), English actor, who was engaged by Christopher RICH for DRURY LANE in 1692 to take the place of William MOUNTFORT, who had just been assassinated. He soon became popular with the public and his fellow-actors, being conscientious and hard-working, and retaining to the end of his life the ability to play high comedy. He was the first to appear as the fine gentlemen of Colley CIBBER's comedies, and one of his best parts was Sir Harry Wildair in FARQUHAR's *The Constant Couple* (1699). He was less good in tragedy, but could always move an audience to tears in parts which, like Macduff in *Macbeth*, demanded pathos. He was also much admired as Hamlet. When Rich left Drury Lane in 1709, Wilks took over the management with Cibber and DOGGETT but, although the theatre prospered, he found himself constantly at odds with the latter, and harmony was not restored until

Doggett was replaced by Barton BOOTH. Even then Wilks's imperious temper often caused trouble and led several actors to migrate to LINCOLN'S INN FIELDS.

WILLARD, EDWARD SMITH (1853–1915), English actor, who was particularly admired for his villains in contemporary MELODRAMA. He made his first appearance on the stage in Weymouth in 1869, and after some years in the provinces was in London, where he made a great success as Clifford Armytage in Sims's *The Lights o' London* (1881), in which year he proved that he could portray tenderness and emotion with equal success by his handling of the title role in *Tom Pinch*, a domestic comedy based by J. J. Dilley and Lewis Clifton on DICKENS's *Martin Chuzzlewit*. He enhanced his reputation with his Captain Skinner in H. A. JONES and Herman's *The Silver King* (1882), James Ralston in Charles Young's *Jim the Penman* (1886), and Jem Dalton in a revival of Tom TAYLOR's *The Ticket-of-Leave Man* in 1888. In 1889 he went into management at the SHAFTESBURY THEATRE, where he played among other parts Cyrus Blenkarn in H. A. Jones's *The Middleman*, and then went to America, where he appeared with such success that he returned there annually for several years, touring the United States and Canada with a repertory of his most successful parts, to which he added the title role in Jones's *Judah* (1890). Among his later roles were the title roles in BARRIE's *The Professor's Love Story* (1894) (which, with *The Middleman*, he revived many times), in Louis N. PARKER's *The Cardinal* (1903), and in Michael Morton's dramatization of Thackeray's *The Newcomes* as *Colonel Newcome* (1906), in which he made his last appearance on the stage.

WILLIAMS, BARNEY (1823–76), American actor, who with his wife Maria Pray, the sister-in-law of W. J. FLORENCE, toured the United States, mainly in Irish comedies. They were both seen in London in 1855 in Samuel LOVER's *Rory O'More*, and remained for four years, returning to New York to continue their Irish impersonations, which became a tradition in the theatre. Williams was for two years manager of WALLACK's old theatre, which he rechristened the Broadway, but soon found that touring was both pleasanter and more profitable, and once more took to the road. He made his first appearance in 1836 and his last in 1875, and to the end remained not so much an actor as an entertainer, a much-loved, jovial, rollicking, drinking, stage Irishman on the boards and off.

WILLIAMS, BERT [Egbert Austin] (*c.* 1876–1922), American Negro entertainer and song writer, born in the Bahamas, who joined a MINSTREL SHOW and in 1895 teamed up with George Walker, also a Negro comedian. Together they perfected a double act on tour, and later appeared in it with outstanding success at KOSTER AND BIAL'S MUSIC HALL in New York, where in 1903 Williams produced a successful all-Negro MUSICAL COMEDY *In Dahomey* which was also seen in London. It was followed by several other shows but after Walker's death in 1909 Williams gave up management and appeared in musical REVUES, chiefly in the *Ziegfeld Follies*, usually writing his own material. A gifted actor, with a rich bass singing voice, he was a serious scholarly man off stage, but for his performances blacked himself and portrayed the shuffling, shiftless Negro of tradition. His pioneering work for his race included the formation of an all-Negro actors' society in 1906.

WILLIAMS, BRANSBY [Bransby William Pharez] (1870–1961), star of the British MUSIC-HALL, who had made a number of appearances as an amateur actor before, encouraged by William TERRISS, he joined a provincial STOCK COMPANY, playing a wide variety of parts which laid the foundations of his excellent technique. In 1896 he deserted the legitimate stage for the 'halls', appearing at the Shoreditch Music-Hall in imitations of Henry IRVING, Beerbohm TREE, and Charles WYNDHAM, with such success that he was immediately engaged for the TIVOLI, and was soon appearing in all the major halls. He had a large repertory of comic characters from DICKENS's novels, and also specialized in dramatic monologues with a musical accompaniment, of which the best-known were Milton Hayes's 'The Green Eye of the Little Yellow God' and 'The Whitest Man I Know'. In 1904 he was commanded to Sandringham to play before Edward VII, and he made many successful appearances in the United States.

WILLIAMS, (George) EMLYN (1905–   ), Welsh actor, dramatist, and theatre director, who first appeared on the stage in FAGAN's *And So To Bed* (1927) in London, and made his New York début later that year in the same play, but in a different role. He had already written several plays before he first attracted attention with his adaptation for London of Sidney HOWARD's translation of *Prenez garde à la peinture* by René Fauchois, which as *The Late Christopher Bean* (1933), with Edith EVANS as Gwenny, ran for over a year. A later production, *Spring 1600* (1934), depicting life backstage in Shakespeare's theatre, failed, in spite of a starry cast directed by John GIELGUD; but Williams achieved an enormous success, as both author and actor, with *Night Must Fall* (1935), in which he played the young homicidal Danny. He repeated his success in New York the following year, the only occasion on which he appeared there in one of his own plays. He also gave an excellent performance in 1937 as Angelo in *Measure for Measure* at the OLD VIC, but for the next few years appeared only in his own work. Another failure, *He Was Born Gay* (1937), was followed by the outstanding success of *The Corn is Green* (1938; N.Y., 1940), probably his best and most popular play. In it he played the young Welsh miner Morgan Evans, with Sybil THORNDIKE as his teacher Miss Moffat. A succession of plays followed: *The Light of Heart* (1940; N.Y., as *Yesterday's Magic*, 1942); *The Morning Star* (1941; N.Y., 1942); *The Druid's Rest* (1944), in which he did not appear; and *The Wind of Heaven* (1945), which postulates Christ's reincarnation as a Welsh boy in a small mining village. Williams then gave a fine performance as Sir Robert Morton in RATTIGAN's *The Winslow Boy* (1946), and after being seen in two more of his own plays, *Trespass* (1947) and *Accolade* (1950), began in 1951 a solo performance as Charles DICKENS. He became famous for his impersonation of the author reading from his own works, touring the world with it and reviving it several times. He also appeared in another of his own plays, *Someone Waiting* (1953; N.Y., 1956), and then began in 1955 another solo performance, *A Boy Growing Up*, in which he read from the works of his fellow Welshman Dylan Thomas. It was also given in New York, and later revived in London and 59 American cities. Williams then appeared in a series of good parts in plays by other authors: Hjalmar Ekdal in IBSEN's *The Wild Duck* in 1955; Shylock in *The Merchant of Venice* and Iago in *Othello* at Stratford-upon-Avon in 1956; and the Author in Robert ARDREY's *Shadow of Heroes* in 1958, in which year he also directed, but did not appear in, his own play *Beth*. He was also seen in a triple bill *Three* (1961), which included John MORTIMER's *Lunch Hour*, N. F. Simpson's *The Form*, and Harold PINTER's *A Slight Ache*. In 1962 he successfully took over the part of Sir Thomas More in BOLT's *A Man For All Seasons* in New York, where he also played Pope Pius XII in HOCHHUTH's *The Deputy* in 1964, in which year his adaptation of Ibsen's *The Master Builder* was produced by the NATIONAL THEATRE company in London. He played in his own adaptation of TURGENEV's *A Month in the Country* in 1965, took over from Gielgud in Alan BENNETT's *Forty Years On* in 1969, and in 1977 gave another solo performance, reading the short stories of Saki (H. H. Munro). He has written two volumes of autobiography, *George* (1961) and *Emlyn* (1973).

WILLIAMS, (Ernest George) HARCOURT (1880–1957), English actor and director, who made his first appearance in Belfast in 1898 with Frank BENSON, appearing also with him in London in 1900. He was with most of the leading players of the day, touring with Ellen TERRY and George ALEXANDER and visiting America with Henry IRVING. In 1907 he was again in London, where he was seen as Valentine in a revival of SHAW's *You Never Can Tell* at the SAVOY THEATRE, and he played Count O'Dowda in the first production of the same author's *Fanny's First Play* (1911) at the LITTLE THEATRE. After service in the First World War he returned to the stage in 1919, as The Chronicler and General Lee in DRINKWATER's *Abraham Lincoln*. Among his later parts were the Stranger in CHESTERTON's *Magic* (1923), the Player King to John BARRYMORE's Hamlet (in 1925), and the Chancellor in Ashley DUKES's adaptation of Neumann's *Der Patriot* as *Such Men Are Dangerous* (1928). In 1929 he was appointed director at the OLD VIC THEATRE, where his innovations in Shakespearian production were first criticized and later hailed as epoch-making, based as they were upon the theories of GRANVILLE-BARKER and Williams's own love of and feeling for the swiftness and splendour of Elizabethan verse. He also introduced Shaw into the repertory with revivals of *Androcles and the Lion* and *The Dark Lady of the Sonnets* in 1930. In 1937 he gave a fine performance as William of Sens in Dorothy L. Sayers's *The Zeal of Thy House*, which he also directed, and some years later returned to the Old Vic company, with which he visited New York in 1946. He celebrated his stage Jubilee in 1948 while playing again in Shaw's *You Never Can Tell*, this time as William the Waiter. His wife, the actress Jean Sterling Mackinlay (1882–1958), whom he married in 1908, was prominently associated with the movement for CHILDREN'S THEATRE in England, and for 27 years staged a unique series of children's matinées in London over the Christmas season.

WILLIAMS, KENNETH, see REVUE.

WILLIAMS, PETER VLADIMIROVICH (1902–47), Soviet theatre designer, creator of some most original settings for *The Pickwick Club* (1934), a dramatization of DICKENS's novel produced at the MOSCOW ART THEATRE. He was equally successful in 1946 with his designs for a revival of Alexei TOLSTOY's *The Death of Ivan the Terrible*, for which, rejecting the usual scenes of grandeur and sumptuousness, he designed a simple wooden structure whose severity well conveyed the harsh spirit of the time. Shortly before his early death he designed the setting for Virta's *Great Days* (1947), his last work for the theatre.

WILLIAMS, TENNESSEE [Thomas Lanier] (1911–83), American dramatist, who after a hard youth began to write plays while still a student. His main concern was the plight of the romantic soul in an unromantic world, and the plays—whose outspokenness has caused much controversy—are characterized by compassion for those who find themselves unable to function in the clear light of American reality. His heroes and heroines are restless creatures, afraid to linger lest they should be corrupted and destroyed, yet unwilling to leave those whom they love and whose brief offers of tender affection seem to constitute the only flicker of light in a world darkened by the soulless pursuit of money and power.

In 1939 Williams was awarded a prize by the THEATRE GUILD for a group of four one-act plays *American Blues*, but after this auspicious start his first full-length play *Battle of Angels* (1940) was a failure and did not reach Broadway. It was later revised as *Orpheus Descending* (1957; London, 1959) with some success. He finally achieved recognition with *The Glass Menagerie* (1945; London, 1948), in which Laurette TAYLOR made her last appearance as the Mother. It was a sensitive study of his own mentally afflicted sister, though in the play her handicap was transmuted into a club-foot which became the image of the arbitrary crippling of individual desires. It was followed by *A Streetcar Named Desire* (1947; London, 1949), a powerful study of the clash between the old and new America—symbolized by the pathetic, self-deceiving Blanche du Bois and her brutish brother-in-law—which was awarded a PULITZER PRIZE. The power revealed by this play, with its characteristic blend of decadence and pathos, was echoed in later works, though Williams is also capable of a disabling sentimentality which tends to undermine his poetic vision. His next plays were *Summer and Smoke* (1948; London, 1951); *The Rose Tattoo* (1951; London, 1959), banned in Dublin in 1957; *Camino Real* (1953; London, 1957), an unsuccessful essay in symbolism based on *Don Quixote*; and *Cat on a Hot Tin Roof* (1955; London, 1958), an intense drama of family relationships set on a plantation in the Mississippi Delta, which also received a Pulitzer Prize. A one-act play *Twenty-Seven Wagons Full of Cotton* was first seen in New York in 1955 as part of a triple bill *All In One*; *Suddenly Last Summer*, produced with *Something Unspoken* in a double bill entitled *Garden District* (N.Y. and London, 1958), featured cannibalism; and *Sweet Bird of Youth* (1959)—which had its British première at the PALACE THEATRE, Watford, in 1968—involved castration. *Period of Adjustment* (1960; London, 1962) was a marital comedy which of all Williams's works probably most resembles an orthodox Broadway offering; but *The Night of the Iguana* (1961; London, 1965), set in a Mexican hotel, returned to more exotic settings. The following decade was not a happy one for Williams, whose talent seemed to be slipping away as he battled through a series of personal crises. His next play *The Milk Train Doesn't Stop Here Anymore* (Spoleto, 1962; N.Y., 1963) was seen at the CITIZENS' THEATRE, Glasgow, in 1969. Of his later works, *Slapstick Tragedy* (1966), made up of two short plays, *The Gnädiges Fräulein* and *The Mutilated, The Seven Descents of Myrtle* (1968), and *In the Bar of a Tokyo Hotel* (1969) had only short runs in New York. The pungent *Small Craft Warnings* (1972; London, 1973) and the much reworked *The Two Character Play*, first produced in London in 1967, which reappeared in New York in 1973 as *Out Cry*, seemed to indicate the return of dramatic vigour; but *The Latter Days of a Celebrated Soubrette* (1974) was simply an unsuccessful revision of *The Gnädiges Fräulein*, and the production of *The Red Devil Battery Sign* destined for New York in 1975 failed to reach there, though the play was seen in London in 1977. Williams's last works were *The Eccentricities of a Nightingale* (1976)—a revision of *Summer and Smoke*—*Vieux Carré* (1977; London, 1978), set in a New Orleans boarding house, *A Lovely Sunday for Creve Coeur* (1979), and a play about Scott and Zelda Fitzgerald, *Clothes for a Summer Hotel* (1980). A volume of *Memoirs* appeared in 1975.

WILLIAMSBURG, Virginia, probably the first English settlement in the New World to have a theatre. In 1716 a merchant named William Levinston, assisted by his indentured servant Charles Stagg and Stagg's wife Mary, formerly actors and teachers of dancing and elocution in England, produced some plays, one of which was possibly given in 1718 before the Governor. Little is known of the building in which they were acted. If it was specially erected it would be the first theatre in America, a distinction claimed also by the SOUTHWARK THEATRE in Philadelphia; but it may have been adapted from an existing building. From the evidence that remains it appears that the auditorium was only 30 ft wide by 86 ft long, with two tiers of boxes on either side and a gallery at the back. It continued in occasional use as a playhouse until 1745, and was demolished about 1769. A second playhouse, located at the back of the Capitol building, was opened in Oct. 1751 by Walter Murray and Thomas KEAN for their theatrical company. They met with little success, perhaps because of the poor quality of their productions, and left Williamsburg after two seasons. Their theatre was later remodelled by the elder Lewis HALLAM, who opened there in 1752 with *The Merchant of Venice* and a farcical afterpiece by RAVENSCROFT, *The Anatomist; or the Sham Doctor*. It is not known how long they stayed, or when the building was demolished.

WILLIAMSON, DAVID (1942–), Australian dramatist, generally considered the leading writer of the Australian dramatic renaissance of the early 1970s, his work being noted for its naturalistic, witty (and bawdy) dialogue. His first play to be produced was *The Removalists* (1971; London, 1973; N.Y., 1974), a fierce attack on the Australian police, after which came *Don's Party* (1972; London, 1973), a study of a group of middle-class liberals at a party on the eve of an Australian Federal Election; *Jugglers Three* (also 1972); *What If You Died Tomorrow?* (1973; London, 1974), in which a man gives up medicine for novel-writing; and *The Department* (1974). More recently he has written *A Handful Of Friends* (1976) and *The Club* (1977; N.Y., as *Players*, 1978; London, 1980), about the battle for control of a rugby club. *Travelling North* (1979; London, 1980), the story of a love affair between a man in his seventies and a woman 20 years younger, was followed by *Celluloid Heroes* (1980).

WILLIAMSON, JAMES CASSIUS (1845–1913), American-born actor-manager, who, with his first wife Maggie MOORE, was appearing at the California Theatre, San Francisco, when in 1874 George COPPIN engaged them both for a tour of Australia. They made their début there in *Struck Oil; or, the Pennsylvania Dutchman*, expanded by Clay Green from an original short play by Sam W. Smith entitled *The German Recruit*, and its immense popularity forced them to revive it constantly, much of it being rewritten by Williamson in performance. It is said to have made a profit of £10,000 in Australia, and was also staged elsewhere, having a run of 100 performances at the ADELPHI in London in 1876. Williamson returned to Australia in 1879 with the Australasian rights to GILBERT and Sullivan's *H.M.S. Pinafore*, in which he played Sir Joseph Porter, and he later obtained the Australasian rights to all the Savoy operas. In 1882 he founded J. C. Williamson Theatres Ltd, and eventually concentrated almost exclusively on its administration, acting only on rare occasions. In 1896 he married as his second wife the Australian dancer Mary Weir.

WILLIAM STREET THEATRE, New York. This was used, probably by amateurs, in 1790 for entertainments and comedies. Among the plays produced there were DODSLEY's *The King and the Miller of Mansfield*, LILLO's *The London Merchant; or, the History of George Barnwell*, and a farce by William DUNLAP, *The Soldier's Return*. There was also a PANTOMIME, *Robinson Crusoe*, displays of transparencies, and 'Italian Shades' or SHADOW SHOWS.

WILLS, IVAH, see COBURN, C. D.

WILL'S COFFEE HOUSE, also known as the Rose Tavern. This was situated at No. 1, Bow Street, Covent Garden, on the west side, at the corner of Russell Street. It was kept by one Will Unwin, and the Restoration wits resorted to it after the play to praise or damn it. They mainly congregated in a room on the first floor. There are many references in contemporary literature to this famous tavern, which was frequented by Dr JOHNSON and often mentioned in *The Spectator*.

WILSON, GEORGES (1921–), French actor, director, and manager, who joined a provincial company at the end of the Second World War. A successful performance for the GRENIER-HUSSENOT company in a stage version of the elder DUMAS's *Les Trois Mousquetaires* led to an invitation in 1952 to join the THÉÂTRE NATIONAL POPULAIRE, where he became an important member of VILAR's troupe, attracting critical acclaim more particularly in the title roles of MUSSET's *Larenzaccio* and JARRY's *Ubu-Roi*. He was already an established director when in 1963 he took over the management of the T.N.P. consequent upon Vilar's resignation. He was instrumental in making BRECHT's plays better known to his audiences (though some Brechtian purists objected to his productions) and in introducing Brendan BEHAN and Edward BOND to the French stage. His last few years at the T.N.P. were overshadowed by administrative and political difficulties, and when the enterprise was finally wound up in 1972 he worked independently, his later roles including James Tyrone in O'NEILL's *Long Day's Journey Into Night* in 1973 and Othello at the AVIGNON FESTIVAL in 1975.

WILSON, JOSEPHINE, see MILES, BERNARD.

WILSON, LANFORD, see CIRCLE REPERTORY COMPANY.

WILTON, MARIE EFFIE, see BANCROFT, SQUIRE.

WILTON'S MUSIC-HALL, London, see MUSIC-HALL.

WINCHESTER MUSIC-HALL, London, in Southwark Bridge Road. This was originally a 'saloon' attached to the Grapes public-house, which became known as the Surrey Music-Hall. When on 15 July 1856 the Surrey (Garden) Music-Hall opened in the Surrey Zoological Gardens, the older hall changed its name to the Winchester. It was run by Richard Preece and his son from its opening until 1878, when it was demolished after a brief period under W. B. Fair, famous as the singer of 'Tommy, Make Room for Your Uncle'. The VOKES family made their first appearance in London at the Winchester, which was also known as the British Saloon and the Grand Philharmonic Hall, and many famous music-hall stars began their careers there. The Surrey (Garden) Music-Hall had a short life, as it was burned down in June 1861. A building on the site, used by St Thomas's Hospital until 1872, was then remodelled as a music-hall and used until 14 Aug. 1877, when it closed. It was demolished a year later.

WIND MACHINE, see SOUND EFFECTS.

WINDMILL THEATRE, London, in Great Windmill Street, near Piccadilly Circus. This intimate theatre, which seats 326 people on two levels, opened on 22 June 1931 in a converted cinema which dated from 1910, its first production being Michael Barrington's *Inquest*. In 1932 Laura Henderson and her manager Vivian Van Damm introduced a policy of continuous variety entitled *Revudeville*, with performances running from 2.30 p.m. to 11 p.m. The Windmill Girls, who appeared in nude tableaux, became a part of London life, and many famous comedians, among them Jimmy Edwards, Harry Secombe, and Tony Hancock, served their apprenticeship there. The Windmill was the only theatre to remain open during the whole of the Second World War, and became renowned under its slogan 'We Never Closed'. After the death of Vivian Van Damm in 1960 *Revudeville* continued under the management of his daughter Sheila until it finally ended on 31 Oct. 1964. The building then reverted to its former role as a cinema until 1974, when it was taken over by Paul Raymond and reopened as a theatre with a comedy entitled *Let's Get Laid* which ran into 1976. It was followed by a nude revue *Rip-Off* which ran until, in 1981, the theatre closed for conversion to a theatre-restaurant.

WINDSOR, Berkshire. Between 1778 and 1793 the only theatre in Windsor, which, since it was situated in a royal residence, operated under licence from the LORD CHAMBERLAIN, was a small barn about a mile outside the town in a muddy field; it was nevertheless referred to as the Theatre Royal, Peascod Street. A new Theatre Royal was built in the High Street in 1793 and attended by George III whenever he was in residence at Windsor Castle. It was turned into a chapel in 1805, but in 1815 a new theatre was opened on the site of the present building, which survived until destroyed by fire in 1908. Rebuilt in 1910, it is now a very beautiful playhouse seating 656. It has good acoustics and excellent sight lines and is charmingly decorated, and is still privately owned and unsubsidized. For a time in the 1920s it housed touring companies, and it was then turned into a cinema, but in 1938, after several unsuccessful attempts, the actor John William Counsell (1905–), now managing director of the Windsor Theatre Company which leases the theatre from its owner, succeeded in establishing a repertory company there, opening with BARRIE's *Dear Brutus*. Six weeks later the theatre received the first of many visits from its royal neighbours when George VI and Queen Elizabeth attended a performance of Clifford Bax's *The Rose Without a Thorn*. It no longer has a resident company, the productions being cast from London. Its programmes tend to be somewhat conservative, consisting mainly of West End successes and revivals of popular classics. About a third of its productions, however, are new works, and over 50 productions have transferred to London, among them John Dighton's *Who Goes There!* (1951), Guy Bolton's *Anastasia* (1953), and the musical *Grab Me a Gondola* (1956). Three farces which later had long runs at the WHITEHALL THEATRE were also first seen at Windsor—John Chapman's *Dry Rot* (1954) and *Simple Spymen* (1958) and Ray Cooney's *Chase Me, Comrade!* (1964).

WINDSOR THEATRE, New York, see FORTY-EIGHTH STREET THEATRE.

WINGS, pairs of FLATS placed at each side of the stage, either facing or obliquely towards the audience. As many as 8 pairs could be used at any one time, though 3 or 4 was the usual number. They may have been used in Italy in the late 16th century by Bernardo Buontalenti (1536–1608), who designed the SCENERY for the Florentine INTERMEZZI of 1589; but it may be that his side-pieces were not true wings but *telari*, three-sided prisms based on the late classical PERIAKTOI. They certainly developed shortly afterwards, and Inigo JONES, on his return from the Continent, introduced them, as Side Scenes or Side Shutters, into his settings for Court MASQUES. From there they descended to the Restoration theatre, and, used in conjunction with a BACKCLOTH and BORDERS, they remained the basic elements of scenery until the introduction in 1832 of the BOX-SET. They are still used for PANTOMIME, big spectacular shows, opera, and ballet. On the Continent they were moved by the CARRIAGE-AND-FRAME method, but in England they usually ran in GROOVES, though in some early Victorian theatres BOOK WINGS were used.

WINSTANLEY, ELIZA [Mrs O'Flaherty] (1818–82), Australian actress, born in England, who went to Australia in 1833 and made her stage début the following year. She was soon playing leading roles in the melodramatic plays popular at the time, but also appeared in a number of classic parts. In 1846 she returned to England, and was seen in Manchester and, in the following year, at the PRINCESS'S THEATRE in London. After a few years on the American stage she was again in London in 1850, and joined the company of Charles KEAN, with whom she remained for eight years. She then ceased to act and embarked on a successful career as a writer, spending her last years in Australia.

WINTER, WILLIAM (1836–1917), American dramatic critic, whose conservatism made him the transatlantic counterpart of Clement SCOTT. In 1861 he joined the staff of the *Albion*, and from 1865 to 1909 worked for the *New York Tribune*. He was the most powerful dramatic critic of his time in the United States, probably because he rarely committed himself to an opinion at variance with that of the great majority of his readers. Although he admired English actors, he was antagonistic to such foreign visitors as DUSE, BERNHARDT, and RÉJANE, dragging irrelevancies concerning their private lives into his notices. He denounced IBSEN ('slimy mush'), MAETERLINCK ('lunacy'), and SHAW. After his retirement from the *Tribune* he contributed articles on theatrical matters to *Harper's Weekly* and other periodicals and wrote biographies of a number of actors, including IRVING, JEFFERSON, BOOTH, MANSFIELD, and BELASCO. His books contain much valuable information, but they are so pompous that it is almost impossible to read them with any pleasure.

WINTER GARDEN THEATRE, London, in Drury Lane, originally the MIDDLESEX MUSIC HALL, later adapted as a theatre with a three-tier auditorium seating 1,581 persons. It opened on 20 May 1919 with *Kissing Time*, starring Leslie HENSON, who appeared in most of the subsequent productions up to 1926, including *A Night Out* (1920),

Sally (1921), *The Beauty Prize* (1923), *To-Night's the Night* (1924), *Primrose* (also 1924), *Tell Me More* (1925), the last two having music by George Gershwin, and *Kid Boots* (1926). Gershwin again provided the music for *Tip-Toes* (also 1926), which starred Dorothy Dickson and Laddie Cliff. In 1927 a musical play *The Vagabond King*, based on Justin McCarthy's *If I Were King*, with Derek Oldham as François Villion, began a long run, and successes were scored by Sophie Tucker in *Follow a Star* (1930) and Gracie Fields in the revue *Walk This Way* (1931), but apart from SHAW's *On the Rocks* (1933) and *Androcles and the Lion* (1934) and a revival of BARRIE's *Peter Pan* in 1942, little of interest occurred until the opening of *It's Time To Dance* (1943), with Jack BUCHANAN and Elsie Randolph. On 3 May 1946 came the first night of Joan Temple's *No Room at the Inn* (previously tried out at the EMBASSY THEATRE), a play about evacuee children in which Freda Jackson made a great success. In 1952 Alec CLUNES transferred the ARTS THEATRE company to the Winter Garden, but with little success, and it was Agatha CHRISTIE's *Witness for the Prosecution* (1953) which gave the theatre another long run. Among later productions were A. P. Herbert's *The Water Gipsies* (1955), FEYDEAU's *Hotel Paradiso* with Alec GUINNESS, Shaw's *The Devil's Disciple* with Tyrone POWER (both 1956), and O'NEILL's *The Iceman Cometh* (1958). After a Christmas production of a musical version of Lewis Carroll's *Alice in Wonderland* (1959), the theatre closed on 23 Jan. 1960 and stood empty until it was demolished in 1965. The NEW LONDON THEATRE now occupies the site.

WINTER GARDEN THEATRE, New York, at 1634 Broadway, between 50th and 51st Streets, on the site of the American Horse Exchange. This theatre, with a seating capacity of 1,479, was opened by the Shuberts on 20 Mar. 1911 with a musical show, and was used mainly for musicals and REVUES, including *The Passing Show* annually from 1912 to 1924. It was then used as a cinema for some years but returned to live theatre in 1933. It has since housed several successful musicals, among them *West Side Story* (1957), with music by Leonard Bernstein, *Funny Girl* (1964), based on the career of the revue star Fanny Brice, played by Barbra Streisand, and *Mame* (1966), from a novel by Patrick Dennis. Stephen SONDHEIM's *Follies* was seen in 1971, and in 1974 the theatre staged one of its rare straight plays when Zero MOSTEL starred in *Ulysses in Nighttown*, based on James JOYCE's *Ulysses*. It returned to musicals later in the year with a revival of *Gypsy*. Sondheim's *Pacific Overtures* (1976) had a disappointingly short run, but *42nd Street* (1980), based on the film musical of 1933, proved enormously popular.

The METROPOLITAN THEATRE under BOUCICAULT in 1858 was renamed the Winter Garden, retaining the name until it was burnt down in 1867.

WINTHROP AMES THEATRE, New York, see LITTLE THEATRE.

WITHERS, GOOGIE [Georgette Lizette] (1917–), English actress, born in India, who studied for the stage under Italia CONTI and made her first appearance in the children's play *The Windmill Man* (1929) by Frederick Bowyer. She was then seen in the chorus of the MUSICAL COMEDY *Nice Goings On* (1933) before appearing in J. B. PRIESTLEY's *Duet in Floodlight* (1935); after playing in N. C. HUNTER's *Ladies and Gentlemen* (1937) she was in another Priestley play *They Came to a City* (1943). In 1945 she took over the part of Amanda in COWARD's *Private Lives* from Kay Hammond, following it with Lee in Ronald MILLAR's farcical comedy *Champagne for Delilah* (1949) and Georgie Elgin in ODETS's *Winter Journey* (1952), giving one of her finest performances as the supportive wife of an alcoholic actor. Later in 1952 she succeeded Peggy ASHCROFT as Hester Collyer in RATTIGAN's *The Deep Blue Sea*, a part in which in 1955 she toured Australia and New Zealand with her Australian-born husband John MCCALLUM, whom she married in 1948. In 1958 she played Gertrude in *Hamlet* and Beatrice in *Much Ado About Nothing* at the SHAKESPEARE MEMORIAL THEATRE. In the same year her husband entered theatrical management in Australia, and most of her later acting has been done in Australia and New Zealand. She made her début in New York in 1961 in Graham GREENE's *The Complaisant Lover* and was seen in Edinburgh and London in 1963 in IONESCO's *Exit the King*, and at the STRAND THEATRE in 1967 in SHAW's *Getting Married*. She then returned to Australia for some years, touring and playing Madame Ranevskaya in CHEKHOV's *The Cherry Orchard* (1972) for the MELBOURNE THEATRE COMPANY. In 1976 she was back in England, appearing with her husband at the CHICHESTER FESTIVAL THEATRE, and later at the HAYMARKET, as Lady Kitty in MAUGHAM's *The Circle*. She was again at Chichester in 1979 to play Lady Bracknell in WILDE's *The Importance of Being Earnest*. Her husband is the author of *Life With Googie* (1979).

WITKIEWICZ [Witkacy], STANISŁAW IGNACY (1885–1939), Polish artist, novelist, and dramatist, who led a stormy and unhappy life, mainly in exile, travelling widely in Russia and even visiting Australia. He committed suicide when the Nazis invaded Poland. His plays, written under the influence of EXPRESSIONISM, were performed only in small experimental theatres during his lifetime and made very little impact. Rediscovered after the Second World War, they were recognized as surrealistic, foreshadowing not only the work of BRECHT but containing also the seeds of BECKETT's and IONESCO's Theatre of the ABSURD and having much in common with EXISTENTIALISM. Most of them were then revived or staged for the first time, and they were widely translated, being everywhere hailed as masterpieces. A volume published in America in 1968, six years after the appearance of his complete works—30 plays in all—in a two-volume Polish edition, contained six plays: *The Shoemakers*, first seen in 1957, revived by Jarocki in 1971; *The Crazy Locomotive* and *The Mother*, both seen in 1964, the former being produced in New York in 1977 and the latter revived by AXER in 1970 and Jarocki in 1972; *They*, seen in 1965; *The Madman and the Nun*, which had its first performance in German in Vienna in 1962 and was staged in 1963 by KANTOR and in New York in 1979; and *The Water Hen*, which Kantor revived in 1967. Among his other works are *The Beelzebub Sonata, or What Really Happened at Mondovar*, not seen until 1966; *Gyubal Wahazar, or The Precipice of Nonsense*, revived by Prus at the Ateneum in Warsaw in 1972; *The Metaphysics of the Two-Headed Calf*; *Lovelies and Dowdies, or The Green Pill* and *The Cuttlefish, or The Hyrcanian World View*, staged by Kantor in 1973 and 1957 respectively; *In a Small Country House*, staged by Kantor in 1961; *Jan Maciej*; *Karol Hellcat*; and a five-act tragedy *The Pragmatists*, written in 1918 and perhaps the most frequently revived of all.

WODEHOUSE, P. G., see MUSICAL COMEDY.

WOFFINGTON, PEG [Margaret] (*c.* 1718–60), Dublin-born actress. She was engaged at the age of 12 by Madame Violante, a famous rope-dancer, to play in a children's company, being seen among other parts as Polly in GAY's *The Beggar's Opera*. In 1732 she made her first appearance in London in the same play, but as Macheath, and then returned to Dublin, where she eventually joined the company at the Smock Alley Theatre. There she played a wide range of parts, including the one in which she was to become famous, Sir Harry Wildair in FARQUHAR's *The Constant Couple*, in which she was first seen in 1740. Later the same year she was engaged by John RICH for COVENT GARDEN, where she played Sir Harry with such spirit and elegance that for a long time no male actor dared attempt the part. A year later she appeared at DRURY LANE under FLEETWOOD, making her first appearance in another BREECHES PART, Silvia in Farquhar's *The Recruiting Officer*, and from then on divided most of her time between the two theatres. Her naturally harsh voice rendered her unfit for tragedy, but in comedy she was outstanding, being much admired as Millamant in CONGREVE's *The Way of the World* and as Lady Townly and Lady Betty Modish in Colley CIBBER's *The Provok'd Husband* and *The Careless Husband* respectively. She was the most beautiful and least vain actress of her day, but her good nature did not extend to her fellow actresses and she was constantly at odds with Kitty CLIVE, Theophilus CIBBER's wife Susanna, and George Anne BELLAMY. She was for some years the mistress of GARRICK, who wrote for her the charming song 'My Lovely Peggy', and played opposite him both in London and in Dublin, where from 1752 to 1754 she appeared triumphantly in many of her most famous roles. She returned to Covent Garden at the end of 1754 as Maria in Cibber's *The Non-Juror* with her usual success, though she suffered somewhat from the rising popularity of Mrs Bellamy. The last breeches part she played was Lothario in ROWE's *The Fair Penitent*, and she was also seen as Lady Randolph in HOME's *Douglas* when it was first seen in London in 1757, with Spranger BARRY as Young Norval. She made her last appearance as Rosalind in *As You Like It* on 17 May 1757, being taken ill at the beginning of the epilogue. She lingered on for three years and gave herself to good works, endowing almshouses at Teddington where she died. She is the subject of the play *Masks and Faces* (1852) by Tom TAYLOR and Charles READE. Her younger sister Mary, known as Polly, was also on the stage for a time, but left to marry Robert, second son of the Earl of Cholmondeley.

WOLF, FRIEDRICH (1888–1953), German dramatist, by profession a doctor, and a dedicated Marxist. He began his literary career as an exponent of EXPRESSIONISM, but soon turned to realistic drama with *Der arme Konrad* (1923), a play about the German Peasants' War, and *Kolonne Hund* (1927). His first successful play was *Cyankali* (1929), an impassioned plea for the abolition of the law against abortion; it was followed by *Die Matrosen von Cattaro* (1930), depicting a mutiny in the Austro-Hungarian navy towards the end of the First World War. Wolf's last play before Hitler's rise to power was *Tai Yang erwacht* (1931), which dealt with a workers' rising in Shanghai in 1927, produced by PISCATOR in Berlin, with the help of film, posters, and mime, as a piece of undisguised political propaganda. Under the impact of Nazi anti-Semitism Wolf then wrote what became one of his best known plays, *Professor Mamlock* (1934), the tragedy of a Jewish surgeon who falls a victim to racial persecution. He went into voluntary exile, producing several more plays on topical events, among them *Floridsdorf*, which dealt with the uprising of the Vienna workers in 1934, and *Patrioten*, about the French Resistance. After the war he returned to East Germany but wrote nothing further of any importance. His early career parallels that of BRECHT in many ways, but his plays, though revolutionary in content, were traditional in form and made no lasting impact.

WOLFENBÜTTEL, capital city of the Duchy of Brunswick, one of whose Dukes, Heinrich Julius (1564–1613), was a patron of the ENGLISH COMEDIANS. They visited him some time in the 1580s and a group of them under Thomas Sackville, the creator of Jan Bouschet, was attached to the ducal household from 1596 until the Duke's death. They appeared in plays written by the Duke, of which eleven are extant, dealing mainly with matrimonial discord, usually with the onus on the female side unless the husband, through sheer stupidity, deserved his punishment. Some of them are written for the old-fashioned MULTIPLE SET, but others are in the new-style single set; they show considerable English influence. The FOOL figuring in all of them is endowed with sound common sense, and in *Vincentius Ladislaus* (1595) gets the better of the braggart soldier or CAPITANO imported from the COMMEDIA DELL'ARTE. After the death of Heinrich Julius theatrical activity ceased, and it was not until 1688 that a theatre was built near the castle by Duke Anton Ulrich (1633–1714), a devotee of French and Italian light opera, who also built an indoor theatre in his new château at Salzdahlum, near Wolfenbüttel, with an open-air theatre in its garden.

WOLFF, PIUS ALEXANDER (1782–1828), German actor, who from 1803 to 1816 was a member of the Court Theatre company in WEIMAR, where he collaborated closely with GOETHE, who regarded him as his leading actor. He married Anna Amelia Becker, also a member of the company, and together they played the leading roles in most of the classical plays produced at the theatre. Tensions within the company led the Wolffs to leave, and in 1816 they settled in BERLIN where they ranked among the best actors of the day.

WOLFIT [Woolfitt], DONALD (1902–68), English actor-manager, knighted in 1957. He made his first appearance on the stage in 1920 and four years later was seen in London as Phirous in Temple Thurston's *The Wandering Jew*. In 1929–30 he was at the OLD VIC, then toured Canada as Robert Browning in Besier's *The Barretts of Wimpole Street*, and was seen in London again as Thomas Mowbray in Gordon Daviot's *Richard of Bordeaux* (1933). He was at Stratford-upon-Avon in 1936, and the following year formed his own company, touring extensively in a Shakespearian repertory in which he also appeared for a season at the KINGSWAY THEATRE in 1940. During the Battle of Britain he gave over 100 lunch-time performances of scenes from Shakespeare, and later continued to tour in Shakespeare and other classics, two of his best performances being the title role in JONSON's *Volpone* and Sir Giles Overreach in MASSINGER's *A New Way to Pay Old Debts*. He was also excellent in the title role of MARLOWE's *Tamburlaine* at the Old Vic in 1951, and as Lord Ogleby in COLMAN and GARRICK's *The Clandestine Marriage*. In 1957 he appeared at the LYRIC THEATRE, Hammersmith, in two plays by MONTHERLANT, *The Master of Santiago* and *Malatesta*. His wife, Rosalind Iden (1911–), the daughter of Ben Iden PAYNE, was an excellent actress who appeared with him in many of his productions on tour and in London. His autobiography, *First Interval*, was published in 1955.

WOMEN'S THEATRE, a theatrical manifestation which developed in the United States in the late 1960s, as part of the Women's Liberation Movement's attempt to define and evaluate the role of woman in society and to create further opportunities for women as directors, playwrights, theatre administrators, and technicians in the professional theatre. The various women's groups, of which there are now some 20 in New York City alone, use improvisation and other theatre techniques, including COLLECTIVE CREATION, to explore various aspects of women's lives and to pinpoint problems which particularly concern them—rape, abortion, divorce, and women's rights among them. The groups vary significantly in style, technique, administrative organization, and degree of permanence, the best known being the It's All Right To Be a Woman Theatre formed in 1970, which plays only for women audiences, putting on 'Cranky Pieces', in which the performers act out an incident previously narrated, and 'Dream Plays', in which they improvize on a dream told by a member of the audience; the more traditional Westbeth Playwrights' Feminist Collective, a mixed company playing to mixed audiences which in 1975 won critical acclaim for its performance of Gertrude Stein's *The Mother of Us All*; the Interart Theatre, run by the Women's Interart Centre; and the New Feminist Theatre.

In England the movement was slower to start, and has not yet been taken seriously by theatre critics. Among its participants are the Women's Theatre Group founded in 1973; The Monstrous Regiment, which followed in 1975; Mrs Worthington's Daughters, which has performed old and new plays by or about women, among them BARRIE's *The Twelve Pound Look* and the one-act *The Oracle*, based on a French play by Theophilus CIBBER's wife Susanna, which was last seen in 1752; Beryl and the Perils; and a cabaret show *Bloomers*, performed by three professional actresses.

WOOD, CHARLES GERALD (1932–), English playwright, noted for his inventive dialogue and for the military background of many of his plays—he served for five years in the regular army. His first work for the theatre was *Cockade* (1963), a triple bill which included *Prisoner and Escort, John Thomas*, and *Spare*, all featuring different aspects of army life. A further one-act play *Don't Make Me Laugh* (1965), depicting the soldier at home, was followed by the farcical fantasy *Meals on Wheels* (also 1965) and *Fill the Stage With Happy Hours* (1966). Wood then wrote two full-length war plays, *Dingo* (1967), set partly in a North African prisoner-of-war camp, seen at the ROYAL COURT THEATRE; and *H, or Monologues at Front of Burning Cities* (1969), a savage indictment of Havelock's march to the relief of Lucknow during the Indian Mutiny, staged by the NATIONAL THEATRE. It was followed by *A Bit of a Holiday* (also 1969) and another triple bill *Welfare* (1970), dealing with the relationship between youth and age pre- and post-Second World War; it included a comic sketch *Tie Up the Ballcock*, a shortened version of *Meals on Wheels*, and *Labour*. *Veterans* (1972), at the Royal Court, which starred John GIELGUD and John Mills as veteran actors, was set on a film location in Turkey, and arose out of Wood's experiences as script writer for the film *The Charge of the Light Brigade*. His next play *Jingo*, presented by the ROYAL SHAKESPEARE COMPANY in 1975, was an attack on the complacency of the British in Singapore in 1941, while *Has 'Washington' Legs?* (1978), seen at the National Theatre, starred Albert FINNEY and again had a film-making background. Latterly he has tended to desert the theatre for films.

WOOD, JOHN, English actor, who was on the stage for some years after leaving OXFORD, where he was President of OUDS, before his excellence was recognized. He was with the OLD VIC company from 1954 to 1956 and made his West End début as Don Quixote in Tennessee WILLIAMS's *Camino Real* (1957). He was seen at the ROYAL COURT in Nigel Dennis's *The Making of Moo* later the same year and in 1967 made his first appearance in New York as Guildenstern in STOPPARD's *Rosencrantz and Guildenstern Are Dead*. His outstanding performance as Richard Rowan in a short run of James JOYCE's *Exiles* (1970) at the MERMAID THEATRE first revealed his true quality to critics and public alike, as he uncovered the layers of meaning hidden within the text and showed a perverse pleasure in listening to his wife's description of her amorous encounters. He then joined the ROYAL SHAKESPEARE COMPANY and displayed his versatility by playing in 1971 Yakov in GORKY's *Enemies* and Sir Fopling Flutter in ETHEREGE's *The Man of Mode*, and in 1972 a Brutus in *Julius Caesar* racked and tormented by his liberal conscience and a psychotic Saturninus in *Titus Andronicus*. He appeared in the West End as the spindly, lecherous, and slightly manic husband in John MORTIMER's *Collaborators* (1973), and returned to the R.S.C. to play the title role in the revival of William GILLETTE's *Sherlock Holmes* and Henry Carr in Stoppard's *Travesties* (both 1974), repeating the last two roles in New York in 1974 and 1975 respectively. In 1976 he was seen with the R.S.C. in SHAW's *The Devil's Disciple* and in the title role of CHEKHOV's *Ivanov*, and in 1978 he made an enormous success in New York in Ira Levin's thriller *Deathtrap*. Back in London in 1979, he gave an electrifying performance as Richard III with the NATIONAL THEATRE company, for which he also appeared in SCHNITZLER's *Undiscovered Country*, Tom Stoppard's adaptation of *Das weite Land*, in the same year, and in VANBRUGH's *The Provoked Wife* in 1980. In 1981 he returned to New York to take over from Ian MCKELLEN in Peter SHAFFER's *Amadeus*.

WOOD, Mrs JOHN [*née* Matilda Charlotte Vining] (1831–1915), English actress and manageress belonging to a well known theatrical family, being first cousin to Fanny Vining, the mother of Fanny DAVENPORT. She made her first appearance in Brighton as a child, and had already made a good reputation in the English provinces as an adult actress when in 1854 she went to America. Widowed in 1863, she took over the OLYMPIC THEATRE in New York and ran it successfully for three years, leaving it to make her first appearance on the London stage in 1866 as Miss Miggs in a dramatization of DICKENS's *Barnaby Rudge*. From 1869 to 1877 she was manageress of the ST JAMES'S THEATRE in London, where she was highly respected by her company and her audiences. She made her last appearance in New York under DALY in 1873, and thereafter was seen only in London. She was at the ROYAL COURT THEATRE from 1883 to 1892, appearing in the first productions of PINERO's farces *The Magistrate* (1885), *The Schoolmistress* (1886), and *Dandy Dick* (1887). A woman of liberal views, she ruled her company kindly but firmly and spared no expense in the running and equipping of her theatre and actors.

WOOD, WILLIAM BURKE (1779–1861), American actor, born in Montreal, Canada, where his parents, English actors who had emigrated to America, had taken refuge during the War of Independence. They later returned to the United States, where their son became the first native-born North American to achieve an important place in American theatre history. In partnership with William WARREN he was for many years manager of the CHESTNUT STREET THEATRE in Philadelphia, and a diary which he kept from 1810 to 1833 contains much interesting material on the theatre. The list of plays given there each season shows in the earlier years a preponderance of Shakespeare and English classic comedies, perhaps because of Wood's own admirable playing of polished comedy, though he was also good in the lighter parts of tragedy. In later years more new American plays were produced. Wood married Juliana, a daughter of Mrs Anthony WESTRAY, who with her two sisters had been at the PARK THEATRE in New York, and she appeared for many years under his management.

WOOD'S MUSEUM, New York, see DALY'S THEATRE.

WOODWARD, HARRY (1717–77), English actor, who made his first appearance on the stage as a child, playing small parts in GAY's *The Beggar's Opera* with a juvenile company in London. A year later he was engaged by John RICH for COVENT GARDEN where, as an excellent dancer and mime, he was known as Lun junior. In 1738 he went to DRURY LANE where he remained for 20 years, writing a number of PANTOMIMES for GARRICK in which he himself played HARLEQUIN. Other parts in which he excelled included Mercutio in *Romeo and Juliet*, Touchstone in *As You Like It*, Petruchio in Garrick's version of *The Taming of the Shrew*, Bobadil in JONSON's *Every Man in His Humour*, and Marplot in his own adaptation of Mrs CENTLIVRE's *The Busybody*. In 1758 he went to Dublin, and with MACKLIN (who soon withdrew) and Spranger BARRY took over the management of the Crow Street Theatre, where they ruined themselves and their rival, MOSSOP at Smock Alley. Returning to London in 1763, Woodward went back to Covent Garden, where in 1775 he created the part of Captain Absolute in SHERIDAN's *The Rivals*. He remained on the stage until he died and was the last of the great Harlequins, before GRIMALDI made CLOWN the chief figure in the pantomime.

WOODWORTH, SAMUEL (1785–1842), American dramatist, author of a domestic drama in the current European style, *The Deed of Gift* (1822); a historical melodrama, *LaFayette*, played before Lafayette himself on his visit to New York in 1824; *The Widow's Son* (1825), a tragedy based on an incident in the War of Independence; and *The Forest Rose; or, American Farmers* (also 1825), featuring the popular Yankee character Jonathan Ploughboy, which later provided an excellent part for Dan MARBLE and Henry PLACIDE. Woodworth, who was also a journalist, edited a Boston newspaper with John Howard PAYNE.

WOOING CEREMONY, see MUMMERS' PLAY.

WOOLGAR, SARAH JANE, see MELLON, MRS ALFRED.

WOOLLCOTT, ALEXANDER HUMPHREYS (1887–1943), American dramatic critic, reputedly the prototype of Sheridan Whiteside in *The Man Who Came to Dinner* (1939) by KAUFMAN and HART. He worked as a police reporter on the *New York Times*, graduating to the drama department and becoming drama critic at the age of 27. In 1922 he went to the *New York Herald* and the following year was transferred to the *Sun*. From 1925 to 1928 he was dramatic critic for the *New York World*; he then retired to devote himself to broadcasting, lecture tours, and writing magazine articles. As a dramatic critic his judgements were capricious. He was more interested in players than in plays and missed or misjudged many important works of his time, including the plays of O'NEILL which he dismissed as completely worthless. He himself turned playwright on two occasions, collaborating with Kaufman on *The Channel Road* (1929) and *The Dark Tower* (1932), and appeared on the stage three times—in BEHRMAN's *Brief Moment* (1932) and *Wine of Choice* (1938), and in the touring company of *The Man Who Came to Dinner*. Though his criticism was ephemeral, he wrote engagingly, and his volumes of fugitive pieces were popular.

WORCESTER'S MEN, see QUEEN ANNE'S MEN.

WORLD THEATRE SEASON, an annual festival of foreign plays in the original, performed by foreign companies invited to London by Peter DAUBENY and housed in the ALDWYCH THEATRE during the absence of the ROYAL SHAKESPEARE COMPANY on tour. It ran from 1964 to 1973, with a final season in 1975. The first season was part of the celebrations for Shakespeare's Quatercentenary, and opened on Mar. 17 with the COMÉDIE-FRANÇAISE in MOLIÈRE's *Tartuffe*, followed by FEYDEAU's *Un Fil à la patte*, with Robert HIRSCH as Bouzin. This was an outstanding success, and the director, Jacques CHARON, was invited to direct Feydeau's *Une Puce à l'oreille* (adapted by John MORTIMER as *A Flea in Her Ear*) for the NATIONAL THEATRE company a year later. The other companies seen during this first season were the Schillertheater from BERLIN in FRISCH's *Andorra* and GOETHE's *Clavigo*; Peppino De Filippo's company in his own play *Metamorphoses of a Wandering Minstrel*; the ABBEY THEATRE company from Dublin in O'CASEY's *Juno and the Paycock* and *The Plough and the Stars*; the Polish Contemporary Theatre in MROŻEK's *What a Lovely Dream!* and *Let's Have Fun!* and FREDRO's *The Life Annuity*; the Greek Art Theatre in ARISTOPHANES' *Birds*; and the MOSCOW ART THEATRE in GOGOL's *Dead Souls*, CHEKHOV's *The Cherry Orchard*, and POGODIN's *Kremlin Chimes*. Several of these companies and others from the same countries were seen in subsequent years, including from Germany a Bremen company in 1967 in WEDEKIND's *Frühlings Erwachen* and the Bochum Schauspielhaus in 1973 in *Kleiner Mann—was nun?*, a cabaret-style re-creation of Berlin in the 1930s based on Hans Fallada's novel; from Italy Romolo VALLI's Compagnia dei Giovani in 1965, 1966, and 1973, as well as companies led by Anna Magnani in 1969 and Eduardo DE FILIPPO in 1972; from Poland, the Cracow Stary in 1972, 1973, and 1975; and from Russia the GORKY THEATRE from Leningrad in 1966 in DOSTOEVSKY's *The Idiot* and Dumbadze and Lovdkipanidze's *Grandma, Uncle Iliko, Hilarion and I*. Other European countries which participated were Austria, Belgium, Czechoslovakia, with Otomar KREJČA's Theatre on the Balustrade in 1968 and Theatre Behind the Gate in 1969, and the Cinoherni Club of Prague in 1970, Spain, with a company led by Núria ESPERT I ROMERO in GARCÍA LORCA's *Yerma* in 1972, Sweden, and Turkey, with the Dormen Theatre in Erol Günaydin's *A Tale of Istanbul* in 1971. From outside Europe came the Kathakali Drama company of India in 1972; the HABIMAH (in 1965) and the CAMERI (in 1967) from Israel; the Nō Theatre in 1967, the Bunraku in 1968, and the Umewaka Nō Troupe in 1973 from Japan; the Natal Theatre Workshop from South Africa in 1972 and 1973 in *Umabatha* (based on *Macbeth*); and the Abafumi Company from Uganda in 1975. From the United States came two companies—the Actors' Studio

Theatre in 1965, in Chekhov's *Three Sisters* and Baldwin's *Blues for Mr Charlie*; and the Negro Ensemble of New York in 1969 in Peter WEISS's *Song of the Lusitanian Bogey* and Ray McIver's *God is a (Guess What?)*.

WORMS, GUSTAVE-HIPPOLYTE (1836–1910), French actor, who made his début at the COMÉDIE-FRANÇAISE in 1858, playing young lovers in comedy and tragedy. He soon proved himself a good actor but, tired of waiting to be received as a member of the company, he left to go to Russia, where he spent several successful and profitable years. Back in Paris in 1877 he returned to the Comédie-Française, making a great success as Don Carlos in a revival of HUGO's *Hernani*, with MOUNET-SULLY in the title role and Sarah BERNHARDT as Doña Sol. He also appeared in several plays by the younger DUMAS, but was at his best in heroic parts or in the elegant heroes of 17th-century comedy. For many years he was a valued professor at the Conservatoire.

WORTH, IRENE (1916–   ), American actress, formerly a teacher, who was not seen on the stage until 1942, when she toured with Elisabeth BERGNER in Margaret Kennedy's *Escape Me Never*. A year later she made her New York début in Martin Vale's *The Two Mrs Carrolls*, and in 1944 she was in London, where she spent most of the next 30 years, making her first appearance in SAROYAN's *The Time of Your Life* (1946) at the LYRIC THEATRE, Hammersmith. She created the role of Celia Coplestone in T. S. ELIOT's *The Cocktail Party* (1949) at the EDINBURGH FESTIVAL, playing it also in New York and London, and then joined the OLD VIC company, where her Helena in *A Midsummer Night's Dream* (1951) showed a strong talent for comedy and her other roles included Desdemona in *Othello* (also 1951) and Portia in *The Merchant of Venice* (1953). After appearing in the first season of the STRATFORD (ONTARIO) FESTIVAL in 1953 she returned to London, where she was seen in N. C. HUNTER's *A Day by the Sea* (also 1953) and in BETTI's *The Queen and the Rebels* (1955). Two contrasting roles—Marcelle in FEYDEAU's *Hotel Paradiso* (1956) and the title role in SCHILLER's *Mary Stuart* (N.Y., 1957)—were followed by Sara Callifer in Graham GREENE's *The Potting Shed* (1958) before she again played Mary at the Old Vic (also 1958). After her Rosalind in *As You Like It* at Stratford, Ontario, in 1959 she returned to New York to star in Lillian HELLMAN's *Toys in the Attic* (1960), and then joined the ROYAL SHAKESPEARE COMPANY in London, playing Lady Macbeth, Goneril in Peter BROOK's production of *King Lear* (both 1962), and Dr Zahnd in DÜRRENMATT's *The Physicists* (1963). During another visit to New York she played the title role in ALBEE's *Tiny Alice* (1964)—in which she was seen with the R.S.C. in 1970—and she reappeared in the West End in three roles in COWARD's *Suite in Three Keys* (1966). After an excellent performance as Hesione Hushabye in SHAW's *Heartbreak House* in 1967, she showed great emotional force as Jocasta in SOPHOCLES' *Oedipus* at the Old Vic in 1968, and returned to Stratford, Ontario, to play the title role in IBSEN's *Hedda Gabler* in 1970. Her gift for comedy gave an unusual piquancy and charm to her Madame Arkadina in CHEKHOV's *The Seagull* at the CHICHESTER FESTIVAL in 1973. In 1975 she returned to New York, where she was seen in Tennessee WILLIAMS's *Sweet Bird of Youth*; as Madame Ranevskaya in Chekhov's *The Cherry Orchard* (1977); as Winnie in BECKETT's *Happy Days* at the PUBLIC THEATRE in 1979; in the brief run of Albee's *The Lady from Dubuque*; and as Ella in Ibsen's *John Gabriel Borkman* (both 1980).

WYCHERLEY, Sir WILLIAM (1640–1716), Restoration dramatist, whose comedies, though coarse and often frankly indecent, show so much strength and savagery in their attack on the vices of the day that their author has been labelled 'a moralist at heart'. He was to some extent influenced by MOLIÈRE, but transmuted his borrowings by his own particular genius, and his style has an individuality seldom found in other writers of his time. His first play, *Love in a Wood; or, St James's Park* (1671), was followed by the somewhat uncharacteristic *The Gentleman Dancing-Master* (1672), based on CALDERÓN's *El Maestro de danzar*. The best of his plays was *The Country Wife* (1675), in which the comedy hinges on the efforts of a jealous husband to keep his young but naturally wanton country wife from the temptations of London. It was a great success, and was revived many times up to 1748. It received a new lease of life when GARRICK adapted it as *The Country Girl* (1766) and produced it at DRURY LANE with Dorothea JORDAN in the title role. The original version has been seen in London several times since the PHOENIX SOCIETY first revived it in 1924, the most recent production being at the NATIONAL THEATRE in 1977. Wycherley's last play, *The Plain-Dealer* (1676), was successful enough for its author to be nicknamed 'Manly' Wycherley, after the name of its hero, with whose outspoken and misanthropic bent he doubtless had much sympathy. It was occasionally revived, and in 1765 was revised by Isaac BICKERSTAFFE for Drury Lane; his version held the stage until 1796, since when the only production (of the original version) has been that at the SCALA THEATRE by the Renaissance Players in 1925.

WYNDHAM [Culverwell], CHARLES (1837–1919), English actor-manager, knighted in 1902. He trained as a doctor, but his success in several amateur productions, under the name of Charles Wyndham which he later adopted legally, caused him to abandon medicine for the theatre, and he made his first appearance on the professional stage at the ROYALTY THEATRE in 1862. When the company disbanded after six months he went to the United States, where the Civil War was at its height, and enlisted in the Federal army as a surgeon. He twice resigned in 1863 to appear on stage, playing Osric to John Wilkes BOOTH's Hamlet in Washington and Thomas Brown to Mrs John WOOD's Pocahontas in a revival of BARKER's *The Indian Princess* in New York. He then returned to the army until the end of 1864. A visit to England, during which he was again seen at the Royalty and also ventured into management without success, proved disappointing, and he returned to New York to play Charles Surface in SHERIDAN's *The School for Scandal* at WALLACK'S THEATRE in 1869. He soon established a reputation as a light comedian, and from 1871 to 1873 led his own comedy company on an extended tour of the Middle West. Back in London in 1874 he made a great success in Bronson HOWARD's *Saratoga* (1870), in which he had appeared in America, renaming it *Brighton*. It opened at the ROYAL COURT THEATRE, but was moved to the CRITERION THEATRE, with which Wyndham was to have a lifelong association, his first outstanding success there being ALBERY's farce *The Pink Dominoes* (1877) in which he played Charles Greyhorne. He made it one of the foremost playhouses of London, and later built and managed the NEW THEATRE and WYNDHAM'S THEATRE with equal success. In 1883 he took his London actors on a long tour of America, the first completely English company to visit California and the Far West. On his return to England he pursued his successful career, being responsible for the production of many interesting new plays. A tall, handsome man, with a mobile, expressive face and a voice which, though at times a little husky, had great charm, he was at his best in high comedy but could play serious roles with conviction; in dramatic scenes, such as his deadly cross-examination of Felicia Hindmarsh in H. A. JONES's *Mrs Dane's Defence* (1900), he had few equals. One of his finest roles was David Garrick in T. W. ROBERTSON's play of that name, which he first revived in 1886, and made his own. He was also outstanding in Jones's *The Case of Rebellious Susan* (1894), in Louis N. PARKER and Murray Carsons *Rosemary* (1896), Jones's *The Liars* (1897), and Hubert Davies's *The Mollusc* (1907). In this last play his leading lady was a fine actress, Mary Moore (1869–1931), the widow of James ALBERY, whom he married as his second wife in 1912. She appeared in many of his productions, and after his death continued to manage his theatres, leaving them on her death to the care of her son Bronson Albery and his step-brother Howard Wyndham, Wyndham's son by his first wife Emma.

WYNDHAM'S THEATRE, London, in Charing Cross Road. Built for Charles WYNDHAM, it seats 770 and has the last complete picture-frame surround in London. It opened on 16 Nov. 1899 with a revival of T. W. ROBERTSON's *David Garrick*, in which Wyndham and his wife Mary Moore had already appeared with great success. The first new play to be given was a translation of ROSTAND's *Cyrano de Bergerac* in Apr. 1900. The years before 1914 were notable for productions of Henry Arthur JONES's *Mrs Dane's Defence* (1900), which gave Wyndham the chance to shine in a strong dramatic role; Charles Marlowe's *When Knights Were Bold* (1907); and the controversial war play *An Englishman's Home* (1909), written by Guy Du Maurier under the pseudonym of 'A Patriot'. In 1910 Gerald DU MAURIER became joint manager of the theatre with Frank Curzon, and in 1913 he enjoyed a considerable success with a revival of SARDOU's *Diplomacy*. He then played Dearth in BARRIE's *Dear Brutus* (1917), and in 1921 made an outstanding success as Bulldog Drummond in the play of that name by 'Sapper' (H. C. McNeile). In 1926 the first of six plays by Edgar WALLACE to be staged at this theatre opened on 1 May; the last, *The Green Pack*, opening on 9 Feb. 1932, the night before the author's death in Hollywood. After the death of Mary Moore in 1931 her sons, the step-brothers Howard Wyndham and Bronson ALBERY, took over the management of the theatre and inaugurated a long series of successful productions, among them Savory's *George and Margaret* (1937), Esther McCracken's *Quiet Wedding* (1938), and its sequel *Quiet Week-End* (1941), the last being the first play at this theatre to have more than 1,000 performances. The first post-war success was BRIDIE's *Daphne Laureola* (1948), with Edith EVANS; Peter USTINOV had a personal triumph in his own play *The Love of Four Colonels* (1951), as did Dorothy TUTIN making her début in Graham GREENE's *The Living Room* (1953). The musical *The Boy Friend* by Sandy Wilson, which had first been seen at the PLAYERS' THEATRE in 1953, began a long run at Wyndham's on 14 Jan. 1954. It closed on 7 Feb. 1959 and was succeeded by the first of four transfers of productions by THEATRE WORKSHOP—*A Taste of Honey* by Shelagh Delaney. Brendan BEHAN's *The Hostage* followed in June, Stephen Lewis's *Sparrers Can't Sing* in 1961, and *Oh, What a Lovely War!* in 1963. Joe ORTON's *Entertaining Mr Sloane* was seen in 1964; John OSBORNE's *Inadmissible Evidence*, with Nicol Williamson, and Jay Presson Allen's *The Prime of Miss Jean Brodie*, with Vanessa REDGRAVE, based on Muriel Spark's novel, both followed in 1965. Alec GUINNESS appeared in a travesty role in Simon GRAY's *Wise Child* in 1967; *The Italian Girl* by Iris Murdoch and James SAUNDERS occupied the theatre for almost the whole of 1968; and a revival of T. S. ELIOT's *The Cocktail Party* was followed by *The Boys in the Band* (1969) by Mart Crowley. Profiting by the abolition of censorship in 1968, it was the first play to portray overtly homosexual characters on the West End stage. Ronald MILLAR's *Abelard and Heloïse* (1970), with Diana RIGG and Keith MICHELL, had a long run, which was followed by *Godspell* (1972), a musical based on the life of Christ, which finally closed on 12 Oct. 1974 after a run of 1,128 performances. In 1975 Paul SCOFIELD returned to the West End as Prospero in

The Tempest, from LEEDS Playhouse, and PINTER'S *No Man's Land* transferred from the OLD VIC. The musical *Side By Side By Sondheim* transferred from the MERMAID THEATRE for a long run in 1976. Unexpectedly successful were Mary O'Malley's *Once a Catholic* (1977) and Dario FO's *Accidental Death of an Anarchist*, presented by the FRINGE group Belt and Braces in 1980.

WYNN, ED, see REVUE and VAUDEVILLE, AMERICAN.

WYNYARD, DIANA [Dorothy Isobel Cox] (1906–64), English actress of great beauty and distinction, who made her first appearance in London in 1925 and was a member of the LIVERPOOL PLAYHOUSE company from 1927 to 1929. She played small parts in London before making her New York début in 1932. Back in London she made a great success as Charlotte Brontë in Clemence DANE'S *Wild Decembers* (1933) and as Belinda Warren in Joyce Carey's *Sweet Aloes* (1934). She played Gilda in Noël COWARD'S *Design for Living* (1939), Linda Easterbrook in S. N. BEHRMAN'S *No Time for Comedy* (1941), and Sara Muller in Lillian HELLMAN'S *Watch on the Rhine* (1942) before touring for ENSA. In 1948 and 1949 she was at the SHAKESPEARE MEMORIAL THEATRE, where her work in Shakespeare's leading women's roles, particularly Lady Macbeth, was much admired. She starred opposite GIELGUD at the PHOENIX THEATRE as Hermione in *The Winter's Tale* (1951) and Beatrice in *Much Ado About Nothing* (1952) and appeared in Moscow as Gertrude to the Hamlet of Paul SCOFIELD in 1955. She was always ready to appear in new plays, however, and was seen in WHITING'S *Marching Song* (1954), Maxwell ANDERSON'S *The Bad Seed* (1955), Tennessee WILLIAMS'S *Camino Real* (1957), N. C. HUNTER'S *A Touch of the Sun* (1958), Lillian Hellman's *Toys in the Attic* (1960), and USTINOV'S *Photo Finish* (1962). She made her last appearances in the NATIONAL THEATRE company's 1963–4 season in *Hamlet* and FRISCH'S *Andorra*.

WYSPIAŃSKI, STANISŁAW (1869–1907), Polish poet, playwright, director, and scene designer. Considered as the successor of MICKIEWICZ, he was the first to co-ordinate and stage the latter's vast poetic drama *Forefathers' Eve* in 1901. In that and in many of his other monumental productions he made innovations in stagecraft which were to have a strong influence on his successors, particularly Leon SCHILLER. Among his own plays the best is usually considered to be *The Wedding* (1901). A revival of this, directed by Hanuszkiewicz, was seen in London during the WORLD THEATRE SEASON of 1966. Other important works are *Akropolis* (1903), in which GROTOWSKI'S company made its first appearance outside Poland in the same year; and *November Night* (1904), seen in London in 1975 in a production by the Stary Theatre, Cracow, under the direction of Wajda and again revived in Warsaw in 1979 under Maciej Prus. Three other plays, *Liberation* (1903), *The Curse* (1905), and *The Judges* (1907), have also been revived in recent years by ŚWINARSKI. Wyspiański had a wide knowledge of Shakespeare, and in an introduction to a Polish version of *Hamlet* he put forward a new view of the play, making the Ghost the central character—a view which resulted in an interesting production, entitled *The Tragical History of Hamlet*, first seen in 1901 and repeated in 1905.

Y

YABLOCHKINA, ALEXANDRA ALEXANDROVNA (1868–1964), Russian actress, who was on the stage at the age of six, her father being manager of the Tbilisi Theatre and later stage director at the ALEXANDRINSKY THEATRE in St Petersburg. She made her adult début at the MALY THEATRE in 1888, where she remained until her death, playing Miss Crawley in a dramatization of Thackeray's *Vanity Fair* at the age of 94. During her long career she appeared in a vast range of parts that included the heroines of Shakespeare, WILDE, and GALSWORTHY, and in Russian plays from GRIBOYEDOV and GORKY to ROMASHOV and KORNEICHUK. From 1916 until her death she was chairman of the ALL-RUSSIAN THEATRICAL SOCIETY.

YACINE, KATEB (1929–), Algerian novelist and dramatist, whose plays are inspired by the struggle of the Algerian people against French colonial oppression and against their own internecine difficulties. Yacine lives in Paris, but his first play, *Le Cadavre encerclé* (1958), directed by SERREAU, had to be performed in Brussels because of the political situation in France. Together with a comic parable, *La Poudre d'intelligence*, and its sequel, *Les Ancêtres redoublent de férocité*—directed by Serreau under the title *La Femme sauvage* (1967)—it forms a trilogy published as *Le Cercle des représailles*. *L'Homme aux sandales de caoutchouc* (*The Man with Rubber Sandals*, 1971), directed by MARÉCHAL, an epic account of the life of Ho-Chi-Minh, caused a political storm in Lyon, where the Théâtre du VIIIe had its subsidy cut by the town council as a result of the production. Although Yacine represents strikingly the massacres and tortures perpetrated by the French army, his plays are pitched on a more ambitious level than the attitude of the authorities might suggest, and his heroes suffer as much from their own deficiencies and from the sorrows of the human condition as from political circumstances. *Mohamed, prends ta valise* (1972), a comic satire on the situation of immigrant workers in France, is bilingual, and was successfully presented to Algerian communities in Paris by an Algerian company.

YAKOVLEV, ALEXEI SEMENOVICH (1773–1817), Russian actor, a member of the St Petersburg Imperial Theatre company, who made his first appearance there in 1794. Helped by his excellent presence and powerful voice, and by the counsels of DMITREVSKY, he was soon playing leading roles. His performances were very unequal, since he relied on intuition, scorned technique, and drank heavily, but he had some perceptions of greatness, and tried to initiate reforms which were later carried out by SHCHEPKIN.

YALE UNIVERSITY and **REPERTORY THEATRE,** New Haven, Connecticut. There were theatrical performances annually at this American university from as early as 1771, among the first productions being STEELE's *The Conscious Lovers*, FARQUHAR's *The Beaux' Stratagem*, and in 1785 an original play, *The Mercenary Match* by 'Mr Bidwell', a student in his senior year. There has been for many years now an amateur dramatic society for the students, the 'Yale Dramat', and in 1925 a Drama Department was inaugurated, with George BAKER as its first director. Its excellent little experimental theatre was built and endowed by Edward Harkness, and its curriculum comprises a wide-ranging schedule of instruction in all branches of professional theatre, including design, direction, administration, playwriting, and dramatic criticism. The library, which houses the personal papers of Eugene O'NEILL, has a theatre collection which includes a vast dossier of photographs of theatrical material collected from all over Europe, begun under Allardyce NICOLL during his term of office as head of the drama department and constantly being extended. In 1966 the then Dean of the School of Drama, Robert Brustein, founded the Yale Repertory Theatre. Originally intended as an adjunct to the School, it left the university premises in 1968 and moved into a church converted into a flexible theatre. In 1975 it was reconstructed to provide four auditoriums, of which the two smaller ones are used exclusively by the School, which also shares a third with the repertory company. The fourth, seating 491, with a thrust stage, is for the sole use of the Repertory Theatre company, among whose more adventurous productions have been an adaptation of AESCHYLUS' *Prometheus Bound* by Robert Lowell, Kenneth Cavender's adaptation of EURIPIDES' *Bacchae*, ARISTOPHANES' *Frogs* (1974) with music by Stephen SONDHEIM, presented in the Yale swimming pool, and *A Midsummer Night's Dream* (1975) incorporating music from Purcell's *The Fairy Queen* for the first time. Modern plays have included the world premières of Eric BENTLEY's *Are You Now Or Have You Ever Been . . .?* (1972) and Arthur KOPIT's *Wings* (1978) and the American premières of several plays by Edward BOND.

The theatre's programme is complementary to that of New Haven's LONG WHARF THEATRE.

YANSHIN, MIKHAIL MIKHAILOVICH (1902–76), Soviet actor and director, who became a member of the MOSCOW ART THEATRE company in 1924. One of the most successful roles of his early career was Lariossik in BULGAKOV's *The Days of the Turbins* (1926), in which he showed warmth, intimacy, and a subtle sense of humour. Among his later outstanding classic parts were Gradoboyev in OSTROVSKY's *Warm Heart*, Telyegin in CHEKHOV's *Uncle Vanya*, and Sir Peter Teazle in SHERIDAN's *The School for Scandal*. He was artistic director of the Gypsy Theatre from 1937 to 1941, and in 1950 became chief producer at the Stanislavsky Theatre in Moscow, where under his direction many new Soviet plays were produced.

YATES, FREDERICK HENRY (1795–1842), English actor, friend of the elder Charles MATHEWS, who persuaded him to go on the stage. He was first seen in Edinburgh, and in 1818 went to London, where he played Iago to the Othello of Charles Mayne YOUNG. In the following year he was seen as Falstaff in *Henry IV, Part I* to the Hotspur of MACREADY. Having no pretensions to be anything more than a useful actor he played everything that was offered to him, but eventually concentrated on comedy where his real talent lay. In 1825, with Daniel TERRY, he took over the ADELPHI THEATRE, and produced there a series of melodramas, farces, and burlesques, being joined on Terry's death in 1829 by Mathews. Some of the best authors of the day wrote for him, and he himself was excellent in such Dickensian parts as Fagin in *Oliver Twist*, Mr Mantalini in *Nicholas Nickleby*, and the grotesque Miss Miggs in *Barnaby Rudge*. He married a member of his company, Elizabeth Brunton (1799–1860), who became the mother of Edmund Yates (1831–94), a well-known journalist and the author of some ephemeral plays, mainly farces.

YATES, RICHARD (1706–96), English comedian, considered almost as good a HARLEQUIN as WOODWARD. He joined GIFFARD's company at the second GOODMAN'S FIELDS THEATRE when it opened in 1732, and remained until it was closed on the passing of the Licensing Act of 1737. He then went to DRURY LANE, where he remained for many years, playing Sir Oliver Surface in the first production of SHERIDAN's *The School for Scandal* (1777). He was also good as Shakespeare's fools, and was considered the only actor of the time to have a just notion of how they should be played. His style in comedy was modelled on that of DOGGETT; fine gentlemen and serious comedy lay outside his range,and he never appeared in tragedy. His second wife, Mary Ann Graham (1728–87), who was at Drury Lane from 1754 to 1785, was considered the finest actress of her day in 18th-century tragedy, for which her dignified presence and powerful though monotonous voice were eminently suitable.

YAVOROV, PEIO, see BULGARIA.

YEATS, WILLIAM BUTLER (1865–1939), Irish poet and dramatist, awarded the NOBEL PRIZE for literature in 1923. With Lady GREGORY he was the founder of the modern Irish dramatic movement, and as a director of the ABBEY THEATRE in Dublin, where all except his earliest plays were first produced, from its foundation in 1904 until his death, he did much to ensure the integrity of the emerging national drama, in spite of political and financial pressures. He also encouraged new playwrights, among them SYNGE and O'CASEY, even when the theatre's repertoire moved away from the aims demonstrated in his own plays. The earliest of these were poetic dramas in the style of MAETERLINCK—*The Countess Cathleen* (1892), *The Land of Heart's Desire* (1894), and *Cathleen ni Houlihan* (1902); the last-named was written in collaboration with Lady Gregory, who was also part-author of *The Pot of Broth* (1904) and of two poetic plays dramatically more powerful than the earlier ones—*The King's Threshold* (also 1904) and *The Unicorn from the Stars* (1908). In the first, the function of poetry and of the poet is the major theme; the second is a bold piece of speculative thinking, based on a mystical experience. Other plays of this period were on subjects taken from the heroic legends of Ireland—*On Baile's Strand* (1904), *Deirdre* (1907), and *The Green Helmet* (1908). *The Shadowy Waters* (1911), again a poetic play in which, as in his earlier works, the theme is remote and the experience it presents, though universal, is not revealed in terms of actual contemporary life, was followed by *The Hour-Glass* (1914), written in verse and so performed at the Abbey; for later productions elsewhere Yeats re-wrote it in prose, owing to the difficulty of finding verse-speakers who met with his approval. There are also two versions of *The Only Jealousy of Emer* (1919), the second, in prose, being entitled *Fighting the Waves*. Together with *At the Hawk's Well* (1917), *The Dreaming of the Bones* (1919), and *Calvary* (1920), the first version formed part of the 'four plays for dancers' in which extreme simplicity of design and setting is matched by a brevity of expression akin to that of the *nō* play of JAPAN. In his last plays this plainness and severity, characteristic also of his poetry after 1916, reaches fulfilment, and the underlying thought makes stricter demands than ever upon the intelligence and imagination of the audience. This is most clearly shown in *The Player Queen* (1922), *The Words Upon the Window Pane* (1934), and *Purgatory* (1938). Yeats also made new versions of SOPHOCLES' *Oedipus the King* (1928) and *Oedipus at Colonus* (1934). In his last play, *The Death of Cuchulain* (1939), he reverted once more to the legends of Celtic Ireland, always one of his main preoccupations.

YEFREMOV, OLEG NIKOLAYEVICH (1927–), Soviet actor and director, who began his career in children's theatre in Moscow before going on to organize a group of young actors, who in 1958 became the resident company at the SOVREMENNIK THEATRE under his directorship. Here he directed and acted in productions of plays by the Soviet dramatists ROZOV, SHWARTZ, Volodin, Shatrov, including a trilogy on Russian revolutionaries —Zorin's *The Decembrists*, Svobodin's *The Populists*, and Shatrov's *The Bolsheviks*—and such modern dramatists as John OSBORNE and William Gibson, making his theatre one of the most popular and original in the capital, with a considerable appeal to the younger generation of Soviet theatregoers. In 1971 he was appointed artistic director of the MOSCOW ART THEATRE in succession to Boris LIVANOV, but his arrival did not produce the expected improvement in the theatre's standards.

YERMOLOVA, MARIA NIKOLAIEVNA (1853–1928), Russian tragic actress, considered by STANISLAVSKY the finest he had ever seen. She made her début in 1870 at the MALY THEATRE in Moscow as Emilia Galotti in LESSING's play of that name, and soon proved herself an outstanding intrepreter of such roles as Lady Macbeth, Joan of Arc in SCHILLER's *Die Jungfrau von Orleans*, and RACINE's Phèdre. In contrast to the somewhat passive and unreal playing of such characters at the time, she made her heroines active, independent members of society. Having weathered the storms of the Russian Revolution, which she welcomed, she celebrated her stage jubilee in 1920 with full Soviet ceremonial and the honour of being named People's Artist of the Republic. A studio named after her in 1930 later became the YERMOLOVA THEATRE.

YERMOLOVA THEATRE, Moscow, a studio named in honour of the actress (above) in 1930, which acquired its present status after its fusion with the Khmelev Studio in 1937 when KHMELEV himself became its director. Its first production was GORKY's *Children of the Sun*, BILL-BELOTSERKOVSKY's *Storm* and OSTROVSKY's *The Poor Bride* were seen in 1938, and in 1939 Khmelev scored a triumph with an outstanding production of *As You Like It*. The theatre later saw the first production of several important Soviet plays, including Surov's war play *Far From Stalingrad* (1946), DOSTOEVSKY's *Crime and Punishment* (1956), Mikhalkov's vaudeville *The Savages* (1958), and Shteyn's *A Game Without Rules* (1962). Modern foreign plays have also been staged, including DE FILIPPO's *Saturday, Sunday, Monday* and DÜRRENMATT's *Play Strindberg*.

YEVREINOV, NIKOLAI, see EVREINOV.

YIDDISH ART THEATRE, New York, see PHOENIX THEATRE and SCHWARTZ, MAURICE.

YORK had its first permanent theatre building, adapted from a TENNIS-COURT in Minster Yard, in 1734. A second theatre was erected on the site of the present Theatre Royal in 1744, a larger theatre, erected on the same site in 1765, obtaining a royal patent in 1769 and thereafter being known correctly as the Theatre Royal. In 1766 the provincial theatre manager Tate WILKINSON took over the York CIRCUIT, which he modified to comprise York, Hull, Leeds, Doncaster, Wakefield, and Pontefract. The York theatre was remodelled in 1822, and further alterations were made to it on later occasions, so that little of the original remains. It became a theatre for touring companies in 1877 and in the 1900s was rebuilt to counteract the appeal of the new Grand Opera House (later the Empire) in New Clifford Street. In 1922 a repertory company was installed for several weeks during the summer, and from 1925 to 1929 a summer season, extending to 10 weeks, was given by a company led by Lena ASHWELL. From 1930 to 1932 the theatre was used for weekly rep. during several months each year, and in 1935 a permanent company was introduced with E. Martin BROWNE as producer. In 1951 he was also responsible for the revival of the York Cycle of MYSTERY PLAYS in the ruins of St Mary's Abbey, since performed regularly, usually at three-year intervals, as part of the York Festival. The Theatre Royal, which is now owned by the local authority and operated by a Trust, was extensively altered and redecorated in 1967 and still functions as a repertory theatre, though some touring companies visit it. It runs a Young People's Theatre Company which works with schools and youth clubs. Seating 930, it has a foyer wing suitable for the presentation of theatrical and other exhibitions.

YOUNG, CHARLES (?–1874), American actor born in Australia, who managed a theatre in Hobart before going to the United States. He appears to have been eccentric and undisciplined, though erratically good in tragedy. He was for a short time the husband of Mrs DUFF, though the marriage was never consummated, and he was also the first husband of Mrs Hermann VEZIN.

YOUNG, CHARLES MAYNE (1777–1856), English actor, who in 1798 made his début in Liverpool as Mr Green. From there he went to Manchester and Edinburgh, where he became an intimate friend of Sir Walter SCOTT. He was also friendly with the elder Charles MATHEWS, and with his help appeared at the HAYMARKET THEATRE in London in 1807 as Hamlet, Mathews playing Polonius. As one of the finest disciples of the KEMBLE school of acting, Young proved himself a tower of strength in support of such players as John Philip Kemble, Mrs SIDDONS, MACREADY, and Miss O'NEILL, and in 1822 was with Edmund KEAN at DRURY LANE. The only new part of any importance which he created was the title role in Miss Mitford's *Rienzi* (1828), but he was good and reliable in both comedy and tragedy, particularly as Hamlet, Macbeth, Iago in *Othello*, Falkland in SHERIDAN's *The Rivals*, Sir Pertinax McSycophant in MACKLIN's *The Man of the World*, Macheath in GAY's *The Beggar's Opera*, and the Stranger in Thompson's adaptation of KOTZEBUE's *Menschenhass und Reue*. He retired in 1832, appearing for the last time as Hamlet, again with Mathews as Polonius, Macready playing the Ghost.

YOUNG AMERICAN ROSCIUS, see COWELL, SAM.

YOUNG ROSCIUS, see BETTY, WILLIAM HENRY WEST.

YOUNG VIC THEATRE, London. The first theatre to be given this name was founded in 1945 as part of the new OLD VIC Drama School, with the idea of presenting adult actors in plays suitable for young people. Under George DEVINE, the company began in Dec. 1946 with GOZZI's *King Stag* (*Il Re Corvo*), and did important work, as did the drama school, developing a pioneering relationship between actor training and professional performance; but lack of influential support and financial stringency caused both projects to be abandoned in 1951.

The second Young Vic was founded in 1970 by Frank DUNLOP as part of the NATIONAL THEATRE, becoming an independent body in 1974. Erected on a bomb-site near the Old Vic, the theatre, which seats 456 on benches and in a narrow balcony running round three sides, with a long thrust stage penetrating deep into the auditorium, provides good theatre for young people at low prices, with emphasis on the text and the acting rather than on elaborate staging. The first production, on 15 Sept. 1970, was *Scapino*, an adaptation of MOLIÈRE's *Les Fourberies de Scapin* by Dunlop and Jim Dale, who played the title role. Later productions, given in an informal atmosphere which encourages audience participation, have included BECKETT's *Waiting for Godot* (1970), JONSON's *The Alchemist*, the musical *Joseph and the Amazing Technicolor Dreamcoat* (both 1972), and STOPPARD's *Rosencrantz and Guildenstern are Dead* (several productions beginning in 1973). The company has toured throughout Great Britain and visited festivals overseas. It also gives shows for younger children, and runs a studio theatre. Frank Dunlop has remained its director except from 1978 to 1980, when Michael Bogdanov took over, founding in 1978 the Young Vic Education Service, which a year later became the Young Vic Education and Community Service.

YOUTH THEATRE, see NATIONAL YOUTH THEATRE.

YOVKOV, YORDAN, see BULGARIA.

YUBANI, ZEF, see ALBANIA.

YUGOSLAVIA as it exists today is not only a relatively new creation, but multi-racial—consisting mainly of Slovenes, Croatians, Serbians, and Macedonians, each with its traditional drama and theatre techniques. There is therefore no single centre of theatrical activity, and important theatrical developments and productions may take place in one of several important towns, in each of which the history of the theatre may date back to Roman times. Among these towns are Belgrade, Zagreb, Ljubljana, Novi Sad, Skopje, Dubrovnik, and Split. Yet the area now covered by Yugoslavia as it was created in 1918 has a continuous and recognizable theatre history. Some theatres of the Graeco-Roman period still stand, and are occasionally used for performances even today. As in other European areas, a LITURGICAL DRAMA developed from the services of the Christian church, and gradually gave way to plays in the vernacular, heavily influenced by the indigenous folk-dramas which have their roots in prehistory. The old MYSTERY and MIRACLE PLAYS were produced side by side with secular comedies; not unnaturally it is the religious plays which are documented rather than the folk-plays and the first extant religious text from Croatia is believed to date back as far as 1492. It may even be considered the first prompt-book, with its text in black and its stage directions in red. Secular plays in the Croat language are known to have been given from 1527 onwards in Dubrovnik, and the pastorals and comedies of Marin Držič (1508–67), in their original form and in various adaptations and translations, are still performed in Yugoslavia, particularly his comedy *Uncle Maroje*, first seen in Dubrovnik in 1550 and revived in 1954. Although the first known dramatic performance in Slovene dates from about the 1660s—*A Play About Paradise*, seen in Ljubljana—there are scattered references to quasi-dramatic recitals before that, and in all parts of the region JESUIT DRAMA flourished. The longest extant dramatic text in the Slovene language dates from this time, a Capuchin PASSION PLAY by Father Romuald, which seems to have achieved its definitive form some time between 1721 and 1733 and was staged in Loka every Good Friday until 1765. It has one remarkable feature: the actors' parts are written in Slovene, the titles of individual scenes in Latin, and the stage-directions in German. There is also a record of a play in Serbian being acted in 1736 at Carlowitz, which then formed part of Austria-Hungary. This appears to have been a secular tragi-comedy by Emanuilo Kazačinski on the life of the last Serbian Tsar, Uroš V. But in that part of the country, as elsewhere under Turkish rule, the development of an indigenous theatre was hampered by the restrictions imposed by the authorities and in many places entertainment was limited to folk song and dance and the activities of strolling players, some of whom are depicted as early as 1317 in a fresco on the walls of a church near Kumanovo, in Macedonia. Nevertheless, all these scattered tendencies were to come to a head in the 19th century when determined efforts were made in several towns to establish some form of national theatre. These were to some extent inspired by the visits of the numerous French and German companies which were not only tolerated by the authorities from the mid-17th century onwards, but even actively encouraged, since it was thought they would help to satisfy and deflect the mimetic impulses which might else have agitated too strongly in favour of a vernacular theatre.

Some of the first performances in the Croat language were given in Zagreb in 1840 by actors from Novi Sad, who formed the first professional indigenous company in the country. They appeared in Belgrade in the following year, but in the absence of a theatre there had to act in a converted hall in the Customs House. The first workers' theatre was established in Idrija in 1850, plays in Slovene being given for the mining community there in a theatre dating from 1796. This is now a cinema, and is the second oldest theatre building in Yugoslavia, the oldest being a charming little playhouse dating from *c.* 1610 on the island of Hvar. It was not until 1861 that the theatre in Zagreb became a national institution, with a state subsidy, and Croatian drama developed rapidly under Dimitrije Demetar (1811–72), author of the historical play *Teuta* (1844), and August Šenoa (1838–81), whose comedy *Ljubica* dates from 1864. Of the Serbian plays of this time the most important are the satirical comedy *The Niggard, or Kir Janja* (1837) and *The Patriots* (1849) by Jovan Sterija Popovič (1806–56) and the patriotic tragedies *Maksim Crnojevič* and *Pera Segedinac* by Laza Kostič (1841–1910), published in 1866 and 1881 respectively. A Slovene tragedy, written jointly by Fran Levstik (1831–87) and Josip Jurčič (1844–81), has also been seen on the modern stage in an adaptation by Bratko Kreft dating from 1946. Many of the plays of this period, though successful in their time, were linked so closely with the struggle for independence that much of their appeal was lost with its achievement. But they helped to lay the foundations of the present theatre, and resulted in the establishment of the present Serbian National Theatre in Belgrade (which opened in 1869, and still stands, though rebuilt and modernized after damage in two world wars) and the Slovene theatre in Ljubljana, where after some 15 years of constant struggle one permanent playhouse was built and opened in 1892.

The years between the two world wars were extremely difficult for the theatre, with centralization in Belgrade causing financial hardships in the other two main centres, Zagreb and Ljubljana; there was also censorship to contend with. Standards of acting and production were however maintained and even improved on, and the repertory was widened to include such European authors as SHAW, IBSEN, and PIRANDELLO. The most important director of this period was undoubtedly Branko Gavella (1885–1962), who first worked in Zagreb in 1914 and later directed productions in all the theatres of Yugoslavia as well as in several towns abroad, among them Sofia, Prague, and Milan. Excellent work was also done in the 1930s by Bojan Stupica, who for the diversity and éclat of his productions was often compared to REINHARDT. The Serbian dramatist Branislav Nušič (1864–1938), who now has a permanent place in the Yugoslav repertory, continued into the 1930s the satiric tradition of Popovič with his comedies, among them the earlier *The Suspicious Person* (1887) and *The National Delegate* (1896) and the post-war *The Lady Minister* (1929) and *The Deceased* (1936). His last play, *Power*, which was left unfinished, was published in 1938, and sums up his satiric view of life, in which bourgeois opportunists scramble for easy jobs which give them disproportionate power, or *Vlast*, as the play is called in the vernacular.

The central figure of the new Slovene theatre was Ivan Cankar (1876–1918), who wrote in the midst of the disintegration of the Austro-Hungarian empire, and whose plays proceed from the realist social criticism of Ibsen to the SYMBOLISM of MAETERLINCK, showing a deeper insight into the emotional complexities of the subconscious than any other Yugoslav dramatist of the time. He wrote only six plays, of which the first was *Jakob Ruda* (1900) and the last *Beautiful Vida* (1912); all are infused with a poetic melancholy which raises them above mere social satire. The symbolism of Cankar was also to be found in the early work of the Croatian dramatist Miroslav Krleža (1893–1981), whose first cycle of plays, collected under the title *Legends*, dealt symbolically with such figures as Columbus and Michelangelo. These were followed by a realistic trilogy, *Golgota* (1920–23), showing the impact of the First World War on all types of Croatian people, and this again by another trilogy, *The Glembays* (1928–30), dealing with the inevitable disintegration of a Zagreb middle-class family from 1913 to 1925. The crowning achievement of Krleža's long career was *Aretheus* (1959), a 'fantasy in five scenes' considered by many Yugoslav critics to be the finest play staged in the country since the end of the Second World War.

Other Croatian dramatists of the early 20th century who deserve mention are Ivo Vojnovič (1857–1929), whose work links the patriotic drama of the mid-19th century with the lyric symbolism of the last years of the century, and whose best known play is *The Lady With the Parasol* (1912); Milan Begovič (1876–1948), a prolific author whose trilogy *God's Man* (1924), *Adventurer at the Door* (1926), and *Without the Third* (1931) has remained in the repertory; and Josip Kosor (1879–1961), whose first play *The Fire of Passion*, an expressionist peasant tragedy, was written in 1910 and seen in Germany and England some ten years before its first production in Yugoslavia in 1922.

In 1941 all theatrical life in Yugoslavia came to a halt with the German invasion, but after the liberation of the country new companies were quickly formed, and there was a rapid expansion, until in 1949 Yugoslavia, which in 1939 had had 12 professional theatres, had no fewer than 66. Many of these were in industrial towns and in remote places which had previously had no, or very little, drama. In Macedonia, for instance, where plays had been seen only sporadically between 1918 and 1941, there were now six theatres, the most important being in Skopje, one of which was Turkish. A Hungarian theatre was founded at Subotica in 1945; another at Bačka Topola a few years later. Other theatres which have opened since the Second World War include an Italian playhouse at Rijeka, a Romanian at Vršac, and a Shiptar at Priština. But the main centres for Serbian plays are still Belgrade and Novi Sad; for Croatian, Zagreb and Dubrovnik, the latter the home of an annual and highly successful summer Festival of the Arts; and for Slovene, Ljubljana. Theatres throughout Yugoslavia are technically well equipped and often reach a high standard in the production of indigenous, European, and American plays, but like so many countries today Yugoslavia is short of good new dramatists.

YURKA [Jurka], BLANCHE (1893–1974), American actress, who made her first appearance in New York in 1907 and was later with stock companies in Buffalo, Dallas and Philadelphia. She already had a long list of leading parts to her credit when in 1922 she played Gertrude to the Hamlet of John BARRYMORE. She later played roles in several plays by IBSEN: Gina in *The Wild Duck* (1928), Hedda in *Hedda Gabler*, Ellida in *The Lady from the Sea* (both 1929), and Hjordis in *The Vikings* (1930), in which she was outstanding. She was also much admired as ARISTOPHANES' Lysistrata later in 1930 and as SOPHOCLES' Electra in 1932. From 1936 to 1938 she toured the United States in a solo programme of scenes from great plays and later toured as Mrs Antrobus in WILDER's *The Skin of Our Teeth*, in a dramatization of Charlotte Brontë's *Jane Eyre* entitled *The Master of Thornfield* (1958), and as Miss Moffat in a revival of Emlyn WILLIAMS's *The Corn is Green* (1961). She made her only appearance in London in 1969 in GIRAUDOUX's *The Madwoman of Chaillot*. She was part-author of *Spring in Autumn* (1933), adapted from a play by MARTÍNEZ SIERRA, and wrote *Dear Audience—A Guide to Greater Enjoyment of the Theatre* (1958) and an autobiography, *Bohemian Girl* (1970).

Z

ZABŁOCKI, FRANCISZEK, see POLAND.

ZACCONI, ERMETE (1857–1948), Italian actor, child of strolling players, who made his first appearance on the stage at the age of seven. While still in his twenties he was recognized as one of the outstanding actors of the naturalistic school—VERISMO—equalled only by NOVELLI. In 1894 he formed his own company and, with DUSE, was instrumental in bringing IBSEN to the notice of the Italian public, being the first to play Oswald Alving in *Ghosts*, which he directed himself. He produced a number of translations of plays by TOLSTOY, HAUPTMANN, and STRINDBERG, and played most of the important roles in Shakespeare, going as far afield as Germany, Austria, and Russia. A most versatile actor, he was considered to be at his best as Hamlet and in the title role of Testoni's *Il Cardinale Lambertini* (1905), a comedy of intrigue based on the life of a 17th-century cardinal of Bologna.

ZACUTO, MOSES, see JEWISH DRAMA.

ZADEK, PETER, see GERMANY.

ZAKHAVA, BORIS EVGENEVICH (1898–1976), Soviet actor and director, pupil and close collaborator of VAKHTANGOV, in whose famous production of GOZZI's *Turandot* (1922) he played Timur. After Vakhtangov's death he joined MEYERHOLD's company, where in 1923 he directed OSTROVSKY's *Truth is Good, but Happiness is Better*. He was also responsible for the first productions of GORKY's *Yegor Bulychov and Others* (1932) and *Dostigayev and Others* (1933). Among his later productions, some of which were considered controversial, were *Hamlet* and DOSTOEVSKY's *The Idiot* in 1958. He then became head of the Shchukin School of Acting in Moscow.

ZAMFIRESCU, GEORGE MIHAIL, see ROMANIA.

ZAMORA, ANTONIO DE, see SPAIN.

ZANGWILL, ISRAEL (1864–1926), Jewish author, who established his reputation with a fine novel, *The Children of the Ghetto* (1892). He also wrote, in English, a number of plays on Jewish subjects, of which *The Melting Pot*, a study of Jewish immigrant life in the United States, was the most important. First produced in New York in 1908, it was a great success, and was several times revived, being first seen in London in 1914. Among his other plays were *The War God* (1911), *Too Much Money* (1918), and *We Moderns* (1923).

ZANNI, term used to describe collectively the servant masks of the COMMEDIA DELL'ARTE; the equivalent English term is Zany. Among them were ARLECCHINO, PANTALONE, and (later) Columbina, the prototypes of the English PANTOMIME players, HARLEQUIN, Pantaloon, and COLUMBINE. The English PUNCH, known in France as POLICHINELLE, also derives from a zany, PULCINELLA, as do three famous French comic characters, Scapin, Scaramouche, and Pasquin—from SCAPINO, SCARAMUCCIA, and PASQUINO respectively. Transported to France, PEDROLINO became a sad, solitary PIERROT, and in England a gregarious member of a seaside concert party, reaching his apotheosis in Pélissier's *Follies* and *The Co-Optimists*. Other minor *zanni* included BRIGHELLA and MEZZETINO. In Elizabethan times the word zany conveyed the idea of a 'clumsy imitator'.

ZAPOLSKA, GABRIELA, see POLAND.

ZARZUELA, Spanish musical play or operetta, which takes its name from the Palacio de la Zarzuela, a hunting-lodge in the woods not far from Madrid. Although many early Spanish plays contained a certain amount of music, the first true *zarzuela* was by CALDERÓN—*El laurel de Apolo* (1658). The music for this and other early *zarzuelas* has been lost, but in 1933 that for the first act of Calderón's *Celos aún del aire matan* (*Jealousy, Even of the Air, Kills*, 1660) was found and published, followed in 1945 by that for the second act, and recently much of the music for Juan Vélez de Guevara's *Los celos hacen estrellas* (*Jealousy Makes Stars*) has been identified. The *zarzuela* flourished as a courtly entertainment from 1660 onwards, its mythological or heroic plot being subtly designed to flatter its royal audience, but in the 18th century it suffered from the popularity of Italian opera. It was revitalized by Ramón de la CRUZ, who with *Las segadoras de Vallecas* (*The Reapers of Vallecas*) introduced a new-style *zarzuela* with a plot drawn from daily life. His work was immensely successful during his lifetime but after his death was again eclipsed by Italian opera, until in the second half of the 19th century a new era began with Barbieri's *Jugar con* *fuego* (*Playing With Fire*, 1851). This led to the opening in 1856 of a Teatro de la Zarzuela in Madrid. The modern *zarzuela*, exemplified by the ever-popular *La verbena de la Paloma* (*The Festival of la Paloma*), by Ricardo de la Vega and BRETÓN, persisted well into the present century and still exists, in a form strongly influenced by foreign revues and musicals.

ZAVADSKY, YURI ALEXEIVICH (1894–1977), Soviet actor and director, who began his career in 1915 as a pupil of VAKHTANGOV, appearing in his last production, GOZZI's *Turandot*, in 1922. In 1924 he opened his own studio, and was later appointed director of the Gorky State Theatre in Rostov. His first productions showed the influence of both Vakhtangov and STANISLAVSKY, but he soon evolved a personal style, combining lyricism with excellent stagecraft, and giving a larger place than usual to music. Many outstanding Soviet actors were his pupils, and GROTOWSKI attended his classes in play production at the Moscow Theatre Institute. During the Second World War he directed TRENEV's *Lyubov Yarovaya* and *On the Banks of the Neva*, GORKY's *Enemies* and *The Philistines*, and *Othello* in Rostov, after going briefly to the MOSSOVIET THEATRE in 1940. He returned there after the war, to contribute not a little to the forward movement of the Soviet theatre during the 1950s. Two of the finest of his later productions were *The Merry Wives of Windsor* in 1947 and a revival of LERMONTOV's *Masquerade* in 1963.

ZEAMI, see JAPAN.

ZEFFIRELLI, FRANCO (1923–), Italian director and scene designer, who first attracted attention in his own country with his sets for VISCONTI's productions of *Troilus and Cressida* in Florence in 1949 and CHEKHOV's *Three Sisters* in Rome in 1951. He was already well known in the world of opera when in 1960 he directed and designed *Romeo and Juliet* at the OLD VIC, the production being seen in New York in 1962. He then directed and designed *Othello* at the ROYAL SHAKESPEARE THEATRE (1961); DUMAS *fils*'s *The Lady of the Camellias* (1963) in New York; *Hamlet* (*Amleto*) (1964), performed by an Italian company at the Old Vic; and *Much Ado About Nothing* (1965) for the NATIONAL THEATRE company at the same theatre. His other work has included MILLER's *After the Fall* (Rome, 1964) and ALBEE's *Who's Afraid of Virginia Woolf?* (Paris, 1964; Milan, 1965) and *A Delicate Balance* (Rome, 1967). He brought the work of Eduardo DE FILIPPO to the notice of a wide audience with his production and décor for the National Theatre in 1973 (N.Y., 1974) of *Saturday, Sunday, Monday*—a piece ideally suited to his distinctive brand of scrupulous but highly decorative realism. His staging of the same author's *Filumena* was seen in the West End in 1977.

ZEMACH, NAHUM, see HABIMAH THEATRE and HEBREW THEATRE, NEW.

ZEYER, JULIUS, see CZECHOSLOVAKIA.

ZIDAROV, KAMEN, see BULGARIA.

ZIEGFELD, FLORENZ (1867–1932), American theatre manager, who perfected the American form of REVUE in a series entitled the *Ziegfeld Follies,* which began in 1907 and continued almost annually until his death, being seen intermittently thereafter until 1957. Ziegfeld based his show on that of the FOLIES-BERGÈRE in Paris, with the emphasis on scenic splendour, comic sketches, VAUDEVILLE specialities, and beautiful girls, many of whom, including Marion Davies, Irene Dunne, and Paulette Goddard, later became film stars. He was also responsible for fostering the talents of such light entertainers as Fanny Brice, W. C. FIELDS, Eddie Cantor, and Bert WILLIAMS, and many famous foreign artistes appeared in New York under his management, including Maurice Chevalier and Evelyn LAYE. Among the many musicals and straight plays which he put on the most successful was *Show Boat* (1927; London, 1928), based on a novel by Edna Ferber and first produced at the ZIEGFELD THEATRE. Ziegfeld's first wife was the actress and singer Anna Held (1873–1918), who made her first appearance in New York in 1896 and was seen mainly in MUSICAL COMEDY. In 1914 he married Billie [Ethelbert Appleton] Burke (1884–1970), who first appeared in London in 1902 and in 1910 went to New York, where she had a long and successful career, mainly in light comedy.

ZIEGFELD THEATRE, New York, on the northwest corner of 6th Avenue and 54th Street, designed for Florenz ZIEGFELD by the scene designer Joseph URBAN, with a handsome auditorium, a fine, stark exterior, and lavish backstage space, and opened successfully on 2 Feb. 1927 with the musical *Rio Rita*. This was followed by *Show Boat*, based by Oscar Hammerstein on Edna Ferber's novel, with music by Jerome Kern. Later successes were COWARD's *Bitter Sweet* in 1929 and the *Ziegfeld Follies* of 1931. After Ziegfeld's death in 1932 his theatre became a cinema, but on 7 Dec. 1944 it was reopened by Billy Rose, under its old name, with the revue *Seven Lively Arts* starring Beatrice LILLIE. Carol Channing appeared there as Lorelei Lee in the musical version of *Gentlemen Prefer Blondes* (1949), and in 1953 the theatre housed a revival of Gershwin's opera *Porgy and Bess*. From 1955 to 1963 it was used for television, but returned to live theatre again on 29 Jan. 1963 with an evening of songs and sketches starring Maurice Chevalier, followed by Bert LAHR in *Foxy* (1964), a musical based on JONSON's *Volpone*. The building was demolished in 1967, to be replaced in 1969 by a cinema erected slightly to the west of the original site.

ZIEGLER, CLARA (1844–1909), German actress, who from 1867 to 1868 was in LEIPZIG and later went to MUNICH for some years as a guest artist. At her best in tragic roles, such as SCHILLER's Joan of Arc, GRILLPARZER's Penthesilea, or EURIPIDES' Medea, she also, in the tradition of her day, played Romeo, and wrote a number of plays. She bequeathed her house and library to the city of Munich as a theatre museum which was formally opened in 1910, its curator for many years being the German theatre historian Franz Rapp.

ZOLA, ÉMILE-EDOUARD-CHARLES-ANTOINE (1840–1902), French novelist and dramatist, who, though mainly remembered for his novels, in particular the twenty volumes of *Les Rougon-Macquart* (1871–93), had an important influence on the development of the French, and so of the European, theatre of his day. As the leader of the naturalistic school of literature, he much disliked the facile, optimistic works of such dramatists as SCRIBE, and thought that a play should be a 'slice of life', thrown on the stage without embellishment or artifice. He set out his theories in a number of critical articles, later republished in two volumes, *Le Naturalisme au théâtre* (1878) and *Nos auteurs dramatiques* (1881), and exemplified them in his own plays, particularly *Thérèse Raquin* (1873), which he based on an earlier novel of the same name. In an English translation, this was seen in London in 1891 and has been several times revived, notably in 1938 as *Thou Shalt Not* . . . In 1945 it was produced in New York as *Therese*. Zola's other plays were less successful. The only other one to have been translated into English is a comedy, *Les Héritiers Rabourdin* (1874), which as *The Heirs of Rabourdin* was produced in 1894 by GREIN's Independent Theatre Club. One of Zola's novels, *L'Assommoir*, was dramatized by Busnach and Gastineau, and in an English version, *Drink*, by Charles READE and Charles WARNER, had a long run in London in 1879, Warner playing the chief part, Coupeau.

ZORRILLA, FRANCISCO DE ROJAS, see ROJAS ZORRILLA.

ZORRILLA Y MORAL, JOSÉ (1817–93), Spanish poet and dramatist of the Romantic movement. His best known play is *Don Juan Tenorio* (1844), which added yet another version to the numerous interpretations of the DON JUAN legend. It is still performed in Spanish theatres, traditionally on All Saints' Day (1 Nov.). Zorrilla is also the author of *El zapatero y el rey* (*The Cobbler and the King*, 1840) and of *Traidor, inconfeso y martír* (1849), based on the life of a pretender to the throne of Portugal. Many of his works are among the foremost literary productions of Spanish Romanticism.

ZUCKMAYER, CARL (1896–1977), German dramatist, who under the influence of EXPRESSIONISM wrote *Kreuzweg* (*Crossroads*, 1920) and *Pankraz erwacht* (*Pankraz Wakes Up*, 1925). These were unsuccessful, and it was only when he turned to REALISM that he had his first resounding success with a comedy, *Der fröhliche Weinberg* (*The Merry Vineyard*, 1926). This was followed by *Schinderhannes* (1927), based on a popular folk-figure of the Napoleonic Wars; *Katharina Knie* (1929), set in a travelling circus; and by the first of his international successes, *Der Hauptmann von Köpenick* (1931). This comedy of a Berlin ex-convict shoemaker who enforces obedience to his orders merely because he is wearing military uniform was seen in London in 1971 at the NATIONAL THEATRE with Paul SCOFIELD. Zuckmayer left Germany in 1933, when because of his outspoken opposition to National Socialism his plays were banned by the authorities, and while in Switzerland wrote *Der Schelm von Bergen* (1934) and *Bellman* (1938). The first play he wrote after settling in the U.S. in 1939 was again an international success, *Des Teufels General*, about Nazi Germany and the problem of resistance, first seen in Zürich in 1946. As *The Devil's General* it was successfully produced in New York and London in 1953. Among Zuckmayer's later plays, none of which achieved a comparable success, were *Der Gesang im Feuerofen* (*The Song in the Fiery Furnace*, 1950), set in occupied France, *Das kalte Licht* (1956), based on the case of the traitorous nuclear physicist Dr Fuchs, *Die Uhr Schlägt eins* (1961), and *Der Rattenfänger* (*The Pied Piper*, 1974).

ZÜRICH, Swiss city whose theatre first emerged from provincial obscurity between 1901 and 1929, mainly through the efforts of Alfred Reucker (1868–1958). To keep abreast of developments in BERLIN and MUNICH, he introduced into the Schauspielhaus a succession of exquisitely mounted revivals of outstanding plays, including those of IBSEN and HOFMANNSTHAL, rather in the manner of REINHARDT, who regularly provided him with guest artists and productions. In 1933 a number of leading actors, fleeing from Hitler's Germany, became available to him, and a gifted company was established which appeared in a mixture of classic plays, notably those of SCHILLER, and such modern works as WOLF's *Professor Mamlock* and BRÜCKNER's *Rassen*. In 1938, when Oscar Wälterlin succeeded Reucker, with Kurt HIRSCHFELD as his assistant, another era of outstanding productions began, with a range of new plays which later formed the basis of many Austrian and West German repertories in the post-war years. It included first productions of BRECHT's *Mutter Courage und ihre Kinder* (1941), *Der gute Mensch von Sezuan* and *Leben des Galilei* (both 1943), and of ZUCKMAYER's *Des Teufels General* (1946). After 1945 many actors returned to West Germany, but the Zürich

Schauspielhaus continued to flourish, giving the first performances of the plays of FRISCH and DÜRRENMATT until Wälterlin's death in 1961 ended an era.

ZWEIG, STEFAN (1881–1942), Austrian poet, novelist, and dramatist. His early plays, which included *Tereites* (1907) and *Das Haus am Meer* (1912), were unremarkable, but were followed by a major theatrical work, *Jeremias* (*Jeremiah*, 1917), a denunciation of war which ends with the destruction of Jerusalem. It was followed by *Legende eines Lebens* (*The Legend of a Life*, 1919) and *Das Lamm des Armen* (*The Ewe-Lamb*, 1929), which exposes the private faces of public figures. A translation of JONSON's *Volpone* in 1926 had a successful run. Zweig committed suicide while living in exile in Brazil.

A GUIDE TO FURTHER READING

Publishers are located in Great Britain unless otherwise stated.

REFERENCE WORKS

Encyclopedias

ADAMS, W. Davenport: *A dictionary of the drama*. Vol. I, A to G. New York: Franklin, 1964. (Never completed)

Enciclopedia dello Spettacolo. 9 vols. Rome: Maschere, 1954–64. *Aggiornamento, 1955–1965*. Rome: Unione Editoriale, 1966.

ESSLIN, Martin, ed.: *Illustrated encyclopaedia of world theatre*. Thames and Hudson, 1977.

GASSNER, John, and QUINN, Edward, eds.: *The reader's encyclopedia of world drama*. New York: Crowell, 1969. Methuen, 1975.

MATLAW, M.: *Modern world drama*. Secker, 1972.

MELCHINGER, Siegfried: *The concise encyclopedia of modern drama*. New York: Horizon, 1966. Vision.

TAYLOR, John Russell: *The Penguin dictionary of the theatre*. Methuen, 1967. New York: Barnes and Noble.

VAUGHN, J. A.: *Drama A to Z*. New York: Ungar, 1979.

General Bibliographies

ARNOTT, J. F., and ROBINSON, J. W.: *English theatrical literature, 1559–1900*. Society for Theatre Research, 1970.

BAKER, Blanch M.: *Theatre and allied arts*. New York: Blom, 1967.

CHESHIRE, David: *Theatre: history, criticism, and reference*. Bingley, 1967. Hamden, Conn.: Archon.

LITTO, Fredric M.: *American dissertations on the drama and theatre*. Kent, Ohio: Kent State U.P., 1969.

LOEWENBERG, A.: *The theatre of the British Isles, excluding London*. Society for Theatre Research, 1950.

STRATMAN, C. J.: *A bibliography of American theatre, excluding New York City*. Chicago: Loyola U.P., 1965. *American theatrical periodicals, 1798–1967: a bibliographical guide*. Durham, N. C.: Duke U. P., 1970. *Britain's theatrical periodicals, 1720–1967: a bibliography*. 2nd edn. New York: New York Public Library, 1972.

Dictionaries and Glossaries

BAND-KUZMANY, Karin R. M., comp.: *Glossary of the theatre in English, French, Italian, and German*. Amsterdam: Elsevier, 1969.

BOWMAN, Walter P., and BALL, Robert Hamilton: *Theatre language. A dictionary of terms in English of the drama and stage from medieval to modern times*. New York: Theatre Arts, 1961.

GRANVILLE, Wilfred: *A dictionary of theatrical terms*. Deutsch, 1952. New York: O.U.P., as *Theatre language*.

RAE, Kenneth, and SOUTHERN, Richard, eds.: *International vocabulary of technical theatre terms in eight languages*. Brussels: Elsevier, 1959. Reinhardt. New York: Theatre Arts.

Biographical Dictionaries

HERBERT, Ian, ed.: *Who's who in the theatre*. 17th edn. 2 vols. Vol. 1 Biographies; Vol. 2 Playbills. Detroit: Gale, 1981. (Earlier edns, 1912–77, pub. by Pitman.)

NUNGEZER, Edwin: *A dictionary of actors and other persons associated with the public representation of plays in England before 1642*. New Haven, Conn.: Yale U.P., 1929. O.U.P.

RIGDON, Walter, ed.: *The biographical encyclopedia and who's who of the American theatre*. New York: J. H. Heineman, 1966.

Who was who in the theatre: 1912–1976. 4 vols. Detroit: Gale, 1978.

GENERAL HISTORIES

ALTMAN, George, *et al*.: *Theater pictorial*. Berkeley: California U.P., 1953.

BERTHOLD, Margot: *A history of world theater*. New York: Ungar, 1972.

BROCKETT, Oscar G.: *History of the theatre*. 3rd edn. Boston, Mass.: Allyn and Bacon, 1977. *The theatre: an introduction*. 4th edn. New York: Holt, Rinehart, 1979.

FREEDLEY, George, and REEVES, John A.: *A history of the theatre*. 3rd edn. New York: Crown, 1968.

GASCOIGNE, Bamber: *World theatre*. Ebury Press, 1968. Boston: Little, Brown.

HARTNOLL, Phyllis: *A concise history of the theatre*. Thames and Hudson, 1968. New York: Abrams.

NAGLER, A. M.: *A source book in theatrical history*. New York: Dover, 1959.

NICOLL, Allardyce: *The development of the theatre*. 5th edn. Harrap, 1966. *World drama from Aeschylus to Anouilh*. 2nd edn. Harrap, 1976.

SOUTHERN, Richard: *Seven ages of the theatre*. 2nd edn. Faber, 1964.

The Modern Drama and Theatre

BENTLEY, Eric: *The playwright as thinker*. New York: Meridian, 1955. Methuen.

BRUSTEIN, Robert: *The theatre of revolt*. Boston: Little, Brown, 1964. Methuen, 1965.

CLARK, Barrett H.: *A study of the modern drama*. Rev. edn. New York: Appleton, 1938.

CLARK, Barrett H., and FREEDLEY, George, eds.: *A history of modern drama*. New York: Appleton, 1947.

ESSLIN, Martin: *The theatre of the absurd*. 3rd edn. Methuen, 1974.

GASSNER, John: *Directions in modern theatre and drama*. Rev. edn. New York: Holt, Rinehart, 1965.

GORELIK, Mordecai: *New theatres for old*. Dobson, 1947. New York: Dutton, 1962.

HAYMAN, Ronald: *Theatre and anti-theatre: new movements since Beckett*. Secker, 1979. New York: O.U.P.

ROOSE-EVANS, James: *Experimental theatre from Stanislavsky to Brecht*. Studio Vista, 1970.

STYAN, J. L.: *Modern drama in theory and practice*. 3 vols. Cambridge U.P., 1981.

VALENCY, M.: *The end of the world: an introduction to contemporary drama*. O.U.P., 1980.

WILLIAMS, Raymond: *Drama from Ibsen to Brecht.* 2nd rev. edn. Penguin, 1976.

CLASSICAL DRAMA AND THEATRE

Ancient Greece

ARNOTT, Peter: *An introduction to the Greek theatre.* Macmillan, 1959. New York: St Martin's.

BIEBER, Margarete: *History of the Greek and Roman theatre.* 2nd edn. O.U.P., 1961. Princeton U.P.

HAIGH, Arthur E. *The Attic theatre: a description of the stage and theatre of the Athenians, and of the dramatic performances in Athens.* New York: Kraus Reprint, 1969.

KITTO, H. D. F.: *Greek tragedy.* Methuen, 1961. New York: Barnes and Noble.

KOTT, Jan: *The eating of the gods: an interpretation of Greek tragedy.* Eyre Methuen, 1974.

PICKARD-CAMBRIDGE, A. W.: *The theatre of Dionysus in Athens.* O.U.P., 1946. *Dramatic festivals of Athens.* 2nd edn. O.U.P., 1968.

TAPLIN, O. P.: *Greek tragedy in action.* Methuen, 1978. Berkeley: California U.P., 1979.

WEBSTER, T. B. L.: *Greek theatre production.* 2nd edn. Methuen, 1970. New York: Barnes and Noble. *Greek tragedy.* O.U.P., 1971.

Ancient Rome

BEARE, W.: *The Roman stage: a short history of Latin drama in the time of the Republic.* 3rd edn. Methuen, 1978. Totowa, N.J.: Rowman and Littlefield.

DUCKWORTH, G. E.: *The nature of Roman comedy.* Princeton U.P., 1952. O.U.P.

GREAT BRITAIN

General Histories and Studies

BRIDGES-ADAMS, W.: *The irresistible theatre* (ending at the Commonwealth). Secker, 1957.

CLUNES, Alec: *The British theatre.* Cassell, 1964. New York: Barnes.

EVANS, Ifor: *A short history of English drama.* 2nd edn. Macgibbon and Kee, 1965.

HARBAGE, A.: *Annals of English drama, 975–1700.* Rev. edn. Methuen, 1964. Philadelphia: Pennsylvania U.P.

MANDER, Raymond, and MITCHENSON, Joe: *Picture history of the British theatre.* Hulton, 1957.

NICOLL, Allardyce: *History of English drama, 1600–1900.* 6 vols. Cambridge U.P., 1952–9. *English drama, 1900–1930.* Cambridge U.P., 1973. *British drama.* 6th edn, rev. by J. C. Trewin. Harrap, 1978. New York: Harper and Row, 1979.

ROOSE-EVANS, James: *London theatres from the Globe to the National.* Phaidon, 1977.

SPEAIGHT, George: *Punch and Judy: a history.* Studio Vista, 1970.

Medieval and Tudor

BOAS, F. S.: *An introduction to Tudor drama.* O.U.P., 1933. New York: A.M.S. Press, 1978.

CHAMBERS, E. K.: *The medieval stage.* 2 vols. O.U.P., 1903

KAHRL, S. J.: *Traditions of medieval English drama.* Hutchinson, 1974.

NAGLER, A. M.: *The medieval religious stage: shapes and phantoms.* New Haven: Yale U.P., 1976.

ROSSITER, A. P.: *English drama from early times to the Elizabethans.* Hutchinson, 1966. Norwood, Pa.: Norwood Edns, 1978.

SOUTHERN, Richard: *The staging of plays before Shakespeare.* Faber, 1973. New York: Theatre Arts. *The medieval theatre in the round.* 2nd edn. Faber, 1975.

STRATMAN, C. J.: *Bibliography of medieval drama.* 2nd edn. New York: Ungar, 1972.

WICKHAM, Glynne: *Early English stages 1300 to 1660: a history of the development of dramatic spectacle and stage convention in England.* 3 vols. Vol. 1 1300–1576 2nd edn, 1980; Vol. 2 1576–1660 Part 1, 1963, Part 2, 1972; Vol. 3, 1980. Routledge. New York: Columbia U.P. *The medieval theatre.* Weidenfeld, 1980.

Elizabethan and Jacobean

BENTLEY, G. E.: *The Jacobean and Caroline stage.* 7 vols. O.U.P., 1941–8.

BRADBROOK, M. C.: *History of Elizabethan drama.* 6 vols. Vol. 1 *Themes and conventions of Elizabethan tragedy*; Vol. 2 *The growth and structure of Elizabethan comedy*; Vol. 3 *The rise of the common player: a study of actor and society in Shakespeare's England*; Vol. 4 *Shakespeare and the Elizabethan poetry: a study of his earlier work in relation to the poetry of the time*; Vol. 5 *Shakespeare the craftsman*; Vol. 6 *The living monument: Shakespeare and the theatre of his time.* Cambridge U.P., 1979.

CHAMBERS, E. K.: *The Elizabethan stage.* 4 vols. O.U.P., 1923.

GREG, W. W.: *Dramatic documents from the Elizabethan playhouses: stage plots, actors' parts, prompt books, reproductions and transcripts.* O.U.P., 1931.

JOSEPH, Bertram: *Elizabethan acting.* 2nd edn. O.U.P., 1964.

KNIGHTS, L. C.: *Drama and society in the age of Jonson.* Chatto, 1957.

LAWRENCE, W. J.: *The Elizabethan playhouse, and other studies.* 2 vols. New York: Russell, 1963.

RIBNER, I.: *Jacobean tragedy.* Methuen, 1979. Totowa, N.J.: Rowman and Littlefield.

WILSON, F. P.: *The English drama, 1485–1585.* O.U.P., 1969.

Shakespearean

BRADBROOK, M. C.: *History of Elizabethan drama,* vols. 4–6 (see above).

CAMPBELL, O. J., and QUINN, Edward, eds.: *The reader's encyclopedia of Shakespeare.* New York: Crowell, 1966. Methuen.

GURR Andrew: *The Shakespearean stage, 1574–1642.* O.U.P., 1970.

HALLIDAY, F. E.: *A Shakespeare companion.* Rev. edn. Duckworth, 1977.

HARBAGE, A.: *Shakespeare's audience.* New York: Columbia U.P., 1941. O.U.P.

KING, T. J.: *Shakespearean staging, 1599–1642.* Cambridge, Mass.: Harvard U.P., 1971.

KOTT, Jan: *Shakespeare, our Contemporary.* 2nd edn. Methuen, 1967.

NAGLER, A. M.: *Shakespeare's stage.* Enlarged edn. New Haven: Yale U.P., 1981.

ODELL, G. C. D.: *Shakespeare from Betterton to Irving.* 2 vols. New York: Blom, 1964. Constable.

PAYNE, W. R. N.: *A Shakespeare bibliography.* Library Association, 1969.

SCHOENBAUM, S.: *Shakespeare's lives.* O.U.P., 1970. *William Shakespeare: a documentary life.* O.U.P., 1975. *Shakespeare, the Globe and the world.* O.U.P., 1979.

SHATTUCK, C. H.: *Shakespeare on the American stage: from the Hallams to Edwin Booth.* Washington, D.C.: Folger, 1976.

SPRAGUE, A. C.: *Shakespeare and the actors: the stage business in his plays.* Cambridge,

Mass.: Harvard U.P., 1944. *Shakespearian players and performances*. Cambridge, Mass.: Harvard U.P., 1953. Black, 1954.

SPRAGUE, A. C., and TREWIN, J. C.: *Shakespeare's plays today: some customs and conventions of the stage*. Sidgwick, 1970.

STYAN, J. L.: *The Shakespeare revolution: criticism and performance in the twentieth century*. Cambridge U.P., 1977.

TREWIN, J. C.: *Shakespeare on the English stage, 1900–1964*. Barrie, 1964. New York: Humanities. *Going to Shakespeare*. Allen and Unwin, 1978.

WAIN, John: *The living world of Shakespeare: a playgoer's guide*. Macmillan, 1980. New York: St Martin's.

Caroline and Restoration

BENTLEY, G. E.: *The Jacobean and Caroline stage*. 7 vols. O.U.P., 1941–68.

CLINTON-BADDELEY, V. C.: *The burlesque tradition in the English theatre after 1660*. Methuen, 1952.

DANCHIN, P.: *The prologues and epilogues of the Restoration (1660–1700). A tentative check-list*. Nancy: 1978.

HOTSON, Leslie: *The Commonwealth and Restoration stage*. New York: Russell, 1962.

HUGHES, Leo: *A century of English farce* (Restoration to mid eighteenth century). Princeton U.P., 1956. O.U.P.

HUME, R. D.: *The development of English drama in the late seventeenth century*. O.U.P., 1976.

ROTHSTEIN, E.: *Restoration tragedy*. Westport, Conn.: Greenwood Press, 1978.

SMITH, Dane Farnsworth: *Plays about the theatre in England, 1671–1737*. O.U.P., 1936.

SUMMERS, Montague: *The Restoration theatre: an account of the staging of plays*. Routledge, 1934. New York: Humanities, 1964. *The playhouse of Pepys, a description of the drama produced in the years 1660–82*. Routledge, 1935. New York: Humanities, 1964.

Eighteenth Century

BATESON, F. W.: *English comic drama, 1700–1750*. O.U.P., 1929. New York: Russell, 1963.

BERNBAUM, E.: *The drama of sensibility: a sketch of the history of English sentimental comedy and domestic tragedy*. Cambridge, Mass.: Harvard U.P., 1925. O.U.P.

CLINTON-BADDELEY, V. C.: *All right on the night*. Putnam, 1954.

KRUTCH, Joseph Wood: *Comedy and conscience after the Restoration*. Rev. edn. New York: Columbia U.P., 1949.

LOFTIS, John: *The politics of drama in Augustan England*. O.U.P., 1963. *Sheridan and the drama of Georgian England*. Blackwell, 1976.

NICOLL, Allardyce: *The Garrick stage: theatres and audience in the eighteenth century* (ed. Sybil Rosenfeld). Manchester U.P., 1980.

PRICE, C.: *Theatre in the age of Garrick*. Blackwell, 1973.

RICHARDS, K., and THOMSON, P.: *Essays on the eighteenth-century English stage*. Methuen, 1972.

ROSENFELD, Sybil: *Strolling players and drama in the provinces, 1660–1765*. Cambridge U.P., 1939. *Temples of Thespis: some private theatricals in England and Wales, 1700–1820*. Society for Theatre Research, 1978.

SOUTHERN, Richard: *The Georgian playhouse*. Pleiades Books, 1948.

VAN LENNEP, William, *et al.*, eds.: *The London stage, 1660–1800*. 11 vols. Part I 1660–

1700; Part II (2 vols.) 1700–1729; Part III (2 vols.) 1729–1747; Part IV (3 vols.) 1747–1776; Part V (3 vols.) 1776–1800. Carbondale, Ill.: Southern Illinois U.P., 1960–8. *Index to the London stage, 1660–1880*. Carbondale, Ill.: Southern Illinois U.P., 1979.

Nineteenth Century

BAKER, M.: *The rise of the Victorian actor*. Croom, 1978. Totowa, N.J.: Rowman and Littlefield.

BOOTH, Michael: *English melodrama*. Jenkins, 1965.

BUSBY, Roy: *British music hall: an illustrated who's who from 1850 to the present day*. Elek, 1976.

CONOLLY, L. W., and WEARING, J. P.: *English drama and theatre, 1800–1900: a guide to information sources*. Detroit: Gale, 1978.

DISHER, M. Willson: *Blood and thunder: mid-Victorian melodrama and its origins*. Rockliff, 1949. *Melodrama*. Rockliff, 1954.

MANDER, Raymond, and MITCHENSON, Joe: *British music hall*. Rev. edn. Gentry Books, 1974.

MAYER, David: *Harlequin in his element: the English pantomime, 1806–1836*. O.U.P., 1969.

REYNOLDS, Ernest: *Early Victorian drama, 1830–70*. Heffer, 1936. New York: Blom, 1965.

RICHARDS, K., and THOMSON, P.: *Essays in nineteenth-century British theatre*. Methuen, 1971.

ROBINSON, Henry Crabb: *The London theatre, 1811–1866*. Society for Theatre Research, 1966.

ROWELL, George: *The Victorian theatre, 1792–1914*. 2nd edn. Cambridge U.P., 1978. *Queen Victoria goes to the theatre*. Elek, 1978. *Theatre in the age of Irving*. Blackwell, 1981.

SCOTT, Clement: *The drama of yesterday and today*. 2 vols. Macmillan, 1899.

SHAW, George Bernard: *Our theatres in the nineties*. 3 vols. Constable, 1932.

SOUTHERN, Richard: *The Victorian theatre: a pictorial survey*. David and Charles, 1970.

WILSON, A. E.: *The story of pantomime*. Home and Van Thal, 1949.

Twentieth Century

AGATE, James: *A short view of the English stage, 1900–1926*. Herbert Jenkins, 1926. New York: Blom, 1969.

ARMSTRONG, William A.: *Experimental theatre*. Bell, 1963.

BEAUMAN, Sally: *The Royal Shakespeare Company: a history of ten decades*. O.U.P., 1982.

ELSOM, J.: *Post-war British theatre*. Rev. edn. Routledge, 1979.

ELSOM, J., and TOMALIN, N.: *The history of the National Theatre*. Cape, 1978.

HAYMAN, Ronald: *British theatre since 1955*. O.U.P., 1979.

KITCHIN, Laurence: *Mid-century drama*. Faber, 1960. *Drama in the sixties*. Faber, 1966.

MAROWITZ, Charles, and TRUSSLER, Simon, eds.: *Theatre at work*. Methuen, 1967. New York: Hill and Wang, 1968.

REYNOLDS, Ernest: *Modern English drama*. Harrap, 1949. Norman: Oklahoma U.P.

TAYLOR, John Russell: *The rise and fall of the well-made play*. Methuen, 1967. New York: Hill and Wang. *Anger and after: a guide to the new British drama*. 2nd edn rev. Eyre Methuen, 1977. New York: Hill and Wang, 1969, as *The angry theatre*. *The second wave: British drama of the sixties*. Eyre Methuen, 1978.

TREWIN, J. C.: *The theatre since 1900*. Dakers, 1951. *The Edwardian theatre*. Blackwell, 1976.

TREWIN, J. C., ed.: *Theatre programme*. Muller, 1954.

TYNAN, Kenneth: *A view of the English stage, 1944–63*. Davis-Poynter, 1975.

WILSON, A. E.: *Edwardian theatre*. Barker, 1951.

IRELAND

BELL, S. H.: *The theatre in Ulster*. Dublin: Gill and Macmillan, 1972.

CLARK, William Smith: *The early Irish stage*. O.U.P., 1955. *The Irish stage in the county towns, 1720–1800*. O.U.P., 1965.

ELLIS-FERMOR, Una: *The Irish dramatic movement*. 2nd edn. Methuen, 1954. New York: Barnes and Noble.

HOGAN, Robert: *After the Irish renaissance: a critical history of the Irish drama since 'The Plough and the Stars'*. Minneapolis: Minnesota U.P., 1967. Macmillan, 1968.

HOGAN, Robert, and KILROY, James: *The modern Irish drama: a documentary history*. 4 vols. Dublin: Dolmen Press, 1975–9.

HUNT, Hugh: *The Abbey: Ireland's National Theatre, 1904–1978*. Dublin: Gill and Macmillan, 1979.

KAVANAGH, Peter: *The Irish theatre*. Tralee: Kerryman, 1946. New York: Blom, 1969.

ROBINSON, Lennox: *Ireland's Abbey Theatre*. Sidgwick, 1951.

EUROPE

France

ARNOTT, Peter: *An introduction to the French theatre*. Macmillan, 1977.

CARLSON, Marvin: *The theatre of the French revolution*. Ithaca: Cornell U.P., 1966.

FOWLIE, Wallace: *Dionysus in Paris* (on contemporary French theatre). Magnolia, Mass.: Smith, 1961. Gollancz.

FRANK, G. *Medieval French drama*. O.U.P., 1954.

GUICHARNAUD, Jacques: *Modern French theatre from Giraudoux to Genet*. Rev. edn. New Haven, Conn.: Yale U.P., 1975.

HAWKINS, Frederick W.: *Annals of the French stage from its origin to the death of Racine*. 2 vols. Chapman and Hall, 1884. New York: Greenwood, 1969. *The French stage in the eighteenth century*. 2 vols. Chapman and Hall, 1888. New York: Greenwood, 1969.

HOBSON, Harold: *The French theatre of today: an English view*. Harrap, 1953. New York: Blom. *French theatre since 1830*. Calder, 1978.

JEFFERY, Brian: *French Renaissance comedy, 1552–1630*. O.U.P., 1969.

KNOWLES, Dorothy: *French drama of the inter-war years, 1918–39*. Harrap, 1967. New York: Barnes and Noble, 1968.

LANCASTER, H. C.: *A history of French dramatic literature, 1610–1700*. 9 vols. Baltimore: Johns Hopkins, 1929–42. New York: Gordian, 1966. *Sunset: a history of Parisian drama in the last years of Louis XIV, 1701–1715*. Baltimore: Johns Hopkins, 1945. *French tragedy in the time of Louis XV and Voltaire, 1715–74*. 2 vols. Baltimore: Johns Hopkins, 1950. *French tragedy in the reign of Louis XVI and the early years of the French Revolution, 1774–92*. Baltimore: Johns Hopkins, 1953.

LAWRENSON, T. E.: *The French stage in the seventeenth century: a study in the advent of the Italian order.* Manchester U.P., 1957.

LOUGH, John: *Paris theatre audiences in the seventeenth and eighteenth centuries.* O.U.P., 1957. *Seventeenth-century French drama: the background.* O.U.P., 1979.

WICKS, C. Beaumont, ed.: *The Parisian stage.* 4 vols. University: Alabama U.P., 1950–67.

WILEY, William: *The early public theatre in France.* Cambridge, Mass.: Harvard U.P., 1960.

Germany

BRAUN, Walter: *Theater in Deutschland.* Munich: Bruckmann, 1956.

BRUFORD, W. H.: *Theatre, drama and audience in Goethe's Germany.* New York: Hillary, 1950. Routledge.

GARTEN, H. F.: *Modern German drama.* 2nd edn. Methuen, 1962. New York: O.U.P., 1964.

GAY, Peter: *Weimar culture: the outsider as insider.* Secker, 1969.

HEITNER, Robert R.: *Modern German drama.* Cambridge U.P., 1979.

INNES, C. D.: *Modern German drama.* Cambridge U.P., 1979.

SHAW, Leroy R., ed.: *German theater today: a symposium.* Austin: Texas U.P., 1964.

SPALTER, Max: *Brecht's tradition.* Baltimore: Johns Hopkins, 1967.

Italy

See also below under DRAMATIC FORMS: Commedia dell'Arte.

HERRICK, Marvin T.: *Italian comedy in the Renaissance.* Urbana: Illinois U.P., 1960. *Italian tragedy in the Renaissance.* Urbana: Illinois U.P., 1965.

KENNARD, Joseph S.: *The Italian theatre.* 2 vols. New York: Blom, 1964.

MCLEOD, Addison: *Plays and players in modern Italy: being a study of the Italian stage as affected by the political life, manners and character of today.* Smith, Elder, 1912. Port Washington, N.Y.: Kennikat, 1970.

NAGLER, A. M.: *Theatre festivals of the Medici, 1539–1637.* New Haven, Conn.: Yale U.P., 1964.

SMITH, Winifred: *Italian actors of the Renaissance.* New York: Blom, 1968.

Russia and the Soviet Union

CARTER, Huntly: *The new spirit in the Russian theatre, 1917–1928.* New York: Brentano's, 1929.

COLEMAN, Arthur P.: *Humour in the Russian comedy from Catherine to Gogol.* New York: Columbia U.P., 1925.

FÜLOP-MILLER, René, and GREGOR, Josef: *The Russian theatre.* New York: Blom, 1930. Harrap.

GORCHAKOV, Nikolai A.: *The theater in Soviet Russia.* New York: Columbia U.P., 1957.

MACLEOD, Joseph: *The new Soviet theatre.* Allen and Unwin, 1943.

SLONIM, Marc: *Russian theater from the Empire to the Soviets.* Cleveland: World, 1961. Methuen, 1963.

Scandinavia

GUSTAFSON, A.: *A history of Swedish literature.* Minneapolis: Minnesota U.P., 1961.

LUCAS, F. L.: *The drama of Ibsen and Strindberg*. New York: Macmillan, 1962. Cassell, 1963.

MCFARLANE, J. W.: *Ibsen and the temper of Norwegian literature*. O.U.P., 1960.

MARKER, F. J. and L.-L.: *The Scandinavian theatre*. Blackwell, 1975.

Spain

CRAWFORD, J. P. W.: *Spanish drama before Lope de Vega*. 2nd edn. Philadelphia: Pennsylvania U.P., 1967.

DONOVAN, R. B.: *The liturgical drama in medieval Spain*. Toronto U.P., 1958.

PEAK, J. Hunter: *Social drama in nineteenth-century Spain*. Chapel Hill: North Carolina U.P., 1965.

RENNERT, Hugo A.: *The Spanish stage in the time of Lope de Vega*. New York: Kraus Reprint, 1963.

SHERGOLD, N. D.: *A history of the Spanish stage, from medieval times until the end of the seventeenth century*. O.U.P., 1967.

UNITED STATES OF AMERICA

ATKINSON, Brooks: *Broadway, 1900–1970*. Rev. edn. New York: Macmillan, 1974.

BIGSBY, C. W. E.: *Confrontation and commitment: a study of contemporary American drama, 1959–1966*. Macgibbon and Kee, 1967. Columbia: University of Missouri Press, 1969.

BLUM, Daniel, and WILLIS, John: *A pictorial history of the American theatre, 1860–1976*. 4th edn. New York: Crown, 1978.

BORDMAN, Gerald: *The American musical theatre*. O.U.P, 1978.

CLURMAN, Harold: *The fervent years* (about the Group Theatre). 2nd edn. New York: Hill and Wang, 1957. Macgibbon and Kee.

DOWNER, Alan S., ed.: *American drama and its critics: a collection of critical essays*. Chicago U.P., 1965.

DOWNER, Alan S.: *The American theatre today*. New York: Basic, 1967.

EWEN, D.: *The story of America's musical theatre*. New York: Chilton, 1968.

GASSNER, John: *Theatre at the crossroads: plays and playwrights on the mid-century American stage*. New York: Holt, Rinehart, 1960.

GILBERT, Douglas: *American vaudeville*. New York: Dover, 1963. Constable.

GOTTFRIED, Martin: *A theater divided: the postwar American stage*. Boston: Little, Brown, 1967.

GREEN, Stanley: *The world of musical comedy*. 3rd edn rev. and enlarged. Cranbury, N.J.: Barnes, 1974. Yoseloff. *Encyclopaedia of the musical*. Cassell, 1977. New York: Dodd, Mead.

GUERNSEY, Otis L., ed.: *The best plays of . . . : the Burns Mantle theater year book*. New York: Dodd, Mead. (First published in 1920; see under Mantle.) *Directory of the American theater 1894–1971*. New York: Dodd, Mead, 1971.

HEWITT, Barnard, ed.: *Theatre U.S.A., 1668–1957*. New York: McGraw-Hill, 1959.

HUGHES, Glenn: *History of the American theatre, 1700–1950*. New York: French, 1951.

KERNAN, Alvin B., ed.: *Modern American theater: a collection of critical essays*. Englewood Cliffs, N.J.: Prentice-Hall, 1967.

KRUTCH, Joseph Wood: *The American drama since 1918*. 2nd edn. New York: Random House, 1957.

LITTLE, Stuart W.: *Off-Broadway: the prophetic theater*. New York: Coward, 1972.

LITTLE, Stuart W., and CANTOR, Arthur: *The playmakers: Broadway from the inside.* New York: Norton, 1970. Reinhardt, 1971.

MCNAMARA, Brooks: *The American playhouse in the eighteenth century*. Cambridge, Mass.: Harvard U.P., 1969.

MANTLE, Robert Burns, ed.: *The best plays of...and the year book of the drama in America.* Boston: Small, 1920–25; New York: Dodd, Mead, 1926–. *The best plays of 1909–19.* New York: Dodd, Mead, 1933. *The best plays of 1899–1909.* New York: Dodd, Mead, 1944. *The best plays of 1894–99.* New York: Dodd, Mead, 1955.

MESERVE, Walter, ed.: *Discussions of modern American drama*. Boston: Heath, 1966.

MORDDEN, Ethan: *The American Theatre*. O.U.P., 1981.

ODELL, G. C. D.: *Annals of the New York stage*. 15 vols. New York: Columbia U.P., 1927–49. A.M.S., 1970.

POGGI, Jack: *Theater in America: the impact of economic forces, 1870–1967.* Ithaca, N.Y.: Cornell U.P., 1968.

QUINN, A. H.: *History of the American drama from the beginnings to the Civil War.* 2nd edn. New York: Appleton, 1943. *History of the American drama from the Civil War to the present day*. 2 vols. in one. 2nd edn. New York: Appleton, 1936. Pitman, 1937.

SHATTUCK, C. H.: *Shakespeare on the American stage: from the Hallams to Edwin Booth*. Washington, D. C.: Folger, 1976.

TAUBMAN, Howard: *Making of the American theatre.* Rev. edn. New York: Coward, 1967. Longmans.

TOLL, Robert C.: *Blacking up: the minstrel show in nineteenth-century America.* O.U.P., 1974. *On with the show: the first century of show business in America.* O.U.P., 1976.

WILLIS, John, ed.: *Theatre World.* New York: Crown (annual). (First published in 1945 by Theatre World.)

WILSON, Garff B.: *A history of American acting*. Bloomington, Ind.: Indiana U.P., 1966.

ASIA

BOWERS, Faubion: *Theatre in the East: a survey of Asian dance and drama.* Nelson, 1956. New York: Grove, 1969.

BRANDON, James R.: *Theatre in Southeast Asia*. Cambridge, Mass.: Harvard U.P., 1967. O.U.P.

PRONKO, Leonard C.: *Theater East and West: perspectives towards a total theater.* Berkeley: California U.P., 1967. Cambridge U.P.

SCOTT, Adolphe C.: *The theatre in Asia.* Weidenfeld, 1972.

TILAKSIRI, J.: *The puppet theatre of Asia*. Colombo, Sri Lanka: 1968.

China

ARLINGTON, L. C.: *The Chinese drama from the earliest times until to-day*. Shanghai: Kelly and Walsh, 1930. New York: Blom.

CHEN, Jack: *The Chinese theatre.* Dobson, 1949. New York: Dufour, 1950.

DOLBY, W.: *A history of Chinese drama*. Elek, 1976.

HOWARD, R.: *Contemporary Chinese theatre.* Hutchinson, 1978.

MACKERRAS, Colin P.: *The rise of the Peking Opera, 1770–1870: social aspects of the theatre in Manchu China.* O.U.P., 1972.

SCOTT, Adolphe C.: *The classical theatre of China.* Unwin, 1957. New York: Barnes and Noble. *An introduction to the Chinese theatre.* New York: Theatre Arts, 1959.

ZUNG, Cecilia S.: *Secrets of the Chinese drama*. Shanghai: Kelly and Walsh, 1937. New York: Blom, 1964.

India

GARGI, Balwant: *Theatre in India*. New York: Theatre Arts, 1962. *Folk theatre of India*. Seattle: Washington U.P., 1966.
KEITH, A. Berriedale: *The Sanskrit drama in its origin, development, theory, and practice*. O.U.P., 1924.
MATHUR, Jagdish C.: *Drama in rural India*. New York: Taplinger, 1964.
SHEKHAR, I.: *Sanskrit drama, its origin and decline*. 2nd edn. New Delhi: 1977.
WELLS, H. W.: *The classical drama of India*. New York: Taplinger, 1963.

Japan

ARNOTT, Peter: *The theatres of Japan*. Macmillan, 1969. New York: St Martin's.
BOWERS, Faubion: *Japanese theatre*. Prentice-Hall, 1974. Rutland, Vt: Tuttle.
ERNST, E.: *The Kabuki theatre*. Secker, 1956. Honolulu: University Press of Hawaii, 1974.
HACHIMONJIYA, Jisho: *The actors' analects*. New York: Columbia U.P., 1969.
IMMOOS, T.: *Japanese theatre*. Studio Vista, 1977.
PRONKO, Leonard C.: *Guide to Japanese drama*. Boston, Mass.: G. K. Hall, 1973.
SCOTT, Adolphe C.: *The Kabuki theatre of Japan*. New York: Collier, 1966.

DRAMATIC FORMS

Tragedy

CORRIGAN, Robert W., ed.: *Tragedy: vision and form*. San Francisco: Chandler, 1965.
LUCAS, F. L.: *Tragedy: serious drama in relation to Aristotle's Poetics*. Hogarth Press, 1927. New York: Macmillan, 1958.
STEINER, George: *The death of tragedy*. Faber, 1961. New York: O.U.P., 1980.
STYAN, J. L.: *The dark comedy: the development of modern comic tragedy*. Cambridge U.P., 1962.
WILLIAMS, Raymond: *Modern tragedy*. Rev. edn. Verso Editions, 1979.

Comedy

BERGSON, Henri: *Laughter*. Garden City, N.Y.: Doubleday, 1956.
CHARNEY, M. M.: *Comedy high and low*. O.U.P., 1978.
CORRIGAN, Robert W., ed.: *Comedy: meaning and form*. San Francisco: Chandler, 1965.
LAUTER, Paul, ed.: *Theories of comedy*. Garden City, N.Y.: Doubleday, 1964.
SEYLER, Athene, and HAGGARD, Stephen: *The craft of comedy*. 2nd edn. New York: Theatre Arts, 1957. Muller, 1958.

Commedia dell'Arte

DUCHARTE, Pierre Louis: *The Italian comedy: the improvisation, scenarios, lives, attributes, portraits and masks of the illustrious characters of the commedia dell'arte*. Harrap, 1929.
NICOLL, Allardyce: *The world of Harlequin: a critical study of the commedia dell'arte*. Cambridge U.P., 1976.

OREGLIA, Giacomo: *The commedia dell'arte.* Methuen, 1968. New York: Hill and Wang.
SMITH, Winifred: *The commedia dell'arte: a study in Italian popular comedy.* New York: Blom, 1964.

PLAYHOUSES AND FORMS OF STAGING

ALOI, Roberto: *Esempi architetture per lo spettacolo.* Milan: Hoepli, 1958. New York: J. H. Heineman, 1959.
BAUR-HEINHOLD, M.: *Baroque theatre.* Thames and Hudson, 1967. New York: McGraw-Hill.
BOYLE, W. P.: *Central and flexible staging.* Berkeley: California U.P., 1956.
BURRIS-MEYER, Harold, and COLE, Edward: *Theatres and auditoriums.* 2nd edn. New York: Reinhold, 1965.
CORRY, Percy: *Planning the stage.* Pitman, 1961.
GLASSTONE, V.: *Victorian and Edwardian theatres.* Thames and Hudson, 1975. Cambridge, Mass.: Harvard U.P.
HOWARD, Diana: *London theatres and music halls, 1850–1950.* Library Association, 1970.
IZENOUR, George: *Theatre design.* New York: McGraw-Hill, 1977.
JOSEPH, Stephen, ed.: *Actor and architect.* Manchester U.P., 1964.
JOSEPH, Stephen: *New theatre forms.* Pitman, 1968. New York: Theatre Arts.
LEACROFT, R.: *The development of the English playhouse.* Eyre Methuen, 1973.
MANDER, Raymond, and MITCHENSON, Joe: *The theatres of London.* New edn. New English Library, 1975. *Lost theatres of London.* New edn. New English Library, 1976.
MIELZINER, Jo: *The shapes of our theatre.* New York: Potter, 1975.
RHODES, E. L.: *Henslowe's Rose: the stage and staging.* Lexington: U.P. of Kentucky, 1976.
SCHLEMMER, Oskar, *et al.*: *The theater of the Bauhaus.* Middletown, Conn.: Wesleyan U.P., 1961.
SILVERMAN, Maxwell, and BOWMAN, M. A.: *Contemporary theatre architecture: an illustrated survey.* New York Public Library, 1965.
SMITH, W. D.: *Shakespeare's playhouse practice.* Hanover, N.H.: U.P. of New England, 1975.
SOUTHERN, Richard: *Proscenium and sight lines.* 2nd edn. Faber, 1964. New York: Theatre Arts.
STODDARD, R.: *Theatre and cinema architecture. A guide to information sources.* De-Detroit: Gale, 1978.
STYAN, J. L.: *Drama, stage and audience.* Cambridge U.P., 1975.

TECHNIQUES OF STAGECRAFT

Comprehensive Studies of the Production Process

CARTER, Conrad, *et al.*: *The production and staging of plays.* New York: Arc Books, 1963.
COTES, Peter: *A handbook for the amateur theatre.* Oldbourne, 1957.
COURTNEY, Richard: *Drama for youth.* Pitman, 1964.
FARBER, Donald C.: *Producing on Broadway: a comprehensive guide.* New York: D.B.S. Publications, 1969.

GASSNER, John: *Producing the play*, including *The new scene technician's handbook* by Philip Barber. Rev. edn. New York: Dryden, 1953.

GRUVER, Bert: *The stage manager's handbook*. New York: D.B.S. Publications, 1961.

HEFFNER, Herbert C., *et al.*: *Modern theatre practice: a handbook of play production, with an appendix on costume and makeup*. 4th edn. New York: Appleton, 1959.

MELVILL, Harald: *Theatrecraft: the A to Z of show business*. Rockliff, 1954.

STODDARD, R.: *Stage scenery, machinery and lighting. A guide to information sources*. Detroit: Gale, 1977.

History and Technique of Directing

COLE, Toby, and CHINOY, Helen Krich, eds.: *Directors on directing*. 2nd edn. Indianapolis: Bobbs-Merrill, 1963.

FERNALD, John: *Sense of direction: the director and his actors*. Secker, 1968. New York: Stein and Day, 1969.

HUNT, Hugh: *The director in the theatre*. Routledge, 1954.

MACGOWAN, Kenneth: *Continental stagecraft*. New York: Blom, 1964.

MCMULLAN, Frank: *The director's handbook: an outline for the teacher and student of play interpretation and direction*. New York: Shoe String Press, 1962.

MARSHALL, Norman: *The producer and the play*. 3rd edn. Davis-Poynter, 1975.

ROOSE-EVANS, James: *Directing the play*. Studio Vista, 1968. New York: Theatre Arts.

History and Techniques of Stage Design

APPIA, Adolphe: *Music and the art of the theatre* (on scenic and lighting design). Coral Gables, Fla.: Miami U.P., 1962.

BURRIS-MEYER, Harold, and COLE, Edward C.: *Scenery for the theatre: the organization, processes, materials, and techniques used to set the stage*. Boston: Little, Brown, 1938. Harrap, 1939.

FRETTE, Guido: *Stage design, 1909–1954*. Milan: G. G. Gorlich, 1955.

HAINAUX, René, ed.: *Stage design throughout the world since 1935*. Harrap, 1957. New York: Theatre Arts, 1965. *Stage design throughout the world since 1950*. Harrap, 1964. New York: Theatre Arts. *Stage design throughout the world since 1960*. Harrap, 1973. New York: Theatre Arts. *Stage design throughout the world, 1970–75*. Harrap, 1976. New York: Theatre Arts.

JOSEPH, Stephen: *Scene painting and design*. Pitman, 1964.

KESLER, J.: *Theatrical costume. A guide to information sources*. Detroit: Gale, 1979.

LAVER, James: *Costume in the theatre*. Harrap, 1964. New York: Hill and Wang.

NEWTON, Stella Mary: *Renaissance theatre costume and the sense of the historic past*. Rapp and Whiting, 1975.

ROSENFELD, Sybil: *A short history of scene design in Great Britain*. Blackwell, 1973. *Georgian scene painters and scene painting*. Cambridge U.P., 1981.

SIMONSON, Lee: *The stage is set*. New York: Theatre Arts, 1963.

SOUTHERN, Richard: *Changeable scenery: its origin and development in the British theatre*. Faber, 1952.

WELKER, D.: *Theatrical set design: the basic techniques*. 2nd edn. Boston: Allyn and Bacon, 1979.

Light and Sound in the Theatre

BENTHAM, Frederick. *The art of stage lighting*. 2nd edn. Pitman, 1976.

BURRIS-MEYER, Harold, and MALLORY, V.: *Sound in the theatre*. New York: Theatre Arts, 1959.

CORRY, Percy: *Lighting the stage*. 3rd edn. Pitman, 1961.

FUCHS, Theodore: *Stage lighting*. New York: Blom, 1963.

HARTMANN, Louis: *Theatre lighting: a manual of the stage switchboard*. New York: D.B.S. Publications, 1970.

MCCANDLESS, Stanley: *A syllabus of stage lighting*. 11th edn. New York: D.B.S. Publications, 1964.

PILBROW, Richard: *Stage lighting*. Rev. edn. Studio Vista, 1979. New York: D.B.S. Publications.

ACTING

BAKER, M.: *The rise of the Victorian actor*. Croom, 1978. Totowa, N.J.: Rowman and Littlefield.

BRADBROOK, M. C.: *The rise of the common player: a study of actor and society in Shakespeare's England*. Cambridge U.P., 1979.

DARLINGTON, W. A.: *The actor and his audience*. Phoenix, 1949.

GILDER, Rosamond: *Enter the actress: the first women in the theatre*. Harrap, 1931. New York: Theatre Arts, 1960.

JOSEPH, Bertram: *The tragic actor*. Routledge, 1959. New York: Theatre Arts. *Elizabethan acting*. 2nd edn. O.U.P., 1964.

WILSON, Garff B.: *A history of American acting*. Bloomington, Ind.: Indiana U.P., 1966.

Theory and Technique

ALBRIGHT, H. and A.: *Acting, the creative process*. 3rd edn. Belmont, Calif.: Wadsworth, 1980.

ARCHER, William: *Masks or faces*? Longmans, 1888. New York: Hill and Wang, 1957.

BLAKELOCK, Denys: *Advice to a player: letters to a young actor*. Heinemann, 1958.

BURTON, Hal: *Great acting*. B.B.C. Publications, 1969. New York: Hill and Wang.

COLE, Toby, ed.: *Acting: a handbook of the Stanislavski Method*. 2nd edn. New York: Crown, 1955.

COLE, Toby, and CHINOY, Helen Krich, eds.: *Actors on acting*. New edn. New York: Crown, 1970.

COQUELIN, Constant: *The art of the actor*. Allen and Unwin, 1954.

CORSON, Richard: *Stage make-up*. 5th edn. Englewood Cliffs, N.J.: Prentice-Hall, 1975.

GIELGUD, John: *Stage directions*. Heinemann Educational, 1979. New York: Random House.

HAYMAN, Ronald: *Techniques of acting*. Methuen, 1969.

HETHMON, Robert H., ed.: *Strasberg at the Actors' Studio: tape recorded sessions*. New York: Viking, 1965. Cape, 1966.

HODGSON, John, and RICHARDS, Ernest: *Improvisation*. Methuen, 1966.

LEWES, G. H.: *On actors and the art of acting*. Smith, Elder, 1857. New York: Grove, 1957.

MAROWITZ, Charles: *The act of being*. Secker, 1978. New York: Taplinger.

MATTHEWS, Brander, ed.: *Papers on acting*. New York: Hill and Wang, 1958.

REDGRAVE, Michael: *Mask or face: reflections in an actor's mirror*. Heinemann, 1958. New York: Theatre Arts. *The actor's ways and means*. Heinemann Educational, 1979. New York: Theatre Arts.

STANISLAVSKI, Constantin: *An actor prepares*. New York: Theatre Arts, 1936. Eyre Methuen, 1980. *Building a character*. New York: Theatre Arts, 1949. Eyre Methuen, 1979. *Creating a role*. New York: Theatre Arts, 1961. Eyre Methuen, 1981.

THEATRE CRITICISM

AGATE, James, ed.: *The English dramatic critics*. Barker, 1932. New York: Hill and Wang, 1958.

DOWNS, Harold: *The critic in the theatre*. Pitman, 1953.

HOBSON, Harold: *Verdict at midnight* (first-night notices put to the test of time). Longmans, 1952.

JONES, T. B., and NICOL, B. de B.: *Neo-classical dramatic criticism, 1560–1770*. Cambridge U.P., 1976.

ROWELL, George: *Victorian dramatic criticism*. Methuen, 1971.

WARD, A. C., ed.: *Specimens of English dramatic criticism*. O.U.P., 1945.

Individual Critics

AGATE, James: *Brief chronicles*. Cape, 1943. *Red letter nights*. Cape, 1944. New York: Blom, 1969. *Immoment toys*. Cape, 1945. New York: Blom, 1969.

ARCHER, William: *The theatrical 'World'*. 5 vols., 1893–7. Walter Scott, 1894–8. New York: Blom, 1969.

BEERBOHM, Max: *Around theatres*. Hart-Davis, 1953. New York: Taplinger, 1969. *More theatres, 1898–1903*. Hart-Davis, 1969. New York: Taplinger. *Last theatres, 1904–1910*. Hart-Davis, 1970.

BENTLEY, Eric: *What is theatre?* New York: Atheneum, 1968. Methuen, 1969.

BROWN, John Mason: *Dramatis personae*. New York: Viking, 1963. Hamilton.

BRUSTEIN, Robert: *Seasons of discontent*. New York: Simon and Schuster, 1965. Cape, 1966. *The third theatre*. New York: Knopf, 1969. Cape, 1970.

CLURMAN, Harold: *Lies like truth*. New York: Macmillan, 1968.

DENNIS, Nigel: *Dramatic essays*. Weidenfeld, 1962.

GREIN, J. T.: *Dramatic criticism*. 5 vols., 1897–1904. Vol. I, J. Long, 1899; Vols. II and III, Greenwood, 1900 and 1902; Vols. IV and V, E. Nash, 1904 and 1905. Vols. I to V, New York: Blom, 1968.

HAZLITT, William: *Hazlitt on theatre*. Walter Scott, 1895. New York: Hill and Wang, 1957.

HOBSON, Harold: *Theatre*. Longmans, 1948. *Theatre 2*. Longmans, 1950.

JAMES, Henry: *The scenic art*. Hart-Davis, 1949. New York: Hill and Wang, 1957.

KERR, Walter: *Pieces at eight*. Reinhardt, 1958. *Thirty plays hath November*. New York: Simon and Schuster, 1969. *Journey to the center of the theater*. New York: Knopf, 1979.

KOTT, Jan: *Theatre notebook, 1947–1967*. Methuen, 1968.

MCCARTHY, Desmond: *Theatre*. Macgibbon, 1954. Westport, Conn.: Greenwood Press, 1977.

MCCARTHY, Mary: *Sights and spectacles, 1937–1958*. New York: Farrar, 1959. Heinemann.

MAROWITZ, Charles: *Confessions of a counterfeit critic: a London theatre notebook, 1958–1971*. Eyre Methuen, 1973.

MONTAGUE, C. E.: *Dramatic values*. Methuen, 1911. New York: Reprint House International.

NATHAN, George Jean: *Passing judgments.* New York: Johnson Reprint, 1969.

SCOTT, Clement: *From The Bells to King Arthur: a critical record of the first-night productions at the Lyceum Theatre from 1871 to 1895.* John Macqueen, 1896. New York: Blom, 1969.

TYNAN, Kenneth: *He that plays the king.* Longmans, 1950. *Curtains.* Longmans, 1961. New York: Atheneum. *The sound of two hands clapping.* Cape, 1975.

WALKLEY, A. B.: *Drama and life.* Methuen, 1907. Freeport, N.Y.: Books for Libraries, 1967.

WHITING, John: *John Whiting on theatre.* Alan Ross, 1966. Chester Springs, Pa.: Dufour.

YOUNG, Stark: *Immortal shadows.* New York: Scribner, 1947.

ACKNOWLEDGEMENTS

Thanks are due to the following for permission to reproduce illustrations on the plates listed:

Anglo-Chinese Educational Institute 96 bottom; Marquess of Bath 17 top; BBC Hulton Picture Library 8 bottom left, 18 bottom, 27 bottom left, 28, 58 top, 60 top, 73 top left, 77, 81 bottom left, 82 top left, 83 top right, bottom left, 84 top right, bottom right, 85, 87 bottom, 88 top left and right, bottom left, 89 top left; Biblioteca Cardinale Giulio Alberoni 9 top left; Bibliothèque Nationale 10 bottom, 34 bottom, 35 bottom, 81 top; Bildarchive Foto Marburg 53 bottom; British Library 12, 19 top left, 33 bottom, 34 top right, 81 bottom right; British Museum 3 top, 13, 94 top, 96 top; John Vere Brown 91 bottom left; Bulloz 41 top; Camera Press 31 bottom, 63, 90 top right, 91 top, bottom left, 95 top; Trustees of the Chatsworth Settlement (photographs Courtauld Institute of Art) 36, 82 top right; Joe Cocks 90 bottom; Comédie-Française 71 top; Courtauld Institute of Art 16 top.

Mike Davis Studios 92 bottom, 95 bottom; Deutsches Archaeologisches Institut 1 top; Documentation Photographique de la Réunion des Musées Nationaux 4 bottom; Dominic 31 top, 67 bottom, 68 bottom, 69 bottom; Dover Publications 74 bottom; Fitzwilliam Museum 41 bottom; Bamber Gascoigne: *World Theatre* 76 top; Giraudon 6 top, 8 bottom right, 9 bottom, 10 top; Greater London Council 17 bottom; Guthrie Theatre, Minneapolis, 72 top; Harvard Theatre Collection (Angus McBean Photographs) 30 top, 66 bottom, 80 bottom; Herzog Anton Ulrich-Museum 16 bottom; Historical Archives of the City of Cologne 15 top; Historische Museum der Stadt Wien 46 top; Kungliga Dramatiska Teatern, Stockholm, 58 bottom, 68 top; the late Professor T. E. Lawrenson 35 top; Lipnitzki-Viollet 52 bottom, 61 top, 62, 70, 89 top right; President and Fellows of Magdalen College, Oxford, 19 bottom; Mander & Mitchenson 22 bottom, 23 top, 24 bottom, 47 bottom, 49 bottom, 53 top, 57, 60 bottom, 76 bottom, 78, 79, 80 top; Mansell Collection 4 top, 7, 8 top, 15 bottom; Martin von Wagner-Museum 2 bottom right; Metropolitan Museum of Art 33 top, 37 top; Museum of London 44, 73 bottom; Museum of the City of New York 52 top.

National Museum of Stockholm at Drottningholm Theatre 34 top left, 43 bottom; National Theatre (photograph Nobby Clark) 71 bottom; National Tourist Office of Greece 2 bottom left, 5 bottom; New York Historical Society 45 bottom; New York Public Library 27 top, 29 bottom, 32 top, 42 bottom, 51, 56, 59, 65 bottom; Novosti Press Agency 25 bottom, 26 top, 27 bottom right, 30 bottom, 50 bottom, 86 top left, right; Österreichische Nationalbibliothek 20 bottom, 37 bottom, 39 top, 42 top, 48; Pat Hodgson Library 9 top right, 18 top left, right, 19 top right, 73 top right; Piccolo Teatro della Città di Milano 66 top; Popperphoto 5 top; RIBA 61 bottom; Richard Southern Accession, University of Bristol Theatre Collection 45 top; Rijksmuseum 38; Governors of the Royal Shakespeare Theatre, Stratford-upon-Avon, 23 bottom, 26 bottom, 29 top, 89 bottom left, right.

Sangeet Natak Akademi, New Delhi, 92 top, 93; Peter Smith 72 bottom; Society for Cultural Relations with the U.S.S.R. 50 top, 55 bottom; SPADEM 83 bottom right; Staatliche Museen zu Berlin 3 bottom; Stratford Shakespeare Festival Foundation of Canada 32 bottom (photograph Robert C. Ragsdale), 90 top left (photograph Douglas Spillane); Gary W. Sweetman 69 top; Swiss Theatre Collection 55 top; Teaterhistorisk Museum, Copenhagen, 46 bottom, 49 top; Theatermuseum, München, 54 bottom; Theatersammlung der Universität Hamburg 47 top; *The Times* 64 top; Verulamium Museum 6 bottom; John Vickers 88 bottom right; Victoria and Albert Museum 11, 14, 20 top, 21, 22 top, 24 top, 25 top, 39 bottom, 40, 43 top, 54 top, 64 bottom, 74 top, 75, 82 bottom left, right, 83 top left. 84 top left, 86 bottom left, right, 87 top left, right; Victoria Theatre, Stoke-on-Trent, 65 top.